Periodic Table of the Elements

Main groups

Transition metals

Main groups

1 1A	2 2A																	18 8A
1 **H** 1.00794																		2 **He** 4.00260
3 **Li** 6.941	4 **Be** 9.01218											13 3A 5 **B** 10.81	14 4A 6 **C** 12.011	15 5A 7 **N** 14.0067	16 6A 8 **O** 15.9994	17 7A 9 **F** 18.998403	10 **Ne** 20.1797
11 **Na** 22.98977	12 **Mg** 24.305	3 3B	4 4B	5 5B	6 6B	7 7B	8	9 8B	10	11 1B	12 2B	13 **Al** 26.98154	14 **Si** 28.0855	15 **P** 30.97376	16 **S** 32.066	17 **Cl** 35.453	18 **Ar** 39.948
19 **K** 39.0983	20 **Ca** 40.078	21 **Sc** 44.9559	22 **Ti** 47.88	23 **V** 50.9415	24 **Cr** 51.996	25 **Mn** 54.9380	26 **Fe** 55.847	27 **Co** 58.9332	28 **Ni** 58.69	29 **Cu** 63.546	30 **Zn** 65.39	31 **Ga** 69.72	32 **Ge** 72.61	33 **As** 74.9216	34 **Se** 78.96	35 **Br** 79.904	36 **Kr** 83.80
37 **Rb** 85.4678	38 **Sr** 87.62	39 **Y** 88.9059	40 **Zr** 91.224	41 **Nb** 92.9064	42 **Mo** 95.94	43 **Tc** (98)	44 **Ru** 101.07	45 **Rh** 102.9055	46 **Pd** 106.42	47 **Ag** 107.8682	48 **Cd** 112.41	49 **In** 114.82	50 **Sn** 118.710	51 **Sb** 121.757	52 **Te** 127.60	53 **I** 126.9045	54 **Xe** 131.29
55 **Cs** 132.9054	56 **Ba** 137.33	57 *La 138.9055	72 **Hf** 178.49	73 **Ta** 180.9479	74 **W** 183.85	75 **Re** 186.207	76 **Os** 190.2	77 **Ir** 192.22	78 **Pt** 195.08	79 **Au** 196.9665	80 **Hg** 200.59	81 **Tl** 204.383	82 **Pb** 207.2	83 **Bi** 208.9804	84 **Po** (20	85 **At** 10)	86 **Rn** (222)
87 **Fr** (223)	88 **Ra** 226.0254	89 †Ac 227.0278	104 **Rf** (261)	105 **Db** (262)	106 **Sg** (266)	107 **Bh** (264)	108 **Hs** (269)	109 **Mt** (268)	110 (271)	111 (272)	112 (277)	114 (289)					118 (293)

*Lanthanide series

| 58 **Ce** 140.12 | 59 **Pr** 140.9077 | 60 **Nd** 144.24 | 61 **Pm** (145) | 62 **Sm** 150.36 | 63 **Eu** 151.96 | 64 **Gd** 157.25 | 65 **Tb** 158.9254 | 66 **Dy** 162.50 | 67 **Ho** 164.9304 | 68 **Er** 167.26 | 69 **Tm** 168.9342 | **Yb** 173. | **Lu** 173. |

†Actinide series

| 90 **Th** 232.0381 | 91 **Pa** 231.0359 | 92 **U** 238.0289 | 93 **Np** 237.048 | 94 **Pu** (244) | 95 **Am** (243) | 96 **Cm** (247) | 97 **Bk** (247) | 98 **Cf** (251) | 99 **Es** (252) | 100 **Fm** (257) | 101 **Md** (258) | 102 **No** (259) |

204
45

KT-433-941

Chemistry

THIRD EDITION

John McMurry

Cornell University

Robert C. Fay

Cornell University

PRENTICE HALL
Upper Saddle River, New Jersey 07458

Library of Congress Cataloging-in-Publication Data

McMurry, John.
    Chemistry / John McMurry, Robert C. Fay-3rd ed.
        p.   cm.
    Includes index.
    ISBN 0-13-087205-9
    1. Chemistry.   I. Fay, Robert C.   II. Title.
QD33.M137   2001
540-dc21   00-032413

Senior Editor: *Kent Porter-Hamann*
Production Editor: *Donna F. Young*
Associate Development Editor: *Nancy Garcia*
Art Director: *Joseph Sengotta*
Art Manager: *Gus Vibal*
Formatter: *Vicki L. Croghan/Envision, Inc.*
Manufacturing Buyer: *Michael Bell*
Media Editor: *Paul Draper*
Assistant Managing Editor, Science Media: *Alison Lorber*
Editorial Director: *Paul F. Corey*
Editor in Chief of Development: *Carol Trueheart*
Senior Marketing Manager: *Steve Sartori*
Vice President of Production and Manufacturing: *David W. Riccardi*
Executive Managing Editor: *Kathleen Schiaparelli*
Creative Director: *Paul Belfanti*
Manufacturing Manager: *Trudy Pisciotti*
Interior Designer: *Judith Matz-Coniglio*
Cover Designer: *Bruce Kenselaar*
Copy Editor: *Luana Richards*
Proofreader: *Jennefer Vecchione*
Illustrators: *Academy Artworks, Inc. and Michael Goodman*
Art Editor: *Karen Branson*
Photo Researcher: *Yvonne Gerin*
Editorial Assistants: *Nancy Bauer, Gillian Buonanno, and Richard Moriarty*
Marketing Assistant: *Matt Redstone*
Art Project Support: *Julie Nazario*
Cover Photo: *Sulfur burning in oxygen,* © *Charles D. Winters*

 © 2001, 1998, 1995 by Prentice-Hall, Inc.
Upper Saddle River, New Jersey 07458

Printed in the United States of America
10 9 8 7 6 5 4 3 2

ISBN 0-13-087205-9

Prentice-Hall International (UK) Limited, *London*
Prentice-Hall of Australia Pty. Limited, *Sydney*
Prentice-Hall Canada Inc., *Toronto*
Prentice-Hall Hispanoamericana, S.A., *Mexico*
Prentice-Hall of India Private Limited, *New Delhi*
Prentice-Hall of Japan, Inc., *Tokyo*
Pearson Education Asia Pte. Ltd.
Editora Prentice-Hall do Brasil, Ltda., *Rio de Janeiro*

Brief Contents

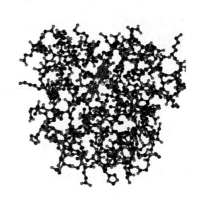

Contents

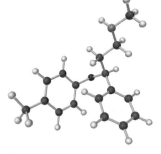

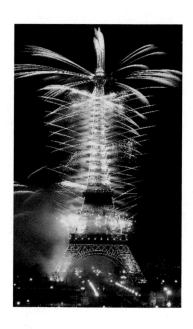

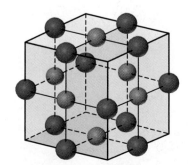

15   Aqueous Equilibria: Acids and Bases   608

16   Applications of Aqueous Equilibria   661

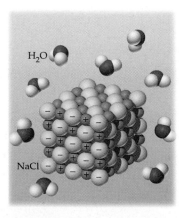

H$_2$O

NaCl

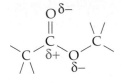

Ester

Interludes

In-text Applications

Preface

The novelist Kurt Vonnegut (a former chemistry major at Cornell University) has a scene in his 1963 book Cat's Cradle in which Francine Pefko, secretary to the famous chemist Dr. Nilsak Horvath, is bemoaning her job:

> "I take dictation from Dr. Horvath and it's just like a foreign language. I don't think I'd understand it—even if I was to go to college."
>
> "If there's something you don't understand, ask Dr.Horvath to explain it. He's very good at explaining. Dr. Hoenikker used to say that any scientist who couldn't explain to an eight-year-old what he was doing was a charlatan."

The case may be overstated, but the underlying sentiment is valid: Scientists should be able to explain their subject in a clear and understandable way to anyone who is willing to learn. And that has been our goal in writing this book—to produce the most readable and effective teaching text possible. The outstanding success of the first two editions suggests that our goal has been met and encourages us to offer this new edition.

About this Book

Our primary goal in writing this book has been to fashion a clear, coherent narrative. Beginning with atomic structure, proceeding next to bonding and molecules, then to bulk physical properties of substances, and ending with a study of chemical properties, we have told a cohesive story about chemistry. Transitions between topics are smooth, explanations are lucid, and tie-ins to earlier material are frequent. Every attempt has been made to explain chemistry in a visual, intuitive way so that it can be understood by those who give it an honest effort.

Insofar as possible, distractions within the text are minimized. Each chapter is broken into numerous sections to provide frequent breathers, and each section has a consistent format. Sections generally begin with an explanation of their subject, move to an Example problem that shows how to work with the material, and end with one or more Practice Problems for the reader to work through. Each chapter ends with a brief Interlude that describes an interesting application or extension of the chapter subject.

About the 3rd Edition

In preparing this third edition, we have again reworked the entire book at the sentence level to make it as easy as possible for a reader to understand and learn chemistry. Among the many changes and improvements, much art has been redrawn, many additional molecular models have been added to enhance the idea of chemistry as a visual, intuitive subject, and even more extensive use has been made of the Periodic Table as a consistent theme throughout the book. In addition, a new Key Concept Summary has been added at the end of each chapter to pull the subject material together in a visual way and make clear the interconnections among topics.

Problems and problem solving have also received a great deal of attention in this third edition. Introduced in the first edition, the use of visual, non-numerical "Key Concept" problems has been much expanded. These problems, which test a real understanding of the material rather than the ability to put numbers into a formula, are increased in number and are now placed within text sections as well as at the end of each chapter just before the Additional Problems. Don't make the mistake of thinking that these Key Concept problems are necessarily simple just because they don't have numbers. Many are real challenges that will test the ability of any student. Other end of-chapter problems are paired, with each odd-numbered problem testing the same concepts as the even-numbered one that precedes it. Finally, we have added new Multi-Concept problems that draw on multiple concepts and that should prove thought-provoking to even the best students.

We hope that this new edition will meet the goals we have set for it and that students will find it to be friendly, accessible, and above all effective in teaching chemistry.

Acknowledgments

Our thanks go to our families and to the many talented people at Prentice Hall who helped bring this book into being. We are especially grateful to Nancy Garcia for her insightful editorial work, Donna Young for her flawless production efforts, Kent Porter-Hamann for her efficient overall project management, Luana Richards for her careful copy editing, and Yvonne Gerin for her fine photo research.

We are particularly pleased to acknowledge the outstanding contributions of several colleagues who created the many important supplements that turn a textbook into a valuable integrated package:

- Robert Pribush at Butler University acted as Consulting Editor to oversee all the supplements. He created the Concept Maps found at the end of each chapter and prepared all annotations for the Instructor's Edition. In addition, he also prepared the Test Bank, the Matter 2001 Instructor CD, and the Instructor's Guide to Media and Print Resources.
- Joseph Topich at Virginia Commonwealth University prepared both the full and partial Solutions Manual as well as the answer appendix at the end of the text.
- DonnaJean Fredeen at Southern Connecticut State University created the Study Guide for students.
- Cheryl Frech at the University of Central Oklahoma wrote the eChapters in the accompanying Chemistry CD.
- Julia Burdge at the University of Akron developed the eMedia Problems and media elements for the accompanying Chemistry CD.

Finally, we want to thank the many colleagues who read, criticized, and improved our work, particularly David Cater who checked the entire final manuscript.

John McMurry
Robert C. Fay

Reviewers of the 3rd edition

Mufeed Basti   *North Carolina A&T State University*
Ronald Bost   *North Central Texas University*
Danielle Brabazon   *Loyola College*
Myron Cherry   *Northeastern State University*
Paul Cohen   *University of New Jersey*
Katherine Covert   *West Virginia University*
Brian Earle   *Cedar Valley College*
Amina El-Ashmawy   *Collin County Community College*
Joanne Follweiler   *Lafayette College*
Wesley Hanson   *John Brown University*
Thomas Herrington   *University of San Diego*
Margaret E. Holzer, *California State University—Northridge*
Narayan Hosmane   *Northern Illinois University*
Jeff Joens   *Florida International University*

John Landrum   *Florida International University*
David Leddy   *Michigan Technological University*
Karen Linscott   *Tri-County Technical College*
Christina Mewhinney   *Eastfield College*
David Miller   *California State University—Northridge*
Abdul Mohammed   *North Carolina A&T State University*
Linda Mona   *United States Naval Academy*
Edward Mottell   *Rose-Hulman Institute*
Gayle Nicoll   *Texas Technological University*
John Schreifels   *George Mason University*
Steven Socol   *McHenry County College*
Kelly Sullivan   *Creighton University*
Erach Talaty   *Wichita State University*
John Vincent   *University of Alabama*
Steve Watton   *Virginia Commonwealth University*

Reviewers of the previous editions

Anneke S. Allen   *Wichita State University*
David Ball   *Cleveland State University*
John Bauman   *University of Missouri-Columbia*
Herbert Beall   *Worcester Polytechnic Institute*
Rathindra N. Bose   *Kent State University*
Virginia Bryan   *Southern Illinois University*
Albert W. Burgstahler   *University of Kansas*
Denise Chauret   *University of Ottawa*
Michael Chetcuti   *University of Notre Dame*
James Coke   *University of North Carolina-Chapel Hill*
Martin Cowie   *University of Alberta*
Robert Crabtree   *Yale University*
Mark Cracolice   *University of Montana*
Michael Denniston   *DeKalb College*
Norman Duffy   *Kent State University*
Royce Engstrom   *University of South Dakota*
Dale D. Ensor   *Tennessee Technological University*
Joanne M. Follweiler   *Lafayette College*
Deanna Franke   *Moorpark College*
Charles Fritchie, Jr.   *Tulane University*
Roy Garvey   *North Dakota State University*
Angela Glisan King   *Wake Forest University*
Mildred V. Hall   *Clark State Community College*
Jane Halverson   *Herkimer County Community College*
Stephen J. Hawkes   *Oregon State University*
Harry G. Hecht   *South Dakota State University*
Werner Horsthemke   *Southern Methodist University*
Colin D. Hubbard   *University of New Hampshire*
James Hutchinson   *Middle Tennessee State University*
Nancy Jones   *La Salle University*
John Luoma   *Cleveland State*
Paul Karr   *Wayne State College*
Nicholas Kildahl   *Worcester Polytechnic Institute*
Leslie Kinsland   *University of Southwestern Louisiana*

Robert Kiser   *University of Kentucky*
Kenneth J. Klabunde   *Kansas State University*
Donald Kleinfelter   *University of Tennessee*
Peter Lykos   *Illinois Institute of Technology*
Jerome Maas   *Oakton Community College*
Kathryn Mansfield Matera   *Baldwin-Wallace College*
William Meena   *Rock Valley College*
Christina Mehwinney   *Brookhaven College*
Gary W. Morrow   *University of Dayton*
Michael Nichols   *John Carroll University*
Deborah Nycz   *Broward Community College*
Scott Perry   *University of Houston*
Robert Pribush   *Butler University*
Heidi Reese   *New York University*
Clyde Riley   *University of Alabama-Huntsville*
B. Ken Robertson   *University of Missouri-Rolla*
Victor Rodwell   *Purdue University*
Steven Ruis   *American River College*
Gene D. Schaumberg   *Sonoma State University*
Richard L. Schowen   *The University of Kansas*
Mary Ellen Scott   *Case Western Reserve University*
Henry Shanfield   *University of Houston*
Charlie Simpson   *Midwestern State University*
Robert Snipp   *Creighton University*
Andrew Sykes   *University of South Dakota*
Klaus H. Theopold   *University of Delaware*
Sidney Toby   *Rutgers University*
Bruno M. Vittimberga   *University of Rhode Island*
Charles A. Wilkie   *Marquette University*
Brenda J. Wojciechowski   *Georgia Southern University*
Shelby D. Worley   *Auburn University*
Charles M. Wynn   *Eastern Connecticut State University*
William H. Zoller   *University of Washington*

Supplements

For the Instructor

- *Annotated Instructor's Edition* **(0-13-089162-2)** with annotations by Robert Pribush, Butler University. This special edition of the text includes the entire student text plus marginal annotations to aid instructors in preparing their lectures. Included are notes on common student misconceptions; suggestions for lecture demonstrations; icons identifying figures reproduced in the transparency pack; cross-references to units in the Student CD-ROM; and cross-references to figures, demonstrations, and animations in the Matter 2001 CD-ROM.

- *Instructor's Guide to Media and Print Resources* **(0-13-088537-1)** by Robert Pribush, Butler University. This package contains teaching tips, lecture demonstrations, lecture outlines, and a grid that organizes over 1,000 still images, animations, and video demonstrations by topic. Also included are complete instructions on how to use the Matter 2001 CD-ROM.

- *Matter 2001 Instructor CD* **(0-13-088524-X)** by Robert Pribush, Butler University. This visual archive CD-ROM contains over 1,000 figures and molecular illustrations, nearly 30 lab demonstration video segments, and over 50 high quality animations. An easy-to-use media asset catalog allows you to use these images and movies with Microsoft PowerPoint or other software, or you can use the prebuilt PowerPoint presentations as a starting point for your lectures. (Free to qualified adopters.)

- *Transparency Pack* **(0-13-088520-7)**. Contains 250 four-color transparencies of images from the text.

- *Solutions Manual* **(0-13-088533-9)** by Joseph Topich, Virginia Commonwealth University. Includes full solutions to all of the problems in the text. Can be made available as a student purchase item with instructor's permission.

- *Test Item File* **(0-13-088534-7)** by Robert Pribush, Butler University. Contains approximately 2,000 multiple-choice questions, all referenced to the text.

- *PH Custom Test* **(Windows: 0-13-088525-8; Mac: 0-13-088526-6)** Computerized versions of the Test Item File allow you to create and tailor exams to your specific needs and includes tools for course management.

- *Companion Website for Chemistry* (http://www.prenhall.com/mcmurry) This free website allows professors to build an online syllabus with calendar-based assignments. Automatically graded homework and quiz sets allow you to evaluate both student preparedness and chapter comprehension efficiently. An enhanced version of this website with additional course management features is also available; see http://www.prenhall.com/demo.

For the Student

- *The Chemistry Media Companion for CW* **(0-13-090567-4) [or** *for CW+* **(0-13-090568-2)]** with the Chemistry Student CD, by Cheryl Frech, University of Central Oklahoma, Julia Burdge, University of Akron, and Thomas Gardner (Tennessee State University). The Student CD is a book-specific CD that allows the student to preview and review the material in an engaging, dynamic way. Difficult concepts are shown in animations, molecules are shown in 3D and can be manipulated, demonstrations are shown in video clips that can be reviewed under student control, and interactive simulations and activities illuminate principles where a static example might not. The CD also serves as the launch point to the Companion Website. The CD is housed in a booklet, the Media Companion, that introduces the Internet and explains how to use online resources to get ahead in your course.

- *The Chemistry Companion Website* (*http://www.prenhall.com/mcmurry*) complements and expands on the material on the CD. Here students can interact with each other and with their instructors via email and an online message board, search for 3D molecules discussed in each chapter, link to interesting and useful websites for each topic, and practice with online quizzes that provide hints and detailed feedback instantly when the student submits answers.

- *Interactive Chemistry Journey* **CD-ROM (0-13-548116-3)** by Steven Gammon, University of Idaho, Lynn R. Hunsberger, University of Louisville, and Sharon Hutchison, University of Idaho. This student tutorial provides an interactive study environment that fosters understanding of a number of core chemical concepts.

- *Math Review ToolKit* **(0-13-088522-3)** by Gary Long, Virginia Polytechnic Institute and State University. Designed to provide assistance to students with weaker math skills, this supplement includes a chapter-by-chapter math review keyed to problems in the text as well as a brief self-assessment test.

- *Student Study Guide* **(0-13-088536-3)** by DonnaJean Fredeen, Southern Connecticut State University. For each chapter, the study guide includes learning goals, an overview, progressive review section with worked examples, and self-tests with answers.

- *Selected Solutions Manual* **(0-13-088532-0)** by Joseph Topich, Virginia Commonwealth University. Contains solutions to all in-chapter problems, and solutions to even-numbered Understanding Key Concepts questions, Additional Problems, and Multi-Concept Problems.

- *The New York Times*/Prentice Hall Themes of the Times. This newspaper-format resource brings together current chemistry-related articles from the award-winning science pages of The New York Times.

- **Prentice Hall Molecular Model Set for General and Organic Chemistry (0-13-955444-0)** This versatile kit allows students to build both ball-and-stick and space-filling models of inorganic and organic compounds. Designed so that it can be used in a subsequent introductory organic chemistry course.

A Guide to Using this Text

We have the same goals. Yours is to learn chemistry; ours is to do everything possible to help you learn. For both of us to succeed, it is important that you enjoy learning about chemistry and come to see its beauty and logic. A good textbook can do much to foster such enjoyment and such insight, and we have worked hard to create such a book. Of course, learning chemistry is going to take some work on your part as well, but we have done all that we can to ensure that the experience is rewarding, not painful. Here are some suggestions for using this book that should prove helpful.

- **Don't read the text immediately.** As you begin each new chapter, scan it over first. Read the chapter outline and introductory paragraphs, and find out what topics will be covered. Then turn to the end of the chapter and read the narrative summary and look at the Key Concept Summary. You'll be in a much better position to learn new material if you first have a general idea of where you're going.

- **Work the problems.** There are no shortcuts here; working problems is an essential part of learning chemistry. The Examples in the text show you how to approach the material. The practice Problems, which accompany all of the Examples, and the Key Concept Problems, which accompany some of the Examples, provide immediate reinforcement. The end-of-chapter Additional Problems and Multi-Concept Problems afford an opportunity for further practice. The Answers to all in-chapter problems and all even-numbered end-of-chapter problems are given at the back of the book. In addition, full solutions to selected problems are given in an accompanying Selected Solutions Manual.

- **Ask questions.** Faculty members and teaching assistants are there for your benefit. They have a sincere interest in helping you learn, so if you don't understand something, ask.

This book has been designed with a view to making the material you will be encountering as clear and as interesting as possible. Our aim has been not just to teach, but to help you learn. Here is a brief overview of its elements, with suggestions as to how you can use them to get the most from your study of chemistry.

Conceptualize

When you clearly understand chemical concepts, you are better equipped to work through quantitative problems and comprehend scientific facts. In the Third Edition of *Chemistry*, the authors expand their signature focus on conceptual understanding to help you succeed.

▶ **PROBLEM 12.5** What are the units of the rate constant for each of the reactions in Table 12.2?

◄ **KEY CONCEPT PROBLEM 12.6** The relative rates of the reaction A + B → products in vessels (a)–(d) are 1:1:4:4. Red spheres represent A molecules, and blue spheres represent B molecules.

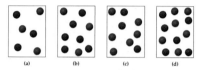

(a)   (b)   (c)   (d)

(a) What is the order of the reaction in A and B, and what is the overall reaction order?
(b) Write the rate law.

12.4 Integrated Rate Law for a First-Order Reaction

Thus far we've focused on the rate law, an equation that tells how a reaction rate depends on reactant concentrations. But we're also interested in how reactant and product concentrations vary with time. For example, it's important to know the rate at which the atmospheric ozone layer is being destroyed, but we also want

◄ Key Concept Problems

Found at the end of chapters in previous editions, the highly praised **Key Concept Problems** have now been expanded and integrated throughout the chapter. They help you focus on learning the most important chemistry concepts.

Key Concept Summary ▶

The **Key Concept Summary** provides a visual outline of the key principles presented in the chapter that (a) includes the major concepts covered in each chapter and (b) presents each concept in the context of associated concepts. It enables you to relate individual concepts and ideas into a meaningful and integrated whole.

Key Concept Summary

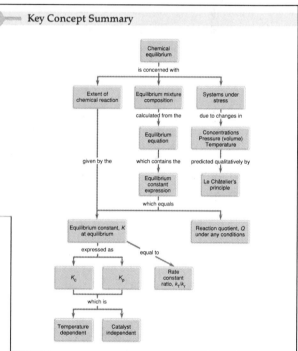

Understanding Key Concepts

Problems 13.1–13.25 appear within the chapter.

13.26 Consider the interconversion of A molecules (red spheres) and B molecules (blue spheres) according to the reaction A ⇌ B. Each of the following series of pictures represents a separate experiment in which time increases from left to right:

Increasing time →

(1)

(2)

(3)

(a) Which of the experiments has resulted in an equilibrium state?
(b) What is the value of the equilibrium constant K_c for the reaction A ⇌ B?

13.28 The reaction A_2 + B ⇌ A + AB has an equilibrium constant K_c = 2. The following pictures represent reaction mixtures that contain A atoms (red), B atoms (blue), and A_2 and AB molecules:

(1)   (2)   (3)

(a) Which reaction mixture is at equilibrium?
(b) For those mixtures that are not at equilibrium, will the reaction go in the forward or reverse direction to reach equilibrium?

13.29 The following pictures represent the initial state and the equilibrium state for the reaction of A_2 molecules (red) with B atoms (blue) to give AB molecules:

Initial state   Equilibrium state

◄ Understanding Key Concepts Problems

Each chapter ends with a set of conceptual questions that encourages you to visualize and think about the actual meaning of key concepts. By solving **Understanding Key Concepts Problems**, you can test your knowledge of fundamental principles before attempting to apply them to more quantitative exercises.

Conceptualize & Solve

Mastering problem-solving techniques is fundamental to the study of chemistry. McMurry and Fay help you learn these essential skills with a wide variety of unique exercises and examples, both conceptual and quantitative.

EXAMPLE 3.13 Stomach acid, a dilute solution of HCl in water, can be neutralized by reaction with sodium hydrogen carbonate, NaHCO₃, according to the equation

$$HCl(aq) + NaHCO_3(aq) \longrightarrow NaCl(aq) + H_2O(l) + CO_2(g)$$

How many milliliters of 0.125 M NaHCO₃ solution are needed to neutralize 18.0 mL of 0.100 M HCl?

BALLPARK SOLUTION The balanced equation shows that HCl and NaHCO₃ react in a 1:1 molar ratio, and we are told that the concentrations of the two solutions are about the same. Thus, the volume of the NaHCO₃ solution must be about the same as that of the HCl solution.

DETAILED SOLUTION Since we need to know the numbers of moles to solve stoichiometry problems, we first have to find how many moles of HCl are in 18.0 mL of a 0.100 M solution by multiplying volume times molarity:

$$\text{Moles of HCl} = 18.0 \text{ mL} \times \frac{1 \text{ L}}{1000 \text{ mL}} \times \frac{0.100 \text{ mol}}{1 \text{ L}} = 1.80 \times 10^{-3} \text{ mol HCl}$$

Next, check the coefficients of the balanced equation to find that each mole of HCl reacts with 1 mol of NaHCO₃, and then calculate how many milliliters of 0.125 M NaHCO₃ solution contains 1.80×10^{-3} mol:

$$1.80 \times 10^{-3} \text{ mol HCl} \times \frac{1 \text{ mol NaHCO}_3}{1 \text{ mol HCl}} \times \frac{1 \text{ L solution}}{0.125 \text{ mol NaHCO}_3} = 0.0144 \text{ L solution}$$

Thus, 14.4 mL of the 0.125 M NaHCO₃ solution is needed to neutralize 18.0 mL of the 0.100 M HCl solution.

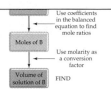

Use coefficients in the balanced equation to find mole ratios

Moles of B

Use molarity as a conversion factor

Volume of solution of B   FIND

FIGURE 3.5 A flow diagram summarizing the use of molarity as a conversion factor between moles and volume in stoichiometry calculations.

◀ Ballpark Solutions

In many examples, the detailed, quantitative solution is preceded by a **Ballpark Solution**—an approximate numerical answer or a qualitative prediction based on conceptual reasoning. This exercise not only provides a check on the reasonableness of your final answers, but also helps you to analyze the problem and devise an effective solution strategy before reaching for a calculator.

Worked Examples ▶

Numerous worked **Examples** allow you to become familiar with the most commonly encountered kinds of problems. Explicit, detailed step-by-step solutions show how to formulate a strategy for solving the **Problem** and arriving at an answer.

Practice Problems ▶

All worked examples are followed by at least one practice **Problem** (usually several) that draws on the same principles. These problems immediately reinforce your grasp of chemical concepts and methods, and provide an opportunity to test your problem-solving skills.

EXAMPLE 3.3 What is the molecular mass of sucrose, C₁₂H₂₂O₁₁? What is the molar mass of sucrose in grams per mole?

SOLUTION The molecular mass of a substance is the sum of the atomic masses of the constituent atoms. First, list the elements present in the molecule, and then look up the atomic mass of each (we'll round off to one decimal place for convenience):

C (12.0 amu); H (1.0 amu); O (16.0 amu)

Next, multiply the atomic mass of each element by the number of times that element appears in the chemical formula, and then total the results.

$$C_{12} (12 \times 12.0 \text{ amu}) = 144.0 \text{ amu}$$
$$H_{22} (22 \times 1.0 \text{ amu}) = 22.0 \text{ amu}$$
$$O_{11} (11 \times 16.0 \text{ amu}) = 176.0 \text{ amu}$$
$$\text{Molec. mass of } C_{12}H_{22}O_{11} = 342.0 \text{ amu}$$

Since one molecule of sucrose has a mass of 342.0 amu, 1 mol of sucrose has a mass of 342.0 grams. Thus, the molar mass of sucrose is 342.0 g/mol.

▶ **PROBLEM 3.4** Calculate the formula mass or molecular mass of the following substances:
(a) Fe₂O₃ (rust) (b) H₂SO₄ (sulfuric acid)
(c) C₆H₈O₇ (citric acid) (d) C₁₆H₁₈N₂O₄S (penicillin G)

▶ **PROBLEM 3.5** The commercial production of iron from iron ore involves the reaction of Fe₂O₃ with CO to yield iron metal plus carbon dioxide:

$$Fe_2O_3(s) + CO(g) \longrightarrow Fe(s) + CO_2(g)$$

Balance the equation, and predict how many moles of CO will react with 0.500 mol of Fe₂O₃.

✦ **KEY CONCEPT PROBLEM 3.6** Methionine, an amino acid used by organisms to make proteins, can be represented by the following ball-and-stick molecular model. Write the formula for methionine, and calculate its molecular mass (red = O, gray = C, blue = N, yellow = S, ivory = H).

Iron is produced commercially by reduction of iron ore with carbon monoxide.

Flowcharts ▶

Commonly used problem-solving procedures are summarized in the form of **flowcharts**. This visual representation makes the logical structure of the solution clearer while helping you to review the specific steps involved.

SOLUTION The problem gives the number of moles of $NaHCO_3$ and asks for a mole-to-mass conversion. First, calculate the formula mass and molar mass of $NaHCO_3$:

$$\text{Form. mass of } NaHCO_3 = 23.0 \text{ amu} + 1.0 \text{ amu} + 12.0 \text{ amu} + (3 \times 16.0 \text{ amu})$$
$$= 84.0 \text{ amu}$$

$$\text{Molar mass of } NaHCO_3 = 84.0 \text{ g/mol}$$

Next, use molar mass as a conversion factor, and set up an equation so that the unwanted unit cancels:

$$0.0626 \text{ mol } NaHCO_3 \times \frac{84.0 \text{ g } NaHCO_3}{1 \text{ mol } NaHCO_3} = 5.26 \text{ g } NaHCO_3$$

EXAMPLE 3.6 Aqueous solutions of sodium hypochlorite (NaOCl), best known as household bleach, are prepared by reaction of sodium hydroxide with chlorine:

$$2 NaOH(aq) + Cl_2(g) \longrightarrow NaOCl(aq) + NaCl(aq) + H_2O(l)$$

How many grams of NaOH are needed to react with 25.0 g of Cl_2?

SOLUTION Finding the relationship between numbers of reactant formula units always requires working in moles. The general strategy was outlined in Figure 3.2 and is reproduced below:

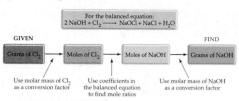

FIGURE 3.2 A summary of conversions between moles and grams for a chemical reaction. The numbers of moles tell how many molecules of each reactant are needed, as given by the balanced equation; the numbers of grams tell how much mass of each reactant is needed.

Multi-Concept Problems

13.111 A 125.4 g quantity of water and an equal molar amount of carbon monoxide were placed in an empty 10.0 L vessel, and the mixture was heated to 700 K. At equilibrium, the partial pressure of CO was 9.80 atm. The reaction is

$$CO(g) + H_2O(g) \rightleftharpoons CO_2(g) + H_2(g)$$

(a) What is the value of K_p at 700 K?
(b) An additional 31.4 g of water was added to the reaction vessel, and a new state of equilibrium was achieved. What are the equilibrium partial pressures of each gas in the mixture? What is the concentration of H_2 in molecules/cm^3?

13.112 A 79.2 g chunk of dry ice (solid CO_2) and 30.0 g of graphite (carbon) were placed in an empty 5.00 L container, and the mixture was heated to achieve equilibrium. The reaction is

$$CO_2(g) + C(s) \rightleftharpoons 2 CO(g)$$

(a) What is the value of K_p at 1000 K if the gas density at 1000 K is 16.3 g/L?
(b) What is the value of K_p at 1100 K if the gas density at 1100 K is 16.9 g/L?
(c) Is the reaction exothermic or endothermic? Explain.

13.113 The amount of carbon dioxide in a gaseous mixture of CO_2 and CO can be determined by passing the gas into an aqueous solution that contains an excess of $Ba(OH)_2$. The CO_2 reacts, yielding a precipitate of $BaCO_3$, but the CO does not react. This method was used to analyze the equilibrium composition of the gas obtained when 1.77 g of CO_2 reacted with 2.0 g of graphite in a 1.000 L container at 1100 K. The analysis yielded 3.41 g of $BaCO_3$. Use these data to calculate K_p at 1100 K for the reaction

$$CO_2(g) + C(s) \rightleftharpoons 2 CO(g)$$

13.114 A 14.58 g quantity of N_2O_4 was placed in a 1.000 L reaction vessel at 400 K. The N_2O_4 decomposed to an equilibrium mixture of N_2O_4 and NO_2 that had a total pressure of 9.15 atm.

(a) What is the value of K_c for the reaction $N_2O_4(g) \rightleftharpoons 2 NO_2(g)$ at 400 K?
(b) How much heat (in kilojoules) was absorbed when the N_2O_4 decomposed to give the equilibrium mixture? (Standard heats of formation may be found in Appendix B.)

◀ Multi-Concept Problems

Appearing at the end of appropriate chapters, **Multi-Concept Problems** connect concepts from the current chapter with those from previous chapters. They serve as an overall review of essential concepts while helping you gain a deeper understanding of how chemistry fits together as an integrated whole.

eMedia Problems ▶

Found at the end of every chapter, these problems are specifically designed to encourage you to use the movies and simulations found on the *Chemistry* Student CD-ROM. These media elements illustrate topics best conveyed in 3-D animations, video demonstrations, or interactive exercises.

 ## eMedia Problems

3.120 The **Balancing Equations** activity (*eChapter 3.1*) allows you to balance three different chemical equations.

(a) In the activity, select the **Decomposition** reaction and balance it.
(b) Rewrite the equation such that ammonium nitrate reacts to yield only elements (N_2, H_2, and O_2) and balance this new equation.
(c) Explain how it is possible for the number of product molecules to be different in parts (a) and (b).
(d) Is it also possible for the number of product atoms to be different in parts (a) and (b)? Explain.

3.121 Calculate the percent composition of each compound in Problem 3.42 (page 111). Use the **Molecular Mass and Percent Mass** activity (*eChapter 3.3*) to verify your answers.

3.122 Using the **Limiting Reagents (Stoichiometry)** simulation (*eChapter 3.6*), perform the redox reaction between magnesium metal and oxygen gas by combining equal masses of both reactants.

(a) Which reactant is the limiting reactant?
(b) Explain how it is possible for there to be a limiting reactant when both are present in equal mass.
(c) Convert the mass you selected to moles for each reactant.
(d) Explain how it is possible for the reactant present in the greatest number of moles to be the limiting reactant.

3.123 Calculate the volume of a solution that

(a) contains 25 g of sulfuric acid and is 0.073 M in sulfuric acid;
(b) contains 40 g of potassium chloride and is 0.101 M in potassium chloride;
(c) contains 17 g of sodium chloride and is 0.646 M in sodium chloride. Use the **Molarity** activity (*eChapter 3.7*) to verify your answers.

3.124 The **Titration** simulation (*eChapter 3.10*) allows you to simulate the titration of five acids, acids A through E. Assuming that the acids can all provide one hydrogen, perform simulated titrations and determine the concentration of all five acids.

Visualize

One of the challenges of studying chemistry is learning to picture atoms and molecules. The carefully designed art program in the Third Edition helps you clearly visualize these abstract concepts.

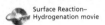 Surface Reaction–
Hydrogenation movie

then move about on the surface until they encounter the C atoms of the adsorbed C_2H_4 molecule. Subsequent stepwise formation of two new C–H bonds gives C_2H_6, which is finally desorbed from the surface.

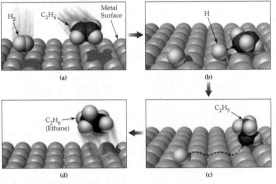

FIGURE 12.17 Proposed mechanism for the catalytic hydrogenation of ethylene (C_2H_4) on a metal surface. **(a)** H_2 and C_2H_4 are adsorbed on the metal surface. **(b)** The H–H bond breaks as H–metal bonds form, and the H atoms move about on the surface. **(c)** One H atom forms a bond to a C atom of the adsorbed C_2H_4 to give a metal-bonded C_2H_5 group. **(d)** A second H atom bonds to the C_2H_5 group, and the resulting C_2H_6 molecule is desorbed from the surface.

Most of the catalysts used in industrial chemical processes are heterogeneous, in part because of the ease with which such catalysts can be separated from the reaction products. Table 12.6 lists some examples of commercial processes that employ heterogeneous catalysts.

◀ Molecular Art

The ability to visualize the three-dimensional structure of molecules is vital to understanding chemistry. Throughout the text, structural formulas are accompanied by newly rendered ball-and-stick and space-filling models that make molecular geometry vividly clear.

Macro-to-Micro Art ▶

Many illustrations in the text are designed to depict processes on the molecular level. Juxtaposed with photos, these **macro-to-micro** illustrations help you to understand the relationship between the seen and unseen chemical worlds. They also make it easier to establish a connection between written descriptions, equations, or graphs and the actual chemical events being represented.

← ENTROPY (S) The amount of randomness, or molecular disorder, in a system.

Entropy has the units J/K (*joules* per kelvin, not kilojoules per kelvin) and is a quantity that can be determined for pure substances, as we'll see in Section 17.5. The larger the value of S, the greater the molecular randomness of the particles in the system. Gases, for example, have more randomness and higher entropy than liquids, and liquids have more randomness and higher entropy than solids.

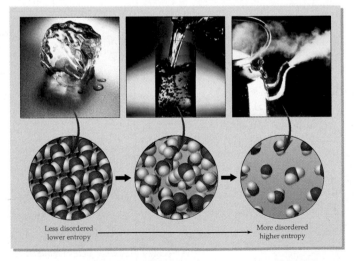

Less disordered
lower entropy

More disordered
higher entropy

A change in entropy is represented as ΔS. When randomness increases, as it does when barium hydroxide octahydrate reacts or ice melts, ΔS has a positive value. The reaction of $Ba(OH)_2 \cdot 8\ H_2O(s)$ with $NH_4Cl(s)$ has $\Delta S° = +428$ J/K, and the melting of ice has $\Delta S° = +22.0$ J/(K · mol). When randomness decreases, ΔS is negative. The freezing of water, for example,

Placement of Art ▶

Art, photographs, and tables are positioned in the text immediately following or adjacent to their text references, making it easier for students to relate text descriptions to visual representations of the same structures or processes.

using an excess amount of one reactant—more than is actually needed according to stoichiometry. Look, for example, at the industrial synthesis of ethylene glycol, $C_2H_6O_2$, a substance used both as automobile antifreeze and as a starting material for the preparation of polyester polymers. More than 2 million tons of ethylene glycol are prepared each year in the United States by reaction of ethylene oxide, C_2H_4O, with water at high temperature:

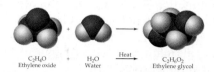

$$C_2H_4O \quad + \quad H_2O \xrightarrow{\text{Heat}} C_2H_6O_2$$
Ethylene oxide   Water   Ethylene glycol

Because water is so cheap and so abundant, it doesn't make sense to worry about using exactly 1 mol of water for each mole of ethylene oxide. It's much easier to use an excess of water to be certain that enough is present to consume entirely the more valuable ethylene oxide reactant. Of course, when an excess of water is present, only the amount required by stoichiometry undergoes reaction. The excess water is only a spectator and is not otherwise involved.

Whenever the ratios of reactant molecules actually used in an experiment are different from those given by the coefficients of the balanced equation, a surplus of one reactant is left over after the reaction is finished. Thus, the extent to which a chemical reaction takes place depends on the reactant that is present in limiting amount—the **limiting reactant**. The other reactant is said to be the *excess reactant*.

Limiting Reactant movie; Limiting Reactant simulation

What happens with excess reactants and limiting reactants is similar to what happens if there are five people in a room but only three chairs. Only three people can sit and the other two stand, because the number of people sitting is limited by the number of available chairs. In the same way, if 5 water molecules come in contact with 3 ethylene oxide molecules, only 3 water molecules can undergo a reaction. The other 2 water molecules are merely spectators, because the number of water molecules that react is limited by the number of available ethylene oxide molecules.

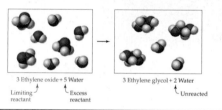

3 Ethylene oxide + 5 Water   3 Ethylene glycol + 2 Water

Limiting reactant   Excess reactant   Unreacted

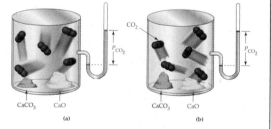

FIGURE 13.3 Thermal decomposition of calcium carbonate: $CaCO_3(s) \rightleftharpoons CaO(s) + CO_2(g)$. At the same temperature, the equilibrium pressure of CO_2 (measured with a closed-end manometer) is the same in **(a)** and **(b)**, independent of how much solid $CaCO_3$ and CaO are present.

CO_2

P_{CO_2}

$CaCO_3$ CaO $CaCO_3$ CaO

(a)   (b)

When liquid mercury is in contact with an aqueous solution that contains $Hg_2(NO_3)_2$ and $Hg(NO_3)_2$, the concentration ratio $[Hg_2^{2+}]/[Hg^{2+}]$ in the aqueous layer (top) has a constant value.

EXAMPLE 13.5 Write the equilibrium equation for each of the following reactions:
(a) $CO_2(g) + C(s) \rightleftharpoons 2\,CO(g)$ **(b)** $Hg(l) + Hg^{2+}(aq) \rightleftharpoons Hg_2^{2+}(aq)$

SOLUTION
(a) Because carbon is a pure solid, its molar concentration is a constant that is incorporated into the equilibrium constant K_c. Therefore,

$$K_c = \frac{[CO]^2}{[CO_2]}$$

Alternatively, because CO and CO_2 are gases, the equilibrium equation can be written using partial pressures:

$$K_p = \frac{(P_{CO})^2}{P_{CO_2}}$$

The relationship between K_p and K_c is $K_p = K_c(RT)^{\Delta n}$, where $\Delta n = 2 - 1 = 1$.
(b) The concentrations of mercury(I) and mercury(II) ions appear in the equilibrium equation, but the concentration of mercury metal is omitted because, as a pure liquid, its concentration is a constant. Therefore,

$$K_c = \frac{[Hg_2^{2+}]}{[Hg^{2+}]}$$

Media Resources for Students

Chemistry is as much a visual, abstract science as it is a mathematical, quantitative one. Prentice Hall offers an array of highly visual tools that transform theoretical concepts into clear mental pictures while reinforcing quantitative applications. These tools are tightly connected to the text and to each other, allowing you to focus on learning chemistry, not correlating and navigating learning tools.

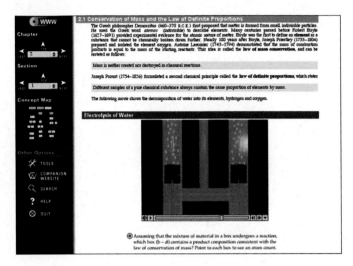

◀ *Chemistry* Student CD-ROM
(0-13-088529-0)

This student CD-ROM, **free** with each new student edition of McMurry/Fay's *Chemistry, Third Edition*, underscores the key chemical concepts found in the text. In context, the *Chemistry* **Student CD-ROM** features rich media elements that explain the material in an engaging, dynamic, and interactive way. It includes simulations, manipulable 3-D molecules, narrated animations, and video clips. From the *Chemistry* **Student CD-ROM**, you can connect directly to the Companion Website at www.prenhall.com/mcmurry.

Chemistry Companion Website ▶
www.prenhall.com/mcmurry

The *Chemistry* **Companion Website** is keyed chapter by chapter to the textbook. On the site, you can explore Key Concept Problems, take interactive quizzes and receive immediate feedback, link to other chemistry-related websites, and access updates and articles from the popular press.

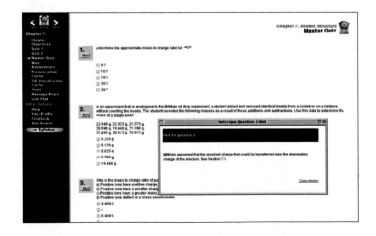

About the Authors

John McMurry *(left)*, educated at Harvard and Columbia, has taught approximately 17,000 students in general and organic chemistry over a 30-year period. A Professor of Chemistry at Cornell University since 1980, Dr. McMurry previously spent 13 years on the faculty at the University of California at Santa Cruz. He has received numerous awards, including the Alfred P. Sloan Fellowship (1969–71), the National Institute of Health Career Development Award (1975–80), the Alexander von Humboldt Senior Scientist Award (1986–87), and the Max Planck Research Award (1991).

Robert C. Fay *(right)*, Professor of Chemistry at Cornell University, has been teaching general and inorganic chemistry at Cornell since 1962. Known for his clear, well-organized lectures, Dr. Fay was the 1980 recipient of the Clark Distinguished Teaching Award. He has also taught as a visiting professor at Harvard University and at the University of Bologna (Italy). A Phi Beta Kappa graduate of Oberlin College, Fay received his Ph.D. from the University of Illinois. He has been an NSF Science Faculty Fellow at the University of East Anglia and the University of Sussex (England) and a NATO/Heineman Senior Fellow at Oxford University.

Chemistry: Matter and Measurement

In both chemistry and daily life, unexpected consequences can result from improperly made measurements.

L ife has changed more in the past two centuries than in all the previously recorded span of human history. The earth's population has increased more than fivefold since 1800, and life expectancy has nearly doubled because of our ability to synthesize medicines, control diseases, and increase crop yields. Methods of transportation have changed from horses and buggies to automobiles and airplanes because of our ability to harness the energy in petroleum. Many goods are now made of polymers and ceramics instead of wood and metal because of our ability to manufacture materials with properties unlike any found in nature.

In one way or another, all these changes involve **chemistry**, the study of the composition, properties, and transformations of matter. Chemistry is deeply involved in both the changes that take place in nature and the profound social changes of the past two centuries. In addition, chemistry is central to the current revolution in molecular biology that is now exploring the details of how life is genetically controlled. No educated person today can understand the modern world without a basic knowledge of chemistry.

1.1   Approaching Chemistry: Experimentation

By opening this book, you have already decided that you need to know more about chemistry. Perhaps you want to learn how medicines are made, how fertilizers and pesticides work, how living organisms function, how new high-temperature ceramics are used in space vehicles, or how microelectronic circuits are etched onto silicon chips. How do you approach chemistry?

Modern computer chips are made by chemical etching of silicon wafers, using "masks" to produce the tiny circuitry.

One way to approach chemistry is to look at what you see around you and try to think of logical explanations for what you see. You would certainly observe, for instance, that different substances have different forms and appearances: Some substances are gases, some are liquids, and some are solids; some are hard and shiny, but others are soft and dull. You'd also observe that different substances behave differently: Iron rusts but gold does not; copper conducts electricity but sulfur doesn't. How can these and a vast number of other observations be explained?

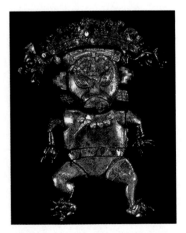

Gold, one of the most valuable of elements, has been prized since antiquity for its beauty and resistance to corrosion.

Iron, though widely used as a structural material, corrodes easily.

Clearly, the natural world is too complex to be understood by looking and thinking alone; a more active approach is needed. Specific questions must be asked, and experiments must be carried out to find their answers. Only when the results of many experiments are known can we then apply logic to devise an interpretation, or *hypothesis*, that explains the results. The hypothesis, in turn, can be used to make more predictions and to suggest more experiments until a consistent explanation, or **theory**, of known observations is finally arrived at.

It's important to keep in mind as you study chemistry or any other science that scientific theories are not laws of nature. All they do is represent the best explanations of experimental results that we can come up with at the present time. Some currently accepted theories will eventually be modified, and others may be replaced altogether if new experiments uncover results that present theories can't explain.

1.2 Chemistry and the Elements

Everything you see around you is formed from one or more of 115 presently known elements. An **element** is a fundamental substance that can't be chemically changed or broken down into anything simpler. Silver, mercury, and sulfur are common examples, as listed in Table 1.1. Note that the names in parentheses in the table are the Latin names from which the symbols for the elements are derived.

Samples of mercury, silver, and sulfur (clockwise from top left).

TABLE 1.1   Names of Some Common Elements and Their Symbols

Aluminum	**Al**	Chlorine	**Cl**	Manganese	**Mn**	Copper (*cuprum*)	**Cu**
Argon	**Ar**	Fluorine	**F**	Nitrogen	**N**	Iron (*ferrum*)	**Fe**
Barium	**Ba**	Helium	**He**	Oxygen	**O**	Lead (*plumbum*)	**Pb**
Boron	**B**	Hydrogen	**H**	Phosphorus	**P**	Mercury (*hydrargyrum*)	**Hg**
Bromine	**Br**	Iodine	**I**	Silicon	**Si**	Potassium (*kalium*)	**K**
Calcium	**Ca**	Lithium	**Li**	Sulfur	**S**	Silver (*argentum*)	**Ag**
Carbon	**C**	Magnesium	**Mg**	Zinc	**Zn**	Sodium (*natrium*)	**Na**

Only about 90 of the 115 presently known elements occur naturally. The remaining ones have been produced artificially by nuclear chemists using high-energy particle accelerators. Furthermore, not all of the 90 or so naturally occurring elements are equally abundant. Hydrogen is thought to account for approximately 75% of the mass in the universe; oxygen and silicon together account for 75% of the earth's crust; and oxygen, carbon, and hydrogen make up more than 90% of the human body (Figure 1.1). By contrast, there is probably less than 20 grams of the element francium (Fr) dispersed over the entire earth at any one time. Francium is an unstable radioactive element, atoms of which are continually being formed and destroyed in natural radiochemical processes. (We'll discuss radioactivity in Chapter 22.)

Chemists refer to specific elements using a shorthand of one- or two-letter symbols. As shown by the examples in Table 1.1, the first letter of an element's symbol is always capitalized, and the second letter, if any, is lowercase. Many of the symbols are just the first one or two letters of the element's English name: H = hydrogen, C = carbon, Al = aluminum, and so forth.

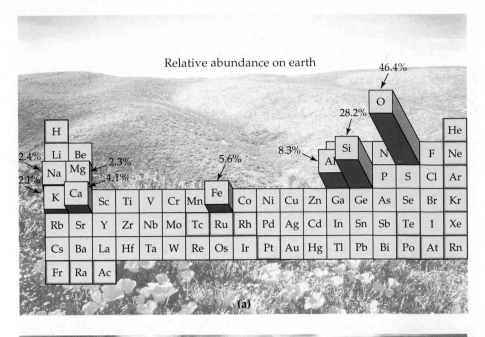

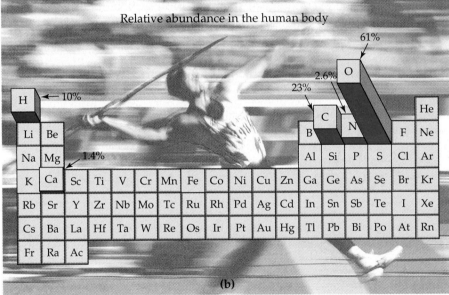

FIGURE 1.1 Estimated elemental composition (by mass percent) of **(a)** the earth's crust and **(b)** the human body. Oxygen is the most abundant element in both. Only the major constituents are shown in each case; small amounts of many other elements are also present.

Other symbols derive from Latin or other languages: Na = sodium (Latin, *natrium*), Pb = lead (Latin, *plumbum*), W = tungsten (German, *wolfram*). The names, symbols, and other information about all 115 elements are given inside the front cover of this book, organized in a form called the *periodic table*.

▶ **PROBLEM 1.1**   Look at the alphabetical list of elements inside the front cover, and find the symbols for the following elements:
(a) Copper (used in electrical wires)
(b) Platinum (used in automobile emission control devices)
(c) Plutonium (used in nuclear weapons)

▶**PROBLEM 1.2**   Look at the alphabetical list of elements inside the front cover, and tell what elements the following symbols represent:
(a) Ag   **(b)** Rh   **(c)** Re   **(d)** Cs   **(e)** Ar   **(f)** As

1.3   Elements and the Periodic Table

Ten elements have been known since the beginning of recorded history: antimony (Sb), carbon (C), copper (Cu), gold (Au), iron (Fe), lead (Pb), mercury (Hg), silver (Ag), sulfur (S), and tin (Sn). The first "new" element to be found in several thousand years was arsenic (As), discovered in about 1250. In fact, only 24 elements were known at the time of the American Revolution in 1776.

The first tabulation of the "chemically simple" substances that we today call elements appeared in a treatise published in 1789 by the French scientist Antoine Lavoisier. As the pace of discovery quickened in the late 1700s and early 1800s, chemists began to look for similarities among elements that might allow general conclusions to be drawn. Particularly important among the early successes was Johann Döbereiner's observation in 1829 that several *triads*, or groups of three elements, appeared to behave similarly. Calcium (Ca), strontium (Sr), and barium (Ba) form one such triad; chlorine (Cl), bromine (Br), and iodine (I) form another; and lithium (Li), sodium (Na), and potassium (K) form a third. By 1843, sixteen such triads were known, and chemists had begun to search for an explanation.

Numerous attempts were made in the mid-1800s to account for the similarities among groups of elements, but the great breakthrough came in 1869 when the Russian chemist Dmitri Mendeleev published the forerunner of the modern **periodic table**, shown in Figure 1.2. In this modern version, elements are placed on a grid that has 7 horizontal rows, called **periods**, and 18 vertical columns, called **groups**. When organized in this way, *the elements in a given group have similar chemical properties*. Lithium, sodium, potassium, and the other metallic elements in group 1A behave similarly. Beryllium, magnesium, calcium, and the other elements in group 2A behave similarly. Fluorine, chlorine, bromine, and the other elements in group 7A behave similarly, and so on throughout the table.

Although the overall form of the modern periodic table is now well established, chemists in different countries have historically used different conventions for labeling the columns. To resolve these difficulties, a newly adopted standard calls for numbering the columns from 1 to 18 going from left to right. This new standard has not yet found complete acceptance, however, and we'll continue to use the U.S. system of numbers and capital letters— group 7A instead of group 17, for example. Labels for the new system are shown in smaller type in Figure 1.2.

One further note: There are actually *32* groups in the table rather than 18, but to make the table fit manageably on a page, the 14 elements following lanthanum (the *lanthanides*) and the 14 following actinium (the *actinides*) are pulled out and shown below the others. These groups are not numbered.

We'll see repeatedly throughout this book that the periodic table of the elements is the most important organizing principle in chemistry. The time you take now to familiarize yourself with the layout and organization of the periodic table will pay you back later on. Notice in Figure 1.2, for example, that there is a regular progression in the size of the seven periods (rows). The first

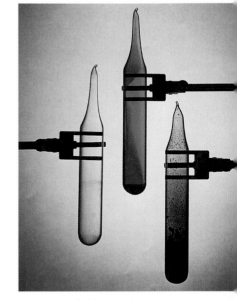

Samples of chlorine, bromine, and iodine, one of Döbereiner's triads of elements with similar chemical properties.

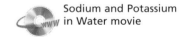

Sodium and Potassium in Water movie

1 H																	2 He
3 Li	4 Be											5 B	6 C	7 N	8 O	9 F	10 Ne
11 Na	12 Mg											13 Al	14 Si	15 P	16 S	17 Cl	18 Ar
19 K	20 Ca	21 Sc	22 Ti	23 V	24 Cr	25 Mn	26 Fe	27 Co	28 Ni	29 Cu	30 Zn	31 Ga	32 Ge	33 As	34 Se	35 Br	36 Kr
37 Rb	38 Sr	39 Y	40 Zr	41 Nb	42 Mo	43 Tc	44 Ru	45 Rh	46 Pd	47 Ag	48 Cd	49 In	50 Sn	51 Sb	52 Te	53 I	54 Xe
55 Cs	56 Ba	57 La	72 Hf	73 Ta	74 W	75 Re	76 Os	77 Ir	78 Pt	79 Au	80 Hg	81 Tl	82 Pb	83 Bi	84 Po	85 At	86 Rn
87 Fr	88 Ra	89 Ac	104 Rf	105 Db	106 Sg	107 Bh	108 Hs	109 Mt	110	111	112		114		116		118

58 Ce	59 Pr	60 Nd	61 Pm	62 Sm	63 Eu	64 Gd	65 Tb	66 Dy	67 Ho	68 Er	69 Tm	70 Yb	71 Lu
90 Th	91 Pa	92 U	93 Np	94 Pu	95 Am	96 Cm	97 Bk	98 Cf	99 Es	100 Fm	101 Md	102 No	103 Lr

FIGURE 1.2   The modern form of the periodic table. Each element is identified by a one- or two-letter symbol and is characterized by an *atomic number*. The table begins with hydrogen (H, atomic number 1) in the upper left-hand corner and continues to the yet unnamed element with atomic number 118. The 14 elements following lanthanum (La, atomic number 57) and the 14 elements following actinium (Ac, atomic number 89) are pulled out and shown below the others.

Interactive Periodic
Table

period has only 2 elements, hydrogen (H) and helium (He); the second and third periods have 8 elements each; the fourth and fifth periods have 18 elements each; and the sixth and (incomplete) seventh periods, which include the lanthanides and actinides, have 32 elements each. We'll see in Chapter 5 that this regular progression in the periodic table reflects a similar regularity in the structure of atoms.

Notice also that not all groups (columns) in the periodic table have the same number of elements. The two larger groups on the left and the six larger groups on the right of the table are called the **main groups**; the ten smaller ones in the middle of the table are called the **transition metal groups**; and the fourteen shown separately at the bottom of the table are called the **inner transition metal groups**.

1.4 Some Chemical Properties of the Elements

Any characteristic that can be used to describe or identify matter is called a **property**. Size, amount, odor, color, and temperature are all well-known examples. Still other properties include such characteristics as melting point, solubility, and chemical behavior. For example, we might list some properties of sodium chloride (table salt) by saying that it melts at 1474°F (or 801°C), dissolves in water, and undergoes a chemical reaction when it comes into contact with a silver nitrate solution.

Properties can be classified as either *intensive* or *extensive*, depending on whether their value changes with the size of the sample. **Intensive properties**, like temperature and melting point, have values that do not depend on the amount of sample. Thus, a small ice cube might have the same temperature as a massive iceberg. **Extensive properties**, like length and volume, have values that *do* depend on the sample size. An ice cube is much smaller than an iceberg.

Properties can also be classified as either *physical* or *chemical* (Table 1.2). **Physical properties** are those characteristics like color, amount, melting point, and temperature that can be determined without changing the chemical makeup of the sample. **Chemical properties** are those that *do* change the chemical makeup of the sample. For example, the rusting that occurs when a bicycle is left out in the rain is due to the chemical combination of oxygen with iron to give the new substance iron oxide. Rusting is therefore a chemical property of iron.

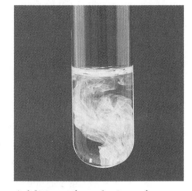

Addition of a solution of sodium chloride to a solution of silver nitrate yields a white precipitate of solid silver chloride.

TABLE 1.2   Some Examples of Physical and Chemical Properties

Physical Properties		Chemical Properties
Temperature	Amount	Rusting (of iron)
Color	Odor	Combustion (of coal)
Melting point	Solubility	Tarnishing (of silver)
Electrical conductivity	Hardness	Hardening (of cement)

As noted in the previous section, groups of elements in the periodic table often show remarkable similarities in their chemical properties. Look at the following groups, for instance, to see some examples of chemical properties:

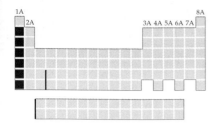

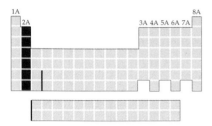

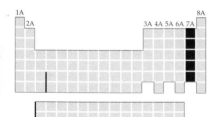

Sodium and Potassium in Water movie
Physical Properties of the Halogens movie

- *Group 1A—Alkali metals:* Lithium (Li), sodium (Na), potassium (K), rubidium (Rb), and cesium (Cs) are shiny, soft metals. All react rapidly (often violently) with water to form products that are highly alkaline, or basic—hence the name *alkali metals*. Because of their high reactivity, the alkali metals are never found in nature in the pure state but only in combination with other elements.

    Note that hydrogen (H) is placed in group 1A even though, as a colorless gas, it is completely different in appearance and behavior from the alkali metals. We'll see the reason for this classification in Section 5.13.

- *Group 2A—Alkaline earth metals:* Beryllium (Be), magnesium (Mg), calcium (Ca), strontium (Sr), barium (Ba), and radium (Ra) are also lustrous, silvery metals, but are less reactive than their neighbors in group 1A. Like the alkali metals, the alkaline earths are never found in nature in the pure state.

Sodium, one of the alkali metals, reacts violently with water to yield hydrogen gas and an alkaline (basic) solution.

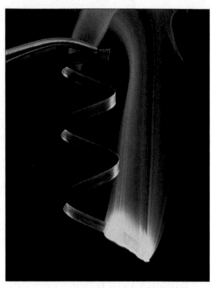

Magnesium, one of the alkaline earth metals, burns in air.

- *Group 7A—Halogens:* Fluorine (F), chlorine (Cl), bromine (Br), and iodine (I), are colorful, corrosive nonmetals. They are found in nature only in combination with other elements, such as with sodium in table salt (sodium chloride, NaCl). In fact, the group name *halogen* is taken from the Greek word *hals*, meaning salt.

Bromine, a halogen, is a corrosive dark red liquid at room temperature.

- **Group 8A—*Noble gases:*** Helium (He), neon (Ne), argon (Ar), krypton (Kr), xenon (Xe), and radon (Rn) are gases of very low reactivity. Helium, neon, and argon don't combine with any other element; krypton and xenon combine with very few.

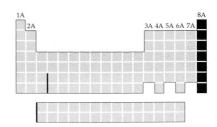

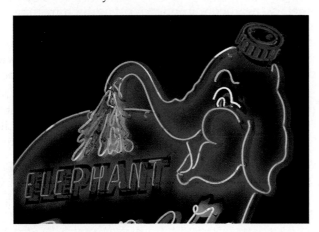

Neon, one of the noble gases, is used in neon lights and signs.

Although the resemblances aren't as pronounced as they are within a single group, *neighboring* groups of elements also behave similarly in some ways. Thus, as indicated in Figure 1.2, the periodic table is often divided into three major classes of elements—metals, nonmetals, and semimetals:

- ***Metals:*** Metals, the largest category of elements, are found on the left side of the periodic table, bounded on the right by a zigzag line running from boron (B) at the top to astatine (At) at the bottom. The metals are easy to characterize by their appearance: All except mercury are solid at room temperature, and most have the silvery shine we normally associate with metals. In addition, metals are generally malleable rather than brittle, can be twisted and drawn into wires without breaking, and are good conductors of heat and electricity.

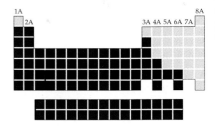

- ***Nonmetals:*** Nonmetals are found on the right side of the periodic table and, like metals, are easy to characterize by their appearance. Eleven of the nonmetals are gases, one is a liquid (bromine), and only five are solids at room temperature (carbon, phosphorus, sulfur, selenium, and iodine). None are silvery in appearance, and several are brightly colored. The solid nonmetals are brittle rather than malleable, and they are poor conductors of heat and electricity.

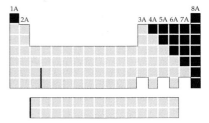

Metals, such as the copper, lead, and aluminum shown here, are solids that conduct electricity and can be formed into wires.

Phosphorus, iodine, and sulfur (clockwise from top left) are typical nonmetals. All are brittle, and none conduct electricity.

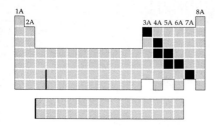

Periodic Table Groups activity

• *Semimetals:* Seven of the nine elements adjacent to the zigzag boundary between metals and nonmetals—boron, silicon, germanium, arsenic, antimony, tellurium, and astatine—are known as semimetals, or *metalloids*, because their properties are intermediate between those of their metallic and nonmetallic neighbors. Though most are silvery in appearance and all are solid at room temperature, semimetals are brittle rather than malleable and tend to be poor conductors of heat and electricity. Silicon, for example, is a widely used *semiconductor*, a substance whose electrical conductivity is intermediate between that of a metal and an insulator.

▶ **PROBLEM 1.3**   Identify the following elements as metals, nonmetals, or semimetals:
(a) Ti   **(b)** Te   **(c)** Se   **(d)** Sc   **(e)** At   **(f)** Ar

➤ **KEY CONCEPT PROBLEM 1.4**   The three so-called "coinage metals" are located near the middle of the periodic table. Use the periodic table to identify them.

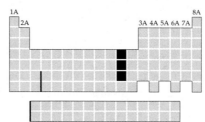

1.5   Experimentation and Measurement

Chemistry is an experimental science. But if our experiments are to be reproducible, we must be able to describe fully the substances we're working with—their amounts, volumes, temperatures, and so forth. Thus, one of the most important requirements in chemistry is that we have a way to *measure* things.

Under an international agreement concluded in 1960, scientists throughout the world now use the **International System of Units** for measurement, abbreviated **SI** for the French *Système Internationale d'Unités*. Based on the well-known metric system, which is used in all industrialized countries of the world except the United States, the SI system has seven fundamental units (Table 1.3). These seven fundamental units, along with others derived from them, suffice for all scientific measurements. We'll look at three of the most common units in this chapter—those for mass, length, and temperature—and will discuss others as the need arises in later chapters.

TABLE 1.3   The Seven Fundamental SI Units of Measure

Physical Quantity	Name of Unit	Abbreviation
Mass	kilogram	kg
Length	meter	m
Temperature	kelvin	K
Amount of substance	mole	mol
Time	second	s
Electric current	ampere	A
Luminous intensity	candela	cd

One problem with any system of measurement is that the sizes of the units sometimes turn out to be inconveniently large or small. For example, a chemist describing the diameter of a sodium atom (0.000 000 000 372 m) would find the meter (m) to be an inconveniently large unit, but an astronomer measuring the average distance from the earth to the sun (150,000,000,000 m) would find the meter to be inconveniently small. For this reason, SI units are modified through the use of prefixes when they refer to either smaller or larger quantities. Thus, the prefix *milli-* means one-thousandth, and a *milli*meter (mm) is 1/1000 of 1 meter. Similarly, the prefix *kilo-* means one thousand, and a *kilo*meter (km) is 1000 meters. (Note that the SI unit for mass (kilogram) already contains the *kilo-* prefix.) A list of prefixes is shown in Table 1.4, with the most commonly used ones given in color.

TABLE 1.4   Some Prefixes for Multiples of SI Units

Factor	Prefix	Symbol	Example
$1{,}000{,}000{,}000 = 10^9$	giga	G	1 gigameter (Gm) $= 10^9$ m
$1{,}000{,}000 = 10^6$	mega	M	1 megameter (Mm) $= 10^6$ m
$1{,}000 = 10^3$	kilo	k	1 kilogram (kg) $= 10^3$ g
$100 = 10^2$	hecto	h	1 hectogram (hg) $= 100$ g
$10 = 10^1$	deka	da	1 dekagram (dag) $= 10$ g
$0.1 = 10^{-1}$	deci	d	1 decimeter (dm) $= 0.1$ m
$0.01 = 10^{-2}$	centi	c	1 centimeter (cm) $= 0.01$ m
$0.001 = 10^{-3}$	milli	m	1 milligram (mg) $= 0.001$ g
*$0.000\ 001 = 10^{-6}$	micro	μ	1 micrometer (μm) $= 10^{-6}$ m
*$0.000\ 000\ 001 = 10^{-9}$	nano	n	1 nanosecond (ns) $= 10^{-9}$ s
*$0.000\ 000\ 000\ 001 = 10^{-12}$	pico	p	1 picosecond (ps) $= 10^{-12}$ s

*For very small numbers, it is becoming common in scientific work to leave a thin space every three digits to the right of the decimal point.

Notice how numbers that are either very large or very small are indicated in Table 1.4 using an exponential format called **scientific notation**. For example, the number 55,000 is written in scientific notation as 5.5×10^4, and the number 0.003 20 as 3.20×10^{-3}. You might want to review Appendix A if you are uncomfortable with scientific notation or if you need to brush up on how to do mathematical manipulations on numbers with exponents.

Note also that all measurements contain both a number and a unit label. A number alone is not much good without a unit to define it. If you asked a friend how far it was to the nearest tennis court, the answer "3" alone wouldn't tell you much. 3 blocks? 3 kilometers? 3 miles?

▶**PROBLEM 1.5**   Express the following quantities in scientific notation:
(a) The diameter of a sodium atom, 0.000 000 000 372 m
(b) The distance from the earth to the sun, 150,000,000,000 m

▶**PROBLEM 1.6**   What units do the following abbreviations represent?
(a) μg   (b) dm   (c) ps   (d) kA   (e) mmol

1.6   Measuring Mass

Mass is defined as the amount of *matter* in an object. **Matter**, in turn, is a catchall term used to describe anything with a physical presence—anything you can touch, taste, or smell. (Stated more scientifically, matter is anything that has mass.) Mass is measured in SI units by the **kilogram** (**kg**; 1 kg = 2.205 U.S. lb). Because the kilogram is too large for many purposes in chemistry, the familiar metric **gram** (**g**; 1 g = 0.001 kg), the **milligram** (**mg**, 1 mg = 0.001 g = 10^{-6} kg), and the **microgram** (**μg**; 1 μg = 0.001 mg = 10^{-6} g = 10^{-9} kg) are more commonly used. One gram is about half the mass of a U.S. penny.

$$1 \text{ kg} = 1,000 \text{ g} = 1,000,000 \text{ mg} = 1,000,000,000 \text{ μg} \quad (2.205 \text{ lb})$$

$$1 \text{ g} = 1,000 \text{ mg} = 1,000,000 \text{ μg} \quad (0.035\ 27 \text{ oz})$$

$$1 \text{ mg} = 1,000 \text{ μg}$$

This pile of 400 pennies weighs about 1 kg.

The standard kilogram, against which all other masses are compared, is defined as the mass of a cylindrical bar of platinum–iridium alloy stored in a vault in a suburb of Paris, France. There are 40 copies of this bar distributed throughout the world, with two (Numbers 4 and 20) now stored at the U.S. National Institute of Standards and Technology near Washington, D.C.

The terms "mass" and "weight," though often used interchangeably, have quite different meanings. *Mass* is a physical property that measures the amount of matter in an object, whereas *weight* measures the pull of gravity on an object by the earth or other celestial body. Clearly, the amount of matter in an object is independent of its location. Your body has the same mass whether you're on earth or on the moon. Just as clearly, though, the weight of an object *does* depend on its location. You might weigh 140 lb on earth, but you would weigh only about 23 lb on the moon, which has a lower gravity than the earth.

At the same location on earth, two objects with identical masses have identical weights; that is, the objects experience an identical pull of the earth's gravity. Thus, the *mass* of an object can be measured by comparing the *weight* of the object to the weight of a reference standard of known mass. Much of the confusion between mass and weight is simply due to a language problem: We speak of "weighing" when we really mean that we are measuring mass by comparing two weights. Figure 1.3 shows two types of balances normally used for measuring mass in the laboratory.

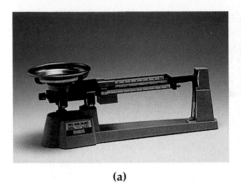

(a)

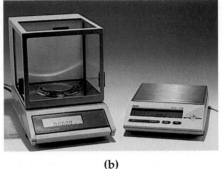

(b)

FIGURE 1.3 **(a)** The single-pan balance has a sliding counterweight that is adjusted until the weight of the object on the pan is just balanced. **(b)** Modern electronic balances.

1.7 Measuring Length

The **meter (m)** is the standard unit of length in the SI system. Although originally defined in 1790 as being 1 ten-millionth of the distance from the equator to the North Pole, the meter was redefined in 1889 as the distance between two thin lines on a bar of platinum–iridium alloy stored outside Paris, France. To accommodate an increasing need for precision, the meter was redefined again in 1983 as equal to the distance traveled by light through a vacuum in 1/299,792,458 second. Although this new definition isn't as easy to grasp as the distance between two scratches on a bar, it has the great advantage that it can't be lost or damaged.

One meter is 39.37 inches, about 10% longer than an English yard and much too large for most measurements in chemistry. Other, more commonly used measures of length are the **centimeter (cm;** 1 cm = 0.01 m, a bit less than half an inch), the **millimeter (mm;** 1 mm = 0.001 m, about the thickness of a U.S. dime), the **micrometer (μm;** 1 μm = 10^{-6} m), the **nanometer (nm;** 1 nm = 10^{-9} m), and the **picometer (pm;** 1 pm = 10^{-12} m). Thus, a chemist might refer to the diameter of a sodium atom as 372 pm (3.72×10^{-10} m).

$$1 \text{ m} = 100 \text{ cm} = 1{,}000 \text{ mm} = 1{,}000{,}000 \text{ } \mu\text{m} = 1{,}000{,}000{,}000 \text{ nm} \qquad (1.0936 \text{ yd})$$

$$1 \text{ cm} = 10 \text{ mm} = 10{,}000 \text{ } \mu\text{m} = 10{,}000{,}000 \text{ nm} \qquad (0.3937 \text{ in.})$$

$$1 \text{ mm} = 1{,}000 \text{ } \mu\text{m} = 1{,}000{,}000 \text{ nm}$$

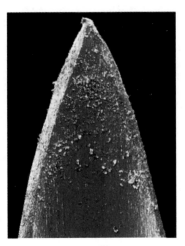

The length of these bacteria on the tip of a pin is about 5×10^{-7} m.

1.8 Measuring Temperature

Just as the kilogram and the meter are slowly replacing the pound and the yard as common units for mass and length measurement in the United States, the **Celsius degree (°C)** is slowly replacing the Fahrenheit degree (°F) as the common unit for temperature measurement. In scientific work, however, the **kelvin (K)** is replacing both. (Note that we say only "kelvin," not "kelvin degree.")

For all practical purposes, the kelvin and the Celsius degree are the same—both are one hundredth of the interval between the freezing point of water and the boiling point of water at atmospheric pressure. The only real difference between the two units is that the numbers assigned to various points on the scales differ. Whereas the Celsius scale assigns a value of 0°C to the freezing point of water and 100°C to the boiling point of water, the Kelvin scale assigns a value of 0 K to the coldest possible temperature, −273.15°C, sometimes called *absolute zero*. Thus, 0 K = −273.15°C and 273.15 K = 0°C. For example, a warm spring day with a Celsius temperature of 25°C has a Kelvin temperature of 25 + 273.15 = 298 K.

$$\text{Temperature in K} = \text{Temperature in °C} + 273.15$$

$$\text{Temperature in °C} = \text{Temperature in K} - 273.15$$

In contrast to the Kelvin and Celsius scales, the common *Fahrenheit scale* specifies an interval of 180° between the freezing point (32°F) and the boiling point (212°F) of water. Thus, it takes 180 Fahrenheit degrees to cover the same range as 100 Celsius degrees (or kelvins), and a Fahrenheit degree is therefore only 100/180 = 5/9 as large as a Celsius degree. Figure 1.4 gives a comparison of the Kelvin, Celsius, and Fahrenheit scales.

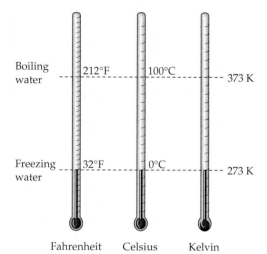

FIGURE 1.4 A comparison of the Kelvin, Celsius, and Fahrenheit temperature scales. One Fahrenheit degree is 5/9 the size of a kelvin or Celsius degree.

Two adjustments need to be made to convert between Fahrenheit and Celsius scales—one to account for the difference in degree size and one to account for the difference in zero point. The size adjustment is made by remembering that a Celsius degree is 9/5 the size of a Fahrenheit degree (or, conversely, that a Fahrenheit degree is 5/9 the size of a Celsius degree). The zero-point adjustment is made by remembering that the freezing point of water is higher by 32 on the Fahrenheit scale than it is on the Celsius scale. Thus, if you want to convert from Celsius to Fahrenheit, you do a size adjustment (multiply °C by 9/5) and then a zero-point adjustment (add 32); if you want to convert from Fahrenheit to Celsius, you find out how many Fahrenheit degrees there are above freezing (by subtracting 32) and then do a size adjustment (multiply by 5/9). The following formulas describe the conversions, and Example 1.1 shows how a conversion can be done.

$$°F = \left(\frac{9°F}{5°C}\right)(°C) + 32°F \qquad °C = \left(\frac{5°C}{9°F}\right)(°F - 32°F)$$

EXAMPLE 1.1 The melting point of table salt is 1474°F. What temperature is this on the Celsius and Kelvin scales?

SOLUTION There are two ways to do this and every other problem in chemistry. One is to think things through to be sure you understand what's going on;

the other is to blindly apply a formula. The thinking approach *always* works; the formula approach works only if you use the right equation. Let's try both ways.

The thinking approach: We're given a temperature in Fahrenheit degrees, and we need to convert to Celsius degrees. Now, a temperature of 1474°F corresponds to (1474 − 32) = 1442 Fahrenheit degrees above the freezing point of water. Since each Fahrenheit degree is only 5/9 as large as a Celsius degree, 1442 Fahrenheit degrees above freezing is equal to (1442 × 5/9) = 801 Celsius degrees above freezing (0°C), or 801°C. The same number of degrees above freezing on the Kelvin scale (273.15 K) corresponds to a temperature of 273.15 + 801 = 1074 K.

The formula approach: Set up an equation using the temperature conversion formula for changing from Fahrenheit to Celsius:

$$°C = \left(\frac{5°C}{9°\cancel{F}} \right)(1474°\cancel{F} - 32°\cancel{F}) = 801°C$$

The melting point of sodium chloride is 1474°F, or 801°C.

Converting to Kelvin gives a temperature of 801° + 273.15° = 1074 K.

Since the answers obtained by the two approaches agree, we can feel confident that our thinking is following the right lines and that we understand the material. (If the answers *did not* agree, we'd be alerted to a misunderstanding somewhere.)

▶ **PROBLEM 1.7** The normal body temperature of a healthy adult is 98.6°F. Express this value on both Celsius and Kelvin scales.

▶ **PROBLEM 1.8** Carry out the indicated temperature conversions.
(a) −78°C = ? K (b) 158°C = ?°F (c) 375 K = ?°F

1.9 Derived Units: Measuring Volume

Look back at the seven fundamental SI units given in Table 1.3 and you'll find that measures for such familiar quantities as area, volume, density, speed, and pressure are missing. All are examples of *derived* quantities rather than fundamental quantities because they can be expressed using one or more of the seven base units (Table 1.5).

TABLE 1.5 Some Derived Quantities

Quantity	Definition	Derived Unit (Name)
Area	Length times length	m^2
Volume	Area times length	m^3
Density	Mass per unit volume	kg/m^3
Speed	Distance per unit time	m/s
Acceleration	Change in speed per unit time	m/s^2
Force	Mass times acceleration	$(kg \cdot m)/s^2$ (newton, N)
Pressure	Force per unit area	$kg/(m \cdot s^2)$ (pascal, Pa)
Energy	Force times distance	$(kg \cdot m^2)/s^2$ (joule, J)

Volume, the amount of space occupied by an object, is measured in SI units by the **cubic meter (m^3)**, defined as the amount of space occupied by a cube 1 meter long on each edge (Figure 1.5).

The cubic meter is equivalent to 264.2 U.S. gallons—much too large a quantity for normal use in chemistry. As a result, smaller, more convenient measures are commonly employed. Both the **cubic decimeter** ($1\ dm^3 = 0.001\ m^3$), equal in size to the more familiar metric **liter (L)**, and the **cubic centimeter** ($1\ cm^3 = 0.001\ dm^3 = 10^{-6}\ m^3$), equal in size to the metric **milliliter (mL)**, are

FIGURE 1.5 A cubic meter is the volume of a cube 1 meter along each edge. Each cubic meter contains 1000 cubic decimeters (liters), and each cubic decimeter contains 1000 cubic centimeters (milliliters).

$$1\ m^3 = 1000\ dm^3$$
$$1\ dm^3 = 1\ L$$
$$= 1000\ cm^3$$
$$1\ cm^3 = 1\ mL$$

particularly convenient. Slightly larger than 1 U.S. quart, a liter has the volume of a cube 1 dm on edge. Similarly, a milliliter has the volume of a cube 1 cm on edge (Figure 1.5).

$$1\ m^3 = 1{,}000\ dm^3 = 1{,}000{,}000\ cm^3 \qquad (264.2\ gal)$$
$$1\ dm^3 = 1\ L = 1{,}000\ mL \qquad (1.057\ qt)$$

Volume measurement is often necessary in laboratory work, particularly for liquids. Figure 1.6 shows some of the more frequently used pieces of equipment.

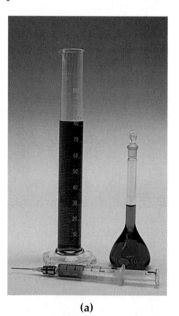

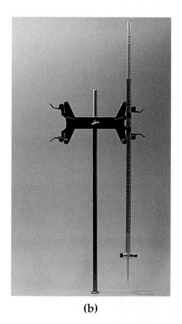

(a)   (b)

FIGURE 1.6 Among the most common items of laboratory equipment used for measuring liquid volume are **(a)** a graduated cylinder, a volumetric flask, a syringe, and **(b)** a buret.

1.10 Derived Units: Measuring Density

The intensive physical property that relates the mass of an object to its volume is called *density*. **Density**, which is simply the mass of an object divided by its volume, is expressed in the SI derived unit g/cm^3 for a solid or g/mL for a liquid. The densities of some common materials are given in Table 1.6.

$$\text{Density} = \frac{\text{Mass (g)}}{\text{Volume (mL or cm}^3)}$$

TABLE 1.6   Densities of Some Common Materials

Substance	Density (g/cm³)	Substance	Density (g/cm³)
Ice (0°C)	0.917	Human fat	0.94
Water (3.98°C)	1.0000	Cork	0.22–0.26
Gold	19.31	Table sugar	1.59
Helium (25°C)	0.000 164	Balsa wood	0.12
Air (25°C)	0.001 185	Earth	5.54

Which weighs more, the brass weight or the pillow? Actually, both have the same mass, but the weight has a higher density because its volume is smaller.

Because most substances change in volume when heated or cooled, densities are temperature-dependent. At 3.98°C, for example, a 1.0000 mL container holds exactly 1.0000 g of water (density = 1.0000 g/mL). As the temperature is raised, however, the volume occupied by the water expands so that only 0.9584 g fits in the 1.0000 mL container at 100°C (density = 0.9584 g/mL). When reporting a density, the temperature must also be specified.

Although most substances expand when heated and contract when cooled, water behaves differently. Water contracts when cooled from 100°C to 3.98°C, but below this temperature it begins to expand again. Thus, the density of liquid water is at its maximum of 1.0000 g/mL at 3.98°C, but decreases to 0.999 87 g/mL at 0°C (Figure 1.7). When freezing occurs, the density drops still further to a value of 0.917 g/cm³ for ice at 0°C. Ice and any other substance with a density less than that of water will float, but any substance with a density greater than that of water will sink.

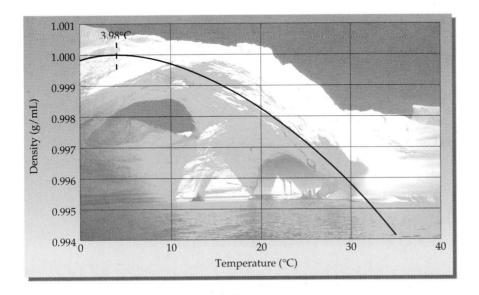

FIGURE 1.7 The density of water at different temperatures. Density reaches a maximum value of 1.000 g/mL at 3.98°C.

Knowing the density of a substance, particularly a liquid, can be very useful because it's often easier to measure a liquid by volume than by mass. Suppose, for example, that you needed 1.5 g of ethyl alcohol. Rather than trying to weigh exactly the right amount, it would be much easier to look up the density of ethyl alcohol (0.7893 g/mL at 20°C) and measure the correct volume with a syringe.

$$\text{Density} = \frac{\text{Mass}}{\text{Volume}} \quad \text{so} \quad \text{Volume} = \frac{\text{Mass}}{\text{Density}}$$

$$\text{Volume} = \frac{1.5 \text{ g ethyl alcohol}}{0.7893 \text{ g/mL}} = 1.9 \text{ mL ethyl alcohol}$$

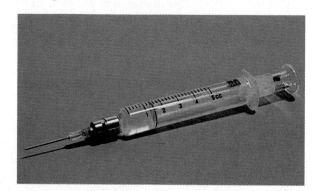

A precise amount of a liquid is easily measured with a syringe if the density of the liquid is known.

EXAMPLE 1.2 What is the density of the element copper (in grams per cubic centimeter) if a sample weighing 324.5 g has a volume of 36.2 cm³?

SOLUTION Density is mass divided by volume:

$$\text{Density} = \frac{\text{Mass}}{\text{Volume}} = \frac{324.5 \text{ g}}{36.2 \text{ cm}^3} = 8.96 \text{ g/cm}^3$$

EXAMPLE 1.3 What is the volume (in cubic centimeters) of 454 g of gold?

SOLUTION Since density is defined as mass divided by volume, volume is mass divided by density:

$$\text{Volume} = \frac{454 \text{ g gold}}{19.31 \text{ g/cm}^3} = 23.5 \text{ cm}^3 \text{ gold}$$

▶ **PROBLEM 1.9** What is the density of glass (in grams per cubic centimeter) if a sample weighing 27.43 g has a volume of 12.40 cm³?

▶ **PROBLEM 1.10** Chloroform, a substance once used as an anesthetic, has a density of 1.483 g/mL at 20°C. How many milliliters would you use if you needed 9.37 g?

1.11 Accuracy, Precision, and Significant Figures in Measurement

Any measurement is only as good as the skill of the person doing the work and the reliability of the equipment being used. You've probably noticed, for instance, that you get slightly different readings when you weigh yourself on a bathroom scale and on a scale at the doctor's office, so there's always some uncertainty about your real weight. The same is true in chemistry—there is always some uncertainty in the value of a measurement.

In talking about the degree of uncertainty in a measurement, we use the words *accuracy* and *precision*. Although most of us use the words interchangeably in daily life, there's actually an important distinction between them. **Accuracy** refers to how close to the true value a given measurement is, whereas **precision** refers to how well a number of independent measurements agree with one another. To see the difference, imagine that you weigh a tennis ball whose true mass is 54.441 78 g. Assume that you take three independent measurements on each of three different types of balance to obtain the data shown in the following table.

Measurement #	Bathroom Scale	Lab Balance	Analytical Balance
1	0.0 kg	54.4 g	54.4419 g
2	0.0 kg	54.7 g	54.4417 g
3	0.1 kg	54.1 g	54.4417 g
(average)	(0.03 kg)	(54.4 g)	(54.4418 g)

This tennis ball has a mass of about 54 g.

If you use a bathroom scale, your measurement (average = 0.03 kg) is neither accurate nor precise. Its accuracy is poor because it measures only to one digit that is far off the true value, and its precision is poor because any two measurements may differ substantially. If you now weigh the ball on an inexpensive single-pan balance, the value you get (average = 54.4 g) has three digits and is fairly accurate, but it is still not very precise because the three readings vary from 54.1 g to 54.7 g, perhaps due to air movements in the room or to a sticky mechanism. Finally, if you weigh the ball on an expensive analytical balance like those found in research laboratories, your measurement (average = 54.4418 g) is both precise and accurate. It's accurate because the measurement is very close to the true value, and it's precise because it has six digits that vary little from one reading to another.

To indicate the uncertainty in a measurement, the value you record should use all the digits you are sure of, plus one additional digit that you estimate. In reading a mercury thermometer that has a mark for each degree, for example, you could be certain about the digits of the nearest mark—say 25°C—but you would have to estimate between two marks—say 25.3°C.

The total number of digits in the measurement is called the number of **significant figures**. For example, the mass of the tennis ball as determined on the single-pan balance (54.4 g) has three significant figures, whereas the mass determined on the analytical balance (54.4418 g) has six significant figures. All digits but the last are certain; the final digit is only a best guess, which we generally assume to have an error of plus or minus one (±1).

Finding the number of significant figures in a measurement is usually easy but can be troublesome if zeros are present. Look at the following four quantities:

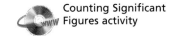

Counting Significant Figures activity

4.803 cm	Four significant figures: 4, 8, 0, 3
0.006 61 g	Three significant figures: 6, 6, 1
55.220 K	Five significant figures: 5, 5, 2, 2, 0
34,200 m	Anywhere from three (3, 4, 2) to five (3, 4, 2, 0, 0) significant figures

The following rules cover the different situations that can arise:

1. *Zeros in the middle of a number are like any other digit; they are always significant.* Thus, 4.803 cm has four significant figures.

2. *Zeros at the beginning of a number are not significant*; they act only to locate the decimal point. Thus, 0.006 61 g has three significant figures. (Note that 0.006 61 g can be rewritten as 6.61×10^{-3} g or as 6.61 mg.)

3. *Zeros at the end of a number and after the decimal point are always significant*. The assumption is that these zeros would not be shown unless they were significant. Thus, 55.220 K has five significant figures. (If the value were known to only four significant figures, we would write 55.22 K.)

4. *Zeros at the end of a number and before the decimal point may or may not be significant*. We can't tell whether they are part of the measurement or whether they only locate the decimal point. Thus, 34,200 m may have three, four, or five significant figures. Often, however, a little common sense is helpful. A temperature reading of 20°C probably has two significant figures rather than one, since one significant figure would imply a temperature anywhere from 10°C to 30°C and would be of little use. Similarly, a volume given as 300 mL probably has three significant figures. On the other hand, a figure of 93,000,000 mi for the distance between the earth and the sun probably has only two or three significant figures.

The fourth rule shows why it's helpful to write numbers in scientific notation rather than ordinary notation. Doing so makes it possible to indicate the number of significant figures. Thus, writing the number 34,200 as 3.42×10^4 indicates three significant figures, but writing it as 3.4200×10^4 indicates five significant figures.

One further point about significant figures: Certain numbers, such as those obtained when counting objects, are *exact* and have an effectively infinite number of significant figures. For example, a week has *exactly* 7 days, not 6.9 or 7.0 or 7.1, and 1 ft has *exactly* 12 in., not 11.9 or 12.0 or 12.1. In addition, the power of 10 used in scientific notation is an exact number. That is, the number 10^3 is exact, but the number 1×10^3 has one significant figure.

EXAMPLE 1.4 How many significant figures does each of the following measurements have?
(a) 0.036 653 m **(b)** 7.2100×10^{-3} g **(c)** 72,100 km **(d)** $25.03

SOLUTION
(a) 5 (by rule 2) **(b)** 5 (by rule 3) **(c)** 3, 4, or 5 (by rule 4)
(d) $25.03 is an exact number

▶**PROBLEM 1.11** A 1.000 mL sample of acetone, a common solvent used as a paint remover, was placed in a small bottle whose mass was known to be 38.0015 g. The following values were obtained when the acetone-filled bottle was weighed: 38.7798 g, 38.7795 g, and 38.7801 g. How would you characterize the precision and accuracy of these measurements if the actual mass of the acetone was 0.7791 g?

▶**PROBLEM 1.12** How many significant figures does each of the following quantities have? Explain your answers.
(a) 76.600 kg **(b)** $4.502\ 00 \times 10^3$ g **(c)** 3000 nm **(d)** 0.003 00 mL
(e) 18 students

1.12 Rounding Numbers

It often happens, particularly when doing arithmetic on a pocket calculator, that a quantity appears to have more significant figures than are really justified. For example, you might calculate the gas mileage of your car by finding that it takes 11.70 gallons of gasoline to drive 278 miles:

$$\text{Mileage} = \frac{\text{Miles}}{\text{Gallons}} = \frac{278 \text{ mi}}{11.70 \text{ gal}} = 23.760\ 684 \text{ mi/gal (mpg)}$$

Calculators often display more figures than are justified by the precision of the data.

Although the answer on the pocket calculator has eight digits, your measurement is really not as precise as it appears. In fact, your answer is precise to only three significant figures and should be **rounded off** to 23.8 mi/gal by removing nonsignificant figures.

How do you decide how many figures to keep and how many to ignore? The full answer to this question involves a mathematical treatment of the data known as *error analysis*. For most purposes, though, a simplified procedure using two easy-to-remember rules is sufficient. These rules give only an approximate value of the actual error, but that approximation is often good enough:

1. *In carrying out a multiplication or division, the answer can't have more significant figures than either of the original numbers.* If you think about it, this rule is just common sense: If you don't know the number of miles you drove to better than three significant figures (278 could mean 277, 278, or 279), you certainly can't calculate your mileage to more than the same number of significant figures.

Three significant figures ↘   *Three significant figures* ↙

$$\frac{278 \text{ mi}}{11.70 \text{ gal}} = 23.8 \text{ mi/gal}$$

↗ *Four significant figures*

2. *In carrying out an addition or subtraction, the answer can't have more digits to the right of the decimal point than either of the original numbers.* For example, if you have 3.18 L of water and you add 0.013 15 L more, you now have 3.19 L. Again, this rule is just common sense. If you don't know the volume you started with past the second decimal place (it could be 3.17, 3.18, or 3.19), you can't know the total of the combined volumes past the same decimal place.

Ends two places past decimal point

3.18? ?? ↙
+ 0.013 15 ← *Ends five places past decimal point*
3.19? ?? ↖

Ends two places past decimal point

Once you decide how many digits to retain for your answer, the rules for rounding off numbers are straightforward:

1. *If the first digit you remove is less than 5, round down by dropping it and all following digits.* Thus, 5.664 **525** becomes 5.66 when rounded to three significant figures because the first of the dropped digits (4) is less than 5.

2. *If the first digit you remove is 6 or greater, round up by adding 1 to the digit on the left.* Thus, 5.664 **525** becomes 5.7 when rounded to two significant figures because the first of the dropped digits (6) is greater than 5.

3. *If the first digit you remove is 5 and there are more nonzero digits following, round up.* Thus, 5.664 **525** becomes 5.665 when rounded to four significant figures because there are nonzero digits (2, 5) after the 5.

4. *If the digit you remove is a 5 with nothing following, round down.* Thus, 5.664 **525** becomes 5.664 **52** when rounded to six significant figures because there is nothing after the 5.

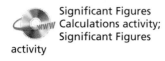

Significant Figures
Calculations activity;
Significant Figures
activity

EXAMPLE 1.5   If it takes 9.25 hours to fly from London, England, to Chicago, a distance of 3952 miles, what is the average speed of the airplane in miles per hour?

SOLUTION   First, set up an equation dividing the number of miles flown by the number of hours:

$$\text{Average speed} = \frac{3952 \text{ mi}}{9.25 \text{ h}} = 427.243\,24 \text{ mi/h}$$

Next, decide how many significant figures should be in your answer. Since the problem involves a division, and since one of the quantities you started with (9.25 h) has only three significant figures, the answer must also have three significant figures.

Finally, round off your answer. Since the first digit to be dropped (2) is less than 5, the answer 427.**243 24** must be rounded off to 427 mi/h. *Notice that in doing a calculation, you use all figures, significant or not, and then round off the final answer.*

▶ **PROBLEM 1.13**   Round off each of the following quantities to the number of significant figures indicated in parentheses:
  (a) 3.774 499 L (4)   **(b)** 255.0974 K (3)   **(c)** 55.265 kg (4)

▶ **PROBLEM 1.14**   Carry out the following calculations, expressing each result with the correct number of significant figures:
  (a) 24.567 g + 0.044 78 g = ? g
  (b) 4.6742 g ÷ 0.003 71 L = ? g/L
  (c) 0.378 mL + 42.3 mL − 1.5833 mL = ? mL

✦ **KEY CONCEPT PROBLEM 1.15**   What is the temperature reading on the following Celsius thermometer? How many significant figures do you have in your answer?

1.13   Calculations: Converting from One Unit to Another

Many scientific activities involve numerical calculations—measuring, weighing, preparing solutions, and so forth—and it's often necessary to convert a quantity from one unit to another. Converting between units isn't difficult; we all do it every day. For example, if you run 7.5 laps around a 200 meter track, you have to convert between the distance unit *lap* and the distance unit *meter* to find that you have run 1500 m (7.5 laps times 200 meters/lap). Converting from one scientific unit to another is just as easy.

$$7.5 \; \text{laps} \times \frac{200 \text{ meters}}{1 \text{ lap}} = 1500 \text{ meters}$$

The simplest way to carry out calculations that involve different units is to use the **dimensional-analysis method**. In this method, a quantity described in one unit is converted into an equivalent quantity with a different unit by using a **conversion factor** to express the relationship between units:

Original quantity × Conversion factor = Equivalent quantity

As an example, we know from Section 1.7 that 1 meter is equal to 39.37 inches. Writing this relationship as a fraction restates it in the form of a conversion factor, either meters per inch or inches per meter:

CONVERSION FACTORS BETWEEN METERS AND INCHES

Since 1 m = 39.37 in., then:
$$\frac{1 \text{ m}}{39.37 \text{ in.}} = \frac{39.37 \text{ in.}}{1 \text{ m}} = 1$$

These runners have to convert from laps to meters to find out how far they have run.

Note that this and all other conversion factors are numerically equal to 1 because the value of the quantity above the division line (the numerator) is equivalent to the value of the quantity below the division line (the denominator). Thus, multiplying by a conversion factor is equivalent to multiplying by 1 and so does not change the value of the quantity.

These two quantities are equivalent
$$\frac{1 \text{ m}}{39.37 \text{ in.}} \quad \text{or} \quad \frac{39.37 \text{ in.}}{1 \text{ m}}$$
These two quantities are equivalent

The key to the dimensional-analysis method of problem-solving is that units are treated like numbers and can thus be multiplied and divided (though not added or subtracted) just as numbers can. The idea when solving a problem is to set up an equation so that all unwanted units cancel, leaving only the desired units. Usually it's best to start by writing what you know and then manipulating that known quantity. For example, if you know your height in inches is 69.5 in. and want to find it in meters, you can write down the height in inches and set up an equation multiplying the height by the conversion factor in meters per inch:

$$69.5 \; \text{in.} \times \frac{1 \text{ m}}{39.37 \text{ in.}} = 1.77 \text{ m}$$

Starting quantity   *Conversion factor*   *Equivalent quantity*

The unit "in." cancels from the left side of the equation because it appears both above and below the division line, and the only unit that remains is "m."

The dimensional-analysis method gives the right answer only if the equation is set up so that the unwanted units cancel. If the equation is set up in any other way, the units won't cancel properly, and you won't get the right answer. Thus, if you were to multiply your height in inches by the incorrect conversion factor inches per meter, you would end up with an incorrect answer expressed in meaningless units:

$$\text{??} \quad 69.5 \text{ in.} \times \frac{39.37 \text{ in.}}{1 \text{ m}} = 2{,}740 \text{ in.}^2/\text{m} \quad \text{??}$$

The main drawback to using the dimensional-analysis method is that it's easy to get the right answer without really understanding what you're doing. It's therefore best to think through a ballpark answer before doing the detailed calculation. If your rough guess isn't close to the calculated answer, there is a misunderstanding somewhere, and you should think through the problem again. Even if you don't estimate the solution, it's important to be sure that your calculated answer makes sense. If, for example, you were trying to calculate the volume of a human cell and you came up with the answer 5.3 cm^3, you should realize that such an answer couldn't possibly be right; cells are too tiny to be distinguished with the naked eye, but a volume of 5.3 cm^3 is about the size of a walnut.

Examples 1.6–1.8 show how to estimate ballpark answers. To conserve space, we'll use this approach routinely in only the next few chapters, but you should make it a standard part of your problem-solving.

EXAMPLE 1.6   Many of the German autobahns have no speed limit, and it's not unusual to see a Porsche or Mercedes flash by at 190 km/h. What is this speed in miles per hour?

BALLPARK SOLUTION   Common sense says that the answer is probably large, perhaps over 100 mi/h. A better estimate is to realize that, since 1 km = 0.6214 mi, a mile is larger than a kilometer and it takes only about 2/3 as many miles as kilometers to measure the same distance. Thus, 190 km is about 120 mi, and 190 km/h is about 120 mi/h.

DETAILED SOLUTION   Use the dimensional-analysis method to set up an equation so that the km units cancel. Place the known quantity first (the speed in km/h), and then multiply by a suitable conversion factor:

$$\frac{190 \text{ ~~km~~}}{1 \text{ h}} \times \frac{0.6214 \text{ mi}}{1 \text{ ~~km~~}} = 118 \frac{\text{mi}}{\text{h}}$$

The ballpark solution and the detailed solution agree. Note that we're not completely sure of the number of significant figures in the answer because the starting value of 190 km/h could have two or three significant figures. Common sense, however, says that three significant figures is more reasonable than two since an automobile speedometer that couldn't distinguish between 180 km/h and 200 km/h wouldn't be of much use.

EXAMPLE 1.7   The Mercedes that just went by at 190 km/h is using gasoline at a rate of 16 L per 100 km. What does this correspond to in miles per gallon?

BALLPARK SOLUTION   Common sense says that the mileage is probably low, perhaps in the range of 10 to 15 mi/gal. This is a harder problem than the previous one to estimate, though, because it requires several different conversions. It's therefore best to think the problem through one step at a time, writing down the intermediate estimates:

A distance of 100 km per 16 L is approximately 7 km/L.
Since 1 km is about 0.6 mi, 7 km/L is about 4 mi/L.
Since 1 L is approximately 1 qt, or 1/4 gal, 4 mi/L is about 16 mi/gal.

DETAILED SOLUTION   It's best to do multiple conversions one step at a time until you get used to dealing with them. You might start by converting the distance from kilometers to miles and the fuel consumption from liters to gallons:

$$100 \text{ km} \times \frac{0.6214 \text{ mi}}{1 \text{ km}} = 62.14 \text{ mi} \qquad 16 \text{ L} \times \frac{1 \text{ gal}}{3.78 \text{ L}} = 4.2328 \text{ gal}$$

Dividing the distance by the amount of fuel gives the mileage:

$$\frac{62.14 \text{ mi}}{4.2328 \text{ gal}} = 14.68 \frac{\text{mi}}{\text{gal}} \qquad \text{Round off to 15 mi/gal}$$

Note that extra digits are carried through the intermediate calculations, and only the final answer is rounded off.

Alternatively, when you become more confident in working conversion problems, you can set up one large equation in which all unwanted units cancel:

$$\frac{100 \text{ km}}{16 \text{ L}} \times \frac{3.78 \text{ L}}{1 \text{ gal}} \times \frac{0.6214 \text{ mi}}{1 \text{ km}} = 14.68 \frac{\text{mi}}{\text{gal}} \qquad \text{Round off to 15 mi/gal}$$

However you do the problem, the ballpark solution and detailed solution agree.

EXAMPLE 1.8   The volcanic explosion that destroyed the Indonesian island of Krakatau on August 27, 1883, released an estimated 4.3 cubic miles (mi^3) of debris into the atmosphere and affected global weather conditions for years. In SI units, how many cubic meters (m^3) were released?

BALLPARK SOLUTION   One meter is much less than 1 mile, so it takes a large number of cubic meters to equal 1 mi^3, and the answer is going to be very large. In fact, 1 mi^3 is the volume of a cube 1 mi on each edge, and 1 km^3 is the volume of a cube 1 km on each edge. Since a kilometer is about 0.6 mi, 1 km^3 is about $(0.6)^3 = 0.2$ times as large as 1 mi^3. Thus, each cubic mile contains about 5 km^3, and 4.3 mi^3 contains about 20 km^3. Each cubic kilometer, in turn, contains $(1000 \text{ m})^3 = 10^9 \text{ m}^3$. Thus, the volume of debris from the Krakatau explosion was about $20 \times 10^9 \text{ m}^3$, or $2 \times 10^{10} \text{ m}^3$.

DETAILED SOLUTION   First convert cubic miles into cubic kilometers:

$$4.3 \text{ mi}^3 \times \left(\frac{1 \text{ km}}{0.6214 \text{ mi}} \right)^3 = 17.92 \text{ km}^3$$

Then convert cubic kilometers into cubic meters:

$$17.92 \text{ km}^3 \times \left(\frac{1000 \text{ m}}{1 \text{ km}} \right)^3 = 1.792 \times 10^{10} \text{ m}^3$$

$$= 1.8 \times 10^{10} \text{ m}^3 \quad \text{rounded off}$$

The ballpark solution and the detailed solution agree.

▶ **PROBLEM 1.16**   First estimate and then calculate answers to the following problems:
  (a) The melting point of gold is 1064°C. What is this temperature in degrees Fahrenheit?
  (b) How large, in cubic centimeters, is the volume of a red blood cell if the cell has a circular shape with a diameter of 6×10^{-6} m and a height of 2×10^{-6} m?

▶ **PROBLEM 1.17**   Gemstones are weighed in *carats*, with 1 carat = 200 mg (exactly). What is the mass in grams of the Hope Diamond, the world's largest blue diamond at 44.4 carats? What is this mass in ounces?

Chemicals, Toxicity, and Risk

Not surprisingly, this bike rider is at much greater risk than the driver of a car.

Life is not risk-free—we all take many risks each day, beginning when we get out of bed in the morning and turn on the lights (200 people are electrocuted in home accidents in the United States each year). We may decide to ride a bike rather than drive, even though there is a 10 times greater likelihood per mile of dying in a bicycle accident than in a car. We may decide to smoke cigarettes, even though smoking increases our chance of getting cancer by 50%. Making judgments that affect our health is something we do every day without thinking about it.

What about risks from "chemicals"? News reports sometimes make it seem that our food is covered with pesticide residues and filled with dangerous additives, that our land is polluted by toxic waste dumps, and that our medicines are unsafe. How bad are the risks from chemicals, and how are the risks evaluated?

First, it's important to realize that *everything* is made of chemicals. There is no such thing as a "chemical-free" food, cleanser, cosmetic, or anything else. Second, it's important to realize that there is no meaningful distinction between a "natural" substance and a "synthetic" one; a chemical is a chemical is a chemical. Many naturally occurring substances—strychnine, for example—are extraordinarily toxic, and many synthetic substances—polyethylene, for example—are harmless.

Risk evaluation of chemicals is carried out by exposing test animals, usually mice or rats, to varying doses of a chemical and then monitoring for signs of harm. To limit the expense and time needed for testing, the doses administered are hundreds or thousands of times larger than those a human might normally encounter. The *acute chemical toxicity* (as opposed to long-term toxicity) observed in animal tests is reported as an *LD_{50} value*, the amount of a substance per kilogram of body weight that is a lethal dose for 50% of the test animals. LD_{50} values of various substances are shown in Table 1.7.

TABLE 1.7 Some LD_{50} Values

Substance	LD_{50} (g/kg)	Substance	LD_{50} (g/kg)
Aflatoxin B_1	4×10^{-4}	Sodium chloride	4
Aspirin	1.5	Sodium cyanide	1.5×10^{-2}
Chloroform	3.2	Sodium cyclamate	17
Ethyl alcohol	10.6	Water	500

Even with an LD_{50} value established in test animals, the risk of human exposure to a given substance is still hard to assess. If a substance is harmful to rats, is it necessarily harmful to humans? How can a large dose for a small animal be translated into a small dose for a large human? All substances are toxic to some organisms to some extent, and the difference between help and harm is often a matter of degree. Common table salt (sodium chloride), for example, is essential to health yet has an LD_{50} of 4 g/kg in rats. Vitamin A is necessary for vision yet causes cancer at high doses. Thus, an important aspect of risk management involves estimating a *threshold* value, or exposure level below which the risk is assumed to be negligible. Unfortunately, the uncertainties involved in finding the threshold are enormous, and the final result may be off by a factor of 10,000 or more.

Even if the threshold value is accurate, the risk must be weighed against potential benefits. Does the benefit of a pesticide that will increase the availability of food outweigh the health risk to 1 person in 1 million who are exposed to it? Do the beneficial effects of a new drug outweigh a potentially dangerous side effect in 0.001% of users? The answers aren't always obvious, but it is the responsibility of legislators and well-informed citizens to keep their responses on a factual level rather than an emotional one.

▶ **PROBLEM 1.18**   What is an LD_{50} value?

▶ **PROBLEM 1.19**   Table salt (sodium chloride) has an LD_{50} of 4 g/kg in rats. Assuming that rats and human have the same LD_{50}, how much salt would a typical 155 lb person have to consume to have a 50% chance of dying?

Key Words

Summary

Chemistry is the study of **matter** and of the changes that matter undergoes. These studies are best approached by posing questions, conducting experiments, and devising **theories** to interpret the experimental results. All matter is formed from one or more of 115 presently known **elements**—fundamental substances that can't be chemically broken down. Elements are symbolized by one- or two-letter abbreviations and are organized into a **periodic table** with **groups** (columns) and **periods** (rows). Elements in the same group of the periodic table show similar chemical properties. The two larger groups on the left and the six larger groups on the right of the table are called the **main groups**; the ten smaller ones in the middle of the table are called the **transition metal groups**; and the fourteen shown separately at the bottom of the table are called the **inner transition metal groups**.

The characteristics, or **properties**, that are used to describe matter can be classified in several ways. **Physical properties** are those that can be determined without changing the chemical composition of the sample, whereas **chemical properties** are those that *do* involve a chemical change in the sample. **Intensive properties** are those whose values do not depend on the size of the sample, whereas **extensive properties** are those whose values *do* depend on sample size.

Accurate measurement is crucial to scientific experimentation. The units used are those of the Système Internationale **(SI units)**. There are seven fundamental SI units, together with other derived units: **Mass**, the amount of matter an object contains, is measured in **kilograms (kg)**; **length** is measured in **meters (m)**; **temperature** is measured in **kelvins (K)**; and **volume** is measured in **cubic meters (m³)**. The more familiar metric **liter (L)** and **milliliter (mL)** are also used for measuring volume, and the **Celsius degree (°C)** is still used for measuring temperature. **Density** is an intensive physical property that relates mass to volume.

Since many experiments involve numerical calculations, it's often necessary to manipulate and convert different units of measure. The simplest way to carry out such conversions is to use the **dimensional-analysis method**, in which an equation is set up so that unwanted units cancel and only the desired units remain. It's also important when measuring physical quantities or carrying out calculations to indicate the precision of the measurement by **rounding off** the result to the correct number of **significant figures**.

Key Concept Summary

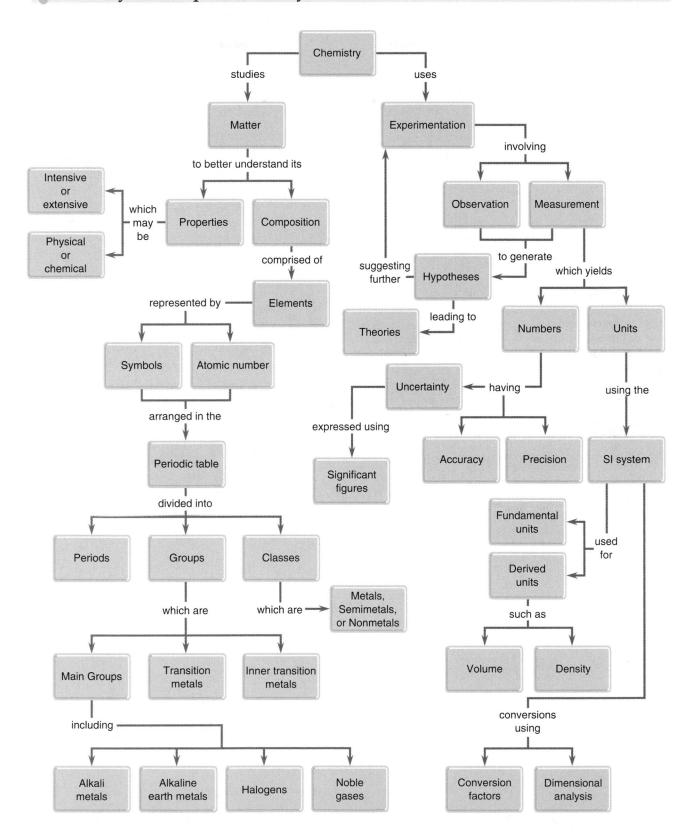

Understanding Key Concepts

The exercises at the end of each chapter begin with a section called "Understanding Key Concepts." The problems in this section are often visual rather than numerical and are intended to probe your understanding rather than your facility with numbers and formulas. Answers to even-numbered problems can be found at the end of the book following the appendixes. Problems 1.1–1.19 appear within the chapter.

1.20 Where on the following outline of a periodic table are the indicated elements or groups of elements?

(a) Alkali metals   (b) Halogens

(c) Alkaline earth metals   (d) Transition metals

(e) Hydrogen   (f) Helium

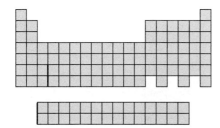

1.21 Where on the following outline of a periodic table does the dividing line between metals and nonmetals fall?

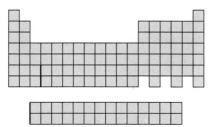

1.22 Is the element marked in red on the following periodic table likely to be a gas, a liquid, or a solid? What is the atomic number of the element in blue? Name at least one other element that is chemically similar to the element in green.

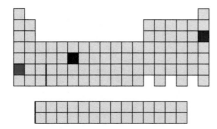

1.23 The radioactive element indicated on the following periodic table is used in smoke detectors. Identify it, give its atomic number, and tell what kind of group it's in.

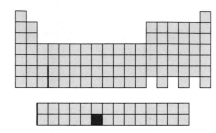

1.24 Characterize each of the following dartboards according to the accuracy and precision of the results.

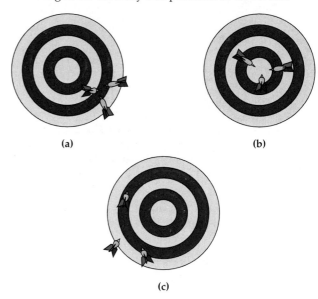

(a)   (b)

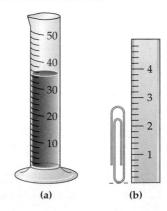

(c)

1.25 How many milliliters of water does the graduated cylinder in (a) contain, and how tall in centimeters is the paperclip in (b)? How many significant figures do you have in each answer?

(a)   (b)

1.26 Assume that you have two graduated cylinders, one with a capacity of 5 mL **(a)** and the other with a capacity of 50 mL **(b)**. Draw a line in each showing how much liquid you would add if you needed to measure 2.64 mL of water. Which cylinder will give more accurate measurements? Explain.

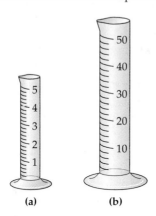

(a)   (b)

1.27 Liquid mercury has a density of 13.546 g/cm^3 at 20°C, and solid lead has a density of 11.35 g/cm^3 at the same temperature. Will lead float or sink in mercury?

Additional Problems

The Additional Problems at the end of each chapter are divided into sections, all but the last two of which deal with specific topics from the chapter. These problems are presented in pairs, with each even-numbered problem followed by an odd-numbered one that requires similar skills. These paired problems are followed by a series of unpaired General Problems that draw on various parts of the chapter and, beginning in Chapter 4, by a section of Multi-Concept Problems that require the use of skills learned over several chapters. All even-numbered problems are answered at the end of the book following the appendixes.

Elements and the Periodic Table

1.28 How many elements are presently known? About how many occur naturally?

1.29 What are the rows called and what are the columns called in the periodic table?

1.30 How many groups are there in the periodic table? How are they labeled?

1.31 What common characteristics do elements within a group of the periodic table have?

1.32 Where in the periodic table are the main-group elements found? Where are the transition metal groups found?

1.33 Where in the periodic table are the metallic elements found? Where are the nonmetallic elements found?

1.34 What is a semimetal, and where in the periodic table are semimetals found?

1.35 List several general properties of the following:
 (a) Alkali metals   **(b)** Noble gases
 (c) Halogens

1.36 Without looking at a periodic table, list as many alkali metals as you can. (There are five common ones.)

1.37 Without looking at a periodic table, list as many alkaline earth metals as you can. (There are five common ones.)

1.38 Without looking at a periodic table, list as many halogens as you can. (There are four common ones.)

1.39 Without looking at a periodic table, list as many noble gases as you can. (There are six common ones.)

1.40 What are the symbols for the following elements?
 (a) Gadolinium (used in color TV screens)
 (b) Germanium (used in semiconductors)
 (c) Technetium (used in biomedical imaging)
 (d) Arsenic (used in pesticides)

1.41 What are the symbols for the following elements?
 (a) Cadmium (used in rechargeable batteries)
 (b) Iridium (used for hardening alloys)
 (c) Beryllium (used in the space shuttle)
 (d) Tungsten (used in light bulbs)

1.42 Give the names corresponding to the following symbols:
 (a) Te   **(b)** Re   **(c)** Be   **(d)** Ar   **(e)** Pu

1.43 Give the names corresponding to the following symbols:
 (a) B   **(b)** Rh   **(c)** Cf   **(d)** Os   **(e)** Ga

1.44 What is wrong with each of the following statements?

(a) The symbol for tin is Ti.

(b) The symbol for manganese is Mg.

(c) The symbol for potassium is Po.

(d) The symbol for helium is HE.

Units and Significant Figures

1.46 Why is accurate measurement crucial in science?

1.47 What is the difference between a number and a physical quantity like length?

1.48 What is the difference between mass and weight?

1.49 What is the difference between a derived SI unit and a fundamental SI unit? Give an example of each.

1.50 What SI units are used for measuring the following quantities?

(a) Mass   **(b)** Length

(c) Temperature   **(d)** Volume

1.51 What SI prefixes correspond to the following multipliers?

(a) 10^3 **(b)** 10^{-6} **(c)** 10^9

(d) 10^{-12} **(e)** 10^{-2}

1.52 Which is larger, a Fahrenheit degree or a Celsius degree? By how much?

1.53 What is the difference between a kelvin and a Celsius degree?

1.54 What is the difference between a cubic decimeter (SI) and a liter (metric)?

1.55 What is the difference between a cubic centimeter (SI) and a milliliter (metric)?

1.56 Which of the following statements use exact numbers?

(a) 1 ft = 12 in.

(b) The population of Mexico City is 8,988,230.

(c) The world record for the 1 mile run, set by Morocco's Hicham el Guerrouj in 1999, is 3 minutes, 43.13 seconds.

1.57 What is the difference in mass between a nickel that weighs 4.8 g and a nickel that weighs 4.8673 g?

1.58 Bottles of wine sometimes carry the notation "Volume = 75 cL." What does the unit cL mean?

1.59 What do the following abbreviations stand for?

(a) dL   **(b)** dm   **(c)** μm   **(d)** nL

1.60 How many picograms are in 1 mg? In 35 ng?

1.61 How many microliters are in 1 L? In 20 mL?

1.62 Carry out the following conversions:

(a) 5 pm = ___ cm = ___ nm

(b) 8.5 cm³ = ___ m³ = ___ mm³

(c) 65.2 mg = ___ g = ___ pg

1.45 What is wrong with each of the following statements?

(a) The symbol for carbon is ca.

(b) The symbol for sodium is So.

(c) The symbol for nitrogen is Ni.

(d) The symbol for chlorine is Cr.

1.63 Which is larger, and by approximately how much?

(a) A liter or a quart   **(b)** A mile or a kilometer

(c) A gram or an ounce   **(d)** A centimeter or an inch

1.64 How many significant figures are in each of the following measurements?

(a) 35.0445 g   **(b)** 59.0001 cm   **(c)** 0.030 03 kg

(d) 0.004 50 m **(e)** 67,000 m²   **(f)** 3.8200×10^3 L

1.65 How many significant figures are in each of the following measurements?

(a) $130.95   **(b)** 2000.003 g   **(c)** 5 ft 3 in.

1.66 The Vehicle Assembly Building at the John F. Kennedy Space Center in Cape Canaveral, Florida, is the largest building in the world, with a volume of 3,666,500 m³. Express this volume in scientific notation.

1.67 The diameter of the earth at the equator is 7926.381 mi. Round off this quantity to four significant figures; to two significant figures. Express the answers in scientific notation.

1.68 Express the following measurements in scientific notation:

(a) 453.32 mg   **(b)** 0.000 042 1 mL   **(c)** 667,000 g

1.69 Convert the following measurements from scientific notation to standard notation:

(a) 3.221×10^{-3} mm   **(b)** 8.940×10^5 m

(c) $1.350\,82 \times 10^{-12}$ m³

1.70 Round off the following quantities to the number of significant figures indicated in parentheses:

(a) 35,670.06 m (4, 6)   **(b)** 68.507 g (2, 3)

(c) 4.995×10^3 cm (3)   **(d)** $2.309\,85 \times 10^{-4}$ kg (5)

1.71 Round off the following quantities to the number of significant figures indicated in parentheses:

(a) 7.0001 kg (4)   **(b)** 1.605 km (3)

(c) 13.2151 g/cm³ (3) **(d)** 2,300,000.1 (7)

1.72 Express the results of the following calculations with the correct number of significant figures:

(a) 4.884×2.05 **(b)** $94.61 \div 3.7$

(c) $3.7 \div 94.61$ **(d)** $5502.3 + 24 + 0.01$

(e) $86.3 + 1.42 - 0.09$ **(f)** 5.7×2.31

1.73 Express the results of the following calculations with the correct number of significant figures:

(a) $\dfrac{3.41 - 0.23}{5.233} \times 0.205$

(b) $\dfrac{5.556 \times 2.3}{4.223 - 0.08}$

Unit Conversions

1.74 Carry out the indicated conversions.

(a) How many grams of meat are in a quarter-pound hamburger (0.25 lb)?

(b) How tall (in meters) is the Sears Tower in Chicago (1454 ft)?

(c) How large (in square meters) is the land area of Australia (2,941,526 mi^2)?

1.75 Convert the following quantities into SI units with the correct number of significant figures:

(a) 5.4 in.   (b) 66.31 lb   (c) 0.5521 gal

(d) 65 mi/h   (e) 978.3 yd^3

1.76 The volume of water used for crop irrigation is measured in acre-feet, where 1 acre-foot is the amount of water needed to cover 1 acre of land to a depth of 1 ft.

(a) If there are 640 acres per square mile, how many cubic feet of water are in 1 acre-foot?

(b) How many acre-feet are in Lake Erie (total volume = 116 mi^3)?

1.77 The height of a horse is usually measured in *hands* instead of in feet, where 1 hand equals 1/3 ft (exactly).

(a) How tall (in centimeters) is a horse of 18.6 hands?

(b) What is the volume (in cubic meters) of a box measuring 6 × 2.5 × 15 hands?

Temperature

1.82 The normal body temperature of a goat is 39.9°C, and that of an Australian spiny anteater is 22.2°C. Express these temperatures in degrees Fahrenheit.

1.83 Of the 90 or so naturally occurring elements, only four are liquid near room temperature: mercury (melting point = −38.87°C), bromine (melting point = −7.2°C), cesium (melting point = 28.40°C), and gallium (melting point = 29.78°C). Convert these melting points to degrees Fahrenheit.

1.84 Tungsten, the element used to make filaments in light bulbs, has a melting point of 6192°F. Convert this temperature to degrees Celsius and to kelvins.

1.85 Suppose that your oven is calibrated in degrees Fahrenheit but a recipe calls for you to bake at 175°C. What oven setting should you use?

1.86 Suppose you were dissatisfied with both Celsius and Fahrenheit units and wanted to design your own temperature scale based on ethyl alcohol (ethanol). On the Celsius scale, ethanol has a melting point of −117.3°C and a boiling point of 78.5°C, but on your new scale calibrated in units of degrees ethanol, °E, you define ethanol to melt at 0°E and boil at 200°E. Answer the following questions.

1.78 Amounts of substances dissolved in solution are often expressed as mass per unit volume. For example, normal human blood has a cholesterol concentration of about 200 mg/100 mL. Express this concentration in the following units:

(a) mg/L   (b) μg/mL   (c) g/L   (d) ng/μL

(e) How much total blood cholesterol (in grams) does a person have if the normal blood volume in the body is 5 L?

1.79 Weights in England are commonly measured in *stones*, where 1 stone = 14 lb. What is the weight (in pounds) of a person who weighs 8.65 stones?

1.80 Among many alternative units that might be considered as a measure of time is the *shake* rather than the second. Based on the expression "faster than a shake of a lamb's tail," we'll define 1 shake as equal to 2.5 × 10^{-4} s. If a car is traveling at 55 mi/h, what is its speed in cm/shake?

1.81 Administration of digitalis, a drug used to control atrial fibrillation in heart patients, must be carefully controlled because even a modest overdosage can be fatal. To take differences between patients into account, drug dosages are prescribed in terms of mg/kg body weight. Thus, a child and an adult differ greatly in weight, but both receive the same dosage per kilogram of body weight. At a dosage of 20 μg/kg body weight, how many milligrams of digitalis should a 160 lb patient receive?

(a) How does your new ethanol degree compare in size with a Celsius degree?

(b) How does an ethanol degree compare in size with a Fahrenheit degree?

(c) What are the melting and boiling points of water on the ethanol scale?

(d) What is normal human body temperature (98.6°F) on the ethanol scale?

(e) If the outside thermometer read 130°E, how would you dress to go out?

1.87 Answer parts (a)–(d) of Problem 1.86 assuming that your new temperature scale is based on ammonia, NH$_3$. On the Celsius scale, ammonia has a melting point of −77.7°C and a boiling point of −33.4°C, but on your new scale calibrated in units of degrees ammonia, °A, you define ammonia to melt at 0°A and boil at 100°A.

Density

1.88 Aspirin has a density of 1.40 g/cm^3. What is the volume (in cubic centimeters) of an aspirin tablet weighing 250 mg? Of a tablet weighing 500 lb?

1.89 Gaseous hydrogen has a density of 0.0899 g/L at $0°C$, and gaseous chlorine has a density of 3.214 g/L at the same temperature. How many liters of each would you need if you wanted 1.0078 g hydrogen and 35.45 g chlorine?

1.90 What is the density of lead (in g/cm^3) if a rectangular bar measuring 0.50 cm in height, 1.55 cm in width, and 25.00 cm in length has a mass of 220.9 g?

1.91 What is the density of lithium metal (in g/cm^3) if a cylindrical wire with a diameter of 2.40 mm and a length of 15.0 cm has a mass of 0.3624 g?

1.92 When an irregularly shaped chunk of silicon weighing 8.763 g was placed in a graduated cylinder containing 25.00 mL of water, the water level in the cylinder rose to 28.76 mL. What is the density of silicon in g/cm^3?

1.93 When the experiment outlined in Problem 1.92 was repeated using a chunk of sodium metal, rather than silicon, an explosion occurred. Was this due to a chemical property or a physical property of sodium?

General Problems

1.94 Give the symbol for each of the following elements:
  (a) Selenium (used in photocopiers)
  (b) Rhenium (used for hardening alloys)
  (c) Cobalt (used in magnets)
  (d) Rhodium (used in catalytic converters)

1.95 Consider the as yet undiscovered elements with atomic numbers 115, 117, and 119.
  (a) Which element is a halogen? Explain.
  (b) Which element should have chemical properties similar to cesium?
  (c) Is element 115 likely to be a metal or a nonmetal? What about element 117? Explain.
  (d) Describe some of the properties you expect for element 119.

1.96 Sodium chloride has a melting point of 1074 K and a boiling point of 1686 K. Convert these temperatures to degrees Celsius and to degrees Fahrenheit.

1.97 The mercury in thermometers freezes at $-38.9°C$. What is this temperature in degrees Fahrenheit?

1.98 The density of chloroform, a widely used organic solvent, is 1.4832 g/mL at $20°C$. How many milliliters would you use if you wanted 112.5 g of chloroform?

1.99 The density of sulfuric acid, H_2SO_4, is 15.28 lb/gal. What is the density of sulfuric acid in g/mL?

1.100 Sulfuric acid (Problem 1.99) is produced in larger amount than any other chemical—9.515×10^{10} lb in 1998. What is the volume of this amount in liters?

1.101 The caliber of a gun is expressed by measuring the diameter of the gun barrel in hundredths of an inch. A ".22" rifle, for example, has a barrel diameter of 0.22 in. What is the barrel diameter of a .22 rifle in millimeters?

1.102 At a certain point, the Celsius and Fahrenheit scales "cross," giving the same numerical value on both. At what temperature does this crossover occur?

1.103 Imagine that you place a cork measuring 1.30 cm $\times$ 5.50 cm $\times$ 3.00 cm in a pan of water and that on top of the cork you place a small cube of lead measuring 1.15 cm on each edge. The density of cork is 0.235 g/cm^3, and the density of lead is 11.35 g/cm^3. Will the combination of cork plus lead float or sink?

1.104 A 125 mL sample of water at 293.2 K was heated for 8 min, 25 s so as to give a constant temperature increase of $3.0°F/\text{min}$. What is the final temperature of the water in degrees Celsius?

1.105 A calibrated flask was filled to the 25.00 mL mark with ethyl alcohol. By weighing the flask before and after adding the alcohol, it was determined that the flask contained 19.7325 g of alcohol. In a second experiment, 25.0920 g of metal beads were added to the flask, and the flask was again filled to the 25.00 mL mark with ethyl alcohol. The total mass of the metal plus alcohol in the flask was determined to be 38.4704 g. What is the density of the metal in g/mL?

1.106 Brass is a copper–zinc alloy. What is the mass in grams of a brass cylinder having a length of 1.62 in. and a diameter of 0.514 in. if the composition of the brass is 67.0% copper and 33.0% zinc by mass? The density of copper is 8.92 g/cm^3, and the density of zinc is 7.14 g/cm^3. Assume that the density of the brass varies linearly with composition.

1.107 Ocean currents are measured in *Sverdrups* (sv) where $1 \text{ sv} = 10^9 \text{ m}^3/\text{s}$. The gulf stream off the tip of Florida, for instance, has a flow of 35 sv.
  (a) What is the flow of the gulf stream in milliliters per minute?
  (b) What mass of water in the gulf stream flows past a given point in 24 hours? The density of water is 1.0 g/mL.
  (c) How much time is required for 1 petaliter (PL; $1 \text{ PL} = 10^{15} \text{ L}$) of water to flow past a given point? The density of water is 1 g/mL.

1.108 The element gallium (Ga) has the second largest liquid range of any element, melting at 29.8°C and boiling at 2204°C at atmospheric pressure.

(a) Is gallium a metal, a nonmetal, or a semimetal?

(b) Name another element whose chemical properties might be similar to those of gallium.

(c) What is the density of gallium in g/cm³ at 25°C if a 1-inch cube has a mass of 0.2133 lb?

(d) Assume that you construct a thermometer using gallium as the fluid instead of mercury, and that you define the melting point of gallium as 0°G and the boiling point of gallium as 1000°G. What is the melting point of sodium chloride (1474°C) on the gallium scale?

---

 eMedia Problems

1.109 Use the **Significant Figures** activity (*eChapter 1.11*) to check your answers to Problems 1.72 and 1.73.

(a) Is it possible for the product of two numbers to have more significant figures than either of the original numbers? Explain.

(b) Is it possible for the sum of two numbers to have more significant figures than one of the original numbers? Explain.

1.110 In Problem 1.93 you were asked to distinguish between chemical and physical properties of sodium metal.

(a) Watch the **Sodium and Potassium in Water** movie (*eChapter 1.4*) and list the physical and chemical properties of sodium illustrated in the movie.

(b) Compare the physical and chemical properties of sodium and potassium. How are they similar? In what ways do they differ?

1.111 Use the **Periodic Table Groups** activity (*eChapter 1.3*) to do the following:

(a) Determine the category in the activity that contains the largest number of elements.

(b) Determine the category that contains the smallest number of elements.

2

Atoms, Molecules, and Ions

If you could take one of these gold coins and cut it into ever smaller and smaller pieces, you would find that it is made of a vast number of tiny fundamental units that we call *atoms*.

People have always been fascinated by changes, particularly those that are dramatic or useful. In the ancient world, the change that occurred when a stick of wood burned, gave off heat, and turned to a small pile of ash was especially important. Similarly, the change that occurred when a reddish lump of rock (iron ore) was heated with charcoal and produced a gray metal (iron) useful for making weapons, tools, and other implements was of enormous value. Observing such changes eventually caused Greek philosophers to think about what different materials might be composed of and led to the idea of fundamental substances that we today call *elements*.

At the same time philosophers were pondering the question of elements, they were also musing about related matters: What is an element made of? Is matter continuously divisible into ever smaller and smaller pieces, or is there an ultimate limit? Can you cut a piece of gold in two, take one of the pieces and cut *it* in two, and so on indefinitely, or is there a point at which you must stop? Although most thinkers, including Plato and Aristotle, believed that matter is continuous, the Greek philosopher Democritus (460–370 B.C.) disagreed. Democritus proposed that elements are composed of tiny particles that we now call *atoms*, from the Greek word *atomos*, meaning indivisible. Little else was learned about elements and atoms until the birth of modern experimental science some 2000 years later.

2.1   Conservation of Mass and the Law of Definite Proportions

The Englishman Robert Boyle (1627–1691) is generally credited with being the first to study chemistry as a separate intellectual discipline and the first to carry out rigorous chemical experiments. Through a careful series of researches into the nature and behavior of gases, Boyle provided clear evidence for the atomic makeup of matter. In addition, Boyle was the first to clearly define an **element** as a substance that cannot be chemically broken down further and to suggest that a substantial number of different elements might exist.

Progress in chemistry was slow in the 100 years following Boyle, and it was not until the work of Joseph Priestley (1733–1804) and Antoine Lavoisier (1743–1794) that the next great leap was made. Priestley prepared and isolated the gas *oxygen* in 1774 by heating mercury(II) oxide, HgO, according to the equation we would now write as $2\,HgO \rightarrow 2\,Hg + O_2$. Lavoisier then showed soon thereafter that oxygen is the key substance involved in combustion.

Heating the red powder HgO causes it to decompose into the silvery liquid mercury and the colorless gas oxygen.

Furthermore, Lavoisier demonstrated by careful measurements that when combustion is carried out in a closed container, the mass of the combustion products is exactly equal to the mass of the starting reactants. For example, when hydrogen gas burns and combines with oxygen to yield water (H_2O), the mass of the water formed is equal to the mass of the hydrogen and oxygen consumed. Called the *law of mass conservation*, this principle is a cornerstone of chemical science.

| ◆— LAW OF MASS CONSERVATION: | Mass is neither created nor destroyed in chemical reactions. |

It's easy to demonstrate the validity of the law of mass conservation by carrying out a simple experiment like that shown in Figure 2.1. If weighed amounts of mercury(II) nitrate [$Hg(NO_3)_2$] and potassium iodide (KI) are dissolved in water and the solutions are mixed, an immediate chemical reaction occurs, leading to formation of the insoluble orange solid mercury(II) iodide (HgI_2). Filtering the reaction mixture removes the mercury(II) iodide, and evaporation of the water leaves a deposit of potassium nitrate (KNO_3). When the two products are weighed, it's found that their combined masses exactly equal the combined masses of the two starting substances.

The combined masses of these two reactants . . .   *. . . equals the combined masses of these two products.*

$$Hg(NO_3)_2 + 2 KI \longrightarrow HgI_2 + 2 KNO_3$$

Mercury(II)   Potassium   Mercury(II)   Potassium
nitrate   iodide   iodide   nitrate

(a)   (b)   (c)

FIGURE 2.1   An illustration of the law of mass conservation. In any chemical reaction, the combined masses of the products formed equals the combined masses of the starting reactants. In this sequence of photos, **(a)** weighed amounts of solid KI and solid $Hg(NO_3)_2$ are dissolved in water. **(b)** The solutions are then mixed to give solid HgI_2, which is removed by filtration, and the solution that remains is evaporated to yield solid KNO_3. **(c)** When the two products are weighed, the combined mass of the reactants KI and $Hg(NO_3)_2$ is found to be equal to the combined mass of the products HgI_2 and KNO_3.

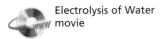
Electrolysis of Water movie

Further investigations in the decades following Lavoisier led the French chemist Joseph Proust (1754–1826) to formulate a second fundamental chemical principle that we now call the *law of definite proportions*.

| ◆— LAW OF DEFINITE PROPORTIONS: | Different samples of a pure chemical substance always contain the same proportion of elements by mass. |

Every sample of water (H_2O) contains 1 part hydrogen and 8 parts oxygen by mass; every sample of carbon dioxide (CO_2) contains 1 part carbon and 2.7 parts oxygen by mass; and so on. *Elements combine in specific proportions, not in random proportions.*

2.2 Dalton's Atomic Theory and the Law of Multiple Proportions

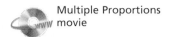
Multiple Proportions movie

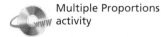
Multiple Proportions activity

How can the law of mass conservation and the law of definite proportions be explained? Why do elements behave as they do? The answers to these questions were provided in 1808 by the English schoolteacher John Dalton (1766–1844), who proposed a new theory of matter. Dalton reasoned as follows:

- *Elements are made of tiny particles called **atoms.*** Although Dalton didn't know what atoms were like, he nevertheless felt that they were necessary to explain why there were so many different elements.

- *Each element is characterized by the mass of its atoms. Atoms of the same element have the same mass, but atoms of different elements have different masses.* Dalton realized that there must be some feature that distinguishes the atoms of one element from those of another. Because Proust's law of definite proportions showed that elements always combine in specific mass ratios, Dalton reasoned that the distinguishing feature between atoms of different elements must be *mass.*

- *Chemical combination of elements to make different substances occurs when atoms join together in small, whole-number ratios.* Only if whole numbers of atoms combine will different samples of a pure compound always contain the same proportion of elements by mass (the law of definite proportions). Fractional parts of atoms are never involved in chemical reactions.

- *Chemical reactions only rearrange the way that atoms are combined; the atoms themselves are not changed.* Dalton realized that atoms must be chemically indestructible for the law of mass conservation to be valid. If the same numbers and kinds of atoms are present in both reactants and products, then the masses of reactants and products must also be the same.

These samples of sulfur and carbon have different masses but contain the same number of atoms.

For any theory to be successful, it must not only explain known observations, it must also lead to the prediction of events or results yet unknown. Dalton realized that his atomic theory does exactly this: It predicts what has come to be called the *law of multiple proportions.*

LAW OF MULTIPLE PROPORTIONS: Elements can combine in different ways to form different substances, whose mass ratios are small, whole-number multiples of each other.

Read again the third statement in Dalton's atomic theory: "Chemical combination . . . occurs when atoms join together in small, whole-number ratios." According to this statement, it might be (and often is) possible for the same elements to combine in *different* ratios to give *different* substances. For example, nitrogen and oxygen can combine either in a 7:8 mass ratio to make a substance we know today as NO, or in a 7:16 mass ratio to make a substance

Copper metal reacts with nitric acid (HNO_3) to yield the brown gas NO_2.

we know as NO_2. Clearly, the second substance contains exactly twice as much oxygen as the first.

NO: 7 g nitrogen per 8 g oxygen   N:O mass ratio = 7:8

NO_2:   7 g nitrogen per 16 g oxygen   N:O mass ratio = 7:16

Comparison of N:O ratios in NO and NO_2

$$\frac{\text{N:O mass ratio in NO}}{\text{N:O mass ratio in NO}_2} = \frac{(7\,\text{g N})/(8\,\text{g O})}{(7\,\text{g N})/(16\,\text{g O})} = 2$$

This result makes sense only if we assume that matter is composed of discrete atoms, which combine with one another in specific and well-defined ways (Figure 2.2). All the atoms of a particular element have the same mass, and atoms of different elements have different characteristic masses.

FIGURE 2.2   An illustration of Dalton's law of multiple proportions. Atoms of nitrogen and oxygen can combine in specific proportions to make either NO or NO_2. NO_2 contains exactly twice as many atoms of oxygen per atom of nitrogen as NO does.

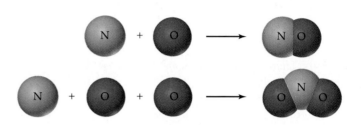

EXAMPLE 2.1   Methane and ethane are both constituents of natural gas. A sample of methane contains 11.40 g of carbon and 3.80 g of hydrogen, whereas a sample of ethane contains 4.47 g of carbon and 1.118 g of hydrogen. Show that the two substances obey the law of multiple proportions.

SOLUTION   First, find the C:H mass ratio in each compound.

$$\text{Methane:}\quad \text{C:H mass ratio} = \frac{11.40\,\text{g C}}{3.80\,\text{g H}} = 3.00$$

$$\text{Ethane:}\quad \text{C:H mass ratio} = \frac{4.47\,\text{g C}}{1.118\,\text{g H}} = 4.00$$

The two C:H ratios are clearly small, whole-number multiples of each other:

$$\frac{\text{C:H mass ratio in methane}}{\text{C:H mass ratio in ethane}} = \frac{3.00}{4.00} = \frac{3}{4}$$

▶**PROBLEM 2.1**   Substances 1 and 2 are colorless gases obtained by combination of sulfur with oxygen. Substance 1 results from combination of 6.00 g of sulfur with 5.99 g of oxygen, and substance 2 results from combination of 8.60 g of sulfur with 12.88 g of oxygen. Show that the mass ratios in the two substances are simple multiples of each other.

Sulfur burns with a bluish flame to yield colorless SO_2 gas.

2.3   The Structure of Atoms: Electrons

Dalton's atomic theory is fine as far as it goes, but it leaves unanswered the obvious question: What is an atom made of? Dalton himself had no way of answering this question, and it was not until nearly a century later that experiments by the English physicist J. J. Thomson (1856–1940) provided some clues.

Thomson's experiments involved the use of *cathode-ray tubes*, early predecessors of today's television and computer displays. As illustrated in Figure 2.3, a cathode-ray tube is a glass tube from which the air has been removed and in which two thin pieces of metal, called *electrodes*, have been sealed. When a sufficiently high voltage is applied across the electrodes, an electric current flows through the tube from the negatively charged electrode (the *cathode*) to the positively charged electrode (the *anode*). If the tube is not fully evacuated but still contains a small amount of air or other gas, the flowing current is visible as a glow called a **cathode ray**. Furthermore, if the anode has a hole in it and the end of the tube is coated with a phosphorescent substance such as zinc sulfide, the rays pass through the hole and strike the end of the tube, where they are visible as a bright spot of light. (In fact, this is exactly what happens in a television set.)

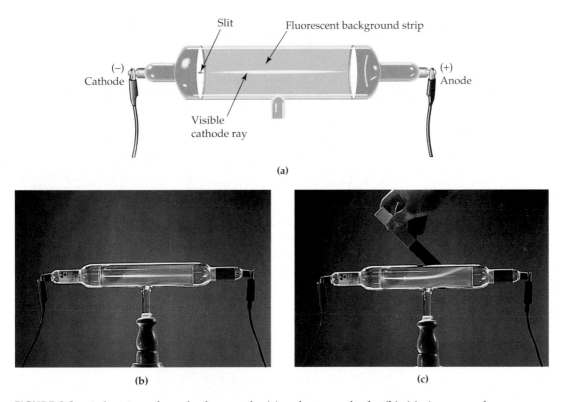

FIGURE 2.3   A drawing of a cathode-ray tube **(a)** and an actual tube **(b), (c)**. A stream of rays (electrons) emitted from the negatively charged cathode passes through a slit, moves toward the positively charged anode, and is detected by a fluorescent strip. The electron beam ordinarily travels in a straight line **(b)**, but it is deflected if either a magnetic field **(c)** or an electric field is present.

Experiments by a number of physicists in the 1890s had shown that cathode rays can be deflected by bringing either a magnet or an electrically charged plate near the tube (Figure 2.3c). Because the beam is produced at a *negative* electrode and is deflected toward a *positive* plate, Thomson proposed that cathode rays must consist of tiny negatively charged particles, which we now call **electrons**. Furthermore, because electrons are emitted from electrodes made of many different metals, all these different substances must contain electrons.

Thomson reasoned that the amount of deflection of an electron beam by a nearby magnetic or electric field must depend on three factors:

- *The strength of the deflecting magnetic or electric field.* The stronger the magnet or the higher the voltage on the charged plate, the greater the deflection.
- *The size of the negative charge on the electron.* The larger the charge on the particle, the greater its interaction with the magnetic or electric field, and the greater the deflection.
- *The mass of the electron.* The lighter the particle, the greater its deflection. (Just as it's easier to deflect a Ping-Pong ball than a bowling ball.)

By carefully measuring the amount of deflection caused by electric and magnetic fields of known strength, Thomson was able to calculate the ratio of the electron's electric charge to its mass—its *charge-to-mass ratio, e/m*. The modern value is

$$\frac{e}{m} = 1.758\ 820 \times 10^8\ \text{C/g}$$

where e is the magnitude of the charge on the electron in *coulombs* (C) and m is the mass of the electron in grams. (We'll say more about coulombs and electrical charge in Chapter 18.) Note that, because e is defined as a *positive* quantity, the actual (negative) charge on the electron is $-e$.

 Millikan Oil Drop Experiment movie

Thomson was able to measure only the *ratio* of charge to mass, not mass itself, and it was left to the American R. A. Millikan (1868–1953) to devise a method for measuring the mass of an electron (Figure 2.4). In Millikan's experiment, a fine mist of oil was sprayed into a box, and the tiny droplets were allowed to fall between two horizontal plates. Observing the spherical drops through a telescopic eyepiece made it possible to determine how rapidly they fell through the air, which allowed their masses to be calculated. The drops were then given a negative charge by irradiation with X rays. By applying a voltage to the plates, with the upper plate positive, it was possible to counteract the downward fall of the charged drops and keep them suspended.

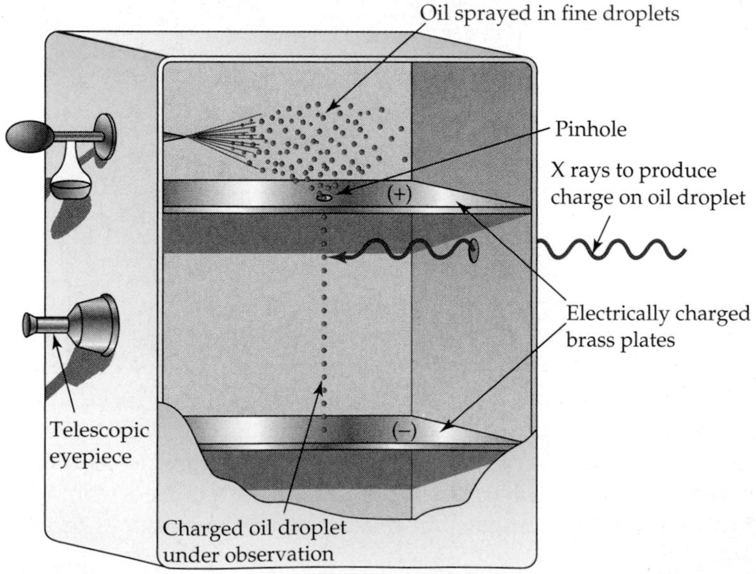

FIGURE 2.4   Millikan's oil-drop experiment. The falling oil droplets are given a negative charge, which makes it possible for them to be suspended between two electrically charged plates. Knowing the mass of the drop and the voltage on the plates makes it possible to calculate the charge on the drop.

Oil sprayed in fine droplets

Pinhole

X rays to produce charge on oil droplet

Electrically charged brass plates

Telescopic eyepiece

Charged oil droplet under observation

With the voltage on the plates and the mass of the drops known, Millikan was able to show that the charge on a given drop was always a small, whole-number multiple of e, whose modern value is $1.602\ 176 \times 10^{-19}$ C. Substituting into Thomson's charge-to-mass ratio then gives the mass m of the electron as $9.109\ 382 \times 10^{-28}$ g:

Because $\qquad \dfrac{e}{m} = 1.758\ 820 \times 10^{8}\ \text{C/g}$

then $\qquad m = \dfrac{e}{1.758\ 820 \times 10^{8}\ \text{C/g}} = \dfrac{1.602\ 176 \times 10^{-19}\ \cancel{\text{C}}}{1.758\ 820 \times 10^{8}\ \dfrac{\cancel{\text{C}}}{\text{g}}}$

$\qquad\qquad = 9.109\ 382 \times 10^{-28}\ g$

2.4   The Structure of Atoms: Protons and Neutrons

Think about the consequences of Thomson's cathode-ray experiments. Because matter is electrically neutral overall, the fact that the atoms in an electrode can give off *negatively* charged particles (electrons) must mean that those same atoms also contain *positively* charged particles. The search for those positively charged particles and for an overall picture of atomic structure led to a landmark experiment published in 1911 by the British physicist Ernest Rutherford (1871–1937).

Rutherford's work involved the use of **alpha (α) particles**, a type of emission previously observed to be given off by a number of naturally occurring radioactive elements, including radium, polonium, and radon. Rutherford knew that alpha particles are about 7000 times more massive than electrons and that they have a positive charge that is twice the magnitude of, but opposite in sign to, the charge on an electron.

When Rutherford directed a beam of alpha particles at a thin gold foil, he found that almost all the particles passed through the foil undeflected. A very small number (about 1 of every 20,000) were deflected at an angle, however, and a few actually bounced back toward the particle source (Figure 2.5).

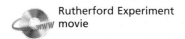

Rutherford Experiment movie

An alpha particle (relative mass = 7000; charge = +2e)

An electron (relative mass = 1; charge = −1e)

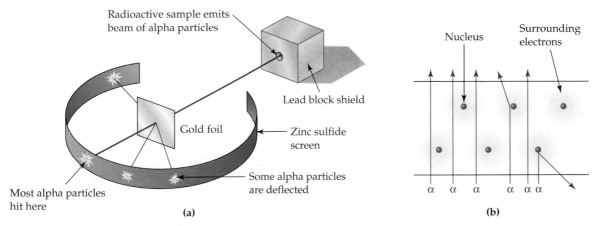

FIGURE 2.5   The Rutherford scattering experiment. **(a)** When a beam of alpha particles is directed at a thin gold foil, most particles pass through the foil undeflected, but a small number are deflected at large angles and a few bounce back toward the particle source. **(b)** A closeup view shows how most of an atom is empty space and only the alpha particles that strike a nucleus are deflected.

Rutherford explained his results by proposing that a metal atom must be almost entirely empty space and have its mass concentrated in a tiny central core that he called the **nucleus**. If the nucleus contains the atom's positive charges and most of its mass, and if the electrons move in space a relatively large distance away, then it is clear why the observed scattering results are obtained: Most alpha particles encounter empty space as they fly through the foil. Only when a positive alpha particle chances to come near a small but massive positive nucleus is it repelled strongly enough to make it bounce backward (Figure 2.5b).

Modern measurements show that an atom has a diameter of roughly 10^{-10} m and that a nucleus has a diameter of about 10^{-15} m. It's difficult to imagine from these numbers alone, though, just how small a nucleus really is. For comparison purposes, if an atom were the size of a large domed stadium, the nucleus would be approximately the size of a small pea in the center of the playing field.

Further experiments by Rutherford and others in the period from 1910 to 1930 showed that a nucleus is composed of two kinds of particles, called *protons* and *neutrons*. **Protons** have a mass of $1.672\,622 \times 10^{-24}$ g (about 1836 times greater than that of an electron) and are positively charged. Because the charge on a proton is opposite in sign but equal in size to that on an electron, the numbers of protons and electrons in a neutral atom are equal. **Neutrons** ($1.674\,927 \times 10^{-24}$ g) are almost identical in mass to protons but carry no charge. Thus, the number of neutrons in a nucleus is not directly related to the numbers of protons and electrons. Table 2.1 compares the three fundamental subatomic particles, and Figure 2.6 gives an overall view of the atom.

The relative size of the nucleus in an atom is roughly the same as that of a pea in the middle of this stadium.

TABLE 2.1 A Comparison of Subatomic Particles

Particle	Mass		Charge	
	grams	amu*	coulombs	e
Electron	$9.109\,382 \times 10^{-28}$	$5.485\,799 \times 10^{-4}$	$-1.602\,176 \times 10^{-19}$	-1
Proton	$1.672\,622 \times 10^{-24}$	$1.007\,276$	$+1.602\,176 \times 10^{-19}$	$+1$
Neutron	$1.674\,927 \times 10^{-24}$	$1.008\,665$	0	0

* The atomic mass unit (amu) is defined in Section 2.6.

FIGURE 2.6 A view of the atom. The protons and neutrons in the nucleus take up very little volume but contain essentially all the atom's mass. A number of electrons equal to the number of protons move around the nucleus and account for most of the atom's volume.

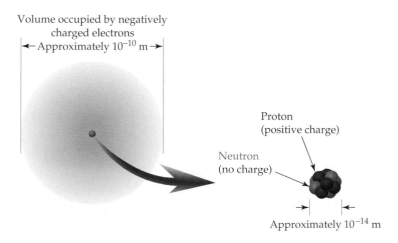

Volume occupied by negatively charged electrons

Approximately 10^{-10} m

Proton (positive charge)

Neutron (no charge)

Approximately 10^{-14} m

EXAMPLE 2.2 Ordinary "lead" pencils actually are made of a form of carbon called graphite. If a pencil line is 0.35 mm wide and the diameter of a carbon atom is 1.5×10^{-10} m, how many atoms wide is the line?

BALLPARK SOLUTION Since a single carbon atom is about 10^{-10} m wide, it takes 10^{10} carbon atoms placed side by side to stretch 1 m, 10^7 carbon atoms to stretch 1 mm, and about 0.3×10^7 (or 3×10^6; 3 *million*) carbon atoms to stretch 0.35 mm.

DETAILED SOLUTION Beginning with a known piece of information, set up an equation using appropriate conversion factors so that the unwanted units cancel. In the present instance, let's start by writing the width of the line in millimeters and then converting to meters. Dividing the line width in meters by the diameter of a single atom in meters then gives the answer:

$$\text{Atoms} = 0.35 \text{ mm} \times \frac{1 \text{ m}}{1000 \text{ mm}} \times \frac{1 \text{ atom}}{1.5 \times 10^{-10} \text{ m}} = 2.3 \times 10^6 \text{ atoms}$$

The ballpark solution and the detailed solution agree.

▶ **PROBLEM 2.2** The gold foil Rutherford used in his scattering experiment had a thickness of approximately 0.0002 in. If a single gold atom has a diameter of 2.9×10^{-8} cm, how many atoms thick was Rutherford's foil?

▶ **PROBLEM 2.3** A small speck of carbon the size of a pinhead contains about 10^{19} atoms, the diameter of a carbon atom is 1.5×10^{-10} m, and the circumference of the earth at the equator is 40,075 km. How many times around the earth would the atoms from this speck of carbon extend if they were laid side by side?

2.5 Atomic Number

Thus far, we have described atoms only in general terms and have not yet answered the most important question: What is it that makes one atom different from another? How, for example, does an atom of gold differ from an atom of carbon? The answer to this question is really quite simple: *Elements differ from one another according to the number of protons in their atoms*, a value called the element's **atomic number (Z)**. All atoms of the same element contain the same number of protons. Hydrogen atoms, atomic number 1, have 1 proton; helium atoms, atomic number 2, have 2 protons; carbon atoms, atomic number 6, have 6 protons; and so on. Of course, every neutral atom also contains a number of electrons equal to its number of protons.

Atomic number (Z) = Number of protons in atom's nucleus

= Number of electrons surrounding atom's nucleus

A hydrogen atom
(1 proton; 1 electron)

A carbon atom
(6 protons; 6 electrons)

A gold atom
(79 protons; 79 electrons)

In addition to protons, the nuclei of most atoms also contain neutrons. The sum of the number of protons (Z) and the number of neutrons (N) in an atom is called the atom's **mass number (A)**. That is, $A = Z + N$.

Mass number (A) = Number of protons (Z) + Number of neutrons (N)

Most hydrogen atoms have 1 proton and no neutrons, so their mass number is 1. Most helium atoms have 2 protons and 2 neutrons, so their mass number is 4. Most carbon atoms have 6 protons and 6 neutrons, so their mass number is 12; and so on. Except for hydrogen, atoms always contain at least as many neutrons as protons, although there is no simple way to predict how many neutrons a given atom will have.

Notice in the previous paragraph that we said *most* hydrogen atoms have mass number 1, *most* helium atoms have mass number 4, and *most* carbon atoms have mass number 12. In fact, different atoms of the same element can have different mass numbers depending on how many neutrons they have. Atoms with identical atomic numbers but different mass numbers are called **isotopes**. Hydrogen, for example, has three isotopes.

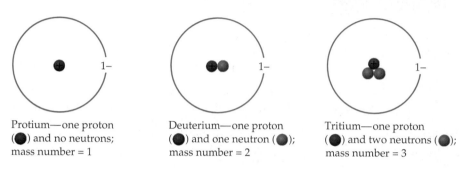

Protium—one proton
(●) and no neutrons;
mass number = 1

Deuterium—one proton
(●) and one neutron (●);
mass number = 2

Tritium—one proton
(●) and two neutrons (●);
mass number = 3

All hydrogen atoms have 1 proton in their nucleus (otherwise they wouldn't be hydrogen), but most (99.985%) have no neutrons. These hydrogen atoms, called *protium*, have mass number 1. In addition, 0.015% of hydrogen atoms, called *deuterium*, have 1 neutron and mass number 2. Still other hydrogen atoms, called *tritium*, have 2 neutrons and mass number 3. An unstable, radioactive isotope, tritium occurs only in trace amounts on earth but is made artificially in nuclear reactors. As other examples, there are 13 known isotopes of carbon, only 2 of which occur commonly, and 19 known isotopes of uranium, only 3 of which occur commonly.

A given isotope is represented by showing its element symbol with mass number as a left superscript and atomic number as a left subscript. Thus, protium is represented as ^1_1H, deuterium as ^2_1H, and tritium as ^3_1H. Similarly, the two naturally occurring isotopes of carbon are represented as $^{12}_6C$ (spoken as "carbon-12") and $^{13}_6C$ (carbon-13). Note that the number of neutrons in an isotope can be calculated simply by subtracting the atomic number (subscript) from the mass number (superscript). For example, subtracting the atomic number 6 from the mass number 12 indicates that a $^{12}_6C$ atom has 6 neutrons.

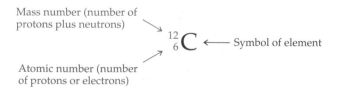

It's important to realize that the number of neutrons in an atom has little effect on the atom's chemical properties. The chemistry of an element is determined almost entirely by the number of electrons it has, which in turn is determined by the number of protons in its nucleus. All three isotopes of hydrogen behave almost identically in their chemical reactions.

EXAMPLE 2.3 The isotope of uranium used in nuclear power plants is $^{235}_{92}$U. How many protons, neutrons, and electrons does an atom of $^{235}_{92}$U have?

Uranium-235 is used as fuel in this nuclear power plant.

SOLUTION The subscript 92 in the symbol $^{235}_{92}$U indicates that a uranium atom has 92 protons and 92 electrons. The number of neutrons is equal to the difference between the mass number (235) and the atomic number (92): $235 - 92 = 143$ neutrons.

EXAMPLE 2.4 Element X is toxic to humans in high concentration but is essential to life at low concentrations. Identify element X, whose atoms contain 24 protons, and write the symbol for the isotope with 28 neutrons.

SOLUTION The atomic number is the number of protons in the atom's nucleus. A look at the periodic table shows that the element with atomic number 24 is chromium, Cr. The particular isotope of chromium in this instance has a mass number of $24 + 28 = 52$ and is written $^{52}_{24}$Cr.

▶ **PROBLEM 2.4** The isotope $^{75}_{34}$Se is used medically for diagnosis of pancreatic disorders. How many protons, neutrons, and electrons does an atom of $^{75}_{34}$Se have?

▶ **PROBLEM 2.5** Chlorine, one of the elements in common table salt (sodium chloride), has two main isotopes, with mass numbers 35 and 37. Look up the atomic number of chlorine, tell how many neutrons each isotope contains, and give the standard symbol for each.

▶ **PROBLEM 2.6** An atom of element X contains 47 protons and 62 neutrons. Identify the element, and write the symbol for the isotope in the standard format.

2.6 Atomic Mass

Atoms are so tiny that even the smallest speck of dust visible to the naked eye contains about 10^{16} atoms. Thus, the mass of a single atom in grams is much too small a number for convenience, and chemists therefore use a unit called an **atomic mass unit (amu)**, also known as a *dalton* (Da). One amu is defined as exactly one-twelfth the mass of an atom of $^{12}_{6}$C and is equal to $1.660\,539 \times 10^{-24}$ g:

$$\text{Mass of one } ^{12}_{6}\text{C atom} = 12 \text{ amu (exactly)}$$

$$1 \text{ amu} = \frac{\text{Mass of one } ^{12}_{6}\text{C atom}}{12} = 1.660\,539 \times 10^{-24} \text{ g}$$

Because the mass of an atom's electrons is negligible compared with the mass of its protons and neutrons, defining 1 amu as 1/12 the mass of a $^{12}_{6}C$ atom means that protons and neutrons each have a mass of almost exactly 1 amu (Table 2.1). Thus, the mass of an atom in atomic mass units—called the atom's *isotopic mass*—is numerically close to the atom's mass number. A $^{1}_{1}H$ atom, for instance, has a mass of 1.007 825 amu; a $^{235}_{92}U$ atom has a mass of 235.043 924 amu; and so forth.

Most elements occur naturally as a mixture of different isotopes. If you look at the periodic table inside the front cover, you'll see listed below the symbol for each element a value called the element's *atomic mass*. (The unit amu is understood but not specified.) An element's **atomic mass** is the weighted average of the isotopic masses of the element's naturally occurring isotopes. (The term *atomic weight* was used formerly, but atomic mass is more correct.)

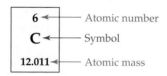

Carbon, for example, occurs on earth as a mixture of two isotopes, $^{12}_{6}C$ (98.89% natural abundance) and $^{13}_{6}C$ (1.11% natural abundance). Although the isotopic mass of any *individual* carbon atom is either 12 amu (a carbon-12 atom) or 13.0034 amu (a carbon-13 atom), the *average* isotopic mass—that is, the atomic mass—of a large collection of carbon atoms is 12.011 amu. (A third carbon isotope, $^{14}_{6}C$, is produced in small amounts in the upper atmosphere when cosmic rays strike $^{14}_{7}N$, but its abundance is so low that it can be ignored when calculating atomic mass.)

$$\text{Atomic mass of C} = (\text{Fraction of } ^{12}_{6}C)(\text{Mass of } ^{12}_{6}C) + (\text{Fraction of } ^{13}_{6}C)(\text{Mass of } ^{13}_{6}C)$$

$$= (0.9889 \times 12 \text{ amu}) + (0.0111 \times 13.0034 \text{ amu})$$

$$= 11.867 \text{ amu} + 0.144 \text{ amu} = 12.011 \text{ amu}$$

The advantage of using atomic masses is that it allows us to *count* a large number of atoms by *weighing* a sample of the element. For example, we can calculate that a small speck of carbon weighing 1.00 mg (1.00×10^{-3} g) contains 5.01×10^{19} carbon atoms:

$$1.00 \times 10^{-3} \text{ g} \times \frac{1 \text{ amu}}{1.6605 \times 10^{-24} \text{ g}} \times \frac{1 \text{ C atom}}{12.011 \text{ amu}} = 5.01 \times 10^{19} \text{ C atoms}$$

We'll have many occasions in future chapters to take advantage of this relationship between mass, atomic mass, and number of atoms.

EXAMPLE 2.5   Chlorine has two naturally occurring isotopes: $^{35}_{17}Cl$ with an abundance of 75.77% and an isotopic mass of 34.969 amu, and $^{37}_{17}Cl$ with an abundance of 24.23% and an isotopic mass of 36.966 amu. What is the atomic mass of chlorine?

SOLUTION   The atomic mass of an element is equal to the sum of the masses of each isotope times the fraction of that isotope:

$$\text{Atomic mass} = (\text{Fraction of } ^{35}_{17}Cl)(\text{Mass of } ^{35}_{17}Cl) + (\text{Fraction of } ^{37}_{17}Cl)(\text{Mass of } ^{37}_{17}Cl)$$

$$= (0.7577 \times 34.969 \text{ amu}) + (0.2423 \times 36.966 \text{ amu}) = 35.45 \text{ amu}$$

An alternative way of looking at the problem is to imagine that you have a large number of chlorine atoms, say 10,000. Of the 10,000 atoms, 7577 of them (75.77%) would have an isotopic mass of 34.969 amu, and 2423 of them (24.23%) would have an isotopic mass of 36.966 amu. Thus, the total mass of the 10,000 atoms would be

$$\left(7577 \ \cancel{atoms} \times 34.969 \ \frac{amu}{\cancel{atom}} \right) + \left(2423 \ \cancel{atoms} \times 36.966 \ \frac{amu}{\cancel{atom}} \right) = 354{,}500 \ amu$$

and the atomic mass would be

$$\frac{354{,}500 \ amu}{10{,}000 \ atoms} = 35.45 \ \frac{amu}{atom}$$

▶ **PROBLEM 2.7**   Copper metal has two naturally occurring isotopes: copper-63 (69.17%; isotopic mass 62.94 amu) and copper-65 (30.83%; isotopic mass 64.93 amu). Calculate the atomic mass of copper, and check your answer in a periodic table.

▶ **PROBLEM 2.8**   Based on your answer to Problem 2.7, how many atoms of copper are in a pure copper penny that weighs 2.15 g? (1 amu = 1.6605×10^{-24} g)

2.7   Compounds and Mixtures

Although only 90 elements occur naturally, there are obviously far more than 90 different kinds of matter on earth. Just look around, and you'll surely find a few hundred. All the many kinds of matter can be classified as either *pure substances* or *mixtures*, as shown in Figure 2.7. Pure substances, in turn, can be either elements or *chemical compounds*.

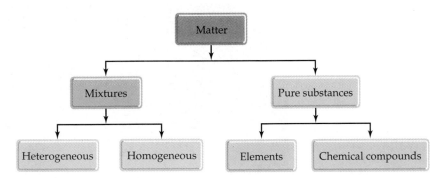

FIGURE 2.7   A scheme for the classification of matter.

Chemical Compounds

A **compound** is a pure substance that is formed when atoms of two or more different elements combine and create a new material with properties completely unlike those of its constituent elements. A compound has a constant composition throughout, and its constituent units are all identical. For example, when atoms of sodium (a soft, lustrous metal) combine with atoms of chlorine (a toxic, yellow-green gas), the familiar white solid called sodium chloride (table salt) is formed. Similarly, when two atoms of hydrogen combine with one atom of oxygen, water is formed. Such transformations are said to be **chemical reactions**.

A compound is written by giving its **chemical formula**, which lists the symbols of the individual constituent elements and indicates the number of atoms of each element with a subscript. If no subscript is given, the number 1 is understood. Thus, sodium chloride is written as NaCl, water as H_2O, and sucrose (table sugar) as $C_{12}H_{22}O_{11}$. A chemical reaction is written in a standard format called a **chemical equation**, in which the starting substances, or *reactants*, are on the left, the final substances, or *products*, are on the right, and an arrow is placed between them to indicate a transformation. Note that the numbers and kinds of atoms are the same on both sides of the reaction arrow, as required by the law of mass conservation.

$$O_2 + 2\,H_2 \longrightarrow 2\,H_2O$$

Two oxygen atoms   Four hydrogen atoms   Two water molecules (H_2O, a chemical compound)

Mixtures

Unlike chemical compounds, whose constituent units are identical, **mixtures** are simply blends of two or more substances added together in some random proportion without chemically changing the individual substances in the mixture. Thus, hydrogen gas and oxygen gas can be mixed in any ratio without changing them (as long as there is no flame nearby to initiate reaction), just as a spoonful of sugar and a spoonful of salt can be mixed.

Mixtures and Compounds movie

The crystalline quartz sand on this beach is a pure compound (SiO_2), but the seawater is a liquid mixture of many compounds dissolved in water.

Mixtures can be further classified as either *heterogeneous* or *homogeneous*. **Heterogeneous mixtures** are those in which the mixing is not uniform and which therefore have regions of different composition. Sand with sugar, water with gasoline, and dust with air, are all heterogeneous mixtures. **Homogeneous mixtures** are those in which the mixing *is* uniform and which therefore have a constant composition throughout. Air is a gaseous mixture of (primarily) oxygen and nitrogen, seawater is a liquid mixture of (primarily) sodium chloride dissolved in water, and brass is a solid mixture of copper and zinc.

With liquids it's often possible to distinguish between a homogeneous mixture and a heterogeneous one simply by looking. Heterogeneous mixtures tend to be cloudy and will separate on standing, whereas homogeneous mixtures are often transparent. We'll look further at the nature and properties of liquid mixtures in Chapter 11.

Bubbles aside, soft drinks are homogeneous mixtures, but the oil and vinegar used as a salad dressing is a heterogeneous mixture.

KEY CONCEPT PROBLEM 2.9   Which of the following drawings represents a mixture, which a pure compound, and which an element?

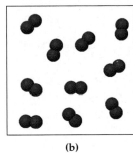

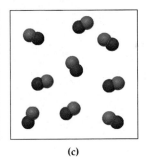

(a)   (b)   (c)

KEY CONCEPT PROBLEM 2.10   Which of the following drawings represents a collection of hydrogen peroxide (H_2O_2) molecules? The red spheres represent oxygen atoms and the ivory spheres represent hydrogen.

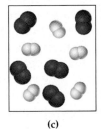

(a)   (b)   (c)   (d)

2.8 Molecules, Ions, and Chemical Bonds

Imagine what must happen when two atoms approach each other at the beginning of a chemical reaction. Because the electrons of an atom occupy a much greater volume than the nucleus, it's the *electrons* that actually make the contact when atoms collide. Thus, it's the electrons that form the connections, or **chemical bonds**, that join atoms together in compounds.

There are three fundamental kinds of chemical bonds between atoms—*covalent bonds*, *ionic bonds*, and *metallic bonds*. As a general rule, covalent bonds occur between two nonmetals, ionic bonds occur between a metal and a nonmetal, and metallic bonds occur between two metals. We'll look briefly at the first two kinds now and defer a discussion of metallic bonds until Chapter 21.

Covalent Bonds: Molecules

A **covalent bond**, the most common kind of chemical bond, results when two atoms *share* some (usually two) electrons. A simple way to think about a covalent bond is to imagine it as a tug-of-war. If two people pull on the same rope, they are effectively joined together. Neither person can escape from the other as long as both hold on. Similarly with atoms: When two atoms both hold on to some shared electrons, the atoms are bonded together.

(a)

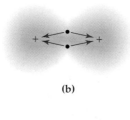

(b)

(a) The two teams are joined together because both are tugging on the same rope. **(b)** Similarly, two atoms are joined together when both nuclei (+) tug on the same electrons, represented here by dots.

The unit of matter that results when two or more atoms are joined by covalent bonds is called a **molecule**. A hydrogen chloride molecule (HCl) results when a hydrogen atom and a chlorine atom share two electrons. A water molecule (H_2O) results when each of two hydrogen atoms shares two electrons with a single oxygen atom. An ammonia molecule (NH_3) results when three hydrogen atoms share two electrons each with a nitrogen atom, and so on. To visualize these and other molecules, it helps to imagine the individual atoms as spheres joined together to form molecules with specific three-dimensional shapes. So-called *ball-and-stick* models

(Figure 2.8a) specifically indicate the covalent bonds between atoms, while *space-filling* models (Figure 2.8b) give an accurate portrait of overall molecular shape.

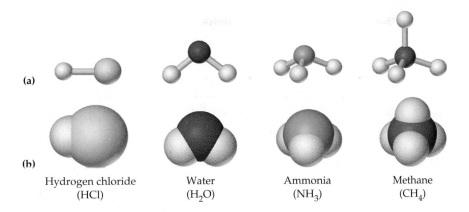

(a)

(b)

| Hydrogen chloride (HCl) | Water (H₂O) | Ammonia (NH₃) | Methane (CH₄) |

FIGURE 2.8 Molecular drawings such as these help in visualizing molecules. **(a)** Ball-and-stick models show individual atoms (spheres) joined together by covalent bonds (sticks). **(b)** Space-filling models accurately portray the overall molecular shape but do not explicitly show covalent bonds.

We sometimes represent a molecule by giving its **structural formula**, which shows the specific connections between atoms and therefore gives more information than the chemical formula alone. Ethyl alcohol, for example, has the chemical formula C_2H_6O and the following structural formula:

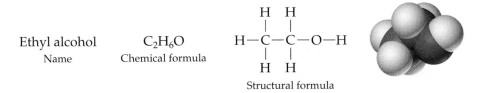

Ethyl alcohol C_2H_6O

Name Chemical formula Structural formula

A structural formula uses lines between atoms to indicate the covalent bonds. Thus, the two carbon atoms in ethyl alcohol are covalently bonded to each other, the oxygen atom is bonded to one of the carbon atoms, and the six hydrogen atoms are distributed three to one carbon, two to the other carbon, and one to the oxygen.

Even some *elements* exist as molecules rather than as atoms. Hydrogen, nitrogen, oxygen, fluorine, chlorine, bromine, and iodine all exist as diatomic (two-atom) molecules whose two atoms are held together by covalent bonds. We therefore have to write them as such when using any of these elements in a chemical equation.

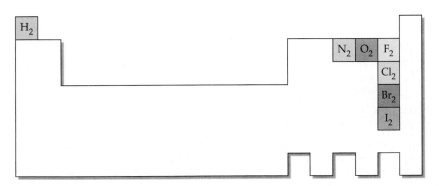

EXAMPLE 2.6   Propane, C_3H_8, has a structure in which the three carbon atoms are bonded in a row, each end carbon is bonded to three hydrogens, and the middle carbon is bonded to two hydrogens. Draw the structural formula, using lines between atoms to represent covalent bonds.

SOLUTION

$$
\begin{array}{c}
\quad\ \ \text{H}\quad\ \ \text{H}\quad\ \ \text{H} \\
\quad\ \ | \qquad | \qquad | \\
\text{H}-\text{C}-\text{C}-\text{C}-\text{H} \qquad \textit{Propane} \\
\quad\ \ | \qquad | \qquad | \\
\quad\ \ \text{H}\quad\ \ \text{H}\quad\ \ \text{H}
\end{array}
$$

▶**PROBLEM 2.11**   Draw the structural formula of methylamine, CH_5N, a substance responsible for the odor of rotting fish. The carbon atom is bonded to the nitrogen atom and to three hydrogens. The nitrogen atom is bonded to the carbon and two hydrogens.

KEY CONCEPT PROBLEM 2.12   Adrenaline, the so-called "flight or fight" hormone, can be represented as the following ball-and-stick model. What is the molecular formula of adrenaline? (Gray = C, ivory = H, red = O, blue = N)

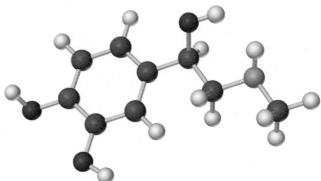

Ionic Bonds

NaCl 3D movie

An **ionic bond**, the second kind of chemical bond, results not from a sharing of electrons but from a *transfer* of one or more electrons from one atom to another. As noted earlier, ionic bonds generally form between a metal and a nonmetal. Metallic elements, such as sodium, magnesium, and zinc, tend to give up electrons, whereas nonmetallic elements, such as oxygen, nitrogen, and chlorine, tend to accept electrons.

Imagine what happens when sodium metal comes in contact with chlorine gas. The sodium atom gives an electron to chlorine, resulting in the formation of two charged particles, called **ions**. Because a sodium atom *loses* one electron, it loses one negative charge and becomes an Na^+ ion with a charge of +1. Such positive ions are called **cations** (**cat**-ions). Conversely, because a chlorine atom *gains* an electron, it gains a negative charge and becomes a Cl^- ion with a charge of −1. Such negative ions are called **anions** (**an**-ions).

Chlorine is a toxic green gas, sodium is a reactive metal, and sodium chloride is a harmless white solid.

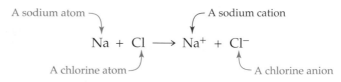

A similar reaction takes place when magnesium and chlorine (Cl_2) come in contact to form $MgCl_2$. Magnesium transfers *two* electrons to two chlorine atoms, yielding the doubly charged Mg^{2+} cation and two Cl^- anions.

$$Mg + Cl_2 \longrightarrow Mg^{2+} + Cl^- + Cl^- \ (MgCl_2)$$

Since opposite charges attract, positively charged cations like Na^+ and Mg^{2+} feel a strong electrical attraction to negatively charged anions like Cl^-, an attraction that we call an ionic bond. Unlike what happens when covalent bonds are formed, though, we can't really talk about discrete Na^+Cl^- *molecules* under normal conditions. We can speak only of an **ionic solid**, in which equal numbers of Na^+ and Cl^- ions are packed together in a regular way (Figure 2.9). In a crystal of table salt, for instance, each Na^+ ion is surrounded by six nearby Cl^- ions, and each Cl^- ion is surrounded by six nearby Na^+ ions, but we can't specify what pairs of ions "belong" to each other as we can with atoms in covalent molecules.

FIGURE 2.9 The arrangement of Na^+ ions and Cl^- ions in a crystal of sodium chloride. Each Na^+ ion is surrounded by six neighboring Cl^- ions, and each Cl^- ion is surrounded by six neighboring Na^+ ions. Thus, there is no discrete "molecule" of NaCl. Instead, the entire crystal is an ionic solid.

Charged, covalently bonded *groups* of atoms, called **polyatomic ions**, also exist—for example, ammonium ion (NH_4^+), hydroxide ion (OH^-), nitrate ion (NO_3^-), and the doubly charged sulfate ion (SO_4^{2-}). You can think of these polyatomic ions as charged molecules because they consist of specific numbers and kinds of atoms joined together by covalent bonds in a definite way, with the overall unit having a positive or negative charge. When writing the formulas of substances that contain more than one of these ions, parentheses are placed around the entire polyatomic unit. The formula $Ba(NO_3)_2$, for instance, indicates a substance made of Ba^{2+} cations and NO_3^- polyatomic anions in a 1:2 ratio. We'll say more about these ions in Section 2.10.

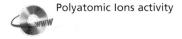
Polyatomic Ions activity

EXAMPLE 2.7 Which of the following compounds would you expect to be ionic and which molecular?
(a) BaF_2 **(b)** SF_4 **(c)** PH_3 **(d)** CH_3OH

SOLUTION Compound **(a)** is composed of a metal (barium) and a nonmetal (fluorine) and is likely to be ionic. Compounds **(b)–(d)** are composed entirely of nonmetals and therefore are probably molecular.

▶ **PROBLEM 2.13** Which of the following compounds would you expect to be ionic and which molecular?
(a) LiBr **(b)** $SiCl_4$ **(c)** BF_3 **(d)** CaO

KEY CONCEPT PROBLEM 2.14 Which of the following drawings is most likely to represent an ionic compound, and which a molecular compound? Explain.

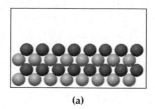

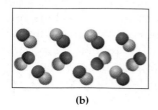

(a)   (b)

2.9 Acids and Bases

Among the many ions we'll be discussing throughout this book, two of the most important are the hydrogen cation (H^+) and the hydroxide anion (OH^-). Since a hydrogen *atom* contains one proton and one electron, a hydrogen *cation* is simply a proton. A hydroxide ion, by contrast, is a polyatomic anion in which an oxygen atom is covalently bonded to a hydrogen atom. Although much of Chapter 15 is devoted to the chemistry of H^+ and OH^- ions, it's worthwhile taking a preliminary look at these two species now.

The importance of the H^+ cation and OH^- anion is that they are fundamental to the concept of *acids* and *bases*. In fact, one useful definition of an **acid** is a substance that provides H^+ ions when dissolved in water, and one definition of a **base** is a substance that provides OH^- ions when dissolved in water.

ACID: A substance that provides H^+ ions in water $HCl, HNO_3, H_2SO_4, H_3PO_4$

BASE: A substance that provides OH^- ions in water $NaOH, KOH, Ba(OH)_2$

Hydrochloric acid (HCl), nitric acid (HNO_3), sulfuric acid (H_2SO_4), and phosphoric acid (H_3PO_4) are among the most common acids. When any of these substances is dissolved in water, H^+ ions are formed along with the corresponding anion. For example, HCl gives H^+ ions and Cl^- ions when it dissolves in water. We sometimes attach the designation (*aq*) to show that the ions are present in aqueous solution. In the same way, we often attach the designation (*g*) for gas, (*l*) for liquid, or (*s*) for solid to indicate the state of other reactants or products. For example, pure HCl is a gas, HCl(*g*), and pure HNO_3 is a liquid, HNO_3(*l*).

Different acids can provide different numbers of H^+ ions, depending on their structure. Hydrochloric acid and nitric acid provide one H^+ ion each per molecule, sulfuric acid can provide two H^+ ions per molecule, and phosphoric acid can provide three H^+ ions per molecule.

Hydrochloric acid:	$HCl(g)$	$\xrightarrow{\text{Dissolve in } H_2O}$	$H^+(aq) + Cl^-(aq)$
Nitric acid:	$HNO_3(l)$	$\xrightarrow{\text{Dissolve in } H_2O}$	$H^+(aq) + NO_3^-(aq)$
Sulfuric acid:	$H_2SO_4(l)$	$\xrightarrow{\text{Dissolve in } H_2O}$	$H^+(aq) + HSO_4^-(aq)$
	$HSO_4^-(aq)$	$\longrightarrow$	$H^+(aq) + SO_4^{2-}(aq)$

Phosphoric acid: $H_3PO_4(l)$ $\xrightarrow{\text{Dissolve in } H_2O}$ $H^+(aq) + H_2PO_4^-(aq)$

$H_2PO_4^-(aq)$ $\longrightarrow$ $H^+(aq) + HPO_4^{2-}(aq)$

$HPO_4^{2-}(aq)$ $\longrightarrow$ $H^+(aq) + PO_4^{3-}(aq)$

Sodium hydroxide (NaOH; also known as *lye* or *caustic soda*), potassium hydroxide (KOH; also known as *caustic potash*), and barium hydroxide [$Ba(OH)_2$] are examples of bases. When any of these compounds dissolves in water, OH^- anions go into solution along with the corresponding metal cation. Sodium hydroxide and potassium hydroxide provide one OH^- ion each, and barium hydroxide provides two OH^- ions, as indicated by its formula.

Sodium hydroxide: $NaOH(s)$ $\xrightarrow{\text{Dissolve in } H_2O}$ $Na^+(aq) + OH^-(aq)$

Potassium hydroxide: $KOH(s)$ $\xrightarrow{\text{Dissolve in } H_2O}$ $K^+(aq) \; + OH^-(aq)$

Barium hydroxide: $Ba(OH)_2(s)$ $\xrightarrow{\text{Dissolve in } H_2O}$ $Ba^{2+}(aq) + 2\,OH^-(aq)$

▶ **PROBLEM 2.15**   Which of the following compounds are acids, and which are bases? Explain.
(a) HF   (b) $Ca(OH)_2$ (c) LiOH   (d) HCN

2.10   Naming Chemical Compounds

In the early days of chemistry, when few pure substances were known, newly discovered compounds were often given fanciful names—morphine, quicklime, muriatic acid, and barbituric acid (named by its discoverer in honor of his friend Barbara)—to cite a few. Today, with well over 18 million pure compounds known, there would be chaos unless a systematic method for naming compounds were used. Ideally, every chemical compound can be given a name that not only defines it uniquely but also allows chemists (and computers) to know the compound's chemical structure.

Morphine, a pain-killing agent found in the opium poppy, was named after Morpheus, the Greek god of dreams.

Different kinds of compounds are named by different rules. Ordinary table salt, for instance, is named *sodium chloride* because of its formula NaCl, but common table sugar ($C_{12}H_{22}O_{11}$) is named *β-D-fructofuranosyl-α-D-glucopyranoside* because of special rules for carbohydrates (don't worry about it). We'll begin in this section by seeing how to name simple binary compounds—those made of only two elements—and then introduce additional rules in later chapters as the need arises.

Naming Binary Ionic Compounds

Binary ionic compounds are named by identifying first the positive ion and then the negative ion. The positive ion takes the same name as the element; the negative ion takes the first part of its name from the element and then adds the ending -*ide*. For example, KBr is named potassium bromide—*potassium* for the K^+ ion, and *bromide* for the negative Br^- ion derived from the element *brom*ine. Figure 2.10 shows some common main-group ions, and Figure 2.11 shows some common transition metal ions.

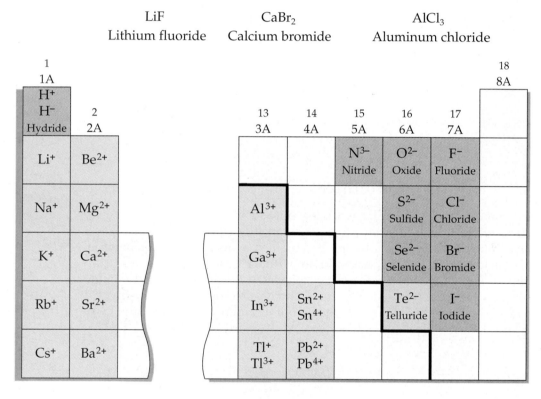

FIGURE 2.10 Main-group cations and anions. A cation bears the same name as the element it is derived from; an anion name has an -*ide* ending.

Main Group Ions activity

There are several interesting features about Figure 2.10. Note, for instance, that metals tend to form cations and nonmetals tend to form anions, as mentioned previously in Section 2.8. Note also that elements within a given group of the periodic table form similar kinds of ions and that the charge on the ion depends on the group number. Main-group metals usually form cations whose charge is equal to the group number. Group 1A elements form monopositive ions (M^+, where M is a metal), group 2A elements form doubly positive ions (M^{2+}), and group 3A elements form triply positive ions (M^{3+}). Main-group nonmetals usually form anions whose charge is equal to the group number minus eight. Thus, group 6A elements form doubly negative ions ($6 - 8 = -2$), group 7A elements form mononegative ions ($7 - 8 = -1$), and group 8A elements form no ions at all ($8 - 8 = 0$).

3	4	5	6	7	8	9	10	11	12
3B	4B	5B	6B	7B		8B		1B	2B
Sc^{3+}	Ti^{3+}	V^{3+}	Cr^{2+} Cr^{3+}	Mn^{2+}	Fe^{2+} Fe^{3+}	Co^{2+}	Ni^{2+}	Cu^{2+}	Zn^{2+}
Y^{3+}					Ru^{3+}	Rh^{3+}	Pd^{2+}	Ag^+	Cd^{2+}
La^{3+}									Hg^{2+}

FIGURE 2.11   Common transition metal ions. Only ions that exist in aqueous solution are shown.

Notice in both Figures 2.10 and 2.11 that some metals form more than one kind of cation. Iron, for example, forms two different cations: the doubly charged Fe^{2+} ion and the triply charged Fe^{3+} ion. In naming these ions, it's necessary to distinguish between them by using a Roman numeral in parentheses to indicate the number of charges. Thus, $FeCl_2$ is iron(II) chloride, and $FeCl_3$ is iron(III) chloride. Alternatively, an older method distinguishes between the ions by using the Latin name of the element (*ferrum*) together with the ending *-ous* for the ion with lower charge and *-ic* for the ion with higher charge. Thus, $FeCl_2$ is sometimes called ferrous chloride, and $FeCl_3$ is called ferric chloride. Though still in use, this older naming system is being phased out.

Transition Metal Ions activity

Ionic Compounds activity

Fe^{2+}	Fe^{3+}	Sn^{2+}	Sn^{4+}
Iron(II) ion	Iron(III) ion	Tin(II) ion	Tin(IV) ion
Ferrous ion	Ferric ion	Stannous ion	Stannic ion
(From the Latin *ferrum* = iron)		(From the Latin *stannum* = tin)	

In any neutral compound, the total positive charge must equal the total negative charge. Thus, in any binary compound, you can always figure out the number of positive charges on a cation by counting the number of negative charges on the associated anion(s). In $FeCl_2$, for example, the iron ion must be Fe(II) because there are two Cl^- ions associated with it. Similarly, in AlF_3 the aluminum ion is Al(III) because there are three F^- anions associated with it. As a general rule, a Roman numeral is needed for transition metal compounds to avoid ambiguity. Metals in group 1A and group 2A form only one cation, though, so Roman numerals are not needed.

Crystals of iron(II) chloride tetrahydrate are greenish; crystals of iron(III) chloride hexahydrate are brownish yellow.

EXAMPLE 2.8   Give systematic names for the following compounds:
(a) $BaCl_2$ **(b)** $CrCl_3$ **(c)** PbS   **(d)** Fe_2O_3

SOLUTION   Refer to Figures 2.10 and 2.11 if you are unsure about charges on the ions.

(a) Barium chloride	No Roman numeral is necessary because barium, a group 2A element, forms only Ba^{2+}.
(b) Chromium(III) chloride	The Roman numeral III is necessary to specify the +3 charge on chromium (a transition metal).
(c) Lead(II) sulfide	The sulfide anion (S^{2-}) has a double negative charge, so the lead cation must be doubly positive.
(d) Iron(III) oxide	The three oxide anions (O^{2-}) have a total negative charge of −6, so the two iron cations must have a total charge of +6. Thus, each is Fe(III).

EXAMPLE 2.9 Write formulas for the following compounds:
(a) Magnesium hydride **(b)** Tin(IV) oxide **(c)** Iron(III) sulfide

SOLUTION Refer to Figures 2.10 and 2.11 to find the charges on ions.

(a) MgH_2 Magnesium (group 2A) forms only a 2+ cation, so there must be two hydride ions (H^-) to balance the charge.

(b) SnO_2 Tin(IV) has a +4 charge, so there must be two oxide ions (O^{2-}) to balance the charge.

(c) Fe_2S_3 Iron(III) has a +3 charge and sulfide ion a −2 charge (S^{2-}), so there must be two irons and three sulfurs.

▶**PROBLEM 2.16** Give systematic names for the following compounds:
(a) CsF **(b)** K_2O **(c)** CuO **(d)** BaS **(e)** $BeBr_2$

▶**PROBLEM 2.17** Write formulas for the following compounds:
(a) Vanadium(III) chloride **(b)** Manganese(IV) oxide
(c) Copper(II) sulfide **(d)** Aluminum oxide

Naming Binary Molecular Compounds

Binary molecular compounds are named by assuming that one of the two elements in the molecule is more cationlike and the other element is more anionlike. As with ionic compounds, the cationlike element takes the name of the element itself, and the anionlike element takes an -*ide* ending. The compound HF, for example, is called *hydrogen fluoride*.

HF Hydrogen is more cationlike because it is farther left in the periodic table, and fluoride is more anionlike because it is farther right. The compound is therefore named *hydrogen fluoride*.

We'll see a quantitative way to decide which element is more cationlike and which is more anionlike in Section 7.4, but we might note for now that it's usually possible to decide by looking at the relative positions of the elements in the periodic table. The farther left and toward the bottom of the periodic table the element occurs, the more likely it is to have cationic character; the farther right and toward the top the element occurs (except for the noble gases), the more likely it is to have anionic character.

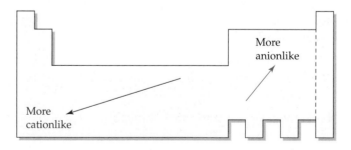

Look at the following examples to see how this generalization applies:

CO Carbon monoxide (C is in group 4A; O is in group 6A)
CO_2 Carbon dioxide
SF_4 Sulfur tetrafluoride (S is in group 6A; F is in group 7A)
N_2O_4 Dinitrogen tetroxide (N is in group 5A; O is in group 6A)

Because nonmetals often combine in different proportions to form different compounds, numerical prefixes are usually included in the names of binary molecular compounds to specify the numbers of each kind of atom present. The compound CO, for example, is called carbon *mon*oxide, and CO_2 is called carbon *di*oxide. Table 2.2 lists the most common prefixes. Note that when the prefix ends in *a* or *o* and the anion name begins with a vowel (*oxide*, for instance), the *a* or *o* on the prefix is dropped to avoid having two vowels together in the name. Thus, we write carbon *mon*oxide rather than carbon *mono*oxide and dinitrogen *tetr*oxide rather than dinitrogen *tetra*oxide. Note also that the *mono* prefix is not used for the atom named first. NO_2, for instance, is called nitrogen dioxide rather than mononitrogen dioxide.

TABLE 2.2 Numerical Prefixes for Naming Compounds	
Prefix	**Meaning**
mono-	1
di-	2
tri-	3
tetra-	4
penta-	5
hexa-	6
hepta-	7
octa-	8

EXAMPLE 2.10   Give systematic names for the following compounds:
(a) PCl_3 **(b)** N_2O_3 **(c)** P_4O_7

SOLUTION   Look at a periodic table to see which element in each compound is more cationlike (farther to the left or lower) and which is more anionlike (farther to the right or higher). Then name the compound using the appropriate numerical prefix.
(a) Phosphorus trichloride   **(b)** Dinitrogen trioxide
(c) Tetraphosphorus heptoxide

▶ **PROBLEM 2.18**   Give systematic names for the following compounds:
  (a) NCl_3 **(b)** P_4O_6 **(c)** S_2F_2

▶ **PROBLEM 2.19**   Write formulas for the following names:
  (a) Disulfur dichloride   **(b)** Iodine monochloride   **(c)** Nitrogen triiodide

Naming Compounds with Polyatomic Ions

Ionic compounds with polyatomic ions (Section 2.8) are named in the same way as binary ionic compounds: First the cation is identified and then the anion. For example, $Ba(NO_3)_2$ is called *barium nitrate* because Ba^{2+} is the cation and the NO_3^- polyatomic anion has the name *nitrate*. Unfortunately, there is no systematic way of naming the polyatomic ions themselves, so it's necessary to memorize the names, formulas, and charges of the most common ones listed in Table 2.3. The ammonium ion NH_4^+ is the only cation on the list; all the others are anions.

Several points about the ions in Table 2.3 need special mention. First, note that the names of most polyatomic anions end in *-ite* or *-ate*; only hydroxide (OH^-), cyanide (CN^-), and peroxide (O_2^{2-}) have the *-ide* ending. Second, note that several of the ions form a series of **oxoanions**, in which an atom of a given element is combined with different numbers of oxygen atoms—hypochlorite (ClO^-), chlorite (ClO_2^-), chlorate (ClO_3^-), and perchlorate (ClO_4^-), for example. When there are only two oxoanions in a series, as with sulfite (SO_3^{2-}) and sulfate (SO_4^{2-}), the ion with fewer oxygens takes the *-ite* ending and the ion with more oxygens takes the *-ate* ending.

These blue crystals of copper sulfate pentahydrate contain the Cu^{2+} cation and the polyatomic SO_4^{2-} anion.

SO_3^{2-} Sulf*ite* ion (fewer oxygens) SO_4^{2-} Sulf*ate* ion (more oxygens)
NO_2^- Nitr*ite* ion (fewer oxygens) NO_3^- Nitr*ate* ion (more oxygens)

TABLE 2.3 Some Common Polyatomic Ions

Formula	Name	Formula	Name
Cation		**Singly charged anions (continued)**	
NH_4^+	Ammonium	NO_2^-	Nitrite
		NO_3^-	Nitrate
Singly charged anions			
$CH_3CO_2^-$	Acetate	**Doubly charged anions**	
CN^-	Cyanide	CO_3^{2-}	Carbonate
ClO^-	Hypochlorite	CrO_4^{2-}	Chromate
ClO_2^-	Chlorite	$Cr_2O_7^{2-}$	Dichromate
ClO_3^-	Chlorate	O_2^{2-}	Peroxide
ClO_4^-	Perchlorate	HPO_4^{2-}	Hydrogen phosphate
$H_2PO_4^-$	Dihydrogen phosphate	SO_3^{2-}	Sulfite
HCO_3^-	Hydrogen carbonate	SO_4^{2-}	Sulfate
	(or bicarbonate)	$S_2O_3^{2-}$	Thiosulfate
HSO_4^-	Hydrogen sulfate		
	(or bisulfate)	**Triply charged anion**	
OH^-	Hydroxide	PO_4^{3-}	Phosphate
MnO_4^-	Permanganate		

When there are more than two oxoanions in the series, the prefix *hypo-* (meaning "less than") is used for the ion with the fewest oxygens, and the prefix *per-* (meaning "more than") is used for the ion with the most oxygens.

ClO^-	*Hypo*chlorite ion (less oxygen than chlorite)
ClO_2^-	Chlorite ion
ClO_3^-	Chlorate ion
ClO_4^-	*Per*chlorate ion (more oxygen than chlorate)

Third, note that several pairs of ions are related by the presence or absence of a hydrogen. The hydrogen carbonate anion (HCO_3^-) differs from the carbonate anion (CO_3^{2-}) by the presence of H^+, and the hydrogen sulfate anion (HSO_4^-) differs from the sulfate anion (SO_4^{2-}) by the presence of H^+. The ion that has the additional hydrogen is sometimes referred to using the prefix *bi-*, although this usage is now discouraged; for example, $NaHCO_3$ is sometimes called sodium bicarbonate.

HCO_3^-	Hydrogen carbonate (*bi*carbonate) ion	CO_3^{2-}	Carbonate ion
HSO_4^-	Hydrogen sulfate (*bi*sulfate) ion	SO_4^{2-}	Sulfate ion

EXAMPLE 2.11 Give systematic names for the following compounds:
(a) $LiNO_3$ **(b)** $KHSO_4$ **(c)** $CuCO_3$ **(d)** $Fe(ClO_4)_3$

SOLUTION Refer to Table 2.3 if you need help with the names of polyatomic ions.

(a) Lithium nitrate	Lithium (group 1A) forms only the Li^+ ion and does not need a Roman numeral.
(b) Potassium hydrogen sulfate	Potassium (group 1A) forms only the K^+ ion.

(c) Copper(II) carbonate   The carbonate ion has a −2 charge, so copper must be +2. A Roman numeral is needed because copper, a transition metal, can form more than one ion.

(d) Iron(III) perchlorate   There are three perchlorate ions, each with a −1 charge, so the iron must have a +3 charge.

EXAMPLE 2.12   Write formulas for the following compounds:
(a) Potassium hypochlorite   **(b)** Silver(I) chromate   **(c)** Iron(III) carbonate

SOLUTION

(a) $KClO$ Potassium forms only the K^+ ion, so only one ClO^- is needed.

(b) Ag_2CrO_4 The polyatomic chromate ion has a −2 charge, so two Ag^+ ions are needed.

(c) $Fe_2(CO_3)_3$ Iron(III) has a +3 charge and the polyatomic carbonate ion has a −2 charge, so there must be two iron ions and three carbonate ions. The polyatomic carbonate ion is set off in parentheses to indicate that there are three of them.

▶ **PROBLEM 2.20**   Give systematic names for the following compounds:
  (a) $Ca(ClO)_2$ **(b)** $Ag_2S_2O_3$ **(c)** NaH_2PO_4 **(d)** $Sn(NO_3)_2$
  (e) $Pb(CH_3CO_2)_4$

▶ **PROBLEM 2.21**   Write formulas for the following compounds:
  (a) Lithium phosphate   **(b)** Magnesium hydrogen sulfate
  (c) Manganese(II) nitrate   **(d)** Chromium(III) sulfate

✦— **KEY CONCEPT PROBLEM 2.22**   The following drawing is that of a solid ionic compound, with green spheres representing the cation and blue spheres representing the anion.

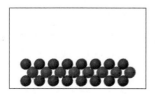

Which of the following formulas are consistent with the drawing?
  (a) $LiBr$ **(b)** $NaNO_2$ **(c)** $CaCl_2$ **(d)** K_2CO_3 **(e)** $Fe_2(SO_4)_3$

Naming Acids

Most acids are **oxoacids**; that is, they contain oxygen in addition to hydrogen and another element. When dissolved in water, an oxoacid yields one or more H^+ ions and a polyatomic oxoanion, such as one of those listed in Table 2.4.

    The names of oxoacids are related to the names of the corresponding oxoanions as described previously, with the *-ite* or *-ate* ending of the anion name replaced by *-ous acid* or *-ic acid*, respectively. In other words, the acid with fewer oxygens has an *-ous* ending, and the acid with more oxygens has an *-ic* ending. The compound HNO_2, for example, is called *nitrous acid* because it has fewer oxygens and yields the nit*rite* ion (NO_2^-) when dissolved in water.

TABLE 2.4 Some Common Oxoacids and Their Anions

Oxoacid		Oxoanion	
HNO_2	Nitr*ous* acid	NO_2^-	Nitr*ite* ion
HNO_3	Nitr*ic* acid	NO_3^-	Nitr*ate* ion
H_3PO_4	Phosphor*ic* acid	PO_4^{3-}	Phosph*ate* ion
H_2SO_3	Sulfur*ous* acid	SO_3^{2-}	Sulf*ite* ion
H_2SO_4	Sulfur*ic* acid	SO_4^{2-}	Sulf*ate* ion
HClO	*Hypo*chlor*ous* acid	ClO^-	*Hypo*chlor*ite* ion
$HClO_2$	Chlor*ous* acid	ClO_2^-	Chlor*ite* ion
$HClO_3$	Chlor*ic* acid	ClO_3^-	Chlor*ate* ion
$HClO_4$	*Per*chlor*ic* acid	ClO_4^-	*Per*chlor*ate* ion

HNO_3 is called *nitric acid* because it has more oxygens and yields the nitr*ate* ion (NO_3^-) when dissolved in water.

Nitr*ous* acid gives nitr*ite ion*

$$HNO_2(l) \xrightarrow{\text{Dissolve in water}} H^+(aq) + NO_2^-(aq)$$

Nitr*ic* acid gives nitr*ate ion*

$$HNO_3(l) \xrightarrow{\text{Dissolve in water}} H^+(aq) + NO_3^-(aq)$$

In a similar way, hypochlor*ous* acid yields the hypochlor*ite* ion, chlor*ous* acid yields the chlor*ite* ion, chlor*ic* acid yields the chlor*ate* ion, and perchlor*ic* acid yields the perchlor*ate* ion (Table 2.4).

In addition to the oxoacids, there are a small number of other acids, such as HCl, that do not contain oxygen. Although the pure, gaseous compound HCl is named hydrogen chloride, the aqueous solution is named *hydrochloric acid*, HCl(*aq*). This example is typical of non-oxygen-containing acids: The prefix *hydro-* and the suffix *-ic acid* are used in such cases.

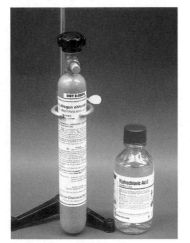

Pure HCl is a colorless gas; hydrochloric acid is an aqueous solution of HCl.

Hydrogen chloride $HCl(g) \xrightarrow{\text{Dissolve in water}} HCl(aq)$ *Hydrochloric acid*

Hydrogen bromide $HBr(g) \xrightarrow{\text{Dissolve in water}} HBr(aq)$ *Hydrobromic acid*

EXAMPLE 2.13 Name the following acids:
(a) HBrO(*aq*)   **(b)** HCN(*aq*)

SOLUTION To name an acid, look at the formula and decide whether the compound is an oxoacid. If so, the name must reflect the number of oxygen atoms, according to Table 2.4. If the compound is not an oxoacid, it is named using the prefix *hydro-* and the suffix *-ic acid*.

(a) This compound is an oxoacid that yields hypobromite ion (BrO^-) when dissolved in water. Its name is hypobromous acid.

(b) This compound is not an oxoacid but yields cyanide ion when dissolved in water. Its name is hydrocyanic acid. (As a pure gas, HCN is named hydrogen cyanide.)

▶ **PROBLEM 2.23** Name the following acids:
   (a) HIO_4 **(b)** $HBrO_2$ **(c)** H_2CrO_4

Are Atoms Real?

The atomic theory of matter lies at the heart of chemistry. Every chemical reaction and every physical law that describes the behavior of matter is explained by chemists in terms of atoms. But how do we know that atoms are real? The best answer to that question is that we can now actually "see" individual atoms with a remarkable device called a *scanning tunneling microscope*, or *STM*. Invented in 1981 by a research team at the IBM Corporation, this special microscope has achieved magnifications of up to *10 million*, allowing chemists to look directly at individual atoms.

The principle behind the operation of an STM is shown in Figure 2.12. A sharp tungsten probe, whose tip is only one or two atoms across, is brought near the surface of the sample, and a small voltage is applied. When the tip comes within a few atomic diameters of the sample, a small electric current flows from the sample to the probe in a process called *electron tunneling*. The strength of the current flow is extremely sensitive to the distance between sample and probe, varying by as much as 1000-fold over a distance of just 100 pm (about one atomic diameter.) Passing the probe across the sample while moving it up and down over individual atoms to keep current flow constant gives a two-dimensional map of the probe's path. By then moving the probe back and forth in a series of closely spaced parallel tracks and storing the data in a computer, a three-dimensional image of the surface can be constructed.

The kind of image produced by scanning tunneling microscopy is quite different from the kind of image we see with our eyes. A normal visual image results when light from the sun or other source reflects off an object, strikes the retina in our eye, and is converted into electrical signals that are processed by the brain. The image obtained with a scanning tunneling microscope, by contrast, is a three-dimensional, computer-generated data plot that uses tunneling current to mimic depth perception. The nature of the computer-generated image depends on the identity of the molecules or atoms on the surface, on the precision with which the probe tip is made, on how the data are manipulated, and on other experimental variables.

This photograph of iron atoms arranged in an oval pattern on a copper surface was taken by a scanning tunneling microscope, a remarkable device that lets scientists actually "see" individual atoms.

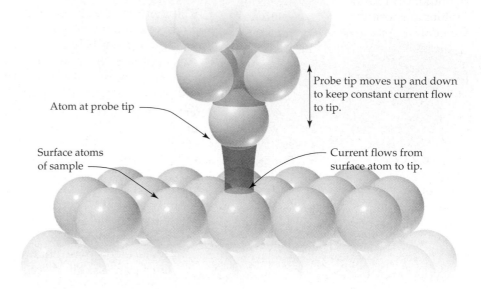

Probe tip moves up and down to keep constant current flow to tip.

Atom at probe tip

Surface atoms of sample

Current flows from surface atom to tip.

FIGURE 2.12 A scanning tunneling microscope works by moving an extremely fine probe along the surface of a sample, applying a small voltage, and measuring current flow between atoms in the sample and the atom at the tip of the probe. By raising and lowering the moving probe to keep current flow constant as the tip passes over atoms, a map of the surface can be obtained.

Most of the present uses of the scanning tunneling microscope involve studies of surface chemistry. Processes such as the deposition of monomolecular layers on smooth surfaces can be studied, the nature of industrial catalysts can be probed, and metal corrosion can be examined. Although the technology is not yet fully developed, the possibility also exists that complex biological structures can be determined with the STM.

▶ **PROBLEM 2.24** How does the image obtained by a scanning tunneling microscope differ from that obtained by the usual optical microscope?

Summary

Elements are made of tiny particles called **atoms**, which can combine in simple numerical ratios. Atoms are composed of three fundamental particles: **Protons** are positively charged, **electrons** are negatively charged, and **neutrons** are neutral. According to the nuclear model of an atom proposed by Ernest Rutherford, protons and neutrons are clustered into a dense core called the **nucleus**, while electrons move around the nucleus at a relatively large distance.

Elements differ from one another according to how many protons their atoms contain, a value called the **atomic number (Z)** of the element. The sum of an atom's protons and neutrons is its **mass number (A)**. Although all atoms of a specific element have the same atomic number, different atoms of an element can have different mass numbers depending on how many neutrons they have. Atoms with identical atomic numbers but different mass numbers are called **isotopes**. Atomic masses are measured using the **atomic mass unit (amu)**, defined as $1/12$ the mass of a ^{12}C atom. Because both protons and neutrons have a mass of approximately 1 amu, the mass of an atom in atomic mass units (the isotopic mass) is numerically close to the atom's mass number. The element's **atomic mass** is a weighted mass average for naturally occurring isotope mixtures.

Most substances on earth are **compounds**, formed when atoms of two or more elements combine in a **chemical reaction**. The atoms in a compound are held together by one of two fundamental kinds of **chemical bonds.** **Covalent bonds** form when two atoms share electrons to give a new unit of matter called a **molecule. Ionic bonds** form when one atom completely transfers one or more electrons to another atom, resulting in the formation of **ions**. Positively charged ions (**cations**) are strongly attracted to negatively charged ions (**anions**) by electrical forces.

The hydrogen ion (H^+) and the **polyatomic** hydroxide ion (OH^-) are among the most important ions in chemistry because they are fundamental to the idea of acids and bases. According to one common definition, **acids** are substances that yield H^+ ions when dissolved in water, and **bases** are substances that yield OH^- ions when dissolved in water.

All chemical compounds can be named systematically by following a series of rules. Binary ionic compounds are named by identifying first the positive ion and then the negative ion. Binary molecular compounds are similarly named by identifying the cationlike and anionlike elements. Naming compounds with polyatomic ions involves memorizing the names and formulas of the most common ones.

Key Words

acid *56*
alpha (α) particle *43*
anion *54*
atom *39*
atomic mass *48*
atomic mass unit (amu) *47*
atomic number (Z) *45*
base *56*
cathode ray *41*
cation *54*
chemical bond *52*
chemical equation *50*
chemical formula *49*
chemical reaction *49*
compound *49*
covalent bond *52*
electron *41*
element *37*
heterogeneous mixture *50*
homogeneous mixture *50*
ion *54*
ionic bond *54*
ionic solid *55*
isotope *46*
law of definite
    proportions *38*
law of mass
    conservation *38*
law of multiple
    proportions *39*
mass number (*A*) *45*
mixture *50*
molecule *52*
neutron *44*
nucleus *44*
oxoacid *63*
oxoanion *61*
polyatomic ion *55*
proton *44*
structural formula *53*

Key Concept Summary

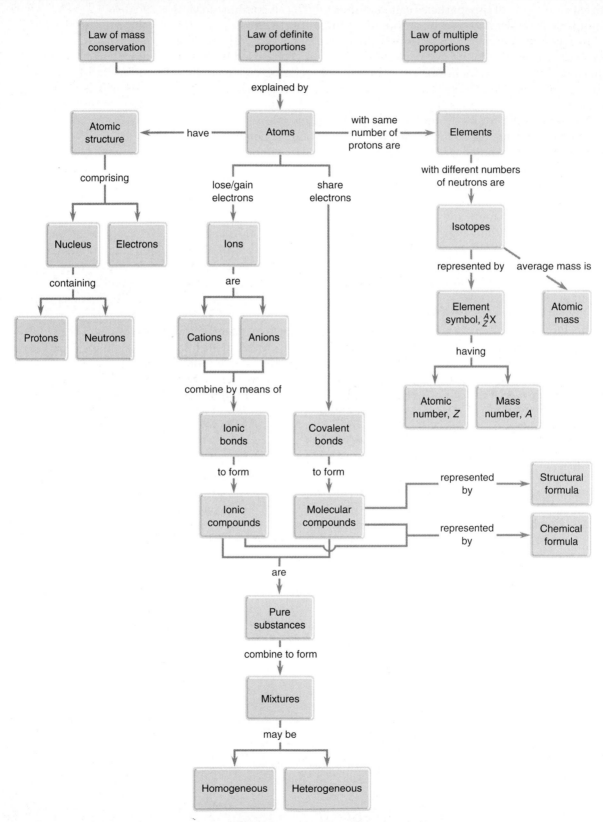

Understanding Key Concepts

Problems 2.1–2.24 appear within the chapter.

2.25 If yellow spheres represent sulfur atoms and red spheres represent oxygen atoms, which of the following drawings depicts a collection of sulfur dioxide molecules?

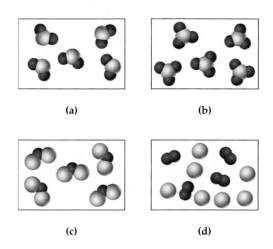

(a)   (b)

(c)   (d)

2.26 Assume that the mixture of substances in drawing **(a)** undergoes a reaction. Which of the drawings **(b)–(d)** represents a product mixture consistent with the law of mass conservation?

2.27 If red and blue spheres represent atoms of different elements, which two of the following drawings illustrate the law of multiple proportions?

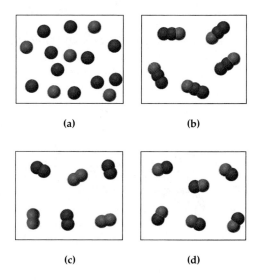

(a)   (b)

(c)   (d)

2.28 Give molecular formulas corresponding to each of the following ball-and-stick molecular representations (red = O, gray = C, blue = N, ivory = H.) In writing the formula, list the atoms in alphabetical order.

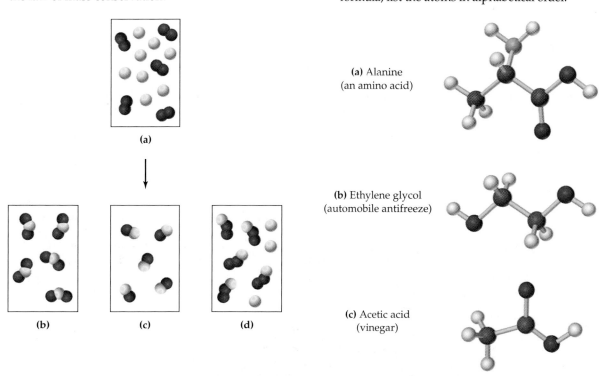

(a)

(b)   (c)   (d)

(a) Alanine (an amino acid)

(b) Ethylene glycol (automobile antifreeze)

(c) Acetic acid (vinegar)

2.29 Which of the following drawings represents an Na atom? A Ca^{2+} ion? An F^- ion?

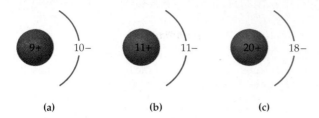

(a) (b) (c)

2.30 Indicate where in the periodic table the following elements are found:

(a) Elements that commonly form anions

(b) Elements that commonly form cations

(c) Elements that commonly form covalent bonds

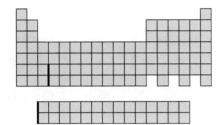

2.31 In the following drawings, red spheres represent cations and blue spheres represent anions. Match each of the drawings **(a)–(d)** with the following ionic compounds:

(i) $Ca_3(PO_4)_2$ (ii) Li_2CO_3

(iii) $FeCl_2$ (iv) $MgSO_4$

(a) (b)

(c) (d)

Additional Problems

Atomic Theory

2.32 How does Dalton's atomic theory account for the law of mass conservation and the law of definite proportions?

2.33 What is the law of multiple proportions, and how is it predicted by Dalton's atomic theory?

2.34 Benzene, ethane, and ethylene are just three of a large number of *hydrocarbons*—compounds that contain only carbon and hydrogen. Show how the following data are consistent with the law of multiple proportions.

Compound	Mass of carbon in 5.00 g sample	Mass of hydrogen in 5.00 g sample
Benzene	4.61 g	0.39 g
Ethane	4.00 g	1.00 g
Ethylene	4.29 g	0.71 g

2.35 In addition to carbon monoxide (CO) and carbon dioxide (CO_2), there is a third compound of carbon and oxygen called *carbon suboxide*. If a 2.500 g sample of carbon suboxide contains 1.32 g of C and 1.18 g of O, show that the law of multiple proportions is followed.

2.36 The atomic mass of carbon (12.011 amu) is approximately 12 times that of hydrogen (1.008 amu).

(a) Show how you can use this knowledge to calculate possible formulas for benzene, ethane, and ethylene (Problem 2.34).

(b) Show how your answer to part **(a)** is consistent with the actual formulas for benzene (C_6H_6), ethane (C_2H_6), and ethylene (C_2H_4).

2.37 What is a possible formula for carbon suboxide (Problem 2.35)?

2.38 **(a)** If the average mass of a single hydrogen atom is 1.67×10^{-24} g, what is the mass in grams of 6.02×10^{23} hydrogen atoms? How does your answer compare numerically with the atomic mass of hydrogen?

(b) If the average mass of a single oxygen atom is 26.558×10^{-24} g, what is the mass in grams of 6.02×10^{23} oxygen atoms? How does your answer compare numerically with the atomic mass of oxygen?

2.39 (a) If the atomic mass of an element is x, what is the mass in grams of 6.02×10^{23} atoms of the element? (See Problem 2.38.)

(b) If 6.02×10^{23} atoms of element Y have a mass of 83.80 g, what is the identity of Y?

2.40 A binary compound of zinc and sulfur contains 67.1% zinc by mass. What is the ratio of zinc and sulfur atoms in the compound?

2.41 There are two binary compounds of titanium and chlorine. One compound contains 31.04% titanium by mass, and the other contains 74.76% chlorine by mass. What are the ratios of titanium and chlorine atoms in the two compounds?

Elements and Atoms

2.42 What is the difference between an atom's atomic number and its mass number?

2.43 What is the difference between an element's atomic number and its atomic mass?

2.44 What is an isotope?

2.45 Carbon-14 and nitrogen-14 both have the same mass number yet they are different elements. Explain.

2.46 The subscript giving the atomic number of an atom is often left off when writing an isotope symbol. For example, $^{13}_{6}C$ is often written simply as ^{13}C. Why is this allowable?

2.47 Iodine has a *lower* atomic mass than tellurium (126.90 for iodine versus 127.60 for tellurium) even though it has a *higher* atomic number (53 for iodine versus 52 for tellurium). Explain.

2.48 Give names and symbols for the following elements:

(a) An element with atomic number 6

(b) An element with 18 protons in its nucleus

(c) An element with 23 electrons

2.49 The radioactive isotope cesium-137 was produced in large amounts in fallout from the 1985 nuclear power plant disaster at Chernobyl, Ukraine. Write the symbol for this isotope in standard format.

2.50 Write standard symbols for the following isotopes:

(a) Radon-220 (b) Polonium-210

(c) Gold-197

2.51 Write symbols for the following isotopes:

(a) $Z = 58$ and $A = 140$ (b) $Z = 27$ and $A = 60$

2.52 How many protons, neutrons, and electrons are in each of the following atoms?

(a) $^{15}_{7}N$ (b) $^{60}_{27}Co$ (c) $^{131}_{53}I$

2.53 How many protons and neutrons are in the nucleus of the following atoms?

(a) ^{27}Al (b) ^{32}S (c) ^{64}Zn (d) ^{207}Pb

2.54 Identify the following elements:

(a) $^{24}_{12}X$ (b) $^{58}_{28}X$

2.55 Identify the following elements:

(a) $^{202}_{80}X$ (b) $^{195}_{78}X$

2.56 Naturally occurring boron consists of two isotopes: ^{10}B (19.9%) with an isotopic mass of 10.0129 amu and ^{11}B (80.1%) with an isotopic mass of 11.009 31 amu. What is the atomic mass of boron? Check your answer by looking at a periodic table.

2.57 Naturally occurring silver consists of two isotopes: ^{107}Ag (51.84%) with an isotopic mass of 106.9051 amu and ^{109}Ag (48.16%) with an isotopic mass of 108.9048 amu. What is the atomic mass of silver? Check your answer in a periodic table.

2.58 Magnesium has three naturally occurring isotopes: ^{24}Mg (23.985 amu) with 78.99% abundance, ^{25}Mg (24.986 amu) with 10.00% abundance, and a third with 11.01% abundance. Look up the atomic mass of magnesium, and then calculate the mass of the third isotope.

2.59 A sample of naturally occurring silicon consists of ^{28}Si (27.9769 amu), ^{29}Si (28.9765 amu), and ^{30}Si (29.9738 amu). If the atomic mass of silicon is 28.0855 amu and the natural abundance of ^{29}Si is 4.67%, what are the natural abundances of ^{28}Si and ^{30}Si?

Compounds and Mixtures, Molecules and Ions

2.60 Which of the following mixtures are homogeneous and which are heterogeneous?

(a) Muddy water (b) Concrete

(c) House paint (d) A soft drink

2.61 Which of the following mixtures are homogeneous?

(a) 18 karat gold (b) Window glass

(c) Tomato juice (d) Liquefied air

2.62 What is the difference between an atom and a molecule? Give an example of each.

2.63 What is the difference between a molecule and an ion? Give an example of each.

2.64 What is the difference between a covalent bond and an ionic bond? Give an example of each.

2.65 Which of the following bonds are likely to be covalent and which ionic? Explain.

(a) B−Br   (b) Na−Br   (c) Br−Cl   (d) O−Br

2.66 The symbol CO stands for carbon monoxide, but the symbol Co stands for the element cobalt. Explain.

2.67 Correct the error in each of the following statements:

(a) The formula of ammonia is NH3.

(b) Molecules of potassium chloride have the formula KCl.

(c) Cl^- is a cation.

(d) CH_4 is a polyatomic ion.

2.68 How many protons and electrons are in each of the following ions?

(a) Be^{2+} (b) Rb^+ (c) Se^{2-} (d) Au^{3+}

2.69 What is the identity of the element X in the following ions?

(a) X^{2+}, a cation that has 36 electrons

(b) X^-, an anion that has 36 electrons

2.70 The structural formula of isopropyl alcohol, better known as "rubbing alcohol," is shown. What is the chemical formula of isopropyl alcohol?

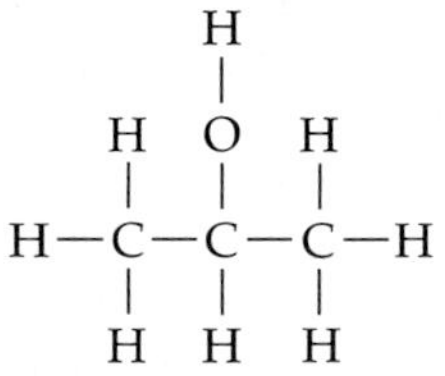

Isopropyl alcohol

2.71 Lactic acid, a compound found both in sour milk and in tired muscles, has the structure shown. What is its chemical formula?

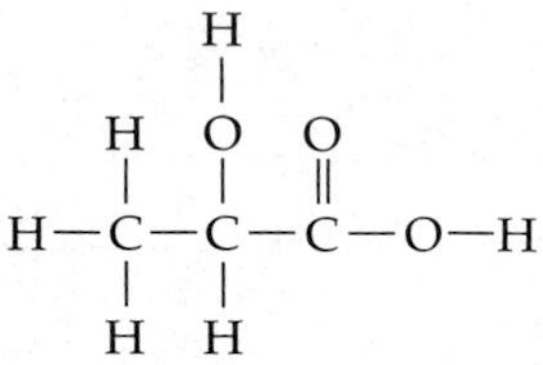

Lactic acid

2.72 Butane, the fuel used in disposable lighters, has the formula C_4H_{10}. The carbon atoms are connected in the sequence C−C−C−C, and each carbon has a total of four covalent bonds. Draw a structural formula for butane.

2.73 Isooctane, the substance in gasoline from which the term *octane rating* derives, has the formula C_8H_{18}. Each carbon has a total of four covalent bonds, and the atoms are connected in the sequence shown. Draw a complete structural formula for isooctane.

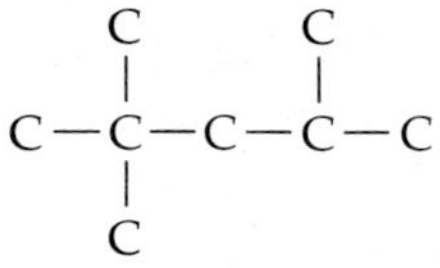

Acids and Bases

2.74 Which of the following compounds are acids and which are bases?

(a) HI   (b) CsOH   (c) H_3PO_4
(d) $Ba(OH)_2$ (e) H_2CO_3

2.75 For each of the acids you identified in Problem 2.74, tell how many H^+ ions can be donated from one molecule of acid.

2.76 Identify the anion that results when each of the acids in Problem 2.74 dissolves in water.

2.77 Identify the cation that results when each of the bases in Problem 2.74 dissolves in water.

Naming Compounds

2.78 Write formulas for the following binary compounds:

(a) Potassium chloride   (b) Tin(II) bromide
(c) Calcium oxide   (d) Barium chloride
(e) Aluminum hydride

2.79 Write formulas for the following compounds:

(a) Calcium acetate   (b) Iron(II) cyanide
(c) Sodium dichromate   (d) Chromium(III) sulfate
(e) Mercury(II) perchlorate

2.80 Name the following ions:
(a) Ba^{2+} (b) Cs^+ (c) V^{3+} (d) HCO_3^-
(e) NH_4^+ (f) Ni^{2+} (g) NO_2^- (h) ClO_2^-
(i) Mn^{2+} (j) ClO_4^-

2.81 Name the following binary molecular compounds:
(a) CCl_4 (b) SiO_2 (c) N_2O (d) N_2O_3

2.82 Give the formulas and charges of the following ions:
(a) Sulfite ion (b) Phosphate ion
(c) Zirconium(IV) ion (d) Chromate ion
(e) Acetate ion (f) Thiosulfate ion

2.83 What are the charges on the positive ions in the following compounds?
(a) $Zn(CN)_2$ (b) $Fe(NO_2)_3$ (c) $Ti(SO_4)_2$
(d) $Sn_3(PO_4)_2$ (e) Hg_2S (f) MnO_2
(g) KIO_4 (h) $Cu(CH_3CO_2)_2$

2.84 Name each of the compounds in Problem 2.83.

2.85 Name each of the following compounds:
(a) $MgSO_3$ (b) $Co(NO_2)_2$ (c) $Mn(HCO_3)_2$
(d) $ZnCrO_4$ (e) $BaSO_4$ (f) $KMnO_4$
(g) $Al_2(SO_4)_3$ (h) $LiClO_3$

2.86 Fill in the missing information to give formulas for the following compounds:
(a) $Na_?SO_4$ (b) $Ba_?(PO_4)_?$ (c) $Ga_?(SO_4)_?$

2.87 Write formulas for each of the following compounds:
(a) Sodium peroxide
(b) Aluminum bromide
(c) Chromium(III) sulfate

General Problems

2.88 Germanium has five naturally occurring isotopes: ^{70}Ge, 20.5%, 69.924 amu; ^{72}Ge, 27.4%, 71.922 amu; ^{73}Ge, 7.8%, 72.923 amu; ^{74}Ge, 36.5%, 73.921 amu; and ^{76}Ge, 7.8%, 75.921 amu. What is the atomic mass of germanium?

2.89 The best balances commonly found in laboratories can weigh amounts as small as 10^{-5} g. If you were to count out carbon atoms at the rate of two per second, how long would it take you to count a pile of atoms large enough to weigh?

2.90 Name the following compounds:
(a) $NaBrO_3$ (b) H_3PO_4 (c) H_3PO_3 (d) V_2O_5

2.91 Write formulas for the following compounds:
(a) Calcium hydrogen sulfate
(b) Tin(II) oxide
(c) Ruthenium(III) nitrate
(d) Ammonium carbonate
(e) Hydriodic acid
(f) Beryllium phosphate

2.92 Ammonia, NH_3, and hydrazine, N_2H_4, are both binary compounds of nitrogen and hydrogen. Based on the law of multiple proportions, how many grams of hydrogen would you expect 2.34 g of nitrogen to combine with to yield ammonia? To yield hydrazine?

2.93 If 3.670 g of nitrogen combines with 0.5275 g of hydrogen to yield compound X, how many grams of nitrogen would combine with 1.575 g of hydrogen to make the same compound? Is X ammonia or hydrazine (Problem 2.92)?

2.94 Tellurium, a group 6A element, forms the oxoanions TeO_4^{2-} and TeO_3^{2-}. What are the likely names of these ions? To what other group 6A oxoanions are they analogous?

2.95 Give the formulas and the likely names of the acids derived from the tellurium-containing oxoanions in Problem 2.94.

2.96 Identify the following atoms or ions:
(a) A halogen anion with 54 electrons
(b) A metal cation with 79 protons and 76 electrons
(c) A noble gas with $A = 84$

2.97 Prior to 1961, the atomic mass unit was defined as 1/16 the mass of the atomic mass of oxygen. That is, the atomic mass of oxygen was defined as exactly 16 amu. What was the mass of a ^{12}C atom prior to 1961 if the atomic mass of oxygen on today's scale is 15.9994 amu?

2.98 What was the mass in atomic mass units of a ^{40}Ca atom prior to 1961 if its mass on today's scale is 39.9626 amu? (See Problem 2.97.)

2.99 Analogous oxoanions of elements in the same group of the periodic table are named similarly. Based on the names of the oxoanions of phosphorus and sulfur, name the following anions:
(a) AsO_4^{3-} (b) SeO_3^{2-}
(c) SeO_4^{2-} (d) $HAsO_4^{2-}$

2.100 Identify the following atoms or ions, and where possible, write the symbol for the specific isotope. If not enough information is given to answer the question, say so.
(a) An alkaline earth metal atom with 20 protons and 20 neutrons
(b) A metal atom with 63 neutrons
(c) A 3+ cation with 23 electrons and 30 neutrons
(d) A 2- anion with 34 protons

2.101 Fluorine occurs naturally as a single isotope. How many protons, neutrons, and electrons are present in one molecule of deuterium fluoride? Is deuterium fluoride an acid or a base? (Deuterium is ^2H.)

2.102 Hydrogen has three isotopes (^{1}H, ^{2}H, and ^{3}H), and chlorine has two isotopes (^{35}Cl and ^{37}Cl). How many kinds of HCl molecules are there? Write the formula for each, and tell how many protons, neutrons, and electrons each contains.

2.103 Cyclohexane, used as a paint and varnish remover, has the formula C_6H_{12}. In this molecule, each C atom is bonded to two H atoms and two other C atoms, and each H atom is bonded to one C atom. Write a structural formula for cyclohexane.

2.104 Pentane, a solvent that is found in petroleum, has the formula C_5H_{12}. In a C_5H_{12} molecule, each C atom forms four bonds, and each H atom forms one bond. Write three possible structural formulas for pentane.

2.105 A 9.520 g quantity of zinc was allowed to react with 40.00 mL of sulfuric acid that had a density of 1.3028 g/mL. The reaction yielded H_2 gas and a solution of zinc sulfate. The density of the gas was determined to be 0.0899 g/L and the mass of the solution was found to be 61.338 g. How many liters of H_2 were formed?

2.106 The *molecular mass* of a molecule is the sum of the atomic masses of all atoms in the molecule. What is the molecular mass of acetaminophen, $C_8H_9NO_2$, the active ingredient in Tylenol?

2.107 The *mass percent* of an element in a compound is the mass of the element (total mass of the element's atoms in the molecule) divided by the mass of the compound (total mass of all atoms in the molecule) times 100%. What is the mass percent of each element in acetaminophen? (See Problem 2.106.)

2.108 Aspirin contains C, H, and O in the following mass percents: C, 60.00%; H, 4.48%; and O, 35.52%. (See Problems 2.106 and 2.107.)

(a) Is aspirin likely to be an ionic compound or a molecular compound?

(b) What is a possible formula for aspirin?

2.109 Element X reacts with element Y to give a product containing X^{3+} ions and Y^{2-} ions.

(a) Is element X likely to be a metal or a nonmetal? Explain.

(b) Is element Y likely to be a metal or a nonmetal? Explain.

(c) What is the formula of the product?

(d) What groups of the periodic table are elements X and Y likely to be in?

 eMedia Problems

2.110 Watch the **Multiple Proportions** movie (*eChapter 2.2*), and describe how the sulfur-to-oxygen mass ratios of SO_2 and SO_3 illustrate the law of multiple proportions. Draw pictures similar to those in problem 2.27 to represent collections of SO_2 and SO_3 molecules.

2.111 After using the **Isotopes of Hydrogen** activity (*eChapter 2.5*), draw pictures of the two naturally occurring boron isotopes referred to in Problem 2.56.

2.112 Give the correct formula and name for the ionic compound formed by each of the following combinations:

(a) Ba^{2+} and OH^-

(b) Ag^+ and CO_3^{2-}

(c) Fe^{2+} and PO_4^{3-}

(d) Fe^{3+} and SO_4^{2-}

(e) Pb^{2+} and ClO_4^-.

Use the **Ionic Compounds** activity (*eChapter 2.10*) to check your answers.

2.113 The **Rutherford Experiment** movie (*eChapter 2.4*) shows alpha particles impinging on a thin gold foil. Describe what happens to the alpha particles and discuss how the results of this experiment shaped the modern view of atomic structure.

2.114 The **Electrolysis of Water** movie (*eChapter 2.7*) shows water being decomposed into hydrogen gas and oxygen gas.

(a) Explain why the result of this experiment indicates that water is a compound and not an element.

(b) How would the result of this experiment be different if the electrolysis apparatus were filled with liquid nitrogen instead of water?

Formulas, Equations, and Moles

The reaction of sodium metal with water is so violent that the hydrogen gas produced bursts into flame.

It's sometimes possible when beginning the study of chemistry to forget that reactions are at the heart of the science. New words, ideas, and principles are sometimes introduced so quickly that the central concern of chemistry—the change of one substance into another—gets lost in the rush.

In this chapter, we'll begin learning about how to describe chemical reactions, starting with a look at the conventions for writing chemical equations and at the necessary mass relationships between reactants and products. Since most chemical reactions are carried out using *solutions* rather than pure materials, we'll also discuss units for describing the concentration of a solution. Finally, we'll see how chemical formulas are determined and how molecular masses are measured.

3.1   Balancing Chemical Equations

Sodium and Potassium in Water movie

The previous two chapters have provided several examples of reactions: hydrogen reacting with oxygen to yield water, sodium reacting with chlorine to yield sodium chloride, mercury(II) nitrate reacting with potassium iodide to yield mercury(II) iodide, and so forth. We can write equations for these reactions in the following format:

4 H and 2 O atoms on this side

4 H and 2 O atoms on this side

$$2\,H_2 + O_2 \longrightarrow 2\,H_2O$$

2 Na and 2 Cl atoms on this side

2 Na and 2 Cl atoms on this side

$$2\,Na + Cl_2 \longrightarrow 2\,NaCl$$

1 Hg, 2 N, 6 O, 2 K, and 2 I atoms on this side

1 Hg, 2 N, 6 O, 2 K, and 2 I atoms on this side

$$Hg(NO_3)_2 + 2\,KI \longrightarrow HgI_2 + 2\,KNO_3$$

Look carefully at how these equations are written. Since we know from Section 2.8 that hydrogen, oxygen, and chlorine exist as covalent H_2, O_2, and Cl_2 *molecules* rather than as isolated atoms, we must write them as such in the chemical equations. Now look at the atoms on each side of the reaction arrow. In each equation, the numbers and kinds of atoms on both sides of the arrow are the same, and the equations are therefore said to be **balanced**.

The requirement that an equation be balanced is a direct consequence of the mass conservation law discussed in Section 2.1: All chemical equations must balance because atoms are neither created nor destroyed in chemical reactions. The numbers and kinds of atoms must be the same in the products as in the reactants.

Balancing a chemical equation involves finding out how many *formula units* of each different substance take part in the reaction. A **formula unit**, as the name implies, is one unit—whether atom, ion, or molecule—corresponding to a given formula. One formula unit of NaCl, for example, is one Na^+ ion and one Cl^- ion. Similarly, one formula unit of $MgBr_2$ is one Mg^{2+} ion and two Br^- ions, and one formula unit of H_2O is one H_2O molecule.

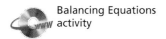

Balancing Equations activity

The balancing process is carried out in four steps using a mixture of common sense and trial-and-error.

1. Write the unbalanced equation using the correct chemical formula for each reactant and product. In the reaction of hydrogen with oxygen to yield water, for example, we begin by writing:

$$H_2 + O_2 \longrightarrow H_2O$$

2. Find suitable **coefficients**—the numbers placed before formulas to indicate how many formula units of each substance are required to balance the equation. *Only these coefficients can be changed when balancing an equation; the formulas themselves can't be changed.* Again taking the reaction of hydrogen with oxygen as an example, we can balance the equation by adding

a coefficient of 2 to both H_2 and H_2O. By so doing, we now have 4 hydrogen atoms and 2 oxygen atoms on each side of the equation:

Add these *coefficients* to balance the equation

$$2\,H_2 + O_2 \longrightarrow 2\,H_2O$$

It might seem easier at first glance to balance the equation simply by adding a subscript 2 to the oxygen atom in water, thereby changing H_2O into H_2O_2. That's not allowed, however, because the resulting equation would no longer describe the same reaction. The substances H_2O (water) and H_2O_2 (hydrogen peroxide) are two entirely different compounds. Water is a substance used for drinking and swimming; hydrogen peroxide is a substance used for bleaching hair and sterilizing wounds.

NOT ALLOWED!
When this subscript is added, we get a completely different reaction.

$$H_2 + O_2 \longrightarrow H_2O_2$$

3. Reduce the coefficients to their smallest whole-number values, if necessary, by dividing them by a common divisor.
4. Check your answer by making sure that the numbers and kinds of atoms are the same on both sides of the equation.

Let's work through some additional examples to see how equations are balanced.

EXAMPLE 3.1   Propane, C_3H_8, is a colorless, odorless gas often used as a heating and cooking fuel in campers and rural homes. Write a balanced equation for the combustion reaction of propane with oxygen to yield carbon dioxide and water.

SOLUTION
Step 1 Write the unbalanced equation using correct chemical formulas for all substances:

$$C_3H_8 + O_2 \longrightarrow CO_2 + H_2O \qquad \text{Unbalanced}$$

Step 2 Find coefficients to balance the equation. It's usually best to start with the most complex substance—in this case C_3H_8—and to deal with one element at a time. Look first at the unbalanced equation, and note that there are 3 carbon atoms on the left side of the equation but only 1 on the right side. If we add a coefficient of 3 to CO_2 on the right, the carbons balance:

$$C_3H_8 + O_2 \longrightarrow 3\,CO_2 + H_2O \qquad \text{Balanced for C}$$

Next, look at the number of hydrogen atoms. There are 8 hydrogens on the left but only 2 (in H_2O) on the right. By adding a coefficient of 4 to the H_2O on the right, the hydrogens balance:

$$C_3H_8 + O_2 \longrightarrow 3\,CO_2 + 4\,H_2O \qquad \text{Balanced for C and H}$$

Finally, look at the number of oxygen atoms. There are 2 on the left but 10 on the right. By adding a coefficient of 5 to the O_2 on the left, the oxygens balance:

$$C_3H_8 + 5\,O_2 \longrightarrow 3\,CO_2 + 4\,H_2O \qquad \text{Balanced for C, H, and O}$$

Step 3 Make sure the coefficients are reduced to their smallest whole-number values. In fact, our answer is already correct, but we might have arrived at a different answer through trial and error:

$$2\,C_3H_8 + 10\,O_2 \longrightarrow 6\,CO_2 + 8\,H_2O$$

Although the preceding equation is balanced, the coefficients are not the smallest whole numbers. It would be necessary to divide all coefficients by 2 to reach the final equation.

Step 4 Check your answer. Count the numbers and kinds of atoms on both sides of the equation to make sure they're the same:

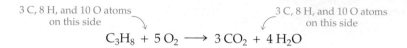

3 C, 8 H, and 10 O atoms on this side   3 C, 8 H, and 10 O atoms on this side

$$C_3H_8 + 5\,O_2 \longrightarrow 3\,CO_2 + 4\,H_2O$$

Potassium chlorate reacts violently with sucrose to yield KCl, CO₂, and H₂O.

EXAMPLE 3.2   The major ingredient in ordinary safety matches is potassium chlorate, $KClO_3$, a substance that can act as a source of oxygen in combustion reactions. Its reaction with ordinary table sugar (sucrose, $C_{12}H_{22}O_{11}$), for example, occurs violently to yield potassium chloride, carbon dioxide, and water. Write a balanced equation for the reaction.

SOLUTION

Step 1 Write the unbalanced equation, making sure the formulas for all substances are correct:

$$KClO_3 + C_{12}H_{22}O_{11} \longrightarrow KCl + CO_2 + H_2O \qquad \text{Unbalanced}$$

Step 2 Find coefficients to balance the equation by starting with the most complex substance (sucrose) and considering one element at a time. Since there are 12 C atoms on the left and only 1 on the right, we can balance for carbon by adding a coefficient of 12 to CO_2:

$$KClO_3 + C_{12}H_{22}O_{11} \longrightarrow KCl + 12\,CO_2 + H_2O \qquad \text{Balanced for C}$$

Since there are 22 H atoms on the left and only 2 on the right, we can balance for hydrogen by adding a coefficient of 11 to H_2O:

$$KClO_3 + C_{12}H_{22}O_{11} \longrightarrow KCl + 12\,CO_2 + 11\,H_2O \quad \text{Balanced for C and H}$$

Since there are now 35 O atoms on the right but only 14 on the left (11 in sucrose and 3 in $KClO_3$), 21 oxygens must be added on the left. We can do this without disturbing the C and H balance by adding 7 more $KClO_3$'s, giving a coefficient of 8 for $KClO_3$:

$$8\,KClO_3 + C_{12}H_{22}O_{11} \longrightarrow KCl + 12\,CO_2 + 11\,H_2O \quad \text{Balanced for C, H, and O}$$

Potassium and chlorine can both be balanced by adding a coefficient of 8 to KCl:

8 K, 8 Cl, 12 C, 22 H, and 35 O atoms   8 K, 8 Cl, 12 C, 22 H, and 35 O atoms

$$8\,KClO_3 + C_{12}H_{22}O_{11} \longrightarrow 8\,KCl + 12\,CO_2 + 11\,H_2O$$

Steps 3 and 4   The coefficients in the balanced equation are already reduced to their smallest whole-number values, and a check shows that the numbers and kinds of atoms are the same on both sides of the equation.

▶ **PROBLEM 3.1** Potassium chlorate, $KClO_3$, decomposes when heated to yield potassium chloride and oxygen, a reaction used to provide oxygen for the emergency breathing masks in airliners. Balance the equation.

Emergency oxygen masks in airliners use $KClO_3$ as the oxygen source.

▶ **PROBLEM 3.2** Balance the following equations:
(a) $C_6H_{12}O_6 \rightarrow C_2H_6O + CO_2$ (fermentation of sugar to yield ethyl alcohol)
(b) $Fe + O_2 \rightarrow Fe_2O_3$ (rusting of iron)
(c) $NH_3 + Cl_2 \rightarrow N_2H_4 + NH_4Cl$ (synthesis of hydrazine for rocket fuel)

◆— **KEY CONCEPT PROBLEM 3.3** Write a balanced equation for the reaction of element A (red spheres) with element B (green spheres) as represented below.

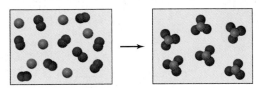

3.2 Chemical Symbols on Different Levels

What does it mean when we write a chemical formula or equation? Answering this question isn't as easy as it sounds because a chemical symbol can have different meanings under different circumstances. Chemists use the same symbols to represent both a small-scale, *microscopic* level and a large-scale, *macroscopic* level, and they tend to slip back and forth between the two levels without realizing the confusion this can cause for newcomers to the field.

On the microscopic level, chemical symbols represent the behavior of individual atoms and molecules. Atoms and molecules are much too small to be seen, but we can nevertheless describe their microscopic behavior if we read the equation $2 H_2 + O_2 \rightarrow 2 H_2O$ to mean "Two molecules of hydrogen react with one molecule of oxygen to yield two molecules of water." It's this microscopic world that we deal with when trying to understand how reactions occur, and it's often helpful in this regard to visualize a molecule as a collection of spheres stuck together. In trying to understand how H_2 reacts with O_2, for example, you might try picturing H_2 and O_2 molecules as made of two spheres pressed together and a water molecule as made of three spheres.

Formation of Water movie

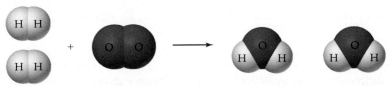

On the macroscopic level, formulas and equations represent the large-scale behaviors of atoms and molecules that give rise to observable properties. In other words, the symbols H_2, O_2, and H_2O can represent not just single molecules but vast numbers of molecules that together have a set of measurable physical properties. A *single* H_2O molecule in isolation is neither solid nor liquid nor gas, but a huge collection of H_2O molecules appears to us as a colorless liquid that freezes at 0°C and boils at 100°C. Clearly, it's this macroscopic behavior we deal with in the laboratory when we weigh out specific amounts of reactants, place them in a flask, and observe visible changes.

What does a chemical formula or equation mean at any particular time? It means whatever you want it to mean depending on the context. The symbol H_2O can mean either one tiny, invisible molecule or a vast collection of molecules large enough to swim in.

3.3   Avogadro's Number and the Mole

Imagine a laboratory experiment—perhaps the reaction of ethylene (C_2H_4) with hydrogen chloride (HCl) to prepare ethyl chloride (C_2H_5Cl), a colorless, low-boiling liquid that doctors and athletic trainers use as a spray-on anesthetic.

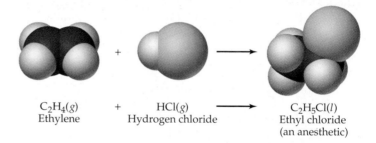

$$C_2H_4(g) \quad + \quad HCl(g) \quad \longrightarrow \quad C_2H_5Cl(l)$$

Ethylene   Hydrogen chloride   Ethyl chloride
(an anesthetic)

Ethyl chloride is often used as a spray-on anesthetic for athletic injuries.

How much ethylene and how much hydrogen chloride should you use for your experiment? According to the coefficients of the balanced equation, a 1:1 number ratio of the two reactants is needed. But because molecules are so small and the number of molecules needed to make a visible sample is so large, your experiment must involve a vast number of ethylene molecules— say 10^{20}—reacting with the same vast number of hydrogen chloride molecules. Unfortunately, though, you can't count the reactant molecules so you have to weigh them. That is, you must convert a *number* ratio of reactant molecules into a *mass* ratio to be sure you are using the right amounts.

Mass ratios are determined by using the *molecular masses* (also called *molecular weights; MW*) of the substances involved in a reaction. Just as the *atomic mass* of an element is the average mass of the element's *atoms*, the **molecular mass** of a substance is the average mass of the substance's *molecules*. Numerically, molecular mass (or more generally **formula mass**) is equal to the sum of the atomic masses of all atoms in the molecule.

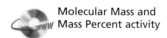

Molecular Mass and
Mass Percent activity

◆—**MOLECULAR MASS**   Sum of atomic masses of all atoms in a molecule.

◆—**FORMULA MASS**   Sum of atomic masses of all atoms in a formula unit of *any* compound, molecular or ionic.

For example, the molecular mass of ethylene is 28.0 amu, the molecular mass of hydrogen chloride is 36.5 amu, the molecular mass of ethyl chloride is

64.5 amu, and the formula mass of sodium chloride is 58.5 amu. (These numbers are rounded off to one decimal place for convenience; the actual values are known more precisely.)

For ethylene, C_2H_4:

at. mass of 2 C	$= 2 \times 12.0 \text{ amu}$	$= 24.0 \text{ amu}$
at. mass of 4 H	$= 4 \times 1.0 \text{ amu}$	$= 4.0 \text{ amu}$
Molec. mass of C_2H_4		$= 28.0 \text{ amu}$

For ethyl chloride, C_2H_5Cl:

at. mass of 2 C	$= 2 \times 12.0 \text{ amu}$	$= 24.0 \text{ amu}$
at. mass of 5 H	$= 5 \times 1.0 \text{ amu}$	$= 5.0 \text{ amu}$
at. mass of Cl		$= 35.5 \text{ amu}$
Molec. mass of C_2H_5Cl		$= 64.5 \text{ amu}$

For hydrogen chloride, HCl:

at. mass of H	$= 1.0 \text{ amu}$
at. mass of Cl	$= 35.5 \text{ amu}$
Molec. mass of HCl	$= 36.5 \text{ amu}$

For sodium chloride, NaCl:

at. mass of Na	$= 23.0 \text{ amu}$
at. mass of Cl	$= 35.5 \text{ amu}$
Form. mass of NaCl	$= 58.5 \text{ amu}$

How are molecular masses used? Since the mass ratio of *one* HCl molecule to *one* ethylene molecule is 36.5 to 28.0, the mass ratio of *any* given number of HCl molecules to the same number of ethylene molecules is also 36.5 to 28.0. In other words, a 36.5 to 28.0 *mass* ratio of HCl and ethylene guarantees a 1:1 *number* ratio. *Samples of different substances contain the same number of molecules (or formula units) whenever their mass ratio is the same as their molecular-mass (or formula-mass) ratio (Figure 3.1).*

(a)   (b)

FIGURE 3.1   **(a)** Because one gumdrop weighs more than one jellybean, you can't get equal numbers merely by taking equal weights. The same is true for different atoms or molecules. **(b)** Equal numbers of HCl and ethylene molecules always have a mass ratio equal to the ratio of their molecular masses, 36.5 to 28.0.

A particularly convenient way to use this mass–number relationship is to measure amounts in grams that are numerically equal to molecular masses. If, for instance, you were to carry out your experiment with 36.5 g of HCl and 28.0 g of ethylene, you could be certain that you have the correct 1:1 number ratio of reactant molecules.

When referring to the enormous numbers of molecules or ions that take part in a visible chemical reaction, it's convenient to use a special unit called a **mole**, abbreviated *mol*. One mole of any substance is the amount whose mass—its **molar mass**—is equal to the molecular or formula mass of the substance in grams. One mole of ethylene has a mass of 28.0 g, one mole of HCl has a mass of 36.5 g, one mole of NaCl has a mass of 58.5 g, and so on. (To be more precise, one mole is formally defined as the amount of a substance that contains the same number of molecules or formula units as there are atoms in exactly 12 g of carbon-12.)

Just how many molecules are there in a mole? Experiments show that one mole of any substance contains 6.022×10^{23} formula units, a value called **Avogadro's number** (abbreviated N_A) after the Italian scientist who first recognized the importance of the mass/number relationship. Avogadro's number of formula units of any substance—that is, one mole—has a mass in grams equal to the molecular or formula mass of the substance.

Molec. mass of HCl = 36.5 amu   Molar mass of HCl = 36.5 g   1 mol HCl = 6.022×10^{23} HCl molecules

Molec. mass of C_2H_4 = 28.0 amu   Molar mass of C_2H_4 = 28.0 g   1 mol C_2H_4 = 6.022×10^{23} C_2H_4 molecules

Molec. mass of C_2H_5Cl = 64.5 amu   Molar mass of C_2H_5Cl = 64.5 g   1 mol C_2H_5Cl = 6.022×10^{23} C_2H_5Cl molecules

Form. mass of NaCl = 58.5 amu   Molar mass of NaCl = 58.5 g   1 mol NaCl = 6.022×10^{23} NaCl formula units

These samples of table sugar, lead shot, potassium dichromate, mercury, water, copper, sodium chloride, and sulfur each contain 1 mol. Do they have the same mass?

Although it's hard to grasp the magnitude of a quantity as large as Avogadro's number, a few comparisons might give you a sense of scale:

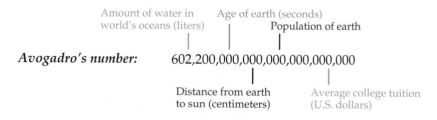

Avogadro's number: 602,200,000,000,000,000,000,000

In effect, molar mass acts as a conversion factor between numbers of molecules and mass. If you know the mass of a sample, you can calculate how many molecules you have; if you know how many molecules you have, you can calculate their mass. Note, though, that it's always necessary when using a molar mass to specify the formula of the substance you're talking about. For example, 1 mol of hydrogen *atoms*, H, has a molar mass of 1.0 g/mol, but 1 mol of hydrogen *molecules*, H_2, has a molar mass of 2.0 g/mol.

Whenever you see a balanced chemical equation, the coefficients tell how many moles of each substance are needed for the reaction. You can then use molar masses to calculate reactant masses. If you saw the following balanced equation for the industrial synthesis of ammonia, for example, you would know that 3 mol of H_2 (3 mol $\times$ 2.0 g/mol = 6.0 g) is needed for reaction with 1 mol of N_2 (28.0 g) to yield 2 mol of NH_3 (2 mol $\times$ 17.0 g/mol = 34.0 g).

This number of moles   . . . reacts with this number   to yield this number of
of hydrogen . . .   of moles of nitrogen . . .   moles of ammonia.

$$3\ H_2 + 1\ N_2 \longrightarrow 2\ NH_3$$

EXAMPLE 3.3   What is the molecular mass of sucrose, $C_{12}H_{22}O_{11}$? What is the molar mass of sucrose in grams per mole?

SOLUTION   The molecular mass of a substance is the sum of the atomic masses of the constituent atoms. First, list the elements present in the molecule, and then look up the atomic mass of each (we'll round off to one decimal place for convenience):

C (12.0 amu); H (1.0 amu); O (16.0 amu)

Next, multiply the atomic mass of each element by the number of times that element appears in the chemical formula, and then total the results.

$$C_{12}\ (12 \times 12.0\ amu) = 144.0\ amu$$
$$H_{22}\ (22 \times 1.0\ amu) = 22.0\ amu$$
$$O_{11}\ (11 \times 16.0\ amu) = 176.0\ amu$$
$$\overline{\text{Molec. mass of } C_{12}H_{22}O_{11} = 342.0\ amu}$$

Since one molecule of sucrose has a mass of 342.0 amu, 1 mol of sucrose has a mass of 342.0 grams. Thus, the molar mass of sucrose is 342.0 g/mol.

▶ **PROBLEM 3.4**   Calculate the formula mass or molecular mass of the following substances:
(a) Fe_2O_3 (rust)   **(b)** H_2SO_4 (sulfuric acid)
(c) $C_6H_8O_7$ (citric acid)   **(d)** $C_{16}H_{18}N_2O_4S$ (penicillin G)

▶ **PROBLEM 3.5**   The commercial production of iron from iron ore involves the reaction of Fe_2O_3 with CO to yield iron metal plus carbon dioxide:

$$Fe_2O_3(s) + CO(g) \longrightarrow Fe(s) + CO_2(g)$$

Balance the equation, and predict how many moles of CO will react with 0.500 mol of Fe_2O_3.

▶— **KEY CONCEPT PROBLEM 3.6**   Methionine, an amino acid used by organisms to make proteins, can be represented by the following ball-and-stick molecular model. Write the formula for methionine, and calculate its molecular mass (red = O, gray = C, blue = N, yellow = S, ivory = H).

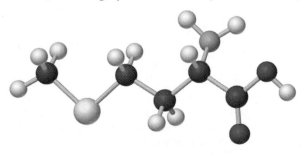

Iron is produced commercially by reduction of iron ore with carbon monoxide.

3.4 Stoichiometry: Chemical Arithmetic

We saw in the previous section that the coefficients in a balanced equation indicate the numbers of moles of substances in a reaction. In actual laboratory work, though, it's necessary to convert between moles and mass to be sure that the correct amounts of reactants are used. In referring to these mole–mass relationships, we use the word **stoichiometry** (stoy-key-**ahm**-uh-tree; from the Greek *stoicheion*, "element," and *metron*, "measure"). Let's look again at the reaction of ethylene with HCl to see how stoichiometric relationships are used.

$$C_2H_4(g) + HCl(g) \longrightarrow C_2H_5Cl(g)$$

Let's assume that we have 15.0 g of ethylene and we need to know how many grams of HCl to use in the reaction. According to the coefficients in the balanced equation, 1 mol of HCl is required for reaction with each mole of ethylene. To find out how many grams of HCl are required to react with 15.0 g of ethylene, we first have to find out how many moles of ethylene are in 15.0 g. We do this gram-to-mole conversion by calculating the molar mass of ethylene and using that value as a conversion factor:

Molecular mass of C_2H_4 = $(2 \times 12.0 \text{ amu}) + (4 \times 1.0 \text{ amu})$ = 28.0 amu

Molar mass of C_2H_4 = 28.0 g/mol

Moles of C_2H_4 = 15.0 $\cancel{\text{g ethylene}} \times \dfrac{1 \text{ mol ethylene}}{28.0 \cancel{\text{ g ethylene}}}$ = 0.536 mol ethylene

Now that we know how many moles of ethylene we have (0.536 mol), we also know from the balanced equation how many moles of HCl we need (0.536 mol), and we have to do a mole-to-gram conversion to find the mass of HCl required. Once again, the conversion is done by calculating the molecular mass of HCl and using molar mass as a conversion factor:

Molecular mass of HCl = 1.0 amu + 35.5 amu = 36.5 amu

Molar mass of HCl = 36.5 g/mol

Grams of HCl = 0.536 $\cancel{\text{mol } C_2H_4} \times \dfrac{1 \cancel{\text{ mol HCl}}}{1 \cancel{\text{ mol } C_2H_4}} \times \dfrac{36.5 \text{ g HCl}}{1 \cancel{\text{ mol HCl}}}$ = 19.6 g HCl

Thus, 19.6 g of HCl is needed to react with 15.0 g of ethylene.

Look carefully at the sequence of steps used in the calculation just completed. The important point is that *moles* (numbers of molecules) are given by the balanced equation but *grams* are used to weigh reactants in the laboratory. Moles tell us *how many molecules* of each reactant we need; grams tell us *how much mass* of each reactant we need.

Moles $\longrightarrow$ Numbers of molecules or formula units

Grams $\longrightarrow$ Mass

The flow diagram in Figure 3.2 illustrates the necessary conversions. Note that you can't go directly from the number of grams of one reactant to the number of grams of another reactant. You *must* first convert to moles.

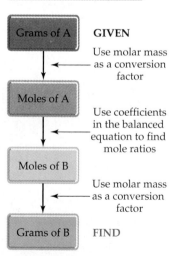

For the balanced equation:
$$a\,A + b\,B \longrightarrow c\,C + d\,D$$

Grams of A **GIVEN**

Use molar mass as a conversion factor

Moles of A

Use coefficients in the balanced equation to find mole ratios

Moles of B

Use molar mass as a conversion factor

Grams of B **FIND**

FIGURE 3.2 A summary of conversions between moles and grams for a chemical reaction. The numbers of moles tell how many molecules of each reactant are needed, as given by the balanced equation; the numbers of grams tell how much mass of each reactant is needed.

EXAMPLE 3.4 How many moles of sucrose are in a tablespoon of sugar that contains 2.85 g?

BALLPARK SOLUTION Since the molecular mass of sucrose (calculated in Example 3.3) is 342.0 amu, 1 mol of sucrose has a mass of 342.0 g. Thus, 2.85 g of sugar is a bit less than one-hundredth of a mole, or 0.01 mol.

DETAILED SOLUTION The problem gives the mass of sucrose and asks for a mass-to-mole conversion. Use the molar mass of sucrose as a conversion factor, and set up an equation so that the unwanted unit cancels:

$$2.85 \ \cancel{\text{g sucrose}} \times \frac{1 \ \text{mol sucrose}}{342.0 \ \cancel{\text{g sucrose}}} = 0.008 \ 33 \ \text{mol sucrose}$$

$$= 8.33 \times 10^{-3} \ \text{mol sucrose}$$

The ballpark answer and the detailed answer agree.

EXAMPLE 3.5 How many grams are there in 0.0626 mol of $NaHCO_3$, the main ingredient in Alka-Seltzer tablets?

SOLUTION The problem gives the number of moles of $NaHCO_3$ and asks for a mole-to-mass conversion. First, calculate the formula mass and molar mass of $NaHCO_3$:

Form. mass of $NaHCO_3$ = 23.0 amu + 1.0 amu + 12.0 amu + (3 × 16.0 amu)

= 84.0 amu

Molar mass of $NaHCO_3$ = 84.0 g/mol

Next, use molar mass as a conversion factor, and set up an equation so that the unwanted unit cancels:

$$0.0626 \ \cancel{\text{mol NaHCO}_3} \times \frac{84.0 \ \text{g NaHCO}_3}{1 \ \cancel{\text{mol NaHCO}_3}} = 5.26 \ \text{g NaHCO}_3$$

EXAMPLE 3.6 Aqueous solutions of sodium hypochlorite (NaOCl), best known as household bleach, are prepared by reaction of sodium hydroxide with chlorine:

$$2 \ NaOH(aq) + Cl_2(g) \longrightarrow NaOCl(aq) + NaCl(aq) + H_2O(l)$$

How many grams of NaOH are needed to react with 25.0 g of Cl_2?

SOLUTION Finding the relationship between numbers of reactant formula units always requires working in moles. The general strategy was outlined in Figure 3.2 and is reproduced below:

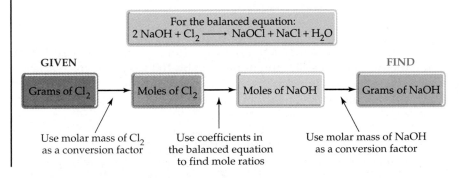

The first job is to find out how many moles of Cl_2 are in 25.0 g of Cl_2. This gram-to-mole conversion is done in the usual way, using the molar mass of Cl_2 (70.9 g/mol) as the conversion factor:

$$25.0 \text{ g } Cl_2 \times \frac{1 \text{ mol } Cl_2}{70.9 \text{ g } Cl_2} = 0.353 \text{ mol } Cl_2$$

Next, look at the coefficients in the balanced equation. Each mole of Cl_2 reacts with 2 mol of NaOH, so 0.353 mol of Cl_2 reacts with $2 \times 0.353 = 0.706$ mol of NaOH. With the number of moles of NaOH known, carry out a mole-to-gram conversion using the molar mass of NaOH (40.0 g/mol) as a conversion factor:

$$\text{Grams of NaOH} = 0.353 \text{ mol } Cl_2 \times \frac{2 \text{ mol NaOH}}{1 \text{ mol } Cl_2} \times \frac{40.0 \text{ g NaOH}}{1 \text{ mol NaOH}}$$

$$= 28.2 \text{ g NaOH}$$

We find that 25.0 g of Cl_2 reacts with 28.2 g of NaOH.

The problem can also be worked by combining the steps and setting up one large equation:

$$\text{Grams of NaOH} = 25.0 \text{ g } Cl_2 \times \frac{1 \text{ mol } Cl_2}{70.9 \text{ g } Cl_2} \times \frac{2 \text{ mol NaOH}}{1 \text{ mol } Cl_2} \times \frac{40.0 \text{ g NaOH}}{1 \text{ mol NaOH}}$$

$$= 28.2 \text{ g NaOH}$$

▶ **PROBLEM 3.7** Aspirin has the formula $C_9H_8O_4$. How many moles of aspirin are in a tablet weighing 500 mg? How many molecules?

▶ **PROBLEM 3.8** Aspirin is prepared by reaction of salicylic acid ($C_7H_6O_3$) with acetic anhydride ($C_4H_6O_3$) according to the following equation:

$$C_7H_6O_3 + C_4H_6O_3 \longrightarrow C_9H_8O_4 + CH_3CO_2H$$

| Salicylic acid | Acetic anhydride | Aspirin | Acetic acid |

How many grams of acetic anhydride are needed to react with 4.50 g of salicylic acid? How many grams of aspirin will result? How many grams of acetic acid are formed as a by-product?

3.5 Yields of Chemical Reactions

In the stoichiometry examples worked out in the preceding section, we made the unstated assumption that all reactions "go to completion." That is, we assumed that *all* reactant molecules are converted to products. In fact, few reactions behave so nicely. Most of the time, a majority of molecules react in the specified way, but other processes, called *side reactions*, also occur. Thus, the amount of product actually formed—the reaction's **yield**—is usually less than the amount that calculations predict.

The amount of product actually formed in a reaction divided by the amount theoretically possible and multiplied by 100% is called the reaction's **percent yield**. For example, if a given reaction *could* provide 6.9 g of a product according to its stoichiometry but actually provides only 4.7 g, then its percent yield is $(4.7/6.9) \times 100\% = 68\%$.

$$\text{Percent yield} = \frac{\text{Actual yield of product}}{\text{Theoretical yield of product}} \times 100\%$$

The following examples show how to calculate and use percent yield.

EXAMPLE 3.7   Methyl *tert*-butyl ether (MTBE, $C_5H_{12}O$), a substance used as an octane booster in gasoline, is made by reaction of isobutylene (C_4H_8) with methanol (CH_4O). What is the percent yield of the reaction if 32.8 g of methyl *tert*-butyl ether is obtained from reaction of 26.3 g of isobutylene with sufficient methanol?

$$C_4H_8(g) + CH_4O(l) \longrightarrow C_5H_{12}O(l)$$

Isobutylene   Methyl *tert*-butyl
                                                   ether (MTBE)

SOLUTION   We need to calculate the amount of methyl *tert*-butyl ether that could theoretically be produced from 26.3 g of isobutylene and compare that theoretical amount to the actual amount (32.8 g).

Always begin stoichiometry problems by calculating the molar masses of the reactants and products:

Isobutylene, C_4H_8: Molec. mass = $(4 \times 12.0 \text{ amu}) + (8 \times 1.0 \text{ amu})$ = 56.0 amu

Molar mass of isobutylene = 56.0 g/mol

MTBE, $C_5H_{12}O$: Molec. mass = $(5 \times 12.0 \text{ amu}) + (12 \times 1.0 \text{ amu}) + 16.0 \text{ amu}$ = 88.0 amu

Molar mass of MTBE = 88.0 g/mol

To calculate the amount of MTBE that could theoretically be produced from 26.3 g of isobutylene, we first have to find the number of moles of reactant, using molar mass as the conversion factor:

$$26.3 \text{ g isobutylene} \times \frac{1 \text{ mol isobutylene}}{56.0 \text{ g isobutylene}} = 0.470 \text{ mol isobutylene}$$

The balanced equation says that 1 mol of product is produced per mol of reactant, so we know that 0.470 mol of isobutylene can theoretically yield 0.470 mol of MTBE. To find the mass of this MTBE, we do a mole-to-mass conversion:

$$0.470 \text{ mol isobutylene} \times \frac{1 \text{ mol MTBE}}{1 \text{ mol isobutylene}} \times \frac{88.0 \text{ g MTBE}}{1 \text{ mol MTBE}} = 41.4 \text{ g MTBE}$$

Dividing the actual amount by the theoretical amount and multiplying by 100% then gives the percent yield:

$$\frac{32.8 \text{ g MTBE}}{41.4 \text{ g MTBE}} \times 100\% = 79.2\%$$

EXAMPLE 3.8   Diethyl ether ($C_4H_{10}O$), the "ether" used medically as an anesthetic, is prepared commercially by treatment of ethyl alcohol (C_2H_6O) with an acid. How many grams of diethyl ether would you obtain from 40.0 g of ethyl alcohol if the percent yield of the reaction is 87%?

$$2 \text{ } C_2H_6O(l) \xrightarrow{\text{Acid}} C_4H_{10}O(l) + H_2O(l)$$

Ethyl   Diethyl
alcohol   ether

SOLUTION   First, calculate the molar masses of the reactant and product:

Ethyl alcohol, C_2H_6O:

Molec. mass = $(2 \times 12$ amu$) + (6 \times 1.0$ amu$) + 16.0$ amu = 46.0 amu

Molar mass of ethyl alcohol = 46.0 /mol

Diethyl ether, $C_4H_{10}O$:

Molec. mass = $(4 \times 12.0$ amu$) + (10 \times 1.0$ amu$) + 16.0$ amu = 74.0 amu

Molar mass of diethyl ether = 74.0 g/mol

Next, find how many moles of ethyl alcohol are in 40.0 g by using molar mass as a conversion factor:

$$40.0 \text{ g ethyl alcohol} \times \frac{1 \text{ mol ethyl alcohol}}{46.0 \text{ g ethyl alcohol}} = 0.870 \text{ mol ethyl alcohol}$$

Because we started with 0.870 mol of ethyl alcohol, and because the balanced equation indicates that 2 mol of ethyl alcohol yield 1 mol of diethyl ether, we can theoretically obtain 0.435 mol of product:

$$0.870 \text{ mol ethyl alcohol} \times \frac{1 \text{ mol diethyl ether}}{2 \text{ mol ethyl alcohol}} = 0.435 \text{ mol diethyl ether}$$

We therefore need to find how many grams of diethyl ether are in 0.435 mol, using molar mass as the conversion factor:

$$0.435 \text{ mol diethyl ether} \times \frac{74.0 \text{ g diethyl ether}}{1 \text{ mol diethyl ether}} = 32.2 \text{ g diethyl ether}$$

Finally, we have to multiply the theoretical amount of product by the observed yield (87% = 0.87) to find how much diethyl ether is actually formed:

$$32.2 \text{ g diethyl ether} \times 0.87 = 28 \text{ g diethyl ether}$$

▶ **PROBLEM 3.9**   Ethyl alcohol is prepared industrially by the reaction of ethylene, C_2H_4, with water. What is the percent yield of the reaction if 4.6 g of ethylene gives 4.7 g of ethyl alcohol?

$$C_2H_4(g) + H_2O(l) \longrightarrow C_2H_6O(l)$$
$$\text{Ethylene} \qquad\qquad\qquad \text{Ethyl alcohol}$$

▶ **PROBLEM 3.10**   Dichloromethane (CH_2Cl_2), sometimes used as a solvent in the decaffeination of coffee beans, is prepared by reaction of methane (CH_4) with chlorine. How many grams of dichloromethane result from reaction of 1.85 kg of methane if the yield is 43.1%?

$$CH_4(g) + 2 Cl_2(g) \longrightarrow CH_2Cl_2(l) + 2 HCl(g)$$
$$\text{Methane} \quad \text{Chlorine} \qquad \text{Dichloromethane}$$

3.6 Reactions with Limiting Amounts of Reactants

Because chemists usually write balanced equations, it's easy to get the impression that reactions are always carried out using exactly the right proportions of reactants. In fact, this is often not the case. Many reactions are carried out using an excess amount of one reactant—more than is actually needed according to stoichiometry. Look, for example, at the industrial synthesis of ethylene glycol, $C_2H_6O_2$, a substance used both as automobile antifreeze and as a starting material for the preparation of polyester polymers. More than 2 million tons of ethylene glycol are prepared each year in the United States by reaction of ethylene oxide, C_2H_4O, with water at high temperature:

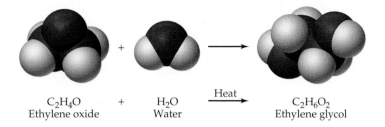

$$C_2H_4O \quad + \quad H_2O \quad \xrightarrow{\text{Heat}} \quad C_2H_6O_2$$
Ethylene oxide + Water → Ethylene glycol

Because water is so cheap and so abundant, it doesn't make sense to worry about using exactly 1 mol of water for each mole of ethylene oxide. It's much easier to use an excess of water to be certain that enough is present to consume entirely the more valuable ethylene oxide reactant. Of course, when an excess of water is present, only the amount required by stoichiometry undergoes reaction. The excess water is only a spectator and is not otherwise involved.

Whenever the ratios of reactant molecules actually used in an experiment are different from those given by the coefficients of the balanced equation, a surplus of one reactant is left over after the reaction is finished. Thus, the extent to which a chemical reaction takes place depends on the reactant that is present in limiting amount—the **limiting reactant**. The other reactant is said to be the *excess reactant*.

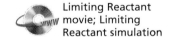

Limiting Reactant movie; Limiting Reactant simulation

What happens with excess reactants and limiting reactants is similar to what happens if there are five people in a room but only three chairs. Only three people can sit and the other two stand, because the number of people sitting is limited by the number of available chairs. In the same way, if 5 water molecules come in contact with 3 ethylene oxide molecules, only 3 water molecules can undergo a reaction. The other 2 water molecules are merely spectators, because the number of water molecules that react is limited by the number of available ethylene oxide molecules.

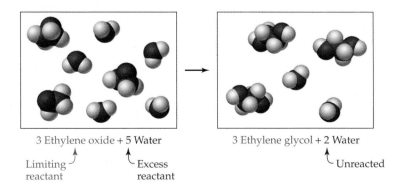

3 Ethylene oxide + 5 Water → 3 Ethylene glycol + 2 Water
Limiting reactant   Excess reactant   Unreacted

The following example shows how to tell if a limiting amount of one reactant is present and how to calculate the amount of the excess reactant that is consumed.

EXAMPLE 3.9   Cisplatin, an anticancer agent used for the treatment of solid tumors, is prepared by the reaction of ammonia with potassium tetrachloroplatinate:

$$K_2PtCl_4 + 2\,NH_3 \longrightarrow Pt(NH_3)_2Cl_2 + 2\,KCl$$

Potassium   Cisplatin
tetrachloroplatinate

Assume that 10.0 g of K_2PtCl_4 and 10.0 g of NH_3 are allowed to react.
(a) Which reactant is limiting, and which is in excess?
(b) How many grams of the excess reactant are consumed, and how many grams remain?
(c) How many grams of cisplatin are formed?

SOLUTION

(a) Complex stoichiometry problems should be worked slowly and carefully, one step at a time. When solving a problem that deals with limiting reactants, the idea is to find how many moles of all reactants are actually present and then compare the mole ratios of those actual amounts to the mole ratios required by the balanced equation. That comparison will identify the reactant there is too much of (the excess reactant) and the reactant there is too little of (the limiting reactant).

Finding the molar amounts of reactants always begins by calculating formula masses and using molar masses as conversion factors:

Form. mass of $K_2PtCl_4 = (2 \times 39.1\ \text{amu}) + 195.1\ \text{amu} + (4 \times 35.5\ \text{amu})$

$$= 415.3\ \text{amu}$$

Molar mass of $K_2PtCl_4 = 415.3$ g/mol

$$\text{Moles of } K_2PtCl_4 = 10.0\ \text{g } K_2PtCl_4 \times \frac{1\ \text{mol } K_2PtCl_4}{415.3\ \text{g } K_2PtCl_4} = 0.0241\ \text{mol } K_2PtCl_4$$

Molec. mass of $NH_3 = 14.0\ \text{amu} + (3 \times 1.0\ \text{amu}) = 17.0\ \text{amu}$

Molar mass of $NH_3 = 17.0$ g/mol

$$\text{Moles of } NH_3 = 10.0\ \text{g } NH_3 \times \frac{1\ \text{mol } NH_3}{17.0\ \text{g } NH_3} = 0.588\ \text{mol } NH_3$$

These calculations tell us that we have 0.588 mol of ammonia and 0.0241 mol of K_2PtCl_4, or $0.588/0.0241 = 24.4$ times as much ammonia as K_2PtCl_4. The coefficients in the balanced equation, however, say that only *two* times as much ammonia as K_2PtCl_4 is needed. Thus, a large excess of NH_3 is present, and K_2PtCl_4 is the limiting reactant.

(b) With the identities of the excess reactant and limiting reactant known, we now have to find how many moles of each undergo reaction and then carry out mole-to-gram conversions to find the mass of each reactant consumed. The entire amount of the limiting reactant (K_2PtCl_4) is used up, but only the amount of the excess reactant (NH_3) required by stoichiometry undergoes reaction:

Moles of K_2PtCl_4 consumed $= 0.0241\ \text{mol } K_2PtCl_4$

$$\text{Moles of } NH_3 \text{ consumed} = 0.0241\ \text{mol } K_2PtCl_4 \times \frac{2\ \text{mol } NH_3}{1\ \text{mol } K_2PtCl_4} = 0.0482\ \text{mol } NH_3$$

$$\text{Grams of } NH_3 \text{ consumed} = 0.0482\ \text{mol } NH_3 \times \frac{17.0\ \text{g } NH_3}{1\ \text{mol } NH_3} = 0.819\ \text{g } NH_3$$

Grams of NH_3 not consumed $= (10.0\ \text{g} - 0.819\ \text{g})\ NH_3 = 9.2\ \text{g } NH_3$

(c) The balanced equation shows that 1 mol of cisplatin is formed for each mole of K_2PtCl_4 consumed. Thus, 0.0241 mol of cisplatin is formed from 0.0241 mol of K_2PtCl_4. To determine the mass of cisplatin produced, we must calculate its molar mass and then carry out a mole-to-gram conversion:

Molec. mass of $Pt(NH_3)_2Cl_2$ − 195.1 amu + (2 × 17.0 amu) + (2 × 35.5 amu)

$$= 300.1 \text{ amu}$$

Molar mass of $Pt(NH_3)_2Cl_2$ = 300.1 g/mol

Grams of $Pt(NH_3)_2Cl_2$ = 0.0241 ~~mol Pt(NH₃)₂Cl₂~~ × $\dfrac{300.1 \text{ g } Pt(NH_3)_2Cl_2}{1 \text{ mol } Pt(NH_3)_2Cl_2}$

$$= 7.23 \text{ g } Pt(NH_3)_2Cl_2$$

▶ **PROBLEM 3.11** Lithium oxide is used aboard the space shuttle to remove water from the air supply according to the equation

$$Li_2O(s) + H_2O(g) \longrightarrow 2\,LiOH(s)$$

If 80.0 kg of water is to be removed and 65 kg of Li_2O is available, which reactant is limiting? How many kilograms of the excess reactant remain?

▶ **PROBLEM 3.12** After lithium hydroxide is produced aboard the space shuttle by reaction of Li_2O with H_2O (Problem 3.11), it is used to remove exhaled carbon dioxide from the air supply according to the equation

$$LiOH(s) + CO_2(g) \longrightarrow LiHCO_3(s)$$

How many grams of CO_2 can 500.0 g of LiOH absorb?

◀ **KEY CONCEPT PROBLEM 3.13** The following diagram represents the reaction of A (red spheres) with B_2 (blue spheres):

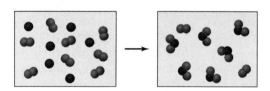

(a) Write a balanced equation for the reaction, and identify the limiting reactant.
(b) How many moles of product can be made from 1.0 mol of A and 1.0 mol of B_2?

3.7 Concentrations of Reactants in Solution: Molarity

For a chemical reaction to occur, the reacting molecules or ions must come into contact. This means that the reactants must have considerable mobility, which in turn means that most chemical reactions are carried out in the liquid state or in solution rather than in the solid state. It's therefore necessary to have a standard means for describing exact quantities of reactants in solution.

Stoichiometry calculations for chemical reactions always require working in *moles*. Thus, the most generally useful means of expressing a solution's concentration is **molarity (M)**, the number of moles of a substance (the *solute*) dissolved in each liter of solution. For example, a solution made by dissolving

Lithium oxide is used aboard the space shuttle to remove water from the air.

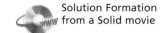

Solution Formation from a Solid movie

1.00 mol (58.5 g) of NaCl in enough water to give 1.00 L of solution has a concentration of 1.00 mol/L, or 1.00 M. The molarity of any solution is found by dividing the number of moles of solute by the number of liters of solution.

Molarity activity

$$\text{Molarity (M)} = \frac{\text{Moles of solute}}{\text{Liters of solution}}$$

Note that it is the final volume of the *solution* that is important, not the starting volume of the *solvent* used. The final volume of the solution might be a bit larger than the volume of the solvent because of the additional volume of the solute. In practice, a solution of known molarity is prepared by weighing an appropriate amount of solute and placing it in a *volumetric flask*, as shown in Figure 3.3. Enough solvent is added to dissolve the solute, and further solvent is added until an accurately calibrated final volume is reached. The solution is then shaken until it is uniformly mixed.

FIGURE 3.3 Preparing a solution of known molarity. **(a)** A measured number of moles of solute is placed in a volumetric flask. **(b)** Enough solvent is added to dissolve the solute by swirling. **(c)** Further solvent is carefully added until the calibration mark on the neck of the flask is reached, and the solution is then shaken until uniform.

(a)   (b)   (c)

Molarity can be used as a conversion factor to relate a solution's volume to the number of moles of solute. If we know the molarity and volume of a solution, we can calculate the number of moles of solute. If we know the number of moles of solute and the molarity of the solution, we can find the solution's volume. Examples 3.10 and 3.11 show how such calculations are done.

$$\text{Molarity} = \frac{\text{Moles of solute}}{\text{Volume of solution (L)}}$$

$$\text{Moles of solute} = \text{Molarity} \times \text{Volume of solution} \qquad \text{Volume of solution} = \frac{\text{Moles of solute}}{\text{Molarity}}$$

EXAMPLE 3.10   What is the molarity of a solution made by dissolving 2.355 g of sulfuric acid (H_2SO_4) in water and diluting to a final volume of 50.0 mL?

SOLUTION   Since molarity is the number of moles of solute per liter of solution, it's first necessary to find the number of moles of sulfuric acid in 2.355 g:

Molec. mass of H_2SO_4 = (2 × 1.0 amu) + 32.1 amu + (4 × 16.0 amu) = 98.1 amu

Molar mass of H_2SO_4 = 98.1 g/mol

$$2.355 \ \cancel{\text{g } H_2SO_4} \times \frac{1 \ \text{mol } H_2SO_4}{98.1 \ \cancel{\text{g } H_2SO_4}} = 0.0240 \ \text{mol } H_2SO_4$$

Next, divide the number of moles of solute by the volume of the solution (in liters) to find the molarity:

$$\frac{0.0240 \text{ mol H}_2\text{SO}_4}{0.0500 \text{ L}} = 0.480 \text{ M}$$

The solution has a sulfuric acid concentration of 0.480 M.

EXAMPLE 3.11 Hydrochloric acid is sold commercially as a 12.0 M solution. How many moles of HCl are in 300.0 mL of 12.0 M solution?

SOLUTION You can calculate the number of moles of solute by multiplying the molarity of the solution by its volume:

$$\text{Moles of HCl} = (\text{Molarity of solution}) \times (\text{Volume of solution})$$

$$= \frac{12.0 \text{ mol HCl}}{1 \cancel{\text{L}}} \times 0.3000 \cancel{\text{L}} = 3.60 \text{ mol HCl}$$

There are 3.60 mol of HCl in 300.0 mL of 12.0 M solution.

▶ **PROBLEM 3.14** How many moles of solute are present in the following solutions?
(a) 125 mL of 0.20 M NaHCO$_3$ (b) 650.0 mL of 2.50 M H$_2$SO$_4$

▶ **PROBLEM 3.15** How many grams of solute would you use to prepare the following solutions?
(a) 500.0 mL of 1.25 M NaOH   (b) 1.50 L of 0.250 M glucose (C$_6$H$_{12}$O$_6$)

▶ **PROBLEM 3.16** How many milliliters of a 0.20 M glucose (C$_6$H$_{12}$O$_6$) solution are needed to provide a total of 25.0 g of glucose?

▶ **PROBLEM 3.17** The concentration of cholesterol (C$_{27}$H$_{46}$O) in normal blood is approximately 0.005 M. How many grams of cholesterol are in 750 mL of blood?

3.8 Diluting Concentrated Solutions

For convenience, chemicals are sometimes bought and stored as concentrated solutions that must be diluted before use. Aqueous hydrochloric acid, for example, is sold commercially as a 12.0 M solution, yet it is most commonly used in the laboratory after dilution with water to a final concentration of 6.0 M or 1.0 M.

$$\text{Concentrated solution} + \text{Solvent} \longrightarrow \text{Dilute solution}$$

The key fact to remember when diluting a concentrated solution is that the number of moles of *solute* remains the same; only the *volume* is changed by adding more solvent. Because the number of moles of solute can be calculated by multiplying molarity times volume, we can set up the following equation:

Solution Formation by Dilution movie

$$\text{Moles of solute} = \text{Molarity} \times \text{Volume}$$

$$= M_i \times V_i = M_f \times V_f$$

where M_i is the initial molarity, V_i is the initial volume, M_f is the final molarity, and V_f is the final volume after dilution. Rearranging this equation into a more useful form shows that the molar concentration after dilution (M_f) can be found by multiplying the initial concentration (M_i) by the ratio of initial and final volumes (V_i/V_f):

$$M_f = M_i \times \frac{V_i}{V_f}$$

Suppose, for example, that we dilute 50.0 mL of a solution of 2.00 M H_2SO_4 to a volume of 200.0 mL. The solution volume *increases* by a factor of four (from 50 mL to 200 mL), so the concentration of the solution must *decrease* by a factor of four (from 2.00 M to 0.500 M):

$$M_f = 2.00 \text{ M} \times \frac{50.0 \text{ mL}}{200.0 \text{ mL}} = 0.500 \text{ M}$$

In practice, dilutions are carried out as shown in Figure 3.4. The volume to be diluted is withdrawn using a calibrated tube called a *pipet*, placed in an empty volumetric flask of the chosen volume, and diluted to the calibration mark on the flask.

FIGURE 3.4   The procedure for diluting a concentrated solution. **(a)** The volume to be diluted is withdrawn and placed in an empty volumetric flask. **(b)** Solvent is added to a level just below the calibration mark, and the flask is shaken. **(c)** More solvent is added to reach the calibration mark, and the solution is inverted a few times to ensure mixing.

(a)   (b)   (c)

EXAMPLE 3.12   How would you prepare 500.0 mL of 0.2500 M NaOH solution starting from a concentration of 1.000 M?

BALLPARK SOLUTION   Because the concentration decreases by a factor of four after dilution (from 1.000 M to 0.2500 M), the volume must increase by a factor of four. Thus, to prepare 500.0 mL of solution, we should start with $500.0/4 = 125.0$ mL.

DETAILED SOLUTION   The problem gives initial and final concentrations (M_i and M_f) and final volume (V_f) and asks for the initial volume (V_i) that we need to dilute. Rewriting the equation $M_i \times V_i = M_f \times V_f$ as $V_i = (M_f/M_i) \times V_f$ gives the answer:

$$V_i = \frac{M_f}{M_i} \times V_f = \frac{0.2500 \text{ M}}{1.000 \text{ M}} \times 500.0 \text{ mL} = 125.0 \text{ mL}$$

We therefore need to place 125.0 mL of 1.000 M NaOH solution in a 500.0 mL volumetric flask and fill to the mark with water.

▶ **PROBLEM 3.18** What is the final concentration if 75.0 mL of a 3.50 M glucose solution is diluted to a volume of 400.0 mL?

▶ **PROBLEM 3.19** Sulfuric acid is normally purchased at a concentration of 18.0 M. How would you prepare 250.0 mL of 0.500 M aqueous H_2SO_4?

3.9 Solution Stoichiometry

We remarked in Section 3.7 that molarity is a conversion factor between numbers of moles of solute and the volume of a solution. If we know the volume and molarity of a solution, we can calculate the number of moles of solute. If we know the number of moles of solute and molarity, we can find the volume.

As indicated by the flow diagram in Figure 3.5, a knowledge of molarity is critical for carrying out stoichiometry calculations on substances in solution. Molarity makes it possible to calculate the volume of one solution needed to react with a given volume of another solution. This sort of calculation is particularly important in acid–base chemistry, as shown in Example 3.13.

EXAMPLE 3.13 Stomach acid, a dilute solution of HCl in water, can be neutralized by reaction with sodium hydrogen carbonate, $NaHCO_3$, according to the equation

$$HCl(aq) + NaHCO_3(aq) \longrightarrow NaCl(aq) + H_2O(l) + CO_2(g)$$

How many milliliters of 0.125 M $NaHCO_3$ solution are needed to neutralize 18.0 mL of 0.100 M HCl?

BALLPARK SOLUTION The balanced equation shows that HCl and $NaHCO_3$ react in a 1:1 molar ratio, and we are told that the concentrations of the two solutions are about the same. Thus, the volume of the $NaHCO_3$ solution must be about the same as that of the HCl solution.

DETAILED SOLUTION Since we need to know the numbers of moles to solve stoichiometry problems, we first have to find how many moles of HCl are in 18.0 mL of a 0.100 M solution by multiplying volume times molarity:

$$\text{Moles of HCl} = 18.0 \; \cancel{mL} \times \frac{1 \; \cancel{L}}{1000 \; \cancel{mL}} \times \frac{0.100 \; mol}{1 \; \cancel{L}} = 1.80 \times 10^{-3} \; mol \; HCl$$

Next, check the coefficients of the balanced equation to find that each mole of HCl reacts with 1 mol of $NaHCO_3$, and then calculate how many milliliters of 0.125 M $NaHCO_3$ solution contains 1.80×10^{-3} mol:

$$1.80 \times 10^{-3} \; \cancel{mol \; HCl} \times \frac{1 \; mol \; \cancel{NaHCO_3}}{1 \; \cancel{mol \; HCl}} \times \frac{1 \; L \; solution}{0.125 \; \cancel{mol \; NaHCO_3}} = 0.0144 \; L \; solution$$

Thus, 14.4 mL of the 0.125 M $NaHCO_3$ solution is needed to neutralize 18.0 mL of the 0.100 M HCl solution.

▶ **PROBLEM 3.20** What volume of 0.250 M H_2SO_4 is needed to react with 50.0 mL of 0.100 M NaOH? The equation is

$$H_2SO_4(aq) + 2\,NaOH(aq) \longrightarrow Na_2SO_4(aq) + 2\,H_2O(l)$$

▶ **PROBLEM 3.21** What is the molarity of an HNO_3 solution if 68.5 mL is needed to react with 25.0 mL of 0.150 M KOH solution?

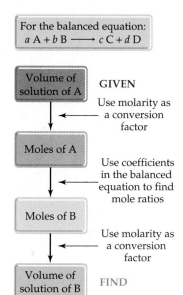

FIGURE 3.5 A flow diagram summarizing the use of molarity as a conversion factor between moles and volume in stoichiometry calculations.

Neutralization of sodium hydrogen carbonate with acid leads to release of CO_2 gas, visible in this fizzing solution.

3.10   Titration

There are two ways to make a solution of known molarity. The first and most obvious way is to make the solution carefully, using an accurately weighed amount of solute dissolved in solvent to an accurately calibrated volume. Often, though, it's more convenient to make up a solution quickly, using an estimated amount of solute and an estimated amount of solvent, and then determine the solution's exact molarity by *titration*.

Titration is a procedure for determining the concentration of a solution by allowing a carefully measured volume to react with a *standard solution* of another substance, whose concentration is known. By finding the volume of the standard solution that reacts with the measured volume of the first solution, the concentration of the first solution can be calculated. (It's necessary, though, that the reaction go to completion and have a yield of 100%.)

To see how titration works, let's imagine that we have an HCl solution (an acid) whose concentration we want to find by allowing it to react with a base such as NaOH in what is called a *neutralization reaction*. (We'll learn more about acid–base neutralization reactions in the next chapter.) The balanced equation is

$$NaOH(aq) + HCl(aq) \longrightarrow NaCl(aq) + H_2O(l)$$

We begin the titration by measuring a known volume of the HCl solution and adding a small amount of an *indicator*, such as phenolphthalein, a compound that is colorless in acidic solution but turns pink in basic solution. Next, we fill a calibrated glass tube called a *buret* with an NaOH standard solution of known concentration, and we slowly add the NaOH to the HCl until the phenolphthalein just begins to turn pink, indicating that all the HCl has reacted and that the solution is starting to become basic. By then reading from the buret to find the volume of the NaOH standard solution that has reacted with the known volume of HCl, we can calculate the concentration of the HCl. The strategy is summarized in Figure 3.6.

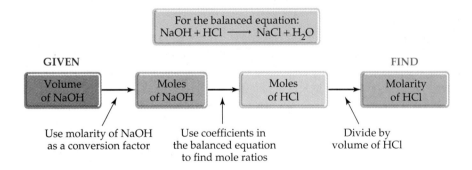

FIGURE 3.6   A flow diagram for an acid–base titration, summarizing the calculations needed to determine the concentration of an HCl solution by titration with an NaOH standard solution.

Let's assume, for example, that we take 20.0 mL of the HCl solution and find that we have to add 48.6 mL of 0.100 M NaOH from a buret to obtain complete reaction. Using the molarity of the NaOH standard solution as a conversion factor, we can calculate the number of moles of NaOH undergoing reaction:

$$\text{Moles of NaOH} = 0.0486 \, \text{L NaOH} \times \frac{0.100 \, \text{mol NaOH}}{1 \, \text{L NaOH}}$$

$$= 0.004\,86 \, \text{mol NaOH}$$

According to the balanced equation, the number of moles of HCl is the same as that of NaOH:

$$\text{Moles of HCl} = 0.004\,86 \; \text{mol NaOH} \times \frac{1 \; \text{mol HCl}}{1 \; \text{mol NaOH}} = 0.004\,86 \; \text{mol HCl}$$

Dividing the number of moles of HCl by the volume then gives the molarity of the HCl:

$$\text{HCl molarity} = \frac{0.004\,86 \; \text{mol HCl}}{0.0200 \; \text{L HCl}} = 0.243 \; \text{M HCl}$$

The titration procedure is shown in Figure 3.7.

(a)   (b)

FIGURE 3.7   Titration of an acid solution of unknown concentration with a base solution of known concentration. **(a)** A measured volume of acid solution is placed in a flask, and phenolphthalein indicator is added. **(b)** A base solution of known concentration is added from a buret until the indicator changes color to signal that all the acid has reacted. Reading the volume of base solution added from the buret makes it possible to calculate the concentration of the acid solution.

▶ **PROBLEM 3.22**   A 25.0 mL sample of vinegar (dilute acetic acid, CH_3CO_2H) is titrated and found to react with 94.7 mL of 0.200 M NaOH. What is the molarity of the acetic acid solution? The reaction is

$$NaOH(aq) + CH_3CO_2H(aq) \longrightarrow CH_3CO_2Na(aq) + H_2O(l)$$

3.11   Percent Composition and Empirical Formulas

Whenever a new compound is made in the laboratory or found in nature, it must be analyzed to find what elements it contains and how much of each element is present—that is, to find its *composition*. The **percent composition** of a compound is expressed by identifying the elements present and giving the mass percent of each. For example, we might express the percent composition of a certain colorless liquid found in gasoline by saying that it contains 84.1% carbon and 15.9% hydrogen by mass. In other words, a 100 g sample of the compound contains 84.1 g of carbon atoms and 15.9 g of hydrogen atoms.

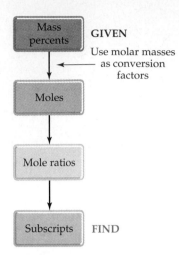

FIGURE 3.8   A flow diagram for calculating the formula of a compound from its percent composition.

Knowing a compound's percent composition makes it possible to calculate the compound's chemical formula. The strategy is to find the relative number of moles of each element in the compound and then use the numbers to establish the mole ratios of the elements, as shown in Figure 3.8.

Let's use for our example the colorless liquid whose composition is 84.1% carbon and 15.9% hydrogen by mass. Arbitrarily taking 100 g of the substance, we find by using molar masses as conversion factors that the 100 g contains:

$$84.1 \; \cancel{g\,C} \times \frac{1 \; mol \; C}{12.01 \; \cancel{g\,C}} = 7.00 \; mol \; C$$

$$15.9 \; \cancel{g\,H} \times \frac{1 \; mol \; H}{1.008 \; \cancel{g\,H}} = 15.8 \; mol \; H$$

With the relative numbers of moles of C and H known, we next find their mole ratio by dividing the larger number of moles (15.8 mol) by the smaller number (7.00 mol):

$$\frac{\text{Moles of H}}{\text{Moles of C}} = \frac{15.8 \; mol \; H}{7.00 \; mol \; C} = \frac{2.26 \; mol \; H}{1 \; mol \; C}$$

The C:H mole ratio of 1:2.26 means that we can write $C_1H_{2.26}$ as a temporary formula for the liquid. Multiplying the subscripts by small integers in a trial-and-error procedure until whole numbers are found then gives the final formula. In the present instance, multiplication of the subscripts by 4 is needed.

$$C_{(1 \times 4)}H_{(2.26 \times 4)} = C_4H_9$$

A formula such as C_4H_9, which is determined from data about percent composition, is called an **empirical formula** because it tells only the *ratios* of atoms in a compound. The **molecular formula**, which tells the actual numbers of atoms in a molecule, can be either the same as the empirical formula or a multiple of it, such as C_8H_{18}. To determine the molecular formula, it's necessary to know the molecular mass of the substance. In the present instance, the molecular mass of our compound (a substance called *octane*) is 114.2 amu, which is a simple multiple of the empirical molecular mass for C_4H_9 (57.1 amu).

To find the multiple, calculate the ratio of the molecular mass to the empirical formula mass:

$$\text{Multiple} = \frac{\text{Molecular mass}}{\text{Empirical formula mass}} = \frac{114.2}{57.1} = 2.00$$

Then multiply the subscripts in the empirical formula by this multiple to obtain the molecular formula. In our example, the molecular formula of octane is $C_{(4 \times 2)}H_{(9 \times 2)}$, or C_8H_{18}.

Just as we can derive the empirical formula of a substance from its percent composition, we can also calculate the percent composition of a substance from its empirical (or molecular) formula. The strategies for the two kinds of calculations are exactly opposite. Aspirin, for example, has the molecular formula $C_9H_8O_4$ and thus has a C:H:O mole ratio of 9:8:4. We can convert this mole ratio into a mass ratio, and thus into percent composition, by carrying out mole-to-gram conversions on a 1 mol sample of compound:

$$1 \text{ mol aspirin} \times \frac{9 \text{ mol C}}{1 \text{ mol aspirin}} \times \frac{12.0 \text{ g C}}{1 \text{ mol C}} = 108 \text{ g C}$$

$$1 \text{ mol aspirin} \times \frac{8 \text{ mol H}}{1 \text{ mol aspirin}} \times \frac{1.01 \text{ g H}}{1 \text{ mol C}} = 8.08 \text{ g H}$$

$$1 \text{ mol aspirin} \times \frac{4 \text{ mol O}}{1 \text{ mol aspirin}} \times \frac{16.0 \text{ g O}}{1 \text{ mol O}} = 64.0 \text{ g O}$$

Dividing the mass of each element by the total mass and multiplying by 100% then gives the percent composition:

Total mass of 1 mol aspirin $= 108 \text{ g} + 8.08 \text{ g} + 64.0 \text{ g} = 180 \text{ g}$

$$\% \text{ C} = \frac{108 \text{ g C}}{180 \text{ g}} \times 100\% = 60.0\%$$

$$\% \text{ H} = \frac{8.08 \text{ g H}}{180 \text{ g}} \times 100\% = 4.49\%$$

$$\% \text{ O} = \frac{64.0 \text{ g O}}{180 \text{ g}} \times 100\% = 35.6\%$$

The answer can be checked by confirming that the sum of the mass percentages is 100%: $60.0\% + 4.49\% + 35.6\% = 100.1\%$.

Examples 3.14 and 3.15 show further conversions between percent composition and empirical formulas.

EXAMPLE 3.14   Vitamin C (ascorbic acid) contains 40.92% C, 4.58% H, and 54.50% O by mass. What is the empirical formula of ascorbic acid?

SOLUTION   Assume that you have 100.00 g of ascorbic acid, and then carry out gram-to-mole conversions to find the number of moles of each element in the sample:

$$40.92 \text{ g C} \times \frac{1 \text{ mol C}}{12.0 \text{ g C}} = 3.41 \text{ mol C}$$

$$4.58 \text{ g H} \times \frac{1 \text{ mol H}}{1.01 \text{ g H}} = 4.53 \text{ mol H}$$

$$54.50 \text{ g O} \times \frac{1 \text{ mol O}}{16.0 \text{ g O}} = 3.41 \text{ mol O}$$

Dividing each of the three numbers by the smallest one (3.41 mol) gives a C:H:O mole ratio of 1:1.33:1 and a temporary formula of $C_1H_{1.33}O_1$. Multiplying the subscripts by small integers in a trial-and-error procedure until whole numbers are found then gives the empirical formula $C_3H_4O_3$.

Multiply subscripts by 2: $C_{(2\times1)}H_{(2\times1.33)}O_{(2\times1)} = C_2H_{2.66}O_2$

Multiply subscripts by 3: $C_{(3\times1)}H_{(3\times1.33)}O_{(3\times1)} = C_3H_4O_3$

EXAMPLE 3.15 Glucose, or blood sugar, has the molecular formula $C_6H_{12}O_6$. What is the empirical formula, and what is the percent composition of glucose?

SOLUTION Dividing the subscripts in the molecular formula $C_6H_{12}O_6$ by 6 gives an empirical formula of CH_2O for glucose.

The percent composition of glucose can be calculated either from the molecular formula or from the empirical formula. Using the molecular formula, for example, the C:H:O mole ratio of 6:12:6 can be converted into a mass ratio by assuming that we have 1 mol of compound and carrying out mole-to-gram conversions:

$$1 \text{ mol glucose} \times \frac{6 \text{ mol C}}{1 \text{ mol glucose}} \times \frac{12.0 \text{ g C}}{1 \text{ mol C}} = 72.0 \text{ g C}$$

$$1 \text{ mol glucose} \times \frac{12 \text{ mol H}}{1 \text{ mol glucose}} \times \frac{1.01 \text{ g H}}{1 \text{ mol H}} = 12.1 \text{ g H}$$

$$1 \text{ mol glucose} \times \frac{6 \text{ mol O}}{1 \text{ mol glucose}} \times \frac{16.0 \text{ g O}}{1 \text{ mol O}} = 96.0 \text{ g O}$$

Dividing the mass of each element by the total mass and multiplying by 100% gives the percent composition:

Total mass of 1 mol glucose = 72.0 g + 12.1 g + 96.0 g = 180.1 g

$$\% \text{ C} = \frac{72.0 \text{ g C}}{180.1 \text{ g}} \times 100\% = 40.0\%$$

$$\% \text{ H} = \frac{12.1 \text{ g H}}{180.1 \text{ g}} \times 100\% = 6.72\%$$

$$\% \text{ O} = \frac{96.0 \text{ g O}}{180.1 \text{ g}} \times 100\% = 53.3\%$$

Note that the sum of the mass percentages is 100%.

▶ **PROBLEM 3.23** What is the empirical formula, and what is the percent composition of dimethylhydrazine, $C_2H_8N_2$, a colorless liquid used as a rocket fuel?

▶ **PROBLEM 3.24** What is the empirical formula of an ingredient in Bufferin tablets that has the percent composition C 14.25%, O 56.93%, Mg 28.83% by mass?

3.12 Determining Empirical Formulas: Elemental Analysis

One of the most common methods used to determine empirical formulas, particularly for compounds containing carbon and hydrogen, is *combustion analysis*. In this method, a compound of unknown composition is burned with oxygen to produce the volatile combustion products CO_2 and H_2O, which are separated and weighed by an automated instrument called a *gas chromatograph*. Methane (CH_4), for instance, burns according to the balanced equation

$$CH_4(g) + 2\,O_2(g) \longrightarrow CO_2(g) + 2\,H_2O(g)$$

With the masses of the carbon-containing product (CO_2) and hydrogen-containing product (H_2O) known, our strategy is to then calculate the number of moles of carbon and hydrogen in the products, from which we can find the C:H mole ratio. This information, in turn, provides the chemical formula, as outlined in Figure 3.9.

As an example of how combustion analysis works, imagine that we have a sample of a pure substance—say, naphthalene (often used for household moth balls). We weigh a known amount of the sample, burn it in pure oxygen, and then analyze the products. Let's say that 0.330 g of naphthalene reacts with O_2 and that 1.133 g of CO_2 and 0.185 g of H_2O are formed. The first thing to find out is the number of moles of carbon and hydrogen in the CO_2 and H_2O products so that we can calculate the number of moles of each element originally present in the naphthalene sample.

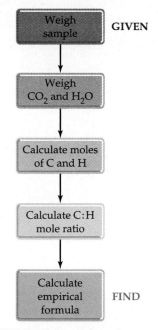

FIGURE 3.9 A flow diagram for determining an empirical formula from combustion analysis of a compound containing C and H.

$$\text{Moles of C in 1.133 g } CO_2 = 1.133 \text{ g } CO_2 \times \frac{1 \text{ mol } CO_2}{44.01 \text{ g } CO_2} \times \frac{1 \text{ mol C}}{1 \text{ mol } CO_2}$$

$$= 0.025\,74 \text{ mol C}$$

$$\text{Moles of H in 0.185 g } H_2O = 0.185 \text{ g } H_2O \times \frac{1 \text{ mol } H_2O}{18.02 \text{ g } H_2O} \times \frac{2 \text{ mol H}}{1 \text{ mol } H_2O}$$

$$= 0.0205 \text{ mol H}$$

Although it's not necessary in this instance since naphthalene contains only carbon and hydrogen, we can make sure that all the mass is accounted for and that no other elements are present. To do so, we carry out mole-to-gram conversions to find the number of grams of C and H in the starting sample:

$$\text{Mass of C} = 0.025\,74 \text{ mol C} \times \frac{12.01 \text{ g C}}{1 \text{ mol C}} = 0.3091 \text{ g C}$$

$$\text{Mass of H} = 0.0205 \text{ mol H} \times \frac{1.01 \text{ g H}}{1 \text{ mol H}} = 0.0207 \text{ g H}$$

$$\text{Total mass of C and H} = 0.3091 \text{ g} + 0.0207 \text{ g} = 0.3298 \text{ g}$$

Because the total mass of the C and H in the products (0.3298 g) is the same as the mass of the starting sample (0.330 g), we know that no other elements are present in naphthalene.

With the relative number of moles of C and H known, divide the larger number of moles by the smaller number to get a C:H mole ratio of 1.26:1 and the temporary formula $C_{1.26}H_1$.

$$\frac{\text{Moles of C}}{\text{Moles of H}} = \frac{0.025\,74 \text{ mol C}}{0.0205 \text{ mol H}} = \frac{1.26 \text{ mol C}}{1 \text{ mol H}}$$

Multiplying the subscripts by small integers in a trial-and-error procedure until whole numbers are found gives the final formula C_5H_4. (Of course, the subscripts may not always be *exact* integers because of small errors in the data, but the discrepancies should be small.)

Multiply subscripts by 2: $C_{(1.26 \times 2)}H_{(1 \times 2)} = C_{2.52}H_2$

Multiply subscripts by 3: $C_{(1.26 \times 3)}H_{(1 \times 3)} = C_{3.78}H_3$

Multiply subscripts by 4: $C_{(1.26 \times 4)}H_{(1 \times 4)} = C_{5.04}H_4$

$$= C_5H_4 \text{ Both subscripts are integers}$$

As mentioned in the previous section, analysis can provide only an empirical formula. To determine the molecular formula, it is also necessary to know the substance's molecular mass. In the present problem, the molecular mass of naphthalene is 128.2 amu, or twice the empirical formula mass of C_5H_4 (64.1 amu).

$$\text{Multiple} = \frac{\text{Molecular mass}}{\text{Empirical formula mass}} = \frac{128.2 \text{ amu}}{64.1 \text{ amu}} = 2.00$$

Thus, the molecular formula of naphthalene is $C_{(2 \times 5)}H_{(2 \times 4)}$, or $C_{10}H_8$.

Example 3.17 shows a combustion analysis when the sample contains oxygen in addition to carbon and hydrogen. Because oxygen yields no combustion products, its presence in a molecule can't be directly detected by this method. Rather, the presence of oxygen must be inferred by subtracting the calculated masses of C and H from the total mass of the sample.

EXAMPLE 3.16   We calculated in Example 3.14 that ascorbic acid (vitamin C) has the empirical formula $C_3H_4O_3$. If the molecular mass of ascorbic acid is 176 amu, what is its molecular formula?

SOLUTION   Calculate a formula mass for the empirical formula, and compare it with the molecular mass of ascorbic acid:

$$\text{Form. mass of } C_3H_4O_3 = (3 \times 12.0 \text{ amu}) + (4 \times 1.0 \text{ amu}) + (3 \times 16.0 \text{ amu})$$
$$= 88.0 \text{ amu}$$

The empirical formula mass (88.0 amu) is half the molecular mass of ascorbic acid (176 amu), so the subscripts in the empirical formula must be multiplied by 2:

$$\text{Molecular mass of ascorbic acid} = 2 \times 88.0 \text{ amu} = 176.0 \text{ amu}$$
$$\text{Molecular formula of ascorbic acid} = C_{(2 \times 3)}H_{(2 \times 4)}O_{(2 \times 3)} = C_6H_8O_6$$

EXAMPLE 3.17   Caproic acid, the substance responsible for the aroma of dirty gym socks and running shoes, contains carbon, hydrogen, and oxygen. On combustion analysis, a 0.450 g sample of caproic acid gives 0.418 g of H_2O and 1.023 g of CO_2. What is the empirical formula of caproic acid? If the molecular mass of caproic acid is 116.2 amu, what is the molecular formula?

The unmistakable aroma of these shoes is due to caproic acid.

SOLUTION First, find the molar amounts of C and H in the sample by using gram-to-mole conversions:

$$\text{Moles of C} = 1.023 \text{ g } CO_2 \times \frac{1 \text{ mol } CO_2}{44.01 \text{ g } CO_2} \times \frac{1 \text{ mol C}}{1 \text{ mol } CO_2} = 0.023 \, 24 \text{ mol C}$$

$$\text{Moles of H} = 0.418 \text{ g } H_2O \times \frac{1 \text{ mol } H_2O}{18.02 \text{ g } H_2O} \times \frac{2 \text{ mol H}}{1 \text{ mol } H_2O} = 0.0464 \text{ mol H}$$

Next, find the number of grams of C and H in the sample by using mole-to-gram conversions:

$$\text{Mass of C} = 0.023 \, 24 \text{ mol C} \times \frac{12.01 \text{ g C}}{1 \text{ mol C}} = 0.2791 \text{ g C}$$

$$\text{Mass of H} = 0.0464 \text{ mol H} \times \frac{1.01 \text{ g H}}{1 \text{ mol H}} = 0.0469 \text{ g H}$$

Subtracting the masses of C and H from the mass of the starting sample indicates that 0.124 g is unaccounted for

$$0.450 \text{ g} - (0.2791 \text{ g} + 0.0469 \text{ g}) = 0.124 \text{ g}$$

Since we are told that oxygen is also present in the sample, the "missing" mass must be due to oxygen, which can't be detected by combustion. We therefore need to carry out a gram-to-mole conversion to find the number of moles of oxygen in the sample:

$$\text{Moles of O} = 0.124 \text{ g O} \times \frac{1 \text{ mol O}}{16.00 \text{ g O}} = 0.007 \, 75 \text{ mol O}$$

Knowing the relative numbers of moles of all three elements, C, H, and O, we next divide the three numbers of moles by the smallest number (0.007 75 mol of oxygen) to arrive at a C:H:O ratio of 3:6:1.

$$\frac{\text{Moles of C}}{\text{Moles of O}} = \frac{0.023 \, 24 \text{ mol C}}{0.007 \, 75 \text{ mol O}} = \frac{3.00 \text{ mol C}}{1 \text{ mol O}}$$

$$\frac{\text{Moles of H}}{\text{Moles of O}} = \frac{0.0464 \text{ mol H}}{0.007 \, 75 \text{ mol O}} = \frac{5.99 \text{ mol H}}{1 \text{ mol O}}$$

The empirical formula of caproic acid is therefore C_3H_6O, and the empirical formula mass is 58.1 amu. Since the molecular mass of caproic acid is 116.2, or twice the empirical formula mass, the molecular formula of caproic acid must be $C_{(2\times3)}H_{(2\times6)}O_{(2\times1)}$, or $C_6H_{12}O_2$.

▶ **PROBLEM 3.25** Menthol, a flavoring agent obtained from peppermint oil, contains carbon, hydrogen, and oxygen. On combustion analysis, 1.00 g of menthol yields 1.161 g of H_2O and 2.818 g of CO_2. What is the empirical formula of menthol?

▶ **PROBLEM 3.26** Ribose, a sugar present in the cells of all living organisms, has a molecular mass of 150 amu and the empirical formula CH_2O. What is the molecular formula of ribose?

▶ **PROBLEM 3.27** Convert the following percent compositions into molecular formulas.
(a) Diborane: H 21.86%, B 78.14%; molec. mass = 27.7 amu
(b) Trioxan: C 40.00%, H 6.71%, O 53.28%; molec. mass = 90.08 amu

3.13 Determining Molecular Masses: Mass Spectrometry

We said in the previous section that elemental analysis gives only an empirical formula. To determine a compound's molecular formula, it's also necessary to know the compound's molecular mass. How is molecular mass determined?

The most common method of determining both atomic and molecular masses is with an instrument called a *mass spectrometer*, shown schematically in Figure 3.10. The sample is first vaporized and is then injected as a dilute gas into an evacuated chamber, where it is bombarded with a beam of high-energy electrons. The electron beam knocks other electrons from the sample molecules, which become positively charged ions. Some of these ionized molecules survive, and others fragment into smaller ions. The various ions of different masses are then accelerated by an electric field and passed between the poles of a strong magnet, which deflects them through a curved, evacuated pipe.

The radius of deflection of a charged ion M^+ as it passes between the magnet poles depends on its mass, with lighter ions deflected more strongly than heavier ones. By varying the strength of the magnetic field, it's possible to focus ions of different masses through a slit and onto a detector assembly. The mass spectrum that results is plotted as a graph of ion mass versus intensity—that is, as the molecular masses of the various ions versus the relative number of those ions produced in the instrument.

Although a typical mass spectrum contains ions of numerous different masses, the heaviest ion is generally due to the ionized molecule itself. By measuring the mass of this ion, the molecular mass of the molecule can be determined. The naphthalene sample discussed in the previous section, for example, gives rise to an intense peak at mass 128 amu in its spectrum, consistent with its molecular formula of $C_{10}H_8$ (Figure 3.10b).

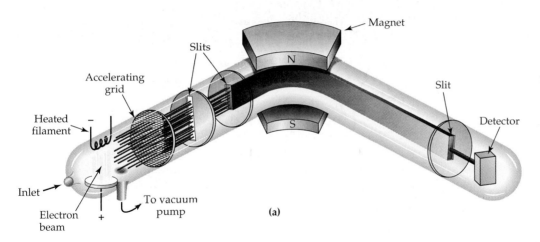

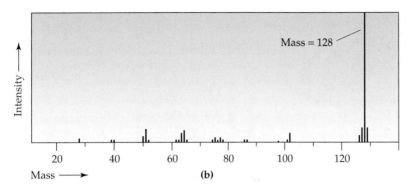

FIGURE 3.10   **(a)** Schematic of a mass spectrometer. Sample molecules are ionized by collision with a high-energy electron beam, and the resulting ions are then passed between the poles of a magnet, where they are deflected according to their mass-to-charge ratio. The deflected ions pass through a slit into a detector assembly. **(b)** A mass spectrum of naphthalene, molec. mass = 128, showing peaks of different masses on the horizontal axis.

Modern mass spectrometers are so precise that molecular masses can often be measured to seven significant figures. A $^{12}C_{10}{}^{1}H_8$ molecule of naphthalene, for example, has a molecular mass of 128.0626 amu as measured by mass spectrometry.

Did Ben Franklin Have Avogadro's Number? A Ballpark Calculation

Benjamin Franklin and Amadeo Avogadro. What did these two have in common?

At length being at Clapham, where there is on the common a large pond . . . I fetched out a cruet of oil and dropped a little of it on the water. I saw it spread itself with surprising swiftness upon the surface. The oil, though not more than a tea-spoonful, produced an instant calm over a space several yards square which spread amazingly and extended itself gradually...making all that quarter of the pond, perhaps half an acre, as smooth as a looking glass.

—*Excerpt from a letter of Benjamin Franklin to William Brownrigg, 1773.*

Benjamin Franklin, author and renowned states-man, was also an inventor and a scientist. Every schoolchild knows of Franklin's experiment with a kite and a key, demonstrating that lightning is electricity. Less well known is that his measurement of the extent to which oil spreads on water makes possible a simple estimate of molecular size and Avogadro's number.

The calculation goes like this: Avogadro's number is the number of molecules in a mole. So, if we can estimate both the number of molecules and the number of moles in Franklin's teaspoon of oil, we can calculate Avogadro's number. Let's start by calculating the number of molecules in the oil.

1. The volume (V) of oil Franklin used was 1 tsp = 4.9 cm^3, and the area (A) covered by the oil was 1/2 acre = 2.0×10^7 cm^2. Let's assume that the oil molecules are tiny cubes that pack closely together and form a layer only one molecule thick. As shown in the accompanying figure, the volume of the oil is equal to the surface area of the layer times the length (l) of the side of one molecule: $V = A \times l$. Rearranging this equation to find the length then gives us an estimate of molecular size:

$$l = \frac{V}{A} = \frac{4.9 \text{ cm}^3}{2.0 \times 10^7 \text{ cm}^2} = 2.4 \times 10^{-7} \text{ cm}$$

2. The area of the oil layer is the area of the side of one molecule (l^2) times the number of molecules (N) of oil: $A = l^2 \times N$. Rearranging this equation gives us the number of molecules:

$$N = \frac{A}{l^2} = \frac{2.0 \times 10^7 \text{ cm}^2}{(2.4 \times 10^{-7} \text{ cm})^2} = 3.5 \times 10^{20} \text{ molecules}$$

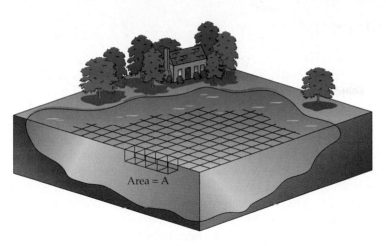

Area = A

3. To calculate the number of moles, we first need to know the mass (m) of the oil. This could have been determined by weighing, but Franklin neglected to do so. Let's therefore estimate the mass by multiplying the volume (V) of the oil by the density (D) of a typical oil, 0.95 g/cm³. [Since oil floats on water, it's not surprising that the density of oil is a bit less than the density of water (1.00 g/cm³).]

$$m = V \times D = 4.9 \text{ cm}^3 \times 0.95 \frac{\text{g}}{\text{cm}^3} = 4.7 \text{ g}$$

4. We now have to make one final assumption about the molecular mass of the oil before completing the calculation. Assuming that a typical oil has molecular mass = 200 amu, then the mass of 1 mol of oil is 200 g. Dividing the mass of the oil by the mass of 1 mol gives the number of moles of oil:

$$\text{Moles of oil} = \frac{4.7 \text{ g}}{200 \text{ g/mol}} = 0.024 \text{ mol}$$

5. Finally, the number of molecules per mole—Avogadro's number—can be obtained:

$$\text{Avogadro's number} = \frac{3.5 \times 10^{20} \text{ molecules}}{0.024 \text{ mol}} = 1.5 \times 10^{22}$$

The calculation is not very accurate, of course, but Ben wasn't really intending for us to calculate Avogadro's number when he made a rough estimate of how much his oil spread out. Nevertheless, the result isn't too bad for such a simple experiment.

▶ **PROBLEM 3.28** What do you think are the main sources of error in calculating Avogadro's number by spreading oil on a pond?

▶ **PROBLEM 3.29** Recalculate Avogadro's number assuming that the oil molecules are tall rectangular boxes rather than cubes, with two edges of equal length and the third edge four times the length of the other two. Assume also that the molecules stand on end in the water.

Key Words

Summary

Because mass is neither created nor destroyed in reactions, all chemical equations must be **balanced**—that is, the numbers and kinds of atoms on both sides of the reaction arrow must be the same. A balanced equation tells the number ratio of reactant and product **formula units** in a reaction.

Just as atomic mass tells the mass of an atom, **molecular mass** tells the mass of a molecule. (The analogous term **formula mass** is used for ionic and other nonmolecular substances.) Molecular mass is the sum of the atomic masses of all atoms in the molecule. When referring to the large numbers of molecules or ions that take part in a visible chemical reaction, it's convenient to use a unit called the **mole**, abbreviated mol. One mole of any object, atom, molecule, or ion contains **Avogadro's number** of formula units, 6.022×10^{23}.

For work in the laboratory, it's necessary to weigh reactants rather than just know mole ratios. Thus, it's necessary to convert between numbers of moles and numbers of grams by using **molar mass** as the conversion factor. The molar mass of any substance is the amount in grams numerically equal to the substance's molecular or formula mass. Carrying out chemical calculations using these relationships is called **stoichiometry**.

The amount of product actually formed in a reaction (the reaction's **yield**) is often less than the amount theoretically possible. Dividing the actual amount by the theoretical amount and multiplying by 100% gives the reaction's **percent yield**. Often, reactions are carried out with an excess of one reactant beyond that called for by the balanced equation. In such cases, the extent to which the reaction takes place depends on the reactant present in limiting amount, the **limiting reactant**.

The concentration of a substance in solution is usually expressed as **molarity (M)**, defined as the number of moles of a substance (the *solute*) dissolved per liter of solution. A solution's molarity acts as a conversion factor between solution volume and number of moles of solute, making it possible to carry out stoichiometry calculations on solutions. Often, chemicals are stored as concentrated aqueous solutions that are diluted before use. When carrying out a dilution, only the volume is changed by adding solvent; the amount of solute is unchanged. A solution's exact concentration can often be determined by **titration**.

The chemical makeup of a substance is described by its **percent composition**—the percentage of the substance's mass due to each of its constituent elements. Elemental analysis is used to calculate a substance's **empirical formula**, which gives the smallest whole-number ratio of atoms of the elements in the compound. To determine the **molecular formula**, which may be a simple multiple of the empirical formula, it's also necessary to know the substance's molecular mass. Molecular masses are usually determined by mass spectrometry.

Key Concept Summary

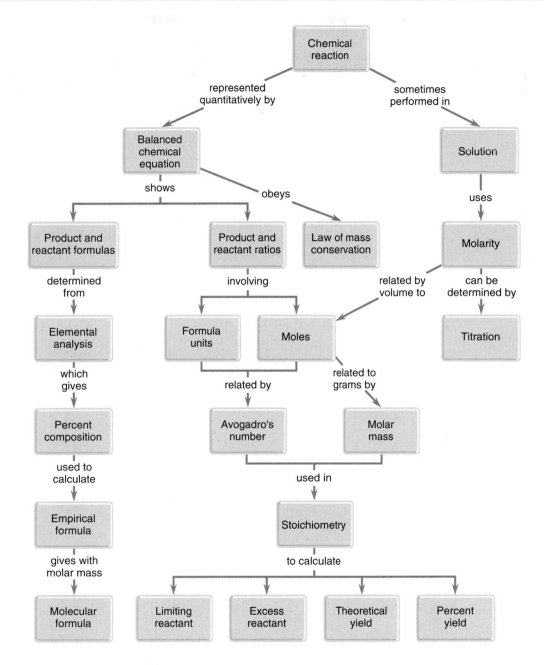

Understanding Key Concepts

Problems 3.1–3.29 appear within the chapter.

3.30 Box **(a)** represents 1.0 mL of a solution of particles at a given concentration. Which of the boxes **(b)–(d)** represents 1.0 mL of the solution that results after **(a)** has been diluted by doubling the volume of its solvent?

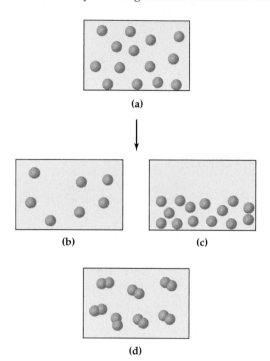

3.31 Reaction of A (green spheres) with B (blue spheres) is shown in the following diagram:

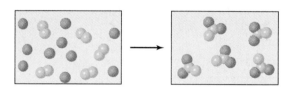

Which equation best describes the stoichiometry of the reaction?
(a) $A_2 + 2 B \rightarrow A_2B_2$
(b) $10 A + 5 B_2 \rightarrow 5 A_2B_2$
(c) $2 A + B_2 \rightarrow A_2B_2$
(d) $5 A + 5 B_2 \rightarrow 5 A_2B_2$

3.32 Balance the equation for the combustion reaction of ethylene with oxygen:

$$C_2H_4 + O_2 \longrightarrow CO_2 + H_2O \qquad \text{Unbalanced}$$

Write the reaction in a format similar to that in Problem 3.31, using red dots to represent O, black dots to represent C, and light green dots to represent H.

3.33 If blue spheres represent nitrogen atoms and red spheres represent oxygen atoms, which box represents reactants and which represents products for the reaction $2 NO(g) + O_2(g) \rightarrow 2 NO_2(g)$?

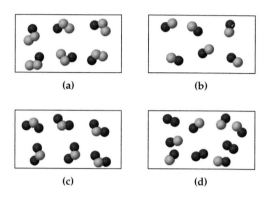

3.34 Fluoxetine, marketed as an antidepressant under the name *Prozac*, can be represented by the following ball-and-stick molecular model. Write the molecular formula for fluoxetine, and calculate its molecular mass (red = O, gray = C, blue = N, yellow-green = F, ivory = H).

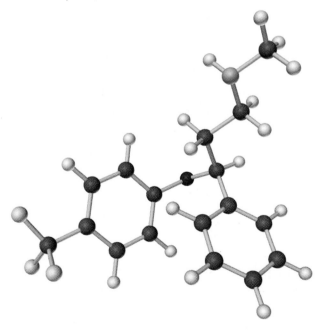

3.35 The unbalanced equation for the reaction of nitrogen with oxygen to yield nitrous oxide is $N_2 + O_2 \rightarrow N_2O$. Why can't this equation be balanced by adding a subscript 2 to the N_2O oxygen: $N_2 + O_2 \rightarrow N_2O_2$?

3.36 Assume that the buret contains H^+ ions, the flask contains OH^- ions, and you are carrying out a titration of the base with the acid. If the volumes in the buret and the flask are identical and the concentration of the acid in the buret is 1.00 M, what is the concentration of base in the flask?

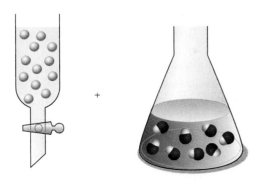

3.37 The following diagram represents the reaction of A_2 (red spheres) with B_2 (blue spheres):

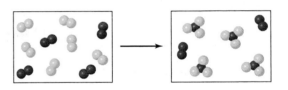

(a) Write a balanced equation for the reaction, and identify the limiting reactant.

(b) How many moles of product can be made from 1.0 mol of A_2 and 1.0 mol of B_2?

Additional Problems

Balancing Equations

3.38 Which of the following equations are balanced?

(a) The development reaction in photography:

$$2\,AgBr + 2\,NaOH + C_6H_6O_2 \longrightarrow$$
$$2\,Ag + H_2O + 2\,NaBr + C_6H_4O_2$$

(b) Preparation of household bleach:

$$2\,NaOH + Cl_2 \longrightarrow NaOCl + NaCl + H_2O$$

3.39 Which of the following equations are balanced? Balance any that need it.

(a) The thermite reaction, used in welding:

$$Al + Fe_2O_3 \longrightarrow Al_2O_3 + Fe$$

(b) The photosynthesis of glucose from CO_2:

$$6\,CO_2 + 6\,H_2O \longrightarrow C_6H_{12}O_6 + 6\,O_2$$

(c) The separation of gold from its ore:

$$Au + 2\,NaCN + O_2 + H_2O \longrightarrow$$
$$NaAu(CN)_2 + 3\,NaOH$$

3.40 Balance the following equations:

(a) $Mg + HNO_3 \rightarrow H_2 + Mg(NO_3)_2$

(b) $CaC_2 + H_2O \rightarrow Ca(OH)_2 + C_2H_2$

(c) $S + O_2 \rightarrow SO_3$

(d) $UO_2 + HF \rightarrow UF_4 + H_2O$

3.41 Balance the following equations:

(a) The explosion of ammonium nitrate:

$$NH_4NO_3 \longrightarrow N_2 + O_2 + H_2O$$

(b) The spoilage of wine into vinegar:

$$C_2H_6O + O_2 \longrightarrow C_2H_4O_2 + H_2O$$

(c) The burning of rocket fuel:

$$C_2H_8N_2 + N_2O_4 \longrightarrow N_2 + CO_2 + H_2O$$

Molecular Masses and Moles

3.42 What are the molecular (formula) masses of the following substances?

(a) Hg_2Cl_2 (calomel, used at one time as a bowel purgative)

(b) $C_4H_8O_2$ (butyric acid, responsible for the odor of rancid butter)

(c) CF_2Cl_2 (a chlorofluorocarbon that destroys the atmospheric ozone layer)

3.43 What are the formulas of the following substances?

(a) $PCl_?$; molec. mass = 137.3 amu

(b) Nicotine, $C_{10}H_{14}N_?$; molec. mass = 162.2 amu

3.44 How many grams are in a mole of each of the following substances?

(a) Ti   **(b)** Br_2 **(c)** Hg   **(d)** H_2O

3.45 How many moles are in a gram of each of the following substances?

(a) Cr   **(b)** Cl_2 **(c)** Au   **(d)** NH_3

3.46 How many moles of ions are in 2.5 mol of NaCl?

3.47 How many moles of positive ions are in 1.45 mol of K_2SO_4?

3.48 How many moles of ions are in 27.5 g of $MgCl_2$?

3.49 How many moles of negative ions are in 35.6 g of AlF_3?

3.50 What is the molecular mass of chloroform if 0.0275 mol weighs 3.28 g?

3.51 What is the molecular mass of cholesterol if 0.5731 mol weighs 221.6 g?

3.52 Iron(II) sulfate, $FeSO_4$, is prescribed for the treatment of anemia. How many moles of $FeSO_4$ are present in a standard 300 mg tablet? How many iron(II) ions?

3.53 The "lead" in lead pencils is actually almost pure carbon, and the mass of a period mark made by a lead pencil is about 0.0001 g. How many carbon atoms are in the period?

3.54 An average cup of coffee contains about 125 mg of caffeine, $C_8H_{10}N_4O_2$. How many moles of caffeine are in a cup? How many molecules of caffeine?

3.55 Let's say that an average egg has a mass of 45 g. What mass would a mole of eggs have?

3.56 How many moles does each of the following samples contain?
  (a) 1.0 g of lithium
  (b) 1.0 g of gold
  (c) 1.0 g of penicillin G potassium, $C_{16}H_{17}N_2O_4SK$

3.57 What is the mass in grams of each of the following samples?
  (a) 0.0015 mol of sodium
  (b) 0.0015 mol of lead
  (c) 0.0015 mol of diazepam (Valium), $C_{16}H_{13}ClN_2O$

Stoichiometry Calculations

3.58 Titanium metal is obtained from the mineral rutile, TiO_2. How many kilograms of rutile are needed to produce 100 kg of Ti?

3.59 Iron metal can be produced from the mineral hematite, Fe_2O_3, by reaction with carbon. How many kilograms of iron are present in 105 kg of hematite?

3.60 In the preparation of iron from hematite (Problem 3.59), Fe_2O_3 reacts with carbon:

$$Fe_2O_3 + C \longrightarrow Fe + CO_2 \qquad \text{Unbalanced}$$

  (a) Balance the equation.
  (b) How many moles of carbon are needed to react with 525 g of hematite?
  (c) How many grams of carbon are needed to react with 525 g of hematite?

3.61 An alternative method for preparing pure iron from Fe_2O_3 (Problem 3.59) is by reaction with carbon monoxide:

$$Fe_2O_3 + CO \longrightarrow Fe + CO_2 \qquad \text{Unbalanced}$$

  (a) Balance the equation.
  (b) How many grams of CO are needed to react with 3.02 g of Fe_2O_3?
  (c) How many grams of CO are needed to react with 1.68 mol of Fe_2O_3?

3.62 Magnesium metal burns in oxygen to form magnesium oxide, MgO.
  (a) Write a balanced equation for the reaction.
  (b) How many grams of oxygen are needed to react with 25.0 g of Mg? How many grams of MgO will result?
  (c) How many grams of Mg are needed to react with 25.0 g of O_2? How many grams of MgO will result?

3.63 Ethylene gas, C_2H_4, reacts with water at high temperature to yield ethyl alcohol, C_2H_6O.
  (a) How many grams of ethylene are needed to react with 0.133 mol of H_2O? How many grams of ethyl alcohol will result?
  (b) How many grams of water are needed to react with 0.371 mol of ethylene? How many grams of ethyl alcohol will result?

3.64 Pure oxygen was first made by heating mercury(II) oxide:

$$HgO \xrightarrow{\text{Heat}} Hg + O_2 \qquad \text{Unbalanced}$$

  (a) Balance the equation.
  (b) How many grams of mercury and how many grams of oxygen are formed from 45.5 g of HgO?
  (c) How many grams of HgO would you need to obtain 33.3 g of O_2?

3.65 Titanium dioxide (TiO_2), the substance used as the pigment in white paint, is prepared industrially by reaction of $TiCl_4$ with O_2 at high temperature.

$$TiCl_4 + O_2 \xrightarrow{\text{Heat}} TiO_2 + 2\,Cl_2$$

  How many kilograms of TiO_2 can be prepared from 5.60 kg of $TiCl_4$?

3.66 Silver metal reacts with chlorine (Cl_2) to yield silver chloride. If 2.00 g of Ag reacts with 0.657 g of Cl_2, what is the empirical formula of silver chloride?

3.67 Aluminum reacts with oxygen to yield aluminum oxide. If 5.0 g of Al reacts with 4.45 g of O_2, what is the empirical formula of aluminum oxide?

Limiting Reactants and Reaction Yield

3.68 Assume that you have 1.39 mol of H_2 and 3.44 mol of N_2. How many grams of ammonia (NH_3) can you make, and how many grams of which reactant will be left over?

$$3 H_2 + N_2 \longrightarrow 2 NH_3$$

3.69 Hydrogen and chlorine react to yield hydrogen chloride: $H_2 + Cl_2 \rightarrow 2 HCl$. How many grams of HCl are formed from reaction of 3.56 g of H_2 with 8.94 g of Cl_2? Which reactant is limiting?

3.70 How many grams of the dry-cleaning solvent ethylene chloride, $C_2H_4Cl_2$, can be prepared by reaction of 15.4 g of ethylene, C_2H_4, with 3.74 g of Cl_2?

$$C_2H_4 + Cl_2 \longrightarrow C_2H_4Cl_2$$

3.71 How many grams of each product result from the following reactions, and how many grams of which reactant is left over?
(a) $(1.3 \text{ g NaCl}) + (3.5 \text{ g AgNO}_3) \rightarrow$
$$(x \text{ g AgCl}) + (y \text{ g NaNO}_3)$$
(b) $(2.65 \text{ g BaCl}_2) + (6.78 \text{ g H}_2\text{SO}_4) \rightarrow$
$$(x \text{ g BaSO}_4) + (y \text{ g HCl})$$

3.72 Limestone ($CaCO_3$) reacts with hydrochloric acid according to the equation $CaCO_3 + 2 HCl \rightarrow CaCl_2 + H_2O + CO_2$. If 1.00 mol of CO_2 has a volume of 22.4 L under the reaction conditions, how many liters of gas can be formed by reaction of 2.35 g of $CaCO_3$ with 2.35 g of HCl? Which reactant is limiting?

3.73 Sodium azide (NaN_3) yields N_2 gas when heated to 300°C, a reaction used in automobile air bags. If 1.00 mol of N_2 has a volume of 47.0 L under the reaction conditions, how many liters of gas can be formed by heating 38.5 g of NaN_3? The reaction is

$$2 NaN_3 \longrightarrow 3 N_2(g) + 2 Na$$

3.74 Acetic acid (CH_3CO_2H) reacts with isopentyl alcohol ($C_5H_{12}O$) to yield isopentyl acetate ($C_7H_{14}O_2$), a fragrant substance with the odor of bananas. If the yield from the reaction of acetic acid with isopentyl alcohol is 45%, how many grams of isopentyl acetate are formed from 3.58 g of acetic acid and 4.75 g of isopentyl alcohol? The reaction is

$$CH_3CO_2H + C_5H_{12}O \longrightarrow C_7H_{14}O_2 + H_2O$$

3.75 Cisplatin [$Pt(NH_3)_2Cl_2$], a compound used in cancer treatment, is prepared by reaction of ammonia with potassium tetrachloroplatinate:

$$K_2PtCl_4 + 2 NH_3 \longrightarrow 2 KCl + Pt(NH_3)_2Cl_2$$

How many grams of cisplatin are formed from 55.8 g of K_2PtCl_4 and 35.6 g of NH_3 if the reaction takes place in 95% yield based on the limiting reactant?

3.76 If 1.87 g of acetic acid reacts with 2.31 g of isopentyl alcohol to give 2.96 g of isopentyl acetate (Problem 3.74), what is the percent yield of the reaction?

3.77 If 3.42 g of K_2PtCl_4 and 1.61 g of NH_3 give 2.08 g of cisplatin (Problem 3.75), what is the percent yield of the reaction?

Molarity, Solution Stoichiometry, Dilution, and Titration

3.78 How many moles of solute are present in each of the following solutions?
(a) 35.0 mL of 1.200 M HNO_3
(b) 175 mL of 0.67 M glucose ($C_6H_{12}O_6$)

3.79 How many grams of solute would you use to prepare each of the following solutions?
(a) 250.0 mL of 0.600 M ethyl alcohol (C_2H_6O)
(b) 167 mL of 0.200 M boric acid (H_3BO_3)

3.80 How many milliliters of a 0.45 M $BaCl_2$ solution contain 15.0 g of $BaCl_2$?

3.81 How many milliliters of a 0.350 M KOH solution contain 0.0171 mol of KOH?

3.82 The sterile saline solution used to rinse contact lenses can be made by dissolving 400 mg of NaCl in sterile water and diluting to 100 mL. What is the molarity of the solution?

3.83 The concentration of glucose ($C_6H_{12}O_6$) in normal blood is approximately 90 mg per 100 mL. What is the molarity of the glucose?

3.84 *Ringer's solution*, used in the treatment of burns and wounds, is prepared by dissolving 4.30 g of NaCl, 0.150 g of KCl, and 0.165 g of $CaCl_2$ in water and diluting to a volume of 500.0 mL. What is the molarity of each of the component ions in the solution?

3.85 Copper reacts with dilute nitric acid according to the equation

$$3 Cu(s) + 8 HNO_3(aq) \longrightarrow$$
$$3 Cu(NO_3)_2(aq) + 2 NO(g) + 4 H_2O(l)$$

If a copper penny weighing 3.045 g is dissolved in a small amount of nitric acid and the resultant solution is diluted to 50.0 mL with water, what is the molarity of the $Cu(NO_3)_2$?

3.86 A bottle of 12.0 M hydrochloric acid has only 35.7 mL left in it. What will the HCl concentration be if the solution is diluted to 250.0 mL?

3.87 What is the volume of the solution that would result by diluting 70.00 mL of 0.0913 M NaOH to a concentration of 0.0150 M?

3.88 A flask containing 450 mL of 0.500 M HBr was accidentally knocked to the floor. How many grams of K_2CO_3 would you need to put on the spill to neutralize the acid according to the equation $2 \, HBr(aq)$ $+ \, K_2CO_3(aq) \rightarrow 2 \, KBr(aq) + CO_2(g) + H_2O(l)$?

3.89 The odor of skunks is caused by chemical compounds called *thiols*. These compounds, of which butanethiol ($C_4H_{10}S$) is a representative example, can be deodorized by reaction with household bleach (NaOCl) according to the following equation:

$$2 \, C_4H_{10}S + NaOCl(aq) \longrightarrow$$
$$C_8H_{18}S_2 + NaCl(aq) + H_2O(aq)$$

How many grams of butanethiol can be deodorized by reaction with 5.00 mL of 0.0985 M NaOCl?

3.90 Potassium permanganate ($KMnO_4$) reacts with oxalic acid ($H_2C_2O_4$) in aqueous sulfuric acid according to the equation

$$2 \, KMnO_4 + 5 \, H_2C_2O_4 + 3 \, H_2SO_4 \longrightarrow$$
$$2 \, MnSO_4 + 10 \, CO_2 + 8 \, H_2O + K_2SO_4$$

How many milliliters of a 0.250 M $KMnO_4$ solution are needed to react completely with 3.225 g of oxalic acid?

3.91 Oxalic acid, $H_2C_2O_4$, is a toxic substance found in spinach leaves. What is the molarity of a solution made by dissolving 12.0 g of oxalic acid in enough water to give 400.0 mL of solution? How many milliliters of 0.100 M KOH would you need to titrate 25.0 mL of the oxalic acid solution according to the following equation?

$$H_2C_2O_4(aq) + 2 \, KOH(aq) \longrightarrow$$
$$K_2C_2O_4(aq) + 2 \, H_2O(l)$$

Formulas and Elemental Analysis

3.92 Urea, a substance commonly used as a fertilizer, has the formula CH_4N_2O. What is its percent composition by mass?

3.93 Calculate the mass percent composition of each of the following substances:

(a) Malachite, a copper-containing mineral: $Cu_2(OH)_2CO_3$

(b) Acetaminophen, a headache remedy: $C_8H_9NO_2$

(c) Prussian blue, an ink pigment: $Fe_4[Fe(CN)_6]_3$

3.94 What is the empirical formula of stannous fluoride, a compound added to toothpaste to protect teeth against decay? Its mass percent composition is 24.25% F, 75.75% Sn.

3.95 What are the empirical formulas of each of the following substances?

(a) Ibuprofen, a headache remedy: 75.69% C, 15.51% O, 8.80% H

(b) Tetraethyllead, the "lead" in gasoline: 29.71% C, 64.06% Pb, 6.23% H

(c) Zircon, a diamondlike mineral: 34.91% O, 15.32% Si, 49.76% Zr

3.96 Combustion analysis of 45.62 mg of toluene, a commonly used solvent, gives 35.67 mg of H_2O and 152.5 mg of CO_2. What is the empirical formula of toluene?

3.97 Coniine, a toxic substance isolated from poison hemlock, contains only carbon, hydrogen, and nitrogen. Combustion analysis of a 5.024 mg sample yields 13.90 mg of CO_2 and 6.048 mg of H_2O. What is the empirical formula of coniine?

3.98 Cytochrome c is an iron-containing enzyme found in the cells of all aerobic organisms. If cytochrome c is 0.43% Fe by mass, what is its minimum molecular mass?

3.99 Nitrogen fixation in the root nodules of peas and other leguminous plants is carried out by the molybdenum-containing enzyme *nitrogenase*. What is the molecular mass of nitrogenase if the enzyme contains two molybdenum atoms and is 0.0872% Mo by mass?

3.100 Disilane, Si_2H_x, is analyzed and found to contain 90.28% by mass silicon. What is the value of x?

3.101 A certain metal sulfide, MS_2, is used extensively as a high-temperature lubricant. If MS_2 is 40.06% by mass sulfur, what is the identity of the metal M?

General Problems

3.102 Give the percent composition of each of the following substances:

(a) Glucose, $C_6H_{12}O_6$

(b) Sulfuric acid, H_2SO_4

(c) Potassium permanganate, $KMnO_4$

(d) Saccharin, $C_7H_5NO_3S$

3.103 What are the empirical formulas of substances with the following mass percent compositions?

(a) Aspirin: 4.48% H, 60.00% C, 35.52% O

(b) Ilmenite (a titanium-containing ore): 31.63% O, 31.56% Ti, 36.81% Fe

(c) Sodium thiosulfate (photographic "fixer"): 30.36% O, 29.08% Na, 40.56% S

3.104 Balance the following equations:

(a) $SiCl_4 + H_2O \rightarrow SiO_2 + HCl$

(b) $P_4O_{10} + H_2O \rightarrow H_3PO_4$

(c) $CaCN_2 + H_2O \rightarrow CaCO_3 + NH_3$

(d) $NO_2 + H_2O \rightarrow HNO_3 + NO$

3.105 Sodium borohydride, $NaBH_4$, a substance used in the synthesis of many pharmaceutical agents, can be prepared by reaction of NaH with B_2H_6 according to the equation $2\,NaH + B_2H_6 \rightarrow 2\,NaBH_4$. How many grams of $NaBH_4$ can be prepared by reaction between 8.55 g of NaH and 6.75 g of B_2H_6? Which reactant is limiting, and how many grams of the excess reactant will be left over?

3.106 Ferrocene, a substance proposed for use as a gasoline additive, has the percent composition 5.42% H, 64.56% C, and 30.02% Fe. What is the empirical formula of ferrocene?

3.107 The molar mass of HCl is 36.5 g/mol, and the average mass per HCl molecule is 36.5 amu. Use the fact that 1 amu $= 1.6605 \times 10^{-24}$ g to calculate Avogadro's number.

3.108 What is the molarity of each ion in a solution prepared by dissolving 0.550 g of Na_2SO_4, 1.188 g of Na_3PO_4, and 0.223 g of Li_2SO_4 in water and diluting to a volume of 100.00 mL?

3.109 Ethylene glycol, commonly used as automobile antifreeze, contains only carbon, hydrogen, and oxygen. Combustion analysis of a 23.46 mg sample yields 20.42 mg of H_2O and 33.27 mg of CO_2. What is the empirical formula of ethylene glycol? What is its molecular formula if it has a molecular mass of 62.0 amu?

3.110 The molecular mass of ethylene glycol (Problem 3.109) is 62.0689 amu when calculated using the atomic masses found in a standard periodic table, yet the molecular mass determined experimentally by high-resolution mass spectrometry is 62.0368 amu. Explain the discrepancy.

3.111 A compound of formula XCl_3 reacts with aqueous $AgNO_3$ to yield a precipitate of solid AgCl according to the following equation:

$$XCl_3(aq) + 3\,AgNO_3(aq) \longrightarrow \\ X(NO_3)_3(aq) + 3\,AgCl(s)$$

When a solution containing 0.634 g of XCl_3 was allowed to react with an excess of aqueous $AgNO_3$, 1.68 g of solid AgCl was formed. What is the identity of the atom X?

3.112 When eaten, dietary carbohydrates are digested to yield glucose ($C_6H_{12}O_6$), which is then metabolized to yield carbon dioxide and water:

$$C_6H_{12}O_6 + O_2 \longrightarrow CO_2 + H_2O \qquad \text{Unbalanced}$$

Balance the equation, and calculate both the mass in grams and the volume in liters of the CO_2 produced from 66.3 g of glucose, assuming that 1 mol of CO_2 has a volume of 25.4 L at normal body temperature.

3.113 Other kinds of titrations are possible in addition to acid–base titrations. For example, the concentration of a solution of potassium permanganate, $KMnO_4$, can be determined by titration against a known amount of oxalic acid, $H_2C_2O_4$, according to the following equation:

$$5\,H_2C_2O_4(aq) + 2\,KMnO_4(aq) + 3\,H_2SO_4(aq) \longrightarrow \\ 10\,CO_2(g) + 2\,MnSO_4(aq) + K_2SO_4(aq) + 8\,H_2O(l)$$

What is the concentration of a $KMnO_4$ solution if 22.35 mL reacts with 0.5170 g of oxalic acid?

3.114 A copper wire having a mass of 2.196 g was allowed to react with an excess of sulfur. The excess sulfur was then burned, yielding SO_2 gas. The mass of the copper sulfide produced was 2.748 g.

(a) What is the percent composition of copper sulfide?

(b) What is its empirical formula?

(c) Calculate the number of copper ions per cubic centimeter if the density of the copper sulfide is 5.6 g/cm^3.

3.115 Element X, a member of group 5A, forms two chlorides, XCl_3 and XCl_5. Reaction of an excess of Cl_2 with 8.729 g of XCl_3 yields 13.233 g of XCl_5. What is the atomic mass and the identity of the element X?

3.116 A mixture of XCl_3 and XCl_5 (see Problem 3.115) weighing 10.00 g contains 81.04% Cl by mass. How many grams of XCl_3 and how many grams of XCl_5 are present in the mixture?

3.117 A 1.268 g sample of a metal carbonate MCO_3 was treated with 100.00 mL of 0.1083 M H_2SO_4, yielding CO_2 gas and an aqueous solution of the metal sulfate. The solution was boiled to remove all the dissolved CO_2 and was then titrated with 0.1241 M NaOH. A 71.02 mL volume of NaOH was required to neutralize the excess H_2SO_4.

(a) What is the identity of the metal M?

(b) How many liters of CO_2 gas were produced if the density of CO_2 is 1.799 g/L?

3.118 Ammonium nitrate, a potential ingredient of terrorist bombs, can be made nonexplosive by addition of diammonium hydrogen phosphate, $(NH_4)_2HPO_4$. Analysis of a such a NH_4NO_3–$(NH_4)_2HPO_4$ mixture showed the mass % of nitrogen to be 30.43%. What is the mass ratio of the two components in the mixture?

3.119 An unidentified metal M reacts with an unidentified halogen X to form a compound MX_2. When heated, the compound decomposes by the reaction

$$2\,MX_2(s) \longrightarrow 2\,MX(s) + X_2(g)$$

When 1.12 g of MX_2 is heated, 0.720 g of MX is obtained, along with 56.0 mL of X_2 gas. Under the conditions used, 1.00 mol of the gas has a volume of 22.41 L.

(a) What is the atomic mass and identity of the halogen X?

(b) What is the atomic mass and identity of the metal M?

 # eMedia Problems

3.120 The **Balancing Equations** activity (*eChapter 3.1*) allows you to balance three different chemical equations.

(a) In the activity, select the **Decomposition** reaction and balance it.

(b) Rewrite the equation such that ammonium nitrate reacts to yield only elements (N_2, H_2, and O_2) and balance this new equation.

(c) Explain how it is possible for the number of product molecules to be different in parts (a) and (b).

(d) Is it also possible for the number of product atoms to be different in parts (a) and (b)? Explain.

3.121 Calculate the percent composition of each compound in Problem 3.42 (page 111). Use the **Molecular Mass and Percent Mass** activity (*eChapter 3.3*) to verify your answers.

3.122 Using the **Limiting Reagents (Stoichiometry)** simulation (*eChapter 3.6*), perform the redox reaction between magnesium metal and oxygen gas by combining equal masses of both reactants.

(a) Which reactant is the limiting reactant?

(b) Explain how it is possible for there to be a limiting reactant when both are present in equal mass.

(c) Convert the mass you selected to moles for each reactant.

(d) Explain how it is possible for the reactant present in the greatest number of moles to be the limiting reactant.

3.123 Calculate the volume of a solution that

(a) contains 25 g of sulfuric acid and is 0.073 M in sulfuric acid;

(b) contains 40 g of potassium chloride and is 0.101 M in potassium chloride;

(c) contains 17 g of sodium chloride and is 0.646 M in sodium chloride. Use the **Molarity** activity (*eChapter 3.7*) to verify your answers.

3.124 The **Titration** simulation (*eChapter 3.10*) allows you to simulate the titration of five acids, acids A through E. Assuming that the acids can all provide one hydrogen, perform simulated titrations and determine the concentration of all five acids.

Reactions in Aqueous Solution

The corrosion evident on parts of this sunken bomber occurs by a typical oxidation–reduction reaction of the kind discussed in this chapter.

Ours is a world based on water. Approximately 71% of the earth's surface is covered by water, and another 3% is covered by ice; 66% of an adult human body is water, and water is needed to sustain all living organisms. It's therefore not surprising that a large amount of important chemistry takes place in water—that is, in **aqueous solution**.

We saw in the previous chapter how chemical reactions are described and how certain mass relationships must be obeyed when reactions occur. In the present chapter, we'll continue our study of chemical reactions by seeing how different reactions can be classified and by learning some of the general ways reactions take place.

4.1 Some Ways that Chemical Reactions Occur

Every chemical reaction occurs for a reason—a reason that involves the energies of reactants and products and that we'll explore in more depth in Chapter 8. Without going into detail at present, it's often convenient to think of reactions as being "driven" from reactants to products by some energetic "driving force" that pushes them along.

Aqueous reactions can be grouped into three general categories, each with its own kind of driving force: *precipitation reactions, acid–base neutralization reactions*, and *oxidation–reduction reactions*. Let's look briefly at an example of each before studying them in more detail in subsequent sections.

Reaction of aqueous $Pb(NO_3)_2$ with aqueous KI gives a yellow precipitate of PbI_2.

1. **Precipitation reactions** are processes in which soluble reactants yield an insoluble solid product that drops out of the solution. Formation of this stable product removes material from the aqueous solution and provides the driving force for the reaction. Most precipitations take place when the anions and cations of two ionic compounds change partners. For example, an aqueous solution of lead(II) nitrate reacts with an aqueous solution of potassium iodide to yield an aqueous solution of potassium nitrate plus an insoluble yellow precipitate of lead iodide:

$$Pb(NO_3)_2(aq) + 2\,KI(aq) \longrightarrow 2\,KNO_3(aq) + PbI_2(s)$$

2. **Acid–base neutralization reactions** are processes in which an acid reacts with a base to yield water plus an ionic compound called a **salt**. You might recall from Section 2.9 that we defined acids as compounds that produce H^+ ions when dissolved in water and bases as compounds that produce OH^- ions when dissolved in water. Thus, the driving force behind a neutralization reaction is the production of the stable covalent water molecule by removal of H^+ and OH^- ions from solution. The reaction between hydrochloric acid and aqueous sodium hydroxide to yield water plus aqueous sodium chloride is a typical example:

$$HCl(aq) + NaOH(aq) \longrightarrow H_2O(l) + NaCl(aq)$$

3. **Oxidation–reduction reactions**, or **redox reactions**, are processes in which one or more electrons are transferred between reaction partners (atoms, molecules, or ions). The driving force is a decrease in electrical potential, analogous to what happens when a live electrical wire is grounded and electrons flow from the wire to the ground. As a result of this transfer of electrons, charges on atoms in the various reactants change. When metallic magnesium reacts with iodine vapor, for instance, a magnesium atom gives an electron to each of two iodine atoms, forming an Mg^{2+} ion and two I^- ions. The charge on the magnesium changes from 0 to +2, and the charge on each iodine changes from 0 to −1:

$$Mg(s) + I_2(g) \longrightarrow MgI_2(s)$$

▶**PROBLEM 4.1** Classify each of the following processes as a precipitation, acid–base neutralization, or redox reaction:
(a) $AgNO_3(aq) + KCl(aq) \rightarrow AgCl(s) + KNO_3(aq)$
(b) $2\,P(s) + 3\,Br_2(l) \rightarrow 2\,PBr_3(l)$
(c) $Ca(OH)_2(aq) + 2\,HNO_3(aq) \rightarrow 2\,H_2O(l) + Ca(NO_3)_2(aq)$

4.2 Electrolytes in Aqueous Solution

We all know from experience that both sugar (sucrose) and table salt (NaCl) dissolve in water. The solutions that result, though, are quite different. When sucrose, a molecular substance, dissolves in water, the solution that results contains neutral sucrose *molecules* surrounded by water. When NaCl, an ionic substance, dissolves in water, the solution contains separate Na^+ and Cl^- *ions* surrounded by water. Because of the presence of the ions, the NaCl solution conducts electricity, but the sucrose solution does not.

$$C_{12}H_{22}O_{11}(s) \xrightarrow{\text{H}_2\text{O}} C_{12}H_{22}O_{11}(aq)$$
$$\text{Sucrose}$$

$$NaCl(s) \xrightarrow{\text{H}_2\text{O}} Na^+(aq) + Cl^-(aq)$$

The electrical conductivity of an aqueous NaCl solution is easy to demonstrate using a battery, a light bulb, and several pieces of wire, connected as shown in Figure 4.1. When the wires are dipped into an aqueous NaCl solution, the positively charged Na^+ ions move through the solution toward the wire connected to the negatively charged terminal of the battery. At the same time, the negatively charged Cl^- ions move toward the wire connected to the positively charged terminal of the battery. The resulting movement of ions allows a current to flow, so the bulb lights. When the wires are dipped into an aqueous sucrose solution, however, there are no ions to carry the current, so the bulb remains dark.

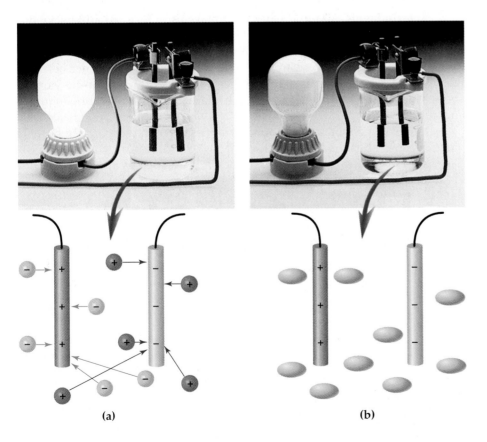

(a) (b)

FIGURE 4.1 A simple device to demonstrate the electrical conductivity of an ionic solution. **(a)** A solution of NaCl conducts electricity because of the movement of charged particles (ions), thereby completing the circuit and allowing the bulb to light. **(b)** A solution of sucrose does not conduct electricity or complete the circuit because it has no charged particles. The bulb therefore remains dark.

Electrolytes and Nonelectrolytes movie

Substances such as NaCl or KBr, which dissolve in water to produce conducting solutions of ions, are called **electrolytes**. Substances such as sucrose or ethyl alcohol, which do not produce ions in aqueous solution, are called **nonelectrolytes**. Most electrolytes are ionic compounds, but some are molecular. Hydrogen chloride, for example, is a molecular compound when pure but **dissociates** (splits apart) to give H^+ and Cl^- ions when it dissolves in water.

$$HCl(g) \xrightarrow{\ H_2O\ } H^+(aq) + Cl^-(aq)$$

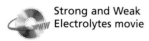

Strong and Weak Electrolytes movie

Compounds that dissociate to a large extent (70–99%) into ions when dissolved in water are said to be **strong electrolytes**, while compounds that dissociate to only a small extent are **weak electrolytes**. Potassium chloride and most other ionic compounds, for instance, are largely dissociated in dilute solution and are thus strong electrolytes. Acetic acid, by contrast, dissociates to the extent of less than 1% in a 0.10 M solution and is a weak electrolyte. As a result, a 0.10 M solution of acetic acid is only weakly conducting.

Dissolution of NaCl in Water movie; Dissolution of KMnO₄ movie

For 0.10 M solutions:

$$KCl(aq) \underset{(2\%)}{\overset{}{\rightleftharpoons}} \underset{(98\%)}{K^+(aq) + Cl^-(aq)} \quad \text{Strong electrolyte}$$

$$\underset{(99\%)}{CH_3CO_2H(aq)} \rightleftharpoons \underset{(1\%)}{H^+(aq) + CH_3CO_2^-(aq)} \quad \text{Weak electrolyte}$$

Note that when we write a dissociation we use a forward-and-backward double arrow $\rightleftharpoons$ to indicate that it takes place simultaneously in both directions. That is, a dissociation is a dynamic process in which an *equilibrium* is established between the forward and reverse reactions. The balance between the two opposing reactions defines the exact concentrations of the various species in solution. We'll learn much more about chemical equilibria in future chapters.

A brief list of some common substances classified according to their electrolyte strength is given in Table 4.1. Note that pure water is classified as a nonelectrolyte because it does not dissociate appreciably into H^+ and OH^- ions. We'll explore the dissociation of water in more detail in Chapter 15.

TABLE 4.1 Electrolyte Classification of Some Common Substances

Strong Electrolytes	Weak Electrolytes	Nonelectrolytes
HCl, HBr, HI	CH_3CO_2H	H_2O
$HClO_4$	HF	CH_3OH (methyl alcohol)
HNO_3		C_2H_5OH (ethyl alcohol)
H_2SO_4		$C_{12}H_{22}O_{11}$ (sucrose)
KBr		Most compounds of carbon
NaCl		(organic compounds)
NaOH, KOH		
Other soluble ionic compounds		

EXAMPLE 4.1 What is the total molar concentration of ions in a 0.350 M solution of the strong electrolyte Na_2SO_4 assuming complete dissociation?

SOLUTION First, we need to know how many ions are produced by dissociation of Na_2SO_4. Writing the equation for dissolving Na_2SO_4 in water shows that 3 mol of ions are formed—2 mol of Na^+ and 1 mol of SO_4^{2-}.

$$Na_2SO_4(s) \xrightarrow{H_2O} 2\,Na^+(aq) + SO_4^{2-}(aq)$$

Thus, the total molar concentration of *ions* is three times the molarity of Na_2SO_4, or 1.05 M.

$$\frac{0.350\ \cancel{\text{mol } Na_2SO_4}}{1\ L} \times \frac{3\ \text{mol ions}}{1\ \cancel{\text{mol } Na_2SO_4}} = 1.05\ M$$

▶ **PROBLEM 4.2** What is the molar concentration of Br^- ions in a 0.225 M aqueous solution of $FeBr_3$ assuming complete dissociation?

KEY CONCEPT PROBLEM 4.3 Three different substances A_2X, A_2Y, and A_2Z were dissolved in water, with the following results. (Water molecules are omitted for clarity.) Which of the substances is the strongest electrolyte, and which the weakest? Explain.

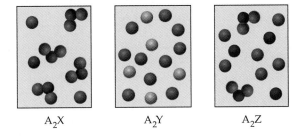

A_2X A_2Y A_2Z

4.3 Aqueous Reactions and Net Ionic Equations

The equations we've been writing up to this point have all been **molecular equations**. That is, all the substances involved in reactions have been written using their full formulas as if they were *molecules*. In Section 4.1, for example, we wrote the precipitation reaction of lead(II) nitrate with potassium iodide to yield solid PbI_2 using only the parenthetical (*aq*) to indicate that the reaction takes place in aqueous solution. Nowhere was it indicated that ions are involved:

A molecular equation

$$Pb(NO_3)_2(aq) + 2\,KI(aq) \longrightarrow 2\,KNO_3(aq) + PbI_2(s)$$

In fact, lead nitrate, potassium iodide, and potassium nitrate are strong electrolytes that dissolve in water to yield solutions of *ions*. Thus, it's more accurate to write the reaction as an **ionic equation**, in which all the ions are explicitly shown:

An ionic equation

$$Pb^{2+}(aq) + 2\,NO_3^-(aq) + 2\,K^+(aq) + 2\,I^-(aq) \longrightarrow 2\,K^+(aq) + 2\,NO_3^-(aq) + PbI_2(s)$$

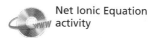

Net Ionic Equation activity

A look at this ionic equation shows that the NO_3^- and K^+ ions undergo no change during the reaction. They appear on both sides of the reaction arrow and act merely as **spectator ions**, whose only role is to balance the charge. The actual reaction, when stripped to its essentials, can be described more simply by writing a **net ionic equation**, in which the spectator ions are removed. A net ionic equation focuses only on the ions undergoing change—the Pb^{2+} and I^- ions in this instance—and ignores everything else:

An ionic equation

$$Pb^{2+}(aq) + 2\,\cancel{NO_3^-}(aq) + 2\,\cancel{K^+}(aq) + 2\,I^-(aq) \longrightarrow 2\,\cancel{K^+}(aq) + 2\,\cancel{NO_3^-}(aq) + PbI_2(s)$$

A net ionic equation

$$Pb^{2+}(aq) + 2\,I^-(aq) \longrightarrow PbI_2(s)$$

Leaving spectator ions out of a net ionic equation does not imply that their presence is irrelevant. Clearly, if a reaction occurs by mixing a solution of Pb^{2+} ions with a solution of I^- ions, then those solutions must also contain additional ions to balance the charge in each: The Pb^{2+} solution must contain an anion, and the I^- solution must contain a cation. Leaving these other ions out of the net ionic equation merely implies that the specific *identity* of the spectator ions is not important; any nonreactive ions could fill the same role.

EXAMPLE 4.2   Aqueous hydrochloric acid reacts with zinc metal to yield hydrogen gas and aqueous zinc chloride. Write a net ionic equation for the process.

$$2\,HCl(aq) + Zn(s) \longrightarrow H_2(g) + ZnCl_2(aq)$$

SOLUTION   First, write the ionic equation, listing all the species present in solution. Both HCl (a molecular compound; Table 4.1) and $ZnCl_2$ (an ionic compound) are strong electrolytes that exist as ions in solution:

$$2\,H^+(aq) + 2\,Cl^-(aq) + Zn(s) \longrightarrow H_2(g) + Zn^{2+}(aq) + 2\,Cl^-(aq)$$

Next, find the ions that are present on both sides of the reaction arrow (the spectator ions) and cancel them, leaving the net ionic equation. In the present instance, the chloride ions cancel from both sides.

$$2\,H^+(aq) + 2\,\cancel{Cl^-}(aq) + Zn(s) \longrightarrow H_2(g) + Zn^{2+}(aq) + 2\,\cancel{Cl^-}(aq)$$

Net ionic equation: $2\,H^+(aq) + Zn(s) \longrightarrow H_2(g) + Zn^{2+}(aq)$

▶ **PROBLEM 4.4**   Write net ionic equations for the following reactions:
(a) $2\,AgNO_3(aq) + Na_2CrO_4(aq) \rightarrow Ag_2CrO_4(s) + 2\,NaNO_3(aq)$
(b) $H_2SO_4(aq) + MgCO_3(s) \rightarrow H_2O(l) + CO_2(g) + MgSO_4(aq)$

Zinc metal reacts with aqueous hydrochloric acid to give hydrogen gas and aqueous Zn^{2+} ions.

4.4   Precipitation Reactions and Solubility Rules

Ionic Compounds activity

To predict whether a precipitation reaction will occur on mixing aqueous solutions of two substances, you must know the **solubilities** of the potential products—that is, how much of each compound will dissolve in a given

Precipitation Reactions
movie

amount of solvent at a given temperature. If a substance has a low solubility in water, it is likely to precipitate from an aqueous solution. If a substance has a high solubility in water, no precipitate will form.

Solubility is a complex matter, and it's not always possible to make correct predictions. As a rule of thumb, though, a compound is probably soluble if it meets either (or both) of the following criteria:

1. **A compound is probably soluble if it contains one of the following** *cations*:
   - Group 1A cation: Li^+, Na^+, K^+, Rb^+, Cs^+
   - Ammonium ion: NH_4^+

2. **A compound is probably soluble if it contains one of the following** *anions*:
   - Halide: Cl^-, Br^-, I^-
     except Ag^+, Hg_2^{2+}, and Pb^{2+} compounds
   - Nitrate (NO_3^-), perchlorate (ClO_4^-), acetate ($CH_3CO_2^-$), sulfate (SO_4^{2-})
     except Ba^{2+}, Hg_2^{2+}, and Pb^{2+} sulfates

If a compound does *not* contain one of the ions listed above, however, it is probably not soluble. Thus, NaOH, which contains a group 1A cation, and $BaCl_2$, which contains a halide anion, are both soluble. A compound such as $CaCO_3$, however, contains neither a group 1A cation nor any of the anions on the list and is therefore not soluble.

Using these guidelines not only makes it possible to predict whether a precipitate will form when solutions of two ionic compounds are mixed but also makes it possible to prepare a specific compound by purposefully carrying out a precipitation. If, for example, you wanted to prepare a sample of silver carbonate, Ag_2CO_3, you could mix a solution of $AgNO_3$ with a solution of Na_2CO_3. Both starting compounds are soluble in water, as is $NaNO_3$. (Why?) Silver carbonate is the only insoluble combination of ions and will therefore precipitate from solution.

Ionic Compounds
activity

$$2\ AgNO_3(aq) + Na_2CO_3(aq) \longrightarrow Ag_2CO_3(s) + 2\ NaNO_3(aq)$$

EXAMPLE 4.3 Predict whether a precipitation reaction will occur when aqueous solutions of $CdCl_2$ and $(NH_4)_2S$ are mixed. Write the net ionic equation.

SOLUTION The idea in this problem is to write the possible reaction, identify the two potential products, and predict the solubility of each. In the present instance, $CdCl_2$ and $(NH_4)_2S$ might give CdS and $2\ NH_4Cl$:

$$??\quad CdCl_2(aq) + (NH_4)_2S(aq) \longrightarrow CdS + 2\ NH_4Cl \quad ??$$

Of the two possible products, the solubility guidelines predict that CdS, a sulfide, is insoluble and that NH_4Cl, an ammonium compound, is soluble. Thus, a precipitation reaction will occur:

$$Cd^{2+}(aq) + S^{2-}(aq) \longrightarrow CdS(s)$$

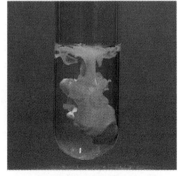

Reaction of aqueous $AgNO_3$ with aqueous Na_2CO_3 gives a white precipitate of Ag_2CO_3.

EXAMPLE 4.4   How might you use a precipitation reaction to prepare a sample of $CuCO_3$? Write the net ionic equation.

SOLUTION   To prepare a precipitate of $CuCO_3$, a soluble Cu^{2+} compound must react with a soluble CO_3^{2-} compound. A look at the solubility guidelines suggests that $Cu(NO_3)_2$ and Na_2CO_3 (among many other possibilities) might work:

$$Cu(NO_3)_2(aq) + Na_2CO_3(aq) \longrightarrow 2\,NaNO_3(aq) + CuCO_3(s)$$
$$Cu^{2+}(aq) + CO_3^{2-}(aq) \longrightarrow CuCO_3(s)$$

▶ **PROBLEM 4.5**   Predict the solubility of each of the following compounds:
(a) $CdCO_3$ (b) MgO (c) Na_2S (d) $PbSO_4$ (e) $(NH_4)_3PO_4$ (f) $HgCl_2$

▶ **PROBLEM 4.6**   Predict whether a precipitation reaction will occur in each of the following situations. Write a net ionic equation for each reaction that does occur.
(a) $NiCl_2(aq) + (NH_4)_2S(aq) \rightarrow$ (b) $Na_2CrO_4(aq) + Pb(NO_3)_2(aq) \rightarrow$
(c) $AgClO_4(aq) + CaBr_2(aq) \rightarrow$

▶ **PROBLEM 4.7**   How might you use a precipitation reaction to prepare a sample of $Ca_3(PO_4)_2$? Write the net ionic equation.

◆— **KEY CONCEPT PROBLEM 4.8**   An aqueous solution of an anion (blue spheres) is added to a solution of a cation (green spheres), and the following result is obtained. Which cations and anions, chosen from the following lists, are compatible with the observed results?

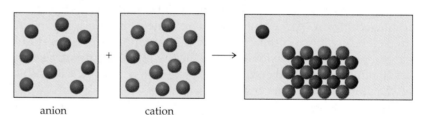

anion   cation

Anions: S^{2-}, PO_4^{3-}, SO_4^{2-}, ClO_4^{-}

Cations: Mg^{2+}, Fe^{3+}, NH_4^{+}, Zn^{2+}

4.5   Acids, Bases, and Neutralization Reactions

We've mentioned the subject of acids and bases on several previous occasions, but now let's look more carefully at what it is that makes an acid an acid, and a base a base. In 1777 the French chemist Antoine Lavoisier proposed that all acids contain a common element, oxygen. In fact, the word "oxygen" is derived from a Greek phrase meaning "acid former." Lavoisier's idea had to be modified, however, when the English chemist Sir Humphrey Davy (1778–1829) showed in 1810 that muriatic acid (now called hydrochloric acid) contains only hydrogen and chlorine but no oxygen. Davy's studies thus suggested that the common element in acids is *hydrogen*, not oxygen.

The relation between acidic behavior and the presence of hydrogen in a compound was clarified in 1887 by the Swedish chemist Svante Arrhenius (1859–1927). Arrhenius proposed that acids are substances that dissociate in

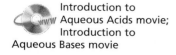
Introduction to
Aqueous Acids movie;
Introduction to
Aqueous Bases movie

water to produce hydrogen ions (H^+) and that bases are substances that dissociate in water to yield hydroxide ions (OH^-):

$$HA(aq) \longrightarrow H^+(aq) + A^-(aq)$$
An acid

$$MOH(aq) \longrightarrow M^+(aq) + OH^-(aq)$$
A base

In these equations, HA is a general formula for an acid—for example, HCl or HNO_3—and MOH is a general formula for a metal hydroxide—for example, NaOH or KOH.

Although convenient to use in equations, the symbol $H^+(aq)$ does not really represent the structure of the ion present in aqueous solution. As a bare proton with no electron nearby, H^+ is much too reactive to exist by itself. Rather, the H^+ attaches to a water molecule, giving the more stable **hydronium ion, H_3O^+**. We'll sometimes write $H^+(aq)$ for convenience, particularly when balancing equations, but will more often write $H_3O^+(aq)$ to represent an aqueous acid solution. Hydrogen chloride, for instance, gives $Cl^-(aq)$ and $H_3O^+(aq)$ when it dissolves in water.

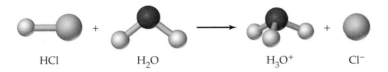

HCl H_2O H_3O^+ Cl^-

Different acids dissociate to different extents in aqueous solution. Those acids that dissociate to a large extent are strong electrolytes and **strong acids;** those acids that dissociate only to a small extent are weak electrolytes and **weak acids**. We've already seen in Table 4.1, for example, that HCl, $HClO_4$, HNO_3, and H_2SO_4 are strong electrolytes, and therefore strong acids, whereas CH_3CO_2H and HF are weak electrolytes, and therefore weak acids.

It should also be reiterated, as pointed out in Section 2.9, that different acids have different numbers of acidic hydrogens and yield different numbers of H_3O^+ ions in solution. Sulfuric acid, for instance, can dissociate twice, and phosphoric acid can dissociate three times. In the case of sulfuric acid, the first dissociation of an H^+ is complete—all H_2SO_4 molecules lose one H^+—but the second dissociation is incomplete, as indicated by the double arrow below. In the case of phosphoric acid, none of the three dissociations is complete.

Sulfuric acid: $H_2SO_4(aq) + H_2O(l) \longrightarrow HSO_4^-(aq) + H_3O^+(aq)$

$HSO_4^-(aq) + H_2O(l) \rightleftharpoons SO_4^{2-}(aq) + H_3O^+(aq)$

Phosphoric acid: $H_3PO_4(aq) + H_2O(l) \rightleftharpoons H_2PO_4^-(aq) + H_3O^+(aq)$

$H_2PO_4^-(aq) + H_2O(l) \rightleftharpoons HPO_4^{2-}(aq) + H_3O^+(aq)$

$HPO_4^{2-}(aq) + H_2O(l) \rightleftharpoons PO_4^{3-}(aq) + H_3O^+(aq)$

Bases, like acids, can also be either strong or weak, depending on the extent to which they dissociate and produce OH^- ions in aqueous solution. Most metal hydroxides, such as NaOH and $Ba(OH)_2$, are strong electrolytes and **strong bases**, but ammonia (NH_3) is a weak electrolyte and a **weak base**. The weakly basic behavior of ammonia is due to the fact that it reacts to a

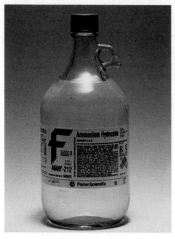

This bottle should be labeled "Aqueous Ammonia" rather than "Ammonium Hydroxide."

small extent with water to yield NH_4^+ and OH^- ions. In fact, aqueous solutions of ammonia are often called "ammonium hydroxide," although this is really a misnomer since the concentrations of NH_4^+ and OH^- ions are low.

$$NH_3(g) + H_2O(l) \rightleftharpoons NH_4^+(aq) + OH^-(aq)$$

As with the dissociation of acetic acid discussed in Section 4.2, the reaction of ammonia with water takes place only to a small extent (about 1%). Most of the ammonia remains unreacted, and we therefore write the reaction with a double arrow to show that a dynamic equilibrium between forward and reverse reactions exists.

Table 4.2 summarizes the names, formulas, and classification of some common acids and bases.

TABLE 4.2   Some Common Acids and Bases

Strong acid	$HClO_4$	Perchloric acid	NaOH	Sodium hydroxide	Strong base
	H_2SO_4	Sulfuric acid	KOH	Potassium hydroxide	
	HBr	Hydrobromic acid	$Ba(OH)_2$	Barium hydroxide	
	HCl	Hydrochloric acid	$Ca(OH)_2$	Calcium hydroxide	
	HNO_3	Nitric acid			
Weak acid	H_3PO_4	Phosphoric acid	NH_3	Ammonia	Weak base
	HF	Hydrofluoric acid			
	CH_3CO_2H	Acetic acid			

When an acid and a base are mixed in the right stoichiometric proportion, both acidic and basic properties disappear because of a **neutralization reaction**, which produces water and a *salt*. The anion of the salt (A^-) comes from the acid, and the cation of the salt (M^+) comes from the base:

A neutralization reaction $$HA(aq) + MOH(aq) \longrightarrow H_2O(l) + MA(aq)$$
$$\text{Acid} \qquad \text{Base} \qquad \text{Water} \qquad \text{A salt}$$

Because salts are generally strong electrolytes in aqueous solution, we can write the neutralization reaction of a strong acid with a strong base as an ionic equation:

$$H^+(aq) + A^-(aq) + M^+(aq) + OH^-(aq) \longrightarrow H_2O(l) + M^+(aq) + A^-(aq)$$

Canceling the ions that appear on both sides of the ionic equation gives the net ionic equation, which holds for the reaction of any strong acid with any strong base in water:

$$H^+(aq) + \cancel{A^-(aq)} + \cancel{M^+(aq)} + OH^-(aq) \longrightarrow H_2O(l) + \cancel{M^+(aq)} + \cancel{A^-(aq)}$$
$$H^+(aq) + OH^-(aq) \longrightarrow H_2O(l)$$
$$\text{or} \quad H_3O^+(aq) + OH^-(aq) \longrightarrow 2 H_2O(l)$$

For the reaction of a weak acid with a strong base, a similar neutralization occurs, but we must write the molecular formula of the acid rather than simply $H^+(aq)$, because the dissociation of the acid in water is incomplete. Instead, the acid exists primarily as the neutral molecule. In the reaction of HF with KOH, for example, we write the net ionic equation as

$$HF(aq) + OH^-(aq) \longrightarrow H_2O(l) + F^-(aq)$$

EXAMPLE 4.5   Write both an ionic equation and a net ionic equation for the neutralization reaction of aqueous HBr and aqueous $Ba(OH)_2$.

SOLUTION   Hydrogen bromide is a strong acid whose aqueous solution contains H^+ ions and Br^- ions. Barium hydroxide is a strong base whose aqueous solution contains Ba^{2+} and OH^- ions. Thus, we have a mixture of four different ions on the reactant side:

$$H^+(aq) + Br^-(aq) + Ba^{2+}(aq) + 2\,OH^-(aq) \longrightarrow$$

The H^+ ions and OH^- ions will react to yield H_2O, so we can write the balanced ionic equation as

$$2\,H^+(aq) + 2\,Br^-(aq) + Ba^{2+}(aq) + 2\,OH^-(aq) \longrightarrow 2\,H_2O(l) + 2\,Br^-(aq) + Ba^{2+}(aq)$$

Canceling the spectator ions $Br^-(aq)$ and $Ba^{2+}(aq)$ from both sides of the ionic equation gives the net ionic equation:

$$2\,H^+(aq) + 2\,\cancel{Br^-(aq)} + \cancel{Ba^{2+}(aq)} + 2\,OH^-(aq) \longrightarrow 2\,H_2O(l) + 2\,\cancel{Br^-(aq)} + \cancel{Ba^{2+}(aq)}$$

$$2\,H^+(aq) + 2\,OH^-(aq) \longrightarrow 2\,H_2O(l) \quad \text{or} \quad H^+(aq) + OH^-(aq) \longrightarrow H_2O(l)$$

The reaction of HBr with $Ba(OH)_2$ thus involves the combination of a proton (H^+) from the acid with OH^- from the base to yield water and a salt $(BaBr_2)$.

▶**PROBLEM 4.9**   Write a balanced ionic equation and net ionic equation for each of the following acid–base reactions:
(a) $2\,CsOH(aq) + H_2SO_4(aq) \rightarrow$
(b) $Ca(OH)_2(aq) + 2\,CH_3CO_2H(aq) \rightarrow$

🔑 **KEY CONCEPT PROBLEM 4.10**   The following pictures represent aqueous solutions of three acids HA (A = X, Y, or Z), with water molecules omitted for clarity. Which of the three is the strongest acid, and which is the weakest?

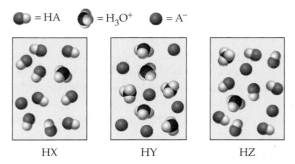

⬤⬤ = HA   🔴 = H_3O^+ ⬤ = A^-

HX   HY   HZ

4.6   Oxidation–Reduction (Redox) Reactions

Magnesium metal burns in air with an intense white light to form solid magnesium oxide. Red phosphorus reacts with liquid bromine to form liquid phosphorus tribromide. Purple aqueous permanganate ion, MnO_4^-, reacts with aqueous Fe^{2+} ion to yield Fe^{3+} and pale pink Mn^{2+}. Although these and many thousands of other reactions appear unrelated, all are examples of *oxidation–reduction* reactions.

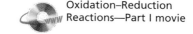

Oxidation–Reduction
Reactions—Part I movie

$$2\,Mg(s) + O_2(g) \longrightarrow 2\,MgO(s)$$

$$2\,P(s) + 3\,Br_2(l) \longrightarrow 2\,PBr_3(l)$$

$$MnO_4^-(aq) + 5\,Fe^{2+}(aq) + 8\,H^+(aq) \longrightarrow Mn^{2+}(aq) + 5\,Fe^{3+}(aq) + 4\,H_2O(l)$$

Magnesium metal burns in air to give MgO.

Elemental phosphorus reacts spectacularly with bromine to give PBr_3.

Aqueous potassium permanganate is frequently used as an *oxidizing agent*, as described in the text.

Historically, the word *oxidation* referred to the combination of an element with oxygen to yield an oxide, and the word *reduction* referred to the removal of oxygen from an oxide to yield the element. Such oxidation–reduction processes have been crucial to the development of human civilization and still have enormous commercial value. The oxidation (rusting) of iron metal by reaction with moist air has been known for millennia and is still a serious problem that causes enormous damage to buildings, bridges, and automobiles. The reduction of iron ore (Fe_2O_3) with charcoal (C) to make iron metal has been carried out since prehistoric times and is still used today in the initial stages of steelmaking.

$$4\,Fe(s) + 3\,O_2(g) \longrightarrow 2\,Fe_2O_3(s) \qquad \text{Rusting of iron: an } \textit{oxidation} \text{ of Fe}$$

$$2\,Fe_2O_3(s) + 3\,C(s) \longrightarrow 4\,Fe(s) + 3\,CO_2(g) \qquad \text{Manufacture of iron: a } \textit{reduction} \text{ of Fe}$$

Today, the words oxidation and reduction have taken on a much broader meaning. An **oxidation** is now defined as the loss of one or more electrons by a substance—element, compound, or ion. Conversely, a **reduction** is the gain of one or more electrons by a substance. Thus, an **oxidation–reduction reaction**, or **redox reaction**, is a process in which *electrons are transferred from one substance to another*.

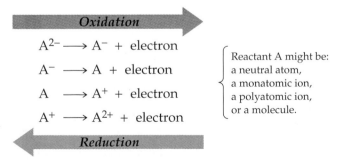

Oxidation

$$A^{2-} \longrightarrow A^- + \text{electron}$$
$$A^- \longrightarrow A + \text{electron}$$
$$A \longrightarrow A^+ + \text{electron}$$
$$A^+ \longrightarrow A^{2+} + \text{electron}$$

Reduction

Reactant A might be: a neutral atom, a monatomic ion, a polyatomic ion, or a molecule.

How can you tell when a redox reaction is taking place? The answer is that we assign to each atom in a substance a value called an **oxidation number** (or *oxidation state*), which provides a measure of whether the atom is neutral, electron-rich, or electron-poor. By comparing the oxidation number of an atom before and after reaction, we can tell whether the atom has gained or lost electrons. Note that *oxidation numbers don't necessarily imply ionic charges*. They are simply a convenient device to help us keep track of electrons in redox reactions.

The rules for assigning oxidation numbers are straightforward:

1. **An atom in its elemental state has an oxidation number of 0.** For example:

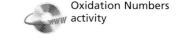

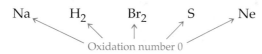

2. **An atom in a monatomic ion has an oxidation number identical to its charge.** (You might want to review Section 2.10 to see the charges on some common ions.) For example:

$$Na^+ \quad Ca^{2+} \quad Al^{3+} \quad Cl^- \quad O^{2-}$$
$$\uparrow \qquad \uparrow \qquad \uparrow \qquad \uparrow \qquad \uparrow$$
$$+1 \qquad +2 \qquad +3 \qquad -1 \qquad -2$$

3. **An atom in a polyatomic ion or in a molecular compound usually has the same oxidation number it would have if it were a monatomic ion.** In the hydroxide ion (OH^-), for example, the oxygen atom has an oxidation number of −2, as if it were a monatomic O^{2-} ion, and the hydrogen atom has an oxidation number of +1, as if it were H^+.

$$
\underset{\substack{\uparrow\qquad\uparrow\qquad\uparrow \\ +1 \quad -2 \quad +1}}{H-O-H}
\qquad
\underset{\substack{\uparrow\qquad\uparrow \\ -2 \quad +1}}{[O-H]^-}
\qquad
\underset{\substack{\uparrow\qquad\uparrow\qquad\uparrow \\ +1 \quad -3 \quad +1}}{H-\overset{\overset{\displaystyle H \leftarrow +1}{|}}{N}-H}
$$

In general, the farther left an element is in the periodic table, the more likely it is that the atom will be "cationlike." Metals, therefore, usually have positive oxidation numbers. The farther right an element is in the periodic table, the more likely it is that the atom will be "anionlike." Non-metals, such as O, N, and the halogens, usually have negative oxidation numbers. We'll see the reasons for this trend in Sections 6.3–6.5.

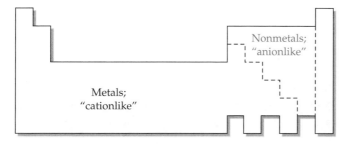

(a) Hydrogen can be either +1 or −1. When bonded to a metal, such as Na or Ca, hydrogen has an oxidation number of −1. When bonded to a nonmetal, such as C, N, O, or Cl, hydrogen has an oxidation number of +1.

$$
\underset{\substack{\nearrow\qquad\searrow \\ +1\qquad -1}}{Na-H}
\qquad
\underset{\substack{\nearrow\quad\uparrow\quad\searrow \\ -1\ +2\ -1}}{H-Ca-H}
\qquad
\underset{\substack{\nearrow\qquad\searrow \\ +1\qquad -1}}{H-Cl}
\qquad
\underset{\substack{\nearrow\quad\uparrow\quad\searrow \\ +1\ -2\ +1}}{H-S-H}
$$

(b) Oxygen usually has an oxidation number of −2. The major exception is in compounds called *peroxides*, which contain either the O_2^{2-} ion or an O−O covalent bond in a molecule. Each oxygen atom in a peroxide has an oxidation number of −1.

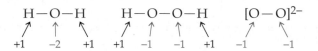

(c) Halogens usually have an oxidation number of −1. The major exception is in compounds of chlorine, bromine, or iodine in which the halogen atom is bonded to oxygen. In such cases, the oxygen has an oxidation number of −2, and the halogen has a positive oxidation number. In Cl_2O, for example, the O atom has an oxidation number of −2 and each Cl atom has an oxidation number of +1.

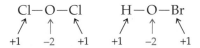

4. **The sum of the oxidation numbers is 0 for a neutral compound and is equal to the net charge for a polyatomic ion.** This rule is particularly useful for finding the oxidation number of an atom in difficult cases. The general idea is to assign oxidation numbers to the "easy" atoms first and then find the oxidation number of the "difficult" atom by subtraction. For example, suppose we need to know the oxidation number of the sulfur atom in sulfuric acid (H_2SO_4). Since each H atom is +1 and each O atom is −2, the S atom must have an oxidation number of +6 for the compound to have no net charge:

$$H_2SO_4 \qquad 2(+1) + (?) + 4(-2) = 0 \text{ net charge}$$
$$+1 \quad ? \quad -2 \qquad ? = 0 - 2(+1) - 4(-2) = +6$$

To find the oxidation number of the chlorine atom in the perchlorate anion (ClO_4^-), we know that each oxygen is −2, so the Cl atom must have an oxidation number of +7 for there to be a net charge of −1:

$$ClO_4^- \qquad ? + 4(-2) = -1 \text{ net charge}$$
$$? \quad -2 \qquad ? = -1 - 4(-2) = +7$$

To find the oxidation number of the nitrogen atom in the ammonium cation (NH_4^+), we know that each H atom is +1, so the N atom must have an oxidation number of −3 for there to be a net charge of +1:

$$NH_4^+ \qquad ? + 4(+1) = +1 \text{ net charge}$$
$$? \quad +1 \qquad ? = +1 - 4(+1) = -3$$

EXAMPLE 4.6 Assign oxidation numbers to each atom in the following substances:

(a) CdS   (b) AlH_3 (c) $S_2O_3^{2-}$ (d) $Na_2Cr_2O_7$

SOLUTION

(a) CdS **(b)** AlH$_3$ **(c)** S$_2$O$_3{}^{2-}$ **(d)** Na$_2$Cr$_2$O$_7$
↑ ↑   ↑ ↑   ↑ ↑   ↑ ↑   ↑
+2 −2   +3 −1   +2 −2   +1 +6 −2

(a) The sulfur atom in S^{2-} has an oxidation number of −2, so Cd must be +2.

(b) H bonded to a metal has the oxidation number −1, so Al must be +3.

(c) O usually has the oxidation number −2, so S must be +2 for the anion to have a net charge of −2: for (2 S^{+2}) (3 O^{-2}), 2(+2) + 3(−2) = −2 net charge.

(d) Na is always +1, and oxygen is −2, so Cr must be +6 for the compound to be neutral: for (2 Na$^+$) (2 Cr^{+6}) (7 O^{-2}), 2(+1) + 2(+6) + 7(−2) = 0 net charge.

▶ **PROBLEM 4.11**   Assign an oxidation number to each atom in the following compounds:

  (a) SnCl$_4$ **(b)** CrO$_3$ **(c)** VOCl$_3$ **(d)** V$_2$O$_3$ **(e)** HNO$_3$ **(f)** FeSO$_4$

4.7   Identifying Redox Reactions

Once oxidation numbers are assigned, it's clear why the reactions mentioned in the previous section are all redox processes. Take the rusting of iron, for example. Two of the reactants, Fe and O$_2$, are elements, and both therefore have an oxidation number of 0. In the product, however, the oxygen atoms have an oxidation number of −2, and the iron atoms have an oxidation number of +3. Thus, Fe has undergone a change from 0 to +3 (a loss of electrons, or oxidation), and O has undergone a change from 0 to −2 (a gain of electrons, or reduction). Note that the total number of electrons given up by the atoms being oxidized (4 Fe × 3 electrons/Fe = 12 electrons) is the same as the number gained by the atoms being reduced (6 O × 2 electrons/O = 12 electrons).

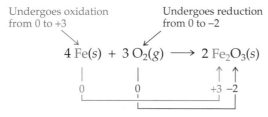

A similar analysis can be carried out for the production of iron metal from its ore. The iron atom is reduced because it goes from an oxidation number of +3 in the reactant (Fe$_2$O$_3$) to 0 in the product (Fe). At the same time, the carbon atom is oxidized because it goes from an oxidation number of 0 in the reactant (C) to +4 in the product (CO$_2$). The oxygen atoms undergo no change because they have an oxidation number of −2 in both reactant and product. The total number of electrons given up by the atoms being oxidized (3 C × 4 electrons/C = 12 electrons) is the same as the number gained by the atoms being reduced (4 Fe × 3 electrons/Fe = 12 electrons).

The iron used for this prehistoric sickle was made by the reduction of iron ore with charcoal.

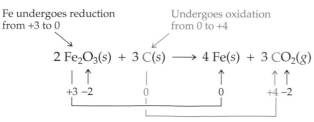

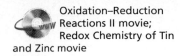

Oxidation–Reduction
Reactions II movie;
Redox Chemistry of Tin
and Zinc movie

As these examples show, oxidations and reductions always occur together. Whenever one atom loses one or more electrons (is oxidized), another atom must gain those electrons (be reduced). The substance that *causes* a reduction by giving up electrons—the iron atom in the reaction of Fe with O_2 and the carbon atom in the reaction of C with Fe_2O_3—is called a **reducing agent**. The substance that causes an oxidation by accepting electrons—the oxygen atom in the reaction of Fe with O_2 and the iron atom in the reaction of C with Fe_2O_3—is called an **oxidizing agent**. The reducing agent is itself oxidized when it gives up electrons, and the oxidizing agent is itself reduced when it accepts electrons.

Reducing agent:
- Causes reduction
- Loses one or more electrons
- Undergoes oxidation
- Oxidation number of atom increases

Oxidizing agent:
- Causes oxidation
- Gains one or more electrons
- Undergoes reduction
- Oxidation number of atom decreases

We'll see in later chapters that redox reactions are common for almost every element in the periodic table except for the noble gas elements of group 8A. In general, metals act as reducing agents, and reactive nonmetals such as O_2 and the halogens act as oxidizing agents.

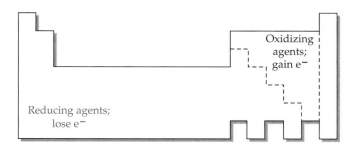

Different metals can give up different numbers of electrons in redox reactions. Lithium, sodium, and the other group 1A elements give up only one electron and become monopositive ions with oxidation numbers of +1. Beryllium, magnesium, and the other group 2A elements, however, typically give up two electrons and become dipositive ions. The transition metals in the middle of the periodic table can give up a variable number of electrons to yield more than one kind of ion depending on the exact reaction. Titanium, for example, can react with chlorine to yield either $TiCl_3$ or $TiCl_4$. Because a chloride ion has a −1 oxidation number, the titanium atom in $TiCl_3$ must have a +3 oxidation number and the titanium atom in $TiCl_4$ must be +4.

EXAMPLE 4.7 Assign oxidation numbers to all atoms, tell in each case which substance is undergoing oxidation and which reduction, and identify the oxidizing and reducing agents.
(a) $Ca(s) + 2\,H^+(aq) \rightarrow Ca^{2+}(aq) + H_2(g)$
(b) $2\,Fe^{2+}(aq) + Cl_2(aq) \rightarrow 2\,Fe^{3+}(aq) + 2\,Cl^-(aq)$

SOLUTION

(a) The neutral elements Ca and H_2 have oxidation numbers of 0; Ca^{2+} is +2 and H^+ is +1:

$$\underset{0}{Ca(s)} + \underset{+1}{2\,H^+(aq)} \longrightarrow \underset{+2}{Ca^{2+}(aq)} + \underset{0}{H_2(g)}$$

Ca is oxidized, because its oxidation number increases from 0 to +2, and H^+ is reduced, because its oxidation number decreases from +1 to 0. The reducing agent is the substance that gives away electrons, thereby going to a higher oxidation number, and the oxidizing agent is the substance that accepts electrons, thereby going to a lower oxidation number. In the present case, calcium is the reducing agent, and H^+ is the oxidizing agent.

(b) Atoms of the neutral element Cl_2 have an oxidation number of 0; the monatomic ions have oxidation numbers equal to their charge:

$$\underset{+2}{2\,Fe^{2+}(aq)} + \underset{0}{Cl_2(aq)} \longrightarrow \underset{+3}{2\,Fe^{3+}(aq)} + \underset{-1}{2\,Cl^-(aq)}$$

Fe^{2+} is oxidized (its oxidation number increases from +2 to +3); Cl_2 is reduced (its oxidation number decreases from 0 to −1). Fe^{2+} is the reducing agent, and Cl_2 is the oxidizing agent.

▶ **PROBLEM 4.12** Aqueous copper(II) ion reacts with aqueous iodide ion to yield solid copper(I) iodide and aqueous iodine. Write the balanced net ionic equation, assign oxidation numbers to all species present, and identify the oxidizing and reducing agents.

▶ **PROBLEM 4.13** Tell in each case which substance is undergoing an oxidation and which a reduction, and identify the oxidizing and reducing agents.
(a) $SnO_2(s) + 2\,C(s) \rightarrow Sn(s) + 2\,CO(g)$
(b) $Sn^{2+}(aq) + 2\,Fe^{3+}(aq) \rightarrow Sn^{4+}(aq) + 2\,Fe^{2+}(aq)$

4.8 The Activity Series of the Elements

The reaction of an aqueous cation, usually a metal ion, with a free element is among the simplest of all redox processes. The products are a different ion and a different element. Iron metal reacts with aqueous copper(II) ion, for example, to give iron(II) ion and copper metal:

$$Fe(s) + Cu^{2+}(aq) \longrightarrow Fe^{2+}(aq) + Cu(s)$$

 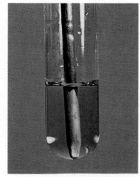

The iron nail reduces the Cu^{2+} ions and becomes coated with metallic Cu. Note the change in intensity of the blue color due to loss of Cu^{2+} ions from solution.

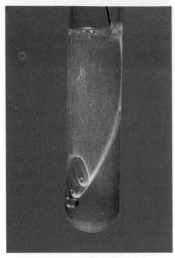

Magnesium metal reacts with aqueous acid to give bubbles of hydrogen gas and Mg^{2+} ion.

Similarly, magnesium metal reacts with aqueous acid to yield magnesium ion and hydrogen gas:

$$Mg(s) + 2\,H^+(aq) \longrightarrow Mg^{2+}(aq) + H_2(g)$$

Whether a reaction occurs between a given ion and a given element depends on the relative ease with which the various species gain or lose electrons—that is, on the relative ease with which the species are reduced or oxidized. By noting the results from a succession of different reactions, it's possible to organize an **activity series**, which ranks the elements in order of their reducing ability in aqueous solution (Table 4.3).

TABLE 4.3   A Partial Activity Series of the Elements

Oxidation Reaction

Strongly reducing	$\begin{aligned} Li &\rightarrow Li^+ + e^- \\ K &\rightarrow K^+ + e^- \\ Ba &\rightarrow Ba^{2+} + 2\,e^- \\ Ca &\rightarrow Ca^{2+} + 2\,e^- \\ Na &\rightarrow Na^+ + e^- \end{aligned}$	These elements react rapidly with aqueous H^+ ions (acid) or with liquid H_2O to release H_2 gas.
	$\begin{aligned} Mg &\rightarrow Mg^{2+} + 2\,e^- \\ Al &\rightarrow Al^{3+} + 3\,e^- \\ Mn &\rightarrow Mn^{2+} + 2\,e^- \\ Zn &\rightarrow Zn^{2+} + 2\,e^- \\ Cr &\rightarrow Cr^{3+} + 3\,e^- \\ Fe &\rightarrow Fe^{2+} + 2\,e^- \end{aligned}$	These elements react with aqueous H^+ ions or with steam to release H_2 gas.
	$\begin{aligned} Co &\rightarrow Co^{2+} + 2\,e^- \\ Ni &\rightarrow Ni^{2+} + 2\,e^- \\ Sn &\rightarrow Sn^{2+} + 2\,e^- \end{aligned}$	These elements react with aqueous H^+ ions to release H_2 gas.
	$\mathbf{H_2 \rightarrow 2\,H^+ + 2\,e^-}$	
Weakly reducing	$\begin{aligned} Cu &\rightarrow Cu^{2+} + 2\,e^- \\ Ag &\rightarrow Ag^+ + e^- \\ Hg &\rightarrow Hg^{2+} + 2\,e^- \\ Pt &\rightarrow Pt^{2+} + 2\,e^- \\ Au &\rightarrow Au^{3+} + 3\,e^- \end{aligned}$	These elements do not react with aqueous H^+ ions to release H_2.

Those elements at the top of Table 4.3 give up electrons readily and are stronger reducing agents, whereas those elements at the bottom give up electrons less readily and are weaker reducing agents. As a result, *any element higher in the activity series will reduce the ion of any element lower in the activity series*. Because copper is above silver, for example, copper metal can give electrons to Ag^+ ions:

$$Cu(s) + 2\,Ag^+(aq) \longrightarrow Cu^{2+}(aq) + 2\,Ag(s)$$

Conversely, because gold is *below* silver in the activity series, gold metal does *not* give electrons to Ag^+ ions:

$$Au(s) + 3\,Ag^+(aq) \xrightarrow{\;\;/\;\;} Au^{3+}(aq) + 3\,Ag(s) \qquad \textit{Does not occur}$$

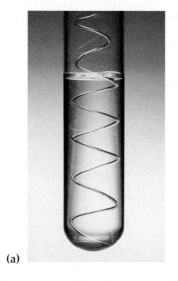
(a)

(b)

(a) The copper wire reduces aqueous Ag^+ ion and (b) becomes coated with metallic silver. Note the blue color due to Cu^{2+} ions in the solution.

The position of hydrogen in the activity series is particularly important because it indicates which metals react with aqueous acid (H^+) to release H_2 gas. The metals at the top of the series—the alkali metals of group 1A and alkaline earth metals of group 2A—are such powerful reducing agents that they react even with pure water, in which the concentration of H^+ is very low:

$$2\,Na(s) + 2\,H_2O(l) \longrightarrow 2\,Na^+(aq) + 2\,OH^-(aq) + H_2(g)$$

Oxidized   Reduced   Unchanged

By contrast, the metals in the middle of the series react with aqueous acid but not with water, and the metals at the bottom of the series react with neither aqueous acid nor water:

$$Fe(s) + 2\,H^+(aq) \longrightarrow Fe^{2+}(aq) + H_2(g)$$

$$Cu(s) + H^+(aq) \longrightarrow \text{No reaction}$$

Notice that the most reactive metals (the top of the activity series) are on the left of the periodic table, whereas the least reactive metals (the bottom of the activity series) are in the transition metal groups closer to the right side of the table. We'll see the reasons for this behavior in Chapter 6.

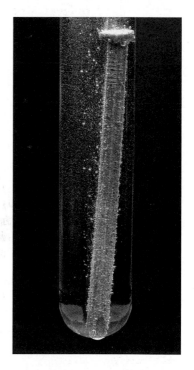
Iron reacts only slowly with aqueous acid at room temperature.

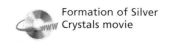
Formation of Silver Crystals movie

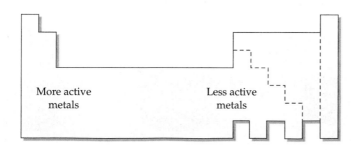
More active metals   Less active metals

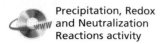

Precipitation, Redox and Neutralization Reactions activity

EXAMPLE 4.8 Predict whether the following reactions will occur:
(a) $Hg^{2+}(aq) + Zn(s) \rightarrow Hg(l) + Zn^{2+}(aq)$
(b) $2\,H^+(aq) + 2\,Ag(s) \rightarrow H_2(g) + 2\,Ag^+(aq)$

SOLUTION Look at Table 4.3 to find the relative activities of the elements.
(a) Zinc is above mercury in the activity series, so this reaction will occur.
(b) Silver is below hydrogen in the activity series, so this reaction will not occur.

▶ **PROBLEM 4.14** Predict whether the following reactions will occur:
(a) $2\,H^+(aq) + Pt(s) \rightarrow H_2(g) + Pt^{2+}(aq)$
(b) $Ca^{2+}(aq) + Mg(s) \rightarrow Ca(s) + Mg^{2+}(aq)$

✦— **KEY CONCEPT PROBLEM 4.15** Element B will reduce the cation of element A (A^+), but will not reduce the cation of element C (C^+). Will element C reduce the cation of element A? Explain.

4.9 Balancing Redox Reactions: The Oxidation-Number Method

Simple redox reactions can be balanced by the common-sense method described in Section 3.1, but other reactions are so complex that a more systematic approach is needed. There are two such systematic approaches often used for balancing redox reactions: the *oxidation-number method* and the *half-reaction method*. Different chemists prefer different methods, so we'll discuss both. The oxidation-number method is useful because it makes you focus on the chemical changes involved; the half-reaction method (discussed in the next section) is useful because it makes you focus on the transfer of electrons, a subject of particular interest when discussing batteries and other aspects of electrochemistry (Chapter 18).

The key to the **oxidation-number method** of balancing redox equations is to realize that the net change in the total of all oxidation numbers must be zero. That is, any *increase* in oxidation number for the oxidized atoms must be matched by a corresponding *decrease* in oxidation number for the reduced atoms. Take the reaction of potassium permanganate, $KMnO_4$, with sodium bromide in aqueous acid, for example. An aqueous acidic solution of the purple permanganate anion, MnO_4^-, is reduced by Br^- to yield the nearly colorless Mn^{2+} ion, while Br^- is oxidized to Br_2. The unbalanced net ionic equation for the process is

$$MnO_4^-(aq) + Br^-(aq) \longrightarrow Mn^{2+}(aq) + Br_2(aq) \qquad \text{Unbalanced}$$

Addition of aqueous Br^- to a solution of MnO_4^- causes a reduction of the permanganate ion, discharging the intense purple color.

The first step is to balance the equation for all atoms other than O and H. In the present instance, a coefficient of 2 is needed for Br^- on the left:

Add this coefficient
to balance for Br.

$$MnO_4^-(aq) + 2 Br^-(aq) \longrightarrow Mn^{2+}(aq) + Br_2(aq)$$

Next, assign oxidation numbers to all atoms, including O and H if present, in both reactants and products:

$$MnO_4^-(aq) + 2 Br^-(aq) \longrightarrow Mn^{2+}(aq) + Br_2(aq)$$

+7 −2   −1   +2   0

Now, decide which of the atoms have changed their oxidation number and have thus been either oxidized or reduced. In the present instance, manganese has been reduced from +7 to +2 (gaining five electrons), and bromine has been oxidized from −1 to 0 (losing one electron).

$$MnO_4^-(aq) + 2 Br^-(aq) \longrightarrow Mn^{2+}(aq) + Br_2(aq)$$

+7   −1   +2   0

Lose $2 \times 1\,e^- = 2\,e^-$

Gain 5 e^-

The next step is to find the net increase in oxidation number for the oxidized atoms and the net decrease in oxidation number for the reduced atoms. Then, multiply the net increase and the net decrease by suitable factors so that the two become equal. In the present instance, the net increase in oxidation number is 2 (as two Br^- ions go from −1 to 0), and the net decrease in oxidation number is 5 (as Mn goes from +7 to +2). Multiplying the net increase by 5 and the net decrease by 2 will make them equal at 10. Thus, we must multiply the coefficients of the manganese species by 2 and the coefficients of the Br species by 5.

Increase in oxidation number: $2\,Br^- \longrightarrow Br_2$ Net increase = +2

−1   0

Decrease in oxidation number: $MnO_4^- \longrightarrow Mn^{2+}$ Net decrease = −5

+7   +2

Multiply by these coefficients to make the net increase
in oxidation number equal to the net decrease.

$$2[MnO_4^-(aq)] + 5[2\,Br^-(aq)] \longrightarrow 2\,Mn^{2+}(aq) + 5\,Br_2(aq)$$

Finally, because we know that the reaction is occurring in acidic solution, we balance the equation for oxygen by adding H_2O to the side with less O and then balance for hydrogen by adding H^+ to the side with less H. In this example, 8 H_2O must be added to the right side to balance for O, and 16 H^+ must then be added to the left side to balance for H. If everything has been done correctly, the final net ionic equation will result. The answer can be checked by noting that the equation is balanced both for atoms and for charge, with +4 on both sides.

First, add these water molecules to balance O.

Next, add these H⁺ ions to balance H.

$$2\,MnO_4^-(aq) + 10\,Br^-(aq) + 16\,H^+(aq) \longrightarrow 2\,Mn^{2+}(aq) + 5\,Br_2(aq) + 8\,H_2O(l)$$

Charge: $(2 \times -1) + (-10) + (+16) = +4$ \qquad\qquad Charge: $(2 \times +2) = +4$

To summarize, balancing a redox reaction in acidic solution by the oxidation-number method is a six-step process, followed by a check of the answer (Figure 4.2.)

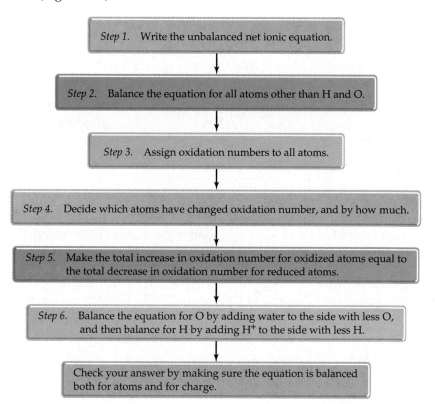

Step 1. Write the unbalanced net ionic equation.

Step 2. Balance the equation for all atoms other than H and O.

Step 3. Assign oxidation numbers to all atoms.

Step 4. Decide which atoms have changed oxidation number, and by how much.

Step 5. Make the total increase in oxidation number for oxidized atoms equal to the total decrease in oxidation number for reduced atoms.

Step 6. Balance the equation for O by adding water to the side with less O, and then balance for H by adding H⁺ to the side with less H.

Check your answer by making sure the equation is balanced both for atoms and for charge.

FIGURE 4.2 The procedure for balancing redox equations by the oxidation-number method.

Example 4.9 shows how the oxidation-number method is used to balance a reaction carried out in basic solution. The procedure is exactly the same as that used for balancing a reaction in acidic solution, but OH⁻ ions are added at the final step to neutralize any H⁺ ions that appear in the equation. This simply reflects the fact that basic solutions contain negligibly small amounts of H⁺ but large amounts of OH⁻.

EXAMPLE 4.9   A purple solution of aqueous potassium permanganate ($KMnO_4$) reacts with aqueous sodium sulfite (Na_2SO_3) in basic solution to yield the green manganate ion (MnO_4^{2-}) and sulfate ion (SO_4^{2-}). The unbalanced net ionic equation is

$$MnO_4^-(aq) + SO_3^{2-}(aq) \longrightarrow MnO_4^{2-}(aq) + SO_4^{2-}(aq) \qquad \text{Unbalanced}$$

Balance the equation by the oxidation-number method.

SOLUTION

Steps 1 and 2. The unbalanced net ionic equation is already balanced for atoms other than O and H.

Step 3. Assign oxidation numbers to all atoms:

$$MnO_4^-(aq) + SO_3^{2-}(aq) \longrightarrow MnO_4^{2-}(aq) + SO_4^{2-}(aq)$$

$$\quad\; \uparrow \uparrow \qquad\quad \uparrow \uparrow \qquad\qquad\quad \uparrow \uparrow \qquad\quad \uparrow \uparrow$$

$$\quad +7\ -2 \qquad +4\ -2 \qquad\qquad +6\ -2 \qquad\; +6\ -2$$

Step 4. Decide which atoms have changed oxidation number and by how much. Manganese has been reduced from +7 to +6 (gaining one electron), and sulfur has been oxidized from +4 to +6 (losing two electrons):

$$MnO_4^-(aq) + SO_3^{2-}(aq) \longrightarrow MnO_4^{2-}(aq) + SO_4^{2-}(aq)$$

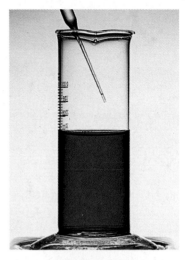

$$\quad\ \ +7 \qquad\quad +4 \qquad\qquad\quad +6 \qquad\qquad +6$$

Lose 2 e⁻

Gain 1 e⁻

Step 5. Find the total net increase in oxidation number for the oxidized atoms and the total net decrease in oxidation number for the reduced atoms:

Increase in oxidation number: $S(+4) \longrightarrow S(+6)$ Net increase = +2

Decrease in oxidation number: $Mn(+7) \longrightarrow Mn(+6)$ Net decrease = −1

Now, choose coefficients that make the net increase equal to the net decrease. In this case, the net decrease must be multiplied by 2, meaning that a coefficient of 2 is needed for both Mn species:

Add this coefficient to make the net increase
and net decrease in oxidation numbers equal.

$$2\,MnO_4^-(aq) + SO_3^{2-}(aq) \longrightarrow 2\,MnO_4^{2-}(aq) + SO_4^{2-}(aq)$$

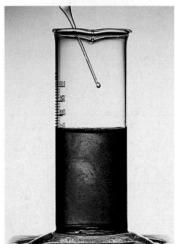

Step 6. Balance the equation for O by adding 1 H_2O on the left, and then balance for H by adding 2 H^+ on the right.

First, add this water molecule
to balance for O.

Next, add these H^+ ions to
balance for H.

$$2\,MnO_4^-(aq) + SO_3^{2-}(aq) + H_2O(l) \longrightarrow 2\,MnO_4^{2-}(aq) + SO_4^{2-}(aq) + 2\,H^+(aq)$$

At this point, the equation is fully balanced. Even so, it's not correct because it assumes an *acidic* solution (the 2 H^+ on the right) but we were told that the reaction occurs in *basic* solution. We can correct the situation by adding 2 OH^- ions to each side of the equation. The 2 OH^- on the right will "neutralize" the 2 H^+ ions, giving 2 H_2O:

2 OH^- ions have been
added to both sides.

These water molecules result from
neutralizing 2 H^+ with 2 OH^-.

$$2\,MnO_4^-(aq) + SO_3^{2-}(aq) + H_2O(l) + 2\,OH^-(aq) \longrightarrow$$
$$2\,MnO_4^{2-}(aq) + SO_4^{2-}(aq) + 2\,H_2O(l)$$

Reduction of the purple MnO_4^- ion with SO_3^{2-} in basic solution yields the deep green MnO_4^{2-} ion.

Finally, we can cancel one H_2O molecule that occurs on both sides of the equation, giving the final net ionic equation that is balanced both for atoms and for charge:

$$2\,MnO_4^-(aq) + SO_3^{2-}(aq) + 2\,OH^-(aq) \longrightarrow 2\,MnO_4^{2-}(aq) + SO_4^{2-}(aq) + H_2O(l)$$

Charge: $(2 \times -1) + (-2) + (2 \times -1) = -6$ Charge: $(2 \times -2) + (-2) = -6$

▶ **PROBLEM 4.16**   Balance the following net ionic equation by the oxidation-number method. The reaction takes place in acidic solution.

$$Cr_2O_7{}^{2-}(aq) + I^-(aq) \longrightarrow Cr^{3+}(aq) + IO_3^-(aq)$$

▶ **PROBLEM 4.17**   Balance the following net ionic equation by the oxidation-number method. The reaction takes place in basic solution.

$$MnO_4^-(aq) + Br^-(aq) \longrightarrow MnO_2(s) + BrO_3^-(aq)$$

4.10 Balancing Redox Reactions: The Half-Reaction Method

An alternative to the oxidation-number method for balancing redox reactions is the **half-reaction method**. The key to this method is to realize that the overall reaction can be broken into two parts, or **half-reactions**. One half-reaction describes the oxidation part of the process, and the other half-reaction describes the reduction part. Each half is balanced separately, and the two halves are then added to obtain the final equation. Let's look at the reaction of aqueous potassium dichromate ($K_2Cr_2O_7$) with aqueous NaCl to see how the method works. The reaction occurs in acidic solution according to the unbalanced net ionic equation

$$Cr_2O_7^{2-}(aq) + Cl^-(aq) \longrightarrow Cr^{3+}(aq) + Cl_2(aq) \qquad \text{Unbalanced}$$

The first step is to decide which atoms have been oxidized and which have been reduced. In the present case, the chloride atom is oxidized (from −1 to 0), and the chromium atom is reduced (from +6 to +3). Thus, we can write two unbalanced half-reactions that show the separate steps:

Oxidation half-reaction: $Cl^-(aq) \longrightarrow Cl_2(aq)$

Reduction half-reaction: $Cr_2O_7^{2-}(aq) \longrightarrow Cr^{3+}(aq)$

With the two half-reactions identified, each is balanced separately. Begin by balancing for all atoms other than H and O. The oxidation half-reaction needs a coefficient of 2 before the Cl^-, and the reduction half-reaction needs a coefficient of 2 before the Cr^{3+}.

The orange dichromate ion is reduced by addition of Cl^- to give the green Cr^{3+} ion.

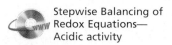
Stepwise Balancing of Redox Equations— Acidic activity

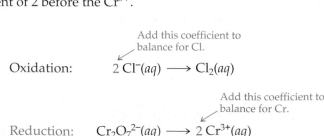

Add this coefficient to balance for Cl.

Oxidation: $2\,Cl^-(aq) \longrightarrow Cl_2(aq)$

Add this coefficient to balance for Cr.

Reduction: $Cr_2O_7^{2-}(aq) \longrightarrow 2\,Cr^{3+}(aq)$

Next, balance both half-reactions for oxygen by adding H_2O to the side with less O, and balance for hydrogen by adding H^+ to the side with less H. The oxidation half-reaction has no H or O, but the reduction reaction needs 7 H_2O on the product side to balance for O and 14 H^+ on the reactant side to balance for H:

Oxidation: $\qquad\qquad\qquad\qquad$ $2\,Cl^-(aq) \longrightarrow Cl_2(aq)$

Then, add 14 H^+ to
balance for H.

First, add 7 H_2O to
balance for O.

Reduction: $\qquad$ $Cr_2O_7^{2-}(aq)\ +\ 14\,H^+(aq) \longrightarrow 2\,Cr^{3+}(aq)\ +\ 7\,H_2O(l)$

Now, balance both half-reactions for charge by adding electrons (e^-) to the side with the greater positive charge. The oxidation half-reaction must have $2\,e^-$ added to the *product* side, and the reduction half-reaction must have $6\,e^-$ added to the *reactant* side:

Add these electrons to
balance for charge.

Oxidation: $\qquad\qquad\qquad\qquad$ $2\,Cl^-(aq) \longrightarrow Cl_2(aq)\ +\ 2\,e^-$

Add these electrons to
balance for charge.

Reduction: $\qquad$ $Cr_2O_7^{2-}(aq)\ +\ 14\,H^+(aq)\ +\ 6\,e^- \longrightarrow 2\,Cr^{3+}(aq)\ +\ 7\,H_2O(l)$

With both half-reactions now balanced, we need to multiply by suitable coefficients so that the number of electrons released in the oxidation half-reaction is the same as the number consumed in the reduction half-reaction. Since the reduction half-reaction has $6\,e^-$ but the oxidation half-reaction has only $2\,e^-$, the entire oxidation half-reaction must be multiplied by 3:

Multiply by this coefficient to
equalize the numbers of electrons
in the two half-reactions.

Oxidation: $\qquad\qquad\qquad$ $3\ \times\ [2\,Cl^-(aq) \longrightarrow Cl_2(aq)\ +\ 2\,e^-]$

$\qquad\qquad\qquad$ or $\quad$ $6\,Cl^-(aq) \longrightarrow 3\,Cl_2(aq)\ +\ 6\,e^-$

Reduction: $\quad$ $Cr_2O_7^{2-}(aq)\ +\ 14\,H^+(aq)\ +\ 6\,e^- \longrightarrow 2\,Cr^{3+}(aq)\ +\ 7\,H_2O(l)$

Adding the two half-reactions together, and canceling the species that occur on both sides (only the electrons in this example), then gives the final balanced equation. Check the answer to make sure it is balanced both for atoms and for charge.

$$6\,Cl^-(aq) \longrightarrow 3\,Cl_2(aq)\ +\ 6\,e^-$$

$$Cr_2O_7^{2-}(aq)\ +\ 14\,H^+(aq)\ +\ 6\,e^- \longrightarrow 2\,Cr^{3+}(aq)\ +\ 7\,H_2O(l)$$

$$\overline{Cr_2O_7^{2-}(aq)\ +\ 14\,H^+(aq)\ +\ 6\,Cl^-(aq) \longrightarrow 3\,Cl_2(aq)\ +\ 2\,Cr^{3+}(aq)\ +\ 7\,H_2O(l)}$$

$\qquad$ Charge: $(-2)+(+14)+(6\times-1)=+6$ $\qquad\qquad\qquad$ Charge: $(2\times+3)=+6$

To summarize, balancing a redox reaction in acidic solution by the half-reaction method is a six-step process, followed by a check of the answer (Figure 4.3.)

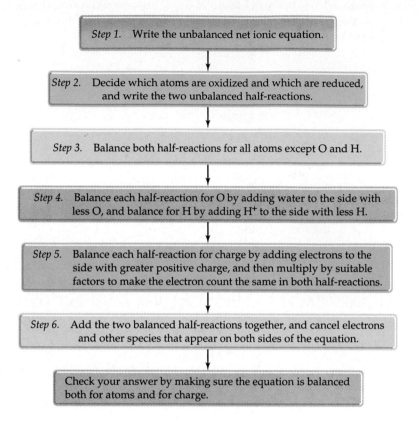

FIGURE 4.3 The procedure for balancing redox equations by the half-reaction method.

Example 4.11 shows how to use the method for balancing a reaction that takes place in basic solution. As in the oxidation-number method, we first balance the reaction for an *acidic* solution and then add OH^- ions in a final step to neutralize H^+.

EXAMPLE 4.10 Write unbalanced half-reactions for the following net ionic equations:
(a) $Mn^{2+}(aq) + ClO_3^-(aq) \rightarrow MnO_2(s) + ClO_2(aq)$
(b) $Cr_2O_7^{2-}(aq) + Fe^{2+}(aq) \rightarrow Cr^{3+}(aq) + Fe^{3+}(aq)$

SOLUTION Look at each equation to see which atoms are being oxidized (increasing in oxidation number) and which are being reduced (decreasing in oxidation number).
(a) Oxidation: $Mn^{2+}(aq) \rightarrow MnO_2(s)$ Mn goes from +2 to +4
    Reduction: $ClO_3^-(aq) \rightarrow ClO_2(aq)$ Cl goes from +5 to +4
(b) Oxidation: $Fe^{2+}(aq) \rightarrow Fe^{3+}(aq)$ Fe goes from +2 to +3
    Reduction: $Cr_2O_7^{2-}(aq) \rightarrow Cr^{3+}(aq)$ Cr goes from + 6 to +3

EXAMPLE 4.11 Aqueous sodium hypochlorite (NaOCl; household bleach) is a strong oxidizing agent that reacts with chromite ion $[Cr(OH)_4^-]$ in basic solution to yield chromate ion (CrO_4^{2-}) and chloride ion. The net ionic equation is

$$ClO^-(aq) + Cr(OH)_4^-(aq) \longrightarrow CrO_4^{2-}(aq) + Cl^-(aq) \qquad \text{Unbalanced}$$

Balance the equation using the half-reaction method.

SOLUTION

Steps 1 and 2. The unbalanced net ionic equation shows that chromium is oxidized (from +3 to +6) and chlorine is reduced (from +1 to −1). Thus, we can write the following two half-reactions:

Oxidation half-reaction: $Cr(OH)_4^-(aq) \longrightarrow CrO_4^{2-}(aq)$

Reduction half-reaction: $ClO^-(aq) \longrightarrow Cl^-(aq)$

Step 3. The half-reactions are already balanced for atoms other than O and H.

Step 4. Balance each half-reaction for O by adding H_2O to the side that has less O, and then balance each for H by adding H^+ to the side that has less H:

Oxidation: $Cr(OH)_4^-(aq) \longrightarrow CrO_4^{2-}(aq) + 4\,H^+(aq)$

Reduction: $ClO^-(aq) + 2\,H^+(aq) \longrightarrow Cl^-(aq) + H_2O(l)$

Step 5. Balance each half-reaction for charge by adding electrons to the side with the greater positive charge:

Oxidation: $Cr(OH)_4^-(aq) \longrightarrow CrO_4^{2-}(aq) + 4\,H^+(aq) + 3\,e^-$

Reduction: $ClO^-(aq) + 2\,H^+(aq) + 2\,e^- \longrightarrow Cl^-(aq) + H_2O(l)$

Next, multiply the half-reactions by factors that make the electron count in each the same. The oxidation half-reaction must be multiplied by 2, and the reduction half-reaction must be multiplied by 3:

Oxidation: $2 \times [Cr(OH)_4^-(aq) \longrightarrow CrO_4^{2-}(aq) + 4\,H^+(aq) + 3\,e^-]$

or $2\,Cr(OH)_4^-(aq) \longrightarrow 2\,CrO_4^{2-}(aq) + 8\,H^+(aq) + 6\,e^-$

Reduction: $3 \times [ClO^-(aq) + 2\,H^+(aq) + 2\,e^- \longrightarrow Cl^-(aq) + H_2O(l)]$

or $3\,ClO^-(aq) + 6\,H^+(aq) + 6\,e^- \longrightarrow 3\,Cl^-(aq) + 3\,H_2O(l)$

Step 6. Add the balanced half-reactions:

$$2\,Cr(OH)_4^-(aq) \longrightarrow 2\,CrO_4^{2-}(aq) + 8\,H^+(aq) + 6\,e^-$$

$$3\,ClO^-(aq) + 6\,H^+(aq) + 6\,e^- \longrightarrow 3\,Cl^-(aq) + 3\,H_2O(l)$$

$$2\,Cr(OH)_4^-(aq) + 3\,ClO^-(aq) + 6\,H^+(aq) + 6\,e^- \longrightarrow$$
$$2\,CrO_4^{2-}(aq) + 3\,Cl^-(aq) + 3\,H_2O(l) + 8\,H^+(aq) + 6\,e^-$$

Now, cancel the species that appear on both sides of the equation:

$$2\,Cr(OH)_4^-(aq) + 3\,ClO^-(aq) \longrightarrow 2\,CrO_4^{2-}(aq) + 3\,Cl^-(aq) + 3\,H_2O(l) + 2\,H^+(aq)$$

Finally, since we know that the reaction takes place in basic solution, we must add 2 OH^- ions to both sides of the equation to neutralize the 2 H^+ ions on the right. The final net ionic equation, balanced for both atoms and charge, is

$$2\,Cr(OH)_4^-(aq) + 3\,ClO^-(aq) + 2\,OH^-(aq) \longrightarrow 2\,CrO_4^{2-}(aq) + 3\,Cl^-(aq) + 5\,H_2O(l)$$

Charge: $(2 \times -1) + (3 \times -1) + (2 \times -1) = -7$ Charge: $(2 \times -2) + (3 \times -1) = -7$

▶ **PROBLEM 4.18** Write unbalanced half-reactions for the following net ionic equations:

(a) $MnO_4^-(aq) + IO_3^-(aq) \rightarrow MnO_2(s) + IO_4^-(aq)$

(b) $NO_3^-(aq) + SO_2(aq) \rightarrow SO_4^{2-}(aq) + NO_2(g)$

▶ **PROBLEM 4.19** Balance the following net ionic equation by the half-reaction method. The reaction takes place in acidic solution.

$$NO_3^-(aq) + Cu(s) \longrightarrow NO(g) + Cu^{2+}(aq)$$

▶ **PROBLEM 4.20** Balance the following equation by the half-reaction method. The reaction takes place in basic solution.

$$Fe(OH)_2(s) + O_2(g) \longrightarrow Fe(OH)_3(s)$$

4.11 Redox Titrations

We saw in Section 3.10 that the concentration of an acid or base solution can be determined by *titration*. A measured volume of acid (or base) solution of unknown concentration is placed in a flask, and a base (or acid) solution of known concentration is slowly added from a buret. By measuring the volume of the added solution needed for a complete reaction, as signaled by an indicator, the unknown concentration can be calculated.

A similar procedure can be used to determine the concentration of many oxidizing or reducing agents using a **redox titration**. All that's necessary is that the substance whose concentration you want to determine undergo an oxidation or reduction reaction in 100% yield and that there be a color change to signal when the reaction is complete. The color change might be due to one of the substances undergoing reaction or to some added redox indicator. Let's imagine, for instance, that we have a potassium permanganate solution whose concentration we want to find. Aqueous $KMnO_4$ reacts with oxalic acid, $H_2C_2O_4$, in acidic solution according to the net ionic equation

$$5\,H_2C_2O_4(aq) + 2\,MnO_4^-(aq) + 6\,H^+(aq) \longrightarrow 10\,CO_2(g) + 2\,Mn^{2+}(aq) + 8\,H_2O(l)$$

The reaction takes place in 100% yield and is accompanied by a sharp color change when the intense purple color of the MnO_4^- ion disappears.

The strategy used for this and other redox titrations is outlined in Figure 4.4. As with acid–base titrations, the general idea is to measure a known amount of one substance—in this case, $H_2C_2O_4$—and use mole ratios from the balanced equation to find the number of moles of the second substance—in this case, $KMnO_4$—necessary for complete reaction. With the molar amount of $KMnO_4$ thus known, titration gives the volume of solution containing that amount. Dividing the number of moles by the volume gives the concentration.

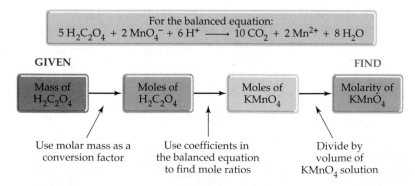

FIGURE 4.4 A flow diagram for a redox titration, summarizing the calculations needed to determine the concentration of a $KMnO_4$ solution by titration of a known mass of $H_2C_2O_4$.

As an example of how this redox titration works, let's carefully weigh an amount of $H_2C_2O_4$—say, 0.2585 g—and dissolve it in approximately 100 mL of 0.5 M H_2SO_4. The exact volume isn't important because we are concerned only with the *amount* of dissolved $H_2C_2O_4$, not with its concentration. Next, we place an aqueous $KMnO_4$ solution of unknown concentration in a buret and slowly add it to the $H_2C_2O_4$ solution. The purple color of MnO_4^- initially disappears as reaction occurs, but we continue the addition until a faint color persists, indicating that all the $H_2C_2O_4$ has reacted and that MnO_4^- ion is no longer being reduced. At this point, we might find that 22.35 mL of the $KMnO_4$ solution has been added (Figure 4.5).

(a) (b) (c)

FIGURE 4.5 The redox titration of oxalic acid, $H_2C_2O_4$, with $KMnO_4$. **(a)** A precise amount of oxalic acid is weighed and dissolved in aqueous sulfuric acid. **(b)** Aqueous $KMnO_4$ of unknown concentration is then added from a buret until **(c)** the purple color persists, indicating that all the $H_2C_2O_4$ has been oxidized.

To calculate the molarity of the $KMnO_4$ solution, we need to find the number of moles of $KMnO_4$ present in the 22.35 mL of solution used for titration. We do this by first calculating the number of moles of oxalic acid that react with the permanganate ion, using a gram-to-mole conversion with the molar mass of $H_2C_2O_4$ as the conversion factor:

$$\text{Moles of } H_2C_2O_4 = 0.2585 \text{ g } H_2C_2O_4 \times \frac{1 \text{ mol } H_2C_2O_4}{90.04 \text{ g } H_2C_2O_4}$$

$$= 2.871 \times 10^{-3} \text{ mol } H_2C_2O_4$$

According to the balanced equation, 5 mol of oxalic acid react with 2 mol of permanganate ion. Thus, we can calculate the number of moles of $KMnO_4$ that react with 2.871×10^{-3} mol of $H_2C_2O_4$:

$$\text{Moles of } KMnO_4 = 2.871 \times 10^{-3} \text{ mol } H_2C_2O_4 \times \frac{2 \text{ mol } KMnO_4}{5 \text{ mol } H_2C_2O_4}$$

$$= 1.148 \times 10^{-3} \text{ mol } KMnO_4$$

Knowing both the number of moles of $KMnO_4$ that react (1.148×10^{-3} mol) and the volume of the $KMnO_4$ solution (22.35 mL) then makes it possible to calculate the molarity:

$$\text{Molarity} = \frac{1.148 \times 10^{-3} \text{ mol } KMnO_4}{22.35 \text{ mL}} \times \frac{1000 \text{ mL}}{1 \text{ L}} = 0.051\ 36 \text{ M}$$

The molarity of the $KMnO_4$ solution as determined by redox titration is 0.051 36 M.

EXAMPLE 4.12 The concentration of an aqueous I_3^- solution can be determined by titration with aqueous sodium thiosulfate, $Na_2S_2O_3$, in the presence of a starch indicator, which turns from deep blue to colorless when all the I_3^- has reacted. What is the molar concentration of I_3^- if 24.55 mL of 0.102 M $Na_2S_2O_3$ is needed for complete reaction with 10.00 mL of the I_3^- solution? The net ionic equation is

$$2\,S_2O_3^{2-}(aq) + I_3^-(aq) \longrightarrow S_4O_6^{2-}(aq) + 3\,I^-(aq)$$

BALLPARK SOLUTION The balanced equation says that the amount of $S_2O_3^{2-}$ needed for the reaction (2 mol) is twice the amount of I_3^- (1 mol). The titration results say that the volume of the $S_2O_3^{2-}$ solution (24.55 mL) is a little over twice the volume of the I_3^- solution (10.00 mL). Thus, the concentrations of the two solutions must be about the same—approximately 0.1 M.

DETAILED SOLUTION We first need to find the number of moles of thiosulfate ion used for the titration:

$$24.55 \text{ mL} \times \frac{1 \text{ L}}{1000 \text{ mL}} \times \frac{0.102 \text{ mol } S_2O_3^{2-}}{1 \text{ L}} = 2.50 \times 10^{-3} \text{ mol } S_2O_3^{2-}$$

According to the balanced equation, 2 mol of $S_2O_3^{2-}$ ion react with 1 mol of I_3^- ion. Thus, we can find the number of moles of I_3^- ion:

$$2.50 \times 10^{-3} \text{ mol } S_2O_3^{2-} \times \frac{1 \text{ mol } I_3^-}{2 \text{ mol } S_2O_3^{2-}} = 1.25 \times 10^{-3} \text{ mol } I_3^-$$

Knowing both the number of moles (1.25×10^{-3} mol) and the volume (10.00 mL) then lets us calculate molarity:

$$\frac{1.25 \times 10 \text{ mol } I_3^-}{10.00 \text{ mL}} \times \frac{10^3 \text{ mL}}{1 \text{ L}} = 0.125 \text{ M}$$

The molarity of the I_3^- solution is 0.125 M.

The red I_3^- solution turns a deep blue color when it is added to a solution containing a small amount of starch.

▶ **PROBLEM 4.21** The concentration of Fe^{2+} ion in aqueous solution can be determined by redox titration with bromate ion, BrO_3^-, according to the net ionic equation

$$6\,Fe^{2+}(aq) + BrO_3^-(aq) + 6\,H^+(aq) \longrightarrow 6\,Fe^{3+}(aq) + Br^-(aq) + 3\,H_2O(l)$$

What is the molar concentration of Fe^{2+} if 31.50 mL of 0.105 M $KBrO_3$ is required for complete reaction with 10.00 mL of the Fe^{2+} solution?

4.12 Some Applications of Redox Reactions

Redox reactions involve almost every element in the periodic table and occur in a vast number of processes throughout nature, biology, and industry. Here are just a few examples:

- **Combustion** is the burning of a fuel by oxidation with oxygen in air. Gasoline, fuel oil, natural gas, wood, paper, and other organic substances of carbon and hydrogen are the most common fuels. Even some metals, such as magnesium and calcium, will burn in air.

$$CH_4(g) + 2\,O_2(g) \longrightarrow CO_2(g) + 2\,H_2O(l)$$

Methane
(natural gas)

- **Bleaching** makes use of redox reactions to decolorize or lighten colored materials. Dark hair is bleached to turn it blond, clothes are bleached to remove stains, wood pulp is bleached to make white paper, and so on. The exact oxidizing agent used depends on the situation: hydrogen peroxide (H_2O_2) is used for hair, sodium hypochlorite (NaOCl) is used for clothes, and elemental chlorine is used for wood pulp, but the principle is always the same. In all cases, colored impurities are destroyed by reaction with a strong oxidizing agent.

- **Batteries**, although they come in many types and sizes, are all based on redox reactions. In a typical redox reaction carried out in the laboratory—say, the reaction of zinc metal with Ag^+ to yield Zn^{2+} and silver metal—the reactants are simply mixed in a flask and electrons are transferred by direct contact between the reactants. In a battery, however, the two reactants are kept in separate compartments and the electrons are transferred through a wire running between them.

   As we'll see in Section 18.9, the common household battery used for flashlights and radios uses a can of zinc metal as one reactant and a paste of solid manganese dioxide as the other. A graphite rod sticks into the MnO_2 paste to provide electrical contact, and a moist paste of ammonium chloride separates the two reactants. When the zinc can and the graphite rod are connected by a wire, zinc sends electrons flowing through the wire toward the MnO_2 in a redox reaction. The resultant electrical current can be used to light a bulb or power a radio. The reaction is

$$Zn(s) + 2\,MnO_2(s) + 2\,NH_4Cl(s) \longrightarrow ZnCl_2(aq) + Mn_2O_3(s) + 2\,NH_3(aq) + H_2O(l)$$

- **Metallurgy**, the science of extracting and purifying metals from their ores, makes use of numerous redox processes. We'll see in Section 21.2, for example, that metallic zinc is prepared by reduction of ZnO with coke, a form of carbon:

$$ZnO(s) + C(s) \longrightarrow Zn(s) + CO(g)$$

- **Corrosion** is the deterioration of a metal by oxidation, such as the rusting of iron in moist air. The economic consequences of rusting are enormous: It has been estimated that up to one-fourth of the iron produced in the United States is used to replace bridges, buildings, and other structures that have been destroyed by corrosion. (The raised dot in the formula $Fe_2O_3 \cdot H_2O$ for rust indicates that one water molecule is associated with each Fe_2O_3 in an undefined way.)

$$4\,Fe(s) + 3\,O_2(g) \xrightarrow{\ H_2O\ } 2\,Fe_2O_3 \cdot H_2O(s)$$
$$\text{Rust}$$

- **Respiration** is the process of breathing and using oxygen for the many biological redox reactions that provide the energy needed by living organisms. We'll see in Chapter 24 that energy is released from food molecules slowly and in complex, multistep pathways, but the overall result of respiration is similar to that of a combustion reaction. For example, the simple sugar glucose ($C_6H_{12}O_6$) reacts with O_2 to give CO_2 and H_2O according to the following equation:

$$C_6H_{12}O_6 + 6\,O_2 \longrightarrow 6\,CO_2 + 6\,H_2O + \text{energy}$$
Glucose
(a carbohydrate)

Photography: A Series of Redox Reactions

Photography is so common that most people never give a moment's thought to how remarkable the process is. Ordinary black-and-white photographic film consists of a celluloid strip that has been coated with a gelatin emulsion containing very tiny crystals, or "grains," of a silver halide, usually AgBr. Making the film involves a considerable amount of art as well as science, and the recipes used by major manufacturers are well-guarded secrets. When exposed to light, the surfaces of the AgBr grains turn dark because of a light-induced redox reaction in which Br^- transfers an electron to Ag^+, producing atoms of elemental silver and Br_2, which reacts with the gelatin emulsion (Figure 4.6). Those areas of the film exposed to the brightest light have the largest number of silver atoms, and those areas exposed to the least light have the smallest number.

$$2 \, AgBr \xrightarrow{\text{Light}} 2 \, Ag + Br_2$$

Perhaps surprisingly in view of what a finished photograph looks like, only a few hundred out of many trillions of Ag^+ ions in each grain are reduced to Ag atoms, and the latent image produced on the film is still invisible at this point. The key to silver halide photography is the *developing* process, in which the latent image is amplified.

By mechanisms still not understood in detail, the presence of a relatively tiny number of Ag atoms on the surface of an AgBr grain sensitizes the remaining Ag^+ ions in the grain toward further reduction when the film is exposed to the organic reducing agent hydroquinone. Those grains that have been exposed to the strongest light—and thus have more Ag atoms—reduce and darken quickly, while those grains with fewer Ag atoms reduce and darken more slowly. By carefully monitoring the amount of time allowed for reduction of the AgBr grains with hydroquinone, it's possible to amplify the latent image on the exposed film and make it visible.

(a)	(b)	(c)

FIGURE 4.6 **(a)** When the paper clip sitting on powdered AgCl crystals is exposed to a bright light, **(b)** the crystals around the clip darken to provide an image of the clip **(c)**.

Once the image is fully formed, the film is *fixed* by washing away the remaining unreduced AgBr so that the film is no longer sensitive to light. Although pure AgBr is insoluble in water, it is made soluble by reaction with a solution of sodium thiosulfate, $Na_2S_2O_3$, called *hypo* by photographers.

$$AgBr(s) + 2\,S_2O_3^{2-}(aq) \longrightarrow Ag(S_2O_3)_2^{3-}(aq) + Br^-(aq)$$

At this point, the film contains a negative image formed by a layer of black, finely divided silver metal, a layer that is denser and darker in those areas exposed to the most light but lighter in those areas exposed to the least light. To convert this negative image into the final printed photograph, the entire photographic procedure is repeated a second time. Light is passed through the negative image onto special photographic paper that is coated with the same kind of gelatin–AgBr emulsion used on the original film. Developing the photographic paper with hydroquinone and fixing the image with sodium thiosulfate reverses the negative image, and a final, positive image is produced. The whole process from film to print is carried out billions of times and consumes over 3 million pounds of silver each year.

▶**PROBLEM 4.22** What is the purpose of $Na_2S_2O_3$, or *hypo*, in the photographic process?

▶**PROBLEM 4.23** The image produced on photographic film is a *negative* one, with dark objects appearing light, and light objects appearing dark. Explain how this negative image is converted to a positive image when printed.

Key Words

Summary

Many reactions, particularly those that involve ionic compounds, take place in **aqueous solution**. Substances whose aqueous solutions contain ions and therefore conduct electricity are called **electrolytes**. Ionic compounds, such as NaCl, and molecular compounds that **dissociate** substantially into ions when dissolved in water are **strong electrolytes**. Substances that dissociate to only a small extent are **weak electrolytes**, and substances that do not produce ions in aqueous solution are **nonelectrolytes**. Acids dissociate in aqueous solutions to yield an anion and a **hydronium ion, H_3O^+**. Those acids that dissociate to a large extent are **strong acids**; those acids that dissociate to a small extent are **weak acids**.

There are three important classes of aqueous reactions. **Precipitation reactions** occur when solutions of two ionic substances are mixed and a precipitate falls from solution. To predict whether a precipitate will form, you must know the **solubilities** of the potential products. **Acid–base neutralization reactions** occur when an acid is mixed with a base, yielding water and a **salt**. The neutralization of a strong acid with a strong base can be written as a **net ionic equation**, in which nonparticipating, **spectator ions** are not specified:

$$H^+(aq) + OH^-(aq) \longrightarrow H_2O(l)$$

Oxidation–reduction reactions, or **redox reactions**, are processes in which one or more electrons are transferred between reaction partners. An **oxidation** is the loss of one or more electrons; a **reduction** is the gain of one or more electrons. Redox reactions can be identified by assigning to each atom in a substance an **oxidation number**, which provides a measure of whether the atom is neutral, electron-rich, or electron-poor. Comparing the oxidation numbers of an atom before and after reaction shows whether the atom has gained or lost electrons.

Oxidations and reductions must occur together. Whenever one substance loses one or more electrons (is oxidized), another substance gains those electrons (is reduced). The substance that causes a reduction by giving up electrons is called a **reducing agent**. The substance that causes an oxidation by accepting electrons is called an **oxidizing agent**. The reducing agent is itself oxidized when it gives up electrons, and the oxidizing agent is itself reduced when it accepts electrons.

Among the simplest of redox processes is the reaction of an aqueous cation, usually a metal ion, with a free element to give a different ion and a different free element. Noting the results from a succession of different reactions makes it possible to organize an **activity series**, which ranks the elements in order of their reducing ability in aqueous solution.

Redox reactions can be balanced using either the **oxidation-number method** or the **half-reaction method**. The concentration of an oxidizing agent or a reducing agent in solution can be determined by a **redox titration**.

Key Concept Summary

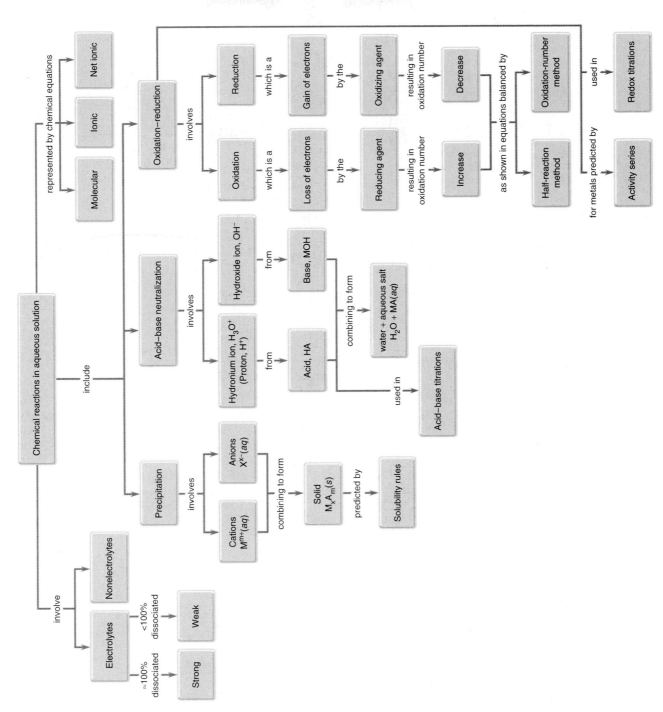

Understanding Key Concepts

Problems 4.1–4.23 appear within the chapter.

4.24 Assume that an aqueous solution of a cation, represented as a red sphere, is allowed to mix with a solution of an anion, represented as a yellow sphere. Three possible outcomes are represented by boxes (1)–(3):

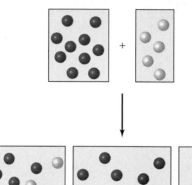

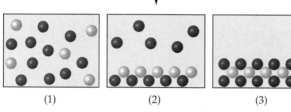

(1)   (2)   (3)

Which outcome corresponds to each of the following reactions?

(a) $2 Na^+(aq) + CO_3^{2-}(aq) \rightarrow$

(b) $Ba^{2+}(aq) + CrO_4^{2-}(aq) \rightarrow$

(c) $2 Ag^+(aq) + SO_3^{2-}(aq) \rightarrow$

4.25 Assume that an aqueous solution of a cation, represented as a blue sphere, is allowed to mix with a solution of an anion, represented as a green sphere, and that the following result is obtained:

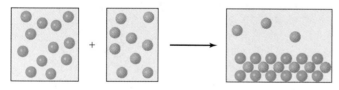

Which combinations of cation and anion, chosen from the following lists, are compatible with the observed results? Explain.

   Cations: Na^+, Ca^{2+}, Ag^+, Ni^{2+}

   Anions: Cl^-, CO_3^{2-}, CrO_4^{2-}, NO_3^-

4.26 Assume that an aqueous solution of OH^-, represented as a blue sphere, is allowed to mix with a solution of an acid H_nA, represented as a red sphere. Three possible outcomes are depicted by boxes (1)–(3), where the green spheres represent A^{n-}, the anion of the acid:

Which outcome corresponds to each of the following reactions?

(a) $HF + OH^- \rightarrow H_2O + F^-$

(b) $H_2SO_3 + 2 OH^- \rightarrow 2 H_2O + SO_3^{2-}$

(c) $H_3PO_4 + 3 OH^- \rightarrow 3 H_2O + PO_4^{3-}$

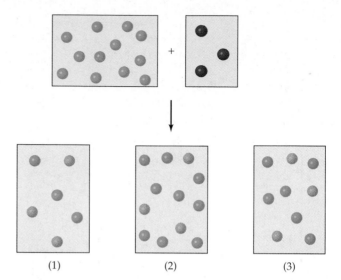

(1)   (2)   (3)

4.27 The concentration of an aqueous solution of $NaOCl$ (sodium hypochlorite; the active ingredient in household bleach) can be determined by a redox titration with iodide ion in acidic solution:

$$OCl^-(aq) + 2 I^-(aq) + 2 H^+(aq) \longrightarrow$$
$$Cl^-(aq) + I_2(aq) + H_2O(l)$$

Assume that the blue spheres in the buret represent I^- ions, the red spheres in the flask represent OCl^- ions, the concentration of the I^- ions in the buret is 0.120 M, and the volumes in the buret and the flask are identical. What is the concentration of $NaOCl$ in the flask? What percentage of the I^- solution in the buret must be added to the flask to react with all the OCl^- ions?

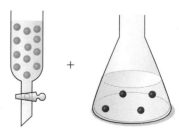

4.28 Use the following reactions to arrange the elements A, B, C, and D in order of their decreasing ability as reducing agents:

$C + B^+ \rightarrow C^+ + B$ $A^+ + D \rightarrow$ No reaction

$C^+ + A \rightarrow$ No reaction $D + B^+ \rightarrow D^+ + B$

4.29 Which of the following reactions would you expect to occur according to the activity series you established in Problem 4.28?

(a) $A^+ + C \rightarrow A + C^+$ **(b)** $A^+ + B \rightarrow A + B^+$

Additional Problems

Aqueous Reactions and Net Ionic Equations

4.30 Classify each of the following reactions as a precipitation, acid–base neutralization, or oxidation–reduction:

(a) $Hg(NO_3)_2(aq) + 2\,NaI(aq) \rightarrow$
$$2\,NaNO_3(aq) + HgI_2(s)$$

(b) $2\,HgO(s) \xrightarrow{\text{Heat}} 2\,Hg(l) + O_2(g)$

(c) $H_3PO_4(aq) + 3\,KOH(aq) \rightarrow K_3PO_4(aq) + 3\,H_2O(l)$

4.31 Classify each of the following reactions as a precipitation, acid–base neutralization, or oxidation–reduction:

(a) $S_8(s) + 8\,O_2(g) \rightarrow 8\,SO_2(g)$

(b) $NiCl_2(aq) + Na_2S(aq) \rightarrow NiS(s) + 2\,NaCl(aq)$

(c) $2\,CH_3CO_2H(aq) + Ba(OH)_2(aq) \rightarrow$
$$(CH_3CO_2)_2Ba(aq) + 2\,H_2O(l)$$

4.32 Write net ionic equations for the reactions listed in Problem 4.30.

4.33 Write net ionic equations for the reactions listed in Problem 4.31.

4.34 Individual solutions of $Ba(OH)_2$ and H_2SO_4 both conduct electricity, but the conductivity disappears when equal molar amounts of the solutions are mixed. Explain.

4.35 A solution of HCl in water conducts electricity but a solution of HCl in chloroform, $CHCl_3$, does not. What does this observation tell you about the way HCl exists in water and the way it exists in chloroform?

4.36 Classify each of the following substances as either a strong or weak electrolyte:

(a) HBr

(b) HF

(c) $NaClO_4$

(d) $(NH_4)_2CO_3$

(e) NH_3

4.37 Is it possible for a molecular substance to be a strong electrolyte? Explain.

4.38 What is the total molar concentration of ions in each of the following solutions, assuming complete dissociation?

(a) A 0.750 M solution of K_2CO_3

(b) A 0.355 M solution of $AlCl_3$

4.39 What is the total molar concentration of ions in each of the following solutions, assuming complete dissociation?

(a) A 1.250 M solution of CH_3OH

(b) A 0.225 M solution of $HClO_4$

Precipitation Reactions and Solubility Rules

4.40 Which of the following substances are likely to be soluble in water?

(a) Ag_2O (b) $Ba(NO_3)_2$ (c) $SnCO_3$ (d) Fe_2O_3

4.41 Which of the following substances are likely to be soluble in water?

(a) ZnS (b) $Au_2(CO_3)_3$ (c) $PbCl_2$ (d) MnO_2

4.42 Predict whether a precipitation reaction will occur when aqueous solutions of the following substances are mixed:

(a) $NaOH + HClO_4$ (b) $FeCl_2 + KOH$

(c) $(NH_4)_2SO_4 + NiCl_2$

4.43 Predict whether a precipitation reaction will occur when aqueous solutions of the following substances are mixed:

(a) $MnCl_2 + Na_2S$ (b) $HNO_3 + CuSO_4$

(c) $Hg(NO_3)_2 + Na_3PO_4$

4.44 How would you prepare the following substances by a precipitation reaction?

(a) $PbSO_4$ (b) $Mg_3(PO_4)_2$ (c) $ZnCrO_4$

4.45 How would you prepare the following substances by a precipitation reaction?

(a) $Al(OH)_3$ (b) FeS (c) $CoCO_3$

4.46 Assume that you have an aqueous mixture of $NaNO_3$ and $AgNO_3$. How could you use a precipitation reaction to separate the two metal ions?

4.47 Assume that you have an aqueous mixture of $BaCl_2$ and $CuCl_2$. How could you use a precipitation reaction to separate the two metal ions?

4.48 Assume that you have an aqueous solution of an unknown salt. Treatment of the solution with dilute $NaOH$, Na_2SO_4, and KCl produces no precipitate. Which of the following cations might the solution contain?

(a) Ag^+ (b) Cs^+ (c) Ba^{2+} (d) NH_4^+

4.49 Assume that you have an aqueous solution of an unknown salt. Treatment of the solution with dilute $BaCl_2$, $AgNO_3$, and $Cu(NO_3)_2$ produces no precipitate. Which of the following anions might the solution contain?

(a) Cl^- (b) NO_3^- (c) OH^- (d) SO_4^{2-}

Acids, Bases, and Neutralization Reactions

4.50 Assume that you are given a solution of an unknown acid or base. How can you tell whether the unknown substance is acidic or basic?

4.51 Why do we use a double arrow ($\rightleftharpoons$) to show the dissociation of a weak acid or weak base in aqueous solution?

4.52 Write balanced ionic equations for the following reactions:

(a) Aqueous perchloric acid is neutralized by aqueous calcium hydroxide.

(b) Aqueous sodium hydroxide is neutralized by aqueous acetic acid.

4.53 Write balanced ionic equations for the following reactions:

(a) Aqueous hydrofluoric acid is neutralized by aqueous calcium hydroxide.

(b) Aqueous magnesium hydroxide is neutralized by aqueous nitric acid.

4.54 Write balanced net ionic equations for the following reactions:

(a) $LiOH(aq) + HI(aq) \rightarrow$

(b) $HBr(aq) + Ca(OH)_2(aq) \rightarrow$

4.55 Write balanced net ionic equations for the following reactions: (Note: $HClO_3$ is a strong acid.)

(a) $Fe(OH)_3(s) + H_2SO_4(aq) \rightarrow$

(b) $HClO_3(aq) + NaOH(aq) \rightarrow$

Redox Reactions and Oxidation Numbers

4.56 Where in the periodic table are the best reducing agents found? The best oxidizing agents?

4.57 Where in the periodic table are the most easily reduced elements found? The most easily oxidized?

4.58 Tell in each of the following instances whether the substance gains electrons or loses electrons in a redox reaction:

(a) An oxidizing agent

(b) A reducing agent

(c) A substance undergoing oxidation

(d) A substance undergoing reduction

4.59 Tell for each of the following substances whether the oxidation number increases or decreases in a redox reaction:

(a) An oxidizing agent

(b) A reducing agent

(c) A substance undergoing oxidation

(d) A substance undergoing reduction

4.60 Assign oxidation numbers to each element in the following compounds:

(a) NO_2　　　(b) SO_3　　　(c) $COCl_2$

(d) CH_2Cl_2　　(e) $KClO_3$　　(f) HNO_3

4.61 Assign oxidation numbers to each element in the following compounds:

(a) $VOCl_3$　　(b) $CuSO_4$　　(c) CH_2O

(d) Mn_2O_7　　(e) OsO_4　　(f) H_2PtCl_6

4.62 Assign oxidation numbers to each element in the following ions:

(a) ClO_3^-　　(b) SO_3^{2-}　　(c) $C_2O_4^{2-}$

(d) NO_2^-　　(e) BrO^-

4.63 Assign oxidation numbers to each element in the following ions:

(a) $Cr(OH)_4^-$　　(b) $S_2O_3^{2-}$　　(c) NO_3^-

(d) MnO_4^{2-}　　(e) HPO_4^{2-}

4.64 Which element is oxidized and which is reduced in each of the following reactions?

(a) $Ca(s) + Sn^{2+}(aq) \rightarrow Ca^{2+}(aq) + Sn(s)$

(b) $ICl(s) + H_2O(l) \rightarrow HCl(aq) + HOI(aq)$

4.65 Which element is oxidized and which is reduced in each of the following reactions?

(a) $Si(s) + 2\,Cl_2(g) \rightarrow SiCl_4(l)$

(b) $Cl_2(g) + 2\,NaBr(aq) \rightarrow Br_2(aq) + 2\,NaCl(aq)$

4.66 Use the activity series of metals (Table 4.3) to predict the outcome of each of the following reactions. If no reaction occurs, write N.R.

(a) $Na^+(aq) + Zn(s) \rightarrow$　　(b) $HCl(aq) + Pt(s) \rightarrow$

(c) $Ag^+(aq) + Au(s) \rightarrow$　　(d) $Au^{3+}(aq) + Ag(s) \rightarrow$

4.67 Neither strontium (Sr) nor antimony (Sb) are shown in the activity series of Table 4.3. Based on their positions in the periodic table, which would you expect to be the better reducing agent? Will the following reaction occur? Explain.

$$2\,Sb^{3+}(aq) + 3\,Sr(s) \longrightarrow 2\,Sb(s) + 3\,Sr^{2+}(aq)$$

4.68 Use the following reactions to arrange the elements A, B, C, and D in order of their redox activity:

$$A + B^+ \rightarrow A^+ + B \qquad C^+ + D \rightarrow No\ reaction$$
$$B + D^+ \rightarrow B^+ + D \qquad B + C^+ \rightarrow B^+ + C$$

4.69 Use the following reactions to arrange the elements A, B, C, and D in order of their redox activity:

$$2\,A + B^{2+} \rightarrow 2\,A^+ + B \qquad B + D^{2+} \rightarrow B^{2+} + D$$
$$A^+ + C \rightarrow No\ reaction \qquad 2\,C + B^{2+} \rightarrow 2\,C^+ + B$$

4.70 Tell which of the following reactions you would expect to occur according to the activity series you established in Problem 4.68:

(a) $A^+ + C \rightarrow A + C^+$ (b) $A^+ + D \rightarrow A + D^+$

4.71 Tell which of the following reactions you would expect to occur according to the activity series you established in Problem 4.69:

(a) $2\,A^+ + D \rightarrow 2\,A + D^{2+}$ (b) $D^{2+} + 2\,C \rightarrow D + 2\,C^+$

Balancing Redox Reactions

4.72 Classify each of the following unbalanced half-reactions as either an oxidation or a reduction:

(a) $NO_3^-(aq) \rightarrow NO(g)$ (b) $Zn(s) \rightarrow Zn^{2+}(aq)$

(c) $Ti^{3+}(aq) \rightarrow TiO_2(s)$ (d) $Sn^{4+}(aq) \rightarrow Sn^{2+}(aq)$

4.73 Classify each of the following unbalanced half-reactions as either an oxidation or a reduction:

(a) $O_2(g) \rightarrow OH^-(aq)$

(b) $H_2O_2(aq) \rightarrow O_2(g)$

(c) $MnO_4^-(aq) \rightarrow MnO_4^{2-}(aq)$

(d) $CH_3OH(aq) \rightarrow CH_2O(aq)$

4.74 Balance the half-reactions in Problem 4.72, assuming that they occur in acidic solution.

4.75 Balance the half-reactions in Problem 4.73, assuming that they occur in basic solution.

4.76 Write unbalanced oxidation and reduction half-reactions for the following processes:

(a) $Te(s) + NO_3^-(aq) \rightarrow TeO_2(s) + NO(g)$

(b) $H_2O_2(aq) + Fe^{2+}(aq) \rightarrow Fe^{3+}(aq) + H_2O(l)$

4.77 Write unbalanced oxidation and reduction half-reactions for the following processes:

(a) $Mn(s) + NO_3^-(aq) \rightarrow Mn^{2+}(aq) + NO_2(g)$

(b) $Mn^{3+}(aq) \rightarrow MnO_2(s) + Mn^{2+}(aq)$

4.78 Balance the following half-reactions:

(a) (acidic) $Cr_2O_7^{2-}(aq) \rightarrow Cr^{3+}(aq)$

(b) (basic) $CrO_4^{2-}(aq) \rightarrow Cr(OH)_4^-(aq)$

(c) (basic) $Bi^{3+}(aq) \rightarrow BiO_3^-(aq)$

(d) (basic) $ClO^-(aq) \rightarrow Cl^-(aq)$

4.79 Balance the following half-reactions:

(a) (acidic) $VO^{2+}(aq) \rightarrow V^{3+}(aq)$

(b) (basic) $Ni(OH)_2(s) \rightarrow Ni_2O_3(s)$

(c) (acidic) $NO_3^-(aq) \rightarrow NO_2(g)$

(d) (basic) $Br_2(aq) \rightarrow BrO_3^-(aq)$

4.80 Write balanced net ionic equations for the following reactions in basic solution:

(a) $MnO_4^-(aq) + IO_3^-(aq) \rightarrow MnO_2(s) + IO_4^-(aq)$

(b) $Cu(OH)_2(s) + N_2H_4(aq) \rightarrow Cu(s) + N_2(g)$

(c) $Fe(OH)_2(s) + CrO_4^{2-}(aq) \rightarrow$
$$Fe(OH)_3(s) + Cr(OH)_4^-(aq)$$

(d) $H_2O_2(aq) + ClO_4^-(aq) \rightarrow ClO_2^-(aq) + O_2(g)$

4.81 Write balanced net ionic equations for the following reactions in basic solution:

(a) $S_2O_3^{2-}(aq) + I_2(aq) \rightarrow S_4O_6^{2-}(aq) + I^-(aq)$

(b) $Mn^{2+}(aq) + H_2O_2(aq) \rightarrow MnO_2(s)$

(c) $Zn(s) + NO_3^-(aq) \rightarrow NH_3(aq) + Zn(OH)_4^{2-}(aq)$

(d) $Bi(OH)_3(s) + Sn(OH)_3^-(aq) \rightarrow$
$$Bi(s) + Sn(OH)_6^{2-}(aq)$$

4.82 Write balanced net ionic equations for the following reactions in acidic solution:

(a) $Zn(s) + VO^{2+}(aq) \rightarrow Zn^{2+}(aq) + V^{3+}(aq)$

(b) $Ag(s) + NO_3^-(aq) \rightarrow Ag^+(aq) + NO_2(g)$

(c) $Mg(s) + VO_4^{3-}(aq) \rightarrow Mg^{2+}(aq) + V^{2+}(aq)$

(d) $I^-(aq) + IO_3^-(aq) \rightarrow I_3^-(aq)$

4.83 Write balanced net ionic equations for the following reactions in acidic solution:

(a) $MnO_4^-(aq) + C_2H_5OH(aq) \rightarrow$
$$Mn^{2+}(aq) + CH_3CO_2H(aq)$$

(b) $H_2O_2(aq) + Cr_2O_7^{2-}(aq) \rightarrow O_2(g) + Cr^{3+}(aq)$

(c) $Sn^{2+}(aq) + IO_4^-(aq) \rightarrow Sn^{4+}(aq) + I^-(aq)$

(d) $PbO_2(s) + Cl^-(aq) \rightarrow PbCl_2(s) + O_2(g)$

Redox Titrations

4.84 Iodine, I_2, reacts with aqueous thiosulfate ion in neutral solution according to the balanced equation

$$I_2(aq) + 2\,S_2O_3^{2-}(aq) \longrightarrow S_4O_6^{2-}(aq) + 2\,I^-(aq)$$

How many grams of I_2 are present in a solution if 35.20 mL of 0.150 M $Na_2S_2O_3$ solution is needed to titrate the I_2 solution?

4.85 How many milliliters of 0.250 M $Na_2S_2O_3$ solution are needed to titrate 2.486 g of I_2 according to the equation in Problem 4.84?

4.86 Titration with solutions of potassium bromate, $KBrO_3$, can be used to determine the concentration of As(III). What is the molar concentration of As(III) in a solution if 22.35 mL of 0.100 M $KBrO_3$ is needed to titrate 50.00 mL of the As(III) solution? The balanced equation is

$$3\,H_3AsO_3(aq) + BrO_3^-(aq) \longrightarrow Br^-(aq) + 3\,H_3AsO_4(aq)$$

4.87 Standardized solutions of $KBrO_3$ are frequently used in redox titrations. The necessary solution can be made by dissolving $KBrO_3$ in water and then titrating it with an As(III) solution. What is the molar concentration of a $KBrO_3$ solution if 28.55 mL of the solution is needed to titrate 1.550 g of As_2O_3? See Problem 4.86 for the balanced equation (As_2O_3 dissolves in aqueous acid solution to yield H_3AsO_3:
$$As_2O_3 + 3 H_2O \rightarrow 2 H_3AsO_3).$$

4.88 The metal content of iron in ores can be determined by a redox procedure in which the sample is first oxidized with Br_2 to convert all the iron to Fe^{3+} and then titrated with Sn^{2+} to reduce the Fe^{3+} to Fe^{2+}. The balanced equation is

$$2 Fe^{3+}(aq) + Sn^{2+}(aq) \rightarrow 2 Fe^{2+}(aq) + Sn^{4+}(aq)$$

What is the mass percent Fe in a 0.1875 g sample if 13.28 mL of a 0.1015 M Sn^{2+} solution is needed to titrate the Fe^{3+}?

4.89 The Sn^{2+} solution used in Problem 4.88 can be standardized by titrating it with a known amount of Fe^{3+}. What is the molar concentration of an Sn^{2+} solution if 23.84 mL is required to titrate 1.4855 g of Fe_2O_3?

4.90 Alcohol levels in blood can be determined by a redox titration with potassium dichromate according to the balanced equation

$$C_2H_5OH(aq) + 2 Cr_2O_7^{2-}(aq) + 16 H^+(aq) \longrightarrow$$
$$2 CO_2(g) + 4 Cr^{3+}(aq) + 11 H_2O(l)$$

What is the blood alcohol level in mass percent if 8.76 mL of 0.049 88 M $K_2Cr_2O_7$ is required for titration of a 10.002 g sample of blood?

4.91 Calcium levels in blood can be determined by adding oxalate ion to precipitate calcium oxalate, CaC_2O_4, followed by dissolving the precipitate in aqueous acid and titrating the oxalic acid ($H_2C_2O_4$) with $KMnO_4$:

$$5 H_2C_2O_4(aq) + 2 MnO_4^-(aq) + 6 H^+(aq) \longrightarrow$$
$$10 CO_2(g) + 2 Mn^{2+}(aq) + 8 H_2O(l)$$

How many milligrams of Ca^{2+} are present in 10.0 mL of blood if 21.08 mL of 0.000 988 M $KMnO_4$ solution is needed for the titration?

General Problems

4.92 Balance the equations for the following reactions in basic solution:

(a) $[Fe(CN)_6]^{3-}(aq) + N_2H_4(aq) \rightarrow$
$$[Fe(CN)_6]^{4-}(aq) + N_2(g)$$
(b) $SeO_3^{2-}(aq) + Cl_2(g) \rightarrow SeO_4^{2-}(aq) + Cl^-(aq)$
(c) $CoCl_2(aq) + HO_2^-(aq) \rightarrow Co(OH)_3(aq) + Cl^-(aq)$

4.93 An alternative procedure to that given in Problem 4.88 for determining the amount of iron in a sample is to convert the iron to Fe^{2+} and then titrate with a solution of $Ce(NH_4)_2(NO_3)_6$:

$$Fe^{2+}(aq) + Ce^{4+}(aq) \longrightarrow Fe^{3+}(aq) + Ce^{3+}(aq)$$

What is the mass percentage of iron in a sample if titration of 1.2284 g of the sample requires 57.91 mL of 0.1018 M $Ce(NH_4)_2(NO_3)_6$?

4.94 Assign oxidation numbers to each atom in the following substances:

(a) Ethane, C_2H_6, a constituent of natural gas
(b) Borax, $Na_2B_4O_7$, a mineral used in laundry detergents
(c) $Mg_2Si_2O_6$, a silicate mineral

4.95 Balance the equations for the following reactions in acidic solution:

(a) $PbO_2(s) + Mn^{2+}(aq) \rightarrow Pb^{2+}(aq) + MnO_4^-(aq)$
(b) $As_2O_3(s) + NO_3^-(aq) \rightarrow H_3AsO_4(aq) + HNO_2(aq)$
(c) $Br_2(aq) + SO_2(g) \rightarrow Br^-(aq) + HSO_4^-(aq)$
(d) $NO_2^-(aq) + I^-(aq) \rightarrow I_2(s) + NO(g)$

4.96 Which of the following ions can be reduced to its elemental form by reaction with iron?
(a) Ni^{2+} (b) Au^{3+} (c) Zn^{2+} (d) Ba^{2+}

4.97 The solubility of an ionic compound can be described quantitatively by a value called the *solubility product constant*, K_{sp}. For the general solubility process $A_aB_b \rightleftharpoons a A^{n+} + b B^{m-}$, $K_{sp} = [A^{n+}]^a [B^{m-}]^b$. The brackets refer to concentrations in moles per liter.

(a) Write the expression for the solubility product constant of Ag_2CrO_4.
(b) If $K_{sp} = 1.1 \times 10^{-12}$ for Ag_2CrO_4, what are the molar concentrations of Ag^+ and CrO_4^{2-} in a saturated solution?

4.98 Write the expression for the solubility product constant of MgF_2 (see Problem 4.97). If $[Mg^{2+}] = 2.6 \times 10^{-4}$ mol/L in a saturated solution, what is the value of K_{sp}?

4.99 Succinic acid, an intermediate in the metabolism of food molecules, has molecular mass = 118.1 amu. When 1.926 g of succinic acid was dissolved in water and titrated, 65.20 mL of 0.5000 M NaOH solution was required to neutralize the acid. How many acidic hydrogens are there in a molecule of succinic acid?

4.100 How could you use a precipitation reaction to separate each of the following pairs of cations? Write the formula for each reactant you would add, and write a balanced net ionic equation for each reaction.
(a) K^+ and Hg_2^{2+} (b) Pb^{2+} and Ni^{2+}
(c) Ca^{2+} and NH_4^+ (d) Fe^{2+} and Ba^{2+}

4.101 How could you use a precipitation reaction to separate each of the following pairs of anions? Write the formula for each reactant you would add, and write a balanced net ionic equation for each reaction.

(a) Cl^- and NO_3^- (b) S^{2-} and SO_4^{2-}

(c) SO_4^{2-} and CO_3^{2-} (d) OH^- and ClO_4^-

4.102 Write a balanced net ionic equation for each of the following reactions in either acidic or basic solution:

(a) $Mn(OH)_2(s) + H_2O_2(aq) \xrightarrow{\text{Base}} Mn(OH)_3(s)$

(b) $MnO_4^{2-}(aq) \xrightarrow{\text{Acid}} MnO_2(s) + MnO_4^-(aq)$

(c) $IO_3^-(aq) + I^-(aq) \xrightarrow{\text{Acid}} I_3^-(aq)$

(d) $P(s) + PO_4^{3-}(aq) \xrightarrow{\text{Base}} HPO_3^{2-}(aq)$

4.103 A 100.0 mL solution containing aqueous HCl and HBr was titrated with 0.1235 M NaOH. The volume of base required to neutralize the acid was 47.14 mL. Aqueous $AgNO_3$ was then added to precipitate the Cl^- and Br^- ions as AgCl and AgBr. The mass of the silver halides obtained was 0.9974 g. What are the molarities of the HCl and HBr in the original solution?

Multi-Concept Problems

4.104 To 100.0 mL of a solution that contains 0.120 M $Cr(NO_3)_2$ and 0.500 M HNO_3 is added 20.0 mL of 0.250 M $K_2Cr_2O_7$. The dichromate and chromium(II) ions react to give chromium(III) ions.

(a) Write a balanced net ionic equation for the reaction.

(b) Calculate the concentrations of all ions in the solution after reaction. Check your concentrations to make sure that the solution is electrically neutral.

4.105 Sodium nitrite, $NaNO_2$, is frequently added to processed meats as a preservative. The amount of nitrite ion in a sample can be determined by acidifying to form nitrous acid (HNO_2), letting the nitrous acid react with an excess of iodide ion, and then titrating the I_3^- ion that results with thiosulfate solution in the presence of a starch indicator. The unbalanced equations are

(1) $HNO_2 + I^- \rightarrow NO + I_3^-$ (in acidic solution)

(2) $I_3^- + S_2O_3^{2-} \rightarrow I^- + S_4O_6^{2-}$

(a) Balance the two redox equations.

(b) When a nitrite-containing sample with a mass of 2.935 g was analyzed, 18.77 mL of 0.1500 M $Na_2S_2O_3$ solution was needed for the titration. What is the mass percentage of NO_2^- ion in the sample?

4.106 Brass is an approximately 4:1 alloy of copper and zinc, along with small amounts of tin, lead, and iron. The mass percents of copper and zinc can be determined by a procedure that begins with dissolving the brass in hot nitric acid. The resulting solution of Cu^{2+} and Zn^{2+} ions is then treated with aqueous ammonia to lower its acidity, followed by addition of sodium thiocyanate (NaSCN) and sulfurous acid (H_2SO_3) to precipitate copper(I) thiocyanate (CuSCN). The solid CuSCN is collected, dissolved in aqueous acid, and treated with potassium iodate (KIO_3) to give iodine, which is then titrated with aqueous sodium thiosulfate ($Na_2S_2O_3$). The filtrate remaining after CuSCN has been removed is neutralized by addition of aqueous ammonia, and a solution of diammonium hydrogen phosphate [$(NH_4)_2HPO_4$] is added to yield a precipitate of zinc ammonium phosphate ($ZnNH_4PO_4$). Heating the precipitate to 900°C converts it to zinc pyrophosphate ($Zn_2P_2O_7$), which is weighed. The equations are

(1) $Cu(s) + NO_3^-(aq) \rightarrow Cu^{2+}(aq) + NO(g)$ (in acid)

(2) $Cu^{2+}(aq) + SCN^-(aq) + HSO_3^-(aq) \rightarrow$
$\qquad\qquad CuSCN(s) + HSO_4^-(aq)$ (in acid)

(3) $Cu^+(aq) + IO_3^-(aq) \rightarrow Cu^{2+}(aq) + I_2(aq)$ (in acid)

(4) $I_2(aq) + S_2O_3^{2-}(aq) \rightarrow I^-(aq) + S_4O_6^{2-}(aq)$ (in acid)

(5) $ZnNH_4PO_4 \rightarrow Zn_2P_2O_7 + H_2O + NH_3$

(a) Balance equations (1)–(5).

(b) When a brass sample with a mass of 0.544 g was subjected to the above analysis, 10.82 mL of 0.1220 M sodium thiosulfate was required for the iodine titration. What is the mass percent copper in the brass?

(c) The brass sample in (b) yielded 0.246 g of $Zn_2P_2O_7$. What is the mass percent zinc in the brass?

eMedia Problems

4.107 After watching the **Strong and Weak Electrolytes** movie (*eChapter 4.2*), draw two pictures (similar to those in Key Concept Problem 4.3, page 121) to represent the two solutions in the movie. Use different colored spheres to represent the hydronium, chloride, and acetate ions.

4.108 The **Dissolution of Mg(OH)₂** movie (*eChapter 4.2*) shows a mixture of magnesium hydroxide and water before and after the addition of hydrochloric acid.

    (a) Do you expect the mixture of magnesium hydroxide to conduct electricity prior to the addition of hydrochloric acid?

    (b) Do you expect the mixture to conduct electricity after the addition of hydrochloric acid?

In each case, explain your answer. Now watch the **Dissolution of Mg(OH)₂** movie (*eChapter 4.2*) that shows a mixture of magnesium hydroxide and water before and after the addition of hydrochloric acid. Which expectation was correct?

4.109 Use the **Ionic Compounds** activity (*eChapter 4.4*) to determine the formula of each of the insoluble iron(III) salts. Then, using your knowledge of the solubility rules, write a molecular, ionic, and net ionic equation for an aqueous reaction that would produce each salt.

4.110 Together, the **Oxidation–Reduction Reactions— Part I** and **Oxidation–Reduction Reactions—Part II** movies (*eChapter 4.6*) show three different reactions in which zinc metal is oxidized.

    (a) Write the molecular, ionic, and net ionic equation corresponding to each of the three reactions.

    (b) Explain how the products of the three reactions are different.

    (c) Explain how they are the same.

4.111 In the **Redox Chemistry of Tin and Zinc** movie (*eChapter 4.6*) zinc metal is being oxidized by an aqueous solution of tin(II) chloride. In actuality, there are two different redox reactions occurring in the movie. Write and balance the redox equation for each reaction.

Periodicity and Atomic Structure

Periodicity—the presence of regularly repeating patterns—is found throughout nature.

The periodic table is the most important organizing principle in chemistry. If you know the properties of any one element in a group, or column, of the periodic table, you can make a good guess at the properties of every other element in the same group and even of the elements in neighboring groups.

To see why it's called the *periodic* table, look at the graph of atomic radius versus atomic number in Figure 5.1. The graph shows a clearly periodic, rise-and-fall pattern. Beginning on the left with atomic number 1 (hydrogen), the size of the atoms increases to a maximum at atomic number 3 (lithium), then decreases to a minimum, then increases again to a maximum at atomic number 11 (sodium), then decreases, and so on. It turns out that all the maxima occur for atoms of group 1A elements—Li (atomic number, $Z = 3$), Na ($Z = 11$), K ($Z = 19$), Rb ($Z = 37$), Cs ($Z = 55$), and Fr ($Z = 87$)—and that the minima occur for atoms of the group 7A elements.

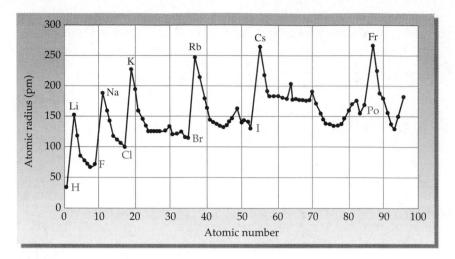

FIGURE 5.1 A graph of atomic radius in picometers (pm) versus atomic number shows a rise-and-fall pattern of periodicity. The maxima occur for atoms of group 1A elements (Li, Na, K, Rb, Cs, Fr); the minima occur for atoms of the group 7A elements. Accurate data are not available for the group 8A elements.

There's nothing unique about the periodicity of atomic radii shown in Figure 5.1. Any of several dozen other physical or chemical properties can be plotted in a similar way with similar results. We'll look at several examples of such periodicity in this chapter and the next.

5.1 Development of the Periodic Table

Interactive Periodic Table

In many ways, the creation of the periodic table by Dmitri Mendeleev in 1869 is an ideal example of how a scientific theory comes into being. At first there is only random information—a large number of elements and many observations about their properties and behavior. As more and more facts become known, people try to organize the data in ways that make sense, until ultimately a consistent hypothesis emerges.

Any good hypothesis must do two things: It must explain known facts, and it must make predictions about phenomena yet unknown. If the predictions are tested and found true, then the hypothesis is a good one and will stand until additional facts require that it be modified or discarded. Mendeleev's hypothesis about how known chemical information could be organized passed all tests. Not only did the periodic table arrange data in a useful and consistent way to explain known facts about chemical reactivity, it also led to several remarkable predictions that were later found to be accurate.

Taking the chemistry of the elements as his primary organizing principle, Mendeleev arranged the known elements by atomic mass and grouped them together according to their chemical reactivity. On so doing, he realized that there were several "holes" in the table, some of which are shown in Figure 5.2. The chemical behavior of aluminum (atomic mass ≈ 27.3) is similar to that of boron (atomic mass ≈ 11), but no element known at the time fit into the slot below aluminum. In the same way, silicon (atomic mass ≈ 28) is similar in many respects to carbon (atomic mass ≈ 12), but no element known fit below silicon.

FIGURE 5.2 A portion of Mendeleev's periodic table, giving the atomic masses known at the time and showing some of the "holes" representing unknown elements. There is an unknown element (which turned out to be gallium, Ga) beneath aluminum (Al) and another one (which turned out to be germanium, Ge) beneath silicon (Si).

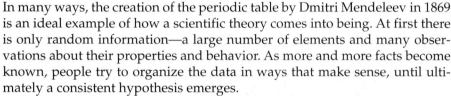

H = 1							
Li = 7	Be = 9.4		B = 11	C = 12	N = 14	O = 16	F = 19
Na = 23	Mg = 24		Al = 27.3	Si = 28	P = 31	S = 32	Cl = 35.5
K = 39	Ca = 40	?, Ti, V, Cr, Mn, Fe, Co, Ni, Cu, Zn	? = 68	? = 72	As = 75	Se = 78	Br = 80

Looking at the holes in the table, Mendeleev predicted that two then-unknown elements existed and might be found at some future time. Furthermore, he predicted with remarkable accuracy what the properties of these unknown elements would be. The element immediately below aluminum, which he called *eka*-aluminum from a Sanskrit word meaning "first," would have an atomic mass near 68 amu, would have a low melting point, and would react with chlorine to form a trichloride XCl_3. Gallium, discovered in 1875, has exactly these predicted properties. The element below silicon, which Mendeleev called *eka*-silicon, would have an atomic mass near 72 amu, would be dark gray in color, and would form an oxide with the formula XO_2. Germanium, discovered in 1886, fits the description perfectly (Table 5.1).

TABLE 5.1 A Comparison of Predicted and Observed Properties for Gallium (*eka*-Aluminum) and Germanium (*eka*-Silicon)

		Mendeleev's Prediction	Observed Property
Gallium	Atomic mass	68 amu	69.72 amu
(*eka*-Aluminum)	Density	5.9 g/cm	5.91 g/cm^3
	Melting point	Low	29.8°C
	Formula of oxide	X_2O_3	Ga_2O_3
	Formula of chloride	XCl_3	$GaCl_3$
Germanium	Atomic mass	72 amu	72.61 amu
(*eka*-Silicon)	Density	5.5 g/cm^3	5.35 g/cm^3
	Color	Dark gray	Light gray
	Formula of oxide	XO_2	GeO_2
	Formula of chloride	XCl_4	$GeCl_4$

Gallium (left) is a shiny, low-melting metal; germanium (right) is a hard, gray semimetal.

The success of these and other predictions convinced chemists of the usefulness of Mendeleev's periodic table and led to its universal acceptance. Even Mendeleev made some mistakes, though. He was completely unaware of the existence of the group 8A elements—He, Ne, Ar, Kr, Xe, and Rn—because none were known at the time. All are colorless, odorless gases with little or no chemical reactivity, and none were discovered until 1894, when argon was first isolated.

5.2 Light and the Electromagnetic Spectrum

What fundamental property of atoms is responsible for the periodic variations we observe in atomic radii and in so many other characteristics of the elements? This question occupied the thoughts of chemists for more than 50 years after Mendeleev, and it was not until well into the 1920s that the answer was established. To understand how the answer slowly emerged, it's necessary to look first at the nature of visible light and other forms of radiant energy. Historically, studies of the interaction of radiant energy with matter have provided immense insight into atomic and molecular structure.

Although they appear quite different to our senses, visible light, infrared radiation, microwaves, radio waves, X rays, and other forms of radiant energy are all different kinds of **electromagnetic radiation**. Collectively, they make up the **electromagnetic spectrum**, shown in Figure 5.3.

Electromagnetic Spectrum activity

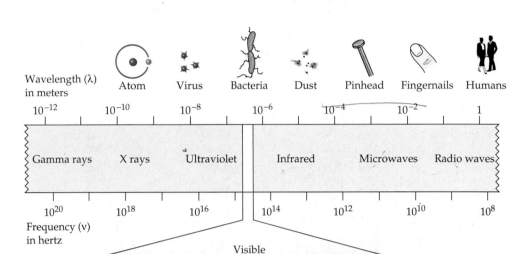

FIGURE 5.3 The electromagnetic spectrum consists of a continuous range of wavelengths and frequencies, from radio waves at the low-frequency end to gamma rays at the high-frequency end. The familiar visible region accounts for only a small portion near the middle of the spectrum. Note that waves in the X-ray region have a length that is approximately the same as the diameter of an atom (10^{-10} m).

Electromagnetic radiation traveling through a vacuum behaves in some ways like ocean waves traveling through water. Like ocean waves, electromagnetic radiation is characterized by a *frequency*, a *wavelength*, and an *amplitude*. If you could stand in one place and look at a sideways, cutaway view of an ocean wave moving through the water, you would see a regular rise-and-fall pattern (Figure 5.4). The **frequency** (ν, Greek nu) of a wave is simply the number of wave peaks that pass by a given point per unit time, usually expressed in units of reciprocal seconds ($1/s$ or s^{-1}), or **hertz** (**Hz**; $1\,Hz = 1\,s^{-1}$). The **wavelength** (λ, Greek lambda) of the wave is the distance from one wave peak to the next, and the **amplitude** of the wave is the height, measured from the center line between peak and trough. Physically, what we perceive as the *intensity* of radiant energy is proportional to the square of the wave amplitude. A very feeble beam and a blinding glare of light may have the same wavelength and frequency, but they will differ greatly in amplitude.

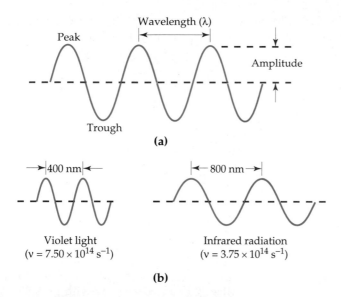

FIGURE 5.4   Electromagnetic waves are characterized by a wavelength, a frequency, and an amplitude. **(a)** Wavelength (λ) is the distance between two successive wave peaks, and frequency (v) is the number of wave peaks that pass a fixed point per unit time. Amplitude is the height of the maximum measured from the center line. **(b)** What we perceive as different kinds of electromagnetic radiation are simply waves with different wavelengths and frequencies.

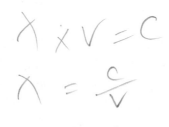

Like electromagnetic waves, ocean waves are characterized by a wavelength, a frequency, and an amplitude.

Multiplying the wavelength of a wave in meters (m) by its frequency in reciprocal seconds (s^{-1}) gives the speed of the wave in meters per second (m/s). The rate of travel of all electromagnetic radiation in a vacuum is a constant value, commonly called the *speed of light* and abbreviated c. Its numerical value is defined as exactly 2.997 924 58 × 10^8 m/s, usually rounded off to 3.00×10^8 m/s:

$$\text{Wavelength} \times \text{Frequency} = \text{Speed}$$
$$\lambda\ (\text{m}) \times v\ (\text{s}^{-1}) = c\ (\text{m/s})$$

which can be rewritten as:

$$\lambda = \frac{c}{v} \quad \text{or} \quad v = \frac{c}{\lambda}$$

This equation says that frequency and wavelength are inversely related: Electromagnetic radiation with a long wavelength has a low frequency, and radiation with a short wavelength has a high frequency.

EXAMPLE 5.1 The light blue glow given off by mercury street lamps has a wavelength of 436 nm. What is its frequency in hertz?

SOLUTION We are given a wavelength and need to find the corresponding frequency. Wavelength and frequency are inversely related by the equation $\lambda \nu = c$, which can be solved for ν. Don't forget to convert from nanometers to meters.

$$\nu = \frac{c}{\lambda} = \frac{\left(3.00 \times 10^8 \frac{\cancel{m}}{s}\right)\left(\frac{10^9 \cancel{nm}}{\cancel{m}}\right)}{436 \cancel{nm}}$$

$$= 6.88 \times 10^{14}\ s^{-1} = 6.88 \times 10^{14}\ Hz$$

The frequency of the light is $6.88 \times 10^{14}\ s^{-1}$, or $6.88 \times 10^{14}\ Hz$.

▶ **PROBLEM 5.1** What is the frequency of a gamma ray with $\lambda = 3.56 \times 10^{-11}$ m? Of a radar wave with $\lambda = 10.3$ cm?

▶ **PROBLEM 5.2** What is the wavelength (in meters) of an FM radio wave with frequency $\nu = 102.5$ MHz? Of a medical X ray with $\nu = 9.55 \times 10^{17}$ Hz?

✦── **KEY CONCEPT PROBLEM 5.3** Two electromagnetic waves are represented below.
(a) Which wave has the higher frequency?
(b) Which wave represents a more intense beam of light?
(c) Which wave represents blue light, and which represents red light?

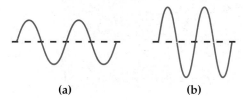

(a)   (b)

5.3 Electromagnetic Radiation and Atomic Spectra

Flame Tests for Metals
www movie

The light that we see from the sun or from a light bulb is "white" light, meaning that it consists of an essentially continuous distribution of all possible wavelengths spanning the entire visible region of the electromagnetic spectrum. When a narrow beam of white light is passed through a glass prism, the different wavelengths travel through the glass at different rates. As a result, the white light is separated into its component colors, ranging from red at the long-wavelength end of the spectrum (700 nm) to violet at the short-wavelength end (400 nm) (Figure 5.5a). This separation into colors is similar to what occurs when light travels through water droplets in the air, forming a rainbow.

What do visible light and other kinds of electromagnetic radiation have to do with atomic structure? The answer involves the fact that atoms give off light when heated or otherwise excited energetically, thereby providing a clue to their atomic makeup. Unlike the white light from the sun, though, the light given off by an energetically excited atom is not a continuous distribution of all possible wavelengths. When passed first through a narrow slit and then through a prism, the light emitted by an excited atom is found to consist of

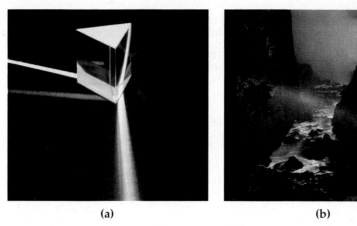

(a)   (b)

FIGURE 5.5 **(a)** When a narrow beam of ordinary white light is passed through a glass prism, different wavelengths travel through the glass at different rates and appear as different colors. **(b)** A similar effect occurs when light passes through water droplets in the air, forming a rainbow.

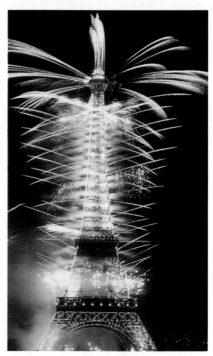

The brilliant colors of these fireworks are due to the emission of light by excited metal atoms.

only a few wavelengths rather than a full rainbow of colors, giving a series of discrete lines separated by blank areas—a **line spectrum**. Excited sodium atoms, produced by heating NaCl or some other sodium salt in the flame of a Bunsen burner, give off yellow light (Figure 5.6a); hydrogen atoms give off a bluish light made of several different colors (Figure 5.6b); and so on. In fact, the brilliant colors of fireworks are produced by adding mixtures of different metal salts to explosive powder.

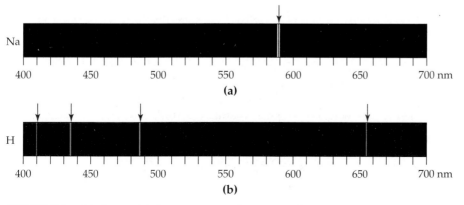

FIGURE 5.6 **(a)** The visible line spectrum of energetically excited sodium atoms consists of a closely spaced pair of yellow lines. **(b)** The visible line spectrum of excited hydrogen atoms consists of four lines, from indigo at 410 nm to red at 656 nm.

Soon after the discovery that energetic atoms emit light of specific wavelengths, chemists began cataloging the line spectra of various elements. They rapidly found that each element has its own unique spectral "signature," and they began using the results to identify the elements present in minerals and other substances. Not until the work of the Swiss schoolteacher Johann Balmer in 1885, though, was a pattern discovered in atomic line spectra. It was known at the time that hydrogen produced a spectrum with four lines, as shown in Figure 5.6b. The wavelengths of the four lines are 656.3 nm (red), 486.1 nm (blue-green), 434.0 nm (blue), and 410.1 nm (indigo).

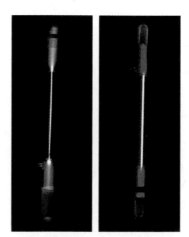

Excited hydrogen atoms give off a bluish light; excited neon atoms emit orange light.

Thinking about the hydrogen spectrum and trying by trial-and-error to organize the data in various ways, Balmer discovered that the wavelengths of the four lines in the hydrogen spectrum can be expressed by the equation

$$\frac{1}{\lambda} = R\left(\frac{1}{2^2} - \frac{1}{n^2}\right) \quad \text{or} \quad \nu = R \cdot c\left(\frac{1}{2^2} - \frac{1}{n^2}\right)$$

where R is a constant (now called the *Rydberg constant*) equal to 1.097×10^{-2} nm^{-1} and n is an integer greater than 2. The red spectral line at 656.3 nm, for example, results from Balmer's equation when $n = 3$:

$$\frac{1}{\lambda} = (1.097 \times 10^{-2} \text{ nm}^{-1})\left(\frac{1}{2^2} - \frac{1}{3^2}\right) = 1.524 \times 10^{-3} \text{ nm}^{-1}$$

$$\lambda = \frac{1}{1.524 \times 10^{-3} \text{ nm}^{-1}} = 656.3 \text{ nm}$$

Similarly, a value of $n = 4$ gives the blue-green line at 486.1 nm, a value of $n = 5$ gives the blue line at 434.0 nm, and so on. Solve Balmer's equation yourself to make sure.

Subsequent to the discovery of the Balmer series of lines in the *visible* region of the electromagnetic spectrum, it was found that many other spectral lines are also present in *nonvisible* regions of the electromagnetic spectrum. Hydrogen, for example, shows a series of spectral lines called the *Lyman series* in the ultraviolet region and still other series (the *Paschen*, *Brackett*, and *Pfund series*) in the infrared region.

By adapting Balmer's equation, the Swedish physicist Johannes Rydberg was able to show that every line in the entire spectrum of hydrogen can be fit by the generalized Balmer–Rydberg equation:

Balmer–Rydberg equation: $\quad \dfrac{1}{\lambda} = R\left(\dfrac{1}{m^2} - \dfrac{1}{n^2}\right) \quad \text{or} \quad \nu = R \cdot c\left(\dfrac{1}{m^2} - \dfrac{1}{n^2}\right)$

where m and n represent integers with $n > m$. If $m = 1$, then the Lyman series of lines results. If $m = 2$, then Balmer's series of visible lines results. If $m = 3$, then the Paschen series is described, and so forth for still larger values of m. Some of these other spectral lines are calculated in Example 5.2.

We'll look further at the Balmer–Rydberg equation and see what the integers m and n represent in Section 5.9.

EXAMPLE 5.2   What are the two longest-wavelength lines (in nanometers) in the Lyman series of the hydrogen spectrum?

SOLUTION   The Lyman series is given by the Balmer–Rydberg equation with $m = 1$ and $n > 1$. The wavelength λ is greatest when n is smallest; that is, when $n = 2$ and $n = 3$.

$$\frac{1}{\lambda} = R\left(\frac{1}{m^2} - \frac{1}{n^2}\right) \quad \text{where } m = 1$$

Solving the equation first for $n = 2$ gives

$$\frac{1}{\lambda} = R\left(\frac{1}{1^2} - \frac{1}{2^2}\right) = (1.097 \times 10^{-2} \text{ nm}^{-1})\left(1 - \frac{1}{4}\right) = 8.228 \times 10^{-3} \text{ nm}^{-1}$$

or $\quad \lambda = \dfrac{1}{8.228 \times 10^{-3} \text{ nm}^{-1}} = 121.5 \text{ nm}$

Solving the equation next for $n = 3$ gives

$$\frac{1}{\lambda} = R\left(\frac{1}{1^2} - \frac{1}{3^2}\right) = (1.097 \times 10^{-2}\,\text{nm}^{-1})\left(1 - \frac{1}{9}\right) = 9.751 \times 10^{-3}\,\text{nm}^{-1}$$

or $$\lambda = \frac{1}{9.751 \times 10^{-3}\,\text{nm}^{-1}} = 102.6\,\text{nm}$$

The two longest-wavelength lines in the Lyman series are at 121.5 nm and 102.6 nm.

EXAMPLE 5.3   What is the shortest-wavelength line (in nanometers) in the Lyman series of the hydrogen spectrum?

SOLUTION   The Lyman series is given by the Balmer–Rydberg equation with $m = 1$ and $n > 1$. The shortest-wavelength line occurs when n is infinitely large so that $1/n^2$ is zero. That is, if $n = \infty$, then $1/n^2 = 0$. Thus, the equation becomes

$$\frac{1}{\lambda} = R\left(\frac{1}{1^2} - \frac{1}{\infty^2}\right) = (1.097 \times 10^{-2}\,\text{nm}^{-1}) \times (1 - 0) = 1.097 \times 10^{-2}\,\text{nm}^{-1}$$

or $$\lambda = \frac{1}{1.097 \times 10^{-2}\,\text{nm}^{-1}} = 91.2\,\text{nm}$$

▶ **PROBLEM 5.4**   The Balmer equation can be extended beyond the visible portion of the electromagnetic spectrum to include lines in the ultraviolet. What is the wavelength (in nanometers) of ultraviolet light in the Balmer series corresponding to a value of $n = 7$?

▶ **PROBLEM 5.5**   What is the longest-wavelength line (in nanometers) in the Paschen series for hydrogen?

▶ **PROBLEM 5.6**   What is the shortest-wavelength line (in nanometers) in the Paschen series for hydrogen?

5.4 Particlelike Properties of Electromagnetic Radiation: The Planck Equation

The existence of atomic line spectra and the fit of the visible hydrogen spectrum to the Balmer–Rydberg equation imply the existence of a general underlying principle about atomic structure, yet it was many years before that principle was found. One of the key discoveries came in 1900, when the German physicist Max Planck (1858–1947) proposed a theory to explain a seemingly unrelated phenomenon called *blackbody radiation*—the visible glow that solid objects give off when heated. The reddish glow from the heating element in an electric stove and the white light emitted by the hot filament in a light bulb are two examples of blackbody radiation.

Experimentally, it's found that the intensity of blackbody radiation varies with the wavelength of the emitted light. When an object such as an iron bar is heated, it first begins to glow a dull red but then changes to a brighter orange and ultimately to a blinding white glare as its temperature increases. Thus, longer wavelengths (red) have a lower intensity, and shorter wavelengths have a higher intensity. If this trend were to continue, the intensity would keep rising indefinitely as the wavelength becomes ever shorter and enters the

The red glow given off by the hot iron is an example of blackbody radiation.

ultraviolet region. In fact, though, the intensity of blackbody radiation does not continue rising indefinitely. Instead, the intensity reaches a maximum and then falls rapidly at wavelengths shorter than about 500 nm (Figure 5.7).

FIGURE 5.7 The dependence of the intensity of blackbody radiation on wavelength at two different temperatures. Intensity increases from right to left on the curve as wavelength decreases. As the wavelength continues to decrease, intensity reaches a maximum and then drops off to zero.

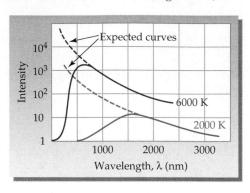

To explain the observation that the intensity of blackbody radiation does not continue to rise indefinitely as the wavelength decreases, Planck concluded that the energy radiated by a heated object can't be continuously variable. Instead the energy is constrained to be emitted only in discrete amounts, or **quanta**. An analogy from daily life is that of stairs versus a ramp. A ramp changes height continuously, but stairs are **quantized**, changing height only in discrete amounts.

The amount of energy, E, associated with each quantum of radiant energy depends on the frequency of the radiation, ν, according to the equation

$$E = h\nu$$

or, since $\nu = \dfrac{c}{\lambda}$,

$$E = \dfrac{hc}{\lambda}$$

A ramp changes height continuously, but stairs are quantized, changing height only in discrete amounts. In the same way, electromagnetic radiation is not continuous but is emitted only in discrete amounts.

The symbol h represents a fundamental physical constant that we now call **Planck's constant**; $h = 6.626 \times 10^{-34}$ J · s. For example, one quantum of red light with a frequency $\nu = 4.62 \times 10^{14}$ s^{-1} (wavelength $\lambda = 649$ nm) has an energy in joules of 3.06×10^{-19} J.

$$E = h\nu = (6.626 \times 10^{-34}\,\text{J} \cdot \text{s}) \times (4.62 \times 10^{14}\,\text{s}^{-1}) = 3.06 \times 10^{-19}\,\text{J}$$

Note that the SI energy unit joule (1 J = 1 kg · m^2/s^2) is a fairly small amount of energy—it takes 100 J to light a 100 watt light bulb for 1 second.

Higher frequencies and shorter wavelengths correspond to higher energy radiation, while lower frequencies and longer wavelengths correspond to lower energy. Blue light ($\lambda \approx 450$ nm), for instance, has a shorter wavelength and is more energetic than red light ($\lambda \approx 650$ nm). Similarly, an X ray ($\lambda \approx 1$ nm) has a shorter wavelength and is more energetic than an FM radio wave ($\lambda \approx 10^{10}$ nm, or 10 m).

Lower frequency; longer wavelength   *Energy*   Higher frequency; shorter wavelength

The idea that electromagnetic energy is quantized rather than continuous received further support in 1905, when Albert Einstein (1879–1955) used it to explain the *photoelectric effect*. Scientists had known since the late 1800s that irradiating a clean metal surface with light causes electrons to be ejected from the metal. Furthermore, the frequency of the light used for the irradiation must be above some threshold value, which is different for every metal. Blue light ($v \approx 6.7 \times 10^{14}$ Hz) causes metallic sodium to emit electrons, for example, but red light ($v \approx 4.0 \times 10^{14}$ Hz) has no effect on sodium.

Photoelectric Effect movie

Einstein explained the photoelectric effect by assuming that a beam of light behaves as if it were composed of a stream of small particles, called **photons**, whose energy, E, is related to their frequency, v, by the Planck equation, $E = hv$. If the frequency (or energy) of the photon striking a metal is below a minimum value, no electron is ejected. Above the threshold level, however, sufficient energy is transferred from the photon for an electron to overcome the attractive forces holding it to the metal (Figure 5.8).

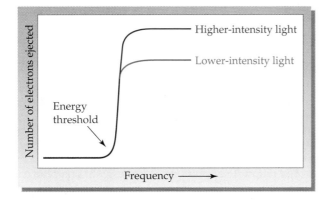

FIGURE 5.8 The photoelectric effect. A plot of the number of electrons ejected from a metal surface versus light frequency shows a threshold value. Increasing the intensity of the light while keeping the frequency the same increases the number of electrons ejected but does not change the threshold value.

Note that the energy of an individual photon depends only on its frequency, not on the intensity of the light beam. The intensity of a light beam is a measure of the *number* of photons in the beam, not of the *energies* of those photons. A low-intensity beam of high-energy photons might easily knock a few electrons loose from a metal, but a high-intensity beam of low-energy photons might not be able to knock loose a single electron.

As a rough analogy, think of throwing balls of different masses at a glass window. A thousand Ping-Pong balls (lower energy) would only bounce off the window, but a single baseball (higher energy) would break the glass. In the same way, low-energy photons bounce off the metal surface, but a single photon at or above a certain threshold energy can "break" the metal and dislodge an electron.

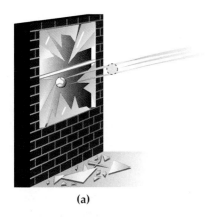

(a)

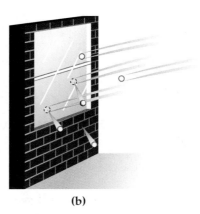

(b)

This glass window can be broken by a single baseball, but a thousand Ping-Pong balls would only bounce off.

The main conclusion from both Planck's and Einstein's work was that the behavior of light and other forms of electromagnetic radiation is more complex than had been formerly believed. *In addition to behaving as waves, light energy can also behave as small particles.* The idea isn't really so strange if you think of light as analogous to matter. The amount of energy corresponding to one quantum of light is almost inconceivably small, just as the amount of matter in one atom is inconceivably small, but the principle is the same: *Both matter and energy occur only in discrete units.* Just as there can be either 1 or 2 hydrogen atoms but not 1.5 or 1.8, there can be 1 or 2 photons of light but not 1.5 or 1.8.

Once the quantized nature of electromagnetic radiation is accepted, part of the puzzle of atomic line spectra is explained. Energetically excited atoms evidently are not able to emit light of continuously varying wavelengths and therefore don't give a continuous spectrum. The atoms are somehow constrained to emit light quanta (photons) of only a few specific energies, and they therefore give a line spectrum. Why atoms should be thus constrained is the next question to answer.

EXAMPLE 5.4   What is the energy (in kilojoules per mole) of photons of radar waves with $v = 3.35 \times 10^8$ Hz?

SOLUTION   The energy of a photon with frequency v can be calculated with the Planck equation $E = hv$:

$$E = hv = (6.626 \times 10^{-34}\ \text{J·s})(3.35 \times 10^8\ \text{s}^{-1}) = 2.22 \times 10^{-25}\ \text{J}$$

To find the energy per mole of photons, the energy of one photon must be multiplied by Avogadro's number:

$$\left(2.22 \times 10^{-25}\ \frac{\text{J}}{\text{photon}}\right)\left(6.022 \times 10^{23}\ \frac{\text{photon}}{\text{mol}}\right) = 0.134\ \text{J/mol}$$

$$= 1.34 \times 10^{-4}\ \text{kJ/mol}$$

▶ **PROBLEM 5.7**   What is the energy (in kilojoules per mole) of photons corresponding to the shortest-wavelength line in the Lyman series for hydrogen (Example 5.3)?

▶ **PROBLEM 5.8**   The biological effects of a given dose of electromagnetic radiation generally become more serious as the energy of the radiation increases: Infrared radiation has a pleasant warming effect; ultraviolet radiation causes tanning and burning; and X rays can cause considerable tissue damage. What energies (in kilojoules per mole) are associated with the following wavelengths: infrared radiation with $\lambda = 1.55 \times 10^{-6}$ m? ultraviolet light with $\lambda = 250$ nm? X rays with $\lambda = 5.49$ nm?

5.5   Wavelike Properties of Matter: The de Broglie Equation

The analogy between matter and radiant energy developed by Planck and Einstein in the early 1900s was further extended in 1924 by the French physicist Louis de Broglie (1892–1987). De Broglie suggested that, if *light* can behave in some respects like *matter*, then perhaps *matter* can behave in some respects like *light*. That is, perhaps *both* light and matter are wavelike as well as particlelike.

To help construct his argument about the wavelike behavior of matter, de Broglie used the now famous equation $E = mc^2$, which had been proposed in 1905 by Einstein to predict the change in mass m of an atom that accompanies the emission of a photon with energy E during radioactive decay

$$\text{Since } E = mc^2 \quad \text{then } m = \frac{E}{c^2}$$

Einstein had seen that, since $E = hc/\lambda$ according to the Planck equation, it is possible to substitute for E to derive a relationship between mass and wavelength:

$$m = \frac{E}{c^2} = \frac{hc/\lambda}{c^2} = \frac{h}{\lambda c}$$

De Broglie suggested that a similar equation might be applied to an *electron* instead of a photon by replacing the speed of the photon (c) by the speed of the electron (v). The resultant **de Broglie equation** thus allows calculation of the "wavelength" of an electron or of any other particle or object of mass m moving at velocity v:

⟵ de Broglie equation: $\quad m = \dfrac{h}{\lambda v} \quad \text{or} \quad \lambda = \dfrac{h}{mv}$

For example, the mass of an electron is 9.11×10^{-31} kg, and the velocity v of an electron in a hydrogen atom is 2.2×10^6 m/s (about 1% of the speed of light). Thus, the de Broglie wavelength of an electron in a hydrogen atom is 3.3×10^{-10} m, or 330 pm:

$$\lambda = \frac{h}{mv} = \frac{6.626 \times 10^{-34} \dfrac{\text{kg} \cdot \text{m}^2}{\text{s}}}{(9.11 \times 10^{-31} \text{ kg})\left(2.2 \times 10^6 \dfrac{\text{m}}{\text{s}}\right)} = 3.3 \times 10^{-10} \text{ m}$$

Note that Planck's constant, which is usually expressed in units of joule seconds (J · s), is expressed for the present purposes in units of $(\text{kg} \cdot \text{m}^2)/\text{s}$ [1 J = 1 $(\text{kg} \cdot \text{m}^2)/\text{s}^2$]. Note also that the calculated de Broglie wavelength of the electron in a hydrogen atom, 330 pm, is larger than the diameter of an isolated H atom itself (about 240 pm).

What does it mean to say that light and matter act both as waves and as particles? The answer is "not much," at least not on the everyday human scale. The problem in trying to understand the dual wave/particle description of light and matter is that our common sense isn't up to the task. Our intuition has been developed from personal experiences, using our eyes and other senses to tell us how light and matter are "supposed" to behave. We have no personal experience on the atomic scale, though, and thus have no common-sense way of dealing with the behavior of light and matter at that level. On the atomic scale, where distances and masses are so tiny, light and matter behave in a manner different from what we're used to.

The dual wave/particle description of light and matter is really just a mathematical *model*. Since we can't see atoms and observe their behavior directly, the best we can do is to construct a set of mathematical equations that correctly account for atomic properties and behavior. The wave/particle description does this extremely well, even though it is not easily understood using day-to-day experience.

The batter has no excuse. He didn't miss because of the ball's de Broglie wavelength.

EXAMPLE 5.5 What is the de Broglie wavelength (in meters) of a pitched baseball with a mass of 120 g and a speed of 100 mph (44.7 m/s)?

SOLUTION The de Broglie relationship says that the wavelength λ of an object with mass m moving at a velocity v can be calculated by the equation $\lambda = h/mv$:

$$\lambda = \frac{h}{mv} = \frac{6.626 \times 10^{-34} \, \dfrac{\text{kg} \cdot \text{m}^2}{\text{s}}}{(0.120 \, \text{kg})\left(44.7 \, \dfrac{\text{m}}{\text{s}}\right)} = 1.24 \times 10^{-34} \text{ m}$$

The de Broglie wavelength of the baseball is 1.24×10^{-34} m—far smaller than the diameter of an atom.

▶ **PROBLEM 5.9** What is the de Broglie wavelength (in meters) of a small car with a mass of 1150 kg traveling at a speed of 55.0 mi/h (24.6 m/s)?

5.6 Quantum Mechanics and the Heisenberg Uncertainty Principle

With the particlelike nature of energy and the wavelike nature of matter now established, let's return to the problem of atomic structure. Several models of atomic structure were proposed in the late nineteenth and early twentieth centuries. A model proposed in 1914 by the Danish physicist Niels Bohr (1885–1962), for example, described the hydrogen atom as a nucleus with an electron circling around it, much as a planet orbits the sun. Furthermore, said Bohr, only certain specific orbits corresponding to certain specific energy levels for the electron are available. The Bohr model was extremely important historically because of its conclusion that electrons have only specific energy levels available to them, but the model fails for atoms with more than one electron.

The breakthrough in understanding atomic structure came in 1926, when the Austrian physicist Erwin Schrödinger (1887–1961) proposed what has come to be called the **quantum mechanical model** of the atom. The fundamental idea behind the model is that it's best to abandon the notion of an electron as a small particle moving around the nucleus in a defined path and to concentrate instead on the electron's wavelike properties. (Recall from the previous section that the de Broglie wavelength of the electron in a hydrogen atom is greater than the diameter of the atom.) In fact, it was shown in 1927 by Werner Heisenberg (1901–1976) that it is *impossible* to know precisely where an electron is and what path it follows—a statement called the **Heisenberg uncertainty principle**.

The Heisenberg uncertainty principle can be understood by imagining what would happen if we tried to determine the position of an electron at a given moment. For us to "see" the electron, light photons of an appropriate frequency would have to interact with and bounce off the electron. But such an interaction would transfer energy from the photon to the electron, thereby increasing the energy of the electron and making it move faster. Thus, the very act of determining the electron's position would make that position change.

In mathematical terms, Heisenberg's principle states that the uncertainty in the electron's position, Δx, times the uncertainty in its momentum, Δmv, is equal to or greater than the quantity $h/4\pi$:

Werner Heisenberg. Where is he? His whereabouts aren't known with certainty.

Heisenberg uncertainty principle: $\quad (\Delta x)(\Delta mv) \geq \dfrac{h}{4\pi}$

The equation says that we can never know both the position and the velocity of an electron (or of any other object) beyond a certain level of precision. If we know the *velocity* with a high degree of certainty (Δmv is small), then the *position* of the electron must be uncertain (Δx must be large). Conversely, if we know the position of the electron exactly (Δx is small), then we can't know its velocity (Δmv must be large). As a result, an electron will always appear as something of a blur whenever we attempt to make any physical measurements of its position and velocity.

A brief calculation can help put the conclusions of the uncertainty principle more clearly. As mentioned in the previous section, the mass m of an electron is 9.11×10^{-31} kg and the velocity v of an electron in a hydrogen atom is 2.2×10^6 m/s. If we assume that the velocity is known to within 10%, or 0.2×10^6 m/s, then we can calculate that the uncertainty in the electron's position as it travels around in a hydrogen atom is greater than 3×10^{-10} m, or 300 pm:

$$\text{If } (\Delta x)(\Delta mv) \geq \frac{h}{4\pi} \quad \text{then } (\Delta x) \geq \frac{h}{(4\pi)(\Delta mv)}$$

$$\Delta x \geq \frac{6.626 \times 10^{-34} \dfrac{\text{kg} \cdot \text{m}^2}{\text{s}}}{(4)(3.1416)(9.11 \times 10^{-31} \text{ kg})\left(0.2 \times 10^6 \dfrac{\text{m}}{\text{s}}\right)}$$

$$\Delta x \geq 3 \times 10^{-10} \text{ m} \quad \text{or} \quad 300 \text{ pm}$$

But since the diameter of a hydrogen atom is only 240 pm, *the uncertainty in the electron's position is similar in size to the atom itself!*

When the mass m of an object is relatively large, as is true in daily life, then Δx and Δv in the Heisenberg relationship can *both* be very small. We therefore have no apparent problem in measuring both position and velocity for visible objects. The problem arises only on the atomic scale. Example 5.6 gives a sample calculation.

EXAMPLE 5.6   Assume that you are traveling at a speed of 90 km/h in a small car with a mass of 1250 kg. If the uncertainty in the velocity of the car is 1% ($\Delta v = 0.9$ km/h), what is the uncertainty (in meters) in the position of the car? How does this compare with the uncertainty in the position of an electron in a hydrogen atom?

SOLUTION   The Heisenberg relationship says that the uncertainty in an object's position, Δx, times the uncertainty in its momentum, Δmv, is equal to or greater than the quantity $h/4\pi$. In the present instance, we need to find Δx when Δv is known:

$$\Delta x \geq \frac{h}{(4\pi)(\Delta mv)}$$

$$\Delta x \geq \frac{6.626 \times 10^{-34} \dfrac{\text{kg} \cdot \text{m}^2}{\text{s}}}{(4)(3.1416)(1250 \text{ kg})\left(0.9 \dfrac{\text{km}}{\text{h}}\right)\left(\dfrac{1 \text{ h}}{3600 \text{ s}}\right)\left(\dfrac{1000 \text{ m}}{1 \text{ km}}\right)}$$

$$\Delta x \geq 2 \times 10^{-37} \text{ m}$$

The uncertainty in the position of the car is far smaller than the uncertainty in the position of an electron in a hydrogen atom (3×10^{-10} m), and far too small a value to have any measurable consequences.

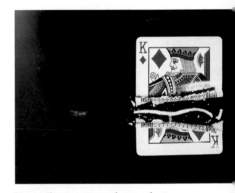

Even the motion of very fast objects can be captured in daily life. On the atomic scale, however, velocity and position can't both be known precisely.

▶ **PROBLEM 5.10**   Calculate the uncertainty (in meters) in the position of a 120 g baseball thrown at a velocity of 45 m/s if the uncertainty in the velocity is 2%.

5.7 Wave Functions and Quantum Numbers

Schrödinger's quantum mechanical model of atomic structure is framed in the form of a *wave equation*, a mathematical equation similar in form to that used for describing the motion of ordinary waves in fluids. The solutions (there are many) to the wave equation are called **wave functions**, or **orbitals**, and are represented by the symbol ψ (Greek psi). The best way to think about a wave function is to regard it as an expression whose square, ψ^2, defines the probability of finding the electron within a specific region of space. As Heisenberg showed, we can never be completely certain about an electron's position. A wave function, however, tells where the electron will most probably be found.

$$\text{Wave equation} \xrightarrow{\text{Solve}} \text{Wave function or orbital } (\psi) \longrightarrow \text{Probability of finding electron in a region of space } (\psi^2)$$

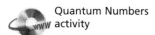

Quantum Numbers activity

A wave function contains three variables called **quantum numbers**, represented as n, l, and m_l, which describe the energy level of the orbital and the three-dimensional shape of the region in space occupied by a given electron.

1. **The principal quantum number (n)** is a positive integer ($n = 1, 2, 3, 4, \ldots$) on which the size and energy level of the orbital primarily depend. For hydrogen and other one-electron atoms, such as He⁺, the energy of an orbital depends *only* on n. For atoms with more than one electron, the energy level of an orbital depends both on n and on the l quantum number.

   As the value of n increases, the number of allowed orbitals increases and the size of those orbitals becomes larger, thus allowing an electron to be farther from the nucleus. Because it takes energy to separate a negative charge from a positive charge, this increased distance between the electron and the nucleus means that the energy of the electron in the orbital increases as the quantum number n increases.

   We often speak of orbitals as being grouped according to the principal quantum number n into successive layers, or **shells**, around the atom. Those orbitals with $n = 3$, for example, are said to be in the third shell.

2. **The angular-momentum quantum number (l)** defines the three-dimensional shape of the orbital. For an orbital whose principal quantum number is n, the angular-momentum quantum number l can have any integral value from 0 to $n - 1$. Thus, within each shell, there are n different shapes for orbitals.

   If $n = 1$, then $l = 0$.

   If $n = 2$, then $l = 0$ or 1.

   If $n = 3$, then $l = 0, 1,$ or 2.

   And so forth.

Just as it's convenient to think of orbitals as being grouped into shells according to the principal quantum number n, we often speak of orbitals as being grouped into **subshells** according to the angular-momentum quantum number l. Different subshells are usually referred to by letter rather than by number, following the order s, p, d, f, g. (Historically, the letters $s, p, d,$ and f arose from the use of the words *sharp, principal, diffuse,* and *fundamental* to describe various lines in atomic spectra. After f, successive subshells are designated alphabetically: $g, h,$ and so on.)

$$\text{Quantum number } l:\quad 0\quad 1\quad 2\quad 3\quad 4\quad \ldots$$
$$\text{Subshell notation:}\quad s\quad p\quad d\quad f\quad g\quad \ldots$$

As an example, we might speak of an orbital with $n = 3$ and $l = 2$ as being a $3d$ orbital: 3 to represent the third shell and d to represent the $l = 2$ subshell.

3. **The magnetic quantum number (m_l)** defines the spatial orientation of the orbital along a standard set of coordinate axes. For an orbital whose angular-momentum quantum number is l, the magnetic quantum number m_l can have any integral value from $-l$ to $+l$. Thus, within each subshell (orbitals with the same shape, or value of l), there are $2l + 1$ different spatial orientations for those orbitals. We'll explore this point further in the next section.

$$\text{If } l = 0, \text{ then } m_l = 0.$$
$$\text{If } l = 1, \text{ then } m_l = -1, 0, \text{ or } +1.$$
$$\text{If } l = 2, \text{ then } m_l = -2, -1, 0, +1, \text{ or } +2.$$
$$\text{And so forth.}$$

A summary of the allowed combinations of quantum numbers for the first four shells is given in Table 5.2.

TABLE 5.2 Allowed Combinations of Quantum Numbers n, l, and m_l for the First Four Shells

n	l	m_l	Orbital Notation	Number of Orbitals in Subshell	Number of Orbitals in Shell
1	0 s	0	$1s$	1	1
2	0 s	0	$2s$	1	4
	1 p	$-1, 0, +1$	$2p$	3	
3	0 s	0	$3s$	1	9
	1 p	$-1, 0, +1$	$3p$	3	
	2 d	$-2, -1, 0, +1, +2$	$3d$	5	
4	0 s	0	$4s$	1	16
	1 p	$-1, 0, +1$	$4p$	3	
	2 d	$-2, -1, 0, +1, +2$	$4d$	5	
	3 f	$-3, -2, -1, 0, +1, +2, +3$	$4f$	7	

The energy levels of various orbitals are shown in Figure 5.9. As noted earlier in this section, the energy levels of different orbitals in a hydrogen atom depend only on the principal quantum number n, but the energy levels of orbitals in multielectron atoms depend on both n and l. In other words, orbitals in a given subshell have the same energy for hydrogen but have slightly different energies for other atoms. In fact, there is even some crossover of energies from one shell to another. A 3d orbital in some multielectron atoms has a higher energy than a 4s orbital, for instance.

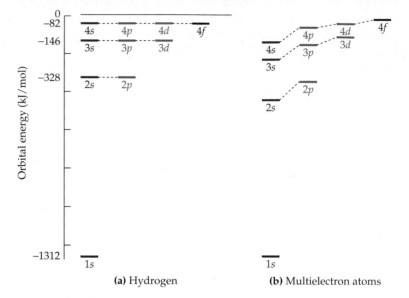

FIGURE 5.9 Orbital energy levels for **(a)** hydrogen and **(b)** a typical multielectron atom. The differences between energies of various subshells in **(b)** are exaggerated for clarity. Note, though, that there is some crossover of energies from one shell to another. In some atoms, a 3d orbital has a higher energy than a 4s orbital, for instance.

EXAMPLE 5.7   Identify the shell and subshell of an orbital with the quantum numbers $n = 3$, $l = 1$, $m_l = 1$.

SOLUTION   The principal quantum number $n = 3$ indicates that the orbital is in the third shell; the angular-momentum quantum number $l = 1$ indicates that the orbital is of the p type. Thus, the orbital has the designation 3p. The magnetic quantum number m_l is related to the spatial orientation of the orbital.

EXAMPLE 5.8   Give the possible combinations of quantum numbers for a 4p orbital.

SOLUTION   The designation 4p indicates that the orbital has a principal quantum number $n = 4$ and an angular-momentum quantum number $l = 1$. The magnetic quantum number m_l can have any of the three values -1, 0, or $+1$. Thus, the allowable combinations are

$$n = 4, l = 1, m_l = -1 \qquad n = 4, l = 1, m_l = 0 \qquad n = 4, l = 1, m_l = +1$$

▶**PROBLEM 5.11**   Extend Table 5.2 to show allowed combinations of quantum numbers when $n = 5$. How many orbitals are there in the fifth shell?

▶**PROBLEM 5.12**   Give orbital notations for electrons in orbitals with the following quantum numbers:
(a) $n = 2, l = 1, m_l = 1$ **(b)** $n = 4, l = 3, m_l = -2$ **(c)** $n = 3, l = 2, m_l = -1$

▶**PROBLEM 5.13**   Give the possible combinations of quantum numbers for the following orbitals:
(a) A 3s orbital   **(b)** A 2p orbital   **(c)** A 4d orbital

5.8 The Shapes of Orbitals

We said in the previous section that a wave function, or orbital, describes the probability of finding the electron within a specific region of space. The shape of a given orbital is defined by the angular-momentum quantum number l, with $l = 0$ called an s orbital, $l = 1$ a p orbital, $l = 2$ a d orbital, and so forth. Of the various possibilities, s, p, d, and f orbitals are the most important because these are the only ones actually occupied in known elements. Let's look at each one individually.

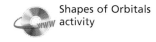

Shapes of Orbitals activity

s Orbitals

All s orbitals are spherical, meaning that the probability of finding an s electron depends only on distance from the nucleus, not on direction. Furthermore, because there is only one possible orientation of a sphere in space, an s orbital has $m_l = 0$ and there is only one s orbital per shell.

As shown in Figure 5.10, the value of ψ^2 for an s orbital is greatest near the nucleus, indicating that this is where the electron is most likely to be found. The probability of finding the electron drops off rapidly as distance from the nucleus increases, although it never goes all the way to zero, even at a large distance. As a result, there is no definite boundary to the atom and no definite "size." For purposes like that of Figure 5.10, however, we usually imagine a boundary surface enclosing the volume where an electron spends *most* (say, 95%) of its time.

Radial Electron Distribution movie

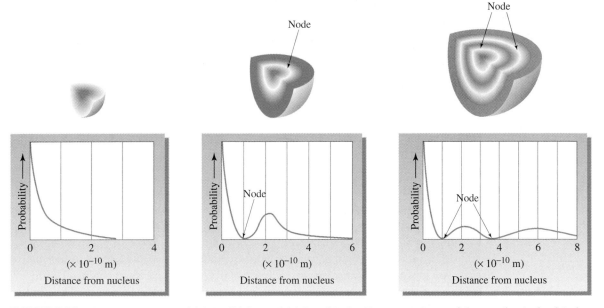

FIGURE 5.10 Representations of **(a)** 1s, **(b)** 2s, and **(c)** 3s orbitals. Cutaway views of these spherical orbitals are shown on the top, with the probability of finding an electron represented by the density of the shading. Electron probability distribution plots of ψ^2 as a function of distance from the nucleus are shown on the bottom. Note that the 2s orbital has buried within it a spherical surface of zero probability (a *node*), and the 3s orbital has within it two spherical surfaces of zero probability. The different colors of different regions in the 2s and 3s orbitals correspond to different algebraic signs of the wave function, analogous to the different phases of a wave, as explained in the text.

Although all s orbitals are spherical, there are significant differences among the s orbitals in different shells. For one thing, the size of the s orbital increases in successively higher shells, implying that an electron in an outer-shell s orbital is farther from the nucleus on average than an electron in an inner-shell *s* orbital. For another thing, the electron distribution in an outer-shell *s* orbital has several regions of maximum probability. As shown in Figure 5.10b, a 2*s* orbital is essentially a sphere within a sphere. There are *two* regions of maximum probability, separated by a surface of zero probability called a **node**. Similarly, a 3*s* orbital has *three* regions of maximum probability and two spherical nodes (Figure 5.10c).

The concept of an orbital node—a surface of zero electron probability separating regions of nonzero probability—is difficult to grasp because it raises the question, How does an electron get from one region of the orbital to another if it's not allowed to be at the node? The question is misleading, though, because it assumes particlelike behavior for the electron rather than wavelike behavior. In fact, *nodes are an intrinsic property of waves*, from moving waves of water in the ocean to the stationary, or standing, wave generated by vibrating a rope (Figure 5.11). A node simply corresponds to the place where the wave has zero amplitude. On either side of the node is a nonzero wave amplitude. Note that a wave has two phases—peaks above the zero line and troughs below—corresponding to different algebraic signs, + and −. Similarly, the different regions of 2*s* and 3*s* orbitals have different phases, + and −, as indicated in Figure 5.10 by different colors.

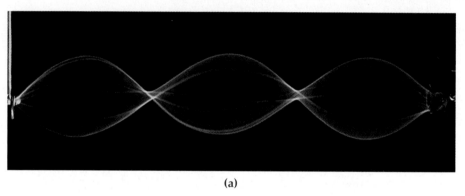

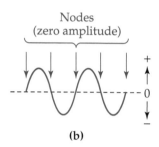

(a)   (b)

FIGURE 5.11 **(a)** When a rope is fixed at one end and vibrated rapidly at the other, a standing wave is generated. **(b)** The wave has two phases of different algebraic sign, + and −, separated by zero-amplitude regions called *nodes*.

p Orbitals

The *p* orbitals are dumbbell-shaped rather than spherical, with their electron distribution concentrated in identical lobes on either side of the nucleus and separated by a planar node cutting through the nucleus. As a result, the probability of finding a *p* electron near the nucleus is zero (Figure 5.12a). As with 2*s* orbitals, the two lobes of a *p* orbital have different phases, as indicated in Figure 5.12b by different shading. We'll see in Chapter 7 that these phases are crucial for bonding, because only lobes of the same phase can interact in forming covalent chemical bonds.

Since there are three allowable values of m_l when $l = 1$, each shell beginning with the second has three *p* orbitals, which are oriented in space at 90° angles to one another along the three coordinate axes *x*, *y*, and *z*. The three *p* orbitals in the second shell, for example, are designated $2p_x$, $2p_y$, and $2p_z$. As you might expect, *p* orbitals in the third and higher shells are larger than those in the second shell and extend farther from the nucleus. Their shape is roughly the same, however.

d and *f* Orbitals

The third and higher shells each contain five *d* orbitals, which differ from their *s* and *p* counterparts because they have two different shapes. Four of the five *d* orbitals are cloverleaf-shaped and have four lobes of maximum electron probability separated by two nodal planes through the nucleus (Figure 5.13a–d). The

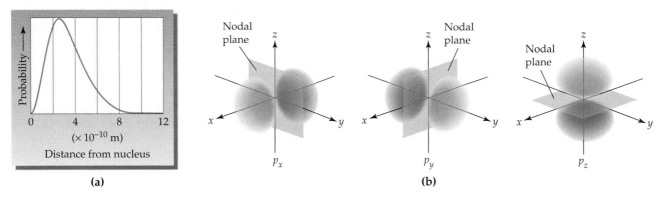

(a)

(b)

FIGURE 5.12 The three $2p$ orbitals. Part **(a)** is a plot giving the probability of finding a $2p$ electron as a function of its distance from the nucleus. Part **(b)** shows representations of the three $2p$ orbitals, each of which is dumbbell-shaped and oriented in space along one of the three coordinate axes x, y, or z. Each p orbital has two lobes of high electron probability separated by a nodal plane passing through the nucleus. The different shadings of the lobes reflect different algebraic signs analogous to the different phases of a wave.

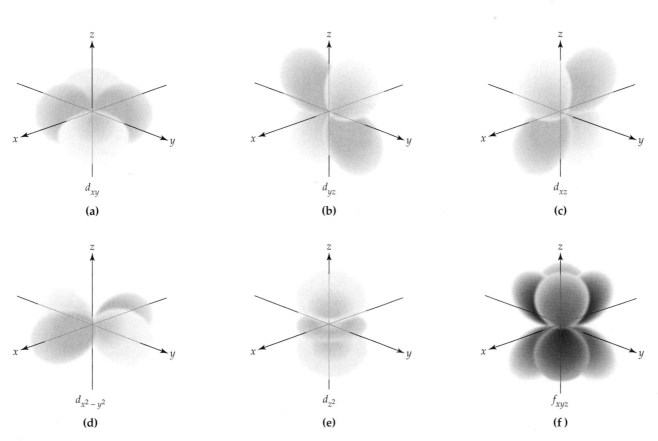

FIGURE 5.13 Representations of the five $3d$ orbitals. Four of the orbitals are shaped like a cloverleaf **(a–d)**, and the fifth is shaped like an elongated dumbbell inside a donut **(e)**. Also shown is one of the seven $4f$ orbitals **(f)**. As with p orbitals in Figure 5.12, the different shadings of the lobes reflect different phases.

fifth *d* orbital is similar in shape to a p_z orbital but has an additional donut-shaped region of electron probability centered in the *xy* plane (Figure 5.13e). In spite of their different shapes, all five *d* orbitals in a given shell have the same energy. As with *p* orbitals, alternating lobes of the *d* orbitals have different phases.

You may have noticed that both the number of nodal planes through the nucleus and the overall geometric complexity of the orbitals increases with the *l* quantum number of the subshell: *s* orbitals have one lobe and no nodal plane through the nucleus; *p* orbitals have two lobes and one nodal plane; *d* orbitals have four lobes and two nodal planes. The seven *f* orbitals are more complex still, having eight lobes of maximum electron probability separated by three nodal planes through the nucleus. (Figure 5.13f shows one of the seven 4*f* orbitals.) Fortunately, most of the elements we'll deal with in the following chapters don't use *f* orbitals in bonding, so we won't worry about them.

▶ **PROBLEM 5.14**   How many nodal planes through the nucleus do you think a g orbital has?

◆― **KEY CONCEPT PROBLEM 5.15**   Give a possible combination of *n* and *l* quantum numbers for the following 4th-shell orbital:

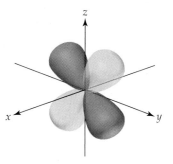

5.9   Quantum Mechanics and Atomic Spectra

Now that we've seen how atoms are structured according to the quantum mechanical model, let's return briefly to the subject of atomic line spectra first mentioned in Section 5.3. How does the quantum mechanical model account for the discrete wavelengths of light found in a line spectrum?

Each electron in an atom occupies an orbital, and each orbital has a specific energy level. Thus, *the energies available to electrons are quantized* and can have only the specific values associated with the orbitals they occupy. When an atom is heated in a flame or electric discharge, the added energy causes an electron to jump from a lower-energy orbital to a higher-energy orbital. In a hydrogen atom, for example, the electron might jump from the 1*s* orbital to a second-shell orbital, to a third-shell orbital, or to *any* higher-shell orbital depending on the amount of energy added. But the energetically excited atom is relatively unstable, and the electron rapidly returns to a lower-energy level accompanied by *emission* of energy equal to the difference between the higher and lower orbitals. *Since the energies of the orbitals are quantized, the amount of energy emitted is also quantized.* Thus, we observe the emission of only specific frequencies of radiation (Figure 5.14). By measuring the frequencies emitted by excited hydrogen atoms, we can calculate the energy differences between orbitals.

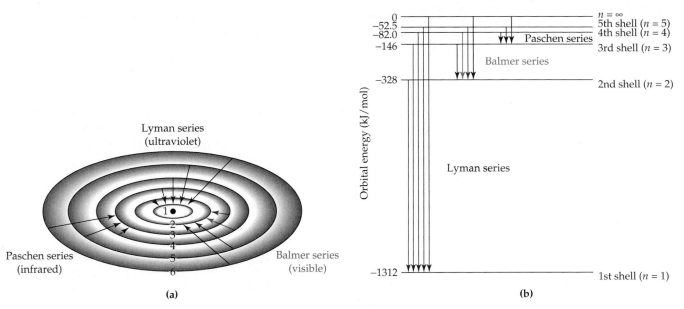

FIGURE 5.14 The origin of atomic line spectra. **(a)** When an electron falls from a higher-energy outer-shell orbital to a lower-energy inner-shell orbital, it emits electromagnetic radiation whose frequency corresponds to the energy difference between the two orbitals. (The orbital radii are not drawn to scale.) **(b)** The different spectral series correspond to electronic transitions from outer-shell orbitals to different inner-shell orbitals.

The variables m and n in the Balmer–Rydberg equation for hydrogen (Section 5.3) represent the principal quantum numbers of the two orbitals involved in the electronic transition. The variable n corresponds to the principal quantum number of the high-energy outer-shell orbital that the transition is *from*, and the variable m corresponds to the principal quantum number of the low-energy inner-shell orbital that the transition is *to*. When $m = 1$ (the Lyman series), for example, the frequencies of emitted light correspond to energy differences between various outer-shell orbitals and the first-shell orbital. When $m = 2$ (the Balmer series), the frequencies correspond to energy differences between outer-shell orbitals and the second-shell orbitals (Figure 5.14b).

$$\frac{1}{\lambda} = R\left(\frac{1}{m^2} - \frac{1}{n^2}\right)$$

Shell the transition is *to* ⟋   ⟍ Shell the transition is *from*
(inner-shell)   (outer-shell)

Notice in Figure 5.14b that as n becomes larger and approaches infinity, the energy difference between the n shell and the first shell converges to a value of 1312 kJ/mol. That is, 1312 kJ is released when electrons come from a great distance (the "infinite" shell) and add to H^+ to give a mole of hydrogen atoms, each with an electron in its first shell:

$$H^+ + e^- \longrightarrow H + \text{Energy (1312 kJ/mol)}$$

Since the energy *released* on adding an electron to H^+ is equal in magnitude to the energy *absorbed* on removing an electron from a hydrogen atom, we can also say that *1312 kJ/mol is required to remove the electron from a hydrogen atom.*

We'll see in the next chapter that the amount of energy necessary to remove an electron from an element provides an important clue about the element's chemical reactivity.

What is true for hydrogen is also true for all other atoms: All atoms show atomic line spectra when energetically excited electrons fall from higher-energy orbitals in outer shells to lower-energy orbitals in inner shells. As you might expect, though, these spectra become very complex for multielectron atoms in which different orbitals within a shell no longer have identical energies and in which a large number of electronic transitions are possible.

EXAMPLE 5.9   What is the energy difference (in kilojoules per mole) between the first and second shells of the hydrogen atom if the first emission in the Lyman series occurs at $\lambda = 121.5$ nm?

SOLUTION   The first line in the Lyman series corresponds to the emission of light as an electron falls from the second shell to the first shell, and the energy of that light is equal to the energy difference between shells. Knowing the wavelength of the light, you can calculate the energy of one photon using the Planck equation, $E = hc/\lambda$:

$$E = \frac{hc}{\lambda} = \frac{(6.626 \times 10^{-34}\text{ J}\cdot\cancel{\text{s}})\left(3.00 \times 10^8\ \frac{\cancel{\text{m}}}{\cancel{\text{s}}}\right)\left(10^9\ \frac{\cancel{\text{nm}}}{\cancel{\text{m}}}\right)}{121.5\ \cancel{\text{nm}}} = 1.64 \times 10^{-18}\text{ J}$$

Multiplying by Avogadro's number gives the answer in joules (or kilojoules) per mole:

$$(1.64 \times 10^{-18}\text{ J})(6.022 \times 10^{23}\text{ mol}^{-1}) = 9.88 \times 10^5\text{ J/mol} = 988\text{ kJ/mol}$$

The energy difference between the first and second shells of the hydrogen atom is 988 kJ/mol.

▶**PROBLEM 5.16**   Calculate (in kilojoules per mole) the energy necessary to remove an electron from the first shell of a hydrogen atom ($R = 1.097 \times 10^{-2}\text{ nm}^{-1}$).

5.10 Electron Spin and the Pauli Exclusion Principle

The three quantum numbers n, l, and m_l discussed in Section 5.7 define the energy, shape, and spatial orientation of orbitals, but they don't quite tell the whole story. When the line spectra of many multielectron atoms are studied in detail, it turns out that some lines actually occur as very closely spaced *pairs*. (You can see this pairing if you look closely at the visible spectrum of sodium in Figure 5.6.) Thus, there are more energy levels than simple quantum mechanics predicts, and a fourth quantum number is required. Denoted m_s, this fourth quantum number is related to a property called *electron spin*.

In certain respects, electrons behave as if they were spinning around an axis, much as the earth spins daily. Unlike the earth, though, electrons are free to spin in either a clockwise or a counterclockwise direction. This spinning charge gives rise to a tiny magnetic field and to a **spin quantum number, m_s,** which can have either of two values, +1/2 or −1/2 (Figure 5.15). A spin of +1/2 is usually represented by an up arrow (↑), and a spin of −1/2 is represented by a down arrow ($\downarrow$). Note that the value of m_s is independent of the other three quantum numbers, unlike the values of n, l, and m_l, which are interrelated.

The importance of the spin quantum number comes when electrons occupy specific orbitals in multielectron atoms. According to the **Pauli exclusion principle**, proposed in 1925 by the Austrian physicist Wolfgang Pauli (1900–1958), no two electrons in an atom can have the same four quantum numbers. In other words, the set of four quantum numbers associated with an electron acts as a unique "address" for that electron in an atom, and no two electrons can have the same address.

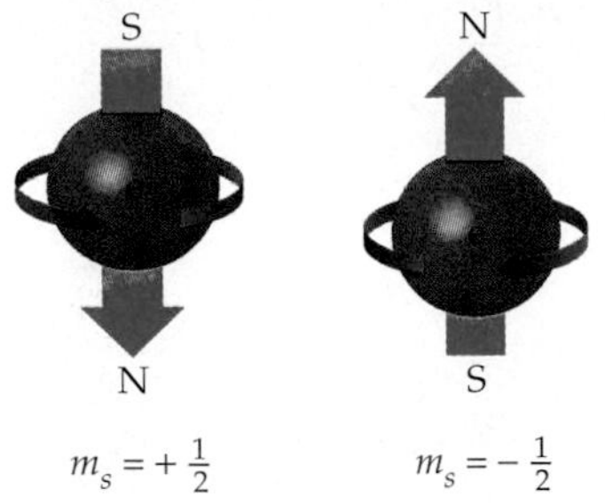

➤ **PAULI EXCLUSION PRINCIPLE:**   No two electrons in an atom can have the same four quantum numbers.

Think about the consequences of the Pauli exclusion principle. Electrons that occupy the same orbital have the same three quantum numbers, n, l, and m_l. But if they have the *same* values for n, l, and m_l, they must have *different* values for the fourth quantum number: either $m_s = +1/2$ or $m_s = -1/2$. Thus, *an orbital can hold only two electrons, which must have opposite spins.* An atom with x number of electrons therefore has at least $x/2$ occupied orbitals (though it might have more if some of its orbitals are only half-filled).

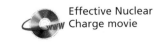

$m_s = +\frac{1}{2}$ $m_s = -\frac{1}{2}$

FIGURE 5.15 Electrons behave in some respects as if they were tiny charged spheres spinning around an axis. This spin (blue arrow) gives rise to a tiny magnetic field (green arrow) and to a fourth quantum number, m_s, which can have a value of either +1/2 or −1/2.

5.11 Orbital Energy Levels in Multielectron Atoms

We said in Section 5.7 that the energy level of an orbital in a hydrogen atom is determined by its principal quantum number n. Within a shell, all hydrogen orbitals have the same energy, independent of their other quantum numbers. The situation is different in multielectron atoms, however, where the energy level of a given orbital depends not only on the shell but also on the subshell. The s, p, d, and f orbitals within a given shell have slightly different energies in a multielectron atom, as shown previously in Figure 5.9, and there is even some crossover of energies between orbitals in different shells.

The difference in energy between subshells in multielectron atoms results from electron–electron repulsions. In hydrogen, the only electrical interaction is the attraction of the positive nucleus for the negative electron, but in multielectron atoms there are many different interactions to consider. Not only are there the *attractions* of the nucleus for each electron, there are also the *repulsions* between every electron and each of its neighboring electrons.

The repulsion of outer-shell electrons by inner-shell electrons is particularly important because the outer-shell electrons are pushed farther away from the nucleus and are thus held less tightly. Part of the attraction of the nucleus for an outer electron is thereby canceled, an effect we describe by saying that the outer electrons are *shielded* from the nucleus by the inner electrons (Figure 5.16). The net nuclear charge actually felt by an electron, called the **effective nuclear charge**, Z_{eff}, is often substantially lower than the actual nuclear charge Z:

Effective Nuclear Charge movie

➤ **Effective nuclear charge:** $Z_{eff} = Z_{actual} -$ Electron shielding

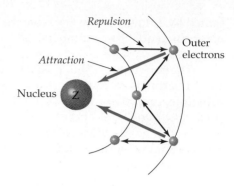

FIGURE 5.16 The origin of electron shielding and Z_{eff}. Outer electrons are attracted toward the nucleus by the nuclear charge but are pushed away by the repulsion of inner electrons. As a result, the nuclear charge actually felt by outer electrons is diminished, and we say that the outer electrons are *shielded* from the full charge of the nucleus by the inner electrons.

How does electron shielding lead to energy differences among orbitals within a shell? The answer is a consequence of the differences in orbital shapes. Compare a 2s orbital with a 2p orbital, for instance. The 2s orbital is spherical and has a large probability density near the nucleus, while the 2p orbitals are dumbbell-shaped and have a node at the nucleus (Section 5.8). Thus, in any given atom, an electron in a 2s orbital spends more time closer to the nucleus and is less shielded than an electron in a 2p orbital. A 2s electron therefore feels a higher Z_{eff}, is more tightly held by the nucleus, and is lower in energy than a 2p electron. In the same way, a 3p electron is closer to the nucleus on average, feels a higher Z_{eff}, and has a lower energy than a 3d electron. More generally, within any given shell n, a lower value of the angular-momentum quantum number l corresponds to a higher Z_{eff} and to a lower energy for the electron.

The idea that electrons in different orbitals are shielded differently and feel different values of Z_{eff} is a very useful one that we'll return to on several occasions to explain various chemical phenomena.

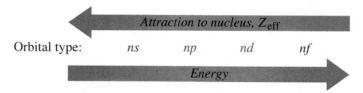

5.12 Electron Configurations of Multielectron Atoms

All the parts are now in place to provide a complete electronic description for every element. Knowing the relative energies of the various orbitals, we can predict for each element which orbitals are occupied by electrons—the element's **electron configuration**.

A set of three rules called the **aufbau principle** (from the German word for "building up") guides the filling order of orbitals. In general, each successive electron added to an atom occupies the lowest-energy orbital available. The resultant lowest-energy electron configuration is called the **ground-state configuration** of the atom. It may happen, of course, that several orbitals will have the same energy level—for example, the three p orbitals or the five d orbitals in a given subshell. Orbitals that have the same energy level are said to be **degenerate**.

Rules of the aufbau principle:

1. Lower-energy orbitals fill before higher-energy orbitals. (The ordering of energy levels for orbitals is shown in Figure 5.9.)

2. An orbital can hold only two electrons, which must have opposite spins. This is just a restatement of the Pauli exclusion principle (Section 5.10),

emphasizing that no two electrons in an atom can have the same four quantum numbers.

3. If two or more degenerate orbitals are available, one electron goes into each until all are half-full, a statement called **Hund's rule**. Only then does a second electron fill one of the orbitals. Furthermore, the electrons in each of the singly occupied orbitals must have the same value for their spin quantum number.

HUND'S RULE: If two or more orbitals with the same energy are available, one electron goes in each until all are half-full. The electrons in the half-filled orbitals all have the same value of their spin quantum number.

Hund's rule is just a matter of common sense. Because electrons repel one another, it makes sense that they remain as far apart as possible. Clearly, they can remain farther apart and be lower in energy if they are in different orbitals describing different spatial regions than if they are in the same orbital occupying the same region.

Let's look at some examples to see how the rules of the aufbau principle are applied.

- **Hydrogen:** Hydrogen has only one electron, which must go into the lowest-energy, 1s orbital. Thus, we say that the ground-state electron configuration of hydrogen is $1s^1$, where the superscript indicates the number of electrons in the specified orbital.

$$\textbf{H: } 1s^1$$

- **Helium:** Helium has two electrons, both of which fit into the lowest-energy, 1s orbital. The two electrons have opposite spins.

$$\textbf{He: } 1s^2$$

- **Lithium and beryllium:** With the 1s orbital full, both the third and fourth electrons go into the next available, 2s orbital.

$$\textbf{Li: } 1s^2 \, 2s^1 \qquad \textbf{Be: } 1s^2 \, 2s^2$$

- **Boron through neon:** The six elements from boron through neon have their three 2p orbitals filled successively. Since these three 2p orbitals have the same energy, they are degenerate and are filled according to Hund's rule. In carbon, for example, the two 2p electrons are in different orbitals, which can be arbitrarily specified as $2p_x$, $2p_y$, or $2p_z$ when writing the electron configuration. Similarly for nitrogen, whose three 2p electrons must be in three different orbitals. Although not usually noted in the written electron configuration, the electrons in each of the singly occupied carbon and nitrogen 2p orbitals must have the same value of the spin quantum number—either +1/2 or −1/2.

For clarity, we sometimes specify electron configurations using orbital-filling diagrams, in which electrons are represented by arrows. The two values of the spin quantum numbers are indicated by having the arrow point either up or down. An up–down pair indicates that an orbital is filled, while a single up (or down) arrow indicates that an orbital is half-filled. Note in the diagrams for carbon and nitrogen that the degenerate $2p$ orbitals are half-filled rather than filled, according to Hund's rule, and that the electron spin is the same in each.

B: $1s^2\,2s^2\,2p^1$ or $\underset{1s}{\uparrow\downarrow}\quad\underset{2s}{\uparrow\downarrow}\qquad\underset{2p}{\uparrow\;\;\underline{}\;\;\underline{}}$

C: $1s^2\,2s^2\,2p_x{}^1\,2p_y{}^1$ or $\underset{1s}{\uparrow\downarrow}\quad\underset{2s}{\uparrow\downarrow}\qquad\underset{2p}{\uparrow\;\;\uparrow\;\;\underline{}}$

N: $1s^2\,2s^2\,2p_x{}^1\,2p_y{}^1\,2p_z{}^1$ or $\underset{1s}{\uparrow\downarrow}\quad\underset{2s}{\uparrow\downarrow}\qquad\underset{2p}{\uparrow\;\;\uparrow\;\;\uparrow}$

From oxygen through neon, the three $2p$ orbitals are successively filled. For fluorine and neon, it's no longer necessary to distinguish among the different $2p$ orbitals, so we can simply write $2p^5$ and $2p^6$.

O: $1s^2\,2s^2\,2p_x{}^2\,2p_y{}^1\,2p_z{}^1$ or $\underset{1s}{\uparrow\downarrow}\quad\underset{2s}{\uparrow\downarrow}\qquad\underset{2p}{\uparrow\downarrow\;\;\uparrow\;\;\uparrow}$

F: $1s^2\,2s^2\,2p^5$ or $\underset{1s}{\uparrow\downarrow}\quad\underset{2s}{\uparrow\downarrow}\qquad\underset{2p}{\uparrow\downarrow\;\;\uparrow\downarrow\;\;\uparrow}$

Ne: $1s^2\,2s^2\,2p^6$ or $\underset{1s}{\uparrow\downarrow}\quad\underset{2s}{\uparrow\downarrow}\qquad\underset{2p}{\uparrow\downarrow\;\;\uparrow\downarrow\;\;\uparrow\downarrow}$

Interactive Periodic Table

- **Sodium and magnesium:** The $3s$ orbital is filled next, giving sodium and magnesium the ground-state electron configurations shown. Note that we often write the configurations in a shorthand version by giving the symbol of the noble gas in the previous row to indicate electrons in filled shells and then specifying only those electrons in partially filled shells.

Neon configuration

Na: $1s^2\,2s^2\,2p^6\,3s^1$ or   [Ne] $3s^1$

Mg: $1s^2\,2s^2\,2p^6\,3s^2$ or   [Ne] $3s^2$

- **Aluminum through argon:** The $3p$ orbitals are now filled, according to the same rules used previously for filling the $2p$ orbitals of boron through neon. Rather than explicitly identify which of the degenerate $3p$ orbitals are occupied in Si, P, and S, we'll simplify the writing by giving just the total number of electrons in the subshell. For example, we'll write $3p^2$ for silicon rather than $3p_x{}^1\,3p_y{}^1$.

Al: [Ne] $3s^2\,3p^1$ **Si:** [Ne] $3s^2\,3p^2$ **P:** [Ne] $3s^2\,3p^3$
S: [Ne] $3s^2\,3p^4$ **Cl:** [Ne] $3s^2\,3p^5$ **Ar:** [Ne] $3s^2\,3p^6$

- **Elements past argon:** Following the filling of the $3p$ subshell in argon, the first crossover in the orbital filling order is encountered. Rather than continue filling the third shell by populating the $3d$ orbitals, the next two electrons in potassium and calcium go into the $4s$ subshell. Only then does filling of the $3d$ subshell occur to give the first transition metal series from scandium through zinc.

K: [Ar] $4s^1$ **Ca:** [Ar] $4s^2$ **Sc:** [Ar] $4s^2\,3d^1 \rightarrow$ **Zn:** [Ar] $4s^2\,3d^{10}$

The experimentally determined ground-state electron configurations of the elements are shown in Figure 5.17.

FIGURE 5.17 Outer-shell, ground-state electron configurations of the elements.

1 H $1s^1$																	2 He $1s^2$
3 Li $2s^1$	4 Be $2s^2$											5 B $2s^2 2p^1$	6 C $2s^2 2p^2$	7 N $2s^2 2p^3$	8 O $2s^2 2p^4$	9 F $2s^2 2p^5$	10 Ne $2s^2 2p^6$
11 Na $3s^1$	12 Mg $3s^2$											13 Al $3s^2 3p^1$	14 Si $3s^2 3p^2$	15 P $3s^2 3p^3$	16 S $3s^2 3p^4$	17 Cl $3s^2 3p^5$	18 Ar $3s^2 3p^6$
19 K $4s^1$	20 Ca $4s^2$	21 Sc $4s^2 3d^1$	22 Ti $4s^2 3d^2$	23 V $4s^2 3d^3$	24 Cr $4s^1 3d^5$	25 Mn $4s^2 3d^5$	26 Fe $4s^2 3d^6$	27 Co $4s^2 3d^7$	28 Ni $4s^2 3d^8$	29 Cu $4s^1 3d^{10}$	30 Zn $4s^2 3d^{10}$	31 Ga $4s^2 3d^{10} 4p^1$	32 Ge $4s^2 3d^{10} 4p^2$	33 As $4s^2 3d^{10} 4p^3$	34 Se $4s^2 3d^{10} 4p^4$	35 Br $4s^2 3d^{10} 4p^5$	36 Kr $4s^2 3d^{10} 4p^6$
37 Rb $5s^1$	38 Sr $5s^2$	39 Y $5s^2 4d^1$	40 Zr $5s^2 4d^2$	41 Nb $5s^1 4d^4$	42 Mo $5s^1 4d^5$	43 Tc $5s^2 4d^5$	44 Ru $5s^1 4d^7$	45 Rh $5s^1 4d^8$	46 Pd $4d^{10}$	47 Ag $5s^1 4d^{10}$	48 Cd $5s^2 4d^{10}$	49 In $5s^2 4d^{10} 5p^1$	50 Sn $5s^2 4d^{10} 5p^2$	51 Sb $5s^2 4d^{10} 5p^3$	52 Te $5s^2 4d^{10} 5p^4$	53 I $5s^2 4d^{10} 5p^5$	54 Xe $5s^2 4d^{10} 5p^6$
55 Cs $6s^1$	56 Ba $6s^2$	57 La $6s^2 5d^1$	72 Hf $6s^2 4f^{14} 5d^2$	73 Ta $6s^2 4f^{14} 5d^3$	74 W $6s^2 4f^{14} 5d^4$	75 Re $6s^2 4f^{14} 5d^5$	76 Os $6s^2 4f^{14} 5d^6$	77 Ir $6s^2 4f^{14} 5d^7$	78 Pt $6s^1 4f^{14} 5d^9$	79 Au $6s^1 4f^{14} 5d^{10}$	80 Hg $6s^2 4f^{14} 5d^{10}$	81 Tl $6s^2 4f^{14} 5d^{10} 6p^1$	82 Pb $6s^2 4f^{14} 5d^{10} 6p^2$	83 Bi $6s^2 4f^{14} 5d^{10} 6p^3$	84 Po $6s^2 4f^{14} 5d^{10} 6p^4$	85 At $6s^2 4f^{14} 5d^{10} 6p^5$	86 Rn $6s^2 4f^{14} 5d^{10} 6p^6$
87 Fr $7s^1$	88 Ra $7s^2$	89 Ac $7s^2 6d^1$	104 Rf $7s^2 5f^{14} 6d^2$	105 Db $7s^2 5f^{14} 6d^3$	106 Sg $7s^2 5f^{14} 6d^4$	107 Bh $7s^2 5f^{14} 6d^5$	108 Hs $7s^2 5f^{14} 6d^6$	109 Mt $7s^2 5f^{14} 6d^7$	110 — 	111 — 	112 — 	114 — 		116 — 			118 —

58 Ce $6s^2 4f^1 5d^1$	59 Pr $6s^2 4f^3$	60 Nd $6s^2 4f^4$	61 Pm $6s^2 4f^5$	62 Sm $6s^2 4f^6$	63 Eu $6s^2 4f^7$	64 Gd $6s^2 4f^7 5d^1$	65 Tb $6s^2 4f^9$	66 Dy $6s^2 4f^{10}$	67 Ho $6s^2 4f^{11}$	68 Er $6s^2 4f^{12}$	69 Tm $6s^2 4f^{13}$	70 Yb $6s^2 4f^{14}$	71 Lu $6s^2 4f^{14} 5d^1$
90 Th $7s^2 6d^2$	91 Pa $7s^2 5f^2 6d^1$	92 U $7s^2 5f^3 6d^1$	93 Np $7s^2 5f^4 6d^1$	94 Pu $7s^2 5f^6$	95 Am $7s^2 5f^7$	96 Cm $7s^2 5f^7 6d^1$	97 Bk $7s^2 5f^9$	98 Cf $7s^2 5f^{10}$	99 Es $7s^2 5f^{11}$	100 Fm $7s^2 5f^{12}$	101 Md $7s^2 5f^{13}$	102 No $7s^2 5f^{14}$	103 Lr $7s^2 5f^{14} 6d^1$

5.13 Electron Configurations and the Periodic Table

TABLE 5.3 Valence-Shell Electron Configurations of Main-Group Elements

Group	Valence-Shell Electron Configuration	
1A	ns^1	(1 total)
2A	ns^2	(2 total)
3A	$ns^2\,np^1$	(3 total)
4A	$ns^2\,np^2$	(4 total)
5A	$ns^2\,np^3$	(5 total)
6A	$ns^2\,np^4$	(6 total)
7A	$ns^2\,np^5$	(7 total)
8A	$ns^2\,np^6$	(8 total)

Why are electron configurations so important, and what do they have to do with the periodic table? The answers become clear when you look closely at Figure 5.17. Focusing only on the electrons in the outermost shell, called the **valence shell**, *all the elements in a given group of the periodic table have similar valence-shell electron configurations* (Table 5.3). The group 1A elements, for example, all have an s^1 valence-shell configuration; the group 2A elements have an s^2 valence-shell configuration; the group 3A elements have an s^2p^1 valence-shell configuration; and so on across every group of the periodic table (except for a small number of anomalies). Furthermore, because the valence-shell electrons are the most loosely held, they are the most important for determining an element's properties, thus explaining why the elements in a given group of the periodic table have similar chemical behavior.

The periodic table can be divided into four regions, or *blocks*, of elements according to the orbitals being filled (Figure 5.18). The group 1A and 2A elements on the left side of the table are called the *s***-block elements** because they result from the filling of an *s* orbital; the group 3A–8A elements on the right side of the table are the *p***-block elements** because they result from the filling of *p* orbitals; the transition metal *d***-block elements** in the middle of the table result from the filling of *d* orbitals; and the lanthanide/actinide *f***-block elements** detached at the bottom of the table result from the filling of *f* orbitals.

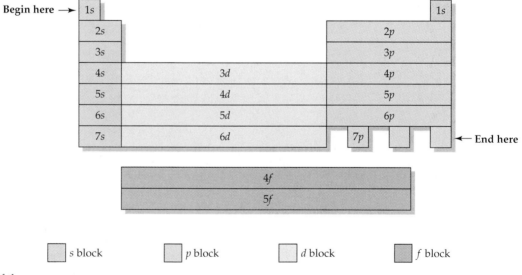

FIGURE 5.18 Blocks of the periodic table, corresponding to filling the different kinds of orbitals. Beginning at the top left and going across successive rows of the periodic table provides a method for remembering the order of orbital filling: $1s \rightarrow 2s \rightarrow 2p \rightarrow 3s \rightarrow 3p \rightarrow 4s \rightarrow 3d \rightarrow 4p$, and so on.

Thinking of the periodic table as outlined in Figure 5.18 provides a simple way to remember the order of orbital filling. Beginning at the top left corner of the periodic table and going across successive rows gives the correct orbital-filling order automatically. The first row of the periodic table, for instance, contains only the two *s*-block elements H and He, so the first available *s* orbital (1*s*) is filled first. The second row begins with two *s*-block elements (Li and Be) and continues with six *p*-block elements (B through Ne), so the next available *s* orbital (2*s*) and then the first available *p* orbitals (2*p*) are filled. The third row is similar to the second row, so the 3*s* and 3*p* orbitals are filled. The fourth row again starts with two *s*-block elements (K and Ca) but is then followed by ten *d*-block elements (Sc through Zn) and six *p*-block elements (Ga through Kr). Thus, the order of orbital filling is $4s$ followed by the first available d orbitals ($3d$) followed by $4p$. Continuing through successive rows of the periodic table gives the entire filling order:

$$1s \longrightarrow 2s \longrightarrow 2p \longrightarrow 3s \longrightarrow 3p \longrightarrow 4s \longrightarrow 3d \longrightarrow 4p \longrightarrow 5s \longrightarrow$$

$$4d \longrightarrow 5p \longrightarrow 6s \longrightarrow 4f \longrightarrow 5d \longrightarrow 6p \longrightarrow 7s \longrightarrow 5f \longrightarrow 6d \longrightarrow 7p$$

EXAMPLE 5.10   Give the ground-state electron configuration of arsenic, $Z = 33$. Draw an orbital-filling diagram, indicating the electrons as up or down arrows.

SOLUTION   Think of the periodic table as having s, p, d, and f blocks of elements, as shown in Figure 5.18. Start with hydrogen at the upper left and fill orbitals until 33 electrons have been added. Remember that only two electrons can go into an orbital and that each one of a set of degenerate orbitals must be half-filled before any one can be completely filled.

$$\text{As: } 1s^2\,2s^2\,2p^6\,3s^2\,3p^6\,4s^2\,3d^{10}\,4p^3 \quad \text{or} \quad [\text{Ar}]\,4s^2\,3d^{10}\,4p^3$$

An orbital-filling diagram indicates the electrons in each orbital as arrows. Note that the three $4p$ electrons all have the same spin:

As: [Ar] ⇅ / $4s$ ⇅ ⇅ ⇅ ⇅ ⇅ / $3d$ ↑ ↑ ↑ / $4p$

▶ **PROBLEM 5.17**   Give expected ground-state electron configurations for the following atoms, and draw orbital-filling diagrams for parts **(a)–(c)**.
(a) Ti ($Z = 22$)   **(b)** Zn ($Z = 30$)   **(c)** Sn ($Z = 50$)   **(d)** Pb ($Z = 82$)

▶ **PROBLEM 5.18**   What do you think is a likely ground-state electron configuration for the sodium *ion*, Na^+, formed by loss of an electron from a neutral sodium atom?

✦ **KEY CONCEPT PROBLEM 5.19**   Identify the atom with the following ground-state electron configuration:

[Ar] ⇅   ⇅ ⇅ ⇅ ↓ ↓   __ __ __

5.14   Some Anomalous Electron Configurations

The rules discussed in the previous section for determining ground-state electron configurations work well but are not completely accurate. As shown in Figure 5.17, 90 electron configurations are correctly accounted for by the rules, but 19 of the predicted configurations are incorrect.

The reasons for the anomalies generally have to do with the unusual stability of both half-filled and fully filled subshells. Chromium, for example, which we would expect to have the configuration [Ar] $4s^2\,3d^4$, actually has the configuration [Ar] $4s^1\,3d^5$. By moving an electron from the $4s$ orbital to an energetically similar $3d$ orbital, chromium trades one filled subshell ($4s^2$) for two half-filled subshells ($4s^1\,3d^5$). In the same way, copper, which we would expect to have the configuration [Ar] $4s^2\,3d^9$, actually has the configuration [Ar] $4s^1\,3d^{10}$. By transferring an electron from the $4s$ orbital to a $3d$ orbital, copper trades one filled subshell ($4s^2$) for a different filled subshell ($3d^{10}$) and gains a half-filled subshell ($4s^1$).

Most of the anomalous electron configurations shown in Figure 5.17 occur in elements with atomic numbers greater than $Z = 40$, where the energy differences between subshells are small. In all cases, the transfer of an electron from one subshell to another lowers the total energy of the atom because of a decrease in electron–electron repulsions.

▶**PROBLEM 5.20** Look at the electron configurations in Figure 5.17, and identify the 19 elements that are anomalous.

5.15 Electron Configurations and Periodic Properties: Atomic Radii

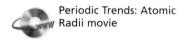
Periodic Trends: Atomic Radii movie

Interactive Periodic Table

One of the many periodic properties of the elements that can be explained by electron configurations is size, or atomic radius. You might wonder, though, how we can talk about a definite "size" for an atom, having said in Section 5.8 that the electron clouds around atoms have no specific boundaries. What is usually done is to assume that the radius of an atom is half the distance between the nuclei of two identical atoms when they are covalently bonded together. In the Cl_2 molecule, for example, the distance between the two chlorine nuclei is 198 pm; in diamond (elemental carbon), the distance between two carbon nuclei is 154 pm. Thus, we say that the atomic radius of chlorine is half the Cl–Cl distance, or 99 pm, and the atomic radius of carbon is half the C–C distance, or 77 pm.

It's possible to check the accuracy of atomic radii by making sure that the assigned values are additive. For instance, since the atomic radius of Cl is 99 pm and the atomic radius of C is 77 pm, the distance between Cl and C nuclei when those two atoms are bonded together ought to be roughly 99 pm + 77 pm, or 176 pm. In fact, the measured distance between chlorine and carbon in the chloromethane molecule (CH_3Cl) is 178 pm, remarkably close to the expected value.

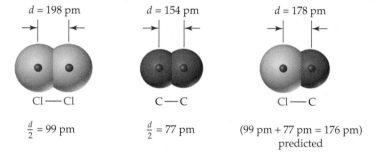

As shown pictorially in Figure 5.19 (and graphically in Figure 5.1 on page 160), a comparison of atomic radius versus atomic number shows a periodic rise-and-fall pattern. Atomic radii *increase* going down a group of the periodic table (Li → Na → K → Rb → Cs, for example) but *decrease* going across a row from left to right (Na → Mg → Al → Si → P → S → Cl, for example).

The increase in atomic radius going down a group of the periodic table occurs because successively larger valence-shell orbitals are occupied. In Li, for example, the outermost occupied shell is the second one ($2s^1$); in Na it's the third one ($3s^1$); in K it's the fourth one ($4s^1$); and so on through Rb ($5s^1$), Cs ($6s^1$), and Fr ($7s^1$). Because larger shells are occupied, the atomic radii are also larger.

The decrease in atomic radius from left to right across the periodic table occurs because of an increase in effective nuclear charge, Z_{eff}, for the valence-shell electrons. As we saw in Section 5.11, Z_{eff}, the net nuclear charge actually felt by an electron, is lower than the true nuclear charge Z because of shielding by other electrons. The amount of shielding felt by an electron depends on both the shell and subshell of the other electrons with which it is interacting.

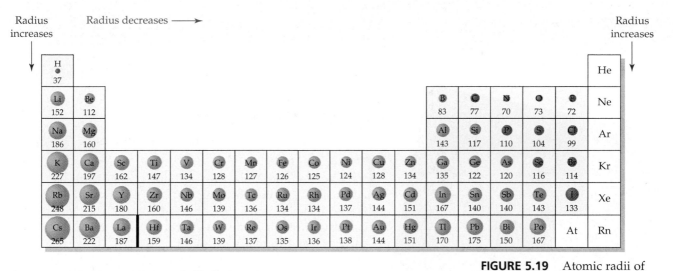

Radius increases

Radius decreases ⟶

Radius increases

FIGURE 5.19 Atomic radii of the elements in picometers.

As a general rule, a valence-shell electron is

- Strongly shielded by electrons in inner shells, which are closer to the nucleus
- Less strongly shielded by other electrons in the same shell, according to the order $s > p > d > f$
- Weakly shielded by other electrons in the same subshell, which are at the same distance from the nucleus

On going across the third period from Na to Ar, for example, each additional electron adds to the same shell (from $3s^1$ for Na to $3s^2\,3p^6$ for Ar). Because electrons in the same shell are at approximately the same distance from the nucleus, they are relatively ineffective at shielding one another. At the same time, though, the nuclear charge Z increases from +11 for Na to +18 for Ar. Thus, the *effective* nuclear charge for the valence-shell electrons increases across the period from 2.20 for Na to 6.75 for Ar, drawing all the valence-shell electrons closer to the nucleus and progressively shrinking the atomic radii (Figure 5.20).

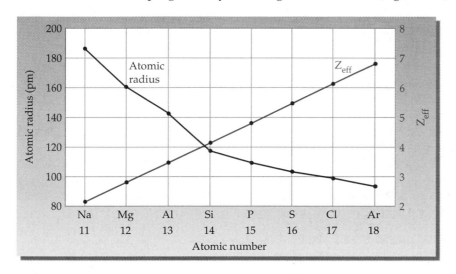

FIGURE 5.20 Plots of Z_{eff} for the highest-energy electron and atomic radius versus atomic number. As Z_{eff} increases, the valence-shell electrons are attracted more strongly to the nucleus, and the atomic radius therefore decreases.

▶ **PROBLEM 5.21** Which atom in each of the following pairs would you expect to be larger? Explain.
(a) Mg or Ba   **(b)** W or Au   **(c)** Si or Sn   **(d)** Ce or Lu

The Aurora Borealis: Atomic Spectra on a Grand Scale

The aurora borealis, or northern lights, are the result of light emission from energetically excited atoms in the earth's atmosphere.

Every so often on a clear evening, residents in Alaska, Canada, and other far-northern parts of the world are treated to a breathtaking display of celestial fireworks— the so-called northern lights, or *aurora borealis*. The lights appear in many different forms—as curtains, arcs, rays, and gauzy patches—and in many different colors. All, however, result from the emission of light by energetically excited atoms in the upper atmosphere, the same phenomenon that gives rise to atomic line spectra.

The aurora borealis is caused by a chain of events that begins on the surface of the sun with a massive solar flare. These flares eject a solar "gas" of energetic protons and electrons that reach earth after about 2 days and are then attracted toward the north and south magnetic poles. (The Southern Hemisphere has its own display of lights called the *aurora australis*.) The energetic electrons are deflected by the earth's magnetic field into a series of sheetlike beams, much as iron filings scattered around a magnet are deflected into a series of lines by the magnet's field of force. The electrons then collide with O_2 and N_2 molecules in the upper atmosphere, exciting them, ionizing them, and breaking them apart into O and N atoms.

The energetically excited atoms, ions, and molecules generated by collisions with electrons emit energy of characteristic wavelengths when they decay to their ground states. The O_2^+ ions emit a red light around 630 nm; N_2^+ ions emit violet and blue light at 391.4 nm and 470.0 nm; and O atoms emit a greenish-yellow light at 557.7 nm and a deep red light at 630.0 nm.

Protons in the solar gas are also responsible for part of the auroral display as they too collide with oxygen atoms when they descend into the upper atmosphere. The protons pull electrons from the oxygen atoms, yielding excited O^+ ions and H atoms that give off still more colors when they return to their ground states. The hydrogen atoms, in particular, emit all the wavelengths of visible light in the Balmer series.

Northern lights are seen almost every night by observers within about 2000 km of the north magnetic pole and are visible in the northern parts of the United States several times a year. They have even been seen as far south as Mexico during times of massive solar disturbances.

▶ **PROBLEM 5.22** Why are the auroras seen primarily near the north and south poles rather than elsewhere on earth?

Summary

Understanding the nature of atoms and molecules begins with an understanding of light and other kinds of **electromagnetic radiation** that make up the **electromagnetic spectrum**. An electromagnetic wave travels through a vacuum at the speed of light (c) and is characterized by its **frequency** (ν), **wavelength** (λ), and **amplitude**. Unlike the white light of the sun, which consists of a nearly continuous distribution of wavelengths, the light emitted by an excited atom consists of only a few discrete wavelengths, a so-called **line spectrum**. The observed wavelengths correspond to the specific energy differences between energies of different orbitals.

Atomic line spectra arise because electromagnetic radiation is **quantized** and occurs only in discrete units, or **quanta**. Just as light behaves in some respects like a stream of small particles (**photons**), so electrons and other tiny units of matter behave in some respects like waves. The wavelength of a particle of mass m traveling at a velocity v is given by the **de Broglie equation**, $\lambda = h/mv$, where h is **Planck's constant**.

The **quantum mechanical model** proposed in 1926 by Erwin Schrödinger describes an atom by a mathematical equation similar to that used to describe wave motion. The behavior of each electron in an atom is characterized by a **wave function**, or **orbital**, the square of which defines the probability of finding the electron in various regions of space. Each wave function has a set of three variables, called **quantum numbers**. The **principal quantum number** n defines the size of the orbital; the **angular-momentum quantum number** l defines the shape of the orbital; and the **magnetic quantum number** m_l defines the spatial orientation of the orbital. In a hydrogen atom, which contains only one electron, the energy of an orbital depends only on n. In a multielectron atom, the energy of an orbital depends on both n and l. In addition, the **spin quantum number** m_s specifies the electron spin as either up or down.

Orbitals can be grouped into successive layers, or **shells**, according to their principal quantum number n. Within a shell, orbitals are grouped into s, p, d, and f **subshells** according to their angular-momentum quantum numbers l. An orbital in an s subshell is spherical, an orbital in a p subshell is dumbbell-shaped, and four of the five orbitals in a d subshell are cloverleaf-shaped.

The **ground-state electron configuration** of a multielectron atom is arrived at by following a series of rules called the **aufbau principle**.

1. The lowest-energy orbitals fill first.
2. Only two electrons of opposite spin go into any one orbital (the **Pauli exclusion principle**).
3. If two or more orbitals are equal in energy (**degenerate**), each is half-filled before any one of them is completely filled (**Hund's rule**).

The periodic table is the most important organizing principle of chemistry. It is successful because elements in each group of the periodic table have similar valence-shell electron configurations and therefore have similar properties. For example, atomic radii of elements show a periodic rise-and-fall pattern according to their position in the table. Atomic radii increase going down a group and decrease from left to right across a period because of an increase in **effective nuclear charge** (Z_{eff}).

Key Words

amplitude *162*
angular-momentum quantum number *l 174*
aufbau principle *184*
Balmer–Rydberg equation *166*
d-block element *188*
de Broglie equation *171*
degenerate *184*
effective nuclear charge (Z_{eff}) *183*
electromagnetic radiation *162*
electromagnetic spectrum *162*
electron configuration *184*
f-block element *188*
frequency (ν) *162*
ground-state electron configuration *184*
Heisenberg uncertainty principle *172*
hertz (Hz) *162*
Hund's rule *185*
line spectrum *165*
magnetic quantum number m_l *175*
node *178*
orbital *174*
Pauli exclusion principle *183*
p-block element *188*
photon *169*
Planck's constant *168*
principal quantum number n *174*
quanta *168*
quantized *168*
quantum mechanical model *172*
quantum number *174*
s-block element *188*
shell *174*
spin quantum number, m_s *182*
subshell *174*
valence shell *188*
wave function *174*
wavelength (λ) *162*

Key Concept Summary

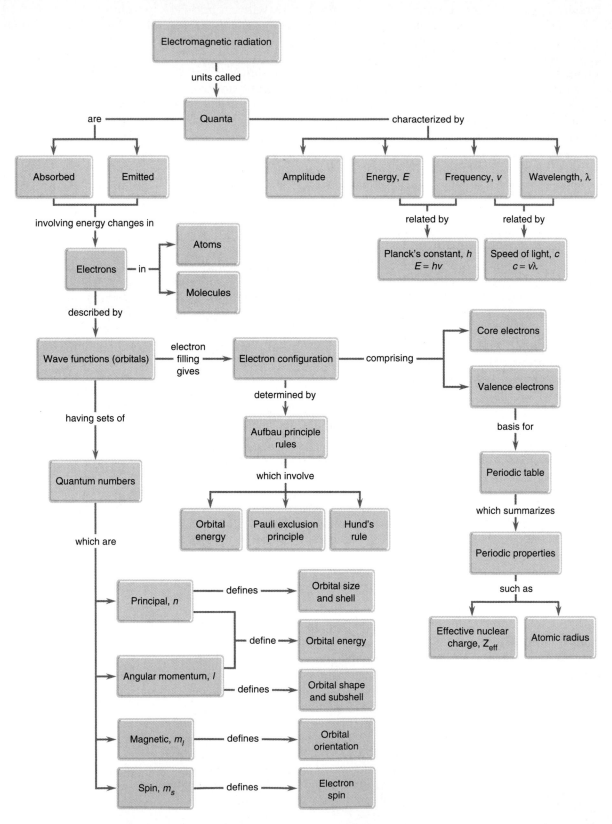

Understanding Key Concepts

Problems 5.1–5.22 appear within the chapter.

5.23 Where on the blank outline of the periodic table do elements that meet the following descriptions appear?

(a) Elements with the valence-shell ground-state electron configuration $ns^2\,np^5$

(b) An element whose fourth shell contains two p electrons

(c) An element with the ground-state electron configuration [Ar] $4s^2\,3d^{10}\,4p^5$

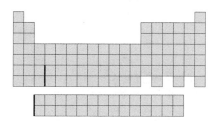

5.24 Where on the periodic table do elements that meet the following descriptions appear?

(a) Elements with electrons whose largest principal quantum number is $n = 4$

(b) Elements with the valence-shell ground-state electron configuration $ns^2\,np^3$

(c) Elements that have only one unpaired p electron

(d) The d-block elements

(e) The p-block elements

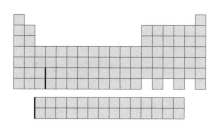

5.25 Two electromagnetic waves are represented below.

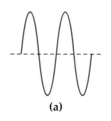

(a) Which wave has the greater intensity?

(b) Which wave corresponds to higher energy radiation?

(c) Which wave represents yellow light, and which represents infrared radiation?

5.26 What atom has the following orbital-filling diagram?

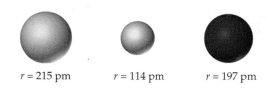

5.27 Draw orbital-filling diagrams of the sort shown in Problem 5.26 for the following atoms. Show each electron as an up or down arrow, and use the abbreviation of the preceding noble gas to represent inner-shell electrons.

(a) K   (b) Mo   (c) Sn   (d) Ir

5.28 Which of the following three spheres represents a Ca atom, which an Sr atom, and which a Br atom?

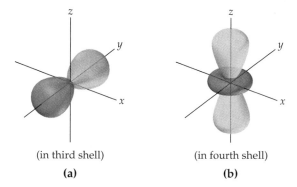

$r = 215$ pm $r = 114$ pm $r = 197$ pm

5.29 Identify each of the following orbitals, and give n and l quantum numbers for each.

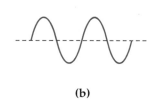

(in third shell)   (in fourth shell)

(a)   (b)

Additional Problems

Electromagnetic Radiation

5.30 Which has the higher frequency, red light or violet light? Which has the longer wavelength? Which has the greater energy?

5.31 Which has the higher frequency, infrared light or ultraviolet light? Which has the longer wavelength? Which has the greater energy?

5.32 What is the wavelength (in meters) of ultraviolet light with $\nu = 5.5 \times 10^{15}$ s^{-1}?

5.33 What is the frequency of a microwave with $\lambda = 4.33 \times 10^{-3}$ m?

5.34 Calculate the energies of the following waves (in kilojoules per mole), and tell which member of each pair has the higher value.

(a) An FM radio wave at 99.5 MHz and an AM radio wave at 115.0 kHz

(b) An X ray with $\lambda = 3.44 \times 10^{-9}$ m and a microwave with $\lambda = 6.71 \times 10^{-2}$ m

5.35 The MRI (magnetic resonance imaging) body scanners used in hospitals operate with 400 MHz radiofrequency energy. How much energy does this correspond to in kilojoules per mole?

5.36 What is the wavelength (in meters) of photons with the following energies? In what region of the electromagnetic spectrum does each appear?

(a) 90.5 kJ/mol

(b) 8.05×10^{-4} kJ/mol

(c) 1.83×10^{3} kJ/mol

5.37 What is the energy of each of the following photons in kilojoules per mole?

(a) $\nu = 5.97 \times 10^{19}$ s^{-1}

(b) $\nu = 1.26 \times 10^{6}$ s^{-1}

(c) $\lambda = 2.57 \times 10^{2}$ m

5.38 Protons and electrons can be given very high energies in particle accelerators. What is the wavelength (in meters) of an electron (mass $= 9.11 \times 10^{-31}$ kg) that has been accelerated to 99% of the speed of light? In what region of the electromagnetic spectrum is this wavelength?

5.39 What is the wavelength (in meters) of a proton (mass $= 1.673 \times 10^{-24}$ g) that has been accelerated to 25% of the speed of light? In what region of the electromagnetic spectrum is this wavelength?

5.40 What is the de Broglie wavelength (in meters) of a baseball weighing 145 g and traveling at 156 km/h?

5.41 What is the de Broglie wavelength (in meters) of a mosquito weighing 1.55 mg and flying at 1.38 m/s?

5.42 At what speed (in meters per second) must a 145 g baseball be traveling to have a de Broglie wavelength of 0.500 nm?

5.43 What velocity would an electron need for its de Broglie wavelength to be that of red light (750 nm)? The mass of an electron is 9.11×10^{-31} kg.

Atomic Spectra

5.44 According to the equation for the Balmer line spectrum of hydrogen, a value of $n = 3$ gives a red spectral line at 656.3 nm, a value of $n = 4$ gives a green line at 486.1 nm, and a value of $n = 5$ gives a blue line at 434.0 nm. Calculate the energy (in kilojoules per mole) of the radiation corresponding to each of these spectral lines.

5.45 According to the values cited in Problem 5.44, the wavelength differences between lines in the Balmer series become smaller as n becomes larger. In other words, the wavelengths converge toward a minimum value as n becomes very large. At what wavelength (in nanometers) do the lines converge?

5.46 The wavelength of light at which the Balmer series converges (Problem 5.45) corresponds to the amount of energy required to completely remove an electron from the second shell of a hydrogen atom. Calculate this energy in kilojoules per mole.

5.47 Lines in the Brackett series of the hydrogen spectrum are caused by emission of energy accompanying the fall of an electron from outer shells to the fourth shell. The lines can be calculated using the Balmer–Rydberg equation:

$$\frac{1}{\lambda} = R\left(\frac{1}{m^2} - \frac{1}{n^2}\right)$$

where $m = 4$, $R = 1.097 \times 10^{-2}$ nm^{-1}, and n is an integer greater than 4. Calculate the wavelengths (in nanometers) and energies (in kilojoules per mole) of the first two lines in the Brackett series. In what region of the electromagnetic spectrum do they fall?

5.48 Sodium atoms emit light with a wavelength of 330 nm when an electron moves from a $4p$ orbital to a $3s$ orbital. What is the energy difference between the orbitals in kilojoules per mole?

5.49 Excited rubidium atoms emit red light with $\lambda = 795$ nm. What is the energy difference (in kilojoules per mole) between orbitals that give rise to this emission?

Orbitals and Quantum Numbers

5.50 What are the four quantum numbers, and what does each specify?

5.51 What is the Heisenberg uncertainty principle, and how does it affect our description of atomic structure?

5.52 Why do we have to use an arbitrary value such as 95% to determine the spatial limitations of an orbital?

5.53 How many nodal surfaces does a $4s$ orbital have? Draw a cutaway representation of a $4s$ orbital showing the nodes and the regions of maximum electron probability.

5.54 What is meant by the term effective nuclear charge, Z_{eff}, and what is it due to?

5.55 How does electron shielding in multielectron atoms give rise to energy differences among $3s$, $3p$, and $3d$ orbitals?

5.56 Give the allowable combinations of quantum numbers for each of the following electrons:
(a) A $4s$ electron **(b)** A $3p$ electron
(c) A $5f$ electron **(d)** A $5d$ electron

5.57 Give the orbital designations of electrons with the following quantum numbers:
(a) $n = 3, l = 0, m_l = 0$ **(b)** $n = 2, l = 1, m_l = -1$
(c) $n = 4, l = 3, m_l = -2$ **(d)** $n = 4, l = 2, m_l = 0$

5.58 Tell which of the following combinations of quantum numbers are not allowed. Explain your answers.
(a) $n = 3, l = 0, m_l = -1$
(b) $n = 3, l = 1, m_l = 1$
(c) $n = 4, l = 4, m_l = 0$

5.59 Which of the following combinations of quantum numbers can refer to an electron in a ground-state cobalt atom ($Z = 27$)?
(a) $n = 3, l = 0, m_l = 2$
(b) $n = 4, l = 2, m_l = -2$
(c) $n = 3, l = 1, m_l = 0$

5.60 What is the maximum number of electrons in an atom whose highest-energy electrons have the principal quantum number $n = 5$?

5.61 What is the maximum number of electrons in an atom whose highest-energy electrons have the principal quantum number $n = 4$ and the angular-momentum quantum number $l = 0$?

5.62 Use the Heisenberg uncertainty principle to calculate the uncertainty (in meters) in the position of a honeybee weighing 0.68 g and traveling at a velocity of 0.85 m/s. Assume that the uncertainty in the velocity is 0.1 m/s.

5.63 The mass of a helium atom is 4.0026 amu, and its average velocity at 25°C is 1.36×10^3 m/s. What is the uncertainty (in meters) in the position of a helium atom if the uncertainty in its velocity is 1%?

Electron Configurations

5.64 Why does the number of elements in successive periods of the periodic table increase by the progression 2, 8, 18, 32?

5.65 Which two of the four quantum numbers determine the energy level of an orbital in a multielectron atom?

5.66 Which orbital in each of the following pairs is higher in energy?
(a) $5p$ or $5d$ **(b)** $4s$ or $3p$ **(c)** $6s$ or $4d$

5.67 Order the orbitals for a multielectron atom in each of the following lists according to increasing energy:
(a) $4d, 3p, 2p, 5s$
(b) $2s, 4s, 3d, 4p$
(c) $6s, 5p, 3d, 4p$

5.68 According to the aufbau principle, which orbital is filled immediately *after* each of the following in a multielectron atom?
(a) $4s$ **(b)** $3d$ **(c)** $5f$ **(d)** $5p$

5.69 According to the aufbau principle, which orbital is filled immediately *before* each of the following?
(a) $3p$ **(b)** $4p$ **(c)** $4f$ **(d)** $5d$

5.70 Give the expected ground-state electron configurations for the following elements:
(a) Ti **(b)** Ru **(c)** Sn **(d)** Sr **(e)** Se

5.71 Give the expected ground-state electron configurations for atoms with the following atomic numbers:
(a) $Z = 55$ **(b)** $Z = 40$
(c) $Z = 80$ **(d)** $Z = 62$

5.72 Draw orbital-filling diagrams for the following atoms. Show each electron as an up or down arrow, and use the abbreviation of the preceding noble gas to represent inner-shell electrons.
(a) Rb **(b)** W **(c)** Ge **(d)** Zr

5.73 Draw orbital-filling diagrams for atoms with the following atomic numbers. Show each electron as an up or down arrow, and use the abbreviation of the preceding noble gas to represent inner-shell electrons.
(a) $Z = 25$ **(b)** $Z = 56$ **(c)** $Z = 28$ **(d)** $Z = 47$

5.74 Order the electrons in the following orbitals according to their shielding ability: $4s$, $4d$, $4f$.

5.75 Order the following elements according to increasing Z_{eff}: Ca, Se, Kr, K.

5.76 What is the expected ground-state electron configuration of the recently discovered element with $Z = 118$?

5.77 What is the atomic number and expected ground-state electron configuration of the yet undiscovered element directly below Fr in the periodic table?

5.78 How many unpaired electrons are present in each of the following ground-state atoms?

(a) O (b) Si (c) K (d) As

5.79 Identify the following atoms:

(a) It has the ground-state electron configuration [Ar] $3d^{10} 4s^2 4p^1$.

(b) It has the ground-state electron configuration [Kr] $4d^{10}$.

5.80 At what atomic number is the filling of a g orbital likely to begin?

5.81 Assuming that g orbitals fill according to Hund's rule, what is the atomic number of first element to have a filled g orbital?

Atomic Radii and Periodic Properties

5.82 Why do atomic radii increase going down a group of the periodic table?

5.83 Why do atomic radii decrease from left to right across a period of the periodic table?

5.84 Order the following atoms according to increasing atomic radius: S, F, O.

5.85 Which atom in each of the following pairs has a larger radius?

(a) Na or K (b) V or Ta

(c) V or Zn (d) Li or Ba

5.86 The amount of energy that must be added to remove an electron from a neutral atom to give a cation is called the atom's *ionization energy*. Which would you expect to have the larger ionization energy, Na or Mg? Explain.

5.87 The amount of energy released when an electron adds to a neutral atom to give an anion is called the atom's *electron affinity*. Which would you expect to have the larger electron affinity, C or F? Explain.

General Problems

5.88 Use the Balmer equation to calculate the wavelength (in nanometers) of the spectral line for hydrogen when $n = 6$. What is the energy (in kilojoules per mole) of the radiation corresponding to this line?

5.89 Lines in the Pfund series of the hydrogen spectrum are caused by emission of energy accompanying the fall of an electron from outer shells to the fifth shell. Use the Balmer–Rydberg equation to calculate the wavelengths (in nanometers) and energies (in kilojoules per mole) of the two longest-wavelength lines in the Pfund series. In what region of the electromagnetic spectrum do they fall?

5.90 What is the shortest wavelength (in nanometers) in the Pfund series (Problem 5.89) for hydrogen?

5.91 What is the wavelength (in meters) of photons with the following energies? In what region of the electromagnetic spectrum does each appear?

(a) 142 kJ/mol

(b) 4.55×10^{-2} kJ/mol

(c) 4.81×10^4 kJ/mol

5.92 What is the energy of each of the following photons in kilojoules per mole?

(a) $\nu = 3.79 \times 10^{11} \, s^{-1}$

(b) $\nu = 5.45 \times 10^4 \, s^{-1}$

(c) $\lambda = 4.11 \times 10^{-5} \, m$

5.93 The *second* in the SI system is defined as the duration of 9,192,631,770 periods of radiation corresponding to the transition between two energy levels of a ground-state cesium-133 atom. What is the energy difference between the two levels in kilojoules per mole?

5.94 Write the symbol, give the ground-state electron configuration, and draw an orbital-filling diagram for each of the following atoms. Use the abbreviation of the preceding noble gas to represent the inner-shell electrons.

(a) The heaviest alkaline earth metal

(b) The lightest transition metal

(c) The heaviest actinide metal

(d) The lightest semimetal

(e) The group 6A element in the fifth period

5.95 Imagine a universe in which the four quantum numbers can have the same possible values as in our universe except that angular-momentum quantum number l can have integral values of $0, 1, 2, \ldots, n + 1$ (instead of $0, 1, 2, \ldots, n - 1$).

(a) How many elements would be in the first two rows of the periodic table in this universe?

(b) What would be the atomic number of the element in the second row and fifth column?

(c) Draw an orbital-filling diagram for the element with atomic number 12.

5.96 Cesium metal is frequently used in photoelectric cells because the amount of energy necessary to eject electrons from a cesium surface is relatively small—only 206.5 kJ/mol. What wavelength of light (in nanometers) does this correspond to?

5.97 The laser light used in compact disc players has $\lambda = 780$ nm. In what region of the electromagnetic spectrum does this light appear? What is the energy of this light in kilojoules per mole?

5.98 Draw orbital-filling diagrams for the following atoms. Show each electron as an up or down arrow, and use the abbreviation of the preceding noble gas to represent inner-shell electrons.

(a) Sr   (b) Cd

(c) Has $Z = 22$ (d) Has $Z = 34$

5.99 The atomic radii of Y (180 pm) and La (187 pm) are significantly different, but the radii of Zr (160 pm) and Hf (159 pm) are essentially identical. Explain.

5.100 One method for calculating Z_{eff} is to use the equation

$$Z_{eff} = \sqrt{\frac{(E)\,(n^2)}{1312 \text{ kJ/mol}}}$$

where E is the energy necessary to remove an electron from an atom and n is the principal quantum number of the electron. Use this equation to calculate Z_{eff} values for the highest-energy electrons in potassium ($E = 418.8$ kJ/mol) and krypton ($E = 1350.7$ kJ/mol).

5.101 One watt (W) is equal to 1 J/s. Assuming that 5.0% of the energy output of a 75 W light bulb is visible light and that the average wavelength of the light is 550 nm, how many photons are emitted by the light bulb each second?

5.102 Microwave ovens work by irradiating food with microwave radiation, which is absorbed and converted into heat. Assuming that radiation with $\lambda = 15.0$ cm is used, that all the energy is converted to heat, and that 4.184 J is needed to raise the temperature of 1.00 g of water by 1.00°C, how many photons are necessary to raise the temperature of a 350 mL cup of water from 20°C to 95°C?

5.103 Photochromic sunglasses, which darken when exposed to light, contain a small amount of colorless $AgCl(s)$ embedded in the glass. When irradiated with light, metallic silver atoms are produced and the glass darkens: $AgCl(s) \rightarrow Ag(s) + Cl$. Escape of the chlorine atoms is prevented by the rigid structure of the glass, and the reaction therefore reverses as soon as the light is removed. If 310 kJ/mol of energy is required to make the reaction proceed, what wavelength of light is necessary?

5.104 The amount of energy necessary to remove an electron from an atom is a quantity called the *ionization energy*, E_i. This energy can be measured by a technique called *photoelectron spectroscopy*, in which light of wavelength λ is directed at an atom, causing an electron to be ejected. The kinetic energy of the ejected electron (E_k) is measured by determining its velocity, v ($E_k = mv^2/2$), and E_i is then calculated using the conservation of energy principle. That is, the energy of the incident light is equal to E_i plus E_k. What is the ionization energy of selenium atoms (in kilojoules per mole) if light with $\lambda = 48.2$ nm produces electrons with a velocity of 2.371×10^6 m/s? The mass, m, of an electron is 9.109×10^{-31} kg.

5.105 X rays with a wavelength of 1.54×10^{-10} m are produced when a copper metal target is bombarded with high-energy electrons that have been accelerated by a voltage difference of 30,000 V. The kinetic energy of the electrons is equal to the product of the voltage difference and the electronic charge in coulombs, where 1 volt-coulomb = 1 J.

(a) What is the kinetic energy (in joules) and the de Broglie wavelength (in meters) of an electron that has been accelerated by a voltage difference of 30,000 V?

(b) What is the energy (in joules) of the X rays emitted by the copper target?

5.106 In the Bohr model of atomic structure, electrons are constrained to orbit a nucleus at specific distances, given by the equation

$$r = \frac{n^2 a_0}{Z}$$

where r is the radius of the orbit, Z is the charge on the nucleus, a_0 is the *Bohr radius* and has value of 5.292×10^{-11} m, and n is a positive integer ($n = 1, 2, 3, \ldots$) like the principal quantum number in the Schrödinger wave equation. Furthermore, Bohr concluded that the energy level E of an electron in a given orbit is

$$E = \frac{-Ze^2}{2r}$$

where e is the charge on an electron. Derive an equation that will let you calculate the difference ΔE between any two energy levels. What relation does your equation have to the Balmer–Rydberg equation?

Multi-Concept Problem

5.107 A 1.000 g sample of alkaline earth metal M reacts completely with 0.8092 g of chlorine gas (Cl_2) to yield an ionic product with the formula MCl_2. In the process, 9.46 kJ of heat is released.

(a) What is the molecular mass and identity of the metal M?

(b) How much heat (in kilojoules) would be released by reaction of 1.000 mol of M with a stoichiometric amount of Cl_2?

 eMedia Problems

5.108 The **Electromagnetic Spectrum** activity (*eChapter 5.2*) allows you to determine the frequency and wavelength of any color of visible light.

(a) Use the activity to determine the range of wavelengths (in nm) for red light, yellow light, and blue light.

(b) For each color in part (a), determine the range of frequencies (in s^{-1}) and the range of photon energies (in kJ/mol).

5.109 Watch the **Photoelectric Effect** movie (*eChapter 5.4*). If it takes 184 kJ/mol to eject electrons from cesium metal via the photoelectric effect, what is the longest wavelength of light that could be used for this purpose?

5.110 View the electron density plots for the noble gases helium, neon, and argon in the **Radial Electron Distribution** movie (*eChapter 5.8*). Then

(a) Explain why the maximum on the helium plot and the first maximum for each of the neon and argon plots occur at such different distances from the nucleus.

(b) Note the relative distances from the nucleus of the outermost maximum on each plot. Why is the outermost maximum on the neon plot farther from the nucleus than the maximum on the helium plot, despite neon's greater nuclear charge?

5.111 In Problem 5.20, you identified elements with anomalous, outer-shell, ground-state electron configurations. Use the **Electron Configuration** activity (*eChapter 5.14*) to identify the electron configurations of these elements. Explain why such configuration anomalies are not observed in the *p*-block elements. In other words, why is carbon's ground-state electron configuration $1s^2 2s^2 2p^2$ instead of $1s^2 2s^1 2p^3$, which would have two half-filled subshells?

5.112 The **Periodic Trends: Atomic Radii** movie (*eChapter 5.15*) shows that atomic radii increase from top to bottom within a group of the periodic table, and decrease from left to right across a row of the periodic table. Explain the factors that account for these trends.

Ionic Bonds and Some Main-Group Chemistry

These varied and fantastic tufa towers growing along the shore of California's Mono Lake are made of calcium carbonate, $CaCO_3$, a simple ionic compound.

What holds atoms together in chemical compounds? Clearly, there must be *some* force holding them together; otherwise, they would simply fly apart. As we saw briefly in Section 2.8, the forces that hold atoms together are called *chemical bonds* and are of two types—ionic bonds and covalent bonds. In this and the next two chapters, we'll look at the nature of chemical bonds and at the energy changes that accompany their formation and breakage. We'll begin in the present chapter with a look at ions and the ionic bonds formed between halogens and main-group metals.

6.1 Ions and Their Electron Configurations

We've seen on numerous occasions that metallic elements on the left side of the periodic table have a tendency to give up electrons to form cations, while the halogens and a few other nonmetallic elements on the right side of the table have a tendency to accept electrons to form anions. What are the ground-state electron configurations of the resultant ions?

For main-group elements, the electrons given up by a metal in forming a cation come from the highest-energy occupied orbital, while the electrons that are accepted by a nonmetal in forming an anion go into the lowest-energy unoccupied orbital according to the aufbau principle (Section 5.12). When a sodium atom ($1s^2\,2s^2\,2p^6\,3s^1$) gives up an electron, for example, the valence-shell $3s$ electron is lost, giving an Na^+ ion with the stable, noble gas electron configuration of neon ($1s^2\,2s^2\,2p^6$). Similarly, when a chlorine atom ($1s^2\,2s^2\,2p^6\,3s^2\,3p^5$) accepts an electron, the electron fills the remaining vacancy in the $3p$ subshell to give a Cl^- ion with the noble gas electron configuration of argon ($1s^2\,2s^2\,2p^6\,3s^2\,3p^6$).

Ion Electron Configuration activity

$$\textbf{Na: } 1s^2\,2s^2\,2p^6\,3s^1 \xrightarrow{-e^-} \textbf{Na}^+\text{: } 1s^2\,2s^2\,2p^6$$

$$\textbf{Cl: } 1s^2\,2s^2\,2p^6\,3s^2\,3p^5 \xrightarrow{+e^-} \textbf{Cl}^-\text{: } 1s^2\,2s^2\,2p^6\,3s^2\,3p^6$$

What is true for sodium is also true for the other elements in group 1A: All form positive ions by losing their valence-shell s electron, and all the resultant ions have noble gas electron configurations. Similarly for the elements in group 2A: All form a doubly positive ion by losing both their valence-shell s electrons. An Mg atom ($1s^2\,2s^2\,2p^6\,3s^2$) goes to an Mg^{2+} ion with the neon configuration $1s^2\,2s^2\,2p^6$ by loss of its two $3s$ electrons, for example.

$$\textbf{Group 1A atom: } [\text{Noble gas}]\,ns^1 \xrightarrow{-e^-} \textbf{Group 1A ion}^+\text{: } [\text{Noble gas}]$$

$$\textbf{Group 2A atom: } [\text{Noble gas}]\,ns^2 \xrightarrow{-2\,e^-} \textbf{Group 2A ion}^{2+}\text{: } [\text{Noble gas}]$$

Just as the group 1A and 2A metals *lose* the appropriate number of electrons to yield ions with noble gas configurations, the group 6A and group 7A nonmetals *gain* the appropriate number of electrons. The halogens in group 7A gain one electron to form singly charged anions with noble gas configurations, and the elements in group 6A gain two electrons to form doubly charged anions with noble gas configurations. Oxygen ($1s^2\,2s^2\,2p^4$), for example, becomes the O^{2-} ion with the neon configuration ($1s^2\,2s^2\,2p^6$).

$$\textbf{Group 6A atom: } [\text{Noble gas}]\,ns^2\,np^4 \xrightarrow{+2\,e^-} \textbf{Group 6A ion}^{2-}\text{: } [\text{Noble gas}]\,ns^2\,np^6$$

$$\textbf{Group 7A atom: } [\text{Noble gas}]\,ns^2\,np^5 \xrightarrow{+e^-} \textbf{Group 7A ion}^-\text{: } [\text{Noble gas}]\,ns^2\,np^6$$

The formulas and electron configurations of the most common main-group ions are shown in Table 6.1.

The situation is a bit different for the formation of ions from the transition metal elements than it is for the main-group elements. Transition metals form cations by first losing their valence-shell s electrons and then losing d electrons. As a result, all the remaining valence electrons in transition metal cations occupy d orbitals. Iron, for example, forms the Fe^{2+} ion by losing its two $4s$ electrons, and forms the Fe^{3+} ion by losing two $4s$ electrons and one $3d$ electron:

TABLE 6.1   Some Common Main-Group Ions and Their Noble Gas Electron Configurations

Group 1A	Group 2A	Group 3A	Group 6A	Group 7A	Electron Configuration
H^+					[None]
H^-					[He]
Li^+	Be^{2+}				[He]
Na^+	Mg^{2+}	Al^{3+}	O^{2-}	F^-	[Ne]
K^+	Ca^{2+}	$^*Ga^{3+}$	S^{2-}	Cl^-	[Ar]
Rb^+	Sr^{2+}	$^*In^{3+}$	Se^{2-}	Br^-	[Kr]
Cs^+	Ba^{2+}	$^*Tl^{3+}$	Te^{2-}	I^-	[Xe]

* These ions do not have a true noble gas electron configuration because they have an additional filled d subshell.

$$\textbf{Fe: } [Ar]\, 4s^2\, 3d^6 \xrightarrow{-2\,e^-} \textbf{Fe}^{2+}\text{: } [Ar]\, 3d^6$$

$$\textbf{Fe: } [Ar]\, 4s^2\, 3d^6 \xrightarrow{-3\,e^-} \textbf{Fe}^{3+}\text{: } [Ar]\, 3d^5$$

It may seem strange that building up the periodic table adds the $3d$ electrons *after* the $4s$ electrons, whereas ion formation from a transition metal removes the $4s$ electrons *before* the $3d$ electrons. Note, though, that the two processes are not the reverse of each other, so they can't be compared directly. Building up the periodic table adds one electron to the valence shell and also adds one positive charge to the nucleus, whereas ion formation removes an electron from the valence shell but does not alter the nucleus.

▶ **PROBLEM 6.1**   Predict the ground-state electron configuration for each of the following ions, and explain your answers.
(a) Ra^{2+} **(b)** La^{3+} **(c)** Ti^{4+} **(d)** N^{3-}

▶ **PROBLEM 6.2**   What doubly positive ion has the following ground-state electron configuration? $1s^2\, 2s^2\, 2p^6\, 3s^2\, 3p^6\, 3d^{10}$

6.2   Ionic Radii

Just as there are systematic differences in the sizes of atoms (Section 5.15), there are also systematic differences in the sizes of ions. As shown in Figure 6.1 for the elements of groups 1A and 2A, atoms shrink dramatically when an electron is removed to form a cation. The radius of an Na atom, for example, is 186 pm, while that of an Na^+ cation is 102 pm. Similarly, the radius of an Mg atom is 160 pm, and that of an Mg^{2+} cation is 72 pm.

FIGURE 6.1   Radii of **(a)** group 1A atoms and their cations; **(b)** group 2A atoms and their cations. The cations are smaller than the neutral atoms both because the principal quantum number of the valence-shell electrons is smaller for the cations and because Z_{eff} is larger.

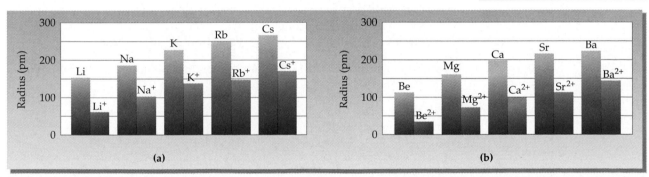

Gain and Loss of
Electrons movie;
Effective Nuclear
Charge movie

The shrinkage that occurs when an electron is removed from an atom is explained both by the fact that the electrons are removed from larger, valence-shell orbitals and by an increase in Z_{eff} (Section 5.11). On going from a neutral Na atom to a charged Na^+ cation, for example, the electron configuration changes from $1s^2\,2s^2\,2p^6\,3s^1$ to $1s^2\,2s^2\,2p^6$. The valence shell of the Na *atom* is the *third* shell, but the valence shell of the Na^+ *cation* is the *second* shell. Thus, the Na^+ ion has a smaller valence shell than the Na atom and therefore a smaller size. Also important in accounting for the smaller size of the ion is that the effective nuclear charge felt by the valence-shell electrons is greater in the Na^+ cation than in the neutral atom. The Na atom has 11 protons and 11 electrons, but the Na^+ cation has 11 protons and only 10 electrons. The smaller number of electrons in the cation means that they shield one another to a lesser extent and therefore feel a stronger pull toward the nucleus.

The same effects felt by the group 1A elements when a single electron is lost are felt by the group 2A elements when two electrons are lost. For example, loss of two valence-shell electrons from an Mg atom ($1s^2\,2s^2\,2p^6\,3s^2$) gives the Mg^{2+} cation ($1s^2\,2s^2\,2p^6$). The smaller valence shell of the Mg^{2+} cation and the increase in effective nuclear charge combine to cause a dramatic shrinkage. Not surprisingly, a similar shrinkage is encountered whenever any of the metal atoms on the left-hand two-thirds of the periodic table is converted into a cation.

Just as atoms shrink when converted to cations by loss of an electron, they expand when converted to anions by gain of an electron. As shown in Figure 6.2 for the group 7A elements (halogens), the expansion is a dramatic one. Chlorine, for example, nearly doubles in size, from 99 pm for the neutral atom to 184 pm for the chloride anion.

FIGURE 6.2 Radii of the group 7A atoms (halogens) and their anions. The anions are larger than the neutral atoms because of additional electron–electron repulsions and a decrease in Z_{eff}.

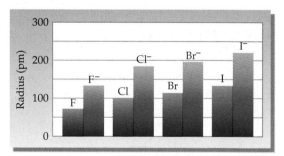

The expansion that occurs when a group 7A atom gains an electron to yield an anion can't be accounted for by a change in the quantum number of the valence shell, since the added electron simply completes an already occupied p subshell: [Ne] $3s^2\,3p^5$ for a Cl atom to [Ne] $3s^2\,3p^6$ for a Cl^- anion, for example. Thus, the expansion is due entirely to the decrease in effective nuclear charge and the increase in electron–electron repulsions that occurs when an extra electron is added.

▶ **PROBLEM 6.3**   Which atom or ion in each of the following pairs would you expect to be larger? Explain.
(a) O or O^{2-} (b) O or S   (c) Fe or Fe^{3+} (d) H or H^-

✦ **KEY CONCEPT PROBLEM 6.4**   Which of the following three spheres represents a K^+ ion, which a K atom, and which a Cl^- ion?

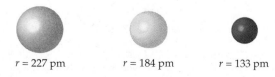

$r = 227$ pm $r = 184$ pm $r = 133$ pm

6.3 Ionization Energy

We saw in the previous chapter that the absorption of light energy by an atom leads to a change in electron configuration. A valence-shell electron is promoted from a lower-energy orbital to a higher-energy one with a larger principal quantum number n. If enough energy is absorbed, the electron can be removed completely from the atom, leaving behind an ion. The amount of energy necessary to remove the outermost, highest-energy electron from an isolated neutral atom in the gaseous state is called the atom's **ionization energy**, abbreviated E_i. For hydrogen atoms, 1312.0 kJ/mol is needed.

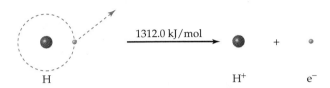

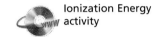

As shown by the plot in Figure 6.3, ionization energies of the various elements differ widely, from a low of 375.7 kJ/mol for cesium to a high of 2372.3 kJ/mol for helium. Furthermore, there is a clear periodicity to the data. The minimum E_i values correspond to the group 1A elements (alkali metals), the maximum E_i values correspond to the group 8A elements (noble gases), and a gradual increase in E_i occurs from left to right across a row of the periodic table—from Na to Ar, for example.

Periodic Trends: Ionization Energies movie

Ionization Energy activity

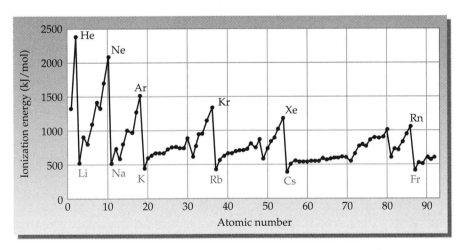

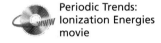

FIGURE 6.3 Ionization energies of the first 92 elements. There is an obvious periodicity to the data, with maximum values for the noble gas elements and minimum values for the alkali metals. Note that all the values are positive, meaning that energy is always required to remove an electron from an atom.

The periodicity evident in Figure 6.3 can be explained by looking at electron configurations. Atoms of the group 8A elements have filled valence subshells, either s (for helium) or both s and p (for the other noble gases). As we saw in Section 5.15, electrons in a filled valence subshell feel a relatively high effective nuclear charge, Z_{eff}, because electrons within the same subshell don't shield one another very strongly. As a result, the electrons are held tightly to the nucleus, the radius of the atom is small, and the energy necessary to remove an electron is relatively large. Atoms of group 1A elements, by contrast, have only a single s electron in their valence shell. This single valence electron is well shielded from the nucleus by all the inner-shell electrons, called the **core electrons**, resulting in a low Z_{eff}. The valence electron is thus held loosely, and the energy necessary to remove it is relatively small.

The three-dimensional display of ionization energies in Figure 6.4 shows other trends in the data beyond the obvious periodicity. One such trend is that ionization energies gradually decrease going down a group in the periodic table, from Li to Fr and from He to Rn, for example. As atomic number increases going down a group, both the principal quantum number of the valence-shell electrons and their average distance from the nucleus also increase. As a result, the valence-shell electrons are less tightly held, and E_i is smaller.

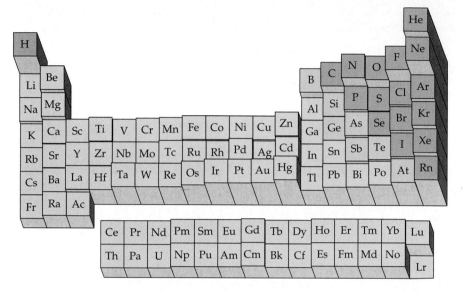

FIGURE 6.4 A three-dimensional display showing how ionization energies increase from left to right across a row and decrease from top to bottom down a group of the periodic table. The elements at the lower left therefore have the smallest E_i values, and the elements at the upper right have the largest.

Still another feature of the E_i data is that minor irregularities occur across a row of the periodic table. As shown by the close-up look at E_i values of the first 20 elements (Figure 6.5), the E_i of beryllium is larger than that of its neighbor boron, and the E_i of nitrogen is larger than that of its neighbor oxygen. Similarly, magnesium has a larger E_i than aluminum, and phosphorus has a slightly larger E_i than sulfur.

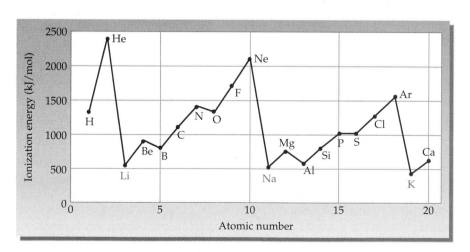

FIGURE 6.5 Ionization energies of the first 20 elements. The group 2A elements (Be, Mg, and Ca) have slightly larger E_i values than might be expected, and the group 6A elements (O and S) have slightly smaller values than might be expected.

The unusually large E_i values for the group 2A elements Be, Mg, and so forth can be explained by looking at their electron configurations. Compare beryllium with boron, for example. An *s* electron is removed on ionization of beryllium, but a *p* electron is removed on ionization of boron:

s electron removed

$$\text{Be } (1s^2\, 2s^2) \longrightarrow \text{Be}^+ (1s^2\, 2s) + e^- \qquad E_i = 899.4 \text{ kJ/mol}$$

p electron removed

$$\text{B } (1s^2\, 2s^2\, 2p^1) \longrightarrow \text{B}^+ (1s^2\, 2s^2) + e^- \qquad E_i = 800.6 \text{ kJ/mol}$$

Because a 2*s* electron is on average closer to the nucleus than a 2*p* electron, it is held more tightly and is harder to remove. Thus, the E_i of beryllium is larger than that of boron.

The unusually low E_i values for atoms of group 6A elements can also be explained by their electron configurations. In comparing nitrogen with oxygen, for example, the nitrogen electron is removed from a half-filled orbital, whereas the oxygen electron is removed from a filled orbital:

Half-filled orbital

$$\text{N } (1s^2\, 2s^2\, 2p_x^1\, 2p_y^1\, 2p_z^1) \longrightarrow \text{N}^+ (1s^2\, 2s^2\, 2p_x^1\, 2p_y^1) + e^- \qquad E_i = 1402.3 \text{ kJ/mol}$$

Filled orbital

$$\text{O } (1s^2\, 2s^2\, 2p_x^2\, 2p_y^1\, 2p_z^1) \longrightarrow \text{O}^+ (1s^2\, 2s^2\, 2p_x^1\, 2p_y^1\, 2p_z^1) + e^- \qquad E_i = 1313.9 \text{ kJ/mol}$$

Because electrons repel one another and tend to stay as far apart as possible, electrons that are forced together in a filled orbital are slightly higher in energy than those in a half-filled orbital, so it is slightly easier to remove one. Thus, oxygen has a smaller E_i than nitrogen.

EXAMPLE 6.1 Arrange the elements Se, Cl, and S in order of increasing ionization energy.

SOLUTION Ionization energy generally increases from left to right across a row of the periodic table and decreases from top to bottom down a group. Chlorine should therefore have a larger E_i than its neighbor sulfur, and selenium should have a smaller E_i than sulfur. Thus, the order is Se < S < Cl.

▶ **PROBLEM 6.5**
Using the periodic table as your guide, predict which element in each of the following pairs has the larger ionization energy:
(a) K or Br **(b)** S or Te **(c)** Ga or Se **(d)** Ne or Sr

6.4 Higher Ionization Energies

Ionization is not limited to the removal of a single electron from an atom. Two, three, or even more electrons can be removed sequentially from an atom, and the amount of energy associated with each step can be measured.

$$\text{M} + \text{Energy} \longrightarrow \text{M}^+ + e^- \qquad \text{First ionization energy } (E_{i1})$$
$$\text{M}^+ + \text{Energy} \longrightarrow \text{M}^{2+} + e^- \qquad \text{Second ionization energy } (E_{i2})$$
$$\text{M}^{2+} + \text{Energy} \longrightarrow \text{M}^{3+} + e^- \qquad \text{Third ionization energy } (E_{i3})$$

And so forth

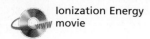

Ionization Energy movie

As you might expect, successively larger amounts of energy are required for each successive ionization step because it is much harder to remove a negatively charged electron from a positively charged ion than from a neutral atom. Interestingly, though, the energy differences between successive steps vary dramatically from one element to another. Removing the second electron from sodium takes nearly ten times as much energy as removing the first one (4562 versus 496 kJ/mol), but removing the second electron from magnesium takes only twice as much energy as removing the first one (1451 versus 738 kJ/mol). Large jumps similar to that for sodium are also found for other elements, as can be seen by following the zigzag line in Table 6.2. Magnesium has a large jump between its second and third ionization energies, aluminum has a large jump between its third and fourth ionization energies, silicon has a large jump between its fourth and fifth ionization energies, and so on.

TABLE 6.2   Successive Ionization Energies (kJ/mol) for Third-Row Elements

E_i Number	Na	Mg	Al	Si	P	S	Cl	Ar
E_{i1}	496	738	578	787	1,012	1,000	1,251	1,520
E_{i2}	4,562	1,451	1,817	1,577	1,903	2,251	2,297	2,665
E_{i3}	6,912	7,733	2,745	3,231	2,912	3,361	3,822	3,931
E_{i4}	9,543	10,540	11,575	4,356	4,956	4,564	5,158	5,770
E_{i5}	13,353	13,630	14,830	16,091	6,273	7,013	6,540	7,238
E_{i6}	16,610	17,995	18,376	19,784	22,233	8,495	9,458	8,781
E_{i7}	20,114	21,703	23,293	23,783	25,397	27,106	11,020	11,995

The large increases in ionization energies that follow the zigzag line in Table 6.2 are yet another consequence of electron configuration. It's relatively easy to remove an electron from a *partially* filled valence shell, where Z_{eff} is lower, but it's relatively difficult to remove an electron from a *filled* valence shell, where Z_{eff} is higher. In other words, a large amount of stability is associated with filled s and p subshells (a noble gas electron configuration), which corresponds to having eight electrons (an *octet*) in the valence shell of an atom or ion. Sodium ([Ne] $3s^1$) loses only one electron easily, magnesium ([Ne] $3s^2$) loses only two electrons easily, aluminum ([Ne] $3s^2 3p^1$) loses only three electrons easily, and so on across the row. We'll further explore the stability of valence-shell electron octets in Section 6.12.

8 electrons in outer (2nd) shell

$$Na\ (1s^2\ 2s^2\ 2p^6\ 3s^1) \longrightarrow Na^+\ (1s^2\ 2s^2\ 2p^6) + e^-$$

$$Mg\ (1s^2\ 2s^2\ 2p^6\ 3s^2) \longrightarrow Mg^{2+}\ (1s^2\ 2s^2\ 2p^6) + 2\ e^-$$

$$Al\ (1s^2\ 2s^2\ 2p^6\ 3s^2\ 3p^1) \longrightarrow Al^{3+}\ (1s^2\ 2s^2\ 2p^6) + 3\ e^-$$

$$\vdots \qquad\qquad\qquad \vdots$$

$$Cl\ (1s^2\ 2s^2\ 2p^6\ 3s^2\ 3p^5) \longrightarrow Cl^{7+}\ (1s^2\ 2s^2\ 2p^6) + 7\ e^-$$

Just as valence-shell electrons rather than core electrons are the most easily lost during ionization, they are also the most easily lost or shared during chemical reactions. We'll see repeatedly in later chapters that *the valence-shell electron configuration of an atom controls the atom's chemistry.*

EXAMPLE 6.2 Which has the larger fifth ionization energy, Ge or As?

SOLUTION   As their positions in the periodic table indicate, the group 4A element germanium has four valence-shell electrons and thus four relatively low ionization energies, whereas the group 5A element arsenic has five valence-shell electrons and five low ionization energies. Germanium therefore has a larger E_{i5} than arsenic.

▶ **PROBLEM 6.6**
(a) Which has the larger third ionization energy, Be or N?
(b) Which has the larger fourth ionization energy, Ga or Ge?

▶ **PROBLEM 6.7**   Three atoms have the following electron configurations:
(a) $1s^2\, 2s^2\, 2p^6\, 3s^2\, 3p^1$ (b) $1s^2\, 2s^2\, 2p^6\, 3s^2\, 3p^5$ (c) $1s^2\, 2s^2\, 2p^6\, 3s^2\, 3p^6\, 4s^1$
Which of the three has the largest E_{i1}? Which has the smallest E_{i4}?

KEY CONCEPT PROBLEM 6.8   Order the indicated three elements according to the ease with which each is likely to lose its third electron:

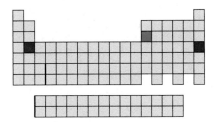

6.5   Electron Affinity

Just as it's possible to measure the energy change on *removing* an electron from an atom to form a cation, it's also possible to measure the energy change on *adding* an electron to an atom to form an anion. An element's **electron affinity**, abbreviated E_{ea}, is the energy change that occurs when an electron is added to an isolated atom in the gaseous state.

Ionization energies (Section 6.3) are always positive because energy must always be added to remove an electron from an atom. Electron affinities, however, are generally negative because energy is usually released when a neutral atom adds an electron.* We'll see in Chapter 8 that this same sign convention is used throughout chemistry: *A positive energy change means energy is absorbed, and a negative energy change means energy is released.*

The more negative the E_{ea}, the greater the tendency of the atom to accept an electron, and the more stable the anion that results. By contrast, an atom that forms an unstable anion by addition of an electron has, in principle, a positive value of E_{ea}, but no experimental measurement can be made in such

Electron Affinity movie; Periodic Trends: Electron Affinity movie

*Some books and reference sources adopt the opposite sign convention for electron affinity, giving it a positive value and defining it as the energy absorbed when an anion loses an electron.

circumstances. All we can say is that the E_{ea} for such an atom is greater than zero. The E_{ea} of hydrogen, for instance, is −72.8 kJ/mol, meaning that energy is released and the H⁻ anion is stable. The E_{ea} of neon, however, is greater than 0 kJ/mol, meaning that Ne does not add an electron and the Ne⁻ anion is unstable.

$$H\ (1s^1) + e^- \longrightarrow H^-\ (1s^2) + 72.8\ \text{kJ/mol} \qquad E_{ea} = -72.8\ \text{kJ/mol}$$

$$Ne\ (1s^2\ 2s^2\ 2p^6) + e^- + \text{Energy} \longrightarrow Ne^-\ (1s^2\ 2s^2\ 2p^6\ 3s^1) \qquad E_{ea} > 0\ \text{kJ/mol}$$

As was true for ionization energies, electron affinities show a periodicity that is related to the electron configurations of the elements. The data in Figure 6.6 indicate that addition of an electron to an atom is energetically favorable, except for the noble gases, some of the alkaline earth metals, and a very few other elements. Group 7A elements have the most negative electron affinities, corresponding to the largest release of energy, while group 2A and group 8A elements have near-zero or positive electron affinities, corresponding to an absorption of energy.

FIGURE 6.6 Measured electron affinities for elements 1–57 and 72–86. A negative value means that energy is released when an electron adds to an atom, while a value of zero means that energy is absorbed but the exact amount can't be measured experimentally. Note that the group 2A elements (alkaline earths) and the group 8A elements (noble gases) have E_{ea} values near zero, while the group 7A elements (halogens) have large negative E_{ea}'s. Accurate electron affinities are not known for elements 58–71.

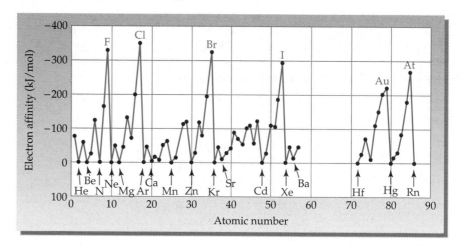

The value of an element's electron affinity is caused by an interplay of several offsetting factors. Attraction between the additional electron and the atomic nucleus favors a negative E_{ea}, but the increase in electron–electron repulsions that result from addition of the extra electron favors a positive E_{ea}.

Large negative E_{ea}'s are found for the halogens (F, Cl, Br, I) because each of these elements has both a high Z_{eff} and room in its valence shell for an additional electron. Halide anions have a noble gas electron configuration with filled s and p sublevels, and the attraction between the additional electron and the atomic nucleus is high. Positive E_{ea}'s are found for the noble gas elements (He, Ne, Ar, Kr, Xe) because the s and p sublevels in these elements are full, and the additional electron must therefore go into the next higher shell, where it is shielded from the nucleus and feels a relatively low Z_{eff}. The attraction of the nucleus for the added electron is therefore small and is outweighed by the additional electron–electron repulsions.

A halogen: $Cl\ (\dots 3s^2\ 3p^5) + e^- \longrightarrow Cl^-(\dots 3s^2\ 3p^6) + \text{Energy} \qquad E_{ea} = -348.6\ \text{kJ/mol}$

A noble gas: $Ar\ (\dots 3s^2\ 3p^6) + e^- + \text{Energy} \longrightarrow Ar^-(\dots 3s^2\ 3p^6\ 4s^1) \qquad E_{ea} > 0\ \text{kJ/mol}$

In looking for other trends in the data of Figure 6.6, the near-zero E_{ea}'s of the alkaline earth metals (Be, Mg, Ca, Sr, Ba) are particularly striking. Atoms of these elements have filled s subshells, which means that the additional electron must go into a p subshell. The higher energy of the p subshell, together with a relatively low Z_{eff} for elements on the left side of the periodic table, means that alkaline earth atoms accept an electron reluctantly and have E_{ea} values near zero.

An alkaline earth: $Mg\,(\ldots 3s^2) + e^- + Energy \longrightarrow Mg^-\,(\ldots 3s^2\,3p^1)$ $E_{ea} \approx 0\ kJ/mol$

EXAMPLE 6.3   Why does nitrogen have a less favorable (more positive) E_{ea} than its neighbors on either side, C and O?

SOLUTION   The magnitude of an element's E_{ea} depends on the element's valence-shell electron configuration. The electron configurations of N, C, and O are

Carbon: $1s^2\,2s^2\,2p_x^1\,2p_y^1$ Nitrogen: $1s^2\,2s^2\,2p_x^1\,2p_y^1\,2p_z^1$

Oxygen: $1s^2\,2s^2\,2p_x^2\,2p_y^1\,2p_z^1$

Carbon has only two electrons in its $2p$ subshell and can readily accept another in its vacant $2p_z$ orbital. Nitrogen, however, has a half-filled $2p$ subshell, and the additional electron must pair up in a $2p$ orbital, where it feels a repulsion from the electron already present. Thus, the E_{ea} of nitrogen is less favorable than that of carbon. Oxygen also must add an electron to an orbital that already has one electron, but the additional stabilizing effect of increased Z_{eff} across the periodic table counteracts the effect of electron repulsion, resulting in a more favorable E_{ea} for O than for N.

▶**PROBLEM 6.9**   Explain why manganese (atomic number 25) has a less favorable E_{ea} than its neighbors on either side.

✦**KEY CONCEPT PROBLEM 6.10**   Which of the indicated three elements has the least favorable E_{ea}, and which has the most favorable E_{ea}?

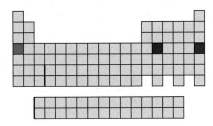

6.6 Ionic Bonds and the Formation of Ionic Solids

What would happen if an element that gives up an electron easily (that is, has a small ionization energy) were to come in contact with an element that accepts an electron easily (that is, has a large negative electron affinity)? The element with the small E_i might transfer an electron to the element with the negative E_{ea}, yielding a cation and an anion. Sodium, for example, reacts with chlorine to give Na^+ ions and Cl^- ions:

$$Na + Cl \longrightarrow Na^+ Cl^-$$

$1s^2\,2s^2\,2p^6\,3s^1$ $1s^2\,2s^2\,2p^6\,3s^2\,3p^5$ $1s^2\,2s^2\,2p^6$ $1s^2\,2s^2\,2p^6\,3s^2\,3p^6$

Sodium metal burns in a chlorine atmosphere to yield solid sodium chloride.

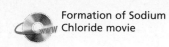

Formation of Sodium Chloride movie

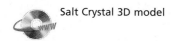

Salt Crystal 3D model

The oppositely charged Na^+ and Cl^- ions that result when sodium transfers an electron to chlorine are attracted to one another by electrostatic forces, and we say that they are joined by an **ionic bond**. The crystalline substance that results is said to be an **ionic solid**. Note, though, that a visible crystal of sodium chloride does not consist of individual pairs of Na^+ and Cl^- ions. Instead, solid NaCl consists of a vast three-dimensional network of ions in which each Na^+ is surrounded by and attracted to *many* Cl^- ions, and each Cl^- is surrounded by and attracted to many Na^+ ions (Figure 6.7).

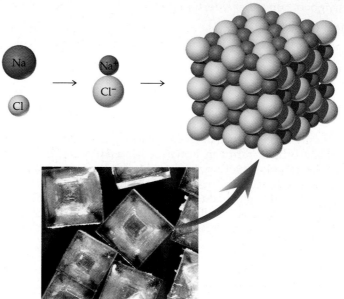

FIGURE 6.7 When sodium atoms transfer electrons to chlorine atoms, crystals of the ionic solid sodium chloride are formed. In the sodium chloride crystal, each Na^+ ion is surrounded by six nearest-neighbor Cl^- ions, and each Cl^- ion is surrounded by six nearest-neighbor Na^+ ions.

What about the energy change, ΔE, that occurs when sodium and chlorine react to yield Na^+ and Cl^- ions? It's apparent from E_i and E_{ea} values that the amount of energy released when a chlorine atom accepts an electron ($E_{ea} = -348.6$ kJ/mol) is not large enough to offset the amount absorbed when a sodium atom loses an electron ($E_i = +495.8$ kJ/mol):

$$E_i \text{ for Na} = +495.8 \text{ kJ/mol} \qquad \text{(Unfavorable)}$$
$$\underline{E_{ea} \text{ for Cl} = -348.6 \text{ kJ/mol} \qquad \text{(Favorable)}}$$
$$\Delta E = +147.2 \text{ kJ/mol} \qquad \text{(Unfavorable)}$$

The net energy change ΔE for the reaction of sodium and chlorine atoms would be unfavorable by +147.2 kJ/mol, and no reaction would occur, unless some other factor were involved. This additional factor, which overcomes the unfavorable energy change of electron transfer, is the large gain in stability due to the formation of ionic bonds. (The Greek capital letter delta, Δ, is used to represent a change in the indicated quantity, in this case, ΔE.)

The actual reaction of sodium with chlorine occurs all at once rather than in a stepwise manner, but energy calculations can be made more easily if we imagine a series of hypothetical steps for which experimentally measured energy values can be obtained. Five contributions must be taken into account to calculate the overall energy change during the formation of solid NaCl from solid sodium metal and gaseous chlorine molecules:

$$Na(s) + 1/2 \, Cl_2(g) \longrightarrow NaCl(s)$$

Step 1. Conversion of solid Na metal into isolated, gaseous Na atoms, a process called *sublimation*. Since energy must be added to disrupt the forces holding atoms together in a solid, the heat required for sublimation has a positive value: 107.3 kJ/mol for Na.

$$Na(s) \longrightarrow Na(g)$$
$$+107.3 \text{ kJ/mol}$$

Step 2. Dissociation of gaseous Cl_2 molecules into individual Cl atoms. Energy must be added to break molecules apart before reaction can occur, and the energy required for bond breaking therefore has a positive value: 243 kJ/mol for Cl_2 (or 122 kJ/mol for $1/2$ Cl_2). We'll look further into the energetics of bond dissociation in Section 8.11.

$$1/2 \, Cl_2(g) \longrightarrow Cl(g)$$
$$+122 \text{ kJ/mol}$$

Step 3. Ionization of isolated Na atoms into Na^+ ions plus electrons. The energy required is simply the first ionization energy of sodium and has a positive value: 495.8 kJ/mol.

$$Na(g) \longrightarrow Na^+(g) + e^-$$
$$+495.8 \text{ kJ/mol}$$

Step 4. Formation of Cl^- ions from Cl atoms by addition of an electron. The energy required is simply the electron affinity of chlorine and has a negative value: −348.6 kJ/mol.

$$Cl(g) + e^- \longrightarrow Cl^-(g)$$
$$-348.6 \text{ kJ/mol}$$

Step 5. Formation of solid NaCl from isolated Na^+ and Cl^- ions. The energy change for this step is a measure of the overall electrostatic attractions between ions in the solid. It is the amount of energy released when isolated ions condense to form a solid, and it has a negative value: −787 kJ/mol for NaCl.

$$Na^+(g) + Cl^-(g) \longrightarrow NaCl(s)$$
$$-787 \text{ kJ/mol}$$

_____

Net reaction: $$Na(s) + 1/2 \, Cl_2(g) \longrightarrow NaCl(s)$$

Net energy change: $$-411 \text{ kJ/mol}$$

A pictorial way of representing the five hypothetical steps in the reaction between sodium and chlorine is shown in Figure 6.8. Called a **Born–Haber cycle**, the figure shows how each step contributes to the overall energy change and how the net process is the sum of the individual steps. As indicated, steps 1, 2, and 3 absorb energy (have positive values), while steps 4 and 5

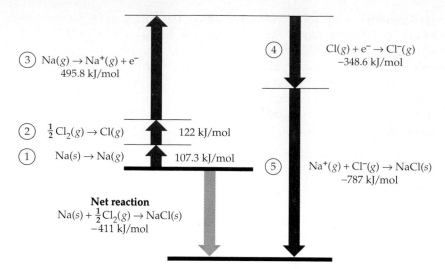

FIGURE 6.8 A Born–Haber cycle for the formation of $NaCl(s)$ from $Na(s)$ and $Cl_2(g)$. The sum of the individual energy changes for the five steps is equal to the net energy change for the overall reaction. Note that the most favorable step is the formation of solid $NaCl$ from gaseous Na^+ and Cl^- ions (step 5).

release energy (have negative values). The largest contribution is step 5, which takes into account the electrostatic interactions between ions in the solid product. Were it not for this large amount of stabilization of the solid due to ionic bonding, no reaction would take place.

Lattice Energy

The sum of the electrostatic interaction energies between ions in a crystal—and thus the measure of the strength of the crystal's ionic bonds—is called the **lattice energy (U)**. By convention, lattice energy is the amount of energy that must be supplied to break an ionic solid into its individual gaseous ions. It therefore has a positive value because energy is required to separate the electrical charges. The formation of a crystal from ions is the reverse of the breakup, however, and step 5 in the Born–Haber cycle of Figure 6.8 therefore has the value $-U$.

$$NaCl(s) \longrightarrow Na^+(g) + Cl^-(g) \qquad U = +787 \text{ kJ/mol} \quad \textbf{(Energy absorbed)}$$

$$Na^+(g) + Cl^-(g) \longrightarrow NaCl(s) \qquad -U = -787 \text{ kJ/mol} \quad \text{(Energy released)}$$

The force F that results from the interaction of electric charges is described by **Coulomb's law** and is equal to a constant k times the charges on the ions, z_1 and z_2, divided by the square of the distance d between their centers:

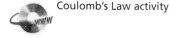

Coulomb's Law activity

$$\textbf{Coulomb's law:} \quad F = k \times \frac{z_1 z_2}{d^2}$$

But because energy is equal to force times distance, the negative of the lattice energy is equal to

$$-U = F \times d = k \times \frac{z_1 z_2}{d}$$

The value of the constant k depends on the geometric arrangement of the ions in the specific compound and is different for different substances.

Lattice energies are largest when the distance d between ions is small and when the charges z_1 and z_2 are large. A small distance d means that the ions are close together, which implies that they have small ionic radii. Thus, if z_1 and z_2 are held constant, the largest lattice energies belong to compounds formed from the smallest ions, as shown in Table 6.3. Within a series of compounds that have the same anion but different cations, lattice energy increases as the cation becomes smaller. Comparing LiF, NaF, and KF, for example, cation size follows the order $K^+ > Na^+ > Li^+$, so lattice energies follow the order LiF > NaF > KF. Similarly, within a series of compounds that have the same cation but different anions, lattice energy increases as anion size decreases. Comparing LiF, LiCl, LiBr, and LiI, for example, anion size follows the order $I^- > Br^- > Cl^- > F^-$, so lattice energies follow the reverse order LiF > LiCl > LiBr > LiI.

TABLE 6.3   Lattice Energies of Some Ionic Solids (kJ/mol)

Cation	Anion				
	F^-	Cl^-	Br^-	I^-	O^{2-}
Li^+	1,036	853	807	757	2,925
Na^+	923	787	747	704	2,695
K^+	821	715	682	649	2,360
Be^{2+}	3,505	3,020	2,914	2,800	4,443
Mg^{2+}	2,957	2,524	2,440	2,327	3,791
Ca^{2+}	2,630	2,258	2,176	2,074	3,401
Al^{3+}	5,215	5,492	5,361	5,218	15,916

Table 6.3 also shows that compounds of ions with higher charges have larger lattice energies than compounds of ions with lower charges. In comparing NaI, MgI_2, and AlI_3, for example, the order of charges on the cations is $Al^{3+} > Mg^{2+} > Na^+$, and the order of lattice energies is $AlI_3 > MgI_2 > NaI$.

EXAMPLE 6.4 Which has the larger lattice energy, NaCl or CsI?

SOLUTION The magnitude of a substance's lattice energy is affected both by the charges on its constituent ions and by the sizes of those ions. The higher the charges on the ions and the smaller the sizes of the ions, the larger the lattice energy. All four ions, Na^+, Cs^+, Cl^-, and I^-, are singly charged, but they differ in size. Since Na^+ is smaller than Cs^+, and Cl^- is smaller than I^-, the distance between ions is smaller in NaCl than in CsI. Thus, NaCl has the larger lattice energy.

▶**PROBLEM 6.11** Which substance in each of the following pairs has the larger lattice energy?
(a) KCl or RbCl   (b) CaF_2 or BaF_2 (c) CaO or KI

▶**PROBLEM 6.12** Calculate the net energy change (in kilojoules per mole) that takes place on formation of KF(s) from the elements: $K(s) + 1/2\ F_2(g) \rightarrow$ KF(s). The following information is needed:

Heat of sublimation for K = 89.2 kJ/mol E_{ea} for F = −328 kJ/mol
Bond dissociation energy for F_2 = 158 kJ/mol E_i for K = 418.8 kJ/mol
Lattice energy for KF = 821 kJ/mol

◆— **KEY CONCEPT PROBLEM 6.13** Which of the following alkaline earth oxides has the larger lattice energy, and which has the smaller lattice energy? Explain.

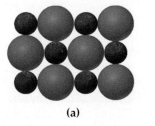

(a)   (b)

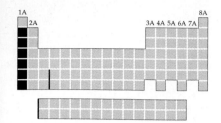

6.7 The Alkali Metals (Group 1A)

Now that we know something about ionization energies, electron affinities, and ionic bonding, let's look at the chemistry of some of the elements that form ionic bonds. The alkali metals of group 1A—Li, Na, K, Rb, Cs, and Fr—have the smallest ionization energies of all the elements because of their valence-shell ns^1 electron configurations (Figure 6.3). They therefore lose this ns^1 electron easily in chemical reactions to yield +1 ions and are thus among the most powerful reducing agents in the periodic table (Sections 4.6–4.8). In fact, the entire chemistry of the alkali metals is dominated by their ability to donate an electron to other elements or compounds.

As their group name implies, the alkali metals are *metallic*. They have a bright, silvery appearance, are malleable, and are good conductors of electricity. Unlike the more common metals such as iron, though, the alkali metals are all soft enough to cut with a dull knife, have low melting points and densities, and are so reactive that they must be stored under oil to prevent their instantaneous reaction with oxygen and water. None is found in the elemental state in nature; they occur only in salts. Their properties are summarized in Table 6.4.

TABLE 6.4 Properties of Alkali Metals

Name	Melting Point (°C)	Boiling Point (°C)	Density (g/cm³)	1st Ionization Energy (kJ/mol)	Abundance on Earth (%)	Atomic Radius (pm)	Ionic (M⁺) Radius (pm)
Lithium	180.5	1342	0.534	520.2	0.0020	152	68
Sodium	97.7	883	0.971	495.8	2.36	186	102
Potassium	63.3	759	0.862	418.8	2.09	227	138
Rubidium	39.3	688	1.532	403.0	0.0090	248	147
Cesium	28.4	671	1.873	375.7	0.000 10	265	167
Francium	—	—	—	≈400	Trace	—	—

Occurrence and Uses of Alkali Metals

Lithium.   First isolated in 1817 from the mineral petalite, $LiAlSi_4O_{10}$, lithium was named from the Greek word *lithos*, meaning stone, because of its common occurrence in rocks. Most lithium today is obtained from the mineral spodumene, $LiAlSi_2O_6$, large deposits of which occur scattered throughout the world in the United States, Canada, Brazil, and the former USSR. The major industrial use of lithium is in all-purpose automotive greases, but lithium salts such as Li_2CO_3 also have a variety of specialized applications, including use as a pharmaceutical agent for the treatment of manic-depressive behavior.

A sample of lithium metal.

Sodium.   Sodium, the sixth most abundant element in the earth's crust, was first prepared in 1807 from caustic soda, NaOH, after which it was named. Sodium occurs throughout the world in vast deposits of NaCl (halite, or rock salt), $NaNO_3$ (saltpeter), $Na_2CO_3 \cdot NaHCO_3 \cdot 2\,H_2O$ (trona), and $Na_2SO_4 \cdot 10\,H_2O$ (mirabilite), all laid down by evaporation of ancient seas. In addition, the world's oceans are approximately 3% by mass NaCl. Uses of sodium and its salts span nearly the entire range of processes in the modern chemical industry. Glass, rubber, pharmaceutical agents, and many other substances use sodium or its salts in their production. [Note that the formulas of trona ($Na_2CO_3 \cdot NaHCO_3 \cdot 2\,H_2O$) and mirabilite ($Na_2SO_4 \cdot 10\,H_2O$) are written with raised dots to specify the overall composition of the substances without indicating exactly how the various parts separated by the dots are bound together.]

A sample of sodium metal.

Potassium.   Potassium, the eighth most abundant element in the earth's crust, was first prepared in 1807 at the same time as sodium. The name of the element is derived from *potash*, a made-up word for K_2CO_3, which had been isolated from the "pot ashes" left over from wood fires. (Today, the term *potash* is used as a general name for all water-soluble, potassium-containing minerals, and the name *caustic potash* is used for KOH.) Potassium is found primarily in deposits of KCl (sylvite) and $KCl \cdot MgCl_2 \cdot 6\,H_2O$ (carnalite), most of which are in Canada and the former USSR. The major use of potassium salts is as a plant fertilizer.

Rubidium, Cesium.   Rubidium and cesium, the two most chemically reactive of the common alkali metals, were both detected as impurities in other substances by chance observation of their characteristic colored spectral lines (Section 5.3). Rubidium, named from the Latin *rubidius* (deepest red), occurs as an impurity in the mineral lepidolite, $K_2Li_3Al_5Si_6O_{20}(OH,F)_4$, and is obtained as a by-product of lithium manufacture. Cesium, named from the Latin *caesius* (sky blue), also occurs with lithium in many minerals and is found in pollucite ($Cs_4Al_4Si_9O_{26} \cdot H_2O$). Neither rubidium nor cesium has any major commercial importance.

A sample of potassium metal.

Francium.   Francium, the heaviest of the group 1A elements, is highly radioactive, and no visible amount of the element has ever been prepared. Little is known about its properties from direct observation, but its behavior would presumably be similar to that of the other alkali metals.

Production of Alkali Metals

Alkali metals are produced commercially by reduction of their chloride salts, although the exact procedure differs for each element. Both lithium metal and sodium metal are produced by *electrolysis*, a process in which an electric current is passed through the molten salt. The details of the process won't be discussed until Sections 18.11 and 18.12, but the fundamental idea is simply to use electrical energy to break down an ionic compound into its elements. A high reaction temperature is necessary to keep the salt liquid.

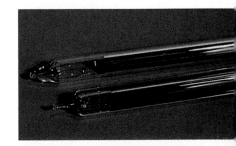

Samples of rubidium and cesium, sealed in glass to protect them from the atmosphere.

$$2\,LiCl(l) \xrightarrow[\substack{450°C}]{\substack{\text{Electrolysis} \\ \text{in KCl}}} 2\,Li(l) + Cl_2(g)$$

$$2\,NaCl(l) \xrightarrow[\substack{580°C}]{\substack{\text{Electrolysis} \\ \text{in } CaCl_2}} 2\,Na(l) + Cl_2(g)$$

Potassium, rubidium, and cesium metals are produced by chemical reduction rather than by electrolysis. Sodium is the reducing agent used in potassium production, and calcium is the reducing agent used for preparing rubidium and cesium.

$$KCl(l) + Na(l) \xrightleftharpoons{850°C} K(g) + NaCl(l)$$

$$2\,RbCl(l) + Ca(l) \xrightleftharpoons{750°C} 2\,Rb(g) + CaCl_2(l)$$

$$2\,CsCl(l) + Ca(l) \xrightleftharpoons{750°C} 2\,Cs(g) + CaCl_2(l)$$

All three of the above reductions appear contrary to the activity series described in Section 4.8, according to which sodium is not a strong enough reducing agent to react with K^+, and calcium is not a strong enough reducing agent to react with either Rb^+ or Cs^+. At high reaction temperatures, however, *equilibria* are established in which small amounts of the products are formed. These products are then removed from the reaction mixture by distillation, thereby driving the reactions toward more product formation. We'll explore the general nature of such chemical equilibria in Chapter 13.

Reactions of Alkali Metals

Reaction with Halogens.   The alkali metals react rapidly with the group 7A elements (halogens) to yield colorless, crystalline ionic salts called *halides*:

$$2\,M(s) + X_2 \longrightarrow 2\,MX(s)$$
A metal halide

where M = Alkali metal (Li, Na, K, Rb, or Cs)

X = Halogen (F, Cl, Br, or I)

The reactivity of the alkali metals increases as their ionization energy decreases, giving a reactivity order $Cs > Rb > K > Na > Li$. Cesium is the most reactive, combining almost explosively with the halogens.

Reaction with Hydrogen and Nitrogen.   Alkali metals react with hydrogen gas to form a series of white crystalline compounds called *hydrides*, MH, in which the hydrogen has an oxidation number of −1. The reaction is slow at room temperature and requires heating to melt the alkali metal before reaction takes place.

$$2\,M(s) + H_2(g) \xrightarrow{\text{Heat}} 2\,MH(s)$$
A metal hydride

where M = Alkali metal (Li, Na, K, Rb, or Cs)

A similar reaction takes place between lithium and nitrogen gas to form lithium nitride, Li_3N, but the other alkali metals do not react with nitrogen.

$$6\,Li(s) + N_2(g) \xrightarrow{\text{Heat}} 2\,Li_3N(s)$$

Reaction with Oxygen.   All the alkali metals react rapidly with oxygen, but they give different kinds of products. Lithium reacts with O_2 to yield the *oxide*, Li_2O; sodium reacts to yield the *peroxide*, Na_2O_2; and the remaining alkali metals, K, Rb, and Cs, form either peroxides or *superoxides*, MO_2, depending on the reaction conditions and on how much oxygen is present. The reasons for the differences have to do largely with the differences in stability of the various products and with the way in which the ions pack together in crystals. Note that the alkali metal cations have a +1 oxidation number in all cases, but the oxidation numbers of the oxygen atoms in the O^{2-}, O_2^{2-}, and O_2^- anions vary from -2 to $-1/2$.

$$4\,Li(s) + O_2(g) \longrightarrow 2\,Li_2O(s)$$ An *oxide*; oxidation number of $O = -2$

$$2\,Na(s) + O_2(g) \longrightarrow Na_2O_2(s)$$ A *peroxide*; oxidation number of $O = -1$

$$K(s) + O_2(g) \longrightarrow KO_2(s)$$ A *superoxide*; oxidation number of $O = -1/2$

Potassium superoxide, KO_2, is a particularly valuable compound because of its use in spacecraft and in self-contained breathing devices to remove moisture and CO_2 from exhaled air, generating oxygen in the process:

$$2\,KO_2(s) + H_2O(g) \longrightarrow KOH(s) + KOOH(s) + O_2(g)$$
$$4\,KO_2(s) + 2\,CO_2(g) \longrightarrow 2\,K_2CO_3(s) + 3\,O_2(g)$$

Reaction with Water. The most well-known and dramatic reaction of the alkali metals is with water to yield hydrogen gas and an alkali metal hydroxide, MOH. In fact, it's this reaction that gives the elements their group name: The solution of metal hydroxide that results from adding an alkali metal to water is *alkaline*, or basic.

Sodium and Potassium in Water movie

$$2\,M(s) + 2\,H_2O(l) \longrightarrow 2\,M^+(aq) + 2\,OH^-(aq) + H_2(g)$$

where $M = $ Li, Na, K, Rb, Cs

Lithium undergoes the reaction with vigorous bubbling as hydrogen is released, sodium reacts rapidly with evolution of heat, and potassium reacts so violently that the hydrogen produced bursts instantly into flame. Rubidium and cesium react almost explosively.

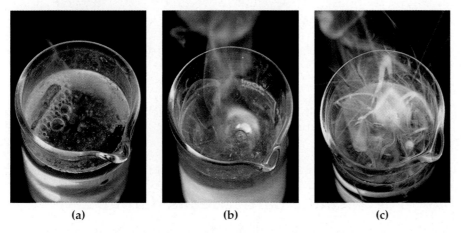

(a)   (b)   (c)

All the alkali metals react with water to generate H_2 gas. **(a)** Lithium reacts vigorously with bubbling, **(b)** sodium reacts violently, and **(c)** potassium reacts almost explosively.

Like all reactions of the alkali metals, the reaction with water is a redox process in which the metal M loses an electron and is oxidized to M^+. At the same time, a hydrogen from water gains an electron and is reduced to H_2 gas, as can be seen by assigning oxidation numbers to the various substances in the usual way. Note that not all hydrogens are reduced; those in the OH⁻ product remain in the +1 oxidation state.

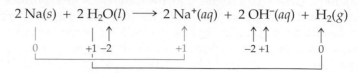

Reaction with Ammonia. The alkali metals react with ammonia, NH_3, to yield H_2 gas plus a metal *amide*, MNH_2. The reaction is exactly analogous to that between an alkali metal and water.

$$2\,M(s) + 2\,NH_3(l) \longrightarrow 2\,M^+(soln) + 2\,NH_2^-(soln) + H_2(g)$$

where M = Li, Na, K, Rb, Cs

The reaction is so slow at low temperature that it's possible for the alkali metals to dissolve in liquid ammonia at −33°C, forming deep blue solutions of metal cations and dissolved electrons. As you might expect, these solutions have extremely powerful reducing properties.

$$M(s) \xrightarrow[\text{solvent}]{\text{Liquid } NH_3} M^+(soln) + e^-(soln)$$

where M = Li, Na, K, Rb, or Cs

Sodium metal dissolves in liquid ammonia to yield a blue solution of Na^+ ions and solvent-surrounded electrons.

▶ **PROBLEM 6.14** Assign oxidation numbers to the oxygen atoms in the following compounds:
(a) Li_2O (b) K_2O_2 (c) CsO_2

▶ **PROBLEM 6.15** Complete and balance the following equations. If no reaction takes place, write N.R.
(a) $Cs(s) + H_2O(l) \rightarrow ?$ (b) $Na(s) + N_2(g) \rightarrow ?$ (c) $Rb(s) + O_2(g) \rightarrow ?$
(d) $K(s) + NH_3(g) \rightarrow ?$ (e) $Rb(s) + H_2(g) \rightarrow ?$

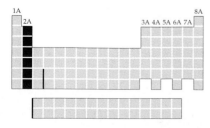

6.8 The Alkaline Earth Metals (Group 2A)

The alkaline earth elements of group 2A—Be, Mg, Ca, Sr, Ba, and Ra—are similar to the alkali metals in many respects. They differ, however, in that they have ns^2 valence-shell electron configurations and can therefore lose *two* electrons in redox reactions. Alkaline earth metals are thus powerful reducing agents and form ions with a +2 charge.

The Born–Haber cycle in Figure 6.9 for the reaction of magnesium with chlorine provides a graphic view of the energy changes involved in the redox reaction of an alkaline earth element. As in the reaction of sodium and chlorine to form NaCl shown previously in Figure 6.8, there are five contributions to the overall energy change in the reaction of magnesium with chlorine to form $MgCl_2$. First, solid magnesium metal must be converted into isolated gaseous magnesium atoms (sublimation). Second, the bond in Cl_2 molecules must be broken to yield two chlorine atoms. Third, magnesium atoms must lose *two* electrons to form the dipositive Mg^{2+} ion. Fourth, the two chlorine atoms formed in step 2 must accept electrons to form two Cl^- ions. Fifth, the gaseous ions must combine to form the ionic solid, $MgCl_2$. As the Born–Haber cycle indicates, it is the large contribution from ionic bonds (the negative of the lattice energy) that releases enough energy to drive the entire process.

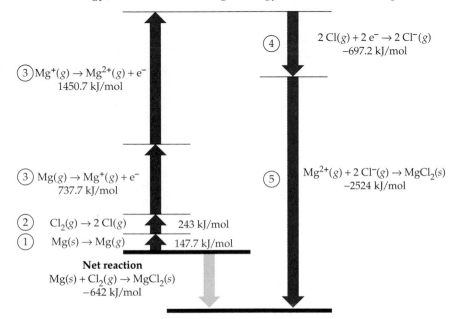

$$4 \quad 2\,Cl(g) + 2\,e^- \rightarrow 2\,Cl^-(g)$$
$$-697.2 \text{ kJ/mol}$$

$$3 \quad Mg^+(g) \rightarrow Mg^{2+}(g) + e^-$$
$$1450.7 \text{ kJ/mol}$$

$$3 \quad Mg(g) \rightarrow Mg^+(g) + e^-$$
$$737.7 \text{ kJ/mol}$$

$$5 \quad Mg^{2+}(g) + 2\,Cl^-(g) \rightarrow MgCl_2(s)$$
$$-2524 \text{ kJ/mol}$$

$$2 \quad Cl_2(g) \rightarrow 2\,Cl(g) \qquad 243 \text{ kJ/mol}$$
$$1 \quad Mg(s) \rightarrow Mg(g) \qquad 147.7 \text{ kJ/mol}$$

Net reaction
$$Mg(s) + Cl_2(g) \rightarrow MgCl_2(s)$$
$$-642 \text{ kJ/mol}$$

FIGURE 6.9   A Born–Haber cycle for the formation of $MgCl_2$ from the elements. The large contribution from ionic bonding in the solid (step 5) provides more than enough energy to remove two electrons from magnesium (step 3).

Though harder than their neighbors in group 1A, the alkaline earth elements are still relatively soft, silvery metals. They tend, however, to have higher melting points and densities than alkali metals, as shown in Table 6.5. Alkaline earth elements are less reactive toward oxygen and water than alkali metals but are nevertheless found in nature only in salts, not in the elemental state.

TABLE 6.5   Properties of Alkaline Earth Metals

Name	Melting Point (°C)	Boiling Point (°C)	Density (g/cm³)	1st Ionization Energy (kJ/mol)	Abundance on Earth (%)	Atomic Radius (pm)	Ionic (M^{2+}) Radius (pm)
Beryllium	1,287	2,471	1.848	899.4	0.000 28	112	44
Magnesium	650	1,090	1.738	737.7	2.33	160	66
Calcium	842	1,484	1.55	589.8	4.15	197	99
Strontium	777	1,382	2.54	549.5	0.038	215	112
Barium	727	1,897	3.51	502.9	0.042	222	134
Radium	700	1,140	≈5.0	509.3	Trace	223	143

Occurrence and Uses of Alkaline Earth Metals

Beryllium.   Beryllium was first detected in 1798 in the gemstones beryl and emerald ($Be_3Al_2Si_6O_{18}$) and was subsequently prepared in pure form in 1828

The gemstones beryl and emerald both contain beryllium.

by the reduction of $BeCl_2$ with potassium. It is obtained today from large commercial deposits of beryl in Brazil and southern Africa. Though beryllium compounds are extremely toxic, particularly when inhaled as dust, the metal is nevertheless useful in forming alloys. Addition of a few percent beryllium to copper or nickel results in hard, corrosion-resistant alloys that are used in airplane engines and precision instruments.

Magnesium.   Compounds of magnesium, the seventh most abundant element in the earth's crust, have been known since ancient times, although the pure metal was not prepared until 1808. The element is named after the Magnesia district in Thessaly, Greece, where large deposits of talc [$Mg_3Si_4O_{10}(OH)_2$] are found. There are many magnesium-containing minerals, including dolomite ($CaCO_3 \cdot MgCO_3$) and magnesite ($MgCO_3$), and the world's oceans provide a nearly infinite supply, since seawater is 0.13% Mg. When alloyed with aluminum, magnesium is used widely as a structural material because of its high strength, low density, and ease in machining. Airplane fuselages, automobile engines, and a great many other products are made of magnesium alloys.

A sample of magnesium.

Calcium.   Calcium is the fifth most abundant element in the earth's crust, owing largely to the presence of huge $CaSO_4 \cdot 2\ H_2O$ (gypsum) and $CaCO_3$ deposits in ancient seabeds. Limestone, marble, chalk, and coral are all slightly different forms of $CaCO_3$. Though the metal was not obtained pure until 1808, compounds of calcium have been known for millennia. Lime (CaO), for example, was prepared by the Romans by heating $CaCO_3$ and was used as a mortar in their constructions. In fact, the name *calcium* is derived from the Latin word *calx*, meaning "lime."

The primary industrial use of calcium metal is as an alloying agent to harden aluminum. Calcium compounds such as lime and gypsum are used for many purposes throughout the chemical and construction industries. Portland cement, for example, contains approximately 70% CaO. In addition, calcium is the primary constituent of teeth and bones.

A sample of calcium.

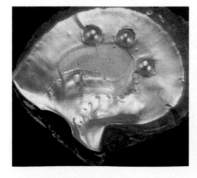

Pearls and coral are two of the many occurrences of calcium carbonate.

Strontium, Barium. Strontium was discovered near, and named after, the small town of Strontian, Scotland, in 1787. There are no commercial uses for the pure metal, but the carbonate salt, $SrCO_3$, is used in the manufacture of glass for color TV picture tubes. Barium is found principally in the minerals witherite ($BaCO_3$) and barite ($BaSO_4$), after which it is named. Though water-soluble salts of barium are extremely toxic, barium sulfate is so insoluble that it is used in medicine as a contrast medium for stomach and intestinal X rays. Like strontium, barium metal has no commercial uses, but various compounds are used in glass manufacture and in drilling oil wells.

Radium. Radium, the heaviest of the group 2A elements, occurs with uranium and was isolated as its chloride salt from the mineral pitchblende by Marie and Pierre Curie in 1898. Radium is highly radioactive, and no more than a few kilograms of the pure metal have ever been produced. Though used for many years as a radiation source for cancer radiotherapy, better sources are now available, and there are no longer any commercial uses for radium.

Production of Alkaline Earth Metals

Like the alkali metals, the pure alkaline earth elements are produced commercially by reduction of their salts, either chemically or through electrolysis. Beryllium is prepared by reduction of BeF_2 with magnesium, and magnesium is prepared by electrolysis of its molten chloride.

$$BeF_2(l) + Mg(l) \xrightarrow{1300°C} Be(l) + MgF_2(l)$$

$$MgCl_2(l) \xrightarrow[750°C]{Electrolysis} Mg(l) + Cl_2(g)$$

Calcium, strontium, and barium are all made by high-temperature reduction of their oxides with aluminum metal.

$$3\,MO(l) + 2\,Al(l) \xrightarrow{High\ temp.} 3\,M(l) + Al_2O_3(s)$$

$$where\ M = Ca,\ Sr,\ Ba$$

Reactions of Alkaline Earth Metals

The alkaline earth metals undergo the same kinds of redox reactions as the alkali metals, but they lose two electrons rather than one to yield dipositive ions, M^{2+}. Because their first ionization energy is larger than that of alkali metals (Figure 6.3), the group 2A metals tend to be somewhat less reactive than alkali metals. The general reactivity trend is $Ba > Sr > Ca > Mg > Be$.

Alkaline earth metals react with halogens to yield ionic halide salts, MX_2, and with oxygen to form oxides, MO:

$$M + X_2 \longrightarrow MX_2 \qquad where\ M = Be,\ Mg,\ Ca,\ Sr,\ or\ Ba$$
$$X = F,\ Cl,\ Br,\ or\ I$$

$$2\,M + O_2 \longrightarrow 2\,MO$$

Beryllium and magnesium are relatively unreactive toward oxygen at room temperature, but both burn with a brilliant white glare when ignited by a flame. Calcium, strontium, and barium are reactive enough that they are best stored under oil to keep them from contact with air. Like the heavier alkali metals, strontium and barium form peroxides, MO_2.

Calcium metal reacts very slowly with water at room temperature.

With the exception of beryllium, the alkaline earth elements react with water to yield metal hydroxides, $M(OH)_2$. Magnesium undergoes reaction only at temperatures above 100°C; calcium and strontium react slowly with liquid water at room temperature. Only barium reacts vigorously.

$$M(s) + 2\,H_2O(l) \longrightarrow M^{2+}(aq) + 2\,OH^-(aq) + H_2(g)$$

$$\text{where } M = Mg, Ca, Sr, \text{ or } Ba$$

EXAMPLE 6.5   Calcium metal reacts with hydrogen gas at high temperature to give calcium hydride. Predict the formula of the product, and write the balanced equation.

SOLUTION   Metal hydrides contain hydrogen with a −1 oxidation number. Since calcium always has a +2 oxidation number, there must be two H^- ions per Ca^{2+} ion, and the formula of calcium hydride must be CaH_2.

$$Ca(s) + H_2(g) \longrightarrow CaH_2(s)$$

▶**PROBLEM 6.16**   Predict the products of the following reactions, and balance the equations:
(a) $Be(s) + Br_2(l) \rightarrow ?$ **(b)** $Sr(s) + H_2O(l) \rightarrow ?$ **(c)** $Mg(s) + O_2(g) \rightarrow ?$

▶**PROBLEM 6.17**   Write a balanced equation for the preparation of beryllium metal by the reduction of beryllium chloride with potassium.

▶**PROBLEM 6.18**   What product do you think is formed by reaction of magnesium with sulfur, a group 6A element? What is the oxidation number of sulfur in the product?

6.9   The Group 3A Elements: Aluminum

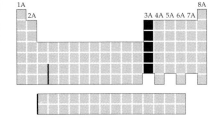

The elements in group 3A—B, Al, Ga, In, and Tl—are the first of the *p*-block elements (Section 5.12) and have the valence-shell electron configuration $ns^2\,np^1$. With the exception of boron, which behaves as a semimetal rather than a metal, the group 3A elements are silvery in appearance, good conductors of electricity, and relatively soft. Gallium, in fact, has a melting point of only 29.8°C. Although properties of the entire group are listed in Table 6.6, we'll concentrate for now on the most common element, aluminum, and look at the others in Chapter 19.

TABLE 6.6   Properties of Group 3A Elements

Name	Melting Point (°C)	Boiling Point (°C)	Density (g/cm³)	1st Ionization Energy (kJ/mol)	Abundance on Earth (%)	Atomic Radius (pm)	Ionic (M^{3+}) Radius (pm)
Boron	2,075	4,000	2.34	800.6	0.001	83	—
Aluminum	660	2,519	2.699	577.6	8.32	143	51
Gallium	29.8	2,204	5.904	578.8	0.001 5	135	62
Indium	157	2,072	7.31	558.3	0.000 01	167	81
Thallium	304	1,473	11.85	589.3	0.000 04	170	95

Aluminum, the most abundant metal in the earth's crust at 8.3%, takes its name from alum, $KAl(SO_4)_2 \cdot 12\,H_2O$, a salt that has been used medicinally since Roman times. In spite of its abundance, the metal nevertheless proved difficult to isolate in pure form. It was such a precious substance in the mid-nineteenth century, in fact, that aluminum cutlery was sometimes used for elegant dinners, and the Washington Monument was capped by a pyramid of pure aluminum. Not until 1886 did an economical manufacturing process become available.

Aluminum occurs in many common minerals and clays, as well as in gemstones. Sapphire and ruby are both impure forms of Al_2O_3 that receive their color from the presence of small amounts of other elements (Cr in ruby; Fe and Ti in sapphire). Most aluminum is currently obtained from bauxite, $Al_2O_3 \cdot x\,H_2O$, which occurs in large deposits in Australia, the United States, Jamaica, and elsewhere. The preparation of Al from ores is extremely energy-intensive, requiring high temperatures and large amounts of electric current to carry out the electrolysis of Al_2O_3. We'll examine the process in more detail in Section 18.12.

$$2\,Al_2O_3(soln) \xrightarrow[\substack{980°C}]{\substack{\text{Electrolysis} \\ \text{in } Na_3AlF_6}} 4\,Al(l) + 3\,O_2(g)$$

Following the trend established by the group 1A and group 2A metals, aluminum is a reducing agent that undergoes redox reactions by losing all three valence-shell electrons to yield Al^{3+} ions. For example, it reacts with the halogens to yield colorless halides, AlX_3, with oxygen to yield an oxide, Al_2O_3, and with nitrogen to yield a nitride, AlN:

$$2\,Al + 3\,X_2 \longrightarrow 2\,AlX_3 \quad \text{where } X = F, Cl, Br, \text{ or } I$$

$$4\,Al + 3\,O_2 \longrightarrow 2\,Al_2O_3$$

$$2\,Al + N_2 \longrightarrow 2\,AlN$$

Reactions with the halogens occur vigorously at room temperature and release large amounts of heat. Reaction with oxygen is also vigorous at room temperature, yet aluminum can be used in a huge array of consumer products without evident corrosion from the air. The explanation for this apparent inconsistency is that aluminum metal reacts rapidly with oxygen only on its *surface*. In so doing, it forms a thin, hard, oxide coating that does not flake off and that protects the underlying metal from contact with air.

Aluminum is less reactive than the group 1A and 2A metals and does not normally react with water because of the oxide coating mentioned previously. It does, however, react with both acidic and basic solutions to give Al^{3+} ions and release H_2 gas.

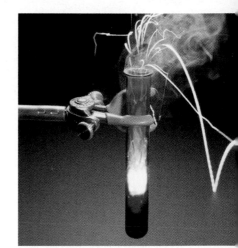

Aluminum metal reacts with liquid bromine in a spectacular display of sparks.

Formation of Aluminum Bromide movie

Acid solution: $2\,Al(s) + 6\,H^+(aq) \longrightarrow 2\,Al^{3+}(aq) + 3\,H_2(g)$

Basic solution: $2\,Al(s) + 2\,OH^-(aq) + 6\,H_2O(l) \longrightarrow 2\,Al(OH)_4^-(aq) + 3\,H_2(g)$

▶ **PROBLEM 6.19**   Identify the oxidizing agent and the reducing agent in the reaction of aluminum metal with $H^+(aq)$.

▶ **PROBLEM 6.20**   Aluminum reacts with sulfur to give a sulfide in the same way that it reacts with oxygen to give an oxide. Identify the product, and write a balanced equation for the reaction.

6.10   The Halogens (Group 7A)

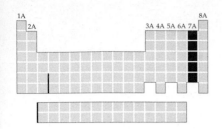

The halogens of group 7A—F, Cl, Br, I, and At—are completely different from the elements we've been discussing up to this point. The halogens are nonmetals rather than metals, they exist as diatomic molecules rather than as individual atoms, and they have a tendency to gain rather than lose electrons when they enter into redox reactions because of their $ns^2\,np^5$ electron configurations. In other words, the halogens are powerful oxidizing agents, characterized by large negative electron affinities and large positive ionization energies. Some of their properties are shown in Table 6.7.

TABLE 6.7   Properties of Halogens

Name	Melting Point (°C)	Boiling Point (°C)	Density (g/cm³)	Electron Affinity (kJ/mol)	Abundance on Earth (%)	Atomic Radius (pm)	Ionic (X⁻) Radius (pm)
Fluorine	−220	−188	1.50 (l)	−328	0.062	72	133
Chlorine	−101	−34	2.03 (l)	−349	0.013	99	181
Bromine	−7	59	3.12 (l)	−325	0.000 3	114	196
Iodine	114	184	4.930 (s)	−295	0.000 05	133	220
Astatine	—	—	—	−270	Trace	—	—

Halogens are too reactive to occur in nature as free elements. Instead, they are found only as their anions in various salts and minerals. Even the name "halogen" implies reactivity, since it comes from the Greek words *hals* (salt) and *gennan* (to form). Thus, a halogen is literally a "salt-former."

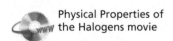
Physical Properties of the Halogens movie

Occurrence and Uses of Halogens

Fluorine.   Fluorine, a corrosive, greenish-yellow gas, is the thirteenth most abundant element in the earth's crust, more common than such well-known elements as sulfur (16th), carbon (17th), and copper (26th). It is found in several common minerals including fluorspar, or fluorite (CaF_2), and fluorapatite [$Ca_5(PO_4)_3F$]. (Note that fluorine is spelled with the *u* first as in *flu* rather than with the *o* first as in *flour*.)

In spite of its toxicity and extreme reactivity, fluorine is widely used for the manufacture of polymers such as Teflon, $(C_2F_4)_n$. Fluorine is also important in the production of UF_6, used in the separation of uranium isotopes for nuclear power plants, and fluoride ion is added to toothpaste in the form of SnF_2 to help prevent tooth decay.

Chlorine.   Chlorine, like fluorine, is a toxic, reactive, greenish-yellow gas that must be handled with great care. Though chlorine is only the twentieth most abundant element in crustal rocks, there are vast additional amounts of chloride ion in the world's oceans (1.9% by mass). The free element was first prepared in 1774 by oxidation of NaCl with MnO_2, though not until 1810 was it recognized that the product of the reaction was indeed a new element rather than a compound.

$$MnO_2(s) + 2\,Cl^-(aq) + 4\,H^+(aq) \longrightarrow Mn^{2+}(aq) + 2\,H_2O(l) + Cl_2(g)$$

Chlorine and chlorine-releasing compounds are frequently added as a disinfectant to the water in swimming pools.

Industrial applications of chlorine center around its uses for the preparation of numerous chlorinated organic chemicals, including PVC [poly(vinyl chloride)] plastic and the solvents chloroform ($CHCl_3$), methylene chloride

(CH$_2$Cl$_2$), and ethylene dichloride (C$_2$H$_4$Cl$_2$). Large amounts of chlorine are also used as a bleach during paper manufacture and as a disinfectant for swimming pools and municipal water supplies.

Bromine.   Bromine, a fourth-row element, is a volatile, reddish liquid rather than a gas like fluorine and chlorine. Its fumes are quite toxic, however, and it causes extremely painful burns when spilled on bare skin. Bromine was first isolated in 1826 by the 23-year-old French chemist A.-J. Balard by oxidation of KBr with MnO$_2$ in a manner similar to that used for the synthesis of chlorine.

$$MnO_2(s) + 2\,Br^-(aq) + 4\,H^+(aq) \longrightarrow Mn^{2+}(aq) + 2\,H_2O(l) + Br_2(aq)$$

The primary uses of bromine are as its silver salt, AgBr, in photographic emulsions and as a reactant for preparing brominated organic compounds. Fuel additives, pesticides, fungicides, and flame retardants are a few of the many kinds of compounds manufactured from bromine.

Iodine.   Iodine is a volatile purple-black solid with a beautiful metallic sheen. As the least reactive halogen, iodine is safe to handle and is widely used as a skin disinfectant. It was first prepared in 1811 from seaweed ash, but commercially useful deposits of the iodine-containing minerals lautarite (CaIO$_3$) and dietzeite [7 Ca(IO$_3$)$_2 \cdot$ 8 CaCrO$_4$] were subsequently found in Chile. Iodine is used in the preparation of numerous organic compounds, including dyes and pharmaceutical agents, but there is no one single use of major importance.

Astatine.   Astatine, like francium in group 1A, is a radioactive element that occurs only in minute amounts in nature. No more than about 5×10^{-8} g has ever been prepared at one time, and little is known about its chemistry.

Production of Halogens

All the free halogens are produced commercially by oxidation of their anions. Fluorine and chlorine are both produced by electrolysis, fluorine from a molten 1:2 mixture of KF and HF, and chlorine from molten NaCl.

$$2\,HF(l) \xrightarrow[100°C]{Electrolysis} H_2(g) + F_2(g)$$

$$2\,NaCl(l) \xrightarrow[580°C]{Electrolysis} 2\,Na(l) + Cl_2(g)$$

Bromine and iodine are both prepared by oxidation of the corresponding halide ion with chlorine. Naturally occurring aqueous solutions of bromide ion with concentrations of up to 5000 ppm are found in Arkansas and in the Dead Sea in Israel. Iodide ion solutions of up to 100 ppm concentration are found in Oklahoma and Michigan.

$$2\,Br^-(aq) + Cl_2(g) \longrightarrow Br_2(l) + 2\,Cl^-(aq)$$

$$2\,I^-(aq) + Cl_2(g) \longrightarrow I_2(s) + 2\,Cl^-(aq)$$

The Dead Sea in Israel has a particularly high concentration of bromide ion.

Reactions of Halogens

Halogens are among the most reactive elements in the periodic table. Fluorine, in fact, forms compounds with every element except the noble gases He, Ne, and Ar. As noted previously, the dominant reaction of halogens is as strong oxidizing agents in redox reactions. That is, their large negative electron affinities allow halogens to accept electrons from other atoms to yield halide anions, X^-.

Reaction with Metals. Halogens react with every metal in the periodic table to yield metal halides. With the alkali and alkaline earth metals, the formula of the halide product is easily predictable. With transition metals, though, more than one product can sometimes form depending on the reaction conditions and the amounts of reactants present. Iron, for example, can react with Cl_2 to form either $FeCl_2$ or $FeCl_3$. Without knowing a good deal more about transition metal chemistry, it's not possible to make predictions at this point. The reaction can be generalized as

$$2\,M + n\,X_2 \longrightarrow 2\,MX_n \qquad \text{where M = Metal}$$
$$X = F, Cl, Br, I$$

Unlike the metallic elements, halogens become *less* reactive going down the periodic table because of their generally decreasing electron affinity. Thus, their reactivity order is $F_2 > Cl_2 > Br_2 > I_2$. Fluorine often reacts violently, chlorine and bromine somewhat less so, and iodine often sluggishly.

Reaction with Hydrogen. Halogens react with hydrogen gas to yield hydrogen halides, HX:

$$H_2(g) + X_2 \longrightarrow 2\,HX(g) \qquad \text{where X = F, Cl, Br, I}$$

Fluorine reacts explosively with hydrogen as soon as the two gases come in contact. Chlorine also reacts explosively once the reaction is initiated by a spark or by ultraviolet light, but the mixture of gases is stable in the dark. Bromine and iodine react more slowly.

The hydrogen halides are valuable because they behave as acids when dissolved in water. Hydrogen fluoride is a weak acid, dissociating only to a small extent in aqueous solution, but the other hydrogen halides are strong acids. As one of the few substances that reacts with glass, HF is frequently used for the etching or fogging of glass. The aqueous solution of HCl, called hydrochloric acid or muriatic acid, is used throughout the chemical industry in a vast number of processes, from pickling steel (removing its iron oxide coating) to dissolving animal bones for producing gelatin.

$$HX \xrightarrow[\text{in } H_2O]{\text{Dissolve}} H^+(aq) + X^-(aq)$$

Reaction with Other Halogens. Since all the halogens exist as diatomic molecules X_2, it's not surprising that a variety of covalent **interhalogen compounds XY** exist, where X and Y are different halogens. Iodine reacts with chlorine, for example, to yield iodine monochloride, ICl, and bromine reacts with fluorine to yield bromine monofluoride, BrF. These reactions can be thought of as redox processes in which the lighter, more reactive element is the oxidizing agent and the heavier, less reactive element is the reducing agent. In the reaction of iodine with chlorine, for example, Cl_2 acts as the oxidizing agent and is reduced to a -1 oxidation state, while I_2 acts as the reducing agent and is oxidized to a $+1$ oxidation state.

$$I_2(s) + Cl_2(g) \longrightarrow 2\ ICl(s)$$

$$\begin{array}{cccc} & & & \quad \uparrow\ \uparrow \\ 0 & 0 & & +1\ -1 \end{array}$$

$$Br_2(l) + F_2(g) \longrightarrow 2\ BrF(g)$$

$$\begin{array}{cccc} & & & \quad \uparrow\ \uparrow \\ 0 & 0 & & +1\ -1 \end{array}$$

As a general rule, the properties of interhalogen compounds are intermediate between those of their parent elements. For example, ICl is a red solid that melts near room temperature, and BrF is a brownish gas that condenses to a liquid near room temperature. All six possible diatomic interhalogen compounds are known, and all act as strong oxidizing agents in redox reactions.

In addition to the diatomic interhalogen compounds, a number of polyatomic substances with formulas XY_3, XY_5, and XY_7 are also known. Typical examples are ClF_3, BrF_5, and IF_7. Once again, the less reactive halogen has a positive oxidation number and the more reactive halogen has a negative oxidation number. In BrF_5, for example, the oxidation number of Br is +5 and that of each F is −1.

EXAMPLE 6.6   The UF_6 used in producing nuclear fuels is prepared by reaction of uranium metal with chlorine trifluoride. Tell which atoms have been oxidized and which reduced, and balance the equation.

$$U(s) + ClF_3(l) \longrightarrow UF_6(l) + ClF(g) \qquad \text{Unbalanced}$$

SOLUTION   First assign oxidation numbers to the various elements, and decide which atoms have undergone a change. Those that have increased in oxidation number have been oxidized, and those that have decreased in oxidation number have been reduced. Uranium is oxidized from 0 to +6, and chlorine is reduced from +3 to +1.

$$U(s) + ClF_3(l) \longrightarrow UF_6(l) + ClF(g)$$

$$\begin{array}{cccccc} & & & & \\ 0 & +3\ -1 & & +6\ -1 & & +1\ -1 \end{array}$$

Balance the equation, either by inspection or by one of the methods discussed in Sections 4.9 and 4.10.

$$U(s) + 3\ ClF_3(l) \longrightarrow UF_6(l) + 3\ ClF(g)$$

▶ **PROBLEM 6.21**   Write the products of the following reactions, and balance the equations:
(a) $Br_2(l) + Cl_2(g) \rightarrow ?$ **(b)** $Al(s) + F_2(g) \rightarrow ?$ **(c)** $H_2(g) + I_2(s) \rightarrow ?$

▶ **PROBLEM 6.22**   Bromine reacts with sodium iodide to yield iodine and sodium bromide. Identify the oxidizing and reducing agents, and write the balanced equation.

6.11 The Noble Gases (Group 8A)

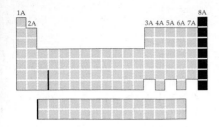

The noble gases of group 8A—He, Ne, Ar, Kr, Xe, and Rn—are completely different from the other elements we've been discussing. They are neither metals like most elements nor reactive nonmetals like the halogens; rather, they are colorless, odorless, unreactive gases. Their $1s^2$ (for He) and $ns^2\, np^6$ (for the others) valence-shell electron configurations make it difficult for the noble gases to either gain or lose electrons, so the elements don't normally enter into redox reactions.

Though sometimes referred to as "rare gases" or "inert gases," these older names are not really accurate because the group 8A elements are neither rare nor completely inert. Argon, for instance, makes up nearly 1% by volume of dry air, and there are several dozen known compounds of krypton and xenon, though none occur naturally. Some properties of the noble gases are shown in Table 6.8.

TABLE 6.8   Properties of Noble Gases

Name	Melting Point (°C)	Boiling Point (°C)	1st Ionization Energy (kJ/mol)	Abundance in Dry Air (vol %)
Helium	−272.2	−268.9	2,372.3	5.2×10^{-4}
Neon	−248.6	−246.1	2,080.6	1.8×10^{-3}
Argon	−189.3	−185.9	1,520.4	0.93
Krypton	−157.4	−153.2	1,350.7	1.1×10^{-4}
Xenon	−111.8	−108.0	1,170.4	9×10^{-6}
Radon	−71	−61.7	1,037	Trace

Occurrence and Uses of Noble Gases

The natural abundance of the noble gas elements depends on where you look. In the universe, helium is the second most abundant element, accounting for about 25% of the total mass (hydrogen accounts for the other 75%). On earth, though, the fact that they're gases means that the abundance of the group 8A elements in crustal rocks is very low. Helium occurs as a minor constituent of natural gas, and argon makes up about 1% of the volume of dry air. The remaining noble gases occur in small amounts in the air (Table 6.8), from which they are obtained by liquefaction and distillation.

Their relative lack of reactivity means that the main commercial uses of the noble gases are in applications that require an inert (unreactive) atmosphere. Argon, for example, is used as the gas in light bulbs and to protect the metal from oxygen during arc welding. Liquid helium is used in scientific research as a cooling agent for extremely low-temperature studies because it has the lowest boiling point (4.2 K) of any substance. Helium is also used in deep-sea diving gas.

Radon, the heaviest of the noble gases, has been much publicized in recent years because of fear that low-level exposures increase the risk of cancer. Like astatine and francium, its neighbors in the periodic table, radon is a radioactive element with only a minute natural abundance. It is produced by radioactive decay of the radium present in small amounts in many granitic rocks, and it can slowly seep into basements, where it remains unless vented. If breathed into the lungs, it can cause radiation damage.

A helium–oxygen mixture is used in diving gas instead of compressed air (nitrogen–oxygen). If air were used, nitrogen would dissolve in the diver's blood at the high underwater pressures and would be released as painful bubbles when the diver returned to the surface.

Reactions of Noble Gases

Helium, neon, and argon undergo no chemical reactions and form no known compounds; krypton and xenon react only with fluorine. Depending on the reaction conditions and on the amounts of reactants present, xenon can form three different fluorides, XeF_2, XeF_4, and XeF_6. All three xenon fluorides are powerful oxidizing agents and undergo a wide variety of redox reactions.

$$Xe(g) + F_2(g) \longrightarrow XeF_2(s)$$
$$Xe(g) + 2\,F_2(g) \longrightarrow XeF_4(s)$$
$$Xe(g) + 3\,F_2(g) \longrightarrow XeF_6(s)$$

The lack of reactivity of the noble gases is a consequence of their unusually large ionization energies (Figure 6.3) and their unusually small electron affinities (Figure 6.6), which result from their valence-shell electron configurations.

▶ **PROBLEM 6.23**   Assign oxidation numbers to the elements in the following compounds of xenon:
(a) XeF_2 (b) XeF_4 (c) $XeOF_4$

6.12 The Octet Rule

Let's list some general conclusions about main-group elements that we can draw from the information in the preceding five sections:

Octet Rule activity

- Group 1A elements tend to lose their ns^1 valence-shell electron, thereby adopting the electron configuration of the noble gas element in the previous row of the periodic table.
- Group 2A elements tend to lose both of their ns^2 valence-shell electrons and adopt a noble gas electron configuration.
- Group 3A elements tend to lose all three of their $ns^2\,np^1$ valence-shell electrons and adopt a noble gas electron configuration.
- Group 7A elements tend to gain one electron, changing from $ns^2\,np^5$ to $ns^2\,np^6$, thereby adopting the configuration of the neighboring noble gas element in the same row.
- Group 8A (noble gas) elements are essentially inert; they rarely gain or lose electrons.

All these observations can be gathered into a single statement called the *octet rule*:

◆— OCTET RULE: *Main-group elements tend to undergo reactions that leave them with eight outer-shell electrons.* That is, main-group elements react so that they attain a noble gas electron configuration with filled *s* and *p* sublevels in their valence electron shell.

There are many exceptions to the octet rule—after all, it's called the octet *rule*, not the octet *law*—but it is nevertheless useful for making predictions and for providing insights about chemical bonding.

Why does the octet rule work? What factors determine whether an atom is likely to gain or to lose electrons? Clearly, electrons are most likely to be lost if they are held loosely in the first place—that is, if they feel a relatively low effective nuclear charge, Z_{eff}, and therefore have small ionization energies. Valence-shell electrons in the group 1A, 2A, and 3A metals, for example, are shielded from the nucleus by core electrons. They feel low values of Z_{eff} (Figure 5.20, page 191), and they are therefore lost relatively easily. Once the next lower noble gas configuration is reached, though, loss of an additional electron is much more difficult because it must come from an inner shell where it feels a high Z_{eff}.

Conversely, electrons are most likely to be gained if they can be held tightly by a high Z_{eff}. Valence-shell electrons in the group 6A and 7A elements, for example, are poorly shielded. They feel high values of Z_{eff} (Figure 5.20) and they aren't lost easily. The high Z_{eff} thus makes possible the gain of one or more additional electrons into vacant valence-shell orbitals. Once the noble gas configuration is reached, though, there are no longer any low-energy orbitals available. An additional electron would have to be placed in a higher-energy orbital, where it would feel only a low Z_{eff}.

Eight is therefore the "magic number" for valence-shell electrons. Taking electrons *from* a filled octet is difficult because they are tightly held by a high Z_{eff}; adding more electrons *to* a filled octet is difficult because, with s and p sublevels full, no low-energy orbital is available.

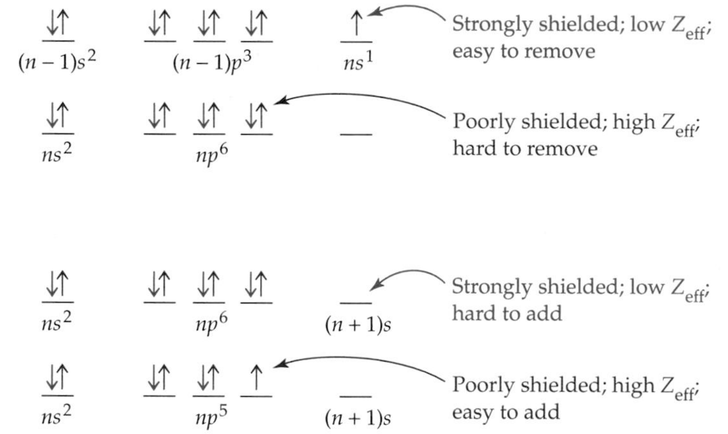

When the octet rule fails, it generally does so for elements toward the right side of the periodic table (groups 3A–8A) that are in the third row and lower (Figure 6.10). The reason is straightforward and has to do with the electron configurations of these elements: With few exceptions, the main-group elements that occasionally break the octet rule have vacant, low-energy d orbitals, which allow them to accommodate additional electrons. Phosphorus, for example, has the electron configuration [Ne] $3s^2\,3p^3$ and has a vacant $3d$ subshell that is only slightly higher in energy than the $3s$ and $3p$ levels. As a result, phosphorus is occasionally able to add more than the three electrons predicted by the octet rule.

1A																	8A
H	2A											3A	4A	5A	6A	7A	He
Li	Be											B	C	N	O	F	Ne
Na	Mg	3B	4B	5B	6B	7B	⎯8B⎯			1B	2B	Al	Si	P	S	Cl	Ar
K	Ca	Sc	Ti	V	Cr	Mn	Fe	Co	Ni	Cu	Zn	Ga	Ge	As	Se	Br	Kr
Rb	Sr	Y	Zr	Nb	Mo	Tc	Ru	Rh	Pd	Ag	Cd	In	Sn	Sb	Te	I	Xe
Cs	Ba	La	Hf	Ta	W	Re	Os	Ir	Pt	Au	Hg	Tl	Pb	Bi	Po	At	Rn
Fr	Ra	Ac	Rf	Db	Sg	Bh	Hs	Mt									

FIGURE 6.10 The octet rule occasionally fails for the shaded main-group elements. These elements, all of which are in the third row or lower, can use low-energy unfilled d orbitals to expand their valence shell beyond the normal octet.

EXAMPLE 6.7   We saw in Section 6.7 that lithium reacts with nitrogen to yield Li_3N. What noble gas configuration does the nitrogen atom in Li_3N have?

SOLUTION   The nitrogen atom in Li_3N has an oxidation number of -3 and has gained three electrons over the neutral atom, giving it a valence-shell octet with the neon configuration:

N configuration: $(1s^2\, 2s^2\, 2p^3)$ N^{3-} configuration: $(1s^2\, 2s^2\, 2p^6)$

▶**PROBLEM 6.24**   What noble gas configurations are the following elements likely to adopt in redox reactions?
(a) Rb   **(b)** Ba   **(c)** Ga   **(d)** F

▶**PROBLEM 6.25**   Although we haven't talked about group 6A elements in this chapter, what are they likely to do in redox reactions—gain or lose electrons? How many?

Salt

Like most people, you probably feel a little guilty about reaching for the salt shaker at mealtime. The notion that high salt intake and high blood pressure go hand in hand is surely among the most highly publicized pieces of nutritional lore to appear in recent decades.

Salt has not always been held in such disrepute. Historically, salt has been prized since the earliest recorded times as a seasoning and a food preservative. Words and phrases in many languages reflect the importance of salt as a life-giving and life-sustaining substance. We refer to a kind and generous person as "the salt of the earth," for instance, and we speak of being "worth one's salt." In Roman times, soldiers were paid in salt; the English word "salary" is derived from the Latin word for paying salt wages (*salarium*).

Salt is perhaps the easiest of all minerals to obtain and purify. The simplest method, used for thousands of years throughout the world in coastal climates where sunshine is abundant and rainfall is scarce, is to evaporate seawater. Though the exact amount varies depending on the source, seawater contains an average of about 3.5% by mass of dissolved substances, most of which is sodium chloride. It has been estimated that evaporation of all the world's oceans would yield approximately *4.5 million cubic miles* of NaCl.

These chandeliers were carved out of salt by miners at the Wieliczka mine.

Only about 10% of current world salt production comes from evaporation of seawater. Most salt is obtained by mining the vast deposits of *halite*, or *rock salt*, formed by evaporation of ancient inland seas. These salt beds can be up to hundreds of meters thick and may occur anywhere from a few meters to thousands of meters below the earth's surface. Salt mining has gone on for at least 3400 years, and the Wieliczka mine in Galicia, Poland, has been worked continuously from A.D. 1000 to the present.

Let's get back now to the dinner table. What about the link between dietary salt intake and high blood pressure? There's no doubt that most people in industrialized nations have a relatively high salt intake, and there's no doubt that high blood pressure among industrialized populations is on the rise. What's not so clear is how the two observations are related. The case against salt has been made largely by comparing widely diverse populations with different dietary salt intakes—by comparing the health of modern Americans with that of inhabitants of the Amazon rain forest, for example. Obviously, though, industrialization brings with it far more changes than simply an increase in dietary salt intake, and many of these other changes may be much more important than salt in contributing to hypertension.

The largest and most definitive study of the connection between salt and high blood pressure was published in the summer of 1991 by the Intersalt Cooperative Research Group, made up of scientists from 32 countries. This group found that there is indeed an increase in blood pressure with increasing salt intake but that the increase is minimal. If Americans were to cut their

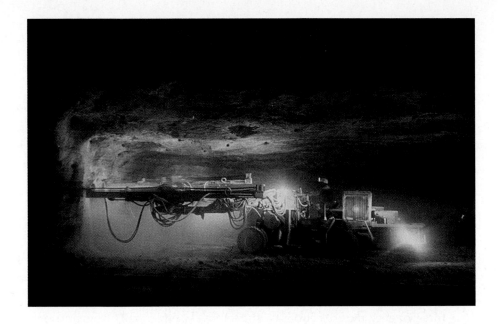

Underground salt mining at the Wieliczka mine in Poland has gone on for nearly 1000 years.

salt intake from the present average of 9 g per day to an average of 3 g per day (a near impossibility), the blood pressure of the average person would decline by 1%. Certain *individuals* would benefit far more substantially, but most people would see no reduction at all.

What should an individual do? The best answer, as in so many things, is to use moderation and common sense. Be aware of your salt intake and reduce it if you can or if you have hypertension, but don't spend a lot of time worrying about it.

▶**PROBLEM 6.26** What are some of the ways that salt is obtained commercially?

Key Words

Summary

Metallic elements on the left side of the periodic table have a tendency to give up electrons to form cations, while the halogens and a few other nonmetallic elements on the right side of the table have a tendency to accept electrons to form anions. The electrons given up by a main-group metal in forming a cation come from the highest-energy occupied orbital, while the electrons that are accepted by a nonmetal in forming an anion go into the lowest-energy unoccupied orbital. Sodium metal, for instance, loses its valence-shell $3s$ electron to form an Na^+ ion with the electron configuration of neon, while chlorine gains a $3p$ electron to form a Cl^- anion with the electron configuration of argon.

The amount of energy necessary to remove a valence electron from an isolated neutral atom is called the atom's **ionization energy, E_i**. Ionization energies are smallest for metallic elements on the left side of the periodic table and largest for nonmetallic elements on the right side, implying that metals can act as electron donors (reducing agents) in chemical reactions.

Ionization is not limited to the removal of a single electron from an atom. Two, three, or even more electrons can be removed sequentially from an atom, although larger amounts of energy are required for each successive ionization step. In general, valence-shell electrons are much more easily removed than core electrons.

The amount of energy released or absorbed when an electron adds to an isolated neutral atom is called the atom's **electron affinity, E_{ea}**. By convention, a negative E_{ea} corresponds to a release of energy and a positive E_{ea} corresponds to an absorption of energy. Electron affinities are most negative for group 7A elements and most positive for group 2A and 8A elements. As a result, the group 7A elements can act as electron acceptors (oxidizing agents) in chemical reactions.

Main-group metallic elements in groups 1A, 2A, and 3A undergo redox reactions with halogens in group 7A, during which the metal loses one or more electrons to the halogen. The product, a metal halide such as NaCl, is an **ionic solid** that consists of metal cations and halide anions electrostatically attracted to one another by **ionic bonds**. The sum of the interaction energies among all ions in a crystal is called the crystal's **lattice energy (U)**, and is derived from **Coulomb's law**.

$$-U = k \times \frac{z_1 z_2}{d}$$

The constant k is characteristic of a specific compound, z_1 and z_2 are the charges on the ions, and d is the distance between their centers.

In general, redox reactions of main-group elements can be described by the **octet rule**, which states that these elements tend to undergo reactions so as to attain a noble gas electron configuration with filled s and p sublevels in their valence shell. Elements on the left side of the periodic table tend to give up electrons until a noble gas configuration is reached; elements on the right side of the table tend to accept electrons until a noble gas configuration is reached; and the noble gases themselves are essentially unreactive.

Key Concept Summary

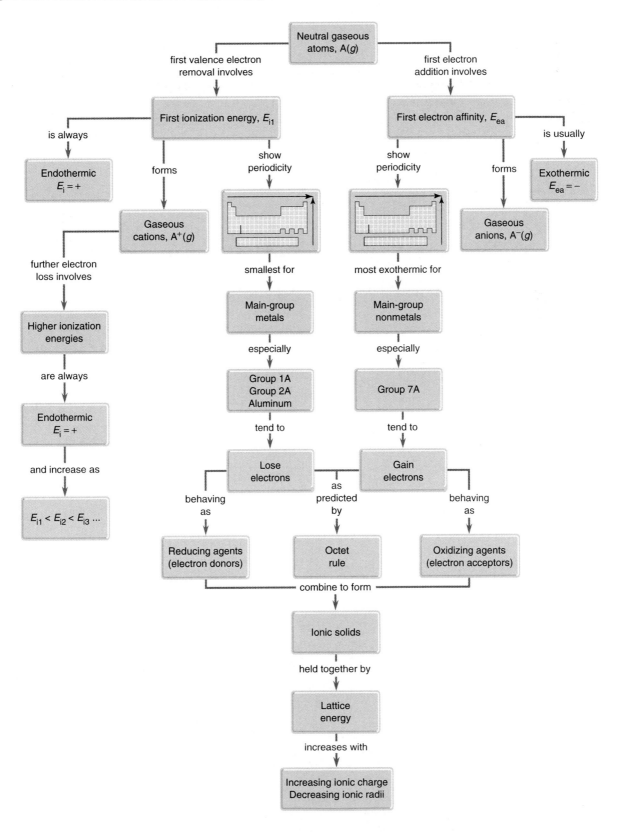

Understanding Key Concepts

Problems 6.1–6.26 appear within the chapter.

6.27 Where on the blank outline of the periodic table do the following elements appear?

(a) Main groups   (b) Halogens

(c) Alkali metals   (d) Noble gases

(e) Alkaline earths (f) Group 3A elements

(g) Lanthanides

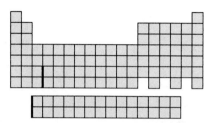

6.28 Which of the following drawings is more likely to represent an ionic compound, and which a covalent compound?

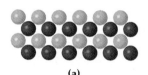   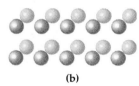

(a)   (b)

6.29 Circle the approximate part or parts of the periodic table where the following elements appear:

(a) Elements with the smallest values of E_{i1}

(b) Elements with the largest atomic radii

(c) Elements with the most negative values of E_{ea}

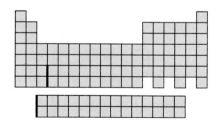

6.30 Where on the periodic table would you find the element that has an ion with each of the following electron configurations? Identify each ion.

(a) 3+ ion: $1s^2 2s^2 2p^6$ (b) 3+ ion: $[Ar] 3d^3$

(c) 2+ ion: $[Kr] 5s^2 4d^{10}$ (d) 1+ ion: $[Kr] 4d^{10}$

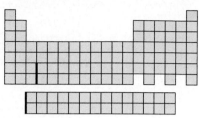

6.31 Which of the following spheres is likely to represent a metal, and which a nonmetal? Which sphere in the products represents a cation, and which an anion?

6.32 Each of the pictures **(a)–(d)** represents one of the following substances at 25°C: sodium, chlorine, iodine, sodium chloride. Which picture corresponds to which substance?

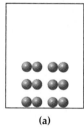 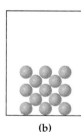  

(a)   (b)   (c)   (d)

6.33 The following pictures represent alkali halide salts. Which salt has the largest lattice energy? Which has the smallest lattice energy?

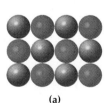   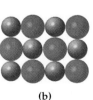

(a)   (b)   (c)

6.34 Three binary compounds are represented on the following drawing—red with red, blue with blue, and green with green. Give a likely formula for each compound, and assign oxidation numbers in each.

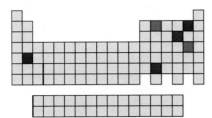

6.35 Given the following values for the formation of LiCl from its elements, draw a Born–Haber cycle similar to that shown in Figure 6.8.

E_{ea} for Cl = −348.6 kJ/mol

Heat of sublimation for Li = +159.4 kJ/mol

E_{i1} for Li = +520 kJ/mol

Bond dissociation energy for Cl_2 = +243 kJ/mol

Lattice energy for LiCl = +853 kJ/mol

Additional Problems

Ions, Ionization Energy, and Electron Affinity

6.36 What are the likely ground-state electron configurations of the following cations?

(a) La^{3+} (b) Ag^+ (c) Sn^{2+}

6.37 What are the likely ground-state electron configurations of the following anions?

(a) Se^{2-} (b) N^{3-}

6.38 There are two elements in the transition metal series Sc through Zn that have four unpaired electrons in their 2+ ions. Identify them.

6.39 Identify the element whose 2+ ion has the ground-state electron configuration [Ar] $3d^{10}$.

6.40 Do ionization energies have a positive sign or a negative sign? Explain.

6.41 Do electron affinities have a positive sign or a negative sign? Explain.

6.42 Which group of elements in the periodic table has the largest E_{i1}, and which group has the smallest? Explain.

6.43 Which element in the periodic table has the smallest ionization energy? Which has the largest?

6.44 (a) Which has the smaller second ionization energy, K or Ca?

(b) Which has the larger third ionization energy, Ga or Ca?

6.45 (a) Which has the smaller fourth ionization energy, Sn or Sb?

(b) Which has the larger sixth ionization energy, Se or Br?

6.46 Three atoms have the following electron configurations:

(a) $1s^2\,2s^2\,2p^6\,3s^2\,3p^3$

(b) $1s^2\,2s^2\,2p^6\,3s^2\,3p^6$

(c) $1s^2\,2s^2\,2p^6\,3s^2\,3p^6\,4s^2$

Which of the three has the largest E_{i2}? Which has the smallest E_{i7}?

6.47 Write the electron configuration of the atom in the third row of the periodic table that has the smallest E_{i4}.

6.48 Which element in each of the following sets has the smallest first ionization energy, and which has the largest?

(a) Li, Ba, K   (b) B, Be, Cl   (c) Ca, C, Cl

6.49 What elements meet the following descriptions?

(a) Has largest E_{i3} (b) Has largest E_{i7}

6.50 What is the relationship between the electron affinity of a monocation such as Na^+ and the ionization energy of the neutral atom?

6.51 What is the relationship between the ionization energy of a monoanion such as Cl^- and the electron affinity of the neutral atom?

6.52 Which has the more negative electron affinity, Na^+ or Na? Na^+ or Cl?

6.53 Which has the more negative electron affinity, Br or Br^-?

6.54 Why is energy usually released when an electron is added to a neutral atom but absorbed when an electron is removed from a neutral atom?

6.55 Why does ionization energy increase regularly across the periodic table from group 1A to group 8A, whereas electron affinity increases irregularly from group 1A to group 7A and then falls dramatically for group 8A?

6.56 Which element in each of the following pairs has the larger (more negative) electron affinity?

(a) F or Fe   (b) Ne or Na   (c) Ba or Br

6.57 According to the data in Figure 6.6, zinc, cadmium, and mercury all have near-zero electron affinities. Explain.

Lattice Energy and Ionic Bonds

6.58 Order the following compounds according to their expected lattice energies: LiCl, KCl, KBr, $MgCl_2$.

6.59 Order the following compounds according to their expected lattice energies: $AlBr_3$, $MgBr_2$, LiBr, CaO.

6.60 Calculate the energy change (in kilojoules per mole) when lithium atoms lose an electron to bromine atoms to form isolated Li^+ and Br^- ions. (The E_i for Li is 520 kJ/mol; the E_{ea} for Br is −325 kJ/mol.)

6.61 Cesium has the smallest ionization energy of all elements (376 kJ/mol) and chlorine has the most negative electron affinity (−349 kJ/mol). Will a cesium atom transfer an electron to a chlorine atom to form isolated Cs^+ and Cl^- ions?

6.62 Find the lattice energy of LiBr in Table 6.3, and calculate the energy change (in kilojoules per mole) for the formation of solid LiBr from the elements. (The sublimation energy for Li is +159.4 kJ/mol, the bond dissociation energy of Br_2 is +224 kJ/mol, and the energy necessary to convert $Br_2(l)$ to $Br_2(g)$ is 30.9 kJ/mol.)

6.63 Look up the lattice energies in Table 6.3, and calculate the energy change (in kilojoules per mole) for the formation of the following substances from their elements:

(a) LiF (The sublimation energy for Li is +159.4 kJ/mol, the E_i for Li is 520 kJ/mol, the E_{ea} for F is −328 kJ/mol, and the bond dissociation energy of F_2 is +158 kJ/mol.)

(b) CaF_2 (The sublimation energy for Ca is +178.2 kJ/mol, $E_{i1} = +589.8$ kJ/mol, and $E_{i2} = +1145$ kJ/mol.)

6.64 Born–Haber cycles, such as that shown in Figure 6.8, are called *cycles* because they form closed loops. If any five of the six energy changes in the cycle are known, the value of the sixth can be calculated. Use the following five values to calculate a lattice energy (in kilojoules per mole) for sodium hydride, NaH:

E_{ea} for H = −72.8 kJ/mol

E_{i1} for Na = +495.8 kJ/mol

Heat of sublimation for Na = +107.3 kJ/mol

Bond dissociation energy for H_2 = +435.9 kJ/mol

Net energy change for the formation of NaH from its elements = −60 kJ/mol

6.65 Calculate a lattice energy for CaH_2 (in kilojoules per mole) using the following information:

E_{ea} for H = −72.8 kJ/mol

E_{i1} for Ca = +589.8 kJ/mol

E_{i2} for Ca = +1145 kJ/mol

Heat of sublimation for Ca = +178.2 kJ/mol

Bond dissociation energy for H_2 = +435.9 kJ/mol

Net energy change for the formation of CaH_2 from its elements = −186.2 kJ/mol

6.66 Calculate the overall energy change (in kilojoules per mole) for the formation of CsF from its elements using the following data:

E_{ea} for F = −328 kJ/mol

E_{i1} for Cs = +375.7 kJ/mol

E_{i2} for Cs = +2422 kJ/mol

Heat of sublimation for Cs = +76.1 kJ/mol

Bond dissociation energy for F_2 = +158 kJ/mol

Lattice energy for CsF = +740 kJ/mol

6.67 The estimated lattice energy for CsF_2 is +2347 kJ/mol. Use the data given in Problem 6.66 to calculate an overall energy change (in kilojoules per mole) for the formation of CsF_2 from its elements. Does the overall reaction absorb energy or release it? In light of your answers to Problem 6.66, which compound is more likely to form in the reaction of cesium with fluorine, CsF or CsF_2?

6.68 Calculate overall energy changes (in kilojoules per mole) for the formation of CaCl from the elements. The following data are needed:

E_{ea} for Cl = −348.6 kJ/mol

E_{i1} for Ca = +589.8 kJ/mol

E_{i2} for Ca = +1145 kJ/mol

Heat of sublimation for Ca = +178.2 kJ/mol

Bond dissociation energy for Cl_2 = +243 kJ/mol

Lattice energy for $CaCl_2$ = +2258 kJ/mol

Lattice energy for CaCl = +717 kJ/mol (estimated)

6.69 Use the data in Problem 6.68 to calculate an overall energy change for the formation of $CaCl_2$ from the elements. Which is more likely to form, CaCl or $CaCl_2$?

6.70 Use the data and the result in Problem 6.64 to draw a Born–Haber cycle for the formation of NaH from its elements.

6.71 Use the data and the result in Problem 6.63a to draw a Born–Haber cycle for the formation of LiF from its elements.

Main-Group Chemistry

6.72 Where in the periodic table do the following elements appear?

(a) Li   (b) Al   (c) Ba   (d) Br

(e) Ne   (f) O   (g) Fr   (h) He

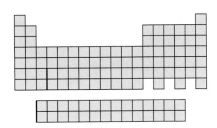

6.73 Where in the periodic table do the following elements appear?

(a) K   (b) Ca   (c) Cl   (d) C

(e) Si   (f) P   (g) B   (h) H

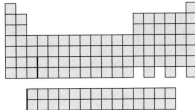

6.74 Which of the elements in groups 7A (F, Cl, Br, I) and 8A (He, Ne, Ar, Kr, Xe) are gases, which are liquids, and which are solids at room temperature?

6.75 Give at least one important use for each of the following elements:

(a) Lithium   (b) Potassium

(c) Strontium   (d) Helium

6.76 Little is known about the chemistry of astatine (At) from direct observation, but reasonable predictions can be made.

(a) Is astatine likely to be a gas, a liquid, or a solid?

(b) What color is astatine likely to have?

(c) Is astatine likely to react with sodium? If so, what is the formula of the product?

6.77 Look at the properties of the alkali metals summarized in Table 6.4, and predict reasonable values for the melting point, boiling point, density, and atomic radius of francium.

6.78 Tell how each of the following elements is produced commercially:

(a) Sodium   (b) Aluminum

(c) Argon   (d) Bromine

6.79 Why does chemical reactivity increase from top to bottom in group 1A but decrease from top to bottom in group 7A?

6.80 Give a brief statement of the octet rule and an explanation of why it works.

6.81 Which main-group elements occasionally break the octet rule?

6.82 Write balanced equations for the reaction of potassium with the following substances. If no reaction occurs, write N.R.

(a) H_2 (b) H_2O (c) NH_3

(d) Br_2 (e) N_2 (f) O_2

6.83 Write balanced equations for the reaction of calcium with the following substances. If no reaction occurs, write N.R.

(a) H_2 (b) H_2O (c) He

(d) Br_2 (e) O_2

6.84 Write balanced equations for the reaction of chlorine with the following substances. If no reaction occurs, write N.R.

(a) H_2 (b) Ar   (c) Br_2 (d) N_2

6.85 As a general rule, more reactive halogens can oxidize the anions of less reactive halogens. Predict the products of the following reactions, and identify the oxidizing and reducing agents in each. If no reaction occurs, write N.R.

(a) $2\,Cl^-(aq) + F_2(g) \rightarrow$?   (b) $2\,Br^-(aq) + I_2(s) \rightarrow$?

(c) $2\,I^-(aq) + Br_2(aq) \rightarrow$?

6.86 Aluminum metal can be prepared by reaction of $AlCl_3$ with Na. Write a balanced equation for the reaction, and tell which atoms have been oxidized and which have been reduced.

6.87 The widely used antacid called *milk of magnesia* is an aqueous suspension of $Mg(OH)_2$. How would you prepare $Mg(OH)_2$ from magnesium metal?

6.88 What is the maximum amount (in grams) of pure iodine that you could obtain from 1.00 kg of the mineral lautarite ($CaIO_3$)?

6.89 Assume that you wanted to prepare a small volume of pure hydrogen by reaction of lithium metal with water. How many grams of lithium would you need to prepare 455 mL of H_2 if the density of hydrogen is 0.0893 g/L?

6.90 How many grams of calcium hydride are formed by reaction of 5.65 g of calcium with 3.15 L of H_2 if the density of hydrogen is 0.0893 g/L and the reaction occurs in 94.3% yield? Which reactant is limiting?

6.91 How many liters of N_2 gas are needed for reaction of 2.87 g of lithium to give lithium nitride if the density of nitrogen is 1.25 g/L?

6.92 Identify the oxidizing agent and the reducing agent in each of the following reactions:

(a) $Mg(s) + 2 H^+(aq) \rightarrow Mg^{2+}(aq) + H_2(g)$

(b) $Kr(g) + F_2(g) \rightarrow KrF_2(s)$

(c) $I_2(s) + 3 Cl_2(g) \rightarrow 2 ICl_3(l)$

6.93 Identify the oxidizing agent and the reducing agent in each of the following reactions:

(a) $2 XeF_2(s) + 2 H_2O(l) \rightarrow 2 Xe(g) + 4 HF(aq) + O_2(g)$

(b) $NaH(s) + H_2O(l) \rightarrow Na^+(aq) + OH^-(aq) + H_2(g)$

(c) $2 TiCl_4(l) + H_2(g) \rightarrow 2 TiCl_3(s) + 2 HCl(g)$

General Problems

6.94 Cu^+ has an ionic radius of 77 pm, but Cu^{2+} has an ionic radius of 73 pm. Explain.

6.95 The following ions all have the same number of electrons: Ti^{4+}, Sc^{3+}, Ca^{2+}, S^{2-}. Order them according to their expected size, and explain your answer.

6.96 Calculate overall energy changes (in kilojoules per mole) for the formation of MgF and MgF_2 from their elements. The following data are needed:

E_{ea} for F = −328 kJ/mol

E_{i1} for Mg = +737.7 kJ/mol

E_{i2} for Mg = +1450.7 kJ/mol

Heat of sublimation for Mg = +147.7 kJ/mol

Bond dissociation energy for F_2 = +158 kJ/mol

Lattice energy for MgF_2 = +2952 kJ/mol

Lattice energy for MgF = 930 kJ/mol (estimated)

6.97 In light of your answers to Problem 6.96, which compound is more likely to form in the reaction of magnesium with fluorine, MgF_2 or MgF?

6.98 Give at least one important use of each of the following elements:

(a) Sodium   (b) Magnesium   (c) Fluorine

6.99 Tell how each of the following elements is produced commercially:

(a) Fluorine   (b) Calcium   (c) Chlorine

6.100 Write balanced equations for the reaction of lithium with the following substances. If no reaction occurs, write N.R.

(a) H_2 (b) H_2O (c) NH_3

(d) Br_2 (e) N_2 (f) O_2

6.101 Write balanced equations for the reaction of fluorine with the following substances. If no reaction occurs, write N.R.

(a) H_2 (b) Na   (c) Br_2 (d) NaBr

6.102 Many early chemists noted a *diagonal relationship* among elements in the periodic table, whereby a given element is sometimes more similar to the element below and to the right than it is to the element directly below. Lithium is more similar to magnesium than to sodium, for example, and boron is more similar to silicon than to aluminum. Use your knowledge about the periodic trends of such properties as atomic radii and Z_{eff} to explain the existence of diagonal relationships.

6.103 We saw in Section 6.6 that the reaction of solid sodium with gaseous chlorine to yield solid sodium chloride (Na^+Cl^-) is favorable by 411 kJ/mol. Calculate the energy change for the alternative reaction that yields chlorine sodide (Cl^+Na^-), and then explain why sodium chloride formation is preferred.

$$2 Na(s) + Cl_2(g) \longrightarrow 2 Cl^+ Na^-(s)$$

Assume that the lattice energy for Cl^+Na^- is the same as that for Na^+Cl^-. The following data are needed in addition to that found in Section 6.6:

E_{ea} for Na = −52.9 kJ/mol

E_{i1} for Cl = +1251 kJ/mol

6.104 Draw a Born–Haber cycle for the reaction of sodium with chlorine to yield chlorine sodide (Problem 6.103).

6.105 One mole of any gas has a volume of 22.4 L at 0°C and 1.00 atmosphere pressure. Assume that 0.719 g of an unknown metal M reacted completely with 94.2 mL of Cl_2 gas at 0°C and 1.00 atmosphere pressure to give a metal halide product. From the data given, several atomic masses for the metal are possible. List several possibilities, and choose the most likely candidate for M.

6.106 Use the following data and data given in Tables 6.2 and 6.3 to calculate the second electron affinity, E_{ea2}, of oxygen. Is the O^{2-} ion stable in the gas phase? Why is it stable in solid MgO?

Heat of sublimation for Mg = +147.7 kJ/mol

Bond dissociation energy for O_2 = +498.4 kJ/mol

E_{ea1} for O = −141.0 kJ/mol

Net energy change for formation of MgO from its elements = −601.7 kJ/mol

Multi-Concept Problems

6.107 The ionization energy of an atom can be measured by photoelectron spectroscopy, in which light of wavelength λ is directed at an atom, causing an electron to be ejected. The kinetic energy of the ejected electron (E_K) is measured by determining its velocity, v, since $E_K = 1/2\ mv^2$. The E_i is then calculated using the relationship that the energy of the incident light is equal to the sum of E_i plus E_K.

    (a) What is the ionization energy of rubidium atoms (in kilojoules per mole) if light with λ = 58.4 nm produces electrons with a velocity of 2.450×10^6 m/s? (The mass of an electron is 9.109×10^{-31} kg.)

    (b) What is the ionization energy of potassium (in kilojoules per mole) if light with λ = 142 nm produces electrons with a velocity of 1.240×10^6 m/s?

6.108 A 1.005 g sample of an unknown alkaline earth metal was allowed to react with a volume of chlorine gas that contains 1.91×10^{22} Cl_2 molecules. The resulting metal chloride was analyzed for chlorine by dissolving a 0.436 g sample in water and adding an excess of $AgNO_3(aq)$. The analysis yielded 1.126 g of AgCl.

    (a) What is the %Cl in the alkaline earth chloride?

    (b) What is the identity of the alkaline earth metal?

    (c) Write balanced equations for all chemical reactions.

    (d) In the reaction of the alkaline earth metal with chlorine, which reactant is in excess, and how many grams of that reactant remain unreacted?

6.109 Element M is prepared industrially from its oxide by a two-step procedure according to the following (unbalanced) equations:

    (i) $M_2O_3(s) + C(s) + Cl_2(g) \longrightarrow MCl_3(l) + CO(g)$

    (ii) $MCl_3(l) + H_2(g) \longrightarrow M(s) + HCl(g)$

    Assume that 0.855 g of M_2O_3 is submitted to the reaction sequence. When the HCl produced in step (ii) is dissolved in water and titrated with 0.511 M NaOH, 144.2 mL of the NaOH solution is required to neutralize the HCl.

    (a) Balance both equations.

    (b) What is the atomic mass of element M, and what is its identity?

    (c) What mass of M (in grams) is produced in the reaction?

6.110 Assume that 20.0 g of Sr is allowed to react with 25.0 g of Cl_2 to give $SrCl_2$.

    (a) What is the net energy change in kJ/mol for the formation of $SrCl_2$ from its elements? Heat of sublimation for Sr = +164.44 kJ/mol; E_{i1} for Sr = +549.5 kJ/mol; E_{i2} for Sr = +1064.2 kJ/mol; bond dissociation energy for Cl_2 = +243 kJ/mol; E_{ea} for Cl = −348.6 kJ/mol; lattice energy for $SrCl_2$ = +2156 kJ/mol.

    (b) Which reactant is limiting, and how many grams of $SrCl_2$ product are formed?

    (c) How much energy in kJ is released during the reaction?

 # eMedia Problems

6.111 According to the **Periodic Trends: Ionization Energies** movie (*eChapter 6.3*), the halogens have some of the highest known first ionization energies. With this in mind, explain why the halogens are almost always found as ions in nature.

6.112 The **Ionization Energy** movie (*eChapter 6.4*) shows a graph of the first, second, third, and fourth ionization energies of aluminum. Explain why each successive ionization energy is greater than the last, and why there is such a big difference between the third and fourth ionization energies.

6.113 The magnitude of the lattice energy of an ionic compound depends on the charges and sizes of the ions that make up the compound. Use the **Coulomb's Law** activity (*eChapter 6.6*) to determine which ionic compound has the larger lattice energy, LiF or SrTe. (Ionic radii in Å for Li^+, F^-, Sr^{2+}, and Te^{2-} are 0.68, 1.33, 1.13, and 2.21, respectively. Note that 1 Å= 100 pm.)

6.114 The **Formation of Sodium Chloride** movie (*eChapter 6.6*) shows the reaction between hot sodium metal and chlorine gas.

    (a) What properties of the elements are responsible for the reaction being so dramatic?

    (b) How would you expect the reaction between rubidium and chlorine to compare to that between sodium and chlorine? Explain.

6.115 Watch the **Sodium and Potassium in Water** movie (*eChapter 6.7*), paying particular attention to the differences between the two reactions shown.

    (a) What property of the alkali metals is responsible for these differences?

    (b) How would you expect the reactions of rubidium and cesium with water to compare to those of sodium and potassium with water?

    (c) How would you expect the reactions of rubidium and cesium with water to compare with each other?

7

Covalent Bonds and Molecular Structure

Shape is crucially important in chemistry, just as it is in architecture.

We saw in the last chapter that the bond formed between a metal and a reactive nonmetal typically involves the transfer of electrons. The metal atom loses one or more electrons and becomes a cation, while the reactive nonmetal atom gains one or more electrons and becomes an anion. The oppositely charged ions are held together by the electrostatic attractions that we call ionic bonds.

How, though, do bonds form between atoms of the same or similar elements? How can we describe the bonds in such substances as H_2, Cl_2, CO_2, and the literally millions of other nonionic compounds? Simply put, the answer is that the bonds in such compounds are formed by the *sharing* of electrons between atoms rather than by the transfer of electrons from one atom to another. The resultant shared-electron bond is called a **covalent bond** and is the most important kind of bond in all of chemistry. We'll explore the nature of covalent bonding in this chapter.

COVALENT BOND A bond that results from the sharing of electrons between atoms.

7.1 The Covalent Bond

To see how the formation of a covalent, shared-electron bond between atoms can be described, let's look at the H−H bond in the H_2 molecule as an example. As two hydrogen atoms come closer together, electrostatic interactions begin to develop between them. The two positively charged nuclei repel each other and the two negatively charged electrons repel each other, but each nucleus attracts both electrons (Figure 7.1). If the attractive forces are stronger than the repulsive forces, a covalent bond is formed, with the two atoms joined together and the two shared electrons occupying the region between the nuclei.

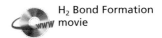
H_2 Bond Formation movie

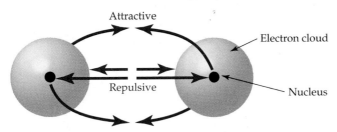

In essence, the shared electrons act as a kind of "glue" to bind the two nuclei together into an H_2 molecule. Both nuclei are simultaneously attracted to the same electrons and are therefore held together, much as two tug-of-war teams pulling on the same rope are held together.

The magnitudes of the various attractive and repulsive forces between nuclei and electrons in a covalent bond depend on how close the atoms are. If the hydrogen atoms are too far apart, the attractive forces are small and no bond exists. If the hydrogen atoms are too close together, the repulsive interaction between the nuclei becomes so strong that it pushes the atoms apart. Thus, there is an optimum distance between nuclei called the **bond length** where net attractive forces are maximized and the H−H molecule is most stable. In the H_2 molecule, the bond length is 74 pm. On a graph of energy versus internuclear distance, the bond length corresponds to the minimum-energy, most stable arrangement (Figure 7.2).

FIGURE 7.1 A covalent H−H bond is the net result of attractive and repulsive electrostatic forces. The nucleus–electron attractions (blue arrows) are greater than the nucleus–nucleus and electron–electron repulsions (red arrows), resulting in a net attractive force that holds the atoms together to form an H_2 molecule.

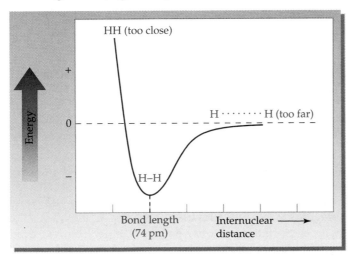

FIGURE 7.2 A graph of potential energy versus internuclear distance for the H_2 molecule. If the hydrogen atoms are too far apart, attractions are weak and no bonding occurs. If the atoms are too close, strong repulsions occur. When the atoms are optimally separated, the energy is at a minimum.

Every covalent bond has its own characteristic length that leads to maximum stability. As you might expect, bond lengths are roughly predictable from a knowledge of atomic radii (Section 5.15). For example, since the atomic radius of hydrogen is 37 pm and the atomic radius of chlorine is 99 pm, the H—Cl bond length in a hydrogen chloride molecule is about 37 pm + 99 pm = 136 pm. (The actual value is 127 pm.)

7.2   Strengths of Covalent Bonds

Look again at Figure 7.2, the graph of energy versus internuclear distance for the H_2 molecule, and note how the H_2 molecule is lower in energy than two separate hydrogen atoms. When pairs of hydrogen atoms bond together, they form lower-energy H_2 molecules and release 436 kJ/mol in the process. Looked at from the other direction, 436 kJ must be *added* per mole of H_2 molecules to split the molecules apart into hydrogen atoms.

The amount of energy that must be supplied to break a chemical bond in an isolated molecule in the gaseous state—and thus the amount of energy that is *released* when the bond forms—is called the **bond dissociation energy, D.** Bond dissociation energies are always positive values because energy must always be supplied to break a bond. Conversely, the amount of energy released on forming a bond is always a negative value.

Every bond in every molecule has its own specific bond dissociation energy. Not surprisingly, though, bonds between the same pairs of atoms usually have similar dissociation energies. For example, carbon–carbon bonds usually have D values of approximately 350–380 kJ/mol regardless of the exact structure of the molecule. (Note in the following examples that all the covalent bonds between atoms are indicated by lines, as described in Section 2.8.)

Ethane

$D = 376$ kJ/mol

Propane

$D = 356$ kJ/mol

Butane

$D = 352$ kJ/mol

Because similar bonds have similar bond dissociation energies, it's possible to construct a useful table of average values (Table 7.1) to compare different kinds of bonds. Keep in mind, though, that the actual value in a specific molecule might vary by ±10% from the average.

The bond dissociation energies listed in Table 7.1 cover a wide range, from a low of 151 kJ/mol for the I—I bond to a high of 570 kJ/mol for the H—F bond. As a rule of thumb, though, most of the bonds commonly encountered

---

TABLE 7.1   Average Bond Dissociation Energies, *D* (kJ/mol)[a]

H—H	436[a]	C—H	410	N—H	390	O—H	460	F—F	159[a]
H—C	410	C—C	350	N—C	300	O—C	350	Cl—Cl	243[a]
H—F	570[a]	C—F	450	N—F	270	O—F	180	Br—Br	193[a]
H—Cl	432[a]	C—Cl	330	N—Cl	200	O—Cl	200	I—I	151[a]
H—Br	366[a]	C—Br	270	N—Br	240	O—Br	210	S—F	310
H—I	298[a]	C—I	240	N—I	—	O—I	220	S—Cl	250
H—N	390	C—N	300	N—N	240	O—N	200	S—Br	210
H—O	460	C—O	350	N—O	200	O—O	180	S—S	225
H—S	340	C—S	260	N—S	—	O—S	—		

Multiple covalent bonds[b]

C=C	611	C≡C	835	C=O	732	O=O	498[a]	N≡N	945[a]

[a]Exact value
[b]We'll discuss multiple covalent bonds in Section 7.5.

---

in naturally occurring molecules (C–H, C–C, C–O) have values in the range 350–400 kJ/mol.

7.3   A Comparison of Ionic and Covalent Compounds

How do the physical properties of ionic compounds and covalent compounds differ? Look at the comparison between NaCl and HCl, shown in Table 7.2. Sodium chloride, an ionic compound, is a white solid with a melting point of 801°C and a boiling point of 1413°C. Hydrogen chloride, a covalent compound, is a colorless gas with a melting point of −115°C and a boiling point of −84.9°C. What accounts for such large differences?

Sodium chloride, an ionic compound, is a white, crystalline solid. Hydrogen chloride, a molecular compound, is a gas at room temperature.

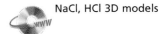

NaCl, HCl 3D models

TABLE 7.2   Some Physical Properties of NaCl and HCl

Property	NaCl	HCl
Formula mass	58.44 amu	36.46 amu
Physical appearance	White solid	Colorless gas
Type of bond	Ionic	Covalent
Melting point	801°C	−115°C
Boiling point	1413°C	−84.9°C

Ionic compounds are high-melting solids because of their ionic bonds. As discussed previously in Section 6.6, a visible sample of sodium chloride consists, not of NaCl molecules, but of a vast three-dimensional network of ions in which each Na⁺ cation is attracted to many surrounding Cl⁻ anions and each Cl⁻ ion is attracted to many surrounding Na⁺ ions. For sodium chloride to melt or boil so that the ions break free of one another, every ionic attraction in the entire crystal must be overcome, a process that requires a large amount of energy.

Covalent compounds, by contrast, are low-melting solids, liquids, or even gases. A sample of a covalent compound such as hydrogen chloride consists of discrete HCl molecules. The covalent bond *within an individual molecule* may be very strong, but the attractive forces *between the different molecules* are fairly weak. As a result, relatively little energy is required to overcome these forces and cause a covalent compound to melt or boil. We'll look at the nature of intermolecular forces and the boiling process in Chapter 10.

7.4 Polar Covalent Bonds: Electronegativity

We've left the impression up to this point that a given bond is either purely ionic, with electrons completely transferred, or purely covalent, with electrons shared equally. In fact, though, ionic and covalent bonds represent only the two extremes of a continuous spectrum of possibilities. Between these two extremes are the large majority of bonds, in which the bonding electrons are shared *unequally* between two atoms but are not completely transferred. Such bonds are said to be **polar covalent bonds** (Figure 7.3). The lowercase Greek letter delta (δ) is used to symbolize the resultant *partial* charges on the atoms, either partial positive ($\delta+$) for the atom that has a smaller share of the bonding electrons or partial negative ($\delta-$) for the atom that has a larger share.

FIGURE 7.3 The bonding continuum from ionic to nonpolar covalent. Polar covalent bonds lie between the two extremes. They are characterized by an unsymmetrical electron distribution in which the bonding electrons are attracted somewhat more strongly by one atom than the other. The symbol δ (Greek delta) means *partial* charge, either partial positive ($\delta+$) or partial negative ($\delta-$).

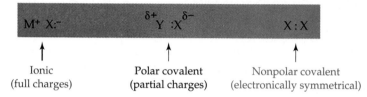

| Ionic | Polar covalent | Nonpolar covalent |
| (full charges) | (partial charges) | (electronically symmetrical) |

As examples of different points along the bonding spectrum, let's compare three substances, NaCl, HCl, and Cl_2:

- **NaCl** The bond in solid sodium chloride is a largely ionic one between Na^+ and Cl^-. In spite of what we've said previously, though, experiments show that the NaCl bond is only about 80% ionic and that the electron transferred from Na to Cl still spends some of its time near sodium. A particularly useful way of visualizing this electron transfer is to represent the compound using what is called an *electrostatic potential map*, which uses color to portray the calculated electron distribution in the molecule. The electron-poor sodium atom is blue, while the electron-rich chlorine is red.

$Na^+ \ Cl^-$ An ionic bond

- **HCl** The bond in a hydrogen chloride molecule is polar covalent. The chlorine atom attracts the bonding electron pair more strongly than hydrogen does, resulting in an unsymmetrical distribution of electrons. Chlorine thus has a partial negative charge (yellow-orange in the electrostatic potential map), and hydrogen has a partial positive charge (blue in the electrostatic potential map). Experimentally, the H—Cl bond has been found to be about 83% covalent and 17% ionic.

$$\delta+ H - Cl^{\delta-}$$

$$[H \ :Cl]$$

A polar covalent bond

The bonding electrons are attracted more strongly by Cl than H.

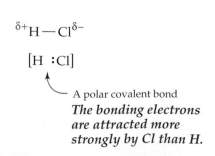

- **Cl₂** The bond in a chlorine molecule is nonpolar covalent, with the bonding electrons attracted equally to the two identical chlorine atoms. A similar situation exists for all such diatomic molecules that contain a covalent bond between two identical atoms.

Cl : Cl   A nonpolar covalent bond

Bond polarity is due to differences in **electronegativity (EN)**, the ability of an atom in a molecule to attract the shared electrons in a covalent bond. As shown graphically in Figure 7.4, metallic elements on the left of the periodic table attract electrons only weakly, whereas the halogens and other reactive nonmetals in the upper right of the table attract electrons strongly. The alkali metals are the least electronegative elements; fluorine, oxygen, nitrogen, and chlorine are the most electronegative. Figure 7.4 also indicates that electronegativity generally decreases down the periodic table within a group.

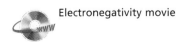

Electronegativity movie

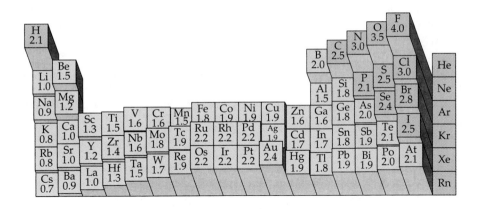

FIGURE 7.4 Electronegativity trends in the periodic table. Electronegativity increases from left to right and generally decreases from top to bottom.

Since electronegativity measures the ability of an atom in a molecule to attract shared electrons, it seems logical that it should be related to electron affinity (E_{ea}, Section 6.5) and ionization energy (E_i, Section 6.3). Electron affinity, after all, is a measure of the tendency of an isolated atom to gain an electron, and ionization energy is a measure of the tendency of an isolated atom to lose an electron. In fact, one of the ways in which electronegativities were first calculated was by taking the average of the absolute values of E_{ea} and E_i and then expressing the results on a unitless scale, with fluorine assigned a value of 4.0.

How can we use a knowledge of electronegativities to make predictions about bond polarity? A general but somewhat arbitrary guideline is that bonds between atoms with the same or similar electronegativities are nonpolar covalent, bonds between atoms whose electronegativities differ by more than about 2 units are substantially ionic, and bonds between atoms whose electronegativities differ by less than 2 units are polar covalent. Thus, we can be reasonably sure that a C−Cl bond in chloroform,

Polarity activity, Interactive Periodic Table

$CHCl_3$, is polar covalent, while an Na^+Cl^- bond in sodium chloride is largely ionic.

Chlorine:	EN = 3.0		Chlorine:	EN = 3.0
Carbon:	EN = 2.5		Sodium:	EN = 0.9
	Difference = 0.5			Difference = 2.1

▶ **PROBLEM 7.1** Use the electronegativity values in Figure 7.4 to predict whether the bonds in the following compounds are polar covalent or ionic: **(a)** $SiCl_4$ **(b)** CsBr **(c)** $FeBr_3$ **(d)** CH_4

▶ **PROBLEM 7.2** Order the following compounds according to the increasing ionic character of their bonds: CCl_4, $BaCl_2$, $TiCl_3$, Cl_2O.

7.5 Electron-Dot Structures

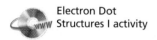

Electron Dot Structures I activity

One of the most convenient ways to picture the sharing of electrons between atoms in covalent or polar covalent bonds is to use what are known as **electron-dot structures**, or *Lewis structures*, named after G. N. Lewis of the University of California at Berkeley. An electron-dot structure represents an atom's valence electrons by dots and indicates by the placement of the dots how the valence electrons are distributed in a molecule. A hydrogen molecule, for example, is drawn showing a pair of dots between the hydrogen atoms, indicating that the hydrogens share the pair of electrons in a covalent bond:

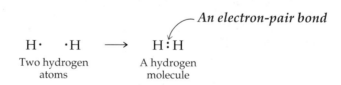

By sharing two electrons in a covalent bond, each hydrogen effectively has one electron pair and the stable, filled-shell electron configuration of helium. As with ions (Section 6.12), a filled valence shell for each atom in a molecule leads to maximum stability.

This hydrogen shares an electron pair ...and this hydrogen shares an electron pair.

Atoms other than hydrogen also form covalent bonds by sharing electron pairs, and the electron-dot structures of the resultant molecules are easily drawn by assigning the correct number of valence electrons to each atom. Group 3A atoms (such as boron) have three valence electrons, group 4A atoms (such as carbon) have four valence electrons, and so on across the periodic table. The group 7A element fluorine has seven valence electrons, and an electron-dot structure for the F_2 molecule shows how a covalent bond can form:

$$:\ddot{F}\cdot \quad \cdot\ddot{F}: \quad \longrightarrow \quad :\ddot{F}:\ddot{F}: \longleftarrow \text{A lone pair}$$

A bonding pair

Two F atoms
(seven valence electrons
per atom)

An F_2 molecule
(each F is surrounded
by eight valence electrons)

Six of the seven valence electrons in a fluorine atom are already paired in three filled atomic orbitals and thus are not shared in bonding. The seventh fluorine valence electron, however, is unpaired and can be used in forming a covalent bond to another fluorine. Each atom in the resultant F_2 molecule thereby achieves a filled valence-shell octet. The three pairs of nonbonding electrons on each fluorine atom are called **lone pairs**, or *nonbonded pairs*, and the shared electrons are called a **bonding pair**.

The tendency of main-group atoms to fill their s and p subshells when they form bonds—the *octet rule* discussed in Section 6.12—is an important guiding principle that makes it possible to predict the formulas and electron-dot structures of a great many molecules. As a general rule, an atom shares as many of its valence-shell electrons as possible, either until it has no more to share or until it reaches an octet configuration. For second-row elements in particular, the following guidelines apply:

- **Group 3A elements**, such as boron, have three valence electrons and can therefore form three electron-pair bonds in neutral molecules such as borane, BH_3. The boron atom in the resultant molecule has only three bonding pairs of electrons and cannot reach an electron octet. (The bonding situation in BH_3 is actually more complicated than suggested here; we'll deal with it in Section 19.4.)

Only six electrons around boron

$$\cdot\dot{B}\cdot + 3\,H\cdot \longrightarrow H:\ddot{B}:H$$

with H above the B

Borane

- **Group 4A elements**, such as carbon, have four valence electrons and form four bonds, as in methane, CH_4. The carbon atom in the resultant molecule has four bonding pairs of electrons.

$$\cdot\dot{C}\cdot + 4\,H\cdot \longrightarrow H:\ddot{C}:H$$

with H above and below the C

Methane

- **Group 5A elements**, such as nitrogen, have five valence electrons and form three bonds, as in ammonia, NH_3. The nitrogen atom in the resultant molecule has three bonding pairs of electrons and one lone pair.

$$\cdot\dot{N}\cdot + 3\,H\cdot \longrightarrow H:\ddot{N}:H$$

with H above the N

Ammonia

- **Group 6A elements**, such as oxygen, have six valence electrons and form two bonds, as in water, H_2O. The oxygen atom in the resultant molecule has two bonding pairs of electrons and two lone pairs.

$$\cdot \overset{\cdot \cdot}{\underset{\cdot \cdot}{O}} \cdot \; + \; 2\,H \cdot \; \longrightarrow \; H \colon \overset{\cdot \cdot}{\underset{\cdot \cdot}{O}} \colon H$$

Water

- **Group 7A elements** (halogens), such as fluorine, have seven valence electrons and form one bond, as in hydrogen fluoride, HF. The fluorine atom in the resultant molecule has one bonding pair of electrons and three lone pairs.

$$\colon \overset{\cdot \cdot}{\underset{\cdot \cdot}{F}} \cdot \; + \; H \cdot \; \longrightarrow \; H \colon \overset{\cdot \cdot}{\underset{\cdot \cdot}{F}} \colon$$

Hydrogen
fluoride

- **Group 8A elements** (noble gases), such as neon, rarely form covalent bonds because they already have valence-shell octets.

$$\colon \overset{\cdot \cdot}{\underset{\cdot \cdot}{Ne}} \colon \qquad \textit{Does not form covalent bonds}$$

These conclusions are summarized in Table 7.3.

TABLE 7.3   Covalent Bonding for Second-Row Elements

Group	Number of Valence Electrons	Number of Bonds	Example
3A	3	3	BH_3
4A	4	4	CH_4
5A	5	3	NH_3
6A	6	2	H_2O
7A	7	1	HF
8A	8	0	Ne

Not all covalent molecules are described as simply as those just discussed. In molecules such as O_2, N_2, and many others, the atoms share more than one pair of electrons, leading to the formation of *multiple* covalent bonds. The oxygen atoms in the O_2 molecule, for example, reach valence-shell octets by sharing four electrons (two pairs), giving a **double bond**. Similarly, the nitrogen atoms in the N_2 molecule share six electrons (three pairs), giving a **triple bond**. In speaking of such molecules, we often use the term **bond order** to refer to the number of electron pairs shared between atoms. Thus, the oxygen–oxygen bond in the O_2 molecule has a bond order of 2, and the nitrogen–nitrogen bond in the N_2 molecule has a bond order of 3. (Although the O_2 molecule does have a double bond, the following electron-dot structure is incorrect in other respects, as we'll see in Section 7.14.)

$$\cdot \overset{\cdot \cdot}{\underset{\cdot \cdot}{O}} \cdot \; + \; \cdot \overset{\cdot \cdot}{\underset{\cdot \cdot}{O}} \cdot \; \longrightarrow \; \overset{\cdot \cdot}{\underset{\cdot \cdot}{O}} \colon\colon \overset{\cdot \cdot}{\underset{\cdot \cdot}{O}}$$

*Two electron pairs
—a double bond*

$$\colon \overset{\cdot}{N} \cdot \; + \; \cdot \overset{\cdot}{N} \colon \; \longrightarrow \; \colon N \vdots\vdots\vdots N \colon$$

*Three electron pairs
—a triple bond*

As you might expect, multiple bonds are both shorter and stronger than their corresponding single-bond counterparts because there are more shared electrons holding the atoms together. Compare, for example, the O=O double bond in O_2 with the O–O single bond in H_2O_2 (hydrogen peroxide), and compare the N≡N triple bond in N_2 with the N–N single bond in N_2H_4 (hydrazine):

$\overset{..}{\underset{..}{O}} = \overset{..}{\underset{..}{O}}$	$H-\overset{..}{\underset{..}{O}}-\overset{..}{\underset{..}{O}}-H$	$:N\equiv N:$	$H-\overset{H}{\underset{..}{N}}-\overset{H}{\underset{..}{N}}-H$
Bond length: 121 pm	148 pm	110 pm	145 pm
Bond strength: 498 kJ/mol	213 kJ/mol	945 kJ/mol	275 kJ/mol

One final point about covalent bonds involves the origin of the bonding electron pair. Although most covalent bonds form when two atoms each contribute one electron, bonds can also form when one atom donates *both* electrons (a lone pair) to another atom that has a vacant valence orbital. The ammonium ion (NH_4^+), for example, forms when the two lone-pair electrons from the nitrogen atom of ammonia, $:NH_3$, bond to H^+. Such bonds are called **coordinate covalent bonds**.

An ordinary covalent bond—each atom donates one electron

$$H\cdot + \cdot H \longrightarrow H:H$$

A coordinate covalent bond—the nitrogen atom donates both electrons

$$H^+ + :\overset{H}{\underset{H}{N}}:H \longrightarrow \left[H:\overset{H}{\underset{H}{N}}:H \right]^+$$

Note that the nitrogen atom in the ammonium ion (NH_4^+) has more than the usual number of bonds—four instead of three—but that it still has an octet of valence electrons. Nitrogen, oxygen, phosphorus, and sulfur form such coordinate covalent bonds frequently.

EXAMPLE 7.1 Draw an electron-dot structure for phosphine, PH_3.

SOLUTION Phosphorus, a group 5A element, has five valence electrons and can achieve a valence-shell octet by forming three bonds and having one lone pair. Each hydrogen supplies one electron.

$$H:\overset{H}{\underset{..}{P}}:H \quad \text{Phosphine}$$

▶ **PROBLEM 7.3** Draw electron-dot structures for the following molecules:
(a) H_2S, a poisonous gas produced by rotten eggs
(b) $CHCl_3$, chloroform

▶ **PROBLEM 7.4** Draw an electron-dot structure for the hydronium ion, H_3O^+, and show how a coordinate covalent bond is formed by the reaction of H_2O with H^+.

7.6 Electron-Dot Structures of Polyatomic Molecules

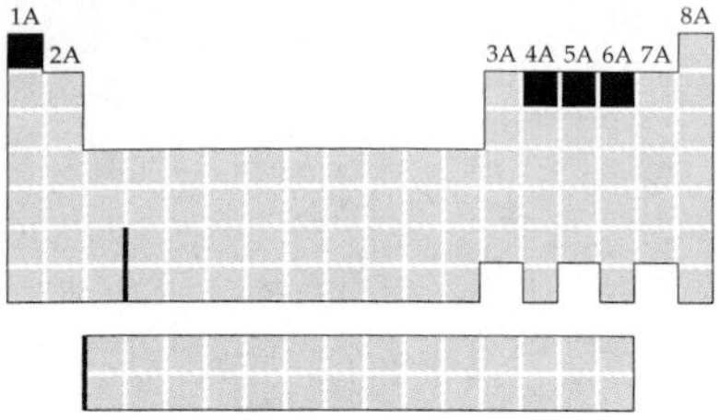

Compounds of Hydrogen and Second-Row Elements: C, N, O

Electron Dot Structures II activity

The vast majority of naturally occurring compounds on which life is based—proteins, fats, carbohydrates, and many others—contain only hydrogen and one or more of the second-row elements carbon, nitrogen, and oxygen. Electron-dot structures are easy to draw for such compounds because the octet rule almost always applies and the number of bonds formed by each element is predictable (Table 7.3).

For relatively small molecules that contain only a few second-row atoms in addition to hydrogen, the second-row atoms are bonded to one another in a central core, with hydrogens on the periphery. In ethane (C_2H_6), for example, two carbon atoms, each of which forms four bonds, combine with six hydrogens, each of which forms one bond. Joining the two carbon atoms and adding the appropriate number of hydrogens to each yields only one possible structure:

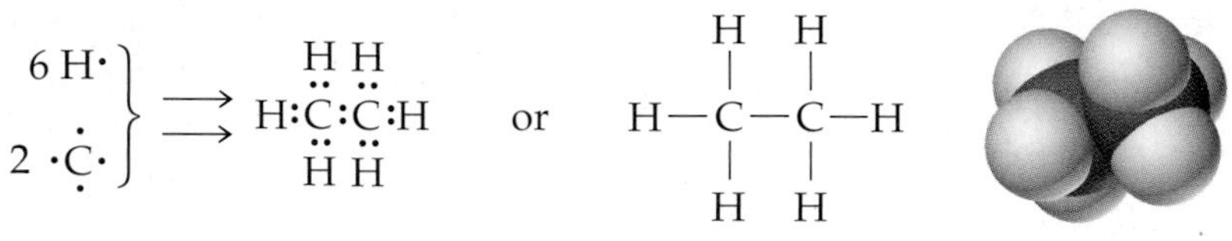

Ethane, C_2H_6

For larger molecules that contain numerous second-row atoms, there is usually more than one possible electron-dot structure. In such cases, some additional knowledge about the order of connections between atoms is necessary before a structure can be drawn.

Note that from now on, we'll follow the usual convention of indicating a two-electron covalent bond by a line. Similarly, we'll use two lines between atoms to represent four shared electrons (two pairs) in a double bond, and three lines to represent six shared electrons (three pairs) in a triple bond. The following examples and problems will give you more practice with electron-dot structures.

EXAMPLE 7.2 Draw an electron-dot structure for hydrazine, N_2H_4.

SOLUTION Join the two nitrogen atoms, and add two hydrogen atoms to each so that each nitrogen forms three bonds and each hydrogen forms one bond:

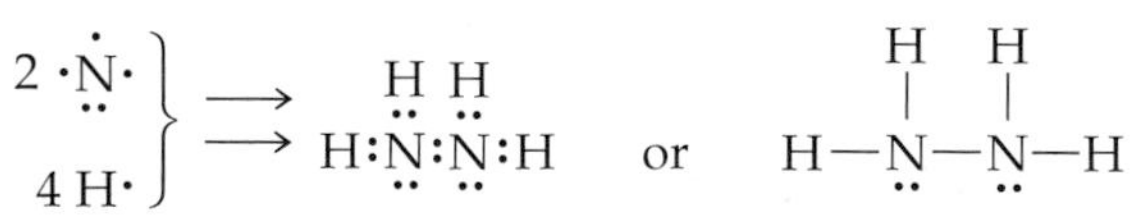

Hydrazine, N_2H_4

EXAMPLE 7.3 Draw an electron-dot structure for carbon dioxide, CO_2.

SOLUTION Connect the atoms so that carbon forms four bonds and each oxygen forms two bonds. The only possible structure contains two carbon–oxygen double bonds.

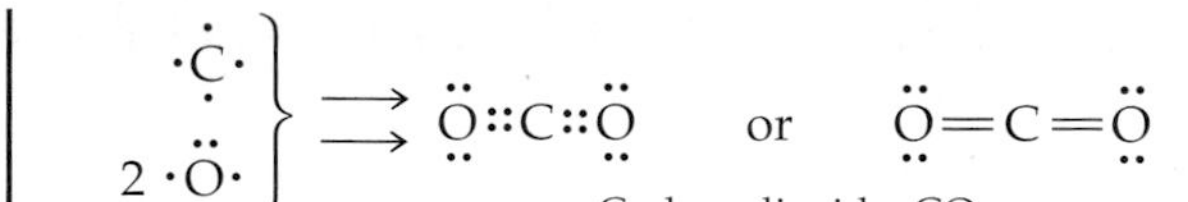

$$\left. \begin{array}{c} \cdot\ddot{C}\cdot \\[2pt] 2\cdot\ddot{O}\cdot \end{array} \right\} \longrightarrow \ddot{O}::C::\ddot{O} \quad \text{or} \quad \ddot{O}=C=\ddot{O}$$

Carbon dioxide, CO_2

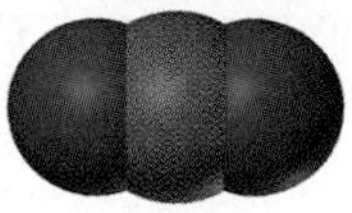

EXAMPLE 7.4 Draw an electron-dot structure for the deadly gas hydrogen cyanide, HCN.

SOLUTION First, connect the carbon and nitrogen atoms. The only way the carbon can form four bonds and the nitrogen can form three bonds is if there is a carbon–nitrogen triple bond.

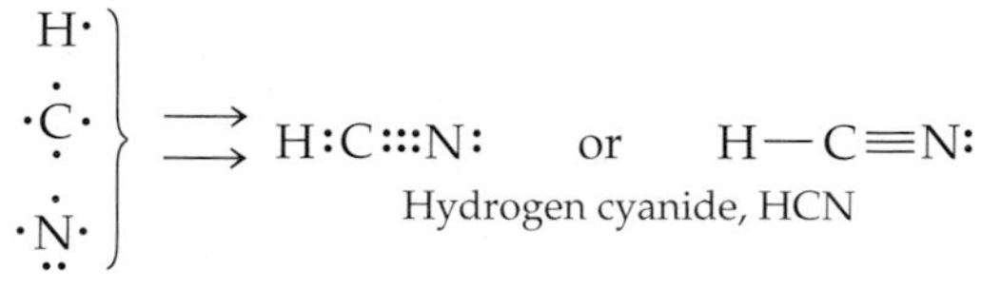

$$\left. \begin{array}{c} H\cdot \\[2pt] \cdot\ddot{C}\cdot \\[2pt] \cdot\ddot{N}\cdot \end{array} \right\} \longrightarrow H:C:::N: \quad \text{or} \quad H-C\equiv N:$$

Hydrogen cyanide, HCN

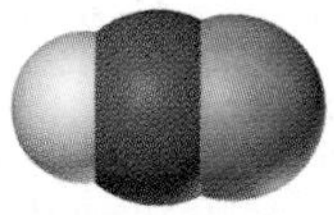

▶ **PROBLEM 7.5** Draw electron-dot structures for the following molecules:
(a) Propane, C_3H_8 (b) Hydrogen peroxide, H_2O_2 (c) Methylamine, CH_5N
(d) Ethylene, C_2H_4 (e) Acetylene, C_2H_2 (f) Phosgene, Cl_2CO

▶ **PROBLEM 7.6** There are two molecules with the formula C_2H_6O. Draw electron-dot structures for both.

━ **KEY CONCEPT PROBLEM 7.7** The following structure is a computer-drawn representation of cytosine, a constituent of the DNA found in all living cells. Only the connections between atoms are shown; multiple bonds are not indicated. Give the molecular formula of cytosine, and complete the structure by showing where the multiple bonds and lone pairs are (red = O, gray = C, blue = N, ivory = H).

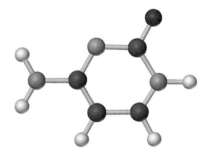

Compounds with Elements Below the Second Row

The simple method of drawing electron-dot structures that works well for most compounds of second-row elements sometimes breaks down for compounds that contain elements below the second row in the periodic table. As we saw in Section 6.12, elements in the third row and lower have unfilled *d* orbitals and are therefore able to expand their valence shell beyond the normal octet of electrons, forming more than the "normal" number of bonds predicted by their group number. In bromine trifluoride, for example, the

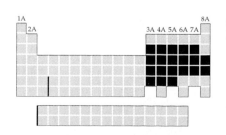

bromine atom forms three electron-pair bonds rather than one and has ten electrons in its valence shell rather than eight:

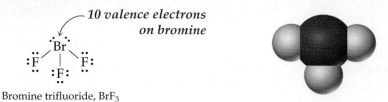

10 valence electrons on bromine

Bromine trifluoride, BrF_3

A general method of drawing electron-dot structures that works for any compound is to use the following steps:

Step 1. Find the total number of valence electrons for all atoms in the molecule. Add one additional electron for each negative charge in an anion and subtract one electron for each positive charge in a cation. In SF_4, for example, the total is 34 (6 from sulfur and 7 from each of 4 fluorines). In OH^-, the total is 8 (6 from oxygen, 1 from hydrogen, and 1 for the negative charge). In NH_4^+, the total is 8 (5 from nitrogen, 1 from each of 4 hydrogens, minus 1 for the positive charge).

SF_4	OH^-	NH_4^+
$:\!\overset{\cdot}{\underset{\cdot\cdot}{S}}\!\cdot$ $4:\!\overset{\cdot\cdot}{\underset{\cdot\cdot}{F}}\!\cdot$	$:\!\overset{\cdot}{\underset{\cdot\cdot}{O}}\!\cdot$ $H\cdot$	$:\!\overset{\cdot}{\underset{\cdot}{N}}\!\cdot$ $4\,H\cdot$
$6e + (4 \times 7e)$	$6e + 1e + 1e$	$5e + (4 \times 1e) - 1e$
$= 34e$	$= 8e$	$= 8e$

Step 2. Decide what the connections are between atoms, and draw a line to represent each covalent bond. Often, you'll be told the connections; other times you'll have to guess. Remember that hydrogen and the halogens usually form only one bond, elements in the second row usually form the number of bonds given in Table 7.3, and elements in the third row and lower often expand their valence shells and occur as the central atom around which other atoms are grouped. Also, it's often the case that the central atom is the least electronegative one (except H). If, for example, you were asked to predict the connections in SF_4, a good guess would be that each fluorine forms one bond to sulfur, which occurs as the central atom and forms more bonds than are predicted by its group number.

$$\begin{matrix} F & & F \\ & \!\diagdown\;\diagup\! & \\ & S & \\ & \!\diagup\;\diagdown\! & \\ F & & F \end{matrix}$$ Sulfur tetrafluoride, SF_4

Step 3. Subtract the number of valence electrons used in bonding from the total number calculated in step 1 to find the number that remain. Assign as many of these remaining electrons as necessary to the terminal atoms (other than hydrogen) so that each has an octet. In SF_4, 8 of the 34 total valence electrons are used in covalent bonding, leaving $34 - 8 = 26$. Twenty-four of the 26 are assigned to the four terminal fluorine atoms to reach an octet configuration for each:

8 + 24 = 32 electrons distributed

Step 4. If unassigned electrons remain after step 3, place them on the central atom. In SF_4, 32 of the 34 electrons have been assigned, leaving the final 2 to be placed on the central S atom:

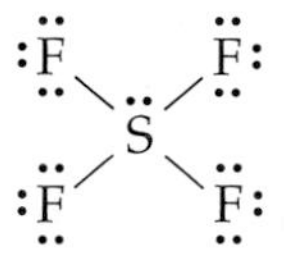

:F̈   F̈:
  \ :S: /
:F̈   F̈:   ***34 electrons distributed***

Step 5. If no unassigned electrons remain after step 3 but the central atom does not yet have an octet, use one or more lone pairs of electrons from a neighboring atom to form a multiple bond (either double or triple). Oxygen, carbon, nitrogen, and sulfur often form multiple bonds. Example 7.6 below shows how to deal with such a case.

EXAMPLE 7.5 Draw an electron-dot structure for phosphorus pentachloride, PCl_5.

SOLUTION First, count the total number of valence electrons. Phosphorus has 5, and each chlorine has 7, for a total of 40. Next, decide on the connections between atoms and draw a line to indicate each bond. Since chlorine normally forms only one bond, it's likely in the case of PCl_5 that all five chlorines are bonded to a central phosphorus atom, which expands its valence shell:

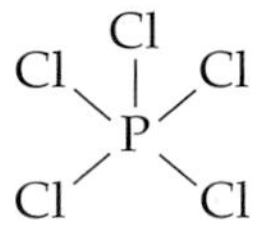

Cl
|
Cl   Cl
 \ | /
  P
 / \
Cl   Cl

Ten of the 40 valence electrons are necessary for the five P—Cl bonds, leaving 30 to be distributed so that each chlorine has an octet. All 30 remaining valence electrons are used in this step, giving the following structure:

:C̈l:
:C̈l | C̈l:
   \ P /
:C̈l: :C̈l:

Phosphorus pentachloride, PCl_5

EXAMPLE 7.6 Draw an electron-dot structure for formaldehyde, CH_2O.

SOLUTION First, count the total number of valence electrons. Carbon has 4, each hydrogen has 1, and the oxygen has 6, for a total of 12. Next, decide on the probable connections between atoms, and draw a line to indicate each bond. In the case of formaldehyde, the less electronegative atom (carbon) is the central atom, and both hydrogens and the oxygen are bonded to carbon:

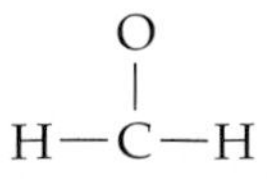

O
|
H—C—H

Formaldehyde-based adhesives are used in the manufacture of plywood and particle board.

Six of the 12 valence electrons are used for bonds, leaving 6 for assignment to the terminal oxygen atom.

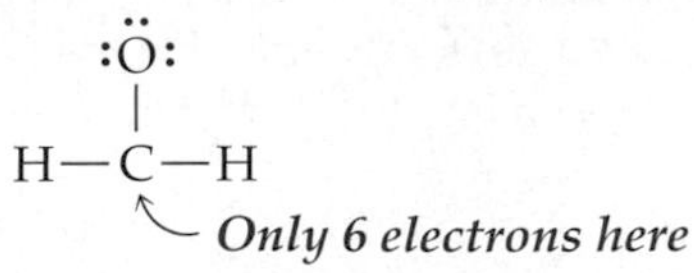

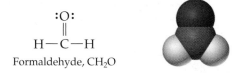

Only 6 electrons here

At this point, all the valence electrons are assigned, but the central carbon atom still does not have an octet. We therefore move two of the oxygen electrons from a lone pair into a bonding pair, generating a carbon–oxygen double bond and satisfying the octet rule for both oxygen and carbon.

Formaldehyde, CH_2O

EXAMPLE 7.7   Draw an electron-dot structure for XeF_5^+.

SOLUTION   Count the total number of valence electrons: Xenon has 8, each fluorine has 7, and 1 is subtracted to account for the positive charge, giving a total of 42. Then, decide on the probable connections between atoms, and draw a line for each bond. In the case of XeF_5^+, it's likely that the five fluorines are bonded to xenon, a fifth-row atom.

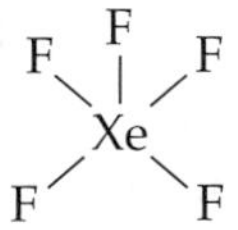

With 10 of the 42 valence electrons used in bonds, distribute as many of the remaining 32 electrons as necessary so that each of the terminal fluorine atoms has an octet. Two electrons still remain, so we assign them to xenon to give the final structure, which has a positive charge.

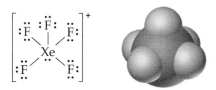

▶**PROBLEM 7.8**   Carbon monoxide, CO, is a deadly gas produced by incomplete combustion of fuels. Draw an electron-dot structure for CO.

▶**PROBLEM 7.9**   Draw an electron-dot structure for each of the following molecules:
(a) $AlCl_3$ **(b)** ICl_3 **(c)** $XeOF_4$ **(d)** HOBr

▶**PROBLEM 7.10**   Draw an electron-dot structure for each of the following ions:
(a) OH^- **(b)** H_3S^+ **(c)** HCO_3^-

7.7   Electron-Dot Structures and Resonance

The steps given in the previous section for drawing electron-dot structures lead to an interesting problem in some cases. Look at ozone, O_3, for instance. Step 1 says that there are 18 valence electrons in the molecule, and steps 2–4 let us draw the following structure:

$$:\ddot{\text{O}} - \ddot{\text{O}} - \ddot{\text{O}}:$$

We find at this point that the central atom does not yet have an octet, and we therefore have to move one of the lone pairs of electrons on a terminal oxygen to become a bonding pair, giving the central oxygen an octet. But do we take a lone pair from the "right-hand" oxygen or the "left-hand" one? Both possibilities lead to acceptable structures:

Move a lone pair from this oxygen?   *Or from this oxygen?*

$$:\ddot{\text{O}} - \ddot{\text{O}} - \ddot{\text{O}}: \quad \longrightarrow \quad \begin{cases} \ddot{\text{O}} = \ddot{\text{O}} - \ddot{\text{O}}: \\ \text{or} \\ :\ddot{\text{O}} - \ddot{\text{O}} = \ddot{\text{O}} \end{cases}$$

Which of the two structures for O_3 is correct? In fact, *neither is correct by itself*. Whenever it's possible to draw more than one valid electron-dot structure for a molecule, the actual electronic structure is an *average* of the different possibilities, called a **resonance hybrid**. Note that *the different resonance forms differ only in the placement of the valence-shell electrons.* The total number of valence electrons remains the same in both structures, the connections between atoms remain the same, and the relative positions of the atoms remain the same.

Ozone doesn't have one O=O double bond and one O–O single bond as the individual structures imply; rather, ozone has two *equivalent* O–O bonds that we can think of as having a bond order of 1.5, midway between pure single bonds and pure double bonds. Both bonds have an identical length of 128 pm.

We can't draw a single electron-dot structure that indicates the equivalence of the two O–O bonds in O_3 because the conventions we use for indicating electron placement aren't good enough. Instead, the idea of resonance is indicated by drawing the two (or more) individual electron-dot structures and using a double-headed *resonance arrow* to show that both contribute to the resonance hybrid. *A straight, double-headed arrow always indicates resonance; it is never used for any other purpose.*

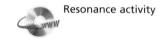

Resonance activity

This double-headed arrow means that the structures on either side are contributors to a resonance hybrid.

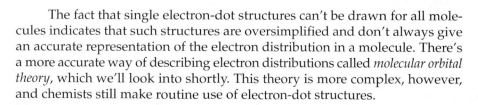

$$:\ddot{\text{O}} - \ddot{\text{O}} = \ddot{\text{O}} \longleftrightarrow \ddot{\text{O}} = \ddot{\text{O}} - \ddot{\text{O}}:$$

The fact that single electron-dot structures can't be drawn for all molecules indicates that such structures are oversimplified and don't always give an accurate representation of the electron distribution in a molecule. There's a more accurate way of describing electron distributions called *molecular orbital theory*, which we'll look into shortly. This theory is more complex, however, and chemists still make routine use of electron-dot structures.

EXAMPLE 7.8   The nitrate ion, NO_3^-, has three equivalent oxygen atoms, and its electronic structure is a resonance hybrid of three electron-dot structures. Draw them.

SOLUTION   There are 24 valence electrons in the nitrate ion: 5 from nitrogen, 6 from each of 3 oxygens, and 1 for the negative charge. The three equivalent oxygens are all bonded to nitrogen, the less electronegative central atom:

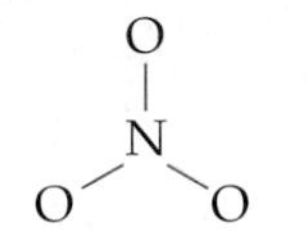

6 of 24 valence electrons assigned

Distributing the remaining 18 valence electrons among the three terminal oxygen atoms completes the octet of each oxygen but leaves nitrogen with only 6 electrons.

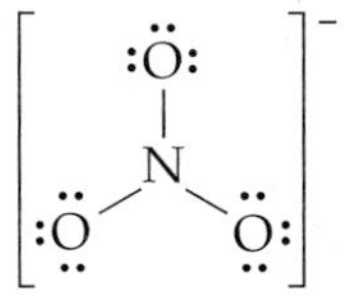

To give nitrogen an octet, one of the oxygen atoms must use a lone pair to form an N—O double bond. But which one? There are three possibilities, and thus three electron-dot resonance structures for the nitrate ion, which differ only in the placement of bonding and lone-pair electrons. The connections between atoms are the same in all three structures, and the atoms have the same positions in all structures.

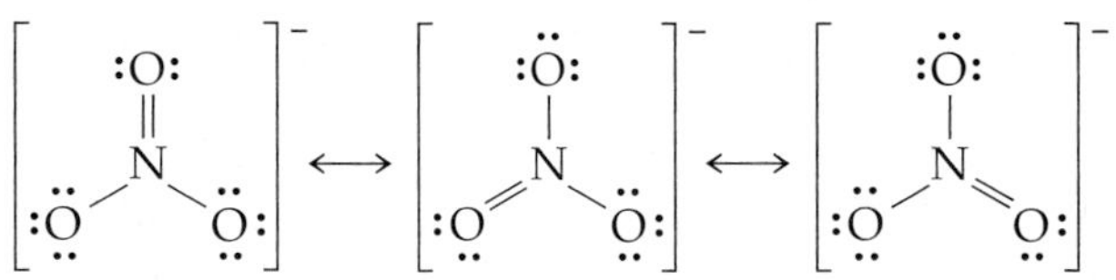

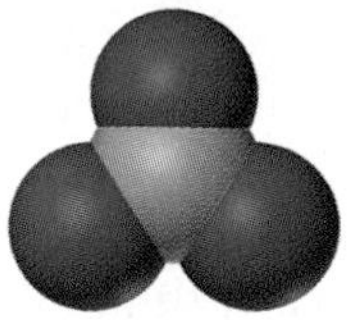

▶ **PROBLEM 7.11**   Called "laughing gas," nitrous oxide (N_2O) is sometimes used by dentists as an anesthetic. Assuming the connections N—N—O, draw two electron-dot structures for N_2O.

▶ **PROBLEM 7.12**   Draw as many electron-dot resonance structures as possible for each of the following molecules or ions, giving all atoms (except H) octets:
(a) SO_2 **(b)** CO_3^{2-} **(c)** HCO_2^-

↤ **KEY CONCEPT PROBLEM 7.13**   The following structure shows the connections between atoms for anisole, a compound used in perfumery. Draw two resonance structures for anisole, showing the positions of the multiple bonds in each (red = O, gray = C, ivory = H).

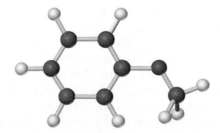

7.8   Formal Charges

Closely related to the ideas of electronegativity and polar covalent bonds discussed in Section 7.4 is the concept of *formal charges* on specific atoms in electron-dot structures. Formal charges result from a kind of electron "bookkeeping" and can be calculated in the following way: Find the number of valence electrons belonging to an atom in a given electron-dot structure and compare that value with the number of valence electrons in the isolated atom. If the numbers aren't the same, then the atom in the molecule has either gained or lost electrons and thus has a formal charge. If the atom in a molecule has more electrons than the isolated atom, it has a negative formal charge; if it has fewer electrons, it has a positive formal charge.

 Formal Charges movie

$$\text{Formal charge} = \begin{pmatrix} \text{Number of} \\ \text{valence electrons} \\ \text{in free atom} \end{pmatrix} - \begin{pmatrix} \textbf{Number of} \\ \textbf{valence electrons} \\ \textbf{in bonded atom} \end{pmatrix}$$

In counting the number of valence electrons in a bonded atom, it's necessary to make a distinction between shared, bonding electrons and unshared, nonbonding electrons. For bookkeeping purposes, an atom can be thought of as "owning" all its nonbonding electrons but only half of its bonding electrons, since the bonding electrons are shared with another atom. Thus, we can rewrite the definition of formal charge as

$$\textbf{Formal charge} = \begin{pmatrix} \text{Number of} \\ \text{valence electrons} \\ \text{in free atom} \end{pmatrix} - \frac{1}{2}\begin{pmatrix} \text{Number of} \\ \text{bonding} \\ \text{electrons} \end{pmatrix} - \begin{pmatrix} \text{Number of} \\ \text{nonbonding} \\ \text{electrons} \end{pmatrix}$$

Look at the atoms in the ammonium ion, NH_4^+, for example. Each of the four equivalent hydrogen atoms has 2 valence electrons (in its covalent bond to nitrogen), and the nitrogen atom has 8 valence electrons (two from each of its four N–H bonds):

$$\begin{bmatrix} \text{H} \\ \text{H}\!:\!\overset{..}{\underset{..}{\text{N}}}\!:\!\text{H} \\ \text{H} \end{bmatrix}^+ \quad \begin{array}{l} \textit{Ammonium ion} \\ \textit{8 valence electrons around nitrogen} \\ \textit{2 valence electrons around each hydrogen} \end{array}$$

For bookkeeping purposes, each hydrogen owns half of its two shared bonding electrons, or 1, while the nitrogen atom owns half of its 8 shared bonding electrons, or 4. Since an isolated hydrogen atom has 1 electron and since the hydrogens in the ammonium ion each still own 1 electron, they have neither gained nor lost electrons and thus have no formal charge. An isolated nitrogen atom, however, has 5 valence electrons. Since the nitrogen atom in NH_4^+ owns only 4 valence electrons, it has a formal charge of +1. The sum of the formal charges on all the atoms (+1 in this example) is, of course, equal to the overall charge on the ion.

$$
\left[\begin{array}{c} H \\ \ddot{} \\ H\!:\!\ddot{N}\!:\!H \\ \ddot{} \\ H \end{array}\right]^{+}
$$

For hydrogen: Isolated hydrogen valence electrons   1
Bound hydrogen bonding electrons   2
Bound hydrogen nonbonding electrons   0

Formal charge $= 1 - \dfrac{1}{2}(2) - 0 = 0$

For nitrogen: Isolated nitrogen valence electrons   5
Bound nitrogen bonding electrons   8
Bound nitrogen nonbonding electrons   0

Formal charge $= 5 - \dfrac{1}{2}(8) - 0 = +1$

The value of formal charge calculations comes from their application to resonance structures, described in the previous section. It often happens that the resonance structures of a given substance are not identical. One of the structures may be "better" than the others, meaning that it approximates the actual electronic structure of the substance more closely. The resonance hybrid in such cases is thus weighted more strongly toward the more favorable structure.

Take the organic substance called acetamide, for instance, a compound related to proteins. We can draw two valid electron-dot structures for acetamide, both of which fulfill the octet rule for the C, N, and O atoms. One of the two structures has no formal charges, while the other has formal charges on the O and N atoms. (Check for yourself that the formal charges are correct.)

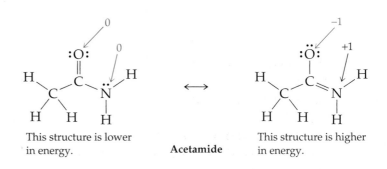

This structure is lower in energy.   **Acetamide**   This structure is higher in energy.

Which of the two structures gives a more accurate representation of the molecule? Since energy is required to separate + and − charges, the structure without formal charges is probably lower in energy than the structure with formal charges. Thus, the actual electronic structure of acetamide is closer to that of the more favorable electron-dot structure.

As another example, the resonance structure for N_2O that places the formal negative charge on the more electronegative oxygen atom rather than on the less electronegative nitrogen atom is probably a more accurate representation of the molecule.

$$
:N\!\equiv\!\overset{+}{N}\!-\!\overset{..}{\underset{..}{O}}:^{-} \quad\longleftrightarrow\quad {}^{-}\overset{..}{\underset{..}{N}}\!=\!\overset{+}{N}\!=\!\overset{..}{\underset{..}{O}}:
$$

This electron-dot structure . . . is more favorable than . . . this one.

EXAMPLE 7.9   Calculate the formal charge on each atom in the following electron-dot structure for SO_2:

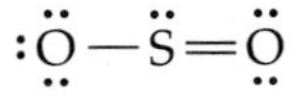

$$:\ddot{O}-\ddot{S}=\ddot{O}$$

SOLUTION

For sulfur:	Isolated sulfur valence electrons	6
	Bound sulfur bonding electrons	6
	Bound sulfur nonbonding electrons	2

$$\text{Formal charge} = 6 - \frac{1}{2}(6) - 2 = +1$$

For singly bonded oxygen:	Isolated oxygen valence electrons	6
	Bound oxygen bonding electrons	2
	Bound oxygen nonbonding electrons	6

$$\text{Formal charge} = 6 - \frac{1}{2}(2) - 6 = -1$$

For doubly bonded oxygen:	Isolated oxygen valence electrons	6
	Bound oxygen bonding electrons	4
	Bound oxygen nonbonding electrons	4

$$\text{Formal charge} = 6 - \frac{1}{2}(4) - 4 = 0$$

These calculations tell us that the sulfur atom of SO_2 has a formal charge of +1 and the singly bonded oxygen atom has a formal charge of −1. We might therefore draw the structure for SO_2 as

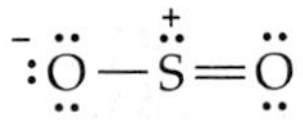

$$\overset{-}{:}\ddot{O}-\overset{+}{\ddot{S}}=\ddot{O}$$

▶**PROBLEM 7.14**   Calculate the formal charge on each atom in the three resonance structures for the nitrate ion in Example 7.8.

▶**PROBLEM 7.15**   Calculate the formal charge on each atom in the following electron-dot structures:

(a) Cyanate ion: $\left[\ddot{N}=C=\ddot{O}\right]^{-}$ **(b)** Ozone: $:\ddot{O}-\ddot{O}=\ddot{O}$

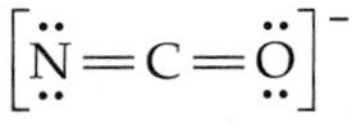

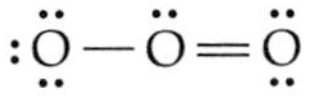

7.9   Molecular Shapes: The VSEPR Model

Look at the following computer-generated ball-and-stick models of water, ammonia, and methane. Each of these molecules—and every other molecule as well—has a specific three-dimensional shape. Often, particularly for biologically important molecules, three-dimensional shape plays a crucial role in determining the molecule's chemistry.

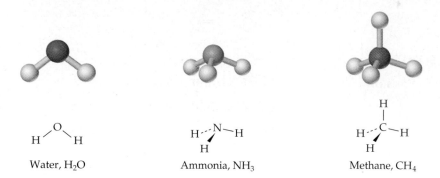

Water, H_2O Ammonia, NH_3 Methane, CH_4

VSEPR movie

VSEPR tutorial

Like so many other molecular properties, shapes are determined by the electronic structure of the bonded atoms. The approximate shape of a molecule can often be predicted by using what is called the **valence-shell electron-pair repulsion (VSEPR) model**. Electrons in bonds and in lone pairs can be thought of as "charge clouds" that repel one another and stay as far apart as possible, thus causing molecules to assume specific shapes. There are only two steps to remember in applying the VSEPR method:

Step 1. Draw an electron-dot structure for the molecule, and count the number of electron charge clouds surrounding the atom of interest. A charge cloud is simply a group of electrons, either in a bond or in a lone pair. Thus, the number of charge clouds is the total number of bonds and lone pairs. Multiple bonds count the same as single bonds because it doesn't matter how many electrons occupy each cloud.

Step 2. Predict the arrangement of charge clouds around each atom by assuming that the clouds are oriented in space as far away from one another as possible. How they achieve this orientation depends on their number. Let's look at the possibilities.

Two Charge Clouds. When there are only two charge clouds, as occurs on the carbon atoms of CO_2 (two double bonds) and HCN (one single bond and one triple bond), the clouds are farthest apart when they point in opposite directions. Thus, CO_2 and HCN are linear molecules with **bond angles** of 180°. Note that the C=O double bonds in CO_2 and the C≡N triple bond in HCN are shown as striped, while the H−C single bond in HCN is gray. *All ball-and-stick models in this book will follow the same convention: Two-electron (single) covalent bonds will be colored a uniform dark gray, and all others, including double bonds, triple bonds, and multiple bonds in resonance hybrid structures, will be striped.*

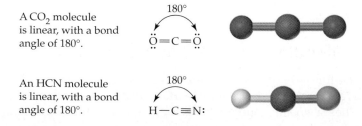

A CO_2 molecule is linear, with a bond angle of 180°.

An HCN molecule is linear, with a bond angle of 180°.

Three Charge Clouds.   When there are three charge clouds, as occurs on the carbon atom of formaldehyde (two single bonds and one double bond) and the sulfur atom of SO_2 (one single bond, one double bond, and one lone pair), the clouds are farthest apart when they lie in the same plane and point to the corners of an equilateral triangle. Thus, a formaldehyde molecule has a *trigonal planar* shape, with H–C–H and H–C=O bond angles near 120°. Similarly, an SO_2 molecule has a trigonal planar arrangement of its three charge clouds on sulfur, but one point of the triangle is occupied by a lone pair and two points by oxygen atoms. The molecule therefore has a *bent* rather than linear shape, with an O–S–O bond angle of approximately 120° rather than 180°.

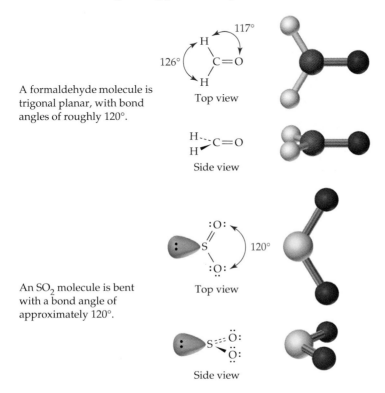

A formaldehyde molecule is trigonal planar, with bond angles of roughly 120°.

An SO_2 molecule is bent with a bond angle of approximately 120°.

Four Charge Clouds.   When there are four charge clouds, as occurs on the central atoms in CH_4 (four single bonds), NH_3 (three single bonds and one lone pair), and H_2O (two single bonds and two lone pairs), the clouds are farthest apart if they extend toward the corners of a *regular tetrahedron*. As illustrated in Figure 7.5, a regular tetrahedron is a geometric solid whose four identical faces are equilateral triangles. The central atom lies in the center of the tetrahedron, the charge clouds point toward the four corners, and the angle between two lines drawn from the center to any two corners is 109.5°.

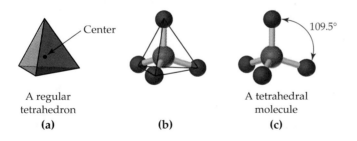

A regular tetrahedron
(a)

A tetrahedral molecule
(c)

(b)

FIGURE 7.5   The tetrahedral geometry of an atom surrounded by four charge clouds. The atom is located in the center of a regular tetrahedron **(a)**, and the four charge clouds point toward the four corners **(b)**. The angle between any two bonds is exactly 109.5° **(c)**.

Because valence electron octets are so common, particularly for second-row elements, the atoms in a great many molecules have shapes based on the tetrahedron. Methane, for example, has a tetrahedral shape, with H−C−H bond angles of 109.5°. In NH_3, the nitrogen atom has a tetrahedral arrangement of its four charge clouds, but one corner of the tetrahedron is occupied by a lone pair, resulting in a *trigonal pyramidal* shape for the molecule. Similarly, H_2O has two corners of the tetrahedron occupied by lone pairs and thus has a *bent* shape.

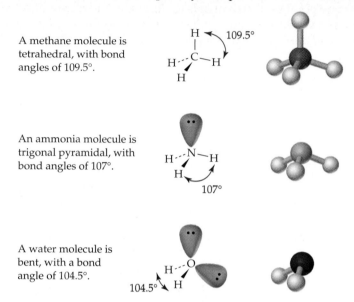

A methane molecule is tetrahedral, with bond angles of 109.5°.

An ammonia molecule is trigonal pyramidal, with bond angles of 107°.

A water molecule is bent, with a bond angle of 104.5°.

Note how the three-dimensional shapes of the molecules in the preceding three structures are indicated. Solid lines are assumed to be in the plane of the paper, dashed lines recede behind the plane of the paper away from the viewer, and heavy, wedged lines protrude out of the paper toward the viewer. Note also that the H−N−H bond angles in ammonia (107°) and the H−O−H bond angle in water (104.5°) are less than the ideal 109.5° tetrahedral value. The angles are diminished somewhat from the tetrahedral value because of the presence of lone pairs. Charge clouds of lone-pair electrons spread out more than charge clouds of bonding electrons because they aren't confined to the space between two atoms. As a result, the somewhat enlarged lone-pair charge clouds tend to compress the bond angles in the rest of the molecule.

Five Charge Clouds. Five charge clouds, such as are found on the central atoms in PCl_5, SF_4, ClF_3, and I_3^-, are oriented toward the corners of a geo-metric figure called a *trigonal bipyramid*. Three clouds lie in a plane and point toward the corners of an equilateral triangle, the fourth cloud points directly up, and the fifth cloud points down:

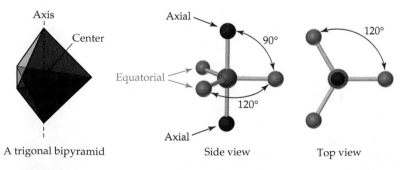

A trigonal bipyramid   Side view   Top view

Trigonal bipyramidal geometry differs from the linear, trigonal planar, and tetrahedral geometries discussed previously because it has two kinds of positions—three *equatorial* positions (around the "equator" of the bipyramid) and two *axial* positions (along the "axis" of the bipyramid). The three equatorial positions are at angles of 120° to one another and at an angle of 90° to the axial positions. The two axial positions are at angles of 180° to each other and at an angle of 90° to the equatorial positions.

Different substances with a trigonal bipyramidal arrangement of charge clouds adopt different shapes depending on whether the five charge clouds are of bonding or nonbonding electrons. Phosphorus pentachloride, for example, has all five positions around phosphorus occupied by chlorine atoms and thus has a trigonal bipyramidal shape:

A PCl$_5$ molecule is trigonal bipyramidal.

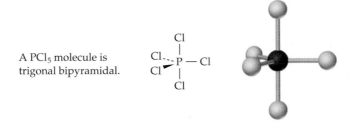

The sulfur atom in SF$_4$ is bonded to four other atoms and has one nonbonding electron lone pair. Because an electron lone pair spreads out and occupies more space than a bonding pair, the nonbonding electrons in SF$_4$ occupy an equatorial position where they are close to (90° away from) only two charge clouds. Were they instead to occupy an axial position, they would be close to three charge clouds. As a result, SF$_4$ has a shape often described as that of a see-saw. The two axial bonds form the board, and the two equatorial bonds form the legs of the see-saw. (You have to tilt your head 90° to see it.)

An SF$_4$ molecule is shaped like a see-saw (turn 90° to see it).

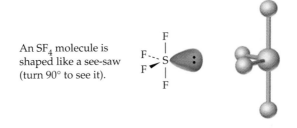

The chlorine atom in ClF$_3$ is bonded to three other atoms and has two nonbonding electron lone pairs. Both lone pairs occupy equatorial positions, resulting in a T shape for the ClF$_3$ molecule. (As with the see-saw, you have to tilt your head 90° to see the T.)

A ClF$_3$ molecule is T-shaped (turn 90° to see it).

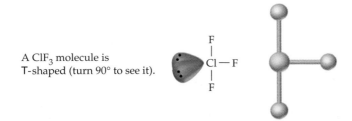

The central iodine atom in the I_3^- ion is bonded to two other atoms and has three lone pairs. All three lone pairs occupy equatorial positions, resulting in a linear shape for I_3^-.

An I_3^- ion is linear.

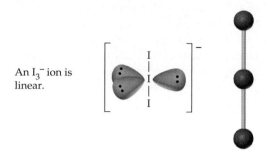

Six Charge Clouds. Six charge clouds around an atom orient toward the six corners of a *regular octahedron*, a geometric solid whose eight faces are equilateral triangles. All six positions are equivalent, and the angle between any two adjacent positions is 90°.

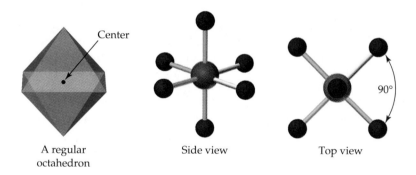

A regular octahedron   Side view   Top view

As was true in the case of five charge clouds, different shapes are possible for molecules having atoms with six charge clouds, depending on whether the clouds are of bonding or nonbonding electrons. Sulfur hexafluoride, for example, has all six positions around sulfur occupied by fluorine atoms:

An SF_6 molecule is octahedral.

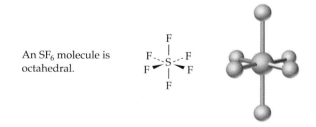

The antimony atom in the $SbCl_5^{2-}$ ion also has six charge clouds, but it is bonded to only five atoms and has one nonbonding electron lone pair. As a result, the ion has a *square pyramidal* shape—a pyramid with a square base:

An $SbCl_5^{2-}$ ion has a square pyramidal shape.

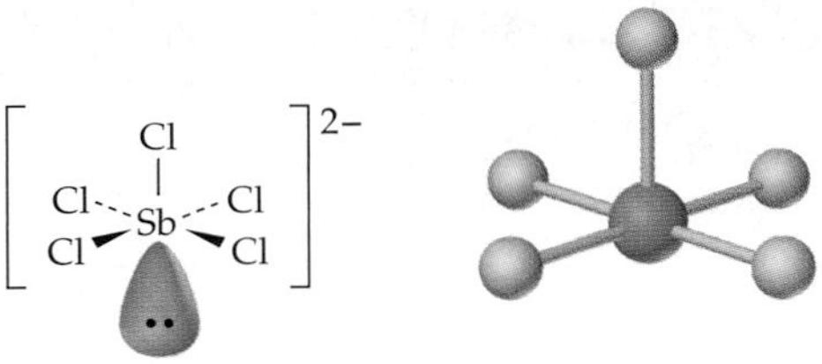

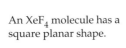

The xenon atom in XeF_4 is bonded to four atoms and has two lone pairs. As you might expect, the lone pairs orient as far away from each other as possible to minimize electronic repulsions, giving the molecule a *square planar* shape:

An XeF_4 molecule has a square planar shape.

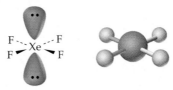

All the geometries discussed above for two to six charge clouds around an atom are summarized in Table 7.4 on pages 270–271.

Shapes of Larger Molecules.   The geometries around individual atoms in larger molecules can also be predicted from the rules summarized in Table 7.4. For example, each of the two carbon atoms in ethylene ($H_2C=CH_2$) has three charge clouds, giving rise to trigonal planar geometry for each carbon. The molecule as a whole is planar, with H−C−C and H−C−H bond angles of approximately 120°.

Each carbon atom in ethylene has trigonal planar geometry. As a result, the entire molecule is planar, with bond angles of 120°.

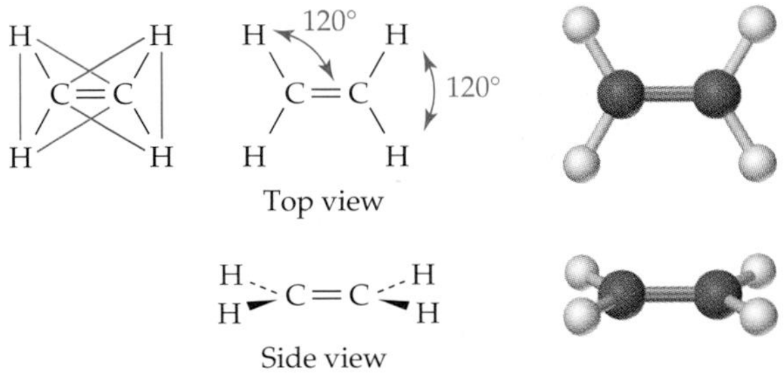

Top view

Side view

Carbon atoms bonded to four other atoms are each at the center of a tetrahedron. As shown below for ethane, $H_3C−CH_3$, the two tetrahedrons are joined so that the central carbon atom of one is a corner atom of the other.

Each carbon atom in ethane has tetrahedral geometry, with bond angles of 109.5°.

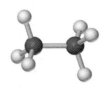

TABLE 7.4   Molecular Geometry Around Atoms with 2, 3, 4, 5, and 6 Charge Clouds

Number of Bonds	Number of Lone Pairs	Number of Charge Clouds	Molecular Geometry		Example
2	0	2		 Linear	O=C=O
3	0	3		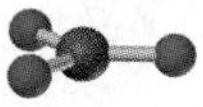 Trigonal planar	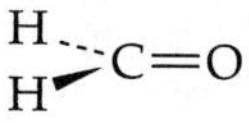 H, H—C=O
2	1			 Bent	 O, O—S
4	0	4		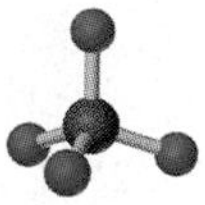 Tetrahedral	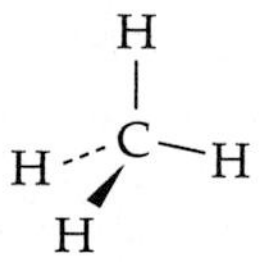 H—C—H (with H above and below)
3	1			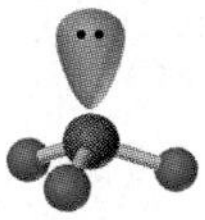 Trigonal pyramidal	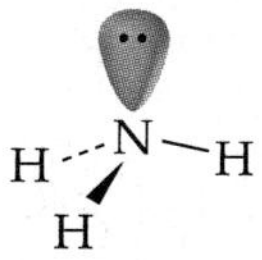 H—N—H (with H below)
2	2			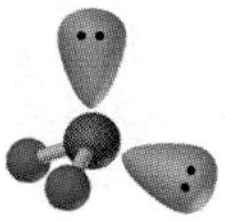 Bent	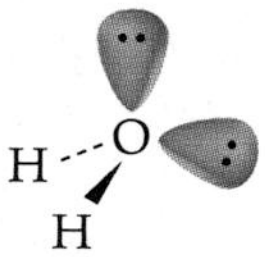 H—O (with H below)
5	0	5		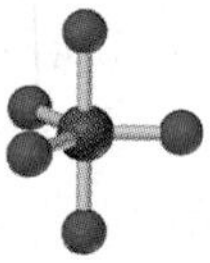 Trigonal bipyramidal	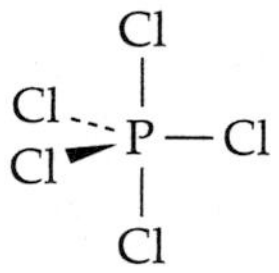 Cl—P—Cl (Cl, Cl above; Cl below)
4	1			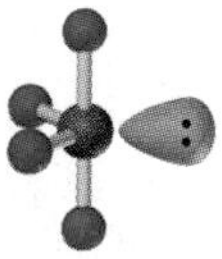 See-saw	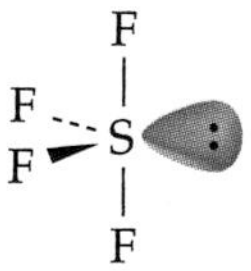 F—S—F (F, F; F below)
3	2			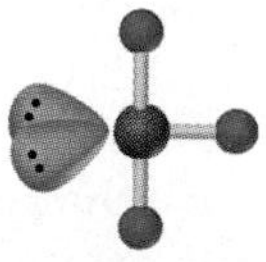 T-shaped	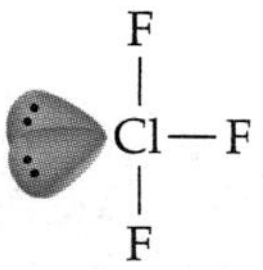 Cl—F (F below)
2	3			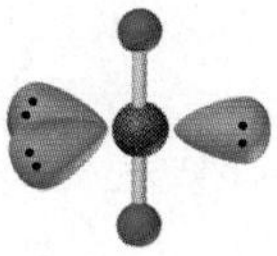 Linear	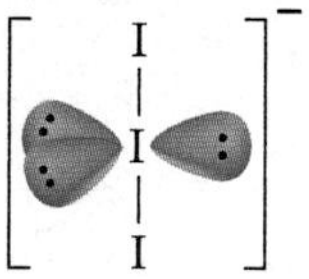 [I—I—I]⁻

Continued

TABLE 7.4 (Continued)

Number of Bonds	Number of Lone Pairs	Number of Charge Clouds	Molecular Geometry		Example
6	0		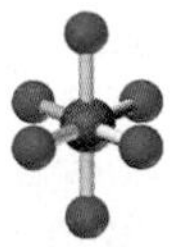	Octahedral	
5	1	6	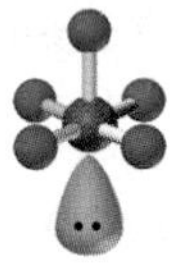	Square pyramidal	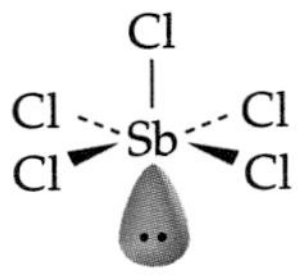
4	2			Square planar	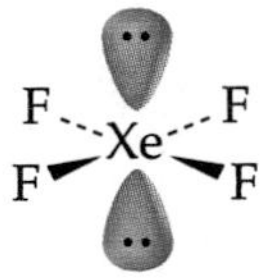

EXAMPLE 7.10 Predict the shape of BrF_5.

SOLUTION First, draw an electron-dot structure for BrF_5 to determine that the central bromine atom has six charge clouds (five bonds and one lone pair):

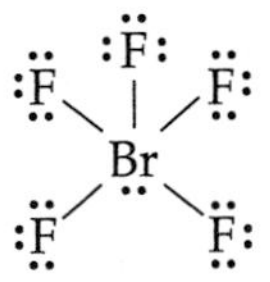

Bromine pentafluoride

Six charge clouds implies an octahedral arrangement; five attached atoms and one lone pair give BrF_5 a square pyramidal shape:

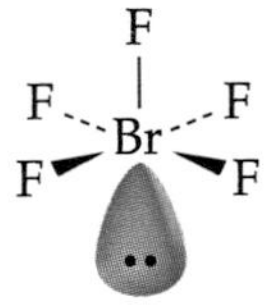

▶ **PROBLEM 7.16** Predict the shapes of the following molecules or ions:
(a) O_3 (b) H_3O^+ (c) XeF_2 (d) PF_6^- (e) $XeOF_4$
(f) AlH_4^- (g) BF_4^- (h) $SiCl_4$ (i) ICl_4^- (j) $AlCl_3$

▶ **PROBLEM 7.17** Acetic acid, CH_3CO_2H, is the main organic constituent of vinegar. Draw an electron-dot structure for acetic acid, and show its overall shape. (The two carbons are connected by a single bond, and both oxygens are connected to the same carbon.)

❯━ **KEY CONCEPT PROBLEM 7.18** What is the geometry around the central atom in each of the following molecular models?

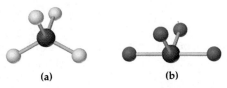

(a)   (b)

7.10 Valence Bond Theory

The electron-dot structures described in Sections 7.6 and 7.7 provide a simple way to predict the distribution of valence electrons in a molecule, and the VSEPR model discussed in Section 7.9 provides a simple way to predict molecular shapes. Neither model, however, says anything about the detailed electronic nature of covalent bonds. To describe bonding, a quantum mechanical model called **valence bond theory** has been developed.

Valence bond theory provides an easily visualized orbital picture of how electron pairs are shared in a covalent bond. In essence, a covalent bond results when two atoms approach each other closely enough so that a singly occupied valence orbital on one atom *overlaps* a singly occupied valence orbital on the other atom. The now-paired electrons in the overlapping orbitals are attracted to the nuclei of both atoms and thus bond the two atoms together. In the H_2 molecule, for instance, the H–H bond results from the overlap of two singly occupied hydrogen $1s$ orbitals:

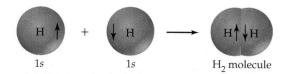

In the valence bond model, the strength of a covalent bond depends on the amount of orbital overlap: The greater the overlap, the stronger the bond. This, in turn, means that bonds formed by overlap of other than s orbitals have a directionality to them. In the F_2 molecule, for instance, each fluorine atom has the electron configuration $[He]\, 2s^2\, 2p_x^2\, 2p_y^2\, 2p_z^1$, meaning that the F–F bond results from the overlap of two singly occupied $2p$ orbitals. The two p orbitals must point directly at each other for optimum overlap to occur, and the F–F bond forms along the internuclear axis. Such bonds that result from head-on orbital overlap and whose shared electrons lie along the internuclear axis are called **sigma (σ) bonds**.

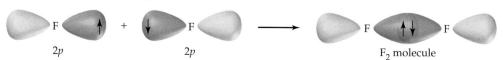

In HCl, the covalent bond involves overlap of a hydrogen $1s$ orbital with a chlorine $3p$ orbital, and forms along the p-orbital axis:

The key ideas of valence bond theory can be summarized as follows:

Key Ideas of Valence Bond Theory

- Covalent bonds are formed by overlap of atomic orbitals, each of which contains one electron of opposite spin.
- Each of the bonded atoms maintains its own atomic orbitals, but the electron pair in the overlapping orbitals is shared by both atoms.
- The greater the amount of orbital overlap, the stronger the bond. This leads to a directional character to the bond when other than s orbitals are involved.

7.11 Hybridization and *sp*³ Hybrid Orbitals

How does valence bond theory describe the electronic structures of poly-atomic molecules, and how does it account for their shapes? Let's look, for example, at a simple tetrahedral molecule such as methane, CH_4. There are several problems to be dealt with.

Carbon has the ground-state electron configuration [He] $2s^2 \, 2p_x^{\,1} \, 2p_y^{\,1}$. It thus has four valence electrons, two of which are paired in a $2s$ orbital and two of which are unpaired in different $2p$ orbitals that we'll arbitrarily designate as $2p_x$ and $2p_y$. But how can carbon form four bonds if two of its valence electrons are already paired and only two unpaired electrons are available for sharing? The answer is that an electron must be promoted from the lower-energy $2s$ orbital to the vacant, higher-energy $2p_z$ orbital, giving an *excited-state configuration* [He] $2s^1 \, 2p_x^{\,1} \, 2p_y^{\,1} \, 2p_z^{\,1}$, which has *four* unpaired electrons.

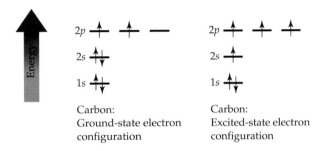

Carbon:
Ground-state electron
configuration

Carbon:
Excited-state electron
configuration

A second problem is more difficult to resolve: If excited-state carbon uses two kinds of orbitals for bonding, $2s$ and $2p$, how can it form four *equivalent* bonds? Furthermore, if the three $2p$ orbitals in carbon are at angles of 90° to one another, and if the $2s$ orbital has no directionality, how can carbon form bonds with angles of 109.5° directed to the corners of a regular tetrahedron? The answers to these questions were provided in 1931 by Linus Pauling, who introduced the idea of *hybrid orbitals*.

Pauling showed that the quantum mechanical wave functions for s and p atomic orbitals derived from the Schrödinger wave equation (Section 5.7) can be mathematically combined to form a new set of equivalent wave functions called **hybrid atomic orbitals**. When one s orbital combines with three p orbitals, as occurs in an excited-state carbon atom, four equivalent orbitals, called *sp*³ **hybrids**, result. (Note that the superscript 3 in the name *sp*³ tells how many p atomic orbitals are combined to construct the hybrid orbitals, not how many electrons occupy each orbital.)

Promotion and
Hybridization tutorial

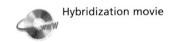

Hybridization movie

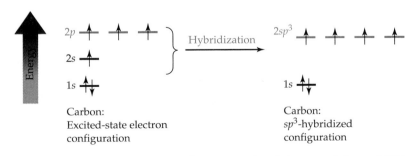

Carbon:
Excited-state electron
configuration

Carbon:
*sp*³-hybridized
configuration

Each of the four equivalent *sp*³ hybrid orbitals has two lobes of different phase (Section 5.8) like an atomic p orbital, but one of the lobes is larger than the other. The four large lobes are oriented toward the four corners of a tetra-hedron (Figure 7.6).

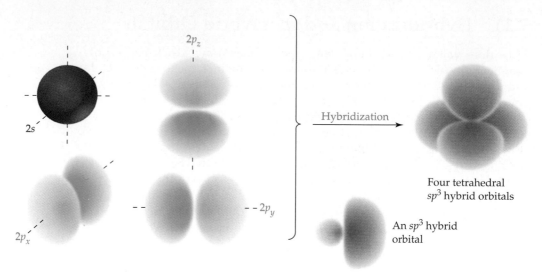

FIGURE 7.6 The formation of four sp^3 hybrid orbitals by combination of an atomic s orbital with three atomic p orbitals. Each sp^3 hybrid orbital has two lobes, one of which is larger than the other. The four large lobes are oriented toward the corners of a tetrahedron at angles of 109.5°.

The shared electrons in a covalent bond made with a strongly directed hybrid orbital spend most of their time in the region between the two bonded nuclei. As a result, covalent bonds made with sp^3 hybrid orbitals are often strong ones. In fact, the energy released on forming the four strong C–H bonds in CH_4 more than compensates for the energy required to produce the excited state of carbon. Figure 7.7 shows how the four C–H sigma bonds in methane can form by head-on overlap of carbon sp^3 hybrid orbitals with hydrogen $1s$ orbitals.

FIGURE 7.7 The bonding in methane. Each of the four C–H bonds results from head-on (σ) overlap of a singly occupied carbon sp^3 hybrid orbital with a singly occupied hydrogen $1s$ orbital.

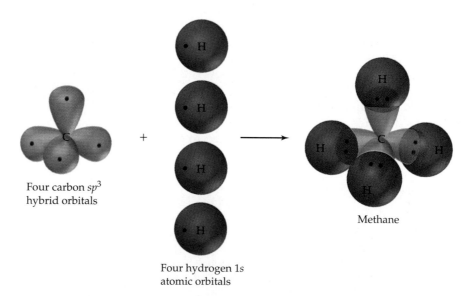

Four carbon sp^3 hybrid orbitals

Four hydrogen $1s$ atomic orbitals

Methane

The same kind of sp^3 hybridization used to describe bonds to carbon in the tetrahedral methane molecule can also be used to describe bonds to nitrogen in the trigonal pyramidal ammonia molecule, to oxygen in the bent water molecule, and to all other atoms that VSEPR theory predicts to have a tetrahedral arrangement of four charge clouds.

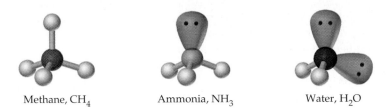

Methane, CH$_4$ Ammonia, NH$_3$ Water, H$_2$O

▶ **PROBLEM 7.19** Describe the bonding in ethane, C$_2$H$_6$. Tell what kinds of orbitals on each atom overlap to form the C–C and C–H bonds.

7.12 Other Kinds of Hybrid Orbitals

Each of the different geometries shown in Table 7.4—whether based on two, three, four, five, or six charge clouds—can be accounted for by a specific kind of orbital hybridization. Let's look at each.

sp² Hybridization

Atoms with three charge clouds undergo hybridization by combination of one atomic *s* orbital with two *p* orbitals, resulting in three *sp²* **hybrid orbitals**. These *sp²* hybrids lie in a plane and are oriented toward the corners of an equilateral triangle at angles of 120° to one another. One *p* orbital remains unchanged and is oriented at a 90° angle to the plane of the *sp²* hybrids, as shown in Figure 7.8.

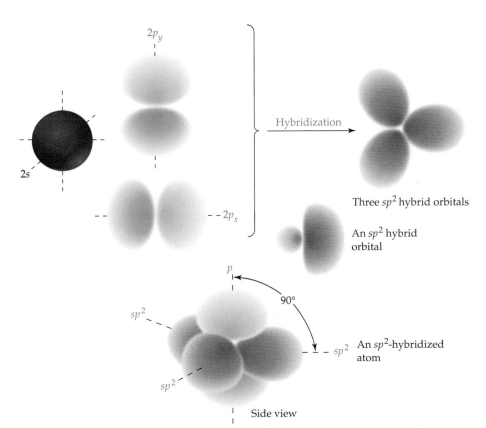

2p_y

Hybridization

2s

2p_x

Three *sp²* hybrid orbitals

An *sp²* hybrid orbital

p

90°

$sp²$

$sp²$

An *sp²*-hybridized atom

$sp²$

Side view

FIGURE 7.8 The formation of *sp²* hybrid orbitals by combination of one *s* orbital and two *p* orbitals. The three hybrids lie in a plane at angles of 120° to one another. One unhybridized *p* orbital remains, oriented at a 90° angle to the plane of the *sp²* orbitals.

The presence of the unhybridized p orbital on an sp^2-hybridized atom has some interesting consequences. Look, for example, at ethylene, $H_2C=CH_2$, a colorless gas used as starting material for the industrial preparation of poly-ethylene. Each carbon atom in ethylene has three charge clouds and is sp^2-hybridized. When two sp^2-hybridized carbon atoms approach each other with sp^2 orbitals aligned head-on for σ bonding, the unhybridized p orbitals on the carbons also approach each other and form a bond, but in a parallel, sideways manner rather than head-on. Such sideways bonding, in which the shared electrons occupy regions above and below a line connecting the nuclei rather than between the nuclei, is called a **pi (π) bond**.

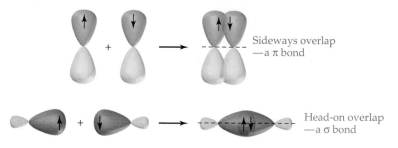

Note that the π bond has two regions of orbital overlap—one above and one below the internuclear axis. Both regions are part of the same bond, and the two shared electrons are spread over both regions. Note also that only lobes of the same phase can overlap properly for bond formation.

The net result of both σ and π overlap is the sharing of four electrons and the formation of a carbon–carbon double bond. The two remaining sp^2 orbitals on each carbon then overlap with hydrogen $1s$ orbitals to form four C–H bonds and complete the $H_2C=CH_2$ structure (Figure 7.9).

FIGURE 7.9 The structure of ethylene. The carbon–carbon double bond consists of one σ bond from the head-on overlap of sp^2 orbitals and one π bond from the sideways overlap of p orbitals. The four C–H σ bonds result from overlap of carbon sp^2 orbitals with hydrogen $1s$ orbitals. The overall shape of the molecule is planar (flat), with H–C–H and H–C–C bond angles of approximately 120°.

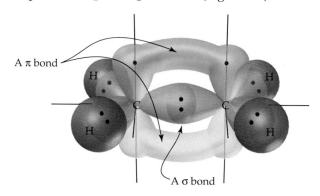

▶ **PROBLEM 7.20** Describe the hybridization of the carbon atom in formaldehyde, $H_2C=O$, and sketch the molecule, showing the orbitals involved in bonding.

sp Hybridization

Atoms with two charge clouds undergo hybridization by combination of one atomic s orbital with one p orbital, resulting in two **sp hybrid orbitals** that are oriented 180° from each other. Since only one p orbital is involved when an atom undergoes sp hybridization, the other two p orbitals are unchanged and are oriented at 90° angles to the sp hybrids, as shown in Figure 7.10.

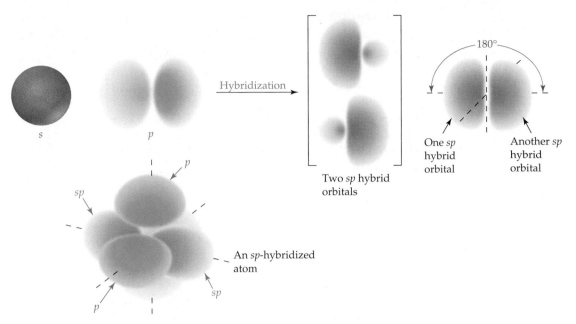

FIGURE 7.10 The combination of one *s* and one *p* orbital gives two *sp* hybrid orbitals oriented 180° apart. Two unhybridized *p* orbitals remain and are oriented at 90° angles to the *sp* hybrids.

A simple example of *sp* hybridization is found in acetylene, H−C≡C−H, a colorless gas used in welding. Both carbon atoms in the acetylene molecule have linear geometry and are *sp*-hybridized. When the two *sp*-hybridized carbon atoms approach each other with their *sp* orbitals aligned head-on for σ bonding, the unhybridized *p* orbitals on each carbon are aligned for π bonding. Two *p* orbitals are aligned in an up/down position, and two are aligned in an in/out position. Thus, there are *two* mutually perpendicular π bonds that form in acetylene by sideways overlap of *p* orbitals, along with one σ bond that forms by head-on overlap of the *sp* orbitals. The net result is the sharing of six electrons and formation of a triple bond (Figure 7.11). In addition, two C−H bonds form by overlap of the remaining two *sp* orbitals with hydrogen 1*s* orbitals.

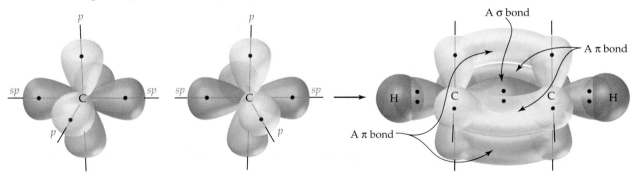

FIGURE 7.11 Formation of a triple bond by two *sp*-hybridized atoms. A σ bond forms by head-on overlap of *sp* orbitals, and two mutually perpendicular π bonds form by sideways overlap of *p* orbitals.

▶ **PROBLEM 7.21** Describe the hybridization of the carbon atom in the hydrogen cyanide molecule, H−C≡N, and sketch the hybrid orbitals it uses for bonding.

sp^3d Hybridization

Atoms with five charge clouds, such as the phosphorus in PCl_5, undergo hybridization by combination of five atomic orbitals. Since a given shell has a total of only four s and p orbitals, the need to use five orbitals implies that a d orbital must be involved. Thus, only atoms in the third or lower rows in the periodic table can form the necessary hybrids.

Hybridization of five atomic orbitals occurs by a combination of one s orbital, three p orbitals, and one d orbital, giving five sp^3d **hybrid orbitals** in a trigonal bipyramidal arrangement (Figure 7.12). Three of the hybrid orbitals lie in a plane at angles of 120°, with the remaining two orbitals perpendicular to the plane, one above and one below.

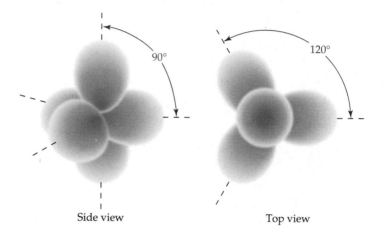

FIGURE 7.12 The five sp^3d hybrid orbitals and their trigonal bipyramidal geometry.

Side view   Top view

sp^3d^2 Hybridization

Atoms with six charge clouds, such as the sulfur in SF_6, undergo hybridization by combination of six atomic orbitals. This again implies that valence-shell d orbitals are involved and that only atoms in the third or lower rows in the periodic table can form the necessary hybrids. Hybridization occurs by a combination of one s orbital, three p orbitals, and two d orbitals, resulting in six sp^3d^2 **hybrid orbitals** with an octahedral arrangement (Figure 7.13). All six orbitals are equivalent, and the angle between any two adjacent orbitals is 90°.

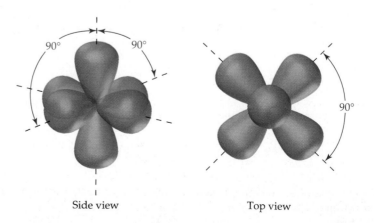

FIGURE 7.13 The six sp^3d^2 hybrid orbitals and their octahedral geometry.

Side view   Top view

A summary of the five most common kinds of hybridization and the geometry that each corresponds to is given in Table 7.5.

TABLE 7.5 Hybrid Orbitals and Their Geometry

Number of Charge Clouds	Geometry of Charge Clouds	Hybridization
2	Linear	sp
3	Trigonal planar	sp^2
4	Tetrahedral	sp^3
5	Trigonal bipyramidal	sp^3d
6	Octahedral	sp^3d^2

EXAMPLE 7.11 Describe the hybridization of the carbon atoms in allene, $H_2C{=}C{=}CH_2$, and sketch the molecule, showing its hybrid orbitals.

SOLUTION Draw an electron-dot structure to find the number of charge clouds on each atom:

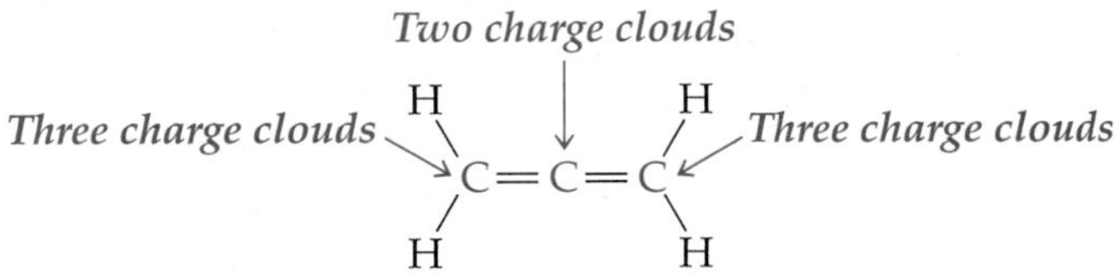

Two charge clouds

Three charge clouds $C{=}C{=}C$ *Three charge clouds*

Because the central carbon atom in allene has two charge clouds (two double bonds), it has a linear geometry and is sp-hybridized. Because the two terminal carbon atoms have three charge clouds each (one double bond and two C−H bonds), they have trigonal planar geometry and are sp^2-hybridized. The central carbon uses its sp orbitals to form two σ bonds at 180° angles and uses its two unhybridized p orbitals to form π bonds, one to each of the terminal carbons. Each terminal carbon atom uses an sp^2 orbital for σ bonding to carbon, a p orbital for π bonding, and its two remaining sp^2 orbitals for C−H bonds.

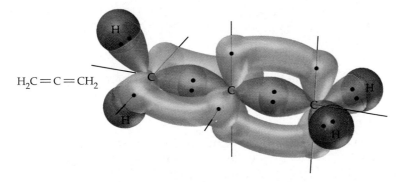

$H_2C{=}C{=}CH_2$

(The carbon orbitals shown in blue are sp and sp^2 hybrids; those shown in green are unhybridized p orbitals.)

▶ **PROBLEM 7.22** Describe the hybridization of the central iodine atom in I_3^-, and sketch the ion, showing the orbitals involved in bonding.

▶ **PROBLEM 7.23** Describe the hybridization of the sulfur atom in SF_2, SF_4, and SF_6 molecules.

KEY CONCEPT PROBLEM 7.24 Identify each of the following sets of hybrid orbitals:

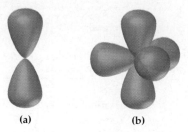

(a)   (b)

7.13 Molecular Orbital Theory: The Hydrogen Molecule

The valence bond model of covalent bonding is easy to visualize and leads to a satisfactory description for most molecules. It does, however, have some problems. Perhaps the most serious flaw in the valence bond model is that it sometimes leads to an incorrect electronic description. For this reason, another bonding description called the **molecular orbital (MO) theory** is often used. The molecular orbital model is more complex than the valence bond model, particularly for larger molecules, but sometimes gives a more satisfactory accounting of chemical and physical properties.

To introduce some of the basic ideas of molecular orbital theory, let's look again at orbitals. The concept of orbitals derives from the quantum mechanical wave equation, in which the square of the wave function gives the probability of finding an electron within a given region of space. The kinds of orbitals that we've been concerned with thus far are called *atomic orbitals* because they are characteristic of individual atoms. Atomic orbitals on the same atom can combine to form hybrids, and atomic orbitals on different atoms can overlap to form covalent bonds, but the orbitals and the electrons in them remain localized on specific atoms.

ATOMIC ORBITAL   A wave function whose square gives the probability of finding an electron within a given region of space *in an atom*.

Molecular orbital theory takes a different approach to bonding by considering the molecule as a whole rather than concentrating on individual atoms. Thus, a *molecular* orbital is to a *molecule* what an *atomic* orbital is to an *atom*.

MOLECULAR ORBITAL   A wave function whose square gives the probability of finding an electron within a given region of space *in a molecule*.

Like atomic orbitals, molecular orbitals have specific energy levels and specific shapes, and they can be occupied by a maximum of two electrons with opposite spins. The energy and shape of a molecular orbital depend on the size and complexity of the molecule and can thus be fairly complicated, but the fundamental analogy between atomic and molecular orbitals remains.

Let's look at the molecular orbital description of the simple diatomic molecule H_2 to see some general features of MO theory. Imagine what might happen when two isolated hydrogen atoms approach each other and begin to interact. The $1s$ orbitals begin to blend together, and the electrons spread out over both atoms. Molecular orbital theory says that there are two ways for the orbital interaction to occur—an additive way and a subtractive way. The additive interaction leads to formation of a molecular orbital that is roughly egg-shaped, whereas the subtractive interaction leads to formation of a molecular orbital that contains a node between atoms (Figure 7.14).

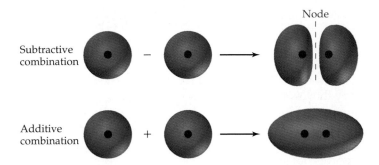

FIGURE 7.14   Formation of molecular orbitals in the H_2 molecule. The additive combination of two atomic $1s$ orbitals leads to formation of a lower-energy, bonding molecular orbital. The subtractive combination leads to formation of a higher-energy, antibonding molecular orbital that has a node between the nuclei.

The additive combination, denoted σ, is lower in energy than the two isolated $1s$ orbitals and is called a **bonding molecular orbital** because any electrons it contains spend most of their time in the region between the two nuclei, helping to bond the atoms together. The subtractive combination, denoted σ^* (spoken as "sigma star"), is higher in energy than the two isolated $1s$ orbitals and is called an **antibonding molecular orbital** because any electrons it contains can't occupy the central region between the nuclei and can't contribute to bonding. Diagrams of the sort shown in Figure 7.15 are used to show the energy relationships of the various orbitals. The two isolated H atomic orbitals are shown on either side, and the two H_2 molecular orbitals are shown in the middle. Each of the starting hydrogen atomic orbitals has one electron, which pair up and occupy the lower-energy bonding MO after covalent bond formation.

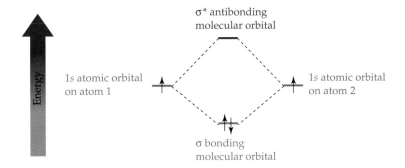

FIGURE 7.15   A molecular orbital diagram for the H_2 molecule. The two electrons are paired in the bonding σ MO, and the antibonding σ^* MO is vacant.

Similar MO diagrams can be drawn, and predictions about stability can be made, for related diatomic species such as H_2^- and He_2. For example, we might imagine constructing the H_2^- ion by bringing together a neutral H· atom with one electron and an H:$^-$ anion with two electrons. Since the resultant H_2^- ion has three electrons, two of them will occupy the lower-energy bonding σ MO and one will occupy the higher-energy antibonding σ^* MO (Figure 7.16a). Two electrons are lowered in energy while only one electron is raised in energy, so a net gain in stability results. We therefore predict (and find experimentally) that the H_2^- ion is stable.

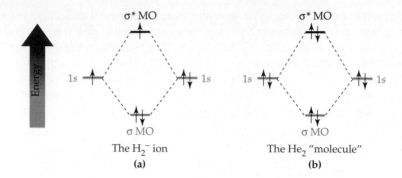

FIGURE 7.16 MO diagrams for **(a)** the stable H_2^- ion and **(b)** the unstable He_2 molecule.

What about He_2? A hypothetical He_2 molecule has four electrons, two of which occupy the lower-energy bonding orbital and two of which occupy the higher-energy antibonding orbital, as shown in Figure 7.16b. Since the decrease in energy for the two bonding electrons is counteracted by the increase in energy for the two antibonding electrons, the He_2 molecule has no net bonding energy and is not stable.

Note that bond orders—the number of electron pairs shared between atoms (Section 7.5)—are easy to calculate from MO diagrams by subtracting the number of antibonding electrons from the number of bonding electrons and dividing by 2:

$$\text{Bond order} = \frac{\left(\begin{array}{c}\text{Number of}\\\text{bonding electrons}\end{array}\right) - \left(\begin{array}{c}\text{Number of}\\\text{antibonding electrons}\end{array}\right)}{2}$$

The H_2 molecule, for example, has a bond order of 1 because it has two bonding electrons and no antibonding electrons. Similarly, the H_2^- ion has a bond order of $1/2$, and the hypothetical He_2 molecule has a bond order of 0, which accounts for the instability of He_2.

The key ideas of the molecular orbital theory of bonding can be summarized as follows:

Key Ideas of Molecular Orbital Theory

- Molecular orbitals are to molecules what atomic orbitals are to atoms. Molecular orbitals describe regions of space in a molecule where electrons are most likely to be found, and they have a specific size, shape, and energy level.
- Molecular orbitals are formed by combining atomic orbitals on different atoms. The number of molecular orbitals formed is the same as the number of atomic orbitals combined.
- Molecular orbitals that are lower in energy than the starting atomic orbitals are bonding; MOs that are higher in energy than the starting atomic orbitals are antibonding.
- Electrons occupy molecular orbitals beginning with the MO of lowest energy. Only two electrons occupy each orbital, and their spins are paired.
- Bond order can be calculated by subtracting the number of electrons in antibonding MOs from the number in bonding MOs and dividing by 2.

▶ **PROBLEM 7.25** Construct an MO diagram for the He_2^+ ion. Is the ion likely to be stable? What is its bond order?

7.14 Molecular Orbital Theory: Other Diatomic Molecules

Now that we've looked at bonding in the H_2 molecule, let's move up a level in complexity and look at the bonding in several second-row diatomic molecules—N_2, O_2, and F_2. The valence bond model developed in Section 7.10 predicts that the nitrogen atoms in N_2 are triply bonded and have one lone pair each, that the oxygen atoms in O_2 are doubly bonded and have two lone pairs each, and that the fluorine atoms in F_2 are singly bonded and have three lone pairs each:

Valence bond theory predicts:

:N≡N:

Ö=Ö

:F̈—F̈:

1 σ bond
and 2 π bonds

1 σ bond
and 1 π bond

1 σ bond

Unfortunately, this simple valence bond picture can't be right because it predicts that the electrons in all three molecules are *spin-paired*. In other words, electron-dot structures indicate that all the occupied atomic orbitals in the three molecules contain two electrons each. It's easy to demonstrate experimentally, however, that the O_2 molecule has two electrons that are *not* spin-paired and that these electrons therefore must be in different, singly occupied orbitals.

Experimental evidence for the electronic structure of O_2 rests on the observation that substances with unpaired electrons are attracted by magnetic fields and are thus said to be **paramagnetic**. The more unpaired electrons the substance has, the stronger the paramagnetic attraction. Substances whose electrons are all spin-paired, by contrast, are weakly repelled by magnetic fields and are said to be **diamagnetic**. Both N_2 and F_2 are diamagnetic, just as predicted by their electron-dot structures, but O_2 is paramagnetic. When liquid O_2 is poured over the poles of a strong magnet, the O_2 sticks to the poles, as shown in Figure 7.17.

FIGURE 7.17 Why does liquid O_2 stick to the poles of a magnet?

Why is O_2 paramagnetic? Although electron-dot structures and valence bond theory fail in their descriptions, MO theory explains the experimental results nicely. In a molecular orbital description of N_2, O_2, and F_2, two atoms come together and their atomic orbitals interact to form molecular orbitals. Four orbital interactions occur, leading to the formation of four bonding MOs and four antibonding MOs, whose relative energies are shown in Figure 7.18. (Note that the relative energies of the σ_{2p} and π_{2p} orbitals in N_2 are different from those in O_2 and F_2.)

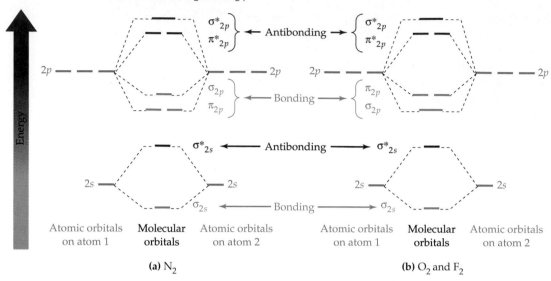

FIGURE 7.18 Molecular orbital energy diagrams for **(a)** N_2 and **(b)** O_2 and F_2. There are eight MOs, four bonding and four antibonding. The two diagrams differ only in the relative energies of the σ_{2p} and π_{2p} orbitals.

The diagrams in Figure 7.18 show the following orbital interactions:

- The $2s$ orbitals interact to give σ_{2s} and σ^*_{2s} MOs.
- The two $2p$ orbitals that lie on the internuclear axis interact head-on to give σ_{2p} and σ^*_{2p} MOs.
- The two remaining pairs of $2p$ orbitals that are perpendicular to the internuclear axis interact in a sideways manner to give two degenerate π_{2p} and two degenerate π^*_{2p} MOs oriented 90° apart. (Recall from Section 5.12 that *degenerate* orbitals have the same energy.)

The shapes of the σ_{2p}, σ^*_{2p}, π_{2p}, and π^*_{2p} MOs are shown in Figure 7.19.

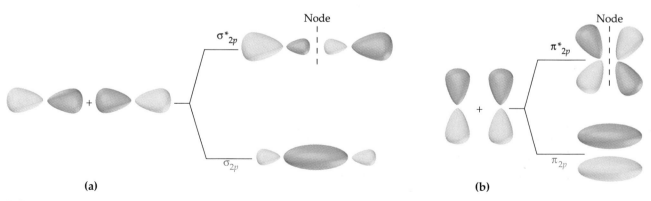

FIGURE 7.19 Formation of **(a)** σ_{2p} and σ^*_{2p} MOs by head-on interaction of two p atomic orbitals, and **(b)** π_{2p} and π^*_{2p} MOs by sideways interaction. Note that the bonding MOs result from interaction of lobes with like phase, while antibonding MOs result from interaction of lobes of opposite phase (Section 5.8). In each case, the bonding MO concentrates electron density between atomic nuclei, whereas the antibonding MO has a node between nuclei.

When appropriate numbers of electrons are added to occupy the molecular orbitals, the results shown in Figure 7.20 are obtained. Both N_2 and F_2 have all their electrons spin-paired, but O_2 has two unpaired electrons in the degenerate π^*_{2p} orbitals. Both N_2 and F_2 are therefore diamagnetic, while O_2 is paramagnetic.

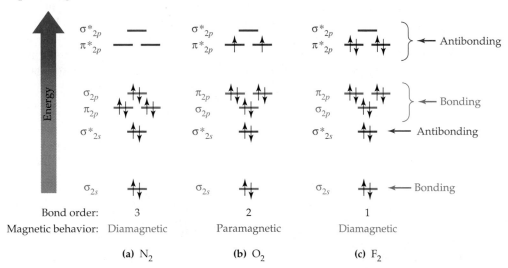

FIGURE 7.20   Molecular orbital diagrams for the second-row diatomic molecules **(a)** N_2, **(b)** O_2, and **(c)** F_2. The O_2 molecule has two unpaired electrons in its two degenerate π^*_{2p} orbitals and is therefore paramagnetic.

It should be pointed out that MO diagrams like that in Figure 7.20 require experience and (often) mathematical calculation to obtain. MO theory is therefore less easy to visualize and understand on an intuitive level than valence bond theory.

▶ **PROBLEM 7.26**   The B_2 and C_2 molecules have MO diagrams similar to that of N_2 in Figure 7.18a. What MOs are occupied in B_2 and C_2, and what is the bond order in each? Would you expect either of these substances to be paramagnetic?

7.15   Combining Valence Bond Theory and Molecular Orbital Theory

Whenever two different theories are used to explain the same concept, the question comes up: Which theory is better? This question isn't easy to answer, because it depends on what is meant by "better." Valence bond theory is better because of its simplicity, but MO theory is better because of its accuracy. Best of all, though, is a blend of the two theories that combines the strengths of both.

Valence bond theory has two main problems: (1) For molecules such as O_2, valence bond theory makes an incorrect prediction about electronic structure. (2) For molecules such as O_3, no single structure is adequate and the concept of resonance involving two or more structures must be added (Section 7.7). The first problem occurs rarely, but the second is much more common. To better deal with resonance, chemists often use a *combination* of bonding theories in which σ bonds are described by valence bond theory and π bonds are described by MO theory.

Take ozone, for instance. Valence bond theory says that ozone is a resonance hybrid of two equivalent structures, both of which have two O—O σ bonds and one O=O π bond (Section 7.7). One structure has a lone pair of electrons in the *p* orbital on the left-hand oxygen atom and has a π bond to the right-hand oxygen. The other structure has a lone pair of electrons in the *p* orbital on the right-hand oxygen and a π bond to the left-hand oxygen. The actual structure of O_3 is an average of the two resonance forms in which four electrons occupy the entire region encompassed by the overlapping set of three *p* orbitals. The only difference between the resonance structures is in the placement of *p* electrons. The atoms themselves are in the same positions in both, and the geometries are the same in both (Figure 7.21).

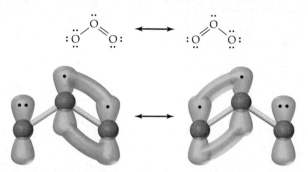

FIGURE 7.21 The ozone molecule is a hybrid of two resonance forms that differ only in the location of *p* electrons. The nuclei and the σ-bond electrons are in the same position in both resonance structures.

Valence bond theory thus gives a good description of the O—O σ bonds but a poor description of the π bonding among *p* atomic orbitals, whose four electrons are spread out, or *delocalized*, over the molecule. Yet this is exactly what MO theory does best—describe bonds in which electrons are delocalized over a molecule. Thus, a combination of valence bond theory and MO theory is used. The σ bonds are best described in valence bond terminology as being localized between pairs of atoms, and the π electrons are best described by MO theory as being delocalized over the entire molecule.

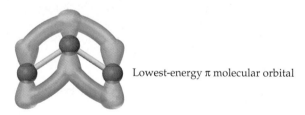

Lowest-energy π molecular orbital

▶ **PROBLEM 7.27** Draw two resonance structures for the formate ion, HCO_2^-, and sketch a π molecular orbital showing how the π electrons are delocalized over both oxygen atoms.

Molecular Shape, Handedness, and Drugs

Why does a right glove fit only on your right hand and not on your left hand? Why do the threads on a light bulb twist only in one direction so that you have to turn the bulb clockwise to screw it in? The reason has to do with the shapes of the glove and the light bulb and the fact that both have a *handedness* to them. When the right-handed glove is held up to a mirror, the reflected image looks like a left-handed glove. (Try it.) When the light bulb with clockwise threads is reflected in a mirror, the threads in the mirror image twist in a counterclockwise direction.

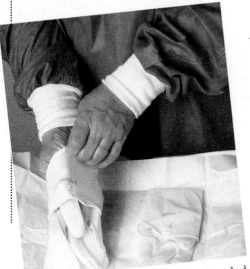

A right hand fits only into a right-handed glove with a complementary shape, not into a left-handed glove.

Molecules too can have shapes that give them a handedness and can thus exist in mirror-image forms, one right-handed and one left-handed. Take, for example, the main classes of biomolecules found in living organisms: carbohydrates (sugars), proteins, fats, and nucleic acids. These and most other biomolecules are handed, and usually only one of the two possible mirror-image forms occurs naturally in a given organism. The other form can often be made in the laboratory but does not occur naturally.

The biological consequences of molecular shape can be dramatic. Look at the structures of dextromethorphan and levomethorphan, for instance. (The Latin prefixes *dextro-* and *levo-* mean "right" and "left," respectively.) One form, dextromethorphan, is a common cough suppressant found in many over-the-counter cold medicines; its mirror-image, levomethorphan, is a powerful narcotic pain-reliever (analgesic) similar in its effects to morphine. The two substances are chemically identical except for their shapes, yet their biological properties are completely different.

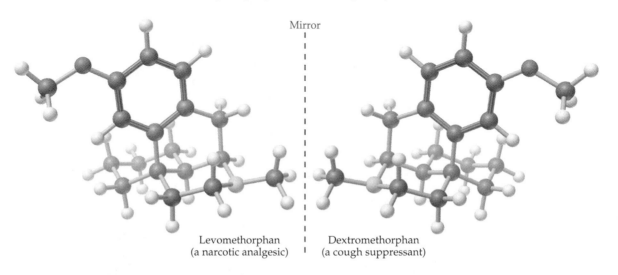

Mirror

Levomethorphan
(a narcotic analgesic)

Dextromethorphan
(a cough suppressant)

The gray spheres in these computer-generated molecular structures represent carbon atoms, the ivory spheres represent hydrogen, the red spheres represent oxygen, and the blue spheres represent nitrogen.

As another example of the effects of shape and molecular handedness, look at the substance called *carvone*. One form of carvone occurs in mint plants and has the characteristic odor of spearmint, while the other form occurs in several herbs and has the odor of caraway seeds. Again, the two structures are the same except for their shapes, yet they have entirely different odors.

Mirror

"Left-handed" carvone
(odor of spearmint)

"Right-handed" carvone
(odor of caraway)

Two plants both produce carvone, but the mint leaves yield the "left-handed" form, while the caraway seeds yield the "right-handed" form.

Why do different mirror-image forms of molecules have different biological properties? The answer goes back to the question about why a right-hand glove fits only on the right hand: A right hand in a right-hand glove is a perfect match because the two shapes are complementary. Putting the same right hand into a left-hand glove produces a mismatch because the two shapes are *not* complementary. In the same way, handed molecules such as dextromethorphan and carvone have specific shapes that match only complementary-shaped receptor sites in the body. The mirror-image forms of the molecules can't fit into the receptor sites and thus don't elicit the same biological response.

Precise molecular shape is of crucial importance to every living organism. Almost every chemical interaction in living systems is governed by complementarity between handed molecules and their receptors.

▶ **PROBLEM 7.28** Why is molecular shape so important in biological chemistry?

◂— **KEY CONCEPT PROBLEM 7.29** One of the following two molecules has a handedness to it and can exist in two mirror-image forms; the other does not. Which is which? Why?

(a)   (b)

Summary

Covalent bonds result from the sharing of electrons between atoms. Every covalent bond has a specific **bond length** that leads to optimum stability and a specific **bond dissociation energy** that describes the strength of the bond. Energy is released when bonds are formed and is absorbed when bonds are broken. As a general rule, an atom will share as many of its valence-shell electrons as possible, either until it has no more to share or until it reaches an octet. Atoms in the third and lower rows of the periodic table can expand their valence shells beyond the normal octet by using *d* orbitals in covalent bond formation.

Electron-dot structures represent an atom's valence electrons by dots and show the two electrons in a **single bond** as a pair of dots shared between atoms or as a single line. In the same way, a **double bond** is represented as four dots or two lines between atoms, and a **triple bond** is represented as six dots or three lines between atoms. Occasionally, a molecule can be represented by more than one satisfactory electron-dot structure. In such cases, no single structure is adequate by itself. The actual electronic structure of the molecule is said to be a **resonance hybrid** of the different individual structures. A double-headed arrow is used to indicate resonance:

$$:\ddot{O}-\ddot{O}=\ddot{O} \longleftrightarrow \ddot{O}=\ddot{O}-\ddot{O}:$$

In a bond between dissimilar atoms, such as that in HCl, one atom often attracts the bonding electrons more strongly than the other, giving rise to a **polar covalent bond**. Bond polarity is due to differences in **electronegativity**, the ability of an atom in a molecule to attract shared electrons. Electronegativity increases from left to right across a row and generally decreases from top to bottom in a group of the periodic table. Fluorine is the most electronegative element; cesium the least.

Molecular shape can often be predicted by the **valence-shell electron-pair repulsion (VSEPR) model**, which treats the electrons around atoms as charge clouds that repel one another and therefore orient themselves as far away from one another as possible. Atoms with two charge clouds adopt a linear arrangement of charge clouds, atoms with three charge clouds adopt a trigonal planar arrangement, and atoms with four charge clouds adopt a tetrahedral arrangement. Similarly, atoms with five charge clouds are trigonal bipyramidal, and atoms with six charge clouds are octahedral.

According to **valence bond theory**, covalent bond formation occurs by the overlap of singly occupied atomic orbitals, either along the internuclear axis to form a **σ bond** or above and below the internuclear axis to form a **π bond**. The observed geometry of covalent bonding is described by assuming that *s*, *p*, and *d* atomic orbitals combine to generate **hybrid orbitals**, which are strongly oriented in specific directions: *sp* **hybrid orbitals** have linear geometry, *sp*² **hybrid orbitals** have trigonal planar geometry, *sp*³ **hybrid orbitals** have tetrahedral geometry, *sp*³*d* **hybrid orbitals** have trigonal bipyramidal geometry, and *sp*³*d*² **hybrid orbitals** have octahedral geometry.

Molecular orbital theory sometimes gives a more accurate picture of electronic structure than the valence bond model. Formed from the mathematical combination of atomic orbitals on different atoms, **molecular orbitals** are wave functions whose square gives the probability of finding an electron in a given region of space in a molecule. Combination of two atomic orbitals gives two molecular orbitals, a **bonding MO** that is lower in energy than the starting atomic orbitals and an **antibonding MO** that is higher in energy than the starting atomic orbitals. Molecular orbital theory is particularly useful for describing delocalized π bonding in molecules.

Key Words

antibonding molecular orbital *281*
atomic orbital *280*
bond angle *264*
bond dissociation energy *246*
bond length *245*
bond order *252*
bonding molecular orbital *281*
bonding pair *251*
coordinate covalent bond *253*
covalent bond *244*
diamagnetic *283*
double bond *252*
electron-dot structure *250*
electronegativity (EN) *249*
formal charge *261*
hybrid atomic orbital *273*
lone pair *251*
molecular orbital *280*
molecular orbital theory (MO) *280*
paramagnetic *283*
pi (π) bond *276*
polar covalent bond *248*
resonance hybrid *259*
sigma (σ) bond *272*
sp hybrid orbital *276*
*sp*² hybrid orbital *275*
*sp*³ hybrid orbital *273*
*sp*³*d* hybrid orbital *278*
*sp*³*d*² hybrid orbital *278*
triple bond *252*
valence bond theory *272*
valence-shell electron-pair repulsion (VSEPR) model *264*

Key Concept Summary

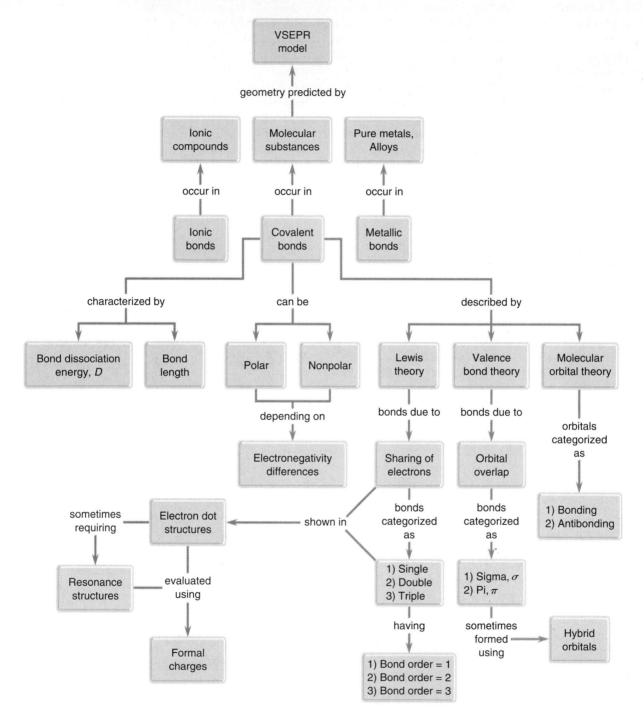

Understanding Key Concepts

Problems 7.1–7.29 appear within the chapter.

7.30 What is the geometry around the central atom in each of the following molecular models?

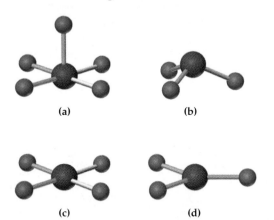

(a)   (b)

(c)   (d)

7.31 What is the geometry around the central atom in each of the following molecular models? (There may be a "hidden" atom in some cases.)

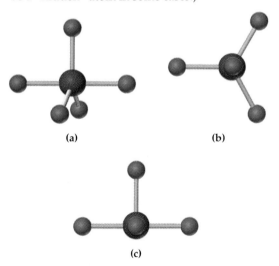

(a)   (b)

(c)

7.32 Three of the following molecular models have a tetrahedral central atom, and one does not. Which is the odd one? (There may be a "hidden" atom in some cases.)

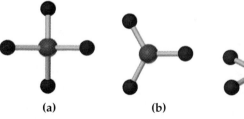

(a)   (b)   (c)   (d)

7.33 Identify each of the following sets of hybrid orbitals:

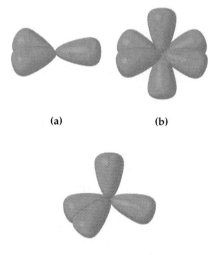

(a)   (b)

(c)

7.34 The following ball-and-stick molecular model is a representation of acetaminophen, the active ingredient in such over-the-counter headache remedies as Tylenol (red = O, gray = C, blue = N, ivory = H):

(a) What is the molecular formula of acetaminophen?

(b) What is the geometry around each carbon and nitrogen? (The lines between atoms indicate connections only, not whether the bonds are single, double, or triple.)

(c) What is the hybridization of each carbon and nitrogen?

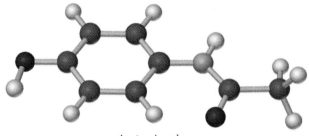

Acetaminophen

7.35 The atoms in the amino acid glycine are connected as shown:

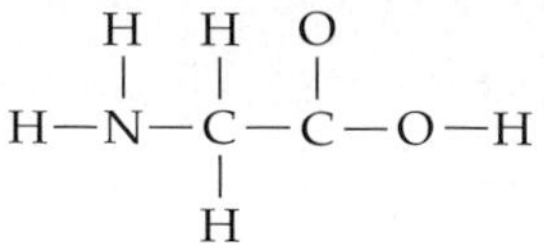

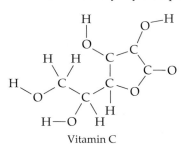

(a) Draw an electron-dot structure for glycine.

(b) Predict approximate values for the H—C—H, O—C—O, and H—N—H bond angles.

(c) Which hybrid orbitals are used by the C and N atoms?

7.36 Vitamin C (ascorbic acid) has the following connections among atoms. Complete the Lewis electron-dot structure for vitamin C, and identify any multiple bonds.

Vitamin C

7.37 The following ball-and-stick molecular model is a representation of thalidomide, a drug that causes birth defects when taken by expectant mothers but that is valuable for its use against leprosy. The lines indicate only the connections between atoms, not whether the bonds are single, double, or triple (red = O, gray = C, blue = N, ivory = H).

(a) What is the molecular formula of thalidomide?

(b) Indicate the positions of the multiple bonds in thalidomide.

(c) What is the geometry around each carbon and each nitrogen?

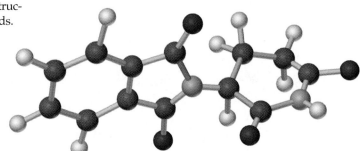

Additional Problems

Electronegativity and Polar Covalent Bonds

7.38 What general trends in electronegativity occur in the periodic table?

7.39 Predict the electronegativity of the undiscovered element with $Z = 119$.

7.40 Order the following elements according to increasing electronegativity: Li, Br, Pb, K, Mg, C.

7.41 Order the following elements according to decreasing electronegativity: C, Ca, Cs, Cl, Cu.

7.42 Which of the following substances are largely ionic and which are covalent?

    (a) HF   **(b)** HI   **(c)** $PdCl_2$

    (d) BBr_3 **(e)** NaOH

7.43 Use the electronegativity data in Figure 7.4 to predict which bond in each of the following pairs is more polar:

    (a) C—H or C—Cl

    (b) Si—Li or Si—Cl

    (c) N—Cl or N—Mg

7.44 Show the direction of polarity for each of the bonds in Problem 7.43 using the δ+/δ− notation.

7.45 Show the direction of polarity for each of the covalent bonds in Problem 7.42 using the δ+/δ− notation.

Electron-Dot Structures and Resonance

7.46 What is the octet rule, and why does it apply primarily to main-group elements, not to transition metals?

7.47 Which of the following substances contains an atom that does not follow the octet rule?
(a) $AlCl_3$ (b) PCl_3
(c) PCl_5 (d) $SiCl_4$

7.48 Draw electron-dot structures for the following molecules or ions:
(a) CBr_4 (b) NCl_3 (c) C_2H_5Cl
(d) BF_4^- (e) O_2^{2-} (f) NO^+

7.49 Draw electron-dot structures for the following molecules, which contain atoms from the third row or lower:
(a) $SbCl_3$ (b) KrF_2 (c) ClO_2
(d) PF_5 (e) H_3PO_4

7.50 Draw as many resonance structures as you can for each of the following molecules or ions:
(a) HN_3 (b) SO_3 (c) SCN^-

7.51 Draw as many resonance structures as you can for the following nitrogen-containing compounds:
(a) N_2O (b) NO
(c) NO_2 (d) N_2O_3 ($ONNO_2$)

7.52 Oxalic acid, $H_2C_2O_4$, is a poisonous substance found in uncooked spinach leaves. If oxalic acid has a C–C single bond and no C–H bond, draw its electron-dot structure.

7.53 Draw an electron-dot structure for carbon disulfide, CS_2. How many double bonds does CS_2 have?

7.54 Which of the following pairs of structures represent resonance forms, and which do not?

(a)  $H-C\equiv N-\ddot{\underset{..}{O}}:$ and $H-C=\ddot{N}-\ddot{\underset{..}{O}}:$

(b) 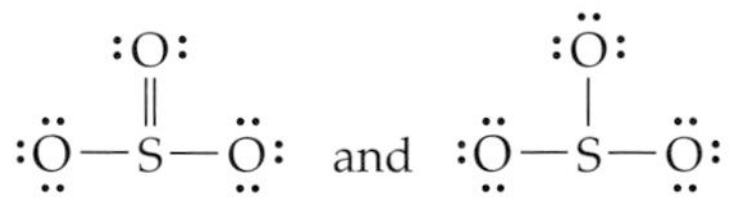 :$\ddot{\underset{..}{O}}-\overset{\overset{:O:}{\|}}{S}-\ddot{\underset{..}{O}}:$ and $:\ddot{\underset{..}{O}}-\overset{\overset{:\ddot{O}:}{\|}}{S}-\ddot{\underset{..}{O}}:$

(c)

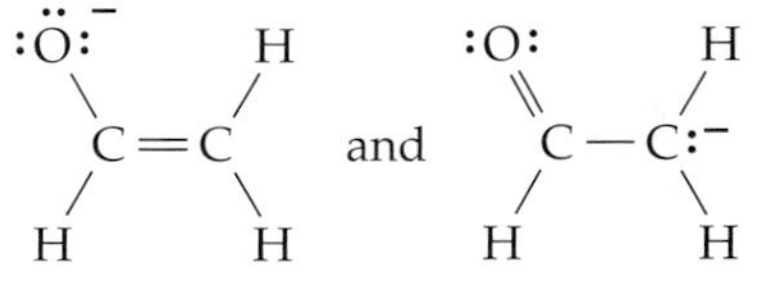

(d)

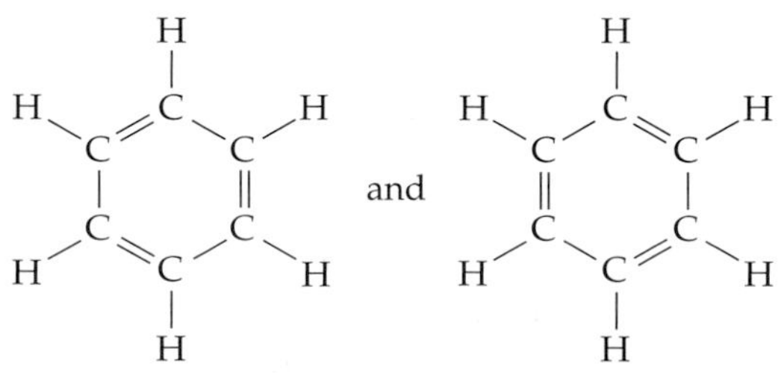

7.55 Which of the following pairs of structures represent resonance forms, and which do not?

(a)

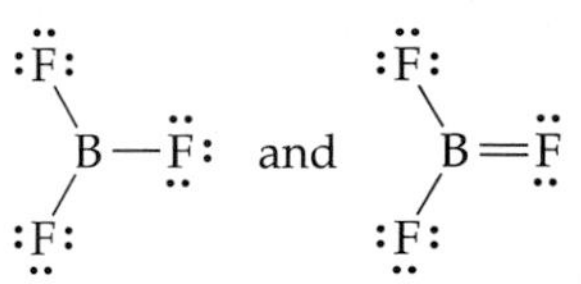

(b)

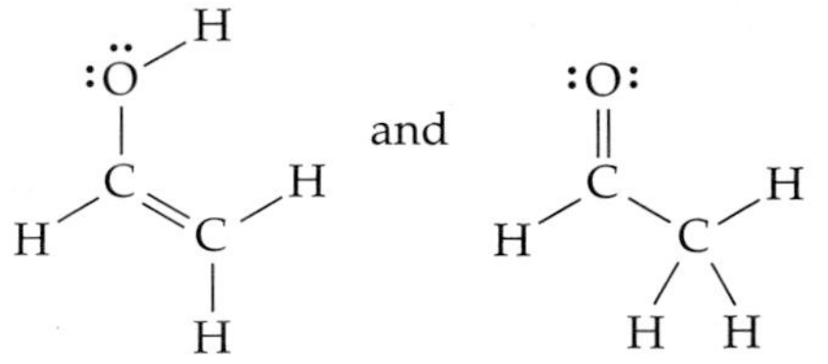

(c)

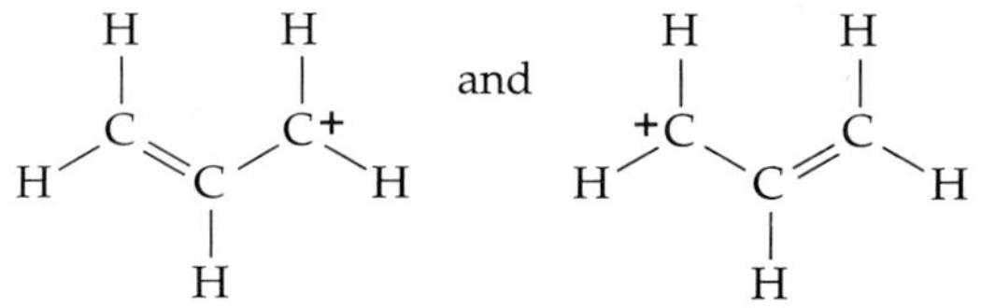

7.56 Identify the third-row elements, X, that form the following ions:

(a) 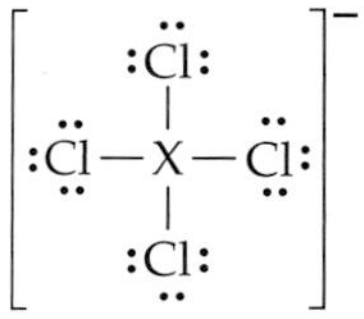 (b)

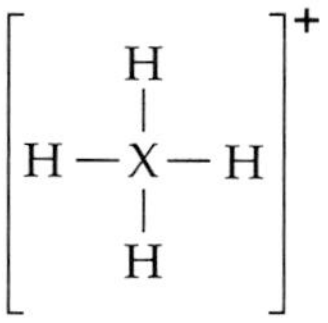

7.57 Identify the fourth-row elements, X, that form the following compounds:

(a) 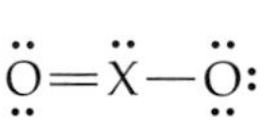 $\ddot{\underset{..}{O}}=\ddot{X}-\ddot{\underset{..}{O}}:$ (b)

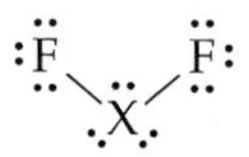

7.58 Draw electron-dot structures for molecules with the following connections:

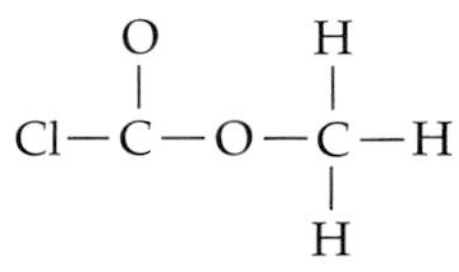

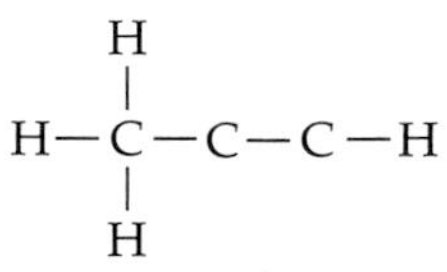

7.59 Draw electron-dot structures for molecules with the following connections:

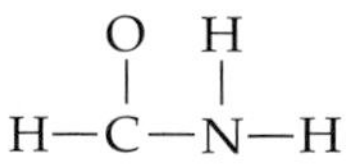

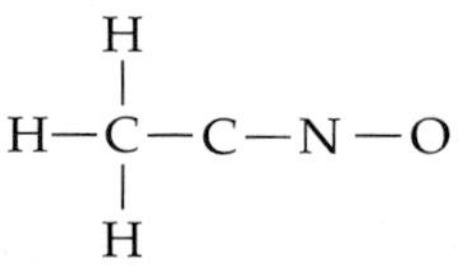

Formal Charges

7.60 Draw an electron-dot structure for carbon monoxide, CO, and assign formal charges to both atoms.

7.61 Assign formal charges to the atoms in the following structures:

(a)

$$H-\overset{\overset{\displaystyle H}{|}}{N}-\overset{..}{\overset{..}{O}}-H$$

(b)

$$\left[H-\overset{..}{N}-\overset{\overset{\displaystyle H}{|}}{\underset{\underset{\displaystyle H}{|}}{C}}-H \right]^{-}$$

(c)

$$:\overset{..}{Cl}-\overset{\overset{\displaystyle :\overset{..}{O}:}{|}}{\underset{\underset{\displaystyle :\overset{..}{Cl}:}{|}}{P}}-\overset{..}{Cl}:$$

7.62 Assign formal charges to the atoms in the following resonance forms of ClO_2^-:

$$\left[:\overset{..}{\underset{..}{O}}-\overset{..}{\underset{..}{Cl}}-\overset{..}{\underset{..}{O}}:\right]^{-} \longleftrightarrow \left[:\overset{..}{\underset{..}{O}}-\overset{..}{\underset{..}{Cl}}=\overset{..}{\underset{..}{O}}:\right]^{-}$$

7.63 Assign formal charges to the atoms in the following resonance forms of H_2SO_3:

$$H\overset{..}{\underset{..}{O}}-\overset{\overset{\displaystyle :O:}{||}}{S}-\overset{..}{\underset{..}{O}}H \longleftrightarrow H\overset{..}{\underset{..}{O}}-\overset{\overset{\displaystyle :\overset{..}{O}:}{|}}{S}-\overset{..}{\underset{..}{O}}H$$

7.64 Assign formal charges to the atoms in the following structures. Which of the two do you think is the more important contributor to the resonance hybrid?

(a)

$$\overset{\overset{\displaystyle H}{\diagdown}}{\underset{\underset{\displaystyle H}{\diagup}}{C}}=N=\overset{..}{N}:$$

(b)

$$\overset{\overset{\displaystyle H}{\diagdown}}{\underset{\underset{\displaystyle H}{\diagup}}{C}}-\overset{..}{N}=\overset{..}{N}:$$

7.65 Calculate formal charges for the C and O atoms in the following structures. Which structure do you think is the more important contributor to the resonance hybrid? Explain.

$$\left[H\overset{\diagup}{\underset{\diagdown}{}}\overset{\overset{\displaystyle :O:}{||}}{\underset{\underset{\displaystyle \overset{\displaystyle ..}{C}}{|}}{C}}\overset{\diagup}{\underset{\diagdown}{}}H \right]^{-} \longleftrightarrow \left[H\overset{\diagup}{\underset{\diagdown}{}}\overset{\overset{\displaystyle :\overset{..}{O}:}{|}}{\underset{\underset{\displaystyle H}{}}{C}}=\overset{}{\underset{}{C}}\overset{\diagup}{\underset{\diagdown}{}}H \right]^{-}$$

The VSEPR Model

7.66 What geometric arrangement of charge clouds do you expect for atoms that have the following number of charge clouds?

(a) 3   (b) 5   (c) 2   (d) 6

7.67 What shape do you expect for molecules that meet the following descriptions?

(a) A central atom with two lone pairs and three bonds to other atoms

(b) A central atom with two lone pairs and two bonds to other atoms

(c) A central atom with two lone pairs and four bonds to other atoms

7.68 How many charge clouds are there around the central atom in molecules that have the following geometry?

(a) Tetrahedral   (b) Octahedral

(c) Bent   (d) Linear

(e) Square pyramidal   (f) Trigonal pyramidal

7.69 How many charge clouds are there around the central atom in molecules that have the following geometry?

(a) See-saw   (b) Square planar

(c) Trigonal bipyramidal (d) T-shaped

(e) Trigonal planar   (f) Linear

7.70 What shape do you expect for each of the following molecules?

(a) H_2Se (b) $TiCl_4$ (c) O_3 (d) GaH_3

7.71 What shape do you expect for each of the following molecules?

(a) XeO_4 (b) SO_2Cl_2 (c) OsO_4 (d) SeO_2

7.72 What shape do you expect for each of the following molecules or ions?

(a) SbF_5 (b) IF_4^+ (c) SeO_3^{2-} (d) CrO_4^{2-}

7.73 Predict the shape of each of the following ions:

(a) NO_3^- (b) NO_2^+ (c) NO_2^-

7.74 What shape do you expect for each of the following anions?

(a) PO_4^{3-} (b) MnO_4^- (c) SO_4^{2-}

(d) SO_3^{2-} (e) ClO_4^-

7.75 What shape do you expect for each of the following cations?
(a) XeF_3^+ (b) SF_3^+ (c) ClF_2^+ (d) CH_3^+

7.76 What bond angles do you expect for each of the following?
(a) The F−S−F angle in SF_2
(b) The H−N−N angle in N_2H_2
(c) The F−Kr−F angle in KrF_4
(d) The Cl−N−O angle in NOCl

7.77 What bond angles do you expect for each of the following?
(a) The Cl−P−Cl angle in PCl_6^-
(b) The Cl−I−Cl angle in ICl_2^-
(c) The O−S−O angle in SO_4^{2-}
(d) The O−B−O angle in BO_3^{3-}

7.78 Acrylonitrile is used as the starting material for manufacturing acrylic fibers. Predict values for all bond angles in acrylonitrile.

$$H_2C=\overset{\overset{\displaystyle H}{|}}{C}-C\equiv N \quad \text{Acrylonitrile}$$

7.79 Predict values for all bond angles in dimethyl sulfoxide, a powerful solvent used in veterinary medicine to treat inflammation.

$$H_3C-\overset{\overset{\displaystyle :\ddot{O}:}{|}}{\underset{..}{S}}-CH_3 \quad \text{Dimethyl sulfoxide}$$

7.80 Explain why cyclohexane, a substance that contains a six-membered ring of carbon atoms, is not flat but instead has a puckered, nonplanar shape. Predict the values of the C−C−C bond angles.

Cyclohexane

Side view

7.81 Like cyclohexane (Problem 7.80), benzene also contains a six-membered ring of carbon atoms, but it is flat rather than puckered. Explain, and predict the values of the C−C−C bond angles.

Benzene

Hybrid Orbitals and Molecular Orbital Theory

7.82 What is the difference in spatial distribution between electrons in a π bond and electrons in a σ bond?

7.83 What is the difference in spatial distribution between electrons in a bonding MO and electrons in an antibonding MO?

7.84 What hybridization do you expect for atoms that have the following numbers of charge clouds?
(a) 2   (b) 5   (c) 6   (d) 4

7.85 What spatial arrangement of charge clouds corresponds to each of the following kinds of hybridization?
(a) sp^3 (b) sp^3d^2 (c) sp

7.86 What hybridization do you expect for the central atom in a molecule that has the following geometry?
(a) Tetrahedral   (b) Octahedral   (c) Bent
(d) Linear   (e) Square pyramidal

7.87 What hybridization do you expect for the central atom in a molecule that has the following geometry?

(a) See-saw
(b) Square planar
(c) Trigonal bipyramidal
(d) T-shaped
(e) Trigonal planar

7.88 What hybridization would you expect for the indicated atom in each of the following?
(a) $H_2C{=}O$ (b) BH_4^- (c) $XeOF_4$ (d) SO_3

7.89 What hybridization would you expect for the indicated atom in each of the following ions?
(a) BrO_3^- (b) HCO_2^- (c) CH_3^+ (d) CH_3^-

7.90 Urea is excreted as a waste product in animal urine. What is the hybridization of the C and N atoms in urea, and what are the approximate values of the various bond angles?

$$H_2N-\overset{\overset{\displaystyle O}{\|}}{C}-NH_2 \quad \text{Urea}$$

7.91 Tell the hybridization of each carbon atom in ascorbic acid (Problem 7.36).

7.92 Use the MO diagram in Figure 7.18b to describe the bonding in O_2^+, O_2, and O_2^-. Which of the three should be stable? What is the bond order of each? Which contain unpaired electrons?

7.93 Use the MO diagram in Figure 7.18a to describe the bonding in N_2^+, N_2, and N_2^-. Which of the three should be stable? What is the bond order of each? Which contain unpaired electrons?

7.94 Make a sketch showing the location and geometry of the p orbitals in the allyl cation. Describe the bonding in this cation using a localized valence bond model for σ bonding and a delocalized MO model for π bonding.

$$H_2C=\overset{\overset{\displaystyle H}{\vert}}{C}-CH_2^+ \quad \text{Allyl cation}$$

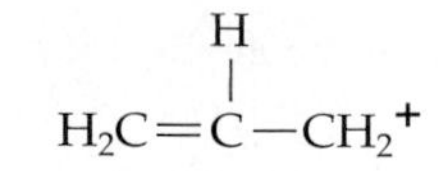

7.95 Make a sketch showing the location and geometry of the p orbitals in the nitrite ion, NO_2^-. Describe the bonding in this ion using a localized valence bond model for σ bonding and a delocalized MO model for π bonding.

General Problems

7.96 The odor of cinnamon oil is due to cinnamaldehyde, C_9H_8O. What is the hybridization of each carbon atom in cinnamaldehyde? How many σ bonds and how many π bonds does cinnamaldehyde have?

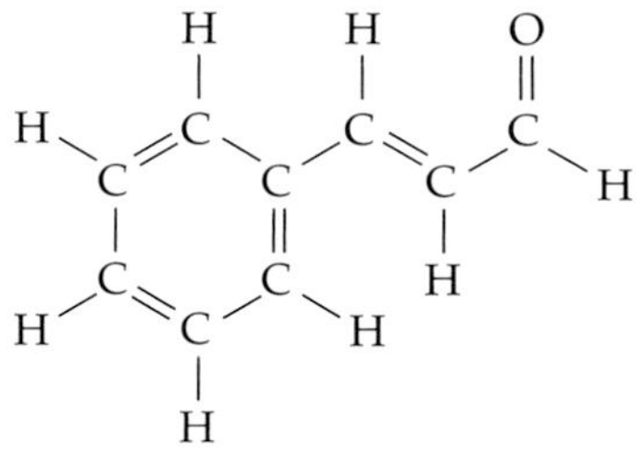

Cinnamaldehyde

7.97 Draw three resonance structures for sulfur tetroxide, SO_4, whose connections are shown below. (This is a neutral molecule; it is not sulfate ion.) Assign formal charges to the atoms in each structure.

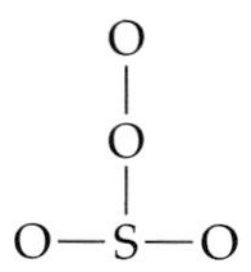

Sulfur tetroxide

7.98 Draw two resonance structures for methyl isocyanate, CH_3NCO, a toxic gas that was responsible for the deaths of at least 3000 people when it was accidentally released into the atmosphere in December 1984 in Bhopal, India. Assign formal charges to the atoms in each resonance structure.

7.99 There are two possible shapes for diimide, $H-N=N-H$. Draw both, and tell if they are resonance forms.

7.100 Boron trifluoride reacts with dimethyl ether to form a compound with a coordinate covalent bond:

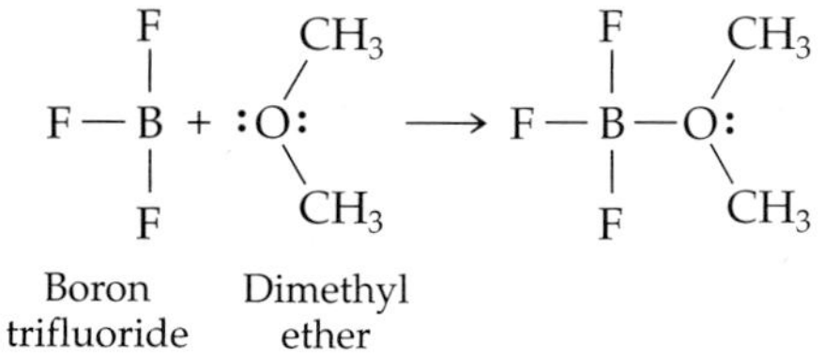

Boron   Dimethyl
trifluoride   ether

(a) Assign formal charges to the B and O atoms in both the reactants and product.

(b) Describe the geometry and hybridization of the B and O atoms in both reactants and product.

7.101 What is the hybridization of the B and N atoms in borazine, what are the values of the B−N−B and N−B−N bond angles, and what is the overall shape of the molecule?

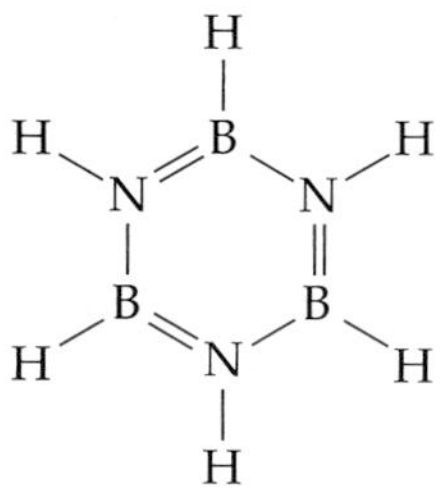

Borazine

7.102 Benzyne, C_6H_4, is a highly energetic and reactive molecule. What hybridization do you expect for the two triply bonded carbon atoms? What are the "theoretical" values for the C−C≡C bond angles? Why do you suppose benzyne is so reactive?

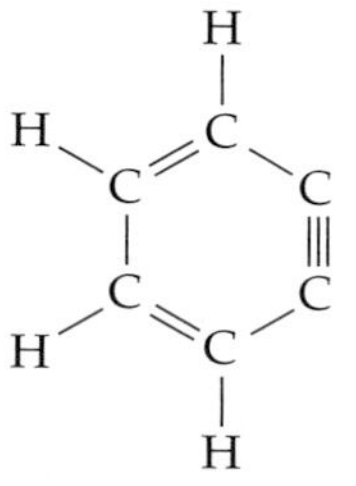

Benzyne

7.103 Propose structures for molecules that meet the following descriptions:
  (a) Contains a C atom that has two π bonds and two σ bonds
  (b) Contains an N atom that has one π bond and two σ bonds
  (c) Contains an S atom that has a coordinate covalent bond

7.104 Draw an electron-dot structure for chloral hydrate, also known in detective novels as "knock-out drops."

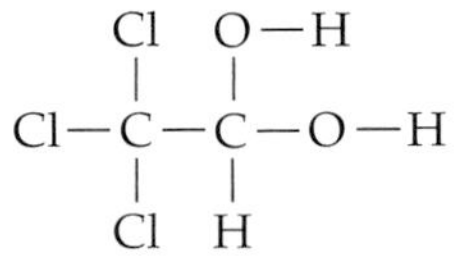

$$
\begin{array}{ccc}
\text{Cl} & \text{O}-\text{H} & \\
| & | & \\
\text{Cl}-\text{C}-\text{C}-\text{O}-\text{H} & & \text{Chloral hydrate}\\
| & | & \\
\text{Cl} & \text{H} &
\end{array}
$$

7.105 Draw a molecular orbital diagram for Li_2. What is the bond order? Is the molecule likely to be stable? Explain.

7.106 Calcium carbide, CaC_2, reacts with water to produce acetylene, C_2H_2, and is sometimes used as a convenient source of that substance. Use the MO diagram in Figure 7.18a to describe the bonding in the carbide anion, C_2^{2-}. What is its bond order?

7.107 There are three substances with the formula $C_2H_2Cl_2$. Draw electron-dot structures for all, and explain how and why they differ.

7.108 The overall energy change during a chemical reaction can be calculated from a knowledge of bond dissociation energies using the following relationship:

Energy change $= D$ (Bonds broken) $- D$ (Bonds formed)

Use the data in Table 7.1 to calculate an energy change for the reaction of methane with chlorine.

$$CH_4(g) + Cl_2(g) \longrightarrow CH_3Cl(g) + HCl(g)$$

7.109 The following structure is a computer-drawn representation of aspartame, $C_{14}H_{18}N_2O_5$, known commercially as NutraSweet. Only the connections between atoms are shown; multiple bonds are not indicated. Complete the structure by indicating the positions of the multiple bonds.

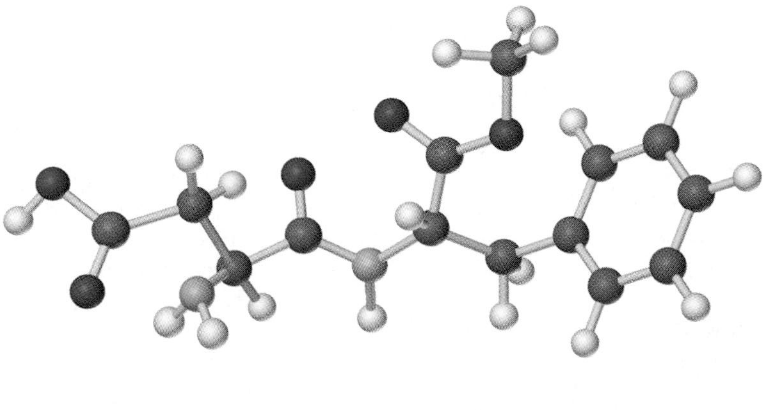

7.110 The N_2O_5 molecule has six N–O σ bonds and two N–O π bonds, but has no N–N bonds and no O–O bonds. Draw eight resonance structures for N_2O_5, and assign formal charges to the atoms in each. Which resonance structures make the more important contributions to the resonance hybrid?

7.111 In the cyanate ion, OCN^-, carbon is the central atom.
  (a) Draw as many resonance structures as you can for OCN^-, and assign formal charges to the atoms in each.
  (b) Which resonance structure makes the greatest contribution to the resonance hybrid? Which makes the least contribution? Explain.
  (c) Is OCN^- linear or bent? Explain.
  (d) Which hybrid orbitals are used by the C atom, and how many π bonds does the C atom form?

7.112 Aspirin has the following connections among atoms. Complete the electron-dot structure for aspirin, tell how many σ bonds and how many π bonds the molecule contains, and tell the hybridization of each carbon atom.

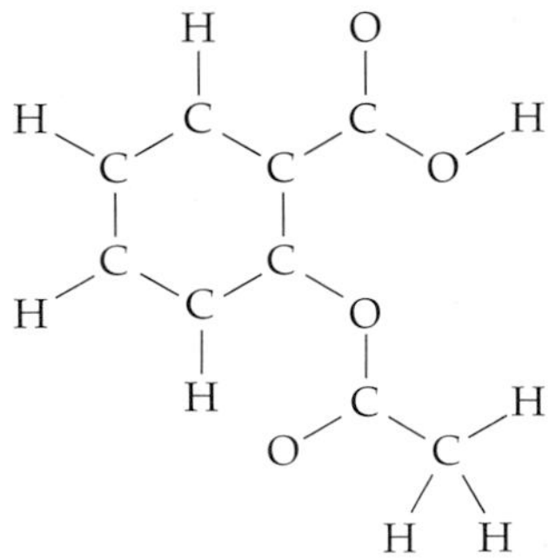

Aspirin

Multi-Concept Problems

7.113 Suppose that the Pauli exclusion principle were somehow changed to allow three electrons per orbital rather than two.

   (a) Instead of an octet, how many outer-shell electrons would be needed for a noble gas electron configuration?

   (b) How many electrons would be shared in a covalent bond?

   (c) Give the electron configuration, and draw an electron-dot structure for element X with $Z = 12$.

   (d) Draw an electron-dot structure for the molecule X_2.

   (e) Assuming that the molecular orbital diagram in Figure 7.20b is valid, tell the bond order for the X_2 molecule.

7.114 The dichromate ion, $Cr_2O_7^{2-}$, has neither Cr–Cr nor O–O bonds.

   (a) Taking both $4s$ and $3d$ electrons into account, draw an electron-dot structure that minimizes the formal charges on the atoms.

   (b) How many outer-shell electrons does each Cr atom have in your electron-dot structure? What is the likely geometry around the Cr atoms?

7.115 A compound with the formula $XOCl_2$, reacts violently with water, yielding HCl and the diprotic acid H_2XO_3. When 0.350 g of $XOCl_2$ was added to 50.0 mL of water and the resultant solution was titrated, 96.1 mL of 0.1225 M NaOH was required to react with all the acid.

   (a) Write a balanced equation for the reaction of $XOCl_2$ with H_2O.

   (b) What is the atomic mass and identity of element X?

   (c) Draw an electron dot structure for $XOCl_2$.

   (d) What is the shape of $XOCl_2$?

 # eMedia Problems

7.116 The **H₂ Bond Formation** movie (*eChapter 7.1*) shows how the internuclear repulsion and attractive forces between the shared electrons and the two nuclei are balanced when the hydrogen nuclei are separated by 74 pm. Although hydrogen atoms can form only a single bond, there are atoms that can form multiple bonds by sharing multiple pairs of electrons. Carbon, for instance, can form a C–C single bond, a C=C double bond, or a C≡C triple bond. Based on the information presented in the movie, which of the carbon–carbon bonds would you expect to be the shortest? Explain.

7.117 The **Electronegativity** movie (*eChapter 7.4*) describes the trends in electronegativity from top to bottom, and from left to right on the periodic table.

   (a) What factors contribute to these trends?

   (b) How is electronegativity different from electron affinity?

7.118 The **VSEPR** movie (*eChapter 7.9*) illustrates the arrangements of charge clouds around a central atom.

   (a) Which of the arrangements does not have all equal bond angles?

   (b) In which of the arrangements are all positions not equivalent with regard to the placement of a single lone pair?

   (c) In which (if any) of the arrangements are all positions equivalent with regard to the placement of a single lone pair, but not equivalent with regard to the placement of a second lone pair?

7.119 In the trigonal bipyramidal arrangement of charge clouds shown in the **VSEPR** movie (*eChapter 7.9*), a single lone pair is shown to occupy an equatorial position rather than an axial position.

   (a) Explain why this is so.

   (b) Is it possible for a molecule with a trigonal bipyramidal arrangement of charge clouds to have linear molecular geometry? If so, how?

   (c) Is it possible for a molecule with a trigonal bipyramidal arrangement of charge clouds to have trigonal planar molecular geometry? Explain.

7.120 In Problem 7.91, you determined the hybridization of each carbon atom in ascorbic acid. After working through the **Promotion and Hybridization** tutorial (*eChapter 7.11*), describe the steps involved in the hybridization of each of ascorbic acid's carbon atoms.

Thermochemistry: Chemical Energy

Oil is where you find it. Much of the energy to power modern society comes from petroleum, forcing oil producers to look in difficult and inhospitable places.

W hy do chemical reactions occur? As mentioned briefly in the previous chapter, the answer involves *stability*. For a reaction to take place spontaneously, the products of the reaction must be more stable than the starting reactants.

But what is "stability," and what does it mean to say that one substance is more stable than another? The most important factor in determining the stability of a substance is the amount of energy it contains. Highly energetic substances are generally less stable and more reactive, while less energetic substances are generally more stable and less reactive. We'll explore some different forms of energy in this chapter and look at the subject of **thermochemistry**—the absorption or release of heat that accompanies chemical reactions.

8.1 Energy

Energy. The word is familiar but is surprisingly hard to define in simple, non-technical terms. The best working definition is that **energy** is the capacity to supply heat or do work:

$$\text{Energy} = \text{Heat} + \text{Work}$$

The water falling over a dam, for example, contains energy that can be used to turn a turbine and generate electricity. A tank of propane gas contains energy that, when released on burning, can heat a house or camper.

Energy can be classified as either *kinetic* or *potential*. **Kinetic energy (E_K)** is the energy of motion. The amount of kinetic energy in a moving object with mass m and velocity v is given by the equation

$$E_K = \frac{1}{2}mv^2$$

The larger the mass of an object and the larger its velocity, the larger the amount of kinetic energy. Thus, water that has fallen over a dam from a great height has a greater velocity and more kinetic energy than the same amount of water that has fallen only a short distance.

Potential energy (E_P), by contrast, is stored energy—perhaps stored in an object because of its height or in a molecule because of its chemical bonds. The water sitting in a reservoir behind the dam contains potential energy because of its height above the stream at the bottom of the dam. When the water is allowed to fall, its potential energy is converted into kinetic energy. Propane and other substances used as fuels contain potential energy in their chemical bonds. When these substances undergo reaction with oxygen during burning, some of this potential energy is released as heat.

The units for energy—$(\text{kg} \cdot \text{m}^2)/\text{s}^2$—follow from the expression $E_K = 1/2\ mv^2$. If, for instance, your body has a mass of 50.0 kg (about 110 lb) and you are riding on a bicycle at a speed of 10.0 m/s (about 22 mi/h), your kinetic energy is 2500 $(\text{kg} \cdot \text{m}^2)/\text{s}^2$:

$$E_K = \frac{1}{2}mv^2 = \frac{1}{2}(50.0\ \text{kg})\left(10.0\ \frac{\text{m}}{\text{s}}\right)^2 = 2500\ \frac{\text{kg} \cdot \text{m}^2}{\text{s}^2} = 2500\ \text{J}$$

The SI energy unit $(\text{kg} \cdot \text{m}^2)/\text{s}^2$ is given the name *joule* (J) after the English physicist James Prescott Joule (1818–1889). The joule is a fairly small amount of energy—it takes roughly 100,000 J to heat a coffee cup full of water from room temperature to boiling—so kilojoules (kJ) are more frequently used in chemistry.

The 75 watt light bulb in this lamp uses energy at the rate of 75 J/s. Only about 5% of that energy appears as light, though; the remainder is given off as heat.

In addition to the SI unit joule, some chemists still use the unit *calorie* (cal, with a lowercase *c*). Originally defined as the amount of energy necessary to raise the temperature of 1 g of water by 1°C (specifically, from 14.5°C to 15.5°C), one calorie is now defined as exactly 4.184 J.

$$1 \text{ cal} = 4.184 \text{ J (exactly)}$$

Nutritionists also use the somewhat confusing unit *Calorie* (Cal, with a capital *C*), which is equal to 1000 calories, or 1 kilocalorie (kcal):

$$1 \text{ Cal} = 1000 \text{ cal} = 1 \text{ kcal} = 4.184 \text{ kJ}$$

The energy value, or caloric content, of food is measured in Calories. Thus, the statement that a banana contains 70 Calories means that 70 Cal (70 kcal, or 290 kJ) of energy is released when the banana is used by the body for fuel.

▶ **PROBLEM 8.1**   What is the kinetic energy (in kilojoules) of a 2300 lb car moving at 55 mi/h?

8.2   Energy Changes and Energy Conservation

Let's pursue a bit further the relationship between potential energy and kinetic energy. According to the *conservation of energy law*, energy can be neither created nor destroyed; it can only be converted from one form into another.

◀— CONSERVATION OF ENERGY LAW   Energy cannot be created or destroyed; it can only be converted from one form into another.

To take the example of falling water again, the water in a reservoir has potential energy because of its height above the outlet stream but has no kinetic energy because it isn't moving ($v = 0$). As the water starts to fall over the dam, though, its height and potential energy decrease while its velocity and kinetic energy increase. The total of potential energy plus kinetic energy always remains constant (Figure 8.1). When the water reaches the bottom and dashes against the rocks or drives the turbine of a generator, its kinetic energy is converted to other forms of energy—perhaps into heat that raises the temperature of the water, or into electrical energy.

FIGURE 8.1   Conservation of energy. The total amount of energy contained by the water in a reservoir is constant. **(a)** At the top of the dam, the energy is potential (E_P). **(b)–(c)** As the water falls over the dam, its velocity increases, and potential energy is converted into kinetic energy (E_K). **(d)** At the bottom of the dam, the kinetic energy gained by the water is largely converted into heat and sound as the water dashes against the rocks.

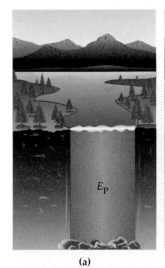

(a)

(b)

(c)

(d)

The conversion of kinetic energy into heat when water falls over a dam and strikes the rocks at the bottom illustrates several other important points about energy. One point is that energy has many forms. Thermal energy, for example, seems different from the kinetic energy of falling water, yet it's really quite similar. *Thermal energy* is simply the kinetic energy of molecular motion, which we measure by finding the **temperature** of an object. An object has a low temperature and we perceive it as "cold" if its atoms or molecules are moving slowly. Conversely, an object has a high temperature and we perceive it as "hot" if its atoms or molecules are moving rapidly and are colliding forcefully with a thermometer or other measuring device. **Heat**, in turn, is the amount of kinetic energy transferred from one object to another as the result of a temperature difference between them.

Chemical energy is another kind of energy that seems different from that of the water in a reservoir, yet again is really quite similar. Chemical energy is a kind of potential energy in which the chemical bonds of molecules act as the "storage" medium. Just as water releases its potential energy when it falls to a more stable position, chemicals release their potential energy in the form of heat or light when they undergo reactions and form more stable products. We'll explore this topic shortly.

The kinetic energy of water falling through the penstocks of a large dam is used to turn huge turbines and generate electricity.

A second point illustrated by the water falling over a dam involves the conservation of energy law. To keep track of all the energy involved, it's necessary to take into account the entire chain of events that ensue from the falling water: the sound of the crashing water, the heating of the rocks at the bottom of the dam, the driving of turbines and electrical generators, the transmission of electrical power, the appliances powered by the electricity, and so on. Carrying the process to its logical extreme, in fact, it's necessary to take the entire universe into account when keeping track of all the energy in the water because the energy lost in one form always shows up elsewhere in another form. So important is the conservation of energy law that it is also known as the **first law of thermodynamics**.

8.3 Internal Energy and State Functions

When keeping track of the energy changes that occur in a chemical reaction, it's helpful to think of the reaction as separate from the world around it. The substances we focus on in an experiment—the starting reactants and the final products—are collectively called the *system*, while everything else—the reaction flask, the room, the building, and so on—is called the *surroundings*. If the system were somehow isolated from its surroundings so that no energy transfer could occur between the two, then the total **internal energy (*E*)** of the system—the sum of all the kinetic and potential energies for every molecule or ion in the system—would be conserved and remain constant throughout the reaction. In fact, this assertion is just a restatement of the first law of thermodynamics:

🔑 FIRST LAW OF THERMODYNAMICS (RESTATED) The total internal energy of an isolated system is constant.

In practice, of course, it's not possible to truly isolate a chemical reaction from its surroundings. In any real situation, the chemicals are in physical contact with the walls of the flask or container, and the container itself is in contact with the surrounding air. What's important, however, is not that the system be isolated but that we be able to measure accurately any energy that enters the system from the surroundings or leaves the system and flows to the surroundings (Figure 8.2). That is, we must be able to measure any *change* in the internal energy of the system, represented by ΔE. The energy change ΔE represents the difference in internal energy between the final and initial states of the system:

$$\Delta E = E_{final} - E_{initial}$$

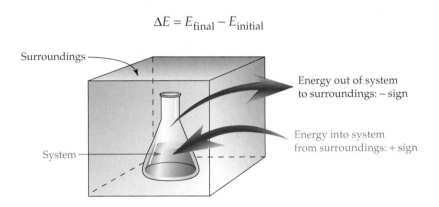

FIGURE 8.2 When energy changes are measured in a chemical reaction, the *system* is the reaction mixture being studied, and the *surroundings* are the flask, the room, and the rest of the universe. The energy change is the difference between final and initial states ($\Delta E = E_{final} - E_{initial}$). Any energy that flows from the system to the surroundings has a negative sign because E_{final} is smaller than $E_{initial}$, and any energy that flows into the system from the surroundings has a positive sign because E_{final} is larger than $E_{initial}$.

By convention, *energy changes are measured from the point of view of the system.* Any energy that flows *from* the system *to* the surroundings has a negative sign because the system has lost it (that is, E_{final} is smaller than $E_{initial}$). Any energy that flows *to* the system *from* the surroundings has a positive sign because the system has gained it (E_{final} is larger than $E_{initial}$). If, for instance, we were to burn 1.00 mol of methane in the presence of 2.00 mol of oxygen, 802 kJ would be released as heat and transferred from the system to the surroundings. The system would have 802 kJ less energy, so $\Delta E = -802$ kJ. This energy flow could be detected and measured by placing the reaction vessel in a water bath and noting the temperature of the bath before and after reaction.

$$CH_4(g) + 2\,O_2(g) \longrightarrow CO_2(g) + 2\,H_2O(g) + 802 \text{ kJ energy}$$

$$\Delta E = -802 \text{ kJ}$$

This experiment tells us that the products of the reaction, $CO_2(g)$ and $2\,H_2O(g)$, have 802 kJ less internal energy than the reactants, $CH_4(g)$ and $2\,O_2(g)$, even though we don't know the exact values at the beginning ($E_{initial}$)

and end (E_{final}) of the reaction. Note that the value $\Delta E = -802$ kJ for the reaction refers to the energy released when reactants are converted to products *in the molar amounts represented by coefficients in the balanced equation.* That is, 802 kJ is released when *1 mol* of methane reacts with *2 mol* of oxygen.

The internal energy of a system depends on many things: chemical identity, sample size, temperature, pressure, physical state (gas, liquid, or solid), and so forth. What the internal energy does *not* depend on is the system's past history. It doesn't matter what the system's temperature or physical state was yesterday or an hour ago, and it doesn't matter how the chemicals were made. All that matters is the present condition of the system. Thus, internal energy is said to be a **state function**—one whose value depends only on the present state of the system. Pressure, volume, and temperature are other examples of state functions, but work and heat are not.

◆— STATE FUNCTION   A function or property whose value depends only on the present state (condition) of the system, not on the path used to arrive at that condition.

Let's illustrate the idea of a state function by imagining a trip from the Artichoke Capitol of the World (Castroville, California) to the Hub of the Universe (Boston, Massachusetts). You are the "system," and your *position* is a state function because how you got to wherever you are is irrelevant. Because your position is a state function, the *change* in your position when you travel from Castroville to Boston (the two cities are about 2720 miles apart) is independent of the path you take, whether through North Dakota or Louisiana (Figure 8.3).

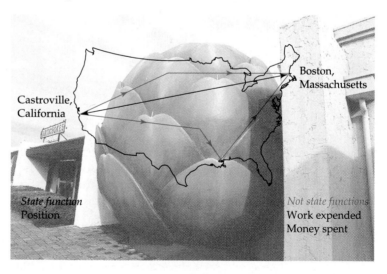

FIGURE 8.3 Since your position is a state function, the change in your position on going from Castroville, California, to Boston, Massachusetts, is independent of the path you take.

The cross-country trip illustrated in Figure 8.3 demonstrates a particularly important point about state functions—their *reversibility*. Imagine that, after traveling from Castroville to Boston, you turn around and go back. Since your final position is now identical to your initial position, the value of the change in position is 0 miles. *The overall change in a state function is zero when the system returns to its original condition.* For a nonstate function, however, the overall change is *not* zero even if the path returns the system to its original condition. Any work you do in making the trip, for instance, is not recovered when you return to your initial position, and any money or time you spent does not reappear.

▶ **PROBLEM 8.2** Which of the following are state functions, and which are not?
(a) The temperature of an ice cube
(b) The volume of an aerosol can
(c) The amount of time required for a 10 mi bike ride

8.4 Expansion Work

Just as energy is hard to define in everyday terms and comes in many forms, so too with *work*. In physics, **work (*w*)** is defined as the force (*F*) that produces the movement of an object times the distance moved (*d*):

$$\textbf{Work} = \text{Force} \times \text{Distance}$$

$$w = F \times d$$

When you run up stairs, for instance, your leg muscles provide a force sufficient to overcome gravity and lift you higher. When you swim, you provide a force sufficient to push water out of the way and pull yourself forward.

This runner going uphill is doing a lot of physical work to overcome gravity.

The most common type of work encountered in chemical systems is the *expansion work* (also called *pressure–volume*, or *PV work*) done as the result of a volume change in the system. Take the reaction of propane (C_3H_8) with oxygen, for instance. The balanced equation says that 7 mol of products come from 6 mol of reactants:

$$\underbrace{C_3H_8(g) + 5\,O_2(g)}_{\text{6 moles of gas}} \longrightarrow \underbrace{3\,CO_2(g) + 4\,H_2O(g)}_{\text{7 moles of gas}}$$

If the reaction takes place inside a container outfitted with a movable piston, the greater volume of gas in the product will force the piston outward against the pressure of the atmosphere (*P*), moving air molecules aside and thereby doing work (Figure 8.4).

A simple calculation gives the exact amount of work done during the expansion. We know from physics that force (*F*) is defined as area times pressure. In Figure 8.4, the force that the expanding gas exerts is the area of the piston (*A*) times the pressure with which the gas pushes against the piston.

FIGURE 8.4 The expansion in volume that occurs during a reaction forces the piston outward against atmospheric pressure P. The amount of work done is equal to the pressure exerted in moving the piston (the opposite of atmospheric pressure, $-P$) times the volume change (ΔV). The volume change is equal to the area of the piston (A) times the distance the piston moves (d). Thus, $w = -P\Delta V$.

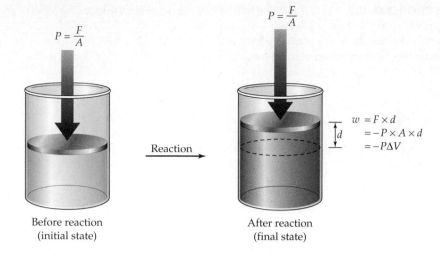

Before reaction
(initial state)

Reaction

After reaction
(final state)

$w = F \times d$
$= -P \times A \times d$
$= -P\Delta V$

Expansion Work activity

This pressure is equal in magnitude but opposite in sign to the external atmospheric pressure (P) that opposes the movement, so it has the value $-P$.

$$F = -P \times A \qquad \text{Where } P \text{ is the external atmospheric pressure}$$

If the piston is moved out a distance d, then the amount of work done is equal to force times distance, or, since $F = -P \times A$, pressure times area times distance:

$$w = F \times d = -P \times A \times d$$

This equation can be put into a more useful form by noticing that the area of the piston times the distance the piston moves is simply the volume change in the system, $\Delta V = A \times d$. Thus, the amount of work done is equal to the pressure the gas exerts against the piston times the volume change (hence the name PV work):

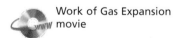
Work of Gas Expansion movie

A negative value ⟍   ⟋ A positive value

$$w = -P\Delta V \qquad\qquad \text{Work done during expansion}$$

What about the sign of the work done during the expansion? Because the work is done *by* the system to move air molecules aside as the piston rises, work energy must be leaving the system. Thus, the negative sign of the work in the above equation is consistent with the convention previously established for ΔE (Section 8.3) whereby we always adopt the point of view of the system. Any energy that flows out of the system has a negative sign because the system has lost it ($E_{\text{final}} < E_{\text{initial}}$).

If the pressure is given in atmospheres (atm) and the volume change is given in liters, then the amount of work done has the unit liter atmosphere (L · atm), where $1 \text{ atm} = 101 \times 10^3 \text{ kg}/(\text{m} \cdot \text{s}^2)$. Thus, $1 \text{ L} \cdot \text{atm} = 101 \text{ J}$:

$$1 \text{ L} \cdot \text{atm} = (1 \text{ L})\left(\frac{10^{-3} \text{ m}^3}{1 \text{ L}}\right)\left(101 \times 10^3 \frac{\text{kg}}{\text{m} \cdot \text{s}^2}\right) = 101 \frac{\text{kg} \cdot \text{m}^2}{\text{s}^2} = 101 \text{ J}$$

When a reaction takes place with a *contraction* in volume rather than an expansion, the ΔV term has a negative sign, and the work has a positive sign.

This is again consistent with adopting the point of view of the system, because the system has now gained work energy ($E_{final} > E_{initial}$). An example is the industrial synthesis of ammonia by reaction of hydrogen with nitrogen. Four moles of gaseous reactants yield only 2 mol of gaseous products, so the volume of the system contracts and work is gained by the system.

$$3\ H_2(g)\ +\ N_2(g)\ \longrightarrow\ 2\ NH_3(g)$$

$$\underbrace{\qquad\qquad\qquad}_{\text{4 moles of gas}}\qquad\underbrace{\qquad\qquad}_{\text{2 moles of gas}}$$

A positive value ↘   ↗ A negative value

$$w\ =\ -P\Delta V \qquad\qquad \text{Work gained during contraction}$$

Of course, if there is no volume change, then $\Delta V = 0$ and there is no work. Such is the case for the combustion of methane where 3 mol of gaseous reactants give 3 mol of gaseous products: $CH_4(g) + 2\ O_2(g) \rightarrow CO_2(g) + 2\ H_2O(g)$.

EXAMPLE 8.1   Calculate the work (in kilojoules) done during a reaction in which the volume expands from 12.0 L to 14.5 L against an external pressure of 5.0 atm.

SOLUTION   Expansion work done during a chemical reaction is calculated by the formula $w = -P\Delta V$, where P is the external pressure opposing the change in volume. In this instance, $P = 5.0$ atm and $\Delta V = (14.5 - 12.0)\ L = 2.5\ L$. Thus:

$$w\ =\ -(5.0\ \text{atm})(2.5\ L)\ =\ -12.5\ L \cdot \text{atm}$$

$$(-12.5\ L \cdot \text{atm})\left(101\ \frac{J}{L \cdot \text{atm}}\right)\ =\ -1.3 \times 10^3\ J\ =\ -1.3\ kJ$$

The negative sign indicates that the expanding system loses work energy and does work on the surroundings.

▶ **PROBLEM 8.3**   Calculate the work (in kilojoules) done during a synthesis of ammonia in which the volume contracts from 8.6 L to 4.3 L at a constant external pressure of 44 atm. In which direction does the work energy flow? What is the sign of the energy change?

KEY CONCEPT PROBLEM 8.4   How much work is done (in kilojoules), and in which direction, as a result of the following reaction?

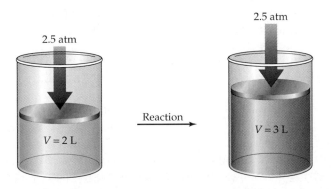

8.5   Energy and Enthalpy

We've seen up to this point that a system can exchange energy with its surroundings either by transferring heat or by doing work. Using the symbol q to represent transferred heat, and remembering from the previous section that $w = -P\Delta V$, we can represent the total energy change of a system, ΔE, as

$$\Delta E = q + w = q - P\Delta V$$

where q has a positive sign if the system gains heat and a negative sign if the system loses it. Rearranging this equation gives the amount of heat transferred:

$$q = \Delta E + P\Delta V$$

Let's look at two ways in which a chemical reaction might be carried out. On the one hand, a reaction might be carried out in a closed container with a constant volume, so that $\Delta V = 0$. In such a case, no PV work is done, and the energy change in the system is due entirely to heat transfer, which we write as q_v to indicate heat at constant volume:

$$q_v = \Delta E \qquad \text{At constant volume; } \Delta V = 0$$

Alternatively, a reaction might be carried out in an open flask or other apparatus that keeps the pressure constant and allows the volume of the system to change freely. In such a case, $\Delta V \neq 0$ and the energy change in the system might be due to both heat transfer and PV work. We indicate the heat transfer at constant pressure by the symbol q_p:

$$q_p = \Delta E + P\Delta V \qquad \text{At constant pressure}$$

Because reactions carried out at constant pressure are so common in chemistry, the heat change for such a process is given a special symbol, ΔH, called the **heat of reaction**, or **enthalpy change** of the reaction. The **enthalpy (H)** of a system is the name given to the quantity $E + PV$.

Many chemical reactions are carried out in open vessels at constant atmospheric pressure.

$$q_p = \Delta E + P\Delta V = \overset{\overset{\text{Enthalpy change}}{\curvearrowleft}}{\Delta H}$$

Note that only the enthalpy *change* during a reaction is important. As with internal energy, E, enthalpy is a state function whose value depends only on the current state of the system, not on the path taken to arrive at that state. Thus, we don't need to know the exact value of the system's enthalpy before and after a reaction. We only need to know the difference between final and initial states:

$$\Delta H = H_{\text{products}} - H_{\text{reactants}}$$

How big a difference is there between ΔE, the heat flow at constant volume, and ΔH, the heat flow at constant pressure? Let's look again at the reaction of propane, C_3H_8, with oxygen as an example. When the reaction is carried out at constant volume, no PV work is possible and all the energy is released as heat: $\Delta E = -2045$ kJ. When the same reaction is carried out at constant pressure, however, only 2043 kJ of heat is released ($\Delta H = -2043$ kJ). The difference, 2 kJ, occurs because at constant pressure, a small amount of expansion work is done against the atmosphere as 6 mol of gaseous reactants are converted into 7 mol of gaseous products.

$$C_3H_8(g) + 5\,O_2(g) \longrightarrow 3\,CO_2(g) + 4\,H_2O(g) \qquad \Delta E = -2045 \text{ kJ}$$

Propane

$$\Delta H = -2043 \text{ kJ}$$

$$P\Delta V = +2 \text{ kJ}$$

What is true of the propane–oxygen reaction is also true of most other reactions: The difference between ΔH and ΔE is usually small, so the two quantities are nearly equal. Chemists usually measure and speak about ΔH, though, because most reactions are carried out at constant atmospheric pressure in loosely covered vessels.

➤ **KEY CONCEPT PROBLEM 8.5** The following reaction has $\Delta E = -186$ kJ/mol.
(a) What is the sign of $P\Delta V$, positive or negative? Explain.
(b) What is the sign and approximate magnitude of ΔH? Explain.

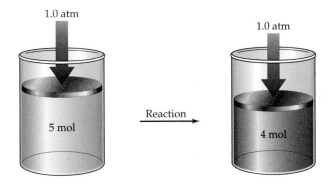

8.6 The Thermodynamic Standard State

The value of the enthalpy change ΔH reported for a reaction is the amount of heat released when reactants are converted to products in the molar amounts represented by coefficients in the balanced equation. In the propane reaction of the previous section, for example, the reaction of 1 mol of propane gas with 5 mol of oxygen gas to give 3 mol of CO_2 gas and 4 mol of water vapor releases 2043 kJ. The actual amount of heat released in a specific reaction, however, depends on the actual amounts of reactants. Thus, reaction of 0.5000 mol of propane with 2.500 mol of O_2 releases 0.5000×2043 kJ = 1022 kJ.

Note that reactants and products must be at the same temperature and that their physical states must be specified as solid (s), liquid (l), gaseous (g), or aqueous (aq) when enthalpy changes are reported. The enthalpy change for the reaction of propane with oxygen is $\Delta H = -2043$ kJ if water is produced as a gas but $\Delta H = -2219$ kJ if water is produced as a liquid.

$$C_3H_8(g) + 5\,O_2(g) \longrightarrow 3\,CO_2(g) + 4\,H_2O(g) \qquad \Delta H = -2043 \text{ kJ}$$
$$C_3H_8(g) + 5\,O_2(g) \longrightarrow 3\,CO_2(g) + 4\,H_2O(l) \qquad \Delta H = -2219 \text{ kJ}$$

The difference of 176 kJ between the values of ΔH for the two reactions arises because the conversion of liquid water to gaseous water requires energy. If liquid water is produced, ΔH is larger (more negative), but if gaseous water is produced, ΔH is smaller (less negative) because 44.0 kJ is needed for the vaporization.

$$H_2O(l) \longrightarrow H_2O(g) \qquad \Delta H = 44.0 \text{ kJ}$$

$$\text{or} \quad 4\,H_2O(l) \longrightarrow 4\,H_2O(g) \qquad \Delta H = 176 \text{ kJ}$$

In addition to specifying the physical state of reactants and products when reporting an enthalpy change, it's also necessary to specify the pressure and temperature. To ensure that all measurements are reported in the same way so that different reactions can be compared, a set of conditions called the **thermodynamic standard state** has been defined.

> **THERMODYNAMIC STANDARD STATE** A pure substance in a specified state, usually its most stable form, at 1 atm pressure* and at a specified temperature, usually 25°C; 1 M concentration for all substances in solution.

Measurements made under these conditions are indicated by addition of the superscript ° to the symbol of the quantity reported. Thus, an enthalpy change measured under standard conditions is called a **standard enthalpy of reaction** and is indicated by the symbol $\Delta H°$. The reaction of propane with oxygen, for example, might be written

$$C_3H_8(g) + 5\,O_2(g) \longrightarrow 3\,CO_2(g) + 4\,H_2O(g) \qquad \Delta H° = -2043 \text{ kJ}$$

EXAMPLE 8.2 The reaction of nitrogen with hydrogen to make ammonia has $\Delta H° = -92.2$ kJ:

$$N_2(g) + 3\,H_2(g) \longrightarrow 2\,NH_3(g) \qquad \Delta H° = -92.2 \text{ kJ}$$

What is the value of ΔE (in kilojoules) if the reaction is carried out at a constant pressure of 40.0 atm and the volume change is -1.12 L?

SOLUTION First rearrange the equation $\Delta H = \Delta E + P\Delta V$ to the form $\Delta E = \Delta H - P\Delta V$, and then substitute the appropriate values for ΔH, P, and ΔV:

$$\Delta E = \Delta H - P\Delta V$$

$$\text{where} \quad \Delta H = -92.2 \text{ kJ}$$

$$P\Delta V = (40.0 \text{ atm})(-1.12 \text{ L}) = -44.8 \text{ L} \cdot \text{atm}$$

$$= (-44.8 \text{ L} \cdot \text{atm})\left(101 \frac{\text{J}}{\text{L} \cdot \text{atm}}\right) = -4520 \text{ J} = -4.52 \text{ kJ}$$

$$\Delta E = (-92.2 \text{ kJ}) - (-4.52 \text{ kJ}) = -87.7 \text{ kJ}$$

---

*The standard pressure, listed here and in most other books as 1 atm, is actually defined to be 1 *bar*, which is equivalent to 0.986 923 atm. The difference is small, however, and has little effect on the accuracy of existing thermodynamic measurements.

Note that ΔE is *smaller* (less negative) than ΔH for this reaction because the volume change is negative. Since the products have less volume than the reactants, a contraction occurs and a small amount of PV work is gained by the system.

▶ **PROBLEM 8.6**   The reaction between hydrogen and oxygen to yield water vapor has $\Delta H° = -484$ kJ:

$$2\,H_2(g) + O_2(g) \longrightarrow 2\,H_2O(g) \qquad \Delta H° = -484\ \text{kJ}$$

How much PV work is done, and what is the value of ΔE (in kilojoules) for the reaction of 0.50 mol of H_2 with 0.25 mol of O_2 at atmospheric pressure if the volume change is -5.6 L?

▶ **PROBLEM 8.7**   The explosion of 2.00 mol of solid trinitrotoluene (TNT) with a volume of approximately 274 mL produces gases with a volume of 448 L at room temperature. How much PV work (in kilojoules) is done during the explosion?

$$2\,C_7H_5N_3O_6(s) \longrightarrow 12\,CO(g) + 5\,H_2(g) + 3\,N_2(g) + 2\,C(s)$$

Explosions cause the essentially instantaneous release of large amounts of gas, thereby doing PV work.

8.7   Enthalpies of Physical and Chemical Change

Almost every change in a system involves either a gain or a loss of enthalpy. The change can be either physical, such as the melting of a solid to a liquid, or chemical, such as the burning of propane. Let's look briefly at examples of both kinds.

 Changes of State movie

Enthalpies of Physical Change

Imagine what would happen if you started with a block of ice at a low temperature, say $-10°C$, and slowly increased its enthalpy by adding heat. The initial input of heat would cause the temperature of the ice to rise until it reached $0°C$. Additional heat would then cause the ice to melt without raising its temperature as the added energy was expended in overcoming the forces that hold H_2O molecules rigidly together in the ice crystal. The amount of heat necessary to melt a substance is called the *enthalpy of fusion*, or *heat of fusion*, ΔH_{fusion}, and has a value of 6.01 kJ/mol for H_2O.

Once the ice has melted, further input of heat raises the water's temperature until it reaches $100°C$, and additional heat then causes the water to boil. Once again, energy is necessary to overcome the forces holding molecules together in the liquid, and the temperature does not rise again until all the liquid has been converted into vapor. The amount of heat required for evaporation is called the *enthalpy of vaporization*, or *heat of vaporization*, ΔH_{vap}, and has a value of 40.7 kJ/mol at $100°C$ for H_2O.

Another kind of physical change is **sublimation**—the direct conversion of a solid to a vapor without going through a liquid state. Solid CO_2 (dry ice), for example, changes directly from solid to vapor at atmospheric pressure without first melting to a liquid. Since enthalpy is a state function, the enthalpy change on going from solid to vapor must be constant regardless of the path taken. Thus, a substance's *enthalpy of sublimation*, or *heat of sublimation*, ΔH_{subl}, is equal to the sum of the heat of fusion and heat of vaporization (Figure 8.5).

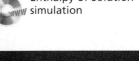 Enthalpy of Solution simulation

Dry ice (solid CO_2) sublimes directly from solid to gas at atmospheric pressure.

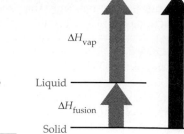

FIGURE 8.5 Because enthalpy is a state function, the enthalpy change from solid to vapor does not depend on the path taken between the two states. Therefore, $\Delta H_{subl} = \Delta H_{fusion} + \Delta H_{vap}$.

Enthalpies of Chemical Change

We saw in Section 8.5 that an enthalpy change is often called a *heat of reaction* because it is a measure of the heat flow into or out of a system at constant pressure. If the products have *more* enthalpy than the reactants, then heat has flowed into the system from the surroundings and ΔH has a positive sign. Such reactions are said to be **endothermic** (*endo* means "within", so heat flows in). The reaction of 1 mol of barium hydroxide octahydrate* with ammonium chloride, for example, absorbs 80.3 kJ from the surroundings ($\Delta H° = +80.3$ kJ). The surroundings, having lost heat, become so cold that water freezes around the outside of the container (Figure 8.6).

FIGURE 8.6 The reaction of barium hydroxide octahydrate with ammonium chloride is so strongly endothermic and draws so much heat from the surroundings that the temperature falls below 0°C and water freezes around the container.

$$Ba(OH)_2 \cdot 8\ H_2O(s) + 2\ NH_4Cl(s) \longrightarrow$$
$$BaCl_2(aq) + 2\ NH_3(aq) + 10\ H_2O(l) \qquad \Delta H° = +80.3\ kJ$$

If the products have *less* enthalpy than the reactants, then heat has flowed from the system to the surroundings and ΔH has a negative sign. Such reactions are said to be **exothermic** (*exo* means "out", so heat flows out). The so-called thermite reaction of aluminum with iron(III) oxide, for example, releases so much heat and the surroundings get so hot ($\Delta H° = -852$ kJ) that it is used in construction work to weld iron (Figure 8.7).

$$2\ Al(s) + Fe_2O_3(s) \longrightarrow 2\ Fe(s) + Al_2O_3(s) \qquad \Delta H° = -852\ kJ$$

As noted previously, the value of $\Delta H°$ given for an equation assumes that the equation is balanced for the number of moles of reactants and products, that all substances are in their standard states, and that the physical state of each substance is as specified. The actual amount of heat released in a specific reaction depends on the amounts of reactants, as illustrated in Example 8.3.

It should also be emphasized that $\Delta H°$ values refer to the reaction going *in the direction written*. For the reverse reaction, the sign of $\Delta H°$ must be changed. Because of the reversibility of state functions (Section 8.3), the enthalpy change for any reaction is equal in magnitude but opposite in sign to that for the reverse reaction. For example, the reaction of iron with aluminum oxide to yield aluminum and iron oxide (the reverse of the thermite reaction) would be endothermic and have $\Delta H° = +852$ kJ:

FIGURE 8.7 The thermite reaction of aluminum with iron(III) oxide is so strongly exothermic, and releases so much heat to the surroundings, that the products become molten.

$$2\ Fe(s) + Al_2O_3(s) \longrightarrow 2\ Al(s) + Fe_2O_3(s) \qquad \Delta H° = +852\ kJ$$
$$2\ Al(s) + Fe_2O_3(s) \longrightarrow 2\ Fe(s) + Al_2O_3(s) \qquad \Delta H° = -852\ kJ$$

---

*Barium hydroxide octahydrate, $Ba(OH)_2 \cdot 8\ H_2O$, is a crystalline compound that contains eight water molecules clustered around the barium atom. We'll learn more about hydrates in Section 14.15.

EXAMPLE 8.3   How much heat (in kilojoules) is evolved when 5.00 g of aluminum reacts with a stoichiometric amount of Fe_2O_3?

$$2\,Al(s)\,+\,Fe_2O_3(s)\,\longrightarrow\,2\,Fe(s)\,+\,Al_2O_3(s)\qquad \Delta H° = -852\text{ kJ}$$

BALLPARK SOLUTION   The balanced equation says that 852 kJ of heat is evolved from the reaction of 2 mol of Al, or about 425 kJ for 1 mol. To find out how much heat is evolved from the reaction of 5.00 g of Al, we have to find out how many moles of aluminum are in 5.00 g. Since the molar mass of Al is about 27 g, 5 g of aluminum is roughly 0.2 mol, and the heat evolved is about 425 kJ/mol × 0.2 mol, or approximately 90 kJ.

DETAILED SOLUTION   The molar mass of Al is 26.98 g/mol, so 5.00 g of Al is equal to 0.185 mol:

$$5.00\text{ g Al} \times \frac{1\text{ mol Al}}{26.98\text{ g Al}} = 0.185\text{ mol Al}$$

Since 2 mol of Al releases 852 kJ of heat, 0.185 mol of Al releases 78.8 kJ of heat:

$$0.185\text{ mol Al} \times \frac{852\text{ kJ}}{2\text{ mol Al}} = 78.8\text{ kJ}$$

▶ **PROBLEM 8.8**   How much heat (in kilojoules) is evolved or absorbed in each of the following reactions?
(a) Burning of 15.5 g of propane:

$$C_3H_8(g)\,+\,5\,O_2(g)\,\longrightarrow\,3\,CO_2(g)\,+\,4\,H_2O(l)\qquad \Delta H° = -2219\text{ kJ}$$

(b) Reaction of 4.88 g of barium hydroxide octahydrate with ammonium chloride:

$$Ba(OH)_2 \cdot 8\,H_2O(s)\,+\,2\,NH_4Cl(s)\,\longrightarrow$$
$$BaCl_2(aq)\,+\,2\,NH_3(aq)\,+\,10\,H_2O(l)\qquad \Delta H° = +80.3\text{ kJ}$$

8.8   Calorimetry and Heat Capacity

The amount of heat transferred during a reaction can be measured with a device called a *calorimeter*, shown schematically in Figure 8.8. At its simplest, a calorimeter is just an insulated vessel with a stirrer, a thermometer, and a loose-fitting lid to keep the contents at atmospheric pressure. The reaction is carried out inside the vessel, and the heat evolved or absorbed is calculated from the temperature change. Since the pressure inside the calorimeter is constant (atmospheric pressure), the temperature measurement makes it possible to calculate the enthalpy change ΔH during a reaction.

A somewhat more complicated device called a *bomb calorimeter* is used to measure the heat released during a combustion reaction, or burning of a flammable substance. (More generally, a *combustion* reaction is any reaction that produces a flame.) The sample is placed in a small cup and sealed under an oxygen atmosphere inside a steel "bomb" that is itself placed in an insulated, water-filled container (Figure 8.9). The reactants are ignited electrically, and the evolved heat is calculated from the temperature change of the surrounding water. Since the reaction takes place at constant volume but not constant pressure, the measurement provides a value for ΔE rather than ΔH.

Thermite movie

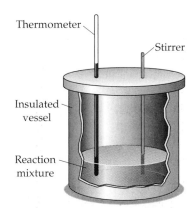

FIGURE 8.8   A calorimeter for measuring the heat flow in a reaction at constant pressure (ΔH). The reaction takes place inside an insulated vessel outfitted with a loose-fitting top, a thermometer, and a stirrer. Measuring the temperature change that accompanies the reaction makes it possible to calculate ΔH.

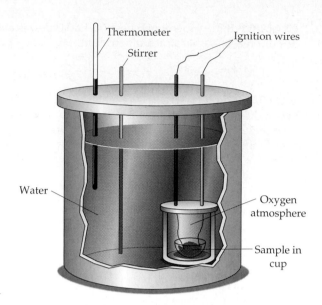

FIGURE 8.9 A bomb calorimeter for measuring the heat evolved at constant volume in a combustion reaction (ΔE). The reaction is carried out inside a steel bomb, and the heat evolved is transferred to the surrounding water, where the temperature rise is measured.

Calorimetry simulation

How can the temperature change inside a calorimeter be used to calculate ΔH (or ΔE) for a reaction? When a calorimeter and its contents absorb a given amount of heat, the temperature rise that results depends on the calorimeter's *heat capacity*. **Heat capacity (C)** is the amount of heat required to raise the temperature of an object or substance a given amount, a relationship that can be expressed by the equation

$$ C = \frac{q}{\Delta T} $$

where q is the quantity of heat transferred and ΔT is the temperature change ($\Delta T = T_{\text{final}} - T_{\text{initial}}$). The greater the heat capacity, the greater the amount of heat needed to produce a given temperature change. A bathtub full of water, for instance, has a greater heat capacity than a coffee cup full, and it therefore takes far more heat to warm the tubful than the cupful. The exact amount of heat absorbed is equal to the heat capacity times the temperature rise:

$$ q = C \times \Delta T $$

Heat capacity is an extensive property (Section 1.4), so its value depends on both the size of an object and its composition. To compare different substances, it's useful to define a quantity called **specific heat**, the amount of heat necessary to raise the temperature of 1 g of a substance by 1°C. The amount of heat necessary to raise the temperature of a given object, then, is the specific heat times the mass of the object times the rise in temperature:

$$ q = (\text{specific heat}) \times (\text{mass of substance}) \times \Delta T $$

Example 8.5 shows how specific heats are used in calorimetry calculations.
Closely related to specific heat is the **molar heat capacity (C_m)**, defined as the amount of heat necessary to raise the temperature of 1 mol of a substance by 1°C. The amount of heat necessary to raise the temperature of a given number of moles of a substance is thus

$$ q = (C_m) \times (\text{Moles of substance}) \times \Delta T $$

Values of specific heats and molar heat capacities for some common substances are given in Table 8.1.

TABLE 8.1   Specific Heats and Molar Heat Capacities for Some Common Substances at 25°C

Substance	Specific Heat J/(g · °C)	Molar Heat Capacity J/(mol · °C)
Air (dry)	1.01	29.1
Aluminum	0.902	24.4
Copper	0.385	24.4
Gold	0.129	25.4
Iron	0.450	25.1
Mercury	0.140	28.0
NaCl	0.864	50.5
Water(s)*	2.03	36.6
Water(l)	4.179	75.3

*At −11°C

As indicated in Table 8.1, the specific heat of liquid water is considerably higher than that of most other substances, and a large transfer of heat is therefore necessary to either cool or warm a given amount of water. One consequence is that large lakes or other bodies of water tend to moderate the air temperature in the surrounding areas. Another consequence is that the human body, which is about 60% water, is able to maintain a steady internal temperature under changing outside conditions.

Large masses of water moderate the temperature of the surroundings because of their high heat capacity.

EXAMPLE 8.4   What is the specific heat of silicon if it takes 192 J to raise the temperature of 45.0 g of Si by 6.0°C?

SOLUTION   The specific heat of a substance is the amount of energy necessary to raise the temperature of 1 g of the substance by 1°C. Since 192 J raises the temperature of 45.0 g of Si by 6.0°C, 192/45.0 = 4.27 J will raise the temperature of 1 g of Si by 6.0°C, and 4.27/6.0 = 0.71 J will raise the temperature of 1 g of Si by 1°C.

$$\text{Specific heat of Si} = \frac{192 \text{ J}}{(45.0 \text{ g})(6.0°C)} = 0.71 \text{ J/(g · °C)}$$

The reaction of aqueous Ag^+ with aqueous Cl^- to yield solid AgCl is an exothermic process.

EXAMPLE 8.5 Aqueous silver ion reacts with aqueous chloride ion to yield a white precipitate of solid silver chloride:

$$Ag^+(aq) + Cl^-(aq) \longrightarrow AgCl(s)$$

When 10.0 mL of 1.00 M $AgNO_3$ solution is added to 10.0 mL of 1.00 M NaCl solution at 25.0°C in a calorimeter, a white precipitate of AgCl forms, and the temperature of the aqueous mixture increases to 32.6°C. Assuming that the specific heat of the aqueous mixture is 4.18 J/(g · °C), that the density of the mixture is 1.00 g/mL, and that the calorimeter itself absorbs a negligible amount of heat, calculate ΔH (in kilojoules) for the reaction.

SOLUTION Since the temperature rises during the reaction, heat must be liberated and ΔH must therefore be negative. The amount of heat evolved during the reaction is equal to the amount of heat absorbed by the mixture and can be calculated in the following way:

Heat evolved = (Specific heat) × (Mass of mixture) × (Temperature change)

where Specific heat = 4.18 J/(g · °C)

$$\text{Mass} = 20.0 \text{ mL} \times 1.00 \frac{g}{mL} = 20.0 \text{ g}$$

Temperature change = 32.6°C − 25.0°C = 7.6°C

$$\text{Heat evolved} = 4.18 \frac{J}{g \cdot °C} \times 20.0 \text{ g} \times 7.6°C = 6.4 \times 10^2 \text{ J}$$

Next, calculate how many moles of AgCl were formed. According to the balanced equation, the number of moles of AgCl produced is equal to the number of moles of Ag^+ (or Cl^-) reacted:

$$\text{Moles of } Ag^+ = 10.0 \text{ mL} \times \frac{1.00 \text{ mol } Ag^+}{1000 \text{ mL}} = 1.00 \times 10^{-2} \text{ mol } Ag^+$$

Moles of AgCl = 1.00×10^{-2} mol AgCl

Finally, calculate ΔH for the reaction. Don't forget that its sign is negative, because heat is evolved.

$$\text{Heat evolved per mole of AgCl} = \frac{6.4 \times 10^2 \text{ J}}{1.00 \times 10^{-2} \text{ mol AgCl}} = 64 \text{ kJ/mol AgCl}$$

Therefore, $\Delta H = -64$ kJ.

▶ **PROBLEM 8.9** Assuming that Coca Cola has the same specific heat as water [4.18 J/(g · °C)], calculate the amount of heat (in kilojoules) transferred when one can (about 350 g) is cooled from 25°C to 3°C.

▶ **PROBLEM 8.10** What is the specific heat of lead if it takes 96 J to raise the temperature of a 75 g block by 10.0°C?

▶ **PROBLEM 8.11** When 25.0 mL of 1.0 M H_2SO_4 is added to 50.0 mL of 1.0 M NaOH at 25.0°C in a calorimeter, the temperature of the aqueous solution increases to 33.9°C. Assuming that the specific heat of the solution is 4.18 J/(g · °C), that its density is 1.00 g/mL, and that the calorimeter itself absorbs a negligible amount of heat, calculate ΔH (in kilojoules) for the reaction.

$$H_2SO_4(aq) + 2 NaOH(aq) \longrightarrow 2 H_2O(l) + Na_2SO_4(aq)$$

8.9 Hess's Law

Having discussed in general terms the energy changes that occur during chemical reactions, let's now look at a specific example in detail. In particular, let's look at the *Haber process*, the industrial method by which approximately 20 million tons of ammonia are produced each year in the United States, primarily for use as fertilizer. The reaction of hydrogen with nitrogen to make ammonia is exothermic, with $\Delta H° = -92.2$ kJ.

$$3\,H_2(g) + N_2(g) \longrightarrow 2\,NH_3(g) \qquad \Delta H° = -92.2\text{ kJ}$$

If we dig into the details of the reaction, we find that it's not as simple as it looks. In fact, the overall reaction occurs by a series of steps, with hydrazine (N_2H_4) produced at an intermediate stage:

$$2\,H_2(g) + N_2(g) \longrightarrow N_2H_4(g) \xrightarrow{\;H_2\;} 2\,NH_3(g)$$
$$\text{Hydrazine} \qquad\qquad \text{Ammonia}$$

The enthalpy change for the conversion of hydrazine to ammonia can be measured as $\Delta H° = -187.6$ kJ, but if we wanted to measure $\Delta H°$ for the formation of hydrazine from hydrogen and nitrogen, we would have difficulty because the reaction doesn't go cleanly. Some of the hydrazine is converted into ammonia while some of the starting nitrogen still remains.

Fortunately, there's a way around the difficulty—a way that makes it possible to measure an energy change indirectly when a direct measurement can't be made. The trick is to realize that, because enthalpy is a state function, ΔH is the same no matter what path is taken between two states. Thus, the sum of the enthalpy changes for the individual steps in a sequence must equal the enthalpy change for the overall reaction, a statement known as **Hess's law**:

➤— Hess's law   The overall enthalpy change for a reaction is equal to the sum of the enthalpy changes for the individual steps in the reaction.

Reactants and products in the individual steps can be added and subtracted like algebraic quantities in determining the overall equation. In the synthesis of ammonia, for example, the sum of steps 1 and 2 is equal to the overall reaction. Thus, the sum of the enthalpy changes for steps 1 and 2 is equal to the enthalpy change for the overall reaction. With this knowledge, we can calculate the enthalpy change for step 1. Figure 8.10 shows the situation pictorially.

Step 1 $2\,H_2(g) + N_2(g) \longrightarrow \cancel{N_2H_4(g)}$ $\Delta H°_1 = ?$

Step 2 $\cancel{N_2H_4(g)} + H_2(g) \longrightarrow 2\,NH_3(g)$ $\Delta H°_2 = -187.6$ kJ

Overall
reaction $3\,H_2(g) + N_2(g) \longrightarrow 2\,NH_3(g)$ $\Delta H°_{reaction} = -92.2$ kJ

Since $\Delta H°_1 + \Delta H°_2 = \Delta H°_{reaction}$

then $\Delta H°_1 = \Delta H°_{reaction} - \Delta H°_2$

$$= (-92.2\text{ kJ}) - (-187.6\text{ kJ}) = +95.4\text{ kJ}$$

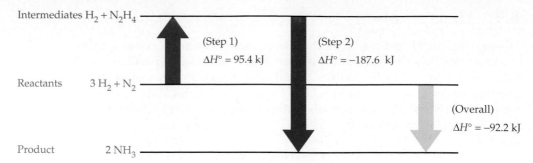

FIGURE 8.10   A representation of the enthalpy changes for steps in the synthesis of ammonia from nitrogen and hydrogen. If $\Delta H°$ values for step 2 and for the overall reaction are known, then $\Delta H°$ for step 1 can be calculated. That is, the enthalpy change for the overall reaction is equal to the sum of the enthalpy changes for the individual steps 1 and 2, a statement known as *Hess's law*.

EXAMPLE 8.6   Methane, the main constituent of natural gas, burns in oxygen to yield carbon dioxide and water:

$$CH_4(g) + 2\,O_2(g) \longrightarrow CO_2(g) + 2\,H_2O(l)$$

Use the following information to calculate $\Delta H°$ (in kilojoules) for the combustion of methane:

$$CH_4(g) + O_2(g) \longrightarrow CH_2O(g) + H_2O(g) \qquad \Delta H° = -284 \text{ kJ}$$
$$CH_2O(g) + O_2(g) \longrightarrow CO_2(g) + H_2O(g) \qquad \Delta H° = -518 \text{ kJ}$$
$$H_2O(l) \longrightarrow H_2O(g) \qquad \Delta H° = 44.0 \text{ kJ}$$

SOLUTION   Though it often takes some trial and error, the idea is to combine the individual reactions so that their sum is the desired reaction. The important points to be aware of are that

- All the reactants [$CH_4(g)$ and $O_2(g)$] must appear on the left.
- All the products [$CO_2(g)$ and $H_2O(l)$] must appear on the right.
- All intermediate products [$CH_2O(g)$ and $H_2O(g)$] must occur on *both* the left and the right so that they cancel.
- Any conversion written in the reverse of the direction given [$H_2O(g) \rightarrow H_2O(l)$] must have the sign of its $\Delta H°$ changed (Section 8.7).
- A reaction can be multiplied by a coefficient as necessary [$H_2O(g) \rightarrow H_2O(l)$ is multiplied by 2], but $\Delta H°$ for the reaction must be multiplied by that same coefficient.

$$CH_4(g) + O_2(g) \longrightarrow \cancel{CH_2O(g)} + \cancel{H_2O(g)} \qquad \Delta H° = -284 \text{ kJ}$$
$$\cancel{CH_2O(g)} + O_2(g) \longrightarrow CO_2(g) + \cancel{H_2O(g)} \qquad \Delta H° = -518 \text{ kJ}$$
$$2\,[\cancel{H_2O(g)} \longrightarrow H_2O(l)] \qquad 2[\Delta H° = -44.0 \text{ kJ}] = -88.0 \text{ kJ}$$

$$\overline{CH_4(g) + 2\,O_2(g) \longrightarrow CO_2(g) + 2\,H_2O(l) \qquad \Delta H° = -890 \text{ kJ}}$$

EXAMPLE 8.7   *Water gas* is the name for the industrially important mixture of CO and H_2 prepared by passing steam over hot charcoal at 1000°C:

$$C(s) + H_2O(g) \longrightarrow CO(g) + H_2(g)$$
$$\text{"Water gas"}$$

The hydrogen is then purified and used as a starting material for preparing ammonia. Use the following information to calculate $\Delta H°$ (in kilojoules) for the water-gas reaction:

$$C(s) + O_2(g) \longrightarrow CO_2(g) \qquad \Delta H° = -393.5 \text{ kJ}$$
$$2\,CO(g) + O_2(g) \longrightarrow 2\,CO_2(g) \qquad \Delta H° = -566.0 \text{ kJ}$$
$$2\,H_2(g) + O_2(g) \longrightarrow 2\,H_2O(g) \qquad \Delta H° = -483.6 \text{ kJ}$$

SOLUTION   As in Example 8.6, the idea is to find a combination of the individual reactions whose sum is the desired reaction. In this instance, it's necessary to reverse the second and third steps and to multiply both by 1/2 to make the overall equation balance. In so doing, the sign of the enthalpy changes for those steps must be changed and also multiplied by 1/2. (Alternatively, we could multiply the first step by 2 and then divide the final result by 2.) Note that $CO_2(g)$ and $O_2(g)$ cancel because they appear on both the right and left sides of reactions.

$$C(s) + \cancel{O_2(g)} \longrightarrow \cancel{CO_2(g)} \qquad\qquad \Delta H° = -393.5 \text{ kJ}$$
$$1/2[2\,\cancel{CO_2(g)} \longrightarrow 2\,CO(g) + \cancel{O_2(g)}] \qquad 1/2[\Delta H° = 566.0 \text{ kJ}] = 283.0 \text{ kJ}$$
$$1/2[2\,H_2O(g) \longrightarrow 2\,H_2(g) + \cancel{O_2(g)}] \qquad 1/2[\Delta H° = 483.6 \text{ kJ}] = 241.8 \text{ kJ}$$
$$\overline{C(s) + H_2O(g) \longrightarrow CO(g) + H_2(g) \qquad\qquad \Delta H° = 131.3 \text{ kJ}}$$

The water-gas reaction is endothermic by 131.3 kJ.

▶ **PROBLEM 8.12**   The industrial degreasing solvent methylene chloride, CH_2Cl_2, is prepared from methane by reaction with chlorine:

$$CH_4(g) + 2\,Cl_2(g) \longrightarrow CH_2Cl_2(g) + 2\,HCl(g)$$

Use the following data to calculate $\Delta H°$ (in kilojoules) for the reaction:

$$CH_4(g) + Cl_2(g) \longrightarrow CH_3Cl(g) + HCl(g) \qquad \Delta H° = -98.3 \text{ kJ}$$
$$CH_3Cl(g) + Cl_2(g) \longrightarrow CH_2Cl_2(g) + HCl(g) \qquad \Delta H° = -104 \text{ kJ}$$

◆— **KEY CONCEPT PROBLEM 8.13**   The reaction of A with B to give D proceeds in two steps and can be represented by the following Hess's law diagram.

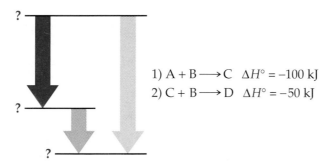

1) $A + B \longrightarrow C \quad \Delta H° = -100 \text{ kJ}$
2) $C + B \longrightarrow D \quad \Delta H° = -50 \text{ kJ}$

(a) What is the equation and $\Delta H°$ for the net reaction?
(b) Which arrow on the diagram corresponds to which step, and which arrow corresponds to the net reaction?
(c) The diagram shows three energy levels. The energies of which substances are represented by each?

◆— **KEY CONCEPT PROBLEM 8.14**   Draw a Hess's law diagram similar to that in Problem 8.13 depicting the energy changes for the reaction in Problem 8.12.

8.10 Standard Heats of Formation

Where do the $\Delta H°$ values we've been using in the previous sections come from? There are so many chemical reactions—several hundred million are known—that it's obviously not possible to measure $\Delta H°$ for them all. A better way is needed.

The most efficient way to manage with the smallest number of experimental measurements is to use what are called **standard heats of formation**, symbolized $\Delta H°_f$.

Formation of Aluminum Bromide movie

◀─ **STANDARD HEAT OF FORMATION** The enthalpy change $\Delta H°_f$ for the formation of 1 mol of a substance in its standard state from its constituent elements in their standard states.

Note several points about this definition. First, the "reaction" to form a substance from its constituent elements can be (and often is) hypothetical. We can't combine carbon and hydrogen in the laboratory to make methane, for instance, yet the heat of formation for methane is $\Delta H°_f = -74.8$ kJ/mol, which corresponds to the standard enthalpy change for the hypothetical reaction

$$C(s) + 2\,H_2(g) \longrightarrow CH_4(g) \qquad \Delta H° = -74.8 \text{ kJ}$$

Second, note that each substance in the reaction must be in its standard state, usually the most stable form at 1 atm pressure and the specified temperature (usually 25°C). Carbon, for example, is most stable as solid graphite (not diamond) under these conditions, and hydrogen is most stable as gaseous H_2 (not H atoms). Table 8.2 gives standard heats of formation for some common substances, and Appendix B gives a more detailed list.

TABLE 8.2 Standard Heats of Formation for Some Common Substances at 25°C

Substance	Formula	$\Delta H°_f$ (kJ/mol)	Substance	Formula	$\Delta H°_f$ (kJ/mol)
Acetylene	$C_2H_2(g)$	226.7	Hydrogen chloride	$HCl(g)$	−92.3
Ammonia	$NH_3(g)$	−46.1	Iron(III) oxide	$Fe_2O_3(s)$	−824.2
Carbon dioxide	$CO_2(g)$	−393.5	Magnesium carbonate	$MgCO_3(s)$	−1095.8
Carbon monoxide	$CO(g)$	−110.5	Methane	$CH_4(g)$	−74.8
Ethanol	$C_2H_5OH(l)$	−277.7	Nitric oxide	$NO(g)$	90.2
Ethylene	$C_2H_4(g)$	52.3	Water (g)	$H_2O(g)$	−241.8
Glucose	$C_6H_{12}O_6(s)$	−1260	Water (l)	$H_2O(l)$	−285.8

No elements are listed in Table 8.2 because, by definition, *the most stable form of any element in its standard state has* $\Delta H°_f = 0$. Remember that all our calculations are based on enthalpy changes, not on actual enthalpy values. Defining $\Delta H°_f$ as zero for all elements thus establishes a kind of thermochemical "sea level," or reference point, from which all changes are measured.

How can standard heats of formation be used for thermochemical calculations? *The standard enthalpy change for any chemical reaction is found by subtracting the sum of the heats of formation of all reactants from the sum of the heats of formation of all products.* (Don't forget, though, to multiply each heat of formation by the coefficient of that substance in the balanced equation.)

$$\Delta H°_{Reaction} = \Delta H°_f(\text{Products}) - \Delta H°_f(\text{Reactants})$$

To find $\Delta H°$ for the reaction

$$\underbrace{a\,A + b\,B + \cdots}_{\substack{\text{Subtract the sum of the} \\ \text{heats of formation for} \\ \text{these reactants} \ldots}} \longrightarrow \underbrace{c\,C + d\,D + \cdots}_{\substack{\ldots \text{from the sum of the} \\ \text{heats of formation for} \\ \text{these products.}}}$$

$$\Delta H°_{Reaction} = [c\,\Delta H°_f(C) + d\,\Delta H°_f(D) + \cdots] - [a\,\Delta H°_f(A) + b\,\Delta H°_f(B) + \cdots]$$

For example, let's calculate $\Delta H°$ for the fermentation of glucose to make ethyl alcohol (ethanol), the reaction that occurs during the production of alcoholic beverages:

$$C_6H_{12}O_6(s) \longrightarrow 2\,C_2H_5OH(l) + 2\,CO_2(g) \qquad \Delta H° = ?$$

Fermentation of the sugar from grapes yields the ethyl alcohol in wine.

Using the data in Table 8.2 gives the following answer:

$$\Delta H° = [2\,\Delta H°_f(\text{Ethanol}) + 2\,\Delta H°_f(CO_2)] - [\Delta H°_f(\text{Glucose})]$$
$$= (2\text{ mol})(-277.7\text{ kJ/mol}) + (2\text{ mol})(-393.5\text{ kJ/mol}) - (1\text{ mol})(-1260\text{ kJ/mol})$$
$$= -82\text{ kJ}$$

The fermentation reaction is exothermic by 82 kJ.

Why does this calculation "work"? It works because fundamentally it's just an application of Hess's law. That is, we can add the individual equations corresponding to the heat of formation for each substance in the reaction to arrive at the enthalpy change for the overall reaction:

Reaction

(1) $C_6H_{12}O_6(s) \longrightarrow 6\,C(s) + 6\,H_2(g) + 3\,O_2(g)$ $-\Delta H°_f = +1260\text{ kJ}$

(2) $2\,[2\,C(s) + 3\,H_2(g) + 1/2\,O_2(g) \longrightarrow C_2H_5OH(l)]$ $2\,[\Delta H°_f = -277.7\text{ kJ}] = -555.4\text{ kJ}$

(3) $2\,[C(s) + O_2(g) \longrightarrow CO_2(g)]$ $2\,[\Delta H°_f = -393.5\text{ kJ}] = -787.0\text{ kJ}$

Net $C_6H_{12}O_6(s) \longrightarrow 2\,C_2H_5OH(l) + 2\,CO_2(g)$ $\Delta H° = -82\text{ kJ}$

Note that reaction (1) represents the formation of glucose from its elements written in reverse, so the sign of $\Delta H°_f$ is reversed. Note also that reactions (2) and (3), which represent the formation of ethyl alcohol and carbon dioxide, respectively, are multiplied by 2 to arrive at the balanced equation.

When we use heats of formation to calculate standard reaction enthalpies, what we're really doing is referencing the enthalpies of both products and reactants to the same point—their constituent elements. By so doing, the product and reactant enthalpies are referenced to one another, and the difference between them is the reaction enthalpy (Figure 8.11). Examples 8.8 and 8.9 give further illustrations of how to use standard heats of formation.

FIGURE 8.11 The standard reaction enthalpy, $\Delta H°$, for the generalized reaction A → B is the difference between the standard heats of formation of products and reactants. Since the different heats of formation are referenced to the same point (the constituent elements), they are referenced to each other.

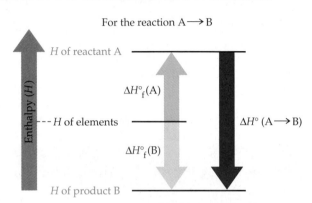

EXAMPLE 8.8 Calculate $\Delta H°$ (in kilojoules) for the synthesis of lime, CaO, from limestone, $CaCO_3$, an important step in the manufacture of cement.

$$CaCO_3(s) \longrightarrow CaO(s) + CO_2(g)$$

$$\Delta H°_f[CaCO_3(s)] = -1206.9 \text{ kJ/mol}$$
$$\Delta H°_f[CaO(s)] = -635.1 \text{ kJ/mol}$$
$$\Delta H°_f[CO_2(g)] = -393.5 \text{ kJ/mol}$$

SOLUTION Subtract the heat of formation of the reactant from the sum of the heats of formation of the products:

$$\Delta H° = [\Delta H°_f(CaO) + \Delta H°_f(CO_2)] - \Delta H°_f(CaCO_3)$$
$$= [(1 \text{ mol})(-635.1 \text{ kJ/mol}) + (1 \text{ mol})(-393.5 \text{ kJ/mol})]$$
$$- (1 \text{ mol})(-1206.9 \text{ kJ/mol})$$
$$= 178.3 \text{ kJ}$$

The reaction is endothermic by 178.3 kJ.

EXAMPLE 8.9 Oxyacetylene welding torches burn acetylene gas, $C_2H_2(g)$. Use the information in Table 8.2 to calculate $\Delta H°$ (in kilojoules) for the combustion reaction of acetylene to yield $CO_2(g)$ and $H_2O(g)$.

SOLUTION First, write the balanced equation:

$$2 C_2H_2(g) + 5 O_2(g) \longrightarrow 4 CO_2(g) + 2 H_2O(g)$$

Next, look up the appropriate heats of formation in Table 8.2:

$$\Delta H°_f[C_2H_2(g)] = 226.7 \text{ kJ/mol} \qquad \Delta H°_f[H_2O(g)] = -241.8 \text{ kJ/mol}$$
$$\Delta H°_f[CO_2(g)] = -393.5 \text{ kJ/mol}$$

Acetylene burns at a high temperature, making the reaction useful for welding.

Finally, carry out the calculation, making sure to multiply each ΔH°_f by the coefficient given in the balanced equation. Remember also that $\Delta H^\circ_f(O_2) = 0\ kJ/mol$.

$$\Delta H^\circ = [4\ \Delta H^\circ_f(CO_2) + 2\ \Delta H^\circ_f(H_2O)] - [2\ \Delta H^\circ_f(C_2H_2)]$$
$$= (4\ mol)(-393.5\ kJ/mol) + (2\ mol)(-241.8\ kJ/mol)$$
$$- (2\ mol)(226.7\ kJ/mol)$$
$$= -2511.0\ kJ$$

The combustion of 2 mol of acetylene according to the balanced equation is exothermic by 2511 kJ!

▶ **PROBLEM 8.15** Use the information in Table 8.2 to calculate ΔH° (in kilojoules) for the reaction of ammonia with O_2 to yield nitric oxide (NO) and $H_2O(g)$, a step in the Ostwald process for the commercial production of nitric acid.

▶ **PROBLEM 8.16** Use the information in Table 8.2 to calculate ΔH° (in kilojoules) for the photosynthesis of glucose from CO_2 and liquid water, a reaction carried out by all green plants.

8.11 Bond Dissociation Energies

The procedure described in the previous section for determining heats of reaction from heats of formation is extremely useful, but it still presents a major problem. To use the method, it's necessary to know ΔH°_f for every substance in a reaction. This implies, in turn, that vast numbers of measurements are needed, since there are over 18 million known chemical compounds. In practice, though, only a few thousand ΔH°_f values have been determined.

For those reactions where insufficient ΔH°_f data are available to allow an exact calculation of ΔH°, it's often possible to get an approximate answer by using the average bond dissociation energies (D) discussed previously in Section 7.2. Although we didn't identify them as such at the time, bond dissociation energies are really just standard enthalpy changes, ΔH°, for the corresponding bond-breaking reactions.

For the reaction $X-Y \longrightarrow X + Y$ $\Delta H^\circ = D = $ Bond dissociation energy

When we say, for example, that Cl_2 has a bond dissociation energy $D = 243\ kJ/mol$, we mean that the standard enthalpy change for the reaction $Cl_2(g) \rightarrow 2\ Cl(g)$ is $\Delta H^\circ = 243\ kJ/mol$. Bond dissociation energies are always positive because energy must always be put into a bond to break it.

Applying Hess's law, we can calculate an approximate enthalpy change for any reaction simply by subtracting the total energy of bonds formed in the products from the total energy of bonds broken in the reactants:

$$\Delta H^\circ = D(\text{Bonds broken}) - D(\text{Bonds formed})$$

In the reaction of H_2 with Cl_2 to yield HCl, for example, one Cl–Cl bond and one H–H bond are broken, while two H–Cl bonds are formed:

Bonds broken   Bond formed

$$H-H + Cl-Cl \longrightarrow 2\ H-Cl$$

According to the data in Table 7.1 (page 247), the bond dissociation energy of Cl_2 is 243 kJ/mol, that of H_2 is 436 kJ/mol, and that of HCl is 432 kJ/mol. We can thus calculate an approximate standard enthalpy change for the reaction.

$$\Delta H° = D(\text{Bonds broken}) - D(\text{Bonds formed})$$
$$= (D_{Cl-Cl} + D_{H-H}) - (2 D_{H-Cl})$$
$$= [(1 \text{ mol})(243 \text{ kJ/mol}) + (1 \text{ mol})(436 \text{ kJ/mol})]$$
$$- (2 \text{ mol})(432 \text{ kJ/mol})$$
$$= -185 \text{ kJ}$$

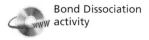
Bond Dissociation
www activity

The reaction is exothermic by approximately 185 kJ.

EXAMPLE 8.10   Use the data in Table 7.1 to find an approximate $\Delta H°$ (in kilo-joules) for the industrial synthesis of chloroform by reaction of methane with Cl_2.

$$CH_4(g) + 3 Cl_2(g) \longrightarrow CHCl_3(g) + 3 HCl(g)$$

SOLUTION   First, draw the structures of the reactants and products to see which bonds are broken and which are formed:

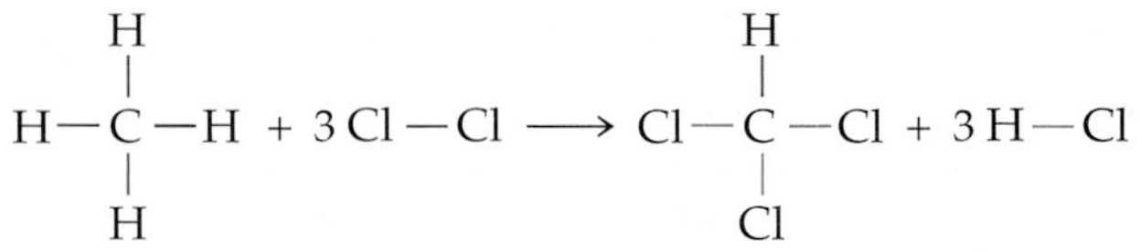

$$
\begin{array}{c}
\quad\quad H \\
\quad\quad | \\
H-C-H + 3\,Cl-Cl \longrightarrow Cl-C-Cl + 3\,H-Cl \\
\quad\quad | \\
\quad\quad H \quad\quad\quad\quad\quad\quad\quad\quad\quad\quad Cl
\end{array}
$$

Three C–H bonds and three Cl–Cl bonds are broken, while three C–Cl and three H–Cl bonds are formed. The bond dissociation energies from Table 7.1 are:

Bonds broken: C–H $D = 410$ kJ/mol   Cl–Cl $D = 243$ kJ/mol
Bonds formed: C–Cl $D = 330$ kJ/mol   H–Cl $D = 432$ kJ/mol

Subtracting the total energies of the bonds formed from the total energies of the bonds broken gives the enthalpy change for the reaction:

$$\Delta H° = D(\text{Bonds broken}) - D(\text{Bonds formed})$$
$$= [3 D_{Cl-Cl} + 3 D_{C-H}] - [3 D_{H-Cl} + 3 D_{C-Cl}]$$
$$= [(3 \text{ mol})(243 \text{ kJ/mol}) + (3 \text{ mol})(410 \text{ kJ/mol})]$$
$$- [(3 \text{ mol})(432 \text{ kJ/mol}) + (3 \text{ mol})(330 \text{ kJ/mol})]$$
$$= -327 \text{ kJ}$$

The reaction is exothermic by approximately 330 kJ.

▶ **PROBLEM 8.17**   Use the data in Table 7.1 to calculate an approximate $\Delta H°$ (in kilojoules) for the industrial synthesis of ethyl alcohol from ethylene: $C_2H_4(g) + H_2O(g) \rightarrow C_2H_5OH(g)$.

▶ **PROBLEM 8.18**   Use the data in Table 7.1 to calculate an approximate $\Delta H°$ (in kilojoules) for the synthesis of hydrazine from ammonia: $2 NH_3(g) + Cl_2(g) \rightarrow N_2H_4(g) + 2 HCl(g)$.

8.12 Fossil Fuels, Fuel Efficiency, and Heats of Combustion

Surely the most familiar of all exothermic reactions is the one that takes place every time we turn up a thermostat, drive a car, or light a match—the burning of a fuel by reaction with oxygen to yield H_2O, CO_2, and heat. The amount of energy released on burning a substance is called its **heat of combustion**, or *combustion enthalpy*, ΔH°_c. This is simply the standard enthalpy change for the reaction of 1 mol of the substance with oxygen. Hydrogen, for instance, has $\Delta H^\circ_c = -285.8$ kJ/mol, and methane has $\Delta H^\circ_c = -890.3$ kJ/mol. Note that the H_2O product is liquid rather than vapor.

$$H_2(g) + 1/2\, O_2(g) \longrightarrow H_2O(l) \qquad\qquad \Delta H^\circ_c = -285.8 \text{ kJ/mol}$$

$$CH_4(g) + 2\, O_2(g) \longrightarrow CO_2(g) + 2\, H_2O(l) \qquad \Delta H^\circ_c = -890.3 \text{ kJ/mol}$$

To compare the efficiency of different fuels, it's more useful to calculate combustion enthalpies per gram or per milliliter of substance rather than per mole (Table 8.3). For applications where weight is important, as in rocket engines, hydrogen is ideal because its combustion enthalpy per gram is the highest of any known fuel. For applications where volume is important, as in automobiles, a mixture of hydrocarbons such as those in gasoline is most efficient because hydrocarbon combustion enthalpies per milliliter are relatively high. Octane and toluene are representative examples.

TABLE 8.3 Thermochemical Properties of Some Fuels

Fuel	Combustion Enthalpy		
	kJ/mol	kJ/g	kJ/mL
Hydrogen, H_2	−285.8	−141.8	−9.9*
Ethanol, C_2H_5OH	−1367	−29.7	−23.4
Graphite, C	−393.5	−32.8	−73.8
Methane, CH_4	−890.3	−55.5	−30.8*
Methanol, CH_3OH	−726.4	−22.7	−17.9
Octane, C_8H_{18}	−5470	−47.9	−33.6
Toluene, C_7H_8	−3910	−42.3	−36.7

*Calculated for the compressed liquid at 0°C

With the exception of hydrogen, all common fuels are organic compounds, whose energy is derived ultimately from the sun through the photosynthesis of carbohydrates in green plants. Though the details are complex, the net result of the photosynthesis reaction is the conversion of carbon dioxide and water into glucose, $C_6H_{12}O_6$. Glucose, once formed, is converted into cellulose and starch, which in turn act as structural materials for plants and as food sources for animals. The conversion is highly endothermic and therefore requires a large input of solar energy. It has been estimated that the total annual amount of solar energy absorbed by the earth's vegetation is approximately 10^{19} kJ, an amount sufficient to synthesize 5×10^{14} kg of glucose per year.

$$6\, CO_2(g) + 6\, H_2O(l) \longrightarrow C_6H_{12}O_6(s) + 6\, O_2(g) \qquad \Delta H^\circ = 2816 \text{ kJ}$$

The so-called *fossil fuels* we use most—coal, natural gas, and petroleum—are the decayed remains of organisms from previous geologic eras. Both coal and petroleum are enormously complex mixtures of compounds. Coal is primarily of vegetable origin, and many of the compounds it contains are structurally similar to graphite (pure carbon). Petroleum is a viscous liquid mixture of *hydrocarbons*—compounds of carbon and hydrogen—that are primarily of marine origin.

Much coal lies near the surface of the earth and is obtained by strip mining.

Coal is burned just as it comes from the mine, but petroleum must be *refined* before use. Refining involves *distillation*—the separation of crude liquid oil into fractions on the basis of their boiling points. So-called straight-run gasoline (bp 30–200°C) consists of compounds with 5–11 carbon atoms per molecule; kerosene (bp 175–300°C) contains compounds in the C_{11}–C_{14} range; gas oil (bp 275–400°C) contains C_{14}–C_{25} substances; and lubricating oils contain whatever remaining compounds will distill. Left over is a tarry residue of asphalt (Figure 8.12).

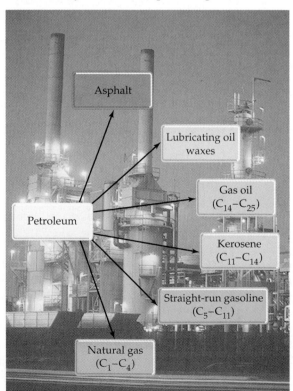

FIGURE 8.12 The products of petroleum refining. The different fractions are grouped according to the number of carbon atoms their molecules contain.

Asphalt

Lubricating oil waxes

Gas oil (C_{14}–C_{25})

Petroleum

Kerosene (C_{11}–C_{14})

Straight-run gasoline (C_5–C_{11})

Natural gas (C_1–C_4)

As the world's petroleum deposits become ever more scarce, other sources of energy will have to be found to replace them. Hydrogen, though it burns cleanly and is relatively nonpolluting, has two drawbacks: low availability and low combustion enthalpy per milliliter. Ethanol and methanol look like good current choices for alternative fuels because both can be produced relatively cheaply and have reasonable combustion enthalpies. Ethanol can be produced from wood by the breakdown of cellulose to glucose and subsequent fermentation. Methanol is produced directly from natural gas by a two-step process:

$$CH_4(g) + H_2O(g) \longrightarrow CO(g) + 3 H_2(g)$$
$$CO(g) + 2 H_2(g) \longrightarrow CH_3OH(l)$$

▶ **PROBLEM 8.19** Liquid butane (C_4H_{10}), the fuel used in many disposable lighters, has $\Delta H°_f = -147.5$ kJ/mol and a density of 0.579 g/mL. Write a balanced equation for the combustion reaction, and use Hess's law to calculate the enthalpy of combustion in kJ/mol, kJ/g, and kJ/mL.

8.13 An Introduction to Entropy

We've said on several occasions that chemical and physical processes occur spontaneously only if they go "downhill" energetically so that the final state is more stable and has less energy than the initial state. In other words, energy must be *released* for a process to occur spontaneously. At the same time, though, we've said that some processes occur perfectly well even though they are endothermic and *absorb* heat. The reaction of barium hydroxide octahydrate with ammonium chloride shown in Figure 8.6, for example, absorbs 80.3 kJ of heat ($\Delta H° = +80.3$ kJ) and leaves the surroundings so cold that water freezes around the outside of the container.

Airbags movie

$$Ba(OH)_2 \cdot 8 H_2O(s) + 2 NH_4Cl(s) \longrightarrow BaCl_2(aq) + 2 NH_3(aq) + 10 H_2O(l)$$
$$\Delta H° = +80.3 \text{ kJ}$$

What's going on? How can the spontaneous reaction of barium hydroxide octahydrate with ammonium chloride *release* energy yet *absorb* heat? The answer is that, in the context of a chemical reaction, the words "energy" and "heat" don't refer to exactly the same thing. There is another factor in addition to heat that determines whether energy is released and thus whether a reaction takes place spontaneously. We'll take only a brief look at this additional factor now and return for a more in-depth study in Chapter 17.

Before exploring the situation further, it's important to understand what the word *spontaneous* means in chemistry, for it's not the same as in everyday language. In chemistry, a **spontaneous process** is one that proceeds on its own without any continuous external influence. The change need not happen quickly, like a spring uncoiling or a rock rolling downhill. It can also happen slowly, like the gradual rusting away of an abandoned car. A **nonspontaneous process**, by contrast, takes place only in the presence of a continuous external influence. Energy must be continuously expended to re-coil a spring or to push a rock uphill. When the external influence stops,

Skiing downhill is a spontaneous process that continues once started. Lifting the skiers back uphill is a nonspontaneous process that requires a continuous input of energy.

the process also stops. The reverse of any spontaneous process is always nonspontaneous.

As another example of a process that takes place spontaneously yet absorbs heat, think about what happens when an ice cube melts. At a temperature of 0°C, ice spontaneously absorbs heat from the surroundings to turn from solid into liquid water without changing temperature.

What do a melting ice cube and the reaction of barium hydroxide octahydrate have in common? *The common feature of these and all other spontaneous processes that absorb heat is an increase in* **entropy (S)***, or molecular randomness, of the system.* The eight water molecules rigidly held in the $Ba(OH)_2 \cdot 8\ H_2O$ crystal break loose and become free to move about in the aqueous liquid product; similarly, the rigidly held H_2O molecules in the ice lose their crystalline ordering and move around more freely in liquid water.

ENTROPY (S) The amount of randomness, or molecular disorder, in a system.

Entropy has the units J/K (*joules* per kelvin, not kilojoules per kelvin) and is a quantity that can be determined for pure substances, as we'll see in Section 17.5. The larger the value of S, the greater the molecular randomness of the particles in the system. Gases, for example, have more randomness and higher entropy than liquids, and liquids have more randomness and higher entropy than solids.

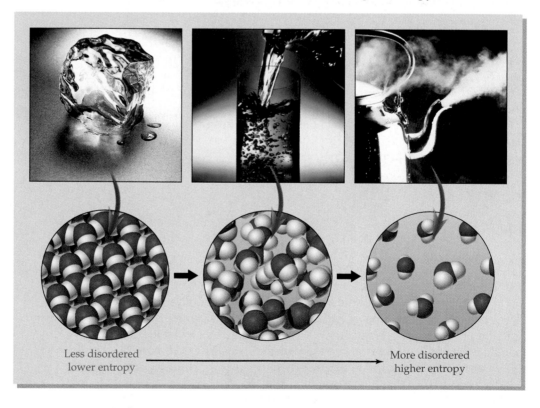

Less disordered
lower entropy
⟶
More disordered
higher entropy

A change in entropy is represented as ΔS. When randomness increases, as it does when barium hydroxide octahydrate reacts or ice melts, ΔS has a positive value. The reaction of $Ba(OH)_2 \cdot 8\ H_2O(s)$ with $NH_4Cl(s)$ has $\Delta S° = +428$ J/K, and the melting of ice has $\Delta S° = +22.0$ J/(K $\cdot$ mol). When randomness decreases, ΔS is negative. The freezing of water, for example, has $\Delta S° = -22.0 \text{ J}/(\text{K} \cdot \text{mol})$. (As with $\Delta H°$, the superscript ° is used in $\Delta S°$ to refer to the standard entropy change in a reaction where all products and reactants are in their standard states.)

$$\Delta S = S_{\text{final}} - S_{\text{initial}}$$

If ΔS is positive, the system has become more random ($S_{\text{final}} > S_{\text{initial}}$).

If ΔS is negative, the system has become less random ($S_{\text{final}} < S_{\text{initial}}$).

Two factors determine the spontaneity of a chemical or physical change in a system: a release or absorption of heat (ΔH) and an increase or decrease in molecular randomness (ΔS). To decide whether a process is spontaneous, both enthalpy and entropy changes must be taken into account:

SPONTANEOUS PROCESS: Favored by decrease in H (negative ΔH)
                         Favored by increase in S (positive ΔS)

NONSPONTANEOUS PROCESS: Favored by increase in H (positive ΔH)
                         Favored by decrease in S (negative ΔS)

Clearly, the two factors don't have to operate in the same direction. It's possible for a process to be *disfavored* by enthalpy (endothermic, positive ΔH) yet still be spontaneous because it is strongly *favored* by entropy (positive ΔS). The melting of ice [$\Delta H° = +6.01 \text{ kJ/mol}$; $\Delta S° = +22.0 \text{ J}/(\text{K} \cdot \text{mol})$] is just such a process, as is the reaction of barium hydroxide octahydrate with ammonium chloride ($\Delta H° = +80.3 \text{ kJ}$; $\Delta S° = +428 \text{ J/K}$). In the latter case, 3 mol of solid reactants produce 10 mol of liquid water, 2 mol of dissolved ammonia, and 3 mol of dissolved ions (1 mol of Ba^{2+} and 2 mol of Cl^-), with a consequent large increase in molecular randomness:

$$\underbrace{Ba(OH)_2 \cdot 8 H_2O(s) + 2 NH_4Cl(s)}_{\text{3 mol solid reactants}} \longrightarrow \underset{\substack{\text{3 mol} \\ \text{dissolved ions}}}{BaCl_2(aq)} + \underset{\substack{\text{2 mol dissolved} \\ \text{molecules}}}{2 NH_3(aq)} + \underset{\substack{\text{10 mol} \\ \text{liquid} \\ \text{water molecules}}}{10 H_2O(l)}$$

$$\Delta H° = +80.3 \text{ kJ} \longleftarrow \text{Unfavorable}$$

$$\Delta S° = +428 \text{ J/K} \longleftarrow \text{Favorable}$$

It's also possible for a process to be favored by enthalpy (exothermic, negative ΔH) yet be nonspontaneous because it is strongly disfavored by entropy (negative ΔS). The conversion of liquid water to ice is nonspontaneous above 0°C, for example, because the process is disfavored by entropy [$\Delta S° = -22.0 \text{ J}/(\text{K} \cdot \text{mol})$] even though it is favored by enthalpy ($\Delta H° = -6.01 \text{ kJ/mol}$).

EXAMPLE 8.11   Predict whether $\Delta S°$ is likely to be positive or negative for each of the following reactions:
(a) $H_2C{=}CH_2(g) + Br_2(g) \rightarrow CH_2BrCH_2Br(l)$
(b) $2 C_2H_6(g) + 7 O_2(g) \rightarrow 4 CO_2(g) + 6 H_2O(g)$

SOLUTION

(a) The amount of molecular randomness in the system decreases when 2 mol of gaseous reactants combine to give 1 mol of liquid product, so the reaction has a negative $\Delta S°$.

(b) The amount of molecular randomness in the system increases when 9 mol of gaseous reactants give 10 mol of gaseous products, so the reaction has a positive $\Delta S°$.

▶ **PROBLEM 8.20** Ethane, C_2H_6, can be prepared by the reaction of acetylene, C_2H_2, with hydrogen:

$$C_2H_2(g) + 2\,H_2(g) \longrightarrow C_2H_6(g)$$

Do you expect $\Delta S°$ for the reaction to be positive or negative? Explain.

◆ **KEY CONCEPT PROBLEM 8.21** Is the reaction represented in the following drawing likely to have a positive or a negative value of $\Delta S°$? Explain.

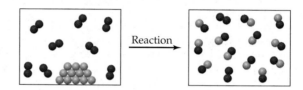

8.14 An Introduction to Free Energy

How do we weigh the contributions of both heat (enthalpy) and randomness (entropy) to the overall spontaneity of a process? To take both factors into account when deciding the spontaneity of a chemical reaction or other process, we define a quantity called the **Gibbs free-energy change (ΔG)**, $\Delta G = \Delta H - T\Delta S$.

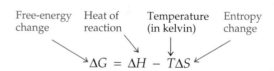

The value of the free-energy change ΔG is a general criterion for the spontaneity of a chemical or physical process. If ΔG has a negative value, the process is spontaneous; if ΔG has a value of 0, the process is neither spontaneous nor nonspontaneous but is instead at equilibrium; and if ΔG has a positive value, the process is nonspontaneous.

$\Delta G < 0$ *Process is spontaneous*

$\Delta G = 0$ *Process is at equilibrium—neither spontaneous nor nonspontaneous*

$\Delta G > 0$ *Process is nonspontaneous*

Figure 8.13 summarizes the possible combinations of ΔH and $-T\Delta S$ to give ΔG.

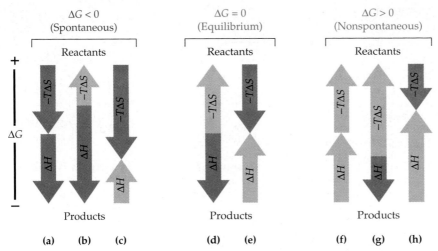

FIGURE 8.13 The possible combinations of ΔH and $-T\Delta S$. A red down arrow indicates that a process is favored, and a blue up arrow indicates that a process is not favored. Processes (a)–(c) are spontaneous: **(a)** Both ΔH and $-T\Delta S$ are favorable; **(b)** a favorable ΔH term outweighs an unfavorable $-T\Delta S$ term; **(c)** a favorable $-T\Delta S$ term outweighs an unfavorable ΔH term. Processes (d)–(e) are at equilibrium: **(d)** A favorable ΔH term is balanced by an unfavorable $-T\Delta S$ term; **(e)** an unfavorable ΔH term is balanced by a favorable $-T\Delta S$ term. Processes (f)–(h) are nonspontaneous: **(f)** Both ΔH and $-T\Delta S$ are unfavorable; **(g)** a favorable ΔH term is outweighed by an unfavorable $-T\Delta S$ term; **(h)** a favorable $-T\Delta S$ term is outweighed by an unfavorable ΔH term.

The fact that the $T\Delta S$ term in the Gibbs free-energy equation is temperature-dependent implies that some processes might be either spontaneous or nonspontaneous depending on the temperature. At low temperature, for instance, an unfavorable (positive) ΔH term might be larger than a favorable (positive) $T\Delta S$ term, but at higher temperature, the $T\Delta S$ term might be larger. Thus, an endothermic process that is nonspontaneous at low temperature can become spontaneous at higher temperature. This, in fact, is exactly what happens in the ice/water transition. At a temperature below 0°C, melting is nonspontaneous because the unfavorable ΔH term outweighs the favorable $T\Delta S$ term. At a temperature above 0°C, however, the melting of ice is spontaneous because the favorable $T\Delta S$ term outweighs the unfavorable ΔH term (Figure 8.14). Exactly at 0°C, the two terms are balanced.

$$\Delta G° = \Delta H° - T\Delta S°$$

At −10°C (263 K): $\Delta G° = 6.01\,\dfrac{kJ}{mol} - (263\,K)\left(0.0220\,\dfrac{kJ}{K\cdot mol}\right) = +0.22\,kJ/mol$

At 0°C (273 K): $\Delta G° = 6.01\,\dfrac{kJ}{mol} - (273\,K)\left(0.0220\,\dfrac{kJ}{K\cdot mol}\right) = 0.00\,kJ/mol$

At +10°C (283 K): $\Delta G° = 6.01\,\dfrac{kJ}{mol} - (283\,K)\left(0.0220\,\dfrac{kJ}{K\cdot mol}\right) = -0.22\,kJ/mol$

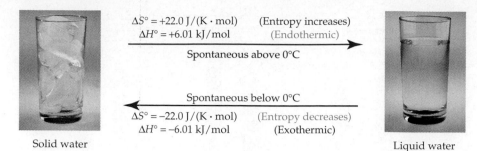

$\Delta S° = +22.0 \text{ J}/(\text{K} \cdot \text{mol})$ (Entropy increases)
$\Delta H° = +6.01 \text{ kJ/mol}$ (Endothermic)

Spontaneous above 0°C

Spontaneous below 0°C

$\Delta S° = -22.0 \text{ J}/(\text{K} \cdot \text{mol})$ (Entropy decreases)
$\Delta H° = -6.01 \text{ kJ/mol}$ (Exothermic)

Solid water   Liquid water

FIGURE 8.14   The melting of ice is disfavored by enthalpy ($+\Delta H$) but favored by entropy ($+\Delta S$). The freezing of water is favored by enthalpy ($-\Delta H$) but disfavored by entropy ($-\Delta S$). Below 0°C, the enthalpy term ΔH dominates the entropy term $T\Delta S$ in the Gibbs free-energy equation, so freezing is spontaneous. Above 0°C, the entropy term dominates the enthalpy term, so melting is spontaneous. At 0°C, the entropy and enthalpy terms are in balance.

An example of a chemical reaction in which temperature controls spontaneity is that of carbon with water to yield carbon monoxide. The reaction has an unfavorable ΔH term (positive) but a favorable $T\Delta S$ term (positive) because randomness increases when a solid and a gas are converted into two gases:

$$C(s) + H_2O(g) \longrightarrow CO(g) + H_2(g) \qquad \begin{array}{ll} \Delta H° = +131 \text{ kJ} & \text{Unfavorable} \\ \Delta S° = +134 \text{ J/K} & \text{Favorable} \end{array}$$

The conversion of reactants to products in their standard states is not spontaneous at room temperature (300 K) because the unfavorable ΔH term is larger than the favorable $T\Delta S$ term. At approximately 978 K (705°C), however, the reaction becomes spontaneous because the favorable $T\Delta S$ term becomes larger than the unfavorable ΔH term. Below 978 K, ΔG has a positive value; at 978 K, $\Delta G = 0$; and above 978 K, ΔG has a negative value:

$$\Delta G° = \Delta H° - T\Delta S°$$

At 695°C (968 K): $\Delta G° = 131 \text{ kJ} - (968 \text{ K})\left(0.134 \dfrac{\text{kJ}}{\text{K}}\right) = +1 \text{ kJ}$

At 705°C (978 K): $\Delta G° = 131 \text{ kJ} - (978 \text{ K})\left(0.134 \dfrac{\text{kJ}}{\text{K}}\right) = 0 \text{ kJ}$

At 715°C (988 K): $\Delta G° = 131 \text{ kJ} - (988 \text{ K})\left(0.134 \dfrac{\text{kJ}}{\text{K}}\right) = -1 \text{ kJ}$

The reaction of carbon with water is, in fact, the first step of an industrial process used to manufacture methanol (CH_3OH). As supplies of natural gas and oil are used up, this reaction may become important for the manufacture of synthetic fuels.

A process is at equilibrium when it is balanced between spontaneous and nonspontaneous—that is, when $\Delta G = 0$ and it is energetically unfavorable to go either from reactants to products or from products to reactants. Thus, at the equilibrium point, we can set up the equation

$$\Delta G = \Delta H - T\Delta S = 0 \qquad \text{At equilibrium}$$

Solving this equation for T gives

$$T = \frac{\Delta H}{\Delta S}$$

which makes it possible to calculate the temperature at which a changeover in behavior between spontaneous and nonspontaneous occurs. Using the known values of ΔH and ΔS for the melting of ice, for instance, we find that the point at which liquid water and solid ice are in equilibrium is

$$T = \frac{\Delta H^\circ}{\Delta S^\circ} = \frac{6.01 \text{ kJ}}{0.0220 \frac{\text{kJ}}{\text{K}}} = 273 \text{ K} = 0°C$$

It's no surprise to find that the ice/water equilibrium point is 273 K, or 0°C, the melting point of ice.

In the same way, the temperature at which the reaction of carbon with water changes between spontaneous and nonspontaneous is 978 K, or 705°C:

$$T = \frac{\Delta H^\circ}{\Delta S^\circ} = \frac{131 \text{ kJ}}{0.134 \frac{\text{kJ}}{\text{K}}} = 978 \text{ K}$$

EXAMPLE 8.12   Quicklime, CaO, is produced by heating limestone, $CaCO_3$, to drive off CO_2 gas. Is the reaction spontaneous under standard conditions at 25°C? Calculate the temperature at which the reaction becomes spontaneous.

$$CaCO_3(s) \longrightarrow CaO(s) + CO_2(g) \qquad \Delta H^\circ = 178.3 \text{ kJ}; \Delta S^\circ = 160 \text{ J/K}$$

SOLUTION   The spontaneity of the reaction at a given temperature can be found by determining whether ΔG is positive or negative. At 25°C (298 K), we have

$$\Delta G^\circ = \Delta H^\circ - T\Delta S^\circ = 178.3 \text{ kJ} - (298 \text{ K})\left(0.160 \frac{\text{kJ}}{\text{K}}\right) = +130.6 \text{ kJ}$$

Since ΔG is positive at this temperature, the reaction is nonspontaneous.

The changeover point between spontaneous and nonspontaneous can be found by setting $\Delta G = 0$ and rearranging the equation:

$$T = \frac{\Delta H^\circ}{\Delta S^\circ} = \frac{178.3 \text{ kJ}}{0.160 \frac{\text{kJ}}{\text{K}}} = 1114 \text{ K}$$

The reaction becomes spontaneous above 1114 K (841°C).

▶ **PROBLEM 8.22**   Which of the following reactions are spontaneous under standard conditions at 25°C, and which are nonspontaneous?
(a) $AgNO_3(aq) + NaCl(aq) \rightarrow AgCl(s) + NaNO_3(aq)$; $\Delta G^\circ = -55.7$ kJ
(b) $2 \text{ C}(s) + 2 \text{ H}_2(g) \rightarrow C_2H_4(g)$; $\Delta G^\circ = 68.1$ kJ

▶ **PROBLEM 8.23**   Is the Haber process for the industrial synthesis of ammonia spontaneous or nonspontaneous under standard conditions at 25°C? At what temperature (°C) does the changeover occur?

$$N_2(g) + 3 \text{ H}_2(g) \longrightarrow 2 \text{ NH}_3(g) \qquad \Delta H^\circ = -92.2 \text{ kJ}; \Delta S^\circ = -199 \text{ J/K}$$

The Evolution of Endothermy

Marlin are one of only a handful of fish in the ocean that are partially warm-blooded.

Most people cringe at the thought of a cold shower. Imagine what it must be like for fish, which spend their lives immersed in bone-chilling water. Water is such an efficient heat conductor and cools bodies so rapidly that most fish don't even try to keep warm: Of the approximately 30,000 known species of bony fishes, all but a handful are cold-blooded. Only tuna, mackerel, billfish (marlin, swordfish), and a few others are warm-blooded.

The active lives of mammals and birds require a high metabolic rate, which these animals achieve by using the energy released during exothermic metabolic reactions to maintain a high body temperature. Fish, however, function at a much lower metabolic rate than mammals and are therefore able to live with a lower, more variable body temperature. Thus, from an evolutionary point of view, the development of warm-bloodedness (*endothermy*) in a few fish species is an extraordinary occurrence, and the reasons for that development are not fully understood. Recent evidence indicates that endothermy has evolved independently in three different fish lineages, probably as a means for the fish to expand their habitat into colder water. The warm-blooded bluefin tuna, for example, migrates annually between tropical and polar waters, and is at home in both.

The different ways in which endothermy has developed in different fishes may provide valuable clues about how it developed in mammals and birds. Most important, it appears that warm-bloodedness is not necessarily an all-or-nothing proposition; it may well have evolved in steps. Both billfish and the butterfly mackerel, for instance, keep only their eyes and brains warm; the rest of their bodies are cold-blooded. Both species accomplish their partial warm-bloodedness through the use of a special heater muscle in the eye. The muscle operates at a high metabolic rate and acts as a heat exchanger to warm the blood passing though it. A special network of blood vessels then distributes this heated blood directly to the eye and brain.

Tunas, by contrast, have evolved a mechanism of endothermy completely different from that of mackerel and billfish. In most fish, the powerful aerobic muscle used for swimming is located just beneath the skin, but in tuna, the muscle is positioned centrally in the body, where it is thermally insulated from the outside water. The heat produced by this muscle during swimming is captured by circulating blood and transported to the cranium through the vascular system. Exactly how all this information about fish relates to the development of endothermy in mammals is unclear, but it appears that warm-bloodedness is a much more complex process than once believed.

Summary

Energy is of two types: *kinetic* and *potential*. **Kinetic energy (E_K)** is the energy of motion; its value depends on both the mass m and velocity v of an object according to the equation $E_K = (1/2)mv^2$. **Potential energy (E_P)** is the energy stored in an object because of its position or in a chemical substance because of its composition. **Heat** is the thermal energy transferred between two objects as the result of a temperature difference, whereas **temperature** is a measure of the kinetic energy of molecular motion.

According to the **conservation of energy law**, also known as the **first law of thermodynamics**, energy can be neither created nor destroyed. Thus, the total energy of an isolated system must remain constant. The total **internal energy (E)** of a system—the sum of all kinetic and potential energies for each particle in the system—is a **state function** because its value depends only on the present condition of the system, not on how that condition was reached.

Work (w) is defined as the distance moved times the force that produces the motion. In chemistry, most work is expansion work (PV work) done as the result of a volume change during a reaction when air molecules are pushed aside. The amount of work done by an expanding gas is given by the equation $w = -P\Delta V$, where P is the pressure against which the system must push and ΔV is the change in volume of the system.

The total internal energy change that takes place during a reaction is the sum of the heat transferred (q) and the work done ($-P\Delta V$). The equation

$$\Delta E = q + (-P\Delta V) \qquad \text{or} \qquad q = \Delta E + P\Delta V = \Delta H$$

where ΔH is called the **enthalpy change** of the system, is one of the fundamental equations of thermochemistry. In general, the $P\Delta V$ term is much smaller than the ΔE term so that the total internal energy change of a reacting system is approximately equal to ΔH, also called the **heat of reaction**. Reactions that have a negative ΔH are said to be **exothermic** because heat is lost by the system, and reactions that have a positive ΔH are said to be **endothermic** because heat is absorbed by the system.

Because enthalpy is a state function, ΔH is the same no matter what path is taken between reactants and products. Thus, the sum of the enthalpy changes for the individual steps in a reaction is equal to the overall enthalpy change for the entire reaction, a relationship known as **Hess's law**. Using this law, it is possible to calculate overall enthalpy changes for individual steps that can't be measured directly. Hess's law also makes it possible to calculate the enthalpy change of any reaction if the standard heats of formation ($\Delta H°_f$) are known for the reactants and products. The **standard heat of formation** is the enthalpy change for the hypothetical formation of 1 mol of a substance in its **thermodynamic standard state** from the most stable forms of the constituent elements in their standard states (1 atm pressure and a specified temperature, usually 25°C).

In addition to enthalpy, **entropy (S)**—a measure of the amount of molecular randomness in a system—is also important in determining whether a given process will occur spontaneously. Together, changes in enthalpy and entropy define a quantity called the **Gibbs free-energy change (ΔG)** according to the equation $\Delta G = \Delta H - T\Delta S$. The value of ΔG is a general criterion for whether a reaction will take place spontaneously. If ΔG is negative, the reaction is **spontaneous**; if ΔG is positive, the reaction is **nonspontaneous**.

Key Words

conservation of energy law *301*

endothermic *312*

energy *300*

enthalpy (H) *308*

enthalpy change (ΔH) *308*

entropy (S) *328*

exothermic *312*

first law of thermodynamics *302*

Gibbs free-energy change (ΔG) *330*

heat *302*

heat capacity (C) *314*

heat of combustion *325*

heat of reaction *308*

Hess's law *317*

internal energy (E) *303*

kinetic energy (E_K) *300*

molar heat capacity (C_m) *314*

nonspontaneous *327*

potential energy (E_P) *300*

specific heat *314*

spontaneous *327*

standard enthalpy of reaction ($\Delta H°$) *310*

standard heat of formation ($\Delta H°_f$) *320*

state function *304*

sublimation *311*

temperature *302*

thermochemistry *299*

thermodynamic standard state *310*

work (w) *305*

Key Concept Summary

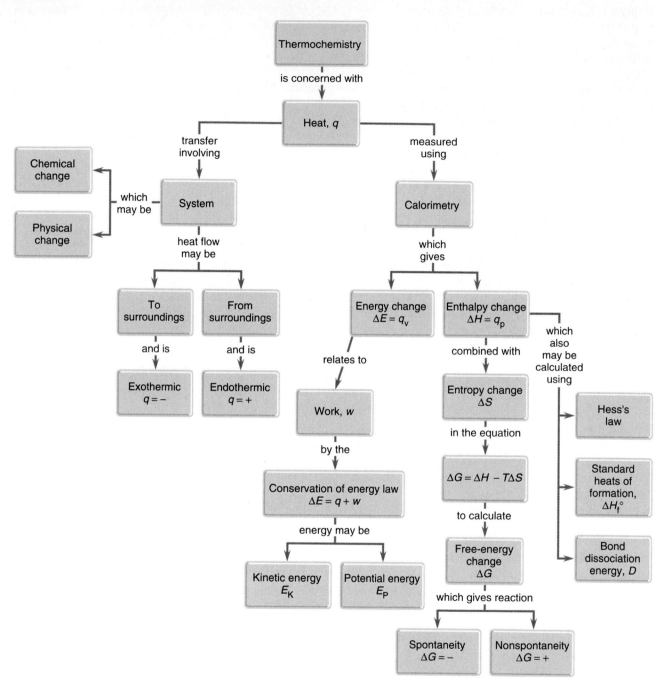

Understanding Key Concepts

Problems 8.1–8.23 appear within the chapter.

8.24 Imagine a reaction that results in a change in both volume and temperature:

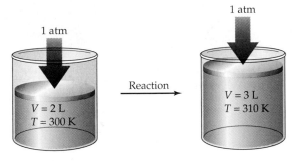

(a) Has any work been done? If so, is its sign positive or negative?

(b) Has there been an enthalpy change? If so, what is the sign of ΔH? Is the reaction exothermic or endothermic?

8.25 Redraw the following diagram to represent the situation (a) when work has been gained by the system and (b) when work has been lost by the system:

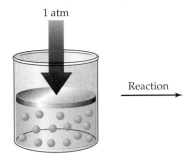

8.26 Acetylene, C_2H_2, reacts with H_2 in two steps to yield ethane, CH_3CH_3:

(1) $HC \equiv CH + H_2 \longrightarrow CH_2 = CH_2$ $\Delta H^\circ = -174.4$ kJ
(2) $CH_2 = CH_2 + H_2 \longrightarrow CH_3CH_3$ $\Delta H^\circ = -137.0$ kJ
Net $HC \equiv CH + 2\,H_2 \longrightarrow CH_3CH_3$ $\Delta H^\circ = -311.4$ kJ

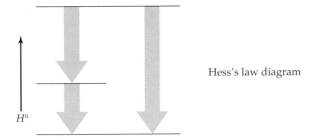

Hess's law diagram

Which arrow in the Hess's law diagram corresponds to which step, and which arrow corresponds to the net reaction? Where are the reactants located on the diagram, and where are the products located?

8.27 Draw a Hess's law diagram similar to the one in Problem 8.26 for the reaction of ethyl alcohol (CH_3CH_2OH) with oxygen to yield acetic acid (CH_3CO_2H).

(1) $CH_3CH_2OH(l) + 1/2\,O_2(g) \longrightarrow$
        $CH_3CHO(g) + H_2O(l)$ $\Delta H^\circ = -174.2$ kJ
(2) $CH_3CHO(g) + 1/2\,O_2(g) \longrightarrow$
        $CH_3CO_2H(l)$ $\Delta H^\circ = -318.4$ kJ
Net $CH_3CH_2OH(l) + O_2(g) \longrightarrow$
        $CH_3CO_2H(l) + H_2O(l)$ $\Delta H^\circ = -492.6$ kJ

8.28 A reaction is carried out in a cylinder fitted with a movable piston, as shown below. The starting volume is $V = 5.00$ L, and the apparatus is held at constant temperature and pressure. Assuming that $\Delta H = -35.0$ kJ and $\Delta E = -34.8$ kJ, redraw the piston to show its position after reaction. Does V increase, decrease, or remain the same?

8.29 The following drawing portrays a reaction of the type $A \longrightarrow B + C$, where the different colored spheres represent different molecular structures. Assume that the reaction has $\Delta H^\circ = +55$ kJ. Is the reaction likely to be spontaneous at all temperatures, nonspontaneous at all temperatures, or spontaneous at some but nonspontaneous at others? Explain.

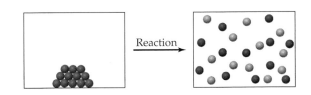

8.30 What are the signs of ΔH, ΔS, and ΔG for the following spontaneous change? Explain.

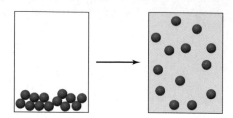

8.31 Consider the following spontaneous reaction of A_3 molecules:

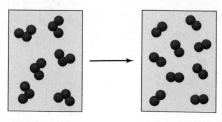

(a) Write a balanced equation for the reaction.

(b) What are the signs of ΔH, ΔS, and ΔG for the reaction? Explain.

Additional Problems

Heat, Work, and Energy

8.32 What is the difference between heat and temperature? Between work and energy? Between kinetic energy and potential energy?

8.33 What is meant by the term *internal energy*?

8.34 Which has more kinetic energy, a 1400 kg car moving at 115 km/h or a 12,000 kg truck moving at 38 km/h?

8.35 Assume that the kinetic energy of a 1400 kg car moving at 115 km/h (Problem 8.34) could be converted entirely into heat. What amount of water could be heated from 20°C to 50°C by the car's energy?

8.36 Calculate the work done (in joules) by a chemical reaction if the volume increases from 3.2 L to 3.4 L against a constant external pressure of 3.6 atm? What is the sign of the energy change?

8.37 The addition of H_2 to C=C double bonds is an important reaction used in the preparation of margarine from vegetable oils. If 50.0 mL of H_2 and 50.0 mL of ethylene (C_2H_4) are allowed to react at 1.5 atm, the product ethane (C_2H_6) has a volume of 50.0 mL. Calculate the amount of PV work done, and tell the direction of the energy flow.

$$C_2H_4(g) + H_2(g) \longrightarrow C_2H_6(g)$$

Energy and Enthalpy

8.38 What is the difference between internal energy change ΔE and enthalpy change ΔH? Which of the two is measured at constant pressure and which at constant volume?

8.39 What is the sign of ΔH for an exothermic reaction? For an endothermic reaction?

8.40 Under what circumstances are ΔE and ΔH essentially equal?

8.41 Which of the following has the highest enthalpy content and which the lowest at a given temperature: $H_2O(s)$, $H_2O(l)$, or $H_2O(g)$? Explain.

8.42 The enthalpy change for the reaction of 50.0 mL of ethylene with 50.0 mL of H_2 at 1.5 atm pressure (Problem 8.37) is $\Delta H = -0.31$ kJ. What is the value of ΔE?

8.43 Assume that a particular reaction evolves 244 kJ of heat and that 35 kJ of PV work is gained by the system. What are the values of ΔE and ΔH for the system? For the surroundings?

8.44 Used in welding metals, the reaction of acetylene with oxygen has $\Delta H° = -1255.5$ kJ:

$$C_2H_2(g) + 5/2\, O_2(g) \longrightarrow H_2O(g) + 2\, CO_2(g)$$
$$\Delta H° = -1255.5 \text{ kJ}$$

How much PV work is done (in kilojoules) and what is the value of ΔE (in kilojoules) for the reaction of 6.50 g of acetylene at atmospheric pressure if the volume change is -2.80 L?

8.45 Ethyl chloride (C_2H_5Cl), a substance used as a topical anesthetic, is prepared by reaction of ethylene with hydrogen chloride:

$$C_2H_4(g) + HCl(g) \longrightarrow C_2H_5Cl(g)$$
$$\Delta H° = -72.3 \text{ kJ}$$

How much PV work is done (in kilojoules) and what is the value of ΔE (in kilojoules) if 89.5 g of ethylene and 125 g of HCl are allowed to react at atmospheric pressure and the volume change is -71.5 L?

8.46 The familiar "ether" used as an anesthetic agent is diethyl ether, $C_4H_{10}O$. Its heat of vaporization is +26.5 kJ/mol at its boiling point. How much energy (in kilojoules) is required to convert 100 mL of diethyl ether at its boiling point from liquid to vapor if its density is 0.7138 g/mL?

8.47 How much energy (in kilojoules) is required to convert 100 mL of water at its boiling point from liquid to vapor, and how does this compare with the result calculated in Problem 8.46 for diethyl ether? [$\Delta H_{vap}(H_2O) = +40.7$ kJ/mol]

8.48 Aluminum metal reacts with chlorine with a spectacular display of sparks:

$$2\,Al(s) + 3\,Cl_2(g) \longrightarrow 2\,AlCl_3(s)$$
$$\Delta H° = -1408.4 \text{ kJ}$$

How much heat (in kilojoules) is released on reaction of 5.00 g of Al?

8.49 How much heat (in kilojoules) is evolved or absorbed in the reaction of 1.00 g of Na with H_2O? Is the reaction exothermic or endothermic?

$$2\,Na(s) + 2\,H_2O(l) \longrightarrow 2\,NaOH(aq) + H_2(g)$$
$$\Delta H° = -368.4 \text{ kJ}$$

8.50 How much heat (in kilojoules) is evolved or absorbed in the reaction of 2.50 g of Fe_2O_3 with enough carbon monoxide to produce iron metal? Is the process exothermic or endothermic?

$$Fe_2O_3(s) + 3\,CO(g) \longrightarrow 2\,Fe(s) + 3\,CO_2(g)$$
$$\Delta H° = -24.8 \text{ kJ}$$

8.51 How much heat (in kilojoules) is evolved or absorbed in the reaction of 233.0 g of calcium oxide with enough carbon to produce calcium carbide? Is the process exothermic or endothermic?

$$CaO(s) + 3\,C(s) \longrightarrow CaC_2(s) + CO(g)$$
$$\Delta H° = 464.8 \text{ kJ}$$

Calorimetry and Heat Capacity

8.52 What is the difference between heat capacity and specific heat?

8.53 Does a measurement carried out in a bomb calorimeter give a value for ΔH or ΔE? Explain.

8.54 Sodium metal is sometimes used as a cooling agent in heat-exchange units because of its relatively high molar heat capacity of 28.2 J/(mol · °C). What is the specific heat of sodium in J/(g · °C)?

8.55 Titanium metal is used as a structural material in many high-tech applications such as in jet engines. What is the specific heat of titanium in J/(g · °C) if it takes 89.7 J to raise the temperature of a 33.0 g block by 5.20°C? What is the molar heat capacity of titanium in J/(mol · °C)?

8.56 When 1.045 g of CaO is added to 50.0 mL of water at 25.0°C in a calorimeter, the temperature of the water increases to 32.3°C. Assuming that the specific heat of the solution is 4.18 J/(g · °C) and that the calorimeter itself absorbs a negligible amount of heat, calculate ΔH (in kilojoules) for the reaction

$$CaO(s) + H_2O(l) \longrightarrow Ca(OH)_2(aq)$$

8.57 When 0.187 g of benzene, C_6H_6, is burned in a bomb calorimeter, the surrounding water bath rises in temperature by 7.48°C. Assuming that the bath contains 250.0 g of water and that the calorimeter itself absorbs a negligible amount of heat, calculate combustion enthalpies for benzene in both kilojoules per gram and kilojoules per mole.

8.58 When a solution containing 8.00 g of NaOH in 50.0 g of water at 25.0°C is added to a solution of 8.00 g of HCl in 250.0 g of water at 25.0°C in a calorimeter, the temperature of the solution increases to 33.5°C. Assuming that the specific heat of the solution is 4.18 J/(g · °C) and that the calorimeter absorbs a negligible amount of heat, calculate ΔH (in kilojoules) for the reaction

$$NaOH(aq) + HCl(aq) \longrightarrow NaCl(aq) + H_2O(l)$$

When the experiment is repeated using a solution of 10.00 g of HCl in 248.0 g of water, the same temperature increase is observed. Explain.

8.59 Instant cold packs used to treat athletic injuries contain solid NH_4NO_3 and a pouch of water. When the pack is squeezed, the pouch breaks and the solid dissolves, lowering the temperature because of the endothermic reaction

$$NH_4NO_3(s) + H_2O(l) \longrightarrow NH_4NO_3(aq)$$

$$\Delta H = +25.7 \text{ kJ}$$

What is the final temperature in a squeezed cold pack that contains 50.0 g of NH_4NO_3 dissolved in 125 mL of water? Assume a specific heat of 4.18 J/(g · °C) for the solution, an initial temperature of 25.0°C, and no heat transfer between the cold pack and the environment.

Hess's Law and Heats of Formation

8.60 How is the standard state of an element defined?

8.61 What is a compound's standard heat of formation?

8.62 What is Hess's law, and why does it "work"?

8.63 Why do elements always have $\Delta H^\circ_f = 0$?

8.64 Sulfuric acid (H_2SO_4), the most widely produced chemical in the world, is made by a two-step oxidation of sulfur to sulfur trioxide, SO_3, followed by reaction with water. Calculate ΔH°_f for SO_3 (in kilojoules per mole), given the following data:

$$S(s) + O_2(g) \longrightarrow SO_2(g) \qquad \Delta H^\circ = -296.8 \text{ kJ}$$
$$SO_2(g) + 1/2\, O_2(g) \longrightarrow SO_3(g) \qquad \Delta H^\circ = -98.9 \text{ kJ}$$

8.65 Calculate ΔH°_f (in kilojoules per mole) for benzene, C_6H_6, from the following data:

$$2\, C_6H_6(l) + 15\, O_2(g) \longrightarrow 12\, CO_2(g) + 6\, H_2O(l)$$
$$\Delta H^\circ = -6534 \text{ kJ}$$
$$\Delta H^\circ_f(H_2O) = -285.8 \text{ kJ/mol}$$
$$\Delta H^\circ_f(CO_2) = -393.5 \text{ kJ/mol}$$

8.66 The standard enthalpy change for the reaction of $SO_3(g)$ with $H_2O(l)$ to yield $H_2SO_4(aq)$ is $\Delta H^\circ = -227.8$ kJ. Use the information in Problem 8.64 to calculate ΔH°_f for $H_2SO_4(aq)$ (in kilojoules per mole). [For $H_2O(l)$, $\Delta H^\circ_f = -285.8$ kJ/mol.]

8.67 Acetic acid (CH_3CO_2H), whose aqueous solutions are known as *vinegar*, is prepared by reaction of ethyl alcohol (CH_3CH_2OH) with oxygen:

$$CH_3CH_2OH(l) + O_2(g) \longrightarrow CH_3CO_2H(l) + H_2O(l)$$

Use the following data to calculate ΔH° (in kilojoules) for the reaction:

$$\Delta H^\circ_f(CH_3CH_2OH) = -277.7 \text{ kJ/mol}$$
$$\Delta H^\circ_f(CH_3CO_2H) = -484.5 \text{ kJ/mol}$$
$$\Delta H^\circ_f(H_2O) = -285.8 \text{ kJ/mol}$$

8.68 Styrene (C_8H_8), the precursor of polystyrene polymers, has a standard heat of combustion of -4395.2 kJ/mol. Write a balanced equation for the combustion reaction, and calculate ΔH°_f for styrene (in kilojoules per mole). [$\Delta H^\circ_f(CO_2) = -393.5$ kJ/mol; $\Delta H^\circ_f(H_2O) = -285.8$ kJ/mol]

8.69 Methyl *tert*-butyl ether (MTBE), $C_5H_{12}O$, a gasoline additive used to boost octane ratings, has $\Delta H^\circ_f = -313.6$ kJ/mol. Write a balanced equation for its combustion reaction, and calculate its standard heat of combustion in kilojoules.

8.70 Methyl *tert*-butyl ether (Problem 8.69) is prepared by reaction of methanol ($\Delta H^\circ_f = -238.7$ kJ/mol) with 2-methylpropene according to the equation

$$\underset{\text{2-Methylpropene}}{CH_3-\underset{\underset{CH_3}{|}}{C}=CH_2} + CH_3OH \longrightarrow$$

$$\underset{\text{Methyl }tert\text{-butyl ether}}{CH_3-\underset{\underset{CH_3}{|}}{\overset{\overset{CH_3}{|}}{C}}-O-CH_3} \qquad \Delta H^\circ = -57.8 \text{ kJ}$$

Calculate ΔH°_f (in kilojoules per mole) for 2-methylpropene.

8.71 One possible use for the cooking fat left over after making french fries is to burn it as fuel. Write a balanced equation, and use the following data to calculate the amount of energy released (in kilojoules per milliliter) from the combustion of cooking fat:

Formula $= C_{51}H_{88}O_6$
Density $= 0.94$ g/mL $\qquad \Delta H^\circ_f = -1310$ kJ/mol

Bond Dissociation Energies

8.72 Use the bond dissociation energies in Table 7.1 to calculate an approximate $\Delta H°$ (in kilojoules) for the reaction of ethylene with hydrogen to yield ethane.

$$H_2C{=}CH_2(g) + H_2(g) \longrightarrow CH_3CH_3(g)$$

8.73 Use the bond dissociation energies in Table 7.1 to calculate an approximate $\Delta H°$ (in kilojoules) for the industrial synthesis of isopropyl alcohol (rubbing alcohol) by reaction of water with propene.

$$\underset{\text{Propene}}{CH_3CH{=}CH_2} + H_2O \longrightarrow \underset{\text{Isopropyl alcohol}}{CH_3\overset{\displaystyle OH}{\underset{\displaystyle |}{C}}HCH_3}$$

8.74 Calculate an approximate heat of combustion for butane (in kilojoules) by using the bond dissociation energies in Table 7.1. (The strength of the O=O bond is 498 kJ/mol, and that of a C=O bond in CO_2 is 804 kJ/mol.)

Butane

8.75 Use the bond dissociation energies in Table 7.1 to calculate an approximate heat of reaction, $\Delta H°$ (in kilojoules), for the industrial reaction of ethanol with acetic acid to yield ethyl acetate (used as nail-polish remover), and water.

Free Energy and Entropy

8.76 What does entropy measure?

8.77 What are the two terms that make up the free-energy change for a reaction, ΔG, and which of the two is usually more important?

8.78 How is it possible for a reaction to be spontaneous yet endothermic?

8.79 Is it possible for a reaction to be nonspontaneous yet exothermic? Explain.

8.80 Tell whether the entropy changes for the following processes are likely to be positive or negative:

(a) The fizzing of a newly opened can of soda

(b) The growth of a plant from seed

8.81 Tell whether the entropy changes, ΔS, for the following processes are likely to be positive or negative:

(a) The conversion of liquid water to water vapor at 100°C

(b) The freezing of liquid water to ice at 0°C

(c) The eroding of a mountain by a glacier

8.82 Tell whether the free-energy changes, ΔG, for the processes listed in Problem 8.81 are likely to be positive, negative, or zero.

8.83 When a bottle of perfume is opened, odorous molecules mix with air and slowly diffuse throughout the entire room. Is ΔG for the diffusion process positive, negative, or zero? What about ΔH and ΔS for the diffusion?

8.84 One of the steps in the cracking of petroleum into gasoline involves the thermal breakdown of large hydrocarbon molecules into smaller ones. For example, the following reaction might occur:

$$C_{11}H_{24} \longrightarrow C_4H_{10} + C_4H_8 + C_3H_6$$

Is ΔS for this reaction likely to be positive or negative? Explain.

8.85 The commercial production of 1,2-dichloroethane, a solvent used in dry cleaning, involves the reaction of ethylene with chlorine:

$$C_2H_4(g) + Cl_2(g) \longrightarrow C_2H_4Cl_2(l)$$

Is ΔS for this reaction likely to be positive or negative? Explain.

8.86 Tell whether reactions with the following values of ΔH and ΔS are spontaneous or nonspontaneous and whether they are exothermic or endothermic:

(a) $\Delta H = -48$ kJ; $\Delta S = +135$ J/K at 400 K

(b) $\Delta H = -48$ kJ; $\Delta S = -135$ J/K at 400 K

(c) $\Delta H = +48$ kJ; $\Delta S = +135$ J/K at 400 K

(d) $\Delta H = +48$ kJ; $\Delta S = -135$ J/K at 400 K

8.87 Tell whether reactions with the following values of ΔH and ΔS are spontaneous or nonspontaneous and whether they are exothermic or endothermic:

(a) $\Delta H = -128$ kJ; $\Delta S = 35$ J/K at 500 K

(b) $\Delta H = +67$ kJ; $\Delta S = -140$ J/K at 250 K

(c) $\Delta H = +75$ kJ; $\Delta S = 95$ J/K at 800 K

8.88 Suppose that a reaction has $\Delta H = -33$ kJ and $\Delta S = -58$ J/K. At what temperature will it change from spontaneous to nonspontaneous?

8.89 Suppose that a reaction has $\Delta H = +41$ kJ and $\Delta S = -27$ J/K. At what temperature, if any, will it change between spontaneous and nonspontaneous?

8.90 Which of the reactions (a)–(d) in Problem 8.86 are spontaneous at all temperatures, which are nonspontaneous at all temperatures, and which have an equilibrium temperature?

8.91 Vinyl chloride ($H_2C=CHCl$), the starting material used in the industrial preparation of poly(vinyl chloride), is prepared by a two-step process that begins with the reaction of Cl_2 with ethylene to yield 1,2-dichloroethane:

$$Cl_2(g) + H_2C=CH_2(g) \longrightarrow ClCH_2CH_2Cl(l)$$

$$\Delta H° = -217.5 \text{ kJ}$$
$$\Delta S° = -233.9 \text{ J/K}$$

(a) Tell whether the reaction is favored by entropy, by enthalpy, by both, or by neither, and then calculate $\Delta G°$ at 298 K.

(b) Tell whether the reaction has an equilibrium temperature between spontaneous and nonspontaneous. If yes, calculate the equilibrium temperature.

8.92 Ethyl alcohol has $\Delta H_{fusion} = 5.02$ kJ/mol and melts at $-114.1°$C. What is the value of ΔS_{fusion} for ethyl alcohol?

8.93 Chloroform has $\Delta H_{vaporization} = 29.2$ kJ/mol and boils at $61.2°$C. What is the value of $\Delta S_{vaporization}$ for chloroform?

General Problems

8.94 When 1.50 g of magnesium metal is allowed to react with 200 mL of 6.00 M aqueous HCl, the temperature rises from 25.0°C to 42.9°C. Calculate ΔH (in kilojoules) for the reaction, assuming that the heat capacity of the calorimeter is 776 J/°C, that the specific heat of the final solution is the same as that of water [(4.18 J/(g · °C)], and that the density of the solution is 1.00 g/mL.

8.95 Use the data in Appendix B to find standard enthalpies of reaction (in kilojoules) for the following processes:

(a) $C(s) + CO_2(g) \rightarrow 2\ CO(g)$

(b) $2\ H_2O_2(aq) \rightarrow 2\ H_2O(l) + O_2(g)$

(c) $Fe_2O_3(s) + 3\ CO(g) \rightarrow 2\ Fe(s) + 3\ CO_2(g)$

8.96 Find $\Delta H°$ (in kilojoules) for the reaction of nitric oxide with oxygen, $2\ NO(g) + O_2(g) \rightarrow N_2O_4(g)$, given the following thermochemical data:

$$N_2O_4(g) \longrightarrow 2\ NO_2(g) \qquad \Delta H° = 57.2 \text{ kJ}$$
$$NO(g) + 1/2\ O_2(g) \longrightarrow NO_2(g) \qquad \Delta H° = -57.0 \text{ kJ}$$

8.97 The boiling point of a substance is defined as the temperature at which liquid and vapor coexist in equilibrium. Use the heat of vaporization ($\Delta H_{vap} = 30.91$ kJ/mol) and entropy of vaporization [$\Delta S_{vap} = 93.2$ J/(K · mol)] to calculate the boiling point (°C) of liquid bromine.

8.98 What is the melting point of benzene (in kelvins) if $\Delta H_{fusion} = 9.95$ kJ/mol and $\Delta S_{fusion} = 35.7$ J/(K · mol)?

8.99 Metallic mercury is obtained by heating the mineral cinnabar (HgS) in air:

$$HgS(s) + O_2(g) \longrightarrow Hg(l) + SO_2(g)$$

(a) Use the data in Appendix B to calculate $\Delta H°$ (in kilojoules) for the reaction.

(b) The entropy change for the reaction is $\Delta S° = +36.7$ J/K. Is the reaction spontaneous at 25°C?

(c) Under what conditions, if any, is the reaction nonspontaneous? Explain.

8.100 Use the average bond dissociation energies in Table 7.1 to calculate approximate reaction enthalpies (in kilojoules) for the following processes:

(a) $2\ CH_4(g) \rightarrow C_2H_6(g) + H_2(g)$

(b) $C_2H_6(g) + F_2(g) \rightarrow C_2H_5F(g) + HF(g)$

(c) $N_2(g) + 3\ H_2(g) \rightarrow 2\ NH_3(g)$

8.101 Methanol (CH_3OH) is made industrially in two steps from CO and H_2. It is so cheap to make that it is being considered for use as a precursor to hydrocarbon fuels such as methane (CH_4):

Step 1. $CO(g) + 2\ H_2(g) \longrightarrow CH_3OH(l)$
$$\Delta S° = -332 \text{ J/K}$$

Step 2. $CH_3OH(l) \longrightarrow CH_4(g) + 1/2\ O_2(g)$
$$\Delta S° = 162 \text{ J/K}$$

(a) Calculate $\Delta H°$ (in kilojoules) for step 1.

(b) Calculate $\Delta G°$ (in kilojoules) for step 1.

(c) Is step 1 spontaneous at 298 K?

(d) Which term is more important, $\Delta H°$ or $\Delta S°$?

(e) In what temperature range is step 1 spontaneous?

(f) Calculate $\Delta H°$ for step 2.

(g) Calculate $\Delta G°$ for step 2.

(h) Is step 2 spontaneous at 298 K?

(i) Which term is more important, $\Delta H°$ or $\Delta S°$?

(j) In what temperature range is step 2 spontaneous?

(k) Calculate an overall $\Delta G°$, $\Delta H°$, and $\Delta S°$ for the formation of CH_4 from CO and H_2.

(l) Is the overall reaction spontaneous?

(m) If you were designing a production facility, would you plan on carrying out the reactions in separate steps or together? Explain.

8.102 Isooctane, C_8H_{18}, is the component of gasoline from which the term *octane rating* derives.

(a) Write a balanced equation for the combustion of isooctane with O_2 to yield $CO_2(g)$ and $H_2O(g)$.

(b) The standard molar heat of combustion for isooctane is -5456.6 kJ/mol. Calculate $\Delta H°_f$ for isooctane.

8.103 We said in Section 8.2 that the potential energy of water at the top of a dam or waterfall is converted into heat when the water dashes against rocks at the bottom. The potential energy of the water at the top is equal to $E_P = mgh$, where m is the mass of the water, g is the acceleration of the falling water due to gravity ($g = 9.81$ m/s^2), and h is the height of the water. Assuming that all the energy is converted to heat, calculate the temperature rise of the water (in degrees Celsius) after falling over California's Yosemite Falls, a distance of 739 m. The specific heat of water is 4.18 J/(g · K).

8.104 For a process to be spontaneous, the total entropy of the system *and its surroundings* must increase. That is,

$$\Delta S_{total} = \Delta S_{system} + \Delta S_{surr} > 0 \quad \text{For a spontaneous process}$$

Furthermore, the entropy change in the surroundings, ΔS_{surr}, is related to the enthalpy change for the process by the equation $\Delta S_{surr} = -\Delta H/T$.

(a) Since both ΔG and ΔS_{total} offer criteria for spontaneity, they must be related. Derive a relationship between them.

(b) What is the value of ΔS_{surr} for the photosynthesis of glucose from CO_2 at 298 K?

$$6\ CO_2(g) + 6\ H_2O(l) \longrightarrow C_6H_{12}O_6(s) + 6\ O_2(g)$$
$$\Delta G° = 2879 \text{ kJ/mol}$$
$$\Delta S° = -210 \text{ J/(K·mol)}$$

8.105 Set up a Hess's law cycle, and use the following information to calculate $\Delta H°_f$ for aqueous nitric acid, $HNO_3(aq)$. You will need to use fractional coefficients for some equations.

$$3\ NO_2(g) + H_2O(l) \longrightarrow 2\ HNO_3(aq) + NO(g)$$
$$\Delta H° = -138.4 \text{ kJ}$$
$$2\ NO(g) + O_2(g) \longrightarrow 2\ NO_2(g)$$
$$\Delta H° = -114.0 \text{ kJ}$$
$$4\ NH_3(g) + 5\ O_2(g) \longrightarrow 4\ NO(g) + 6\ H_2O(l)$$
$$\Delta H° = -1169.6 \text{ kJ}$$

$NH_3(g) \quad \Delta H°_f = -46.1 \text{ kJ/mol}$
$H_2O(l) \quad \Delta H°_f = -285.8 \text{ kJ/mol}$

8.106 Hess's law can be used to calculate reaction enthalpies for hypothetical processes that can't be carried out in the laboratory. Set up a Hess's law cycle that will let you calculate $\Delta H°$ for the conversion of methane to ethylene:

$$2\ CH_4(g) \longrightarrow C_2H_4(g) + 2\ H_2(g)$$

You can use the following information:

$$2\ C_2H_6(g) + 7\ O_2(g) \longrightarrow 4\ CO_2(g) + 6\ H_2O(l)$$
$$\Delta H° = -3119.4 \text{ kJ}$$
$$CH_4(g) + 2\ O_2(g) \longrightarrow CO_2(g) + 2\ H_2O(l)$$
$$\Delta H° = -890.3 \text{ kJ}$$
$$C_2H_4(g) + H_2(g) \longrightarrow C_2H_6(g)$$
$$\Delta H° = -137.0 \text{ kJ}$$

$H_2O(l) \quad \Delta H°_f = -285.8 \text{ kJ/mol}$

Multi-Concept Problems

8.107 Acid spills are often neutralized with sodium carbonate or sodium hydrogen carbonate. For neutralization of acetic acid, the unbalanced equations are

(1) $2\ CH_3CO_2H(l) + Na_2CO_3(s) \longrightarrow$
$\qquad\qquad 2\ CH_3CO_2Na(aq) + CO_2(g) + H_2O(l)$

(2) $CH_3CO_2H(l) + NaHCO_3(s) \longrightarrow$
$\qquad\qquad CH_3CO_2Na(aq) + CO_2(g) + H_2O(l)$

(a) Balance both reactions.

(b) How many kilograms of each substance is needed to neutralize a 1.000 gallon spill of pure acetic acid (density = 1.049 g/mL)?

(c) How much heat (in kilojoules) is absorbed or liberated in each reaction? See Appendix B for standard heats of formation; $\Delta H°_f = -726.1$ kJ/mol for $CH_3CO_2Na(aq)$.

8.108 (a) Write a balanced equation for the reaction of potassium metal with water.

(b) Use the data in Appendix B to calculate $\Delta H°$ for the reaction of potassium metal with water.

(c) Assume that a chunk of potassium weighing 7.55 g is dropped into 400.0 g of water at 25.0°C. What is the final temperature of the water if all the heat released is used to warm the water?

(d) What is the molarity of the KOH solution prepared in part (c), and how many milliliters of 0.554 M H_2SO_4 are required to neutralize it?

8.109 Hydrazine, a component of rocket fuel, undergoes combustion to yield N_2 and H_2O:

$$N_2H_4(l) + O_2(g) \longrightarrow N_2(g) + 2 H_2O(l)$$

(a) Draw an electron-dot structure for hydrazine, predict the geometry about each nitrogen atom, and tell the hybridization of each nitrogen.

(b) Use the following information to set up a Hess's law cycle, and then calculate $\Delta H°$ for the combustion reaction. You will need to use fractional coefficients for some equations.

$$2 NH_3(g) + 3 N_2O(g) \longrightarrow 4 N_2(g) + 3 H_2O(l)$$
$$\Delta H° = -1011.2 \text{ kJ}$$

$$N_2O(g) + 3 H_2(g) \longrightarrow N_2H_4(l) + H_2O(l)$$
$$\Delta H° = -317 \text{ kJ}$$

$$4 NH_3(g) + O_2(g) \longrightarrow 2 N_2H_4(l) + 2 H_2O(l)$$
$$\Delta H° = -286 \text{ kJ}$$

$$H_2O(l) \qquad \Delta H°_f = -285.8 \text{ kJ/mol}$$

(c) How much heat is released on combustion of 100.0 g of hydrazine?

 ## eMedia Problems

8.110 Use the **Calorimetry** simulation (*eChapter 8.8*) to determine the molar heat of combustion of citric acid (molec. mass = 192.1 amu). Suppose you performed this experiment in the lab and later realized when you opened the calorimeter that not all of the citric acid was consumed in the combustion. Discuss the effects of this error on the calculated molar heat of combustion.

8.111 The **Formation of Aluminum Bromide** movie (*eChapter 8.10*) shows the reaction that takes place when aluminum metal and bromine liquid are combined. Write the balanced equation for the reaction shown. Is the $\Delta H°$ for this balanced reaction the standard heat of formation for aluminum bromide? Explain.

8.112 Watch the **Airbags** movie (*eChapter 8.13*) and determine the signs of ΔH, ΔS, and ΔG for the decomposition of sodium azide. Could the decomposition of sodium azide be used to inflate airbags if the reaction were endothermic? Explain.

8.113 Practice determining the signs of ΔH, ΔS, and ΔG in the **Spontaneous Reactions** simulation (*eChapter 8.14*). (a) Describe a case in which the sign of one of the parameters can be determined from the signs of the other two. (b) Explain why it is not always possible to determine the sign of the last parameter even if the other two are known. Describe two such cases.

Gases: Their Properties and Behavior

The life of this parasailer depends on an invisible blanket of air. Without it, he would sink like a stone.

CONTENTS

A quick look around tells you that matter takes different forms. Most of the things around you are *solids*, substances whose constituent atoms, molecules, or ions are held rigidly together in a definite way, giving the solid a definite volume and shape. Other substances are *liquids*, whose constituent atoms or molecules are held together less strongly, giving the liquid a definite volume but a changeable and indefinite shape. Still other substances are *gases*, whose constituent atoms or molecules have little attraction for one another and are therefore free to move about in whatever volume is available.

Though gases are few in number—only about a hundred substances are gases at room temperature—their study was enormously important in the historical development of chemical theories. We'll look briefly at this historical development in the present chapter, and we'll see how the behavior of gases can be described.

9.1 Gases and Gas Pressure

We live surrounded by a blanket of air—the mixture of gases that make up the earth's atmosphere. As shown in Table 9.1, nitrogen and oxygen account for more than 99% by volume of dry air. The remaining 1% is largely argon, with trace amounts of several other substances also present. Carbon dioxide, about which there is so much current concern because of the so-called greenhouse effect, is present in air only to the extent of about 0.037%, or 370 parts per million (ppm). Though small, this value has risen in the past 150 years from an estimated 290 ppm in 1850, as the burning of fossil fuels and the deforestation of tropical rain forests have increased.

TABLE 9.1   Composition of Dry Air at Sea Level

Constituent	% Volume	% Mass
N_2	78.08	75.52
O_2	20.95	23.14
Ar	0.93	1.29
CO_2	0.037	0.05
Ne	1.82×10^{-3}	1.27×10^{-3}
He	5.24×10^{-4}	7.24×10^{-5}
CH_4	1.7×10^{-4}	9.4×10^{-5}
Kr	1.14×10^{-4}	3.3×10^{-4}

Air is typical of gases in many respects, and its behavior illustrates several important points about gases. For instance, gas mixtures are always *homogeneous*. Unlike liquids, which often fail to mix with one another and which may separate into distinct layers—oil and water, for example—gases always mix thoroughly. Furthermore, gases are *compressible*. When pressure is applied, the volume of a gas contracts proportionately. Solids and liquids, however, are nearly incompressible, and even the application of great pressure changes their volume only slightly.

Homogeneous mixing and compressibility both result from the fact that the molecules in gases are far apart (Figure 9.1). Mixing occurs because individual gas molecules have little interaction with their neighbors and the chemical identities of those neighbors are therefore irrelevant. In solids and liquids, by contrast, molecules are packed closely together, where they are affected by various attractive and repulsive forces that can inhibit their mixing. Compressibility is possible in gases because less than 0.1% of the volume of a typical gas is taken up by the molecules themselves under normal circumstances; the remaining 99.9% is empty space. By contrast, approximately 70% of a solid's or liquid's volume is taken up by the molecules.

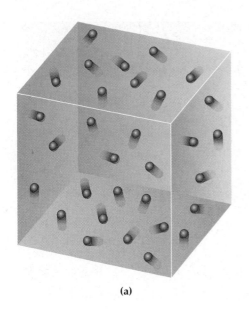
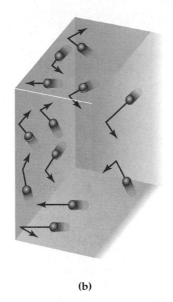

(a)   (b)

FIGURE 9.1 **(a)** A gas is a large collection of particles moving at random throughout a volume that is primarily empty space. **(b)** Collisions of randomly moving particles with the walls of the container exert a force per unit area that we perceive as gas pressure.

One of the most obvious characteristics of gases is that they exert a measurable *pressure* on the walls of their container (Figure 9.1b). We're all familiar with pumping up a tire or inflating a balloon and feeling the hardness that results from the pressure inside. In scientific terms, pressure (P) is defined as a force F exerted per unit area A. Force, in turn, is defined as mass (m) times acceleration (a), which, on earth, is usually the acceleration due to gravity, $a = 9.81 \text{ m/s}^2$.

$$\text{Pressure } (P) = \frac{F}{A} = \frac{m \times a}{A}$$

The SI unit for force is the **newton**, where $1 \text{ N} = 1 \text{ (kg} \cdot \text{m)/s}^2$, and the SI unit for pressure is the **pascal**, where $1 \text{ Pa} = 1 \text{ N/m}^2 = 1 \text{ kg/(m} \cdot \text{s}^2)$. Expressed in more familiar units, a pascal is actually a very small amount—the pressure exerted by a mass of 10.2 mg resting on an area of 1.00 cm^2:

Forcing more air into the tire increases the pressure and makes the tire feel "hard."

$$P = \frac{m \times a}{A} = \frac{(10.2 \text{ mg})\left(\dfrac{1 \text{ kg}}{10^6 \text{ mg}}\right)\left(9.81 \dfrac{\text{m}}{\text{s}^2}\right)}{(1.00 \text{ cm}^2)\left(\dfrac{1 \text{ m}^2}{10^4 \text{ cm}^2}\right)} = \frac{1.00 \times 10^{-4} \dfrac{\text{kg} \cdot \text{m}}{\text{s}^2}}{1.00 \times 10^{-4} \text{ m}^2} = 1.00 \text{ Pa}$$

In rough terms, a penny sitting on the tip of your finger exerts a pressure of about 250 Pa.

Just as the air in a tire and a penny on your fingertip exert pressure, the mass of the atmosphere pressing down on the earth's surface exerts what we call *atmospheric pressure*. In fact, a 1.00 m^2 column of air extending from the earth's surface through the upper atmosphere has a mass of about 10,300 kg, producing an atmospheric pressure of approximately 101,000 Pa, or 101 kPa (Figure 9.2).

$$P = \frac{m \times a}{A} = \frac{10,300 \text{ kg} \times 9.81 \dfrac{\text{m}}{\text{s}^2}}{1.00 \text{ m}^2} = 101,000 \text{ Pa} = 101 \text{ kPa}$$

FIGURE 9.2 A column of air 1.00 m² in cross-sectional area extending from the earth's surface through the upper atmosphere has a mass of about 10,300 kg, producing an atmospheric pressure of approximately 101,000 Pa.

As is frequently the case with SI units, which must serve many scientific disciplines, the pascal is an inconvenient size for most chemical measurements. Thus, the alternative pressure units *millimeter of mercury (mm Hg)* and *atmosphere (atm)* are more frequently used.

The millimeter of mercury, also called a *torr* after the seventeenth-century Italian scientist Evangelista Torricelli (1608–1647), is based on atmospheric pressure measurements using a mercury *barometer*. As shown in Figure 9.3, a barometer consists simply of a long, thin tube that is sealed at one end, filled with mercury, and then inverted into a dish of mercury. Some mercury runs from the tube into the dish until the downward pressure of the mercury in the column is exactly balanced by the outside atmospheric pressure, which presses on the mercury in the dish and pushes it up the column. The height of the mercury column varies slightly from day to day depending on the altitude and weather conditions, but standard atmospheric pressure at sea level is defined to be exactly 760 mm Hg.

Knowing the density of mercury ($1.359\ 51 \times 10^4$ kg/m³ at 0°C) and the acceleration due to gravity ($9.806\ 65$ m/s²), it's possible to calculate the pressure exerted by the column of mercury 760 mm (0.76 m) in height. Thus, 1 standard atmosphere of pressure (1 atm), or 760 mm Hg, is equal to 101,325 Pa:

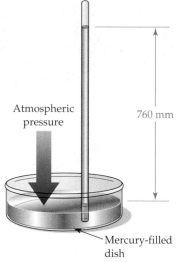

Atmospheric pressure

760 mm

Mercury-filled dish

FIGURE 9.3 A mercury barometer is used to measure atmospheric pressure by determining the height of a mercury column supported in a sealed glass tube. The downward pressure of the mercury in the column is exactly balanced by the outside atmospheric pressure that presses down on the mercury in the dish and pushes it up the column.

$$P = (0.76\ \text{m})\left(1.359\ 51 \times 10^4\ \frac{\text{kg}}{\text{m}^3}\right)\left(9.806\ 65\ \frac{\text{m}}{\text{s}^2}\right) = 101{,}325\ \text{Pa}$$

1 atm = 760 mm Hg = 101,325 Pa

Gas pressure inside a container is often measured using an open-end *manometer*, a simple instrument similar in principle to the mercury barometer. As shown in Figure 9.4, an open-end manometer consists of a U-tube filled with mercury, with one end connected to a gas-filled container and the other end open to the atmosphere. The difference between the pressure of the gas and the pressure of the atmosphere is equal to the difference between the heights of the mercury levels in the two arms of the U-tube. If the gas pressure inside the container is less than atmospheric, the mercury level is higher in the

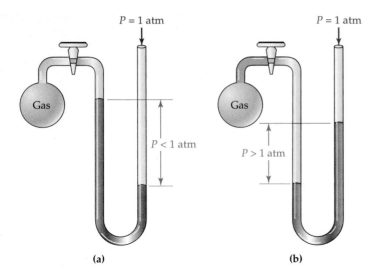

FIGURE 9.4 Open-end manometers for measuring pressure in a gas-filled bulb. In **(a)**, the pressure in the bulb is lower than atmospheric, so the mercury level is higher in the arm open to the bulb; in **(b)**, the pressure in the bulb is higher than atmospheric, so the mercury level is higher in the arm open to the atmosphere.

arm connected to the container (Figure 9.4a). If the gas pressure inside the container is greater than atmospheric, the mercury level is higher in the arm open to the atmosphere (Figure 9.4b).

EXAMPLE 9.1 Typical atmospheric pressure on top of Mt. Everest (29,035 ft) is 265 mm Hg. Convert this value to pascals and to atmospheres.

BALLPARK SOLUTION One atmosphere is defined as 760 mm Hg pressure. Since 265 mm Hg is about 1/3 of 760 mm Hg, the air pressure on Mt. Everest is about 1/3 of standard atmospheric pressure—approximately 30,000 Pa, or 0.3 atm.

DETAILED SOLUTION Use the conversion factors 1 atm/760 mm Hg and 101,325 Pa/760 mm Hg to carry out the necessary calculations:

$$265 \text{ mm Hg} \times \frac{101,325 \text{ Pa}}{760 \text{ mm Hg}} = 3.53 \times 10^4 \text{ Pa}$$

$$265 \text{ mm Hg} \times \frac{1 \text{ atm}}{760 \text{ mm Hg}} = 0.349 \text{ atm}$$

EXAMPLE 9.2 Assume that you are using an open-end manometer (Figure 9.4) filled with mineral oil rather than mercury. What is the gas pressure (in millimeters of mercury) if the level of mineral oil in the arm connected to the bulb is 237 mm higher than the level in the arm connected to the atmosphere and atmospheric pressure is 746 mm Hg? The density of mercury is 13.6 g/mL, and the density of mineral oil is 0.822 g/mL.

SOLUTION The pressure of the gas is less than atmospheric because the liquid level is higher on the side connected to the sample. Because mercury is more dense than mineral oil by a factor of 13.6/0.822, or 16.5, a given pressure will hold a column of mercury only 1/16.5 times the height of a column of mineral oil. Thus, a pressure of 237 mm mineral oil corresponds to a pressure of 14.3 mm Hg:

$$237 \text{ mm mineral oil} \times \frac{0.822 \text{ mm Hg}}{13.6 \text{ mm mineral oil}} = 14.3 \text{ mm Hg}$$

Atmospheric pressure decreases as altitude increases. On the top of Mt. Everest, typical atmospheric pressure is 265 mm Hg.

The gas pressure in the bulb is equal to the difference between outside pressure and the manometer reading:

$$P_{gas} = 746 \text{ mm Hg} - 14.3 \text{ mm Hg} = 732 \text{ mm Hg}$$

▶ **PROBLEM 9.1** Pressures are often given in the familiar unit pounds per square inch (psi). How many pounds per square inch correspond to 1.00 atm? To 1.00 mm Hg?

▶ **PROBLEM 9.2** If the density of water is 1.00 g/mL and the density of mercury is 13.6 g/mL, how high a column of water (in meters) can be supported by standard atmospheric pressure?

▶ **PROBLEM 9.3** What is the pressure (in atmospheres) in a container of gas connected to a mercury-filled, open-end manometer if the level in the arm connected to the container is 24.7 cm higher than in the arm open to the atmosphere and atmospheric pressure is 0.975 atm?

⊷ **KEY CONCEPT PROBLEM 9.4** What is the pressure of the gas inside the following apparatus (in mm Hg) if outside pressure is 750 mm Hg??

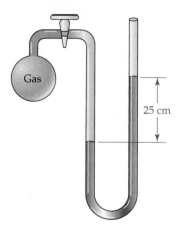

9.2 The Gas Laws

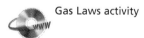

Unlike solids and liquids, different gases show remarkably similar physical behavior regardless of their chemical makeup. Helium and fluorine, for example, are vastly different in their chemical properties yet are almost identical in much of their physical behavior. Numerous observations made in the late 1600s showed that the physical properties of any gas can be defined by four variables: pressure (P), temperature (T), volume (V), and amount, or number of moles (n). The specific relationships among these four variables are called the **gas laws**.

Boyle's Law: The Relationship Between Volume and Pressure

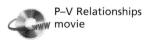

Imagine that you have a sample of gas inside a cylinder with a movable piston at one end (Figure 9.5). What would happen if you were to decrease the volume of the gas by pushing the piston partway down? Experience tells you that you would feel a resistance to pushing the piston because the pressure of the gas in the cylinder would increase. According to **Boyle's law**, the volume

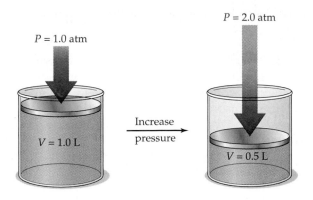

FIGURE 9.5   Boyle's law. At constant n and T, the volume of a gas decreases proportionately as its pressure increases. If the pressure is doubled, the volume is halved.

of a fixed amount of gas at a constant temperature varies inversely with its pressure. If the gas volume is halved, the gas pressure doubles; if the volume is doubled, the pressure is halved.

BOYLE'S LAW   The volume of a gas varies inversely with pressure. That is, P times V is constant when n and T are kept constant. (The symbol $\propto$ means "is proportional to," and k denotes a constant.)

$$V \propto 1/P \quad \text{or} \quad PV = k \text{ at constant } n \text{ and } T$$

The validity of Boyle's law is easy to demonstrate by making a simple series of pressure–volume measurements on a gas sample (Table 9.2) and plotting them as in Figure 9.6. When V is plotted versus P, as in Figure 9.6a, the result is a curve in the form of a hyperbola. When V is plotted versus $1/P$, as in Figure 9.6b, the result is a straight line. Such graphical behavior is characteristic of mathematical equations of the form $y = mx + b$. In this case, $y = V$, $m = $ the slope of the line (the constant k in the present instance), $x = 1/P$, and $b = $ the y-intercept (a constant; 0 in the present instance). (See Appendix A.3 for a review of linear equations.)

$$V = k\left(\frac{1}{P}\right) + 0 \quad (\text{or } PV = k)$$
$$\uparrow \quad \uparrow \uparrow \qquad \uparrow$$
$$y = m \ x \ + \ b$$

TABLE 9.2   Pressure–Volume Measurements on a Gas Sample

Pressure (mm Hg)	Volume (L)
760	1
380	2
253	3
190	4
152	5
127	6
109	7
95	8
84	9
76	10

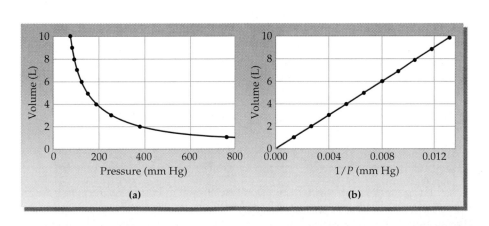

(a)   **(b)**

FIGURE 9.6   Boyle's law. **(a)** A plot of V versus P for a gas sample is a hyperbola, but **(b)** a plot of V versus $1/P$ is a straight line. Such a straight-line graph is characteristic of equations having the form $y = mx + b$.

Charles' Law: The Relationship Between Volume and Temperature

Imagine that you again have a gas sample inside a cylinder with a movable piston at one end (Figure 9.7). What would happen if you were to raise the temperature of the sample while letting the piston move freely to keep the pressure constant? Common sense tells you that the piston would move up because the volume of the gas in the cylinder would expand. According to **Charles' law**, the volume of a fixed amount of gas at a constant pressure varies directly with its absolute temperature. If the gas temperature in kelvins is doubled, the volume is doubled; if the gas temperature is halved, the volume is halved.

◆— CHARLES' LAW   The volume of a gas varies directly with absolute temperature. That is, V divided by T is constant when n and P are held constant.

$$V \propto T \quad \text{or} \quad V/T = k \text{ at constant } n \text{ and } P$$

FIGURE 9.7 Charles' law. At constant n and P, the volume of a gas increases proportionately as its absolute temperature increases. If the absolute temperature is doubled, the volume is doubled.

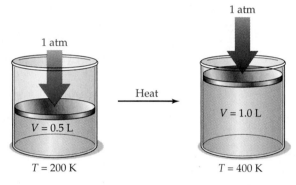

The validity of Charles' law can be demonstrated by making a series of temperature–volume measurements on a gas sample, giving the results listed in Table 9.3. Like Boyle's law, Charles' law also takes the mathematical form $y = mx + b$, where $y = V$, $m =$ the slope of the line (the constant k in the present instance), $x = T$, and $b =$ the y-intercept (0 in the present instance). A plot of V versus T is therefore a straight line whose slope is equal to the constant k (Figure 9.8).

$$V = kT + 0 \qquad \text{(or } \frac{V}{T} = k\text{)}$$
$$y = mx + b$$

TABLE 9.3   Temperature – Volume Measurements on a Gas Sample at Constant n, P

Temperature (K)	Volume (L)
123	0.45
173	0.63
223	0.82
273	1.00
323	1.18
373	1.37

FIGURE 9.8 Charles' law. A plot of V versus T for a gas sample is a straight line that can be extrapolated to absolute zero.

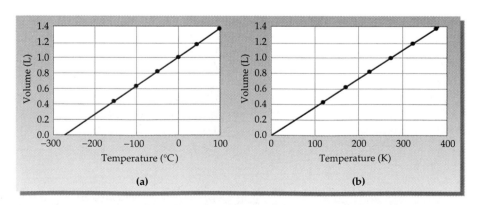

The plots of volume versus temperature shown in Figure 9.8 demonstrate an interesting point. When temperature is plotted on the Celsius scale, as in Figure 9.8a, the straight line can be extrapolated to $V = 0$ at $T = -273°C$. Because matter can't have a negative volume, this extrapolation suggests that −273°C must be the lowest possible temperature, or *absolute zero* on the Kelvin scale. In fact, the approximate value of absolute zero was first determined using this simple method.

Avogadro's Law: The Relationship Between Volume and Amount

Imagine that you have two more gas samples inside cylinders with movable pistons (Figure 9.9). One cylinder contains 1 mol of a gas and the other cylinder contains 2 mol of the gas at the same temperature and pressure as the first. Common sense tells you that the gas in the second cylinder will have twice the volume of the gas in the first cylinder because there is twice as much of it. According to **Avogadro's law**, the volume of a gas at a fixed pressure and temperature depends on its molar amount. If the amount of the gas is halved, the gas volume is halved; if the amount is doubled, the volume is doubled.

▸ AVOGADRO'S LAW   The volume of a gas varies directly with its molar amount. That is, V divided by n is constant when T and P are held constant.

$$V \propto n \quad \text{or} \quad V/n = k \text{ at constant } T \text{ and } P$$

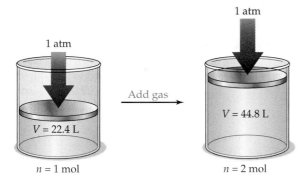

FIGURE 9.9   Avogadro's law. At constant T and P, the volume of a gas increases proportionately as its molar amount increases. If the molar amount is doubled, the volume is doubled.

The total volume of these three basketballs is almost exactly 22.4 L, the standard molar volume.

Put another way, Avogadro's law also says that equal volumes of different gases at the same temperature and pressure contain the same molar amounts. A 1 L container of oxygen contains the same number of moles as a 1 L container of helium, fluorine, argon, or any other gas at the same T and P. Experiments show that 1 mol of a gas occupies a volume (the **standard molar volume**) of 22.414 L at 0°C and 1.0000 atm pressure. (For comparison, the standard molar volume is almost exactly the same as the volume of three basketballs.)

▸ **KEY CONCEPT PROBLEM 9.5**   Show the approximate level of the movable piston in drawings **(a)** and **(b)** after the indicated changes have been made to the initial gas sample.

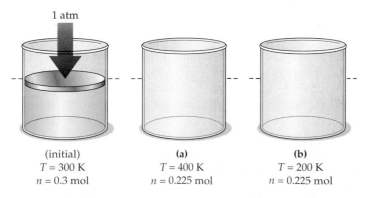

(initial)	(a)	(b)
$T = 300$ K	$T = 400$ K	$T = 200$ K
$n = 0.3$ mol	$n = 0.225$ mol	$n = 0.225$ mol

9.3 The Ideal Gas Law

All three of the gas laws discussed in the previous section can be combined into a single statement called the **ideal gas law**, which describes how the volume of a gas is affected by changes in pressure, temperature, and amount. When the values of any three of the variables P, V, T, and n are known, the ideal gas law lets us calculate the value of the fourth. The constant R in the equation is called the **gas constant** and has the same value for all gases.

➤— IDEAL GAS LAW $\qquad V = \dfrac{nRT}{P} \quad$ or $\quad PV = nRT$

Notice how the ideal gas law can be rearranged in different ways to take the form of Boyle's law, Charles' law, or Avogadro's law.

Boyle's law: $\qquad PV = nRT = k \qquad$ (When n and T are constant)

Charles' law: $\qquad \dfrac{V}{T} = \dfrac{nR}{P} = k \qquad$ (When n and P are constant)

Avogadro's law: $\qquad \dfrac{V}{n} = \dfrac{RT}{P} = k \qquad$ (When T and P are constant)

Notice also that the value of the gas constant R can be calculated from a knowledge of the molar volume of a gas. Since 1 mol of an ideal gas occupies a volume of 22.414 L at 0°C (273.15 K) and 1 atm pressure, the gas constant R is equal to 0.082 058 (L · atm)/(K · mol), or 8.3145 J/(K · mol) in SI units:

$$R = \frac{P \cdot V}{n \cdot T} = \frac{(1 \text{ atm})(22.414 \text{ L})}{(1 \text{ mol})(273.15 \text{ K})} = 0.082 \, 058 \, \frac{\text{L} \cdot \text{atm}}{\text{K} \cdot \text{mol}}$$

$$= 8.3145 \text{ J/(K·mol)} \qquad \text{When } P \text{ is in pascals and } V \text{ is in cubic meters}$$

The specific conditions used for the calculation, 1 atm and 0°C (273.15 K), are said to represent **standard temperature and pressure** (abbreviated **STP**). These standard conditions are generally used when reporting measurements on gases. Note that the standard temperature for gas measurements (0°C, or 273.15 K) is different from that usually assumed for thermodynamic measurements (25°C, or 298.15 K; Section 8.6). [It should also be noted that the standard pressure for gas measurements has been redefined to be 1 *bar*, or 100,000 Pa, rather than 1 atm (101,325 Pa), although this change is not yet in wide use. Thus, the new standard molar volume is 22.711 L rather than 22.414 L.]

➤— STANDARD TEMPERATURE AND $\qquad T = 0°\text{C} \qquad P = 1 \text{ atm}$
PRESSURE (**STP**) FOR GASES

The name *ideal gas* law implies that there must be some gases whose behavior is *nonideal*. In fact, there is no such thing as an ideal gas that obeys the equation perfectly under all circumstances; all real gases deviate slightly from the behavior predicted by the law. As Table 9.4 shows, for example, the actual molar volume of a real gas often differs slightly from the 22.414 L ideal volume. Under most conditions, though, the deviations from ideal behavior are so slight as to make little difference. We'll discuss circumstances in Section 9.8 where the deviations are greater.

TABLE 9.4   Molar Volumes of Some Real Gases at STP

22.40	22.09	22.40	22.06	22.38	22.40	22.43	22.41
NH_3	Ar	CO_2	Cl_2	F_2	N_2	H_2	He

Molar volume (L): 30, 20, 10, 0

EXAMPLE 9.3   How many moles of air are in the lungs of an average adult with a lung capacity of 3.8 L? Assume that the person is at 1.00 atm pressure and has a normal body temperature of 37°C.

BALLPARK SOLUTION   A lung volume of 4 L is about 1/6 of 22.4 L, the standard molar volume of an ideal gas. Thus, the lungs should have a capacity of about 1/6 mol, or 0.17 mol.

DETAILED SOLUTION   This problem asks for a value of n when V, P, and T are given. Rearranging the ideal gas law to the form $n = PV/RT$, converting the temperature from degrees Celsius to Kelvin, and substituting the given values of P, V, and T into the equation gives

$$n = \frac{PV}{RT} = \frac{(1.00 \text{ atm})(3.8 \text{ L})}{\left(0.082\ 06\ \dfrac{\text{L} \cdot \text{atm}}{\text{K} \cdot \text{mol}}\right)(310 \text{ K})} = 0.15 \text{ mol}$$

The lungs of an average adult hold 0.15 mol of air.

EXAMPLE 9.4   In a typical automobile engine, the mixture of gasoline and air in a cylinder is compressed from 1.0 atm to 9.5 atm. If the uncompressed volume of the cylinder is 410 mL, what is the volume (in milliliters) when the mixture is fully compressed?

BALLPARK SOLUTION   Since the pressure in the cylinder increases about 10-fold, the volume must decrease about 10-fold according to Boyle's law, from approximately 400 mL to 40 mL.

DETAILED SOLUTION   This is a Boyle's law problem, since only P and V are changing while n and T remain fixed. We can therefore set up the following equality:

$$nRT = (PV)_{\text{initial}} = (PV)_{\text{final}}$$

Solving for V_{final} gives

$$V_{\text{final}} = \frac{(PV)_{\text{initial}}}{P_{\text{final}}} = \frac{1.0 \text{ atm} \times 410 \text{ mL}}{9.5 \text{ atm}} = 43 \text{ mL}$$

▶ **PROBLEM 9.6**   How many moles of methane gas, CH_4, are in a storage tank with a volume of 1.000×10^5 L at STP? How many grams?

How many moles of methane are in these tanks?

▶ **PROBLEM 9.7** An aerosol spray can with a volume of 350 mL contains 3.2 g of propane gas (C_3H_8) as propellant. What is the pressure (in atmospheres) of gas in the can at 20°C?

▶ **PROBLEM 9.8** A helium gas cylinder of the sort used to fill balloons has a volume of 43.8 L and a pressure of 1.51×10^4 kPa at 25.0°C. How many moles of helium are in the tank?

▶ **PROBLEM 9.9** What final temperature (°C) is required for the pressure inside an automobile tire to increase from 2.15 atm at 0°C to 2.37 atm, assuming the volume remains constant?

9.4 Stoichiometric Relationships with Gases

Many chemical reactions, including some of the most important processes in the chemical industry, involve gases. Twenty million tons of ammonia, for example, are manufactured each year by reaction of hydrogen with nitrogen according to the equation $3 H_2 + N_2 \rightarrow 2 NH_3$. Thus, it's necessary to be able to calculate amounts of gaseous reactants just as it's necessary to calculate amounts of solids, liquids, and solutions (Sections 3.4–3.9).

Most gas calculations are just applications of the ideal gas law in which three of the variables P, V, T, and n are known, and the fourth variable must be calculated. For example, the reaction used in the deployment of automobile air bags is the high-temperature decomposition of sodium azide, NaN_3, to produce N_2 gas. (The sodium is then removed by a subsequent reaction.) How many liters of N_2 at 1.15 atm and 30°C are produced by decomposition of 145 g of NaN_3?

Automobile air bags are inflated with N_2 gas produced by decomposition of sodium azide.

Airbags movie

$$2 NaN_3(s) \longrightarrow 2 Na(s) + 3 N_2(g)$$

Values for P and T are given in this problem, the value of n can be calculated, and the ideal gas law will then let us find V. To find n, the number of moles of N_2 gas produced, we first need to find how many moles of NaN_3 are in 145 g:

Molar mass of NaN_3 = 65.0 g/mol

$$\text{Moles of } NaN_3 = 145 \text{ g } NaN_3 \times \frac{1 \text{ mol } NaN_3}{65.0 \text{ g } NaN_3} = 2.23 \text{ mol } NaN_3$$

Next, find how many moles of N_2 are produced in the decomposition reaction. The balanced equation says that 2 mol of NaN_3 yields 3 mol of N_2. Thus, 2.23 mol of NaN_3 yields 3.35 mol of N_2:

$$\text{Moles of } N_2 = 2.23 \text{ mol } NaN_3 \times \frac{3 \text{ mol } N_2}{2 \text{ mol } NaN_3} = 3.35 \text{ mol } N_2$$

Finally, use the ideal gas law to calculate the volume of N_2. Remember to use the Kelvin temperature (303 K) rather than the Celsius temperature (30°C) in the calculation:

$$V = \frac{nRT}{P} = \frac{3.35 \text{ mol} \times 0.082\,06\, \frac{\text{L} \cdot \text{atm}}{\text{K} \cdot \text{mol}} \times 303 \text{ K}}{1.15 \text{ atm}} = 72.4 \text{ L}$$

Example 9.5 illustrates another gas stoichiometry calculation.

Still other applications of the ideal gas law make it possible to calculate such properties as density and molar mass. Densities are calculated simply by weighing a known volume of a gas at a known temperature and pressure, as shown in Figure 9.10. Using the ideal gas law to find the volume at STP and then dividing the measured mass by the volume gives the density at STP. Example 9.6 gives a sample calculation.

Molar masses, and therefore molecular masses, can also be calculated using the ideal gas law. Imagine, for example, that an unknown gas bubbling up in a swamp is collected, placed in a sample bulb, and found to have a density of 0.714 g/L at STP. What is the molecular mass of the gas?

Let's assume that we have 1.00 L of sample, which has a mass of 0.714 g. Since the density is measured at STP, we know T, P, and V, and we need to find n, the molar amount of gas that has a mass of 0.714 g:

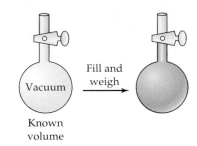

FIGURE 9.10 Determining the density of an unknown gas. A bulb of known volume is evacuated, weighed when empty, filled with gas at a known pressure and temperature, and weighed again. Dividing the mass by the volume gives the density.

$$n = \frac{PV}{RT} = \frac{(1.00\ \text{atm})(1.00\ \text{L})}{\left(0.082\ 06\ \dfrac{\text{L}\cdot\text{atm}}{\text{K}\cdot\text{mol}}\right)(273\ \text{K})} = 0.0446\ \text{mol}$$

Dividing the mass of the sample by the number of moles then gives the molar mass:

$$\text{Molar mass} = \frac{0.714\ \text{g}}{0.0446\ \text{mol}} = 16.0\ \text{g/mol}$$

Thus, the molar mass of the unknown gas (actually methane, CH_4) is 16.0 g/mol, and the molecular mass is 16.0 amu.

It's often true in chemistry, particularly in gas-law calculations, that a problem can be solved in more than one way. As an alternative method for calculating the molar mass of the unknown swamp gas, you might recognize that 1 mol of an ideal gas has a volume of 22.4 L at STP. Since 1 L of the unknown gas has a mass of 0.714 g, 22.4 L of the gas (1 mol) has a mass of 16.0 g:

$$\text{Molar mass} = \left(0.714\ \frac{\text{g}}{\text{L}}\right)\left(22.4\ \frac{\text{L}}{\text{mol}}\right) = 16.0\ \text{g/mol}$$

Example 9.7 illustrates another calculation of the molar mass of an unknown gas.

EXAMPLE 9.5   A typical high-pressure tire on a racing bicycle might have a volume of 365 mL and a pressure of 7.80 atm at 25°C. Suppose the rider filled the tire with helium to minimize weight. What is the mass of the helium in the tire?

SOLUTION   We are given V, P, and T, and we need to use the ideal gas law to calculate n, the number of moles of helium in the tire. With n known, we can then do a mole-to-mass conversion.

$$n = \frac{PV}{RT} = \frac{(7.80\ \text{atm})(0.365\ \text{L})}{\left(0.082\ 06\ \dfrac{\text{L}\cdot\text{atm}}{\text{K}\cdot\text{mol}}\right)(298\ \text{K})} = 0.116\ \text{mol}$$

$$\text{Grams of helium} = 0.116\ \text{mol He} \times \frac{4.00\ \text{g He}}{1\ \text{mol He}} = 0.464\ \text{g}$$

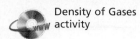
Density of Gases
www activity

EXAMPLE 9.6 What is the density (in grams per liter) of ammonia at STP if the gas in a 1.000 L bulb weighs 0.672 g at 25°C and 733.4 mm Hg pressure?

SOLUTION The density of any substance is mass divided by volume. For the ammonia sample, the mass is 0.672 g, but the volume of the gas is given under nonstandard conditions and must first be converted to STP. Since the amount of sample n is constant, we can set the quantity PV/RT measured under non-standard conditions equal to PV/RT at STP and then solve for V at STP:

$$n = \left(\frac{PV}{RT}\right)_{measured} = \left(\frac{PV}{RT}\right)_{STP} \quad \text{or} \quad V_{STP} = \left(\frac{PV}{RT}\right)_{measured} \times \left(\frac{RT}{P}\right)_{STP}$$

$$V_{STP} = \left(\frac{733.4 \text{ mm Hg} \times 1.000 \text{ L}}{298 \text{ K}}\right)\left(\frac{273 \text{ K}}{760 \text{ mm Hg}}\right) = 0.884 \text{ L}$$

Thus, the amount of gas in the 1.000 L bulb under the measured nonstandard conditions would have a volume of only 0.884 L at STP. Dividing the given mass by this volume gives the density of ammonia at STP:

$$\text{Density} = \frac{\text{Mass}}{\text{Volume}} = \frac{0.672 \text{ g}}{0.884 \text{ L}} = 0.760 \text{ g/L}$$

EXAMPLE 9.7 To identify the contents of an unlabeled cylinder of gas, a sample was collected and found to have a density of 5.380 g/L at 15°C and 736 mm Hg pressure. What is the molar mass of the gas?

SOLUTION Let's assume we have a 1.000 L sample of the gas, which weighs 5.380 g. We know the temperature, volume, and pressure of the gas and can therefore use the ideal gas law to find n, the number of moles in the sample:

$$PV = nRT \quad \text{or} \quad n = \frac{PV}{RT}$$

$$n = \frac{\left(736 \text{ mm Hg} \times \dfrac{1 \text{ atm}}{760 \text{ mm Hg}}\right)(1.000 \text{ L})}{\left(0.082\ 06 \dfrac{\text{L} \cdot \text{atm}}{\text{K} \cdot \text{mol}}\right)(288 \text{ K})} = 0.0410 \text{ mol}$$

Dividing the number of grams in the sample by the number of moles gives the molar mass:

$$\frac{5.380 \text{ g}}{0.0410 \text{ mol}} = 131 \text{ g/mol}$$

The gas is probably xenon (atomic mass = 131.3 amu).

▶ **PROBLEM 9.10** Carbonate-bearing rocks like limestone ($CaCO_3$) react with dilute acids such as HCl to produce carbon dioxide according to the equation $CaCO_3(s) + 2 \text{ HCl}(aq) \rightarrow CaCl_2(aq) + CO_2(g) + H_2O(l)$. How many grams of CO_2 would be formed by complete reaction of 33.7 g of limestone? What is the volume (in liters) of this CO_2 at STP?

▶ **PROBLEM 9.11** Propane gas (C_3H_8) is used as a fuel in rural areas. How many liters of CO_2 are formed at STP by complete combustion of the propane in a container with a volume of 15.0 L and a pressure of 4.5 atm at 25°C? The unbalanced equation is

$$C_3H_8(g) + O_2(g) \longrightarrow CO_2(g) + H_2O(l)$$

Carbonate-bearing rocks like limestone ($CaCO_3$) react with dilute acids such as HCl to produce bubbles of carbon dioxide.

▶**PROBLEM 9.12**   A foul-smelling gas produced by reaction of HCl with Na_2S was collected, and a 1.00 L sample was found to have a mass of 1.52 g at STP. What is the molecular mass of the gas? What is its likely formula and name?

9.5   Partial Pressure and Dalton's Law

Just as the gas laws apply to all *pure* gases, regardless of chemical identity, they also apply to *mixtures* of gases, such as air. The pressure, volume, temperature, and amount of a gas mixture are all related by the ideal gas law.

What is responsible for the pressure in a gas mixture? Since the pressure of a pure gas at constant temperature and volume is proportional to its amount ($P = nRT/V$), the pressure contribution from each individual gas in a mixture is also proportional to *its* amount in the mixture. In other words, the total pressure exerted by a mixture of gases in a container at constant V and T is equal to the sum of the pressures of each individual gas in the container, a statement known as **Dalton's law of partial pressures**.

◆— **DALTON'S LAW OF** $P_{total} = P_1 + P_2 + P_3 + \dots$ At constant V and T
PARTIAL PRESSURES   where $P_1, P_2, \dots$ refer to the pressures each
individual gas would have if it were alone.

The individual pressures of the various gases in the mixture, P_1, P_2, and so forth, are called *partial pressures* and refer to the pressure each individual gas would exert if it were alone in the container. That is,

$$P_1 = n_1\left(\frac{RT}{V}\right) \qquad P_2 = n_2\left(\frac{RT}{V}\right) \qquad P_3 = n_3\left(\frac{RT}{V}\right) \qquad \dots \text{and so forth}$$

But since all the gases in the mixture have the same temperature and volume, we can rewrite Dalton's law to indicate that the total pressure depends only on the total molar amount of gas present and not on the chemical identities of the individual gases:

$$P_{total} = (n_1 + n_2 + n_3 + \cdots)\left(\frac{RT}{V}\right)$$

The concentration of any individual component in a gas mixture is usually expressed as a **mole fraction (X)**, which is defined simply as the number of moles of the component divided by the total number of moles in the mixture:

$$\textbf{Mole fraction } (X) = \frac{\text{Moles of component}}{\text{Total moles in mixture}}$$

The mole fraction of component 1, for example, is

$$X_1 = \frac{n_1}{n_1 + n_2 + n_3 + \cdots} = \frac{n_1}{n_{total}}$$

But since $n = PV/RT$, we can also write

$$X_1 = \frac{P_1\left(\dfrac{V}{RT}\right)}{P_{total}\left(\dfrac{V}{RT}\right)} = \frac{P_1}{P_{total}}$$

which can be rearranged to solve for P_1, the partial pressure of component 1:

$$P_1 = X_1 \cdot P_{total}$$

This equation says that *the partial pressure exerted by each component in a gas mixture is equal to the mole fraction of that component times the total pressure.* In air, for example, the mole fractions of N_2, O_2, Ar, and CO_2 are 0.7808, 0.2095, 0.0093, and 0.000 37, respectively (Table 9.1), and the total pressure of the air is the sum of the individual partial pressures:

$$P_{air} = P_{N_2} + P_{O_2} + P_{Ar} + P_{CO_2} + \cdots$$

Thus, at a total air pressure of 1 atm (760 mm Hg), the partial pressures of the individual components are

$$
\begin{aligned}
P_{N_2} &= 0.780\,8 \text{ atm } N_2 &= 593.4 \text{ mm Hg} \\
P_{O_2} &= 0.209\,5 \text{ atm } O_2 &= 159.2 \text{ mm Hg} \\
P_{Ar} &= 0.009\,3 \text{ atm Ar} &= 7.1 \text{ mm Hg} \\
P_{CO_2} &= 0.000\,37 \text{ atm } CO_2 &= 0.3 \text{ mm Hg} \\
\hline
P_{air} &= 1.000\,0 \text{ atm air} &= 760.0 \text{ mm Hg}
\end{aligned}
$$

There are numerous practical applications of Dalton's law, ranging from the use of anesthetic agents in hospital operating rooms, where partial pressures of both oxygen and anesthetic in the patient's lungs must be constantly monitored, to the composition of diving gases used for underwater exploration. Example 9.8 gives an illustration.

The partial pressure of oxygen in the scuba tanks must be the same underwater as in air at atmospheric pressure.

EXAMPLE 9.8 At an underwater depth of 250 ft, the pressure is 8.38 atm. What should the mole percent of oxygen in the diving gas be for the partial pressure of oxygen in the gas to be 0.21 atm, the same as it is in air at 1.0 atm?

SOLUTION The partial pressure of a gas in a mixture is equal to the mole fraction of the gas times the total pressure. Rearranging this equation lets us solve for the mole fraction of O_2:

$$\text{Since} \quad P_{O_2} = X_{O_2} \cdot P_{total} \quad \text{then} \quad X_{O_2} = \frac{P_{O_2}}{P_{total}}$$

$$X_{O_2} = \frac{0.21 \text{ atm}}{8.38 \text{ atm}} = 0.025$$

$$\text{Percent } O_2 = 0.025 \times 100\% = 2.5\% \; O_2$$

The diving gas should contain 2.5% O_2 for the partial pressure of O_2 to be the same at 8.38 atm as it is in air at 1.0 atm.

▶ **PROBLEM 9.13** What is the mole fraction of each component in a mixture of 12.45 g of H_2, 60.67 g of N_2, and 2.38 g of NH_3?

▶ **PROBLEM 9.14** What is the total pressure (in atmospheres) and what is the partial pressure of each component if the gas mixture in Problem 9.13 is in a 10.00 L steel container at 90°C?

▶ **PROBLEM 9.15** On a humid day in summer, the mole fraction of gaseous H_2O (water vapor) in the air at 25°C can be as high as 0.0287. Assuming a total pressure of 0.977 atm, what is the partial pressure (in atmospheres) of H_2O in the air?

✦ **KEY CONCEPT PROBLEM 9.16** What is the partial pressure of each gas—red, yellow, and green—if the total pressure inside the following container is 600 mm Hg?

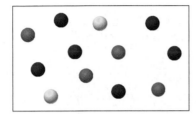

9.6 The Kinetic–Molecular Theory of Gases

Thus far, we've concentrated on *describing* the behavior of gases rather than on understanding the reasons for that behavior. Actually, the reasons are straightforward and were explained more than a century ago using a model called the **kinetic–molecular theory**. The theory is based on the following assumptions:

1. A gas consists of tiny particles, either atoms or molecules, moving about at random.
2. The volume of the particles themselves is negligible compared with the total volume of the gas; most of the volume of a gas is empty space.
3. The gas particles act independently of one another; there are no attractive or repulsive forces between particles.
4. Collisions of the gas particles, either with other particles or with the walls of a container, are elastic; that is, the total kinetic energy of the gas particles is constant at constant T.
5. The average kinetic energy of the gas particles is proportional to the Kelvin temperature of the sample.

Kinetic Energy in a Gas movie

Beginning with these assumptions, it's possible not only to understand the behavior of gases but also to derive quantitatively the ideal gas law (though we'll not do so here). For example, let's look at how the individual gas laws follow from the five postulates of kinetic–molecular theory:

• *Boyle's law* ($P \propto 1/V$): Gas pressure is a measure of the number and forcefulness of collisions between gas particles and the walls of their container. The smaller the volume at constant n and T, the more crowded together the particles are and the greater the frequency of collisions. Thus, pressure increases as volume decreases (Figure 9.11a).

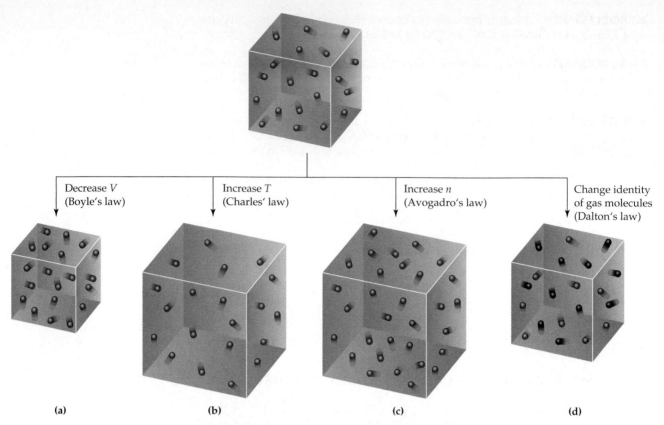

Decrease V
(Boyle's law)

Increase T
(Charles' law)

Increase n
(Avogadro's law)

Change identity
of gas molecules
(Dalton's law)

(a)   (b)   (c)   (d)

FIGURE 9.11   **(a)** Decreasing the volume of the gas at constant n and T increases the frequency of collisions with the container walls and therefore increases the pressure (Boyle's law). **(b)** Increasing the temperature (kinetic energy) at constant n and P increases the volume of the gas (Charles' law). **(c)** Increasing the amount of gas at constant T and P increases the volume (Avogadro's law). **(d)** Changing the identity of some gas molecules at constant T and V has no effect on the pressure (Dalton's law).

- *Charles' law* ($V \propto T$): Temperature is a measure of the average kinetic energy of the gas particles. The higher the temperature at constant n and P, the faster the gas particles move and the more room they require to move around in to avoid increasing their collisions with the walls of the container. Thus, volume increases as temperature increases (Figure 9.11b).

- *Avogadro's law* ($V \propto n$): The more particles there are in a gas sample, the more volume the particles need at constant P and T to avoid increasing their collisions with the walls of the container. Thus, volume increases as amount increases (Figure 9.11c).

- *Dalton's law* ($P_{total} = P_1 + P_2 + \cdots$): The chemical identity of the particles in a gas is irrelevant. Total pressure of a fixed volume of gas depends only on the temperature T and the total number of moles of gas n. The pressure exerted by a specific kind of particle thus depends on the mole fraction of that kind of particle in the mixture (Figure 9.11d).

One of the more important conclusions from kinetic–molecular theory involves the relationship between temperature and E_K, the kinetic energy of molecular motion (assumption 5). It can be shown by a somewhat complex derivation that the total kinetic energy of a mole of gas particles is equal to

$3RT/2$ and that the average kinetic energy per particle is thus $3RT/2N_A$, where N_A is Avogadro's number. Knowing this relationship makes it possible to calculate the average speed u of a gas particle. To take a helium atom at room temperature (298 K), for example, we can write

$$E_K = \frac{3RT}{2N_A} = \frac{1}{2}mu^2$$

which can be rearranged to give

$$u^2 = \frac{3RT}{mN_A}$$

$$\text{or} \quad u = \sqrt{\frac{3\,RT}{mN_A}} = \sqrt{\frac{3\,RT}{M}} \quad \text{where } M \text{ is the molar mass}$$

Substituting appropriate values for R [8.314 J/(K · mol)] and for M, the molar mass of helium (4.00×10^{-3} kg/mol), we have

$$u = \sqrt{\frac{3 \times 8.314 \dfrac{J}{K \cdot mol} \times 298\ K}{4.00 \times 10^{-3} \dfrac{kg}{mol}}} = \sqrt{1.86 \times 10^6 \frac{J}{kg}}$$

$$= \sqrt{1.86 \times 10^6 \frac{\dfrac{kg \cdot m^2}{s^2}}{kg}} = 1.36 \times 10^3\ m/s$$

Thus, the average speed of a helium atom at room temperature is more than 1.3 km/s, or about 3000 mi/h! Average speeds of some other molecules at 25°C are given in Table 9.5. The heavier the molecule, the slower the average speed.

TABLE 9.5   Average Speeds (m/s) of Some Molecules at 25°C

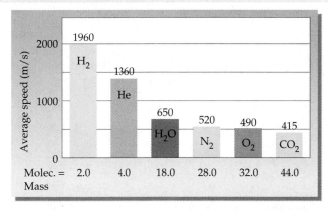

Just because the *average* speed of helium atoms at 298 K is 1.36 km/s doesn't mean that all helium atoms are moving at that speed or that a specific atom will travel from Maine to California in 1 h. As shown in Figure 9.12, there is a broad distribution of speeds among particles in a gas, a distribution that flattens out and moves higher as the temperature increases. Furthermore, an individual gas particle is likely to travel only a very short distance before it collides with another particle and bounces off in a different direction. Thus, the actual path followed by a gas particle is a random zigzag.

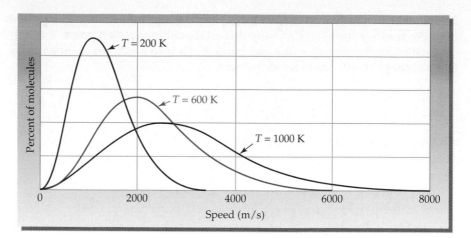

FIGURE 9.12 The distribution of speeds for helium atoms at different temperatures.

For helium at room temperature and 1 atm pressure, the average distance between collisions (the *mean free path*) is only about 2×10^{-7} m, or 1000 atomic diameters, and there are approximately 10^{10} collisions per second. For a larger O_2 molecule, the mean free path is about 6×10^{-8} m.

▶ **PROBLEM 9.17** Calculate the average speed of a nitrogen molecule (in meters per second) on a hot day in summer ($T = 37°C$) and on a cold day in winter ($T = -25°C$).

9.7 Graham's Law: Diffusion and Effusion of Gases

The constant motion and high velocities of gas particles lead to some important practical consequences. One such consequence is that gases mix rapidly when they come in contact. Take the stopper off a bottle of perfume, for instance, and the odor will spread rapidly through the room as perfume molecules mix with the molecules in the air. This mixing of different gases by random molecular motion and with frequent collisions is called **diffusion**. A similar process in which gas molecules escape without collisions through a tiny hole into a vacuum is called **effusion** (Figure 9.13).

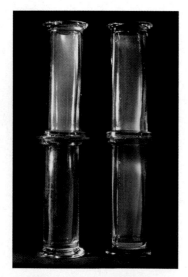

An illustration of diffusion. Glass lids initially separate the lower jars, which contain red bromine vapor, from the upper jars, which contain only air. When the lid on the right is removed, diffusion causes the gases to mix.

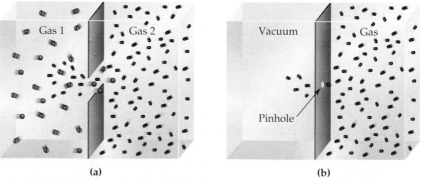

FIGURE 9.13 **(a)** Diffusion is the mixing of gas molecules by random motion under conditions where molecular collisions occur. **(b)** Effusion is the escape of a gas through a pinhole without molecular collisions.

According to **Graham's law**, formulated in the mid-1800s by the Scottish chemist Thomas Graham (1805–1869), the rate of effusion of a gas is inversely proportional to the square root of its molar mass. In other words, the lighter the molecule, the more rapidly it effuses.

⌐ GRAHAM'S LAW   The rate of effusion of a gas is inversely proportional to the square root of its molar mass, M.

$$\text{Rate} \propto \frac{1}{\sqrt{M}}$$

In comparing two gases at the same temperature and pressure, we can set up an equation saying that the ratio of the effusion rates of the two gases is inversely proportional to the ratio of the square roots of their molar masses:

$$\frac{\text{Rate}_1}{\text{Rate}_2} = \frac{\sqrt{M_2}}{\sqrt{M_1}} = \sqrt{\frac{M_2}{M_1}}$$

The inverse relationship between rate of effusion and square root of the molar mass follows directly from the connection between temperature and kinetic energy described in the previous section. Since temperature is a measure of average kinetic energy and is independent of the gas's chemical identity, different gases at the same temperature have the same average kinetic energy:

Since $\dfrac{1}{2}\,mu^2 = \dfrac{3RT}{2N_A}$ for any gas

then $\left(\dfrac{1}{2}\,mu^2\right)_{\text{gas 1}} = \left(\dfrac{1}{2}\,mu^2\right)_{\text{gas 2}}$ at the same T

Canceling the factor of 1/2 from both sides and rearranging, we find that the average speeds of the molecules in two gases vary as the inverse ratio of the square roots of their masses:

Since $\left(\dfrac{1}{2}\,mu^2\right)_{\text{gas 1}} = \left(\dfrac{1}{2}\,mu^2\right)_{\text{gas 2}}$

then $(mu^2)_{\text{gas 1}} = (mu^2)_{\text{gas 2}}$ and $\dfrac{(u_{\text{gas 1}})^2}{(u_{\text{gas 2}})^2} = \dfrac{m_2}{m_1}$

so $\dfrac{u_{\text{gas 1}}}{u_{\text{gas 2}}} = \dfrac{\sqrt{m_2}}{\sqrt{m_1}} = \sqrt{\dfrac{m_2}{m_1}}$

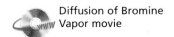

Diffusion of Bromine Vapor movie

If, as seems reasonable, the rate of effusion of a gas is proportional to the average speed of the gas molecules, then Graham's law results.

One of the most important practical consequences of Graham's law is that mixtures of gases can be separated into their pure components by taking advantage of the different rates of diffusion of the components. (Diffusion is more complex than effusion because of the molecular collisions that occur, but Graham's law usually works as a good approximation.) For example, naturally occurring uranium is a mixture of isotopes, primarily ^{235}U (0.72%) and ^{238}U (99.28%). In uranium enrichment plants that purify the fissionable uranium-235 used for fuel in nuclear reactors, elemental uranium is converted into volatile uranium hexafluoride, bp 56°C, and UF_6 gas is allowed to diffuse from one chamber to another through a permeable membrane. The $^{235}\text{UF}_6$ and $^{238}\text{UF}_6$ molecules diffuse through the

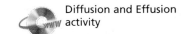

Diffusion and Effusion activity

membrane at slightly different rates according to the square root of the ratio of their masses:

For $^{235}UF_6$, $m = 349.03$ amu

For $^{238}UF_6$, $m = 352.04$ amu

so $\quad \dfrac{\text{Rate of } ^{235}UF_6 \text{ diffusion}}{\text{Rate of } ^{238}UF_6 \text{ diffusion}} = \sqrt{\dfrac{352.04 \text{ amu}}{349.03 \text{ amu}}} = 1.0043$

The UF_6 gas that passes through the membrane is thus very slightly enriched in the lighter, faster-moving isotope. After repeating the process many thousands of times, a separation of isotopes can be achieved. Approximately 85% of the Western world's nuclear fuel supply—some 5000 tons per year—is produced by this gas diffusion method.

EXAMPLE 9.9 Assume that you have a sample of hydrogen gas containing H_2, HD, and D_2 that you want to separate into pure components (H = ^{1}H and D = ^{2}H). What are the relative rates of diffusion of the three molecules according to Graham's law?

SOLUTION First, find the masses of the three molecules: for H_2, $m = 2.0$ amu; for HD, $m = 3.0$ amu; for D_2, $m = 4.0$ amu.

Next, apply Graham's law to different pairs of gas molecules. Since D_2 is the heaviest of the three molecules, it will diffuse most slowly, and we'll call its relative rate 1.00. We can then compare HD and H_2 with D_2:

Comparing HD with D_2, we have

$$\frac{\text{Rate of HD diffusion}}{\text{Rate of } D_2 \text{ diffusion}} = \sqrt{\frac{\text{mass of } D_2}{\text{mass of HD}}} = \sqrt{\frac{4.0 \text{ amu}}{3.0 \text{ amu}}} = 1.15$$

Comparing H_2 with D_2, we have

$$\frac{\text{Rate of } H_2 \text{ diffusion}}{\text{Rate of } D_2 \text{ diffusion}} = \sqrt{\frac{\text{mass of } D_2}{\text{mass of } H_2}} = \sqrt{\frac{4.0 \text{ amu}}{2.0 \text{ amu}}} = 1.41$$

Thus, the relative rates of diffusion are H_2 (1.41) > HD (1.15) > D_2 (1.00).

▶ **PROBLEM 9.18** Which gas in each of the following pairs diffuses more rapidly, and what are the relative rates of diffusion?
(a) Kr and O_2 (b) N_2 and acetylene, C_2H_2

▶ **PROBLEM 9.19** What are the relative rates of diffusion of the three naturally occurring isotopes of neon, ^{20}Ne, ^{21}Ne, and ^{22}Ne?

9.8 The Behavior of Real Gases

Before ending this discussion of gases, it's worthwhile expanding on a point made earlier: The behavior of a real gas is often a bit different from that of the generalized ideal gas we've been discussing up to this point. For instance, kinetic–molecular theory assumes that the volume of the gas particles themselves is negligible compared with the total gas volume. The assumption is valid at STP, where the volume taken up by molecules of a typical gas is only about 0.05% of the total volume, but at 500 atm and 0°C, the volume of the molecules is about 20% of the total volume (Figure 9.14) so the assumption is invalid. As a result, the volume of a real gas at high pressure is larger than predicted by the ideal gas law.

(a)

(b)

FIGURE 9.14 The volume taken up by the gas particles themselves is less important at lower pressure **(a)** than at higher pressure **(b)**. As a result, the volume of a real gas at high pressure is somewhat larger than the ideal value.

A second problem with real gases is the assumption that there are no attractive forces between particles. At lower pressures, this assumption is a good one because the gas particles are far apart and the attractive forces between them are negligible. At higher pressures, however, the particles are much closer together and the attractive forces between them become more important. In general, intermolecular attractions become significant at a distance of about 10 molecular diameters and increase rapidly as the distance diminishes (Figure 9.15). The result is that molecules draw together slightly, decreasing the volume at a given pressure (or decreasing the pressure for a given volume).

Attraction

FIGURE 9.15 Molecules attract one another at distances up to about 10 molecular diameters. The result of the attraction is a decrease in the actual volume of most real gases when compared with ideal gases at pressures up to 300 atm.

Note that the effect of molecular volume (to increase V) is opposite that of intermolecular attractions (to decrease V). The two factors therefore tend to cancel at intermediate pressures, but the effect of molecular volume is dominant above about 350 atm.

Both problems can be dealt with by a modification of the ideal gas law called the *van der Waals equation*, which uses two correction factors, a and b. The increase in V caused by the effect of molecular volume is corrected by subtracting an amount nb from the observed volume. For reasons we won't go into, the decrease in V (or, equivalently, the decrease in P) caused by the effect of intermolecular attractions is best corrected by adding an amount an^2/V^2 to the pressure.

Correction for intermolecular attractions.

Correction for molecular volume.

van der Waals equation: $\left(P + \dfrac{an^2}{V^2}\right)(V - nb) = nRT$

or $P = \dfrac{nRT}{V - nb} - \dfrac{an^2}{V^2}$

▶ **PROBLEM 9.20** Assume that you have 0.500 mol of N_2 in a volume of 0.600 L at 300 K. Calculate the pressure in atmospheres using both the ideal gas law and the van der Waals equation. For N_2, $a = 1.35$ (L$^2 \cdot$ atm)/mol^2, and $b = 0.0387$ L/mol.

Real Gases activity

9.9   The Earth's Atmosphere

The mantle of gases surrounding the earth is far from the uniform mixture you might expect. Although atmospheric pressure decreases in a regular way at higher altitudes (Figure 9.16a), the profile of temperature versus altitude is much more complex (Figure 9.16b). Four regions of the atmosphere have been defined based on this temperature curve. The temperature in the *troposphere*, the region nearest the earth's surface, decreases regularly up to about 12 km altitude, where it reaches a minimum value, and then increases in the *stratosphere*, up to about 50 km. Above the stratosphere, in the *mesosphere*, (50–85 km), the temperature again decreases but then again increases in the *thermosphere*. To give you a feeling for these altitudes, passenger jets normally fly near the top of the troposphere at altitudes of 10–12 km, and the world altitude record for aircraft is 37.65 km—roughly in the middle of the stratosphere.

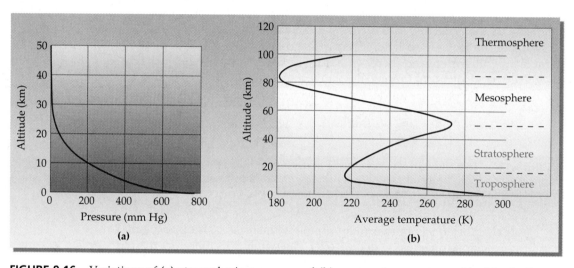

FIGURE 9.16   Variations of **(a)** atmospheric pressure and **(b)** average temperature with altitude. Four regions of the earth's atmosphere can be defined based on the temperature variations.

The earth and its atmosphere as seen from space.

Chemistry of the Troposphere

Not surprisingly, it's the layer nearest the earth's surface—the troposphere—that is the most easily disturbed by human activities and has the greatest effect on the earth's surface conditions. Among those effects, air pollution, acid rain, and the greenhouse effect are particularly important.

Air Pollution.   Air pollution has appeared in the last two centuries as an unwanted by-product of our industrialized societies. Its causes are relatively straightforward; its control is difficult. The main causes of air pollution are the release of unburned hydrocarbon molecules and the production of nitric oxide, NO, during combustion of petroleum products in the more than 500 million automobile engines presently in use worldwide. The NO is further oxidized by reaction with air to yield nitrogen dioxide, NO_2, which splits into NO plus free oxygen atoms in the presence of sunlight (symbolized by $h\nu$). Reaction of the oxygen atoms with O_2 molecules then yields ozone, O_3, a highly reactive substance that can further combine with unburned hydrocarbons in the air. The end result is the production of so-called photochemical smog, the hazy, brownish layer lying over many cities.

$$NO_2(g) + h\nu \longrightarrow NO(g) + O(g)$$

$$O(g) + O_2(g) \longrightarrow O_3(g)$$

Acid Rain. Acid rain, a second major environmental problem, results primarily from the production of sulfur dioxide, SO_2, that accompanies the burning of sulfur-containing coal in power-generating plants. Sulfur dioxide is slowly converted to SO_3 by reaction with oxygen in air, and SO_3 dissolves in rainwater to yield dilute sulfuric acid, H_2SO_4.

$$S(\text{in coal}) + O_2(g) \longrightarrow SO_2(g)$$

$$2\,SO_2(g) + O_2(g) \longrightarrow 2\,SO_3(g)$$

$$SO_3(g) + H_2O(l) \longrightarrow H_2SO_4(aq)$$

Among the many dramatic effects of acid rain are the extinction of fish from acidic lakes throughout parts of the northeastern United States, Canada, and Scandinavia, the damage to forests throughout much of central and eastern Europe, and the deterioration everywhere of marble buildings and statuary. Marble is a form of calcium carbonate, $CaCO_3$, and, like all metal carbonates, reacts with acid to produce CO_2. The result is a slow eating away of the stone.

$$CaCO_3(s) + H_2SO_4(aq) \longrightarrow CaSO_4(aq) + H_2O(l) + CO_2(g)$$

The photochemical smog over many cities is the end result of pollution from automobile exhausts.

Greenhouse Effect. The third major atmospheric problem, the so-called greenhouse effect and the global warming that could result, is less well documented and not as well understood as either air pollution or acid rain. The basis of concern about the greenhouse problem is the fear that human activities over the past century may have disturbed the earth's delicate thermal balance. One component of that balance is the radiant energy the earth's surface receives from the sun, a certain amount of which is radiated back into space as infrared energy. Although much of this radiation passes out through the atmosphere, some is absorbed by atmospheric gases, particularly water vapor, carbon dioxide, and methane. This absorbed radiation warms the atmosphere and acts to maintain a relatively stable temperature at the earth's surface. Should increasing amounts of radiation be absorbed, increased atmospheric heating could result and global temperatures could rise.

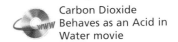

Carbon Dioxide Behaves as an Acid in Water movie

Measurements show that concentrations of atmospheric carbon dioxide have been rising in the last 150 years, largely because of the increased use of fossil fuels, from an estimated 290 parts per million (ppm) in 1850 to 367 ppm in early 1999 (Figure 9.17). Thus, there is concern among many atmospheric scientists that increased absorption of infrared radiation and widespread global warming might follow. Thus far, there have been no major consequences of a greenhouse effect, and current worries stem largely from the predictions of sophisticated computer models. These models predict a potential warming by as much as 3°C by the year 2050, an amount that could result in an increased melting of polar ice caps and a resultant rise in ocean levels.

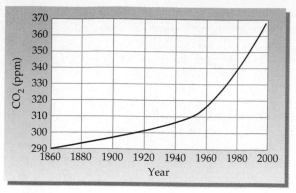

FIGURE 9.17   Concentrations of atmospheric CO_2 have increased dramatically in the last century as a result of increased fossil fuel use. Though the consequences to date have been small, atmospheric scientists worry that global atmospheric warming may soon occur.

Chemistry of the Upper Atmosphere

Relatively little of the atmosphere's mass is located above the troposphere, but the chemistry that occurs there is nonetheless crucial to maintaining life on earth. Particularly important is what takes place in the *ozone layer*, an atmospheric band stretching from about 20 to 40 km above the earth's surface. Ozone (O_3) is a severe pollutant at low altitudes but is critically important in the upper atmosphere because it absorbs intense ultraviolet radiation from the sun. Even though it is present in very small amounts in the stratosphere, ozone acts as a shield to prevent high-energy solar radiation from reaching the earth's surface, where it can cause such problems as eye cataracts and skin cancer.

Around 1976, a disturbing decrease in the amount of ozone present over the South Pole began showing up (Figure 9.18), and more recently a similar phenomenon has been found over the North Pole. Ozone levels drop to below 50% of normal in the polar spring before returning to near normal in the autumn.

FIGURE 9.18   A false color satellite image of the ozone hole over Antarctica in October, 1999. The lowest ozone concentrations are represented by the black and violet regions, where ozone levels are up to 50% lower than normal.

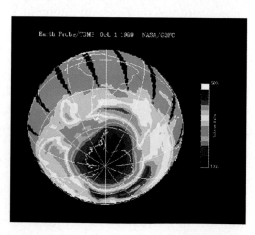

The principal cause of ozone depletion appears to be the presence in the stratosphere of *chlorofluorocarbons* (CFCs), such as CF_2Cl_2 and $CFCl_3$. Widely used as aerosol propellants and as refrigerants in air-conditioners, CFCs have been released in large amounts in the past several decades and have ultimately found their way into the stratosphere.

There are several different mechanisms of ozone destruction that predominate under different stratospheric conditions. All are multistep processes that begin when ultraviolet light ($h\nu$) strikes a CFC molecule, breaking a carbon–chlorine bond and generating a chlorine atom:

$$CFCl_3 + h\nu \longrightarrow CFCl_2 + Cl$$

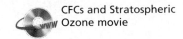

CFCs and Stratospheric
Ozone movie

The resultant chlorine atom reacts with ozone to yield O_2 and ClO, and two ClO molecules then give Cl_2O_2. Further reaction occurs when Cl_2O_2 is struck by more ultraviolet light to generate O_2 and two more chlorine atoms.

$$
\begin{array}{ll}
(1) & 2\,[Cl + O_3 \longrightarrow O_2 + ClO] \\
(2) & 2\,ClO \longrightarrow Cl_2O_2 \\
(3) & Cl_2O_2 + h\nu \longrightarrow 2\,Cl + O_2 \\
\hline
\text{Net:} & 2\,O_3 + h\nu \longrightarrow 3\,O_2
\end{array}
$$

Look at the overall result of the above reaction sequence. Chlorine atoms are used in the first step but are regenerated in the third step, so they do not appear in the net equation. Thus, the net sequence is a *chain reaction*, in which the generation of just a few chlorine atoms from a few CFC molecules leads to the destruction of a great many ozone molecules.

Recognition of the problem led the U.S. government in 1980 to ban the use of CFCs for aerosol propellants and, more recently, for refrigerants. Worldwide action to reduce CFC use began in September 1987, and an international agreement calling for a total ban on the release of CFCs took effect in 1996. The amounts of CFCs in the stratosphere appear to be peaking and are expected to decline slowly until their eventual disappearance sometime in the middle of the century.

▶ **PROBLEM 9.21** The ozone layer is about 20 km thick, has an average total pressure of 10 mm Hg (1.3×10^{-2} atm), and has an average temperature of 230 K. The partial pressure of ozone in the layer is only about 1.2×10^{-6} mm Hg (1.6×10^{-9} atm). How many meters thick would the layer be if all the ozone contained in it were compressed into a thin layer of pure O_3 at STP?

Inhaled Anesthetics

William Morton's demonstration in 1846 of ether-induced anesthesia during dental surgery ranks as one of the most important medical breakthroughs of all time. Before that date, all surgery had been carried out with the patient conscious. Use of chloroform as an anesthetic quickly followed Morton's work, made popular by Queen Victoria of England, who in 1853 gave birth to a child while anesthetized by chloroform.

Literally hundreds of substances in addition to ether ($CH_3CH_2OCH_2CH_3$) and chloroform ($CHCl_3$) have subsequently been shown to act as inhaled anesthetics. Halothane, enflurane, isoflurane, and methoxyflurane are at present the most commonly used agents in hospital operating rooms. All four are potent at relatively low doses, are nontoxic, and are nonflammable, an important safety feature.

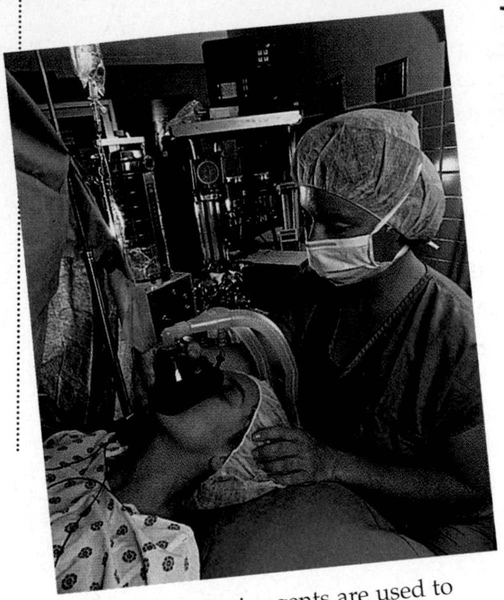

Inhaled anesthetic agents are used to prepare patients for surgery.

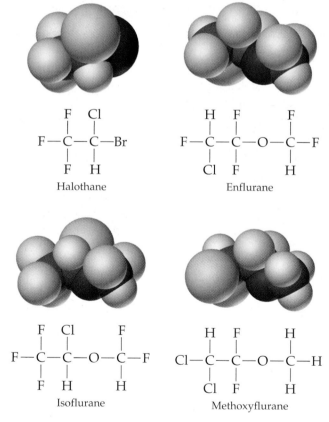

Halothane

Enflurane

Isoflurane

Methoxyflurane

Despite their great value, surprisingly little is known about how inhaled anesthetics work in the body. Even the definition of anesthesia as a behavioral state is imprecise, and the nature of the changes in brain function leading to anesthesia are unknown. Remarkably enough, the potency of different inhaled anesthetics correlates well with their solubility in olive oil: The more soluble in olive oil, the more potent as an anesthetic. This unusual observation has led many scientists to believe that anesthetics act by dissolving in the fatty membranes surrounding nerve cells. The resultant changes in the fluidity and shape of the membranes apparently decrease the ability of sodium ions to pass into the nerve cells, thereby blocking the firing of nerve impulses.

Depth of anesthesia is determined by the concentration of anesthetic agent that reaches the brain. Brain concentration, in turn, depends on the solubility and transport of the anesthetic agent in the bloodstream and on its partial pressure in inhaled air. Anesthetic potency is usually expressed as a *minimum alveolar concentration (MAC)*, defined as the percent concentration of anesthetic in inhaled air that results in anesthesia in 50% of patients. As shown in Table 9.6, nitrous oxide, N_2O, is the least potent of the common anesthetics. Fewer than 50% of patients are immobilized by breathing an 80/20 mix of nitrous oxide and oxygen. Methoxyflurane is the most potent agent; a partial pressure of only 1.2 mm Hg is sufficient to anesthetize 50% of patients, and a partial pressure of 1.4 mm Hg will anesthetize 95%.

TABLE 9.6 Relative Potency of Inhaled Anesthetics

Anesthetic	MAC (%)	MAC (partial pressure, mm Hg)
Nitrous oxide	—	>760
Enflurane	1.7	13
Isoflurane	1.4	11
Halothane	0.75	5.7
Methoxyflurane	0.16	1.2

▶ **PROBLEM 9.22** For ether, a partial pressure of 15 mm Hg results in anesthesia in 50% of patients. What is the MAC for ether?

▶ **PROBLEM 9.23** Chloroform has an MAC of 0.77%.
(a) What partial pressure of chloroform is required to anesthetize 50% of patients?
(b) What mass of chloroform in 10.0 L of air at STP will produce the appropriate MAC?

Key Words

Summary

A gas is a large collection of atoms or molecules moving independently through a volume that is largely empty space. Collisions of the randomly moving particles with the walls of their container exert a force per unit area that we perceive as gas pressure. The SI unit for pressure is the **pascal**, but the *atmosphere* and the *millimeter of mercury* are more commonly used. The physical condition of any gas is defined by four variables: pressure (P), temperature (T), volume (V), and molar amount (n). The specific relationships among these variable are called the **gas laws:**

Boyle's law:	The volume of a gas varies inversely with its pressure. That is, $V \propto 1/P$ or $PV = k$ at constant n, T.
Charles' law:	The volume of a gas varies directly with Kelvin temperature. That is, $V \propto T$ or $V/T = k$ at constant n, P.
Avogadro's law:	The volume of a gas varies directly with its molar amount. That is, $V \propto n$ or $V/n = k$ at constant T, P.

The three individual gas laws can be combined into a single **ideal gas law**, $PV = nRT$. The constant R in the equation is called the **gas constant** and has the same value for all gases. At **standard temperature and pressure (STP;** 1 atm and 0°C), the **standard molar volume** of an ideal gas is 22.414 L. The ideal gas law simplifies stoichiometric calculations for reactions where one of the components is a gas. If any three of the four variables P, V, T and n are known, the fourth can be calculated.

The gas laws apply to mixtures of gases as well as to pure gases. According to **Dalton's law of partial pressures**, the total pressure exerted by a mixture of gases in a container is equal to the sum of the pressures each individual gas would exert alone.

The behavior of gases can be accounted for using a model called the **kinetic–molecular theory**, a group of five postulates from which the ideal gas law can be derived:

1. A gas consists of tiny particles moving about at random.
2. The volume of the gas particles is negligible compared with the total volume occupied by the gas.
3. There are no forces between particles, either attractive or repulsive.
4. Collisions of gas particles are elastic.
5. The average kinetic energy of the gas particles is proportional to the absolute temperature.

The connection between temperature and kinetic energy obtained from the kinetic–molecular theory makes it possible to calculate the average speed of a gas particle at any temperature. An important practical consequence of this relationship is **Graham's law**, which states that the rate of a gas's **effusion**, or spontaneous passage through a pinhole in a membrane, depends inversely on the square root of the gas's mass.

Real gases differ in their behavior from that predicted by the ideal gas law, particularly at high pressure, where gas particles are forced close together and intermolecular attractions become significant.

Key Concept Summary

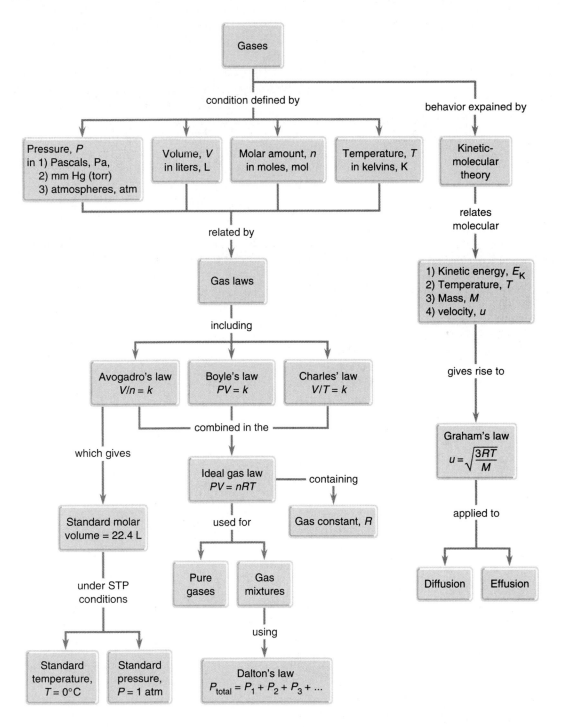

Understanding Key Concepts

Problems 9.1–9.23 appear within the chapter.

9.24 Assume that you have a sample of gas in a cylinder with a movable piston, as shown in the following drawing:

Redraw the apparatus to show what the sample will look like after **(a)** the temperature is increased from 300 K to 450 K at constant pressure, **(b)** the pressure is increased from 1 atm to 2 atm at constant temperature, and **(c)** the temperature is decreased from 300 K to 200 K and the pressure is decreased from 3 atm to 2 atm.

9.25 Assume that you have a sample of gas at 350 K in a sealed container, as represented in **(a)**. Which of the drawings **(b)**–**(d)** represents the gas after the temperature is lowered from 350 K to 150 K?

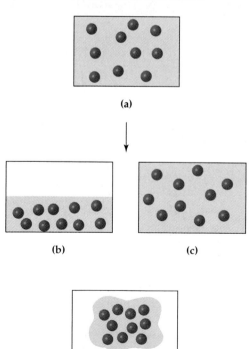

(a)

(b)   **(c)**

(d)

9.26 Assume that you have a mixture of He (atomic mass = 4 amu) and Xe (atomic mass = 131 amu) at 300 K. Which of the drawings best represents the mixture (blue = He; green = Xe)?

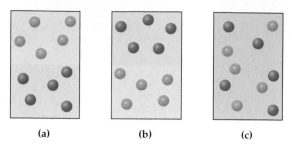

(a)   **(b)**   **(c)**

9.27 Three bulbs, two of which contain different gases and one of which is empty, are connected as shown in the following drawing:

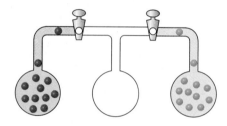

Redraw the apparatus to represent the gases after the stopcocks are opened and the system is allowed to come to equilibrium.

9.28 The apparatus shown is called a *closed-end* manometer because the arm not connected to the gas sample is closed to the atmosphere and is under vacuum. Explain how you can read the gas pressure in the bulb.

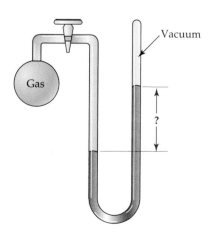

9.29 Redraw the following open-end manometer to show what it would look like when stopcock A is opened.

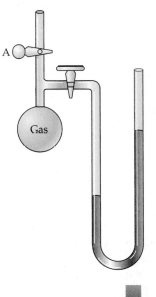

9.30 Effusion of a 1:1 mixture of two gases through a small pinhole produces the results shown below.
   (a) Which gas molecules—yellow or blue—have a higher average speed?
   (b) If the yellow molecules have a molecular mass of 25 amu, what is the molecular mass of the blue molecules?

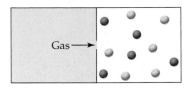

9.31 Show the approximate level of the movable piston in drawings **(a)**, **(b)**, and **(c)** after the indicated changes have been made to the gas.

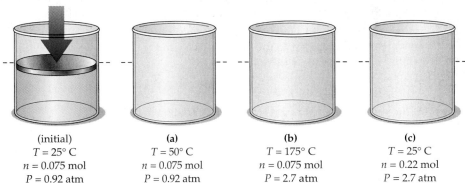

(initial)	(a)	(b)	(c)
$T = 25°$ C	$T = 50°$ C	$T = 175°$ C	$T = 25°$ C
$n = 0.075$ mol	$n = 0.075$ mol	$n = 0.075$ mol	$n = 0.22$ mol
$P = 0.92$ atm	$P = 0.92$ atm	$P = 2.7$ atm	$P = 2.7$ atm

Additional Problems

Gases and Gas Pressure

9.32 What is temperature a measure of?

9.33 Why are gases so much more compressible than solids or liquids?

9.34 Atmospheric pressure at the top of Pike's Peak in Colorado is approximately 480 mm Hg. Convert this value to atmospheres and to pascals.

9.35 Carry out the following conversions:
   (a) 352 torr to kPa
   (b) 0.255 atm to mm Hg
   (c) 0.0382 mm Hg to Pa

9.36 What is the pressure (in millimeters of mercury) inside a container of gas connected to a mercury-filled, open-end manometer of the sort shown in Figure 9.4 when the level in the arm connected to the container is 17.6 cm lower than the level in the arm open to the atmosphere, and the atmospheric pressure reading outside the apparatus is 754.3 mm Hg?

9.37 What is the pressure (in atmospheres) inside a container of gas connected to a mercury-filled, open-end manometer when the level in the arm connected to the container is 28.3 cm higher than the level in the arm open to the atmosphere, and the atmospheric pressure reading outside the apparatus is 1.021 atm?

9.38 Assume that you have an open-end manometer filled with ethyl alcohol (density = 0.7893 g/mL at 20°C) rather than mercury (density = 13.546 g/mL at 20°C). What is the pressure (in pascals) if the level in the arm open to the atmosphere is 55.1 cm higher than the level in the arm connected to the gas sample and the atmospheric pressure reading is 752.3 mm Hg?

9.39 Assume that you have an open-end manometer filled with chloroform (density = 1.4832 g/mL at 20°C) rather than mercury (density = 13.546 g/mL at 20°C). What is the difference in height between the liquid in the two arms if the pressure in the arm connected to the gas sample is 0.788 atm and the atmospheric pressure reading is 0.849 atm? In which arm is the chloroform level higher?

9.40 Calculate the average molecular mass of air from the data given in Table 9.1.

9.41 What is the average molecular mass of a diving-gas mixture that contains 2.0% by volume O_2 and 98.0% by volume He?

The Gas Laws

9.42 Assume that you have a cylinder with a movable piston. What would happen to the gas pressure inside the cylinder if you were to do the following?
(a) Triple the Kelvin temperature while holding the volume constant
(b) Reduce the amount of gas by 1/3 while holding the temperature and volume constant
(c) Decrease the volume by 45% at constant T
(d) Halve the Kelvin temperature and triple the volume

9.43 Assume that you have a cylinder with a movable piston. What would happen to the gas volume of the cylinder if you were to do the following?
(a) Halve the Kelvin temperature while holding the pressure constant
(b) Increase the amount of gas by 1/4 while holding the temperature and pressure constant
(c) Decrease the pressure by 75% at constant T
(d) Double the Kelvin temperature and double the pressure

9.44 Which sample contains the most molecules: 1.00 L of O_2 at STP, 1.00 L of air at STP, or 1.00 L of H_2 at STP?

9.45 Which sample contains more molecules: 2.50 L of air at 50°C and 750 mm Hg pressure or 2.16 L of CO_2 at −10°C and 765 mm Hg pressure?

9.46 Oxygen gas is commonly sold in 49.0 L steel containers at a pressure of 150 atm. What volume (in liters) would the gas occupy at a pressure of 1.02 atm if its temperature remained unchanged? If its temperature was raised from 20.0°C to 35.0°C at constant $P = 150$ atm?

9.47 A compressed-air tank carried by scuba divers has a volume of 8.0 L and a pressure of 140 atm at 20°C. What is the volume of air in the tank (in liters) at STP?

9.48 If 15.0 g of CO_2 gas has a volume of 0.30 L at 300 K, what is its pressure in millimeters of mercury?

9.49 If 20.0 g of N_2 gas has a volume of 0.40 L and a pressure of 6.0 atm, what is its Kelvin temperature?

9.50 The matter in interstellar space consists almost entirely of hydrogen atoms at a temperature of 100 K and a density of approximately 1 atom/cm^3. What is the gas pressure in millimeters of mercury?

9.51 Methane gas, CH_4, is sold in a 43.8 L cylinder containing 5.54 kg. What is the pressure inside the cylinder (in kilopascals) at 20°C?

9.52 Many laboratory gases are sold in steel cylinders with a volume of 43.8 L. What mass (in grams) of argon is inside a cylinder whose pressure is 17,180 kPa at 20°C?

9.53 A small cylinder of helium gas used for filling balloons has a volume of 2.30 L and a pressure of 13,800 kPa at 25°C. How many balloons can you fill if each one has a volume of 1.5 L and a pressure of 1.25 atm at 25°C?

Gas Stoichiometry

9.54 Which sample contains more molecules, 15.0 L of steam (gaseous H_2O) at 123.0°C and 0.93 atm pressure or a 10.5 g ice cube at −5.0°C?

9.55 Which sample contains more molecules, 3.14 L of Ar at 85.0°C and 1111 mm Hg pressure or 11.07 g of Cl_2?

9.56 Imagine that you have two identical flasks, one containing hydrogen at STP and the other containing oxygen at STP. How can you tell which is which without opening them?

9.57 Imagine that you have two identical flasks, one containing chlorine gas and the other containing argon. How can you tell which is which without opening them?

9.58 What is the total mass (in grams) of oxygen in a room measuring 4.0 m by 5.0 m by 2.5 m? Assume that the gas is at STP and that air contains 20.95% oxygen by volume.

9.59 The average oxygen content of arterial blood is approximately 0.25 g of O_2 per liter. Assuming a body temperature of 37°C, how many moles of oxygen are transported by each liter of arterial blood? How many milliliters?

9.60 One mole of any gas has a volume of 22.414 L at STP. What are the densities of the following gases (in grams per liter) at STP?
(a) CH_4 (b) CO_2 (c) O_2 (d) UF_6

9.61 What is the density (in grams per liter) of a gas mixture that contains 27.0% F_2 and 73.0% He by volume at 714 mm Hg and 27.5°C?

9.62 An unknown gas is placed in a 1.500 L bulb at a pressure of 356 mm Hg and a temperature of 22.5°C, and is found to weigh 0.9847 g. What is the molecular mass of the gas?

9.63 What are the molecular masses of the gases with the following densities:

(a) 1.342 g/L at STP

(b) 1.053 g/L at 25°C and 752 mm Hg

9.64 Pure oxygen gas was first prepared by heating mercury(II) oxide, HgO:

$$2 HgO(s) \longrightarrow 2 Hg(l) + O_2(g)$$

What volume (in liters) of oxygen at STP is released by heating 10.57 g of HgO?

9.65 How many grams of HgO would you need to heat if you wanted to prepare 0.0155 mol of O_2 according to the equation in Problem 9.64?

9.66 Hydrogen gas can be prepared by reaction of zinc metal with aqueous HCl:

$$Zn(s) + 2 HCl(aq) \longrightarrow ZnCl_2(aq) + H_2(g)$$

(a) How many liters of H_2 would be formed at 742 mm Hg and 15°C if 25.5 g of zinc was allowed to react?

(b) How many grams of zinc would you start with if you wanted to prepare 5.00 L of H_2 at 350 mm Hg and 30.0°C?

9.67 Ammonium nitrate can decompose explosively when heated according to the equation

$$2 NH_4NO_3(s) \longrightarrow 2 N_2(g) + 4 H_2O(g) + O_2(g)$$

How many liters of gas would be formed at 450°C and 1.00 atm pressure by explosion of 450 g of NH_4NO_3?

9.68 The reaction of sodium peroxide (Na_2O_2) with CO_2 is used in space vehicles to remove CO_2 from the air and generate O_2 for breathing:

$$2 Na_2O_2(s) + 2 CO_2(g) \longrightarrow 2 Na_2CO_3(s) + O_2(g)$$

(a) Assuming that air is breathed at an average rate of 4.50 L/min (25°C; 735 mm Hg) and that the concentration of CO_2 in expelled air is 3.4% by volume, how many grams of CO_2 are produced in 24 h?

(b) How many days would a 3.65 kg supply of Na_2O_2 last?

9.69 Titanium(III) chloride, a substance used in catalysts for preparing polyethylene, is made by high-temperature reaction of $TiCl_4$ vapor with H_2:

$$2 TiCl_4(g) + H_2(g) \longrightarrow 2 TiCl_3(s) + 2 HCl(g)$$

(a) How many grams of $TiCl_4$ are needed for complete reaction with 155 L of H_2 at 435°C and 795 mm Hg pressure?

(b) How many liters of HCl gas at STP will result from the reaction described in part (a)?

Dalton's Law and Mole Fraction

9.70 Use the information in Table 9.1 to calculate the partial pressure (in atmospheres) of each gas in dry air at STP.

9.71 Natural gas is a mixture of many substances, primarily CH_4, C_2H_6, C_3H_8, and C_4H_{10}. Assuming that the total pressure of the gases is 1.48 atm and that their mole ratio is 94 : 4.0 : 1.5 : 0.50, calculate the partial pressure (in atmospheres) of each gas.

9.72 A special gas mixture used in bacterial growth chambers contains 1.00% by weight CO_2 and 99.0% O_2. What is the partial pressure (in atmospheres) of each gas at a total pressure of 0.977 atm?

9.73 A gas mixture for use in some lasers contains 5.00% by weight HCl, 1.00% H_2, and 94% Ne. The mixture is sold in cylinders that have a volume of 49.0 L and a pressure of 13,800 kPa at 21.0°C. What is the partial pressure (in kilopascals) of each gas in the mixture?

9.74 What is the mole fraction of each gas in the mixture described in Problem 9.73?

9.75 A mixture of Ar and N_2 gases has a density of 1.413 g/L at STP. What is the mole fraction of each gas?

9.76 Magnesium metal reacts with aqueous HCl to yield H_2 gas:

$$Mg(s) + 2 HCl(aq) \longrightarrow MgCl_2(aq) + H_2(g)$$

The gas that forms is found to have a volume of 3.557 L at 25°C and a pressure of 747 mm Hg. Assuming that the gas is saturated with water vapor at a partial pressure of 23.8 mm Hg, what is the partial pressure (in millimeters of mercury) of the H_2? How many grams of magnesium metal were used in the reaction?

9.77 Chlorine gas was first prepared in 1774 by oxidation of NaCl with MnO_2:

$$2 NaCl(s) + 2 H_2SO_4(l) + MnO_2(s) \longrightarrow$$
$$Na_2SO_4(s) + MnSO_4(s) + 2 H_2O(g) + Cl_2(g)$$

Assume that the gas produced is saturated with water vapor at a partial pressure of 28.7 mm Hg and that it has a volume of 0.597 L at 27°C and 755 mm Hg pressure.

(a) What is the mole fraction of Cl_2 in the gas?

(b) How many grams of NaCl were used in the experiment, assuming complete reaction?

Kinetic–Molecular Theory and Graham's Law

9.78 What are the basic assumptions of the kinetic–molecular theory?

9.79 What is the difference between effusion and diffusion?

9.80 What is the difference between heat and temperature?

9.81 Why does a helium-filled balloon lose pressure faster than an air-filled balloon?

9.82 The average temperature at an altitude of 20 km is 220 K. What is the average speed (in meters per second) of an N_2 molecule at this altitude?

9.83 At what temperature (°C) will xenon atoms have the same average speed that Br_2 molecules have at 20°C?

9.84 Which has a higher average speed, H_2 at 150 K or He at 375°C?

9.85 Which has a higher average speed, a Ferrari at 145 mph or a UF_6 molecule at 25°C?

9.86 An unknown gas is found to diffuse through a porous membrane 2.92 times more slowly than H_2. What is the molecular mass of the gas?

9.87 What is the molecular mass of a gas that diffuses through a porous membrane 1.86 times faster than Xe? What might the gas be?

9.88 Rank the following gases in order of their speed of diffusion through a membrane, and calculate the ratio of their diffusion rates: HCl, F_2, Ar

9.89 Which will diffuse through a membrane more rapidly, CO or N_2? Assume that the samples contain only the most abundant isotopes of each element, ^{12}C, ^{16}O, and ^{14}N.

9.90 A big-league fastball travels at about 45 m/s. At what temperature (°C) do helium atoms have this same average speed?

9.91 Traffic on the German autobahns reaches speeds of up to 230 km/h. At what temperature (°C) do oxygen molecules have this same average speed?

General Problems

9.92 Chlorine occurs as a mixture of two isotopes, ^{35}Cl and ^{37}Cl. What is the ratio of the diffusion rates of the three species $(^{35}Cl)_2$, $^{35}Cl^{37}Cl$, and $(^{37}Cl)_2$?

9.93 What would the atmospheric pressure be (in millimeters of mercury) if our atmosphere were composed of pure CO_2 gas?

9.94 The surface temperature of Venus is about 1050 K, and the pressure is about 75 earth atmospheres. Assuming that these conditions represent a Venusian "STP," what is the standard molar volume (in liters) of a gas on Venus?

9.95 When you look directly up at the sky, you are actually looking through a very tall, transparent column of air that extends from the surface of the earth thousands of kilometers into space. If the air in this column were liquefied, how tall would it be? The density of liquid air is 0.89 g/mL.

9.96 Assume that you take a flask, evacuate it to remove all the air, and find its mass to be 478.1 g. You then fill the flask with argon to a pressure of 2.15 atm and reweigh it. What would the balance read (in grams) if the flask has a volume of 7.35 L and the temperature is 20.0°C?

9.97 The apparatus shown consists of three bulbs connected by stopcocks. What is the pressure inside the system when the stopcocks are opened? Assume that the lines connecting the bulbs have zero volume and that the temperature remains constant.

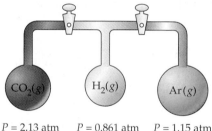

| P = 2.13 atm | P = 0.861 atm | P = 1.15 atm |
| V = 1.50 L | V = 1.00 L | V = 2.00 L |

9.98 The apparatus shown consists of three temperature-jacketed 1.000 L bulbs connected by stopcocks. Bulb A contains a mixture of $H_2O(g)$, $CO_2(g)$, and $N_2(g)$ at 25°C and a total pressure of 564 mm Hg. Bulb B is empty and is held at a temperature of −70°C. Bulb C is also empty and is held at a temperature of −190°C. The stopcocks are closed, and the volume of the lines connecting the bulbs is zero. CO_2 sublimes at −78°C, and N_2 boils at −196°C.

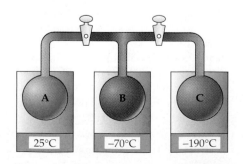

(a) The stopcock between A and B is opened, and the system is allowed to come to equilibrium. The pressure in A and B is now 219 mm Hg. What do bulbs A and B contain?

(b) How many moles of H_2O are in the system?

(c) Both stopcocks are opened, and the system is again allowed to come to equilibrium. The pressure throughout the system is 33.5 mm Hg. What do bulbs A, B, and C contain?

(d) How many moles of N_2 are in the system?

(e) How many moles of CO_2 are in the system?

9.99 Assume that you have 1.00 g of nitroglycerin in a 500.0 mL steel container at 20.0°C and 1.00 atm pressure. An explosion occurs, raising the temperature of the container and its contents to 425°C. The balanced equation is

$$4\ C_3H_5N_3O_9(l) \longrightarrow$$
$$12\ CO_2(g) + 10\ H_2O(g) + 6\ N_2(g) + O_2(g)$$

(a) How many moles of nitroglycerin and how many moles of gas (air) were in the container originally?

(b) How many moles of gas are in the container after the explosion?

(c) What is the pressure (in atmospheres) inside the container after the explosion according to the ideal gas law?

9.100 Use both the ideal gas law and the van der Waals equation to calculate the pressure (in atmospheres) of 45.0 g of NH_3 gas in a 1.000 L container at 0°C, 50°C, and 100°C. For NH_3, $a = 4.17$ ($L^2 \cdot$ atm)/mol^2 and $b = 0.0371$ L/mol.

9.101 When solid mercury(I) carbonate, Hg_2CO_3, is added to nitric acid, HNO_3, a reaction occurs to give mercury(II) nitrate, $Hg(NO_3)_2$, water, and two gases A and B:

$$Hg_2CO_3(s) + HNO_3(aq) \longrightarrow$$
$$Hg(NO_3)_2(aq) + H_2O(l) + A(g) + B(g)$$

(a) When the gases are placed in a 500.0 mL bulb at 20°C, the pressure is 258 mm Hg. How many moles of gas are present?

(b) When the gas mixture is passed over CaO(s), gas A reacts, forming $CaCO_3(s)$:

$$CaO(s) + A(g) + B(g) \longrightarrow CaCO_3(s) + B(g)$$

The remaining gas B is collected in a 250.0 mL container at 20°C and found to have a pressure of 344 mm Hg. How many moles of B are present?

(c) The mass of gas B collected in part (b) was found to be 0.218 g. What is the density of B (in grams per liter)?

(d) What is the molecular mass of B, and what is its formula?

(e) Write a balanced equation for the reaction of mercury(I) carbonate with nitric acid.

9.102 Dry ice (solid CO_2) has occasionally been used as an "explosive" in mining. A hole is drilled, dry ice and a small amount of gunpowder are placed in the hole, a fuse is added, and the hole is plugged. When lit, the exploding gunpowder rapidly vaporizes the dry ice, building up an immense pressure. Assume that 500.0 g of dry ice is placed in a cavity with a volume of 0.800 L and the ignited gunpowder heats the CO_2 to 700 K. What is the final pressure inside the hole?

9.103 Consider the combustion reaction of 0.148 g of a hydrocarbon having formula C_nH_{2n+2} with an excess of O_2 in a 400.0 mL steel container. Before reaction, the gaseous mixture had a temperature of 25.0°C and a pressure of 2.000 atm. After complete combustion and loss of considerable heat, the mixture of products and excess O_2 had a temperature of 125.0°C and a pressure of 2.983 atm.

(a) What is the formula and molar mass of the hydrocarbon?

(b) What are the partial pressures (in atmospheres) of the reactants?

(c) What are the partial pressures (in atmospheres) of the products and the excess O_2?

9.104 Natural gas is a mixture of hydrocarbons, primarily methane (CH_4) and ethane (C_2H_6). A typical mixture might have $X_{methane} = 0.915$ and $X_{ethane} = 0.085$. Let's assume that we have a 15.50 g sample of natural gas in a volume of 15.00 L at a temperature of 20.00°C.

(a) How many total moles of gas are in the sample?

(b) What is the pressure of the sample (in atmospheres)?

(c) What is the partial pressure of each component in the sample (in atmospheres)?

(d) When the sample is burned in an excess of oxygen, how much heat (in kilojoules) is liberated?

Multi-Concept Problems

9.105 When a gaseous compound X containing only C, H, and O is burned in O_2, 1 volume of the unknown gas reacts with 3 volumes of O_2 to give 2 volumes of CO_2 and 3 volumes of gaseous H_2O. Assume all volumes are measured at the same temperature and pressure.

    (a) Calculate a formula for the unknown gas, and write a balanced equation for the combustion reaction.

    (b) Is the formula you calculated an empirical formula or a molecular formula? Explain.

    (c) Draw two different possible electron-dot structures for the compound X.

    (d) Combustion of 5.000 g of X releases 144.2 kJ of heat. Look up ΔH_f° values for $CO_2(g)$ and $H_2O(g)$ in Appendix B, and calculate ΔH_f° for compound X.

9.106 Isooctane, C_8H_{18}, is the component of gasoline from which the term *octane rating* derives.

    (a) Write a balanced equation for the combustion of isooctane to yield CO_2 and H_2O.

    (b) Assuming that gasoline is 100% isooctane, that isooctane burns to produce only CO_2 and H_2O, and that the density of isooctane is 0.792 g/mL, what mass of CO_2 (in kilograms) is produced each year by the annual U.S. gasoline consumption of 4.6×10^{10} L?

    (c) What is the volume (in liters) of this CO_2 at STP?

    (d) How many moles of air are necessary for the combustion of 1 mol of isooctane, assuming that air is 21.0% O_2 by volume? What is the volume (in liters) of this air at STP?

9.107 The *Rankine* temperature scale used in engineering is to the Fahrenheit scale as the Kelvin scale is to the Celsius scale. That is, 1 Rankine degree is the same size as 1 Fahrenheit degree, and $0°R$ = absolute zero.

    (a) What temperature corresponds to the freezing point of water on the Rankine scale?

    (b) What is the value of the gas constant R on the Rankine scale in $(L \cdot atm)/(°R \cdot mol)$?

    (c) Use the van der Waals equation to determine the pressure inside a 400.0 mL vessel that contains 2.50 mol of CH_4 at a temperature of $525°R$. For CH_4, $a = 2.253 \ (L^2 \cdot atm)/mol^2$; $b = 0.04278 \ L/mol$.

9.108 Chemical explosions are characterized by the instantaneous release of large quantities of hot gases, which set up a shock wave of enormous pressure (up to 700,000 atm) and velocity (up to 20,000 mi/h). For example, explosion of nitroglycerin ($C_3H_5N_3O_9$) releases four gases, A, B, C, and D:

$$n \ C_3H_5N_3O_9(l) \longrightarrow a \ A(g) + b \ B(g) + c \ C(g) + d \ D(g)$$

Assume that the explosion of 1 mol (227 g) of nitroglycerin releases gases with a temperature of 1950°C and a volume of 1323 L at 1.00 atm pressure.

    (a) How many moles of hot gas are released by explosion of 0.004 00 mol of nitroglycerin?

    (b) When the products released by an explosion of 0.004 00 mol of nitroglycerin were placed in a 500.0 mL flask and the flask was cooled to −10°C, product A solidified, and the pressure inside the flask was 623 mm Hg. How many moles of A were present, and what is its likely identity?

    (c) When gases B, C, and D were passed through a tube of powdered Li_2O, gas B reacted to form Li_2CO_3. The remaining gases, C and D, were collected in another 500.0 mL flask and found to have a pressure of 260 mm Hg at 25°C. How many moles of B were present, and what is its likely identity?

    (d) When gases C and D were passed through a hot tube of powdered copper, gas C reacted to form CuO. The remaining gas, D, was collected in a third 500.0 mL flask and found to have a mass of 0.168 g and a pressure of 223 mm Hg at 25°C. How many moles each of C and D were present, and what are their likely identities?

    (e) Write a balanced equation for the explosion of nitroglycerin.

eMedia Problems

9.109 The **P-V Relationships** movie (*eChapter 9.2*) shows the relationship between pressure and volume of a gas at constant temperature. Under less controlled conditions, compressing a gas will cause a temperature increase. Under these conditions, how would the plot of volume versus pressure be different from the one in the movie?

9.110 The **Gas Laws** activity (*eChapter 9.2*) allows you to compare the volumes of several gases as a function of pressure and temperature. Compare equal masses of nitrogen and carbon dioxide at various temperatures and pressures. Under identical conditions, do equal masses of N_2 and CO_2 have the same volume? Explain.

9.111 As shown in the **Airbags** movie (*eChapter 9.4*), automobile airbags are inflated by the explosive decomposition of sodium azide. If the nitrogen has an initial pressure of 1.05 atm and an initial temperature of 100°C, what mass of sodium azide would be required to inflate a 20.0-L bag?

9.112 Use the **Density of Gases** activity (*eChapter 9.4*) to compare the densities of two different gases at the same pressure and temperature. Use the kinetic-molecular theory to explain why the molar mass of a gas is necessary to determine the density of a gas, but not necessary to determine the pressure exerted by the gas.

9.113 The relative speeds of helium and neon atoms are illustrated in the **Kinetic Energy in a Gas** movie (*eChapter 9.6*). Given that the average kinetic energies of all gases are equal at a given temperature, determine how much faster helium atoms move (on average) than neon atoms. How does this particular aspect of kinetic molecular theory ($E_K \propto$ to absolute temperature) explain Charles's observation that pressure increases with increasing temperature at constant volume?

10

Liquids, Solids, and Phase Changes

Water in all three of its phases—solid, liquid, and vapor—is visible on Mt. Erebus, the world's southernmost active volcano.

The kinetic–molecular theory developed in the previous chapter accounts for the properties of gases by assuming that gas particles act independently of one another. Because the attractive forces between them are so weak, the particles in gases are free to move about at random and occupy whatever space is available. The same is not true in liquids and solids, however. Liquids and solids are distinguished from gases by the presence of strong attractive forces between particles. In liquids, these attractive forces are strong enough to hold the particles in close contact while still letting them slip and slide over one another. In solids, the forces are so strong that they hold the particles rigidly in place and prevent their movement (Figure 10.1).

In this chapter, we'll examine the nature of the forces responsible for the properties of liquids and solids, paying particular attention to the ordering of particles in solids and to the different kinds of solids that result. In addition, we'll look at what happens during transitions between solid, liquid, and gaseous states and at the effects of temperature and pressure on these transitions.

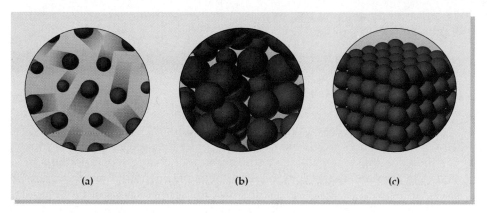

FIGURE 10.1 A molecular comparison of gases, liquids, and solids. **(a)** In gases, the particles feel little attraction for one another and are free to move about randomly. **(b)** In liquids, the particles are held close together by attractive forces but are free to move over one another. **(c)** In solids, the particles are rigidly held in an ordered arrangement.

10.1 Polar Covalent Bonds and Dipole Moments

Before looking at the forces between molecules, it's first necessary to develop the ideas of *bond dipoles* and *dipole moments*. We saw in Section 7.4 that polar covalent bonds form between atoms of different electronegativity. Chlorine is more electronegative than carbon, for example, and the chlorine atom in chloromethane (CH_3Cl) thus attracts the electrons in the C–Cl bond toward itself. The C–Cl bond is therefore polarized so that the chlorine atom is slightly electron-rich ($\delta-$) and the carbon atom is slightly electron-poor ($\delta+$).

Because the polar C–Cl bond in chloromethane has two charged ends—a positive end and a negative end—we describe it as being a bond **dipole**, and we often represent the dipole using an arrow with a cross at one end ($\longmapsto$) to indicate the direction of electron displacement. The point of the arrow represents the negative end of the dipole ($\delta-$), and the crossed end (which looks like a plus sign) represents the positive end ($\delta+$):

Chlorine is at negative end of bond dipole.

Carbon is at positive end of bond dipole.

Chloromethane, CH_3Cl

Just as individual bonds in molecules are often polar, molecules as a whole are also often polar because of the net sum of individual bond polarities and lone-pair contributions in the molecule. The resultant *molecular dipoles* can be looked at in the following way: Assume that there is a center of mass of all positive charges (nuclei) in a molecule and a center of mass of all negative charges (electrons). If these two centers don't coincide, then the molecule has a net polarity.

The measure of net molecular polarity is a quantity called the **dipole moment, μ** (Greek mu), which is defined as the magnitude of the charge Q at either end of the molecular dipole times the distance r between the charges: $\mu = Q \times r$. Dipole moments are expressed in *debyes* (D), where 1 D = 3.336×10^{-30} coulomb meters (C · m) in SI units. To help calibrate your thinking, the charge on an electron is 1.60×10^{-19} coulomb. Thus, if one proton and one electron were separated by 100 pm (a bit less than the length of an average covalent bond), then the dipole moment would be 1.60×10^{-29} C · m, or 4.80 D:

$$\mu = Q \times r$$

$$\mu = (1.60 \times 10^{-19} \text{ C})(100 \times 10^{-12} \text{ m})\left(\frac{1 \text{ D}}{3.336 \times 10^{-30} \text{ C·m}}\right) = 4.80 \text{ D}$$

It's relatively easy to measure dipole moments, and values for some common substances are given in Table 10.1. Once the dipole moment is known, it's then possible to get an idea of the amount of charge separation in a molecule. In chloromethane, for example, the experimentally measured dipole moment is $\mu = 1.87$ D. If we assume that the contributions of the nonpolar C–H bonds are small, then most of the chloromethane dipole moment is due to the C–Cl bond. Since the C–Cl bond distance is 178 pm, we can calculate that the dipole moment of chloromethane would be 1.78×4.80 D $= 8.54$ D if the C–Cl bond were ionic (that is, if a full negative charge on chlorine were separated from a full positive charge on carbon by a distance of 178 pm). But because the measured dipole moment of chloromethane is only 1.87 D, we can conclude that the C–Cl bond is only about $(1.87/8.54)(100\%) = 22\%$ ionic. Thus, the chlorine atom in chloromethane has an excess of about 0.2 electron, and the carbon atom has a deficiency of about 0.2 electron.

This polarity is clearly visible in an electrostatic potential map (Section 7.4), which shows the electron-rich chlorine as red and the electron-poor remainder of the molecule as blue-green.

TABLE 10.1 Dipole Moments of Some Common Compounds

Compound	Dipole Moment (D)
NaCl*	9.0
CH$_3$Cl	1.87
H$_2$O	1.85
NH$_3$	1.47
CO$_2$	0
CCl$_4$	0

* Measured in the gas phase.

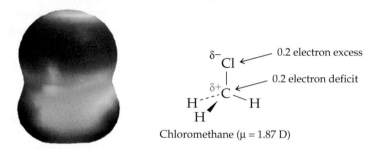

Chloromethane ($\mu = 1.87$ D)

Not surprisingly, the largest dipole moment listed in Table 10.1 belongs to the ionic compound NaCl. Water and ammonia also have substantial dipole moments because both oxygen and nitrogen are electronegative relative to hydrogen and because both O and N have lone pairs of electrons that make substantial contributions to net molecular polarity. In keeping with these polarities, electrostatic potential maps show that the hydrogens in both molecules are electron-poor (blue), while the oxygen and nitrogen atoms are electron-rich (red.)

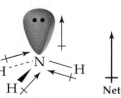

Ammonia ($\mu = 1.47$ D)

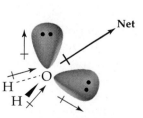

Water ($\mu = 1.85$ D)

In contrast with water and ammonia, carbon dioxide and tetrachloromethane (CCl_4) have zero dipole moments. Molecules of both substances contain *individual* polar covalent bonds, but because of the symmetry of their structures, the individual bond polarities exactly cancel. Thus, the electrostatic potential maps of both substances are symmetrical:

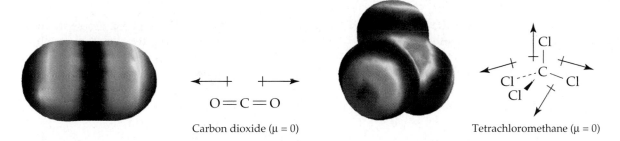

Carbon dioxide ($\mu = 0$)

Tetrachloromethane ($\mu = 0$)

EXAMPLE 10.1   The dipole moment of HCl is 1.03 D, and the distance between atoms is 127 pm. What is the percent ionic character of the HCl bond?

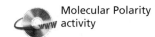

Molecular Polarity activity

SOLUTION   If HCl were 100% ionic, a negative charge (Cl^-) would be separated from a positive charge (H^+) by 127 pm, and we could calculate an expected dipole moment of 6.09 D:

$$\mu = Q \times r$$

$$\mu = (1.60 \times 10^{-19} \, C)(127 \times 10^{-12} \, m)\left(\frac{1 \, D}{3.336 \times 10^{-30} \, C \cdot m}\right) = 6.09 \, D$$

The fact that the observed dipole moment of HCl is 1.03 D implies that the H–Cl bond is only about 17% ionic:

$$\frac{1.03 \, D}{6.09 \, D} \times 100\% = 16.9\%$$

EXAMPLE 10.2   Would you expect vinyl chloride ($H_2C{=}CHCl$), the starting material used for preparation of poly(vinyl chloride) polymer, to have a dipole moment? If yes, indicate the direction.

SOLUTION   First, use the VSEPR model described in Section 7.9 to predict the molecular shape of vinyl chloride. Since both carbon atoms have three charge clouds, each has trigonal planar geometry, and the molecule as a whole is planar:

Vinyl chloride

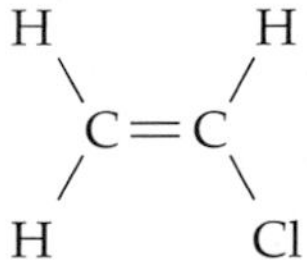

Top view 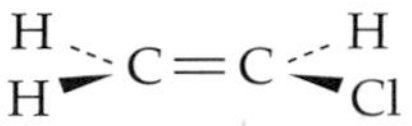   Side view

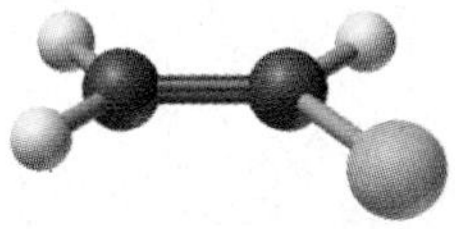

Next, assign polarities to the individual bonds according to the differences in electronegativity of the bonded atoms (Figure 7.4), and make a reasonable guess about the overall polarity that would result by summing the individual contributions. In vinyl chloride, only the C–Cl bond has a substantial polarity, giving the molecule a net polarity:

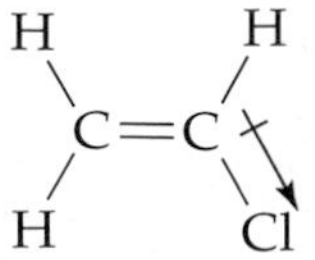

▶ **PROBLEM 10.1**   The dipole moment of HF is $\mu = 1.82$ D, and the bond length is 92 pm. Calculate the percent ionic character of the H–F bond. Is HF more ionic or less ionic than HCl (Example 10.1)?

▶ **PROBLEM 10.2**   Tell which of the following compounds is likely to have a dipole moment, and show the direction of each.
(a) SF_6 (b) $H_2C=CH_2$ (c) $CHCl_3$ (d) CH_2Cl_2

✦ **KEY CONCEPT PROBLEM 10.3**   Account for the observed dipole moment of methanol ($\mu = 1.70$ D) by using arrows to indicate the direction in which electrons are displaced.

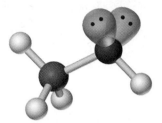

10.2   Intermolecular Forces

Now that we know a bit about molecular polarities, let's see how they give rise to some of the forces that occur between molecules. The existence of such forces is easy to show. Take H_2O, for example. An individual H_2O molecule consists of two hydrogen atoms and one oxygen atom joined together in a specific way by the *intra*molecular forces that we call covalent bonds. But a visible sample of H_2O exists either as solid ice, liquid water, or gaseous steam, depending on its temperature. Thus, there must also be some *inter***molecular forces** between molecules that hold the molecules together at certain temperatures (Figure 10.2).

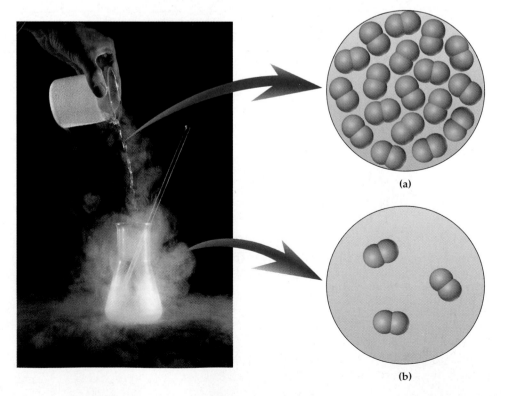

FIGURE 10.2   **(a)** In an individual N_2 molecule, the atoms are held together by the strong intramolecular force we call a covalent bond. Different N_2 molecules are weakly attracted to one another at low temperature by intermolecular forces, causing nitrogen to become liquid. **(b)** At a higher temperature, intermolecular forces are no longer able to keep molecules close together, so nitrogen becomes a gas.

(Strictly speaking, the term *intermolecular* refers only to molecular substances, but we'll use it generally to refer to interactions among all kinds of particles, including molecules, ions, and atoms.)

Intermolecular forces as a whole are usually called **van der Waals forces** after the Dutch scientist Johannes van der Waals (1837–1923). These forces are of several different types, including *dipole–dipole forces, London dispersion forces,* and *hydrogen bonds.* In addition, *ion–dipole forces* operate between ions and molecules. All these intermolecular forces are electrical in origin and result from the mutual attraction of unlike charges or the mutual repulsion of like charges. If the particles are ions, then full charges are present and the ion–ion attractions are so strong (energies on the order of 500–1000 kJ/mol) that they give rise to what we call *ionic bonds* (Section 6.6). If the particles are neutral, then only partial charges are present at best, but even so the attractive forces can be substantial.

Ion–Dipole Forces

We saw in the previous section that a molecule has a net polarity and an overall dipole moment if the sum of its individual bond dipoles is nonzero. One side of the molecule has a net excess of electrons and a partial negative charge ($\delta-$), while the other side has a net deficiency of electrons and a partial positive charge ($\delta+$). An **ion–dipole force** is the result of electrical interactions between an ion and the partial charges on a polar molecule (Figure 10.3).

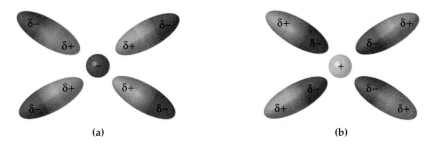

(a)   (b)

FIGURE 10.3 Polar molecules orient toward ions so that **(a)** the positive end of the dipole is near an anion and **(b)** the negative end of the dipole is near a cation.

As you might expect, the favored orientation of a polar molecule in the presence of ions is one where the positive end of the dipole is near an anion and the negative end of the dipole is near a cation. The magnitude of the interaction energy E depends on the charge on the ion z, on the strength of the dipole as measured by its dipole moment μ, and on the inverse square of the distance r from the ion to the dipole: $E = z \cdot \mu/r^2$. Ion–dipole forces are particularly important in aqueous solutions of ionic substances such as NaCl, in which polar water molecules surround the ions. We'll explore this point in more detail in the next chapter.

Dipole–Dipole Forces

Polar molecules experience **dipole–dipole forces** as the result of electrical interactions among dipoles on neighboring molecules. The forces can be either attractive or repulsive, depending on the orientation of the molecules (Figure 10.4), and the net force in a large collection of molecules results from many individual interactions of both types. The forces are generally weak, with energies on the order of 3–4 kJ/mol, and are significant only when molecules are in close contact.

(a) **(b)**

FIGURE 10.4 **(a)** Polar molecules attract one another and approach closely when oriented with unlike charges together, but **(b)** they repel one another and push apart when oriented with like charges together.

Not surprisingly, the strength of a given dipole–dipole interaction depends on the sizes of the dipole moments involved. The more polar the substance, the greater the strength of its dipole–dipole interactions. Table 10.2 lists several substances with similar molecular masses but different dipole moments and shows that there is a rough correlation between dipole moment and boiling point. The higher the dipole moment, the stronger the intermolecular forces, and the greater the amount of heat that must be added to overcome those forces. Thus, substances with higher dipole moments generally have higher boiling points.

TABLE 10.2   Comparison of Molecular Masses, Dipole Moments, and Boiling Points

Substance	Mol Mass (amu)	Dipole Moment (D)	bp (K)
$CH_3CH_2CH_3$	44.10	0.1	231
CH_3OCH_3	46.07	1.3	248
CH_3Cl	50.49	1.9	249
CH_3CN	41.05	3.9	355

London Dispersion Forces

The causes of intermolecular forces among charged and polar particles are easy to understand, but it's less obvious how such forces arise among nonpolar molecules or the individual atoms of a noble gas. Benzene (C_6H_6), for example, has zero dipole moment, as indicated by the symmetry of its electrostatic potential map, and therefore experiences no dipole–dipole forces. Nevertheless, there must be *some* intermolecular forces present among benzene molecules because the substance is a liquid rather than a gas at room temperature, with a melting point of 5.5°C and a boiling point of 80.1°C.

Benzene
$\mu = 0$
mp = 5.5°C
bp = 80.1°C

All atoms and molecules experience **London dispersion forces**, which result from the motion of electrons. Take atoms of the noble gas helium, for example. Averaged over time, the electron distribution around a helium atom is spherically symmetrical. At any given *instant*, though, the electron distribution in an atom may be unsymmetrical, giving the atom a short-lived dipole moment. This instantaneous dipole on one atom can affect the electron distributions in neighboring atoms and *induce* temporary dipoles in those neighbors (Figure 10.5). As a result, weak attractive forces develop.

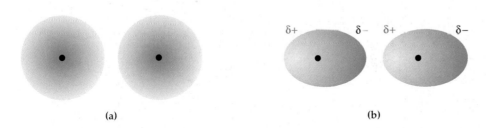

(a) (b)

FIGURE 10.5 **(a)** Averaged over time, the electron distribution in a helium atom is spherically symmetrical. **(b)** At any given instant, however, the electron distribution in an atom may be unsymmetrical, resulting in a temporary dipole moment in that atom and inducing a complementary attractive dipole in neighboring atoms.

London forces are generally small—their energies are in the range 1–10 kJ/mol—and their exact magnitude depends on the ease with which a molecule's electron cloud can be distorted by a nearby electric field, a property referred to as *polarizability*. A smaller molecule or lighter atom is less polarizable and has smaller dispersion forces because it has only a few, tightly held electrons. A larger molecule or heavier atom, however, is more polarizable and has larger dispersion forces because it has many electrons, some of which are less tightly held and are farther from a nucleus. Among the halogens, for example, the F_2 molecule is small and less polarizable, while I_2 is larger and more polarizable. As a result, F_2 has smaller dispersion forces and is a gas at room temperature, while I_2 has larger dispersion forces and is a solid (Table 10.3).

Shape is also important in determining the magnitude of the dispersion forces affecting a molecule. More spread-out shapes, which maximize molecular surface area, allow greater contact between molecules and give rise to higher dispersion forces than do more compact shapes, which minimize molecular contact. Pentane, for example, boils at 309.4 K, whereas 2,2-dimethylpropane boils at 282.7 K. Both substances have the same molecular formula, C_5H_{12}, but pentane is longer and somewhat spread out, whereas 2,2-dimethylpropane is more spherical and compact (Figure 10.6).

TABLE 10.3 Melting Points and Boiling Points of the Halogens

Halogen	mp (K)	bp (K)
F_2	53.5	85.0
Cl_2	172.2	238.6
Br_2	265.9	331.9
I_2	386.7	457.5

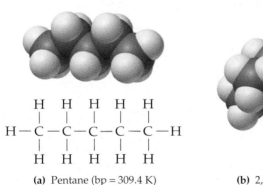

H H H H H
| | | | |
H—C—C—C—C—C—H
| | | | |
H H H H H

(a) Pentane (bp = 309.4 K)

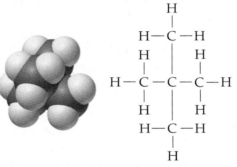

H
|
H—C—H
H | H
| | |
H—C—C—C—H
| | |
H | H
H—C—H
|
H

(b) 2,2-Dimethylpropane (bp = 282.7 K)

FIGURE 10.6 **(a)** Longer, less compact molecules like pentane feel stronger dispersion forces and have consequently higher boiling points than **(b)** more compact molecules like 2,2-dimethylpropane.

Hydrogen Bonds

A **hydrogen bond** is an attractive interaction between a hydrogen atom bonded to a very electronegative atom (O, N, or F) and an unshared electron pair on another electronegative atom. For example, hydrogen bonds occur in both water and ammonia:

Hydrogen bonds arise because O–H, N–H, and F–H bonds are highly polar, with a partial positive charge on the hydrogen and a partial negative charge on the electronegative atom. In addition, the hydrogen atom has no core electrons to shield its nucleus, and it has a small size, so it can be approached closely by other molecules. As a result, the dipole–dipole attraction between the hydrogen and a nearby electronegative atom is unusually strong, giving rise to a hydrogen bond. Water, in particular, is able to form a vast three-dimensional network of hydrogen bonds because each H_2O molecule has two hydrogens and two electron pairs (Figure 10.7).

FIGURE 10.7 Liquid water contains a vast three-dimensional network of hydrogen bonds resulting from the attraction between positively polarized hydrogens and electron pairs on negatively polarized oxygens. Each oxygen can form two hydrogen bonds, represented by dotted lines.

Hydrogen bonds can be quite strong, with energies up to 40 kJ/mol. To see one effect of hydrogen bonding, look at Table 10.4, which plots the boiling points of the covalent binary hydrides for the group 4A–7A elements. As you might expect, the boiling points generally increase with molecular mass down a group of the periodic table as a result of increased London dispersion forces—for example, $CH_4 < SiH_4 < GeH_4 < SnH_4$. Three substances, however, are clearly anomalous: NH_3, H_2O, and HF. All three have higher boiling points than might be expected because of the hydrogen bonds they contain.

TABLE 10.4   Boiling Points of the Binary Hydrides of Groups 4A, 5A, 6A, and 7A

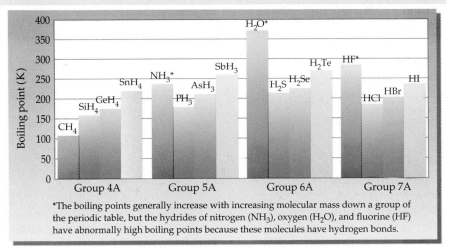

*The boiling points generally increase with increasing molecular mass down a group of the periodic table, but the hydrides of nitrogen (NH_3), oxygen (H_2O), and fluorine (HF) have abnormally high boiling points because these molecules have hydrogen bonds.

Hydrogen bonding has enormous consequences for all living organisms. Hydrogen bonds are largely responsible for defining the crucial three-dimensional shapes of large protein molecules throughout our bodies, for making possible the storage of genetic information in a cell's DNA, and for affecting the solubility and transport through the body of many thousands of vital substances. It's no exaggeration to say that life as we know it could not exist without hydrogen bonds. We'll have more to say about this in Chapter 24.

A comparison of the various kinds of intermolecular forces discussed in this section is shown in Table 10.5.

TABLE 10.5   A Comparison of Intermolecular Forces

Force	Strength	Characteristics
Ion–dipole	Moderate (10–50 kJ/mol)	Occurs between ions and polar solvents
Dipole–dipole	Weak (3–4 kJ/mol)	Occurs between polar molecules
London dispersion	Weak (1–10 kJ/mol)	Occurs between all molecules; strength depends on size, polarizability
Hydrogen bond	Moderate (10–40 kJ/mol)	Occurs between molecules with O — H, N — H, and F — H bonds

EXAMPLE 10.3   Identify the likely kinds of intermolecular forces in the following substances:

(a) HCl   (b) CH_3CH_3 (c) CH_3NH_2 (d) Kr

SOLUTION

(a) HCl is a polar molecule but can't form hydrogen bonds. It has dipole–dipole forces and dispersion forces.

(b) CH_3CH_3 is a nonpolar molecule and has only dispersion forces.

(c) CH_3NH_2 is a polar molecule that can form hydrogen bonds. In addition, it has dipole–dipole forces and dispersion forces.

(d) Kr is nonpolar and has only dispersion forces.

▶**PROBLEM 10.4**   Of the substances Ar, Cl_2, CCl_4, and HNO_3, which has:
(a) the largest dipole–dipole forces?
(b) the largest hydrogen-bond forces?
(c) the smallest dispersion forces?

▶**PROBLEM 10.5**   Consider the kinds of intermolecular forces present in the following compounds, and then rank the substances in likely order of increasing boiling point: H_2S (34 amu), CH_3OH (32 amu), C_2H_6 (30 amu), Ar (40 amu).

10.3   Some Properties of Liquids

Many familiar and observable properties of liquids can be explained by the intermolecular forces just discussed. We all know, for instance, that some liquids, such as water or gasoline, flow easily when poured, whereas others, such as motor oil or maple syrup, flow sluggishly.

The measure of a liquid's resistance to flow is called its **viscosity**. Not surprisingly, viscosity is related to the ease with which individual molecules move around in the liquid and thus to the intermolecular forces present. Substances with small, nonpolar molecules, such as pentane and benzene, experience only weak intermolecular forces and have relatively low viscosities, whereas more polar substances, such as glycerol [$C_3H_5(OH)_3$], experience stronger intermolecular forces and so have higher viscosities.

Another familiar property of liquids is **surface tension**, the resistance of a liquid to spread out and increase its surface area. Surface tension is caused by the different intermolecular forces experienced by molecules in the interior of a liquid and those on the surface. Molecules in the interior of a liquid are surrounded and experience maximum intermolecular forces, while molecules at the surface have fewer neighbors and feel weaker forces. Surface molecules are therefore less stable, and the liquid tends to minimize their number by minimizing the surface area (Figure 10.8). The ability of a water strider to walk on water and the beading-up of water on a newly waxed car are both due to surface tension.

Viscous liquids like motor oil flow sluggishly when poured, while less viscous liquids like water flow more freely.

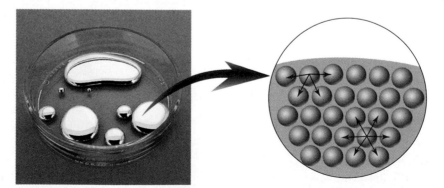

FIGURE 10.8   Surface tension, which causes these drops of liquid mercury to form beads, is due to the different forces experienced by atoms on the surface and those in the interior. Atoms on the surface are less stable because they have fewer neighbors and feel fewer attractive forces than atoms in the interior, so the liquid acts to minimize their number by minimizing the area of the surface.

Surface tension, like viscosity, is generally higher in liquids that have stronger intermolecular forces. Both properties are also temperature-dependent because molecules at higher temperatures have more kinetic energy to counteract the attractive forces holding them together. Data for some common substances are given in Table 10.6. Note that mercury has a particularly large surface tension, causing droplets to form beads (Figure 10.8) and giving the top of the mercury column in a barometer a particularly rounded appearance.

Surface tension allows a water strider to walk on a pond without penetrating the surface.

TABLE 10.6   Viscosities and Surface Tensions of Some Common Substances at 20°C

Name	Formula	Viscosity ($N \cdot s/m^2$)	Surface Tension (J/m^2)
Pentane	C_5H_{12}	2.4×10^{-4}	1.61×10^{-2}
Benzene	C_6H_6	6.5×10^{-4}	2.89×10^{-2}
Water	H_2O	1.00×10^{-3}	7.29×10^{-2}
Ethanol	C_2H_5OH	1.20×10^{-3}	2.23×10^{-2}
Mercury	Hg	1.55×10^{-3}	4.6×10^{-1}
Glycerol	$C_3H_5(OH)_3$	1.49	6.34×10^{-2}

10.4   Phase Changes

Solid ice melts to liquid water; liquid water freezes to solid ice or evaporates to gaseous steam; gaseous steam condenses to liquid water. Such processes, in which the physical form but not the chemical identity of a substance changes, are called **phase changes**, or *changes of state*. Matter in any one state, or **phase**, can change into either of the other two. Solids can even change directly into gases, as occurs when dry ice (solid CO_2) *sublimes*. The names of the various phase changes are

Changes of State movie

Fusion (melting)	solid $\rightarrow$ liquid
Freezing	liquid $\rightarrow$ solid
Vaporization	liquid $\rightarrow$ gas
Condensation	gas $\rightarrow$ liquid
Sublimation	solid $\rightarrow$ gas
Deposition	gas $\rightarrow$ solid

Like all naturally occurring processes, every phase change has associated with it a free-energy change, ΔG. This free-energy change ΔG is made up of two contributions, an enthalpy part (ΔH) and a temperature-dependent entropy part ($T\Delta S$), according to the equation $\Delta G = \Delta H - T\Delta S$ that we saw in Section 8.14. The enthalpy part is the heat flow associated with making or breaking the intermolecular attractions that hold liquids and solids together, while the entropy part is associated with the change in disorder or randomness between the various phases. Gases are more random and have more entropy than liquids, which in turn are more random and have more entropy than solids.

The melting of a solid to a liquid, the sublimation of a solid to a gas, and the vaporization of a liquid to a gas all involve a change from a less random phase to a more random one, and all *absorb* heat energy to overcome the intermolecular forces holding particles together. Thus, both ΔS and ΔH are positive for these phase changes. By contrast, the freezing of a liquid to a solid, the deposition of a gas to a solid, and the condensation of a gas to a liquid all involve a change from a more random phase to a less random one, and all *release* heat energy as intermolecular attractions increase to hold particles more tightly together. Thus, both ΔS and ΔH have negative values for these phase changes. The situations are summarized in Figure 10.9. Note that the energy required for sublimation of a solid is the sum of the energies required for melting and vaporization, as we saw in Section 8.7.

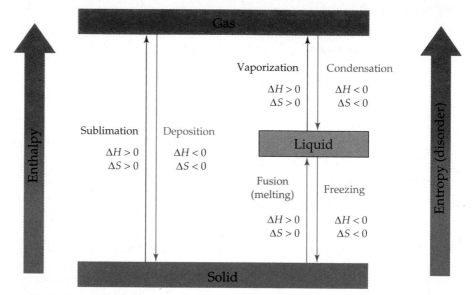

FIGURE 10.9   Changes from a less random phase to a more random one have positive values of ΔH and ΔS. Changes from a more random phase to a less random one have negative values of ΔH and ΔS.

Evaporation of perspiration carries away heat and cools the body after exertion.

Why do citrus growers spray their trees with water on cold nights?

Let's look at the transitions of solid ice to liquid water and liquid water to vapor to see some examples of energy relationships during phase changes. For the melting, or fusion, of ice to water, $\Delta H° = +6.01$ kJ/mol and $\Delta S° = +22.0$ J/(K · mol); for the vaporization of water to steam, $\Delta H° = +40.67$ kJ/mol and $\Delta S° = +109$ J/(K · mol). Notice that both $\Delta H°$ and $\Delta S°$ are larger for the liquid → vapor change than for the solid → liquid change because many more intermolecular attractions need to be overcome and much more randomness is gained in the change of liquid to vapor. This highly endothermic conversion of liquid water to gaseous water vapor is used by many organisms as a cooling mechanism. When our bodies perspire on a warm day, evaporation of the perspiration absorbs heat and leaves the skin feeling cooler.

For phase changes in the opposite direction, the numbers have the same absolute values but opposite signs. That is, $\Delta H° = -6.01$ kJ/mol and $\Delta S° = -22.0$ J/(K · mol) for the freezing of liquid water to ice, and $\Delta H° = -40.67$ kJ/mol and $\Delta S° = -109$ J/(K · mol) for the condensation of steam to liquid water. Citrus growers take advantage of the exothermic freezing of water when they spray their trees with water on cold nights to prevent frost damage. As water freezes on the leaves, it releases enough heat to protect the tree.

Knowing the values of $\Delta H°$ and $\Delta S°$ for a phase transition makes it possible to calculate the temperature at which the change occurs. Recall from Section 8.14 that $\Delta G°$ is negative for a spontaneous process, positive for a nonspontaneous process, and zero for a process at equilibrium. Thus, by setting $\Delta G° = 0$ and solving for T in the free-energy equation, we can calculate the temperature at which two phases are in equilibrium under standard conditions. For the solid → liquid phase change in water, for instance, we have

$$\Delta G° = \Delta H° - T\Delta S° = 0$$

or
$$T = \Delta H°/\Delta S°$$

where $\Delta H° = +6.01$ kJ/mol and $\Delta S° = +22.0$ J/(K · mol).

So
$$T = \frac{6.01 \dfrac{kJ}{mol}}{0.0220 \dfrac{kJ}{K \cdot mol}} = 273 \text{ K}$$

In other words, ice turns into liquid water, and liquid water turns into ice, at 273 K, or 0°C, at 1 atm pressure—hardly a surprise. (In practice, the calculation is done in the opposite direction. That is, the temperature at which a phase change occurs is measured and then used to calculate $\Delta S°$ for the process: $\Delta S° = \Delta H°/T$.)

The results of continuously adding heat to a substance can be displayed on a *heating curve* like that shown in Figure 10.10 for H_2O. Beginning with solid H_2O at an arbitrary temperature, say −25.0°C, addition of heat raises the ice's temperature until it reaches 0°C. Since the molar heat capacity of ice (Section 8.8) is 36.57 J/(mol · °C), and since we need to raise the temperature 25.0°C, 914 J/mol is required:

Heating Curves activity

$$\text{Energy to heat ice from } -25°C \text{ to } 0°C = 36.57 \frac{J}{mol \cdot °C} \times 25.0°C$$

$$= 914 \text{ J/mol}$$

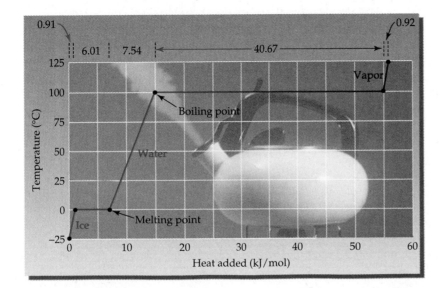

FIGURE 10.10 A heating curve for H_2O, showing the temperature changes and phase transitions that occur when heat is added. The plateau at 0°C represents the melting of solid ice, and the plateau at 100°C represents the boiling of liquid water.

Once the temperature of the ice reaches 0°C, addition of further heat goes into disrupting hydrogen bonds and other intermolecular forces rather than into increasing the temperature, as indicated by the plateau at 0°C on the heating curve in Figure 10.10. At this temperature—the *melting point*—solid and liquid coexist in equilibrium as molecules break free from their positions in the ice crystals and enter the liquid phase. Not until the solid turns completely to liquid does heat again cause the temperature to rise. The amount of energy required to overcome enough intermolecular forces to convert a solid into a liquid is the *enthalpy of fusion*, more commonly called the **heat of fusion**, ΔH_{fusion}. For ice, ΔH_{fusion} = +6.01 kJ/mol.

Continued addition of heat to liquid water raises the temperature until it reaches 100°C. We can calculate from the molar heat capacity of liquid water [75.4 J/(mol · °C)] that 7.54 kJ/mol is required:

$$\text{Energy to heat water from 0°C to 100°C} = 75.4 \frac{J}{mol \cdot °C} \times 100°C$$

$$= 7.54 \times 10^3 \text{ J/mol}$$

Once the temperature of the water reaches 100°C, addition of further heat again goes into overcoming intermolecular forces rather than into increasing the temperature, as indicated by the second plateau at 100°C on the heating curve. At this temperature—the *boiling point*—liquid and vapor coexist in equilibrium as molecules break free from the surface of the liquid and enter the gas phase. The amount of energy necessary to convert a liquid into a gas is called the *enthalpy of vaporization*, or **heat of vaporization**, ΔH_{vap}. For water, ΔH_{vap} = +40.67 kJ/mol. Only after the liquid has been completely vaporized does the temperature again rise.

Notice that the largest part (40.67 kJ/mol) of the 56.05 kJ/mol required to convert solid ice at −25°C to gaseous steam at 125°C is used for vaporization. The heat of vaporization for water is large because all the hydrogen bonds must be broken before molecules can escape from the liquid. Table 10.7 gives further data on both heat of fusion and heat of vaporization for some common compounds. Note that what is true for water is also true for other compounds: The heat of vaporization of a compound is always larger than its heat of fusion, because almost all intermolecular forces must be overcome before vaporization can occur, but relatively fewer intermolecular forces must be overcome for a solid to change to a liquid.

TABLE 10.7 Heats of Fusion and Heats of Vaporization for Some Common Compounds

Name	Formula	ΔH_{fusion} (kJ/mol)	ΔH_{vap} (kJ/mol)
Ammonia	NH_3	5.97	23.4
Benzene	C_6H_6	9.95	30.8
Ethanol	C_2H_5OH	5.02	38.6
Helium	He	0.02	0.10
Mercury	Hg	2.33	56.9
Water	H_2O	6.01	40.67

EXAMPLE 10.4   The boiling point of water is 100°C, and the enthalpy change for the conversion of water to steam is $\Delta H_{vap} = 40.67$ kJ/mol. What is the entropy change for vaporization, ΔS_{vap}, in J/(K · mol)?

SOLUTION   At the temperature where a phase change occurs, the two phases coexist in equilibrium and ΔG, the free-energy difference between the phases, is zero: $\Delta G = \Delta H - T\Delta S = 0$.

Rearranging this equation gives $\Delta S = \Delta H/T$, where both ΔH and T are known. Remember that T must be expressed in kelvins:

$$\Delta S_{vap} = \frac{\Delta H_{vap}}{T} = \frac{40.67 \dfrac{kJ}{mol}}{373.15 \text{ K}} = 0.1090 \text{ kJ}/(\text{K} \cdot \text{mol}) = 109.0 \text{ J}/(\text{K} \cdot \text{mol})$$

As you would expect, there is a large positive entropy change, corresponding to a large increase in randomness, on converting water from a liquid to a gas.

▶ **PROBLEM 10.6**   Which of the following processes would you expect to have a positive value of ΔS and which a negative value?
(a) Sublimation of dry ice
(b) Formation of dew on a cold morning
(c) Mixing of cigarette smoke with air in a closed room

▶ **PROBLEM 10.7**   Chloroform has $\Delta H_{vap} = 29.2$ kJ/mol and has $\Delta S_{vap} = 87.5$ J/(K · mol). What is the boiling point of chloroform in kelvins?

10.5   Evaporation, Vapor Pressure, and Boiling Point

The conversion of a liquid to a vapor takes place in a visible way when the liquid boils, but it takes place under other conditions as well. Let's imagine the two experiments illustrated in Figure 10.11. In experiment (a), we place a liquid in an open container; in experiment (b), we place the liquid in a closed container connected to a mercury manometer (Section 9.1). After a certain amount of time has passed, the liquid in the first container has evaporated, while the liquid in the second container remains but the pressure has risen. At equilibrium and at a constant temperature, the pressure has a constant value called the **vapor pressure** of the liquid.

FIGURE 10.11   Liquids after sitting for a length of time in **(a)** an open container and **(b)** a closed container. The liquid in the open container has evaporated, but the liquid in the closed container has brought about a rise in pressure.

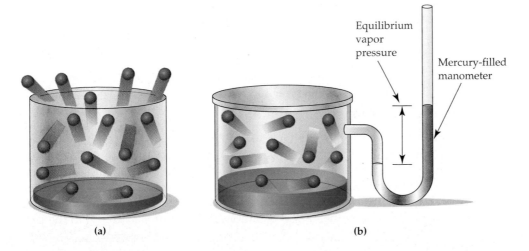

Equilibrium vapor pressure

Mercury-filled manometer

(a)   (b)

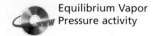

Evaporation and vapor pressure are both explained on a molecular level by the kinetic–molecular theory developed in Section 9.6 to account for the behavior of gases. The molecules in a liquid are in constant motion, but at a variety of speeds depending on the amount of kinetic energy they have. In considering a large sample, molecular kinetic energies follow a distribution curve like that shown in Figure 10.12. The exact shape of the curve depends on the temperature. The higher the temperature and the lower the boiling point of the substance, the greater the fraction of molecules in the sample that have sufficient kinetic energy to break free from the surface of the liquid and escape into the vapor.

FIGURE 10.12 The distribution of molecular kinetic energies in a liquid at two temperatures. Only the faster-moving molecules have sufficient kinetic energy to escape from the liquid and enter the vapor. The higher the temperature, the larger the number of molecules with enough energy to escape.

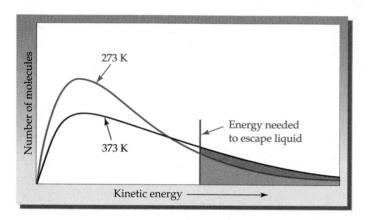

Molecules that enter the vapor phase in an open container can escape the liquid and drift away until the liquid evaporates entirely, but molecules in a closed container are trapped. As more and more molecules pass from the liquid to the vapor, chances increase that random motion will cause some of them to return occasionally to the liquid. Ultimately, the number of molecules returning to the liquid and the number escaping become equal, at which point a dynamic equilibrium exists. Although *individual* molecules are constantly passing back and forth from one phase to the other, the *total numbers* of molecules in both liquid and vapor phases remain constant.

The numerical value of a liquid's vapor pressure depends on the magnitude of the intermolecular forces present and on the temperature. The smaller the intermolecular forces, the higher the vapor pressure because loosely held molecules escape more easily. The higher the temperature, the higher the vapor pressure because a larger fraction of molecules have sufficient kinetic energy to escape.

The Clausius–Clapeyron Equation

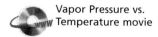

As indicated by the plots in Figure 10.13a, the vapor pressure of a liquid rises with temperature in a nonlinear way. A linear relationship *is* found, however, when the logarithm of the vapor pressure, $\ln P_{vap}$, is plotted against the inverse of the Kelvin temperature, $1/T$. Table 10.8 gives the appropriate data for water, and Figure 10.13b shows the plot. As noted in Section 9.2, a linear graph is characteristic of a mathematical equation of the form $y = mx + b$. In the present instance, $y = \ln P_{vap}$, $x = 1/T$, m is the slope of the line $(-\Delta H_{vap}/R)$, and b

TABLE 10.8   Vapor Pressure of Water at Various Temperatures

Temp (K)	P_{vap} (mm Hg)	ln P_{vap}	1/T
273	4.58	1.522	0.003 66
283	9.21	2.220	0.003 53
293	17.5	2.862	0.003 41
303	31.8	3.459	0.003 30
313	55.3	4.013	0.003 19
323	92.5	4.527	0.003 10
333	149.4	5.007	0.003 00
343	233.7	5.454	0.002 92
353	355.1	5.872	0.002 83
363	525.9	6.265	0.002 75
373	760.0	6.633	0.002 68
378	906.0	6.809	0.002 65

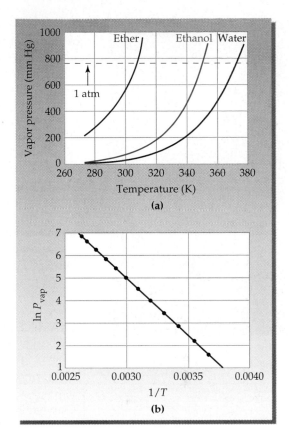

FIGURE 10.13   (a) The vapor pressures of water, ethanol, and diethyl ether show a nonlinear rise when plotted as a function of temperature. **(b)** A plot of ln P_{vap} versus 1/T (Kelvin) for water, prepared from the data in Table 10.8, shows a linear relationship.

is the y-intercept (a constant, C). Thus, the data fit an expression known as the *Clausius–Clapeyron equation*:

Clausius–Clapeyron equation:

$$\ln P_{vap} = \left(-\frac{\Delta H_{vap}}{R}\right)\frac{1}{T} + C$$

$$y \quad = \quad m \quad x \; + \; b$$

where ΔH_{vap} is the heat of vaporization of the liquid, R is the familiar gas constant (Section 9.3), and C is a constant characteristic of each specific substance.

The Clausius–Clapeyron equation makes it possible to calculate the heat of vaporization of a liquid by measuring its vapor pressure at several temperatures and then plotting the results to obtain the slope of the line. Alternatively, once the heat of vaporization and the vapor pressure at one temperature are known, the vapor pressure of the liquid at any other temperature can be calculated, as shown in Example 10.5.

When the vapor pressure of a liquid rises to the point where it becomes equal to the external pressure, the liquid changes completely into vapor and the liquid boils. On a molecular level, you might picture boiling in the following way: Imagine that a few molecules in the interior of the liquid momentarily break free from their neighbors and form a microscopic bubble. If the external pressure from the atmosphere is greater than the vapor pressure inside the bubble, the bubble is immediately crushed. At the temperature where the external pressure and the vapor pressure in the bubble are the same,

What is the vapor pressure of the liquid at its boiling point?

however, the bubble is not crushed. Instead, it rises through the denser liquid, grows larger as more molecules join it, and appears as part of the vigorous action we associate with boiling.

The temperature at which boiling occurs when the external pressure is exactly 1 atm is called the **normal boiling point** of the liquid. On the plots in Figure 10.13a, the normal boiling points of the three liquids are reached when the curves cross the dashed line representing 760 mm Hg—for ether, 34.6°C (307.8 K); for ethanol, 78.3°C (351.5 K); and for water, 100.0°C (373.15 K).

If the external pressure is *less* than 1 atm, then the vapor pressure necessary for boiling is reached earlier than 1 atm and the liquid boils at a lower than normal temperature. On top of Mt. Everest, for example, where the atmospheric pressure is only about 260 mm Hg, water boils at approximately 71°C. Conversely, if the external pressure on a liquid is *greater* than 1 atm, the vapor pressure necessary for boiling is reached later and the liquid boils at a greater than normal temperature. Pressure cookers take advantage of this effect by causing water to boil at a higher temperature, thereby allowing food to cook more rapidly.

EXAMPLE 10.5   The vapor pressure of ethanol at 34.7°C is 100.0 mm Hg, and the heat of vaporization of ethanol is 38.6 kJ/mol. What is the vapor pressure of ethanol (in millimeters of mercury) at 65.0°C?

SOLUTION   There are several ways to do this problem. One way is to use the vapor pressure at $T = 307.9$ K (34.7°C) to find a value for C in the Clausius–Clapeyron equation:

$$C = \ln P_{vap} + \frac{\Delta H_{vap}}{RT}$$

Then you could use that value to solve for $\ln P_{vap}$ at $T = 338.2$ K (65.0°C).

Alternatively, because C is a constant, its value is the same at any two pressures and temperatures. That is,

$$C = \ln P_1 + \frac{\Delta H_{vap}}{RT_1} = \ln P_2 + \frac{\Delta H_{vap}}{RT_2}$$

This equation can be rearranged to solve for the desired quantity, $\ln P_2$:

$$\ln P_2 = \ln P_1 + \left(\frac{\Delta H_{vap}}{R}\right)\left(\frac{1}{T_1} - \frac{1}{T_2}\right)$$

where $P_1 = 100.0$ mm Hg and $\ln P_1 = 4.6052$, $\Delta H_{vap} = 38.6$ kJ/mol, $R = 8.3145$ J/(K · mol), $T_2 = 338.2$ K (65.0°C), and $T_1 = 307.9$ K (34.7°C). Thus,

$$\ln P_2 = 4.6052 + \left(\frac{38{,}600 \dfrac{J}{mol}}{8.3145 \dfrac{J}{K \cdot mol}}\right)\left(\frac{1}{307.9 \text{ K}} - \frac{1}{338.2 \text{ K}}\right)$$

$$\ln P_2 = 4.6052 + 1.3509 = 5.9561$$

$$P_2 = \text{antiln}(5.9561) = 386.1 \text{ mm Hg}$$

Antilogarithms are reviewed in Appendix A.2.

EXAMPLE 10.6   Ether has $P_{vap} = 400$ mm Hg at 17.9°C and a normal boiling point of 34.6°C. What is the heat of vaporization, ΔH_{vap}, for ether in kilojoules per mole?

SOLUTION   The heat of vaporization, ΔH_{vap}, of a liquid can be obtained either graphically from the slope of a plot of $\ln P_{vap}$ versus $1/T$, or algebraically from the Clausius–Clapeyron equation. As derived in Example 10.5,

$$\ln P_2 = \ln P_1 + \left(\frac{\Delta H_{vap}}{R} \right)\left(\frac{1}{T_1} - \frac{1}{T_2} \right)$$

which can be solved for ΔH_{vap}:

$$\Delta H_{vap} = \frac{(\ln P_2 - \ln P_1)(R)}{\dfrac{1}{T_1} - \dfrac{1}{T_2}}$$

where $P_1 = 400$ mm Hg and $\ln P_1 = 5.991$, $P_2 = 760$ mm Hg at the normal boiling point and $\ln P_2 = 6.633$, $R = 8.3145$ J/(K · mol), $T_1 = 291.1$ K (17.9°C), and $T_2 = 307.8$ K (34.6°C). Thus,

$$\Delta H_{vap} = \frac{(6.633 - 5.991)\left(8.3145 \, \dfrac{J}{K \cdot mol} \right)}{\dfrac{1}{291.1 \ K} - \dfrac{1}{307.8 \ K}} = 28{,}600 \ J/mol = 28.6 \ kJ/mol$$

▶ **PROBLEM 10.8**   The normal boiling point of benzene is 80.1°C, and the heat of vaporization is $\Delta H_{vap} = 30.8$ kJ/mol. What is the boiling point of benzene (in °C) on top of Mt. Everest, where $P = 260$ mm Hg?

▶ **PROBLEM 10.9**   Bromine has $P_{vap} = 400$ mm at 41.0°C and a normal boiling point of 331.9 K. What is the heat of vaporization, ΔH_{vap}, of bromine in kilojoules per mole?

10.6   Kinds of Solids

It's obvious from a brief look around that most substances are solids rather than liquids or gases. It's also obvious that there are many different kinds of solids. Some solids, such as iron and aluminum, are hard and metallic; others, such as sugar and table salt, are crystalline and easily broken; and still others, such as rubber and many plastics, are soft and amorphous.

The most fundamental distinction between kinds of solids is that some are crystalline and some are amorphous. **Crystalline solids** are those whose atoms, ions, or molecules have an ordered arrangement extending over a long range. This order on the atomic level is also seen on the visible level, because crystalline solids usually have flat faces and distinct angles (Figure 10.14). **Amorphous solids**, by contrast, are those whose constituent particles are randomly arranged and have no ordered long-range structure. Rubber is an example.

Crystalline solids can be further categorized as *ionic, molecular, covalent network,* or *metallic*. **Ionic solids** are those like sodium chloride, whose constituent particles are ions. A crystal of sodium chloride is composed of

FIGURE 10.14 **(a)** A crystalline solid, such as the gypsum ($CaSO_4 \cdot 2\ H_2O$) shown here, has flat faces and distinct angles. These regular macroscopic features reflect a similarly ordered arrangement of particles at the atomic level. **(b)** An amorphous solid, such as rubber, has a disordered arrangement of its constituent particles.

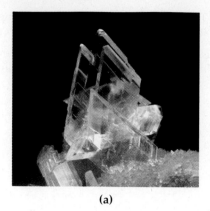

(a) **(b)**

Cesium Chloride, Ice, Diamond, Gold 3D models

alternating Na^+ and Cl^- ions ordered in a regular three-dimensional arrangement and held together by ionic bonds, as discussed in Section 6.6. **Molecular solids** are those like sucrose or ice, whose constituent particles are molecules held together by the intermolecular forces discussed in Section 10.2. A crystal of ice, for example, is composed of H_2O molecules held together in a regular way by hydrogen bonding (Figure 10.15a). **Covalent network solids** are those like quartz (Figure 10.15b) or diamond, whose atoms are linked together by covalent bonds into a giant three-dimensional array. In effect, a covalent network solid is one *very* large molecule. **Metallic solids**, such as silver or iron, also consist of large arrays of atoms, but their crystals have metallic properties such as electrical conductivity. We'll discuss metals in Chapter 21.

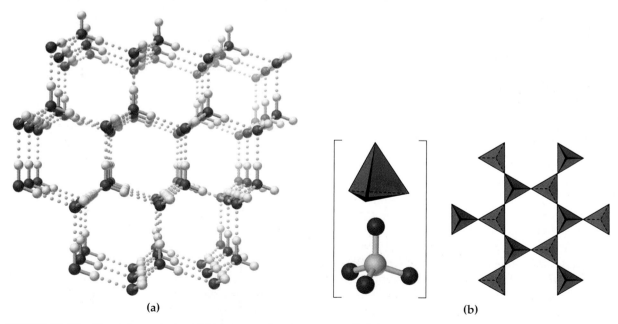

(a) **(b)**

FIGURE 10.15 Crystal structures of **(a)** ice, a molecular solid, and **(b)** quartz, a covalent network solid. Ice consists of individual H_2O molecules held together in a regular manner by hydrogen bonds. Quartz (SiO_2) is essentially one very large molecule whose Si and O atoms are linked by covalent bonds. Each silicon atom has tetrahedral geometry and is bonded to four oxygens; each oxygen has approximately linear geometry and is bonded to two silicons. The shorthand representation on the right shows how SiO_4 tetrahedra join at their corners to share oxygen atoms.

A summary of the different types of crystalline solids and their characteristics is given in Table 10.9.

TABLE 10.9 Types of Crystalline Solids and Their Characteristics

Type of Solid	Intermolecular Forces	Properties	Examples
Ionic	Ion–ion forces	Brittle, hard, high-melting	NaCl, KBr, $MgCl_2$
Molecular	Dispersion forces, dipole–dipole forces, hydrogen bonds	Soft, low-melting, nonconducting	H_2O, Br_2, CO_2, CH_4
Covalent network	Covalent bonds	Hard, high-melting	Quartz (SiO_2), C (diamond, graphite)
Metallic	Metallic bonds	Variable hardness and melting point, conducting	Na, Zn, Cu, Fe

10.7 Probing the Structure of Solids: X-Ray Crystallography

A fundamental principle of optics says that the wavelength of the light used to observe an object must be smaller than the object itself. Since atoms have diameters of around 2×10^{-10} m and the visible light detected by our eyes has wavelengths of 4–7×10^{-7} m, it is impossible to see atoms using even the finest optical microscope. To "see" atoms, we must use "light" with a wavelength of approximately 10^{-10} m, which is in the X-ray region of the electromagnetic spectrum (Section 5.2).

The origins of X-ray crystallography go back to the work of Max von Laue in 1912. On passing X rays through a crystal of sodium chloride and letting them strike a photographic plate, Laue noticed that a pattern of spots was produced on the plate, indicating that the X rays were being *diffracted* by the atoms in the crystal. A typical diffraction pattern is shown in Figure 10.16.

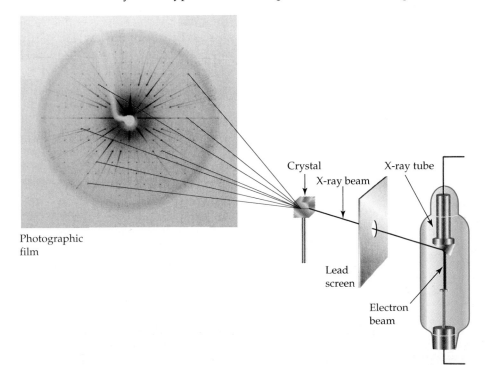

Photographic film

Crystal

X-ray beam

X-ray tube

Lead screen

Electron beam

FIGURE 10.16 An X-ray diffraction experiment. A beam of X rays is passed through a crystal and allowed to strike a photographic film. The rays are diffracted by atoms in the crystal, giving rise to a regular pattern of spots on the film.

Diffraction of electromagnetic radiation occurs when a beam is scattered by an object containing regularly spaced lines (such as those in a diffraction grating) or points (such as the atoms in a crystal). This scattering can happen only if the spacing between the lines or points is comparable to the wavelength of the radiation. As shown in Figure 10.17, diffraction is due to *interference* between two waves passing through the same region of space at the same time. If the waves are in-phase, peak to peak and trough to trough, the interference is constructive and the combined wave is increased in intensity. If the waves are out-of-phase, however, the interference is destructive and the wave is canceled. Constructive interference gives rise to the intense spots observed on Laue's photographic plate, while destructive interference causes the surrounding light areas.

FIGURE 10.17 Interference of electromagnetic waves. **(a)** Constructive interference occurs if the waves are in-phase and produces a wave with increased intensity. **(b)** Destructive interference occurs if the waves are out-of-phase and results in cancellation.

How does the diffraction of X rays by atoms in a crystal give rise to the observed pattern of spots on a photographic plate? According to an explanation advanced in 1913 by the English physicist William H. Bragg and his 22-year-old son William L. Bragg, the X rays are diffracted by different layers of atoms in the crystal, leading to constructive interference in some instances but destructive interference in others.

To understand the Bragg analysis, imagine that incoming X rays with wavelength λ strike a crystal face at an angle θ and then bounce off at the same angle, just as light bounces off a mirror (Figure 10.18). Those rays that strike an atom in the top layer are reflected at the same angle θ, and those rays that strike an atom in the second layer are also reflected at the angle θ. But because the second layer of atoms is farther from the X-ray source, the distance the X rays travel to reach the second layer is farther than the distance they travel to reach the first layer by an amount indicated as *BC* in Figure 10.18. Using trigonometry, you can show that the extra distance *BC* is equal to the distance between atomic layers *d* (= *AC*) times the sine of the angle θ:

$$\sin \theta = \frac{BC}{d} \quad \text{so} \quad BC = d \sin \theta$$

Of course, the extra distance *BC* = *CB'* must also be traveled again by the *reflected* rays as they exit the crystal, making the total extra distance traveled equal to 2*d* sin θ.

FIGURE 10.18 Diffraction of X rays of wavelength λ from atoms in the top two layers of a crystal. Rays striking atoms in the second layer travel a distance equal to *BC* + *CB'* farther than rays striking atoms in the first layer. If this distance is a whole number of wavelengths, the reflected rays are in-phase and interfere constructively. Knowing the angle θ then makes it possible to calculate the distance *d* between the layers.

$$BC + CB' = 2d \sin \theta$$

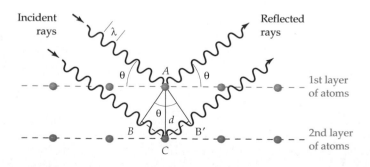

The key to the Bragg analysis is the realization that the different rays striking the two layers of atoms are in-phase initially but can be in-phase after reflection only if the extra distance $BC + CB'$ is equal to a whole number of wavelengths $n\lambda$, where n is an integer (1, 2, 3, . . .). If the extra distance is not a whole number of wavelengths, then the reflected rays will be out-of-phase and will cancel. Setting the extra distance $2d \sin \theta = n\lambda$ and rearranging to solve for d gives the *Bragg equation*:

$$BC + CB' = 2d \sin \theta = n\lambda$$

Bragg equation: $$d = \frac{n\lambda}{2 \sin \theta}$$

Of the variables in the Bragg equation, the value of the wavelength λ is known, the value of $\sin \theta$ can be measured, and the value of n is a small integer, usually assumed to be 1. Thus, the distance d between layers of atoms in a crystal can be calculated. For their work, the Braggs shared the 1915 Nobel Prize in physics; the younger Bragg was 25 at the time.

Computer-controlled *X-ray diffractometers* that automatically rotate a crystal and measure the diffraction from all angles are now available. Analysis of the X-ray diffraction pattern then makes it possible to measure the interatomic distance between any two atoms in a crystal. For molecular substances, this knowledge of interatomic distances indicates which atoms are close enough to form a bond. X-ray analysis thus provides a means for determining the structures of molecules (Figure 10.19).

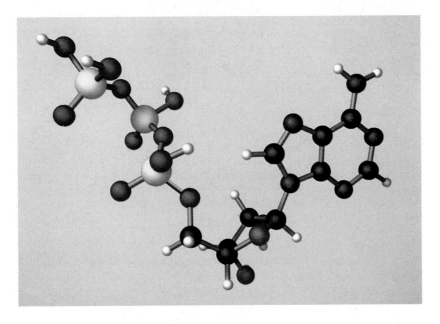

FIGURE 10.19   A computer-generated structure of adenosine triphosphate (ATP), the biological molecule that has been called "the energy currency of the living cell," as determined by X-ray crystallography.

10.8   Unit Cells and the Packing of Spheres in Crystalline Solids

How do particles—whether atoms, ions, or molecules—pack together in crystals? Let's look at metals, which are the simplest examples of crystal packing because the individual atoms can be treated as spheres. Not surprisingly, metal atoms (and other kinds of particles as well) generally pack together in crystals so that they can be as close as possible and maximize intermolecular attractions.

If you were to take a large number of uniformly sized marbles and arrange them in a box in some orderly way, there are four possibilities you might come up with. One way to arrange the marbles is in orderly rows and stacks, with the spheres in one layer sitting directly on top of those in the previous layer so that all layers are identical (Figure 10.20a). Called **simple cubic packing**, each sphere is touched by six neighbors—four in its own layer, one above, and one below—and is thus said to have a **coordination number** of 6. Only 52% of the available volume is occupied by the spheres in simple cubic packing, making inefficient use of space and minimizing attractive forces. Of all the metals in the periodic table, only polonium crystallizes in this way.

Alternatively, space could be used more efficiently if, instead of stacking the spheres directly on top of one another, you slightly separate the spheres in a given layer and offset alternating layers in an *a-b-a-b* arrangement so that the spheres in the *b* layers fit into the depressions between spheres in the *a* layers, and vice versa (Figure 10.20b). Called **body-centered cubic packing**, each sphere has a coordination number of 8—four neighbors above and four below—and space is used quite efficiently: 68% of the available volume is occupied. Iron, sodium, and 14 other metals crystallize in this way.

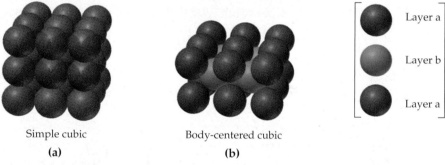

Simple cubic

(a)

Body-centered cubic

(b)

Layer a

Layer b

Layer a

FIGURE 10.20 **(a)** In simple cubic packing of spheres, all the layers are identical and all atoms are lined up in stacks and rows. Each sphere is touched by six neighbors, four in the same layer, one directly above, and one directly below. **(b)** In body-centered cubic packing of spheres, the spheres in layer *a* are separated slightly and the spheres in layer *b* are offset so that they fit into the depressions between atoms in layer *a*. Each sphere is touched by eight neighbors, four in the layer below and four in the layer above.

The remaining two packing arrangements of spheres are both said to be *closest-packed*. The **hexagonal closest-packed** arrangement (Figure 10.21a) has two alternating layers, *a-b-a-b*. Each layer has a hexagonal arrangement of touching spheres, which are offset so that spheres in a *b* layer fit into the small triangular depressions between spheres in an *a* layer. Zinc, magnesium, and 19 other metals crystallize in this way.

The **cubic closest-packed** arrangement (Figure 10.21b) has *three* alternating layers, *a-b-c-a-b-c*. The *a-b* layers are identical to those in the hexagonal closest-packed arrangement, but the third layer is offset from both *a* and *b* layers. Silver, copper, and 16 other metals crystallize with this arrangement.

In both kinds of closest-packed arrangements, each sphere has a coordination number of 12—six neighbors in the same layer, three above, and three

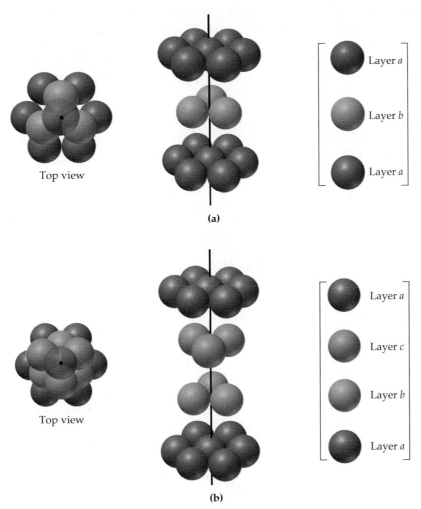

(a)

(b)

FIGURE 10.21 (a) In hexagonal closest-packing of spheres, there are two alternating hexagonal layers *a* and *b* offset from each other so that the spheres in one layer sit in the small triangular depressions of neighboring layers. (b) In cubic closest-packing of spheres, there are three alternating hexagonal layers, *a*, *b*, and *c*, offset from one another so that the spheres in one layer sit in the small triangular depressions of neighboring layers. In both kinds of closest-packing, each sphere is touched by 12 neighbors, 6 in the same layer, 3 in the layer above, and 3 in the layer below.

What kind of packing arrangement do these oranges have?

below—and 74% of the available volume is filled. The next time you're in a grocery store, look to see how the oranges are stacked in their display box. Chances are good they'll have a closest-packed arrangement.

Unit Cells

Having now taken a bulk view of how spheres can pack in a crystal, it's also useful to take a close-up view. Just as a large wall might be made up of many identical bricks stacked together in a repeating pattern, a crystal is made up of many small repeat units called **unit cells** stacked together in three dimensions.

Fourteen different unit-cell geometries occur in crystalline solids. All are parallelepipeds: six-sided geometric solids whose faces are

Just as this wall is made of many bricks stacked together in a regular way, a crystal is made of many small repeating units called unit cells.

parallelograms. We'll be concerned here only with those unit cells that have cubic symmetry—cells whose edges are equal in length and whose angles are 90°.

There are three kinds of cubic unit cells: *primitive-cubic, body-centered cubic,* and *face-centered cubic.* As shown in Figure 10.22a, a **primitive-cubic unit cell** for a metal has an atom at each of its eight corners, where it is shared with seven neighboring cubes that come together at the same point. As a result, only 1/8 of each corner atom "belongs to" a given cubic unit. This primitive-cubic unit cell, with all atoms arranged in orderly rows and stacks, is the repeat unit found in simple cubic packing.

Sodium 3D model

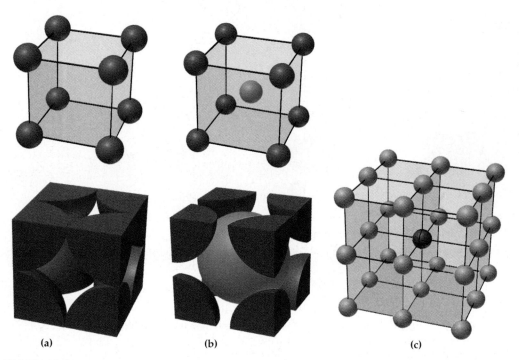

(a)   (b)   (c)

FIGURE 10.22 Geometries of **(a)** primitive-cubic and **(b)** body-centered cubic unit cells in both a skeletal view (top) and a space-filling view (bottom). Part **(c)** shows how eight primitive-cubic unit cells stack together to share a common corner where they meet.

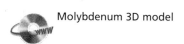
Molybdenum 3D model

A **body-centered cubic unit cell** has eight corner atoms plus an additional atom in the center of the cube (Figure 10.22b). This body-centered cubic unit cell, with two repeating offset layers and with the spheres in a given layer slightly separated, is the repeat unit found in body-centered cubic packing.

A **face-centered cubic unit cell** has eight corner atoms plus an additional atom on each of its six faces, where it is shared with one other neighboring cube. Thus, 1/2 of each face atom belongs to a given unit cell. This face-centered cubic unit cell is the repeat unit found in cubic closest packing. Though it's a bit difficult to visualize, the faces of the unit-cell cube are at 54.7° angles to the layers of the atoms (Figure 10.23).

Gold 3D model

A summary of stacking patterns, coordination numbers, amount of space used, and unit cells for the four kinds of packing of spheres is given in

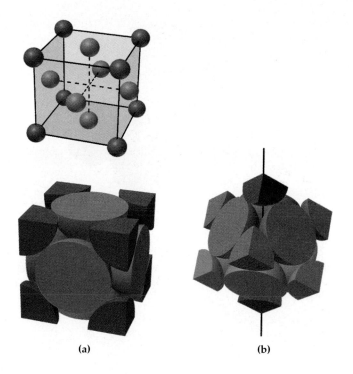

(a)

(b)

FIGURE 10.23 **(a)** Geometry of a face-centered cubic unit cell, and **(b)** a view showing how this unit cell is found in cubic closest-packing. The faces are tilted at 54.7° angles to the three repeating atomic layers.

Table 10.10. Note that hexagonal closest-packing is the only one of the four with a noncubic unit cell.

TABLE 10.10   Summary of the Four Kinds of Packing for Spheres

Structure	Stacking Pattern	Coordination Number	Space Used (%)	Unit Cell
Simple cubic	a-a-a-a-	6	52	Primitive cubic
Body-centered cubic	a-b-a-b-	8	68	Body-centered cubic
Hexagonal closest-packed	a-b-a-b-	12	74	(Noncubic)
Cubic closest-packed	a-b-c-a-b-c-	12	74	Face-centered cubic

EXAMPLE 10.7   How many atoms are in one primitive-cubic unit cell of a metal?

SOLUTION   As shown in Figure 10.22a, there is an atom at each of the eight corners of the primitive-cubic unit cell. When unit cells are stacked together, however, each corner atom is shared by eight cubes so that only 1/8 of each atom "belongs" to a given unit cell. Thus, there is $1/8 \times 8 = 1$ atom per unit cell.

EXAMPLE 10.8   Silver metal crystallizes in a cubic closest-packed arrangement with the edge of the unit cell having a length $d = 407$ pm. What is the radius (in picometers) of a silver atom?

SOLUTION   Cubic closest-packing uses a face-centered cubic unit cell. Looking at any one face of the cube head-on shows that the face atoms touch the corner atoms along the diagonal of the face but that corner atoms do not touch one another along the edges. Each diagonal is therefore equal to four atomic radii, $4r$:

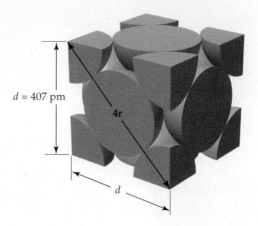

Since the diagonal and two edges of the cube form a right triangle, we can use the Pythagorean theorem to set the sum of the squares of the two edges equal to the square of the diagonal, $d^2 + d^2 = (4r)^2$, and then solve for r, the radius of one atom:

$$d^2 + d^2 = (4r)^2$$

$$2d^2 = 16r^2 \quad \text{and} \quad r^2 = \frac{d^2}{8}$$

$$\text{thus} \quad r = \sqrt{\frac{d^2}{8}} = \sqrt{\frac{(407 \text{ pm})^2}{8}} = 144 \text{ pm}$$

The radius of a silver atom is 144 pm.

EXAMPLE 10.9   Nickel has a face-centered cubic unit cell with a length of 352.4 pm along an edge. What is the density of nickel in g/cm^3?

SOLUTION   Density is equal to mass divided by volume. The volume of a single cubic unit cell with edge d is simply $d^3 = (3.524 \times 10^{-8} \text{ cm})^3 = 4.376 \times 10^{-23} \text{ cm}^3$. The mass of a single unit cell can be calculated by counting the number of atoms in the cell and multiplying by the mass of a single atom.

Each of the eight corner atoms in a face-centered cubic unit cell is shared by eight unit cells so that only $1/8 \times 8 = 1$ atom belongs to a single cell. In addition, each of the six face atoms is shared by two unit cells, so that $1/2 \times 6 = 3$ atoms belong to a single cell. Thus, a single cell has 1 corner atom and 3 face atoms, for a total of 4, and each atom has a mass equal to the molar mass of nickel (58.69 g/mol) divided by Avogadro's number (6.022×10^{23} atoms/mol). We can now calculate the density:

$$\text{Density} = \frac{\text{Mass}}{\text{Volume}} = \frac{(4 \text{ atoms}) \left(\dfrac{58.69 \dfrac{\text{g}}{\text{mol}}}{6.022 \times 10^{23} \dfrac{\text{atoms}}{\text{mol}}} \right)}{4.376 \times 10^{-23} \text{ cm}^3} = 8.909 \text{ g/cm}^3$$

The calculated density of nickel is 8.909 g/cm^3. (The measured value is 8.90 g/cm^3.)

EXAMPLE 10.10   Imagine a tiled floor in the following pattern. Identify the smallest repeating rectangular unit, analogous to a two-dimensional unit cell.

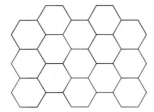

SOLUTION   Some trial-and-error is required, but the idea is to draw two perpendicular sets of parallel lines that define a repeating rectangular unit. (There may be more than one possibility.)

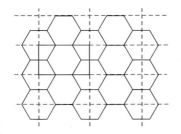   or

▶ **PROBLEM 10.10**   How many atoms are in
(a) One body-centered cubic unit cell of a metal?
(b) One face-centered cubic unit cell of a metal?

▶ **PROBLEM 10.11**   Polonium metal crystallizes in a simple cubic arrangement, with the edge of a unit cell having a length $d = 334$ pm. What is the radius (in picometers) of a polonium atom?

▶ **PROBLEM 10.12**   What is the density of polonium (Problem 10.11) in g/cm^3?

✦— **KEY CONCEPT PROBLEM 10.13**   Imagine a tiled floor in the following pattern. Identify the smallest repeating rectangular unit, analogous to a two-dimensional unit cell.

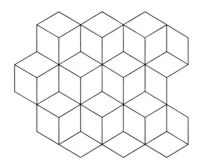

10.9   Structures of Some Ionic Solids

Simple ionic solids such as NaCl and KBr are similar to metals in that the individual ions are spheres that pack together in a regular way. They differ from metals, however, in that the spheres are not all the same size—anions are generally larger than cations (Section 6.2). As a result, ionic solids adopt a variety of different unit cells depending on the size and charge of the ions. NaCl, KCl, and a number of other salts have a face-centered cubic unit cell in which the larger Cl^- anions occupy corners and faces, while the smaller Na^+ cations fit into the holes between adjacent anions (Figure 10.24).

Salt Crystal 3D model

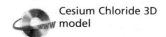
Cesium Chloride 3D model

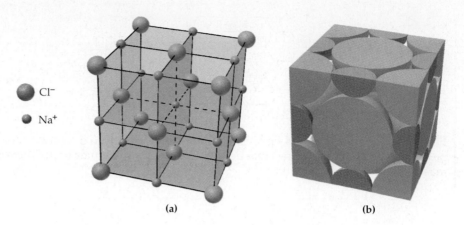

FIGURE 10.24 The unit cell of NaCl in both **(a)** a skeletal view and **(b)** a space-filling view in which one face of the unit cell is viewed head-on. The larger chloride anions adopt a face-centered cubic unit cell, with the smaller sodium cations fitting into the holes between adjacent anions.

(a)

(b)

It's necessary, of course, that the unit cell of an ionic substance be electrically neutral, with equal numbers of positive and negative charges. In the NaCl unit cell, for instance, there are four Cl^- anions ($1/8 \times 8 = 1$ corner atom, plus $1/2 \times 6 = 3$ face atoms) and also four Na^+ cations ($1/4 \times 12 = 3$ edge atoms, plus 1 center atom). (Remember that each corner atom in a cubic unit cell is shared by eight cells, each face atom is shared by two cells, and each edge atom is shared by four cells.)

Two other common ionic unit cells are shown in Figure 10.25. Copper(I) chloride has a face-centered cubic arrangement of the larger Cl^- anions, with

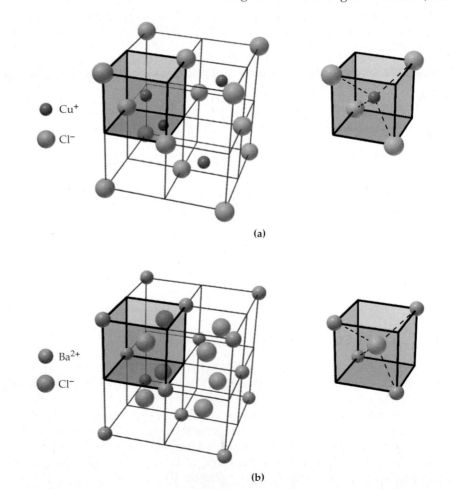

(a)

(b)

FIGURE 10.25 Unit cells of **(a)** CuCl and **(b)** BaCl$_2$. Both are based on a face-centered cubic arrangement of one ion, with the other ion tetrahedrally surrounded in holes. In CuCl, only alternating holes are filled, while in BaCl$_2$, all holes are filled.

the smaller Cu^+ cations in holes so that each is surrounded by a tetrahedron of four anions. Barium chloride, by contrast, has a face-centered cubic arrangement of the smaller Ba^{2+} *cations*, with the larger Cl^- anions surrounded tetrahedrally. As required for charge neutrality, there are twice as many Cl^- anions as there are Ba^{2+} cations.

▶ **PROBLEM 10.14**   Count the numbers of + and − charges in the CuCl and $BaCl_2$ unit cells (Figure 10.25), and show that both cells are electrically neutral.

✦ **KEY CONCEPT PROBLEM 10.15**   Rhenium oxide crystallizes in the following cubic unit cell:

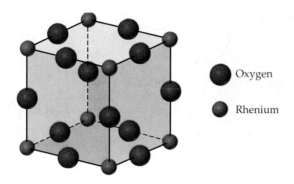

Oxygen

Rhenium

(a) How many rhenium atoms and how many oxygen atoms are in each unit cell?
(b) What is the formula of rhenium oxide?
(c) What is the oxidation state of rhenium?
(d) What is the geometry around each oxygen atom?
(e) What is the geometry around each rhenium atom?

10.10   Structures of Some Covalent Network Solids

Carbon

Carbon exists in more than 40 known structural forms, or **allotropes**, several of which are crystalline but most of which are amorphous. Graphite, the most common and stable form of elemental carbon under normal conditions, is a crystalline covalent network solid that consists of two-dimensional sheets of fused six-membered rings (Figure 10.26a). Each carbon atom is sp^2-hybridized and is connected to three other carbons. The diamond form of elemental carbon is a covalent network solid in which each carbon atom is sp^3-hybridized and is covalently bonded with tetrahedral geometry to four other carbons (Figure 10.26b).

In addition to graphite and diamond, a third crystalline allotrope of carbon called *fullerene* was discovered in 1990 as a constituent of soot. Fullerene consists of spherical C_{60} molecules arranged in the extraordinary shape of a soccer ball. The C_{60} ball has 12 pentagonal and 20 hexagonal faces, with each atom sp^2-hybridized and bonded to three other atoms (Figure 10.27).

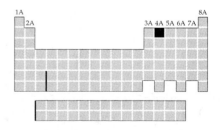

Graphite, Diamond, Fullerene 3D models

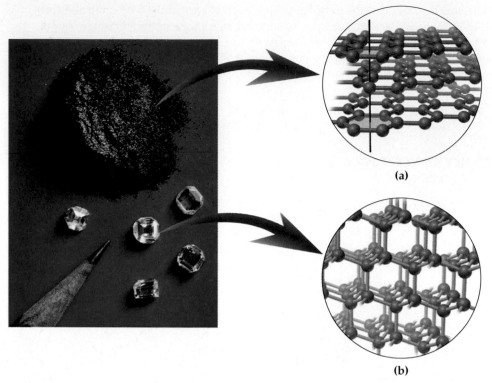

FIGURE 10.26 Two crystalline allotropes of carbon: **(a)** Graphite is a covalent network solid consisting of two-dimensional sheets of six-membered rings. The atoms in each sheet are offset slightly from the atoms in the neighboring sheets. **(b)** Diamond is a vast, three-dimensional array of sp^3-hybridized carbon atoms, each of which is bonded with tetrahedral geometry to four other carbons.

FIGURE 10.27 Fullerene, C_{60}, is a molecular solid whose molecules have the shape of a soccer ball. The ball has 12 pentagonal and 20 hexagonal faces, and each carbon atom is sp^2-hybridized.

The different structures of the carbon allotropes lead to widely different properties. Because of its three-dimensional network of strong single bonds that tie all atoms in a crystal together, diamond is the hardest known substance. In addition to its use in jewelry, diamond is widely used industrially for the tips of saw blades and drilling bits. It is an electrical insulator and has a melting point of over 3550°C. Clear, colorless, and highly crystalline, diamonds are very rare and are found in only a few places in the world, primarily in South Africa.

Graphite is the black, slippery substance used as the "lead" in pencils, as an electrode material in batteries, and as a lubricant in locks. All these properties result from its sheetlike structure. Since the layers are held together only by London dispersion forces rather than by covalent bonds, they can slide over one another, giving graphite its greasy feeling and lubricating properties. Graphite is more stable than diamond at normal pressures but can be converted into diamond at very high pressure. Some industrial diamonds are now made from graphite by applying 150,000 atm pressure at high temperature.

Fullerene, black and shiny like graphite, is yet too newly discovered to have any known useful properties, though research is continuing.

Silica

Just as living organisms are based on carbon compounds, most rocks and minerals are based on silicon compounds. Quartz and much sand, for instance, are nearly pure *silica*, SiO_2; silicon and oxygen together make up nearly 75% of the mass of the earth's crust. Considering that silicon and carbon are both in group 4A of the periodic table, you might expect SiO_2 to be similar in its properties to CO_2. In fact, though, CO_2 is a molecular substance and a gas at room temperature, whereas SiO_2 is a covalent network solid (Figure 10.15b) with a melting point over 1600°C.

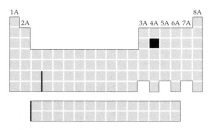

The dramatic difference in properties between CO_2 and SiO_2 is due primarily to the difference in electronic structure between carbon and silicon. The π part of a *carbon*–oxygen double bond is formed by sideways overlap of a carbon 2*p* orbital with an oxygen 2*p* orbital (Section 7.12). If a similar *silicon*–oxygen double bond were to form, it would require overlap of an oxygen 2*p* orbital and a silicon 3*p* orbital. But because the Si–O bond distance is longer than the C–O distance and a 3*p* orbital is larger than a 2*p* orbital, overlap between the two is not as favorable. As a result, silicon forms four single bonds to four oxygens in a covalent network structure rather than two double bonds to two oxygens in a molecular structure.

Colored glasses contain transition metal ions; borosilicate glass is used in cooking utensils because of its resistance to thermal shock.

Heating silica above about 1600°C breaks many of its Si–O bonds and turns it from a crystalline solid into a viscous liquid. When this fluid is cooled, some of the Si–O bonds re-form in a random arrangement, and a noncrystalline, amorphous solid called *quartz glass* is formed. If additives are mixed in before cooling, a wide variety of glasses can be prepared. Common window glass, for instance, is prepared by adding $CaCO_3$ and Na_2CO_3. Addition of various transition metal ions results in the preparation of colored glasses, and addition of B_2O_3 produces a high-melting *borosilicate glass* that is sold under the trade name Pyrex. Borosilicate glass is particularly useful for cooking utensils and laboratory glassware because it expands very little when heated and is thus unlikely to crack.

10.11 Phase Diagrams

Now that we've looked at the three phases of matter individually, let's take an overall view. As noted previously, any one phase of matter can change spontaneously into either of the other two, depending on the temperature and pressure. A particularly convenient way to picture these pressure and temperature dependencies of a pure substance in a closed system is to use what are called **phase diagrams**. As illustrated for water in Figure 10.28, a typical phase diagram shows which phase is stable at different combinations of pressure and temperature. When a boundary line between phases

Phase Diagram movie

is crossed by changing either the temperature or the pressure, a phase change occurs.

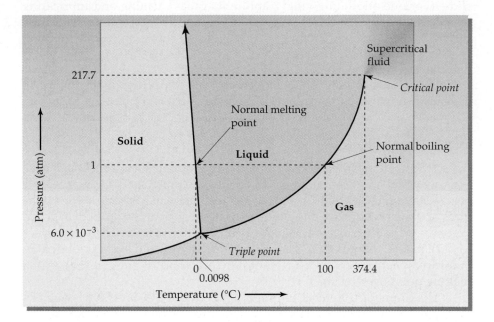

FIGURE 10.28 A phase diagram for H_2O, showing a negative slope for the solid/liquid boundary. Various features of the diagram are discussed in the text. Note that the pressure and temperature axes are not drawn to scale.

The simplest way to understand a phase diagram is to begin at the origin in the lower left corner of Figure 10.28 and travel up and right along the boundary line between solid on the left and gas on the right. Points on this line represent pressure/temperature combinations at which the two phases are in equilibrium in a closed system and a phase transition (sublimation) between solid ice and gaseous water vapor occurs. At some point along the solid/gas line, an intersection is reached where two lines diverge to form the bounds of the liquid region. The solid/liquid boundary for H_2O goes up and slightly left, while the liquid/gas boundary continues curving up and to the right. Called the **triple point**, this three-way intersection represents a unique combination of pressure and temperature at which all three phases coexist in equilibrium. For water, the triple-point temperature T_t is 0.0098°C, and the triple-point pressure P_t is 6.0×10^{-3} atm.

Continuing up and slightly left from the triple point, the solid/liquid boundary line represents the melting point of solid ice (or the freezing point of liquid water) at various pressures. When the pressure is 1 atm, the melting point (called the **normal melting point**) is exactly 0°C. There is a slight negative slope to the line, indicating that the melting point of ice decreases as pressure increases. Water is unusual in this respect, since most substances have a positive slope to their solid/liquid line, indicating that their melting points *increase* with pressure. We'll say more about this behavior shortly.

Continuing up and right from the triple point, the liquid/gas boundary line represents those pressure/temperature combinations at which liquid and gas coexist and water boils (or steam condenses). In fact, the part of the curve up to 1 atm pressure is simply the vapor pressure curve we saw previously in Figure 10.13a. When the pressure is 1 atm, water is at its **normal boiling point** of 100°C. Continuing along the liquid/gas boundary line, we suddenly reach the **critical point**, where the line abruptly ends. The critical temperature T_c is the temperature beyond which a gas cannot be liquefied, no matter how great

At the triple point, solid exists in the boiling liquid. That is, solid, liquid, and gas coexist in equilibrium.

the pressure; the critical pressure P_c is the pressure beyond which a liquid cannot be vaporized, no matter how high the temperature. For water, $T_c = 374.4°C$ and $P_c = 217.7$ atm.

We're all used to seeing solid/liquid and liquid/gas phase transitions, but behavior at the critical point lies so far outside our normal experiences that it's hard to imagine. A gas at the critical point is under such high pressure, and its molecules are so close together, that it becomes indistinguishable from a liquid. A liquid at the critical point is at such a high temperature, and its molecules so relatively far apart, that it becomes indistinguishable from a gas. The two phases simply blend into each other to form a **supercritical fluid** that is neither true liquid nor true gas. No distinct physical phase change occurs on going beyond the critical point. Rather, a whitish, pearly sheen momentarily appears, and the visible boundary between liquid and gas suddenly vanishes. Frankly, you have to see it to believe it.

The phase diagram of CO_2 shown in Figure 10.29 has many of the same features as that of water but differs in several interesting respects. First, the triple point is at $P_t = 5.11$ atm, meaning that CO_2 can't be a liquid below this pressure, no matter what the temperature. At 1 atm pressure, CO_2 is a solid below $-78.5°C$ but a gas above this temperature. Second, the slope of the solid/liquid boundary is positive, meaning that the solid phase is favored as the pressure rises and that the melting point of solid CO_2 therefore increases with pressure.

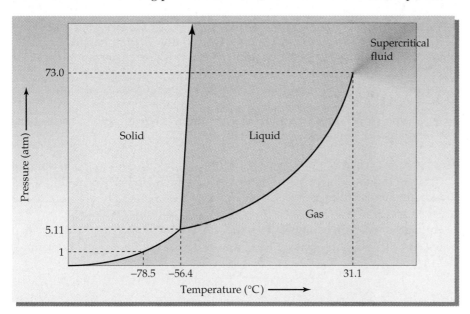

FIGURE 10.29 A phase diagram for CO_2, showing a positive slope for the solid/liquid boundary. The pressure and temperature axes are not to scale.

The effect of pressure on the slope of the solid/liquid boundary line—negative for H_2O but positive for CO_2 and most other substances—depends on the relative densities of the solid and liquid phases. For CO_2 and most other substances, the solid phase is denser than the liquid because particles are packed closer together in the solid. Increasing the pressure pushes the molecules even closer together, thereby favoring the solid phase even more and giving the solid/liquid boundary line a positive slope. Water, however, reaches its maximum density as a liquid at 3.98°C (Section 1.10). Water then becomes *less* dense when it freezes to a solid because large empty spaces are left between molecules due to the ordered three-dimensional network of hydrogen bonds in ice (Figure 10.15a). As a result, increasing the pressure favors the liquid phase, giving the solid/liquid boundary a negative slope.

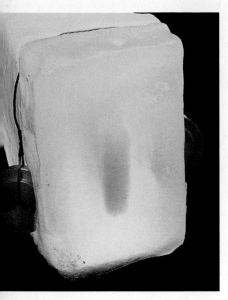

FIGURE 10.30   Why does the weighted wire cut through this block of ice?

Figure 10.30 shows a simple demonstration of the effect of pressure on melting point. If a thin wire with heavy weights at each end is draped over a block of ice near 0°C, the wire rapidly cuts through the block because the increased pressure lowers the melting point of the ice under the wire, causing the ice to liquefy. If the same experiment is tried with a block of dry ice (solid CO_2), however, nothing happens. The dry ice is unaffected because increased pressure makes melting more difficult rather than less difficult.

EXAMPLE 10.11   Freeze-dried foods are prepared by freezing the food and removing water by sublimation of ice at low pressure. Look at the phase diagram of water in Figure 10.28, and tell the maximum pressure (in mm Hg) at which ice and water vapor are in equilibrium.

SOLUTION   Ice and water vapor are in equilibrium only below the triple-point pressure, $P_t = 6.0 \times 10^{-3}$ atm. Converting to millimeters of mercury gives

$$6.0 \times 10^{-3} \text{ atm} \times \frac{760 \text{ mm Hg}}{1 \text{ atm}} = 4.6 \text{ mm Hg}$$

▶ **PROBLEM 10.16**   Look at the phase diagram of CO_2 in Figure 10.29, and tell the minimum pressure (in atmospheres) at which liquid CO_2 can exist.

▶ **PROBLEM 10.17**   Look at the phase diagram of CO_2 in Figure 10.29, and describe what happens to a CO_2 sample when the following changes are made:
(a) The temperature is increased from −100°C to 0°C at a constant pressure of 2 atm.
(b) The pressure is reduced from 72 atm to 5.0 atm at a constant temperature of 30°C.
(c) The pressure is first increased from 3.5 atm to 76 atm at −10°C, and the temperature is then increased from −10°C to 45°C.

KEY CONCEPT PROBLEM 10.18   Gallium metal has the following phase diagram (the pressure axis is not to scale). In the region shown, gallium has two different solid phases.
(a) Where on the diagram are the solid, liquid, and vapor regions?
(b) How many triple points does gallium have? Circle each on the diagram.
(c) At 1 atm pressure, which phase is more dense, solid or liquid? Explain.

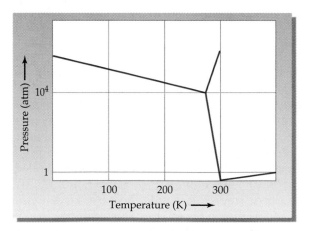

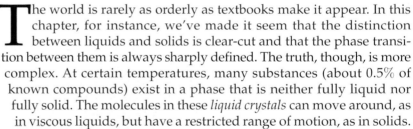

Liquid Crystals

The world is rarely as orderly as textbooks make it appear. In this chapter, for instance, we've made it seem that the distinction between liquids and solids is clear-cut and that the phase transition between them is always sharply defined. The truth, though, is more complex. At certain temperatures, many substances (about 0.5% of known compounds) exist in a phase that is neither fully liquid nor fully solid. The molecules in these *liquid crystals* can move around, as in viscous liquids, but have a restricted range of motion, as in solids.

The molecules in most liquid crystals have a rigid, rodlike shape with a length four to eight times greater than their diameter. When packed together, the molecules tend to orient with their long axes roughly parallel, like logs in a stack of firewood. Individual molecules can migrate through the fluid and can spin around their long axis, but they can't rotate end over end. Several different liquid crystalline phases exist, depending on the amount of ordering. Two of the most common are the *nematic* phase, in which the ends of the molecules are randomly arranged, and the *smectic* phase, in which the molecules are arranged in layers.

The screen in this lap-top computer uses liquid crystals for its display.

Nematic

Smectic

The widespread use of liquid crystals for displays in digital watches, pocket calculators, and computer screens hinges on the fact that the orientation of liquid-crystal molecules is extremely sensitive to the presence of small electric fields and to the nature of nearby surfaces. As shown in Figure 10.31, a typical liquid-crystal display (LCD) contains a thin layer of nematic liquid-crystal molecules sandwiched between two glass sheets that have been rubbed in different directions with a thin nylon brush and then layered with tiny transparent electrode strips made of indium/tin oxide. The outside of each glass sheet is coated with a *polarizer* oriented parallel to the rubbing direction, and one of the sheets is further coated with a reflecting mirror. Because the molecules in the liquid crystal align parallel to the direction of rubbing, and

421

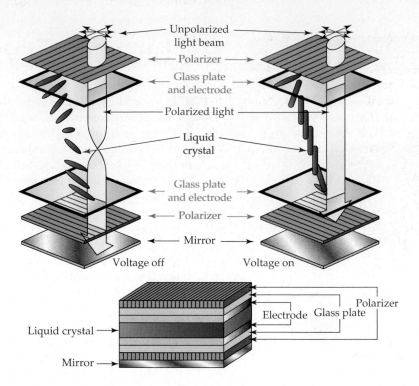

FIGURE 10.31 A liquid-crystal display (LCD), whose operation is explained in the text.

because the two glass sheets are rubbed at 90° angles to each other, the molecules undergo a gradual 90° twist in orientation between the two surfaces.

An LCD works in the following way: A beam of ordinary light consists of electromagnetic waves vibrating in various planes perpendicular to the direction of travel. When light passes through a polarizer, such as the lenses of Polaroid sunglasses, only those waves vibrating in a single plane pass through, resulting in a beam of *plane-polarized light*. The light then passes through a layer of liquid crystal that twists the plane of polarization by 90°, thereby allowing the beam to pass through the second polarizer, strike the mirror, and reflect back to be seen as a white background.

To form an LCD image, a voltage is applied to an appropriate pattern of tiny electrodes, causing the molecules in the liquid crystal to reorient parallel to the electric field. As a result, the molecules are no longer able to twist the plane of the polarized light, and the light can no longer pass through the second polarizer to reflect off the mirror. The areas under the electrodes are therefore seen as dark.

▶ **PROBLEM 10.19** How does a liquid crystal differ from a typical solid crystal, such as that of sodium chloride?

▶ **PROBLEM 10.20** What structural feature do liquid crystal molecules share?

Summary

The presence of polar covalent bonds in a molecule can cause the molecule as a whole to have a net polarity, a property measured by the **dipole moment**.

Intermolecular forces, known collectively as **van der Waals forces**, are the attractions responsible for holding particles together in the liquid and solid phases. There are several kinds of intermolecular forces, all of which arise from electrical attractions: **Dipole–dipole forces** occur between two polar molecules. **London dispersion forces** are characteristic of all molecules and result from the presence of temporary dipole moments caused by the presence of momentarily unsymmetrical electron distributions. A **hydrogen bond** is the attraction between a positively polarized hydrogen atom bonded to O, N, or F and a lone pair of electrons on an O, N, or F atom of another molecule. In addition, **ion–dipole forces** occur between an ion and a polar molecule.

Matter in any one phase—solid, liquid, or gas—can undergo a **phase change** to any of the other two phases. Like all naturally occurring processes, a phase change has associated with it a free-energy change, $\Delta G = \Delta H - T\Delta S$, that has both an enthalpy term and a temperature-dependent entropy term. The enthalpy component, ΔH, is a measure of the change in intermolecular forces; the entropy component, ΔS, is a measure of the change in molecular randomness accompanying the phase transition. The enthalpy change for the solid–liquid transition is called the **heat of fusion**, and the enthalpy change for the liquid–vapor transition is the **heat of vaporization**.

The effects of temperature and pressure on phase changes can be displayed graphically on **phase diagrams**. A typical phase diagram has three regions—solid, liquid, and gas—separated by three boundary lines: solid/gas, solid/liquid, and liquid/gas. The boundary lines represent pressure/temperature combinations at which two phases are in equilibrium and phase changes occur. At exactly 1 atm pressure, the temperature at the solid/liquid boundary corresponds to the **normal melting point** of the substance, and the temperature at the liquid/gas boundary corresponds to the **normal boiling point**. The three lines meet at the **triple point**, a unique combination of temperature and pressure at which all three phases coexist in equilibrium. The liquid/gas line runs from the triple point to the **critical point**, a pressure/temperature combination beyond which liquid and gas phases form a **supercritical fluid** that is neither liquid nor gas.

Solids can be characterized as **amorphous** if their particles are randomly arranged or **crystalline** if their particles are ordered. Crystalline solids can be further characterized as **ionic solids** if their particles are ions, **molecular solids** if their particles are molecules, **covalent network solids** if they consist of a covalently bonded array of atoms without discrete molecules, or **metallic solids** if their particles are metal atoms.

The regular three-dimensional network of particles in a crystal is made up of small repeating units called **unit cells**. There are 14 kinds of unit cells, three of which have cubic symmetry and are commonly used by metals and simple ionic compounds. **Simple cubic packing** uses a **primitive-cubic unit cell**, with an atom at each corner of the cube. **Body-centered cubic packing** uses a **body-centered cubic unit cell**, with an atom at the center and at each corner of the cube. **Cubic closest-packing** uses a **face-centered cubic unit cell**, with an atom at the center of each face and at each corner of the cube. A fourth kind of packing, called **hexagonal closest-packing**, uses a noncubic unit cell.

Key Words

allotrope *415*
amorphous solid *403*
body-centered cubic packing *408*
body-centered cubic unit cell *410*
Bragg equation *407*
Clausius–Clapeyron equation *401*
coordination number *408*
covalent network solid *404*
critical point *418*
crystalline solid *403*
cubic closest-packing *408*
diffraction *406*
dipole *385*
dipole moment (μ) *385*
dipole–dipole force *389*
face-centered cubic unit cell *410*
heat of fusion (ΔH_{fusion}) *398*
heat of vaporization (ΔH_{vap}) *398*
hexagonal closest-packing *408*
hydrogen bond *392*
intermolecular force *388*
ion–dipole force *389*
ionic solid *403*
London dispersion force *391*
metallic solid *404*
molecular solid *404*
normal boiling point *402*
normal melting point *418*
phase *395*
phase change *395*
phase diagram *417*
primitive-cubic unit cell *410*
simple cubic packing *408*
supercritical fluid *419*
surface tension *394*
triple point *418*
unit cell *409*
van der Waals force *389*
vapor pressure *399*
viscosity *394*

Key Concept Summary

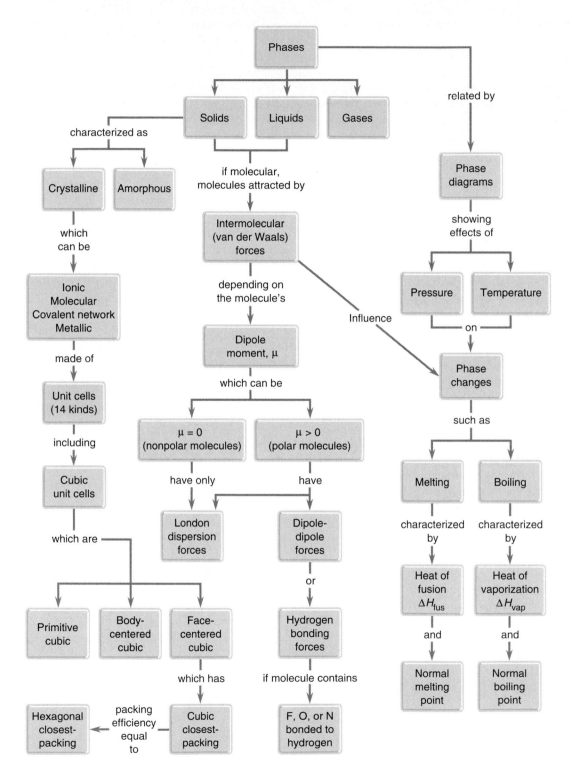

Understanding Key Concepts

Problems 10.1–10.20 appear within the chapter.

10.21 Identify each of the following kinds of packing.

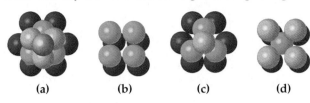

(a)   (b)   (c)   (d)

10.22 Zinc sulfide, or sphalerite, crystallizes in the following cubic unit cell:

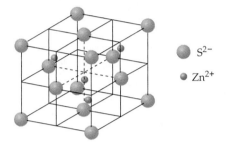

S²⁻
Zn²⁺

(a) What kind of packing do the sulfide ions adopt?

(b) How many S^{2-} ions and how many Zn^{2+} ions are in the unit cell?

10.23 Perovskite, a mineral containing calcium, oxygen, and titanium, crystallizes in the following cubic unit cell:

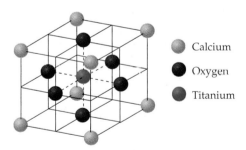

Calcium

Oxygen

Titanium

(a) What is the formula of perovskite?

(b) What is the oxidation number of titanium in perovskite?

10.24 The phase diagram of a substance is shown below.

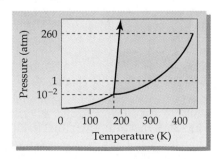

(a) Approximately what is the normal boiling point, and what is the normal melting point of the substance?

(b) What is the physical state of the substance under the following conditions?

   (i) $T = 150$ K, $P = 0.5$ atm

   (ii) $T = 325$ K, $P = 0.9$ atm

   (iii) $T = 450$ K, $P = 165$ atm

10.25 Boron nitride, BN, is a covalent network solid with a structure similar to that of graphite. Sketch a small portion of the boron nitride structure.

10.26 Imagine a tiled floor made of square and octagonal tiles in the following pattern. Identify the smallest repeating rectangular unit, analogous to a unit cell.

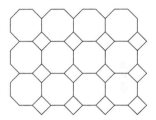

10.27 The following phase diagram of elemental carbon has three different solid phases in the region shown.

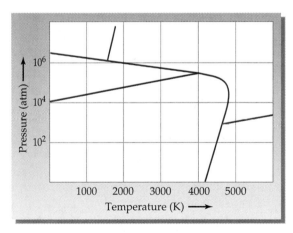

(a) Show where the solid, liquid, and vapor regions are on the diagram.

(b) How many triple points does carbon have? Circle each on the diagram.

(c) Graphite is the most stable solid phase under normal conditions. Identify the graphite phase on the diagram.

(d) On heating graphite to 2500 K at a pressure of 100,000 atm, graphite can be converted into diamond. Identify the diamond phase on the graph.

(e) Which phase is more dense, graphite or diamond? Explain.

Additional Problems

Dipole Moments and Intermolecular Forces

10.28 Why do some molecules have dipole moments while others do not?

10.29 What is the difference between London dispersion forces and dipole–dipole forces?

10.30 What are the most important kinds of intermolecular forces present in each of the following substances?

(a) Chloroform, $CHCl_3$ (b) Oxygen, O_2

(c) Polyethylene, C_nH_{2n+2} (d) Methanol, CH_3OH

10.31 Of the substances Xe, CH_3Cl, HF, which has:

(a) The smallest dipole–dipole forces?

(b) The largest hydrogen bond forces?

(c) The largest dispersion forces?

10.32 Methanol (CH_3OH; bp = 65°C) boils nearly 230° higher than methane (CH_4; bp = −164°C), but 1-decanol ($C_{10}H_{21}OH$; bp = 229°C) boils only 55° higher than decane ($C_{10}H_{22}$; bp = 174°C). Explain.

10.33 Which substance in each of the following pairs would you expect to have larger dispersion forces?

(a) Ethane, C_2H_6, or octane, C_8H_{18}

(b) HCl or HI

(c) H_2O or H_2Se

10.34 Which of the following substances would you expect to have a nonzero dipole moment? Explain, and show the direction of each.

(a) Cl_2O (b) XeF_4

(c) Chloroethane, CH_3CH_2Cl (d) BF_3

10.35 Which of the following substances would you expect to have a nonzero dipole moment? Explain, and show the direction of each.

(a) NF_3 (b) CH_3NH_2 (c) XeF_2 (d) PCl_5

10.36 How can you account for the fact that the dipole moment of SO_2 is 1.63 D, but that of CO_2 is zero?

10.37 Draw three-dimensional structures of PCl_3 and PCl_5, and then explain why one of the molecules has a dipole moment and one does not.

10.38 Draw a picture that shows how hydrogen bonding takes place between two ammonia molecules.

10.39 1,3-Propanediol, $HOCH_2CH_2CH_2OH$, can form *intra*molecular hydrogen bonds as well as intermolecular hydrogen bonds. Draw a structure of 1,3-propanediol that shows an intramolecular hydrogen bond.

Vapor Pressure and Phase Changes

10.40 Why is ΔH_{vap} usually larger than ΔH_{fusion}?

10.41 Why is the heat of sublimation, ΔH_{subl}, equal to the sum of ΔH_{vap} and ΔH_{fusion}?

10.42 Mercury has mp = −38.9°C and bp = 356.6°C. What, if any, phase changes take place under the following conditions at 1.0 atm pressure?

(a) The temperature of a sample is raised from −30°C to 365°C.

(b) The temperature of a sample is lowered from 291 K to 238 K.

(c) The temperature of a sample is lowered from 638 K to 231 K.

10.43 Iodine has mp = 113.5°C and bp = 184.4°C. What, if any, phase changes take place under the following conditions at 1.0 atm pressure?

(a) The temperature of a solid sample is held at 113.5°C while heat is added.

(b) The temperature of a sample is lowered from 452 K to 389 K.

10.44 Water at room temperature is placed in a flask connected by rubber tubing to a vacuum pump, and the pump is turned on. After several minutes, the volume of the water has decreased and what remains has turned to ice. Explain.

10.45 Ether at room temperature is placed in a flask connected by a rubber tube to a vacuum pump, the pump is turned on, and the ether begins boiling. Explain.

10.46 How much energy (in kilojoules) is needed to heat 5.00 g of ice from −10.0°C to 30.0°C? The heat of fusion of water is 6.01 kJ/mol, and the molar heat capacity is 36.6 J/(K · mol) for ice and 75.3 J/(K · mol) for liquid water.

10.47 How much energy (in kilojoules) is released when 15.3 g of steam at 115.0°C is condensed to give liquid water at 75.0°C? The heat of vaporization of liquid water is 40.67 kJ/mol, and the molar heat capacity is 75.3 J/(K · mol) for the liquid and 33.6 J/(K · mol) for the vapor.

10.48 How much energy (in kilojoules) is released when 7.55 g of water at 33.5°C is cooled to −10.0°C? (See Problem 10.46 for the necessary data.)

10.49 How much energy (in kilojoules) is released when 25.0 g of ethanol vapor at 93.0°C is cooled to −10.0°C? Ethanol has mp = −114.5°C, bp = 78.4°C, ΔH_{vap} = 38.56 kJ/mol, and ΔH_{fusion} = 4.60 kJ/mol. The molar heat capacity is 113 J/(K · mol) for the liquid and 65.7 J/(K · mol) for the vapor.

10.50 Draw a molar heating curve for ethanol, C_2H_5OH, similar to that shown for water in Figure 10.10. Begin with solid ethanol at its melting point, and raise the temperature to 100°C. The necessary data are given in Problem 10.49.

10.51 Draw a molar heating curve for sodium similar to that shown for water in Figure 10.10. Begin with solid sodium at its melting point, and raise the temperature to 1000°C. The necessary data are mp = 97.8°C, bp = 883°C, ΔH_{vap} = 89.6 kJ/mol, and ΔH_{fusion} = 2.64 kJ/mol. Assume that the molar heat capacity is 28.2 J/(K · mol) for both liquid and vapor phases and does not change with temperature.

10.52 Naphthalene, better known as "mothballs," has bp = 218°C and ΔH_{vap} = 43.3 kJ/mol. What is the entropy of vaporization, ΔS_{vap} [in J/(K · mol)] for naphthalene?

10.53 What is the entropy of fusion, ΔS_{fusion} [in J/(K · mol)] for sodium? The necessary data are given in Problem 10.51.

10.54 Carbon disulfide, CS_2, has P_{vap} = 100 mm at −5.1°C and a normal boiling point of 46.5°C. What is ΔH_{vap} for carbon disulfide in kJ/mol?

10.55 The vapor pressure of $SiCl_4$ is 100 mm Hg at 5.4°C, and the normal boiling point is 56.8°C. What is ΔH_{vap} for $SiCl_4$ in kJ/mol?

10.56 What is the vapor pressure of CS_2 (in mm Hg) at 20.0°C? (See Problem 10.54.)

10.57 What is the vapor pressure of $SiCl_4$ (in mm Hg) at 30.0°C? (See Problem 10.55.)

10.58 Dichloromethane, CH_2Cl_2, is an organic solvent used for removing caffeine from coffee beans. The following table gives the vapor pressure of dichloromethane at various temperatures. Fill in the rest of the table, and use the data to plot curves of P_{vap} versus T and ln P_{vap} versus $1/T$.

Temp (K)	P_{vap} (mm Hg)	ln P_{vap}	1/T
263	80.1	?	?
273	133.6	?	?
283	213.3	?	?
293	329.6	?	?
303	495.4	?	?
313	724.4	?	?

10.59 The following table gives the vapor pressure of mercury at various temperatures. Fill in the rest of the table, and use the data to plot curves of P_{vap} versus T and ln P_{vap} versus $1/T$.

Temp (K)	P_{vap} (mm Hg)	ln P_{vap}	1/T
500	39.3	?	?
520	68.5	?	?
540	114.4	?	?
560	191.6	?	?
580	286.4	?	?
600	432.3	?	?

10.60 Use the plot you made in Problem 10.58 to find a value (in kJ/mol) for ΔH_{vap} for dichloromethane.

10.61 Use the plot you made in Problem 10.59 to find a value (in kJ/mol) for ΔH_{vap} for mercury. The normal boiling point of mercury is 630 K.

10.62 Choose any two temperatures and corresponding vapor pressures in the table given in Problem 10.58, and use those values to calculate ΔH_{vap} for dichloromethane (in kJ/mol). How does the value you calculated compare to the value you read from your plot in Problem 10.60?

10.63 Choose any two temperatures and corresponding vapor pressures in the table given in Problem 10.59, and use those values to calculate ΔH_{vap} for mercury (in kJ/mol). How does the value you calculated compare to the value you read from your plot in Problem 10.61?

Structures of Solids

10.64 List the four main classes of crystalline solids, and give a specific example of each.

10.65 What kinds of particles are present in each of the four main classes of crystalline solids?

10.66 What is a unit cell?

10.67 Which of the four kinds of packing used by metals makes the most efficient use of space, and which makes the least efficient use?

10.68 Copper crystallizes in a face-centered cubic unit cell with an edge length of 362 pm. What is the radius of a copper atom (in picometers)? What is the density of copper in g/cm³?

10.69 Lead crystallizes in a face-centered cubic unit cell with an edge length of 495 pm. What is the radius of a lead atom (in picometers)? What is the density of lead in g/cm³?

10.70 Aluminum has a density of 2.699 g/cm³ and crystallizes with a face-centered cubic unit cell. What is the edge length of a unit cell (in picometers)?

10.71 Tungsten crystallizes in a body-centered cubic unit cell with an edge length of 317 pm. What is the length (in picometers) of a unit-cell diagonal that passes through the center atom?

10.72 In light of your answer to Problem 10.71, what is the radius (in picometers) of a tungsten atom?

10.73 Sodium has a density of 0.971 g/cm^3 and crystallizes with a body-centered cubic unit cell. What is the radius of a sodium atom, and what is the edge length of the cell (both in picometers)?

10.74 Titanium metal has a density of 4.54 g/cm^3 and an atomic radius of 144.8 pm. In what cubic unit cell does titanium crystallize?

10.75 Calcium metal has a density of 1.55 g/cm^3 and crystallizes in a cubic unit cell with an edge length of 558.2 pm.

    (a) How many Ca atoms are in one unit cell?

    (b) In which of the three cubic unit cells does calcium crystallize?

10.76 Sodium hydride, NaH, crystallizes in a face-centered cubic unit cell similar to that of NaCl (Figure 10.24). How many Na^+ ions touch each H^- ion, and how many H^- ions touch each Na^+ ion?

10.77 Cesium chloride crystallizes in a cubic unit cell with Cl^- ions at the corners and a Cs^+ ion in the center. Count the numbers of $+$ and $-$ charges, and show that the unit cell is electrically neutral.

10.78 If the edge length of an NaH unit cell is 488 pm, what is the length (in picometers) of an Na–H bond? (See Problem 10.76.)

10.79 The edge length of a CsCl unit cell (Problem 10.77) is 412.3 pm. What is the length (in picometers) of the Cs–Cl bond? If the ionic radius of a Cl^- ion is 181 pm, what is the ionic radius (in picometers) of a Cs^+ ion?

Phase Diagrams

10.80 Look at the phase diagram of CO_2 in Figure 10.29, and tell what phases are present under the following conditions:

    (a) $T = -60°C$, $P = 0.75$ atm

    (b) $T = -35°C$, $P = 18.6$ atm

    (c) $T = -80°C$, $P = 5.42$ atm

10.81 Look at the phase diagram of H_2O in Figure 10.28, and tell what happens to an H_2O sample when the following changes are made:

    (a) The temperature is reduced from 48°C to −4.4°C at a constant pressure of 6.5 atm.

    (b) The pressure is increased from 85 atm to 226 atm at a constant temperature of 380°C.

10.82 Bromine has $T_t = -7.3°C$, $P_t = 44$ mm Hg, $T_c = 315°C$, and $P_c = 102$ atm. The density of the liquid is 3.1 g/cm^3, and the density of the solid is 3.4 g/cm^3. Sketch a phase diagram for bromine, and label all points of interest.

10.83 Oxygen has $T_t = 54.3$ K, $P_t = 1.14$ mm Hg, $T_c = 154.6$ K, and $P_c = 49.77$ atm. The density of the liquid is 1.14 g/cm^3, and the density of the solid is 1.33 g/cm^3. Sketch a phase diagram for oxygen, and label all points of interest.

10.84 Refer to the bromine phase diagram you sketched in Problem 10.82, and tell what phases are present under the following conditions:

    (a) $T = -10°C$, $P = 0.0075$ atm

    (b) $T = 25°C$, $P = 16$ atm

10.85 Refer to the oxygen phase diagram you sketched in Problem 10.83, and tell what phases are present under the following conditions:

    (a) $T = -210°C$, $P = 1.5$ atm

    (b) $T = -100°C$, $P = 66$ atm

10.86 Does solid oxygen (Problem 10.83) melt when pressure is applied, as water does? Explain.

10.87 Assume that you have samples of the following three gases at 25°C. Which of the three can be liquefied by applying pressure and which cannot? Explain.

    Ammonia: $T_c = 132.5°C$ and $P_c = 112.5$ atm

    Methane: $T_c = -82.1°C$ and $P_c = 45.8$ atm

    Sulfur dioxide: $T_c = 157.8°C$ and $P_c = 77.7$ atm

10.88 Benzene has a melting point of 5.53°C and a boiling point of 80.09°C at atmospheric pressure. Its density is 0.8787 g/cm^3 when liquid and 0.899 g/cm^3 when solid; it has $T_c = 289.01°C$, $P_c = 48.34$ atm, $T_t = 5.52°C$, and $P_t = 0.0473$ atm. Starting from a point at 200 K and 66.5 atm, trace the following path on a phase diagram:

    (1) First, increase T to 585 K while keeping P constant.

    (2) Next, decrease P to 38.5 atm while keeping T constant.

    (3) Then, decrease T to 278.66 K while keeping P constant.

    (4) Finally, decrease P to 0.0025 atm while keeping T constant.

    What is your starting phase and what is your final phase?

10.89 Refer to the oxygen phase diagram you drew in Problem 10.83, and trace the following path starting from a point at 0.0011 atm and −225°C:

    (1) First, increase P to 35 atm while keeping T constant.

    (2) Next, increase T to −150°C while keeping P constant.

    (3) Then, decrease P to 1.0 atm while keeping T constant.

    (4) Finally, decrease T to −215°C while keeping P constant.

    What is your starting phase and what is your final phase?

10.90 How many phase transitions did you pass through in Problem 10.88, and what are they?

10.91 What phase transitions did you pass through in Problem 10.89?

General Problems

10.92 Fluorine is more electronegative than chlorine (Figure 7.4), yet fluoromethane (CH_3F; $\mu = 1.85$ D) has a smaller dipole moment than chloromethane (CH_3Cl; $\mu = 1.87$ D). Explain.

10.93 What is the atomic radius (in picometers) of an argon atom if solid argon has a density of 1.623 g/cm^3 and crystallizes at low temperature in a face-centered cubic unit cell?

10.94 Mercury has mp $= -38.9°C$, a molar heat capacity of 27.9 J/(K · mol) for the liquid and 28.2 J/(K · mol) for the solid, and $\Delta H_{fusion} = 2.33$ kJ/mol. Assuming that the heat capacities don't change with temperature, how much energy (in kilojoules) is needed to heat 7.50 g of Hg from a temperature of $-50.0°C$ to $+50.0°C$?

10.95 Silicon carbide, SiC, is a covalent network solid with a structure similar to that of diamond. Sketch a small portion of the SiC structure.

10.96 In Denver, the Mile-High City, water boils at 95°C. What is atmospheric pressure (in atmospheres) in Denver? ΔH_{vap} for H_2O is 40.67 kJ/mol.

10.97 Acetic acid, the principal nonaqueous constituent of vinegar, exists as a dimer in the liquid phase, with two acetic acid molecules joined together by two hydrogen bonds. Sketch the structure you would expect this dimer to have.

$$\text{Acetic acid}$$

10.98 Magnesium metal has $\Delta H_{fusion} = 9.037$ kJ/mol and $\Delta S_{fusion} = 9.79$ J/(K · mol). What is the melting point (in °C) of magnesium?

10.99 Titanium tetrachloride, $TiCl_4$, has a melting point of $-23.2°C$ and has $\Delta H_{fusion} = 9.37$ kJ/mol. What is the entropy of fusion, ΔS_{fusion} [in J/(K · mol)], for $TiCl_4$?

10.100 Dichlorodifluoromethane, CCl_2F_2, one of the chlorofluorocarbon refrigerants responsible for destroying part of the earth's ozone layer, has $P_{vap} = 40.0$ mm Hg at $-81.6°C$ and $P_{vap} = 400$ mm Hg at $-43.9°C$. What is the normal boiling point of CCl_2F_2 (in °C)?

10.101 Trichlorofluoromethane, CCl_3F, another chlorofluorocarbon refrigerant, has $P_{vap} = 100.0$ mm Hg at $-23°C$ and $\Delta H_{vap} = 24.77$ kJ/mol.

   (a) What is the normal boiling point of trichlorofluoromethane (in °C)?

   (b) What is ΔS_{vap} for trichlorofluoromethane?

10.102 Nitrous oxide, N_2O, occasionally used by dentists under the name "laughing gas," has $P_{vap} = 100$ mm Hg at $-110.3°C$ and a normal boiling point of $-88.5°C$. What is the heat of vaporization of nitrous oxide (in kJ/mol)?

10.103 Acetone, a common laboratory solvent, has $\Delta H_{vap} = 29.1$ kJ/mol and a normal boiling point of 56.2°C. At what temperature (in °C) does acetone have $P_{vap} = 105$ mm Hg?

10.104 Use the following data to sketch a phase diagram for krypton: $T_t = -169°C$, $P_t = 133$ mm Hg, $T_c = -63°C$, $P_c = 54$ atm, mp $= -156.6°C$, bp $= -152.3°C$. The density of solid krypton is 2.8 g/cm^3, and the density of the liquid is 2.4 g/cm^3. Can a sample of gaseous krypton at room temperature be liquefied by raising the pressure?

10.105 What is the physical phase of krypton (Problem 10.104) under the following conditions:

   (a) $P = 5.3$ atm, $T = -153°C$

   (b) $P = 65$ atm, $T = 250$ K

10.106 Calculate the percent volume occupied by the spheres in a body-centered cubic unit cell.

10.107 Iron crystallizes in a body-centered cubic unit cell with an edge length of 287 pm. What is the radius of an iron atom (in picometers)?

10.108 Iron metal has a density of 7.86 g/cm^3 and a molar mass of 55.85 g. Use this information together with the data in Problem 10.107 to calculate a value for Avogadro's number.

10.109 Silver metal crystallizes in a face-centered cubic unit cell, with an edge length of 408 pm. The molar mass of silver is 107.9 g/mol, and its density is 10.50 g/cm^3. Use these data to calculate a value for Avogadro's number.

10.110 The NaCl unit cell is shown in Figure 10.24.

   (a) What is the edge length (in picometers) of the NaCl unit cell? The ionic radius of Na^+ is 97 pm, and the ionic radius of Cl^- is 181 pm.

   (b) What is the density of NaCl in g/cm^3?

10.111 Niobium oxide crystallizes in the following cubic unit cell:

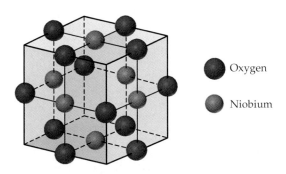

   (a) How many niobium atoms and how many oxygen atoms are in each unit cell?

   (b) What is the formula of niobium oxide?

   (c) What is the oxidation state of niobium?

Multi-Concept Problems

10.112 The mineral *magnetite* is an iron oxide ore with a density of 5.20 g/cm³. At high temperature, magnetite reacts with carbon monoxide to yield iron metal and carbon dioxide. When 2.660 g of magnetite is allowed to react with sufficient carbon monoxide, the CO_2 product is found to have a volume of 1.136 L at 298 K and 751 mm Hg pressure.

   (a) What mass of iron (in grams) is formed in the reaction?

   (b) What is the formula of magnetite?

   (c) Magnetite has a somewhat complicated cubic unit cell with an edge length of 839 pm. How many Fe and O atoms are present in each unit cell?

10.113 A group 3A metal has a density of 2.70 g/cm³ and a cubic unit cell with an edge length of 404 pm. Reaction of a 1.07 cm³ chunk of the metal with an excess of hydrochloric acid gives a colorless gas that occupies 4.00 L at 23.0°C and a pressure of 740 mm Hg.

   (a) Identify the metal.

   (b) Is the unit cell primitive, body-centered, or face-centered?

   (c) What is the atomic radius of the metal atom in picometers?

10.114 A cube-shaped crystal of an alkali metal, 1.62 mm on an edge, was vaporized in a 500.0 mL evacuated flask. The pressure of the resulting vapor was 12.5 mm Hg at 802°C. The structure of the solid metal is known to be body-centered cubic.

   (a) What is the atomic radius of the metal atom in picometers?

   (b) Use the data in Figure 5.21 to identify the alkali metal.

   (c) What are the densities of the solid and the vapor (in g/cm³)?

10.115 Assume that 1.588 g of an alkali metal undergoes complete reaction with the amount of gaseous halogen contained in a 0.500 L flask at 298 K and 755 mm Hg pressure. In the reaction, 22.83 kJ is released ($\Delta H = -22.83$ kJ). The product, a binary ionic compound, crystallizes in a unit cell with anions in a face-centered cubic arrangement and with cations centered along each edge between anions. In addition, there is a cation in the center of the cube.

   (a) What is the identity of the alkali metal?

   (b) The edge length of the unit cell is 535 pm. Find the radius of the alkali metal cation from the data in Figure 6.1 and then calculate the radius of the halide anion. Identify the anion from the data in Figure 6.2.

   (c) Sketch a space-filling, head-on view of the unit cell, labeling the ions. Are the anions in contact with one another?

   (d) What is the density of the compound in g/cm³?

   (e) What is the standard heat of formation for the compound?

 # eMedia Problems

10.115 The **Molecular Polarity** activity (*eChapter 10.1*) allows you to experiment with various polyatomic molecules. Use geometry to explain how a molecule such as BF_3 can contain individual bond dipoles and yet have a zero dipole moment overall.

10.116 **(a)** Using the information in Figure 10.29, sketch a heating curve for carbon dioxide (at 1 atm) similar to the one in the **Changes of State** movie (*eChapter 10.4*). Begin the curve at -85°C and end it at 30°C.

   (b) Sketch the heating curve for carbon dioxide again as it would appear at 25 atm. Use the same initial and final temperatures.

   (c) Under what conditions would the heating curve for water resemble the one you drew for carbon dioxide in part (a)? (See Figure 10.28, p. 418.)

10.117 Using the **Equilibrium Vapor Pressure** activity (*eChapter 10.5*), compare the vapor pressures of methanol, ethanol, acetic acid, water, and benzene.

   (a) Arrange the compounds in order of decreasing equilibrium vapor pressure at 100°C and at 50°C.

   (b) If one compound has a higher equilibrium vapor pressure than another at a particular temperature, will it necessarily have a higher equilibrium vapor pressure at all temperatures? If not, give an example of two compounds whose vapor pressure curves cross, and the temperature at which their vapor pressures are roughly equal.

10.118 Acetic acid, CH_3CO_2H, and ethanol, C_2H_5OH, have very similar molar masses and both exhibit hydrogen bonding. For which of these two compounds does hydrogen bonding contribute more significantly to the overall intermolecular forces? Use data from the **Equilibrium Vapor Pressure** activity (*eChapter 10.5*) to support your answer.

Solutions and Their Properties

These pearls are a type of *colloid*, a mixture of compounds with particles substantially larger than those of normal molecules.

Thus far, we've been concerned primarily with pure substances, both elements and compounds. If you look around, though, it's clear that most of the substances you see in day-to-day life are *mixtures*. Air is a gaseous mixture of (primarily) oxygen and nitrogen, blood is a liquid mixture of many different components, and rocks are solid mixtures of different minerals.

We saw in Section 2.7 that a mixture is any intimate combination of two or more pure substances and that mixtures are classified as either *heterogeneous* or *homogeneous*, depending on their appearance. Heterogeneous mixtures are those in which the mixing of components is visually nonuniform and which therefore have regions of different composition. Homogeneous mixtures are those in which the mixing *is* uniform, at least to the naked eye, and which therefore have the same composition throughout. We'll explore the properties of some homogeneous mixtures in this chapter, with particular emphasis on the mixtures we call *solutions*.

(a)

(b)

(a) Wine, a solution, contains dissolved molecules and is clear, but **(b)** milk, a colloid, contains larger particles and is murky.

11.1 Solutions

Homogeneous mixtures can be classified according to the size of their constituent particles as either *solutions* or *colloids*. **Solutions**, the most important class of homogeneous mixtures, contain particles with diameters in the range 0.1–2 nm—the size of a typical ion or small molecule. They are transparent, though they may be colored, and they do not separate on standing. **Colloids**, such as milk and fog, contain particles with diameters in the range 2–500 nm. Although they are often murky or opaque to light, they do not separate on standing. Mixtures called *suspensions* also exist, having even larger particles than colloids. These are not truly homogeneous, however, because their particles separate on standing and are visible with a low-power microscope. Blood, paint, and aerosol sprays, are examples.

We usually think of a solution as a solid dissolved in a liquid or as a mixture of liquids, but there are many other kinds of solutions as well. In fact, any one state of matter can form a solution with any other state, and seven different kinds of solutions are possible (Table 11.1). Even solutions of one solid with another and solutions of a gas in a solid are well known. Metal alloys, such as stainless steel (4–30% chromium in iron) and brass (10–40% zinc in copper) are examples of solid/solid solutions, and hydrogen in palladium is an example of a gas/solid solution. Metallic palladium, in fact, is able to absorb up to 935 times its own volume of H_2 gas.

TABLE 11.1 Some Different Kinds of Solutions

Kind of Solution	Example
Gas in gas	Air (O_2, N_2, Ar, and other gases)
Gas in liquid	Carbonated water (CO_2 in water)
Gas in solid	H_2 in palladium metal
Liquid in liquid	Gasoline (mixture of hydrocarbons)
Liquid in solid	Dental amalgam (mercury in silver)
Solid in liquid	Seawater (NaCl and other salts in water)
Solid in solid	Metal alloys, such as sterling silver (92.5% Ag, 7.5% Cu)

For solutions in which a gas or solid is dissolved in a liquid, the dissolved substance is called the **solute** and the liquid is called the **solvent**. When one liquid is dissolved in another, the minor component is usually considered the solute and the major component is the solvent. Thus, ethyl alcohol is the solute and water the solvent in a mixture of 10% ethyl alcohol and 90% water, but water is the solute and ethyl alcohol the solvent in a mixture of 90% ethyl alcohol and 10% water.

11.2 Energy Changes and the Solution Process

With the exception of gas/gas mixtures, such as air, the different kinds of solutions listed in Table 11.1 involve *condensed phases*, either liquid or solid. Thus, all the intermolecular forces described in Chapter 10 to explain the properties of pure liquids and solids are also important for explaining the properties of solutions. The situation is more complex for solutions than for pure substances, though, because there are three types of interactions among particles that have to be taken into account: solvent–solvent interactions, solvent–solute interactions, and solute–solute interactions.

A good rule of thumb, often summarized in the phrase "like dissolves like," is that solutions will form when the three types of interactions are similar in kind and in magnitude. Thus, ionic solids like NaCl dissolve in polar solvents like water because the strong ion–dipole attractions between Na^+ and Cl^- ions and polar H_2O molecules are similar in magnitude to the strong dipole–dipole attractions between water molecules and to the strong ion–ion attractions between Na^+ and Cl^- ions. In the same way, nonpolar organic substances like cholesterol, $C_{27}H_{46}O$, dissolve in nonpolar organic solvents like benzene, C_6H_6, because of the similar London dispersion forces present among both kinds of molecules. Oil, however, does not dissolve in water because the two liquids have different kinds of intermolecular forces.

Why don't oil and water mix?

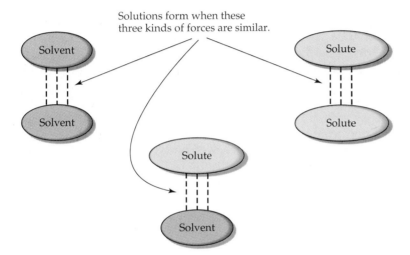

The dissolution of a solid in a liquid can be visualized as shown in Figure 11.1 for NaCl. When solid NaCl is placed in water, those ions that are less tightly held because of their position at a corner or an edge of the crystal

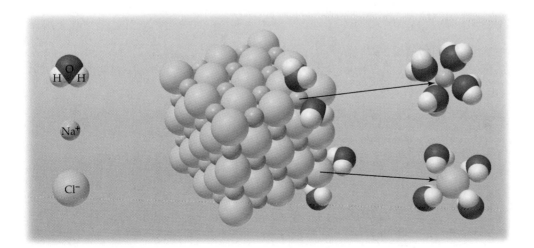

FIGURE 11.1   Dissolution of NaCl crystals in water. Water molecules surround an accessible edge or corner ion in a crystal and collide with it until the ion breaks free. Additional water molecules then surround the ion and stabilize it by means of ion–dipole attractions.

are exposed to water molecules, which collide with them until an ion happens to break free. More water molecules then cluster around the ion, stabilizing it by means of ion–dipole attractions. A new edge or corner is thereby exposed on the crystal, and the process continues until the entire crystal has dissolved. The ions in solution are said to be *solvated*—more specifically, *hydrated* when water is the solvent—meaning that they are surrounded and stabilized by an ordered shell of solvent molecules.

Like all chemical and physical processes, the dissolution of a substance in a solvent has associated with it a free-energy change, ΔG (Section 8.14). If ΔG is negative, the process is spontaneous and the substance dissolves; if ΔG is positive, the process is nonspontaneous and the substance does not dissolve. The free-energy change has two terms: $\Delta G = \Delta H - T\Delta S$. The enthalpy term ΔH measures the heat flow into or out of the system during dissolution, and the temperature-dependent entropy term $T\Delta S$ measures the change in the amount of molecular disorder or randomness in the system. The enthalpy change is called the *enthalpy of solution*, or **heat of solution** (ΔH_{soln}), and the entropy change is called the **entropy of solution** (ΔS_{soln}).

What values might we expect for ΔH_{soln} and ΔS_{soln}? Let's take the entropy change first. Entropies of solution are usually positive because molecular randomness usually increases during dissolution: $+43.4 \ J/(K \cdot mol)$ for NaCl in water, for example. When a solid dissolves in a liquid, randomness increases on going from a well-ordered crystal to a less-ordered state in which solvated ions or molecules are able to move freely in solution. When one liquid dissolves in another, randomness increases as the different molecules intermingle. Table 11.2 lists values of ΔS_{soln} for some common ionic substances.

Enthalpy of Solution activity

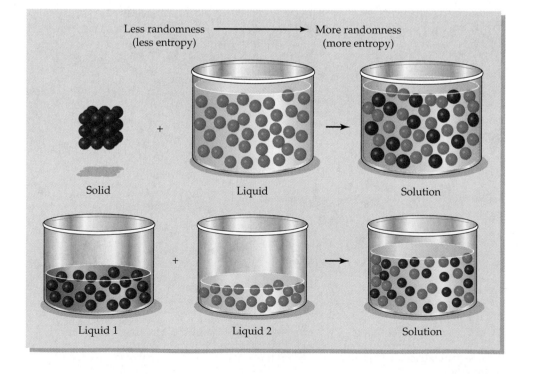

Less randomness (less entropy) ⟶ More randomness (more entropy)

Solid + Liquid → Solution

Liquid 1 + Liquid 2 → Solution

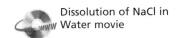
Dissolution of NaCl in
Water movie

**TABLE 11.2   Some Enthalpies and Entropies of
Solution in Water at 25°C**

Substance	ΔH_{soln} (kJ/mol)	ΔS_{soln} [J/(K · mol)]
LiCl	−37.0	10.5
NaCl	3.9	43.4
KCl	17.2	75.0
LiBr	−48.8	21.5
NaBr	−0.6	54.6
KBr	19.9	89.0
KOH	−57.6	12.9

Values for the enthalpy of solution, ΔH_{soln}, are difficult to predict (Table 11.2). Some solids dissolve exothermically and have negative values of ΔH_{soln} (−37.0 kJ/mol for LiCl in water), but others dissolve endothermically and have positive values of ΔH_{soln} (+17.2 kJ/mol for KCl in water). Athletes take advantage of both situations when they use instant hot packs or cold packs to treat injuries. Both kinds of instant packs consist of a pouch of water and a dry chemical, either $CaCl_2$ or $MgSO_4$ for hot packs, and NH_4NO_3 for cold packs. When the pack is squeezed, the pouch breaks and the solid dissolves, either raising or lowering the temperature (Figure 11.2).

Hot packs:	$CaCl_2(s)$	$\Delta H_{soln} = -81.3 \text{ kJ/mol}$
	$MgSO_4(s)$	$\Delta H_{soln} = -91.2 \text{ kJ/mol}$
Cold packs:	$NH_4NO_3(s)$	$\Delta H_{soln} = +25.7 \text{ kJ/mol}$

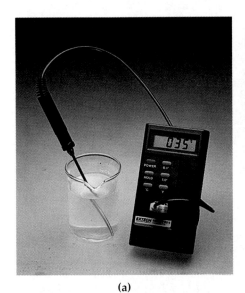

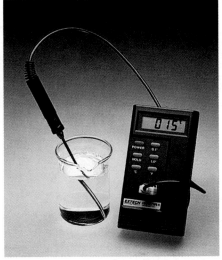

(a)	(b)

FIGURE 11.2   **(a)** Dissolution of $CaCl_2$ in water is exothermic, causing the temperature of the water to rise from its initial value of 25°C. **(b)** Dissolution of NH_4NO_3 is endothermic, causing the temperature of the water to fall from its initial value.

The value of the heat of solution for a substance results from an interplay of the three kinds of interactions mentioned earlier:

- **Solvent–solvent interactions:** Energy is required (positive ΔH) to overcome intermolecular forces between solvent molecules because the molecules must be separated and pushed apart to make room for solute particles.
- **Solute–solute interactions:** Energy is required (positive ΔH) to overcome intermolecular forces holding solute particles together in a crystal. For an ionic solid, this is the lattice energy (Section 6.6). Substances with higher lattice energies therefore tend to be less soluble than substances with lower lattice energies.
- **Solvent–solute interactions:** Energy is released (negative ΔH) when solvent molecules cluster around solute particles and solvate them. For ionic substances in water, the amount of hydration energy released is generally greater for smaller cations than for larger ones because water molecules can approach the positive nuclei of smaller ions more closely and thus bind more tightly. In addition, hydration energy generally increases as the charge on the ion increases.

The first two kinds of interactions are endothermic, requiring an input of energy to spread apart solvent molecules and to break apart crystals. Only the third interaction is exothermic, as attractive intermolecular forces develop between solvent and solute. The sum of the three interactions determines whether ΔH_{soln} is endothermic or exothermic. For some substances, the one exothermic interaction is sufficiently large to outweigh the two endothermic interactions, but for other substances, the reverse is true (Figure 11.3).

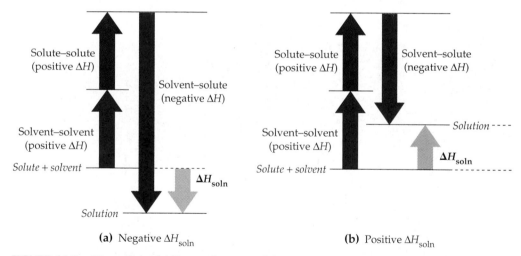

(a) Negative ΔH_{soln} **(b)** Positive ΔH_{soln}

FIGURE 11.3 The value of ΔH_{soln} is the sum of three terms: solvent–solvent, solute–solute, and solvent–solute. It can be either **(a)** negative or **(b)** positive.

EXAMPLE 11.1 Pentane (C_5H_{12}) and 1-butanol (C_4H_9OH) are organic liquids with similar molecular masses, but their solubility properties differ substantially. Which of the two would you expect to be more soluble in water? Explain.

SOLUTION Pentane is a nonpolar molecule and is unlikely to have strong intermolecular interactions with polar water molecules. 1-Butanol, however, has an –OH group just as water does, and is therefore a polar molecule that can form hydrogen bonds with water. 1-Butanol is more soluble in water.

▶**PROBLEM 11.1** Arrange the following compounds in order of their expected increasing solubility in water: Br_2, KBr, toluene (C_7H_8, a constituent of gasoline).

▶**PROBLEM 11.2** Which would you expect to have the larger (more negative) hydration energy?
(a) Na^+ or Cs^+ **(b)** K^+ or Ba^{2+}

11.3   Units of Concentration

In daily life, it's often sufficient to describe a solution as either *dilute* or *concentrated*. In scientific work, though, it's usually necessary to know the exact concentration of a solution—that is, to know the exact amount of solute dissolved in a given amount of solvent. There are many ways of expressing concentration, each of which has its own advantages and disadvantages. We'll look briefly at four of the most common methods: *molarity, mole fraction, mass percent,* and *molality*.

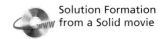
Solution Formation
from a Solid movie

Molarity (M)

The most common way of expressing concentration in a chemistry laboratory is to use *molarity*. As discussed previously in Section 3.7, a solution's molarity is given by the number of moles of solute per liter of solution (mol/L, abbreviated M). If, for example, you dissolve 0.500 mol (20.0 g) of NaOH in enough water to give 1.000 L of solution, then the solution has a concentration of 0.500 M.

Expressions of
Concentration activity

$$\text{Molarity (M)} = \frac{\text{Moles of solute}}{\text{Liters of solution}}$$

The advantages of using molarity are twofold: (1) Stoichiometry calculations are simplified because numbers of moles are used rather than mass, and (2) amounts of solution (and therefore of solute) are measured by volume rather than by mass. As a result, titrations are particularly easy (Section 3.10).

The disadvantages of using molarity are also twofold: (1) The exact concentration depends on the temperature, because the volume of a solution expands or contracts as the temperature changes, and (2) the exact amount of solvent in a given volume can't be determined unless the density of the solution is known. (Remember from Section 3.7 that solutions of a given molarity are prepared by dissolving a solute in a small amount of solvent and then diluting with solvent *to* the desired volume. The solution is not made by dissolving the solute *in* the desired volume of solvent.)

Mole fraction (*X*)

As discussed in Section 9.5, the mole fraction (*X*) of any component in a solution is given by the number of moles of the component divided by the total number of moles making up the solution (including solvent):

$$\text{Mole fraction } (X) = \frac{\text{Moles of component}}{\text{Total moles making up the solution}}$$

For example, a solution prepared by dissolving 1.00 mol (32.0 g) of methyl alcohol (CH_3OH) in 5.00 mol (90.0 g) of water has a methyl alcohol concentration $X = 1.00 \text{ mol}/(1.00 \text{ mol} + 5.00 \text{ mol}) = 0.167$. Note that mole fractions are dimensionless because the units cancel.

Mole fractions are independent of temperature and are particularly useful for calculations involving gas mixtures. Except in special situations, mole fractions are not often used for liquid solutions because other methods are generally more convenient.

Mass percent (mass %)

As the name suggests, the *mass percent* of any component in a solution is the mass of that component divided by the total mass of the solution times 100%:

$$\textbf{Mass percent} = \frac{\text{Mass of component}}{\text{Total mass of solution}} \times 100\%$$

For example, a solution prepared by dissolving 10.0 g of glucose in 100.0 g of water has a glucose concentration of 9.09 mass %:

$$\text{Mass \% glucose} = \frac{10.0 \text{ g}}{10.0 \text{ g} + 100.0 \text{ g}} \times 100\% = 9.09 \text{ mass \%}$$

Closely related to mass percent, and particularly useful for very dilute solutions, are the concentration units *parts per million (ppm)* and *parts per billion (ppb)*:

$$\textbf{Parts per million (ppm)} = \frac{\text{Mass of component}}{\text{Total mass of solution}} \times 10^6$$

$$\textbf{Parts per billion (ppb)} = \frac{\text{Mass of component}}{\text{Total mass of solution}} \times 10^9$$

Dissolving a quarter teaspoon of table sugar in this swimming pool would give a concentration of about 1 ppb.

A concentration of 1 ppm for a substance means that each kilogram of solution contains 1 mg of the solute. (For dilute aqueous solutions near room temperature, where 1 kg has a volume of 1 L, 1 ppm means that each liter of solution contains 1 mg of solute.) Similarly, a concentration of 1 ppb means that each liter of an aqueous solution contains 0.001 mg of solute. Values in ppm and ppb are frequently used for expressing the concentrations of trace amounts of impurities in air or water. Thus, you might express the maximum allowable concentration of lead in drinking water as 50 ppb, or about 1 g per 20,000 L.

The advantage of using mass percent (or ppm) for expressing concentration is that the values are independent of temperature because masses don't change when substances are heated or cooled. The disadvantage of using mass percent is that it's generally more difficult when working with liquid solutions to measure amounts by mass rather than by volume. Furthermore, the density of a solution must be known before a concentration in mass percent can be converted into molarity. Example 11.3 shows how to make the conversion.

EXAMPLE 11.2   Assume that you have a 5.75 mass % solution of LiCl in water. What mass of solution (in grams) contains 1.60 g of LiCl?

SOLUTION Describing a solution as 5.75 mass % LiCl in water means that 100.0 g of solution contains 5.75 g of LiCl (and 94.25 g of water). The mass of solution per 1.60 g of LiCl is therefore

$$\text{Mass of soln} = 1.60 \text{ g LiCl} \times \frac{100.0 \text{ g soln}}{5.75 \text{ g LiCl}} = 27.8 \text{ g soln}$$

Thus, 27.8 g of solution contains 1.60 g of LiCl.

EXAMPLE 11.3 The density of a 25.0 mass % solution of sulfuric acid (H_2SO_4) in water is 1.1783 g/mL at 25.0°C. What is the molarity of the solution?

SOLUTION Describing a solution as 25.0 mass % sulfuric acid in water means that 100.0 g of solution contains 25.0 g of H_2SO_4 and 75.0 g of water. Since we want to calculate the concentration in molarity, we need to find the number of moles of sulfuric acid dissolved in a specific volume of solution. First, convert the 25.0 g of H_2SO_4 into moles:

$$\frac{\text{Moles } H_2SO_4}{100.0 \text{ g solution}} = \frac{25.0 \text{ g } H_2SO_4}{100.0 \text{ g solution}} \times \frac{1 \text{ mol } H_2SO_4}{98.1 \text{ g } H_2SO_4} = \frac{0.255 \text{ mol } H_2SO_4}{100.0 \text{ g solution}}$$

Next, find the volume of 100.0 g of solution, using density as the conversion factor:

$$\text{Volume} = 100.0 \text{ g soln} \times \frac{1 \text{ mL}}{1.1783 \text{ g soln}} = 84.87 \text{ mL} = 0.084 \, 87 \text{ L}$$

Finally, calculate the molarity of the solution:

$$\text{Molarity} = \frac{\text{Moles } H_2SO_4}{\text{Liters of solution}} = \frac{0.255 \text{ mol } H_2SO_4}{0.084 \, 87 \text{ L}} = 3.00 \text{ M}$$

The molarity of the 25.0 mass % sulfuric acid solution is 3.00 M.

▶ **PROBLEM 11.3** What is the mass percent concentration of a saline solution prepared by dissolving 1.00 mol of NaCl in 1.00 L of water?

▶ **PROBLEM 11.4** The legal limit for human exposure to carbon monoxide in the workplace is 35 ppm. Assuming that the density of air is 1.3 g/L, how many grams of carbon monoxide are in 1.0 L of air at the maximum allowable concentration?

▶ **PROBLEM 11.5** Assuming that seawater is an aqueous solution of NaCl, what is its molarity? The density of seawater is 1.025 g/mL at 20°C, and the NaCl concentration is 3.50 mass %.

Molality (*m*)

The *molality* of a solution is defined as the number of moles of solute per kilogram of solvent (mol/kg):

$$\text{Molality } (m) = \frac{\text{Moles of solute}}{\text{Mass of solvent (kg)}}$$

To prepare a 1.000 m solution of KBr in water, for example, you would dissolve 1.000 mol of KBr (119.0 g) in 1.000 kg (1000 mL) of water. You can't say for sure what the final volume of the solution will be, although it will probably be a bit larger than 1000 mL. Although the names sound similar, note the differences between molarity and molality. Molarity is the number of moles of solute per *volume* (liter) of *solution*, whereas molality is the number of moles of solute per *mass* (kilogram) of *solvent*.

The main advantage of using molality is that it is temperature independent because masses don't change when substances are heated or cooled. Thus, it is well suited for calculating certain properties of solutions that we'll discuss later in this chapter. The disadvantages of using molality are that amounts of solution must be measured by mass rather than by volume and that the density of the solution must be known to convert molality into molarity (see Example 11.5).

A summary of the four methods of expressing concentration, together with a comparison of their relative advantages and disadvantages, is given in Table 11.3.

TABLE 11.3   A Comparison of Various Concentration Units

Name	Units	Advantages	Disadvantages
Molarity (M)	$\dfrac{\text{mol solute}}{\text{L solution}}$	Useful in stoichiometry; measure by volume	Temperature-dependent; must know density to find solvent mass
Mole fraction (X)	none	Temperature-independent; useful in special applications	Measure by mass; must know density to convert to molarity
Mass %	%	Temperature-independent; useful for small amounts	Measure by mass; must know density to convert to molarity
Molality (m)	$\dfrac{\text{mol solute}}{\text{kg solvent}}$	Temperature-independent; useful in special applications	Measure by mass; must know density to convert to molarity

EXAMPLE 11.4   What is the molality of a solution made by dissolving 1.45 g of table sugar (sucrose, $C_{12}H_{22}O_{11}$) in 30.0 mL of water?

SOLUTION   The molar mass of sucrose, $C_{12}H_{22}O_{11}$, is 342.3 g/mol, so 1.45 g of sucrose is 4.24×10^{-3} mol:

$$1.45 \text{ g sucrose} \times \frac{1 \text{ mol sucrose}}{342.3 \text{ g sucrose}} = 4.24 \times 10^{-3} \text{ mol sucrose}$$

The molality of the solution is obtained by dividing the number of moles of sucrose by the mass of the solvent in kilograms. Since the density of water is 1.00 g/mL, 30.0 mL of water has a mass of 30.0 g, or 0.0300 kg. Thus, the molality of the solution is

$$\text{Molality} = \frac{4.24 \times 10^{-3} \text{ mol}}{0.0300 \text{ kg}} = 0.141 \text{ } m$$

EXAMPLE 11.5   Ethylene glycol, $C_2H_4(OH)_2$, is a colorless liquid used as automobile antifreeze. If the density at 20°C of a 4.028 m solution of ethylene glycol in water is 1.0241 g/mL, what is the molarity of the solution? The molar mass of ethylene glycol is 62.07 g/mol.

SOLUTION A 4.028 m solution of ethylene glycol in water contains 4.028 mol of ethylene glycol per kilogram of water. To find the solution's molarity, we need to find the number of moles of solute per volume (liter) of solution. The volume, in turn, can be found from the mass of the solution by using density as a conversion factor. Thus, we first need to find the mass of the solution.

The mass of the solution is the sum of the masses of solute and solvent. Assuming that 1.000 kg of solvent is used to dissolve 4.028 mol of ethylene glycol, we need to find the mass of the ethylene glycol:

$$\text{Mass of ethylene glycol} = 4.028 \text{ mol} \times 62.07 \frac{g}{mol} = 250.0 \text{ g}$$

Dissolving this 250.0 g of ethylene glycol in 1.000 kg (or 1000 g) of water gives a total mass of the solution of 1250 g:

$$\text{Mass of solution} = 250.0 \text{ g} + 1000 \text{ g} = 1250 \text{ g}$$

The volume of the solution is obtained from its mass by using its density as a conversion factor:

$$\text{Volume of solution} = 1250 \text{ g} \times \frac{1 \text{ mL}}{1.0241 \text{ g}} = 1221 \text{ mL} = 1.221 \text{ L}$$

The molarity of the solution is the number of moles of solute divided by the volume of solution:

$$\text{Molarity of solution} = \frac{4.028 \text{ mol}}{1.221 \text{ L}} = 3.299 \text{ M}$$

EXAMPLE 11.6 A 0.750 M solution of H_2SO_4 in water has a density of 1.046 g/mL at 20°C. What is the concentration of this solution in **(a)** mole fraction, **(b)** mass percent, and **(c)** molality? The molar mass of H_2SO_4 is 98.1 g/mol.

SOLUTION
(a) Let's pick an arbitrary amount of the solution that will make the calculations easy, say 1.00 L. Since the concentration of the solution is 0.750 mol/L and the density is 1.046 g/mL (or 1.046 kg/L), 1.00 L of the solution contains 0.750 mol (73.6 g) of H_2SO_4 and has a mass of 1.046 kg:

$$\text{Moles of } H_2SO_4 \text{ in 1.00 L soln} = 0.750 \frac{mol}{L} \times 1.00 \text{ L} = 0.750 \text{ mol}$$

$$\text{Mass of } H_2SO_4 \text{ in 1.00 L soln} = 0.750 \text{ mol} \times 98.1 \frac{g}{mol} = 73.6 \text{ g}$$

$$\text{Mass of 1.00 L soln} = 1.00 \text{ L} \times 1.046 \frac{kg}{L} = 1.046 \text{ kg}$$

Subtracting the mass of H_2SO_4 from the total mass of the solution gives 0.972 kg of water, or 54.0 mol in 1.00 L of solution:

$$\text{Mass of } H_2O \text{ in 1.00 L soln} = (1.046 \text{ kg}) - (0.0736 \text{ kg}) = 0.972 \text{ kg } H_2O$$

$$\text{Moles of } H_2O \text{ in 1.00 L of soln} = 972 \text{ g} \times \frac{1 \text{ mol}}{18.0 \text{ g}} = 54.0 \text{ mol } H_2O$$

Thus, the mole fraction of H_2SO_4 is

$$X_{H_2SO_4} = \frac{0.750 \text{ mol } H_2SO_4}{0.750 \text{ mol } H_2SO_4 + 54.0 \text{ mol } H_2O} = 0.0137$$

(b) The mass percent concentration can be determined from the calculations in part (a):

$$\text{Mass \% of } H_2SO_4 = \frac{0.0736 \text{ kg } H_2SO_4}{1.046 \text{ kg total}} \times 100\% = 7.04\%$$

(c) The molality of the solution can also be determined from the calculations in part (a). Since 0.972 kg of water has 0.750 mol of H_2SO_4 dissolved in it, 1.00 kg of water would have 0.772 mol of H_2SO_4 dissolved in it:

$$1.00 \text{ kg } H_2O \times \frac{0.750 \text{ mol } H_2SO_4}{0.972 \text{ kg } H_2O} = 0.772 \text{ mol } H_2SO_4$$

Thus, the molality of the sulfuric acid solution is 0.772 m.

▶ **PROBLEM 11.6** What is the molality of a solution prepared by dissolving 0.385 g of cholesterol, $C_{27}H_{46}O$, in 40.0 g of chloroform, $CHCl_3$? What is the mole fraction of cholesterol in the solution?

▶ **PROBLEM 11.7** What mass (in grams) of a 0.500 m solution of sodium acetate, CH_3CO_2Na, in water would you use to obtain 0.150 mol of sodium acetate?

▶ **PROBLEM 11.8** The density at 20°C of a 0.258 m solution of glucose in water is 1.0173 g/mL, and the molar mass of glucose is 180.2 g. What is the molarity of the solution?

▶ **PROBLEM 11.9** The density at 20°C of a 0.500 M solution of acetic acid in water is 1.0042 g/mL. What is the concentration of this solution in molality? The molar mass of acetic acid, CH_3CO_2H, is 60.05 g.

▶ **PROBLEM 11.10** Assuming that seawater is a 3.50 mass % aqueous solution of NaCl, what is the molality of seawater?

11.4 Some Factors Affecting Solubility

If you take solid NaCl and add it to water, dissolution occurs rapidly at first but then slows down as more and more NaCl is added. Eventually the dissolution stops because a dynamic equilibrium is reached where the number of Na^+ and Cl^- ions leaving a crystal to go into solution is equal to the number of ions returning from solution to the crystal. At this point, the solution is said to be **saturated** in that solute.

$$\text{Solute } + \text{ Solvent } \underset{\text{Crystallize}}{\overset{\text{Dissolve}}{\rightleftharpoons}} \text{ Solution}$$

Note that this definition requires a saturated solution to be at *equilibrium* with undissolved solid. Substances that are more soluble at high temperature than at low temperature can sometimes form what are called **supersaturated** solutions, which contain a greater-than-equilibrium amount of solute. For example, when a saturated solution of sodium acetate is prepared at high temperature and then cooled slowly, a supersaturated solution results, as shown in Figure 11.4. Such a solution is unstable, however, and precipitation occurs when a tiny seed crystal of sodium acetate is added to initiate crystallization.

FIGURE 11.4 A supersaturated solution of sodium acetate in water. When a tiny seed crystal is added, larger crystals begin to grow and precipitate from the solution until equilibrium is reached.

Effect of Temperature on Solubility

The amount of solute per unit of solvent needed to form a saturated solution is called the solute's **solubility**. Like melting point and boiling point, the solubility of a substance in a given solvent is a physical property characteristic of that substance. Different substances can have greatly different solubilities, as shown in Figure 11.5. Sodium chloride, for instance, has a solubility of 36.0 g/100 mL of water at 20°C, and sodium nitrate has a solubility of 87.6 g/100 mL of water at 20°C. Sometimes, particularly when two liquids are involved, the solvent and solute are **miscible**, meaning that they are mutually soluble in all proportions. Ethyl alcohol and water is an example.

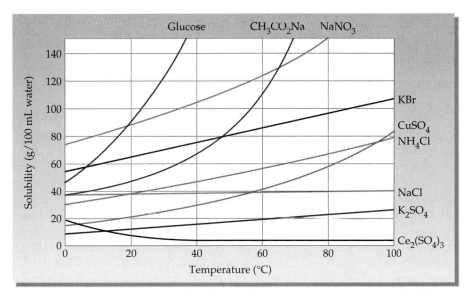

FIGURE 11.5 Solubilities of some common solids in water as a function of temperature. Most substances become more soluble as temperature rises, although the exact relationship is often complex and nonlinear.

Solubilities are temperature-dependent, and the temperature at which a specific measurement is made must be reported. As Figure 11.5 shows, there is no obvious correlation between structure and solubility or between solubility and temperature. The solubilities of most molecular and ionic solids increase with increasing temperature, though the solubilities of some (NaCl) are almost unchanged, and the solubilities of others [$Ce_2(SO_4)_3$] decrease.

The effect of temperature on the solubility of gases is more predictable than its effect on the solubility of solids. Most gases (helium is the only common exception) become less soluble in water as the temperature increases (Figure 11.6). One consequence of this decreased solubility is that carbonated drinks bubble continuously as they warm up to room temperature after being refrigerated. Soon, they lose so much dissolved CO_2 that they become "flat." A much more important consequence is the damage to aquatic life that can result from the decrease in concentration of dissolved oxygen when hot water is discharged from an industrial plant into lakes and rivers, an effect known as thermal pollution.

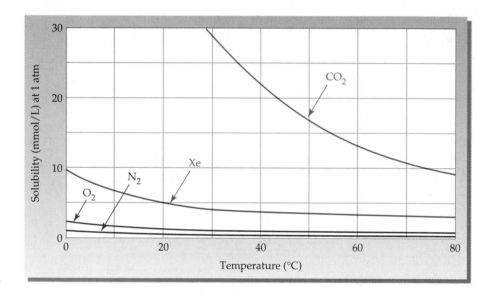

FIGURE 11.6 Solubilities of some gases in water as a function of temperature. Most gases become less soluble in water as the temperature rises. The concentration units are millimoles per liter (mmol/L) at a gas pressure of 1 atm.

Effect of Pressure on Solubility

Pressure has practically no effect on the solubility of liquids and solids but has a profound effect on the solubility of gases. According to *Henry's law*, the solubility of a gas in a liquid at a given temperature is directly proportional to the partial pressure of the gas over the solution:

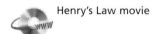
Henry's Law movie

⬥— Henry's law: Solubility = $k \cdot P$

The constant k in this expression is the Henry's-law constant characteristic of a specific gas, and P is the partial pressure of the gas over the solution. Doubling the partial pressure doubles the solubility, tripling the partial pressure triples the solubility, and so forth. Henry's-law constants are usually given in units of mol/(L · atm), and measurements are reported at 25°C. Note that at a gas partial pressure P of 1 atm the Henry's-law constant k is numerically equal to the solubility of the gas in moles per liter.

The most common example of Henry's-law behavior occurs when you open a can of soda or other carbonated drink. Bubbles of gas come fizzing out of solution because the pressure of CO_2 in the can drops and CO_2 suddenly becomes less soluble. A more serious example of Henry's-law behavior occurs when a deep-sea diver surfaces too quickly and develops a painful and life-threatening condition called the "bends." Bends occur because large amounts of nitrogen dissolve in the blood at high underwater pressures. When the diver ascends and pressure decreases too rapidly, bubbles of nitrogen form in the blood, blocking capillaries and inhibiting blood flow. The condition can be prevented by using an oxygen/helium mixture for breathing rather than air (oxygen/nitrogen), because helium has a much lower solubility in blood than nitrogen.

On a molecular level, the increase in gas solubility with increasing pressure occurs because of a change in the position of the equilibrium between dissolved and undissolved gas. At a given pressure, an equilibrium is established in which equal numbers of gas particles enter and leave the solution (Figure 11.7a). When the pressure is increased, however, more particles are forced into solution than leave it, and gas solubility therefore increases until a new equilibrium is established (Figure 11.7b–c).

Divers who ascend too quickly can develop the bends, a condition caused by formation of nitrogen bubbles in the blood. Treatment involves placement in this high-pressure tank called a hyperbaric chamber, and a slow, controlled change to atmospheric pressure.

 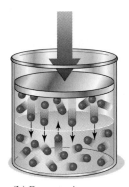 

(a) Equilibrium   **(b)** Pressure increase   **(c)** Equilibrium restored

FIGURE 11.7 A molecular view of Henry's law. **(a)** At a given pressure, an equilibrium exists in which equal numbers of gas particles enter and leave the solution. **(b)** When pressure is increased by pushing on the piston, more gas particles are temporarily forced into solution than are able to leave, so solubility increases until a new equilibrium is reached **(c)**.

EXAMPLE 11.7 The Henry's-law constant of methyl bromide (CH_3Br), a gas used as a soil fumigating agent, is $k = 0.159$ mol/(L · atm) at 25°C. What is the solubility (in mol/L) of methyl bromide in water at 25°C and a partial pressure of 125 mm Hg?

 Henry's Law activity

SOLUTION Henry's law says that the solubility of a gas in water is equal to $k \cdot P$. In the present instance,

$$k = 0.159 \text{ mol/(L} \cdot \text{atm)}$$

$$P = 125 \text{ mm Hg} \times \frac{1 \text{ atm}}{760 \text{ mm Hg}} = 0.164 \text{ atm}$$

$$\text{Solubility} = k \cdot P = 0.159 \frac{\text{mol}}{\text{L} \cdot \text{atm}} \times 0.164 \text{ atm} = 0.0261 \text{ M}$$

The solubility of methyl bromide in water at a partial pressure of 125 mm Hg is 0.0261 M.

Why does this carbonated drink fizz when the bottle is opened?

▶**PROBLEM 11.11**   The solubility of CO_2 in water is 3.2×10^{-2} M at 25°C and 1 atm pressure. What is the Henry's-law constant for CO_2 in mol/(L · atm)?

▶**PROBLEM 11.12**   The partial pressure of CO_2 in air is approximately 4.0×10^{-4} atm. Use the Henry's-law constant you calculated in Problem 11.11 to find the concentration of CO_2 in
(a) A can of soda under a CO_2 pressure of 2.5 atm at 25°C
(b) A can of soda open to the atmosphere at 25°C

11.5   Physical Behavior of Solutions: Colligative Properties

The behavior of solutions is qualitatively similar to that of pure solvents but is quantitatively different. Pure water boils at 100.0°C and freezes at 0.0°C, for example, but a 1.00 *m* (molal) solution of NaCl in water boils at 101.0°C and freezes at −3.7°C.

The elevation of boiling point and the lowering of freezing point that are observed on comparing a pure solvent with a solution are examples of **colligative properties**—properties that depend on the *amount* of dissolved solute but not on its chemical identity. (The word *colligative* means "bound together in a collection" and is used because a "collection" of solute particles is responsible for the observed effects.) Other colligative properties are a decrease in vapor pressure of a solution compared with the pure solvent and *osmosis*, the migration of solvent and other small molecules through a semipermeable membrane.

In comparing the properties of a pure solvent with those of a solution...

Colligative properties
{
Vapor pressure of solution is lower.
Boiling point of solution is higher.
Freezing point of solution is lower.
Osmosis, the migration of solvent molecules through a semipermeable membrane, occurs when solvent and solution are separated by the membrane.
}

We'll look at each of the four colligative properties in more detail in Sections 11.6–11.8.

11.6   Vapor-Pressure Lowering of Solutions: Raoult's Law

We said in Section 10.5 that a liquid in a closed container is in equilibrium with its vapor and that the amount of pressure exerted by the vapor is called the *vapor pressure*. When you compare the vapor pressure of a pure solvent with that of a solution at the same temperature, however, you find that the two values are different. If the solute is nonvolatile and has no appreciable vapor pressure of its own, as occurs when a solid is dissolved, then the vapor pressure of the solution is always lower than that of the pure solvent. If the solute *is* volatile and has a significant vapor pressure of its own, as often occurs in a mixture of two liquids, then the vapor pressure of the mixture is intermediate between the vapor pressures of the two pure substances.

Solutions with a Nonvolatile Solute

It's easy to demonstrate with manometers that a solution of a nonvolatile solute has a lower vapor pressure than the pure solvent (Figure 11.8). Alternatively, you can show the same effect by comparing the evaporation rate of pure solvent with the evaporation rate of a solution. A solution always evaporates more slowly than a pure solvent does because its vapor pressure is lower and its molecules therefore escape less readily.

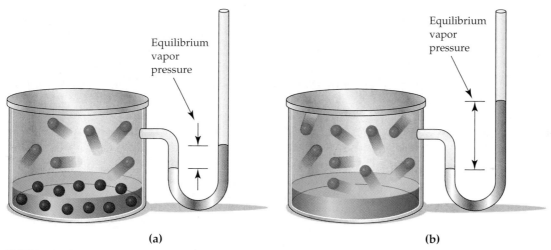

Equilibrium vapor pressure

Equilibrium vapor pressure

(a)   (b)

FIGURE 11.8   The equilibrium vapor pressure of a solution with **(a)** a nonvolatile solute is always lower than that of **(b)** the pure solvent by an amount that depends on the mole fraction of the solvent.

According to **Raoult's law**, the vapor pressure of a solution containing a nonvolatile solute is equal to the vapor pressure of pure solvent times the mole fraction of the solvent. That is,

Raoult's law: $P_{soln} = P_{solv} \times X_{solv}$

where P_{soln} is the vapor pressure of the solution, P_{solv} is the vapor pressure of pure solvent at the same temperature, and X_{solv} is the mole fraction of the solvent in the solution.

Take a solution of 1.00 mol of glucose in 15.0 mol of water at 25°C, for instance. The vapor pressure of pure water at 25°C is 23.76 mm Hg, and the mole fraction of water in the solution is $15.0 \text{ mol}/(1.0 \text{ mol} + 15.0 \text{ mol}) = 0.938$. Thus, Raoult's law predicts a vapor pressure for the solution of 23.76 mm Hg $\times 0.938 = 22.3$ mm Hg, which corresponds to a vapor-pressure lowering, ΔP_{soln}, of 1.5 mm Hg:

$$P_{soln} = P_{solv} \times X_{solv} = 23.76 \text{ mm Hg} \times \frac{15.0 \text{ mol}}{1.00 \text{ mol} + 15.0 \text{ mol}} = 22.3 \text{ mm Hg}$$

$$\Delta P_{soln} = P_{solv} - P_{soln} = 23.76 \text{ mm Hg} - 22.3 \text{ mm Hg} = 1.5 \text{ mm Hg}$$

Alternatively, the amount of vapor-pressure lowering can be calculated by multiplying the mole fraction of the *solute* times the vapor pressure of the pure solvent. That is,

$$\Delta P_{\text{soln}} = P_{\text{solv}} \times X_{\text{solute}} = 23.76 \text{ mm Hg} \times \frac{1.00 \text{ mol}}{1.00 \text{ mol} + 15.0 \text{ mol}}$$

$$= 1.5 \text{ mm Hg}$$

If an ionic substance such as NaCl is the solute, we have to calculate mole fractions based on the total concentration of solute *particles* (ions) rather than NaCl formula units. A solution of 1.00 mol NaCl in 15.0 mol water at 25°C, for example, contains 2.00 mol of dissolved particles (assuming complete dissociation), resulting in a mole fraction for water of 0.882:

$$X_{\text{water}} = \frac{15.0 \text{ mol H}_2\text{O}}{1.00 \text{ mol Na}^+ + 1.00 \text{ mol Cl}^- + 15.0 \text{ mol H}_2\text{O}} = 0.882$$

Since the mole fraction of water is smaller in the 1.00 mol NaCl solution than in the 1.00 mol glucose solution, the vapor pressure of the NaCl solution is lower: 21.0 mm Hg for NaCl versus 22.3 mm Hg for glucose at 25°C.

$$P_{\text{soln}} = P_{\text{solv}} \times X_{\text{solv}} = 23.76 \text{ mm Hg} \times 0.882 = 21.0 \text{ mm Hg}$$

What accounts for the lowering of the vapor pressure when a nonvolatile solute is dissolved in a solvent? As we've noted on numerous prior occasions, a physical process such as the vaporization of a liquid to a gas is accompanied by a free-energy change, $\Delta G_{\text{vap}} = \Delta H_{\text{vap}} - T\Delta S_{\text{vap}}$. The more negative the value of ΔG_{vap}, the more favored the vaporization process. Thus, if we want to compare the ease of vaporization of a pure solvent with that of the solvent in a solution, we have to compare the signs and relative magnitudes of the ΔH_{vap} and ΔS_{vap} terms in the two cases.

The vaporization of a liquid to a gas is *disfavored* by enthalpy (positive ΔH_{vap}) because energy is required to overcome intermolecular attractions between liquid molecules. At the same time, however, vaporization is *favored* by entropy (positive ΔS_{vap}) because molecular randomness increases when molecules go from a semiordered liquid state to a disordered gaseous state.

The heats of vaporization for a pure solvent and for a solution are similar because similar intermolecular forces must be overcome in both cases for solvent molecules to escape from the liquid. The entropies of vaporization for a pure solvent and for a solution are *not* similar, however. Because a solvent in a solution has more molecular disorder and higher entropy than a pure solvent does, the entropy *change* on going from liquid to vapor is smaller for the solution than for the pure solvent. Subtracting a *smaller* $T\Delta S_{\text{vap}}$ from ΔH_{vap} thus results in a *larger* (more positive) ΔG_{vap} for the solution. As a result, vaporization is less favored for the solution, and vapor pressure of the solution at equilibrium is lower (Figure 11.9).

FIGURE 11.9 The lower vapor pressure of a solution relative to that of a pure solvent is due to the difference in their entropies of vaporization, ΔS_{vap}. Because the entropy of the solution is higher to begin with, ΔS_{vap} is smaller for the solvent in the solution than for the pure solvent. As a result, vaporization of the solution is less favored (less negative ΔG_{vap}), and the vapor pressure of the solution is lower.

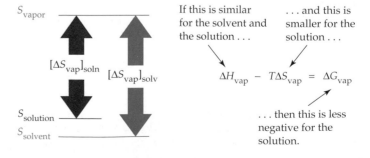

It should be pointed out that, just as the ideal gas law discussed in Section 9.3 applies only to "ideal" gases, Raoult's law applies only to ideal solutions. Raoult's law approximates the behavior of most real solutions, but significant deviations from ideality occur as solute concentration increases. The law works best when solute concentrations are low and when solute and solvent particles have similar intermolecular forces.

If the intermolecular forces between solute particles and solvent molecules are weaker than the forces between solvent molecules alone, then the solvent molecules are less tightly held in the solution, and the vapor pressure is higher than Raoult's law predicts. Conversely, if the intermolecular forces between solute and solvent molecules are stronger than the forces between solvent molecules alone, then the solvent molecules are more tightly held in the solution, and the vapor pressure is lower than predicted. Solutions of ionic substances, in particular, often have a vapor pressure significantly lower than predicted because the ion–dipole forces between dissolved ions and polar water molecules are so strong. In addition, ionic substances rarely dissociate to the extent of 100%, so a solution of an ionic compound usually contains fewer particles than the formula of the compound suggests.

EXAMPLE 11.8   What is the vapor pressure (in mm Hg) of a solution made by dissolving 18.3 g of NaCl in 500.0 g of H_2O at 70°C? The vapor pressure of pure water at 70°C is 233.7 mm Hg.

SOLUTION   Raoult's law says that the vapor pressure of the solution is equal to the vapor pressure of pure solvent times the mole fraction of the solvent in the solution. Thus, we have to find a value for the mole fraction of solvent. First, calculate the number of moles of NaCl and H_2O. The molar mass of NaCl is 58.44 g/mol, and the molar mass of water is 18.02 g/mol.

$$\text{Moles of NaCl} = 18.3 \text{ g NaCl} \times \frac{1 \text{ mol NaCl}}{58.44 \text{ g NaCl}} = 0.313 \text{ mol NaCl}$$

$$\text{Moles of } H_2O = 500.0 \text{ g } H_2O \times \frac{1 \text{ mol } H_2O}{18.02 \text{ g } H_2O} = 27.75 \text{ mol } H_2O$$

Next, calculate the mole fraction of water in the solution. Since NaCl is an ionic substance that dissociates into two particles when dissolved in water, the solution contains 0.626 mol of dissolved particles: 0.313 mol Na^+ ions and 0.313 mol Cl^- ions. Thus, the mole fraction of water is

$$\text{Mole fraction of } H_2O = \frac{27.75 \text{ mol}}{0.626 \text{ mol} + 27.75 \text{ mol}} = 0.9779$$

From Raoult's law, the vapor pressure of the solution is

$$P_{soln} = P_{solv} \times X_{solv} = 233.7 \text{ mm Hg} \times 0.9779 = 228.5 \text{ mm Hg}$$

EXAMPLE 11.9   How many grams of sucrose must be added to 320 g of water to lower the vapor pressure by 1.5 mm Hg at 25°C? The vapor pressure of water at 25°C is 23.8 mm Hg, and the molar mass of sucrose is 342.3 g/mol.

SOLUTION   The vapor pressure of the solution, P_{soln}, can be calculated by subtracting the amount of vapor-pressure lowering from the vapor pressure of the pure solvent, P_{solv}:

$$P_{soln} = 23.8 \text{ mm Hg} - 1.5 \text{ mm Hg} = 22.3 \text{ mm Hg}$$

According to Raoult's law, the vapor pressure of the solution is equal to the vapor pressure of pure solvent times the mole fraction of the solvent, X_{solv}. Thus, X_{solv} is equal to P_{soln} divided by P_{solv}:

$$\text{Since} \quad P_{soln} = P_{solv} \times X_{solv}$$

$$\text{then} \quad X_{solv} = \frac{P_{soln}}{P_{solv}} = \frac{22.3 \text{ mm Hg}}{23.8 \text{ mm Hg}} = 0.937$$

The mole fraction of water just calculated, X_{solv}, is the number of moles of water divided by the total number of moles of sucrose plus water:

$$X_{solv} = \frac{\text{Moles of water}}{\text{Total moles}}$$

Since the number of moles of water is

$$\text{Moles of water} = 320 \text{ g} \times \frac{1 \text{ mol}}{18.0 \text{ g}} = 17.8 \text{ mol}$$

then the total number of moles of sucrose plus water is

$$\text{Total moles} = \frac{\text{Moles of water}}{X_{solv}} = \frac{17.8 \text{ mol}}{0.937} = 19.0 \text{ mol}$$

Subtracting the number of moles of water from the total number of moles gives the number of moles of sucrose needed:

$$\text{Moles of sucrose} = 19.0 \text{ mol} - 17.8 \text{ mol} = 1.2 \text{ mol}$$

Converting moles into grams then gives the mass of sucrose needed:

$$\text{Grams of sucrose} = 1.2 \text{ mol} \times 342.3 \frac{\text{g}}{\text{mol}} = 4.1 \times 10^2 \text{ g}$$

▶ **PROBLEM 11.13**   What is the vapor pressure (in mm Hg) of a solution prepared by dissolving 5.00 g of benzoic acid ($C_7H_6O_2$) in 100.00 g of ethyl alcohol (C_2H_5OH) at 35°C? The vapor pressure of pure ethyl alcohol at 35°C is 100.5 mm Hg.

▶ **PROBLEM 11.14**   How many grams of NaBr must be added to 250 g of water to lower the vapor pressure by 1.30 mm Hg at 40°C? The vapor pressure of water at 40°C is 55.3 mm Hg.

← KEY CONCEPT PROBLEM 11.15   The following diagram shows a close-up view of part of the vapor-pressure curve for a pure solvent and a solution. Which curve represents the pure solvent, and which the solution?

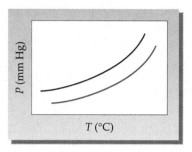

Solutions with a Volatile Solute

As you might expect from Dalton's law (Section 9.5), the overall vapor pressure P_{total} of a mixture of two volatile liquids A and B is the sum of the vapor-pressure contributions of the individual components, P_A and P_B:

$$P_{total} = P_A + P_B$$

The individual vapor pressures P_A and P_B are calculated by Raoult's law. That is, the vapor pressure of A is equal to the mole fraction of A (X_A) times the vapor pressure of pure A ($P°_A$), and the vapor pressure of B is equal to the mole fraction of B (X_B) times the vapor pressure of pure B ($P°_B$). Thus, the total vapor pressure of the solution is

$$P_{total} = P_A + P_B = (P°_A \cdot X_A) + (P°_B \cdot X_B)$$

Take a mixture of the two similar organic liquids benzene (C_6H_6, bp = 80.1°C) and toluene (C_7H_8, bp = 110.6°C), for example. Pure benzene has a vapor pressure $P° = 96.0$ mm Hg at 25°C, and pure toluene has $P° = 30.3$ mm Hg at the same temperature. In a 1:1 molar mixture of the two, where the mole fraction of each is $X = 0.500$, the vapor pressure of the solution is 63.2 mm Hg:

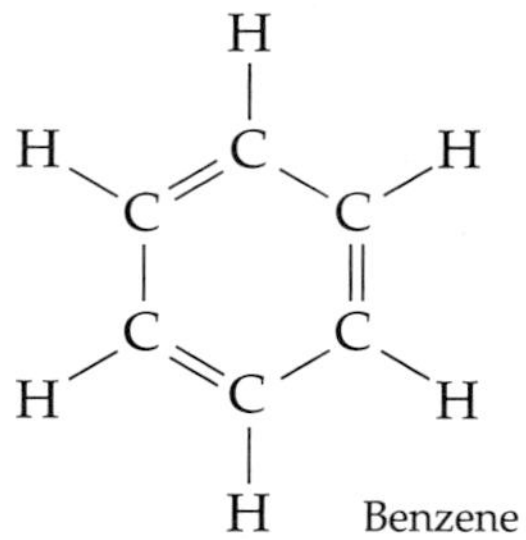

Benzene 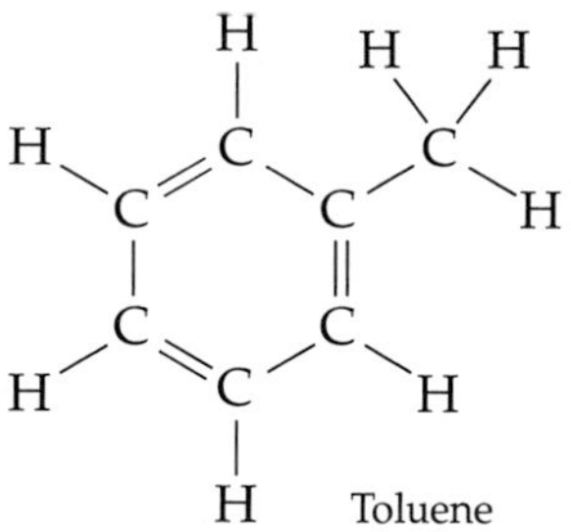   Toluene

$$
\begin{aligned}
P_{total} &= (P°_{benzene})(X_{benzene}) + (P°_{toluene})(X_{toluene})\\
&= (96.0 \text{ mm Hg} \times 0.500) + (30.3 \text{ mm Hg} \times 0.500)\\
&= 48.0 \text{ mm Hg} + 15.2 \text{ mm Hg} = 63.2 \text{ mm Hg}
\end{aligned}
$$

Note that the vapor pressure of the mixture has a value intermediate between the vapor pressures of the two pure liquids (Figure 11.10). In addition, you might note that the vapor pressure of the lower-boiling component is greater than that of the higher-boiling component.

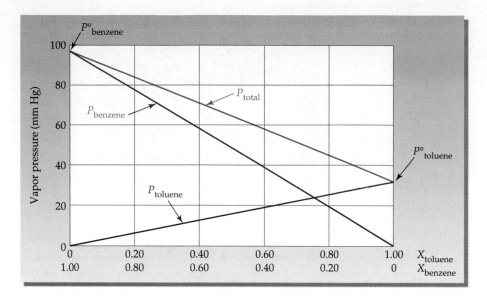

FIGURE 11.10 The vapor pressure of a solution of the two volatile liquids benzene and toluene at 25°C is the sum of the two individual vapor pressures, each calculated by Raoult's law.

As with nonvolatile solutes, the Raoult's-law behavior shown in Figure 11.10 for a mixture of benzene and toluene applies only to ideal solutions. Most real solutions show behavior that deviates slightly from the ideal in either a positive or negative way, depending on the kinds and strengths of intermolecular forces present in the solution.

▶ **PROBLEM 11.16**

(a) What is the vapor pressure (in mm Hg) of a solution prepared by dissolving 25.0 g of ethyl alcohol (C_2H_5OH) in 100.0 g of water at 25°C? The vapor pressure of pure water is 23.8 mm Hg, and the vapor pressure of ethyl alcohol is 61.2 mm Hg at 25°C.

(b) What is the vapor pressure of the solution if 25.0 g of water is dissolved in 100.0 g of ethyl alcohol at 25°C?

KEY CONCEPT PROBLEM 11.17 The following diagram shows a close-up view of part of the vapor-pressure curves for two pure liquids and a mixture of the two. Which curves represent pure liquids, and which the mixture?

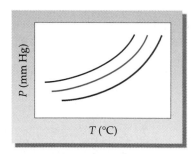

11.7 Boiling-Point Elevation and Freezing-Point Depression of Solutions

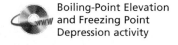

Boiling-Point Elevation and Freezing Point Depression activity

We said in Section 10.5 that the vapor pressure of a liquid rises with increasing temperature and that the liquid boils when its vapor pressure equals atmospheric pressure. Since a solution of a nonvolatile solute has a lower vapor pressure than a pure solvent has at a given temperature, it follows that the solution must be heated to a higher temperature to cause it to boil. Furthermore, because the solution has a lower vapor pressure than the pure solvent at a given temperature, its liquid/vapor phase transition line on a phase diagram is always lower than that of the pure solvent. As a result, the triple-point temperature T_t is lower for the solution than for the solvent, and the solid/liquid phase transition line is shifted to a lower temperature. In other words, the solution must be cooled to a lower temperature to freeze. Figure 11.11 shows the situation.

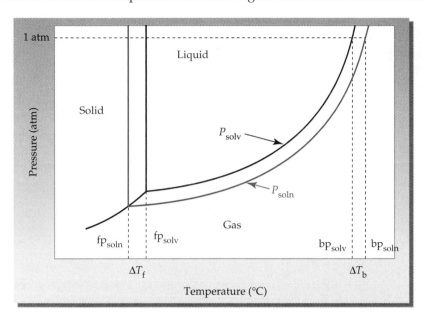

FIGURE 11.11 Phase diagrams for a pure solvent (red line) and a solution (green line). Because the vapor pressure of the solution is lower than that of the pure solvent at a given temperature, the temperature at which the vapor pressure reaches atmospheric pressure is higher for the solution than for the solvent. Thus, the boiling point of the solution is higher by an amount ΔT_b. Furthermore, because the liquid/vapor phase transition line is lower for the solution than for the solvent, the triple-point temperature T_t is lower and the solid/liquid phase transition line is shifted to a lower temperature. As a result, the freezing point of the solution is lower than that of the pure solvent by an amount ΔT_f.

The boiling-point elevation of a solution relative to that of a pure solvent depends on the concentration of dissolved particles, just as vapor-pressure lowering does. Thus, a 1.00 m solution of glucose in water boils at 100.51°C at 1 atm pressure (0.51°C above normal), but a 1.00 m solution of NaCl in water boils at 101.02°C (1.02°C above normal) because there are twice as many particles (ions) dissolved in the NaCl solution as there are in the glucose solution.

The change in boiling point ΔT_b for a solution is

$$\Delta T_b = K_b \cdot m$$

where m is the *molal* (*not* molar) concentration of solute particles and K_b is the **molal boiling-point-elevation constant** characteristic of each liquid. The concentration must be expressed in molality rather than molarity so that the solute concentration is independent of temperature. Molal boiling-point-elevation constants are given in Table 11.4 for some common substances.

TABLE 11.4 Molal Boiling-Point-Elevation Constants (K_b) and Molal Freezing-Point-Depression Constants (K_f) for Some Common Substances

Substance	K_b [(°C · kg)/mol]	K_f [(°C · kg)/mol]
Benzene (C_6H_6)	2.53	5.12
Camphor ($C_{10}H_{16}O$)	5.95	37.7
Chloroform ($CHCl_3$)	3.63	4.70
Diethyl ether ($C_4H_{10}O$)	2.02	1.79
Ethyl alcohol (C_2H_5OH)	1.22	1.99
Water (H_2O)	0.51	1.86

The freezing-point depression of a solution relative to that of a pure solvent depends on the concentration of solute particles, just as boiling-point elevation does. For example, a 1.00 m solution of glucose in water freezes at $-1.86°C$, and a 1.00 m solution of NaCl in water freezes at $-3.72°C$. The change in freezing point ΔT_f for a solution is

$$\Delta T_f = K_f \cdot m$$

where m is the molal concentration of solute particles and K_f is the **molal freezing-point-depression constant** characteristic of each solvent. Some molal freezing-point-depression constants are also given in Table 11.4.

The fundamental cause of boiling-point elevation and freezing-point depression in solutions is the same as the cause of vapor-pressure lowering (Section 11.6): the entropy difference between the pure solvent and the solvent in the solution. Let's take boiling-point elevations first. We know that liquid and vapor phases are in equilibrium at the boiling point (T_b) and that the free-energy difference between the two phases (ΔG_{vap}) is therefore zero (Section 8.14).

Since $\quad \Delta G_{vap} = \Delta H_{vap} - T_b\Delta S_{vap} = 0$

then $\quad \Delta H_{vap} = T_b\Delta S_{vap} \quad$ and $\quad T_b = \dfrac{\Delta H_{vap}}{\Delta S_{vap}}$

In comparing the heats of vaporization (ΔH_{vap}) for a pure solvent and for a solution, we find that the two values are similar because similar intermolecular forces holding the solvent molecules together must be overcome in both cases. In comparing the *entropies* of vaporization, however, the two values are not similar. Because the solvent in a solution has more molecular randomness than a pure solvent has, the entropy change between solution and vapor is smaller than the entropy change between pure solvent and vapor. But if ΔS_{vap} is *smaller* for the solution, then T_b must be correspondingly *larger*. In other words, the boiling point of the solution (T_b) is higher than that of the pure solvent (Figure 11.12).

FIGURE 11.12 The higher boiling point of a solution relative to that of a pure solvent is due to a difference in their entropies of vaporization, ΔS_{vap}. Because the solvent in the solution has a higher entropy to begin with, ΔS_{vap} is smaller for the solution than for the pure solvent. As a result, the boiling point of the solution T_b is higher than that of the pure solvent.

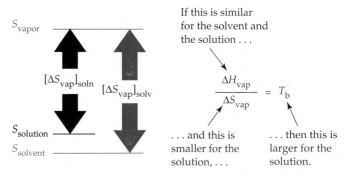

A similar explanation explains freezing-point depression. We know that liquid and solid phases are in equilibrium at the freezing point, so the free-energy difference between the phases (ΔG_{fusion}) is zero.

Since $\quad \Delta G_{fusion} = \Delta H_{fusion} - T_f\Delta S_{fusion} = 0$

then $\quad \Delta H_{fusion} = T_f\Delta S_{fusion} \quad$ and $\quad T_f = \dfrac{\Delta H_{fusion}}{\Delta S_{fusion}}$

In comparing a solution with a pure solvent, the heats of fusion (ΔH_{fusion}) are similar because similar intermolecular forces between solvent molecules are involved. The entropies of fusion (ΔS_{fusion}), however, are not similar. Because the solvent in a solution has more molecular randomness than a pure solvent has, the entropy change between solution and solid is larger than the entropy change between pure solvent and solid. With ΔS_{fusion} *larger* for the solution, T_f must be correspondingly *smaller*, meaning that the freezing point of the solution (T_f) is lower than that of the pure solvent (Figure 11.13).

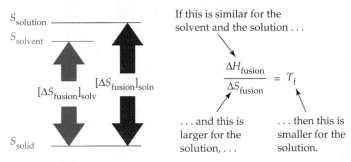

FIGURE 11.13   The lower freezing point of a solution relative to that of a pure solvent is due to a difference in their entropies of fusion, ΔS_{fusion}. Because the solvent in the solution has a higher entropy level to begin with, ΔS_{fusion} is larger for the solution than for the pure solvent. As a result, the freezing point of the solution T_f is lower than that of the pure solvent.

EXAMPLE 11.10   What is the molality of an aqueous glucose solution if the boiling point of the solution at 1 atm pressure is 101.27°C? The molal boiling-point-elevation constant for water is given in Table 11.4.

SOLUTION   Rearrange the equation for molal boiling-point elevation to solve for m:

$$\Delta T_b = K_b \cdot m \qquad \text{so} \qquad m = \frac{\Delta T_b}{K_b}$$

where $K_b = 0.51$ (°C · kg)/mol and $\Delta T_b = 101.27°C - 100.00°C = 1.27°C$. Then

$$m = \frac{1.27\ °C}{0.51\ \dfrac{°C \cdot kg}{mol}} = 2.5\ \frac{mol}{kg} = 2.5\ m$$

The molality of the solution is 2.5 m.

▶**PROBLEM 11.18**   What is the normal boiling point (in °C) of a solution prepared by dissolving 1.50 g of aspirin (acetylsalicylic acid, $C_9H_8O_4$) in 75.00 g of chloroform ($CHCl_3$)? The normal boiling point of chloroform is 61.7°C, and K_b for chloroform is given in Table 11.4.

▶**PROBLEM 11.19**   What is the freezing point (in °C) of a solution prepared by dissolving 7.40 g of K_2SO_4 in 110 g of water? (Assume 100% dissociation for K_2SO_4.) The value of K_f for water is given in Table 11.4.

▶**PROBLEM 11.20**   What is the molality of an aqueous solution of KBr whose freezing point is −2.95°C? (Assume 100% dissociation for KBr.) The molal freezing-point-depression constant of water is given in Table 11.4.

← **KEY CONCEPT PROBLEM 11.21** The following phase diagram shows a close-up view of the liquid–vapor phase transition boundaries for pure chloroform and a solution of a nonvolatile solute in chloroform.

(a) What is the approximate boiling point of pure chloroform?

(b) What is the approximate molal concentration of the nonvolatile solute? (See Table 11.4 to find K_b for chloroform.)

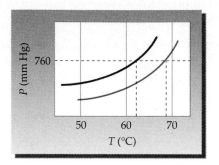

11.8 Osmosis and Osmotic Pressure

Certain materials, including those that make up the membranes around living cells, are *semipermeable*. That is, they allow water or other small molecules to pass through, but they block the passage of large solute molecules or ions. When a solution and a pure solvent (or two solutions of different concentration) are separated by the right kind of semipermeable membrane, solvent molecules pass through the membrane in a process called **osmosis**. Although the passage of solvent through the membrane takes place in both directions, passage from the pure solvent side to the solution side is more favored and occurs faster. As a result, the amount of liquid on the pure solvent side decreases, the amount of liquid on the solution side increases, and the concentration of the solution decreases.

Osmosis can be demonstrated with the experimental setup shown in Figure 11.14, in which a solution in the bulb is separated by a semipermeable membrane from pure solvent in the container. Solvent passes through the membrane from the beaker to the bulb, causing the liquid level in the attached tube to rise. The increased weight of liquid in the tube creates an increased pressure that pushes solvent back through the membrane until the rates of forward and reverse passage become equal and the liquid level stops rising. The amount of pressure necessary to achieve this equilibrium is called the **osmotic pressure, Π** (Greek capital pi), of the solution. Osmotic pressures can be extremely high, even for relatively dilute solutions. The osmotic pressure of a 0.15 M NaCl solution at 25°C, for example, is 7.3 atm, a value that will support a difference in water level of approximately 250 ft!

The amount of osmotic pressure between solution and pure solvent depends on the concentration of solute particles in the solution according to the equation

$$\Pi = MRT$$

A cucumber (top) shrivels into a pickle (bottom) when immersed in salt water because osmotic pressure drives water from the cucumber's cells.

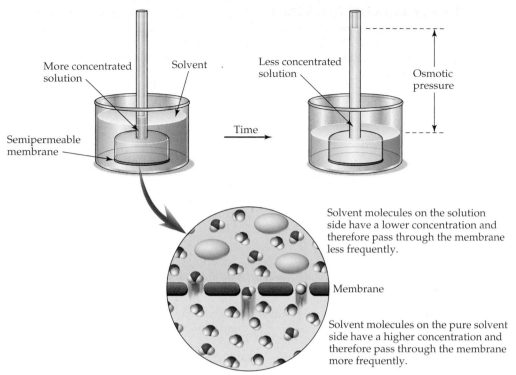

Solvent molecules on the solution side have a lower concentration and therefore pass through the membrane less frequently.

Membrane

Solvent molecules on the pure solvent side have a higher concentration and therefore pass through the membrane more frequently.

FIGURE 11.14   The phenomenon of osmosis. A solution inside the bulb is separated from pure solvent in the container by a semipermeable membrane. Net passage of solvent from the container through the membrane occurs, and the liquid in the tube rises until an equilibrium is reached. At equilibrium, the osmotic pressure exerted by the column of liquid in the tube is sufficient to prevent further net passage of solvent.

where M is the molar concentration of solute particles, R is the gas constant [0.082 06 (L · atm)/(K · mol)], and T is the temperature in kelvins. For example, a 1.00 M solution of glucose in water at 300 K has an osmotic pressure of 24.6 atm:

$$\Pi = MRT = 1.00 \, \frac{mol}{L} \times 0.082 \, 06 \, \frac{L \cdot atm}{K \cdot mol} \times 300 \text{ K} = 24.6 \text{ atm}$$

Note that the solute concentration is given in *molarity* when calculating osmotic pressure rather than in molality as with other colligative properties. Because osmotic-pressure measurements are made at the specific temperature given in the equation $\Pi = MRT$, it's not necessary to express concentration in a temperature-independent unit such as molality.

Osmosis, like all colligative properties, results from an increase in entropy as pure solvent passes through the membrane and mixes with the solution. Perhaps the simplest explanation of osmosis is seen by looking at the molecular level (Figure 11.14). Solvent molecules on the solvent side of the membrane, because of their somewhat greater concentration, approach the membrane a bit more frequently than molecules on the solution side, thereby passing through more often.

Osmotic pressure is responsible for the rise of sap in tall trees.

EXAMPLE 11.11 The total concentration of dissolved particles inside red blood cells is approximately 0.30 M, and the membrane surrounding the cells is semipermeable. What would the osmotic pressure (in atmospheres) inside the cells become if the cells were removed from blood plasma and placed in pure water at 298 K?

SOLUTION If red blood cells were removed from the body and placed in pure water, water would pass through the cell membrane, causing an increase in pressure inside the cells. The amount of this pressure would be

$$\Pi = MRT$$

where M = 0.30 mol/L, R = 0.082 06 (L · atm)/(K · mol), T = 298 K. Then

$$\Pi = 0.30 \frac{mol}{L} \times 0.082\ 06 \frac{L \cdot atm}{K \cdot mol} \times 298\ K = 7.3\ atm$$

The buildup of internal pressure would cause the blood cells to burst.

EXAMPLE 11.12 A solution of an unknown substance in water at 293 K gives rise to an osmotic pressure of 5.66 atm. What is the molarity of the solution?

SOLUTION We are given values for Π and T, and we need to solve for M in the equation $\Pi = MRT$. Rearranging this equation, we get

$$M = \frac{\Pi}{RT}$$

where Π = 5.66 atm, R = 0.082 06 (L · atm)/(K · mol), and T = 293 K. Thus,

$$M = \frac{5.66\ atm}{0.082\ 06 \frac{L \cdot atm}{K \cdot mol} \times 293\ K} = 0.235\ M$$

▶**PROBLEM 11.22** What osmotic pressure (in atmospheres) would you expect for a solution of 0.125 M $CaCl_2$ that is separated from pure water by a semipermeable membrane at 310 K? Assume 100% dissociation for $CaCl_2$.

▶**PROBLEM 11.23** A solution of an unknown substance in water at 300 K gives rise to an osmotic pressure of 3.85 atm. What is the molarity of the solution?

11.9 Some Uses of Colligative Properties

We use colligative properties in many ways, both in the chemical laboratory and in day-to-day life. Motorists in winter, for example, take advantage of freezing-point lowering when they drive on streets where the snow has been melted by a sprinkling of salt. The antifreeze added to automobile radiators and the de-icer solution sprayed on airplane wings also work by lowering the freezing point of water. That same automobile antifreeze keeps radiator water from boiling over in summer by raising its boiling point.

One of the more interesting uses of colligative properties is the desalination of seawater by **reverse osmosis**. When pure water and seawater are separated by a suitable membrane, the passage of water molecules from the pure side to the solution side is faster than passage in the reverse direction. As osmotic pressure builds up, though, the rates of forward and reverse water passage eventually become equal at an osmotic pressure of about 30 atm at 25°C. If a pressure even *greater* than 30 atm is now applied to the solution side, the reverse passage of water becomes favored. As a result, pure water can be obtained from seawater (Figure 11.15).

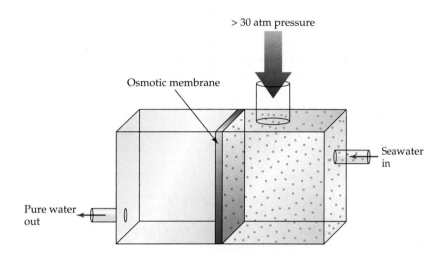

> 30 atm pressure

Osmotic membrane

Seawater in

Pure water out

FIGURE 11.15 The desalination of seawater by reverse osmosis. By applying a pressure on the seawater that is greater than osmotic pressure, water is forced through the osmotic membrane from the seawater side to the pure water side.

The most important use of colligative properties in the laboratory is for determining the molecular mass of an unknown substance. Any of the four colligative properties we've discussed can be used, but the most accurate values are obtained from osmotic pressure measurements because the magnitude of the osmosis effect is so great. For example, a solution of 0.0200 M glucose in water at 300 K will give an osmotic pressure reading of 374.2 mm Hg, a value that can easily be read to four significant figures. The same solution, however, will lower the freezing point by only 0.04°C, a value that can be read easily to only one significant figure. Example 11.13 shows how osmotic pressure can be used to find molecular mass.

EXAMPLE 11.13   A solution prepared by dissolving 20.0 mg of insulin in water and diluting to a volume of 5.00 mL gives an osmotic pressure of 12.5 mm Hg at 300 K. What is the molecular mass of insulin?

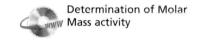
Determination of Molar Mass activity

SOLUTION   To determine molecular mass, we need to know the number of moles of insulin represented by the 20.0 mg sample. We can do this by first rearranging the equation for osmotic pressure to find the molar concentration of the insulin solution:

$$\text{Since} \quad \Pi = MRT, \quad \text{then} \quad M = \frac{\Pi}{RT}$$

$$M = \frac{12.5 \text{ mm Hg} \times \dfrac{1 \text{ atm}}{760 \text{ mm Hg}}}{0.082\ 06 \dfrac{\text{L} \cdot \text{atm}}{\text{K} \cdot \text{mol}} \times 300 \text{ K}} = 6.68 \times 10^{-4} \text{ M}$$

Since the volume of the solution is 5.00 mL, the number of moles of insulin is

$$\text{Moles insulin} = 6.68 \times 10^{-4} \frac{\text{mol}}{\text{L}} \times \frac{1 \text{ L}}{1000 \text{ mL}} \times 5.00 \text{ mL} = 3.34 \times 10^{-6} \text{ mol}$$

Knowing both the mass and the number of moles of insulin, we can now calculate the molar mass, and hence molecular mass:

$$\text{Molar mass} = \frac{\text{mass of insulin}}{\text{moles of insulin}} = \frac{0.0200 \text{ g insulin}}{3.34 \times 10^{-6} \text{ mol insulin}} = 5990 \text{ g/mol}$$

The molecular mass of insulin is 5990 amu.

▶**PROBLEM 11.24** A solution of 0.250 g of naphthalene (mothballs) in 35.00 g of camphor lowers the freezing point by 2.10°C. What is the molar mass of naphthalene? The freezing-point-depression constant for camphor is 37.7 (°C · kg)/mol.

▶**PROBLEM 11.25** What is the molar mass of sucrose (table sugar) if a solution prepared by dissolving 0.822 g of sucrose in 300.0 mL of water has an osmotic pressure of 149 mm Hg at 298 K?

11.10 Fractional Distillation of Liquid Mixtures

Distillation of petroleum into fractions according to boiling point is carried out in the large towers of this refinery.

Perhaps the most commercially important of all applications of colligative properties is in the refining of petroleum to make gasoline. Petroleum refineries appear as a vast maze of pipes, towers, and tanks. The pipes, though, are just for transferring the petroleum or its products, and the tanks are just for storage. It's in the towers that the important separation of crude petroleum into usable fractions takes place. As we saw in Section 8.12, petroleum is a complex mixture of hydrocarbon molecules that are refined by distillation into different fractions: straight-run gasoline (bp 30–200°C), kerosene (bp 175–300°C), and gas oil (bp 275–400°C).

Called **fractional distillation**, this separation of petroleum into fractions occurs when a mixture of volatile liquids is boiled, and the vapors are condensed. Because the vapor is enriched in the component with the higher vapor pressure according to Raoult's law (Section 11.6), the condensed vapors are also enriched in that component, and a partial purification can be effected. If the boil/condense cycle is then repeated a large number of times, complete purification of the more volatile liquid component can be achieved.

Let's look at the separation by fractional distillation of a 1:1 molar mixture of benzene and toluene. If we begin by heating the mixture, boiling occurs when the sum of the vapor pressures equals atmospheric pressure—that is, when $X \cdot P°_{\text{benzene}} + X \cdot P°_{\text{toluene}} = 760$ mm Hg. Reading from the vapor-pressure curves in Figure 11.16 (or calculating values with the Clausius–Clapeyron equation as discussed in Section 10.5), we find that boiling occurs at 365.3 K (92.2°C), where $P°_{\text{benzene}} = 1084$ mm Hg and $P°_{\text{toluene}} = 436$ mm Hg:

$$P_{\text{mixt}} = X \cdot P°_{\text{benzene}} + X \cdot P°_{\text{toluene}}$$

$$= (0.5000)(1084 \text{ mm Hg}) + (0.5000)(436 \text{ mm Hg})$$

$$= 542 \text{ mm Hg} + 218 \text{ mm Hg}$$

$$= 760 \text{ mm Hg}$$

Although the starting *liquid* mixture of benzene and toluene has a 1:1 molar composition, the composition of the *vapor* is not 1:1. Of the 760 mm Hg total vapor pressure for the boiling mixture, 542/760 = 71.3% is due to benzene and 218/760 = 28.7% is due to toluene. If we now condense the vapor, the liquid we get has this 71.3:28.7 composition. On boiling this new liquid mixture, the composition of the

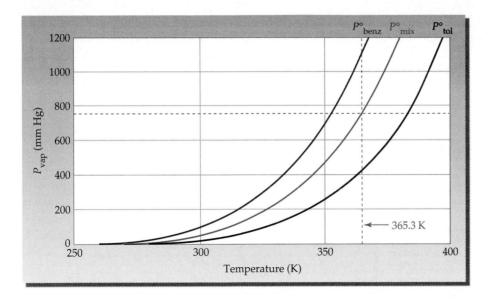

FIGURE 11.16 Vapor-pressure curves for pure benzene (blue), pure toluene (red), and a 1:1 mixture of the two (green). The mixture boils at 92.2°C (365.3 K) at atmospheric pressure.

vapor now becomes 86.4% benzene and 13.6% toluene. A third condense/boil cycle brings the composition of the vapor to 94.4% benzene/5.6% toluene, and so on through further cycles until the desired level of purity is reached.

Fractional distillation can be represented on a liquid/vapor phase diagram by plotting temperature versus composition, as shown in Figure 11.17. The lower region of the diagram represents the liquid phase, and the upper region represents the vapor phase. Between the two is a thin equilibrium region where liquid and vapor coexist.

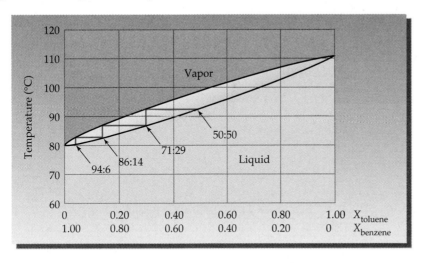

FIGURE 11.17 A phase diagram of temperature versus composition (mole fraction) for a mixture of benzene and toluene. Liquid composition is given by the lower curve, and vapor composition is given by the top curve. The thin region between curves represents an equilibrium between phases. Liquid and vapor compositions at a given temperature are connected by a horizontal tie line, as explained in the text.

To understand how the diagram works, let's imagine starting with the 50:50 benzene/toluene mixture and heating it to its boiling point (92.2°C on the diagram). The lower curve represents the liquid composition (50:50), but the upper curve represents the vapor composition (approximately 71:29 at 92.2°C). The two points are connected by a short horizontal line called a *tie line* to indicate that the temperature is the same at both points. Condensing the 71:29 vapor mixture by lowering the temperature gives a 71:29 liquid mixture that, when heated to *its* boiling point (86.6°C), has an 86:14 vapor composition, as represented by another tie line. In essence, fractional distillation is simply a walk across successive tie lines in whatever number of steps is necessary to reach the desired purity.

In practice, the successive boil/condense cycles occur naturally in the distillation column, and there is no need to isolate liquid mixtures at intermediate stages of purification. Fractional distillation is therefore relatively simple to carry out and is routinely used on a daily basis in chemical plants and laboratories throughout the world (Figure 11.18).

FIGURE 11.18   A simple fractional distillation column used in a chemistry laboratory. The vapors from a boiling mixture of liquids rise inside the column, where they condense on contact with the cool column walls, drip back, and are reboiled by contact with more hot vapor. Numerous boil/condense cycles occur before vapors finally pass out the top of the column, reach the water-cooled condenser, and drip into the receiver.

Thermometer

Water out

Water in

Distillation column

Receiver

Mixture to be separated

KEY CONCEPT PROBLEM 11.26   The following graph is a phase diagram of temperature versus composition for mixtures of the two liquids chloroform and dichloromethane.

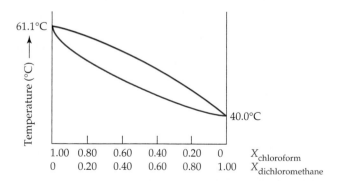

(a) Label the regions on the diagram corresponding to liquid and vapor.

(b) Assume that you begin with a mixture of 60% chloroform and 40% dichloromethane. At what approximate temperature will the mixture begin to boil? Mark as point *a* on the diagram the liquid composition at the boiling point, and mark as point *b* the vapor composition at the boiling point.

(c) Assume that the vapor at point *b* condenses and is reboiled. Mark as point *c* on the diagram the liquid composition of the condensed vapor and as point *d* the vapor composition of the reboiled material.

(d) What will the approximate composition of the liquid be after carrying out two cycles of boiling and condensing?

Dialysis

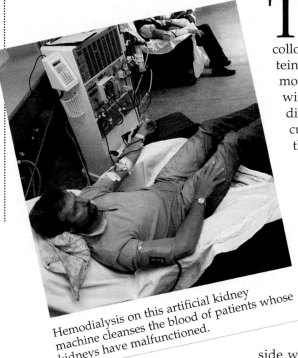

Hemodialysis on this artificial kidney machine cleanses the blood of patients whose kidneys have malfunctioned.

The process of *dialysis* is analogous to osmosis, except that both solvent molecules and small solute particles can pass through the semipermeable dialysis membrane. Only large colloidal particles such as cells and large molecules such as proteins can't pass. (The exact dividing line between a "small" molecule and a "large" one is imprecise, and dialysis membranes with a variety of pore sizes can be obtained.) Because they don't dialyze, proteins can be separated from small ions and molecules, making dialysis a valuable procedure for purification of the proteins needed in laboratory studies.

Perhaps the most important medical use of dialysis is in artificial kidney machines, where *hemodialysis* is used to cleanse the blood of patients whose kidneys have malfunctioned. Blood is diverted from the body and pumped through a cellophane dialysis tube suspended in a solution formulated to contain many of the same components as blood plasma. These substances—glucose, NaCl, NaHCO$_3$, and KCl—have the same concentrations in the dialysis solution as they do in blood so that they have no net passage through the cellophane membrane.

Small waste products such as urea pass through the dialysis membrane from the blood to the solution side where they are washed away, but cells, proteins, and other important blood components are prevented by their size from passing through the membrane. The wash solution is changed every 2 hours, and a typical hemodialysis treatment lasts for approximately 4–7 hours.

▶ **PROBLEM 11.27** What is the difference between a dialysis membrane and the typical semipermeable membrane used for osmosis?

Key Words

Summary

Solutions are homogeneous mixtures that contain particles the size of a typical ion or small molecule. Any one state of matter can mix with any other state, leading to seven possible kinds of solutions. For solutions in which a gas or solid is dissolved in a liquid, the dissolved substance is called the **solute** and the liquid is called the **solvent**.

The dissolution of a solute in a solvent has associated with it a free-energy change, ΔG, which has two terms, an enthalpy term ΔH and a temperature-dependent entropy term $T\Delta S$. The enthalpy change is the **heat of solution** (ΔH_{soln}), and the entropy change is the **entropy of solution** (ΔS_{soln}). Heats of solution can be either positive or negative depending on the relative strengths of solvent–solvent, solute–solute, and solvent–solute intermolecular forces. Entropies of solution are usually positive because disorder increases when a pure solute dissolves in a pure solvent.

The concentration of a solution can be expressed in many ways, including **molarity** (moles of solute per liter of solution), **mole fraction** (moles of solute per mole of solution), **mass percent** (mass of solute per mass of solution times 100%), and **molality** (moles of solute per kilogram of solvent). When equilibrium is reached and no further solute dissolves in a given amount of solvent, a solution is said to be **saturated**. The concentration at this point represents the **solubility** of the solute. Solubilities are usually temperature-dependent, though often not in a simple way. Gas solubilities usually decrease with increasing temperature, but the solubilities of solids can either increase or decrease. The solubilities of gases also depend on pressure. According to **Henry's law**, the solubility of a gas in a liquid at a given temperature is proportional to the partial pressure of the gas over the solution.

In comparison with a pure solvent, a solution has a lower vapor pressure at a given temperature, a lower freezing point, and a higher boiling point. In addition, a solution that is separated from solvent by a semipermeable membrane gives rise to the phenomenon of **osmosis**. All four of these properties of solutions depend only on the concentration of dissolved solute rather than on the chemical identity of the solute and are therefore called **colligative properties**. The fundamental cause of all colligative properties is the same: the higher entropy of the solvent in the solution relative to that of the pure solvent.

Colligative properties have many practical uses, including the melting of snow by salt, the desalination of seawater by reverse osmosis, the separation and purification of volatile liquids by fractional distillation, and the determination of molecular mass by osmotic pressure measurement.

Key Concept Summary

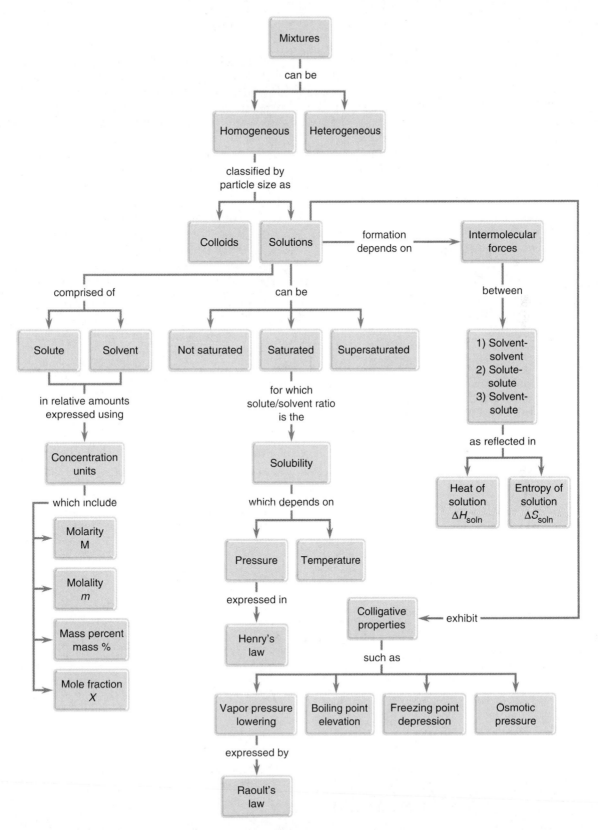

Understanding Key Concepts

Problems 11.1–11.27 appear within the chapter.

11.28 Rank the situations represented by the following drawings according to increasing entropy.

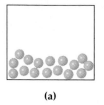

(a)

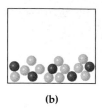

(b)

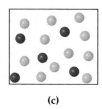

(c)

11.29 The following diagram shows a close-up view of part of the vapor-pressure curves for a solvent (red curve) and a solution of the solvent with a second liquid (green curve). Is the second liquid more volatile or less volatile than the solvent?

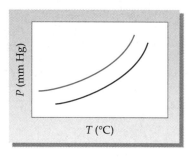

11.30 Assume that two liquids are separated by a semi-permeable membrane. Draw the situation after equilibrium is reached.

Before equilibrium

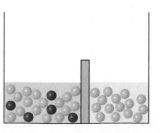

11.31 When 1 mol of NaCl is added to 1 L of water, the boiling point increases. When 1 mol of methyl alcohol is added to 1 L of water, the boiling point decreases. Explain.

11.32 When 100 mL of 9 M H_2SO_4 at 0°C is added to 100 mL of liquid water at 0°C, the temperature rises to 12°C. When 100 mL of 9 M H_2SO_4 at 0°C is added to 100 g of solid ice at 0°C, the temperature falls to −12°C. Explain the difference in behavior.

11.33 The enthalpy of solution (ΔH_{soln}) for HBr(g) in water is −85.1 kJ/mol and that for $AgNO_3$(s) is +22.6 kJ/mol. Assuming that you begin at room temperature and make a 0.10 M solution of each, which solution will be warm to the touch and which will be cool?

11.34 Two beakers, one with pure water (blue) and the other with a solution of NaCl in water (green), are placed in a closed container as represented by drawing **(a)**. Which of the drawings **(b)–(d)** represents what the beakers will look like after a substantial amount of time has passed?

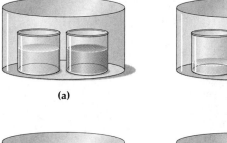

(a)

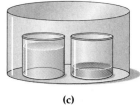

(b)

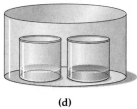

(c)

(d)

11.35 A phase diagram of temperature versus composition for a mixture of the two volatile liquids octane (bp = 69°C) and decane (bp = 126°C) is shown. Assume that you begin with a mixture containing 0.60 mol of decane and 0.40 mol of octane.

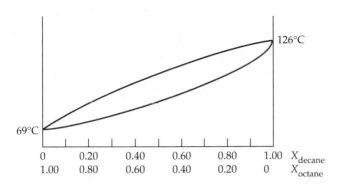

(a) What region on the diagram corresponds to vapor, and what region corresponds to liquid?

(b) At what approximate temperature will the mixture begin to boil? Mark as point *a* on the diagram the liquid composition at the boiling point, and mark as point *b* the vapor composition at the boiling point.

(c) Assume that the vapor at point *b* condenses and is reboiled. Mark as point *c* on the diagram the liquid composition of the condensed vapor and as point *d* on the diagram the vapor composition of the reboiled material.

Additional Problems

Solutions and Energy Changes

11.36 If a single 5 g block of NaCl is placed in water, it dissolves slowly, but if 5 g of powdered NaCl is placed in water, it dissolves rapidly. Explain.

11.37 Give an example of each of the following kinds of solutions:

(a) A gas in a liquid

(b) A solid in a solid

(c) A liquid in a solid

11.38 Explain the solubility rule of thumb "like dissolves like" in terms of the intermolecular forces that occur in solutions.

11.39 Br_2 is much more soluble in tetrachloromethane, CCl_4, than in water. Explain.

11.40 Why do ionic substances with higher lattice energies tend to be less soluble in water than substances with lower lattice energies?

11.41 Which would you expect to have the larger hydration energy, SO_4^{2-} or ClO_4^-? Explain.

11.42 Ethyl alcohol, CH_3CH_2OH, is miscible with water at 20°C, but pentyl alcohol, $CH_3CH_2CH_2CH_2CH_2OH$, is soluble in water only to the extent of 2.7 g/100 mL. Explain.

11.43 Pentyl alcohol (Problem 11.42) is miscible with octane, C_8H_{18}, but methyl alcohol, CH_3OH, is insoluble in octane. Explain.

11.44 The dissolution of $CaCl_2(s)$ in water is exothermic, with $\Delta H_{soln} = -81.3$ kJ/mol. If you were to prepare a 1.00 *m* solution of $CaCl_2$ beginning with water at 25.0°C, what would the final temperature of the solution be (in °C)? Assume that the specific heats of both pure H_2O and the solution are the same, 4.18 J/(K · g).

11.45 The dissolution of $NH_4ClO_4(s)$ in water is endothermic, with $\Delta H_{soln} = +33.5$ kJ/mol. If you prepare a 1.00 *m* solution of NH_4ClO_4 beginning with water at 25.0°C, what is the final temperature of the solution (in °C)? Assume that the specific heats of both pure H_2O and the solution are the same, 4.18 J/(K · g).

Units of Concentration

11.46 What is the difference between molarity and molality?

11.47 What is the difference between a saturated and a supersaturated solution?

11.48 How would you prepare each of the following solutions?

(a) A 0.150 M solution of glucose in water

(b) A 1.135 *m* solution of KBr in water

(c) A solution of methyl alcohol and water in which $X_{CH_3OH} = 0.15$ and $X_{H_2O} = 0.85$

11.49 How would you prepare each of the following solutions?

(a) 100 mL of a 155 ppm solution of urea, CH_4N_2O, in water

(b) 100 mL of an aqueous solution whose K^+ concentration is 0.075 M

11.50 How would you prepare 165 mL of a 0.0268 M solution of benzoic acid ($C_7H_6O_2$) in chloroform ($CHCl_3$)?

11.51 How would you prepare 165 mL of a 0.0268 *m* solution of benzoic acid ($C_7H_6O_2$) in chloroform ($CHCl_3$)?

11.52 Which of the following solutions is more concentrated?

(a) 0.500 M KCl or 0.500 mass % KCl in water

(b) 1.75 M glucose or 1.75 *m* glucose in water

11.53 Which of the following solutions has the higher molarity?

(a) 10 ppm KI in water or 10,000 ppb KBr in water

(b) 0.25 mass % KCl in water or 0.25 mass % citric acid ($C_6H_8O_7$) in water

11.54 What is the mass percent concentration of the following solutions?

(a) Dissolve 0.655 mol of citric acid, $C_6H_8O_7$, in 1.00 kg of water.

(b) Dissolve 0.135 mg of KBr in 5.00 mL of water.

(c) Dissolve 5.50 g of aspirin, $C_9H_8O_4$, in 145 g of dichloromethane, CH_2Cl_2.

11.55 What is the molality of each solution prepared in Problem 11.54?

11.56 The so-called *ozone layer* in the earth's stratosphere has an average total pressure of 10 mm Hg (1.3×10^{-2} atm). The partial pressure of ozone in the layer is about 1.2×10^{-6} mm Hg (1.6×10^{-9} atm). What is the concentration of ozone (in parts per million), assuming that the average molar mass of air is 29 g/mol?

11.57 Persons are medically considered to have lead poisoning if they have a concentration of greater than 10 micrograms of lead per deciliter of blood. What is this concentration in parts per billion?

11.58 What is the concentration of each of the following solutions?

(a) The molality of a solution prepared by dissolving 25.0 g of H_2SO_4 in 1.30 L of water

(b) The mole fraction of each component of a solution prepared by dissolving 2.25 g of nicotine, $C_{10}H_{14}N_2$, in 80.0 g of CH_2Cl_2

11.59 Household bleach is a 5.0 mass % aqueous solution of sodium hypochlorite, NaOCl. What is the molality of the bleach? What is the mole fraction of NaOCl in the bleach?

11.60 The density of a 16.0 mass % solution of sulfuric acid in water is 1.1094 g/mL at 25.0°C. What is the molarity of the solution?

11.61 Ethylene glycol, $C_2H_6O_2$, is the principal constituent of automobile antifreeze. If the density of a 40.0 mass % solution of ethylene glycol in water is 1.0514 g/mL at 20°C, what is the molarity?

11.62 What is the molality of the 40.0 mass % ethylene glycol solution used for automobile antifreeze (Problem 11.61)?

11.63 What is the molality of the 16.0 mass % solution of sulfuric acid in Problem 11.60?

11.64 Nalorphine ($C_{19}H_{21}NO_3$), a relative of morphine, is used to combat withdrawal symptoms in narcotics users. How many grams of a 1.3×10^{-3} *m* aqueous solution of nalorphine are needed to obtain a dose of 1.5 mg?

11.65 How many grams of water should you add to 32.5 g of sucrose, $C_{12}H_{22}O_{11}$, to get a 0.850 *m* solution?

11.66 A 0.944 M solution of glucose, $C_6H_{12}O_6$, in water has a density of 1.0624 g/mL at 20°C. What is the concentration of this solution in:

(a) Mole fraction?

(b) Mass percent?

(c) Molality?

11.67 Lactose, $C_{12}H_{22}O_{11}$, is a naturally occurring sugar found in mammalian milk. A 0.335 M solution of lactose in water has a density of 1.0432 g/mL at 20°C. What is the concentration of this solution in:

(a) Mole fraction?

(b) Mass percent?

(c) Molality?

Solubility and Henry's Law

11.68 Vinyl chloride ($CH_2{=}CHCl$), the starting material from which PVC polymer is made, has a Henry's-law constant of 0.091 mol/(L · atm) at 25°C. What is the solubility of vinyl chloride in water (in mol/L) at 25°C and a partial pressure of 0.75 atm?

11.69 Hydrogen sulfide, H_2S, is a toxic gas responsible for the odor of rotten eggs. The solubility of $H_2S(g)$ in water at STP is 0.195 M. What is the Henry's-law constant of H_2S at 0°C? What is the solubility of H_2S in water at 0°C and a partial pressure of 25.5 mm Hg?

11.70 Fish generally need an O_2 concentration in water of at least 4 mg/L for survival. What partial pressure of oxygen above the water (in atmospheres at 0°C)

is needed to obtain this concentration? The solubility of O_2 in water at 0°C and 1 atm partial pressure is 2.21×10^{-3} mol/L.

11.71 At an altitude of 10,000 ft, the partial pressure of oxygen in the lungs is about 68 mm Hg. What is the concentration (in mg/L) of dissolved O_2 in blood (or water) at this partial pressure and a normal body temperature of 37°C? The solubility of O_2 in water at 37°C and 1 atm partial pressure is 1.93×10^{-3} mol/L.

11.72 Look at the solubility graph in Figure 11.6, and estimate an approximate Henry's-law constant for xenon at STP.

11.73 Ammonia, NH_3, is one of the few gases that does not obey Henry's law. Suggest a reason.

Colligative Properties

11.74 What factor is responsible for all colligative properties?

11.75 What is osmotic pressure?

11.76 Draw a phase diagram showing how the phase boundaries differ for a pure solvent compared with a solution.

11.77 A solution concentration must be expressed in molality when considering boiling-point elevation or freezing-point depression but can be expressed in molarity when considering osmotic pressure. Why?

11.78 What is the vapor pressure (in mm Hg) of the following solutions, each of which contains a nonvolatile solute? The vapor pressure of water at 45.0°C is 71.93 mm Hg.

    (a) A solution of 10.0 g of urea, CH_4N_2O, in 150.0 g of water at 45.0°C

    (b) A solution of 10.0 g of LiCl in 150.0 g of water at 45.0°C

11.79 What is the vapor pressure (in mm Hg) of a solution of 16.0 g of glucose ($C_6H_{12}O_6$) in 80.0 g of methanol (CH_3OH) at 27°C? The vapor pressure of pure methanol at 27°C is 140 mm Hg.

11.80 What is the boiling point (in °C) of each of the solutions in Problem 11.78? The molal boiling-point-elevation constant for water is given in Table 11.4.

11.81 What is the freezing point (in °C) of each of the solutions in Problem 11.78? The molal freezing-point-depression constant for water is given in Table 11.4.

11.82 Acetone, C_3H_6O, and ethyl acetate, $C_4H_8O_2$, are organic liquids often used as solvents. At 30°C, the vapor pressure of acetone is 285 mm Hg and the vapor pressure of ethyl acetate is 118 mm Hg. What is the vapor pressure (in mm Hg) at 30°C of a solution prepared by dissolving 25.0 g of acetone in 25.0 g of ethyl acetate?

11.83 The industrial solvents chloroform, $CHCl_3$, and dichloromethane, CH_2Cl_2, are prepared commercially by reaction of methane with chlorine, followed by fractional distillation of the product mixture. At 25°C, the vapor pressure of $CHCl_3$ is 205 mm Hg and the vapor pressure of CH_2Cl_2 is 415 mm Hg. What is the vapor pressure (in mm Hg) at 25°C of a mixture of 15.0 g of $CHCl_3$ and 37.5 g of CH_2Cl_2?

11.84 What is the mole fraction of each component in the liquid mixture in Problem 11.82, and what is the mole fraction of each component in the vapor at 30°C?

11.85 What is the mole fraction of each component in the liquid mixture in Problem 11.83, and what is the mole fraction of each component in the vapor at 25°C?

11.86 A solution prepared by dissolving 5.00 g of aspirin, $C_9H_8O_4$, in 215 g of chloroform has a normal boiling point that is elevated by $\Delta T = 0.47°C$ over that of pure chloroform. What is the value of the molal boiling-point-elevation constant for chloroform?

11.87 A solution prepared by dissolving 3.00 g of ascorbic acid (vitamin C, $C_6H_8O_6$) in 50.0 g of acetic acid has a freezing point that is depressed by $\Delta T = 1.33°C$ below that of pure acetic acid. What is the value of the molal freezing-point-depression constant for acetic acid?

11.88 A solution of citric acid, $C_6H_8O_7$, in 50.0 g of acetic acid has a boiling point elevation of $\Delta T = 1.76°C$. What is the molality of the solution if the molal boiling-point-elevation constant for acetic acid is $K_b = 3.07\ (°C \cdot kg)/mol$.

11.89 What is the normal boiling point (in °C) of ethyl alcohol if a solution prepared by dissolving 26.0 g of glucose ($C_6H_{12}O_6$) in 285 g of ethyl alcohol has a boiling point of 79.1°C? See Table 11.4 to find K_b for ethyl alcohol.

11.90 What osmotic pressure (in atmospheres) would you expect for each of the following solutions?

    (a) 5.00 g of NaCl in 350.0 mL of aqueous solution at 50°C

    (b) 6.33 g of sodium acetate, CH_3CO_2Na, in 55.0 mL of aqueous solution at 10°C

11.91 What osmotic pressure (in mm Hg) would you expect for an aqueous solution of 11.5 mg of insulin (mol. mass = 5990 amu) in 6.60 mL of solution at 298 K? What would the height of the water column be (in meters)? The density of mercury is 13.534 g/mL at 298 K.

11.92 A solution of an unknown molecular substance in water at 300 K gives rise to an osmotic pressure of 4.85 atm. What is the molarity of the solution?

11.93 Human blood gives rise to an osmotic pressure of approximately 7.7 atm at body temperature, 37.0°C. What must the molarity of an intravenous glucose solution be to have the same osmotic pressure as blood?

Uses of Colligative Properties

11.94 Which of the four colligative properties is most often used for molecular mass determination, and why?

11.95 If cost per gram were not a concern, which of the following substances would be the most efficient per unit mass for melting snow from sidewalks and roads: glucose ($C_6H_{12}O_6$), LiCl, NaCl, $CaCl_2$? Explain.

11.96 Cellobiose is a sugar obtained by degradation of cellulose. If 200.0 mL of aqueous solution containing 1.500 g of cellobiose at 25.0°C gives rise to an osmotic pressure of 407.2 mm Hg, what is the molecular mass of cellobiose?

11.97 Met-enkephalin is one of the so-called endorphins, a class of naturally occurring morphinelike chemicals in the brain. What is the molecular mass of met-enkephalin if 20.0 mL of an aqueous solution containing 15.0 mg met-enkephalin at 298 K supports a column of water 32.9 cm high? The density of mercury at 298 K is 13.534 g/mL.

11.98 The freezing point of a solution prepared by dissolving 1.00 mol of hydrogen fluoride, HF, in 500 g

water is −3.8°C, but the freezing point of a solution prepared by dissolving 1.00 mol of hydrogen chloride, HCl, in 500 g of water is −7.4°C. Explain.

11.99 The boiling point of a solution prepared by dissolving 71 g of Na_2SO_4 in 1.00 kg of water is 100.8°C. Explain.

11.100 Elemental analysis of β-carotene, a dietary source of vitamin A, shows that it contains 10.51% H and 89.49% C. Dissolving 0.0250 g of β-carotene in 1.50 g of camphor gives a freezing-point depression of 1.17°C. What are the molecular mass and formula of β-carotene? [K_f for camphor is 37.7 (°C · kg)/mol.]

11.101 Lysine, one of the amino acid building blocks found in proteins, contains 49.29% C, 9.65% H, 19.16% N, and 21.89% O by elemental analysis. A solution prepared by dissolving 30.0 mg of lysine in 1.200 g of the organic solvent biphenyl, gives a freezing-point depression of 1.37°C. What are the molecular mass and formula of lysine? [K_f for biphenyl is 8.00 (°C · kg)/mol.]

General Problems

11.102 When salt is spread on snow-covered roads at −2°C, the snow melts. When salt is spread on snow-covered roads at −30°C, nothing happens. Explain.

11.103 How many grams of KBr dissolved in 125 g of water is needed to raise the boiling point of water to 103.2°C?

11.104 How many grams of ethylene glycol (automobile antifreeze, $C_2H_6O_2$) dissolved in 3.55 kg of water is needed to lower the freezing point of water in an automobile radiator to −22.0°C?

11.105 When 1 mL of toluene is added to 100 mL of benzene (bp 80.1°C), the boiling point of the benzene solution rises, but when 1 mL of benzene is added to 100 mL of toluene (bp 110.6°C), the boiling point of the toluene solution falls. Explain.

11.106 When solid $CaCl_2$ is added to liquid water, the temperature rises. When solid $CaCl_2$ is added to ice at 0°C, the temperature falls. Explain.

11.107 Silver chloride has a solubility of 0.007 mg/mL in water at 5°C. What is the osmotic pressure (in atmospheres) of a saturated solution of AgCl?

11.108 How many grams of naphthalene, $C_{10}H_8$ (commonly used as household mothballs), should be added to 150.0 g of benzene to depress its freezing point by 0.35°C? See Table 11.4 to find K_f for benzene.

11.109 Bromine is sometimes used as a solution in tetrachloromethane, CCl_4. What is the vapor pressure (in mm Hg) of a solution of 1.50 g of Br_2 in 145.0 g of

CCl_4 at 300 K? The vapor pressure of pure bromine at 300 K is 30.5 kPa, and the vapor pressure of CCl_4 is 16.5 kPa.

11.110 Assuming that seawater is a 3.5 mass % solution of NaCl and that its density is 1.00 g/mL, calculate both its boiling point and its freezing point in °C.

11.111 There's actually much more in seawater than just dissolved NaCl. Major ions present include 19,000 ppm Cl^-, 10,500 ppm Na^+, 2650 ppm SO_4^{2-}, 1350 ppm Mg^{2+}, 400 ppm Ca^{2+}, 380 ppm K^+, 140 ppm HCO_3^-, and 65 ppm Br^-.

(a) What is the total molality of all ions present in seawater?

(b) Assuming molality and molarity to be equal, what amount of osmotic pressure (in atmospheres) would seawater give rise to at 300 K?

11.112 Rubbing alcohol is a 90 mass % solution of isopropyl alcohol, C_3H_8O, in water.

(a) How many grams of rubbing alcohol contains 10.5 g of isopropyl alcohol?

(b) How many moles of isopropyl alcohol are in 50.0 g of rubbing alcohol?

11.113 Although inconvenient, it's possible to use osmotic pressure to measure temperature. What is the temperature (in kelvins) if a solution prepared by dissolving 17.5 mg of glucose ($C_6H_{12}O_6$) in 50.0 mL of aqueous solution gives rise to an osmotic pressure of 37.8 mm Hg?

11.114 The steroid hormone *estradiol* contains only C, H, and O; combustion analysis of a 3.47 mg sample yields 10.10 mg CO_2 and 2.76 mg H_2O. On dissolving 7.55 mg of estradiol in 0.500 g of camphor, the melting point of camphor is depressed by 2.10°C. What is the molecular mass of estradiol, and what is a probable formula? [For camphor, $K_f = 37.7$ (°C · kg)/mol.]

11.115 Many acids are partially dissociated into ions in aqueous solution. Trichloroacetic acid (CCl_3CO_2H), for instance, is partially dissociated in water according to the equation

$$CCl_3CO_2H(aq) \rightleftharpoons H^+(aq) + CCl_3CO_2^-(aq)$$

What is the percentage of molecules dissociated if the freezing point of a 1.00 m solution of trichloroacetic acid in water is −2.53°C?

11.116 Addition of 50.00 mL of 2.238 m H_2SO_4 (solution density = 1.1243 g/mL) to 50.00 mL of 2.238 M $BaCl_2$ gives a white precipitate.

(a) What is the mass of the precipitate in grams?

(b) If you filter the mixture and add more H_2SO_4 solution to the filtrate, would you obtain more precipitate? Explain.

11.117 A solid mixture of KCl, KNO_3, and $Ba(NO_3)_2$ is 20.92 mass % chlorine, and a 1.000 g sample of the mixture in 500.0 mL of aqueous solution at 25°C has an osmotic pressure of 744.7 mm Hg. What are the mass percents of KCl, KNO_3, and $Ba(NO_3)_2$ in the mixture?

11.118 A solution of LiCl in a mixture of water and methanol (CH_3OH) has a vapor pressure of 39.4 mm Hg at 17°C and 68.2 mm Hg at 27°C. The vapor pressure of pure water is 14.5 mm Hg at 17°C and 26.8 mm Hg at 27°C, and the vapor pressure of pure methanol is 82.5 mm Hg at 17°C and 140.3 mm Hg at 27°C. What is the composition of the solution (in mass %)?

Multi-Concept Problems

11.119 Treatment of 1.385 g of an unknown metal M with an excess of aqueous HCl evolved a gas that was found to have a volume of 382.6 mL at 20.0°C and 755 mm Hg pressure. Heating the reaction mixture to evaporate the water and remaining HCl then gave a white crystalline compound, MCl_x. After dissolving the compound in 25.0 g of water, the melting point of the resulting solution was −3.53°C.

(a) How many moles of H_2 gas are evolved?

(b) What mass of MCl_x is formed?

(c) What is the molality of particles (ions) in the solution of MCl_x?

(d) How many moles of ions are in solution?

(e) What is the formula and molecular mass of MCl_x?

(f) What is the identity of the metal M?

11.120 A compound that contains only C and H was burned in excess O_2 to give CO_2 and H_2O. When 0.270 g of the compound was burned, the amount of CO_2 formed reacted completely with 20.0 mL of 2.00 M NaOH solution according to the equation

$$2 OH^-(aq) + CO_2(g) \longrightarrow CO_3^{2-}(aq) + H_2O(l)$$

When 0.270 g of the compound was dissolved in 50.0 g of camphor, the resulting solution had a freezing point of 177.9°C. [Pure camphor freezes at 179.8°C and has $K_f = 37.7$(°C · kg)/mol.]

(a) What is the empirical formula of the compound?

(b) What is the molecular mass of the compound?

(c) What is the molecular formula of the compound?

 ## www eMedia Problems

11.121 Watch the **Solution Formation from a Solid** movie (*eChapter 11.3*). Assuming that the density of the resulting solution is 1.23 g/mL, express its concentration in:

(a) mole fraction

(b) mole percent

(c) molality

11.122 The solution prepared in the **Solution Formation from a Solid** movie (*eChapter 11.3*) has a concentration of 1.00 M. Discuss the errors in calculation of concentration that would result from:

(a) neglecting to subtract the mass of the paper

(b) using the molar mass of *anhydrous* $CuSO_4$ (without water molecules) instead of that of $CuSO_4 \cdot 5 H_2O$

11.123 The **Enthalpy of Solution** activity (*eChapter 11.2*) allows you to experiment with the dissolution of several different solids, some of which dissolve endothermically and some of which dissolve exothermically. Examine Figure 11.3, and explain why the dissolution of a gas in liquid is virtually always exothermic.

11.124 For which of the four solids in the **Enthalpy of Solution** activity (*eChapter 11.2*) would solubility increase as temperature increased? For which would the solubility *decrease* as temperature increased? Using the activity, compare data for sodium chloride and ammonium nitrate. Which solubility would you expect to change by the greatest amount (in g/L) for a given change in temperature? Explain your reasoning.

11.125 The **Boiling-Point Elevation and Freezing-Point Depression** activity (*eChapter 11.7*) illustrates how the boiling point and freezing point of water are affected by the addition of a solute.

(a) If you were to dissolve 25.0 g of sodium chloride in 750 g of water, what would be the boiling point of the resulting solution?

(b) If you heated the solution from part (a) to its boiling point and continued adding heat to keep the solution at a boil, would the boiling temperature remain constant? If not, explain why not.

Chemical Kinetics

12

The speed of these airplanes is defined as the change in location per unit time (meters per second, m/s). Similarly, the speed, or rate, of a chemical reaction is defined as a change in concentration per unit time (molar per second, M/s).

Chemists ask three fundamental questions when they study chemical reactions: What happens? To what extent does it happen? How fast does it happen? The answer to the first question is given by the balanced chemical equation, which identifies the reactants, the products, and the stoichiometry of the reaction. The answer to the second question will be addressed in the next chapter, which deals with chemical equilibrium. In this chapter, we'll look at the answer to the third question—the speeds, or rates, at which chemical reactions occur. The area of chemistry concerned with reaction rates and the sequence of steps by which reactions occur is called **chemical kinetics**.

Chemical kinetics is a subject of crucial environmental and economic importance. In the upper atmosphere, for example, maintenance or depletion of the ozone layer, which protects us from the sun's harmful ultraviolet radiation, depends on the relative rates of reactions that produce and destroy O_3 molecules. In the chemical industry, the profitability of the process for

473

Liquid ammonia is used as a fertilizer.

synthesis of ammonia, which is used as a fertilizer, depends on the rate at which gaseous N_2 and H_2 can be converted to NH_3.

In this chapter, we'll describe reaction rates and examine how they are affected by variables such as reactant concentrations and temperature. We'll also see how chemists use rate data to propose a *mechanism*, or pathway, by which a reaction takes place. By understanding reaction mechanisms, we can control known reactions and predict new ones.

12.1 Reaction Rates

The rates of chemical reactions differ greatly. Some reactions, such as the combination of sodium and bromine, occur instantly. Other reactions, such as the rusting of iron, are imperceptibly slow. To describe the rate of a reaction quantitatively, we must specify how fast the concentration of a reactant or a product changes per unit time.

$$\text{Rate} = \frac{\text{Concentration change}}{\text{Time change}}$$

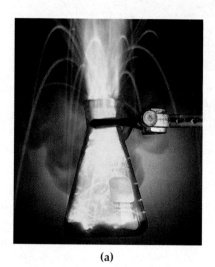

(a)

(b)

(a) The reaction between sodium and bromine is very fast, whereas (b) the rusting of iron is very slow.

One example of a reaction that has been studied in detail is the thermal decomposition of gaseous dinitrogen pentoxide, N_2O_5, to give the brown gas nitrogen dioxide and molecular oxygen:

$$2\,N_2O_5(g) \longrightarrow 4\,NO_2(g) + O_2(g)$$

Colorless   Brown   Colorless

Changes in concentration as a function of time, and thus the reaction rate, can be determined by measuring the increase in pressure as 2 gas molecules are converted to 5 gas molecules. Alternatively, concentration changes can be monitored by measuring the intensity of the brown color due to NO_2. Reactant and product concentrations as a function of time at 55°C are listed in Table 12.1. Note that the concentrations of NO_2 and O_2 increase as the concentration of N_2O_5 decreases.

The **reaction rate** is defined either as the *increase* in the concentration of a *product* per unit time or as the *decrease* in the concentration of a *reactant* per

TABLE 12.1 Concentrations as a Function of Time at 55°C for the Reaction $2 N_2O_5(g) \rightarrow 4 NO_2(g) + O_2(g)$

Time (s)	Concentration (M)		
	N_2O_5	NO_2	O_2
0	0.0200	0	0
100	0.0169	0.0063	0.0016
200	0.0142	0.0115	0.0029
300	0.0120	0.0160	0.0040
400	0.0101	0.0197	0.0049
500	0.0086	0.0229	0.0057
600	0.0072	0.0256	0.0064
700	0.0061	0.0278	0.0070

unit time (the two definitions are equivalent). Let's look first at product formation. In the decomposition of N_2O_5, the rate of formation of O_2 is given by the equation

$$\text{Rate of formation of } O_2 = \frac{\Delta[O_2]}{\Delta t}$$

$$= \frac{\text{Conc of } O_2 \text{ at time } t_2 - \text{Conc of } O_2 \text{ at time } t_1}{t_2 - t_1}$$

where the square brackets surrounding O_2 denote molar concentration; $\Delta[O_2]$ is the change in the molar concentration of O_2; Δt is the change in time; and $\Delta[O_2]/\Delta t$ is the change in the molar concentration of O_2 during the interval from time t_1 to t_2. During the time period 300 to 400 s, for example, the average rate of formation of O_2 is 9×10^{-6} M/s:

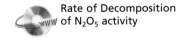 Rate of Decomposition of N_2O_5 activity

$$\text{Rate of formation of } O_2 = \frac{\Delta[O_2]}{\Delta t} = \frac{0.0049 \text{ M} - 0.0040 \text{ M}}{400 \text{ s} - 300 \text{ s}} = 9 \times 10^{-6} \text{ M/s}$$

The units of reaction rate are molar per second, M/s, or, equivalently, moles per liter second, mol/(L · s). We define reaction rate in terms of concentration (moles per liter) rather than amount (moles) because we want the rate to be independent of the scale of the reaction. When twice as much 0.0200 M N_2O_5 decomposes in a vessel of twice the volume, twice the number of moles of O_2 form per second, but the number of moles of O_2 *per liter* that form per second is unchanged.

Plotting the data of Table 12.1 to give the three curves that appear in Figure 12.1 gives additional insight into the concept of reaction rate. Looking at the time period 300 to 400 s on the O_2 line, $\Delta[O_2]$ and Δt are represented, respectively, by the vertical and horizontal sides of a right triangle. The slope of the third side, the hypotenuse of the triangle, is $\Delta[O_2]/\Delta t$, the average rate of O_2 formation during that time period. The steeper the slope of the hypotenuse, the faster the rate. Look, for example, at the triangle defined by $\Delta[NO_2]$ and Δt. The average rate of formation of NO_2 during the time period 300 to 400 s is 3.7×10^{-5} M/s, which is four times the rate of formation of O_2, in accord with the 4:1 ratio of the coefficients of NO_2 and O_2 in the chemical equation for decomposition of N_2O_5.

FIGURE 12.1
Concentrations measured as a function of time when gaseous N_2O_5 at an initial concentration of 0.0200 M decomposes to gaseous NO_2 and O_2 at 55°C. Note that the concentrations of O_2 and NO_2 go up as the concentration of N_2O_5 goes down. The slope of the hypotenuse of each triangle gives the average rate of change of the product or reactant concentration during the indicated time interval. The rate of formation of O_2 is one-fourth the rate of formation of NO_2 and one-half the rate of decomposition of N_2O_5.

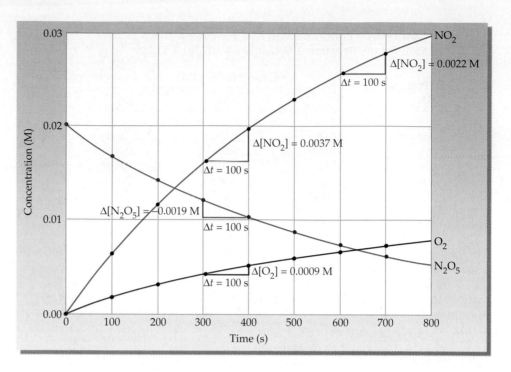

$$\text{Rate of formation of } NO_2 = \frac{\Delta[NO_2]}{\Delta t} = \frac{0.0197 \text{ M} - 0.0160 \text{ M}}{400 \text{ s} - 300 \text{ s}}$$

$$= 3.7 \times 10^{-5} \text{ M/s}$$

As O_2 and NO_2 form, N_2O_5 disappears. Consequently, $\Delta[N_2O_5]/\Delta t$ is negative, in accord with the negative slope of the hypotenuse of the triangle defined by $\Delta[N_2O_5]$ and Δt in Figure 12.1. Because *reaction rate is defined as a positive quantity*, we must always introduce a minus sign in calculating the rate of disappearance of a reactant. During the time period 300 to 400 s, for example, the average rate of decomposition of N_2O_5 is 1.9×10^{-5} M/s:

$$\text{Rate of decomposition of } N_2O_5 = \frac{-\Delta[N_2O_5]}{\Delta t} = \frac{-(0.0101 \text{ M} - 0.0120 \text{ M})}{400 \text{ s} - 300 \text{ s}}$$

$$= 1.9 \times 10^{-5} \text{ M/s}$$

It's important to specify the reactant or product when quoting a rate because the relative rates of product formation and reactant consumption depend on the coefficients in the balanced equation. For the decomposition of N_2O_5, 4 mol of NO_2 form and 2 mol of N_2O_5 disappear for each mole of O_2 that forms. Therefore, the rate of formation of O_2 is one-fourth the rate of formation of NO_2 and one-half the rate of decomposition of N_2O_5:

$$\begin{pmatrix} \text{Rate of formation} \\ \text{of } O_2 \end{pmatrix} = \frac{1}{4} \begin{pmatrix} \text{Rate of formation} \\ \text{of } NO_2 \end{pmatrix} = \frac{1}{2} \begin{pmatrix} \text{Rate of decomposition} \\ \text{of } N_2O_5 \end{pmatrix}$$

or $\qquad \dfrac{\Delta[O_2]}{\Delta t} \qquad = \qquad \dfrac{1}{4}\left(\dfrac{\Delta[NO_2]}{\Delta t}\right) \qquad = \qquad -\dfrac{1}{2}\left(\dfrac{\Delta[N_2O_5]}{\Delta t}\right)$

It's also important to specify the time when quoting a rate because the rate changes as the reaction proceeds. For example, the average rate of formation of NO_2 is 3.7×10^{-5} M/s during the time period 300 to 400 s but is considerably slower during the period 600 to 700 s. During the latter period, $\Delta[NO_2]/\Delta t = 0.0022$ M/100 s $= 2.2 \times 10^{-5}$ M/s. Ordinarily, reaction rates decrease as the reaction mixture runs out of reactants, as indicated by the decreasing slopes of the curves in Figure 12.1 as time passes.

Often, chemists want to know the rate of a reaction at a specific time t rather than the rate averaged over a time interval Δt. For example, what is the rate of formation of NO_2 at time $t = 350$ s? If we make our measurements at shorter and shorter time intervals, the triangle defined by $\Delta[NO_2]$ and Δt will shrink to a point, and the slope of the hypotenuse of the triangle will approach the slope of the tangent to the curve, as shown in Figure 12.2. The slope of the tangent to a concentration-versus-time curve at a time t is called the **instantaneous rate** of the reaction at that particular time. The instantaneous rate at the beginning of a reaction ($t = 0$) is called the **initial rate**.

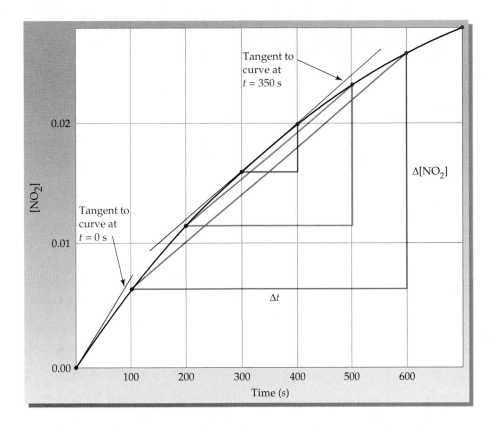

FIGURE 12.2 Concentration of NO_2 versus time when N_2O_5 decomposes at 55°C. The average rate of formation of NO_2 during a time interval Δt equals the slope of the hypotenuse of the triangle defined by $\Delta[NO_2]$ and Δt. As the time interval about the time $t = 350$ s gets smaller, the triangle shrinks to a point, and the slope of the hypotenuse approaches the slope of the tangent to the curve at time t. The slope of the tangent at time t is defined as the instantaneous rate of the reaction at that particular time. The initial rate is the slope of the tangent to the curve at $t = 0$.

EXAMPLE 12.1 Ethanol (C_2H_5OH), the active ingredient in alcoholic beverages and an octane booster in gasoline, is produced by fermentation of glucose. The balanced equation is

$$C_6H_{12}O_6(aq) \longrightarrow 2\,C_2H_5OH(aq) + 2\,CO_2(g)$$

(a) How is the rate of formation of ethanol related to the rate of consumption of glucose?

(b) Write this relationship in terms of $\Delta[C_2H_5OH]/\Delta t$ and $\Delta[C_6H_{12}O_6]/\Delta t$.

SOLUTION

(a) According to the balanced equation, 2 mol of ethanol are produced for each mole of glucose that reacts. Therefore, the rate of formation of ethanol is twice the rate of consumption of glucose.

(b) Since the rate of formation of ethanol is $\Delta[C_2H_5OH]/\Delta t$ and the rate of consumption of glucose is $-\Delta[C_6H_{12}O_6]/\Delta t$ (note the minus sign), we can write

$$\frac{\Delta[C_2H_5OH]}{\Delta t} = -2\frac{\Delta[C_6H_{12}O_6]}{\Delta t}$$

▶ **PROBLEM 12.1** The oxidation of iodide ion by arsenic acid, H_3AsO_4, is described by the balanced equation

$$3\,I^-(aq) + H_3AsO_4(aq) + 2\,H^+(aq) \longrightarrow I_3^-(aq) + H_3AsO_3(aq) + H_2O(l)$$

(a) If $-\Delta[I^-]/\Delta t = 4.8 \times 10^{-4}$ M/s, what is the value of $\Delta[I_3^-]/\Delta t$ during the same time interval?

(b) What is the average rate of consumption of H^+ during that time interval?

▶ **PROBLEM 12.2** Use the data in Table 12.1 to calculate the average rate of decomposition of N_2O_5 and the average rate of formation of O_2 during the time interval 200–300 s.

12.2 Rate Laws and Reaction Order

We noted in the previous section that the rate of decomposition of N_2O_5 depends on its concentration, slowing down as the N_2O_5 concentration decreases. Ordinarily, the rate of a chemical reaction depends on the concentrations of at least some of the reactants.

Let's consider the general reaction

$$a\,A + b\,B \longrightarrow \text{Products}$$

where A and B are the reactants, and a and b are stoichiometric coefficients in the balanced chemical equation. The dependence of the reaction rate on the concentration of each reactant is given by an equation called the **rate law**.

The rate law is usually written in the form

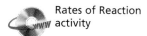
Rates of Reaction activity

$$\text{Rate} = -\frac{\Delta[A]}{\Delta t} = k[A]^m[B]^n$$

where k is a proportionality constant called the **rate constant**. We have arbitrarily expressed the rate as the rate of disappearance of A ($-\Delta[A]/\Delta t$), but we could equally well have written it as the rate of disappearance of any reactant (say, $-\Delta[B]/\Delta t$) or as the rate of appearance of any product. The exponents m and n in the rate law indicate how sensitive the rate is to changes in [A] and [B], and they are usually unrelated to the coefficients a and b in the balanced equation. For the simple reactions discussed in this book, the exponents are usually small positive integers. For more complex reactions, however, the exponents can be negative, zero, or even fractions.

An exponent of 1 means that the rate depends linearly on the concentration of the corresponding reactant. For example, if $m = 1$ and [A] is doubled, the rate doubles. If $m = 2$ and [A] is doubled, $[A]^2$ quadruples and the rate increases by a factor of 4. Figure 12.3 shows how the rate changes depend on the value of the exponent. When m is zero, the rate is independent of the concentration of A because any number raised to the zeroth power is unity ($[A]^0 = 1$). When m is negative, the rate *decreases* as [A] increases. For example, if $m = -1$ and [A] is doubled, $[A]^{-1}$ is halved and the rate decreases by a factor of 2.

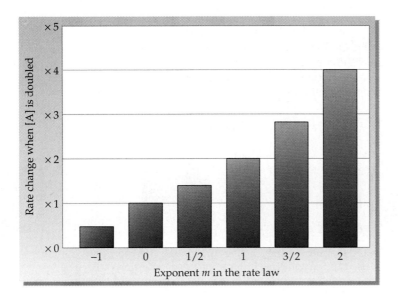

FIGURE 12.3 Change in reaction rate when the concentration of a reactant A is doubled for different values of the exponent m in the rate law, rate $= k[A]^m[B]^n$. Note that the rate change increases as m increases.

The values of the exponents m and n determine the **reaction order** with respect to A and B, respectively. The sum of the exponents ($m + n$) defines the *overall* reaction order. Thus, if the rate law is

$$\text{Rate} = k[A]^2[B] \qquad m = 2; n = 1; \text{and } m + n = 3$$

we say that the reaction is second order in A, first order in B, and third order overall.

The values of the exponents in a rate law must be determined by experiment; they cannot be deduced from the stoichiometry of the reaction. As Table 12.2 shows, there is no general relationship between the stoichiometric coefficients in the balanced chemical equation and the exponents in the rate law. In the first reaction in Table 12.2, for example, the coefficients of $(CH_3)_3CBr$ and H_2O in the balanced equation are both 1, but the exponents in the rate law are 1 for $(CH_3)_3CBr$ and 0 for H_2O: Rate $= k[(CH_3)_3CBr]^1[H_2O]^0 = k[(CH_3)_3CBr]$. In Section 12.8, we'll see that the exponents in the rate law depend on the reaction mechanism.

TABLE 12.2   Balanced Chemical Equations and Experimentally Determined Rate Laws for Some Reactions

Reaction	Rate Law
$(CH_3)_3CBr(aq) + H_2O(aq) \rightarrow (CH_3)_3COH(aq) + H^+(aq) + Br^-(aq)$	Rate $= k[(CH_3)_3CBr]$
$HCO_2H(aq) + Br_2(aq) \rightarrow 2\,H^+(aq) + 2\,Br^-(aq) + CO_2(g)$	Rate $= k[Br_2]$
$BrO_3^-(aq) + 5\,Br^-(aq) + 6\,H^+(aq) \rightarrow 3\,Br_2(aq) + 3\,H_2O(l)$	Rate $= k[BrO_3^-][Br^-][H^+]^2$
$H_2(g) + I_2(g) \rightarrow 2\,HI(g)$	Rate $= k[H_2][I_2]$
$CH_3CHO(g) \rightarrow CH_4(g) + CO(g)$	Rate $= k[CH_3CHO]^{3/2}$

EXAMPLE 12.2 Look at the second reaction in Table 12.2, shown in progress in Figure 12.4. What is the order of the reaction with respect to each of the reactants, and what is the overall reaction order?

SOLUTION To find the reaction order, look at the exponents in the rate law for this reaction in Table 12.2:

$$\text{Rate} = k[\text{Br}_2]$$

Because HCO_2H (formic acid) does not appear in the rate law, the rate is independent of the concentration of HCO_2H, and the reaction is zero order in HCO_2H. Because the exponent on $[\text{Br}_2]$ (understood) is 1, the reaction is first order in Br_2. The reaction is first order overall because the sum of the exponents is 1.

FIGURE 12.4 The reaction of formic acid (HCO_2H) and bromine (Br_2). As time passes (left to right), the red color of bromine disappears because Br_2 is reduced to the colorless Br^- ion. The concentration of Br_2 as a function of time, and thus the reaction rate, can be determined by measuring the intensity of the color.

▶ **PROBLEM 12.3** Consider the last three reactions in Table 12.2. What is the order of each reaction in the various reactants, and what is the overall reaction order for each?

12.3 Experimental Determination of a Rate Law

One method of determining the values of the exponents in a rate law is to carry out a series of experiments in which the initial rate of a reaction is measured as a function of different sets of initial concentrations. Consider, for example, the air oxidation of nitric oxide, one of the reactions that contributes to the formation of acid rain:

$$2\,\text{NO}(g) + \text{O}_2(g) \longrightarrow 2\,\text{NO}_2(g)$$

Look at the initial rate data in Table 12.3.

TABLE 12.3 Initial Concentration and Rate Data for the Reaction $2\,\text{NO}(g) + \text{O}_2(g) \rightarrow 2\,\text{NO}_2(g)$

Experiment	Initial [NO]	Initial [O$_2$]	Initial Rate of Formation of NO$_2$ (M/s)
1	0.015	0.015	0.048
2	0.030	0.015	0.192
3	0.015	0.030	0.096
4	0.030	0.030	0.384

Note that pairs of experiments are designed to investigate the effect on the initial rate of a change in the initial concentration of a single reactant. In the first two experiments, for example, the concentration of NO is doubled from 0.015 M to 0.030 M while the concentration of O_2 is held constant. The initial rate increases by a factor of 4, from 0.048 M/s to 0.192 M/s, indicating that the rate depends on the concentration of NO squared, $[NO]^2$. When [NO] is held constant and $[O_2]$ is doubled (experiments 1 and 3), the initial rate doubles from 0.048 M/s to 0.096 M/s, showing that the rate depends on the concentration of O_2 to the first power, $[O_2]^1$. Therefore, the rate law is

$$\text{Rate} = k[NO]^2[O_2]$$

In accord with this rate law, which is second order in NO, first order in O_2, and third order overall, the initial rate increases by a factor of 8 when the concentrations of both NO and O_2 are doubled (experiments 1 and 4).

The preceding method uses initial rates rather than rates at a later stage of the reaction because we want to avoid complications from the reverse reaction: reactants $\leftarrow$ products. As the product concentrations build up, the rate of the reverse reaction increases. If the reverse rate becomes comparable to the forward rate, the measured reaction rate will depend on the concentrations of both reactants and products. At the beginning of the reaction, however, the product concentrations are zero, and therefore the products can't affect the measured rate. When we measure an initial rate, we are measuring the rate of only the forward reaction, so only reactants (and catalysts; see Section 12.11) can appear in the rate law.

One aspect of determining a rate law is to establish the reaction order. Another is to evaluate the numerical value of the rate constant k. Each reaction has its own characteristic value of the rate constant, which depends on temperature but does not depend on concentrations. To evaluate k for the formation of NO_2, for instance, we can use the data from any one of the experiments in Table 12.3. Solving the rate law for k and substituting the initial rate and concentrations from the first experiment, we obtain

$$k = \frac{\text{Rate}}{[NO_2]^2[O_2]} = \frac{0.048\,\dfrac{M}{s}}{(0.015\,M)^2(0.015\,M)} = 1.4 \times 10^4/(M^2 \cdot s)$$

Try repeating the calculation for experiments 2–4, and show that you get the same value of k. Note that the units of k in this example are $1/(M^2 \cdot s)$, read as "one over molar squared second." Because rates have units of M/s, the units of k depend on the number of concentration terms in the rate law and on the values of the exponents. Units for some common cases are given below:

Rate Law	Overall Reaction Order	Units for k
Rate = $k[A]$	First order	$1/s$ or s^{-1}
Rate = $k[A][B]$	Second order	$1/(M \cdot s)$
Rate = $k[A][B]^2$	Third order	$1/(M^2 \cdot s)$

Be careful not to confuse the rate of a reaction and the rate constant. The *rate* depends on concentrations, whereas the rate *constant* does not (it is a constant). The rate always has units of M/s, whereas the units of the rate constant depend on the overall reaction order.

Example 12.3 gives another instance of how a rate law can be determined from initial rates.

EXAMPLE 12.3 Initial rate data for decomposition of gaseous N_2O_5 at 55°C are as follows:

Experiment	Initial [N_2O_5]	Initial Rate of Decomposition of N_2O_5 (M/s)
1	0.020	3.4×10^{-5}
2	0.050	8.5×10^{-5}

(a) What is the rate law?

(b) What is the value of the rate constant?

SOLUTION

(a) The rate law for decomposition of N_2O_5 can be written as

$$\text{Rate} = -\frac{\Delta[N_2O_5]}{\Delta t} = k[N_2O_5]^m$$

where m is both the order of the reaction in N_2O_5 and the overall reaction order. Comparing experiments 1 and 2 shows that an increase in the initial concentration of N_2O_5 by a factor of 2.5 increases the initial rate by a factor of 2.5:

$$\frac{[N_2O_5]_2}{[N_2O_5]_1} = \frac{0.050 \text{ M}}{0.020 \text{ M}} = 2.5 \qquad \frac{(\text{Rate})_2}{(\text{Rate})_1} = \frac{8.5 \times 10^{-5} \text{ M/s}}{3.4 \times 10^{-5} \text{ M/s}} = 2.5$$

The rate depends linearly on the concentration of N_2O_5, and therefore the rate law is

$$\text{Rate} = -\frac{\Delta[N_2O_5]}{\Delta t} = k[N_2O_5]$$

Note that the reaction is first order in N_2O_5. If the rate had increased by a factor of $(2.5)^2 = 6.25$, the reaction would have been second order in N_2O_5. If the rate had increased by a factor of $(2.5)^3 = 15.6$, the reaction would have been third order in N_2O_5; and so on.

A more formal way to approach this problem is to write the rate law for each experiment:

$$(\text{Rate})_1 = k[N_2O_5]_1 = k(0.020 \text{ M})^m \qquad (\text{Rate})_2 = k[N_2O_5]_2 = k(0.050 \text{ M})^m$$

If we then divide the second equation by the first, we obtain

$$\frac{(\text{Rate})_2}{(\text{Rate})_1} = \frac{k(0.050 \text{ M})^m}{k(0.020 \text{ M})^m} = (2.5)^m$$

Comparing this ratio to the ratio of the experimental rates,

$$\frac{(\text{Rate})_2}{(\text{Rate})_1} = \frac{8.5 \times 10^{-5} \text{ M/s}}{3.4 \times 10^{-5} \text{ M/s}} = 2.5$$

shows that the exponent m must have a value of 1. Therefore, the rate law is

$$\text{Rate} = -\frac{\Delta[N_2O_5]}{\Delta t} = k[N_2O_5]$$

(b) The value of the rate constant k can be found by solving the rate law for k and then substituting in the data from either experiment. Using the data from the first experiment, we obtain

$$k = \frac{\text{Rate}}{[N_2O_5]} = \frac{3.4 \times 10^{-5}\,\dfrac{M}{s}}{0.020\ M} = 1.7 \times 10^{-3}\ s^{-1}$$

Note that the units of k, $1/s$ or s^{-1}, are given by the ratio of the units for the rate and the concentration. These are the expected units for a first-order reaction.

EXAMPLE 12.4   What is the initial rate of decomposition of N_2O_5 at 55°C when its concentration is 0.030 M? (See Example 12.3.)

BALLPARK SOLUTION   Because the reaction is first order in N_2O_5 and the concentration of N_2O_5 is 3/2 times that in experiment 1 of Example 12.3, the decomposition rate will increase by a factor of 3/2 from 3.4×10^{-5} M/s to about 5×10^{-5} M/s.

DETAILED SOLUTION   Calculate the rate by substituting the rate constant found in Example 12.3 ($1.7 \times 10^{-3}\ s^{-1}$) and the given concentration (0.030 M) into the rate law:

$$\text{Rate} = -\frac{\Delta[N_2O_5]}{\Delta t} = k[N_2O_5] = \left(\frac{1.7 \times 10^{-3}}{s}\right)(0.030\ M) = 5.1 \times 10^{-5}\ M/s$$

▶ **PROBLEM 12.4**   The oxidation of iodide ion by hydrogen peroxide in an acidic solution is described by the balanced equation

$$H_2O_2(aq) + 3\,I^-(aq) + 2\,H^+(aq) \longrightarrow I_3^-(aq) + 2\,H_2O(l)$$

The rate of formation of the red-colored triiodide ion, $\Delta[I_3^-]/\Delta t$, can be determined by measuring the rate of appearance of the color (Figure 12.5). Following are initial rate data at 25°C:

Experiment	Initial [H_2O_2]	Initial [I^-]	Initial Rate of Formation of I_3^- (M/s)
1	0.100	0.100	1.15×10^{-4}
2	0.100	0.200	2.30×10^{-4}
3	0.200	0.100	2.30×10^{-4}
4	0.200	0.200	4.60×10^{-4}

(a) What is the rate law for formation of I_3^-?

(b) What is the value of the rate constant?

(c) What is the initial rate of formation of I_3^- when the concentrations are $[H_2O_2] = 0.300$ M and $[I^-] = 0.400$ M?

FIGURE 12.5   A sequence of photographs showing the progress of the reaction of hydrogen peroxide (H_2O_2) and iodide ion (I^-). As time passes (left to right), the red color due to triiodide ion (I_3^-) increases in intensity.

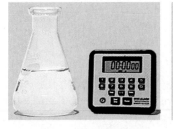

▶ **PROBLEM 12.5** What are the units of the rate constant for each of the reactions in Table 12.2?

⌐ **KEY CONCEPT PROBLEM 12.6** The relative rates of the reaction A + B → products in vessels (a)–(d) are 1:1:4:4. Red spheres represent A molecules, and blue spheres represent B molecules.

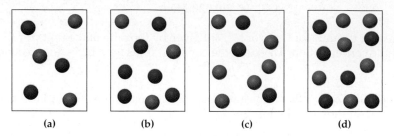

| (a) | (b) | (c) | (d) |

(a) What is the order of the reaction in A and B, and what is the overall reaction order?

(b) Write the rate law.

12.4 Integrated Rate Law for a First-Order Reaction

Thus far we've focused on the rate law, an equation that tells how a reaction rate depends on reactant concentrations. But we're also interested in how reactant and product concentrations vary with time. For example, it's important to know the rate at which the atmospheric ozone layer is being destroyed, but we also want to know what the ozone concentration will be 50 years from now and how long it will take for the concentration to be reduced by a given amount, say, 10%.

Because the kinetics of the pollution-induced decomposition of ozone is a very complicated problem, let's consider instead a simple, general first-order reaction

$$a \, A \longrightarrow \text{Products}$$

A **first-order reaction** is one whose rate depends on the concentration of a single reactant raised to the first power.

$$\text{Rate} = -\frac{\Delta[A]}{\Delta t} = k[A]$$

An example is the decomposition of hydrogen peroxide in basic solution, which we'll discuss in Example 12.5 at the end of this section:

$$2 \, H_2O_2(aq) \longrightarrow 2 \, H_2O(l) + O_2(g)$$

Using calculus, it's possible to convert the rate law to another form, called the **integrated rate law**:

$$\ln \frac{[A]_t}{[A]_0} = -kt$$

In this equation, ln denotes the natural logarithm, $[A]_0$ designates the concentration of A at some initial time, arbitrarily considered to be $t = 0$, and $[A]_t$ is the concentration of A at any time t thereafter. (See Appendix A.2 for a review of logarithms.) The ratio $[A]_t/[A]_0$ is the fraction of A that remains at time t. Thus, the integrated rate law is a *concentration–time equation* that makes it possible to calculate the concentration of A at any time t or the fraction of A that remains at any time t. The integrated rate law can also be used to calculate the time required for the initial concentration of A to drop to any particular value or to any particular fraction of its initial concentration (Figure 12.6a). Example 12.5 shows how to use the integrated rate law.

Since $\ln([A]_t/[A]_0) = \ln[A]_t - \ln[A]_0$, we can rewrite the integrated rate law as

$$\ln[A]_t = -kt + \ln[A]_0$$

This equation is of the form $y = mx + b$, which says that $\ln[A]_t$ is a linear function of time:

$$\ln[A]_t = (-k)t + \ln[A]_0$$

$$\underset{y}{\uparrow} \qquad \underset{m\ \ x}{\uparrow\ \uparrow} \quad \underset{b}{\uparrow}$$

If we graph $\ln[A]$ versus time, we obtain a straight line having a slope $m = -k$ and an intercept $b = \ln[A]_0$ (Figure 12.6b). The value of the rate constant is simply equal to minus the slope of the straight line:

$$k = -(\text{Slope})$$

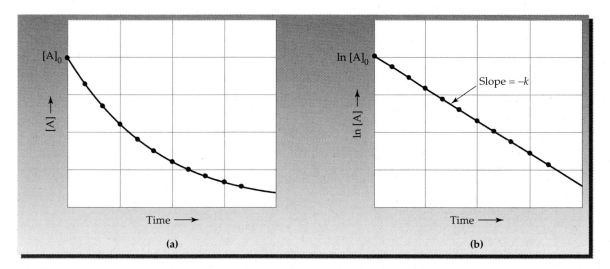

FIGURE 12.6   Plots of **(a)** reactant concentration versus time and **(b)** natural logarithm of reactant concentration versus time for a first-order reaction. A first-order reaction exhibits an exponential decay of the reactant concentration **(a)** and a linear decay of the logarithm of the reactant concentration **(b)**. The slope of the plot of $\ln[A]$ versus time gives the rate constant.

This graphical method of determining a rate constant, illustrated in Example 12.6, is an alternative to the method of initial rates used in Example 12.3. It should be emphasized, however, that a plot of $\ln[A]$ versus time will give a straight line only if the reaction is first order in A. Indeed, a good way of testing whether a reaction is first order is to examine the appearance of such a plot.

EXAMPLE 12.5 The decomposition of hydrogen peroxide in dilute sodium hydroxide solution is described by the equation

$$2\,H_2O_2(aq) \longrightarrow 2\,H_2O(l) + O_2(g)$$

The reaction is first order in H_2O_2, the rate constant at 20°C is $1.8 \times 10^{-5}\,s^{-1}$, and the initial concentration of H_2O_2 is 0.30 M.
(a) What is the concentration of H_2O_2 after 4.00 h?
(b) How long will it take for the H_2O_2 concentration to drop to 0.12 M?
(c) How long will it take for 90% of the H_2O_2 to decompose?

SOLUTION
(a) Since this reaction has a first-order rate law, $-\Delta[H_2O_2]/\Delta t = k[H_2O_2]$, we can use the corresponding concentration–time equation for a first-order reaction:

$$\ln \frac{[H_2O_2]_t}{[H_2O_2]_0} = -kt$$

Because k has units of s^{-1}, we must first convert the time from hours to seconds:

$$t = (4.00\,h)\left(\frac{60\,min}{h}\right)\left(\frac{60\,s}{min}\right) = 14{,}400\,s$$

Then, substitute the values of $[H_2O_2]_0$, k, and t into the concentration–time equation:

$$\ln \frac{[H_2O_2]_t}{0.30\,M} = -(1.8 \times 10^{-5}\,s^{-1})(1.44 \times 10^4\,s) = -0.259$$

Taking the antilogarithm (antiln) of both sides gives

$$\frac{[H_2O_2]_t}{0.30\,M} = e^{-0.259} = 0.772$$

$$[H_2O_2]_t = (0.772)(0.30\,M) = 0.23\,M$$

(b) First, solve the concentration–time equation for time:

$$t = -\frac{1}{k}\ln\frac{[H_2O_2]_t}{[H_2O_2]_0}$$

Then evaluate the time by substituting the concentrations and the value of k:

$$t = -\left(\frac{1}{1.8 \times 10^{-5}\,s^{-1}}\right)\ln\frac{0.12\,M}{0.30\,M} = -\left(\frac{1}{1.8 \times 10^{-5}\,s^{-1}}\right)(-0.916) = 5.1 \times 10^4\,s$$

Thus, the H_2O_2 concentration reaches 0.12 M at a time of $5.1 \times 10^4\,s$ (14 h).
(c) When 90% of the H_2O_2 has decomposed, 10% remains. Therefore,

$$\frac{[H_2O_2]_t}{[H_2O_2]_0} = \frac{(0.10)(0.30\,M)}{0.30\,M} = 0.10$$

The time required for 90% decomposition is

$$t = -\left(\frac{1}{1.8 \times 10^{-5}\,\text{s}^{-1}}\right)\ln 0.10 = -\left(\frac{1}{1.8 \times 10^{-5}\,\text{s}^{-1}}\right)(-2.30) = 1.3 \times 10^5\,\text{s (36 h)}$$

EXAMPLE 12.6 Experimental concentration-versus-time data for the decomposition of gaseous N_2O_5 at 55°C are listed in Table 12.1 and are plotted in Figure 12.1. Use these data to confirm that the decomposition of N_2O_5 is a first-order reaction. What is the value of the rate constant?

SOLUTION To confirm that the reaction is first order, check to see if a plot of $\ln[N_2O_5]$ versus time gives a straight line. Values of $\ln[N_2O_5]$ are listed in the following table and are plotted versus time in the graph:

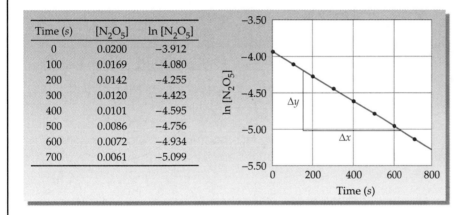

Time (s)	$[N_2O_5]$	$\ln[N_2O_5]$
0	0.0200	-3.912
100	0.0169	-4.080
200	0.0142	-4.255
300	0.0120	-4.423
400	0.0101	-4.595
500	0.0086	-4.756
600	0.0072	-4.934
700	0.0061	-5.099

Because the data points lie on a straight line, the reaction is first order in N_2O_5. The slope of the line can be determined from the coordinates of any two widely separated points on the line, and the rate constant k can be calculated from the slope:

$$\text{Slope} = \frac{\Delta y}{\Delta x} = \frac{(-5.02) - (-4.17)}{650\,\text{s} - 150\,\text{s}} = \frac{-0.85}{500\,\text{s}} = -1.7 \times 10^{-3}\,\text{s}^{-1}$$

$$k = -(\text{Slope}) = 1.7 \times 10^{-3}\,\text{s}^{-1}$$

Note that the slope is negative, k is positive, and the value of k agrees with the value obtained earlier in Example 12.3 by the method of initial rates.

▶ **PROBLEM 12.7** In acidic aqueous solution, the complex ion $Co(NH_3)_5Br^{2+}$ undergoes a slow reaction in which the bromide ion is replaced by a water molecule:

$$Co(NH_3)_5Br^{2+}(aq) + H_2O(l) \longrightarrow Co(NH_3)_5(H_2O)^{3+}(aq) + Br^-(aq)$$

The reaction is first order in $Co(NH_3)_5Br^{2+}$, the rate constant at 25°C is $6.3 \times 10^{-6}\,\text{s}^{-1}$, and the initial concentration of $Co(NH_3)_5Br^{2+}$ is 0.100 M.
(a) What is its molarity after a reaction time of 10.0 h?
(b) How many hours are required for 75% of the $Co(NH_3)_5Br^{2+}$ to react?

The purple complex ion $Co(NH_3)_5Br^{2+}$ (left) reacts with water in acidic solution, yielding the pinkish-orange complex ion $Co(NH_3)_5(H_2O)^{3+}$ (right).

▶ **PROBLEM 12.8** At high temperatures, cyclopropane is converted to propene, the material from which polypropylene is made:

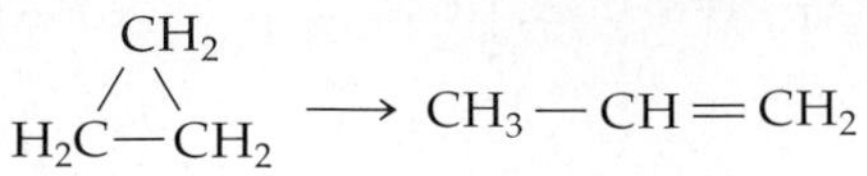

$$\underset{\text{Cyclopropane}}{\overset{\displaystyle CH_2}{\underset{H_2C-CH_2}{\diagup \diagdown}}} \longrightarrow \underset{\text{Propene}}{CH_3-CH=CH_2}$$

Given the following concentration data, test whether the reaction is first order, and calculate the value of the rate constant:

Time (min)	0	5.0	10.0	15.0	20.0
[Cyclopropane]	0.098	0.080	0.066	0.054	0.044

12.5 Half-Life of a First-Order Reaction

The **half-life** of a reaction, symbolized by $t_{1/2}$, is the time required for the reactant concentration to drop to one-half of its initial value. Consider the first-order reaction

$$a\,A \longrightarrow \text{Products}$$

To relate the reaction's half-life to the rate constant, let's begin with the integrated rate law:

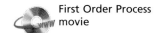

First Order Process movie

$$\ln \frac{[A]_t}{[A]_0} = -kt$$

When $t = t_{1/2}$, the fraction of A that remains, $[A]_t/[A]_0$, is 1/2. Therefore,

$$\ln \frac{1}{2} = -kt_{1/2}$$

$$\text{so} \quad t_{1/2} = \frac{-\ln \frac{1}{2}}{k} = \frac{\ln 2}{k}$$

$$\text{or} \quad t_{1/2} = \frac{0.693}{k}$$

Rates of Reaction activity

Thus, the half-life of a first-order reaction is readily calculated from the rate constant, and vice versa.

It's evident from the equation $t_{1/2} = 0.693/k$ that the half-life of a first-order reaction is a constant because it depends only on the rate constant and not on the reactant concentration. This point is worth noting because reactions that are not first order have half-lives that *do* depend on concentration; that is, the amount of time in one half-life changes as the reactant concentration changes for a non-first-order reaction.

The constancy of the half-life for a first-order reaction is illustrated in Figure 12.7. Each successive half-life is an equal period of time in which the concentration decreases by a factor of 2. We'll see in Chapter 22 that half-lives are widely used in describing radioactive decay rates.

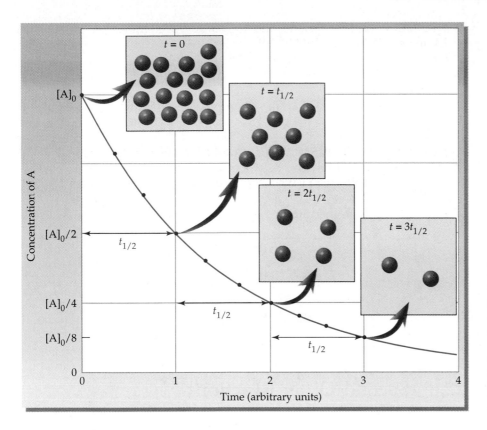

FIGURE 12.7   Concentration of a reactant A as a function of time for a first-order reaction. The concentration falls from its initial value, $[A]_0$, to $[A]_0/2$ after one half-life, to $[A]_0/4$ after a second half-life, to $[A]_0/8$ after a third half-life, and so on. For a first-order reaction, each half-life represents an equal amount of time.

EXAMPLE 12.7

(a) Estimate the half-life for the decomposition of gaseous N_2O_5 at 55°C from the concentration-versus-time plot in Figure 12.1.

(b) Calculate the half-life from the rate constant (1.7×10^{-3} s^{-1}).

(c) If the initial concentration of N_2O_5 is 0.020 M, what is the concentration of N_2O_5 after five half-lives?

(d) How long will it take for the N_2O_5 concentration to fall to 12.5% of its initial value?

Radioactive Decay simulation

SOLUTION

(a) Figure 12.1 shows that the concentration of N_2O_5 falls from 0.020 M to 0.010 M during a time period of approximately 400 s. At 800 s, $[N_2O_5]$ has decreased by another factor of 2, to 0.0050 M. Therefore, $t_{1/2} \approx 400$ s.

(b) Based on the value of the rate constant,

$$t_{1/2} = \frac{0.693}{k} = \frac{0.693}{1.7 \times 10^{-3} \text{ s}^{-1}} = 4.1 \times 10^2 \text{ s} \ (6.8 \text{ min})$$

(c) At $5t_{1/2}$, $[N_2O_5]$ will be $(1/2)^5 = 1/32$ of its initial value. Therefore,

$$[N_2O_5] = \frac{0.020 \text{ M}}{32} = 0.000\ 62 \text{ M}$$

(d) Since 12.5% of the initial concentration corresponds to 1/8 or $(1/2)^3$ of the initial concentration, the time required is three half-lives:

$$t = 3t_{1/2} = 3(4.1 \times 10^2 \text{ s}) = 1.2 \times 10^3 \text{ s} \ (20 \text{ min})$$

▶ **PROBLEM 12.9** Consider the first-order decomposition of H_2O_2 in Example 12.5.

(a) What is the half-life (in hours) of the reaction at 20°C?

(b) What is the molarity of H_2O_2 after four half-lives if the initial concentration of H_2O_2 is 0.30 M?

(c) How many hours will it take for the concentration to drop to 25% of its initial value?

◆━ **KEY CONCEPT PROBLEM 12.10** Consider the first-order reaction A → B in which A molecules (red spheres) are converted to B molecules (blue spheres).

(a) Given the following pictures at $t = 0$ min and $t = 10$ min, what is the half-life of the reaction?

(b) Draw a picture that shows the number of A and B molecules present at $t = 15$ min.

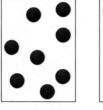

$t = 0$ min $t = 10$ min

12.6 Second-Order Reactions

A **second-order reaction** is one whose rate depends on the concentration of a single reactant raised to the second power or on the concentrations of two different reactants, each raised to the first power. For the simpler type, a A → products, the rate law is

$$\text{Rate} = -\frac{\Delta[A]}{\Delta t} = k[A]^2$$

An example is the thermal decomposition of nitrogen dioxide to yield NO and O_2:

$$2\,NO_2(g) \longrightarrow 2\,NO(g) + O_2(g)$$

Using calculus, it's possible to convert the rate law to the integrated rate law

$$\frac{1}{[A]_t} = kt + \frac{1}{[A]_0}$$

This integrated rate law then allows us to calculate the concentration of A at any time t if the initial concentration $[A]_0$ is known.

Since the integrated rate law has the form $y = mx + b$, a graph of $1/[A]$ versus time is a straight line if the reaction is second order:

$$\underset{y}{\frac{1}{[A]_t}} = \underset{mx}{kt} + \underset{b}{\frac{1}{[A]_0}}$$

The slope of the straight line is the rate constant k, and the intercept is $1/[A]_0$. Thus, by plotting $1/[A]$ versus time, we can test whether the reaction is second order and can determine the value of the rate constant (see Example 12.8).

We can obtain an expression for the half-life of a second-order reaction by substituting $[A]_t = [A]_0/2$ and $t = t_{1/2}$ into the integrated rate law:

$$t_{1/2} = \frac{1}{k[A]_0}$$

In contrast with a first-order reaction, the time required for the concentration of A to drop to one-half of its initial value in a second-order reaction depends on both the rate constant and the initial concentration. Thus, the value of $t_{1/2}$ increases as the reaction proceeds because the value of $[A]_0$ at the beginning of each successive half-life is smaller by a factor of 2. Consequently, each half-life for a second-order reaction is twice as long as the preceding one (Figure 12.8).

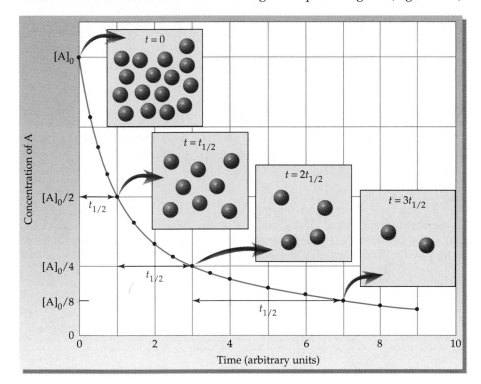

FIGURE 12.8   Concentration of a reactant A as a function of time for a second-order reaction. Note that each half-life is twice as long as the preceding one because $t_{1/2} = 1/k[A]_0$ and the concentration of A at the beginning of each successive half-life is smaller by a factor of 2.

Table 12.4 summarizes some important differences between first-order and second-order reactions of the type $a\, A \rightarrow$ products.

TABLE 12.4   Characteristics of First- and Second-Order Reactions of the Type $a\, A \rightarrow$ Products

	First-Order	Second-Order
Rate Law	$-\dfrac{\Delta[A]}{\Delta t} = k[A]$	$-\dfrac{\Delta[A]}{\Delta t} = k[A]^2$
Concentration–time equation	$\ln[A]_t = -kt + \ln[A]_0$	$\dfrac{1}{[A]_t} = kt + \dfrac{1}{[A]_0}$
Linear graph	$\ln[A]$ versus t	$\dfrac{1}{[A]}$ versus t
Graphical determination of k	$k = -(\text{Slope})$	$k = \text{Slope}$
Half-life	$t_{1/2} = \dfrac{0.693}{k}$ (constant)	$t_{1/2} = \dfrac{1}{k[A]_0}$ (not constant)

EXAMPLE 12.8 At elevated temperatures, nitrogen dioxide decomposes to nitric oxide and molecular oxygen:

$$2\,NO_2(g) \longrightarrow 2\,NO(g) + O_2(g)$$

Concentration–time data for this reaction at 300°C are as follows:

Time (s)	[NO₂]	Time (s)	[NO₂]
0	8.00×10^{-3}	200	4.29×10^{-3}
50	6.58×10^{-3}	300	3.48×10^{-3}
100	5.59×10^{-3}	400	2.93×10^{-3}
150	4.85×10^{-3}	500	2.53×10^{-3}

(a) Is the reaction first order or second order?

(b) What is the value of the rate constant?

(c) What is the concentration of NO_2 at $t = 20.0$ min?

(d) What is the half-life of the reaction when the initial concentration of NO_2 is 6.00×10^{-3} M?

(e) What is $t_{1/2}$ when $[NO_2]_0$ is 3.00×10^{-3} M?

SOLUTION

(a) To determine whether the reaction is first order or second order, calculate values of ln $[NO_2]$ and $1/[NO_2]$, and then graph these values versus time. The plot of ln $[NO_2]$ versus time is curved, but the plot of $1/[NO_2]$ versus time is a straight line. The reaction is therefore second order in NO_2.

Time (s)	[NO₂]	ln [NO₂]	1/[NO₂]
0	8.00×10^{-3}	−4.828	125
50	6.58×10^{-3}	−5.024	152
100	5.59×10^{-3}	−5.187	179
150	4.85×10^{-3}	−5.329	206
200	4.29×10^{-3}	−5.451	233
300	3.48×10^{-3}	−5.661	287
400	2.93×10^{-3}	−5.833	341
500	2.53×10^{-3}	−5.980	395

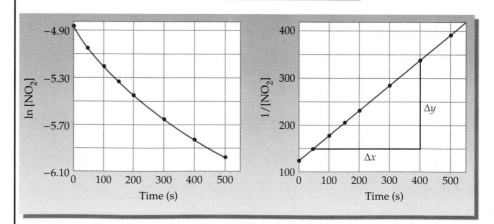

(b) The rate constant is equal to the slope of the line in the plot of $1/[NO_2]$ versus time, which we can estimate from the coordinates of two widely separated points on the graph:

$$k = \text{Slope} = \frac{\Delta y}{\Delta x} = \frac{340\ M^{-1} - 150\ M^{-1}}{400\ s - 50\ s} = \frac{190\ M^{-1}}{350\ s} = 0.54/(M \cdot s)$$

(c) The concentration of NO_2 at $t = 20.0$ min (1200 s) can be calculated using the integrated rate law:

$$\frac{1}{[NO_2]_t} = kt + \frac{1}{[NO_2]_0}$$

Substituting the values of k, t, and $[NO_2]_0$ gives

$$\frac{1}{[NO_2]_t} = \left(\frac{0.54}{M \cdot s}\right)(1200\ s) + \frac{1}{8.00 \times 10^{-3}\ M}$$

$$= \frac{648}{M} + \frac{125}{M} = \frac{773}{M}$$

$$[NO_2]_t = 1.3 \times 10^{-3}\ M$$

(d) The half-life of a second-order reaction can be calculated from the rate constant and the initial concentration of NO_2 (6.00×10^{-3} M):

$$t_{1/2} = \frac{1}{k[NO_2]_0} = \frac{1}{\left(\dfrac{0.54}{M \cdot s}\right)(6.00 \times 10^{-3}\ M)} = 3.1 \times 10^2\ s$$

(e) When $[NO_2]_0$ is 3.00×10^{-3} M, $t_{1/2} = 6.2 \times 10^2$ s (twice as long as when $[NO_2]_0$ is 6.00×10^{-3} M because $[NO_2]_0$ is now smaller by a factor of 2).

▶ **PROBLEM 12.11** Hydrogen iodide gas decomposes at 410°C:

$$2\ HI(g) \longrightarrow H_2(g) + I_2(g)$$

The following data describe this decomposition.

Time (min)	0	20	40	60	80
[HI]	0.500	0.382	0.310	0.260	0.224

(a) Is the reaction first order or second order?
(b) What is the value of the rate constant?
(c) At what time (in minutes) does the HI concentration reach 0.100 M?
(d) How many minutes does it take for the HI concentration to drop from 0.400 M to 0.200 M?

12.7 Reaction Mechanisms

Thus far, our discussion of chemical kinetics has centered on reaction rates. We've seen that the rate of a reaction depends on both reactant concentrations and the value of the rate constant. An equally important issue in chemical kinetics is the **reaction mechanism**, the sequence of molecular events, or reaction steps, that defines the pathway from reactants to products. Chemists want to know the sequence in which the various reaction steps take place so they can better control known reactions and predict new ones.

◆— REACTION MECHANISM   The sequence of reaction steps that defines the pathway from reactants to products.

A single step in a reaction mechanism is called an **elementary reaction** or an **elementary step**. To clarify the crucial distinction between an elementary reaction and an overall reaction, let's consider the gas-phase reaction of nitrogen dioxide and carbon monoxide to give nitric oxide and carbon dioxide:

$$NO_2(g) + CO(g) \longrightarrow NO(g) + CO_2(g) \qquad \text{Overall reaction}$$

Experimental evidence suggests that this reaction takes place by a two-step mechanism:

Step 1. $\quad NO_2(g) + NO_2(g) \longrightarrow NO(g) + NO_3(g) \qquad$ Elementary reaction

Step 2. $\quad NO_3(g) + CO(g) \longrightarrow NO_2(g) + CO_2(g) \qquad$ Elementary reaction

In the first elementary step, two NO_2 molecules collide with enough energy to break one N–O bond and form another, resulting in the transfer of an oxygen atom from one NO_2 molecule to the other. In the second step, the NO_3 molecule formed in the first step collides with a CO molecule, and the transfer of an oxygen atom from NO_3 to CO yields an NO_2 molecule and a CO_2 molecule (Figure 12.9).

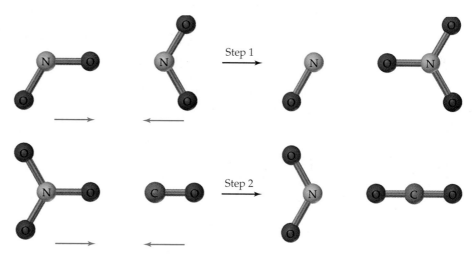

FIGURE 12.9   Elementary steps in the reaction of NO_2 with CO.

The chemical equation for an elementary reaction is a description of an individual molecular event that involves the breaking and/or making of chemical bonds. By contrast, the balanced equation for an *overall* reaction describes only the stoichiometry of the overall process but provides no information about how the reaction occurs. The equation for the reaction of NO_2 with CO, for example, does not tell us that the reaction occurs by direct transfer of an oxygen atom from an NO_2 molecule to a CO molecule.

Elementary reaction—describes behavior of individual molecules.

Overall reaction—describes reaction stoichiometry.

Of course, the elementary steps in a proposed reaction mechanism must sum to give the overall reaction. When we sum the elementary steps in the reaction of NO_2 with CO and then cancel the molecules that appear on both sides of the resulting equation, we obtain the overall reaction:

Step 1.	$NO_2(g) + NO_2(g) \longrightarrow NO(g) + NO_3(g)$	Elementary reaction
Step 2.	$NO_3(g) + CO(g) \longrightarrow NO_2(g) + CO_2(g)$	Elementary reaction

$$NO_2(g) + \cancel{NO_2(g)} + \cancel{NO_3(g)} + CO(g) \longrightarrow NO(g) + \cancel{NO_3(g)} + \cancel{NO_2(g)} + CO_2(g)$$

$$NO_2(g) + CO(g) \longrightarrow NO(g) + CO_2(g) \qquad \text{Overall reaction}$$

A species that is formed in one step of a reaction mechanism and consumed in a subsequent step, such as NO_3 in our example, is called a **reaction intermediate**. Note that reaction intermediates do not appear in the net equation for the overall reaction, and it's only by looking at the elementary steps that their presence is noticed.

Elementary reactions are classified on the basis of their **molecularity**, the number of molecules (or atoms) on the reactant side of the chemical equation for the elementary reaction. A **unimolecular reaction** is an elementary reaction that involves a single reactant molecule—for example, the unimolecular decomposition of ozone in the upper atmosphere:

$$O_3{}^*(g) \longrightarrow O_2(g) + O(g)$$

(The asterisk on O_3 indicates that the ozone molecule is in an energetically excited state as a result of previous absorption of ultraviolet light from the sun. The absorbed energy causes one of the two $O-O$ bonds to break, with loss of an oxygen atom.)

The beautiful northern lights, or aurora borealis, are often observed in the Northern Hemisphere at high latitudes. The light is produced in part by excited O atoms in the upper atmosphere.

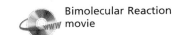

Bimolecular Reaction movie

A **bimolecular reaction** is an elementary reaction that results from an energetic collision between two reactant molecules. An example taken from the chemistry of the upper atmosphere is the reaction of an ozone molecule with an oxygen atom to yield two O_2 molecules:

$$O_3(g) + O(g) \longrightarrow 2\,O_2(g)$$

Both unimolecular and bimolecular reactions are common, but **termolecular reactions**, which involve three atoms or molecules, are rare. As any pool player knows, three-body collisions are much less probable than two-body collisions. There are some reactions, however, that require a three-body collision, notably the combination of two atoms to form a diatomic molecule. For example, oxygen atoms in the upper atmosphere combine as a result of collisions involving some third molecule M:

$$O(g) + O(g) + M(g) \longrightarrow O_2(g) + M(g)$$

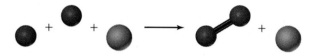

In the atmosphere, M is most likely N_2, but in principle it could be any atom or molecule. The role of M is to carry away the energy that is released when the O—O bond is formed. If M were not involved in the collision, the two oxygen atoms would simply bounce off each other, and no reaction would occur.

EXAMPLE 12.9   The following two-step mechanism has been proposed for the gas-phase decomposition of nitrous oxide (N_2O):

Step 1. $N_2O(g) \longrightarrow N_2(g) + O(g)$

Step 2. $N_2O(g) + O(g) \longrightarrow N_2(g) + O_2(g)$

(a) Write the chemical equation for the overall reaction.
(b) Identify any reaction intermediates.
(c) What is the molecularity of each of the elementary reactions?
(d) What is the molecularity of the overall reaction?

SOLUTION
(a) The overall reaction is the sum of the two elementary steps:

Step 1. $N_2O(g) \longrightarrow N_2(g) + O(g)$ Elementary reaction

Step 2. $N_2O(g) + O(g) \longrightarrow N_2(g) + O_2(g)$ Elementary reaction

---

$2\,N_2O(g) + \cancel{O(g)} \longrightarrow 2\,N_2(g) + \cancel{O(g)} + O_2(g)$

$2\,N_2O(g) \longrightarrow 2\,N_2(g) + O_2(g)$ Overall reaction

(b) The oxygen atom is a reaction intermediate because it is formed in the first elementary step and consumed in the second step.

(c) The first elementary reaction is unimolecular because it involves a single reactant molecule. The second elementary step is bimolecular because it involves two reactant molecules.

(d) It's not appropriate to use the word *molecularity* in connection with the overall reaction because the overall reaction does not describe an individual molecular event. Only an elementary reaction can have a molecularity.

◆— **KEY CONCEPT PROBLEM 12.12** A suggested mechanism for the reaction of nitrogen dioxide and molecular fluorine is

Step 1. $NO_2(g) + F_2(g) \longrightarrow NO_2F(g) + F(g)$

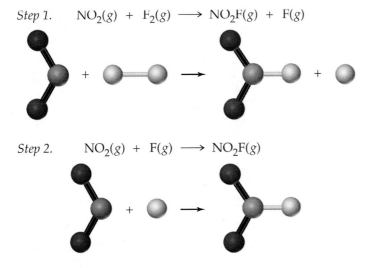

Step 2. $NO_2(g) + F(g) \longrightarrow NO_2F(g)$

(a) Give the chemical equation for the overall reaction, and identify any reaction intermediates.

(b) What is the molecularity of each of the elementary reactions?

12.8 Rate Laws and Reaction Mechanisms

We emphasized in Section 12.2 that the rate law for an overall chemical reaction must be determined by experiment. It can't be deduced from the stoichiometric coefficients in the balanced equation for the overall reaction. By contrast, the rate law for an elementary reaction follows directly from its molecularity. The concentration of each reactant in an elementary reaction appears in the rate law, with an exponent equal to its coefficient in the chemical equation for the elementary reaction.

Consider, for example, the unimolecular decomposition of ozone:

$$O_3(g) \longrightarrow O_2(g) + O(g)$$

The number of moles of O_3 per liter that decompose per unit time is directly proportional to the molar concentration of O_3:

$$\text{Rate} = -\frac{\Delta[O_3]}{\Delta t} = k[O_3]$$

The rate of a unimolecular reaction is always first order in the concentration of the reactant molecule.

Bimolecular Reaction movie

For a bimolecular elementary reaction of the type $A + B \rightarrow$ products, the reaction rate depends on the frequency of collisions between A and B molecules. The frequency of AB collisions involving any *particular* A molecule is proportional to the molar concentration of B, while the total frequency of AB collisions involving *all* A molecules is proportional to the molar concentration of A times the molar concentration of B (Figure 12.10). Therefore, the reaction obeys the second-order rate law

$$\text{Rate} = -\frac{\Delta[A]}{\Delta t} = -\frac{\Delta[B]}{\Delta t} = k[A][B]$$

FIGURE 12.10 The effect of concentration on the frequency of collisions between A molecules (blue) and B molecules (red). **(a)** The frequency of AB collisions involving any one A molecule is proportional to the concentration of B molecules. **(b)** Doubling the concentration of A molecules (from 1 to 2 per unit volume) doubles the total frequency of AB collisions. **(c)** Doubling the concentration of B molecules doubles the frequency of AB collisions involving any one A molecule. **(d)** Doubling the concentration of A molecules *and* doubling the concentration of B molecules *quadruples* the total frequency of AB collisions. Thus, the total frequency of AB collisions is proportional to the concentration of A molecules times the concentration of B molecules.

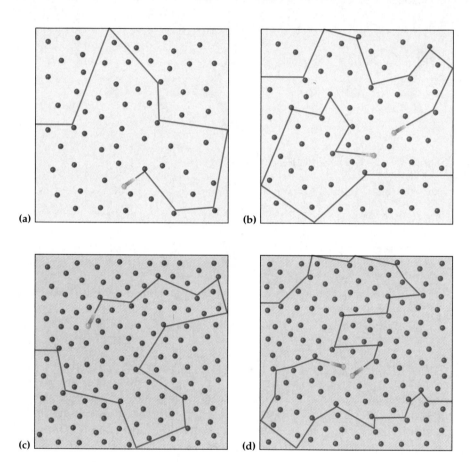

(a) (b) (c) (d)

A similar line of reasoning indicates that a bimolecular reaction of the type

$$A + A \longrightarrow \text{Products} \quad \text{or} \quad 2\,A \longrightarrow \text{Products}$$

has the second-order rate law

$$\text{Rate} = -\frac{\Delta[A]}{\Delta t} = k[A][A] = k[A]^2$$

Rate laws for elementary reactions are summarized in Table 12.5. Note that the overall reaction order for an elementary reaction is always equal to its molecularity.

TABLE 12.5   Rate Laws for Elementary Reactions

Elementary Reaction	Molecularity	Rate Law
A $\rightarrow$ Products	Unimolecular	Rate = $k[A]$
A + A $\rightarrow$ Products	Bimolecular	Rate = $k[A]^2$
A + B $\rightarrow$ Products	Bimolecular	Rate = $k[A][B]$
A + A + B $\rightarrow$ Products	Termolecular	Rate = $k[A]^2[B]$
A + B + C $\rightarrow$ Products	Termolecular	Rate = $k[A][B][C]$

The experimentally observed rate law for an *overall* reaction depends on the reaction mechanism—that is, on the sequence of elementary steps and their relative rates. If the overall reaction occurs in a single elementary step, the experimental rate law is the same as the rate law for the elementary reaction. An example is the conversion of bromomethane to methanol in basic solution:

$$CH_3Br(aq) + OH^-(aq) \longrightarrow Br^-(aq) + CH_3OH(aq)$$

Bromomethane   Hydroxide ion   Bromide ion   Methanol

This reaction occurs in a single bimolecular step in which a new C−O bond forms at the same time as the C−Br bond breaks. The experimental rate law is

$$\text{Rate} = -\frac{\Delta[CH_3Br]}{\Delta t} = k[CH_3Br][OH^-]$$

When an overall reaction occurs in two or more elementary steps, one of the steps is often much slower than the others. The slowest step in a reaction mechanism is called the **rate-determining step** because that step acts as a bottleneck, limiting the rate at which reactants can be converted to products. A chemical reaction is somewhat like the cafeteria line in a dining hall in this respect. The rate at which the line moves is determined not by the faster steps, perhaps picking up a salad or a beverage, but by the slowest step, perhaps waiting for a well-done hamburger. The overall reaction can occur no faster than the speed of the rate-determining step.

In the reaction of nitrogen dioxide with carbon monoxide, for example, the first step in the mechanism is slower and rate-determining, whereas the second step occurs more rapidly:

Which is more likely to slow down this line: picking up a beverage or waiting for a hamburger?

$$NO_2(g) + NO_2(g) \xrightarrow{k_1} NO(g) + NO_3(g) \qquad \text{Slower, rate-determining}$$

$$NO_3(g) + CO(g) \xrightarrow{k_2} NO_2(g) + CO_2(g) \qquad \text{Faster}$$

$$\overline{NO_2(g) + CO(g) \longrightarrow NO(g) + CO_2(g)} \qquad \text{Overall reaction}$$

The constants k_1 and k_2, written above the arrows in the preceding equations, are the rate constants for the elementary reactions. The rate of the overall reaction is determined by the rate of the first, slower step. In the second step, the unstable intermediate (NO_3) reacts as soon as it is formed in the first step.

Because the rate law for an overall reaction depends on the reaction mechanism, it provides important clues to the mechanism. A plausible mechanism must meet two criteria: (1) The elementary steps must sum up to give the overall reaction, and (2) the mechanism must be consistent with the observed rate law for the overall reaction.

The procedure that chemists use in establishing a reaction mechanism goes something like this: First, the rate law is determined by experiment. Then, a series of elementary steps is devised, and the rate law predicted by the proposed mechanism is worked out. If the observed and predicted rate laws don't agree, the mechanism must be discarded and another one must be devised. If the observed and predicted rate laws *do* agree, the proposed mechanism is a plausible (though not necessarily correct) pathway for the reaction. Figure 12.11 summarizes the procedure.

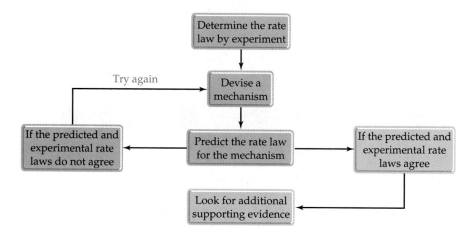

FIGURE 12.11 Flowchart illustrating the logic used in studies of reaction mechanisms.

In the reaction of NO_2 with CO, for example, the experimental rate law is

$$\text{Rate} = -\frac{\Delta[NO_2]}{\Delta t} = k[NO_2]^2$$

The rate law predicted by the proposed mechanism is that for the rate-determining step, where k_1 is the rate constant for that step. Note that the predicted rate law follows directly from the molecularity of the rate-determining step:

$$\text{Rate} = -\frac{\Delta[NO_2]}{\Delta t} = k_1[NO_2]^2$$

Because the observed and predicted rate laws have the same form (second-order dependence on $[NO_2]$), the proposed mechanism is consistent with the experimental rate law. The observed rate constant k is the same as k_1, the rate constant for the first elementary step.

The case for a particular mechanism is strengthened considerably if a reaction intermediate can be isolated or if an unstable intermediate (such as NO_3) can be detected. It's easy to disprove a mechanism, but it's never possible to finally "prove" a mechanism because there may be an alternative reaction pathway, not yet imagined, that also fits the experimental facts. The best that we can do to establish a mechanism is to accumulate a convincing body of experimental evidence that supports it. Proving a reaction mechanism is more like proving a case in a law court than proving a theorem in mathematics.

EXAMPLE 12.10   The following reaction has a second-order rate law:

$$H_2(g) + 2\,ICl(g) \longrightarrow I_2(g) + 2\,HCl(g) \qquad Rate = k[H_2][ICl]$$

Devise a possible reaction mechanism.

SOLUTION   The reaction doesn't occur in a single elementary step because, if it did, the rate law would be third order: rate $= k[H_2][ICl]^2$. The observed rate law will be obtained if the rate-determining step involves the bimolecular reaction of H_2 and ICl. A plausible sequence of elementary steps is

$$H_2(g) + ICl(g) \xrightarrow{\ k_1\ } HI(g) + HCl(g) \qquad \text{Slower, rate-determining}$$

$$\underline{HI(g) + ICl(g) \xrightarrow{\ k_2\ } I_2 + HCl(g) \qquad \text{Faster}}$$

$$H_2(g) + 2\,ICl(g) \longrightarrow I_2(g) + 2\,HCl(g) \qquad \text{Overall reaction}$$

The rate law predicted by this mechanism, rate $= k_1[H_2][ICl]$, is in accord with the observed rate law.

▶ **PROBLEM 12.13**   Write the rate law for each of the following elementary reactions:
(a) $O_3(g) + O(g) \rightarrow 2\,O_2(g)$
(b) $Br(g) + Br(g) + Ar(g) \rightarrow Br_2(g) + Ar(g)$
(c) $Co(CN)_5(H_2O)^{2-}(aq) \rightarrow Co(CN)_5^{2-}(aq) + H_2O(l)$

▶ **PROBLEM 12.14**   The following substitution reaction has a first-order rate law:

$$Co(CN)_5(H_2O)^{2-}(aq) + I^-(aq) \longrightarrow Co(CN)_5I^{3-}(aq) + H_2O(l)$$
$$Rate = k[Co(CN)_5(H_2O)^{2-}]$$

Suggest a possible reaction mechanism, and show that your mechanism is in accord with the observed rate law.

12.9   Reaction Rates and Temperature: The Arrhenius Equation

Everyday experience tells us that the rates of chemical reactions increase with increasing temperature. Familiar fuels such as gas, oil, and coal are relatively inert at room temperature but burn rapidly at elevated temperatures. Many foods last almost indefinitely when stored in a freezer but spoil quickly at room temperature. Metallic magnesium is inert in cold water but reacts with hot water (Figure 12.12). As a rule of thumb, reaction rates tend to double when the temperature is increased by 10°C.

To understand why reaction rates depend on temperature, we need a picture of how reactions take place. According to the **collision theory** model, a bimolecular reaction occurs when two properly oriented reactant molecules come together in a sufficiently energetic collision. To be specific, let's consider one of the simplest possible reactions, the reaction of an atom A with a diatomic molecule BC to give a diatomic molecule AB and an atom C:

$$A + BC \longrightarrow AB + C$$

FIGURE 12.12   Magnesium is inert in cold water (left) but reacts with hot water (right). Evidence of reaction is the pink color of phenolphthalein, which indicates formation of an alkaline solution. The reaction is $Mg(s) + 2\,H_2O(l) \rightarrow Mg^{2+}(aq) + 2\,OH^-(aq) + H_2(g)$.

An example from atmospheric chemistry is the reaction of an oxygen atom with an HCl molecule to give an OH molecule and a chlorine atom:

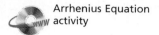
Arrhenius Equation
activity

$$O(g) + HCl(g) \longrightarrow OH(g) + Cl(g)$$

If the reaction occurs in a single step, the electron distribution about the three nuclei must change in the course of the collision such that a new bond, A–B, develops at the same time the old bond, B–C, breaks. Between the reactant and product stages, the nuclei pass through a configuration in which all three atoms are weakly linked together. We can picture the progress of the reaction as

$$A + B–C \longrightarrow A\text{-}\text{-}\text{-}B\text{-}\text{-}\text{-}C \longrightarrow A–B + C$$

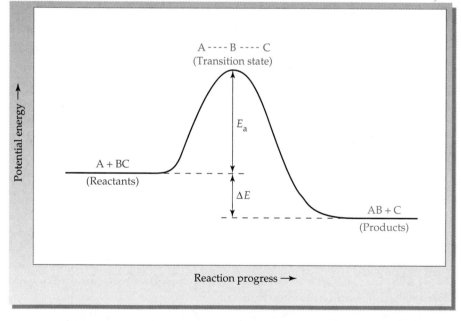

If A and BC have filled shells of electrons (no unpaired electrons or vacant, low-energy orbitals), they will repel each other. To achieve the configuration A---B---C, then, the atoms require energy to overcome this repulsion. The energy comes from the kinetic energy of the colliding particles and is converted to potential energy in A---B---C. In fact, A---B---C has more potential energy than either the reactants or the products. Thus, there is a potential energy barrier that must be surmounted before reactants can be converted to products, as depicted graphically on the potential energy profile in Figure 12.13.

FIGURE 12.13 Potential energy profile for the reaction A + BC → AB + C, showing the energy barrier between the reactants and products. As the reaction progresses, kinetic energy of the reactants is first converted into potential energy of the transition state and is then transformed into kinetic energy of the products. At each point along the profile, the *total* energy is conserved. The profile is drawn for an exothermic reaction, so ΔE, the energy of reaction, is negative.

The height of the barrier is called the **activation energy**, E_a, and the configuration of atoms at the maximum in the potential energy profile is called the **transition state**, or the *activated complex*. Since energy is conserved in the collision, all the energy needed to climb the potential energy hill must come from the kinetic energy of the colliding molecules. If the collision energy is less than E_a, the reactant molecules can't surmount the barrier and they simply bounce apart. If the collision energy is at least as great as E_a, however, the reactants can climb over the barrier and be converted to products.

Experimental evidence for the notion of an activation energy barrier comes from a comparison of collision rates and reaction rates. Collision rates in gases can be calculated from kinetic–molecular theory (Section 9.6). For a gas at room temperature (298 K) and 1 atm pressure, each molecule undergoes approximately 10^9 collisions per second, or 1 collision every 10^{-9} s. Thus, if every collision resulted in reaction, every gas-phase reaction would be complete in about 10^{-9} s. By contrast, observed reactions often have half-lives of minutes or hours, so it's clear that only a tiny fraction of the collisions lead to reaction.

Very few collisions are productive because very few occur with a collision energy as large as the activation energy. The fraction of collisions with an energy equal to or greater than the activation energy E_a is represented in Figure 12.14 at two different temperatures by the areas under the curves to the right of E_a. When E_a is large compared with RT, this fraction f is approximated by the equation

$$f = e^{-E_a/RT}$$

where R is the gas constant [8.314 J/(K · mol)] and T is the absolute temperature in kelvins. Note that f is a very small number. For example, for a reaction having an activation energy of 75 kJ/mol, the value of f at 298 K is 7×10^{-14}:

$$f = \exp\left[\dfrac{-75{,}000\ \dfrac{\text{J}}{\text{mol}}}{\left(8.314\ \dfrac{\text{J}}{\text{K} \cdot \text{mol}}\right)(298\ \text{K})} \right] = e^{-30.3} = 7 \times 10^{-14}$$

Thus, only *7 collisions in 100 trillion* are sufficiently energetic to convert reactants to products.

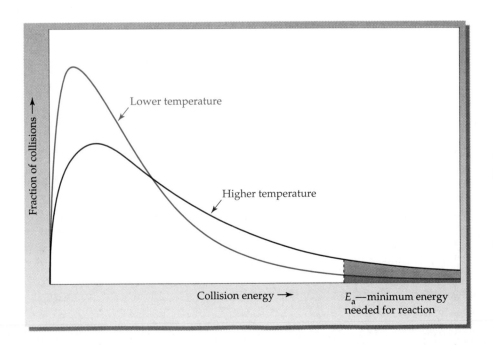

FIGURE 12.14 Plots of the fraction of collisions with a particular energy at two different temperatures. For each plot, the total area under the curve is unity, and the area to the right of E_a is the fraction f of the collisions with an energy greater than or equal to E_a. The fraction of collisions that are sufficiently energetic to result in reaction increases rapidly with increasing temperature.

As the temperature increases, the distribution of collision energies broadens and shifts to higher energies (Figure 12.14), resulting in a rapid increase in the fraction of collisions that lead to products. At 308 K, for example, the calculated value of f for the reaction with $E_a = 75$ kJ/mol is 2×10^{-13}. Thus, a temperature increase of just 3%, from 298 K to 308 K, increases the value of f by a factor of 3. Collision theory therefore accounts nicely for the exponential dependence of reaction rates on reciprocal temperature. As T increases ($1/T$ decreases), $f = e^{-E_a/RT}$ increases exponentially. Collision theory also explains why reaction rates are so much lower than collision rates. (Collision rates also increase with increasing temperature, but only by a small amount—less than 2% on going from 298 K to 308 K.)

The fraction of collisions that lead to products is further reduced by an orientation requirement. Even if the reactants collide with sufficient energy, they won't react unless the orientation of the reaction partners is correct for formation of the transition state. For example, a collision of A with the C end of the molecule BC can't result in formation of AB:

$$A + C\text{-}B \longrightarrow A\text{---}C\text{---}B \;\not\longrightarrow\; A\text{-}B + C$$

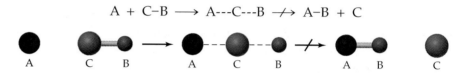

The reactant molecules would simply collide and then separate without reaction:

$$A + C\text{-}B \longrightarrow A\text{---}C\text{---}B \longrightarrow A + C\text{-}B$$

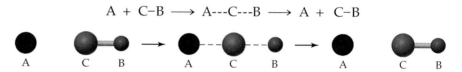

The fraction of collisions having proper orientation for conversion of reactants to products is called the **steric factor**, p. For the reaction, A + BC → AB + C, the value of p is expected to be about 0.5 because A has a nearly 50:50 probability of colliding with each of the B and C ends of BC. (This assumes that B and C have similar sizes and electronic properties.) For reactions of complex molecules, p is a fraction considerably less than 0.5.

Now let's see how the two parameters p and f come into the rate law. Since bimolecular collisions between any two molecules—say, A and B—occur at a rate that is proportional to their concentrations, we can write

$$\text{Collision rate} = Z[A][B]$$

where Z is a constant related to the collision frequency and has units of a second-order rate constant, $1/(\text{M} \cdot \text{s})$ or $\text{M}^{-1}\text{s}^{-1}$. The reaction rate is lower than the collision rate by a factor $p \times f$ because only a fraction of the colliding molecules have the correct orientation and the minimum energy needed for reaction:

$$\text{Reaction rate} = p \times f \times \text{Collision rate} = pfZ[A][B]$$

Since the rate law is

$$\text{Reaction rate} = k[A][B]$$

the rate constant predicted by collision theory is $k = pfZ$, or $k = pZe^{-E_a/RT}$.

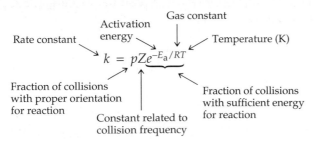

This expression is usually written in a form called the **Arrhenius equation**, named after Svante Arrhenius, the Swedish chemist who proposed it in 1889 on the basis of experimental studies of reaction rates:

Arrhenius equation: $k = Ae^{-E_a/RT}$

The parameter $A (= pZ)$ is called the **frequency factor** (or pre-exponential factor). In accord with the minus sign in the exponent, the rate constant decreases as E_a increases and increases as T increases.

⊷ KEY CONCEPT PROBLEM 12.15 The potential energy profile for the one-step reaction $AB + CD \rightarrow AC + BD$ is shown below. The energies are in kJ/mol relative to an arbitrary zero of energy.

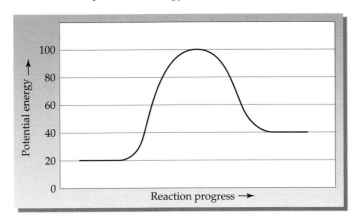

(a) What is the value of the activation energy for this reaction?
(b) Is the reaction exothermic or endothermic?
(c) Suggest a plausible structure for the transition state.

12.10   Using the Arrhenius Equation

As we saw in the previous section, the activation energy E_a is one of the most important factors affecting the rate of a chemical reaction. Its value can be determined using the Arrhenius equation if values of the rate constant are known at different temperatures. Taking the natural logarithm of both sides of the Arrhenius equation, we obtain the logarithmic form

$$\ln k = \ln A - \frac{E_a}{RT}$$

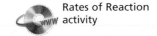 Rates of Reaction activity

This equation can be rearranged into the form $y = mx + b$, so a graph of ln k versus $1/T$ (called an Arrhenius plot) gives a straight line with slope $m = -E_a/R$ and intercept $b = \ln A$:

$$\underset{y}{\ln k} = \underset{m}{\left(\frac{-E_a}{R}\right)}\underset{x}{\left(\frac{1}{T}\right)} + \underset{b}{\ln A}$$

The experimental value of the activation energy is easily determined from the slope of the straight line, as shown in Example 12.11.

$$E_a = -R(\text{Slope})$$

Still another form of the Arrhenius equation can be derived that allows us to estimate the activation energy from rate constants at just two temperatures. At temperature T_1,

$$\ln k_1 = \left(\frac{-E_a}{R}\right)\left(\frac{1}{T_1}\right) + \ln A$$

and at temperature T_2,

$$\ln k_2 = \left(\frac{-E_a}{R}\right)\left(\frac{1}{T_2}\right) + \ln A$$

Subtracting the first equation from the second, and remembering that $(\ln k_2 - \ln k_1) = \ln (k_2/k_1)$, we obtain

$$\ln\left(\frac{k_2}{k_1}\right) = \left(\frac{-E_a}{R}\right)\left(\frac{1}{T_2} - \frac{1}{T_1}\right)$$

This equation can be used to calculate E_a from rate constants k_1 and k_2 at temperatures T_1 and T_2. By the same token, if we know E_a and the rate constant k_1 at one temperature T_1, we can calculate the rate constant k_2 at another temperature T_2. Example 12.11 shows how this is done.

EXAMPLE 12.11   Rate constants for the gas-phase decomposition of hydrogen iodide, $2\,HI(g) \rightarrow H_2(g) + I_2(g)$, are listed in the following table:

Temperature (°C)	k [1/(M · s)]	Temperature (°C)	k [1/(M · s)]
283	3.52×10^{-7}	427	1.16×10^{-3}
356	3.02×10^{-5}	508	3.95×10^{-2}
393	2.19×10^{-4}		

(a) Find the activation energy (in kJ/mol) using all five data points.
(b) Calculate E_a from the rate constants at 283°C and 508°C.
(c) Given the rate constant at 283°C and the value of E_a obtained in part (b), what is the rate constant at 293°C?

SOLUTION

(a) The activation energy E_a can be determined from the slope of a linear plot of $\ln k$ versus $1/T$. Since the temperature in the Arrhenius equation is expressed in kelvins, we must first convert the Celsius temperatures to absolute temperatures. Then calculate values of $1/T$ and $\ln k$, and plot $\ln k$ versus $1/T$. The results are shown in the following table and graph:

t (°C)	T (K)	k [1/(M · s)]	$1/T$ (1/K)	$\ln k$
283	556	3.52×10^{-7}	0.001 80	−14.860
356	629	3.02×10^{-5}	0.001 59	−10.408
393	666	2.19×10^{-4}	0.001 50	−8.426
427	700	1.16×10^{-3}	0.001 43	−6.759
508	781	3.95×10^{-2}	0.001 28	−3.231

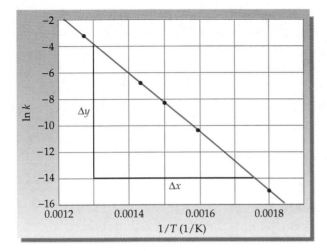

The slope of the straight-line plot can be determined from the coordinates of two widely separated points:

$$\text{Slope} = \frac{\Delta y}{\Delta x} = \frac{(-14.0) - (-3.9)}{(0.001\ 75\ \text{K}^{-1}) - (0.001\ 30\ \text{K}^{-1})} = \frac{-10.1}{0.000\ 45\ \text{K}^{-1}}$$

$$= -2.24 \times 10^4\ \text{K}$$

Finally, calculate the activation energy from the slope:

$$E_a = -R(\text{Slope}) = -\left(8.314\ \frac{\text{J}}{\text{K} \cdot \text{mol}}\right)(-2.24 \times 10^4\ \text{K})$$

$$= 1.9 \times 10^5\ \text{J/mol} = 190\ \text{kJ/mol}$$

Note that the slope of the Arrhenius plot is negative and the activation energy is positive. The greater the activation energy for a particular reaction, the steeper the slope of the $\ln k$ versus $1/T$ plot and the greater the increase in the rate constant for a given increase in temperature.

(b) To calculate E_a from values of the rate constant at two temperatures, use the equation

$$\ln\left(\frac{k_2}{k_1}\right) = \left(\frac{-E_a}{R}\right)\left(\frac{1}{T_2} - \frac{1}{T_1}\right)$$

Substituting the values of $k_1 = 3.52 \times 10^{-7}\,\text{M}^{-1}\text{s}^{-1}$ at $T_1 = 556\,\text{K}$ (283°C) and $k_2 = 3.95 \times 10^{-2}\,\text{M}^{-1}\text{s}^{-1}$ at $T_2 = 781\,\text{K}$ (508°C) gives

$$\ln\left(\frac{3.95 \times 10^{-2}\,\text{M}^{-1}\text{s}^{-1}}{3.52 \times 10^{-7}\,\text{M}^{-1}\text{s}^{-1}}\right) = \left(\frac{-E_a}{8.314\,\dfrac{\text{J}}{\text{K} \cdot \text{mol}}}\right)\left(\frac{1}{781\,\text{K}} - \frac{1}{556\,\text{K}}\right)$$

Simplifying this equation gives

$$11.628 = \left(\frac{-E_a}{8.314\,\dfrac{\text{J}}{\text{K} \cdot \text{mol}}}\right)\left(\frac{-5.18 \times 10^{-4}}{\text{K}}\right)$$

$$E_a = 1.87 \times 10^5\,\text{J/mol} = 187\,\text{kJ/mol}$$

(c) Use the same equation as in part (b), but now the known values are $k_1 = 3.52 \times 10^{-7}\,\text{M}^{-1}\text{s}^{-1}$ at $T_1 = 556\,\text{K}$ (283°C) and $E_a = 1.87 \times 10^5\,\text{J/mol}$, and k_2 at $T_2 = 566\,\text{K}$ (293°C) is the unknown:

$$\ln\left(\frac{k_2}{3.52 \times 10^{-7}\,\text{M}^{-1}\text{s}^{-1}}\right) = \left(\frac{-1.87 \times 10^5\,\dfrac{\text{J}}{\text{mol}}}{8.314\,\dfrac{\text{J}}{\text{K} \cdot \text{mol}}}\right)\left(\frac{1}{566\,\text{K}} - \frac{1}{556\,\text{K}}\right) = 0.715$$

Taking the antiln of both sides gives

$$\frac{k_2}{3.52 \times 10^{-7}\,\text{M}^{-1}\text{s}^{-1}} = e^{0.715} = 2.04$$

$$k_2 = 7.18 \times 10^{-7}\,\text{M}^{-1}\text{s}^{-1}$$

In this temperature range, a rise in temperature of 10 K doubles the rate constant.

▶ **PROBLEM 12.16** Rate constants for decomposition of gaseous dinitrogen pentoxide are $3.7 \times 10^{-5}\,\text{s}^{-1}$ at 25°C and $1.7 \times 10^{-3}\,\text{s}^{-1}$ at 55°C.

$$2\,\text{N}_2\text{O}_5(g) \longrightarrow 4\,\text{NO}_2(g) + \text{O}_2(g)$$

(a) What is the activation energy for this reaction in kJ/mol?

(b) What is the rate constant at 35°C?

12.11 Catalysis

Reaction rates are affected not only by reactant concentrations and temperature but also by the presence of *catalysts*. A **catalyst** is a substance that increases the rate of a reaction without being consumed in the reaction. An example is manganese dioxide, a black powder that speeds up the thermal decomposition of potassium chlorate:

$$2\,\text{KClO}_3(s) \xrightarrow[\text{Heat}]{\text{MnO}_2\text{ catalyst}} 2\,\text{KCl}(s) + 3\,\text{O}_2(g)$$

In the absence of a catalyst, $KClO_3$ decomposes very slowly, even when heated, but when a small amount of MnO_2 is mixed with the $KClO_3$ before heating, rapid evolution of oxygen ensues. The MnO_2 can be recovered unchanged after the reaction is complete.

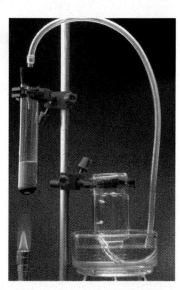

MnO_2 catalyzes the thermal decomposition of $KClO_3$ to KCl and O_2 gas.

Catalysts are of enormous importance, both in the chemical industry and in living organisms. Nearly all industrial processes for the manufacture of essential chemicals use catalysts to favor formation of specific products and to lower reaction temperatures, thus reducing energy costs. In environmental chemistry, catalysts such as nitric oxide play a role in the formation of air pollutants, while other catalysts, such as platinum in automobile catalytic converters, are potent weapons in the battle to control air pollution.

In living organisms, chemical reactions are catalyzed by large molecules called *enzymes*, which facilitate specific reactions of crucial biological importance (see Section 24.6). For example, nitrogenase, an enzyme present in bacteria on the root nodules of leguminous plants such as peas and beans, catalyzes the conversion of atmospheric nitrogen to ammonia. The ammonia serves as a nitrogen fertilizer for plant growth. In the human body, the enzyme carbonic anhydrase catalyzes the reaction of carbon dioxide with water:

$$CO_2(g) + H_2O(l) \rightleftharpoons H^+(aq) + HCO_3^-(aq)$$

The forward reaction occurs when the blood takes up CO_2 in the tissues, and the reverse reaction occurs when the blood releases CO_2 in the lungs. Remarkably, carbonic anhydrase increases the rate of these reactions by a factor of about 10^6.

How does a catalyst work? A catalyst accelerates the rate of a reaction by making available a new and more efficient mechanism for conversion of reactants to products. Let's consider the decomposition of hydrogen peroxide in a basic, aqueous solution:

$$2\,H_2O_2(aq) \longrightarrow 2\,H_2O(l) + O_2(g)$$

Although unstable with respect to water and oxygen, hydrogen peroxide decomposes only very slowly at room temperature because the reaction has a high activation energy (76 kJ/mol). In the presence of iodide ion, however, the reaction is appreciably faster (Figure 12.15) because it can proceed by a different, lower-energy pathway:

Step 1. $H_2O_2(aq) + I^-(aq) \longrightarrow H_2O(l) + IO^-(aq)$ Slower, rate-determining

Step 2. $H_2O_2(aq) + IO^-(aq) \longrightarrow H_2O(l) + O_2(g) + I^-(aq)$ Faster

_____

$2\,H_2O_2(aq) \longrightarrow 2\,H_2O(l) + O_2(g)$ Overall reaction

The H_2O_2 first oxidizes the catalyst (I^-) to hypoiodite ion (IO^-) and then reduces the intermediate IO^- back to I^-. Note that the catalyst does not appear in the overall reaction because it is consumed in one step and regenerated in a later step. The catalyst is, however, intimately involved in the reaction, as is evident from the presence of I^- in the observed rate law:

$$\text{Rate} = k[H_2O_2][I^-]$$

The rate law is consistent with the reaction of H_2O_2 and I^- as the rate-determining step.

(a)

(b)

FIGURE 12.15 The rate of decomposition of aqueous hydrogen peroxide can be monitored qualitatively by collecting the evolved oxygen gas in a balloon. **(a)** In the absence of a catalyst, little O_2 is produced. **(b)** After addition of aqueous sodium iodide, the balloon rapidly inflates with O_2.

The catalyzed pathway for a reaction might have a faster rate than the uncatalyzed pathway either because of a larger frequency factor A or a smaller activation energy E_a in the Arrhenius equation. Usually, however, catalysts function by making available a reaction pathway with a lower activation energy (Figure 12.16). In the decomposition of hydrogen peroxide, for example, catalysis by I^- lowers E_a for the overall reaction by 19 kJ/mol.

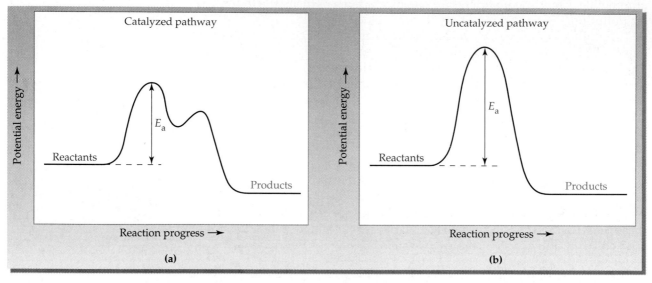

FIGURE 12.16 Typical potential energy profiles for a reaction whose activation energy is lowered by the presence of a catalyst: **(a)** the catalyzed pathway; **(b)** the uncatalyzed pathway. The shape of the barrier for the catalyzed pathway describes the decomposition of H_2O_2: The first of the two maxima is higher because the first step is rate-determining.

 Catalysis movie

Because the reaction occurs in two steps, the energy profile in Figure 12.16 exhibits two maxima (two transition states), with a minimum in between that represents the energy of the intermediate species present after the first step. The first maximum is higher than the second because the first step is rate-determining, and the activation energy for the overall reaction is E_a for the first step. Maxima for both steps, though, are lower than the top of the barrier for the uncatalyzed pathway. Note that a catalyst does not affect the energies of the reactants and products.

◆— **KEY CONCEPT PROBLEM 12.17**   The relative rates of the reaction A + B → AB in vessels (a)–(d) are 1:2:1:2. Red spheres represent A molecules, blue spheres represent B molecules, and yellow spheres represent molecules of a third substance C.

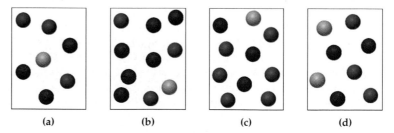

(a)   (b)   (c)   (d)

(a) What is the order of the reaction in A, B, and C?
(b) Write the rate law.
(c) Write a mechanism that agrees with the rate law.
(d) Why doesn't C appear in the equation for the overall reaction?

12.12   Homogeneous and Heterogeneous Catalysts

Catalysts are commonly classified as either *homogeneous* or *heterogeneous*. A **homogeneous catalyst** is one that exists in the same phase as the reactants. For example, iodide ion is a homogeneous catalyst for the decomposition of aqueous hydrogen peroxide because both I^- and H_2O_2 are present in the same aqueous solution phase.

In the atmosphere, nitric oxide is a gas-phase homogeneous catalyst for the conversion of molecular oxygen to ozone, a process described by the following series of reactions:

$$1/2\,O_2(g) + NO(g) \longrightarrow NO_2(g)$$

$$NO_2(g) \xrightarrow{\text{Sunlight}} NO(g) + O(g)$$

$$\underline{O(g) + O_2(g) \longrightarrow O_3(g)}$$

$$3/2\,O_2(g) \longrightarrow O_3(g) \qquad \text{Overall reaction}$$

Nitric oxide first reacts with atmospheric O_2 to give nitrogen dioxide, a poisonous brown gas. Subsequent absorption of sunlight by NO_2 results in dissociation to give an oxygen atom, which then reacts with O_2 to form ozone. As usual, the catalyst (NO) and the intermediates (NO_2 and O) do not appear in the chemical equation for the overall reaction.

A **heterogeneous catalyst** is one that exists in a different phase from that of the reactants. Ordinarily, the catalyst is a solid, and the reactants are either gases or liquids. For example, in the Fischer–Tropsch process for manufacturing synthetic gasoline, tiny particles of a metal such as iron or cobalt coated on alumina (Al_2O_3) catalyze the conversion of gaseous carbon monoxide and hydrogen to hydrocarbons such as octane (C_8H_{18}):

$$8\,CO(g) + 17\,H_2(g) \xrightarrow[\text{catalyst}]{Co/Al_2O_3} C_8H_{18}(l) + 8\,H_2O(l)$$

The mechanism of heterogeneous catalysis is often complex and not well understood. Important steps, however, involve (1) adsorption of reactants onto the surface of the catalyst, (2) conversion of reactants to products on the surface, and (3) desorption of products from the surface. The adsorption step is thought to involve chemical bonding of reactants to the highly reactive metal atoms on the surface with accompanying breaking, or at least weakening, of bonds in the reactants.

To illustrate, consider the catalytic hydrogenation of compounds with C=C double bonds, a reaction used in the food industry to convert unsaturated vegetable oils to solid fats. The simplest reaction of this type is the hydrogenation of ethylene, which is catalyzed by finely divided metals such as Ni, Pd, or Pt:

$$\underset{\text{Ethylene}}{H_2C = CH_2(g)} + H_2(g) \xrightarrow[\text{catalyst}]{\text{Metal}} \underset{\text{Ethane}}{H_3C - CH_3(g)}$$

As shown in Figure 12.17, the function of the metal surface is to adsorb the reactants and facilitate the rate-determining step, breaking the strong H–H bond in the H_2 molecule. Because the H–H bond breaking is accompanied by simultaneous formation of bonds from the separating H atoms to the surface metal atoms, the activation energy for the process is lowered. The H atoms

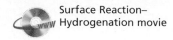

Surface Reaction–
Hydrogenation movie

then move about on the surface until they encounter the C atoms of the adsorbed C_2H_4 molecule. Subsequent stepwise formation of two new C–H bonds gives C_2H_6, which is finally desorbed from the surface.

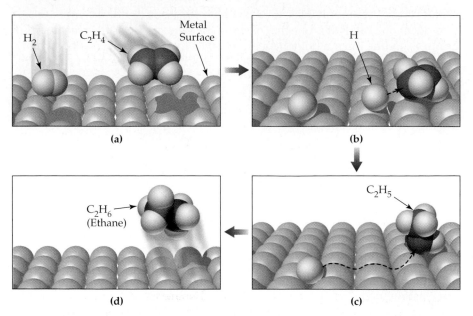

FIGURE 12.17 Proposed mechanism for the catalytic hydrogenation of ethylene (C_2H_4) on a metal surface. **(a)** H_2 and C_2H_4 are adsorbed on the metal surface. **(b)** The H–H bond breaks as H–metal bonds form, and the H atoms move about on the surface. **(c)** One H atom forms a bond to a C atom of the adsorbed C_2H_4 to give a metal-bonded C_2H_5 group. **(d)** A second H atom bonds to the C_2H_5 group, and the resulting C_2H_6 molecule is desorbed from the surface.

Most of the catalysts used in industrial chemical processes are heterogeneous, in part because of the ease with which such catalysts can be separated from the reaction products. Table 12.6 lists some examples of commercial processes that employ heterogeneous catalysts.

TABLE 12.6   Some Heterogeneous Catalysts Used in Commercially Important Reactions

Reaction	Catalyst	Commercial process	End product: Commercial uses
$2\,SO_2 + O_2 \rightarrow 2\,SO_3$	Pt or V_2O_5	Intermediate step in the contact process for synthesis of sulfuric acid	H_2SO_4: Manufacture of fertilizers, chemicals; oil refining
$4\,NH_3 + 5\,O_2 \rightarrow 4\,NO + 6\,H_2O$	Pt and Rh	First step in the Ostwald process for synthesis of nitric acid	HNO_3: Manufacture of explosives, fertilizers, plastics, dyes, lacquers
$N_2 + 3\,H_2 \rightarrow 2\,NH_3$	Fe, K_2O, and Al_2O_3	Haber process for synthesis of ammonia	NH_3: Manufacture of fertilizers, nitric acid
$H_2O + CH_4 \rightarrow CO + 3\,H_2$	Ni	Steam–hydrocarbon re-forming process for synthesis of hydrogen	H_2: Manufacture of ammonia, methanol
$CO + H_2O \rightarrow CO_2 + H_2$	ZnO and CuO	Gas shift reaction to improve yield in the synthesis of H_2	H_2: Manufacture of ammonia, methanol
$CO + 2\,H_2 \rightarrow CH_3OH$	ZnO and Cr_2O_3	Industrial synthesis of methanol	CH_3OH: Manufacture of plastics, adhesives, gasoline additives; industrial solvent
$\underset{/}{\overset{\backslash}{C}}\!\!=\!\!\underset{\backslash}{\overset{/}{C}} + H_2 \longrightarrow \underset{/|}{\overset{H\backslash}{C}}\!-\!\underset{|\backslash}{\overset{/H}{C}}$	Ni, Pd, or Pt	Catalytic hydrogenation of compounds with C=C bonds, as in conversion of unsaturated vegetable oils to solid fats	Food products: margarine, shortening

Another important application of heterogeneous catalysts is in automobile catalytic converters. Despite much work on engine design and fuel composition, automotive exhaust emissions contain air pollutants such as unburned hydrocarbons (C_xH_y), carbon monoxide, and nitric oxide. Carbon monoxide results from incomplete combustion of hydrocarbon fuels, and nitric oxide is produced when atmospheric nitrogen and oxygen combine at the high temperatures present in an automobile engine. Catalytic converters promote conversion of the offending pollutants to carbon dioxide, water, nitrogen, and oxygen (Figure 12.18):

$$C_xH_y(g) + (x + y/4)\,O_2(g) \longrightarrow x\,CO_2(g) + (y/2)\,H_2O(g)$$

$$2\,CO(g) + O_2(g) \longrightarrow 2\,CO_2(g)$$

$$2\,NO(g) \longrightarrow N_2(g) + O_2(g)$$

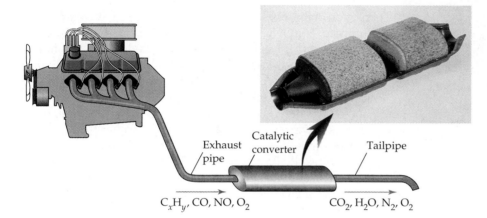

Exhaust pipe

Catalytic converter

Tailpipe

C_xH_y, CO, NO, O_2

CO_2, H_2O, N_2, O_2

FIGURE 12.18 The gases exhausted from an automobile engine pass through a catalytic converter where air pollutants such as unburned hydrocarbons (C_xH_y), CO, and NO are converted to CO_2, H_2O, N_2, and O_2. The photo shows a cutaway view of a catalytic converter. The beads are impregnated with the heterogeneous catalyst.

Typical catalysts for these reactions are the so-called noble metals Pt and Pd, and transition metal oxides V_2O_5, Cr_2O_3, and CuO. Because the surface of the catalyst is rendered ineffective ("poisoned") by adsorption of lead, automobiles with catalytic converters must use unleaded gasoline.

Explosives

Some chemical reactions are slow and some are fast, but *explosions* are in a class by themselves. Chemical explosions are characterized by the nearly instantaneous release of large quantities of hot gases such as N_2, CO_2, and H_2O. The rapidly expanding gases set up a shock wave of enormous pressure—up to 700,000 atm—that propagates through the surroundings at speeds of up to 9000 m/s (20,000 mi/h), causing the physical destruction we've all seen in movies and television newscasts.

Explosives are generally categorized as either *primary* or *secondary*, depending on their sensitivity to shock. Primary explosives are the most sensitive to heat and shock. They are generally used in detonators, blasting caps, and military fuses to initiate the explosion of the less-sensitive secondary explosives. Mercury(II) fulminate, $Hg(ONC)_2$, was the first initiator to be used commercially, but it has been largely replaced by lead(II) azide, $Pb(N_3)_2$, which is more stable when stored under hot conditions.

Secondary explosives, or *high explosives*, are generally less sensitive to heat and shock than primary explosives and are therefore safer to manufacture, transport, and handle. Most secondary explosives will simply burn rather than explode when ignited in air, and most can be detonated only by the nearby explosion of a primary initiator. Among the most common secondary explosives are nitroglycerin, trinitrotoluene (TNT), pentaerythritol tetranitrate (PETN), and RDX.

The high-pressure shock wave caused by a chemical explosion is powerful enough to demolish this building.

As is clear from looking at their structures, most explosives are rich in oxygen and nitrogen, and most contain *nitro groups*, $-NO_2$. Because the chemical bonds in nitro groups are relatively weak (about 200 kJ/mol), and because the explosion products (CO_2, N_2, H_2O, and others) are extremely stable, a great deal of energy is released within a few microseconds during an explosion. One mole (227 g) of nitroglycerin, for example, releases 1427 kJ when it explodes. The actual mix of reaction products is complex, but the reaction can be approximated by the balanced equation

$$4\,C_3H_5N_3O_9(l) \longrightarrow 12\,CO_2(g) + 10\,H_2O(g) + 6\,N_2(g) + O_2(g)$$

Note that 29 mol of gaseous products with a volume of 650 L at STP (Section 9.3) are produced from just 4 mol of liquid nitroglycerin.

The first commercially important high explosive was nitroglycerin, prepared in 1847 by reaction of glycerin with nitric acid in the presence of sulfuric acid:

$$\begin{array}{c}
\text{H} \\
| \\
\text{H—C—OH} \\
| \\
\text{H—C—OH} \\
| \\
\text{H—C—OH} \\
| \\
\text{H}
\end{array}
\quad + \; 3\,\text{HONO}_2 \quad \xrightarrow{\text{H}_2\text{SO}_4} \quad
\begin{array}{c}
\text{H} \\
| \\
\text{H—C—ONO}_2 \\
| \\
\text{H—C—ONO}_2 \\
| \\
\text{H—C—ONO}_2 \\
| \\
\text{H}
\end{array}
\quad + \; 3\,\text{H}_2\text{O}$$

Glycerin   Nitroglycerin

As you might expect, the reaction is extremely hazardous to carry out. In fact, it wasn't until 1865 that the Swedish chemist and industrialist Alfred Nobel succeeded in finding a relatively safe method of producing nitroglycerin and of incorporating it into a reliable commercial blasting product known as *dynamite*. (Nobel's discovery resulted in a large personal fortune, which he used to establish the Nobel Prizes.) Modern industrial dynamite used for quarrying stone and blasting roadbeds is a mixture of ammonium nitrate and nitroglycerin absorbed onto diatomaceous earth.

Military explosives are generally used as fillings for bombs or shells and must therefore have a very low sensitivity to impact shock on firing. In addition, they must have good stability for long-term storage under adverse conditions, and they should have a low mass-to-energy ratio. TNT and RDX are the most commonly used military high explosives. PETN and RDX are also often compounded with waxes or synthetic polymers to make so-called plastic explosives.

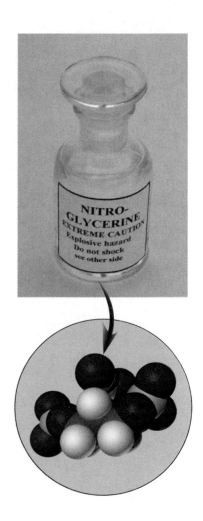

▶ **PROBLEM 12.18** The explosive decomposition of nitroglycerin is highly exothermic. Explain.

▶ **PROBLEM 12.19** Would you expect a higher activation energy for the decomposition of a primary or a secondary explosive? Explain.

▶ **PROBLEM 12.20** The reaction that occurs on detonation of pentaerythritol tetranitrate (PETN) can be approximated by the equation

$$\text{C}_5\text{H}_8\text{N}_4\text{O}_{12}(s) \longrightarrow 4\,\text{CO}_2(g) + 4\,\text{H}_2\text{O}(g) + 2\,\text{N}_2(g) + \text{C}(s)$$

Use the standard heats of formation given in Appendix B to calculate the standard heat of reaction in kilojoules. For PETN, $\Delta H°_f = 537$ kJ/mol. (For a review of $\Delta H°_f$, see Section 8.10.)

▶ **PROBLEM 12.21** How many liters of gas are produced from the decomposition of 1.54 kg of PETN according to the equation in Problem 12.20? Assume that the gas has a temperature of 800°C and a pressure of 0.975 atm.

Key Words

Summary

Chemical kinetics is the area of chemistry concerned with reaction rates. A **reaction rate** is defined as the increase in the concentration of a product, or the decrease in the concentration of a reactant, per unit time. It can be expressed as the average rate during a given time interval, the **instantaneous rate** at a particular time, or the **initial rate** at the beginning of the reaction.

Reaction rates depend on reactant concentrations, temperature, and the presence of catalysts. The concentration dependence is given by the **rate law**, rate = $k[A]^m[B]^n$, where k is the **rate constant**, m and n specify the **reaction order** with respect to reactants A and B, and $m + n$ is the overall reaction order. The values of m and n must be determined by experiment; they can't be deduced from the stoichiometry of the overall reaction.

The **integrated rate law** is a concentration–time equation that allows us to calculate concentrations at any time t or the time required for an initial concentration to reach any particular value. For a **first-order reaction**, the integrated rate law is $\ln [A]_t = -kt + \ln [A]_0$. A graph of $\ln [A]$ versus time is a straight line with a slope equal to $-k$. For a **second-order reaction**, the integrated rate law is $1/[A]_t = kt + 1/[A]_0$. A graph of $1/[A]$ versus time is linear with a slope equal to k. The **half-life** ($t_{1/2}$) of a reaction is the time required for the reactant concentration to drop to one-half its initial value.

A **reaction mechanism** is the sequence of **elementary reactions**, or **elementary steps**, that defines the pathway from reactants to products. Elementary reactions are classified as **unimolecular**, **bimolecular**, or **termolecular**, depending on the number of reactant molecules. The rate law for an elementary reaction follows directly from its **molecularity**: rate = $k[A]$ for a unimolecular reaction, and rate = $k[A]^2$ or rate = $k[A][B]$ for a bimolecular reaction. The observed rate law for an overall reaction depends on the sequence of elementary steps and their relative rates. The slowest step in a reaction mechanism is called the **rate-determining step**. A chemical species that is formed in one elementary step and consumed in a subsequent step is termed a **reaction intermediate**. An acceptable mechanism must meet two criteria: (1) The elementary steps must sum to give the overall reaction, and (2) the mechanism must be consistent with the observed rate law.

The temperature dependence of rate constants is described by the **Arrhenius equation**, $k = Ae^{-E_a/RT}$, where A is the **frequency factor** and E_a is the **activation energy**. The value of E_a can be determined from the slope of a linear plot of $\ln k$ versus $1/T$, and it can be interpreted as the height of the potential energy barrier between reactants and products. The configuration of atoms at the top of the barrier is called the **transition state**. According to **collision theory**, the rate constant is given by $k = pZe^{-E_a/RT}$, where p is a **steric factor** (the fraction of collisions in which the molecules have the proper orientation for reaction), Z is a constant related to the collision frequency, and $e^{-E_a/RT}$ is the fraction of collisions with energy equal to or greater than E_a.

A **catalyst** is a substance that increases the rate of a reaction without being consumed in the reaction. It functions by making available an alternative reaction pathway that has a lower activation energy. A **homogeneous catalyst** is present in the same phase as the reactants, whereas a **heterogeneous catalyst** is present in a different phase.

Key Concept Summary

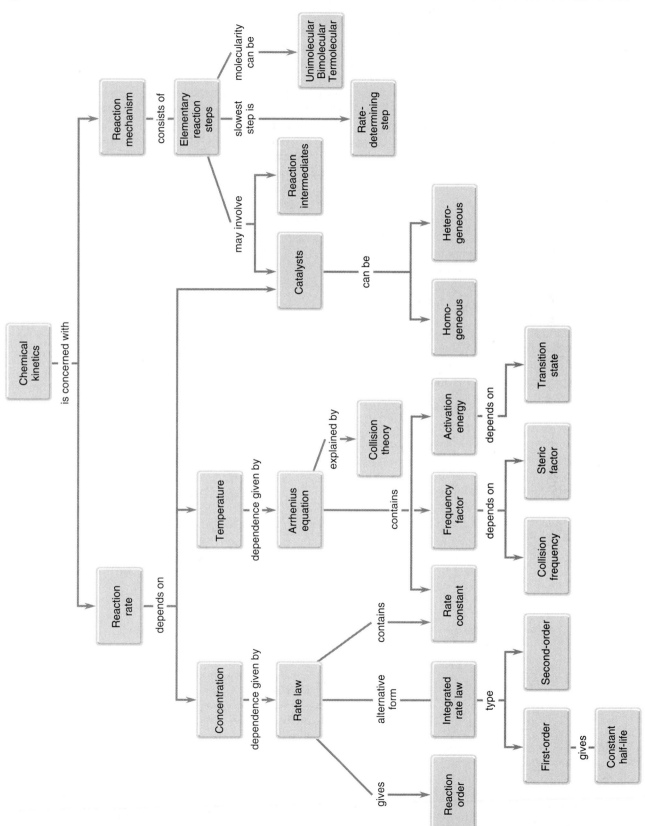

Understanding Key Concepts

Problems 12.1–12.21 appear within the chapter.

12.22 The following reaction is first order in A and first order in B:

$$A + B \longrightarrow Products \qquad Rate = k[A][B]$$

What are the relative rates of this reaction in vessels (a)–(d)? Each vessel has the same volume. Red spheres represent A molecules, and blue spheres represent B molecules.

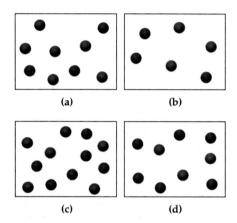

(a)   (b)

(c)   (d)

12.23 What are the relative values of the rate constant k for cases (a)–(d) in Problem 12.22?

12.24 Consider the first-order decomposition of A molecules (red spheres) in three vessels of equal volume.

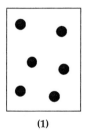      

(1)   (2)   (3)

(a) What are the relative rates of decomposition in vessels (1)–(3)?

(b) What are the relative half-lives of the reactions in vessels (1)–(3)?

(c) How will the rates and half-lives be affected if the volume of each vessel is decreased by a factor of 2?

12.25 Consider the first-order reaction $A \rightarrow B$ in which A molecules (red spheres) are converted to B molecules (blue spheres).

(a) Given the following pictures at $t = 0$ min and $t = 1$ min, draw pictures that show the number of A and B molecules present at $t = 2$ min and $t = 3$ min.

(b) What is the half-life of the reaction?

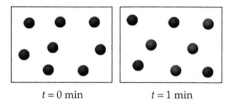

$t = 0$ min $t = 1$ min

12.26 The following pictures represent the progress of a reaction in which two A molecules combine to give a more complex molecule A_2, $2\,A \rightarrow A_2$.

(a) Is the reaction first order or second order in A?

(b) What is the rate law?

(c) Draw an appropriate picture in the last box, and specify the time.

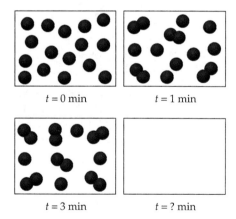

$t = 0$ min $t = 1$ min

$t = 3$ min $t = ?$ min

12.27 What is the molecularity of each of the following elementary reactions?

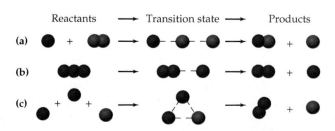

12.28 You wish to determine the reaction order and rate constant for the following thermal decomposition reaction:

$$2\,AB_2 \longrightarrow A_2 + 2\,B_2$$

(a) What data would you collect?

(b) How would you use these data to determine whether the reaction is first order or second order?

(c) Describe how you would determine the value of the rate constant.

12.29 You wish to determine the activation energy for the following first-order reaction:

$$A \longrightarrow B + C$$

(a) What data would you collect?

(b) How would you use these data to determine the activation energy?

12.30 The rate of the reaction $A + B_2 \rightarrow AB + B$ is directly proportional to the concentration of B_2, independent of the concentration of A, and directly proportional to the concentration of a substance C.

(a) What is the rate law?

(b) Write a mechanism that agrees with the experimental facts.

(c) What is the role of C in this reaction, and why doesn't C appear in the chemical equation for the overall reaction?

12.31 Consider a reaction that occurs by the following mechanism:

$$A + BC \longrightarrow AC + B$$
$$AC + D \longrightarrow A + CD$$

The potential energy profile for this reaction is shown below:

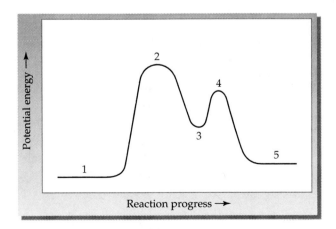

(a) What is the equation for the overall reaction?

(b) Write structural formulas for the species present at reaction stages 1–5. Identify each species as a reactant, product, catalyst, intermediate, or transition state.

(c) Which of the two steps in the mechanism is the rate-determining step? Write the rate law for the overall reaction.

(d) Is the reaction endothermic or exothermic? Add labels to the diagram that show the values of the energy of reaction ΔE and the activation energy E_a for the overall reaction.

12.32 Consider three reactions with different values of E_a and ΔE:

Reaction 1. $E_a = 20$ kJ/mol; $\Delta E = -60$ kJ/mol
Reaction 2. $E_a = 10$ kJ/mol; $\Delta E = -20$ kJ/mol
Reaction 3. $E_a = 40$ kJ/mol; $\Delta E = +15$ kJ/mol

(a) Sketch a potential energy profile for each reaction that shows the potential energy of reactants, products, and the transition state. Include labels that define E_a and ΔE.

(b) Assuming that all three reactions are carried out at the same temperature and that all three have the same frequency factor A, which reaction is the fastest and which is the slowest?

(c) Which reaction is the most endothermic, and which is the most exothermic?

12.33 Draw a plausible transition state for the bimolecular reaction of nitric oxide with ozone. Use dashed lines to indicate the atoms that are weakly linked together in the transition state.

$$NO(g) + O_3(g) \longrightarrow NO_2(g) + O_2(g)$$

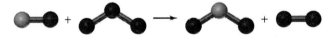

Additional Problems

Reaction Rates

12.34 What are the usual units of reaction rate?

12.35 Concentrations of trace constituents of the atmosphere are sometimes expressed in molecules/cm³. If those units are used for concentrations, what are the units of reaction rate?

12.36 Given the concentration–time data in Problem 12.8, calculate the average rate of decomposition of cyclopropane during the following time intervals:
   (a) 0–5.0 min   **(b)** 15.0–20.0 min

12.37 Concentration–time data for decomposition of nitrogen dioxide are given in Example 12.8. What is the average rate of decomposition of NO_2 during the following time periods?
   (a) 50–100 s   **(b)** 100–150 s

12.38 From the plot of concentration–time data in Figure 12.1, estimate:
   (a) The instantaneous rate of decomposition of N_2O_5 at $t = 200$ s
   (b) The initial rate of decomposition of N_2O_5

12.39 From a plot of the concentration–time data in Example 12.8, estimate:
   (a) The instantaneous rate of decomposition of NO_2 at $t = 100$ s
   (b) The initial rate of decomposition of NO_2

12.40 Ammonia is manufactured in large amounts by the reaction

$$N_2(g) + 3 H_2(g) \longrightarrow 2 NH_3(g)$$

(a) How is the rate of consumption of H_2 related to the rate of consumption of N_2?

(b) How is the rate of formation of NH_3 related to the rate of consumption of N_2?

12.41 In the first step of the Ostwald process for synthesis of nitric acid, ammonia is converted to nitric oxide by the high-temperature reaction

$$4 NH_3(g) + 5 O_2(g) \longrightarrow 4 NO(g) + 6 H_2O(g)$$

(a) How is the rate of consumption of O_2 related to the rate of consumption of NH_3?

(b) How are the rates of formation of NO and H_2O related to the rate of consumption of NH_3?

12.42 For the reaction $N_2(g) + 3 H_2(g) \rightarrow 2 NH_3(g)$, what is the relationship among $\Delta[N_2]/\Delta t$, $\Delta[H_2]/\Delta t$, and $\Delta[NH_3]/\Delta t$?

12.43 The oxidation of iodide ion by peroxydisulfate ion is described by the equation

$$3 I^-(aq) + S_2O_8{}^{2-}(aq) \longrightarrow I_3{}^-(aq) + 2 SO_4{}^{2-}(aq)$$

(a) If $-\Delta[S_2O_8{}^{2-}]/\Delta t = 1.5 \times 10^{-3}$ M/s for a particular time interval, what is the value of $\Delta[I^-]/\Delta t$ for the same time interval?

(b) What is the average rate of formation of $SO_4{}^{2-}$ during that time interval?

Rate Laws

12.44 The gas-phase reaction of nitric oxide and bromine yields nitrosyl bromide:

$$2 NO(g) + Br_2(g) \longrightarrow 2 NOBr(g)$$

The rate law is rate = $k[NO]^2[Br_2]$. What is the reaction order with respect to each of the reactants, and what is the overall reaction order?

12.45 The reaction of gaseous chloroform and chlorine is described by the equation

$$CHCl_3(g) + Cl_2(g) \longrightarrow CCl_4(g) + HCl(g)$$

The rate law is rate = $k[CHCl_3][Cl_2]^{1/2}$. What is the order of the reaction with respect to $CHCl_3$ and Cl_2? What is the overall reaction order?

12.46 The gas-phase reaction of hydrogen and iodine monochloride,

$$H_2(g) + 2 ICl(g) \longrightarrow 2 HCl(g) + I_2(g)$$

is first order in H_2 and first order in ICl. What is the rate law, and what are the units of the rate constant?

12.47 The reaction $2 NO(g) + 2 H_2(g) \rightarrow N_2(g) + 2 H_2O(g)$ is first order in H_2 and second order in NO. Write the rate law, and specify the units of the rate constant.

12.48 Bromomethane is converted to methanol in an alkaline solution:

$$CH_3Br(aq) + OH^-(aq) \longrightarrow CH_3OH(aq) + Br^-(aq)$$

The reaction is first order in each reactant.
   (a) Write the rate law.
   (b) How does the reaction rate change if the OH^- concentration is decreased by a factor of 5?
   (c) What is the change in rate if the concentrations of both reactants are doubled?

12.49 The oxidation of Br^- by BrO_3^- in acidic solution is described by the equation

$$5\,Br^-(aq) + BrO_3^-(aq) + 6\,H^+(aq) \longrightarrow$$
$$3\,Br_2(aq) + 3\,H_2O(l)$$

The reaction is first order in Br^-, first order in BrO_3^-, and second order in H^+.

(a) Write the rate law.

(b) What is the overall reaction order?

(c) How does the reaction rate change if the H^+ concentration is tripled?

(d) What is the change in rate if the concentrations of both Br^- and BrO_3^- are halved?

12.50 At 600°C, acetone (CH_3COCH_3) decomposes to ketene ($CH_2{=}C{=}O$) and various hydrocarbons. Initial rate data are given in the table:

Experiment	Initial [CH$_3$COCH$_3$]	Initial Rate of Decomposition of CH$_3$COCH$_3$ (M/s)
1	6.0×10^{-3}	5.2×10^{-5}
2	9.0×10^{-3}	7.8×10^{-5}

(a) Determine the rate law.

(b) Calculate the rate constant.

(c) Calculate the rate of decomposition when the acetone concentration is 1.8×10^{-3} M.

12.51 The initial rates listed in the table were determined for the thermal decomposition of azomethane (CH_3NNCH_3):

$$CH_3NNCH_3(g) \longrightarrow C_2H_6(g) + N_2(g)$$

Experiment	Initial [CH$_3$NNCH$_3$]	Initial Rate of Decomposition of CH$_3$NNCH$_3$ (M/s)
1	2.4×10^{-2}	6.0×10^{-6}
2	8.0×10^{-3}	2.0×10^{-6}

(a) What is the rate law?

(b) What is the value of the rate constant?

(c) What is the rate of decomposition when the concentration of azomethane is 0.020 M?

12.52 Initial rate data at 25°C are listed in the table for the reaction

$$NH_4^+(aq) + NO_2^-(aq) \longrightarrow N_2(g) + 2\,H_2O(l)$$

Experiment	Initial [NH$_4$$^+$]	Initial [NO$_2$$^-$]	Initial Rate of Consumption of NH$_4$$^+$ (M/s)
1	0.24	0.10	7.2×10^{-6}
2	0.12	0.10	3.6×10^{-6}
3	0.12	0.15	5.4×10^{-6}

(a) What is the rate law?

(b) What is the value of the rate constant?

(c) What is the reaction rate when the concentrations are $[NH_4^+] = 0.39$ M and $[NO_2^-] = 0.052$ M?

12.53 The initial rates listed in the table were determined for the reaction

$$2\,NO(g) + Cl_2(g) \longrightarrow 2\,NOCl(g)$$

Experiment	Initial [NO]	Initial [Cl$_2$]	Initial Rate of Consumption of Cl$_2$ (M/s)
1	0.13	0.20	1.0×10^{-2}
2	0.26	0.20	4.0×10^{-2}
3	0.13	0.10	5.0×10^{-3}

(a) What is the rate law?

(b) What is the value of the rate constant?

(c) What is the reaction rate when both reactant concentrations are 0.12 M?

Integrated Rate Law; Half-Life

12.54 At 500°C, cyclopropane (C_3H_6) rearranges to propene (CH_3–$CH{=}CH_2$). The reaction is first order, and the rate constant is 6.7×10^{-4} s^{-1}. If the initial concentration of C_3H_6 is 0.0500 M:

(a) What is the molarity of C_3H_6 after 30 min?

(b) How many minutes does it take for the C_3H_6 concentration to drop to 0.0100 M?

(c) How many minutes does it take for 25% of the C_3H_6 to react?

12.55 The rearrangement of methyl isonitrile (CH_3NC) to acetonitrile (CH_3CN) is a first-order reaction and has a rate constant of 5.11×10^{-5} s^{-1} at 472 K.

$$CH_3{-}N{\equiv}C \longrightarrow CH_3{-}C{\equiv}N$$

Methyl isonitrile   Acetonitrile

If the initial concentration of CH_3NC is 0.0340 M:

(a) What is the molarity of CH_3NC after 2.00 h?

(b) How many minutes does it take for the CH_3NC concentration to drop to 0.0300 M?

(c) How many minutes does it take for 20% of the CH_3NC to react?

12.56 What is the half-life (in minutes) of the reaction in Problem 12.54? How many minutes will it take for the concentration of cyclopropane to drop to 6.25% of its initial value?

12.57 What is the half-life (in hours) of the reaction in Problem 12.55? How many hours will it take for the concentration of methyl isonitrile to drop to 12.5% of its initial value?

12.58 The decomposition of aqueous hydrogen peroxide to gaseous O_2 and water is a first-order reaction. If it takes 8.0 h for the concentration of H_2O_2 to decrease from 0.80 M to 0.40 M, how many hours are required for the concentration to decrease from 0.60 M to 0.15 M?

12.59 Sucrose (table sugar) reacts with water in acidic solution to give glucose and fructose, two simpler sugars that have the same molecular formulas but different structures.

$$C_{12}H_{22}O_{11}(aq) + H_2O(l) \longrightarrow C_6H_{12}O_6(aq) + C_6H_{12}O_6(aq)$$
$$\text{Sucrose} \qquad\qquad\qquad \text{Glucose} \qquad \text{Fructose}$$

At 25°C, it takes 3.33 h for the concentration of sucrose to drop from 1.20 M to 0.600 M. How many hours are required for the concentration to drop from 0.800 M to 0.0500 M?

12.60 Butadiene (C_4H_6) reacts with itself to form a dimer with the formula C_8H_{12}. The reaction is second order in C_4H_6. If the rate constant at a particular temperature is $4.0 \times 10^{-2}\ M^{-1}s^{-1}$ and the initial concentration of C_4H_6 is 0.0200 M:

(a) What is its molarity after a reaction time of 1.00 h?

(b) What is the time (in hours) when the C_4H_6 concentration reaches a value of 0.0020 M?

12.61 Hydrogen iodide decomposes slowly to H_2 and I_2 at 600 K. The reaction is second order in HI and the rate constant is $9.7 \times 10^{-6}\ M^{-1}s^{-1}$. If the initial concentration of HI is 0.100 M:

(a) What is its molarity after a reaction time of 6.00 days?

(b) What is the time (in days) when the HI concentration reaches a value of 0.020 M?

12.62 What is the half-life (in minutes) of the reaction in Problem 12.60 when the initial C_4H_6 concentration is 0.0200 M? How many minutes does it take for the concentration of C_4H_6 to drop from 0.0100 M to 0.0050 M?

12.63 What is the half-life (in days) of the reaction in Problem 12.61 when the initial HI concentration is 0.100 M? How many days does it take for the concentration of HI to drop from 0.0250 M to 0.0125 M?

12.64 At elevated temperatures, nitrous oxide decomposes according to the equation

$$2\ N_2O(g) \longrightarrow 2\ N_2(g) + O_2(g)$$

Given the following data, plot the appropriate graphs to determine whether the reaction is first order or second order. What is the value of the rate constant?

Time (min)	0	60	90	120	180
[N_2O]	0.250	0.218	0.204	0.190	0.166

12.65 Nitrosyl bromide decomposes at 10°C:

$$2\ NOBr(g) \longrightarrow 2\ NO(g) + Br_2(g)$$

Use the following kinetic data to determine the order of the reaction and the value of the rate constant.

Time (s)	0	10	20	30	40
[NOBr]	0.0400	0.0303	0.0244	0.0204	0.0175

12.66 At 25°C, the half-life of a certain first-order reaction is 248 s. What is the value of the rate constant at this temperature?

12.67 At 100°C, a certain substance exhibits second-order decomposition with a half-life of 25 min. If the initial concentration of the substance is 0.036 M, what is the value of the rate constant at 100°C?

Reaction Mechanisms

12.68 What is the difference between an elementary reaction and an overall reaction?

12.69 What is the difference between molecularity and reaction order?

12.70 What is the relationship between the coefficients in a balanced chemical equation for an overall reaction and the exponents in the rate law?

12.71 What distinguishes the rate-determining step from the other steps in a reaction mechanism? How does the rate-determining step affect the observed rate law?

12.72 Consider the following mechanism for the reaction of hydrogen and iodine monochloride:

Step 1. $H_2(g) + ICl(g) \longrightarrow HI(g) + HCl(g)$
Step 2. $HI(g) + ICl(g) \longrightarrow I_2(g) + HCl(g)$

(a) Write the equation for the overall reaction.

(b) Identify any reaction intermediates.

(c) What is the molecularity of each elementary step?

12.73 The following mechanism has been proposed for the reaction of nitric oxide and chlorine:

Step 1. $NO(g) + Cl_2(g) \longrightarrow NOCl_2(g)$
Step 2. $NOCl_2(g) + NO(g) \longrightarrow 2\ NOCl(g)$

(a) What is the overall reaction?

(b) Identify any reaction intermediates.

(c) What is the molecularity of each elementary step?

12.74 Give the molecularity and the rate law for each of the following elementary reactions:

(a) $O_3(g) + Cl(g) \rightarrow O_2(g) + ClO(g)$

(b) $NO_2(g) \rightarrow NO(g) + O(g)$

(c) $ClO(g) + O(g) \rightarrow Cl(g) + O_2(g)$

(d) $Cl(g) + Cl(g) + N_2(g) \rightarrow Cl_2(g) + N_2(g)$

12.75 Identify the molecularity and write the rate law for each of the following elementary reactions:

(a) $I_2(g) \rightarrow 2\ I(g)$

(b) $2\ NO(g) + Br_2(g) \rightarrow 2\ NOBr(g)$

(c) $CH_3Br(aq) + OH^-(aq) \rightarrow CH_3OH(aq) + Br^-(aq)$

(d) $N_2O_5(g) \rightarrow NO_2(g) + NO_3(g)$

12.76 The thermal decomposition of nitryl chloride, NO_2Cl, is believed to occur by the following mechanism:

$$NO_2Cl(g) \xrightarrow{k_1} NO_2(g) + Cl(g)$$

$$Cl(g) + NO_2Cl(g) \xrightarrow{k_2} NO_2(g) + Cl_2(g)$$

(a) What is the overall reaction?

(b) What is the molecularity of each of the elementary steps?

(c) What rate law is predicted by this mechanism if the first step is the rate-determining step?

12.77 The substitution reactions of molybdenum hexacarbonyl, $Mo(CO)_6$, with a variety of other molecules L are believed to occur by the following mechanism:

$$Mo(CO)_6 \xrightarrow{k_1} Mo(CO)_5 + CO$$

$$Mo(CO)_5 + L \xrightarrow{k_2} Mo(CO)_5L$$

(a) What is the overall reaction?

(b) What is the molecularity of each of the elementary steps?

(c) Write the rate law, assuming that the first step is the rate-determining step.

12.78 The reaction $2\ NO_2(g) + F_2(g) \rightarrow 2\ NO_2F(g)$ has a second-order rate law, rate $= k[NO_2][F_2]$. Suggest a mechanism that is consistent with this rate law.

12.79 The decomposition of ozone in the upper atmosphere is facilitated by nitric oxide. The overall reaction and the rate law are

$$O_3(g) + O(g) \longrightarrow 2\ O_2(g) \qquad \text{Rate} = k[O_3][NO]$$

Write a mechanism that is consistent with the rate law.

The Arrhenius Equation

12.80 Why don't all collisions between reactant molecules lead to chemical reaction?

12.81 Two reactions have the same activation energy, but their rates at the same temperature differ by a factor of 10. Explain.

12.82 Rate constants for the reaction $2\ N_2O_5(g) \rightarrow 4\ NO_2(g) + O_2(g)$ exhibit the following temperature dependence:

Temperature (°C)	k (1/s)	Temperature (°C)	k (1/s)
25	3.7×10^{-5}	55	1.7×10^{-3}
45	5.1×10^{-4}	65	5.2×10^{-3}

Plot an appropriate graph of the data, and determine the activation energy (in kJ/mol) for this reaction.

12.83 The following rate constants describe the thermal decomposition of nitrogen dioxide:

$$2\ NO_2(g) \longrightarrow 2\ NO(g) + O_2(g)$$

Temperature (°C)	k [1/(M · s)]	Temperature (°C)	k [1/(M · s)]
330	0.77	378	4.1
354	1.8	383	4.7

Plot an appropriate graph of the data, and calculate the value of E_a for this reaction in kJ/mol.

12.84 Rate constants for the reaction $NO_2(g) + CO(g) \rightarrow NO(g) + CO_2(g)$ are 1.3/(M · s) at a temperature of 700 K and 23.0/(M · s) at 800 K.

(a) What is the value of the activation energy in kJ/mol?

(b) What is the rate constant at 750 K?

12.85 A certain first-order reaction has a rate constant of $1.0 \times 10^{-3}\ s^{-1}$ at 25°C.

(a) If the reaction rate doubles when the temperature is increased to 35°C, what is the activation energy for this reaction in kJ/mol?

(b) What is the E_a (in kJ/mol) if the same temperature change causes the rate to triple?

12.86 Reaction of the anti-cancer drug cisplatin, $Pt(NH_3)_2Cl_2$, with water is described by the equation

$$Pt(NH_3)_2Cl_2(aq) + H_2O(l) \longrightarrow$$
$$Pt(NH_3)_2(H_2O)Cl^+(aq) + Cl^-(aq)$$

The rate of this reaction increases by a factor of 15 on raising the temperature from 25°C to 50°C. What is the value of the activation energy in kJ/mol?

12.87 The widely used solvent ethyl acetate undergoes the following reaction in basic solution:

$$CH_3CO_2C_2H_5(aq) + OH^-(aq) \longrightarrow$$
$$CH_3CO_2^-(aq) + C_2H_5OH(aq)$$

The rate of this reaction increases by a factor of 6.37 on raising the temperature from 15°C to 45°C. Calculate the value of the activation energy in kJ/mol.

12.88 Values of $E_a = 183$ kJ/mol and $\Delta E = 9$ kJ/mol have been measured for the reaction

$$2\,HI(g) \longrightarrow H_2(g) + I_2(g)$$

(a) Sketch a potential energy profile for this reaction that shows the potential energy of reactants, products, and the transition state. Include labels that define E_a and ΔE.

(b) Considering the geometry of the reactants and products, suggest a plausible structure for the transition state.

12.89 Values of $E_a = 248$ kJ/mol and $\Delta E = 41$ kJ/mol have been measured for the reaction

$$H_2(g) + CO_2(g) \longrightarrow H_2O(g) + CO(g)$$

(a) Sketch a potential energy profile for this reaction that shows the potential energy of reactants, products, and the transition state. Include labels that define E_a and ΔE.

(b) Considering the geometry of the reactants and products, suggest a plausible structure for the transition state.

Catalysis

12.90 Comment on the following statement: "A catalyst increases the rate of a reaction, but it is not consumed because it does not participate in the reaction."

12.91 Why doesn't a catalyst appear in the overall chemical equation for a reaction?

12.92 What effect does a catalyst have on the rate, mechanism, and activation energy of a chemical reaction?

12.93 Distinguish between a homogeneous catalyst and a heterogeneous catalyst, and give an example of each.

12.94 In the upper atmosphere, chlorofluorocarbons such as $CFCl_3$ absorb sunlight, and subsequent fragmentation produces Cl atoms. The Cl atoms participate in the following mechanism for destruction of ozone:

$$Cl(g) + O_3(g) \longrightarrow ClO(g) + O_2(g)$$
$$ClO(g) + O(g) \longrightarrow Cl(g) + O_2(g)$$

(a) Write the chemical equation for the overall reaction.

(b) What is the role of the Cl atoms in this reaction?

(c) Is ClO a catalyst or a reaction intermediate?

(d) What distinguishes a catalyst from an intermediate?

12.95 Sulfur dioxide is oxidized to sulfur trioxide in the following sequence of reactions:

$$2\,SO_2(g) + 2\,NO_2(g) \longrightarrow 2\,SO_3(g) + 2\,NO(g)$$
$$2\,NO(g) + O_2(g) \longrightarrow 2\,NO_2(g)$$

(a) Write the chemical equation for the overall reaction.

(b) Identify any molecule that acts as a catalyst or intermediate in this reaction.

12.96 Consider the following mechanism for the decomposition of nitramide (NH_2NO_2) in aqueous solution:

$$NH_2NO_2(aq) + OH^-(aq) \longrightarrow NHNO_2^-(aq) + H_2O(l)$$
$$NHNO_2^-(aq) \longrightarrow N_2O(g) + OH^-(aq)$$

(a) Write the chemical equation for the overall reaction.

(b) Identify the catalyst and the reaction intermediate.

(c) How will the rate of the overall reaction be affected if HCl is added to the solution?

12.97 In Problem 12.79, you wrote a mechanism for the nitric oxide-facilitated decomposition of ozone. Does your mechanism involve a catalyst or a reaction intermediate? Explain.

General Problems

12.98 Consider the potential energy profile in Figure 12.16a for the iodide ion-catalyzed decomposition of H_2O_2. What point on the profile represents the potential energy of the transition state for the first step in the reaction? What point represents the potential energy of the transition state for the second step? What point represents the potential energy of the intermediate products $H_2O(l) + IO^-(aq)$?

12.99 Decomposition of N_2O_5 is a first-order reaction. At 25°C, it takes 5.2 h for the concentration to drop from 0.120 M to 0.060 M. How many hours does it take for the concentration to drop from 0.030 M to 0.015 M? From 0.480 M to 0.015 M?

12.100 Consider the reaction $H_2(g) + I_2(g) \rightarrow 2\,HI(g)$. The reaction of a fixed amount of H_2 and I_2 is studied in a cylinder fitted with a movable piston. Indicate the effect of each of the following changes on the rate of the reaction:

(a) An increase in temperature at constant volume

(b) An increase in volume at constant temperature

(c) Addition of a catalyst

(d) Addition of argon (an inert gas) at constant volume

12.101 When the temperature of a gas is raised by 10°C, the collision frequency increases by only about 2%, but the reaction rate increases by 100% (factor of 2) or more. Explain.

12.102 The initial rates listed in the table were measured in methanol solution for the reaction

$$C_2H_4Br_2 + 3I^- \longrightarrow C_2H_4 + 2Br^- + I_3^-$$

Experiment	Initial $[C_2H_4Br_2]$	Initial $[I^-]$	Initial Rate of Formation of I_3^- (M/s)
1	0.127	0.102	6.45×10^{-5}
2	0.343	0.102	1.74×10^{-4}
3	0.203	0.125	1.26×10^{-4}

(a) What is the rate law?

(b) What is the value of the rate constant?

(c) What is the reaction rate when both reactant concentrations are 0.150 M?

12.103 Concentration–time data for the conversion of A and B to D are listed below.

(a) Write a balanced equation for the reaction.

(b) What is the reaction order with respect to A, B, and C, and what is the overall reaction order?

(c) What is the rate law?

(d) Is a catalyst involved in this reaction? Explain.

(e) Suggest a mechanism that is consistent with the data.

(f) Calculate the rate constant for the formation of D.

Experiment	Time (s)	[A]	[B]	[C]	[D]
1	0	5.00	2.00	1.00	0.00
	60	4.80	1.90	1.00	0.10
2	0	10.00	2.00	1.00	0.00
	60	9.60	1.80	1.00	0.20
3	0	5.00	4.00	1.00	0.00
	60	4.80	3.90	1.00	0.10
4	0	5.00	2.00	2.00	0.00
	60	4.60	1.80	2.00	0.20

12.104 What fraction of the molecules in a gas at 300 K collide with an energy equal to or greater than E_a when E_a is equal to 50 kJ/mol? What is the value of this fraction when E_a is 100 kJ/mol?

12.105 If the rate of a reaction increases by a factor of 2.5 for a temperature rise from 20°C to 30°C, what is the value of the activation energy in kJ/mol? By what factor does the rate of this reaction increase when the temperature is raised from 120°C to 130°C?

12.106 A two-step mechanism has been suggested for the reaction of nitric oxide and bromine:

$$NO(g) + Br_2(g) \xrightarrow{k_1} NOBr_2(g)$$

$$NOBr_2(g) + NO(g) \xrightarrow{k_2} 2NOBr(g)$$

(a) What is the overall reaction?

(b) What is the role of $NOBr_2$ in this reaction?

(c) What is the predicted rate law if the first step is much slower than the second step?

(d) The observed rate law is rate = $k[NO]^2[Br_2]$. What can you conclude about the rate-determining step?

12.107 Consider the following mechanism for the reaction of nitric oxide and hydrogen:

$$2NO(g) \underset{k_{-1}}{\overset{k_1}{\rightleftharpoons}} N_2O_2(g) \qquad \text{Fast}$$

$$N_2O_2(g) + H_2(g) \xrightarrow{k_2} N_2O(g) + H_2O(g) \quad \text{Slow, rate-determining}$$

$$N_2O(g) + H_2(g) \xrightarrow{k_3} N_2(g) + H_2O(g) \quad \text{Fast}$$

The first step, which is reversible, has a rate constant of k_1 for the forward reaction and k_{-1} for the reverse reaction. The forward and reverse reactions in step 1 occur at the same rate, and both are fast compared with step 2, the rate-determining step.

(a) Write an equation for the overall reaction.

(b) Identify all reaction intermediates.

(c) Write the rate law for the rate-determining step.

(d) Derive an equation for the rate of the overall reaction in terms of the concentrations of the reactants. (The concentrations of intermediates should not appear in the rate law for the overall reaction.)

12.108 Consider the following concentration–time data for the reaction of iodide ion and hypochlorite ion (OCl^-). The products are chloride ion and hypoiodite ion (OI^-).

Experiment	Time (s)	$[I^-]$	$[OCl^-]$	$[OH^-]$
1	0	2.40×10^{-4}	1.60×10^{-4}	1.00
	10	2.17×10^{-4}	1.37×10^{-4}	1.00
2	0	1.20×10^{-4}	1.60×10^{-4}	1.00
	10	1.08×10^{-4}	1.48×10^{-4}	1.00
3	0	2.40×10^{-4}	4.00×10^{-5}	1.00
	10	2.34×10^{-4}	3.40×10^{-5}	1.00
4	0	1.20×10^{-4}	1.60×10^{-4}	2.00
	10	1.14×10^{-4}	1.54×10^{-4}	2.00

(a) Write a balanced equation for the reaction.

(b) Determine the rate law, and calculate the value of the rate constant.

(c) Does the reaction occur by a single-step mechanism? Explain.

(d) Propose a mechanism that is consistent with the rate law. (*Hint*: Transfer of an H^+ ion between H_2O and OCl^- is a rapid reversible reaction.)

12.109 For the thermal decomposition of nitrous oxide, $2\,N_2O(g) \rightarrow 2\,N_2(g) + O_2(g)$, values of the parameters in the Arrhenius equation are $A = 4.2 \times 10^9\ s^{-1}$ and $E_a = 222\ kJ/mol$. If a stream of N_2O is passed through a tube 25 mm in diameter and 20 cm long at a flow rate of 0.75 L/min, at what temperature should the tube be maintained to have a partial pressure of 1.0 mm of O_2 in the exit gas? Assume that the total pressure of the gas in the tube is 1.50 atm.

12.110 Consider the reversible, first-order interconversion of two molecules A and B:

$$A \underset{k_r}{\overset{k_f}{\rightleftharpoons}} B$$

where $k_f = 3.0 \times 10^{-3}\ s^{-1}$ is the rate constant for the forward reaction and $k_r = 1.0 \times 10^{-3}\ s^{-1}$ is the rate constant for the reverse reaction. We'll see in the next chapter that a reaction does not go to completion, but instead reaches a state of equilibrium with comparable concentrations of reactants and products if the rate constants k_f and k_r have comparable values.

(a) What are the rate laws for the forward and reverse reactions?

(b) Draw a qualitative graph that shows how the rates of the forward and reverse reactions vary with time.

(c) What are the relative concentrations of B and A when the rates of the forward and reverse reactions become equal?

12.111 Radioactive decay exhibits a first-order rate law, rate $= -\Delta N/\Delta t = kN$, where N denotes the number of radioactive nuclei present at time t. The half-life of strontium-90, a dangerous nuclear fission product, is 29 yr.

(a) What fraction of the strontium-90 remains after three half-lives?

(b) What is the value of the rate constant for the decay of strontium-90?

(c) How many years are required for 99% of the strontium-90 to disappear?

12.112 The age of any remains from a once-living organism can be determined by *radiocarbon dating*, a procedure that works by determining the concentration of radioactive ^{14}C in the remains. All living organisms contain an equilibrium concentration of radioactive ^{14}C that gives rise to an average of 15.3 nuclear decay events per minute per gram of carbon. At death, however, no additional ^{14}C is taken in, so the concentration slowly drops as radioactive decay occurs. What is the age of a bone fragment from an archaeological dig if the bone shows an average of 2.3 radioactive events per minute per gram of carbon? Radioactive decay is kinetically a first-order process, and $t_{1/2}$ for ^{14}C is 5730 years.

Multi-Concept Problems

12.113 Values of $E_a = 6.3\ kJ/mol$ and $A = 6.0 \times 10^8/(M \cdot s)$ have been measured for the bimolecular reaction:

$$NO(g) + F_2(g) \longrightarrow NOF(g) + F(g)$$

(a) Calculate the rate constant at 25°C.

(b) The product of the reaction is nitrosyl fluoride. Its formula is usually written as NOF, but its structure is actually ONF. Is the ONF molecule linear or bent?

(c) Draw a plausible transition state for the reaction. Use dashed lines to indicate the atoms that are weakly linked together in the transition state.

(d) Why does the reaction have such a low activation energy?

12.114 A 1.50 L sample of gaseous HI having a density of $0.0101\ g/cm^3$ is heated at 410°C. As time passes, the HI decomposes to gaseous H_2 and I_2. The rate law is $-\Delta[HI]/\Delta t = k[HI]^2$, where $k = 0.031/(M \cdot min)$ at 410°C.

(a) What is the initial rate of production of I_2 in molecules/min?

(b) What is the partial pressure of H_2 after a reaction time of 8.00 h?

12.115 The rate constant for the decomposition of gaseous NO_2 to NO and O_2 is $4.7/(M \cdot s)$ at 383°C. Consider the decomposition of a sample of pure NO_2 having an initial pressure of 746 mm Hg in a 5.00 L reaction vessel at 383°C.

(a) What is the order of the reaction?

(b) What is the initial rate of formation of O_2 in g/(L·s)?

(c) What is the mass of O_2 in the vessel after a reaction time of 1.00 min?

12.116 The rate constant for the first-order decomposition of gaseous N_2O_5 to NO_2 and O_2 is $1.7 \times 10^{-3}\ s^{-1}$ at 55°C. If 2.70 g of gaseous N_2O_5 is introduced into an evacuated 2.00 L container maintained at a constant temperature of 55°C, what is the total pressure in the container after a reaction time of 13.0 minutes?

12.117 Consider the decomposition of N_2O_5 under the conditions described in Problem 12.116, and assume that the heat of the reaction is independent of temperature.

(a) Use the data in Appendix B to calculate the initial rate at which the reaction mixture absorbs heat (in J/s).

(b) What is the total amount of heat absorbed (in kilojoules) after a reaction time of 10.0 min?

12.118 A 0.500 L reaction vessel equipped with a movable piston is filled completely with a 3.00% aqueous solution of hydrogen peroxide. The H_2O_2 decomposes to water and O_2 gas in a first-order reaction that has a half-life of 10.7 h. As the reaction proceeds, the gas formed pushes the piston against a constant external atmospheric pressure of 738 mm Hg. Calculate the *PV* work done (in joules) after a reaction time of 4.02 h. (You may assume that the density of the solution is 1.00 g/mL and that the temperature of the system is maintained at 20°C.)

 # eMedia Problems

12.119 Use the **Rate of Decomposition of N_2O_5** simulation (*eChapter 12.1*) to determine the average rate of appearance of NO_2 between 100 and 500 s. The simulation begins at time zero with only N_2O_5 and no products, and NO_2 is produced at four times the rate that O_2 is produced. If you were able to carry out the reaction by starting with equal concentrations of N_2O_5 and NO_2, would it change the relative rates of appearance of NO_2 and O_2? Explain.

12.120 The **Rates of Reaction** simulation (*eChapter 12.3*) allows you to vary the activation energy, overall energy change, temperature, and initial reactant concentration for a chemical reaction. Use the simulation to determine what effect each of the following actions has on the initial reaction rate, the rate law, the rate constant, and the half-life of the reaction:

(a) Increasing the initial reactant concentration

(b) Increasing the activation energy

(c) Increasing the overall energy change

(d) Increasing the temperature

12.121 (a) Watch the **First-Order Process** movie (*eChapter 12.5*) and determine the value of the rate constant for the process shown.

(b) Assume that the process in the movie takes place at room temperature (25°C) and has a frequency factor of 3.86×10^8. Calculate the activation energy.

(c) Using your answers from parts (a) and (b), determine what the value of the rate constant and that of the half-life would be at 100°C.

12.122 Radioactive decay is a first-order process. In Problem 12.111 (page 526) you determined the rate constant and the number of years necessary for 99% of a sample of strontium-90 to decay. Use the **Radioactive Decay** simulation (*eChapter 12.5*) to determine the same two things for uranium-238, thorium-232, uranium-235, and potassium-40.

12.123 (a) Using the **Rates of Reaction** simulation (*eChapter 12.3*), select an activation energy and determine the rate constant for the reaction at 0°C.

(b) With the Arrhenius equation and your answer from part (a), determine what the rate constant will be at 30° and at 50°C. Verify your answers with the simulation.

(c) Based on your results, by approximately how much must the temperature increase in order for the rate of reaction to increase by a factor of 10?

13

Chemical Equilibrium

CONTENTS

Fish cannot survive without dissolved O_2. The dynamic equilibrium between atmospheric O_2 and dissolved O_2 gives seawater an O_2 concentration of 0.21 mmol/L at 20°C.

At the beginning of the previous chapter, we raised three key questions about chemical reactions: What happens? How fast does it happen? To what extent does it happen? The answer to the first question is given by the stoichiometry of the balanced chemical equation, and the answer to the second question is given by the kinetics of the reaction. In this chapter, we'll look at the answer to the third question: How far does a reaction proceed toward completion before it reaches a state of *chemical equilibrium*—a state in which the concentrations of reactants and products no longer change?

◆— CHEMICAL EQUILIBRIUM   The state that is reached when the concentrations of reactants and products remain constant over time

We're already familiar with the concept of equilibrium from our study of the evaporation of liquids (Section 10.5). When a liquid evaporates in a closed container, it soon gives rise to a constant vapor pressure because of a dynamic equilibrium in which the number of molecules leaving the liquid becomes equal to the number returning from the vapor. Chemical reactions behave similarly. They can occur in both forward and reverse directions, and when the rates of the forward and reverse reactions become equal, the concentrations of reactants and products remain constant.

Chemical equilibria are important in numerous biological and environmental processes. For example, equilibria involving O_2 molecules and the protein hemoglobin play a crucial role in the transport and delivery of oxygen from our lungs to our muscles. Similar equilibria involving CO molecules and hemoglobin account for the toxicity of carbon monoxide.

A mixture of reactants and products in the equilibrium state is called an **equilibrium mixture**. In this chapter, we'll address a number of important questions about the composition of equilibrium mixtures: What is the relationship between the concentrations of reactants and products in an equilibrium mixture? How can we determine equilibrium concentrations from initial concentrations? What factors can be exploited to alter the composition of an equilibrium mixture? This last question is particularly important when choosing conditions for the synthesis of industrial chemicals such as hydrogen, ammonia, and lime (CaO).

13.1 The Equilibrium State

In previous chapters, we've generally assumed that chemical reactions result in complete conversion of reactants to products. Many reactions, however, do not go to completion. Take, for example, the decomposition of the colorless gas dinitrogen tetroxide (N_2O_4) to the dark brown gas nitrogen dioxide (NO_2).

$$N_2O_4(g) \rightleftharpoons 2 NO_2(g)$$

Colorless   Brown

Figure 13.1 shows the results of two experiments at 25°C that illustrate the interconversion of N_2O_4 and NO_2. In the first experiment (Figure 13.1a), 0.0400 mol of N_2O_4 is placed in a 1.000 L flask to give an initial N_2O_4 concentration of 0.0400 M. The formation of NO_2 is indicated by the appearance of a brown color, and its concentration can be monitored by measuring the intensity of the color. According to the balanced equation, 2.0 mol of NO_2 forms for each mole of N_2O_4 that disappears, so the concentration of N_2O_4 at any time is equal to the initial concentration of N_2O_4 minus half the concentration of NO_2. As time passes, the concentration of N_2O_4 decreases and the concentration of NO_2 increases until both concentrations level off at constant, equilibrium values: $[N_2O_4] = 0.0337$ M; $[NO_2] = 0.0125$ M.

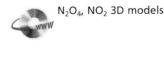

N_2O_4, NO_2 3D models

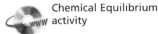

Chemical Equilibrium activity

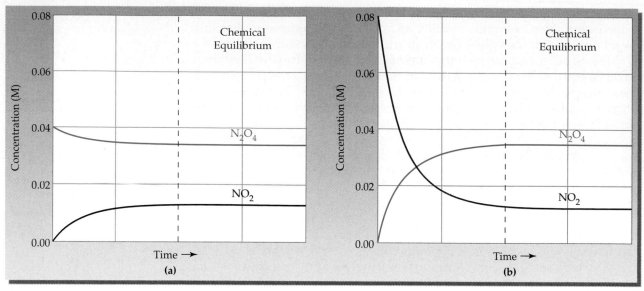

FIGURE 13.1 Change in the concentrations of N_2O_4 and NO_2 with time in two experiments at 25°C: **(a)** Only N_2O_4 is present initially; **(b)** only NO_2 is present initially. In experiment **(a)**, $[NO_2]$ increases as $[N_2O_4]$ decreases. In experiment **(b)**, $[N_2O_4]$ increases as $[NO_2]$ decreases. In both experiments, a state of chemical equilibrium is reached when the concentrations level off at constant values: $[N_2O_4] = 0.0337$ M; $[NO_2] = 0.0125$ M.

In the second experiment, shown in Figure 13.1b, we begin with NO_2, rather than N_2O_4, at a concentration of 0.0800 M. The conversion of NO_2 to N_2O_4 proceeds until the concentrations level off at the same values as obtained in the first experiment. Taken together, the two experiments demonstrate that the interconversion of N_2O_4 and NO_2 is *reversible* and that the same equilibrium state is reached starting from either substance.

$$N_2O_4(g) \rightleftharpoons 2\,NO_2(g) \qquad \text{Reaction occurs in both directions}$$

To indicate that the reaction can proceed in both forward and reverse directions, we write the balanced equation with two arrows, one pointing from reactants to products, and the other pointing from products to reactants. (The terms "reactants" and "products" could be confusing in this context because the products of the forward reaction are reactants in the reverse reaction. To avoid confusion, we'll restrict the term *reactants* to the substances on the left side of the chemical equation and the term *products* to the substances on the right side of the equation.)

Strictly speaking, *all* chemical reactions are reversible. What we sometimes call irreversible reactions are simply those that proceed *nearly* to completion, so that the equilibrium mixture contains almost all products and almost no reactants. For such reactions, the reverse reaction is often too slow to be detected.

Why do the reactions of N_2O_4 and NO_2 appear to stop after the concentrations reach their equilibrium values? We'll explore that question in more detail in Section 13.11, but will note for now that the concentrations reach constant values, not because the reactions stop, but because the *rates* of the forward and reverse reactions become equal. Take, for example, the experiment in which N_2O_4 is converted to an equilibrium mixture of NO_2 and N_2O_4. Because reaction rates depend on concentrations (Section 12.2), the rate of the forward reaction ($N_2O_4 \rightarrow 2\ NO_2$) decreases as the concentration of N_2O_4 decreases, while the rate of the reverse reaction ($N_2O_4 \leftarrow 2\ NO_2$) increases as the concentration of NO_2 increases. Eventually, the decreasing rate of the forward reaction and the increasing rate of the reverse reaction become equal. At that point, there are no further changes in concentrations because N_2O_4 and NO_2 both disappear as fast as they're formed. Thus, chemical equilibrium is a *dynamic state* in which forward and reverse reactions continue at equal rates so that there is no net conversion of reactants to products (Figure 13.2).

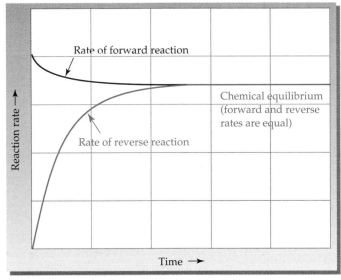

FIGURE 13.2 Rates of the forward and reverse reactions for decomposition of N_2O_4 to NO_2. As N_2O_4 is consumed, the rate of the forward reaction decreases; as NO_2 is formed, the rate of the reverse reaction increases. When the two rates become equal, an equilibrium state is attained and there are no further changes in concentrations.

13.2 The Equilibrium Constant K_c

Table 13.1 lists concentration data for the experiments in Figure 13.1 along with data for three additional experiments. In experiments 1 and 2, the equilibrium mixtures have identical compositions because the initial concentration of N_2O_4 in experiment 1 is half the initial concentration of NO_2 in experiment 2 (that is, the total number of N and O atoms is the same in both experiments). In experiments 3–5, different initial concentrations of N_2O_4 and/or NO_2 give different equilibrium concentrations. In all the experiments, however, the equilibrium concentrations are related. The last column of Table 13.1 shows that, at equilibrium, the expression $[NO_2]^2/[N_2O_4]$ has a constant value of approximately 4.64×10^{-3} M.

It's interesting to note that the expression $[NO_2]^2/[N_2O_4]$ appears to be related to the balanced equation for the reaction, $N_2O_4(g) \rightleftarrows 2\ NO_2(g)$: The concentration of the product is in the numerator, raised to the power of its coefficient in the balanced equation, and the concentration of the reactant is in the denominator. You might wonder if there is an analogous expression with a constant value for every chemical reaction. If so, how is the form of that expression related to the balanced equation for the reaction?

If the rate at which people move from the first floor to the second equals the rate at which people move from the second floor to the first, the number of people on each floor remains constant, and the two populations are in dynamic equilibrium.

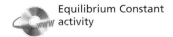
Equilibrium Constant activity

TABLE 13.1   Concentration Data at 25°C for the Reaction $N_2O_4(g) \rightleftharpoons 2\,NO_2(g)$

Experiment Number	Initial Concentrations (M)		Equilibrium Concentrations (M)		Equilibrium Constant Expression
	$[N_2O_4]$	$[NO_2]$	$[N_2O_4]$	$[NO_2]$	$[NO_2]^2/[N_2O_4]$
1	0.0400	0.0000	0.0337	0.0125	4.64×10^{-3}
2	0.0000	0.0800	0.0337	0.0125	4.64×10^{-3}
3	0.0600	0.0000	0.0522	0.0156	4.66×10^{-3}
4	0.0000	0.0600	0.0246	0.0107	4.65×10^{-3}
5	0.0200	0.0600	0.0429	0.0141	4.63×10^{-3}

To answer those questions, let's consider a general reversible reaction:

$$a\,A + b\,B \rightleftharpoons c\,C + d\,D$$

where A and B are the reactants, C and D are the products, and a, b, c, and d are stoichiometric coefficients in the balanced chemical equation. On the basis of experimental studies of many reversible reactions, the Norwegian chemists Cato Maximilian Guldberg and Peter Waage proposed in 1864 that the concentrations in an equilibrium mixture are related by the following **equilibrium equation,** where K_c is the *equilibrium constant* and the expression on the right side is called the *equilibrium constant expression*.

Equilibrium equation: $K_c = \dfrac{[C]^c[D]^d}{[A]^a[B]^b}$

Equilibrium constant   *Equilibrium constant expression*

As usual, square brackets indicate the molar concentration of the substance within the brackets (hence the subscript c for "concentration" in K_c). The equilibrium equation is also known as the *law of mass action* because in the early days of chemistry, concentration was called "active mass."

The **equilibrium constant K_c** is the number obtained by multiplying the equilibrium concentrations of all the products and dividing by the product of the equilibrium concentrations of all the reactants, with the concentration of each substance raised to the power of its coefficient in the balanced chemical equation. No matter what the individual equilibrium concentrations may be in a particular experiment, *the equilibrium constant for a reaction at a particular temperature always has the same value.* Thus, the equilibrium equation for the decomposition reaction of N_2O_4 to give $2\,NO_2$ is

$$K_c = \frac{[NO_2]^2}{[N_2O_4]} = 4.64 \times 10^{-3} \qquad \text{At } 25°C$$

where the equilibrium constant expression is $[NO_2]^2/[N_2O_4]$ and the equilibrium constant K_c has a value of 4.64×10^{-3} at 25°C (Table 13.1).

The units of K_c depend on the particular equilibrium equation. For the decomposition of N_2O_4, K_c has units of $M^2/M = M$. It's customary, however, to omit the units when quoting values of equilibrium constants.

Equilibrium constants are temperature-dependent, so the temperature must be given when citing a value of K_c. For example, K_c for the decomposition of N_2O_4 increases from 4.64×10^{-3} at 25°C to 1.53 at 127°C.

The form of the equilibrium constant expression and the numerical value of the equilibrium constant depend on the form of the balanced chemical equation. Consider again the chemical equation and the equilibrium equation for a general reaction:

$$a\,A + b\,B \rightleftharpoons c\,C + d\,D \qquad K_c = \frac{[C]^c[D]^d}{[A]^a[B]^b}$$

If we write the chemical equation in the reverse direction, the new equilibrium constant expression is the reciprocal of the original expression, and the new equilibrium constant K_c' is the reciprocal of the original equilibrium constant K_c:

$$c\,C + d\,D \rightleftharpoons a\,A + b\,B \qquad K_c' = \frac{[A]^a[B]^b}{[C]^c[D]^d} = \frac{1}{K_c}$$

(The prime distinguishes K_c' from K_c.) Because the equilibrium constants K_c and K_c' have different numerical values, it's important to specify the form of the balanced chemical equation when quoting the value of an equilibrium constant.

EXAMPLE 13.1 Write the equilibrium equation for each of the following reactions:
(a) $N_2(g) + 3\,H_2(g) \rightleftharpoons 2\,NH_3(g)$ **(b)** $2\,NH_3(g) \rightleftharpoons N_2(g) + 3\,H_2(g)$

SOLUTION
(a) Put the concentration of the reaction product, NH_3, in the numerator of the equilibrium constant expression and the concentrations of the reactants, N_2 and H_2, in the denominator. Then raise the concentration of each substance to the power of its coefficient in the balanced chemical equation:

$$K_c = \frac{[NH_3]^2}{[N_2][H_2]^3}$$

Coefficient of NH_3

Coefficient of H_2

(b) Because the balanced equation is the reverse of that in part (a), the equilibrium constant expression is the reciprocal of the expression in part (a) and the equilibrium constant K_c' is the reciprocal of K_c.

$$K_c' = \frac{[N_2][H_2]^3}{[NH_3]^2} \qquad K_c' = \frac{1}{K_c}$$

EXAMPLE 13.2 The following concentrations were measured for an equilibrium mixture at 500 K: $[N_2] = 3.0 \times 10^{-2}$ M; $[H_2] = 3.7 \times 10^{-2}$ M; $[NH_3] = 1.6 \times 10^{-2}$ M. Calculate the equilibrium constant at 500 K for each of the reactions in Example 13.1.

SOLUTION
(a) Calculate the value of K_c by substituting the equilibrium concentrations into the equilibrium equation:

$$K_c = \frac{[NH_3]^2}{[N_2][H_2]^3} = \frac{(1.6 \times 10^{-2})^2}{(3.0 \times 10^{-2})(3.7 \times 10^{-2})^3} = 1.7 \times 10^2$$

(b) The value of the equilibrium constant K_c' is

$$K_c' = \frac{[N_2][H_2]^3}{[NH_3]^2} = \frac{(3.0 \times 10^{-2})(3.7 \times 10^{-2})^3}{(1.6 \times 10^{-2})^2} = 5.9 \times 10^{-3}$$

Note that K_c' is the reciprocal of K_c. That is,

$$5.9 \times 10^{-3} = \frac{1}{1.7 \times 10^2}$$

▶ **PROBLEM 13.1** The oxidation of sulfur dioxide to give sulfur trioxide is an important step in the industrial process for synthesis of sulfuric acid. Write the equilibrium equation for each of the following reactions:
(a) $2\,SO_2(g) + O_2(g) \rightleftarrows 2\,SO_3(g)$ **(b)** $2\,SO_3(g) \rightleftarrows 2\,SO_2(g) + O_2(g)$

▶ **PROBLEM 13.2** The following equilibrium concentrations were measured at 800 K: $[SO_2] = 3.0 \times 10^{-3}$ M; $[O_2] = 3.5 \times 10^{-3}$ M; $[SO_3] = 5.0 \times 10^{-2}$ M. Calculate the equilibrium constant at 800 K for each of the reactions in Problem 13.1.

✦— **KEY CONCEPT PROBLEM 13.3** The following pictures represent mixtures of A molecules (red spheres) and B molecules (blue spheres), which interconvert according to the equation $A \rightleftarrows B$. If mixture (1) is at equilibrium, which of the other mixtures are also at equilibrium? Explain.

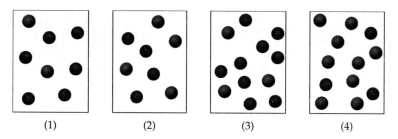

(1)   (2)   (3)   (4)

13.3 The Equilibrium Constant K_p

Because gas pressures are easily measured, equilibrium equations for gas-phase reactions are often written using partial pressures rather than molar concentrations. For example, the equilibrium equation for the decomposition of N_2O_4 can be written as

$$K_p = \frac{(P_{NO_2})^2}{P_{N_2O_4}} \quad \text{For the reaction } N_2O_4(g) \rightleftharpoons 2\,NO_2(g)$$

where $P_{N_2O_4}$ and P_{NO_2} are the partial pressures (in atmospheres) of reactants and products at equilibrium, and the subscript p on K reminds us that the **equilibrium constant K_p** is defined using partial pressures. In this particular example, the units of K_p are $atm^2/atm = atm$, but, as for K_c, it's common practice to omit the units. Note that the equilibrium equations for K_p and K_c have the same form except that the expression for K_p contains partial pressures instead of molar concentrations.

The constants K_p and K_c for the general gas-phase reaction $a\,A + b\,B \rightleftarrows c\,C + d\,D$ are related because the pressure of each component in a mixture of ideal gases is directly proportional to its molar concentration. For component A, for example,

$$P_A V = n_A RT$$

so

$$P_A = \frac{n_A}{V}RT = [A]RT$$

Similarly, $P_B = [B]RT$, $P_C = [C]RT$, and $P_D = [D]RT$. The equilibrium equation for K_p is therefore given by

$$K_p = \frac{(P_C)^c(P_D)^d}{(P_A)^a(P_B)^b} = \frac{([C]RT)^c([D]RT)^d}{([A]RT)^a([B]RT)^b} = \frac{[C]^c[D]^d}{[A]^a[B]^b} \times (RT)^{(c+d)-(a+b)}$$

Because the first term on the right side is equal to K_c, the values of K_p and K_c are related by the equation

$$K_p = K_c(RT)^{\Delta n} \qquad \text{For the reaction } a\,A + b\,B \rightleftharpoons c\,C + d\,D$$

Here, R is the gas constant, 0.082 06 (L · atm)/(K · mol), T is the absolute temperature, and $\Delta n = (c+d) - (a+b)$ is the number of moles of gaseous products minus the number of moles of gaseous reactants.

For the decomposition of 1 mol of N_2O_4 to 2 mol of NO_2, $\Delta n = 2 - 1 = 1$, and $K_p = K_c(RT)$:

$$N_2O_4(g) \rightleftharpoons 2\,NO_2(g) \qquad K_p = K_c(RT)$$

For the reaction of 1 mol of hydrogen with 1 mol of iodine to give 2 mol of hydrogen iodide, $\Delta n = 2 - (1 + 1) = 0$, and $K_p = K_c(RT)^0 = K_c$:

$$H_2(g) + I_2(g) \rightleftharpoons 2\,HI(g) \qquad K_p = K_c$$

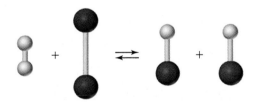

In general, K_p is equal to K_c only if the same number of moles of gases appear on both sides of the balanced chemical equation so that $\Delta n = 0$.

EXAMPLE 13.3 Methane (CH_4) reacts with hydrogen sulfide to yield H_2 and carbon disulfide, a solvent used in manufacturing rayon and cellophane:

$$CH_4(g) + 2\,H_2S(g) \rightleftharpoons CS_2(g) + 4\,H_2(g)$$

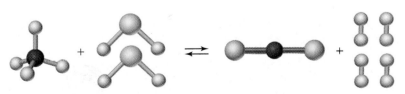

What is the value of K_p at 1000 K if the partial pressures in an equilibrium mixture at 1000 K are 0.20 atm of CH_4, 0.25 atm of H_2S, 0.52 atm of CS_2, and 0.10 atm of H_2?

SOLUTION Write the equilibrium equation by setting K_p equal to the equilibrium constant expression using partial pressures:

$$K_p = \frac{(P_{CS_2})(P_{H_2})^4}{(P_{CH_4})(P_{H_2S})^2}$$

Coefficient of H_2

Coefficient of H_2S

The partial pressures of products are in the numerator and the partial pressures of reactants are in the denominator, with the pressure of each substance raised to the power of its coefficient in the balanced chemical equation. Then substitute the partial pressures into the equilibrium equation and solve for K_p:

$$K_p = \frac{(P_{CS_2})(P_{H_2})^4}{(P_{CH_4})(P_{H_2S})^2} = \frac{(0.52)(0.10)^4}{(0.20)(0.25)^2} = 4.2 \times 10^{-3}$$

EXAMPLE 13.4   Hydrogen is produced industrially by the steam–hydrocarbon re-forming process. The reaction that takes place in the first step of this process is

$$H_2O(g) + CH_4(g) \rightleftharpoons CO(g) + 3\,H_2(g)$$

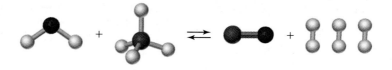

(a) If $K_c = 3.8 \times 10^{-3}$ at 1000 K, what is the value of K_p at the same temperature?
(b) If $K_p = 6.1 \times 10^4$ at 1125°C, what is the value of K_c at 1125°C?

SOLUTION
(a) For this reaction, $\Delta n = (1+3) - (1+1) = 2$. Therefore,

$$K_p = K_c(RT)^{\Delta n} = K_c(RT)^2 = (3.8 \times 10^{-3})[(0.082\,06)(1000)]^2 = 26$$

(b) Solving the equation $K_p = K_c(RT)^2$ for K_c gives

$$K_c = \frac{K_p}{(RT)^2} = \frac{6.1 \times 10^4}{[(0.082\,06)(1398)]^2} = 4.6$$

Note that the temperature in these equations is the absolute temperature; 1125°C corresponds to $1125 + 273 = 1398$ K.

▶ **PROBLEM 13.4**   In the industrial synthesis of hydrogen, mixtures of CO and H_2 are enriched in H_2 by allowing the CO to react with steam. The chemical equation for this so-called *water-gas shift reaction* is

$$CO(g) + H_2O(g) \rightleftharpoons CO_2(g) + H_2(g)$$

What is the value of K_p at 700 K if the partial pressures in an equilibrium mixture at 700 K are 1.31 atm of CO, 10.0 atm of H_2O, 6.12 atm of CO_2, and 20.3 atm of H_2?

▶ **PROBLEM 13.5**   Nitric oxide reacts with oxygen to give nitrogen dioxide, an important reaction in the Ostwald process for the industrial synthesis of nitric acid:

$$2\,NO(g) + O_2(g) \rightleftharpoons 2\,NO_2(g)$$

If $K_c = 6.9 \times 10^5$ at 227°C, what is the value of K_p at this temperature? If $K_p = 1.3 \times 10^{-2}$ at 1000 K, what is the value of K_c at 1000 K?

13.4   Heterogeneous Equilibria

Thus far we've been discussing **homogeneous equilibria**, in which all reactants and products are in a single phase, usually either gaseous or solution. **Heterogeneous equilibria**, by contrast, are those in which reactants and products are present in more than one phase. Take, for example, the thermal decomposition of solid calcium carbonate, a reaction used in manufacturing cement:

$$CaCO_3(s) \rightleftharpoons CaO(s) + CO_2(g)$$

When the reaction is carried out in a closed container, three phases are present at equilibrium: solid calcium carbonate, solid calcium oxide, and gaseous carbon dioxide. If we were to write the usual equilibrium equation for the reaction, including all the reactants and products, we would have

$$"K_c" = \frac{[CaO][CO_2]}{[CaCO_3]}$$

The manufacture of cement begins with the thermal decomposition of limestone, $CaCO_3$, in large kilns.

But because both CaO and $CaCO_3$ are pure solids, their molar "concentrations" are constants that can be calculated from their densities and molar masses:

$$[CaO] = \frac{3.25\ g}{cm^3} \times \frac{1\ mol}{56.1\ g} \times \frac{1000\ cm^3}{L} = 57.9\ mol/L$$

$$[CaCO_3] = \frac{2.71\ g}{cm^3} \times \frac{1\ mol}{100.1\ g} \times \frac{1000\ cm^3}{L} = 27.1\ mol/L$$

In general, the concentration of any pure solid is independent of its amount because its concentration is the *ratio* of its amount (in moles) to its volume (in liters). If, for example, you double the amount of $CaCO_3$, you also double its volume, but the ratio of the two (the concentration) remains constant.

Rearranging the equilibrium equation for the decomposition of $CaCO_3$ to combine the constants $[CaCO_3]$, $[CaO]$, and "K_c", we obtain

$$"K_c" \times \frac{[CaCO_3]}{[CaO]} = [CO_2]$$

We can rewrite this equation as $K_c = [CO_2]$, where the new equilibrium constant is $K_c = "K_c"[CaCO_3]/[CaO]$. The analogous equilibrium equation in terms of pressure is $K_p = P_{CO_2}$, where P_{CO_2} is the equilibrium pressure of CO_2 in atmospheres:

$$K_c = [CO_2] \qquad K_p = P_{CO_2}$$

As a general rule, the concentrations of pure solids and pure liquids are not included when writing an equilibrium equation because their concentrations are constants that are incorporated into the value of the equilibrium constant. We include only the concentrations of gases and the concentrations of solutes in solutions because only those concentrations can be varied.

To establish equilibrium between solid $CaCO_3$, solid CaO, and gaseous CO_2, all three components must be present. It follows from the equations $K_c = [CO_2]$ and $K_p = P_{CO_2}$, however, that the concentration and pressure of CO_2 at equilibrium are constant, independent of how much solid CaO and $CaCO_3$ is present (Figure 13.3). Of course, if the temperature is changed, the concentration and pressure of CO_2 will also change because the values of K_c and K_p depend on temperature.

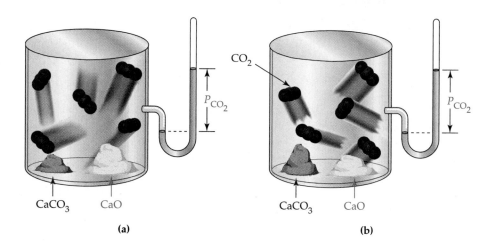

FIGURE 13.3 Thermal decomposition of calcium carbonate: $CaCO_3(s) \rightleftarrows CaO(s) + CO_2(g)$. At the same temperature, the equilibrium pressure of CO_2 (measured with a closed-end manometer) is the same in **(a)** and **(b)**, independent of how much solid $CaCO_3$ and CaO are present.

(a) (b)

When liquid mercury is in contact with an aqueous solution that contains $Hg_2(NO_3)_2$ and $Hg(NO_3)_2$, the concentration ratio $[Hg_2^{2+}]/[Hg^{2+}]$ in the aqueous layer (top) has a constant value.

EXAMPLE 13.5 Write the equilibrium equation for each of the following reactions:
(a) $CO_2(g) + C(s) \rightleftarrows 2\,CO(g)$ **(b)** $Hg(l) + Hg^{2+}(aq) \rightleftarrows Hg_2^{2+}(aq)$

SOLUTION
(a) Because carbon is a pure solid, its molar concentration is a constant that is incorporated into the equilibrium constant K_c. Therefore,

$$K_c = \frac{[CO]^2}{[CO_2]}$$

Alternatively, because CO and CO_2 are gases, the equilibrium equation can be written using partial pressures:

$$K_p = \frac{(P_{CO})^2}{P_{CO_2}}$$

The relationship between K_p and K_c is $K_p = K_c(RT)^{\Delta n}$, where $\Delta n = 2 - 1 = 1$.
(b) The concentrations of mercury(I) and mercury(II) ions appear in the equilibrium equation, but the concentration of mercury metal is omitted because, as a pure liquid, its concentration is a constant. Therefore,

$$K_c = \frac{[Hg_2^{2+}]}{[Hg^{2+}]}$$

▶ **PROBLEM 13.6**   Write the equilibrium constant expressions for K_c and K_p for each of the following reactions:
(a) $2\ Fe(s) + 3\ H_2O(g) \rightleftarrows Fe_2O_3(s) + 3\ H_2(g)$
(b) $2\ H_2O(l) \rightleftarrows 2\ H_2(g) + O_2(g)$

13.5 Using the Equilibrium Constant

Knowing the value of the equilibrium constant for a chemical reaction is important in many ways. For example, it lets us judge the extent of the reaction, predict the direction of the reaction, and calculate equilibrium concentrations from any initial concentrations. Let's look at each possibility.

Judging the Extent of Reaction

The numerical value of the equilibrium constant for a reaction indicates the extent to which reactants are converted to products; that is, it measures how far the reaction proceeds before the equilibrium state is reached. Consider, for example, the reaction of H_2 with O_2, which has a very large equilibrium constant ($K_c = 2.4 \times 10^{47}$ at 500 K):

$$2\ H_2(g) + O_2(g) \rightleftharpoons 2\ H_2O(g)$$

$$K_c = \frac{[H_2O]^2}{[H_2]^2[O_2]} = 2.4 \times 10^{47} \qquad \text{At 500 K}$$

Because products appear in the numerator of the equilibrium constant expression and reactants are in the denominator, a very large value of K_c means that the equilibrium ratio of products to reactants is very large. In other words, the reaction proceeds nearly to completion. For example, if stoichiometric amounts of H_2 and O_2 are allowed to react and $[H_2O] = 1.0$ M at equilibrium, then the concentrations of H_2 and O_2 that remain at equilibrium are negligibly small: $[H_2] = 2.0 \times 10^{-16}$ M and $[O_2] = 1.0 \times 10^{-16}$ M. (Try substituting these concentrations into the equilibrium equation to show that they satisfy the equation.)

By contrast, if a reaction has a very *small* value of K_c, the equilibrium ratio of products to reactants is very small and the reaction proceeds hardly at all before equilibrium is reached. For example, the reverse of the reaction of H_2 with O_2 gives the same equilibrium mixture as obtained from the forward reaction ($[H_2] = 2.0 \times 10^{-16}$ M, $[O_2] = 1.0 \times 10^{-16}$ M, $[H_2O] = 1.0$ M). The reverse reaction does not occur to any appreciable extent, however, because its equilibrium constant is so small: $K_c' = 1/K_c = 1/(2.4 \times 10^{47}) = 4.1 \times 10^{-48}$.

$$2\ H_2O(g) \rightleftharpoons 2\ H_2(g) + O_2(g)$$

$$K_c' = \frac{[H_2]^2[O_2]}{[H_2O]^2} = 4.1 \times 10^{-48} \qquad \text{At 500 K}$$

If a reaction has an intermediate value of K_c—say, a value in the range of 10^3 to 10^{-3}—then appreciable concentrations of both reactants and products are present in the equilibrium mixture. Take the reaction of hydrogen with iodine, which has $K_c = 57.0$ at 700 K:

$$H_2(g) + I_2(g) \rightleftharpoons 2\ HI(g)$$

$$K_c = \frac{[HI]^2}{[H_2][I_2]} = 57.0 \qquad \text{At 700 K}$$

If the equilibrium concentrations of H_2 and I_2 are both 0.010 M, then the concentration of HI at equilibrium is 0.075 M:

$$[HI]^2 = K_c[H_2][I_2]$$
$$[HI] = \sqrt{K_c[H_2][I_2]} = \sqrt{(57.0)(0.010)(0.010)} = 0.075 \text{ M}$$

Thus, the concentrations of both reactants and products—0.010 M and 0.075 M—are appreciable.

The gas-phase decomposition of N_2O_4 to NO_2 is another example of a reaction with a value of K_c that is neither large nor small: $K_c = 4.64 \times 10^{-3}$ at 25°C. Accordingly, equilibrium mixtures contain appreciable concentrations of both N_2O_4 and NO_2, as shown previously in Table 13.1.

We can make the following generalizations concerning the composition of equilibrium mixtures:

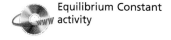
Equilibrium Constant activity

1. If $K_c > 10^3$, products predominate over reactants. If K_c is very large, the reaction proceeds nearly to completion.
2. If $K_c < 10^{-3}$, reactants predominate over products. If K_c is very small, the reaction proceeds hardly at all.
3. If K_c is in the range 10^{-3} to 10^3, appreciable concentrations of both reactants and products are present.

These points are illustrated in Figure 13.4.

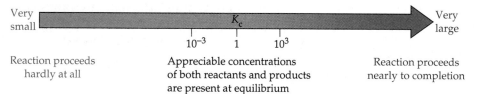

Very small

K_c

Very large

10^{-3} 1 10^3

Reaction proceeds hardly at all

Appreciable concentrations of both reactants and products are present at equilibrium

Reaction proceeds nearly to completion

FIGURE 13.4   Judging the extent of a reaction. The larger the value of the equilibrium constant K_c, the farther the reaction proceeds to the right before reaching the equilibrium state.

▶ **PROBLEM 13.7**   The value of K_c for the dissociation reaction $H_2(g) \rightleftharpoons 2\,H(g)$ is 1.2×10^{-42} at 500 K. Does the equilibrium mixture contain mainly H_2 molecules or H atoms? Explain in light of the position of hydrogen in the periodic table.

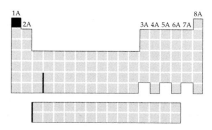

Predicting the Direction of Reaction

Consider again the gaseous reaction of hydrogen with iodine:

$$H_2(g) + I_2(g) \rightleftharpoons 2\,HI(g) \qquad K_c = 57.0 \text{ at } 700\,K$$

Suppose that we have a mixture of $H_2(g)$, $I_2(g)$, and $HI(g)$ at 700 K with concentrations $[H_2]_t = 0.10$ M, $[I_2]_t = 0.20$ M, and $[HI]_t = 0.40$ M. (The subscript t on the concentration symbols means that the concentrations were measured at some arbitrary time t, not necessarily at equilibrium.) If we substitute these concentrations into the equilibrium constant expression, we obtain a value called the **reaction quotient** Q_c.

$$\textbf{Reaction quotient:} \quad Q_c = \frac{[HI]_t^2}{[H_2]_t[I_2]_t} = \frac{(0.40)^2}{(0.10)(0.20)} = 8.0$$

The reaction quotient Q_c is defined in the same way as the equilibrium constant K_c except that the concentrations in Q_c are not necessarily equilibrium values.

For the case at hand, the numerical value of Q_c (8.0) is not equal to K_c (57.0), which means that the mixture of $H_2(g)$, $I_2(g)$, and $HI(g)$ is not at equilibrium—hardly surprising since the concentrations were chosen at random. As time passes, though, reaction will occur, changing the concentrations and thus changing the value of Q_c in the direction of K_c. After a sufficiently long time, an equilibrium state will be reached, and $Q_c = K_c$.

The usefulness of the reaction quotient Q_c is that it lets us predict the direction of reaction by comparing the values of Q_c and K_c. If Q_c is less than K_c, movement toward equilibrium increases Q_c by converting reactants to products, meaning that the reaction proceeds from left to right. If Q_c is greater than K_c, movement toward equilibrium decreases Q_c by converting products to reactants, meaning that the reaction proceeds from right to left. If Q_c equals K_c, the reaction mixture is already at equilibrium, and no net reaction occurs.

Thus, we can make the following generalizations concerning the direction of the reaction:

1. If $Q_c < K_c$, the reaction goes from left to right.
2. If $Q_c > K_c$, the reaction goes from right to left.
3. If $Q_c = K_c$, no net reaction occurs.

These points are illustrated in Figure 13.5.

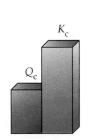
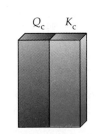

| Reactants → Products | Reactants and products are at equilibrium | Reactants ← Products |

FIGURE 13.5 Predicting the direction of reaction. The direction of reaction depends on the relative values of Q_c and K_c.

EXAMPLE 13.6   A mixture of 1.57 mol of N_2, 1.92 mol of H_2, and 8.13 mol of NH_3 is introduced into a 20.0 L reaction vessel at 500 K. At this temperature, the equilibrium constant K_c for the reaction $N_2(g) + 3\,H_2(g) \rightleftarrows 2\,NH_3(g)$ is 1.7×10^2. Is the reaction mixture at equilibrium? If not, what is the direction of the reaction?

BALLPARK SOLUTION   Approximate initial concentrations are calculated by dividing rounded values of the number of moles of each substance by the volume; $[N_2] \approx (1.6\ \text{mol})/(20\ \text{L}) \approx 0.08\ \text{M}$, $[H_2] \approx (2\ \text{mol})/(20\ \text{L}) \approx 0.1\ \text{M}$, and $[NH_3] \approx (8\ \text{mol})/(20\ \text{L}) \approx 0.4\ \text{M}$. An approximate value of the reaction quotient is obtained by substituting these concentrations into the equilibrium constant expression:

$$Q_c = \frac{[NH_3]_t^2}{[N_2]_t[H_2]_t^3} \approx \frac{(0.4)^2}{(0.08)(0.1)^3} \approx 2 \times 10^3$$

The numerator (0.16) divided by the denominator (0.08×10^{-3}) gives a value of 2×10^3 for Q_c. Because this approximate value of Q_c is about 10 times larger than K_c, we can be fairly confident that the reaction mixture is not at equilibrium and that the reaction will proceed from right to left, decreasing the concentration of NH_3 and increasing the concentrations of N_2 and H_2 until $Q_c = K_c = 1.7 \times 10^2$.

DETAILED SOLUTION   The initial concentration of N_2 is $(1.57\ \text{mol})/(20.0\ \text{L}) = 0.0785\ \text{M}$. Similarly, $[H_2] = 0.0960\ \text{M}$ and $[NH_3] = 0.406\ \text{M}$. Substituting these concentrations into the equilibrium constant expression gives

$$Q_c = \frac{[NH_3]_t^2}{[N_2]_t[H_2]_t^3} = \frac{(0.406)^2}{(0.0785)(0.0960)^3} = 2.37 \times 10^3$$

Again, we find that Q_c is greater than K_c (1.7×10^2), and so the reaction will proceed from right to left to reach equilibrium.

▶ **PROBLEM 13.8**   The equilibrium constant K_c for the reaction $2\ NO(g) + O_2(g) \rightleftharpoons 2\ NO_2(g)$ is 6.9×10^5 at 500 K. A 5.0 L reaction vessel at this temperature was filled with 0.060 mol of NO, 1.0 mol of O_2, and 0.80 mol of NO_2.
  (a) Is the reaction mixture at equilibrium? If not, in which direction does the reaction proceed?
  (b) What is the direction of the reaction if the initial amounts are 5.0×10^{-3} mol of NO, 0.20 mol of O_2, and 4.0 mol of NO_2?

◀━ **KEY CONCEPT PROBLEM 13.9**   The reaction $A_2 + B_2 \rightleftharpoons 2\ AB$ has an equilibrium constant $K_c = 4$. The following pictures represent reaction mixtures that contain A_2 molecules (red), B_2 molecules (blue), and AB molecules:

(1)   (2)   (3)

  (a) Which reaction mixture is at equilibrium?
  (b) For those reaction mixtures that are not at equilibrium, will the reaction go in the forward or reverse direction to reach equilibrium?

Calculating Equilibrium Concentrations

If the equilibrium constant and all the equilibrium concentrations but one are known, the unknown concentration can be calculated directly from the equilibrium equation. To illustrate, let's consider the following problem: What is the concentration of NO in an equilibrium mixture of gaseous NO, O_2, and NO_2 at 500 K that contains 1.0×10^{-3} M O_2 and 5.0×10^{-2} M NO_2? At this temperature, the equilibrium constant K_c for the reaction $2\ NO(g) + O_2(g) \rightleftharpoons 2\ NO_2(g)$ is 6.9×10^5.

In this problem, K_c and all the equilibrium concentrations except one are known, and we're asked to calculate the unknown equilibrium concentration. First, we write the equilibrium equation for the reaction, and solve for the unknown concentration:

$$K_c = \frac{[NO_2]^2}{[NO]^2[O_2]} \qquad [NO] = \sqrt{\frac{[NO_2]^2}{[O_2]K_c}}$$

Then we substitute the known values of K_c, $[O_2]$, and $[NO_2]$ into the expression for $[NO]$, and calculate its equilibrium concentration:

$$[NO] = \sqrt{\frac{(5.0 \times 10^{-2})^2}{(1.0 \times 10^{-3})(6.9 \times 10^5)}} = \sqrt{3.6 \times 10^{-6}} = \pm 1.9 \times 10^{-3}\,M$$

(Pressing the $\sqrt{x}$ key on a calculator, gives a positive number. Remember, though, that the square root of a positive number can be positive or negative.) Of the two roots, we choose the positive one ($[NO] = 1.9 \times 10^{-3}\,M$) because the concentration of a chemical substance is always a positive quantity.

To be sure that we haven't made any errors, it's a good idea to check the result by substituting it into the equilibrium equation:

$$K_c = 6.9 \times 10^5 = \frac{[NO_2]^2}{[NO]^2[O_2]} = \frac{(5.0 \times 10^{-2})^2}{(1.9 \times 10^{-3})^2(1.0 \times 10^{-3})} = 6.9 \times 10^5$$

Another type of problem is one in which we know the initial concentrations but do not know any of the equilibrium concentrations. To solve this kind of problem, follow the series of steps summarized in Figure 13.6 and illustrated in Examples 13.7 and 13.8.

Step 1. Write the balanced equation for the reaction.

Step 2. Under the balanced equation, make a table that lists for each substance involved in the reaction:
   (a) The initial concentration
   (b) The change in concentration on going to equilibrium
   (c) The equilibrium concentration
In constructing the table, define x as the concentration (mol/L) of one of the substances that reacts on going to equilibrium.

Step 3. Substitute the equilibrium concentrations into the equilibrium equation for the reaction and solve for x. If you must solve a quadratic equation, choose the mathematical solution that makes chemical sense.

Step 4. Calculate the equilibrium concentrations from the calculated value of x.

Step 5. Check your results by substituting them into the equilibrium equation.

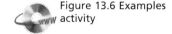

Figure 13.6 Examples activity

FIGURE 13.6 Steps to follow in calculating equilibrium concentrations from initial concentrations.

EXAMPLE 13.7 The equilibrium constant K_c for the reaction of H_2 with I_2 is 57.0 at 700 K:

$$H_2(g) + I_2(g) \rightleftharpoons 2\,HI(g) \qquad K_c = 57.0 \text{ at } 700 \text{ K}$$

If 1.00 mol of H_2 is allowed to react with 1.00 mol of I_2 in a 10.0 L reaction vessel at 700 K, what are the concentrations of H_2, I_2, and HI at equilibrium? What is the composition of the equilibrium mixture in moles?

SOLUTION This problem involves calculation of equilibrium concentrations from initial concentrations, so we should use the method outlined in Figure 13.6.

Step 1 The balanced equation is given: $H_2(g) + I_2(g) \rightleftharpoons 2\,HI(g)$.

Step 2 The initial concentrations are $[H_2] = [I_2] = (1.00 \text{ mol})/(10.0 \text{ L}) = 0.100$ M. For convenience, define an unknown, x, as the concentration (mol/L) of H_2 that reacts. According to the balanced equation for the reaction, x mol/L of H_2 reacts with x mol/L of I_2 to give $2x$ mol/L of HI. This reduces the initial concentrations of H_2 and I_2 from 0.100 mol/L to $(0.100 - x)$ mol/L at equilibrium. Let's summarize these results in a table under the balanced equation:

	$H_2(g)$	+	$I_2(g)$	$\rightleftharpoons$	$2\,HI(g)$
Initial concentration (M)	0.100		0.100		0
Change (M)	$-x$		$-x$		$+2x$
Equilibrium Concentration (M)	$(0.100 - x)$		$(0.100 - x)$		$2x$

Step 3 Substitute the equilibrium concentrations into the equilibrium equation for the reaction:

$$K_c = 57.0 = \frac{[HI]^2}{[H_2][I_2]} = \frac{(2x)^2}{(0.100 - x)(0.100 - x)} = \left(\frac{2x}{0.100 - x}\right)^2$$

Because the right side of this equation is a perfect square, we can take the square root of both sides:

$$\sqrt{57.0} = \pm 7.55 = \frac{2x}{0.100 - x}$$

Solving for x, we obtain two solutions. The equation with the positive square root of 57.0 gives

$$+7.55(0.100 - x) = 2x$$
$$0.755 = 2x + 7.55x$$
$$x = \frac{0.755}{9.55} = 0.0791 \text{ M}$$

The equation with the negative square root of 57.0 gives

$$-7.55(0.100 - x) = 2x$$
$$-0.755 = 2x - 7.55x$$
$$x = \frac{-0.755}{-5.55} = 0.136 \text{ M}$$

Because the initial concentrations of H_2 and I_2 are 0.100 M, x can't exceed 0.100 M. Therefore, discard $x = 0.136$ M as chemically unreasonable and choose the first solution, $x = 0.0791$ M.

Step 4 Calculate the equilibrium concentrations from the calculated value of **x**:

$$[H_2] = [I_2] = 0.100 - x = 0.100 - 0.0791 = 0.021 \text{ M}$$
$$[HI] = 2x = (2)(0.0791) = 0.158 \text{ M}$$

Step 5 Check the results by substituting them into the equilibrium equation:

$$K_c = 57.0 = \frac{[HI]^2}{[H_2][I_2]} = \frac{(0.158)^2}{(0.021)(0.021)} = 57$$

The number of moles of each substance in the equilibrium mixture can be obtained by multiplying each concentration by the volume of the reaction vessel:

$$\text{Moles of } H_2 = \text{Moles of } I_2 = (0.021 \text{ mol/L})(10.0 \text{ L}) = 0.21 \text{ mol}$$
$$\text{Moles of HI} = (0.158 \text{ mol/L})(10.0 \text{ L}) = 1.58 \text{ mol}$$

EXAMPLE 13.8   Calculate the equilibrium concentrations of H_2, I_2, and HI at 700 K if the initial concentrations are $[H_2] = 0.100$ M and $[I_2] = 0.200$ M. The equilibrium constant K_c for the reaction $H_2(g) + I_2(g) \rightleftharpoons 2 \text{ HI}(g)$ is 57.0 at 700 K.

SOLUTION   This problem is similar to Example 13.7 except that the initial concentrations of H_2 and I_2 are unequal.

Step 1 The balanced equation is $H_2(g) + I_2(g) \rightleftharpoons 2 \text{ HI}(g)$.

Step 2 Again, define x as the concentration of H_2 that reacts. Set up a table of concentrations under the balanced equation:

	$H_2(g)$	$+$	$I_2(g)$	$\rightleftharpoons$	$2 \text{ HI}(g)$
Initial concentration (M)	0.100		0.200		0
Change (M)	$-x$		$-x$		$+2x$
Equilibrium concentration (M)	$(0.100 - x)$		$(0.200 - x)$		$2x$

Step 3 Substitute the equilibrium concentrations into the equilibrium equation:

$$K_c = 57.0 = \frac{[HI]^2}{[H_2][I_2]} = \frac{(2x)^2}{(0.100 - x)(0.200 - x)}$$

Because the right side of this equation is not a perfect square, we must put the equation into the standard quadratic form, $ax^2 + bx + c = 0$, and then solve for x using the quadratic formula (Appendix A.4):

$$x = \frac{-b \pm \sqrt{b^2 - 4ac}}{2a}$$

Rearranging the equilibrium equation gives

$$(57.0)(0.0200 - 0.300x + x^2) = 4x^2$$

or

$$53.0x^2 - 17.1x + 1.14 = 0$$

Substituting the values of a, b, and c into the quadratic formula gives two solutions:

$$x = \frac{17.1 \pm \sqrt{(-17.1)^2 - 4(53.0)(1.14)}}{2(53.0)} = \frac{17.1 \pm 7.1}{106} = 0.228 \quad \text{and} \quad 0.0943$$

Discard the solution that uses the positive square root ($x = 0.228$) because the H_2 concentration can't change by more than its initial value (0.100 M). Therefore, choose the solution that uses the negative square root ($x = 0.0943$).

Step 4 Calculate the equilibrium concentrations from the calculated value of x:

$$[H_2] = 0.100 - x = 0.100 - 0.0943 = 0.006 \text{ M}$$
$$[I_2] = 0.200 - x = 0.200 - 0.0943 = 0.106 \text{ M}$$
$$[HI] = 2x = (2)(0.0943) = 0.189 \text{ M}$$

Step 5 Check the results by substituting them into the equilibrium equation:

$$K_c = 57.0 = \frac{[HI]^2}{[H_2][I_2]} = \frac{(0.189)^2}{(0.006)(0.106)} = 56.2$$

The calculated value of K_c (56.2), which should be rounded to one significant figure (6×10^1), agrees with the value given in the problem (57.0).

▶ **PROBLEM 13.10** In Problem 13.7, we found that an equilibrium mixture of H_2 molecules and H atoms at 500 K contains mainly H_2 molecules because the equilibrium constant for the dissociation reaction $H_2(g) \rightleftarrows 2 H(g)$ is very small ($K_c = 1.2 \times 10^{-42}$).
(a) What is the molar concentration of H atoms if $[H_2] = 0.10$ M?
(b) How many H atoms and H_2 molecules are present in 1.0 L of 0.10 M H_2 at 500 K?

▶ **PROBLEM 13.11** The H_2/CO ratio in mixtures of carbon monoxide and hydrogen (called *synthesis gas*) is increased by the water-gas shift reaction $CO(g) + H_2O(g) \rightleftarrows CO_2(g) + H_2(g)$, which has an equilibrium constant $K_c = 4.24$ at 800 K. Calculate the equilibrium concentrations of CO_2, H_2, CO, and H_2O at 800 K if only CO and H_2O are present initially at concentrations of 0.150 M.

▶ **PROBLEM 13.12** Calculate the equilibrium concentrations of N_2O_4 and NO_2 at 25°C in a vessel that contains an initial N_2O_4 concentration of 0.0500 M. The equilibrium constant K_c for the reaction $N_2O_4(g) \rightleftarrows 2 NO_2(g)$ is 4.64×10^{-3} at 25°C.

▶ **PROBLEM 13.13** Calculate the equilibrium concentrations at 25°C for the reaction in Problem 13.12 for initial concentrations of $[N_2O_4] = 0.0200$ M and $[NO_2] = 0.0300$ M.

13.6 Factors That Alter the Composition of an Equilibrium Mixture

One of the principal goals of chemical synthesis is to effect a maximum conversion of reactants to products with a minimum expenditure of energy. This objective is achieved easily if the reaction goes nearly to completion at mild temperature and pressure. But if the reaction gives an equilibrium mixture that is rich in reactants and poor in products, then the experimental conditions must be adjusted. For example, in the Haber process for the synthesis of ammonia from N_2 and H_2 (Figure 13.7), the choice of experimental conditions is of real economic importance. Annual U.S. production of ammonia is about 20 million tons, primarily for use as fertilizer.

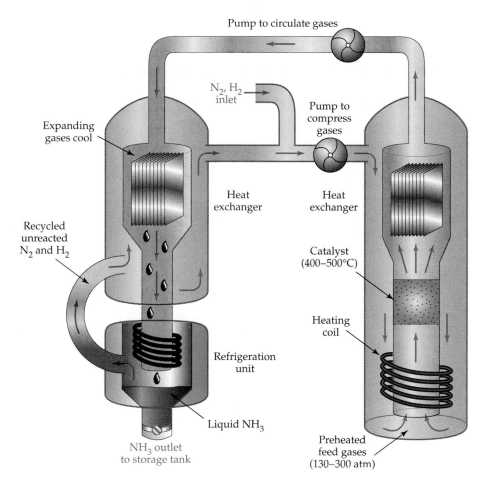

FIGURE 13.7 Representation of the Haber process for the industrial production of ammonia. A mixture of gaseous N_2 and H_2 at 130–300 atm pressure is passed over a catalyst at 400–500°C, and ammonia is produced by the reaction $N_2(g) + 3\,H_2(g) \rightleftharpoons 2\,NH_3(g)$. The NH_3 in the gaseous mixture of reactants and products is liquefied, and the unreacted N_2 and H_2 are recycled.

Several factors can be exploited to alter the composition of an equilibrium mixture:

1. The concentration of reactants or products can be changed.
2. The pressure and volume can be changed.
3. The temperature can be changed.

(A possible fourth factor, addition of a catalyst, increases only the rate at which equilibrium is reached. As we'll see in Section 13.10, a catalyst does not affect the equilibrium concentrations.)

The qualitative effect of the listed changes on the composition of an equilibrium mixture can be predicted using a principle first described by the French chemist Henri-Louis Le Châtelier:

◆—— LE CHÂTELIER'S PRINCIPLE If a stress is applied to a reaction mixture at equilibrium, reaction occurs in the direction that relieves the stress.

The word "stress" in this context means a change in concentration, pressure, volume, or temperature that disturbs the original equilibrium. Reaction then occurs to change the composition of the mixture until a new state of equilibrium is reached. The direction that the reaction takes (reactants to products, or vice versa) is the one that reduces the stress. In the next three sections, we'll look at the different kinds of stress that can change the composition of an equilibrium mixture.

13.7 Altering an Equilibrium Mixture: Changes in Concentration

Let's consider the equilibrium that occurs in the Haber process for synthesis of ammonia:

$$N_2(g) + 3 H_2(g) \rightleftharpoons 2 NH_3(g) \quad K_c = 0.291 \text{ at } 700 \text{ K}$$

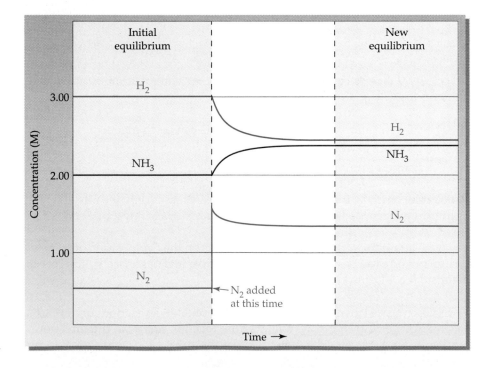

Suppose that we have an equilibrium mixture of 0.50 M N_2, 3.00 M H_2, and 1.98 M NH_3 at 700 K, and that we disturb the equilibrium by increasing the N_2 concentration to 1.50 M. Le Châtelier's principle tells us that reaction will occur to relieve the stress of the increased concentration of N_2 by converting some of the N_2 to NH_3. Of course, as the N_2 concentration decreases, the H_2 concentration must also decrease and the NH_3 concentration must increase in accord with the stoichiometry of the balanced equation. These changes are illustrated in Figure 13.8.

FIGURE 13.8   Changes in concentrations when N_2 is added to an equilibrium mixture of N_2, H_2, and NH_3. Net conversion of N_2 and H_2 to NH_3 occurs until a new equilibrium is established. That is, N_2 and H_2 concentrations decrease, while the NH_3 concentration increases.

In general, when an equilibrium is disturbed by the addition or removal of any reactant or product, Le Châtelier's principle predicts that

1. The concentration stress of an *added* reactant or product is relieved by reaction in the direction that *consumes* the added substance.
2. The concentration stress of a *removed* reactant or product is relieved by reaction in the direction that *replenishes* the removed substance.

Application of these rules to the equilibrium $N_2(g) + 3 H_2(g) \rightleftarrows 2 NH_3(g)$ indicates that the yield of ammonia is increased by an increase in the N_2 or H_2 concentration or by a decrease in the NH_3 concentration (Figure 13.9). In the industrial production of ammonia, the concentration of gaseous NH_3 is decreased by liquefying the ammonia (bp $-33°C$) as it's formed, and so more ammonia is produced.

Le Châtelier's principle is a handy rule for predicting changes in the composition of an equilibrium mixture, but it doesn't explain *why* those changes occur. To see why Le Châtelier's principle works, let's look again at the reaction quotient Q_c. For the initial equilibrium mixture of 0.50 M N_2, 3.00 M H_2, and 1.98 M NH_3 at 700 K, Q_c is equal to the equilibrium constant K_c (0.291) because the system is at equilibrium:

$$Q_c = \frac{[NH_3]_t^2}{[N_2]_t[H_2]_t^3} = \frac{(1.98)^2}{(0.50)(3.00)^3} = 0.29 = K_c$$

When we disturb the equilibrium by increasing the N_2 concentration to 1.50 M, the denominator of the equilibrium constant expression increases and Q_c decreases to a value less than K_c:

$$Q_c = \frac{[NH_3]_t^2}{[N_2]_t[H_2]_t^3} = \frac{(1.98)^2}{(1.50)(3.00)^3} = 0.0968 < K_c$$

For the system to move to a new state of equilibrium, Q_c must increase; that is, the numerator of the equilibrium constant expression must increase and the denominator must decrease. This implies net conversion of N_2 and H_2 to NH_3, just as predicted by Le Châtelier's principle. When the new equilibrium is established (Figure 13.8), the concentrations are 1.31 M N_2, 2.43 M H_2, and 2.36 M NH_3, and Q_c is again equal to K_c:

$$Q_c = \frac{[NH_3]_t^2}{[N_2]_t[H_2]_t^3} = \frac{(2.36)^2}{(1.31)(2.43)^3} = 0.296 = K_c$$

As another example of how a change in concentration affects an equilibrium, let's consider the reaction in aqueous solution of iron(III) and thiocyanate (SCN^-) ions to give an equilibrium mixture that contains the Fe–N bonded red complex ion $FeNCS^{2+}$:

$$Fe^{3+}(aq) + SCN^-(aq) \rightleftarrows FeNCS^{2+}(aq)$$
Pale yellow   Colorless   Red

Shifts in the position of this equilibrium can be detected by observing how the color of the solution changes when we add various reagents (Figure 13.10). If we add aqueous $FeCl_3$, the red color gets darker, as

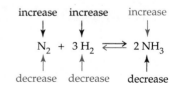

increase   increase   increase

$N_2 + 3 H_2 \rightleftarrows 2 NH_3$

decrease   decrease   decrease

FIGURE 13.9 Effect of concentration changes on the equilibrium $N_2(g) + 3 H_2(g) \rightleftarrows 2 NH_3(g)$. An increase in the N_2 or H_2 concentration or a decrease in the NH_3 concentration shifts the equilibrium from left to right. A decrease in the N_2 or H_2 concentration or an increase in the NH_3 concentration shifts the equilibrium from right to left.

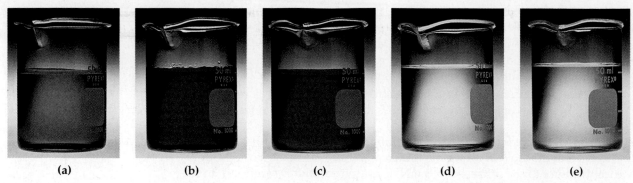

(a) (b) (c) (d) (e)

FIGURE 13.10 Color changes produced by adding various reagents to an equilibrium mixture of Fe^{3+} (pale yellow), SCN^- (colorless), and $FeNCS^{2+}$ (red): **(a)** The original solution. **(b)** After adding $FeCl_3$ to the original solution, the red color is darker because of an increase in $[FeNCS^{2+}]$. **(c)** After adding KSCN to the original solution, the red color again deepens. **(d)** After adding $H_2C_2O_4$ to the original solution, the red color disappears because of a decrease in $[FeNCS^{2+}]$; the yellow color is due to $Fe(C_2O_4)_3^{3-}$. **(e)** After adding $HgCl_2$ to the original solution, the red color again vanishes.

Le Châtelier's Principle movie

predicted by Le Châtelier's principle. The concentration stress of added Fe^{3+} is relieved by reaction from left to right, which consumes some of the Fe^{3+} and increases the concentration of $FeNCS^{2+}$. (Note that the Cl^- ions are not involved in the reaction.) Similarly, if we add aqueous KSCN, the stress of added SCN^- shifts the equilibrium from left to right, and again the red color gets darker.

The equilibrium can be shifted in the opposite direction by adding reagents that remove Fe^{3+} or SCN^- ions. For example, oxalic acid ($H_2C_2O_4$), a poisonous substance present in plants such as rhubarb, reacts with Fe^{3+} to form the stable complex ion $Fe(C_2O_4)_3^{3-}$, thus decreasing the concentration of free $Fe^{3+}(aq)$. In accord with Le Châtelier's principle, the concentration stress of removed Fe^{3+} is relieved by dissociation of $FeNCS^{2+}$ to replenish the Fe^{3+} ions. Because the concentration of $FeNCS^{2+}$ decreases, the red color disappears.

Addition of aqueous $HgCl_2$ also eliminates the red color because Hg^{2+} reacts with SCN^- ions to form the stable Hg–S bonded complex ion $Hg(SCN)_4^{2-}$. Removal of free $SCN^-(aq)$ shifts the equilibrium $Fe^{3+}(aq) + SCN^-(aq) \rightleftarrows FeNCS^{2+}(aq)$ from right to left to replenish the SCN^- ions.

EXAMPLE 13.9 The reaction of iron(III) oxide with carbon monoxide occurs in a blast furnace when iron ore is reduced to iron metal:

$$Fe_2O_3(s) + 3\,CO(g) \rightleftharpoons 2\,Fe(l) + 3\,CO_2(g)$$

Use Le Châtelier's principle to predict the direction of reaction when an equilibrium mixture is disturbed by:
(a) Adding Fe_2O_3
(b) Removing CO_2
(c) Removing CO; also account for the change using the reaction quotient Q_c.

SOLUTION
(a) Because Fe_2O_3 is a solid, its "concentration" doesn't change when more Fe_2O_3 is added. Therefore, there is no concentration stress, and the original equilibrium is undisturbed.

(b) Le Châtelier's principle predicts that the concentration stress of removed CO_2 will be relieved by reaction from left to right to replenish the CO_2.

(c) Le Châtelier's principle predicts that the concentration stress of removed CO will be relieved by reaction from right to left to replenish the CO. The reaction quotient is

$$Q_c = \frac{[CO_2]_t^3}{[CO]_t^3}$$

When the equilibrium is disturbed by reducing [CO], Q_c increases, so that $Q_c > K_c$. For the system to move to a new state of equilibrium, Q_c must decrease— that is, $[CO_2]$ must decrease and [CO] must increase. Therefore, the reaction goes from right to left, as predicted by Le Châtelier's principle.

▶ **PROBLEM 13.14** Consider the equilibrium for the water-gas shift reaction:

$$CO(g) + H_2O(g) \rightleftharpoons CO_2(g) + H_2(g)$$

Use Le Châtelier's principle to predict how the concentration of H_2 will change when the equilibrium is disturbed by:
(a) Adding CO
(b) Adding CO_2
(c) Removing H_2O
(d) Removing CO_2; also account for the change using the reaction quotient Q_c.

13.8 Altering an Equilibrium Mixture: Changes in Pressure and Volume

To illustrate how an equilibrium mixture is affected by a change in pressure as a result of a change in the volume, let's return to the Haber synthesis of ammonia. The balanced equation for the reaction has 4 mol of gas on the left side of the equation and 2 mol on the right side:

$$N_2(g) + 3 H_2(g) \rightleftharpoons 2 NH_3(g) \qquad K_c = 0.291 \text{ at } 700 \text{ K}$$

What happens to the composition of the equilibrium mixture if we increase the pressure by decreasing the volume? (Recall from Sections 9.2 and 9.3 that the pressure of an ideal gas is inversely proportional to the volume at constant temperature and constant number of moles of gas; $P = nRT/V$.) According to Le Châtelier's principle, reaction will occur in the direction that relieves the stress of the increased pressure, which means that the number of moles of gas must decrease. Therefore, we predict that the reaction will proceed from left to right because the forward reaction converts 4 mol of gaseous reactants to 2 mol of gaseous products.

In general, Le Châtelier's principle predicts that

1. An *increase* in pressure by reducing the volume will bring about net reaction in the direction that *decreases* the number of moles of gas.

2. A *decrease* in pressure by enlarging the volume will bring about net reaction in the direction that *increases* the number of moles of gas.

To see why Le Châtelier's principle works for pressure (volume) changes, let's look again at the reaction quotient for the equilibrium mixture of 0.50 M N_2, 3.00 M H_2, and 1.98 M NH_3 at 700 K:

$$Q_c = \frac{[NH_3]_t^2}{[N_2]_t[H_2]_t^3} = \frac{(1.98)^2}{(0.50)(3.00)^3} = 0.29 = K_c$$

If we disturb the equilibrium by reducing the volume by a factor of 2, we not only double the total pressure, we also double the molar concentration of each reactant and product (because molarity $= n/V$). Because the balanced equation has more moles of gaseous reactants than gaseous products, the increase in the denominator of the equilibrium constant expression is greater than the increase in the numerator, and the new value of Q_c is less than the equilibrium constant K_c:

$$Q_c = \frac{[NH_3]_t^2}{[N_2]_t[H_2]_t^3} = \frac{(3.96)^2}{(1.00)(6.00)^3} = 0.0726 < K_c$$

For the system to move to a new state of equilibrium, Q_c must increase, which means that the reaction must go from left to right, as predicted by Le Châtelier's principle (Figure 13.11). In practice, the yield of ammonia in the Haber process is increased by running the reaction at high pressure, typically 130–300 atm.

FIGURE 13.11 Qualitative effect of pressure and volume on the equilibrium $N_2(g) + 3 H_2(g) \rightleftarrows 2 NH_3(g)$. **(a)** A mixture of gaseous N_2, H_2, and NH_3 at equilibrium. **(b)** When the pressure is increased by decreasing the volume, the mixture is no longer at equilibrium ($Q_c < K_c$). **(c)** Reaction occurs from left to right, decreasing the total number of gaseous molecules until equilibrium is re-established ($Q_c = K_c$).

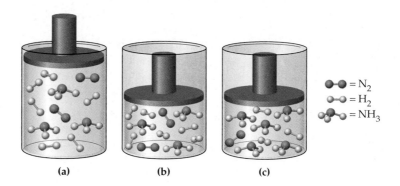

(a)   (b)   (c)

$\bullet\!\!-\!\!\bullet$ = N_2
$\circ\!\!-\!\!\circ$ = H_2
$\circ\!\!\bullet\!\!\circ$ = NH_3

Of course, the composition of an equilibrium mixture is unaffected by a change in pressure if the reaction involves no change in the number of moles of gas. For example, the reaction of hydrogen with iodine has 2 mol of gas on both sides of the balanced equation:

$$H_2(g) + I_2(g) \rightleftarrows 2 HI(g)$$

If we double the pressure by halving the volume, the numerator and denominator of the reaction quotient change by the same factor, and Q_c remains unchanged:

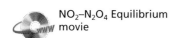

NO$_2$–N$_2$O$_4$ Equilibrium movie

$$Q_c = \frac{[HI]_t^2}{[H_2]_t[I_2]_t}$$

In applying Le Châtelier's principle to a heterogeneous equilibrium, the effect of pressure changes on solids and liquids can be ignored because the volume (and concentration) of a solid or a liquid is nearly independent of pressure. Consider, for example, the high-temperature reaction of carbon with steam, the first step in converting coal to gaseous fuels:

$$C(s) + H_2O(g) \rightleftarrows CO(g) + H_2(g)$$

Ignoring the carbon because it's a solid, we predict that a decrease in volume (increase in pressure) will shift the equilibrium from right to left because the reverse reaction decreases the amount of gas from 2 mol to 1 mol.

Throughout this section, we've been careful to limit the application of Le Châtelier's principle to pressure changes that result from a change *in volume*. What happens, though, if we keep the volume constant but increase the total pressure by adding a gas that is not involved in the reaction—say, an inert gas such as argon? In that case, the equilibrium remains undisturbed because adding an inert gas at constant volume does not change the partial pressures or the molar concentrations of the substances involved in the reaction. Only if the added gas is a reactant or product does the reaction quotient change.

EXAMPLE 13.10   Does the number of moles of reaction products increase, decrease, or remain the same when each of the following equilibria is subjected to a decrease in pressure by increasing the volume?
(a) $PCl_5(g) \rightleftharpoons PCl_3(g) + Cl_2(g)$
(b) $CaO(s) + CO_2(g) \rightleftharpoons CaCO_3(s)$
(c) $3 Fe(s) + 4 H_2O(g) \rightleftharpoons Fe_3O_4(s) + 4 H_2(g)$

SOLUTION
(a) According to Le Châtelier's principle, the stress of a decrease in pressure is relieved by net reaction in the direction that increases the number of moles of gas. Since the forward reaction converts 1 mol of gas to 2 mol of gas, the reaction will go from left to right, thus increasing the number of moles of PCl_3 and Cl_2.

(b) Because there is 1 mol of gas on the left side of the balanced equation and none on the right side, the stress of a decrease in pressure is relieved by reaction from right to left. The number of moles of $CaCO_3$ therefore decreases.

(c) Because there are 4 mol of gas on both sides of the balanced equation, the composition of the equilibrium mixture is unaffected by a change in pressure. The number of moles of Fe_3O_4 and H_2 remains the same.

▶ **PROBLEM 13.15**   Does the number of moles of products increase, decrease, or remain the same when each of the following equilibria is subjected to an increase in pressure by decreasing the volume?
(a) $CO(g) + H_2O(g) \rightleftharpoons CO_2(g) + H_2(g)$
(b) $2 CO(g) \rightleftharpoons C(s) + CO_2(g)$
(c) $N_2O_4(g) \rightleftharpoons 2 NO_2(g)$

KEY CONCEPT PROBLEM 13.16   The following picture represents the equilibrium mixture for the gas-phase reaction $A_2 \rightleftharpoons 2 A$.

Draw a picture that shows how the concentrations change when the pressure is increased by reducing the volume.

13.9 Altering an Equilibrium Mixture: Changes in Temperature

When an equilibrium is disturbed by a change in concentration, pressure, or volume, the composition of the equilibrium mixture changes because the reaction quotient Q_c is no longer equal to the equilibrium constant K_c. As long as the temperature remains constant, however, concentration, pressure, or volume changes don't change the *value* of the equilibrium constant.

By contrast, a change in temperature nearly always changes the value of the equilibrium constant. For the Haber synthesis of ammonia, which is an exothermic reaction, the equilibrium constant K_c decreases by a factor of 10^{11} over the temperature range 300–1000 K (Figure 13.12).

$$N_2(g) + 3\,H_2(g) \rightleftharpoons 2\,NH_3(g) + 92.2\ kJ \qquad \Delta H° = -92.2\ kJ$$

At low temperatures, the equilibrium mixture is rich in NH_3 because K_c is large. At high temperatures, the equilibrium shifts in the direction of N_2 and H_2.

FIGURE 13.12 Temperature dependence of the equilibrium constant for the reaction $N_2(g) + 3\,H_2(g) \rightleftharpoons 2\,NH_3(g)$. Note that K_c is plotted on a logarithmic scale and decreases by a factor of 10^{11} on raising the temperature from 300 K to 1000 K.

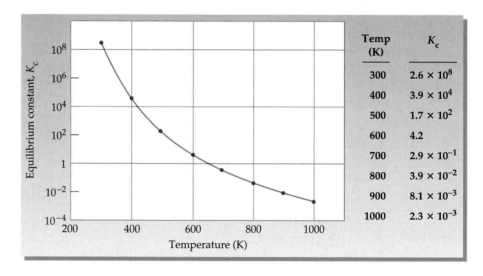

Temp (K)	K_c
300	2.6×10^8
400	3.9×10^4
500	1.7×10^2
600	4.2
700	2.9×10^{-1}
800	3.9×10^{-2}
900	8.1×10^{-3}
1000	2.3×10^{-3}

In general, the temperature dependence of the equilibrium constant depends on the sign of $\Delta H°$ for the reaction.

1. The equilibrium constant for an exothermic reaction (negative $\Delta H°$) decreases as the temperature increases.
2. The equilibrium constant for an endothermic reaction (positive $\Delta H°$) increases as the temperature increases.

Le Châtelier's Principle movie

If you forget the way in which K_c depends on temperature, you can predict it using Le Châtelier's principle. Take the endothermic decomposition of N_2O_4, for example:

$$N_2O_4(g) + 57.2\ kJ \rightleftharpoons 2\,NO_2(g) \qquad \Delta H° = +57.2\ kJ$$

Colorless   Brown

Le Châtelier's principle says that if heat is added to an equilibrium mixture (thus increasing its temperature), net reaction occurs in the direction that relieves the stress of the added heat. For an endothermic reaction, such as the decomposition of N_2O_4, heat is absorbed by reaction in the forward direction. The equilibrium therefore shifts to the right at the higher temperature, which means that K_c increases with increasing temperature.

Because N_2O_4 is colorless and NO_2 has a brown color, the effect of temperature on the N_2O_4–NO_2 equilibrium is readily apparent from the color of the mixture (Figure 13.13). For an exothermic reaction, such as the Haber synthesis of NH_3, heat is absorbed by reaction in the reverse direction, and consequently K_c decreases with increasing temperature.

FIGURE 13.13 Sample tubes containing an equilibrium mixture of N_2O_4 and NO_2 immersed in ice water (left) and hot water (right). The darker brown color of the sample at the higher temperature indicates that the equilibrium $N_2O_4(g) \rightleftharpoons 2 NO_2(g)$ shifts from left to right (K_c increases) with increasing temperature, as expected for an endothermic reaction.

EXAMPLE 13.11   In the first step of the Ostwald process for synthesis of nitric acid, ammonia is oxidized to nitric oxide by the reaction

$$4 NH_3(g) + 5 O_2(g) \rightleftharpoons 4 NO(g) + 6 H_2O(g) \qquad \Delta H° = -905.6 \text{ kJ}$$

How does the equilibrium amount of NO vary with an increase in temperature?

SOLUTION   Le Châtelier's principle predicts that a stress of added heat when the temperature is increased will be relieved by net reaction in the direction that absorbs the heat. Because the reaction is exothermic, heat is absorbed when the reaction proceeds from right to left. The equilibrium will therefore shift to the left (K_c will decrease) with an increase in temperature. Consequently, the equilibrium mixture will contain less NO at higher temperatures.

▶ **PROBLEM 13.17**   When air is heated at very high temperatures in an automobile engine, the air pollutant nitric oxide is produced by the reaction

$$N_2(g) + O_2(g) \rightleftharpoons 2 NO(g) \qquad \Delta H° = +180.5 \text{ kJ}$$

How does the equilibrium amount of NO vary with an increase in temperature?

▶ **PROBLEM 13.18**   Ethyl acetate, a solvent used as fingernail-polish remover, is made by reaction of acetic acid with ethanol:

$$\underset{\text{Acetic acid}}{CH_3CO_2H(soln)} + \underset{\text{Ethanol}}{C_2H_5OH(soln)} \rightleftharpoons \underset{\text{Ethyl acetate}}{CH_3CO_2C_2H_5(soln)} + H_2O(soln)$$

$$\Delta H = -2.9 \text{ kJ}$$

In this equation, *soln* denotes a largely organic solution that also contains water. Does the amount of ethyl acetate in an equilibrium mixture increase or decrease when the temperature is increased? How does K_c change when the temperature is decreased? Justify your answers using Le Châtelier's principle.

Fingernail polish can be removed by dissolving it in ethyl acetate.

⊶ KEY CONCEPT PROBLEM 13.19 The following pictures represent the composition of the equilibrium mixture for the reaction $A(g) + B(s) \rightleftarrows AB(g)$ at 400 K and 500 K:

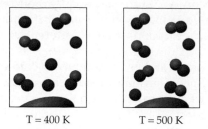

T = 400 K   T = 500 K

Is the reaction endothermic or exothermic? Explain using Le Châtelier's principle.

13.10 The Effect of a Catalyst on Equilibrium

We saw in Section 12.11 that a catalyst increases the rate of a chemical reaction by making available a new, lower-energy pathway for conversion of reactants to products. Because the forward and reverse reactions pass through the same transition state, a catalyst lowers the activation energy for the forward and reverse reactions by exactly the same amount. As a result, the rates of the forward and reverse reactions increase by the same factor (Figure 13.14).

FIGURE 13.14 Potential energy profiles for a reaction whose activation energy is lowered by the presence of a catalyst. The activation energy for the catalyzed pathway (red curve) is lower than that for the uncatalyzed pathway (blue curve) by an amount ΔE_a. The catalyst lowers the activation energy barrier for the forward and reverse reactions by exactly the same amount. The catalyst therefore accelerates the forward and reverse reactions by the same factor, and the composition of the equilibrium mixture is unchanged.

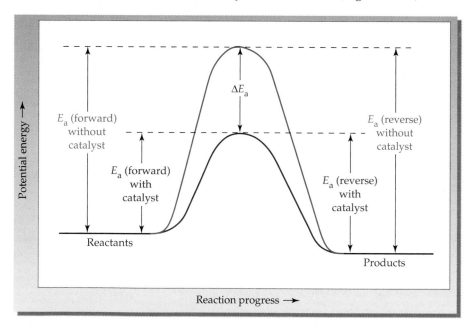

If a reaction mixture is at equilibrium in the absence of a catalyst, meaning that the forward and reverse rates are equal, it will still be at equilibrium after addition of a catalyst because the forward and reverse rates, though faster, remain equal. If a reaction mixture is not at equilibrium, a catalyst accelerates the rate at which equilibrium is reached, but it does not affect the composition of the equilibrium mixture. Because a catalyst has no effect on the equilibrium concentrations, it does not appear in the balanced chemical equation or in the equilibrium constant expression.

Even though it doesn't change the position of an equilibrium, a catalyst can nevertheless have an important influence on the choice of optimum conditions for a reaction. Consider again the Haber synthesis of ammonia. Because the reaction $N_2(g) + 3\,H_2(g) \rightleftharpoons 2\,NH_3(g)$ is exothermic, its equilibrium constant decreases with increasing temperature, and optimum yields of NH_3 are obtained at low temperatures. At those low temperatures, however, the rate at which equilibrium is reached is too slow for the reaction to be practical. We thus have what appears to be a no-win situation: Low temperatures give good yields but slow rates, whereas high temperatures give satisfactory rates but poor yields. The answer to the dilemma is to find a catalyst.

In the early 1900s, the German chemist Fritz Haber discovered that a catalyst consisting of iron mixed with certain metal oxides causes the reaction to occur at a satisfactory rate at temperatures where the equilibrium concentration of NH_3 is reasonably favorable. The yield of NH_3 can be improved further by running the reaction at high pressures. Typical reaction conditions for the industrial synthesis of ammonia are 400–500°C and 130–300 atm.

▶ **PROBLEM 13.20** A platinum catalyst is used in automobile catalytic converters to hasten the oxidation of carbon monoxide:

$$2\,CO(g) \;+\; O_2(g) \;\overset{Pt}{\rightleftharpoons}\; 2\,CO_2(g) \qquad \Delta H° = -566\ kJ$$

Suppose that you have a reaction vessel containing an equilibrium mixture of $CO(g)$, $O_2(g)$, and $CO_2(g)$. Will the amount of CO increase, decrease, or remain the same when:

(a) A platinum catalyst is added?

(b) The temperature is increased?

(c) The pressure is increased by decreasing the volume?

(d) The pressure is increased by adding argon gas?

(e) The pressure is increased by adding O_2 gas?

13.11 The Link Between Chemical Equilibrium and Chemical Kinetics

We emphasized in Section 13.1 that the equilibrium state is a dynamic one in which reactant and product concentrations remain constant, not because the reaction stops, but because the rates of the forward and reverse reactions are equal. To explore this idea further, let's consider the general, reversible reaction

$$A + B \rightleftharpoons C + D$$

Let's assume that the forward and reverse reactions occur in a single bimolecular step; that is, they are elementary reactions (Section 12.7). We can then write the following rate laws:

$$\text{Rate of forward reaction} = k_f[A][B]$$

$$\text{Rate of reverse reaction} = k_r[C][D]$$

If we begin with a mixture that contains all reactants and no products, the initial rate of the reverse reaction is zero because $[C] = [D] = 0$. As A and B are converted to C and D by the forward reaction, the rate of the forward reaction decreases because [A] and [B] are getting smaller. At the same time, the rate of the reverse reaction increases because [C] and [D] are getting larger. Eventually, the rates of the forward and reverse reactions become equal, and thereafter the concentrations remain constant; that is, the system is at chemical equilibrium.

Because the forward and reverse rates are equal at equilibrium, we can write

$$k_f[A][B] = k_r[C][D] \qquad \text{At equilibrium}$$

which can be rearranged to give

$$\frac{k_f}{k_r} = \frac{[C][D]}{[A][B]}$$

The right side of this equation is the equilibrium constant expression for the forward reaction, which is equal to the equilibrium constant K_c since the reaction mixture is at equilibrium.

$$K_c = \frac{[C][D]}{[A][B]}$$

Therefore, the equilibrium constant is simply the ratio of the rate constants for the forward and reverse reactions:

$$K_c = \frac{k_f}{k_r}$$

Note that, in deriving this equation for K_c, we have assumed a single-step mechanism. For a multistep mechanism, each step has a characteristic rate constant ratio, k_f/k_r. When equilibrium is reached, each step in the mechanism must be at equilibrium, and K_c for the overall reaction is equal to the product of the rate constant ratios for the individual steps.

The equation relating K_c to k_f and k_r provides a fundamental link between chemical equilibrium and chemical kinetics: The relative values of the rate constants for the forward and reverse reactions determine the composition of the equilibrium mixture. When k_f is much larger than k_r, K_c is very large and the reaction goes almost to completion. Such a reaction is said to be irreversible because the reverse reaction is often too slow to be detected. When k_f and k_r have comparable values, K_c has a value near unity, and comparable concentrations of both reactants and products are present at equilibrium. This is the usual situation for a reversible reaction.

Addition of a catalyst to a reaction mixture increases both rate constants k_f and k_r because the reaction takes place by a different, lower-energy mechanism. Because k_f and k_r increase by the same factor, though, the ratio k_f/k_r is unaffected, and the value of the equilibrium constant $K_c = k_f/k_r$ remains unchanged. Thus, addition of a catalyst does not alter the composition of an equilibrium mixture.

The equation $K_c = k_f/k_r$ also helps us understand why equilibrium constants depend on temperature. In Section 12.9, we saw that rate constants increase as the temperature increases, in accord with the Arrhenius equation $k = Ae^{-E_a/RT}$. In general, the forward and reverse reactions have different values of the activation energy, so k_f and k_r increase by different amounts as the temperature increases. The ratio $k_f/k_r = K_c$ is therefore temperature dependent. For an exothermic reaction, which has $\Delta E = E_a(\text{forward}) - E_a(\text{reverse}) < 0$, $E_a(\text{reverse})$ is greater than $E_a(\text{forward})$. Consequently, k_r increases by more than k_f increases as the temperature increases, and so $K_c = k_f/k_r$ for an exothermic reaction *decreases* as the temperature increases. Conversely, K_c for an endothermic reaction *increases* as the temperature increases.

EXAMPLE 13.12   The equilibrium constant K_c for the reaction of hydrogen with iodine is 57.0 at 700 K, and the reaction is endothermic ($\Delta E = 9$ kJ).

$$H_2(g) \ + \ I_2(g) \ \underset{k_r}{\overset{k_f}{\rightleftharpoons}} \ 2\,HI(g) \qquad K_c \ = \ 57.0 \text{ at } 700 \text{ K}$$

(a) Assuming that the reaction occurs in a single step, is the rate constant k_f for the formation of HI larger or smaller than the rate constant k_r for the decomposition of HI?
(b) The value of k_r at 700 K is 1.16×10^{-3} $M^{-1}s^{-1}$. What is the value of k_f at the same temperature?
(c) How are the values of k_f, k_r, and K_c affected by the addition of a catalyst?
(d) How are the values of k_f, k_r, and K_c affected by an increase in temperature?

SOLUTION
(a) Because $K_c = k_f/k_r = 57.0$, the rate constant for the formation of HI (forward reaction) is larger than the rate constant for the decomposition of HI (reverse reaction) by a factor of 57.0.
(b) Since $K_c = k_f/k_r$,

$$k_f = (K_c)(k_r) = (57.0)(1.16 \times 10^{-3}\,M^{-1}s^{-1}) = 6.61 \times 10^{-2}\,M^{-1}s^{-1}$$

(c) A catalyst lowers the activation energy barrier for the forward and reverse reactions by the same amount, thus increasing the rate constants k_f and k_r by the same factor. Because the equilibrium constant K_c equals the ratio of k_f to k_r, the value of K_c is unaffected by the addition of a catalyst.
(d) Because the reaction is endothermic, $E_a(\text{forward})$ is greater than $E_a(\text{reverse})$. Consequently, as the temperature increases, k_f increases by more than k_r increases, and therefore $K_c = k_f/k_r$ increases. Note that this result is in accord with Le Châtelier's principle.

▶ **PROBLEM 13.21**   Nitric oxide emitted from the engines of supersonic transport planes can contribute to destruction of stratospheric ozone:

$$NO(g) \ + \ O_3(g) \ \underset{k_r}{\overset{k_f}{\rightleftharpoons}} \ NO_2(g) \ + \ O_2(g)$$

This single-step, bimolecular reaction is highly exothermic ($\Delta E = -200$ kJ) and its equilibrium constant K_c is 3.4×10^{34} at 300 K.
(a) Which rate constant is larger, k_f or k_r?
(b) The value of k_f at 300 K is 8.5×10^6 $M^{-1}s^{-1}$. What is the value of k_r at the same temperature?
(c) A typical temperature in the stratosphere is 230 K. Do the values of k_f, k_r, and K_c increase or decrease when the temperature is lowered from 300 K to 230 K?

Nitric oxide emissions from supersonic aircraft can contribute to destruction of the ozone layer.

Breathing and Oxygen Transport

TABLE 13.2   Partial Pressure of Oxygen in the Lungs and Blood at Sea Level

Source	P_{O_2} (mm Hg)
Dry air	159
Alveolar air	100
Arterial blood	95
Venous blood	40

Humans, like all animals, need oxygen. The oxygen comes, of course, from breathing: About 500 mL of air is drawn into the lungs of an average person with each breath. When the freshly inspired air travels through the bronchial passages and enters the approximately 150 million alveolar sacs of the lungs, it picks up moisture and mixes with air remaining from the previous breath. As it mixes, the concentrations of both water vapor and carbon dioxide increase. These gas concentrations are measured by their partial pressures (Section 9.5), with the partial pressure of oxygen in the lungs usually around 100 mm Hg (Table 13.2). Oxygen then diffuses through the delicate walls of the lung alveoli and into arterial blood, which transports it to all body tissues.

Only about 3% of the oxygen in blood is dissolved; the rest is chemically bound to *hemoglobin* molecules (Hb), large proteins that contain *heme* groups embedded in them. Each hemoglobin molecule contains four heme groups, and each heme group contains an iron atom that is able to bind to one O_2 molecule. Thus, a single hemoglobin molecule can bind four molecules of oxygen.

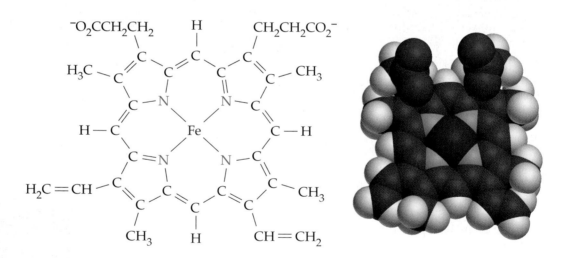

Heme — an O_2 molecule binds
to the central iron atom.

The entire system of oxygen transport and delivery in the body depends on the pickup and release of O_2 by hemoglobin according to the following series of equilibria:

$$Hb + O_2 \rightleftharpoons Hb(O_2)$$

$$Hb(O_2) + O_2 \rightleftharpoons Hb(O_2)_2$$

$$Hb(O_2)_2 + O_2 \rightleftharpoons Hb(O_2)_3$$

$$Hb(O_2)_3 + O_2 \rightleftharpoons Hb(O_2)_4$$

The positions of the different equilibria depend on the partial pressures of O_2 (P_{O_2}) in the various tissues. In hard-working, oxygen-starved muscles, where P_{O_2} is low, oxygen is released from hemoglobin as the equilibria shift toward the left, according to Le Châtelier's principle. In the lungs, where P_{O_2} is high, oxygen is absorbed by hemoglobin as the equilibria shift toward the right.

The amount of oxygen carried by hemoglobin at any given value of P_{O_2} is usually expressed as a percent saturation and can be found from the curve shown in Figure 13.15. The saturation is 97.5% in the lungs, where $P_{O_2} = 100$ mm Hg, meaning that each hemoglobin is carrying close to its maximum possible amount of 4 O_2 molecules. When $P_{O_2} = 26$ mm Hg, however, the saturation drops to 50%.

What about people who live at high altitudes? In Leadville, Colorado, for example, where the altitude is 10,156 ft, the partial pressure of O_2 in the lungs is only about 68 mm Hg. Hemoglobin is only 90% saturated with O_2 at this pressure, meaning that less oxygen is available for delivery to the tissues. People who climb suddenly from sea level to high altitude thus experience a feeling of oxygen deprivation, or hypoxia, as their bodies are unable to supply enough oxygen to tissues. The body soon copes with the situation, though, by producing more hemoglobin molecules, which both provide more capacity for O_2 transport and also drive the $Hb + O_2$ equilibria to the right. The time required to adapt to the lower O_2 pressures is typically days to weeks, so athletes and hikers must train at high altitudes for some time.

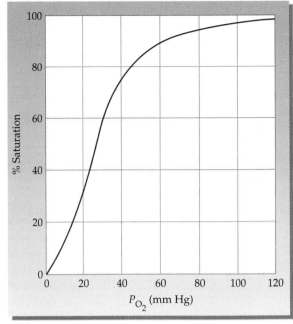

FIGURE 13.15 An oxygen-carrying curve for hemoglobin. The percent saturation of the oxygen-binding sites on hemoglobin depends on the partial pressure of oxygen (P_{O_2}).

▶ **PROBLEM 13.22** The affinity of hemoglobin (Hb) for CO is greater than its affinity for O_2. Use Le Châtelier's principle to predict how CO affects the equilibrium $Hb + O_2 \rightleftarrows Hb(O_2)$. Suggest a reason for the toxicity of CO.

▶ **PROBLEM 13.23** In which direction does the equilibrium $Hb + O_2 \rightleftarrows Hb(O_2)$ shift on taking an elevator to the top of the Empire State Building? Explain.

▶ **PROBLEM 13.24** The inner core of the heme group, which consists of 20 C atoms and 4 N atoms, is planar. Just as the π electrons in the ozone molecule are delocalized over all three O atoms, so the π electrons of the heme group are delocalized over all 24 atoms in the inner core. How many π electrons are there in the inner core? (Be careful in counting the π electrons on the N atoms.)

▶ **PROBLEM 13.25** How many O_2 molecules are drawn into the lungs of an average person with each breath? Assume that the ambient air temperature is 25°C.

The bodies of mountain dwellers produce increased amounts of hemoglobin to cope with the low O_2 pressures at high altitudes.

Key Words

Summary

Chemical equilibrium is a dynamic state in which the concentrations of reactants and products remain constant because the rates of the forward and reverse reactions are equal. For the general reaction $a\,A + b\,B \rightleftharpoons c\,C + d\,D$, concentrations in the **equilibrium mixture** are related by the **equilibrium equation**:

$$K_c = \frac{[C]^c[D]^d}{[A]^a[B]^b}$$

The quotient on the right side of the equation is called the *equilibrium constant expression*. The **equilibrium constant K_c** is the number obtained when equilibrium concentrations (in mol/L) are substituted into the equilibrium constant expression. The value of K_c varies with temperature and depends on the form of the balanced chemical equation.

The **equilibrium constant K_p** can be used for gas-phase reactions. It is defined in the same way as K_c except that the equilibrium constant expression contains partial pressures (in atmospheres) instead of molar concentrations. The constants K_p and K_c are related by the equation $K_p = K_c(RT)^{\Delta n}$, where $\Delta n = (c + d) - (a + b)$ for the reaction $a\,A + b\,B \rightleftharpoons c\,C + d\,D$.

Homogeneous equilibria are those in which all reactants and products are in a single phase; **heterogeneous equilibria** are those in which reactants and products are present in more than one phase. The equilibrium equation for a heterogeneous equilibrium does not include concentrations of pure solids or pure liquids.

The value of the equilibrium constant for a reaction makes it possible to judge the extent of reaction, predict the direction of reaction, and calculate equilibrium concentrations from initial concentrations. The farther the reaction proceeds toward completion, the larger the value of K_c. The direction of a reaction not at equilibrium depends on the relative values of K_c and the **reaction quotient Q_c**, which is defined in the same way as K_c except that the concentrations in the equilibrium constant expression are not necessarily equilibrium concentrations. If $Q_c < K_c$, the reaction goes from left to right to attain equilibrium; if $Q_c > K_c$, the reaction goes from right to left; if $Q_c = K_c$, the system is at equilibrium.

The composition of an equilibrium mixture can be altered by changes in concentration, pressure (volume), or temperature. The qualitative effect of these changes is predicted by **Le Châtelier's principle**, which says that if a stress is applied to a reaction mixture at equilibrium, reaction occurs in the direction that relieves the stress. Temperature changes affect equilibrium concentrations because K_c is temperature-dependent. As the temperature increases, K_c for an exothermic reaction decreases, and K_c for an endothermic reaction increases.

A catalyst increases the rate at which chemical equilibrium is reached, but it does not affect the equilibrium constant or the equilibrium concentrations. The equilibrium constant for a single-step reaction is equal to the ratio of the rate constants for the forward and reverse reactions; $K_c = k_f/k_r$.

Key Concept Summary

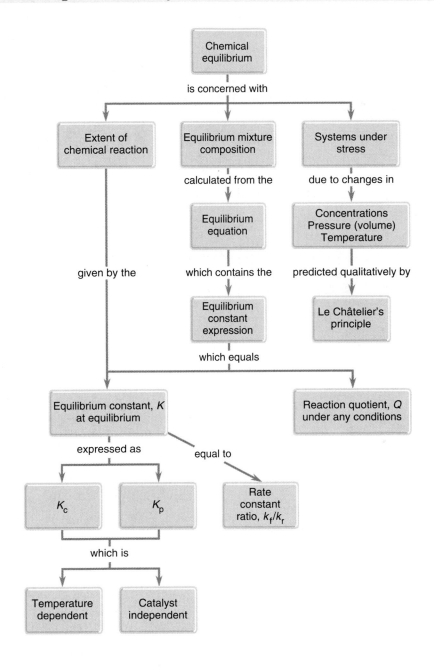

Chemical equilibrium

is concerned with

Extent of chemical reaction

Equilibrium mixture composition

Systems under stress

calculated from the

due to changes in

Equilibrium equation

Concentrations Pressure (volume) Temperature

given by the

which contains the

predicted qualitatively by

Equilibrium constant expression

Le Châtelier's principle

which equals

Equilibrium constant, K at equilibrium

Reaction quotient, Q under any conditions

expressed as

equal to

K_c

K_p

Rate constant ratio, k_f/k_r

which is

Temperature dependent

Catalyst independent

Understanding Key Concepts

Problems 13.1–13.25 appear within the chapter.

13.26 Consider the interconversion of A molecules (red spheres) and B molecules (blue spheres) according to the reaction $A \rightleftharpoons B$. Each of the following series of pictures represents a separate experiment in which time increases from left to right:

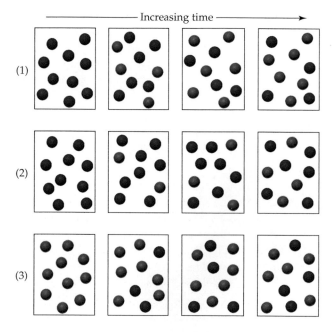

(a) Which of the experiments has resulted in an equilibrium state?

(b) What is the value of the equilibrium constant K_c for the reaction $A \rightleftharpoons B$?

(c) Explain why you can calculate K_c without knowing the volume of the reaction vessel.

13.27 The following pictures represent the equilibrium state for three different reactions of the type $A_2 + X_2 \rightleftharpoons 2\,AX$ ($X = B$, C, or D):

  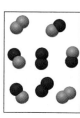

$$A_2 + B_2 \rightleftharpoons 2\,AB \qquad A_2 + C_2 \rightleftharpoons 2\,AC \qquad A_2 + D_2 \rightleftharpoons 2\,AD$$

(a) Which reaction has the largest equilibrium constant?

(b) Which reaction has the smallest equilibrium constant?

13.28 The reaction $A_2 + B \rightleftharpoons A + AB$ has an equilibrium constant $K_c = 2$. The following pictures represent reaction mixtures that contain A atoms (red), B atoms (blue), and A_2 and AB molecules:

(1)   (2)   (3)

(a) Which reaction mixture is at equilibrium?

(b) For those mixtures that are not at equilibrium, will the reaction go in the forward or reverse direction to reach equilibrium?

13.29 The following pictures represent the initial state and the equilibrium state for the reaction of A_2 molecules (red) with B atoms (blue) to give AB molecules:

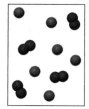 

Initial state   Equilibrium state

(a) Write a balanced chemical equation for the reaction.

(b) If the volume of the equilibrium mixture is decreased, will the number of AB molecules increase, decrease, or remain the same? Explain.

13.30 Consider the reaction $A + B \rightleftharpoons AB$. The vessel on the right contains an equilibrium mixture of A molecules (red spheres), B molecules (blue spheres), and AB molecules. If the stopcock is opened and the contents of the two vessels are allowed to mix, will the reaction go in the forward or reverse direction? Explain.

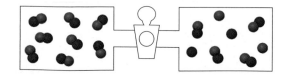

13.31 The following pictures represent the composition of the equilibrium mixture for the reaction $A + B \rightleftarrows AB$ at 300 K and at 400 K:

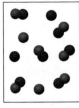 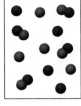

$T = 300 \text{ K}$　　　$T = 400 \text{ K}$

Is the reaction exothermic or endothermic? Explain using Le Châtelier's principle.

13.32 The following pictures represent equilibrium mixtures at 325 K and 350 K for a reaction involving A atoms (red), B atoms (blue), and AB molecules:

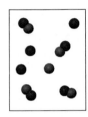 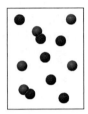

$T = 325 \text{ K}$　　　$T = 350 \text{ K}$

(a) Write a balanced equation for the reaction that occurs on raising the temperature.

(b) Is the reaction exothermic or endothermic? Explain using Le Châtelier's principle.

(c) If the volume of the container is increased, will the number of A atoms increase, decrease, or remain the same? Explain.

13.33 The following picture represents an equilibrium mixture of solid $BaCO_3$, solid BaO, and gaseous CO_2.

(a) Draw a picture that represents the equilibrium mixture after addition of four more CO_2 molecules.

(b) Draw a picture that represents the equilibrium mixture at a higher temperature. The decomposition of $BaCO_3$ is endothermic.

13.34 The following picture represents the composition of the equilibrium mixture for the endothermic reaction $A_2 \rightleftarrows 2 A$ at 500 K:

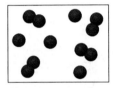

Draw a picture that represents the equilibrium mixture after each of the following changes:

(a) Adding a catalyst

(b) Increasing the volume

(c) Decreasing the temperature

13.35 The following picture represents the equilibrium state for the reaction $2 AB \rightleftarrows A_2 + B_2$:

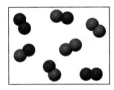

Which rate constant is larger, k_f or k_r? Explain.

Additional Problems

Equilibrium Expressions and Equilibrium Constants

13.36 For each of the following equilibria, write the equilibrium constant expression for K_c:

(a) $CH_4(g) + H_2O(g) \rightleftarrows CO(g) + 3 H_2(g)$

(b) $3 F_2(g) + Cl_2(g) \rightleftarrows 2 ClF_3(g)$

(c) $H_2(g) + F_2(g) \rightleftarrows 2 HF(g)$

13.37 For each of the following equilibria, write the equilibrium constant expression for K_c:

(a) $2 C_2H_4(g) + O_2(g) \rightleftarrows 2 CH_3CHO(g)$

(b) $2 NO(g) \rightleftarrows N_2(g) + O_2(g)$

(c) $4 NH_3(g) + 5 O_2(g) \rightleftarrows 4 NO(g) + 6 H_2O(g)$

13.38 For each of the equilibria in Problem 13.36, write the equilibrium constant expression for K_p and give the equation that relates K_p and K_c.

13.39 For each of the equilibria in Problem 13.37, write the equilibrium constant expression for K_p and give the equation that relates K_p and K_c.

13.40 Diethyl ether, used as an anesthetic, is synthesized by heating ethanol with concentrated sulfuric acid. Write the equilibrium constant expression for K_c.

$$2 C_2H_5OH(soln) \rightleftarrows C_2H_5OC_2H_5(soln) + H_2O(soln)$$
　　Ethanol　　　　　　Diethyl ether

13.41 Ethylene glycol, used as antifreeze in automobile radiators, is manufactured by hydration of ethylene oxide. Write the equilibrium constant expression for K_c.

$$H_2C{-}CH_2(soln) + H_2O(soln) \rightleftharpoons HOCH_2CH_2OH(soln)$$

Ethylene oxide   Ethylene glycol

13.42 Write the equilibrium constant expression for K_c for the following metabolic reaction:

Citrate $\rightleftharpoons$ Isocitrate

13.43 Write the equilibrium constant expression for K_c for the following metabolic reaction:

$$CH_3CO_2H(aq) + HO_2CCCH_2CO_2H(aq) \rightleftharpoons$$

Acetic acid   Oxaloacetic acid

Citric acid

13.44 If $K_c = 7.5 \times 10^{-9}$ at 1000 K for the reaction $N_2(g) + O_2(g) \rightleftharpoons 2\,NO(g)$, what is K_c at 1000 K for the reaction $2\,NO(g) \rightleftharpoons N_2(g) + O_2(g)$?

13.45 At 400 K, $K_p = 50.2$ for the reaction $N_2O_4(g) \rightleftharpoons 2\,NO_2(g)$. What is K_p at 400 K for the reaction $2\,NO_2(g) \rightleftharpoons N_2O_4(g)$?

13.46 An equilibrium mixture of PCl_5, PCl_3, and Cl_2 at a certain temperature contains 8.3×10^{-3} M PCl_5, 1.5×10^{-2} M PCl_3, and 3.2×10^{-2} M Cl_2. Calculate the equilibrium constant K_c for the reaction $PCl_5(g) \rightleftharpoons PCl_3(g) + Cl_2(g)$.

13.47 The partial pressures in an equilibrium mixture of NO, Cl_2, and NOCl at 500 K are as follows: $P_{NO} = 0.240$ atm; $P_{Cl_2} = 0.608$ atm; $P_{NOCl} = 1.35$ atm. What is K_p at 500 K for the reaction $2\,NO(g) + Cl_2(g) \rightleftharpoons 2\,NOCl(g)$?

13.48 A sample of HI (9.30×10^{-3} mol) was placed in an empty 2.00 L container at 1000 K. After equilibrium was reached, the concentration of I_2 was 6.29×10^{-4} M. Calculate the value of K_c at 1000 K for the reaction $H_2(g) + I_2(g) \rightleftharpoons 2\,HI(g)$.

13.49 Vinegar contains acetic acid, a weak acid that is partially dissociated in aqueous solution:

$$CH_3CO_2H(aq) \rightleftharpoons H^+(aq) + CH_3CO_2^-(aq)$$

(a) Write the equilibrium constant expression for K_c.
(b) What is the value of K_c if the extent of dissociation in 1.0 M CH_3CO_2H is 0.42%?

13.50 The industrial solvent ethyl acetate is produced by reaction of acetic acid with ethanol:

$$CH_3CO_2H(soln) + C_2H_5OH(soln) \rightleftharpoons$$

Acetic acid   Ethanol

$$CH_3CO_2C_2H_5(soln) + H_2O(soln)$$

Ethyl acetate

(a) Write the equilibrium constant expression for K_c.
(b) A solution prepared by mixing 1.00 mol of acetic acid and 1.00 mol of ethanol contains 0.65 mol of ethyl acetate at equilibrium. Calculate the value of K_c. Explain why you can calculate K_c without knowing the volume of the solution.

13.51 A characteristic reaction of ethyl acetate is hydrolysis, the reverse of the reaction in Problem 13.50. Write the equilibrium equation for hydrolysis of ethyl acetate, and use the data in Problem 13.50 to calculate K_c for the hydrolysis reaction.

13.52 At 500 K, $K_c = 0.575$ for the reaction $PCl_5(g) \rightleftharpoons PCl_3(g) + Cl_2(g)$. What is K_p at the same temperature?

13.53 One step in the manufacture of sulfuric acid involves the oxidation of sulfur dioxide, $2\,SO_2(g) + O_2(g) \rightleftharpoons 2\,SO_3(g)$. If K_p for this reaction is 3.30 at 1000 K, what is the value of K_c at the same temperature?

13.54 The vapor pressure of water at 25°C is 0.0313 atm. Calculate the values of K_p and K_c at 25°C for the equilibrium $H_2O(l) \rightleftharpoons H_2O(g)$.

13.55 Naphthalene, a white solid used to make mothballs, has a vapor pressure of 0.10 mm Hg at 27°C. Calculate the values of K_p and K_c at 27°C for the equilibrium $C_{10}H_8(s) \rightleftharpoons C_{10}H_8(g)$.

13.56 For each of the following equilibria, write the equilibrium constant expression for K_c. Where appropriate, also write the equilibrium constant expression for K_p.
(a) $Fe_2O_3(s) + 3\,CO(g) \rightleftharpoons 2\,Fe(l) + 3\,CO_2(g)$
(b) $4\,Fe(s) + 3\,O_2(g) \rightleftharpoons 2\,Fe_2O_3(s)$
(c) $BaSO_4(s) \rightleftharpoons BaO(s) + SO_3(g)$
(d) $BaSO_4(s) \rightleftharpoons Ba^{2+}(aq) + SO_4^{2-}(aq)$

13.57 For each of the following equilibria, write the equilibrium constant expression for K_c. Where appropriate, also write the equilibrium constant expression for K_p.
(a) $WO_3(s) + 3\,H_2(g) \rightleftharpoons W(s) + 3\,H_2O(g)$
(b) $Ag^+(aq) + Cl^-(aq) \rightleftharpoons AgCl(s)$
(c) $2\,FeCl_3(s) + 3\,H_2O(g) \rightleftharpoons Fe_2O_3(s) + 6\,HCl(g)$
(d) $MgCO_3(s) \rightleftharpoons MgO(s) + CO_2(g)$

Using the Equilibrium Constant

13.58 When the following reactions come to equilibrium, does the equilibrium mixture contain mostly reactants or mostly products?

(a) $H_2(g) + S(s) \rightleftarrows H_2S(g)$; $K_c = 7.8 \times 10^5$

(b) $N_2(g) + 2\,H_2(g) \rightleftarrows N_2H_4(g)$; $K_c = 7.4 \times 10^{-26}$

13.59 Which of the following reactions goes almost all the way to completion, and which proceeds hardly at all?

(a) $N_2(g) + O_2(g) \rightleftarrows 2\,NO(g)$; $K_c = 2.7 \times 10^{-18}$

(b) $2\,NO(g) + O_2(g) \rightleftarrows 2\,NO_2(g)$; $K_c = 6.0 \times 10^{13}$

13.60 For which of the following reactions will the equilibrium mixture contain an appreciable concentration of both reactants and products?

(a) $Cl_2(g) \rightleftarrows 2\,Cl(g)$; $K_c = 6.4 \times 10^{-39}$

(b) $Cl_2(g) + 2\,NO(g) \rightleftarrows 2\,NOCl(g)$; $K_c = 3.7 \times 10^8$

(c) $Cl_2(g) + 2\,NO_2(g) \rightleftarrows 2\,NO_2Cl(g)$; $K_c = 1.8$

13.61 Which of the following reactions yield appreciable equilibrium concentrations of both reactants and products?

(a) $2\,Cu(s) + O_2(g) \rightleftarrows 2\,CuO(s)$; $K_c = 4 \times 10^{45}$

(b) $H_3PO_4(aq) \rightleftarrows H^+(aq) + H_2PO_4^-(aq)$; $K_c = 7.5 \times 10^{-3}$

(c) $2\,HBr(g) \rightleftarrows H_2(g) + Br_2(g)$; $K_c = 2 \times 10^{-19}$

13.62 When wine spoils, ethanol is oxidized to acetic acid as O_2 from the air reacts with the wine:

$$\underset{\text{Ethanol}}{C_2H_5OH(aq)} + O_2(aq) \rightleftarrows \underset{\text{Acetic acid}}{CH_3CO_2H(aq)} + H_2O(l)$$

The value of K_c for this reaction at 25°C is 1.2×10^{82}. Will much ethanol remain when the reaction has reached equilibrium? Explain.

13.63 The value of K_c for the reaction $3\,O_2(g) \rightleftarrows 2\,O_3(g)$ is 1.7×10^{-56} at 25°C. Do you expect pure air at 25°C to contain much O_3 (ozone) when O_2 and O_3 are in equilibrium? If the equilibrium concentration of O_2 in air at 25°C is 8×10^{-3} M, what is the equilibrium concentration of O_3?

13.64 At 1400 K, $K_c = 2.5 \times 10^{-3}$ for the reaction $CH_4(g) + 2\,H_2S(g) \rightleftarrows CS_2(g) + 4\,H_2(g)$. A 10.0 L reaction vessel at 1400 K contains 2.0 mol of CH_4, 3.0 mol of CS_2, 3.0 mol of H_2, and 4.0 mol of H_2S. Is the reaction mixture at equilibrium? If not, in which direction does the reaction proceed to reach equilibrium?

13.65 The first step in the industrial synthesis of hydrogen is the reaction of steam and methane to give synthesis gas, a mixture of carbon monoxide and hydrogen:

$$H_2O(g) + CH_4(g) \rightleftarrows CO(g) + 3\,H_2(g) \quad K_c = 4.7 \text{ at } 1400 \text{ K}$$

A mixture of reactants and products at 1400 K contains 0.035 M H_2O, 0.050 M CH_4, 0.15 M CO, and 0.20 M H_2. In which direction does the reaction proceed to reach equilibrium?

13.66 An equilibrium mixture of N_2, H_2, and NH_3 at 700 K contains 0.036 M N_2 and 0.15 M H_2. At this temperature, K_c for the reaction $N_2(g) + 3\,H_2(g) \rightleftarrows 2\,NH_3(g)$ is 0.29. What is the concentration of NH_3?

13.67 An equilibrium mixture of O_2, SO_2, and SO_3 contains equal concentrations of SO_2 and SO_3. Calculate the concentration of O_2 if $K_c = 2.7 \times 10^2$ for the reaction $2\,SO_2(g) + O_2(g) \rightleftarrows 2\,SO_3(g)$.

13.68 The air pollutant NO is produced in automobile engines because of the high-temperature reaction $N_2(g) + O_2(g) \rightleftarrows 2\,NO(g)$; $K_c = 1.7 \times 10^{-3}$ at 2300 K. If the initial concentrations of N_2 and O_2 at 2300 K are both 1.40 M, what are the concentrations of NO, N_2, and O_2 when the reaction mixture reaches equilibrium?

13.69 Recalculate the equilibrium concentrations in Problem 13.68 if the initial concentrations are 2.24 M N_2 and 0.56 M O_2. (This N_2/O_2 concentration ratio is the ratio found in air.)

13.70 At a certain temperature, the reaction $PCl_5(g) \rightleftarrows PCl_3(g) + Cl_2(g)$ has an equilibrium constant $K_c = 5.8 \times 10^{-2}$. Calculate the equilibrium concentrations of PCl_5, PCl_3, and Cl_2 if only PCl_5 is present initially, at a concentration of 0.160 M.

13.71 Recalculate the equilibrium concentrations in Problem 13.70 if the initial concentrations are $[PCl_5] = 0.200$ M, $[PCl_3] = 0.100$ M, and $[Cl_2] = 0.040$ M.

13.72 The value of K_c for the reaction of acetic acid with ethanol is 3.4 at 25°C:

$$\underset{\text{Acetic acid}}{CH_3CO_2H(soln)} + \underset{\text{Ethanol}}{C_2H_5OH(soln)} \rightleftarrows$$
$$\underset{\text{Ethyl acetate}}{CH_3CO_2C_2H_5(soln)} + H_2O(soln) \quad K_c = 3.4$$

(a) How many moles of ethyl acetate are present in an equilibrium mixture that contains 4.0 mol of acetic acid, 6.0 mol of ethanol, and 12.0 mol of water at 25°C?

(b) Calculate the number of moles of all reactants and products in an equilibrium mixture prepared by mixing 1.00 mol of acetic acid and 10.00 mol of ethanol.

13.73 In a basic aqueous solution, chloromethane undergoes a substitution reaction in which Cl^- is replaced by OH^-:

$$\underset{\text{Chloromethane}}{CH_3Cl(aq)} + OH^-(aq) \rightleftarrows \underset{\text{Methanol}}{CH_3OH(aq)} + Cl^-(aq)$$

The equilibrium constant K_c is 1×10^{16}. Calculate the equilibrium concentrations of CH_3Cl, CH_3OH, OH^-, and Cl^- in a solution prepared by mixing equal volumes of 0.1 M CH_3Cl and 0.2 M NaOH. (*Hint*: In defining x, assume that the reaction goes 100% to completion, and then take account of a small amount of the reverse reaction.)

13.74 Equilibrium pressures can be calculated from initial pressures and K_p in the same way that equilibrium concentrations can be calculated from initial concentrations and K_c (Figure 13.6). At 700 K, $K_p = 0.140$ for the reaction $ClF_3(g) \rightleftharpoons ClF(g) + F_2(g)$. Calculate the equilibrium partial pressures of ClF_3, ClF, and F_2 if only ClF_3 is present initially, at a partial pressure of 1.47 atm.

13.75 At 1000 K, $K_p = 19.9$ for the reaction $Fe_2O_3(s) + 3\ CO(g) \rightleftharpoons 2\ Fe(s) + 3\ CO_2(g)$. What are the equilibrium partial pressures of CO and CO_2 if CO is the only gas present initially, at a partial pressure of 0.978 atm?

Le Châtelier's Principle

13.76 Consider the following equilibrium: $Ag^+(aq) + Cl^-(aq) \rightleftharpoons AgCl(s)$. Use Le Châtelier's principle to predict how the amount of solid silver chloride will change when the equilibrium is disturbed by:

(a) Adding NaCl

(b) Adding $AgNO_3$

(c) Adding NH_3, which reacts with Ag^+ to form the complex ion, $Ag(NH_3)_2^+$

(d) Removing Cl^-; also account for the change using the reaction quotient Q_c.

13.77 Will the concentration of NO_2 increase, decrease, or remain the same when the equilibrium $NO_2Cl(g) + NO(g) \rightleftharpoons NOCl(g) + NO_2(g)$ is disturbed by:

(a) Adding NOCl

(b) Adding NO

(c) Removing NO

(d) Adding NO_2Cl; also account for the change using the reaction quotient Q_c.

13.78 When each of the following equilibria is disturbed by increasing the pressure as a result of decreasing the volume, does the number of moles of reaction products increase, decrease, or remain the same?

(a) $2\ CO_2(g) \rightleftharpoons 2\ CO(g) + O_2(g)$

(b) $N_2(g) + O_2(g) \rightleftharpoons 2\ NO(g)$

(c) $Si(s) + 2\ Cl_2(g) \rightleftharpoons SiCl_4(g)$

13.79 For each of the following equilibria, use Le Châtelier's principle to predict the direction of reaction when the volume is increased.

(a) $C(s) + H_2O(g) \rightleftharpoons CO(g) + H_2(g)$

(b) $2\ H_2(g) + O_2(g) \rightleftharpoons 2\ H_2O(g)$

(c) $2\ Fe(s) + 3\ H_2O(g) \rightleftharpoons Fe_2O_3(s) + 3\ H_2(g)$

13.80 For the water-gas shift reaction $CO(g) + H_2O(g) \rightleftharpoons CO_2(g) + H_2(g)$, $\Delta H° = -41.2$ kJ, does the amount of H_2 in an equilibrium mixture increase or decrease when the temperature is increased? How does K_c change when the temperature is decreased? Justify your answers using Le Châtelier's principle.

13.81 The value of $\Delta H°$ for the reaction $3\ O_2(g) \rightleftharpoons 2\ O_3(g)$ is +285 kJ. Does the equilibrium constant for this reaction increase or decrease when the temperature increases? Justify your answer using Le Châtelier's principle.

13.82 Consider the exothermic reaction $CoCl_4^{2-}(aq) + 6\ H_2O(l) \rightleftharpoons Co(H_2O)_6^{2+}(aq) + 4\ Cl^-(aq)$. Will the equilibrium concentration of $CoCl_4^{2-}$ increase or decrease when the following changes occur?

(a) HCl is added.

(b) $Co(NO_3)_2$ is added.

(c) The solution is diluted with water.

(d) The temperature is increased.

13.83 Consider the endothermic reaction $Fe^{3+}(aq) + Cl^-(aq) \rightleftharpoons FeCl^{2+}(aq)$. Use Le Châtelier's principle to predict how the equilibrium concentration of the complex ion $FeCl^{2+}$ will change when:

(a) $Fe(NO_3)_3$ is added.

(b) Cl^- is precipitated as AgCl by addition of $AgNO_3$.

(c) The temperature is increased.

(d) A catalyst is added.

13.84 Methanol (CH_3OH) is manufactured by reaction of carbon monoxide with hydrogen in the presence of a ZnO/Cr_2O_3 catalyst:

$$CO(g) + 2\ H_2(g) \xrightleftharpoons{ZnO/Cr_2O_3} CH_3OH(g) \quad \Delta H° = -91\ kJ$$

Does the amount of methanol increase, decrease, or remain the same when an equilibrium mixture of reactants and products is subjected to the following changes?

(a) The temperature is increased.

(b) The volume is decreased.

(c) Helium is added.

(d) CO is added.

(e) The catalyst is removed.

13.85 In the gas phase at 400°C, isopropyl alcohol (rubbing alcohol) decomposes to acetone, an important industrial solvent:

$$(CH_3)_2CHOH(g) \rightleftharpoons (CH_3)_2CO(g) + H_2(g)$$

Isopropyl alcohol   Acetone

$$\Delta H° = +57.3\ kJ$$

Does the amount of acetone increase, decrease, or remain the same when an equilibrium mixture of reactants and products is subjected to the following changes?

(a) The temperature is increased.

(b) The volume is increased.

(c) Argon is added.

(d) H_2 is added.

(e) A catalyst is added.

Chemical Equilibrium and Chemical Kinetics

13.86 Consider a general, single-step reaction of the type $A + B \rightleftarrows C$. Show that the equilibrium constant is equal to the ratio of the rate constants for the forward and reverse reactions, $K_c = k_f/k_r$.

13.87 Which of the following relative values of k_f and k_r results in an equilibrium mixture that contains large amounts of reactants and small amounts of products?

(a) $k_f > k_r$ (b) $k_f = k_r$ (c) $k_f < k_r$

13.88 Consider the gas-phase hydration of hexafluoroacetone, $(CF_3)_2CO$:

$$(CF_3)_2CO(g) + H_2O(g) \underset{k_r}{\overset{k_f}{\rightleftarrows}} (CF_3)_2C(OH)_2(g)$$

At 76°C, the forward and reverse rate constants are $k_f = 0.13\ M^{-1}s^{-1}$ and $k_r = 6.2 \times 10^{-4}\ s^{-1}$. What is the value of the equilibrium constant K_c?

13.89 Consider the reaction of chloromethane with OH^- in aqueous solution:

$$CH_3Cl(aq) + OH^-(aq) \underset{k_r}{\overset{k_f}{\rightleftarrows}} CH_3OH(aq) + Cl^-(aq)$$

At 25°C, the rate constant for the forward reaction is $6 \times 10^{-6}\ M^{-1}s^{-1}$, and the equilibrium constant K_c is 1×10^{16}. Calculate the rate constant for the reverse reaction at 25°C.

13.90 Listed in the table are forward and reverse rate constants for the reaction $2\ NO(g) \rightleftarrows N_2(g) + O_2(g)$.

Temperature (K)	k_f (M^{-1}s^{-1})	k_r (M^{-1}s^{-1})
1400	0.29	1.1×10^{-6}
1500	1.3	1.4×10^{-5}

Is the reaction endothermic or exothermic? Explain in terms of kinetics.

13.91 Forward and reverse rate constants for the reaction $CO_2(g) + N_2(g) \rightleftarrows CO(g) + N_2O(g)$ exhibit the following temperature dependence:

Temperature (K)	k_f (M^{-1}s^{-1})	k_r (M^{-1}s^{-1})
1200	9.1×10^{-11}	1.5×10^5
1300	2.7×10^{-9}	2.6×10^5

Is the reaction endothermic or exothermic? Explain in terms of kinetics.

General Problems

13.92 When 0.500 mol of N_2O_4 is placed in a 4.00 L reaction vessel and heated at 400 K, 79.3% of the N_2O_4 decomposes to NO_2.

(a) Calculate K_c and K_p at 400 K for the reaction $N_2O_4(g) \rightleftarrows 2\ NO_2(g)$.

(b) Draw an electron-dot structure for NO_2, and rationalize the structure of N_2O_4.

13.93 At 25°C, $K_c = 1.6 \times 10^{24}$ for the reaction $CaO(s) + CO_2(g) \rightleftarrows CaCO_3(s)$. Use the value of K_c to make an estimate of the extent of this reaction. Verify your estimate by calculating the concentration of CO_2 gas that is in equilibrium with solid CaO and solid $CaCO_3$ at 25°C.

13.94 What concentration of NH_3 is in equilibrium with $1.0 \times 10^{-3}\ M\ N_2$ and $2.0 \times 10^{-3}\ M\ H_2$ at 700 K? At this temperature, $K_c = 0.291$ for the reaction $N_2(g) + 3\ H_2(g) \rightleftarrows 2\ NH_3(g)$.

13.95 The F–F bond in F_2 is relatively weak because the lone pairs of electrons on one F atom repel the lone pairs on the other F atom; $K_p = 7.83$ at 1500 K for the reaction $F_2(g) \rightleftarrows 2\ F(g)$.

(a) If the equilibrium partial pressure of F_2 molecules at 1500 K is 0.200 atm, what is the equilibrium partial pressure of F atoms in atm?

(b) What fraction of the F_2 molecules dissociate at 1500 K?

(c) Why is the F–F bond in F_2 weaker than the Cl–Cl bond in Cl_2?

13.96 The equilibrium concentrations in a gas mixture at a particular temperature are 0.13 M H_2, 0.70 M I_2, and 2.1 M HI. What equilibrium concentrations are obtained at the same temperature when 0.20 mol of HI is injected into an empty 500.0 mL container?

13.97 A 5.00 L reaction vessel is filled with 1.00 mol of H_2, 1.00 mol of I_2, and 2.50 mol of HI. Calculate the equilibrium concentrations of H_2, I_2, and HI at 500 K. The equilibrium constant K_c at 500 K for the reaction $H_2(g) + I_2(g) \rightleftarrows 2\ HI(g)$ is 129.

13.98 At 1000 K, the value of K_c for the reaction $C(s) + H_2O(g) \rightleftarrows CO(g) + H_2(g)$ is 3.0×10^{-2}. Calculate the equilibrium concentrations of H_2O, CO, and H_2 in a reaction mixture obtained by heating 6.00 mol of steam and an excess of solid carbon in a 5.00 L container. What is the molar composition of the equilibrium mixture?

13.99 The equilibrium constant K_p for the reaction $PCl_5(g)$ $\rightleftharpoons PCl_3(g) + Cl_2(g)$ is 3.81×10^2 at 600 K and 2.69×10^3 at 700 K.

(a) Is the reaction endothermic or exothermic?

(b) How are the equilibrium amounts of reactants and products affected by (i) an increase in volume, (ii) addition of an inert gas, or (iii) addition of a catalyst?

13.100 Consider the following gas-phase reaction: $2\,A(g) + B(g) \rightleftharpoons C(g) + D(g)$. An equilibrium mixture of reactants and products is subjected to the following changes:

(a) A decrease in volume

(b) An increase in temperature

(c) Addition of reactants

(d) Addition of a catalyst

(e) Addition of an inert gas

Which of these changes affect the composition of the equilibrium mixture, but leave the value of the equilibrium constant K_c unchanged? Which of the changes affect the value of K_c? Which affect neither the composition of the equilibrium mixture nor K_c?

13.101 Baking soda (sodium hydrogen carbonate) decomposes when it is heated:

$$2\,NaHCO_3(s) \rightleftharpoons Na_2CO_3(s) + CO_2(g) + H_2O(g)$$

$$\Delta H° = +136 \text{ kJ}$$

Consider an equilibrium mixture of reactants and products in a closed container. How does the number of moles of CO_2 change when the mixture is disturbed by:

(a) Adding solid $NaHCO_3$?

(b) Adding water vapor?

(c) Decreasing the volume of the container?

(d) Increasing the temperature?

13.102 Acetic acid tends to form dimers, $(CH_3CO_2H)_2$, because of hydrogen bonding:

Monomer   Dimer

The equilibrium constant K_c for this reaction is 1.51×10^2 in benzene solution, but only 3.7×10^{-2} in water solution.

(a) Calculate the ratio of dimers to monomers for 0.100 M acetic acid in benzene.

(b) Calculate the ratio of dimers to monomers for 0.100 M acetic acid in water.

(c) Why is K_c for the water solution so much smaller than K_c for the benzene solution?

13.103 Consider the reaction $C(s) + CO_2(g) \rightleftharpoons 2\,CO(g)$. When 1.50 mol of CO_2 and an excess of solid carbon are heated in a 20.0 L container at 1100 K, the equilibrium concentration of CO is 7.00×10^{-2} M. Calculate:

(a) The equilibrium concentration of CO_2

(b) The value of the equilibrium constant K_c at 1100 K

13.104 When 1.000 mol of PCl_5 is introduced into a 5.000 L container at 500 K, 78.50% of the PCl_5 dissociates to give an equilibrium mixture of PCl_5, PCl_3, and Cl_2:

$$PCl_5(g) \rightleftharpoons PCl_3(g) + Cl_2(g)$$

(a) Calculate the values of K_c and K_p.

(b) If the initial concentrations in a particular mixture of reactants and products are $[PCl_5] = 0.500$ M, $[PCl_3] = 0.150$ M, and $[Cl_2] = 0.600$ M, in which direction does the reaction proceed to reach equilibrium? What are the concentrations when the mixture reaches equilibrium?

13.105 At a certain temperature, $K_p = 1.42$ for the reaction $PCl_5(g) \rightleftharpoons PCl_3(g) + Cl_2(g)$. Calculate the equilibrium partial pressures and the total pressure if the initial partial pressures are $P_{PCl_5} = 3.00$ atm, $P_{PCl_3} = 2.00$ atm, and $P_{Cl_2} = 1.50$ atm.

13.106 Refining petroleum involves cracking large hydrocarbon molecules into smaller, more volatile pieces. A simple example of hydrocarbon cracking is the gas-phase thermal decomposition of butane to give ethane and ethylene:

Butane, C_4H_{10} Ethane, C_2H_6 Ethylene, C_2H_4

(a) Write the equilibrium constant expressions for K_p and K_c.

(b) The value of K_p at 500°C is 12. What is the value of K_c?

(c) A sample of butane having a pressure of 50 atm is heated at 500°C in a closed container at constant volume. When equilibrium is reached, what percentage of the butane has been converted to ethane and ethylene? What is the total pressure at equilibrium?

(d) How would the percent conversion in part (c) be affected by a decrease in volume?

13.107 The equilibrium constant K_c for the gas-phase thermal decomposition of cyclopropane to propene is 1.0×10^5 at 500 K:

$$\underset{\text{Cyclopropane}}{\overset{\displaystyle CH_2}{\underset{\displaystyle H_2C-CH_2}{\diagup\diagdown}}} \rightleftarrows \underset{\text{Propene}}{CH_3-CH=CH_2} \qquad K_c = 1.0 \times 10^5$$

(a) What is the value of K_p at 500 K?

(b) What is the equilibrium partial pressure of cyclopropane at 500 K when the partial pressure of propene is 5.0 atm?

(c) Can you alter the ratio of the two concentrations at equilibrium by adding cyclopropane or by decreasing the volume of the container? Explain.

(d) Which has the larger rate constant, the forward reaction or the reverse reaction?

(e) Why is cyclopropane so reactive? (*Hint*: Consider the hybrid orbitals used by the C atoms.)

13.108 The equilibrium constant K_p for the gas-phase thermal decomposition of *tert*-butyl chloride is 3.45 at 500 K:

$$\underset{\text{\textit{tert}-Butyl chloride}}{(CH_3)_3CCl(g)} \rightleftarrows \underset{\text{Isobutylene}}{(CH_3)_2C=CH_2(g)} + HCl(g)$$

(a) Calculate the value of K_c at 500 K.

(b) Calculate the molar concentrations of reactants and products in an equilibrium mixture obtained by heating 1.00 mol of *tert*-butyl chloride in a 5.00 L vessel at 500 K.

(c) A mixture of isobutylene (0.400 atm partial pressure at 500 K) and HCl (0.600 atm partial pressure at 500 K) is allowed to reach equilibrium at 500 K. What are the equilibrium partial pressures of *tert*-butyl chloride, isobutylene, and HCl?

13.109 As shown in Figure 13.14, a catalyst lowers the activation energy for the forward and reverse reactions by the same amount, ΔE_a.

(a) Apply the Arrhenius equation, $k = Ae^{-E_a/RT}$, to the forward and reverse reactions, and show that a catalyst increases the rates of both reactions by the same factor.

(b) Use the relation between the equilibrium constant and the forward and reverse rate constants, $K_c = k_f/k_r$, to show that a catalyst does not affect the value of the equilibrium constant.

13.110 Given the Arrhenius equation, $k = Ae^{-E_a/RT}$, and the relation between the equilibrium constant and the forward and reverse rate constants, $K_c = k_f/k_r$, explain why K_c for an exothermic reaction decreases with increasing temperature.

Multi-Concept Problems

13.111 A 125.4 g quantity of water and an equal molar amount of carbon monoxide were placed in an empty 10.0 L vessel, and the mixture was heated to 700 K. At equilibrium, the partial pressure of CO was 9.80 atm. The reaction is

$$CO(g) + H_2O(g) \rightleftarrows CO_2(g) + H_2(g)$$

(a) What is the value of K_p at 700 K?

(b) An additional 31.4 g of water was added to the reaction vessel, and a new state of equilibrium was achieved. What are the equilibrium partial pressures of each gas in the mixture? What is the concentration of H_2 in molecules/cm^3?

13.112 A 79.2 g chunk of dry ice (solid CO_2) and 30.0 g of graphite (carbon) were placed in an empty 5.00 L container, and the mixture was heated to achieve equilibrium. The reaction is

$$CO_2(g) + C(s) \rightleftarrows 2 CO(g)$$

(a) What is the value of K_p at 1000 K if the gas density at 1000 K is 16.3 g/L?

(b) What is the value of K_p at 1100 K if the gas density at 1100 K is 16.9 g/L?

(c) Is the reaction exothermic or endothermic? Explain.

13.113 The amount of carbon dioxide in a gaseous mixture of CO_2 and CO can be determined by passing the gas into an aqueous solution that contains an excess of $Ba(OH)_2$. The CO_2 reacts, yielding a precipitate of $BaCO_3$, but the CO does not react. This method was used to analyze the equilibrium composition of the gas obtained when 1.77 g of CO_2 reacted with 2.0 g of graphite in a 1.000 L container at 1100 K. The analysis yielded 3.41 g of $BaCO_3$. Use these data to calculate K_p at 1100 K for the reaction

$$CO_2(g) + C(s) \rightleftarrows 2 CO(g)$$

13.114 A 14.58 g quantity of N_2O_4 was placed in a 1.000 L reaction vessel at 400 K. The N_2O_4 decomposed to an equilibrium mixture of N_2O_4 and NO_2 that had a total pressure of 9.15 atm.

(a) What is the value of K_c for the reaction $N_2O_4(g) \rightleftarrows 2 NO_2(g)$ at 400 K?

(b) How much heat (in kilojoules) was absorbed when the N_2O_4 decomposed to give the equilibrium mixture? (Standard heats of formation may be found in Appendix B.)

13.115 Consider the sublimation of moth balls at 27°C in a room having dimensions 8.0 ft × 10.0 ft × 8.0 ft. Assume that the moth balls are pure solid naphthalene (density 1.16 g/cm³) and that they are spheres with a diameter of 12.0 mm. The equilibrium constant K_c for sublimation of naphthalene is 5.40×10^{-6} at 27°C.

$$C_{10}H_8(s) \rightleftharpoons C_{10}H_8(g)$$

(a) When excess moth balls are present, how many gaseous naphthalene molecules are in the room at equilibrium?

(b) How many moth balls are required to saturate the room with gaseous naphthalene?

13.116 Ozone is unstable with respect to decomposition to ordinary oxygen:

$$2 O_3(g) \rightleftharpoons 3 O_2(g) \qquad K_p = 1.3 \times 10^{57}$$

How many O_3 molecules are present at equilibrium in 10 million cubic meters of air at 25°C and 720 mm Hg pressure?

---

 eMedia Problems

13.117 Use the **Chemical Equilibrium** simulation (*eChapter 13.2*) to explore the reaction between the iron(III) ion and the thiocyanate ion.

(a) Write the equilibrium constant expression for the reaction as it appears in the simulation.

(b) Does the equilibrium lie to the right or to the left?

(c) Would it be possible to establish equilibrium starting with only the product and none of the reactants? Explain.

(d) Write the equilibrium constant expression for the reverse reaction.

13.118 Calculate the equilibrium concentrations of all species in the **Chemical Equilibrium** simulation (*eChapter 13.2*) when equilibrium is established from the following initial concentrations:

(a) 0.0010 M $Fe^{3+}(aq)$ and 0.0005 M $SCN^-(aq)$

(b) 0.0005 M $Fe^{3+}(aq)$ and 0.0005 M $SCN^-(aq)$

(c) 0.0006 M $Fe^{3+}(aq)$ and 0.0010 M $SCN^-(aq)$

(d) 0.0001 M $Fe(SCN)^{2+}(aq)$

(e) 0.0010 M $Fe(SCN)^{2+}(aq)$ and 0.0005 M $SCN^-(aq)$.

Use the simulation to check your answers for (a), (b), and (c).

13.119 Using the **Equilibrium Constant** simulation (*eChapter 13.2*), compare the reactions A $\rightleftharpoons$ B and A $\rightleftharpoons$ 2 B in Key Concept Problem 13.3 (page 534). (Each picture represents the contents of a 1.00×10^{-24} L vessel.)

(a) If figure 1 represents an equilibrium mixture for the reaction A $\rightleftharpoons$ B, what is the value of K_c for the reaction?

(b) If picture 1 represents an equilibrium mixture for the reaction A $\rightleftharpoons$ 2 B, what is the value of K_c for the reaction?

(c) Which (if any) of the other pictures also represents an equilibrium mixture for the reaction A $\rightleftharpoons$ 2 B? Explain your answer.

13.120 Consider the reaction A $\rightleftharpoons$ 2 B. Equilibrium is achieved beginning with 0.55 M A and no B. The equilibrium concentration of B is 0.10 M. What is the value of K_c for the reaction? Use the **Equilibrium Constant** simulation (*eChapter 13.2*) to verify your answer.

Hydrogen, Oxygen, and Water

Approximately 2.0% of the earth's water is in the form of ice.

Hydrogen is the most abundant element in the universe, and oxygen is the most abundant element on the earth's surface. When chemically combined, they yield water, perhaps the most important and familiar of all chemical compounds.

Hydrogen combines with every element in the periodic table except the noble gases and forms more compounds than any other element. Industrially, large amounts of elemental hydrogen are produced for use in the synthesis of such chemicals as ammonia and methanol. Oxygen is essential for respiration and is the oxidizing agent in the energy-generating combustion processes that maintain our industrialized civilization. Approximately 31 million tons of oxygen are produced annually in the United States, largely for use in making steel.

In this chapter, we'll take a detailed look at the chemistry of hydrogen and oxygen, and we'll discuss some of the properties of water, the solvent for all the reactions to be discussed in Chapters 15 and 16.

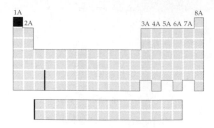

14.1 Hydrogen

Henry Cavendish (1731–1810), an English chemist and physicist, was the first person to isolate hydrogen in pure form. Cavendish showed that the action of acids on metals, such as zinc, iron, and tin, produces a flammable gas that can be distinguished from other gases by its unusually low density:

$$2\,H^+(aq) + Zn(s) \longrightarrow H_2(g) + Zn^{2+}(aq)$$

The French chemist Lavoisier called the gas "hydrogen," meaning "water former," because it combines with oxygen to produce water.

At ordinary temperatures and pressures, hydrogen is a colorless, odorless, and tasteless gas comprised of diatomic H_2 molecules. Because H_2 is a nonpolar molecule that contains only two electrons, intermolecular forces are extremely weak (Section 10.2). As a result, hydrogen has a very low melting point ($-259.2°C$) and a very low boiling point ($-252.8°C$). The bonding forces within the H_2 molecule are exceptionally strong, however: The H–H bond dissociation energy is 436 kJ/mol, greater than that for any other single bond between two atoms of the same element (Section 7.2):

$$H_2(g) \longrightarrow 2\,H(g) \quad D = 436\ \text{kJ/mol}$$

By comparison, the bond dissociation energies of the halogens range from 151 kJ/mol for I_2 to 243 kJ/mol for Cl_2. Because of the strong H–H bond, H_2 is thermally stable. Even at 2000 K, only 1 of every 2500 H_2 molecules is dissociated into H atoms at 1 atm pressure.

Hydrogen is thought to account for approximately 75% of the mass of the universe. Our sun and other stars, for instance, are composed mainly of hydrogen, which serves as their nuclear fuel. On earth, though, hydrogen is rarely found in uncombined form because the earth's gravity is too weak to hold such a light molecule. Nearly all the H_2 originally present in the earth's atmosphere has been lost to space, so that the atmosphere now contains only 0.53 ppm of H_2 by volume. In the earth's crust and oceans, hydrogen is the ninth most abundant element on a mass basis (0.9 mass %) and the third most abundant on an atom basis (15.4 atom %). Hydrogen is found in water, petroleum, proteins, carbohydrates, fats, and literally millions of other compounds.

▶ **PROBLEM 14.1** Hydrogen is used to inflate weather balloons because it is much less dense than air. Calculate the density of gaseous H_2 at 25°C and 1 atm pressure. Compare your result with the density of dry air under the same conditions (1.185×10^{-3} g/cm³).

14.2 Isotopes of Hydrogen

As mentioned in Section 2.5, there are three isotopes of hydrogen: *protium*, or ordinary hydrogen (^1_1H); *deuterium*, or heavy hydrogen (^2_1H or D); and *tritium* (^3_1H or T). Nearly all (99.985%) the atoms in naturally occurring hydrogen are protium. The terrestrial abundance of deuterium is only 0.015 atom %, and tritium is present only in trace amounts ($\sim 10^{-16}$ atom %).

The properties of protium, deuterium, and tritium are similar (Table 14.1) because chemical behavior is determined primarily by electronic structure and all three isotopes have the same electron configuration ($1s^1$). There are,

The Lagoon Nebula in the constellation Sagittarius. These interstellar gas clouds consist largely of hydrogen, the most abundant element in the universe. The gas is heated by radiation from nearby stars. Can you explain its characteristic red glow? (Recall Section 5.3.)

TABLE 14.1 Properties of Hydrogen Isotopes

Property	Protium	Deuterium	Tritium
Atomic hydrogen (H)			
Mass, amu	1.0078	2.0141	3.0160
Ionization energy, kJ/mol	1311.7	1312.2	
Nuclear stability	Stable	Stable	Radioactive
Molecular hydrogen (H_2)			
Melting point, K	13.96	18.73	20.62
Boiling point, K	20.39	23.67	25.04
Bond dissociation energy , kJ/mol	435.9	443.4	446.9
Water (H_2O)			
Melting point, °C	0.00	3.81	4.48
Boiling point, °C	100.00	101.42	101.51
Density at 25°C, g/mL	0.997	1.104	1.214
Dissociation constant at 25°C	1.01×10^{-14}	0.195×10^{-14}	$\sim 0.06 \times 10^{-14}$

however, quantitative differences in properties, known as **isotope effects**, that arise from the differences in the masses of the isotopes. For example, D_2 has a higher melting point, a higher boiling point, and a greater bond dissociation energy than H_2. Similarly, D_2O has a higher melting point and a higher boiling point than H_2O, and the equilibrium constant for dissociation of D_2O is about one-fifth that for H_2O:

$$H_2O(l) \rightleftharpoons H^+(aq) + OH^-(aq) \qquad K = 1.01 \times 10^{-14}$$

$$D_2O(l) \rightleftharpoons D^+(aq) + OD^-(aq) \qquad K = 0.195 \times 10^{-14}$$

Isotope effects are much greater for hydrogen than for any other element because the percentage differences between the masses of the various isotopes are considerably larger for hydrogen than for heavier elements. For example, a deuterium (^2H) atom is 100% heavier than a protium (^1H) atom, whereas a ^{37}Cl atom is only 6% heavier than a ^{35}Cl atom.

The effect of isotopic mass on the rate of a chemical reaction is called a *kinetic-isotope effect*. For example, deuterium can be separated from protium by passing an electric current through a solution of an inert electrolyte in ordinary water. (*Electrolysis*, the process of using an electric current to bring about chemical change, will be discussed in Section 18.11.) Because the heavier D atom forms stronger bonds than the lighter H atom, O—D bonds break more slowly than O—H bonds. As a result, D_2 evolves from D_2O more slowly than H_2 evolves from H_2O, so the remaining water is enriched in D_2O as the electrolysis proceeds.

$$2\,H_2O(l) \xrightarrow{\text{Electrolysis}} 2\,H_2(g) + O_2(g) \qquad \text{Faster}$$

$$2\,D_2O(l) \xrightarrow{\text{Electrolysis}} 2\,D_2(g) + O_2(g) \qquad \text{Slower}$$

In a typical experiment, reduction of the water's volume from 2400 L to 83 mL yields 99% pure D_2O. Large amounts of D_2O (about 160 tons per year in the United States) are manufactured by this method for use as a coolant and a moderator in nuclear reactors.

Electrolysis of water gives H_2 gas at one electrode and O_2 gas at the other electrode. Which gas is at which electrode?

▶**PROBLEM 14.2** The most abundant elements (by mass) in the body of a healthy human adult are oxygen (61.4%), carbon (22.9%), hydrogen (10.0%), and nitrogen (2.6%).
(a) Calculate the percent D if all the hydrogen atoms in a human were deuterium atoms.
(b) Calculate the percent C if all the carbon atoms were atoms of the isotope having a mass of 13 amu ($^{13}_{6}C$).
(c) Are isotope effects larger for hydrogen or for carbon? Explain.

14.3 Preparation and Uses of Hydrogen

The purest hydrogen (>99.95% pure) is made by electrolysis of water. However, this process requires a large amount of energy—286 kJ per mole of H_2 produced—and thus is not economical for large-scale production.

$$2 H_2O(l) \longrightarrow 2 H_2(g) + O_2(g) \qquad \Delta H° = +572 \text{ kJ}$$

Small amounts of hydrogen are conveniently prepared in the laboratory by reaction of dilute acid with an active metal such as zinc:

$$\text{Zn}(s) + 2 H^+(aq) \longrightarrow H_2(g) + \text{Zn}^{2+}(aq)$$

Oxidation $\quad\quad\uparrow\quad\quad\quad\uparrow\quad\quad\quad\quad\uparrow\quad\quad\quad\uparrow$
numbers $\quad\quad 0 \quad\quad\quad +1 \quad\quad\quad\quad 0 \quad\quad\quad +2$

This is a redox reaction (Section 4.6) in which hydrogen in H^+ is reduced from the +1 to the 0 oxidation state, while zinc is oxidized from the 0 to the +2 oxidation state. A typical apparatus for generating and collecting hydrogen is shown in Figure 14.1.

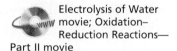

Electrolysis of Water movie; Oxidation–Reduction Reactions—Part II movie

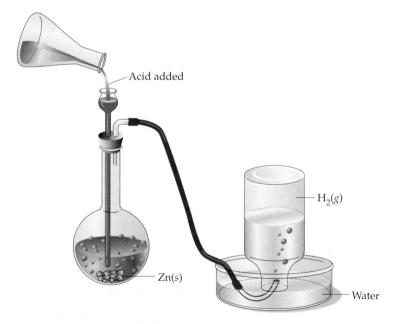

FIGURE 14.1 Preparation of hydrogen by reaction of zinc metal with dilute acid. The H_2 gas, which is nearly insoluble in water, displaces the water in the collection vessel.

Because water is the cheapest and most readily available source of hydrogen, all large-scale, industrial methods for producing hydrogen use an inexpensive reducing agent such as hot iron, carbon, or methane (natural gas) to extract the oxygen from steam:

$$4\ H_2O(g) + 3\ Fe(s) \xrightarrow{\text{Heat}} Fe_3O_4(s) + 4\ H_2(g) \qquad \Delta H° = -151\ kJ$$

$$H_2O(g) + C(s) \xrightarrow{1000°C} CO(g) + H_2(g) \qquad \Delta H° = +131\ kJ$$

At present, the most important industrial method for producing hydrogen is the three-step, **steam–hydrocarbon re-forming process**. The first step in the process is the conversion of steam and methane to a mixture of carbon monoxide and hydrogen known as *synthesis gas* (so-called because it can be used as the starting material for the synthesis of liquid fuels). The reaction requires high temperature, moderately high pressure, and a nickel catalyst:

$$H_2O(g) + CH_4(g) \xrightarrow[\text{Ni catalyst}]{1100°C} CO(g) + 3\ H_2(g) \qquad \Delta H° = +206\ kJ$$

In the second step, the synthesis gas and additional steam are passed over a metal oxide catalyst at about 400°C. Under these conditions, the carbon monoxide component of the synthesis gas and the steam are converted to carbon dioxide and more hydrogen. This reaction of CO with H_2O is called the **water-gas shift reaction** because it shifts the composition of synthesis gas by removing the toxic carbon monoxide and producing more of the economically important hydrogen:

$$CO(g) + H_2O(g) \xrightarrow[\text{Catalyst}]{400°C} CO_2(g) + H_2(g) \qquad \Delta H° = -41\ kJ$$

Finally, the unwanted carbon dioxide is removed in a third step by passing the H_2/CO_2 mixture through a basic aqueous solution. This treatment converts the carbon dioxide to carbonate ion, which remains in the aqueous phase:

$$CO_2(g) + 2\ OH^-(aq) \longrightarrow CO_3^{2-}(aq) + H_2O(l)$$

Approximately 95% of the H_2 produced in industry is synthesized and consumed in industrial plants that manufacture other chemicals. The largest single consumer of hydrogen is the Haber process for synthesizing ammonia (Sections 13.6–13.10):

$$N_2(g) + 3\ H_2(g) \rightleftharpoons 2\ NH_3(g)$$

Large amounts of hydrogen are also used for the synthesis of methanol, CH_3OH, from carbon monoxide:

$$CO(g) + 2\ H_2(g) \xrightarrow[\text{catalyst}]{\text{Cobalt}} CH_3OH(l)$$

Methanol is an industrial solvent, a precursor to additives in unleaded gasolines, and a starting material for the manufacture of formaldehyde, CH_2O, used in making plastics. Annual U.S. production of methanol, about 1.7 billion gallons, consumes about 700,000 tons of hydrogen.

▶**PROBLEM 14.3**   Write a balanced net ionic equation for the reaction of gallium metal with dilute acid.

14.4   Reactivity of Hydrogen

The hydrogen atom is the simplest of all atoms, containing only a single $1s$ electron and a single proton. In most versions of the periodic table, hydrogen is located in group 1A above the alkali metals because they too have just one valence electron. Alternatively, hydrogen could be placed in group 7A above the halogens because, like the halogens, it is just one electron short of a noble gas configuration (He). Thus, hydrogen has properties similar to those of both alkali metals and halogens. A hydrogen atom can lose an electron to form a hydrogen cation, H^+, or it can gain an electron to yield a hydride anion, H^-:

Alkali-metal-like reaction: $H(g) \longrightarrow H^+(g) + e^-$ $E_i = +1312$ kJ/mol

Halogen-like reaction: $H(g) + e^- \longrightarrow H^-(g)$ $E_{ea} = -73$ kJ/mol

Because the amount of energy needed to ionize a hydrogen atom is so large ($E_i = 1312$ kJ/mol), hydrogen doesn't completely transfer its valence electron in chemical reactions. Instead, it shares this electron with a nonmetallic element to give a covalent compound such as CH_4, NH_3, H_2O, or HF. In this regard, hydrogen differs markedly from the alkali metals, which have much smaller ionization energies (ranging from 520 kJ/mol for Li to 376 kJ/mol for Cs) and which form ionic compounds with nonmetals.

Complete ionization of a hydrogen atom is possible in the gas phase, where the hydrogen ion is present as a bare proton of radius ~1.5×10^{-3} pm, about 100,000 times smaller than the radius of a hydrogen atom. In liquids and solids, however, the bare proton is too reactive to exist by itself. Instead, the proton bonds to a molecule that has a lone pair of electrons. In water, for example, the proton bonds to an H_2O molecule to give a hydronium ion, $H_3O^+(aq)$.

Although adding an electron to hydrogen ($E_{ea} = -73$ kJ/mol) releases less energy than adding an electron to the halogens ($E_{ea} = -295$ to -349 kJ/mol), hydrogen will accept an electron from an active metal to give an ionic hydride, such as NaH or CaH_2. In this regard, the behavior of hydrogen parallels that of the halogens, which form ionic halides such as NaCl and $CaCl_2$.

At room temperature, H_2 is relatively unreactive because of its strong H–H bond, although it does react with F_2 to give HF and with Cl_2 in the presence of light to give HCl. Reactions of H_2 with O_2, N_2, or C, however, require high temperatures, the presence of a catalyst, or both. Catalysts such as metallic iron, nickel, palladium, or platinum facilitate the dissociation of H_2 into highly reactive H atoms (Section 12.12). The reaction of hydrogen and oxygen is highly exothermic, and gas mixtures that contain as little as 4% by volume hydrogen in air are highly flammable and potentially explosive.

$$2\,H_2(g) + O_2(g) \longrightarrow 2\,H_2O(l) \qquad \Delta H° = -572\text{ kJ}$$

Formation of Water movie

Explosive burning of the hydrogen-filled dirigible *Hindenburg* during landing at Lakehurst, New Jersey, on May 6, 1937 killed 36 of the 97 persons aboard.

14.5 Binary Hydrides

The **binary hydrides** are compounds that contain hydrogen and just one other element. Formulas and melting points of the simplest hydrides of the main-group elements are listed in Figure 14.2. Binary hydrides can be classified as ionic, covalent, or metallic.

(1) 1A	(2) 2A		(13) 3A	(14) 4A	(15) 5A	(16) 6A	(17) 7A	(18) 8A
LiH 692	BeH$_2$ d 250		B$_2$H$_6$ −165	CH$_4$ −182	NH$_3$ −78	H$_2$O 0	HF −83	
NaH d 800	MgH$_2$ d 280		AlH$_3$ d 150	SiH$_4$ −185	PH$_3$ −134	H$_2$S −86	HCl −115	
KH d	CaH$_2$ 816		GaH$_3$ −15	GeH$_4$ −165	AsH$_3$ −116	H$_2$Se −66	HBr −88	
RbH d	SrH$_2$ d 675		InH$_3$ (?)	SnH$_4$ −146	SbH$_3$ −88	H$_2$Te −51	HI −51	
CsH d	BaH$_2$ d 675		TlH$_3$ (?)	PbH$_4$	BiH$_3$	H$_2$Po	HAt	

FIGURE 14.2 Formulas and melting points (°C) of the simplest hydrides of the main-group elements. The group 1A and the heavier group 2A hydrides, shown in blue, are ionic, while the other main-group hydrides, shown in red, are covalent. The change in bond type, however, is gradual and continuous. Transition metal hydrides (not shown) are classified as metallic. The letter "d" indicates decomposition rather than melting on heating to the indicated temperature. The existence of InH$_3$ and TlH$_3$ is uncertain.

Ionic Hydrides

Ionic hydrides are saltlike, high-melting, white, crystalline compounds formed by the alkali metals and the heavier alkaline earth metals Ca, Sr, and Ba. They can be prepared by direct reaction of the elements at about 400°C:

$$2\,\text{Na}(l) + \text{H}_2(g) \longrightarrow 2\,\text{NaH}(s) \qquad \Delta H° = -112.6 \text{ kJ}$$

$$\text{Ca}(s) + \text{H}_2(g) \longrightarrow \text{CaH}_2(s) \qquad \Delta H° = -181.5 \text{ kJ}$$

The alkali metal hydrides contain alkali metal cations and H$^-$ anions in a face-centered cubic crystal structure like that of sodium chloride (Section 10.9). Alkali metal hydrides are also ionic in the liquid state, as shown by the fact that the molten compounds conduct electricity.

The H$^-$ anion is a good proton acceptor, and ionic hydrides therefore react with water to give H$_2$ gas and OH$^-$ ions:

$$\text{CaH}_2(s) + 2\,\text{H}_2\text{O}(l) \longrightarrow 2\,\text{H}_2(g) + \text{Ca}^{2+}(aq) + 2\,\text{OH}^-(aq)$$

This reaction of an ionic hydride with water is a redox reaction because the hydride reduces the water (+1 oxidation state for H) to H$_2$ (0 oxidation state). In turn, the hydride (−1 oxidation state for H) is oxidized to H$_2$. In general, ionic hydrides are good reducing agents. Some, such as potassium hydride, catch fire in air because of a rapid redox reaction with oxygen:

$$2\,\text{KH}(s) + \text{O}_2(g) \longrightarrow \text{H}_2\text{O}(g) + \text{K}_2\text{O}(s)$$

Calcium hydride reacts with water to give bubbles of H$_2$ gas and OH$^-$ ions. The red color is due to added phenolphthalein, which turns from colorless to red in the presence of a base.

Covalent Hydrides

Covalent hydrides, as their name implies, are compounds in which hydrogen is attached to another element by a covalent bond. The most common examples are hydrides of nonmetallic elements, such as diborane (B_2H_6), methane (CH_4), ammonia (NH_3), water (H_2O), and the hydrogen halides (HX). Only the simplest covalent hydrides are listed in Figure 14.2, though more complex examples, such as hydrogen peroxide (H_2O_2) and hydrazine (N_2H_4), are also known. Because most covalent hydrides consist of discrete, small molecules that have relatively weak intermolecular forces, they are gases or volatile liquids at ordinary temperatures.

Metallic Hydrides

Metallic hydrides are formed by reaction of the lanthanide and actinide metals and certain of the d-block transition metals with variable amounts of hydrogen. These hydrides have the general formula MH_x, where the x subscript represents the number of H atoms in the simplest formula. They are often called **interstitial hydrides** because they are thought to consist of a crystal lattice of metal atoms with the smaller hydrogen atoms occupying holes, or *interstices*, between the larger metal atoms (Figure 14.3).

The nature of the bonding in metallic hydrides is not well understood, and it's not known whether the hydrogens are present as neutral H atoms, H^+ cations, or H^- anions. Because the hydrogen atoms can fill a variable number of interstices, many metallic hydrides are **nonstoichiometric compounds**, meaning that their atomic composition can't be expressed as a ratio of small whole numbers. Examples are $TiH_{1.7}$, $ZrH_{1.9}$, and PdH_x ($x < 1$). Other metallic hydrides, however, are stoichiometric compounds—for example, TiH_2 and UH_3.

FIGURE 14.3 One plane of the structure of an interstitial metallic hydride. The metal atoms (larger spheres) have a face-centered cubic structure, and the hydrogen atoms (smaller spheres) occupy interstices (holes) between the metal atoms.

The properties of metallic hydrides depend on their composition, which is a function of the partial pressure of H_2 gas in the surroundings. For example, PdH_x behaves as a metallic conductor for small values of x but becomes a semiconductor when x reaches about 0.5. (Semiconductors are discussed in Section 21.5.) The H atoms in PdH_x are highly mobile, and H_2 can pass through a membrane of palladium metal. The process probably involves dissociation of H_2 into H atoms on one surface of the membrane, diffusion of H atoms through the membrane as they jump from one interstice to another, and recombination to form H_2 on the opposite surface of the membrane. Because other gases don't penetrate palladium, this process can be used to separate H_2 or D_2 from other components of gas mixtures.

Interstitial hydrides are of current interest as potential hydrogen-storage devices because they can contain a remarkably large amount of hydrogen. Palladium, for example, absorbs up to 935 times its own volume of H_2. This amount corresponds to a density of hydrogen comparable to that in liquid hydrogen. For use as a fuel, hydrogen could be stored as PdH_x and then liberated when needed simply by heating the PdH_x.

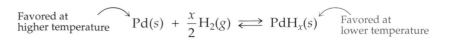

Favored at higher temperature $\quad Pd(s) + \dfrac{x}{2} H_2(g) \rightleftharpoons PdH_x(s) \quad$ Favored at lower temperature

EXAMPLE 14.1 Write a balanced net ionic equation for reaction of each of the following hydrides with water:
(a) Lithium hydride **(b)** Barium hydride

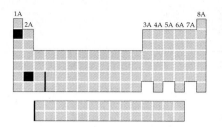

SOLUTION
(a) Because lithium is a group 1A metal, lithium hydride has formula LiH and is an ionic hydride. It therefore reacts with water to give H_2 gas and OH^- ions. Balance the equation either by inspection or by using the method of oxidation numbers.

$$LiH(s) + H_2O(l) \longrightarrow H_2(g) + Li^+(aq) + OH^-(aq)$$

Oxidation numbers: LiH -1, H_2O $+1$, H_2 0

The H^- ion reduces water to H_2 gas, and in the process H^- is oxidized to H_2 gas.
(b) Because barium is a group 2A metal, barium hydride has the formula BaH_2 and is an ionic hydride. It also reacts with water to give H_2 gas and OH^- ions. Balance the equation either by inspection or by using the method of oxidation numbers.

$$BaH_2(s) + 2 H_2O(l) \longrightarrow 2 H_2(g) + Ba^{2+}(aq) + 2 OH^-(aq)$$

Oxidation numbers: BaH_2 -1, H_2O $+1$, H_2 0

EXAMPLE 14.2 How many grams of barium hydride must be treated with water to obtain 4.36 L of hydrogen at 20°C and 0.975 atm pressure?

BALLPARK SOLUTION The balanced equation for the reaction is given in Example 14.1(b). Although the 4.36 L of H_2 is not at STP (273 K and 1 atm pressure), the actual temperature (293 K) and pressure (0.975 atm) of the gas differ from STP by less than 10%. Therefore, we will ignore the difference. Since 1 mol of an ideal gas occupies a volume of about 22 L at STP (Section 9.3), the number of moles of H_2 in the 4.36 L volume of gas is approximately 4.4 L divided by 22 L/mol, or 0.20 mol of H_2. The balanced equation for the reaction states that 2 mol of H_2 are obtained for every 1 mol of BaH_2 that reacts. Therefore, the amount of BaH_2 needed is approximately 0.20/2 = 0.10 mol, or about 14 g since the molar mass of BaH_2 is 139.3 g/mol.

DETAILED SOLUTION First, convert the volume of H_2 to moles by using the ideal gas law:

$$n = \frac{PV}{RT} = \frac{(0.975 \text{ atm})(4.36 \text{ L})}{\left(0.082\,06 \dfrac{\text{L} \cdot \text{atm}}{\text{mol} \cdot \text{K}}\right)(293 \text{ K})} = 0.177 \text{ mol } H_2$$

Next, use the balanced equation to calculate the number of moles of BaH_2 required to produce 0.177 mol of H_2:

$$\text{Moles of } BaH_2 = 0.177 \text{ mol } H_2 \times \frac{1 \text{ mol } BaH_2}{2 \text{ mol } H_2} = 0.0885 \text{ mol } BaH_2$$

Finally, use the molar mass of BaH_2 (139.3 g/mol) to convert moles of BaH_2 to grams of BaH_2:

$$\text{Grams of } BaH_2 = 0.0885 \text{ mol } BaH_2 \times \frac{139.3 \text{ g } BaH_2}{1 \text{ mol } BaH_2} = 12.3 \text{ g } BaH_2$$

▶ **PROBLEM 14.4** Write a balanced net ionic equation for reaction of each of the following hydrides with water:
(a) Strontium hydride **(b)** Potassium hydride

▶ **PROBLEM 14.5** Calcium hydride is a convenient, portable source of hydrogen that is used, for example, to inflate weather balloons. If the reaction of CaH_2 with water is used to inflate a balloon with 2.0×10^5 L of H_2 gas at 25°C and 1.00 atm pressure, how many kilograms of CaH_2 is needed?

◄ **KEY CONCEPT PROBLEM 14.6** The following pictures represent binary hydrides AH_x, where A = K, Ti, C, or F. Ivory spheres represent H atoms or ions, and burgundy spheres represent atoms or ions of the element A.

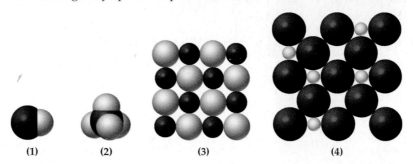

(1)   (2)   (3)   (4)

(a) Write the formula of each hydride, and classify each as ionic, covalent, or interstitial.
(b) Which hydride has the lowest melting point? Explain.
(c) Which hydride reacts with water at 25°C to give H_2 gas?

▶ **PROBLEM 14.7** If palladium metal (density 12.0 g/cm³) dissolves 935 times its own volume of H_2 at STP, what is the value of x in the formula PdH_x? What is the density of hydrogen in PdH_x? What is the molarity of H atoms in PdH_x? Assume that the volume of palladium is unchanged when the H atoms go into the interstices.

14.6 Oxygen

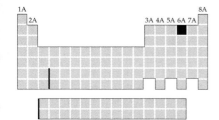

Oxygen was first isolated and characterized in the period 1771–1774 by the English chemist Joseph Priestley and the Swedish chemist Karl Wilhelm Scheele. Priestley and Scheele found that heating certain compounds such as mercury(II) oxide generates a colorless, odorless, tasteless gas that supports combustion better than air does:

$$2\,HgO(s) \xrightarrow{\text{Heat}} 2\,Hg(l) + O_2(g)$$

Priestley called the gas "dephlogisticated air," but Lavoisier soon recognized it as an element and named it "oxygen."

Gaseous O_2 condenses at −183°C to form a pale blue liquid and freezes at −219°C to give a pale blue solid. In all three phases—gas, liquid, and solid—O_2 is paramagnetic, as illustrated previously in Figure 7.17. The bond length in O_2 is 121 pm, appreciably shorter than the O–O single bond in H_2O_2 (148 pm), and the bond dissociation energy of O_2 (498 kJ) is intermediate between that for the single bond in F_2 (158 kJ) and the triple bond in N_2 (945 kJ). These properties are consistent with the presence of a double bond in O_2 (Section 7.14).

Oxygen is the most abundant element on the surface of our planet and is crucial to human life. It's in the air we breathe, the water we drink, and the

Liquid oxygen has a pale blue color.

food we eat. It's the oxidizing agent in the metabolic "burning" of foods, and it's an important component of biological molecules: Approximately one-fourth of the atoms in living organisms are oxygen. Moreover, oxygen is the oxidizing agent in the combustion processes that provide thermal and electrical energy for maintaining our industrialized civilization.

On a mass basis, oxygen constitutes 23% of the atmosphere (21% by volume), 46% of the lithosphere (the earth's crust), and more than 85% of the hydrosphere. In the atmosphere, oxygen is found primarily as O_2, sometimes called *dioxygen*. The oxygen in the hydrosphere is, of course, in the form of H_2O, but enough dissolved O_2 is present to maintain aquatic life. In the lithosphere, oxygen is combined with other elements in crustal rocks composed of silicates, carbonates, oxides, and other oxygen-containing minerals.

The amount of oxygen in the atmosphere remains fairly constant at about 1.18×10^{18} kg because the combustion and respiration processes that remove O_2 are balanced by photosynthesis, the complex process in which green plants use solar energy to produce O_2 and glucose from carbon dioxide and water:

$$6\,CO_2 + 6\,H_2O \xrightarrow{h\nu} 6\,O_2 + \underset{\text{Glucose}}{C_6H_{12}O_6}$$

Oxygen in the earth's atmosphere is produced by photosynthesis in plants, such as this underwater *elodea*.

The metabolism of carbohydrates in our bodies to give carbon dioxide and water is essentially the reverse of the photosynthesis reaction. The energy from the sun that is absorbed in the endothermic photosynthetic process is released when organic matter is burned or metabolized to carbon dioxide and water. This cycling of oxygen between the atmosphere and the biosphere acts as the mechanism for converting solar energy to the chemical energy needed for metabolic processes. Ultimately, nearly all our energy comes from the sun.

14.7 Preparation and Uses of Oxygen

Small amounts of O_2 can be prepared in the laboratory by electrolysis of water, by decomposition of aqueous hydrogen peroxide in the presence of a catalyst such as Fe^{3+}, or by thermal decomposition of an oxoacid salt, such as potassium chlorate, $KClO_3$:

$$2\,H_2O(l) \xrightarrow{\text{Electrolysis}} 2\,H_2(g) + O_2(g)$$

$$2\,H_2O_2(aq) \xrightarrow{\text{Catalyst}} 2\,H_2O(l) + O_2(g)$$

$$2\,KClO_3(s) \xrightarrow[\text{MnO}_2\text{ catalyst}]{\text{Heat}} 2\,KCl(s) + 3\,O_2(g)$$

Because oxygen is relatively insoluble in water, it can be collected by water displacement using the same apparatus employed to collect hydrogen (Figure 14.1). Oxygen is seldom prepared in the laboratory, though, because it is commercially available as a compressed gas in high-pressure steel cylinders.

Oxygen is produced on an industrial scale, along with nitrogen and argon, by fractional distillation of liquefied air. When liquid air warms in a suitable distilling column, the more volatile components—nitrogen (bp −196°C) and argon (bp −186°C)—can be removed as gases from the top of the column. The less volatile oxygen (bp −183°C) remains as a liquid at the bottom. Oxygen is the third-ranking industrial chemical produced in the United States (31 million tons); only sulfuric acid (48 million tons) and nitrogen (33 million tons) are produced in greater quantities.

Crude iron is converted to steel by oxidizing impurities with O_2 gas.

More than two-thirds of the oxygen produced industrially is used in making steel. The crude iron obtained from a blast furnace contains impurities, such as carbon, silicon, and phosphorus, which adversely affect the mechanical properties of the metal. In refining crude iron to strong steels, the impurity levels are lowered by treating the iron with a controlled amount of O_2, thus converting the impurities to the corresponding oxides.

Among its other uses, oxygen is used in sewage treatment to destroy malodorous compounds and in paper bleaching to oxidize compounds that impart unwanted colors. In the oxyacetylene torch, the highly exothermic reaction between O_2 and acetylene provides the high temperatures ($>3000°C$) needed for cutting and welding metals:

$$2\ HC \equiv CH(g) + 5\ O_2(g) \longrightarrow 4\ CO_2(g) + 2\ H_2O(g) \qquad \Delta H° = -2511\ kJ$$
Acetylene

In all its applications, O_2 serves as an inexpensive and readily available oxidizing agent.

EXAMPLE 14.3   How many milliliters of O_2 gas at 25°C and 1.00 atm pressure can be obtained by thermal decomposition of 0.200 g of $KClO_3$?

$$2\ KClO_3(s) \longrightarrow 2\ KCl(s) + 3\ O_2(g)$$

SOLUTION   Follow a three-step procedure. First, use the molar mass of $KClO_3$ (122.6 g/mol) to calculate the number of moles of $KClO_3$ available:

$$\text{Moles of } KClO_3 = 0.200\ g\ KClO_3 \times \frac{1\ mol\ KClO_3}{122.6\ g\ KClO_3} = 1.63 \times 10^{-3}\ mol\ KClO_3$$

Next, use the balanced equation and the number of moles of $KClO_3$ to calculate the number of moles of O_2 produced:

$$\text{Moles of } O_2 = \left(\frac{3\ mol\ O_2}{2\ mol\ KClO_3}\right)(1.63 \times 10^{-3}\ mol\ KClO_3) = 2.44 \times 10^{-3}\ mol\ O_2$$

Finally, use the ideal gas law and the number of moles of O_2 to calculate the volume of O_2:

$$V = \frac{nRT}{P} = \frac{(2.44 \times 10^{-3}\ mol)\left(0.082\ 06\ \dfrac{L \cdot atm}{K \cdot mol}\right)(298\ K)}{1.00\ atm}$$

$$= 5.97 \times 10^{-2}\ L = 59.7\ mL$$

▶ **PROBLEM 14.8**   How many milliliters of O_2 gas at 25°C and 1.00 atm pressure are obtained by thermal decomposition of 0.200 g of $KMnO_4$? The balanced equation for the reaction is $2\ KMnO_4(s) \rightarrow K_2MnO_4(s) + MnO_2(s) + O_2(g)$.

14.8   Reactivity of Oxygen

We can anticipate the reactivity of oxygen on the basis of the electron configuration of an oxygen atom ($1s^2\ 2s^2\ 2p^4$) and its high electronegativity (3.5). With six valence electrons, oxygen is just two electrons short of the stable octet configuration of neon, the next noble gas. Oxygen can therefore achieve an octet configuration either by accepting two electrons from an active metal or by gaining a share in two additional electrons through covalent bonding. Thus, oxygen reacts with active metals, such as lithium and magnesium, to give *ionic oxides*:

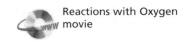

Reactions with Oxygen movie

$$4 \, \text{Li}(s) + \text{O}_2(g) \longrightarrow 2 \, \text{Li}_2\text{O}(s)$$

$$2 \, \text{Mg}(s) + \text{O}_2(g) \longrightarrow 2 \, \text{MgO}(s)$$

On the other hand, with nonmetals, such as hydrogen, carbon, sulfur, and phosphorus, oxygen forms *covalent oxides*:

$$2 \, \text{H}_2(g) + \text{O}_2(g) \longrightarrow 2 \, \text{H}_2\text{O}(l)$$

$$\text{C}(s) + \text{O}_2(g) \longrightarrow \text{CO}_2(g)$$

$$\text{S}_8(s) + 8 \, \text{O}_2(g) \longrightarrow 8 \, \text{SO}_2(g)$$

$$\text{P}_4(s) + 5 \, \text{O}_2(g) \longrightarrow \text{P}_4\text{O}_{10}(s)$$

In covalent compounds, oxygen generally achieves an octet configuration either by forming two single bonds, as in H_2O, or one double bond, as in CO_2. Oxygen often forms a double bond to small atoms such as carbon and nitrogen because there is good π overlap between the relatively compact p orbitals of second-row atoms. With larger atoms such as silicon, however, there is less efficient π overlap, and double bond formation is therefore less common (Figure 14.4).

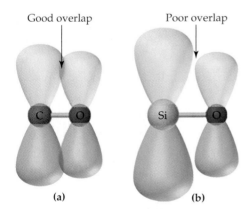

FIGURE 14.4 π overlap between the p orbitals of oxygen and other atoms. **(a)** With a second-row atom, such as carbon, oxygen forms a strong π bond. **(b)** With a larger atom, such as silicon, oxygen tends not to form π bonds because the longer Si–O distance and the larger, more diffuse silicon orbitals result in poor π overlap.

Oxygen reacts directly with all the elements in the periodic table except the noble gases and a few inactive metals, such as platinum and gold. Fortunately, most of these reactions are slow at room temperature; otherwise, cars would rust faster and many elements would spontaneously burst into flame in air. At higher temperatures, however, oxygen is extremely reactive. Figure 14.5 illustrates reactions of oxygen with magnesium, sulfur, and white phosphorus. To initiate burning of magnesium and sulfur requires heat, but white phosphorus spontaneously ignites in oxygen at room temperature. The brilliant white light emitted by burning magnesium makes this element particularly useful in fireworks.

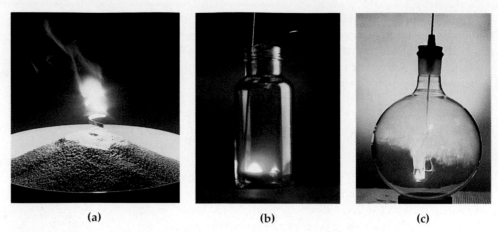

(a)   (b)   (c)

FIGURE 14.5   Some reactions of oxygen. **(a)** Magnesium metal burns in oxygen with a bright white flame, producing a white smoke of solid magnesium oxide, MgO. **(b)** Hot, molten sulfur burns in oxygen with a blue flame, forming gaseous sulfur dioxide, SO_2. **(c)** White phosphorus spontaneously inflames in oxygen, yielding an incandescent white smoke of solid P_4O_{10}.

14.9   Oxides

FIGURE 14.6   Formulas, acid–base properties, and the covalent–ionic character of the oxides of main-group elements in their highest oxidation states. Basic oxides are shown in blue, acidic oxides are shown in red, and amphoteric oxides are shown in violet.

It's convenient to classify binary oxygen-containing compounds on the basis of oxygen's oxidation state. Binary compounds with oxygen in the −2 oxidation state are called **oxides**, compounds with oxygen in the −1 oxidation state are **peroxides**, and compounds with oxygen in the −1/2 oxidation state are **superoxides**. We'll look first at oxides and then consider peroxides and superoxides in the next section.

We can categorize oxides in terms of their acid–base character as basic, acidic, or *amphoteric* (both basic and acidic), as shown in Figure 14.6.

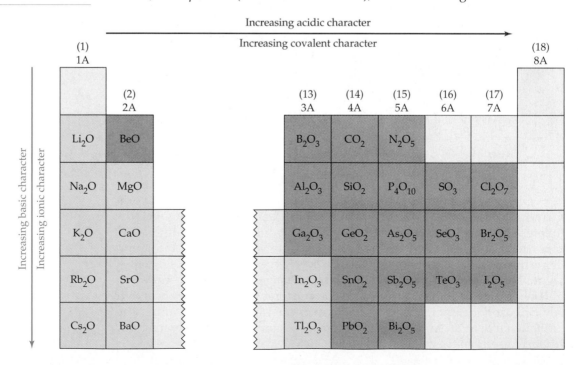

Basic oxides, also called base anhydrides, are ionic and are formed by metals on the left side of the periodic table. Water-soluble basic oxides, such as Na_2O, dissolve by reacting with water to produce OH^- ions:

$$Na_2O(s) + H_2O(l) \longrightarrow 2\,Na^+(aq) + 2\,OH^-(aq)$$

Water-insoluble basic oxides, such as MgO, can dissolve in strong acids because H^+ ions from the acid combine with the O^{2-} ion to produce water:

$$MgO(s) + 2\,H^+(aq) \longrightarrow Mg^{2+}(aq) + H_2O(l)$$

Acidic oxides, also called acid anhydrides, are covalent and are formed by the nonmetals on the right side of the periodic table. Water-soluble acidic oxides, such as N_2O_5, dissolve by reacting with water to produce aqueous H^+ ions:

$$N_2O_5(s) + H_2O(l) \longrightarrow 2\,H^+(aq) + 2\,NO_3^-(aq)$$

Water-insoluble acidic oxides, such as SiO_2, can dissolve in strong bases:

$$SiO_2(s) + 2\,OH^-(aq) \longrightarrow SiO_3^{2-}(aq) + H_2O(l)$$

Amphoteric oxides exhibit both acidic and basic properties. (The term **amphoteric** [am-fo-**tare**-ic] comes from the Greek word *amphoteros*, meaning "in both ways.") For example, Al_2O_3 is insoluble in water, but it dissolves both in strong acids and in strong bases. Al_2O_3 behaves as a base when it reacts with acids, giving the Al^{3+} ion, but it behaves as an acid when it reacts with bases, yielding the aluminate ion, $Al(OH)_4^-$.

Basic behavior: $Al_2O_3(s) + 6\,H^+(aq) \longrightarrow 2\,Al^{3+}(aq) + 3\,H_2O(l)$

Acidic behavior: $Al_2O_3(s) + 2\,OH^-(aq) + 3\,H_2O(l) \longrightarrow 2\,Al(OH)_4^-(aq)$

The elements that form amphoteric oxides have intermediate electronegativities, and the bonds in their oxides have intermediate ionic–covalent character.

The acid–base properties and the ionic–covalent character of an element's oxide depend on both the element's position in the periodic table and its oxidation number. As Figure 14.6 shows, both the acidic character and the covalent character of an oxide increase across the periodic table from the active metals on the left to the electronegative nonmetals on the right. In the third row, for example, Na_2O and MgO are basic, Al_2O_3 is amphoteric, and SiO_2, P_4O_{10}, SO_3, and Cl_2O_7 are acidic. Within a group in the periodic table, both the basic character and the ionic character of an oxide increase from the more electronegative elements at the top to the less electronegative ones at the bottom. In group 3A, for example, B_2O_3 is acidic, Al_2O_3 and Ga_2O_3 are amphoteric, and In_2O_3 and Tl_2O_3 are basic. Combining the horizontal and vertical trends in acidity, we find the most acidic oxides in the upper right of the periodic table, the most basic oxides in the lower left, and the amphoteric oxides in a roughly diagonal band stretching across the middle.

Both the acidic character and the covalent character of different oxides of the same element increase with increasing oxidation number of the element.

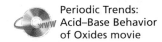

Periodic Trends: Acid–Base Behavior of Oxides movie

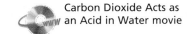

Carbon Dioxide Acts as an Acid in Water movie

Thus, sulfur(VI) oxide (SO_3) is more acidic than sulfur(IV) oxide (SO_2). Reaction of SO_3 with water gives a strong acid (sulfuric acid, H_2SO_4), whereas reaction of SO_2 with water yields a weak acid (sulfurous acid, H_2SO_3). The oxides of chromium exhibit the same trend. Chromium(VI) oxide (CrO_3) is acidic, chromium(III) oxide (Cr_2O_3) is amphoteric, and chromium(II) oxide (CrO) is basic.

As the bonding in oxides gradually changes from ionic to covalent, a corresponding change in structure, from extended three-dimensional structures to discrete molecular structures, occurs. This results in a change in physical properties, as illustrated by the melting points in Figure 14.7. MgO, an ionic oxide with the face-centered cubic NaCl crystal structure, has a very high melting point because of its high lattice energy (Section 6.6). Al_2O_3 and SiO_2 also have extended three-dimensional structures (Section 10.10), but they exhibit increasingly covalent character and have increasingly lower melting points. The trend continues to P_4O_{10}, SO_3, and Cl_2O_7, which are molecular substances with strong covalent bonds within each molecule but only weak intermolecular forces between molecules. Thus, P_4O_{10} is a volatile, relatively low-melting solid, and SO_3 and Cl_2O_7 are volatile liquids.

FIGURE 14.7   Melting points (in kelvins) of oxides of the third-period elements in their highest oxidation states. (No melting point is given for Na_2O because it sublimes and has a vapor pressure of 1 atm at 1548 K.)

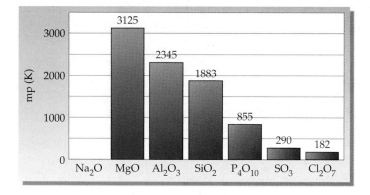

The uses of oxides are determined by their properties. Because of their high thermal stability, mechanical strength, and electrical resistance, MgO and Al_2O_3 are used as high-temperature electrical insulators in products such as electrical heaters and spark plugs. SiO_2 is the main component of the optical fibers used for communications. The acidic oxides of the nonmetals are important as precursors to industrial acids, such as HNO_3, H_3PO_4, and H_2SO_4.

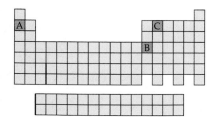

KEY CONCEPT PROBLEM 14.9   Look at the location of elements A, B, and C in the accompanying periodic table.

(a) Write the formula of the oxide of each element.

(b) Which oxide is the most ionic and which is the most covalent?

(c) Which oxide is the most acidic and which is the most basic?

(d) Which oxide can react with both $H^+(aq)$ and $OH^-(aq)$?

▶**PROBLEM 14.10**   Write balanced net ionic equations for the following reactions:

(a) Dissolution of solid Li_2O in water

(b) Dissolution of SO_3 in water

(c) Dissolution of the amphoteric oxide Cr_2O_3 in strong acid

(d) Dissolution of Cr_2O_3 in strong base to give $Cr(OH)_4^-$ ions

14.10 Peroxides and Superoxides

When the heavier group 1A and 2A metals are heated in an excess of oxygen, they form either *peroxides*, such as Na_2O_2 and BaO_2, or *superoxides*, such as KO_2, RbO_2, and CsO_2. Under the same conditions, however, the lighter group 1A and 2A metals form normal oxides, such as Li_2O, MgO, CaO, and SrO. All these compounds are ionic solids, and the nature of the product obtained depends on the amount of O_2 present, the temperature, the sizes of the ions and how they pack together, and the resultant lattice energies of the various crystalline solids (Section 6.6).

The metal peroxides contain the peroxide ion, O_2^{2-}, and the metal superoxides contain the superoxide ion, O_2^-:

$$\left[:\ddot{O}:\ddot{O}: \right]^{2-} \qquad \left[:\ddot{O}:\ddot{O}: \right]^- \longleftrightarrow \left[:\ddot{O}:\ddot{O}: \right]^-$$

Peroxide ion   Superoxide ion

The peroxide ion is diamagnetic and has an O—O single bond length of 149 pm. The superoxide ion, with one unpaired electron, is paramagnetic and has an O—O bond length of 133 pm, intermediate between the peroxide ion's bond length and the O=O double bond length of 121 pm in O_2. This suggests a bond order of 1.5 for O_2^-.

The trend in bond lengths and the magnetic properties of O_2, O_2^-, and O_2^{2-} are all nicely explained by the molecular orbital (MO) theory discussed in Section 7.14. Recall that the occupied MOs of highest energy in O_2 are the two, degenerate, antibonding π^*_{2p} orbitals. Each of these orbitals contains one electron, and the two electrons have the same spin, in accord with Hund's rule. Thus, the O_2 molecule is paramagnetic and has a bond order of 2. The O_2^- ion has one additional electron (a total of 3) in the π^*_{2p} orbitals, and the O_2^{2-} ion has two additional electrons (a total of 4) in the π^*_{2p} orbitals. As the electron population of the antibonding π^*_{2p} orbitals increases, the bond order decreases and the bond length increases (Table 14.2).

TABLE 14.2   Bond and Magnetic Properties of Diatomic Oxygen Species

Species	Number of π^*_{2p} Electrons	Number of Unpaired Electrons	Bond Order	Bond Length (pm)	Magnetic Properties
O_2	2	2	2	121	Paramagnetic
O_2^-	3	1	1.5	133	Paramagnetic
O_2^{2-}	4	0	1	149	Diamagnetic

Like the oxide ion, the peroxide ion is a basic anion. When a metal peroxide dissolves in water, the O_2^{2-} ion picks up a proton from water, forming HO_2^- and OH^- ions. The dissolution of Na_2O_2 is typical:

$$Na_2O_2(s) + H_2O(l) \longrightarrow 2\,Na^+(aq) + HO_2^-(aq) + OH^-(aq)$$

In the presence of a strong acid, the O_2^{2-} ion combines with two protons, yielding hydrogen peroxide, H_2O_2. For example, aqueous solutions of

Dissolution of sodium peroxide in water that contains phenolphthalein gives a red color due to formation of OH^- ions.

A self-contained breathing apparatus used by fire fighters. The source of O_2 is the reaction of KO_2 with exhaled water vapor.

H_2O_2 are conveniently prepared in the laboratory by reaction of barium peroxide with a stoichiometric amount of sulfuric acid:

$$BaO_2(s) + H_2SO_4(aq) \longrightarrow BaSO_4(s) + H_2O_2(aq)$$

The byproduct, barium sulfate, is insoluble and easily removed by filtration.

When metal superoxides, such as KO_2, dissolve in water, they decompose with evolution of oxygen:

$$2\ KO_2(s) + H_2O(l) \longrightarrow O_2(g) + 2\ K^+(aq) + HO_2^-(aq) + OH^-(aq)$$

Oxidation numbers: $-1/2$ (for KO_2), 0 (for O_2), -1 (for HO_2^-)

This decomposition is a redox reaction in which the oxygen in KO_2 is simultaneously oxidized from the $-1/2$ oxidation state in O_2^- to the 0 oxidation state in O_2 and reduced from the $-1/2$ oxidation state in O_2^- to the -1 oxidation state in HO_2^-. Such a reaction, in which a substance is both oxidized and reduced, is called a **disproportionation** reaction. One useful consequence of the KO_2 disproportionation reaction in water is that potassium superoxide can serve as a convenient source of oxygen in masks worn by fire fighters. The oxygen results from reaction of KO_2 with exhaled water vapor.

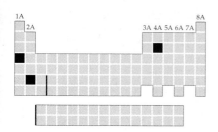

EXAMPLE 14.4   What is the oxidation number of oxygen in each of the following compounds? Tell whether the compound is an oxide, a peroxide, or a superoxide.
(a) KO_2 **(b)** BaO_2 **(c)** SiO_2

SOLUTION   First determine the oxidation number of the metal, which is generally equal to the periodic group number. Then assign an oxidation number to oxygen so that the sum of the oxidation numbers is zero.
(a) Because K in group 1A has an oxidation number of $+1$, the oxidation number of oxygen in KO_2 must be $-1/2$. Thus, KO_2 is a superoxide.
(b) Because Ba in group 2A has an oxidation number of $+2$, the oxidation number of oxygen in BaO_2 must be -1. Thus, BaO_2 is a peroxide.
(c) Because Si in group 4A has an oxidation number of $+4$, the oxidation number of oxygen in SiO_2 must be -2. Thus, SiO_2 is an oxide.

▶ **PROBLEM 14.11**   What is the oxidation number of oxygen in each of the following compounds? Tell whether the compound is an oxide, a peroxide, or a superoxide.
(a) Rb_2O_2 **(b)** CaO **(c)** CsO_2 **(d)** SrO_2 **(e)** CO_2

▶ **PROBLEM 14.12**   Write a balanced net ionic equation for the reaction of water with each of the oxygen compounds listed in Problem 14.11.

◀ **KEY CONCEPT PROBLEM 14.13**   Draw a molecular orbital (MO) energy-level diagram for O_2^-, including the MOs derived from the oxygen $2s$ and $2p$ orbitals. Show the electron population of the MOs, and verify that O_2^- is paramagnetic and has a bond order of 1.5.

14.11   Hydrogen Peroxide

H_2O_2 3D model

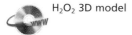

Hydrogen peroxide (HOOH) is sold in drugstores as a 3% aqueous solution for domestic use and is marketed as a 30% aqueous solution for industrial and laboratory use. Because of its oxidizing properties, hydrogen peroxide is used as a mild antiseptic and as a bleach for textiles, paper pulp, and hair. In the chemical industry, hydrogen peroxide is a starting material for synthesis of other peroxide compounds, some of which are used in the manufacture of plastics.

Pure hydrogen peroxide is an almost colorless, syrupy liquid that freezes at $-0.4°C$ and boils at an estimated $150°C$. The exact boiling point is not known with certainty, however, because pure H_2O_2 explodes when heated. The relatively high estimated boiling point indicates strong hydrogen bonding between H_2O_2 molecules in the liquid (Section 10.2). In aqueous solutions, hydrogen peroxide behaves as a weak acid, partially dissociating to give H^+ and HO_2^- ions.

Hydrogen peroxide is both a strong oxidizing agent and a reducing agent. When hydrogen peroxide acts as an oxidizing agent, oxygen is reduced from the -1 oxidation state in H_2O_2 to the -2 oxidation state in H_2O (or OH^-). For example, hydrogen peroxide oxidizes Br^- to Br_2 in acidic solution:

The two H–O–O planes in the H_2O_2 molecule are nearly perpendicular to each other.

$$H_2O_2(aq) + 2 H^+(aq) + 2 Br^-(aq) \longrightarrow 2 H_2O(l) + Br_2(aq)$$

Oxidation
numbers -1 -1 -2 0

When hydrogen peroxide acts as a reducing agent, oxygen is oxidized from the -1 oxidation state in H_2O_2 to the 0 oxidation state in O_2. A typical example is the reaction of hydrogen peroxide with permanganate ion in acidic solution:

$$5 H_2O_2(aq) + 2 MnO_4^-(aq) + 6 H^+(aq) \longrightarrow 5 O_2(g) + 2 Mn^{2+}(aq) + 8 H_2O(l)$$

-1 $+7$ 0 $+2$

The reduction of manganese in this reaction from the $+7$ oxidation state in MnO_4^- to the $+2$ oxidation state in the manganese(II) ion, Mn^{2+}, is accompanied by a beautiful color change (Figure 14.8).

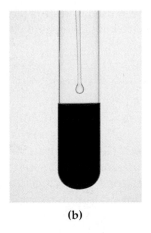

(a)   (b)   (c)

FIGURE 14.8 Some reactions of hydrogen peroxide. **(a)** Addition of a colorless aqueous solution of H_2O_2 to a colorless acidic solution of NaBr produces a yellow-orange color due to the formation of bromine, Br_2. **(b)** Addition of aqueous H_2O_2 to a violet acidic solution of $KMnO_4$ produces **(c)** bubbles of O_2 gas and decolorizes the solution as the violet MnO_4^- ion is converted to the nearly colorless Mn^{2+} ion.

Because it is both an oxidizing agent and a reducing agent, H_2O_2 can oxidize and reduce itself. Thus, hydrogen peroxide is unstable and undergoes disproportionation to water and oxygen:

$$2 H_2O_2(l) \longrightarrow 2 H_2O(l) + O_2(g) \qquad \Delta H° = -196 \text{ kJ}$$

Oxidation
numbers -1 -2 0

In the absence of a catalyst, the disproportionation is too slow to be observed at room temperature. Rapid, exothermic, and potentially explosive decomposition of hydrogen peroxide is initiated, however, by heat and by a broad range of catalysts, including transition metal ions, certain anions (such as I^-), metal surfaces, blood (Figure 14.9), and even tiny particles of dust. Because decomposition is accelerated by light, hydrogen peroxide is stored in dark bottles. It is best handled in dilute aqueous solutions; concentrated solutions and the pure liquid are extremely hazardous materials.

FIGURE 14.9 When a few drops of blood are added to aqueous hydrogen peroxide (left), the hydrogen peroxide decomposes rapidly, evolving bubbles of oxygen that produce a thick foam (right). The reaction is catalyzed by *catalase*, an enzyme present in blood.

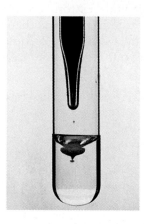

▶ **PROBLEM 14.14** Draw an electron-dot structure for H_2O_2. Is your structure consistent with its O—O bond length of 148 pm? (Look at Table 14.2 to see how bond length and bond order are related.)

▶ **PROBLEM 14.15** The discoloration and restoration of old oil paintings involves some interesting chemistry. On exposure to polluted air containing H_2S, white lead carbonate pigments are converted to PbS, a black solid. Hydrogen peroxide has been used to restore the original white color. Write a balanced equation for the reaction, which involves oxidation of black PbS to white $PbSO_4$.

14.12 Ozone

Oxygen exists in two allotropes (Section 10.10): ordinary dioxygen, O_2, and ozone, O_3. Ozone is a toxic, pale blue gas with a characteristic sharp, penetrating odor that you can detect at concentrations as low as 0.01 ppm. You've probably noticed the odor of ozone around sparking electric motors or after a severe electrical storm. Ozone is produced when an electric discharge passes through O_2, providing the energy needed to bring about the endothermic reaction:

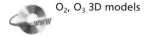
O_2, O_3 3D models

$$3\,O_2(g) \xrightarrow[\text{discharge}]{\text{Electric}} 2\,O_3(g) \qquad \Delta H^\circ = +285 \text{ kJ}$$

Although ozone can be prepared in the laboratory by passing O_2 through an electrical device like that shown in Figure 14.10, it is unstable and decomposes exothermically to O_2. Decomposition of the dilute gas is slow, but the concentrated gas, liquid ozone (bp $-112°C$), or solid ozone (mp $-192°C$) can decompose explosively.

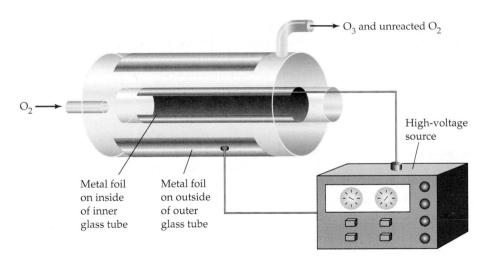

→ O₃ and unreacted O₂

O₂ →

High-voltage source

Metal foil on inside of inner glass tube

Metal foil on outside of outer glass tube

FIGURE 14.10 Generator for preparing ozone. A stream of O_2 gas flows through the ozonizer tube and is partially converted to O_3 when an electric discharge passes through the gas.

As discussed in Section 7.7, two resonance structures are required to explain the structure of ozone because the two O–O bonds have equal lengths:

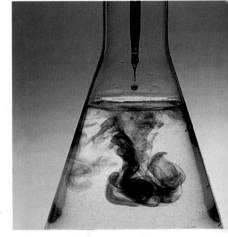

The bent structure is in accord with the VSEPR model described in Section 7.9, which predicts a bond angle near 120° for a triatomic molecule having a central atom surrounded by three charge clouds. The π electrons are shared by the three oxygen atoms, giving a net bond order of 1.5 between each pair of oxygen atoms (1 σ bond + 0.5 π bond). This bonding description is in agreement with the O–O bond length of 128 pm, intermediate between the lengths of an O–O single bond (148 pm) and an O=O double bond (121 pm).

Ozone is an extremely powerful oxidizing agent. In fact, of the common oxidizing agents, only F_2 is more potent. A standard method for detecting ozone in polluted air is to pass the air through a basic solution of potassium iodide that contains a starch indicator. The ozone oxidizes iodide ion to iodine, I_2, which combines with the starch to give the deep blue starch–iodine complex:

Addition of aqueous iodine to a starch solution gives the blue starch–iodine complex. The complex forms when I^- is oxidized to I_2 in the presence of starch.

$$O_3(g) + 2\ I^-(aq) + H_2O(l) \longrightarrow O_2(g) + I_2(aq) + 2\ OH^-(aq)$$

Oxidation numbers: $0 \quad\quad -1 \quad\quad\quad\quad\quad\quad\quad 0 \quad\quad -2$

Because of its oxidizing properties, ozone is sometimes used to kill bacteria in drinking water.

14.13 Water

Water, the most important compound of hydrogen and oxygen, is the most familiar and abundant compound on earth. Nearly three-fourths of the earth's surface is covered with water, and an estimated 1.35×10^{18} m³ of water is present in the oceans. (It's interesting to note that the volume of the oceans in milliliters [1.35×10^{24} mL] is roughly twice Avogadro's number.) Water accounts for nearly two-thirds of the mass of the adult human body and 93% of the mass of the human embryo in the first month.

H₂O 3D model

Approximately 97.3% of the world's vast supply of water is in the oceans. Most of the rest is in the form of polar ice caps and glaciers (2.0%) and underground fresh water (0.6%). Freshwater lakes and rivers account for less than 0.01% of the total, yet they nevertheless contain an enormous amount of water (1.26×10^{14} m³).

Seawater is unfit for drinking or agriculture because each kilogram contains about 35 g of dissolved salts. The most abundant salt in seawater is sodium chloride, but more than 60 different elements are present in small amounts. Table 14.3 lists the ions that account for more than 99% of the mass of the dissolved salts. Although the oceans represent an almost unlimited source of chemicals, ion concentrations are so low that recovery costs are high. Only three substances are obtained from seawater commercially: sodium chloride, magnesium, and bromine.

Water for use in homes, agriculture, and industry is generally obtained from freshwater lakes, rivers, or underground sources. The water you drink must be purified to remove solid particles, colloidal material, bacteria, and other harmful impurities. Important steps in a typical purification process include preliminary filtration, sedimentation, sand filtration, aeration, and sterilization (Figure 14.11).

TABLE 14.3   Major Ionic Constituents of Seawater

Ion	g/kg of Seawater
Cl^-	19.0
Na^+	10.5
SO_4^{2-}	2.65
Mg^{2+}	1.35
Ca^{2+}	0.40
K^+	0.38
HCO_3^-	0.14
Br^-	0.065

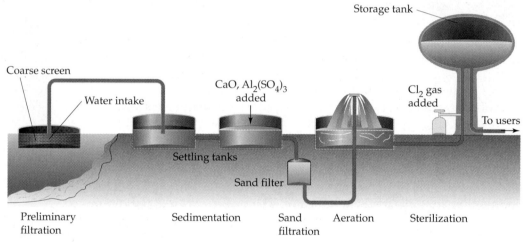

FIGURE 14.11   Purification of drinking water.

The sedimentation, or settling, of suspended matter takes place in large tanks and is accelerated by the addition of lime, CaO, and aluminum sulfate, $Al_2(SO_4)_3$. The lime makes the water slightly basic, which precipitates the added Al^{3+} ions as aluminum hydroxide:

$$CaO(s) + H_2O(l) \longrightarrow Ca^{2+}(aq) + 2\,OH^-(aq)$$

$$Al^{3+}(aq) + 3\,OH^-(aq) \longrightarrow Al(OH)_3(s)$$

As the gelatinous precipitate of aluminum hydroxide slowly settles, it carries with it suspended solids, colloidal material, and most of the bacteria. The water is then filtered through a bed of sand and subsequently sprayed into the air to oxidize dissolved organic impurities. Finally, the water is sterilized by adding chlorine or ozone, which kills the remaining bacteria. The water still contains up to 0.5 g/L of inorganic ions such as Na^+, K^+, Mg^{2+}, Ca^{2+}, Cl^-, F^-, SO_4^{2-}, and HCO_3^-, but in such low concentrations that they aren't harmful.

Water that contains appreciable concentrations of doubly charged cations such as Ca^{2+}, Mg^{2+}, and Fe^{2+} is called **hard water**. These cations combine with long-chain organic anions in soaps to give the undesirable, insoluble precipitates (soap scum) often observed in sinks and bathtubs. They also form the unwanted metal carbonate precipitates known as *boiler scale* that deposit in boilers, hot water heaters, and tea kettles, thus reducing the efficiency of heat transfer. You can understand the formation of boiler scale in terms of Le Châtelier's principle (Section 13.6). When hard water containing HCO_3^- anions is heated, the equilibrium involving decomposition of HCO_3^- to CO_2 and CO_3^{2-} shifts to the right as CO_2 gas escapes from the solution. The resulting CO_3^{2-} ions then combine with cations, such as Ca^{2+}, to form insoluble metal carbonates.

$$2\,HCO_3^-(aq) \xrightleftharpoons{\text{Heat}} CO_2(g) + H_2O(l) + CO_3^{2-}(aq)$$

$$Ca^{2+}(aq) + CO_3^{2-}(aq) \longrightarrow CaCO_3(s)$$

Hard water can be softened by **ion exchange**, a process in which the Ca^{2+} and Mg^{2+} ions are replaced by Na^+. The exchange occurs when hard water is passed through a resin that has ionic $SO_3^-Na^+$ groups attached to it. The more highly charged Ca^{2+} and Mg^{2+} cations bond to the negative SO_3^- groups more strongly than Na^+, and so the Na^+ ions are replaced as the hard water passes through the resin. The ion exchange process can be represented by the following equation, where RSO_3^- represents the ionic groups on the resin:

$$2\,RSO_3^-Na^+(s) + Ca^{2+}(aq) \longrightarrow (RSO_3^-)_2Ca^{2+}(s) + 2\,Na^+(aq)$$
$$\qquad\qquad \text{Hard water} \qquad\qquad\qquad\qquad\qquad \text{Soft water}$$

Ion exchange resins make convenient household water softeners because the Na^+ ion form of the resin is easily regenerated by treating it with a concentrated solution of NaCl.

14.14 Reactivity of Water

Water reacts with the alkali metals, the heavier alkaline earth metals (Ca, Sr, Ba, and Ra), and the halogens. With most other elements, water is unreactive at room temperature. Water is reduced to hydrogen by the alkali and alkaline earth metals, which are oxidized to aqueous metal hydroxides:

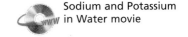 Sodium and Potassium in Water movie

$$2\,Na(s) + 2\,H_2O(l) \longrightarrow H_2(g) + 2\,Na^+(aq) + 2\,OH^-(aq)$$

$$Ca(s) + 2\,H_2O(l) \longrightarrow H_2(g) + Ca^{2+}(aq) + 2\,OH^-(aq)$$

Oxidation numbers: $\quad 0 \qquad +1 \qquad\qquad 0 \qquad +2$

Only fluorine is more electronegative than oxygen and thus is the only element able to oxidize water to oxygen. In the process, fluorine is reduced to hydrofluoric acid:

$$2\,F_2(g) + 2\,H_2O(l) \longrightarrow O_2(g) + 4\,HF(aq)$$

Oxidation numbers: $\quad 0 \qquad -2 \qquad\qquad 0 \qquad -1$

Chlorine doesn't oxidize water, but instead disproportionates to a limited extent. The products are hypochlorous acid, HOCl, a weak acid in which chlorine is in the +1 oxidation state, and hydrochloric acid, a strong acid in which chlorine is in the −1 oxidation state:

$$Cl_2(g) + H_2O(l) \rightleftharpoons HOCl(aq) + H^+(aq) + Cl^-(aq)$$

Oxidation numbers: $0 \qquad +1 \qquad -1$

Bromine and iodine behave similarly, but the extent of disproportionation decreases markedly in the series $Cl > Br > I$.

▶ **PROBLEM 14.16** Write a balanced net ionic equation for the reaction of water with each of the following elements:
(a) Li (b) Sr (c) Br_2

14.15 Hydrates

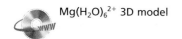
$Mg(H_2O)_6{}^{2+}$ 3D model

Solid compounds that contain water molecules are called **hydrates**. Hydrated salts, such as magnesium perchlorate hexahydrate, $Mg(ClO_4)_2 \cdot 6\,H_2O$, and aluminum chloride hexahydrate, $AlCl_3 \cdot 6\,H_2O$ are examples. Because the structures of hydrates are sometimes complex or unknown, a dot is used in the formula of a hydrate to specify the composition without indicating how the water is bound. If the structure is known, a more informative formula can be given. The formulas $[Mg(H_2O)_6](ClO_4)_2$ and $[Al(H_2O)_6]Cl_3$, for instance, indicate that the six water molecules in each compound are attached to the metal ion. As shown in Figure 14.12, the negative (oxygen) end of each dipolar water molecule bonds to the positive metal cation, and the six water molecules are located at the vertices of an octahedron. Because bonding interactions between water and a metal cation increase with increasing charge on the cation, hydrate formation is common for salts that contain +2 and +3 cations.

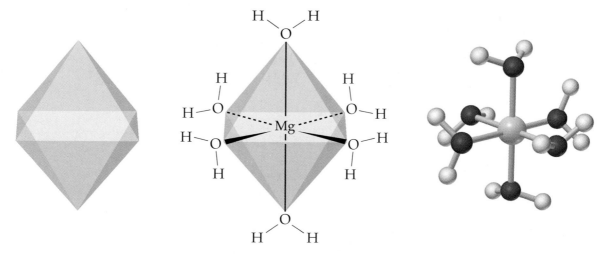

FIGURE 14.12 Octahedral structure of the $Mg(H_2O)_6{}^{2+}$ cation. **(a)** A regular octahedron is a polyhedron that has eight equilateral triangular faces and six vertices. **(b)** Octahedral structure of the hydrated metal cation in $[Mg(H_2O)_6](ClO_4)_2$. The six Mg–O bonds point toward the six vertices of a regular octahedron. **(c)** A view of the $Mg(H_2O)_6{}^{2+}$ cation showing only the location of the atoms and the octahedral arrangement of the bonds to the six H_2O molecules.

When hydrates are heated, the water is driven off. If you heat blue crystals of $CuSO_4 \cdot 5\ H_2O$ above 350°C, for example, you'll obtain anhydrous copper sulfate, $CuSO_4$. This transformation is readily observed, since the anhydrous compound is white (Figure 14.13).

Some anhydrous compounds are so prone to forming hydrates that they absorb water from the atmosphere. Anhydrous $Mg(ClO_4)_2$, for example, picks up water from air, yielding $Mg(ClO_4)_2 \cdot 6\ H_2O$. Compounds that absorb water from the air are said to be **hygroscopic** and are often useful as drying agents.

EXAMPLE 14.5   When 1.823 g of the hydrate $CaCrO_4 \cdot x\ H_2O$ was heated at 200°C, 1.479 g of anhydrous $CaCrO_4$ was obtained. What is the formula of the hydrate?

SOLUTION   The value of x in the formula of the hydrate is the number of moles of H_2O per mole of $CaCrO_4$, so we need to use the masses and the molar masses of H_2O (18.02 g/mol) and $CaCrO_4$ (156.1 g/mol) to calculate their mole ratio:

$$\text{Grams of } H_2O = 1.823\ g - 1.479\ g = 0.344\ g\ H_2O$$

$$\text{Moles of } H_2O = 0.344\ g\ H_2O \times \frac{1\ mol\ H_2O}{18.02\ g\ H_2O} = 1.91 \times 10^{-2}\ mol\ H_2O$$

$$\text{Moles of } CaCrO_4 = 1.479\ g\ CaCrO_4 \times \frac{1\ mol\ CaCrO_4}{156.1\ g\ CaCrO_4}$$

$$= 9.475 \times 10^{-3}\ mol\ CaCrO_4$$

$$x = \frac{\text{Moles of } H_2O}{\text{Moles of } CaCrO_4} = \frac{1.91 \times 10^{-2}\ mol}{9.475 \times 10^{-3}\ mol} = 2.02$$

Since $x = 2$ within experimental uncertainty, the formula of the hydrate is $CaCrO_4 \cdot 2\ H_2O$.

▶ **PROBLEM 14.17**   A 3.10 g sample of anhydrous $NiSO_4$ was exposed to moist air. If 5.62 g of a hydrate of nickel sulfate was obtained, what is the formula of the hydrate?

FIGURE 14.13   Blue crystals of $CuSO_4 \cdot 5\ H_2O$ are converted to white anhydrous $CuSO_4$ when the sample is heated with a Bunsen flame.

A "Hydrogen Economy"

ydrogen is an enormously attractive fuel because it's environmentally clean, giving only water as a combustion product. If hydrogen is burned in air, small amounts of nitrogen oxides can be produced because of the high-temperature combination of nitrogen and oxygen, but the combustion products are free of CO, CO_2, SO_2, unburned hydrocarbons, and other environmental pollutants that result from the combustion of petroleum fuels. In addition, the amount of heat liberated when hydrogen burns is 242 kJ/mol (121 kJ/g), more than twice that of gasoline, oil, or natural gas on a mass basis.

$$H_2(g) + 1/2\, O_2(g) \longrightarrow H_2O(g) \qquad \Delta H° = -242 \text{ kJ}$$

A BMW test car powered by liquid hydrogen. The fuel tank located in the trunk of the car is receiving a fill up of liquid hydrogen fuel.

As a result, some people envision what they call a "hydrogen economy" in which our energy needs are met by gaseous, liquid, and solid hydrogen. For heating homes, gaseous hydrogen could be conveyed through underground pipes, while liquid hydrogen could be shipped by truck or by rail in large vacuum-insulated tanks. Automobiles might be powered by liquid hydrogen or by "solid hydrogen" in the form of solid interstitial hydrides or hydrogen stored in the recently discovered tube-shaped molecules called *carbon nanotubes*. Prototype cars have already been built with their engines modified to run on hydrogen.

What is keeping us from reaching a hydrogen economy? Before a hydrogen economy can become a reality, cheaper ways of producing hydrogen must be found. Since hydrogen is not a naturally occurring energy source like coal, oil, or natural gas, energy must first be expended to produce the hydrogen before it can be used. Current research therefore focuses on finding cheaper methods for extracting hydrogen from its compounds.

One approach for producing hydrogen is to use solar energy to "split" water into H_2 and O_2. The feasibility of this scheme depends on the development of catalysts that absorb sunlight and then use the energy to reduce water to hydrogen. Another strategy employs thermal energy to effect a series of reactions that bring about the net conversion of water to hydrogen and oxygen. One such reaction series uses the following high-temperature reactions with iron compounds in which iron is shuttled between different oxidation states:

$$3\,FeCl_2(s) + 4\,H_2O(g) \xrightarrow{500°C} Fe_3O_4(s) + 6\,HCl(g) + H_2(g)$$

$$Fe_3O_4(s) + 3/2\,Cl_2(g) + 6\,HCl(g) \xrightarrow{100°C} 3\,FeCl_3(s) + 3\,H_2O(g) + 1/2\,O_2(g)$$

$$3\,FeCl_3(s) \xrightarrow{300°C} 3\,FeCl_2(s) + 3/2\,Cl_2(g)$$

$$\text{Net:} \quad H_2O(g) \longrightarrow H_2(g) + 1/2\,O_2(g)$$

The use of liquid hydrogen as a fuel in the U.S. space program is well known. Hydrogen powered the *Saturn V* rocket that carried the first astronauts to the moon, and it fuels the rocket engines of the space shuttle (Figure 14.14a). Although liquid hydrogen has been handled safely for many years, it is an extremely dangerous substance. The disastrous breakup of the *Challenger* space shuttle (Figure 14.14b), which took the lives of seven astronauts in 1986, resulted from a leak in the O-ring of the solid-fuel rocket boosters and subsequent explosive burning of massive amounts of hydrogen. Before liquid hydrogen can come into more general use as a fuel, the hazards of storing and distributing this flammable and explosive material must be solved.

(a)   (b)

FIGURE 14.14 (a) The space shuttle, consisting of the orbiter, two solid-fuel rocket boosters, and the huge external fuel tank. The fuel tank, which is 47.0 m high and 8.4 m in diameter, contains 1.45×10^6 L of liquid hydrogen and 5.41×10^5 L of liquid oxygen at liftoff. (b) Breakup of the *Challenger* space shuttle. The rocket boosters are at the top of the photo, and the orbiter is at the bottom left, with its rocket engines still firing. Wreckage of the external fuel tank is obscured by vapor and smoke.

▶ **PROBLEM 14.18** Hydrogen is a gas at ordinary temperatures. Explain how it can be stored as a solid.

▶ **PROBLEM 14.19** The space shuttle fuel tank contains 1.45×10^6 L of liquid hydrogen, which has a density of 0.088 g/L. How much heat (in kilojoules) is liberated when the hydrogen burns in an excess of oxygen? How many kilograms of oxygen is needed to oxidize the hydrogen?

Key Words

amphoteric *587*
binary hydride *579*
covalent hydride *580*
disproportionation *590*
hard water *595*
hydrate *596*
hygroscopic *597*
interstitial hydride *580*
ion exchange *595*
ionic hydride *579*
isotope effect *575*
metallic hydride *580*
nonstoichiometric
   compound *580*
oxide *586*
peroxide *586*
steam–hydrocarbon
   re-forming process *577*
superoxide *586*
water-gas shift reaction *577*

Summary

Hydrogen, the most abundant element in the universe, has three isotopes: protium ($_1^1H$), deuterium ($_1^2H$), and tritium ($_1^3H$). The isotopes of hydrogen exhibit small differences in properties, known as **isotope effects**. A hydrogen atom, which has the electron configuration $1s^1$, can lose its electron, forming a hydrogen cation (H^+), or it can gain an electron, yielding a hydride anion (H^-). At ordinary temperatures, hydrogen exists as diatomic H_2 molecules, which are thermally stable and unreactive because of the strong H–H bond.

Hydrogen for industrial purposes is produed by the **steam–hydrocarbon re-forming process**, and is used in the synthesis of ammonia and methanol. In the laboratory, hydrogen is prepared by reaction of dilute acid with an active metal, such as zinc.

Hydrogen forms three types of **binary hydrides**. Active metals give **ionic hydrides**, such as LiH and CaH_2; nonmetals give **covalent hydrides**, such as NH_3, H_2O, and HF; and transition metals give **metallic**, or **interstitial, hydrides**, such as PdH_x. Interstitial hydrides are often **nonstoichiometric compounds**.

Oxygen is the most abundant element in the earth's crust. Dioxygen (O_2) can be prepared in the laboratory by electrolysis of water, by catalytic decomposition of hydrogen peroxide, or by thermal decomposition of $KClO_3$. Oxygen is manufactured by fractional distillation of liquefied air, and is used in making steel. The O_2 molecule is paramagnetic and has an O=O double bond. Ozone (O_3), an allotrope of oxygen, is a powerful oxidizing agent.

Oxygen forms ionic **oxides**, such as Li_2O and MgO, with active metals, and covalent oxides, such as P_4O_{10} and SO_3, with nonmetals. Oxides can also be classified in terms of their acid–base properties. Basic oxides are ionic, and acidic oxides are covalent. **Amphoteric** oxides, such as Al_2O_3, exhibit both acidic and basic properties.

Metal **peroxides**, such as Na_2O_2, are ionic compounds that contain the O_2^{2-} anion and have oxygen in the -1 oxidation state. Metal **superoxides**, such as KO_2, contain the O_2^- anion and have oxygen in the $-1/2$ oxidation state. Hydrogen peroxide (H_2O_2), a strong oxidizing agent and also a reducing agent, is unstable with respect to **disproportionation** to H_2O and O_2. A disproportionation reaction is one in which a substance is simultaneously oxidized and reduced.

Water is the most abundant compound on earth. Seawater, which accounts for 97.3% of the world's water supply, contains 3.5 mass % of dissolved salts. Purification of drinking water involves preliminary filtration, sedimentation, sand filtration, aeration, and sterilization. **Hard water**, which contains appreciable concentrations of doubly charged cations such as Ca^{2+}, Mg^{2+}, and Fe^{2+}, can be softened by **ion exchange**. Water is reduced to H_2 by the alkali metals and heavier alkaline earth metals, and is oxidized to O_2 by fluorine. Solid compounds that contain water are known as **hydrates**.

Key Concept Summary

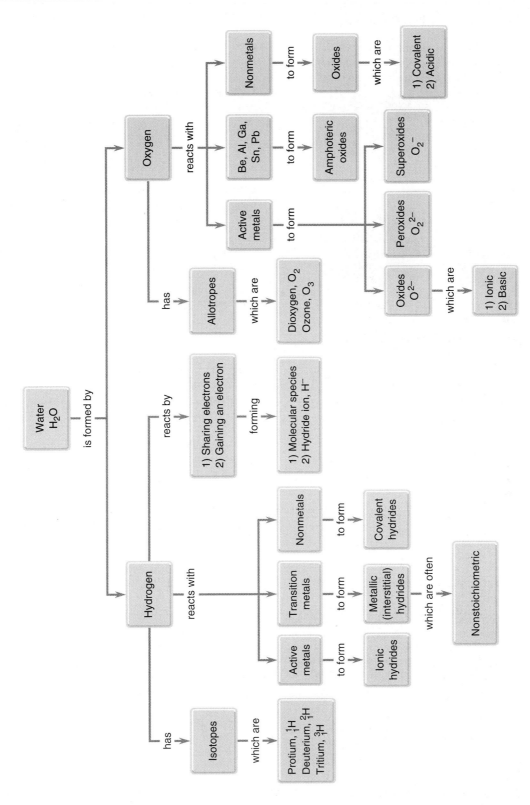

Understanding Key Concepts

Problems 14.1–14.19 appear within the chapter.

14.20 In the following pictures of binary hydrides, ivory spheres represent H atoms or ions, and burgundy spheres represent atoms or ions of the other element.

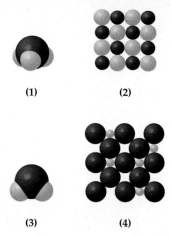

(1)   (2)

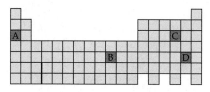

(3)   (4)

(a) Identify each binary hydride as ionic, covalent, or interstitial.

(b) What is the oxidation state of hydrogen in compounds (1), (2), and (3)? What is the oxidation state of the other element?

14.21 Look at the location of elements A, B, C, and D in the following periodic table:

(a) Write the formula of the simplest binary hydride of each element.

(b) Classify each binary hydride as ionic, covalent, or interstitial.

(c) Which of these hydrides are molecular? Which are solids with an infinitely extended three-dimensional crystal structure?

(d) What are the oxidation states of hydrogen and the other element in the hydrides of A, C, and D?

14.22 Which of the following molecules have similar properties, and which have quite different properties?

(a) $H_2{}^{16}O$ (b) $D_2{}^{17}O$ (c) $H_2{}^{16}O_2$

(d) $H_2{}^{17}O$ (e) $D_2{}^{17}O_2$

Draw the structure of each molecule.

14.23 In the following pictures of oxides, red spheres represent O atoms or ions, and green spheres represent atoms or ions of a second- or third-row element in its highest oxidation state.

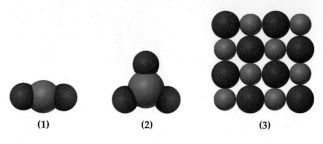

(1)   (2)   (3)

(a) What is the oxidation state of oxygen in each oxide? What is the oxidation state of the other element?

(b) Identify each oxide as ionic or covalent.

(c) Identify each oxide as acidic or basic.

(d) What is the identity of the other element in (1) and (2)?

14.24 In the following pictures of oxides, red spheres represent O atoms or ions, and green spheres represent atoms or ions of a first- or second-row element in its highest oxidation state.

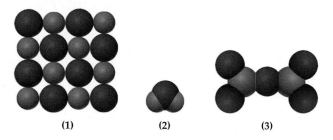

(1)   (2)   (3)

(a) What is the oxidation state of oxygen in each oxide? What is the oxidation state of the other element?

(b) Which of these oxides is (are) molecular, and which has (have) an infinitely extended three-dimensional structure?

(c) Which of these oxides is (are) likely to be a gas or a liquid, and which is (are) likely to be a high-melting solid?

(d) Identify the other element in (2) and (3).

14.25 Look at the location of elements A, B, C, and D in the following periodic table:

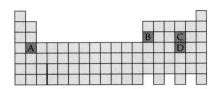

(a) Write the formula of the oxide that has each of these elements in its highest oxidation state.

(b) Classify each oxide as basic, acidic, or amphoteric.

(c) Which oxide is the most ionic? Which is the most covalent?

(d) Which of these oxides are molecular? Which are solids with an infinitely extended three-dimensional crystal structure?

(e) Which of these oxides has the highest melting point? Which has the lowest melting point?

14.26 The following pictures represent structures of the hydrides of four second-row elements.

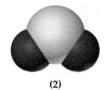

(1)   (2)   (3)   (4)

(a) Which compound has the highest melting point?

(b) Which compound has the lowest boiling point?

(c) Which compounds yield H_2 gas when they are mixed together?

14.27 The following pictures represent structures of the oxides of carbon and sulfur. Which has the stronger bonds? Explain.

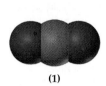

(1)   (2)

Additional Problems

Chemistry of Hydrogen

14.28 Explain what is meant by an isotope effect, and give two examples.

14.29 How can protium and deuterium be separated?

14.30 Calculate the percentage mass difference between:

(a) ^1H and ^2H (b) ^2H and ^3H

Would you expect the differences in properties for H_2O and D_2O to be larger or smaller than the differences in properties for D_2O and T_2O? Do the data in Table 14.1 support your prediction?

14.31 (a) If the volume of the oceans is 1.35×10^{18} m^3 and the abundance of deuterium is 0.0156 atom %, how many kilograms of deuterium are present in the oceans? (Neglect the presence of dissolved substances and assume that the density of water is 1.00 g/cm^3.)

(b) Do the same calculation for tritium, assuming that the abundance of tritium is approximately 10^{-16} atom %.

14.32 There are three isotopes of hydrogen and three naturally occurring isotopes of oxygen (^{16}O, ^{17}O, and ^{18}O). How many kinds of water are possible? Draw their structures.

14.33 There are three isotopes of hydrogen and just one naturally occurring isotope of phosphorus (^{31}P). How many kinds of phosphine (PH_3) are possible? Draw their structures.

14.34 Write a balanced equation for the synthesis of hydrogen using each of the following starting materials.

(a) Zn   (b) C   (c) CH_4 (d) H_2O

14.35 Complete and balance the equation for each of the following reactions:

(a) $Fe(s) + H^+(aq) \rightarrow$

(b) $Ca(s) + H_2O(l) \rightarrow$

(c) $Al(s) + H^+(aq) \rightarrow$

(d) $C_2H_6(g) + H_2O(g) \xrightarrow[\text{Catalyst}]{\text{Heat}}$

14.36 What is the most important method for the industrial production of hydrogen? Write balanced equations for the reactions involved.

14.37 Write a balanced equation for each of the following reactions:

(a) Reduction of steam by hot iron

(b) Production of synthesis gas from propane, C_3H_8

(c) The water-gas shift reaction

14.38 Ionic metal hydrides react with water to give hydrogen gas and an aqueous solution of the metal hydroxide.

(a) On reaction of equal masses of LiH and CaH_2 with water, which compound gives more hydrogen?

(b) How many kilograms of CaH_2 is needed to fill a 100 L tank with compressed H_2 gas at 150 atm pressure and 25°C?

14.39 The hydrogen-filled dirigible *Hindenburg* had a volume of 1.99×10^8 L. If the hydrogen used was produced by reaction of carbon with steam, how many kilograms of carbon would have been needed to produce enough hydrogen to fill the dirigible at 20°C and 740 mm pressure?

$$C(s) + H_2O(g) \longrightarrow CO(g) + H_2(g)$$

14.40 In the following compounds, is hydrogen present as H^+, H^-, or as a covalently bound H atom?

(a) MgH_2 (b) PH_3 (c) KH   (d) HBr

14.41 In the following compounds, is hydrogen present as H^+, H^-, or as a covalently bound H atom?

(a) H_2Se (b) RbH   (c) CaH_2 (d) GeH_4

14.42 Compare some of the physical properties of H_2S, NaH, and PdH_x.

14.43 Compare some of the physical properties of $TiH_{1.7}$, HCl, and CaH_2.

14.44 Describe the bonding in:

(a) CH_4 (b) NaH

14.45 Describe the bonding in:

(a) CaH_2 (b) NH_3

14.46 Predict the molecular structure of:

(a) H_2Se (b) AsH_3 (c) SiH_4

14.47 Describe the molecular geometry of:

(a) GeH_4 (b) H_2S (c) NH_3

14.48 What is a nonstoichiometric compound? Give an example, and account for its lack of stoichiometry in terms of structure.

14.49 Explain why the hydrogen atoms in interstitial hydrides are mobile.

14.50 Titanium hydride, TiH_2, has a density of 3.9 g/cm^3.

(a) Calculate the density of hydrogen in TiH_2, and compare it with that in liquid H_2 (0.070 g/cm^3).

(b) How many cubic centimeters of H_2 at STP are absorbed in making 1.00 cm^3 of TiH_2?

14.51 The density of LiH is 0.82 g/cm^3.

(a) Calculate the density of hydrogen in LiH and the number of H atoms in 5.0 cm^3 of LiH.

(b) How many liters of H_2 at 20°C and 740 mm Hg are required to make 5.0 cm^3 of LiH?

Chemistry of Oxygen

14.52 How is O_2 prepared (a) in industry and (b) in the laboratory? Write balanced equations for the reactions involved.

14.53 In what forms is oxygen commonly found in nature?

14.54 How many liters of O_2 gas at 25°C and 0.985 atm pressure can be obtained by catalytic decomposition of 20.4 g of hydrogen peroxide?

14.55 In the oxyacetylene torch, how many grams of acetylene and how many liters of O_2 at STP are needed to generate 1000 kJ of heat?

14.56 Write a balanced equation for the reaction of an excess of O_2 with each of the following elements:

(a) Li   (b) P   (c) Al   (d) Si

14.57 Write a balanced equation for the reaction of an excess of O_2 with each of the following elements:

(a) Ca   (b) C   (c) As   (d) B

14.58 Draw some electron-dot structures for O_2, and explain why they are inconsistent with the paramagnetism of O_2 and its O=O double bond.

14.59 Use molecular orbital theory to account for the paramagnetism of O_2 and its O=O double bond.

14.60 Arrange the following oxides in order of increasing covalent character: B_2O_3, BeO, CO_2, Li_2O, N_2O_5.

14.61 Arrange the following oxides in order of increasing ionic character: SiO_2, K_2O, P_4O_{10}, Ga_2O_3, GeO_2.

14.62 Arrange the following oxides in order of increasing basic character: Al_2O_3, Cs_2O, K_2O, N_2O_5.

14.63 Arrange the following oxides in order of increasing acidic character: BaO, Cl_2O_7, SO_3, SnO_2.

14.64 Which is more acidic?

(a) Cr_2O_3 or CrO_3 (b) N_2O_5 or N_2O_3

(c) SO_2 or SO_3

14.65 Which is more basic?

(a) CrO or Cr_2O_3 (b) SnO_2 or SnO

(c) As_2O_3 or As_2O_5

14.66 Write a balanced net ionic equation for reaction of each of the following oxides with water:

(a) Cl_2O_7 (b) K_2O (c) SO_3

14.67 Write a balanced net ionic equation for reaction of each of the following oxides with water:

(a) BaO (b) Cs$_2$O (c) N$_2$O$_5$

14.68 Write a balanced net ionic equation for reaction of the amphoteric oxide ZnO with:

(a) Hydrochloric acid

(b) Aqueous sodium hydroxide; the product is Zn(OH)$_4$$^{2-}$

14.69 Write a balanced net ionic equation for reaction of the amphoteric oxide Ga$_2$O$_3$ with:

(a) Aqueous sulfuric acid

(b) Aqueous potassium hydroxide; the product is Ga(OH)$_4$$^-$

14.70 Distinguish between a peroxide and a superoxide, and give an example of each.

14.71 Classify each of the following compounds as an oxide, a peroxide, or a superoxide:

(a) Na$_2$O$_2$ (b) GeO$_2$ (c) RbO$_2$ (d) BaO$_2$

14.72 What products are formed when the following metals are burned in an excess of air?

(a) Ba (b) Ca (c) Cs (d) Li (e) Na

14.73 Write balanced net ionic equations for the reaction of water with:

(a) BaO$_2$ (b) RbO$_2$

14.74 Draw MO energy-level diagrams for O$_2$, O$_2$$^-$, and O$_2$$^{2-}$, including only MOs derived from the oxygen 2p atomic orbitals. Show the electron population of the MOs. (See Section 7.14.)

(a) Why does the O–O bond length increase in the series O$_2$, O$_2$$^-$, O$_2$$^{2-}$?

(b) Why is O$_2$$^-$ paramagnetic, whereas O$_2$$^{2-}$ is diamagnetic?

14.75 Draw an MO energy-level diagram for O$_2$$^+$, including only MOs derived from the oxygen 2p atomic orbitals. Show the electron population of the MOs.

(a) Predict the bond order in O$_2$$^+$, and tell whether the O–O bond should be longer or shorter than that in O$_2$.

(b) Is O$_2$$^+$ paramagnetic or diamagnetic?

14.76 Write a balanced net ionic equation for each of the following reactions:

(a) Oxidation by H$_2$O$_2$ of I$^-$ to I$_2$ in acidic solution

(b) Reduction by H$_2$O$_2$ of Cr$_2$O$_7$$^{2-}$ to Cr^{3+} in acidic solution

14.77 Write a balanced net ionic equation for each of the following reactions:

(a) Oxidation by H$_2$O$_2$ of Fe^{2+} to Fe^{3+} in acidic solution

(b) Reduction by H$_2$O$_2$ of IO$_4$$^-$ to IO$_3$$^-$ in basic solution

14.78 Give a description of the electronic structure of ozone that is consistent with the fact that the two O–O bond lengths are equal.

14.79 What experiment could you perform to distinguish O$_3$ from O$_2$?

14.80 How is ozone made in the laboratory?

14.81 How many kilojoules must be supplied to convert 10.0 g of O$_2$ to O$_3$?

Chemistry of Water

14.82 Write a balanced net ionic equation for the reaction of water with each of the following:

(a) F$_2$ (b) Cl$_2$ (c) I$_2$ (d) Ba

14.83 Can water undergo a disproportionation reaction? Explain.

14.84 Give an example of a hydrate, and indicate how the water is bound.

14.85 Describe the structure of the cation in [Fe(H$_2$O)$_6$](NO$_3$)$_3$.

14.86 What is the mass percent water in plaster of paris, CaSO$_4$ · 1/2 H$_2$O?

14.87 Calculate the mass percent water in CuSO$_4$ · 5 H$_2$O.

14.88 When 3.44 g of the mineral gypsum, CaSO$_4$ · x H$_2$O, is heated to 128°C, 2.90 g of CaSO$_4$ · 1/2 H$_2$O is obtained. What is the value of x in the formula of gypsum?

14.89 Anhydrous, hygroscopic, blue CoCl$_2$ forms red-violet CoCl$_2$ · x H$_2$O on exposure to moist air. If the color change is accompanied by an 83.0% increase in mass, what is the formula of the hydrate?

14.90 If seawater contains 3.5 mass % of dissolved salts, how many kilograms of salts are present in 1.0 mi^3 of seawater? (1 mi = 1609 m; density of seawater = 1.025 g/cm^3.)

14.91 How many kilograms of magnesium are present in a cubic meter of seawater? Assume the Mg^{2+} ion concentration listed in Table 14.3 and a density for seawater of 1.025 g/cm^3.

General Problems

14.92 To prepare H_2 from water, would you allow water to react with an oxidizing agent or a reducing agent? Which of the following metals could be used in the reaction? (*Hint:* Recall the activity series, Section 4.8.)

(a) Ag   (b) Al   (c) Au   (d) Ca

14.93 How many tons of hydrogen are required for the annual U.S. production of ammonia (20 million tons)?

14.94 How many liters of H_2 at STP are required for the hydrogenation of 2.7 kg of butadiene?

$$H_2C{=}CH{-}CH{=}CH_2(g) + 2\,H_2(g) \longrightarrow$$

1,3-Butadiene

$$CH_3{-}CH_2{-}CH_2{-}CH_3(g)$$

Butane

14.95 Give the formula and the name of a compound that has oxygen in each of the following oxidation states: $-1/2, -1, -2$.

14.96 Name each of the following compounds:

(a) B_2O_3 (b) H_2O_2 (c) SrH_2

(d) CsO_2 (e) $HClO_4$ (f) BaO_2

14.97 Give the chemical formula for each of the following compounds:

(a) Calcium hydroxide

(b) Chromium(III) oxide

(c) Rubidium superoxide

(d) Sodium peroxide

(e) Barium hydride

(f) Hydrogen selenide

14.98 Three isotopes of oxygen exist (^{16}O, ^{17}O, and ^{18}O).

(a) How many kinds of dioxygen (O_2) molecules are possible? Draw their structures.

(b) How many kinds of ozone (O_3) molecules are possible? Draw their structures.

14.99 What is the oxidation number of oxygen in each of the following compounds?

(a) Al_2O_3 (b) SrO_2 (c) SnO_2 (d) CsO_2

14.100 Write a balanced equation for a reaction in which each of the following acts as an oxidizing agent:

(a) O_2 (b) O_3 (c) H_2O_2 (d) H_2 (e) H_2O

14.101 Write a balanced equation for a reaction in which each of the following acts as a reducing agent:

(a) H_2 (b) H_2O_2 (c) H_2O

14.102 Which of the following elements are oxidized by water? Which are reduced by water? Which undergo a disproportionation reaction when treated with water?

(a) Cl_2 (b) F_2 (c) K   (d) Br_2

14.103 How many liters of seawater (density 1.025 g/cm^3) must be processed to obtain 2.0 million kg of bromine? Assume the Br^- ion concentration listed in Table 14.3 and a recovery rate of 20%.

14.104 Use the standard heats of formation in Appendix B to calculate $\Delta H°$ (in kilojoules) for each of the following reactions:

(a) $CO(g) + 2\,H_2(g) \rightarrow CH_3OH(l)$

(b) $CO(g) + H_2O(g) \rightarrow CO_2(g) + H_2(g)$

(c) $2\,KClO_3(s) \rightarrow 2\,KCl(s) + 3\,O_2(g)$

(d) $6\,CO_2(g) + 6\,H_2O(l) \rightarrow 6\,O_2(g) + C_6H_{12}O_6(s)$

14.105 One model of an acid–base reaction involves proton transfer from the acid to a solvent water molecule to give an H_3O^+ ion. Write balanced chemical equations that show how aqueous solutions of acidic oxides such as SO_2 and SO_3 can yield H_3O^+ ions. Which of these two oxides gives a higher concentration of H_3O^+ ions? Explain.

Multi-Concept Problems

14.106 How much heat (in joules) is liberated when 1.000 L of ozone at 20°C and 63.6 mm Hg decomposes to ordinary oxygen, O_2?

14.107 Sodium hydride, which has the NaCl crystal structure, has a density of 0.92 g/cm^3. If the ionic radius of Na^+ is 102 pm, what is the ionic radius of H^- in NaH?

14.108 A 1.84 g sample of an alkaline earth metal hydride was treated with an excess of dilute hydrochloric acid, and the resulting gas was collected in a 1.000 L container at 20°C. The measured pressure of the gas

was 750.0 mm Hg. Identify the alkaline earth metal, and write the formula for the metal hydride.

14.109 A 250.0 mL gaseous sample of a sulfur oxide at 77°C and 720.0 mm Hg pressure was allowed to react completely with an excess of water. Prior to reaction, the density of the gas was determined to be 2.64 g/L.

(a) What is the formula for the sulfur oxide?

(b) How much heat (in kilojoules) is released when the gas reacts with the water?

(c) How many milliliters of 0.160 M NaOH is needed to neutralize the aqueous solution?

14.110 A 300.0 mL sample of ordinary water was allowed to react with 5.4 g of N_2O_5, and an excess of zinc metal was then added. The resulting gas was collected in a 500.0 mL container at 25°C.

(a) What is the partial pressure (in mm Hg) of HD in the container?

(b) How many HD molecules are present in the container?

(c) How many D_2 molecules are present in the container?

14.111 Sodium amalgam is an alloy of sodium and mercury. The percent sodium in the alloy can be determined by reacting the amalgam with an excess of hydrochloric acid and collecting the liberated H_2 gas.

(a) When a 5.26 g sample of amalgam was treated with 250.0 mL of 0.2000 M HCl and the liberated H_2 was collected in a 500.0 mL container at 22°C, the gas had a pressure of 434 mm Hg. What is the mass % sodium in the amalgam?

(b) After the reaction in part (a), a 50.00 mL portion of the solution was titrated with 0.1000 M NaOH. How many milliliters of the NaOH solution is needed to neutralize the excess HCl?

 # eMedia Problems

14.112 As shown in the **Formation of Water** movie (*eChapter 14.4*), hydrogen and oxygen react explosively to form water. Is the reaction between hydrogen and oxygen an oxidation–reduction reaction? If so, which substance is the reducing agent and which is the oxidizing agent?

14.113 Compare the reaction of white phosphorus and the reaction of sulfur in the **Reactions with Oxygen** movie (*eChapter 14.8*). Which of these two reactions do you think has the higher activation energy? For each reaction, sketch an energy profile similar to the one in Figure 12.13 (page 502).

14.114 Watch the **Carbon Dioxide Behaves as an Acid in Water** movie (*eChapter 14.9*), and write the equation for the reaction that occurs between carbon dioxide and water. Carbon dioxide is known as the anhydride of carbonic acid. What is meant by the term anhydride? What are the acid anhydrides of sulfuric and nitric acids?

14.115 The **Acid–Base Behavior of Oxides** movie (*eChapter 14.9*) shows oxides of strongly metallic elements forming basic solutions with water. Explain how a compound such as Na_2O can produce a basic solution with water when the compound does not contain the hydroxide ion.

14. 116 The reaction between sodium metal and water is shown in the **Sodium and Potassium in Water** movie (*eChapter 14.14*). Write and balance the equation for this reaction. Is this an oxidation–reduction reaction? If so, identify the oxidizing and reducing agents.

15

Aqueous Equilibria: Acids and Bases

The sour taste of citrus fruits is due to acids such as citric acid and ascorbic acid (vitamin C).

Acids and bases are among the most familiar of all chemical compounds. Acetic acid in vinegar, citric acid in lemons and other citrus fruits, magnesium hydroxide (milk of magnesia) in commercial antacids, and ammonia in household cleaning products are among the acids and bases that we encounter every day. Hydrochloric acid is the acid in gastric juice; it is essential to digestion and is secreted by the lining of our stomachs in quantities of 1.2–1.5 L per day.

The characteristic properties of acids and bases have been known for centuries. Acids have a sour taste,* they react with metals such as iron and zinc to yield H_2 gas, and they change the color of the plant dye *litmus* from blue to red. By contrast, bases have a bitter taste and slippery feel, and they change

*Although many early chemists tasted the substances they worked with and survived, you should never taste any laboratory chemical.

the color of litmus from red to blue. When acids and bases are mixed in the right proportion, the characteristic acidic and basic properties disappear, and new substances known as *salts* are obtained.

What is it that makes an acid an acid and a base a base? We first raised those questions in Section 4.5, and we'll now take a closer look at some of the concepts that chemists have developed to describe the chemical behavior of acids and bases. We'll also apply the principles of chemical equilibrium discussed in Chapter 13 to determine the concentrations of the substances present in aqueous solutions of acids and bases. An enormous amount of chemistry can be understood in terms of acid–base reactions, perhaps the most important reaction type in all of chemistry.

15.1   Acid–Base Concepts: The Brønsted–Lowry Theory

Thus far we've been using the Arrhenius theory of acids and bases (Section 4.5). According to Arrhenius, acids are substances that dissociate in water to produce hydrogen ions (H^+), and bases are substances that dissociate in water to yield hydroxide ions (OH^-). Thus, HCl and H_2SO_4 are acids, and NaOH and $Ba(OH)_2$ are bases.

◆— A GENERALIZED ARRHENIUS ACID: $HA(aq) \rightleftharpoons H^+(aq) + A^-(aq)$

◆— A GENERALIZED ARRHENIUS BASE: $MOH(aq) \rightleftharpoons M^+(aq) + OH^-(aq)$

The Arrhenius theory accounts for the properties of many common acids and bases, but it has important limitations. For one thing, the Arrhenius theory is restricted to aqueous solutions; for another, it doesn't account for the basicity of substances like ammonia (NH_3) that don't contain OH groups. In 1923, a more general theory of acids and bases was proposed independently by the Danish chemist Johannes Brønsted and the English chemist Thomas Lowry. According to the *Brønsted–Lowry theory*, an acid is any substance (molecule or ion) that can transfer a proton (H^+ ion) to another substance, and a base is any substance that can accept a proton. In short, acids are proton donors, bases are proton acceptors, and acid–base reactions are proton-transfer reactions:

◆— BRØNSTED–LOWRY ACID   A substance that can transfer H^+

◆— BRØNSTED–LOWRY BASE   A substance that can accept H^+

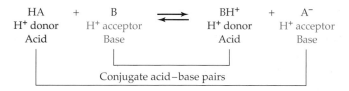

It follows from this equation that the products of a Brønsted–Lowry acid–base reaction, BH^+ and A^-, are themselves acids and bases. The species BH^+ produced when the base B accepts a proton from HA can itself donate a proton back to A^-, meaning that it is a Brønsted–Lowry acid. Similarly, the species A^- produced when HA loses a proton can itself accept a proton back from BH^+, meaning that it is a Brønsted–Lowry base. Chemical species whose formulas differ only by one proton are said to be **conjugate acid–base pairs**.

Thus, A^- is the **conjugate base** of the acid HA, and HA is the **conjugate acid** of the base A^-. Similarly, B is the conjugate base of the acid BH^+, and BH^+ is the conjugate acid of the base B.

To see what's going on in an acid–base reaction, keep your eye on the proton. For example, when a Brønsted–Lowry acid HA is placed in water, it reacts reversibly with water in an *acid-dissociation equilibrium*. The acid transfers a proton to the solvent, which acts as a base (a proton acceptor). The products are the **hydronium ion**, H_3O^+ (the conjugate acid of H_2O), and A^- (the conjugate base of HA):

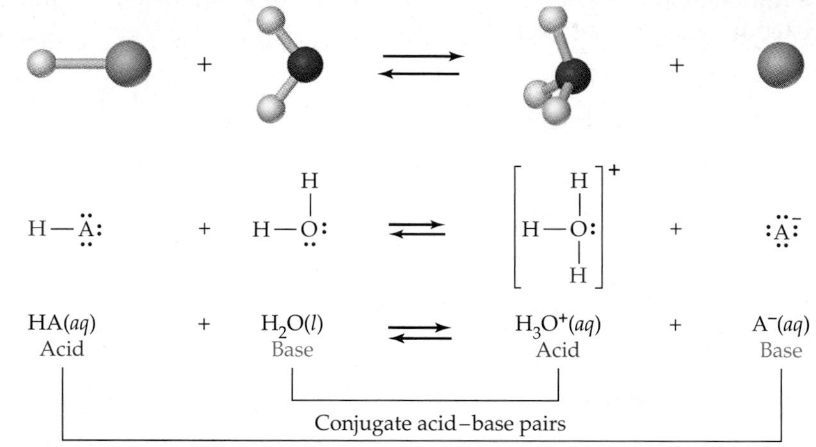

In the reverse reaction, H_3O^+ acts as the proton donor (acid) and A^- acts as the proton acceptor (base). Typical examples of Brønsted–Lowry acids include not only electrically neutral molecules, such as HCl, HNO_3, and HF, but also cations and anions that contain transferable protons, such as NH_4^+, HSO_4^-, and HCO_3^-.

When a Brønsted–Lowry base such as NH_3 dissolves in water, it accepts a proton from the solvent, which acts as an acid. The products are the hydroxide ion, OH^- (the conjugate base of water), and the ammonium ion, NH_4^+ (the conjugate acid of NH_3). In the reverse reaction, NH_4^+ acts as the proton donor, and OH^- acts as the proton acceptor:

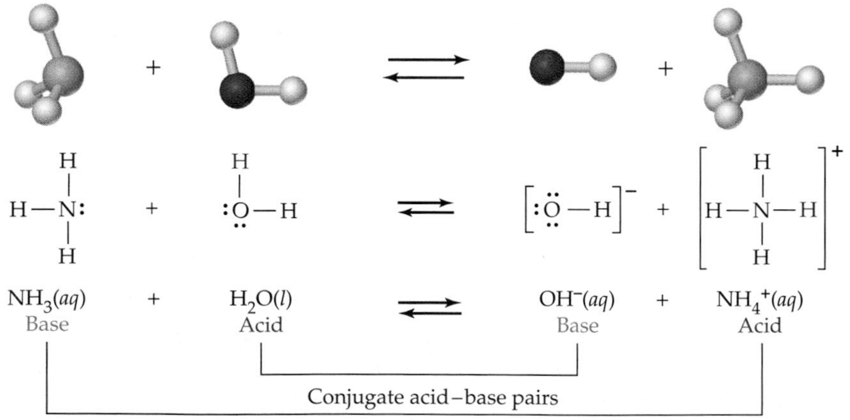

For a molecule or ion to accept a proton, it must have at least one unshared pair of electrons that it can use for bonding to the proton. As shown by the following electron-dot structures, all Brønsted–Lowry bases have one or more lone pairs of electrons:

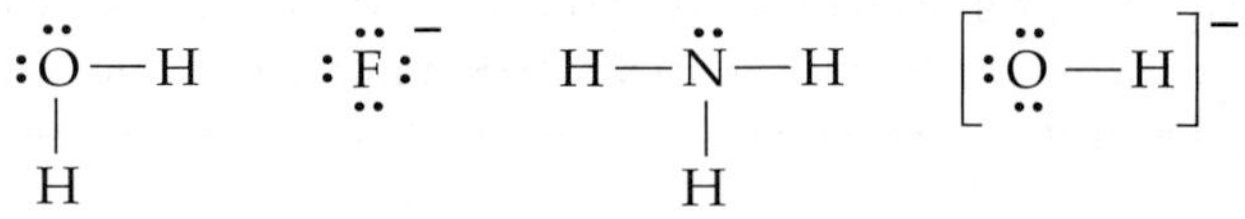

$$:\overset{..}{\underset{|}{O}}-H \qquad :\overset{..}{\underset{..}{F}}:^{-} \qquad H-\overset{..}{\underset{|}{N}}-H \qquad \left[:\overset{..}{\underset{..}{O}}-H\right]^{-}$$

Some Brønsted–Lowry bases

EXAMPLE 15.1 Account for the acidic properties of nitrous acid (HNO_2) in terms of the Arrhenius theory and the Brønsted–Lowry theory, and identify the conjugate base of HNO_2.

SOLUTION HNO_2 is an Arrhenius acid because it undergoes dissociation in water to produce H^+ ions:

$$HNO_2(aq) \rightleftharpoons H^+(aq) + NO_2^-(aq)$$

Nitrous acid is a Brønsted–Lowry acid because it acts as a proton donor when it dissociates, transferring a proton to water to give the hydronium ion:

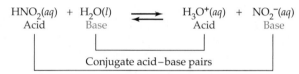

$$\underset{\text{Acid}}{HNO_2(aq)} + \underset{\text{Base}}{H_2O(l)} \rightleftharpoons \underset{\text{Acid}}{H_3O^+(aq)} + \underset{\text{Base}}{NO_2^-(aq)}$$

Conjugate acid–base pairs

The conjugate base of HNO_2 is NO_2^-, the species that remains after HNO_2 has lost a proton.

▶ **PROBLEM 15.1** Write a balanced equation for the dissociation of each of the following Brønsted–Lowry acids in water:
(a) H_2SO_4 **(b)** HSO_4^- **(c)** H_3O^+
What is the conjugate base of the acid in each case?

▶ **PROBLEM 15.2** What is the conjugate acid of each of the following Brønsted–Lowry bases?
(a) HCO_3^- **(b)** CO_3^{2-} **(c)** OH^-

↤ **KEY CONCEPT PROBLEM 15.3** For the following reaction, identify the Brønsted–Lowry acids, bases, and conjugate acid–base pairs:

○—◯ + ○—◯ ⇌ ◯◯○ + ○

○ = H ◯ = S ○ = F

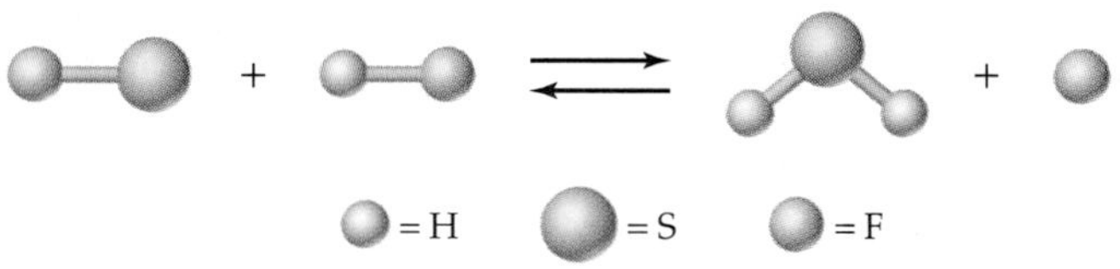

15.2 Acid Strength and Base Strength

A helpful way of viewing an acid-dissociation equilibrium is to realize that the two bases, H_2O and A^-, are competing for protons:

Introduction to Aqueous Acids movie; Introduction to Aqueous Bases movie

$$\underset{\text{Acid}}{HA(aq)} + \underset{\text{Base}}{H_2O(l)} \rightleftharpoons \underset{\text{Acid}}{H_3O^+(aq)} + \underset{\text{Base}}{A^-(aq)}$$

If H_2O is a stronger base (a stronger proton acceptor) than A^-, the H_2O molecules will get the protons, and the solution will contain mainly H_3O^+ and A^-. If A^- is a stronger base than H_2O, the A^- ions will get the protons, and the solution will contain mainly HA and H_2O. When beginning with equal concentrations of reactants and products, *the proton is always transferred to the stronger base.* This means that the direction of reaction to reach equilibrium is proton transfer from the stronger acid to the stronger base to give the weaker acid and the weaker base:

Stronger acid + Stronger base ⟶ Weaker acid + Weaker base

$$\underset{\text{donor}}{\text{Stronger H}^+} + \underset{\text{acceptor}}{\text{Stronger H}^+} \longrightarrow \underset{\text{donor}}{\text{Weaker H}^+} + \underset{\text{acceptor}}{\text{Weaker H}^+}$$

Different acids differ in their ability to donate protons. A **strong acid** is one that is almost completely dissociated in water (Section 4.5). Thus, the acid-dissociation equilibrium of a strong acid lies nearly 100% to the right, and the solution contains almost entirely H_3O^+ and A^- ions with only a negligible amount of undissociated HA molecules. Typical examples of strong acids are perchloric acid ($HClO_4$), hydrochloric acid (HCl), hydrobromic acid (HBr), hydroiodic acid (HI), nitric acid (HNO_3), and sulfuric acid (H_2SO_4). It follows from this definition that *strong acids have very weak conjugate bases.* The ions ClO_4^-, Cl^-, Br^-, I^-, NO_3^-, and HSO_4^- have only a negligible tendency to combine with a proton in aqueous solution, and they are therefore much weaker bases than H_2O.

A **weak acid** is one that is only partially dissociated in water. Only a small fraction of the weak acid molecules transfer a proton to water, and the solution therefore contains mainly undissociated HA molecules along with small amounts of H_3O^+ and the conjugate base A^-. Typical examples of weak acids are nitrous acid (HNO_2), hydrofluoric acid (HF), and acetic acid (CH_3CO_2H). In the case of very weak acids such as NH_3, OH^-, and H_2, the acid has practically no tendency to transfer a proton to water, and the acid-dissociation equilibrium lies essentially 100% to the left. It follows from this definition that *very weak acids have strong conjugate bases.* For example, the NH_2^-, O^{2-}, and H^- ions are essentially 100% protonated in aqueous solution and are much stronger bases than H_2O.

The equilibrium concentrations of HA, H_3O^+, and A^- for strong acids, weak acids, and very weak acids are represented graphically in Figure 15.1.

FIGURE 15.1   Dissociation of HA involves H^+ transfer to H_2O, yielding H_3O^+ and A^-. The extent of dissociation is **(a)** nearly 100% for a strong acid, **(b)** considerably less than 100% for a weak acid, and **(c)** nearly 0% for a very weak acid.

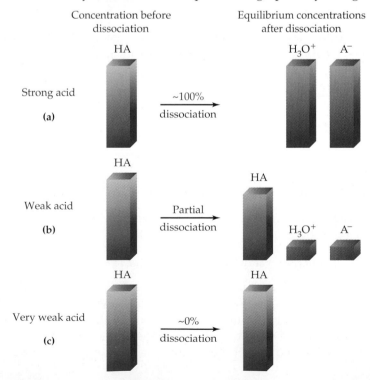

The inverse relationship between the strength of an acid and the strength of its conjugate base is illustrated in Table 15.1.

TABLE 15.1   Relative Strengths of Conjugate Acid–Base Pairs

	Acid, HA		Base, A$^-$		
Stronger acid	$HClO_4$ HCl H_2SO_4 HNO_3	Strong acids. 100% dissociated in aqueous solution.	$ClO_4{}^-$ Cl^- $HSO_4{}^-$ $NO_3{}^-$	Very weak bases. Negligible tendency to be protonated in aqueous solution.	Weaker base
	H_3O^+		H_2O		
	$HSO_4{}^-$ H_3PO_4 HNO_2 HF CH_3CO_2H H_2CO_3 H_2S $NH_4{}^+$ HCN $HCO_3{}^-$	Weak acids. Exist in solution as a mixture of HA, A$^-$, and H_3O^+.	$SO_4{}^{2-}$ $H_2PO_4{}^-$ $NO_2{}^-$ F^- $CH_3CO_2{}^-$ $HCO_3{}^-$ HS^- NH_3 CN^- $CO_3{}^{2-}$	Weak bases. Moderate tendency to be protonated in aqueous solution.	
	H_2O		OH^-		
Weaker acid	NH_3 OH^- H_2	Very weak acids. Negligible tendency to dissociate.	$NH_2{}^-$ O^{2-} H^-	Strong bases. 100% protonated in aqueous solution.	Stronger base

EXAMPLE 15.2   If you mix equal concentrations of reactants and products, which of the following reactions proceed to the right and which proceed to the left?
(a) $H_2SO_4(aq) + NH_3(aq) \rightleftharpoons NH_4{}^+(aq) + HSO_4{}^-(aq)$
(b) $HCO_3{}^-(aq) + SO_4{}^{2-}(aq) \rightleftharpoons HSO_4{}^-(aq) + CO_3{}^{2-}(aq)$

SOLUTION
(a) In this reaction, H_2SO_4 and $NH_4{}^+$ are the proton donors (acids), and NH_3 and $HSO_4{}^-$ are the proton acceptors (bases). According to Table 15.1, H_2SO_4 is a stronger acid than $NH_4{}^+$, and NH_3 is therefore a stronger base than $HSO_4{}^-$. Because proton transfer occurs from the stronger acid to the stronger base, the reaction proceeds from left to right.

$$H_2SO_4(aq) \; + \; NH_3(aq) \; \longrightarrow \; NH_4^+(aq) \; + \; HSO_4^-(aq)$$

Stronger acid   Stronger base   Weaker acid   Weaker base

(b) $HCO_3{}^-$ and $HSO_4{}^-$ are the acids, and $SO_4{}^{2-}$ and $CO_3{}^{2-}$ are the bases. Table 15.1 indicates that $HSO_4{}^-$ is the stronger acid, and $CO_3{}^{2-}$ is the stronger base. Therefore, $CO_3{}^{2-}$ gets the proton, and the reaction proceeds from right to left.

$$HCO_3^-(aq) \; + \; SO_4^{2-}(aq) \; \longleftarrow \; HSO_4^-(aq) \; + \; CO_3^{2-}(aq)$$

Weaker acid   Weaker base   Stronger acid   Stronger base

▶ **PROBLEM 15.4**   If you mix equal concentrations of reactants and products, which of the following reactions proceed to the right and which proceed to the left?
(a) $HF(aq) + NO_3{}^-(aq) \rightleftharpoons HNO_3(aq) + F^-(aq)$
(b) $NH_4{}^+(aq) + CO_3{}^{2-}(aq) \rightleftharpoons HCO_3{}^-(aq) + NH_3(aq)$

⊷ KEY CONCEPT PROBLEM 15.5 The following pictures represent aqueous solutions of two acids HA (A = X or Y); water molecules have been omitted for clarity.

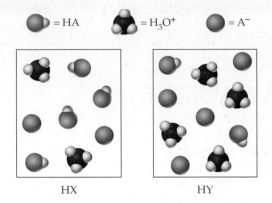

HX   HY

(a) Which is the stronger acid, HX or HY?
(b) Which is the stronger base, X^- or Y^-?
(c) If you mix equal concentrations of reactants and products, will the following reaction proceed to the right or to the left?

$$HX + Y^- \rightleftharpoons HY + X^-$$

15.3 Hydrated Protons and Hydronium Ions

The proton is fundamental to both the Arrhenius and the Brønsted–Lowry definitions of an acid. Dissociation of an Arrhenius acid HA gives an aqueous hydrogen ion, or hydrated proton, written as $H^+(aq)$:

$$HA(aq) \rightleftharpoons H^+(aq) + A^-(aq)$$

As a bare proton, the positively charged H^+ ion is too reactive to exist in aqueous solution, and so it bonds to the oxygen atom of a solvent water molecule to give the trigonal pyramidal hydronium ion, H_3O^+. The H_3O^+ ion, which can be regarded as the simplest hydrate of the proton, $[H(H_2O)]^+$, can associate through hydrogen bonding with additional water molecules to give higher hydrates with the general formula $[H(H_2O)_n]^+$ ($n = 2, 3,$ or 4), for example, $H_5O_2^+$, $H_7O_3^+$, and $H_9O_4^+$. It's likely that acidic aqueous solutions contain a distribution of $[H(H_2O)_n]^+$ ions having different values of n. In this book, though, we'll use the symbols $H^+(aq)$ and $H_3O^+(aq)$ to mean the same thing—namely, a proton hydrated by an unspecified number of water molecules. Ordinarily, we use H_3O^+ in acid–base reactions to emphasize the proton transfer character of these reactions.

H_3O^+ — the hydronium ion, or hydrated H^+

15.4 Dissociation of Water

One of the most important properties of water is its ability to act both as an acid and as a base. In the presence of an acid, water acts as a base, whereas in the presence of a base, water acts as an acid. It's not surprising, therefore, that in pure water one molecule can donate a proton to another in a reaction in which water acts as both an acid and a base *at the same time*:

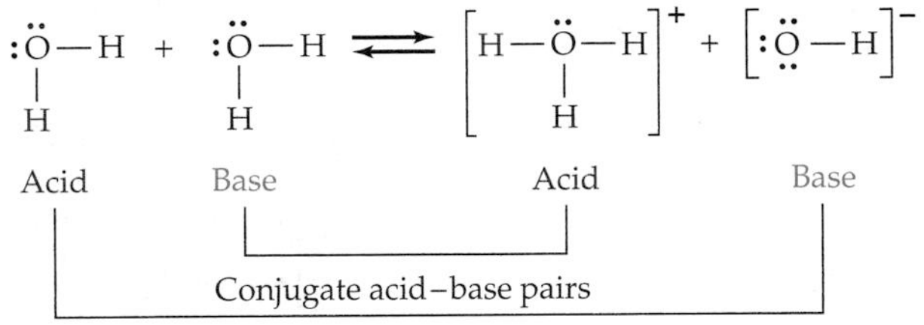

Called the **dissociation of water**, this reaction is characterized by the usual kind of equilibrium equation:

$$2\,H_2O(l) \rightleftharpoons H_3O^+(aq) + OH^-(aq)$$

$$"K_c" = \frac{[H_3O^+][OH^-]}{[H_2O]^2}$$

Because the molar concentration of water is a constant (water is a pure liquid), we can combine $[H_2O]^2$ and the equilibrium constant $"K_c"$ into a single constant, which we denote by K_w and call the **ion-product constant for water:**

$$K_w = "K_c" \times [H_2O]^2 = [H_3O^+][OH^-]$$

There are two important aspects of the dynamic equilibrium in the dissociation of water. First, the forward and reverse reactions are rapid; H_2O molecules, H_3O^+ ions, and OH^- ions continually interconvert as protons transfer quickly from one species to another. Second, the position of the equilibrium lies far to the left; at any given instant, only a tiny fraction of the water molecules are dissociated into H_3O^+ and OH^- ions. The vast majority of the molecules are undissociated.

We can calculate the extent of dissociation of the water molecules starting from experimental measurements that show the H_3O^+ concentration in pure water to be 1.0×10^{-7} M at 25°C:

$$[H_3O^+] = 1.0 \times 10^{-7}\,\text{M} \qquad \text{At 25°C}$$

Since the dissociation reaction of water produces equal concentrations of H_3O^+ and OH^- ions, the OH^- concentration in pure water is also 1.0×10^{-7} M at 25°C:

$$[H_3O^+] = [OH^-] = 1.0 \times 10^{-7}\,\text{M} \qquad \text{At 25°C}$$

Furthermore, we know that the molar concentration of pure water, calculated from its density and molar mass, is 55.4 M at 25°C:

$$[H_2O] = \left(\frac{997\,\text{g}}{\text{L}}\right)\left(\frac{1\,\text{mol}}{18.0\,\text{g}}\right) = 55.4\,\text{mol/L} \qquad \text{At 25°C}$$

From these facts, we can conclude that the ratio of dissociated to undissociated water molecules is about 2 in 10^9, a very small number indeed:

$$\frac{1.0 \times 10^{-7}\,M}{55.4\,M} = 1.8 \times 10^{-9} \qquad \text{About 2 in } 10^9$$

In addition, we can calculate that the numerical value of K_w at 25°C is 1.0×10^{-14}. As is common practice for equilibrium constants, the units of K_w (mol^2/L^2) are omitted:

$$K_w = [H_3O^+][OH^-] = (1.0 \times 10^{-7})(1.0 \times 10^{-7})$$
$$= 1.0 \times 10^{-14} \qquad \text{At 25°C}$$

In very dilute solutions, the product of the H_3O^+ and OH^- concentrations is not affected by the presence of solutes. This is not true in more concentrated solutions, but we'll neglect that complication and assume that *the product of the H_3O^+ and OH^- concentrations is always 1.0×10^{-14} at 25°C in any aqueous solution.*

We can distinguish acidic, neutral, and basic aqueous solutions by the relative values of the H_3O^+ and OH^- concentrations:

$$\text{Acidic:} \quad [H_3O^+] > [OH^-]$$
$$\text{Neutral:} \quad [H_3O^+] = [OH^-]$$
$$\text{Basic:} \quad [H_3O^+] < [OH^-]$$

At 25°C, $[H_3O^+] > 1.0 \times 10^{-7}$ M in an acidic solution, $[H_3O^+] = [OH^-] = 1.0 \times 10^{-7}$ M in a neutral solution, and $[H_3O^+] < 1.0 \times 10^{-7}$ M in a basic solution (Figure 15.2). If one of the concentrations, $[H_3O^+]$ or $[OH^-]$, is known, the other is readily calculated:

$$\text{Since} \quad [H_3O^+][OH^-] = K_w = 1.0 \times 10^{-14}$$

$$\text{then} \quad [H_3O^+] = \frac{1.0 \times 10^{-14}}{[OH^-]} \quad \text{and} \quad [OH^-] = \frac{1.0 \times 10^{-14}}{[H_3O^+]}$$

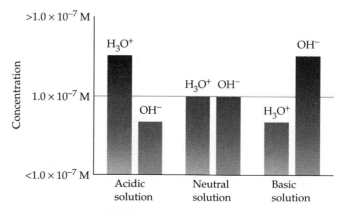

FIGURE 15.2 Values of the H_3O^+ and OH^- concentrations at 25°C in acidic, neutral, and basic solutions.

In the previous discussion, we were careful to emphasize that the value of $K_w = 1.0 \times 10^{-14}$ applies only at 25°C. Because K_w, like all equilibrium constants, is affected by temperature, the H_3O^+ and OH^- concentrations in neutral aqueous solutions at temperatures other than 25°C deviate from 1.0×10^{-7} M (see Problem 15.7). Unless otherwise indicated, we'll always assume a temperature of 25°C.

EXAMPLE 15.3 The concentration of H_3O^+ ions in a sample of lemon juice is 2.5×10^{-3} M. Calculate the concentration of OH^- ions, and classify the solution as acidic, neutral, or basic.

BALLPARK SOLUTION Since the product of the H_3O^+ and OH^- concentrations must equal 10^{-14}, and since the H_3O^+ concentration is in the range 10^{-3} to 10^{-2} M, the OH^- concentration must be in the range 10^{-11} to 10^{-12} M. Because $[H_3O^+] > [OH^-]$, the solution is acidic.

DETAILED SOLUTION When $[H_3O^+]$ is known, the OH^- concentration can be found from the expression

$$[OH^-] = \frac{K_w}{[H_3O^+]} = \frac{1.0 \times 10^{-14}}{2.5 \times 10^{-3}} = 4.0 \times 10^{-12} \text{ M}$$

▶ **PROBLEM 15.6** The concentration of OH^- in a sample of seawater is 5.0×10^{-6} M. Calculate the concentration of H_3O^+ ions, and classify the solution as acidic, neutral, or basic.

▶ **PROBLEM 15.7** At 50°C the value of K_w is 5.5×10^{-14}. What are the concentrations of H_3O^+ and OH^- in a neutral solution at 50°C?

15.5 The pH Scale

Rather than write hydronium ion concentrations in molarity, it's more convenient to express them on a logarithmic scale known as the *pH scale*. The term **pH** is derived from the French *puissance d'hydrogène* ("power of hydrogen") and refers to the power of 10 (the exponent) used to express the molar H_3O^+ concentration. The pH of a solution is defined as the negative base-10 logarithm (log) of the molar hydronium ion concentration:

$$pH = -\log[H_3O^+] \quad \text{or} \quad [H_3O^+] = \text{antilog}(-pH) = 10^{-pH}$$

Thus, an acidic solution having $[H_3O^+] = 10^{-2}$ M has a pH of 2, a basic solution having $[OH^-] = 10^{-2}$ M and $[H_3O^+] = 10^{-12}$ M has a pH of 12, and a neutral solution having $[H_3O^+] = 10^{-7}$ M has a pH of 7. Note that although we express $[H_3O^+]$ in moles per liter, we take the log of the number only, not the units. (The pOH can be defined in the same way as the pH. Just as $pH = -\log[H_3O^+]$, so $pOH = -\log[OH^-]$. It follows from the equation $[H_3O^+][OH^-] = 1.0 \times 10^{-14}$ that $pH + pOH = 14.00$.)

If you use a calculator to find the pH from the H_3O^+ concentration, your answer will have more decimal places than the proper number of significant figures. For example, the pH of the lemon juice in Example 15.3 ($[H_3O^+] = 2.5 \times 10^{-3}$ M) is found on a calculator to be

$$pH = -\log(2.5 \times 10^{-3}) = 2.602\ 06$$

This result should be rounded to pH 2.60 (two significant figures) because $[H_3O^+]$ has only two significant figures. Note that the only significant figures in a logarithm are the digits to the right of the decimal point; the number to the left of the decimal point is an exact number related to the integral power of 10 in the exponential expression for $[H_3O^+]$:

$$pH = -\log(2.5 \times 10^{-3}) = -\log 10^{-3} - \log 2.5 = 3 - 0.40 = 2.60$$

2 significant figures (2 SF's)   Exact number   Exact number   2 SF's   Exact number   2 SF's

Because the pH scale is logarithmic, the pH changes by 1 unit when $[H_3O^+]$ changes by a factor of 10, by 2 units when $[H_3O^+]$ changes by a factor of 100, and by 6 units when $[H_3O^+]$ changes by a factor of 1,000,000. To appreciate the extent to which the pH scale is a compression of the $[H_3O^+]$ scale, compare the amounts of 12 M HCl required to change the pH of the water in a backyard swimming pool. Only about 100 mL of 12 M HCl is needed to change the pH from 7 to 6, but a 10,000 L truckload of 12 M HCl is needed to change the pH from 7 to 1.

The pH scale and pH values for some common substances are shown in Figure 15.3. Because the pH is the *negative* log of $[H_3O^+]$, the pH decreases as $[H_3O^+]$ increases. Thus, when $[H_3O^+]$ increases from 10^{-7} M to 10^{-6} M, the pH decreases from 7 to 6. As a result, acidic solutions have pH less than 7, and basic solutions have pH greater than 7.

Acidic solution:	pH < 7
Neutral solution:	pH = 7
Basic solution:	pH > 7

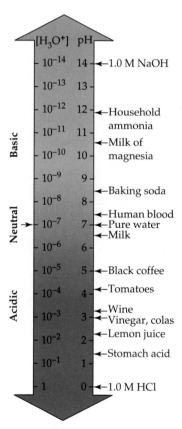

FIGURE 15.3 The pH scale and pH values for some common substances.

EXAMPLE 15.4 Calculate the pH of an aqueous ammonia solution that has an OH^- concentration of 1.9×10^{-3} M.

SOLUTION First calculate the H_3O^+ concentration from the OH^- concentration, and then take the negative of the logarithm of $[H_3O^+]$ to convert to pH:

$$[H_3O^+] = \frac{K_w}{[OH^-]} = \frac{1.0 \times 10^{-14}}{1.9 \times 10^{-3}} = 5.3 \times 10^{-12} \text{ M}$$

$$pH = -\log[H_3O^+] = -\log(5.3 \times 10^{-12}) = 11.28$$

Note that the pH is quoted to two significant figures (.28) because $[H_3O^+]$ is known to two significant figures (5.3).

EXAMPLE 15.5 Acid rain is a matter of serious concern because most species of fish die in waters having a pH lower than 4.5–5.0. Calculate the H_3O^+ concentration in a lake that has a pH of 4.5.

BALLPARK SOLUTION Because a pH of 4.5 is between pH 4 and pH 5, $[H_3O^+]$ is between 10^{-4} M and 10^{-5} M.

DETAILED SOLUTION Calculate the H_3O^+ concentration by taking the antilogarithm of the negative of the pH:

$$[H_3O^+] = \text{antilog}(-pH) = 10^{-pH} = 10^{-4.5} = 3 \times 10^{-5} \text{ M}$$

[H$_3$O$^+$] is reported to only one significant figure because the pH has only one digit beyond the decimal point. (If you need help in finding the antilog of a number, see Appendix A.2.)

▶**PROBLEM 15.8** Calculate the pH of each of the following solutions:
(a) A sample of seawater that has an OH$^-$ concentration of 1.58×10^{-6} M
(b) A sample of acid rain that has an H$_3$O$^+$ concentration of 6.0×10^{-5} M

▶**PROBLEM 15.9** Human blood has a pH of 7.40. Calculate the concentrations of H$_3$O$^+$ and OH$^-$.

15.6 Measuring pH

The approximate pH of a solution can be determined by using an **acid–base indicator**, a substance that changes color in a specific pH range (Figure 15.4). Indicators (abbreviated HIn) exhibit pH-dependent color changes because they are weak acids and have different colors in their acid (HIn) and conjugate base (In$^-$) forms:

$$\underset{\text{Color A}}{\text{HIn}(aq)} + \text{H}_2\text{O}(l) \rightleftharpoons \text{H}_3\text{O}^+(aq) + \underset{\text{Color B}}{\text{In}^-(aq)}$$

Bromthymol blue, for example, changes color in the pH range 6.0–7.6 from yellow in its acid form to blue in its base form. Phenolphthalein changes in the pH range 8.2–9.8 from colorless in its acid form to pink in its base form.

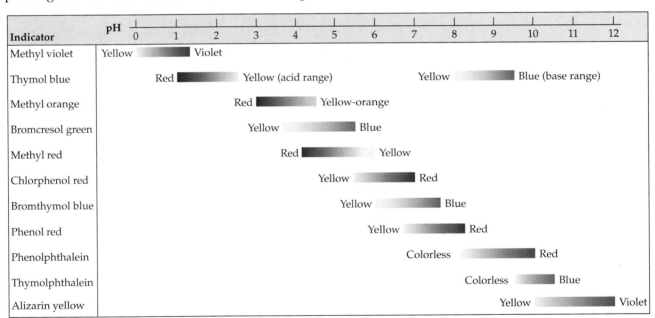

FIGURE 15.4 Some common acid–base indicators and their color changes. Note that the color of an indicator changes over a range of about 2 pH units.

Because indicators change color over a range of about 2 pH units, it's possible to determine the pH of a solution to within approximately ±1 unit simply by adding to the solution a few drops of an indicator that changes color in the appropriate pH range. To make the determination particularly easy, a mixture of indicators known as *universal indicator* is commercially available for making approximate pH measurements in the range 3–10 (Figure 15.5). More

Natural Indicators movie

(b)

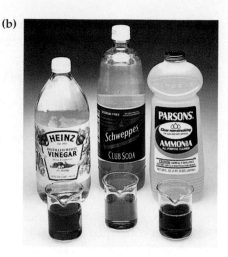

(a)

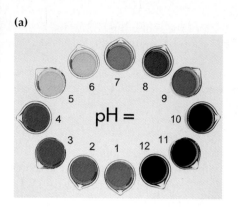

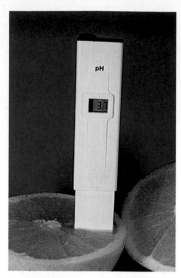

FIGURE 15.5 **(a)** The color of universal indicator in solutions of known pH from 1 to 12. **(b)** The color of universal indicator in some familiar products gives the following approximate pH values: vinegar, pH 3; club soda, pH 4–5; household ammonia, pH ≥10.

FIGURE 15.6 A pH meter with its electrical probe dipping into a grapefruit. An accurate value of the pH (3.7) is shown on the meter.

accurate values can be determined with an electronic instrument called a *pH meter* (Figure 15.6), a device that measures the pH-dependent electrical potential of the test solution. We'll learn more about pH meters in Section 18.7.

15.7 The pH in Solutions of Strong Acids and Strong Bases

The commonly encountered strong acids listed in Table 15.1 include three **monoprotic acids** ($HClO_4$, HCl, and HNO_3), which contain a single dissociable proton, and one **diprotic acid** (H_2SO_4), which has two dissociable protons. Because strong monoprotic acids are 100% dissociated in aqueous solution, the H_3O^+ and A^- concentrations are equal to the initial concentration of the acid, and the concentration of undissociated HA molecules is essentially zero.

$$HA(aq) + H_2O(l) \xrightarrow{100\%} H_3O^+(aq) + A^-(aq)$$

The pH of a solution of a strong monoprotic acid is easily calculated from the H_3O^+ concentration, as we'll illustrate in Example 15.6. Calculation of the pH of an H_2SO_4 solution is more complicated because 100% of the H_2SO_4 molecules dissociate to give H_3O^+ and HSO_4^- ions, but less than 100% of the resulting HSO_4^- ions dissociate to give H_3O^+ and SO_4^{2-} ions (we'll have more to say about diprotic acids in Section 15.11).

The most familiar examples of strong bases are alkali metal hydroxides, MOH, such as NaOH (*caustic soda*) and KOH (*caustic potash*). These compounds are water-soluble ionic solids that exist in aqueous solution as alkali metal cations (M^+) and OH^- anions:

$$MOH(s) \xrightarrow{H_2O} MOH(aq) \xrightarrow{100\%} M^+(aq) + OH^-(aq)$$

Thus, 0.10 M NaOH contains 0.10 M Na^+ and 0.10 M OH^-, and the pH is readily calculated from the OH^- concentration, as we'll show in Example 15.7.

The alkaline earth metal hydroxides $M(OH)_2$ (M = Mg, Ca, Sr, or Ba) are also strong bases (~100% dissociated), but they give lower OH^- concentrations because they are less soluble. Their solubility at room temperature varies from 38 g/L for the relatively soluble $Ba(OH)_2$ to ~10^{-2} g/L for the relatively insoluble $Mg(OH)_2$. Aqueous suspensions of $Mg(OH)_2$, called *milk of magnesia*, are used as an antacid. The most common and least expensive alkaline earth hydroxide is $Ca(OH)_2$, which is used in making mortars and cements. It is called *slaked lime* because it is made by treating *lime* (CaO) with water. Aqueous solutions of the slightly soluble $Ca(OH)_2$ (solubility ~1.3 g/L) are known as *limewater*.

Alkaline earth oxides, such as CaO, are even stronger bases than the corresponding hydroxides because the oxide ion (O^{2-}) is a stronger base than OH^- (Table 15.1). In fact, the O^{2-} ion can't exist in aqueous solutions because it is immediately and completely protonated by water, yielding OH^- ions:

$$O^{2-}(aq) + H_2O(l) \xrightarrow{\ 100\%\ } OH^-(aq) + OH^-(aq)$$

Thus, when lime is dissolved in water, it gives 2 OH^- ions per CaO formula unit:

$$CaO(s) + H_2O(l) \longrightarrow Ca^{2+}(aq) + 2\,OH^-(aq)$$

Lime is the world's most important strong base. It is produced in enormous quantities (22 million tons per year in the United States) for use in steelmaking, water purification, and chemical manufacture. Lime is made by decomposition of limestone, $CaCO_3$, at temperatures of 800–1000°C:

$$CaCO_3(s) \xrightarrow{\ Heat\ } CaO(s) + CO_2(g)$$

Lime is spread on lawns to raise the pH of acidic soils.

EXAMPLE 15.6   Calculate the pH of a 0.025 M HNO_3 solution.

SOLUTION   Since nitric acid is a strong acid, it is almost completely dissociated in aqueous solution. Therefore, $[H_3O^+] = 0.025$ M, and pH = 1.60:

$$HNO_3(aq) + H_2O(l) \xrightarrow{\ 100\%\ } H_3O^+(aq) + NO_3^-(aq)$$
$$pH = -\log[H_3O^+] = -\log(2.5 \times 10^{-2}) = 1.60$$

EXAMPLE 15.7   Calculate the pH of each of the following solutions:
(a) A 0.10 M solution of NaOH
(b) A 0.0050 M solution of slaked lime [$Ca(OH)_2$]
(c) A solution prepared by dissolving 0.28 g of lime (CaO) in enough water to make 1.00 L of limewater [$Ca(OH)_2(aq)$]

SOLUTION
(a) Because NaOH is a strong base, it is 100% dissociated. Therefore, $[OH^-] = 0.10$ M, $[H_3O^+] = 1.0 \times 10^{-13}$ M, and pH = 13.00.

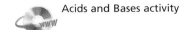

Acids and Bases activity

$$[H_3O^+] = \frac{K_w}{[OH^-]} = \frac{1.0 \times 10^{-14}}{0.10} = 1.0 \times 10^{-13} \text{ M}$$
$$pH = -\log(1.0 \times 10^{-13}) = 13.00$$

(b) Because slaked lime is a strong base, it is 100% dissociated, providing 2 OH^- per $Ca(OH)_2$ formula unit. Therefore, $[OH^-] = 2(0.0050\ M) = 0.010\ M$, $[H_3O^+] = 1.0 \times 10^{-12}\ M$, and pH = 12.00:

$$[H_3O^+] = \frac{K_w}{[OH^-]} = \frac{1.0 \times 10^{-14}}{0.010} = 1.0 \times 10^{-12}\ M$$

$$pH = -\log(1.0 \times 10^{-12}) = 12.00$$

(c) First calculate the number of moles of CaO dissolved from the given mass of CaO and its molar mass (56.1 g/mol):

$$\text{Moles of CaO} = 0.28\ \text{g CaO} \times \frac{1\ \text{mol CaO}}{56.1\ \text{g CaO}} = 0.0050\ \text{mol CaO}$$

Protonation of the O^{2-} ion produces 2 mol of OH^- per mole of CaO dissolved:

$$CaO(s) + H_2O(l) \longrightarrow Ca^{2+}(aq) + 2\ OH^-(aq)$$

$$\text{Moles of } OH^- \text{ produced} = 2(0.0050\ \text{mol}) = 0.010\ \text{mol}$$

Since the solution volume is 1.00 L,

$$[OH^-] = \frac{0.010\ \text{mol}}{1.00\ \text{L}} = 0.010\ M$$

The $[OH^-]$ happens to be identical to that in part (b). Therefore, pH = 12.00.

▶ **PROBLEM 15.10**   Calculate the pH of:
(a) 0.050 M $HClO_4$ (b) 6.0 M HCl
(c) 0.020 M KOH   (d) 0.010 M $Ba(OH)_2$

▶ **PROBLEM 15.11**   Calculate the pH of a solution prepared by dissolving 0.25 g of BaO in enough water to make 0.500 L of solution.

15.8   Equilibria in Solutions of Weak Acids

A weak acid is not the same thing as a dilute solution of a strong acid. Whereas a strong acid is 100% dissociated in aqueous solution, a weak acid is only partially dissociated. It might therefore happen by chance that the H_3O^+ concentration from complete dissociation of a dilute strong acid is the same as that from partial dissociation of a concentrated weak acid.

Like the equilibrium reactions discussed in Chapter 13, the dissociation of a weak acid in water is characterized by an equilibrium equation. The equilibrium constant for the dissociation reaction, denoted K_a, is called the **acid-dissociation constant**:

$$HA(aq) + H_2O(l) \rightleftharpoons H_3O^+(aq) + A^-(aq)$$

$$"K_c" = \frac{[H_3O^+][A^-]}{[HA][H_2O]} \qquad K_a = "K_c" \times [H_2O] = \frac{[H_3O^+][A^-]}{[HA]}$$

$$K_a = \frac{[H_3O^+][A^-]}{[HA]}$$

Note that $[H_2O]$, which is essentially constant in dilute aqueous solutions, has been incorporated into the equilibrium constant K_a and is therefore omitted from the equilibrium constant expression for K_a.

Values of K_a and $pK_a = -\log K_a$ for some typical weak acids are listed in Table 15.2. (Just as the pH is defined as $-\log[H^+]$, so the pK_a of an acid is

TABLE 15.2   Acid-Dissociation Constants at 25°C

	Acid	Molecular Formula	Structural Formula*	K_a	pK_a[†]
Stronger acid	Hydrochloric	HCl	H—Cl	2×10^6	-6.3
	Nitrous	HNO_2	H—O—N=O	4.5×10^{-4}	3.35
	Hydrofluoric	HF	H—F	3.5×10^{-4}	3.46
	Acetylsalicylic (aspirin)	$C_9H_8O_4$	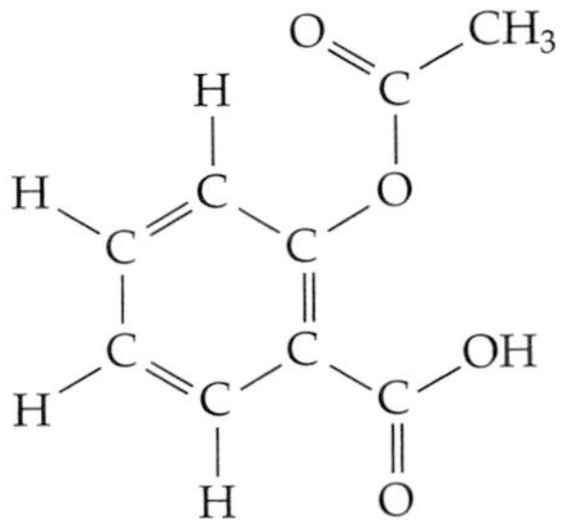	3.0×10^{-4}	3.52
	Formic	HCO_2H	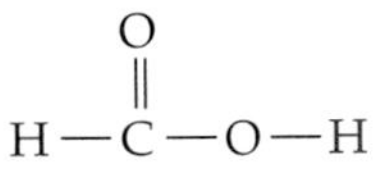	1.8×10^{-4}	3.74
	Ascorbic (vitamin C)	$C_6H_8O_6$	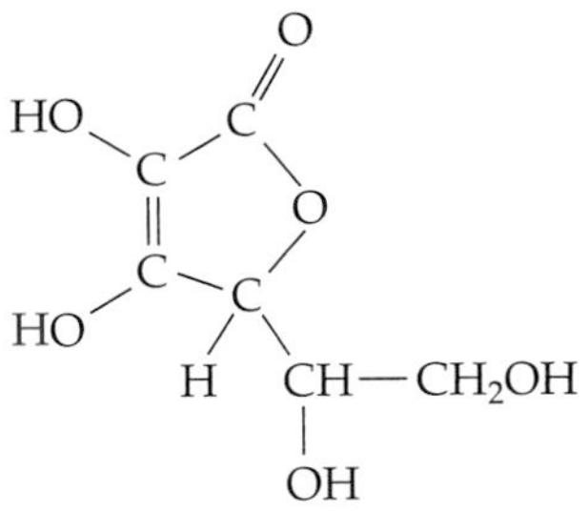	8.0×10^{-5}	4.10
	Benzoic	$C_6H_5CO_2H$	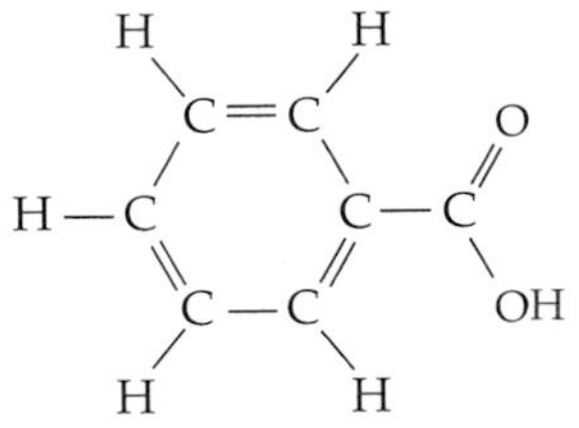	6.5×10^{-5}	4.19
	Acetic	CH_3CO_2H	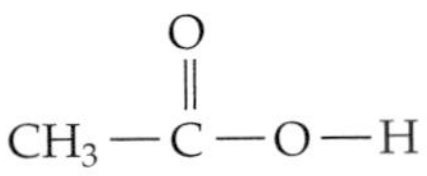	1.8×10^{-5}	4.74
	Hypochlorous	HOCl	H—O—Cl	3.5×10^{-8}	7.46
Weaker acid	Hydrocyanic	HCN	H—C≡N	4.9×10^{-10}	9.31
	Methanol	CH_3OH	CH_3—O—H	2.9×10^{-16}	15.54

*The proton that is transferred to water when the acid dissociates is shown in color.   [†]$pK_a = -\log K_a$.

defined as $-\log K_a$.) Also included in Table 15.2 for comparison are values for HCl, a typical strong acid. As indicated by the equilibrium equation, the larger the value of K_a, the stronger the acid. Thus, methanol ($K_a = 2.9 \times 10^{-16}$) is the weakest of the acids listed in Table 15.2, and nitrous acid ($K_a = 4.5 \times 10^{-4}$) is the strongest of the weak acids. Strong acids, such as HCl, have K_a values that are much greater than 1. A more complete list of K_a values for weak acids is given in Appendix C.

Numerical values of acid-dissociation constants are determined from pH measurements, as shown in Example 15.8.

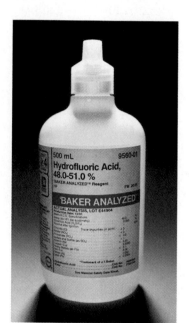

Of the hydrohalic acids HF, HCl, HBr, and HI, hydrofluoric acid is the only weak acid.

EXAMPLE 15.8 The pH of 0.250 M HF is 2.036. What is the value of K_a for hydrofluoric acid?

SOLUTION First write the balanced equation for the dissociation equilibrium and the equilibrium equation that defines K_a:

$$HF(aq) + H_2O(l) \rightleftharpoons H_3O^+(aq) + F^-(aq)$$

$$K_a = \frac{[H_3O^+][F^-]}{[HF]}$$

To obtain a value for K_a, we need to calculate the concentrations of the various species in the equilibrium mixture. Let's begin by calculating the H_3O^+ concentration from the pH:

$$[H_3O^+] = \text{antilog}(-pH) = 10^{-pH} = 10^{-2.036} = 9.20 \times 10^{-3} \text{ M}$$

Since dissociation of one HF molecule gives one H_3O^+ ion and one F^- ion, the H_3O^+ and F^- concentrations are equal:

$$[F^-] = [H_3O^+] = 9.20 \times 10^{-3} \text{ M}$$

Furthermore, the HF concentration at equilibrium is equal to the initial concentration (0.250 M) minus whatever dissociates (9.20×10^{-3} M):

$$[HF] = 0.250 - 0.00920 = 0.241 \text{ M}$$

Substituting the equilibrium concentrations into the equilibrium equation gives the value of K_a:

$$K_a = \frac{[H_3O^+][F^-]}{[HF]} = \frac{(9.20 \times 10^{-3})(9.20 \times 10^{-3})}{0.241} = 3.52 \times 10^{-4}$$

▶ **PROBLEM 15.12** The pH of 0.10 M HOCl is 4.23. Calculate K_a for hypochlorous acid, and check your answer against the value given in Table 15.2.

✦ **KEY CONCEPT PROBLEM 15.13** The following pictures represent aqueous solutions of three acids HA (A = X, Y, or Z); water molecules have been omitted for clarity:

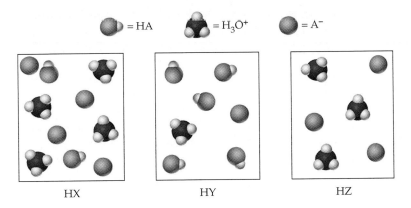

(a) Arrange the three acids in order of increasing value of K_a.
(b) Which acid, if any, is a strong acid?
(c) Which solution has the highest pH, and which has the lowest pH?

15.9 Calculating Equilibrium Concentrations in Solutions of Weak Acids

Once the K_a value for a weak acid has been measured, it can be used to calculate equilibrium concentrations and the pH in a solution of the acid. We'll illustrate the approach to such a problem by calculating the concentrations of all species present (H_3O^+, CN^-, HCN, and OH^-) and the pH in a 0.10 M HCN solution. The approach we'll take is quite general and will be useful on numerous later occasions.

The key to solving acid–base equilibrium problems is to think about the chemistry—that is, to consider the possible proton-transfer reactions that can take place between Brønsted–Lowry acids and bases.

Step 1. Let's begin by listing the species present initially before any dissociation reactions and identifying them as acids or bases. We'll include in our list the acid HCN and the solvent H_2O, but won't include the species that are produced by dissociation reactions (H_3O^+, CN^-, and OH^-) because they are present only in small concentrations. Since water can behave either as an acid or a base, our list of species present initially is

$$\begin{array}{cc} \text{HCN} & \text{H}_2\text{O} \\ \text{Acid} & \text{Acid or base} \end{array}$$

Step 2. Because we have two acids (HCN and H_2O) and just one base (H_2O), two proton-transfer reactions are possible:

$$\text{HCN}(aq) + \text{H}_2\text{O}(l) \rightleftharpoons \text{H}_3\text{O}^+(aq) + \text{CN}^-(aq) \qquad K_a = 4.9 \times 10^{-10}$$
$$\text{H}_2\text{O}(l) + \text{H}_2\text{O}(l) \rightleftharpoons \text{H}_3\text{O}^+(aq) + \text{OH}^-(aq) \qquad K_w = 1.0 \times 10^{-14}$$

The value of K_a for HCN comes from Table 15.2.

Step 3. The proton-transfer reaction that proceeds farther to the right—the one that has the larger equilibrium constant—is called the **principal reaction**. Any other proton-transfer reactions are termed **subsidiary reactions**. Since K_a for HCN is more than 1000 times greater than K_w, the principal reaction in this case is dissociation of HCN, and dissociation of water is a subsidiary reaction. Although the principal reaction and the subsidiary reaction both produce H_3O^+ ions, there is only one H_3O^+ concentration in the solution, which must simultaneously satisfy the equilibrium equations for both reactions. To make life simple, we'll assume that essentially all the H_3O^+ comes from the principal reaction:

$$[H_3O^+]\,(\text{total}) = [H_3O^+]\,(\text{from principal reaction}) + [H_3O^+]\,(\text{from subsidiary reaction})$$
$$\approx [H_3O^+]\,(\text{from principal reaction})$$

In other words, we'll assume that the equilibrium concentration of H_3O^+ is established entirely by the dissociation of the stronger acid, HCN, while dissociation of the weaker acid, H_2O, makes a negligible contribution.

$$[H_3O^+]\,(\text{total}) \approx [H_3O^+]\,(\text{from HCN})$$

Step 4. Next, we express the concentrations of the species involved in the principal reaction in terms of the concentration of HCN that dissociates—say, x mol/L. According to the balanced equation for the dissociation of HCN, if x mol/L of HCN dissociates, then x mol/L of H_3O^+ and x mol/L of CN^- are formed, and the initial concentration of HCN before dissociation (0.10 mol/L in our example) is reduced to $(0.10 - x)$ mol/L at equilibrium. Let's summarize these considerations in a table under the principal reaction:

Principal reaction:	HCN(aq) + H$_2$O(l) $\rightleftharpoons$	H$_3$O$^+$(aq) +	CN$^-$(aq)
Initial conc (M)	0.10	~0	0
Change (M)	$-x$	$+x$	$+x$
Equilibrium conc (M)	$0.10 - x$	x	x

Step 5. Substituting the equilibrium concentrations into the equilibrium equation for the principal reaction gives

$$K_a = 4.9 \times 10^{-10} = \frac{[H_3O^+][CN^-]}{[HCN]} = \frac{(x)(x)}{(0.10 - x)}$$

Because K_a is very small, the principal reaction will not proceed very far to the right, and x will be negligibly small in comparison with 0.10. Therefore, we can make the approximation that $(0.10 - x) \approx 0.10$, which greatly simplifies the solution:

$$4.9 \times 10^{-10} = \frac{(x)(x)}{(0.10 - x)} \approx \frac{x^2}{0.10}$$

$$x^2 = 4.9 \times 10^{-11}$$

$$x = 7.0 \times 10^{-6}$$

Step 6. Next, we use the calculated value of x to obtain the equilibrium concentration of all species involved in the principal reaction:

$$[H_3O^+] = [CN^-] = x = 7.0 \times 10^{-6} \, M$$

$$[HCN] = 0.10 - x = 0.10 - (7.0 \times 10^{-6}) = 0.10 \, M$$

Note that our simplifying approximation, $0.10 - x \approx 0.10$, is valid because x is only 7.0×10^{-6} and the initial [HCN] is 0.10. *It's important to check the validity of the simplifying approximation in every problem* because x is not always negligible in comparison with the initial concentration of the acid. Example 15.9 illustrates such a case.

Step 7. The concentrations of the species involved in the principal reaction are the "big" concentrations. The species involved in the subsidiary reaction(s) are present in smaller concentrations that can be calculated from equilibrium equations for the subsidiary reaction(s) and the big concentrations already determined. In the present problem, only the OH^- concentration remains to be calculated. It is determined from the subsidiary equilibrium equation, $[H_3O^+][OH^-] = K_w$, and the H_3O^+ concentration (7.0×10^{-6} M) already calculated from the principal reaction:

$$[OH^-] = \frac{K_w}{[H_3O^+]} = \frac{1.0 \times 10^{-14}}{7.0 \times 10^{-6}} = 1.4 \times 10^{-9} \, M$$

Note that $[OH^-]$ is 5000 times smaller than $[H_3O^+]$.

At this point, we can check the initial assumption that essentially all the H_3O^+ comes from the principal reaction. Because dissociation of water gives one H_3O^+ ion for each OH^- ion and because water is the only source of OH^-, $[H_3O^+]$ *from the dissociation of water* is equal to $[OH^-]$, which we just calculated to be 1.4×10^{-9} M. This value is negligible compared with $[H_3O^+]$ from the dissociation of HCN (7.0×10^{-6} M).

$$[H_3O^+] \, (\text{total}) = [H_3O^+] \, (\text{from HCN}) + [H_3O^+] \, (\text{from } H_2O)$$

$$= (7.0 \times 10^{-6} \, M) + (1.4 \times 10^{-9} \, M) = 7.0 \times 10^{-6} \, M$$

Step 8. Finally, we can calculate the pH:

$$pH = -\log(\text{total} \, [H_3O^+]) = -\log(7.0 \times 10^{-6}) = 5.15$$

Figure 15.7 summarizes the steps followed in solving this problem. We'll apply the same systematic approach to all aqueous equilibrium problems in this and the next chapter.

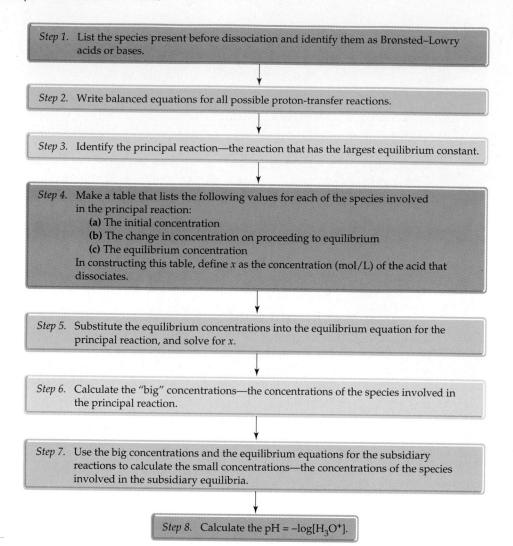

FIGURE 15.7 Steps to follow in solving problems involving weak acids.

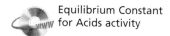

Equilibrium Constant for Acids activity

EXAMPLE 15.9   Calculate the pH and the concentrations of all species present (H_3O^+, F^-, HF, and OH^-) in 0.050 M HF.

SOLUTION   Follow the eight-step sequence outlined in Figure 15.7.

Step 1.   The species present initially are

$$\underset{\text{Acid}}{HF} \qquad \underset{\text{Acid or base}}{H_2O}$$

Step 2.   The possible proton-transfer reactions are

$$HF(aq) + H_2O(l) \rightleftharpoons H_3O^+(aq) + F^-(aq) \qquad K_a = 3.5 \times 10^{-4}$$
$$H_2O(l) + H_2O(l) \rightleftharpoons H_3O^+(aq) + OH^-(aq) \qquad K_w = 1.0 \times 10^{-14}$$

Step 3.   Since $K_a \gg K_w$, the principal reaction is dissociation of HF.

Step 4.

Principal reaction:	$HF(aq) + H_2O(l) \rightleftharpoons H_3O^+(aq) + F^-(aq)$		
Initial conc (M)	0.050	~0	0
Change (M)	$-x$	$+x$	$+x$
Equilibrium conc (M)	$0.050 - x$	x	x

Step 5. Substituting the equilibrium concentrations into the equilibrium equation for the principal reaction gives

$$K_a = 3.5 \times 10^{-4} = \frac{[H_3O^+][F^-]}{[HF]} = \frac{(x)(x)}{(0.050 - x)}$$

Making the usual approximation that x is negligible compared with the initial concentration of the acid, we assume that $(0.050 - x) \approx 0.050$ and then solve for an approximate value of x:

$$x^2 \approx (3.5 \times 10^{-4})(0.050)$$
$$x \approx 4.2 \times 10^{-3}$$

Since the initial concentration of HF (0.050 M) is known to the third decimal place, x is negligible compared with the initial [HF] only if x is less than 0.001 M. Our approximate value of x (0.0042 M) is not negligible compared with 0.050 M, and so our approximation, $0.050 - x \approx 0.050$, is invalid. We must therefore solve the quadratic equation without making approximations:

$$3.5 \times 10^{-4} = \frac{x^2}{(0.050 - x)}$$
$$x^2 + (3.5 \times 10^{-4})x - (1.75 \times 10^{-5}) = 0$$

We use the standard quadratic formula (Appendix A.4):

$$x = \frac{-b \pm \sqrt{b^2 - 4ac}}{2a}$$

$$x = \frac{-(3.5 \times 10^{-4}) \pm \sqrt{(3.5 \times 10^{-4})^2 - 4(-1.75 \times 10^{-5})}}{2}$$

$$= \frac{-(3.5 \times 10^{-4}) \pm (8.37 \times 10^{-3})}{2}$$

$$= +4.0 \times 10^{-3} \quad \text{or} \quad -4.4 \times 10^{-3}$$

Of the two solutions for x, only the positive value has physical meaning, since x is the H_3O^+ concentration. Therefore,

$$x = 4.0 \times 10^{-3}$$

Step 6. The big concentrations are

$$[H_3O^+] = [F^-] = x = 4.0 \times 10^{-3} \text{ M}$$
$$[HF] = (0.050 - x) = (0.050 - 0.0040) = 0.046 \text{ M}$$

Step 7. The small concentration, $[OH^-]$, is obtained from the subsidiary equilibrium, the dissociation of water:

$$[OH^-] = \frac{K_w}{[H_3O^+]} = \frac{1.0 \times 10^{-14}}{4.0 \times 10^{-3}} = 2.5 \times 10^{-12} \text{ M}$$

Step 8. $pH = -\log[H_3O^+] = -\log(4.0 \times 10^{-3}) = 2.40$

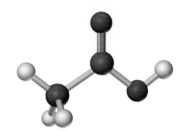

▶ **PROBLEM 15.14** Acetic acid, CH_3CO_2H, is the solute that gives vinegar its characteristic odor and sour taste. Calculate the pH and the concentrations of all species present (H_3O^+, $CH_3CO_2^-$, CH_3CO_2H, and OH^-) in:
(a) 1.00 M CH_3CO_2H **(b)** 0.0100 M CH_3CO_2H

▶ **PROBLEM 15.15** A vitamin C tablet containing 250 mg of ascorbic acid ($C_6H_8O_6$; $K_a = 8.0 \times 10^{-5}$) is dissolved in a 250 mL glass of water. What is the pH of the solution?

15.10 Percent Dissociation in Solutions of Weak Acids

In addition to K_a, another useful measure of the strength of a weak acid is the **percent dissociation**, defined as the concentration of the acid that dissociates divided by the initial concentration of the acid times 100%:

$$\text{Percent dissociation} = \frac{[HA] \text{ dissociated}}{[HA] \text{ initial}} \times 100\%$$

Take, for example, the 1.00 M acetic acid solution in Problem 15.14a. If you solved that problem correctly, you found that 1.00 M CH_3CO_2H has an H_3O^+ concentration of 4.2×10^{-3} M. Because $[H_3O^+]$ equals the concentration of CH_3CO_2H that dissociates, the percent dissociation in 1.00 M CH_3CO_2H is 0.42%:

Equilibrium Constant for Acids activity

$$\text{Percent dissociation} = \frac{[CH_3CO_2H] \text{ dissociated}}{[CH_3CO_2H] \text{ initial}} \times 100\%$$

$$= \frac{4.2 \times 10^{-3} \text{ M}}{1.00 \text{ M}} \times 100\% = 0.42\%$$

In general, the percent dissociation depends on the acid and increases with increasing value of K_a. For a given weak acid, the percent dissociation increases with increasing dilution, as shown in Figure 15.8. The 0.0100 M CH_3CO_2H solution in Problem 15.14b, for example, has $[H_3O^+] = 4.2 \times 10^{-4}$ M, and the percent dissociation is 4.2%:

$$\text{Percent dissociation} = \frac{[CH_3CO_2H] \text{ dissociated}}{[CH_3CO_2H] \text{ initial}} \times 100\%$$

$$= \frac{4.2 \times 10^{-4} \text{ M}}{0.0100 \text{ M}} \times 100\% = 4.2\%$$

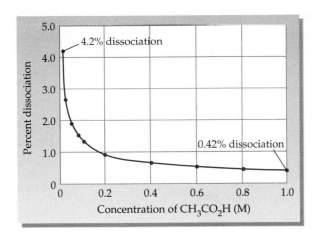

FIGURE 15.8 The percent dissociation of acetic acid increases as the concentration of the acid decreases. A 100-fold *decrease* in [CH$_3$CO$_2$H] results in a 10-fold *increase* in the percent dissociation.

▶ **PROBLEM 15.16**   Calculate the percent dissociation of HF ($K_a = 3.5 \times 10^{-4}$) in:
(a) 0.050 M HF   **(b)** 0.50 M HF

15.11 Polyprotic Acids

Acids that contain more than one dissociable proton are called **polyprotic acids**. Polyprotic acids dissociate in a stepwise manner, and each dissociation step is characterized by its own acid-dissociation constant, K_{a1}, K_{a2}, and so forth. For example, carbonic acid (H$_2$CO$_3$), the diprotic acid that forms when gaseous carbon dioxide dissolves in water, and is important in maintaining a constant pH in human blood, undergoes the following dissociation reactions:

$$\text{H}_2\text{CO}_3(aq) + \text{H}_2\text{O}(l) \rightleftharpoons \text{H}_3\text{O}^+(aq) + \text{HCO}_3^-(aq) \qquad K_{a1} = \frac{[\text{H}_3\text{O}^+][\text{HCO}_3^-]}{[\text{H}_2\text{CO}_3]} = 4.3 \times 10^{-7}$$

$$\text{HCO}_3^-(aq) + \text{H}_2\text{O}(l) \rightleftharpoons \text{H}_3\text{O}^+(aq) + \text{CO}_3^{2-}(aq) \qquad K_{a2} = \frac{[\text{H}_3\text{O}^+][\text{CO}_3^{2-}]}{[\text{HCO}_3^-]} = 5.6 \times 10^{-11}$$

As shown in Table 15.3, the values of stepwise dissociation constants of polyprotic acids decrease, typically by a factor of 10^4 to 10^6, in the order $K_{a1} > K_{a2} > K_{a3}$. Because of electrostatic forces, it's more difficult to remove a positively charged proton from a negative ion such as HCO$_3^-$ than from an uncharged molecule such as H$_2$CO$_3$, so $K_{a2} < K_{a1}$. In the case of triprotic acids (such as H$_3$PO$_4$), it's more difficult to remove H$^+$ from an anion with a double negative charge (such as HPO$_4^{2-}$), than from an anion with a single negative charge (such as H$_2$PO$_4^-$), so $K_{a3} < K_{a2}$.

TABLE 15.3   Stepwise Dissociation Constants for Polyprotic Acids at 25°C

Name	Formula	K_{a1}	K_{a2}	K_{a3}
Carbonic acid	H$_2$CO$_3$	4.3×10^{-7}	5.6×10^{-11}	
Hydrogen sulfide*	H$_2$S	1.0×10^{-7}	$\sim 10^{-19}$	
Oxalic acid	H$_2$C$_2$O$_4$	5.9×10^{-2}	6.4×10^{-5}	
Phosphoric acid	H$_3$PO$_4$	7.5×10^{-3}	6.2×10^{-8}	4.8×10^{-13}
Sulfuric acid	H$_2$SO$_4$	Very large	1.2×10^{-2}	
Sulfurous acid	H$_2$SO$_3$	1.5×10^{-2}	6.3×10^{-8}	

*Because of its very small size, K_{a2} for H$_2$S is difficult to measure and its value is uncertain.

Polyprotic acid solutions contain a mixture of acids—H_2A, HA^-, and H_2O in the case of a diprotic acid. Because H_2A is by far the stronger acid, the principal reaction is dissociation of H_2A, and essentially all the H_3O^+ in the solution comes from the first dissociation step. Example 15.10 shows how calculations are done.

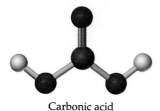

Carbonic acid

EXAMPLE 15.10   Calculate the pH and the concentrations of all species present (H_2CO_3, HCO_3^-, CO_3^{2-}, H_3O^+, and OH^-) in a 0.040 M carbonic acid solution.

SOLUTION   Use the eight-step procedure summarized in Figure 15.7.

Steps 1–3. The species present initially are H_2CO_3 (acid) and H_2O (acid or base). Because $K_{a1} \gg K_w$, the principal reaction is dissociation of H_2CO_3.

Step 4.

Principal reaction:	$H_2CO_3(aq) + H_2O(l) \rightleftharpoons H_3O^+(aq) + HCO_3^-(aq)$		
Initial conc (M)	0.040	~0	0
Change (M)	$-x$	$+x$	$+x$
Equilibrium conc (M)	$0.040 - x$	x	x

Step 5. Substituting the equilibrium concentrations into the equilibrium equation for the principal reaction gives

$$K_{a1} = 4.3 \times 10^{-7} = \frac{[H_3O^+][HCO_3^-]}{[H_2CO_3]} = \frac{(x)(x)}{(0.040 - x)}$$

where the K_{a1} value and the K_{a2} value below come from Table 15.3. Assuming that $(0.040 - x) \approx 0.040$,

$$x^2 = (4.3 \times 10^{-7})(0.040)$$

$$x = 1.3 \times 10^{-4} \quad \text{Approximation } (0.040 - x) \approx 0.040 \text{ is justified.}$$

Step 6. The big concentrations are

$$[H_3O^+] = [HCO_3^-] = x = 1.3 \times 10^{-4} \text{ M}$$

$$[H_2CO_3] = 0.040 - x = 0.040 - 0.000\,13 = 0.040 \text{ M}$$

Step 7. The small concentrations are obtained from the subsidiary equilibria—(1) dissociation of HCO_3^- and (2) dissociation of water—and from the big concentrations already determined:

$$(1) \quad HCO_3^-(aq) + H_2O(l) \rightleftharpoons H_3O^+(aq) + CO_3^{2-}(aq)$$

$$K_{a2} = 5.6 \times 10^{-11} = \frac{[H_3O^+][CO_3^{2-}]}{[HCO_3^-]} = \frac{(1.3 \times 10^{-4})[CO_3^{2-}]}{1.3 \times 10^{-4}}$$

$$[CO_3^{2-}] = K_{a2} = 5.6 \times 10^{-11} \text{ M}$$

(In general, for a solution of a weak diprotic acid H_2A, $[A^{2-}] = K_{a2}$.)

$$(2) \quad [OH^-] = \frac{K_w}{[H_3O^+]} = \frac{1.0 \times 10^{-14}}{1.3 \times 10^{-4}} = 7.7 \times 10^{-11} \text{ M}$$

The second dissociation of H_2CO_3 produces a negligible amount of H_3O^+ compared with the H_3O^+ obtained from the first dissociation. Of the 1.3×10^{-4} mol/L of HCO_3^- produced by the first dissociation, only 5.6×10^{-11} mol/L dissociates to form H_3O^+ and CO_3^{2-}.

Step 8. $pH = -\log[H_3O^+] = -\log(1.3 \times 10^{-4}) = 3.89$

▶ **PROBLEM 15.17**   Calculate the pH and the concentrations of all species present in 0.10 M H_2SO_3. Values of K_a are in Table 15.3.

▶ **PROBLEM 15.18**   Calculate the pH and the concentrations of all species present in 0.50 M H_2SO_4. (*Hint*: All the H_2SO_4 molecules dissociate to give H_3O^+ and HSO_4^- ions, but less than 100% of the resulting HSO_4^- ions dissociate to give H_3O^+ and SO_4^{2-} ions. The second dissociation step takes place in the presence of H_3O^+ from the first step.) Values of K_a are given in Table 15.3.

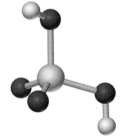

Sulfuric acid

15.12 Equilibria in Solutions of Weak Bases

Weak bases, such as ammonia, accept a proton from water to give the conjugate acid of the base and OH^- ions:

$$NH_3(aq) + H_2O(l) \rightleftharpoons NH_4^+(aq) + OH^-(aq)$$

The equilibrium reaction of any base B with water is characterized by an equilibrium equation similar in form to that for the dissociation of a weak acid. The equilibrium constant K_b is called the **base-dissociation constant**:

$$B(aq) + H_2O(l) \rightleftharpoons BH^+(aq) + OH^-(aq)$$

$$K_b = \frac{[BH^+][OH^-]}{[B]}$$

As usual, $[H_2O]$ is omitted from the equilibrium constant expression. Table 15.4 lists some typical weak bases and gives their K_b values. (The term *base-protonation constant* might be a more descriptive name for K_b, but the term *base-dissociation constant* is still widely used.)

TABLE 15.4 K_b Values for Some Weak Bases and K_a Values for Their Conjugate Acids

Base	Formula, B	K_b	Conjugate Acid, BH^+	K_a
Ammonia	NH_3	1.8×10^{-5}	NH_4^+	5.6×10^{-10}
Aniline	$C_6H_5NH_2$	4.3×10^{-10}	$C_6H_5NH_3^+$	2.3×10^{-5}
Dimethylamine	$(CH_3)_2NH$	5.4×10^{-4}	$(CH_3)_2NH_2^+$	1.9×10^{-11}
Hydrazine	N_2H_4	8.9×10^{-7}	$N_2H_5^+$	1.1×10^{-8}
Hydroxylamine	NH_2OH	9.1×10^{-9}	NH_3OH^+	1.1×10^{-6}
Methylamine	CH_3NH_2	3.7×10^{-4}	$CH_3NH_3^+$	2.7×10^{-11}

Many weak bases are organic compounds called *amines*, derivatives of ammonia in which one or more hydrogen atoms are replaced by another

Ethylamine 3D model

group. Methylamine, for example, is an organic amine responsible for the odor of rotting fish.

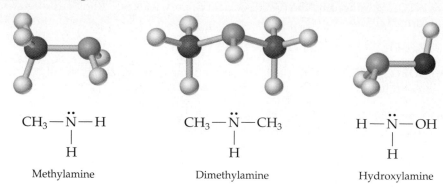

$$CH_3 - \overset{\cdot\cdot}{N} - H$$
$$|$$
$$H$$

Methylamine

$$CH_3 - \overset{\cdot\cdot}{N} - CH_3$$
$$|$$
$$H$$

Dimethylamine

$$H - \overset{\cdot\cdot}{N} - OH$$
$$|$$
$$H$$

Hydroxylamine

Many over-the-counter drugs contain salts formed from amines and hydrochloric acid.

The basicity of an amine is due to the lone pair of electrons on the nitrogen atom, which can be used for bonding to a proton.

Equilibria in solutions of weak bases are treated by the same procedure used for solving problems involving weak acids. Example 15.11 illustrates the procedure.

EXAMPLE 15.11   Codeine ($C_{18}H_{21}NO_3$), a drug used in painkillers and cough medicines, is a naturally occurring amine that has $K_b = 1.6 \times 10^{-6}$. Calculate the pH and the concentrations of all species present in a 0.0012 M solution of codeine.

SOLUTION   Use the procedure outlined in Figure 15.7.

Step 1. Let's use Cod as an abbreviation for codeine and $CodH^+$ for its conjugate acid. The species present initially are Cod (base) and H_2O (acid or base).

Step 2. There are two possible proton-transfer reactions:

$$Cod(aq) + H_2O(l) \rightleftharpoons CodH^+(aq) + OH^-(aq) \qquad K_b = 1.6 \times 10^{-6}$$
$$H_2O(l) + H_2O(l) \rightleftharpoons H_3O^+(aq) + OH^-(aq) \qquad K_w = 1.0 \times 10^{-14}$$

Step 3. Since Cod is a much stronger base than H_2O ($K_b \gg K_w$), the principal reaction involves protonation of codeine.

Step 4.

Principal reaction:	$Cod(aq) + H_2O(l) \rightleftharpoons CodH^+(aq) + OH^-(aq)$		
Initial conc (M)	0.0012	0	~0
Change (M)	$-x$	$+x$	$+x$
Equilibrium conc (M)	$0.0012 - x$	x	x

Step 5. The value of x is obtained from the equilibrium equation:

$$K_b = 1.6 \times 10^{-6} = \frac{[CodH^+][OH^-]}{[Cod]} = \frac{(x)(x)}{(0.0012 - x)}$$

Assuming that $(0.0012 - x) \approx 0.0012$,

$$x^2 = (1.6 \times 10^{-6})(0.0012)$$

$$x = 4.4 \times 10^{-5} \qquad \text{Approximation } (0.0012 - x) \approx 0.0012 \text{ is justified.}$$

Step 6. The big concentrations are

$$[CodH^+] = [OH^-] = x = 4.4 \times 10^{-5}\ M$$
$$[Cod] = 0.0012 - x = 0.0012 - 0.000\,044 = 0.0012\ M$$

Step 7. The small concentration is obtained from the subsidiary equilibrium, the dissociation of water:

$$[H_3O^+] = \frac{K_w}{[OH^-]} = \frac{1.0 \times 10^{-14}}{4.4 \times 10^{-5}} = 2.3 \times 10^{-10}\ M$$

Step 8. $pH = -\log[H_3O^+] = -\log(2.3 \times 10^{-10}) = 9.64$

Note that the pH is greater than 7, as expected for a solution of a weak base.

▶ **PROBLEM 15.19**   Calculate the pH and the concentrations of all species present in 0.40 M NH_3 ($K_b = 1.8 \times 10^{-5}$).

▶ **PROBLEM 15.20**   Strychnine ($C_{21}H_{22}N_2O_2$), a deadly poison used for killing rodents, is a weak base having $K_b = 1.8 \times 10^{-6}$. Calculate the pH of a saturated solution of strychnine (16 mg/100 mL).

15.13   Relation Between K_a and K_b

We've seen in previous sections that the strength of an acid can be expressed by its K_a, and the strength of a base can be expressed by its K_b. For a conjugate acid–base pair, the two equilibrium constants are related in a simple way that makes it possible to calculate either one from the other. Let's consider the conjugate acid–base pair NH_4^+ and NH_3, for example, where K_a refers to proton transfer from the acid NH_4^+ to water, and K_b refers to proton transfer from water to the base NH_3. The sum of the two reactions is simply the dissociation of water:

$$\text{NH}_4^+(aq) + H_2O(l) \rightleftharpoons H_3O^+(aq) + \text{NH}_3(aq) \qquad K_a = \frac{[H_3O^+][NH_3]}{[NH_4^+]} = 5.6 \times 10^{-10}$$

$$\text{NH}_3(aq) + H_2O(l) \rightleftharpoons \text{NH}_4^+(aq) + OH^-(aq) \qquad K_b = \frac{[NH_4^+][OH^-]}{[NH_3]} = 1.8 \times 10^{-5}$$

$$\text{Net:}\quad 2\,H_2O(l) \rightleftharpoons H_3O^+(aq) + OH^-(aq) \qquad K_w = [H_3O^+][OH^-] = 1.0 \times 10^{-14}$$

Notice that the equilibrium constant for the net reaction is equal to the *product* of the equilibrium constants for the reactions added:

$$K_a \times K_b = \frac{[H_3O^+][\text{NH}_3]}{[\text{NH}_4^+]} \times \frac{[\text{NH}_4^+][OH^-]}{[\text{NH}_3]} = [H_3O^+][OH^-] = K_w$$

$$= (5.6 \times 10^{-10})(1.8 \times 10^{-5}) = 1.0 \times 10^{-14}$$

What we've shown in this particular case is true in general. *Whenever chemical equations for two (or more) reactions are added to get the equation for a net reaction, the equilibrium constant for the net reaction is equal to the product of the equilibrium constants for the individual reactions:*

$$K_{net} = K_1 \times K_2 \times \cdots$$

For any conjugate acid–base pair, the product of the acid-dissociation constant for the acid and the base-dissociation constant for the base is always equal to the ion-product constant for water:

$$K_a \times K_b = K_w$$

As the strength of an acid increases (larger K_a), the strength of its conjugate base decreases (smaller K_b) because the product $K_a \times K_b$ remains constant at 1.0×10^{-14}. This inverse relationship between the strength of an acid and the strength of its conjugate base was illustrated qualitatively in Table 15.1.

Compilations of equilibrium constants, such as Appendix C, generally list either K_a or K_b, but not both, because K_a is easily calculated from K_b and vice versa:

$$K_a = \frac{K_w}{K_b} \quad \text{and} \quad K_b = \frac{K_w}{K_a}$$

EXAMPLE 15.12

(a) K_b for trimethylamine is 6.5×10^{-5}. Calculate K_a for the trimethylammonium ion, $(CH_3)_3NH^+$.

(b) K_a for HCN is 4.9×10^{-10}. Calculate K_b for CN^-.

SOLUTION

(a) K_a for $(CH_3)_3NH^+$ is the equilibrium constant for the acid-dissociation reaction

$$(CH_3)_3NH^+(aq) + H_2O(l) \rightleftharpoons H_3O^+(aq) + (CH_3)_3N(aq)$$

Because $K_a = K_w/K_b$, we can find K_a for $(CH_3)_3NH^+$ from K_b for its conjugate base $(CH_3)_3N$:

$$K_a = \frac{K_w}{K_b} = \frac{1.0 \times 10^{-14}}{6.5 \times 10^{-5}} = 1.5 \times 10^{-10}$$

(b) K_b for CN^- is the equilibrium constant for the base-protonation reaction

$$CN^-(aq) + H_2O(l) \rightleftharpoons HCN(aq) + OH^-(aq)$$

Because $K_b = K_w/K_a$, we can find K_b for CN^- from K_a for its conjugate acid HCN:

$$K_b = \frac{K_w}{K_a} = \frac{1.0 \times 10^{-14}}{4.9 \times 10^{-10}} = 2.0 \times 10^{-5}$$

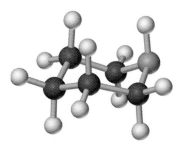

Piperidine

▶ **PROBLEM 15.21**

(a) Piperidine $(C_5H_{11}N)$ is an amine found in black pepper. Find K_b for piperidine in Appendix C, and then calculate K_a for the $C_5H_{11}NH^+$ cation.

(b) Find K_a for HOCl in Appendix C, and then calculate K_b for OCl^-.

15.14 Acid–Base Properties of Salts

When an acid neutralizes a base (Section 4.1), an ionic compound called a *salt* is formed. Salt solutions can be neutral, acidic, or basic, depending on the acid–base properties of the constituent cations and anions (Figure 15.9). As a general rule, salts formed by reaction of a strong acid with a strong base are neutral, salts formed by reaction of a strong acid with a weak base are acidic, and salts formed by reaction of a weak acid with a strong base are basic. It's as if, in an acid–base reaction, the influence of the stronger partner is dominant:

$$\textbf{Strong acid } + \textbf{ Strong base} \longrightarrow \text{Neutral solution}$$

$$\textbf{Strong Acid } + \text{Weak base} \longrightarrow \textbf{Acidic} \text{ solution}$$

$$\text{Weak acid } + \textbf{Strong base} \longrightarrow \textbf{Basic} \text{ solution}$$

FIGURE 15.9 Some 0.10 M aqueous salt solutions (left to right): NaCl, NH_4Cl, $AlCl_3$, NaCN, and $(NH_4)_2CO_3$. A few drops of universal indicator have been added to each solution. The color of the indicator shows that the NaCl solution is neutral, the NH_4Cl and $AlCl_3$ solutions are acidic, and the NaCN and $(NH_4)_2CO_3$ solutions are basic.

Salts that Yield Neutral Solutions

Salts such as NaCl that are derived from a strong base (NaOH) and a strong acid (HCl) yield neutral solutions because neither the cation nor the anion reacts appreciably with water to produce H_3O^+ or OH^- ions. As the conjugate base of a strong acid, Cl^- has no tendency to make the solution basic by picking up a proton from water. As the cation of a strong base, the hydrated Na^+ ion has only a negligible tendency to make the solution acidic by transferring a proton to a solvent water molecule.

The following ions do not react appreciably with water to produce either H_3O^+ or OH^- ions:

1. Cations from strong bases:
     Alkali metal cations of group 1A (Li^+, Na^+, K^+)
     Alkaline earth cations of group 2A (Ca^{2+}, Sr^{2+}, Ba^{2+}), except for Be^{2+}
2. Anions from strong monoprotic acids:
     Cl^-, Br^-, I^-, NO_3^-, and ClO_4^-

Salts that contain only these ions give neutral solutions in pure water (pH = 7). (Usually, water is slightly acidic because of dissolved atmospheric CO_2. To prepare a neutral salt solution, you would first have to remove the dissolved CO_2.)

Salts that Yield Acidic Solutions

Salts such as NH_4Cl that are derived from a weak base (NH_3) and a strong acid (HCl) produce acidic solutions. In such a case, the anion is neither an acid nor a base, but the cation is a weak acid:

$$NH_4^+(aq) + H_2O(l) \rightleftharpoons H_3O^+(aq) + NH_3(aq)$$

Related ammonium salts derived from amines, such as [CH₃NH₃]Cl, [(CH₃)₂NH₂]Cl, and [(CH₃)₃NH]Cl, also give acidic solutions because they too have cations with at least one dissociable proton. The pH of a solution that contains an acidic cation can be calculated by the standard procedure outlined in Figure 15.7. For a 0.10 M NH₄Cl solution, the pH is 5.12. Although the reaction of a cation or anion of a salt with water to produce H_3O^+ or OH^- ions is sometimes called a *salt hydrolysis reaction*, there is no fundamental difference between a salt hydrolysis reaction and any other Brønsted–Lowry acid–base reaction.

Another type of acidic cation is a hydrated cation of a small, highly charged metal ion, such as Al^{3+}. In aqueous solution, the Al^{3+} ion bonds to six water molecules to give the hydrated cation $Al(H_2O)_6^{3+}$. All metal ions exist in aqueous solution as hydrated cations, but their acidity varies greatly depending on the charge and size of the unhydrated metal ion. Because of the high (3+) charge on the Al^{3+} ion, electrons in the O—H bonds of the bound water molecules are attracted toward the Al^{3+} ion. The attraction is strong because the Al^{3+} ion is small and the electrons in the O—H bonds are relatively close to the center of positive charge. As a result, electron density shifts from the O—H bonds toward the Al^{3+} ion, thus weakening the O—H bonds and increasing their polarity, which in turn eases the transfer of a proton to a solvent water molecule:

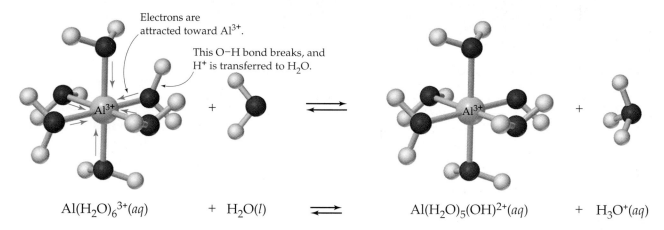

Electrons are attracted toward Al^{3+}.

This O—H bond breaks, and H^+ is transferred to H_2O.

$$Al(H_2O)_6^{3+}(aq) \quad + \quad H_2O(l) \quad \rightleftharpoons \quad Al(H_2O)_5(OH)^{2+}(aq) \quad + \quad H_3O^+(aq)$$

The acid-dissociation constant for $Al(H_2O)_6^{3+}$, $K_a = 1.4 \times 10^{-5}$, is much larger than $K_w = 1.0 \times 10^{-14}$, which means that the water molecules in the hydrated cation are much stronger proton donors than are free solvent water molecules. In fact, the acid strength of $Al(H_2O)_6^{3+}$ is comparable to that of acetic acid, which has $K_a = 1.8 \times 10^{-5}$. In general, the acidity of hydrated main-group cations increases from left to right in the periodic table as the metal ion charge increases and the metal ion size decreases ($Li^+ < Be^{2+}$; $Na^+ < Mg^{2+} < Al^{3+}$). Transition metal cations, such as Zn^{2+}, Cr^{3+}, and Fe^{3+}, also give acidic solutions; their K_a values are listed in Table C.2 of Appendix C.

EXAMPLE 15.13 Calculate the pH of a 0.10 M solution of AlCl₃; K_a for $Al(H_2O)_6^{3+}$ is 1.4×10^{-5}.

SOLUTION Because this problem is similar to others done earlier, we'll abbreviate the procedure in Figure 15.7.

Steps 1–4. The species present initially are $Al(H_2O)_6^{3+}$ (acid), Cl^- (inert), and H_2O (acid or base). Because $Al(H_2O)_6^{3+}$ is a much stronger acid than water ($K_a \gg K_w$), the principal reaction is dissociation of $Al(H_2O)_6^{3+}$:

Principal reaction:	$Al(H_2O)_6{}^{3+}(aq)$	$+ H_2O(l)$	$\rightleftharpoons H_3O^+(aq) +$	$Al(H_2O)_5(OH)^{2+}(aq)$
Equilibrium conc (M)	$0.10 - x$		x	x

Step 5. The value of x is obtained from the equilibrium equation:

$$K_a = 1.4 \times 10^{-5} = \frac{[H_3O^+][Al(H_2O)_5(OH)^{2+}]}{[Al(H_2O)_6{}^{3+}]} = \frac{(x)(x)}{(0.10 - x)} \approx \frac{x^2}{0.10}$$

$$x = [H_3O^+] = 1.2 \times 10^{-3} \text{ M}$$

Step 8. $\text{pH} = -\log(1.2 \times 10^{-3}) = 2.92$

Note that $Al(H_2O)_6{}^{3+}$ is a much stronger acid than $NH_4{}^+$, in accord with the colors of the indicator in Figure 15.9.

▶ **PROBLEM 15.22** Predict whether the following salt solutions are neutral or acidic, and calculate the pH of each:
(a) 0.25 M NH_4Br **(b)** 0.40 M $ZnCl_2$; K_a for $Zn(H_2O)_6{}^{2+}$ is 2.5×10^{-10}.

Salts that Yield Basic Solutions

Salts such as NaCN that are derived from a strong base (NaOH) and a weak acid (HCN) yield basic solutions. In this case, the cation is neither an acid nor a base, but the anion is a weak base:

$$CN^-(aq) + H_2O(l) \rightleftharpoons HCN(aq) + OH^-(aq)$$

Other anions that exhibit basic properties are listed in Table 15.1 and include $NO_2{}^-$, F^-, $CH_3CO_2{}^-$, and $CO_3{}^{2-}$. The pH of a basic salt solution can be calculated by the standard procedure, as shown in Example 15.14.

EXAMPLE 15.14 Calculate the pH of a 0.10 M solution of NaCN; K_a for HCN is 4.9×10^{-10}.

SOLUTION

Step 1. The species present initially are Na^+ (inert), CN^- (base), and H_2O (acid or base).

Step 2. There are two possible proton-transfer reactions:

$$CN^-(aq) + H_2O(l) \rightleftharpoons HCN(aq) + OH^-(aq) \qquad K_b$$
$$H_2O(l) + H_2O(l) \rightleftharpoons H_3O^+(aq) + OH^-(aq) \qquad K_w$$

Step 3. As shown in Example 15.12b, $K_b = K_w/(K_a \text{ for HCN}) = 2.0 \times 10^{-5}$. Because $K_b \gg K_w$, CN^- is a stronger base than H_2O, and the principal reaction is proton transfer from H_2O to CN^-.

Step 4.

Principal reaction:	$CN^-(aq) + H_2O(l)$	$\rightleftharpoons HCN(aq) +$	$OH^-(aq)$
Equilibrium conc (M)	$0.10 - x$	x	x

Step 5. The value of x is obtained from the equilibrium equation:

$$K_b = 2.0 \times 10^{-5} = \frac{[HCN][OH^-]}{[CN^-]} = \frac{(x)(x)}{(0.10 - x)} \approx \frac{x^2}{0.10}$$

$$x = [OH^-] = 1.4 \times 10^{-3} \, M$$

Step 7.

$$[H_3O^+] = \frac{K_w}{[OH^-]} = \frac{1.0 \times 10^{-14}}{1.4 \times 10^{-3}} = 7.1 \times 10^{-12}$$

Step 8. $pH = -\log(7.1 \times 10^{-12}) = 11.15$

In accord with the color of the indicator in Figure 15.9, the solution is basic.

▶ **PROBLEM 15.23** Calculate the pH of 0.20 M $NaNO_2$; K_a for HNO_2 is 4.6×10^{-4}.

Salts that Contain Acidic Cations and Basic Anions

Finally, let's look at a salt such as $(NH_4)_2CO_3$ in which both the cation and the anion can undergo proton-transfer reactions. Because NH_4^+ is a weak acid and CO_3^{2-} is a weak base, the pH of an $(NH_4)_2CO_3$ solution depends on the relative acid strength of the cation and base strength of the anion:

$$NH_4^+(aq) + H_2O(l) \rightleftharpoons H_3O^+(aq) + NH_3(aq) \qquad \text{Acid strength } (K_a)$$

$$CO_3^{2-}(aq) + H_2O(l) \rightleftharpoons HCO_3^-(aq) + OH^-(aq) \qquad \text{Base strength } (K_b)$$

We can distinguish three possible cases:

(a) $K_a > K_b$. If K_a for the cation is greater than K_b for the anion, the solution will contain an excess of H_3O^+ ions (pH < 7).

(b) $K_a < K_b$. If K_a for the cation is less than K_b for the anion, the solution will contain an excess of OH^- ions (pH > 7).

(c) $K_a \approx K_b$. If K_a for the cation and K_b for the anion are comparable, the solution will contain pproximately equal concentrations of H_3O^+ and OH^- ions (pH ≈ 7).

To determine whether an $(NH_4)_2CO_3$ solution is acidic, basic, or neutral, let's work out the values of K_a for NH_4^+ and K_b for CO_3^{2-}:

$$K_a \text{ for } NH_4^+ = \frac{K_w}{K_b \text{ for } NH_3} = \frac{1.0 \times 10^{-14}}{1.8 \times 10^{-5}} = 5.6 \times 10^{-10}$$

$$K_b \text{ for } CO_3^{2-} = \frac{K_w}{K_a \text{ for } HCO_3^-} = \frac{K_w}{K_{a2} \text{ for } H_2CO_3} = \frac{1.0 \times 10^{-14}}{5.6 \times 10^{-11}} = 1.8 \times 10^{-4}$$

Because $K_a < K_b$, the solution is basic (pH > 7), in accord with the color of the indicator in Figure 15.9.

A summary of the acid–base properties of salts is given in Table 15.5.

TABLE 15.5 Acid–Base Properties of Salts

Type of Salt	Examples	Ions That React with Water	pH of Solution
Cation from strong base; anion from strong acid	$NaCl$, KNO_3, BaI_2	None	~ 7
Cation from weak base; anion from strong acid	NH_4Cl, NH_4NO_3, $[(CH_3)_3NH]Cl$	Cation	< 7
Small, highly charged cation; anion from strong acid	$AlCl_3$, $Cr(NO_3)_3$, $Fe(ClO_4)_3$	Hydrated cation	< 7
Cation from strong base; anion from weak acid	$NaCN$, KF, Na_2CO_3	Anion	> 7
Cation from weak base; anion from weak acid	NH_4CN, NH_4F, $(NH_4)_2CO_3$	Cation and anion	< 7 if $K_a > K_b$ > 7 if $K_a < K_b$ ~ 7 if $K_a \approx K_b$

▶ **PROBLEM 15.24** Calculate K_a for the cation and K_b for the anion in an aqueous NH_4CN solution. Is the solution acidic, basic, or neutral?

▶ **PROBLEM 15.25** Classify each of the following salt solutions as acidic, basic, or neutral:
(a) KBr **(b)** $NaNO_2$ **(c)** NH_4Br **(d)** $ZnCl_2$ **(e)** NH_4F

15.15 Factors That Affect Acid Strength

Why is one acid stronger than another? Although a complete analysis of the factors that determine the strength of an acid is a complex business, the extent of dissociation of an acid HA is often determined by the strength and polarity of the H–A bond. The strength of the H–A bond, as we saw in Section 8.11, is the enthalpy required to dissociate HA into an H atom and an A atom. The polarity of the H–A bond increases with an increase in the electronegativity of A and is related to the ease of electron transfer from an H atom to an A atom to give an H^+ cation and an A^- anion. In general, the weaker the H–A bond, the stronger the acid, and the more polar the H–A bond, the stronger the acid.

Let's look first at the hydrohalic acids HF, HCl, HBr, and HI. Electrostatic potential maps (Section 7.4) show that all these molecules are polar, with the halogen atom being electron rich (red) and the H atom being electron poor (blue).

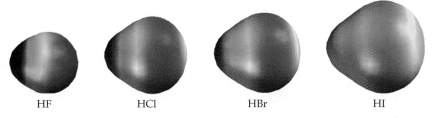

HF   HCl   HBr   HI

The variation in polarity in this series, however, is much less important than the variation in bond strength, which decreases markedly from 567 kJ/mol for HF to 299 kJ/mol for HI.

In general, for binary acids of elements in the same *group* of the periodic table, the H–A bond strength is the most important determinant of acidity.

The H−A bond strength generally decreases with increasing size of element A down a group, so acidity increases. For HA (A = F, Cl, Br, or I), for example, the size of A increases from F to I, so bond strength decreases and acidity increases from HF to HI. Hydrofluoric acid is a weak acid ($K_a = 3.5 \times 10^{-4}$), whereas HCl, HBr, and HI are strong acids.

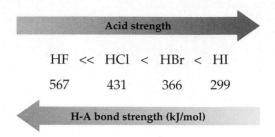

$$\text{HF} \ll \text{HCl} < \text{HBr} < \text{HI}$$

| 567 | 431 | 366 | 299 |

As a further example of this effect, H₂S ($K_{a1} = 1.0 \times 10^{-7}$) is a stronger acid than H_2O.

For binary acids of elements in the same *row* of the periodic table, changes in the H−A bond strength are smaller, and the polarity of the H−A bond is the most important determinant of acid strength. The strengths of binary acids of the second-row elements, for example, increase with increasing electronegativity of A:

Acid strength

$$\text{CH}_4 < \text{NH}_3 \ll \text{H}_2\text{O} < \text{HF}$$

| 2.5 | 3.0 | 3.5 | 4.0 |

Electronegativity of A

The C−H bond is relatively nonpolar, and methane has no tendency to dissociate in water into H_3O^+ and CH_3^- ions. The N−H bond is more polar, but dissociation of NH_3 into H_3O^+ and NH_2^- ions is still negligibly small. Water and hydrofluoric acid, however, are increasingly stronger acids. Periodic trends in the strength of binary acids are summarized in Figure 15.10.

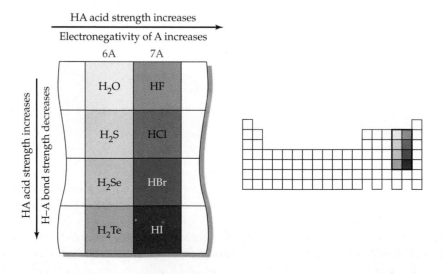

FIGURE 15.10 The acid strength of a binary acid HA increases from left to right in the periodic table, with increasing electronegativity of A , and from top to bottom, with decreasing H−A bond strength.

Another familiar class of acids consists of oxoacids, such as H_2CO_3, HNO_3, H_2SO_4, and $HClO$. These compounds have the general formula H_nYO_m, where Y is a nonmetallic atom, such as C, N, S, or Cl, and n and m are integers. The atom Y is always bonded to one or more hydroxyl (OH) groups and can be bonded, in addition, to one or more oxygen atoms:

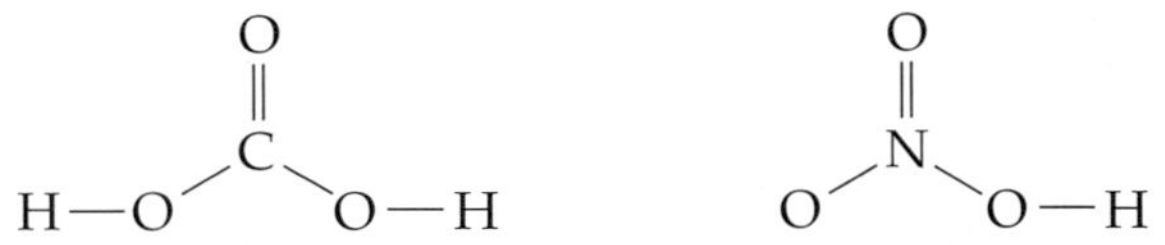

Carbonic acid

Nitric acid

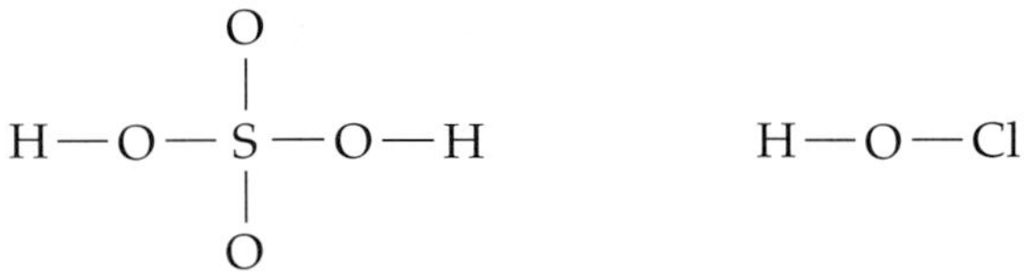

Sulfuric acid

Hypochlorous acid

Because dissociation of an oxoacid involves breaking an O—H bond, any factor that weakens the O—H bond or increases its polarity increases the strength of the acid. Two such factors are the electronegativity of Y and the oxidation number of Y in the general reaction

$$-\overset{|}{\underset{|}{Y}}-O-H \; + \; H_2O \; \rightleftharpoons \; H_3O^+ \; + \; -\overset{|}{\underset{|}{Y}}-O^-$$

1. For oxoacids that contain the same number of OH groups and the same number of O atoms, acid strength increases with increasing electronegativity of Y. For example, the acid strength of the hypohalous acids HOY (Y = Cl, Br, or I) increases with increasing electronegativity of the halogen:

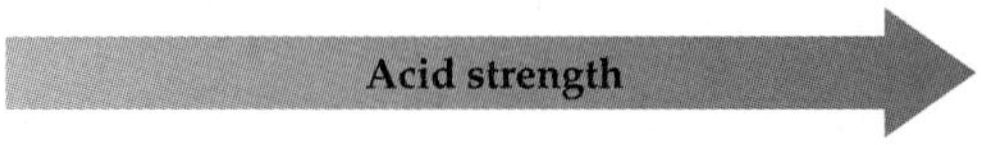

Acid strength

	H—O—I	<	H—O—Br	<	H—O—Cl
Dissociation constant, K_a	2.3×10^{-11}		2.0×10^{-9}		3.5×10^{-8}
Electronegativity	2.5		2.8		3.0

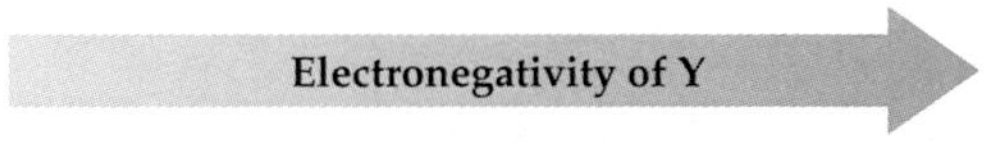

Electronegativity of Y

As the halogen becomes more electronegative, an increasing amount of electron density shifts from the O—H bond toward the halogen, thus weakening the O—H bond and increasing its polarity. As a result, the proton is more easily transferred to a solvent water molecule, and so the acid strength increases.

2. For oxoacids that contain the same atom Y but different numbers of oxygen atoms, acid strength increases with increasing oxidation number

of Y, which increases, in turn, with an increasing number of oxygen atoms. This effect is illustrated by the oxoacids of chlorine:

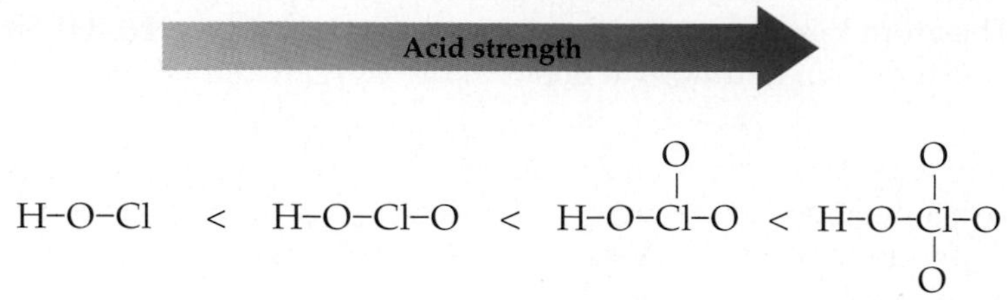

Acid strength

$$H-O-Cl \quad < \quad H-O-Cl-O \quad < \quad H-O-\overset{\overset{\displaystyle O}{|}}{Cl}-O \quad < \quad H-O-\overset{\overset{\displaystyle O}{|}}{\underset{\underset{\displaystyle O}{|}}{Cl}}-O$$

Name of acid	Hypochlorous	Chlorous	Chloric	Perchloric
Dissociation constant, K_a	3.5×10^{-8}	1.2×10^{-2}	~1	Very large
Oxidation number of Cl	+1	+3	+5	+7

Oxidation number of Cl

As the number of O atoms in $HClO_m$ increases, an increasing amount of electron density shifts from the Cl atom toward the more electronegative O atoms. The amount of positive charge on the Cl atom therefore increases as its oxidation number increases. The increased positive charge on the Cl atom in turn attracts an increasing amount of electron density from the O–H bond, thus weakening the O–H bond and increasing its polarity. [Recall that this same charge effect was discussed in the previous section for $Al(H_2O)_6{}^{3+}$.] As a result, the proton is more easily transferred to a solvent water molecule.

Another factor that affects the acid strength of oxoacids is the relative stability of the corresponding oxoanions. The $ClO_m{}^-$ anion becomes more stable as the number of O atoms increases in the series $ClO^- < ClO_2{}^- < ClO_3{}^- < ClO_4{}^-$ because a larger number of electronegative O atoms can better accommodate the anion's negative charge. As the stability of the anion increases, the corresponding acid has a greater tendency to dissociate. The increase in acid strength with increasing number of O atoms is further illustrated by the oxoacids of sulfur: H_2SO_4 is a stronger acid than H_2SO_3.

▶ **PROBLEM 15.26** Identify the stronger acid in each of the following pairs:
(a) H_2S or H_2Se (b) HI or H_2Te
(c) HNO_2 or HNO_3 (d) H_2SO_3 or H_2SeO_3

15.16 Lewis Acids and Bases

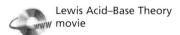
Lewis Acid–Base Theory
www movie

In 1923, the same year in which Brønsted and Lowry defined acids and bases in terms of their proton donor/acceptor properties, the American chemist G. N. Lewis proposed an even more general concept of acids and bases. Lewis noticed that when a base accepts a proton, it shares a lone pair of electrons with the proton to form a new covalent bond. Using ammonia as an example, the reaction can be written in the following format, in which the curved arrow represents donation of the nitrogen lone pair to form a bond with H^+:

$$
H^+ \; + \; :N-H \longrightarrow \left[H-\underset{\underset{H}{|}}{\overset{\overset{H}{|}}{N}}-H \right]^+
$$

In this reaction, the proton behaves as an electron-pair acceptor, and the ammonia molecule behaves as an electron-pair donor. Consequently, the Lewis definition of acids and bases states that a *Lewis acid* is an electron-pair acceptor, and a *Lewis base* is an electron-pair donor.

>— **LEWIS ACID**   An electron-pair acceptor

>— **LEWIS BASE**   An electron-pair donor

Since all proton acceptors have an unshared pair of electrons, and since all electron-pair donors can accept a proton, the Lewis and the Brønsted–Lowry definitions of a base are simply different ways of looking at the same property. All Lewis bases are Brønsted–Lowry bases, and all Brønsted–Lowry bases are Lewis bases. The Lewis definition of an acid, however, is considerably more general than the Brønsted–Lowry definition. Lewis acids include not only H^+ but also other cations and neutral molecules having vacant valence orbitals that can accept a share in a pair of electrons donated by a Lewis base.

Common examples of cationic Lewis acids are metal ions, such as Al^{3+} and Cu^{2+}. Hydration of the Al^{3+} ion, for example, is a Lewis acid–base reaction in which each of six H_2O molecules donates a pair of electrons to Al^{3+} to form the hydrated cation $Al(H_2O)_6{}^{3+}$:

$$
Al^{3+} \; + \; 6 \; :\!\ddot{O}\!-H \longrightarrow \left[Al \left(\ddot{O}\!-H \right)_6 \right]^{3+}
$$

Lewis acid   Lewis base

Similarly, the reaction of Cu^{2+} ion with ammonia is a Lewis acid–base reaction in which each of four NH_3 molecules donates a pair of electrons to Cu^{2+} to form the deep blue ion $Cu(NH_3)_4{}^{2+}$ (Figure 15.11).

$$
Cu^{2+} \; + \; 4 \; :NH_3 \longrightarrow Cu(NH_3)_4{}^{2+}
$$

Lewis acid   Lewis base

FIGURE 15.11 Addition of aqueous ammonia to a solution of the light blue $Cu^{2+}(aq)$ ion (left) gives a light blue precipitate of $Cu(OH)_2$ (center). Addition of an excess of ammonia yields the deep blue ion $Cu(NH_3)_4{}^{2+}$ (right).

Examples of neutral Lewis acids are halides of group 3A elements, such as BF_3. Boron trifluoride, a colorless gas, is an excellent Lewis acid because the boron atom in the trigonal planar BF_3 molecule is surrounded by only six valence electrons (Figure 15.12). The boron atom uses three sp^2 hybrid orbitals to bond to the three F atoms and has a vacant $2p$ valence orbital that can accept a share in a pair of electrons from a Lewis base, such as NH_3. The Lewis acid and base sites are evident in electrostatic potential maps, which show the electron poor B atom (blue) and the electron rich N atom (red):

Lewis acid   Lewis base   Acid–base adduct

In the product, called an *acid–base adduct*, the boron atom has acquired the usual stable octet of electrons.

(a)   (b)   (c)

FIGURE 15.12 (a) Electron-dot structure of BF_3. (b) Trigonal planar structure of BF_3 and vacant $2p$ orbital perpendicular to the molecular plane. (c) Structure of the F_3B–NH_3 adduct. The geometry about boron changes from trigonal planar in BF_3 to tetrahedral in the adduct. Both boron and nitrogen use sp^3 hybrid orbitals in the adduct.

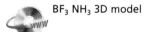 BF$_3$ NH$_3$ 3D model

Additional examples of neutral Lewis acids are oxides of nonmetals, such as CO_2, SO_2, and SO_3. The reaction of SO_3 with water, for example, can be viewed as a Lewis acid–base reaction in which SO_3 accepts a lone pair of electrons from a water molecule:

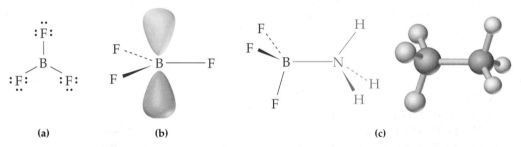

Lewis acid   Lewis base   Sulfuric acid

Because the S=O bond is polar, with a partial positive charge (δ+) on the less electronegative S atom, the S atom attracts an electron pair from H_2O. Formation of a bond from the water O atom to the S atom in the first step is helped along by a shift of a shared pair of electrons to oxygen. In the second step, a proton shifts from one oxygen atom to another, yielding sulfuric acid (H_2SO_4).

EXAMPLE 15.15 For each of the following reactions, identify the Lewis acid and the Lewis base.
(a) $CO_2 + OH^- \rightarrow HCO_3^-$
(b) $B(OH)_3 + OH^- \rightarrow B(OH)_4^-$
(c) $6\,CN^- + Fe^{3+} \rightarrow Fe(CN)_6^{3-}$

SOLUTION
(a) The carbon atom of O=C=O bears a partial positive charge (δ+) because oxygen is more electronegative than carbon. Therefore, the carbon atom attracts an electron pair from OH^-. Formation of a covalent bond from OH^- to CO_2 is helped along by a shift of a shared electron pair to oxygen:

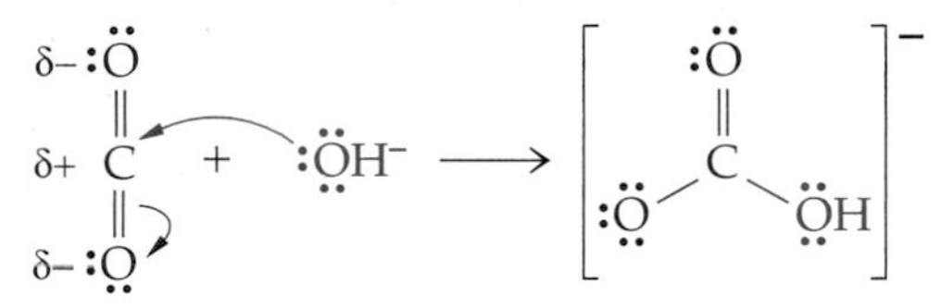

The Lewis acid (electron-pair acceptor) is CO_2; the Lewis base (electron-pair donor) is OH^-.

(b) The Lewis acid is boric acid, $B(OH)_3$, a weak acid and mild antiseptic used in eyewash. The boron atom completes its octet by accepting a pair of electrons from the Lewis base, OH^-.

(c) The Lewis acid is Fe^{3+}, and the Lewis base is CN^-.

▶ **PROBLEM 15.27** For each of the following reactions, identify the Lewis acid and the Lewis base.
(a) $AlCl_3 + Cl^- \rightarrow AlCl_4^-$ (b) $2\,NH_3 + Ag^+ \rightarrow Ag(NH_3)_2^+$
(c) $SO_2 + OH^- \rightarrow HSO_3^-$ (d) $6\,H_2O + Cr^{3+} \rightarrow Cr(OH_2)_6^{3+}$

◀ **KEY CONCEPT PROBLEM 15.28** For the following Lewis acid-base reaction, draw electron dot structures for the reactants and products, and use the curved arrow notation to represent the donation of a lone pair of electrons from the Lewis base to the Lewis acid.

$$BeCl_2 + 2\,Cl^- \longrightarrow BeCl_4^{2-}$$

Acid Rain

Plants that burn sulfur-containing coal and oil release large quantities of sulfur oxides into the atmosphere, ultimately leading to acid rain.

The problem of acid rain has emerged as one of the more important environmental issues of recent times. Both the causes and the effects of acid rain are well understood. The problem is what to do about it.

As the water that has evaporated from oceans and lakes condenses into raindrops, it dissolves small quantities of gases from the atmosphere. Under normal conditions, rain is slightly acidic, with a pH close to 5.6, because of dissolved CO_2. In recent decades, however, the acidity of rainwater in many industrialized areas of the world has increased by a factor of over 100, to a pH between 3 and 3.5.

The primary cause of acid rain is industrial and automotive pollution. Each year in industrialized countries, large power plants and smelters that burn sulfur-containing fossil fuels pour millions of tons of sulfur dioxide (SO_2) gas into the atmosphere, where some is oxidized by air to produce sulfur trioxide (SO_3). Sulfur oxides then dissolve in rain to form dilute sulfurous acid and sulfuric acid:

$$SO_2(g) + H_2O(l) \longrightarrow H_2SO_3(aq) \qquad \text{Sulfurous acid}$$

$$SO_3(g) + H_2O(l) \longrightarrow H_2SO_4(aq) \qquad \text{Sulfuric acid}$$

Nitrogen oxides produced by the high-temperature reaction of N_2 with O_2 in coal-burning plants and in automobile engines further contribute to the problem. Nitrogen dioxide (NO_2) dissolves in water to form dilute nitric acid (HNO_3) and nitric oxide (NO):

$$3 NO_2(g) + H_2O(l) \longrightarrow 2 HNO_3(aq) + NO(g)$$

Oxides of both sulfur and nitrogen have always been present in the atmosphere, produced by such natural sources as volcanoes and lightning bolts, but their amounts have increased dramatically over the last century because of industrialization.

Many processes in nature require such a fine pH balance that they are dramatically upset by the shift that has occurred in the pH of rain. Thousands of lakes in the Adirondack region of upper New York State and in southeastern Canada have become so acidic that all fish life has disappeared. Massive tree die-offs have occurred throughout central and eastern Europe as acid rain has lowered the pH of the soil and has leached nutrients from leaves. Countless marble statues have been slowly dissolved away as their calcium carbonate has been attacked by acid rain.

$$CaCO_3(s) + 2 H^+(aq) \longrightarrow Ca^{2+}(aq) + H_2O(l) + CO_2(g)$$

Fortunately, acidic emissions from automobiles and power plants have been greatly reduced in recent years. Nitrogen oxide emissions have been lowered by equipping automobiles with catalytic converters (Section 12.12), which catalyze the decomposition of nitrogen oxides to N_2 and O_2. Sulfur dioxide emissions from power plants have been reduced by *scrubbing* combustion products before they are emitted from plant smoke stacks. The process involves addition of an aqueous suspension of lime (CaO) to the combustion chamber and the stack. The lime reacts with SO_2 to give calcium sulfite ($CaSO_3$):

$$CaO(s) + SO_2(g) \longrightarrow CaSO_3(s)$$

Unfortunately, scrubbers are expensive, and the $CaSO_3$, which has no commercial uses, must be disposed of in land fills. Much more work on methods to control acidic emissions remains to be done because the problem will grow more serious as sources of low-sulfur coal are exhausted and power plants are forced to rely on more abundant sources of high-sulfur coal.

▶ **PROBLEM 15.29** The reaction of lime (CaO) with SO_2 in the scrubber of a power plant can be regarded as a Lewis acid-base reaction. Explain.

▶ **PROBLEM 15.30** What is the pH of 1.00 L of rainwater that has dissolved 5.47 mg of NO_2? Assume that all of the NO_2 has reacted with water to give nitric acid.

A marble statue is being slowly dissolved by reaction of calcium carbonate with acid rain (top). A researcher examines tree branches damaged by acid rain on Mount Mitchell in North Carolina (bottom).

Key Words

Summary

According to the Arrhenius theory, acids (HA) are substances that dissociate in water to produce $H^+(aq)$. Bases (MOH) are substances that dissociate to yield $OH^-(aq)$. The more general **Brønsted–Lowry theory** defines an acid as a proton donor, a base as a proton acceptor, and an acid–base reaction as a proton-transfer reaction. Examples of Brønsted–Lowry acids are HCl, NH_4^+, and HSO_4^-; examples of Brønsted–Lowry bases are OH^-, F^-, and NH_3.

A **strong acid** HA is nearly 100% dissociated, whereas a **weak acid** HA is only partially dissociated, existing as an equilibrium mixture of HA, H_3O^+, and A^-. The extent of dissociation of a weak acid HA is measured by the **acid-dissociation constant, K_a**:

$$HA(aq) + H_2O(l) \rightleftharpoons H_3O^+(aq) + A^-(aq) \qquad K_a = \frac{[H_3O^+][A^-]}{[HA]}$$

The strength of an acid (HA) and the strength of its **conjugate base** (A^-) are inversely related. Strong acids, such as HCl, have very weak conjugate bases (Cl^-). Strong bases, such as OH^-, have very weak conjugate acids (H_2O). The H_3O^+ ion, a hydrated proton, is called the **hydronium ion**.

Water, which can act both as an acid and as a base, undergoes the **dissociation** reaction $H_2O + H_2O \rightleftharpoons H_3O^+ + OH^-$. In pure water at 25°C, $[H_3O^+] = [OH^-] = 1.0 \times 10^{-7}$ M. The **ion-product constant for water, K_w**, is given by $K_w = [H_3O^+][OH^-] = 1.0 \times 10^{-14}$. The acidity of an aqueous solution is expressed on the **pH** scale, where $pH = -\log[H_3O^+]$. Acidic solutions have $pH < 7$, basic solutions have $pH > 7$, and neutral solutions have $pH = 7$. The pH of a solution can be determined with an **acid–base indicator** or a pH meter.

Polyprotic acids contain more than one dissociable proton and dissociate in a stepwise manner. Because the stepwise dissociation constants decrease in the order $K_{a1} \gg K_{a2} \gg K_{a3}$, nearly all the H_3O^+ in a polyprotic acid solution comes from the first dissociation step.

The extent of dissociation of a weak base B is measured by the **base-dissociation constant, K_b**:

$$B(aq) + H_2O(l) \rightleftharpoons BH^+(aq) + OH^-(aq) \qquad K_b = \frac{[BH^+][OH^-]}{[B]}$$

Examples of weak bases are NH_3 and derivatives of NH_3 called *amines*. For any conjugate acid–base pair, (K_a for the acid) $\times$ (K_b for the base) = K_w.

Aqueous solutions of salts can be neutral, acidic, or basic, depending on the acid–base properties of the constituent ions. Group 1A and 2A cations (except Be^{2+}) and anions that are conjugate bases of strong acids, such as Cl^-, do not react appreciably with water to produce H_3O^+ or OH^- ions. Cations that are conjugate acids of weak bases, such as NH_4^+, and hydrated cations of small, highly charged metal ions, such as Al^{3+}, yield acidic solutions, whereas anions that are conjugate bases of weak acids, such as CN^-, yield basic solutions.

The acid strength of a binary acid HA increases with decreasing strength of the H–A bond and increasing polarity of the H–A bond. The acid strength of an oxoacid, H_nYO_m (Y = C, N, S, Cl), increases with increasing electronegativity and increasing oxidation number of the atom Y.

A **Lewis acid** is an electron-pair acceptor and a **Lewis base** is an electron-pair donor. Lewis acids include not only H^+ but also other cations and neutral molecules that can accept a share in a pair of electrons from a Lewis base. Examples of Lewis acids are Al^{3+}, Cu^{2+}, BF_3, SO_3, and CO_2.

Key Concept Summary

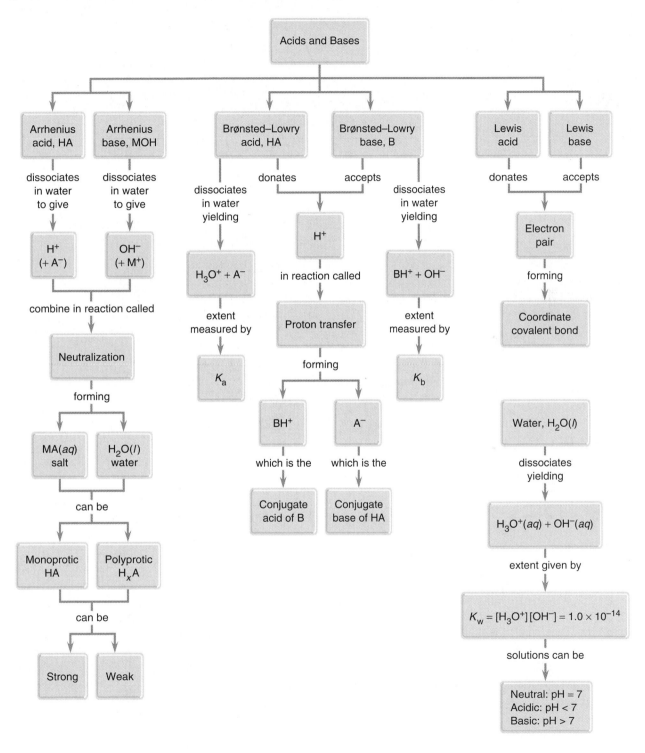

Acids and Bases

Arrhenius acid, HA — dissociates in water to give → H⁺ (+ A⁻)

Arrhenius base, MOH — dissociates in water to give → OH⁻ (+ M⁺)

H⁺ (+ A⁻) and OH⁻ (+ M⁺) combine in reaction called → **Neutralization** — forming → **MA(aq) salt** and **H₂O(l) water** — can be → **Monoprotic HA** and **Polyprotic HₓA** — can be → **Strong** and **Weak**

Brønsted–Lowry acid, HA — donates → H⁺; dissociates in water yielding → $H_3O^+ + A^-$ — extent measured by → K_a

Brønsted–Lowry base, B — accepts → H⁺; dissociates in water yielding → $BH^+ + OH^-$ — extent measured by → K_b

H⁺ in reaction called → **Proton transfer** — forming → **BH⁺** which is the → **Conjugate acid of B**; and **A⁻** which is the → **Conjugate base of HA**

Lewis acid — donates → Electron pair
Lewis base — accepts → Electron pair
— forming → **Coordinate covalent bond**

Water, H₂O(l) — dissociates yielding → $H_3O^+(aq) + OH^-(aq)$ — extent given by → $K_w = [H_3O^+][OH^-] = 1.0 \times 10^{-14}$ — solutions can be →

Neutral: pH = 7
Acidic: pH < 7
Basic: pH > 7

Understanding Key Concepts

Problems 15.1–15.30 appear within the chapter.

15.31 For each of the following reactions, identify the Brøn-sted–Lowry acids and bases:

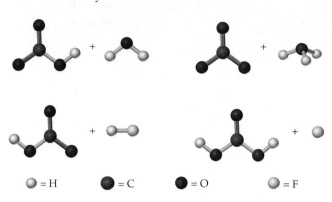

  = H   = C   = O   = F

15.32 The following pictures represent aqueous solutions of three acids HA (A = X, Y, or Z); water molecules have been omitted for clarity:

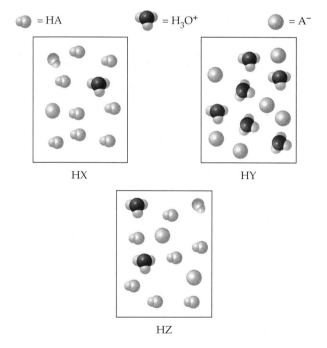

= HA   = H₃O⁺   = A⁻

HX   HY

HZ

(a) What is the conjugate base of each acid?
(b) Arrange the three acids in order of increasing acid strength.
(c) Which acid, if any, is a strong acid?
(d) Which acid has the smallest value of K_a?
(e) What is the percent dissociation in the solution of HZ?

15.33 Which of the following pictures represents a solution of a weak diprotic acid H_2A? (Water molecules are omitted for clarity.) Which pictures represent an impossible situation? Explain.

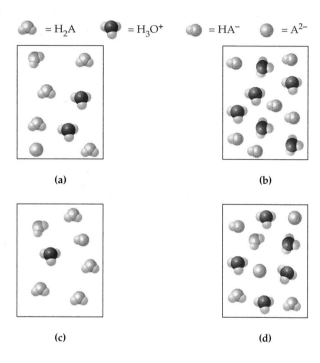

  = H_2A = H_3O^+ = HA^- = A^{2-}

(a)   (b)

(c)   (d)

15.34 Which of the following pictures best represents an aqueous solution of sulfuric acid? Explain. (Water molecules have been omitted for clarity.)

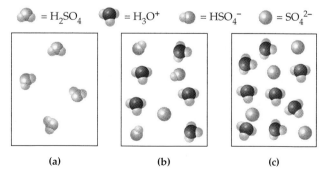

  = H_2SO_4 = H_3O^+ = HSO_4^- = SO_4^{2-}

(a)   (b)   (c)

15.35 The following pictures represent aqueous solutions of three acids HA (A = X, Y, or Z); water molecules have been omitted for clarity:

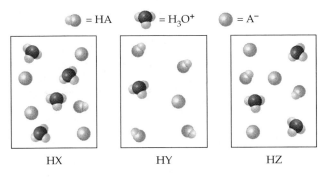

(a) Which conjugate base (A⁻ = X⁻, Y⁻, or Z⁻) has the largest value of K_b?

(b) Which A⁻ ion is the weakest base?

15.36 The following pictures represent solutions of three salts NaA (A⁻ = X⁻, Y⁻, or Z⁻); water molecules and Na⁺ ions have been omitted for clarity:

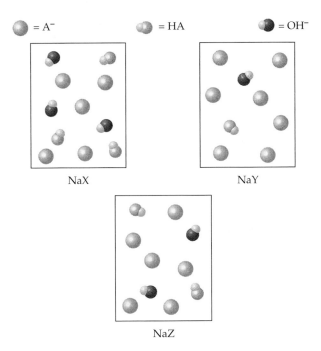

(a) Arrange the three A⁻ anions in order of increasing base strength.

(b) Which A⁻ anion has the strongest conjugate acid?

(c) Why does each box contain the same number of HA molecules and OH⁻ anions?

15.37 Locate sulfur, selenium, chlorine, and bromine in the periodic table:

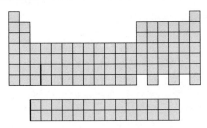

(a) Which binary acid (H_2S, H_2Se, HCl, or HBr) is the strongest? Which is the weakest? Explain.

(b) Which oxoacid (H_2SO_3, H_2SeO_3, $HClO_3$, or $HBrO_3$) is the strongest? Which is the weakest? Explain.

15.38 Look at the electron-dot structures of the following molecules and ions:

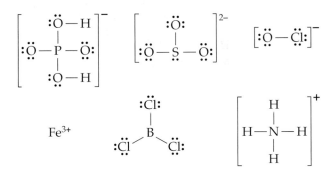

(a) Which of these molecules and ions can behave as a Brønsted–Lowry acid? Which can behave as a Brønsted–Lowry base?

(b) Which can behave as a Lewis acid? Which can behave as a Lewis base?

15.39 For each of the following Lewis acid–base reactions, draw electron-dot structures for the reactants and products, and use the curved arrow notation (Section 15.16) to represent the donation of a lone pair of electrons from the Lewis base to the Lewis acid.

(a) $NH_3 + B(CH_3)_3 \rightarrow H_3NB(CH_3)_3$

(b) $AlCl_3 + Cl^- \rightarrow AlCl_4^-$

(c) $F^- + PF_5 \rightarrow PF_6^-$

(d) $H_2O + SO_2 \rightarrow H_2SO_3$

Additional Problems

Acid–Base Concepts

15.40 Give three examples of molecules or ions that are Brønsted–Lowry bases but not Arrhenius bases.

15.41 Give an example of an anion that can behave both as a Brønsted–Lowry acid and as a Brønsted–Lowry base.

15.42 Give the formula for the conjugate base of each of the following Brønsted–Lowry acids:

(a) HSO_4^- (b) H_2SO_3 (c) $H_2PO_4^-$

(d) NH_4^+ (e) H_2O (f) NH_3

15.43 Give the formula for the conjugate acid of each of the following Brønsted–Lowry bases:

(a) SO_3^{2-} (b) H_2O (c) CH_3NH_2

(d) OH^- (e) HCO_3^- (f) H^-

15.44 For each of the following reactions, identify the Brønsted–Lowry acids and bases and the conjugate acid–base pairs:

(a) $CH_3CO_2H(aq) + NH_3(aq) \rightleftharpoons$
$$NH_4^+(aq) + CH_3CO_2^-(aq)$$

(b) $CO_3^{2-}(aq) + H_3O^+(aq) \rightleftharpoons H_2O(l) + HCO_3^-(aq)$

(c) $HSO_3^-(aq) + H_2O(l) \rightleftharpoons H_3O^+(aq) + SO_3^{2-}(aq)$

(d) $HSO_3^-(aq) + H_2O(l) \rightleftharpoons H_2SO_3(aq) + OH^-(aq)$

15.45 For each of the following reactions, identify the Brønsted–Lowry acids and bases and the conjugate acid–base pairs:

(a) $CN^-(aq) + H_2O(l) \rightleftharpoons OH^-(aq) + HCN(aq)$

(b) $H_2PO_4^-(aq) + H_2O(l) \rightleftharpoons H_3O^+(aq) + HPO_4^{2-}(aq)$

(c) $HPO_4^{2-}(aq) + H_2O(l) \rightleftharpoons OH^-(aq) + H_2PO_4^-(aq)$

(d) $NH_4^+(aq) + NO_2^-(aq) \rightleftharpoons HNO_2(aq) + NH_3(aq)$

15.46 Which of the following species behave as strong acids or as strong bases in aqueous solution? See Table 15.1 to check your answers.

(a) HNO_2 (b) HNO_3 (c) NH_4^+ (d) Cl^-

(e) H^- (f) O^{2-} (g) H_2SO_4

15.47 Which acid in each of the following pairs has the stronger conjugate base? See Table 15.1 for help with parts (c) and (d).

(a) H_2CO_3 or H_2SO_4 (b) HCl or HF

(c) HF or NH_4^+ (d) HCN or HSO_4^-

15.48 If you mix equal concentrations of reactants and products, which of the following reactions proceed to the right and which proceed to the left? Use the data in Table 15.1, and remember that the stronger base gets the proton.

(a) $H_2CO_3(aq) + HSO_4^-(aq) \rightleftharpoons$
$$H_2SO_4(aq) + HCO_3^-(aq)$$

(b) $HF(aq) + Cl^-(aq) \rightleftharpoons HCl(aq) + F^-(aq)$

(c) $HF(aq) + NH_3(aq) \rightleftharpoons NH_4^+(aq) + F^-(aq)$

(d) $HSO_4^-(aq) + CN^-(aq) \rightleftharpoons HCN(aq) + SO_4^{2-}(aq)$

15.49 If you mix equal concentrations of reactants and products, which of the following reactions proceed to the right and which proceed to the left? Use the data in Table 15.1.

(a) $HSO_4^-(aq) + CH_3CO_2^-(aq) \rightleftharpoons$
$$CH_3CO_2H(aq) + SO_4^{2-}(aq)$$

(b) $HNO_2(aq) + NO_3^-(aq) \rightleftharpoons HNO_3(aq) + NO_2^-(aq)$

(c) $HCO_3^-(aq) + F^-(aq) \rightleftharpoons HF(aq) + CO_3^{2-}(aq)$

(d) $NH_4^+(aq) + CN^-(aq) \rightleftharpoons HCN(aq) + NH_3(aq)$

Dissociation of Water; pH

15.50 For each of the following solutions, calculate $[OH^-]$ from $[H_3O^+]$, or $[H_3O^+]$ from $[OH^-]$. Classify each solution as acidic, basic, or neutral.

(a) $[H_3O^+] = 3.4 \times 10^{-9}$ M

(b) $[OH^-] = 0.010$ M

(c) $[OH^-] = 1.0 \times 10^{-10}$ M

(d) $[H_3O^+] = 1.0 \times 10^{-7}$ M

(e) $[H_3O^+] = 8.6 \times 10^{-5}$ M

15.51 For each of the following solutions, calculate $[OH^-]$ from $[H_3O^+]$, or $[H_3O^+]$ from $[OH^-]$. Classify each solution as acidic, basic, or neutral.

(a) $[H_3O^+] = 2.5 \times 10^{-4}$ M

(b) $[H_3O^+] = 2.0$ M

(c) $[OH^-] = 5.6 \times 10^{-9}$ M

(d) $[OH^-] = 1.5 \times 10^{-3}$ M

(e) $[OH^-] = 1.0 \times 10^{-7}$ M

15.52 Calculate the pH to the correct number of significant figures for solutions with the following concentrations of H_3O^+ or OH^-:

(a) $[H_3O^+] = 2.0 \times 10^{-5}$ M

(b) $[OH^-] = 4 \times 10^{-3}$ M

(c) $[H_3O^+] = 3.56 \times 10^{-9}$ M

(d) $[H_3O^+] = 10^{-3}$ M

(e) $[OH^-] = 12$ M

15.53 What is the pH to the correct number of significant figures for solutions with the following concentrations of H_3O^+ or OH^-?

(a) $[OH^-] = 7.6 \times 10^{-3}$ M

(b) $[H_3O^+] = 10^{-8}$ M

(c) $[H_3O^+] = 5.0$ M

(d) $[OH^-] = 1.0 \times 10^{-7}$ M

(e) $[H_3O^+] = 2.18 \times 10^{-10}$ M

15.54 Calculate the H_3O^+ concentration to the correct number of significant figures for solutions with the following pH values:

(a) 4.1   (b) 10.82   (c) 0.00

(d) 14.25   (e) −1.0

15.55 What is the H_3O^+ concentration to the correct number of significant figures for solutions with the following pH values?

(a) 9.0   (b) 7.00   (c) −0.3

(d) 15.18   (e) 2.63

15.56 What is the change in pH if $[H_3O^+]$ changes by each of the following factors?

(a) 1000   (b) 1.0×10^5 (c) 2.0

15.57 By what factor must $[H_3O^+]$ change to produce the following pH changes?

(a) 1.0 unit   (b) 10.00 units   (c) 0.10 unit

15.58 Given the following approximate concentrations of H_3O^+ or OH^- for various biological fluids, calculate the pH.

(a) Gastric juice, $[H_3O^+] = 10^{-2}$ M

(b) Spinal fluid, $[H_3O^+] = 4 \times 10^{-8}$ M

(c) Bile, $[OH^-] = 8 \times 10^{-8}$ M

(d) Urine, $[OH^-] = 6 \times 10^{-10}$ M to 2×10^{-6} M

15.59 What is the pH of each of the following foods?

(a) Sauerkraut, $[H_3O^+] = 3 \times 10^{-4}$ M

(b) Eggs, $[OH^-] = 6 \times 10^{-7}$ M

(c) Apples, $[OH^-] = 10^{-11}$ M

(d) Limes, $[H_3O^+] = 1.3 \times 10^{-2}$ M

Strong Acids and Strong Bases

15.60 Calculate the pH of each of the following solutions:

(a) 0.40 M HBr

(b) 3.7×10^{-4} M KOH

(c) 5.0×10^{-5} M $Ca(OH)_2$

15.61 What is the pH of each of the following solutions?

(a) 1.8 M $HClO_4$

(b) 1.2 M LiOH

(c) 5.3×10^{-3} M $Ba(OH)_2$

15.62 Calculate the pH of solutions prepared by:

(a) Dissolving 4.8 g of lithium hydroxide in water to give 250 mL of solution

(b) Dissolving 0.93 g of hydrogen chloride in water to give 0.40 L of solution

(c) Diluting 50.0 mL of 0.10 M HCl to a volume of 1.00 L

(d) Mixing 100.0 mL of 2.0×10^{-3} M HCl and 400.0 mL of 1.0×10^{-3} M $HClO_4$ (assume that volumes are additive)

15.63 Calculate the pH of solutions prepared by:

(a) Dissolving 0.20 g of sodium oxide in water to give 100.0 mL of solution

(b) Dissolving 1.26 g of pure nitric acid in water to give 0.500 L of solution

(c) Diluting 40.0 mL of 0.075 M $Ba(OH)_2$ to a volume of 300.0 mL

(d) Mixing equal volumes of 0.20 M HCl and 0.50 M HNO_3 (assume that volumes are additive)

Weak Acids

15.64 Write a balanced net ionic equation and the corresponding equilibrium equation for dissociation of the following weak acids:

(a) Chlorous acid, $HClO_2$

(b) Hypobromous acid, HOBr

(c) Formic acid, HCO_2H

15.65 Write a balanced net ionic equation and the corresponding equilibrium equation for dissociation of the following weak acids:

(a) Hydrazoic acid, HN_3

(b) Benzoic acid, $C_6H_5CO_2H$

(c) Hydrogen peroxide, H_2O_2

15.66 Using values of K_a in Appendix C, arrange the following acids in order of (a) increasing acid strength and (b) decreasing percent dissociation: C_6H_5OH, HNO_3, CH_3CO_2H, HOCl. Also estimate $[H_3O^+]$ in a 1.0 M solution of each acid.

15.67 Use the values of K_a in Appendix C to arrange the following acids in order of (a) increasing acid strength and (b) decreasing percent dissociation: HCO_2H, HCN, $HClO_4$, HOBr. Also estimate $[H_3O^+]$ in a 1.0 M solution of each acid.

15.68 The pH of 0.040 M hypobromous acid (HOBr) is 5.05. Set up the equilibrium equation for dissociation of HOBr, and calculate the value of the acid-dissociation constant.

15.69 Lactic acid ($C_3H_6O_3$), which occurs in sour milk and foods such as sauerkraut, is a weak monoprotic acid. The pH of a 0.10 M solution of lactic acid is 2.43. What is the value of K_a for lactic acid?

15.70 Phenol (C_6H_5OH) is a weak acid used as a general disinfectant and in the manufacture of plastics. Calculate the pH and the concentrations of all species present (H_3O^+, $C_6H_5O^-$, C_6H_5OH, and OH^-) in a 0.10 M solution of phenol ($K_a = 1.3 \times 10^{-10}$). Also calculate the percent dissociation.

15.71 Formic acid (HCO_2H) is an organic acid secreted by ants and stinging nettles. Calculate the pH and the concentrations of all species present (HCO_2H, HCO_2^-, H_3O^+, and OH^-) in 0.20 M HCO_2H ($K_a = 1.8 \times 10^{-4}$). Also calculate the percent dissociation.

15.72 Calculate the pH and the percent dissociation in 1.5 M HNO_2 ($K_a = 4.5 \times 10^{-4}$).

15.73 A typical aspirin tablet contains 324 mg of aspirin (acetylsalicylic acid, $C_9H_8O_4$), a monoprotic acid having $K_a = 3.0 \times 10^{-4}$. If you dissolve two aspirin tablets in a 300 mL glass of water, what is the pH of the solution and the percent dissociation?

Polyprotic Acids

15.74 Write balanced net ionic equations and the corresponding equilibrium equations for stepwise dissociation of the diprotic acid H_2SeO_4.

15.75 Write balanced net ionic equations and the corresponding equilibrium equations for stepwise dissociation of the triprotic acid H_3PO_4.

15.76 Calculate the pH and the concentrations of all species present (H_2CO_3, HCO_3^-, CO_3^{2-}, H_3O^+, and OH^-) in 0.010 M H_2CO_3 ($K_{a1} = 4.3 \times 10^{-7}$; $K_{a2} = 5.6 \times 10^{-11}$).

15.77 Calculate the pH and the concentrations of H_2SO_3, HSO_3^-, SO_3^{2-}, H_3O^+, and OH^- in 0.025 M H_2SO_3 ($K_{a1} = 1.5 \times 10^{-2}$; $K_{a2} = 6.3 \times 10^{-8}$).

15.78 Oxalic acid ($H_2C_2O_4$) is a diprotic acid that occurs in plants such as rhubarb and spinach. Calculate the pH and the concentration of $C_2O_4^{2-}$ ions in 0.20 M $H_2C_2O_4$ ($K_{a1} = 5.9 \times 10^{-2}$; $K_{a2} = 6.4 \times 10^{-5}$).

15.79 Calculate the concentrations of H_3O^+ and SO_4^{2-} in a solution prepared by mixing equal volumes of 0.2 M HCl and 0.6 M H_2SO_4 (K_{a2} for H_2SO_4 is 1.2×10^{-2}).

Weak Bases; Relation Between K_a and K_b

15.80 Write a balanced net ionic equation and the corresponding equilibrium equation for reaction of the following weak bases with water:
(a) Dimethylamine, $(CH_3)_2NH$
(b) Aniline, $C_6H_5NH_2$
(c) Cyanide ion, CN^-

15.81 Write a balanced net ionic equation and the corresponding equilibrium equation for reaction of the following weak bases with water:
(a) Pyridine, C_5H_5N
(b) Ethylamine, $C_2H_5NH_2$
(c) Acetate ion, $CH_3CO_2^-$

15.82 Morphine ($C_{17}H_{19}NO_3$), a narcotic used in painkillers, is a weak organic base. If the pH of a 7.0×10^{-4} M solution of morphine is 9.5, what is the value of K_b?

15.83 A 1.00×10^{-3} M solution of quinine, a drug used in treating malaria, has a pH of 9.75. What is the value of K_b for quinine?

15.84 Using the values of K_b in Appendix C, calculate $[OH^-]$ and the pH for each of the following solutions:
(a) 0.24 M methylamine
(b) 0.040 M pyridine
(c) 0.075 M hydroxylamine

15.85 Aniline ($C_6H_5NH_2$) is an organic base used in the manufacture of dyes. Calculate the pH and the concentrations of all species present ($C_6H_5NH_2$, $C_6H_5NH_3^+$, OH^-, and H_3O^+) in a 0.15 M solution of aniline ($K_b = 4.3 \times 10^{-10}$).

15.86 Using values of K_b in Appendix C, calculate values of K_a for each of the following ions:
(a) Propylammonium ion, $C_3H_7NH_3^+$
(b) Hydroxylammonium ion, NH_3OH^+
(c) Anilinium ion, $C_6H_5NH_3^+$
(d) Pyridinium ion, $C_5H_5NH^+$

15.87 Using values of K_a in Appendix C, calculate values of K_b for each of the following ions:
(a) Fluoride ion, F^-
(b) Hypobromite ion, OBr^-
(c) Hydrogen sulfide ion, HS^-
(d) Sulfide ion, S^{2-}

Acid–Base Properties of Salts

15.88 Write a balanced net ionic equation for the reaction of each of the following ions with water. In each case, identify the Brønsted–Lowry acids and bases and the conjugate acid–base pairs.

(a) $CH_3NH_3^+$ (b) $Cr(H_2O)_6^{3+}$

(c) $CH_3CO_2^-$ (d) PO_4^{3-}

15.89 Write a balanced net ionic equation for the principal reaction in solutions of each of the following salts. In each case, identify the Brønsted–Lowry acids and bases and the conjugate acid–base pairs.

(a) Na_2CO_3 (b) NH_4NO_3

(c) $NaCl$ (d) $ZnCl_2$

15.90 Classify each of the following ions according to whether they react with water to give a neutral, acidic, or basic solution:

(a) F^- (b) Br^- (c) NH_4^+

(d) $K(H_2O)_6^+$ (e) SO_3^{2-} (f) $Cr(H_2O)_6^{3+}$

15.91 Classify each of the following salt solutions as neutral, acidic, or basic. See Appendix C for values of equilibrium constants.

(a) $Fe(NO_3)_3$ (b) $Ba(NO_3)_2$ (c) $NaOCl$

(d) NH_4I (e) NH_4NO_2 (f) $(CH_3NH_3)Cl$

15.92 Calculate the concentrations of all species present and the pH in 0.10 M solutions of the following substances. See Appendix C for values of equilibrium constants.

(a) Ethylammonium nitrate, $(C_2H_5NH_3)NO_3$

(b) Sodium acetate, $Na(CH_3CO_2)$

(c) Sodium nitrate, $NaNO_3$.

15.93 Calculate the pH and the percent dissociation of the hydrated cation in 0.020 M solutions of the following substances. See Appendix C for values of equilibrium constants.

(a) $Fe(NO_3)_2$ (b) $Fe(NO_3)_3$

Factors That Affect Acid Strength

15.94 Arrange each group of compounds in order of increasing acid strength. Explain your reasoning.

(a) HCl, H_2S, PH_3

(b) NH_3, PH_3, AsH_3

(c) $HBrO$, $HBrO_2$, $HBrO_3$

15.95 Arrange each group of compounds in order of decreasing acid strength. Explain your reasoning.

(a) H_2O, H_2S, H_2Se

(b) $HClO_3$, $HBrO_3$, HIO_3

(c) PH_3, H_2S, HCl

15.96 Identify the strongest acid in each of the following sets. Explain your reasoning.

(a) H_2O, HF, or HCl

(b) $HClO_2$, $HClO_3$, or $HBrO_3$

(c) HBr, H_2S, or H_2Se

15.97 Identify the weakest acid in each of the following sets. Explain your reasoning.

(a) H_2SO_3, $HClO_3$, $HClO_4$

(b) NH_3, H_2O, H_2S

(c) $B(OH)_3$, $Al(OH)_3$, $Ga(OH)_3$

15.98 Identify the stronger acid in each of the following pairs. Explain your reasoning.

(a) H_2Se or H_2Te (b) H_3PO_4 or H_3AsO_4

(c) $H_2PO_4^-$ or HPO_4^{2-} (d) CH_4 or NH_4^+

15.99 Identify the stronger base in each of the following pairs. Explain your reasoning.

(a) ClO_2^- or ClO_3^- (b) HSO_4^- or $HSeO_4^-$

(c) HS^- or OH^- (d) HS^- or Br^-

Lewis Acids and Bases

15.100 For each of the following reactions, identify the Lewis acid and the Lewis base:

(a) $SiF_4 + 2\,F^- \rightarrow SiF_6^{2-}$

(b) $4\,NH_3 + Zn^{2+} \rightarrow Zn(NH_3)_4^{2+}$

(c) $2\,Cl^- + HgCl_2 \rightarrow HgCl_4^{2-}$

(d) $CO_2 + H_2O \rightarrow H_2CO_3$

15.101 For each of the following reactions, identify the Lewis acid and the Lewis base:

(a) $2\,Cl^- + BeCl_2 \rightarrow BeCl_4^{2-}$

(b) $Mg^{2+} + 6\,H_2O \rightarrow Mg(H_2O)_6^{2+}$

(c) $SO_3 + OH^- \rightarrow HSO_4^-$

(d) $F^- + BF_3 \rightarrow BF_4^-$

15.102 For each of the Lewis acid–base reactions in Problem 15.100, draw electron-dot structures for the reactants and products, and use the curved arrow notation (Section 15.16) to represent the donation of a lone pair of electrons from the Lewis base to the Lewis acid.

15.103 For each of the Lewis acid–base reactions in Problem 15.101, draw electron-dot structures for the reactants and products, and use the curved arrow notation (Section 15.16) to represent the donation of a lone pair of electrons from the Lewis base to the Lewis acid.

15.104 Classify each of the following as a Lewis acid or a Lewis base:

(a) CN^-　(b) H^+　(c) H_2O　(d) Fe^{3+}

(e) OH^-　(f) CO_2　(g) $P(CH_3)_3$　(h) $B(CH_3)_3$

15.105 Which would you expect to be the stronger Lewis acid in each of the following pairs? Explain.

(a) BF_3 or BH_3　　(b) SO_2 or SO_3

(c) Sn^{2+} or Sn^{4+}　(d) CH_3^+ or CH_4

General Problems

15.106 Aqueous solutions of hydrogen sulfide contain H_2S, HS^-, S^{2-}, H_3O^+, OH^-, and H_2O in varying concentrations. Which of these species can act only as an acid? Which can act only as a base? Which can act both as an acid and as a base?

15.107 Given the following pH values for some common foods, calculate $[H_3O^+]$ and $[OH^-]$.

(a) Dill pickles, 3.2　　(b) Eggs, 7.8

(c) Apples, 3.1　　(d) Milk, 6.4

(e) Tomatoes, 4.2　　(f) Limes, 1.9

15.108 Draw an electron-dot structure for H_3O^+, and explain how H_3O^+ can form higher hydrates such as $H_5O_2^+$, $H_7O_3^+$, and $H_9O_4^+$.

15.109 The hydronium ion H_3O^+ is the strongest acid that can exist in aqueous solution because stronger acids dissociate by transferring a proton to water. What is the strongest base that can exist in aqueous solution?

15.110 Baking powder contains baking soda ($NaHCO_3$) and an acidic substance such as sodium alum, $NaAl(SO_4)_2 \cdot 12\ H_2O$. These components react in an aqueous medium to produce CO_2 gas, which "raises" the dough. Write a balanced net ionic equation for the reaction.

15.111 Arrange the following substances in order of increasing $[H_3O^+]$ for a 0.10 M solution of each:

(a) $Zn(NO_3)_2$　(b) Na_2O　(c) $NaOCl$

(d) $NaClO_4$　(e) $HClO_4$

15.112 At 0°C, the density of liquid water is 0.9998 g/mL and the value of K_w is 1.14×10^{-15}. What fraction of the molecules in liquid water are dissociated at 0°C? What is the percent dissociation at 0°C? What is the pH of a neutral solution at 0°C?

15.113 Use the conjugate acid–base pair HCN and CN^- to derive the relationship between K_a and K_b.

15.114 Nicotine ($C_{10}H_{14}N_2$) can accept two protons because it has two basic N atoms ($K_{b1} = 1.0 \times 10^{-6}$; $K_{b2} = 1.3 \times 10^{-11}$). Calculate the values of K_a for the conjugate acids $C_{10}H_{14}N_2H^+$ and $C_{10}H_{14}N_2H_2^{2+}$.

15.115 Sodium benzoate ($C_6H_5CO_2Na$) is used as a food preservative. Calculate the pH and the concentrations of all species present (Na^+, $C_6H_5CO_2^-$, $C_6H_5CO_2H$, H_3O^+, and OH^-) in 0.050 M sodium benzoate; K_a for benzoic acid ($C_6H_5CO_2H$) is 6.5×10^{-5}.

15.116 The hydrated cation $M(H_2O)_6^{3+}$ has $K_a = 10^{-4}$, and the acid HA has $K_a = 10^{-5}$. Identify the principal reaction in an aqueous solution of each of the following salts, and classify each solution as acidic, basic, or neutral:

(a) NaA　　(b) $M(NO_3)_3$

(c) $NaNO_3$　(d) MA_3

15.117 Calculate the pH and the concentrations of all species present (H_3O^+, F^-, HF, Cl^-, and OH^-) in a solution that contains 0.10 M HF ($K_a = 3.5 \times 10^{-4}$) and 0.10 M HCl.

15.118 Classify each of the following salt solutions as neutral, acidic, or basic. See Appendix C for values of equilibrium constants.

(a) NH_4F　　(b) $(NH_4)_2SO_3$

15.119 Calculate the percent dissociation in each of the following solutions. What is the quantitative relationship between the percent dissociation and the concentration of the acid? What is the quantitative relationship between the percent dissociation and the value of K_a?

(a) 2.0 M HOCl ($K_a = 3.5 \times 10^{-8}$)

(b) 0.020 M HOCl

(c) 2.0 M HF ($K_a = 3.5 \times 10^{-4}$)

15.120 Beginning with the equilibrium equation for the dissociation of a weak acid, show that the percent dissociation varies directly as the square root of K_a and inversely as the square root of the concentration of the acid.

15.121 Calculate the pH and the concentrations of all species present in 0.25 M solutions of each of the salts in Problem 15.118. (*Hint:* The principal reaction is proton transfer from the cation to the anion.)

15.122 For a solution of two weak acids with comparable values of K_a, there is no single principal reaction. The two acid-dissociation equilibrium equations must therefore be solved simultaneously. Calculate the pH in a solution that is 0.10 M in acetic acid (CH_3CO_2H, $K_a = 1.8 \times 10^{-5}$) and 0.10 M in benzoic acid ($C_6H_5CO_2H$, $K_a = 6.5 \times 10^{-5}$). (*Hint*: Let $x = [CH_3CO_2H]$ that dissociates and $y = [C_6H_5CO_2H]$ that dissociates; then $[H_3O^+] = x + y$.)

15.123 What is the pH and the principal source of H_3O^+ ions in 1.0×10^{-10} M HCl? (*Hint*: The pH of an acid solution can't exceed 7.) What is the pH of 1.0×10^{-7} M HCl?

Multi-Concept Problems

15.124 A 7.0 wt % solution of H_3PO_4 in water has a density of 1.0353 g/mL. Calculate the pH and the concentrations of all species present (H_3PO_4, $H_2PO_4^-$, HPO_4^{2-}, PO_4^{3-}, H_3O^+, and OH^-) in the solution. Values of equilibrium constants are listed in Appendix C.

15.125 In the case of very weak acids, $[H_3O^+]$ from the dissociation of water is significant compared with $[H_3O^+]$ from dissociation of the weak acid. The sugar substitute saccharin ($C_7H_5NO_3S$), for example, is a very weak acid having $K_a = 2.1 \times 10^{-12}$ and a solubility in water of 348 mg/100 mL. Calculate $[H_3O^+]$ in a saturated solution of saccharin. (*Hint*: Equilibrium equations for the dissociation of saccharin and water must be solved simultaneously.)

15.126 In aqueous solution, sodium acetate behaves as a strong electrolyte, yielding Na^+ cations and $CH_3CO_2^-$ anions. A particular solution of sodium acetate has a pH of 9.07 and a density of 1.0085 g/mL. What is the molality of this solution, and what is its freezing point?

15.127 During a certain time period, 4.0 million tons of SO_2 was released into the atmosphere and was subsequently oxidized to SO_3. As explained in the Interlude, the acid rain produced when the SO_3 dissolves in water can damage marble statues:

$$CaCO_3(s) + H_2SO_4(aq) \longrightarrow$$
$$CaSO_4(aq) + CO_2(g) + H_2O(l)$$

(a) How many 500 pound marble statues could be damaged by the acid rain? (Assume that the statues are pure $CaCO_3$ and that a statue is damaged when 3.0% of its mass is dissolved.)

(b) How many liters of CO_2 gas at 20°C and 735 mm Hg is produced as a by-product?

(c) The cation in aqueous H_2SO_4 is trigonal pyramidal rather than trigonal planar. Explain.

15.128 Neutralization reactions involving either a strong acid or a strong base go essentially to completion, and therefore we must take such neutralizations into account before calculating concentrations in mixtures of acids and bases. Consider a mixture of 3.28 g of Na_3PO_4 and 300.0 mL of 0.180 M HCl. Write balanced net ionic equations for the neutralization reactions, and calculate the pH of the solution.

15.129 We've said that alkali metal cations do not react appreciably with water to produce H_3O^+ ions, but in fact, all cations are acidic to some extent. The most acidic alkali metal cation is the smallest one, Li^+, which has $K_a = 2.5 \times 10^{-14}$ for the reaction

$$Li(H_2O)_4^+(aq) + H_2O(l) \rightleftharpoons$$
$$H_3O^+(aq) + Li(H_2O)_3(OH)(aq)$$

This reaction and the dissociation of water must be considered simultaneously in calculating the pH of Li^+ solutions, which nevertheless have pH $\approx$ 7. Check this by calculating the pH of 0.10 M LiCl.

 eMedia Problems

15.130 The **Introduction to Aqueous Acids** movie (*eChapter 15.2*) shows the dissociation of two strong acids and one weak acid. Given that all three of the acids dissociate to give hydronium ion in water, what is it about the behavior of the weak acid that makes it different from the strong acids? What is the effect of this difference in terms of pH?

15.131 The K_a of HCN is 4.9×10^{-10}. Calculate the concentration of an aqueous HCN solution that would have pH $= 5.5$. Use the **Acids and Bases** activity (*eChapter 15.7*) to verify your answer. What is the percent dissociation of HCN at this concentration?

15.132 Measure the pH of a 0.10 M solution of each of the compounds available in the **Acids and Bases** activity (*eChapter 15.7*) and classify each of the compounds as a strong acid, a weak acid, a strong base, or a weak base.

15.133 Use data from the **Acids and Bases** activity (*eChapter 15.7*) to determine the K_b of methylamine, CH_3NH_2. Determine the percent dissociation of methylamine at 0.10 M, 0.0010 M, and 1.0×10^{-5} M concentrations.

15.134 Use the **Equilibrium Constant** activity (*eChapter 15.9*) to experiment with the dissociation of a weak acid. What would be the hydronium ion concentration and pH of 0.10 M solutions of weak acids with $K_a = 1 \times 10^{-5}$; $K_a = 1 \times 10^{-10}$; and $K_a = 1 \times 10^{-15}$?

Applications of Aqueous Equilibria

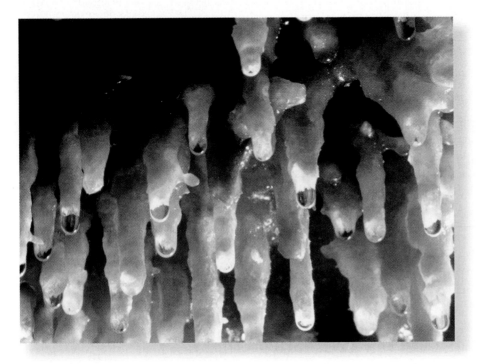

These downward growing, icicle-shaped structures called *stalactites* are formed in limestone caves by the slow precipitation of calcium carbonate from dripping water.

Aqueous equilibria play a crucial role in many biological and environmental processes. The pH of human blood, for example, is carefully controlled at a value of 7.4 by equilibria involving, primarily, the conjugate acid–base pair H_2CO_3 and HCO_3^-. The pH of many lakes and streams must remain near 5.5 for plant and aquatic life to flourish.

We began a study of aqueous equilibria in Chapter 15, where we examined the dissociation of weak acids and weak bases. In this chapter, we'll continue the study. First we'll see how to calculate the pH of mixtures of acids and bases. Then we'll look at the dissolution and precipitation of slightly soluble salts and the factors that affect solubility. Aqueous equilibria involving the dissolution and precipitation of salts are important in a great many natural processes, from tooth decay to the formation of limestone caves.

16.1   Neutralization Reactions

We've seen on numerous occasions that the neutralization reaction of an acid with a base produces water and a salt. But to what extent does a neutralization reaction go to completion? We must answer that question before we can make pH calculations on mixtures of acids and bases. Let's look at four types of neutralization reactions: (1) strong acid–strong base, (2) weak acid–strong base, (3) strong acid–weak base, and (4) weak acid–weak base.

Strong Acid–Strong Base

Strong Acid–Strong Base Neutralization activity

Let's consider the reaction of hydrochloric acid with aqueous sodium hydroxide to give water and an aqueous solution of sodium chloride:

$$HCl(aq) + NaOH(aq) \longrightarrow H_2O(l) + NaCl(aq)$$

Because $HCl(aq)$, $NaOH(aq)$, and $NaCl(aq)$ are all completely dissociated, the net ionic equation for the neutralization reaction is

$$H_3O^+(aq) + OH^-(aq) \rightleftharpoons 2\,H_2O(l)$$

If we mix equal numbers of moles of $HCl(aq)$ and $NaOH(aq)$, the concentrations of H_3O^+ and OH^- remaining in the NaCl solution after neutralization will be the same as those in pure water, $[H_3O^+] = [OH^-] = 1.0 \times 10^{-7}$ M. In other words, the reaction of HCl with NaOH proceeds far to the right.

We come to the same conclusion by looking at the equilibrium constant for the reaction. Because the neutralization reaction of any strong acid with a strong base is the reverse of the dissociation of water, its equilibrium constant, K_n ("n" for neutralization), is just the reciprocal of the ion-product constant for water, $K_n = 1/K_w$:

$$H_3O^+(aq) + OH^-(aq) \rightleftharpoons 2\,H_2O(l)$$

$$K_n = \frac{1}{[H_3O^+][OH^-]} = \frac{1}{K_w} = \frac{1}{1.0 \times 10^{-14}} = 1.0 \times 10^{14}$$

The value of K_n (1.0×10^{14}) for a strong acid–strong base reaction is a large number, which means that the neutralization reaction proceeds essentially 100% to completion. After neutralization of equal molar amounts of acid and base, the solution contains a salt derived from a strong base and a strong acid. Because neither the cation nor the anion of the salt has acidic or basic properties, the pH is 7 (Section 15.14).

Weak Acid–Strong Base

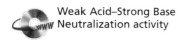

Weak Acid–Strong Base Neutralization activity

Because a weak acid HA is largely undissociated, the net ionic equation for the neutralization reaction involves proton transfer from HA to the strong base, OH^-:

$$HA(aq) + OH^-(aq) \rightleftharpoons H_2O(l) + A^-(aq)$$

Acetic acid (CH_3CO_2H), for example, reacts with aqueous NaOH to give water and aqueous sodium acetate (CH_3CO_2Na):

$$CH_3CO_2H(aq) + OH^-(aq) \rightleftharpoons H_2O(l) + CH_3CO_2^-(aq)$$

Note that Na^+ ions do not appear in the net ionic equation because both NaOH and CH_3CO_2Na are completely dissociated.

To obtain the equilibrium constant K_n, we multiply known equilibrium constants for reactions that add to give the net ionic equation for the neutralization reaction. Because CH_3CO_2H is on the left side of the equation and $CH_3CO_2^-$ is on the right side, one of the reactions needed is the dissociation of CH_3CO_2H. Since H_2O is on the right side of the equation and OH^- is on the left side, the other reaction needed is the reverse of the dissociation of H_2O. Note that H_3O^+ and one H_2O molecule cancel when the two equations are added:

$$CH_3CO_2H(aq) + H_2O(l) \rightleftharpoons H_3O^+(aq) + CH_3CO_2^-(aq) \qquad K_a = 1.8 \times 10^{-5}$$

$$H_3O^+(aq) + OH^-(aq) \rightleftharpoons 2\,H_2O(l) \qquad 1/K_w = 1.0 \times 10^{14}$$

Net: $\quad CH_3CO_2H(aq) + OH^-(aq) \rightleftharpoons H_2O(l) + CH_3CO_2^-(aq) \qquad K_n = (K_a)(1/K_w) = (1.8 \times 10^{-5})(1.0 \times 10^{14})$

$$= 1.8 \times 10^9$$

As we saw in Section 15.13, the equilibrium constant for a net reaction is always equal to the *product* of the equilibrium constants for the reactions added. Therefore, we multiply K_a for CH_3CO_2H by the reciprocal of K_w to get K_n for the neutralization reaction. (We use $1/K_w$ because the H_2O dissociation reaction is written in the reverse direction.) The resulting large value of K_n (1.8×10^9) means that the neutralization reaction proceeds nearly 100% to completion.

As a general rule, the neutralization of any weak acid by a strong base will go 100% to completion because OH^- has a great affinity for protons. After neutralization of equal molar amounts of CH_3CO_2H and NaOH, the solution contains Na^+, which has no acidic or basic properties, and $CH_3CO_2^-$, which is a weak base. Therefore, the pH is greater than 7 (Section 15.14).

Strong Acid–Weak Base

A strong acid HA is completely dissociated into H_3O^+ and A^- ions, and the neutralization reaction therefore involves proton transfer from H_3O^+ to the weak base B:

$$H_3O^+(aq) + B(aq) \rightleftharpoons H_2O(l) + BH^+(aq)$$

For example, the net ionic equation for neutralization of hydrochloric acid with aqueous ammonia is

$$H_3O^+(aq) + NH_3(aq) \rightleftharpoons H_2O(l) + NH_4^+(aq)$$

When NaOH(aq) is added to $CH_3CO_2H(aq)$ containing the acid–base indicator phenolphthalein, the color of the indicator changes from colorless to pink in the pH range 8.2–9.8 due to neutralization of the acetic acid.

As in the weak acid–strong base case, we can obtain the equilibrium constant for the neutralization reaction by multiplying known equilibrium constants for reactions that add to give the net ionic equation:

$$NH_3(aq) + H_2O(l) \rightleftharpoons NH_4^+(aq) + OH^-(aq) \qquad K_b = 1.8 \times 10^{-5}$$

$$H_3O^+(aq) + OH^-(aq) \rightleftharpoons 2\,H_2O(l) \qquad 1/K_w = 1.0 \times 10^{14}$$

Net: $\quad H_3O^+(aq) + NH_3(aq) \rightleftharpoons H_2O(l) + NH_4^+(aq) \qquad K_n = (K_b)(1/K_w) = (1.8 \times 10^{-5})(1.0 \times 10^{14})$

$$= 1.8 \times 10^9$$

Again, the neutralization reaction proceeds nearly 100% to the right because its equilibrium constant K_n is a very large number (1.8×10^9). (It's

When hydrochloric acid is added to aqueous ammonia containing the acid–base indicator methyl red, the color of the indicator changes from yellow to red in the pH range 6.0–4.2 due to neutralization of the NH_3.

purely coincidental that the neutralization reactions of CH_3CO_2H with NaOH and of HCl with NH_3 have the same value of K_n. The two K_n values are the same because K_a for CH_3CO_2H happens to have the same value as K_b for NH_3.)

The neutralization of any weak base with a strong acid generally goes 100% to completion because H_3O^+ is a powerful proton donor. After neutralization of equal molar amounts of NH_3 and HCl, the solution contains NH_4^+, which is a weak acid, and Cl^-, which has no acidic or basic properties. Therefore, the pH is less than 7 (Section 15.14).

Weak Acid–Weak Base

Both a weak acid HA and a weak base B are largely undissociated, and the neutralization reaction therefore involves proton transfer from the weak acid to the weak base. For example, the net ionic equation for neutralization of acetic acid with aqueous ammonia is

$$CH_3CO_2H(aq) + NH_3(aq) \rightleftharpoons NH_4^+(aq) + CH_3CO_2^-(aq)$$

The equilibrium constant K_n can be obtained by adding equations for (1) the acid dissociation of acetic acid, (2) the base protonation of ammonia, and (3) the reverse of the dissociation of water:

$$CH_3CO_2H(aq) + H_2O(l) \rightleftharpoons H_3O^+(aq) + CH_3CO_2^-(aq) \qquad K_a = 1.8 \times 10^{-5}$$

$$NH_3(aq) + H_2O(l) \rightleftharpoons NH_4^+(aq) + OH^-(aq) \qquad K_b = 1.8 \times 10^{-5}$$

$$\underline{H_3O^+(aq) + OH^-(aq) \rightleftharpoons 2\,H_2O(l) \qquad\qquad\qquad 1/K_w = 1.0 \times 10^{14}}$$

Net: $CH_3CO_2H(aq) + NH_3(aq) \rightleftharpoons NH_4^+(aq) + CH_3CO_2^-(aq) \qquad K_n = (K_a)(K_b)(1/K_w)$

$$K_n = (K_a)(K_b)\left(\frac{1}{K_w}\right) = (1.8 \times 10^{-5})(1.8 \times 10^{-5})(1.0 \times 10^{14}) = 3.2 \times 10^4$$

Note that the value of K_n in this case is smaller than it is for the preceding three cases, so the neutralization does not proceed as far toward completion.

In general, weak acid–weak base neutralizations have less tendency to proceed to completion than neutralizations involving strong acids or strong bases. The neutralization of HCN with aqueous ammonia, for example, has a value of K_n less than unity, which means that the reaction proceeds less than halfway to completion:

$$HCN(aq) + NH_3(aq) \rightleftharpoons NH_4^+(aq) + CN^-(aq) \qquad K_n = 0.88$$

EXAMPLE 16.1 Write a balanced net ionic equation for neutralization of equal molar amounts of nitric acid and methylamine (CH_3NH_2). Indicate whether the pH after neutralization is greater than, equal to, or less than 7.

SOLUTION Because HNO_3 is a strong acid and CH_3NH_2 is a weak base, the net ionic equation is

$$H_3O^+(aq) + CH_3NH_2(aq) \rightleftharpoons H_2O(l) + CH_3NH_3^+(aq)$$

After neutralization, the solution contains $CH_3NH_3^+$, a weak acid, and NO_3^-, which has no acidic or basic properties. Therefore, the pH is less than 7.

▶**PROBLEM 16.1** Write a balanced net ionic equation for neutralization of equal molar amounts of the following acids and bases. Indicate whether the pH after neutralization is greater than, equal to, or less than 7. Values of K_a and K_b are listed in Appendix C.
(a) HNO_2 and KOH   (b) HBr and NH_3 (c) KOH and $HClO_4$

▶**PROBLEM 16.2** Write a balanced net ionic equation for neutralization of the following acids and bases, calculate the value of K_n for each neutralization reaction, and arrange the reactions in order of increasing tendency to proceed to completion. Values of K_a and K_b are listed in Appendix C.
(a) HF and NaOH   (b) HCl and KOH   (c) HF and NH_3

16.2 The Common-Ion Effect

A solution of a weak acid and its conjugate base is an important acid–base mixture because such mixtures regulate the pH in biological systems. To illustrate pH calculations for weak acid–conjugate base mixtures, let's calculate the pH of a solution prepared by dissolving 0.10 mol of acetic acid and 0.10 mol of sodium acetate in water and then diluting the solution to a volume of 1.00 L. This problem, like those discussed in Chapter 15, can be solved by thinking about the chemistry involved. First identify the acid–base properties of the various species in solution, and then consider the possible proton-transfer reactions these species can undergo. We'll follow the procedure outlined in Figure 15.7.

Step 1. Since acetic acid is largely undissociated in aqueous solution, and since the salt sodium acetate is essentially 100% dissociated, the species present initially are

$$CH_3CO_2H \quad Na^+ \quad CH_3CO_2^- \quad H_2O$$
$$\text{Acid} \qquad \text{Inert} \qquad \text{Base} \qquad \text{Acid or base}$$

Steps 2–3. Because we have two acids and two bases, there are four possible proton-transfer reactions. We know, however, that acetic acid is a stronger acid than water and that the principal reaction therefore involves proton transfer from CH_3CO_2H to either $CH_3CO_2^-$ or H_2O:

$$CH_3CO_2H(aq) + CH_3CO_2^-(aq) \rightleftharpoons CH_3CO_2^-(aq) + CH_3CO_2H(aq) \qquad K = 1$$
$$CH_3CO_2H(aq) + H_2O(l) \rightleftharpoons H_3O^+(aq) + CH_3CO_2^-(aq) \qquad K_a = 1.8 \times 10^{-5}$$

Although the first of these reactions has the larger equilibrium constant, we can't consider it to be the principal reaction because the reactants and products are identical. Proton transfer from acetic acid to its conjugate base is constantly occurring, but that reaction doesn't change any concentrations and therefore can't be used to calculate equilibrium concentrations. Consequently, the principal reaction is dissociation of acetic acid.

Step 4. Now we can set up a table of concentrations for the species involved in the principal reaction. As usual (Section 15.9), we define x as the concentration of acid that dissociates—here, acetic acid—but we need to take into account the fact that the acetate ions come from two sources: 0.10 mol/L of acetate comes from the sodium acetate present initially, and x mol/L comes from dissociation of acetic acid.

Principal reaction:	$CH_3CO_2H(aq) + H_2O(l) \rightleftharpoons H_3O^+(aq) + CH_3CO_2^-(aq)$		
Initial conc (M)	0.10	~0	0.10
Change (M)	$-x$	$+x$	$+x$
Equilibrium conc (M)	$0.10 - x$	x	$0.10 + x$

Step 5. Substituting the equilibrium concentrations into the equilibrium equation for the principal reaction, we obtain

$$K_a = 1.8 \times 10^{-5} = \frac{[H_3O^+][CH_3CO_2^-]}{[CH_3CO_2H]} = \frac{(x)(0.10 + x)}{0.10 - x}$$

Because K_a is small, x is small compared with 0.10, and we can make the approximation that $(0.10 + x) \approx (0.10 - x) \approx 0.10$, which simplifies the solution of the equation to

$$1.8 \times 10^{-5} = \frac{(x)(0.10 + x)}{0.10 - x} \approx \frac{(x)(0.10)}{0.10}$$

$$x = [H_3O^+] = 1.8 \times 10^{-5} \text{ M}$$

Step 8. $pH = -\log(1.8 \times 10^{-5}) = 4.74$

It's interesting to compare the $[H_3O^+]$ in a 0.10 M acetic acid–0.10 M sodium acetate solution with the $[H_3O^+]$ in a 0.10 M solution of pure acetic acid. The principal reaction is the same in both cases, but in 0.10 M acetic acid all the acetate ions come from dissociation of acetic acid:

Principal reaction:	$CH_3CO_2H(aq) + H_2O(l) \rightleftharpoons H_3O^+(aq) + CH_3CO_2^-(aq)$		
Equilibrium conc (M)	$0.10 - x$	x	x

The calculated $[H_3O^+]$ of 0.10 M acetic acid is 1.3×10^{-3} M, and the pH is 2.89 versus a pH of 4.74 for the acetic acid–sodium acetate solution. The difference in pH is illustrated in Figure 16.1.

FIGURE 16.1   The 0.10 M acetic acid–0.10 M sodium acetate solution on the left has a lower H_3O^+ concentration ($[H_3O^+] = 1.8 \times 10^{-5}$ M; pH 4.74) than the 0.10 M acetic acid solution on the right ($[H_3O^+] = 1.3 \times 10^{-3}$ M; pH 2.89). The difference in pH is revealed by the color of the indicator methyl orange, which changes from yellow to red in the pH range 4.4–3.2.

The decrease in $[H_3O^+]$ on adding acetate ions to an acetic acid solution is an example of the **common-ion effect**, the shift in an equilibrium on adding a substance that provides more of an ion already involved in the equilibrium. Thus, added acetate ions shift the acetic acid dissociation equilibrium to the left, as shown in Figure 16.2.

$$CH_3CO_2H(aq) + H_2O(l) \rightleftharpoons H_3O^+(aq) + CH_3CO_2^-(aq)$$

The common-ion effect is just another example of Le Châtelier's principle (Section 13.6), in which the stress on the equilibrium that results from raising one of the product concentrations is relieved by shifting the equilibrium to the left. Another case is found in Example 16.2.

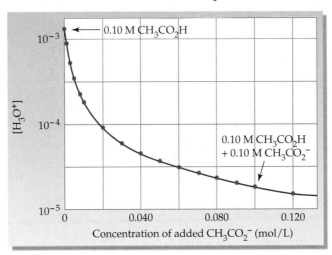

FIGURE 16.2 The common-ion effect. The concentration of H_3O^+ in a 0.10 M acetic acid solution decreases as the concentration of added sodium acetate increases because added acetate ions shift the acid-dissociation equilibrium to the left. Note that $[H_3O^+]$ is plotted on a logarithmic scale.

EXAMPLE 16.2   In 0.15 M NH_3, the pH is 11.21 and the percent dissociation is 1.1%. Calculate the concentrations of all species present, the pH, and the percent dissociation of ammonia in a solution that is 0.15 M in NH_3 and 0.45 M in NH_4Cl.

BALLPARK SOLUTION   The principal reaction is proton transfer to NH_3 from H_2O.

$$NH_3(aq) + H_2O(l) \rightleftharpoons NH_4^+(aq) + OH^-(aq)$$

In the NH_3–NH_4Cl solution, NH_4^+ is the common ion, and raising its concentration shifts the equilibrium for the principal reaction to the left. Thus, the percent dissociation in the 0.15 M NH_3–0.45 M NH_4Cl solution will be less than the 1.1% in 0.15 M NH_3. Also, the $[OH^-]$ will be less than that in 0.15 M NH_3, and so the pH will be less than 11.21.

DETAILED SOLUTION   Since NH_4^+ ions come both from the NH_4Cl present initially (0.45 M) and from reaction of NH_3 with H_2O, the concentrations of the species involved in the principal reaction are as follows:

Principal reaction:	$NH_3(aq)$ + $H_2O(l)$ $\rightleftharpoons$	$NH_4^+(aq)$ +	$OH^-(aq)$
Initial conc (M)	0.15	0.45	~0
Change (M)	$-x$	$+x$	$+x$
Equilibrium conc (M)	$0.15 - x$	$0.45 + x$	x

The equilibrium equation for the principal reaction is

$$K_b = 1.8 \times 10^{-5} = \frac{[NH_4^+][OH^-]}{[NH_3]} = \frac{(0.45 + x)(x)}{0.15 - x} \approx \frac{(0.45)(x)}{0.15}$$

We assume x is negligible in comparison with 0.45 and 0.15 because (1) the equilibrium constant K_b is small and (2) the equilibrium is shifted to the left by the common-ion effect. Therefore,

$$x = [OH^-] = \frac{(1.8 \times 10^{-5})(0.15)}{0.45} = 6.0 \times 10^{-6} \text{ M}$$

$$[NH_3] = 0.15 - x = 0.15 - (6.0 \times 10^{-6}) = 0.15 \text{ M}$$

$$[NH_4^+] = 0.45 + x = 0.45 + (6.0 \times 10^{-6}) = 0.45 \text{ M}$$

Note that the assumption concerning the size of x is justified. The H_3O^+ concentration and the pH are

$$[H_3O^+] = \frac{K_w}{[OH^-]} = \frac{1.0 \times 10^{-14}}{6.0 \times 10^{-6}} = 1.7 \times 10^{-9} \text{ M}$$

$$pH = -\log(1.7 \times 10^{-9}) = 8.77$$

The percent dissociation of ammonia is

$$\text{Percent dissociation} = \frac{[NH_3]_{\text{dissociated}}}{[NH_3]_{\text{initial}}} \times 100\%$$

$$= \frac{6.0 \times 10^{-6}}{0.15} \times 100\% = 0.0040\%$$

▶ **PROBLEM 16.3** Calculate the concentrations of all species present, the pH, and the percent dissociation of HCN ($K_a = 4.9 \times 10^{-10}$) in a solution that is 0.025 M in HCN and 0.010 M in NaCN.

▶ **PROBLEM 16.4** Calculate the pH in a solution prepared by dissolving 0.10 mol of solid NH_4Cl in 0.500 L of 0.40 M NH_3. Assume that there is no volume change.

KEY CONCEPT PROBLEM 16.5 The following pictures represent solutions of a weak acid HA that may also contain the sodium salt NaA. Which solution has the highest pH, and which has the largest percent dissociation of HA? (Na^+ ions and solvent water molecules have been omitted for clarity.)

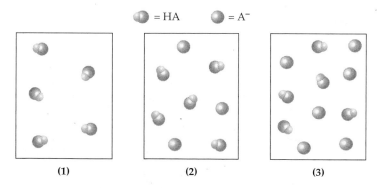

= HA   = A⁻

(1)   (2)   (3)

16.3 Buffer Solutions

Solutions like those discussed in the previous section, which contain a weak acid and its conjugate base, are called **buffer solutions** because they resist drastic changes in pH. If a small amount of OH^- is added to a buffer solution, the pH increases, but not by much because the acid component of the buffer solution neutralizes the added OH^-. If a small amount of H_3O^+ is added to a buffer solution, the pH decreases, but again not by much because the base component of the buffer solution neutralizes the added H_3O^+.

Buffer solution: Weak acid + Conjugate base   For example
$$
\begin{cases}
CH_3CO_2H + CH_3CO_2^- \\
HF + F^- \\
NH_4^+ + NH_3 \\
H_2PO_4^- + HPO_4^{2-}
\end{cases}
$$

Buffer solutions are very important in biological systems. Blood, for example, is a buffer solution that can soak up the acids and bases produced in biological reactions. The pH of human blood is carefully controlled at a value very close to 7.4 by conjugate acid–base pairs, primarily H_2CO_3 and its conjugate base HCO_3^-. The oxygen-carrying ability of blood depends on control of the pH to within 0.1 pH unit.

To see how a buffer solution works, let's return to the 0.10 M acetic acid–0.10 M sodium acetate solution discussed in the previous section. The principal reaction and the equilibrium concentrations for the solution are

Principal reaction:	$CH_3CO_2H(aq) + H_2O(l) \rightleftharpoons H_3O^+(aq) + CH_3CO_2^-(aq)$		
Equilibrium conc (M)	$0.10 - x$	x	$0.10 + x$

If we solve the equilibrium equation for $[H_3O^+]$, we obtain

$$
K_a = \frac{[H_3O^+][CH_3CO_2^-]}{[CH_3CO_2H]}
$$

$$
[H_3O^+] = K_a \frac{[CH_3CO_2H]}{[CH_3CO_2^-]}
$$

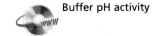
Buffer pH activity

Thus, the H_3O^+ concentration in a buffer solution has a value close to the value of K_a for the weak acid but differs by a factor equal to the concentration ratio [weak acid]/[conjugate base]. In the 0.10 M acetic acid–0.10 M sodium acetate solution where the concentration ratio is unity, $[H_3O^+]$ equals K_a:

$$
[H_3O^+] = K_a \frac{(0.10 - x)}{(0.10 + x)} = K_a\left(\frac{0.10}{0.10}\right) = K_a = 1.8 \times 10^{-5} \text{ M}
$$

$$
pH = -\log(1.8 \times 10^{-5}) = 4.74
$$

Note that in calculating this result we have set the *equilibrium* concentrations, $(0.10 - x)$ and $(0.10 + x)$, equal to the *initial* concentrations, 0.10, because x is negligible compared with the initial concentrations. For commonly used buffer solutions, K_a is small and the initial concentrations are relatively large. As a result, x is generally negligible compared with the initial concentrations, and we can use initial concentrations in the calculations.

Addition of OH⁻ to a Buffer

Now let's consider what happens when we add H_3O^+ or OH^- to a buffer solution. First, suppose that we add 0.01 mol of solid NaOH to 1.00 L of the 0.10 M acetic acid–0.10 M sodium acetate solution. Because neutralization reactions involving strong acids or strong bases go essentially 100% to completion (Section 16.1), we must take account of neutralization before calculating $[H_3O^+]$. Initially, we have $(1.00 \text{ L})(0.10 \text{ mol/L}) = 0.10$ mol of acetic acid and an equal amount of acetate ion. When we add 0.01 mol of NaOH, the neutralization reaction will alter the numbers of moles:

Neutralization reaction:	$CH_3CO_2H(aq) + OH^-(aq)$ $\xrightarrow{100\%}$	$H_2O(l) + CH_3CO_2^-(aq)$
Before reaction (mol)	0.10	0.01	0.10
Change (mol)	−0.01	−0.01	+0.01
After reaction (mol)	0.09	~0	0.11

If we assume that the solution volume remains constant at 1.00 L, the concentrations of the buffer components after neutralization are

$$[CH_3CO_2H] = \frac{0.09 \text{ mol}}{1.00 \text{ L}} = 0.09 \text{ M}$$

$$[CH_3CO_2^-] = \frac{0.11 \text{ mol}}{1.00 \text{ L}} = 0.11 \text{ M}$$

Substituting these concentrations into the expression for $[H_3O^+]$, we can then calculate the pH:

$$[H_3O^+] = K_a \frac{[CH_3CO_2H]}{[CH_3CO_2^-]}$$

$$= (1.8 \times 10^{-5})\left(\frac{0.09}{0.11}\right) = 1.5 \times 10^{-5} \text{ M}$$

$$pH = 4.82$$

Adding 0.01 mol of NaOH changes $[H_3O^+]$ by only a small amount because the concentration ratio [weak acid]/[conjugate base] changes by only a small amount, from unity to 9/11 (Figure 16.3a). The corresponding change in pH, from 4.74 to 4.82, is only 0.08 pH unit.

FIGURE 16.3 **(a)** When OH^- is added to a buffer solution, some of the weak acid is neutralized and thus converted to the conjugate base. **(b)** When H_3O^+ is added to a buffer solution, some of the conjugate base is neutralized and thus converted to the weak acid. However, as long as the concentration ratio [weak acid]/[conjugate base] stays close to its original value, $[H_3O^+]$ and the pH won't change very much.

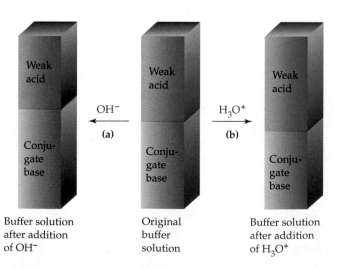

Addition of H_3O^+ to a Buffer

Now suppose that we add 0.01 mol of HCl to 1.00 L of the 0.10 M acetic acid–0.10 M sodium acetate buffer solution. The added strong acid will convert 0.01 mol of acetate ions to 0.01 mol of acetic acid because of the neutralization reaction

$$H_3O^+(aq) + CH_3CO_2^-(aq) \xrightarrow{100\%} H_2O(l) + CH_3CO_2H(aq)$$

The concentrations after neutralization will be $[CH_3CO_2H] = 0.11$ M and $[CH_3CO_2^-] = 0.09$ M, and the pH of the solution will be 4.66:

$$[H_3O^+] = K_a \frac{[CH_3CO_2H]}{[CH_3CO_2^-]}$$

$$= (1.8 \times 10^{-5})\left(\frac{0.11}{0.09}\right) = 2.2 \times 10^{-5} \text{ M}$$

$$pH = 4.66$$

Again, the change in pH, from 4.74 to 4.66, is small because the concentration ratio [weak acid]/[conjugate base] remains close to its original value (Figure 16.3b).

Buffer Capacity

To appreciate the ability of a buffer solution to maintain a nearly constant pH, let's contrast the behavior of the 0.10 M acetic acid–0.10 M sodium acetate buffer with that of a 1.8×10^{-5} M HCl solution. This very dilute HCl solution has the same pH (4.74) as the buffer solution, but it doesn't have the capacity to soak up added acid or base. For example, if we add 0.01 mol of solid NaOH to 1.00 L of 1.8×10^{-5} M HCl, a negligible amount of OH^- (1.8×10^{-5} mol) is neutralized, and the concentration of OH^- after neutralization is 0.01 mol/1.00 L = 0.01 M. As a result, the pH rises from 4.74 to 12.0:

$$[H_3O^+] = \frac{K_w}{[OH^-]} = \frac{(1.0 \times 10^{-14})}{(0.01)} = 1 \times 10^{-12} \text{ M}$$

$$pH = 12.0$$

The abilities of the HCl solution and the buffer solution to absorb added base are contrasted in Figure 16.4.

FIGURE 16.4 The color of each solution is due to the presence of a few drops of methyl red, an acid–base indicator that is red at pH less than about 5.4 and yellow at pH greater than about 5.4. **(a)** 1.00 L of 1.8×10^{-5} M HCl (pH = 4.74); **(b)** the solution from part (a) turns yellow (pH > 5.4) after addition of only a few drops of 0.10 M NaOH; **(c)** 1.00 L of a 0.10 M acetic acid–0.10 M sodium acetate buffer solution (pH = 4.74); **(d)** the solution from part (c) is still red (pH < 5.4) after addition of 100 mL of 0.10 M NaOH.

(a)

(b)

(c)

(d)

We sometimes talk about the buffering ability of a solution using the term **buffer capacity** as a measure of the amount of acid or base that the solution can absorb without a significant change in pH. Buffer capacity is also a measure of how little the pH changes with the addition of a given amount of acid or base. Buffer capacity depends on how many moles of weak acid and conjugate base are present. For equal volumes of solution, the more concentrated the solution, the greater the buffer capacity. For solutions having the same concentration, the greater the volume, the greater the buffer capacity.

◆— KEY CONCEPT PROBLEM 16.6   The following pictures represent solutions that contain a weak acid HA and/or its sodium salt NaA. (Na$^+$ ions and solvent water molecules have been omitted for clarity.)

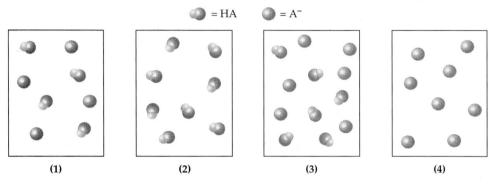

(1)   (2)   (3)   (4)

(a) Which of the solutions are buffer solutions?

(b) Which solution has the greatest buffer capacity?

▶PROBLEM 16.7   Calculate the pH of 0.100 L of a buffer solution that is 0.25 M in HF and 0.50 M in NaF. What is the change in pH on addition of
(a) 0.002 mol of HNO$_3$?   **(b)** 0.004 mol of KOH?

▶PROBLEM 16.8   Calculate the change in pH when 0.002 mol of HNO$_3$ is added to 0.100 L of a buffer solution that is 0.050 M in HF and 0.100 M in NaF. Does this solution have more or less buffer capacity than the one in Problem 16.7?

16.4   The Henderson–Hasselbalch Equation

We saw in the previous section that the H$_3$O$^+$ concentration in a buffer solution depends on the dissociation constant of the weak acid and on the concentration ratio [weak acid]/[conjugate base]:

$$[H_3O^+] = K_a \frac{[\text{Acid}]}{[\text{Base}]}$$

This equation can be rewritten in logarithmic form by taking the negative base-10 logarithm of both sides:

$$pH = -\log[H_3O^+] = -\log K_a - \log \frac{[\text{Acid}]}{[\text{Base}]}$$

If we then define pK_a = −log K_a by analogy with pH = −log [H$_3$O$^+$], and take account of the fact that

$$-\log \frac{[\text{Acid}]}{[\text{Base}]} = \log \frac{[\text{Base}]}{[\text{Acid}]}$$

we obtain an expression called the **Henderson–Hasselbalch equation**:

Henderson–Hasselbalch equation: $pH = pK_a + \log \dfrac{[\text{Base}]}{[\text{Acid}]}$

The Henderson–Hasselbalch equation says that the pH of a buffer solution has a value close to the pK_a of the weak acid, differing only by the amount log [base]/[acid]. When [base]/[acid] = 1, then log [base]/[acid] = 0, and the pH equals the pK_a.

The real importance of the Henderson–Hasselbalch equation, particularly in biochemistry, is that it tells us how the pH affects the percent dissociation of a weak acid. Suppose, for example, that you have a solution containing the amino acid glycine, one of the molecules from which proteins are made, and that the pH of the solution is 2.00 pH units greater than the pK_a of glycine:

$$\overset{+}{H_3}NCH_2\overset{\displaystyle O}{\overset{\|}{C}}O^-(aq) + H_2O(l) \rightleftharpoons H_2NCH_2\overset{\displaystyle O}{\overset{\|}{C}}O^-(aq) + H_3O^+(aq) \qquad K_a = 2.5 \times 10^{-10}$$
$$\text{Glycine} \qquad\qquad\qquad\qquad\qquad\qquad\qquad\qquad pK_a = 9.60$$

$$pH = pK_a + 2.00 = 11.60$$

Since $pH = pK_a + 2.00$, then log [base]/[acid] = 2.00, and [base]/[acid] = $1.0 \times 10^2 = 100/1$. Therefore, the Henderson–Hasselbalch equation tells us that 100 of every 101 glycine molecules are dissociated, which corresponds to 99% dissociation:

$$\log \dfrac{[\text{Base}]}{[\text{Acid}]} = pH - pK_a = 2.00$$

$$\dfrac{[\text{Base}]}{[\text{Acid}]} = 1.0 \times 10^2 = \dfrac{100}{1} \qquad 99\% \text{ dissociation}$$

The Henderson–Hasselbalch equation thus gives the following relationships:

At $pH = pK_a + 2.00$: $\dfrac{[\text{Base}]}{[\text{Acid}]} = 1.0 \times 10^2 = \dfrac{100}{1}$ 99% dissociation

At $pH = pK_a + 1.00$: $\dfrac{[\text{Base}]}{[\text{Acid}]} = 1.0 \times 10^1 = \dfrac{10}{1}$ 91% dissociation

At $pH = pK_a + 0.00$: $\dfrac{[\text{Base}]}{[\text{Acid}]} = 1.0 \times 10^0 = \dfrac{1}{1}$ 50% dissociation

At $pH = pK_a - 1.00$: $\dfrac{[\text{Base}]}{[\text{Acid}]} = 1.0 \times 10^{-1} = \dfrac{1}{10}$ 9% dissociation

At $pH = pK_a - 2.00$: $\dfrac{[\text{Base}]}{[\text{Acid}]} = 1.0 \times 10^{-2} = \dfrac{1}{100}$ 1% dissociation

The Henderson–Hasselbalch equation also tells us how to prepare a buffer solution with a given pH. The general idea is to select a weak acid whose pK_a is close to the desired pH and then adjust the [base]/[acid] ratio to the value specified by the Henderson–Hasselbalch equation. For example, to prepare a buffer having pH near 7, we might use the $H_2PO_4^-$–HPO_4^{2-} conjugate acid–base pair because pK_a for $H_2PO_4^-$ is $-\log(6.2 \times 10^{-8}) = 7.21$. Similarly, a mixture of NH_4Cl and NH_3 would be a good choice for a buffer having pH near 9 because pK_a for NH_4^+ is $-\log(5.6 \times 10^{-10}) = 9.25$. As a rule of thumb, the pK_a of the weak acid component of a buffer should be within ± 1 pH unit of the desired pH.

Because buffer solutions are widely used in the laboratory and in medicine, prepackaged buffers having a variety of precisely known pH values are commercially available (Figure 16.5). The manufacturer prepares these buffers by choosing a buffer system having an appropriate pK_a value and then adjusting the amounts of the ingredients so that the [base]/[acid] ratio has the proper value.

It's important to realize that the pH of a buffer solution does not depend on the volume of the solution. Because a change in solution volume changes the concentrations of the acid and base by the same amount, the [base]/[acid] ratio and the pH remain unchanged. As a result, the volume of water used to prepare a buffer solution is not critical, and you can dilute a buffer without a change in pH. The pH depends only on pK_a and on the relative molar amounts of weak acid and conjugate base.

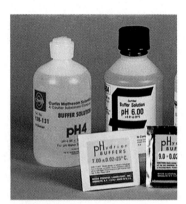

FIGURE 16.5 Prepackaged buffer solutions of known pH, and solid ingredients for preparing buffer solutions of known pH.

 Calculating pH Using Henderson–Hasselbalch Equation activity

EXAMPLE 16.3

(a) Use the Henderson–Hasselbalch equation to calculate the pH of a buffer solution that is 0.45 M in NH_4Cl and 0.15 M in NH_3.

(b) How would you prepare an NH_4Cl–NH_3 buffer that has a pH of 9.00?

SOLUTION

(a) We've already solved this problem by another method in Example 16.2. Now that we've discussed equilibria in buffer solutions, though, we can use the Henderson–Hasselbalch equation as a shortcut. Since NH_4^+ is the weak acid in an NH_4^+–NH_3 buffer solution, we need to find the pK_a for NH_4^+. Calculate it from the tabulated K_b value for NH_3 (Appendix C):

$$K_a = \frac{K_w}{K_b} = \frac{1.0 \times 10^{-14}}{1.8 \times 10^{-5}} = 5.6 \times 10^{-10}$$

$$pK_a = -\log K_a = -\log(5.6 \times 10^{-10}) = 9.25$$

Since [base] = $[NH_3]$ = 0.15 M and [acid] = $[NH_4^+]$ = 0.45 M,

$$pH = pK_a + \log \frac{[Base]}{[Acid]} = 9.25 + \log\left(\frac{0.15}{0.45}\right) = 9.25 - 0.48 = 8.77$$

The pH of the buffer solution is 8.77.

A common error in using the Henderson–Hasselbalch equation is to invert the [base]/[acid] ratio. It is therefore wise to check that your answer makes chemical sense. If the concentrations of the acid and its conjugate base are equal, the pH will equal the pK_a. If the acid predominates, the pH will be less than the pK_a, and if the conjugate base predominates, the pH will be greater than the pK_a. In this example, [acid] = $[NH_4^+]$ is greater than [base] = $[NH_3]$, and so the calculated pH (8.77) should be less than the pK_a (9.25).

(b) Rearrange the Henderson–Hasselbalch equation to obtain an expression for the relative amounts of NH_3 and NH_4^+ in a solution having pH = 9.00:

$$\log \frac{[\text{Base}]}{[\text{Acid}]} = \text{pH} - \text{p}K_a = 9.00 - 9.25 = -0.25$$

Therefore,

$$\frac{[NH_3]}{[NH_4^+]} = \text{antilog}(-0.25) = 10^{-0.25} = 0.56$$

The solution must contain 0.56 mol of NH_3 for every 1.00 mol of NH_4Cl, but the volume of the solution isn't critical. One way of preparing the buffer would be to combine 1.00 mol of NH_4Cl (53.5 g) with 0.56 mol of NH_3 (say, 560 mL of 1.00 M NH_3).

EXAMPLE 16.4   What $[NH_3]/[NH_4^+]$ ratio is required for a buffer solution that has pH = 7.00? Why is a mixture of NH_3 and NH_4Cl a poor choice for a buffer having pH = 7.00?

SOLUTION   We can calculate the required $[NH_3]/[NH_4^+]$ ratio from the Henderson–Hasselbalch equation:

$$\log \frac{[NH_3]}{[NH_4^+]} = \text{pH} - \text{p}K_a = 7.00 - 9.25 = -2.25$$

$$\frac{[NH_3]}{[NH_4^+]} = \text{antilog}(-2.25) = 10^{-2.25} = 5.6 \times 10^{-3}$$

For a typical value of $[NH_4^+]$—say, 1.0 M—the NH_3 concentration would have to be very small (0.0056 M). Such a solution is a poor buffer because it has little capacity to absorb added acid. Also, because the $[NH_3]/[NH_4^+]$ ratio is far from unity, addition of a small amount of H_3O^+ or OH^- will result in a large change in pH.

▶ **PROBLEM 16.9**   Use the Henderson–Hasselbalch equation to calculate the pH of a buffer solution prepared by mixing equal volumes of 0.20 M $NaHCO_3$ and 0.10 M Na_2CO_3. (K_a values are given in Appendix C.)

▶ **PROBLEM 16.10**   Give a recipe for preparing a $NaHCO_3$–Na_2CO_3 buffer solution that has pH = 10.40.

▶ **PROBLEM 16.11**   Suppose you are performing an experiment that requires a constant pH of 7.50. Suggest an appropriate buffer system based on the K_a values in Appendix C.

▶ **PROBLEM 16.12**   The pK_a of the amino acid serine is 9.15. At what pH is serine:
(a) 66% dissociated?   **(b)** 5% dissociated?

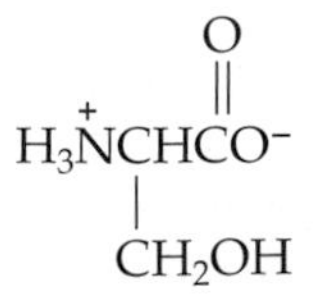

Serine
pK_a = 9.15

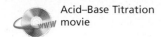

Acid–Base Titration movie

16.5 pH Titration Curves

In a typical acid–base titration (Section 3.10), a solution containing a known concentration of base (or acid) is added slowly from a buret to a second solution containing an unknown concentration of acid (or base). The progress of the titration is monitored, either by using a pH meter or by observing the color of a suitable acid–base indicator. With a pH meter, you can record data to produce a **pH titration curve**, a plot of the pH of the solution as a function of the volume of added titrant (Figure 16.6).

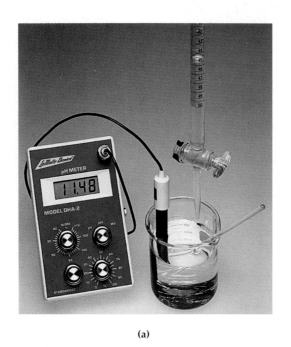

(a)

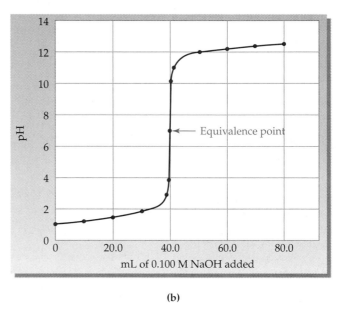

(b)

FIGURE 16.6 A strong acid–strong base titration curve. **(a)** In this pH titration, 0.100 M NaOH is added slowly from a buret to an HCl solution of unknown concentration. The pH of the solution is measured with a pH meter and is recorded as a function of the volume of NaOH added. **(b)** The pH titration curve for titration of 40.0 mL of 0.100 M HCl with 0.100 M NaOH. The pH increases gradually in the regions before and after the equivalence point, but increases rapidly in the region near the equivalence point. The equivalence point comes after addition of 40.0 mL of 0.100 M NaOH. The pH at the equivalence point is 7.00.

Why study titration curves? The shape of a pH titration curve makes it possible to identify the **equivalence point** in a titration, the point at which stoichiometrically equivalent quantities of acid and base have been mixed together. Knowing the shape of the titration curve is also useful in selecting a suitable indicator to signal the equivalence point. We'll explore both these points later.

We can calculate pH titration curves using the principles of aqueous solution equilibria. To understand why titration curves have certain characteristic shapes, let's calculate these curves for four important types of titration: (1) strong acid–strong base, (2) weak acid–strong base, (3) weak base–strong acid, and (4) polyprotic acid–strong base. For convenience, we'll express amounts of solute in millimoles (mmol) and solution volumes in milliliters (mL). Molar concentration can thus be expressed in mmol/mL, a unit that is equivalent to mol/L:

$$\text{Molarity} = \frac{\text{mmol of solute}}{\text{mL of solution}} = \frac{10^{-3} \text{ mol of solute}}{10^{-3} \text{ L of solution}} = \frac{\text{mol of solute}}{\text{L of solution}}$$

16.6   Strong Acid–Strong Base Titrations

As an example of a strong acid–strong base titration, let's consider the titration of 40.0 mL of 0.100 M HCl with 0.100 M NaOH. We'll calculate the pH at selected points in the course of the titration to illustrate the procedures we use to calculate the entire curve.

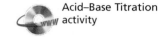

Acid–Base Titration activity

1. **Before Addition of Any NaOH**   Since HCl is a strong acid, the initial concentration of H_3O^+ is 0.100 M, and the pH is 1.00. (We've rounded the value of the pH to two significant figures.)

2. **Before the Equivalence Point**   Let's calculate the pH after addition of 10.0 mL of 0.100 M NaOH. The added OH^- ions will decrease $[H_3O^+]$ because of the neutralization reaction

$$H_3O^+(aq) + OH^-(aq) \xrightarrow{100\%} 2\,H_2O(l)$$

The number of millimoles of H_3O^+ present initially is the product of the initial volume of HCl and its molarity:

$$\text{mmol } H_3O^+ \text{ initial} = (40.0 \text{ mL})(0.100 \text{ mmol/mL}) = 4.00 \text{ mmol}$$

Similarly, the number of millimoles of OH^- added is the product of the volume of NaOH added and its molarity:

$$\text{mmol } OH^- \text{ added} = (10.0 \text{ mL})(0.100 \text{ mmol/mL}) = 1.00 \text{ mmol}$$

For each mmol of OH^- added, an equal amount of H_3O^+ will disappear because of the neutralization reaction. The number of millimoles of H_3O^+ remaining after neutralization is therefore

$$\text{mmol } H_3O^+ \text{ after neutralization} = \text{mmol } H_3O^+_{\text{initial}} - \text{mmol } OH^-_{\text{added}}$$
$$= 4.00 \text{ mmol} - 1.00 \text{ mmol} = 3.00 \text{ mmol}$$

We've carried out this calculation using *amounts* of acid and base (mmol) rather than *concentrations* (molarity) because the volume changes as the titration proceeds. If we divide the number of millimoles of H_3O^+ after neutralization by the total volume (now $40.0 + 10.0 = 50.0$ mL), we obtain $[H_3O^+]$ after neutralization:

$$[H_3O^+] \text{ after neutralization} = \frac{3.00 \text{ mmol}}{50.0 \text{ mL}} = 6.00 \times 10^{-2} \text{ M}$$

$$pH = -\log(6.00 \times 10^{-2}) = 1.22$$

This same procedure can be used to calculate the pH at other points prior to the equivalence point, giving the results summarized in Table 16.1.

TABLE 16.1 Sample Results for pH Calculations at Various Points in the Titration of 40.0 mL of 0.100 M HCl with 0.100 M NaOH

mL NaOH added	mmol OH⁻ added	mmol H₃O⁺ after neutr.	Total volume (mL)		[H₃O⁺] after neutr.	pH
Before the equivalence point:						
0.0	0.0	4.00	40.0		1.00×10^{-1}	1.00
10.0	1.00	3.00	50.0		6.00×10^{-2}	1.22
20.0	2.00	2.00	60.0		3.33×10^{-2}	1.48
30.0	3.00	1.00	70.0		1.43×10^{-2}	1.84
39.0	3.90	0.10	79.0		1.27×10^{-3}	2.90
39.9	3.99	0.01	79.9		1.3×10^{-4}	3.9
At the equivalence point:						
40.0	4.00	0.00	80.0		1.0×10^{-7}	7.00
Beyond the equivalence point:						
		mmol OH⁻ after neutr.		[OH⁻] after neutr.		
40.1	4.01	0.01	80.1	1.2×10^{-4}	8.3×10^{-11}	10.1
41.0	4.10	0.10	81.0	1.2×10^{-3}	8.3×10^{-12}	11.08
50.0	5.00	1.00	90.0	1.11×10^{-2}	9.0×10^{-13}	12.05
60.0	6.00	2.00	100.0	2.00×10^{-2}	5.0×10^{-13}	12.30
70.0	7.00	3.00	110.0	2.73×10^{-2}	3.7×10^{-13}	12.43
80.0	8.00	4.00	120.0	3.33×10^{-2}	3.0×10^{-13}	12.52

3. **At the Equivalence Point** After addition of 40.0 mL of 0.100 M NaOH, we have added (40.0 mL)(0.100 mmol/mL) = 4.00 mmol of NaOH, which is just enough OH⁻ to neutralize all the 4.00 mmol of HCl initially present. This is the equivalence point of the titration, and the pH is 7.00 because the solution contains only water and NaCl, a salt derived from a strong base and a strong acid.

4. **Beyond the Equivalence Point** After addition of 60.0 mL of 0.100 M NaOH, we have added (60.0 mL)(0.100 mmol/mL) = 6.00 mmol of NaOH, which is more than enough to neutralize the 4.00 mmol of HCl initially present. Consequently, an excess of OH⁻ (6.00 − 4.00 = 2.00 mmol) is present. Since the total volume is now 40.0 + 60.0 = 100.0 mL, the concentration of OH⁻ is

$$[\text{OH}^-] \text{ after neutralization} = \frac{2.00 \text{ mmol}}{100.0 \text{ mL}} = 2.00 \times 10^{-2} \text{ M}$$

The H₃O⁺ concentration and the pH are

$$[\text{H}_3\text{O}^+] = \frac{K_w}{[\text{OH}^-]} = \frac{1.0 \times 10^{-14}}{2.00 \times 10^{-2}} = 5.0 \times 10^{-13} \text{ M}$$

$$\text{pH} = -\log(5.0 \times 10^{-13}) = 12.30$$

Sample results for pH calculations at other places beyond the equivalence point are also included in Table 16.1.

Plotting the pH data in Table 16.1 as a function of milliliters of NaOH added gives the pH titration curve in Figure 16.6b (page 676). This curve exhibits a gradual increase in pH in the regions before and after the equivalence point but a very sharp increase in pH in the region near the equivalence point. Thus, when the volume of added NaOH increases from 39.9 to 40.1 mL (0.2 mL is only about 4 drops from a buret), the pH increases from 3.9 to 10.1 (Table 16.1). This very sharp increase in pH in the region of the equivalence point is characteristic of the titration curve for any strong acid–strong base reaction. It is this feature that allows us to identify the equivalence point when the concentration of the acid is unknown.

The pH curve for titration of a strong base with a strong acid is similar except that the initial pH is high and then decreases as acid is added.

▶ **PROBLEM 16.13** Calculate the pH of the solution resulting from titration of 40.0 mL of 0.100 M HCl with **(a)** 35.0 mL and **(b)** 45.0 mL of 0.100 M NaOH. Are your results consistent with the pH data in Table 16.1?

▶ **PROBLEM 16.14** A 40.0 mL volume of 0.100 M NaOH is titrated with 0.0500 M HCl. Calculate the pH after addition of the following volumes of acid:
(a) 60.0 mL   **(b)** 80.2 mL   **(c)** 100.0 mL

16.7 Weak Acid–Strong Base Titrations

As an example of a weak acid–strong base titration, let's consider the titration of 40.0 mL of 0.100 M acetic acid with 0.100 M NaOH. Calculation of the pH at selected points along the titration curve is straightforward because we've already met all the equilibrium problems that arise.

1. **Before Addition of any NaOH** The equilibrium problem at this point is the familiar one of calculating the pH of a solution of a weak acid. The calculated pH of 0.100 M acetic acid is 2.89.

2. **Before the Equivalence Point** Since acetic acid is largely undissociated and NaOH is completely dissociated, the neutralization reaction is

$$CH_3CO_2H(aq) + OH^-(aq) \xrightarrow{100\%} H_2O(l) + CH_3CO_2^-(aq)$$

After addition of 20.0 mL of 0.100 M NaOH, we have added (20.0 mL) (0.100 mmol/mL) = 2.00 mmol of NaOH, which is enough OH$^-$ to neutralize exactly half the 4.00 mmol of CH$_3$CO$_2$H present initially. Neutralization gives a buffer solution that contains 2.00 mmol of CH$_3$CO$_2^-$ and 4.00 − 2.00 = 2.00 mmol of CH$_3$CO$_2$H. Consequently, the [base]/[acid] ratio is unity, and pH = pK_a:

$$pH = pK_a + \log\frac{[\text{Base}]}{[\text{Acid}]} = pK_a = 4.74$$

3. **At the Equivalence Point** The equivalence point is reached after adding 40.0 mL of 0.100 M NaOH (4.00 mmol), which is just enough OH$^-$ to neutralize all the 4.00 mmol of CH$_3$CO$_2$H initially present. After neutralization, the solution contains 0.0500 M CH$_3$CO$_2^-$:

$$[\text{Na}^+] = [CH_3CO_2^-] = \frac{4.00 \text{ mmol}}{40.0 \text{ mL} + 40.0 \text{ mL}} = 0.0500 \text{ M}$$

Because Na$^+$ is neither an acid nor a base and CH$_3$CO$_2^-$ is a weak base, we have a basic salt solution (Section 15.14), whose pH can be calculated as 8.72 by the method outlined in Example 15.14. For a weak monoprotic acid–strong base titration, the pH at the equivalence point is always greater than 7 because the anion of the weak acid is a base.

4. **After the Equivalence Point** After addition of 60.0 mL of 0.100 M NaOH, we have added $(60.0 \text{ mL})(0.100 \text{ mmol/mL}) = 6.00$ mmol of NaOH, which is more than enough OH^- to neutralize the 4.00 mmol of CH_3CO_2H present initially. The total volume is $40.0 + 60.0 = 100.0$ mL, and the concentrations after neutralization are

$$[CH_3CO_2^-] = \frac{4.00 \text{ mmol}}{100.0 \text{ mL}} = 0.0400 \text{ M}$$

$$[OH^-] = \frac{6.00 \text{ mmol} - 4.00 \text{ mmol}}{100.0 \text{ mL}} = 0.0200 \text{ M}$$

The principal reaction is the same as that at the equivalence point:

$$CH_3CO_2^-(aq) + H_2O(l) \rightleftharpoons CH_3CO_2H(aq) + OH^-(aq)$$

In this case, however, $[OH^-]$ from the principal reaction is negligible compared with $[OH^-]$ from the excess NaOH. The hydronium ion concentration and the pH can be calculated from that $[OH^-]$:

$$[H_3O^+] = \frac{K_w}{[OH^-]} = \frac{1.0 \times 10^{-14}}{0.0200} = 5.0 \times 10^{-13} \text{ M}$$

$$pH = 12.30$$

In general, $[OH^-]$ from reaction of the anion of a weak acid with water is negligible beyond the equivalence point, and the pH is determined by the concentration of OH^- from the excess NaOH.

The results of pH calculations for titration of 0.100 M CH_3CO_2H with 0.100 M NaOH are plotted in Figure 16.7. Comparison of the titration curves

FIGURE 16.7 **(a)** A weak acid–strong base titration curve compared with **(b)** a strong acid–strong base curve. The curves shown are for titration of **(a)** 40.0 mL of 0.100 M CH_3CO_2H with 0.100 M NaOH (blue curve) and **(b)** 40.0 mL of 0.100 M HCl with 0.100 M NaOH (red curve). The pH ranges in which the acid–base indicators phenolphthalein and methyl red change color are indicated. Note that phenolphthalein is an excellent indicator for the weak acid–strong base titration because the equivalence point **(a)** is at pH 8.72; methyl red is an unsatisfactory indicator because it changes color well before the equivalence point. Either phenolphthalein or methyl red can be used for the strong acid–strong base titration because the curve rises very steeply in the region of the equivalence point **(b)** at pH 7.00.

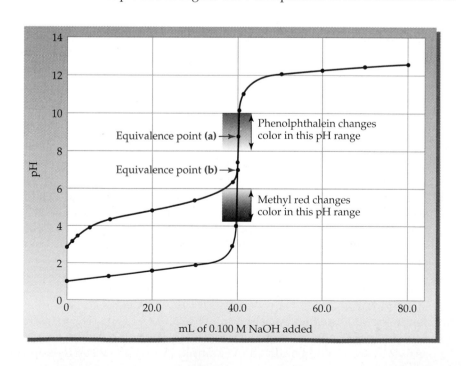

for the weak acid–strong base titration and the strong acid–strong base case shows several significant differences:

1. The initial rise in pH is greater for titration of the weak acid, but the curve then becomes more level in the region midway to the equivalence point. Both effects are due to the buffering action of the weak acid–conjugate base mixture. The curve has minimum slope exactly halfway to the equivalence point, where the buffering action is maximized and the $pH = pK_a$ for the weak acid.

2. The increase in pH in the region near the equivalence point is smaller than in the strong acid–strong base case. The weaker the acid, the smaller the increase in the pH, as illustrated in Figure 16.8.

3. The pH at the equivalence point is greater than 7 because the anion of a weak acid is a base.

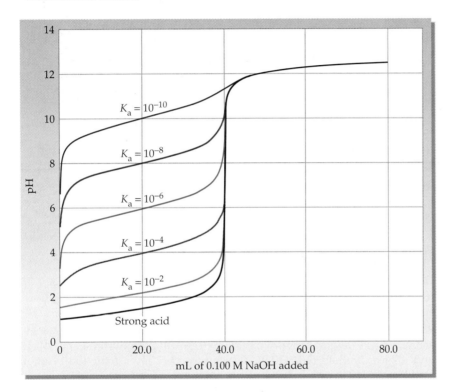

FIGURE 16.8 Various weak acid–strong base pH titration curves. The curves shown are for titration of 40.0 mL of 0.100 M solutions of various weak acids with 0.100 M NaOH. In each case, the equivalence point comes after addition of 40.0 mL of 0.100 M NaOH, but the increase in pH at the equivalence point gets smaller and the equivalence point gets more difficult to detect as the K_a value of the weak acid decreases.

Beyond the equivalence point, the curves for the weak acid–strong base and strong acid–strong base titrations are identical because the pH in both cases is determined by the concentration of OH^- from the excess NaOH.

Figure 16.7 shows how knowing the shape of a pH titration curve makes possible selection of a suitable acid–base indicator to signal the equivalence point of a titration. Phenolphthalein is an excellent indicator for the CH_3CO_2H–NaOH titration because the pH at the equivalence point (8.72) falls within the pH range (8.2–9.8) in which phenolphthalein changes color. Methyl red, however, is an unacceptable indicator for this titration because the pH range in which it changes color (4.2–6.0) corresponds to a pH well before the equivalence point. Anyone who tried to determine the acetic acid content in a solution of unknown concentration would badly underestimate the amount of acid present if methyl red were used as the indicator.

Either phenolphthalein or methyl red can be used as the indicator for a strong acid–strong base titration. The increase in pH in the region of the equivalence point is so steep that any indicator changing color in the pH range 4–10 can be used without making a significant error in locating the equivalence point.

KEY CONCEPT PROBLEM 16.15   The following pictures represent solutions at various points in the titration of a weak acid HA with aqueous NaOH. (Na$^+$ ions and solvent water molecules have been omitted for clarity.)

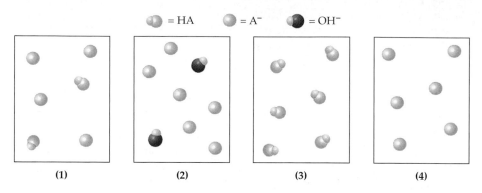

To which of the following points do solutions (1)–(4) correspond?

(a) Before addition of any NaOH

(b) Before the equivalence point

(c) At the equivalence point

(d) After the equivalence point

PROBLEM 16.16   Consider the titration of 100.0 mL of 0.016 M HOCl ($K_a = 3.5 \times 10^{-8}$) with 0.0400 M NaOH. How many milliliters of 0.0400 M NaOH are required to reach the equivalence point? Calculate the pH:

(a) After addition of 10.0 mL of 0.0400 M NaOH

(b) Halfway to the equivalence point

(c) At the equivalence point

PROBLEM 16.17   The following acid–base indicators change color in the indicated pH ranges: bromthymol blue (6.0–7.6), thymolphthalein (9.4–10.6), and alizarin yellow (10.1–12.0). Which indicator is best for the titration in Problem 16.16? Which indicator is unacceptable? Explain.

16.8   Weak Base–Strong Acid Titrations

Figure 16.9 shows the pH titration curve for a typical weak base–strong acid titration, the titration of 40.0 mL of 0.100 M NH$_3$ with 0.100 M HCl. The pH calculations are simply outlined to save space; you should verify the results yourself.

1. **Before Addition of Any HCl**   The equilibrium problem at the start of the titration is the familiar one of calculating the pH of a solution of a weak base (Section 15.12). The principal reaction at this point is the reaction of ammonia with water:

$$NH_3(aq) + H_2O(l) \rightleftharpoons NH_4^+(aq) + OH^-(aq) \qquad K_b = 1.8 \times 10^{-5}$$

The initial pH is 11.12.

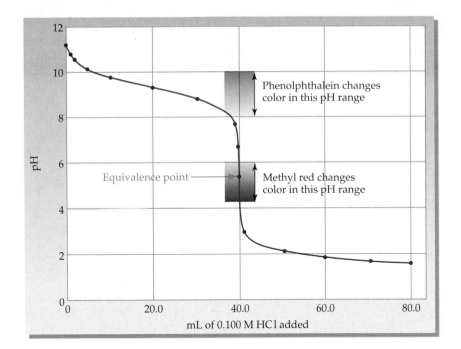

FIGURE 16.9 A weak base–strong acid titration curve. The curve shown is for titration of 40.0 mL of 0.100 M NH_3 with 0.100 M HCl. The pH is 11.12 at the start of the titration, 9.25 (the pK_a value for NH_4^+) in the buffer region halfway to the equivalence point, and 5.28 at the equivalence point. Note that methyl red is a good indicator for this titration, but phenolphthalein is unacceptable.

2. **Before the Equivalence Point** As HCl is added to the NH_3 solution, NH_3 is converted to NH_4^+ because of the neutralization reaction

$$NH_3(aq) \ + \ H_3O^+(aq) \ \xrightarrow{100\%} \ NH_4^+(aq) \ + \ H_2O(l)$$

The neutralization reaction goes to completion, but the amount of H_3O^+ added before the equivalence point is not sufficient to convert all the NH_3 to NH_4^+. We therefore have an NH_4^+–NH_3 buffer solution, which accounts for the leveling of the titration curve in the buffer region between the start of the titration and the equivalence point. The pH at any specific point can be calculated from the Henderson–Hasselbalch equation. After addition of 20.0 mL of 0.100 M HCl (halfway to the equivalence point), $[NH_3] = [NH_4^+]$ and the pH is equal to pK_a for NH_4^+ (9.25).

3. **At the Equivalence Point** The equivalence point is reached after adding 40.0 mL of 0.100 M HCl (4.00 mmol). At this point the 4.00 mmol of NH_3 present initially has been converted to 4.00 mmol of NH_4^+; $[NH_4^+] =$ (4.00 mmol)/(80.0 mL) = 0.0500 M. Since NH_4^+ is a weak acid and Cl^- is neither an acid nor a base, we have an acidic salt solution (Section 15.14). The principal reaction is

$$NH_4^+(aq) \ + \ H_2O(l) \ \rightleftharpoons \ H_3O^+(aq) \ + \ NH_3(aq) \qquad K_a = K_w/K_b = 5.6 \times 10^{-10}$$

The pH at the equivalence point is 5.28.

4. **Beyond the Equivalence Point** In this region, all the NH_3 has been converted to NH_4^+, and an excess of H_3O^+ is present from the excess HCl. Since acid dissociation of NH_4^+ produces a negligible $[H_3O^+]$ compared with $[H_3O^+]$ from the excess HCl, the pH can be calculated directly from the concentration of excess HCl. For example, after addition of 60.0 mL of 0.100 M HCl:

$$[H_3O^+] = \frac{6.00 \text{ mmol} \ - \ 4.00 \text{ mmol}}{100.0 \text{ mL}} = 0.0200 \text{ M}$$

$$pH = 1.70$$

16.9 Polyprotic Acid–Strong Base Titrations

As a final example of an acid–base titration, let's consider the gradual addition of NaOH to the protonated form of the amino acid alanine (H_2A^+), a substance that acts as a diprotic acid. Amino acids (about which we'll have more to say in Chapter 24) are both acidic and basic, and can be protonated by strong acids. The resultant protonated amino acid thus has two dissociable protons and can react with two molar amounts of OH^- to give first the neutral form and then the anionic form:

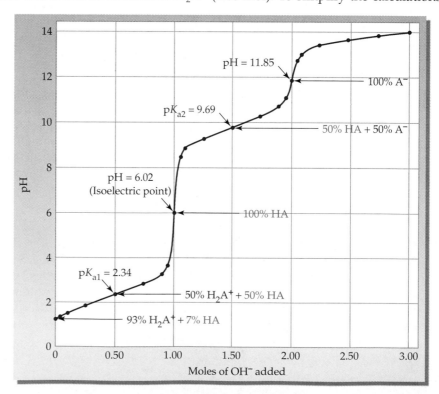

The proton of the $-CO_2H$ group is more acidic than the proton of the $-NH_3^+$ group and is neutralized first. The neutral form of alanine (HA) that results has a plus charge on the $-NH_3^+$ group and a minus charge on the $-CO_2^-$ group but is electrically neutral overall. In a second step, the proton of the $-NH_3^+$ group is neutralized, yielding the anionic form of alanine (A^-). The dissociation equilibria and their K_a values are

$$H_2A^+(aq) + H_2O(l) \rightleftharpoons H_3O^+(aq) + HA(aq) \qquad K_{a1} = 4.6 \times 10^{-3}; pK_{a1} = 2.34$$

$$HA(aq) + H_2O(l) \rightleftharpoons H_3O^+(aq) + A^-(aq) \qquad K_{a2} = 2.0 \times 10^{-10}; pK_{a2} = 9.69$$

Figure 16.10 shows the pH titration curve for addition of NaOH to 1.00 L of a 1.00 M solution of H_2A^+ (1.00 mol). To simplify the calculations,

FIGURE 16.10 Change in the pH of 1.00 L of a 1.00 M solution of H_2A^+ on addition of solid NaOH. The protonated form of alanine, H_2A^+, is a diprotic acid, so the titration curve exhibits two equivalence points, at pH 6.02 and pH 11.85, and two buffer regions, near pH 2.34 and pH 9.69.

we've assumed that the added base is solid NaOH so that we can neglect volume changes in the course of the titration. Let's calculate the pH at some of the more important points.

1. **Before Addition of Any NaOH**   The equilibrium problem at the start of the titration is the familiar one of calculating the pH of a diprotic acid (Section 15.11). The principal reaction is dissociation of H_2A^+, and $[H_3O^+]$ can be calculated from the equilibrium equation

$$K_{a1} = 4.6 \times 10^{-3} = \frac{[H_3O^+][HA]}{[H_2A^+]} = \frac{(x)(x)}{1.00 - x}$$

Solving the quadratic equation gives $[H_3O^+] = 0.066$ M and pH $= 1.18$.

2. **Halfway to the First Equivalence Point**   As NaOH is added, H_2A^+ is converted to HA because of the neutralization reaction

$$H_2A^+(aq) + OH^-(aq) \xrightarrow{100\%} H_2O(l) + HA(aq)$$

Halfway to the first equivalence point, we have an H_2A^+–HA buffer solution with $[H_2A^+] = [HA]$. The Henderson–Hasselbalch equation gives pH $= pK_{a1} = 2.34$.

3. **At the First Equivalence Point**   At this point, we have added just enough NaOH to convert all the H_2A^+ to HA. The principal reaction at the first equivalence point is proton transfer between HA molecules:

$$2\,HA(aq) \rightleftharpoons H_2A^+(aq) + A^-(aq) \qquad K = K_{a2}/K_{a1} = 4.3 \times 10^{-8}$$

It can be shown (see Problem 16.132) that the pH at this point is equal to the average of the two pK_a values:

$$\text{pH (at first equivalence point)} = \frac{pK_{a1} + pK_{a2}}{2}$$

Since H_2A^+ has $pK_{a1} = 2.34$ and $pK_{a2} = 9.69$, the pH at the first equivalence point is $(2.34 + 9.69)/2 = 6.02$. The same situation holds for most polyprotic acids: The pH at the first equivalence point is equal to the average of pK_{a1} and pK_{a2}.

For an amino acid, the pH value, $(pK_{a1} + pK_{a2})/2$, is called the *isoelectric point* (Figure 16.10). At that point the concentration of the neutral HA is at a maximum, and the very small concentrations of H_2A^+ and A^- are equal. Biochemists take advantage of isoelectric points in separating mixtures of amino acids and proteins.

4. **Halfway Between the First and Second Equivalence Points**   At this point, half the HA has been converted to A^- because of the neutralization reaction

$$HA(aq) + OH^-(aq) \xrightarrow{100\%} H_2O(l) + A^-(aq)$$

We thus have an HA–A^- buffer solution with $[HA] = [A^-]$. Therefore, pH $= pK_{a2} = 9.69$.

5. **At the Second Equivalence Point** At this point, we have added enough NaOH to convert all the HA to A⁻, and we have a 1.00 M solution of a basic salt (Section 15.14). The principal reaction is

$$A^-(aq) + H_2O(l) \rightleftharpoons HA(aq) + OH^-(aq)$$

and its equilibrium constant is

$$K_b = \frac{K_w}{K_a \text{ for HA}} = \frac{K_w}{K_{a2}} = \frac{1.0 \times 10^{-14}}{2.0 \times 10^{-10}} = 5.0 \times 10^{-5}$$

We can obtain [OH⁻] from the equilibrium equation for the principal reaction and then calculate [H₃O⁺] and pH in the usual way:

$$K_b = 5.0 \times 10^{-5} = \frac{[HA][OH^-]}{[A^-]} = \frac{(x)(x)}{1.00 - x}; \quad x = [OH^-] = 7.1 \times 10^{-3} \text{ M}$$

$$[H_3O^+] = \frac{K_w}{[OH^-]} = \frac{1.0 \times 10^{-14}}{7.1 \times 10^{-3}} = 1.4 \times 10^{-12} \text{ M}$$

$$pH = 11.85$$

Since the initial solution of H₂A⁺ contained 1.00 mol of H₂A⁺, the amount of NaOH required to reach the second equivalence point is 2.00 mol. Beyond the second equivalence point, the pH is determined by [OH⁻] from the excess NaOH.

EXAMPLE 16.5 Consider the titration of 30.0 mL of a 0.0600 M solution of the protonated form of the amino acid methionine (H₂A⁺) with 0.0900 M NaOH. Calculate the pH after addition of 20.0 mL of base.

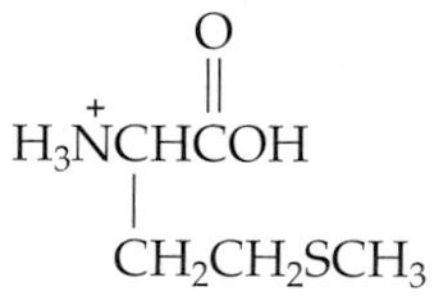

$$\begin{array}{c} \overset{+}{H_3}\text{NCHCOH} \\ | \\ CH_2CH_2SCH_3 \end{array} \quad \begin{array}{l} \overset{O}{\parallel} \\ \\ \end{array}$$

Methionine cation (H₂A⁺)
$K_{a1} = 5.2 \times 10^{-3}$
$K_{a2} = 6.2 \times 10^{-10}$

SOLUTION The number of millimoles of H₂A⁺ present initially and of NaOH added are

mmol H₂A⁺ initial = (30.0 mL)(0.0600 mmol/mL) = 1.80 mmol

mmol NaOH added = (20.0 mL)(0.0900 mmol/mL) = 1.80 mmol

The added base is just enough to reach the first equivalence point, converting all the H₂A⁺ to HA because of the neutralization reaction

$$H_2A^+(aq) + OH^-(aq) \xrightarrow{100\%} H_2O(l) + HA(aq)$$

Therefore, the pH is equal to the average of pK_{a1} and pK_{a2}:

$$pH = \frac{pK_{a1} + pK_{a2}}{2} = \frac{[-\log(5.2 \times 10^{-3})] + [-\log(6.2 \times 10^{-10})]}{2}$$

$$= \frac{2.28 + 9.21}{2} = 5.74$$

▶ **PROBLEM 16.18** Consider the titration of 40.0 mL of 0.0800 M H_2SO_3 ($K_{a1} = 1.5 \times 10^{-2}$; $K_{a2} = 6.3 \times 10^{-8}$) with 0.160 M NaOH. Calculate the pH after addition of the following volumes of 0.160 M NaOH:
(a) 20.0 mL (b) 30.0 mL (c) 35.0 mL

▶ **PROBLEM 16.19** Consider the titration of 40.0 mL of a 0.0250 M solution of the protonated form of the amino acid valine (H_2A^+) with 0.100 M NaOH. Calculate the pH after addition of the following volumes of 0.100 M NaOH:
(a) 10.0 mL (b) 15.0 mL (c) 20.0 mL

$$\begin{array}{c} O \\ \parallel \\ H_3\overset{+}{N}CHCOH \\ | \\ CHCH_3 \\ | \\ CH_3 \end{array}$$

Valine cation (H_2A^+)

$K_{a1} = 4.8 \times 10^{-3}$
$K_{a2} = 2.4 \times 10^{-10}$

16.10 Solubility Equilibria

Many biological and environmental processes involve the dissolution or precipitation of a sparingly soluble ionic compound. Tooth decay, for example, begins when tooth enamel, composed of the mineral hydroxyapatite, $Ca_5(PO_4)_3OH$, dissolves on reaction with organic acids produced by bacterial decomposition of foods rich in sugar. Kidney stones form when moderately insoluble calcium salts, such as calcium oxalate, CaC_2O_4, precipitate slowly over a long period of time. To understand the quantitative aspects of such solubility and precipitation phenomena, we must examine the principles of solubility equilibria.

Let's consider the solubility equilibrium in a saturated solution of calcium fluoride in contact with an excess of solid calcium fluoride. Like most sparingly soluble ionic solutes, calcium fluoride is a strong electrolyte in water and exists in the aqueous phase as dissociated hydrated ions, $Ca^{2+}(aq)$ and $F^-(aq)$. At equilibrium, the ion concentrations remain constant because the rate at which solid CaF_2 dissolves to give $Ca^{2+}(aq)$ and $F^-(aq)$ exactly equals the rate at which the ions crystallize to form solid CaF_2:

$$CaF_2(s) \rightleftharpoons Ca^{2+}(aq) + 2\,F^-(aq)$$

The equilibrium equation for the dissolution reaction is

$$K_{sp} = [Ca^{2+}][F^-]^2$$

where the equilibrium constant K_{sp} is called the **solubility product constant**, or simply the **solubility product**. As usual for a heterogeneous equilibrium, the concentration of the solid, CaF_2, has been omitted from the equilibrium equation because it is constant (Section 13.4).

For the general solubility equilibrium

$$M_mX_x(s) \rightleftharpoons m\,M^{n+}(aq) + x\,X^{y-}(aq)$$

the equilibrium constant expression for K_{sp} is

$$K_{sp} = [M^{n+}]^m[X^{y-}]^x$$

Most kidney stones consist of insoluble calcium salts, such as calcium oxalate.

A saturated solution of calcium fluoride in contact with solid CaF_2 contains constant equilibrium concentrations of $Ca^{2+}(aq)$ and $F^-(aq)$ because at equilibrium the ions crystallize at the same rate as the solid dissolves.

Thus, K_{sp} always equals the product of the equilibrium concentrations of all the ions on the right side of the chemical equation, with the concentration of each ion raised to the power of its coefficient in the balanced equation.

EXAMPLE 16.6   Write the expression for the solubility product of silver chromate, Ag_2CrO_4.

SOLUTION   First write the balanced equation for the solubility equilibrium:

$$Ag_2CrO_4(s) \rightleftharpoons 2\,Ag^+(aq) + CrO_4^{2-}(aq)$$

The exponents in the equilibrium constant expression for K_{sp} are the coefficients in the balanced equation. Therefore,

$$K_{sp} = [Ag^+]^2[CrO_4^{2-}]$$

▶ **PROBLEM 16.20**   Write the expression for K_{sp} of:
(a) $AgCl$ (b) PbI_2 (c) $Ca_3(PO_4)_2$

16.11   Measuring K_{sp} and Calculating Solubility from K_{sp}

The numerical value of a solubility product K_{sp} is measured by experiment. For example, we could determine K_{sp} for CaF_2 by adding an excess of solid CaF_2 to water, stirring the mixture to give a saturated solution of CaF_2, and then measuring the concentrations of Ca^{2+} and F^- in the saturated solution. To make sure that the concentrations had reached constant equilibrium values, we would want to stir the mixture for an additional period of time and then repeat the measurements. Suppose that we found $[Ca^{2+}] = 3.3 \times 10^{-4}$ M and $[F^-] = 6.7 \times 10^{-4}$ M. (The value of $[F^-]$ is twice the value of $[Ca^{2+}]$ because each mole of CaF_2 that dissolves yields 1 mol of Ca^{2+} ions and 2 mol of F^- ions.) We could then calculate K_{sp} for CaF_2:

$$K_{sp} = [Ca^{2+}][F^-]^2 = (3.3 \times 10^{-4})(6.7 \times 10^{-4})^2 = 1.5 \times 10^{-10}$$

Another way to measure K_{sp} for CaF_2 is to approach the equilibrium from the opposite direction—that is, by mixing sources of Ca^{2+} and F^- ions to give a precipitate of solid CaF_2 and a saturated solution of CaF_2. Suppose, for example, that we mix solutions of $CaCl_2$ and NaF, allow time for equilibrium to be reached, and then measure $[Ca^{2+}] = 1.5 \times 10^{-4}$ M and $[F^-] = 1.0 \times 10^{-3}$ M. These ion concentrations yield the same value of K_{sp}:

$$K_{sp} = [Ca^{2+}][F^-]^2 = (1.5 \times 10^{-4})(1.0 \times 10^{-3})^2 = 1.5 \times 10^{-10}$$

Values of K_{sp} are not affected by the presence of other ions in solution, such as Na^+ from NaF and Cl^- from $CaCl_2$, as long as the solution is very dilute. As ion concentrations increase, K_{sp} values are somewhat modified because of electrostatic interactions between ions. We'll ignore that complication here.

If the saturated solution is prepared by a method other than dissolution of CaF_2 in pure water, there are no separate restrictions on $[Ca^{2+}]$ and $[F^-]$; the only restriction on the ion concentrations is that the equilibrium constant expression $[Ca^{2+}][F^-]^2$ must equal the K_{sp}. That condition is satisfied by an infinite number of combinations of $[Ca^{2+}]$ and $[F^-]$, and therefore we can prepare many different solutions that are saturated with respect to CaF_2. For example, if $[F^-]$ is 1.0×10^{-2} M, then $[Ca^{2+}]$ must be 1.5×10^{-6} M:

$$[Ca^{2+}] = \frac{K_{sp}}{[F^-]^2} = \frac{1.5 \times 10^{-10}}{(1.0 \times 10^{-2})^2} = 1.5 \times 10^{-6} \text{ M}$$

Selected values of K_{sp} for various ionic compounds at 25°C are listed in Table 16.2, and additional values can be found in Appendix C. Like all equilibrium constants, values of K_{sp} depend on temperature (Section 13.9).

TABLE 16.2 K_{sp} Values for Some Ionic Compounds at 25°C

Name	Formula	K_{sp}
Aluminum hydroxide	$Al(OH)_3$	1.9×10^{-33}
Barium carbonate	$BaCO_3$	2.6×10^{-9}
Calcium carbonate	$CaCO_3$	5.0×10^{-9}
Calcium fluoride	CaF_2	1.5×10^{-10}
Lead(II) chloride	$PbCl_2$	1.2×10^{-5}
Lead(II) chromate	$PbCrO_4$	2.8×10^{-13}
Silver chloride	$AgCl$	1.8×10^{-10}
Silver sulfate	Ag_2SO_4	1.2×10^{-5}

Once the K_{sp} value for a compound has been measured, you can use it to calculate the solubility of the compound—the amount of compound that dissolves per unit volume of saturated solution. Because of two complications, however, calculated solubilities are often approximate. First, K_{sp} values can be difficult to measure, and values listed in different sources might differ by a factor of 10 or more. Second, calculated solubilities can be less than observed solubilities because of side reactions. For example, dissolution of $PbCl_2$ gives both Pb^{2+} and $PbCl^+$ because of some ion association between Pb^{2+} and Cl^- ions:

$$(1) \quad PbCl_2(s) \rightleftharpoons Pb^{2+}(aq) + 2\,Cl^-(aq)$$

$$(2) \quad Pb^{2+}(aq) + Cl^-(aq) \rightleftharpoons PbCl^+(aq)$$

In this book, we will calculate approximate solubilities assuming that ionic solutes are completely dissociated [reaction (1)]. In the case of $PbCl_2$, ignoring the second equilibrium gives a calculated solubility that is too low by a factor of about 2.

EXAMPLE 16.7   A particular saturated solution of silver chromate, Ag_2CrO_4, has $[Ag^+] = 5.0 \times 10^{-5}$ M and $[CrO_4^{2-}] = 4.4 \times 10^{-4}$ M. What is the value of K_{sp} for Ag_2CrO_4?

SOLUTION   Substituting the equilibrium concentrations into the expression for K_{sp} of Ag_2CrO_4 (Example 16.6) gives the value of K_{sp}:

$$K_{sp} = [Ag^+]^2[CrO_4^{2-}] = (5.0 \times 10^{-5})^2(4.4 \times 10^{-4}) = 1.1 \times 10^{-12}$$

Addition of aqueous K_2CrO_4 to aqueous $AgNO_3$ gives a red precipitate of Ag_2CrO_4 and a saturated solution of Ag_2CrO_4.

EXAMPLE 16.8   A saturated solution of Ag_2CrO_4 prepared by dissolving solid Ag_2CrO_4 in water has $[CrO_4^{2-}] = 6.5 \times 10^{-5}$ M. Calculate K_{sp} for Ag_2CrO_4.

SOLUTION   Because both the Ag^+ and CrO_4^{2-} ions come from dissolution of solid Ag_2CrO_4, $[Ag^+]$ must be twice $[CrO_4^{2-}]$. Therefore, $[Ag^+] = (2)(6.5 \times 10^{-5}) = 1.3 \times 10^{-4}$ M. The value of K_{sp} is

$$K_{sp} = [Ag^+]^2[CrO_4^{2-}] = (1.3 \times 10^{-4})^2(6.5 \times 10^{-5}) = 1.1 \times 10^{-12}$$

EXAMPLE 16.9   Calculate the solubility of MgF_2 in water at 25°C in units of:
(a) Moles per liter   **(b)** Grams per liter

SOLUTION   Write the balanced equation for the solubility equilibrium assuming complete dissociation of MgF_2, and then look up K_{sp} for MgF_2 in Appendix C:

$$MgF_2(s) \rightleftharpoons Mg^{2+}(aq) + 2\,F^-(aq) \qquad K_{sp} = 7.4 \times 10^{-11}$$

(a) If we define x as the number of moles per liter of MgF_2 that dissolves, then the saturated solution contains x mol/L of Mg^{2+} and $2x$ mol/L of F^-. Substituting these equilibrium concentrations into the expression for K_{sp} gives

$$K_{sp} = 7.4 \times 10^{-11} = [Mg^{2+}][F^-]^2 = (x)(2x)^2$$
$$4x^3 = 7.4 \times 10^{-11}$$
$$x^3 = 1.8 \times 10^{-11}$$
$$x = 2.6 \times 10^{-4}\,\text{mol/L}$$

Thus, the molar solubility of MgF_2 in water at 25°C is 2.6×10^{-4} M. (Taking account of the side reaction that produces MgF^+ would increase the calculated solubility by about 6%.)

Note that the number 2 appears twice in the expression $(x)(2x)^2$. The exponent 2 is required because of the equilibrium equation, $K_{sp} = [Mg^{2+}][F^-]^2$. The coefficient 2 in $2x$ is required because each mole of MgF_2 that dissolves gives 2 mol of $F^-(aq)$.

(b) To convert the solubility from units of moles per liter to units of grams per liter, multiply the molar solubility of MgF_2 by its molar mass (62.3 g/mol):

$$\text{Solubility (in g/L)} = \frac{2.6 \times 10^{-4}\,\text{mol}}{L} \times \frac{62.3\,\text{g}}{\text{mol}} = 1.6 \times 10^{-2}\,\text{g/L}$$

▶**PROBLEM 16.21**   A saturated solution of $Ca_3(PO_4)_2$ has $[Ca^{2+}] = 2.01 \times 10^{-8}$ M and $[PO_4^{3-}] = 1.6 \times 10^{-5}$ M. Calculate K_{sp} for $Ca_3(PO_4)_2$.

▶**PROBLEM 16.22**   Prior to having an X-ray exam of the upper gastrointestinal tract, a patient drinks an aqueous suspension of solid $BaSO_4$. (Scattering of X rays by barium greatly enhances the quality of the photograph.) Although Ba^{2+} is toxic, ingestion of $BaSO_4$ is safe because it is quite insoluble. If a saturated solution prepared by dissolving solid $BaSO_4$ in water has $[Ba^{2+}] = 1.05 \times 10^{-5}$ M, what is the value of K_{sp} for $BaSO_4$?

▶**PROBLEM 16.23**   Which has the greater molar solubility: AgCl with $K_{sp} = 1.8 \times 10^{-10}$ or Ag_2CrO_4 with $K_{sp} = 1.1 \times 10^{-12}$? Which has the greater solubility in grams per liter?

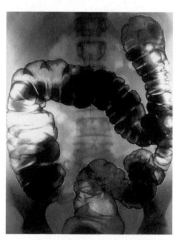

This X-ray photograph of a normal colon was taken soon after the patient drank a barium sulfate "cocktail."

KEY CONCEPT PROBLEM 16.24 The following pictures represent saturated solutions of three silver salts: AgX, AgY, and AgZ. (Other ions and solvent water molecules have been omitted for clarity.)

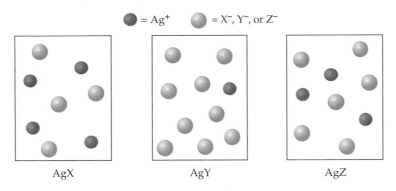

● = Ag^+ ○ = X^-, Y^-, or Z^-

AgX   AgY   AgZ

(a) Which salt has the largest value of K_{sp}?
(b) Which salt has the smallest value of K_{sp}?

16.12 Factors That Affect Solubility

The Common-Ion Effect

We've already discussed the common-ion effect in connection with the dissociation of weak acids and bases (Section 16.2). To see how a common ion affects the position of a solubility equilibrium, let's look again at the solubility of MgF_2:

Common Ion Effect movie

$$MgF_2(s) \rightleftharpoons Mg^{2+}(aq) + 2\,F^-(aq)$$

In Example 16.9, we found that the molar solubility of MgF_2 in pure water at 25°C is 2.6×10^{-4} M. Thus,

$$[Mg^{2+}] = 2.6 \times 10^{-4}\ M \qquad [F^-] = 5.2 \times 10^{-4}\ M$$

When MgF_2 dissolves in a solution that contains a common ion from another source—say, F^- from NaF—the position of the solubility equilibrium is shifted to the left by the common-ion effect. If $[F^-]$ is larger than 5.2×10^{-4} M, then $[Mg^{2+}]$ must be correspondingly smaller than 2.6×10^{-4} M to maintain the equilibrium expression $[Mg^{2+}][F^-]^2$ at a constant value of $K_{sp} = 7.4 \times 10^{-11}$. A smaller value of $[Mg^{2+}]$ thus means that MgF_2 is less soluble in a sodium fluoride solution than it is in pure water. Similarly, the presence of Mg^{2+} from another source—say, $MgCl_2$—shifts the solubility equilibrium to the left and decreases the solubility of MgF_2.

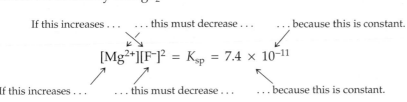

If this increases . . .   . . . this must decrease . . .   . . . because this is constant.

$$[Mg^{2+}][F^-]^2 = K_{sp} = 7.4 \times 10^{-11}$$

If this increases . . .   . . . this must decrease . . .   . . . because this is constant.

In general, the solubility of a slightly soluble ionic compound is decreased by the presence of a common ion in the solution, as illustrated in

Figure 16.11. The quantitative aspects of the common-ion effect are explored in Example 16.10.

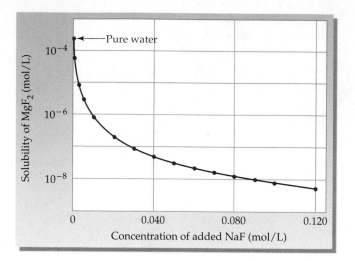

FIGURE 16.11 The common-ion effect. The solubility of MgF_2 at 25°C decreases markedly on addition of F^- ions. Note that the calculated solubility is plotted on a logarithmic scale.

EXAMPLE 16.10 Calculate the molar solubility of MgF_2 in 0.10 M NaF at 25°C.

SOLUTION Once again, we define x as the molar solubility of MgF_2. Dissolution of x mol/L of MgF_2 provides x mol/L of Mg^{2+} and $2x$ mol/L of F^-, but the total concentration of F^- is $(0.10 + 2x)$ mol/L because the solution already contains 0.10 mol/L of F^- from the NaF. It's helpful to summarize the equilibrium concentrations under the balanced equation:

Solubility equilibrium:	$MgF_2(s) \rightleftharpoons Mg^{2+}(aq) + 2\,F^-(aq)$	
Equilibrium conc (M)	x	$0.10 + 2x$

Substituting the equilibrium concentrations into the expression for K_{sp} gives

$$K_{sp} = 7.4 \times 10^{-11} = [Mg^{2+}][F^-]^2 = (x)(0.10 + 2x)^2$$

Because K_{sp} is small, $2x$ will be small compared with 0.10, and we can make the approximation that $(0.10 + 2x) \approx 0.10$. Therefore,

$$7.4 \times 10^{-11} = (x)(0.10 + 2x)^2 \approx (x)(0.10)^2$$

$$x = \frac{7.4 \times 10^{-11}}{(0.10)^2} = 7.4 \times 10^{-9} \text{ M}$$

Note that the calculated solubility of MgF_2 in 0.10 M NaF is less than that in pure water by a factor of about 35,000! (See Figure 16.11.) (Taking account of the equilibrium $Mg^{2+} + F^- \rightleftharpoons MgF^+$ would increase the calculated solubility by a factor of about 10. In this case, quite a bit of the Mg^{2+} is converted to MgF^+ because of the relatively high F^- concentration in 0.10 M NaF.)

▶**PROBLEM 16.25** Calculate the molar solubility of MgF_2 in 0.10 M $MgCl_2$ at 25°C.

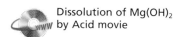

Dissolution of $Mg(OH)_2$ by Acid movie

The pH of the Solution

An ionic compound that contains a basic anion becomes more soluble as the acidity of the solution increases. The solubility of $CaCO_3$, for example, increases with decreasing pH (Figure 16.12) because the CO_3^{2-} ions combine

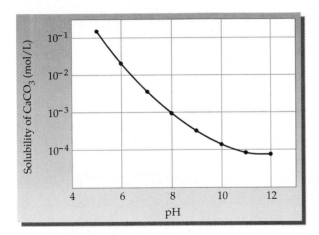

FIGURE 16.12 The solubility of $CaCO_3$ at 25°C increases as the solution becomes more acidic because the CO_3^{2-} ions combine with protons, thus driving the solubility equilibrium to the right. Note that the solubility is plotted on a logarithmic scale.

with protons to give HCO_3^- ions. As CO_3^{2-} ions are removed from the solution, the solubility equilibrium shifts to the right, as predicted by Le Châtelier's principle. The net reaction is dissolution of $CaCO_3$ in acidic solution to give Ca^{2+} ions and HCO_3^- ions:

$$CaCO_3(s) \rightleftharpoons Ca^{2+}(aq) + CO_3^{2-}(aq)$$

$$H_3O^+(aq) + CO_3^{2-}(aq) \rightleftharpoons HCO_3^-(aq) + H_2O(l)$$

$$\text{Net: } CaCO_3(s) + H_3O^+(aq) \rightleftharpoons Ca^{2+}(aq) + HCO_3^-(aq) + H_2O(l)$$

Other salts that contain basic anions, such as CN^-, PO_4^{3-}, S^{2-}, or F^-, behave similarly. By contrast, pH has no effect on the solubility of salts that contain anions of strong acids (Cl^-, Br^-, I^-, NO_3^-, and ClO_4^-) because these anions are not protonated by H_3O^+.

The effect of pH on the solubility of $CaCO_3$ has important environmental consequences. For instance, the formation of limestone caves, such as Mammoth Cave in Kentucky, is due to the slow dissolution of limestone ($CaCO_3$) in the slightly acidic natural water of underground streams. Marble, another form of $CaCO_3$, also dissolves in acid, which accounts for the deterioration of marble monuments on exposure to acid rain.

The effect of pH on solubility is also important in understanding how fluoride ion reduces tooth decay. When tooth enamel comes in contact with F^- ions in drinking water or fluoride-containing toothpaste, OH^- ions in hydroxyapatite, $Ca_5(PO_4)_3OH$, are replaced by F^- ions, giving the mineral fluorapatite, $Ca_5(PO_4)_3F$. Because F^- is a much weaker base than OH^-, $Ca_5(PO_4)_3F$ is much more resistant than $Ca_5(PO_4)_3OH$ to dissolving in acids.

▶ **PROBLEM 16.26** Which of the following compounds are more soluble in acidic solution than in pure water?
(a) AgCN   **(b)** PbI_2 **(c)** $Al(OH)_3$ **(d)** ZnS

Formation of Complex Ions

The solubility of an ionic compound increases dramatically if the solution contains a Lewis base that can form a coordinate covalent bond (Section 7.5) to the metal cation. Silver chloride, for example, is insoluble in water and in acid, but it dissolves in an excess of aqueous ammonia, forming the *complex ion* $Ag(NH_3)_2^+$. A **complex ion** is an ion that contains a metal cation bonded to one or more small molecules or ions, such as NH_3, CN^-, or OH^-. In accord with Le Châtelier's principle, ammonia shifts the solubility equilibrium to the right by tying up the Ag^+ ion in the form of the complex ion:

$$AgCl(s) \rightleftharpoons Ag^+(aq) + Cl^-(aq)$$

$$Ag^+(aq) + 2\,NH_3(aq) \rightleftharpoons Ag(NH_3)_2{}^+(aq)$$

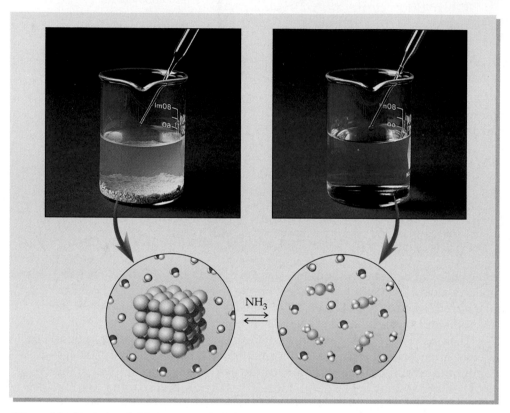

Silver chloride is insoluble in water (left) but dissolves on addition of an excess of aqueous ammonia (right).

The formation of a complex ion is a stepwise process, and each step has its own characteristic equilibrium constant. For formation of $Ag(NH_3)_2{}^+$, the reactions are

$$Ag^+(aq) + NH_3(aq) \rightleftharpoons Ag(NH_3)^+(aq) \qquad K_1 = 2.1 \times 10^3$$

$$Ag(NH_3)^+(aq) + NH_3(aq) \rightleftharpoons Ag(NH_3)_2{}^+(aq) \qquad K_2 = 8.1 \times 10^3$$

Net: $\quad Ag^+(aq) + 2\,NH_3(aq) \rightleftharpoons Ag(NH_3)_2{}^+(aq) \qquad K_f = 1.7 \times 10^7$

$$K_1 = \frac{[Ag(NH_3)^+]}{[Ag^+][NH_3]} = 2.1 \times 10^3 \qquad K_2 = \frac{[Ag(NH_3)_2{}^+]}{[Ag(NH_3)^+][NH_3]} = 8.1 \times 10^3$$

$$K_f = K_1K_2 = \frac{[Ag(NH_3)_2{}^+]}{[Ag^+][NH_3]^2} = 1.7 \times 10^7 \qquad \text{At } 25°C$$

The stability of a complex ion is measured by its **formation constant K_f** (or **stability constant**), the equilibrium constant for formation of the complex ion from the hydrated metal cation. The large value of K_f for $Ag(NH_3)_2^+$ means that this complex ion is quite stable, and nearly all the Ag^+ ion in an aqueous ammonia solution is therefore present in the form of $Ag(NH_3)_2^+$ (see Example 16.11).

The net reaction for dissolution of AgCl in aqueous ammonia is the sum of the equations for the dissolution of AgCl in water and the reaction of $Ag^+(aq)$ with $NH_3(aq)$ to give $Ag(NH_3)_2^+$:

$$AgCl(s) \rightleftharpoons Ag^+(aq) + Cl^-(aq)$$

$$Ag^+(aq) + 2\,NH_3(aq) \rightleftharpoons Ag(NH_3)_2^+(aq)$$

$$\text{Net:} \quad AgCl(s) + 2\,NH_3(aq) \rightleftharpoons Ag(NH_3)_2^+(aq) + Cl^-(aq)$$

Its equilibrium constant K is the product of the equilibrium constants for the reactions added:

$$K = \frac{[Ag(NH_3)_2^+][Cl^-]}{[NH_3]^2} = (K_{sp})(K_f) = (1.8 \times 10^{-10})(1.7 \times 10^7) = 3.1 \times 10^{-3}$$

Because K is much larger than K_{sp}, the solubility equilibrium for AgCl in the presence of ammonia lies much farther to the right than it does in the absence of ammonia. The increase in the solubility of AgCl on addition of ammonia is shown graphically in Figure 16.13. In general, the solubility of an ionic compound increases when the metal cation is tied up in the form of a complex ion. The quantitative effect of complex formation on the solubility of AgCl is explored in Example 16.12.

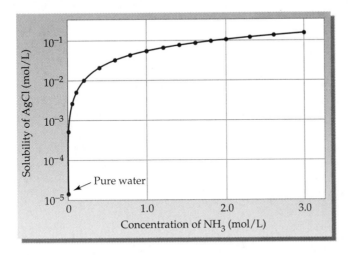

FIGURE 16.13 The solubility of AgCl in aqueous ammonia at 25°C increases with increasing ammonia concentration owing to formation of the complex ion $Ag(NH_3)_2^+$. Note that the solubility is plotted on a logarithmic scale.

EXAMPLE 16.11   What are the concentrations of Ag^+, $Ag(NH_3)^+$, and $Ag(NH_3)_2^+$ in a solution prepared by adding 0.10 mol of $AgNO_3$ to 1.0 L of 3.0 M NH_3?

SOLUTION   Because K_1, K_2, and K_f for $Ag(NH_3)_2^+$ are all large numbers, nearly all the Ag^+ from $AgNO_3$ will be converted to $Ag(NH_3)_2^+$:

$$Ag^+(aq) + 2\,NH_3(aq) \rightleftharpoons Ag(NH_3)_2^+(aq) \qquad K_f = 1.7 \times 10^7$$

To calculate the concentrations, it's convenient to introduce the fiction of 100% conversion of Ag^+ to $Ag(NH_3)_2^+$ followed by a tiny amount of back-reaction [dissociation of $Ag(NH_3)_2^+$] to give a small equilibrium concentration of Ag^+. Conversion of 0.10 mol/L of Ag^+ to $Ag(NH_3)_2^+$ consumes 0.20 mol/L of NH_3. Assuming 100% conversion to $Ag(NH_3)_2^+$, the following concentrations are obtained:

$$[Ag^+] = 0 \text{ M}$$
$$[Ag(NH_3)_2^+] = 0.10 \text{ M}$$
$$[NH_3] = 3.0 - 0.20 = 2.8 \text{ M}$$

Dissociation of x mol/L of $Ag(NH_3)_2^+$ in the back-reaction produces x mol/L of Ag^+ and $2x$ mol/L of NH_3. Therefore, the equilibrium concentrations (in mol/L) are

$$[Ag(NH_3)_2^+] = 0.10 - x$$
$$[Ag^+] = x$$
$$[NH_3] = 2.8 + 2x$$

Let's summarize our reasoning in a table under the balanced equation:

	$Ag^+(aq)$ +	2 $NH_3(aq)$	$\rightleftharpoons$	$Ag(NH_3)_2^+(aq)$
Initial conc (M)	0.10	3.0		0
After 100% reaction (M)	0	2.8		0.10
Equilibrium conc (M)	x	$2.8 + 2x$		$0.10 - x$

Substituting the equilibrium concentrations into the expression for K_f, and making the approximation that x is negligible compared with 0.10 (and with 2.8) gives

$$K_f = 1.7 \times 10^7 = \frac{[Ag(NH_3)_2^+]}{[Ag^+][NH_3]^2} = \frac{0.10 - x}{(x)(2.8 + 2x)^2} \approx \frac{0.10}{(x)(2.8)^2}$$

$$[Ag^+] = x = \frac{0.10}{(1.7 \times 10^7)(2.8)^2} = 7.5 \times 10^{-10} \text{ M}$$

$$[Ag(NH_3)_2^+] = 0.10 - x = 0.10 - (7.5 \times 10^{-10}) = 0.10 \text{ M}$$

The concentration of $Ag(NH_3)^+$ can be calculated from either of the stepwise equilibria. Let's use the equilibrium equation for formation of $Ag(NH_3)^+$ from Ag^+:

$$K_1 = \frac{[Ag(NH_3)^+]}{[Ag^+][NH_3]} = 2.1 \times 10^3$$

$$[Ag(NH_3)^+] = K_1[Ag^+][NH_3] = (2.1 \times 10^3)(7.5 \times 10^{-10})(2.8) = 4.4 \times 10^{-6} \text{ M}$$

Note that nearly all the Ag^+ is in the form of $Ag(NH_3)_2^+$.

EXAMPLE 16.12 Calculate the molar solubility of AgCl at 25°C in:
(a) pure water   (b) 3.0 M NH_3

SOLUTION
(a) In pure water, the solubility equilibrium is

$$AgCl(s) \rightleftharpoons Ag^+(aq) + Cl^-(aq)$$

Substituting the equilibrium concentrations (x mol/L) into the expression for K_{sp} gives

$$K_{sp} = 1.8 \times 10^{-10} = [Ag^+][Cl^-] = (x)(x)$$

$$x = \text{Molar solubility} = \sqrt{1.8 \times 10^{-10}} = 1.3 \times 10^{-5} \text{ M}$$

(b) The balanced equation for dissolution of AgCl in aqueous NH_3 is

$$AgCl(s) + 2\,NH_3(aq) \rightleftharpoons Ag(NH_3)_2{}^+(aq) + Cl^-(aq) \qquad K = 3.1 \times 10^{-3}$$

If we define x as the number of moles per liter of AgCl that dissolves, then the saturated solution contains x mol/L of $Ag(NH_3)_2{}^+$, x mol/L of Cl^-, and $(3.0 - 2x)$ mol/L of NH_3. (We're assuming that essentially all the Ag^+ is in the form of $Ag(NH_3)_2{}^+$, as proved in Example 16.11.) Substituting the equilibrium concentrations into the equilibrium equation gives

$$K = 3.1 \times 10^{-3} = \frac{[Ag(NH_3)_2{}^+][Cl^-]}{[NH_3]^2} = \frac{(x)(x)}{(3.0 - 2x)^2}$$

Taking the square root of both sides, we obtain

$$5.6 \times 10^{-2} = \frac{x}{3.0 - 2x}$$

$$x = (5.6 \times 10^{-2})(3.0 - 2x) = 0.17 - 0.11x$$

$$x = \frac{0.17}{1.11} = 0.15 \text{ M}$$

The molar solubility of AgCl in 3.0 M NH_3 is 0.15 M. Note that AgCl is much more soluble in aqueous NH_3 than in pure water, as shown in Figure 16.13.

▶ **PROBLEM 16.27**   In an excess of $NH_3(aq)$, Cu^{2+} ion forms a deep blue complex ion, $Cu(NH_3)_4{}^{2+}$, which has a formation constant $K_f = 5.6 \times 10^{11}$. Calculate the Cu^{2+} concentration in a solution prepared by adding 5.0×10^{-3} mol of $CuSO_4$ to 0.500 L of 0.40 M NH_3.

▶ **PROBLEM 16.28**   The "fixing" of photographic film involves dissolving unexposed silver bromide in a thiosulfate ($S_2O_3{}^{2-}$) solution:

$$AgBr(s) + 2\,S_2O_3{}^{2-}(aq) \rightleftharpoons Ag(S_2O_3)_2{}^{3-}(aq) + Br^-(aq)$$

Using $K_{sp} = 5.4 \times 10^{-13}$ for AgBr and $K_f = 4.7 \times 10^{13}$ for $Ag(S_2O_3)_2{}^{3-}$, calculate the equilibrium constant K for the dissolution reaction, and calculate the molar solubility of AgBr in 0.10 M $Na_2S_2O_3$.

This photographic film was immersed in a thiosulfate solution to dissolve the unexposed silver bromide. Because the film is light sensitive, the light used in photographic darkrooms is low-intensity red light, which has a wavelength in the least energetic region of the visible spectrum.

Amphoterism

We saw in Section 14.9 that amphoteric oxides, such as aluminum oxide, are soluble both in strongly acidic and in strongly basic solutions:

In acid: $Al_2O_3(s) + 6\,H_3O^+(aq) \rightleftharpoons 2\,Al^{3+}(aq) + 9\,H_2O(l)$

In base: $Al_2O_3(s) + 2\,OH^-(aq) + 3\,H_2O(l) \rightleftharpoons 2\,Al(OH)_4{}^-(aq)$

The corresponding hydroxides behave similarly (Figure 16.14):

In acid: $Al(OH)_3(s) + 3\,H_3O^+(aq) \rightleftharpoons Al^{3+}(aq) + 6\,H_2O(l)$

In base: $Al(OH)_3(s) + OH^-(aq) \rightleftharpoons Al(OH)_4{}^-(aq)$

FIGURE 16.14
(a) Aluminum hydroxide, a gelatinous white precipitate, forms on addition of aqueous NaOH to $Al^{3+}(aq)$. **(b)** The precipitate dissolves on addition of excess aqueous NaOH, yielding the colorless $Al(OH)_4^-$ ion . (The precipitate also dissolves in aqueous HCl, yielding the colorless Al^{3+} ion.)

(a)   **(b)**

Dissolution of $Al(OH)_3$ in excess base is just a special case of the effect of complex-ion formation on solubility: $Al(OH)_3$ dissolves because excess OH^- ions convert it to the soluble complex ion $Al(OH)_4^-$ (aluminate ion). The effect of pH on the solubility of $Al(OH)_3$ is shown in Figure 16.15.

FIGURE 16.15   A plot of solubility versus pH shows that $Al(OH)_3$ is an amphoteric hydroxide. $Al(OH)_3$ is essentially insoluble between pH 4 and 10, but it dissolves both in strongly acidic and in strongly basic solutions.

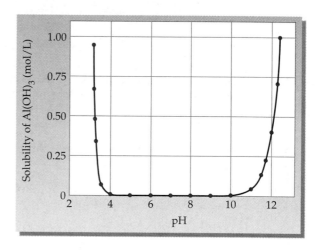

Other examples of amphoteric hydroxides include $Zn(OH)_2$, $Cr(OH)_3$, $Sn(OH)_2$, and $Pb(OH)_2$, which react with excess OH^- ions to form the soluble complex ions $Zn(OH)_4^{2-}$ (zincate ion), $Cr(OH)_4^-$ (chromite ion), $Sn(OH)_3^-$ (stannite ion), and $Pb(OH)_3^-$ (plumbite ion), respectively. By contrast, basic hydroxides, such as $Mn(OH)_2$, $Fe(OH)_2$, and $Fe(OH)_3$, dissolve in strong acid, but not in strong base.

The dissolution of Al_2O_3 in base is important in the production of aluminum metal from its ore. Aluminum is mined as bauxite ($Al_2O_3 \cdot x\ H_2O$), a hydrated oxide that is always contaminated with Fe_2O_3 and SiO_2. In the **Bayer process**, Al_2O_3 is purified by treating bauxite with hot aqueous NaOH. Aluminum oxide and SiO_2 dissolve, forming $Al(OH)_4^-$ and SiO_3^{2-} (silicate ion), but Fe_2O_3 remains undissolved. Subsequent filtration removes Fe_2O_3, and treatment of the filtrate with a weak acid (CO_2 in air) precipitates $Al(OH)_3$, leaving SiO_3^{2-} in solution. Removal of water from $Al(OH)_3$ by heating gives pure Al_2O_3, which is then converted to aluminum metal by electrolysis (see Section 18.12).

16.13   Precipitation of Ionic Compounds

A common problem in chemistry is to decide whether a precipitate of an ionic compound will form when solutions that contain the constituent ions are mixed. For example, will CaF_2 precipitate on mixing solutions of $CaCl_2$ and NaF? In other words, will the dissolution reaction proceed in the reverse direction, from right to left?

$$CaF_2(s) \rightleftharpoons Ca^{2+}(aq) + 2\,F^-(aq)$$

We touched on this question briefly in Section 4.4 when we looked at solubility rules, but we can now get a more quantitative view. The answer depends on the value of the **ion product (IP)**, a number defined by the expression

$$IP = [Ca^{2+}]_t[F^-]_t^2$$

The IP is defined in the same way as K_{sp}, except that the concentrations in the expression for IP are arbitrary concentrations at time t, not necessarily equilibrium concentrations. Thus, the IP is actually a reaction quotient Q_c (Section 13.5), but the term *ion product* is more descriptive because the equilibrium constant expression isn't a quotient.

If the value of IP is greater than K_{sp}, the solution is supersaturated with respect to CaF_2—a nonequilibrium situation. In that case, CaF_2 will precipitate, thus reducing the ion concentrations until IP equals K_{sp}. At that point, solubility equilibrium is reached, and the solution is saturated.

In general, we only need to calculate the value of IP and then compare it with K_{sp} to decide whether an ionic compound will precipitate. Three cases arise:

1. If $IP > K_{sp}$, the solution is supersaturated, and precipitation will occur.
2. If $IP = K_{sp}$, the solution is saturated, and equilibrium exists already.
3. If $IP < K_{sp}$, the solution is unsaturated, and precipitation will not occur.

EXAMPLE 16.13   Will a precipitate form when 0.150 L of 0.10 M $Pb(NO_3)_2$ and 0.100 L of 0.20 M NaCl are mixed?

SOLUTION   After the two solutions are mixed, the combined solution contains Pb^{2+}, NO_3^-, Na^+, and Cl^- ions and has a volume of 0.150 L + 0.100 L = 0.250 L. Because sodium salts and nitrate salts are soluble in water, the only compound that might precipitate is $PbCl_2$, which has $K_{sp} = 1.2 \times 10^{-5}$ (Appendix C). To calculate the value of IP for $PbCl_2$, first calculate the number of moles of Pb^{2+} and Cl^- in the combined solution:

$$\text{Moles } Pb^{2+} = (0.150\ \text{L})(0.10\ \text{mol/L}) = 1.5 \times 10^{-2}\ \text{mol}$$
$$\text{Moles } Cl^- = (0.100\ \text{L})(0.20\ \text{mol/L}) = 2.0 \times 10^{-2}\ \text{mol}$$

Then convert moles to molar concentrations:

$$[Pb^{2+}] = \frac{1.5 \times 10^{-2}\ \text{mol}}{0.250\ \text{L}} = 6.0 \times 10^{-2}\ \text{M}$$

$$[Cl^-] = \frac{2.0 \times 10^{-2}\ \text{mol}}{0.250\ \text{L}} = 8.0 \times 10^{-2}\ \text{M}$$

The ion product is

$$IP = [Pb^{2+}]_t[Cl^-]_t^2 = (6.0 \times 10^{-2})(8.0 \times 10^{-2})^2 = 3.8 \times 10^{-4}$$

Since $K_{sp} = 1.2 \times 10^{-5}$, IP is greater than K_{sp}, and $PbCl_2$ will precipitate.

When 0.150 L of 0.10 M $Pb(NO_3)_2$ and 0.100 L of 0.20 M NaCl are mixed, a white precipitate of $PbCl_2$ forms because the ion product is greater than K_{sp}.

▶ **PROBLEM 16.29**   Will a precipitate form on mixing equal volumes of the following solutions?
(a) 3.0×10^{-3} M $BaCl_2$ and 2.0×10^{-3} M Na_2CO_3
(b) 1.0×10^{-5} M $Ba(NO_3)_2$ and 4.0×10^{-5} M Na_2CO_3

▶ **PROBLEM 16.30**   Will a precipitate form on mixing 25 mL of 1.0×10^{-3} M $MnSO_4$, 25 mL of 1.0×10^{-3} M $FeSO_4$, and 200 mL of a buffer solution that is 0.20 M in NH_4Cl and 0.20 M in NH_3? Values of K_{sp} can be found in Appendix C.

16.14   Separation of Ions by Selective Precipitation

A convenient method for separating a mixture of ions is to add a solution that will precipitate some of the ions but not others. The anions SO_4^{2-} and Cl^-, for example, can be separated by addition of a solution of $Ba(NO_3)_2$. Insoluble $BaSO_4$ precipitates, but Cl^- remains in solution because $BaCl_2$ is soluble. Similarly, the cations Ag^+ and Zn^{2+} can be separated by addition of dilute HCl. Silver chloride, AgCl, precipitates, but Zn^{2+} stays in solution because $ZnCl_2$ is soluble.

In the next section, we'll see that mixtures of metal cations, M^{2+}, can be separated into two groups by selective precipitation of metal sulfides, MS. For example, Pb^{2+}, Cu^{2+}, and Hg^{2+}, which form very insoluble sulfides, can be separated from Mn^{2+}, Fe^{2+}, Co^{2+}, Ni^{2+}, and Zn^{2+}, which form more soluble sulfides. The separation is carried out in an acidic solution and makes use of the following solubility equilibrium:

$$MS(s) + 2 H_3O^+(aq) \rightleftharpoons M^{2+}(aq) + H_2S(aq) + 2 H_2O(l)$$

The equilibrium constant for this reaction, called the *solubility product in acid*, is given the symbol K_{spa}:

$$K_{spa} = \frac{[M^{2+}][H_2S]}{[H_3O^+]^2}$$

The separation depends on adjusting the H_3O^+ concentration so that the reaction quotient Q_c exceeds K_{spa} for the very insoluble sulfides but not for the more soluble ones (Table 16.3).

TABLE 16.3 Solubility Products in Acid (K_{spa}) at 25°C for Metal Sulfides

Metal Sulfide, MS	K_{spa}	Metal Sulfide, MS	K_{spa}
MnS	3×10^{10}	ZnS	3×10^{-2}
FeS	6×10^2	PbS	3×10^{-7}
CoS	3	CuS	6×10^{-16}
NiS	8×10^{-1}	HgS	2×10^{-32}

We use K_{spa} for metal sulfides rather than K_{sp} for two reasons. First, the ion separations are carried out in acidic solution, so use of K_{spa} is more convenient. Second, the K_{sp} values given in many reference books for the reaction $MS(s) \rightleftharpoons M^{2+} + S^{2-}$ are incorrect because they are based on a K_{a2} value for H_2S (1.3×10^{-14}) that is greatly in error. The correct value of K_{a2} for H_2S ($\sim 10^{-19}$) is very small, which means that S^{2-}, like O^{2-}, is highly basic and is not an important species in aqueous solutions. The principal sulfide-containing species in aqueous solutions are H_2S in acidic solutions and HS^- in basic solutions. The relation between K_{spa} and the traditional K_{sp}, K_{a1}, and K_{a2} values is $K_{spa} = K_{sp}/(K_{a1}K_{a2})$.

In a typical experiment, the M^{2+} concentrations are about 0.01 M, and the H_3O^+ concentration is adjusted to about 0.3 M by adding HCl. The solution is then saturated with H_2S gas, which gives an H_2S concentration of about 0.10 M. Substituting these concentrations into the equilibrium constant expression, we find that the reaction quotient Q_c is 1×10^{-2}:

$$Q_c = \frac{[M^{2+}]_t[H_2S]_t}{[H_3O^+]_t^2} = \frac{(0.01)(0.10)}{(0.3)^2} = 1 \times 10^{-2}$$

This value of Q_c exceeds K_{spa} for PbS, CuS, and HgS (Table 16.3) but does not exceed K_{spa} for MnS, FeS, CoS, NiS, or ZnS. As a result, PbS, CuS, and HgS precipitate under these acidic conditions, but Mn^{2+}, Fe^{2+}, Co^{2+}, Ni^{2+}, and Zn^{2+} remain in solution.

Adding H_2S to an acidic solution of Hg^{2+} and Ni^{2+} precipitates Hg^{2+} as black HgS but leaves green Ni^{2+} in solution.

▶ **PROBLEM 16.31** Determine whether Cd^{2+} can be separated from Zn^{2+} by bubbling H_2S through a 0.3 M HCl solution that contains 0.005 M Cd^{2+} and 0.005 M Zn^{2+}. (K_{spa} for CdS is 8×10^{-7}.)

16.15 Qualitative Analysis

Qualitative analysis is a procedure for identifying the ions present in an unknown solution. The ions are identified by specific chemical tests, but because one ion can interfere with the test for another, the ions must first be

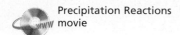

separated. In the traditional scheme of analysis for metal cations, some 20 cations are separated initially into five groups by selective precipitation (Figure 16.16).

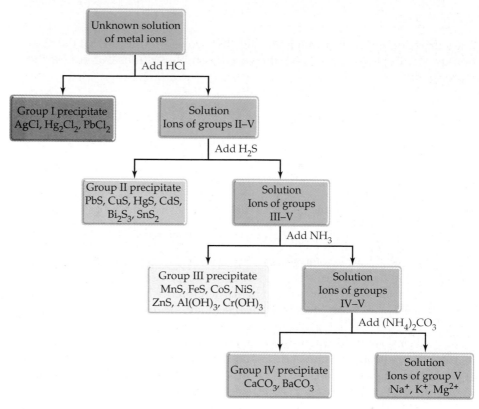

FIGURE 16.16 Flowchart for separation of metal cations in qualitative analysis.

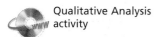

- **Group I: Ag^+, Hg_2^{2+}, and Pb^{2+}**   When aqueous HCl is added to the unknown solution, the cations of group I precipitate as insoluble chlorides—AgCl, Hg_2Cl_2, and $PbCl_2$. The cations of groups II–V, which form soluble chlorides, remain in solution. A small amount of the Pb^{2+} also remains in solution because $PbCl_2$ is slightly soluble.

- **Group II: Pb^{2+}, Cu^{2+}, Hg^{2+}, Cd^{2+}, Bi^{3+}, and Sn^{4+}**   After the insoluble chlorides have been removed, the solution is treated with H_2S to precipitate the cations of group II as insoluble sulfides—PbS, CuS, HgS, CdS, Bi_2S_3, and SnS_2. Because the solution is strongly acidic at this point ($[H_3O^+]$ ≈ 0.3 M), only the most insoluble sulfides precipitate. The acid-insoluble sulfides are then removed from the solution.

- **Group III: Mn^{2+}, Fe^{2+}, Co^{2+}, Ni^{2+}, Zn^{2+}, Al^{3+}, and Cr^{3+}**   At this point aqueous NH_3 is added, neutralizing the acidic solution and giving an NH_4^+–NH_3 buffer that is slightly basic (pH ≈ 8). The decrease in $[H_3O^+]$ shifts the metal sulfide solubility equilibrium to the left, thus precipitating the 2+ cations of group III as insoluble sulfides—MnS, FeS, CoS, NiS, and ZnS. The 3+ cations precipitate from the basic solution, not as sulfides, but as insoluble hydroxides—$Al(OH)_3$ and $Cr(OH)_3$.

- **Group IV: Ca^{2+} and Ba^{2+}** After the base-insoluble sulfides and the insoluble hydroxides have been removed, the solution is treated with $(NH_4)_2CO_3$ to precipitate group IV cations as insoluble carbonates—$CaCO_3$ and $BaCO_3$. Magnesium carbonate does not precipitate at this point because $[CO_3^{2-}]$ in the NH_4^+–NH_3 buffer is maintained at a low value.

- **Group V: Na^+, K^+, and Mg^{2+}** The only ions remaining in solution at this point are those whose chlorides, sulfides, and carbonates are soluble under conditions of the previous reactions. Magnesium ion is separated and identified by addition of a solution of $(NH_4)_2HPO_4$; if Mg^{2+} is present, a white precipitate of $Mg(NH_4)PO_4$ forms. The alkali metal ions are usually identified by the characteristic colors that they impart to a Bunsen flame (Figure 16.17).

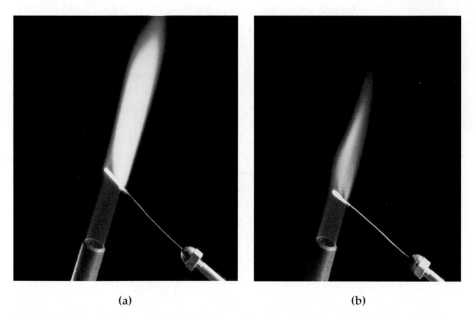

(a)	(b)

FIGURE 16.17 Flame tests for **(a)** sodium (persistent yellow) and **(b)** potassium (fleeting violet).

Once the cations have been separated into groups, further separations and specific tests are carried out to determine the presence or absence of the ions in each group. In group I, for example, lead can be separated from silver and mercury by treating the precipitate with hot water. The more soluble $PbCl_2$ dissolves, but the less soluble $AgCl$ and Hg_2Cl_2 do not. To test for Pb^{2+}, the solid chlorides are removed, and the solution is treated with a solution of K_2CrO_4. If Pb^{2+} is present, a yellow precipitate of $PbCrO_4$ forms.

Detailed procedures for separating and identifying all the ions can be found in general chemistry laboratory manuals. Although modern methods of metal-ion analysis employ sophisticated analytical instruments, qualitative analysis is still included in many general chemistry laboratory courses because it is an excellent vehicle for developing laboratory skills and for learning about acid–base, solubility, and complex-ion equilibria.

When aqueous potassium chromate is added to a solution that contains Pb^{2+}, a yellow precipitate of $PbCrO_4$ forms.

Analyzing Proteins by Electrophoresis

It has been estimated that there are as many as 140,000 different proteins in the human body. Many of these proteins act as *enzymes*, biological catalysts that regulate the tens of thousands of different reactions going on in your body each minute. Other proteins protect against disease, act as transport agents to move ions and small molecules through the body, or provide mechanical support for muscle, skin, and bone. Regardless of the exact details, virtually all biological processes involve proteins in some way.

As you might imagine, a biochemist faced with the need to purify one specific protein out of a mixture of thousands of substances with similar structures has a very difficult task. Using some knowledge of acid–base chemistry, however, makes the job easier. Proteins have various acidic and basic sites spread throughout their structures. Each of these acidic or basic sites in a given protein has its own pK_a or pK_b, and each site is therefore either charged or uncharged at a given pH. A protein with an abundance of acidic sites (AH), for example, will be uncharged at a low pH, where dissociation is suppressed, but will gain successively more negative charges as the pH is raised and dissociations begin to occur. Conversely, a protein with an abundance of basic sites (B) will have many positive charges at a low pH, where the basic sites are protonated, but will become successively less charged as the pH is raised and the basic sites are deprotonated. Different proteins have different numbers of acidic and basic sites and therefore have different overall charges at a given pH. By taking advantage of these different overall charges, biochemists can separate protein molecules from one another.

Muscle tissue is made of protein filaments.

AH AH AH AH →Raise pH→ A⁻ A⁻ A⁻ A⁻

Low pH; acidic sites are uncharged | High pH; acidic sites are dissociated

⁺BH ⁺BH ⁺BH ⁺BH →Raise pH→ B B B B

Low pH; basic sites are protonated | High pH; basic sites are uncharged

When an aqueous solution of a protein is placed in an electric field between two electrodes, a positively charged protein migrates toward the negative electrode, and a negatively charged protein migrates toward the positive electrode (Figure 16.18). The amount of this movement, called *electrophoresis*, varies with the size and shape of the protein, with the strength of the electric field, and with the number of charges on the protein. The number of charges, in turn, is determined by the number of acidic and basic sites on the protein and by the pH of the aqueous solution through which the protein moves.

Electrophoresis is routinely used both in research laboratories for isolating new proteins and in clinical laboratories for determining protein concentrations in blood serum. In the protein separation shown in Figure 16.19, blood serum is separated by electrophoresis into a pattern of five or six different

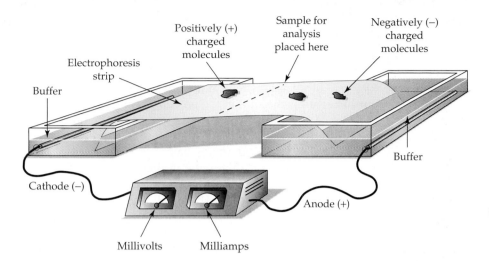

FIGURE 16.18 An electrophoresis apparatus for protein separation. An aqueous mixture of proteins is buffered to a given pH and placed in an electric field. Different proteins migrate toward one or the other of the electrodes at a rate dependent on the protein's overall charge.

protein fractions—albumin, two α-globulins, one or two β-globulins, and the γ-globulins. An abnormally high or low reading for a particular fraction can indicate a specific clinical condition.

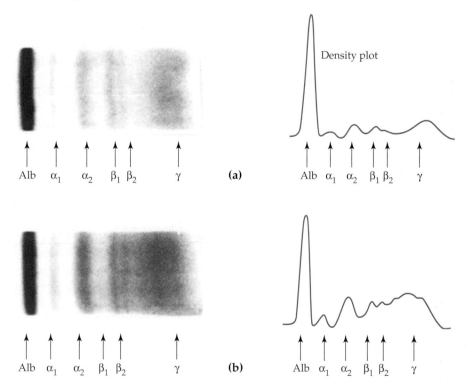

FIGURE 16.19 **(a)** A normal electrophoresis pattern of blood serum. **(b)** An abnormal pattern, with elevated γ-globulin, indicating the possibility of liver disease, collagen disorder, or infection.

▶ **PROBLEM 16.32** In an electrophoresis experiment, a particular protein migrates toward the negative electrode in the presence of an H_3PO_4–$H_2PO_4^-$ buffer but migrates toward the positive electrode in the presence of an H_3BO_3–$H_2BO_3^-$ buffer. Explain.

▶ **PROBLEM 16.33** Consider the separation of proteins by electrophoresis in the presence of an $H_2PO_4^-$–HPO_4^{2-} buffer. If you want to increase the rate at which the proteins migrate toward the negative electrode, should you increase or decrease the concentration ratio $[HPO_4^{2-}]/[H_2PO_4^-]$? Explain.

Key Words

Summary

Neutralization reactions involving a strong acid and/or a strong base have very large equilibrium constants (K_n) and proceed nearly 100% to completion. Weak acid–weak base neutralizations tend not to go to completion.

The **common-ion effect** is the shift in the position of an equilibrium that occurs when a substance is added that provides more of an ion already involved in the equilibrium. An example is the decrease in percent dissociation of a weak acid on addition of its conjugate base.

A solution of a weak acid and its conjugate base is called a **buffer solution** because it resists drastic changes in pH. The ability of a buffer solution to absorb small amounts of added H_3O^+ or OH^- without a significant change in pH (**buffer capacity**) increases with increasing amounts of weak acid and conjugate base. The pH of a buffer solution has a value close to pK_a ($-\log K_a$) of the weak acid and can be calculated from the **Henderson–Hasselbalch equation**:

$$pH = pK_a + \log \frac{[\text{Conjugate base}]}{[\text{Weak acid}]}$$

A **pH titration curve** is a plot of the pH of a solution as a function of the volume of base (or acid) added in the course of an acid–base titration. For a strong acid–strong base titration, the titration curve exhibits a sharp change in pH in the region of the **equivalence point**, the point at which stoichiometrically equivalent amounts of acid and base have been mixed together. For weak acid–strong base and weak base–strong acid titrations, the titration curves display a relatively flat region midway to the equivalence point, a smaller change in pH in the region of the equivalence point, and a pH at the equivalence point that is not equal to 7.00.

The **solubility product, K_{sp},** for an ionic compound is the equilibrium constant for dissolution of the compound in water. The solubility of the compound and K_{sp} are related by the equilibrium equation for the dissolution reaction. The solubility of an ionic compound is (1) suppressed by the presence of a common ion in the solution; (2) increased by decreasing the pH if the compound contains a basic anion, such as OH^-, S^{2-}, or CO_3^{2-}; and (3) increased by the presence of a Lewis base, such as NH_3, CN^-, or OH^-, that can bond to the metal cation to form a **complex ion**. The stability of a complex ion is measured by its **formation constant, K_f.**

When solutions of soluble ionic compounds are mixed, an insoluble compound will precipitate if the **ion product (IP)** for the insoluble compound exceeds its K_{sp}. The IP is defined in the same way as K_{sp}, except that the concentrations in the expression for IP are not necessarily equilibrium concentrations. Certain metal cations can be separated by selective precipitation of metal sulfides. Selective precipitation is important in **qualitative analysis**, a procedure for identifying the ions present in an unknown solution.

Key Concept Summary

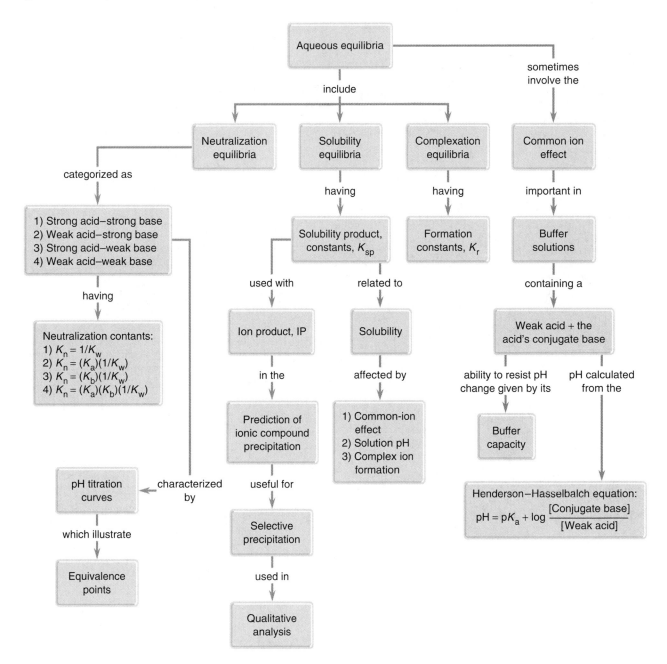

Understanding Key Concepts

Problems 16.1–16.33 appear within the chapter.

16.34 The following pictures represent solutions that contain one or more of the compounds H_2A, NaHA, and Na_2A, where H_2A is a weak diprotic acid. (Na^+ ions and solvent water molecules have been omitted for clarity.)

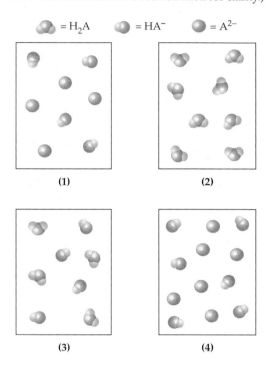

(1)   (2)

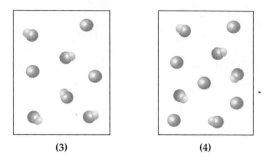

(3)   (4)

(a) Which solution has the highest pH? Which has the lowest pH?

(b) Draw a picture that represents the equilibrium state of solution (1) after addition of two H_3O^+ ions.

(c) Draw a picture that represents the equilibrium state of solution (1) after addition of two OH^- ions.

16.36 The strong acid HA is mixed with an equal molar amount of aqueous NaOH. Which of the following pictures represents the equilibrium state of the solution? (Na^+ ions and solvent water molecules have been omitted for clarity.)

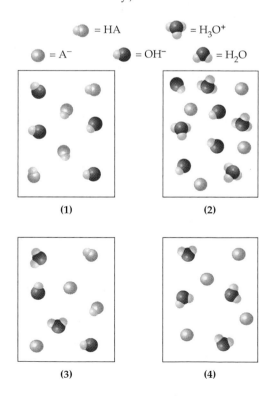

(3)   (4)

(a) Which of the solutions are buffer solutions?

(b) Which solution has the greatest buffer capacity?

16.35 The following pictures represent solutions that contain a weak acid HA ($pK_a = 6.0$) and its sodium salt NaA. (Na^+ ions and solvent water molecules have been omitted for clarity.)

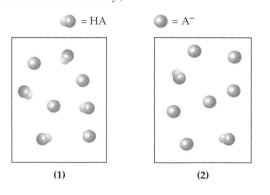

(1)   (2)

16.37 The following pictures represent solutions at various stages in the titration of a weak diprotic acid H_2A with aqueous NaOH. (Na^+ ions and solvent water molecules have been omitted for clarity.)

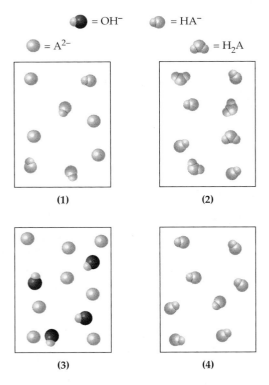

(1)   (2)

(3)   (4)

(a) To which of the following stages do solutions (1)–(4) correspond?
   (i)   Halfway to the first equivalence point
   (ii) At the first equivalence point
   (iii) Halfway between the first and second equivalence points
   (iv) Beyond the second equivalence point
(b) Which solution has the highest pH? Which has the lowest pH?

16.38 The following pictures represent solutions at various stages in the titration of a weak base B with aqueous HCl. (Cl^- ions and solvent water molecules have been omitted for clarity.)

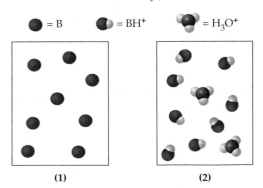

(1)   (2)

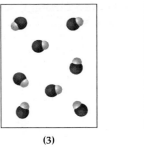

(3)   (4)

(a) To which of the following stages do solutions (1)–(4) correspond?
   (i)   The initial solution before addition of any HCl
   (ii) Halfway to the equivalence point
   (iii) At the equivalence point
   (iv) Beyond the equivalence point
(b) Is the pH at the equivalence point more or less than 7?

16.39 The following pictures represent solutions of AgCl, which also may contain ions other than Ag^+ and Cl^- that are not shown. If solution (1) is a saturated solution of AgCl, classify solutions (2)–(4) as unsaturated, saturated, or supersaturated.

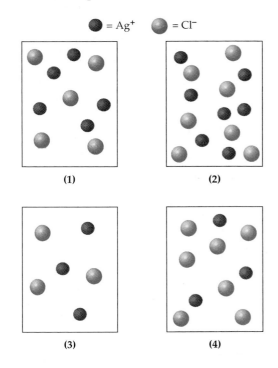

(1)   (2)

(3)   (4)

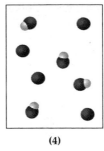

16.40 The following pictures represent solutions of Ag_2CrO_4, which also may contain ions other than Ag^+ and CrO_4^{2-} that are not shown. Solution (1) is in equilibrium with solid Ag_2CrO_4. Will a precipitate of solid Ag_2CrO_4 form in solutions (2)–(4)? Explain.

$\bullet = Ag^+$ $\bigcirc = CrO_4^{2-}$

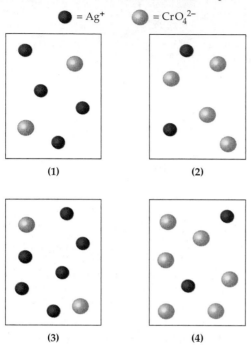

(1)   (2)

(3)   (4)

16.41 Consider the titration of 50.0 mL of 0.010 M HA ($K_a = 1.0 \times 10^{-4}$) with 0.010 M NaOH.

(a) Sketch the pH titration curve, and label the equivalence point.

(b) How many milliliters of 0.010 M NaOH are required to reach the equivalence point?

(c) Is the pH at the equivalence point greater than, equal to, or less than 7?

(d) What is the pH exactly halfway to the equivalence point?

16.42 The following plot shows two titration curves, each representing the titration of 50.0 mL of 0.100 M acid with 0.100 M NaOH:

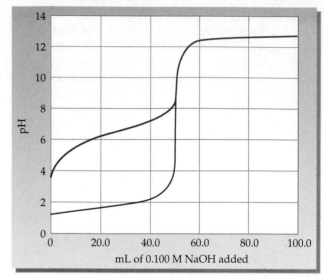

(a) Which of the two curves represents the titration of a strong acid, and which a weak acid?

(b) What is the approximate pH at the equivalence point for each of the acids?

(c) What is the approximate pK_a of the weak acid?

16.43 Consider saturated solutions of the slightly soluble salts AgBr and $BaCO_3$.

(a) Is the solubility of AgBr increased, decreased, or unaffected by addition of each of the following substances?

(i) HBr   (ii) HNO_3
(iii) $AgNO_3$ (iv) NH_3

(b) Is the solubility of $BaCO_3$ increased, decreased, or unaffected by addition of each of the following substances?

(i) HNO_3 (ii) $Ba(NO_3)_2$
(iii) Na_2CO_3 (iv) CH_3CO_2H

Additional Problems

Neutralization Reactions

16.44 Is the pH greater than, equal to, or less than 7 after neutralization of each of the following pairs of acids and bases?

(a) HI and NaOH

(b) HOCl and $Ba(OH)_2$

(c) HNO_3 and aniline ($C_6H_5NH_2$)

(d) Benzoic acid ($C_6H_5CO_2H$) and KOH

16.45 Is the pH greater than, equal to, or less than 7 after neutralization of each of the following pairs of acids and bases?

(a) CsOH and HNO_2 (b) NH_3 and HBr

(c) KOH and $HClO_4$ (d) NH_3 and HOBr

16.46 Calculate the equilibrium constant, K_n, for each neutralization reaction in Problem 16.44, and arrange the four reactions in order of increasing tendency to proceed to completion. See Appendix C for values of K_a and K_b.

16.47 Calculate the equilibrium constant, K_n, for each neutralization reaction in Problem 16.45, and arrange the four reactions in order of increasing tendency to proceed to completion. See Appendix C for values of K_a and K_b.

16.48 Which of the following mixtures has the higher pH?

(a) 0.10 M HF and 0.10 M NaOH

(b) 0.10 M HCl and 0.10 M NaOH

16.49 Which of the following mixtures has the lower pH?

    (a) 0.10 M $HClO_4$ and 0.10 M NaOH

    (b) 0.10 M $HClO_4$ and 0.10 M NH_3

16.50 Phenol (C_6H_5OH, $K_a = 1.3 \times 10^{-10}$) is a weak acid used in mouthwashes, and pyridine (C_5H_5N, $K_b = 1.8 \times 10^{-9}$) is a weak base used as a solvent. Calculate the value of K_n for neutralization of phenol by pyridine. Does the neutralization reaction proceed very far toward completion?

16.51 Aniline ($C_6H_5NH_2$, $K_b = 4.3 \times 10^{-10}$) is a weak base used in the manufacture of dyes. Calculate the value of K_n for neutralization of aniline by vitamin C (ascorbic acid, $C_6H_8O_6$, $K_a = 8.0 \times 10^{-5}$). Does much aniline remain at equilibrium?

The Common-Ion Effect

16.52 Which of the following substances affects the percent dissociation of HNO_2?

    (a) $NaNO_2$ **(b)** NaCl

    (c) HCl   **(d)** $Ba(NO_2)_2$

16.53 Which of the following substances affects the pH of an aqueous NH_3 solution?

    (a) KOH   **(b)** NH_4NO_3

    (c) NH_4Br **(d)** KBr

16.54 Does the pH increase, decrease, or remain the same on addition of each of the following?

    (a) LiF to an HF solution

    (b) KI to an HI solution

    (c) NH_4Cl to an NH_3 solution

16.55 Does the pH increase, decrease, or remain the same on addition of each of the following?

    (a) NH_4NO_3 to an NH_3 solution

    (b) Na_2CO_3 to an $NaHCO_3$ solution

    (c) $NaClO_4$ to an NaOH solution

16.56 Calculate the pH of a solution that is 0.25 M in HF and 0.10 M in NaF.

16.57 Calculate the pH of a solution prepared by mixing equal volumes of 0.20 M methylamine (CH_3NH_2, $K_b = 3.7 \times 10^{-4}$) and 0.60 M CH_3NH_3Cl.

16.58 Calculate the percent dissociation of 0.10 M hydrazoic acid (HN_3, $K_a = 1.9 \times 10^{-5}$). Recalculate the percent dissociation of 0.10 M HN_3 in the presence of 0.10 M HCl, and explain the change.

16.59 Calculate the pH of 100 mL of 0.30 M NH_3 before and after addition of 4.0 g of NH_4NO_3, and account for the change. Assume that the volume remains constant.

Buffer Solutions

16.60 Which of the following gives a buffer solution when equal volumes of the two solutions are mixed?

    (a) 0.10 M HF and 0.10 M NaF

    (b) 0.10 M HF and 0.10 M NaOH

    (c) 0.20 M HF and 0.10 M NaOH

    (d) 0.10 M HCl and 0.20 M NaF

16.61 Which of the following gives a buffer solution when equal volumes of the two solutions are mixed?

    (a) 0.10 M NH_3 and 0.10 M HCl

    (b) 0.20 M NH_3 and 0.10 M HCl

    (c) 0.10 M NH_4Cl and 0.10 M NH_3

    (d) 0.20 M NH_4Cl and 0.10 M NaOH

16.62 Which of the following solutions has the greater buffer capacity: **(a)** 100 mL of 0.30 M HNO_2–0.30 M $NaNO_2$ or **(b)** 100 mL of 0.10 M HNO_2–0.10 M $NaNO_2$? Explain.

16.63 Which of the following solutions has the greater buffer capacity: **(a)** 50 mL of 0.20 M NH_4Br–0.30 M NH_3 or **(b)** 50 mL of 0.40 M NH_4Br–0.60 M NH_3? Explain.

16.64 The following reaction is important in maintaining the pH of blood at a nearly constant value of about 7.4: $H_2CO_3 + H_2O \rightleftharpoons H_3O^+ + HCO_3^-$. What happens to the position of this equilibrium and the pH when blood absorbs added acid or base?

16.65 Explain how the $H_2PO_4^-$–HPO_4^{2-} buffer system can help to maintain the pH of intracellular fluid at a value close to 7.4.

16.66 Calculate the pH of a buffer solution that is 0.20 M in HCN and 0.12 M in NaCN. Will the pH change if the solution is diluted by a factor of 2? Explain.

16.67 Calculate the pH of a buffer solution prepared by dissolving 4.2 g of $NaHCO_3$ and 5.3 g of Na_2CO_3 in 0.20 L of water. Will the pH change if the solution volume is increased by a factor of 10? Explain.

16.68 Calculate the pH of 0.500 L of a 0.200 M NH_4Cl–0.200 M NH_3 buffer before and after addition of **(a)** 0.0050 mol of NaOH and **(b)** 0.020 mol of HCl. Assume that the volume remains constant.

16.69 Calculate the pH of 0.300 L of a 0.500 M $NaHSO_3$–0.300 M Na_2SO_3 buffer before and after addition of **(a)** 5.0 mL of 0.20 M HCl and **(b)** 5.0 mL of 0.10 M NaOH.

16.70 What is the pK_a value for each of the following acids? (K_a values are listed in Appendix C.)

    (a) Boric acid (H_3BO_3)

    (b) Formic acid (HCO_2H)

    (c) Hypochlorous acid (HOCl)

    How does pK_a vary as acid strength increases?

16.71 What is the value of K_a for an acid that has:

(a) $pK_a = 5.00$?   (b) $pK_a = 8.70$?

Which of the two acids is weaker?

16.72 Use the Henderson–Hasselbalch equation to calculate the pH of a buffer solution that is 0.25 M in formic acid (HCO_2H) and 0.50 M in sodium formate (HCO_2Na).

16.73 Use the Henderson–Hasselbalch equation to calculate the ratio of H_2CO_3 to HCO_3^- in blood having a pH of 7.40.

16.74 In what volume ratio should you mix 1.0 M solutions of NH_4Cl and NH_3 to produce a buffer solution having pH = 9.80?

16.75 Give a recipe for preparing a CH_3CO_2H–CH_3CO_2Na buffer solution that has pH = 4.44.

16.76 You need a buffer solution that has pH = 7.00. Which of the following buffer systems should you choose? Explain.

(a) H_3PO_4 and $H_2PO_4^-$

(b) $H_2PO_4^-$ and HPO_4^{2-}

(c) HPO_4^{2-} and PO_4^{3-}

16.77 Which of the following conjugate acid–base pairs should you choose to prepare a buffer solution that has pH = 4.50? Explain.

(a) HSO_4^- and SO_4^{2-}

(b) $HOCl$ and OCl^-

(c) $C_6H_5CO_2H$ and $C_6H_5CO_2^-$

pH Titration Curves

16.78 Consider the titration of 60.0 mL of 0.150 M HNO_3 with 0.450 M NaOH.

(a) How many millimoles of HNO_3 are present at the start of the titration?

(b) How many milliliters of NaOH are required to reach the equivalence point?

(c) What is the pH at the equivalence point?

(d) Sketch the general shape of the pH titration curve.

16.79 Make a rough plot of pH versus milliliters of acid added for the titration of 50.0 mL of 1.0 M NaOH with 1.0 M HCl. Indicate the pH at the following points, and tell how many milliliters of acid are required to reach the equivalence point.

(a) At the start of the titration

(b) At the equivalence point

(c) After addition of a large excess of acid

16.80 A 60.0 mL sample of a monoprotic acid is titrated with 0.150 M NaOH. If 20.0 mL of base is required to reach the equivalence point, what is the concentration of the acid?

16.81 A 25.0 mL sample of a diprotic acid is titrated with 0.240 M KOH. If 60.0 mL of base is required to reach the second equivalence point, what is the concentration of the acid?

16.82 A 50.0 mL sample of 0.120 M HBr is titrated with 0.240 M NaOH. Calculate the pH after addition of the following volumes of base, and construct a plot of pH versus milliliters of NaOH added.

(a) 0.0 mL   (b) 20.0 mL   (c) 24.9 mL

(d) 25.0 mL   (e) 25.1 mL   (f) 40.0 mL

16.83 A 40.0 mL sample of 0.150 M $Ba(OH)_2$ is titrated with 0.400 M HNO_3. Calculate the pH after addition of the following volumes of acid, and plot the pH versus milliliters of HNO_3 added.

(a) 0.0 mL   (b) 10.0 mL   (c) 20.0 mL

(d) 30.0 mL   (e) 40.0 mL

16.84 Consider the titration of 40.0 mL of 0.250 M HF with 0.200 M NaOH. How many milliliters of base are required to reach the equivalence point? Calculate the pH at each of the following points:

(a) After addition of 10.0 mL of base

(b) Halfway to the equivalence point

(c) At the equivalence point

(d) After addition of 80.0 mL of base

16.85 A 100.0 mL sample of 0.100 M methylamine (CH_3NH_2, $K_b = 3.7 \times 10^{-4}$) is titrated with 0.250 M HNO_3. Calculate the pH after addition of each of the following volumes of acid:

(a) 0.0 mL   (b) 20.0 mL   (c) 40.0 mL   (d) 60.0 mL

16.86 Consider the titration of 50.0 mL of a 0.100 M solution of the protonated form of the amino acid alanine (H_2A^+; $K_{a1} = 4.6 \times 10^{-3}$, $K_{a2} = 2.0 \times 10^{-10}$) with 0.100 M NaOH. Calculate the pH after addition of each of the following volumes of base:

(a) 10.0 mL   (b) 25.0 mL   (c) 50.0 mL

(d) 75.0 mL   (e) 100.0 mL

16.87 Consider the titration of 25.0 mL of 0.0200 M H_2CO_3 with 0.0250 M KOH. Calculate the pH after addition of each of the following volumes of base:

(a) 10.0 mL   (b) 20.0 mL   (c) 30.0 mL

(d) 40.0 mL   (e) 50.0 mL

16.88 What is the pH at the equivalence point for titration of 0.10 M solutions of the following acids and bases, and which of the indicators in Figure 15.4 would be suitable for each titration?

(a) HNO_2 and NaOH

(b) HI and NaOH

(c) CH_3NH_2 (methylamine) and HCl

16.89 What is the pH at the equivalence point for titration of 0.20 M solutions of the following acids and bases, and which of the indicators in Figure 15.4 would be suitable for each titration?

(a) $C_5H_{11}N$ (piperidine) and HNO_3

(b) $NaHSO_3$ and NaOH

(c) $Ba(OH)_2$ and HBr

Solubility Equilibria

16.90 For each of the following compounds, write a balanced net ionic equation for dissolution of the compound in water, and write the equilibrium expression for K_{sp}:

    (a) Ag_2CO_3 **(b)** $PbCrO_4$

    (c) $Al(OH)_3$ **(d)** Hg_2Cl_2

16.91 For each of the following, write the equilibrium expression for K_{sp}:

    (a) $Ca(OH)_2$ **(b)** Ag_3PO_4

    (c) $BaCO_3$ **(d)** $Ca_5(PO_4)_3OH$

16.92 A particular saturated solution of PbI_2 has $[Pb^{2+}] = 5.0 \times 10^{-3}$ M and $[I^-] = 1.3 \times 10^{-3}$ M.

    (a) What is the value of K_{sp} for PbI_2?

    (b) What is $[I^-]$ in a saturated solution of PbI_2 that has $[Pb^{2+}] = 2.5 \times 10^{-4}$ M?

    (c) What is $[Pb^{2+}]$ in a saturated solution that has $[I^-] = 2.5 \times 10^{-4}$ M?

16.93 A particular saturated solution of $Ca_3(PO_4)_2$ has $[Ca^{2+}] = [PO_4^{3-}] = 2.9 \times 10^{-7}$ M.

(a) What is the value of K_{sp} for $Ca_3(PO_4)_2$?

(b) What is $[Ca^{2+}]$ in a saturated solution of $Ca_3(PO_4)_2$ that has $[PO_4^{3-}] = 0.010$ M?

(c) What is $[PO_4^{3-}]$ in a saturated solution that has $[Ca^{2+}] = 0.010$ M?

16.94 If a saturated solution prepared by dissolving Ag_2CO_3 in water has $[Ag^+] = 2.56 \times 10^{-4}$ M, what is the value of K_{sp} for Ag_2CO_3?

16.95 Use the following solubility data to calculate a value of K_{sp} for each compound:

    (a) $CdCO_3$; 2.5×10^{-6} M

    (b) $Ca(OH)_2$; 1.06×10^{-2} M

    (c) $PbBr_2$; 4.34 g/L

    (d) $BaCrO_4$; 2.8×10^{-3} g/L

16.96 Use the values of K_{sp} in Appendix C to calculate the molar solubility of the following compounds:

    (a) $BaCrO_4$ **(b)** $Mg(OH)_2$ **(c)** Ag_2SO_3

16.97 Use the values of K_{sp} in Appendix C to calculate the solubility of the following compounds (in g/L):

    (a) Ag_2CO_3 **(b)** $CuBr$ **(c)** $Cu_3(PO_4)_2$

Factors That Affect Solubility

16.98 Use Le Châtelier's principle to explain the following changes in the solubility of Ag_2CO_3 in water:

    (a) Decrease on addition of $AgNO_3$

    (b) Increase on addition of HNO_3

    (c) Decrease on addition of Na_2CO_3

    (d) Increase on addition of NH_3

16.99 Use Le Châtelier's principle to predict whether the solubility of BaF_2 will increase, decrease, or remain the same on addition of each of the following substances:

    (a) HCl   **(b)** KF

    (c) $NaNO_3$ **(d)** $Ba(NO_3)_2$

16.100 Calculate the molar solubility of $PbCrO_4$ in:

    (a) Pure water   **(b)** 1.0×10^{-3} M K_2CrO_4

16.101 Calculate the molar solubility of SrF_2 in:

    (a) 0.010 M $Sr(NO_3)_2$ **(b)** 0.010 M NaF

16.102 Which of the following compounds are more soluble in acidic solution than in pure water? Write a balanced net ionic equation for each dissolution reaction.

    (a) AgBr   **(b)** $CaCO_3$

    (c) $Ni(OH)_2$ **(d)** $Ca_3(PO_4)_2$

16.103 Which of the following compounds are more soluble in acidic solution than in pure water? Write a balanced net ionic equation for each dissolution reaction.

    (a) MnS   **(b)** $Fe(OH)_3$

    (c) AgCl   **(d)** $BaCO_3$

16.104 Silver ion reacts with excess CN^- to form a colorless complex ion, $Ag(CN)_2^-$, which has a formation constant $K_f = 3.0 \times 10^{20}$. Calculate the concentration of Ag^+ in a solution prepared by mixing equal volumes of 2.0×10^{-3} M $AgNO_3$ and 0.20 M NaCN.

16.105 Dissolution of 5.0×10^{-3} mol of $Cr(OH)_3$ in 1.0 L of 1.0 M NaOH gives a solution of the complex ion $Cr(OH)_4^-$ ($K_f = 8 \times 10^{29}$). What fraction of the chromium in such a solution is present as uncomplexed Cr^{3+}?

16.106 Write a balanced net ionic equation for each of the following dissolution reactions, and use the appropriate K_{sp} and K_f values in Appendix C to calculate the equilibrium constant for each.

    (a) AgI in aqueous NaCN to form $Ag(CN)_2^-$

    (b) $Al(OH)_3$ in aqueous NaOH to form $Al(OH)_4^-$

    (c) $Zn(OH)_2$ in aqueous NH_3 to form $Zn(NH_3)_4^{2+}$

16.107 Write a balanced net ionic equation for each of the following dissolution reactions, and use the appropriate K_{sp} and K_f values in Appendix C to calculate the equilibrium constant for each.

    (a) $Zn(OH)_2$ in aqueous NaOH to form $Zn(OH)_4^{2-}$

    (b) $Cu(OH)_2$ in aqueous NH_3 to form $Cu(NH_3)_4^{2+}$

    (c) AgBr in aqueous NH_3 to form $Ag(NH_3)_2^+$

16.108 Calculate the molar solubility of AgI in:

    (a) Pure water

    (b) 0.10 M NaCN; K_f for $Ag(CN)_2^-$ is 3.0×10^{20}

16.109 Calculate the molar solubility of $Cr(OH)_3$ in 0.50 M NaOH; K_f for $Cr(OH)_4^-$ is 8×10^{29}.

Precipitation; Qualitative Analysis

16.110 Will a precipitate of $BaSO_4$ form when 100 mL of 4.0×10^{-3} M $BaCl_2$ and 300 mL of 6.0×10^{-4} M Na_2SO_4 are mixed? Explain.

16.111 Will a precipitate of $PbCl_2$ form on mixing equal volumes of 0.010 M $Pb(NO_3)_2$ and 0.010 M HCl? Explain. What minimum Cl^- concentration is required to begin precipitation of $PbCl_2$ from 5.0×10^{-3} M $Pb(NO_3)_2$?

16.112 What compound, if any, will precipitate when 80 mL of 1.0×10^{-5} M $Ba(OH)_2$ is added to 20 mL of 1.0×10^{-5} M $Fe_2(SO_4)_3$?

16.113 "Hard" water contains alkaline earth cations such as Ca^{2+}, which reacts with CO_3^{2-} to form insoluble deposits of $CaCO_3$. Will a precipitate of $CaCO_3$ form if a 250 mL sample of hard water having $[Ca^{2+}] = 8.0 \times 10^{-4}$ M is treated with:

(a) 0.10 mL of 2.0×10^{-3} M Na_2CO_3?

(b) 10 mg of solid Na_2CO_3?

16.114 The pH of a sample of hard water (Problem 16.113) having $[Mg^{2+}] = 2.5 \times 10^{-4}$ M is adjusted to pH 10.80. Will $Mg(OH)_2$ precipitate?

16.115 In qualitative analysis, Al^{3+} and Mg^{2+} are separated in an NH_4^+–NH_3 buffer having pH ≈ 8. Assuming cation concentrations of 0.010 M, show why $Al(OH)_3$ precipitates but $Mg(OH)_2$ does not.

16.116 Can Fe^{2+} be separated from Sn^{2+} by bubbling H_2S through a 0.3 M HCl solution that contains 0.01 M Fe^{2+} and 0.01 M Sn^{2+}? A saturated solution of H_2S has $[H_2S] \approx 0.10$ M. Values of K_{spa} are 6×10^2 for FeS and 1×10^{-5} for SnS.

16.117 Will CoS precipitate in a solution that is 0.10 M in $Co(NO_3)_2$, 0.5 M in HCl, and 0.10 M in H_2S? Will CoS precipitate if the pH of the solution is adjusted to pH 8 with an NH_4^+–NH_3 buffer? $K_{spa} = 3$ for CoS.

16.118 Using the qualitative analysis flowchart in Figure 16.16, tell how you could separate the following pairs of ions:

(a) Ag^+ and Cu^{2+} **(b)** Na^+ and Ca^{2+}

(c) Mg^{2+} and Mn^{2+} **(d)** K^+ and Cr^{3+}

16.119 Give a method for separating the following pairs of ions by addition of no more than two substances:

(a) Hg_2^{2+} and Co^{2+} **(b)** Na^+ and Mg^{2+}

(c) Fe^{2+} and Hg^{2+} **(d)** Ba^{2+} and Pb^{2+}

General Problems

16.120 Assume that you have three white solids: NaCl, KCl, and $MgCl_2$. What tests could you do to tell which is which?

16.121 Which of the following pairs of substances, when mixed in any proportion you wish, can be used to prepare a buffer solution?

(a) NaCN and HCN

(b) NaCN and NaOH

(c) HCl and NaCN

(d) HCl and NaOH

(e) HCN and NaOH

16.122 Which of the following pairs gives a buffer solution when equal volumes of the two solutions are mixed?

(a) 0.10 M $NaHCO_3$ and 0.10 M H_2CO_3

(b) 0.10 M $NaHCO_3$ and 0.10 M Na_2CO_3

(c) 0.10 M $NaHCO_3$ and 0.10 M HCl

(d) 0.20 M $NaHCO_3$ and 0.10 M NaOH

16.123 On the same graph, sketch pH titration curves for titration of (1) a strong acid with a strong base and (2) a weak acid with a strong base. How do the two curves differ with respect to:

(a) The initial pH?

(b) The pH in the region between the start of the titration and the equivalence point?

(c) The pH at the equivalence point?

(d) The pH beyond the equivalence point?

(e) The volume of base required to reach the equivalence point?

16.124 How many milliliters of 3.0 M NH_4Cl must be added to 250 mL of 0.20 M NH_3 to obtain a buffer solution having pH = 9.40?

16.125 Consider a buffer solution that contains equal concentrations of $H_2PO_4^-$ and HPO_4^{2-}. Will the pH increase, decrease, or remain the same when each of the following substances is added?

(a) Na_2HPO_4 **(b)** HBr   **(c)** KOH

(d) KI   **(e)** H_3PO_4 **(f)** Na_3PO_4

16.126 A saturated solution of $Mg(OH)_2$ in water has pH = 10.35. Calculate K_{sp} for $Mg(OH)_2$.

16.127 The mercurous ion, Hg_2^{2+}, reacts with Cl^- to give a white precipitate of Hg_2Cl_2. How much Hg_2^{2+} remains in solution after addition of 1 drop (about 0.05 mL) of 6 M HCl to 1.0 mL of 0.010 M $Hg_2(NO_3)_2$? Express your answer in:

(a) mol/L   **(b)** g/L

16.128 Calculate the concentrations of NH_4^+ and NH_3 and the pH in a solution prepared by mixing 20.0 g of NaOH and 0.500 L of 1.5 M NH_4Cl. Assume that the volume remains constant.

16.129 In qualitative analysis, Ag^+, Hg_2^{2+}, and Pb^{2+} are separated from other cations by addition of HCl. Calculate the concentration of Cl^- required to just begin precipitation of **(a)** AgCl, **(b)** Hg_2Cl_2, and **(c)** $PbCl_2$ in a solution having metal ion concentrations of 0.030 M. What fraction of the Pb^{2+} remains in solution when the Ag^+ just begins to precipitate?

16.130 Calculate the molar solubility of MnS in a 0.30 M NH_4Cl–0.50 M NH_3 buffer solution that is saturated with H_2S ($[H_2S] \approx 0.10$ M). What is the solubility of MnS (in g/L)? K_{spa} for MnS is 3×10^{10}.

16.131 What is the molar solubility of $Mg(OH)_2$ in a buffer solution that has a pH of 9.00?

Multi-Concept Problems

16.132 When a typical diprotic acid H_2A ($K_{a1} = 10^{-4}$; $K_{a2} = 10^{-10}$) is titrated with NaOH, the principal A-containing species at the first equivalence point is HA^-.

(a) By considering all four proton-transfer reactions that can occur in an aqueous solution of HA^-, show that the principal reaction is $2\ HA^- \rightleftharpoons H_2A + A^{2-}$.

(b) Assuming that this is the principal reaction, show that the pH at the first equivalence point is equal to the average of pK_{a1} and pK_{a2}.

(c) How many A^{2-} ions are present in 50.0 mL of 1.0 M NaHA?

16.133 Ethylenediamine ($NH_2CH_2CH_2NH_2$, abbreviated en) is an organic base that can accept two protons:

$$en(aq) + H_2O(l) \rightleftharpoons enH^+(aq) + OH^-(aq)$$
$$K_{b1} = 5.2 \times 10^{-4}$$
$$enH^+(aq) + H_2O(l) \rightleftharpoons enH_2^{2+}(aq) + OH^-(aq)$$
$$K_{b2} = 3.7 \times 10^{-7}$$

(a) Consider the titration of 30.0 mL of 0.100 M ethylenediamine with 0.100 M HCl. Calculate the pH after addition of the following volumes of acid, and construct a qualitative plot of pH versus milliliters of HCl added:

(i)   0.0 mL   (ii) 15.0 mL   (iii) 30.0 mL
(iv) 45.0 mL   (v) 60.0 mL   (vi) 75.0 mL

(b) Draw the structure of ethylenediamine, and explain why it can accept two protons.

(c) What hybrid orbitals do the N atoms use for bonding?

16.134 A 40.0 mL sample of a mixture of HCl and H_3PO_4 was titrated with 0.100 M NaOH. The first equivalence point was reached after 88.0 mL of base, and the second equivalence point was reached after 126.4 mL of base.

(a) What is the concentration of H_3O^+ at the first equivalence point?

(b) What are the initial concentrations of HCl and H_3PO_4 in the mixture?

(c) What percent of the HCl is neutralized at the first equivalence point?

(d) What is the pH of the mixture before addition of any base?

(e) Sketch the pH titration curve, and label the buffer regions and equivalence points.

(f) What indicators would you select to signal the equivalence points?

16.135 A 1.000 L sample of HCl gas at 25°C and 732.0 mm Hg was absorbed completely in an aqueous solution that contained 6.954 g of Na_2CO_3 and 250.0 g of water.

(a) What is the pH of the solution?

(b) What is the freezing point of the solution?

(c) What is the vapor pressure of the solution? (The vapor pressure of pure water at 25°C is 23.76 mm Hg.)

16.136 A saturated solution of an ionic salt MX exhibits an osmotic pressure of 74.4 mm Hg at 25°C. Assuming that MX is completely dissociated in solution, what is the value of its K_{sp}?

16.137 Consider the reaction that occurs on mixing 50.0 mL of 0.560 M $NaHCO_3$ and 50.0 mL of 0.400 M NaOH at 25°C.

(a) Write a balanced net ionic equation for the reaction.

(b) What is the pH of the resulting solution?

(c) How much heat (in joules) is liberated by the reaction? (Standard heats of formation are given in Appendix B.)

(d) What is the final temperature of the solution to the nearest 0.1°C? You may assume that all the heat liberated is absorbed by the solution, the mass of the solution is 100.0 g, and its specific heat is 4.18 J/(g · °C).

eMedia Problems

16.138 Use the **Calculating pH Using the Henderson–Hasselbalch Equation** activity (*eChapter 16.4*) to explore the properties of buffers.

(a) Use data from the activity to determine the pK_a of hypochlorous acid, HClO.

(b) Using the pK_a from part (a) to calculate the pH of a 0.025 M aqueous solution of hypochlorous acid.

(c) Will the pH of the 0.025 M solution change if it is diluted with water to twice its original volume?

(d) Will the pH of a buffer change when it is diluted with water to twice its original volume? Explain.

16.139 Prepare 1.00 L of a buffer containing 5.00 g of acetic acid and 10.0 g of sodium acetate using the **Calculating pH Using the Henderson–Hasselbalch Equation** activity (*eChapter 16.4*).

(a) Describe what the term *buffer capacity* means.

(b) Without a dramatic change in pH, does this buffer have a greater capacity to absorb a strong acid or a strong base?

16.140 Use the **Acid–Base Titration** activity (*eChapter 16.5*) to determine the concentration of an unknown acid by adding 0.40 M NaOH in increments of 1.0 mL. Repeat the titration adding increments of 0.10 mL of base as you approach the equivalence point. Once more, repeat the titration adding increments of 0.05 mL near the equivalence point. If your acid is dilute enough, repeat the titration three more times using 0.10 M NaOH in 1.0 mL, 0.10 mL, and 0.05 mL increments.

(a) Tabulate the acid concentrations that you calculate from each titration experiment.

(b) Are the values all exactly the same? If not, explain why.

(c) Which concentration do you think is the most precise, and why?

16.141 The **Common Ion Effect** movie (*eChapter 16.2*) illustrates how the solubility of silver iodide in water is reduced by the addition of sodium iodide. Explain how the solubility of silver iodide would be affected by the addition of silver nitrate, the addition of sodium chloride, and the addition of hydrochloric acid.

16.142 The **Dissolution of Mg(OH)$_2$ by Acid** movie (*eChapter 16.12*) shows how the relatively insoluble solid, magnesium hydroxide, can be made more soluble in water by the addition of acid.

(a) Write the net ionic equation for the process by which magnesium hydroxide dissolves in water.

(b) Write the net ionic equation for the combination of two hydronium ions and two hydroxide ions.

(c) Show that the net ionic equations from parts (a) and (b) sum to give the overall net ionic equation shown in the movie.

(d) Determine the equilibrium constant for the process represented by the overall net ionic equation.

(e) Calculate the solubility of magnesium hydroxide in 0.0010 M HCl.

Thermodynamics:
Entropy, Free Energy,
and Equilibrium

If the second law of thermodynamics states that all spontaneous processes lead to increased disorder, how can living things grow and evolve, creating ever more complex and highly ordered structures? The answer is in this chapter.

What factors determine the direction and extent of a chemical reaction? Some reactions, such as the combustion of hydrocarbon fuels, go almost to completion. Others, such as the combination of gold and oxygen, hardly occur at all. Still others—for example, the industrial synthesis of ammonia from N_2 and H_2 at 400–500°C—result in an equilibrium mixture that contains appreciable amounts of both reactants and products.

As we saw in Section 13.5, the extent of any particular reaction is described by the value of its equilibrium constant K: A value of K much larger than 1 indicates that the reaction goes far toward completion, and a value of K much smaller than 1 means that the reaction does not proceed very far before reaching an equilibrium state. But what determines the value of the equilibrium constant, and can we predict its value without measuring it? Put another way, what fundamental properties of nature determine the direction and extent of a particular chemical reaction? For answers to these questions, we turn to **thermodynamics**, the area of science that deals with the interconversion of heat and other forms of energy.

17.1 Spontaneous Processes

We have defined a **spontaneous process** as one that proceeds on its own without any external influence (Section 8.13). The reverse of a spontaneous process is always nonspontaneous and takes place only in the presence of some continuous external influence. Consider, for example, the expansion of a gas into a vacuum. When the stopcock in the apparatus shown in Figure 17.1 is opened, the gas in bulb A expands spontaneously into the evacuated bulb B until the gas pressure in the two bulbs is the same. The reverse process, migration of all the gas molecules into one bulb, does not occur spontaneously. To compress a gas from a larger to a smaller volume, we would have to push on the gas with a piston.

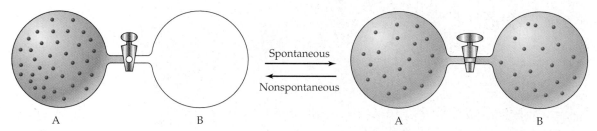

FIGURE 17.1 When the stopcock is opened, the gas in bulb A expands spontaneously into evacuated bulb B to fill all the available volume. The reverse process, compression of the gas, is nonspontaneous.

As a second example, consider the combination of hydrogen and oxygen in the presence of a platinum catalyst:

$$2 H_2(g) + O_2(g) \xrightarrow{\text{Catalyst}} 2 H_2O(l)$$

The forward reaction occurs spontaneously, but the reverse reaction, decomposition of water into its elements, does not occur no matter how long we wait. We'll see in the next chapter that we can force the reverse reaction to occur by electrolysis, but that reverse process is nonspontaneous and requires a continuous input of electrical energy.

In general, whether the forward or reverse reaction is spontaneous depends on the temperature, pressure, and composition of the reaction mixture. Consider the Haber synthesis of ammonia:

$$N_2(g) + 3 H_2(g) \xrightarrow{\text{Catalyst}} 2 NH_3(g)$$

A mixture of gaseous N_2, H_2, and NH_3, each at a partial pressure of 1 atm, reacts spontaneously at 300 K to convert some of the N_2 and H_2 to NH_3. We can predict the direction of spontaneous reaction from the relative values of the equilibrium constant K and the reaction quotient Q (Section 13.5). Since $K_p = 4.4 \times 10^5$ at 300 K and $Q_p = 1$ for partial pressures of 1 atm, the reaction will proceed in the forward direction because Q_p is less than K_p. Under these conditions, the reverse reaction is nonspontaneous. At 700 K, however, $K_p = 8.8 \times 10^{-5}$, and the reverse reaction is spontaneous because Q_p is greater than K_p.

A spontaneous reaction always moves a reaction mixture toward equilibrium. By contrast, a nonspontaneous reaction moves the composition of a mixture away from the equilibrium composition. Remember, though, that the word "spontaneous" doesn't mean the same thing as "fast." A spontaneous reaction can be either fast or slow—for example, the gradual rusting of iron metal is a slow spontaneous reaction. Thermodynamics tells us where a reaction is headed, but it says nothing about how long it takes to get there. As discussed in Section 12.9, the rate at which equilibrium is achieved depends on kinetics, especially on the height of the activation energy barrier between the reactants and products (Figure 17.2).

(a)

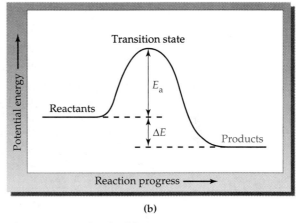

(b)

FIGURE 17.2   **(a)** The rusting of this ship is a spontaneous reaction, but it occurs slowly. **(b)** A spontaneous reaction occurs slowly if it has a high activation energy E_a.

▶ **PROBLEM 17.1**   Which of the following processes are spontaneous, and which are nonspontaneous?
(a) Diffusion of perfume molecules from one side of a room to the other
(b) Heat flow from a cold object to a hot object
(c) Decomposition of rust ($Fe_2O_3 \cdot H_2O$) to iron metal, oxygen, and water
(d) Decomposition of solid $CaCO_3$ to solid CaO and gaseous CO_2 at 25°C and 1 atm pressure ($K_p = 1.4 \times 10^{-23}$)

Nitrogen Triiodide movie

17.2   Enthalpy, Entropy, and Spontaneous Processes: A Brief Review

Let's look more closely at spontaneous processes and at the thermodynamic driving forces that cause them to occur. We saw in Chapter 8 that most spontaneous chemical reactions are accompanied by the conversion of potential energy to heat. For example, when methane burns in air, the potential energy stored in the chemical bonds of CH_4 and O_2 is partly converted to heat, which flows from the system (reactants plus products) to the surroundings:

$$CH_4(g) + 2\,O_2(g) \longrightarrow CO_2(g) + 2\,H_2O(l) \quad \Delta H° = -890.3\ \text{kJ}$$

Because heat is lost by the system, the reaction is exothermic and the standard enthalpy of reaction is negative ($\Delta H° = -890.3$ kJ). Of course, the total energy is conserved, so all the energy lost by the system shows up as heat gained by the surroundings.

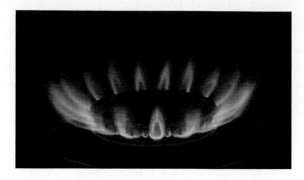

The combustion of CH_4 in air is a spontaneous, exothermic reaction.

Because spontaneous reactions so often give off heat, the nineteenth-century French scientist Marcellin Berthelot proposed that spontaneous chemical or physical changes are *always* exothermic. But Berthelot's proposal can't be correct. Common sense tells you, for instance, that ice spontaneously absorbs heat and melts at temperatures above 0°C. Similarly, liquid water absorbs heat and spontaneously boils at temperatures above 100°C. As further examples, gaseous N_2O_4 absorbs heat when it decomposes to NO_2 at 400 K, and table salt absorbs heat when it dissolves in water at room temperature:

$$H_2O(s) \longrightarrow H_2O(l) \qquad\qquad \Delta H_{fusion} = +6.01 \text{ kJ}$$

$$H_2O(l) \longrightarrow H_2O(g) \qquad\qquad \Delta H_{vap} = +40.7 \text{ kJ}$$

$$N_2O_4(g) \longrightarrow 2\,NO_2(g) \qquad\qquad \Delta H° = +57.1 \text{ kJ}$$

$$NaCl(s) \longrightarrow Na^+(aq) + Cl^-(aq) \qquad \Delta H° = +3.88 \text{ kJ}$$

All these processes are endothermic, yet all are spontaneous. In all cases, the system moves spontaneously to a state of *higher* potential energy by absorbing heat from the surroundings.

Since some spontaneous reactions are exothermic and others are endothermic, it's clear that enthalpy alone can't account for the direction of spontaneous change; a second factor must be involved. This second thermodynamic driving force is nature's tendency to move to a condition of maximum randomness or disorder (Section 8.13).

The tendency of things to get "messed up" is common in everyday life. You may rake the leaves on your lawn into an orderly pile, but after a few windy days the leaves are again scattered randomly. The reverse process is nonspontaneous; the wind never blows the randomly disordered leaves into a neatly arranged pile. Molecular systems behave similarly: *Molecular systems tend to move spontaneously to a state of maximum randomness or disorder.*

Molecular randomness, or disorder, is called **entropy** and is denoted by the symbol S. Entropy is a state function (Section 8.3), and the entropy change ΔS for a process thus depends only on the initial and final states of the system:

$$\Delta S = S_{final} - S_{initial}$$

When the randomness or disorder of a system increases, ΔS has a positive value; when randomness decreases, ΔS is negative.

Why aren't these leaves ever blown into a neat pile?

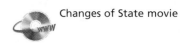
Changes of State movie

If you analyze the four spontaneous endothermic processes mentioned previously, you'll see that each involves an increase in the randomness of the system. When ice melts, for example, randomness increases because the highly ordered crystalline arrangement of rigidly held water molecules collapses and the molecules become free to move about in the liquid. When liquid water vaporizes, randomness further increases because the molecules can now move independently in the much larger volume of the gas. In general, processes that convert a solid to a liquid or a liquid to a gas involve an increase in randomness and thus an increase in entropy (Figure 17.3).

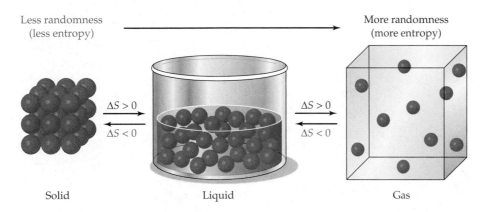

FIGURE 17.3 Molecular randomness—and thus entropy—increases when a solid melts and when a liquid vaporizes. Conversely, randomness and entropy decrease when a vapor condenses and when a liquid freezes. Note the sign of ΔS for each process.

The decomposition of N_2O_4 ($O_2N–NO_2$) is accompanied by an increase in randomness because breaking the N—N bond allows the two gaseous NO_2 fragments to move independently. Whenever a molecule breaks into two or more pieces, the amount of molecular randomness increases. More specifically, randomness—and thus entropy—increases whenever a reaction results in an increase in the number of gaseous molecules (Figure 17.4).

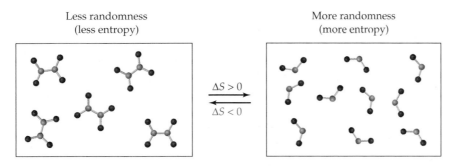

FIGURE 17.4 Molecular randomness—and thus entropy—increases when a reaction results in an increase in the number of gaseous particles. For example, ΔS is positive for decomposition of N_2O_4 to NO_2 and is negative for formation of N_2O_4 from NO_2.

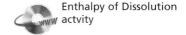
Enthalpy of Dissolution actvity

The increase in entropy on dissolving sodium chloride in water involves disruption of the crystal structure of solid NaCl and hydration of the Na^+ and Cl^- ions (Section 11.2). Disruption of the crystal increases randomness, since the Na^+ and Cl^- ions are rigidly held in the solid but are free to move about in the liquid. The hydration process, however, *decreases* randomness because it puts the polar, hydrating water molecules into an orderly arrangement about the Na^+ and Cl^- ions. It turns out that the overall dissolution process results in a net increase in randomness, and ΔS is thus positive (Figure 17.5). This is usually the case for dissolution of molecular solids, such as $HgCl_2$, and salts that contain +1 cations and −1 anions. For salts such as $CaSO_4$, which contain more highly charged ions, the hydrating water molecules are more strongly ordered about the ions, and the dissolution process often results in a net decrease in entropy. The following dissolution reactions illustrate the point:

$$HgCl_2(s) \longrightarrow HgCl_2(aq) \qquad \Delta S = +9 \, J/(K \cdot mol)$$

$$NaCl(s) \longrightarrow Na^+(aq) + Cl^-(aq) \qquad \Delta S = +43 \, J/(K \cdot mol)$$

$$CaSO_4(s) \longrightarrow Ca^{2+}(aq) + SO_4^{2-}(aq) \qquad \Delta S = -140 \, J/(K \cdot mol)$$

FIGURE 17.5 When NaCl dissolves in water, the crystal breaks up, and the Na^+ and Cl^- ions are surrounded by hydrating water molecules. The polar H_2O molecules are oriented such that the partially negative O atoms are near the cations and the partially positive H atoms are near the anions. Disruption of the crystal increases the entropy, but the hydration process decreases the entropy. For dissolution of NaCl, the net effect is an entropy increase.

Less randomness
(less entropy)

More randomness
(more entropy)

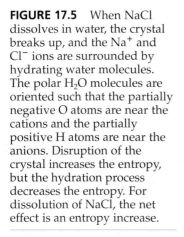
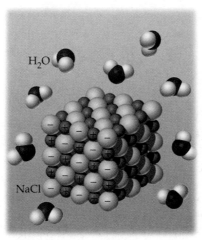

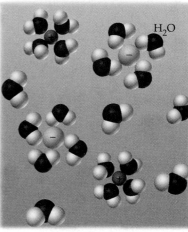

$\Delta S > 0$
$\Delta S < 0$

H_2O

NaCl

H_2O

$NaCl(s) + H_2O(l)$

$Na^+(aq) + Cl^-(aq)$

EXAMPLE 17.1   Predict the sign of ΔS in the system for each of the following processes:
(a) $CO_2(s) \rightarrow CO_2(g)$ (sublimation of dry ice)
(b) $CaSO_4(s) \rightarrow CaO(s) + SO_3(g)$
(c) $N_2(g) + 3 H_2(g) \rightarrow 2 NH_3(g)$
(d) $I_2(s) \rightarrow I_2(aq)$ (dissolution of iodine in water)

SOLUTION
(a) A gas is much more disordered than a solid. Therefore, ΔS is positive.

(b) One mole of gaseous molecules appears on the product side of the equation and none on the reactant side. Since the reaction increases the number of gaseous molecules, the entropy change is positive.

(c) The entropy change is negative because the reaction decreases the number of gaseous molecules from 4 mol to 2 mol. Fewer particles can move independently after reaction than before.

(d) Iodine molecules are electrically neutral and form a molecular solid. The dissolution process destroys the order of the crystal and enables the iodine molecules to move about randomly in the liquid. Therefore, ΔS is positive.

▶**PROBLEM 17.2**   Predict the sign of ΔS in the system for each of the following processes:
(a) $H_2O(g) \rightarrow H_2O(l)$ (formation of rain droplets)
(b) $I_2(g) \rightarrow 2 I(g)$
(c) $CaCO_3(s) \rightarrow CaO(s) + CO_2(g)$
(d) $Ag^+(aq) + Br^-(aq) \rightarrow AgBr(s)$

KEY CONCEPT PROBLEM 17.3   Consider the gas-phase reaction of A_2 molecules (red) with B atoms (blue):

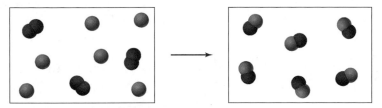

(a) Write a balanced equation for the reaction.
(b) Predict the sign of ΔS for the reaction.

17.3   Entropy and Probability

Why do systems tend to move spontaneously to a state of maximum randomness or disorder? The answer is that a disordered state is more probable than an ordered state because the disordered state can be achieved in more ways. A familiar example illustrates the point: Suppose that you shake a box containing 20 identical coins and then count the number of heads (H) and tails (T). Not surprisingly, it's very unlikely that all 20 coins will come up heads; that is, a perfectly ordered arrangement is much less probable than the totally disordered state in which heads and tails come up randomly.

The probabilities of the ordered and disordered states are proportional to the number of ways that the states can be achieved. The perfectly ordered state (20 H) can be achieved in only one way because it consists of a single configuration. In how many ways, though, can the totally disordered state be achieved? The totally disordered state of a collection of coins includes all possible configurations in which heads and tails are arranged at random. Thus, if there were just two coins in the box, each of them could come up in two ways (H or T), and the two together could come up in $2 \times 2 = 2^2 = 4$ ways (HH, HT, TH, or TT). Three coins could come up in $2 \times 2 \times 2 = 2^3 = 8$ ways (HHH, HTH, THH, TTH, HHT, HTT, THT, or TTT), and so on. For the case of 20 coins, the number of possible arrangements is $2^{20} = 1,048,576$.

Because the perfectly ordered state of 20 heads can be achieved in only one way, and the totally disordered state can be achieved in 2^{20} ways, the totally disordered state is 2^{20} times more probable than the perfectly ordered state. If you begin with the ordered state of 20 heads and shake the box, the system will move to the disordered state of higher probability.

An analogous chemical example is a crystal containing diatomic molecules such as carbon monoxide in which the two distinct ends of the CO molecule correspond to the heads and tails of a coin. Let's suppose that the long dimensions of the molecules are oriented vertically (Figure 17.6) and that the temperature is 0 K, so that the molecules are locked into a fixed arrangement. The state in which the molecules pack together in a perfectly ordered

Shaking a box that contains 20 quarters gives a random arrangement of heads and tails.

(a)

(b)

FIGURE 17.6   A hypothetical crystal containing 20 CO molecules. In **(a)**, the molecules are arranged in a perfectly ordered "heads-up" structure. In **(b)**, the molecules are arranged randomly in one of the 2^{20} ways in which the disordered structure can be obtained.

"heads-up" arrangement (Figure 17.6a) can be achieved in only one way, whereas the state in which the molecules are arranged randomly with respect to the vertical direction can be achieved in many ways—2^{20} ways for a hypothetical crystal containing 20 CO molecules (Figure 17.6b). Therefore, the disordered state of the crystal is 2^{20} times more probable than the perfectly ordered "heads-up" structure.

The Austrian physicist Ludwig Boltzmann proposed in 1896 that the entropy of a particular state is related to the number of ways that the state can be achieved, according to the formula

$$S = k \ln W$$

where S is the entropy of the state, $\ln W$ is the natural logarithm of the number of ways that the state can be achieved, and k, now known as *Boltzmann's constant*, is a universal constant equal to the gas constant R divided by Avogadro's number ($k = R/N_A = 1.38 \times 10^{-23}$ J/K). Because a logarithm is dimensionless, you can see from the Boltzmann equation that entropy has the same units as the constant k, namely, joules per kelvin.

Now let's apply Boltzmann's formula to our hypothetical crystal containing 20 CO molecules. Because the perfectly ordered state can be achieved in only one way ($W = 1$ in the Boltzmann equation) and because $\ln 1 = 0$, the entropy of the perfectly ordered state is zero:

$$S = k \ln W = k \ln 1$$
$$= 0$$

The more probable disordered state, however, can be achieved in 2^{20} ways and thus has a higher entropy:

$$S = k \ln W = k \ln 2^{20}$$
$$= (1.38 \times 10^{-23} \text{ J/K})(20)(\ln 2)$$
$$= 1.91 \times 10^{-22} \text{ J/K}$$

where we have made use of the relation $\ln x^a = a \ln x$ (Appendix A.2).

If our crystal contained 1 mol of CO molecules, the entropy of the perfectly ordered state (6.02×10^{23} C atoms up) would still be zero, but the entropy of the totally disordered state would be much higher because Avogadro's number of molecules can be arranged randomly in a huge number of ways ($W = 2^{N_A} = 2^{6.02 \times 10^{23}}$). According to Boltzmann's formula, the entropy of the disordered state is

$$S = k \ln W = k \ln 2^{N_A} = k N_A \ln 2$$

Because $k = R/N_A$,

$$S = R \ln 2 = (8.314 \text{ J/K})(0.693)$$
$$= 5.76 \text{ J/K}$$

Based on experimental measurements, the entropy of 1 mol of solid carbon monoxide near 0 K is about 5 J/K, indicating that the CO molecules adopt a nearly random arrangement.

The nearly random arrangement of CO molecules in crystalline carbon monoxide is unusual but is easily understood in terms of molecular structure. Because CO molecules have a dipole moment of only 0.11 D, intermolecular dipole–dipole forces are unusually weak (Sections 10.1 and 10.2), and the molecules therefore have little preference for the slightly lower energy, completely ordered arrangement. By contrast, HCl, with a dipole moment of 1.03 D, forms an ordered crystalline solid, and so the entropy of 1 mol of solid HCl at 0 K is 0 J/K.

Boltzmann's formula also explains why a gas expands into a vacuum. If the two bulbs in Figure 17.1 have equal volumes, each molecule has one chance in two of being in bulb A (heads, in our coin example) and one chance in two of being in bulb B (tails) when the stopcock is opened. It's exceedingly unlikely that all the molecules in 1 mol of gas will be in bulb A, since that state can be achieved in only one way. The state in which Avogadro's number of molecules are randomly distributed between bulbs A and B can be achieved in $2^{6.02 \times 10^{23}}$ ways, and the entropy of the disordered state is therefore higher than the entropy of the ordered state by the now familiar amount, $R \ln 2 = 5.76$ J/K. Thus, a gas expands spontaneously because the state of greater volume is more probable.

In general, when the volume of 1 mol of an ideal gas changes from $V_{initial}$ to V_{final} at constant temperature, the entropy of the gas changes by an amount

$$\Delta S = R \ln \frac{V_{final}}{V_{initial}}$$

Because the pressure and volume of an ideal gas are related inversely ($P = nRT/V$), we can also write

$$\Delta S = R \ln \frac{P_{initial}}{P_{final}}$$

Thus, the entropy of a gas *increases* when its pressure *decreases* at constant temperature, and the entropy *decreases* when pressure *increases*. Common sense tells us that the more we squeeze the gas, the less space the gas molecules have and so the more ordered they will be.

▶ **PROBLEM 17.4** Which state has the higher entropy? Explain in terms of probability.
(a) A perfectly ordered crystal of solid nitrous oxide (N≡N–O) or a disordered crystal in which the molecules are oriented randomly
(b) Silica glass or a quartz crystal
(c) 1 mol of N_2 gas at STP or 1 mol of N_2 gas at 273 K in a volume of 11.2 L
(d) 1 mol of N_2 gas at STP or 1 mol of N_2 gas at 273 K and 0.25 atm

17.4 Entropy and Temperature

Thus far, we've seen that entropy is associated with the orientation and distribution of molecules in space. Disordered crystals have higher entropy than ordered crystals, and diffuse gases have higher entropy than compressed gases.

Entropy is also associated with molecular motion. As the temperature of a substance increases, random molecular motion increases, and there is a corresponding increase in the average kinetic energy of the molecules. But not all the molecules have the same energy. As we saw in Section 9.6, there is a distribution of molecular speeds in a gas, a distribution that broadens and shifts to higher speeds with increasing temperature (Figure 9.12, page 364). In solids, liquids, and gases, the total energy of a substance can be distributed among the individual molecules in a number of ways that increases as the total energy increases. According to Boltzmann's formula, the more ways that the energy can be distributed, the greater the randomness of the state and the higher its entropy. Therefore, the entropy of a substance increases with increasing temperature (Figure 17.7).

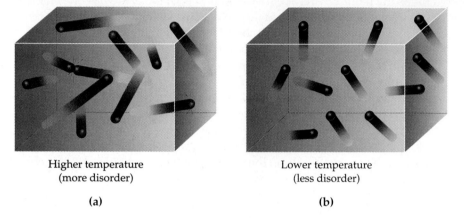

Higher temperature
(more disorder)

Lower temperature
(less disorder)

(a)   **(b)**

FIGURE 17.7   **(a)** A substance at a higher temperature has greater molecular motion, more disorder, and greater entropy than **(b)** the same substance at a lower temperature.

A typical plot of entropy versus temperature is shown in Figure 17.8. At absolute zero, every substance is a solid whose particles are rigidly fixed in a crystalline structure. If there is no residual orientational disorder,

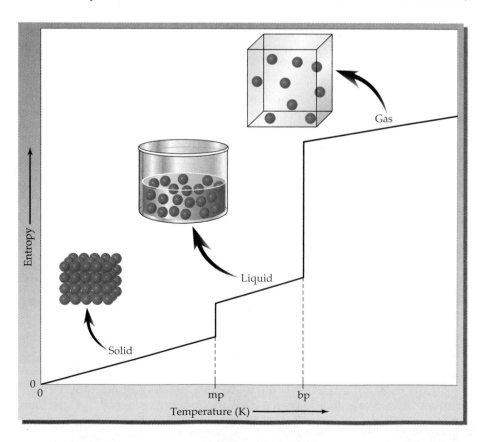

FIGURE 17.8   The entropy of a pure substance, equal to zero at 0 K, shows a steady increase with rising temperature, punctuated by discontinuous jumps in entropy at the temperatures of the phase transitions.

like that in carbon monoxide (Figure 17.6b), the entropy of the substance at 0 K will be zero, a general result summarized in the **third law of thermodynamics**:

◆━ THIRD LAW OF THERMODYNAMICS   The entropy of a perfectly ordered crystalline substance at 0 K is zero.

(The first law of thermodynamics was discussed in Section 8.3. We'll review the first law and discuss the second law in Section 17.6.)

As the temperature of a solid is raised, the added energy causes the molecules to vibrate. The number of ways in which the vibrational energy can be distributed increases with rising temperature, and the entropy of the solid thus increases steadily as the temperature goes up.

At the melting point, there is a discontinuous jump in entropy because there are many more ways of arranging the molecules in the liquid than in the solid. Furthermore, the molecules in the liquid can undergo translational and rotational as well as vibrational motion, and so there are many more ways of distributing the total energy in the liquid. An even greater jump in entropy is observed at the boiling point because molecules in the gas are free to occupy a much larger volume. Between the melting point and the boiling point, the entropy of a liquid increases steadily as molecular motion increases and the number of ways of distributing the total energy among the individual molecules increases. For the same reason, the entropy of a gas rises steadily as its temperature increases.

17.5 Standard Molar Entropies and Standard Entropies of Reaction

We won't describe how the entropy of a substance is determined, except to note that two approaches are available: (1) calculations based on Boltzmann's formula and (2) experimental measurements of heat capacities (Section 8.8) down to very low temperatures. Suffice it to say that *standard molar entropies*, denoted by $S°$, are known for many substances.

The Thermite Reaction
www movie

◆━ STANDARD MOLAR   The entropy of 1 mol of the pure substance at
ENTROPY, $S°$:   1 atm pressure and a specified temperature, usually 25°C.

Values of $S°$ for some common substances at 25°C are listed in Table 17.1, and additional values are given in Appendix B. Note that the units of $S°$ are *joules* (not kilojoules) per kelvin mole [J/(K · mol)]. Standard molar entropies are often called *absolute entropies* because they are measured with respect to an absolute reference point—the entropy of the perfectly ordered crystalline substance at 0 K [$S° = 0$ J/(K · mol) at $T = 0$ K].

Standard molar entropies make it possible to compare the entropies of different substances under the same conditions of temperature and pressure. It's apparent from Table 17.1, for example, that the entropies of gaseous substances tend to be larger than those of liquids, which, in turn, tend to be larger than those of solids. Table 17.1 also shows that $S°$ values increase with increasing molecular complexity. Compare, for example, CH_3OH, which has $S° = 127$ J/(K · mol), to CH_3CH_2OH, which has $S° = 161$ J/(K · mol).

TABLE 17.1 Standard Molar Entropies for Some Common Substances at 25°C

Substance	Formula	$S°$ [J/(K · mol)]	Substance	Formula	$S°$ [J/(K · mol)]
Gases			*Liquids*		
Acetylene	C_2H_2	200.8	Acetic acid	CH_3CO_2H	160
Ammonia	NH_3	192.3	Ethanol	CH_3CH_2OH	161
Carbon dioxide	CO_2	213.6	Methanol	CH_3OH	127
Carbon monoxide	CO	197.6	Water	H_2O	69.9
Ethylene	C_2H_4	219.5	*Solids*		
Hydrogen	H_2	130.6	Calcium carbonate	$CaCO_3$	92.9
Methane	CH_4	186.2	Calcium oxide	CaO	39.7
Nitrogen	N_2	191.5	Diamond	C	2.4
Nitrogen dioxide	NO_2	240.0	Graphite	C	5.7
Dinitrogen tetroxide	N_2O_4	304.2	Iron	Fe	27.3
Oxygen	O_2	205.0	Iron(III) oxide	Fe_2O_3	87.4

Once we have values for standard molar entropies, it's easy to calculate the entropy change for a chemical reaction. The **standard entropy of reaction, $\Delta S°$**, can be obtained simply by subtracting the standard molar entropies of all the reactants from the standard molar entropies of all the products:

$$\Delta S° = S°(\text{products}) - S°(\text{reactants})$$

Because $S°$ values are quoted on a per-mole basis, the $S°$ value for each substance must be multiplied by the stoichiometric coefficient of that substance in the balanced chemical equation. Thus, for the general reaction

$$a\,A + b\,B \longrightarrow c\,C + d\,D$$

the standard entropy of reaction is

$$\Delta S° = [c\,S°(C) + d\,S°(D)] - [a\,S°(A) + b\,S°(B)]$$

where the units of the coefficients are moles, the units of $S°$ are J/(K · mol), and the units of $\Delta S°$ are J/K.

As an example, let's calculate the standard entropy change for the reaction

$$N_2O_4(g) \longrightarrow 2\,NO_2(g)$$

Using the appropriate $S°$ values obtained from Table 17.1, we find that $\Delta S° = 175.8$ J/K:

$$\Delta S° = 2\,S°(NO_2) - S°(N_2O_4)$$

$$= (2\,\text{mol})\left(240.0\,\frac{J}{K \cdot mol}\right) - (1\,\text{mol})\left(304.2\,\frac{J}{K \cdot mol}\right)$$

$$= 175.8\,\text{J/K}$$

Although the standard molar entropy of N_2O_4 is larger than that of NO_2, as expected for a more complex molecule, $\Delta S°$ for the reaction is positive because

1 mol of N_2O_4 is converted to 2 mol of NO_2. As noted earlier, we expect an increase in entropy whenever a molecule breaks into two or more pieces.

EXAMPLE 17.2   Calculate the standard entropy of reaction at 25°C for the synthesis of ammonia:

$$N_2(g) + 3\,H_2(g) \longrightarrow 2\,NH_3(g)$$

SOLUTION   The standard entropy change for the reaction is

$$\Delta S° = 2\,S°(NH_3) - [S°(N_2) + 3\,S°(H_2)]$$

Substituting into this equation the appropriate $S°$ values from Table 17.1, we obtain

$$\Delta S° = (2\text{ mol})\left(192.3\,\frac{J}{K\cdot mol}\right) - \left[(1\text{ mol})\left(191.5\,\frac{J}{K\cdot mol}\right) + (3\text{ mol})\left(130.6\,\frac{J}{K\cdot mol}\right)\right]$$

$$= -198.7\text{ J/K}$$

Note that the entropy change is negative because the number of gaseous molecules decreases from 4 mol to 2 mol. (This result was predicted in Example 17.1c.)

▶**PROBLEM 17.5**   Calculate the standard entropy of reaction at 25°C for the decomposition of calcium carbonate:

$$CaCO_3(s) \longrightarrow CaO(s) + CO_2(g)$$

17.6   Entropy and the Second Law of Thermodynamics

We've seen thus far that molecular systems tend to move spontaneously toward a state of minimum enthalpy and maximum entropy. In any particular reaction, however, the enthalpy of the system can either increase or decrease. Similarly, the entropy of the system can either increase or decrease. How, then, can we decide whether a reaction will occur spontaneously? In Section 8.14, we said that it is the value of the *free-energy change*, ΔG, that is the criterion for spontaneity, where $\Delta G = \Delta H - T\Delta S$. If $\Delta G < 0$, the reaction is spontaneous; if $\Delta G > 0$, the reaction is nonspontaneous; and if $\Delta G = 0$, the reaction is at equilibrium. In this section and the next, we'll see how that conclusion was reached. Let's begin by looking at the first two laws of thermodynamics:

◆━ **FIRST LAW OF THERMODYNAMICS**   In any process, spontaneous or nonspontaneous, the total energy of a system and its surroundings is constant.

◆━ **SECOND LAW OF THERMODYNAMICS**   In any *spontaneous* process, the total entropy of a system and its surroundings always increases.

The first law (Section 8.3) is simply a statement of the conservation of energy. It says that energy (or enthalpy) can flow between a system and its surroundings but that the total energy of the system plus the surroundings always remains constant. In an exothermic reaction, the system loses enthalpy to the surroundings; in an endothermic reaction, the system gains enthalpy from the surroundings. Since energy is conserved in *all* processes, spontaneous and nonspontaneous, the first law helps us keep track of energy flow between the system and the surroundings, but it doesn't tell us whether a particular reaction will be spontaneous or nonspontaneous.

The second law, however, provides a clear-cut criterion of spontaneity. It says that the direction of spontaneous change is always determined by the sign of the total entropy change:

$$\Delta S_{total} = \Delta S_{system} + \Delta S_{surroundings}$$

Specifically,

> If $\Delta S_{total} > 0$, the reaction is spontaneous.
>
> If $\Delta S_{total} < 0$, the reaction is nonspontaneous
>
> If $\Delta S_{total} = 0$, the reaction mixture is at equilibrium.

All reactions proceed spontaneously in the direction that increases the entropy of the system plus surroundings. A reaction that is nonspontaneous in the forward direction is spontaneous in the reverse direction because ΔS_{total} for the reverse reaction equals $-\Delta S_{total}$ for the forward reaction. If ΔS_{total} is zero, the reaction doesn't go spontaneously in either direction, and the reaction mixture is at equilibrium.

To determine the value of ΔS_{total}, we need values for the entropy changes in the system and the surroundings. The entropy change in the system, ΔS_{sys}, is just the entropy of reaction, which can be calculated from standard molar entropies (Table 17.1), as described in the previous section. For a reaction that occurs at constant pressure, the entropy change in the surroundings is directly proportional to the enthalpy change for the reaction (ΔH) and inversely proportional to the kelvin temperature (T) of the surroundings, according to the equation

$$\Delta S_{surr} = \frac{-\Delta H}{T}$$

Although we won't derive this equation to calculate ΔS_{surr}, we can nevertheless justify its form. To see why ΔS_{surr} is proportional to $-\Delta H$, recall that for an exothermic reaction ($\Delta H < 0$), the system loses heat to the surroundings (Figure 17.9a). As a result, the random, chaotic motion of the molecules in the surroundings increases, and the entropy of the surroundings also increases ($\Delta S_{surr} > 0$). Conversely, for an endothermic reaction ($\Delta H > 0$), the system gains

FIGURE 17.9 **(a)** When an exothermic reaction occurs in the system ($\Delta H < 0$), the surroundings gain heat and their entropy increases ($\Delta S_{surr} > 0$). **(b)** When an endothermic reaction occurs in the system ($\Delta H > 0$), the surroundings lose heat and their entropy decreases ($\Delta S_{surr} < 0$).

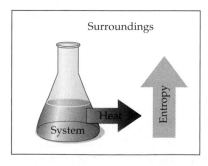

(a)

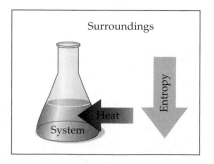

(b)

heat from the surroundings (Figure 17.9b), and the entropy of the surroundings therefore decreases ($\Delta S_{surr} < 0$). Because ΔS_{surr} is positive when ΔH is negative, and vice versa, ΔS_{surr} is proportional to $-\Delta H$:

$$\Delta S_{surr} \propto -\Delta H$$

The reason why ΔS_{surr} is inversely proportional to the absolute temperature T is more subtle. We can think of the surroundings as an infinitely large constant-temperature bath to which heat can be added without changing its temperature. If the surroundings have a low temperature, they have only a small amount of disorder, in which case addition of a given quantity of heat results in a substantial increase in the amount of disorder (a relatively large value of ΔS_{surr}). If the surroundings have a high temperature, they already have a large amount of disorder, and addition of the same quantity of heat produces only a marginal increase in the amount of disorder (a relatively small value of ΔS_{surr}). Thus, we might expect ΔS_{surr} to vary inversely with temperature.

$$\Delta S_{surr} \propto \frac{1}{T}$$

Adding heat to the surroundings is somewhat analogous to tossing a rock into a lake. If the lake has little disorder (calm, smooth surface), the rock's impact produces considerable disorder, evident in a circular pattern of waves. If the lake is already appreciably disordered (rough, choppy surface), the additional disorder produced when the rock hits the water is hardly noticeable.

Adding heat to cold surroundings is analogous to tossing a rock into calm waters. Both processes produce a considerable increase in disorder and thus a relatively large increase in entropy.

Adding heat to hot surroundings is analogous to tossing a rock into rough waters. Both processes produce a relatively small increase in disorder and thus a relatively small increase in entropy.

EXAMPLE 17.3   Consider the oxidation of iron metal:

$$4\,\text{Fe}(s) + 3\,\text{O}_2(g) \longrightarrow 2\,\text{Fe}_2\text{O}_3(s)$$

By determining the sign of ΔS_{total}, show that the reaction is spontaneous at 25°C.

The oxidation (rusting) of bulk iron, though spontaneous, is slow at 25°C. When heated, finely divided iron such as steel wool burns rapidly in air.

SOLUTION   To determine the sign of $\Delta S_{\text{total}} = \Delta S_{\text{sys}} + \Delta S_{\text{surr}}$, we need to calculate the values of ΔS_{sys} and ΔS_{surr}. The entropy change in the system is equal to the standard entropy of reaction and can be calculated using the standard molar entropies in Table 17.1:

$$\Delta S_{\text{sys}} = \Delta S° = 2\, S°(\text{Fe}_2\text{O}_3) - [4\, S°(\text{Fe}) + 3\, S°(\text{O}_2)]$$

$$= (2 \text{ mol})\left(87.4 \, \frac{\text{J}}{\text{K} \cdot \text{mol}}\right) - \left[(4 \text{ mol})\left(27.3 \, \frac{\text{J}}{\text{K} \cdot \text{mol}}\right) + \right.$$

$$\left. (3 \text{ mol})\left(205.0 \, \frac{\text{J}}{\text{K} \cdot \text{mol}}\right)\right]$$

$$= -549.5 \text{ J/K}$$

Thus, ΔS_{sys} is negative, as expected for a reaction that consumes 3 mol of gas.

To obtain $\Delta S_{\text{surr}} = -\Delta H°/T$, first calculate $\Delta H°$ for the reaction from standard enthalpies of formation (Section 8.10):

$$\Delta H° = 2\, \Delta H°_f(\text{Fe}_2\text{O}_3) - [4\, \Delta H°_f(\text{Fe}) + 3\, \Delta H°_f(\text{O}_2)]$$

Because $\Delta H°_f = 0$ for elements and $\Delta H°_f = -824.2$ kJ/mol for Fe_2O_3 (Appendix B), $\Delta H°$ for the reaction is

$$\Delta H° = 2\, \Delta H°_f(\text{Fe}_2\text{O}_3) = (2 \text{ mol})(-824.2 \text{ kJ/mol}) = -1648.4 \text{ kJ}$$

The reaction is highly exothermic, and accordingly there is a large positive entropy change in the surroundings:

$$\Delta S_{\text{surr}} = \frac{-\Delta H°}{T} = \frac{-(-1{,}648{,}400 \text{ J})}{298.15 \text{ K}} = 5529 \text{ J/K}$$

The total entropy change is positive, and the reaction is therefore spontaneous under standard-state conditions at 25°C:

$$\Delta S_{\text{total}} = \Delta S_{\text{sys}} + \Delta S_{\text{surr}} = -549.5 \text{ J/K} + 5529 \text{ J/K} = 4980 \text{ J/K}$$

▶ **PROBLEM 17.6**   By determining the sign of ΔS_{total}, show whether the decomposition of calcium carbonate is spontaneous under standard-state conditions at 25°C.

$$\text{CaCO}_3(s) \longrightarrow \text{CaO}(s) + \text{CO}_2(g)$$

17.7   Free Energy

Chemists are generally more interested in the system (the reaction mixture) than the surroundings, and it's therefore convenient to restate the second law in terms of the thermodynamic properties of the system, without regard to the surroundings. For this purpose, we use the thermodynamic property called **free energy**, denoted by G in honor of J. Willard Gibbs (1839–1903), the American mathematical physicist who laid the foundations of chemical thermodynamics. As discussed in Section 8.14, the free energy G of a system is defined as

◆— **FREE ENERGY:** $G = H - TS$

where H is the enthalpy, T is the temperature in kelvins, and S is the entropy. As you would expect from its name, free energy has units of energy (J or kJ).

Why is G called the *free* energy? If you think of TS as the part of the system's energy that is already disordered, then $H - TS (= G)$ is the part of the system's energy that is still ordered and therefore free (available) to cause spontaneous change by becoming disordered.

Free energy, like enthalpy and entropy, is a state function, and the change in free energy (ΔG) for a process is therefore independent of path. For a reaction at constant temperature, ΔG is equal to the change in enthalpy minus the product of temperature times the change in entropy:

$$\Delta G = \Delta H - T\Delta S$$

To see what this equation for free-energy change has to do with spontaneity, let's return to the relationship

$$\Delta S_{total} = \Delta S_{sys} + \Delta S_{surr} = \Delta S + \Delta S_{surr}$$

where we have now dropped the subscript "sys." (It's generally understood that symbols without a subscript refer to the system, not the surroundings.) Since $\Delta S_{surr} = -\Delta H/T$, where ΔH is the heat gained by the system at constant pressure, we can also write

$$\Delta S_{total} = \Delta S - \frac{\Delta H}{T}$$

Multiplying both sides by $-T$ gives

$$-T\Delta S_{total} = \Delta H - T\Delta S$$

But the right-hand side of this equation is just ΔG, the change in the free energy of the system at constant temperature and pressure. Therefore,

$$-T\Delta S_{total} = \Delta G$$

Note that ΔG and ΔS_{total} have opposite signs because the absolute temperature T is always positive.

According to the second law of thermodynamics, a reaction is spontaneous if ΔS_{total} is positive, nonspontaneous if ΔS_{total} is negative, and at equilibrium if ΔS_{total} is zero. Since $-T\Delta S_{total} = \Delta G$ and since ΔG and ΔS_{total} have opposite signs, we can restate the thermodynamic criterion for the spontaneity of a reaction carried out at constant temperature and pressure in the following way:

> **If $\Delta G < 0$, the reation is spontaneous.**
>
> If $\Delta G > 0$, the reaction is nonspontaneous.
>
> If $\Delta G = 0$, the reaction mixture is at equilibrium.

In other words, *in any spontaneous process at constant temperature and pressure, the free energy of the system always decreases.*

As discussed in Section 8.14, the temperature T acts as a weighting factor that determines the relative importance of the enthalpy and entropy contributions to ΔG in the free-energy equation $\Delta G = \Delta H - T\Delta S$. If ΔH and ΔS are either both negative or both positive, the sign of ΔG (and therefore the spontaneity of the reaction) depends on the temperature (Table 17.2; see also Figure 8.13). If ΔH and ΔS are both negative, the reaction will be spontaneous only if the absolute value of ΔH is larger than the absolute value of $T\Delta S$. This is most likely at low temperatures, where the weighting factor T in $T\Delta S$ is small. If ΔH and ΔS are both positive, the reaction will be spontaneous only if $T\Delta S$ is larger than ΔH, which is most likely at high temperatures. We've already seen how these considerations apply to phase changes (Section 10.4).

TABLE 17.2   Signs of Enthalpy, Entropy, and Free-Energy Changes and Reaction Spontaneity for a Reaction at Constant Temperature and Pressure

ΔH	ΔS	$\Delta G = \Delta H - T\Delta S$	Reaction Spontaneity
$-$	$+$	$-$	Spontaneous at all temperatures
$-$	$-$	$-$ or $+$	Spontaneous at low temperatures where ΔH outweighs $T\Delta S$
			Nonspontaneous at high temperatures where $T\Delta S$ outweighs ΔH
$+$	$-$	$+$	Nonspontaneous at all temperatures
$+$	$+$	$-$ or $+$	Spontaneous at high temperatures where $T\Delta S$ outweighs ΔH
			Nonspontaneous at low temperatures where ΔH outweighs $T\Delta S$

EXAMPLE 17.4   Iron metal can be produced by reducing iron(III) oxide with hydrogen:

$$Fe_2O_3(s) + 3\,H_2(g) \longrightarrow 2\,Fe(s) + 3\,H_2O(g)$$

$$\Delta H° = +98.8 \text{ kJ}; \Delta S° = +141.5 \text{ J/K}$$

(a) Is this reaction spontaneous at 25°C?

(b) At what temperature will the reaction become spontaneous?

(a) BALLPARK SOLUTION   Let's use rounded values of ΔH (100 kJ), T (300 K), and ΔS (0.14 kJ/K) to estimate the relative values of ΔH and $T\Delta S$. Because $T\Delta S$ = 300 K × 0.14 kJ/K = 42 kJ is smaller than ΔH (100 kJ), $\Delta G = \Delta H - T\Delta S$ is positive and the reaction is nonspontaneous at 300 K.

(a) DETAILED SOLUTION   At 25°C (298 K), ΔG for the reaction is

$$\Delta G = \Delta H - T\Delta S = (98.8 \text{ kJ}) - (298 \text{ K})(0.1415 \text{ kJ/K})$$

$$= (98.8 \text{ kJ}) - (42.2 \text{ kJ})$$

$$= 56.6 \text{ kJ}$$

Because the positive ΔH term is larger than the positive $T\Delta S$ term, ΔG is positive and the reaction is nonspontaneous at 298 K.

(b) SOLUTION   At sufficiently high temperatures, $T\Delta S$ becomes larger than ΔH, ΔG becomes negative, and the reaction becomes spontaneous. We can estimate the temperature at which ΔG changes from a positive to a negative value by setting $\Delta G = \Delta H - T\Delta S = 0$. Solving for T, we find that the reaction becomes spontaneous at 698 K:

$$T = \frac{\Delta H}{\Delta S} = \frac{98.8 \text{ kJ}}{0.1415 \text{ kJ/K}} = 698 \text{ K}$$

Note that this calculation assumes the values of ΔH and ΔS are unchanged on going from 298 K to 698 K. In general, the enthalpies and entropies of both reactants and products increase with increasing temperature, but the increases for the products tend to cancel the increases for the reactants. As a result, values of ΔH and ΔS for a reaction are relatively independent of temperature, at least over a relatively small temperature range. In this example, the temperature range is quite large (400 K), and so the calculated value of T is only an estimate.

▶ **PROBLEM 17.7**   Consider the decomposition of gaseous N_2O_4:

$$N_2O_4(g) \longrightarrow 2\ NO_2(g) \qquad \Delta H° = +57.1\ \text{kJ};\ \Delta S° = +175.8\ \text{J/K}$$

(a) Is this reaction spontaneous at 25°C?
(b) Estimate the temperature at which the reaction becomes spontaneous.

▶ **PROBLEM 17.8**   The following data apply to the vaporization of mercury: $\Delta H_{vap} = 58.5\ \text{kJ/mol};\ \Delta S_{vap} = 92.9\ \text{J/(K · mol)}$.
(a) Does mercury boil at 325°C and 1 atm pressure?
(b) What is the normal boiling point of mercury?

✦ **KEY CONCEPT PROBLEM 17.9**   What are the signs (+, −, or 0) of ΔH, ΔS, and ΔG for the following spontaneous reaction of A atoms (red) and B atoms (blue)?

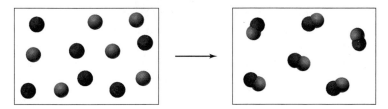

17.8   Standard Free-Energy Changes for Reactions

The free energy of a substance, like its enthalpy and entropy, depends on temperature, pressure, the physical state of the substance (solid, liquid, or gas), and its concentration (in the case of solutions). As a result, free-energy changes for chemical reactions must be compared under a well-defined set of standard-state conditions:

✦ **STANDARD-STATE CONDITIONS:** $\begin{cases} \text{Solids, liquids, and gases in pure form at 1 atm pressure} \\ \text{Solutes at 1 M concentration} \\ \text{A specified temperature, usually 25°C} \end{cases}$

The **standard free-energy change, $\Delta G°$,** for a reaction is the change in free energy that occurs when reactants in their standard states are converted to products in their standard states. As with $\Delta H°$ (Section 8.10), the value of $\Delta G°$ is an extensive property that refers to the number of moles indicated in the chemical equation. For example, $\Delta G°$ at 25°C for the reaction

$$Na(s) + H_2O(l) \longrightarrow 1/2\ H_2(g) + Na^+(aq) + OH^-(aq)$$

is the change in free energy that occurs when 1 mol of solid sodium reacts completely with 1 mol of liquid water to give 0.5 mol of hydrogen gas at 1 atm pressure, along with an aqueous solution that contains 1 mol of Na^+ ions and 1 mol of OH^- ions at concentrations of 1 M, with all reactants and products at a temperature of 25°C. For this reaction, $\Delta G° = -182$ kJ.

Because the free-energy change for any process at constant temperature and pressure is $\Delta G = \Delta H - T\Delta S$, we can calculate the standard free-energy change $\Delta G°$ for a reaction from the standard enthalpy change $\Delta H°$ and the standard entropy change $\Delta S°$. Consider again the Haber synthesis of ammonia:

$$N_2(g) + 3\,H_2(g) \longrightarrow 2\,NH_3(g) \qquad \Delta H° = -92.2 \text{ kJ}; \Delta S° = -198.7 \text{ J/K}$$

The standard free-energy change at 25°C (298 K) is

$$\Delta G° = \Delta H° - T\Delta S° = (-92.2 \times 10^3 \text{ J}) - (298 \text{ K})(-198.7 \text{ J/K})$$
$$= (-92.2 \times 10^3 \text{ J}) - (-59.2 \times 10^3 \text{ J})$$
$$= -33.0 \text{ kJ}$$

Because the negative $\Delta H°$ term is larger than the negative $T\Delta S°$ term at 25°C, $\Delta G°$ is negative and the reaction is spontaneous under standard-state conditions.

Gibbs Free Energy activity

Note that a standard free-energy change applies to a *hypothetical* process rather than an actual process. In a hypothetical process, separate reactants in their standard states are completely converted to separate products in their standard states. In an actual process, however, reactants and products are mixed together, and the reaction may not go to completion.

Take the Haber synthesis, for example. The hypothetical process is

1 mol $N_2(g)$ (1 atm, 25°C)	+	3 mol $H_2(g)$ (1 atm, 25°C)	$\xrightarrow{\Delta G° = -33.0 \text{ kJ}}$	2 mol $NH_3(g)$ (1 atm, 25°C)

State 1 (under first two boxes) ... State 2 (under last box)

where $\Delta G° = -33.0$ kJ is the change in the free energy of the system on going from state 1 to state 2. In an actual synthesis of ammonia, however, the reactants N_2 and H_2 are not separate but are mixed together. Moreover, the reaction doesn't go to completion; it reaches an equilibrium state in which both reactants and products are present together.

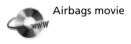
Airbags movie

How, then, should we think about the meaning of $\Delta G°$ in the context of an actual reaction? One way would be to suppose that we have a mixture of N_2, H_2, and NH_3, with each substance present at a partial pressure of 1 atm. Suppose further that the mixture behaves as an ideal gas so that the free energy of each component in the mixture is the same as the free energy of the pure substance. Finally, suppose that the number of moles of each component— say x, y, and z—is very large so that the partial pressures don't change appreciably when 1 mol of N_2 and 3 mol of H_2 are converted to 2 mol of NH_3. In other words, we are imagining the following real process:

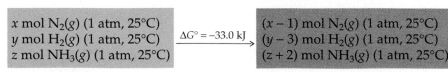

x mol $N_2(g)$ (1 atm, 25°C) y mol $H_2(g)$ (1 atm, 25°C) z mol $NH_3(g)$ (1 atm, 25°C)	$\xrightarrow{\Delta G° = -33.0 \text{ kJ}}$	$(x-1)$ mol $N_2(g)$ (1 atm, 25°C) $(y-3)$ mol $H_2(g)$ (1 atm, 25°C) $(z+2)$ mol $NH_3(g)$ (1 atm, 25°C)

State 1 ... State 2

The free-energy change for this process is the standard free-energy change $\Delta G°$, since each reactant and product is present at 1 atm pressure. If $\Delta G°$ is negative, the reaction will proceed spontaneously to give more products. If $\Delta G°$ is positive, the reaction will proceed in the reverse direction to give more reactants. As always, the value of $\Delta G°$ provides no information concerning the rate of the reaction.

EXAMPLE 17.5   Iron metal is produced commercially by reducing iron(III) oxide in iron ore with carbon monoxide:

$$Fe_2O_3(s) + 3\ CO(g) \longrightarrow 2\ Fe(s) + 3\ CO_2(g)$$

(a) Calculate the standard free-energy change for this reaction at 25°C.

(b) Is the reaction spontaneous under standard-state conditions at 25°C?

(c) Does the reverse reaction become spontaneous at higher temperatures? Explain.

SOLUTION

(a) We can calculate the standard free-energy change from the relation $\Delta G° = \Delta H° - T\Delta S°$, but first we must find $\Delta H°$ and $\Delta S°$ from standard enthalpies of formation ($\Delta H°_f$) and standard molar entropies ($S°$). The following values are found in Appendix B:

	$Fe_2O_3(s)$	$CO(g)$	$Fe(s)$	$CO_2(g)$
$\Delta H°_f$ (kJ/mol)	−824.2	−110.5	0	−393.5
$S°$ [J/(K · mol)]	87.4	197.6	27.3	213.6

So we have

$$\Delta H° = [2\ \Delta H°_f(Fe) + 3\ \Delta H°_f(CO_2)] - [\Delta H°_f(Fe_2O_3) + 3\ \Delta H°_f(CO)]$$
$$= [(2\ mol)(0\ kJ/mol) + (3\ mol)(-393.5\ kJ/mol)]$$
$$\qquad - [(1\ mol)(-824.2\ kJ/mol) + (3\ mol)(-110.5\ kJ/mol)]$$
$$\Delta H° = -24.8\ kJ$$

and

$$\Delta S° = [2\ S°(Fe) + 3\ S°(CO_2)] - [S°(Fe_2O_3) + 3\ S°(CO)]$$
$$= \left[(2\ mol)\left(27.3\ \frac{J}{K \cdot mol}\right) + (3\ mol)\left(213.6\ \frac{J}{K \cdot mol}\right)\right]$$
$$\qquad - \left[(1\ mol)\left(87.4\ \frac{J}{K \cdot mol}\right) + (3\ mol)\left(197.6\ \frac{J}{K \cdot mol}\right)\right]$$
$$\Delta S° = +15.0\ J/K \quad or \quad 0.0150\ kJ/K$$

Therefore,

$$\Delta G° = \Delta H° - T\Delta S°$$
$$= (-24.8\ kJ) - (298\ K)(0.0150\ kJ/K)$$
$$\Delta G° = -29.3\ kJ$$

(b) Because $\Delta G°$ is negative, the reaction is spontaneous at 25°C. This means that a mixture of $Fe_2O_3(s)$, $CO(g)$, $Fe(s)$, and $CO_2(g)$, with each gas at a partial pressure of 1 atm, will react at 25°C to produce more iron metal.

(c) Because $\Delta H°$ is negative and $\Delta S°$ is positive, $\Delta G°$ will be negative at all temperatures. The forward reaction is therefore spontaneous at all temperatures, and the reverse reaction does not become spontaneous at higher temperatures.

▶ **PROBLEM 17.10** Consider the thermal decomposition of calcium carbonate:

$$CaCO_3(s) \longrightarrow CaO(s) + CO_2(g)$$

(a) Using the data in Appendix B, calculate the standard free-energy change for this reaction at 25°C.
(b) Will a mixture of solid $CaCO_3$, solid CaO, and gaseous CO_2 at 1 atm pressure react spontaneously at 25°C to produce more CaO and CO_2?
(c) Assuming that $\Delta H°$ and $\Delta S°$ are independent of temperature, estimate the temperature at which the reaction becomes spontaneous.

✦ **KEY CONCEPT PROBLEM 17.11** Consider the following endothermic decomposition of AB_2 molecules:

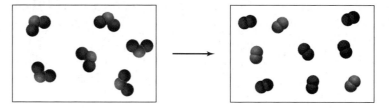

(a) What is the sign (+, −, or 0) of $\Delta S°$ for the reaction?
(b) Is the reaction more likely to be spontaneous at high temperatures or at low temperatures? Explain.

17.9 Standard Free Energies of Formation

The **standard free energy of formation, $\Delta G°_f$,** of a substance is the free-energy change for formation of 1 mol of the substance in its standard state from the most stable form of its constituent elements in their standard states. For example, we found in the previous section that the standard free-energy change $\Delta G°$ for synthesis of 2 mol of NH_3 from its constituent elements is −33.0 kJ:

$$N_2(g) + 3 H_2(g) \longrightarrow 2 NH_3(g) \qquad \Delta G° = -33.0 \text{ kJ}$$

Therefore, $\Delta G°_f$ for ammonia is −33.0 kJ/2 mol, or −16.5 kJ/mol.

Values of $\Delta G°_f$ at 25°C for some common substances are listed in Table 17.3, and additional values are given in Appendix B. Note that $\Delta G°_f$ for an element in its most stable form at 25°C is defined to be zero. Thus, solid graphite has $\Delta G°_f = 0$ kJ/mol, but diamond, a less stable form of solid carbon at 25°C, has $\Delta G°_f = 2.9$ kJ/mol. As with standard enthalpies of formation, $\Delta H°_f$, a zero value of $\Delta G°_f$ for elements in their most stable form establishes a thermochemical "sea level," or reference point, with respect to which the standard free energies of other substances are measured. We can't measure the absolute value of a substance's free energy (as we can the entropy), but that's not a problem because we are interested only in free-energy *differences* between reactants and products.

TABLE 17.3   Standard Free Energies of Formation for Some Common Substances at 25°C

Substance	Formula	$\Delta G°_f$ (kJ/mol)	Substance	Formula	$\Delta G°_f$ (kJ/mol)
Gases			*Liquids*		
Acetylene	C_2H_2	209.2	Acetic acid	CH_3CO_2H	−390
Ammonia	NH_3	−16.5	Ethanol	C_2H_5OH	−174.9
Carbon dioxide	CO_2	−394.4	Methanol	CH_3OH	−166.4
Carbon monoxide	CO	−137.2	Water	H_2O	−237.2
Ethylene	C_2H_4	68.1	*Solids*		
Hydrogen	H_2	0	Calcium carbonate	$CaCO_3$	−1128.8
Methane	CH_4	−50.8	Calcium oxide	CaO	−604.0
Nitrogen	N_2	0	Diamond	C	2.9
Nitrogen dioxide	NO_2	51.3	Graphite	C	0
Dinitrogen tetroxide	N_2O_4	97.8	Iron(III) oxide	Fe_2O_3	−742.2

The standard free energy of formation of a substance measures its thermodynamic stability with respect to its constituent elements. Substances that have a negative value of $\Delta G°_f$, such as carbon dioxide and water, are stable and do not decompose to their constituent elements under standard-state conditions. Substances that have a positive value of $\Delta G°_f$, such as ethylene and nitrogen dioxide, are thermodynamically unstable with respect to their constituent elements. Once prepared, though, such substances can exist for long periods of time if the rate of their decomposition is slow.

Clearly, there's no point in trying to synthesize a substance from its elements under standard-state conditions if the substance has a positive value of $\Delta G°_f$. Such a substance would have to be prepared at other temperatures and/or pressures, or it would have to be made from alternative starting materials using a reaction that has a negative free-energy change. Thus, a knowledge of thermodynamics can save considerable time in chemical synthesis.

In the previous section, we calculated standard free-energy changes for reactions from the equation $\Delta G° = \Delta H° - T\Delta S°$, using tabulated values of $\Delta H°_f$ and $S°$ to find $\Delta H°$ and $\Delta S°$. Alternatively, we can calculate $\Delta G°$ more directly by subtracting the standard free energies of formation of all the reactants from the standard free energies of formation of all the products:

$$\Delta G° = \Delta G°_f(\text{products}) - \Delta G°_f(\text{reactants})$$

For the general reaction

$$a\,A + b\,B \longrightarrow c\,C + d\,D$$

the standard free-energy change is

$$\Delta G° = [c\,\Delta G°_f(C) + d\,\Delta G°_f(D)] - [a\,\Delta G°_f(A) + b\,\Delta G°_f(B)]$$

To illustrate, let's calculate the standard free-energy change for the reaction in Example 17.5—the reduction of iron(III) oxide with carbon monoxide:

$$Fe_2O_3(s) + 3\,CO(g) \longrightarrow 2\,Fe(s) + 3\,CO_2(g)$$

Using the $\Delta G°_f$ values in Table 17.3, we obtain

$$\Delta G° = [2 \, \Delta G°_f(Fe) + 3 \, \Delta G°_f(CO_2)] - [\Delta G°_f(Fe_2O_3) + 3 \, \Delta G°_f(CO)]$$

$$= [(2 \, mol)(0 \, kJ/mol) + (3 \, mol)(-394.4 \, kJ/mol)]$$

$$- [(1 \, mol)(-742.2 \, kJ/mol) + (3 \, mol)(-137.2 \, kJ/mol)]$$

$$\Delta G° = -29.4 \, kJ$$

This result agrees well with the value of −29.3 kJ calculated from the equation $\Delta G° = \Delta H° - T\Delta S°$ in Example 17.5.

EXAMPLE 17.6

(a) Calculate the standard free-energy change for the oxidation of ammonia to give nitric oxide (NO) and water. Is it worth trying to find a catalyst for this reaction?

$$4 \, NH_3(g) + 5 \, O_2(g) \longrightarrow 4 \, NO(g) + 6 \, H_2O(l)$$

(b) Is it worth trying to find a catalyst for the synthesis of NO from gaseous N_2 and O_2 at 25°C?

SOLUTION

(a) We can calculate $\Delta G°$ most easily from tabulated standard free energies of formation (Appendix B):

$$\Delta G° = [4 \, \Delta G°_f(NO) + 6 \, \Delta G°_f(H_2O)] - [4 \, \Delta G°_f(NH_3) + 5 \, \Delta G°_f(O_2)]$$

$$= [(4 \, mol)(86.6 \, kJ/mol) + (6 \, mol)(-237.2 \, kJ/mol)]$$

$$- [(4 \, mol)(-16.5 \, kJ/mol) + (5 \, mol)(0 \, kJ/mol)]$$

$$\Delta G° = -1010.8 \, kJ$$

It is worth looking for a catalyst because the negative value of $\Delta G°$ indicates that the reaction is spontaneous. (This reaction is the first step in the *Ostwald process* for production of nitric acid. In industry, the reaction is carried out using a platinum–rhodium catalyst.)

(b) It's not worth looking for a catalyst for the reaction $N_2(g) + O_2(g) \rightarrow 2 \, NO(g)$ because the standard free energy of formation of NO is positive ($\Delta G°_f = 86.6$ kJ/mol). This means that NO is unstable with respect to decomposition to N_2 and O_2 at 25°C. A catalyst could only *increase the rate of decomposition*. It can't affect the composition of the equilibrium mixture (Section 13.10), and so it can't affect the direction of the reaction.

▶ **PROBLEM 17.12**

(a) Using values of $\Delta G°_f$ in Appendix B, calculate the standard free-energy change for the reaction of calcium carbide (CaC_2) with water. Might this reaction be used for synthesis of acetylene (C_2H_2)?

$$CaC_2(s) + 2 \, H_2O(l) \longrightarrow C_2H_2(g) + Ca(OH)_2(s)$$

(b) Is it possible to synthesize acetylene from solid graphite and gaseous H_2 at 25°C and 1 atm pressure?

17.10 Free-Energy Changes and Composition of the Reaction Mixture

The sign of the standard free-energy change $\Delta G°$ tells the direction of spontaneous reaction when both reactants and products are present at standard-state conditions. In actual reactions, however, the composition of the reaction mixture seldom corresponds to standard-state pressures and concentrations. Moreover, the partial pressures and concentrations change as a reaction proceeds. So how do we calculate the free-energy change ΔG for a reaction when the reactants and products are present at nonstandard-state pressures and concentrations?

The answer is given by the relation

$$\Delta G = \Delta G° + RT \ln Q$$

where ΔG is the free-energy change under nonstandard-state conditions, $\Delta G°$ is the free-energy change under standard-state conditions, R is the gas constant, T is the absolute temperature in kelvins, and Q is the reaction quotient (Q_p for reactions involving gases or Q_c for reactions involving solutes in solution). Recall from Section 13.5 that the reaction quotient Q_c is an expression having the same form as the equilibrium constant expression K_c except that the concentrations do not necessarily have equilibrium values. Similarly, Q_p has the same form as K_p except that the partial pressures have arbitrary values. For example, for the Haber synthesis of ammonia:

$$N_2(g) + 3\,H_2(g) \rightleftharpoons 2\,NH_3(g) \qquad Q_p = \frac{(P_{NH_3})^2}{(P_{N_2})(P_{H_2})^3}$$

(For reactions that involve both gases and solutes in solution, the reaction quotient Q contains partial pressures of gases and molar concentrations of solutes.)

We won't derive the equation for ΔG under nonstandard-state conditions. It's hardly surprising, however, that ΔG and Q should turn out to be related because both predict the direction of a reaction. Example 17.7 shows how to use this equation.

EXAMPLE 17.7   Calculate the free-energy change for ammonia synthesis at 25°C (298 K) given the following sets of partial pressures:
(a) 1.0 atm N_2, 3.0 atm H_2, 0.020 atm NH_3
(b) 0.010 atm N_2, 0.030 atm H_2, 2.0 atm NH_3

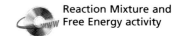

Reaction Mixture and
Free Energy activity

$$N_2(g) + 3\,H_2(g) \rightleftharpoons 2\,NH_3(g) \qquad \Delta G° = -33.0 \text{ kJ}$$

(a) BALLPARK SOLUTION   Based on Le Châtelier's principle (Section 13.6), we expect that the reaction will have a greater tendency to occur under the cited conditions than under standard-state conditions because one of the reactant partial pressures is greater than 1 atm and the product partial pressure is less than 1 atm. We therefore predict that ΔG will be more negative than $\Delta G°$, which means that the reaction has a greater thermodynamic driving force under the cited conditions than it does under standard-state conditions.

(a) DETAILED SOLUTION   We can calculate ΔG from the relation $\Delta G = \Delta G° + RT \ln Q$, where Q is Q_p for the reaction $N_2(g) + 3\,H_2(g) \rightleftharpoons 2\,NH_3(g)$:

$$Q_p = \frac{(P_{NH_3})^2}{(P_{N_2})(P_{H_2})^3} = \frac{(0.020)^2}{(1.0)(3.0)^3} = 1.5 \times 10^{-5}$$

Substituting this value of Q_p into the equation for ΔG gives

$$\Delta G = \Delta G° + RT \ln Q$$
$$= (-33.0 \times 10^3 \text{ J/mol}) + [8.314 \text{ J/(K} \cdot \text{mol)}](298 \text{ K})(\ln 1.5 \times 10^{-5})$$
$$= (-33.0 \times 10^3 \text{ J/mol}) + (-27.5 \times 10^3 \text{ J/mol})$$
$$\Delta G = -60.5 \text{ kJ/mol}$$

To maintain consistent units in this calculation, we have expressed ΔG and $\Delta G°$ in units of kJ/mol because R has units of J/(K · mol). The "per mole" in this context means per molar amounts of reactants and products indicated by the coefficients in the balanced equation. Thus, the free energy change is −60.5 kJ when 1 mole of N_2 and 3 moles of H_2 are converted to 2 moles of NH_3 under the specified conditions.

Note that ΔG is more negative than $\Delta G°$ because Q_p is less than 1 and $\ln Q_p$ is therefore a negative number. Note also that when each reactant and product is present at a partial pressure of 1 atm, $Q_p = 1$, $\ln Q_p = 0$, and $\Delta G = \Delta G°$.

(b) BALLPARK SOLUTION   In this case, the reaction mixture is rich in the product and poor in the reactants. Therefore, Q_p is expected to be greater than 1, and ΔG should be more positive than $\Delta G°$.

(b) DETAILED SOLUTION   The value of Q_p for the reaction $N_2(g) + 3\,H_2(g) \rightleftharpoons 2\,NH_3(g)$ is

$$Q_p = \frac{(P_{NH_3})^2}{(P_{N_2})(P_{H_2})^3} = \frac{(2.0)^2}{(0.010)(0.030)^3} = 1.5 \times 10^7$$

The corresponding value of ΔG is

$$\Delta G = \Delta G° + RT \ln Q_p$$
$$= (-33.0 \times 10^3\ \text{J/mol}) + [8.314\ \text{J/(K·mol)}](298\ \text{K})(\ln 1.5 \times 10^7)$$
$$= (-33.0 \times 10^3\ \text{J/mol}) + (40.9 \times 10^3\ \text{J/mol})$$
$$\Delta G = 7.9\ \text{kJ/mol}$$

Because Q_p is large enough to give a positive value for ΔG, the reaction is nonspontaneous in the forward direction but spontaneous in the reverse direction. Thus, as we saw in Section 13.7, the direction in which a reaction proceeds spontaneously depends on the composition of the reaction mixture.

▶ **PROBLEM 17.13**   Calculate ΔG for the formation of ethylene (C_2H_4) from carbon and hydrogen at 25°C when the partial pressures are 100 atm H_2 and 0.10 atm C_2H_4.

$$2\,C(s) + 2\,H_2(g) \longrightarrow C_2H_4(g) \qquad \Delta G° = 68.1\ \text{kJ}$$

Is the reaction spontaneous in the forward or the reverse direction?

◆— **KEY CONCEPT PROBLEM 17.14**   Consider the following gas-phase reaction of A_2 (red) and B_2 (blue) molecules:

$$A_2 + B_2 \rightleftharpoons 2\,AB \qquad \Delta G° = 15\ \text{kJ}$$

(a) Which of the following reaction mixtures has the largest ΔG of reaction, and which has the smallest?

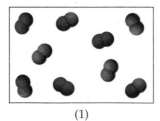
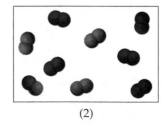
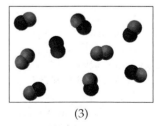

(1)   (2)   (3)

(b) If the partial pressure of each reactant and product in reaction mixture (1) is equal to 1 atm, what is the value of ΔG for the reaction in mixture (1)?

17.11   Free Energy and Chemical Equilibrium

Now that we've seen how ΔG for a reaction depends on composition, we can understand how the total free energy of a reaction mixture changes as the reaction progresses toward equilibrium. Look again at the expression for calculating ΔG:

$$\Delta G = \Delta G° + RT \ln Q$$

If the reaction mixture contains mainly reactants and almost no products, Q will be much less than 1 and $RT \ln Q$ will be a very large negative number (minus infinity when $Q = 0$). Consequently, no matter what the value of $\Delta G°$ (positive or negative), the negative $RT \ln Q$ term will dominate the $\Delta G°$ term, and ΔG will be negative. This means that the forward reaction is always spontaneous when the concentration of products is very small. Conversely, if the reaction mixture contains mainly products and almost no reactants, Q will be much greater than 1, and $RT \ln Q$ will be a very large positive number (plus infinity when no reactants are present). Consequently, the positive $RT \ln Q$ term will dominate the $\Delta G°$ term, and ΔG will be positive. Thus, the reverse reaction is always spontaneous when the concentration of reactants is very small. These conditions are summarized as follows:

- When the reaction mixture is mostly reactants:

$$Q << 1 \qquad RT \ln Q << 0 \qquad \Delta G < 0$$

  The total free energy decreases as the reaction proceeds spontaneously in the forward direction.

- When the reaction mixture is mostly products:

$$Q >> 1 \qquad RT \ln Q >> 0 \qquad \Delta G > 0$$

  The total free energy decreases as the reaction proceeds spontaneously in the reverse direction.

Figure 17.10 shows how the total free energy of a reaction mixture changes as the reaction progresses. Because the free energy decreases as pure reactants form products and also decreases as pure products form reactants, the free-energy curve must go through a minimum somewhere between pure reactants and pure products. At that minimum free-energy composition, the system is at equilibrium because conversion of either reactants to products or products to reactants would involve an increase in free energy. The equilibrium composition persists indefinitely unless the system is disturbed by an external influence.

The sign of ΔG for the reaction is the same as the sign of the slope of the free-energy curve (Figure 17.10). To the left of the equilibrium composition, ΔG and the slope of the curve are negative, and the free energy decreases as reactants are converted to products. To the right of the equilibrium composition, ΔG and the slope of the curve are positive. Exactly at the equilibrium composition, ΔG and the slope of the curve are zero, and no reaction occurs.

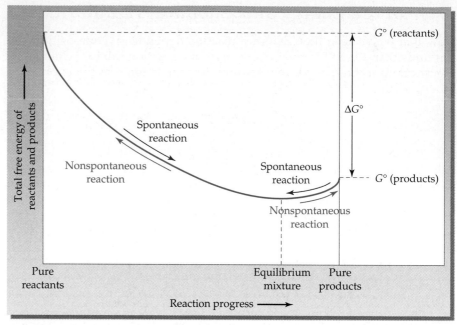

FIGURE 17.10   The total free energy of a reaction mixture as a function of the progress of the reaction. Beginning with either pure reactants or pure products, the free energy decreases (ΔG is negative) as the system moves toward equilibrium. The graph is drawn assuming that the pure reactants and pure products are in their standard states and that $\Delta G°$ for the reaction is negative so the equilibrium composition is rich in products.

We can now derive a relationship between free energy and the equilibrium constant. At equilibrium, ΔG for a reaction is zero and the reaction quotient Q is equal to the equilibrium constant K. Substituting $\Delta G = 0$ and $Q = K$ into the equation

$$\Delta G = \Delta G° + RT \ln Q$$

gives

$$0 = \Delta G° + RT \ln K$$

or

$$\Delta G° = -RT \ln K$$

(Here, K is K_p for reactions involving gases because the standard state for gases is 1 atm pressure. K is K_c for reactions involving solutes in solution because the standard-state for solutes is 1 M concentration. For reactions involving both gases and solutes in solution, the equilibrium constant expression contains partial pressures of gases and molar concentrations of solutes.)

The equation $\Delta G° = -RT \ln K$ is one of the most important relationships in chemical thermodynamics because it allows us to calculate the equilibrium constant for a reaction from the standard free-energy change, or vice versa. This relationship is especially useful when K is difficult to measure. Consider a reaction so slow that it takes more than an experimenter's lifetime to reach equilibrium, or a reaction that goes essentially to completion, so that the equilibrium concentrations of the reactants are extremely small and hard to measure. We can't measure K directly in such cases, but we can calculate its value from $\Delta G°$.

The relationship between $\Delta G°$ and the equilibrium constant K is summarized in Table 17.4. A reaction with a negative value of $\Delta G°$ has an equilibrium constant greater than 1, which corresponds to a minimum in the

TABLE 17.4   Relationship Between the Standard Free-Energy Change and the Equilibrium Constant for a Reaction: $\Delta G° = -RT \ln K$

$\Delta G°$	$\ln K$	K	Comment
$\Delta G° < 0$	$\ln K > 0$	$K > 1$	The equilibrium mixture is mainly products.
$\Delta G° > 0$	$\ln K < 0$	$K < 1$	The equilibrium mixture is mainly reactants.
$\Delta G° = 0$	$\ln K = 0$	$K = 1$	The equilibrium mixture contains comparable amounts of reactants and products ($K = 1$ for 1 M concentrations and 1 atm partial pressures).

free-energy curve of Figure 17.10 at a composition rich in products. Conversely, a reaction that has a positive value of $\Delta G°$ has an equilibrium constant less than 1, and a minimum in the free-energy curve at a composition rich in reactants. Try redrawing Figure 17.10 for the case where $\Delta G° > 0$.

We have now answered the fundamental question posed at the beginning of this chapter: What determines the value of the equilibrium constant—that is, what properties of nature determine the direction and extent of a particular chemical reaction? The answer is that the value of the equilibrium constant is determined by the standard free-energy change, $\Delta G°$, for the reaction, which depends, in turn, on the standard heats of formation and the standard molar entropies of the reactants and products.

EXAMPLE 17.8   Methanol (CH_3OH), an important alcohol used in the manufacture of adhesives, fibers, and plastics, is synthesized industrially by the reaction

$$CO(g) + 2\,H_2(g) \rightleftharpoons CH_3OH(g)$$

Use the thermodynamic data in Appendix B to calculate the equilibrium constant for this reaction at 25°C.

SOLUTION   First calculate $\Delta G°$ for the reaction from the tabulated values of $\Delta G°_f$ for reactants and products:

$$\Delta G° = \Delta G°_f(CH_3OH) - [\Delta G°_f(CO) + 2\,\Delta G°_f(H_2)]$$
$$= (1\text{ mol})(-161.9\text{ kJ/mol}) - [(1\text{ mol})(-137.2\text{ kJ/mol}) + (2\text{ mol})(0\text{ kJ/mol})]$$
$$\Delta G° = -24.7\text{ kJ}$$

Solving the equation

$$\Delta G° = -RT \ln K$$

for $\ln K$ gives

$$\ln K = \frac{-\Delta G°}{RT} = \frac{-(-24.7 \times 10^3\text{ J/mol})}{[8.314\text{ J/K}\cdot\text{mol}](298\text{ K})} = 9.97$$

Therefore,

$$K = K_p = \text{antiln } 9.97 = e^{9.97} = 2.1 \times 10^4$$

The equilibrium constant obtained by this procedure is K_p because the reactants and products are gases and their standard states are defined in terms of pressure. If we want the value of K_c, we must calculate it from the relation $K_p = K_c(RT)^{\Delta n}$ (Section 13.3), where R must be expressed in the proper units $[R = 0.08206 \ (L \cdot atm)/(K \cdot mol)]$.

EXAMPLE 17.9   The value of $\Delta G°_f$ at 25°C for gaseous mercury is 31.85 kJ/mol. What is the vapor pressure of mercury at 25°C?

SOLUTION   The vapor pressure (in atm) is equal to K_p for the reaction

$$Hg(l) \rightleftharpoons Hg(g) \qquad K_p = P_{Hg}$$

Note that $Hg(l)$ is omitted from the equilibrium constant expression because it is a pure liquid. Because the standard state for elemental mercury is the pure liquid, $\Delta G°_f = 0$ for $Hg(l)$, and $\Delta G°$ for the vaporization reaction is simply equal to $\Delta G°_f$ for $Hg(g)$ (31.85 kJ/mol). We can calculate K_p from $\Delta G°$ as in Example 17.8:

$$\ln K_p = \frac{-\Delta G°}{RT} = \frac{-(31.85 \times 10^3 \ J/mol)}{[8.314 \ J/(K \cdot mol)](298 \ K)} = -12.86$$

$$K_p = \text{antiln}\,(-12.86) = e^{-12.86} = 2.6 \times 10^{-6}$$

Since K_p is defined in units of atmospheres, the vapor pressure of mercury at 25°C is 2.6×10^{-6} atm (0.0020 mm Hg). Because the vapor pressure is appreciable and mercury is toxic in the lungs, mercury should not be handled without adequate ventilation.

Mercury has an appreciable vapor pressure at room temperature, and its handling requires adequate ventilation.

EXAMPLE 17.10   At 25°C, K_{sp} for $PbCrO_4$ is 2.8×10^{-13}. Calculate the standard free-energy change at 25°C for the reaction $PbCrO_4(s) \rightleftharpoons Pb^{2+}(aq) + CrO_4^{2-}(aq)$.

SOLUTION   We can calculate $\Delta G°$ directly from the equilibrium constant K_{sp}:

$$\Delta G° = -RT \ln K_{sp} = -[8.314 \ J/(K \cdot mol)](298 \ K)(\ln 2.8 \times 10^{-13})$$
$$= 71.6 \times 10^3 \ J/mol = 71.6 \ kJ/mol$$

Note that $\Delta G°$ is a large positive number, in accord with a K_{sp} value much less than 1.

▶ **PROBLEM 17.15**   Given the data in Appendix B, calculate K_p at 25°C for the reaction $CaCO_3(s) \rightleftharpoons CaO(s) + CO_2(g)$.

▶ **PROBLEM 17.16**   Use the data in Appendix B to calculate the vapor pressure of water at 25°C.

▶ **PROBLEM 17.17**   At 25°C, K_w for the dissociation of water is 1.0×10^{-14}. Calculate $\Delta G°$ for the reaction $2 \ H_2O(l) \rightleftharpoons H_3O^+(aq) + OH^-(aq)$.

Some Random Thoughts About Entropy

The idea of entropy has intrigued thinkers for more than a century. Poets, philosophers, physicists, and biologists have all struggled to understand the consequences of entropy.

Poets and philosophers have spoken of entropy as "time's arrow," a metaphor that arises out of the second law of thermodynamics. According to the second law, all spontaneously occurring processes are accompanied by an increase in the disorder of the universe. At some far distant time, when all is disorder and there is no order or available energy left, the universe as we know it must end.

At some distant time, when all is disorder, the universe as we know it must end.

Physicists speak of entropy as giving a directionality to time. There is a symmetry to basic physical laws that makes them equally valid when the signs of the quantities are reversed. The attraction between a positively charged proton and a negatively charged electron, for example, is exactly the same as the attraction between a negatively charged proton (a so-called *antiproton*) and a positively charged electron (a *positron*). Time, however, cannot be reversed because of entropy. Any process that takes place spontaneously over time must increase the entropy of the universe. The reverse process, which would have to go backward over time, can't occur because it would decrease the entropy of the universe. Thus, there is a one-way nature to time that is not shared by other physical quantities.

Biologists, too, have been intrigued by entropy and its consequences. Their problem is that, if the disorder of the universe is always increasing, how is it possible for enormously complex and increasingly sophisticated life forms to evolve? After all, the more complex the organism, the greater the amount of order required and the *lower* the entropy. The answer to the biologists' question again arises from the second law of thermodynamics: All spontaneous processes increase the disorder of the universe—that is, the total disorder of both system and surroundings. It's perfectly possible, however, for the disorder of any *system* to decrease spontaneously as long as the disorder of the *surroundings* increases by an even greater amount.

The energy used to power all living organisms on earth comes from sunlight, caused ultimately by nuclear reactions in the sun. Photosynthetic cells in plants use the sun's energy to make glucose, which is then used by animals as their primary source of energy. The energy an animal obtains from glucose is then used to build and organize complex molecules, resulting in a *decrease* in entropy for the animal. At the same time, however, the entropy of the surroundings *increases* as the animal releases small, simple waste products such as CO_2 and H_2O. Furthermore, heat is released by the animal, further increasing the entropy of the surroundings. Thus, an organism pays for its decrease in entropy by increasing the entropy of the rest of the universe.

747

Plants capture solar energy and convert it to chemical potential energy by using it to make energy-rich molecules such as glucose. Animals tap this energy by eating plants (or other animals that have eaten plants) and use it to build even more complex molecular structures. These processes involve a local decrease in entropy, but produce a net increase in the entropy of the universe as a whole.

▶ **PROBLEM 17.18** Consider the growth of a human adult from a single cell. Does this process violate the second law of thermodynamics? Explain.

▶ **PROBLEM 17.19** If you watched a movie run backwards, would you expect to see violations of the second law? Explain.

Summary

Thermodynamics deals with the interconversion of heat and other forms of energy and allows us to predict the direction and extent of chemical reactions and other spontaneous processes. A **spontaneous process** proceeds on its own without any external influence. All spontaneous reactions move toward equilibrium.

Entropy, denoted by S, is a state function that measures molecular disorder, or randomness. The entropy of a system (reactants plus products) increases (ΔS is positive) for the following processes: phase transitions that convert a solid to a liquid or a liquid to a gas; reactions that increase the number of gaseous molecules; dissolution of molecular solids and certain salts in water; raising the temperature of a substance; expansion of a gas at constant temperature.

A disordered state of a system (state of high entropy) can be achieved in more ways (W) than an ordered state and is therefore more probable. The entropy of a state can be calculated from Boltzmann's formula, $S = k \ln W$. According to the **third law of thermodynamics**, the entropy of a pure, perfectly ordered crystalline substance at 0 K is zero.

The **standard molar entropy, $S°$,** of a substance is the absolute entropy of 1 mol of the pure substance at 1 atm pressure and a specified temperature, usually 25°C. The **standard entropy of reaction, $\Delta S°$,** can be calculated from the relation $\Delta S° = S°(\text{products}) - S°(\text{reactants})$.

The **first law of thermodynamics** states that in any process the total energy of a system and its surroundings remains constant. The **second law of thermodynamics** says that in any *spontaneous* process, the total entropy of a system and its surroundings ($\Delta S_{\text{total}} = \Delta S_{\text{sys}} + \Delta S_{\text{surr}}$) always increases. A chemical reaction is spontaneous if $\Delta S_{\text{total}} > 0$, nonspontaneous if $\Delta S_{\text{total}} < 0$, and at equilibrium if $\Delta S_{\text{total}} = 0$. Reactions that are nonspontaneous in the forward direction are spontaneous in the reverse direction. For a reaction at constant pressure, $\Delta S_{\text{surr}} = -\Delta H/T$, and ΔS_{sys} is $\Delta S°$ for the reaction.

Free energy, $G = H - TS$, is a state function that indicates whether a reaction is spontaneous or nonspontaneous. A reaction at constant temperature and pressure is spontaneous if $\Delta G < 0$, nonspontaneous if $\Delta G > 0$, and at equilibrium if $\Delta G = 0$. In the equation $\Delta G = \Delta H - T\Delta S$, temperature is a weighting factor that determines the relative importance of the enthalpy and entropy contributions to ΔG.

The **standard free-energy change, $\Delta G°$,** for a reaction is the change in free energy that occurs when reactants in their standard states are converted to products in their standard states. The **standard free energy of formation, $\Delta G°_f$,** of a substance is the free-energy change for formation of 1 mol of the substance in its standard state from the most stable form of the constituent elements in their standard states. Substances with a negative value of $\Delta G°_f$ are thermodynamically stable with respect to the constituent elements. We can calculate $\Delta G°$ for a reaction in either of two ways: (1) $\Delta G° = \Delta G°_f(\text{products}) - \Delta G°_f(\text{reactants})$ or (2) $\Delta G° = \Delta H° - T\Delta S°$.

The free-energy change, ΔG, for a reaction under nonstandard-state conditions is given by $\Delta G = \Delta G° + RT \ln Q$, where Q is the reaction quotient. At equilibrium, $\Delta G = 0$ and $Q = K$. As a result, $\Delta G° = -RT \ln K$, which allows us to calculate the equilibrium constant from $\Delta G°$ and vice versa.

Key Words

entropy (S) *720*

first law of
  thermodynamics *729*

free energy (S) *732*

second law of
  thermodynamics *729*

spontaneous process *718*

standard entropy of
  reaction ($\Delta S°$) *728*

standard free-energy
  change ($\Delta G°$) *735*

standard free energy of
  formation ($\Delta G°_f$) *738*

standard molar entropy
  ($S°$) *727*

standard-state
  conditions *735*

thermodynamics *718*

third law of
  thermodynamics *727*

Key Concept Summary

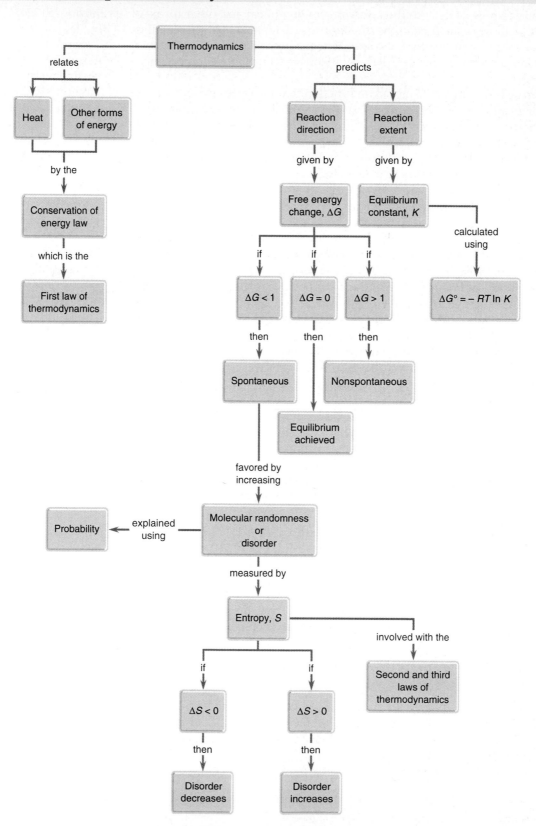

Understanding Key Concepts

Problems 17.1–17.19 appear within the chapter.

17.20 Ideal gases A (red spheres) and B (blue spheres) occupy two separate bulbs. The contents of both bulbs constitute the initial state of an isolated system. Consider the process that occurs when the stopcock is opened.

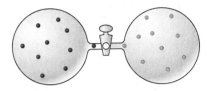

    (a) Sketch the final (equilibrium) state of the system.

    (b) What are the signs (+, −, or 0) of ΔH, ΔS, and ΔG for this process? Explain.

    (c) How does this process illustrate the second law of thermodynamics?

    (d) Is the reverse process spontaneous or nonspontaneous? Explain.

17.21 What are the signs (+, −, or 0) of ΔH, ΔS, and ΔG for the spontaneous sublimation of a crystalline solid? Explain.

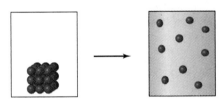

17.22 What are the signs (+, −, or 0) of ΔH, ΔS, and ΔG for the spontaneous condensation of a vapor to a liquid? Explain.

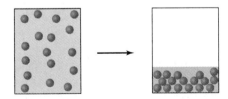

17.23 An ideal gas is compressed at constant temperature. What are the signs (+, −, or 0) of ΔH, ΔS, and ΔG for the process? Explain.

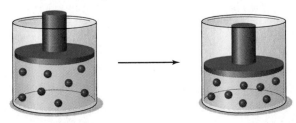

17.24 Consider the following spontaneous reaction of A_2 molecules (red) and B_2 molecules (blue):

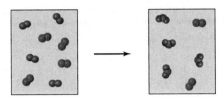

    (a) Write a balanced equation for the reaction.

    (b) What are the signs (+, −, or 0) of ΔH, ΔS, and ΔG for the reaction? Explain.

17.25 Consider the dissociation reaction $A_2(g) \rightleftharpoons 2\,A(g)$. The following pictures represent two possible initial states and the equilibrium state of the system:

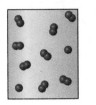

Initial state 1   Initial state 2

Equilibrium state

    (a) Is the reaction quotient Q_p for initial state 1 greater than, less than, or equal to the equilibrium constant K_p? Is Q_p for initial state 2 greater than, less than, or equal to K_p?

    (b) What are the signs (+, −, or 0) of ΔH, ΔS, and ΔG when the system goes from initial state 1 to the equilibrium state? Explain. Is this a spontaneous process?

    (c) What are the signs (+, −, or 0) of ΔH, ΔS, and ΔG when the system goes from initial state 2 to the equilibrium state? Explain. Is this a spontaneous process?

    (d) Relate each of the pictures to the graph in Figure 17.10.

17.26 Consider again the dissociation reaction $A_2(g) \rightleftharpoons 2A(g)$ (Problem 17.25).

(a) What are the signs (+, −, or 0) of the standard enthalpy change, $\Delta H°$, and the standard entropy change, $\Delta S°$, for the forward reaction?

(b) Distinguish between the meaning of $\Delta S°$ for the dissociation reaction and ΔS for the process in which the system goes from initial state 1 to the equilibrium state (pictured in Problem 17.25).

(c) Can you say anything about the sign of $\Delta G°$ for the dissociation reaction? How does $\Delta G°$ depend on temperature? Will $\Delta G°$ increase, decrease, or remain the same if the temperature increases?

(d) Will the equilibrium constant K_p increase, decrease, or remain the same if the temperature increases? How will the picture for the equilibrium state (Problem 17.25) change if the temperature increases?

(e) What is the value of ΔG for the dissociation reaction when the system is at equilibrium?

17.27 Make a qualitative plot of free energy versus reaction progress for a reaction that has a positive value of $\Delta G°$. Account for the shape of the curve, and identify the point at which $\Delta G = 0$. What is the significance of that point?

17.28 The following pictures represent equilibrium mixtures for the interconversion of A molecules (red) and X, Y, or Z molecules (blue):

(1) $A \rightleftharpoons X$ (2) $A \rightleftharpoons Y$

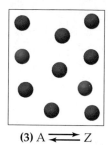

(3) $A \rightleftharpoons Z$

What is the sign of $\Delta G°$ for each of the three reactions?

17.29 Consider reaction (1) $A \rightleftharpoons X$ in Problem 17.28.

(a) What is the sign of ΔG for the reaction after 4 molecules of A have been added to the reaction mixture?

(b) Sketch the reaction mixture after a new state of equilibrium has been reached.

Additional Problems

Spontaneous Processes

17.30 Distinguish between a spontaneous process and a nonspontaneous process, and give an example of each.

17.31 Decomposition of hydrogen peroxide to gaseous O_2 and water is a spontaneous process, yet H_2O_2 is stable for long periods of time. Explain.

17.32 Which of the following processes are spontaneous, and which are nonspontaneous?

(a) Freezing of water at 2°C

(b) Corrosion of iron metal

(c) Expansion of a gas to fill the available volume

(d) Separation of an unsaturated aqueous solution of potassium chloride into solid KCl and liquid water

17.33 Tell whether the following processes are spontaneous or nonspontaneous:

(a) Dissolving sugar in hot coffee

(b) Decomposition of NaCl to solid sodium and gaseous chlorine at 25°C and 1 atm pressure

(c) Uniform mixing of bromine vapor and nitrogen gas

(d) Boiling of gasoline at 25°C and 1 atm pressure

17.34 Assuming that gaseous reactants and products are present at 1 atm partial pressure, which of the following reactions are spontaneous in the forward direction?

(a) $N_2(g) + 2H_2(g) \rightarrow N_2H_4(l)$; $K_p = 7 \times 10^{-27}$

(b) $2Mg(s) + O_2(g) \rightarrow 2MgO(s)$; $K_p = 2 \times 10^{198}$

(c) $MgCO_3(s) \rightarrow MgO(s) + CO_2(g)$; $K_p = 9 \times 10^{-10}$

(d) $2CO(g) + O_2(g) \rightarrow 2CO_2(g)$; $K_p = 1 \times 10^{90}$

17.35 Assuming that dissolved reactants and products are present at 1 M concentrations, which of the following reactions are nonspontaneous in the forward direction?

(a) $HCN(aq) + H_2O(l) \rightarrow H_3O^+(aq) + CN^-(aq)$;
$K = 4.9 \times 10^{-10}$

(b) $H_3O^+(aq) + OH^-(aq) \rightarrow 2H_2O(l)$; $K = 1.0 \times 10^{14}$

(c) $Ba^{2+}(aq) + CO_3^{2-}(aq) \rightarrow BaCO_3(s)$; $K = 3.8 \times 10^8$

(d) $AgCl(s) \rightarrow Ag^+(aq) + Cl^-(aq)$; $K = 1.8 \times 10^{-10}$

Entropy

17.36 Define entropy, and give an example of a process in which the entropy of a system increases.

17.37 Comment on the following statement: Exothermic reactions are spontaneous, but endothermic reactions are nonspontaneous.

17.38 Predict the sign of the entropy change in the system for each of the following processes:

(a) A solid sublimes

(b) A liquid freezes

(c) AgI precipitates from a solution containing Ag^+ and I^- ions

(d) Gaseous CO_2 bubbles out of a carbonated beverage

17.39 Predict the sign of ΔS in the system for each of the following reactions:

(a) $PCl_5(s) \rightarrow PCl_3(l) + Cl_2(g)$

(b) $CH_4(g) + 2\,O_2(g) \rightarrow CO_2(g) + 2\,H_2O(l)$

(c) $2\,H_3O^+(aq) + CO_3^{2-}(aq) \rightarrow CO_2(g) + 3\,H_2O(l)$

(d) $Mg(s) + Cl_2(g) \rightarrow MgCl_2(s)$

17.40 Predict the sign of ΔS for each process in Problem 17.32.

17.41 Predict the sign of ΔS for each process in Problem 17.33.

17.42 Consider a disordered crystal of monodeuteriomethane in which each tetrahedral CH_3D molecule is oriented randomly in one of four possible ways. Use Boltzmann's formula to calculate the entropy of the disordered state of the crystal if the crystal contains:

(a) 12 molecules

(b) 120 molecules

(c) 1 mol of molecules

What is the entropy of the crystal if the C–D bond of each of the CH_3D molecules points in the same direction?

17.43 Consider the distribution of ideal gas molecules among three bulbs (A, B, and C) of equal volume. For each of the following states, determine the number of ways (W) that the state can be achieved, and use Boltzmann's formula to calculate the entropy of the state:

(a) 2 molecules in bulb A

(b) 2 molecules randomly distributed among bulbs A, B, and C

(c) 3 molecules in bulb A

(d) 3 molecules randomly distributed among bulbs A, B, and C

(e) 1 mol of molecules in bulb A

(f) 1 mol of molecules randomly distributed among bulbs A, B, and C

What is ΔS on going from state (e) to state (f)? Compare your result with ΔS calculated from the equation $\Delta S = R \ln(V_{final}/V_{initial})$.

17.44 Which state in each of the following pairs has the higher entropy per mole of substance?

(a) H_2 at 25°C in a volume of 10 L or H_2 at 25°C in a volume of 50 L

(b) O_2 at 25°C and 1 atm or O_2 at 25°C and 10 atm

(c) H_2 at 25°C and 1 atm or H_2 at 100°C and 1 atm

(d) CO_2 at STP or CO_2 at 100°C and 0.1 atm

17.45 Which state in each of the following pairs has the higher entropy per mole of substance?

(a) Ice at −40°C or ice at 0°C

(b) N_2 at STP or N_2 at 0°C and 10 atm

(c) N_2 at STP or N_2 at 0°C in a volume of 50 L

(d) Water vapor at 150°C and 1 atm or water vapor at 100°C and 2 atm

Standard Molar Entropies and Standard Entropies of Reaction

17.46 What is meant by the standard molar entropy of a substance? How are standard molar entropies used to calculate standard entropies of reaction?

17.47 What are the units of (a) standard molar entropies and (b) standard entropies of reaction? Why are standard molar entropies sometimes called absolute entropies?

17.48 Which substance in each of the following pairs would you expect to have the higher standard molar entropy? Explain.

(a) $C_2H_2(g)$ or $C_2H_6(g)$

(b) $CO_2(g)$ or $CO(g)$

(c) $I_2(s)$ or $I_2(g)$

(d) $CH_3OH(g)$ or $CH_3OH(l)$

17.49 Which substance in each of the following pairs would you expect to have the higher standard molar entropy? Explain.

(a) $NO(g)$ or $NO_2(g)$

(b) $CH_3CO_2H(l)$ or $HCO_2H(l)$

(c) $Br_2(l)$ or $Br_2(s)$

(d) $S(s)$ or $SO_3(g)$

17.50 Use the standard molar entropies in Appendix B to calculate $\Delta S°$ at 25°C for each of the following reactions. Account for the sign of the entropy change in each case.

(a) $2\,H_2O_2(l) \rightarrow 2\,H_2O(l) + O_2(g)$

(b) $2\,Na(s) + Cl_2(g) \rightarrow 2\,NaCl(s)$

(c) $2\,O_3(g) \rightarrow 3\,O_2(g)$

(d) $4\,Al(s) + 3\,O_2(g) \rightarrow 2\,Al_2O_3(s)$

17.51 Use the $S°$ values in Appendix B to calculate $\Delta S°$ at 25°C for each of the following reactions. Suggest a reason for the sign of $\Delta S°$ in each case.

(a) $2 S(s) + 3 O_2(g) \rightarrow 2 SO_3(g)$

(b) $SO_3(g) + H_2O(l) \rightarrow H_2SO_4(aq)$

(c) $AgCl(s) \rightarrow Ag^+(aq) + Cl^-(aq)$

(d) $NH_4NO_3(s) \rightarrow N_2O(g) + 2 H_2O(g)$

Entropy and the Second Law of Thermodynamics

17.52 State the second law of thermodynamics.

17.53 An isolated system is one that exchanges neither matter nor energy with the surroundings. What is the entropy criterion for spontaneous change in an isolated system? Give an example of a spontaneous process in an isolated system.

17.54 Give an equation that relates the entropy change in the surroundings to the enthalpy change in the system. What is the sign of ΔS_{surr} for:

(a) An exothermic reaction?

(b) An endothermic reaction?

17.55 When heat is added to the surroundings, the entropy of the surroundings increases. How does ΔS_{surr} depend on the temperature of the surroundings? Explain.

17.56 Use the data in Appendix B to calculate ΔS_{sys}, ΔS_{surr}, and ΔS_{total} at 25°C for the reaction

$$N_2(g) + 2 O_2(g) \longrightarrow N_2O_4(g)$$

Is this reaction spontaneous under standard-state conditions at 25°C?

17.57 Copper metal is obtained by smelting copper(I) sulfide ores:

$$Cu_2S(s) + O_2(g) \longrightarrow 2 Cu(s) + SO_2(g)$$

Use the data in Appendix B to calculate ΔS_{sys}, ΔS_{surr}, and ΔS_{total} at 25°C for this reaction. Is the reaction spontaneous under standard-state conditions at 25°C?

17.58 For vaporization of benzene, $\Delta H_{vap} = 30.7$ kJ/mol and $\Delta S_{vap} = 87.0$ J/(K · mol). Calculate ΔS_{surr} and ΔS_{total} at:

(a) 70°C   (b) 80°C   (c) 90°C

Does benzene boil at 70°C and 1 atm pressure? Calculate the normal boiling point of benzene.

17.59 For melting of sodium chloride, $\Delta H_{fusion} = 30.2$ kJ/mol and $\Delta S_{fusion} = 28.1$ J/(K · mol). Calculate ΔS_{surr} and ΔS_{total} at:

(a) 1050 K   (b) 1075 K   (c) 1100 K

Does NaCl melt at 1100 K? Calculate the melting point of NaCl.

Free Energy

17.60 Describe how the signs of ΔH and ΔS determine whether a reaction is spontaneous or nonspontaneous at constant temperature and pressure.

17.61 What determines the direction of spontaneous reaction when ΔH and ΔS are both positive or both negative? Explain.

17.62 The melting point of benzene is 5.5°C. Predict the signs of ΔH, ΔS, and ΔG for melting of benzene at:

(a) 0°C   (b) 15°C

17.63 Consider a twofold expansion of an ideal gas at 25°C in the isolated system shown in Figure 17.1.

(a) What are the values of ΔH, ΔS, and ΔG for the process?

(b) How does this process illustrate the second law of thermodynamics?

17.64 Given the data in Problem 17.58, calculate ΔG for vaporization of benzene at:

(a) 70°C   (b) 80°C   (c) 90°C

Predict whether benzene will boil at each of these temperatures and 1 atm pressure.

17.65 Given the data in Problem 17.59, calculate ΔG for melting of sodium chloride at:

(a) 1050 K   (b) 1075 K   (c) 1100 K

Predict whether NaCl will melt at each of these temperatures and 1 atm pressure.

17.66 Calculate the melting point of benzoic acid $(C_6H_5CO_2H)$, given the following data: $\Delta H_{fusion} = 17.3$ kJ/mol and $\Delta S_{fusion} = 43.8$ J/(K · mol).

17.67 Calculate the enthalpy of fusion of naphthalene $(C_{10}H_8)$, given that its melting point is 128°C and its entropy of fusion is 47.7 J/(K · mol).

Standard Free-Energy Changes and Standard Free Energies of Formation

17.68 Define **(a)** the standard free-energy change, $\Delta G°$, for a reaction and **(b)** the standard free energy of formation, $\Delta G°_f$, of a substance.

17.69 What is meant by the standard state of a substance?

17.70 Use the data in Appendix B to calculate $\Delta H°$ and $\Delta S°$ for each of the following reactions. From the values of $\Delta H°$ and $\Delta S°$, calculate $\Delta G°$ at 25°C, and predict whether each reaction is spontaneous under standard-state conditions.
   (a) $N_2(g) + 2\, O_2(g) \rightarrow 2\, NO_2(g)$
   (b) $2\, KClO_3(s) \rightarrow 2\, KCl(s) + 3\, O_2(g)$
   (c) $CH_3CH_2OH(l) + O_2(g) \rightarrow CH_3CO_2H(l) + H_2O(l)$

17.71 Use the data in Appendix B to calculate $\Delta H°$ and $\Delta S°$ for each of the following reactions. From the values of $\Delta H°$ and $\Delta S°$, calculate $\Delta G°$ at 25°C, and predict whether each reaction is spontaneous under standard-state conditions.
   (a) $2\, SO_2(g) + O_2(g) \rightarrow 2\, SO_3(g)$
   (b) $N_2(g) + 2\, H_2(g) \rightarrow N_2H_4(l)$
   (c) $CH_3OH(l) + O_2(g) \rightarrow HCO_2H(l) + H_2O(l)$

17.72 Use the standard free energies of formation in Appendix B to calculate $\Delta G°$ at 25°C for each reaction in Problem 17.70.

17.73 Use the standard free energies of formation in Appendix B to calculate $\Delta G°$ at 25°C for each reaction in Problem 17.71.

17.74 Use the data in Appendix B to tell which of the following compounds are thermodynamically stable with respect to their constituent elements at 25°C:
   (a) $BaCO_3(s)$ **(b)** $HBr(g)$
   (c) $N_2O(g)$ **(d)** $C_2H_4(g)$

17.75 Use the data in Appendix B to decide whether synthesis of the following compounds from their constituent elements is thermodynamically feasible at 25°C:
   (a) $C_6H_6(l)$ **(b)** $NO(g)$
   (c) $PH_3(g)$ **(d)** $FeO(s)$

17.76 Ethanol is manufactured in industry by hydration of ethylene:

$$CH_2{=}CH_2(g) + H_2O(l) \longrightarrow CH_3CH_2OH(l)$$

Using the data in Appendix B, calculate $\Delta G°$ and show that this reaction is spontaneous at 25°C. Why does this reaction become nonspontaneous at higher temperatures? Estimate the temperature at which the reaction becomes nonspontaneous.

17.77 Sulfur dioxide in the effluent gases from coal-burning electric power plants is one of the principal causes of acid rain. One method for reducing SO_2 emissions involves partial reduction of SO_2 to H_2S, followed by catalytic conversion of the H_2S and the remaining SO_2 to elemental sulfur:

$$2\, H_2S(g) + SO_2(g) \longrightarrow 3\, S(s) + 2\, H_2O(g)$$

Using the data in Appendix B, calculate $\Delta G°$ and show that this reaction is spontaneous at 25°C. Why does this reaction become nonspontaneous at high temperatures? Estimate the temperature at which the reaction becomes nonspontaneous.

17.78 Consider the conversion of acetylene to benzene:

$$3\, C_2H_2(g) \longrightarrow C_6H_6(l)$$

Is a catalyst for this reaction possible? Is it possible to synthesize benzene from graphite and gaseous H_2 at 25°C and 1 atm pressure?

17.79 Consider the conversion of dichloroethane to vinyl chloride, the starting material for manufacturing poly(vinyl chloride) (PVC) plastics:

$$CH_2ClCH_2Cl(l) \longrightarrow CH_2{=}CHCl(g) + HCl(g)$$
Dichloroethane   Vinyl chloride

Is this reaction spontaneous under standard-state conditions? Would it help to carry out the reaction in the presence of base to remove HCl? Explain. Is it possible to synthesize vinyl chloride from graphite, gaseous H_2, and gaseous Cl_2 at 25°C and 1 atm pressure?

Free Energy, Composition, and Chemical Equilibrium

17.80 What is the relationship between the free-energy change under nonstandard-state conditions, ΔG, the free-energy change under standard-state conditions, $\Delta G°$, and the reaction quotient, Q?

17.81 Compare the values of ΔG and $\Delta G°$ when:
   (a) $Q < 1$ **(b)** $Q = 1$ **(c)** $Q > 1$
   Does the thermodynamic driving force increase or decrease as Q increases?

17.82 Sulfuric acid is produced in larger amounts by weight than any other chemical. It is used in manufacturing fertilizers, oil refining, and hundreds of other processes. An intermediate step in the industrial process for synthesis of H_2SO_4 involves catalytic oxidation of sulfur dioxide:

$$2\, SO_2(g) + O_2(g) \longrightarrow 2\, SO_3(g) \qquad \Delta G° = -141.8\ kJ$$

Calculate ΔG at 25°C given the following sets of partial pressures:
   (a) 100 atm SO_2, 100 atm O_2, 1.0 atm SO_3
   (b) 2.0 atm SO_2, 1.0 atm O_2, 10 atm SO_3
   (c) Each reactant and product at a partial pressure of 1.0 atm

17.83 Urea (NH_2CONH_2), an important nitrogen fertilizer, is produced industrially by the reaction

$$2\,NH_3(g) + CO_2(g) \longrightarrow NH_2CONH_2(aq) + H_2O(l)$$

Given that $\Delta G° = -13.6$ kJ, calculate ΔG at 25°C for the following sets of conditions:

(a) 10 atm NH_3, 10 atm CO_2, 1.0 M NH_2CONH_2

(b) 0.10 atm NH_3, 0.10 atm CO_2, 1.0 M NH_2CONH_2

Is the reaction spontaneous for the conditions in part (a) and/or part (b)?

17.84 What is the relationship between the standard free-energy change, $\Delta G°$, for a reaction and the equilibrium constant, K? What is the sign of $\Delta G°$ when:

(a) $K > 1$ **(b)** $K = 1$ **(c)** $K < 1$

17.85 Do you expect a large or small value of the equilibrium constant for a reaction when:

(a) $\Delta G°$ is positive?   **(b)** $\Delta G°$ is negative?

17.86 Calculate the equilibrium constant K_p at 25°C for the reaction in Problem 17.82.

17.87 Calculate the equilibrium constant at 25°C for the reaction in Problem 17.83.

17.88 Given values of $\Delta G°_f$ at 25°C for liquid ethanol (-174.9 kJ/mol) and gaseous ethanol (-168.6 kJ/mol), calculate the vapor pressure of ethanol at 25°C.

17.89 At 25°C, K_a for acid dissociation of aspirin ($C_9H_8O_4$) is 3.0×10^{-4}. Calculate $\Delta G°$ for the reaction $C_9H_8O_4(aq) + H_2O(l) \rightleftharpoons H_3O^+(aq) + C_9H_7O_4^-(aq)$

17.90 Ethylene oxide, C_2H_4O, is used to make antifreeze (ethylene glycol, $HOCH_2CH_2OH$). It is produced industrially by catalyzed air oxidation of ethylene:

$$2\,CH_2{=}CH_2(g) + O_2(g) \longrightarrow 2\,H_2C{-}CH_2(g)$$
$$\diagdown\!\!\diagup$$
$$O$$

Ethylene oxide

Use the data in Appendix B to calculate $\Delta G°$ and K_p for this reaction at 25°C.

17.91 Use the data in Appendix B to calculate K_p at 25°C for the reaction

$$CO(g) + 2\,H_2(g) \rightleftharpoons CH_3OH(g)$$

What is ΔG for this reaction at 25°C when each reactant and product is present at a partial pressure of 20 atm?

General Problems

17.92 Sorbitol ($C_6H_{14}O_6$), a substance used as a sweetener in foods, is prepared by reaction of glucose with hydrogen in the presence of a catalyst:

$$C_6H_{12}O_6(aq) + H_2(g) \longrightarrow C_6H_{14}O_6(aq)$$

Which of the following quantities are affected by the catalyst?

(a) Rate of the forward reaction

(b) Rate of the reverse reaction

(c) Spontaneity of the reaction

(d) $\Delta H°$

(e) $\Delta S°$

(f) $\Delta G°$

(g) The equilibrium constant

(h) Time required to reach equilibrium

17.93 Indicate whether the following processes are spontaneous or nonspontaneous:

(a) Heat transfer from a block of ice to a room maintained at 25°C

(b) Evaporation of water from an open beaker

(c) Conversion of iron(III) oxide to iron metal and oxygen

(d) Uphill motion of an automobile

17.94 Do you agree with the following statements? If not, explain.

(a) Spontaneous reactions are always fast.

(b) In any spontaneous process, the entropy of the system always increases.

(c) An endothermic reaction is always nonspontaneous.

(d) A reaction that is nonspontaneous in the forward direction is always spontaneous in the reverse direction.

17.95 When rolling a pair of dice, there are two ways to get a point total of 3 (1 + 2; 2 + 1) but only one way to get a point total of 2 (1 + 1). How many ways are there of getting point totals of 4–12? What is the most probable point total?

17.96 Make a rough, qualitative plot of standard molar entropy versus temperature for methane from 0 K to 298 K. Incorporate the following data into your plot: mp = -182°C; bp = -164°C; $S° = 186.2$ J/(K · mol) at 25°C.

17.97 The standard free-energy change at 25°C for dissociation of water is 79.9 kJ:

$$2\,H_2O(l) \rightleftharpoons H_3O^+(aq) + OH^-(aq) \qquad \Delta G° = 79.9\ \text{kJ}$$

For each of the following sets of concentrations, calculate ΔG at 25°C, and indicate whether the reaction is spontaneous in the forward or reverse direction.

(a) $[H_3O^+] = [OH^-] = 1.0$ M

(b) $[H_3O^+] = [OH^-] = 1.0 \times 10^{-7}$ M

(c) $[H_3O^+] = 1.0 \times 10^{-7}$ M, $[OH^-] = 1.0 \times 10^{-10}$ M

Are your results consistent with Le Châtelier's principle? Use the thermodynamic data to calculate the equilibrium constant for the reaction.

17.98 Calculate the normal boiling point of ethanol (CH_3CH_2OH) given that its enthalpy of vaporization is 38.6 kJ/mol and its entropy of vaporization is 110 J/(K · mol).

17.99 Chloroform ($CHCl_3$) has a normal boiling point of 61°C and an enthalpy of vaporization of 29.24 kJ/mol. What are its values of ΔG_{vap} and ΔS_{vap} at 61°C?

17.100 The entropy change for a certain nonspontaneous reaction at 50°C is 104 J/K.

(a) Is the reaction endothermic or exothermic?

(b) What is the minimum value of ΔH (in kJ) for the reaction?

17.101 Ammonium nitrate is dangerous because it decomposes (sometimes explosively) when heated:

$$NH_4NO_3(s) \longrightarrow N_2O(g) + 2\,H_2O(g)$$

(a) Using the data in Appendix B, show that this reaction is spontaneous at 25°C.

(b) How does $\Delta G°$ for the reaction change when the temperature is raised?

(c) Calculate the equilibrium constant K_p at 25°C.

(d) Calculate ΔG for the reaction when the partial pressure of each gas is 30 atm.

17.102 Use the data in Appendix B to calculate $\Delta H°$, $\Delta S°$, and $\Delta G°$ at 25°C for each of the following reactions:

(a) $2\,Mg(s) + O_2(g) \rightarrow 2\,MgO(s)$

(b) $MgCO_3(s) \rightarrow MgO(s) + CO_2(g)$

(c) $Fe_2O_3(s) + 2\,Al(s) \rightarrow Al_2O_3(s) + 2\,Fe(s)$

(d) $2\,NaHCO_3(s) \rightarrow Na_2CO_3(s) + CO_2(g) + H_2O(g)$

Are these reactions spontaneous or nonspontaneous at 25°C and 1 atm pressure? How does $\Delta G°$ change when the temperature is raised?

17.103 *Trouton's rule* says that the ratio of the molar heat of vaporization of a liquid to its normal boiling point (in kelvins) is approximately the same for all liquids: $\Delta H_{vap}/T_{bp} \approx 88$ J/(K · mol).

(a) Check the reliability of Trouton's rule for the liquids listed in the following table:

(b) Explain why liquids tend to have the same value of $\Delta H_{vap}/T_{bp}$.

(c) Which of the liquids in the table deviate(s) from Trouton's rule? Explain.

Liquid	bp (°C)	ΔH_{vap} (kJ/mol)
Ammonia	−77.7	23.4
Benzene	80.1	30.8
Carbon tetrachloride	76.8	29.8
Chloroform	61.1	29.2
Mercury	356.6	56.9

17.104 Just as we can define a standard enthalpy of formation ($\Delta H°_f$) and a standard free energy of formation ($\Delta G°_f$), we can define an analogous standard entropy of formation ($\Delta S°_f$) as being the entropy change for formation of a substance in its standard state from its constituent elements in their standard states. Use the standard molar entropies given in Appendix B to calculate $\Delta S°_f$ for the following substances:

(a) Benzene, $C_6H_6(l)$

(b) $CaSO_4(s)$

(c) Ethanol, $C_2H_5OH(l)$

Check your answers by calculating $\Delta S°_f$ from the values given in Appendix B for $\Delta H°_f$ and $\Delta G°_f$.

17.105 Use the data in Appendix B to calculate the equilibrium pressure of CO_2 in a closed vessel that contains each of the following samples:

(a) 15 g of $MgCO_3$ and 1.0 g of MgO at 25°C

(b) 15 g of $MgCO_3$ and 1.0 g of MgO at 280°C

(c) 30 g of $MgCO_3$ and 1.0 g of MgO at 280°C

You may assume that $\Delta H°$ and $\Delta S°$ are independent of temperature.

17.106 The equilibrium constant K_b for dissociation of aqueous ammonia is 1.710×10^{-5} at 20°C and 1.892×10^{-5} at 50°C. What are the values of $\Delta H°$ and $\Delta S°$ for the reaction?

$$NH_3(aq) + H_2O(l) \rightleftharpoons NH_4^+(aq) + OH^-(aq)$$

17.107 The temperature dependence of the equilibrium constant is given by the equation

$$\ln K = \frac{-\Delta H°}{R}\left(\frac{1}{T}\right) + \frac{\Delta S°}{R}$$

where $\Delta H°$ and $\Delta S°$ are assumed to be independent of temperature.

(a) Derive this equation from equations given in this chapter.

(b) Explain how this equation can be used to determine experimental values of $\Delta H°$ and $\Delta S°$ from values of K at several different temperatures.

(c) Use this equation to predict the sign of $\Delta H°$ for a reaction whose equilibrium constant increases with increasing temperature. Is the reaction endothermic or exothermic? Is your prediction in accord with Le Châtelier's principle?

17.108 The normal boiling point of bromine is 58.8°C, and the standard entropies of the liquid and vapor are $S°[Br_2(l)] = 152.2$ J/(K · mol); $S°[Br_2(g)] = 245.4$ J/(K · mol). At what temperature does bromine have a vapor pressure of 227 mm Hg?

Multi-Concept Problems

17.109 A mixture of NO_2 and N_2O_4, each at an initial partial pressure of 1.00 atm, is heated to 100°C. Use the data in Appendix B to calculate the partial pressure of each gas at equilibrium. You may assume that $\Delta H°$ and $\Delta S°$ are independent of temperature.

$$N_2O_4(g) \rightleftharpoons 2\,NO_2(g)$$

17.110 A mixture of 14.0 g of N_2 and 3.024 g of H_2 in a 5.00 L container is heated to 400°C. Use the data in Appendix B to calculate the molar concentrations of N_2, H_2, and NH_3 at equilibrium. Assume that $\Delta H°$ and $\Delta S°$ are independent of temperature, and remember that the standard state of a gas is defined in terms of pressure.

$$N_2(g) + 3\,H_2(g) \rightleftharpoons 2\,NH_3(g)$$

17.111 One step in the commercial synthesis of sulfuric acid involves catalytic oxidation of sulfur dioxide:

$$2\,SO_2(g) + O_2(g) \rightleftharpoons 2\,SO_3(g)$$

(a) A mixture of 192 g of SO_2, 48.0 g of O_2, and a V_2O_5 catalyst is heated to 800 K in a 15.0 L vessel. Use the data in Appendix B to calculate the partial pressures of SO_3, SO_2, and O_2 at equilibrium. You may assume that $\Delta H°$ and $\Delta S°$ are independent of temperature.

(b) Does the percent yield of SO_3 increase or decrease on raising the temperature from 800 K to 1000 K? Explain.

(c) Does the total pressure increase or decrease on raising the temperature from 800 K to 1000 K? Calculate the total pressure (in atm) at 1000 K.

17.112 The lead storage battery uses the reaction

$$Pb(s) + PbO_2(s) + 2\,H^+(aq) + 2\,HSO_4^-(aq) \longrightarrow$$
$$2\,PbSO_4(s) + 2\,H_2O(l)$$

(a) Use the data in Appendix B to calculate $\Delta G°$ for this reaction.

(b) Calculate ΔG for this reaction on a cold winter's day (10°F) in a battery that has "run down" to the point where the sulfuric acid concentration is only 0.100 M.

---

 eMedia Problems

17.113 The **Enthalpy of Solution** activity (*eChapter 17.2*) allows you to measure temperature changes for the spontaneous dissolution of several different compounds in water at 25°C.

(a) Using data from this simulation determine the signs of $\Delta G°$, $\Delta H°$, and $\Delta S°$ for the dissolution of sodium chloride.

(b) Is the dissolution of sodium chloride spontaneous at all temperatures? Explain.

17.114 The spontaneous decomposition of sodium azide is shown in the **Airbags** movie (*eChapter 17.8*).

(a) What are the signs of $\Delta G°$, $\Delta H°$, and $\Delta S°$ for this decomposition reaction?

(b) Is there any temperature at which this reaction is not spontaneous? Explain.

17.115 When a balloon filled with hydrogen is ignited, hydrogen and oxygen react explosively to form water. The narration in the **Formation of Water** movie (*eChapter 14.4*) makes the statement: "The equilibrium lies far to the right."

(a) Explain what this statement means in terms of products and reactants.

(b) Explain the relationship between the equilibrium constant, K, and the sign of $\Delta G°$.

17.116 Explore the relationship between spontaneity and temperature using the **Gibbs Free Energy** activity (*eChapter 17.8*).

(a) Which of the five reactions available in the activity are spontaneous at all temperatures?

(b) Which of the reactions are nonspontaneous at all temperatures?

(c) Under what conditions will a reaction for which $\Delta S°$ and $\Delta H°$ are both negative be spontaneous?

(d) Under what conditions will a reaction for which $\Delta S°$ and $\Delta H°$ are both positive be spontaneous?

17.117 Using the **Gibbs Free Energy** activity (*eChapter 17.8*), determine the value of the equilibrium constant for the decomposition of dinitrogen tetroxide at 0 K, 330 K, and 1000 K. Describe what conditions of temperature and pressure favor the formation of nitrogen dioxide.

Electrochemistry

Chrome plating is an electrochemical process.

Batteries are everywhere in modern societies. They provide the electric current to start our automobiles and to power a host of products such as pocket calculators, digital watches, heart pacemakers, radios, and tape recorders. A battery is an example of an **electrochemical cell**, a device for inter-converting chemical and electrical energy. A battery takes the energy released by a spontaneous chemical reaction and uses it to produce electricity.

Electrochemistry, the area of chemistry concerned with the inter-conversion of chemical and electrical energy, is enormously important in modern science and technology, not only because of batteries, but also because it makes possible the manufacture of essential industrial chemicals and materials. Sodium hydroxide, for example, which is used in the manufacture of paper, textiles, soaps, and detergents, is produced by passing an electric current through an aqueous solution of sodium chloride. Chlorine, essential to the manufacture of plastics such as poly(vinyl chloride) (PVC), is obtained in the same process. Aluminum metal is also produced in an electrochemical process, as is pure copper for use in electrical wiring.

In this chapter, we'll look at the principles involved in the design and operation of electrochemical cells. In addition, we'll explore some important connections between electrochemistry and thermodynamics.

18.1 Galvanic Cells

FIGURE 18.1 **(a)** A strip of zinc metal is immersed in an aqueous copper sulfate solution. The redox reaction takes place at the metal–solution interface and involves direct transfer of two electrons from Zn atoms to Cu^{2+} ions. **(b)** As time passes, a dark-colored deposit of copper metal appears on the zinc, and the blue color due to $Cu^{2+}(aq)$ fades from the solution.

Electrochemical cells are of two basic types: **galvanic cells** (also called **voltaic cells**) and **electrolytic cells**. The names "galvanic" and "voltaic" honor the Italian scientists Luigi Galvani (1737–1798) and Alessandro Volta (1745–1827), who conducted pioneering work in the field of electrochemistry. In a galvanic cell, a spontaneous chemical reaction generates an electric current. In an electrolytic cell, an electric current drives a nonspontaneous reaction. The two types are therefore the reverse of each other. We'll take up galvanic cells in this section and will examine electrolytic cells later. First, though, let's review some of the basics of oxidation–reduction, or redox, reactions (Section 4.6).

If you immerse a strip of zinc metal in an aqueous solution of copper sulfate, you find that a dark-colored solid deposits on the surface of the zinc and that the blue color characteristic of the Cu^{2+} ion slowly disappears from the solution (Figure 18.1). Chemical analysis shows that the dark-colored

Oxidation–
Reduction
Reactions movie

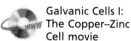
Galvanic Cells I:
The Copper–Zinc
Cell movie

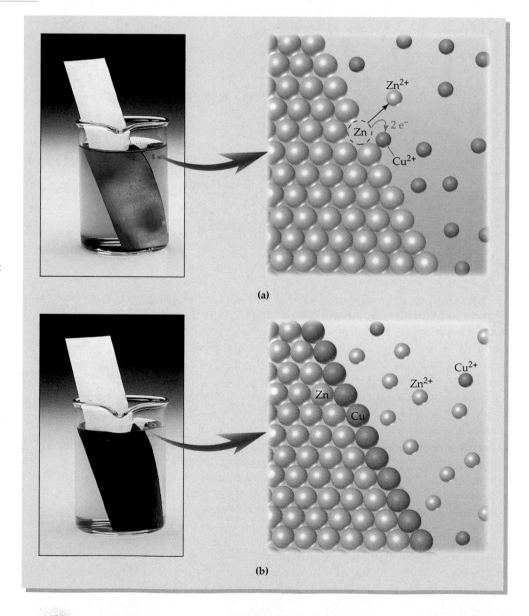

(a)

(b)

deposit is finely divided copper metal and that the solution now contains zinc ions. Therefore, the reaction is

$$Zn(s) + Cu^{2+}(aq) \longrightarrow Zn^{2+}(aq) + Cu(s)$$

This is a redox reaction in which Zn is oxidized to Zn^{2+} and Cu^{2+} is reduced to Cu. Recall that an *oxidation* involves a loss of electrons (an increase in oxidation number) and a *reduction* involves a gain of electrons (a decrease in oxidation number).

We can represent the oxidation and reduction aspects of the reaction by separating the overall process into **half-reactions**, one representing the oxidation reaction and the other representing the reduction:

Oxidation half-reaction: $\quad Zn(s) \longrightarrow Zn^{2+}(aq) + 2\,e^-$

Reduction half-reaction: $\quad Cu^{2+}(aq) + 2\,e^- \longrightarrow Cu(s)$

We say that Cu^{2+} is the *oxidizing agent* because, in gaining electrons from Zn, it causes the oxidation of Zn to Zn^{2+}. Similarly, we say that Zn is the *reducing agent* because, in losing electrons to Cu^{2+}, it causes the reduction of Cu^{2+} to Cu.

If the reaction is carried out as shown in Figure 18.1, electrons are transferred directly from Zn to Cu^{2+}, and the enthalpy of reaction is lost to the surroundings as heat. If, however, the reaction is carried out using the electrochemical cell depicted in Figure 18.2, then some of the chemical energy released by the reaction is converted to electrical energy, which can be used to light a light bulb or run an electric motor.

The apparatus shown in Figure 18.2 is a type of galvanic cell called a *Daniell cell*, after John Frederick Daniell, the English chemist who first constructed it in 1836. It consists of two *half-cells*, a beaker containing a strip of zinc that dips into

FIGURE 18.2 **(a)** A galvanic cell that uses oxidation of zinc metal to Zn^{2+} ions and reduction of Cu^{2+} ions to copper metal. Note that the negative particles (electrons in the wire and anions in solution) travel around the circuit in the same direction. The resulting electric current can be used to light a light bulb. **(b)** An operating Daniell cell. The salt bridge in part (a) is replaced by a porous glass disk that allows ion flow between the anode and cathode compartments but prevents bulk mixing, which would bring Cu^{2+} ions into direct contact with zinc and short-circuit the cell. The light bulb in part (a) is replaced with a digital voltmeter (more about this in Section 18.3).

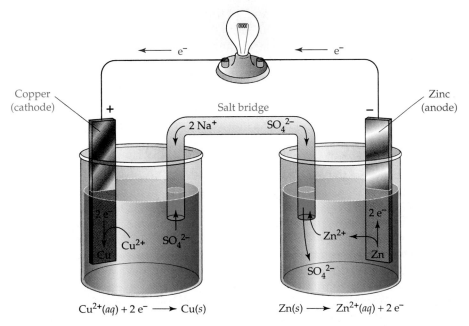

Copper (cathode)

Salt bridge

$2\,Na^+$ $\quad SO_4^{2-}$

Zinc (anode)

$2\,e^-$

Cu^{2+} $\quad SO_4^{2-}$

Zn^{2+}

Zn

SO_4^{2-}

$Cu^{2+}(aq) + 2\,e^- \longrightarrow Cu(s)$ $\qquad$ $Zn(s) \longrightarrow Zn^{2+}(aq) + 2\,e^-$

(a)

(b)

an aqueous solution of zinc sulfate, and a second beaker containing a strip of copper that dips into aqueous copper sulfate. The strips of zinc and copper are called **electrodes** and are connected by an electrically conducting wire. In addition, the two solutions are connected by a **salt bridge**, a U-shaped tube that contains a gel permeated with a solution of an inert electrolyte, such as Na_2SO_4. The ions of the inert electrolyte do not react with the other ions in the solutions, and they are not oxidized or reduced at the electrodes.

The reaction that occurs in the Daniell cell is the same one that occurs when Zn reacts directly with Cu^{2+}, but now, because the Zn metal and Cu^{2+} ions are in separate compartments, the electrons are transferred from Zn to Cu^{2+} through the wire. Consequently, the oxidation and reduction half-reactions occur at separate electrodes and an electric current flows through the wire. Note that electrons are not transferred through the solution because the metal wire is a much better conductor of electrons than is water. In fact, free electrons react rapidly with water and are therefore unstable in aqueous solutions.

The electrode at which oxidation takes place is called the **anode** (the zinc strip in this example), and the electrode at which reduction takes place is called the **cathode** (the copper strip). Of course, the anode and cathode half-reactions must add to give the overall cell reaction:

Anode (oxidation) half-reaction: $$Zn(s) \longrightarrow Zn^{2+}(aq) + 2\,e^-$$

Cathode (reduction) half-reaction: $$Cu^{2+}(aq) + 2\,e^- \longrightarrow Cu(s)$$

Overall cell reaction: $$Zn(s) + Cu^{2+}(aq) \longrightarrow Zn^{2+}(aq) + Cu(s)$$

The salt bridge is necessary to complete the electrical circuit. Without it, the solution in the anode compartment would become positively charged as Zn^{2+} ions appeared in it, and the solution in the cathode compartment would become negatively charged as Cu^{2+} ions were removed from it. Because of the charge imbalance, the electrode reactions would quickly come to a halt, and electron flow through the wire would cease.

With the salt bridge in place, electrical neutrality is maintained in both compartments by a flow of ions. Anions (in this case SO_4^{2-}) flow through the salt bridge from the cathode compartment to the anode compartment, and cations migrate through the salt bridge from the anode compartment to the cathode compartment. For the cell shown in Figure 18.2, Na^+ ions move out of the salt bridge into the cathode compartment and Zn^{2+} ions move into the salt bridge from the anode compartment. (It's interesting to note that the anode and cathode get their names from the direction of ion flow between the two compartments: *An*ions move toward the *an*ode, and *cat*ions move toward the *cat*hode.)

The electrodes of commercial galvanic cells (batteries) are generally labeled with plus (+) and minus (−) signs, although the magnitude of the actual charge on the electrodes is infinitesimally small and the sign of the charge associated with each electrode depends on the point of view. From the perspective of the wire, the anode looks negative because a stream of negatively charged electrons comes from it. From the perspective of the solution, the anode looks positive because positively charged Zn^{2+} ions move from it. Because galvanic cells are used to supply electric current to an external circuit, it makes sense to adopt the perspective of the wire. Consequently, we regard the anode as the negative (−) electrode and the cathode as the positive (+) electrode. Thus, electrons move through the external circuit from the negative electrode, where they are produced by the anode half-reaction, to the positive electrode, where they are consumed by the cathode half-reaction.

Anode: $\begin{cases} \text{Is where oxidation occurs} \\ \text{Is where electrons are produced} \\ \text{Is what anions migrate toward} \\ \text{Has a negative sign} \end{cases}$ **Cathode:** $\begin{cases} \text{Is where reduction occurs} \\ \text{Is where electrons are consumed} \\ \text{Is what cations migrate toward} \\ \text{Has a positive sign} \end{cases}$

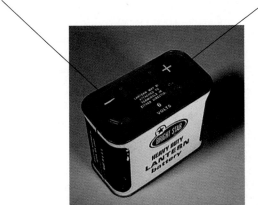

Why do anions move toward the anode? Shouldn't the negative ions be repelled by the negative charge of the anode? The answer is that the anode's negative charge is shielded by the surrounding Zn^{2+} cations, which enter the solution from the surface of the electrode when zinc is oxidized. From the perspective of the anions, the solution around the anode "looks" positive, and so the anions move toward the anode to neutralize the positive charge of the Zn^{2+} cations (Figure 18.3).

EXAMPLE 18.1 Design a galvanic cell that uses the redox reaction

$$Fe(s) + 2\,Fe^{3+}(aq) \longrightarrow 3\,Fe^{2+}(aq)$$

Identify the anode and cathode half-reactions, and sketch the experimental setup. Label the anode and cathode, indicate the direction of electron and ion flow, and identify the sign of each electrode.

SOLUTION In the overall cell reaction, iron metal is oxidized to iron(II) ions, and iron(III) ions are reduced to iron(II) ions. Therefore, the cell half-reactions are

Anode (oxidation):	$Fe(s) \longrightarrow Fe^{2+}(aq) + 2\,e^-$
Cathode (reduction):	$2 \times [Fe^{3+}(aq) + e^- \longrightarrow Fe^{2+}(aq)]$
Overall cell reaction:	$Fe(s) + 2\,Fe^{3+}(aq) \longrightarrow 3\,Fe^{2+}(aq)$

The cathode half-reaction has been multiplied by a factor of 2 so that the two half-reactions will add to give the overall cell reaction. Whenever half-reactions are added, the electrons must cancel. No electrons can appear in the overall reaction because all of the electrons lost by the reducing agent are gained by the oxidizing agent.

A possible experimental setup is shown on the next page. The anode compartment consists of an iron metal electrode dipping into an aqueous solution of $Fe(NO_3)_2$. Note, though, that *any inert electrolyte* can be used to carry the current in the anode compartment; Fe^{2+} does not need to be present initially because it's not a reactant in the anode half-reaction.

Since Fe^{3+} is a reactant in the cathode half-reaction, $Fe(NO_3)_3$ would be a good electrolyte for the cathode compartment. The cathode can be any electrical conductor that doesn't react with the ions in the solution. A platinum wire is a

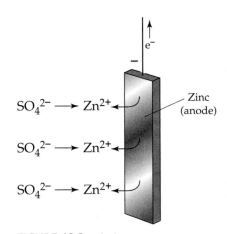

FIGURE 18.3 Anions move toward the anode to neutralize the positive charge of the cations produced in solution when zinc metal is oxidized.

common inert electrode. (Iron metal can't be used because it would react directly with Fe^{3+}, thus short-circuiting the cell.) The salt bridge contains $NaNO_3$, but any inert electrolyte would do. Electrons flow through the wire from the iron anode (−) to the platinum cathode (+). Anions move from the cathode compartment toward the anode while cations migrate from the anode compartment toward the cathode.

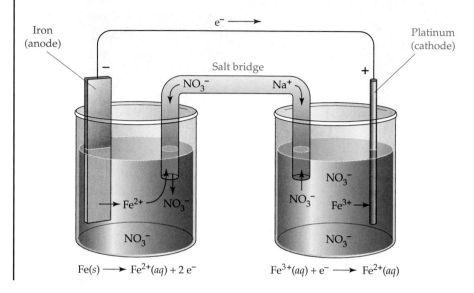

$$Fe(s) \longrightarrow Fe^{2+}(aq) + 2\,e^- \qquad\qquad Fe^{3+}(aq) + e^- \longrightarrow Fe^{2+}(aq)$$

▶ **PROBLEM 18.1** Describe a galvanic cell that uses the reaction

$$2\,Ag^+(aq) + Ni(s) \longrightarrow 2\,Ag(s) + Ni^{2+}(aq)$$

Identify the anode and cathode half-reactions, and sketch the experimental setup. Label the anode and cathode, indicate the direction of electron and ion flow, and identify the sign of each electrode.

18.2 Shorthand Notation for Galvanic Cells

Rather than describing a galvanic cell in words, it's convenient to use a shorthand notation for representing the cell. For the Daniell cell in Figure 18.2, which uses the reaction

$$Zn(s) + Cu^{2+}(aq) \longrightarrow Zn^{2+}(aq) + Cu(s)$$

we can write the following expression:

$$Zn(s)\,|\,Zn^{2+}(aq)\,\|\,Cu^{2+}(aq)\,|\,Cu(s)$$

In this notation, a single vertical line (|) represents a phase boundary, such as that between a solid electrode and an aqueous solution, and the double vertical line (‖) denotes a salt bridge. The shorthand for the anode half-cell is always written on the left of the salt-bridge symbol, followed on the right by the shorthand for the cathode half-cell. The electrodes are written on the extreme left (anode) and on the extreme right (cathode), and the reactants in each half-cell are written first, followed by the products. With these arbitrary conventions, electrons move through the external circuit from left to right (from anode to cathode). Reading the shorthand thus suggests the overall cell reaction: Zn is oxidized to Zn^{2+}, and Cu^{2+} is reduced to Cu.

Salt bridge

Anode half-cell $\mid$ Cathode half-cell

$$Zn(s) \mid Zn^{2+}(aq) \parallel Cu^{2+}(aq) \mid Cu(s)$$

Phase boundary   Electrons flow this way   Phase boundary

For the galvanic cell in Example 18.1, based on the reaction

$$Fe(s) + 2 Fe^{3+}(aq) \longrightarrow 3 Fe^{2+}(aq)$$

the shorthand notation is

$$Fe(s) \mid Fe^{2+}(aq) \parallel Fe^{3+}(aq), Fe^{2+}(aq) \mid Pt(s)$$

The shorthand for the cathode half-cell includes both reactant (Fe^{3+}) and product (Fe^{2+}) as well as the electrode (Pt). The two ions $Fe^{3+}(aq)$ and $Fe^{2+}(aq)$ are separated by a comma rather than a vertical line because they are in the same phase.

The notation for a cell involving a gas has an additional vertical line because an additional phase is present. Thus, the notation

$$Cu(s) \mid Cu^{2+}(aq) \parallel Cl_2(g) \mid Cl^-(aq) \mid C(s)$$

specifies a cell in which copper is oxidized to Cu^{2+} at a copper anode and Cl_2 gas is reduced to Cl^- at a graphite (carbon) cathode. The cell reaction is

$$Cu(s) + Cl_2(g) \longrightarrow Cu^{2+}(aq) + 2 Cl^-(aq)$$

A more detailed notation would include ion concentrations and gas pressures, for example:

$$Cu(s) \mid Cu^{2+}(1.0 \text{ M}) \parallel Cl_2(1 \text{ atm}) \mid Cl^-(1.0 \text{ M}) \mid C(s)$$

EXAMPLE 18.2   Given the following cell notation

$$Pt(s) \mid Sn^{2+}(aq), Sn^{4+}(aq) \parallel Ag^+(aq) \mid Ag(s)$$

write a balanced equation for the cell reaction, and give a brief description of the cell.

SOLUTION   Because the anode always appears at the left in the shorthand notation, the anode (oxidation) half-reaction is

$$Sn^{2+}(aq) \longrightarrow Sn^{4+}(aq) + 2 e^-$$

The cathode (reduction) half-reaction is

$$2 \times [Ag^+(aq) + e^- \longrightarrow Ag(s)]$$

Note that we multiply the cathode half-reaction by a factor of 2 so that the electrons will cancel when we sum the two half-reactions to give the cell reaction:

$$Sn^{2+}(aq) + 2\,Ag^+(aq) \longrightarrow Sn^{4+}(aq) + 2\,Ag(s)$$

The cell consists of a platinum wire anode dipping into an Sn^{2+} solution—say, $Sn(NO_3)_2(aq)$—and a silver cathode dipping into an Ag^+ solution—say, $AgNO_3(aq)$. As usual, the anode and cathode half-cells must be connected by a wire and a salt bridge containing inert ions.

Although the anode half-cell always appears on the left in the shorthand notation, its location in a cell drawing is arbitrary. This means that you can't infer which electrode is the anode and which is the cathode from the location of the electrodes in a cell drawing. You must identify the electrodes based on whether each electrode half-reaction is an oxidation or a reduction.

▶ **PROBLEM 18.2** Write the shorthand notation for a galvanic cell that uses the reaction

$$Fe(s) + Sn^{2+}(aq) \longrightarrow Fe^{2+}(aq) + Sn(s)$$

▶ **PROBLEM 18.3** Write a balanced equation for the overall cell reaction, and give a brief description of a galvanic cell represented by the following shorthand notation:

$$Pb(s)\,|\,Pb^{2+}(aq)\,\|\,Br_2(l)\,|\,Br^-(aq)\,|\,Pt(s)$$

◆— **KEY CONCEPT PROBLEM 18.4** Consider the following galvanic cell:

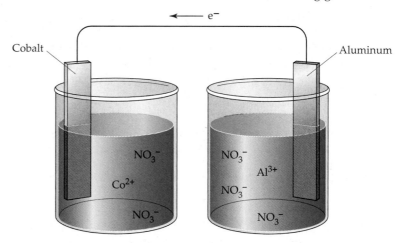

(a) Complete the drawing by adding any components essential for a functioning cell.
(b) Label the anode and cathode, and indicate the direction of ion flow.
(c) Write a balanced equation for the cell reaction.
(d) Write the shorthand notation for the cell.

18.3 Cell Potentials and Free-Energy Changes for Cell Reactions

Let's return to the Daniell cell shown in Figure 18.2 to find an electrical measure of the driving force of a cell reaction. Electrons move through the external circuit from the zinc anode to the copper cathode because they have lower energy when on copper than on zinc. The driving force that pushes the negatively charged electrons away from the anode (– electrode) and pulls them toward the cathode (+ electrode) is an electrical potential called the **electromotive force (emf)**, also known as the **cell potential (*E*)** or the **cell voltage**. The SI unit of electrical potential is the volt (V), and the potential of a galvanic cell is defined as a positive quantity.

The relationship between the volt and the SI units of energy (joule, J) and electric charge (coulomb, C) is given by the equation

$$1\,J = 1\,C \times 1\,V$$

where 1 C is the amount of charge transferred when a current of 1 ampere (A) flows for 1 second (s). (The current passing through a 100 W household light bulb is about 1 A, which means that the electric charge of the electrons passing through the bulb in 1 s is 1 C.) When 1 C of charge moves between two electrodes that differ in electrical potential by 1 V, 1 J of energy is released by the cell and can be used to do electrical work.

A cell potential is measured with an electronic instrument called a *voltmeter* (Figure 18.2b), which is designed to give a positive reading when the + and − terminals of the voltmeter are connected to the + (cathode) and − (anode) electrodes of the cell, respectively. Thus, the voltmeter–cell connections required to get a positive reading on the voltmeter indicate which electrode is the anode and which is the cathode.

We've now seen two quantitative measures of the driving force of a chemical reaction: the cell potential *E* (an electrochemical quantity) and the free-energy change ΔG (a thermochemical quantity, Section 17.7). The values of ΔG and *E* are directly proportional and are related by the equation

$$\Delta G = -nFE$$

where *n* is the number of moles of electrons transferred in the reaction and *F* is the **faraday** (or **Faraday constant**), the electric charge on 1 mol of electrons (96,485 C/mol e⁻). In our calculations, we'll round the value of *F* to three significant figures:

$$F = 96{,}500\,C/mol\,e^{-}$$

The faraday is named in honor of Michael Faraday (1791–1867), the nineteenth-century English scientist who laid the foundations for our current understanding of electricity.

Two features of the equation $\Delta G = -nFE$ are worth noting: the units and the minus sign. When we multiply the charge transferred (*nF*) in coulombs by the cell potential (*E*) in volts, we obtain an energy (ΔG) in joules, in accord with the relationship $1\,J = 1\,C \times 1\,V$. The minus sign is required because *E* and ΔG have opposite signs: The spontaneous reaction in a galvanic cell has a positive cell potential but a negative free-energy change (Section 17.7).

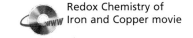 Redox Chemistry of Iron and Copper movie

Later in this chapter, we'll see that cell potentials, like free-energy changes, depend on the composition of the reaction mixture. The **standard cell potential *E*°** is the cell potential when both reactants and products are in their standard states— solutes at 1 M concentrations, gases at a partial pressure of 1 atm, solids and liquids in pure form, with all at a specified temperature, usually 25°C. For example, *E*° for the reaction

$$Zn(s) + Cu^{2+}(aq) \longrightarrow Zn^{2+}(aq) + Cu(s)$$

is the cell potential measured at 25°C for a cell that has pure Zn and Cu metal electrodes and 1 M concentrations of Zn^{2+} and Cu^{2+}.

The standard free-energy change and the standard cell potential are related by the equation

$$\Delta G° = -nFE°$$

Because $\Delta G°$ and $E°$ are directly proportional, a voltmeter can be regarded as a "free-energy meter." When a voltmeter measures $E°$, it also indirectly measures $\Delta G°$.

EXAMPLE 18.3   Calculate the standard free-energy change at 25°C for the following reaction. The standard cell potential is 1.10 V at 25°C.

$$Zn(s) + Cu^{2+}(aq) \longrightarrow Zn^{2+}(aq) + Cu(s)$$

SOLUTION   Two moles of electrons are transferred from Zn to Cu^{2+} in this reaction, and the standard free-energy change is therefore

$$\Delta G° = -nFE° = -(2 \text{ mol } e^-)\left(\frac{96,500 \text{ C}}{\text{mol } e^-}\right)(1.10 \text{ V})\left(\frac{1 \text{ J}}{1 \text{ C} \cdot \text{V}}\right)$$

$$= -212,000 \text{ J} = -212 \text{ kJ}$$

▶ **PROBLEM 18.5**   The standard cell potential at 25°C is 0.92 V for the reaction

$$Al(s) + Cr^{3+}(aq) \longrightarrow Al^{3+}(aq) + Cr(s)$$

What is the standard free-energy change for this reaction at 25°C?

18.4   Standard Reduction Potentials

The standard potential of any galvanic cell is the sum of the standard half-cell potentials for oxidation at the anode and reduction at the cathode:

$$E°_{cell} = E°_{ox} + E°_{red}$$

Consider, for example, a cell in which H_2 gas is oxidized to H^+ ions at the anode and Cu^{2+} ions are reduced to copper metal at the cathode (Figure 18.4):

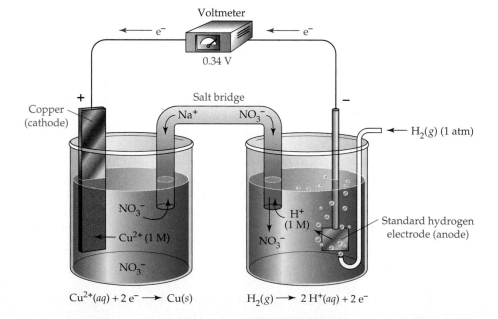

FIGURE 18.4   A galvanic cell consisting of a Cu^{2+}(1 M)/Cu half-cell and a standard hydrogen electrode (S.H.E.). The S.H.E. is a piece of platinum foil that is in contact with bubbles of $H_2(g)$ at 1 atm pressure and with $H^+(aq)$ at 1 M concentration. Electrons flow from the S.H.E. (anode) to the copper cathode. The measured standard cell potential at 25°C is 0.34 V.

Anode (oxidation): $H_2(g) \longrightarrow 2\,H^+(aq) + 2\,e^-$

Cathode (reduction): $Cu^{2+}(aq) + 2\,e^- \longrightarrow Cu(s)$

Overall cell reaction: $H_2(g) + Cu^{2+}(aq) \longrightarrow 2\,H^+(aq) + Cu(s)$

[We represent the hydrated proton as $H^+(aq)$ rather than $H_3O^+(aq)$ because we're interested here in electron transfer, not proton transfer as in Chapter 15.] The standard potential for this cell, 0.34 V at 25°C, is a measure of the combined driving forces of the oxidation and reduction half-reactions:

$$E^\circ_{\text{cell}} = E^\circ_{\text{ox}} + E^\circ_{\text{red}} = E^\circ_{H_2 \to H^+} + E^\circ_{Cu^{2+} \to Cu} = 0.34\text{ V}$$

If we could determine E° values for individual half-reactions, we could combine those values to obtain E° values for a host of cell reactions. Unfortunately, it's not possible to measure the potential of a single electrode; we can measure only a potential *difference* by placing a voltmeter between *two* electrodes. Nevertheless, we can develop a set of standard half-cell potentials by choosing an arbitrary standard half-cell as a reference point, assigning it an arbitrary potential, and then expressing the potential of all other half-cells relative to the reference half-cell. Recall that this same approach was used in Section 8.10 for determining standard enthalpies of formation, ΔH°_f.

To define an electrochemical "sea level," chemists have chosen a reference half-cell called the **standard hydrogen electrode (S.H.E.,** shown in Figure 18.4). It consists of a platinum electrode in contact with H_2 gas and aqueous H^+ ions at standard-state conditions [1 atm $H_2(g)$, 1 M $H^+(aq)$, 25°C]. The corresponding half-reaction, written in either direction, is assigned an arbitrary potential of exactly 0 V:

$$2\,H^+(aq, 1\text{ M}) + 2\,e^- \longrightarrow H_2(g, 1\text{ atm}) \qquad E^\circ = 0\text{ V}$$
$$H_2(g, 1\text{ atm}) \longrightarrow 2\,H^+(aq, 1\text{ M}) + 2\,e^- \qquad E^\circ = 0\text{ V}$$

With this choice of standard reference electrode, the entire potential of the cell

$$Pt(s)\,|\,H_2(1\text{ atm})\,|\,H^+(1\text{ M})\,\|\,Cu^{2+}(1\text{ M})\,|\,Cu(s)$$

can be attributed to the Cu^{2+}/Cu half-cell:

$$E^\circ_{\text{cell}} = E^\circ_{H_2 \to H^+} + E^\circ_{Cu^{2+} \to Cu} = 0.34\text{ V}$$
$$\uparrow \qquad\qquad \uparrow \qquad\qquad \uparrow$$
$$0.34\text{ V} \qquad 0\text{ V} \qquad 0.34\text{ V}$$

Because the Cu^{2+}/Cu half-reaction is a reduction, the corresponding half-cell potential, $E^\circ = 0.34$ V, is called a **standard reduction potential:**

$$Cu^{2+}(aq) + 2\,e^- \longrightarrow Cu(s) \qquad \text{Standard reduction potential: } E^\circ = 0.34\text{ V}$$

In a cell in which this half-reaction occurs in the opposite direction, the corresponding half-cell potential has the same magnitude but opposite sign:

$$Cu(s) \longrightarrow Cu^{2+}(aq) + 2\,e^- \qquad E^\circ = -0.34\text{ V}$$

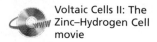

Whenever the direction of a half-reaction is reversed, the sign of $E°$ must also be reversed. Thus, the standard potential for an oxidation half-reaction is the negative of the standard reduction potential.

We can determine standard potentials for other half-cells simply by constructing galvanic cells in which each half-cell of interest is paired up with the S.H.E. For example, to find the potential of a half-cell consisting of a zinc electrode dipping into a 1 M Zn^{2+} solution, we would build the cell shown in Figure 18.5. The voltmeter–electrode connections required to get a positive reading on the voltmeter (0.76 V) tell us that the zinc electrode is the anode and the S.H.E. is the cathode. Therefore, the half-cell reactions involve oxidation of Zn and reduction of H^+. Alternatively, we could identify the direction of the half-reactions by noting that the H^+ concentration decreases as the reaction progresses.

Anode (oxidation):	$Zn(s) \longrightarrow Zn^{2+}(aq) + 2\,e^-$	$E° = ?$
Cathode (reduction):	$2\,H^+(aq) + 2\,e^- \longrightarrow H_2(g)$	$E° = 0\,V$
Overall cell reaction:	$Zn(s) + 2\,H^+(aq) \longrightarrow Zn^{2+}(aq) + H_2(g)$	$E° = 0.76\,V$

Since the anode and cathode half-cell potentials must sum to give the overall cell potential, the $E°$ value for oxidation of Zn to Zn^{2+} must be 0.76 V and the standard reduction potential for the Zn^{2+}/Zn half-cell is therefore -0.76 V.

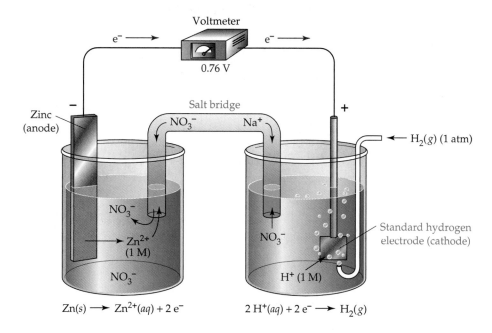

FIGURE 18.5 A galvanic cell consisting of a Zn/Zn^{2+} (1 M) half-cell and a standard hydrogen electrode. Electrons flow from the zinc anode to the S.H.E. (cathode). The measured standard cell potential at 25°C is 0.76 V.

We could also have calculated the standard reduction potential for the Zn^{2+}/Zn half-cell from the observed standard potential of the Daniell cell (1.10 V) and the standard reduction potential for the Cu^{2+}/Cu half-cell (0.34 V):

Anode (oxidation):	$Zn(s) \longrightarrow Zn^{2+}(aq) + 2\,e^-$	$E° = ?$
Cathode (reduction):	$Cu^{2+}(aq) + 2\,e^- \longrightarrow Cu(s)$	$E° = 0.34\,V$
Overall cell reaction:	$Zn(s) + Cu^{2+}(aq) \longrightarrow Zn^{2+}(aq) + Cu(s)$	$E° = 1.10\,V$

Since 1.10 V = 0.76 V + 0.34 V, we would again conclude that the $E°$ value for oxidation of Zn to Zn^{2+} is 0.76 V and the standard reduction potential for the Zn^{2+}/Zn half-cell is therefore -0.76 V.

From experiments of the sort just described, hundreds of half-cell potentials have been determined. A short list is presented in Table 18.1, and a more complete tabulation is given in Appendix D. The following conventions are observed in constructing a table of half-cell potentials:

1. *The half-reactions are written as reductions* rather than as oxidations. This means that oxidizing agents and electrons are on the left side of each half-reaction and reducing agents are on the right side.

2. *The listed half-cell potentials are standard reduction potentials*, also known as **standard electrode potentials**.

3. *The half-reactions are listed in order of decreasing standard reduction potential* (decreasing tendency to occur in the forward direction; increasing tendency to occur in the reverse direction). Consequently, the strongest oxidizing agents are located in the upper left of the table (F_2, H_2O_2, MnO_4^-, and so forth), and the strongest reducing agents are found in the lower right of the table (Li, Na, Mg, and so forth).

By choosing $E° = 0$ V for the standard hydrogen electrode, we obtain standard reduction potentials that range from about $+3$ V to -3 V.

TABLE 18.1   Standard Reduction Potentials at 25°C

	Reduction Half-Reaction		$E°$ (V)	
Stronger oxidizing agent	$F_2(g) + 2\,e^-$	$\longrightarrow 2\,F^-(aq)$	2.87	Weaker reducing agent
	$H_2O_2(aq) + 2\,H^+(aq) + 2\,e^-$	$\longrightarrow 2\,H_2O(l)$	1.78	
	$MnO_4^-(aq) + 8\,H^+(aq) + 5\,e^-$	$\longrightarrow Mn^{2+}(aq) + 4\,H_2O(l)$	1.51	
	$Cl_2(g) + 2\,e^-$	$\longrightarrow 2\,Cl^-(aq)$	1.36	
	$Cr_2O_7^{2-}(aq) + 14\,H^+(aq) + 6\,e^-$	$\longrightarrow 2\,Cr^{3+}(aq) + 7\,H_2O(l)$	1.33	
	$O_2(g) + 4\,H^+(aq) + 4\,e^-$	$\longrightarrow 2\,H_2O(l)$	1.23	
	$Br_2(l) + 2\,e^-$	$\longrightarrow 2\,Br^-(aq)$	1.09	
	$Ag^+(aq) + e^-$	$\longrightarrow Ag(s)$	0.80	
	$Fe^{3+}(aq) + e^-$	$\longrightarrow Fe^{2+}(aq)$	0.77	
	$O_2(g) + 2\,H^+(aq) + 2\,e^-$	$\longrightarrow H_2O_2(aq)$	0.70	
	$I_2(s) + 2\,e^-$	$\longrightarrow 2\,I^-(aq)$	0.54	
	$O_2(g) + 2\,H_2O(l) + 4\,e^-$	$\longrightarrow 4\,OH^-(aq)$	0.40	
	$Cu^{2+}(aq) + 2\,e^-$	$\longrightarrow Cu(s)$	0.34	
	$Sn^{4+}(aq) + 2\,e^-$	$\longrightarrow Sn^{2+}(aq)$	0.15	
	$2\,H^+(aq) + 2\,e^-$	$\longrightarrow H_2(g)$	0	
	$Pb^{2+}(aq) + 2\,e^-$	$\longrightarrow Pb(s)$	-0.13	
	$Ni^{2+}(aq) + 2\,e^-$	$\longrightarrow Ni(s)$	-0.26	
	$Cd^{2+}(aq) + 2\,e^-$	$\longrightarrow Cd(s)$	-0.40	
	$Fe^{2+}(aq) + 2\,e^-$	$\longrightarrow Fe(s)$	-0.45	
	$Zn^{2+}(aq) + 2\,e^-$	$\longrightarrow Zn(s)$	-0.76	
	$2\,H_2O(l) + 2\,e^-$	$\longrightarrow H_2(g) + 2\,OH^-(aq)$	-0.83	
	$Al^{3+}(aq) + 3\,e^-$	$\longrightarrow Al(s)$	-1.66	
	$Mg^{2+}(aq) + 2\,e^-$	$\longrightarrow Mg(s)$	-2.37	Stronger reducing agent
Weaker oxidizing agent	$Na^+(aq) + e^-$	$\longrightarrow Na(s)$	-2.71	
	$Li^+(aq) + e^-$	$\longrightarrow Li(s)$	-3.04	

Note how the ordering of the half-reactions in Table 18.1 corresponds to their ordering in the activity series in Table 4.3 (page 134). The more active metals at the top of the activity series have the more positive oxidation potentials and therefore the more negative standard reduction potentials.

▶ **PROBLEM 18.6** The standard potential for the following galvanic cell is 0.92 V:

$$Al(s)\,|\,Al^{3+}(aq)\,\|\,Cr^{3+}(aq)\,|\,Cr(s)$$

Look up the standard reduction potential for the Al^{3+}/Al half-cell in Table 18.1, and calculate the standard reduction potential for the Cr^{3+}/Cr half-cell.

18.5 Using Standard Reduction Potentials

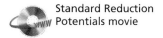
Standard Reduction Potentials movie

A table of standard reduction potentials summarizes an enormous amount of chemical information in a very small space. It enables us to arrange oxidizing or reducing agents in order of increasing strength, and it permits us to predict the spontaneity or nonspontaneity of thousands of redox reactions. Suppose, for example, that the table contains just 100 half-reactions. We can pair each reduction half-reaction with any one of the remaining 99 oxidation half-reactions to give a total of $100 \times 99 = 9900$ cell reactions. By calculating the $E°$ values for these cell reactions we would find that half of them are spontaneous and the other half are nonspontaneous. (Can you see why?)

To illustrate how to use the tabulated $E°$ values, let's calculate $E°$ for the oxidation of Zn(s) by $Ag^+(aq)$:

$$2\,Ag^+(aq)\,+\,Zn(s)\,\longrightarrow\,2\,Ag(s)\,+\,Zn^{2+}(aq)$$

First, we find the relevant half-reactions in Table 18.1 and write them in the appropriate direction for reduction of Ag^+ and oxidation of Zn. Next, we add the half-reactions to get the overall reaction. Before adding, though, we must multiply the Ag^+/Ag half-reaction by a factor of 2 so that the electrons will cancel:

Reduction:	$2 \times [Ag^+(aq)\,+\,e^-\,\longrightarrow\,Ag(s)]$	$E° = 0.80\ \text{V}$
Oxidation:	$Zn(s)\,\longrightarrow\,Zn^{2+}(aq)\,+\,2\,e^-$	$E° = -(-0.76\ \text{V})$
Overall reaction:	$2\,Ag^+(aq)\,+\,Zn(s)\,\longrightarrow\,2\,Ag(s)\,+\,Zn^{2+}(aq)$	$E° = 1.56\ \text{V}$

Then we tabulate the $E°$ values for the half-reactions, remembering that $E°$ for oxidation of zinc is the negative of the standard reduction potential (-0.76 V). We do *not* multiply the $E°$ value for reduction of Ag^+ (0.80 V) by a factor of 2, however, because an electrical potential does not depend on how much reaction occurs.

The reason why $E°$ values are independent of the amount of reaction can be understood by looking at the equation $\Delta G° = -nFE°$. Free energy is an *extensive* property (Section 1.4) because it depends on the amount of substance. If we double the amount of Ag^+ reduced, the free-energy change, $\Delta G°$, doubles. But, of course, the number of electrons transferred, n, also doubles, and the ratio $E° = -\Delta G°/nF$ is therefore constant. Electrical potential is therefore an *intensive* property, which does not depend on the amount of substance.

The $E°$ value for the overall reaction of Zn(s) with $Ag^+(aq)$ is the sum of the $E°$ values for the two half-reactions: 0.80 V + 0.76 V = 1.56 V. Because $E°$ is positive (and $\Delta G°$ is negative), oxidation of zinc by Ag^+ is a spontaneous reaction under standard-state conditions. Just as Ag^+ can oxidize Zn, it's evident from Table 18.1 that Ag^+ can oxidize any reducing agent that lies below it in the table (Fe^{2+}, H_2O_2, I^-, and so forth). The sum of $E°$ for the Ag^+/Ag reduction (0.80 V) and $-E°$ for any half-reaction that lies below the Ag^+/Ag half-reaction always gives a positive $E°$ for the overall reaction.

In general, an oxidizing agent can oxidize any reducing agent that lies below it in the table but can't oxidize a reducing agent that appears above it in the table. Thus, Ag^+ can't oxidize Br^-, H_2O, Cr^{3+}, and so forth because $E°$ for the overall reaction is negative. Simply by glancing at the locations of the oxidizing and reducing agents in the table, we can predict whether a reaction is spontaneous or nonspontaneous.

EXAMPLE 18.4

(a) Arrange the following oxidizing agents in order of increasing strength under standard-state conditions: $Br_2(l)$, $Fe^{3+}(aq)$, $Cr_2O_7{}^{2-}(aq)$.

(b) Arrange the following reducing agents in order of increasing strength under standard-state conditions: Al(s), Na(s), Zn(s).

SOLUTION

(a) Pick out the half-reactions in Table 18.1 that involve Br_2, Fe^{3+}, and $Cr_2O_7{}^{2-}$, and list them in the order in which they occur in the table:

$$Cr_2O_7{}^{2-}(aq) + 14\,H^+(aq) + 6\,e^- \longrightarrow 2\,Cr^{3+}(aq) + 7\,H_2O(l) \qquad E° = 1.33\text{ V}$$

$$Br_2(l) + 2\,e^- \longrightarrow 2\,Br^-(aq) \qquad E° = 1.09\text{ V}$$

$$Fe^{3+}(aq) + e^- \longrightarrow Fe^{2+}(aq) \qquad E° = 0.77\text{ V}$$

We can see that $Cr_2O_7{}^{2-}$ has the greatest tendency to be reduced (largest $E°$), and Fe^{3+} has the least tendency to be reduced (smallest $E°$). The species that has the greatest tendency to be reduced is the strongest oxidizing agent, so oxidizing strength increases in the order $Fe^{3+} < Br_2 < Cr_2O_7{}^{2-}$. As a shortcut, simply note that the strength of the oxidizing agents, listed on the left side of Table 18.1, increases on moving up in the table.

(b) List the half-reactions that involve Al(s), Na(s), and Zn(s) in the order in which they occur in Table 18.1:

$$Zn^{2+}(aq) + 2\,e^- \longrightarrow Zn(s) \qquad E° = -0.76\text{ V}$$

$$Al^{3+}(aq) + 3\,e^- \longrightarrow Al(s) \qquad E° = -1.66\text{ V}$$

$$Na^+(aq) + e^- \longrightarrow Na(s) \qquad E° = -2.71\text{ V}$$

The last half-reaction has the least tendency to occur in the forward direction (most negative $E°$) and the greatest tendency to occur in the reverse direction. Therefore, Na is the strongest reducing agent, and reducing strength increases in the order Zn < Al < Na. As a shortcut, note that the strength of the reducing agents, listed on the right side of Table 18.1, increases on moving down the table.

EXAMPLE 18.5   Predict from Table 18.1 whether $Pb^{2+}(aq)$ can oxidize Al(s) or Cu(s) under standard-state conditions. Calculate $E°$ for each reaction at 25°C.

SOLUTION   We find that $Pb^{2+}(aq)$ is above Al(s) in the table but below Cu(s). Therefore, $Pb^{2+}(aq)$ can oxidize Al(s) but can't oxidize Cu(s). To confirm these predictions, calculate $E°$ values for the overall reactions.

For oxidation of Al by Pb^{2+}, $E°$ is positive (1.53 V), and the reaction is therefore spontaneous:

$$3 \times [Pb^{2+}(aq) + 2\,e^- \longrightarrow Pb(s)] \qquad\qquad E° = -0.13\ V$$
$$\underline{2 \times [Al(s) \longrightarrow Al^{3+}(aq) + 3\,e^-] \qquad\qquad E° = 1.66\ V}$$
$$3\,Pb^{2+}(aq) + 2\,Al(s) \longrightarrow 3\,Pb(s) + 2\,Al^{3+}(aq) \qquad E° = 1.53\ V$$

Note that we have multiplied the Pb^{2+}/Pb half-reaction by a factor of 3 and the Al/Al^{3+} half-reaction by a factor of 2, so that the electrons will cancel, but we do not multiply the $E°$ values by these factors because electrical potential is an intensive property.

For oxidation of Cu by Pb^{2+}, $E°$ is negative (-0.47 V), and the reaction is therefore nonspontaneous:

$$Pb^{2+}(aq) + 2\,e^- \longrightarrow Pb(s) \qquad\qquad E° = -0.13\ V$$
$$\underline{Cu(s) \longrightarrow Cu^{2+}(aq) + 2\,e^- \qquad\qquad E° = -0.34\ V}$$
$$Pb^{2+}(aq) + Cu(s) \longrightarrow Pb(s) + Cu^{2+}(aq) \qquad E° = -0.47\ V$$

▶ **PROBLEM 18.7** Which is the stronger oxidizing agent, $Cl_2(g)$ or $Ag^+(aq)$? Which is the stronger reducing agent, Fe(s) or Mg(s)?

▶ **PROBLEM 18.8** Predict from Table 18.1 whether each of the following reactions can occur under standard-state conditions:
(a) $2\,Fe^{3+}(aq) + 2\,I^-(aq) \rightarrow 2\,Fe^{2+}(aq) + I_2(s)$
(b) $3\,Ni(s) + 2\,Al^{3+}(aq) \rightarrow 3\,Ni^{2+}(aq) + 2\,Al(s)$

Confirm your predictions by calculating the value of $E°$ for each reaction. Which reaction(s) can occur in the reverse direction under standard-state conditions?

✦ **KEY CONCEPT PROBLEM 18.9** Consider the following table of standard reduction potentials:

Reduction Half-Reaction	$E°$ (V)
$A^{3+} + 2\,e^- \rightarrow A^+$	1.47
$B^{2+} + 2\,e^- \rightarrow B$	0.60
$C^{2+} + 2\,e^- \rightarrow C$	-0.21
$D^+ + e^- \rightarrow D$	-1.38

(a) Which substance is the strongest reducing agent? Which is the strongest oxidizing agent?
(b) Which substances can be oxidized by B^{2+}? Which can be reduced by C?
(c) Write a balanced equation for the overall cell reaction that delivers the highest voltage, and calculate $E°$ for the reaction.

18.6 Cell Potentials and Composition of the Reaction Mixture: The Nernst Equation

Cell potentials, like free-energy changes (Section 17.10), depend on temperature and on the composition of the reaction mixture—that is, on the concentrations of solutes and the partial pressures of gases. This dependence can be derived from the equation

$$\Delta G = \Delta G° + RT \ln Q$$

Recall from Section 17.10 that ΔG is the free-energy change for a reaction under nonstandard-state conditions, $\Delta G°$ is the free-energy change under standard-state conditions, and Q is the reaction quotient. Since $\Delta G = -nFE$ and $\Delta G° = -nFE°$, we can rewrite the equation for ΔG in the form

$$-nFE = -nFE° + RT \ln Q$$

Dividing by $-nF$, we obtain the **Nernst equation**, named after Walther Nernst (1864–1941), the German chemist who first derived it:

◆— NERNST EQUATION: $E = E° - \dfrac{RT}{nF} \ln Q$ or $E = E° - \dfrac{2.303RT}{nF} \log Q$

Because of an intimate connection between the cell voltage and pH (Section 18.7), we will write the Nernst equation in terms of base-10 logarithms. At 25°C, $2.303RT/F$ has a value of 0.0592 V, and therefore

$$E = E° - \frac{0.0592 \text{ V}}{n} \log Q \qquad \text{In volts, at 25°C}$$

In actual galvanic cells, the concentrations and partial pressures of reactants and products seldom have standard-state values, and the values change as the cell reaction proceeds. The Nernst equation is useful because it enables us to calculate cell potentials under nonstandard-state conditions, as shown in Example 18.6.

EXAMPLE 18.6   Consider a galvanic cell that uses the reaction

Nernst Equation activity

$$Zn(s) + 2 H^+(aq) \longrightarrow Zn^{2+}(aq) + H_2(g)$$

Calculate the cell potential at 25° when $[H^+] = 1.0$ M, $[Zn^{2+}] = 0.0010$ M, and $P_{H_2} = 0.10$ atm.

BALLPARK SOLUTION   Qualitatively, we expect that the reaction will have a greater tendency to occur under the cited conditions than under standard-state conditions because the product concentrations are lower than standard-state values. We therefore predict that the cell potential E will be greater than the standard cell potential $E°$.

DETAILED SOLUTION   Quantitatively, the cell potential at 25°C is

$$E = E° - \frac{0.0592 \text{ V}}{n} \log Q$$

$$= E° - \left(\frac{0.0592 \text{ V}}{n} \right) \left(\log \frac{[Zn^{2+}](P_{H_2})}{[H^+]^2} \right)$$

where the reaction quotient contains both molar concentrations of solutes and the partial pressure of a gas (in atm). As usual, zinc has been omitted from the reaction quotient because it's a pure solid. We can calculate the value of $E°$ from the standard potentials in Table 18.1:

$$E° = E°_{Zn \to Zn^{2+}} + E°_{H^+ \to H_2} = -(-0.76 \text{ V}) + 0 \text{ V} = 0.76 \text{ V}$$

For this reaction, 2 mol of electrons are transferred, so $n = 2$. Substituting into the Nernst equation the appropriate values of $E°$, n, $[H^+]$, $[Zn^{2+}]$, and P_{H_2} gives

$$E = (0.76 \text{ V}) - \left(\frac{0.0592 \text{ V}}{2}\right)\left(\log \frac{(0.0010)(0.10)}{(1.0)^2}\right) = (0.76 \text{ V}) - \left(\frac{0.0592 \text{ V}}{2}\right)(-4.0)$$

$$= 0.76 \text{ V} + 0.12 \text{ V}$$

$$= 0.88 \text{ V} \qquad \text{At } 25°C$$

▶ **PROBLEM 18.10** Consider a galvanic cell that uses the reaction

$$Cu(s) + 2 \text{ Fe}^{3+}(aq) \longrightarrow Cu^{2+}(aq) + 2 \text{ Fe}^{2+}(aq)$$

What is the potential of a cell at 25°C that has the following ion concentrations?

$$[\text{Fe}^{3+}] = 1.0 \times 10^{-4} \text{ M} \qquad [\text{Cu}^{2+}] = 0.25 \text{ M} \qquad [\text{Fe}^{2+}] = 0.20 \text{ M}$$

◆— **KEY CONCEPT PROBLEM 18.11** Consider the following galvanic cell:

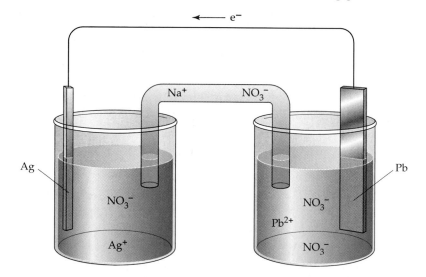

(a) What is the quantitative change in the cell voltage on increasing the ion concentrations in the anode compartment by a factor of 10?

(b) What is the quantitative change in the cell voltage on increasing the ion concentrations in the cathode compartment by a factor of 10?

18.7 Electrochemical Determination of pH

The electrochemical determination of pH using a pH meter is a particularly important application of the Nernst equation. Consider, for example, a cell with a hydrogen electrode as the anode and a second reference electrode as the cathode:

$$Pt \mid H_2(1 \text{ atm}) \mid H^+(? \text{ M}) \parallel \text{Reference cathode}$$

The hydrogen electrode consists of a platinum wire that is in contact with H_2 at 1 atm and dips into a solution of unknown pH. The potential of this cell is

$$E_{\text{cell}} = E_{H_2 \rightarrow H^+} + E_{\text{ref}}$$

We can calculate the potential for the hydrogen electrode half-reaction

$$H_2(g) \longrightarrow 2\,H^+(aq) + 2\,e^-$$

by applying the Nernst equation to this half-reaction:

$$E_{H_2 \rightarrow H^+} = (E^\circ{}_{H_2 \rightarrow H^+}) - \left(\frac{0.0592\ V}{n}\right)\left(\log \frac{[H^+]^2}{P_{H_2}}\right)$$

Since $E^\circ = 0$ V for the standard hydrogen electrode, $n = 2$, and $P_{H_2} = 1$ atm, we can rewrite this equation as

$$E_{H_2 \rightarrow H^+} = -\left(\frac{0.0592\ V}{2}\right)(\log [H^+]^2)$$

Further, because $\log [H^+]^2 = 2 \log [H^+]$ and $-\log [H^+] = pH$, the half-cell potential for the hydrogen electrode is directly proportional to the pH:

$$E_{H_2 \rightarrow H^+} = -\left(\frac{0.0592\ V}{2}\right)(2)(\log [H^+]) = (0.0592\ V)(-\log [H^+])$$

$$= (0.0592\ V)(pH)$$

The overall cell potential is

$$E_{cell} = (0.0592\ V)(pH) + E_{ref}$$

and the pH is therefore a linear function of the cell potential:

$$pH = \frac{E_{cell} - E_{ref}}{0.0592\ V}$$

A higher cell potential indicates a higher pH, meaning that we can measure the pH of a solution simply by measuring E_{cell}.

In actual pH measurements, a *glass* electrode replaces the cumbersome hydrogen electrode and a *calomel* electrode is used as the reference. A glass electrode consists of a silver wire coated with silver chloride that dips into a reference solution of dilute hydrochloric acid (Figure 18.6). The hydrochloric acid is separated from the test solution of unknown pH by a thin glass membrane. For rugged applications, the exterior of the electrode is made of an epoxy resin, which protects the glass tip. A calomel electrode consists of mercury(I) chloride (Hg_2Cl_2, calomel) in contact with liquid mercury and aqueous KCl. The cell half-reactions are

$$2 \times [Ag(s) + Cl^-(aq) \longrightarrow AgCl(s) + e^-] \qquad E^\circ = -0.22\ V$$

$$Hg_2Cl_2(s) + 2\,e^- \longrightarrow 2\,Hg(l) + 2\,Cl^-(aq) \qquad E^\circ = 0.28\ V$$

The overall cell potential, E_{cell}, depends not only on the potentials of these two half-reactions but also on the boundary potential that develops across the thin glass membrane separating the reference HCl solution from the test solution. Because the boundary potential depends linearly on the difference in the pH of the solutions on the two sides of the membrane, the pH of the test solution can be determined by measuring E_{cell}. The cell potential is measured with a pH meter, a voltage-measuring device that electronically converts E_{cell} to pH and displays the result in pH units.

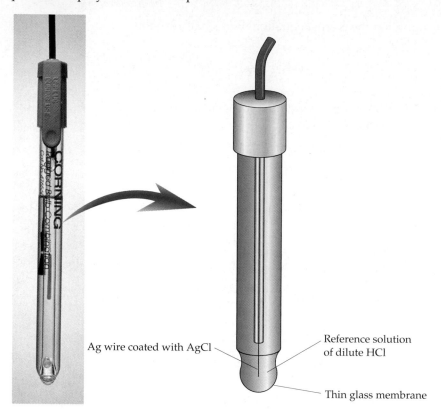

FIGURE 18.6   A glass electrode consists of a silver wire coated with silver chloride that dips into a reference solution of dilute hydrochloric acid. The hydrochloric acid is separated from the test solution of unknown pH by a thin glass membrane. When a glass electrode is immersed in the test solution, its electrical potential depends linearly on the difference in the pH of the solutions on the two sides of the membrane.

Ag wire coated with AgCl

Reference solution of dilute HCl

Thin glass membrane

EXAMPLE 18.7   The following cell has a potential of 0.55 V at 25°C:

$$Pt(s)\,|\,H_2(1\ atm)\,|\,H^+(?\ M)\,\|\,Cl^-(1\ M)\,|\,Hg_2Cl_2(s)\,|\,Hg(l)$$

What is the pH of the solution in the anode compartment?

SOLUTION   The cell reaction is

$$H_2(g)\ +\ Hg_2Cl_2(s)\ \longrightarrow\ 2\,H^+(aq)\ +\ 2\,Hg(l)\ +\ 2\,Cl^-(aq)$$

and the cell potential is

$$E_{cell}\ =\ E_{H_2 \to H^+}\ +\ E_{Hg_2Cl_2 \to Hg}\ =\ 0.55\ V$$

Because the reference electrode is the standard calomel electrode, which has $E = E° = 0.28$ V (Appendix D), the half-cell potential for the hydrogen electrode is 0.27 V:

$$E_{H_2 \to H^+}\ =\ E_{cell}\ -\ E_{Hg_2Cl_2 \to Hg}\ =\ 0.55\ V\ -\ 0.28\ V\ =\ 0.27\ V$$

We can then apply the Nernst equation to the half-reaction $H_2(g) \to 2\,H^+(aq) + 2\,e^-$:

$$E_{H_2 \to H^+}\ =\ (E°_{H_2 \to H^+})\ -\ \left(\frac{0.0592\ V}{n}\right)\left(\log \frac{[H^+]^2}{P_{H_2}}\right)$$

Substituting in the values of E, $E°$, n, and P_{H_2} gives

$$0.27 \text{ V} = (0 \text{ V}) - \left(\frac{0.0592 \text{ V}}{2}\right)\left(\log \frac{[H^+]^2}{1}\right) = (0.0592 \text{ V})(\text{pH})$$

Therefore, the pH is

$$\text{pH} = \frac{0.27 \text{ V}}{0.0592 \text{ V}} = 4.6$$

▶ **PROBLEM 18.12**   What is the pH of the solution in the anode compartment of the following cell if the measured cell potential at 25°C is 0.28 V?

$$Pt(s)\,|\,H_2(1 \text{ atm})\,|\,H^+(? \text{ M})\,\|\,Pb^{2+}(1 \text{ M})\,|\,Pb(s)$$

18.8 Standard Cell Potentials and Equilibrium Constants

We saw in Section 18.3 that the standard free-energy change for a reaction is related to the standard cell potential by the equation

$$\Delta G° = -nFE°$$

In addition, we showed in Section 17.11 that the standard free-energy change is also related to the equilibrium constant for the reaction:

$$\Delta G° = -RT \ln K$$

Combining these two equations, we obtain

$$-nFE° = -RT \ln K$$

or

$$E° = \frac{RT}{nF} \ln K = \frac{2.303 \, RT}{nF} \log K$$

Since $2.303 \, RT/F$ has a value of 0.0592 V at 25°C, we can rewrite this equation in the simplified form

$$E° = \frac{0.0592 \text{ V}}{n} \log K \qquad \text{In volts, at 25°C}$$

Very small concentrations are difficult to measure, so the determination of an equilibrium constant from concentration measurements is not feasible when K is either very large or very small. Standard cell potentials, however, are relatively easy to measure. Consequently, the most common use of this equation is in calculating equilibrium constants from standard cell potentials.

As an example, let's calculate the value of K for the reaction in the Daniell cell:

$$Zn(s) + Cu^{2+}(aq) \longrightarrow Zn^{2+}(aq) + Cu(s)$$

Solving the equation for log K and substituting in the appropriate values of $E°$ and n, we obtain

$$E° = \frac{0.0592 \text{ V}}{n} \log K$$

$$\log K = \frac{nE°}{0.0592 \text{ V}} = \frac{(2)(1.10 \text{ V})}{0.0592 \text{ V}} = 37.2$$

Therefore, K is the antilog of 37.2 (Appendix A.2):

$$K = \text{antilog } 37.2 = 10^{37.2} = 2 \times 10^{37} \qquad \text{At 25°C}$$

Because the equilibrium constant is a very large number, the reaction goes essentially to completion. When $[Zn^{2+}] = 1$ M, for example, $[Cu^{2+}]$ is less than 10^{-37} M.

The preceding calculation shows that even a relatively small value of $E°$ (+1.10 V) corresponds to a huge value of K (2×10^{37}). A positive value of $E°$ corresponds to a positive value of log K and therefore $K > 1$, and a negative value of $E°$ corresponds to a negative value of log K and therefore $K < 1$. Because the standard reduction potentials in Table 18.1 span a range of about 6 V, $E°$ for a redox reaction can range from +6 V for reaction of the strongest oxidizing agent with the strongest reducing agent to −6 V for reaction of the weakest oxidizing agent with the weakest reducing agent. However, $E°$ values outside the range +3 V to −3 V are uncommon. For the case of $n = 2$, the correspondence between the values of $E°$ and K is indicated in Figure 18.7. Equilibrium constants for redox reactions tend to be either very large or very small in comparison with equilibrium constants for acid–base reactions, which are in the range of 10^{14} to 10^{-14}. Redox reactions typically go either essentially to completion (K is very large) or almost not at all (K is very small).

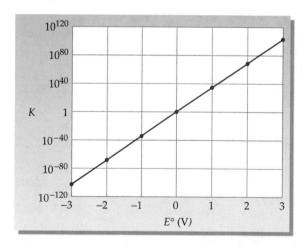

FIGURE 18.7   The relationship between the equilibrium constant K for a redox reaction with $n = 2$ and the standard cell potential $E°$. Note that K is plotted on a logarithmic scale.

In previous chapters, we discussed two different ways to determine the value of an equilibrium constant K: from concentration data (Section 13.2) and from thermochemical data (Section 17.11). In this section, we've added a third way: from electrochemical data. The following are the key relationships needed for each approach:

1. *K* from concentration data for solutes: $K = \dfrac{[C]^c[D]^d}{[A]^a[B]^b}$

2. *K* from thermochemical data: $\Delta G° = -RT \ln K; \quad \ln K = \dfrac{-\Delta G°}{RT}$

3. *K* from electrochemical data: $E° = \dfrac{RT}{nF} \ln K; \quad \ln K = \dfrac{nFE°}{RT}$

EXAMPLE 18.8 Use the standard reduction potentials in Table 18.1 to calculate the equilibrium constant at 25°C for the reaction

$$6\,Br^-(aq) + Cr_2O_7^{2-}(aq) + 14\,H^+(aq) \longrightarrow 3\,Br_2(l) + 2\,Cr^{3+}(aq) + 7\,H_2O(l)$$

SOLUTION Find the relevant half-reactions in Table 18.1, and write them in the proper direction for oxidation of Br^- and reduction of $Cr_2O_7^{2-}$. Before adding the half-reactions to get the overall reaction, multiply the Br^-/Br_2 half-reaction by a factor of 3 so that the electrons will cancel:

$3 \times [2\,Br^-(aq) \longrightarrow Br_2(l) + 2\,e^-]$	$E° = -1.09\ V$
$Cr_2O_7^{2-}(aq) + 14\,H^+(aq) + 6\,e^- \longrightarrow 2\,Cr^{3+}(aq) + 7\,H_2O(l)$	$E° = 1.33\ V$
$6\,Br^-(aq) + Cr_2O_7^{2-}(aq) + 14\,H^+(aq) \longrightarrow 3\,Br_2(l) + 2\,Cr^{3+}(aq) + 7\,H_2O(l)$	$E° = 0.24\ V$

Note that $E°$ for the Br^-/Br_2 *oxidation* is the negative of the tabulated standard *reduction* potential (1.09 V), and remember that we don't multiply this $E°$ value by a factor of 3 because electrical potential is an intensive property. The $E°$ value for the overall reaction is the sum of the $E°$ values for the half-reactions: $-1.09\ V + 1.33\ V = 0.24\ V$. To calculate the equilibrium constant, use the relation between log K and $nE°$, with $n = 6$:

$$\log K = \frac{nE°}{0.0592\ V} = \frac{(6)(0.24\ V)}{0.0592\ V} = 24 \qquad K = 1 \times 10^{24} \qquad \text{At 25°C}$$

▶ **PROBLEM 18.13** Use the data in Table 18.1 to calculate the equilibrium constant at 25°C for the reaction

$$4\,Fe^{2+}(aq) + O_2(g) + 4\,H^+(aq) \longrightarrow 4\,Fe^{3+}(aq) + 2\,H_2O(l)$$

▶ **PROBLEM 18.14** What is the value of $E°$ for a redox reaction involving the transfer of 2 mol of electrons if its equilibrium constant is 1.8×10^{-5} (the value of the acid-dissociation constant K_a for acetic acid)?

18.9 Batteries

By far the most important practical application of galvanic cells is their use as *batteries*. In multicell batteries, such as those in automobiles, the individual galvanic cells are linked in series, with the anode of each cell connected to the cathode of the adjacent cell. The voltage provided by the battery is the sum of the individual cell voltages. The features required in a battery depend on the application. In general, however, a commercially successful battery should be compact, lightweight, physically rugged, and inexpensive, and it must provide a stable source of power for relatively long periods of time. Battery design is an active area of research that requires considerable ingenuity as well as a solid understanding of electrochemistry.

Let's look at several of the most common types of commercial batteries.

Lead Storage Battery

The *lead storage battery* is perhaps the most familiar of all galvanic cells because it has been used as a reliable source of power for starting automobiles for more than three-quarters of a century. A typical 12 V battery consists of six cells connected in series, each cell providing a potential of about 2 V. The cell design is illustrated in Figure 18.8. The anode, a series of lead grids packed with spongy lead, and the cathode, a second series of grids packed with lead dioxide, dip into the electrolyte, an aqueous solution of sulfuric acid (38% by weight). When the cell is discharging (providing current), the electrode half-reactions and the overall cell reaction are:

Anode: $Pb(s) + HSO_4^-(aq) \longrightarrow PbSO_4(s) + H^+(aq) + 2\,e^-$ $E° = 0.296$ V

Cathode: $PbO_2(s) + 3\,H^+(aq) + HSO_4^-(aq) + 2\,e^- \longrightarrow PbSO_4(s) + 2\,H_2O(l)$ $E° = 1.628$ V

Overall: $Pb(s) + PbO_2(s) + 2\,H^+(aq) + 2\,HSO_4^-(aq) \longrightarrow 2\,PbSO_4(s) + 2\,H_2O(l)$ $E° = 1.924$ V

(These equations contain HSO_4^- ions because SO_4^{2-} is protonated in strongly acidic solutions.)

Lead is oxidized to lead sulfate at the anode, and lead dioxide is reduced to lead sulfate at the cathode. The cell doesn't need to have separate anode and cathode compartments because the oxidizing and reducing agents are both solids (PbO_2 and Pb) that are kept from coming in contact by the presence of insulating spacers between the grids.

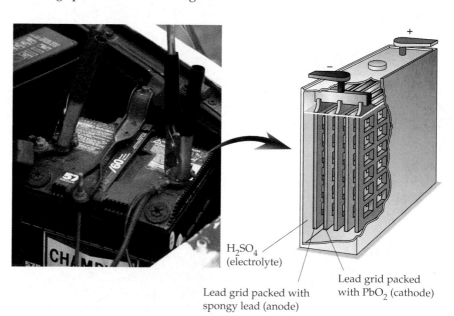

FIGURE 18.8   A lead storage battery and a cutaway view of one cell. Each electrode consists of several grids with a large surface area so that the battery can deliver the high currents required to start an automobile engine. The electrolyte is aqueous sulfuric acid.

H_2SO_4
(electrolyte)

Lead grid packed with spongy lead (anode)

Lead grid packed with PbO_2 (cathode)

Because the reaction product (solid $PbSO_4$) adheres to the surface of the electrodes, a "run-down" lead storage battery can be recharged by using an external source of direct current to drive the cell reaction in the reverse, nonspontaneous direction. In an automobile, the battery is continuously recharged by a device called an alternator, which is driven by the engine.

A lead storage battery typically provides good service for several years, but eventually mechanical shock from driving over rough roads dislodges the $PbSO_4$ from the electrodes. Then it's no longer possible to recharge the battery, and it must be replaced.

Dry-Cell Batteries

The common household batteries used as power sources for flashlights, portable radios, and tape recorders use the *dry cell*, or *Leclanché cell*, patented in 1866 by the Frenchman George Leclanché. A dry-cell battery has a zinc metal can, which serves as the anode, and an inert graphite rod surrounded by a paste of solid manganese dioxide and carbon black, which functions as the cathode (Figure 18.9). Surrounding the MnO_2-containing paste is the electrolyte, a moist paste of ammonium chloride and zinc chloride in starch. The dry cell is not completely dry but gets its name from the fact that the electrolyte is a viscous, aqueous paste rather than a liquid solution. The electrode reactions, which are rather complicated, can be represented in simplified form by the following equations:

Anode: $$Zn(s) \longrightarrow Zn^{2+}(aq) + 2\,e^-$$

Cathode: $$2\,MnO_2(s) + 2\,NH_4{}^+(aq) + 2\,e^- \longrightarrow Mn_2O_3(s) + 2\,NH_3(aq) + H_2O(l)$$

This cell provides a potential of about 1.5 V, but the potential deteriorates to about 0.8 V as the cell is used.

Insulator

Graphite rod (cathode)

MnO_2 and carbon black paste

NH_4Cl and $ZnCl_2$ paste (electrolyte)

Zinc metal can (anode)

FIGURE 18.9 Leclanché dry cell and a cutaway view.

The *alkaline dry cell* is a modified version of the Leclanché cell in which the acidic NH_4Cl electrolyte of the Leclanché cell is replaced by a basic electrolyte, either NaOH or KOH. As in the Leclanché cell, the electrode reactions involve oxidation of zinc and reduction of manganese dioxide, but the oxidation product is zinc oxide, as is appropriate to the basic conditions:

Anode: $$Zn(s) + 2\,OH^-(aq) \longrightarrow ZnO(s) + H_2O(l) + 2\,e^-$$

Cathode: $$2\,MnO_2(s) + H_2O(l) + 2\,e^- \longrightarrow Mn_2O_3(s) + 2\,OH^-(aq)$$

Corrosion of the zinc anode is a significant side reaction under acidic conditions because zinc reacts with $H^+(aq)$ to give $Zn^{2+}(aq)$ and $H_2(g)$. Under basic conditions, however, the cell has a longer life because zinc corrodes more slowly. The alkaline cell also produces higher power and more stable current and voltage because of more efficient ion transport in the alkaline electrolyte.

Closely related to the alkaline dry cell is the *mercury battery*, often used in watches, heart pacemakers, and other devices where a battery of small size is required (Figure 18.10). The anode of the mercury battery is zinc, as in the alkaline dry cell, but the cathode is steel in contact with mercury(II) oxide (HgO) in an alkaline medium of KOH and $Zn(OH)_2$. Zinc is oxidized at the anode, and HgO is reduced at the cathode:

Anode: $\quad\quad\quad Zn(s) + 2\,OH^-(aq) \longrightarrow ZnO(s) + H_2O(l) + 2\,e^-$

Cathode: $\quad HgO(s) + H_2O(l) + 2\,e^- \longrightarrow Hg(l) + 2\,OH^-(aq)$

Even though mercury batteries can be made very small, they still produce a stable potential (about 1.3 V) for long periods of time. When possible, used mercury batteries should be recycled to recover the mercury because of its toxicity.

Steel (cathode)

Insulator

HgO in KOH and $Zn(OH)_2$

Zinc container (anode)

FIGURE 18.10 A small mercury battery and a cutaway view.

Nickel–Cadmium Batteries

Nickel–cadmium, or "ni–cad," batteries are popular for use in calculators and portable power tools because, unlike most other dry-cell batteries, they are rechargeable (Figure 18.11). The anode of a ni–cad battery is cadmium metal, and the cathode is the nickel(III) compound NiO(OH) supported on nickel metal. The electrode reactions are

Anode: $\quad\quad\quad\quad Cd(s) + 2\,OH^-(aq) \longrightarrow Cd(OH)_2(s) + 2\,e^-$

Cathode: $\quad NiO(OH)(s) + H_2O(l) + e^- \longrightarrow Ni(OH)_2(s) + OH^-(aq)$

Ni–cad batteries can be recharged many times because the solid products of the electrode reactions adhere to the surface of the electrodes.

FIGURE 18.11 Rechargeable nickel–cadmium, or "ni–cad," storage batteries.

Lithium Batteries

Rechargeable lithium batteries may prove to be the batteries of the future because of their light weight and high voltage (about 3.0 V). More than 400 million were sold in 1998, with production doubling annually. Lithium has a higher $E°$ value for oxidation than any other metal, and only 6.94 g of lithium is needed to provide 1 mol of electrons. Lithium batteries consist of a lithium anode—either lithium metal itself or lithium atoms inserted into a graphite electrode—a metal oxide or metal sulfide cathode that can incorporate Li^+

ions, and an electrolyte that contains a lithium salt, such as $LiClO_4$, in an organic solvent. Solid-state polymer electrolytes that can transport Li^+ ions have also been used. When the cathode material is MnO_2, for example, the electrode reactions are

Anode: $$Li(s) \longrightarrow Li^+ + e^-$$

Cathode: $$MnO_2(s) + Li^+ + e^- \longrightarrow LiMnO_2(s)$$

Lithium batteries are now used in cell phones, laptop computers, and cameras.

Fuel Cells

A **fuel cell** is a galvanic cell in which one of the reactants is a traditional fuel such as methane or hydrogen. A fuel cell differs from an ordinary battery in that the reactants are not contained within the cell but instead are continuously supplied from an external reservoir. Perhaps the best-known example is the hydrogen–oxygen fuel cell, which is used as a source of electric power in space vehicles (Figure 18.12). This cell contains porous carbon electrodes impregnated with metallic catalysts and an electrolyte consisting of hot, aqueous KOH. The fuel (gaseous H_2) and the oxidizing agent (gaseous O_2) don't react directly but instead flow into separate cell compartments where H_2 is oxidized at the anode and O_2 is reduced at the cathode. The overall cell reaction is simply the conversion of hydrogen and oxygen to water:

Anode: $$2\,H_2(g) + 4\,OH^-(aq) \longrightarrow 4\,H_2O(l) + 4\,e^-$$

Cathode: $$O_2(g) + 2\,H_2O(l) + 4\,e^- \longrightarrow 4\,OH^-(aq)$$

Overall: $$2\,H_2(g) + O_2(g) \longrightarrow 2\,H_2O(l)$$

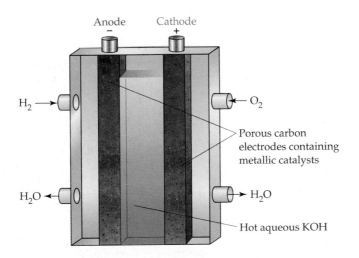

FIGURE 18.12 A hydrogen–oxygen fuel cell. Gaseous H_2 is oxidized to water at the anode, and gaseous O_2 is reduced to hydroxide ion at the cathode. The net reaction is the conversion of H_2 and O_2 to water.

To date, large-scale applications of fuel cells have been limited because of cost, although the Tokyo Electric Power Company in Japan is now operating an 11 MW fuel-cell power plant capable of supplying power to about 4000 households. Future applications include use as the power source in environmentally clean electric vehicles. Several fuel-cell powered buses have been tested in Chicago, and DaimlerChrysler has recently introduced NECar 4, its fourth-generation new electric car, which is powered by a hydrogen-oxygen fuel cell.

This environmentally clean, fuel-cell powered electric bus runs on compressed hydrogen. Its only exhaust gas is water.

▶ **PROBLEM 18.15** Write a balanced equation for the overall cell reaction in each of the following batteries:

(a) Leclanché dry cell   **(b)** Alkaline dry cell

(c) Mercury battery   **(d)** Nickel–cadmium battery

18.10   Corrosion

The steel support column of this bridge has corroded because the iron has been oxidized by moist air, yielding $Fe_2O_3 \cdot H_2O$ (rust).

Corrosion is the oxidative deterioration of a metal, such as the conversion of iron to rust, a hydrated iron(III) oxide of approximate composition $Fe_2O_3 \cdot H_2O$. The rusting of iron has enormous economic consequences. It has been estimated that as much as one-fourth of the steel produced in the United States goes to replace steel structures and products that have been destroyed by corrosion.

To prevent corrosion, we first have to understand how it occurs. One important fact is that the rusting of iron requires both oxygen and water; it doesn't occur in oxygen-free water or in dry air. Another clue is the observation that rusting involves pitting of the metal surface, but the rust is deposited at a location physically separated from the pits. This suggests that rust does not form by direct reaction of iron and oxygen, but rather by an electrochemical process in which iron is oxidized in one region of the surface and oxygen is reduced in another region.

A possible mechanism for rusting, consistent with the known facts, is illustrated in Figure 18.13. The surface of the iron and a droplet of surface water constitute a tiny galvanic cell in which different regions of the surface act as anode and cathode while the aqueous phase serves as the electrolyte. Iron is oxidized more readily in some regions (anode regions) than in others (cathode regions) because the composition of the metal is somewhat inhomogeneous and the surface is irregular. Factors such as impurities, phase boundaries, and mechanical stress may influence the ease of oxidation in a particular region of the surface.

At an anode region, iron is oxidized to Fe^{2+} ions,

$$Fe(s) \longrightarrow Fe^{2+}(aq) + 2\,e^- \qquad E° = 0.45 \text{ V}$$

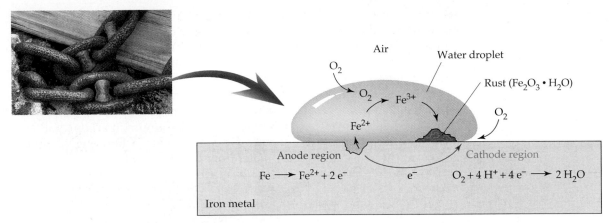

FIGURE 18.13 An electrochemical mechanism for corrosion of iron. The metal and a surface water droplet constitute a tiny galvanic cell in which iron is oxidized to Fe^{2+} in a region of the surface (anode region) remote from atmospheric O_2, and O_2 is reduced near the edge of the droplet at another region of the surface (cathode region). Electrons flow from anode to cathode through the metal, while ions flow through the water droplet. Dissolved O_2 oxidizes Fe^{2+} further to Fe^{3+} before it is deposited as rust ($Fe_2O_3 \cdot H_2O$).

while at a cathode region, oxygen is reduced to water:

$$O_2(g) + 4\,H^+(aq) + 4\,e^- \longrightarrow 2\,H_2O(l) \qquad E° = 1.23\ \text{V}$$

The actual potential for the reduction half-reaction is less than the standard potential (1.23 V) because the water droplet is not 1 M in H^+ ions. (In fact, the water is only slightly acidic because the main source of H^+ ions is the reaction of water with dissolved atmospheric carbon dioxide.) Even at pH 7, however, the potential for the reduction half-reaction is 0.81 V, which means that the cell potential is highly positive, indicative of a spontaneous reaction.

The electrons required for reduction of O_2 at the cathode region are supplied by a current that flows through the metal from the more easily oxidized anode region (Figure 18.13). The electrical circuit is completed by migration of ions in the water droplet. When Fe^{2+} ions migrate away from the pitted, anode region, they come in contact with O_2 dissolved in the surface portion of the water droplet and are further oxidized to Fe^{3+} ions:

$$4\,Fe^{2+}(aq) + O_2(g) + 4\,H^+(aq) \longrightarrow 4\,Fe^{3+}(aq) + 2\,H_2O(l)$$

Iron(III) forms a very insoluble hydrated oxide even in moderately acidic solutions, and the iron(III) is deposited as the familiar red-brown material that we call rust:

$$2\,Fe^{3+}(aq) + 4\,H_2O(l) \longrightarrow \underset{\text{Rust}}{Fe_2O_3 \cdot H_2O(s)} + 6\,H^+(aq)$$

An electrochemical mechanism for corrosion also explains nicely why automobiles rust more rapidly in parts of the country where road salt is used to melt snow and ice. Dissolved salts in the water droplet greatly increase the conductivity of the electrolyte, thus accelerating the pace of corrosion.

A glance at a table of standard reduction potentials indicates that the O_2/H_2O half-reaction lies above the M^{n+}/M half-reaction for nearly all metals, meaning that O_2 can oxidize all metals except a few, such as gold and platinum. Aluminum, for example, has $E° = -1.66$ V for the Al^{3+}/Al half-reaction and is oxidized more readily than iron. In other words, the corrosion of aluminum products such as aircraft and automobile parts, window frames, cooking utensils, and soda cans should be a serious problem. Fortunately it isn't, because oxidation of aluminum gives a very hard, almost impenetrable film of Al_2O_3 that adheres to the surface of the metal and protects it from further attack. Other metals such as magnesium, chromium, titanium, and zinc form similar protective oxide coatings. In the case of iron, however, rust is too porous to shield the underlying metal from further oxidation.

This titanium bicycle doesn't corrode because of a hard, impenetrable layer of TiO_2 that adheres to the surface and protects the metal from further oxidation.

Prevention of Corrosion

Corrosion of iron can be prevented, or at least minimized, by shielding the metal surface from oxygen and moisture. A coat of paint is effective for a while, but rust begins to form as soon as the paint is scratched or chipped. Metals such as chromium, tin, or zinc afford a more durable surface coating for iron. The steel used in making automobiles, for example, is coated by dipping into a bath of molten zinc, a process known as **galvanizing**. As the potentials indicate, zinc is oxidized more easily than iron, and therefore, when the metal is oxidized, zinc is oxidized instead of iron. Any incipient oxidation of iron would be reversed immediately because Zn can reduce Fe^{2+} to Fe. As long as the zinc and iron are in contact, the zinc protects the iron from oxidation even if the zinc layer becomes scratched (Figure 18.14).

$$Fe^{2+}(aq) + 2\,e^- \longrightarrow Fe(s) \qquad E° = -0.45\text{ V}$$

$$Zn^{2+}(aq) + 2\,e^- \longrightarrow Zn(s) \qquad E° = -0.76\text{ V}$$

FIGURE 18.14 A layer of zinc protects iron from oxidation, even when the zinc layer becomes scratched. The zinc (anode), iron (cathode), and water droplet (electrolyte) constitute a tiny galvanic cell. Oxygen is reduced at the cathode, and zinc is oxidized at the anode, thus protecting the iron from oxidation.

Air

Water droplet

Zn^{2+}

O_2

Scratch in zinc layer

Zinc (anode) Zn $\longrightarrow$ Zn^{2+} + 2 e^-

Zinc metal

e^-

Iron (cathode)

$O_2 + 4\,H^+ + 4\,e^- \longrightarrow 2\,H_2O$

Iron metal

The technique of protecting a metal from corrosion by connecting it to a second metal that is more easily oxidized is called **cathodic protection**. It's not necessary to cover the entire surface of the metal with a second metal, as in galvanizing iron. All that's required is electrical contact with the second metal. An underground steel pipeline, for example, can be protected by connecting it through an insulated wire to a stake of magnesium, which acts as a **sacrificial anode** and corrodes instead of iron. In effect, the arrangement is a galvanic cell in which the easily oxidized magnesium acts as the anode, the pipeline behaves as the cathode, and moist soil is the electrolyte. The cell half-reactions are

Anode: $\qquad\qquad\qquad Mg(s) \longrightarrow Mg^{2+}(aq) + 2\,e^- \qquad E° = 2.37\ V$

Cathode: $\quad O_2(g) + 4\,H^+(aq) + 4\,e^- \longrightarrow 2\,H_2O(l) \qquad E° = 1.23\ V$

For large steel structures such as pipelines, storage tanks, bridges, and ships, cathodic protection is the best defense against premature rusting.

18.11 Electrolysis and Electrolytic Cells

Thus far, we've been concerned only with *galvanic cells*—electrochemical cells in which a spontaneous redox reaction produces an electric current. A second important kind of electrochemical cell is the **electrolytic cell**, in which an electric current is used to drive a nonspontaneous reaction. Thus, the processes occurring in galvanic and electrolytic cells are the reverse of each other: A galvanic cell converts chemical energy to electrical energy when a reaction with a positive value of E (and a negative value of ΔG) proceeds toward equilibrium; an electrolytic cell converts electrical energy to chemical energy when an electric current drives a reaction with a negative value of E (and a positive value of ΔG) in a direction away from equilibrium.

The process of using an electric current to bring about chemical change is called **electrolysis**. The opposite signs of E and $\Delta G = -nFE$ for the two kinds of cells are summarized in Table 18.2, along with the situation for a reaction that has reached equilibrium—a dead battery!

TABLE 18.2 Relationship Between Cell Potentials E and Free-Energy Changes ΔG

Reaction Type	E	ΔG	Cell Type
Spontaneous	+	−	Galvanic
Nonspontaneous	−	+	Electrolytic
Equilibrium	0	0	Dead battery

Electrolysis of Molten Sodium Chloride

An electrolytic cell has two electrodes that dip into an electrolyte and are connected to a battery or some other source of direct electric current. A cell for electrolysis of molten sodium chloride, for example, is illustrated in Figure 18.15. The battery serves as an electron pump, pushing electrons into one electrode and pulling them out of the other. The negative electrode attracts Na^+ cations, which combine with the electrons supplied by the battery and are thereby reduced to liquid sodium metal. Similarly, the positive electrode attracts Cl^- anions, which replenish the electrons removed by the battery and are thereby oxidized to chlorine gas. The electrode reactions and overall cell reaction are

Anode (oxidation): $\qquad$ $2\,Cl^-(l) \longrightarrow Cl_2(g) + 2\,e^-$

Cathode (reduction): $\quad$ $2\,Na^+(l) + 2\,e^- \longrightarrow 2\,Na(l)$

Overall cell reaction: $\quad$ $2\,Na^+(l) + 2\,Cl^-(l) \longrightarrow 2\,Na(l) + Cl_2(g)$

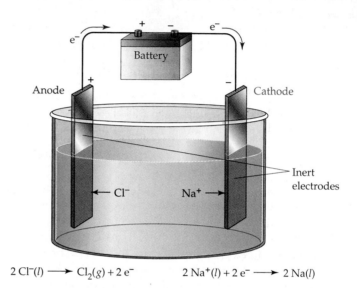

$2\,Cl^-(l) \longrightarrow Cl_2(g) + 2\,e^- \qquad\qquad 2\,Na^+(l) + 2\,e^- \longrightarrow 2\,Na(l)$

FIGURE 18.15 Electrolysis of molten sodium chloride. Chloride ions are oxidized to Cl_2 gas at the anode, and Na^+ ions are reduced to sodium metal at the cathode.

As in a galvanic cell, the anode is the electrode where oxidation takes place, and the cathode is the electrode where reduction takes place. The signs of the electrodes, however, are opposite for the two kinds of cells. In a galvanic cell, the anode is considered negative because it supplies electrons to the external circuit, but in an electrolytic cell, the anode is considered positive because electrons are pulled out of it by the battery. Note that the sign of each electrode in the electrolytic cell is the same as the sign of the battery electrode to which it is attached.

Electrolysis of Aqueous Sodium Chloride

When an aqueous salt solution is electrolyzed, the electrode reactions may differ from those for electrolysis of the molten salt because water may be involved. In the electrolysis of aqueous sodium chloride, for example, the cathode half-reaction might involve either reduction of Na^+ to sodium metal, as in the case of molten sodium chloride, or reduction of water to hydrogen gas:

Electrolysis activity

$$Na^+(aq) + e^- \longrightarrow Na(s) \qquad\qquad E° = -2.71\ V$$
$$2\,H_2O(l) + 2\,e^- \longrightarrow H_2(g) + 2\,OH^-(aq) \qquad E° = -0.83\ V$$

Because the standard potential is much less negative for reduction of water than for reduction of Na^+, water is reduced preferentially and bubbles of hydrogen gas are produced at the cathode.

The anode half-reaction might involve either oxidation of Cl^- to Cl_2 gas, as in the case of molten sodium chloride, or oxidation of water to oxygen gas:

$$2\,Cl^-(aq) \longrightarrow Cl_2(g) + 2\,e^- \qquad\qquad E° = -1.36\text{ V}$$

$$2\,H_2O(l) \longrightarrow O_2(g) + 4\,H^+(aq) + 4\,e^- \qquad E° = -1.23\text{ V}$$

Based on the $E°$ values, we might expect a slight preference for oxidation of water in a solution having 1 M ion concentrations. For a neutral solution ($[H^+]$ = 10^{-7} M), the preference for water oxidation will be even greater because its oxidation potential at pH 7 is −0.81 V. The observed product at the anode, however, is Cl_2, not O_2, because of a phenomenon called *overvoltage*.

Experiments indicate that the applied voltage required for an electrolysis is always greater than the voltage calculated from standard reduction potentials. The additional voltage required is the **overvoltage**. The overvoltage is needed because the magnitude of the current that passes through an electrolytic cell is often limited by the rate of electron transfer at the electrode–solution interface for one or both of the cell half-reactions. If a half-reaction has a substantial activation energy barrier for electron transfer, and a correspondingly slow rate (Section 12.9), then an additional voltage, or overvoltage, must be applied to surmount the barrier and cause the reaction to proceed at a satisfactory rate. For electrode half-reactions involving solution or deposition of metals, the overvoltage is quite small, but for half-reactions involving formation of O_2 (or H_2) gas, the overvoltage can be as large as 1 V. Present theory is unable to predict the magnitude of the overvoltage, but it's known that the overvoltage for formation of O_2 is much larger than that for formation of Cl_2. Because of overvoltage, it's sometimes difficult to predict which half-reaction will occur when $E°$ values for the competing half-reactions are similar. In such cases, only experiment can tell what actually happens.

The observed electrode reactions and overall cell reaction for electrolysis of aqueous sodium chloride are

Anode (oxidation):	$2\,Cl^-(aq) \longrightarrow Cl_2(g) + 2\,e^-$	$E° = -1.36\text{ V}$
Cathode (reduction):	$2\,H_2O(l) + 2\,e^- \longrightarrow H_2(g) + 2\,OH^-(aq)$	$E° = -0.83\text{ V}$
Overall cell reaction:	$2\,Cl^-(aq) + 2\,H_2O(l) \longrightarrow Cl_2(g) + H_2(g) + 2\,OH^-(aq)$	$E° = -2.19\text{ V}$

Sodium ion acts as a spectator ion and is not involved in the electrode reactions. Thus, the sodium chloride solution is converted to a sodium hydroxide solution as the electrolysis proceeds. The minimum potential required to force this nonspontaneous reaction to occur under standard-state conditions is 2.19 V plus the overvoltage.

Electrolysis of Water

The electrolysis of any aqueous solution requires the presence of an electrolyte to carry the current in solution. But if the ions of the electrolyte are less easily oxidized and reduced than water, then water will react at both electrodes. Consider, for example, the electrolysis of an aqueous solution of the inert electrolyte Na_2SO_4. Water is oxidized at the anode in preference to SO_4^{2-} ions and is reduced at the cathode in preference to Na^+ ions. The electrode and overall cell reactions are

Electrolysis of Water movie

Anode (oxidation):	$2\,H_2O(l) \longrightarrow O_2(g) + 4\,H^+(aq) + 4\,e^-$
Cathode (reduction):	$4\,H_2O(l) + 4\,e^- \longrightarrow 2\,H_2(g) + 4\,OH^-(aq)$
Overall cell reaction:	$6\,H_2O(l) \longrightarrow 2\,H_2(g) + O_2(g) + 4\,H^+(aq) + 4\,OH^-(aq)$

Electrolysis of water gives O_2 gas at the anode (on the left) and H_2 gas at the cathode in a 1:2 mole ratio.

If the anode and cathode solutions are mixed, the H^+ and OH^- ions react to form water:

$$4\,H^+(aq) + 4\,OH^-(aq) \longrightarrow 4\,H_2O(l)$$

The net electrolysis reaction is therefore the decomposition of water, a process sometimes used in the laboratory to produce small amounts of pure H_2 and O_2:

$$2\,H_2O(l) \longrightarrow 2\,H_2(g) + O_2(g)$$

KEY CONCEPT PROBLEM 18.16   Metallic potassium was first prepared by Humphrey Davy in 1807 by electrolysis of molten potassium hydroxide:

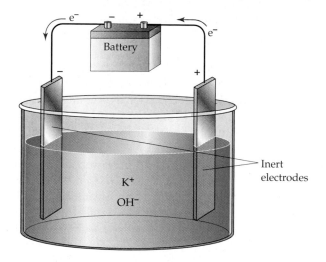

(a) Label the anode and cathode, and show the direction of ion flow.

(b) Write balanced equations for the anode, cathode, and overall cell reactions.

▶ **PROBLEM 18.17**   Predict the half-cell reactions that occur when aqueous solutions of the following salts are electrolyzed in a cell with inert electrodes. What is the overall cell reaction in each case?
(a) LiCl   **(b)** $CuSO_4$

18.12   Commercial Applications of Electrolysis

Electrolysis is used in the manufacture of many important chemicals and in numerous processes for purification and electroplating of metals. Let's look at some examples.

Manufacture of Sodium

Sodium metal is produced commercially in a *Downs cell* by electrolysis of a molten mixture of sodium chloride and calcium chloride (Figure 18.16). The presence of $CaCl_2$ allows the cell to be operated at a lower temperature, because the melting point of the $NaCl$–$CaCl_2$ mixture (about 580°C) is depressed well below that of pure $NaCl$ (801°C). The liquid sodium produced at the cylindrical steel cathode is less dense than the molten salt and thus

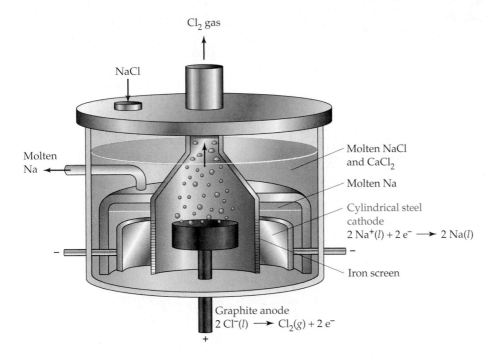

Cl$_2$ gas

NaCl

Molten
Na

Molten NaCl
and CaCl$_2$

Molten Na

Cylindrical steel
cathode
$2\,Na^+(l) + 2\,e^- \longrightarrow 2\,Na(l)$

Iron screen

Graphite anode
$2\,Cl^-(l) \longrightarrow Cl_2(g) + 2\,e^-$

FIGURE 18.16 Cross-sectional view of a Downs cell for commercial production of sodium metal by electrolysis of molten sodium chloride. The cell design keeps the sodium and chlorine apart so that they can't react with each other.

floats to the top part of the cell, where it is drawn off into a suitable container. Chlorine gas forms at the graphite anode, which is separated from the cathode by an iron screen. The cell design keeps the highly reactive sodium and chlorine away from each other and out of contact with air. Because the Downs process requires high currents (typically 25,000–40,000 A), plants for production of sodium are located near sources of inexpensive hydroelectric power, such as Niagara Falls, New York.

Manufacture of Chlorine and Sodium Hydroxide

Production of chlorine and sodium hydroxide by electrolysis of aqueous sodium chloride is the basis of the *chlor-alkali industry*, a business that generates annual sales of approximately $4 billion in the United States alone. Both chlorine and sodium hydroxide rank among the top chemicals in terms of production: Annual output of each in the United States is 12–13 million tons. Chlorine is used in water and sewage treatment and in the manufacture of plastics such as poly(vinyl chloride) (PVC). Sodium hydroxide is employed in making paper, textiles, soaps, and detergents.

Figure 18.17 shows the essential features of a membrane cell for commercial production of chlorine and sodium hydroxide. A saturated aqueous solution of sodium chloride (*brine*) flows into the anode compartment, where Cl^- is oxidized to Cl_2 gas, and water enters the cathode compartment, where it is converted to H_2 gas and OH^- ions. Between the anode and cathode compartments is a special plastic membrane that is permeable to cations but not to anions or water. The membrane keeps the Cl_2 and OH^- ions apart but allows a current of Na^+ ions to flow into the cathode compartment, thus carrying the current in solution and maintaining electrical neutrality in both compartments. The Na^+ and OH^- ions flow out of the cathode compartment as an aqueous solution of NaOH.

FIGURE 18.17 A membrane cell for electrolytic production of Cl_2 and NaOH. Chloride ion is oxidized to Cl_2 gas at the anode, and water is converted to H_2 gas and OH^- ions at the cathode. Sodium ions move from the anode compartment to the cathode compartment through a cation-permeable membrane. Reactants (brine and water) enter the cell, and products (Cl_2 gas, H_2 gas, aqueous NaOH, and depleted brine) leave through appropriately placed pipes.

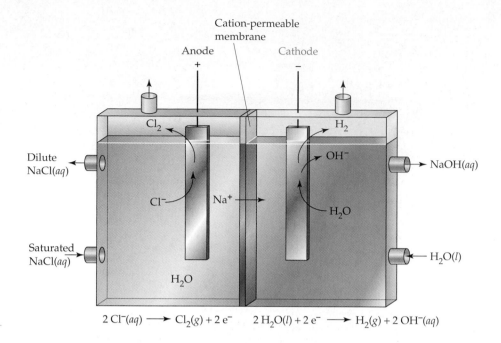

$$2\ Cl^-(aq) \longrightarrow Cl_2(g) + 2\ e^- \qquad 2\ H_2O(l) + 2\ e^- \longrightarrow H_2(g) + 2\ OH^-(aq)$$

Manufacture of Aluminum

Although aluminum is the third most abundant element in the earth's crust (8.3% by weight), it remained a rare and expensive metal until 1886, when a 22-year-old American, Charles Martin Hall, and a 23-year-old Frenchman, Paul Heroult, independently devised a practical process for electrolytic production of aluminum. Still used today, the **Hall–Heroult process** involves electrolysis of a molten mixture of aluminum oxide (Al_2O_3) and cryolite (Na_3AlF_6) at about 1000°C in a cell with graphite electrodes (Figure 18.18). Electrolysis of pure Al_2O_3 is impractical because it melts at a very high temperature (2045°C), and electrolysis of aqueous Al^{3+} solutions is not feasible because water is reduced in preference to Al^{3+} ions. Thus, the use of cryolite as a solvent for Al_2O_3 is the key to the success of the Hall–Heroult process.

FIGURE 18.18 An electrolytic cell for production of aluminum by the Hall–Heroult process. Molten aluminum metal forms at the graphite cathode that lines the cell. Because molten aluminum is more dense than the Al_2O_3–Na_3AlF_6 mixture, it collects at the bottom of the cell and is drawn off periodically.

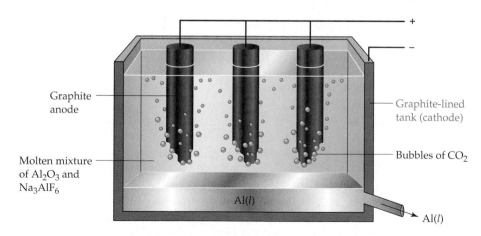

The electrode reactions are still not fully understood, but they probably involve complex anions of the type $AlF_xO_y^{+3-x-2y}$, formed by reaction of Al_2O_3 and Na_3AlF_6. The complex anions are reduced at the cathode to molten aluminum metal and are oxidized at the anode to O_2 gas, which reacts with the graphite anodes to give CO_2 gas. As a result, the anodes are chewed up rapidly and must be replaced frequently. The cell operates at a low voltage (5–6 V) but with very high currents (up to 250,000 A) because 1 mol of electrons produces only 9.0 g of aluminum. Electrolytic production of aluminum is the largest single consumer of electricity in the United States today, making recycling of aluminum products highly desirable.

Electrorefining and Electroplating

The purification of a metal by means of electrolysis is called **electrorefining**. For example, impure copper obtained from ores is converted to pure copper in an electrolytic cell that has impure copper as the anode and pure copper as the cathode (Figure 18.19). The electrolyte is an aqueous solution of copper sulfate.

Electroplating movie

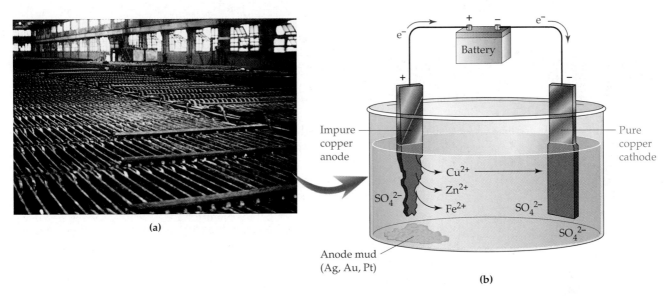

(a)

(b)

FIGURE 18.19 Electrorefining of copper metal. **(a)** Alternating slabs of impure copper and pure copper serve as the electrodes in electrolytic cells for the refining of copper. **(b)** Copper is transferred through the $CuSO_4$ solution from the impure Cu anode to the pure Cu cathode. More easily oxidized impurities (Zn, Fe) remain in solution as cations, but noble metal impurities (Ag, Au, Pt) are not oxidized and collect as anode mud.

At the impure Cu anode, copper is oxidized along with more easily oxidized metallic impurities such as zinc and iron. Less easily oxidized impurities such as silver, gold, and platinum fall to the bottom of the cell as *anode mud*, which is reprocessed to recover the precious metals. At the pure Cu cathode, Cu^{2+} ions are reduced to pure copper metal, but the less easily reduced metal ions (Zn^{2+}, Fe^{2+}, and so forth) remain in the solution.

Anode (oxidation): $M(s) \longrightarrow M^{2+}(aq) + 2\,e^-$ (M = Cu, Zn, Fe)

Cathode (reduction): $Cu^{2+}(aq) + 2\,e^- \longrightarrow Cu(s)$

Thus, the net cell reaction simply involves transfer of copper metal from the impure anode to the pure cathode. The process typically takes about 4 weeks, and the copper obtained is 99.95% pure.

Closely related to electrorefining is **electroplating**, the coating of one metal on the surface of another using electrolysis. For example, steel automobile bumpers are plated with chromium to protect them from corrosion, and silver-plating is commonly used in making items of fine table service. The object to be plated is carefully cleaned and then set up as the cathode of an electrolytic cell that contains a solution of ions of the metal to be deposited.

18.13 Quantitative Aspects of Electrolysis

Michael Faraday showed in the 1830s that the amount of substance produced at an electrode by electrolysis depends on the quantity of charge passed through the cell. For example, passage of 1 mol of electrons through a Downs cell yields 1 mol (23.0 g) of sodium at the cathode:

$$Na^+(l) + e^- \longrightarrow Na(l)$$

$$\text{1 mol} \qquad \text{1 mol} \qquad \text{1 mol (23.0 g)}$$

Similarly, passage of 1 mol of electrons in the Hall–Heroult process produces 1/3 mol (9.0 g) of aluminum, because 3 mol of electrons are required to reduce 1 mol of Al^{3+} to aluminum metal:

$$Al^{3+}(l) + 3\,e^- \longrightarrow Al(l)$$

$$\text{1/3 mol} \qquad \text{1 mol} \qquad \text{1/3 mol (9.0 g)}$$

In general, the amount of product formed in an electrode reaction follows directly from the stoichiometry of the reaction and the molar mass of the product.

To find out how many moles of electrons pass through a cell in a particular experiment, we need to measure the electric current and the time that the current flows. The number of coulombs of charge passed through the cell is equal to the product of the current in amperes (coulombs per second) and the time in seconds:

$$\text{Charge (C)} = \text{Current (A)} \times \text{Time (s)}$$

Because the charge on 1 mol of electrons is 96,500 C (Section 18.3), the number of moles of electrons passed through the cell is

$$\text{Moles of e}^- = \text{Charge (C)} \times \frac{1 \text{ mol e}^-}{96,500 \text{ C}}$$

The sequence of conversions in Figure 18.20 is used to calculate the mass or volume of product produced by passing a known current through a cell for a fixed period of time. The key is to think of the electrons as a "reactant" in a balanced chemical equation and then to proceed as with any other stoichiometry problem. Example 18.9 illustrates the calculations. Alternatively, we can calculate the current (or time) required to produce a given amount of product by working through the sequence in Figure 18.20 in the reverse direction, as shown in Example 18.10.

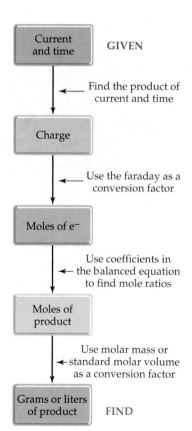

FIGURE 18.20 Sequence of conversions used to calculate the amount of product produced by passing a current through an electrolytic cell for a fixed period of time.

EXAMPLE 18.9 A constant current of 30.0 A is passed through an aqueous solution of NaCl for a time of 1.00 h. How many grams of NaOH and how many liters of Cl_2 gas at STP are produced?

SOLUTION   Because electrons can be thought of as a reactant in the electrolysis process, the first step is to calculate the charge and the number of moles of electrons passed through the cell:

$$\text{Charge} = \left(30.0 \,\frac{\text{C}}{\text{s}}\right)(1.00 \text{ h})\left(\frac{60 \text{ min}}{\text{h}}\right)\left(\frac{60 \text{ s}}{\text{min}}\right) = 1.08 \times 10^5 \text{ C}$$

$$\text{Moles of } e^- = (1.08 \times 10^5 \text{ C})\left(\frac{1 \text{ mol } e^-}{96{,}500 \text{ C}}\right) = 1.12 \text{ mol } e^-$$

The cathode reaction yields 2 mol of OH^- per 2 mol of electrons (Section 18.11), so 1.12 mol of NaOH will be obtained:

$$2 \text{ H}_2\text{O}(l) + 2 \text{ e}^- \longrightarrow \text{H}_2(g) + 2 \text{ OH}^-(aq)$$

$$\text{Moles of NaOH} = (1.12 \text{ mol } e^-)\left(\frac{2 \text{ mol NaOH}}{2 \text{ mol } e^-}\right) = 1.12 \text{ mol NaOH}$$

Converting the number of moles of NaOH to grams of NaOH gives

$$\text{Grams of NaOH} = (1.12 \text{ mol NaOH})\left(\frac{40.0 \text{ g NaOH}}{\text{mol NaOH}}\right) = 44.8 \text{ g NaOH}$$

The anode reaction gives 1 mol of Cl_2 per 2 mol of electrons, so 0.560 mol of Cl_2 will be obtained:

$$2 \text{ Cl}^-(aq) \longrightarrow \text{Cl}_2(g) + 2 \text{ e}^-$$

$$\text{Moles of Cl}_2 = (1.12 \text{ mol } e^-)\left(\frac{1 \text{ mol Cl}_2}{2 \text{ mol } e^-}\right) = 0.560 \text{ mol Cl}_2$$

Since 1 mol of an ideal gas occupies 22.4 L at STP, the volume of Cl_2 obtained is

$$\text{Liters of Cl}_2 = (0.560 \text{ mol Cl}_2)\left(\frac{22.4 \text{ L Cl}_2}{\text{mol Cl}_2}\right) = 12.5 \text{ L Cl}_2$$

As a shortcut, the entire sequence of conversions can be carried out in one step. For example, the volume of Cl_2 produced at the anode is

$$\left(30.0 \,\frac{\text{C}}{\text{s}}\right)(1.00 \text{ h})\left(\frac{3600 \text{ s}}{\text{h}}\right)\left(\frac{1 \text{ mol } e^-}{96{,}500 \text{ C}}\right)\left(\frac{1 \text{ mol Cl}_2}{2 \text{ mol } e^-}\right)\left(\frac{22.4 \text{ L Cl}_2}{\text{mol Cl}_2}\right) = 12.5 \text{ L Cl}_2$$

EXAMPLE 18.10   How many amperes must be passed through a Downs cell to produce sodium metal at a rate of 30.0 kg/h?

SOLUTION   Proceed through a sequence of conversions similar to that in Example 18.9, but in reverse order. Because the molar mass of sodium is 23.0 g/mol, the number of moles of sodium produced per hour is

$$\text{Moles of Na} = (30.0 \text{ kg Na})\left(\frac{1000 \text{ g}}{1 \text{ kg}}\right)\left(\frac{1 \text{ mol Na}}{23.0 \text{ g Na}}\right) = 1.30 \times 10^3 \text{ mol Na}$$

To produce each mole of sodium, 1 mol of electrons must be passed through the cell:

$$\text{Na}^+(l) + \text{e}^- \longrightarrow \text{Na}(l)$$

Therefore, the charge passed per hour is

$$\text{Charge} = (1.30 \times 10^3 \text{ mol Na})\left(\frac{1 \text{ mol e}^-}{1 \text{ mol Na}}\right)\left(\frac{96,500 \text{ C}}{\text{mol e}^-}\right) = 1.25 \times 10^8 \text{ C}$$

Since there are 3600 s in 1 h, the current required is

$$\text{Current} = \frac{1.25 \times 10^8 \text{ C}}{3600 \text{ s}} = 3.47 \times 10^4 \text{ C/s} = 34,700 \text{ A}$$

▶ **PROBLEM 18.18**   How many kilograms of aluminum can be produced in 8.00 h by passing a constant current of 1.00×10^5 A through a molten mixture of aluminum oxide and cryolite?

▶ **PROBLEM 18.19**   A layer of silver is electroplated on a coffee server using a constant current of 0.100 A. How much time is required to deposit 3.00 g of silver?

Electrochemical Art

I f aluminum is a silvery metal, then why are some aluminum objects brightly colored? Aluminum bicycle parts, for instance, come in a spectrum of colors including red, blue, purple, yellow, and black.

Aluminum, chromium, titanium, and several other metals can be colored by an electrochemical process called *anodizing*. Unlike electroplating, in which a metal ion in the electrolyte is *reduced* and the metal is coated onto the surface of the cathode, anodizing involves *oxidation* of a metal anode to yield a metal oxide coat. In the oxidation of aluminum, for instance, the electrode reactions are

Cathode (reduction): $\quad 6\,H^+(aq) + 6\,e^- \longrightarrow 3\,H_2(g)$

Anode (oxidation): $\quad 2\,Al(s) + 3\,H_2O(l) \longrightarrow Al_2O_3(s) + 6\,H^+(aq) + 6\,e^-$

Overall cell reaction: $\quad 2\,Al(s) + 3\,H_2O(l) \longrightarrow Al_2O_3(s) + 3\,H_2(g)$

The thickness of the aluminum oxide coating that forms on the anode can be controlled by varying the current flow during electrolysis. Typically, the coating is about 0.01 mm thick, which corresponds to about 4×10^4 atomic layers of Al_2O_3 on top of the underlying aluminum metal. The porosity of the Al_2O_3 coating can also be controlled by varying the electrolysis conditions, and it's this porosity that makes coloring possible. When an organic dye is added to the electrolyte, dye molecules soak into the spongy surface coating as it forms and become trapped as the surface hardens.

Titanium anodizing proceeds much like that of aluminum, but the resultant coat of TiO_2 is much thinner (10^{-4} mm) than the corresponding coat of Al_2O_3 (10^{-2} mm). Furthermore, the iridescent colors of anodized titanium result not from the absorption of organic dyes but from interference of light as it is reflected by the anodized surface. When a beam of white light strikes the anodized surface, part of the light is reflected from the outer TiO_2, while part penetrates through the semitransparent TiO_2 and is reflected from the inner metal. If the two reflections of a particular wavelength are out of phase, they interfere destructively and that wavelength is canceled from the reflected light. As a result, the light that remains is colored. (Similar effects are responsible for the colors in oil slicks and peacock feathers.)

The exact color of anodized titanium depends on the thickness of the TiO_2 layer, which in turn varies with the voltage used to produce the layer. A voltage of 5 V, for example, gives a TiO_2 coating 3×10^{-5} mm thick with a yellow appearance, and a voltage of 30 V gives a coating 6×10^{-5} mm thick with a light blue appearance. Artists use these effects to produce striking metal sculptures whose vibrant colors appear to change depending on the viewing angle.

These colored containers are made by anodizing aluminum, a silvery metal.

The iridescent colors on this piece of metal are produced by anodizing different parts of the surface at different voltages.

▶ **PROBLEM 18.20** The reflection of light from anodized titanium is somewhat similar to the diffraction of X rays from a crystal (Section 10.7). Explain why the color of anodized titanium depends on the thickness of the TiO_2 layer.

▶ **PROBLEM 18.21** How many minutes are required to produce a 0.0100 mm thick coating of Al_2O_3 (density 3.97 g/cm³) on a square piece of aluminum metal 10.0 cm on an edge if the current passed through the piece is 0.600 A?

Key Words

Summary

Electrochemistry is the area of chemistry concerned with the interconversion of chemical and electrical energy. Chemical energy is converted to electrical energy in a **galvanic cell**, a device in which a spontaneous redox reaction is used to produce an electric current. Electrical energy is converted to chemical energy in an **electrolytic cell**, a cell in which an electric current drives a nonspontaneous reaction. It's convenient to separate cell reactions into **half-reactions** because oxidation and reduction occur at separate **electrodes**. The electrode at which oxidation occurs is called the **anode**, and the electrode at which reduction occurs is called the **cathode**.

The **cell potential** *E* (also called the **cell voltage** or **electromotive force**) is an electrical measure of the driving force of the cell reaction. Cell potentials depend on temperature, ion concentrations, and gas pressures. The **standard cell potential** *E*° is the cell potential when reactants and products are in their standard states. Cell potentials are related to free-energy changes by the equations $\Delta G = -nFE$ and $\Delta G° = -nFE°$, where $F = 96,500 \text{ C/mol e}^-$ is the **faraday**, the charge on 1 mol of electrons.

The **standard reduction potential** for the a half-reaction is defined relative to an arbitrary value of 0 V for the **standard hydrogen electrode (S.H.E.)**:

$$2 \, H^+(aq, 1 \, M) + 2 \, e^- \longrightarrow H_2(g, 1 \, atm) \qquad E° = 0 \, V$$

Tables of standard reduction potentials—also called **standard electrode potentials**—are used to arrange oxidizing and reducing agents in order of increasing strength, to calculate *E*° values for cell reactions, and to decide whether a particular redox reaction is spontaneous.

Cell potentials under nonstandard-state conditions can be calculated using the **Nernst equation**,

$$E = E° - \frac{0.0592 \, V}{n} \log Q \qquad \text{In volts, at 25°C}$$

where *Q* is the reaction quotient. The equilibrium constant *K* and the standard cell potential *E*° are related by the equation

$$E° = \frac{0.0592 \, V}{n} \log K \qquad \text{In volts, at 25°C}$$

A battery consists of one or more galvanic cells. A **fuel cell** differs from a battery in that the reactants are continuously supplied to the cell.

Corrosion of iron (rusting) is an electrochemical process in which iron is oxidized in an anode region of the metal surface and oxygen is reduced in a cathode region. Corrosion can be prevented by covering iron with another metal, such as zinc in the process called **galvanizing**, or simply by putting the iron in electrical contact with a second metal that is more easily oxidized, called **cathodic protection**.

Electrolysis, the process of using an electric current to bring about chemical change, is employed to produce sodium, chlorine, sodium hydroxide, and aluminum (**Hall–Heroult process**), and is used in **electrorefining** and **electroplating**.

The product obtained at an electrode depends on reduction potentials and **overvoltage**. The amount of product obtained is related to the number of moles of electrons passed through the cell, which depends on the current and the time that the current flows.

Key Concept Summary

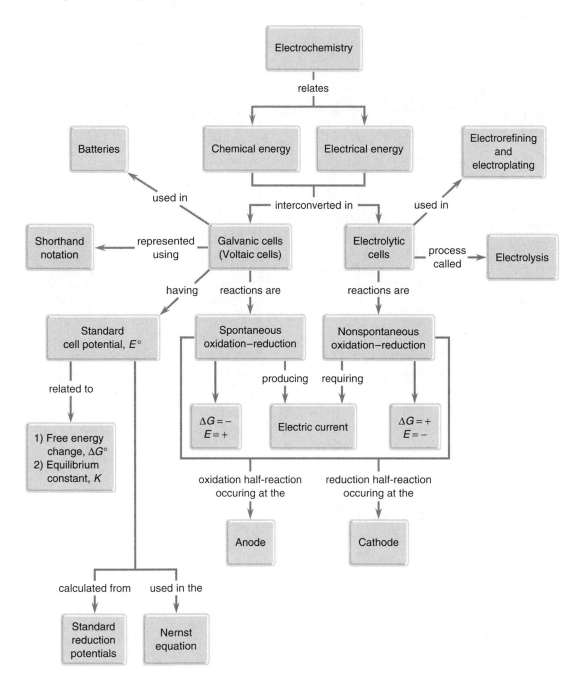

Understanding Key Concepts

Problems 18.1–18.21 appear within the chapter.

18.22 The following picture of a galvanic cell has lead and zinc electrodes:

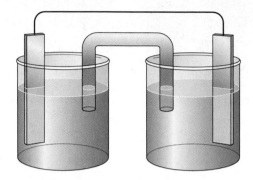

(a) Label the electrodes, and identify the ions present in the solutions.

(b) Label the anode and cathode.

(c) Indicate the direction of electron flow in the wire and ion flow in the solutions.

(d) Tell what electrolyte could be used in the salt bridge, and indicate the direction of ion flow.

(e) Write balanced equations for the electrode and overall cell reactions.

18.23 Consider the following galvanic cell:

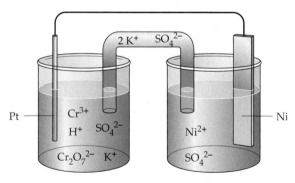

(a) Identify the anode and cathode.

(b) Write a balanced equation for the cell reaction.

(c) Write the shorthand notation for the cell.

18.24 Consider the following galvanic cells:

(1) $Cu(s)|Cu^{2+}(1\ M)\|Fe^{3+}(1\ M), Fe^{2+}(1\ M)|Pt(s)$

(2) $Cu(s)|Cu^{2+}(1\ M)\|Fe^{3+}(1\ M), Fe^{2+}(5\ M)|Pt(s)$

(3) $Cu(s)|Cu^{2+}(0.1\ M)\|Fe^{3+}(0.1\ M), Fe^{2+}(0.1\ M)|Pt(s)$

(a) Write a balanced equation for each cell reaction.

(b) Sketch each cell. Label the anode and cathode, and indicate the direction of electron and ion flow.

(c) Which of the three cells has the largest cell potential? Which has the smallest cell potential? Explain.

18.25 Sketch a cell with inert electrodes suitable for electrolysis of aqueous $CuBr_2$.

(a) Label the anode and cathode.

(b) Indicate the direction of electron and ion flow.

(c) Write balanced equations for the anode, cathode, and overall cell reactions.

18.26 Consider the following electrochemical cell:

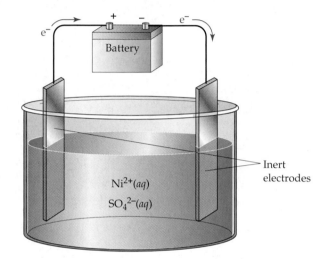

(a) Is the cell a galvanic or an electrolytic cell? Explain.

(b) Label the anode and cathode, and show the direction of ion flow.

(c) Write balanced equations for the anode, cathode, and overall cell reactions.

18.27 The following cell reactions occur spontaneously:

$$B + A^+ \longrightarrow B^+ + A$$
$$C + A^+ \longrightarrow C^+ + A$$
$$B + C^+ \longrightarrow B^+ + C$$

(a) Arrange the following reduction half-reactions in order of decreasing tendency to occur: $A^+ + e^- \rightarrow A$, $B^+ + e^- \rightarrow B$, and $C^+ + e^- \rightarrow C$.

(b) Which of these substances (A, A^+, B, B^+, C, C^+) is the strongest oxidizing agent? Which is the strongest reducing agent?

(c) Which of the three cell reactions delivers the highest voltage?

18.28 Consider the following substances: Fe(s), PbO$_2$(s), H$^+$(aq), Al(s), Ag(s), Cr$_2$O$_7^{2-}$(aq).

    (a) Look at the $E°$ values in Appendix D, and classify each substance as an oxidizing agent or reducing agent.

    (b) Which is the strongest oxidizing agent? Which is the weakest oxidizing agent?

    (c) Which is the strongest reducing agent? Which is the weakest reducing agent?

    (d) Which substances can be oxidized by Cu^{2+}(aq)? Which can be reduced by H$_2$O$_2$(aq)?

18.29 Consider a Daniell cell with 1.0 M ion concentrations:

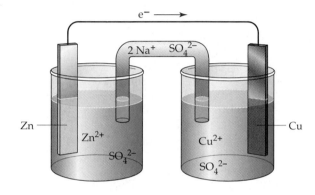

Does the cell voltage increase, decrease, or remain the same when each of the following changes is made? Explain.

    (a) 5.0 M CuSO$_4$ is added to the cathode compartment.

    (b) 5.0 M H$_2$SO$_4$ is added to the cathode compartment.

    (c) 5.0 M Zn(NO$_3$)$_2$ is added to the anode compartment.

    (d) 1.0 M Zn(NO$_3$)$_2$ is added to the anode compartment.

Additional Problems

Galvanic Cells

18.30 Define anode and cathode with reference to a specific galvanic cell.

18.31 Does the oxidizing agent react at the anode or at the cathode in a galvanic cell? Explain.

18.32 Why is the cathode of a galvanic cell considered to be the positive electrode?

18.33 What is the function of a salt bridge in a galvanic cell?

18.34 Describe galvanic cells that use the following reactions. In each case, write the anode and cathode half-reactions, and sketch the experimental setup. Label the anode and cathode, identify the sign of each electrode, and indicate the direction of electron and ion flow.

    (a) Cd(s) + Sn^{2+}(aq) → Cd^{2+}(aq) + Sn(s)

    (b) 2 Al(s) + 3 Cd^{2+}(aq) → 2 Al^{3+}(aq) + 3 Cd(s)

    (c) Cr$_2$O$_7^{2-}$(aq) + 6 Fe^{2+}(aq) + 14 H$^+$(aq) →
                            2 Cr^{3+}(aq) + 6 Fe^{3+}(aq) + 7 H$_2$O(l)

18.35 Describe galvanic cells that use the following reactions. In each case, write the anode and cathode half-reactions, and sketch the experimental setup. Label the anode and cathode, identify the sign of each electrode, and indicate the direction of electron and ion flow.

    (a) 3 Cu^{2+}(aq) + 2 Cr(s) → 3 Cu(s) + 2 Cr^{3+}(aq)

    (b) Pb(s) + 2 H$^+$(aq) → Pb^{2+}(aq) + H$_2$(g)

    (c) Cl$_2$(g) + Sn^{2+}(aq) → Sn^{4+}(aq) + 2 Cl$^-$(aq)

18.36 Write the standard shorthand notation for each cell in Problem 18.34.

18.37 Write the standard shorthand notation for each cell in Problem 18.35.

18.38 An H$_2$/H$^+$ half-cell (anode) and an Ag$^+$/Ag half-cell (cathode) are connected by a wire and a salt bridge.

    (a) Sketch the cell, indicating the direction of electron and ion flow.

    (b) Write balanced equations for the electrode and overall cell reactions.

    (c) Give the shorthand notation for the cell.

18.39 A galvanic cell is constructed from a Zn/Zn^{2+} half-cell (anode) and a Cl$_2$/Cl$^-$ half-cell (cathode).

    (a) Sketch the cell, indicating the direction of electron and ion flow.

    (b) Write balanced equations for the electrode and overall cell reactions.

    (c) Give the shorthand notation for the cell.

18.40 Write balanced equations for the electrode and overall cell reactions in the following galvanic cells. Sketch each cell, labeling the anode and cathode and showing the direction of electron and ion flow.

(a) $Co(s)\,|\,Co^{2+}(aq)\,\|\,Cu^{2+}(aq)\,|\,Cu(s)$

(b) $Fe(s)\,|\,Fe^{2+}(aq)\,\|\,O_2(g)\,|\,H^+(aq), H_2O(l)\,|\,Pt(s)$

18.41 Write balanced equations for the electrode and overall cell reactions in the following galvanic cells. Sketch each cell, labeling the anode and cathode and showing the direction of electron and ion flow.

(a) $Mn(s)\,|\,Mn^{2+}(aq)\,\|\,Pb^{2+}(aq)\,|\,Pb(s)$

(b) $Pt(s)\,|\,H_2(g)\,|\,H^+(aq)\,\|\,Cl^-(aq)\,|\,AgCl(s)\,|\,Ag(s)$

Cell Potentials and Free-Energy Changes; Standard Reduction Potentials

18.42 What are the SI units of electrical potential, electric charge, and energy? How are they related?

18.43 Define all terms in the equation $\Delta G = -nFE$.

18.44 What conditions must be met for a cell potential E to qualify as a standard cell potential $E°$?

18.45 How are standard reduction potentials defined?

18.46 The silver oxide–zinc battery used in hearing aids delivers a voltage of 1.60 V. Calculate the free-energy change (in kilojoules) for the cell reaction

$$Zn(s) + Ag_2O(s) \longrightarrow ZnO(s) + 2\,Ag(s)$$

18.47 The standard cell potential for a lead storage battery is 1.924 V. Calculate $\Delta G°$ (in kilojoules) for the cell reaction

$$Pb(s) + PbO_2(s) + 2\,H^+(aq) + 2\,HSO_4^-(aq) \longrightarrow$$
$$2\,PbSO_4(s) + 2\,H_2O(l)$$

18.48 Using the standard free energies of formation in Appendix B, calculate the standard cell potential for the reaction in the hydrogen–oxygen fuel cell:

$$2\,H_2(g) + O_2(g) \longrightarrow 2\,H_2O(l)$$

18.49 Consider a fuel cell that uses the reaction

$$CH_4(g) + 2\,O_2(g) \longrightarrow CO_2(g) + 2\,H_2O(l)$$

Given the standard free energies of formation in Appendix B, what is the value of $E°$ for the cell reaction?

18.50 The standard potential for the following galvanic cell is 0.40 V:

$$Zn(s)\,|\,Zn^{2+}(aq)\,\|\,Eu^{3+}(aq), Eu^{2+}(aq)\,|\,Pt(s)$$

(Europium, Eu, is one of the lanthanide elements.) Use the data in Table 18.1 to calculate the standard reduction potential for the Eu^{3+}/Eu^{2+} half-cell.

18.51 The following reaction has an $E°$ value of 0.27 V:

$$Cu^{2+}(aq) + 2\,Ag(s) + 2\,Br^-(aq) \longrightarrow$$
$$Cu(s) + 2\,AgBr(s)$$

Use the data in Table 18.1 to calculate the standard reduction potential for the half-reaction

$$AgBr(s) + e^- \longrightarrow Ag(s) + Br^-(aq)$$

18.52 Arrange the following oxidizing agents in order of increasing strength under standard-state conditions: $Br_2(l)$, $MnO_4^-(aq)$, $Sn^{4+}(aq)$.

18.53 List the following reducing agents in order of increasing strength under standard-state conditions: $Al(s)$, $Pb(s)$, $Fe(s)$.

18.54 Consider the following substances: $I_2(s)$, $Fe^{2+}(aq)$, $Cr_2O_7^{2-}(aq)$. Which is the strongest oxidizing agent? Which is the weakest oxidizing agent?

18.55 Consider the following substances: $Fe^{2+}(aq)$, $Sn^{2+}(aq)$, $I^-(aq)$. Identify the strongest reducing agent and the weakest reducing agent.

18.56 Calculate the standard cell potential and the standard free-energy change (in kilojoules) for each reaction in Problem 18.34. (See Appendix D for standard reduction potentials.)

18.57 Calculate $E°$ and $\Delta G°$ (in kilojoules) for the cell reactions in Problem 18.35. (See Appendix D for standard reduction potentials.)

18.58 Calculate $E°$ for each of the following reactions, and tell which are spontaneous under standard-state conditions:

(a) $2\,Fe^{2+}(aq) + Pb^{2+}(aq) \rightarrow 2\,Fe^{3+}(aq) + Pb(s)$

(b) $Mg(s) + Ni^{2+}(aq) \rightarrow Mg^{2+}(aq) + Ni(s)$

18.59 Calculate $E°$ for each of the following reactions, and tell which are spontaneous under standard-state conditions:

(a) $5\,Ag^+(aq) + Mn^{2+}(aq) + 4\,H_2O(l) \rightarrow$
$$5\,Ag(s) + MnO_4^-(aq) + 8\,H^+(aq)$$

(b) $2\,H_2O_2(aq) \rightarrow O_2(g) + 2\,H_2O(l)$

18.60 Use the data in Table 18.1 to predict whether the following reactions can occur under standard-state conditions:

(a) Oxidation of $Sn^{2+}(aq)$ by $Br_2(l)$

(b) Reduction of $Ni^{2+}(aq)$ by $Sn^{2+}(aq)$

(c) Oxidation of $Ag(s)$ by $Pb^{2+}(aq)$

(d) Reduction of $I_2(s)$ by $Cu(s)$

18.61 What reaction can occur, if any, when the following experiments are carried out under standard-state conditions?

(a) A strip of zinc is dipped into an aqueous solution of $Pb(NO_3)_2$.

(b) An acidic solution of $FeSO_4$ is exposed to oxygen.

(c) A silver wire is immersed in an aqueous solution of $NiCl_2$.

(d) Hydrogen gas is bubbled through aqueous $Cd(NO_3)_2$.

The Nernst Equation

18.62 Consider a galvanic cell that uses the reaction

$$2\,Ag^+(aq) + Sn(s) \longrightarrow 2\,Ag(s) + Sn^{2+}(aq)$$

Calculate the potential at 25°C for a cell that has the following ion concentrations: $[Ag^+] = 0.010$ M, $[Sn^{2+}] = 0.020$ M.

18.63 Consider a galvanic cell based on the reaction

$$2\,Fe^{2+}(aq) + Cl_2(g) \longrightarrow 2\,Fe^{3+}(aq) + 2\,Cl^-(aq)$$

Calculate the cell potential at 25°C when $[Fe^{2+}] = 1.0$ M, $[Fe^{3+}] = 1.0 \times 10^{-3}$ M, $[Cl^-] = 3.0 \times 10^{-3}$ M, and $P_{Cl_2} = 0.50$ atm.

18.64 What is the cell potential at 25°C for the following galvanic cell?

$$Pb(s)\,|\,Pb^{2+}(1.0\text{ M})\,\|\,Cu^{2+}(1.0 \times 10^{-4}\text{ M})\,|\,Cu(s)$$

If the Pb^{2+} concentration is maintained at 1.0 M, what is the Cu^{2+} concentration when the cell potential drops to zero?

18.65 A galvanic cell has an iron electrode in contact with 0.10 M $FeSO_4$ and a copper electrode in contact with a $CuSO_4$ solution. If the measured cell potential at 25°C is 0.67 V, what is the concentration of Cu^{2+} in the $CuSO_4$ solution?

18.66 The Nernst equation applies to both cell reactions and half-reactions. For the conditions specified, calculate the potential for the following half-reactions at 25°C:

(a) $I_2(s) + 2\,e^- \rightarrow 2\,I^-(aq)$; $[I^-] = 0.020$ M

(b) $Fe^{3+}(aq) + e^- \rightarrow Fe^{2+}(aq)$;
$$[Fe^{3+}] = [Fe^{2+}] = 0.10\text{ M}$$

(c) $Sn^{2+}(aq) \rightarrow Sn^{4+}(aq) + 2\,e^-$;
$$[Sn^{2+}] = 1.0 \times 10^{-3}\text{ M}, [Sn^{4+}] = 0.40\text{ M}$$

(d) $2\,Cr^{3+}(aq) + 7\,H_2O(l) \rightarrow$
$$Cr_2O_7^{2-}(aq) + 14\,H^+(aq) + 6\,e^-;$$
$$[Cr^{3+}] = [Cr_2O_7^{2-}] = 1.0\text{ M}, [H^+] = 0.010\text{ M}$$

18.67 What is the reduction potential at 25°C for the hydrogen electrode in each of the following solutions? The half-reaction is

$$2\,H^+(aq) + 2\,e^- \longrightarrow H_2(g, 1\text{ atm})$$

(a) 1.0 M HCl

(b) A solution having pH 4.00

(c) Pure water

(d) 1.0 M NaOH

18.68 The following cell has a potential of 0.27 V at 25°C:

$$Pt(s)\,|\,H_2(1\text{ atm})\,|\,H^+(?\text{ M})\,\|\,Ni^{2+}(1\text{ M})\,|\,Ni(s)$$

What is the pH of the solution in the anode compartment?

18.69 What is the pH of the solution in the cathode compartment of the following cell if the measured cell potential at 25°C is 0.58 V?

$$Zn(s)\,|\,Zn^{2+}(1\text{ M})\,\|\,H^+(?\text{ M})\,|\,H_2(1\text{ atm})\,|\,Pt(s)$$

Standard Cell Potentials and Equilibrium Constants

18.70 Beginning with the equations that relate $E°$, $\Delta G°$, and K, show that $\Delta G°$ is negative and $K > 1$ for a reaction that has a positive value of $E°$.

18.71 If a reaction has an equilibrium constant $K < 1$, is $E°$ positive or negative? What is the value of K when $E° = 0$ V?

18.72 Use the data in Table 18.1 to calculate the equilibrium constant at 25°C for the reaction

$$Ni(s) + 2\,Ag^+(aq) \longrightarrow Ni^{2+}(aq) + 2\,Ag(s)$$

18.73 From standard reduction potentials, calculate the equilibrium constant at 25°C for the reaction

$$2\,MnO_4^-(aq) + 10\,Cl^-(aq) + 16\,H^+(aq) \longrightarrow$$
$$2\,Mn^{2+}(aq) + 5\,Cl_2(g) + 8\,H_2O(l)$$

18.74 Calculate the equilibrium constant at 25°C for each reaction in Problem 18.34.

18.75 Calculate the equilibrium constant at 25°C for each reaction in Problem 18.35.

18.76 Calculate the equilibrium constant at 25°C for disproportionation of Hg_2^{2+}:

$$Hg_2^{2+}(aq) \longrightarrow Hg(l) + Hg^{2+}(aq)$$

See Appendix D for standard reduction potentials.

18.77 Use standard reduction potentials to calculate the equilibrium constant at 25°C for disproportionation of hydrogen peroxide:

$$2\,H_2O_2(l) \longrightarrow 2\,H_2O(l) + O_2(g)$$

Batteries; Corrosion

18.78 What is rust? What causes it to form? What can be done to prevent its formation?

18.79 The standard oxidation potential for the reaction $Cr(s) \rightarrow Cr^{3+}(aq) + 3 e^-$ is 0.74 V. Despite the large, positive oxidation potential, chromium is used as a protective coating on steel automobile bumpers. Why doesn't the chromium corrode?

18.80 What is meant by cathodic protection? Which of the following metals can offer cathodic protection to iron?

$$Zn, \quad Ni, \quad Al, \quad Sn$$

18.81 What is a sacrificial anode? Give an example.

18.82 For a lead storage battery:
   (a) Sketch one cell that shows the anode, cathode, electrolyte, direction of electron and ion flow, and sign of the electrodes.
   (b) Write the anode, cathode, and overall cell reactions.

(c) Calculate the equilibrium constant for the cell reaction ($E° = 1.924$ V).

(d) What is the cell voltage when the cell reaction reaches equilibrium?

18.83 Calculate the values of $E°$, $\Delta G°$ (in kilojoules), and K at 25°C for the cell reaction in a hydrogen–oxygen fuel cell: $2 H_2(g) + O_2(g) \rightarrow 2 H_2O(l)$. What is the cell voltage at 25°C if the partial pressure of each gas is 25 atm?

18.84 How many grams of HgO react at the cathode of a mercury battery when 2.00 g of zinc is consumed at the anode?

18.85 Write a balanced equation for the reaction that occurs when a nickel–cadmium battery is recharged. If 10.0 g of $Ni(OH)_2$ is oxidized in the charging process, how many grams of cadmium are formed?

Electrolysis

18.86 Magnesium metal is produced by electrolysis of molten magnesium chloride using inert electrodes.
   (a) Sketch the cell, label the anode and cathode, indicate the sign of the electrodes, and show the direction of electron and ion flow.
   (b) Write balanced equations for the anode, cathode, and overall cell reactions.

18.87 **(a)** Sketch a cell with inert electrodes suitable for electrolysis of an aqueous solution of sulfuric acid. Label the anode and cathode, and indicate the direction of electron and ion flow. Identify the positive and negative electrodes.
   (b) Write balanced equations for the anode, cathode, and overall cell reactions.

18.88 List the anode and cathode half-reactions that might occur when an aqueous solution of $MgCl_2$ is electrolyzed in a cell having inert electrodes. Predict which half-reactions will occur, and justify your answers.

18.89 What products should be formed when the following reactants are electrolyzed in a cell having inert electrodes? Account for any differences.
   (a) molten KCl   **(b)** aqueous KCl

18.90 Predict the anode, cathode, and overall cell reactions when an aqueous solution of each of the following salts is electrolyzed in a cell having inert electrodes:
   (a) NaBr   **(b)** $CuCl_2$ **(c)** LiOH

18.91 Predict the anode, cathode, and overall cell reactions when an aqueous solution of each of the following salts is electrolyzed in a cell having inert electrodes:
   (a) Ag_2SO_4 **(b)** $Ca(OH)_2$ **(c)** KI

18.92 How many grams of silver will be obtained when an aqueous silver nitrate solution is electrolyzed for 20.0 min with a constant current of 2.40 A?

18.93 A constant current of 100.0 A is passed through an electrolytic cell having an impure copper anode, a pure copper cathode, and an aqueous $CuSO_4$ electrolyte. How many kilograms of copper are refined by transfer from the anode to the cathode in a 24.0 h period?

18.94 How many hours are required to produce 1.00×10^3 kg of sodium by electrolysis of molten NaCl with a constant current of 3.00×10^4 A? How many liters of Cl_2 at STP will be obtained as a by-product?

18.95 What constant current (in amperes) is required to produce aluminum by the Hall–Heroult process at a rate of 40.0 kg/h?

18.96 How many grams of $PbSO_4$ are reduced at the cathode if you charge a lead storage battery for 1.50 h with a constant current of 10.0 A?

18.97 A layer of chromium is electroplated on an automobile bumper by passing a constant current of 200.0 A through a cell that contains $Cr^{3+}(aq)$. How many minutes are required to deposit 125 g of chromium?

General Problems

18.98 Consider a galvanic cell that uses the following half-reactions:

$$MnO_4^-(aq) + 8\,H^+(aq) + 5\,e^- \longrightarrow$$
$$Mn^{2+}(aq) + 4\,H_2O(l)$$
$$Sn^{4+}(aq) + 2\,e^- \longrightarrow Sn^{2+}(aq)$$

(a) Write a balanced equation for the overall cell reaction.

(b) What is the oxidizing agent, and what is the reducing agent?

(c) Calculate the standard cell potential.

18.99 Given the following half-reactions and $E°$ values,

$$Mn^{3+}(aq) + e^- \longrightarrow Mn^{2+}(aq) \quad E° = 1.54\,V$$
$$MnO_2(s) + 4\,H^+(aq) + e^- \longrightarrow Mn^{3+}(aq) + 2\,H_2O(l)$$
$$E° = 0.95\,V$$

write a balanced equation for disproportionation of $Mn^{3+}(aq)$, and calculate the value of $E°$ for this reaction. Is the reaction spontaneous under standard-state conditions?

18.100 Consider the following half-reactions and $E°$ values:

$$Ag^+(aq) + e^- \longrightarrow Ag(s) \qquad E° = 0.80\,V$$
$$Cu^{2+}(aq) + 2\,e^- \longrightarrow Cu(s) \qquad E° = 0.34\,V$$
$$Pb^{2+}(aq) + 2\,e^- \longrightarrow Pb(s) \qquad E° = -0.13\,V$$

(a) Which of these metals or ions is the strongest oxidizing agent? Which is the strongest reducing agent?

(b) The half-reactions can be used to construct three different galvanic cells. Tell which cell delivers the highest voltage, identify the anode and cathode, and tell the direction of electron and ion flow.

(c) Write the cell reaction for part (b), and calculate the values of $E°$, $\Delta G°$ (in kilojoules), and K for this reaction.

(d) Calculate the voltage for the cell in part (b) if both ion concentrations are 0.010 M.

18.101 *Standard* reduction potentials for the Pb^{2+}/Pb and Cd^{2+}/Cd half-reactions are -0.13 V and -0.40 V, respectively. At what relative concentrations of Pb^{2+} and Cd^{2+} will these half-reactions have the same reduction potential?

18.102 Consider a galvanic cell that uses the following half-reactions:

$$2\,H^+(aq) + 2\,e^- \longrightarrow H_2(g)$$
$$Al^{3+}(aq) + 3\,e^- \longrightarrow Al(s)$$

(a) What materials are used for the electrodes? Identify the anode and cathode, and indicate the direction of electron and ion flow.

(b) Write a balanced equation for the cell reaction, and calculate the standard cell potential.

(c) Calculate the cell potential at 25°C if the ion concentrations are 0.10 M and the partial pressure of H_2 is 10.0 atm.

(d) Calculate $\Delta G°$ (in kilojoules) and K for the cell reaction at 25°C.

(e) Calculate the mass change (in grams) of the aluminum electrode after the cell has supplied a constant current of 10.0 A for 25.0 min.

18.103 A Daniell cell delivers a constant current of 0.100 A for 200.0 h. How many grams of zinc are oxidized at the anode?

18.104 Approximately 13 million tons of Cl_2 are produced annually in the United States by electrolysis of brine. How many kilowatt-hours (kWh) of energy are required for the electrolysis if the cells operate at a potential of 4.5 V? (1 ton $= 2000$ lb; 1 kWh $= 3.6 \times 10^6$ J.)

18.105 The standard half-cell potential for reduction of quinone to hydroquinone is 0.699 V:

Quinone, $C_6H_4O_2$

Hydroquinone, $C_6H_4(OH)_2$

When a photographic film is developed, silver bromide is reduced by hydroquinone (the developer) in a basic aqueous solution to give quinone and tiny black particles of silver metal:

$$2\,AgBr(s) + C_6H_4(OH)_2(aq) + 2\,OH^-(aq) \longrightarrow$$
$$2\,Ag(s) + 2\,Br^-(aq) + C_6H_4O_2(aq) + 2\,H_2O(l)$$

(a) What is the value of $E°$ for this reaction under acidic conditions ($[H^+] = 1.0$ M)? Is the reaction spontaneous in 1.0 M acid?

(b) What is the value of $E°$ for the reaction under basic conditions ($[OH^-] = 1.0$ M)?

18.106 When suspected drunk drivers are tested with a Breathalyzer, the alcohol (ethanol) in the exhaled breath is oxidized to acetic acid with an acidic solution of potassium dichromate:

$$3\,CH_3CH_2OH(aq) + 2\,Cr_2O_7^{2-}(aq) + 16\,H^+(aq) \longrightarrow$$

Ethanol

$$3\,CH_3CO_2H(aq) + 4\,Cr^{3+}(aq) + 11\,H_2O(l)$$

Acetic acid

The color of the solution changes because some of the orange $Cr_2O_7^{2-}$ is converted to the green Cr^{3+}. The Breathalyzer measures the color change and produces a meter reading calibrated in terms of blood alcohol content.

(a) What is $E°$ for the reaction if the standard half-cell potential for reduction of acetic acid to ethanol is 0.058 V?

(b) What is the value of E for the reaction when the concentrations of ethanol, acetic acid, $Cr_2O_7^{2-}$, and Cr^{3+} are 1.0 M and the pH is 4.00?

18.107 Consider addition of the following half-reactions:

(1) $Fe^{3+}(aq) + 3\,e^- \longrightarrow Fe(s)$	$E°_1 = -0.04$ V	
(2) $Fe(s) \longrightarrow Fe^{2+}(aq) + 2\,e^-$	$E°_2 = 0.45$ V	
(3) $Fe^{3+}(aq) + e^- \longrightarrow Fe^{2+}(aq)$	$E°_3 = ?$	

Because half-reactions (1) and (2) contain a different number of electrons, the net reaction (3) is another half-reaction, and $E°_3$ can't be obtained simply by adding $E°_1$ and $E°_2$. The free-energy changes, however, are additive because G is a state function:

$$\Delta G°_3 = \Delta G°_1 + \Delta G°_2$$

(a) Starting with the relationship between $\Delta G°$ and $E°$, derive a general equation that relates the $E°$ values for half-reactions (1), (2), and (3).

(b) Calculate the value of $E°_3$ for the Fe^{3+}/Fe^{2+} half-reaction.

(c) Explain why the $E°$ values would be additive ($E°_3 = E°_1 + E°_2$) if reaction (3) were an overall cell reaction rather than a half-reaction.

18.108 The following galvanic cell has a potential of 0.578 V at 25°C:

$$Ag(s)\,|\,AgCl(s)\,|\,Cl^-(1.0\ M)\,\|\,Ag^+(1.0\ M)\,|\,Ag(s)$$

Use this information to calculate K_{sp} for AgCl at 25°C.

18.109 A galvanic cell has a silver electrode in contact with 0.050 M $AgNO_3$ and a copper electrode in contact with 1.0 M $Cu(NO_3)_2$.

(a) Write a balanced equation for the cell reaction, and calculate the cell potential at 25°C.

(b) Excess NaBr(aq) is added to the $AgNO_3$ solution to precipitate AgBr. What is the cell potential at 25°C after precipitation of AgBr if the concentra-

tion of excess Br^- is 1.0 M? Write a balanced equation for the cell reaction under these conditions. (K_{sp} for AgBr at 25°C is 5.4×10^{-13}.)

(c) Use the result in part (b) to calculate the standard reduction potential $E°$ for the half-reaction $AgBr(s) + e^- \rightarrow Ag(s) + Br^-(aq)$.

18.110 The following galvanic cell has a potential of 1.214 V at 25°C:

$$Hg(l)\,|\,Hg_2Br_2(s)\,|\,Br^-(0.10\ M)\,\|\,MnO_4^-(0.10\ M),$$
$$Mn^{2+}(0.10\ M),\,H^+(0.10\ M)\,|\,Pt(s)$$

Calculate the value of K_{sp} for Hg_2Br_2 at 25°C.

18.111 For the following half-reaction, $E° = 1.103$ V:

$$Cu^{2+}(aq) + 2\,CN^-(aq) + e^- \longrightarrow Cu(CN)_2^-(aq)$$

Calculate the formation constant K_f for $Cu(CN)_2^-$.

18.112 Accidentally chewing on a stray fragment of aluminum foil can cause a sharp tooth pain if the aluminum comes in contact with an amalgam filling. The filling, an alloy of silver, tin, and mercury, acts as the cathode of a tiny galvanic cell, the aluminum behaves as the anode, and saliva serves as the electrolyte. When the aluminum and the filling come in contact, an electric current passes from the aluminum to the filling, which is sensed by a nerve in the tooth. Aluminum is oxidized at the anode, and O_2 gas is reduced to water at the cathode.

(a) Write balanced equations for the anode, cathode, and overall cell reactions.

(b) Write the Nernst equation in a form that applies at body temperature (37°C).

(c) Calculate the cell voltage at 37°C. You may assume that $[Al^{3+}] = 1.0 \times 10^{-9}$ M, $P_{O_2} = 0.20$ atm, and that saliva has a pH of 7.0. Also assume that the $E°$ values in Appendix D apply at 37°C.

18.113 Copper reduces dilute nitric acid to nitric oxide (NO) but reduces concentrated nitric acid to nitrogen dioxide (NO_2):

(1) $3\,Cu(s) + 2\,NO_3^-(aq) + 8\,H^+(aq) \longrightarrow$	
$3\,Cu^{2+}(aq) + 2\,NO(g) + 4\,H_2O(l)$	$E° = 0.62$ V
(2) $Cu(s) + 2\,NO_3^-(aq) + 4\,H^+(aq) \longrightarrow$	
$Cu^{2+}(aq) + 2\,NO_2(g) + 2\,H_2O(l)$	$E° = 0.45$ V

Assuming that $[Cu^{2+}] = 0.10$ M and that the partial pressures of NO and NO_2 are 1.0×10^{-3} atm, calculate the potential (E) for reactions (1) and (2) at 25°C and show which reaction has the greater thermodynamic tendency to occur when the concentration of HNO_3 is

(a) 1.0 M

(b) 10.0 M

(c) At what HNO_3 concentration do reactions 1 and 2 have the same value of E?

Multi-Concept Problems

18.114 Adiponitrile, a key intermediate in the manufacture of nylon, is made industrially by an electrolytic process involving reduction of acrylonitrile:

Anode (oxidation): $2\,H_2O \longrightarrow O_2 + 4\,H^+ + 4\,e^-$

Cathode (reduction):

$2\,CH_2{=}CHCN + 2\,H^+ + 2\,e^- \longrightarrow NC(CH_2)_4CN$

Acrylonitrile   Adiponitrile

(a) Write a balanced equation for the overall cell reaction.

(b) How many kilograms of adiponitrile are produced in 10.0 h in a cell that has a constant current of 3.00×10^3 A?

(c) How many liters of O_2 at 740 mm Hg and 25°C are produced as a by-product?

18.115 The reaction of MnO_4^- with oxalic acid ($H_2C_2O_4$) in acidic solution is widely used to determine the concentration of permanganate solutions:

(a) Write a balanced net ionic equation for the reaction.

(b) Use the data in Appendix D to calculate $E°$ for the reaction.

(c) Show that the reaction goes to completion by calculating the values of $\Delta G°$ and K at 25°C.

(d) A 1.200 g sample of sodium oxalate ($Na_2C_2O_4$) is dissolved in dilute H_2SO_4 and then titrated with a $KMnO_4$ solution. If 32.50 mL of the $KMnO_4$ solution is required to reach the equivalence point, what is the molarity of the $KMnO_4$ solution?

18.116 Consider the redox titration of 120.0 mL of 0.100 M $FeSO_4$ with 0.120 M $K_2Cr_2O_7$ at 25°C, assuming that the pH of the solution is maintained at 2.00 with a suitable buffer. The solution is in contact with a platinum electrode and constitutes one half-cell of an electrochemical cell. The other half-cell is a standard hydrogen electrode. The two half-cells are connected with a wire and a salt bridge, and the progress of the titration is monitored by measuring the cell potential with a voltmeter.

(a) Write a balanced net ionic equation for the titration reaction.

(b) What is the cell potential at the equivalence point?

18.117 Consider the reaction that occurs in the hydrogen–oxygen fuel cell:

$$2\,H_2(g) + O_2(g) \longrightarrow 2\,H_2O(l)$$

(a) Use the thermodynamic data in Appendix B to calculate the values of $\Delta G°$ and $E°$ at 95°C, assuming that $\Delta H°$ and $\Delta S°$ are independent of temperature.

(b) Calculate the cell voltage at 95°C when the partial pressures of H_2 and O_2 are 25 atm.

18.118 Consider a galvanic cell that utilizes the following half-reactions:

Anode: $Zn(s) + H_2O(l) \longrightarrow$
$$ZnO(s) + 2\,H^+(aq) + 2\,e^-$$

Cathode: $Ag^+(aq) + e^- \longrightarrow Ag(s)$

(a) Write a balanced equation for the cell reaction, and use the thermodynamic data in Appendix B to calculate the values of $\Delta H°$, $\Delta S°$, and $\Delta G°$ for the reaction.

(b) What are the values of $E°$ and the equilibrium constant K for the cell reaction?

(c) What happens to the cell voltage if aqueous ammonia is added to the cathode compartment? Calculate the cell voltage assuming that the solution in the cathode compartment was prepared by mixing 50.0 mL of 0.100 M $AgNO_3$ and 50.0 mL of 4.00 M NH_3.

(d) Will AgCl precipitate if 10.0 mL of 0.200 M NaCl is added to the solution in part (c)? Will AgBr precipitate if 10.0 mL of 0.200 M KBr is added to the resulting solution?

eMedia Problems

18.119 The **Oxidation–Reduction Reactions** movie (*eChapter 18.1*) and the **Galvanic Cells I** movie (*eChapter 18.1*) both illustrate the same reaction, oxidation of zinc metal by copper(II) ions. Explain why this reaction as it is shown in the **Oxidation–Reduction Reactions** movie cannot be used to generate a voltage.

18.120 The **Redox Chemistry of Iron and Copper** movie (*eChapter 18.3*) shows an iron nail in a solution of copper sulfate.

(a) Write the half-reaction for the formation of the copper metal on the surface of the nail.

(b) What species is being oxidized? Write the half-reaction for the oxidation.

(c) Write the equation for the oxidation–reduction reaction taking place in the movie.

(d) What is the sign of $\Delta G°$ for the overall process?

18.121 Chromium metal can be electroplated onto a copper electrode by electrolysis of a solution containing $H_2Cr_2O_7$, as is shown in the **Electroplating** movie (*eChapter 18.12*).

(a) Write the half-reaction for the reduction of dichromate ion to chromium metal.

(b) What mass of chromium metal could be plated out by application of a 5.0 A current for a period of 1.0 hour?

18.122 Calculate what mass of aluminum metal would be plated out from a 1.5 M solution of aluminum nitrate if a current of 45 A is applied for a period of 1.0 minute. Use the **Electrolysis** activity (*eChapter 18.13*) to verify your answer.

18.123 For a given current, which of the four metals in the **Electrolysis** activity (*eChapter 18.13*) will deposit the greatest number of grams per second? Which will deposit the greatest number of moles per second? If your answer is different for these two questions, explain why.

The Main-Group Elements

Lightning is a major source of fixed nitrogen. It provides the energy needed to convert atmospheric N_2 to nitrogen oxides.

The main-group elements are the 47 elements that occupy groups 1A–8A of the periodic table. They are subdivided into the *s*-block elements of groups 1A and 2A, with valence electron configuration ns^1 or ns^2, and the *p*-block elements of groups 3A–8A, with valence configurations $ns^2 np^{1-6}$. Main-group elements are important because of their high natural abundance and their presence in commercially valuable chemicals. Eight of the 10 most abundant elements in the earth's crust and all 10 of the most abundant elements in the human body are main-group elements (Figure 19.1). In addition, the 10 most widely used industrial chemicals contain only main-group elements (Table 19.1).

In Sections 6.7–6.11 we surveyed the alkali and alkaline earth metals (groups 1A and 2A), aluminum (group 3A), the halogens (group 7A), and the noble gases (group 8A). In Chapter 14 we looked in detail at the chemistry of

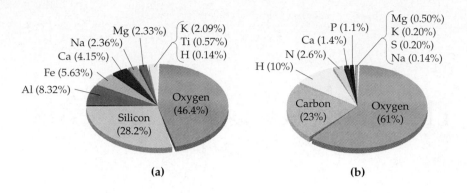

FIGURE 19.1 The 10 most abundant elements by mass **(a)** in the earth's crust and **(b)** in the human body. All are main-group elements except iron and titanium.

(a)   (b)

TABLE 19.1   The Top 10 Chemicals (1998 U.S. Production)

Chemical	Millions of Tons	Principal Uses
Sulfuric acid (H_2SO_4)	47.6	Fertilizers, chemicals, oil refining
Nitrogen (N_2)	32.9	Inert atmospheres, low temperatures
Oxygen (O_2)	31.1	Steelmaking, welding, medical uses
Ethylene ($CH_2{=}CH_2$)	26.0	Plastics, antifreeze
Lime (CaO)	22.5	Steelmaking, chemicals, water treatment
Ammonia (NH_3)	19.8	Fertilizers, nitric acid
Phosphoric acid (H_3PO_4)	14.4	Fertilizers, detergents
Propylene ($CH_3CH{=}CH_2$)	14.3	Plastics, fibers, solvents
Chlorine (Cl_2)	12.8	Chemicals, plastics, water treatment
Ethylene dichloride (CH_2ClCH_2Cl)	12.3	Plastics, solvent

hydrogen and oxygen. Now we'll examine the remaining main-group elements, paying special attention to boron (group 3A), carbon and silicon (group 4A), nitrogen and phosphorus (group 5A), and sulfur (group 6A).

19.1 A Review of General Properties and Periodic Trends

Interactive Periodic Table

Let's begin our survey of the main-group elements by reviewing the periodic trends that make it possible to classify these elements as metals, nonmetals, or semimetals. Figure 19.2 shows the main-group regions of the periodic table, with metals to the left of the heavy stair-step line, nonmetals to the right of the line, and semimetals—elements with intermediate properties—along the line. The elements usually classified as semimetals are boron (group 3A), silicon and germanium (group 4A), arsenic and antimony (group 5A), tellurium (group 6A), and astatine (group 7A).

From left to right across the periodic table, the effective nuclear charge Z_{eff} increases, because each additional valence electron does not completely shield the additional nuclear charge (Section 5.11). As a result, the atom's electrons are more strongly attracted to the nucleus, ionization energy generally increases, atomic radius decreases, and electronegativity increases. The elements on the left side of the table tend to form cations by losing electrons, and those on the right tend to form anions by gaining electrons. Thus, metallic character decreases and nonmetallic character increases across the table

Z_{eff}, ionization energy, electronegativity, and nonmetallic character increase →

Atomic radius and metallic character decrease →

(left vertical axis, top arrow) Ionization energy, electronegativity, and nonmetallic character decrease

(left vertical axis, bottom arrow) Atomic radius and metallic character increase

1 1A				13 3A	14 4A	15 5A	16 6A	17 7A	18 8A
H	2 2A								He
Li	Be			B	C	N	O	F	Ne
Na	Mg			Al	Si	P	S	Cl	Ar
K	Ca			Ga	Ge	As	Se	Br	Kr
Rb	Sr			In	Sn	Sb	Te	I	Xe
Cs	Ba			Tl	Pb	Bi	Po	At	Rn
Fr	Ra				114		116		118

FIGURE 19.2 Periodic trends in the properties of the main-group elements. The metallic elements (green) and the nonmetallic elements (lavender) are separated by the heavy stair-step line. The semimetals, shown in blue, lie along the line.

from left to right. In the third row, for example, sodium, magnesium, and aluminum are metals, silicon is a semimetal, and phosphorus, sulfur, and chlorine are nonmetals.

From the top to the bottom of a group in the periodic table, additional shells of electrons are occupied, and atomic radius therefore increases. Because the valence electrons are farther from the nucleus, though, ionization energy and electronegativity generally decrease. As a result, metallic character increases and nonmetallic character decreases down a group from top to bottom. In group 4A, for example, carbon is a nonmetal, silicon and germanium are semimetals, and tin and lead are metals. The horizontal and vertical periodic trends combine to locate the element with the most metallic character (francium) in the lower left of the periodic table, the element with the most nonmetallic character (fluorine) in the upper right, and the semimetals along the diagonal stair-step that stretches across the middle.

In earlier chapters, we saw examples of how the metallic or nonmetallic character of an element affects its chemistry. Metals tend to form ionic compounds with nonmetals, whereas nonmetals tend to form covalent, molecular compounds with one another. Thus, binary metallic hydrides, such as NaH and CaH_2, are ionic solids with high melting points, and binary nonmetallic hydrides, such as CH_4, NH_3, H_2O, and HF, are covalent, molecular compounds that exist at room temperature as gases or volatile liquids (Section 14.5).

Oxides exhibit similar trends. In the third row, for example, Na_2O and MgO are typical high-melting, ionic solids, and P_4O_{10}, SO_3, and Cl_2O_7 are volatile, covalent, molecular compounds (Section 14.9). Of course, the metallic or nonmetallic character of an oxide also affects its acid–base properties: Na_2O and MgO are basic, but P_4O_{10}, SO_3, and Cl_2O_7 are acidic. Table 19.2 summarizes some of the properties that distinguish metallic and nonmetallic elements.

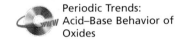

Periodic Trends: Acid–Base Behavior of Oxides

TABLE 19.2   Properties of Metallic and Nonmetallic Elements

Metals	Nonmetals
All are solids at 25°C except Hg, which is a liquid	Eleven are gases at 25°C, one is a liquid (Br), and five are solids (C, P, S, Se, and I)
Most have a silvery shine	Most lack a metallic luster
Malleable and ductile	Nonmalleable and brittle
Good conductors of heat and electricity	Poor conductors of heat and electricity, except graphite
Relatively low ionization energies	Relatively high ionization energies
Relatively low electronegativities	Relatively high electronegativities
Lose electrons to form cations	Gain electrons to form anions; share electrons to form oxoanions
Hydrides are ionic (or interstitial)	Hydrides are covalent and molecular
Oxides are ionic and basic	Oxides are covalent, molecular, and acidic

EXAMPLE 19.1   Use the periodic table to predict which element in each of the following pairs has more metallic character:
(a) Ga or As   **(b)** P or Bi   **(c)** Sb or S

SOLUTION
(a) Ga and As are in the same row of the periodic table, but Ga (group 3A) lies to the left of As (group 5A). Therefore, Ga is more metallic.
(b) Bi lies below P in group 5A and is therefore more metallic.
(c) Sb (group 5A) has more metallic character because it lies below and to the left of S (group 6A).

▶**PROBLEM 19.1**   Predict which element in each of the following pairs has more nonmetallic character:
(a) B or Al   **(b)** Ge or Br   **(c)** In or Se

19.2   Distinctive Properties of the Second-Row Elements

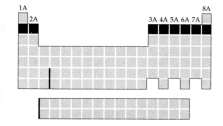

The properties of elements in the second row of the periodic table differ markedly from those of heavier elements in the same periodic group. The second-row atoms have especially small sizes and especially high electronegativities. In group 5A, for example, the electronegativity of N is 3.0, whereas the electronegativities of P, As, Sb, and Bi are all in the range 2.1–1.9. Figure 7.4 (page 249) shows graphically the discontinuity in electronegativity that distinguishes the second-row elements from other elements of the same periodic group.

The small sizes and high electronegativities of the second-row elements accentuate their nonmetallic behavior. Thus, BeO is amphoteric, but the oxides of the other group 2A elements are basic. Boron differs from the metallic elements of group 3A in forming mainly covalent, molecular compounds. For example, BF_3 (bp −100°C) is a gaseous, molecular halide, but AlF_3 (mp 1290°C) is a typical high-melting, ionic solid. Furthermore, hydrogen bonding interactions are generally restricted to compounds of the highly electronegative second-row elements N, O, and F (Section 10.2). Recall also

Interactive Periodic Table

that HF contrasts with HCl, HBr, and HI in being the only weak hydrohalic acid (Section 15.2).

Another factor that distinguishes the second-row elements from the heavier elements is their lack of valence d orbitals. Because the second-row elements have only four valence orbitals ($2s$, $2p_x$, $2p_y$, and $2p_z$), they form a maximum of four covalent bonds. By contrast, the third-row elements can use d orbitals to form more than four bonds. Thus, nitrogen forms only NCl_3, but phosphorus forms both PCl_3 and PCl_5:

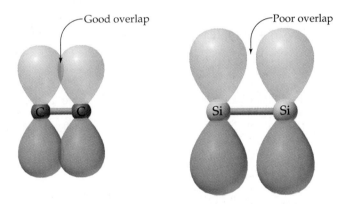

A further consequence of the small size of the second-row atoms C, N, and O is their ability to form multiple bonds involving π overlap of the $2p$ orbitals. By contrast, the $3p$ orbitals of the corresponding third-row atoms Si, P, and S are more diffuse, and the longer bond distances for these larger atoms result in poor π overlap.

As a result of this poor overlap, π bonds involving p orbitals are rare for elements of the third and higher rows. Although compounds with C=C double bonds are common, molecules with Si=Si double bonds are uncommon and have been synthesized only recently. In group 5A, elemental nitrogen contains triply bonded N_2 molecules, whereas white phosphorus contains tetrahedral P_4 molecules, in which each P atom forms three single bonds rather than one triple bond. Similarly, O_2 contains an O=O double bond, whereas elemental sulfur contains crown-shaped S_8 rings, in which each S atom forms two single bonds rather than one double bond.

EXAMPLE 19.2   Account for the following observations:

(a) CO_2 is a gaseous molecular substance, but SiO_2 is a covalent network solid in which SiO_4 tetrahedra are linked to four neighboring SiO_4 tetrahedra by shared oxygen atoms.

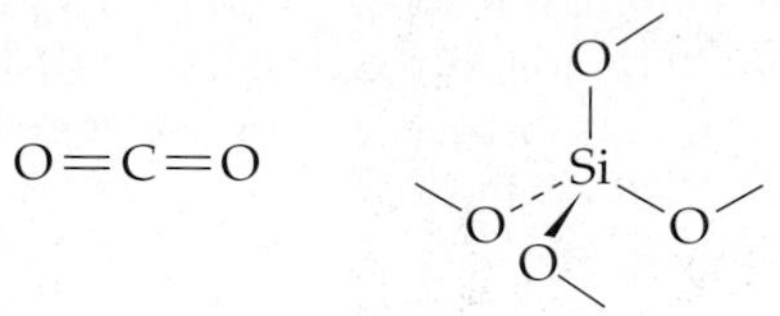

$$O=C=O$$

(b) Glass made of SiO_2 is attacked by hydrofluoric acid with formation of $SiF_6{}^{2-}$ anions. The analogous $CF_6{}^{2-}$ anion does not exist.

SOLUTION

(a) Because of its small size and good π overlap with other small atoms, carbon forms strong double bonds with two oxygens to give discrete CO_2 molecules. Because the larger Si atom does not have good π overlap with other atoms, it uses its four valence electrons to form four single bonds rather than two double bonds.

(b) Silicon has $3d$ orbitals and can use octahedral sp^3d^2 hybrid orbitals to bond to six F^- ions. With just $2s$ and $2p$ valence orbitals, carbon can form a maximum of only four bonds.

▶**PROBLEM 19.2**

    (a) Draw electron-dot structures for HNO_3 and H_3PO_4, and suggest a reason for the difference in the formulas of these acids.

    (b) Sulfur forms SF_6, but oxygen bonds to a maximum of two F atoms, yielding OF_2. Explain.

◀— **KEY CONCEPT PROBLEM 19.3**   The organic solvent acetone has the molecular formula $(CH_3)_2CO$. The silicon analogue, a thermally stable lubricant, is a polymer, $[(CH_3)_2SiO]_n$. Account for the difference in structure:

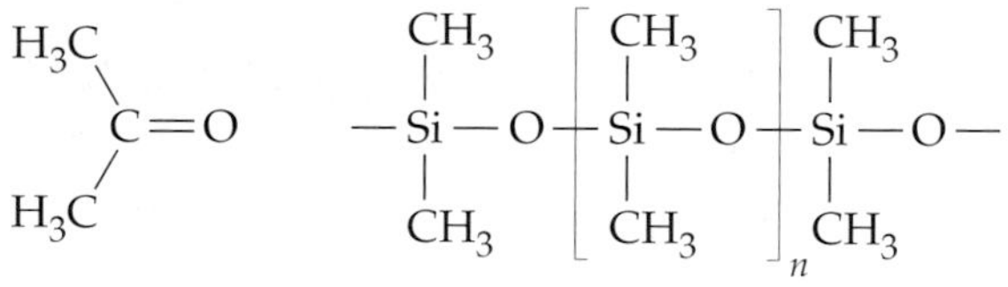

Acetone   Polydimethylsiloxane (silicone oil)

19.3   The Group 3A Elements

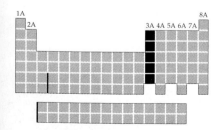

The group 3A elements are boron, aluminum, gallium, indium, and thallium. We discussed the most important of these, aluminum, in Section 6.9, and we'll take up boron in the next section. Gallium is remarkable for its unusually low melting point (29.7°C) and unusually large liquid range (29.7–2204°C). Its most important use is in making gallium arsenide (GaAs), a semiconductor material employed in the manufacture of diode lasers for laser printers, compact-disc players, and fiber-optic communication devices. (We'll say more about semiconductors in Section 21.5.) Indium is also used in making semiconductor devices, such as transistors and electrical resistance thermometers called *thermistors*. Thallium is extremely toxic and has no commercial uses.

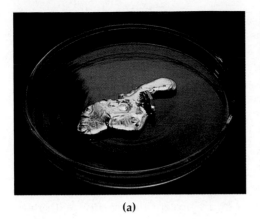
(a)

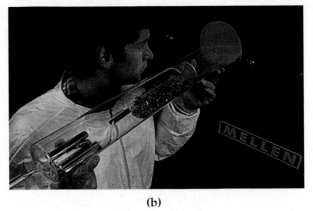
(b)

(a) A sample of gallium metal. **(b)** The semiconductor material gallium arsenide is prepared by heating gallium and arsenic in a furnace. The tube contains bars of gallium metal (left) and pieces of arsenic (right).

The valence electron configuration of the group 3A elements is $ns^2 np^1$, and their primary oxidation state is +3. In addition, the heavier elements exhibit a +1 state, which is uncommon for gallium and indium but is the most stable oxidation state for thallium.

Despite some irregularities, the properties of the group 3A elements are generally consistent with increasing metallic character down the group (Tables 6.6 and 19.3). All the group 3A elements are metals except boron. Boron has a much smaller atomic radius and a higher electronegativity than the other elements of the group; it therefore shares its valence electrons in covalent bonds rather than transferring them to another element. Accordingly, boron has nonmetallic properties.

TABLE 19.3   Properties of the Group 3A Elements

Property	Boron	Aluminum	Gallium	Indium	Thallium
Valence electron configuration	$2s^2 2p^1$	$3s^2 3p^1$	$4s^2 4p^1$	$5s^2 5p^1$	$6s^2 6p^1$
Common oxidation states	+3	+3	+3	+3	+3, +1
Atomic radius (pm)	83	143	135	167	170
M^{3+} ionic radius (pm)		51	62	81	95
First ionization energy (kJ/mol)	801	578	579	558	589
Electronegativity	2.0	1.5	1.6	1.7	1.8
Redox potential, $E°$ (V) for $M^{3+}(aq) + 3 e^- \rightarrow M(s)$	−0.87*	−1.66	−0.56	−0.34	−0.34†

*$E°$ for the reaction $B(OH)_3(aq) + 3 H^+(aq) + 3 e^- \rightarrow B(s) + 3 H_2O(l)$.
†$E°$ for the reaction $Tl^+(aq) + e^- \rightarrow Tl(s)$.

19.4 Boron

Boron is a relatively rare element, accounting for only about 0.001% of the earth's crust by mass. Nevertheless, boron is readily available because it occurs in concentrated deposits of borate minerals such as borax, $Na_2B_4O_7 \cdot 10 H_2O$.

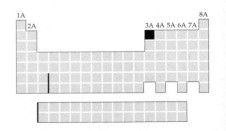

Elemental boron can be prepared by high-temperature reduction of B_2O_3 with magnesium, but the product is impure and amorphous:

$$B_2O_3(l) + 3\,Mg(s) \longrightarrow 2\,B(s) + 3\,MgO(s)$$

High-purity, crystalline boron is best obtained by reaction of boron tribromide and hydrogen on a heated tantalum filament at high temperatures:

$$2\,BBr_3(g) + 3\,H_2(g) \xrightarrow[1200°C]{Ta\ wire} 2\,B(s) + 6\,HBr(g)$$

Crystalline boron is a strong, hard, high-melting substance (mp 2075°C) that is chemically inert at room temperature, except for reaction with fluorine. These properties make boron fibers a desirable component in high-strength composite materials used in making sports equipment and military aircraft (see Section 21.8). Unlike Al, Ga, In, and Tl, which are metallic conductors, boron is a semiconductor.

The borate minerals in this open-pit borax mine near Boron, California, are believed to have been formed by evaporation of water from hot springs that were once present.

Boron Compounds

Boron Halides   The boron halides are highly reactive, volatile, covalent compounds that consist of trigonal planar BX_3 molecules. At room temperature, BF_3 and BCl_3 are gases, BBr_3 is a liquid, and BI_3 is a low-melting solid (mp 43°C). In their most important reactions, the boron halides behave as Lewis acids. For example, BF_3 reacts with ammonia to give the Lewis acid–base adduct F_3B-NH_3 (Section 15.16); it reacts with metal fluorides, yielding salts that contain the tetrahedral BF_4^- anion; and it acts as a catalyst in many industrially important organic reactions. In all these reactions, the boron atom uses its vacant $2p$ orbital in accepting a share in a pair of electrons from a Lewis base.

Boron Hydrides   The boron hydrides, or **boranes**, are volatile, molecular compounds with formulas B_nH_m. The simplest is diborane (B_2H_6), the dimer of the unstable BH_3. Diborane can be prepared by reaction of sodium borohydride, $NaBH_4$, and iodine in an appropriate organic solvent:

$$2\,NaBH_4 + I_2 \longrightarrow B_2H_6 + H_2 + 2\,NaI$$

Because boranes have high heats of combustion, they were once considered as potential lightweight, high-energy rocket fuels, but they proved to

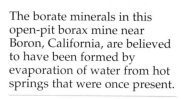

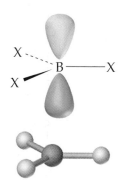

The boron halides are Lewis acids because they have a vacant $2p$ orbital.

Boron Halides activity; BF_3 3D model

be too expensive to prepare. Nevertheless, boranes continue to be of interest to chemists because of their unusual structures and bonding.

The diborane molecule has a structure in which two BH_2 groups are connected by two bridging H atoms. The geometry about the B atoms is roughly tetrahedral, and the bridging B—H bonds are significantly longer than the terminal B—H bonds, 133 pm versus 119 pm. The structure differs from that of ethane (C_2H_6) and is unusual because hydrogen normally forms only one bond.

Diborane

B_2H_6 3D model

Ethane

If each line in the structural formula of diborane represented an ordinary 2-electron covalent bond between 2 atoms (a *two-center, two-electron,* or 2c–2e bond), there would be 8 pairs, or 16 valence electrons. But diborane has a total of only 12 valence electrons—3 electrons from each boron and 1 from each hydrogen. Thus, diborane is said to be *electron-deficient*: It doesn't have enough electrons to form a 2c–2e bond between each pair of bonded atoms.

Because the geometry about the B atoms is roughly tetrahedral, we can assume that each boron uses sp^3 hybrid orbitals to bond to the four neighboring H atoms. The four terminal B—H bonds are assumed to be ordinary 2c–2e bonds, formed by overlap of a boron sp^3 hybrid orbital and a hydrogen $1s$ orbital, thereby using 4 of the 6 pairs of electrons. Each of the 2 remaining pairs of electrons forms a **three-center, two-electron bond** (3c–2e bond), which joins each bridging H atom to *both* B atoms. Each electron pair occupies a three-center molecular orbital formed by overlap of three atomic orbitals— one sp^3 hybrid orbital from each B atom and the $1s$ orbital on one bridging H atom (Figure 19.3). Because the 2 electrons in the B—H—B bridge are spread out over 3 atoms, the electron density between adjacent atoms is less than in an ordinary 2c–2e bond. The bridging B—H bonds are therefore weaker and correspondingly longer than the terminal B—H bonds.

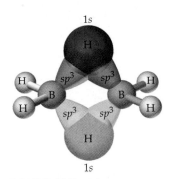

FIGURE 19.3 Three-center bonding molecular orbitals in diborane. Each of the two MOs (one shown darker than the other) is formed by overlap of an sp^3 hybrid orbital on each B atom and the $1s$ orbital on one bridging H atom. Each MO contains one pair of electrons and accounts for the bonding in one B—H—B bridge.

19.5 The Group 4A Elements

The group 4A elements—carbon, silicon, germanium, tin, and lead—are especially important, both in industry and in living organisms. Carbon is present in all plants and animals, accounts for 23% of the mass of the human body, and is an essential constituent of the molecules on which life is based. Silicon is equally important in the mineral world: It is present in numerous silicate minerals and is the second most abundant element in the earth's crust. Both silicon

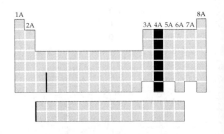

and germanium are used in making modern solid-state electronic devices. Tin and lead have been known and used since ancient times.

The group 4A elements further illustrate the increase in metallic character down a group in the periodic table: Carbon is a nonmetal; silicon and germanium are semimetals; and tin and lead are metals. The usual periodic trends in atomic size, ionization energy, and electronegativity are evident in the data of Table 19.4.

TABLE 19.4   Properties of the Group 4A Elements

Property	Carbon	Silicon	Germanium	Tin	Lead
Valence electron configuration	$2s^2\,2p^2$	$3s^2\,3p^2$	$4s^2\,4p^2$	$5s^2\,5p^2$	$6s^2\,6p^2$
Melting point (°C)	>3550*	1414	938	232†	327
Boiling point (°C)		3265	2833	2602	1749
Density (g/cm^3)	3.51*	2.33	5.32	7.26†	11.3
Abundance in earth's crust (mass %)	0.020	28.2	0.0005	0.0002	0.0013
Common oxidation states	+2, +4	+4	+4	+2, +4	+2, +4
Atomic radius (pm)	77	117	122	140	175
First ionization energy (kJ/mol)	1086	786	762	709	716
Electronegativity	2.5	1.8	1.8	1.8	1.9
Redox potential, $E°$ (V) for $M^{2+}(aq) + 2\,e^- \rightarrow M(s)$				−0.14	−0.13

*Diamond. †White Sn.

Because the group 4A elements have the valence electron configuration $ns^2\,np^2$, their most common oxidation state is +4, as in CCl_4, $SiCl_4$, $GeCl_4$, $SnCl_4$, and $PbCl_4$. These compounds are volatile, molecular liquids in which the group 4A atom uses tetrahedral sp^3 hybrid orbitals to form covalent bonds to the Cl atoms. The +2 oxidation state occurs for tin and lead and is the most stable oxidation state for lead. Both $Sn^{2+}(aq)$ and $Pb^{2+}(aq)$ are common solution species, but there are no simple $M^{4+}(aq)$ ions for any of the group 4A elements. Instead, M(IV) species exist in solution as covalently bonded complex ions—for example, $SiF_6{}^{2-}$, $GeCl_6{}^{2-}$, $Sn(OH)_6{}^{2-}$, and $Pb(OH)_6{}^{2-}$. In general, the +4 oxidation-state compounds are covalent, and the compounds with tin and lead in the +2 oxidation state are largely ionic.

19.6   Carbon

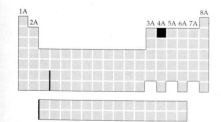

Carbon, although the second most abundant element in living organisms, accounts for only 0.02% of the mass of the earth's crust. It is present in carbonate minerals, such as limestone ($CaCO_3$), and in fossil fuels, such as coal, petroleum, and natural gas. In uncombined form, carbon is found as diamond and graphite.

Recall from Section 10.10 that diamond has a covalent network structure in which each C atom uses sp^3 hybrid orbitals to form a tetrahedral array of σ bonds, with bond lengths of 154 pm. (See Figure 10.26b, page 416.) The interlocking, three-dimensional network of strong bonds makes diamond the hardest known substance and gives it the highest melting point for an element (>3550°C). Because the valence electrons are localized in the σ bonds, they are not free to move when an electrical potential is applied and diamond is therefore an electrical insulator.

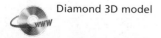

Diamond 3D model

Graphite has a two-dimensional sheetlike structure (Figure 10.26a, page 416) in which each C atom uses sp^2 hybrid orbitals to form trigonal planar σ bonds to three neighboring C atoms. In addition, each C atom uses its remaining p orbital, which is perpendicular to the plane of the sheet, to form a π bond. Because each C atom must share its π bond with its three neighbors, the π electrons are delocalized and are free to move in the plane of the sheet. As a result, the electrical conductivity of graphite in a direction parallel to the sheets is about 10^{20} times greater than the conductivity of diamond, which makes graphite useful as an electrode material.

Graphite 3D model

The carbon sheets in graphite are separated by a distance of 335 pm and are held together by only London dispersion forces. Atmospheric gases can be absorbed between the sheets, thus enabling the sheets to easily slide over one another. As a result, graphite has a slippery feel and can be used as a lubricant. Because the sheets are so far apart, it's relatively difficult for an electron to hop from one sheet to the next, and the electrical conductivity in the direction perpendicular to the sheets is therefore about 10^4 times smaller than the conductivity parallel to the sheets.

A third allotrope of carbon, called *fullerene*, was discovered in 1985 by Robert F. Curl Jr., Harold Kroto, and Richard E. Smalley, who were awarded the 1996 Nobel Prize in chemistry for their pioneering work. It consists of nearly spherical C_{60} molecules with the shape of a soccer ball (Figure 10.27, page 416). Fullerene can be prepared in relatively large amounts by electrically heating a graphite rod in an atmosphere of helium. Unlike diamond and graphite, C_{60} is a molecular substance, and it is therefore soluble in nonpolar organic solvents. Numerous derivatives of fullerene have been prepared, including compounds such as $C_{60}F_{36}$, in which other atoms are attached to the outside of the C_{60} cage, and compounds such as La@C_{60}, in which a metal atom is trapped within the cage. (The symbol @ means that the lanthanum atom is located inside the C_{60} cage.) Closely related to C_{60} are other carbon clusters, such as the egg-shaped molecule C_{70} and the recently discovered tube-shaped molecules called *carbon nanotubes*.

Fullerene 3D model

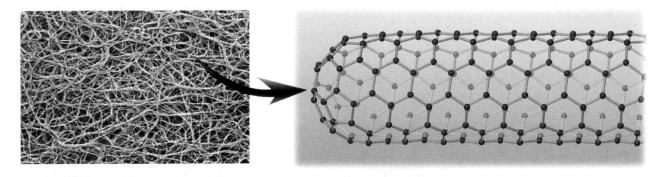

A carbon nanotube has walls made of graphite sheets rolled into a cylinder and hemispherical tips made of a portion of a fullerene molecule.

Carbon also exists in more than 40 amorphous (noncrystalline) forms that resemble graphite; *coke*, *charcoal*, and *carbon black* are a few. Coke is made by heating coal in the absence of air and is used as a reducing agent in the manufacture of steel (as we'll see in Section 21.3). Charcoal is formed when wood is heated in the absence of air. Because charcoal has a porous, sponge-like structure with an immense surface area (≈ 25 m^2/g), it has strong adsorbent properties and is widely used in filters for removing foul-smelling molecules from air and water. Carbon black, used in the manufacture of printing inks and automobile tires, is made by heating hydrocarbons such as natural gas (CH_4) in a limited supply of oxygen:

$$CH_4(g) + O_2(g) \longrightarrow C(s) + 2 H_2O(g)$$

Carbon Compounds

Carbon forms many millions of compounds, most of which are classified as *organic*; only CO, CO_2, $CaCO_3$, HCN, CaC_2, and a handful of others are considered to be *inorganic*. The distinction is a historical one rather than a scientific one, though, and we'll deal with it in more detail in Chapter 23. For the present, we'll look only at some simple inorganic compounds of carbon.

Oxides of Carbon   The most important oxides of carbon are carbon monoxide (CO) and carbon dioxide (CO_2). Carbon monoxide is a colorless, odorless, toxic gas that forms when carbon or hydrocarbon fuels are burned in a limited supply of oxygen. In an excess of oxygen, CO burns to give CO_2:

$$2 C(s) + O_2(g) \longrightarrow 2 CO(g) \qquad \Delta H° = -221 \text{ kJ}$$
$$2 CO(g) + O_2(g) \longrightarrow 2 CO_2(g) \qquad \Delta H° = -566 \text{ kJ}$$

Carbon monoxide is synthesized industrially, along with H_2, by heating hydrocarbons or coke with steam (Section 14.3). One of its main uses is in the industrial synthesis of methanol, CH_3OH:

$$CO(g) + 2 H_2(g) \xrightarrow[\text{ZnO/Cr}_2\text{O}_3 \text{ catalyst}]{400°C} CH_3OH(g)$$

The high toxicity of CO results from its ability to bond strongly to the iron(II) atom of hemoglobin, the oxygen-carrying protein in red blood cells. Because hemoglobin has a greater affinity for CO than for O_2 by a factor of 200, even small concentrations of CO in the blood can convert a substantial fraction of the O_2-bonded hemoglobin, called *oxyhemoglobin*, to the CO-bonded form, called *carboxyhemoglobin*, thus impairing the ability of hemoglobin to carry O_2 to the tissues:

$$\underset{\text{Oxyhemoglobin}}{\text{Hb–O}_2 + \text{CO}} \rightleftharpoons \underset{\text{Carboxyhemoglobin}}{\text{Hb–CO} + \text{O}_2}$$

A CO concentration in air of only 200 ppm can produce symptoms such as headache, dizziness, and nausea, and a concentration of 1000 ppm can cause death within 4 hours. One hazard of cigarette smoking is chronic exposure to low levels of CO. Because CO reduces the blood's ability to carry O_2, the heart must work harder to supply O_2 to the tissues, thus increasing the risk of heart attack.

Carbon dioxide is a colorless, odorless, nonpoisonous gas. It is produced when fuels are burned in an excess of oxygen and is an end product of food metabolism in humans and animals. Commercially, carbon dioxide is obtained as a by-product of the yeast-catalyzed fermentation of sugar in the manufacture of alcoholic beverages:

$$\underset{\text{Glucose}}{C_6H_{12}O_6(aq)} \xrightarrow{\text{Yeast}} \underset{\text{Ethanol}}{2 \text{ CH}_3\text{CH}_2\text{OH}(aq)} + 2 CO_2(g)$$

Carbon dioxide can also be obtained by heating metal carbonates, and it is produced in the laboratory when metal carbonates are treated with strong acids:

$$CaCO_3(s) \xrightarrow{\text{Heat}} CaO(s) + CO_2(g)$$

$$Na_2CO_3(s) + 2\,H^+(aq) \longrightarrow 2\,Na^+(aq) + CO_2(g) + H_2O(l)$$

Carbon dioxide is used in beverages and in fire extinguishers. The "bite" of carbonated beverages is due to the mild acidity of CO_2 solutions (pH $\approx$ 4), which results from approximately 0.3% conversion of the dissolved CO_2 to carbonic acid (H_2CO_3), a weak diprotic acid (Section 15.11):

Carbon Dioxide Behaves as an Acid in Water movie

$$CO_2(aq) + H_2O(l) \rightleftharpoons H_2CO_3(aq) \rightleftharpoons H^+(aq) + HCO_3^-(aq)$$

CO_2 is useful in fighting fires because it is nonflammable and is about 1.5 times more dense than air. It therefore settles over a small fire like a blanket, separating the fire from its source of oxygen. Solid CO_2 (dry ice), which sublimes at $-78°C$, is used primarily as a refrigerant.

These cattle were asphyxiated by an enormous bubble of carbon dioxide that welled up from the depths of Lake Nyos, Cameroon, in 1986. The gas, although not toxic in itself, is heavier than air. Released suddenly from solution, possibly as a result of volcanic activity, it blanketed the shores and overflowed into lower-lying surrounding regions, displacing oxygen and suffocating some 1700 people.

Carbonates Carbonic acid, H_2CO_3, has never been isolated as a pure substance, but it forms two series of salts: carbonates, which contain the trigonal planar CO_3^{2-} ion, and hydrogen carbonates (bicarbonates), which contain the HCO_3^- ion. Several million tons of soda ash, Na_2CO_3, are used annually in making glass, and $Na_2CO_3 \cdot 10\,H_2O$, known as washing soda, is used in laundering textiles. The carbonate ion removes cations such as Ca^{2+} and Mg^{2+} from hard water, and it acts as a base to give OH^- ions, which help in removing grease from fabrics:

$$Ca^{2+}(aq) + CO_3^{2-}(aq) \longrightarrow CaCO_3(s)$$

$$CO_3^{2-}(aq) + H_2O(l) \rightleftharpoons HCO_3^-(aq) + OH^-(aq)$$

Sodium hydrogen carbonate, $NaHCO_3$, is called baking soda because it reacts with acidic substances in food to yield bubbles of CO_2 gas that cause dough to rise:

$$NaHCO_3(s) + H^+(aq) \longrightarrow Na^+(aq) + CO_2(g) + H_2O(l)$$

Hydrogen Cyanide and Cyanides   Hydrogen cyanide is a highly toxic, volatile substance (bp 26°C) produced when metal cyanide solutions are acidified:

$$CN^-(aq) + H^+(aq) \longrightarrow HCN(aq)$$

Aqueous solutions of HCN, known as hydrocyanic acid, are very weakly acidic ($K_a = 4.9 \times 10^{-10}$).

The cyanide ion is called a *pseudohalide* ion because it behaves like Cl^- in forming an insoluble, white silver salt, AgCN. In complex ions such as $Fe(CN)_6^{3-}$, CN^- acts as a Lewis base (Section 15.16), bonding to transition metals through the lone pair of electrons on carbon. In fact, the toxicity of HCN and other cyanides is due to the strong bonding of CN^- to iron(III) in cytochrome oxidase, an important enzyme involved in the oxidation of food molecules. With CN^- attached to the iron, the enzyme is unable to function. Cellular energy production thus comes to a halt, and rapid death follows.

The bonding of CN^- to gold and silver is exploited in the extraction of these metals from their ores. The crushed rock containing small amounts of the precious metals is treated with an aerated cyanide solution, and the metals are then recovered from their $M(CN)_2^-$ complex ions by reduction with zinc. For gold, the reactions are

$$4\,Au(s) + 8\,CN^-(aq) + O_2(g) + 2\,H_2O(l) \longrightarrow 4\,Au(CN)_2^-(aq) + 4\,OH^-(aq)$$

$$2\,Au(CN)_2^-(aq) + Zn(s) \longrightarrow 2\,Au(s) + Zn(CN)_4^{2-}(aq)$$

Carbides   Carbon forms a number of binary compounds called *carbides*, in which the carbon atom has a negative oxidation state. Examples include ionic carbides of active metals such as CaC_2 and Al_4C_3, interstitial carbides of transition metals such as Fe_3C, and covalent network carbides such as SiC. Calcium carbide is a high-melting, colorless solid that has an NaCl type of structure with Ca^{2+} ions in place of Na^+ and C_2^{2-} ions in place of Cl^-. It is prepared by heating lime (CaO) and coke (C) at high temperatures and is used to prepare acetylene (C_2H_2) for oxyacetylene welding:

$$CaO(s) + 3\,C(s) \xrightarrow{2200°} CaC_2(s) + CO(g)$$

$$CaC_2(s) + 2\,H_2O(l) \longrightarrow C_2H_2(g) + Ca(OH)_2(s)$$

Iron carbide (Fe_3C) is an important constituent of steel, and silicon carbide (SiC) is the industrial abrasive called carborundum. Almost as hard as diamond, SiC has a diamondlike structure with alternating Si and C atoms.

▶**PROBLEM 19.4**   Hydrogen cyanide, HCN, is a linear triatomic molecule. Draw its electron-dot structure, and indicate which hybrid orbitals are used by the carbon atom.

▶**PROBLEM 19.5**   The equilibrium between oxyhemoglobin and carboxyhemoglobin suggests an approach to treating mild cases of carbon monoxide poisoning. Explain.

19.7   Silicon

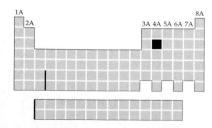

Silicon is a hard, gray, semiconducting solid that melts at 1410°C. It crystallizes in a diamondlike structure but does not form a graphitelike allotrope because of the relatively poor overlap of silicon π orbitals. In nature, silicon is generally found combined with oxygen in SiO_2 and in various silicate minerals. It is obtained in elemental form by reduction of silica sand (SiO_2) with coke (C) in an electric furnace:

$$SiO_2(l) + 2\,C(s) \xrightarrow{\text{Heat}} Si(l) + 2\,CO(g)$$

The silicon used for making solid-state semiconductor devices such as transistors, computer chips, and solar cells must be ultrapure, with impurities at a level of less than $10^{-7}\%$ (1 ppb). For electronic applications, silicon is purified by converting it to $SiCl_4$, a volatile liquid (bp 58°C) that can be separated from impurities by fractional distillation and then converted back to elemental silicon by reduction with hydrogen:

$$Si(s) + 2\,Cl_2(g) \longrightarrow SiCl_4(l)$$
$$SiCl_4(g) + 2\,H_2(g) \xrightarrow{\text{Heat}} Si(s) + 4\,HCl(g)$$

The silicon is purified further by a process called **zone refining** (Figure 19.4a), in which a heater melts a narrow zone of a silicon rod. Because the impurities are more soluble in the liquid phase than in the solid, they concentrate in the molten zone. As the heater sweeps slowly down the rod, ultrapure silicon crystallizes at the trailing edge of the molten zone, and the impurities are dragged to the rod's lower end. Figure 19.4b shows some samples of ultrapure silicon.

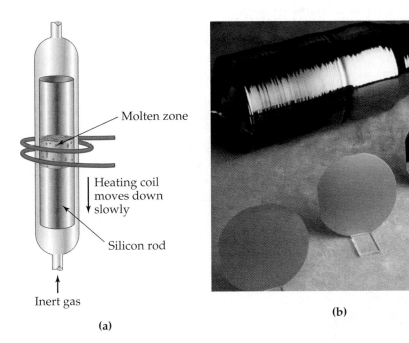

Molten zone

Heating coil moves down slowly

Silicon rod

Inert gas

(a)

(b)

FIGURE 19.4   **(a)** Purification of silicon by zone refining. The heater coil sweeps the molten zone and the impurities to the lower end of the rod. After the rod has cooled, the impurities are removed by cutting off the rod's lower end. **(b)** A rod of ultrapure silicon and silicon wafers cut from the rod. Silicon wafers are used to produce the integrated-circuit chips found in solid-state electronic devices.

Silicates   Approximately 90% of the earth's crust consists of **silicates**, ionic compounds that contain silicon oxoanions along with cations such as Na^+, K^+, Mg^{2+}, or Ca^{2+} to balance the negative charge of the anions. As shown in Figure 19.5, the basic structural building block in silicates is the SiO_4 tetrahedron, a unit that occurs as the simple orthosilicate ion (SiO_4^{4-}) in the mineral

The mineral zircon ($ZrSiO_4$) is a relatively inexpensive gemstone.

zircon, $ZrSiO_4$. If two SiO_4 tetrahedra share a common O atom, the disilicate anion $Si_2O_7^{6-}$, found in $Sc_2Si_2O_7$, results.

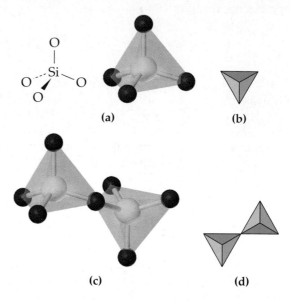

FIGURE 19.5 **(a)** A view of the SiO_4^{4-} anion showing the tetrahedral SiO_4 structural unit. **(b)** A tetrahedron is used as a shorthand representation of the SiO_4^{4-} anion. An O atom is located at each corner of the tetrahedron and the Si atom is at the center. **(c)** A view of the $Si_2O_7^{6-}$ anion. (d) A shorthand representation of the $Si_2O_7^{6-}$ anion. The corner shared by the two tetrahedra represents a shared O atom.

Simple anions such as SiO_4^{4-} and $Si_2O_7^{6-}$ are relatively rare in silicate minerals. More common are larger anions in which two or more O atoms bridge between Si atoms to give rings, chains, layers, and extended three-dimensional structures. The sharing of two O atoms per SiO_4 tetrahedron gives either cyclic anions, such as $Si_6O_{18}^{12-}$, or infinitely extended chain anions with repeating $Si_2O_6^{4-}$ units (Figure 19.6). The $Si_6O_{18}^{12-}$ cyclic anion is present in the mineral beryl ($Be_3Al_2Si_6O_{18}$) and in the gemstone emerald, a beryl in which about 2% of the Al^{3+} is replaced by green Cr^{3+} cations. Chain anions are found in minerals such as diopside, $CaMgSi_2O_6$.

FIGURE 19.6 Samples of the silicate minerals: **(a)** emerald, a green beryl ($Be_3Al_2Si_6O_{18}$) with about 2% Cr^{3+} ions substituting for Al^{3+}, and **(b)** diopside ($CaMgSi_2O_6$). A shorthand representation of the structure of **(c)** the cyclic anion $Si_6O_{18}^{12-}$ in beryl and **(d)** the infinitely extended chain anion ($Si_2O_6^{4-}$)$_n$ in diopside. Note that the number of negative charges on the $Si_2O_6^{4-}$ repeating unit equals the number of terminal (unshared) O atoms in that unit (four).

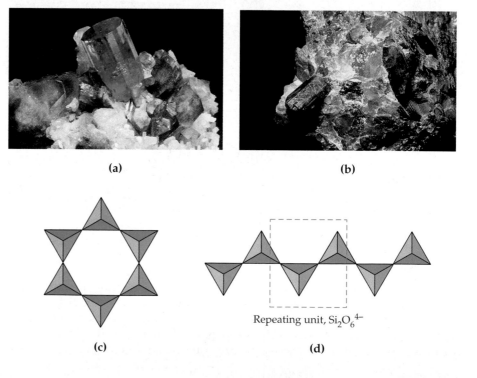

As shown in Figure 19.7, additional sharing of O atoms gives the double-stranded chain anions $(Si_4O_{11}^{6-})_n$ found in asbestos minerals such as tremolite, $Ca_2Mg_5(Si_4O_{11})_2(OH)_2$, and the infinitely extended two-dimensional layer anions $(Si_4O_{10}^{4-})_n$ found in clay minerals, micas, and talc, $Mg_3(OH)_2(Si_4O_{10})$. Asbestos is fibrous, as shown in the figure, because the ionic bonds between the silicate chain anions and the Ca^{2+} and Mg^{2+} cations that lie between the chains and hold them together are relatively weak and easily broken. Similarly, mica is sheetlike because the ionic bonds between the two-dimensional layer anions and the interposed metal cations are much weaker than the Si–O covalent bonds within the layer anions.

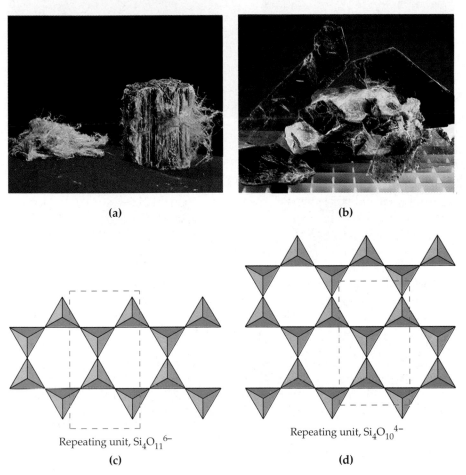

(a) (b)

Repeating unit, $Si_4O_{11}^{6-}$

(c)

Repeating unit, $Si_4O_{10}^{4-}$

(d)

FIGURE 19.7 **(a)** Asbestos is a fibrous material because of its chain structure. **(b)** Mica cleaves into thin sheets because of its two-dimensional layer structure. **(c)** A shorthand representation of the double-stranded chain anion $(Si_4O_{11}^{6-})_n$ in asbestos minerals. Two of the single-stranded chains of Figure 19.6d are laid side by side, and half of the SiO_4 tetrahedra share an additional O atom. **(d)** The layer anion $(Si_4O_{10}^{4-})_n$ in mica is formed by the sharing of three O atoms per SiO_4 tetrahedron. Note again that the number of negative charges on each repeating unit equals the number of terminal O atoms in that unit.

If the layer anions of Figure 19.7d are stacked on top of one another and the terminal O atoms are shared, an infinitely extended three-dimensional structure is obtained in which all four O atoms of each SiO_4 tetrahedron are shared between two Si atoms, resulting in *silica* (SiO_2). The mineral quartz is one of many crystalline forms of SiO_2.

Partial substitution of the Si^{4+} of SiO_2 with Al^{3+} gives **aluminosilicates** called *feldspars*, the most abundant of all minerals. An example is orthoclase, $KAlSi_3O_8$, which has a three-dimensional structure like that of SiO_2. The $(Si_3O_8^-)_n$ framework consists of SiO_4 and AlO_4 tetrahedra that share all four of their corners with neighboring tetrahedra. The K^+ cation balances the negative charge. In the aluminosilicates known as *zeolites*, the SiO_4 and AlO_4 tetrahedra are joined together in an open structure that has a three-dimensional network of cavities linked by channels. Because only small molecules can enter these channels, zeolites act as molecular sieves for separating small molecules from larger ones. They are also used as catalysts in the manufacture of gasoline.

(a) Pure crystalline quartz, one form of SiO_2, is colorless.
(b) Orthoclase, $KAlSi_3O_8$, has a structure similar to that of SiO_2.

(a)   (b)

KEY CONCEPT PROBLEM 19.6 The following pictures represent silicate anions. What is the formula and charge of each anion?

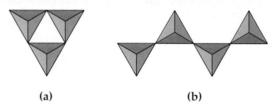

(a)   (b)

19.8 Germanium, Tin, and Lead

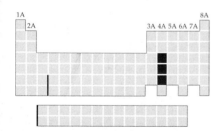

Germanium, tin, and lead have relatively low abundances in the earth's crust (Table 19.4, page 820), but tin and lead are concentrated in workable deposits and are readily extracted from their ores. Tin is obtained from the mineral cassiterite (SnO_2) by reduction of the purified oxide with carbon:

$$SnO_2(s) + 2\,C(s) \xrightarrow{\text{Heat}} Sn(l) + 2\,CO(g)$$

Tin is used as a protective coating over steel in making tin cans, and is an important component of alloys such as bronze (10% Sn, 90% Cu), pewter (90% Sn, with Cu and Sb), and lead solder (33% Sn, 67% Pb).

Lead is obtained from its ore, galena (PbS), by roasting the sulfide in air and reducing the resulting PbO with carbon monoxide in a blast furnace:

$$2\,PbS(s) + 3\,O_2(g) \longrightarrow 2\,PbO(s) + 2\,SO_2(g)$$
$$PbO(s) + CO(g) \longrightarrow Pb(l) + CO_2(g)$$

Sample of galena (PbS).

Lead is used in making pipes, cables, pigments, and, most importantly, electrodes for storage batteries (Section 18.9).

Although tin and lead have been known for over 5000 years, germanium was not discovered until 1886, by C. A. Winkler in Germany. In fact, germanium was one of the "holes" in Mendeleev's periodic table (Figure 5.2, page 160). Germanium is used in making transistors and special glasses for infrared devices.

The physical properties of the heavier group 4A elements nicely illustrate the gradual transition from semimetal to metallic character. Germanium is a relatively high-melting, brittle semiconductor that has the same crystal structure as diamond and silicon. Tin exists in two allotropic forms, the usual silvery white metallic form called *white tin* and a brittle, semiconducting form with the diamond structure called *gray tin*. White tin is the stable form at room temperature, but when kept for long periods of time below the transition temperature of 13°C, it slowly crumbles to gray tin, a phenomenon known as "tin disease":

$$\text{White tin} \xrightleftharpoons{13°C} \text{Gray tin}$$

Only the metallic form occurs for lead. Both white tin and lead are soft, malleable, low-melting metals.

  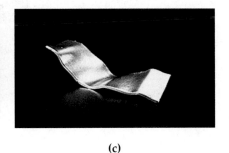

(a)   (b)   (c)

Samples of **(a)** germanium, **(b)** tin, and **(c)** lead.

19.9 The Group 5A Elements

The group 5A elements are nitrogen, phosphorus, arsenic, antimony, and bismuth. As shown in Table 19.5, these elements exhibit the expected trends of increasing atomic size, decreasing ionization energy, and decreasing electronegativity down the periodic group from N to Bi. Accordingly, metallic character increases in the same order: N and P are typical nonmetals, As and Sb are semimetals, and Bi is a metal. Thus, nitrogen is a gaseous substance made up of N_2 molecules, but bismuth is a silvery solid having an extended three-dimensional structure. The increasing metallic character of the heavier

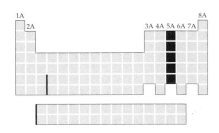

TABLE 19.5 Properties of the Group 5A Elements

Property	Nitrogen	Phosphorus	Arsenic	Antimony	Bismuth
Valence electron configuration	$2s^2\,2p^3$	$3s^2\,3p^3$	$4s^2\,4p^3$	$5s^2\,5p^3$	$6s^2\,6p^3$
Melting point (°C)	−210	44*	614†	631	271
Boiling point (°C)	−196	280		1587	1564
Atomic radius (pm)	70	110	120	140	150
First ionization energy (kJ/mol)	1402	1012	947	834	703
Electronegativity	3.0	2.1	2.0	1.9	1.9

*White phosphorus. †Sublimes.

elements is also evident in the acid–base properties of their oxides: Most nitrogen and phosphorus oxides are acidic, arsenic and antimony oxides are amphoteric, and Bi_2O_3 is basic.

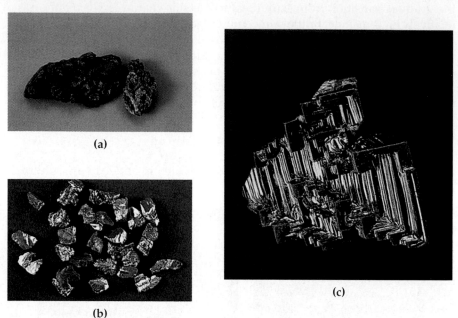

(a)

(c)

(b)

Samples of **(a)** arsenic, **(b)** antimony, and **(c)** bismuth.

The valence electron configuration of the group 5A elements is $ns^2\,np^3$. They exhibit a maximum oxidation state of +5 in compounds such as HNO_3 and PF_5, in which they share all five valence electrons with a more electronegative element. They show a minimum oxidation state of −3 in compounds such as NH_3 and PH_3, where they share three valence electrons with a less electronegative element. The −3 state also occurs in ionic compounds such as Li_3N and Mg_3N_2, which contain the N^{3-} anion.

Nitrogen and phosphorus are unusual in that they exhibit all oxidation states between −3 and +5. For arsenic and antimony, the most important oxidation states are +3, as in $AsCl_3$, As_2O_3, and H_3AsO_3, and +5, as in AsF_5, As_2O_5, and H_3AsO_4. The +5 state becomes increasingly less stable from As to Sb to Bi. Another indication of increasing metallic character down the group is the existence of Sb^{3+} and Bi^{3+} cations in salts such as $Sb_2(SO_4)_3$ and $Bi(NO_3)_3$. By contrast, no simple cations are found in compounds of N or P.

The natural abundances of As, Sb, and Bi in the earth's crust are relatively low—about 0.0002% for As and about 0.000 02% for Sb and Bi. All three elements are found in sulfide ores and are used in making various metal alloys. Arsenic is also used in making pesticides and semiconductors, such as GaAs. Bismuth compounds are present in some pharmaceuticals, such as Pepto-Bismol.

19.10   Nitrogen

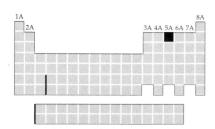

Elemental nitrogen is a colorless, odorless, tasteless gas that makes up 78% of the earth's atmosphere by volume. Because nitrogen (bp −196°C) is the most volatile component of liquid air, it is readily separated from the less volatile oxygen (bp −183°C) and argon (bp −186°C) by fractional distillation. As shown in Table 19.1 on page 812, the annual U.S. production of nitrogen is 32.9 million tons, greater than that of any other industrial chemical except sulfuric acid. Nitrogen gas is used as a protective inert atmosphere in manufacturing processes, and the liquid is used as a refrigerant. By far the most important use of nitrogen, however, is in the Haber process for the manufacture of ammonia, used in nitrogen fertilizers.

Under most conditions, the N_2 molecule is unreactive because a large amount of energy is required to break its strong nitrogen–nitrogen triple bond:

$$:N\equiv N: \longrightarrow 2 \; :\dot{N}\cdot \qquad \Delta H° = 945 \text{ kJ}$$

As a result, reactions involving N_2 often have a high activation energy and/or an unfavorable equilibrium constant. For example, N_2 and O_2 do not combine to form nitric oxide at 25°C because the equilibrium constant for the reaction is 4.5×10^{-31}:

$$N_2(g) + O_2(g) \rightleftharpoons 2 \text{ NO}(g) \qquad \Delta H° = 180.5 \text{ kJ}; K_c = 4.5 \times 10^{-31} \text{ at } 25°C$$

At higher temperatures, however, the reaction does occur because it is endothermic and the equilibrium shifts to the right with increasing temperature (Section 13.9). Indeed, high-temperature formation of NO from air in automobile engines is a major source of air pollution. Atmospheric N_2 and O_2 also react to form NO during electrical storms, where lightning discharges provide the energy required for the highly endothermic reaction.

Nitrogen reacts with H_2 in the Haber process for synthesis of NH_3, but the reaction requires high temperatures, high pressures, and a catalyst (Sections 13.8–13.10):

$$N_2(g) + 3 \text{ H}_2(g) \xrightarrow[\text{Fe/K}_2\text{O/Al}_2\text{O}_3 \text{ catalyst}]{400–500°C, \; 130–300 \text{ atm}} 2 \text{ NH}_3(g)$$

It's convenient to classify nitrogen compounds by oxidation state, as shown in Table 19.6.

TABLE 19.6 Oxidation States of Nitrogen and Representative Compounds

Oxidation State	Compound	Formula	Electron-Dot Structure
−3	Ammonia	NH_3	H—N̈—H with H below
−2	Hydrazine	N_2H_4	H—N̈—N̈—H with H, H below
−1	Hydroxylamine	NH_2OH	H—N̈—Ö—H with H below
+1	Nitrous oxide	N_2O	:N≡N—Ö:
+2	Nitric oxide	NO	:N̈=Ö:
+3	Nitrous acid	HNO_2	H—Ö—N̈=Ö:
+4	Nitrogen dioxide	NO_2	:Ö—N̈=Ö:
+5	Nitric acid	HNO_3	H—Ö—N=Ö: with :Ö: below

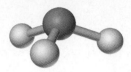

Nitrogen Compounds

Ammonia Ammonia and its synthesis by the Haber process serve as the gateway to nitrogen chemistry because ammonia is the starting material for the industrial synthesis of other important nitrogen compounds, such as nitric acid. Used in agriculture as a fertilizer, ammonia is the most commercially important compound of nitrogen.

Ammonia is a colorless, pungent-smelling gas, consisting of polar, trigonal pyramidal NH_3 molecules that have a lone pair of electrons on the N atom. Because of hydrogen bonding (Section 10.2), gaseous NH_3 is extremely soluble in water and is easily condensed to liquid NH_3, which boils at $-33°C$. Like water, liquid ammonia is an excellent solvent for ionic compounds. It also dissolves alkali metals, as mentioned in Section 6.7.

Because ammonia is a Brønsted–Lowry base (Section 15.1), its aqueous solutions are weakly alkaline:

$$NH_3(aq) + H_2O(l) \rightleftharpoons NH_4^+(aq) + OH^-(aq) \qquad K_b = 1.8 \times 10^{-5}$$

Neutralization of aqueous ammonia with acids yields ammonium salts, which resemble alkali metal salts in their solubility.

Hydrazine Hydrazine (H_2NNH_2) can be regarded as a derivative of NH_3 in which one H atom is replaced by an amino (NH_2) group. It can be prepared by reaction of ammonia with a basic solution of sodium hypochlorite (NaOCl):

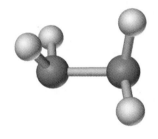

$$2\,NH_3(aq) + OCl^-(aq) \longrightarrow N_2H_4(aq) + H_2O(l) + Cl^-(aq)$$

No doubt you've heard that household cleaners should never be mixed because of possible exothermic reactions or formation of dangerous products. Formation of hydrazine on mixing household ammonia and hypochlorite-containing chlorine bleaches is a case in point.

N_2H_4 3D model

Pure hydrazine is a poisonous, colorless liquid that smells like ammonia, freezes at 2°C, and boils at 114°C. It is violently explosive in the presence of air or other oxidizing agents and is used as a rocket fuel. For example, the *Apollo* lunar-landing module used a fuel composed of hydrazine and a derivative of hydrazine, along with dinitrogen tetroxide (N_2O_4) as the oxidizer. The highly exothermic reaction is

$$2\,N_2H_4(l) + N_2O_4(l) \longrightarrow 3\,N_2(g) + 4\,H_2O(g) \qquad \Delta H° = -1049 \text{ kJ}$$

Hydrazine can be handled safely in aqueous solutions, where it behaves as a weak base ($K_b = 8.9 \times 10^{-7}$) and a versatile reducing agent. It reduces Fe^{3+} to Fe^{2+}, I_2 to I^-, and Ag^+ to metallic Ag, for example.

Oxides of Nitrogen Nitrogen forms a large number of oxides, but here we'll discuss only three: nitrous oxide (dinitrogen monoxide, N_2O), nitric oxide (nitrogen monoxide, NO), and nitrogen dioxide (NO_2). We discussed N_2O_5 and N_2O_4 in Chapters 12 and 13.

Nitrous oxide (N_2O) is a colorless, sweet-smelling gas obtained when molten ammonium nitrate is heated gently at about 270°C:

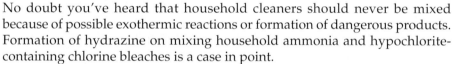

$$NH_4NO_3(l) \longrightarrow N_2O(g) + 2\,H_2O(g)$$

(Strong heating can cause an explosion.) Known as "laughing gas" because small doses are mildly intoxicating, nitrous oxide is used as a dental anesthetic and as a propellant for dispensing whipped cream.

Nitric oxide (NO) is a colorless gas, produced in the laboratory when copper metal is treated with dilute nitric acid:

$$3\,Cu(s) + 2\,NO_3^-(aq) + 8\,H^+(aq) \longrightarrow 3\,Cu^{2+}(aq) + 2\,NO(g) + 4\,H_2O(l)$$

As we'll discuss later, NO is prepared in large quantities by catalytic oxidation of ammonia, the first step in the industrial synthesis of nitric acid. Recently, it has been found that nitric oxide is important in biological processes: It plays a role in transmitting messages between nerve cells and in killing harmful microorganisms. It also helps to protect the heart from insufficient oxygen levels by dilating blood vessels.

Nitrogen dioxide (NO_2) is the highly toxic, reddish brown gas that forms rapidly when nitric oxide is exposed to air (Figure 19.8):

$$2\,NO(g) + O_2(g) \longrightarrow 2\,NO_2(g)$$

NO_2 is also produced when copper reacts with concentrated nitric acid:

$$Cu(s) + 2\,NO_3^-(aq) + 4\,H^+(aq) \longrightarrow Cu^{2+}(aq) + 2\,NO_2(g) + 2\,H_2O(l)$$

NO 3D model
www

    (a)   **(b)**   **(c)**

FIGURE 19.8 **(a)** Nitric oxide is a colorless gas. **(b)** It turns brown on contact with air because NO is rapidly oxidized to NO_2. **(c)** Copper reacts with concentrated HNO_3 yielding noxious, red-brown fumes of NO_2. The blue color of the solution is due to Cu^{2+} ions.

Nitrogen Dioxide and Dinitrogen Tetraoxide movie

Because NO_2 has an odd number of electrons (23), it is paramagnetic. It tends to dimerize, forming colorless, diamagnetic N_2O_4, in which the unpaired electrons of two NO_2 molecules pair up to give an N–N bond:

$$O_2N\cdot + \cdot NO_2 \rightleftharpoons O_2N\text{–}NO_2 \qquad \Delta H° = -57.2\ kJ$$
     Brown   Colorless

In the gas phase, NO_2 and N_2O_4 are present in equilibrium. Because dimer formation is exothermic, N_2O_4 predominates at lower temperatures, and NO_2 predominates at higher temperatures. Thus, the color of the mixture fades on cooling and darkens on warming, as was illustrated in Figure 13.13 (page 555). It's interesting to note that gaseous NO, which also has an odd number of electrons, has little tendency to dimerize.

Nitrous Acid Nitrogen dioxide reacts with water to yield a mixture of nitrous acid (HNO_2) and nitric acid (HNO_3):

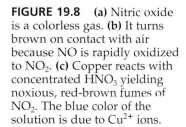

$$2\,NO_2(g) + H_2O(l) \longrightarrow HNO_2(aq) + H^+(aq) + NO_3^-(aq)$$

This is a disproportionation reaction in which nitrogen goes from the +4 oxidation state in NO_2 to the +3 state in HNO_2 and the +5 state in HNO_3. Nitrous acid is a weak acid ($K_a = 4.5 \times 10^{-4}$) that tends to disproportionate to nitric oxide and nitric acid. It has not been isolated as a pure compound:

$$3 \, HNO_2(aq) \rightleftharpoons 2 \, NO(g) + H_3O^+(aq) + NO_3^-(aq)$$

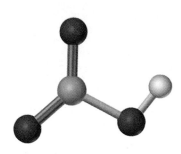

Nitric Acid Nitric acid, one of the most important inorganic acids, is used mainly in making ammonium nitrate for fertilizers, but it is also used in manufacturing explosives, plastics, and dyes. Annual U.S. production of HNO_3 is approximately 9 million tons.

Nitric acid is produced industrially by the multistep **Ostwald process**, which involves (1) air oxidation of ammonia to nitric oxide at about 850°C over a platinum–rhodium catalyst, (2) rapid oxidation of the nitric oxide to nitrogen dioxide, and (3) disproportionation of NO_2 in water:

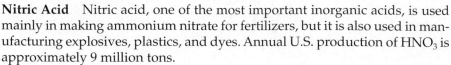

(1) $\quad 4 \, NH_3(g) + 5 \, O_2(g) \xrightarrow[\text{Pt/Rh catalyst}]{850°C} 4 \, NO(g) + 6 \, H_2O(g)$

(2) $\quad 2 \, NO(g) + O_2(g) \longrightarrow 2 \, NO_2(g)$

(3) $\quad 3 \, NO_2(g) + H_2O(l) \longrightarrow 2 \, HNO_3(aq) + NO(g)$

Distillation of the resulting aqueous HNO_3 removes some of the water and gives concentrated (15 M) nitric acid, an HNO_3–H_2O mixture that is 68.5% HNO_3 by mass.

Further removal of water is required to obtain pure nitric acid, a colorless liquid that boils at 83°C. In the laboratory, you've probably noticed that concentrated nitric acid often has a yellow-brown color. The color is due to NO_2, produced by a slight amount of decomposition:

$$4 \, HNO_3(aq) \longrightarrow 4 \, NO_2(aq) + O_2(g) + 2 \, H_2O(l)$$

Nitric acid is a strong acid, essentially 100% dissociated in water. It's also a strong oxidizing agent, as indicated by large, positive $E°$ values for reduction to lower oxidation states:

Why does freshly prepared concentrated nitric acid (left) turn yellow-brown on standing (right)?

$$NO_3^-(aq) + 2 \, H^+(aq) + e^- \longrightarrow NO_2(g) + H_2O(l) \qquad E° = 0.79 \text{ V}$$
$$NO_3^-(aq) + 4 \, H^+(aq) + 3 \, e^- \longrightarrow NO(g) + 2 \, H_2O(l) \qquad E° = 0.96 \text{ V}$$

Thus, nitric acid is a stronger oxidizing agent than $H^+(aq)$ and can oxidize relatively inactive metals like copper and silver that are not oxidized by aqueous HCl. The reduction product of HNO_3 in a particular reaction depends on the nature of the reducing agent and the reaction conditions. We've already seen, for example, that copper reduces dilute HNO_3 to NO, but it reduces concentrated HNO_3 to NO_2.

An even more potent oxidizing agent than HNO_3 is *aqua regia*, a mixture of concentrated HCl and concentrated HNO_3 in a 3:1 ratio by volume. Aqua regia can oxidize even inactive metals like gold, which do not react with either HCl or HNO_3 separately:

$$Au(s) + 3 \, NO_3^-(aq) + 6 \, H^+(aq) + 4 \, Cl^-(aq) \longrightarrow AuCl_4^-(aq) + 3 \, NO_2(g) + 3 \, H_2O(l)$$

In this reaction, NO_3^- serves as the oxidizing agent, and Cl^- provides an additional driving force by converting the Au(III) oxidation product to the $AuCl_4^-$ complex ion.

19.11 Phosphorus

Phosphorus is the most abundant element of group 5A, accounting for 0.10% of the mass of the earth's crust. It is found in phosphate rock, which is mostly calcium phosphate, $Ca_3(PO_4)_2$, and in fluorapatite, $Ca_5(PO_4)_3F$. The apatites are phosphate minerals with the formula $3 \, Ca_3(PO_4)_2 \cdot CaX_2$, where X^- is usually F^- or OH^-. Phosphorus is also important in living systems and is the sixth most abundant element in the human body (Figure 19.1). Our bones are mostly $Ca_3(PO_4)_2$, and tooth enamel is almost pure $Ca_5(PO_4)_3OH$. Phosphate groups are an integral part of the nucleic acids DNA and RNA, the molecules that pass genetic information from generation to generation (see Section 24.12).

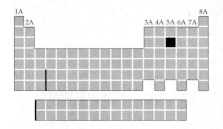

Elemental phosphorus is produced industrially by heating phosphate rock, coke, and silica sand at about 1500°C in an electric furnace. The reaction can be represented by the simplified equation

$$2 \, Ca_3(PO_4)_2(s) \; + \; 10 \, C(s) \; + \; 6 \, SiO_2(s) \longrightarrow P_4(g) \; + \; 10 \, CO(g) \; + \; 6 \, CaSiO_3(l)$$

To condense the phosphorus, the gaseous reaction products are passed through water. Phosphorus is used to make phosphoric acid, one of the top 10 industrial chemicals (Table 19.1).

Phosphorus exists in two common allotropic forms: white phosphorus and red phosphorus (Figure 19.9). White phosphorus, the form produced in the industrial synthesis, is a toxic, waxy, white solid that contains discrete tetrahedral P_4 molecules. Red phosphorus, by contrast, is essentially nontoxic and has a polymeric structure.

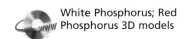
White Phosphorus; Red Phosphorus 3D models

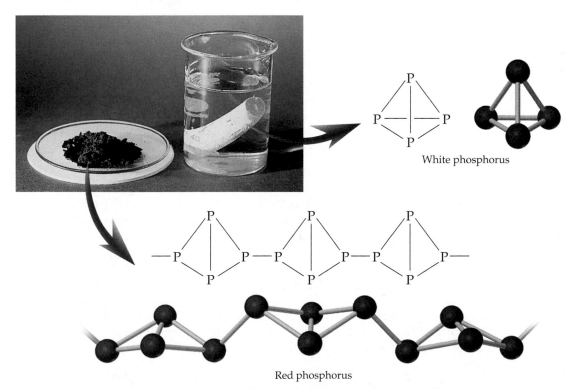

White phosphorus

Red phosphorus

FIGURE 19.9 Red phosphorus (left), and white phosphorus stored under water (right).

Finely divided white phosphorus, deposited on a piece of filter paper by evaporation of a carbon disulfide solution of P_4, bursts into flame in air.

As expected for a molecular solid that contains small, nonpolar molecules, white phosphorus has a low melting point (44°C) and is soluble in nonpolar solvents such as carbon disulfide, CS_2. It is highly reactive, bursting into flames when exposed to air, and is thus stored under water. When white phosphorus is heated in the absence of air at about 300°C, it is converted to the more stable red form. Consistent with its polymeric structure, red phosphorus is higher melting (mp ≈ 600°C), less soluble, and less reactive than white phosphorus, and it does not ignite on contact with air (Figure 19.9).

The high reactivity of white phosphorus is due to an unusual bonding that produces considerable strain in the P_4 molecules. If each P atom uses three $3p$ orbitals to form three P–P bonds, all the bond angles should be 90°. The geometry of P_4, however, requires that all the bonds have 60° angles, which means that the p orbitals can't overlap in a head-on fashion. As a result, the P–P bonds are "bent," relatively weak, and highly reactive (Figure 19.10).

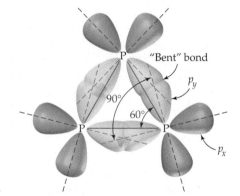

FIGURE 19.10 One equilateral triangular face of a tetrahedral P_4 molecule showing the 60° bond angles and the 90° angles between the p orbitals. The relatively poor orbital overlap in the bent bonds accounts for the high reactivity of white phosphorus.

Phosphorus Compounds

Like nitrogen, phosphorus forms compounds in all oxidation states between −3 and +5, but the +3 state, as in PCl_3, P_4O_6, and H_3PO_3, and the +5 state, as in PCl_5, P_4O_{10}, and H_3PO_4, are the most important. In comparison with nitrogen, phosphorus is more likely to be found in a positive oxidation state because of its lower electronegativity (Table 19.5, page 829).

Phosphine Phosphine (PH_3), a colorless, extremely poisonous gas, is the most important hydride of phosphorus. Like NH_3, phosphine has a trigonal pyramidal structure and has the group 5A atom in the −3 oxidation state. Unlike NH_3, however, its aqueous solutions are neutral, indicating that PH_3 is a poor proton acceptor. In accord with the low electronegativity of phosphorus, phosphine is easily oxidized, burning in air to form phosphoric acid:

$$PH_3(g) + 2\,O_2(g) \longrightarrow H_3PO_4(l)$$

Phosphorus Halides Phosphorus reacts with all the halogens, forming phosphorus(III) halides, PX_3, or phosphorus(V) halides, PX_5 (X = F, Cl, Br, or I), depending on the relative amounts of the reactants:

Limited amount of X_2: $P_4 + 6\,X_2 \longrightarrow 4\,PX_3$

Excess amount of X_2: $P_4 + 10\,X_2 \longrightarrow 4\,PX_5$

All these halides are gases, volatile liquids, or low-melting solids. For example, phosphorus trichloride is a colorless liquid that boils at 76°C, and phosphorus pentachloride is an off-white solid that melts at 167°C. Both fume on contact with moist air because of a reaction with water that breaks the P—Cl bonds, converting PCl_3 to phosphorous acid (H_3PO_3), and PCl_5 to phosphoric acid (H_3PO_4):

$$PCl_3(l) + 3\,H_2O(l) \longrightarrow H_3PO_3(aq) + 3\,HCl(aq)$$
$$PCl_5(s) + 4\,H_2O(l) \longrightarrow H_3PO_4(aq) + 5\,HCl(aq)$$

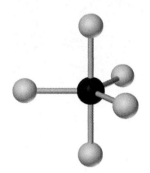

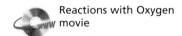

PH_3; PCl_3; H_3PO_3
www 3D models

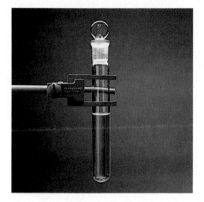

Samples of PCl_3 (left) and PCl_5 (right).

Oxides and Oxoacids of Phosphorus   When phosphorus burns in air or oxygen, it yields tetraphosphorus hexoxide (P_4O_6, mp 24°C) or tetraphosphorus decoxide (P_4O_{10}, mp 420°C), depending on the amount of oxygen present:

Limited amount of O_2: $P_4(s) + 3\,O_2(g) \longrightarrow P_4O_6(s)$

Excess amount of O_2: $P_4(s) + 5\,O_2(g) \longrightarrow P_4O_{10}(s)$

Reactions with Oxygen
www movie

Both oxides are molecular compounds and have structures with a tetrahedral array of P atoms, as in white phosphorus. One O atom bridges each of the six edges of the P_4 tetrahedron, and an additional, terminal O atom is bonded to each P atom in P_4O_{10}.

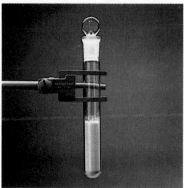

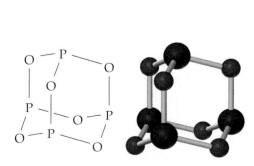

P_4O_6

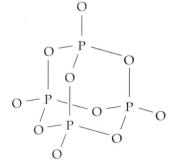

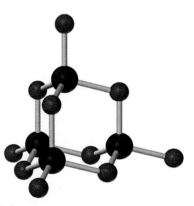

P_4O_{10}

The reaction of P_4O_{10} with water produces enough heat to convert some of the water to steam.

Both P_4O_6 and P_4O_{10} are acidic oxides, and they react with water to form aqueous solutions of phosphorous acid and phosphoric acid, respectively:

$$P_4O_6(s) + 6\,H_2O(l) \longrightarrow 4\,H_3PO_3(aq)$$

$$P_4O_{10}(s) + 6\,H_2O(l) \longrightarrow 4\,H_3PO_4(aq)$$

Because P_4O_{10} has a great affinity for water, it is widely used as a drying agent for gases and organic solvents.

It's interesting to note that phosphorous acid (H_3PO_3) is a weak *diprotic* acid because only two of its three H atoms are bonded to oxygen. The H atom bonded directly to phosphorus is not acidic because the P–H bond is nonpolar. In phosphoric acid, however, all three hydrogens are attached to oxygen, and thus phosphoric acid is a weak triprotic acid. The geometry about the P atom in both molecules is tetrahedral, as expected. Note that the successive dissociation constants decrease by a factor of about 10^5 (Section 15.11).

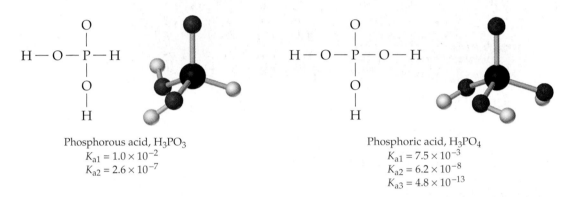

Phosphorous acid, H_3PO_3
$K_{a1} = 1.0 \times 10^{-2}$
$K_{a2} = 2.6 \times 10^{-7}$

Phosphoric acid, H_3PO_4
$K_{a1} = 7.5 \times 10^{-3}$
$K_{a2} = 6.2 \times 10^{-8}$
$K_{a3} = 4.8 \times 10^{-13}$

Pure phosphoric acid is a low-melting, colorless, crystalline solid (mp 42°C), but the commercially available phosphoric acid used in the laboratory is a syrupy, aqueous solution containing about 82% H_3PO_4 by mass. The method used to manufacture phosphoric acid depends on its intended application. For use as a food additive—for example, as the tart ingredient in various soft drinks—pure phosphoric acid is made by burning molten phosphorus in a mixture of air and steam:

$$P_4(l) \xrightarrow{\;O_2\;} P_4O_{10}(s) \xrightarrow{\;H_2O\;} H_3PO_4(aq)$$

For use in making fertilizers, an impure form of phosphoric acid is produced by treating phosphate rock with sulfuric acid:

$$Ca_3(PO_4)_2(s) + 3\,H_2SO_4(aq) \longrightarrow 2\,H_3PO_4(aq) + 3\,CaSO_4(aq)$$

Reaction of the phosphoric acid with phosphate rock gives $Ca(H_2PO_4)_2$, a water-soluble fertilizer known as triple superphosphate:

$$Ca_3(PO_4)_2(s) + 4\,H_3PO_4(aq) \longrightarrow 3\,Ca^{2+}(aq) + 6\,H_2PO_4^-(aq)$$

Phosphoric acid is sometimes called *orthophosphoric acid* to distinguish it from other phosphoric acids that are obtained when H_3PO_4 is heated. For example, *diphosphoric acid*, $H_4P_2O_7$, also called *pyrophosphoric acid*, is obtained when two molecules of H_3PO_4 combine, with elimination of one water molecule:

$$H-O-P(=O)(OH)-O-H + H-O-P(=O)(OH)-O-H \longrightarrow H-O-P(=O)(OH)-O-P(=O)(OH)-O-H + H_2O$$

H₂O eliminated

Orthophosphoric acid, H_3PO_4
(Phosphoric acid)

Diphosphoric acid, $H_4P_2O_7$
(Pyrophosphoric acid)

The next in the series is triphosphoric acid, $H_5P_3O_{10}$, and the one with an indefinitely long chain of phosphate groups is called polymetaphosphoric acid, $(HPO_3)_n$:

$$H-O-P(=O)(OH)-O-P(=O)(OH)-O-P(=O)(OH)-O-H$$

Triphosphoric acid, $H_5P_3O_{10}$

$$-O-P(=O)(OH)-O-P(=O)(OH)-O-P(=O)(OH)-O-$$

Repeating unit,
HPO_3
Polymetaphosphoric acid, $(HPO_3)_n$

All these acids have phosphorus in the +5 oxidation state, and all have structures in which PO_4 tetrahedra share bridging O atoms.

The sodium salt of triphosphoric acid, $Na_5P_3O_{10}$, is a component of some synthetic detergents. It is made by heating a stoichiometric mixture of the powdered orthophosphate salts Na_2HPO_4 and NaH_2PO_4:

$$2\,Na_2HPO_4(s) + NaH_2PO_4(s) \longrightarrow Na_5P_3O_{10}(s) + 2\,H_2O(g)$$

The triphosphate ion $P_3O_{10}^{5-}$ acts as a water softener, bonding to ions such as Ca^{2+} and Mg^{2+} that are responsible for formation of soap scum in hard water. Unfortunately, $P_3O_{10}^{5-}$ can also contribute to excessive fertilization and rampant growth of algae, a process called *eutrophication*, when phosphate-rich waste water is discharged into lakes. Subsequent decomposition of the algae can deplete a lake of oxygen and cause fish to die off.

Sodium triphosphate, $Na_5P_3O_{10}$, in synthetic detergents can contribute to the eutrophication of rivers and lakes.

EXAMPLE 19.3 Write balanced equations for conversion of orthophosphoric acid to:

(a) Triphosphoric acid   **(b)** Polymetaphosphoric acid

SOLUTION Both reactions involve combination of H_3PO_4 molecules, with elimination of water. For a phosphate chain of any length, the number of water molecules eliminated is one less than the number of P atoms in the chain. For a very long chain as in $(HPO_3)_n$, the number of water molecules eliminated is essentially the same as the number of P atoms in the chain.

(a) $3 H_3PO_4(aq) \rightarrow H_5P_3O_{10}(aq) + 2 H_2O(l)$

(b) $n H_3PO_4(aq) \rightarrow (HPO_3)_n(aq) + n H_2O(l)$

▶ **PROBLEM 19.7** Elimination of water from H_3PO_4 leads to the formation of tetraphosphoric acid, $H_6P_4O_{13}$. Write a balanced equation for the reaction. Draw structural formulas for H_3PO_4 and $H_6P_4O_{13}$, and show how elimination of water gives $H_6P_4O_{13}$.

19.12 The Group 6A Elements

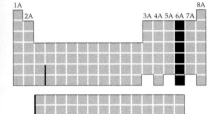

The group 6A elements are oxygen, sulfur, selenium, tellurium, and polonium. As shown in Table 19.7, their properties exhibit the usual periodic trends. Both oxygen and sulfur are typical nonmetals. Selenium and tellurium are primarily nonmetallic in character, though the most stable allotrope of selenium, gray selenium, is a lustrous semiconducting solid. Tellurium is also a semiconductor and is usually classified as a semimetal. Polonium, a radioactive element that occurs in trace amounts in uranium ores, is a silvery white metal.

TABLE 19.7   Properties of the Group 6A Elements

Property	Oxygen	Sulfur	Selenium	Tellurium	Polonium
Valence electron configuration	$2s^2\,2p^4$	$3s^2\,3p^4$	$4s^2\,4p^4$	$5s^2\,5p^4$	$6s^2\,6p^4$
Melting point (°C)	−219	113*	221†	450	254
Boiling point (°C)	−183	445	685	988	962
Atomic radius (pm)	66	104	116	143	167
X^{2-} ionic radius (pm)	132	184	191	211	
First ionization energy (kJ/mol)	1314	1000	941	869	812
Electron affinity (kJ/mol)	−141	−200	−195	−190	−183
Electronegativity	3.5	2.5	2.4	2.1	2.0
Redox potential, $E°$ (V) for $X + 2 H^+ + 2 e^- \rightarrow H_2X$	1.23	0.14	−0.40	−0.79	—

*Rhombic S.   †Gray Se.

Elemental tellurium.

Elemental selenium.

With valence electron configuration $ns^2\,np^4$, the group 6A elements are just two electrons short of an octet configuration, and the −2 oxidation state is therefore a common one. The stability of the −2 state decreases, however, with increasing metallic character, as indicated by the $E°$ values in Table 19.7. Thus, oxygen is a powerful oxidizing agent, but $E°$ values for reduction of Se and Te are negative, which means that H_2Se and H_2Te are reducing agents. Because S, Se, and Te are much less electronegative than oxygen, they are commonly found in positive oxidation states, especially +4, as in SF_4, SO_2, and H_2SO_3, and +6, as in SF_6, SO_3, and H_2SO_4.

Commercial uses of Se, Te, and Po are limited, though selenium is used in making red-colored glass and in photocopiers (see the Interlude at the end of this chapter). Tellurium is used in alloys to improve their machinability, and polonium (^{210}Po) has been used as a heat source in space equipment and as a source of alpha particles in scientific research.

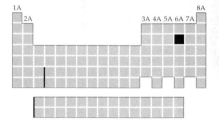

The color of the red glass in these traffic signals is due to cadmium selenide, CdSe.

19.13 Sulfur

Sulfur is the sixteenth most abundant element in the earth's crust—0.026% by mass. It occurs in elemental form in large underground deposits and is present in numerous minerals such as pyrite (FeS_2, which contains the S_2^{2-} ion), galena (PbS), cinnabar (HgS), and gypsum ($CaSO_4 \cdot 2\,H_2O$). Sulfur is also present in natural gas as H_2S and in crude oil as organic sulfur compounds. In plants and animals, sulfur occurs in various proteins, and it is one of the 10 most abundant elements in the human body (Figure 19.1).

Pyrite (FeS_2) is often called fool's gold because of its golden yellow color. It contains the disulfide ion (S_2^{2-}), the sulfur analog of the peroxide ion (O_2^{2-}).

Elemental sulfur is obtained from underground deposits and recovered from natural gas and crude oil. Sulfur is removed from these fuels prior to burning in order to prevent pollution of the air with SO_2 and subsequent formation of acid rain (see the Interlude in Chapter 15). The sulfur compounds in gas and oil are first converted to H_2S, one-third of which is then burned to give SO_2. Subsequent reaction of the SO_2 with the remaining H_2S yields elemental sulfur:

$$2\,H_2S(g) + 3\,O_2(g) \longrightarrow 2\,SO_2(g) + 2\,H_2O(g)$$

$$SO_2(g) + 2\,H_2S(g) \xrightarrow[\text{Fe}_2\text{O}_3\text{ catalyst}]{300°C} 3\,S(g) + 2\,H_2O(g)$$

H_2S; SO_2; SO_3
3D models

A sample of rhombic sulfur, the most stable allotrope of sulfur.

S_8 3D model

In the United States, 88% of the sulfur produced is used to manufacture sulfuric acid.

Sulfur exists in many allotropic forms, but the most stable at 25°C is rhombic sulfur, a yellow crystalline solid (mp 113°C) that contains crown-shaped S_8 rings:

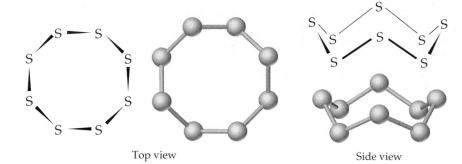

Top view   Side view

Above 95°C, rhombic sulfur is less stable than monoclinic sulfur (mp 119°C), an allotrope in which the cyclic S_8 molecules pack differently in the crystal. The phase transition from rhombic to monoclinic sulfur is very slow, however, and rhombic sulfur simply melts at 113°C when heated at an ordinary rate.

As shown in Figure 19.11, molten sulfur exhibits some striking changes when its temperature is increased. Just above its melting point, sulfur is a fluid, straw-colored liquid, but between 160°C and 195°C its color becomes dark reddish brown, and its viscosity increases by a factor of more than 10,000. At still higher temperatures, the liquid becomes more fluid again and then boils at 445°C to give a vapor that contains mostly S_8 molecules along with smaller amounts of other S_n molecules ($2 \le n \le 10$). If the liquid is cooled rapidly by pouring it into water, the sulfur forms an amorphous, rubbery material called plastic sulfur.

(a)

(b)

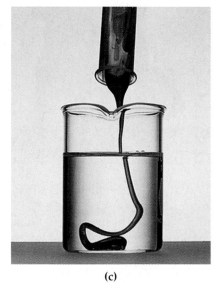

(c)

FIGURE 19.11 Effect of temperature on the properties of sulfur. **(a)** Fluid, straw-colored liquid sulfur at about 120°C. **(b)** Viscous, reddish brown liquid sulfur at about 180°C. **(c)** Plastic sulfur, obtained by pouring liquid sulfur into water. Plastic sulfur is unstable and reverts to rhombic sulfur on standing at room temperature.

The dramatic increase in the viscosity of molten sulfur at 160–195°C is due to the opening of the S_8 rings, yielding S_8 chains that form long polymers with more than 200,000 S atoms in the chain:

$$S_8 \text{ rings} \xrightarrow{\text{Heat}} \cdot S\text{—}S_6\text{—}S \cdot \text{ chains}$$

$$\cdot S\text{—}S_6\text{—}S \cdot \text{ chains } + S_8 \text{ rings} \longrightarrow \cdot S\text{—}S_{14}\text{—}S \cdot \text{ chains}$$

$$\xrightarrow{\boxed{\text{Etc.}}} S_n \text{ chains}$$

$$n > 200,000$$

Whereas the small S_8 rings easily slide over one another in the liquid, the long polymer chains become entangled, thus accounting for the increase in viscosity. Above 200°C, the polymer chains begin to fragment into smaller pieces, and the viscosity therefore decreases. On rapid cooling, the chains are temporarily frozen in a disordered, tangled arrangement, which accounts for the elastic properties of plastic sulfur.

Sulfur Compounds

Hydrogen Sulfide   Hydrogen sulfide is a colorless gas (bp −61°C) with the strong, foul odor we associate with rotten eggs, in which it occurs as a result of bacterial decomposition of sulfur-containing proteins. Hydrogen sulfide is extremely toxic, causing headaches and nausea at concentrations of 10 ppm, and sudden paralysis and death at 100 ppm. On initial exposure, the odor of H_2S can be detected at about 0.02 ppm, but unfortunately the gas tends to dull the sense of smell. It is thus an extremely insidious poison, even more dangerous than HCN.

In the laboratory, H_2S can be prepared by treating iron(II) sulfide with dilute sulfuric acid:

$$\text{FeS}(s) + 2\,\text{H}^+(aq) \longrightarrow \text{H}_2\text{S}(g) + \text{Fe}^{2+}(aq)$$

For use in qualitative analysis (Section 16.15), H_2S is usually generated in solution by hydrolysis of thioacetamide:

$$\underset{\text{Thioacetamide}}{\overset{\displaystyle S \atop \displaystyle \|}{\text{CH}_3\text{—C—NH}_2(aq)}} + \text{H}_2\text{O}(l) \longrightarrow \underset{\text{Acetamide}}{\overset{\displaystyle O \atop \displaystyle \|}{\text{CH}_3\text{—C—NH}_2(aq)}} + \text{H}_2\text{S}(aq)$$

Hydrogen sulfide is a very weak diprotic acid ($K_{a1} = 1.0 \times 10^{-7}$; $K_{a2} \approx 10^{-19}$) and a mild reducing agent. In reactions with mild oxidizing agents, it is oxidized to a milky white suspension of elemental sulfur:

$$\text{H}_2\text{S}(aq) + 2\,\text{Fe}^{3+}(aq) \longrightarrow \text{S}(s) + 2\,\text{Fe}^{2+}(aq) + 2\,\text{H}^+(aq)$$

Oxides and Oxoacids of Sulfur   Sulfur dioxide (SO_2) and sulfur trioxide (SO_3) are the most important of the various oxides of sulfur. Sulfur dioxide, a colorless, toxic gas (bp −10°C) with a pungent, choking odor, is formed when sulfur burns in air:

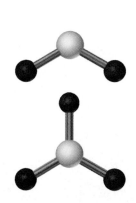

$$\text{S}(s) + \text{O}_2(g) \longrightarrow \text{SO}_2(g)$$

Sulfur burns in air with a blue flame, yielding SO_2 gas.

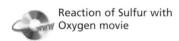

Reaction of Sulfur with Oxygen movie

Sulfur dioxide is slowly oxidized in the atmosphere to SO_3, which dissolves in rainwater to give sulfuric acid. The burning of sulfur-containing fuels is thus a major cause of acid rain (Section 9.9). In the laboratory, SO_2 is conveniently prepared by treating sodium sulfite with dilute acid:

$$Na_2SO_3(s) + 2\,H^+(aq) \longrightarrow SO_2(g) + H_2O(l) + 2\,Na^+(aq)$$

Because SO_2 is toxic to microorganisms, it is used for sterilizing wine and dried fruit.

Both SO_2 and SO_3 are acidic, though aqueous solutions of SO_2 contain mainly dissolved SO_2 and little, if any, sulfurous acid (H_2SO_3):

$$SO_2(aq) + H_2O(l) \rightleftharpoons H_2SO_3(aq)$$

Nevertheless, it's convenient to write the acidic species as H_2SO_3, a weak diprotic acid:

$$H_2SO_3(aq) + H_2O(l) \rightleftharpoons H_3O^+(aq) + HSO_3^-(aq) \qquad K_{a1} = 1.5 \times 10^{-2}$$

$$HSO_3^-(aq) + H_2O(l) \rightleftharpoons H_3O^+(aq) + SO_3^{2-}(aq) \qquad K_{a2} = 6.3 \times 10^{-8}$$

Like carbonic acid, sulfurous acid has never been isolated in pure form.

H_2SO_4 3D model

Sulfuric acid (H_2SO_4), the world's most important industrial chemical, is manufactured by the **contact process**, a three-step reaction sequence in which (1) sulfur burns in air to give SO_2, (2) SO_2 is oxidized to SO_3 in the presence of a vanadium(V) oxide catalyst, and (3) SO_3 reacts with water to give H_2SO_4:

(1) $S(s) + O_2(g) \longrightarrow SO_2(g)$

(2) $2\,SO_2(g) + O_2(g) \xrightarrow[\text{V_2O_5 catalyst}]{\text{Heat}} 2\,SO_3(g)$

(3) $SO_3(g) + H_2O \text{ (in conc } H_2SO_4) \longrightarrow H_2SO_4(l)$

In the third step, the SO_3 is absorbed in concentrated sulfuric acid rather than in water because dissolution of SO_3 in water is slow. Water is then added to achieve the desired concentration. Commercial concentrated sulfuric acid is 98% H_2SO_4 by mass (18 M H_2SO_4). Anhydrous (100%) H_2SO_4 is a viscous, colorless liquid that freezes at 10.4°C and boils above 300°C. The H_2SO_4 molecule is tetrahedral, as predicted by the VSEPR model (Section 7.9)

Sulfuric acid is a strong acid in the dissociation of its first proton and has $K_{a2} = 1.2 \times 10^{-2}$ for the dissociation of its second proton. As a diprotic acid, it forms two series of salts: hydrogen sulfates (bisulfates), such as $NaHSO_4$, and sulfates, such as Na_2SO_4.

The oxidizing properties of sulfuric acid depend on its concentration and temperature. In dilute solutions at room temperature, H_2SO_4 behaves like HCl, oxidizing metals that stand above hydrogen in the activity series (Table 4.3):

$$Fe(s) + 2 H^+(aq) \longrightarrow Fe^{2+}(aq) + H_2(g)$$

Hot, concentrated H_2SO_4 is a better oxidizing agent than the dilute, cold acid and can oxidize metals like copper, which are not oxidized by $H^+(aq)$. In the process, H_2SO_4 is reduced to SO_2:

$$Cu(s) + 2 H_2SO_4(l) \longrightarrow Cu^{2+}(aq) + SO_4^{2-}(aq) + SO_2(g) + 2 H_2O(l)$$

U.S. production of sulfuric acid in 1998 was 47.6 million tons, far exceeding that of any other chemical (Table 19.1). It is used mostly in manufacturing soluble phosphate and ammonium sulfate fertilizers, but it is essential to many other industries (Figure 19.12). So widespread is the use of sulfuric acid in industrial countries that the amount produced is sometimes regarded as an indicator of economic activity.

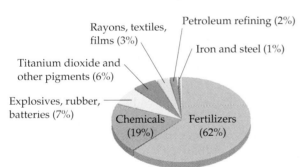

Rayons, textiles, films (3%)
Petroleum refining (2%)
Iron and steel (1%)
Titanium dioxide and other pigments (6%)
Explosives, rubber, batteries (7%)
Chemicals (19%)
Fertilizers (62%)

FIGURE 19.12 Uses of sulfuric acid in the United States.

KEY CONCEPT PROBLEM 19.8 Consider the following sulfur-containing oxoanions:

(a) Write the formula of each oxoanion, including its charge.
(b) Which oxoanion is the strongest acid?
(c) Which is the strongest base?
(d) Which is the weakest base?

▶**PROBLEM 19.9** Write electron-dot structures for each of the following molecules, and use VSEPR theory to predict the structure of each:
(a) H_2S **(b)** SO_2 **(c)** SO_3

19.14   The Halogens: Oxoacids and Oxoacid Salts

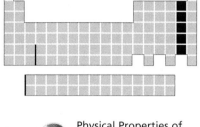

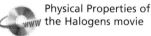
Physical Properties of the Halogens movie

The halogens (group 7A) have valence electron configuration $ns^2\, np^5$ and are the most electronegative group of elements in the periodic table. We saw in Section 6.10 that halogen atoms tend to achieve an octet configuration by gaining one electron in reactions with metals, thus forming ionic compounds such as NaCl. They also share one electron with nonmetals to give molecular compounds such as HCl, BCl_3, PF_5, and SF_6. In all these compounds, the halogen is in the −1 oxidation state.

Among the most important compounds of halogens in positive oxidation states are the oxoacids of Cl, Br, and I (Table 19.8) and the corresponding oxoacid salts. In these compounds, the halogen shares its valence electrons with oxygen, a more electronegative element. (Electronegativities are O, 3.5; Cl, 3.0; Br, 2.8; I, 2.5.) The general formula for a halogen oxoacid is HXO_n, and the oxidation state of the halogen is +1, +3, +5, or +7, depending on the value of n.

TABLE 19.8   Oxoacids of the Halogens

Oxidation State	Generic Name (Formula)	Chlorine	Bromine	Iodine
+1	Hypohalous acid (HXO)	HClO	HBrO	HIO
+3	Halous acid (HXO_2)	$HClO_2$	—	—
+5	Halic acid (HXO_3)	$HClO_3$	$HBrO_3$	HIO_3
+7	Perhalic acid (HXO_4)	$HClO_4$	$HBrO_4$	HIO_4, H_5IO_6

Only four of the acids listed in Table 19.8 have been isolated in pure form: perchloric acid ($HClO_4$), iodic acid (HIO_3), and the two periodic acids, metaperiodic acid (HIO_4) and paraperiodic acid (H_5IO_6). The others are stable only in aqueous solution or in the form of their salts. Note that chlorous acid ($HClO_2$) is the only known halous acid.

The acid strength of the halogen oxoacids increases with the increasing oxidation state of the halogen (Section 15.15). For example, acid strength increases from HClO, a weak acid ($K_a = 3.5 \times 10^{-8}$), to $HClO_4$, a very strong acid ($K_a \gg 1$). Note that the acidic proton is bonded to oxygen, not to the halogen, even though we usually write the molecular formula of these acids as HXO_n. All the halogen oxoacids and their salts are strong oxidizing agents.

A hypohalous acid is formed when Cl_2, Br_2, or I_2 dissolves in cold water:

$$X_2(g, l, \text{or } s)\ +\ H_2O(l)\ \rightleftharpoons\ HOX(aq)\ +\ X^-(aq)$$

In this reaction, the halogen disproportionates, going to the +1 oxidation state in HOX and the −1 state in X^-. The equilibrium lies to the left but is shifted to the right in basic solution:

$$X_2(g, l, \text{or } s)\ +\ 2\,OH^-(aq)\ \rightleftharpoons\ OX(aq)\ +\ X^-(aq)\ +\ H_2O(l)$$

Large amounts of aqueous sodium hypochlorite (NaOCl) are produced in the chlor-alkali industry (Section 18.12) when the Cl_2 gas and aqueous NaOH from electrolysis of aqueous NaCl are allowed to mix. Aqueous NaOCl is a strong oxidizing agent and is sold in a 5% solution as chlorine bleach.

Further disproportionation of OCl^- to ClO_3^- and Cl^- is slow at room temperature but becomes fast at higher temperatures. Thus, when Cl_2 gas reacts with hot aqueous NaOH, it gives a solution that contains sodium chlorate ($NaClO_3$) rather than NaOCl:

$$3\,Cl_2(g) + 6\,OH^-(aq) \longrightarrow ClO_3^-(aq) + 5\,Cl^-(aq) + 3\,H_2O(l)$$

Chlorate salts are used as weed killers and as strong oxidizing agents. Potassium chlorate, for example, is an oxidizer in matches, fireworks, and explosives. It also reacts vigorously with organic matter.

Sodium perchlorate ($NaClO_4$) is produced commercially by electrolytic oxidation of aqueous sodium chlorate and is converted to perchloric acid by reaction with concentrated HCl:

$$ClO_3^-(aq) + H_2O(l) \xrightarrow{\text{Electrolysis}} ClO_4^-(aq) + H_2(g)$$

$$NaClO_4(s) + HCl(aq) \longrightarrow HClO_4(aq) + NaCl(s)$$

The $HClO_4$ is then concentrated by distillation at reduced pressure.

Pure, anhydrous perchloric acid is a colorless, shock-sensitive liquid that decomposes explosively on heating. It is a powerful and dangerous oxidizing agent, violently oxidizing organic matter and rapidly oxidizing even silver and gold. Perchlorate salts are also strong oxidants, and they too must be handled with caution. Ammonium perchlorate (NH_4ClO_4), in fact, is the oxidizer in the solid booster rockets used to propel the space shuttle.

Iodine differs from the other halogens in forming more than one perhalic acid. Paraperiodic acid (H_5IO_6) is obtained as white crystals (mp 128°C) when periodic acid solutions are evaporated. When heated to 100°C at reduced pressure, these crystals lose water and are converted to metaperiodic acid (HIO_4):

$$H_5IO_6(s) \xrightarrow[\text{12 mm Hg}]{100°C} HIO_4(s) + 2\,H_2O(g)$$

Metaperiodic acid is a strong monoprotic acid, whereas paraperiodic acid is a weak polyprotic acid ($K_{a1} = 5.1 \times 10^{-4}$; $K_{a2} = 4.9 \times 10^{-9}$). It has an octahedral structure in which a central iodine atom is bonded to one O atom and five OH groups:

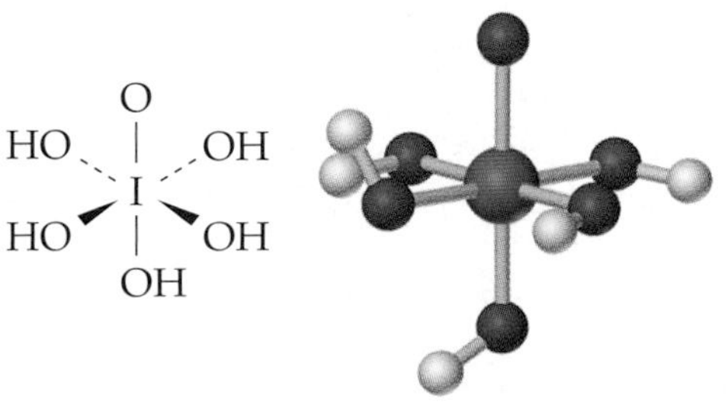

Paraperiodic acid, H_5IO_6 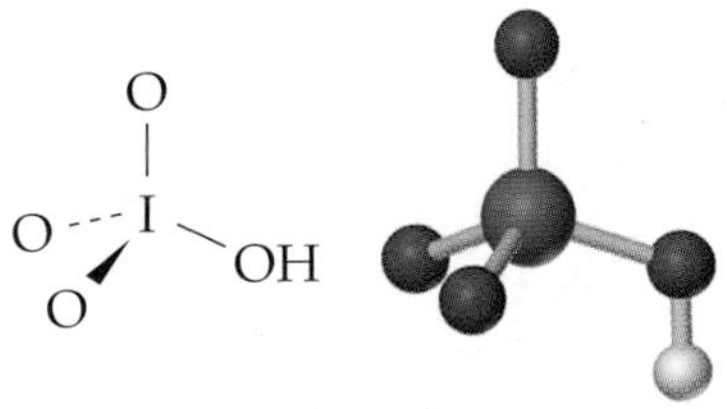   Metaperiodic acid, HIO_4

Chlorine and bromine do not form perhalic acids of the type H_5XO_6 because their smaller sizes favor a tetrahedral structure over an octahedral one.

Photocopiers

Not too many years ago, copies of documents were made either by using various wet photographic methods or by typing the original using "carbon paper." The introduction of the plain-paper photocopy machine in the mid-1950s revolutionized the way offices handle paper. Now, the most widely used method of document copying is based on *xerography*, a term derived from the Greek words for "dry writing."

Dry photocopiers make use of an unusual property of selenium, the group 6A element below sulfur in the periodic table. Selenium is a *photoconductor*, a substance that is a poor electrical conductor when dark but whose conductivity increases (by a factor of 1000) when exposed to light. When the light is removed, the conductivity again drops.

As illustrated in Figure 19.13, the xerographic process begins when a selenium-coated drum is given a uniform positive charge and is then exposed through a lens to a brightly illuminated document. Those areas on the drum that correspond to a light part of the document become conducting when exposed and lose their charge, while those areas on the drum that correspond to a dark part of the document remain nonconducting and retain their charge. Thus, an image of the document is formed on the drum as an array of positive electrical charges.

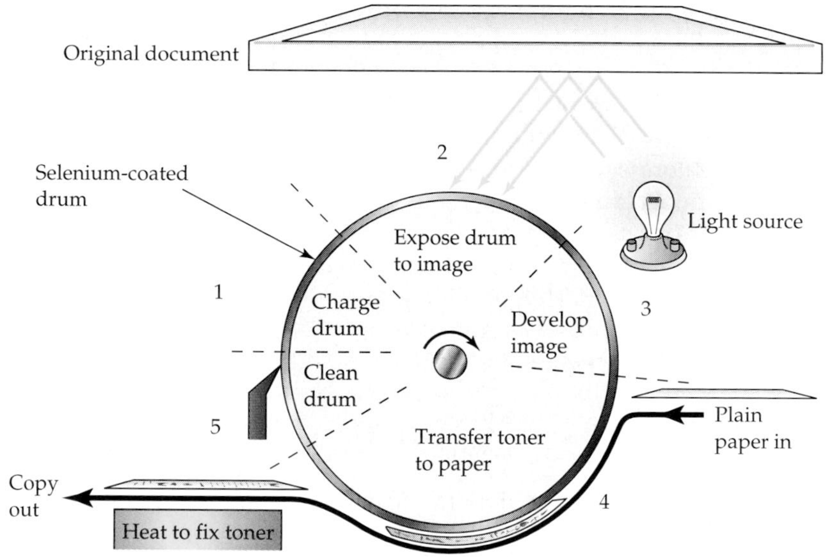

FIGURE 19.13 The photocopying process. A selenium-coated rotating drum is given a uniform positive charge (step 1) and is then exposed to an image (step 2). Negatively charged toner particles are attracted to the charged areas of the drum (step 3), and the image is transferred from the drum to a sheet of paper (step 4). Heating then fixes the image, and the drum is flooded with light and cleaned to ready the machine for another cycle (step 5).

Following its formation, the image is developed by exposing the drum to negatively charged dry ink particles (*toner*), which are attracted to the positively charged areas of the drum. The developed image is then transferred to paper by passing a sheet of paper between the drum and a positively charged development electrode, which induces the negatively charged toner particles to jump from the drum to the paper.

The toner particles, in addition to serving as pigment, are made of a resinous plastic material that fuses to the paper when heated, thereby fixing the image. The final copy then rolls out of the machine, and the drum is restored to its original condition by flooding it with light to remove all remaining charges and gently scraping off any bits of excess toner.

▶ **PROBLEM 19.10** What would be the effect of coating the photocopier drum with copper instead of selenium? Explain.

Summary

The main-group elements are the s-block elements of groups 1A and 2A and the p-block elements of groups 3A–8A. From left to right across the periodic table, ionization energy, electronegativity, and nonmetallic character generally increase, while atomic radius and metallic character decrease. From top to bottom of a group in the periodic table, ionization energy, electronegativity, and nonmetallic character generally decrease, while atomic radius and metallic character increase. The second-row elements form strong multiple bonds but are unable to form more than four bonds because they lack valence d orbitals.

The group 3A elements—B, Al, Ga, In, and Tl—are metals except for boron, which is a semimetal. Boron is a semiconductor and forms molecular compounds. **Boranes**, such as diborane (B_2H_6), are electron-deficient molecules that contain **three-center, two-electron bonds** (B–H–B).

The group 4A elements—C, Si, Ge, Sn, and Pb—exhibit the usual increase in metallic character down the group. Their most common oxidation state is +4, but the +2 state becomes increasingly more stable from Ge to Sn to Pb. In elemental form, carbon exists as diamond, graphite, fullerene, coke, charcoal, and carbon black.

Silicon, the second most abundant element in the earth's crust, is obtained by reducing silica sand (SiO_2) with coke. It is purified for use in the semiconductor industry by **zone refining**. In the **silicates**, SiO_4 tetrahedra share common O atoms to give silicon oxoanions with ring, chain, layer, and extended three-dimensional structures. In **aluminosilicates**, such as $KAlSi_3O_8$, Al^{3+} replaces some of the Si^{4+}.

Molecular nitrogen (N_2) is unreactive because of its strong N≡N triple bond. Nitrogen exhibits all oxidation states between −3 and +5. Nitric acid is manufactured by the **Ostwald process**.

Phosphorus, the most abundant group 5A element, exists in two common allotropic forms—white phosphorus, which contains highly reactive tetrahedral P_4 molecules, and red phosphorus, which is polymeric. The most common oxidation states of P are −3, as in phosphine (PH_3); +3, as in PCl_3, P_4O_6, and H_3PO_3; and +5, as in PCl_5, P_4O_{10}, and H_3PO_4. On heating, H_3PO_4 molecules combine, with elimination of water molecules, yielding polyphosphoric acids (for example, $H_5P_3O_{10}$) or polymetaphosphoric acid $(HPO_3)_n$.

Sulfur is obtained from underground deposits and is recovered from natural gas and crude oil. The properties of sulfur change dramatically on heating as the S_8 rings of rhombic sulfur open and polymerize to give long chains, which then fragment at higher temperatures. The most common oxidation states of S are −2, as in H_2S; +4, as in SO_2 and H_2SO_3; and +6, as in SO_3 and H_2SO_4. Sulfuric acid, the world's most important industrial chemical, is manufactured by the **contact process**.

Chlorine, bromine, and iodine form a series of oxoacids: hypohalous acid (HXO), halous acid (HXO_2 for X = Cl), halic acid (HXO_3), and perhalic acid (HXO_4). Acid strength increases as the oxidation state of the halogen increases from +1 to +7. Iodine forms two perhalic acids, HIO_4 and H_5IO_6. Halogen oxoacids and their salts are strong oxidizing agents.

Key Words

Key Concept Summary

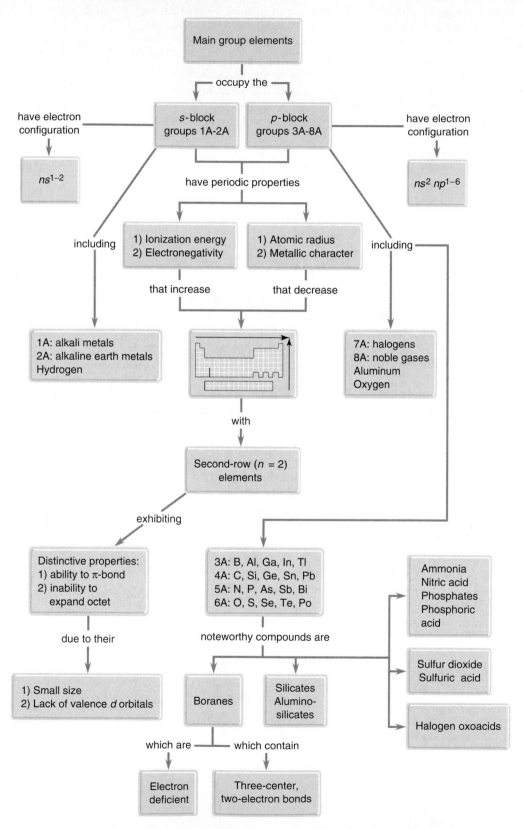

Understanding Key Concepts

Problems 19.1–19.10 appear within the chapter.

19.11 Locate each of the following groups of elements on the periodic table.

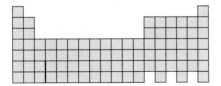

    (a) Main-group elements
    (b) *s*-Block elements
    (c) *p*-Block elements
    (d) Main-group metals
    (e) Nonmetals
    (f) Semimetals

19.12 Locate each of the following elements on the periodic table.

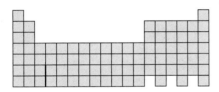

    (a) Element with the lowest ionization energy
    (b) Most electronegative element
    (c) Group 4A element with the largest atomic radius
    (d) Group 6A element with the smallest atomic radius
    (e) Group 3A element that is a semiconductor
    (f) Group 5A element that forms the strongest π bonds

19.13 Locate the following elements on the periodic table.

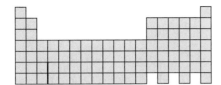

    (a) Elements that are gases at room temperature (25°C)
    (b) An element that is a liquid at 25°C
    (c) Nonmetals that are solids at 25°C
    (d) Elements that exist as diatomic molecules at 25°C

19.14 Consider the six second- and third-row elements in groups 5A–7A of the periodic table:

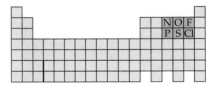

Possible molecular structures for common allotropes of these elements are shown below:

    (a) What is the molecular structure of each of the six elements?
    (b) Using electron-dot structures, explain why each element has its particular molecular structure.
    (c) Explain why nitrogen and phosphorus have different molecular structures and why oxygen and sulfur have different molecular structures, but fluorine and chlorine have the same molecular structure.

19.15 The following models represent the structures of binary oxides of second- and third-row elements in their highest oxidation states:

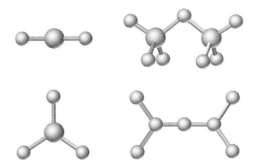

    (a) Identify the nonoxygen atom in each case, and write the molecular formula for each oxide.
    (b) Draw an electron-dot structure for each oxide. For which oxides are resonance structures needed?

19.16 The following models represent the structures of binary hydrides of second-row elements:

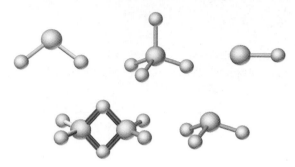

(a) Identify the nonhydrogen atom in each case, and write the molecular formula for each hydride.

(b) Draw an electron-dot structure for each hydride. For which hydride is there a problem in drawing the structure? Explain.

19.17 The following pictures represent various silicate anions. Give the formula and charge of each anion.

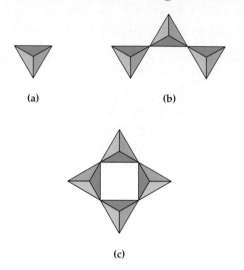

(a)　　　　　(b)

(c)

Additional Problems

General Properties and Periodic Trends

19.18 Which element in each of the following pairs has the higher ionization energy?

(a) S or Cl　　　(b) Si or Ge　　　(c) In or O

19.19 Arrange the following elements in order of increasing ionization energy:

(a) P　　(b) K　　(c) Al　　(d) F

19.20 Which element in each of the following pairs has the larger atomic radius?

(a) B or Al　　　(b) P or S　　　(c) Pb or Br

19.21 Arrange the following elements in order of increasing atomic radius:

(a) As　　(b) O　　(c) Sn　　(d) S

19.22 Which element in each of the following pairs has the higher electronegativity?

(a) Te or I　　　(b) N or P　　　(c) In or F

19.23 Arrange the following elements in order of increasing electronegativity:

(a) N　　(b) Ge　　(c) O　　(d) P

19.24 Which element in each of the following pairs has more metallic character?

(a) Si or Sn　　　(b) Ge or Se　　　(c) Bi or I

19.25 Which element in each of the following pairs has more nonmetallic character?

(a) S or Te　　　(b) Cl or P　　　(c) Bi or Br

19.26 Which compound in each of the following pairs is more ionic?

(a) CaH_2 or NH_3

(b) P_4O_6 or Ga_2O_3

(c) $SiCl_4$ or KCl

19.27 Which compound in each of the following pairs is more covalent?

(a) PCl_3 or AlF_3

(b) CaO or NO

(c) NH_3 or KH

19.28 Which of the following compounds are molecular, and which have an extended three-dimensional structure?

(a) B_2H_6

(b) $KAlSi_3O_8$

(c) SO_3

(d) $GeCl_4$

19.29 Which of the following compounds are molecular, and which have an extended three-dimensional structure?

(a) KF

(b) P_4O_{10}

(c) $SiCl_4$

(d) $CaMgSi_2O_6$

19.30 Which oxide in each of the following pairs is more acidic?

(a) Al_2O_3 or P_4O_{10}

(b) B_2O_3 or Ga_2O_3

(c) SO_2 or SnO_2

19.31 Which oxide in each of the following pairs is more basic?

(a) SO_2 or SnO_2

(b) In_2O_3 or Ga_2O_3

(c) Al_2O_3 or N_2O_5

19.32 Consider the following list of elements: C, Se, B, Sn, Cl. Identify the element on this list that:

(a) Has the largest atomic radius

(b) Is the most electronegative

(c) Is the best electrical conductor

(d) Has a maximum oxidation state of +6

(e) Forms a hydride with the empirical formula XH_3

19.33 Consider the following list of elements: N, Si, Al, S, F. Identify the element on this list that:

(a) Has the highest ionization energy

(b) Has the most metallic character

(c) Forms the strongest π bonds

(d) Is a semiconductor

(e) Forms a 2− anion

19.34 BF_3 reacts with F^- to give BF_4^-, but AlF_3 reacts with F^- to give AlF_6^{3-}. Explain.

19.35 $GeCl_4$ reacts with Cl^- to give $GeCl_6^{2-}$, but CCl_4 does not react with excess Cl^-. Explain.

19.36 At ordinary temperatures sulfur exists as S_8, but oxygen exists as O_2. Explain.

19.37 Elemental nitrogen exists as N_2, but white phosphorus exists as P_4. Explain.

The Group 3A Elements

19.38 What is the most common oxidation state for each of the group 3A elements?

19.39 What is the oxidation state of the group 3A element in each of the following compounds?

(a) $NaBF_4$ **(b)** $GaCl_3$ **(c)** $TlCl$ **(d)** B_2H_6

19.40 List three ways in which the properties of boron differ from those of the other group 3A elements.

19.41 Explain why the properties of boron differ so markedly from the properties of the other group 3A elements.

19.42 How is crystalline boron prepared? Write a balanced equation for the reaction.

19.43 Write a balanced equation for the reduction of boron oxide by magnesium.

19.44 Tell what is meant by:

(a) An electron-deficient molecule

(b) A three-center two-electron bond

Illustrate each definition with an example.

19.45 Describe the structure of diborane (B_2H_6), and explain why the bridging B−H bonds are longer than the terminal B−H bonds.

19.46 Identify the group 3A element that best fits each of the following descriptions:

(a) Is the most abundant element of the group

(b) Is stable in the +1 oxidation state

(c) Is a semiconductor

(d) Forms a molecular fluoride

19.47 Identify the group 3A element that best fits each of the following descriptions:

(a) Has an unusually low melting point

(b) Is the most electronegative

(c) Is extremely toxic

(d) Forms an acidic oxide

The Group 4A Elements

19.48 Identify the group 4A element that best fits each of the following descriptions:

(a) Prefers the +2 oxidation state

(b) Forms the strongest π bonds

(c) Is the second most abundant element in the earth's crust

(d) Forms the most acidic oxide

19.49 Select the group 4A element that best fits each of the following descriptions:

(a) Forms the most basic oxide

(b) Is the least dense semimetal

(c) Is the second most abundant element in the human body

(d) Is the most electronegative

19.50 Describe the geometry of each of the following molecules or ions, and tell which hybrid orbitals are used by the central atom:

(a) $GeBr_4$ **(b)** CO_2 **(c)** CO_3^{2-} **(d)** $Sn(OH)_6^{2-}$

19.51 What is the geometry of each of the following molecules or ions, and which hybrid orbitals are used by the central atom?

(a) SiO_4^{4-} **(b)** PF_6^- **(c)** $SnCl_2$ **(d)** N_2O

19.52 List three properties of diamond, and account for them in terms of structure and bonding.

19.53 Describe the structure and bonding in graphite, and explain why graphite is a good lubricant and a good electrical conductor.

19.54 Give the name and formula of a compound in which carbon exhibits an oxidation state of:

(a) +4   (b) +2   (c) −4

19.55 Give an example of an ionic carbide. What is the oxidation state of carbon in this substance?

19.56 List three commercial uses for carbon dioxide, and relate each use to one of carbon dioxide's properties.

19.57 Why are CO and CN^- so toxic to humans?

19.58 Describe the preparation of silicon from silica sand, and tell how silicon is purified for use in semiconductor devices. Write balanced equations for all reactions.

19.59 What minerals are the starting materials for preparation of:

(a) Silicon?   b) Tin?   (c) Lead?

Give the name and chemical formula for each mineral, and write a balanced equation for the preparation of each element.

19.60 Using the notation of Figure 19.5 (page 826), draw the structure of the silicate anion in:

(a) K_4SiO_4 (b) $Ag_{10}Si_4O_{13}$

What is the relationship between the charge on the anion and the number of terminal O atoms?

19.61 Using the notation of Figure 19.5 (page 826), draw the structure of the cyclic silicate anion in which four SiO_4 tetrahedra share O atoms to form an eight-membered ring of alternating Si and O atoms. Give the formula and charge of the anion.

19.62 The silicate anion in the mineral kinoite is a chain of three SiO_4 tetrahedra that share corners with adjacent tetrahedra. The mineral also contains Ca^{2+} ions, Cu^{2+} ions, and water molecules in a 1:1:1 ratio.

(a) Give the formula and charge of the silicate anion.

(b) Give the complete formula for the mineral.

19.63 Suggest a plausible structure for the silicate anion in each of the following minerals:

(a) Spodumene, $LiAlSi_2O_6$

(b) Wollastonite, $Ca_3Si_3O_9$

(c) Thortveitite, $Sc_2Si_2O_7$

(d) Albite, $NaAlSi_3O_8$

The Group 5A Elements

19.64 Identify the group 5A element(s) that best fits each of the following descriptions:

(a) Makes up part of bones and teeth

(b) Forms stable salts containing M^{3+} ions

(c) Is the most abundant element in the atmosphere

(d) Forms a basic oxide

19.65 Identify the group 5A element that best fits each of the following descriptions:

(a) Forms strong π bonds

(b) Is a metal

(c) Is the most abundant group 5A element in the earth's crust

(d) Forms oxides with the group 5A element in the +1, +2, and +4 oxidation states

19.66 Give the chemical formula for each of the following compounds, and indicate the oxidation state of the group 5A element:

(a) Nitrous oxide   (b) Hydrazine

(c) Calcium phosphide   (d) Phosphorous acid

(e) Arsenic acid

19.67 Give the chemical formula for each of the following compounds, and indicate the oxidation state of the group 5A element:

(a) Nitric oxide

(b) Nitrous acid

(c) Phosphine

(d) Tetraphosphorus decoxide

(e) Triphosphoric acid

19.68 Draw an electron-dot structure for N_2, and explain why this molecule is so unreactive.

19.69 Draw electron-dot structures for:

(a) Nitrous oxide

(b) Nitric oxide

(c) Nitrogen dioxide

Predict the molecular geometry of each, and indicate which are expected to be paramagnetic.

19.70 Predict the geometry of each of the following molecules or ions, and tell which hybrid orbitals are used by the central atom:

(a) NO_2^- (b) PH_3

(c) PF_5 (d) PCl_4^+

19.71 Predict the geometry of each of the following molecules or ions, and tell which hybrid orbitals are used by the central atom:

(a) PCl_6^- (b) N_2O

(c) H_3PO_3 (d) NO_3^-

19.72 Describe the structures of the white and red allotropes of phosphorus, and explain why white phosphorus is so reactive.

19.73 Draw the structure of each of the following molecules:

(a) Tetraphosphorus hexoxide

(b) Tetraphosphorus decoxide

(c) Phosphorous acid

(d) Orthophosphoric acid

(e) Polymetaphosphoric acid

19.74 Account for each of the following observations:

(a) Phosphorous acid is a diprotic acid.

(b) Nitrogen doesn't exist as a four-atom molecule like P_4.

19.75 Account for each of the following observations:

(a) Nitric acid is a strong oxidizing agent, but phosphoric acid is not.

(b) Phosphorus, arsenic, and antimony form trichlorides and pentachlorides, but nitrogen forms only NCl_3.

19.76 Write a balanced equation to account for each of the following observations:

(a) Nitric oxide turns brown when exposed to air.

(b) Nitric acid turns yellow-brown on standing.

(c) Silver dissolves in dilute HNO_3, yielding a colorless gas.

(d) Hydrazine reduces iodine to I^- and in the process is oxidized to N_2 gas.

19.77 Describe the process used for industrial production of the following chemicals:

(a) Nitrogen   (b) Ammonia

(c) Nitric acid   (d) Phosphoric acid

Write balanced equations for all chemical reactions.

The Group 6A Elements

19.78 Identify the group 6A element that best fits each of the following descriptions:

(a) Is the most electronegative

(b) Is a semimetal

(c) Is radioactive

(d) Is the most abundant element in the earth's crust

19.79 Identify the group 6A element that best fits each of the following descriptions:

(a) Is a metal

(b) Is the most abundant element in the human body

(c) Is the strongest oxidizing agent

(d) Has the most negative electron affinity

19.80 Describe the structure of the sulfur molecules in:

(a) Rhombic sulfur

(b) Monoclinic sulfur

(c) Plastic sulfur

(d) Liquid sulfur above 160°C

19.81 The viscosity of liquid sulfur increases sharply at about 160°C and then decreases again above 200°C. Explain.

19.82 Give the name and formula of two compounds in which sulfur exhibits an oxidation state of:

(a) −2   (b) +4   (c) +6

19.83 What is the oxidation state of sulfur in each of the following compounds?

(a) HgS (b) $Ca(HSO_4)_2$ (c) H_2SO_3

(d) FeS_2 (e) SF_4

19.84 Describe the contact process for manufacture of sulfuric acid, and write balanced equations for all reactions.

19.85 Describe a convenient laboratory method for preparing each of the following compounds, and write balanced equations for all reactions:

(a) Sulfur dioxide   (b) Hydrogen sulfide

(c) Sodium hydrogen sulfate

19.86 Write a balanced net ionic equation for each of the following reactions:

(a) $Zn(s) + $ dilute $H_2SO_4(aq) \rightarrow$

(b) $BaSO_3(s) + HCl(aq) \rightarrow$

(c) $Cu(s) + $ hot, conc $H_2SO_4(l) \rightarrow$

(d) $H_2S(aq) + I_2(aq) \rightarrow$

19.87 Write a balanced net ionic equation for each of the following reactions:

(a) $ZnS(s) + HCl (aq) \rightarrow$

(b) $H_2S(aq) + Fe(NO_3)_3(aq) \rightarrow$

(c) $Fe(s) + $ dilute $H_2SO_4(aq) \rightarrow$

(d) $BaO(s) + H_2SO_4(aq) \rightarrow$

19.88 Account for each of the following observations:

(a) H_2SO_4 is a stronger acid than H_2SO_3.

(b) SF_4 exists, but OF_4 does not.

(c) The S_8 ring is nonplanar.

19.89 Account for each of the following observations:

(a) Oxygen is more electronegative than sulfur.

(b) Sulfur forms long S_n chains, but oxygen does not.

(c) The SO_3 molecule is trigonal planar, but the SO_3^{2-} ion is trigonal pyramidal.

Halogen Oxoacids and Oxoacid Salts

19.90 Write the formula for each of the following compounds, and indicate the oxidation state of the halogen:

(a) Bromic acid

(b) Hypoiodous acid

(c) Sodium chlorite

(d) Potassium metaperiodate

19.91 Write the formula for each of the following compounds, and indicate the oxidation state of the halogen:

(a) Potassium hypobromite

(b) Paraperiodic acid

(c) Sodium bromate

(d) Chlorous acid

19.92 Name each of the following compounds:

(a) HIO_3 (b) $HClO_2$

(c) $NaOBr$ (d) $LiClO_4$

19.93 Name each of the following compounds:

(a) $KClO_2$ (b) HIO_4

(c) $HOBr$ (d) $NaBrO_3$

19.94 Write an electron-dot structure for each of the following molecules or ions, and predict the molecular geometry:

(a) HIO_3 (b) ClO_2^-

(c) $HOCl$ (d) IO_6^{5-}

19.95 Write an electron-dot structure for each of the following molecules or ions, and predict the molecular geometry:

(a) BrO_4^- (b) ClO_3^-

(c) HIO_4 (d) $HOBr$

19.96 Explain why acid strength increases in the order $HClO < HClO_2 < HClO_3 < HClO_4$.

19.97 Explain why acid strength increases in the order $HIO < HBrO < HClO$.

19.98 Write a balanced net ionic equation for each of the following reactions:

(a) $Br_2(l) + $ cold $NaOH(aq) \rightarrow$

(b) $Cl_2(g) + $ cold $H_2O(l) \rightarrow$

(c) $Cl_2(g) + $ hot $NaOH(aq) \rightarrow$

19.99 Write a balanced equation for the reaction of potassium chlorate and sucrose. The products are $KCl(s)$, $CO_2(g)$, and $H_2O(g)$.

General Problems

19.100 What is the chemical formula for the most important mineral sources of each of the following elements?

(a) B   (b) P   (c) S

19.101 What are the principal uses of the following industrial chemicals?

(a) H_2SO_4 (b) N_2 (c) NH_3

(d) HNO_3 (e) H_3PO_4

19.102 Which compound in each of the following pairs has the higher melting point?

(a) $LiCl$ or PCl_3

(b) CO_2 or SiO_2

(c) P_4O_{10} or NO_2

19.103 Which element in each of the following pairs is the better electrical conductor?

(a) B or Ga

(b) In or S

(c) Pb or P

19.104 The structures of the various phosphate and silicate anions have many similarities. List some.

19.105 Compare and contrast the properties of ammonia and phosphine.

19.106 How many of the four most abundant elements in the earth's crust and in the human body can you identify without consulting Figure 19.1?

19.107 Identify as many of the 10 most important industrial chemicals as you can without consulting Table 19.1.

19.108 Which of the group 4A elements have allotropes with the diamond structure? Which have metallic allotropes? How does the variation in the structure of the group 4A elements illustrate how metallic character varies down a periodic group?

19.109 Write a balanced chemical equation for a laboratory preparation of each of the following compounds:

(a) NH_3 (b) CO_2

(c) B_2H_6 (diborane)   (d) C_2H_2 (acetylene)

(e) N_2O (f) NO_2

19.110 Write balanced equations for the reactions of (a) H_3PO_4 and (b) $B(OH)_3$ with water. Classify each acid as a Brønsted–Lowry acid or a Lewis acid.

19.111 What sulfur-containing compound is present in acid rain, and how is it formed by the burning of sulfur-containing fossil fuels?

19.112 Account for each of the following observations:

(a) Diamond is extremely hard and high melting, whereas graphite is very soft and high melting.

(b) Chlorine does not form a perhalic acid, H_5ClO_6.

19.113 Give one example from main-group chemistry that illustrates each of the following descriptions:

(a) Covalent network solid

(b) Disproportionation reaction

(c) Paramagnetic oxide

(d) Polar molecule that violates the octet rule

(e) Lewis acid

(f) Amphoteric oxide

(g) Semiconductor

(h) Strong oxidizing agent

(i) Allotropes

19.114 Could the strain in the P_4 molecule be reduced by using sp^3 hybrid orbitals in bonding instead of pure p orbitals? Explain.

19.115 Carbon is an essential element in the molecules on which life is based. Would silicon be equally satisfactory? Explain.

Multi-Concept Problems

19.116 An important physiological reaction of nitric oxide (NO) is its interaction with the superoxide ion (O_2^-) to form the peroxynitrite ion ($ONOO^-$).

(a) Write electron-dot structures for NO, O_2^-, and $ONOO^-$, and predict the O–N–O bond angle in $ONOO^-$.

(b) The bond length in NO (115 pm) is intermediate between the length of an NO triple bond and an NO double bond. Account for the bond length and the paramagnetism of NO using molecular orbital theory.

19.117 Consider phosphorous acid, a polyprotic acid with formula H_3PO_3.

(a) Draw two plausible structures for H_3PO_3. For each one, predict the shape of the pH titration curve for titration of the H_3PO_3 ($K_{a1} = 1.0 \times 10^{-2}$) with aqueous NaOH.

(b) For the structure with the H atoms in two different environments, calculate the pH at the first and second equivalence points assuming that 30.00 mL of 0.1240 M H_3PO_3 ($K_{a2} = 2.6 \times 10^{-7}$) is titrated with 0.1000 M NaOH.

19.118 We've said that the +1 oxidation state is uncommon for indium but is the most stable state for thallium. Verify this statement by calculating $E°$ and $\Delta G°$ (in kilojoules) for the disproportionation reaction

$$3\,M^+(aq) \longrightarrow M^{3+}(aq) + 2\,M(s) \quad M = In \text{ or } Tl$$

Is disproportionation a spontaneous reaction for In^+ and/or Tl^+? Standard reduction potentials for the relevant half-reactions are

$$In^{3+}(aq) + 2\,e^- \longrightarrow In^+(aq) \quad E° = -0.44 \text{ V}$$
$$In^+(aq) + e^- \longrightarrow In(s) \quad E° = -0.14 \text{ V}$$
$$Tl^{3+}(aq) + 2\,e^- \longrightarrow Tl^+(aq) \quad E° = +1.25 \text{ V}$$
$$Tl^+(aq) + e^- \longrightarrow Tl(s) \quad E° = -0.34 \text{ V}$$

19.119 Terrorists often use ammonium nitrate fertilizer as an ingredient in car bombs. When ammonium nitrate explodes, it decomposes to gaseous nitrogen, oxygen, and water vapor. The force of the explosion results from sudden production of a huge volume of hot gas.

(a) Write a balanced equation for the reaction.

(b) What volume of gas (in liters) is produced from explosion of 1.80 m^3 of solid NH_4NO_3? Assume that the gas has a temperature of 500°C and a pressure of 1.00 atm. The density of NH_4NO_3 is 1.725 g/cm^3.

(c) Use the thermodynamic data in Appendix B to calculate the amount of heat (in kilojoules) released in the reaction.

19.120 It has been claimed that NH_4NO_3 fertilizer can be rendered unexplodable (see Problem 19.119) by a process that involves additives such as diammonium hydrogen phosphate, $(NH_4)_2HPO_4$. Analysis of such a "desensitized" sample of NH_4NO_3 showed the mass % nitrogen to be 33.81%.

(a) Assuming that the mixture contains only NH_4NO_3 and $(NH_4)_2HPO_4$, what is the mass % of each of these two components?

(b) A 0.965 g sample of the mixture was dissolved in enough water to make 50.0 mL of solution. What is the pH of the solution? (Hint: The strongest base present is HPO_4^{2-}.)

19.121 A 5.00 g quantity of white phosphorus was burned in an excess of oxygen, and the product was dissolved in enough water to make 250.0 mL of solution.

(a) Write balanced equations for the reactions.

(b) What is the pH of the solution?

(c) When the solution was treated with an excess of aqueous $Ca(NO_3)_2$, a white precipitate was obtained. Write a balanced equation for the reaction, and calculate the mass of the precipitate in grams.

(d) The precipitate in part (c) was removed, and the solution that remained was treated with an excess of zinc, yielding a colorless gas that was collected at 20°C and 742 mm Hg. Identify the gas, and determine its volume.

 eMedia Problems

19.122 The **Physical Properties of the Halogens** movie (*eChapter 19.14*) shows three of the halogens and indicates the physical state of each at room temperature. Given that all of the halogens are nonpolar, diatomic molecules, explain in terms of intermolecular forces how the halogens can exist in three different states at room temperature.

19.123 Watch the **Nitrogen Dioxide and Dinitrogen Tetroxide** movie (*eChapter 19.10*) and answer the following questions:

(a) What is the equation corresponding to the equilibrium between the two species?

(b) Based on the information in the movie, which species would predominate at high temperature? Which species would predominate at low temperature?

(c) Is the reaction in part (a) endothermic or exothermic? Explain your reasoning.

19.124 Many nonmetals, including nitrogen, can form several different oxides. The **Nitrogen Dioxide and Dinitrogen Tetroxide** movie (*eChapter 19.10*) illustrates the equilibrium between NO_2 and N_2O_4. Like nitrogen, sulfur forms several different oxides including SO_2. Why is it that NO_2 molecules can combine to produce N_2O_4 molecules but SO_2 molecules cannot combine to give S_2O_4 molecules? Draw electron-dot structures to support your answer.

Transition Elements and Coordination Chemistry

Colorful transition metal compounds are pigments used in many paints.

The **transition elements** occupy the central part of the periodic table, bridging the gap between the active *s*-block metals of groups 1A and 2A on the left and the *p*-block metals, semimetals, and nonmetals of groups 3A–8A on the right (Figure 20.1). Because the *d* subshells are being filled in this region of the periodic table, the transition elements are also called the **d-block elements**.

Each *d* subshell consists of five orbitals and can accommodate 10 electrons, so each transition series consists of 10 elements. The first series extends from scandium through zinc and includes many familiar metals, such as chromium, iron, and copper. The second series runs from yttrium through cadmium, and the third series runs from lanthanum through mercury. In addition, there is a fourth transition series made up of actinium through the recently discovered element 112.

Main groups

Main groups

1 1A																	18 8A
1 H	2 2A											13 3A	14 4A	15 5A	16 6A	17 7A	2 He
3 Li	4 Be			Transition-metal groups								5 B	6 C	7 N	8 O	9 F	10 Ne
11 Na	12 Mg	3 3B	4 4B	5 5B	6 6B	7 7B	8	9 8B	10	11 1B	12 2B	13 Al	14 Si	15 P	16 S	17 Cl	18 Ar
19 K	20 Ca	21 Sc	22 Ti	23 V	24 Cr	25 Mn	26 Fe	27 Co	28 Ni	29 Cu	30 Zn	31 Ga	32 Ge	33 As	34 Se	35 Br	36 Kr
37 Rb	38 Sr	39 Y	40 Zr	41 Nb	42 Mo	43 Tc	44 Ru	45 Rh	46 Pd	47 Ag	48 Cd	49 In	50 Sn	51 Sb	52 Te	53 I	54 Xe
55 Cs	56 Ba	57 La	72 Hf	73 Ta	74 W	75 Re	76 Os	77 Ir	78 Pt	79 Au	80 Hg	81 Tl	82 Pb	83 Bi	84 Po	85 At	86 Rn
87 Fr	88 Ra	89 Ac	104 Rf	105 Db	106 Sg	107 Bh	108 Hs	109 Mt	110	111	112		114		116		118

Lanthanides	58 Ce	59 Pr	60 Nd	61 Pm	62 Sm	63 Eu	64 Gd	65 Tb	66 Dy	67 Ho	68 Er	69 Tm	70 Yb	71 Lu
Actinides	90 Th	91 Pa	92 U	93 Np	94 Pu	95 Am	96 Cm	97 Bk	98 Cf	99 Es	100 Fm	101 Md	102 No	103 Lr

FIGURE 20.1 The transition elements (*d*-block elements, shown in yellow) are located in the central region of the periodic table between the *s*-block and *p*-block main-group elements. The two series of inner transition elements (*f*-block elements, shown in green) follow lanthanum and actinium.

Tucked into the periodic table between lanthanum (atomic number 57) and hafnium (atomic number 72) are the **lanthanides**. In this series of 14 metallic elements, the seven 4*f* orbitals are progressively filled, as shown in Figure 5.18 (page 188). Following actinium (atomic number 89) is a second series of 14 elements, the **actinides**, in which the 5*f* subshell is progressively filled. The lanthanides and actinides together comprise the *f*-block elements, or **inner transition elements**.

The transition metals iron and copper have been known since antiquity and have played an important role in the development of civilization. Iron, the main constituent of steel, is still important as a structural material. Worldwide production of steel amounts to some 800 million tons per year. In newer technologies, other transition elements are useful. For example, the strong, lightweight metal titanium is a major component in modern jet aircraft. Transition metals are also used as heterogeneous catalysts in automobile catalytic converters and in the industrial synthesis of essential chemicals such as sulfuric acid, nitric acid, and ammonia.

The role of the transition elements in living systems is equally important. Iron is present in biomolecules such as hemoglobin, which transports oxygen from our lungs to other parts of the body. Cobalt is an essential component of vitamin B_{12}. Nickel, copper, and zinc are vital constituents of many enzymes, the large protein molecules that act as catalysts for biochemical reactions.

In this chapter, we'll look at the properties and chemical behavior of transition metal compounds, paying special attention to *coordination compounds*, in which a central metal ion (or atom)—usually a transition metal—is attached to a group of surrounding molecules or ions by coordinate covalent bonds (Section 7.5).

20.1 Electron Configurations

Look at the electron configurations of potassium and calcium, the *s*-block elements immediately preceding the first transition series. These atoms have 4*s* valence electrons, but no *d* electrons:

$$K \ (Z = 19): [Ar] \, 3d^0 \, 4s^1 \qquad Ca \ (Z = 20): [Ar] \, 3d^0 \, 4s^2$$

The filling of the $3d$ subshell begins at atomic number 21 (scandium) and continues until the subshell is completely filled at atomic number 30 (zinc):

$$\text{Sc } (Z = 21): [\text{Ar}] \, 3d^1 \, 4s^2 \longrightarrow \text{Zn } (Z = 30): [\text{Ar}] \, 3d^{10} \, 4s^2$$

(It's convenient to list the valence electrons in order of principal quantum number because the $4s$ electrons are lost first in forming ions.)

The valence electrons are generally considered to be those in the outermost shell (Section 5.12) because they are the ones that are involved in chemical bonding. For transition elements, however, both the $(n-1)d$ and the ns electrons are involved in bonding and are considered valence electrons.

The filling of the $3d$ subshell generally proceeds according to Hund's rule (Section 5.12) with one electron adding to each of the five $3d$ orbitals before a second electron adds to any one of them. There are just two exceptions to the expected regular filling pattern, chromium and copper:

	Cr ($Z = 24$)	Cu ($Z = 29$)
Expected configuration	↑ ↑ ↑ ↑ ___ ↑↓ $\underbrace{}_{3d^4}$ $\underbrace{}_{4s^2}$	↑↓ ↑↓ ↑↓ ↑↓ ↑ ↑↓ $\underbrace{}_{3d^9}$ $\underbrace{}_{4s^2}$
Observed configuration	↑ ↑ ↑ ↑ ↑ ↑ $\underbrace{}_{3d^5}$ $\underbrace{}_{4s^1}$	↑↓ ↑↓ ↑↓ ↑↓ ↑↓ ↑ $\underbrace{}_{3d^{10}}$ $\underbrace{}_{4s^1}$

Electron configurations depend on both orbital energies and electron–electron repulsions. Consequently, it's not always possible to predict configurations when two valence subshells have similar energies. It's often found, however, that exceptions from the expected orbital filling pattern result in either half-filled or completely filled subshells. In the case of chromium, for example, the $3d$ and $4s$ subshells have similar energies. It's evidently advantageous to shift one electron from the $4s$ to the $3d$ subshell, which decreases electron–electron repulsions and gives two half-filled subshells. Because each valence electron is in a separate orbital, the electron–electron repulsion that would otherwise occur between the two $4s$ electrons in the expected configuration is eliminated. A similar shift of one electron from $4s$ to $3d$ in copper gives a completely filled $3d$ subshell and a half-filled $4s$ subshell.

In contrast to the situation for the neutral atoms, the electron configurations of transition metal cations are easy to predict: The valence s orbital is vacant, and all the valence electrons occupy the d orbitals. Iron, for example, which forms 2+ and 3+ cations, has the following valence electron configurations:

Fe: ↑↓ ↑ ↑ ↑ ↑   ↑↓   Fe^{2+}: ↑↓ ↑ ↑ ↑ ↑   Fe^{3+}: ↑ ↑ ↑ ↑ ↑
$\quad\underbrace{}_{3d^6}\ \underbrace{}_{4s^2}$ $\qquad\underbrace{}_{3d^6}$ $\qquad\quad\underbrace{}_{3d^5}$

When a neutral atom loses one or more electrons, the remaining electrons are less shielded, and the effective nuclear charge (Z_{eff}) increases. Consequently, the remaining electrons are more strongly attracted to the nucleus, and their orbital energies decrease. It turns out that the $3d$ orbitals experience a steeper drop in energy with increasing Z_{eff} than does the $4s$ orbital, making the $3d$ orbitals in cations lower in energy than the $4s$ orbital. As a result, all the valence electrons occupy the $3d$ orbitals, and the $4s$ orbital is vacant.

In neutral molecules and complex anions, the metal atom usually has a positive oxidation state. It therefore has a partial positive charge and a higher Z_{eff} than that of the neutral atom. As a result, the $3d$ orbitals are again lower in energy than the $4s$ orbital, and so all the metal's valence electrons occupy the d orbitals. The metal atom in both VCl_4 and MnO_4^{2-}, for example, has the valence configuration $3d^1$. Electron configurations and other properties for atoms and common ions of first-series transition elements are summarized in Table 20.1.

TABLE 20.1   Selected Properties of First-Series Transition Elements

Group:	3B	4B	5B	6B	7B		8B		1B	2B
Element:	Sc	Ti	V	Cr	Mn	Fe	Co	Ni	Cu	Zn
Valence electron configuration										
M atom	$3d^14s^2$	$3d^24s^2$	$3d^34s^2$	$3d^54s^1$	$3d^54s^2$	$3d^64s^2$	$3d^74s^2$	$3d^84s^2$	$3d^{10}4s^1$	$3d^{10}4s^2$
M^{2+} ion		$3d^2$	$3d^3$	$3d^4$	$3d^5$	$3d^6$	$3d^7$	$3d^8$	$3d^9$	$3d^{10}$
M^{3+} ion	$3d^0$	$3d^1$	$3d^2$	$3d^3$	$3d^4$	$3d^5$	$3d^6$			
Elec. conductivity*	3	4	8	11	1	17	24	24	96	27
Melting point (°C)	1541	1668	1910	1907	1246	1538	1495	1455	1085	420
Boiling point (°C)	2836	3287	3407	2671	2061	2861	2927	2913	2562	907
Density (g/cm³)	2.99	4.51	6.0	7.15	7.3	7.87	8.86	8.90	8.96	7.14
Atomic radius (pm)	162	147	134	128	127	126	125	124	128	134
E_i (kJ/mol)†										
First	631	658	650	653	717	759	758	737	745	906
Second	1235	1310	1414	1592	1509	1561	1646	1753	1958	1733
Third	2389	2652	2828	2987	3248	2957	3232	3393	3554	3833

*Electrical conductivity relative to an arbitrary value of 100 for silver.
†Ionization energy.

Transition elements are placed in groups of the periodic table designated 1B–8B because their valence electron configurations are similar to those of analogous elements in the main groups 1A–8A. Thus, copper in group 1B ([Ar] $3d^{10}\,4s^1$) and zinc in group 2B ([Ar] $3d^{10}\,4s^2$) have valence electron configurations similar to those of potassium in group 1A ([Ar] $4s^1$) and calcium in group 2A ([Ar] $4s^2$). Similarly, scandium in group 3B ([Ar] $3d^1\,4s^2$) through iron in group 8B ([Ar] $3d^6\,4s^2$) have the same number of valence electrons as the p-block elements aluminum in group 3A ([Ne] $3s^2\,3p^1$) through argon in group 8A ([Ne] $3s^2\,3p^6$). Cobalt ([Ar] $3d^7\,4s^2$) and nickel ([Ar] $3d^8\,4s^2$) are also assigned to group 8B although there are no main-group elements with 9 or 10 valence electrons.

EXAMPLE 20.1   Give the electron configuration of the metal in each of the following atoms or ions:
(a) Ni   (b) Cr^{3+} (c) FeO_4^{2-} (ferrate ion)

SOLUTION
(a) Nickel (Z = 28) has a total of 28 electrons, including the argon core of 18. In neutral atoms of the first transition series, the $4s$ orbital is usually filled with 2 electrons, and the remaining electrons occupy the $3d$ orbitals. The electron configuration is therefore [Ar] $3d^8\,4s^2$.
(b) A neutral Cr atom (Z = 24) has a total of 24 electrons; a Cr^{3+} ion has 24 − 3 = 21 electrons. In transition metal ions, all the valence electrons occupy the d orbitals. Thus, the electron configuration is [Ar] $3d^3$.

(c) First, determine the oxidation number of iron in FeO_4^{2-}. Because each of the four oxygens has an oxidation number of -2 and the overall charge on the oxo-anion is -2, the oxidation number of the iron must be $+6$. An iron(VI) atom has a total of 20 electrons, 6 less than a neutral iron atom ($Z = 26$). Because the valence electrons occupy the $3d$ orbitals, the electron configuration of Fe(VI) is [Ar] $3d^2$.

▶ **PROBLEM 20.1**   Give the electron configuration of the metal in each of the following:
(a) V   **(b)** Co^{2+} **(c)** MnO_2

◀— **KEY CONCEPT PROBLEM 20.2**   On the periodic table below, locate the transition metal atom or ion with the following electron configurations. Identify each atom or ion.
(a) An atom: [Ar] $3d^5\,4s^2$ **(b)** A 2+ ion: [Ar] $3d^8$
(c) An atom: [Kr] $4d^{10}\,4s^1$ **(d)** A 3+ ion: [Kr] $3d^3$

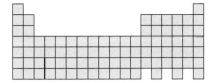

20.2   Properties of Transition Elements

Let's look at some trends in the properties of the transition elements shown in Table 20.1 and try to understand them in terms of electron configurations.

Metallic Properties

All the transition elements are metals. Like the metals of groups 1A and 2A, the transition metals are malleable, ductile, lustrous, and good conductors of heat and electricity. Silver has the highest electrical conductivity of any element at room temperature, with copper a close second. The transition metals are harder, have higher melting and boiling points, and are more dense than the group 1A and 2A metals, largely because the sharing of d, as well as s, electrons results in stronger bonding.

    From left to right across the first transition series, melting points increase from 1541°C for Sc to a maximum of 1910°C for V in group 5B, then decrease to 1085°C for Cu (Table 20.1, Figure 20.2). The second- and third-series

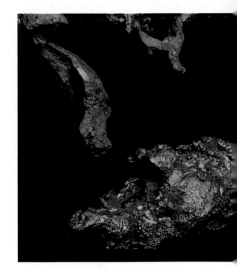

Native copper.

FIGURE 20.2   Relative melting points of the transition elements. Melting points reach a maximum value in the middle of each series.

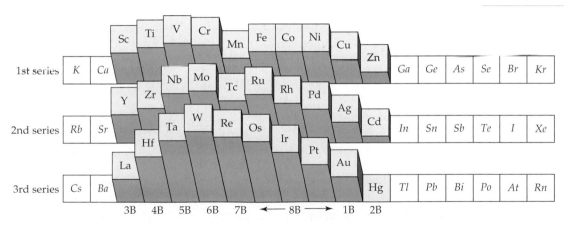

Interactive Periodic Table

elements exhibit a similar maximum in melting point, but at group 6B: 2623°C for Mo and 3422°C for W, the metal with the highest melting point. The melting points increase as the number of unpaired *d* electrons available for bonding increases and then decrease as the *d* electrons pair up and become less available for bonding. Note that zinc ($3d^{10} 4s^2$), in which all the *d* and *s* electrons are paired, has a relatively low melting point (420°C) and that mercury ($4f^{14} 5d^{10} 6s^2$) is a liquid at room temperature (mp −39°C). We'll look at bonding in metals in more detail in Section 21.4.

Atomic Radii and Densities

Atomic radii are given in Figure 20.3. From left to right across a transition series, the atomic radii decrease, at first markedly and then more gradually after group 6B. Toward the end of each series, the radii increase again. The decrease in radii with increasing atomic number is due to the fact that the added *d* electrons only partially shield the added nuclear charge (Section 5.15). As a result, the effective nuclear charge Z_{eff} increases. With increasing Z_{eff}, the electrons are more strongly attracted to the nucleus, and atomic size decreases. The upturn in radii toward the end of each series is probably due to more effective shielding and to increasing electron–electron repulsion as double occupation of the *d* orbitals is completed. In contrast to the large variation in radii for main-group elements, all transition metal atoms have quite similar radii, which accounts for their ability to blend together in forming alloys such as brass (mostly copper and zinc).

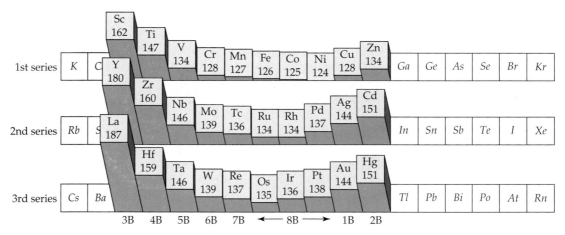

FIGURE 20.3 Atomic radii (in pm) of the transition elements. The radii decrease with increasing atomic number and then increase again toward the end of each transition series. Note that the second- and third-series transition elements have nearly identical radii.

It's interesting to note that the atomic radii of the second- and third-series transition elements from group 4B on are nearly identical, though we would expect an increase in size on adding an entire principal quantum shell of electrons. The small sizes of the third-series atoms are associated with what is called the **lanthanide contraction**, the general decrease in atomic radii of the *f*-block lanthanide elements between the second and third transition series (Figure 20.4).

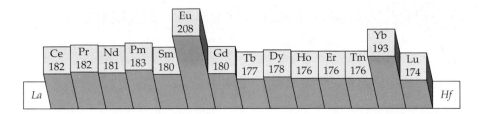

FIGURE 20.4   Atomic radii (in pm) of the lanthanide elements. The radii generally decrease with increasing atomic number.

The lanthanide contraction is due to the increase in effective nuclear charge with increasing atomic number as the $4f$ subshell is filled. By the end of the lanthanides, the size *decrease* due to a larger Z_{eff} almost exactly compensates for the expected size *increase* due to an added quantum shell of electrons. Consequently, atoms of the third transition series have radii very similar to those of the second transition series.

As you might expect, the densities of the transition metals are inversely related to their atomic radii (Figure 20.5). The densities initially increase from left to right across each transition series and then decrease toward the end of each series. Because the second- and third-series elements have nearly the same atomic volume, the much heavier third-series elements have unusually high densities: 22.6 g/cm^3 for osmium and 22.5 g/cm^3 for iridium, the most dense elements.

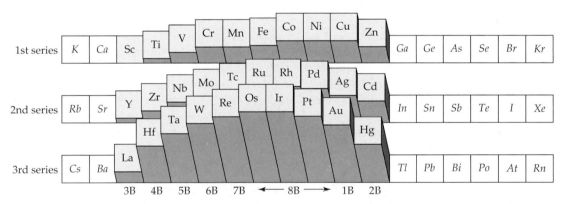

FIGURE 20.5   Relative densities of the transition metals. Density initially increases across each series and then decreases.

Ionization Energies and Oxidation Potentials

Ionization energies generally increase from left to right across a transition series, though there are some irregularities, as indicated in Table 20.1 for the atoms of the first transition series. The general trend correlates with an increase in effective nuclear charge and a decrease in atomic radius.

Table 20.2 lists standard potentials $E°$ for oxidation of first-series transition metals. Note that these potentials are the negative of the corresponding standard reduction potentials (Table 18.1, page 771). Except for copper, all the $E°$ values are positive, which means that the solid metal is oxidized to its aqueous cation more readily than H_2 gas is oxidized to $H^+(aq)$.

$$M(s) \longrightarrow M^{2+}(aq) + 2\,e^- \qquad E° > 0\ V \qquad \text{Product is } M^{3+} \text{ for } M = Sc$$

$$H_2(g) \longrightarrow 2\,H^+(aq) + 2\,e^- \qquad E° = 0\ V$$

TABLE 20.2   Standard Potentials for Oxidation of First-Series Transition Metals

Oxidation Half-Reaction	$E°$ (V)	Oxidation Half-Reaction	$E°$ (V)
$Sc(s) \rightarrow Sc^{3+}(aq) + 3\ e^-$	2.08	$Fe(s) \rightarrow Fe^{2+}(aq) + 2\ e^-$	0.45
$Ti(s) \rightarrow Ti^{2+}(aq) + 2\ e^-$	1.63	$Co(s) \rightarrow Co^{2+}(aq) + 2\ e^-$	0.28
$V(s) \rightarrow V^{2+}(aq) + 2\ e^-$	1.18	$Ni(s) \rightarrow Ni^{2+}(aq) + 2\ e^-$	0.26
$Cr(s) \rightarrow Cr^{2+}(aq) + 2\ e^-$	0.91	$Cu(s) \rightarrow Cu^{2+}(aq) + 2\ e^-$	−0.34
$Mn(s) \rightarrow Mn^{2+}(aq) + 2\ e^-$	1.18	$Zn(s) \rightarrow Zn^{2+}(aq) + 2\ e^-$	0.76

In other words, the first-series metals, except for copper, are stronger reducing agents than H_2 gas and can therefore be oxidized by the H^+ ion in acids like HCl that lack an oxidizing anion:

$$M(s) + 2\ H^+(aq) \longrightarrow M^{2+}(aq) + H_2(g) \quad E° > 0\ V \text{ (except for M = Cu)}$$

Oxidation of copper requires a stronger oxidizing agent, such as HNO_3.

The standard potential for oxidation of a metal is a composite property that depends on $\Delta G°$ for sublimation of the metal, the ionization energies of the metal atom, and $\Delta G°$ for hydration of the metal ion:

$$M(s) \xrightarrow{\Delta G°_{subl}} M(g) \xrightarrow{E_i\ (-2\ e^-)} M^{2+}(g) \xrightarrow{\Delta G°_{hydr}} M^{2+}(aq)$$

Nevertheless, the general trend in the $E°$ values shown in Table 20.2 correlates with the general trend in the ionization energies in Table 20.1. The ease of oxidation of the metal decreases as the ionization energies increase across the transition series from Sc to Zn. (Only Mn and Zn deviate from the trend of decreasing $E°$ values.) Thus, the so-called *early* transition metals, those on the left side of the d block (Sc through Mn), are oxidized most easily and are the strongest reducing agents.

20.3   Oxidation States of Transition Elements

The transition elements differ from most main-group metals in that they exhibit a variety of oxidation states. Sodium, magnesium, and aluminum, for example, have a single oxidation state equal to their periodic group number (Na^+, Mg^{2+}, and Al^{3+}), but the transition elements frequently have oxidation states less than their group number. For example, manganese in group 7B shows oxidation states of +2 in $Mn^{2+}(aq)$, +3 in $Mn(OH)_3(s)$, +4 in $MnO_2(s)$, +6 in $MnO_4^{2-}(aq)$ (manganate ion), and +7 in $MnO_4^-(aq)$ (permanganate ion). Figure 20.6 summarizes the common oxidation states for elements of the first transition series, with the most frequently encountered ones indicated in red.

FIGURE 20.6   Common oxidation states for first-series transition elements. The states encountered most frequently are shown in red. The highest oxidation state for the group 3B–7B metals is their periodic group number, but the group 8B transition metals have a maximum oxidation state less than their group number. Most transition elements have more than one common oxidation state.

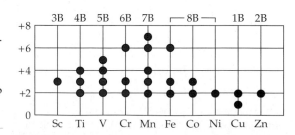

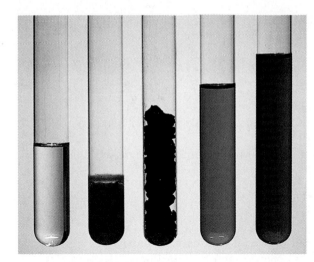

Manganese has different oxidation states and different colors in $Mn^{2+}(aq)$, $Mn(OH)_3(s)$, $MnO_2(s)$, $MnO_4^{2-}(aq)$, and $MnO_4^-(aq)$ (from left to right).

All the first-series transition elements except scandium form a 2+ cation, corresponding to loss of the two $4s$ valence electrons. Because the $3d$ and $4s$ orbitals have similar energies, loss of a $3d$ electron is also possible, yielding 3+ cations such as $V^{3+}(aq)$, $Cr^{3+}(aq)$, and $Fe^{3+}(aq)$. Additional energy is required to remove the third electron, but this is more than compensated for by the larger (more negative) $\Delta G°$ of hydration of the more highly charged 3+ cation. Still higher oxidation states result from loss or sharing of additional d electrons. In their highest oxidation states, the transition elements are combined with the most electronegative elements (F and O): for example, $VF_5(l)$ and $V_2O_5(s)$ for vanadium in group 5B; CrO_4^{2-} (chromate ion) and $Cr_2O_7^{2-}$ (dichromate ion) for chromium in group 6B; MnO_4^- for manganese in group 7B.

Note in Figure 20.6 that the highest oxidation state for the group 3B–7B metals is the group number, corresponding to loss or sharing of all the valence s and d electrons. For the group 8B transition metals, though, loss or sharing of all the valence electrons is energetically prohibitive because of the increasing value of Z_{eff}. Consequently, only lower oxidation states are accessible for these transition metals—for example, +6 in FeO_4^{2-} and +3 in Co^{3+}. Even these species have a great tendency to be reduced to still lower oxidation states. For example, the aqueous Co^{3+} ion oxidizes water to O_2 gas and is thereby reduced to Co^{2+}:

$$4\,Co^{3+}(aq) + 2\,H_2O(l) \longrightarrow 4\,Co^{2+}(aq) + O_2(g) + 4\,H^+(aq) \qquad E° = +0.58\ V$$

In general, ions that have the transition metal in a high oxidation state tend to be good oxidizing agents—for example, $Cr_2O_7^{2-}$, MnO_4^-, and FeO_4^{2-}. Conversely, early transition metal ions with the metal in a low oxidation state are good reducing agents —for example, V^{2+} and Cr^{2+}. Divalent ions of the *later* transition metals on the right side of the d block, such as Co^{2+}, Ni^{2+}, Cu^{2+}, and Zn^{2+}, are poor reducing agents because of the larger value of Z_{eff}. In fact, zinc has only one oxidation state (+2).

The elements of the second and third transition series also exhibit a variety of oxidation states. In general, the stability of the higher oxidation states increases down a periodic group. In group 8B, for example, the oxide of iron with the highest oxidation state is iron(III) oxide, Fe_2O_3. Ruthenium and osmium, though, form volatile tetroxides, RuO_4 and OsO_4, in which the metals have an oxidation state of +8.

← **KEY CONCEPT PROBLEM 20.3** How does the effective nuclear charge, Z_{eff}, vary from left to right across the first transition series?

Sc	Ti	V	Cr	Mn	Fe	Co	Ni	Cu	Zn

Based on the variation in Z_{eff},

(a) Which M^{2+} ion (M = Ti–Zn) should be the strongest reducing agent? Which should be the weakest?

(b) Which oxoanion (VO_4^{3-}, CrO_4^{2-}, MnO_4^{2-}, or FeO_4^{2-}) should be the strongest oxidizing agent? Which should be the weakest?

20.4 Chemistry of Selected Transition Elements

Experimental work in transition metal chemistry is particularly enjoyable because most transition metal compounds have brilliant colors. In this section, we'll look at the chemistry of some representative elements commonly encountered in the laboratory.

Chromium

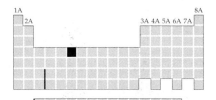

Chromium, which gets its name from the Greek word for color (*chroma*), is obtained from the ore chromite, a mixed metal oxide with the formula $FeO \cdot Cr_2O_3$, or $FeCr_2O_4$. Reduction of chromite with carbon gives the alloy ferrochrome, which is used in making stainless steel, a hard, corrosion-resistant steel that contains up to 30% chromium.

$$FeCr_2O_4(s) + 4\,C(s) \longrightarrow \underbrace{Fe(s) + 2\,Cr(s)}_{\text{Ferrochrome}} + 4\,CO(g)$$

Pure chromium is obtained by reducing chromium(III) oxide with aluminum:

$$Cr_2O_3(s) + 2\,Al(s) \longrightarrow 2\,Cr(s) + Al_2O_3(s)$$

In addition to its use in making steels, chromium is widely used to electroplate metallic objects with an attractive, protective coating (Section 18.12). Chromium is hard and lustrous, takes a high polish, and is resistant to corrosion because an invisible, microscopic film of chromium(III) oxide shields the surface from further oxidation.

The aqueous solution chemistry of chromium can be systematized according to its oxidation states and the species that exist under acidic and basic conditions (Table 20.3). The common oxidation states are +2, +3, and +6, with the +3 state the most stable.

Chromium metal reacts with aqueous acids in the absence of oxygen to give $Cr^{2+}(aq)$, the beautiful blue chromium(II) (chromous) ion, in which Cr^{2+} is bound to six water molecules, $Cr(H_2O)_6^{2+}$ (Figure 20.7a):

$$Cr(s) + 2\,H^+(aq) \longrightarrow Cr^{2+}(aq) + H_2(g) \qquad E^\circ = +0.91\text{ V}$$

In air, the reaction of chromium metal with acids yields chromium(III) because chromium(II) is rapidly oxidized by atmospheric oxygen:

$$4\,Cr^{2+}(aq) + O_2(g) + 4\,H^+(aq) \longrightarrow 4\,Cr^{3+}(aq) + 2\,H_2O(l) \qquad E^\circ = +1.64\text{ V}$$

This sculpture, *Herakles in Ithaka* by Jason Seley, was made from chromium-plated automobile bumpers.
(Jason Seley [1919–1983], Herakles in Ithaka I, 1980–81. Sculpture, welded steel, 342.9 cm. Gift of the artist, Herbert F. Johnson Museum of Art, Cornell University. Photo by Robert Barker, Cornell University Photography.)

TABLE 20.3   Chromium Species in Common Oxidation States

	Oxidation State		
	+2	**+3**	**+6**
Acidic solution:	$Cr^{2+}(aq)$ $\xrightarrow{+0.41\ V}$	$Cr^{3+}(aq)$ $\xleftarrow{+1.33\ V}$	$Cr_2O_7^{2-}(aq)$
	Chromium(II) ion (Chromous ion) Blue	Chromium(III) ion (Chromic ion) Violet	Dichromate ion Orange
Basic solution:	$Cr(OH)_2(s)$ Light blue	$Cr(OH)_3(s)$ Pale green $\xleftarrow{-0.13\ V}$	$CrO_4^{2-}(aq)$ Chromate ion Yellow
		$Cr(OH)_4^{-}(aq)$ Chromite ion Deep green	

Although the $Cr(H_2O)_6^{3+}$ ion is violet, chromium(III) solutions are often green because anions replace some of the bound water molecules to give green complex ions such as $Cr(H_2O)_5Cl^{2+}$ and $Cr(H_2O)_4Cl_2^+$ (Figure 20.7b).

In basic solution, chromium(III) precipitates as chromium(III) hydroxide, a pale green solid that dissolves both in acid and in excess base (Figure 20.8):

In acid: $Cr(OH)_3(s) + 3\ H_3O^+(aq) \rightleftharpoons Cr^{3+}(aq) + 6\ H_2O(l)$

In excess base: $Cr(OH)_3(s) + OH^-(aq) \rightleftharpoons Cr(OH)_4^-(aq)$

Recall that this behavior is typical of amphoteric metal hydroxides (Section 16.12).

(a)   **(b)**

FIGURE 20.7 **(a)** Reaction of chromium metal with 6 M H_2SO_4 in the absence of air gives bubbles of H_2 gas and a solution containing the blue chromium(II) ion. **(b)** Reaction of chromium metal with 6 M HCl in the presence of air yields a green solution containing chromium(III) species.

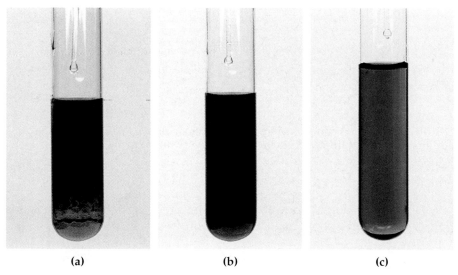

(a)   **(b)**   **(c)**

FIGURE 20.8 **(a)** Slow addition of aqueous NaOH to a solution of the violet $Cr(H_2O)_6^{3+}$ ion gives a pale green precipitate of chromium(III) hydroxide, $Cr(OH)_3$. **(b)** The $Cr(OH)_3$ dissolves on addition of H_2SO_4, reforming the $Cr(H_2O)_6^{3+}$ ion. **(c)** $Cr(OH)_3$ also dissolves on addition of excess NaOH, yielding the deep green chromite ion, $Cr(OH)_4^-$.

In contrast to the amphoteric $Cr(OH)_3$, chromium(II) hydroxide is a typical basic hydroxide. It dissolves in acid, but not in excess base. Conversely, the chromium(VI) compound, $CrO_2(OH)_2$, is a strong acid (chromic acid, H_2CrO_4). Recall from Section 15.15 that acid strength increases with increasing polarity of the O–H bonds, which increases, in turn, with increasing oxidation number of the chromium atom.

$$
\begin{array}{ccc}
+2 & +3 & +6 \\
Cr(OH)_2 & Cr(OH)_3 & CrO_2(OH)_2 \\
\text{Basic} & \text{Amphoteric} & \text{Acidic}
\end{array}
$$

⟶ Increasing strength as a proton donor ⟶

In the +6 oxidation state, the most important solution species are the yellow chromate ion (CrO_4^{2-}) and the orange dichromate ion ($Cr_2O_7^{2-}$). These ions are interconverted by the rapid equilibrium reaction

$$2\, CrO_4^{2-}(aq) + 2\, H^+(aq) \rightleftharpoons Cr_2O_7^{2-}(aq) + H_2O(l) \qquad K \approx 10^{14}$$

Because the equilibrium constant is about 10^{14}, CrO_4^{2-} ions predominate in basic solutions and $Cr_2O_7^{2-}$ ions predominate in acidic solutions (Figure 20.9).

(a)   (b)

FIGURE 20.9 **(a)** An aqueous solution of sodium chromate (Na_2CrO_4). **(b)** On addition of dilute sulfuric acid to the solution in (a), the yellow CrO_4^{2-} ion is converted to the orange $Cr_2O_7^{2-}$ ion.

The $Cr_2O_7^{2-}$ ion is a powerful oxidizing agent in acidic solution and is widely used as an oxidant in analytical chemistry.

$$Cr_2O_7^{2-}(aq) + 14\, H^+(aq) + 6\, e^- \longrightarrow 2\, Cr^{3+}(aq) + 7\, H_2O(l) \qquad E° = +1.33\ V$$

In basic solution, where CrO_4^{2-} is the predominant species, chromium(VI) is a much weaker oxidizing agent:

$$CrO_4^{2-}(aq) + 4\, H_2O(l) + 3\, e^- \longrightarrow Cr(OH)_3(s) + 5\, OH^-(aq) \qquad E° = -0.13\ V$$

Iron

Iron, the fourth most abundant element in the earth's crust (5.6% by mass), is immensely important both in human civilization and in living systems. Because iron is relatively soft and easily corroded, it is combined with carbon and other metals, such as vanadium, chromium, and manganese, to make alloys (steels) that are harder and less reactive than pure iron. Iron is obtained from its most important ores, hematite (Fe_2O_3) and magnetite (Fe_3O_4), by reduction with coke in a blast furnace (see Section 21.3). In living systems, iron is an essential constituent of numerous biomolecules. The body of a healthy human adult contains about 4 g of iron, 65% of which is present in the oxygen-carrying protein hemoglobin (Chapter 13 Interlude).

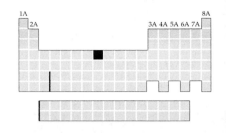

The most common oxidation states of iron are +2 (ferrous) and +3 (ferric). When iron metal reacts in the absence of air with an acid such as HCl, which lacks an oxidizing anion, the product is the light green iron(II) ion, $Fe(H_2O)_6^{2+}$:

$$Fe(s) + 2\,H^+(aq) \longrightarrow Fe^{2+}(aq) + H_2(g) \qquad E° = +0.45\,V$$

The oxidation stops at the iron(II) stage because the standard potential for oxidation of the iron(II) ion is negative:

$$Fe^{2+}(aq) \longrightarrow Fe^{3+}(aq) + e^- \qquad E° = -0.77\,V$$

In air, the iron(II) ion is slowly oxidized by atmospheric oxygen to the iron(III) ion, $Fe(H_2O)_6^{3+}$:

$$4\,Fe^{2+}(aq) + O_2(g) + 4\,H^+(aq) \longrightarrow$$
$$4\,Fe^{3+}(aq) + 2\,H_2O(l) \qquad E° = +0.46\,V$$

When iron is treated with an acid that has an oxidizing anion—for example, dilute nitric acid—the metal is oxidized directly to iron(III):

$$Fe(s) + NO_3^-(aq) + 4\,H^+(aq) \longrightarrow$$
$$Fe^{3+}(aq) + NO(g) + 2\,H_2O(l) \qquad E° = +1.00\,V$$

Addition of base to iron(III) solutions precipitates the gelatinous, red-brown hydrous oxide, $Fe_2O_3 \cdot x\,H_2O$, usually written as $Fe(OH)_3$ (Figure 20.10):

$$Fe^{3+}(aq) + 3\,OH^-(aq) \longrightarrow Fe(OH)_3(s)$$

FIGURE 20.10 When dilute NaOH is added to a solution of iron(III) sulfate, $Fe_2(SO_4)_3$, a gelatinous, red-brown precipitate of $Fe(OH)_3$ forms.

Because $Fe(OH)_3$ is very insoluble ($K_{sp} = 2.6 \times 10^{-39}$), it forms as soon as the pH rises above pH 2. The red-brown rust stains that you've seen in sinks and bathtubs are due to air oxidation of $Fe^{2+}(aq)$ followed by deposition of hydrous iron(III) oxide. Unlike $Cr(OH)_3$, $Fe(OH)_3$ is not appreciably amphoteric. It dissolves in acid, but not in excess base.

Copper

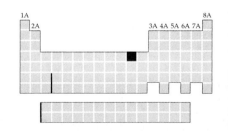

Copper, a reddish colored metal, is a relatively rare element, accounting for only 0.0068% of the earth's crust by mass. Like the other group 1B elements silver and gold, copper is found in the elemental state. Its most important ores are sulfides, such as chalcopyrite, $CuFeS_2$. In a multistep process, copper sulfides are concentrated, separated from iron, and converted to molten copper(I) sulfide, which is then reduced to elemental copper by blowing air through the hot liquid:

$$Cu_2S(l) + O_2(g) \longrightarrow 2\,Cu(l) + SO_2(g)$$

The product, containing about 99% copper, is purified by electrolysis (Section 18.12).

Because of its high electrical conductivity and negative oxidation potential, copper is widely used to make electrical wiring and corrosion-resistant water pipes. Copper is also used in coins and is combined with other metals to make alloys such as brass (mostly copper and zinc) and bronze (mostly copper and tin). Though less reactive than other first-series transition metals, copper is oxidized on prolonged exposure to O_2, CO_2, and water in moist air, forming basic copper(II) carbonate, $Cu_2(OH)_2CO_3$. Subsequent reaction with dilute H_2SO_4 in acid rain then forms $Cu_2(OH)_2SO_4$, the green patina seen on bronze monuments.

$$2\,Cu(s) + O_2(g) + CO_2(g) + H_2O(g) \longrightarrow Cu_2(OH)_2CO_3(s)$$
$$Cu_2(OH)_2CO_3(s) + H_2SO_4(aq) \longrightarrow Cu_2(OH)_2SO_4(s) + CO_2(g) + H_2O(l)$$

Why do monuments made of copper or bronze turn green with age?

In its compounds, copper exists in two common oxidation states, +1 (cuprous) and +2 (cupric). Because $E°$ for the Cu^+/Cu^{2+} half-reaction is less negative than that for the Cu/Cu^+ half-reaction, any oxidizing agent strong enough to oxidize copper to the copper(I) ion is also able to oxidize the copper(I) ion to the copper(II) ion.

$$Cu(s) \longrightarrow Cu^+(aq) + e^- \qquad E° = -0.52\text{ V}$$
$$Cu^+(aq) \longrightarrow Cu^{2+}(aq) + e^- \qquad E° = -0.15\text{ V}$$

Dilute nitric acid, for example, oxidizes copper to the +2 oxidation state:

$$3\,Cu(s) + 2\,NO_3^-(aq) + 8\,H^+(aq) \longrightarrow$$
$$3\,Cu^{2+}(aq) + 2\,NO(g) + 4\,H_2O(l) \qquad E° = +0.62\text{ V}$$

It follows from the $E°$ values that $Cu^+(aq)$ can undergo a disproportionation reaction, oxidizing and reducing itself:

$$
\begin{array}{ll}
Cu^+(aq) + e^- \longrightarrow Cu(s) & E° = +0.52\text{ V} \\
\underline{Cu^+(aq) \longrightarrow Cu^{2+}(aq) + e^-} & \underline{E° = -0.15\text{ V}} \\
2\,Cu^+(aq) \longrightarrow Cu(s) + Cu^{2+}(aq) & E° = +0.37\text{ V}
\end{array}
$$

The positive $E°$ value for the disproportionation corresponds to a large equilibrium constant, indicating that the reaction proceeds far toward completion:

$$2\,Cu^+(aq) \rightleftharpoons Cu(s) + Cu^{2+}(aq) \qquad K_c = 1.8 \times 10^6$$

Thus, the copper(I) ion is not an important species in aqueous solution, though copper(I) does exist in solid compounds such as CuCl. In the presence of Cl^-

ions, the disproportionation equilibrium is reversed because precipitation of the insoluble, white copper(I) chloride drives the following reaction to the right:

$$Cu(s) + Cu^{2+}(aq) + 2\,Cl^-(aq) \longrightarrow 2\,CuCl(s)$$

The more common +2 oxidation state is found in the blue aqueous ion, $Cu(H_2O)_6^{2+}$, and in numerous solid compounds and complex ions. Addition of base (aqueous ammonia) to a solution of a copper(II) salt gives a blue precipitate of copper(II) hydroxide, which dissolves in excess aqueous ammonia, yielding the dark blue complex ion $Cu(NH_3)_4^{2+}$ (Figure 20.11):

$$Cu^{2+}(aq) + 2\,OH^-(aq) \longrightarrow Cu(OH)_2(s)$$
$$Cu(OH)_2(s) + 4\,NH_3(aq) \longrightarrow Cu(NH_3)_4^{2+}(aq) + 2\,OH^-(aq)$$

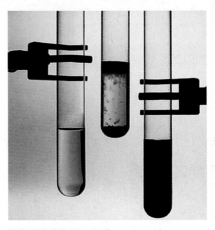

FIGURE 20.11   When an aqueous solution of $CuSO_4$ (left) is treated with aqueous ammonia, a blue precipitate of $Cu(OH)_2$ forms (center). On addition of excess ammonia, the precipitate dissolves, yielding the deep blue $Cu(NH_3)_4^{2+}$ ion (right).

Perhaps the most common of all copper compounds is the blue-colored copper(II) sulfate pentahydrate, $CuSO_4 \cdot 5\,H_2O$. Four of the five water molecules are bound to the copper(II) ion, and the fifth is hydrogen bonded to the sulfate ion. When heated, $CuSO_4 \cdot 5\,H_2O$ loses its water and its color (Figure 14.13, page 597), suggesting that the blue color of the pentahydrate is due to bonding of Cu^{2+} to the water molecules.

20.5   Coordination Compounds

A **coordination compound** is a compound in which a central metal ion (or atom) is attached to a group of surrounding molecules or ions by coordinate covalent bonds. A good example is the anticancer drug cisplatin, $Pt(NH_3)_2Cl_2$, in which two NH_3 molecules and two Cl^- ions use lone pairs of electrons to bond to the platinum(II) ion:

$$\begin{array}{c} :\ddot{C}l: \\ :\ddot{C}l:\ddot{P}t:NH_3 \qquad \text{Cisplatin} \\ NH_3 \end{array}$$

The molecules or ions that surround the central metal ion in a coordination compound are called **ligands**, and the atoms that are attached directly to the metal are called **ligand donor atoms**. In cisplatin, for example, the ligands are NH_3 and Cl^-, and the ligand donor atoms are N and Cl. Note that formation of a coordination compound is a Lewis acid–base interaction (Section 15.16) in which the ligands act as Lewis bases (electron-pair donors) and the central metal ion behaves as a Lewis acid (an electron-pair acceptor).

Not all coordination compounds are neutral molecules like $Pt(NH_3)_2Cl_2$. Many are salts, such as $[Ni(NH_3)_6]Cl_2$ and $K_3[Fe(CN)_6]$, which contain a complex cation or anion along with enough ions of opposite charge to give a compound that is electrically neutral overall. To emphasize that the complex ion is a discrete structural unit, it is always enclosed in brackets in the formula of the salt. Thus, $[Ni(NH_3)_6]Cl_2$ contains $[Ni(NH_3)_6]^{2+}$ cations and Cl^- anions. The term **metal complex** (or simply complex) refers both to neutral molecules, such as $Pt(NH_3)_2Cl_2$, and to complex ions, such as $[Ni(NH_3)_6]^{2+}$ and $[Fe(CN)_6]^{3-}$.

The number of ligand donor atoms that surround a central metal ion in a complex is called the **coordination number** of the metal. Thus, platinum(II) has a coordination number of 4 in $Pt(NH_3)_2Cl_2$, and iron(III) has a coordination number of 6 in $[Fe(CN)_6]^{3-}$. The most common coordination numbers are 4 and 6, but others are well known (Table 20.4). The coordination number of a metal ion in a particular complex depends on the metal ion's size, charge, and electron configuration, and on the size and shape of the ligands.

TABLE 20.4   Examples of Complexes with Various Coordination Numbers

Coordination Number	Complex
2	$[Ag(NH_3)_2]^+$, $[CuCl_2]^-$
3	$[HgI_3]^-$
4	$[Zn(NH_3)_4]^{2+}$, $[Ni(CN)_4]^{2-}$
5	$[Ni(CN)_5]^{3-}$, $Fe(CO)_5$
6	$[Cr(H_2O)_6]^{3+}$, $[Fe(CN)_6]^{3-}$
7	$[ZrF_7]^{3-}$
8	$[Mo(CN)_8]^{4-}$

Metal complexes have characteristic shapes, depending on the metal ion's coordination number. Two-coordinate complexes, such as $[Ag(NH_3)_2]^+$, are linear. Four-coordinate complexes are either tetrahedral or square planar; for example, $[Zn(NH_3)_4]^{2+}$ is tetrahedral, and $[Ni(CN)_4]^{2-}$ is square planar. Nearly all six-coordinate complexes are octahedral. The more common coordination geometries are illustrated in Figure 20.12. Coordination geometries were first deduced by the Swiss chemist Alfred Werner, who was awarded the 1913 Nobel Prize in chemistry for his pioneering studies.

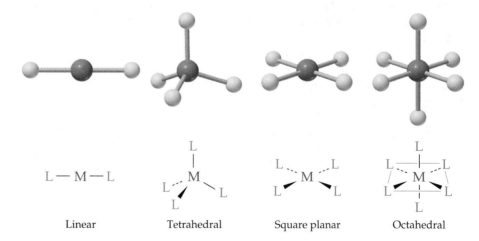

FIGURE 20.12 The arrangement of ligand donor atoms (L) in ML_n complexes with coordination numbers 2, 4, and 6. In the octahedral arrangement, four ligands are at the corners of a square, with one more above and one below the plane of the square.

Linear   Tetrahedral   Square planar   Octahedral

The charge on a metal complex is equal to the charge on the metal ion (its oxidation state) plus the sum of the charges on the ligands. Thus, if we know the charge on each ligand and the charge on the complex, we can easily find the oxidation state of the metal.

EXAMPLE 20.2   A cobalt(III) ion forms a complex with four ammonia molecules and two chloride ions. What is the formula of the complex?

SOLUTION   The charge on the complex is the sum of the charges on the Co^{3+} ion, the four NH_3 ligands, and the two Cl^- ligands:

$$Co^{3+} \quad 4\,NH_3 \quad 2\,Cl^-$$
$$+3 \;+\; (4 \times 0) \;+\; [2 \times (-1)] = +1$$

Therefore, the formula of the complex is $[Co(NH_3)_4Cl_2]^+$.

EXAMPLE 20.3 What is the oxidation state of platinum in the coordination compound $K[Pt(NH_3)Cl_5]$?

SOLUTION Because the compound is electrically neutral overall and contains one K^+ cation per complex anion, the anion must be $[Pt(NH_3)Cl_5]^-$. Since ammonia is neutral, and chloride has a charge of -1, the oxidation state of platinum must be $+4$:

$$K^+ \quad Pt^{n+} \quad NH_3 \quad 5\,Cl^-$$
$$+1 \;+\; n \;+\; 0 \;+\; [5 \times (-1)] = 0; \quad n = +4$$

▶**PROBLEM 20.4** What is the formula of the chromium(III) complex that contains two ammonia and four thiocyanate (SCN^-) ligands?

▶**PROBLEM 20.5** What is the oxidation state of iron in $Na_4[Fe(CN)_6]$?

20.6 Ligands

Structures of some typical ligands are shown in Figure 20.13. Because all ligands are Lewis bases, they have at least one lone pair of electrons that can be used to form a coordinate covalent bond to a metal ion. They can be classified as *monodentate* or *polydentate*, depending on the number of ligand donor atoms that bond to the metal. Ligands such as H_2O, NH_3, or Cl^-, which bond using the electron pair of a single donor atom, are called **monodentate** ligands (literally, "one-toothed" ligands). Those that bond through electron pairs on more than one donor atom are termed **polydentate** ligands ("many-toothed" ligands). For example, ethylenediamine ($NH_2CH_2CH_2NH_2$, abbreviated en) is a **bidentate** ligand because it bonds to a metal using an electron pair on each of its two nitrogen atoms. The **hexadentate** ligand ethylenediaminetetraacetate ion ($EDTA^{4-}$) bonds to a metal ion through electron pairs on six donor atoms (two N atoms and four O atoms).

Ethylenediamine; Glycinate Ion; Oxalate Ion 3D models

Polydentate ligands are known as **chelating agents** (pronounced **key**-late-ing, from the Greek word *chele*, meaning "claw") because their multipoint attachment to a metal ion resembles the grasping of an object by the claws of a crab. For example, ethylenediamine holds a cobalt(III) ion with two claws, its two nitrogen donor atoms (Figure 20.14). The resulting five-membered ring consisting of the Co(III) ion, two N atoms, and two C atoms of the ligand is called a **chelate ring**. A complex such as $[Co(en)_3]^{3+}$ or $[Co(EDTA)]^-$ that contains one or more chelate rings is known as a metal **chelate**. Because $EDTA^{4-}$ bonds to a metal ion through six donor atoms, it forms especially stable complexes and is often used to hold metal ions in solution. For example, in the treatment of lead poisoning, $EDTA^{4-}$ bonds to Pb^{2+}, which is then excreted by the kidneys as the soluble chelate $[Pb(EDTA)]^{2-}$. $EDTA^{4-}$ is commonly added to food products such as commercial salad dressings to complex any metal cations that might be present in trace amounts. The free metal ions might otherwise catalyze the oxidation of oils, thus causing the dressing to become rancid.

$EDTA^{4-}$ Ion 3D models

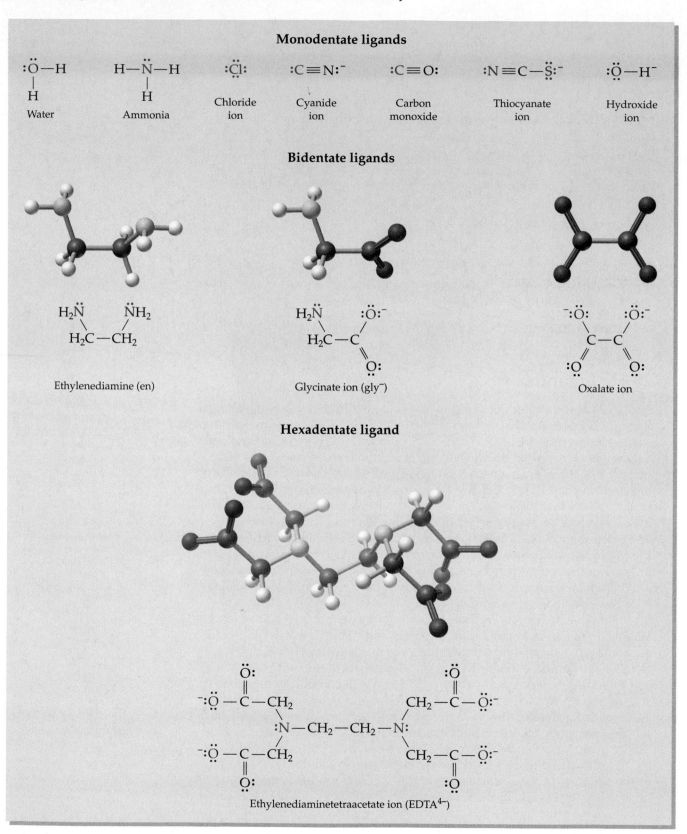

FIGURE 20.13 Structures of some common ligands. Ligand donor atoms are in color. The thiocyanate ion can bond to a metal through either the S atom or the N atom.

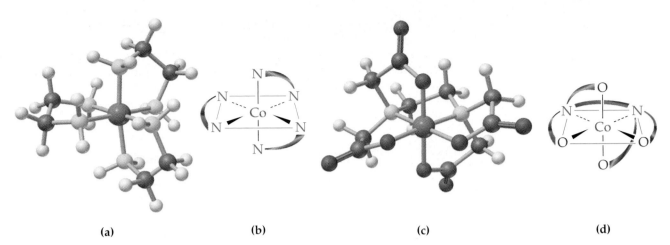

(a)   (b)   (c)   (d)

FIGURE 20.14 **(a)** Molecular model and **(b)** shorthand representation of the $[Co(en)_3]^{3+}$ ion. The complex contains three cobalt–ethylenediamine chelate rings. In (b), the symbol N⌒N represents a bidentate $NH_2CH_2CH_2NH_2$ ligand, which spans adjacent corners of the octahedron. **(c)** Molecular model and **(d)** shorthand representation of the $[Co(EDTA)]^-$ ion. The hexadentate $EDTA^{4-}$ ligand uses its two N atoms and four O atoms to bond to the metal, thus forming five chelate rings.

Heme 3D model

Naturally occurring chelating ligands are essential constituents of many important biomolecules. For example, the *heme* group in hemoglobin contains a planar, tetradentate ligand that uses the lone pair of electrons on each of its four N atoms to bond to an iron(II) ion (Chapter 13 Interlude). The ligand in heme is an example of a *porphyrin*, a derivative of the porphine molecule (Figure 20.15a) in which the porphine's peripheral H atoms are replaced by various substituent groups (CH_3, $CH_2{=}CH$, and $CH_2CH_2CO_2^-$ in heme). Bonding of the porphyrin to the Fe(II) ion involves prior loss of the two NH protons, which makes room for the Fe(II) to occupy the cavity between the four N atoms.

In hemoglobin (and in the iron storage protein myoglobin), the heme is linked to a protein (the globin) through an additional Fe–N bond, as shown in Figure 20.15b. In addition, the Fe(II) can bond to an O_2 molecule to give the six-coordinate, octahedral complex present in oxyhemoglobin and oxymyoglobin. The three-dimensional shape of the protein part of the molecule makes possible the reversible binding of O_2.

FIGURE 20.15 **(a)** The structure of the porphine molecule. Loss of the two NH protons gives a planar, tetradentate 2− ligand that can bond to a metal cation. The porphyrins are derivatives of porphine in which the peripheral H atoms are replaced by various substituent groups. **(b)** Schematic of the planar heme group, the attached protein chain, and the bound O_2 molecule in oxyhemoglobin and oxymyoglobin. The Fe(II) ion has a six-coordinate, octahedral environment, and the O_2 acts as a monodentate ligand.

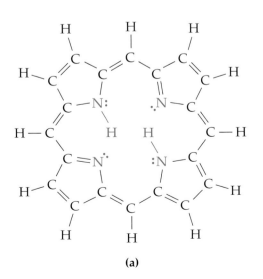

(a)

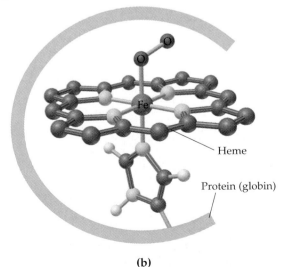

(b)

20.7 Naming Coordination Compounds

In the early days of coordination chemistry, coordination compounds were named after their discoverer or according to their color. Now we use systematic names that specify the number of ligands of each particular type, the metal, and its oxidation state. Before listing the rules used to name coordination compounds, let's consider a few examples that will illustrate how to apply the rules:

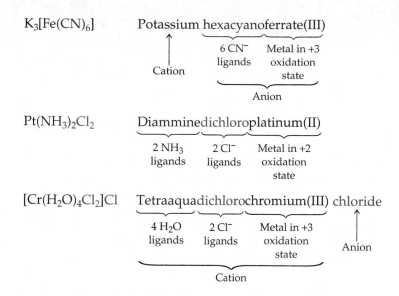

The following list summarizes the nomenclature rules recommended by the International Union of Pure and Applied Chemistry:

1. If the compound is a salt, name the cation first and then the anion, just as in naming simple salts (Section 2.10). For example, $K_3[Fe(CN)_6]$ is potassium hexacyanoferrate(III).
2. In naming a complex ion or a neutral complex, name the ligands first and then the metal. The names of anionic ligands end in -*o*. As shown in Table 20.5, they are usually obtained by changing the anion endings -*ide* to -*o* and -*ate* to -*ato*. Neutral ligands are specified by their usual names,

TABLE 20.5   Names of Some Common Ligands

Anionic Ligand	Ligand Name	Neutral Ligand	Ligand Name
Bromide, Br^-	Bromo	Ammonia, NH_3	Ammine
Carbonate, CO_3^{2-}	Carbonato	Water, H_2O	Aqua
Chloride, Cl^-	Chloro	Carbon monoxide, CO	Carbonyl
Cyanide, CN^-	Cyano	Ethylenediamine, en	Ethylenediamine
Fluoride, F^-	Fluoro		
Glycinate, gly^-	Glycinato		
Hydroxide, OH^-	Hydroxo		
Oxalate, $C_2O_4^{2-}$	Oxalato		
Thiocyanate, SCN^-	Thiocyanato*		
	Isothiocyanato[†]		

* Ligand donor atom is S.   [†]Ligand donor atom is N.

except for H_2O, NH_3, and CO, which are called aqua, ammine (note spelling), and carbonyl, respectively. The name of a complex is one word, with no space between the various ligand names and no space between the names of the last ligand and the metal.

3. If the complex contains more than one ligand of a particular type, indicate the number with the appropriate Greek prefix: *di-*, *tri-*, *tetra-*, *penta-*, *hexa-*, and so forth. The ligands are listed in alphabetical order, and the prefixes are ignored in determining the order. Thus, tetra*a*qua precedes *di*chloro in the name for $[Cr(H_2O)_4Cl_2]Cl$: tetraaquadichlorochromium(III) chloride.

4. If the name of a ligand itself contains a Greek prefix—for example, ethylene*di*amine—put the ligand name in parentheses and use one of the following alternative prefixes to specify the number of ligands: *bis-* (2), *tris-* (3), *tetrakis-* (4), and so forth. Thus, the name of $[Co(en)_3]Cl_3$ is tris(ethylenediamine)cobalt(III) chloride.

5. Use a Roman numeral in parentheses, immediately following the name of the metal, to indicate the metal's oxidation state. As shown by the preceding examples, there is no space between the name of the metal and the parenthesis.

6. In naming the metal, use the ending *-ate* if the metal is in an anionic complex. Thus, $[Fe(CN)_6]^{3-}$ is the hexacyanoferrate(III) anion. There are no simple rules for going from the name of the metal to the name of the metallate anion, partly because some of the anions have Latin names. Some common examples are given in Table 20.6.

The rules for naming coordination compounds make it possible to go from a formula to the systematic name or from a systematic name to the appropriate formula. The following examples give some practice.

EXAMPLE 20.4 Name each of the following:
(a) $[Co(NH_3)_6]Cl_3$, prepared in 1798 by B. M. Tassaert and generally considered to be the first coordination compound
(b) $[Rh(NH_3)_5I]I_2$, a yellow compound obtained by heating $[Rh(NH_3)_5(H_2O)]I_3$ at 100°C
(c) $Fe(CO)_5$, a highly toxic, volatile liquid
(d) $[Fe(C_2O_4)_3]^{3-}$, the ion formed when Fe_2O_3 rust stains are dissolved in oxalic acid $(H_2C_2O_4)$

SOLUTION
(a) Because the chloride ion has a charge of −1 and ammonia is neutral, the oxidation state of cobalt is +3. Use the prefix *hexa-* to indicate that the cation contains six NH_3 ligands, and use a Roman numeral III in parentheses to indicate the oxidation state of cobalt. The name of $[Co(NH_3)_6]Cl_3$ is hexaamminecobalt(III) chloride.
(b) Because the iodide ion has a charge of −1, the complex cation is $[Rh(NH_3)_5I]^{2+}$ and the rhodium has an oxidation state of +3. List the *a*mmine ligands before the *i*odo ligand, and use the prefix *penta-* to indicate the presence of five NH_3 ligands. The name of $[Rh(NH_3)_5I]I_2$ is pentaammineiodorhodium(III) iodide.
(c) Because the carbonyl ligand is neutral, the oxidation state of iron is zero. The systematic name of $Fe(CO)_5$ is pentacarbonyliron(0), but the common name iron pentacarbonyl is often used.
(d) Because each oxalate ligand $(C_2O_4{}^{2-})$ has a charge of −2, and because $[Fe(C_2O_4)_3]^{3-}$ has an overall charge of −3, iron must have an oxidation state of +3. Use the name ferrate(III) for the metal because the complex is an anion. The name of $[Fe(C_2O_4)_3]^{3-}$ is the trioxalatoferrate(III) ion.

A sample of tris(ethylenediamine)cobalt(III) chloride, $[Co(en)_3]Cl_3$.

TABLE 20.6 Names of Some Common Metallate Anions

Metal	Anion Name
Aluminum	Aluminate
Chromium	Chromate
Cobalt	Cobaltate
Copper	Cuprate
Gold	Aurate
Iron	Ferrate
Manganese	Manganate
Nickel	Nickelate
Platinum	Platinate
Zinc	Zincate

The compound $K_3[Fe(C_2O_4)_3] \cdot 3 H_2O$ contains the trioxalatoferrate(III) ion.

EXAMPLE 20.5 Write the formula for each of the following:
(a) Potassium tetracyanonickelate(II)
(b) Aquachlorobis(ethylenediamine)cobalt(III) chloride
(c) Sodium hexafluoroaluminate
(d) Diamminesilver(I) ion

SOLUTION

(a) Bonding of four CN^- ligands to Ni^{2+} gives an $[Ni(CN)_4]^{2-}$ anion, which must be balanced by two K^+ cations. The compound's formula is therefore $K_2[Ni(CN)_4]$. This compound is obtained when excess KCN is added to a solution of a nickel(II) salt.

(b) Because the complex cation contains one H_2O, one Cl^-, and two neutral en ligands, and because the metal is Co^{3+}, the cation is $[Co(en)_2(H_2O)Cl]^{2+}$. The 2+ charge of the cation must be balanced by two Cl^- anions, so the formula of the compound is $[Co(en)_2(H_2O)Cl]Cl_2$. The cation is the first product formed when $[Co(en)_2Cl_2]^+$ reacts with water.

(c) Sodium hexafluoroaluminate, also called cryolite, is used in the electrolytic production of aluminum metal (Section 18.12) and is an example of a coordination compound of a main-group element. The oxidation state of aluminum is omitted from the name because aluminum has only one oxidation state (+3). Because F^- has a charge of −1, the anion is $[AlF_6]^{3-}$. The charge of the anion must be balanced by three Na^+ cations; therefore, the formula for the compound is $Na_3[AlF_6]$.

(d) Diamminesilver(I) ion, formed when silver chloride dissolves in an excess of aqueous ammonia, has the formula $[Ag(NH_3)_2]^+$.

▶ **PROBLEM 20.6** Name each of the following:
  (a) $[Cu(NH_3)_4]SO_4$, a deep blue compound obtained when $CuSO_4$ is treated with an excess of ammonia
  (b) $Na[Cr(OH)_4]$, the compound formed when $Cr(OH)_3$ is dissolved in an excess of aqueous NaOH
  (c) $Co(gly)_3$, a complex that contains the anion of the amino acid glycine
  (d) $[Fe(H_2O)_5(NCS)]^{2+}$, the red complex ion formed in a qualitative analysis test for iron

▶ **PROBLEM 20.7** Write the formula for each of the following:
  (a) Tetraamminezinc(II) nitrate, the compound formed when zinc nitrate is treated with an excess of ammonia
  (b) Tetracarbonylnickel(0), the first metal carbonyl (prepared in 1888) and an important compound in the industrial refining of nickel metal
  (c) Potassium amminetrichloroplatinate(II), a compound that contains a square planar anion
  (d) The dicyanoaurate(I) ion, an ion important in the extraction of gold from its ores

20.8 Isomers

Isomerism movie

One of the more interesting aspects of coordination chemistry is the existence of **isomers**, compounds that have the same formula but a different arrangement of their constituent atoms. Because their atoms are arranged differently, isomers are different compounds with different chemical reactivity and different physical properties (such as color, solubility, and melting point). Figure 20.16 shows a scheme for classifying some of the kinds of isomers in coordination chemistry. As we'll see in Chapters 23 and 24, isomers are also important in organic chemistry and biochemistry.

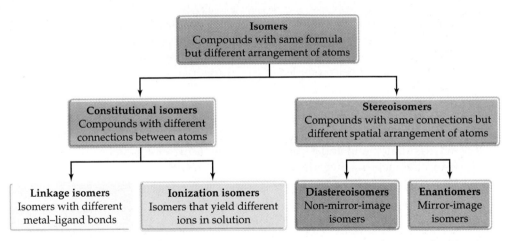

FIGURE 20.16   Classification scheme for the kinds of isomers in coordination chemistry. See the text for examples.

Constitutional Isomers

Isomers that have different connections among their constituent atoms are called **constitutional isomers**. Of the various types of constitutional isomers, we'll mention just two: *linkage isomers* and *ionization isomers*.

   Linkage isomers arise when a ligand can bond to a metal through either of two different donor atoms. For example, the nitrite (NO_2^-) ion forms two different pentaamminecobalt(III) complexes: a yellow *nitro* complex that contains a Co−N bond, $[Co(NH_3)_5(NO_2)]^{2+}$, and a red *nitrito* complex that contains a Co−O bond, $[Co(NH_3)_5(ONO)]^{2+}$ (Figure 20.17). The ligand in the nitrito

FIGURE 20.17   Samples and structures of **(a)** the nitro complex $[Co(NH_3)_5(NO_2)]^{2+}$, which contains an N-bonded NO_2^- ligand, and **(b)** the nitrito complex $[Co(NH_3)_5(ONO)]^{2+}$, which contains an O-bonded NO_2^- ligand.

(a)

(b)

complex is written as ONO to emphasize that it's linked to the cobalt through the oxygen atom. The thiocyanate (SCN^-) ion is another ligand that gives linkage isomers because it can bond to a metal through either the sulfur atom to give a thiocyanato complex or the nitrogen atom to give an isothiocyanato complex.

Ionization isomers differ in the anion that is bonded to the metal ion. An example is the pair $[Co(NH_3)_5Br]SO_4$, a violet compound that has a Co–Br bond and a free sulfate anion, and $[Co(NH_3)_5SO_4]Br$, a red compound that has a Co–sulfate bond and a free bromide ion. Ionization isomers get their name from the fact that they yield different ions in solution.

Stereoisomers

Isomers that have the same connections among atoms but a different arrangement of the atoms in space are called **stereoisomers**. In coordination chemistry, there are two kinds of stereoisomers: *diastereoisomers* and *enantiomers* (we'll leave the study of the latter to Section 20.9).

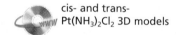
cis- and trans- Pt(NH₃)₂Cl₂ 3D models

Diastereoisomers, also called **geometric isomers**, have the same connections among atoms but different spatial orientations of their metal–ligand bonds. For example, in the square planar complex $Pt(NH_3)_2Cl_2$ the two Pt–Cl bonds can be oriented either adjacent at a 90° angle or opposite at a 180° angle, as shown in Figure 20.18. The isomer in which identical ligands occupy adjacent corners of the square is called the **cis isomer**, and that in which identical ligands are across from each other is called the **trans isomer**. (The Latin word *cis* means next to; *trans* means across.) Cis and trans isomers are different compounds with different properties. Thus, *cis*-$Pt(NH_3)_2Cl_2$ is a polar molecule and is more soluble in water than *trans*-$Pt(NH_3)_2Cl_2$. The trans isomer is nonpolar because the two Pt–Cl and the two Pt–NH₃ bond dipoles point in opposite directions and therefore cancel. It's also interesting that *cis*-$Pt(NH_3)_2Cl_2$ (cisplatin) is an effective anticancer drug, whereas the trans isomer is physiologically inactive.

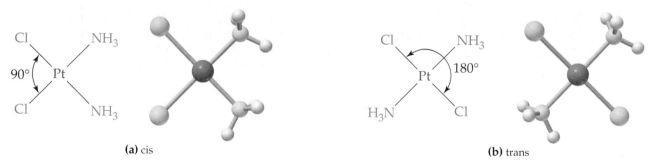

(a) cis   **(b)** trans

FIGURE 20.18   Diastereoisomers of the square planar complex $Pt(NH_3)_2Cl_2$. The two compounds have the same connections among atoms but different arrangements of the atoms in space.

In general, square planar complexes of the type MA_2B_2 and MA_2BC—where M is a metal ion and A, B, and C are ligands—can exist as cis–trans isomers. No cis–trans isomers are possible, however, for four-coordinate *tetrahedral* complexes because all four corners of a tetrahedron are adjacent to one another.

Octahedral complexes of the type MA_4B_2 can also exist as diastereoisomers because the two B ligands can be on either adjacent or opposite corners of the octahedron. Examples are the violet compound *cis*-[Co(NH$_3$)$_4$Cl$_2$]Cl and the green compound *trans*-[Co(NH$_3$)$_4$Cl$_2$]Cl. As Figure 20.19 shows, there are several ways of drawing the cis and trans isomers because each complex can be rotated in space, changing the perspective but not the identity of the isomer.

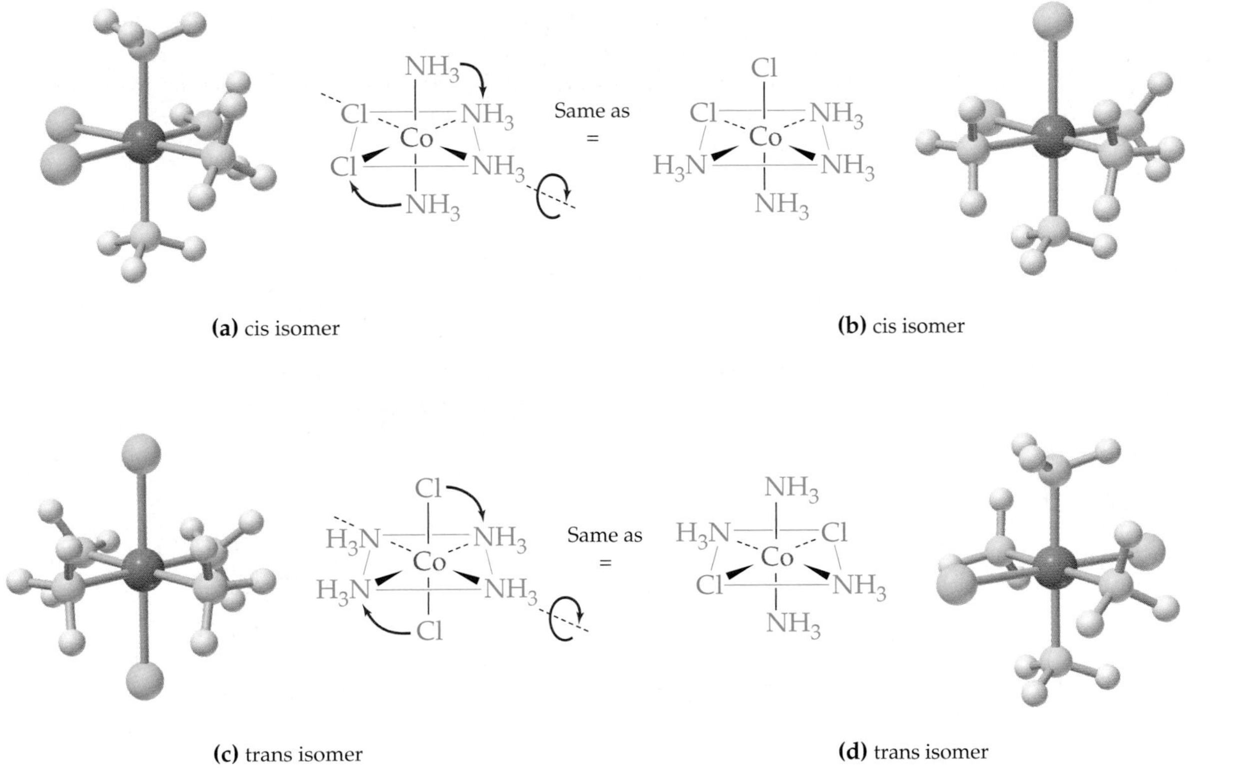

(a) cis isomer

(b) cis isomer

(c) trans isomer

(d) trans isomer

How can we be sure that there are only two diastereoisomers of the [Co(NH$_3$)$_4$Cl$_2$]$^+$ ion? The first Cl$^-$ ligand can be located at any one of the six corners of the octahedron. Once one Cl$^-$ is present, however, the five corners remaining are no longer equivalent. The second Cl$^-$ can be located either on one of the four corners adjacent to the first Cl$^-$, which gives the cis isomer, or on the unique corner opposite the first Cl$^-$, which gives the trans isomer. Thus, only two diastereoisomers are possible for complexes of the type MA$_4$B$_2$ (and MA$_4$BC).

FIGURE 20.19
Diastereoisomers of the [Co(NH$_3$)$_4$Cl$_2$]$^+$ ion. Although they may appear different, cis isomers **(a)** and **(b)** are identical, as can be seen by rotating the entire complex by 90° about a Cl–Co–NH$_3$ axis. Similarly, trans isomers **(c)** and **(d)** are identical.

EXAMPLE 20.6 Platinum(II) forms square planar complexes, and platinum(IV) gives octahedral complexes. How many diastereoisomers are possible for each of the following complexes? Describe their structures.
(a) [Pt(NH$_3$)$_3$Cl]$^+$ **(b)** [Pt(NH$_3$)Cl$_5$]$^-$
(c) Pt(NH$_3$)$_2$Cl(NO$_2$)   **(d)** [Pt(NH$_3$)$_4$ClBr]$^{2+}$

SOLUTION
(a) No isomers are possible for a square planar complex of the type MA$_3$B.

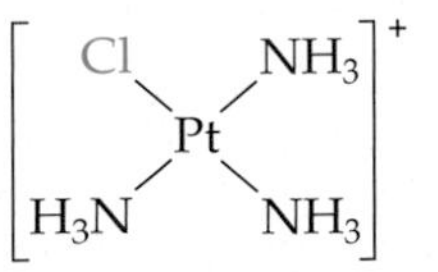

(b) No isomers are possible for an octahedral complex of the type MA_5B.

(c) Cis and trans isomers are possible for a square planar complex of the type MA_2BC. The Cl^- and NO_2^- ligands can be on either adjacent or opposite corners of the square.

(d) Cis and trans isomers are possible for an octahedral complex of the type MA_4BC. The Cl^- and Br^- ligands can be on either adjacent or opposite corners of the octahedron.

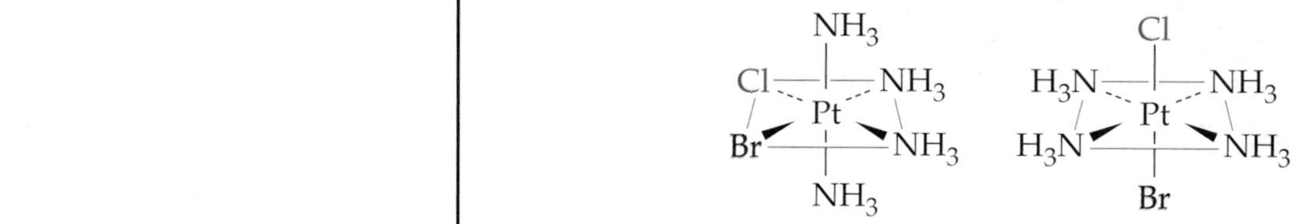

EXAMPLE 20.7 Draw the structures of all possible diastereoisomers of $Co(NH_3)_3Cl_3$.

SOLUTION Two diastereoisomers are possible for an octahedral complex of the type MA_3B_3. Isomer (a) shown below has the three Cl^- ligands in adjacent positions on one triangular face of the octahedron; isomer (b) has all three Cl^- ligands in a plane that contains the Co(III) ion. In isomer (a), all three Cl–Co–Cl bond angles are 90°, whereas in isomer (b) two Cl–Co–Cl angles are 90° and the third is 180°.

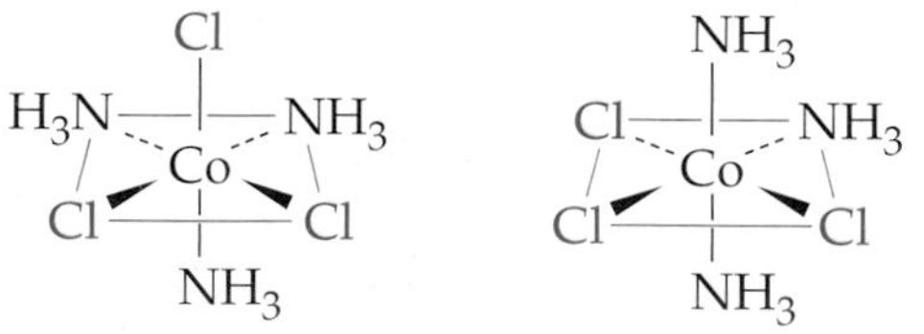

To convince yourself that only two diastereoisomers of $Co(NH_3)_3Cl_3$ exist, look back at Figure 20.19 and consider the products obtained from *cis*- and *trans*-$[Co(NH_3)_4Cl_2]^+$ when one of the four NH_3 ligands is replaced by a Cl^-. There are two kinds of NH_3 ligands in *cis*-$[Co(NH_3)_4Cl_2]^+$: two NH_3 ligands that are trans to each other and two NH_3 ligands that are trans to a Cl^-. Replacing an NH_3 trans to another NH_3 with Cl^- gives $Co(NH_3)_3Cl_3$ isomer (a), while replacing an NH_3 trans to a Cl^- with another Cl^- gives isomer (b). All four NH_3 ligands in *trans*-$[Co(NH_3)_4Cl_2]^+$ are equivalent, and replacing any one of them with a Cl^- gives isomer (b). [If you rotate the resulting $Co(NH_3)_3Cl_3$ complex, you can see that it is identical to the previous drawing of isomer (b).] Thus, there are only two diastereoisomers for a complex of the type MA_3B_3.

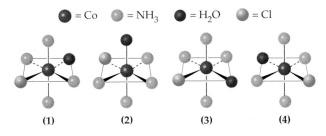

▶ **PROBLEM 20.8**

How many diastereoisomers are possible for the following complexes? Draw the structure of each diastereoisomer.

(a) $Pt(NH_3)_2(SCN)_2$ (b) $[CoCl_2Br_2]^{2-}$ (tetrahedral)

(c) $Co(NH_3)_3(NO_2)_3$ (d) $Pt(en)Cl_2$

(e) $[Cr(en)_2Br_2]^+$ (f) $[Rh(en)_3]^{3+}$

✦━ **KEY CONCEPT PROBLEM 20.9**   Consider the following isomers of the $[Co(NH_3)_4(H_2O)Cl]^{2+}$ ion:

● = Co   ● = NH_3 ● = H_2O ● = Cl

(1)   (2)   (3)   (4)

(a) Label the isomers as cis or trans.

(b) Which isomers are identical, and which are different?

20.9 Enantiomers and Molecular Handedness

Diastereoisomers are easy to distinguish because the various bonds in the cis and trans isomers point in different directions. *Enantiomers*, however, are stereoisomers that differ in a more subtle way: **Enantiomers** are *molecules or ions that are nonidentical mirror images of each other*; they differ because of their handedness.

Handedness affects almost everything we do. Anyone who has played much softball knows that the last available glove always fits the wrong hand. Any left-handed person sitting next to a right-handed person in a lecture knows that taking notes sometimes means bumping elbows. The reason for these difficulties is that our hands aren't identical—they're mirror images. When you hold your left hand up to a mirror, the image you see looks like your right hand. (Try it.)

Not all objects are handed, of course. There's no such thing as a "right-handed" tennis ball or a "left-handed" coffee mug. When a tennis ball or a coffee mug is held up to a mirror, the image reflected is identical to the ball or mug itself. Objects that do have a handedness to them are said to be **chiral**

Chirality movie

(pronounced **ky**-ral, from the Greek *cheir*, meaning hand), and objects that lack handedness are said to be nonchiral, or **achiral**.

Why is it that some objects are chiral but others aren't? In general, an object is not chiral if, like the coffee mug, it has a **symmetry plane** cutting through its middle so that one half of the object is a mirror image of the other half. If you were to cut the mug in half, one half of the mug would be the mirror image of the other half. A hand, however, has no symmetry plane and is therefore chiral. If you were to cut a hand in two, one "half" of the hand would not be a mirror image of the other half (Figure 20.20).

FIGURE 20.20 The meaning of a symmetry plane: An achiral object like the coffee mug has a symmetry plane passing through it, making the two halves mirror images. A chiral object like the hand has no symmetry plane because the two "halves" of the hand are not mirror images.

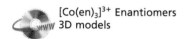

$[Co(en)_3]^{3+}$ Enantiomers 3D models

Just as certain objects like a hand are chiral, certain molecules and ions are also chiral. Consider the tris(ethylenediamine)cobalt(III) ion, $[Co(en)_3]^{3+}$, for example (Figure 20.21). The $[Co(en)_3]^{3+}$ cation has no symmetry plane because its two halves aren't mirror images, but it does have a threefold symmetry axis, which passes through the Co(III) ion and is nearly perpendicular to the plane of the figure. Like a hand, $[Co(en)_3]^{3+}$ is chiral and can exist in two nonidentical mirror-image forms—a "right-handed" enantiomer in which the three ethylenediamine ligands spiral to the right (clockwise) about the threefold axis and a "left-handed" enantiomer in which the ethylenediamine ligands spiral to the left (counterclockwise) about the threefold axis. The direction of spiral is indicated in the figure by the red arrows. By contrast, the analogous ammonia complex $[Co(NH_3)_6]^{3+}$ is achiral because, like a coffee mug, it has a symmetry plane. (Actually, $[Co(NH_3)_6]^{3+}$ has several symmetry planes, though only one is shown in Figure 20.21.) Thus, $[Co(NH_3)_6]^{3+}$ exists in a single form and does not have enantiomers.

Optical Activity movie

Enantiomers have identical properties except for their reactions with other chiral substances and their effect on *plane-polarized light*. In ordinary light, the electric vibrations occur in all planes parallel to the direction in which the light wave is traveling, but in plane-polarized light, the electric vibrations of the light wave are restricted to a single plane, as shown in Figure 20.22. Plane-polarized light is obtained by passing ordinary light through a polarizing filter, like that found in certain kinds of sunglasses. If the plane-polarized light is then passed through a solution of one enantiomer, the plane of polarization is rotated, either to the right or to the left. If the light is passed through a solution of the other enantiomer, its plane of polarization is rotated through an equal angle, but in the opposite direction. Enantiomers are sometimes called *optical isomers* because of their effect on plane-polarized light.

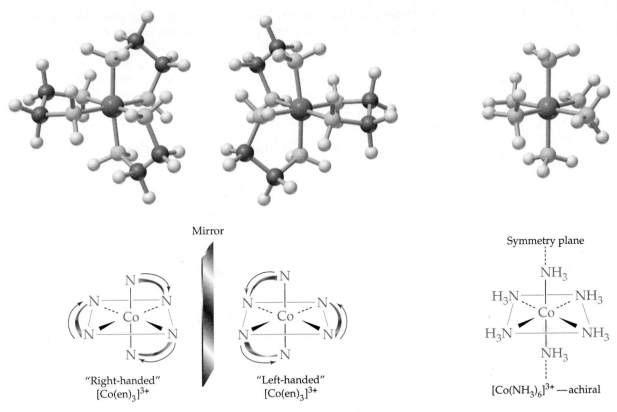

FIGURE 20.21 The structures of the $[Co(en)_3]^{3+}$ enantiomers and the achiral $[Co(NH_3)_6]^{3+}$ ion. $[Co(en)_3]^{3+}$ has a helical structure in which the three en ligands lie along the threads of a screw. As the red arrows show, one enantiomer is "right-handed" in the sense that the screw would advance into the page as you rotated it to the right. The other enantiomer is "left-handed" because it would advance into the page as you rotated it to the left. In contrast, $[Co(NH_3)_6]^{3+}$ has several symmetry planes and is achiral. (Only one of the symmetry planes is identified.)

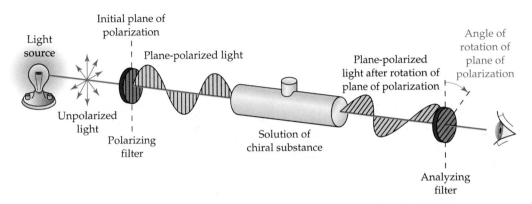

FIGURE 20.22 The essential features of a polarimeter. The polarimeter measures the angle through which the plane of plane-polarized light is rotated when the light is passed through a solution of a chiral substance.

Enantiomers are labeled (+) or (−), depending on the direction of rotation of the plane of polarization. For example, the isomer of $[Co(en)_3]^{3+}$ that rotates the plane of polarization to the right is labeled (+)-$[Co(en)_3]^{3+}$, and the isomer that rotates the plane of polarization to the left is designated (−)-$[Co(en)_3]^{3+}$. A 50:50 mixture of the (+) and (−) isomers, called a **racemic mixture**, produces no net optical rotation because the rotations produced by the individual enantiomers exactly cancel.

EXAMPLE 20.8 Draw the structures of all possible diastereoisomers and enantiomers of $[Co(en)_2Cl_2]^+$.

SOLUTION Ethylenediamine always spans adjacent corners of an octahedron, but the Cl^- ligands can be on either adjacent or opposite corners. Therefore, there are two diastereoisomers, cis and trans. Because the trans isomer has several symmetry planes—one cuts through the Co and the en ligands—it is achiral and has no enantiomers. The cis isomer, however, is chiral and exists as a pair of enantiomers that are nonidentical mirror images. You can see that the cis enantiomer on the right is not the same as the one on the left if you rotate it by 180° about the vertical N−Co−N axis.

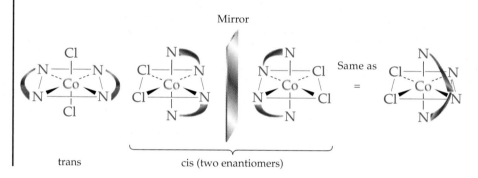

trans   cis (two enantiomers)

▶ **PROBLEM 20.10** Which of the following objects are chiral?
(a) A chair (b) A foot (c) A pencil
(d) A corkscrew (e) A banana

✦ **KEY CONCEPT PROBLEM 20.11** Consider the following ethylenediamine complexes of rhodium:

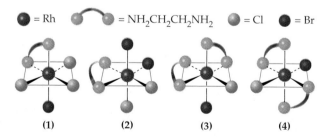

● = Rh   ⌒ = $NH_2CH_2CH_2NH_2$ ○ = Cl   ● = Br

(1)   (2)   (3)   (4)

(a) Which complexes are chiral, and which are achiral?
(b) Draw the enantiomer of each chiral complex.

▶ **PROBLEM 20.12** Which of the following complexes can exist as enantiomers? Draw the structure of each enantiomer.
(a) $[Fe(C_2O_4)_3]^{3-}$ (b) $[Co(NH_3)_4(en)]^{3+}$
(c) $[Co(NH_3)_2(en)_2]^{3+}$ (d) $[Cr(H_2O)_4Cl_2]^+$

20.10   Color of Transition Metal Complexes

Most transition metal complexes have beautiful colors that depend on the identity of the metal and the ligands. The color of an aqua complex, for example, depends on the metal: $[Co(H_2O)_6]^{2+}$ is pink, $[Ni(H_2O)_6]^{2+}$ is green, $[Cu(H_2O)_6]^{2+}$ is blue, but $[Zn(H_2O)_6]^{2+}$ is colorless (Figure 20.23). If we keep the metal constant but vary the ligand, the color also changes. For example, $[Ni(H_2O)_6]^{2+}$ is green, $[Ni(NH_3)_6]^{2+}$ is blue, and $[Ni(en)_3]^{2+}$ is violet (Figure 20.24).

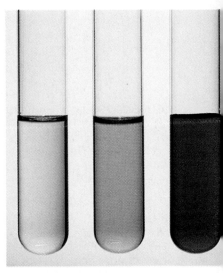

FIGURE 20.23   Aqueous solutions of the nitrate salts of cobalt(II), nickel(II), copper(II), and zinc(II) (from left to right).

FIGURE 20.24   Aqueous solutions that contain $[Ni(H_2O)_6]^{2+}$, $[Ni(NH_3)_6]^{2+}$, and $[Ni(en)_3]^{2+}$ (from left to right). The two solutions on the right were prepared by adding ammonia and ethylene-diamine, respectively, to aqueous nickel(II) nitrate.

How can we account for the color of transition metal complexes? Let's begin by recalling that white light consists of a continuous spectrum of wavelengths corresponding to different colors (Section 5.2). When white light strikes a colored substance, some wavelengths are transmitted while others are absorbed. Just as atoms can absorb light by undergoing electronic transitions between atomic energy levels, thereby giving rise to atomic spectra (Section 5.3), so a metal complex can absorb light by undergoing an electronic transition from its lowest (ground) energy state to a higher (excited) energy state.

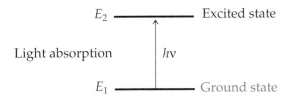

The wavelength λ of the light absorbed by a metal complex depends on the energy separation $\Delta E = E_2 - E_1$ between the two states, as given by the Planck equation $\Delta E = h\nu = hc/\lambda$ (Section 5.4), where h is Planck's constant, ν is the frequency of the light, and c is the speed of light:

$$\Delta E = E_2 - E_1 = h\nu = \frac{hc}{\lambda}$$

or

$$\lambda = \frac{hc}{\Delta E}$$

The measure of the amount of light absorbed by a substance is called the *absorbance*, and a plot of absorbance versus wavelength is called an **absorption spectrum**. For example, the absorption spectrum of the red-violet $[Ti(H_2O)_6]^{3+}$ ion has a broad absorption band at about 500 nm, a wavelength in the blue-green part of the visible spectrum (Figure 20.25). Because the absorbance is smaller in the red and violet regions of the spectrum, these

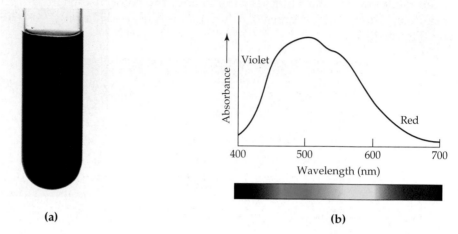

FIGURE 20.25 **(a)** A solution that contains the $[Ti(H_2O)_6]^{3+}$ ion. **(b)** Visible absorption spectrum of the $[Ti(H_2O)_6]^{3+}$ ion.

(a)

(b)

colors are largely transmitted, and we perceive the color of $[Ti(H_2O)_6]^{3+}$ to be red-violet. In general, the color that we see is complementary to the color absorbed (Figure 20.26).

FIGURE 20.26 Using an artist's color wheel, we can determine the observed color of a substance from the color of the light absorbed. Complementary colors are shown on opposite sides of the color wheel, and an approximate wavelength range for each color is indicated. Observed and absorbed colors are generally complementary. For example, if a complex absorbs only red light of 720 nm wavelength, it has a green color.

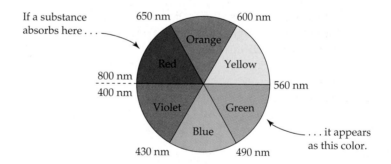

20.11 Bonding in Complexes: Valence Bond Theory

Color Wheel activity

According to the **valence bond theory** (Section 7.10), the bonding in metal complexes arises when a filled ligand orbital containing a pair of electrons overlaps a vacant hybrid orbital on the metal ion to give a coordinate covalent bond:

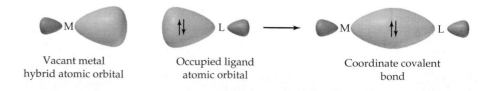

Vacant metal hybrid atomic orbital

Occupied ligand atomic orbital

Coordinate covalent bond

The hybrid orbitals (Sections 7.11 and 7.12) that a metal ion uses to accept a share in the ligand electrons are those that point in the directions of the ligands. Remember that geometry and hybridization go together. Once you know the geometry of a complex, you automatically know which hybrid orbitals the metal ion uses. The relationship between the geometry of a complex and the hybrid orbitals used by the metal ion is summarized in Table 20.7.

TABLE 20.7   Hybrid Orbitals for Common Coordination Geometries

Coordination Number	Geometry	Hybrid Orbitals	Example
2	Linear	sp	$[Ag(NH_3)_2]^+$
4	Tetrahedral	sp^3	$[CoCl_4]^{2-}$
4	Square planar	dsp^2	$[Ni(CN)_4]^{2-}$
6	Octahedral	d^2sp^3 or sp^3d^2	$[Cr(H_2O)_6]^{3+}$, $[Co(H_2O)_6]^{2+}$

To illustrate the relationship between geometry and hybridization, let's consider the tetrahedral complex $[CoCl_4]^{2-}$. A free Co^{2+} ion has the electron configuration [Ar] $3d^7$, and its orbital diagram is

$$Co^{2+}: \text{[Ar]} \quad \uparrow\downarrow \;\; \uparrow\downarrow \;\; \uparrow \;\; \uparrow \;\; \uparrow \qquad \underline{} \qquad \underline{} \; \underline{} \; \underline{}$$
$$\qquad\qquad\qquad\qquad 3d \qquad\qquad\qquad 4s \qquad\qquad 4p$$

Because the geometry of $[CoCl_4]^{2-}$ is tetrahedral, the hybrid orbitals that Co^{2+} uses to accept a share in the four pairs of ligand electrons must be sp^3 hybrids formed from the vacant $4s$ and $4p$ orbitals. It's convenient to represent the bonding in the complex by an orbital diagram that shows the hybridization of the metal orbitals and the four pairs of ligand electrons, now shared in the bonds between the metal and the ligands:

$$[CoCl_4]^{2-}: \text{[Ar]} \quad \uparrow\downarrow \;\; \uparrow\downarrow \;\; \uparrow \;\; \uparrow \;\; \uparrow \qquad \boxed{\uparrow\downarrow \qquad \uparrow\downarrow \;\; \uparrow\downarrow \;\; \uparrow\downarrow}$$
$$\qquad\qquad\qquad\qquad\quad 3d \qquad\qquad\qquad 4s \qquad\qquad 4p$$

Four sp^3 bonds to the ligands

In accord with this description, $[CoCl_4]^{2-}$ has three unpaired electrons and is paramagnetic. Recall from Section 7.14 that paramagnetic substances contain unpaired electrons and are attracted by magnetic fields, whereas diamagnetic substances contain only paired electrons and are weakly repelled by magnetic fields. The number of unpaired electrons in a transition metal complex can be determined experimentally by quantitative measurement of the force exerted on the complex by a magnetic field.

As an example of a square planar complex, consider $[Ni(CN)_4]^{2-}$. A free Ni^{2+} ion has eight $3d$ electrons, two of which are unpaired in accord with Hund's rule:

$$Ni^{2+}: \text{[Ar]} \quad \uparrow\downarrow \;\; \uparrow\downarrow \;\; \uparrow\downarrow \;\; \uparrow \;\; \uparrow \qquad \underline{} \qquad \underline{} \; \underline{} \; \underline{}$$
$$\qquad\qquad\qquad\qquad\quad 3d \qquad\qquad\qquad 4s \qquad\qquad 4p$$

In square planar complexes, the metal uses a set of four hybrid orbitals called dsp^2 hybrids, which point toward the four corners of a square. By pairing up the two unpaired d electrons in one d orbital, we obtain a vacant $3d$ orbital that can be hybridized with the $4s$ orbital and two of the $4p$ orbitals to give the square planar dsp^2 hybrids. These hybrids form bonds to the ligands by accepting a share in the four pairs of ligand electrons:

$$[Ni(CN)_4]^{2-}: [Ar]\ \underset{3d}{\uparrow\downarrow\ \uparrow\downarrow\ \uparrow\downarrow\ \uparrow\downarrow}\ \boxed{\underset{4s}{\uparrow\downarrow}\quad \underset{4p}{\uparrow\downarrow\ \uparrow\downarrow\ \uparrow\downarrow}}\ \_\_$$

Four dsp^2 bonds to the ligands

In agreement with this description, $[Ni(CN)_4]^{2-}$ is diamagnetic.

In octahedral complexes, the metal ion uses either sp^3d^2 or d^2sp^3 hybrid orbitals. To see the difference between these two kinds of hybrids, let's consider the cobalt(III) complexes $[CoF_6]^{3-}$ and $[Co(CN)_6]^{3-}$. A free Co^{3+} ion has six $3d$ electrons, four of which are unpaired:

$$Co^{3+}: [Ar]\ \underset{3d}{\uparrow\downarrow\ \uparrow\ \uparrow\ \uparrow\ \uparrow}\quad \underset{4s}{\_}\quad \underset{4p}{\_\ \_\ \_}\quad \underset{4d}{\_\ \_\ \_\ \_\ \_}$$

Magnetic measurements indicate that $[CoF_6]^{3-}$ is paramagnetic and contains four unpaired electrons. Evidently, none of the $3d$ orbitals is available to accept a share in the ligand electrons because each is already at least partially occupied. Consequently, the octahedral hybrids that Co^{3+} uses are formed from the vacant $4s$, $4p$, and $4d$ orbitals. These orbitals, called sp^3d^2 hybrids, share in the six pairs of ligand electrons, as shown in the following orbital diagram:

$$[CoF_6]^{3-}: [Ar]\ \underset{3d}{\uparrow\downarrow\ \uparrow\ \uparrow\ \uparrow\ \uparrow}\ \boxed{\underset{4s}{\uparrow\downarrow}\quad \underset{4p}{\uparrow\downarrow\ \uparrow\downarrow\ \uparrow\downarrow}\quad \uparrow\downarrow\ \uparrow\downarrow}\ \underset{4d}{\_\ \_\ \_}$$

Six sp^3d^2 bonds to the ligands

In contrast to $[CoF_6]^{3-}$, magnetic measurements indicate that $[Co(CN)_6]^{3-}$ is diamagnetic. All six $3d$ electrons are therefore paired and occupy just three of the five $3d$ orbitals. That leaves two vacant $3d$ orbitals, which combine with the vacant $4s$ and $4p$ orbitals to give a set of six octahedral hybrid orbitals called d^2sp^3 hybrids. The d^2sp^3 hybrids form bonds to the ligands by accepting a share in the six pairs of ligand electrons:

$$[Co(CN)_6]^{3-}: [Ar]\ \underset{3d}{\uparrow\downarrow\ \uparrow\downarrow\ \uparrow\downarrow}\ \boxed{\uparrow\downarrow\ \uparrow\downarrow\quad \underset{4s}{\uparrow\downarrow}\quad \underset{4p}{\uparrow\downarrow\ \uparrow\downarrow\ \uparrow\downarrow}}\ \underset{4d}{\_\ \_\ \_\ \_\ \_}$$

Six d^2sp^3 bonds to the ligands

Note that the difference between d^2sp^3 and sp^3d^2 hybrids lies in the principal quantum number of the d orbitals. In d^2sp^3 hybrids, the principal quantum number of the d orbitals is one less than the principal quantum number of the s and p orbitals. In sp^3d^2 hybrids, the s, p, and d orbitals have the same principal quantum number. To determine which set of hybrids is used in any given complex, we must know the magnetic properties of the complex.

Complexes of metals like Co^{3+} that exhibit more than one spin state are classified as *high-spin* or *low-spin*. A **high-spin complex**, such as $[CoF_6]^{3-}$, is one in which the d electrons are arranged according to Hund's rule to give the maximum number of unpaired electrons. A **low-spin complex**, such as $[Co(CN)_6]^{3-}$, is one in which the d electrons are paired up to give a maximum number of doubly occupied d orbitals and a minimum number of unpaired electrons.

EXAMPLE 20.9 Give a valence bond description of the bonding in $[V(NH_3)_6]^{3+}$. Include orbital diagrams for the free metal ion and the metal ion in the complex. Tell which hybrid orbitals the metal ion uses and the number of unpaired electrons present.

SOLUTION The free V^{3+} ion has the electron configuration [Ar] $3d^2$ and the orbital diagram

$$V^{3+}: [Ar] \; \underset{3d}{\uparrow \; \uparrow \; \underline{} \; \underline{} \; \underline{}} \quad \underset{4s}{\underline{}} \quad \underset{4p}{\underline{} \; \underline{} \; \underline{}}$$

Because $[V(NH_3)_6]^{3+}$ is octahedral, the V^{3+} ion must use either d^2sp^3 or sp^3d^2 hybrid orbitals in accepting a share in six pairs of electrons from the six NH_3 ligands. The preferred hybrids are d^2sp^3 because several $3d$ orbitals are vacant and d^2sp^3 hybrids have lower energy than sp^3d^2 hybrids (because the $3d$ orbitals have lower energy than the $4d$ orbitals). Thus, $[V(NH_3)_6]^{3+}$ has the following orbital diagram:

$$[V(NH_3)_6]^{3+}: [Ar] \; \underset{3d}{\uparrow \; \uparrow \; \underline{}} \; \boxed{\underset{}{\uparrow\downarrow \; \uparrow\downarrow} \quad \underset{4s}{\uparrow\downarrow} \quad \underset{4p}{\uparrow\downarrow \; \uparrow\downarrow \; \uparrow\downarrow}}$$

Six d^2sp^3 bonds to the ligands

The complex has two unpaired electrons and is therefore paramagnetic.

▶ **PROBLEM 20.13** Give a valence bond description of the bonding in each of the following complexes. Include orbital diagrams for the free metal ion and the metal ion in the complex. Tell which hybrid orbitals the metal ion uses and the number of unpaired electrons in each complex.
  (a) $[Fe(CN)_6]^{3-}$ (low-spin)   **(b)** $[Co(H_2O)_6]^{2+}$ (high-spin)
  (c) $[VCl_4]^-$ (tetrahedral)   **(d)** $[PtCl_4]^{2-}$ (square planar)

20.12 Crystal Field Theory

Valence bond theory helps us visualize the bonding in complexes, but it doesn't account for their color or explain why some complexes are high-spin whereas others are low-spin. To explain these properties, we turn to the **crystal field theory**, a model that views the bonding in complexes as arising from electrostatic interactions and considers the effect of the ligand charges on the energies of the metal ion d orbitals. This model was first applied to transition metal ions in ionic crystals—hence the name crystal field theory—but it is also applicable to metal complexes where the "crystal field" is the electric field due to the charges or dipoles of the ligands.

Octahedral Complexes

Let's first consider an octahedral complex such as $[TiF_6]^{3-}$ (Figure 20.27). According to crystal field theory, the bonding is ionic and involves electrostatic attraction between the positively charged Ti^{3+} ion and the negatively charged F^- ligands. Of course, the F^- ligands repel one another, which is why they adopt the geometry (octahedral) that locates them as far apart from one another as possible. Because the metal–ligand attractions are greater than the ligand–ligand repulsions, the complex is more stable than the separated ions, thus accounting for the bonding.

Note the differences between crystal field theory and valence bond theory. In crystal field theory, there are no covalent bonds, no shared electrons, and no hybrid orbitals—just electrostatic interactions within an array of ions.

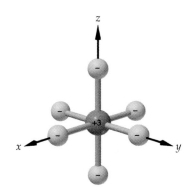

FIGURE 20.27 Crystal field model of the octahedral $[TiF_6]^{3-}$ complex. The metal ion and ligands are regarded as charged particles held together by electrostatic attraction. The ligands lie along the $\pm x$, $\pm y$, and $\pm z$ directions.

In complexes that contain neutral dipolar ligands, such as H_2O or NH_3, the electrostatic interactions are of the ion–dipole type (Section 10.2). For example, in $[Ti(H_2O)_6]^{3+}$, the Ti^{3+} ion attracts the negative end of the water dipoles.

To explain why complexes are colored, we need to look at the effect of the ligand charges on the energies of the d orbitals. Recall that four of the d orbitals are shaped like a cloverleaf, and the fifth (d_{z^2}) is shaped like a dumbbell inside a donut (Section 5.8). Figure 20.28 shows the spatial orientation of the d orbitals with respect to an octahedral array of charged ligands located along the x, y, and z coordinate axes.

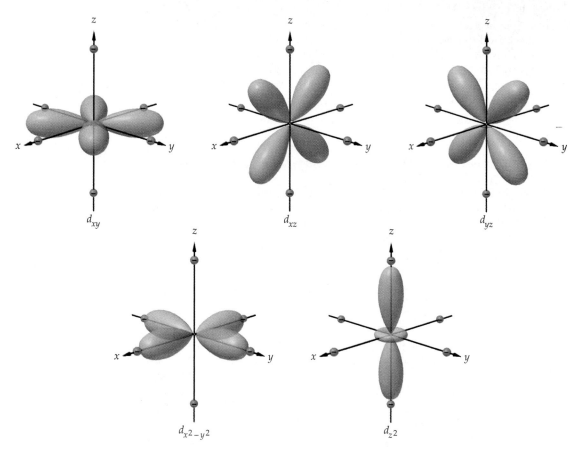

FIGURE 20.28 The shapes of the five d orbitals and their orientation with respect to an octahedral array of charged ligands.

Because the d electrons are negatively charged, they are repelled by the negatively charged ligands. Thus, their orbital energies are higher in the complex than in the free metal ion. But not all the d orbitals are raised in energy by the same amount. As shown in Figure 20.29, the d_{z^2} and $d_{x^2-y^2}$ orbitals, which point directly at the ligands, are raised in energy to a greater extent than the d_{xy}, d_{xz}, and d_{yz} orbitals, which point between the ligands. This energy splitting between the two sets of d orbitals is called the **crystal field splitting** and is represented by the Greek letter Δ.

In general, the crystal field splitting energy Δ corresponds to wavelengths of light in the visible region of the spectrum, and the colors of complexes can therefore be attributed to electronic transitions between the lower- and higher-energy sets of d orbitals. Consider, for example, $[Ti(H_2O)_6]^{3+}$, a complex that contains a single d electron (Ti^{3+} has the electron configuration [Ar] $3d^1$). In the ground-energy state, the d electron occupies one of the lower-energy orbitals—xy, xz, or yz (from now on we'll denote the d orbitals by their

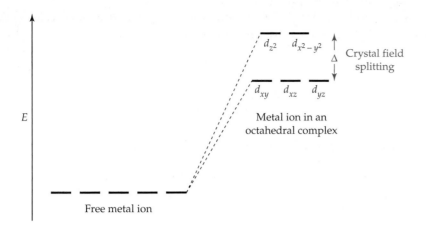

FIGURE 20.29 A *d*-orbital energy level diagram for a free metal ion and a metal ion in an octahedral complex. In the absence of ligands, the five *d* orbitals have the same energy. When the metal ion is surrounded by an octahedral array of ligands, the *d* orbitals increase in energy and split into two sets that are separated in energy by the crystal field splitting, Δ. The *d* orbitals whose lobes point directly toward the ligands (d_{z^2} and $d_{x^2-y^2}$) are higher in energy than the *d* orbitals whose lobes point between the ligands (d_{xy}, d_{xz}, and d_{yz}).

subscripts). When $[Ti(H_2O)_6]^{3+}$ absorbs blue-green light with a wavelength of about 500 nm, the absorbed energy is used to promote the *d* electron to one of the higher-energy orbitals, z^2 or $x^2 - y^2$:

We can calculate the value of Δ from the wavelength of the absorbed light, about 500 nm for $[Ti(H_2O)_6]^{3+}$:

$$\Delta = h\nu = \frac{hc}{\lambda} = \frac{(6.626 \times 10^{-34}\,J\cdot s)(3.00 \times 10^8\,m/s)}{500 \times 10^{-9}\,m} = 3.98 \times 10^{-19}\,J$$

This is the energy needed to excite a single $[Ti(H_2O)_6]^{3+}$ ion. To express Δ on a per-mole basis, we multiply by Avogadro's number:

$$\Delta = \left(3.98 \times 10^{-19}\,\frac{J}{ion}\right)\left(6.02 \times 10^{23}\,\frac{ions}{mol}\right)$$

$$= 2.40 \times 10^5\,J/mol = 240\,kJ/mol$$

The absorption spectra of different complexes indicate that the size of the crystal field splitting depends on the nature of the ligands. For example, Δ for Ni^{2+} ([Ar] $3d^8$) complexes increases as the ligand varies from H_2O to NH_3 to ethylenediamine (en). Accordingly, the electronic transitions shift to higher energy (shorter wavelength) as the ligand varies from H_2O to NH_3 to en, thus accounting for the observed variation in color (Figure 20.24):

In general, the crystal field splitting increases as the ligand varies in the following order, known as the **spectrochemical series**:

Weak-field ligands $I^- < Br^- < Cl^- < F^- < H_2O < NH_3 < en < CN^-$ Strong-field ligands

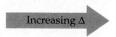

Increasing Δ

Ligands such as halides and H_2O, which give a relatively small value of Δ, are called **weak-field ligands**. Ligands such as NH_3, en, and CN^-, which produce a relatively large value of Δ, are known as **strong-field ligands**. Different metal ions have different values of Δ, which explains why their complexes with the same ligand have different colors (Figure 20.23). Because d^0 ions such as Ti^{4+}, d^{10} ions such as Zn^{2+}, and main-group ions don't have partially filled d subshells, they can't undergo d–d electronic transitions, and most of their compounds are therefore colorless.

Crystal field theory accounts for the magnetic properties of complexes as well as for their color. It explains, for example, why complexes with weak-field ligands, such as $[CoF_6]^{3-}$, are high-spin, whereas related complexes with strong-field ligands, such as $[Co(CN)_6]^{3-}$, are low-spin. In $[CoF_6]^{3-}$, the six d electrons of Co^{3+} ([Ar] $3d^6$) occupy both the higher- and lower-energy d orbitals, whereas in $[Co(CN)_6]^{3-}$, all six d electrons are spin-paired in the lower-energy orbitals:

$[CoF_6]^{3-}$ (high-spin) $[Co(CN)_6]^{3-}$ (low-spin)

What determines which of the two spin states has the lower energy? In general, when an electron moves from a z^2 or $x^2 - y^2$ orbital to one of the lower-energy orbitals, the orbital energy decreases by Δ. But because of electron–electron repulsion, it costs energy to put the electron into an orbital that already contains another electron. The energy required is called the spin-pairing energy P. If Δ is greater than P, as it is for $[Co(CN)_6]^{3-}$, then the low-spin arrangement has lower energy. If Δ is less than P, as it is for $[CoF_6]^{3-}$, the high-spin arrangement has lower energy. Thus, the observed spin state depends on the relative values of Δ and P. In general, strong-field ligands give low-spin complexes, and weak-field ligands give high-spin complexes.

Note that a choice between high-spin and low-spin electron configurations arises only for complexes of metal ions with four to seven d electrons, so-called d^4–d^7 complexes. For d^1–d^3 and d^8–d^{10} complexes, only one ground-state electron configuration is possible. In d^1–d^3 complexes, all the electrons occupy the lower-energy d orbitals, independent of the value of Δ. In d^8–d^{10} complexes, the lower-energy set of d orbitals is filled with three pairs of electrons, while the higher-energy set contains two, three, or four electrons, again independent of the value of Δ.

EXAMPLE 20.10   Draw a crystal field orbital energy level diagram, and predict the number of unpaired electrons for each of the following complexes:
(a) $[Cr(en)_3]^{3+}$ **(b)** $[Mn(CN)_6]^{3-}$ **(c)** $[Co(H_2O)_6]^{2+}$

SOLUTION

(a) Cr^{3+} ([Ar] $3d^3$) has three unpaired electrons. In the complex, they occupy the lower-energy set of d orbitals.

(b) Mn^{3+} ([Ar] $3d^4$) can have a high-spin or low-spin configuration. Because CN^- is a strong-field ligand, all four d electrons go into the lower-energy d orbitals. The complex is low-spin, with two unpaired electrons.

(c) Co^{2+} ([Ar] $3d^7$) has a high-spin configuration with three unpaired electrons because H_2O is a weak-field ligand.

In the following orbital energy level diagrams, the relative values of the crystal field splitting Δ are in accord with the positions of the ligands in the spectrochemical series ($H_2O < en < CN^-$):

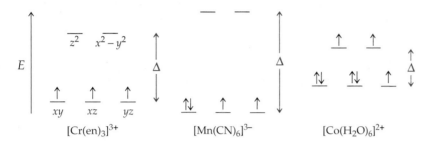

▶ **PROBLEM 20.14**   Draw a crystal field d-orbital energy level diagram, and predict the number of unpaired electrons for each of the following complexes:
(a) $[Fe(H_2O)_6]^{2+}$ **(b)** $[Fe(CN)_6]^{4-}$ **(c)** $[VF_6]^{3-}$

Tetrahedral and Square Planar Complexes

As you might expect, different geometric arrangements of the ligands give different energy splittings for the d orbitals. Figure 20.30 shows d-orbital energy level diagrams for tetrahedral and square planar complexes.

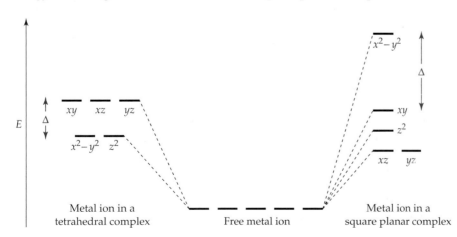

FIGURE 20.30   Energies of the d orbitals in tetrahedral and square planar complexes relative to their energy in the free metal ion. The crystal field splitting energy Δ is small in tetrahedral complexes but much larger in square planar complexes.

The splitting pattern in tetrahedral complexes is just the opposite of that in octahedral complexes; that is, the xy, xz, and yz orbitals have higher energy than the $x^2 - y^2$ and z^2 orbitals. (As with octahedral complexes, the energy ordering follows from the relative orientation of the orbital lobes and the ligands, but we won't try to derive the result.) Because none of the orbitals point directly at the ligands in tetrahedral geometry, and because there are only four ligands instead of six, the crystal field splitting in tetrahedral complexes is only about half of that in octahedral complexes. Consequently, Δ is almost always smaller than the spin-pairing energy P, and nearly all tetrahedral complexes are high-spin.

Square planar complexes look like octahedral ones (Figure 20.28) except that the two trans ligands along the z axis are missing. In square planar complexes, the $x^2 - y^2$ orbital is high in energy (Figure 20.30) because it points directly at all four ligands, which lie along the x and y axes. The splitting pattern is more complicated here than for the octahedral and tetrahedral cases, but the main point to remember is that a large energy gap exists between the $x^2 - y^2$ orbital and the four lower-energy orbitals. Square planar geometry is most common for metal ions with electron configuration d^8 because this configuration favors low-spin complexes in which all four lower-energy orbitals are filled and the higher-energy $x^2 - y^2$ orbital is vacant. Common examples are $[Ni(CN)_4]^{2-}$, $[PdCl_4]^{2-}$, and $Pt(NH_3)_2Cl_2$.

EXAMPLE 20.11 Draw crystal field energy level diagrams and predict the number of unpaired electrons for the following complexes:
(a) $[FeCl_4]^-$ (tetrahedral)   **(b)** $[PtCl_4]^{2-}$ (square planar)

SOLUTION
(a) Take the tetrahedral energy level diagram from Figure 20.30. Because nearly all tetrahedral complexes are high-spin (small Δ), the five d electrons of Fe^{3+} ($[Ar]\,3d^5$) are distributed between the higher- and lower-energy orbitals as shown below. The complex is high-spin with five unpaired electrons.
(b) Pt^{2+} ($[Xe]\,4f^{14}\,5d^8$) has eight d electrons. Because Δ is large in square planar complexes, all the electrons occupy the four lower-energy d orbitals. There are no unpaired electrons, and the complex is diamagnetic.

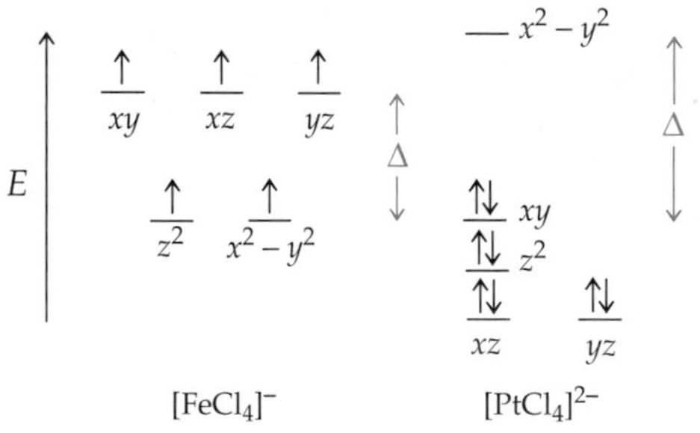

$[FeCl_4]^-$ $[PtCl_4]^{2-}$

▶ **PROBLEM 20.15** Draw a crystal field energy level diagram, and predict the number of unpaired electrons for the following complexes:
(a) $[NiCl_4]^{2-}$ (tetrahedral)   **(b)** $[Ni(CN)_4]^{2-}$ (square planar)

Titanium: A High-Tech Metal

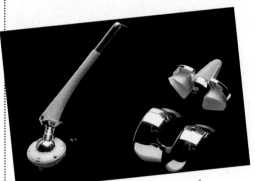

This artificial knee joint is made of titanium, a nontoxic, noncorrosive, high-tech metal.

Named after the Titans, Greek mythological figures symbolic of power and strength, titanium is the ninth most abundant element in the earth's crust (0.57% by mass). It occurs in rutile (TiO_2) and in the mineral ilmenite ($FeTiO_3$), found in the United States, Australia, Canada, Malaysia, and elsewhere. Despite its abundance, titanium is difficult and expensive to produce in its pure form, and it remained a curiosity until about 1950, when its potential applications in aerospace technology were recognized.

Titanium is a superb structural material because of its hardness, strength, heat resistance (mp 1668°C), and relatively low density (4.51 g/cm³). Titanium is just as strong as steel, but 45% lighter; titanium is twice as strong as aluminum but only 60% heavier. When alloyed with a few percent aluminum and vanadium, titanium has a higher strength-to-weight ratio than any other engineering metal. These properties make titanium an ideal choice in aerospace applications, such as airframes and jet engines, as well as for recreational use. (Those who can afford one rave about the ride and feel of a titanium bike frame.)

Although titanium has a large positive $E°$ for oxidation, and Ti dust will burn in air, the bulk metal is remarkably immune to corrosion because its surface becomes coated with a thin, protective oxide film. Titanium objects are inert to seawater, nitric acid, hot aqueous NaOH, and even to aqueous chlorine gas. Titanium is therefore used in chemical plants, in desalination equipment, and in numerous other industrial processes that demand inert, noncorrosive materials. Because it is nontoxic and inert to body fluids, titanium is even used for manufacturing artificial joints and dental implants.

Pure titanium is obtained commercially from rutile (TiO_2) by an indirect route that involves initial reaction of TiO_2 with Cl_2 gas and coke to yield liquid $TiCl_4$ (bp 136°C), which is purified by fractional distillation. Subsequent reduction to Ti metal is then carried out by reaction with molten magnesium at 900°C, and further purification is effected by melting the titanium in an electric arc under an atmosphere of argon.

$$TiO_2(s) + 2\,Cl_2(g) + 2\,C(s) \longrightarrow TiCl_4(l) + 2\,CO(g)$$

$$TiCl_4(g) + 2\,Mg(l) \longrightarrow Ti(s) + 2\,MgCl_2(l)$$

Although the process is extremely expensive and energy-intensive, the cost of producing titanium is justified because of its unique properties. Worldwide production of titanium now exceeds 100,000 tons per year.

▶ **PROBLEM 20.16** $E°$ values for oxidation of the metal to the $M^{3+}(aq)$ ion are +1.37 V for M = Ti and +0.04 V for M = Fe. Why then doesn't titanium corrode?

▶ **PROBLEM 20.17** How many liters of Cl_2 at 20°C and 740 mm Hg pressure would be needed to produce 1.00×10^5 tons of titanium metal from TiO_2 if the overall yield is 93.5%?

Key Words

Summary

Transition elements (*d*-block elements) are the metallic elements in the central part of the periodic table. Most of the neutral atoms have valence electron configuration $(n-1)d^{1-10}\,ns^2$, and the cations have configuration $(n-1)d^{1-10}$. Transition metals exhibit a variety of oxidation states. Ions with the metal in a high oxidation state tend to be good oxidizing agents ($Cr_2O_7^{2-}$, MnO_4^-), and ions with the metal in a low oxidation state are good reducing agents (V^{2+}, Cr^{2+}). The **lanthanides** and **actinides** comprise the *f*-block, or **inner transition, elements**.

Coordination compounds (metal complexes) are compounds in which a central metal ion is attached to a group of surrounding **ligands** by coordinate covalent bonds. Ligands can be **monodentate** or **polydentate**, depending on the number of **donor atoms** attached to the metal. Polydentate ligands are also called **chelating agents**. They form complexes (metal **chelates**) that contain rings of atoms known as **chelate rings**.

The number of ligand donor atoms bonded to a metal is called the **coordination number** of the metal. Common coordination numbers and geometries are 2 (linear), 4 (tetrahedral or square planar), and 6 (octahedral). Systematic names for complexes specify the number of ligands of each particular type, the metal, and its oxidation state.

Many complexes exist as **isomers**, compounds that have the same formula but a different arrangement of the constituent atoms. **Constitutional isomers**, such as **linkage isomers** and **ionization isomers**, have different connections between their constituent atoms. **Stereoisomers** (**diastereoisomers** and **enantiomers**) have the same connections but a different arrangement of the atoms in space. The most common diastereoisomers are **cis** and **trans isomers** of square planar MA_2B_2 and octahedral MA_4B_2 complexes. Enantiomers, such as "right-handed" and "left-handed" $[Co(en)_3]^{3+}$, are nonidentical mirror images. One isomer rotates the plane of plane-polarized light to the right, and the other rotates this plane through an equal angle but in the opposite direction. A 50:50 mixture of the two isomers is called a **racemic mixture**. Molecules that have handedness are said to be **chiral**.

Valence bond theory describes the bonding in complexes in terms of two-electron, coordinate covalent bonds resulting from overlap of filled ligand orbitals with vacant metal hybrid orbitals that point in the direction of the ligands: sp (linear), sp^3 (tetrahedral), dsp^2 (square planar), and d^2sp^3 or sp^3d^2 (octahedral).

Crystal field theory assumes that the metal–ligand bonding is entirely ionic. Because of electrostatic repulsions between the *d* electrons and the ligands, the *d* orbitals are raised in energy and are differentiated by an energy separation called the **crystal field splitting**, Δ. In octahedral complexes, the z^2 and x^2-y^2 orbitals have higher energy than the xy, xz, and yz orbitals. Tetrahedral and square planar complexes exhibit different splitting patterns. The colors of complexes are due to electronic transitions from one set of *d* orbitals to another, and the transition energies depend on the position of the ligand in the **spectrochemical series**. **Weak-field ligands** give small Δ values, and **strong-field ligands** give large Δ values. Crystal field theory accounts for the magnetic properties of complexes in terms of the relative values of Δ and the spin-pairing energy P. Small Δ values favor **high-spin complexes** (those with a maximum number of unpaired *d* electrons), and large Δ values favor **low-spin complexes**.

Key Concept Summary

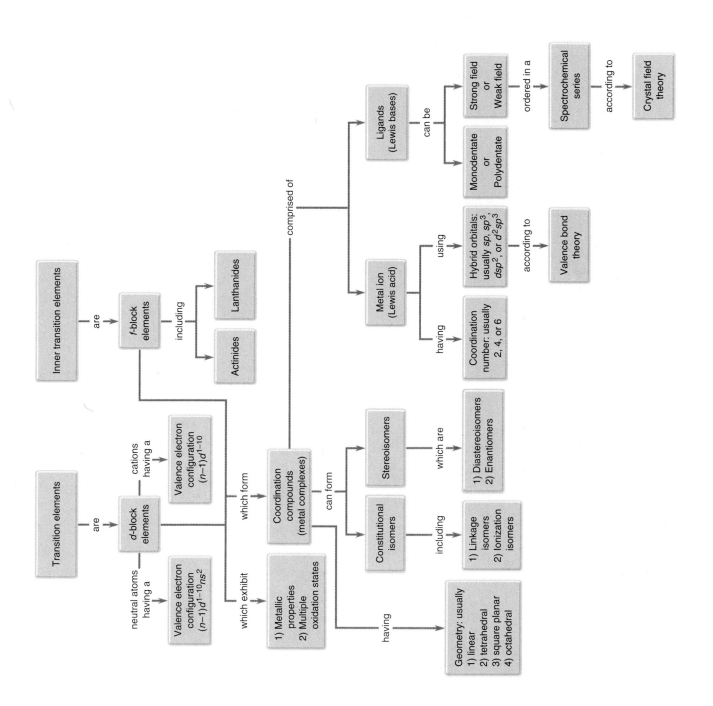

Understanding Key Concepts

Problems 20.1–20.17 appear within the chapter.

20.18 Locate on the periodic table below the transition elements with the following electron configurations. Identify each element.

(a) [Ar] $3d^7\ 4s^2$ **(b)** [Ar] $3d^5\ 4s^1$

(c) [Kr] $4d^2\ 5s^2$ **(d)** [Xe] $4f^3\ 6s^2$

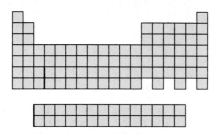

20.19 What is the general trend in the following properties from left to right across the first transition series (Sc to Zn)? Explain each trend.

(a) Atomic radius

(b) Density

(c) Ionization energy

(d) Standard oxidation potential

Sc	Ti	V	Cr	Mn	Fe	Co	Ni	Cu	Zn

20.20 Classify the following ligands as monodentate, bidentate, or tridentate. Which can form chelate rings?

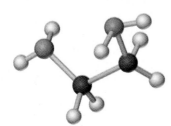

(a) $NH_2CH_2CH_2NH_2$

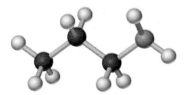

(b) $CH_3CH_2CH_2NH_2$

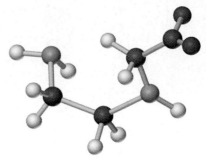

(c) $NH_2CH_2CH_2NHCH_2CO_2^-$

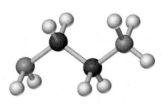

(d) $NH_2CH_2CH_2NH_3^+$

20.21 Draw the structure of the following complexes. What is the oxidation state, coordination number, and coordination geometry of the metal in each?

(a) $Na[Au(CN)_2]$

(b) $[Co(NH_3)_5Br]SO_4$

(c) $Pt(en)Cl_2$

(d) $(NH_4)_2[PtCl_2(C_2O_4)_2]$

20.22 Consider the following isomers of $[Cr(NH_3)_2Cl_4]^-$:

(1) (2) (3) (4)

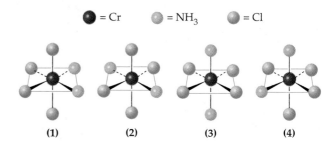

(a) Label the isomers as cis or trans.

(b) Which isomers are identical, and which are different?

(c) Do any of these isomers exist as enantiomers? Explain.

20.23 Consider the following ethylenediamine complexes:

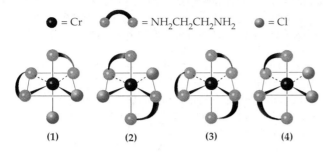

(1)   (2)   (3)   (4)

(a) Which complexes are chiral, and which are achiral?

(b) Draw the enantiomer of each chiral complex.

(c) Which, if any, of the chiral complexes are enantiomers of one another?

20.24 Predict the crystal field energy level diagram for a square pyramidal ML_5 complex that has two ligands along the $\pm x$ and $\pm y$ axes but only one ligand along the z axis. Your diagram should be intermediate between those for an octahedral ML_6 complex and a square planar ML_4 complex.

20.25 Imagine two complexes, one tetrahedral and one square planar, in which the central atom is bonded to four different ligands (shown here in four different colors). Is either complex chiral? Explain.

Additional Problems

Electron Configurations and Properties of Transition Elements

20.26 Use the periodic table to write the electron configuration for each of the following atoms:

(a) Ni   (b) Cr   (c) Zr   (d) Zn

20.27 What is the electron configuration for each of the following atoms?

(a) Mn   (b) Cd   (c) Co   (d) Fe

20.28 Use the periodic table to give the electron configuration for each of the following:

(a) Co^{2+} (b) Fe^{3+}

(c) Mo^{3+} (d) Cr(VI) in CrO_4^{2-}

20.29 Specify the electron configuration for each of the following:

(a) Cr^{3+} (b) Ag^+

(c) Rh^{3+} (d) Mn(VI) in MnO_4^{2-}

20.30 Predict the number of unpaired electrons for each of the following:

(a) Cu^{2+} (b) Ti^{2+} (c) Zn^{2+} (d) Cr^{3+}

20.31 Predict the number of unpaired electrons for each of the following:

(a) Sc^{3+} (b) Co^{2+} (c) Mn^{3+} (d) Cr^{2+}

20.32 Titanium, used in making jet aircraft engines, is much harder than potassium or calcium. Explain.

20.33 Molybdenum (mp 2623°C) has a higher melting point than yttrium (mp 1522°C) or cadmium (321°C). Explain.

20.34 Briefly account for each of the following observations:

(a) Atomic radii decrease in the order Sc > Ti > V.

(b) Densities increase in the order Ti < V < Cr.

20.35 Arrange the following atoms in order of decreasing atomic radii, and account for the trend:

(a) Cr   (b) Ti   (c) Mn   (d) V

20.36 What is the lanthanide contraction, and why does it occur?

20.37 The atomic radii of zirconium and hafnium are nearly identical. Explain.

20.38 Calculate the sum of the first two ionization energies for the first-series transition elements, and account for the general trend.

20.39 What is the general trend in standard potentials for oxidation of first-series transition metals from Sc to Zn? What is the reason for the trend?

20.40 Write a balanced net ionic equation for reaction of each of the following metals with hydrochloric acid in the absence of air. If no reaction occurs, indicate N.R.

(a) Cr   (b) Zn   (c) Cu   (d) Fe

20.41 Write a balanced net ionic equation for reaction of each of the following metals with dilute sulfuric acid in the absence of air. If no reaction occurs, indicate N.R.

(a) Mn   (b) Ag   (c) Sc   (d) Ni

Oxidation States

20.42 Which of the following metals have more than one oxidation state?

(a) Zn   (b) Mn   (c) Sr   (d) Cu

20.43 Which of the following metals have only one oxidation state?

(a) V   (b) Al   (c) Co   (d) Sc

20.44 What is the highest oxidation state for each of the elements from Sc to Zn?

20.45 The highest oxidation state for the early transition metals Sc, Ti, V, Cr, and Mn is the periodic group number. The highest oxidation state for the later transition elements Fe, Co, and Ni is less than the periodic group number. Explain.

20.46 Which is the stronger oxidizing agent, Cr^{2+} or Cu^{2+}? Explain.

20.47 Which is the stronger reducing agent, Ti^{2+} or Ni^{2+}? Explain.

20.48 Which is more easily oxidized, Cr^{2+} or Ni^{2+}? Explain.

20.49 Which is more easily reduced, V^{3+} or Fe^{3+}? Explain.

20.50 Arrange the following substances in order of increasing strength as an oxidizing agent, and account for the trend:

(a) Mn^{2+} (b) MnO_2 (c) MnO_4^-

20.51 Arrange the following ions in order of increasing strength as a reducing agent, and account for the trend:

(a) Cr^{2+} (b) Cr^{3+} (c) $Cr_2O_7^{2-}$

Chemistry of Selected Transition Elements

20.52 Write a balanced equation for the industrial production of:

(a) Chromium metal from chromium(III) oxide

(b) Copper metal from copper(I) sulfide

20.53 Write a balanced net ionic equation for reaction of nitric acid with:

(a) Iron metal

(b) Copper metal

(c) Chromium metal

20.54 What product is formed when dilute H_2SO_4, a nonoxidizing acid, is added to each of the following ions or compounds?

(a) CrO_4^{2-} (b) $Cr(OH)_3$ (c) $Cr(OH)_2$

(d) $Fe(OH)_2$ (e) $Cu(OH)_2$

20.55 What product is formed when excess aqueous NaOH is added to a solution of each of the following ions?

(a) Cr^{2+} (b) Cr^{3+} (c) $Cr_2O_7^{2-}$

(d) Fe^{2+} (e) Fe^{3+}

20.56 Arrange the following hydroxy compounds in order of increasing acid strength, and account for the trend:

(a) $CrO_2(OH)_2$ (b) $Cr(OH)_2$ (c) $Cr(OH)_3$

20.57 Explain how $Cr(OH)_3$ can act both as an acid and as a base.

20.58 Which of the following compounds is amphoteric?

(a) $Cr(OH)_2$ (b) $Fe(OH)_2$

(c) $Cr(OH)_3$ (d) $Fe(OH)_3$

20.59 Which of the following ions disproportionates in aqueous solution? Write a balanced net ionic equation for the reaction.

(a) Cr^{3+} (b) Fe^{3+} (c) Cu^+ (d) Cu^{2+}

20.60 How might you separate the following pairs of ions by addition of a single reagent? Include formulas for the major products of the reactions.

(a) Fe^{3+} and Na^+

(b) Cr^{3+} and Fe^{3+}

(c) Fe^{3+} and Cu^{2+}

20.61 Give a method for separating the following pairs of ions by addition of a single reagent. Include formulas for the major products of the reactions.

(a) K^+ and Cu^{2+}

(b) Cu^{2+} and Cr^{3+}

(c) Fe^{3+} and Al^{3+}

20.62 Complete and balance the net ionic equation for each of the following reactions in acidic solution:

(a) $Cr_2O_7^{2-}(aq) + Fe^{2+}(aq) \rightarrow$

(b) $Fe^{2+}(aq) + O_2(g) \rightarrow$

(c) $Cu_2O(s) + H^+(aq) \rightarrow$

(d) $Fe(s) + H^+(aq) \rightarrow$

20.63 Complete and balance the net ionic equation for each of the following reactions in acidic solution:

(a) $Cr^{2+}(aq) + Cr_2O_7^{2-}(aq) \rightarrow$

(b) $Cu(s) + $ conc $HNO_3(aq) \rightarrow$

(c) $Cu^{2+}(aq) + $ excess $NH_3(aq) \rightarrow$

(d) $Cr(OH)_4^-(aq) + $ excess $H^+(aq) \rightarrow$

20.64 Write a balanced net ionic equation for each of the following reactions:

(a) A CrO_4^{2-} solution turns from yellow to orange upon addition of acid.

(b) $Fe^{3+}(aq)$ reacts with aqueous KSCN to give a deep red solution.

(c) Copper metal reacts with nitric acid to give NO gas and a blue solution.

(d) A deep green solution of $Cr(OH)_3$ in excess base turns yellow on addition of hydrogen peroxide.

20.65 Write a balanced net ionic equation for each of the following reactions:

(a) A $CuSO_4$ solution becomes dark blue when excess ammonia is added.

(b) A solution of $Na_2Cr_2O_7$ turns from orange to yellow on addition of base.

(c) When base is added to a solution of $Fe(NO_3)_3$, a red-brown precipitate forms.

(d) Dissolution of CuS in hot HNO_3 gives NO gas, a blue solution, and a yellow solid.

Coordination Compounds; Ligands

20.66 Forming $[Ni(en)_3]^{2+}$ from Ni^{2+} and ethylenediamine is a Lewis acid–base reaction. Explain.

20.67 Identify the Lewis acid and the Lewis base in the reaction of oxalate ions ($C_2O_4^{2-}$) with Fe^{3+} to give $[Fe(C_2O_4)_3]^{3-}$.

20.68 Give an example of a coordination compound in which the metal exhibits a coordination number of:

(a) 2   (b) 4   (c) 6.

20.69 What is the coordination number of the metal in each of the following complexes?

(a) $AgCl_2^-$ (b) $[Cr(H_2O)_5Cl]^{2+}$

(c) $[Co(NCS)_4]^{2-}$ (d) $[ZrF_8]^{4-}$

(e) $Co(NH_3)_3(NO_2)_3$

20.70 What is the oxidation state of the metal in each of the complexes in Problem 20.69?

20.71 Identify the oxidation state of the metal in each of the following complexes:

(a) $[Ni(CN)_5]^{3-}$ (b) $Ni(CO)_4$

(c) $[Co(en)_2(H_2O)Br]^{2+}$ (d) $[Cu(H_2O)_2(C_2O_4)_2]^{2-}$

20.72 What is the formula of a complex that has each of the following geometries?

(a) Tetrahedral   (b) Linear

(c) Octahedral   (d) Square planar

20.73 What is the formula for each of the following complexes?

(a) An iridium(III) complex with three ammonia and three chloride ligands

(b) A chromium(III) complex with two water and two oxalate ligands

(c) A platinum(IV) complex with two ethylenediamine and two thiocyanate ligands

20.74 Draw the structure of the iron oxalate complex $[Fe(C_2O_4)_3]^{3-}$. Describe the coordination geometry, and identify any chelate rings. What is the coordination number and the oxidation number of the iron?

20.75 Draw the structure of the platinum ethylenediamine complex $[Pt(en)_2]^{2+}$. Describe the coordination geometry, and identify any chelate rings. What is the coordination number and the oxidation number of the platinum?

20.76 Identify the oxidation state of the metal in each of the following compounds:

(a) $Co(NH_3)_3(NO_2)_3$ (b) $[Ag(NH_3)_2]NO_3$

(c) $K_3[Cr(C_2O_4)_2Cl_2]$ (d) $Cs[CuCl_2]$

20.77 What is the oxidation state of the metal in each of the following compounds?

(a) $(NH_4)_3[RhCl_6]$ (b) $[Cr(NH_3)_4(SCN)_2]Br$

(c) $[Cu(en)_2]SO_4$ (d) $Na_2[Mn(EDTA)]$

Naming Coordination Compounds

20.78 What is the systematic name for each of the following ions?

(a) $[MnCl_4]^{2-}$ (b) $[Ni(NH_3)_6]^{2+}$

(c) $[Co(CO_3)_3]^{3-}$ (d) $[Pt(en)_2(SCN)_2]^{2+}$

20.79 Assign a systematic name to each of the following ions:

(a) $[AuCl_4]^-$ (b) $[Fe(CN)_6]^{4-}$

(c) $[Fe(H_2O)_5NCS]^{2+}$ (d) $[Cr(NH_3)_2(C_2O_4)_2]^-$

20.80 What is the systematic name for each of the following coordination compounds?

(a) $Cs[FeCl_4]$ (b) $[V(H_2O)_6](NO_3)_3$

(c) $[Co(NH_3)_4Br_2]Br$ (d) $Cu(gly)_2$

20.81 What is the systematic name for each of the following compounds?

(a) $[Cu(NH_3)_4]SO_4$ (b) $Cr(CO)_6$

(c) $K_3[Fe(C_2O_4)_3]$ (d) $[Co(en)_2(NH_3)CN]Cl_2$

20.82 Write the formula for each of the following compounds:

(a) Tetraammineplatinum(II) chloride

(b) Sodium hexacyanoferrate(III)

(c) Tris(ethylenediamine)platinum(IV) sulfate

(d) Triamminetrithiocyanatorhodium(III)

20.83 Write the formula for each of the following compounds:

(a) Diamminesilver(I) nitrate

(b) Potassium diaquadioxalatocobaltate(III)

(c) Hexacarbonylmolybdenum(0)

(d) Diamminebis(ethylenediamine)chromium(III) chloride

Isomers

20.84 Draw all possible constitutional isomers of $[Ru(NH_3)_5(NO_2)]Cl$. Label the isomers as linkage isomers or ionization isomers.

20.85 There are six possible isomers for a square planar palladium(II) complex that contains two Cl^- and two SCN^- ligands. Sketch the structures of all six, and label them according to the isomer classification scheme in Figure 20.16.

20.86 Which of the following complexes can exist as diastereoisomers?
   (a) $[Cr(NH_3)_2Cl_4]^-$
   (b) $[Co(NH_3)_5Br]^{2+}$
   (c) $[FeCl_2(NCS)_2]^{2-}$ (tetrahedral)
   (d) $[PtCl_2Br_2]^{2-}$ (square planar)

20.87 Tell how many diastereoisomers are possible for each of the following complexes, and draw their structures:
   (a) $Pt(NH_3)_2(CN)_2$ (b) $[Co(en)(SCN)_4]^-$
   (c) $[Cr(H_2O)_4Cl_2]^+$ (d) $Ru(NH_3)_3I_3$

20.88 Which of the following complexes are chiral?
   (a) $[Pt(en)Cl_2]$ (b) cis-$[Co(NH_3)_4Br_2]^+$
   (c) cis-$[Cr(en)_2(H_2O)_2]^{3+}$ (d) $[Cr(C_2O_4)_3]^{3-}$

20.89 Which of the following complexes can exist as enantiomers? Draw their structures.
   (a) $[Cr(en)_3]^{3+}$
   (b) cis-$[Co(en)_2(NH_3)Cl]^{2+}$
   (c) $trans$-$[Co(en)_2(NH_3)Cl]^{2+}$
   (d) $[Pt(NH_3)_3Cl_3]^+$

20.90 Draw all possible diastereoisomers and enantiomers of each of the following complexes:
   (a) $Ru(NH_3)_4Cl_2$
   (b) $[Pt(en)_3]^{4+}$
   (c) $[Pt(en)_2ClBr]^{2+}$

20.91 Draw all possible diastereoisomers and enantiomers of each of the following complexes:
   (a) $[Rh(C_2O_4)_2I_2]^{3-}$
   (b) $[Cr(NH_3)_2Cl_4]^-$
   (c) $[Co(EDTA)]^-$

20.92 How does plane-polarized light differ from ordinary light? Draw the structure of a chromium complex that rotates the plane of plane-polarized light.

20.93 What is a racemic mixture? Does it affect plane-polarized light? Explain.

Color of Complexes; Valence Bond and Crystal Field Theories

20.94 What is an absorption spectrum? If the absorption spectrum of a complex has just one band at 455 nm, what is the color of the complex?

20.95 A red-colored complex has just one absorption band in the visible region of the spectrum. Predict the approximate wavelength of this band.

20.96 Give a valence bond description of the bonding in each of the following complexes. Include orbital diagrams for the free metal ion and the metal ion in the complex. Indicate which hybrid orbitals the metal ion uses for bonding, and specify the number of unpaired electrons.
   (a) $[Ti(H_2O)_6]^{3+}$
   (b) $[NiBr_4]^{2-}$ (tetrahedral)
   (c) $[Fe(CN)_6]^{3-}$ (low-spin)
   (d) $[MnCl_6]^{3-}$ (high-spin)

20.97 For each of the following complexes, describe the bonding using valence bond theory. Include orbital diagrams for the free metal ion and the metal ion in the complex. Indicate which hybrid orbitals the metal ion uses for bonding, and specify the number of unpaired electrons.
   (a) $[AuCl_4]^-$ (square planar)
   (b) $[Ag(NH_3)_2]^+$
   (c) $[Fe(H_2O)_6]^{2+}$ (high-spin)
   (d) $[Fe(CN)_6]^{4-}$ (low-spin)

20.98 Draw a crystal field energy level diagram for the $3d$ orbitals of titanium in $[Ti(H_2O)_6]^{3+}$. Indicate what is meant by the crystal field splitting, and explain why $[Ti(H_2O)_6]^{3+}$ is colored.

20.99 Use a sketch to explain why the d_{xy} and $d_{x^2-y^2}$ orbitals have different energies in an octahedral complex. Which of the two orbitals has higher energy?

20.100 The $[Ti(NCS)_6]^{3-}$ ion exhibits a single absorption band at 544 nm. Calculate the crystal field splitting energy Δ (in kJ/mol). Is NCS^- a stronger or weaker field ligand than water? Predict the color of $[Ti(NCS)_6]^{3-}$.

20.101 The $[Cr(H_2O)_6]^{3+}$ ion is violet in color, and $[Cr(CN)_6]^{3-}$ is yellow. Explain this difference using crystal field theory. Use the colors to order H_2O and CN^- in the spectrochemical series.

20.102 For each of the following complexes, draw a crystal field energy level diagram, assign the electrons to orbitals, and predict the number of unpaired electrons:
   (a) $[CrF_6]^{3-}$ (b) $[V(H_2O)_6]^{3+}$ (c) $[Fe(CN)_6]^{3-}$

20.103 Draw a crystal field energy level diagram, assign the electrons to orbitals, and predict the number of unpaired electrons for each of the following:
   (a) $[Cu(en)_3]^{2+}$
   (b) $[FeF_6]^{3-}$
   (c) $[Co(en)_3]^{3+}$ (low-spin)

20.104 The $Ni^{2+}(aq)$ cation is green in color, but $Zn^{2+}(aq)$ is colorless. Explain.

20.105 The $Cr^{3+}(aq)$ cation is violet in color, but $Y^{3+}(aq)$ is colorless. Explain.

20.106 Weak-field ligands tend to give high-spin complexes, but strong-field ligands tend to give low-spin complexes. Explain.

20.107 Explain why nearly all tetrahedral complexes are high-spin.

20.108 Draw a crystal field energy level diagram for a square planar complex, and explain why square planar geometry is especially common for d^8 complexes.

General Problems

20.110 Which of the following complexes are paramagnetic?
- **(a)** $[Mn(CN)_6]^{3-}$
- **(b)** $[Zn(NH_3)_4]^{2+}$
- **(c)** $[Fe(CN)_6]^{4-}$
- **(d)** $[FeF_6]^{4-}$

20.111 Which of the following complexes are diamagnetic?
- **(a)** $[Ni(H_2O)_6]^{2+}$
- **(b)** $[Co(CN)_6]^{3-}$
- **(c)** $[HgI_4]^{2-}$
- **(d)** $[Cu(NH_3)_4]^{2+}$

20.112 For each of the following reactions in acidic solution, predict the products and write a balanced net ionic equation:
- **(a)** $Co^{3+}(aq) + H_2O(l) \rightarrow$
- **(b)** $Cr^{2+}(aq) + O_2(g) \rightarrow$
- **(c)** $Cu(s) + Cr_2O_7^{2-}(aq) \rightarrow$
- **(d)** $CrO_4^{2-}(aq) + H^+(aq) \rightarrow$

20.113 Write a balanced net ionic equation for each of the following reactions in acidic, basic, or neutral solution:
- **(a)** $MnO_4^-(aq) + C_2O_4^{2-}(aq) \xrightarrow{\text{Acidic}}$
  $$Mn^{2+}(aq) + CO_2(g)$$
- **(b)** $Cr_2O_7^{2-}(aq) + Ti^{3+}(aq) \xrightarrow{\text{Acidic}}$
  $$Cr^{3+}(aq) + TiO^{2+}(aq)$$
- **(c)** $MnO_4^-(aq) + SO_3^{2-}(aq) \xrightarrow{\text{Basic}}$
  $$MnO_4^{2-}(aq) + SO_4^{2-}(aq)$$
- **(d)** $Fe(OH)_2(s) + O_2(g) \xrightarrow{\text{Neutral}} Fe(OH)_3(s)$

20.114 Calculate the concentrations of Fe^{3+}, Cr^{3+}, and $Cr_2O_7^{2-}$ in a solution prepared by mixing 100.0 mL of 0.100 M $K_2Cr_2O_7$ and 100.0 mL of 0.400 M $FeSO_4$. The initial solutions are strongly acidic.

20.115 In basic solution, $Cr(OH)_4^-$ is oxidized to CrO_4^{2-} by hydrogen peroxide. Calculate the concentration of CrO_4^{2-} in a solution prepared by mixing 40.0 mL of 0.030 M $Cr_2(SO_4)_3$, 10.0 mL of 0.20 M H_2O_2, and 50.0 mL of 1.0 M NaOH.

20.116 Name each of the following compounds:
- **(a)** $Na[Pt(H_2O)Br(C_2O_4)_2]$
- **(b)** $[Cr(NH_3)_6][Co(C_2O_4)_3]$
- **(c)** $[Co(NH_3)_6][Cr(C_2O_4)_3]$
- **(d)** $[Rh(en)_2(NH_3)_2]_2(SO_4)_3$

20.109 For each of the following complexes, draw a crystal field energy level diagram, assign the electrons to orbitals, and predict the number of unpaired electrons:
- **(a)** $[Pt(NH_3)_4]^{2+}$ (square planar)
- **(b)** $[MnCl_4]^{2-}$ (tetrahedral)
- **(c)** $[Co(NCS)_4]^{2-}$ (tetrahedral)
- **(d)** $[Cu(en)_2]^{2+}$ (square planar)

20.117 Describe the bonding in $[Mn(CN)_6]^{3-}$ using both crystal field theory and valence bond theory. Include the appropriate crystal field d orbital energy level diagram and the valence bond orbital diagram. Which model allows you to predict the number of unpaired electrons? How many do you expect?

20.118 The complex $[FeCl_6]^{3-}$ is more paramagnetic than $[Fe(CN)_6]^{3-}$. Explain.

20.119 Although Cl^- is a weak-field ligand and CN^- is a strong-field ligand, $[CrCl_6]^{3-}$ and $[Cr(CN)_6]^{3-}$ exhibit approximately the same amount of paramagnetism. Explain.

20.120 In octahedral complexes, the choice between high-spin and low-spin electron configurations arises only for d^4–d^7 complexes. Explain.

20.121 Draw a crystal field energy level diagram and predict the number of unpaired electrons for each of the following:
- **(a)** $[Mn(H_2O)_6]^{2+}$
- **(b)** $Pt(NH_3)_2Cl_2$
- **(c)** $[FeO_4]^{2-}$
- **(d)** $[Ru(NH_3)_6]^{2+}$ (low-spin)

20.122 Explain why $[CoCl_4]^{2-}$ (blue) and $[Co(H_2O)_6]^{2+}$ (pink) have different colors. Which complex has its absorption bands at longer wavelengths?

20.123 The complex $[Rh(en)_2(NO_2)(SCN)]^+$ can exist in 12 isomeric forms, including constitutional isomers and stereoisomers. Sketch the structures of all 12 isomers.

20.124 The glycinate anion, $gly^- = NH_2CH_2CO_2^-$, bonds to metal ions through the N atom and one of the O atoms. Using N⌒O to represent gly^-, sketch the structures of the four stereoisomers of $Co(gly)_3$.

20.125 Sketch the five possible diastereoisomers of $Pt(gly)_2Cl_2$, where gly^- is the bidentate ligand described in Problem 20.124. Which of the isomers has a dipole moment? Which can exist as a pair of enantiomers?

20.126 Draw the structures of all possible diastereoisomers of an octahedral complex with the formula $MA_2B_2C_2$. Which of the diastereoisomers, if any, can exist as enantiomers?

20.127 Predict the crystal field energy level diagram for a linear ML_2 complex that has two ligands along the $\pm z$ axis:

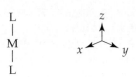

Multi-Concept Problems

20.128 Formation constants for the ammonia and ethylene-diamine complexes of nickel(II) indicate that $Ni(en)_3^{2+}$ is much more stable than $Ni(NH_3)_6^{2+}$:

(1) $Ni(H_2O)_6^{2+}(aq) + 6\,NH_3(aq) \rightleftharpoons$
$\quad Ni(NH_3)_6^{2+}(aq) + 6\,H_2O(l)$ $\qquad K_f = 2.0 \times 10^8$

(2) $Ni(H_2O)_6^{2+}(aq) + 3\,en(aq) \rightleftharpoons$
$\quad Ni(en)_3^{2+}(aq) + 6\,H_2O(l)$ $\qquad K_f = 4 \times 10^{17}$

The enthalpy changes for the two reactions, $\Delta H°_1$ and $\Delta H°_2$, should be about the same because both complexes have six Ni–N bonds.

(a) Which of the two reactions should have the larger entropy change, $\Delta S°$? Explain.

(b) Account for the greater stability of $Ni(en)_3^{2+}$ in terms of the relative values of $\Delta S°$ for the two reactions.

(c) Assuming that $\Delta H°_2 - \Delta H°_1$ is zero, calculate the value of $\Delta S°_2 - \Delta S°_1$.

20.129 The percent iron in iron ore can be determined by dissolving the ore in acid, then reducing the iron to Fe^{2+}, and finally titrating the Fe^{2+} with aqueous $KMnO_4$. The reaction products are Fe^{3+} and Mn^{2+}.

(a) Write a balanced net ionic equation for the titration reaction.

(b) Use the $E°$ values in Appendix D to calculate $\Delta G°$ (in kilojoules) and the equilibrium constant for the reaction.

(c) Draw a crystal field energy level diagram for the reactants and products, MnO_4^-, $[Fe(H_2O)_6]^{2+}$, $[Fe(H_2O)_6]^{3+}$, and $[Mn(H_2O)_6]^{2+}$, and predict the number of unpaired electrons for each.

(d) Does the paramagnetism of the solution increase or decrease as the reaction proceeds? Explain.

(e) What is the mass % Fe in the iron ore if titration of the Fe^{2+} from a 1.265 g sample of ore requires 34.83 mL of 0.05132 M $KMnO_4$ to reach the equivalence point?

20.130 Complete reaction of 2.60 g of chromium metal with 50.00 mL of 1.200 M H_2SO_4 in the absence of air gave a blue solution and a colorless gas that was collected at 25°C and a pressure of 735 mm Hg.

(a) Write a balanced net ionic equation for the reaction.

(b) How many liters of gas were produced?

(c) What is the pH of the solution?

(d) Describe the bonding in the blue colored ion using both the crystal field theory and the valence bond theory. Include the appropriate crystal field d-orbital energy level diagram and the valence bond orbital diagram. Identify the hybrid orbitals used in the valence bond description.

(e) When an excess of KCN is added to the solution, the color changes and the paramagnetism of the solution decreases. Explain.

20.131 In acidic aqueous solution, *trans*-$[Co(en)_2Cl_2]^+$ undergoes the following substitution reaction:

trans-$[Co(en)_2Cl_2]^+(aq) + H_2O(l) \longrightarrow$
$\quad$ *trans*-$[Co(en)_2(H_2O)Cl]^{2+}(aq) + Cl^-(aq)$

The reaction is first order in *trans*-$[Co(en)_2Cl_2]^+$, and the rate constant at 25°C is $3.2 \times 10^{-5}\ s^{-1}$.

(a) What is the half-life of the reaction in hours?

(b) If the initial concentration of *trans*-$[Co(en)_2Cl_2]^+$ is 0.138 M, what is its molarity after a reaction time of 16.5 h?

(c) Devise a possible reaction mechanism with a unimolecular rate-determining step.

(d) Is the reaction product chiral or achiral? Explain.

(e) Draw a crystal field energy level diagram for *trans*-$[Co(en)_2Cl_2]^+$ that takes account of the fact that Cl^- is a weaker field ligand than ethylenediamine.

eMedia Problems

20.132 Watch the **Isomerism** movie (*eChapter 20.8*), and answer the following questions. Draw structures to support your answers.

    (a) Which of the three geometries can exhibit cis–trans isomerism?

    (b) Which of the three geometries can exhibit optical isomerism?

20.133 The **Isomerism** movie (*eChapter 20.8*) shows the cis and trans isomers of the tetraamminedichlorocobalt(III) ion. Which if either of these isomers is chiral? Is it possible for an octahedral complex not to exhibit cis–trans isomerism and yet still be chiral? Explain.

20.134 The **Optical Activity** movie (*eChapter 20.9*) illustrates the interaction of plane-polarized light with a solution containing a single enantiomer of a chiral compound.

    (a) Define the term *enantiomer*.

    (b) How do the chemical, physical, and optical properties of enantiomers differ?

    (c) What is a *racemic* mixture?

    (d) How does a solution containing a racemic mixture interact with plane-polarized light?

20.135 Enantiomers are designated either + or −, depending on whether they rotate the plane of plane-polarized light clockwise or counterclockwise, respectively. Watch the **Optical Activity** movie (*eChapter 20.9*), and draw both enantiomers for each of the following optically active complexes.

    (a) cis-$[Cr(en)_2Cl_2]^+$

    (b) cis-$[Cr(C_2O_4)_2(H_2O)_2]^-$

    (c) $[Ni(en)_3]^{2+}$

20.136 Based on information in the **Optical Activity** movie (*eChapter 20.9*), which of the following solutions would you expect to rotate the plane of plane-polarized light? Explain each prediction.

    (a) An equimolar mixture of (+)-cis-$[Cr(en)_2Cl_2]^+$ and (−)-cis-$[Cr(en)_2Cl_2]^+$

    (b) An equimolar mixture of (+)-cis-$[Cr(en)_2Cl_2]^+$ and trans-$[Cr(en)_2Cl_2]^+$

    (c) An equimolar mixture of (+)-$[Co(en)_3]^{3+}$ and (+)-cis-$[Cr(en)_2Cl_2]^+$

21 Metals and Solid-State Materials

An experimental magnetically levitated train (Maglev) in Japan is suspended above superconducting magnets that are cooled with liquid helium. On April 14, 1999, this five-car train surpassed the speed record of its three-car predecessor, attaining a maximum speed of 552 km/h in a manned vehicle run.

The materials available for making tools and weapons, houses and sky-scrapers, computers and lasers have had a profound effect on the development of human civilization. Indeed, archaeologists organize early human history in terms of materials—the Stone Age, in which only natural materials such as wood and stone were available; the Bronze Age, in which implements were made of copper alloyed with tin; and the Iron Age, in which ornaments, weapons, and tools were made of iron. Copper and iron are still of enormous importance. Copper is used in making electrical wiring, and iron is the main constituent of steel. Today, metals unknown in ancient times, such as aluminum and titanium, play a leading role in modern technology. These metals are widely used, for instance, in the aircraft industry because of their low densities and high resistance to corrosion.

Modern technology is made possible by a host of *solid-state materials* such as *semiconductors*, *superconductors*, *advanced ceramics*, and *composites*. Semiconductors are used in the miniature solid-state electronic devices found in computers. Super-conductors are used to make the powerful magnets found in the magnetic resonance imaging (MRI) instruments employed in medical diagnosis. Advanced ceramics and composites have numerous engineering, electronic, and biomed-ical applications and are likely to be among the more important materials in future technologies.

The MRI instruments used in hospitals contain magnets made from superconductors.

In this chapter we'll look at both metals and solid-state materials. We'll examine the natural sources of the metallic elements, the methods used to obtain metals from their ores, and the models used to describe the bonding in metals. We'll also look at the structure, bonding, properties, and applications of semiconductors, superconductors, ceramics, and composites.

21.1 Sources of the Metallic Elements

Most metals occur in nature as **minerals**, the crystalline, inorganic constituents of the rocks that make up the earth's crust. Silicates and aluminosilicates (Section 19.7) are the most abundant minerals, but they are difficult to concentrate and reduce and are therefore generally unimportant as commercial sources of metals. More important are oxides and sulfides, such as hematite (Fe_2O_3), rutile (TiO_2), and cinnabar (HgS) (Figure 21.1), which yield iron, titanium, and

FIGURE 21.1 Samples of **(a)** hematite (Fe_2O_3), **(b)** star-shaped needles of rutile (TiO_2) on a quartz matrix, and **(c)** cinnabar (HgS).

(a)   (b)   (c)

mercury, respectively. Mineral deposits from which metals can be produced economically are called **ores** (Table 21.1).

TABLE 21.1   Principal Ores of Some Important Metals

Metal	Ore	Formula	Location of Important Deposits
Aluminum	Bauxite	$Al_2O_3 \cdot x\ H_2O$	Australia, Brazil, Jamaica
Chromium	Chromite	$FeCr_2O_4$	Russia, South Africa
Copper	Chalcopyrite	$CuFeS_2$	U.S., Chile, Canada
Iron	Hematite	Fe_2O_3	Australia, Ukraine, U.S.
Lead	Galena	PbS	U.S., Australia, Canada
Manganese	Pyrolusite	MnO_2	Russia, Gabon, South Africa
Mercury	Cinnabar	HgS	Spain, Algeria, Mexico
Tin	Cassiterite	SnO_2	Malaysia, Bolivia
Titanium	Rutile	TiO_2	Australia
	Ilmenite	$FeTiO_3$	Canada, U.S., Australia
Zinc	Sphalerite	ZnS	U.S., Canada, Australia

It's interesting to note how the chemical compositions of the most common ores correlate with the locations of the metals in the periodic table (Figure 21.2). The early transition metals on the left side of the *d* block generally occur as oxides, and the more electronegative, late transition metals on the right side of the *d* block occur as sulfides. This pattern makes sense because the less electronegative metals tend to form ionic compounds by losing electrons to highly electronegative nonmetals such as oxygen. By contrast, the more electronegative metals tend to form compounds with more covalent character by bonding to the less electronegative nonmetals such as sulfur.

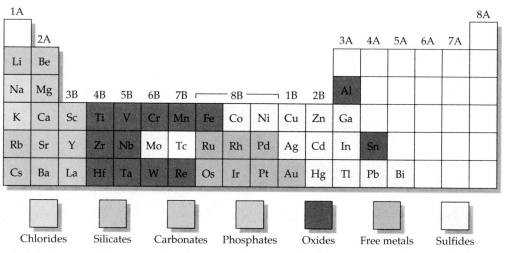

FIGURE 21.2   Primary mineral sources of metals. The *s*-block metals occur as chlorides, silicates, and carbonates. The *d*- and *p*-block metals are found as oxides and sulfides, except for the group 3B metals, which occur as phosphates, and the platinum-group metals and gold, which occur in uncombined form. There is no mineral source of technetium (Tc in group 7B), a radioactive element that is made in nuclear reactors.

We might also expect to find oxide ores for the *s*-block metals and sulfide ores for the more electronegative *p*-block metals. In fact, sulfide ores are common for the *p*-block metals, except for Al and Sn, but oxides of the *s*-block metals are strongly basic and far too reactive to exist in an environment that contains acidic oxides such as CO_2 and SiO_2. Consequently, s-block metals are found in nature as carbonates, as silicates, and, in the case of Na and K, as chlorides (Sections 6.7 and 6.8). Only gold and the platinum-group metals (Ru, Os, Rh, Ir, Pd, and Pt) are sufficiently unreactive to occur commonly in uncombined form as the free metals.

21.2 Metallurgy

An ore is a complex mixture of a metal-containing mineral and economically worthless material called **gangue** (pronounced "gang"), consisting of sand, clay, and other impurities. The extraction of a metal from its ore requires several steps: (1) concentration of the ore and, if necessary, chemical treatment prior to reduction; (2) reduction of the mineral to the free metal; and (3) refining or purification of the metal. These processes are a part of **metallurgy**, the science and technology of extracting metals from their ores. Another aspect of metallurgy is the making of **alloys**, metallic solutions composed of two or more elements. Steels and bronze are examples of alloys.

Concentration and Chemical Treatment of Ores

Ores are concentrated by separating the mineral from the gangue. The mineral and the gangue have different properties, which are exploited in various separation methods. Density differences, for example, are important in panning for gold, a procedure in which prospectors flush water over gold-bearing earth in a pan. The less dense gangue is washed away, and the more dense gold particles remain at the bottom of the pan. Differences in magnetic properties are used in concentrating the iron ore magnetite (Fe_3O_4). The Fe_3O_4 is strongly attracted by magnets, but the gangue is unaffected.

Metal sulfide ores are concentrated by **flotation**, a process that exploits differences in the ability of water and oil to wet the surfaces of the mineral and the gangue. A powdered ore such as chalcopyrite ($CuFeS_2$) is mixed with water, oil, and a detergent, and the mixture is vigorously agitated in a tank by blowing air through the liquid (Figure 21.3). The gangue, which contains ionic silicates, is moistened by the polar water molecules and sinks to the bottom of the tank. The mineral particles, which contain the less polar metal sulfide, are coated by the nonpolar oil molecules and become attached to the soapy air bubbles created by the detergent. The metal sulfide particles are

In panning for gold, prospectors use density differences to separate gold from the gangue.

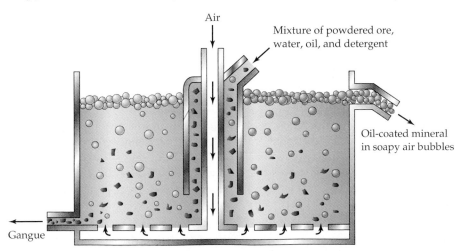

FIGURE 21.3 The flotation process for concentrating metal sulfide ores. Mineral particles float to the top of the tank along with the soapy air bubbles, while the gangue sinks to the bottom.

thus carried to the surface in the soapy froth, which is skimmed off at the top of the tank.

Sometimes, ores are concentrated by chemical treatment. In the *Bayer process*, for instance, the Al_2O_3 in bauxite ($Al_2O_3 \cdot x\ H_2O$) is separated from Fe_2O_3 impurities by treating the ore with hot aqueous NaOH. The amphoteric Al_2O_3 dissolves as the aluminate ion, $Al(OH)_4^-$, but the basic Fe_2O_3 does not:

$$Al_2O_3(s) + 2\ OH^-(aq) + 3\ H_2O(l) \longrightarrow 2\ Al(OH)_4^-(aq)$$

The Bayer process is described in more detail in Section 16.12.

Chemical treatment is also used to convert minerals to compounds that are more easily reduced to the metal. For example, sulfide minerals, such as sphalerite (ZnS), are converted to oxides by **roasting**, a process that involves heating the mineral in air:

$$2\ ZnS(s) + 3\ O_2(g) \xrightarrow{\text{Heat}} 2\ ZnO(s) + 2\ SO_2(g)$$

Formerly, the sulfur dioxide by-product was a source of acid rain because it is oxidized in the atmosphere to SO_3, which reacts with water vapor to yield sulfuric acid. In modern roasting facilities, however, the SO_2 is converted to sulfuric acid rather than being vented to the atmosphere.

Reduction

Once an ore has been concentrated, it is reduced to the free metal, either by chemical reduction or by electrolysis. The method used (Table 21.2) depends on the activity of the metal as measured by its standard reduction potential (Table 18.1). The most active metals have the most negative standard reduction potentials and are the most difficult to reduce; the least active metals have the most positive standard reduction potentials and are the easiest to reduce.

TABLE 21.2   Reduction Methods for Producing Some Common Metals

	Metal	Reduction Method
Least active	Au, Pt	None; found in nature as the free metal
	Cu, Ag, Hg	Roasting of the metal sulfide
	V, Cr, Mn, Fe, Ni, Zn, W, Pb	Reduction of the metal oxide with carbon, hydrogen, or a more active metal
	Al	Electrolysis of molten Al_2O_3 in cryolite
Most active	Li, Na, Mg	Electrolysis of the molten metal chloride

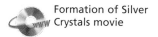
Formation of Silver Crystals movie

Gold and platinum are so inactive that they are usually found in nature in uncombined form, but copper and silver, which are slightly more active, are found in both combined and uncombined form. Copper, silver, and mercury commonly occur in sulfide ores that are easily reduced by roasting. Cinnabar (HgS), for example, yields elemental mercury when the ore is heated at 600°C in a stream of air:

$$HgS(s) + O_2(g) \xrightarrow{600°C} Hg(g) + SO_2(g)$$

Although it may seem strange that heating a substance in oxygen reduces it to the metal, both oxidation and reduction of HgS occur in this reaction: Sulfide ions are oxidized and mercury(II) is reduced.

More active metals, such as chromium, zinc, and tungsten, are obtained by reducing their oxides with a chemical reducing agent such as carbon, hydrogen, or a more active metal (Na, Mg, or Al). Pure chromium, for example, is produced by reducing Cr_2O_3 with aluminum (Section 20.4). Zinc, used in the automobile industry for galvanizing steel (Section 18.10), is obtained by reducing ZnO with coke, a form of carbon produced by heating coal in the absence of air:

$$ZnO(s) + C(s) \xrightarrow{\text{Heat}} Zn(g) + CO(g)$$

Although carbon is the cheapest available reducing agent, it is unsatisfactory for reducing oxides of metals like tungsten that form very stable carbides. (Tungsten carbide, WC, is an extremely hard material used to make high-speed cutting tools.) The preferred method for producing tungsten involves reducing tungsten(VI) oxide with hydrogen:

$$WO_3(s) + 3\,H_2(g) \xrightarrow{850°C} W(s) + 3\,H_2O(g)$$

Because of its high strength, high melting point (3422°C), low volatility, and high efficiency for converting electrical energy into light, tungsten is used to make filaments for electric light bulbs (Figure 21.4).

There are no chemical reducing agents strong enough to reduce compounds of the most active metals, so these metals are produced by electrolytic reduction (Section 18.12). Lithium, sodium, and magnesium, for example, are obtained by electrolysis of their molten chlorides. Aluminum is manufactured by electrolysis of purified Al_2O_3 in molten cryolite (Na_3AlF_6).

Refining

For most applications, the metals obtained from processing and reducing ores need to be purified. The methods used include distillation, chemical purification, and electrorefining. Zinc (bp 907°C), for example, is volatile enough to be separated from cadmium, lead, and other impurities by distillation. In this way, the purity of the zinc obtained from reduction of ZnO is increased from 99% to 99.99%. Nickel, used as a catalyst and as a battery material, is purified by the **Mond process**, a chemical method involving formation and subsequent decomposition of the volatile compound nickel tetracarbonyl, $Ni(CO)_4$ (bp 43°C). Carbon monoxide is passed over impure nickel at about 150°C and 20 atm pressure, forming $Ni(CO)_4$ and leaving metal impurities behind. The $Ni(CO)_4$ is then decomposed at higher temperatures (about 230°C) on pellets of pure nickel. The process works because the equilibrium shifts to the left with increasing temperature:

$$Ni(s) + 4\,CO(g) \underset{\text{Higher temp.}}{\overset{\text{Lower temp.}}{\rightleftharpoons}} Ni(CO)_4(g)$$

$$\Delta H° = -160.8 \text{ kJ}; \ \Delta S° = -410 \text{ J/K}$$

A similar strategy is employed to purify zirconium, which is used as cladding for fuel rods in nuclear reactors. The crude metal is heated at about

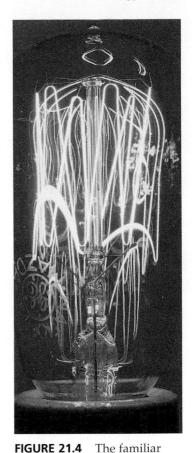

FIGURE 21.4 The familiar incandescent light bulb emits white light when an electric current passes through the tungsten wire filament, heating it to a high temperature. The glass bulb contains gases such as argon and nitrogen, which carry away heat from the filament. Although tungsten has the highest boiling point of any element (5555°C), the tungsten slowly evaporates from the hot filament and condenses as the black spot of tungsten metal usually visible on the inside surface of a burned-out bulb.

Zirconium is used as cladding for uranium dioxide fuel rods in nuclear reactors.

200°C with a small amount of iodine in an evacuated container to form the volatile ZrI_4. The ZrI_4 is then decomposed to pure zirconium by letting the vapor come in contact with an electrically heated tungsten or zirconium filament at about 1300°C:

$$Zr(s) + 4 I_2(g) \underset{1300°C}{\overset{200°C}{\rightleftharpoons}} ZrI_4(g)$$

Copper from the reduction of ores must be purified for use in making electrical wiring because impurities increase its electrical resistance. The method used is electrorefining, an electrolytic process in which copper is oxidized to Cu^{2+} at an impure copper anode, and Cu^{2+} from an aqueous copper sulfate solution is reduced to copper at a pure copper cathode. The process is described in Section 18.12.

▶ **PROBLEM 21.1**   Write a balanced equation for production of each of the following metals:
(a) Chromium by reducing Cr_2O_3 with aluminum
(b) Copper by roasting Cu_2S
(c) Lead by reducing PbO with coke
(d) Potassium by electrolysis of molten KCl

21.3   Iron and Steel

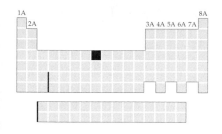

The metallurgy of iron is of special technological importance because iron is the major constituent of steel, the most widely used of all metals. Worldwide production of steel amounts to some 750 million tons per year. Iron is produced by carbon monoxide reduction of iron ore, usually hematite (Fe_2O_3), in a huge reactor called a **blast furnace** (Figure 21.5). A charge of iron ore, coke, and limestone ($CaCO_3$) is introduced at the top of the furnace, and a blast of hot air is sent in at the bottom, burning coke and yielding carbon monoxide at a temperature of about 2000°C:

$$2 C(s) + O_2(g) \longrightarrow 2 CO(g) \qquad \Delta H° = -221.0 \text{ kJ}$$

As the charge descends and the hot carbon monoxide rises, a complex series of high-temperature reactions occurs in the various regions of the furnace, as shown in Figure 21.5. The key overall reaction is the reduction of Fe_2O_3 to iron metal (mp 1538°C), which is obtained as an impure liquid at the bottom of the furnace:

Thermite movie

$$Fe_2O_3(s) + 3 CO(g) \longrightarrow 2 Fe(l) + 3 CO_2(g)$$

The purpose of the limestone is to remove the gangue from the iron ore. At the high temperatures of the furnace, the limestone decomposes yielding lime (CaO), a basic oxide that reacts with SiO_2 and other acidic oxides present in the gangue. The product, called **slag**, is a molten material consisting mainly of calcium silicate:

$$CaCO_3(s) \longrightarrow CaO(s) + CO_2(g)$$

$$\underset{\text{Lime}}{CaO(s)} + \underset{\text{Sand}}{SiO_2(s)} \longrightarrow \underset{\text{Slag}}{CaSiO_3(l)}$$

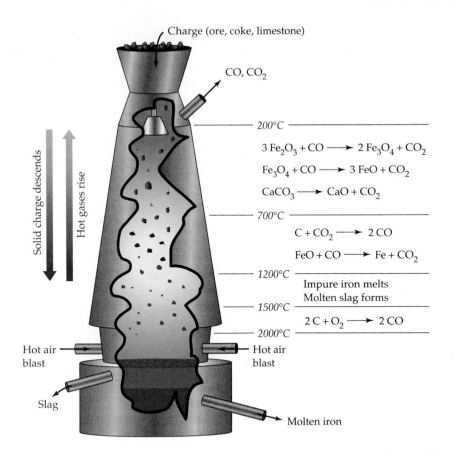

Charge (ore, coke, limestone)

CO, CO$_2$

Solid charge descends

Hot gases rise

200°C

$$3 Fe_2O_3 + CO \longrightarrow 2 Fe_3O_4 + CO_2$$
$$Fe_3O_4 + CO \longrightarrow 3 FeO + CO_2$$
$$CaCO_3 \longrightarrow CaO + CO_2$$

700°C

$$C + CO_2 \longrightarrow 2 CO$$
$$FeO + CO \longrightarrow Fe + CO_2$$

1200°C

Impure iron melts
Molten slag forms

1500°C

$$2 C + O_2 \longrightarrow 2 CO$$

2000°C

Hot air blast

Hot air blast

Slag

Molten iron

FIGURE 21.5 A diagram of a blast furnace for reduction of iron ore. Modern blast furnaces are as large as 60 m in height and 14 m in diameter. They are designed for continuous operation and produce up to 10,000 tons of iron per day. Note the approximate temperatures and the chemical reactions that occur in the various regions of the furnace.

The slag, which is less dense than the molten iron, floats on the surface of the iron, thus allowing the iron and the slag to be removed from the bottom of the furnace through separate taps.

The iron obtained from a blast furnace is a brittle material called *cast iron*, or *pig iron*. It contains about 4% elemental carbon and smaller amounts of other impurities such as elemental silicon, phosphorus, sulfur, and manganese, which are formed from their compounds in the reducing atmosphere of the furnace. The most important of several methods for purifying the iron and converting it to steel is the **basic oxygen process**. Molten iron from the blast furnace is exposed to a jet of pure oxygen gas for about 20 min in a furnace that is lined with basic oxides such as CaO. The impurities in the iron are oxidized, and the acidic oxides that form react with the basic CaO to yield a molten slag that can be poured off. Phosphorus, for example, is oxidized to P$_4$O$_{10}$, which then reacts with CaO to give molten calcium phosphate:

$$P_4(l) + 5 O_2(g) \longrightarrow P_4O_{10}(l)$$
$$6 CaO(s) + P_4O_{10}(l) \longrightarrow 2 Ca_3(PO_4)_2(l)$$
Basic oxide   Acidic oxide   Slag

Manganese also passes into the slag because its oxide is basic and reacts with added SiO$_2$, yielding molten manganese silicate:

$$2 Mn(l) + O_2(g) \longrightarrow 2 MnO(s)$$
$$MnO(s) + SiO_2(s) \longrightarrow MnSiO_3(l)$$
Basic oxide   Acidic oxide   Slag

Using the basic oxygen process to make steel.

The basic oxygen process produces steels that contain about 1% carbon but only very small amounts of phosphorus and sulfur. Usually, the composition of the liquid steel is monitored by chemical analysis, and the amounts of oxygen and impure iron used are adjusted to achieve the desired concentrations of carbon and other impurities. The hardness, strength, and malleability of the steel depend on its chemical composition, on the rate at which the liquid steel is cooled, and on subsequent heat treatment of the solid. The mechanical and chemical properties of a steel can also be altered by adding other metals. Stainless steel, for example, is a corrosion-resistant iron alloy that contains up to 30% chromium along with smaller amounts of nickel.

21.4   Bonding in Metals

Thus far, we've discussed the sources, production, and properties of some important metals. Some properties, such as hardness and melting point, vary considerably among metals, but other properties are characteristic of metals in general. For instance, all metals can be drawn into wires (ductility) or beaten into sheets (malleability) without breaking into pieces like glass or an ionic crystal. Furthermore, all metals have a high thermal and electrical conductivity. When you touch a metal, it feels cold because the metal efficiently conducts heat away from your hand, and when you connect a metal wire to the terminals of a battery, the wire conducts an electric current.

To understand those properties, we need to look at the bonding in metals. We'll consider two theoretical models that are commonly used: the *electron-sea model* and the *molecular orbital theory*.

Because of its ductility, aluminum can be drawn into wires.

Electron-Sea Model of Metals

If you try to draw an electron-dot structure for a metal, you'll quickly realize that there aren't enough valence electrons available to form an electron-pair bond between every pair of adjacent atoms. Sodium, for example, which has just one valence electron per atom ($3s^1$), crystallizes in a body-centered cubic structure in which each Na atom is surrounded by eight nearest neighbors (Section 10.8). Consequently, the valence electrons can't be localized in a bond between any particular pair of atoms. Instead, they are delocalized and belong to the crystal as a whole.

In the **electron-sea model**, a metal crystal is viewed as a three-dimensional array of metal cations immersed in a sea of delocalized electrons that are free to move throughout the crystal (Figure 21.6). The continuum of delocalized, mobile valence electrons acts as an electrostatic glue that holds the metal cations together.

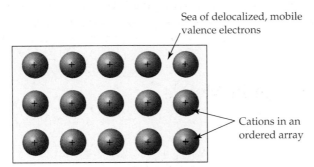

Sea of delocalized, mobile valence electrons

Cations in an ordered array

FIGURE 21.6   Two-dimensional representation of the electron-sea model of a metal. An ordered array of cations is immersed in a continuous distribution of delocalized, mobile valence electrons. The valence electrons do not belong to any particular metal ion but to the crystal as a whole.

The electron-sea model affords a simple qualitative explanation for the electrical and thermal conductivity of metals. Because the electrons are mobile, they are free to move away from a negative electrode and toward a positive electrode when a metal is subjected to an electrical potential. The mobile electrons can also conduct heat by carrying kinetic energy from one part of the crystal to another. The malleability and ductility of metals follows from the fact that the delocalized bonding extends in all directions; that is, it is not confined to oriented bond directions as in covalent network solids like SiO_2. When a metallic crystal is deformed, no localized bonds are broken. Instead, the electron sea simply adjusts to the new distribution of cations, and the energy of the deformed structure is similar to that of the original. Thus, the energy required to deform a metal like sodium is relatively small. Of course, the energy required to deform a transition metal like iron is greater because iron has more valence electrons ($4s^2\,3d^6$), and the electrostatic glue is therefore more dense.

Molecular Orbital Theory for Metals

A more detailed understanding of the bonding in metals is provided by the molecular orbital theory, a model that is a logical extension of the molecular orbital description of small molecules discussed in Sections 7.13–7.15. Recall that in the H_2 molecule the $1s$ orbitals of the two H atoms overlap to give a σ bonding MO and a higher-energy σ^* antibonding MO. The bonding in the gaseous Na_2 molecule is similar: The $3s$ orbitals of the two Na atoms combine to give a σ and a σ^* MO. Because each Na atom has just one $3s$ valence electron, the lower-energy bonding orbital is filled and the higher-energy antibonding orbital is empty:

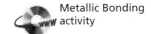

Metallic Bonding activity

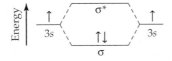

Now consider what happens if we bring together an increasingly larger number of Na atoms to build up a crystal of sodium metal. The key idea to remember from Section 7.13 is that the number of molecular orbitals formed is the same as the number of atomic orbitals combined. Thus, there will be three MOs for a triatomic Na_3 molecule, four MOs for Na_4, and so on. A cubic crystal of sodium metal, 1.5 mm on an edge, contains about 10^{20} Na atoms and therefore has about 10^{20} MOs, each of which is delocalized over all the atoms in the crystal. As shown in Figure 21.7, the difference in energy between successive MOs in an Na_n molecule decreases as the number of Na atoms increases, so that the MOs merge into an almost continuous band of energy levels. Consequently, MO theory for metals is often called **band theory**. The bottom half of the band consists of bonding MOs and is filled, whereas the top half of the band consists of antibonding MOs and is empty.

FIGURE 21.7 Molecular orbital energy levels for Na_n molecules. A crystal of sodium metal can be regarded as a giant Na_n molecule, where n has a value of about 10^{20}. As the value of n increases, the energy levels merge into an almost continuous band. Because each Na atom has one 3s valence electron and each MO can hold two electrons, the 3s band is half-filled. In this and subsequent figures, the deep red color denotes the filled portion of a band.

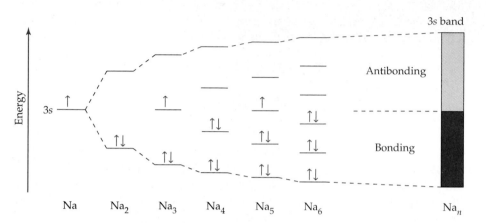

How does band theory account for the electrical conductivity of metals? Because each of the MOs in a metal has a definite energy (Figure 21.7), each electron in a metal has a specific kinetic energy and a specific velocity. These values depend on the particular MO energy level and increase from the bottom of a band to the top. For a one-dimensional metal wire, electrons traveling in opposite directions at the same speed have the same kinetic energy. Thus, the energy levels within a band occur in degenerate pairs; one set of energy levels applies to electrons moving to the right, and the other set applies to electrons moving to the left. (Recall from Section 5.12 that degenerate energy levels are levels that have the same energy.)

In the absence of an electrical potential, the two sets of levels are equally populated. That is, for each electron moving to the right, another electron moves to the left with exactly the same speed (Figure 21.8). As a result, there

FIGURE 21.8 Half-filled 3s band of MO energy levels for sodium metal. The direction of electron motion for the two degenerate sets of energy levels is indicated by the horizontal arrows. **(a)** In the absence of an electrical potential, the two sets of levels are equally populated, and no electric current flows through the wire. **(b)** In the presence of an electrical potential (positive electrode on the right), some of the electrons shift from one set of energy levels to the other, and there is a net current of electrons from left to right.

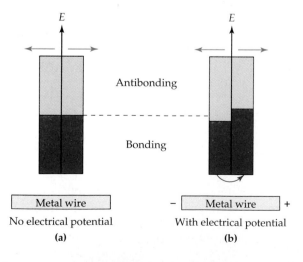

is no net electric current in either direction. In the presence of an electrical potential, however, those electrons moving to the right (toward the positive terminal of a battery) are accelerated, those moving to the left (toward the negative terminal) are slowed down, and those moving to the left with very slow speeds undergo a change of direction. Thus, the number of electrons moving to the right is now greater than the number moving to the left, and so there is a net electric current.

It's evident from Figure 21.8 that an electrical potential can shift electrons from one set of energy levels to the other only if the band is partially filled. If the band is completely filled, there are no available vacant energy levels to which electrons can be excited, and therefore the two sets of levels must remain equally populated, even in the presence of an electrical potential. This means that an electrical potential can't accelerate the electrons in a completely filled band. *Materials that have only completely filled bands are therefore electrical insulators.* By contrast, *materials that have partially filled bands are metals.*

Based on the preceding analysis, we would predict that magnesium should be an insulator because it has the electron configuration [Ar] $3s^2$ and should therefore have a completely filled 3s band. This prediction is wrong, however, because we have not yet considered the 3p valence orbitals. Just as the 3s orbitals combine to form a 3s band, so the 3p orbitals can combine to form a 3p band. If the 3s and 3p bands were widely separated in energy, the 3s band would be filled, the 3p band would be empty, and magnesium would be an insulator. In fact, though, the 3s and 3p bands overlap in energy, and the resulting composite band is only partially filled (Figure 21.9). Thus, magnesium is a conductor.

Transition metals have a d band that can overlap the s band to give a composite band consisting of six MOs per metal atom. Half of these MOs are bonding and half are antibonding. We might therefore expect maximum bonding for metals that have six valence electrons per metal atom because six electrons will just fill the bonding MOs and leave the antibonding MOs empty. In accord with this picture, maximum bonding near group 6B causes the melting points of the transition metals to also be at a maximum near group 6B (Section 20.2).

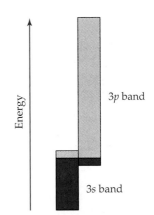

FIGURE 21.9   In magnesium metal, the 3s and 3p bands have similar energies and overlap to give a composite band consisting of four MOs per Mg atom. The composite band can accommodate eight electrons per Mg atom but is only partially filled, since each Mg atom has just two valence electrons. (In this and subsequent figures, the separate sets of energy levels for the right- and left-moving electrons aren't shown.)

EXAMPLE 21.1   The melting points of chromium and zinc are 1907°C and 420°C, respectively. Use band theory to account for the difference.

SOLUTION   The electron configurations are [Ar] $3d^5 4s^1$ for Cr and [Ar] $3d^{10} 4s^2$ for Zn. Assume that the 3d and 4s bands overlap. The composite band, which can accommodate 12 valence electrons per metal atom, will be half-filled for Cr and completely filled for Zn. Strong bonding and consequently a high melting point are expected for Cr because all the bonding MOs are occupied and all the antibonding MOs are empty. Weak bonding and a low melting point are expected for Zn because both the bonding and the antibonding MOs are occupied. (The fact that Zn is a metal suggests that the 4p orbitals also contribute to the composite band.)

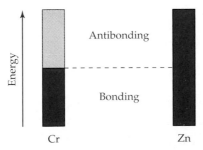

▶ **PROBLEM 21.2** Mercury metal is a liquid at room temperature. Use band theory to suggest a reason for its low melting point (−39°C).

◀ **KEY CONCEPT PROBLEM 21.3** The following pictures represent the electron population of the composite s–d band for three metals—Ag, Mo, and Y:

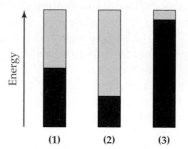

(a) Which picture corresponds to which metal? Explain.

(b) Which metal has the highest melting point, and which has the lowest? Explain.

(c) Molybdenum is very hard, whereas silver is relatively soft. Explain.

21.5 Semiconductors

A **semiconductor**, such as silicon or germanium, is a material that has an electrical conductivity intermediate between that of a metal and that of an insulator. To understand the electrical properties of semiconductors, let's look first at the bonding in insulators. Consider diamond, for example, a covalent network solid in which each C atom is bonded tetrahedrally to four other C atoms (Figure 10.26). In a localized description of the bonding, C–C electron-pair bonds result from overlap of sp^3 hybrid orbitals. In a delocalized description, the 2s and 2p valence orbitals of all the C atoms combine to give bands of bonding and antibonding MOs—a total of four MOs per C atom. As is generally the case for insulators, the bonding MOs, called the **valence band**, and the higher-energy antibonding MOs, called the **conduction band**, are separated in energy by a large **band gap**. The band gap in diamond is about 520 kJ/mol.

Each of the two bands in diamond can accommodate four electrons per C atom. Because carbon has just four valence electrons ($2s^2\ 2p^2$), the valence band is completely filled and the conduction band is completely empty. Diamond is therefore an insulator because there are no vacant MOs in the valence band to which electrons can be excited by an electrical potential, and because population of the vacant MOs of the conduction band is prevented by the large band gap. By contrast, metallic conductors have no energy gap between the highest occupied and lowest unoccupied MOs. This fundamental difference between the energy levels of metals and insulators is illustrated in Figure 21.10.

FIGURE 21.10 Bands of MO energy levels for **(a)** a metallic conductor, **(b)** an electrical insulator, and **(c)** a semiconductor. A metallic conductor has a partially filled band. An electrical insulator has a completely filled valence band and a completely empty conduction band, which are separated in energy by a large band gap. In a semiconductor, the band gap is smaller. As a result, the conduction band is partially occupied with a few electrons, and the valence band is partially empty. Electrical conductivity in metals and semiconductors results from the presence of partially filled bands.

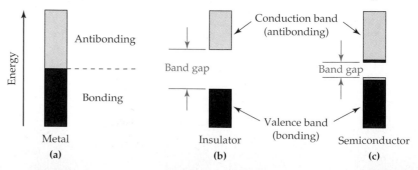

The MOs of a semiconductor are similar to those of an insulator, but the band gap in a semiconductor is smaller (Figure 21.10). As a result, a few electrons have enough energy to jump the gap and occupy the higher-energy conduction band. The conduction band is thus partially filled, and the valence band is partially empty because it now contains a few unoccupied MOs. When an electrical potential is applied to a semiconductor, it conducts a small amount of current because the potential can accelerate the electrons in the partially filled bands. Table 21.3 shows how the electrical properties of the group 4A elements vary with the size of the band gap.

TABLE 21.3   Band Gaps for the Group 4A Elements

Element*	Band Gap (kJ/mol)	Type of Material
C (diamond)	520	Insulator
Si	107	Semiconductor
Ge	65	Semiconductor
Sn (gray tin)	8	Semiconductor
Sn (white tin)	0	Metal
Pb	0	Metal

* Si, Ge, and gray Sn have the same structure as diamond.

The electrical conductivity of a semiconductor *increases* with increasing temperature because the number of electrons with sufficient energy to occupy the conduction band increases as the temperature rises. At higher temperatures, there are more charge carriers (electrons) in the conduction band and more vacancies in the valence band. By contrast, the electrical conductivity of a metal *decreases* with increasing temperature. At higher temperatures, the metal cations undergo increased vibrational motion about their lattice sites, and vibration of the cations disrupts the flow of electrons through the crystal.

The conductivity of a semiconductor can be greatly increased by adding certain impurities in small (ppm) amounts, a process called **doping**. Consider, for example, the addition of a group 5A element such as phosphorus to a group 4A semiconductor such as silicon. Like diamond, silicon has a structure in which each Si atom is surrounded tetrahedrally by four others. The added P atoms occupy normal Si positions in the structure, but each P atom has five valence electrons and therefore introduces an extra electron not needed for bonding. In the MO picture, the extra electrons occupy the conduction band, as shown in Figure 21.11a. The number of electrons in the conduction band of the doped silicon is much greater than in pure silicon, and the conductivity of the doped semiconductor is therefore correspondingly higher. Because the charge carriers are *negative* electrons, the silicon doped with a group 5A element is called an **n-type semiconductor**.

FIGURE 21.11 MO energy levels for doped semiconductors. **(a)** An *n*-type semiconductor, such as silicon doped with phosphorus, has more electrons than needed for bonding and thus has *negative* electrons in the partially filled conduction band. **(b)** A *p*-type semiconductor, such as silicon doped with boron, has fewer electrons than needed for bonding and thus has vacancies—*positive* holes—in the valence band.

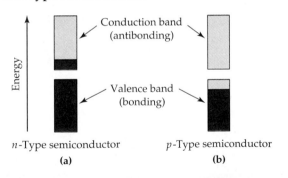
n-Type semiconductor
(a)

p-Type semiconductor
(b)

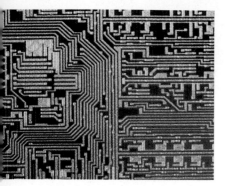

A computer microprocessor chip.

Now let's consider a semiconductor in which silicon is doped with a group 3A element such as boron. Each B atom has just three valence electrons and therefore does not have enough electrons to form bonds to its four Si neighbors. In the MO picture, the bonding MOs of the valence band are only partially filled, as shown in Figure 21.11b. The vacancies in the valence band can be regarded as *positive* holes in a filled band, and the silicon doped with a group 3A element is therefore called a ***p*-type semiconductor**.

Doped semiconductors are essential components in the modern solid-state electronic devices found in radios, television sets, pocket calculators, and computers. Devices such as transistors, which control electrical signals in these products, are made from *n*-type and *p*-type semiconductors. In modern integrated circuits, an amazing number of extremely small devices can be packed into a small space, thus decreasing the size and increasing the speed of electrical equipment. For example, computer microprocessors now contain up to 28 million transistors on a silicon chip with a surface area of about 1 cm^2 and are able to execute as many as 1 billion instructions per second.

EXAMPLE 21.2 Consider a crystal of germanium that has been doped with a small amount of aluminum. Is the doped crystal an *n*-type or *p*-type semiconductor? Compare the conductivity of the doped crystal with that of pure germanium.

SOLUTION Germanium, like silicon, is a group 4A semiconductor, and aluminum, like boron, is a group 3A element. The doped germanium is therefore a *p*-type semiconductor because each Al atom has one less valence electron than needed for bonding to the four neighboring Ge atoms. (Like silicon, germanium has the diamond structure.) The valence band is thus partially filled, which accounts for the electrical conductivity. The conductivity is greater than that of pure germanium because the doped germanium has many more positive holes in the valence band. That is, it has more vacant MOs available to which electrons can be excited by an electrical potential.

▶**PROBLEM 21.4** Is germanium doped with arsenic an *n*-type or *p*-type semiconductor? Why is its conductivity greater than that of pure germanium?

✦**KEY CONCEPT PROBLEM 21.5** The following pictures show the electron population of the bands of MO energy levels for four materials—diamond, silicon, silicon doped with aluminum, and white tin:

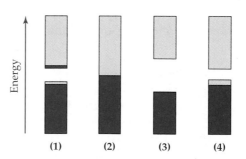

(a) Which picture corresponds to which material?
(b) Arrange the four materials in order of increasing electrical conductivity. Explain.

21.6 Superconductors

The discovery of high-temperature superconductors is surely one of the most exciting scientific developments in the last 20 years. It has stimulated an enormous amount of research in chemistry, physics, and materials science that could some day lead to a world of superfast computers, magnetically levitated trains, and power lines that carry electric current without loss of energy.

A **superconductor** is a material that loses all electrical resistance below a characteristic temperature called the **superconducting transition temperature, T_c.** This phenomenon was discovered in 1911 by the Dutch physicist Heike Kamerlingh Onnes, who found that mercury abruptly loses its electrical resistance when it is cooled with liquid helium to 4.2 K (Figure 21.12). Below its T_c, a material becomes a perfect conductor, and an electric current, once started, flows indefinitely without loss of energy.

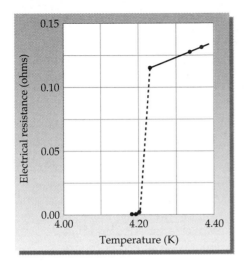

FIGURE 21.12 The electrical resistance of mercury falls to zero at its superconducting transition temperature, $T_c = 4.2$ K. Above T_c, mercury is a metallic conductor: Its resistance increases (conductivity decreases) with increasing temperature. Below T_c, mercury is a superconductor.

 YBa$_2$Cu$_3$O$_7$ 3D model

Since 1911, scientists have been searching for materials that superconduct at higher temperatures, and more than 6000 superconductors are now known. Until 1986, however, the record value of T_c was only 23.2 K (for the compound Nb$_3$Gc). The situation changed dramatically in 1986 when K. Alex Müller and J. Georg Bednorz of the IBM Zürich Research Laboratory reported a T_c of 35 K for the nonstoichiometric barium lanthanum copper oxide Ba$_x$La$_{2-x}$CuO$_4$, where x has a value of about 0.1. Soon thereafter, scientists found even higher values of T_c for other copper-containing oxides: 90 K for YBa$_2$Cu$_3$O$_7$, 125 K for Tl$_2$Ca$_2$Ba$_2$Cu$_3$O$_{10}$, and 133 K for HgCa$_2$Ba$_2$Cu$_3$O$_{8+x}$, the record holder as of this writing. High values of T_c for these compounds were completely unexpected because most metal oxides—nonmetallic inorganic solids called *ceramics*—are electrical insulators. Within just one year of discovering the first ceramic superconductor, Müller and Bednorz were awarded the 1987 Nobel Prize in physics.

The crystal structure of YBa$_2$Cu$_3$O$_7$, the so-called 1-2-3 compound (1 yttrium, 2 bariums, and 3 coppers), is illustrated in Figure 21.13, and one unit cell of the structure is shown in Figure 21.14. The crystal contains parallel planes of Y, Ba, and Cu atoms. Two-thirds of the Cu atoms are surrounded by a square pyramid of five O atoms, some of which are shared with neighboring CuO$_5$ groups to give two-dimensional layers of square pyramids. The remaining Cu atoms are surrounded by a square of four O atoms, two of which are shared with neighboring CuO$_4$ squares to give chains of CuO$_4$ groups. It's

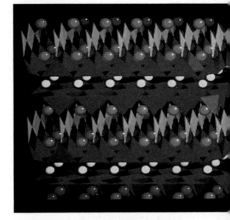

FIGURE 21.13 A computer graphics representation of the crystal structure of YBa$_2$Cu$_3$O$_7$. Layers of Y atoms (yellow) and Ba atoms (blue) are sandwiched between layers of CuO$_5$ square pyramids (red) and chains of square planar CuO$_4$ groups (green).

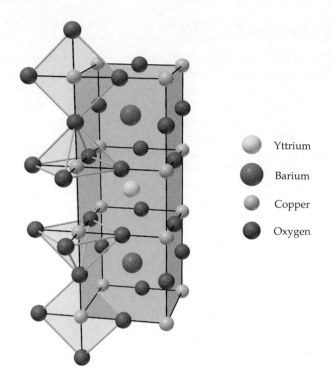

FIGURE 21.14 One unit cell of the crystal structure of $YBa_2Cu_3O_7$ contains one Y atom, two Ba atoms, three Cu atoms, and seven O atoms. In counting the Cu and O atoms, recall that the unit cell contains one-eighth of each corner atom, one-fourth of each edge atom, and one-half of each face atom (Section 10.8). The figure includes eight O atoms from neighboring unit cells to show the square pyramidal CuO_5 groups and the square planar CuO_4 groups.

Yttrium

Barium

Copper

Oxygen

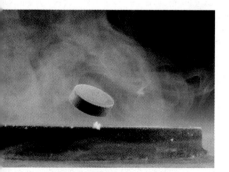

FIGURE 21.15 Levitation of a magnet above a pellet of $YBa_2Cu_3O_7$ cooled to 77 K with liquid nitrogen. $YBa_2Cu_3O_7$ becomes a superconductor at approximately 90 K.

interesting to note that the Cu atoms have a fractional oxidation number of +2.33, based on the usual oxidation numbers of +3 for Y, +2 for Ba, and −2 for O. Both the infinitely extended layers of Cu and O atoms and the fractional oxidation number of Cu appear to play a role in the current flow, but a generally accepted theory of superconductivity in ceramic superconductors is not yet available. This is a field where experiment is far ahead of theory.

One of the most dramatic properties of a superconductor is its ability to levitate a magnet (Figure 21.15). When a superconductor is cooled below its T_c and a magnet is lowered toward it, the superconductor and the magnet repel each other, and the magnet hovers above the superconductor as though suspended in midair.

The repulsive force arises in the following way: When the magnet moves toward the superconductor, it induces a supercurrent in the surface of the superconductor that continues to flow even after the magnet stops moving. The supercurrent, in turn, induces a magnetic field in the superconductor that exactly cancels the field from the magnet. Thus, the net magnetic field within the bulk of the superconductor is zero, a phenomenon called the *Meissner effect*. Outside the superconductor, however, the magnetic fields due to the magnet and the supercurrent repel each other, just as the north poles of two bar magnets do. The magnet therefore experiences an upward magnetic force as well as the usual downward gravitational force, and it remains suspended above the superconductor at the point where the two forces are equal. Potential applications of the Meissner effect include high-speed, magnetically levitated trains.

Some applications of superconductors already exist. For example, powerful superconducting magnets are essential components in the magnetic resonance imaging (MRI) instruments widely used in medical diagnosis (see Section 22.10). Superconductors are also used to make the magnets that bend the path of charged particles in high-energy particle accelerators. All present applications, however, use conventional superconductors ($T_c \leq 20$ K) that are cooled to 4.2 K with liquid helium, an expensive substance that requires sophisticated cryogenic (cooling) equipment. Much of the excitement surrounding the new, high-temperature superconductors arises because their T_c values are above the boiling point of liquid nitrogen (bp 77 K), an abundant refrigerant that is cheaper than milk. Of course, the search goes on for materials with still higher values of T_c. For applications such as long-distance electric power transmission, the goal is a material that superconducts at room temperature.

Several serious problems must be surmounted before applications of high-temperature superconductors can become a reality. Presently known ceramic superconductors are brittle powders with high melting points, so they are not easily fabricated into the wires and coils needed for electrical equipment. Also, the currents that these materials are able to carry at 77 K are still too low for practical applications. Thus, applications are likely in the future but are not right around the corner.

In 1991, scientists at AT&T Bell Laboratories discovered a new class of high-temperature superconductors based on fullerene, the allotrope of carbon that contains C_{60} molecules (Sections 10.10 and 19.6). Called "buckyballs," after the architect R. Buckminster Fuller, these soccer ball-shaped C_{60} molecules react with potassium to give K_3C_{60}. This stable crystalline solid contains a face-centered cubic array of buckyballs, with K^+ ions in the cavities between the C_{60} molecules (Figure 21.16). At room temperature K_3C_{60} is a metallic conductor, but it becomes a superconductor at 18 K. The rubidium fulleride, Rb_3C_{60}, and a rubidium–thallium–C_{60} compound of unknown stoichiometry have higher T_c values of 30 K and 45–48 K, respectively.

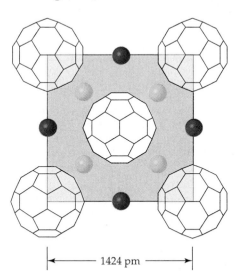

1424 pm

FIGURE 21.16 A portion of one unit cell of the face-centered cubic structure of K_3C_{60} viewed perpendicular to a cube face. The C_{60} "buckyballs" are located at the cube corners and face centers, and the K^+ ions (red and blue spheres) are in two kinds of holes between the $C_{60}{}^{3-}$ ions. The K^+ ions shown in red lie in the plane of the cube face (the plane of the paper) and are surrounded octahedrally by six $C_{60}{}^{3-}$ ions. Those shown in blue lie in a plane one-fourth of a cell edge length below the plane of the paper and are surrounded tetrahedrally by four $C_{60}{}^{3-}$ ions.

The copper oxide ceramic superconductors are two-dimensional conductors: They can conduct a current parallel to the layers of Cu and O atoms but not between the layers. The fullerides, by contrast, are three-dimensional conductors that conduct equally in all directions. Because of this property, they may prove to be superior materials for making superconducting wires.

▶**PROBLEM 21.6** Show that one unit cell of $YBa_2Cu_3O_7$ (Figure 21.14) contains three Cu atoms and seven O atoms.

21.7 Ceramics

Ceramics are inorganic, nonmetallic, nonmolecular solids, including both crystalline materials such as quartz (SiO_2) and amorphous materials such as glasses. Known since ancient times, traditional silicate ceramics, such as pottery and porcelain, are made by heating aluminosilicate clays to high temperatures. Modern, so-called **advanced ceramics**—materials that have high-tech engineering, electronic, and biomedical applications—include *oxide ceramics*, such as alumina (Al_2O_3), and *nonoxide ceramics*, such as silicon carbide (SiC) and silicon nitride (Si_3N_4). Additional examples are listed in Table 21.4, which compares properties of ceramics with those of aluminum and steel. (Note that oxide ceramics are named by replacing the *-um* ending of the element name with an *-a*; thus, BeO is beryllia, ZrO_2 is zirconia, and so forth.)

TABLE 21.4   Properties of Some Ceramic and Metallic Materials

Material	Melting Point (°C)	Density (g/cm³)	Elastic Modulus (GPa)*	Hardness (mohs)[†]
Oxide ceramics				
Alumina, Al_2O_3	2054	3.97	380	9
Beryllia, BeO	2578	3.01	370	8
Zirconia, ZrO_2	2710	5.68	210	8
Nonoxide ceramics				
Boron carbide, B_4C	2350	2.50	280	9
Silicon carbide, SiC	2830	3.16	400	9
Silicon nitride, Si_3N_4	1900	3.17	310	9
Metals				
Aluminum	660	2.70	70	3
Plain carbon steel	1515	7.86	205	5

* The elastic modulus, measured in units of pressure (1 gigapascal = 1 GPa = 10^9 Pa) indicates the stiffness of a material when it is subjected to a load. The larger the value, the stiffer the material.

[†] Numbers on the Mohs hardness scale range from 1 for talc, a very soft material, to 10 for diamond, the hardest known substance.

In many respects, the properties of ceramics are superior to those of metals: Ceramics have higher melting points, and they are stiffer, harder, and more resistant to wear and corrosion. Moreover, they maintain much of their strength at high temperatures, where metals either melt or corrode because of oxidation. Silicon nitride and silicon carbide, for example, are stable to oxidation in air up to 1400–1500°C, and oxide ceramics don't react with oxygen because they are already fully oxidized. Because ceramics are less dense than steel, they are attractive lightweight, high-temperature materials for replacing metal components in aircraft, space vehicles, and automotive engines.

Unfortunately, ceramics are brittle, as anyone who has dropped a coffee cup well knows. The brittleness, hardness, stiffness, and high melting points of ceramics are due to strong chemical bonding. Take silicon carbide, for example, a covalent network solid that crystallizes in the diamond structure (Figure 21.17). Each Si atom is bonded tetrahedrally to four C atoms, and each C atom is bonded tetrahedrally to four Si atoms. The strong, highly directional covalent bonds ($D_{Si-C} = 435$ kJ/mol) prevent the planes of atoms from sliding over one another when the solid is subjected to the stress of a

Silicon nitride rotor for use in gas-turbine engines.

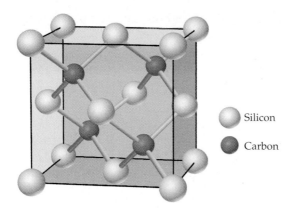

FIGURE 21.17 One unit cell of the cubic form of silicon carbide, SiC. The Si atoms are located at the corners and face centers of a face-centered cubic unit cell, while the C atoms occupy cavities (tetrahedral holes) between four Si atoms. Each C atom is bonded tetrahedrally to four Si atoms, and each Si atom is bonded tetrahedrally to four C atoms. The crystal can't deform under stress because the bonds are strong and highly directional.

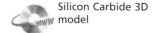 Silicon Carbide 3D model

load or an impact. As a result, the solid can't deform to relieve the stress. It maintains its shape up to a point, but then the bonds give way suddenly, and the material fails catastrophically when the stress exceeds a certain threshold value. Oxide ceramics, in which the bonding is largely ionic, behave similarly. By contrast, metals are able to deform under stress because their planes of metal cations can slide easily in the electron sea (Section 21.4). As a result, metals dent but ceramics shatter.

Ceramic processing, the series of steps that leads from raw material to the finished ceramic object, is important in determining the strength and the resistance to fracture of the product. Processing often begins with a fine powder, which is combined with an organic binder, shaped, compacted, and finally *sintered* at temperatures of 1300–2000°C. **Sintering**, which occurs below the melting point, is a process in which the particles of the powder are "welded" together without completely melting. During sintering, the crystal grains grow larger and the density of the material increases as the void spaces between particles disappear. Unfortunately, impurities and remaining voids can lead to microscopic cracks that cause the material to fail under stress. It is therefore important to minimize impurities and voids by beginning with high-purity, fine powders that can be tightly compacted prior to sintering.

One approach to making such powders is the **sol–gel method**, which involves synthesis of a metal oxide powder from a metal alkoxide, a compound derived from a metal and an alcohol. In the synthesis of titania (TiO_2) from titanium ethoxide, $Ti(OCH_2CH_3)_4$, for example, the $Ti(OCH_2CH_3)_4$ starting material is made by the reaction of titanium(IV) chloride with ethanol and ammonia in a benzene solution:

$$TiCl_4 + 4\ \underset{\text{Ethanol}}{HOCH_2CH_3} + 4\ NH_3(g) \xrightarrow{C_6H_6} \underset{\text{Titanium ethoxide}}{Ti(OCH_2CH_3)_4} + 4\ NH_4Cl(s)$$

Pure $Ti(OCH_2CH_3)_4$ is then dissolved in an appropriate organic solvent, and water is added to bring about a *hydrolysis* reaction, represented by the following simplified equation:

$$Ti(OCH_2CH_3)_4 + 4\ H_2O \xrightarrow{\text{Solution}} Ti(OH)_4(s) + 4\ HOCH_2CH_3$$

In this reaction, $Ti–OCH_2CH_3$ bonds are broken, $Ti–OH$ bonds are formed, and ethanol is regenerated. The $Ti(OH)_4$ forms in the solution as a colloidal dispersion called a *sol*, consisting of extremely fine particles having a diameter of only 0.001–0.1 μm. Subsequent reactions eliminate water molecules and form oxygen bridges between Ti atoms:

$$(HO)_3Ti—O—H + H—O—Ti(OH)_3 \longrightarrow (HO)_3Ti—O—Ti(OH)_3 + H_2O$$

Because all the OH groups can undergo this reaction, the particles of the sol link together through a three-dimensional network of oxygen bridges, and the sol is converted to a more rigid, gelatinlike material called a *gel*. The remaining water and solvent are then removed by heating the gel, and TiO_2 is obtained as a fine powder consisting of high-purity particles with a diameter less than 1 μm (Figure 21.18).

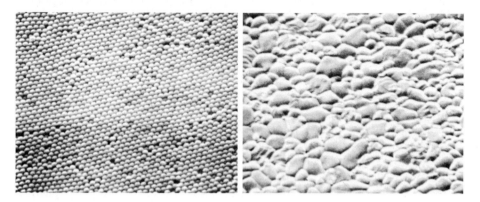

FIGURE 21.18 Electron micrographs of a titania powder consisting of tightly packed particles of TiO_2 with a diameter less than 1 μm (left) and the dense ceramic produced by sintering the powder (right). The powder was made by the sol–gel method.

Oxide ceramics have many important uses. Alumina, for example, is the material of choice for making spark-plug insulators because of its high electrical resistance, high strength, high thermal stability, and chemical inertness. Because alumina is nontoxic and essentially inert in biological systems, it is used in constructing dental crowns and the heads of artificial hips. Alumina is also used as a substrate material for electronic circuit boards.

EXAMPLE 21.3 Alumina (Al_2O_3) powders can be prepared from aluminum ethoxide by the sol–gel method. Write a balanced equation for the hydrolysis of aluminum ethoxide.

SOLUTION First, determine the chemical formula of aluminum ethoxide. Because aluminum is a group 3A element, it has an oxidation number of +3. The ethoxide ligand is the anion of ethanol, $HOCH_2CH_3$, and has a charge of −1. Since aluminum ethoxide is a neutral compound, its formula must be $Al(OCH_2CH_3)_3$. The hydrolysis reaction, which breaks the $Al–OCH_2CH_3$ bonds and forms $Al–OH$ bonds, requires one H_2O molecule for each ethoxide ligand:

$$Al(OCH_2CH_3)_3 + 3 H_2O \longrightarrow Al(OH)_3 + 3 HOCH_2CH_3$$

Subsequent heating converts the aluminum hydroxide to alumina.

EXAMPLE 21.4 The 1-2-3 ceramic superconductor $YBa_2Cu_3O_7$ has been synthesized by the sol–gel method from a stoichiometric mixture of yttrium ethoxide, barium ethoxide, and copper(II) ethoxide in an appropriate organic solvent. The oxide product, before being heated in oxygen, has the formula $YBa_2Cu_3O_{6.5}$. Write a balanced equation for hydrolysis of the stoichiometric mixture of metal ethoxides.

SOLUTION   First, determine the chemical formulas of the metal ethoxides. Because yttrium is a group 3B element and has an oxidation number of +3, the formula of yttrium ethoxide must be $Y(OCH_2CH_3)_3$. Similarly, because both barium (in group 2A) and copper(II) have an oxidation number of +2, the formulas of barium ethoxide and copper ethoxide must be $Ba(OCH_2CH_3)_2$ and $Cu(OCH_2CH_3)_2$, respectively. If the three metal ethoxides were present separately, hydrolysis would give $Y(OH)_3$, $Ba(OH)_2$, and $Cu(OH)_2$. Because they are present together in a 1:2:3 ratio, we write the product as $Y(OH)_3 \cdot 2\,Ba(OH)_2 \cdot 3\,Cu(OH)_2$, or $YBa_2Cu_3(OH)_{13}$. Thus, the hydrolysis reaction requires 13 H_2O molecules for 13 OCH_2CH_3 ligands, and the balanced equation is

$$Y(OCH_2CH_3)_3 + 2\,Ba(OCH_2CH_3)_2 + 3\,Cu(OCH_2CH_3)_2 + 13\,H_2O \longrightarrow$$
$$YBa_2Cu_3(OH)_{13} + 13\,HOCH_2CH_3$$

Subsequent heating removes water, converting the mixed-metal hydroxide to the oxide $YBa_2Cu_3O_{6.5}$, which is then oxidized to $YBa_2Cu_3O_7$ by heating in O_2 gas.

▶ **PROBLEM 21.7**   Silica glasses used in lenses, laser mirrors, and other optical components can be made by the sol–gel method. One step in the process involves hydrolysis of $Si(OCH_3)_4$. Write a balanced equation for the reaction.

▶ **PROBLEM 21.8**   Crystals of the oxide ceramic barium titanate, $BaTiO_3$, have an unsymmetrical arrangement of ions, which gives the crystals an electric dipole moment. Such materials are called ferroelectrics and are used in making various electronic devices. Barium titanate can be made by the sol–gel method, which involves hydrolysis of a mixture of metal alkoxides. Write a balanced equation for the hydrolysis of a 1:1 mixture of barium isopropoxide and titanium isopropoxide, and explain how the resulting sol is converted to $BaTiO_3$. (The isopropoxide ligand is the anion of isopropyl alcohol, $HOCH(CH_3)_2$, also known as rubbing alcohol.)

21.8   Composites

We saw in the previous section that ceramics are brittle and prone to fracture. They can be strengthened and toughened, though, by mixing the ceramic powder prior to sintering with fibers of a second ceramic material, such as carbon, boron, or silicon carbide. The resulting hybrid material, called a **ceramic composite**, combines the advantageous properties of both components. An example is the composite consisting of fine grains of alumina reinforced with *whiskers* of silicon carbide. Whiskers are tiny, fiber-shaped particles, about 0.5 μm in diameter and 50 μm long, that are very strong because they are single crystals. Silicon carbide-reinforced alumina possesses high strength and high shock resistance, even at high temperatures, and has therefore been used to make high-speed cutting tools for machining very hard steels.

How do fibers and whiskers increase the strength and fracture toughness of a composite material? First, fibers have great strength along the fiber axis because most of the chemical bonds are aligned in that direction. Second, there are several ways in which fibers can prevent microscopic cracks from propagating to the point that they lead to fracture of the material. The fibers can deflect cracks, thus preventing them from moving cleanly in one direction, and they can bridge cracks, thus holding the two sides of a crack together.

Silicon carbide-reinforced alumina is an example of a composite in which both the fibers and the surrounding matrix are ceramics. There are other composites, however, in which the two phases are different types of materials.

Examples are **ceramic–metal composites**, or **cermets**, such as aluminum metal reinforced with boron fiber, and **ceramic–polymer composites**, such as boron/epoxy and carbon/epoxy. (Epoxy is a resin consisting of long-chain organic molecules.) These materials are popular for aerospace and military applications because of their high strength-to-weight ratios. Boron-reinforced aluminum, for example, is used as a lightweight structural material in the space shuttle, and boron/epoxy and carbon/epoxy skins are used on military aircraft. Increased use of composite materials in commercial aircraft could result in weight savings of 20–30% and corresponding savings in fuel (which accounts for 10–15% of the cost of operating an airline).

The skin of the B-2 advanced technology aircraft is a strong, lightweight composite material that contains carbon fibers.

Ceramic fibers used in composites are usually made by high-temperature methods. Carbon (graphite) fiber, for example, can be made by thermal decomposition of fibers of polyacrylonitrile, a long-chain organic molecule also used to make the textile Orlon:

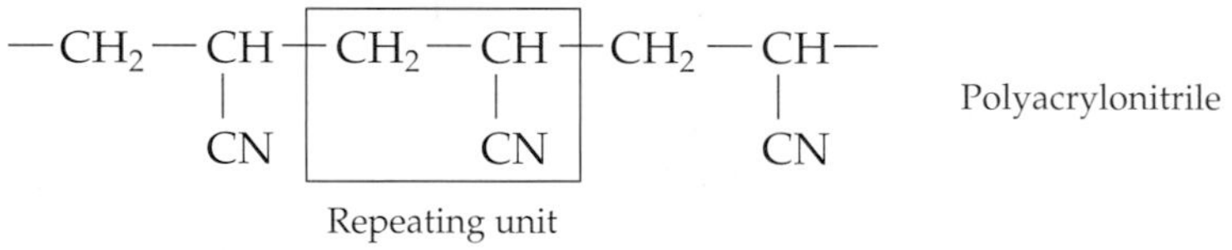

$$-CH_2-CH-\boxed{CH_2-CH}-CH_2-CH-$$
$$CN\boxed{CN}CN$$

Polyacrylonitrile

Repeating unit

In the final step of the multistep process, the carbon in the fiber is converted to graphite by heating at 1400–2500°C. Similarly, silicon carbide fiber can be made by heating fibers that contain long-chain molecules with alternating silicon and carbon atoms:

$$-SiH_2-CH_2-\boxed{SiH_2-CH_2}-SiH_2-CH_2-\xrightarrow[-H_2]{Heat} SiC\ fiber$$

Repeating unit

▶ **PROBLEM 21.9** Classify each of these composites as ceramic–ceramic, ceramic–metal, or ceramic–polymer:
(a) Cobalt/tungsten carbide   (b) Silicon carbide/zirconia
(c) Boron nitride/epoxy   (d) Boron carbide/titanium

Diamonds and Diamond Films

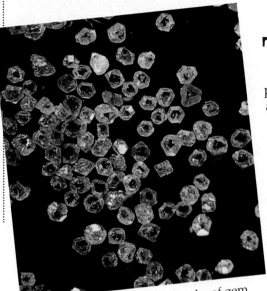

These diamonds, fine enough to be of gem quality, were made from graphite at high temperature and high pressure.

Their sparkling beauty has made diamonds an object of fascination and desire for millennia. Far more important—at least industrially—is their hardness. Diamond powder is unsurpassed for polishing, diamond-tipped bits are unequalled for drilling, and diamond-tipped sawblades are unparalleled for cutting. In fact, the industrial demand for diamonds dwarfs that of the jewelry trade and far outstrips the supply of natural stones. As a result of this demand, an intensive effort to produce synthetic diamonds began in the 1940s and culminated in the mid-1950s with independent announcements by the General Electric company in the United States and the Allemanna Svenska Elektriska company in Sweden of a method for preparing diamond from graphite.

The original General Electric method for producing diamond from graphite used a combination of high temperature (2500°C) and pressure (100,000 atm) and yielded a black gritty material suitable for use as an industrial abrasive. Subsequent technical improvements have made possible the synthesis of gem-quality stones of up to 2 carats in mass (about 0.4 g) at a price competitive with that of natural diamond.

The most exciting recent advance in diamond-producing technology was the development by several research groups in the 1980s of methods for depositing diamond *films* on a variety of surfaces at low pressure. When a dilute mixture (about 1% by volume) of CH_4 in H_2 gas at a pressure of about 40 mm Hg is heated in a microwave discharge to a temperature near 1000°C, free carbon and hydrogen atoms are produced. The hydrogen atoms evidently impede the formation of graphite, and the carbon atoms therefore deposit on a nearby surface in the form of diamond. The thin, filmlike coating that results is made up of tiny diamond crystals about 10–20 nm in diameter.

Imagine the possibilities, some of which are just now becoming reality: Tools, knifeblades, and scalpels coated with diamond remain forever sharp. Eyeglass lenses and wristwatches coated with diamond remain scratch-free. High-fidelity loudspeakers coated with diamond give a nearly perfect, undistorted sound at high frequencies. Hip joints and other biological implants coated with diamond are not rejected by the body's immune system.

Further improvements in deposition technology are still needed—only substrates stable at a temperature of 1000°C can be coated at present—but the time may not be too far off before diamond coatings on a variety of consumer products become commonplace.

▶ **PROBLEM 21.10** Is diamond a ceramic? Explain.

▶ **PROBLEM PROBLEM 21.11** Draw one unit cell of a diamond crystal, and explain why diamond is so hard.

Key Words

Summary

Most metals occur in nature as **minerals**—the silicates, carbonates, oxides, sulfides, phosphates, and chlorides that make up the rocks of the earth's crust. Minerals from which metals can be produced economically are called **ores**.

Metallurgy, the science and technology of extracting metals from their ores, involves three steps: (1) concentration and chemical treatment to separate the mineral from impurities called **gangue**; (2) reduction of the mineral to the free metal; and (3) refining or purification of the metal. The first step uses processes such as **flotation** and **roasting**. The reduction method (roasting, chemical reduction or electrolysis) depends on the activity of the metal. The refining step involves distillation, chemical purification, or electrorefining.

Iron is produced in a blast furnace by reducing iron oxide ores with CO. Added limestone ($CaCO_3$) decomposes to CaO, which reacts with SiO_2 and other acidic oxide impurities to yield **slag** (mainly $CaSiO_3$) that is separated from the molten iron. The iron is converted to steel by the **basic oxygen process**, which removes impurities such as elemental Si, P, S, and C.

Two bonding models are used for metals. The **electron-sea model** pictures a metal as an array of metal cations immersed in a sea of delocalized, mobile valence electrons that act as an electrostatic glue. In the molecular orbital theory for metals, also called **band theory**, the delocalized valence electrons occupy a vast number of MO energy levels that are so closely spaced that they merge into an almost continuous band. Both theories account for properties such as malleability, ductility, and high thermal and electrical conductivity, but band theory better explains how the number of valence electrons affects properties such as melting point and hardness.

Band theory also accounts for the electrical properties of metals, insulators, and semiconductors. Because electrical conductivity involves excitation of valence electrons to readily accessible higher-energy MOs, materials with partially filled bands are metallic conductors, and materials with only completely filled bands are electrical insulators. In insulators, the bonding MOs, called the **valence band**, and the antibonding MOs, called the **conduction band**, are separated in energy by a large **band gap**. In **semiconductors**, the band gap is smaller, and a few electrons have enough energy to occupy the conduction band. The resulting partially filled valence and conduction bands give rise to a small conductivity. The conductivity can be increased by **doping**—adding a group 5A impurity to a group 4A element, which gives an *n*-**type semiconductor**, or adding a group 3A impurity to a group 4A element, which gives a *p*-**type semiconductor**.

A **superconductor** is a material that loses all electrical resistance below a characteristic temperature called the **superconducting transition temperature,** T_c. An example is $YBa_2Cu_3O_7$ ($T_c = 90$ K). Below T_c, a superconductor is able to levitate a magnet, a consequence of the Meissner effect.

Ceramics are inorganic, nonmetallic, nonmolecular solids. Modern **advanced ceramics** include oxide ceramics such as Al_2O_3 and $YBa_2Cu_3O_7$ and nonoxide ceramics such as SiC and Si_3N_4. Ceramics are generally higher-melting, lighter, stiffer, harder, and more resistant to wear and corrosion than metals. One approach to ceramic processing is the **sol–gel method**, involving hydrolysis of a metal alkoxide. The strength and fracture toughness of ceramics can be increased by making **ceramic composites**, hybrid materials such as Al_2O_3/SiC. Other types of composites include **ceramic–metal composites** and **ceramic–polymer composites**.

Key Concept Summary

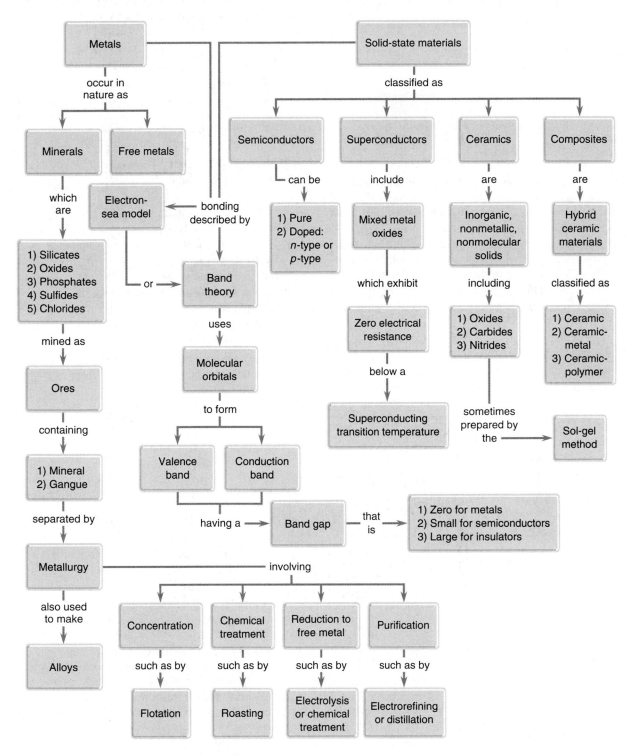

Understanding Key Concepts

Problems 21.1–21.11 appear within the chapter.

21.12 Look at the location of elements A, B, C, and D in the following periodic table:

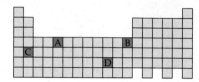

Without looking at Figure 21.2, predict whether these elements are likely to be found in nature as carbonates, oxides, sulfides, or in uncombined form. Explain.

21.13 Among the methods for extracting metals from their ores are (i) roasting a metal sulfide, (ii) chemical reduction of a metal oxide, and (iii) electrolysis. The preferred method depends on the $E°$ value for the reduction half-reaction $M^{n+}(aq) + n\,e^- \rightarrow M(s)$.

(a) Which method is appropriate for the metals with the most negative $E°$ values?

(b) Which method is appropriate for the metals with the most positive $E°$ values?

(c) Which method is appropriate for metals A, B, C, and D in the following periodic table?

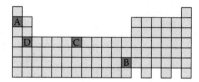

21.14 The following pictures show the electron populations of the bands of MO energy levels for four different materials:

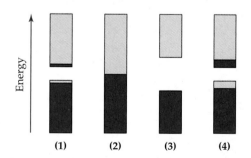

(a) Classify each material as an insulator, a semiconductor, or a metal.

(b) Arrange the four materials in order of increasing electrical conductivity. Explain.

(c) Tell whether the conductivity of each material increases or decreases when the temperature increases.

21.15 The following pictures show the electron populations of the composite s–d bands for three different transition metals:

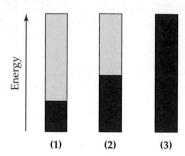

(a) Which metal has the highest melting point? Explain.

(b) Which metal has the lowest melting point? Explain.

(c) Arrange the metals in order of increasing hardness. Explain.

21.16 The following picture represents the electron population of the bands of MO energy levels for elemental silicon:

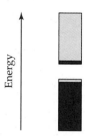

(a) Identify the valence band, conduction band, and band gap.

(b) In a drawing, show how the electron population changes when the silicon is doped with gallium.

(c) In a drawing, show how the electron population changes when the silicon is doped with arsenic.

(d) Compare the electrical conductivity of the doped silicon samples with that of pure silicon. Account for the differences.

21.17 Sketch the electron populations of the bands of MO energy levels for elemental carbon (diamond), silicon, germanium, gray tin, and white tin. (Band gap data are given in Table 21.3.) Your sketches should show how the populations of the different bands vary with a change in the group 4A element.

Additional Problems

Sources of the Metallic Elements

21.18 List three metals that are found in nature as oxide ores.

21.19 List three metals that are found in nature in uncombined form.

21.20 Locate the following metals in the periodic table, and predict whether they are likely to be found in nature as oxides, sulfides, or in uncombined form:

(a) Copper   (b) Zirconium

(c) Palladium   (d) Bismuth

21.21 Locate the following metals in the periodic table, and predict whether they are likely to be found in nature as oxides, sulfides, or in uncombined form:

(a) Vanadium   (b) Silver

(c) Rhodium   (d) Hafnium

21.22 Explain why early transition metals occur in nature as oxides but late transition metals occur as sulfides.

21.23 Explain why iron occurs in nature as an oxide but calcium occurs as a carbonate.

21.24 Name the following minerals:

(a) Fe_2O_3 (b) PbS

(c) TiO_2 (d) $CuFeS_2$

21.25 Write chemical formulas for the following minerals:

(a) Cinnabar   (b) Bauxite

(c) Sphalerite   (d) Chromite

Metallurgy

21.26 Describe the flotation process for concentrating a metal sulfide ore, and explain why this process would not work well for concentrating a metal oxide ore.

21.27 Gallium is a minor constituent in bauxite, $Al_2O_3 \cdot x\, H_2O$, and like aluminum, forms an amphoteric oxide, Ga_2O_3. Explain how the Bayer process makes it possible to separate Ga_2O_3 from Fe_2O_3, which is also present in bauxite. Write a balanced net ionic equation for the reaction.

21.28 Zinc, cadmium, and mercury are found in nature as sulfide ores. Values of $E°$ for the reduction half-reaction $M^{2+}(aq) + 2\,e^- \rightarrow M(s)$ are -0.76 V for M = Zn, -0.40 V for M = Cd, and $+0.85$ V for M = Hg. Explain why roasting sphalerite (ZnS) gives ZnO but roasting cinnabar (HgS) gives elemental mercury. What product would you expect from roasting CdS?

21.29 Zinc metal is produced by reducing ZnO with coke, and magnesium metal is produced by electrolysis of molten $MgCl_2$. Look at the standard reduction potentials in Appendix D, and explain why magnesium can't be prepared by the method used for zinc.

21.30 Complete and balance each of the following equations:

(a) $V_2O_5(s) + Ca(s) \rightarrow$ (b) $PbS(s) + O_2(g) \rightarrow$

(c) $MoO_3(s) + H_2(g) \rightarrow$ (d) $MnO_2(s) + Al(s) \rightarrow$

(e) $MgCl_2(l) \xrightarrow{\text{Electrolysis}}$

21.31 Complete and balance each of the following equations:

(a) $Fe_2O_3(s) + H_2(g) \rightarrow$ (b) $Cr_2O_3(s) + Al(s) \rightarrow$

(c) $Ag_2S(s) + O_2(g) \rightarrow$ (d) $TiCl_4(g) + Mg(l) \rightarrow$

(e) $LiCl(l) \xrightarrow{\text{Electrolysis}}$

21.32 Use the data in Appendix B to calculate $\Delta H°$ and $\Delta G°$ for the roasting of sphalerite, ZnS:

$$2\,ZnS(s) + 3\,O_2(g) \longrightarrow 2\,ZnO(s) + 2\,SO_2(g)$$

Explain why $\Delta H°$ and $\Delta G°$ have different values, and account for the sign of $(\Delta H° - \Delta G°)$.

21.33 Use the thermodynamic data in Appendix B to calculate $\Delta H°$ and $\Delta G°$ for the roasting of cinnabar, HgS:

$$HgS(s) + O_2(g) \longrightarrow Hg(l) + SO_2(g)$$

Explain why $\Delta H°$ and $\Delta G°$ have different values, and account for the sign of $(\Delta H° - \Delta G°)$.

21.34 Ferrochrome, an iron–chromium alloy used in making stainless steel, is produced by reducing chromite ($FeCr_2O_4$) with coke:

$$FeCr_2O_4(s) + 4\,C(s) \longrightarrow \underbrace{Fe(s) + 2\,Cr(s)}_{\text{Ferrochrome}} + 4\,CO(g)$$

(a) How many kilograms of chromium can be obtained by the reaction of 236 kg of chromite with an excess of coke?

(b) How many liters of carbon monoxide at 25°C and 740 mm Hg are obtained as a by-product?

21.35 One step in the industrial process for producing copper from chalcopyrite ($CuFeS_2$) involves reducing molten copper(I) sulfide with a blast of hot air:

$$Cu_2S(l) + O_2(g) \longrightarrow 2\,Cu(l) + SO_2(g)$$

(a) How many kilograms of Cu_2S must be reduced to account for the world's annual copper production of about 8×10^9 kg?

(b) How many liters of SO_2 at STP are produced as a by-product?

(c) If all the SO_2 escaped into the atmosphere and was converted to sulfuric acid in acid rain, how many kilograms of H_2SO_4 would there be in the rain?

21.36 Nickel, used in making stainless steel, can be purified by electrorefining. The electrolysis cell has an impure nickel anode, a pure nickel cathode, and an aqueous solution of nickel sulfate as the electrolyte. How many kilograms of nickel can be refined in 8.00 h if the current passed through the cell is held constant at 52.5 A?

21.37 Pure copper for use in electrical wiring is obtained by electrorefining. How many hours are required to transfer 7.50 kg of copper from an impure copper anode to a pure copper cathode if the current passed through the electrolysis cell is held constant at 40.0 A?

Iron and Steel

21.38 Write a balanced equation for the overall reaction that occurs in a blast furnace when iron ore is reduced to iron metal. Identify the oxidizing agent and the reducing agent.

21.39 What is the role of coke in the commercial process for producing iron? Write balanced equations for the relevant chemical reactions.

21.40 What is slag, and what is its role in the commercial process for producing iron?

21.41 Why is limestone added to the blast furnace in the commercial process for producing iron? Write balanced equations for the relevant chemical reactions.

21.42 Briefly describe the basic oxygen process. Write balanced equations for the relevant reactions.

21.43 Why does slag form in the basic oxygen process? Write balanced equations for the relevant reactions.

21.44 When iron ore is reduced in a blast furnace, some of the SiO_2 impurity is also reduced by reaction with carbon to give elemental silicon and carbon monoxide. The silicon is subsequently reoxidized in the basic oxygen process, and the resulting SiO_2 reacts with CaO, yielding slag, which is then separated from the molten steel. Write balanced equations for the three reactions involving SiO_2.

21.45 In a blast furnace, some of the $CaSO_4$ impurity in iron ore is reduced by carbon, yielding elemental sulfur and carbon monoxide. The sulfur is subsequently oxidized in the basic oxygen process, and the product reacts with CaO to give a molten slag. Write balanced equations for the reactions.

Bonding in Metals

21.46 Potassium metal crystallizes in a body-centered cubic structure. Draw one unit cell, and try to draw an electron-dot structure for bonding of the central K atom to its nearest-neighbor K atoms. What is the problem?

21.47 Describe the electron-sea model of the bonding in cesium metal. Cesium has a body-centered cubic structure.

21.48 How does the electron-sea model account for the malleability and ductility of metals?

21.49 How does the electron-sea model account for the electrical and thermal conductivity of metals?

21.50 Cesium metal is very soft, and tungsten metal is very hard. Explain the difference using the electron-sea model.

21.51 Sodium melts at 98°C, and magnesium melts at 649°C. Account for the higher melting point of magnesium using the electron-sea model.

21.52 Why is the molecular orbital theory for metals called band theory?

21.53 Draw an MO energy level diagram that shows the population of the 4s band for potassium metal.

21.54 How does band theory account for the electrical conductivity of metals?

21.55 Materials with partially filled bands are metallic conductors, and materials with only completely filled bands are electrical insulators. Explain why the population of the bands affects the conductivity.

21.56 Draw an MO energy level diagram for beryllium metal, and show the population of the MOs for the following two cases:

(a) The 2s and 2p bands are well separated in energy.

(b) The 2s and 2p bands overlap in energy.

Which diagram agrees with the fact that beryllium has a high electrical conductivity? Explain.

21.57 Draw an MO energy level diagram for calcium metal, and show the population of the MOs for the following two cases:

(a) The 4s and 3d bands are well separated in energy.

(b) The 4s and 3d bands overlap in energy.

Which diagram agrees with the fact that calcium has a high electrical conductivity? Explain.

21.58 The melting points for the second-series transition elements increase from 1522°C for yttrium to 2623°C for molybdenum and then decrease again to 321°C for cadmium. Account for the trend using band theory.

21.59 Copper has a Mohs hardness value of 3, and iron has a Mohs hardness value of 5. Use band theory to explain why copper is softer than iron.

Semiconductors

21.60 Define a semiconductor, and give three examples.

21.61 Explain each of the following terms:
(a) Valence band   (b) Conduction band
(c) Band gap   (d) Doping

21.62 Draw the bands of MO energy levels and the electron population for:
(a) A semiconductor   (b) An electrical insulator
Explain why a semiconductor has the higher electrical conductivity.

21.63 Draw the bands of MO energy levels and the electron population for:
(a) A semiconductor   (b) A metallic conductor
Explain why a semiconductor has the lower electrical conductivity.

21.64 How does the electrical conductivity of a semiconductor change as the size of the band gap increases? Explain.

21.65 How does the electrical conductivity of a semiconductor change as the temperature increases? Explain.

21.66 Explain what an n-type semiconductor is, and give an example. Draw an MO energy level diagram, and show the population of the valence band and the conduction band for an n-type semiconductor.

21.67 Explain what a p-type semiconductor is, and give an example. Draw an MO energy level diagram, and show the population of the valence band and the conduction band for a p-type semiconductor.

21.68 Explain why germanium doped with phosphorus has a higher electrical conductivity than pure germanium.

21.69 Explain why silicon doped with gallium has a higher electrical conductivity than pure silicon.

21.70 Classify the following semiconductors as n-type or p-type:
(a) Si doped with In
(b) Ge doped with Sb
(c) Gray Sn doped with As

21.71 Classify the following semiconductors as n-type or p-type:
(a) Ge doped with As
(b) Ge doped with B
(c) Si doped with Sb

21.72 Arrange the following materials in order of increasing electrical conductivity:
(a) Cu   (b) Al_2O_3 (c) Fe
(d) Pure Ge   (e) Ge doped with In

21.73 Arrange the following materials in order of increasing electrical conductivity:
(a) Pure gray Sn
(b) Gray Sn doped with Sb
(c) NaCl
(d) Ag
(e) Pure Si

Superconductors

21.74 What are the two most characteristic properties of a superconductor?

21.75 $YBa_2Cu_3O_7$ is a superconductor below its T_c of 90 K and a metallic conductor above 90 K. Make a rough plot of electrical resistance versus temperature for $YBa_2Cu_3O_7$.

21.76 Compare the structure and properties of ceramic superconductors such as $YBa_2Cu_3O_7$ with those of fullerene-based superconductors such as Rb_3C_{60}.

21.77 Look at Figure 21.14, and identify the coordination numbers of the Cu, Y, and Ba atoms.

Ceramics and Composites

21.78 What is a ceramic, and what properties distinguish a ceramic from a metal?

21.79 Contrast the bonding in ceramics with the bonding in metals.

21.80 Why are ceramics more wear-resistant than metals?

21.81 Why are oxide ceramics more corrosion-resistant than metals?

21.82 Silicon nitride (Si_3N_4), a high-temperature ceramic useful for making engine components, is a covalent network solid in which each Si atom is bonded to four N atoms and each N atom is bonded to three Si atoms. Explain why silicon nitride is more brittle than a metal like copper.

21.83 Magnesia (MgO), used as an insulator for electrical heating devices, has a face-centered cubic structure like that of NaCl. Draw one unit cell of the structure of MgO, and explain why MgO is more brittle than magnesium metal.

21.84 What is ceramic processing?

21.85 Describe what happens when a ceramic powder is sintered.

21.86 Describe the differences between a sol and a gel.

21.87 What is the difference between a hydrolysis reaction and the reaction that occurs when a sol is converted to a gel?

21.88 Zirconia (ZrO_2), an unusually tough oxide ceramic, has been used to make very sharp table knives. Write a balanced equation for the hydrolysis of zirconium isopropoxide in the sol–gel method for making zirconia powders. The isopropoxide ligand is the anion of isopropyl alcohol, $HOCH(CH_3)_2$.

21.89 Zinc oxide is a semiconducting ceramic used to make *varistors* (variable resistors). Write a balanced equation for the hydrolysis of zinc ethoxide in the sol–gel method for making ZnO powders.

21.90 Describe the reactions that occur when an $Si(OH)_4$ sol becomes a gel. What is the formula of the ceramic obtained when the gel is dried and sintered?

21.91 Describe the reactions that occur when a $Y(OH)_3$ sol becomes a gel. What is the chemical formula of the ceramic obtained when the gel is dried and sintered?

21.92 Silicon nitride powder can be made by the reaction of silicon tetrachloride vapor with gaseous ammonia. The by-product is gaseous hydrogen chloride. Write a balanced equation for the reaction.

21.93 Boron, which is used in making composites, is deposited on a tungsten wire when the wire is heated electrically in the presence of boron trichloride vapor and gaseous hydrogen. Write a balanced equation for the reaction.

21.94 Explain why graphite/epoxy composites are good materials for making tennis rackets and golf clubs.

21.95 Explain why silicon carbide-reinforced alumina is stronger and tougher than pure alumina.

General Problems

21.96 Imagine a planet with an atmosphere that contains O_2 and SO_2 but no CO_2. Give the chemical composition of the minerals you would expect to find for the alkaline earth metals on such a planet.

21.97 Superconductors with values of T_c above 77 K are of special interest. What's so special about 77 K?

21.98 What properties of metals are better explained by band theory than by the electron-sea model?

21.99 Tungsten is hard and has a very high melting point (3422°C), and gold is soft and has a relatively low melting point (1064°C). Are these facts in better agreement with the electron-sea model or the MO model (band theory)? Explain.

21.100 Explain why the enthalpy of vaporization of vanadium (460 kJ/mol) is much larger than that of zinc (114 kJ/mol).

21.101 Classify each of the following materials as a metallic conductor, an *n*-type semiconductor, a *p*-type semiconductor, or an electrical insulator:

(a) MgO   (b) Si doped with Sb

(c) White tin   (d) Ge doped with Ga

(e) Stainless steel

21.102 Gallium arsenide, a material used in the manufacture of laser printers and compact disc players, has a band gap of 130 kJ/mol. Is GaAs a metallic conductor, a semiconductor, or an electrical insulator? With what group 4A element is GaAs isoelectronic? (Isoelectronic substances have the same number of electrons.) Draw an MO energy level diagram for GaAs, and show the population of the bands.

21.103 A 3.4×10^3 kg batch of cast iron contains 0.45% by mass of phosphorus as an impurity. In the conversion of cast iron to steel by the basic oxygen process, the phosphorus is oxidized to P_4O_{10}, which then reacts with CaO and is removed as slag.

(a) Write balanced equations for the oxidation of P_4 and for the formation of slag.

(b) How many kilograms of CaO are required to react with all the P_4O_{10}?

21.104 The $YBa_2Cu_3O_7$ superconductor can be synthesized by the sol–gel method from a stoichiometric mixture of metal ethoxides. How many grams of $Y(OCH_2CH_3)_3$ and how many grams of $Ba(OCH_2CH_3)_2$ are required for reaction with 75.4 g of $Cu(OCH_2CH_3)_2$ and an excess of water? Assuming a 100% yield, how many grams of $YBa_2Cu_3O_7$ are obtained?

21.105 A 0.3249 g sample of stainless steel was analyzed for iron by dissolving the sample in sulfuric acid and titrating the Fe^{2+} in the resulting solution with 0.018 54 M $K_2Cr_2O_7$. If 38.89 mL of the $K_2Cr_2O_7$ solution was required to reach the equivalence point, what is the mass percentage of iron in the steel?

21.106 Mullite, $3\,Al_2O_3 \cdot 2\,SiO_2$, a high-temperature ceramic being considered for use in engines, can be made by the sol–gel method.

(a) Write a balanced equation for formation of the sol by hydrolysis of a stoichiometric mixture of aluminum ethoxide and tetraethoxysilane, $Si(OCH_2CH_3)_4$.

(b) Describe the reactions that convert the sol to a gel.

(c) What additional steps are required to convert the gel to the ceramic product?

21.107 The glass in photochromic sunglasses is called a nanocomposite because it contains tiny silver halide particles with a diameter of less than 10 nm. This glass darkens and becomes less transparent outdoors in sunlight but returns to its original transparency indoors, where there is less light. How might these reversible changes occur?

21.108 Small molecules with C=C double bonds, called monomers, can join with one another to form long chain molecules called polymers. Thus, acrylonitrile, $H_2C=CHCN$, polymerizes under appropriate conditions to give polyacrylonitrile, a common starting material for production of the carbon fibers used in composites.

$$-CH_2-CH\underbrace{-CH_2-CH}-CH_2-CH-$$
$$\hspace{1.6cm}|\hspace{2.0cm}|\hspace{2.0cm}|$$
$$\hspace{1.5cm}CN\hspace{1.9cm}CN\hspace{1.9cm}CN$$

Repeating unit

Polyacrylonitrile

(a) Write electron-dot structures for acrylonitrile and polyacrylonitrile, and show how rearranging the electrons can lead to formation of the polymer.

(b) Use the bond dissociation energies in Table 7.1 to calculate ΔH per CH_2CHCN unit for conversion of acrylonitrile to polyacrylonitrile. Is the reaction endothermic or exothermic?

Multi-Concept Problems

21.109 At high temperatures, coke reduces silica impurities in iron ore to elemental silicon:

$$SiO_2(s) + 2\,C(s) \longrightarrow Si(s) + 2\,CO(g)$$

(a) Use the data in Appendix B to calculate values of $\Delta H°$, $\Delta S°$, and $\Delta G°$ for this reaction at 25°C.

(b) Is the reaction endothermic or exothermic?

(c) Account for the sign of the entropy change.

(d) Is the reaction spontaneous at 25°C and 1 atm pressure of CO?

(e) Assuming that $\Delta H°$ and $\Delta S°$ do not depend on temperature, estimate the temperature at which the reaction becomes spontaneous at 1 atm pressure of CO.

21.110 The Mond process for purifying nickel involves the reaction of impure nickel with carbon monoxide at about 150°C to give nickel tetracarbonyl. The nickel tetracarbonyl then decomposes to pure nickel at about 230°C:

$$Ni(s) + 4\,CO(g) \rightleftharpoons Ni(CO)_4(g)$$
$$\Delta H° = -160.8\ kJ;\ \Delta S° = -410\ J/K$$

The values of $\Delta H°$ and $\Delta S°$ apply at 25°C, but they are relatively independent of temperature and can be used at 150°C and 230°C.

(a) Calculate $\Delta G°$ and the equilibrium constant K_p at 150°C.

(b) Calculate $\Delta G°$ and the equilibrium constant K_p at 230°C.

(c) Why does the reaction have a large negative value for $\Delta S°$? Show that the change in $\Delta G°$ with increasing temperature is consistent with a negative value of $\Delta S°$.

(d) Show that the change in K_p with increasing temperature is consistent with a negative value of $\Delta H°$.

21.111 Figure 21.5 indicates that carbon reacts with carbon dioxide to form carbon monoxide at high temperatures:

$$C(s) + CO_2(g) \longrightarrow 2\,CO(g)$$

(a) Use the thermodynamic data in Appendix B to calculate the total pressure and the molar concentrations of CO and CO_2 at 500°C in a 50.00 L vessel that initially contains 100.0 g of CO_2 and an excess of solid carbon. Assume that the reaction mixture has reached equilibrium and that $\Delta H°$ and $\Delta S°$ are independent of temperature.

(b) Calculate the total pressure and the equilibrium concentrations at 1000°C.

(c) Use $\Delta H°$ and $\Delta S°$ to explain why the concentrations increase or decrease on raising the temperature.

21.112 Zinc chromite ($ZnCr_2O_4$), which is used in making magnetic tape for cassette recorders, can be prepared by thermal decomposition of $(NH_4)_2Zn(CrO_4)_2$. The other reaction products are N_2 and water vapor.

(a) Write a balanced equation for the reaction.

(b) How many grams of $ZnCr_2O_4$ can be prepared from 10.36 g of $(NH_4)_2Zn(CrO_4)_2$?

(c) How many liters of gaseous by-products are obtained at 292°C and 745 mm pressure?

(d) $ZnCr_2O_4$ has the cubic spinel structure consisting of a cubic closest-packed (face-centered cubic) array of O^{2-} ions with Zn^{2+} ions in the tetrahedral holes and Cr^{3+} ions in the octahedral holes. The tetrahedral and octahedral holes are like those in K_3C_{60} (see Figure 21.16). What fraction of each of the two types of holes is occupied?

(e) Draw a crystal field d-orbital energy level diagram for the Cr^{3+} ion in $ZnCr_2O_4$. Would you expect this compound to be colored? Explain. Does the Zn^{2+} ion contribute to the color? Explain.

21.113 The alkali metal fulleride superconductors M_3C_{60} have a cubic closest-packed (face-centered cubic) arrangement of nearly spherical C_{60}^{3-} anions with M^+ cations in the holes between the larger C_{60}^{3-} ions. The holes are of two types: octahedral holes, which are surrounded octahedrally by six C_{60}^{3-} ions, and tetrahedral holes, which are surrounded tetrahedrally by four C_{60}^{3-} ions.

(a) Sketch the three-dimensional structure of one unit cell.

(b) How many C_{60}^{3-} ions, octahedral holes, and tetrahedral holes are present per unit cell?

(c) Specify fractional coordinates for all the octahedral and tetrahedral holes.

(Fractional coordinates are fractions of the unit cell edge lengths. For example, a hole at the center of the cell has fractional coordinates $\frac{1}{2}, \frac{1}{2}, \frac{1}{2}$.)

(d) The radius of a C_{60}^{3-} ion is about 500 pm. Assuming that the C_{60}^{3-} ions are in contact along the face diagonals of the unit cell, calculate the radii of the octahedral and tetrahedral holes.

(e) The ionic radii of Na^+, K^+, and Rb^+ are 97, 133, and 147 pm, respectively. Which of these ions will fit into the octahedral and tetrahedral holes? Which ions will fit only if the framework of C_{60}^{3-} ions expands?

---

 eMedia Problems

21.114 The **Electroplating** movie (*eChapter 18.12*) shows chromium metal being produced by the electrolysis of an aqueous solution of chromic acid. Using standard reduction potentials in Appendix D, explain why sodium metal isn't produced by electrolysis of an aqueous sodium chloride solution.

21.115 Write the equation for the reaction between iron(III) oxide and aluminum metal shown in the **Thermite** movie (*eChapter 21.3*). What is the purpose of adding potassium chlorate, sugar, and concentrated sulfuric acid?

21.116 Use the **Metallic Bonding** activity (*eChapter 21.4*) to determine the melting points of the transition elements in row 4, 5, or 6 of the periodic table.

(a) What is the relationship between the melting point of a transition metal and the relative numbers of electrons in bonding and antibonding molecular orbitals?

(b) Explain why the electron-sea model of bonding in metals does not accurately predict the trends in transition metal melting points.

21.117 The **1-2-3 Superconductor** $YBa_2Cu_3O_7$ (*eChapter 21.6*) has one of the highest known T_c values, at 90 K. By turning the structure, identify the CuO_5 group and the CuO_4 group. What are the geometries of these groups? What properties of this and other high temperature superconductors keep them from being widely used?

21.118 What advantages do ceramics such as **Silicon Carbide** (*eChapter 21.7*) have relative to metals? What are the disadvantages of using ceramics? Turn the 3D model and see if you can find a special orientation where planes of atoms are separated by largely empty space, devoid of bonds. What does the result of this investigation say about the physical properties of ceramics?

Nuclear Chemistry

The Z machine at Sandia National Laboratories has reached temperatures of 1.8 million degrees, close to the 2 million degrees needed for nuclear fusion.

W hen methane (CH_4) burns in oxygen, the C, H, and O atoms recombine to yield CO_2 and H_2O, but they still remain C, H, and O. When aqueous NaCl reacts with aqueous $AgNO_3$, solid AgCl precipitates from solution, but the Ag^+ and Cl^- ions themselves remain the same. In fact, in all the reactions we've discussed thus far, only the *bonds* between atoms have changed; the chemical identities of the atoms themselves have remained unchanged. But anyone who reads the paper or watches television knows that atoms *can* change, often resulting in the conversion of one element into another. Atomic weapons, nuclear energy, and radioactive radon gas in our homes are all topics of societal importance, and all involve **nuclear chemistry**—the study of the properties and reactions of atomic nuclei.

22.1   Nuclear Reactions and Their Characteristics

Recall from Section 2.5 that an atom is characterized by its *atomic number*, Z, and its *mass number*, A. The atomic number, written as a subscript to the left of the element symbol, gives the number of protons in the nucleus. The mass number, written as a superscript to the left of the element symbol, gives the total number of **nucleons**, a general term for both protons (p) and neutrons (n). The most common isotope of carbon, for example, has 12 nucleons: 6 protons and 6 neutrons.

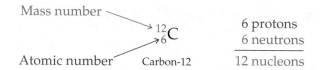

Mass number	6 protons
$^{12}_{6}\text{C}$	6 neutrons
Atomic number Carbon-12	12 nucleons

Atoms with identical atomic numbers but different mass numbers are called *isotopes*, and the nucleus of a specific isotope is called a **nuclide**. There are 13 known isotopes of carbon, two of which occur commonly (^{12}C and ^{13}C) and one of which (^{14}C) is produced in small amounts in the upper atmosphere by the action of neutrons from cosmic rays on ^{14}N. The remaining 10 carbon isotopes have been produced artificially. Only the two commonly occurring ones are indefinitely stable; the other 11 undergo spontaneous **nuclear reactions**, which change their nuclei. Carbon-14, for example, slowly decomposes to give nitrogen-14 plus an electron, a process we can write as

$$^{14}_{6}\text{C} \longrightarrow \, ^{14}_{7}\text{N} + \, ^{0}_{-1}\text{e}$$

The electron is often written as $^{0}_{-1}\text{e}$, where the superscript 0 indicates that the mass of an electron is essentially zero when compared with that of a proton or neutron, and the subscript -1 indicates that the charge is -1. (The subscript in this instance is not a true atomic number.)

Nuclear reactions, such as the spontaneous decay of ^{14}C, are distinguished from chemical reactions in several ways:

- A *nuclear* reaction involves a change in an atom's nucleus, usually producing a different element. A *chemical* reaction, by contrast, involves only a change in distribution of the outer-shell electrons around the atom and never changes the nucleus itself or produces a different element.
- Different isotopes of an element have essentially the same behavior in chemical reactions but often have completely different behavior in nuclear reactions.
- The rate of a nuclear reaction is unaffected by a change in temperature or pressure or by the addition of a catalyst.
- The nuclear reaction of an atom is essentially the same, regardless of whether the atom is in a chemical compound or in uncombined elemental form.
- The energy change accompanying a nuclear reaction is far greater than that accompanying a chemical reaction. The nuclear transformation of 1.0 g of uranium-235 releases 8.2×10^{7} kJ, for example, whereas the chemical combustion of 1.0 g of methane releases only 56 kJ.

More than 250 ships on the world's oceans are powered by nuclear reactors, including this Russian icebreaker.

22.2 Nuclear Reactions and Radioactivity

Scientists have known since 1896 that many nuclides are **radioactive**—that is, they spontaneously emit radiation. Early studies of radioactive nuclei, or **radionuclides**, by the British physicist Ernest Rutherford in 1897 showed that there are three common types of radiation with markedly different properties: *alpha* (α), *beta* (β), and *gamma* (γ) *radiation*, named after the first three letters of the Greek alphabet.

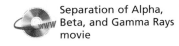
Separation of Alpha, Beta, and Gamma Rays movie

Alpha (α) Radiation

Using the simple experiment shown in Figure 22.1, Rutherford found that **α radiation** consists of a stream of particles that are repelled by a positively charged electrode, are attracted by a negatively charged electrode, and have a mass-to-charge ratio identifying them as helium nuclei, $_2^4\text{He}^{2+}$. Alpha particles thus consist of two protons and two neutrons.

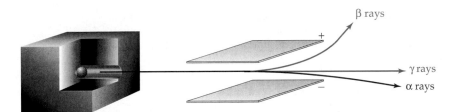

FIGURE 22.1   The effect of an electric field on α, β, and γ radiation. The radioactive source in the shielded box emits radiation, which passes between two electrodes. Alpha radiation is deflected toward the negative electrode, β radiation is strongly deflected toward the positive electrode, and γ radiation is undeflected.

Because the emission of an alpha particle from a nucleus results in a loss of two protons and two neutrons, it reduces the mass number of the nucleus by 4 and the atomic number by 2. Alpha emission is particularly common for heavy radioactive isotopes, or **radioisotopes**: Uranium-238, for example, spontaneously emits an alpha particle and forms thorium-234.

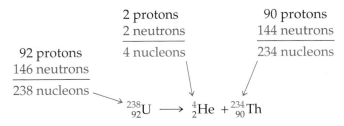

Note how the **nuclear equation** for the radioactive decay of uranium-238 into lighter elements is written. The equation is not balanced in the usual chemical sense because the kinds of nuclei are not the same on both sides of the arrow. Instead, we say that a nuclear equation is balanced when the sums of the nucleons are the same on both sides of the equation and when the sums of the charges on the nuclei and any elementary particles (protons, neutrons, and electrons) are the same on both sides. In the decay of $_{92}^{238}\text{U}$ to give $_2^4\text{He}$ and $_{90}^{234}\text{Th}$, for example, there are 238 nucleons and 92 nuclear charges on both sides of the nuclear equation.

Note also that we are concerned only with charges on elementary particles and on nuclei when we write nuclear equations, not with ionic charges on atoms. The alpha particle is thus written as $_2^4\text{He}$ rather than $_2^4\text{He}^{2+}$, and the thorium resulting from radioactive decay of $_{92}^{238}\text{U}$ is written as $_{90}^{234}\text{Th}$ rather than $_{90}^{234}\text{Th}^{2-}$. We ignore the ionic charges both because they are irrelevant to nuclear disintegration and because they soon disappear.

The $_2^4\text{He}^{2+}$ ion immediately picks up two electrons from whatever it strikes, yielding a neutral helium atom, and the $_{90}^{234}\text{Th}^{2-}$ ion immediately gives two electrons to whatever it is in electrical contact with, yielding a neutral thorium atom.

Beta (β) Radiation

Further work by Rutherford in the late 1800s showed that **β radiation** consists of a stream of particles that are attracted to a positive electrode (Figure 22.1), are repelled by a negative electrode, and have a mass-to-charge ratio identifying them as electrons, $_{-1}^0\text{e}$ or β^-. Beta emission occurs when a neutron in the nucleus spontaneously decays into a proton plus an electron, which is then ejected. The product nucleus has the same mass number as the starting nucleus because a neutron has turned into a proton, but it has a higher atomic number because of the newly created proton. The reaction of ^{131}I to give ^{131}Xe is an example:

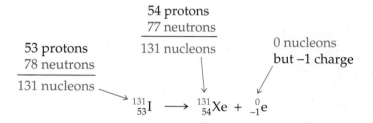

Writing the emitted β particle as $_{-1}^0\text{e}$ in the nuclear equation makes clear the charge balance of the nuclear reaction: The subscript in the $_{53}^{131}\text{I}$ nucleus on the left (53) is balanced by the sum of the two subscripts on the right ($54 - 1 = 53$).

Gamma (γ) Radiation

Gamma radiation is unaffected by electric fields (Figure 22.1), has no mass, and is simply electromagnetic radiation of very high energy and thus of very short wavelength ($\lambda = 10^{-11}$–10^{-14} m). In modern terms, we would say that **γ rays** are a stream of high-energy photons. Gamma radiation almost always accompanies α and β emission as a mechanism for the release of energy, but it is often not shown when writing nuclear equations because it does not change either the mass number or the atomic number of the product nucleus.

Positron Emission and Electron Capture

In addition to α, β, and γ radiation, two other common types of radioactive decay processes also occur: *positron emission* and *electron capture*. **Positron emission** involves the conversion of a proton in the nucleus into a neutron plus an ejected *positron*, $_1^0\text{e}$ or β^+, a particle that can be thought of as a "positive electron." A positron has the same mass as an electron but an opposite charge. The result of positron emission is a decrease in the atomic number of the product nucleus but no change in the mass number. Potassium-40, for example, undergoes positron emission to yield argon-40, a nuclear reaction important in geology for dating rocks. Note once again that the sum of the two subscripts on the right of the nuclear equation ($18 + 1 = 19$) is equal to the subscript in the $_{19}^{40}\text{K}$ nucleus on the left.

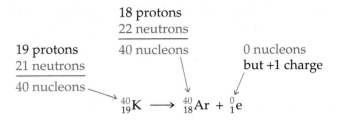

$$^{40}_{19}\text{K} \longrightarrow {}^{40}_{18}\text{Ar} + {}^{0}_{1}\text{e}$$

Electron capture is a process in which the nucleus captures an inner-shell electron, thereby converting a proton into a neutron. The mass number of the product nucleus is unchanged, but the atomic number decreases by 1, just as in positron emission. The conversion of mercury-197 into gold-197 is an example:

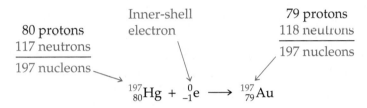

$$^{197}_{80}\text{Hg} + {}^{0}_{-1}\text{e} \longrightarrow {}^{197}_{79}\text{Au}$$

Characteristics of the different kinds of radioactive decay processes are summarized in Table 22.1.

TABLE 22.1   A Summary of Radioactive Decay Processes

Process	Symbol	Change in Atomic Number	Change in Mass Number	Change in Neutron Number
α emission	${}^{4}_{2}\text{He}$ or α	−2	−4	−2
β emission	${}^{0}_{-1}\text{e}$ or β⁻	+1	0	−1
γ emission	${}^{0}_{0}\gamma$ or γ	0	0	0
Positron emission	${}^{0}_{1}\text{e}$ or β⁺	−1	0	+1
Electron capture	E. C.	−1	0	+1

EXAMPLE 22.1   Write balanced nuclear equations for each of the following processes:
(a) Alpha emission from curium-242: ${}^{242}_{96}\text{Cm} \rightarrow {}^{4}_{2}\text{He} + ?$
(b) Beta emission from magnesium-28: ${}^{28}_{12}\text{Mg} \rightarrow {}^{0}_{-1}\text{e} + ?$
(c) Positron emission from xenon-118: ${}^{118}_{54}\text{Xe} \rightarrow {}^{0}_{1}\text{e} + ?$

SOLUTION   The key to writing nuclear equations is to make sure that the number of nucleons is the same on both sides of the equation and that the number of elementary and nuclear charges is the same.
(a) In alpha emission, the mass number decreases by 4, and the atomic number decreases by 2, giving plutonium-238: ${}^{242}_{96}\text{Cm} \rightarrow {}^{4}_{2}\text{He} + {}^{238}_{94}\text{Pu}$
(b) In beta emission, the mass number is unchanged, and the atomic number increases by 1, giving aluminum-28: ${}^{28}_{12}\text{Mg} \rightarrow {}^{0}_{-1}\text{e} + {}^{28}_{13}\text{Al}$
(c) In positron emission, the mass number is unchanged, and the atomic number decreases by 1, giving iodine-118: ${}^{118}_{54}\text{Xe} \rightarrow {}^{0}_{1}\text{e} + {}^{118}_{53}\text{I}$

▶ **PROBLEM 22.1**　Write balanced nuclear equations for each of the following processes:

(a) Beta emission from ruthenium-106: $^{106}_{44}\text{Ru} \rightarrow {}^{0}_{-1}\text{e} + ?$

(b) Alpha emission from bismuth-189: $^{189}_{83}\text{Bi} \rightarrow {}^{4}_{2}\text{He} + ?$

(c) Electron capture by polonium-204: $^{204}_{84}\text{Po} + {}^{0}_{-1}\text{e} \rightarrow ?$

▶ **PROBLEM 22.2**　What particle is produced by decay of thorium-214 to radium-210?

$$^{214}_{90}\text{Th} \longrightarrow {}^{210}_{88}\text{Ra} + ?$$

22.3　Radioactive Decay Rates

First-Order Process
movie

The rates of different radioactive decay processes vary enormously. Some radionuclides, such as uranium-238, decay at a barely perceptible rate over billions of years; others, such as carbon-17, decay within milliseconds.

Radioactive decay is kinetically a first-order process (Section 12.3), whose rate is proportional to the number of radioactive nuclei N in a sample times the first-order rate constant k, called the **decay constant**:

$$\text{Decay rate} = k \times N$$

As we saw in Section 12.4, a first-order rate law can be converted into an integrated rate law of the form

$$\ln\left(\frac{N}{N_0}\right) = -kt$$

where N_0 is the number of radioactive nuclei originally present in a sample and N is the number remaining at time t. Note that the decay constant k has units of time^{-1}, so the quantity kt is unitless.

Like all first-order processes, radioactive decay is characterized by a half-life, $t_{1/2}$, the time required for the number of radioactive nuclei in a sample to drop to half its initial value (Section 12.5). For example, the half-life of iodine-131, a radioisotope used in thyroid testing, is 8.02 days. If today you have 1.000 g of $^{131}_{53}\text{I}$, then 8.02 days from now you will have only 0.500 g of $^{131}_{53}\text{I}$ remaining because one-half of the sample will have decayed (by beta emission), yielding 0.500 g of $^{131}_{54}\text{Xe}$, After 8.02 more days (16.04 total), only 0.250 g of $^{131}_{53}\text{I}$ will remain; after a further 8.02 days (24.06 total), only 0.125 g will remain; and so on. Each passage of a half-life causes the decay of one-half of whatever sample remains, as shown graphically by the curve in Figure 22.2. The half-life is the same no matter what the size of the sample, the temperature, or any other external condition.

$$^{131}_{53}\text{I} \longrightarrow {}^{131}_{54}\text{Xe} + {}^{0}_{-1}\text{e} \qquad t_{1/2} = 8.02 \text{ days}$$

Mathematically, the value of $t_{1/2}$ can be calculated from the integrated rate law by setting $N = 1/2\,N_0$ at time $t_{1/2}$:

$$\ln\left(\frac{\frac{1}{2}N_0}{N_0}\right) = -kt_{1/2} \quad \text{and} \quad \ln\frac{1}{2} = -\ln 2 = -kt_{1/2}$$

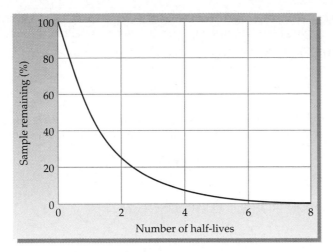

FIGURE 22.2 The decay of a radionuclide over time. No matter what the value of the half-life, 50% of the sample remains after one half-life, 25% remains after two half-lives, 12.5% remains after three half-lives, and so on.

$$\text{so} \qquad t_{1/2} = \frac{\ln 2}{k} \quad \text{and} \quad k = \frac{\ln 2}{t_{1/2}}$$

These equations say that if we know the value of either the decay constant k or the half-life $t_{1/2}$, we can calculate the value of the other. Furthermore, if we know the value of $t_{1/2}$, we can calculate the ratio of remaining and initial amounts of radioactive sample N/N_0 at any time t by substituting the expression for k into the integrated rate law:

$$\text{Since} \qquad \ln\left(\frac{N}{N_0}\right) = -kt \quad \text{and} \quad k = \frac{\ln 2}{t_{1/2}}$$

$$\text{then} \qquad \ln\left(\frac{N}{N_0}\right) = (-\ln 2)\left(\frac{t}{t_{1/2}}\right)$$

Example 22.2 shows how to calculate a half-life from a decay constant, and Example 22.4 shows how to determine the percentage of radioactive sample remaining at time t.

The half-lives of some useful radioisotopes are given in Table 22.2. As you might expect, radioisotopes used internally in medical applications have fairly short half-lives so that they decay rapidly and don't cause long-term health hazards.

TABLE 22.2   Half-lives of Some Useful Radioisotopes

Radioisotope	Symbol	Radiation	Half-life	Use
Tritium	$^{3}_{1}\text{H}$	β^-	12.33 years	Biochemical tracer
Carbon-14	$^{14}_{6}\text{C}$	β^-	5730 years	Archaeological dating
Phosphorus-32	$^{32}_{15}\text{P}$	β^-	14.26 days	Leukemia therapy
Potassium-40	$^{40}_{19}\text{K}$	β^-	1.28×10^9 years	Geological dating
Cobalt-60	$^{60}_{27}\text{Co}$	β^-, γ	5.27 years	Cancer therapy
Technetium-99m*	$^{99m}_{43}\text{Tc}$	γ	6.01 hours	Brain scans
Iodine-123	$^{123}_{53}\text{I}$	γ	13.27 hours	Thyroid therapy
Uranium-235	$^{235}_{92}\text{U}$	α, γ	7.04×10^8 years	Nuclear reactors

*The m in technetium-99m stands for *metastable*, meaning that it undergoes gamma emission but does not change its mass number or atomic number.

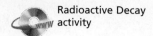
Radioactive Decay
activity

EXAMPLE 22.2   The decay constant for sodium-24, a radioisotope used medically in blood studies, is 4.63×10^{-2} h^{-1}. What is the half-life of ^{24}Na?

SOLUTION   Half-life can be calculated from the decay constant by using the equation

$$t_{1/2} = \frac{\ln 2}{k}$$

Substituting the values $k = 4.63 \times 10^{-2}$ h^{-1} and $\ln 2 = 0.693$ into the equation gives

$$t_{1/2} = \frac{0.693}{4.63 \times 10^{-2} \text{ h}^{-1}} = 15.0 \text{ h}$$

The half-life of sodium-24 is 15.0 h.

EXAMPLE 22.3   The half-life of radon-222, a radioactive gas of concern as a health hazard in some homes, is 3.823 days. What is the decay constant of ^{222}Rn?

SOLUTION   A decay constant can be calculated from the half-life by using the equation

$$k = \frac{\ln 2}{t_{1/2}}$$

Substituting the values $t_{1/2} = 3.823$ days and $\ln 2 = 0.693$ into the equation gives

$$k = \frac{0.693}{3.823 \text{ days}} = 0.181 \text{ day}^{-1}$$

EXAMPLE 22.4   Phosphorus-32, a radioisotope used in leukemia therapy, has a half-life of 14.26 days. What percent of a sample remains after 35.0 days?

SOLUTION   The ratio of remaining (N) and initial (N_0) amounts of a radioactive sample at time t is given by the equation

$$\ln\left(\frac{N}{N_0}\right) = (-\ln 2)\left(\frac{t}{t_{1/2}}\right)$$

Substituting values for t and for $t_{1/2}$ into the equation gives

$$\ln\left(\frac{N}{N_0}\right) = -0.693\left(\frac{35.0 \text{ days}}{14.26 \text{ days}}\right) = -1.70$$

Taking the natural antilog of -1.70 then gives the ratio N/N_0:

$$\frac{N}{N_0} = \text{antiln } (-1.70) = 0.183$$

Since the initial amount of ^{32}P was 100%, we can set $N_0 = 100\%$ and solve for N:

$$\frac{N}{100\%} = 0.183 \quad \text{so} \quad N = 0.183 \times 100\% = 18.3\%$$

After 35.0 days, 18.3% of a ^{32}P sample remains, and $100\% - 18.3\% = 81.7\%$ has decayed.

EXAMPLE 22.5   A sample of ^{41}Ar, a radioisotope used to measure the flow of gases from smokestacks, decays initially at a rate of 34,500 disintegrations/min, but the decay rate falls to 21,500 disintegrations/min after 75.0 min. What is the half-life of ^{41}Ar?

The flow of gases from a smokestack can be measured by releasing ^{41}Ar and monitoring its passage.

SOLUTION   The half-life of a radioactive decay process is given by finding $t_{1/2}$ in the equation

$$\ln\left(\frac{N}{N_0}\right) = (-\ln 2)\left(\frac{t}{t_{1/2}}\right)$$

In the present instance, though, we are given decay rates at two different times rather than values of N and N_0. Nevertheless, for a first-order process like radioactive decay, in which rate $= kN$, the ratio of the decay rate at any time t to the decay rate at time $t = 0$ is the same as the ratio of N to N_0:

$$\frac{\text{Decay rate at time } t}{\text{Decay rate at time } t = 0} = \frac{kN}{kN_0} = \frac{N}{N_0}$$

Substituting the proper values into the equation gives

$$\ln\left(\frac{21,500}{34,500}\right) = -0.693\left(\frac{75.0 \text{ min}}{t_{1/2}}\right) \quad \text{or} \quad -0.473 = \frac{-52.0 \text{ min}}{t_{1/2}}$$

$$\text{so} \quad t_{1/2} = \frac{-52.0 \text{ min}}{-0.473} = 110 \text{ min}$$

The half-life of ^{41}Ar is 110 min.

▶ **PROBLEM 22.3**   The decay constant for mercury-197, a radioisotope used medically in kidney scans, is $1.08 \times 10^{-2} \text{ h}^{-1}$. What is the half-life of mercury-197?

▶ **PROBLEM 22.4**   The half-life of carbon-14 is 5730 years. What is its decay constant?

▶**PROBLEM 22.5** What percentage of $^{14}_{6}C$ ($t_{1/2}$ = 5730 years) remains in a sample estimated to be 16,230 years old?

▶**PROBLEM 22.6** What is the half-life of iron-59, a radioisotope used medically in the diagnosis of anemia, if a sample with an initial decay rate of 16,800 disintegrations/min decays at a rate of 10,860 disintegrations/min after 28.0 days?

✦ **KEY CONCEPT PROBLEM 22.7** What is the half-life of the radionuclide that shows the following decay curve?

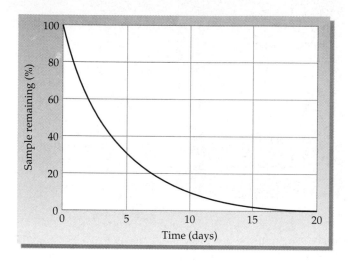

22.4 Nuclear Stability

Why do some nuclei undergo radioactive decay while others do not? Why, for instance, does a carbon-*14* nucleus, with six protons and *eight* neutrons, spontaneously emit a β particle, whereas a carbon-*13* nucleus, with six protons and *seven* neutrons, is stable indefinitely? Before answering these questions, it's important to define what we mean by "stable." In the context of nuclear chemistry, we'll use the word *stable* to refer to isotopes whose half-lives can be measured, even if that half-life is only a fraction of a second. We'll call those isotopes that decay too rapidly for their half-lives to be measured *unstable*, and those isotopes that do not undergo radioactive decay *nonradioactive*, or *stable indefinitely*.

The answer to the question about why some nuclei are radioactive while others are not has to do with the neutron/proton ratio in the nucleus and the forces holding the nucleus together. To see the effect of the neutron/proton ratio on nuclear stability, look at the grid pictured in Figure 22.3. Along the side of the grid are divisions representing the number of neutrons in nuclei, a number arbitrarily cut off at 200. Along the bottom of the grid are divisions representing the number of protons in nuclei—the first 92 divisions represent the naturally occurring elements, the next 17 represent the artificially produced **transuranium elements**, and the divisions beyond 109 represent elements about which little or nothing is known. (Actually, only 90 of the first 92 elements occur naturally. Technetium and promethium do not occur naturally because all their isotopes are radioactive and have very short half-lives. Francium and astatine occur on the earth only in very tiny amounts.)

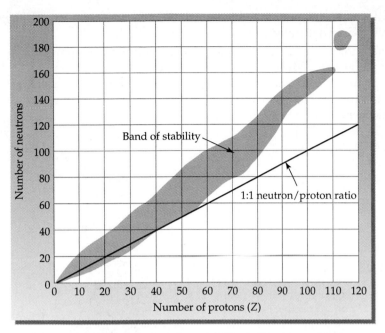

FIGURE 22.3   The band of nuclear stability indicating various neutron/proton combinations that give rise to observable nuclei with measurable half-lives. Combinations outside the band are not stable. The "island of stability" near 114 protons and 184 neutrons corresponds to a group of superheavy nuclei that are predicted to be stable. The first members of this group were reported in 1999.

When the more than 3600 known nuclides are plotted on the neutron/proton grid in Figure 22.3, they fall in a curved band sometimes called the "band of nuclear stability." Even within the band, all but 264 of the nuclides disintegrate spontaneously, though their rates of decay vary enormously. On either side of the band is a "sea of instability" representing the large number of unstable neutron/proton combinations that have never been seen. Particularly interesting is the "island of stability" predicted to exist for a few superheavy nuclides near 114 protons and 184 neutrons. The first members of this group—^{287}114, ^{288}114, ^{289}114, ^{289}116, and ^{293}118—were prepared in 1999 and do indeed seem to be unusually stable. Isotope ^{289}114, for example, has a half-life of 30.4 seconds.

Several generalizations can be made about the data in Figure 22.3:

- Every element in the periodic table has at least one radioactive isotope.
- Hydrogen is the only element whose most abundant stable isotope ($^1_1$H) contains more protons (1) than neutrons (0). The most abundant stable isotopes of other elements lighter than calcium ($Z = 20$) usually have either the same number of protons and neutrons ($^4_2$He, $^{12}_6$C, $^{16}_8$O, and $^{28}_{14}$Si, for example), or have only one more neutron than protons (($^{11}_5$B, $^{19}_9$F, and $^{23}_{11}$Na, for example).
- The ratio of neutrons to protons gradually increases for elements heavier than calcium, giving a curved appearance to the band of stability. The most abundant stable isotope of bismuth, for example, has 126 neutrons and 83 protons ($^{209}_{83}$Bi).

954   Chapter 22   Nuclear Chemistry

- All isotopes heavier than bismuth-209 are radioactive, even though they may occur naturally.
- Of the 264 nonradioactive isotopes, 207 have an even number of neutrons in their nuclei. Most nonradioactive isotopes (156) have even numbers of both protons and neutrons, 51 have an even number of neutrons but an odd number of protons, and only 4 have an odd number of both protons and neutrons (Figure 22.4).

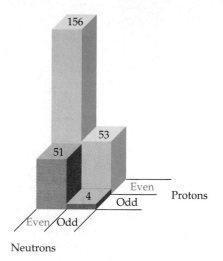

FIGURE 22.4   Numbers of nonradioactive isotopes with various even/odd combinations of neutrons and protons. The majority of nonradioactive isotopes have both an even number of protons and an even number of neutrons. Only four nonradioactive isotopes have both an odd number of protons and an odd number of neutrons.

Observations such as those in Figure 22.3 suggest that neutrons function as a kind of nuclear "glue" that holds nuclei together by overcoming proton–proton repulsions. The more protons there are in the nucleus, the more glue is needed. Furthermore, there appear to be certain "magic numbers" of protons or neutrons—2, 8, 20, 28, 50, 82, 126—that give rise to particularly stable nuclei. For instance, there are 10 naturally occurring isotopes of tin, which has a magic number of protons ($Z = 50$), but there are only 2 naturally occurring isotopes of its neighbors on either side, indium ($Z = 49$) and antimony ($Z = 51$). Lead-208 is especially stable because it has a *double* magic number of nucleons: 126 neutrons and 82 protons. We'll see in the next section how the relative stability of different nuclei can be measured quantitatively.

The correlation of nuclear stability with special numbers of nucleons is reminiscent of the correlation of chemical stability with special numbers of electrons—the octet rule discussed in Section 6.12. In fact, a shell model of nuclear structure has been proposed, analogous to the shell model of electronic structure. The magic numbers of nucleons correspond to filled nuclear-shell configurations, although the details are relatively complex.

A close-up look at a segment of the band of nuclear stability reveals some more interesting trends (Figure 22.5). One trend is that elements with an even atomic number have a larger number of nonradioactive isotopes than do elements with an odd atomic number. Tungsten ($Z = 74$) has 5 nonradioactive isotopes and osmium ($Z = 76$) has 7, for example, while their neighbor rhenium ($Z = 75$) has only 1.

Another trend is that radioactive nuclei with higher neutron/proton ratios (top side of the band) tend to emit beta particles, while nuclei with lower

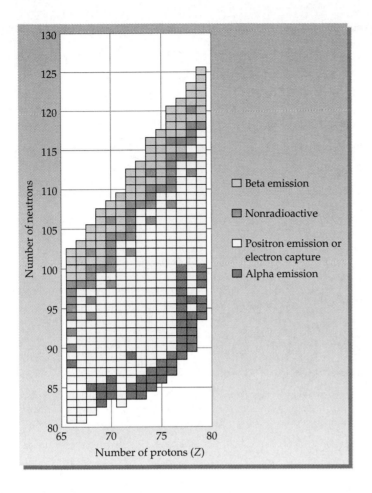

FIGURE 22.5   A close-up look at the band of nuclear stability in the region from $Z = 66$ (dysprosium) through $Z = 79$ (gold) shows the types of radioactive processes undergone by various nuclides. Nuclides with lower neutron/proton ratios tend to undergo positron emission, electron capture, or alpha emission, whereas nuclides with higher neutron/proton ratios tend to undergo beta emission.

neutron/proton ratios (bottom side of the band) tend to undergo nuclear disintegration by positron emission, electron capture, or alpha emission. This makes sense if you think about it: The nuclei on the top side of the band are neutron-rich and therefore undergo a process that *decreases* the neutron/proton ratio. The nuclei on the bottom side of the band, by contrast, are neutron-poor and therefore undergo processes that *increase* the neutron/proton ratio. (Take a minute to convince yourself that alpha emission does, in fact, increase the neutron/proton ratio for heavy nuclei in which n > p.)

This process decreases the neutron/proton ratio:

$\begin{cases} \text{Beta emission:} \qquad \text{Neutron} \longrightarrow \text{Proton} + \beta^- \end{cases}$

These processes increase the neutron/proton ratio:

$\begin{cases} \text{Positron emission:} \quad \text{Proton} \longrightarrow \text{Neutron} + \beta^+ \\ \text{Electron capture:} \quad \text{Proton} + \text{Electron} \longrightarrow \text{Neutron} \\ \text{Alpha emission:} \quad {}^{A}_{Z}X \longrightarrow {}^{A-4}_{Z-2}Y + {}^{4}_{2}He \end{cases}$

One further point about nuclear decay is that some nuclides, particularly those of the heavy elements above bismuth, can't reach a nonradioactive decay product by a single emission. The product nucleus resulting from the first decay is itself radioactive and therefore undergoes a further disintegration. In fact, such nuclides must often undergo a whole series of nuclear disintegrations—a **decay series**—before they ultimately reach a nonradioactive product. Uranium-238, for example, undergoes a series of 14 sequential nuclear reactions, ultimately ending at lead-206 (Figure 22.6).

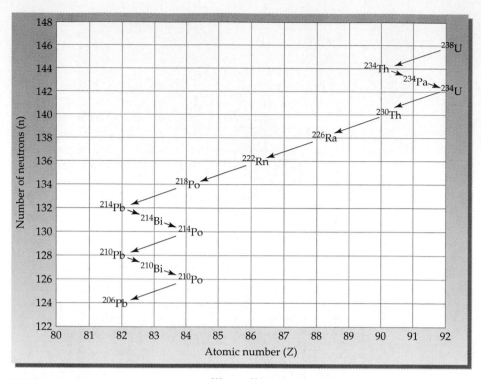

FIGURE 22.6   The decay series from $^{238}_{92}$U to $^{206}_{82}$Pb. Each nuclide except for the last is radioactive and undergoes nuclear decay. The left-pointing, longer arrows (red) represent alpha emissions, and the right-pointing, shorter arrows (blue) represent beta emissions.

▶ **PROBLEM 22.8**   Of the two isotopes ^{173}Au and ^{199}Au, one decays by beta emission and one decays by alpha emission. Which is which?

✦— **KEY CONCEPT PROBLEM 22.9**   The following series has two kinds of processes: one represented by the shorter arrows pointing right and the other represented by the longer arrows pointing left. Tell what kind of nuclear decay process each arrow corresponds to, and identify each nuclide A–E in the series.

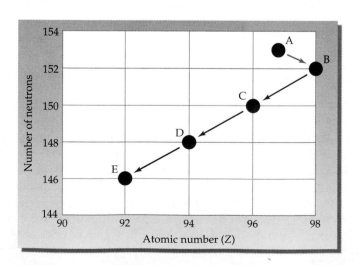

22.5 Energy Changes During Nuclear Reactions

We noted in the previous section that neutrons appear to act as a kind of nuclear glue by overcoming the proton–proton repulsions that would otherwise cause the nucleus to fly apart. In principle, it should be possible to measure the strength of the forces holding a nucleus together by measuring the amount of heat released on forming the nucleus from isolated protons and neutrons. For example, the energy change associated with combining two neutrons and two protons to yield a helium-4 nucleus should be a direct measure of nuclear stability just as the energy change associated with formation of a chemical bond is a measure of the bond's stability.

$$2\,{}^{1}_{1}\text{H} + 2\,{}^{1}_{0}\text{n} \longrightarrow {}^{4}_{2}\text{He} \qquad \Delta E = ?$$

Unfortunately, there is a problem with actually carrying out the measurement: Temperatures rivaling those in the interior of the sun (10^7 K) are necessary for the reaction to occur! That is, the activation energy (Section 12.9) required to force the elementary particles close enough for reaction is extremely high. Nevertheless, the energy change for the process can be calculated by using the Einstein equation $\Delta E = \Delta mc^2$, which relates the energy change (ΔE) of a nuclear process to a corresponding mass change Δm (Section 5.5).

Take a helium-4 nucleus, for example. We know from Table 2.1 that the mass of two neutrons and two protons is 4.031 88 amu:

$$\begin{aligned}
\text{Mass of 2 neutrons} &= (2)(1.008\ 66\ \text{amu}) = 2.017\ 32\ \text{amu}\\
\text{Mass of 2 protons} &= (2)(1.007\ 28\ \text{amu})\ \ = 2.014\ 56\ \text{amu}\\
\hline
\text{Total mass of 2 n} + \text{2 p} &\qquad\qquad\qquad\quad = 4.031\ 88\ \text{amu}
\end{aligned}$$

Furthermore, we can subtract the mass of two electrons from the experimentally measured mass of a helium-4 *atom* to find that the mass of a helium-4 *nucleus* is 4.001 50 amu:

$$\begin{aligned}
\text{Mass of helium-4 atom} &\qquad\qquad\qquad\quad = \ \ 4.002\ 60\ \text{amu}\\
-\ \text{Mass of 2 electrons} &= -(2)(5.486 \times 10^{-4}\ \text{amu}) = -0.001\ 10\ \text{amu}\\
\hline
\text{Mass of helium-4 nucleus} &\qquad\qquad\qquad\qquad\ = \ \ 4.001\ 50\ \text{amu}
\end{aligned}$$

Subtracting the mass of the helium nucleus from the combined mass of its constituent neutrons and protons shows a difference of 0.030 38 amu. That is, 0.030 38 amu (or 0.030 38 g/mol) is *lost* when two protons and two neutrons combine to form a helium-4 nucleus:

$$\begin{aligned}
\text{Mass of 2 n} + \text{2 p} &= \ \ 4.031\ 88\ \text{amu}\\
-\ \text{Mass of } {}^{4}\text{He nucleus} &= -4.001\ 50\ \text{amu}\\
\hline
\text{Mass difference} &= \ \ 0.030\ 38\ \text{amu} \quad (\text{or } 0.030\ 38\ \text{g/mol})
\end{aligned}$$

The loss in mass that occurs when protons and neutrons combine to form a nucleus is called the **mass defect** of the nucleus. This lost mass is converted into energy that is released during the nuclear reaction and is thus a direct measure of the **binding energy** holding the nucleons together. The larger the binding energy, the more stable the nucleus. Using the Einstein equation, we can calculate this binding energy for a helium-4 nucleus:

$$\Delta E = \Delta mc^2 \qquad \text{where } c = 3.00 \times 10^8 \text{ m/s}$$
$$= (3.038 \times 10^{-5} \text{ kg/mol})(3.00 \times 10^8 \text{ m/s})^2$$
$$= 2.73 \times 10^{12} \text{ kg} \cdot \text{m}^2/(\text{mol} \cdot \text{s}^2)$$
$$= 2.73 \times 10^{12} \text{ J/mol} = 2.73 \times 10^9 \text{ kJ/mol}$$

The binding energy for helium-4 nuclei is 2.73×10^9 kJ/mol. In other words, 2.73×10^9 kJ/mol is released when helium-4 nuclei are formed, and 2.73×10^9 kJ/mol must be supplied to cause disintegration of helium-4 nuclei into isolated protons and neutrons. This enormous amount of energy is more than *10 million times* the energy change associated with a typical chemical process!

To make comparisons among different nuclides easier, binding energies are usually expressed on a per-nucleon basis using *electron volts* (*eV*) as the energy unit, where $1 \text{ eV} = 1.60 \times 10^{-19}$ J and 1 million electron volts (1 MeV) = 1.60×10^{-13} J. Thus, the helium-4 binding energy is 7.08 MeV/nucleon:

$$\text{Helium-4 binding energy} = \left(\frac{2.73 \times 10^{12} \text{ J/mol}}{6.022 \times 10^{23} \text{ nuclei/mol}} \right) \left(\frac{1 \text{ MeV}}{1.60 \times 10^{-13} \text{ J}} \right) \left(\frac{1 \text{ nucleus}}{4 \text{ nucleons}} \right)$$
$$= 7.08 \text{ MeV/nucleon}$$

A plot of binding energy per nucleon for the most stable isotope of each element is shown in Figure 22.7. Since a higher binding energy per nucleon corresponds to greater stability, the most stable nuclei are at the top of the curve. Iron-56, with a binding energy of 8.79 MeV/nucleon, is the most stable isotope known. If all the mass in the universe were somehow converted to its most stable form, the universe would become a chunk of iron.

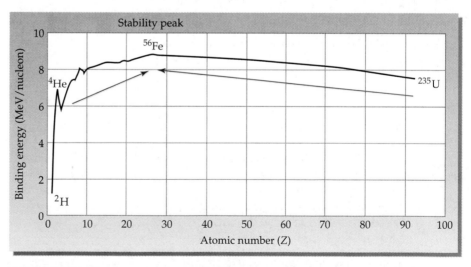

FIGURE 22.7 The binding energy per nucleon for the most stable isotope of each naturally occurring element. Binding energy reaches a maximum of 8.79 MeV/nucleon at ^{56}Fe. As a result, there is an increase in stability when much lighter elements fuse together to yield heavier elements up to ^{56}Fe and when much heavier elements split apart to yield lighter elements down to ^{56}Fe, as indicated by the arrows.

The idea that mass and energy are interconvertible is a potentially disturbing one because it seems to overthrow two of the fundamental principles on which chemistry is based—the law of mass conservation and the law of energy conservation. In fact, what the mass/energy interconversion means is that the two individual laws must be *combined*. Neither mass nor energy is conserved separately; only the combination of the two is conserved. Every time we do a reaction, whether nuclear *or* chemical, mass and energy are interconverted, but the combination of the two is conserved. For the energy changes involved in typical chemical reactions, however, the effects are so small that the mass change can't be detected by even the best analytical balance. Example 22.7 illustrates a mass–energy calculation for a chemical reaction.

EXAMPLE 22.6 Helium-6 is a radioactive isotope with $t_{1/2} = 0.861$ s. Calculate the mass defect (in g/mol) for the formation of a ^{6}He nucleus, and calculate the binding energy in MeV/nucleon. Is a ^{6}He nucleus more stable or less stable than a ^{4}He nucleus? (The mass of a ^{6}He atom is 6.018 89 amu.)

SOLUTION First, calculate the total mass of the nucleons (4 n + 2 p):

$$\text{Mass of 4 neutrons} = (4)(1.008\ 66\ \text{amu}) = 4.034\ 64\ \text{amu}$$

$$\text{Mass of 2 protons} = (2)(1.007\ 28\ \text{amu}) = 2.014\ 56\ \text{amu}$$

$$\overline{\text{Mass of 4 n} + 2 \qquad\qquad\qquad\qquad = 6.049\ 20\ \text{amu}}$$

Next, calculate the mass of a ^{6}He nucleus by subtracting the mass of two electrons from the mass of a ^{6}He atom:

$$\text{Mass of helium-6 atom} \qquad\qquad\qquad = 6.018\ 89\ \text{amu}$$

$$-\text{Mass of 2 electrons} = -(2)(5.486 \times 10^{-4}\ \text{amu}) = -0.001\ 10\ \text{amu}$$

$$\overline{\text{Mass of helium-6 nucleus} \qquad\qquad\qquad = 6.017\ 79\ \text{amu}}$$

Then subtract the mass of the ^{6}He nucleus from the mass of the nucleons to find the mass defect:

$$\text{Mass defect} = \text{Mass of nucleons} - \text{Mass of nucleus}$$
$$= (6.049\ 20\ \text{amu}) - (6.017\ 79\ \text{amu})$$
$$= 0.031\ 41\ \text{amu, or } 0.031\ 41\ \text{g/mol}$$

Now, use the Einstein equation to convert the mass defect into binding energy:

$$\Delta E = \Delta mc^2 = \left(0.031\ 41\ \frac{\text{g}}{\text{mol}} \right)\left(10^{-3}\ \frac{\text{kg}}{\text{g}} \right)\left(3.00 \times 10^8\ \frac{\text{m}}{\text{s}} \right)^2$$
$$= 2.83 \times 10^{12}\ \text{J/mol} = 2.83 \times 10^9\ \text{kJ/mol}$$

This value can be expressed in units of MeV/nucleon by dividing by Avogadro's number, converting to MeV, and dividing by the number of nucleons:

$$\left(\frac{2.83 \times 10^{12}\ \dfrac{\text{J}}{\text{mol}}}{6.022 \times 10^{23}\ \dfrac{\text{nuclei}}{\text{mol}}} \right)\left(\frac{1\ \text{MeV}}{1.60 \times 10^{-13}\ \text{J}} \right)\left(\frac{1\ \text{nucleus}}{6\ \text{nucleons}} \right) = 4.89\ \text{MeV/nucleon}$$

The binding energy of a radioactive ^{6}He nucleus is 4.89 MeV/nucleon, making it less stable than a ^{4}He nucleus, whose binding energy is 7.08 MeV/nucleon.

EXAMPLE 22.7 What is the change in mass (in grams) when 2 mol of hydrogen atoms combine to form 1 mol of hydrogen molecules?

$$2\,H \longrightarrow H_2 \qquad \Delta E = -436\ kJ$$

SOLUTION The problem asks us to calculate a mass defect Δm when the energy change ΔE is known. To do this, we have to rearrange the Einstein equation to solve for mass, remembering that $1\,J = 1\ kg \cdot m^2/s^2$:

$$\Delta m = \frac{\Delta E}{c^2} = \frac{(-436\ kJ)\left(10^3\ \dfrac{J}{kJ}\right)\left(\dfrac{1\ kg \cdot m^2/s^2}{1\ J}\right)}{\left(3.00 \times 10^8\ \dfrac{m}{s}\right)^2}$$

$$= -4.84 \times 10^{-12}\ kg = -4.84 \times 10^{-9}\ g$$

The loss in mass accompanying formation of 1 mol of H_2 molecules from its constituent atoms is 4.84×10^{-9} g, far too small an amount to be detectable by any balance currently available.

▶ **PROBLEM 22.10** Calculate the mass defect (in g/mol) for the formation of an oxygen-16 nucleus, and calculate the binding energy in MeV/nucleon. The mass of an ^{16}O atom is 15.994 92 amu.

▶ **PROBLEM 22.11** What is the mass change (in g/mol) for the thermite reaction of aluminum with iron(III) oxide?

$$2\,Al(s) + Fe_2O_3(s) \longrightarrow Al_2O_3(s) + 2\,Fe(s) \qquad \Delta E = -852\ kJ$$

22.6 Nuclear Fission and Fusion

A careful look at the plot of atomic number versus binding energy per nucleon in Figure 22.7 leads to some interesting and enormously important conclusions. The fact that binding energy per nucleon begins at a relatively low value for $_1^2H$, reaches a maximum at $_{26}^{56}Fe$, and then gradually tails off implies that both lighter and heavier elements are less stable than midweight elements near iron-56. Very heavy elements can therefore gain stability and release energy if they fragment to yield midweight elements, while very light elements can gain stability and release energy if they fuse together. The two resultant processes—**fission** for the fragmenting of heavy nuclei and **fusion** for the joining together of light nuclei—have changed the world since their discovery in the late 1930s and early 1940s.

Nuclear Fission

Certain nuclei—uranium-233, uranium-235, and plutonium-239, for example—do more than undergo simple radioactive decay; they break into fragments when struck by neutrons. As illustrated in Figure 22.8, an incoming neutron causes the nucleus to split into two smaller pieces of roughly similar size.

The fission of a nucleus does not occur in exactly the same way each time: Nearly 400 different fission pathways have been identified for uranium-235, yielding nearly 800 different fission products. One of the more frequently occurring pathways generates barium-142 and krypton-91,

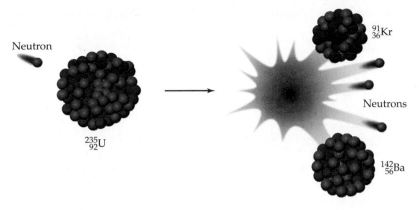

FIGURE 22.8 A representation of nuclear fission. A uranium-235 nucleus fragments when struck by a neutron, yielding two smaller nuclei and releasing a large amount of energy.

along with two additional neutrons plus the one neutron that initiated the fission:

$$^1_0n + ^{235}_{92}U \longrightarrow ^{142}_{56}Ba + ^{91}_{36}Kr + 3\,^1_0n$$

The three neutrons released by fission of a ^{235}U nucleus can induce three more fissions yielding nine neutrons, which can induce nine more fissions yielding 27 neutrons, and so on indefinitely. The result is a **chain reaction** that continues to occur even if the supply of neutrons from outside is cut off. If the sample size is small, many of the neutrons escape before initiating additional fission events, and the chain reaction soon stops. If there is a sufficient amount of ^{235}U, though—an amount called the **critical mass**—enough neutrons remain for the chain reaction to become self-sustaining. Under high-pressure conditions that confine the ^{235}U to a small volume, the chain reaction may even occur so rapidly that a nuclear explosion results. For ^{235}U, the critical mass is about 56 kg, although the amount can be reduced to 15 kg by placing a coating of ^{238}U around the ^{235}U to reflect back some of the escaping neutrons.

The amount of energy released during nuclear fission can be calculated as in Example 22.8 by finding the accompanying mass change and then using the Einstein mass–energy relationship discussed in the previous section. When calculating the mass change, it's simplest to use the masses of the *atoms* corresponding to the relevant nuclei, rather than the masses of the nuclei themselves, because the number of electrons is the same in both reactants and products and thus cancels from the calculation.

An enormous amount of energy is released in the explosion that accompanies an uncontrolled nuclear chain reaction.

EXAMPLE 22.8   How much energy (in kJ/mol) is released by the fission of uranium-235 to form barium-142 and krypton-91? The fragment masses are ^{235}U (235.0439 amu), ^{142}Ba (141.9164 amu), ^{91}Kr (90.9234 amu), and n (1.008 66 amu).

$$^1_0n + ^{235}_{92}U \longrightarrow ^{142}_{56}Ba + ^{91}_{36}Kr + 3\,^1_0n$$

SOLUTION   First calculate the mass change by subtracting the masses of the products from the mass of the ^{235}U reactant:

Mass of ^{235}U	=	235.0439 amu
− Mass of ^{142}Ba	=	−141.9164 amu
− Mass of ^{91}Kr	=	−90.9234 amu
− Mass of 2 n = −(2)(1.008 66 amu) =		−2.0173 amu
Mass change: =		0.1868 amu (or 0.1868 g/mol)

Next, use the Einstein equation to convert the mass loss into an energy change:

$$\Delta E = \Delta mc^2$$

$$= \left(0.1868\,\frac{\text{g}}{\text{mol}}\right)\left(1 \times 10^{-3}\,\frac{\text{kg}}{\text{g}}\right)\left(3.00 \times 10^8\,\frac{\text{m}}{\text{s}}\right)^2$$

$$= 1.68 \times 10^{13}\,\text{kg} \cdot \text{m}^2/(\text{s}^2 \cdot \text{mol}) = 1.68 \times 10^{10}\,\text{kJ/mol}$$

Nuclear fission of 1 mol of ^{235}U releases 1.68×10^{10} kJ.

▶ **PROBLEM 22.12**   An alternative pathway for the nuclear fission of ^{235}U produces tellurium-137 and zirconium-97. How much energy (in kJ/mol) is released in this fission pathway?

$$^{1}_{0}\text{n} + \,^{235}_{92}\text{U} \longrightarrow \,^{137}_{52}\text{Te} + \,^{97}_{40}\text{Zr} + 2\,^{1}_{0}\text{n}$$

The masses are ^{235}U (235.0439 amu), ^{137}Te (136.9254 amu), ^{97}Zr (96.9110 amu), and n (1.008 66 amu).

Nuclear Reactors

The same fission process that can lead to a nuclear explosion under some conditions can be used for the generation of electric power when carried out in a controlled manner in a nuclear reactor (Figure 22.9). The principle behind a nuclear reactor is simple: Uranium fuel is placed in a containment vessel surrounded by circulating coolant, and *control rods* are added. Made of substances such as boron and cadmium, which absorb and thus regulate the flow of neutrons, the control rods are raised and lowered as necessary to maintain the fission at a barely self-sustainable rate so that overheating is prevented. Energy from the controlled fission heats the circulating coolant, which in turn produces steam to drive a turbine and produce electricity.

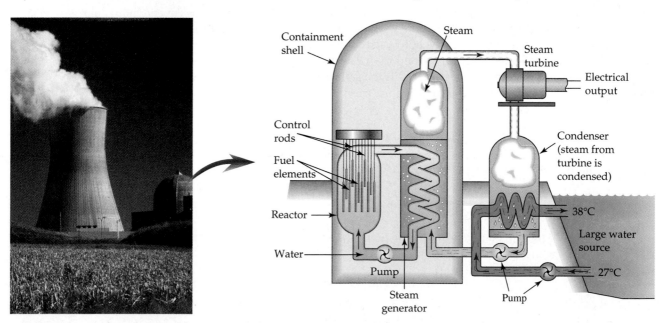

FIGURE 22.9   A nuclear power plant. Heat produced in the reactor core is transferred by coolant circulating in a closed loop to a steam generator, and the steam then drives a turbine to generate electricity.

Naturally occurring uranium is a mixture of two isotopes. The nonfissionable ^{238}U isotope has a natural abundance of 99.3%, while the fissionable ^{235}U isotope is present only to the extent of 0.7%. The fuel used in nuclear reactors is typically made of compressed pellets of UO_2 that have been isotopically enriched to a 3% concentration of ^{235}U and then encased in zirconium rods. The rods are placed in a pressure vessel filled with water, which acts as a moderator to slow the neutrons so they can be captured more readily. No nuclear explosion can occur in a reactor because the amount and concentration of fissionable fuel is too low and because the fuel is not confined by pressure into a small volume. In a worst-case accident, however, uncontrolled fission could lead to enormous overheating that could melt the reactor and surrounding containment vessel, thereby releasing large amounts of radioactivity to the environment.

Thirty countries around the world now obtain some of their electricity from nuclear energy (Figure 22.10). Lithuania leads with 87%, followed by a number of other European countries that have also made a substantial commitment to the technology. The United States has been more cautious, with only 22% of its power coming from nuclear plants. Worldwide, 434 nuclear plants were in operation in 1999, with an additional 36 under construction, most of them in Asia. Approximately 17% of the world's electrical power is currently generated by nuclear reactors.

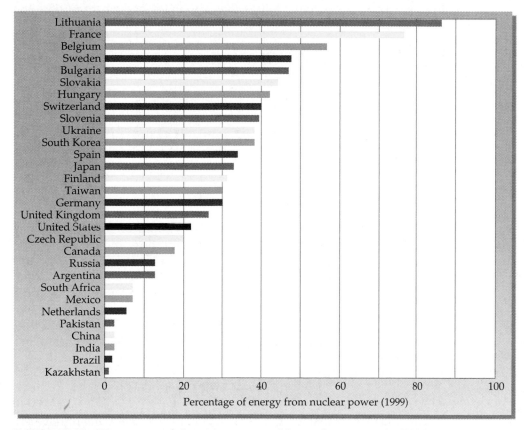

FIGURE 22.10   Percentage of electricity generated by nuclear power in 1999.

The primary problem holding back future development is the yet unsolved matter of how to dispose of the radioactive wastes generated by the plants. It will take at least 600 years for waste strontium-90 to decay to safe levels, and at least 20,000 years for plutonium-239 to decay.

Nuclear Fusion

Just as heavy nuclei such as ^{235}U release energy when they undergo *fission*, very light nuclei such as the isotopes of hydrogen release enormous amounts of energy when they undergo *fusion*. In fact, it's just this fusion reaction of hydrogen nuclei to produce helium that powers our sun and other stars. Among the processes thought to occur in the sun are those in the following sequence leading to helium-4:

$$^1_1H + {}^1_1H \longrightarrow {}^2_1H + {}^0_1e$$

$$^1_1H + {}^2_1H \longrightarrow {}^3_2He$$

$$^3_2He + {}^3_2He \longrightarrow {}^4_2He + 2\,{}^1_1H$$

$$^3_2He + {}^1_1H \longrightarrow {}^4_2He + {}^0_1e$$

The main appeal of nuclear fusion as a power source is that the hydrogen isotopes used as fuel are cheap and plentiful and that the fusion products are nonradioactive and nonpolluting. The technical problems that must be solved before achieving a practical and controllable fusion method are staggering, however. Not the least of the problems is that a temperature of approximately 40 *million* kelvins is needed to initiate the fusion process.

▶ **PROBLEM 22.13** Calculate the amount of energy released (in kJ/mol) for the fusion reaction of ^{1}H and ^{2}H atoms to yield a ^{3}He atom:

$$^1_1H + {}^2_1H \longrightarrow {}^3_2He$$

The atomic masses are ^{1}H (1.007 83 amu), ^{2}H (2.014 10 amu), and ^{3}He (3.016 03 amu).

22.7 Nuclear Transmutation

Only about 300 of the more than 3600 known isotopes occur naturally. The remainder have been made by **nuclear transmutation**, the change of one element into another. Such transmutation is often brought about by bombardment of an atom with a high-energy particle such as a proton, neutron, or α particle. In the ensuing collision between particle and atom, an unstable nucleus is momentarily created, a nuclear change occurs, and a different element is produced. The first nuclear transmutation was accomplished in 1917 by Rutherford, who bombarded ^{14}N nuclei with α particles and found that ^{17}O was produced:

$$^{14}_7N + {}^4_2He \longrightarrow {}^{17}_8O + {}^1_1H$$

Other nuclear transmutations can lead to the synthesis of entirely new elements never before seen on earth. In fact, all the transuranium elements—those elements with atomic numbers greater than 92—have been produced by bombardment reactions. Plutonium, for example, can be made by bombardment of uranium-238 with α particles:

$$^{238}_{92}U + {}^4_2He \longrightarrow {}^{241}_{94}Pu + {}^1_0n$$

The plutonium-241 that results from uranium-238 bombardment is itself radioactive with a half-life of 14.4 years, decaying by β emission to yield americium-241. (If the name sounds familiar, it's because americium is used commercially in making smoke detectors.) Americium-241 is also radioactive, decaying by α emission with a half-life of 432 years.

$$^{241}_{94}\text{Pu} \longrightarrow \,^{241}_{95}\text{Am} + \,^{0}_{-1}\text{e}$$

$$^{241}_{95}\text{Am} \longrightarrow \,^{237}_{93}\text{Np} + \,^{4}_{2}\text{He}$$

Still other nuclear transmutations are carried out using neutrons, protons, or other particles for bombardment. The cobalt-60 used in radiation therapy for cancer patients can be prepared by neutron bombardment of iron-58. Iron-58 first absorbs a neutron to yield iron-59, the iron-59 undergoes β decay to yield cobalt-59, and the cobalt-59 then absorbs a second neutron to yield cobalt-60:

$$^{58}_{26}\text{Fe} + \,^{1}_{0}\text{n} \longrightarrow \,^{59}_{26}\text{Fe}$$

$$^{59}_{26}\text{Fe} \longrightarrow \,^{59}_{27}\text{Co} + \,^{0}_{-1}\text{e}$$

$$^{59}_{27}\text{Co} + \,^{1}_{0}\text{n} \longrightarrow \,^{60}_{27}\text{Co}$$

The overall change can be written as

$$^{58}_{26}\text{Fe} + 2\,^{1}_{0}\text{n} \longrightarrow \,^{60}_{27}\text{Co} + \,^{0}_{-1}\text{e}$$

The Fermi National Accelerator Laboratory has a particle accelerator 4 miles in circumference that is able to accelerate protons to energies of 1 trillion eV.

EXAMPLE 22.9   The element berkelium, first prepared at the University of California at Berkeley in 1949, is made by α bombardment of $^{241}_{95}\text{Am}$. Two neutrons are also produced during the reaction. What isotope of berkelium results from this transmutation? Write a balanced nuclear equation.

SOLUTION   According to the periodic table, berkelium has Z = 97. Since the sum of the reactant mass numbers is 241 + 4 = 245, and since 2 neutrons are produced, the berkelium isotope must have a mass number of 243.

$$^{241}_{95}\text{Am} + \,^{4}_{2}\text{He} \longrightarrow \,^{243}_{97}\text{Bk} + 2\,^{1}_{0}\text{n}$$

▶**PROBLEM 22.14**   Write a balanced nuclear equation for the reaction of argon-40 with a proton: $^{40}_{18}\text{Ar} + \,^{1}_{1}\text{H} \rightarrow ? + \,^{1}_{0}\text{n}$

▶**PROBLEM 22.15**   Write a balanced nuclear equation for the reaction of uranium-238 with a deuteron ($^{2}_{1}\text{H}$): $^{238}_{92}\text{U} + \,^{2}_{1}\text{H} \rightarrow ? + 2\,^{1}_{0}\text{n}$

22.8 Detecting and Measuring Radioactivity

Radioactive emissions are invisible. We can't see, hear, smell, touch, or taste them, no matter how high the dose. We can, however, detect radiation by measuring its *ionizing* properties. High-energy radiation of all kinds is usually grouped under the name **ionizing radiation** because interaction of the radiation with a molecule knocks an electron from the molecule, thereby ionizing it.

$$\text{Molecule} \xrightarrow[\text{radiation}]{\text{Ionizing}} \text{Ion} + e^-$$

Ionizing radiation includes not only α particles, β particles, and γ rays, but also X rays and *cosmic rays*. X rays, like γ rays, are high-energy photons ($\lambda = 10^{-8}$–10^{-11} m) rather than particles; **cosmic rays** are energetic particles coming from interstellar space. They consist primarily of protons, along with some α and β particles.

The simplest device for detecting radiation is the photographic film badge worn by people who routinely work with radioactive materials. Any radiation striking the badge causes it to fog. Perhaps the best known method for measuring radiation is the *Geiger counter*, an argon-filled tube containing two electrodes (Figure 22.11). The inner walls of the tube are coated with an electrically conducting material and given a negative charge, and a wire in the center of the tube is given a positive charge. As radiation enters the tube through a thin window, it strikes and ionizes argon atoms, releasing electrons that briefly conduct a tiny electric current between the electrodes. The passage of the current is detected, amplified, and used to produce a clicking sound or to register on a meter. The more radiation that enters the tube, the more frequent the clicks.

This photographic film badge is a common device for monitoring radiation exposure.

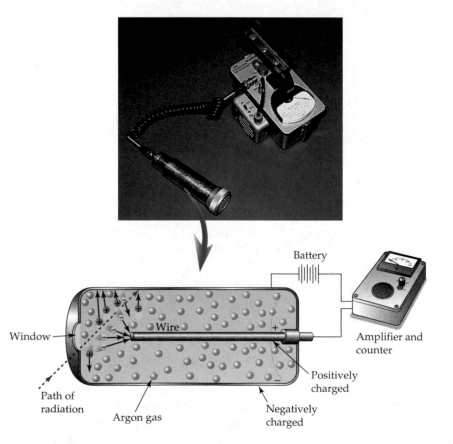

FIGURE 22.11 A Geiger counter for measuring radiation. As radiation enters the tube through a thin window, it ionizes argon atoms and produces electrons that conduct a tiny electric current from the negatively charged walls to the positively charged center electrode. The current flow then registers on the meter.

The most versatile method for measuring radiation in the laboratory is the *scintillation counter*, in which a substance called a *phosphor* (often a sodium iodide crystal containing a small amount of thallium iodide) emits a flash of light when struck by radiation. The number of flashes is counted electronically and converted into an electrical signal.

Radiation intensity is expressed in different ways, depending on what is being measured. Some units measure the number of nuclear decay events; others measure the amount of exposure to radiation or the biological consequences of radiation (Table 22.3).

TABLE 22.3 Units for Measuring Radiation

Unit	Quantity Measured	Description
Becquerel (Bq)	Decay events	Amount of sample that undergoes 1 disintegration/s
Curie (Ci)	Decay events	Amount of sample that undergoes 3.7×10^{10} disintegrations/s
Gray (Gy)	Energy absorbed per kilogram of tissue	1 Gy = 1 J/kg tissue
Rad	Energy absorbed per kilogram of tissue	1 rad = 0.01 Gy
Sievert (Sv)	Tissue damage	1 Sv = 1 J/kg
Rem	Tissue damage	1 rem = 0.01 Sv

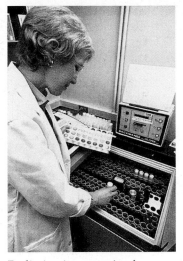

Radiation is conveniently detected and measured using this scintillation counter, which electronically counts the flashes produced when radiation strikes a phosphor.

- The *becquerel* (Bq) is the SI unit for measuring the number of radioactive disintegrations occurring per second in a sample: 1 Bq = 1 disintegration/s. The *curie* (Ci) and *millicurie* (mCi) also measure disintegrations per unit time, but they are far larger units than the becquerel and are more often used, particularly in medicine and biochemistry. One curie is the decay rate of 1 g of radium, equal to 3.7×10^{10} Bq:

$$1 \text{ Bq} = 1 \text{ disintegration/s}$$
$$1 \text{ Ci} = 3.7 \times 10^{10} \text{ Bq} = 3.7 \times 10^{10} \text{ disintegrations/s}$$

For example, a 1.5 mCi sample of tritium is equal to 5.6×10^7 Bq, meaning that it undergoes 5.6×10^7 disintegrations/s.

$$(1.5 \text{ mCi})\left(10^{-3} \frac{\text{Ci}}{\text{mCi}} \right)\left(3.7 \times 10^{10} \frac{\text{Bq}}{\text{Ci}} \right) = 5.6 \times 10^7 \text{ Bq}$$

- The *gray* (Gy) is the SI unit for measuring the amount of energy absorbed per kilogram of tissue exposed to a radiation source: 1 Gy = 1 J/kg. The *rad* (*radiation absorbed dose*) also measures tissue exposure and is more often used in medicine.

$$1 \text{ Gy} = 1 \text{ J/kg} \qquad 1 \text{ rad} = 0.01 \text{ Gy}$$

- The *sievert* (Sv) is the SI unit that measures the amount of tissue damage caused by radiation. It takes into account not just the energy absorbed per kilogram of tissue but also the different biological effects of different kinds of radiation. For example, 1 Gy of α radiation causes 20 times more tissue damage than 1 Gy of γ rays, but 1 Sv of α radiation and 1 Sv of γ rays cause the same amount of damage. The *rem* (*roentgen equivalent for man*) is an analogous non-SI unit that is more frequently used in medicine.

$$1 \text{ rem} = 0.01 \text{ Sv}$$

22.9 Biological Effects of Radiation

The effects of ionizing radiation on the human body vary with the kind and energy of the radiation, the length of exposure, and whether the radiation is from an external or internal source. When the source is external, X rays and γ radiation are more harmful than α and β particles because they penetrate through clothing and skin. When the source is internal, however, α and β particles are particularly dangerous because all their radiation energy is given up to the surrounding tissue. Alpha emitters are especially hazardous internally and are almost never used in medical applications.

Because of their relatively large mass, α particles move slowly (up to only one-tenth the speed of light) and can be stopped by a few sheets of paper or by the top layer of skin. Beta particles, because they are much lighter, move at up to nine-tenths the speed of light and have about 100 times the penetrating power of α particles. A block of wood or heavy protective clothing is necessary to stop beta radiation, which would otherwise penetrate and burn the skin. Gamma rays and X rays move at the speed of light and have about 1000 times the penetrating power of α particles. A lead block several inches thick is needed to stop γ and X radiation, which would otherwise penetrate and damage the body's internal organs. Some properties of different kinds of ionizing radiation are summarized in Table 22.4.

TABLE 22.4 Some Properties of Ionizing Radiation

Type of Radiation	Energy Range	Penetrating Distance in Water*
α	3–9 MeV	0.02–0.04 mm
β	0–3 MeV	0–4 mm
X	100 eV–10 keV	0.01–1 cm
γ	10 keV–10 MeV	1–20 cm

* Distances at which one-half of the radiation has been stopped.

The biological effects of different radiation doses are given in Table 22.5. Although the effects sound fearful, the average radiation dose received annually by most people is only about 120 mrem. About 70% of this radiation comes

TABLE 22.5 Biological Effects of Short-Term Radiation on Humans

Dose (rem)	Biological Effects
0–25	No detectable effects
25–100	Temporary decrease in white blood cell count
100–200	Nausea, vomiting, longer-term decrease in white blood cells
200–300	Vomiting, diarrhea, loss of appetite, listlessness
300–600	Vomiting, diarrhea, hemorrhaging, eventual death in some cases
Above 600	Eventual death in nearly all cases

from natural sources (rocks and cosmic rays); the remaining 30% comes from medical procedures such as X rays. The amount due to emissions from nuclear power plants and to fallout from atmospheric testing of nuclear weapons in the 1950s is barely detectable.

22.10 Applications of Nuclear Chemistry

Dating with Radioisotopes

Biblical scrolls are found in a cave near the Dead Sea. Are they authentic? A mummy is discovered in an Egyptian tomb. How old is it? The burned bones of a man are dug up near Lubbock, Texas. How long have humans lived in the area? These and many other questions can be answered by archaeologists using a technique called *radiocarbon dating*. (The Dead Sea Scrolls are 1900 years old and authentic, the mummy is 3100 years old, and the human remains found in Texas are 9900 years old.)

Radiocarbon dating of archaeological artifacts depends on the slow and constant production of radioactive carbon-14 in the upper atmosphere by neutron bombardment of nitrogen atoms. (The neutrons come from the bombardment of other atoms by cosmic rays.)

$$^{14}_{7}N + {}^{1}_{0}n \longrightarrow {}^{14}_{6}C + {}^{1}_{1}H$$

Radiocarbon dating places the age of this Egyptian mummy at 3100 years.

Carbon-14 atoms produced in the upper atmosphere combine with oxygen to yield $^{14}CO_2$, which slowly diffuses into the lower atmosphere, where it mixes with ordinary $^{12}CO_2$ and is taken up by plants during photosynthesis. When these plants are eaten, carbon-14 enters the food chain and is ultimately distributed evenly throughout all living organisms.

As long as a plant or animal is living, a dynamic equilibrium exists in which an organism excretes or exhales the same amount of ^{14}C that it takes in. As a result, the ratio of ^{14}C to ^{12}C in the living organism is the same as that in the atmosphere—about 1 part in 10^{12}. When the plant or animal dies, however, it no longer takes in more ^{14}C, and the $^{14}C/^{12}C$ ratio in the organism slowly decreases as ^{14}C undergoes radioactive decay by beta emission, with $t_{1/2} = 5730$ years.

$$^{14}_{6}C \longrightarrow {}^{14}_{7}N + {}^{0}_{-1}e$$

At 5730 years (one ^{14}C half-life) after the death of the organism, the $^{14}C/^{12}C$ ratio has decreased by a factor of 2; at 11,460 years after death, the $^{14}C/^{12}C$ ratio has decreased by a factor of 4; and so on. By measuring the present $^{14}C/^{12}C$ ratio in the traces of any once-living organism, archaeologists can determine how long ago the organism died. Human or animal hair from well-preserved remains, charcoal or wood fragments from once-living trees, and cotton or linen from once-living plants are all useful sources for radiocarbon dating. The technique becomes less accurate as samples get older and the amount of ^{14}C they contain diminishes, but artifacts with an age of 1000–20,000 years can be dated with reasonable accuracy. The outer limit of the technique is about 60,000 years.

Just as radiocarbon measurements allow dating of once-living organisms, similar measurements on other radioisotopes make possible the dating of rocks. Uranium-238, for example, has a half-life of 4.47×10^9 years and decays through the series of events shown previously in Figure 22.6 to yield lead-206. The age of a uranium-containing rock can therefore be determined by measuring the $^{238}U/^{206}Pb$ ratio. Similarly, potassium-40 has a half-life of

1.28×10^9 years and decays through electron capture and positron emission to yield argon-40. (Both processes yield the same product.)

$$^{40}_{19}\text{K} + ^{\ \ 0}_{-1}\text{e} \longrightarrow ^{40}_{18}\text{Ar}$$

$$^{40}_{19}\text{K} \longrightarrow ^{40}_{18}\text{Ar} + ^{0}_{1}\text{e}$$

The age of a rock can be found by crushing a sample, measuring the amount of ^{40}Ar gas that escapes, and comparing the amount of argon-40 with the amount of ^{40}K remaining in the sample. It is through techniques such as these that the age of the earth has been estimated at approximately 4.5 billion years.

EXAMPLE 22.10 Radiocarbon measurements made in 1988 on the Shroud of Turin, a religious artifact thought by some to be the burial shroud of Christ, showed a ^{14}C decay rate of 14.2 disintegrations/min per gram of carbon. What is the age of the shroud if currently living organisms decay at the rate of 15.3 disintegrations/min per gram of carbon? The half-life of ^{14}C is 5730 years.

SOLUTION As we saw in Example 22.5, the ratio of the decay rate at any time t to the decay rate at time $t = 0$ is the same as the ratio of N to N_0:

$$\frac{\text{Decay rate at time } t}{\text{Decay rate at time } t = 0} = \frac{kN}{kN_0} = \frac{N}{N_0}$$

To date the shroud, we need to calculate the time t that corresponds to the observed decay rate. This can be done by using the equation

$$\ln\left(\frac{N}{N_0}\right) = (-\ln 2)\left(\frac{t}{t_{1/2}}\right)$$

Substituting the proper values into the equation gives

$$\ln\left(\frac{14.2}{15.3}\right) = -0.693\left(\frac{t}{5730 \text{ years}}\right) \quad \text{or} \quad -0.0746 = -0.693\left(\frac{t}{5730 \text{ years}}\right)$$

so
$$t = \frac{(-0.0746)(5730 \text{ years})}{-0.693} = 617 \text{ years}$$

The Shroud of Turin is approximately 617 years old, showing it to be from medieval times.

▶ **PROBLEM 22.16** Charcoal found in the Lascaux cave in France, site of many prehistoric cave paintings, was observed to decay at a rate of 2.4 disintegrations/min per gram of carbon. What is the age of the charcoal if currently living organisms decay at the rate of 15.3 disintegrations/min per gram of carbon? The half-life of ^{14}C is 5730 years.

Medical Uses of Radioactivity

The origins of nuclear medicine date to 1901, when the French physician Henri Danlos first used radium in the treatment of a tuberculous skin lesion. Since that time, uses of radioactivity have become a crucial part of modern medical care, both diagnostic and therapeutic. Current nuclear techniques can be grouped into three classes: (1) in vivo procedures, (2) therapeutic procedures, and (3) imaging procedures.

What is the age of these cave paintings?

In Vivo Procedures   In vivo studies—those that take place *inside* the body—are carried out to assess the functioning of a particular organ or body system. A radiopharmaceutical agent is administered, and its path in the body—whether it is absorbed, excreted, diluted, or concentrated—is determined by analysis of blood or urine samples.

An example of the many in vivo procedures using radioactive agents is the determination of whole-blood volume by injecting a known quantity of red blood cells labeled with radioactive chromium-51. After a suitable interval to allow the labeled cells to be distributed evenly throughout the body, a blood sample is taken, the amount of dilution of the ^{51}Cr is measured, and the blood volume is calculated. Recall from Section 3.8 that when a concentrated solution is diluted, the amount of solute (^{51}Cr in the present instance) remains the same and only the volume changes. That is,

$$\text{Amount of } ^{51}Cr = C_0 \times V_0 = C_{blood} \times V_{blood}$$

or
$$V_{blood} = \frac{C_0 V_0}{C_{blood}}$$

where:

C_0 = Concentration of labeled cells injected ($\mu Ci/mL$)

V_0 = Volume of labeled cells injected (mL)

C_{blood} = Concentration of labeled cells in blood ($\mu Ci/mL$)

V_{blood} = Blood volume

Therapeutic Procedures   Therapeutic procedures—those in which radiation is used as a weapon to kill diseased tissue—can involve either external or internal sources of radiation. External radiation therapy for the treatment of cancer is often carried out with gamma rays from a cobalt-60 source. The highly radioactive source is shielded by a thick lead container and has a small opening directed toward the site of the tumor. By focusing the radiation beam on the tumor and rotating the patient's body, the tumor receives the full exposure while the exposure of surrounding parts of the body is minimized. Nevertheless, sufficient exposure occurs so that most patients suffer some effects of radiation sickness.

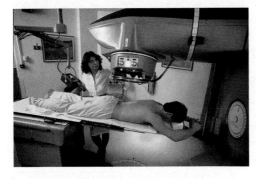

Cancerous tumors can be treated by irradiation with gamma rays from this cobalt-60 source.

Internal radiation therapy is a much more selective technique than external therapy. In the treatment of thyroid disease, for example, iodine-131, a powerful beta emitter known to localize in the target tissue, is administered internally. Since β particles penetrate no farther than several millimeters, the localized ^{131}I produces a high radiation dose that destroys only the surrounding diseased tissue.

Imaging Procedures Imaging procedures give diagnostic information about the health of body organs by analyzing the distribution pattern of radioisotopes introduced into the body. A radiopharmaceutical agent that is known to concentrate in a specific tissue or organ is injected into the body, and its distribution pattern is monitored by external radiation detectors. Depending on the disease and organ, a diseased organ might concentrate more of the radiopharmaceutical than a normal organ and thus show up as a radioactive "hot" spot against a "cold" background. Alternatively, the diseased organ might concentrate less of the radiopharmaceutical than a normal organ and thus show up as a cold spot on a hot background.

The radioisotope most widely used today is technetium-99m, whose short half-life of 6.01 hours minimizes a patient's exposure to harmful effects. Bone scans using Tc-99m, such as that shown in Figure 22.12a, are an important tool in the diagnosis of cancer and other pathological conditions.

Another kind of imaging procedure makes use of a technique called *magnetic resonance imaging* (MRI). MRI uses no radioisotopes and has no known side-effects. Instead, MRI uses radio waves to stimulate certain nuclei in the presence of a powerful magnetic field. The stimulated nuclei (normally the hydrogen nuclei in H_2O molecules) then give off a signal that can be measured, interpreted, and correlated with their environment in the body. Figure 22.12b shows a brain scan carried out by MRI and indicates the position of a tumor.

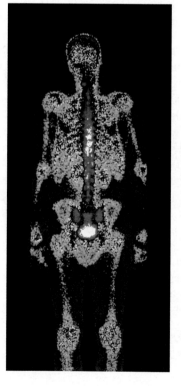

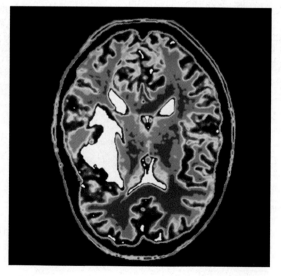

FIGURE 22.12 (a) A bone scan using radioactive technetium-99m. (b) An MRI brain scan, showing the position of a tumor (the large white area at left).

(a)

(b)

The Origin of Chemical Elements

The stars in the Milky Way galaxy condensed from gas clouds under gravitational attraction.

Cosmologists tell us that the universe began some 15 billion years ago in an extraordinary event they call the "big bang." Initially, the temperature must have been inconceivably high, but after 1 second, the temperature dropped to about 10^{10} K and elementary particles began to form: protons, neutrons, and electrons, as well as positrons and *neutrinos*—neutral particles with a mass much less than that of an electron. After 3 minutes, the temperature dropped to 10^9 K, and protons began fusing with neutrons to form helium nuclei, $^4_2\text{He}^{2+}$. Matter remained in this form for many millions of years until the expanding universe cooled to about 10,000 K. Electrons were then able to bind to protons and to helium nuclei, forming stable hydrogen and helium atoms.

The attractive force of gravity acting on regions of higher-than-average density slowly produced massive local concentrations of matter and ultimately formed billions of galaxies, each with many billions of stars. As the gas clouds of hydrogen and helium condensed under gravitational attraction and stars formed, their temperatures reached 10^7 K, and their densities reached 100 g/cm³. Protons and neutrons again fused to yield helium nuclei, generating vast amounts of heat and light—about 6×10^8 kJ per mole of protons undergoing fusion.

Most of these early stars probably burned out after a few billion years, but a few were so massive that, as their nuclear fuel diminished, gravitational attraction caused a rapid contraction leading to still higher core temperatures and higher densities—up to 5×10^8 K and 5×10^5 g/cm³. Much larger nuclei were now formed, including carbon, oxygen, silicon, magnesium, and iron. Ultimately, the stars underwent a gravitational collapse resulting in the synthesis of still heavier elements and an explosion visible throughout the universe as a *supernova*.

Matter from exploding supernovas was blown throughout the galaxy, forming a new generation of stars and planets. Our own sun and solar system formed only about 4.5 billion years ago from matter released by former supernovas. Except for hydrogen and helium, all the atoms in our bodies, our planet, and our solar system were created more than 5 billion years ago in exploding stars.

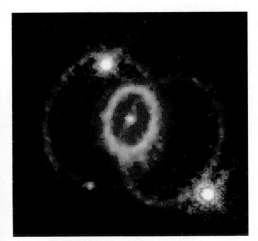

The instantaneous gravitational collapse of a massive star resulted in this supernova explosion observed in 1987.

▶ **PROBLEM 22.17** How do elements heavier than iron arise?

Key Words

Summary

Nuclear chemistry is the study of the properties and reactions of atomic nuclei. **Nuclear reactions** differ from chemical reactions in that they involve a change in an atom's nucleus, often producing a different element. The rate of a nuclear reaction is unaffected by the addition of a catalyst or by a change in temperature or pressure, and the energy change accompanying a nuclear reaction is far greater than that accompanying any chemical reaction. Of the more than 3600 known isotopes, most have been made by **nuclear transmutation**, the change of one element into another. Such transmutation is often brought about by bombardment of an atom with a high-energy particle such as a proton, neutron, or α particle.

Radioactivity is the spontaneous emission of radiation from an unstable nucleus. **Alpha (α) radiation** consists of helium nuclei, small particles containing two protons and two neutrons ($_2^4He$). **Beta (β) radiation** consists of electrons ($_{-1}^0e$), and **gamma (γ) radiation** consists of high-energy photons that have no mass. **Positron emission** is the conversion of a proton in the nucleus into a neutron plus an ejected *positron*, $_1^0e$ or $β^+$, a particle that has the same mass as an electron but an opposite charge. **Electron capture** is the capture of an inner-shell electron by a proton in the nucleus. The process is accompanied by emission of gamma rays and results in the conversion of a proton in the nucleus into a neutron. Every element in the periodic table has at least one radioactive isotope, or **radioisotope**. Radioactive decay is characterized kinetically by a first-order **decay constant** and by a half-life, $t_{1/2}$, the time required for the number of radioactive nuclei in a sample to drop to half its initial value.

The stability of a given nucleus is related to its neutron/proton ratio. Neutrons act as a kind of nuclear "glue" that holds nuclei together by overcoming proton–proton repulsions. The strength of the forces involved can be measured by calculating an atom's **mass defect**—the difference in mass between a given nucleus and the total mass of its individual **nucleons** (protons and neutrons). Applying the Einstein equation $\Delta E = \Delta mc^2$ then allows calculation of the nuclear **binding energy**.

Certain heavy nuclei such as uranium-235 undergo **nuclear fission** when struck by neutrons, breaking apart into fragment nuclei and releasing enormous amounts of energy. Light nuclei such as the isotopes of hydrogen undergo **nuclear fusion** when heated to sufficiently high temperatures, forming heavier nuclei and releasing energy.

High-energy radiation of all types—α particles, β particles, γ rays, X rays, and cosmic rays—is known collectively as **ionizing radiation**. When ionizing radiation strikes a molecule, it dislodges an electron and leaves an ion. Radiation intensity is expressed in different ways according to what property is being measured. The becquerel (Bq) and the curie (Ci) measure the number of radioactive disintegrations per second in a sample. The gray (Gy) and the rad measure the amount of radiation absorbed per kilogram of tissue. The sievert (Sv) and the rem measure the amount of tissue damage caused by radiation. Radiation effects become noticeable with a human exposure of 25 rem and become lethal at an exposure above 600 rem.

Key Concept Summary

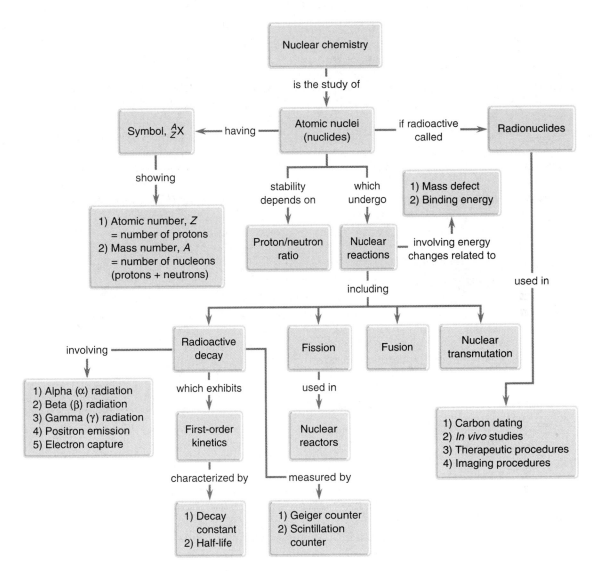

Understanding Key Concepts

Problems 22.1–22.17 appear within the chapter.

22.18 $^{40}_{19}$K decays by positron emission to give $^{40}_{18}$Ar. If yellow spheres represent $^{40}_{19}$K atoms and blue spheres represent $^{40}_{18}$Ar atoms, how many half-lives have passed in the following sample?

22.19 Write the symbol of the isotope represented by the following drawing. Blue spheres represent neutrons, and red spheres represent protons.

22.20 The nuclide shown in Problem 22.19 is radioactive. Would you expect it to decay by positron emission or β emission? Explain.

22.21 Of the two isotopes of iodine, ^{136}I and ^{122}I, one decays by β emission and one decays by positron emission. Which is which? Explain.

22.22 Of the two isotopes of tungsten, ^{160}W and ^{185}W, one decays by β emission and one decays by α emission. Which is which? Explain.

22.23 Identify the isotopes involved, and tell the type of decay process occurring in the following nuclear reaction:

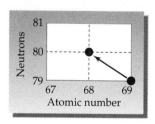

22.24 Isotope A decays to isotope E through the series of steps shown below. The series has two kinds of processes: one represented by the shorter arrows pointing right and the other represented by the longer arrows pointing left.

(a) What kind of nuclear decay process does each kind of arrow correspond to?

(b) Identify and write the symbol $^{A}_{Z}$X for each isotope in the decay series:

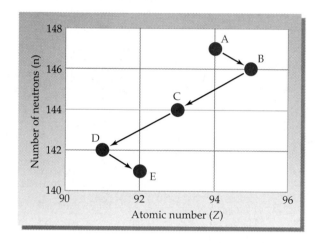

22.25 What is the half-life of the radionuclide that shows the following decay curve?

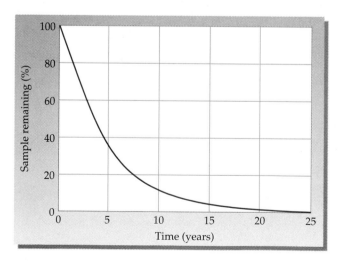

Additional Problems

Nuclear Reactions and Radioactivity

22.26 Positron emission and electron capture both give a product nuclide whose atomic number is 1 less than the starting nuclide. Explain.

22.27 What is the difference between an α particle and a helium atom?

22.28 Why is an α emitter more hazardous to an organism internally than externally, whereas a gamma emitter is equally hazardous internally and externally?

22.29 Why do nuclides that are "neutron-rich" emit β particles, but nuclides that are "neutron-poor" emit α particles or positrons or undergo electron capture?

22.30 Harmful chemical spills can often be cleaned up by treatment with another chemical. A spill of H_2SO_4, for example, can be neutralized by addition of $NaHCO_3$. Why can't harmful radioactive wastes from nuclear power plants be cleaned up just as easily?

22.31 The chemical reactions of a ^{24}Na atom and a $^{24}Na^+$ ion are completely different, but the nuclear reactions of a ^{24}Na atom and a $^{24}Na^+$ ion are identical (both emit β particles). Explain.

22.32 Complete and balance the following nuclear equations:

(a) $^{126}_{50}Sn \rightarrow {}^{0}_{-1}e + ?$

(b) $^{210}_{88}Ra \rightarrow {}^{4}_{2}He + ?$

(c) $^{77}_{37}Rb \rightarrow {}^{0}_{1}e + ?$

(d) $^{76}_{36}Kr + {}^{0}_{-1}e \rightarrow ?$

22.33 Complete and balance the following nuclear equations:

(a) $^{90}_{38}Sr \rightarrow {}^{0}_{-1}e + ?$

(b) $^{247}_{100}Fm \rightarrow {}^{4}_{2}He + ?$

(c) $^{49}_{25}Mn \rightarrow {}^{0}_{1}e + ?$

(d) $^{37}_{18}Ar + {}^{0}_{-1}e \rightarrow ?$

22.34 What particle is produced in each of the following decay reactions?

(a) $^{188}_{80}Hg \rightarrow {}^{188}_{79}Au + ?$

(b) $^{218}_{85}At \rightarrow {}^{214}_{83}Bi + ?$

(c) $^{234}_{90}Th \rightarrow {}^{234}_{91}Pa + ?$

22.35 What particle is produced in each of the following decay reactions?

(a) $^{24}_{11}Na \rightarrow {}^{24}_{12}Mg + ?$

(b) $^{135}_{60}Nd \rightarrow {}^{135}_{59}Pr + ?$

(c) $^{170}_{78}Pt \rightarrow {}^{166}_{76}Os + ?$

22.36 Write balanced nuclear equations for the following processes:

(a) α emission of ^{162}Re

(b) Electron capture of ^{138}Sm

(c) β emission of ^{188}W

(d) Positron emission of ^{165}Ta

22.37 Write balanced nuclear equations for the following processes:

(a) β emission of ^{157}Eu

(b) Electron capture of ^{126}Ba

(c) α emission of ^{146}Sm

(d) Positron emission of ^{125}Ba

22.38 Americium-241, a radioisotope used in smoke detectors, decays by a series of 12 reactions involving sequential loss of α, α, β, α, α, β, α, α, α, β, α, and β particles. Identify each intermediate nuclide and the final stable product nucleus.

22.39 Radon-222 decays by a series of three α emissions and two β emissions. What is the final stable nuclide?

22.40 Thorium-232 decays by a 10-step series, ultimately yielding lead-208. How many α particles and how many β particles are emitted?

22.41 How many α particles and how many β particles are emitted in the 11-step decay of ^{235}U into ^{207}Pb?

Radioactive Decay Rates

22.42 What does it mean when we say that the half-life of iron-59 is 44.5 days?

22.43 What is the difference between a half-life and a decay constant, and what is the relationship between them?

22.44 The half-life of indium-111, a radioisotope used in studying the distribution of white blood cells, is $t_{1/2} = 2.805$ days. What is the decay constant of ^{111}In?

22.45 What is the decay constant of gallium-67, a radioisotope used for imaging soft-tissue tumors? The half-life of ^{67}Ga is $t_{1/2} = 78.25$ hours.

22.46 The decay constant of thallium-201, a radioisotope used for parathyroid imaging, is 0.228 day^{-1}. What is the half-life of ^{201}Tl?

22.47 The decay constant of plutonium-239, a waste product from nuclear reactors, is 2.88×10^{-5} year^{-1}. What is the half-life of ^{239}Pu?

22.48 The half-life of ^{241}Am is 432.2 years. What percentage of a ^{241}Am sample remains after 65 days? After 65 years? After 650 years?

22.49 Fluorine-18 has $t_{1/2} = 109.8$ min. What percentage of an ^{18}F sample remains after 24 min? After 24 hours? After 24 days?

22.50 How old is a sample of wood whose ^{14}C content is found to be 43% that of a living tree? The half-life of ^{14}C is 5730 years.

22.51 What is the age of a rock whose $^{40}Ar/^{40}K$ ratio is 1.15? The half-life of ^{40}K is 1.28×10^9 years.

22.52 The decay constant of ^{35}S is 7.89×10^{-3} day^{-1}. What percentage of a ^{35}S sample remains after 185 days?

22.53 Plutonium-239 has a decay constant of 2.88×10^{-5} year^{-1}. What percentage of a ^{239}Pu sample remains after 1000 years? After 25,000 years? After 100,000 years?

22.54 Polonium-209, an α emitter, has a half-life of 102 years. How many α particles are emitted in 1.0 s from a 1.0 ng sample of ^{209}Po?

22.55 Chlorine-36 is a β emitter, with a half-life of 3.0×10^5 years. How many β particles are emitted in 1.0 min from a 5.0 mg sample of ^{36}Cl? How many curies of radiation does the 5.0 mg sample represent?

22.56 A 1.0 mg sample of ^{79}Se decays initially at a rate of 1.5×10^5 disintegrations/s. What is the half-life of ^{79}Se in years?

22.57 What is the half-life (in years) of ^{44}Ti if a 1.0 ng sample decays initially at a rate of 4.8×10^3 disintegrations/s?

22.58 A sample of ^{37}Ar undergoes 8540 disintegrations/min initially but undergoes 6990 disintegrations/min after 10.0 days. What is the half-life of ^{37}Ar in days?

22.59 A sample of ^{28}Mg decays initially at a rate of 53,500 disintegrations/min, but the decay rate falls to 10,980 disintegrations/min after 48.0 hours. What is the half-life of ^{28}Mg in hours?

Energy Changes During Nuclear Reactions

22.60 Why does a given nucleus have less mass than the sum of its constituent protons and neutrons?

22.61 In an endothermic chemical reaction, do the products have more mass, less mass, or the same mass as the reactants? Explain.

22.62 What is the wavelength (in nm) of gamma rays whose energy is 1.50 MeV?

22.63 What is the frequency (in Hz) of X rays whose energy is 6.82 keV?

22.64 Calculate the mass defect (in g/mol) for the following nuclides:
(a) ^{52}Fe (atomic mass = 51.948 11 amu)
(b) ^{92}Mo (atomic mass = 91.906 81 amu)

22.65 Calculate the mass defect (in g/mol) for the following nuclides:
(a) ^{32}S (atomic mass = 31.972 07 amu)
(b) ^{40}Ca (atomic mass = 39.962 59 amu)

22.66 Calculate the binding energy (in MeV/nucleon) for the following nuclides:
(a) ^{58}Ni (atomic mass = 57.935 35 amu)
(b) ^{84}Kr (atomic mass = 83.911 51 amu)

22.67 Calculate the binding energy (in MeV/nucleon) for the following nuclides:
(a) ^{63}Cu (atomic mass = 62.939 60 amu)
(b) ^{84}Sr (atomic mass = 83. 913 43 amu)

22.68 What is the energy change ΔE (in kJ/mol) when an α particle is emitted from ^{174}Ir? The atomic mass of ^{174}Ir is 173.966 66 amu, the atomic mass of ^{170}Re is 169.958 04 amu, and the atomic mass of a ^4He atom is 4.002 60 amu.

$$^{174}_{77}Ir \longrightarrow {}^{170}_{75}Re + {}^4_2He \qquad \Delta E = ?$$

22.69 Magnesium-28 is a β emitter that decays to aluminum-28. How much energy is released in kJ/mol? The atomic mass of ^{28}Mg is 27.983 88 amu, and the atomic mass of ^{28}Al is 27.981 91 amu.

22.70 What is the mass change (in grams) accompanying the formation of NH_3 from H_2 and N_2?

$$N_2(g) + 3\,H_2(g) \longrightarrow 2\,NH_3(g) \quad \Delta H° = -92.2 \text{ kJ}$$

22.71 What is the mass change (in grams) accompanying the formation of CO and H_2 in the water gas reaction?

$$C(s) + H_2O(g) \longrightarrow CO(g) + H_2(g) \quad \Delta H° = +131 \text{ kJ}$$

22.72 A positron has the same mass as an electron (9.109×10^{-31} kg) but an opposite charge. When the two particles encounter each other, annihilation occurs and only gamma rays are produced. How much energy (in kJ/mol) is produced?

22.73 How much energy is released (in kJ) in the fusion reaction of ^2H to yield 1 mol of ^3He? The atomic mass of ^2H is 2.0141 amu, and the atomic mass of ^3He is 3.0160 amu.

$$2\,{}^2_1H \longrightarrow {}^3_2He + {}^1_0n$$

Nuclear Transmutation

22.74 Give the products of the following nuclear reactions:

(a) $^{109}_{47}\text{Ag} + ^4_2\text{He} \rightarrow$?

(b) $^{10}_5\text{B} + ^4_2\text{He} \rightarrow$? $+ ^1_0\text{n}$

22.75 Balance the following equations for the nuclear fission of ^{235}U:

(a) $^{235}_{92}\text{U} \rightarrow ^{160}_{62}\text{Sm} + ^{72}_{30}\text{Zn} + ?^1_0\text{n}$

(b) $^{235}_{92}\text{U} \rightarrow ^{87}_{35}\text{Br} + ? + 2^1_0\text{n}$

22.76 Element 109 ($^{266}_{109}\text{Mt}$) was prepared in 1982 by bombardment of ^{209}Bi atoms with ^{58}Fe atoms. Identify the other product that must have formed, and write a balanced nuclear equation.

22.77 Molybdenum-99 is formed by neutron bombardment of a naturally occurring isotope. If one neutron is absorbed and no by-products are formed, what is the starting isotope?

22.78 Californium-246 is formed by bombardment of uranium-238 atoms. If four neutrons are formed as by-products, what particle is used for the bombardment?

22.79 Balance the following transmutation reactions:

(a) $^{246}_{96}\text{Cm} + ^{12}_6\text{C} \rightarrow ? + 4^1_0\text{n}$

(b) $^{253}_{99}\text{Es} + ? \rightarrow ^{256}_{101}\text{Md} + ^1_0\text{n}$

(c) $^{250}_{98}\text{Cf} + ^{11}_5\text{B} \rightarrow ? + 4^1_0\text{n}$

General Problems

22.80 How much energy (in kJ/mol) is released during the overall decay of ^{232}Th to ^{208}Pb (Problem 22.40)? The relevant masses are ^{232}Th = 232.038 054 amu; ^{208}Pb = 207.976 627 amu; electron = 5.485 799 × 10^{-4} amu; α particle = 4.001 506 amu.

22.81 Potassium ion, K^+, is present in most foods and is an essential nutrient in the human body. Potassium-40, however, which has a natural abundance of 0.0117%, is radioactive with $t_{1/2} = 1.28 \times 10^9$ years. What is the decay constant of ^{40}K? How many $^{40}\text{K}^+$ ions are present in 1.00 g of KCl? How many disintegrations/s does 1.00 g of KCl undergo?

22.82 How much energy (in kJ/mol) is released during the overall decay process shown in Problem 22.24? The atomic mass of A = 241.056 845 amu; atomic mass of E = 233.039 628 amu; electron mass = 5.485 799 × 10^{-4} amu; α particle mass = 4.001 506 amu.

22.83 Three new elements were created in a single experiment in 1999 when ^{208}Pb was bombarded with ^{86}Kr. Identify the isotopes X, Y, and Z.

$$^{208}_{82}\text{Pb} + ^{86}_{36}\text{Kr} \xrightarrow{-\text{n}} X \xrightarrow{-\alpha} Y \xrightarrow{-\alpha} Z$$

22.84 Chlorine-34 has a half-life of only 1.53 s. How long does it take for 99.99% of a ^{34}Cl sample to decay?

22.85 The decay constant for ^{20}F is 0.063 s^{-1}. How long does it take for 99.99% of a ^{20}F sample to decay?

22.86 Calculate the mass defect (in g/mol) and the binding energy (in MeV/nucleon) for the following nuclides. Which of the two is more stable?

(a) ^{50}Cr (atomic mass = 49.946 05 amu)

(b) ^{64}Zn (atomic mass = 63.929 15 amu)

22.87 What is the age of a bone fragment that shows an average of 2.9 disintegrations/min per gram of carbon? The carbon in living organisms undergoes an average of 15.3 disintegrations/min per gram, and the half-life of ^{14}C is 5730 years.

22.88 How much energy (in kJ/mol) is released in the fusion reaction of ^2H with ^3He?

$$^2_1\text{H} + ^3_2\text{He} \longrightarrow ^4_2\text{He} + ^1_1\text{H}$$

The relevant masses are ^2H (2.0141 amu), ^3He (3.0160 amu), ^4He (4.0026 amu), ^1H (1.0078 amu).

22.89 The longest half-life yet measured for radioactive decay is the double β emission of selenium-82, for which $t_{1/2} = 1.1 \times 10^{20}$ years.

$$^{82}_{34}\text{Se} \longrightarrow ^{82}_{36}\text{Kr} + 2^{\ 0}_{-1}\text{e}$$

How many disintegrations per day occur initially in a 1.0 mol sample of ^{82}Se?

22.90 The most abundant isotope of uranium, ^{238}U, does not undergo fission. In a *breeder reactor*, however, a ^{238}U atom captures a neutron and emits two β particles to make a fissionable isotope of plutonium, which can then be used as fuel in a nuclear reactor. Write a balanced nuclear equation.

22.91 It has been estimated that 3.9×10^{23} kJ/s is radiated into space by the sun. What is the rate of the sun's mass loss in kg/s?

Multi-Concept Problems

22.92 A small sample of wood from an archeological site in Clovis, New Mexico, was burned in O_2, and the CO_2 produced was bubbled through a solution of $Ba(OH)_2$ to produce a precipitate of $BaCO_3$. When the $BaCO_3$ was collected by filtration, a 1.000 g sample was found to have a radioactivity of 4.0×10^{-3} Bq. The half-life of ^{14}C is 5730 years, and living organisms have a radioactivity due to ^{14}C of 15.3 disintegrations/min per gram of carbon. What is the age of the Clovis site?

22.93 Polonium-210, a naturally occurring radioisotope, is an α emitter, with $t_{1/2} = 138$ days. Assume that a sample of ^{210}Po with a mass of 0.700 mg was placed in a 250.0 mL flask, which was evacuated, sealed, and allowed to sit undisturbed. What would the pressure be inside the flask (in mm Hg) at 20°C after 365 days if all the α particles emitted had become helium atoms?

22.94 A blood-volume determination was carried out on patient by injection with 20.0 mL of blood that had been radioactively labeled with Cr-51 to an activity of 4.10 μCi/mL. After a brief period to allow for mixing in the body, blood was drawn from the patient for analysis. Unfortunately, a mixup in the laboratory prevented an immediate analysis, and it was not until 17.0 days later that a scintillation measurement on the blood was made. The radiation level was then determined to be 0.00935 μCi/mL. If ^{51}Cr has $t_{1/2} = 27.7$ days, what is the volume of blood in the patient?

 eMedia Problems

22.95 The **Separation of Alpha, Beta, and Gamma Rays** movie (*eChapter 22.2*) shows how alpha particles are deflected away from the positive pole of an electric field indicating that they are positively charged. Write the balanced nuclear equation for alpha emission from each of the following nuclei.

(a) ^{177}Ir

(b) ^{156}Yb

(c) ^{154}Dy

(d) ^{161}Re

22.96 The **Separation of Alpha, Beta, and Gamma Rays** movie (*eChapter 22.2*) shows how beta particles are deflected away from the negative pole of an electric field indicating that they are negatively charged. Write the balanced nuclear equation for beta emission from each of the following nuclei.

(a) ^{197}Ir

(b) ^{195}Os

(c) ^{172}Er

(d) ^{182}Hf

22.97 ^{238}U has a half-life of 4.47×10^9 years. Use the **Radioactive Decay** activity (*eChapter 22.3*) to determine the half-life of ^{235}U. Which isotope is more abundant in nature? Comment on the apparent relationship between half-life and natural abundance of these two isotopes.

22.98 In Problem 22.51, you determined the age of a rock using the $^{40}Ar/^{40}K$ ratio. Using the rate constant that you calculated for that problem, calculate how much of a 20.00 kg sample of ^{40}K would remain after 1.6 billion years. Use the **Radioactive Decay** activity (*eChapter 22.3*) to verify your answer. By what process(es) does ^{40}K decay to ^{40}Ar?

22.99 Determine the half-life of ^{232}Th to three significant figures using the **Radioactive Decay** activity (*eChapter 22.3*). What percentage of a 20.00 kg sample of ^{232}Th would remain after 5.05×10^9 years?

Organic Chemistry

The Neotronics Olfactory Sensing Equipment (NOSE) may complement or replace human quality control testers for analyzing odors and aromas in the food, beverage, and fragrance industries.

F rom the very beginning of chemical studies in the mid-1700s, people noticed that substances obtained from plants and animals were more difficult to purify and work with than those from minerals. To express this difference, the term *organic chemistry* was used to mean the study of compounds from living organisms, while *inorganic chemistry* was used to refer to the study of compounds from nonliving sources. Today we know that there are no fundamental differences between organic and inorganic compounds—the same principles apply to both. The only common characteristic of compounds from living sources is that all contain the element *carbon*. Thus, **organic chemistry** is now defined as the study of carbon compounds.

Why is carbon special, and why do chemists still treat organic chemistry as a separate branch of science? The answers to these questions involve the ability of carbon atoms to bond together, forming long chains and rings. Of all the elements, only carbon is able to form such an immense array of compounds, from methane, with one carbon atom, to deoxyribonucleic acid (DNA), with tens of billions of carbon atoms. More than 17 million organic compounds have been made, and thousands of new ones are made each week in chemical laboratories throughout the world.

23.1   The Nature of Organic Molecules

Let's review what we've seen in earlier chapters about organic molecules:

- Carbon is *tetravalent* (Section 7.5); it has four outer-shell electrons ($1s^2\ 2s^2\ 2p^2$) and forms four bonds. In methane, for example, carbon is bonded to four hydrogen atoms:

$$H - \underset{\displaystyle H}{\overset{\displaystyle H}{\underset{|}{\overset{|}{C}}}} - H \qquad \text{Methane, CH}_4$$

- Organic molecules have *covalent bonds* (Section 7.1). In ethane, for example, all bonds result from the sharing of two electrons, either between C and C or between C and H:

$$\begin{array}{c} H\ H \\ H\!:\!\ddot{C}\!:\!\ddot{C}\!:\!H \\ H\ H \end{array} = H - \underset{\displaystyle H\ \ H}{\overset{\displaystyle H\ \ H}{\underset{|\ \ |}{\overset{|\ \ |}{C - C}}}} - H \qquad \text{Ethane, C}_2\text{H}_6 \text{ or CH}_3\text{CH}_3$$

- Organic molecules have *polar covalent bonds* when carbon bonds to an element on the right or left side of the periodic table (Section 7.4). In chloromethane, the electronegative chlorine atom attracts electrons more strongly than carbon does, resulting in polarization of the C–Cl bond so that carbon has a partial positive charge, δ+. In methyllithium, the lithium attracts electrons less strongly than carbon does, resulting in polarization of the carbon–lithium bond so that carbon has a partial negative charge, δ−. These polarity patterns are seen clearly in the following electrostatic potential maps, where electron-rich regions appear in red and electron-poor regions in blue.

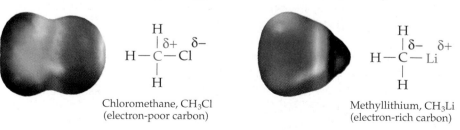

Chloromethane, CH$_3$Cl
(electron-poor carbon)

Methyllithium, CH$_3$Li
(electron-rich carbon)

- Carbon can form *multiple covalent bonds* by sharing more than two electrons with a neighboring atom (Section 7.5). In ethylene, the two carbon atoms share four electrons in a double bond. In acetylene, the two carbons share six electrons in a triple bond:

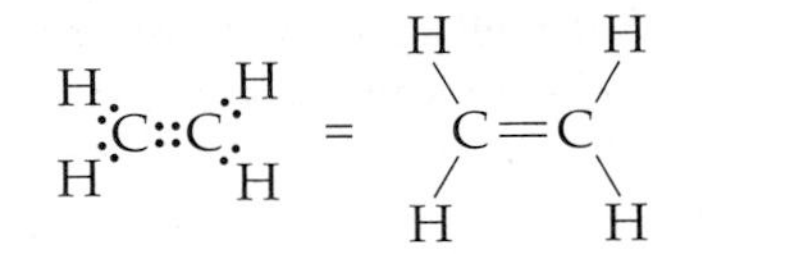

H₂C::C H $=$ C=C Ethylene, C_2H_4

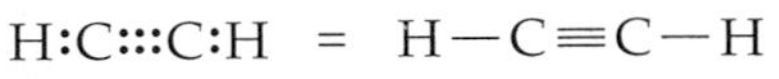

H:C:::C:H $=$ H—C≡C—H   Acetylene, C_2H_2

- Organic molecules have specific three-dimensional shapes, which can be predicted by the VSEPR model (Section 7.9). When carbon is bonded to four atoms, as in methane, the bonds are oriented toward the four corners of a tetrahedron with carbon in the center and with H–C–H angles near 109.5°:

109.5°

When carbon bonds to three atoms, as in ethylene, the bonds are at angles of approximately 120° to one another. When carbon bonds to two atoms, as in acetylene, the bonds are at angles of 180°. Note that in the models of these molecules, the multiple bonds are shown with stripes.

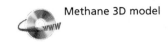

Methane 3D model
WWW

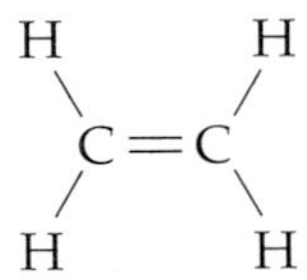

Ethylene (120° angles)

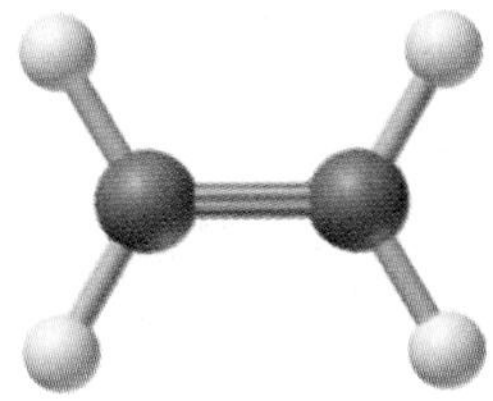

Acetylene (180° angles)

- Carbon uses *hybrid atomic orbitals* for bonding (Sections 7.11 and 7.12). A carbon that bonds to four atoms uses sp^3 orbitals formed by the combination of an atomic s orbital with three atomic p orbitals. These sp^3 orbitals point toward the corners of a tetrahedron, accounting for the observed geometry of carbon.

Doubly bonded carbons are sp^2-hybridized. Carbon has three sp^2 hybrid orbitals, which lie in a plane and point toward the corners of an equilateral triangle, and one unhybridized p orbital, which is oriented at a 90° angle to the plane of the sp^2 hybrids. When two sp^2-hybridized carbon atoms approach each other with sp^2 orbitals aligned head-on for sigma bonding, the unhybridized p orbitals on each carbon overlap to form a pi bond, resulting in a net carbon–carbon double bond.

Triply bonded carbons are sp-hybridized. Carbon has two sp hybrid orbitals, which are at 180° to each other, and two unhybridized p orbitals, which are oriented 90° from the sp hybrids and 90° from each other. When two sp-hybridized carbon atoms approach each other with sp orbitals aligned head-on for sigma bonding, the *p* orbitals on each carbon overlap to form two pi bonds, resulting in a net carbon–carbon triple bond.

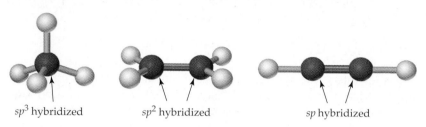

sp³ hybridized   *sp² hybridized*   *sp hybridized*

Covalent bonding gives organic compounds properties that are quite different from those of ionic compounds. Intermolecular forces between individual organic molecules are relatively weak, and organic compounds therefore have lower melting and boiling points than do ionic compounds. In fact, many simple organic compounds are liquid at room temperature. In addition, most organic compounds are insoluble in water and don't conduct electricity. Only a few small polar organic molecules such as glucose, acetic acid, and ethyl alcohol dissolve in water.

23.2   Alkanes and Their Isomers

Why are there so many organic compounds? The answer is that a relatively small number of atoms can bond together in a great many ways. Take molecules that contain only carbon and hydrogen (**hydrocarbons**) and have only single bonds. Such compounds belong to the family of organic molecules called **saturated hydrocarbons**, or **alkanes**.

If you imagine ways that one carbon and four hydrogens can combine, only methane, CH_4, is possible. If you imagine ways that two carbons and six hydrogens can combine, only ethane, C_2H_6, is possible; and if you imagine the combination of three carbons with eight hydrogens, only propane, C_3H_8, is possible.

The paraffin wax coating that makes these apples so shiny is a mixture of alkanes.

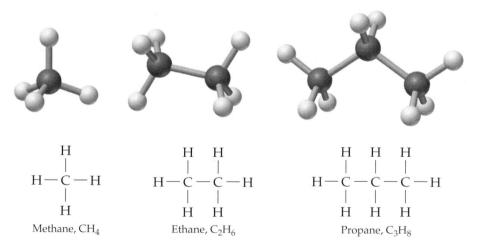

Methane, CH_4 Ethane, C_2H_6 Propane, C_3H_8

When larger numbers of carbons and hydrogens combine, though, *more than one structure can result*. There are two structures with the formula C_4H_{10}, for instance: One has the four carbons in a row, and the other has them in a branched arrangement. Similarly, there are three structures with the formula C_5H_{12}, and even more for larger alkanes. Compounds with all their carbons connected in a row are called **straight-chain alkanes**, and those with a branching connection of carbons are called **branched-chain alkanes**.

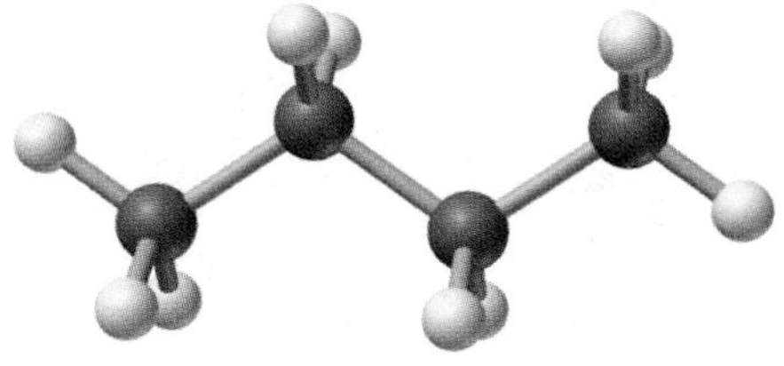

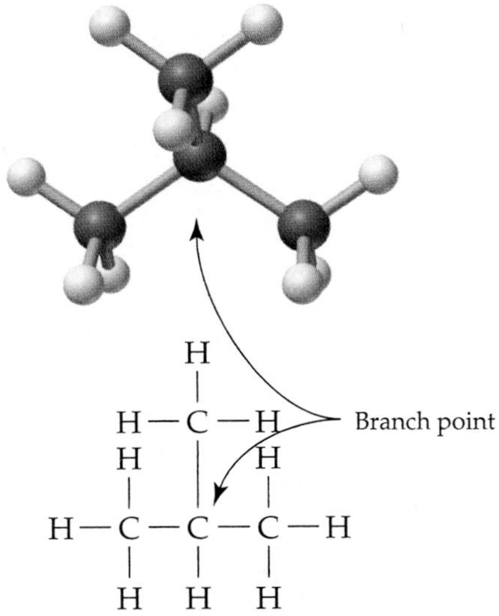

C_4H_{10}

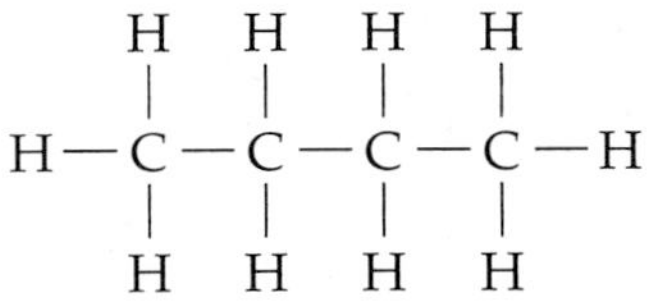

Butane (straight chain)

2-Methylpropane (branched chain)

Branch point

C_5H_{12}

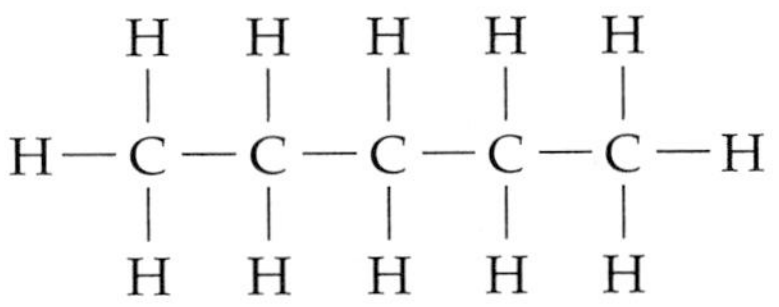

Pentane
(straight chain)

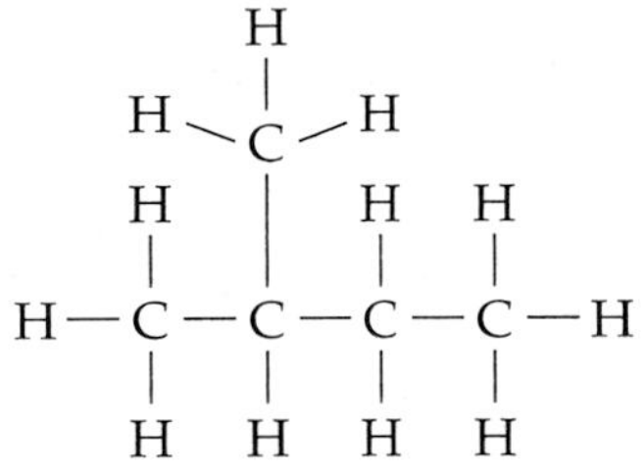

2-Methylbutane
(branched chain)

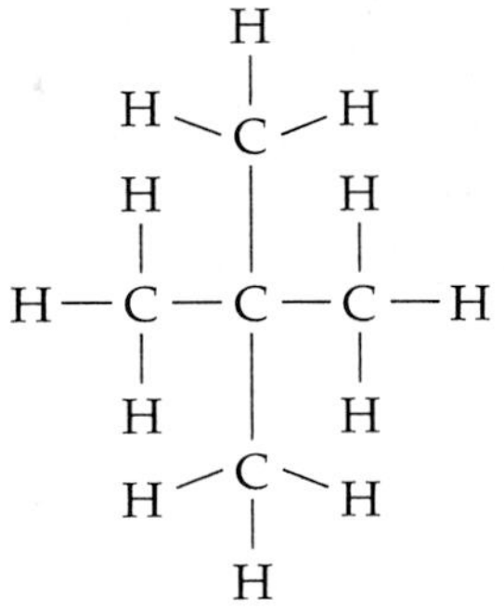

2,2-Dimethylpropane
(branched chain)

Compounds like the two different C_4H_{10} molecules and the three different C_5H_{12} molecules, which have the same molecular formula but different structures, are called *isomers*, as we saw in Section 20.8. The number of possible alkane isomers grows rapidly as the number of carbon atoms increases (Table 23.1).

TABLE 23.1 Number of Possible Alkane Isomers

Formula	No. of Isomers	Formula	No. of Isomers
C_6H_{14}	5	$C_{10}H_{22}$	75
C_7H_{16}	9	$C_{20}H_{42}$	366,319
C_8H_{18}	18	$C_{30}H_{62}$	4,111,846,763
C_9H_{20}	35	$C_{40}H_{82}$	62,491,178,805,831

It's important to realize that different isomers are different chemical compounds. They have different structures, different chemical properties, and different physical properties, such as melting point and boiling

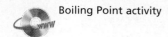

point. For example, ethyl alcohol (ethanol, or grain alcohol) and dimethyl ether both have the formula C_2H_6O, yet ethyl alcohol is a liquid with a boiling point of 78.5°C, whereas dimethyl ether is a gas with a boiling point of −23°C.

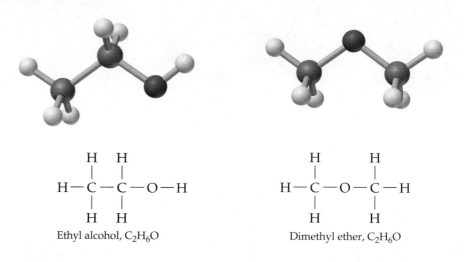

Ethyl alcohol, C_2H_6O Dimethyl ether, C_2H_6O

▶ **PROBLEM 23.1**   Draw the straight-chain isomer with the formula C_7H_{16}.

▶ **PROBLEM 23.2**   Draw the five alkane isomers with the formula C_6H_{14}.

23.3   Drawing Organic Structures

It's both awkward and time-consuming to draw all the bonds and all the atoms in an organic molecule, even for a relatively small one like C_4H_{10}. Thus, a shorthand way of drawing **condensed structures** is often used. In condensed structures, carbon–hydrogen and most carbon–carbon single bonds aren't shown; rather, they're "understood." If a carbon atom has three hydrogens bonded to it, we write CH_3; if the carbon has two hydrogens bonded to it, we write CH_2; and so on. For example, the four-carbon, straight-chain alkane (called *butane*) and its branched-chain isomer (called *2-methylpropane*, or *isobutane*) can be written in the following way:

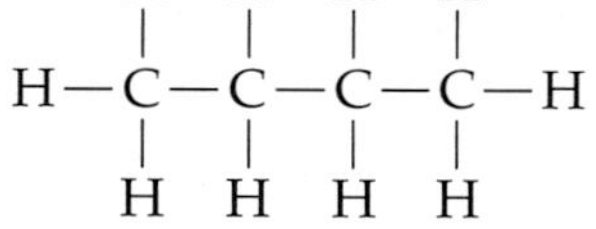

$$
\begin{array}{c}
\overset{\displaystyle H \;\; H \;\; H \;\; H}{\underset{\displaystyle H \;\; H \;\; H \;\; H}{H-C-C-C-C-H}}
\end{array}
\;=\; CH_3CH_2CH_2CH_3 \qquad \text{Butane}
$$

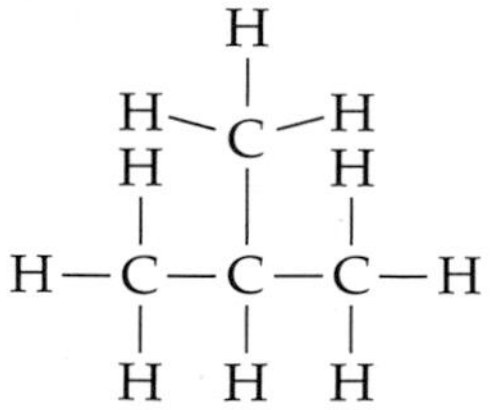

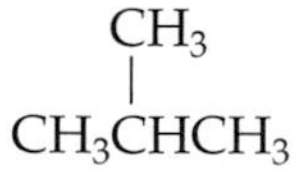

$$
\begin{array}{c}
H-C-C-C-H
\end{array}
\;=\; CH_3CHCH_3 \qquad \begin{array}{c}\text{2- Methylpropane}\\ \text{(Isobutane)}\end{array}
$$

Note that the horizontal bonds between carbons aren't shown—the CH_3 and CH_2 units are simply placed next to each other—but the vertical bond in 2-methylpropane is shown for clarity.

▶**PROBLEM 23.3**   Draw the three isomers of C_5H_{12} as condensed structures.

✦ **KEY CONCEPT PROBLEM 23.4**   Give the formula of the following molecular
model, and convert the model into a condensed structure.

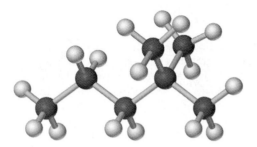

23.4 The Shapes of Organic Molecules

The condensed structure of an organic molecule implies nothing about three-dimensional shape; it only indicates the connections among atoms. Thus, a molecule can be arbitrarily drawn in many different ways. The branched-chain alkane called 2-methylbutane, for instance, might be represented by any of the following structures. All have four carbons connected in a row, with a $-CH_3$ branch on the second carbon from the end.

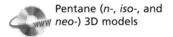

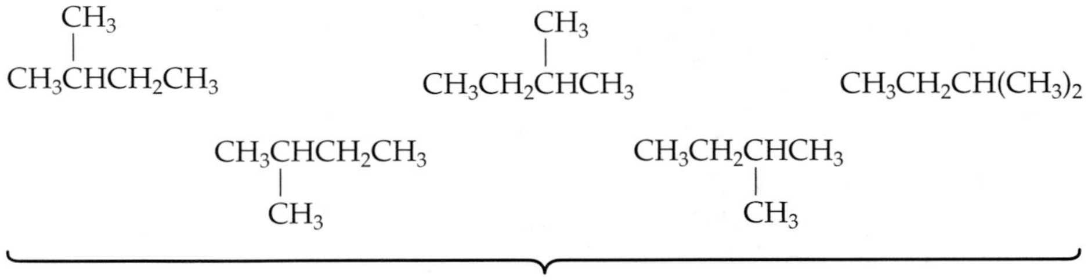

Pentane (*n*-, *iso*-, and
neo-) 3D models

$$\overset{\displaystyle CH_3}{\underset{\displaystyle |}{CH_3CHCH_2CH_3}}\qquad\qquad \overset{\displaystyle CH_3}{\underset{\displaystyle |}{CH_3CH_2CHCH_3}}\qquad\qquad CH_3CH_2CH(CH_3)_2$$

$$\underset{\displaystyle \underset{\displaystyle |}{CH_3}}{CH_3CHCH_2CH_3}\qquad\qquad\qquad \underset{\displaystyle \underset{\displaystyle |}{CH_3}}{CH_3CH_2CHCH_3}$$

Some representations of 2-methylbutane

In fact, 2-methylbutane has no one single shape because *rotation* occurs around carbon–carbon single bonds. The two parts of a molecule joined by a carbon–carbon single bond are free to spin around the bond, giving rise to an infinite number of possible three-dimensional structures, or **conformations**. Thus, a large sample of 2-methylbutane contains a great many molecules that are constantly changing their shape. At any given instant, though, most of the molecules have an extended, zigzag conformation, which is slightly more stable than other possibilities. The same is true for other alkanes.

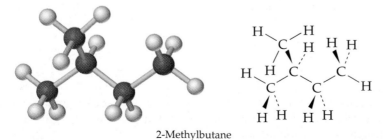

2-Methylbutane

EXAMPLE 23.1   The following condensed structures have the same formula, C_8H_{18}. Which of them represent the same molecule?

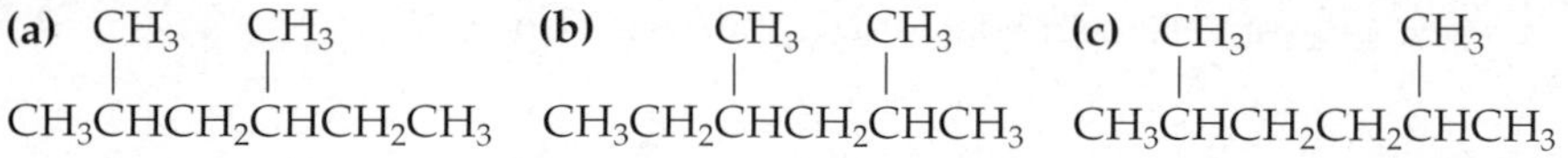

(a) CH₃   CH₃
     | |
CH₃CHCH₂CHCH₂CH₃

(b)   CH₃   CH₃
          | |
CH₃CH₂CHCH₂CHCH₃

(c) CH₃   CH₃
     | |
CH₃CHCH₂CH₂CHCH₃

SOLUTION   Pay attention to the order of connection between atoms. Don't get confused by the apparent differences caused by writing a structure right-to-left versus left-to-right. In this example, structure (a) has a straight chain of six carbons with −CH₃ branches on the second and fourth carbons from the end. Structure (b) also has a straight chain of six carbons with −CH₃ branches on the second and fourth carbons from the end and is therefore identical to (a). The only difference between (a) and (b) is that one is written "forward" and one is written "backward." Structure (c) has a straight chain of six carbons with −CH₃ branches on the second and *fifth* carbons from the end, so it is an isomer of (a) and (b).

▶ **PROBLEM 23.5**   Which of the following structures are identical?

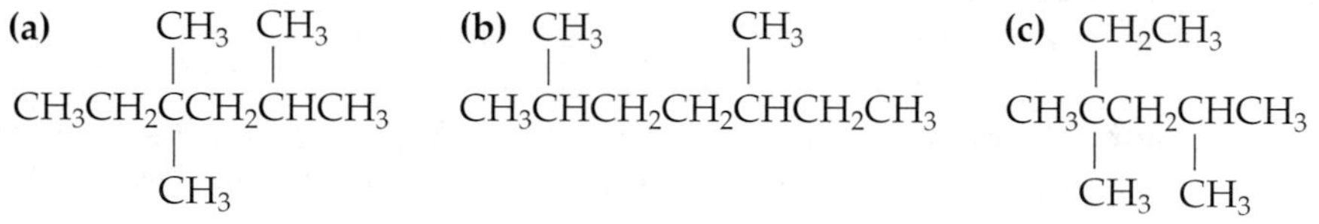

(a)   CH₃ CH₃
         | |
CH₃CH₂CCH₂CHCH₃
         |
         CH₃

(b) CH₃   CH₃
     | |
CH₃CHCH₂CH₂CHCH₂CH₃

(c) CH₂CH₃
     |
CH₃CCH₂CHCH₃
     | |
     CH₃ CH₃

23.5   Naming Alkanes

In earlier times, when relatively few pure organic chemicals were known, new compounds were named at the whim of their discoverer. Thus, urea is a crystalline substance first isolated from urine, and the barbiturates are a group of tranquilizing agents named by their discoverer in honor of his friend Barbara. As more and more compounds became known, however, the need for a systematic method of naming organic compounds became apparent.

The system of naming now used was devised by the International Union of Pure and Applied Chemistry, abbreviated IUPAC. In the IUPAC system, a chemical name has three parts: prefix, parent, and suffix. The parent name tells how many carbon atoms are present in the longest continuous chain; the suffix identifies what family the molecule belongs to; and the prefix (if needed) specifies the location of various substituent groups attached to the parent chain:

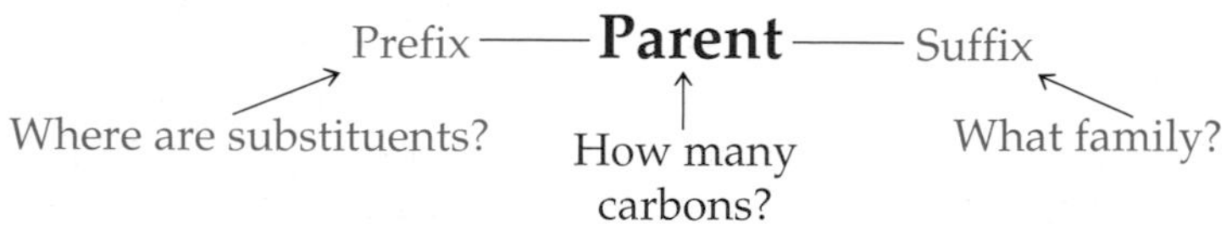

Prefix —— **Parent** —— Suffix

Where are substituents?   How many carbons?   What family?

Straight-chain alkanes are named by counting the number of carbon atoms in the chain and adding the family suffix *-ane*. With the exception of the first four compounds—methane, ethane, propane, and butane—whose names have historical origins, the alkanes are named from Greek numbers according to the number of carbons present. Thus, *pent*ane is the five-carbon alkane, *hex*ane is the six-carbon alkane, and so on, as shown in Table 23.2.

TABLE 23.2 Names of Straight-Chain Alkanes

Number of Carbons	Structure	Name	Number of Carbons	Structure	Name
1	CH_4	*Methane*	6	$CH_3CH_2CH_2CH_2CH_2CH_3$	*Hexane*
2	CH_3CH_3	*Ethane*	7	$CH_3CH_2CH_2CH_2CH_2CH_2CH_3$	*Heptane*
3	$CH_3CH_2CH_3$	*Propane*	8	$CH_3CH_2CH_2CH_2CH_2CH_2CH_2CH_3$	*Octane*
4	$CH_3CH_2CH_2CH_3$	*Butane*	9	$CH_3CH_2CH_2CH_2CH_2CH_2CH_2CH_2CH_3$	*Nonane*
5	$CH_3CH_2CH_2CH_2CH_3$	*Pentane*	10	$CH_3CH_2CH_2CH_2CH_2CH_2CH_2CH_2CH_2CH_3$	*Decane*

Branched-chain alkanes are named by following four steps:

Step 1. Name the main chain. Find the longest continuous chain of carbons in the molecule, and use the name of that chain as the parent name. The longest chain may not always be obvious from the manner of writing; you may have to "turn corners" to find it:

$$CH_3-CH_2$$
$$CH_3-CH-CH_2-CH_2-CH_3$$

Named as hexane, not as a pentane, because the longest chain has six carbons.

If you prefer, you can redraw the structure so that the longest chain is on one line:

$$CH_3-CH_2$$
$$CH_3-CH-CH_2-CH_2-CH_3$$

same as

$$CH_3$$
$$CH_3-CH_2-CH-CH_2-CH_2-CH_3$$

Step 2. Number the carbon atoms in the main chain. Beginning at the end nearer the first branch point, number each carbon atom in the parent chain:

$$CH_3$$
$$CH_3-CH_2-CH-CH_2-CH_2-CH_3$$
$$1\quad\ 2\quad\ 3\quad\ 4\quad\ 5\quad\ 6$$
$$[\,6\quad\ 5\quad\ 4\quad\ 3\quad\ 2\quad\ 1\,]$$

Wrong numbering

The first (and only) branch occurs at the third carbon, C3, if we start numbering from the left, but would occur at C4 if we started from the right by mistake.

Step 3. Identify and number the branching substituent. Assign a number to each branching substituent group on the parent chain according to its point of attachment.

$$CH_3$$
$$CH_3-CH_2-CH-CH_2-CH_2-CH_3$$
$$1\quad\ 2\quad\ 3\quad\ 4\quad\ 5\quad\ 6$$

The main chain is a hexane. There is a $-CH_3$ substituent group connected to C3 of the chain.

If there are two substituent groups on the same carbon, assign the same number to both. There must always be as many numbers in the name as there are substituents.

$$CH_2-CH_3$$
$$CH_3-CH_2-C-CH_2-CH_2-CH_3$$
$$1\quad\ 2\quad\ 3\ |\ \ 4\quad\ 5\quad\ 6$$
$$CH_3$$

The main chain is hexane. There are two substituents, a $-CH_3$ and a $-CH_2CH_3$, both connected to C3 of the chain.

The $-CH_3$ and $-CH_2CH_3$ substituents that branch off the main chain in this compound are called **alkyl groups**. You can think of an alkyl group as the part of an alkane that remains when a hydrogen is removed. For example, removing a hydrogen from methane, CH_4, gives the *methyl group*, $-CH_3$, and removing a hydrogen from ethane, CH_3CH_3, gives the *ethyl group*, $-CH_2CH_3$. Alkyl groups are named by replacing the *-ane* ending of the parent alkane with an *-yl* ending.

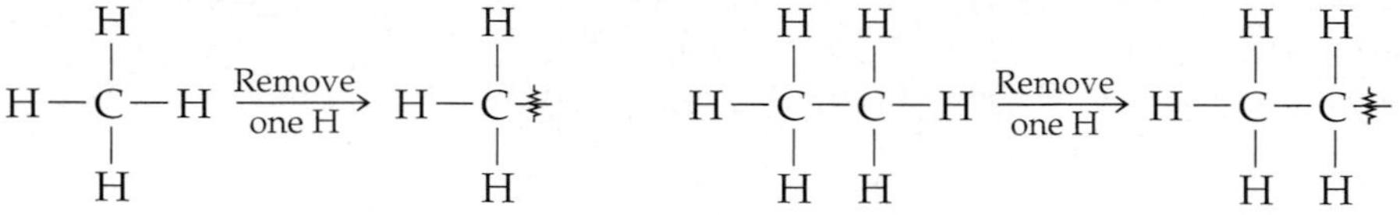

Methane   A methyl group   Ethane   An ethyl group

Step 4. Write the name as a single word. Use hyphens to separate the different prefixes, and use commas to separate numbers when there are more than one. If two or more different substituent groups are present, list them in alphabetical order. If two or more identical substituents are present, use one of the prefixes *di-*, *tri-*, *tetra-*, and so forth, but don't use these prefixes for alphabetizing purposes.

Look at the following examples to see how names are written:

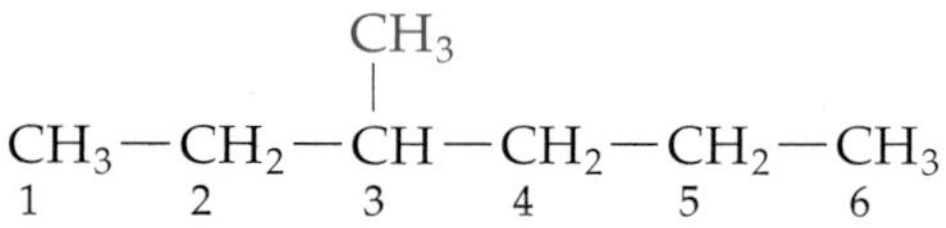

3-Methylhexane—a six-carbon main chain with a 3-methyl substituent

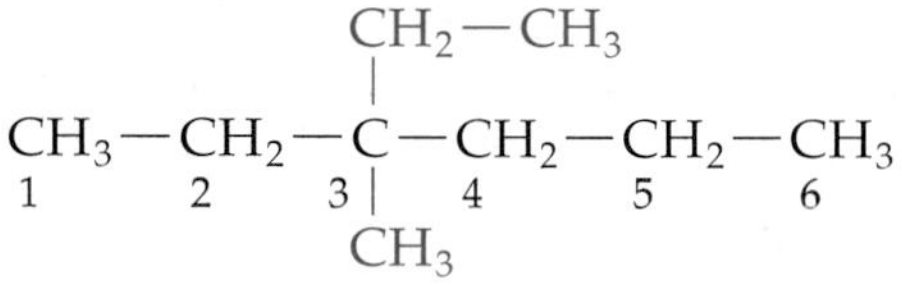

3-Ethyl-3-methylhexane—a six-carbon main chain with 3-ethyl and 3-methyl substituents

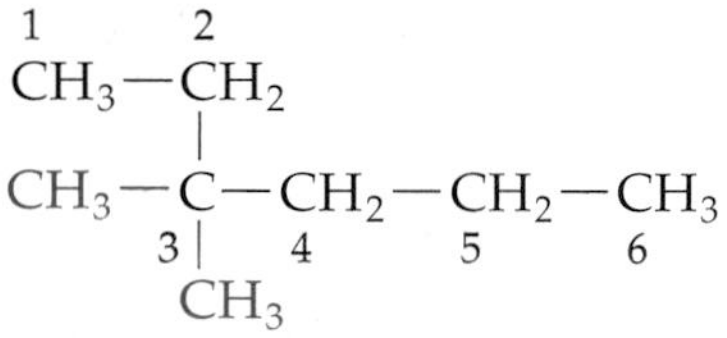

3,3-Dimethylhexane—a six-carbon main chain with two 3-methyl substituents

More About Alkyl Groups

It doesn't matter which hydrogen is removed from CH_4 to form a methyl group or which hydrogen is removed from CH_3CH_3 to form an ethyl group because all the hydrogen atoms in each molecule are equivalent. The eight hydrogens in $CH_3CH_2CH_3$, however, are not all equivalent. Propane has two "kinds" of hydrogens—six on the end carbons and two on the middle carbon. Depending on which kind of hydrogen is removed, two different propyl groups can result. Removal of one of the six hydrogens attached to an end carbon yields a straight-chain group called *propyl*. Removal of one of the two hydrogens attached to the middle carbon yields a branched-chain group called *isopropyl*.

Similarly, there are four different kinds of butyl groups. Two (butyl and *sec*-butyl) are derived from straight-chain butane, and two (isobutyl and *tert*-butyl) are derived from branched-chain isobutane. The prefixes *sec*- (for *secondary*) and *tert*- (for *tertiary*) refer to the number of other carbon atoms attached to the branching carbon. There are two other carbons attached to the branch point in a *sec*-butyl group and three other carbons attached to the branch point in a *tert*-butyl group.

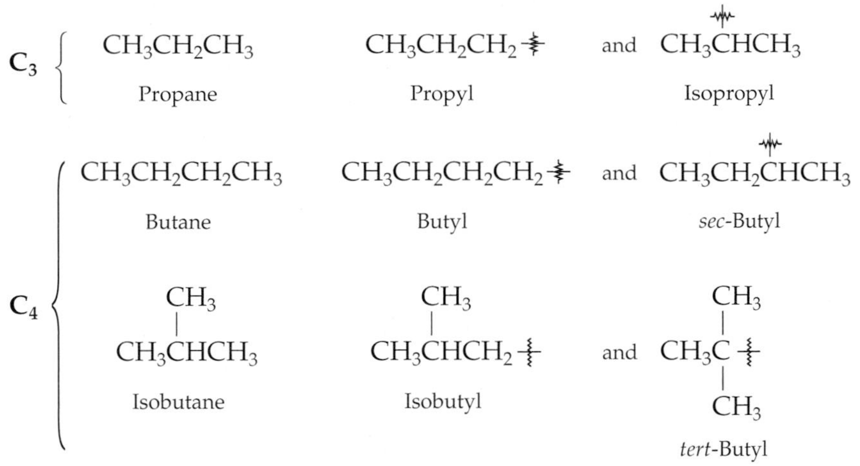

C_3 {

CH₃CH₂CH₃   CH₃CH₂CH₂⫶ and CH₃CHCH₃

Propane   Propyl   Isopropyl

C_4 {

CH₃CH₂CH₂CH₃   CH₃CH₂CH₂CH₂⫶ and CH₃CH₂CHCH₃

Butane   Butyl   *sec*-Butyl

$$CH_3 \qquad\qquad CH_3 \qquad\qquad CH_3$$
$$| \qquad\qquad\quad | \qquad\qquad\quad |$$
CH₃CHCH₃   CH₃CHCH₂⫶ and CH₃C⫶

Isobutane   Isobutyl   CH₃

                                                   tert-Butyl

It's important to realize that alkyl groups themselves are not stable compounds and that the "removal" of a hydrogen from an alkane is just a useful way of looking at things, not a chemical reaction. Alkyl groups are simply parts of molecules that help us to name compounds.

EXAMPLE 23.2   What is the IUPAC name of the following alkane?

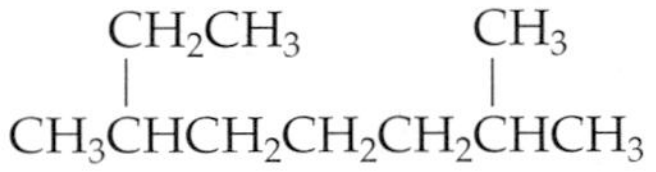

$$CH_2CH_3 \qquad\qquad CH_3$$
$$| \qquad\qquad\qquad |$$
CH₃CHCH₂CH₂CH₂CHCH₃

SOLUTION   The molecule has a chain of eight carbons (octane) with two methyl substituents. Numbering from the end nearer the first methyl substituent indicates that the methyls are at C2 and C6, giving the name 2,6-dimethyloctane. The numbers are separated by a comma and are set off from the rest of the name by a hyphen.

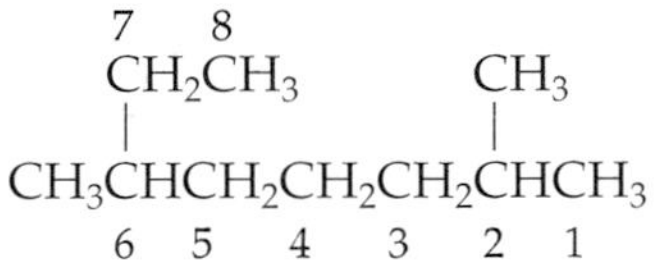

$$\overset{7}{C}H_2\overset{8}{C}H_3 \qquad\qquad CH_3$$
$$| \qquad\qquad\qquad\qquad |$$
CH₃CHCH₂CH₂CH₂CHCH₃   2,6-Dimethyloctane
6   5   4   3   2   1

EXAMPLE 23.3   Draw the structure of 3-isopropyl-2-methylhexane.

SOLUTION   First, look at the parent name (hexane) and draw its carbon structure:

C — C — C — C — C — C   Hexane

Next, find the substituents (3-isopropyl and 2-methyl), and place them on the proper carbons:

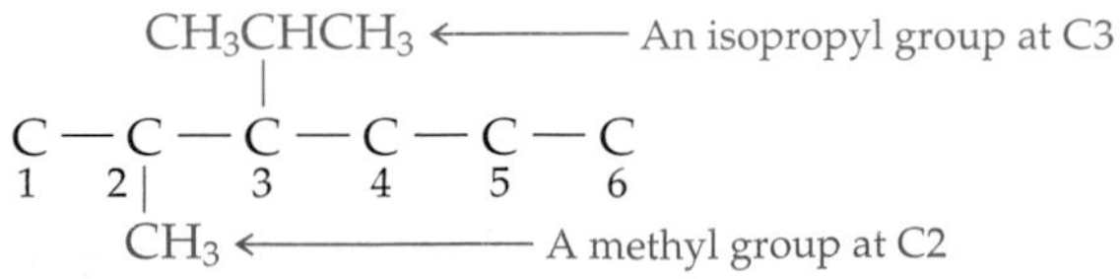

CH₃CHCH₃ ⟵———— An isopropyl group at C3
         |
C — C — C — C — C — C
1   2| 3   4   5   6
     CH₃ ⟵————— A methyl group at C2

Finally, add hydrogens to complete the structure:

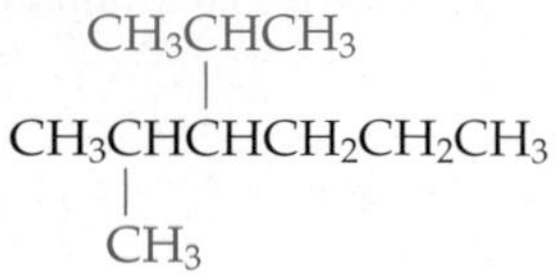

$$CH_3CHCH_3$$
$$|$$
$$CH_3CHCHCH_2CH_2CH_3$$ 3-Isopropyl-2-methylhexane
$$|$$
$$CH_3$$

▶ **PROBLEM 23.6**   What are the IUPAC names of the following alkanes?
(a) The three isomers of C_5H_{12}

(b) CH_3
              $|$
$CH_3CH_2CHCHCH_3$
              $|$
            CH_2CH_3

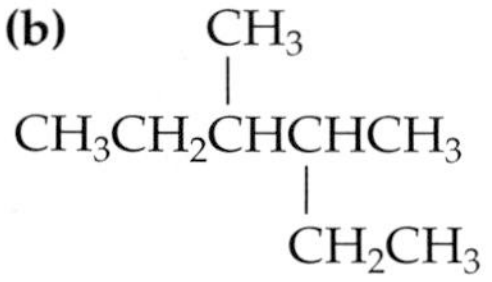

(c) CH_3 CH_3
       $|$ $|$
$CH_3CHCH_2CHCH_3$

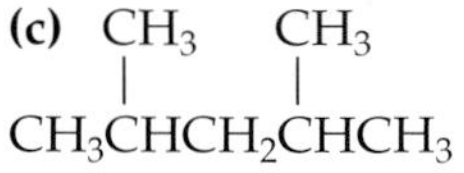

(d) CH_3 CH_2CH_3
       $|$ $|$
$CH_3CCH_2CH_2CHCH_3$
       $|$
     CH_3

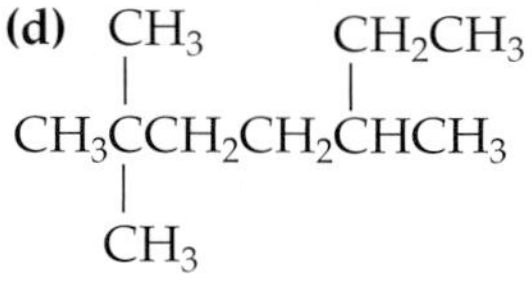

▶ **PROBLEM 23.7**   Draw condensed structures corresponding to the following IUPAC names:
(a) 3,4-Dimethylnonane
(b) 3-Ethyl-4,4-dimethylheptane
(c) 2,2-Dimethyl-4-propyloctane
(d) 2,2,4-Trimethylpentane

↞ **KEY CONCEPT PROBLEM 23.8**   What is the IUPAC name of the following alkane?

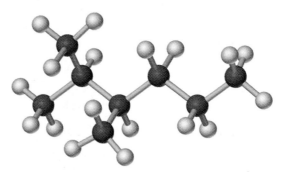

23.6   Cycloalkanes

The compounds we've been dealing with thus far have all been open-chain, or *acyclic*, alkanes. **Cycloalkanes**, which contain rings of carbon atoms, are also well known and are widespread throughout nature. Compounds of all ring sizes from 3 through 30 carbons and beyond have been prepared. The four simplest cycloalkanes having 3 carbons (cyclopropane), 4 carbons (cyclobutane), 5 carbons (cyclopentane), and 6 carbons (cyclohexane) are shown below.

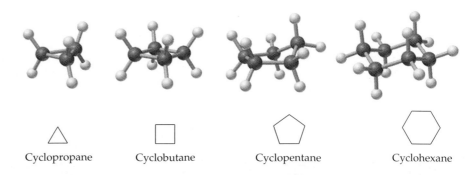

Cyclopropane   Cyclobutane   Cyclopentane   Cyclohexane

Even condensed structures are awkward for cyclic molecules, and a streamlined way of drawing structures is often used in which cycloalkanes are represented by polygons. A triangle represents cyclopropane, a square represents cyclobutane, and so on. Notice that carbon and hydrogen atoms aren't shown in these **line structures**. A carbon atom is simply understood to be at every junction of lines, and the proper number of hydrogen atoms needed to give each carbon four bonds is supplied mentally. Methylcyclohexane, for instance looks like this in a line structure:

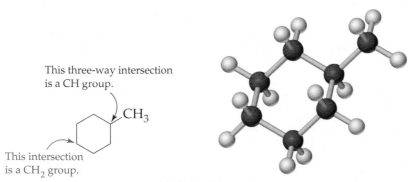

This three-way intersection is a CH group.

CH$_3$

This intersection is a CH$_2$ group.

Methylcyclohexane

As you might expect, the C–C bonds in cyclopropane and cyclobutane are considerably distorted from the ideal 109.5° value. Cyclopropane, for example, has the shape of an equilateral triangle, with C–C–C angles of 60°. As a result, the bonds in three- and four-membered rings are weaker than normal, and the molecules are more reactive than other alkanes. Cyclopentane, cyclohexane, and larger cycloalkanes, however, pucker into shapes that allow bond angles to be near their normal tetrahedral value, as shown in the computer-generated models at the beginning of this section.

Cyclopropane; Cyclobutane; Cyclopentane; Cyclohexane 3D models

Substituted cycloalkanes are named using the cycloalkane as the parent name and identifying the positions on the ring where substituents are attached. Start numbering at the group that has alphabetical priority, and proceed in the direction that gives the second substituent the lower possible number. For example,

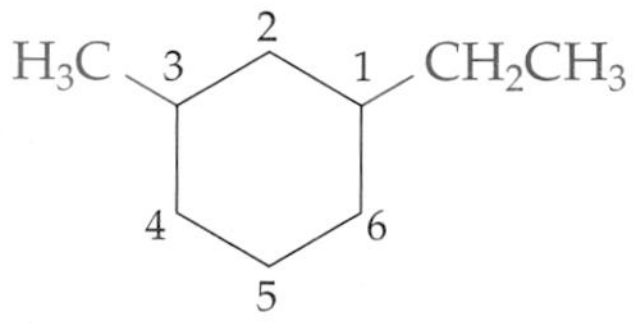

H$_3$C $\overset{2}{}$ $\overset{1}{}$ CH$_2$CH$_3$

$\overset{3}{}$

$\overset{4}{}$ $\overset{6}{}$

$\overset{5}{}$

1-Ethyl-3-methylcyclohexane

Not 1-methyl-3-ethylcyclohexane or 1-ethyl-5-methylcyclohexane or 1-methyl-5-ethylcyclohexane

EXAMPLE 23.4   What is the IUPAC name of the following cycloalkane?

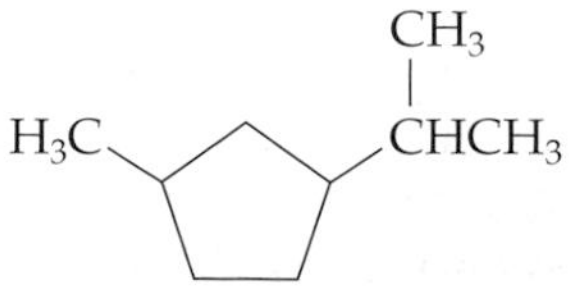

CH$_3$
|
H$_3$C   CHCH$_3$

SOLUTION   First, identify the parent cycloalkane (cyclopentane) and the two substituents (a methyl group and an isopropyl group). Then, number the ring beginning at the group having alphabetical priority (isopropyl rather than methyl), and proceed in a direction that gives the second group the lower possible number.

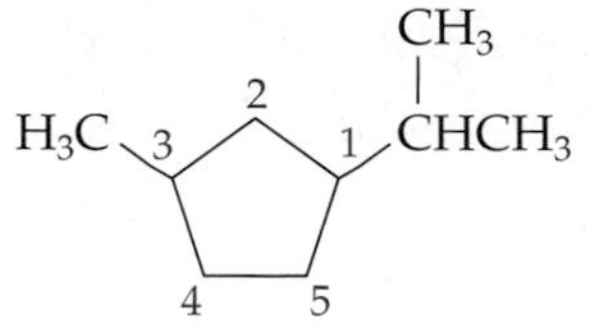

1-Isopropyl-3-methylcyclopentane

▶ **PROBLEM 23.9**   Give IUPAC names for the following cycloalkanes:

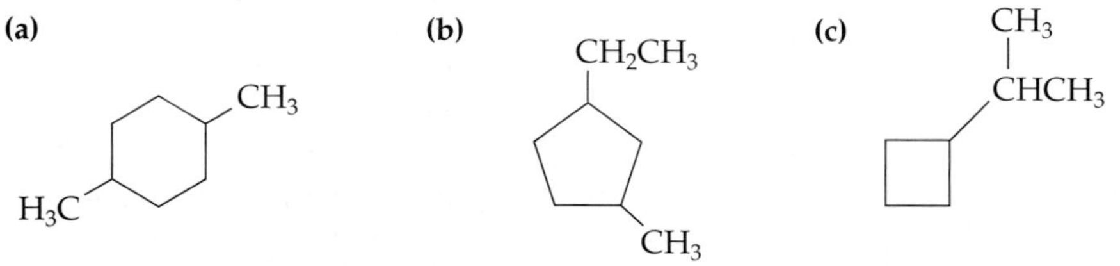

(a)

(b)

(c)

▶ **PROBLEM 23.10**   Draw structures corresponding to the following IUPAC names. Use simplified line structures rather than condensed structures for the rings.
(a) 1,1-Dimethylcyclobutane
(b) 1-*tert*-Butyl-2-methylcyclopentane
(c) 1,3,5-Trimethylcycloheptane

23.7   Reactions of Alkanes

Methane gas is burned off on these oil wells.

Alkanes have relatively low chemical reactivity and are inert to acids, bases, and most other common laboratory reagents. They do, however, react with oxygen and with halogens under appropriate conditions. The chemical reaction of alkanes with oxygen occurs during combustion in an engine or furnace when the alkane is burned as fuel. Carbon dioxide and water are formed as products, and a large amount of heat is released. For example, methane, the main component of natural gas, reacts with oxygen to release 890 kJ per mole of methane burned:

$$CH_4(g) + 2\,O_2(g) \longrightarrow CO_2(g) + 2\,H_2O(l) \qquad \Delta H° = -890 \text{ kJ}$$

Propane (the LP gas used in campers and rural homes), gasoline (a mixture of C_5–C_{11} alkanes), kerosene (a mixture of C_{11}–C_{14} alkanes), and other alkanes burn similarly.

Reaction of alkanes with Cl_2 or Br_2 occurs when a mixture of the two reactants is irradiated with ultraviolet light, denoted by *hv*. Depending on the relative amounts of the two reactants and on the time allowed for reaction, a sequential substitution of the alkane hydrogen atoms by halogen occurs, leading to a mixture of halogenated products. Methane, for example, reacts with chlorine to yield a mixture of chloromethane (CH_3Cl), dichloromethane (methylene chloride; CH_2Cl_2), trichloromethane (chloroform; $CHCl_3$), and tetrachloromethane (carbon tetrachloride; CCl_4):

$$CH_4 + Cl_2 \xrightarrow{h\nu} CH_3Cl + HCl$$

$$\overset{Cl_2}{\longrightarrow} CH_2Cl_2 + HCl$$

$$\overset{Cl_2}{\longrightarrow} CHCl_3 + HCl$$

$$\overset{Cl_2}{\longrightarrow} CCl_4 + HCl$$

▶ **PROBLEM 23.11**   Draw all the monochloro substitution products you would expect to obtain from chlorination of 2-methylbutane.

23.8   Families of Organic Molecules: Functional Groups

Chemists have learned through experience that organic compounds can be classified into families according to their structural features and that the chemical behavior of a family is often predictable. The structural features that make it possible to class compounds together are called *functional groups*. A **functional group** is composed of an atom or group of atoms within a molecule and has a characteristic chemical behavior. A given functional group undergoes the same kinds of reactions in every molecule it's a part of. Look at the carbon–carbon double-bond functional group, for example. Ethylene (C_2H_4), the simplest compound with a double bond, undergoes chemical reactions similar to those of α-pinene ($C_{10}H_{16}$), a larger and more complex compound (and major constituent of turpentine). Both, for example, react with hydrogen in the same way (Figure 23.1).

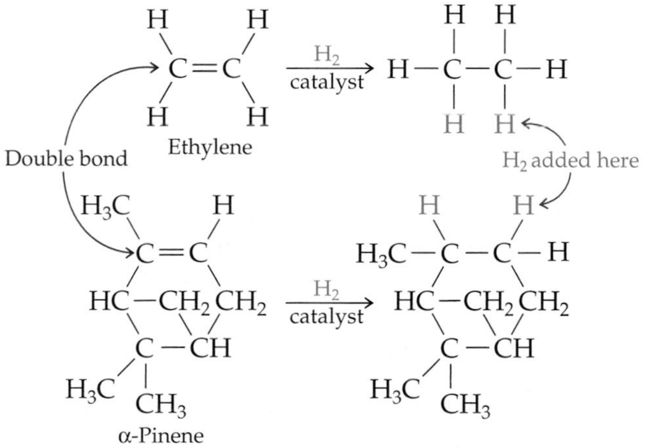

FIGURE 23.1   The reactions of ethylene and α-pinene with hydrogen. The carbon–carbon double-bond functional group adds two hydrogen atoms in each case, regardless of the complexity of the rest of the molecule.

The example shown in Figure 23.1 is typical: *The chemistry of an organic molecule, regardless of its size and complexity, is largely determined by the functional groups it contains.* Look carefully at Table 23.3, which lists some of the most common functional groups and gives examples of their occurrence. Some functional groups, such as alkenes, alkynes, and aromatic rings, have only carbon–carbon double or triple bonds; others contain single bonds to oxygen, nitrogen, or halogen atoms; and still others have carbon–oxygen double bonds.

TABLE 23.3 Some Important Families of Organic Molecules

Family Name	Functional Group Structure	Simple Example	Name	Name Ending
Alkane	(contains only C—H and C—C single bonds)	CH_3CH_3	Ethane	-ane
Alkene	$\diagdown C=C \diagup$	$H_2C=CH_2$	Ethene (Ethylene)	-ene
Alkyne	$-C\equiv C-$	$H-C\equiv C-H$	Ethyne (Acetylene)	-yne
Arene (aromatic)	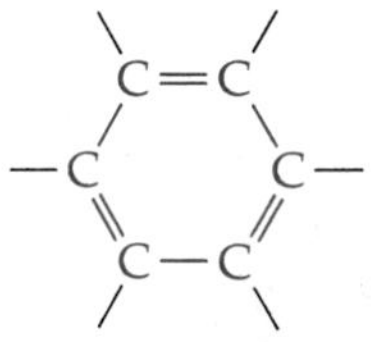 (benzene ring structure)	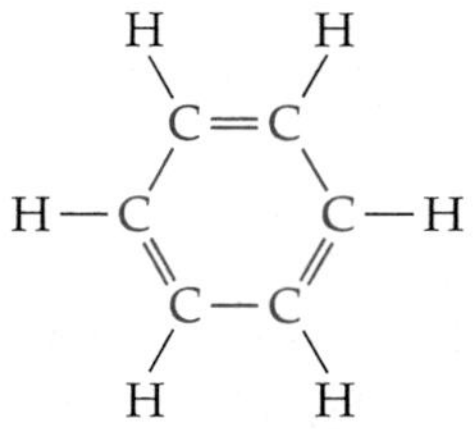 (benzene ring structure)	Benzene	None
Alcohol	$-\overset{\mid}{\underset{\mid}{C}}-O-H$	CH_3OH	Methanol	-ol
Ether	$-\overset{\mid}{\underset{\mid}{C}}-O-\overset{\mid}{\underset{\mid}{C}}-$	CH_3OCH_3	Dimethyl ether	ether
Amine	$-\overset{\mid}{\underset{\mid}{C}}-\overset{\mid}{N}-$	CH_3NH_2	Methylamine	-amine
Aldehyde	$-\overset{\mid}{\underset{\mid}{C}}-\overset{\overset{O}{\parallel}}{C}-H$	CH_3CH with O double bond	Ethanal (Acetaldehyde)	-al
Ketone	$-\overset{\mid}{\underset{\mid}{C}}-\overset{\overset{O}{\parallel}}{C}-\overset{\mid}{\underset{\mid}{C}}-$	CH_3CCH_3 with O double bond	Propanone (Acetone)	-one
Carboxylic acid	$-\overset{\mid}{\underset{\mid}{C}}-\overset{\overset{O}{\parallel}}{C}-O-H$	CH_3COH with O double bond	Ethanoic acid (Acetic acid)	-oic acid
Ester	$-\overset{\mid}{\underset{\mid}{C}}-\overset{\overset{O}{\parallel}}{C}-O-\overset{\mid}{\underset{\mid}{C}}-$	CH_3COCH_3 with O double bond	Methyl ethanoate (Methyl acetate)	-oate
Amide	$-\overset{\mid}{\underset{\mid}{C}}-\overset{\overset{O}{\parallel}}{C}-\overset{\mid}{N}-$	CH_3CNH_2 with O double bond	Ethanamide (Acetamide)	-amide

*The bonds whose connections aren't specified are assumed to be attached to carbon or hydrogen atoms in the rest of the molecule.

▶**PROBLEM 23.12**
Locate and identify the functional groups in the following molecules:
(a) Lactic acid, from sour milk   (b) Styrene, used to make polystyrene

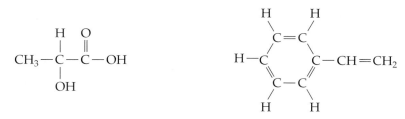

▶**PROBLEM 23.13**   Propose structures for molecules that fit the following descriptions:
(a) C_2H_4O, containing an aldehyde functional group
(b) $C_3H_6O_2$, containing a carboxylic acid functional group

23.9   Alkenes and Alkynes

In contrast to alkanes, which have only single bonds, alkenes and alkynes have *multiple* bonds: **Alkenes** are hydrocarbons that contain a carbon–carbon double bond, and **alkynes** are hydrocarbons that contain a carbon–carbon triple bond. Both groups of compounds are **unsaturated**, meaning that they have fewer hydrogens per carbon than the related alkanes. Ethylene ($H_2C{=}CH_2$), for example, has the formula C_2H_4, whereas ethane (CH_3CH_3) has the formula C_2H_6.

Alkenes are named by counting the longest chain of carbons that contains the double bond and adding the family suffix *-ene*. Thus, ethylene, the simplest alkene, is followed by propene, butene, pentene, hexene, and so on. Note that ethylene should properly be called *ethene*, but the name ethylene has been used for so long that it is accepted by IUPAC. Similarly, the name *propylene* is often used for propene.

$H_2C{=}CH_2$ $CH_3CH{=}CH_2$ $CH_3CH_2CH{=}CH_2$ $CH_3CH{=}CHCH_3$

Ethene   Propene   1-Butene   2-Butene
(Ethylene)   (Propylene)

Isomers are possible for butene and higher alkenes, depending on the position of the double bond in the chain, which must be specified by a numerical prefix. Numbering starts from the chain end nearer the double bond, and only the first of the double-bond carbons is cited. If a substituent is present on the chain, its identity is noted and its position of attachment is given. If the double bond is equidistant from both ends of the chain, numbering starts from the end nearer the substituent.

$$CH_3$$
$$|$$
$$CH_3CH{=}CHCH_2CHCH_3$$
1   2   3   4   5   6

5-Methyl-2-hexene

(numbered to give double
bond the lower number)

$$CH_3$$
$$|$$
$$CH_3CH_2CH{=}CHCHCH_3$$
6   5   4   3   2   1

2-Methyl-3-hexene

(numbered to give substituent
the lower number when the double
bond is equidistant from both ends)

In addition to the alkene isomers that exist because of double-bond *position*, alkene isomers can also exist because of double-bond *geometry*. For example, there are two geometrical, or **cis–trans isomers**, of 2-butene, which differ in their geometry about the double bond. The cis isomer has its two $-CH_3$ groups on the same side of the double bond, and the trans isomer has its two $-CH_3$ groups on opposite sides. Like the other kinds of isomers we've discussed, the individual cis and trans isomers of an alkene are different substances with different physical properties and different (though often similar) chemical behavior. *cis*-2-Butene boils at 4°C, for example, but *trans*-2-butene boils at 0.9°C.

cis- and *trans*-2-Butene
wWw 3D models

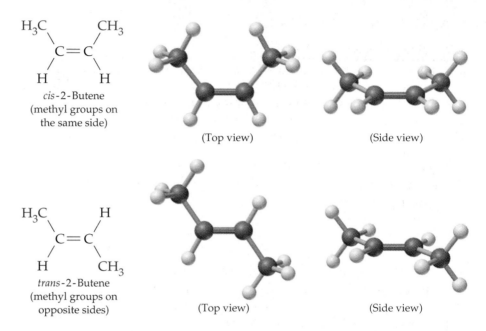

cis-2-Butene
(methyl groups on
the same side)

(Top view)

(Side view)

trans-2-Butene
(methyl groups on
opposite sides)

(Top view)

(Side view)

Cis–trans isomerism in alkenes arises because the electronic structure of the carbon–carbon double bond makes bond rotation energetically unfavorable at normal temperatures. Were it to occur, rotation would break the pi part of the double bond by disrupting the sideways overlap of two parallel p orbitals (Figure 23.2). In fact, an energy input of 240 kJ/mol is needed to cause bond rotation.

FIGURE 23.2 Rotation around a carbon–carbon double bond requires a large amount of energy because p orbital overlap is destroyed. Cis–trans alkene isomers are stable and do not interchange as a result of this barrier to rotation.

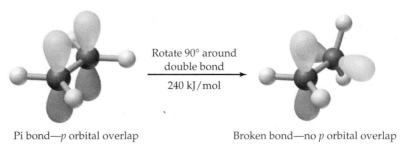

Rotate 90° around
double bond

240 kJ/mol

Pi bond—p orbital overlap

Broken bond—no p orbital overlap

Alkynes are similar in many respects to alkenes and are named using the family suffix -*yne*. The simplest alkyne, HC≡CH, is often called by its alternative name *acetylene* rather than by its systematic name *ethyne*.

Ethyne 3D model
wWw

HC≡CH

CH₃C≡CH

$\overset{1}{C}H_3\overset{2}{C}\equiv\overset{3}{C}\overset{4}{C}H_3$

$\overset{4}{C}H_3\overset{3}{C}H_2\overset{2}{C}\equiv\overset{1}{C}H$

Ethyne
(Acetylene)

Propyne

2-Butyne

1-Butyne

As with alkenes, isomers are possible for butyne and higher alkynes, depending on the position of the triple bond in the chain. Unlike the alkenes, however, no cis–trans isomers are possible for alkynes because of their linear geometry.

EXAMPLE 23.5   Draw the structure of *cis*-3-heptene.

SOLUTION   The name 3-heptene indicates that the molecule has seven carbons (hept-) and has a double bond between carbons 3 and 4:

$$\overset{1}{C}H_3\overset{2}{C}H_2\overset{3}{C}H=\overset{4}{C}H\overset{5}{C}H_2\overset{6}{C}H_2\overset{7}{C}H_3 \quad \text{A 3-Heptene}$$

The prefix *cis*- indicates that the two alkyl groups attached to the double-bond carbons lie on the same side of the double bond:

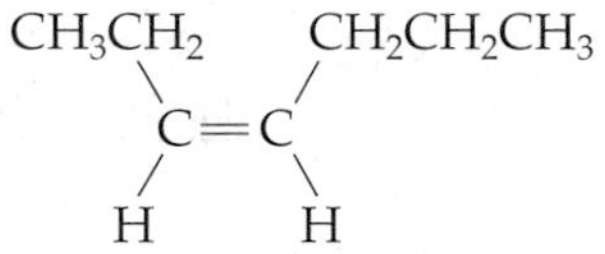

$$\begin{array}{ccc} CH_3CH_2 & & CH_2CH_2CH_3 \\ \diagdown & & \diagup \\ & C=C & \\ \diagup & & \diagdown \\ H & & H \end{array} \qquad \textit{cis}\text{-3-Heptene}$$

▶ **PROBLEM 23.14**   Give IUPAC names for the following alkenes and alkynes:

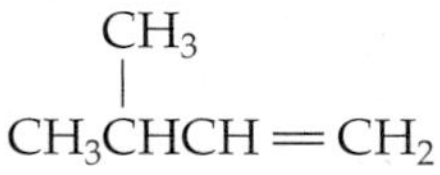

$$\begin{array}{c} CH_3 \\ | \\ CH_3CHCH=CH_2 \end{array} \qquad \begin{array}{c} CH_3CH_2CH_2 \\ | \\ CH_3C=CHCH_2CH_3 \end{array} \qquad \begin{array}{c} CH_2CH_3 \\ | \\ HC\equiv CCHCH_2CH_2CH_3 \end{array}$$

(a) 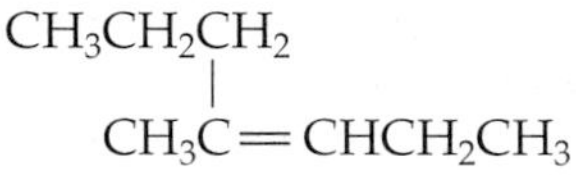   (b) 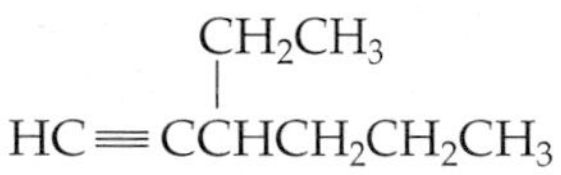   (c)

▶ **PROBLEM 23.15**   Draw structures corresponding to the following IUPAC names:
(a) 2,2-Dimethyl-3-hexene
(b) 4-Isopropyl-2-heptyne
(c) *trans*-3-Heptene

23.10   Reactions of Alkenes and Alkynes

The most important transformations of alkenes and alkynes are **addition reactions**. That is, a reagent we might write in a general way as X—Y adds to the multiple bond of the unsaturated reactant to yield a saturated product. Alkenes and alkynes react similarly, but we'll look only at alkenes because they're more common.

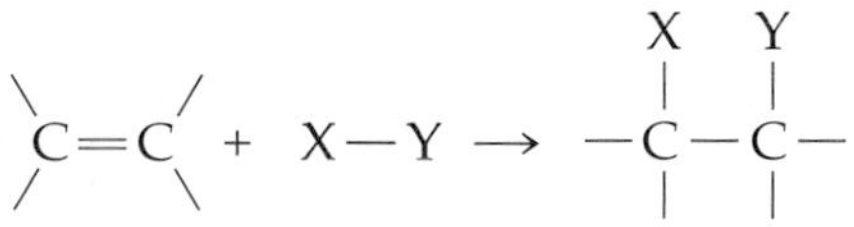

$$\begin{array}{c} \diagdown \quad \diagup \\ C=C \\ \diagup \quad \diagdown \end{array} + \; X-Y \; \longrightarrow \; \begin{array}{c} X \quad Y \\ | \quad | \\ -C-C- \\ | \quad | \end{array} \qquad \textbf{An addition reaction}$$

Addition of Hydrogen   Alkenes react with hydrogen gas in the presence of a platinum or palladium catalyst to yield the corresponding alkane product. For example,

$$CH_3CH_2CH=CH_2 + H_2 \xrightarrow[\text{catalyst}]{\text{Pd}} CH_3CH_2CH_2CH_3$$

1-Butene   Butane

This addition of hydrogen to an alkene, often called **hydrogenation**, is used commercially to convert unsaturated vegetable oils to the saturated fats used in margarine and cooking fats.

$$
\overset{O}{\overset{\|}{\text{≠OCCH}_2\text{CH}_2\text{CH}_2\text{CH}_2\text{CH}_2\text{CH}=\text{CHCH}_2\text{CH}_2\text{CH}_2\text{CH}_2\text{CH}_2\text{CH}_3}}
$$

Partial structure of a vegetable oil

$$
\downarrow \quad \text{H}_2, \text{ Pd catalyst}
$$

$$
\overset{O}{\overset{\|}{\text{≠OCCH}_2\text{CH}_2\text{CH}_2\text{CH}_2\text{CH}_2\text{CH} - \text{CHCH}_2\text{CH}_2\text{CH}_2\text{CH}_2\text{CH}_2\text{CH}_3}}
$$

Partial structure of a saturated cooking fat

Testing for Unsaturated Hydrocarbons with Bromine movie; Surface Reaction–Hydrogenation movie

Addition of Cl₂ and Br₂ Alkenes react with the halogens Cl_2 and Br_2 to give dihaloalkane addition products, a process called **halogenation**. For example,

$$
\text{H}_2\text{C}=\text{CH}_2 + \text{Cl}_2 \longrightarrow \text{H}-\overset{\text{Cl}}{\underset{\text{H}}{\overset{|}{\underset{|}{\text{C}}}}}-\overset{\text{Cl}}{\underset{\text{H}}{\overset{|}{\underset{|}{\text{C}}}}}-\text{H}
$$

Ethene (Ethylene)

1,2-Dichloroethane

More than 12 million tons of 1,2-dichloroethane are manufactured each year in the United States by reaction between ethylene and chlorine as the first step in making PVC [poly(vinyl chloride)] plastics.

Addition of Water Alkenes don't react with pure water, but in the presence of a strong acid catalyst such as sulfuric acid, a **hydration** reaction takes place to yield an *alcohol*. An −H from water adds to one carbon, and an −OH adds to the other. For example, nearly 300 million gallons of ethyl alcohol (ethanol) are produced each year in the United States by the acid-catalyzed addition of water to ethylene:

$$
\text{H}_2\text{C}=\text{CH}_2 + \text{H}_2\text{O} \xrightarrow[\text{catalyst}]{\text{H}_2\text{SO}_4} \text{H}-\overset{\text{H}}{\underset{\text{H}}{\overset{|}{\underset{|}{\text{C}}}}}-\overset{\text{OH}}{\underset{\text{H}}{\overset{|}{\underset{|}{\text{C}}}}}-\text{H}
$$

Ethene (Ethylene)

Ethanol (an alcohol)

▶ **PROBLEM 23.16** Show the products of the reaction of 2-butene with the following reagents:
(a) H₂, Pd catalyst   (b) Br₂   (c) H₂O, H₂SO₄ catalyst

▶ **PROBLEM 23.17** Reaction of 2-pentene with H₂O in the presence of H₂SO₄ yields a mixture of two alcohol products. Draw their structures.

✦ **KEY CONCEPT PROBLEM 23.18**   Draw the structure of the alcohol you would expect to obtain by acid-catalyzed reaction of the following cyclic alkene with H_2O:

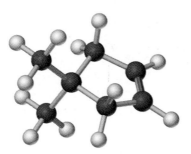

23.11   Aromatic Compounds and Their Reactions

In the early days of organic chemistry, the word *aromatic* was used to describe certain fragrant substances from fruits, trees, and other natural sources. Chemists soon realized, however, that substances grouped as aromatic behaved in a chemically different manner from most other organic compounds. Today, the term **aromatic** refers to the class of compounds that can be represented as having a six-membered ring with three double bonds. Benzene is the simplest aromatic compound, but aspirin, the steroid sex hormone estradiol, and many other important compounds also contain aromatic rings.

Benzene   Aspirin   Estradiol

Benzaldehyde, a close structural relative of benzene, is an aromatic compound responsible for the odor of cherries.

Benzene is a flat, symmetrical molecule that is often represented as a six-membered ring with three double bonds. The problem with this representation, however, is that it gives the wrong impression about benzene's chemical reactivity. Since benzene appears to have three double bonds, we might expect it to react with H_2, Br_2, and H_2O to give the same kinds of addition products that alkenes do. But this expectation is wrong. Benzene and other aromatic compounds are much less reactive than alkenes and don't normally undergo addition reactions.

Benzene's relative lack of reactivity is a consequence of its electronic structure. As shown by the orbital picture in Figure 23.3b, each of the six carbons in benzene is sp^2-hybridized and has a p orbital perpendicular to the ring. When these p orbitals overlap to form pi bonds, there are two possibilities, shown in Figure 23.3c.

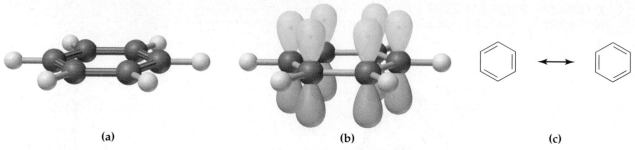

FIGURE 23.3   Some representations of benzene: **(a)** a molecular model; **(b)** an orbital picture; and **(c)** two equivalent resonance structures, which differ only in the positions of the double bonds.

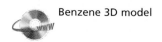
Benzene 3D model

Neither of the two equivalent structures in Figure 23.3c is correct by itself. Rather, each represents one *resonance* form of the true benzene structure, which is a resonance hybrid of the two. (For a review of resonance, you might want to reread Section 7.7). The stability of benzene is thus explained by the fact that its six pi electrons are spread equally around the entire ring. Because the six electrons aren't confined to specific double bonds in the normal way, benzene doesn't react to give addition products in the normal way. Such an idea is hard to represent using lines for covalent bonds, so chemists sometimes indicate the situation by representing the six electrons as a circle inside the six-membered ring. In this book, though, we'll indicate aromatic rings by showing just one of the individual resonance structures, because it's easier to keep track of electrons that way.

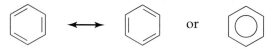

Substituted aromatic compounds are named using the suffix -*benzene*. Thus, C_6H_5Br is bromobenzene, $C_6H_5CH_3$ is methylbenzene (also called *toluene*), $C_6H_5NO_2$ is nitrobenzene, and so on. Disubstituted aromatic compounds are named using one of the prefixes *ortho-*, *meta-*, or *para-*. An *ortho-* or *o*-disubstituted benzene has its two substituents in a 1,2 relationship on the ring; a *meta-* or *m*-disubstituted benzene has its two substituents in a 1,3 relationship; and a *para-* or *p*-disubstituted benzene has its substituents in a 1,4 relationship. When the benzene ring itself is a substituent, the name **phenyl** (pronounced **fen**-nil) is used.

ortho-Dimethylbenzene   *meta*-Dibromobenzene   *para*-Dinitrobenzene   A phenyl group

Unlike alkenes, which undergo addition reactions, aromatic compounds usually undergo **substitution reactions**. That is, a group Y *substitutes* for one of the hydrogen atoms on the aromatic ring without changing the ring itself. It doesn't matter which of the six ring hydrogens in benzene is replaced because all six are equivalent.

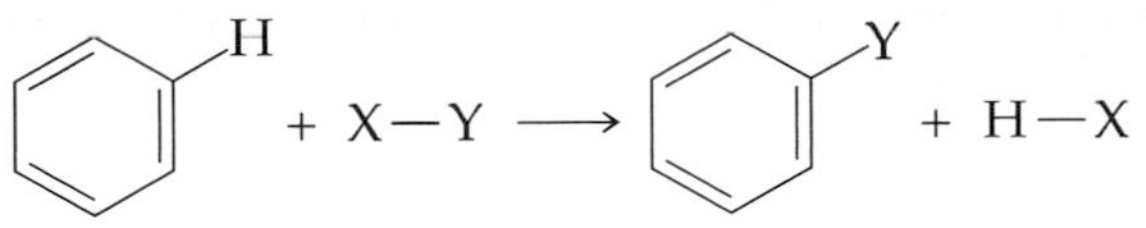

The dyes used to add the bright colors to clothing are made by a process that begins with an aromatic nitration reaction.

A substitution reaction

Nitration Substitution of a nitro group ($-NO_2$) for a ring hydrogen occurs when benzene reacts with nitric acid in the presence of sulfuric acid as catalyst:

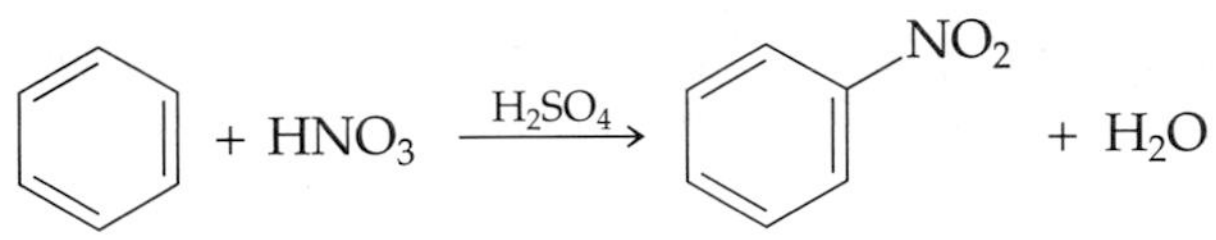

Benzene $+$ HNO$_3$ $\xrightarrow{\text{H}_2\text{SO}_4}$ Nitrobenzene $+$ H$_2$O

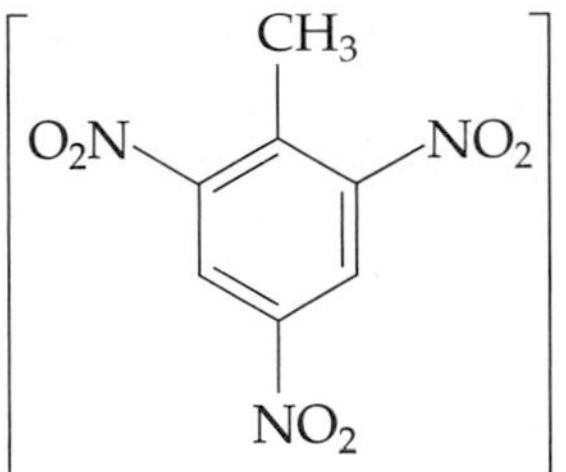

Trinitrotoluene (TNT)

Nitration of aromatic rings is a key step in the synthesis of explosives such as TNT (trinitrotoluene) and many important pharmaceutical agents. Nitrobenzene itself is a starting material for preparing many of the brightly colored dyes used in clothing.

Halogenation Substitution of a bromine or chlorine for a ring hydrogen occurs when benzene reacts with Br$_2$ or Cl$_2$ in the presence of FeBr$_3$ or FeCl$_3$ as catalyst:

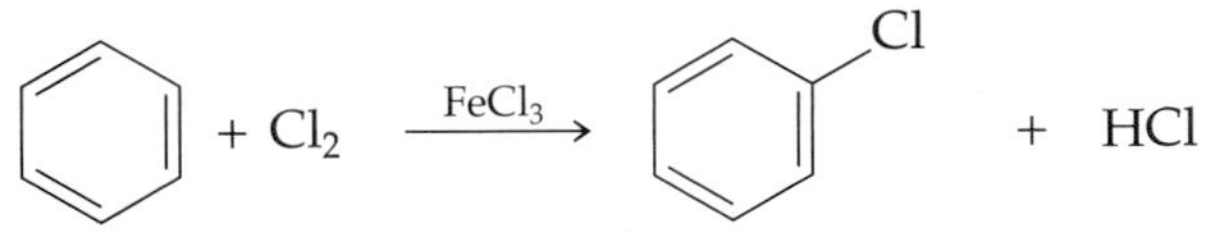

Benzene $+$ Cl$_2$ $\xrightarrow{\text{FeCl}_3}$ Chlorobenzene $+$ HCl

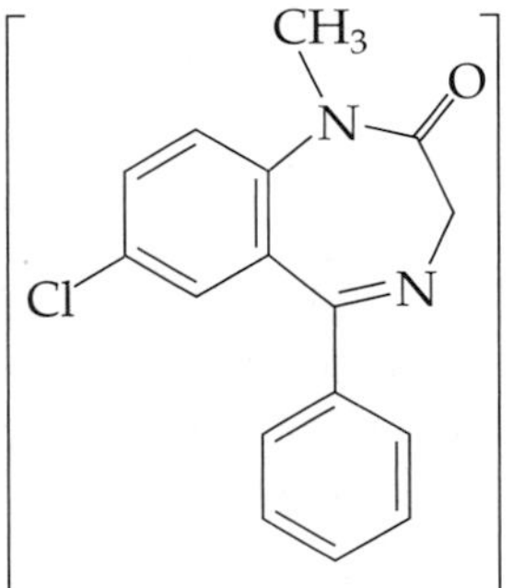

Diazepam (Valium)

The chlorination of an aromatic ring is a step used in the synthesis of numerous pharmaceutical agents, such as the tranquilizer Valium.

Sulfonation Substitution of a sulfonic acid group ($-SO_3H$) for a ring hydrogen occurs when benzene reacts with concentrated sulfuric acid and sulfur trioxide:

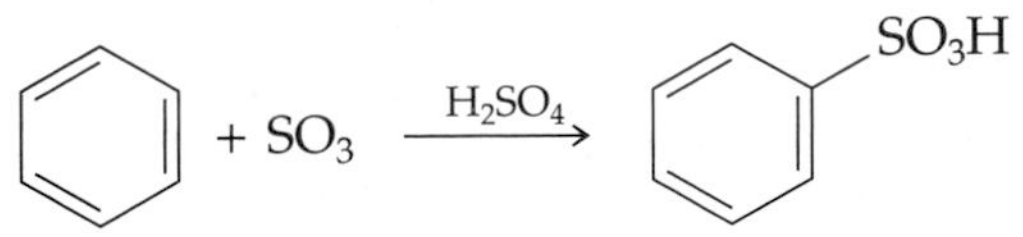

Benzene $+$ SO$_3$ $\xrightarrow{\text{H}_2\text{SO}_4}$ Benzenesulfonic acid

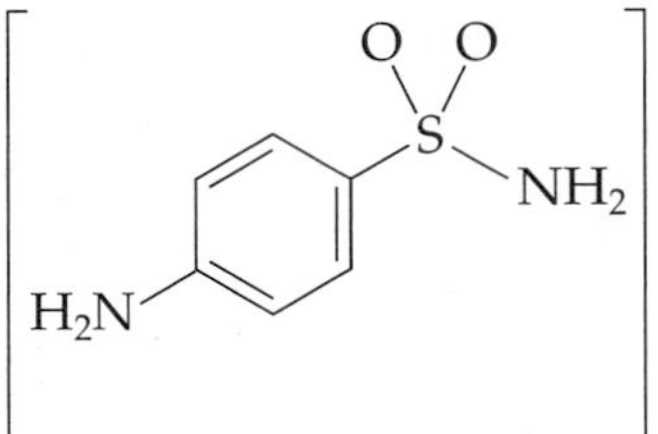

Sulfanilamide
(a sulfa antibiotic)

Aromatic-ring sulfonation is a key step in the synthesis of such compounds as aspirin and the sulfa-drug family of antibiotics.

▶ **PROBLEM 23.19**   Draw structures corresponding to the following names:
(a) *o*-Dibromobenzene   **(b)** *p*-Chloronitrobenzene
(c) *m*-Diethylbenzene

▶ **PROBLEM 23.20**   Write the products from reaction of the following reagents with *p*-dimethylbenzene (also called *p*-xylene):
(a) Br_2, $FeBr_3$ **(b)** HNO_3, H_2SO_4 **(c)** SO_3, H_2SO_4

▶ **PROBLEM 23.21**   Reaction of $Br_2/FeBr_3$ with toluene (methylbenzene) can lead to a mixture of *three* substitution products. Show the structure of each.

23.12   Alcohols, Ethers, and Amines

Alcohols

Alcohols can be thought of either as derivatives of water, in which one of the hydrogens is replaced by an organic substituent, or as derivatives of alkanes, in which one of the hydrogens is replaced by a hydroxyl group (−OH).

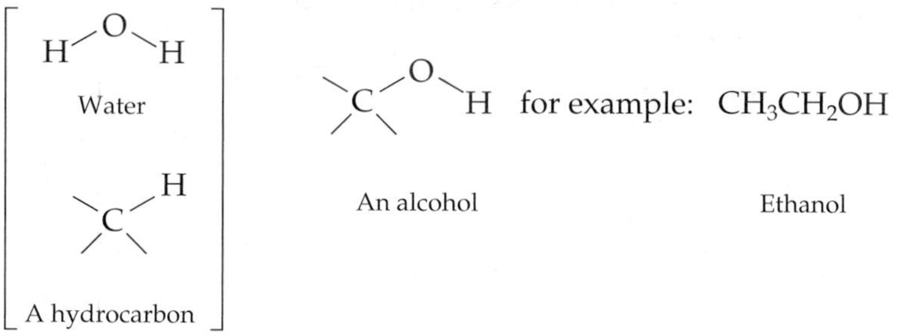

Water

A hydrocarbon

An alcohol   for example: CH_3CH_2OH

Ethanol

Like water, alcohols form hydrogen bonds (Section 10.2), which affect many of their chemical and physical properties. Alcohols are generally higher-boiling than alkanes of similar size, and simple alcohols are often soluble in water because of their ability to form hydrogen bonds to the solvent (Figure 23.4).

FIGURE 23.4   Alcohols, like water, form intermolecular hydrogen bonds. As a result, alcohols are relatively high-boiling and are often soluble in water.

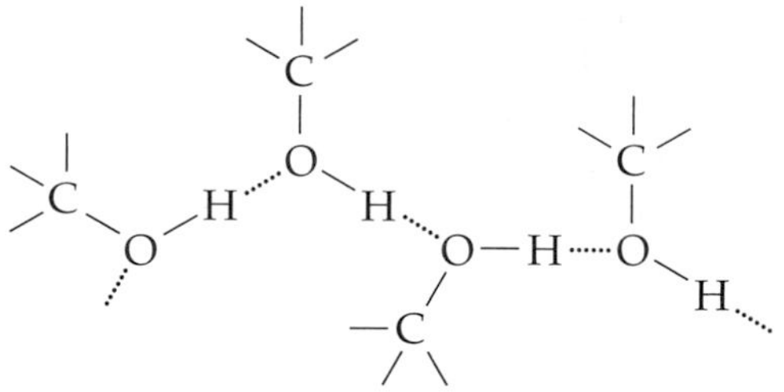

Alcohols are named by specifying the point of attachment of the −OH group to the hydrocarbon chain and using the family suffix *-ol* to replace the terminal *-e* in the alkane name. Numbering of the chain begins at the end nearer the −OH group. For example,

$$CH_3CH_2CH_2OH$$
3   2   1

1-Propanol

$$CH_3\overset{OH}{\underset{}{C}}HCH_3$$
1   2   3

2-Propanol

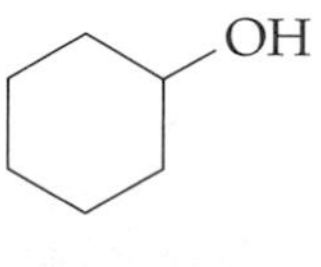

$$CH_3\overset{OH}{\underset{\underset{CH_3}{|}}{C}}CH_3$$
1   2| 3

2-Methyl-2-propanol

Cyclohexanol

Alcohols are among the most important and commonly encountered of all organic chemicals. Methanol (CH_3OH), the simplest member of the family, was once known as *wood alcohol* because it was prepared by heating wood in the absence of air. Approximately 1.7 billion gallons of methanol are manufactured each year in the United States by catalytic reduction of carbon monoxide with hydrogen gas:

$$CO + 2 H_2 \xrightarrow[\text{ZnO/Cr}_2\text{O}_3 \text{ catalyst}]{400°C} CH_3OH$$

Though toxic to humans, causing blindness in low doses (15 mL) and death in larger amounts (100–200 mL), methanol is an important industrial starting material for preparing formaldehyde (CH_2O), acetic acid (CH_3CO_2H), and other chemicals.

Ethanol (CH_3CH_2OH) is one of the oldest known pure organic chemicals. Its production by fermentation of grains and sugars, and its subsequent purification by distillation, go back at least as far as the twelfth century A.D. Sometimes called *grain alcohol*, ethanol is the "alcohol" present in all wines (10–13%), beers (3–5%), and distilled liquors (35–90%). Fermentation is carried out by adding yeast to an aqueous sugar solution and allowing enzymes in the yeast to break down carbohydrates into ethanol and CO_2:

$$C_6H_{12}O_6 \xrightarrow{\text{Yeast}} 2 CO_2 + 2 CH_3CH_2OH$$

Glucose   Ethanol

Only about 5% of the ethanol produced industrially comes from fermentation. Most is obtained by acid-catalyzed addition of water to ethylene (Section 23.10).

2-Propanol [($CH_3)_2CHOH$], commonly called isopropyl alcohol or *rubbing alcohol*, is used primarily as a solvent. It is prepared industrially by addition of water to propene:

$$CH_3CH{=}CH_2 + H_2O \xrightarrow[\text{catalyst}]{\text{Acid}} \overset{\displaystyle OH}{\underset{}{CH_3\overset{|}{C}HCH_3}}$$

Propene   2-Propanol
(Propylene)

Still other important alcohols include 1,2-ethanediol (ethylene glycol), 1,2,3-propanetriol (glycerol), and the aromatic alcohol phenol. Ethylene glycol is the principal constituent of automobile antifreeze (Section 11.7), glycerol is used as a moisturizing agent in many foods and cosmetics, and phenol is used for preparing nylon, epoxy adhesives, and heat-setting resins.

$$HOCH_2CH_2OH \qquad HOCH_2\overset{\displaystyle OH}{\underset{\displaystyle |}{C}}HCH_2OH$$

1,2-Ethanediol
(Ethylene glycol)

1,2,3-Propanetriol
(Glycerol)

Phenol

Propylene glycol (1,2-propane-diol) is sprayed on airplanes in winter to de-ice the wings prior to takeoff.

Ethers

Ethers are compounds that have two organic groups bonded to the same oxygen atom. They are fairly inert chemically and are often used as reaction solvents for that reason. Diethyl ether, the most common member of the ether family, was used for many years as a surgical anesthetic agent but has now been replaced by safer nonflammable alternatives.

Diethyl ether 3D model

$$CH_3CH_2OCH_2CH_3$$
Diethyl ether

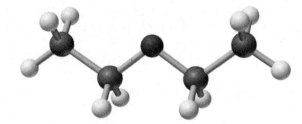

Amines

Amines are organic derivatives of ammonia in the same way that alcohols and ethers are organic derivatives of water. That is, one or more of the ammonia hydrogens is replaced in amines by an organic substituent. As the following examples indicate, the suffix *-amine* is used in naming these compounds:

The characteristic aroma of ripe fish is due to methylamine, CH_3NH_2.

Like ammonia, amines are bases because they can use the lone pair of electrons on nitrogen to accept H^+ from an acid and give ammonium salts (Section 15.12). Because they're ionic, ammonium salts are much more soluble in water than are neutral amines. Thus, a water-insoluble amine such as triethylamine dissolves readily in water when converted to its ammonium salt by reaction with HCl:

Ethylamine 3D model

$$CH_3CH_2-\overset{\displaystyle ..}{\underset{\displaystyle |}{N}}-CH_2CH_3 \; + \; HCl(aq) \longrightarrow CH_3CH_2-\overset{\displaystyle H}{\underset{\displaystyle |}{N^+}}-CH_2CH_3 \;\; Cl^-(aq)$$
$$\qquad\qquad | \qquad\qquad\qquad\qquad\qquad\qquad\qquad\qquad |$$
$$\qquad\;\; CH_2CH_3 \qquad\qquad\qquad\qquad\qquad\qquad CH_2CH_3$$

Triethylamine
(water-insoluble)

Triethylammonium chloride
(water-soluble)

This increase in water solubility on conversion of an amine to its protonated salt has enormous practical consequences in drug delivery. Many important amine-containing drugs, such as morphine (a painkiller) and tetracycline (an antibiotic), are insoluble in aqueous body fluids and are thus difficult to deliver to the appropriate site within the body. Converting these drugs to their ammonium salts, however, increases their solubility to the point where delivery through the bloodstream becomes possible.

▶ **PROBLEM 23.22** Write the structures of the ammonium salts produced by reaction of the following amines with HCl:

(a)

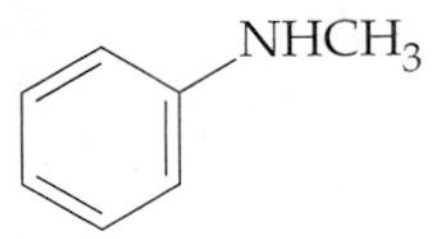

NHCH$_3$

(b)

CH$_3$CH$_2$CH$_2$NH$_2$

23.13 Aldehydes and Ketones

Look back at the functional groups listed in Table 23.3 and you'll see that many of them have a carbon–oxygen double bond (C=O), called a **carbonyl group** (pronounced car-bo-**neel**). Carbonyl-containing compounds are everywhere. Carbohydrates, fats, proteins, and nucleic acids all contain carbonyl groups; most pharmaceutical agents contain carbonyl groups; and many of the synthetic polymers used for clothing and other applications contain carbonyl groups.

As shown in the following electrostatic potential maps, the C=O bond in carbonyl compounds is polar because the electronegative oxygen atom attracts electrons away from the carbon atom. Nevertheless, some carbonyl compounds are more polar than others because of what else is bonded to the carbonyl carbon atom.

It's useful to classify carbonyl compounds into two categories based on the nature of the groups bonded to the C=O and on the chemical properties that result. In one category are *aldehydes* and *ketones*; in the other are *carboxylic acids*, *esters*, and *amides*. In aldehydes and ketones, the carbonyl carbon is bonded to atoms (H and C) that are not strongly electronegative and thus contribute no additional polarity to the molecule. In carboxylic acids, esters, and amides, however, the carbonyl-group carbon *is* bonded to an atom (O or N) that is strongly electronegative, giving these compounds even greater polarity and greater chemical reactivity.

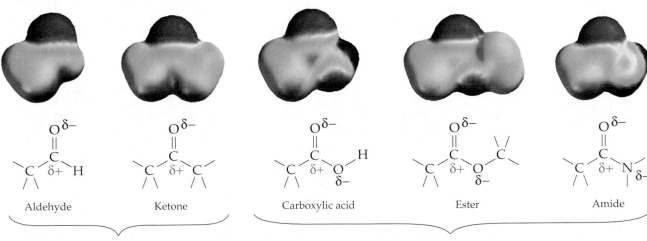

The adhesive binders used in plywood are made starting with formaldehyde.

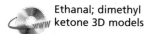

Ethanal; dimethyl ketone 3D models

Aldehydes, which have a hydrogen atom bonded to the carbonyl group, and **ketones**, which have two carbon atoms bonded to the carbonyl group, are used throughout chemistry and biology. For example, an aqueous solution of formaldehyde (properly named *methanal*) is used under the name *formalin* as a biological sterilant and preservative. Formaldehyde is also used in the chemical industry as a starting material for the manufacture of the plastics Bakelite and melamine, as a component of the adhesives used to bind plywood, and as a part of the foam insulation used in houses. Note that formaldehyde differs from other aldehydes in having two hydrogens attached to the carbonyl group rather than one. Acetone (properly named *propanone*) is perhaps the most widely used of all organic solvents. You might have seen cans of acetone sold in paint stores for general-purpose cleanup work. When naming these groups of compounds, aldehydes take the suffix *-al* and ketones take the suffix *-one*.

$$\underset{\substack{\text{Formaldehyde} \\ \text{(Methanal)}}}{\overset{\overset{\displaystyle O}{\|}}{HCH}} \qquad \underset{\substack{\text{Acetaldehyde} \\ \text{(Ethanal)}}}{\overset{\overset{\displaystyle O}{\|}}{CH_3CH}} \qquad \underset{\substack{\text{Acetone} \\ \text{(Propanone)}}}{\overset{\overset{\displaystyle O}{\|}}{CH_3CCH_3}} \qquad \underset{\substack{\text{2-Butanone}}}{\overset{\overset{\displaystyle O}{\|}}{\underset{1 \quad 2\,3 \quad 4}{CH_3CCH_2CH_3}}} \qquad \underset{\text{Cyclohexanone}}{}$$

Aldehyde and ketone functional groups are also present in many biologically important compounds. Glucose and most other sugars contain aldehyde groups, for instance. Testosterone and many other steroid hormones contain ketone groups.

Glucose—a pentahydroxyhexanal

Testosterone—a steroid hormone

The industrial preparation of simple aldehydes and ketones usually involves an oxidation reaction of the related alcohol. Thus, formaldehyde is prepared by oxidation of methanol, and acetone is prepared by oxidation of 2-propanol.

Methanol $\xrightarrow{\text{Air, 300°C}}$ Formaldehyde

2-Propanol $\xrightarrow{\text{Air, 300°C}}$ Acetone

23.14 Carboxylic Acids, Esters, and Amides

Carboxylic acids, esters, and amides have their carbonyl groups bonded to an atom (O or N) that strongly attracts electrons. All three families undergo carbonyl-group substitution reactions, in which a group we can represent as −Y substitutes for the −OH, −OC, or −N group of the carbonyl reactant.

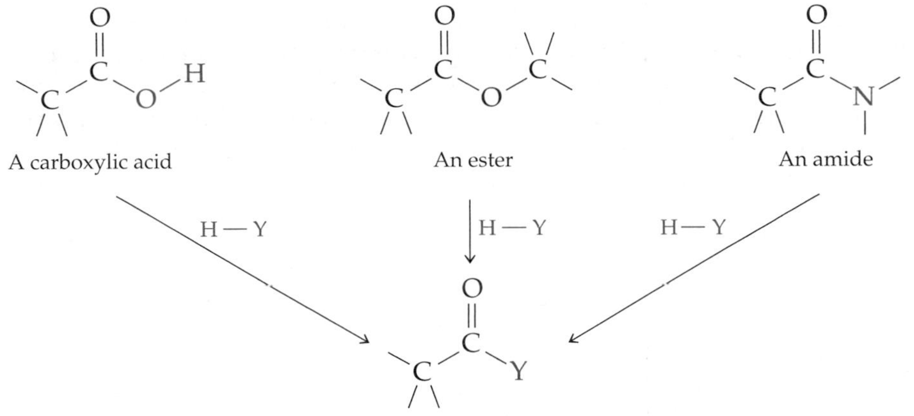

A carboxylic acid   An ester   An amide

H—Y   H—Y   H—Y

A carbonyl-group substitution reaction

Carboxylic Acids

Carboxylic acids, which contain the $-\overset{\overset{\displaystyle O}{\|}}{C}-OH$ functional group, occur widely throughout the plant and animal kingdoms. Acetic acid (ethanoic acid), for example, is the principal organic constituent of vinegar, and butanoic acid is responsible for the odor of rancid butter. In addition, long-chain carboxylic acids such as stearic acid are components of all animal fats and vegetable oils. Although many carboxylic acids have common names—*acetic* acid instead of *ethanoic* acid, for instance—systematic names are derived by replacing the final -*e* of the corresponding alkane with -*oic acid*.

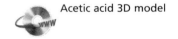

Acetic acid 3D model

$$CH_3\overset{\overset{\displaystyle O}{\|}}{C}OH$$

Acetic acid
(Ethanoic acid)

$$CH_3CH_2CH_2\overset{\overset{\displaystyle O}{\|}}{C}OH$$

Butanoic acid

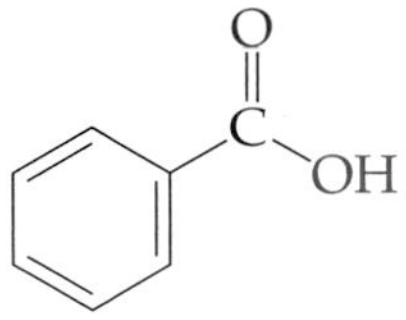

Benzoic acid

$$CH_3CH_2CH_2CH_2CH_2CH_2CH_2CH_2CH_2CH_2CH_2CH_2CH_2CH_2CH_2CH_2CH_2\overset{\overset{\displaystyle O}{\|}}{C}OH$$

Stearic acid
(Octadecanoic acid)

As their name implies, carboxylic acids are *acidic*; they dissociate slightly in aqueous solution to give H_3O^+ and a **carboxylate anion**. Carboxylic acids are much weaker than inorganic acids like HCl or H_2SO_4, however. The K_a of acetic acid, for example, is 1.78×10^{-5} (p$K_a = 4.75$), meaning that only about 1% of acetic acid molecules dissociate in a 1.0 M aqueous solution. Note in the following electrostatic potential map of acetic acid that the acidic −OH hydrogen is positively polarized (blue).

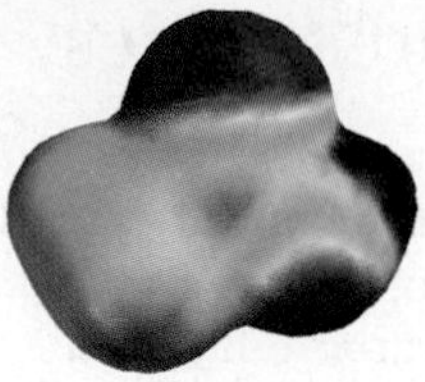

$$CH_3\overset{\displaystyle O}{\overset{\|}{C}}OH + H_2O \rightleftharpoons CH_3\overset{\displaystyle O}{\overset{\|}{C}}O^- + H_3O^+$$

Acetic acid
(Ethanoic acid)
$pK_a = 4.75$

Acetate ion

One of the most important chemical transformations of carboxylic acids is their acid-catalyzed reaction with an alcohol to yield an ester. Acetic acid, for example, reacts with ethanol in the presence of H_2SO_4 to yield ethyl acetate, a widely used solvent. The reaction is a typical carbonyl-group substitution, with $-OCH_2CH_3$ from the alcohol replacing $-OH$ from the acid.

$$CH_3-\overset{\displaystyle O}{\overset{\|}{C}}\boxed{-OH + H-}OCH_2CH_3 \xrightarrow[\text{catalyst}]{H^+} CH_3-\overset{\displaystyle O}{\overset{\|}{C}}-OCH_2CH_3 + H_2O$$

Acetic acid   Ethanol   Ethyl acetate

Esters

Esters, which contain the $-\overset{\displaystyle O}{\overset{\|}{C}}-O-C$ functional group, have many uses in medicine, industry, and living systems. In medicine, a number of important pharmaceutical agents, including aspirin and the local anesthetic benzocaine, are esters. In industry, polyesters such as Dacron and Mylar are used to make synthetic fibers and films. In nature, many simple esters are responsible for the fragrant odors of fruits and flowers. For example, pentyl acetate is found in bananas, and octyl acetate is found in oranges.

Ethyl acetate 3D model

The odor of these bananas is due to pentyl acetate, a simple ester.

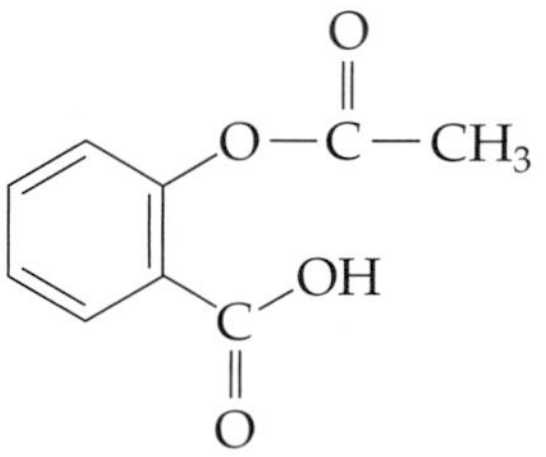

Aspirin

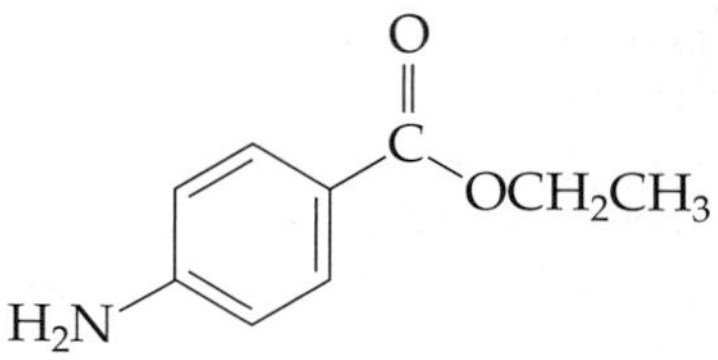

Benzocaine

$$CH_3\overset{\displaystyle O}{\overset{\|}{C}}OCH_2CH_2CH_2CH_2CH_3$$

Pentyl acetate

The most important reaction of esters is their conversion by a carbonyl-group substitution reaction into carboxylic acids. Both in the laboratory and in the body, esters undergo a reaction with water—a **hydrolysis**—that splits the ester molecule into a carboxylic acid and an alcohol. The net effect is a substitution of −OC by −OH. Although the reaction is slow in pure water, it is catalyzed by both acid and base. Base-catalyzed ester hydrolysis is often called **saponification**, from the Latin word *sapo* meaning "soap." Soap, in fact, is a mixture of sodium salts of long-chain carboxylic acids and is produced by hydrolysis with aqueous NaOH of the naturally occurring esters in animal fat.

$$CH_3-\overset{\overset{\displaystyle O}{\|}}{C}\boxed{-OCH_2CH_3 + H}-O-H \xrightarrow[\text{catalyst}]{\text{H}^+ \text{ or OH}^-} CH_3-\overset{\overset{\displaystyle O}{\|}}{C}-OH + H-OCH_2CH_3$$

Ethyl acetate   Acetic acid   Ethanol

Since esters are derived from carboxylic acids and alcohols, they are named by first identifying the alcohol-related part and then the acid-related part using the *-ate* ending. Ethyl acetate, for example, is the ester derived from ethanol and acetic acid.

Amides

Amides are compounds with the $-\overset{\overset{\displaystyle O}{\|}}{C}-N$ functional group. Without amides, there would be no life. As we'll see in the next chapter, the amide bond between nitrogen and a carbonyl-group carbon is the fundamental link used by organisms to form proteins. In addition, some synthetic polymers such as nylon contain amide groups, and important pharmaceutical agents such as acetaminophen, the aspirin substitute found in Tylenol and Excedrin, are amides.

$$\underset{\text{Repeating unit of nylon 66}}{\xi CCH_2CH_2CH_2CH_2\overset{\overset{\displaystyle O}{\|}}{C}-\overset{\overset{\displaystyle H}{|}}{N}CH_2CH_2CH_2CH_2CH_2CH_2\overset{\overset{\displaystyle H}{|}}{N}\xi}$$

Acetaminophen

Unlike amines, which also contain nitrogen (Section 23.12), amides are neutral rather than basic. Amides do not act as proton acceptors and do not form ammonium salts when treated with acid. The neighboring carbonyl group causes the unshared pair of electrons on nitrogen to be held tightly, thus preventing the electrons from bonding to H⁺.

Although better methods are normally used, amides can be prepared by the reaction of a carboxylic acid with ammonia or an amine, just as esters are prepared by the reaction of a carboxylic acid with an alcohol. In both cases, water is a by-product, and the −OH part of the carboxylic acid is replaced. Amides are named by first citing the *N*-alkyl group on the amine part (*N* because the group is attached to nitrogen) and then identifying the carboxylic acid part using the *-amide* ending.

Acetamide 3D model

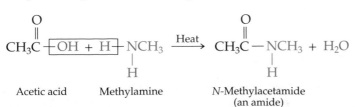

$$\underset{\text{Acetic acid}}{CH_3\overset{\overset{\displaystyle O}{\|}}{C}}\boxed{-OH + H}\underset{\overset{\displaystyle |}{H}}{-N}CH_3 \xrightarrow{\text{Heat}} \underset{\substack{\textit{N}\text{-Methylacetamide} \\ \text{(an amide)}}}{CH_3\overset{\overset{\displaystyle O}{\|}}{C}-\underset{\overset{\displaystyle |}{H}}{N}CH_3} + H_2O$$

Acetic acid   Methylamine

Amides undergo an acid- or base-catalyzed hydrolysis reaction with water in the same way that esters do. Just as an ester yields a carboxylic acid and an alcohol, an amide yields a carboxylic acid and an amine (or ammonia). The net effect is a substitution of −N by −OH. This hydrolysis of amides is the key process that occurs in the stomach during digestion of proteins.

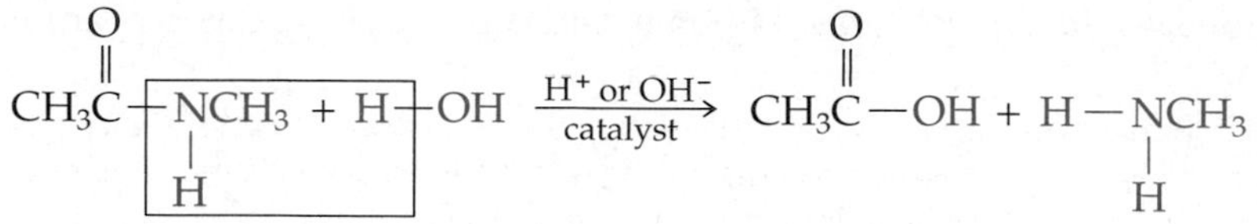

$$CH_3C{-}NCH_3 + H{-}OH \xrightarrow[\text{catalyst}]{H^+ \text{ or } OH^-} CH_3C{-}OH + H{-}NCH_3$$

N-Methylacetamide   Acetic acid   Methylamine

EXAMPLE 23.6   Give the systematic names of the following compounds:

(a)
$$CH_3CH_2CH_2COCH_2CH_2CH_3$$

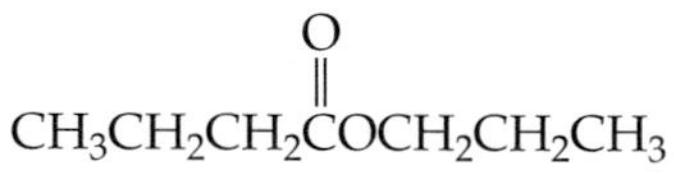

(b)

$$\text{(benzene ring)}{-}\underset{\text{O}}{\overset{\|}{C}}{-}N(CH_3)_2$$

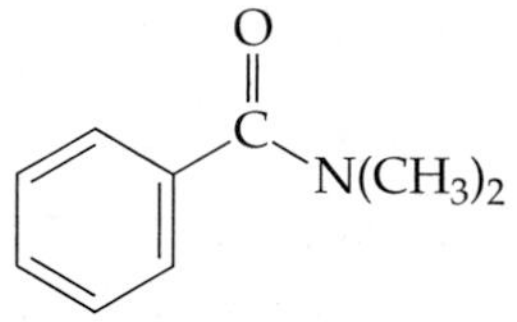

SOLUTION
(a) First identify the alcohol-derived part (propanol) and the acid-derived part (butanoic acid), and then assign the name: propyl butanoate.

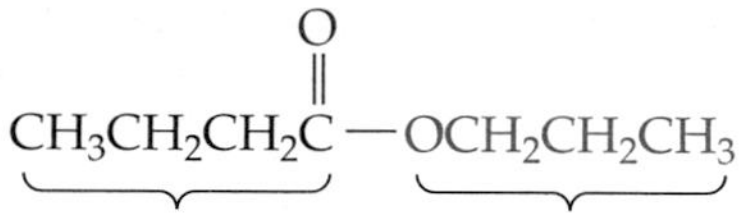

$$CH_3CH_2CH_2\overset{O}{\overset{\|}{C}}{-}OCH_2CH_2CH_3 \quad \text{Propyl butanoate}$$

Butanoic acid   Propanol

(b) First identify the amine-derived part (dimethylamine) and the acid-derived part (benzoic acid), and then assign the name: N,N-dimethylbenzamide.

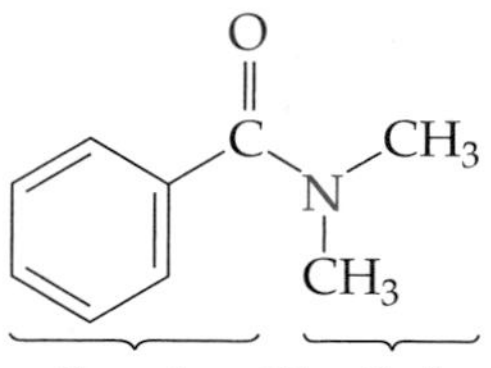

$$\text{(benzene ring)}{-}\underset{\underset{CH_3}{|}}{\overset{O}{\overset{\|}{C}}}{-}\underset{}{N}{-}CH_3 \qquad \text{N,N-Dimethylbenzamide}$$

Benzoic   Dimethylamine
acid

EXAMPLE 23.7   Write the products of the following reactions:

(a) $$CH_3\overset{CH_3}{\overset{|}{C}}HCH_2\overset{O}{\overset{\|}{C}}OH + HOCHCH_3 \longrightarrow ?$$

(b) $$CH_3CH_2\overset{Br}{\overset{|}{C}}HCH_2\overset{O}{\overset{\|}{C}}NH_2 + H_2O \longrightarrow ?$$

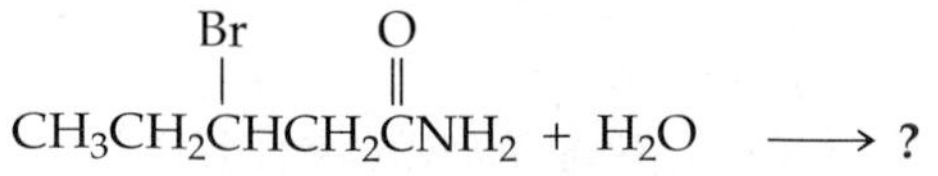

SOLUTION

(a) The reaction of a carboxylic acid with an alcohol yields an ester plus water. Write the reagents to show how H_2O is removed, and then connect the remaining fragments to complete the substitution reaction:

$$\underset{\overset{|}{CH_3CHCH_2C}}{\overset{CH_3}{|}}\overset{O}{\overset{\|}{}}-OH + H-OCHCH_3 \longrightarrow \underset{\overset{|}{CH_3CHCH_2C}}{\overset{CH_3}{|}}\overset{O}{\overset{\|}{}}-\overset{CH_3}{\overset{|}{OCHCH_3}} + H_2O$$

(b) The reaction of an amide with water yields a carboxylic acid and an amine (or ammonia). Write the reagents to show how NH_3 is removed, and then connect the remaining fragments to complete the substitution reaction:

$$\underset{\overset{|}{CH_3CH_2CHCH_2C}}{\overset{Br}{|}}\overset{O}{\overset{\|}{}}-NH_2 + H-OH \longrightarrow \underset{\overset{|}{CH_3CH_2CHCH_2}}{\overset{Br}{|}}\overset{O}{\overset{\|}{COH}} + NH_3$$

▶ **PROBLEM 23.23** Draw structures corresponding to the following names:
(a) 4-Methylpentanoic acid   (b) Isopropyl benzoate
(c) N-Ethylpropanamide

▶ **PROBLEM 23.24** Write the products of the following reactions:

(a)

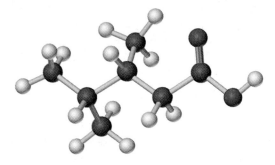

$$+ NH_3 \xrightarrow{\text{Heat}} ?$$

(b) $\underset{\overset{|}{CH_3CHCH_2}}{\overset{Cl}{|}}\overset{O}{\overset{\|}{COH}} + \underset{\overset{|}{CH_3CH_2CHOH}}{\overset{CH_3}{|}} \xrightarrow[\text{catalyst}]{H^+} ?$

◀ **KEY CONCEPT PROBLEM 23.25** Draw the structure of the ester you would obtain by acid-catalyzed reaction of the following carboxylic acid with 2-propanol:

23.15 Synthetic Polymers

Polymers are large molecules formed by the repetitive bonding together of many smaller molecules, called **monomers**. As we'll see in the next chapter, biological polymers occur throughout nature. Cellulose and starch are polymers built from small sugar monomers, proteins are polymers built from

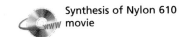

Synthesis of Nylon 610 movie

These striking examples of suburban yard art are made of polypropylene, an alkene polymer.

Styrene Monomer, Polymer 3D models

amino acid monomers, and nucleic acids are polymers built from nucleotide monomers. The basic idea is the same, but synthetic polymers are much less complex than biopolymers because the starting monomer units are usually smaller and simpler.

Many simple alkenes, called **vinyl monomers**, undergo polymer-forming (*polymerization*) reactions: Ethylene yields polyethylene, propylene (propene) yields polypropylene, styrene yields polystyrene, and so forth. The polymer molecules that result may have anywhere from a few hundred to many thousand monomer units incorporated into a long chain. Some commercially important polymers are shown in Table 23.4.

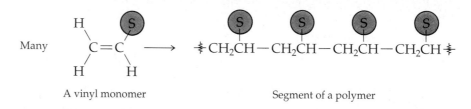

A vinyl monomer   Segment of a polymer

where (S) represents a substituent, such as H, CH_3, Cl, OH, or phenyl

TABLE 23.4   Some Alkene Polymers and Their Uses

Monomer Name	Structure	Polymer Name	Uses
Ethylene	$H_2C=CH_2$	Polyethylene	Packaging, bottles
Propylene	$H_2C=CH-CH_3$	Polypropylene	Bottles, rope, pails medical tubing
Vinyl chloride	$H_2C=CH-Cl$	Poly(vinyl chloride)	Insulation, plastic pipe
Styrene	$H_2C=CH-\bigcirc$	Polystyrene	Foams and molded plastics
Acrylonitrile	$H_2C=CH-C\equiv N$	Orlon, Acrilan	Fibers, outdoor carpeting

The fundamental process in alkene polymerization is a double-bond addition reaction similar to those discussed in Section 23.10. A species called an *initiator*, In·, first adds to the double bond of an alkene, yielding a reactive intermediate that in turn adds to a second alkene molecule to produce another reactive intermediate, and so on.

A second kind of polymerization process occurs when molecules with *two* functional groups react. We've seen, for example, that a carboxylic acid reacts with an amine to yield an amide (Section 23.14). If a molecule with two carboxylic acid groups reacts with a molecule having two amino groups, an initial reaction joins the two molecules together, and further reactions then link more and more molecules together until a giant polyamide chain results. Nylon 66, one of the most important such polymers, is prepared by heating adipic acid (hexanedioic acid) with 1,6-hexanediamine at 280°C.

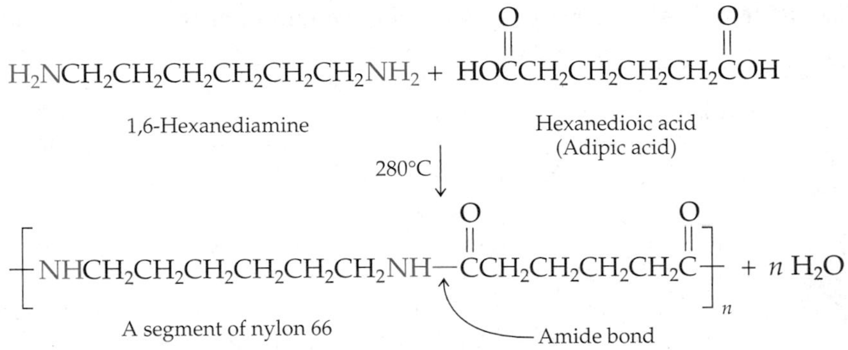

$$H_2NCH_2CH_2CH_2CH_2CH_2CH_2NH_2 + HOCCH_2CH_2CH_2CH_2COH$$

1,6-Hexanediamine   Hexanedioic acid (Adipic acid)

280°C

$$\left[NHCH_2CH_2CH_2CH_2CH_2CH_2NH-CCH_2CH_2CH_2CH_2C \right]_n + n\ H_2O$$

A segment of nylon 66   Amide bond

Nylons have many uses, both in engineering applications and in fibers. High impact strength and resistance to abrasion make nylon an excellent material for bearings and gears. High tensile strength makes it suitable as fibers for a range of applications from clothing to mountaineering ropes to carpets.

Just as diacids and diamines react to give *polyamides*, diacids and dialcohols react to give *polyesters*. The most industrially important polyester, made from reaction of terephthalic acid (1,4-benzenedicarboxylic acid) with ethylene glycol (1,2-ethanediol), is used under the trade name Dacron to make clothing fiber and under the name Mylar to make plastic film and recording tape.

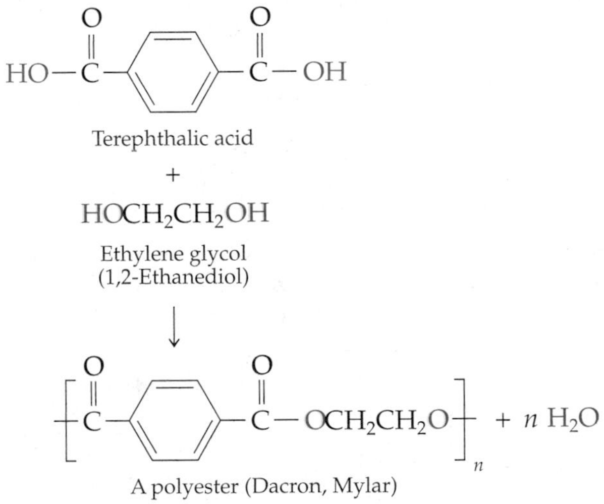

$$HO-C-\langle\ \rangle-C-OH$$

Terephthalic acid

+

$$HOCH_2CH_2OH$$

Ethylene glycol (1,2-Ethanediol)

$$\left[C-\langle\ \rangle-C-OCH_2CH_2O \right]_n + n\ H_2O$$

A polyester (Dacron, Mylar)

▶ **PROBLEM 23.26** Identify the monomer units used to make the following polymers:

(a)
$$\left[\begin{array}{c} CO_2CH_3 \quad\quad CO_2CH_3 \\ CH-CH_2-CH-CH_2 \end{array} \right]$$

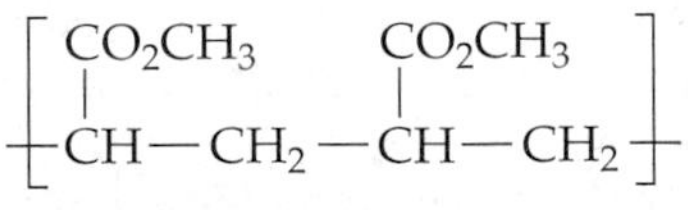

(b)
$$\left[OCH_2CH_2CH_2O-CCH_2CH_2C-O \right]$$

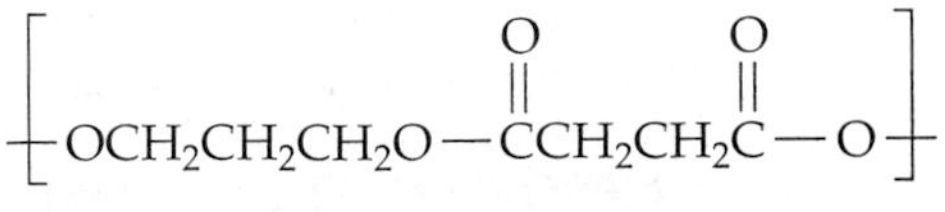

Natural or Synthetic?

Prior to the development of organic chemistry in the 20th century, only substances from natural sources were available for treating our diseases, dying our clothes, cleansing and perfuming our bodies, and so forth. Extracts of the opium poppy, for instance, have been used since the seventeenth century for the relief of pain. The prized purple dye called *Tyrian purple*, obtained from a Middle Eastern mollusk, has been known since antiquity. Oils distilled from bergamot, sweet bay, rose, and lavender, have been employed for centuries in making perfume.

Many of these so-called *natural products* were first used without any knowledge of their chemical composition. As organic chemistry developed, though, chemists learned how to work out the structures of the compounds in natural products. The disease-curing properties of limes and other citrus fruits, for example, were known for centuries but the chemical structure of vitamin C, the active ingredient, was not determined until 1933. Today there is a renewed interest in folk remedies, and a large effort is being made to identify medicinally important chemical compounds found in plants.

Whether from the laboratory or from food, the vitamin C is the same.

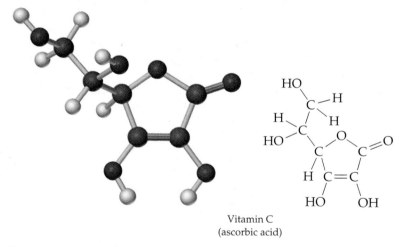

Vitamin C
(ascorbic acid)

Once a structure is known, organic chemists try to *synthesize* the compound in the laboratory. If the starting materials are inexpensive and the synthesis process is simple enough, it may become more economical to manufacture a compound than to isolate it from a plant or bacterium. In the case of vitamin C, a complete synthesis was achieved in 1933, and it is now much cheaper to synthesize it from glucose than to extract it from citrus or other natural sources. Worldwide, more than 80 million pounds are synthesized each year.

But is the "synthetic" vitamin C as good as the "natural" one? Some people still demand vitamins only from natural sources, assuming that natural is somehow better. Although eating an orange is probably better than taking a tablet, the difference lies in the many other substances present in the orange. The vitamin C itself is exactly the same, just as the NaCl produced by reacting sodium and chlorine in the laboratory is exactly the same as the NaCl found in the ocean. Natural and synthetic compounds are identical in all ways.

▶ **PROBLEM 23.27** Identify the functional groups present in vitamin C.

Summary

Organic chemistry is the study of carbon compounds. More than 17 million organic compounds are known and can be organized into families according to the functional groups they contain. A **functional group** is an atom or group of atoms within a molecule that has characteristic chemical behavior and undergoes the same kinds of reactions in every molecule it's a part of.

The simplest family of compounds is comprised of the **saturated hydrocarbons**, or **alkanes**, which contain only carbon and hydrogen and have only single bonds. **Straight-chain alkanes** have all their carbons connected in a row, **branched-chain alkanes** have a branched connection of atoms in their chain, and **cycloalkanes** have a ring of carbon atoms. Isomerism is possible in alkanes having four or more carbons. Straight-chain alkanes are named in the IUPAC system by adding the family ending *-ane* to the Greek number that tells how many carbon atoms are present. Branched-chain alkanes are named by identifying the longest continuous chain of carbon atoms and then telling what **alkyl groups** are present as branches off the main chain. Alkanes are chemically rather inert, although they undergo combustion with oxygen and undergo a reaction with chlorine.

Alkenes are hydrocarbons that contain a carbon–carbon double bond, and **alkynes** are hydrocarbons that contain a carbon–carbon triple bond. **Cis–trans isomers** are possible for substituted alkenes because of the lack of rotation about the carbon–carbon double bond. The cis isomer has two substituents on the same side of the double bond, and the trans isomer has two substituents on opposite sides. The most important transformations of alkenes and alkynes are **addition reactions**, in which a substance adds to the multiple bond to yield a saturated product.

Aromatic compounds are often represented as having a six-membered ring with three double bonds. These compounds usually undergo **substitution reactions**, in which a group substitutes for one of the hydrogen atoms on the aromatic ring. **Alcohols** and **ethers** can be thought of as derivatives of water in which one or both of the hydrogens are replaced by an organic substituent. Similarly, **amines** are derivatives of ammonia in which one or more of the ammonia hydrogens are replaced by an organic substituent. Amines are bases and can be protonated by acids to yield ammonium salts.

Compounds that contain a **carbonyl group**, C=O, can be classified into two categories based on their chemical properties. In **aldehydes** and **ketones**, the carbonyl-group carbon is bonded to atoms (H and C) that don't attract electrons strongly. In **carboxylic acids**, **esters**, and **amides**, the carbonyl-group carbon is bonded to an atom (O or N) that *does* attract electrons strongly. As a result, these latter three families of compounds undergo carbonyl-group substitution reactions, in which a group −Y substitutes for the −OH, −OC, or −N group of the carbonyl reactant.

Polymers are large molecules formed by the repetitive bonding together of many smaller **monomers**. Alkene polymers such as polyethylene result from the polymerization of simple alkenes. Nylons and polyesters result from the sequential reaction of two difunctional molecules.

Key Words

addition reaction *999*
alcohol *1004*
aldehyde *1008*
alkane *984*
alkene *997*
alkyl group *990*
alkyne *997*
amide *1011*
amine *1006*
aromatic *1001*
branched-chain alkane *985*
carbonyl group *1007*
carboxylate anion *1009*
carboxylic acid *1009*
cis–trans isomers *998*
condensed structure *986*
conformation *987*
cycloalkane *992*
ester *1010*
ether *1006*
functional group *995*
halogenation *1000*
hydration *1000*
hydrocarbon *984*
hydrogenation *1000*
hydrolysis *1011*
ketone *1008*
line structure *993*
monomer *1013*
organic chemistry *981*
phenyl *1002*
polymer *1013*
saponification *1011*
saturated hydrocarbon *984*
straight-chain alkane *985*
substitution reaction *1002*
unsaturated *997*
vinyl monomer *1014*

Key Concept Summary

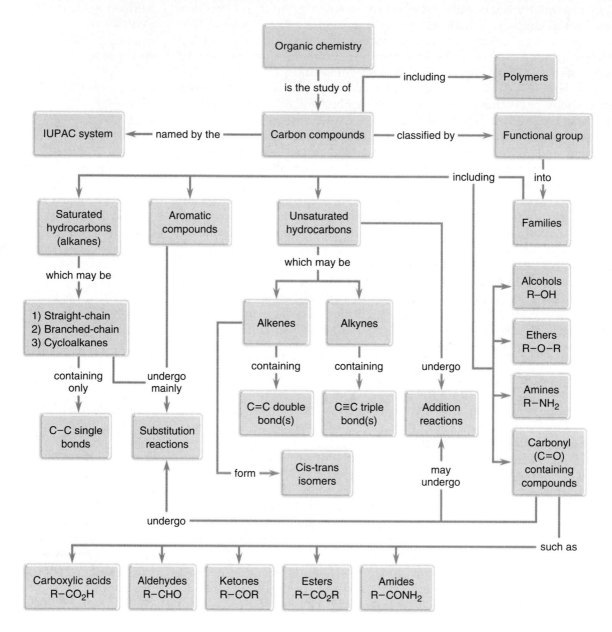

Understanding Key Concepts

Problems 23.1–23.27 appear within the chapter.

23.28 Convert each of the following models into a condensed formula:

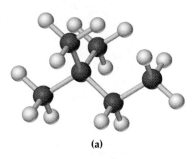

(a)

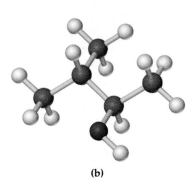

(b)

23.29 Convert each of the following models into a line drawing:

(a)

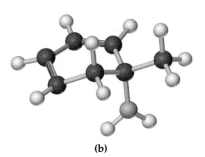

(b)

23.30 Identify the functional groups in each of the following compounds:

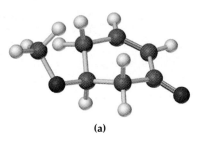

(a)

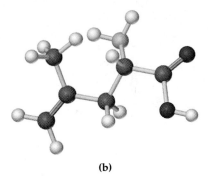

(b)

23.31 Give a systematic name for each of the following compounds:

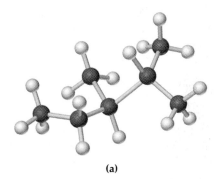

(a)

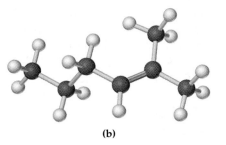

(b)

23.32 The following structure represents a segment of an alkene polymer. Identify the monomer from which the polymer was made.

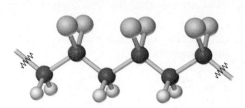

23.33 Identify the carboxylic acid and alcohol from which the following ester was made:

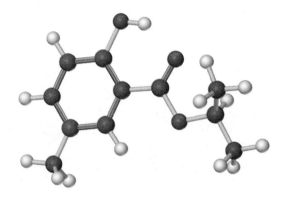

23.34 Draw two isomers of the following compound:

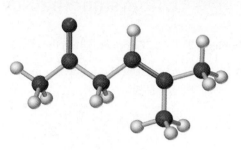

23.35 Draw three resonance forms for naphthalene, showing the positions of the double bonds.

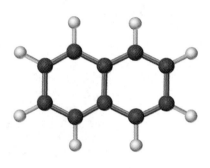

Additional Problems

Functional Groups and Isomers

23.36 What are functional groups, and why are they important?

23.37 Describe the structure of the functional group in each of the following families:

(a) Alkene   (b) Alcohol

(c) Ester   (d) Amine

23.38 Propose structures for molecules that meet the following descriptions:

(a) A ketone with the formula $C_5H_{10}O$

(b) An ester with the formula $C_6H_{12}O_2$

(c) A compound with formula $C_2H_5NO_2$ that is both an amine and a carboxylic acid

23.39 Write structures for each of the following molecular formulas. You may have to use rings and/or multiple bonds in some instances.

(a) C_2H_7N (b) C_4H_8

(c) C_2H_4O (d) CH_2O_2

23.40 There are three isomers with the formula C_3H_8O. Draw their structures.

23.41 Write as many isomers as you can that fit the following descriptions:

(a) Alcohols with formula $C_4H_{10}O$

(b) Amines with formula C_3H_9N

(c) Ketones with formula $C_5H_{10}O$

(d) Aldehydes with formula $C_5H_{10}O$

23.42 Identify the functional groups in the following molecules:

(a)

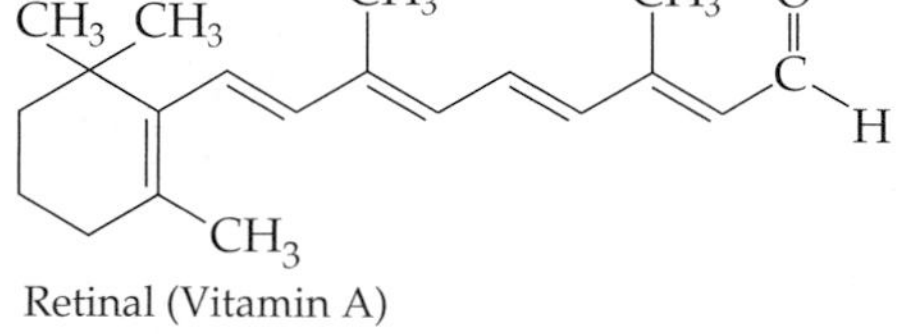

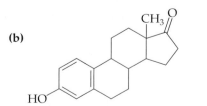

Retinal (Vitamin A)

(b)

Estrone, a female sex hormone

23.43 Identify the functional groups in cocaine.

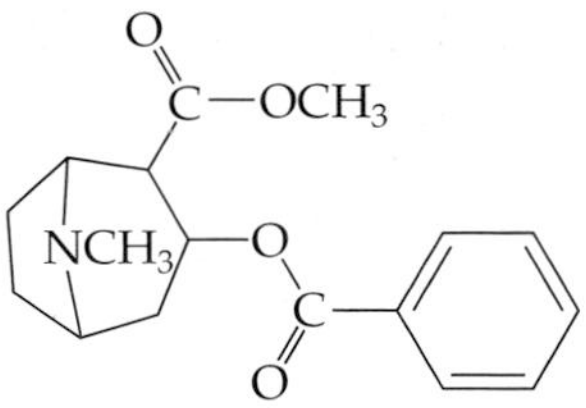

Cocaine

Alkanes

23.44 What is the difference between a straight-chain alkane and a branched-chain alkane?

23.45 What is the difference between an alkane and an alkyl group?

23.46 What kind of hybrid orbitals does carbon use in forming alkanes?

23.47 Why are alkanes said to be saturated?

23.48 If someone reported the preparation of a compound with the formula C_3H_9, most chemists would be skeptical. Why?

23.49 What is wrong with each of the following structures?

(a) $CH_3=CHCH_2CH_2OH$

(b) 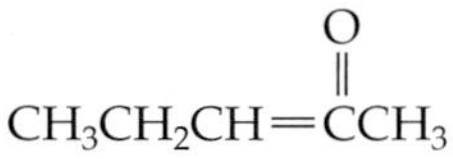 $CH_3CH_2CH=\overset{\overset{\displaystyle O}{\|}}{C}CH_3$

(c) $CH_3CH_2C\equiv CH_2CH_3$

23.50 What are the IUPAC names of the following alkanes?

(a) 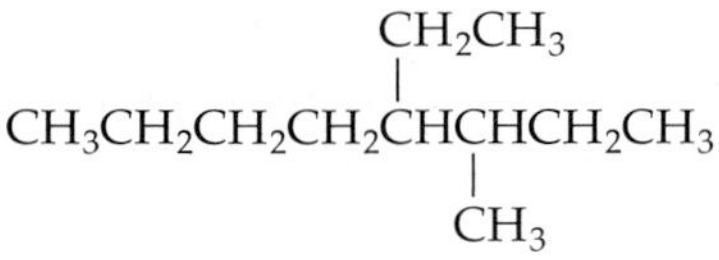 $CH_3CH_2CH_2CH_2\overset{\overset{\displaystyle CH_2CH_3}{|}}{C}HCHCH_2CH_3$
     $\quad\quad\quad\quad\quad\quad\quad\quad\quad |$
     $\quad\quad\quad\quad\quad\quad\quad\quad\quad CH_3$

(b) $CH_3CH_2CH_2\overset{\overset{\displaystyle CH_3CHCH_3}{|}}{C}HCH_2\overset{}{C}HCH_3$
     $\quad\quad\quad\quad\quad\quad\quad\quad\quad\quad |$
     $\quad\quad\quad\quad\quad\quad\quad\quad\quad\quad CH_3$

(c) $CH_3\overset{\overset{\displaystyle CH_3}{|}}{C}CH_2CH_2CH_2\overset{\overset{\displaystyle CH_3}{|}}{C}HCH_3$
     $\quad\quad |$
     $\quad\quad CH_3$

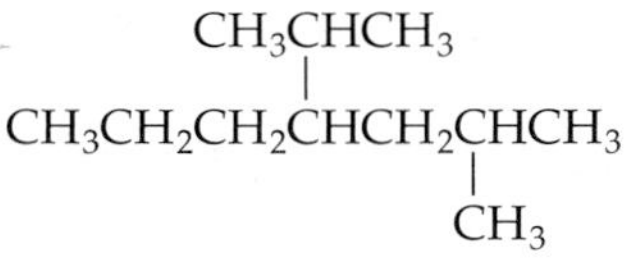

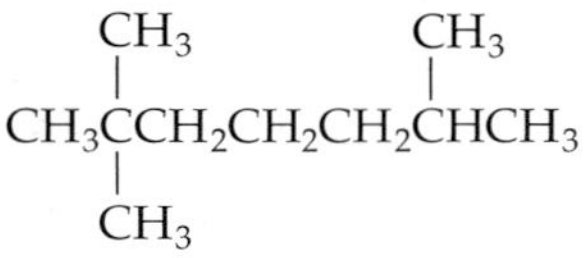

(d) 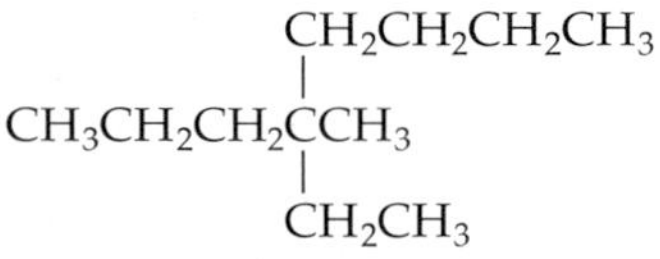 $CH_3CH_2CH_2\overset{\overset{\displaystyle CH_2CH_2CH_2CH_3}{|}}{C}CH_3$
     $\quad\quad\quad\quad\quad\quad\quad |$
     $\quad\quad\quad\quad\quad\quad\quad CH_2CH_3$

23.51 The following compound, known commonly as isooctane, is important as a reference substance for determining the octane rating of gasoline. What is the IUPAC name of isooctane?

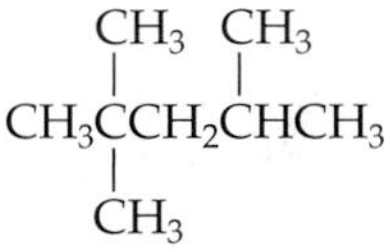

$$CH_3\overset{\overset{\displaystyle CH_3}{|}}{C}CH_2\overset{\overset{\displaystyle CH_3}{|}}{C}HCH_3 \quad \text{Isooctane}$$
$$\quad |$$
$$\quad CH_3$$

23.52 Write condensed structures for each of the following compounds:

(a) 3-Ethylhexane
(b) 2,2,3-Trimethylpentane
(c) 3-Ethyl-3,4-dimethylheptane
(d) 5-Isopropyl-2-methyloctane

23.53 Draw structures corresponding to the following IUPAC names:

(a) Cyclooctane
(b) 1,1-Dimethylcyclopentane
(c) 1,2,3,4-Tetramethylcyclobutane
(d) 4-Ethyl-1,1-dimethylcyclohexane

23.54 Give IUPAC names for each of the following cycloalkanes:

(a) 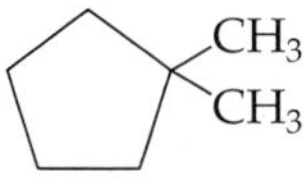 (b) 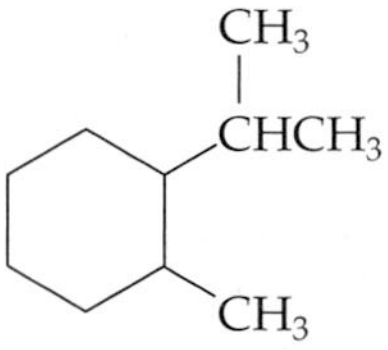 (c)

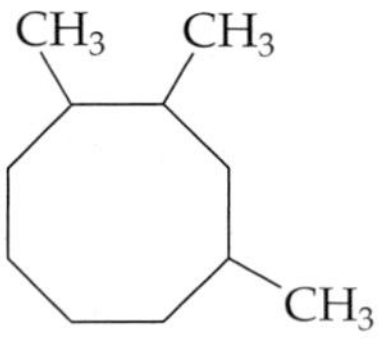

23.55 The following names are incorrect. What is wrong with each, and what are the correct names?

(a) 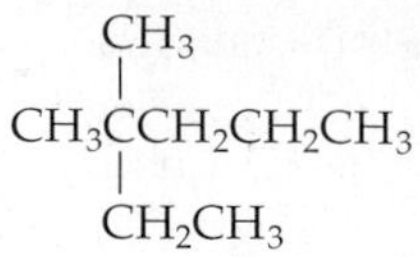
$$CH_3$$
$$|$$
$$CH_3CCH_2CH_2CH_3$$
$$|$$
$$CH_2CH_3$$
4-Ethyl-4-methylpentane

(b) 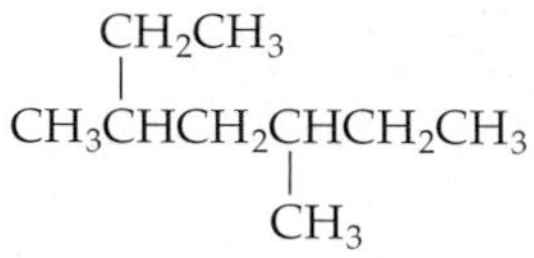
$$CH_2CH_3$$
$$|$$
$$CH_3CHCH_2CHCH_2CH_3$$
$$|$$
$$CH_3$$
5-Ethyl-3-methylhexane

(c) 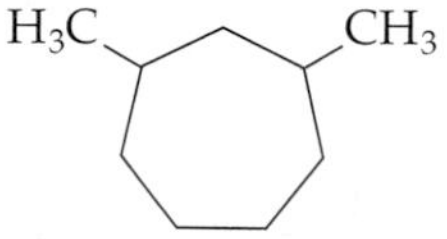
$$H_3C \qquad CH_3$$

1,4-Dimethylcyclooctane

23.56 Give IUPAC names for each of the five isomers with the formula C_6H_{14}.

23.57 Draw structures and give IUPAC names for the nine isomers of C_7H_{16}.

23.58 Write the formulas of all monochlorinated substitution products that might result from a substitution reaction of the following substances with Cl_2:

(a) Hexane

(b) 3-Methylpentane

(c) Methylcyclohexane

23.59 Which of the following reactions is likely to have a higher yield? Explain.

(a) 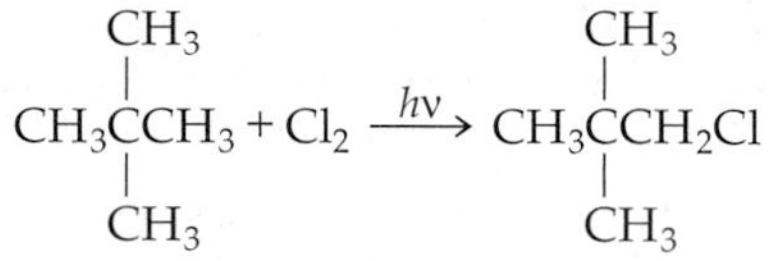
$$CH_3 \qquad\qquad CH_3$$
$$| \qquad\qquad\qquad |$$
$$CH_3CCH_3 + Cl_2 \xrightarrow{h\nu} CH_3CCH_2Cl$$
$$| \qquad\qquad\qquad |$$
$$CH_3 \qquad\qquad CH_3$$

(b) 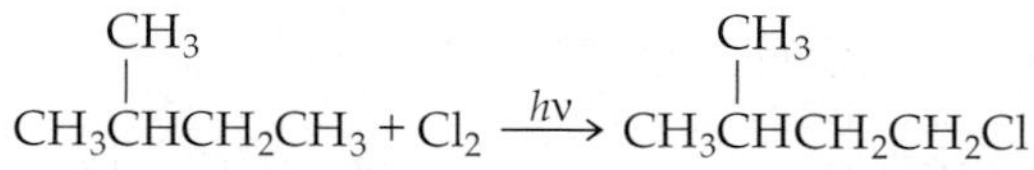
$$CH_3 \qquad\qquad CH_3$$
$$| \qquad\qquad\qquad |$$
$$CH_3CHCH_2CH_3 + Cl_2 \xrightarrow{h\nu} CH_3CHCH_2CH_2Cl$$

Alkenes, Alkynes, and Aromatic Compounds

23.60 What kind of hybrid orbitals does carbon use in forming:

(a) double bonds?

(b) triple bonds?

(c) aromatic rings?

23.61 Why are alkenes, alkynes, and aromatic compounds said to be unsaturated?

23.62 Not all compounds that smell nice are called "aromatic," and not all compounds called "aromatic" smell nice. Explain.

23.63 What is meant by the term *addition reaction*?

23.64 Write structural formulas for compounds that meet the following descriptions:

(a) An alkene with five carbons

(b) An alkyne with four carbons

(c) A substituted aromatic compound with eight carbons

23.65 How many dienes (compounds with two double bonds) are there with the formula C_5H_8? Draw structures of as many as you can.

23.66 Give IUPAC names for the following compounds:

(a) 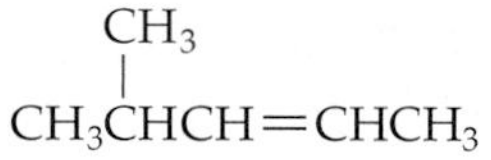
$$CH_3$$
$$|$$
$$CH_3CHCH=CHCH_3$$

(b) 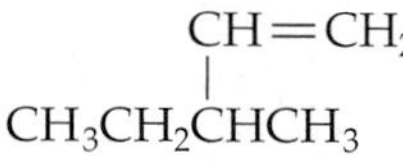
$$CH=CH_2$$
$$|$$
$$CH_3CH_2CHCH_3$$

(c) 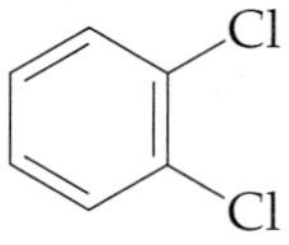
(structure: benzene ring with two Cl substituents in ortho position)

(d) 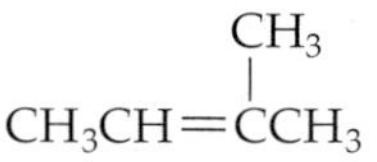
$$CH_3$$
$$|$$
$$CH_3CH=CCH_3$$

(e) 
$$CH_3$$
$$|$$
$$CH_3CH_2C\equiv CCH_2CH_2CHCH_3$$

23.67 Draw structures corresponding to the following IUPAC names:

(a) *cis*-2-Hexene

(b) 2-Methyl-3-hexene

(c) 2-Methyl-1,3-butadiene

23.68 Ignoring cis–trans isomers, there are five alkenes with the formula C_5H_{10}. Draw structures for as many as you can, and give their IUPAC names. Which can exist as cis–trans isomers?

23.69 There are three alkynes with the formula C_5H_8. Draw and name them.

23.70 Which of the following compounds are capable of cis–trans isomerism?

(a) 1-Hexene   **(b)** 2-Hexene   **(c)** 3-Hexene

23.71 Which of the following compounds are capable of cis–trans isomerism?

$$\begin{matrix} & CH_3 \\ & | \\ \textbf{(a)} \; CH_3CHCH\!=\!CHCH_3 \end{matrix}$$

$$\begin{matrix} & CH\!=\!CH_2 \\ & | \\ \textbf{(b)} \; CH_3CH_2CHCH_3 \end{matrix}$$

$$\begin{matrix} & Cl \\ & | \\ \textbf{(c)} \; CH_3CH\!=\!CHCHCH_2CH_3 \end{matrix}$$

23.72 Draw structures of the following compounds, indicating the cis or trans geometry of the double bond if necessary:

(a) *cis*-3-Heptene

(b) *cis*-4-Methyl-2-pentene

(c) *trans*-2,5-Dimethyl-3-hexene

23.73 The following names are incorrect by IUPAC rules. Draw the structures represented and give the correct names.

(a) 2-Methyl-4-hexene

(b) 5,5-Dimethyl-3-hexyne

(c) 2-Butyl-1-propene

(d) 1,5-Diethylbenzene

23.74 Why is cis–trans isomerism possible for alkenes but not for alkanes or alkynes?

23.75 Why do you suppose small-ring cycloalkenes such as cyclohexene don't exist as cis–trans isomers?

23.76 Write equations for the reaction of 2,3-dimethyl-2-butene with each of the following reagents:

(a) H_2 and Pd catalyst

(b) Br_2

(c) H_2O and H_2SO_4 catalyst

23.77 Write equations for the reaction of 2-methyl-2-butene with the reagents given in Problem 23.76.

23.78 Write equations for the reaction of *p*-dichlorobenzene with the following reagents:

(a) Br_2 and $FeBr_3$ catalyst

(b) HNO_3 and H_2SO_4 catalyst

(c) H_2SO_4 and SO_3

(d) Cl_2 and $FeCl_3$ catalyst

23.79 Benzene and other aromatic compounds don't normally react with hydrogen in the presence of a palladium catalyst. If very high pressures (200 atm) and high temperatures are used, however, benzene will add three molecules of H_2 to give an addition product. What is a likely structure for the product?

Alcohols, Amines, and Carbonyl Compounds

23.80 Draw structures corresponding to the following names:

(a) 2,4-Dimethyl-2-pentanol

(b) 2,2-Dimethylcyclohexanol

(c) 5,5-Diethyl-1-heptanol

(d) 3-Ethyl-3-hexanol

23.81 Draw structures corresponding to the following names:

(a) Propylamine

(b) Diethylamine

(c) *N*-Methylpropylamine

23.82 Assume that you have samples of quinine (an amine) and menthol (an alcohol). What simple chemical test could you do to distinguish between them?

23.83 Assume that you're given samples of pentanoic acid and methyl butanoate, both of which have the formula $C_5H_{10}O_2$. Describe how you can tell them apart.

23.84 What is the structural difference between an aldehyde and a ketone?

23.85 How do aldehydes and ketones differ from carboxylic acids, esters, and amides?

23.86 How are industrially important ketones and aldehydes usually prepared?

23.87 What general kind of reaction do carboxylic acids, esters, and amides undergo?

23.88 Identify the kinds of carbonyl groups in the following molecules (aldehyde, amide, ester, or ketone):

(a)   **(b)** $CH_3CH_2CH_2CHO$

$$\begin{matrix} & CH_3 \\ & | \\ \textbf{(c)} \; CH_3CHCH_2COCH_3 \end{matrix}$$ **(d)**

$$\begin{matrix} & CH_3 \\ & | \\ \textbf{(e)} \; CH_3CHCH_2COOCH_3 \end{matrix}$$

23.89 Draw and name compounds that meet the following descriptions:

(a) Three different amides with the formula $C_5H_{11}NO$

(b) Three different esters with the formula $C_6H_{12}O_2$

23.90 Write the equation for the dissociation of benzoic acid in water. If the K_a of benzoic acid is 6.5×10^{-5}, what is its percent dissociation in a 1.0 M solution?

23.91 Assume that you have a sample of acetic acid ($pK_a = 4.75$) dissolved in water.

(a) Draw the structure of the major species present in the water solution.

(b) Now assume that aqueous HCl is added to the acetic acid solution until pH 2 is reached. Draw the structure of the major species present.

(c) Finally, assume that aqueous NaOH is added to the acetic acid solution until pH 12 is reached. Draw the structure of the major species present.

23.92 Give the IUPAC names of the following compounds:

(a) $\underset{\underset{CH_3}{|}}{CH_3CH}CH_2CH_2\overset{\overset{O}{||}}{C}OCH_3$ **(b)** $\underset{\underset{CH_3}{|}}{CH_3\overset{\overset{CH_3}{|}}{C}}CH_2CH_2\overset{\overset{O}{||}}{C}OH$

(c) $CH_3CH_2CH_2\underset{\underset{CH_3}{|}}{CH}\overset{\overset{O}{||}}{C}NH_2$

23.93 Give the IUPAC names of the following compounds:

(a) $\underset{\underset{CH_2CH_3}{|}}{CH_3CH}CH_2CH_2\overset{\overset{O}{||}}{C}\underset{\underset{CH_3}{|}}{N}CH_3$ **(b)** $\underset{\underset{CH_3}{|}}{CH_3CH}\overset{\overset{O}{||}}{C}O\underset{\underset{CH_3}{|}}{CH}CH_3$

(c) (benzene ring with $CONHCH_2CH_3$ and Cl substituents)

23.94 Draw structures corresponding to the following IUPAC names:

(a) Methyl pentanoate

(b) Isopropyl 2-methylbutanoate

(c) Cyclohexyl acetate

23.95 Draw structures corresponding to the following IUPAC names:

(a) 3-Methylpentanamide

(b) N-Phenylacetamide

(c) N-Ethyl-N-methylbenzamide

23.96 Write equations showing how you could prepare each of the esters in Problem 23.94 from the appropriate alcohols and carboxylic acids.

23.97 Write equations showing how you could prepare each of the amides in Problem 23.95 from the appropriate amines and carboxylic acids.

23.98 Novocaine, a local anesthetic, has the following structure. Identify the functional groups present in novocaine, and show the structures of the alcohol and carboxylic acid you would use to prepare it.

$$H_2N-\text{(benzene ring)}-\overset{\overset{O}{||}}{C}-OCH_2CH_2\underset{\underset{CH_2CH_3}{|}}{N}CH_2CH_3$$

Novocaine

23.99 Ordinary soap is a mixture of the sodium or potassium salts of long-chain carboxylic acids that arise from saponification of animal fat. Draw the structures of soap molecules produced in the following reaction:

$$\begin{array}{l} CH_2O\overset{\overset{O}{||}}{C}(CH_2)_{14}CH_3 \\ | \\ CHO\overset{\overset{O}{||}}{C}(CH_2)_7CH{=}CH(CH_2)_7CH_3 + 3\ KOH \longrightarrow \\ | \\ CH_2O\overset{\overset{O}{||}}{C}(CH_2)_{16}CH_3 \\ \text{A fat} \end{array}$$

Polymers

23.100 What is the difference between a monomer and a polymer?

23.101 What is the difference between a polymer like polyethylene and a polymer like nylon?

23.102 Show the structure of poly(vinyl chloride) by drawing several repeating units. Vinyl chloride is $H_2C{=}CHCl$.

23.103 Show the structures of the polymers you would obtain from the following monomers:

(a) $F_2C{=}CF_2$ gives Teflon

(b) $H_2C{=}CHCOOCH_3$ gives Lucite

23.104 Show the monomer units you would use to prepare the following alkene polymers:

(a)
$$\begin{array}{ccc} & \text{CN} & \text{CN} & \text{CN} \\ & | & | & | \\ \text{⧧ CH}_2\text{CHCH}_2\text{CHCH}_2\text{CH ⧧} \end{array}$$

(b)
$$\begin{array}{ccc} & \text{CH}_3 & \text{CH}_3 & \text{CH}_3 \\ & | & | & | \\ \text{⧧ CH}_2\text{CHCH}_2\text{CHCH}_2\text{CH ⧧} \end{array}$$

(c)
$$\begin{array}{ccc} & \text{Cl} & \text{Cl} & \text{Cl} \\ & | & | & | \\ \text{⧧ CH}_2\text{CCH}_2\text{CCH}_2\text{C ⧧} \\ & | & | & | \\ & \text{Cl} & \text{Cl} & \text{Cl} \end{array}$$

23.105 What monomer unit is used to prepare poly(methylcyanoacrylate), also known as "superglue"?

$$\begin{array}{ccc} & \text{CN} & \text{CN} & \text{CN} \\ & | & | & | \\ \text{⧧ CH}_2\text{CCH}_2\text{CCH}_2\text{C ⧧} \\ & | & | & | \\ & \text{CH}_3 & \text{CH}_3 & \text{CH}_3 \end{array}$$

Poly(methylcyanoacrylate)

23.106 Kevlar, a nylon polymer prepared by reaction of 1,4-benzenedicarboxylic acid (terephthalic acid) with *p*-diaminobenzene, is so strong that it's used to make bulletproof vests. Draw the structure of a segment of Kevlar.

23.107 Draw the structure of a segment of the polyester that results from reaction of the dialcohol ethylene glycol (HOCH$_2$CH$_2$OH) with the diacid butanedioic acid (HO$_2$CCH$_2$CH$_2$CO$_2$H).

General Problems

23.108 Draw structural formulas for the following compounds:
(a) 2-Methylheptane
(b) 4-Ethyl-2-methylhexane
(c) 4-Ethyl-3,4-dimethyloctane
(d) 2,4,4-Trimethylheptane
(e) 1,1-Dimethylcyclopentane
(f) 4-Isopropyl-3-methylheptane

23.109 Give IUPAC names for the following alkanes:

(a)
$$\begin{array}{c} \text{CH}_3 \\ | \\ \text{CH}_3\text{CH}_2\text{CH}_2\text{CHCHCH}_3 \\ | \\ \text{CH}_3 \end{array}$$

(b)
$$\begin{array}{c} \text{CH}_3 \\ | \\ \text{CH}_3\text{CH}_2\text{CH}_2\text{CHCHCH}_3 \\ | \\ \text{CH}_2\text{CH}_2\text{CH}_2\text{CH}_3 \end{array}$$

(c)
$$\begin{array}{c} \text{CH}_3 \quad\quad \text{CH}_2\text{CH}_3 \\ | \quad\quad\quad | \\ \text{CH}_3\text{CHCH}_2\text{CCH}_3 \\ | \\ \text{CH}_2\text{CH}_3 \end{array}$$

(d)
$$\begin{array}{c} \text{CH}_2\text{CH}_3 \\ | \\ \text{CH}_3\text{CH}_2\text{CCH}_2\text{CH}_3 \\ | \\ \text{CH}_2\text{CH}_3 \end{array}$$

23.110 Assume that you have two unlabeled bottles, one with cyclohexane and one with cyclohexene. How could you tell them apart by doing chemical reactions?

23.111 Assume you have two unlabeled bottles, one with cyclohexene and one with benzene. How could you tell them apart by doing chemical reactions?

23.112 Write the products of the following reactions:

(a)
$$\begin{array}{c} \text{CH}_3 \\ | \\ \text{CH}_3\text{CH}_2\text{CH}=\text{CHCHCH}_3 \end{array} \xrightarrow[\text{Pd catalyst}]{\text{H}_2}$$

(b)
$$\text{Br}-\!\!\!\bigcirc\!\!\!-\text{Br} \xrightarrow[\text{H}_2\text{SO}_4]{\text{HNO}_3}$$

(c)
a cyclohexene ring with CH$_3$ groups on both double-bond carbons $\xrightarrow[\text{H}_2\text{SO}_4]{\text{H}_2\text{O}}$

23.113 Show the structure of the nylon polymer that results from heating 6-aminohexanoic acid.

$$\text{H}_2\text{NCH}_2\text{CH}_2\text{CH}_2\text{CH}_2\text{CH}_2\overset{\overset{\displaystyle O}{\|}}{\text{C}}\text{OH} \xrightarrow{\text{Heat}}$$

6-Aminohexanoic acid

Multi-Concept Problems

23.114 Fumaric acid is an organic substance widely used as a food additive. Its elemental composition is 41.4% C, 3.5% H, and 55.1% O. A solution made by dissolving 0.1500 g of fumaric acid in water and diluting to a volume of 100.0 mL gave rise to an osmotic pressure of 240.3 mm Hg at 298 K. On titration of a sample weighing 0.573 g , 94.1 mL of 0.105 M NaOH was required to reach an equivalence point. Fumaric acid reacts with 1 mol of HCl to give an addition product and with 1 mol of H_2 to give a hydrogenation product.

(a) What is the empirical formula of fumaric acid?

(b) What is the molecular mass of fumaric acid?

(c) Draw three possible structures for fumaric acid.

(d) If fumaric acid contains a trans double bond, which of your structures is correct?

23.115 When 0.0552 g of an unknown liquid containing only C, H, and O was subjected to combustion analysis, 0.1213 g of CO_2 and 0.0661 g of H_2O were formed.

(a) Calculate a formula for the unknown, and write a balanced equation for the combustion reaction.

(b) Is the formula you calculated an empirical formula or a molecular formula? Explain.

(c) Draw three possible structures for the compound, and identify the functional groups in each.

(d) Reaction of the compound with CrO_3 yields acetone. Which of the three structures you drew in part (c) is most likely to be correct?

(e) Combustion of 5.000 g of the compound releases 166.9 kJ heat. Look up ΔH_f° values for $CO_2(g)$ and $H_2O(l)$ in Appendix B, and calculate ΔH_f° for the compound.

 # eMedia Problems

23.116 Use the **Boiling Point** activity (*eChapter 23.2*) to get the boiling points of the first six straight-chain alkanes (methane, ethane, propane, butane, pentane, and hexane), and the corresponding alcohols derived by replacing one of the CH_3 hydrogens at the end of each chain with an OH group.

(a) For a given number of carbons, which has the higher boiling point, the alkane or the alcohol? Explain this observation in terms of intermolecular forces.

(b) Does the difference in boiling point for an alkane and its corresponding alcohol increase or decrease with increasing number of carbons?

(c) Using your answer from part (b), describe how the relative importance of hydrogen bonding and London dispersion forces changes with increasing length of a carbon chain. Which type of interaction contributes more significantly to the overall intermolecular forces for small molecules, and which for large molecules?

23.117 Determine the boiling points of at least five different alcohols and their corresponding amines using the **Boiling Point** activity (*eChapter 23.2*).

(a) Based on the boiling points, which functional group, OH or NH_2, appears to exhibit more significant hydrogen bonding?

(b) Draw electron-dot structures of alcohol and amine functional groups. Use the electron-dot structures to explain the observation that one group exhibits more significant hydrogen bonding than the other.

23.118 What is the hybridization of the carbons in the **Cycloalkanes structures** (*eChapter 23.6*)? Which of these structures do you expect to be the most unstable and why?

23.119 Watch the **Synthesis of Nylon 6,10** movie (*eChapter 23.15*) and draw a segment of the polymer product of this condensation polymerization process. (HCl is the other product.)

24

Biochemistry

The genetic details of every living organism, no matter how fantastic they may appear, are encoded in the organism's DNA.

I f the ultimate goal of chemistry is to understand the world around us on a molecular level, then a knowledge of **biochemistry**—the chemistry of living organisms—is clearly a crucial part of that goal. In this chapter we'll look at the main classes of biomolecules: proteins, carbohydrates, lipids, and nucleic acids. First, though, we'll take a brief look at where the energy to drive biological reactions comes from and at how biochemical energy is used by living organisms.

24.1 Biochemical Energy

All living things do mechanical work. Microorganisms engulf food, plants bend toward the sun, animals move about. Organisms also do the chemical work of synthesizing biomolecules needed for energy storage, growth, and repair. Even individual cells do work when they move molecules and ions across cell membranes.

In animals, it is the energy extracted from food and released in the exquisitely interconnected reactions of *metabolism* that allows work to be done. All animals are powered by the cellular oxidation of food molecules containing mainly carbon, hydrogen, and oxygen. The end products of this biological oxidation are carbon dioxide, water, and energy, just as they are when an organic fuel such as methane is burned with oxygen in a furnace.

$$\text{C, H, O (food molecules)} + O_2 \longrightarrow CO_2 + H_2O + \text{Energy}$$

The energy used by the vast majority of living organisms comes ultimately from the sun. Photosynthesis in plants converts sunlight to potential energy that is stored mainly in the chemical bonds of carbohydrates. Plant-eating animals use some of this energy for living and store the rest of it, mainly in the chemical bonds of fats. Other animals then eat smaller animals, and chemical energy thus moves higher up the food chain.

The energy used by almost all living organisms ultimately comes from the sun. Plants use solar energy for the photosynthesis of glucose from CO_2, and animals then eat the plants.

The sum of the many organic reactions that take place in cells is called **metabolism**. These reactions usually occur in long sequences, which may be either linear or cyclic. In a linear sequence, the product of one reaction serves as the starting material for the next. In a cyclic sequence, a series of reactions regenerates the first reactant and produces other products along the way.

$$A \longrightarrow B \longrightarrow C \longrightarrow D \longrightarrow \ldots$$

A linear sequence

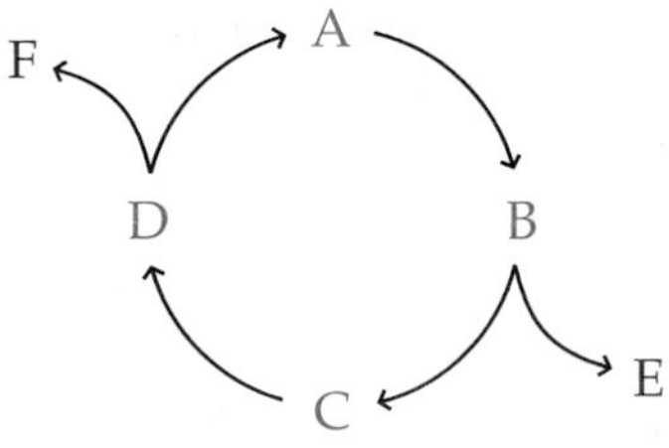

A cyclic sequence

where E and F are products

Those reaction sequences that break molecules apart are known collectively as **catabolism**, while those that put building blocks back together to assemble larger molecules are known as **anabolism**. Catabolic reactions generally release energy that is used to power living organisms, whereas anabolic reactions generally absorb energy. The overall picture of catabolism and energy production can be roughly divided into the four stages shown in Figure 24.1.

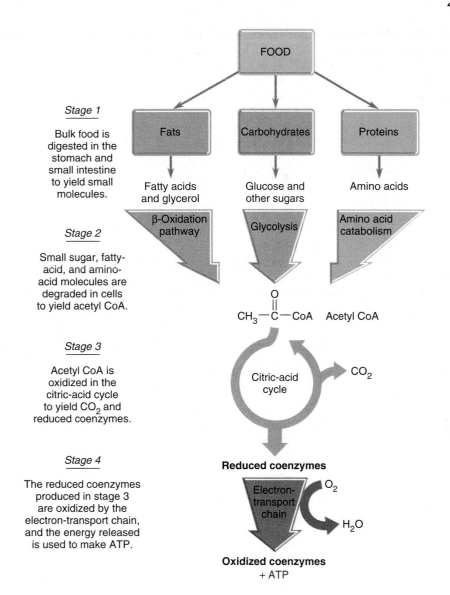

Stage 1

Bulk food is digested in the stomach and small intestine to yield small molecules.

Stage 2

Small sugar, fatty-acid, and amino-acid molecules are degraded in cells to yield acetyl CoA.

Stage 3

Acetyl CoA is oxidized in the citric-acid cycle to yield CO_2 and reduced coenzymes.

Stage 4

The reduced coenzymes produced in stage 3 are oxidized by the electron-transport chain, and the energy released is used to make ATP.

FIGURE 24.1 An overview of catabolic pathways in the four stages of food degradation and the production of biochemical energy.

The first stage of catabolism, *digestion*, takes place in the stomach and small intestine when bulk food is broken down into small molecules such as simple sugars, long-chain carboxylic acids (called *fatty acids*), and amino acids. In stage 2, these small molecules are further degraded inside cells to yield two-carbon acetyl groups ($CH_3C=O$) attached to a large carrier molecule called *coenzyme A*. The resultant compound, *acetyl coenzyme A* (*acetyl CoA*), is an intermediate in the breakdown of all main classes of food molecules.

Acetyl groups are oxidized in the third stage of catabolism, the *citric acid cycle*, to yield carbon dioxide and water. This stage releases a great deal of energy that is used in stage 4, the *electron-transport chain*, to make molecules of **adenosine triphosphate (ATP)**. ATP, the final result of food catabolism, plays a pivotal role in the production of biological energy. As the crucial molecule for driving many metabolic reactions, ATP has been called the "energy currency of the living cell." Catabolic reactions "pay off" in ATP by synthesizing it from *adenosine diphosphate* (ADP) plus hydrogen phosphate ion, HPO_4^{2-}. Anabolic reactions "spend" ATP by transferring a phosphate group

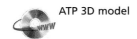

ATP 3D model

to other molecules, thereby regenerating ADP. The entire process of energy production revolves around the ATP $\rightleftharpoons$ ADP interconversion.

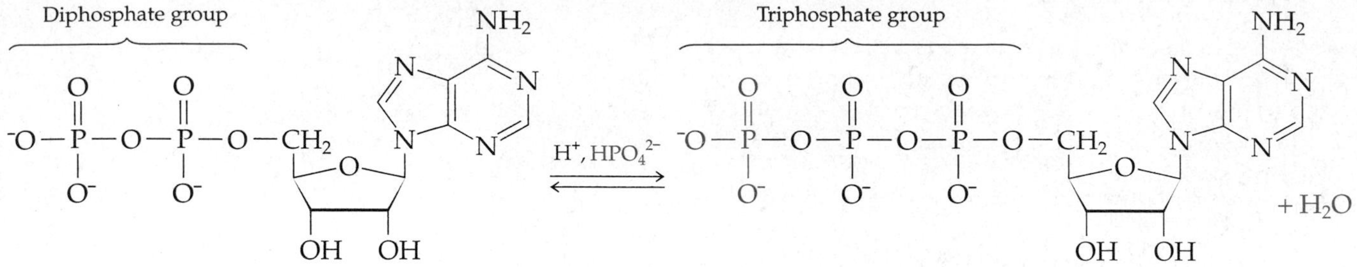

Diphosphate group

Adenosine diphosphate (ADP)

H^+, HPO_4^{2-}

Triphosphate group

Adenosine triphosphate (ATP)

$+ H_2O$

Because the primary metabolic function of ATP is to drive reactions, biochemists often refer to it as a "high-energy molecule" or an "energy store-house." These terms don't mean that ATP is somehow different from other compounds; they mean only that ATP releases a large amount of energy when its P–O–P (*phosphoric anhydride*) bonds are broken and a phosphate group is transferred.

What does the body do with ATP? Recall from Sections 8.14 and 17.7 that a chemical reaction is favorable, or spontaneous, if the free-energy change, ΔG, for the reaction is negative. Conversely, a reaction is unfavorable if the free-energy change is positive. The change in free energy depends on two factors: the release or absorption of heat (ΔH) and the increase or decrease in entropy (ΔS). The larger the amount of heat released and the greater the increase in entropy, the more favorable the reaction.

$$\Delta G = \Delta H - T\Delta S$$

Reactions in living organisms are no different from laboratory reactions in flasks. Both follow the same laws, and both have the same energy requirements. For any biochemical reaction to occur spontaneously, ΔG must be negative. For example, complete oxidation of 1 mol of glucose, the principal source of energy for animals, has $\Delta G = -2870$ kJ.

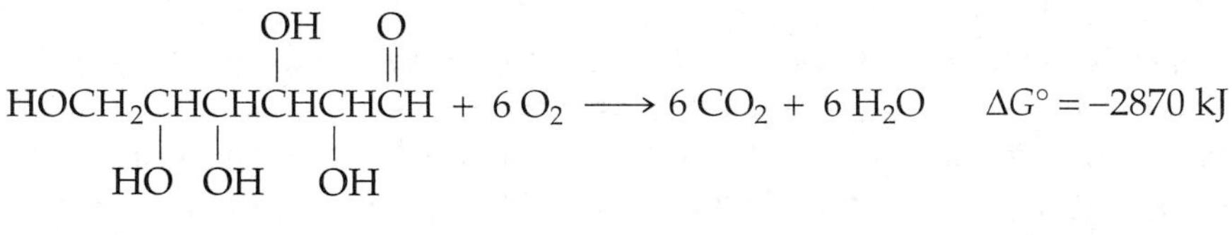

$$HOCH_2CHCHCHCHCH + 6\,O_2 \longrightarrow 6\,CO_2 + 6\,H_2O \qquad \Delta G° = -2870 \text{ kJ}$$

Glucose ($C_6H_{12}O_6$)

Reactions in which the free-energy change is positive can also occur, but such reactions can't be spontaneous. An example is the conversion of glucose to glucose-6-phosphate, an important step in the breakdown of dietary carbohydrates.

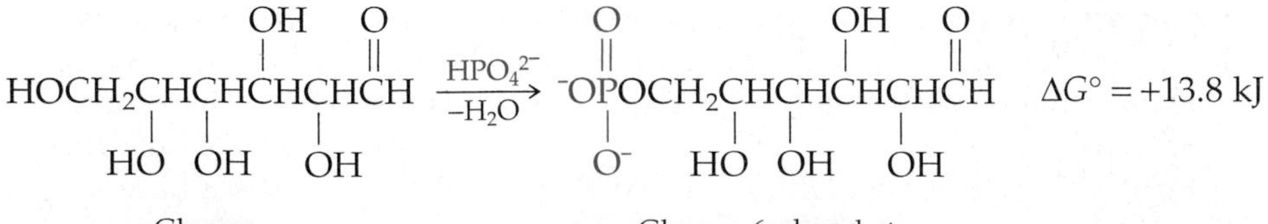

$$HOCH_2CHCHCHCHCH \xrightarrow[-H_2O]{HPO_4^{2-}} {}^-OPOCH_2CHCHCHCHCH \qquad \Delta G° = +13.8 \text{ kJ}$$

Glucose

Glucose-6-phosphate

What usually happens when an energetically unfavorable reaction occurs is that it is "coupled" to an energetically favorable reaction so that the *overall* free-energy change for the combined reactions is favorable. For example, the reaction of glucose with hydrogen phosphate ion (HPO_4^{2-}) to yield glucose-6-phosphate plus water is energetically unfavorable by 13.8 kJ and thus does not take place spontaneously. At the same time, however, the reaction of ATP with water to yield ADP plus hydrogen phosphate ion is energetically *favorable* by about 30.5 kJ. Thus, if the two reactions are coupled, the overall process for the synthesis of glucose-6-phosphate is favorable by about 16.7 kJ. That is, the free-energy change for the reaction of ATP with water is so favorable that it can "drive" the unfavorable reaction of glucose with hydrogen phosphate ion.

$$\text{Glucose} + HPO_4^{2-} \longrightarrow \text{Glucose-6-phosphate} + H_2O \qquad \Delta G° = +13.8 \text{ kJ}$$

$$\text{ATP} + H_2O \longrightarrow \text{ADP} + HPO_4^{2-} + H^+ \qquad \Delta G° = -30.5 \text{ kJ}$$

$$\text{Net: Glucose} + \text{ATP} \longrightarrow \text{Glucose-6-phosphate} + \text{ADP} + H^+ \qquad \Delta G° = -16.7 \text{ kJ}$$

It's this ability to drive otherwise unfavorable reactions that makes ATP so useful. In fact, most of the thousands of reactions going on in your body every minute are somehow coupled to ATP. It's no exaggeration to say that the transfer of a phosphate group from ATP is the single most important chemical reaction in making life possible.

▶ **PROBLEM 24.1** One of the steps in fat metabolism is the reaction of glycerol [$HOCH_2-CH(OH)-CH_2OH$] with ATP to yield glycerol-1-phosphate. Write the reaction, and draw the structure of glycerol-1-phosphate.

24.2 Amino Acids and Peptides

Taken from the Greek *proteios*, meaning "primary," the name *protein* aptly describes a group of biological molecules that are of primary importance to all living organisms. Approximately 50% of the body's dry weight is protein, and almost all the reactions that occur in the body are catalyzed by proteins. In fact, a human body contains over *100,000* different kinds of proteins.

Proteins have many different biological functions. Some, such as the keratin in skin, hair, and fingernails, serve a structural purpose. Others, such as the insulin that controls glucose metabolism, act as *hormones*—chemical messengers that coordinate the activities of different cells in an organism. And still other proteins, such as DNA polymerase, are *enzymes*, the biological catalysts that carry out body chemistry. All **proteins** are biological polymers made up of many amino acids linked together to form a long chain. As their name implies, *amino acids* are molecules that contain two functional groups, a basic amino group ($-NH_2$) and an acidic $-COOH$ group. Alanine is a simple example.

Bird feathers are made largely of the protein *keratin*.

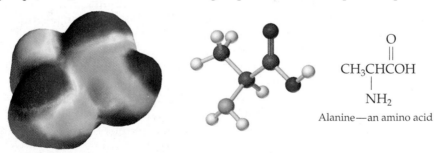

$$\begin{array}{c} \quad\quad\quad O \\ \quad\quad\quad \| \\ CH_3CHCOH \\ \quad | \\ \quad NH_2 \end{array}$$

Alanine—an amino acid

Two or more amino acids can link together by forming amide bonds (Section 23.14), usually called **peptide bonds**. A *dipeptide* results from the linking together of two amino acids by formation of a peptide bond between the $-NH_2$ group of one and the $-COOH$ group of the other. Similarly, a *tripeptide* results from linking three amino acids by two peptide bonds, and so on. Short chains of up to 100 amino acids are often called **polypeptides**, while the term *protein* is reserved for longer chains.

$$H-\underset{\underset{H}{|}}{\overset{\overset{H}{|}}{N}}-\underset{\underset{R}{|}}{\overset{}{C}}-\overset{\overset{O}{||}}{C}\boxed{OH + H}\underset{\underset{H}{|}}{\overset{\overset{H}{|}}{N}}-\underset{\underset{R'}{|}}{\overset{}{C}}-\overset{\overset{O}{||}}{C}-OH \longrightarrow H-\underset{\underset{H}{|}}{\overset{\overset{H}{|}}{N}}-\underset{\underset{R}{|}}{\overset{}{C}}-\overset{\overset{O}{||}}{C}-\underset{\underset{H}{|}}{\overset{\overset{H}{|}}{N}}-\underset{\underset{R'}{|}}{\overset{}{C}}-\overset{\overset{O}{||}}{C}-OH + H_2O$$

α Amino acids—The groups symbolized by R and R' represent different amino acid side chains.

A peptide bond

Twenty different amino acids are commonly found in proteins, as shown in Figure 24.2. Each amino acid is referred to by a three-letter shorthand code, such as Ala (alanine), Gly (glycine), Pro (proline), and so on. All 20 are called **alpha- (α-) amino acids** because the amino group in each is connected to the carbon atom *alpha to* (next to) the carboxylic acid group. Nineteen of the 20 have an $-NH_2$ amino group, and one (proline) has an $-NH-$ amino group as part of a ring.

The 20 amino acids differ only in the nature of the group attached to the α carbon. Usually called the **side chain**, this group can be symbolized in a general way by the letter R. (In a broader context, the symbol R is used throughout organic chemistry to refer to an organic fragment of unspecified structure.) Our bodies can synthesize only 10 of the 20 amino acids. The remaining 10, highlighted in Figure 24.2, are called *essential amino acids* because they must be obtained from the diet.

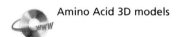

Amino Acid 3D models

α carbon

Side chain

$$R-\underset{\underset{NH_2}{|}}{CH}-\overset{\overset{O}{||}}{C}OH$$

Generalized structure of an α-amino acid

R

The 20 common amino acids can be classified as *neutral*, *basic*, or *acidic*, depending on the structure of their side chains. Fifteen of the 20 have neutral side chains. Two (aspartic acid and glutamic acid) have an additional carboxylic acid group in their side chains and are classified as acidic amino acids. Three (lysine, arginine, and histidine) have an additional amine function in their side chains and are classified as basic amino acids. The 15 neutral amino acids can be further divided into those with nonpolar side chains and those with polar functional groups such as amide or hydroxyl groups. Nonpolar side chains are often described as *hydrophobic* (water fearing) because they are repelled by water, while polar side chains are described as *hydrophilic* (water loving) because they are attracted to water.

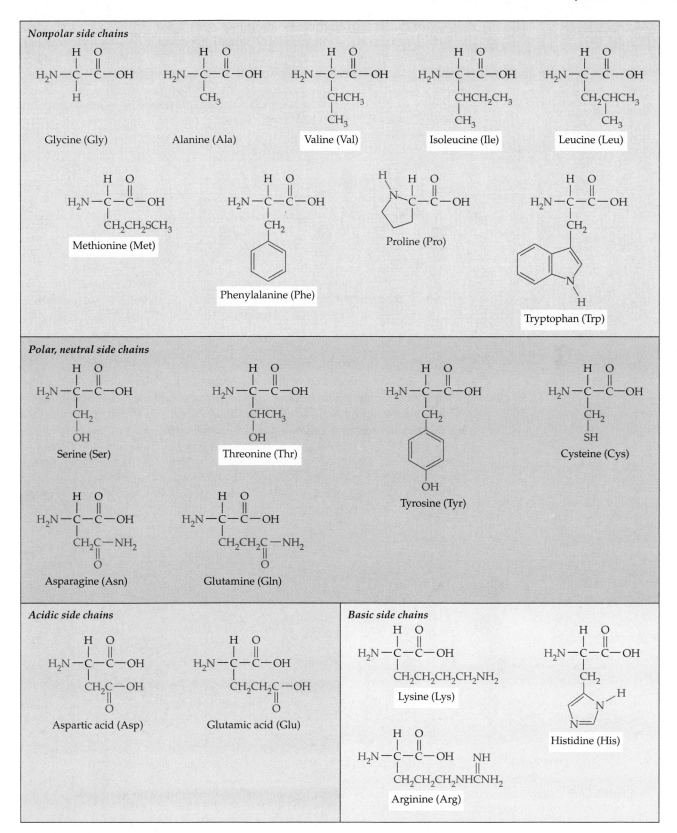

FIGURE 24.2 Structures of the 20 α-amino acids commonly found in proteins. Fifteen of the 20 have neutral side chains, 2 have acidic side chains, and 3 have basic side chains. The names of the 10 essential amino acids are highlighted.

▶ **PROBLEM 24.2** Which common amino acids contain an aromatic (benzene-like) ring? Which contain sulfur? Which are alcohols? Which have alkyl-group side chains?

▶ **PROBLEM 24.3** Draw alanine, using wedges and dashes to show the tetrahedral geometry of its α carbon.

✎ **KEY CONCEPT PROBLEM 24.4** Identify the following amino acids, and tell whether each is acidic, basic, or neutral:

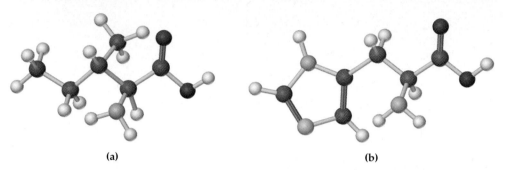

(a)   (b)

24.3 Amino Acids and Molecular Handedness

Chirality movie

We saw in Section 20.9 that certain molecules lack a plane of symmetry and are therefore *chiral*. When held up to a mirror, a chiral molecule is not identical to its reflected image. Instead, the molecule and its mirror image have a right-hand/left-hand relationship. Compare alanine and propane, for example (Figure 24.3). An alanine molecule is chiral; it has no symmetry plane and can exist in two forms—a "right-handed" form called D-alanine (from *dextro*, Latin for "right") and a "left-handed" form called L-alanine (from *levo*, Latin for "left"). Propane, however, is achiral. It has a symmetry plane cutting through the three carbons and thus exists in only a single form.

L-Alanine; D-Alanine 3D models

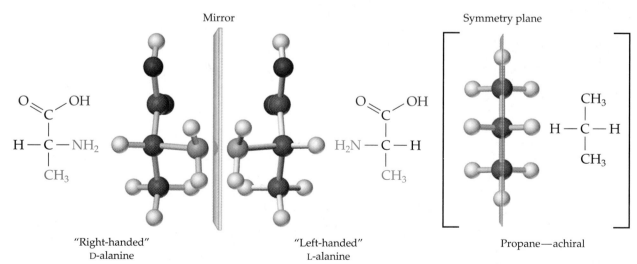

FIGURE 24.3 Alanine (2-aminopropanoic acid) has no symmetry plane and can therefore exist in two forms—a "right-handed" form and a "left-handed" form. Propane, however, has a symmetry plane and is achiral.

Why are some organic molecules chiral while others are not? The answer has to do with the three-dimensional nature of molecules. We've seen that carbon forms four bonds that are oriented toward the four corners of a regular tetrahedron. Whenever a carbon atom is bonded to four *different* atoms or groups of atoms, chirality results. If a carbon is bonded to two or more of the same groups, however, no chirality is possible. In alanine, for example, carbon 2 is bonded to four different groups: a −COOH group, an −H atom, an −NH_2 group, and a −CH_3 group. Thus, alanine is chiral. In propane, however, each carbon is bonded to at least two groups—the −H atoms—that are identical. Thus, propane is achiral.

Groups attached to C2

1. −COOH
2. −H
3. −NH_2
4. −CH_3
} Different

Alanine—chiral

Groups attached to C2

1. −CH_3
2. −CH_3
} Identical
3. −H
4. −H
} Identical

Propane—achiral

As mentioned in Section 20.9, the two mirror-image forms of a chiral molecule are called *enantiomers*, or *optical isomers*. The mirror-image relationship of the enantiomers of a molecule with four different groups on a chiral carbon atom is shown in Figure 24.4.

Mirror

FIGURE 24.4 A molecule with a carbon atom that is bonded to four different groups is chiral and is not identical to its mirror image. It thus exists in two enantiomeric forms.

Of the 20 common amino acids, 19 are chiral because they have four different groups bonded to their α carbons, −H, −NH_2, −COOH, and −R (the side chain). Only glycine, H_2NCH_2COOH, is achiral. Even though chiral α-amino acids can exist as pairs of enantiomers, only L-amino acids are found in proteins. When drawn with the −COOH group at the top and the side-chain R group at the bottom, an L-amino acid has its −NH_2 group coming out of the plane of the paper on the left side of the structure:

An L-amino acid

EXAMPLE 24.1 Lactic acid can be isolated from sour milk. Is lactic acid chiral?

$$
\begin{array}{cc}
\text{OH} & \text{O} \\
| & \| \\
\text{CH}_3-\text{CH}-\text{COH} \\
\phantom{\text{CH}}3 \quad 2 \quad 1
\end{array}
\qquad \text{Lactic acid}
$$

SOLUTION To find out if lactic acid is chiral, list the groups attached to each carbon:

Groups on carbon 1	Groups on carbon 2	Groups on carbon 3
1. −OH	1. −COOH	1. −CH(OH)COOH
2. =O	2. −OH	2. −H
3. −CH(OH)CH$_3$	3. −H	3. −H
	4. −CH$_3$	4. −H

Next, look at the lists to see if any carbon is attached to four different groups. Of the three carbons, carbon 2 has four different groups, and lactic acid is therefore chiral.

▶ **PROBLEM 24.5** Which of the following objects are chiral?
(a) A glove   (b) A baseball   (c) A screw   (d) A nail

▶ **PROBLEM 24.6** 2-Aminopropane is an achiral molecule, but 2-amino-butane is chiral. Explain.

▶ **PROBLEM 24.7** Which of the following molecules are chiral?
(a) 3-Chloropentane   (b) 2-Chloropentane

$$
\begin{array}{cc}
\text{CH}_3 & \text{CH}_3 \\
| & | \\
\end{array}
$$
(c) CH$_3$CHCH$_2$CHCH$_2$CH$_3$

▶ **PROBLEM 24.8** Two of the 20 common amino acids have two chiral carbon atoms in their structures. Identify them.

✦— **KEY CONCEPT PROBLEM 24.9** Two of the following three molecules are identical, and one is the enantiomer of the other two. Which one is the enantiomer?

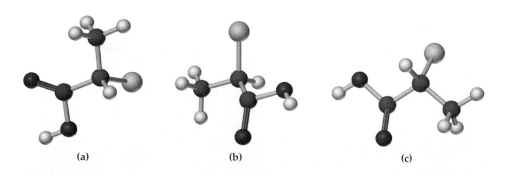

(a)   (b)   (c)

24.4 Proteins

Proteins are amino acid polymers in which the individual amino acids, often called *residues*, are linked together by peptide (amide) bonds. The repeating chain of amide linkages to which the side chains are attached is called the *backbone*.

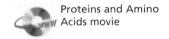
Proteins and Amino Acids movie

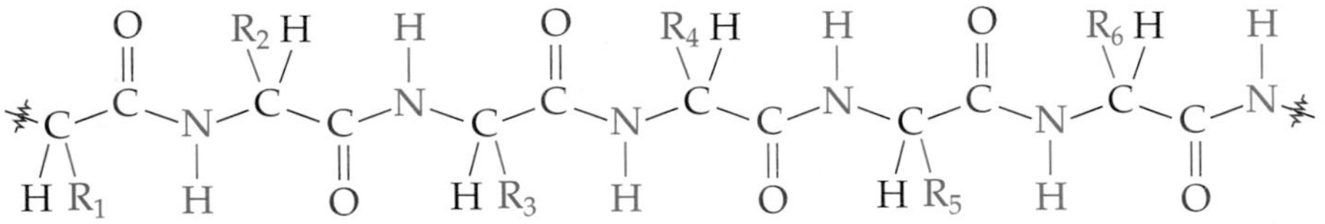

A segment of a protein backbone. The side-chain R groups
of the individual amino acids are substituents on the backbone.

Since amino acids can be assembled in any order, depending on which −COOH group joins with which −NH$_2$ group, the number of possible isomeric peptides increases rapidly as the number of amino acid residues increases. There are six ways in which three different amino acids can be joined, more than 40,000 ways in which the eight amino acids present in the blood pressure-regulating hormone angiotensin II can be joined (Figure 24.5), and a staggering number of ways in which the *1800* amino acids in myosin, the major component of muscle filaments, can be arranged.

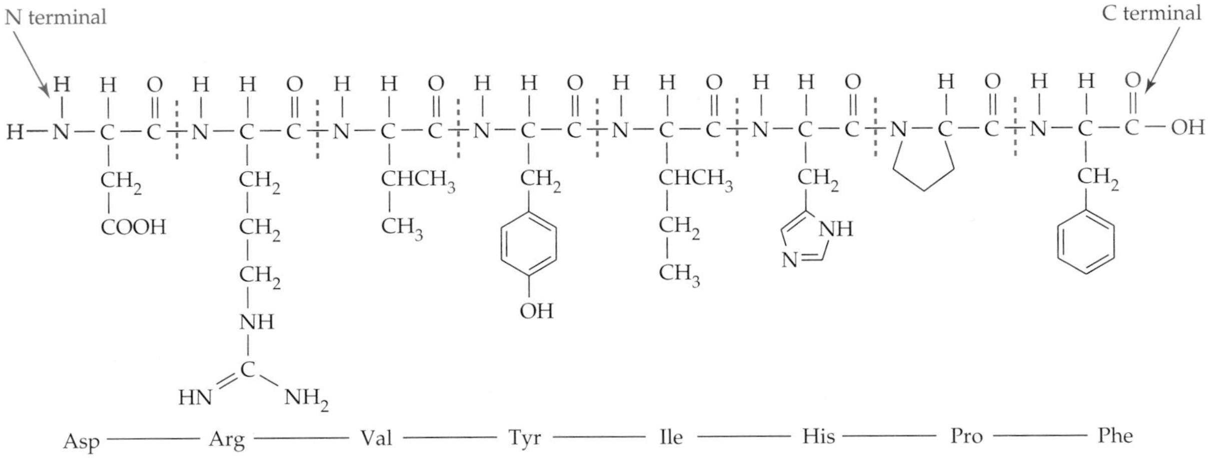

Asp ——— Arg ——— Val ——— Tyr ——— Ile ——— His ——— Pro ——— Phe

FIGURE 24.5 The structure of angiotensin II, an octapeptide in blood plasma that regulates blood pressure.

No matter how long the chain, all noncyclic proteins have an **N-terminal amino acid** with a free −NH$_2$ group on one end and a **C-terminal amino acid** with a free −COOH group on the other end. By convention, a protein is written with the N-terminal residue on the left and the C-terminal residue on the right, and its name is indicated with the three-letter abbreviations listed in Figure 24.2. Thus, angiotensin II (Figure 24.5) is abbreviated Asp-Arg-Val-Tyr-Ile-His-Pro-Phe.

Proteins are classified according to their three-dimensional shape as either *fibrous* or *globular*. Fibrous proteins, such as collagen and the keratins, consist of polypeptide chains arranged side by side in long filaments. Because these proteins are tough and insoluble in water, they are used in nature for structural materials like skin, tendons, hair, ligaments, and muscle. Globular

FIGURE 24.6   A computer-generated view of the enzyme pepsin, a typical globular protein.

proteins, by contrast, are usually coiled into compact, nearly spherical shapes, as seen in the computer-generated picture of the digestive enzyme pepsin in Figure 24.6. Most of the 2000 or so known enzymes present inside cells are globular proteins.

Another way to classify proteins is according to biological function. As indicated in Table 24.1, proteins have a remarkable diversity of roles.

TABLE 24.1   Some Biological Functions of Proteins

Type	Function	Example
Enzymes	Catalyze biological processes	Pepsin
Hormones	Regulate body processes	Insulin
Storage proteins	Store nutrients	Ferritin
Transport proteins	Transport oxygen and other substances through the body	Hemoglobin
Structural proteins	Form an organism's structure	Collagen
Protective proteins	Help fight infection	Antibodies
Contractile proteins	Form muscles	Actin, myosin
Toxic proteins	Serve as a defense for the plant or animal	Snake venoms

EXAMPLE 24.2   Draw the structure of the dipeptide Ala-Ser.

SOLUTION   First, look up the structures of the two amino acids, Ala (alanine) and Ser (serine). Since alanine is N-terminal and serine is C-terminal, Ala-Ser must have an amide bond between the alanine $-COOH$ and the serine $-NH_2$.

$$\text{N terminal} \quad\quad \overset{\displaystyle O}{\overset{\|}{\quad}} \quad\quad \overset{\displaystyle O}{\overset{\|}{\quad}} \quad \text{C terminal}$$

$$H_2NCHC - NHCHCOH$$
$$\underset{CH_3}{|} \quad\quad \underset{CH_2OH}{|}$$

Alanine   Serine

Ala-Ser

▶ **PROBLEM 24.10**   Use the three-letter shorthand notations to name the two isomeric dipeptides that can be made from valine and cysteine. Draw both structures.

▶ **PROBLEM 24.11**   Name the six tripeptides that contain valine, tyrosine, and glycine.

24.5   Levels of Protein Structure

With molecular masses of up to one-half *million* amu, many proteins are so large that the word *structure* takes on a broader meaning than it does with simpler molecules. In fact, chemists usually speak about four levels of structure when describing proteins. The **primary structure** of a protein specifies the sequence in which the various amino acids are linked together. **Secondary structure** refers to how segments of the protein chain are oriented into a regular pattern. **Tertiary structure** refers to how the entire protein chain is coiled and folded into a specific three-dimensional shape, and **quaternary structure** refers to how several protein chains aggregate to form a larger unit.

Primary Structure

Primary structure is the most fundamental of the four structural levels because it is the protein's amino acid sequence that determines its overall shape and function. So crucial is primary structure to function that the change of only one amino acid out of several hundred can drastically alter biological properties. The disease sickle-cell anemia, for example, is caused by a genetic defect in blood hemoglobin whereby valine is substituted for glutamic acid at only one position in a chain of 146 amino acids.

Secondary Protein Structure

When looking at the primary structure of a protein like angiotensin II in Figure 24.5, you might get the idea that the molecule is long and threadlike. In fact, though, most proteins fold in such a way that segments of the protein chain orient into regular patterns, called *secondary structures*. There are two common patterns: the *alpha-helix* and the *beta-pleated sheet*.

Keratin, a fibrous protein found in wool, hair, fingernails, and feathers, wraps into a helical coil, much like the cord on a telephone (Figure 24.7). Called an **alpha- (α-) helix**, this secondary structure is stabilized by the formation of hydrogen bonds (Section 10.2) between the N–H group of one amino acid and the C=O group of another amino acid four residues away. Each turn of the helix contains 3.6 amino acid residues, with a distance between turns of 0.54 nm.

Lamb's wool is made largely of keratin, a fibrous protein with an α-helical secondary structure.

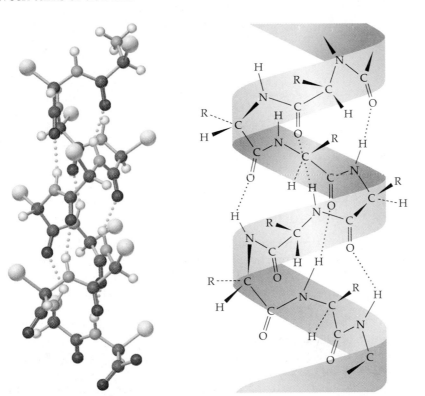

FIGURE 24.7 The α-helical secondary structure of keratin. The amino acid backbone winds in a right-handed spiral, much like that of a telephone cord.

Fibroin, the fibrous protein found in silk, has a secondary structure called a **beta- (β-) pleated sheet**, in which a polypeptide chain doubles back on itself after a hairpin bend. The two sections of the chain on either side of the bend line up in a parallel arrangement held together by hydrogen bonds

(Figure 24.8). Although not as common as the α-helix, small pleated-sheet regions are often found in proteins.

Silk is made of fibroin, a fibrous protein with a pleated-sheet secondary structure.

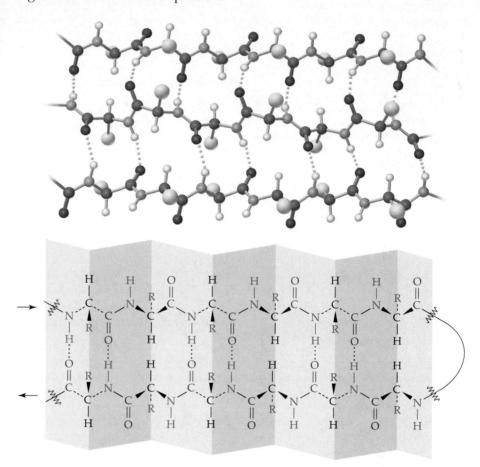

FIGURE 24.8 The β-pleated-sheet secondary structure of silk fibroin. The amino acid side chains are above and below the rough plane of the sheet. (Dotted lines indicate hydrogen bonds between chains.)

Tertiary Protein Structure

Secondary protein structures result primarily from hydrogen bonding between amide linkages along the protein backbone, but higher levels of structure result primarily from interactions of side-chain R groups in the protein. Myoglobin, for example, is a globular protein found in the skeletal muscles of sea mammals, where it stores oxygen needed to sustain the animals during long dives. With a single chain of 153 amino acid residues, myoglobin consists of eight straight segments, each of which adopts an α-helical secondary structure. These helical sections then fold further to form a compact, nearly spherical, tertiary structure (Figure 24.9).

The most important force stabilizing a protein's tertiary structure results from the hydrophobic interactions of hydrocarbon side chains on amino acids. Those amino acids with neutral, nonpolar side chains have a strong tendency to congregate on the hydrocarbon-like interior of a protein molecule, away from the aqueous medium. Those amino acids with polar side chains, by contrast, are usually found on the exterior of the protein

The muscles of sea mammals contain myoglobin, a globular protein that stores oxygen.

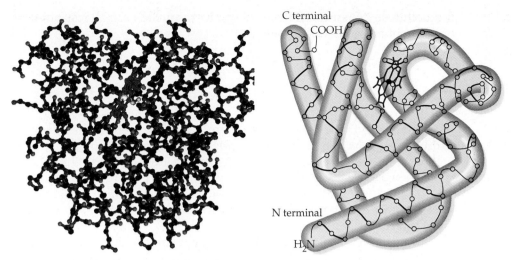

FIGURE 24.9 Secondary and tertiary structure of myoglobin, a globular protein found in the muscles of sea mammals. Myoglobin has eight helical sections.

where they can be solvated by water. Also important for stabilizing a protein's tertiary structure are *disulfide bridges* (covalent S−S bonds formed between nearby cysteine residues), *salt bridges* (ionic attractions between positively and negatively charged sites on the protein), and hydrogen bonds between nearby amino acids.

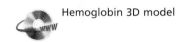
Hemoglobin 3D model

$$O=C \quad\quad C=O \quad O=C \quad\quad\quad C=O$$
$$|\quad\quad\quad\quad\quad | \quad\quad | \quad\quad\quad\quad\quad\quad |$$
$$CHCH_2SH + HSCH_2CH \xrightarrow{[O]} CHCH_2S-SCH_2CH$$
$$|\quad\quad\quad\quad\quad | \quad\quad | \quad\quad\quad\quad\quad\quad |$$
$$NH \quad\quad\quad\quad NH \quad\quad NH \quad\quad\quad\quad NH$$

Two cysteine residues   A disulfide bridge

24.6 Enzymes

Enzymes are large proteins that act as catalysts for biological reactions. A *catalyst*, as we saw in Section 12.11, is an agent that speeds up the rate of a chemical reaction without itself undergoing change. For example, sulfuric acid catalyzes the reaction of a carboxylic acid with an alcohol to yield an ester (Section 23.14).

Enzymes, too, catalyze reactions that might otherwise occur very slowly, but they differ from catalysts like sulfuric acid in two important respects. First, enzymes are much larger, more complicated molecules than simple inorganic catalysts. Second, enzymes are far more specific in their action. Sulfuric acid catalyzes the reaction of nearly *every* carboxylic acid with nearly *every* alcohol, but enzymes often catalyze only a *single* reaction of a single compound, called the enzyme's **substrate**. For example, the enzyme *amylase* found in human digestive systems catalyzes the breakdown of starch to yield glucose but has no effect on cellulose, even though starch and cellulose are structurally similar. Thus, humans can digest potatoes (starch) but not grass (cellulose).

$$\text{Starch} + H_2O \xrightarrow{\text{Amylase}} \text{Many glucose molecules}$$

$$\text{Cellulose} + H_2O \xrightarrow{\text{Amylase}} \text{No reaction}$$

The catalytic activity of an enzyme is measured by its **turnover number**, which is defined as the number of substrate molecules acted on by one molecule of enzyme per unit time. As indicated in Table 24.2, enzymes vary greatly in their turnover number. Most enzymes have values around a few thousand, but carbonic anhydrase, which catalyzes the reaction of CO_2 with water to yield bicarbonate ion, acts on *600,000* substrate molecules per second.

In addition to their protein part, many enzymes contain small, nonprotein parts called **cofactors**. In such enzymes, the protein part is called an **apoenzyme**, and the entire assembly of apoenzyme plus cofactor is called a **holoenzyme**. Only holoenzymes are active as catalysts; neither apoenzyme nor cofactor alone can catalyze a reaction.

Holoenzyme = Apoenzyme + Cofactor

An enzyme cofactor can be either an inorganic ion (usually a metal cation) or a small organic molecule called a **coenzyme**. (The requirement of many enzymes for metal-ion cofactors is the main reason behind our dietary need for trace minerals. Iron, zinc, copper, manganese, molybdenum, cobalt, nickel, and selenium are all essential trace elements that function as enzyme cofactors.) A large number of different organic molecules serve as coenzymes. Often, though not always, the coenzyme is a *vitamin*. Thiamine (vitamin B_1), for example, is a coenzyme required in the metabolism of carbohydrates.

Thiamine (vitamin B_1); an enzyme cofactor

How Enzymes Work: The Lock-and-Key Model

Our current picture of how enzymes work uses the **lock-and-key model** to explain both specificity and catalytic activity. In this model, an enzyme is pictured as a large, irregularly shaped molecule with a cleft, or crevice, in its middle. Inside the crevice is an **active site**, a small region with the shape and chemical composition necessary to bind the substrate and catalyze the appropriate reaction. In other words, the active site acts like a lock into which only a specific key can fit (Figure 24.10). An enzyme's active site is lined by various acidic, basic, and neutral amino acid side chains, all properly positioned for maximum interaction with the substrate.

Enzyme-catalyzed reactions begin when the substrate migrates into the active site to form an enzyme–substrate complex. No covalent bonds are formed; the enzyme and substrate are held together by hydrogen bonds and by weak intermolecular attractions between functional groups. With enzyme

TABLE 24.2 Turnover Numbers of Some Enzymes	
Enzyme	**Turnover Number (s^{-1})**
Carbonic anhydrase	600,000
Acetylcholinesterase	25,000
β-Amylase	18,000
Penicillinase	2,000
DNA polymerase I	15

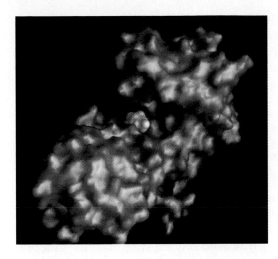

FIGURE 24.10 According to the lock-and-key model, an enzyme is a large, three-dimensional molecule containing a crevice with an active site. Only a substrate whose shape and structure are complementary to those of the active site can fit into the enzyme. The active site of the enzyme hexose kinase is clearly visible as the cleft at the top in this computer-generated structure, as is the fit of the substrate (blue) in the active site.

and substrate thus held together in a precisely defined arrangement, the appropriately positioned functional groups in the active site carry out a chemical reaction on the substrate molecule, and enzyme plus product then separate:

$$\text{E} \; + \; \text{S} \longrightarrow \text{E–S} \longrightarrow \text{E} \; + \; \text{P}$$

Enzyme   Substrate   Complex   Enzyme   Product

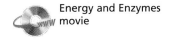

Energy and Enzymes movie

As an example of enzyme action, look in Figure 24.11 at the enzyme hexose kinase, which catalyzes the reaction of adenosine triphosphate (ATP) with glucose to yield glucose-6-phosphate and adenosine diphosphate (ADP). The enzyme first binds a molecule of ATP cofactor at a position near the active site, and glucose then bonds to the active site with its C6 hydroxyl group held rigidly in position next to the ATP molecule. Reaction ensues, and the two products are released from the enzyme.

Hexose Kinase 3D model

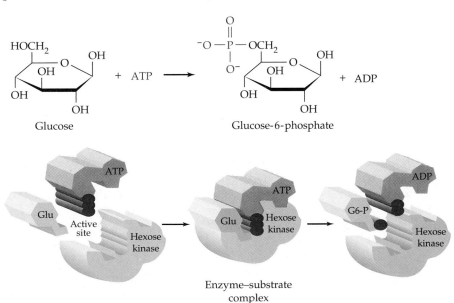

FIGURE 24.11 The hexose kinase–catalyzed reaction of glucose with ATP. Glucose enters the cleft of the enzyme and binds to the active site, where it reacts with a molecule of ATP cofactor held nearby.

Compare the enzyme-catalyzed reaction shown in Figure 24.11 with the same reaction in the absence of enzyme. Without enzyme present, the two reactants would spend most of their time surrounded by solvent molecules, far away from each other and only occasionally bumping together. With an enzyme present, however, the two reactants are forced into close contact. Enzymes act as catalysts because of their ability to bring reactants together, hold them at the exact distance and with the exact orientation necessary for reaction, and provide acidic or basic sites as required.

24.7   Carbohydrates

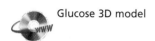

Glucose 3D model

Carbohydrates occur in every living organism. The starch in food and the cellulose in grass are nearly pure carbohydrate. Modified carbohydrates form part of the coating around all living cells, and other carbohydrates are found in the DNA that carries genetic information from one generation to the next. The word *carbohydrate* was used originally to describe glucose, which has the formula $C_6H_{12}O_6$ and was once thought to be a "hydrate of carbon," $C_6(H_2O)_6$. Although this view was soon abandoned, the word persisted and is now used to refer to a large class of polyhydroxylated aldehydes and ketones. Glucose, for example, is a six-carbon aldehyde with five hydroxyl groups.

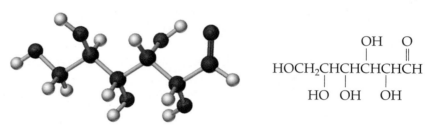

Glucose—a pentahydroxy aldehyde

Carbohydrates are classified as either *simple* or *complex*. Simple sugars, or **monosaccharides**, are carbohydrates such as glucose and fructose that can't be broken down into smaller molecules by hydrolysis with aqueous acid. Complex carbohydrates, or **polysaccharides**, are compounds such as cellulose and starch that are made of many simple sugars linked together. On hydrolysis, polysaccharides are cleaved to yield many molecules of simple sugars.

Monosaccharides are further classified as either aldoses or ketoses. An *aldose* contains an aldehyde carbonyl group; a *ketose* contains a ketone carbonyl group (Section 23.13). The *-ose* suffix indicates a sugar, and the number of carbon atoms in the sugar is specified by using one of the prefixes *tri-*, *tetr-*, *pent-*, or *hex-*. Thus, glucose is an aldohexose (a six-carbon aldehyde sugar), fructose is a ketohexose (a six-carbon ketone sugar), and ribose is an aldopentose (a five-carbon aldehyde sugar). Most commonly occurring sugars are either aldopentoses or aldohexoses.

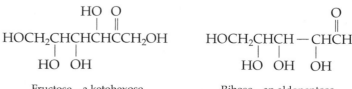

Fructose—a ketohexose   Ribose—an aldopentose

▶**PROBLEM 24.12** Classify each of the following monosaccharides:

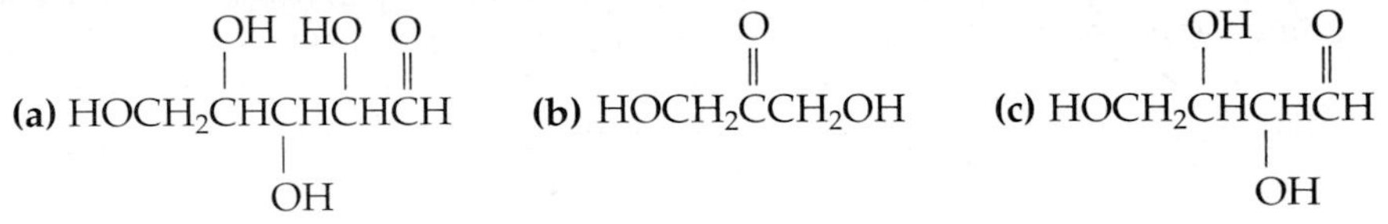

(a)
$$\text{HOCH}_2\overset{\displaystyle \text{OH}}{\underset{\displaystyle \text{OH}}{\text{CH}}}\overset{\displaystyle \text{HO}}{\text{CH}}\overset{\displaystyle \text{O}}{\overset{\|}{\text{CH}}}\text{CH}$$

(b)
$$\text{HOCH}_2\overset{\displaystyle \text{O}}{\overset{\|}{\text{C}}}\text{CH}_2\text{OH}$$

(c)
$$\text{HOCH}_2\overset{\displaystyle \text{OH}}{\text{CH}}\overset{}{\underset{\displaystyle \text{OH}}{\text{CH}}}\overset{\displaystyle \text{O}}{\overset{\|}{\text{CH}}}$$

24.8 Handedness of Carbohydrates

We saw in Section 24.3 that compounds are chiral if they have a carbon atom bonded to four different atoms or groups of atoms. Such compounds lack a plane of symmetry and can exist as a pair of enantiomers in either a "right-handed" D form or a "left-handed" L form. For instance, the simple triose glyceraldehyde is chiral because it has four different groups bonded to C2: −CHO, −H, −OH, and −CH$_2$OH. Of the two enantiomers, only D-glyceraldehyde occurs naturally.

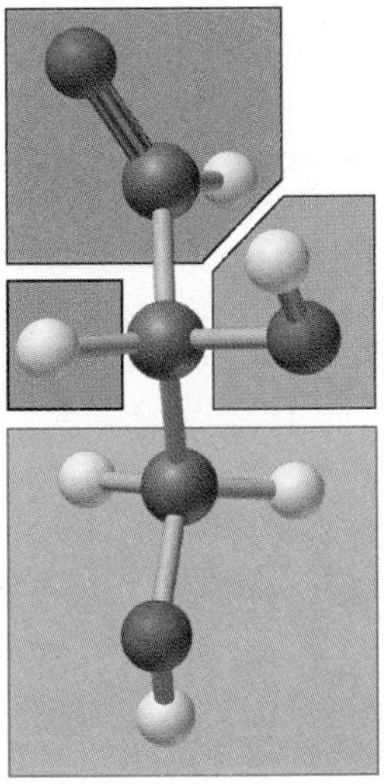

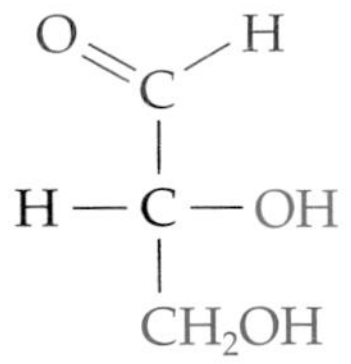

$$\overset{\displaystyle O}{\underset{\displaystyle}{\overset{\diagdown}{\underset{\displaystyle}{C}}}}\overset{\displaystyle H}{\diagup}$$
$$\text{H}-\overset{\displaystyle}{\underset{\displaystyle}{C}}-\text{OH}$$
$$\text{CH}_2\text{OH}$$

D-Glyceraldehyde

Groups bonded to C2
1.—CHO
2.—H
3.—OH
4.—CH$_2$OH

Glyceraldehyde has only one chiral carbon atom and can exist as two enantiomers, but other sugars have two, three, four, or even more chiral carbons. In general, a compound with n chiral carbon atoms has a maximum of 2^n possible forms. Glucose, for example, has four chiral carbon atoms, so a total of $2^4 = 16$ isomers are possible, differing in the spatial arrangements of the substituents around the chiral carbon atoms.

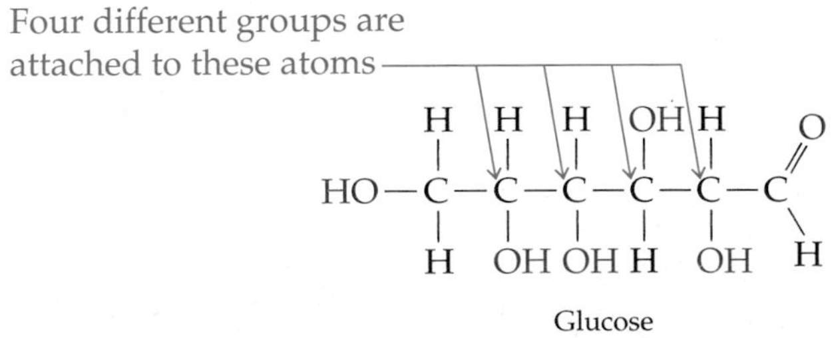

Four different groups are
attached to these atoms ⎯⎯

$$\text{HO}-\overset{\displaystyle \text{H}}{\underset{\displaystyle \text{H}}{\text{C}}}-\overset{\displaystyle \text{H}}{\underset{\displaystyle \text{OH}}{\text{C}}}-\overset{\displaystyle \text{H}}{\underset{\displaystyle \text{OH}}{\text{C}}}-\overset{\displaystyle \text{OH}}{\underset{\displaystyle \text{H}}{\text{C}}}-\overset{\displaystyle \text{H}}{\underset{\displaystyle \text{OH}}{\text{C}}}-\overset{\displaystyle \text{O}}{\underset{\displaystyle \text{H}}{\overset{\diagup}{\underset{\diagdown}{\text{C}}}}}$$

Glucose

▶ **PROBLEM 24.13** Draw tetrahedral representations of the two glyceraldehyde enantiomers using wedged, dashed, and normal lines to show three-dimensionality.

▶ **PROBLEM 24.14** Ribose has three chiral carbon atoms. What is the maximum number of isomers?

✦ **KEY CONCEPT PROBLEM 24.15** Classify the following monosaccharide, and identify each chiral carbon in its structure:

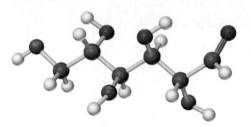

24.9 Cyclic Structures of Monosaccharides

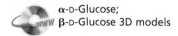
α-D-Glucose;
β-D-Glucose 3D models

Glucose and other monosaccharides are often shown for convenience as having open-chain structures. They actually exist, however, primarily as cyclic molecules in which an −OH group near the bottom of the chain adds to the carbonyl group near the top of the chain to form a ring. In glucose itself, ring formation occurs between the −OH group on C5 and the C=O group at C1 (Figure 24.12).

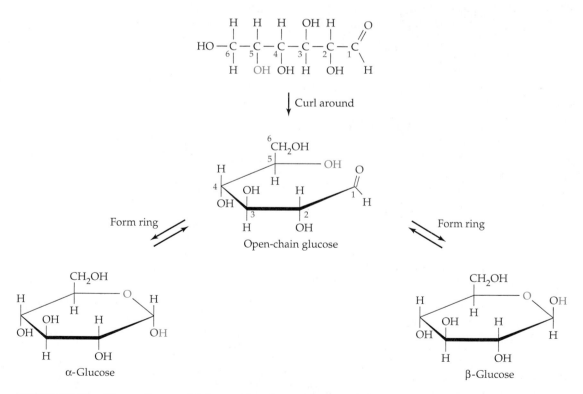

FIGURE 24.12 The cyclic α and β forms of D-glucose that result from ring formation between the −OH group at C5 and the C=O group at C1. The α form has the −OH group at C1 on the bottom side of the ring; the β form has the −OH group on the top.

Note that two cyclic forms of glucose can result from ring formation, depending on whether the newly formed −OH group at C1 is on the bottom or top side of the ring. The ordinary crystalline glucose you might take from a bottle is entirely the cyclic α form, in which the C1 −OH group is on the bottom side of the ring. At equilibrium in water solution, however, all three forms are present in the proportion 0.02% open-chain form, 36% α form, and 64% β form.

24.10   Some Common Disaccharides and Polysaccharides

Lactose

Lactose, or *milk sugar*, is the major carbohydrate present in mammalian milk. Human milk, for example, is about 7% lactose. Structurally, lactose is a disaccharide whose hydrolysis with aqueous acid yields one molecule of glucose and one molecule of another monosaccharide called galactose. The two sugars are bonded together by what is called a *1,4 link*, a bridging oxygen atom between C1 of β-galactose and C4 of β-glucose.

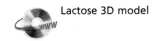

Lactose 3D model

Sucrose

Sucrose, or plain table sugar, is probably the most common pure organic chemical in the world. Although sucrose is found in many plants, sugar beets (20% by mass) and sugar cane (15% by mass) are the most common sources. Hydrolysis of sucrose yields one molecule of glucose and one molecule of fructose. The 50:50 mixture of glucose and fructose that results, often called *invert sugar*, is commonly used as a food additive.

Cellulose and Starch

Cellulose, the fibrous substance that forms the structural material in grasses, leaves, and stems, consists of several thousand β-glucose molecules joined together by 1,4 links to form an immense polysaccharide.

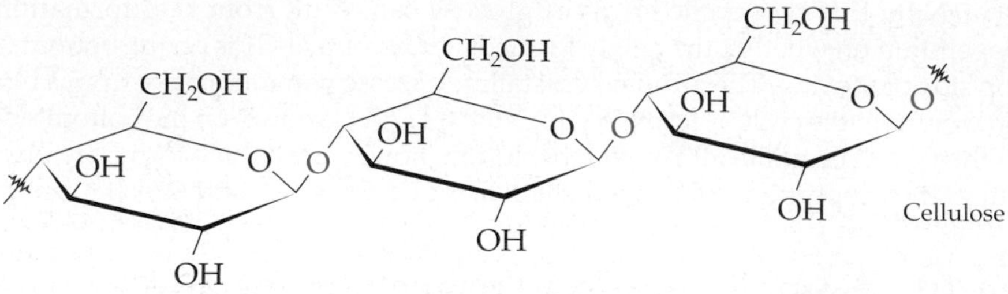

Cellulose

β-Glucose units

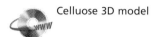

Celluose 3D model

Starch, like cellulose, is also a polymer of glucose. Unlike cellulose, though, starch is edible. Indeed, the starch in such vegetables as beans, rice, and potatoes is an essential part of the human diet. Structurally, starch differs from cellulose in that it contains α- rather than β-glucose units. Starch is also more structurally complex than cellulose and is of two types: *amylose* and *amylopectin*. Amylose, which accounts for about 20% of starch, consists of several hundred to 1000 α-glucose units joined together in a long chain by 1,4 links (Figure 24.13a). Amylopectin, which accounts for about 80% of starch, is much larger than amylose (up to 100,000 glucose units per molecule) and has *branches* approximately every 25 units along its chain. A glucose molecule at a branch point uses two of its hydroxyl groups (those at C4 and C6) to form links to two other sugars (Figure 24.13b).

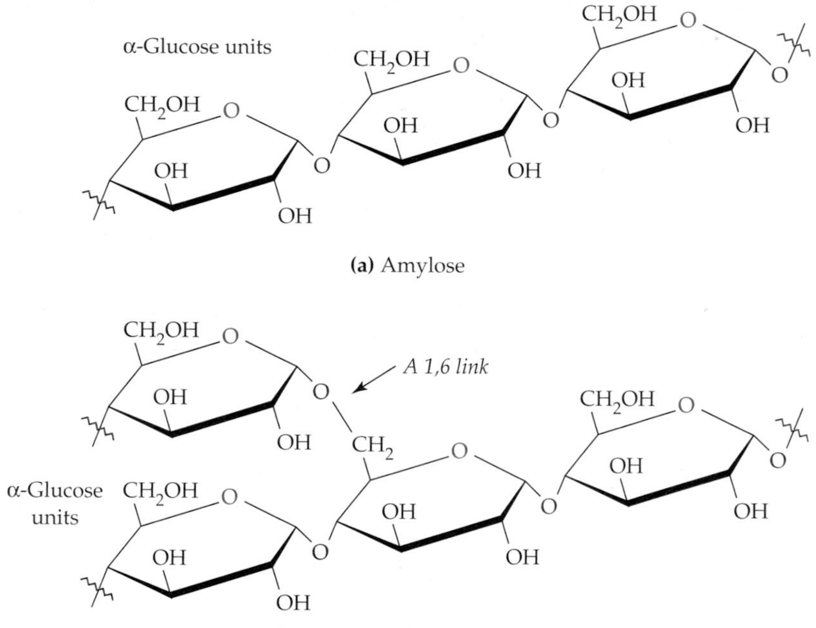

α-Glucose units

(a) Amylose

FIGURE 24.13 Glucose polymers in starch.
(a) Amylose consists only of linear chains of α-glucose units linked by 1,4 bonds.
(b) Amylopectin has branch points about every 25 sugars in the chain. A glucose unit at a branch point uses two of its hydroxyls (at C4 and C6) to form 1,4 and 1,6 links to two other sugars.

A 1,6 link

α-Glucose units

(b) Amylopectin

Starch molecules are digested in the stomach by enzymes called α-*glycosidases*, which break down the polysaccharide chain and release individual glucose molecules. As is usually the case with enzyme-catalyzed reactions, α-glycosidases are highly specific in their action. They hydrolyze only the links between α units while leaving the links between β units untouched. Thus, starch is easily digested but cellulose is not.

Glycogen

Glycogen, sometimes called *animal starch*, serves the same food storage role in animals that starch serves in plants. After we eat starch and the body breaks it down into simple glucose units, some of the glucose is used immediately as fuel and some is stored in the body as glycogen for later use. Structurally, glycogen is similar to amylopectin in being a long polymer of α-glucose units with branch points in its chain. Glycogen has many more branches than amylopectin, however, and is much larger—up to 1 million glucose units per glycogen molecule.

Glycogen provides the stored energy that lets these marathoners finish the race.

24.11 Lipids

Lipids are less well known to most people than proteins or carbohydrates, yet they are just as essential to life. Lipids have many important biological roles, serving as sources of fuel storage, as protective coatings around many plants and insects, and as major components of the membranes that surround every living cell.

Chemically, **lipids** are the naturally occurring organic molecules that dissolve in nonpolar organic solvents when a sample of plant or animal tissue is crushed or ground. Because they're defined by solubility, a physical property, rather than by chemical structure, it's not surprising that there are a great many different kinds of lipids (Figure 24.14). Note that all the lipids in Figure 24.14 contain large hydrocarbon portions, which accounts for their solubility behavior.

Duck feathers are protected by a waxy coating of lipids.

$$\begin{array}{c}\quad\quad\;\; O\\ \quad\quad\;\; \|\\ CH_2OCCH_2CH_2CH_2CH_2CH_2CH_2CH_2CH_2CH_2CH_2CH_2CH_2CH_2CH_2CH_3\\ \quad\; O\\ \quad\; \|\\ CHOCCH_2CH_2CH_2CH_2CH_2CH_2CH_2CH=CHCH_2CH_2CH_2CH_2CH_2CH_2CH_2CH_3\\ \quad\quad\;\; O\\ \quad\quad\;\; \|\\ CH_2OCCH_2CH_2CH_2CH_2CH_2CH_2CH_2CH_2CH_2CH_2CH_2CH_2CH_3\end{array}$$

An animal fat or vegetable oil

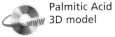

Palmitic Acid
3D model

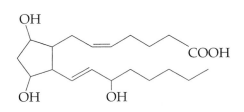

Cholesterol—a steroid $PGF_{2\alpha}$—a prostaglandin

FIGURE 24.14 Structures of some representative lipids isolated from plant and animal tissue by extraction with nonpolar organic solvents. All have large hydrocarbon portions.

Fats and Oils

Animal fats and vegetable oils are the most plentiful lipids in nature. Although they appear physically different—animal fats like butter and lard are usually solid, whereas vegetable oils like corn and peanut oil are liquid—their structures are similar. All fats and oils are **triacylglycerols**, or *triglycerides*, esters of glycerol (1,2,3-propanetriol) with three long-chain carboxylic acids called **fatty acids**. The fatty acids are usually unbranched and have an even number of carbon atoms in the range 12–22.

As shown by the triacylglycerol structure in Figure 24.14, the three fatty acids of a given molecule need not be the same. Furthermore, the fat or oil from a given source is a complex mixture of many different triacylglycerols. Table 24.3 shows the structures of some commonly occurring fatty acids, and Table 24.4 lists the composition of several fats and oils. Note that vegetable oils are largely unsaturated, but animal fats contain a high percentage of saturated fatty acids.

TABLE 24.3 Structures of Some Common Fatty Acids

Name	No. of Carbons	No. of Double Bonds	Structure	Melting Point (°C)
Saturated				
Myristic	14	0	$CH_3(CH_2)_{12}COOH$	58
Palmitic	16	0	$CH_3(CH_2)_{14}COOH$	63
Stearic	18	0	$CH_3(CH_2)_{16}COOH$	70
Unsaturated				
Oleic	18	1	$CH_3(CH_2)_7CH=CH(CH_2)_7COOH$	16
Linoleic	18	2	$CH_3(CH_2)_4CH=CHCH_2CH=CH(CH_2)_7COOH$	−5
Linolenic	18	3	$CH_3CH_2CH=CHCH_2CH=CHCH_2CH=CH(CH_2)_7COOH$	−11

TABLE 24.4 Approximate Composition of Some Common Fats and Oils

Source	Saturated Fatty Acids (%) C_{14} Myristic	C_{16} Palmitic	C_{18} Stearic	Unsaturated Fatty Acids (%) C_{18} Oleic	C_{18} Linoleic
Animal Fat					
Butter	10	25	10	25	5
Human fat	3	25	8	46	10
Whale blubber	8	12	3	35	10
Vegetable Oil					
Corn	1	10	4	35	45
Olive	1	5	5	80	7
Peanut	—	7	5	60	20

About 40 different fatty acids occur naturally. Palmitic acid (C_{16}) and stearic acid (C_{18}) are the most abundant saturated acids; oleic and linoleic acids (both C_{18}) are the most abundant unsaturated ones. Oleic acid is *monounsaturated* because it has only one double bond, but linoleic and linolenic acids are *polyunsaturated fatty acids* (called *PUFAs*) because they have more than one carbon–carbon double bond. Although the reasons are not yet clear, it appears that a diet rich in saturated fats leads to a higher level of blood cholesterol and consequent higher risk of heart attack than a diet rich in unsaturated fats.

The data in Table 24.3 show that unsaturated fatty acids generally have lower melting points than their saturated counterparts, a trend that also holds true for triacylglycerols. Since vegetable oils have a higher proportion of unsaturated fatty acids than do animal fats, they have lower melting points and appear as liquids rather than solids. This behavior arises because the carbon–carbon double bonds in unsaturated vegetable oils introduce bends and kinks into the hydrocarbon chains, making it difficult for the chains to nestle closely together to become solid crystals.

The carbon–carbon double bonds in vegetable oils can be hydrogenated to yield saturated fats in the same way that any alkene can react with hydrogen to yield an alkane (Section 23.10). By carefully controlling the extent of hydrogenation, the final product can have any desired consistency. Margarine, for example, is prepared so that only about two-thirds of the double bonds present in the starting vegetable oil are hydrogenated.

$$\overset{O}{\underset{\|}{}}$$
‡OCCH₂CH₂CH₂CH₂CH₂CH₂CH₂CH=CHCH₂CH=CHCH₂CH=CHCH₂CH₃

Partial structure of a vegetable oil

| H₂, Pd catalyst

$$\overset{O}{\underset{\|}{}}$$
‡OCCH₂CH₂CH₂CH₂CH₂CH₂CH₂CH₂CH₂CH₂CH₂CH₂CH₂CH₂CH₂CH₂CH₃

Saturated product

▶ **PROBLEM 24.16** Show the structure of glyceryl trioleate, a fat molecule whose components are glycerol and three oleic acid units.

Steroids

A **steroid** is a lipid whose structure is based on the tetracyclic (four-ring) system shown in the following examples. Three of the rings are six-membered, while the fourth is five-membered. Steroids have many diverse roles throughout both the plant and animal kingdoms. Some steroids, such as digitoxigenin, isolated from the purple foxglove *Digitalis purpurea*, are used in medicine as heart stimulants. Others, such as hydrocortisone, are hormones, and still others have a wide variety of different physiological functions.

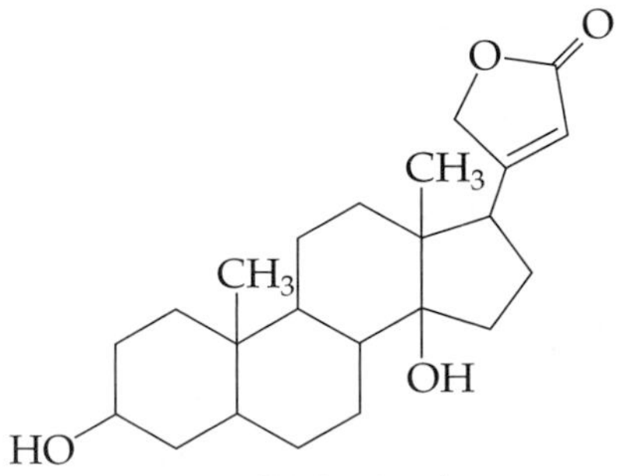

Digitoxigenin

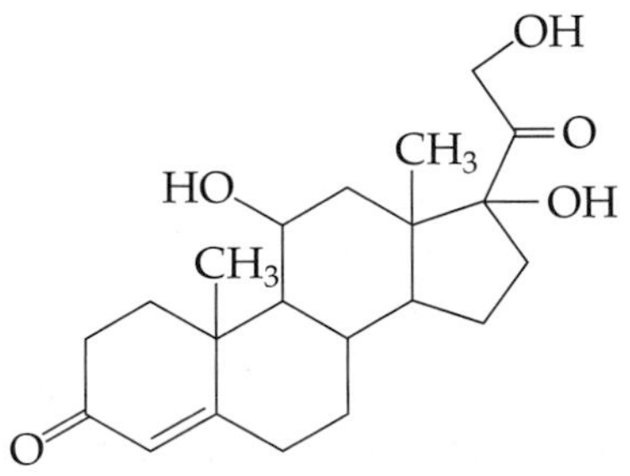

Hydrocortisone

The purple foxglove, a common backyard plant contains the steroidal heart stimulant digitoxigenin.

Cholesterol, an unsaturated alcohol whose structure was shown in Figure 24.14, is the most abundant animal steroid. It has been estimated that a 60 kg (130 lb) person has a total of about 175 g of cholesterol distributed throughout the body. Much of this cholesterol is bonded through ester links to fatty acids, but some is found as the free alcohol. Gallstones, for example, are nearly pure cholesterol.

Cholesterol serves two important functions in the body. First, it is a minor component of cell membranes, where it helps to keep the membranes fluid. Second, it serves as the body's starting material for the synthesis of all other steroids, including the sex hormones. Although news reports often make cholesterol sound dangerous, there would be no life without it. The human body obtains its cholesterol both by synthesis in the liver and by ingestion of food. Even on a strict no-cholesterol diet, an adult is able to synthesize approximately 800 mg per day.

24.12 Nucleic Acids

How does a seed "know" what kind of plant to become? How does a fertilized ovum know how to grow into a human being? How does a cell know what part of the body it's in so that it can carry out the right reactions? The answers to such questions involve the biological molecules called **nucleic acids**. **Deoxyribonucleic acid (DNA)** and **ribonucleic acid (RNA)** are the chemical carriers of an organism's genetic information. Coded in an organism's DNA is all the information that determines the nature of the organism and all the directions needed for producing the thousands of different proteins required by the organism.

Just as proteins are polymers made of amino acid units, nucleic acids are polymers made up of **nucleotide** units linked together to form a long chain. Each nucleotide is composed of a **nucleoside** plus phosphoric acid, H_3PO_4, and each nucleoside is composed of an aldopentose sugar plus an amine base.

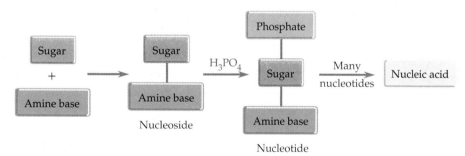

The sugar component in RNA is ribose, and the sugar in DNA is 2-deoxyribose (2-deoxy means that oxygen is missing from C2 of ribose).

Ribose   2-Deoxyribose

There are four different cyclic amine bases in DNA: adenine, guanine, cytosine, and thymine. Adenine, guanine, and cytosine also occur in RNA, but thymine is replaced in RNA by a related base called uracil.

Adenine (A)
DNA
RNA

Guanine (G)
DNA
RNA

Cytosine (C)
DNA
RNA

Thymine (T)
DNA

Uracil (U)
RNA

In both DNA and RNA, the cyclic amine base is bonded to C1′ of the sugar, and the phosphoric acid is bonded to the C5′ sugar position. Thus, nucleosides and nucleotides have the general structures shown in Figure 24.15. (In discussions of RNA and DNA, numbers with a prime superscript refer to positions on the sugar component of a nucleotide; numbers without a prime refer to positions on the cyclic amine base.)

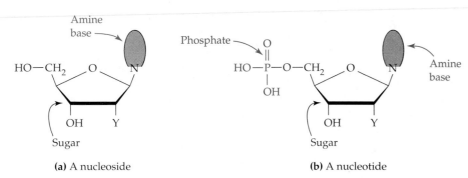

(a) A nucleoside

(b) A nucleotide

FIGURE 24.15 General structures of **(a)** a nucleoside and **(b)** a nucleotide. When Y = H, the sugar is deoxyribose; when Y = OH, the sugar is ribose.

Nucleotides join together in nucleic acids by forming a phosphate ester bond between the phosphate group at the 5′ end of one nucleotide and the hydroxyl group on the sugar component at the 3′ end of another nucleotide (Figure 24.16).

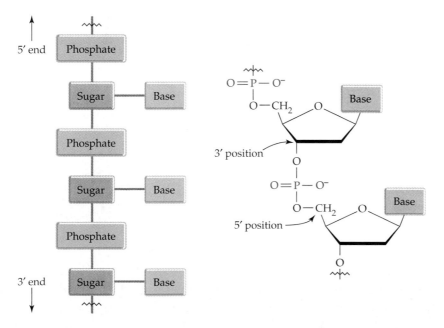

FIGURE 24.16 Generalized structure of a nucleic acid.

Just as the structure of a protein depends on its sequence of individual amino acids, the structure of a nucleic acid depends on its sequence of individual nucleotides. To carry the analogy further, just as a protein has a polyamide backbone with different side chains attached to it, a nucleic acid has an alternating sugar–phosphate backbone with different amine base side chains attached.

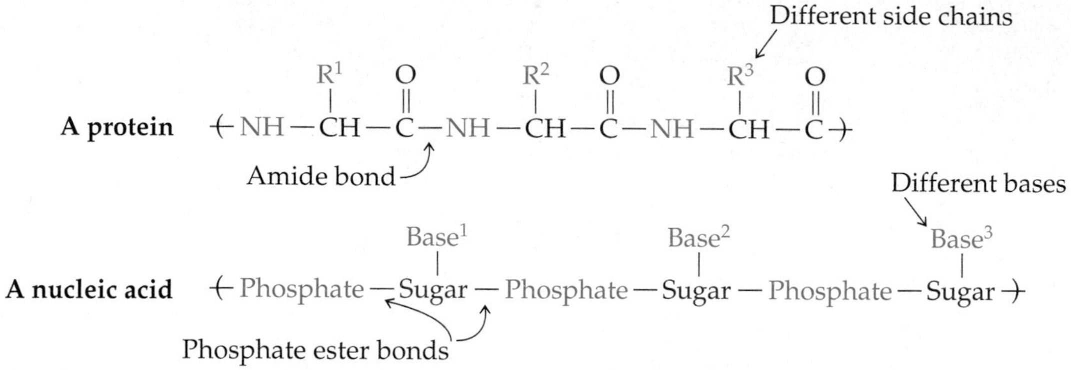

The sequence of nucleotides is described by starting at the 5' phosphate end and identifying the bases in order. Abbreviations are used for each nucleotide: A for adenosine, G for guanosine, C for cytidine, T for thymidine, and U for uracil. Thus, a typical DNA sequence might be written as -T-A-G-G-C-T-.

EXAMPLE 24.3 Draw the full structure of the DNA dinucleotide C-T.

SOLUTION

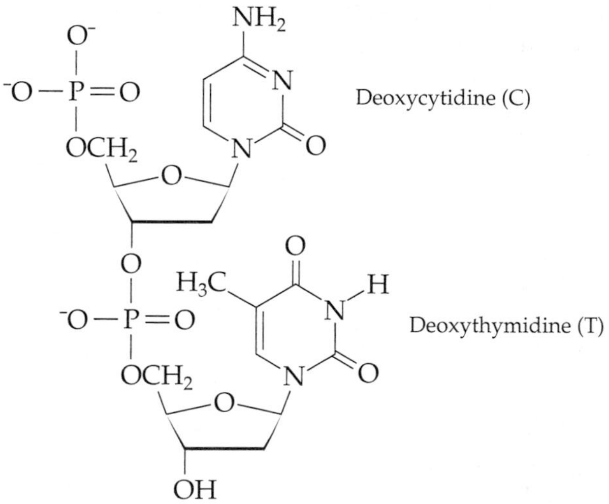

▶**PROBLEM 24.17** Draw the full structure of the DNA dinucleotide A-G.

▶**PROBLEM 24.18** Draw the full structure of the RNA dinucleotide U-A.

►KEY CONCEPT PROBLEM 24.19   Identify the following bases, and tell whether each is found in DNA, in RNA, or in both:

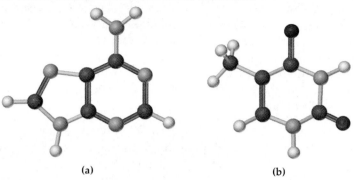

(a)   (b)

24.13   Base Pairing in DNA: The Watson–Crick Model

Molecules of DNA isolated from different tissues of the same species have the same proportions of nucleotides, but molecules from different species can have quite different proportions. For example, human DNA contains about 30% each of A and T and about 20% each of G and C, but the bacterium *Clostridium perfringens* contains about 37% each of A and T and only 13% each of G and C. Note that in both cases, the bases occur in pairs. Adenine and thymine are usually present in equal amounts, as are guanine and cytosine. Why should this be?

According to the **Watson–Crick model**, DNA consists of two polynucleotide strands coiled around each other in a *double helix*. The sugar–phosphate backbone is on the outside of the helix, and the amine bases are on the inside, so that a base on one strand points directly in toward a base on the second strand. The two strands run in opposite directions and are held together by hydrogen bonds between pairs of bases. Adenine and thymine form two strong hydrogen bonds to each other, but not to G or C; G and C form three strong hydrogen bonds to each other, but not to A or T.

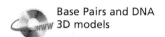

Base Pairs and DNA
www 3D models

Adenine   Thymine   Guanine   Cytosine

The two strands of the DNA double helix aren't identical; rather, they're complementary. Whenever a G base occurs in one strand, a C base occurs opposite it in the other strand. When an A base occurs in one strand, a T base occurs in the other strand. This complementary pairing of bases explains why A and T are always found in equal amounts, as are G and C. Figure 24.17 illustrates this base pairing, showing how the two complementary strands coil into the double helix. X-ray measurements show that the DNA double helix is 2.0 nm wide, that there are exactly 10 base pairs in each full turn, and that each turn is 3.4 nm high.

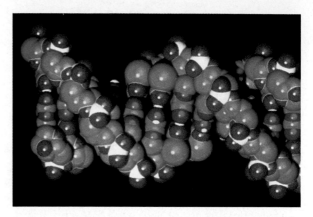

FIGURE 24.17 The coil of the sugar–phosphate backbone is visible on the outside of the DNA double helix in this computer-generated structure, while the hydrogen-bonded pairs of amine bases lie flat on the inside.

EXAMPLE 24.4   What sequence of bases on one strand of DNA is complementary to the sequence G-C-A-T-T-A-T on another strand?

SOLUTION   Since A and G form complementary pairs with T and C, respectively, go through the given sequence replacing A with T, G with C, T with A, and C with G:

Original: G-C-A-T-T-A-T

Complement: C-G-T-A-A-T-A

▶**PROBLEM 24.20**   What sequence of bases on one strand of DNA is complementary to the following sequence on another strand?

G-G-C-C-C-G-T-A-A-T

24.14 Nucleic Acids and Heredity

Most DNA of higher organisms, both plant and animal, is found in the nucleus of cells in the form of threadlike strands that are coated with proteins and wound into complex assemblies called **chromosomes**. Each chromosome is made up of several thousand **genes**, where a gene is a segment of a DNA chain that contains the instructions necessary to make a specific protein. By decoding the right genes at the right time, an organism uses genetic information to synthesize the thousands of proteins needed for living. Thus, the function of DNA is to act as a storage medium for an organism's genetic information. The function of RNA is to read, decode, and use the information received from DNA to make proteins.

Three processes take place in the transfer and use of genetic information: **Replication** is the means by which identical copies of DNA are made, forming additional molecules and preserving genetic information for passing on to offspring. **Transcription** is the means by which information in the DNA is transferred to and decoded by RNA. **Translation** is the means by which RNA uses the information to build proteins.

Replication (DNA $\xrightarrow{\text{Transcription}}$ RNA $\xrightarrow{\text{Translation}}$ Proteins

Replication

DNA replication is an enzyme-catalyzed process that begins with a partial unwinding of the double helix. As the DNA strands separate and bases are exposed, new nucleotides line up on each strand in a complementary manner, A to T and C to G, and two new strands begin to grow. Each new strand is complementary to its old template strand, and two new, identical DNA double helixes are produced (Figure 24.18).

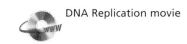
DNA Replication movie

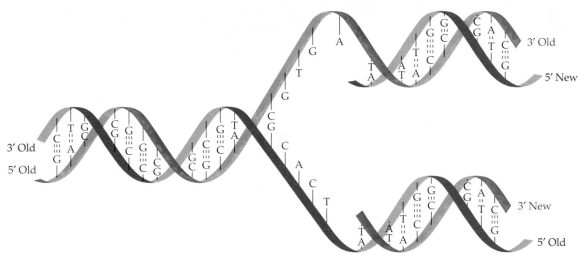

FIGURE 24.18   DNA replication. A portion of the DNA double helix unwinds, and complementary nucleotides line up for linking to yield two new DNA molecules. Each of the new DNA molecules contains one of the original strands and one new strand.

The magnitude of the replication process is extraordinary. The nucleus of a human cell contains 46 chromosomes (23 pairs), each of which consists of one large DNA molecule. Each chromosome, in turn, is made up of several thousand genes, and the sum of all genes in a human cell (the *genome*) is approximately 3 billion base pairs. This immense base sequence is faithfully copied during replication, with an error occurring only about once each 10–100 billion bases.

Transcription

The genetic instructions contained in DNA are transcribed into RNA when a small portion of the DNA double helix unwinds and one of the two DNA strands acts as a template for complementary *ribonucleotides* to line up, a process similar to that of DNA replication (Figure 24.19). The only difference is that uracil (U) rather than thymine lines up opposite adenine. Once completed, the RNA molecule separates from the DNA template, and the DNA rewinds to its stable double-helix conformation.

FIGURE 24.19 Transcription of DNA to synthesize RNA. A small portion of the DNA double helix unwinds, and one of the two DNA strands acts as a template on which ribonucleotides line up. The RNA produced is complementary to the DNA strand from which it is transcribed.

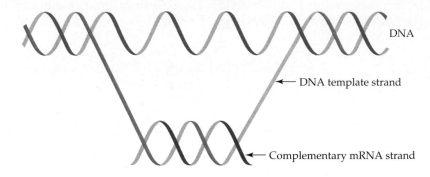

DNA

DNA template strand

Complementary mRNA strand

Translation

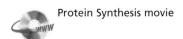

Protein biosynthesis is directed by a special kind of RNA called *messenger RNA*, or *mRNA*. It takes place on knobby protuberances within a cell called *ribosomes*. The specific ribonucleotide sequence in mRNA acts like a long coded sentence to specify the order in which different amino acid residues are to be joined. Each of the estimated 100,000 proteins in the human body is synthesized from a different mRNA that has been transcribed from a specific gene segment on DNA.

Each "word" along the mRNA chain consists of a series of three ribonucleotides that is specific for a given amino acid. For example, the series cytosine-uracil-guanine (C-U-G) on mRNA is a three-letter word directing that the amino acid leucine be incorporated into the growing protein. The words are read by another kind of RNA called *transfer RNA*, or *tRNA*. Each of the 60 or so different tRNAs contains a complementary base sequence that allows it to recognize a three-letter word on mRNA and act as a carrier to bring a specific amino acid into place for transfer to the growing peptide chain (Figure 24.20). When synthesis of the protein is completed, a "stop" word signals the end, and the protein is released from the ribosome.

FIGURE 24.20 Protein biosynthesis. Messenger RNA is read by tRNA that contains complementary three-base sequences. Transfer RNA then assembles the proper amino acids (AA_1, AA_2, and so on) into position for incorporation into the peptide.

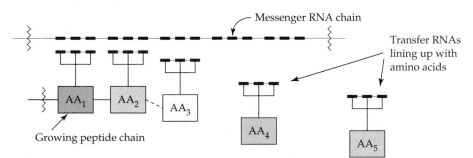

Messenger RNA chain

Transfer RNAs lining up with amino acids

AA_1 AA_2 AA_3 AA_4 AA_5

Growing peptide chain

EXAMPLE 24.5 What RNA base sequence is complementary to the following DNA base sequence?

G-C-C-T-A-A-G-T-G

SOLUTION Go through the DNA sequence and replace A with U, G with C, T with A, and C with G:

Original DNA: G-C-C-T-A-A-G-T-G

Complementary RNA: C-G-G-A-U-U-C-A-C

▶ **PROBLEM 24.21** Show how uracil can form strong hydrogen bonds to adenine.

▶ **PROBLEM 24.22** What RNA sequence is complementary to the following DNA sequence?

C-G-T-G-A-T-T-A-C-A

▶ **PROBLEM 24.23** From what DNA sequence was the following RNA sequence transcribed?

U-G-C-A-U-C-G-A-G-U

DNA Fingerprinting

DNA fingerprinting requires determining the sequence of nucleotides in long DNA chains.

Being a criminal is a lot harder now than it was in the good old days, and part of the reason is the development of *DNA fingerprinting*. The technique of DNA fingerprinting arose from the discovery in 1984 that human genes contain short, repeating sequences of noncoding DNA, called *short tandem repeat* (STR) *loci*. The base sequences in these STR loci are slightly different for every individual (except identical twins), so a pattern unique to each person can be obtained by determining the DNA sequences.

Forensic scientists have agreed on 13 specific STR loci that are most accurate for identification of an individual. Based on these 13 loci, a Combined DNA Index System (CODIS) has been established to serve as a registry of convicted offenders. If the profile of sequences from a known individual and the profile from DNA obtained at a crime scene match, the probability is approximately 82 billion to 1 that the DNA is from the same individual. In paternity cases, where the DNA of father and offspring are related but not fully identical, the identity of the father can be established with a probability of 100,000 to 1.

Of course, determining a DNA sequence is easier said than done, and several steps are needed. When a DNA sample is obtained from a crime scene—from blood, hair follicles, skin, or semen, for example—the sample is first subjected to cleavage by special enzymes that cut out the fragments containing the STR loci. The fragments are then isolated and purified by electrophoresis, a technique similar to that discussed in the Chapter 16 Interlude *Analyzing Proteins by Electrophoresis*.

Once pure, the fragments are amplified using the *polymerase chain reaction* (PCR), a process that has been described as being to DNA what Gutenberg's invention of the printing press was to the written word. Just as a printing press produces multiple copies of a book, PCR produces multiple copes of a given DNA sequence. Starting with less than 1 picogram of DNA with a chain length of 10,000 nucleotides (about 100,000 molecules), PCR can produce several micrograms (about 10^{11} molecules) in just a few hours.

Finally, the sequence of nucleotides in the DNA fragments is determined, using a technique called the *Sanger dideoxy method*. This method is particularly amenable to automation, and robotic sequencing machines are now able to analyze sequences up to 1000 nucleotides in length in the space of just a few hours. If your DNA is found at a crime scene, there's no way to hide it.

▶ **PROBLEM 24.24** What characteristic of human DNA makes fingerprinting possible?

▶ **PROBLEM 24.25** What are the steps involved in DNA fingerprinting?

Summary

Fats, carbohydrates, and proteins are **metabolized** in the body to yield acetyl CoA, which is further degraded in the citric acid cycle to yield two molecules of CO_2 plus a large amount of energy. The energy output of the various steps in the citric acid cycle is coupled to the electron-transport chain, a series of enzyme-catalyzed reactions whose ultimate purpose is to synthesize **adenosine triphosphate (ATP)**.

Proteins are large biomolecules consisting of **α-amino acids** linked together by amide, or peptide, bonds. Twenty amino acids are commonly found in proteins, and all except glycine have a handedness. In general, any carbon atom bonded to four different groups has a handedness and is said to be chiral. Proteins can be classified either by shape or biological function. Fibrous proteins are tough, threadlike, and water insoluble; globular proteins are compact, water soluble, and mobile within cells. Some proteins are enzymes, some are hormones, and some act as structural or transport agents.

A protein's **primary structure** is its amino acid sequence. Its **secondary structure** is the orientation of segments of the protein chain into a regular pattern, such as an **α-helix** or a **β-pleated sheet**. Its **tertiary structure** is the three-dimensional shape into which the entire protein molecule is coiled.

Enzymes are large proteins that function as biological catalysts and whose specificity is due to a **lock-and-key** fit between enzyme and **substrate**. Enzymes contain a crevice, inside which is an **active site**, a small region of the enzyme with the specific shape necessary to bind the proper substrate.

Carbohydrates are polyhydroxy aldehydes and ketones. Simple carbohydrates such as glucose can't be hydrolyzed to smaller molecules; complex carbohydrates such as starch contain many simple sugars linked together. **Monosaccharides** such as glucose exist as a mixture of an open-chain form and two cyclic forms called the α form and the β form. Disaccharides such sucrose contain two simple sugars joined by a linking oxygen atom.

Lipids are the naturally occurring organic molecules that dissolve in a non-polar solvent. There are a great many different kinds of lipids. Animal fats and vegetable oils are **triacylglycerols**—esters of glycerol with three long-chain **fatty acids**. The fatty acids are unbranched, have an even number of carbon atoms, and may be either saturated or unsaturated.

Deoxyribonucleic acid (DNA) and **ribonucleic acid (RNA)** are the chemical carriers of an organism's genetic information. Nucleic acids are made up of many individual building blocks, called **nucleotides**, linked together to form a long chain. Each nucleotide consists of a cyclic amine base linked to C1 of a sugar, with the sugar in turn linked to phosphoric acid. The sugar component in RNA is ribose; the sugar in DNA is 2-deoxyribose. The bases in DNA are adenine, guanine, cytosine, and thymine; the bases in RNA are adenine, guanine, cytosine, and uracil. Molecules of DNA consist of two complementary polynucleotide strands held together by hydrogen bonds between bases on the two strands and coiled into a double helix.

Three processes take place in the transfer of genetic information: **Replication** is the process by which identical copies of DNA are made and genetic information is preserved. **Transcription** is the process by which messenger RNA is produced. **Translation** is the process by which mRNA directs protein synthesis.

Key Words

active site *1042*
adenosine triphosphate
  (ATP) *1029*
alpha- (α-) amino acid *1032*
alpha- (α-) helix *1039*
anabolism *1028*
apoenzyme *1042*
beta- (β-) pleated sheet *1039*
biochemistry *1027*
carbohydrate *1044*
catabolism *1028*
chromosome *1056*
coenzyme *1042*
cofactor *1042*
C-terminal amino acid *1037*
deoxyribonucleic acid
  (DNA) *1052*
enzyme *1041*
fatty acid *1050*
gene *1056*
holoenzyme *1042*
lipid *1049*
lock-and-key model *1042*
metabolism *1028*
monosaccharide *1044*
N-terminal amino acid *1037*
nucleic acid *1052*
nucleoside *1052*
nucleotide *1052*
peptide bond *1032*
polypeptide *1032*
polysaccharide *1044*
primary structure *1038*
protein *1031*
quaternary structure *1038*
replication *1057*
ribonucleic acid (RNA) *1052*
secondary structure *1038*
side chain *1032*
steroid *1051*
substrate *1041*
tertiary structure *1038*
transcription *1057*
translation *1057*
triacylglycerol *1050*
turnover number *1042*
Watson–Crick model *1055*

Key Concept Summary

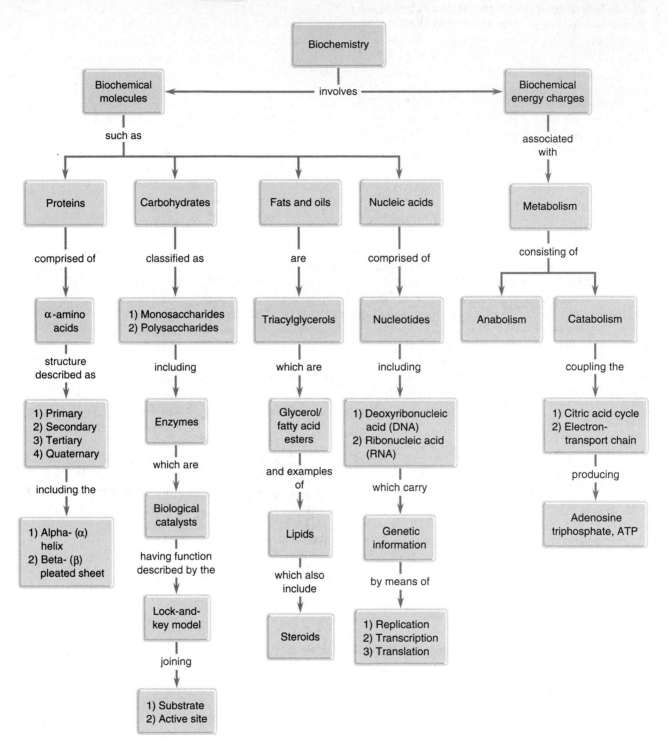

Understanding Key Concepts

Problems 24.1–24.25 appear within the chapter.

24.26 Identify the following amino acids:

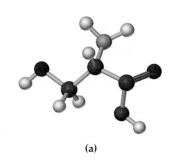

(a)

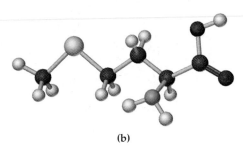

(b)

24.27 Does the following model represent a D-amino acid or an L-amino acid? Identify it.

24.28 Is the following model of glucose in the α form or the β form?

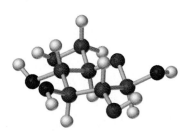

24.29 The following model represents D-ribose in its open-chain form. Is ribose an aldose or a ketose? How many chiral carbon atoms does ribose have?

24.30 Identify the following dipeptide:

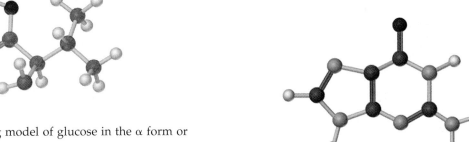

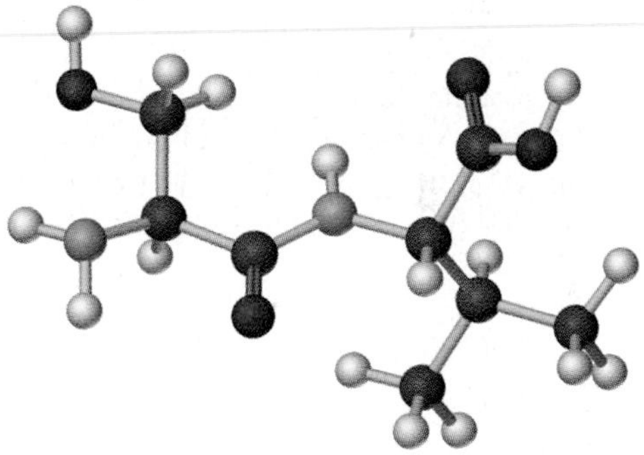

24.31 Identify the following amine bases found in nucleic acids:

(a)

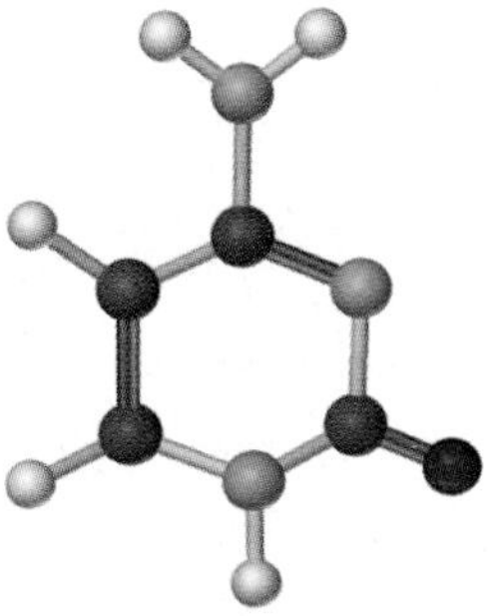

(b)

Additional Problems

Amino Acids, Peptides, and Proteins

24.32 What does the prefix "α" mean when referring to α-amino acids?

24.33 Why are the naturally occurring amino acids called L-amino acids?

24.34 What amino acids do the following abbreviations stand for?

(a) Ser   (b) Thr   (c) Pro

(d) Phe   (e) Cys

24.35 Name and draw the structures of amino acids that fit the following descriptions:

(a) Contains an isopropyl group

(b) Contains an alcohol group

(c) Contains a thiol (–SH) group

(d) Contains an aromatic ring

24.36 Much of the chemistry of amino acids is the familiar chemistry of carboxylic acid and amine functional groups. What products would you expect to obtain from the following reactions of glycine?

(a) H_2NCH_2COH $+$ CH_3OH $\xrightarrow[\text{catalyst}]{H_2SO_4}$

(b) H_2NCH_2COH $+$ HCl $\longrightarrow$

24.37 *Aspartame*, marketed under the trade name Nutra-Sweet for use as a nonnutritive sweetener, is the methyl ester of a simple dipeptide. Identify the two amino acids present in aspartame, and show all the products of digestion, assuming that both amide and ester bonds are hydrolyzed in the stomach.

$$H_2N-CH-\overset{\overset{\displaystyle O}{\|}}{C}-NH-CH-\overset{\overset{\displaystyle O}{\|}}{C}-OCH_3$$

with CH₂–COOH on first CH and CH₂–phenyl on second CH

Aspartame

24.38 Identify the amino acids present in the following hexapeptide:

$$H_2N-CH-\overset{O}{\overset{\|}{C}}-NH-CH-\overset{O}{\overset{\|}{C}}-NH-CH-\overset{O}{\overset{\|}{C}}-NH-CH-\overset{O}{\overset{\|}{C}}-NH-CH-\overset{O}{\overset{\|}{C}}-NH-CH-\overset{O}{\overset{\|}{C}}-OH$$

side chains: $CH(CH_3)_2$ CH_2OH CH_2–phenyl $CH_2CH_2SCH_3$ $CHOH$ / CH_3 CH_3

24.39 Look at the structure of angiotensin II in Figure 24.5, and identify both the N-terminal and C-terminal amino acids.

24.40 What is meant by the following terms as they apply to proteins?

(a) Primary structure

(b) Secondary structure

(c) Tertiary structure

24.41 What is the difference between fibrous and globular proteins?

24.42 What kinds of intramolecular interactions are important in stabilizing a protein's tertiary structure?

24.43 What kind of bonding stabilizes helical and β-pleated-sheet secondary protein structures?

24.44 Why is cysteine such an important amino acid for defining the tertiary structure of proteins?

24.45 Which of the following amino acids are most likely to be found on the outside of a globular protein and which on the inside? Explain.

(a) Valine   (b) Leucine

(c) Aspartic acid   (d) Asparagine

24.46 Use the three-letter abbreviations to name all tripeptides containing methionine, isoleucine, and lysine.

24.47 How many tetrapeptides containing alanine, serine, leucine, and glutamic acid do you think there are? Use the three-letter abbreviations to name three.

24.48 Write structural formulas for the two dipeptides containing phenylalanine and glutamic acid.

24.49 Write the structural formula of Asp-Met-Pro-Gly.

Molecular Handedness

24.50 Which of the following objects are chiral?
    (a) A shoe   **(b)** A bed
    (c) A light bulb   **(d)** A flower pot

24.51 Give two examples of chiral objects and two examples of achiral objects.

24.52 Which of the following compounds are chiral?
    (a) 2,4-Dimethylheptane
    (b) 5-Ethyl-3,3-dimethylheptane

24.53 Draw chiral molecules that meet the following descriptions:
    (a) A chloroalkane, $C_5H_{11}Cl$
    (b) An alcohol, $C_6H_{14}O$
    (c) An alkene, C_6H_{12}
    (d) An alkane, C_8H_{18}

24.54 There are eight alcohols with the formula $C_5H_{12}O$. Draw them and tell which are chiral.

24.55 Propose structures for compounds that meet the following descriptions:
    (a) A chiral alcohol with four carbons
    (b) A chiral aldehyde
    (c) A compound with two chiral carbons

Carbohydrates

24.56 What is the structural difference between an aldose and a ketose?

24.57 Classify each of the following carbohydrates by indicating the nature of its carbonyl group and the number of carbon atoms present. For example, glucose is an aldohexose.

    (a)
$$\underset{\underset{OH}{|}}{HOCH_2CH}\underset{\underset{OH}{|}}{CH}\overset{\overset{OH}{|}}{CH}\overset{\overset{O}{\parallel}}{CH}$$

    (b)
$$HOCH_2CHCHCHCHCH$$

    (c)
$$HOCH_2CHCHCHCCH_2OH$$

24.58 Starch and cellulose are both polymers of glucose. What is the main structural difference between them, and what different roles do they serve in nature?

24.59 Starch and glycogen are both α-linked polymers of glucose. What is the structural difference between them, and what different roles do they serve in nature?

24.60 Write the open-chain structure of a ketotetrose.

24.61 Write the open-chain structure of a four-carbon deoxy sugar.

24.62 D-Mannose, an aldohexose found in orange peels, has the following structure in open-chain form. Coil mannose around, and draw it in cyclic α and β forms.

$$HOCH_2CHCHCHCH{-}CH \quad \text{D-Mannose}$$

24.63 Draw D-galactose in its cyclic α and β forms.

$$HOCH_2CHCHCHCH{-}CH \quad \text{D-Galactose}$$

24.64 Show two D-mannose molecules (Problem 24.62) attached by an α-1,4 link.

24.65 Show two D-galactose molecules (Problem 24.63) attached by a β-1,4 link.

Lipids

24.66 What is a fatty acid?

24.67 What does it mean to say that fats and oils are triacylglycerols?

24.68 Draw the structure of glycerol myristate, a fat made from glycerol and three myristic acid molecules (see Table 24.3).

24.69 Spermaceti, a fragrant substance isolated from sperm whales, was a common ingredient in cosmetics until its use was banned in 1976 to protect the whales from extinction. Chemically, spermaceti is cetyl palmitate, the ester of palmitic acid (see Table 24.3) with cetyl alcohol (the straight-chain C_{16} alcohol). Show the structure of spermaceti.

24.70 There are two isomeric fat molecules whose components are glycerol, one palmitic acid, and two stearic acids (see Table 24.3). Draw the structures of both, and explain how they differ.

24.71 One of the two molecules you drew in Problem 24.70 is chiral. Which molecule is chiral, and why?

24.72 Draw the structures of all products you would obtain by reaction of the following lipid with aqueous KOH. What are the names of the products?

$$CH_2-O-\overset{\overset{\textstyle O}{\|}}{C}(CH_2)_{16}CH_3$$

$$CH-O-\overset{\overset{\textstyle O}{\|}}{C}(CH_2)_7CH=CH(CH_2)_7CH_3$$

$$CH_2-O-\overset{\overset{\textstyle O}{\|}}{C}(CH_2)_7CH=CHCH_2CH=CHCH_2CH=CHCH_2CH_3$$

24.73 Draw the structure of the product you would obtain on hydrogenation of the lipid in Problem 24.72. What is its name? Would the product have a higher or lower melting point than the original lipid? Why?

24.74 What products would you obtain by treating oleic acid with the following reagents?
(a) Br_2 (b) H_2, Pd catalyst
(c) CH_3OH, HCl catalyst

24.75 Look up the structure of linoleic acid in Table 24.3 and draw all potential products of its reaction with 2 mol of HCl.

Nucleic Acids

24.76 What is a nucleotide, and what three kinds of components does it contain?

24.77 What are the names of the sugars in DNA and RNA, and how do they differ?

24.78 Where in the cell is most DNA found?

24.79 What is meant by the following terms as they apply to nucleic acids?
(a) Base pairing   (b) Replication
(c) Translation   (d) Transcription

24.80 What is the difference between a gene and a chromosome?

24.81 What genetic information does a single gene contain?

24.82 Draw structures to show how the phosphate and sugar components of a nucleic acid are joined.

24.83 Draw structures to show how the sugar and amine base components of a nucleic acid are joined.

24.84 Draw the complete structure of deoxycytidine 5'-phosphate, one of the four deoxyribonucleotides.

24.85 Draw the complete structure of guanosine 5'-phosphate, one of the four ribonucleotides.

24.86 If the sequence T-A-C-C-G-A appeared on one strand of DNA, what sequence would appear opposite it on the other strand?

24.87 What sequence would appear on the mRNA molecule transcribed from the DNA in Problem 24.86?

24.88 Human insulin is composed of two polypeptide chains. One chain contains 21 amino acids, and the other contains 30 amino acids. How many nucleotides are present in the DNA to code for each chain?

24.89 The DNA from sea urchins contains about 32% A and about 18% G. What percentages of T and C would you expect in sea urchin DNA? Explain.

General Problems

24.90 One of the constituents of the carnauba wax used in floor and furniture polish is an ester of a C_{32} straight-chain alcohol with a C_{20} straight-chain carboxylic acid. Draw the structure of this ester.

24.91 Cytochrome c is an important enzyme found in the cells of all aerobic organisms. Elemental analysis of cytochrome c shows that it contains 0.43% iron. What is the minimum molecular mass of this enzyme?

24.92 The catabolism of glucose to yield carbon dioxide and water has $\Delta G° = -2870$ kJ/mol. What is the value of $\Delta G°$ for the photosynthesis of glucose from carbon dioxide and water in green plants?

24.93 The *endorphins* are a group of naturally occurring neuroproteins that act in a manner similar to morphine to control pain. Research has shown that the biologically active part of the endorphin molecule is a simple pentapeptide called an *enkephalin*, with the structure Tyr-Gly-Gly-Phe-Met. Draw the complete structure of this enkephalin.

24.94 Write full structures for the following peptides, and indicate the positions of the peptide bonds:
  (a) Val-Phe-Cys
  (b) Glu-Pro-Ile-Leu

24.95 The α-helical segments of myoglobin and other proteins stop when a proline residue is encountered in the chain. Why is proline never encountered in a protein α-helix?

24.96 Jojoba wax, used in candles and cosmetics, contains the ester of stearic acid and a straight-chain C_{22} alcohol. Draw the structure of this ester.

24.97 Write representative structures for the following:
  (a) A fat
  (b) A vegetable oil
  (c) A steroid

24.98 What DNA sequence is complementary to the following sequence?

A-G-T-T-C-A-T-C-G

24.99 Protonation of the side chain in arginine occurs on the doubly bonded nitrogen atom. Draw three resonance structures of protonated product.

Multi-Concept Problem

24.100 The relative degree of unsaturation in a fat or oil is expressed as an *iodine number*. Olive oil, for instance, is highly unsaturated and has an iodine number of 172, while butter is much less unsaturated and has an iodine number of 37. Defined as the number of grams of I_2 absorbed per 100 grams of fat, iodine number is based on the fact that the carbon–carbon double bonds in fats or oils undergo an addition reaction with I_2. The greater the number of double bonds, the larger the amount of I_2 that reacts.

$$\underset{/}{\overset{\backslash}{C}}=\underset{\backslash}{\overset{/}{C}} \quad \xrightarrow{I_2} \quad \overset{I}{\underset{I}{\overset{|}{C}-\overset{|}{C}}}$$

To determine an iodine number, a known amount of fat is treated with a known amount of I_2. When reaction is complete (about 1 h), the amount of excess I_2 remaining is determined by titration with $Na_2S_2O_3$ according to the equation

$$2\,Na_2S_2O_3(aq) + I_2(aq) \longrightarrow Na_2S_4O_6(aq) + 2\,NaI(aq)$$

Knowing both the amount of I_2 originally added and the amount remaining after reaction, the iodine number can be calculated.

Assume that 0.500 g of human milkfat is treated with 25.0 mL of 0.200 M I_2 solution and that 81.99 mL of 0.100 M $Na_2S_2O_3$ is required for complete reaction with the excess I_2.

  (a) What amount (in grams) of I_2 was added initially?
  (b) How many grams of I_2 reacted with the milkfat, and how many grams were in excess?
  (c) What is the iodine number of human milkfat?
  (d) Assuming a molecular mass of 800 amu, how many double bonds does an average molecule of milkfat contain?

 ## eMedia Problems

24.101 Watch the **Chirality** movie (*eChapter 24.3*) and answer the following questions:
  (a) What conditions are required for a carbon in a molecule to be chiral?
  (b) How many chiral carbons are there in the structure in Problem 24.7(c)?

24.102 Problem 24.46 asks you to identify all of the tripeptides that contain the amino acids methionine, isoleucine, and lysine. Using information in the **Proteins and Amino Acids** movie (*eChapter 24.4*) and in Figure 24.2, draw structures for the tripeptides you identified in Problem 24.46.

24.103 Watch the **Energy and Enzymes** movie (*eChapter 24.6*) and answer the following questions:
  (a) What is meant by the term *inhibition*?
  (b) Name two different ways that enzyme inhibition can occur.

Mathematical Operations

A.1 Scientific Notation

The numbers that you encounter in chemistry are often either very large or very small. For example, there are about 33,000,000,000,000,000,000,000 H_2O molecules in 1.0 mL of water, and the distance between the H and O atoms in an H_2O molecule is 0.000 000 000 095 7 m. These quantities are more conveniently written in scientific notation as 3.3×10^{22} molecules and 9.57×10^{-11} m, respectively. In scientific notation, numbers are written in the exponential format $A \times 10^n$, where A is a number between 1 and 10, and the exponent n is a positive or negative integer.

How do you convert a number from ordinary notation to scientific notation? If the number is greater than or equal to 10, shift the decimal point to the *left* by n places until you obtain a number between 1 and 10. Then, multiply the result by 10^n. For example, the number 8137.6 is written in scientific notation as 8.1376×10^3:

$$8137.6 = 8.1376 \times 10^3$$

Shift decimal point to the left
by 3 places to get a number
between 1 and 10

Number of places decimal
point was shifted to the left

When you shift the decimal point to the left by three places, you are in effect dividing the number by $10 \times 10 \times 10 = 1000 = 10^3$. Therefore, you must multiply the result by 10^3 so that the value of the number is unchanged.

To convert a number less than 1 to scientific notation, shift the decimal point to the *right* by n places until you obtain a number between 1 and 10. Then, multiply the result by 10^{-n}. For example, the number 0.012 is written in scientific notation as 1.2×10^{-2}:

$$0.012 = 1.2 \times 10^{-2}$$

Shift decimal point to the right
by 2 places to get a number
between 1 and 10

Number of places decimal
point was shifted to the right

When you shift the decimal point to the right by two places, you are in effect multiplying the number by $10 \times 10 = 100 = 10^2$. Therefore, you must multiply the result by 10^{-2} so that the value of the number is unchanged. ($10^2 \times 10^{-2} = 10^0 = 1$.)

The following table gives some additional examples. To convert from scientific notation to ordinary notation, simply reverse the preceding process. Thus, to write the number 5.84×10^4 in ordinary notation, drop the factor of 10^4 and move the decimal point by 4 places to the *right* ($5.84 \times 10^4 = 58{,}400$). To write the number 3.5×10^{-1} in ordinary notation, drop the factor of 10^{-1} and move the decimal point by 1 place to the *left* ($3.5 \times 10^{-1} = 0.35$). Note that you don't need scientific notation for numbers between 1 and 10 because $10^0 = 1$.

Number	Scientific notation
58,400	5.84×10^4
0.35	3.5×10^{-1}
7.296	$7.296 \times 10^0 = 7.296$

Addition and Subtraction

To add or subtract two numbers expressed in scientific notation, both numbers must have the same exponent. Thus, to add 7.16×10^3 and 1.32×10^2, first write the latter number as 0.132×10^3 and then add:

$$\begin{array}{r} 7.16 \ \times 10^3 \\ + \ 0.132 \ \times 10^3 \\ \hline 7.29 \ \times 10^3 \end{array}$$

The answer has three significant figures. (Significant figures are discussed in Section 1.11.) Alternatively, you can write the first number as 71.6×10^2 and then add:

$$\begin{array}{r} 71.6 \ \times 10^2 \\ + \ \ \ 1.32 \ \times 10^2 \\ \hline 72.9 \ \times 10^2 = 7.29 \times 10^3 \end{array}$$

Multiplication and Division

To multiply two numbers expressed in scientific notation, multiply the factors in front of the powers of 10 and then add the exponents:

$$(A \times 10^n)(B \times 10^m) = AB \times 10^{n+m}$$

For example,

$$(2.5 \times 10^4)(4.7 \times 10^7) = (2.5)(4.7) \times 10^{4+7} = 12 \times 10^{11} = 1.2 \times 10^{12}$$
$$(3.46 \times 10^5)(2.2 \times 10^{-2}) = (3.46)(2.2) \times 10^{5+(-2)} = 7.6 \times 10^3$$

Both answers have two significant figures.

To divide two numbers expressed in scientific notation, divide the factors in front of the powers of 10 and then subtract the exponent in the denominator from the exponent in the numerator:

$$\frac{A \times 10^n}{B \times 10^m} = \frac{A}{B} \times 10^{n-m}$$

For example,

$$\frac{3 \times 10^6}{7.2 \times 10^2} = \frac{3}{7.2} \times 10^{6-2} = 0.4 \times 10^4 = 4 \times 10^3 \qquad \text{(1 significant figure)}$$

$$\frac{7.50 \times 10^{-5}}{2.5 \times 10^{-7}} = \frac{7.50}{2.5} \times 10^{-5-(-7)} = 3.0 \times 10^2 \qquad \text{(2 significant figures)}$$

Powers and Roots

To raise a number $A \times 10^n$ to a power m, raise the factor A to the power m and then multiply the exponent n by the power m:

$$(A \times 10^n)^m = A^m \times 10^{n \times m}$$

For example, 3.6×10^2 raised to the 3rd power is 4.7×10^7:

$$(3.6 \times 10^2)^3 = (3.6)^3 \times 10^{2 \times 3} = 47 \times 10^6 = 4.7 \times 10^7 \qquad \text{(2 significant figures)}$$

To take the mth root of a number $A \times 10^n$, raise the number to the power $1/m$. That is, raise factor A to the power $1/m$ and then divide the exponent n by the root m:

$$\sqrt[m]{A \times 10^n} = (A \times 10^n)^{1/m} = A^{1/m} \times 10^{n/m}$$

For example, the square root of 9.0×10^8 is 3.0×10^4:

$$\sqrt[2]{9.0 \times 10^8} = (9.0 \times 10^8)^{1/2} = (9.0)^{1/2} \times 10^{8/2} = 3.0 \times 10^4 \qquad \text{(2 significant figures)}$$

Because the exponent in the answer (n/m) is an integer we must sometimes rewrite the original number by shifting the decimal point so that the exponent n is an integral multiple of the root m. For example, to take the cube root of 6.4×10^{10}, we first rewrite this number as 64×10^9 so that the exponent (9) is an integral multiple of the root 3:

$$\sqrt[3]{6.4 \times 10^{10}} = \sqrt[3]{64 \times 10^9} = (64)^{1/3} \times 10^{9/3} = 4.0 \times 10^3$$

Scientific Notation and Electronic Calculators

With a scientific calculator you can carry out calculations in scientific notation. You should consult the instruction manual for your particular calculator to learn how to enter and manipulate numbers expressed in an exponential format. On most calculators, you enter the number $A \times 10^n$ by (i) entering the number A, (ii) pressing a key labeled EXP or EE, and (iii) entering the exponent n. If the exponent is negative, you press a key labeled $+/-$ before entering the value of n. (Note that you do not enter the number 10.) The calculator displays the number $A \times 10^n$ with the number A on the left followed by some space and then the exponent n. For example,

$$4.625 \times 10^2 \quad \text{is displayed as} \quad 4.625 \ \ 02$$

To add, subtract, multiply, or divide exponential numbers, use the same sequence of key strokes as you would in working with ordinary numbers. When you add or subtract on a calculator, the numbers need not have the same exponent; the calculator automatically takes account of the different exponents. Remember, though, that the calculator often gives more digits in the answer than the allowed number of significant figures. It's sometimes helpful to outline the calculation on paper, as in the preceding examples, in order to keep track of the number of significant figures.

Most calculators have x^2 and $\sqrt{x}$ keys for squaring a number and finding its square root. Just enter the number, and press the appropriate key. You probably have a y^x (or a^x) key for raising a number to a power. To raise 4.625×10^2 to the 3rd power, for example, use the following keystrokes: (i) enter the number 4.625×10^3 in the usual way, (ii) press the y^x key, (iii) enter the power 3, and (iv) press the = key. The result is displayed as 9.8931641 07, but it must be rounded to 4 significant figures. Therefore, $(4.625 \times 10^2)^3 = 9.893 \times 10^7$.

To take the mth root of a number, raise the number to the power $1/m$. For example, to take the 5th root of 4.52×10^{11}, use the following keystrokes: (i) enter the number 4.52×10^{11}, (ii) press the y^x key, (iii) enter the number 5 (for the 5th root), (iv) press the $1/x$ key (to convert the 5th root to the power $1/5$), and (v) press the = key. The result is

$$\sqrt[5]{4.52 \times 10^{11}} = (4.52 \times 10^{11})^{1/5} = 2.14 \times 10^2$$

The calculator is able to handle the nonintegral exponent 11/5, and there is therefore no need to enter the number as 45.2×10^{10} so that the exponent is an integral multiple of the root 5.

▶ **PROBLEM A.1**   Perform the following calculations, expressing the result in scientific notation with the correct number of significant figures. (You don't need a calculator for these.)
(a) $(1.50 \times 10^4) + (5.04 \times 10^3)$ **(b)** $(2.5 \times 10^{-2}) - (5.0 \times 10^{-3})$ **(c)** $(4.0 \times 10^4)^2$
(d) $\sqrt[3]{8 \times 10^{12}}$ **(e)** $\sqrt{2.5 \times 10^5}$

ANSWERS:
(a) 2.00×10^4 **(b)** 2.0×10^{-2} **(c)** 1.6×10^9 **(d)** 2×10^4 **(e)** 5.0×10^2

▶ **PROBLEM A.2**   Perform the following calculations, expressing the result in scientific notation with the correct number of significant figures. (Use a calculator for these.)
(a) $(9.72 \times 10^{-1}) + (3.4823 \times 10^2)$ **(b)** $3.772 \times 10^3) - (2.891 \times 10^4)$
(c) $(7.62 \times 10^{-3})^4$ **(d)** $\sqrt[3]{8.2 \times 10^7}$ **(e)** $\sqrt[5]{3.47 \times 10^{-12}}$

ANSWERS:
(a) 3.4920×10^2 **(b)** -2.514×10^4 **(c)** 3.37×10^{-9} **(d)** 4.3×10^2
(e) 5.11×10^{-3}

A.2   Logarithms

Common Logarithms

Any positive number x can be written as 10 raised to some power z—that is, $x = 10^z$. The exponent z is called the *common*, or *base 10*, *logarithm* of the number x and is denoted $\log_{10} x$, or simply $\log x$:

$$x = 10^z \qquad \log x = z$$

For example, 100 can be written as 10^2, and $\log 100$ is therefore equal to 2:

$$100 = 10^2 \qquad \log 100 = 2$$

Similarly,

$$10 = 10^1 \qquad \log 10 = 1$$
$$1 = 10^0 \qquad \log 1 = 0$$
$$0.1 = 10^{-1} \qquad \log 0.1 = -1$$

In general, the logarithm of a number x is the power z to which 10 must be raised to equal the number x.

As Figure A.1 shows, the logarithm of a number greater than 1 is positive, the logarithm of 1 is zero, and the logarithm of a positive number less than 1 is negative. The logarithm of a *negative* number is undefined because 10 raised to any power is always positive ($x = 10^z > 0$).

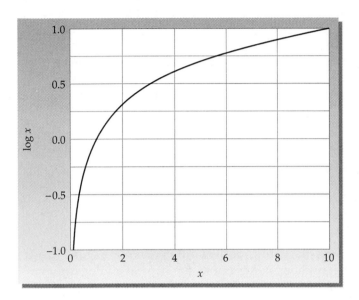

FIGURE A.1   Values of log x for values of x in the range 0.1 to 10.

You can use a calculator to find the logarithm of a number that is not an integral power of 10. For example, to find the logarithm of 60.2, simply enter 60.2, and press the LOG key. The logarithm should be between 1 and 2 because 60.2 is between 10^1 and 10^2. The calculator gives a value of 1.7795965, which must be rounded to 1.780 because 60.2 has three significant figures. The only significant figures in a logarithm are the digits beyond the decimal point, as explained in Section 15.5.

Antilogarithms

The antilogarithm, denoted antilog, is the inverse of the common logarithm. If z is the logarithm of x, then x is the antilogarithm of z. But since x can be written as 10^z, the antilogarithm of z is 10^z:

$$\text{If } z = \log x \qquad \text{then} \qquad x = \text{antilog } z = 10^z$$

In other words, the antilog of a number is 10 raised to a power equal to that number. For example, the antilog of 2 is $10^2 = 100$, and the antilog of 0.70 is $10^{0.70}$.

To find the value of antilog 0.70, use your calculator. If you have a 10^x key, enter 0.70 and press the 10^x key. If you have a y^x key, use the following keystrokes: (i) enter 10, (ii) press the y^x key, (iii) enter the exponent 0.70, and (iv) press the = key. If you have an INV (inverse) key, enter 0.70, press the INV key, and then press the LOG key. The result is

$$\text{antilog } 0.70 = 5.0 \text{ (2 significant figures)}.$$

Natural Logarithms

The number $e = 2.718\ 28\ldots$, like $\pi = 3.141\ 59\ldots$, turns up in many scientific problems. It is therefore convenient to define a logarithm based on e, just as we defined a logarithm based on 10. Just as a number x can be written as 10^z, it can also be written as e^u. The exponent u is called the *natural*, or *base e, logarithm* of the number x and is denoted $\log_e x$, or more commonly, $\ln x$:

$$x = e^u \qquad \ln x = u$$

The natural logarithm of a number x is the power u to which e must be raised to equal the number x. For example, the number 10 can be written as $e^{2.303}$, and the natural logarithm of 10 is therefore equal to 2.303:

$$10 = e^{2.303} = (2.718\ 28\ldots)^{2.303} \qquad \ln 10 = 2.303$$

To find the natural logarithm of a number on your calculator, simply enter the number, and press the LN key.

The natural antilogarithm, denoted antiln, is the inverse of the natural logarithm. If u is the natural logarithm of x, then $x \, (= e^u)$ is the natural antilogarithm of u:

$$\text{If } u = \ln x \qquad \text{then} \qquad x = \text{antiln } u = e^u$$

In other words, the natural antilogarithm of a number is e raised to a power equal to that number. For example, the natural antilogarithm of 0.70 is $e^{0.70}$, which equals 2.0:

$$\text{antiln } 0.70 = e^{0.70} = 2.0$$

Your calculator probably has an INV (inverse) key or an e^x key. To find the natural antilogarithm of a number—say, 0.70—enter 0.70, press the INV key, and then press the LN key. Alternatively, you can enter 0.70, and press the e^x key.

Some Mathematical Properties of Logarithms

Because logarithms are exponents, the algebraic properties of exponents can be used to derive the following useful relationships involving logarithms:

1. The logarithm (either common or natural) of a product xy equals the sum of the logarithm of x and the logarithm of y:

$$\log xy = \log x + \log y \qquad \ln xy = \ln x + \ln y$$

2. The logarithm of a quotient x/y equals the difference between the logarithm of x and the logarithm of y:

$$\log \frac{x}{y} = \log x - \log y \qquad \ln \frac{x}{y} = \ln x - \ln y$$

It follows from these relationships that

$$\log \frac{y}{x} = -\log \frac{x}{y} \qquad \ln \frac{y}{x} = -\ln \frac{x}{y}$$

Because $\log 1 = \ln 1 = 0$, it also follows that

$$\log \frac{1}{x} = -\log x \qquad \ln \frac{1}{x} = -\ln x$$

3. The logarithm of x raised to a power a equals a times the logarithm of x:

$$\log x^a = a \log x \qquad \ln x^a = a \ln x$$

Similarly,

$$\log x^{1/a} = \frac{1}{a} \log x \qquad \ln x^{1/a} = \frac{1}{a} \ln x$$

where

$$x^{1/a} = \sqrt[a]{x}$$

What is the numerical relationship between the common logarithm and the natural logarithm? To derive it, we begin with the definitions of $\log x$ and $\ln x$:

$$\log x = z \qquad \text{where} \qquad x = 10^z$$
$$\ln x = u \qquad \text{where} \qquad x = e^u$$

We then write $\ln x$ in terms of 10^z and make use of the property that $\ln x^a = a \ln x$:

$$\ln x = \ln 10^z = z \ln 10$$

Because $z = \log x$ and $\ln 10 = 2.303$, we find that the natural logarithm is 2.303 times the common logarithm:

$$\ln x = 2.303 \log x$$

▶ **PROBLEM A.3**

Use a calculator to evaluate the following expressions:
(a) $\log 705$ (b) $\ln (3.4 \times 10^{-6})$ (c) antilog (-2.56) (d) antiln 8.1

ANSWERS:
(a) 2.848   (b) -12.59 (c) 2.8×10^{-3} (d) 3×10^3

A.3   Straight-Line Graphs and Linear Equations

The results of a scientific experiment are often summarized in the form of a graph. Consider an experiment in which some property y is measured as a function of some variable x. (A real example would be measurement of the volume of a gas as a function of its temperature, but we'll use y and x

to keep the discussion general.) Suppose that we obtain the following experimental data:

x	y
−1	−5
1	1
3	7
5	13

The graph in Figure A.2 shows values of x, called the independent variable, along the horizontal axis and values of y, the dependent variable, along the vertical axis. Each pair of experimental values of x and y is represented by a point on the graph. For this particular experiment, the four data points lie on a straight line.

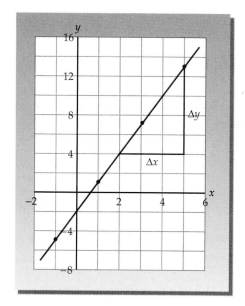

FIGURE A.2   A straight-line y versus x plot of the data in the table.

The equation of a straight line can be written as

$$y = mx + b$$

where m is the slope of the line and b is the intercept, the value of y at the point where the line crosses the y axis—that is, the value of y when $x = 0$. The slope of the line is the change in y (Δy) for a given change in x (Δx):

$$m = \text{slope} = \frac{\Delta y}{\Delta x}$$

The right-triangle in Figure A.2 shows that y changes from 4 to 13 when x changes from 2 to 5. Therefore, the slope of the line is 3:

$$m = \text{slope} = \frac{\Delta y}{\Delta x} = \frac{13 - 4}{5 - 2} = \frac{9}{3} = 3$$

The graph shows a y intercept of -2 ($b = -2$), and the equation of the line is therefore

$$y = 3x - 2$$

An equation of the form $y = mx + b$ is called a *linear equation* because values of x and y that satisfy such an equation are the coordinates of points that lie on a straight line. We also say that y is a *linear function* of x, or that y is *directly proportional* to x. In our example, the rate of change of y is 3 times that of x.

A.4 Quadratic Equations

A quadratic equation is an equation that can be written in the form

$$ax^2 + bx + c = 0$$

where a, b, and c are constants. The equation contains only powers of x and is called quadratic because the highest power of x is 2. The solutions to a quadratic equation (values of x that satisfy the equation) are given by the *quadratic formula*:

$$x = \frac{-b \pm \sqrt{b^2 - 4ac}}{2a}$$

The $\pm$ indicates that there are two solutions, one given by the $+$ sign, and the other given by the $-$ sign.

As an example, let's solve the equation

$$x^2 = \frac{2 - 6x}{3}$$

First, we put the equation into the form $ax^2 + bx + c = 0$ by multiplying it by 3 and moving $2 - 6x$ to the left side. The result is

$$3x^2 + 6x - 2 = 0$$

Then we apply the quadratic formula with $a = 3$, $b = 6$, and $c = -2$:

$$x = \frac{-6 \pm \sqrt{(6)^2 - 4(3)(-2)}}{2(3)}$$

$$= \frac{-6 \pm \sqrt{36 + 24}}{6} = \frac{-6 \pm \sqrt{60}}{6} = \frac{-6 \pm 7.746}{6}$$

The two solutions are

$$x = \frac{-6 + 7.746}{6} = \frac{1.746}{6} = 0.291 \quad \text{and} \quad x = \frac{-6 - 7.746}{6} = \frac{-13.746}{6} = -2.291$$

B Thermodynamic Properties at 25°C

TABLE B.1 Inorganic Substances

Substance and State	$\Delta H°_f$ (kJ/mol)	$\Delta G°_f$ (kJ/mol)	$S°$ [J/(K · mol)]	Substance and State	$\Delta H°_f$ (kJ/mol)	$\Delta G°_f$ (kJ/mol)	$S°$ [J/(K · mol)]
Aluminum				Calcium			
Al(s)	0	0	28.3	Ca(s)	0	0	41.4
Al(g)	330.0	289.4	164.5	Ca(g)	177.8	144.0	154.8
AlCl₃(s)	−704.2	−628.9	110.7	Ca²⁺(aq)	−542.8	−553.6	−53.1
Al₂O₃(s)	−1676	−1582	50.9	CaF₂(s)	−1228.0	−1175.6	68.5
Barium				CaCl₂(s)	−795.4	−748.8	108.4
Ba(s)	0	0	62.8	CaH₂(s)	−181.5	−142.5	41.4
Ba(g)	180.0	146.0	170.1	CaC₂(s)	−59.8	−64.8	70.0
Ba²⁺(aq)	−537.6	−560.8	9.6	CaO(s)	−635.1	−604.0	39.7
BaCl₂(s)	−855.0	−806.7	123.7	Ca(OH)₂(s)	−986.1	−898.6	83.4
BaO(s)	−548.0	−520.3	72.1	CaCO₃(s)	−1206.9	−1128.8	92.9
BaCO₃(s)	−1216	−1138	112	CaSO₄(s)	−1434.1	−1321.9	107
BaSO₄(s)	−1473	−1362	132	Ca₃(PO₄)₂(s)	−4120.8	−3884.7	236.0
Beryllium				Carbon			
Be(s)	0	0	9.5	C(s, graphite)	0	0	5.7
BeO(s)	−609.6	−580.3	14.1	C(s, diamond)	1.9	2.9	2.4
Be(OH)₂(s)	−902.5	−815.0	45.2	C(g)	716.7	671.3	158.0
Boron				CO(g)	−110.5	−137.2	197.6
B(s)	0	0	5.9	CO₂(g)	−393.5	−394.4	213.6
BF₃(g)	−1136.0	−1119.4	254.3	CO₂(aq)	−413.8	−386.0	117.6
BCl₃(g)	−403.8	−388.7	290.0	CO₃²⁻(aq)	−677.1	−527.8	−56.9
B₂H₆(g)	35.6	86.7	232.0	HCO₃⁻(aq)	−692.0	−586.8	91.2
B₂O₃(s)	−1273.5	−1194.3	54.0	H₂CO₃(aq)	−699.7	−623.2	187.4
H₃BO₃(s)	−1094.3	−968.9	90.0	HCN(l)	108.9	125.0	112.8
Bromine				HCN(g)	135.1	124.7	201.7
Br(g)	111.9	82.4	174.9	CS₂(l)	89.0	64.6	151.3
Br⁻(aq)	−121.5	−104.0	82.4	CS₂(g)	116.7	67.1	237.7
Br₂(l)	0	0	152.2	COCl₂(g)	−219.1	−204.9	283.4
Br₂(g)	30.9	3.14	245.4	Cesium			
HBr(g)	−36.4	−53.4	198.6	Cs(s)	0	0	85.2
Cadmium				Cs(g)	76.5	49.6	175.6
Cd(s)	0	0	51.8	Cs⁺(aq)	−258.3	−292.0	133.1
Cd(g)	111.8	77.3	167.6	CsF(s)	−553.5	−525.5	92.8
Cd²⁺(aq)	−75.9	−77.6	−73.2	CsCl(s)	−443.0	−414.5	101.2
CdCl₂(s)	−391.5	−343.9	115.3	CsBr(s)	−405.8	−391.4	113.1
CdO(s)	−258.4	−228.7	54.8	CsI(s)	−346.6	−340.6	123.1
CdS(s)	−161.9	−156.5	64.9				
CdSO₄(s)	−933.3	−822.7	123.0				

TABLE B.1   Inorganic Substances (*continued*)

Substance and State	$\Delta H°_f$ (kJ/mol)	$\Delta G°_f$ (kJ/mol)	$S°$ [J/(K · mol)]	Substance and State	$\Delta H°_f$ (kJ/mol)	$\Delta G°_f$ (kJ/mol)	$S°$ [J/(K · mol)]
Chlorine				**Iron**			
$Cl(g)$	121.7	105.7	165.1	$Fe(s)$	0	0	27.3
$Cl^-(aq)$	−167.2	−131.3	56.5	$Fe(g)$	416.3	370.3	180.5
$Cl_2(g)$	0	0	223.0	$FeCl_2(s)$	−341.8	−302.3	118.0
$HCl(g)$	−92.3	−95.3	186.8	$FeCl_3(s)$	−399.5	−334.0	142.3
$HCl(aq)$	−167.2	−131.2	56.5	$FeO(s)$	−272	−255	61
$ClO_2(g)$	102.5	120.5	256.7	$Fe_2O_3(s)$	−824.2	−742.2	87.4
$Cl_2O(g)$	80.3	97.9	266.1	$Fe_3O_4(s)$	−1118	−1015	146
Chromium				$FeS_2(s)$	−178.2	−166.9	52.9
$Cr(s)$	0	0	23.8	**Lead**			
$Cr(g)$	396.6	351.8	174.4	$Pb(s)$	0	0	64.8
$Cr_2O_3(s)$	−1140	−1058	81.2	$Pb(g)$	195.2	162.2	175.3
Cobalt				$PbCl_2(s)$	−359.4	−314.1	136.0
$Co(s)$	0	0	30.0	$PbBr_2(s)$	−278.7	−261.9	161.5
$Co(g)$	424.7	380.3	179.4	$PbO(s)$	−217.3	−187.9	68.7
$CoO(s)$	−237.9	−214.2	53.0	$PbO_2(s)$	−277	−217.4	68.6
Copper				$PbS(s)$	−100	−98.7	91.2
$Cu(s)$	0	0	33.1	$PbCO_3(s)$	−699.1	−625.5	131.0
$Cu(g)$	337.4	297.7	166.3	$PbSO_4(s)$	−919.9	−813.2	148.6
$Cu^{2+}(aq)$	64.8	65.5	−99.6	**Lithium**			
$CuCl(s)$	−137.2	−119.9	86.2	$Li(s)$	0	0	29.1
$CuCl_2(s)$	−220.1	−175.7	108.1	$Li(g)$	159.3	126.6	138.7
$CuO(s)$	−157	−130	42.6	$Li^+(aq)$	−278.5	−293.3	13
$Cu_2O(s)$	−168.6	−146.0	93.1	$LiF(s)$	−616.0	−587.7	35.7
$CuS(s)$	−53.1	−53.6	66.5	$LiCl(s)$	−408.6	−384.4	59.3
$Cu_2S(s)$	−79.5	−86.2	120.9	$LiBr(s)$	−351.2	−342.0	74.3
$CuSO_4(s)$	−771.4	−662.2	109.2	$LiI(s)$	−270.4	−270.3	86.8
Fluorine				$Li_2O(s)$	−597.9	−561.2	37.6
$F(g)$	79.0	61.9	158.6	$LiOH(s)$	−484.9	−439.0	42.8
$F^-(aq)$	−332.6	−278.8	−13.8	**Magnesium**			
$F_2(g)$	0	0	202.7	$Mg(s)$	0	0	32.7
$HF(g)$	−271	−273	173.7	$Mg(g)$	147.1	112.5	148.6
Hydrogen				$MgCl_2(s)$	−641.6	−591.8	89.6
$H(g)$	218.0	203.3	114.6	$MgO(s)$	−601.7	−569.4	26.9
$H^+(aq)$	0	0	0	$MgCO_3(s)$	−1096	−1012	65.7
$H_2(g)$	0	0	130.6	$MgSO_4(s)$	−1284.9	−1170.6	91.6
$OH^-(aq)$	−230.0	−157.3	−10.8	**Manganese**			
$H_2O(l)$	−285.8	−237.2	69.9	$Mn(s)$	0	0	32.0
$H_2O(g)$	−241.8	−228.6	188.7	$Mn(g)$	280.7	238.5	173.6
$H_2O_2(l)$	−187.8	−120.4	110	$MnO(s)$	−385.2	−362.9	59.7
$H_2O_2(g)$	−136.3	−105.6	232.6	$MnO_2(s)$	−520.0	−465.1	53.1
$H_2O_2(aq)$	−191.2	−134.1	144	**Mercury**			
Iodine				$Hg(l)$	0	0	76.0
$I(g)$	106.8	70.3	180.7	$Hg(g)$	61.32	31.85	174.8
$I^-(aq)$	−55.2	−51.6	111	$Hg^{2+}(aq)$	171.1	164.4	−32.2
$I_2(s)$	0	0	116.1	$Hg_2^{2+}(aq)$	172.4	153.5	84.5
$I_2(g)$	62.4	19.4	260.6	$HgCl_2(s)$	−224	−179	146
$HI(g)$	26.5	1.7	206.5	$Hg_2Cl_2(s)$	−265.2	−210.8	192
				$HgO(s)$	−90.8	−58.6	70.3
				$HgS(s)$	−58.2	−50.6	82.4

TABLE B.1   Inorganic Substances (continued)

Substance and State	$\Delta H°_f$ (kJ/mol)	$\Delta G°_f$ (kJ/mol)	$S°$ [J/(K · mol)]	Substance and State	$\Delta H°_f$ (kJ/mol)	$\Delta G°_f$ (kJ/mol)	$S°$ [J/(K · mol)]
Nickel				Potassium (cont.)			
Ni(s)	0	0	29.9	KBr(s)	−393.8	−380.7	95.9
Ni(g)	429.7	384.5	182.1	KI(s)	−327.9	−324.9	106.3
NiCl$_2$(s)	−305.3	−259.1	97.7	K$_2$O(s)	−361.5		
NiO(s)	−240	−212	38.0	K$_2$O$_2$(s)	−494.1	−425.1	102.1
NiS(s)	−82.0	−79.5	53.0	KO$_2$(s)	−284.9	−239.4	116.7
Nitrogen				KOH(s)	−424.8	−379.1	78.9
N(g)	472.7	455.6	153.2	KOH(aq)	−482.4	−440.5	91.6
N$_2$(g)	0	0	191.5	KClO$_3$(s)	−397.7	−296.3	143
NH$_3$(g)	−46.1	−16.5	192.3	KClO$_4$(s)	−432.8	−303.1	151.0
NH$_3$(aq)	−80.3	−26.6	111	KNO$_3$(s)	−494.6	−394.9	133.1
NH$_4$$^+$(aq)	−132.5	−79.4	113	Rubidium			
N$_2$H$_4$(l)	50.6	149.2	121.2	Rb(s)	0	0	76.8
N$_2$H$_4$(g)	95.4	159.3	238.4	Rb(g)	80.9	53.1	170.0
NO(g)	90.2	86.6	210.7	Rb$^+$(aq)	−251.2	−284.0	121.5
NO$_2$(g)	33.2	51.3	240.0	RbF(s)	−557.7		
N$_2$O(g)	82.0	104.2	219.7	RbCl(s)	−435.4	−407.8	95.9
N$_2$O$_4$(g)	9.16	97.8	304.2	RbBr(s)	−394.6	−381.8	110.0
N$_2$O$_5$(g)	11	115	356	RbI(s)	−333.8	−328.9	118.4
NOCl(g)	51.7	66.1	261.6	Selenium			
NO$_2$Cl(g)	12.6	54.4	272.1	Se(s, black)	0	0	42.44
HNO$_3$(l)	−174.1	−80.8	155.6	H$_2$Se(g)	29.7	15.9	219.0
HNO$_3$(g)	−133.9	−73.5	266.8	Silicon			
HNO$_2$(aq)	−119	−50.6	136	Si(s)	0	0	18.8
HNO$_3$(aq)	−207.4	−111.3	146	Si(g)	450.0	405.5	167.9
NO$_3$$^-$(aq)	−207.4	−111.3	146.4	SiF$_4$(g)	−1615.0	−1572.8	282.7
NH$_4$Cl(s)	−314.4	−202.9	94.6	SiCl$_4$(l)	−687.0	−619.9	240
NH$_4$NO$_3$(s)	−365.6	−184.0	151.1	SiO$_2$(s, quartz)	−910.9	−856.7	41.8
Oxygen				Silver			
O(g)	249.2	231.7	160.9	Ag(s)	0	0	42.6
O$_2$(g)	0	0	205.0	Ag(g)	284.9	246.0	173.0
O$_3$(g)	143	163	238.8	Ag$^+$(aq)	105.6	77.1	72.7
Phosphorus				AgF(s)	−204.6		
P(s, white)	0	0	41.1	AgCl(s)	−127.1	−109.8	96.2
P(s, red)	−18	−12	22.8	AgBr(s)	−100.4	−96.9	107
P$_4$(g)	58.9	24.5	279.9	AgI(s)	−61.8	−66.2	115
PH$_3$(g)	5.4	13	210.1	Ag$_2$O(s)	−31.0	−11.2	121
PCl$_3$(l)	−320	−272	217	Ag$_2$S(s)	−32.6	−40.7	144.0
PCl$_3$(g)	−287.0	−267.8	311.7	AgNO$_3$(s)	−124.4	−33.4	140.9
PCl$_5$(s)	−443.5			Sodium			
PCl$_5$(g)	−374.9	−305.0	364.5	Na(s)	0	0	51.2
P$_4$O$_{10}$(s)	−2984	−2698	228.9	Na(g)	107.3	76.8	153.6
PO$_4$$^{3-}$(aq)	−1277.4	−1018.7	−220.5	Na$^+$(aq)	−240.1	−261.9	59.0
HPO$_4$$^{2-}$(aq)	−1292.1	−1089.2	−33.5	NaF(s)	−576.6	−546.3	51.1
H$_2$PO$_4$$^-$(aq)	−1296.3	−1130.2	90.4	NaCl(s)	−411.2	−384.2	72.1
H$_3$PO$_4$(s)	−1279	−1119	110.5	NaBr(s)	−361.1	−349.0	86.8
Potassium				NaI(s)	−287.8	−286.1	98.5
K(s)	0	0	64.2	NaH(s)	−56.3	−33.5	40.0
K(g)	89.2	60.6	160.2	Na$_2$O(s)	−414.2	−375.5	75.1
K$^+$(aq)	−252.4	−283.3	103	Na$_2$O$_2$(s)	−510.9	−447.7	95.0
KF(s)	−567.3	−537.8	66.6	NaO$_2$(s)	−260.2	−218.4	115.9
KCl(s)	−436.7	−409.2	82.6	NaOH(s)	−425.6	−379.5	64.5

TABLE B.1   Inorganic Substances (*continued*)

Substance and State	$\Delta H°_f$ (kJ/mol)	$\Delta G°_f$ (kJ/mol)	$S°$ [J/(K · mol)]	Substance and State	$\Delta H°_f$ (kJ/mol)	$\Delta G°_f$ (kJ/mol)	$S°$ [J/(K · mol)]
Sodium (*cont.*)				Titanium			
NaOH(*aq*)	−470.1	−419.2	48.2	Ti(*s*)	0	0	30.6
Na$_2$CO$_3$(*s*)	−1130.7	−1044.5	135.0	Ti(*g*)	473.0	428.4	180.2
NaHCO$_3$(*s*)	−950.8	−851.0	102	TiCl$_4$(*l*)	−804.2	−737.2	252.3
NaNO$_3$(*s*)	−467.9	−367.0	116.5	TiCl$_4$(*g*)	−763.2	−726.3	353.2
NaNO$_3$(*aq*)	−447.5	−373.2	205.4	TiO$_2$(*s*)	−944.7	−889.5	50.3
Na$_2$SO$_4$(*s*)	−1387.1	−1270.2	149.6	Tungsten			
Sulfur				W(*s*)	0	0	32.6
S(*s*, rhombic)	0	0	31.8	W(*g*)	849.4	807.1	174.0
S(*s*, monoclinic)	0.3			WO$_3$(*s*)	−842.9	−764.0	75.9
S(*g*)	277.2	236.7	167.7	Zinc			
S$_2$(*g*)	128.6	79.7	228.2	Zn(*s*)	0	0	41.6
H$_2$S(*g*)	−20.6	−33.6	205.7	Zn(*g*)	130.7	95.2	160.9
H$_2$S(*aq*)	−39.7	−27.9	121	Zn^{2+}(*aq*)	−153.9	−147.1	−112.1
HS$^-$(*aq*)	−17.6	12.1	62.8	ZnCl$_2$(*s*)	−415.1	−369.4	111.5
SO$_2$(*g*)	−296.8	−300.2	248.1	ZnO(*s*)	−348.3	−318.3	43.6
SO$_3$(*g*)	−395.7	−371.1	256.6	ZnS(*s*)	−206.0	−201.3	57.7
H$_2$SO$_4$(*l*)	−814.0	−690.1	156.9	ZnSO$_4$(*s*)	−982.8	−871.5	110.5
H$_2$SO$_4$(*aq*)	−909.3	−744.6	20				
HSO$_4^-$(*aq*)	−887.3	−756.0	132				
SO$_4^{2-}$(*aq*)	−909.3	−744.6	20				
Tin							
Sn(*s*, white)	0	0	51.5				
Sn(*s*, gray)	−2.1	0.1	44.1				
Sn(*g*)	301.2	266.2	168.4				
SnCl$_4$(*l*)	−511.3	−440.1	258.6				
SnCl$_4$(*g*)	−471.5	−432.2	365.8				
SnO(*s*)	−286	−257	56.5				
SnO$_2$(*s*)	−580.7	−519.7	52.3				

TABLE B.2   Organic Substances

Substance and State	Formula	$\Delta H°_f$ (kJ/mol)	$\Delta G°_f$ (kJ/mol)	$S°$ [J/(K · mol)]
Acetaldehyde(g)	CH_3CHO	−166.2	−133.0	263.8
Acetic acid(l)	CH_3CO_2H	−484.5	−390	160
Acetylene(g)	C_2H_2	226.7	209.2	200.8
Benzene(l)	C_6H_6	49.0	124.5	172.8
Butane(g)	C_4H_{10}	−126	−17	310
Carbon tetrachloride(l)	CCl_4	−135.4	−65.3	216.4
Dichloroethane(l)	CH_2ClCH_2Cl	−165.2	−79.6	208.5
Ethane(g)	C_2H_6	−84.7	−32.9	229.5
Ethanol(l)	C_2H_5OH	−277.7	−174.9	161
Ethanol(g)	C_2H_5OH	−235.1	−168.6	282.6
Ethylene(g)	C_2H_4	52.3	68.1	219.5
Ethylene oxide(g)	C_2H_4O	−52.6	−13.1	242.4
Formaldehyde(g)	$HCHO$	−108.6	−102.5	218.8
Formic acid(l)	HCO_2H	−424.7	−361.4	129.0
Glucose(s)	$C_6H_{12}O_6$	−1260	−910	212.1
Methane(g)	CH_4	−74.8	−50.8	186.2
Methanol(l)	CH_3OH	−238.7	−166.4	127
Methanol(g)	CH_3OH	−201.2	−161.9	238
Propane(g)	C_3H_8	−105	−25	270
Vinyl chloride(g)	$CH_2=CHCl$	35	51.9	263.9

Equilibrium Constants at 25°C

TABLE C.1 Acid-Dissociation Constants at 25°C

Acid	Formula	K_{a1}	K_{a2}	K_{a3}
Acetic	CH_3CO_2H	1.8×10^{-5}		
Acetylsalicylic	$C_9H_8O_4$	3.0×10^{-4}		
Arsenic	H_3AsO_4	5.6×10^{-3}	1.7×10^{-7}	4.0×10^{-12}
Arsenious	H_3AsO_3	6×10^{-10}		
Ascorbic	$C_6H_8O_6$	8.0×10^{-5}		
Benzoic	$C_6H_5CO_2H$	6.5×10^{-5}		
Boric	H_3BO_3	5.8×10^{-10}		
Carbonic	H_2CO_3	4.3×10^{-7}	5.6×10^{-11}	
Chloroacetic	CH_2ClCO_2H	1.4×10^{-3}		
Citric	$C_6H_8O_7$	7.1×10^{-4}	1.7×10^{-5}	4.1×10^{-7}
Formic	HCO_2H	1.8×10^{-4}		
Hydrazoic	HN_3	1.9×10^{-5}		
Hydrocyanic	HCN	4.9×10^{-10}		
Hydrofluoric	HF	3.5×10^{-4}		
Hydrogen peroxide	H_2O_2	2.4×10^{-12}		
Hydrosulfuric	H_2S	1.0×10^{-7}	$\approx 10^{-19}$	
Hypobromous	$HOBr$	2.0×10^{-9}		
Hypochlorous	$HOCl$	3.5×10^{-8}		
Hypoiodous	HOI	2.3×10^{-11}		
Iodic	HIO_3	1.7×10^{-1}		
Lactic	$HC_3H_5O_3$	1.4×10^{-4}		
Nitrous	HNO_2	4.5×10^{-4}		
Oxalic	$H_2C_2O_4$	5.9×10^{-2}	6.4×10^{-5}	
Phenol	C_6H_5OH	1.3×10^{-10}		
Phosphoric	H_3PO_4	7.5×10^{-3}	6.2×10^{-8}	4.8×10^{-13}
Phosphorous	H_3PO_3	1.0×10^{-2}	2.6×10^{-7}	
Saccharin	$C_7H_5NO_3S$	2.1×10^{-12}		
Selenic	H_2SeO_4	Very large	1.2×10^{-2}	
Selenious	H_2SeO_3	3.5×10^{-2}	5×10^{-8}	
Sulfuric	H_2SO_4	Very large	1.2×10^{-2}	
Sulfurous	H_2SO_3	1.5×10^{-2}	6.3×10^{-8}	
Tartaric	$C_4H_6O_6$	1.0×10^{-3}	4.6×10^{-5}	
Water	H_2O	1.8×10^{-16}		

TABLE C.2 Acid-Dissociation Constants at 25°C for Hydrated Metal Cations

Cation	K_a	Cation	K_a
$Fe^{2+}(aq)$	3.2×10^{-10}	$Be^{2+}(aq)$	3×10^{-7}
$Co^{2+}(aq)$	1.3×10^{-9}	$Al^{3+}(aq)$	1.4×10^{-5}
$Ni^{2+}(aq)$	2.5×10^{-11}	$Cr^{3+}(aq)$	1.6×10^{-4}
$Zn^{2+}(aq)$	2.5×10^{-10}	$Fe^{3+}(aq)$	6.3×10^{-3}

Note: As an example, K_a for $Fe^{2+}(aq)$ is the equilibrium constant for the reaction

$$Fe(H_2O)_6^{2+}(aq) + H_2O(l) \rightleftharpoons H_3O^+(aq) + Fe(H_2O)_5(OH)^+(aq)$$

TABLE C.3   Base-Dissociation Constants at 25°C

Base	Formula	K_b
Ammonia	NH_3	1.8×10^{-5}
Aniline	$C_6H_5NH_2$	4.3×10^{-10}
Codeine	$C_{18}H_{21}NO_3$	1.6×10^{-6}
Dimethylamine	$(CH_3)_2NH$	5.4×10^{-4}
Ethylamine	$C_2H_5NH_2$	6.4×10^{-4}
Hydrazine	N_2H_4	8.9×10^{-7}
Hydroxylamine	NH_2OH	9.1×10^{-9}
Methylamine	CH_3NH_2	3.7×10^{-4}
Morphine	$C_{17}H_{19}NO_3$	1.6×10^{-6}
Piperidine	$C_5H_{11}N$	1.3×10^{-3}
Propylamine	$C_3H_7NH_2$	5.1×10^{-4}
Pyridine	C_5H_5N	1.8×10^{-9}
Strychnine	$C_{21}H_{22}N_2O_2$	1.8×10^{-6}
Trimethylamine	$(CH_3)_3N$	6.5×10^{-5}

TABLE C.4   Solubility Product Constants at 25°C

Compound	Formula	K_{sp}
Aluminum hydroxide	$Al(OH)_3$	1.9×10^{-33}
Barium carbonate	$BaCO_3$	2.6×10^{-9}
Barium chromate	$BaCrO_4$	1.2×10^{-10}
Barium fluoride	BaF_2	1.8×10^{-7}
Barium hydroxide	$Ba(OH)_2$	5.0×10^{-3}
Barium sulfate	$BaSO_4$	1.1×10^{-10}
Cadmium carbonate	$CdCO_3$	6.2×10^{-12}
Cadmium hydroxide	$Cd(OH)_2$	5.3×10^{-15}
Calcium carbonate	$CaCO_3$	5.0×10^{-9}
Calcium fluoride	CaF_2	1.5×10^{-10}
Calcium hydroxide	$Ca(OH)_2$	4.7×10^{-6}
Calcium phosphate	$Ca_3(PO_4)_2$	2.1×10^{-33}
Calcium sulfate	$CaSO_4$	7.1×10^{-5}
Chromium(III) hydroxide	$Cr(OH)_3$	6.7×10^{-31}
Cobalt(II) hydroxide	$Co(OH)_2$	1.1×10^{-15}
Copper(I) bromide	$CuBr$	6.3×10^{-9}
Copper(I) chloride	$CuCl$	1.7×10^{-7}
Copper(II) carbonate	$CuCO_3$	2.5×10^{-10}
Copper(II) hydroxide	$Cu(OH)_2$	1.6×10^{-19}
Copper(II) phosphate	$Cu_3(PO_4)_2$	1.4×10^{-37}
Iron(II) hydroxide	$Fe(OH)_2$	4.9×10^{-17}
Iron(III) hydroxide	$Fe(OH)_3$	2.6×10^{-39}
Lead(II) bromide	$PbBr_2$	6.6×10^{-6}
Lead(II) chloride	$PbCl_2$	1.2×10^{-5}
Lead(II) chromate	$PbCrO_4$	2.8×10^{-13}
Lead(II) iodide	PbI_2	8.5×10^{-9}
Lead(II) sulfate	$PbSO_4$	1.8×10^{-8}
Magnesium carbonate	$MgCO_3$	6.8×10^{-6}
Magnesium fluoride	MgF_2	7.4×10^{-11}
Magnesium hydroxide	$Mg(OH)_2$	5.6×10^{-12}
Manganese(II) carbonate	$MnCO_3$	2.2×10^{-11}
Manganese(II) hydroxide	$Mn(OH)_2$	2.1×10^{-13}
Mercury(I) bromide	Hg_2Br_2	6.4×10^{-23}

TABLE C.4   Solubility Product Constants at 25°C (*continued*)

Compound	Formula	K_{sp}
Mercury(I) chloride	Hg_2Cl_2	1.4×10^{-18}
Mercury(I) iodide	Hg_2I_2	5.3×10^{-29}
Mercury(II) hydroxide	$Hg(OH)_2$	3.1×10^{-26}
Nickel(II) hydroxide	$Ni(OH)_2$	5.5×10^{-16}
Silver bromide	$AgBr$	5.4×10^{-13}
Silver carbonate	Ag_2CO_3	8.4×10^{-12}
Silver chloride	$AgCl$	1.8×10^{-10}
Silver chromate	Ag_2CrO_4	1.1×10^{-12}
Silver cyanide	$AgCN$	6.0×10^{-17}
Silver iodide	AgI	8.5×10^{-17}
Silver sulfate	Ag_2SO_4	1.2×10^{-5}
Silver sulfite	Ag_2SO_3	1.5×10^{-14}
Strontium carbonate	$SrCO_3$	5.6×10^{-10}
Tin(II) hydroxide	$Sn(OH)_2$	5.4×10^{-27}
Zinc carbonate	$ZnCO_3$	1.2×10^{-10}
Zinc hydroxide	$Zn(OH)_2$	4.1×10^{-17}

TABLE C.5   Solubility Products in Acid (K_{spa}) at 25°C

Compound	Formula	K_{spa}
Cadmium sulfide	CdS	8×10^{-7}
Cobalt(II) sulfide	CoS	3
Copper(II) sulfide	CuS	6×10^{-16}
Iron(II) sulfide	FeS	6×10^{2}
Lead(II) sulfide	PbS	3×10^{-7}
Manganese(II) sulfide	MnS	3×10^{10}
Mercury(II) sulfide	HgS	2×10^{-32}
Nickel(II) sulfide	NiS	8×10^{-1}
Silver sulfide	Ag_2S	6×10^{-30}
Tin(II) sulfide	SnS	1×10^{-5}
Zinc sulfide	ZnS	3×10^{-2}

Note: K_{spa} for MS is the equilibrium constant for the reaction

$$MS(s) + 2 H_3O^+(aq) \rightleftharpoons M^{2+}(aq) + H_2S(aq) + 2 H_2O(l)$$

We use K_{spa} for metal sulfides rather than K_{sp} because the traditional values of K_{sp} are now known to be inccorrect since they are based on a K_{a2} value for H_2S that is greatly in error (see R. J. Myers, J. Chem. Ed., **1986**, *63*, 687–690).

TABLE C.6   Formation Constants for Complex Ions at 25°C

Complex Ion	K_f	Complex Ion	K_f
$Ag(CN)_2^-$	3.0×10^{20}	$Ga(OH)_4^-$	3×10^{39}
$Ag(NH_3)_2^+$	1.7×10^{7}	$Ni(CN)_4^{2-}$	1.7×10^{30}
$Ag(S_2O_3)_2^{3-}$	4.7×10^{13}	$Ni(NH_3)_6^{2+}$	2.0×10^{8}
$Al(OH)_4^-$	3×10^{33}	$Ni(en)_3^{2+}$	4×10^{17}
$Be(OH)_4^{2-}$	4×10^{18}	$Pb(OH)_3^-$	8×10^{13}
$Cr(OH)_4^-$	8×10^{29}	$Sn(OH)_3^-$	3×10^{25}
$Cu(NH_3)_4^{2+}$	5.6×10^{11}	$Zn(CN)_4^{2-}$	4.7×10^{19}
$Fe(CN)_6^{4-}$	3×10^{35}	$Zn(NH_3)_4^{2+}$	7.8×10^{8}
$Fe(CN)_6^{3-}$	4×10^{43}	$Zn(OH)_4^{2-}$	3×10^{15}

D Standard Reduction Potentials at 25°C

Half-Reaction	$E°$ (V)
$F_2(g) + 2\ e^- \rightarrow 2\ F^-(aq)$	2.87
$O_3(g) + 2\ H^+(aq) + 2\ e^- \rightarrow O_2(g) + H_2O(l)$	2.08
$S_2O_8^{2-}(aq) + 2\ e^- \rightarrow 2\ SO_4^{2-}(aq)$	2.01
$Co^{3+}(aq) + e^- \rightarrow Co^{2+}(aq)$	1.81
$H_2O_2(aq) + 2\ H^+(aq) + 2\ e^- \rightarrow 2\ H_2O(l)$	1.78
$Ce^{4+}(aq) + e^- \rightarrow Ce^{3+}(aq)$	1.72
$MnO_4^-(aq) + 4\ H^+(aq) + 3\ e^- \rightarrow MnO_2(s) + 2\ H_2O(l)$	1.68
$PbO_2(s) + 3\ H^+(aq) + HSO_4^-(aq) + 2\ e^- \rightarrow PbSO_4(s) + 2\ H_2O(l)$	1.628
$2\ HClO\ (aq) + 2\ H^+(aq) + 2\ e^- \rightarrow Cl_2(g) + 2\ H_2O(l)$	1.61
$MnO_4^-(aq) + 8\ H^+(aq) + 5\ e^- \rightarrow Mn^{2+}(aq) + 4\ H_2O(l)$	1.51
$2\ BrO_3^-(aq) + 12\ H^+(aq) + 10\ e^- \rightarrow Br_2(l) + 6\ H_2O(l)$	1.48
$Cl_2(g) + 2\ e^- \rightarrow 2\ Cl^-(aq)$	1.36
$Cr_2O_7^{2-}(aq) + 14\ H^+(aq) + 6\ e^- \rightarrow 2\ Cr^{3+}(aq) + 7\ H_2O(l)$	1.33
$O_2(g) + 4\ H^+(aq) + 4\ e^- \rightarrow 2\ H_2O(l)$	1.23
$MnO_2(s) + 4\ H^+(aq) + 2\ e^- \rightarrow Mn^{2+}(aq) + 2\ H_2O(l)$	1.22
$2\ IO_3^-(aq) + 12\ H^+(aq) + 10\ e^- \rightarrow I_2(s) + 6\ H_2O(l)$	1.20
$Br_2(l) + 2\ e^- \rightarrow 2\ Br^-(aq)$	1.09
$HNO_2(aq) + H^+(aq) + e^- \rightarrow NO(g) + H_2O(l)$	0.98
$NO_3^-(aq) + 4\ H^+(aq) + 3\ e^- \rightarrow NO(g) + 2\ H_2O(l)$	0.96
$2\ Hg^{2+}(aq) + 2\ e^- \rightarrow Hg_2^{2+}(aq)$	0.92
$HO_2^-(aq) + H_2O(l) + 2\ e^- \rightarrow 3\ OH^-$	0.88
$Hg^{2+}(aq) + 2\ e^- \rightarrow Hg(l)$	0.85
$ClO^-(aq) + H_2O(l) + 2\ e^- \rightarrow Cl^-(aq) + 2\ OH^-(aq)$	0.81
$Ag^+(aq) + e^- \rightarrow Ag(s)$	0.80
$Hg_2^{2+}(aq) + 2\ e^- \rightarrow 2\ Hg(l)$	0.80
$Fe^{3+}(aq) + e^- \rightarrow Fe^{2+}(aq)$	0.77
$O_2(g) + 2\ H^+(aq) + 2\ e^- \rightarrow H_2O_2(aq)$	0.70
$MnO_4^-(aq) + e^- \rightarrow MnO_4^{2-}(aq)$	0.56
$I_2(s) + 2\ e^- \rightarrow 2\ I^-(aq)$	0.54
$Cu^+(aq) + e^- \rightarrow Cu(s)$	0.52
$H_2SO_3(aq) + 4\ H^+(aq) + 4\ e^- \rightarrow S(s) + 3\ H_2O(l)$	0.45
$O_2(g) + 2\ H_2O(l) + 4\ e^- \rightarrow 4\ OH^-(aq)$	0.40
$Cu^{2+}(aq) + 2\ e^- \rightarrow Cu(s)$	0.34
$Hg_2Cl_2(s) + 2\ e^- \rightarrow 2\ Hg(l) + 2\ Cl^-(aq)$	0.28
$AgCl(s) + e^- \rightarrow Ag(s) + Cl^-(aq)$	0.22
$SO_4^{2-}(aq) + 4\ H^+(aq) + 2\ e^- \rightarrow H_2SO_3(aq) + H_2O(l)$	0.17

Half-Reaction	$E°$ (V)
$Cu^{2+}(aq) + e^- \rightarrow Cu^+(aq)$	0.15
$Sn^{4+}(aq) + 2\,e^- \rightarrow Sn^{2+}(aq)$	0.15
$S(s) + 2\,H^+(aq) + 2\,e^- \rightarrow H_2S(g)$	0.14
$AgBr(s) + e^- \rightarrow Ag(s) + Br^-(aq)$	0.07
$2\,H^+(aq) + 2\,e^- \rightarrow H_2(g)$	0
$Fe^{3+}(aq) + 3\,e^- \rightarrow Fe(s)$	−0.04
$Pb^{2+}(aq) + 2\,e^- \rightarrow Pb(s)$	−0.13
$Sn^{2+}(aq) + 2\,e^- \rightarrow Sn(s)$	−0.14
$AgI(s) + e^- \rightarrow Ag(s) + I^-(aq)$	−0.15
$Ni^{2+}(aq) + 2\,e^- \rightarrow Ni(s)$	−0.26
$Co^{2+}(aq) + 2\,e^- \rightarrow Co(s)$	−0.28
$PbSO_4(s) + H^+(aq) + 2\,e^- \rightarrow Pb(s) + HSO_4^-(aq)$	−0.296
$Cd^{2+}(aq) + 2\,e^- \rightarrow Cd(s)$	−0.40
$Cr^{3+}(aq) + e^- \rightarrow Cr^{2+}(aq)$	−0.41
$Fe^{2+}(aq) + 2\,e^- \rightarrow Fe(s)$	−0.45
$2\,CO_2(g) + 2\,H^+(aq) + 2\,e^- \rightarrow H_2C_2O_4(aq)$	−0.49
$Cr^{3+}(aq) + 3\,e^- \rightarrow Cr(s)$	−0.74
$Zn^{2+}(aq) + 2\,e^- \rightarrow Zn(s)$	−0.76
$2\,H_2O(l) + 2\,e^- \rightarrow H_2(g) + 2\,OH^-(aq)$	−0.83
$Cr^{2+}(aq) + 2\,e^- \rightarrow Cr(s)$	−0.91
$Mn^{2+}(aq) + 2\,e^- \rightarrow Mn(s)$	−1.18
$Al^{3+}(aq) + 3\,e^- \rightarrow Al(s)$	−1.66
$Mg^{2+}(aq) + 2\,e^- \rightarrow Mg(s)$	−2.37
$Na^+(aq) + e^- \rightarrow Na(s)$	−2.71
$Ca^{2+}(aq) + 2\,e^- \rightarrow Ca(s)$	−2.87
$K^+(aq) + e^- \rightarrow K(s)$	−2.93
$Li^+(aq) + e^- \rightarrow Li(s)$	−3.04

Answers to Selected Problems

Chapter 1

1.1 (a) Cu; (b) Pt; (c) Pu   **1.2** (a) silver; (b) rhodium; (c) rhenium; (d) cesium; (e) argon; (f) arsenic   **1.3** (a) Ti, metal; (b) Te, semimetal; (c) Se, nonmetal; (d) Sc, metal; (e) At, semimetal; (f) Ar, nonmetal   **1.4** copper (Cu), silver (Ag), and gold (Au)   **1.5** (a) 3.72×10^{-10} m; (b) 1.5×10^{11} m   **1.6** (a) microgram; (b) decimeter; (c) picosecond; (d) kiloampere; (e) millimole   **1.7** 37.0°C; 310.2 K   **1.8** (a) 195 K; (b) 316°F; (c) 215°F   **1.9** 2.212 g/cm^3 **1.10** 6.32 mL   **1.11** The results are both precise and accurate.   **1.12** (a) 5 significant figures; (b) 6 significant figures; (c) 1, 2, 3, or 4 significant figures; (d) 3 significant figures; (e) 18 is an exact number.   **1.13** (a) 3.774 L; (b) 255 K; (c) 55.26 kg   **1.14** (a) 24.612 g; (b) 1.26×10^3 g/L; (c) 41.1 mL   **1.15** 32.6°C   **1.16** (a) Calculation: 1947°F; (b) Calculation: 6×10^{-11} cm^3 **1.17** 8.88 g; 0.313 ounces   **1.18** An LD$_{50}$ value is the amount of a substance per kilogram of body weight that is a lethal dose for 50% of the test animals.   **1.19** 300 g   *Understanding Key Concepts*   **1.20**

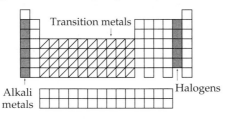

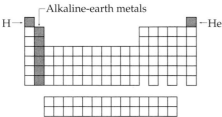

1.22 red—gas; blue—42; green—sodium   **1.24** (a) good precision, poor accuracy; (b) good precision, good accuracy; (c) poor precision, poor accuracy
1.26

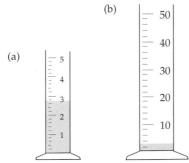

The 5 mL graduated cylinder will give more accurate measurements.   ***Additional Problems***   **1.28** 115 elements are presently known. About 90 elements occur naturally.   **1.30** There are 18 groups in the periodic table: 1A, 2A, 3B, 4B, 5B, 6B, 7B, 8B (3 groups), 1B, 2B, 3A, 4A, 5A, 6A, 7A, 8A   **1.32**

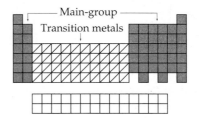

1.34

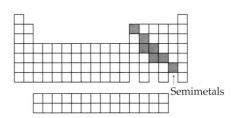

1.36 Li, Na, K, Rb, and Cs   **1.38** F, Cl, Br, and I   **1.40** (a) Gd; (b) Ge; (c) Tc; (d) As   **1.42** (a) tellurium; (b) rhenium; (c) beryllium; (d) argon; (e) plutonium   **1.44** (a) Tin is Sn: Ti is titanium; (b) Manganese is Mn: Mg is magnesium; (c) Potassium is K: Po is polonium; (d) The symbol for helium is He.   **1.46** Accurate measurement is crucial in science because, if our experiments are to be reproducible, we must be able to describe fully the substances we are working with—their amounts, sizes, temperatures, etc.   **1.48** Mass measures the amount of matter in an object, whereas weight measures the pull of gravity on an object by the earth or other celestial body.   **1.50** (a) kilogram, kg; (b) meter, m; (c) kelvin, K; (d) cubic meter, m^3 **1.52** A Celsius degree is larger than a Fahrenheit degree by a factor of 9/5.   **1.54** The volume of a cubic decimeter (dm^3) and a liter (L) are the same.   **1.56** (a) and (b) are exact; (c) is not exact.   **1.58** cL is centiliter (10^{-2} L)   **1.60** 1×10^9 pg/mg; 3.5×10^4 pg/35 ng   **1.62** (a) 5 pm = 5×10^{-10} cm = 5×10^{-3} nm; (b) 8.5 cm^3 = 8.5×10^{-6} m^3 = 8.5×10^3 mm^3; (c) 65.2 mg = 0.0652 g = 6.52×10^{10} pg   **1.64** (a) 6 significant figures; (b) 6 significant figures; (c) 4 significant figures; (d) 3 significant figures; (e) 2, 3, 4, or 5 significant figures; (f) 5 significant figures   **1.66** 3.6665×10^6 **1.68** (a) 4.5332×10^2 mg; (b) 4.21×10^{-5} mL; (c) 6.67×10^5 g   **1.70** (a) 3.567×10^4 or 35,670 m (4 significant figures), 35,670.1 m (6 significant figures); (b) 69 g (2 significant figures), 68.5 g (3 significant figures); (c) 4.99×10^3 cm (3 significant figures); (d) 2.3098×10^{-4} kg (5 significant figures)   **1.72** (a) 10.0; (b) 26; (c) 0.039; (d) 5526;

(e) 87.6; (f) 13 **1.74** (a) 110 g; (b) 443.2 m; (c) 7.6181×10^{12} m^2 **1.76** (a) 43,560 ft^3; (b) 3.92×10^8 acre-ft **1.78** (a) 2000 mg/L; (b) 2000 µg/mL; (c) 2 g/L; (d) 2000 ng/µL; (e) 10 g **1.80** 0.61 cm/shake **1.82** 103.8°F; 72.0°F **1.84** 3422°C; 3695 K **1.86** (a) 1.021 °E/°C; (b) 0.5675 °E/°F; (c) H$_2$O melting point = 119.8°E, H$_2$O boiling point = 220.0°E; (d) 157.6°E; (e) Since the outside temperature is 50.0°F, I would wear a sweater or light jacket. **1.88** 0.18 cm^3; 162,000 cm^3 **1.90** 11 g/cm^3 **1.92** 2.33 g/cm^3 **1.94** (a) Se; (b) Re; (c) Co; (d) Rh **1.96** mp = 801°C = 1474°F; bp = 1413°C = 2575°F **1.98** 75.85 mL **1.100** 2.357×10^{10} L **1.102** The Celsius and Fahrenheit scales "cross" at –40°C (–40°F). **1.104** 34.1°C **1.106** 45.9 g **1.108** (a) a metal; (b) indium; (c) 5.904 g/cm^3 (d) 664.2°G.

Chapter 2

2.1 3/2 **2.2** 2×10^4 Au atoms **2.3** ~40 times **2.4** 34 p, 34 e⁻, and 41 n **2.5** $^{35}_{17}$Cl has 18 n; $^{37}_{17}$Cl has 20 n **2.6** $^{109}_{47}$Ag **2.7** 63.55 amu **2.8** 2.04×10^{22} Cu atoms **2.9** (a) represents a mixture; (c) represents a pure compound; (b) represents a diatomic element. **2.10** (b) represents a collection of H$_2$O$_2$ molecules.
2.11

$$H-\underset{\underset{H}{|}}{\overset{\overset{H}{|}}{C}}-\underset{}{\overset{\overset{H}{|}}{N}}-H$$

2.12 C$_9$H$_{13}$NO$_3$ **2.13** (a) ionic; (b) molecular; (c) molecular; (d) ionic. **2.14** (a) represents an ionic compound because there are no discrete molecules. (b) represents a molecular compound because discrete molecules are present. **2.15** (a) HF is an acid. In water, HF dissociates to produce H⁺(aq); (b) Ca(OH)$_2$ is a base. In water, Ca(OH)$_2$ dissociates to produce OH⁻(aq); (c) LiOH is a base. In water, LiOH dissociates to produce OH⁻(aq); (d) HCN is an acid. In water, HCN dissociates to produce H⁺(aq). **2.16** (a) cesium fluoride; (b) potassium oxide; (c) copper(II) oxide; (d) barium sulfide; (e) beryllium bromide **2.17** (a) VCl$_3$; (b) MnO$_2$; (c) CuS; (d) Al$_2$O$_3$ **2.18** (a) nitrogen trichloride; (b) tetraphosphorus hexoxide; (c) disulfur difluoride **2.19** (a) S$_2$Cl$_2$; (b) ICl; (c) NI$_3$ **2.20** (a) calcium hypochlorite; (b) silver(I) thiosulfate or silver thiosulfate; (c) sodium dihydrogen phosphate; (d) tin(II) nitrate; (e) lead(IV) acetate **2.21** (a) Li$_3$PO$_4$; (b) Mg(HSO$_4$)$_2$; (c) Mn(NO$_3$)$_2$; (d) Cr$_2$(SO$_4$)$_3$ **2.22** CaCl$_2$ **2.23** (a) periodic acid; (b) bromous acid; (c) chromic acid **2.24** The image obtained with a scanning tunneling microscope is a three-dimensional, computer-generated data plot that uses tunneling current to mimic depth perception. *Understanding Key Concepts* **2.26** (d) **2.28** (a) C$_3$H$_7$NO$_2$; (b) C$_2$H$_6$O$_2$; (c) C$_2$H$_4$O$_2$
2.30

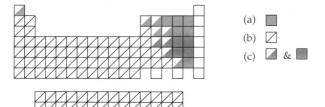

(a) ▨
(b) ▱
(c) ◩ & ▨

Additional Problems **2.32** The law of mass conservation states that chemical reactions only rearrange the way that atoms are combined; the atoms themselves are not changed. The law of definite proportions states that the chemical combination of elements to make different substances occurs when atoms join together in small, whole-number ratios.
2.34

$$\frac{\text{C:H mass ratio in benzene}}{\text{C:H mass ratio in ethane}} = \frac{12}{4.0} = \frac{3}{1}$$

$$\frac{\text{C:H mass ratio in benzene}}{\text{C:H mass ratio in ethylene}} = \frac{12}{6.0} = \frac{2}{1}$$

$$\frac{\text{C:H mass ratio in ethylene}}{\text{C:H mass ratio in ethane}} = \frac{6.0}{4.0} = \frac{3}{2}$$

2.36 (a) benzene, CH; ethane, CH$_3$; ethylene, CH$_2$; (b) These ratios are consistent with their modern formulas. **2.38** (a) 1.01 g. This result is numerically equal to the atomic mass of H in grams. (b) 16.0 g. This result is numerically equal to the atomic mass of O in grams. **2.40** 1 Zn/1 S **2.42** The atomic number is equal to the number of protons. The mass number is equal to the sum of the number of protons and the number of neutrons. **2.44** Atoms of the same element that have different numbers of neutrons are called isotopes. **2.46** The subscript giving the atomic number of an atom is often left off of an isotope symbol because one can readily look up the atomic number in the periodic table. **2.48** (a) carbon, C; (b) argon, Ar; (c) vanadium, V **2.50** (a) $^{220}_{86}$Rn; (b) $^{210}_{84}$Po; (c) $^{197}_{79}$Au **2.52** (a) 7 p, 7 e⁻, 8 n; (b) 27 p, 27 e⁻, 33 n; (c) 53 p, 53 e⁻, 78 n **2.54** (a) $^{24}_{12}$Mg, magnesium; (b) $^{58}_{28}$Ni, nickel **2.56** 10.8 amu **2.58** 25.982 (^{26}Mg) **2.60** (a) heterogeneous; (b) heterogeneous; (c) homogeneous; (d) homogeneous **2.62** An atom is the smallest particle that retains the chemical properties of an element. A molecule is matter that results when two or more atoms are joined by covalent bonds. H and O are atoms, H$_2$O is a water molecule. **2.64** A covalent bond results when two atoms share several (usually two) of their electrons. An ionic bond results from a complete transfer of one or more electrons from one atom to another. The C–H bonds in methane (CH$_4$) are covalent bonds. The bond in NaCl (Na⁺Cl⁻) is an ionic bond. **2.66** Element symbols are composed of one or two letters. If the element symbol is two letters, the first letter is uppercase and the second letter is lowercase. CO stands for carbon and oxygen in carbon monoxide. **2.68** (a) 4 p, 2 e⁻; (b) 37 p, 36 e⁻; (c) 34 p, 36 e⁻; (d) 79 p, 76 e⁻ **2.70** C$_3$H$_8$O
2.72

$$H-\underset{\underset{H}{|}}{\overset{\overset{H}{|}}{C}}-\underset{\underset{H}{|}}{\overset{\overset{H}{|}}{C}}-\underset{\underset{H}{|}}{\overset{\overset{H}{|}}{C}}-\underset{\underset{H}{|}}{\overset{\overset{H}{|}}{C}}-H$$

2.74 (a) acid; (b) base; (c) acid; (d) base; (e) acid **2.76** (a) I⁻; (b) mainly H$_2$PO$_4$⁻; (c) mainly HCO$_3$⁻ **2.78** (a) KCl; (b) SnBr$_2$; (c) CaO; (d) BaCl$_2$; (e) AlH$_3$ **2.80** (a) barium ion; (b) cesium ion; (c) vanadium(III) ion; (d) hydrogen carbonate ion; (e) ammonium ion; (f) nickel(II) ion; (g) nitrite ion; (h) chlorite ion; (i) manganese(II) ion; (j) perchlorate ion **2.82** (a) SO$_3$²⁻; (b) PO$_4$³⁻; (c) Zr⁴⁺; (d) CrO$_4$²⁻;

(e) $CH_3CO_2^-$; (f) $S_2O_3^{2-}$ **2.84** (a) zinc(II) cyanide; (b) iron(III) nitrite; (c) titanium(IV) sulfate; (d) tin(II) phosphate; (e) mercury(I) sulfide; (f) manganese(IV) oxide; (g) potassium periodate; (h) copper(II) acetate   **2.86** (a) Na_2SO_4; (b) $Ba_3(PO_4)_2$; (c) $Ga_2(SO_4)_3$ **2.88** 72.6 amu   **2.90** (a) sodium bromate; (b) phosphoric acid; (c) phosphorous acid; (d) vanadium(V) oxide   **2.92** For NH_3, 0.505 g H; for N_2H_4, 0.337 g H   **2.94** TeO_4^{2-}, tellurate; TeO_3^{2-}, tellurite. TeO_4^{2-} and TeO_3^{2-} are analogous to SO_4^{2-} and SO_3^{2-}.   **2.96** (a) I^-; (b) Au^{3+}; (c) Kr   **2.98** 39.9641 amu   **2.100** (a) ^{40}Ca; (b) Not enough information, several different isotopes can have 63 neutrons. (c) $^{56}Fe^{3+}$; (d) Se^{2-} **2.102** $^1H^{35}Cl$ has 18 p, 18 n, and 18 e$^-$; $^1H^{37}Cl$ has 18 p, 20 n, and 18 e$^-$; $^2H^{35}Cl$ has 18 p, 19 n, and 18 e$^-$; $^2H^{37}Cl$ has 18 p, 21 n, and 18 e$^-$; $^3H^{35}Cl$ has 18 p, 20 n, and 18 e$^-$; $^3H^{37}Cl$ has 18 p, 22 n, and 18 e$^-$

2.104

2.106 151.165 amu   **2.108** (a) a molecular compound; (b) $C_9H_8O_4$

Chapter 3

3.1 $2 KClO_3 \rightarrow 2 KCl + 3 O_2$ **3.2** (a) $C_6H_{12}O_6 \rightarrow 2 C_2H_6O + 2 CO_2$; (b) $4 Fe + 3 O_2 \rightarrow 2 Fe_2O_3$; (c) $4 NH_3 + Cl_2 \rightarrow N_2H_4 + 2 NH_4Cl$ **3.3** $3 A_2 + 2 B \rightarrow 2 BA_3$ **3.4** (a) 159.7 amu; (b) 98.1 amu; (c) 192.1 amu; (d) 334.4 amu   **3.5** $Fe_2O_3(s) + 3 CO(g) \rightarrow 2 Fe(s) + 3 CO_2(g)$; 1.50 mol CO   **3.6** $C_5H_{11}NO_2S$; 149.24 amu   **3.7** 2.77×10^{-3} mol; 1.67×10^{21} molecules   **3.8** 3.33 g $C_4H_6O_3$; 5.87 g $C_9H_8O_4$; 1.96 g $C_2H_4O_2$ **3.9** 63%   **3.10** 4220 g   **3.11** Li_2O is the limiting reactant; 41 kg H_2O **3.12** 921 g CO_2 **3.13** (a) $A + B_2 \rightarrow AB_2$. A is the limiting reactant. (b) 1.0 mol of AB_2 **3.14** (a) 0.025 mol; (b) 1.62 mol   **3.15** (a) 25.0 g; (b) 67.6 g   **3.16** 690 mL   **3.17** 1 g   **3.18** 0.656 M   **3.19** Dilute 6.94 mL of 18.0 M H_2SO_4 with enough water to make 250.0 mL of solution.   **3.20** 10.0 mL   **3.21** 5.47×10^{-2} M   **3.22** 0.758 M   **3.23** CH_4N; 39.9% C, 13.4% H, 46.6% N   **3.24** $MgCO_3$ **3.25** $C_{10}H_{20}O$ **3.26** $C_5H_{10}O_5$ **3.27** (a) B_2H_6; (b) $C_3H_6O_3$ **3.28** the assumptions that (i) the oil molecules are tiny cubes, (ii) the oil layer is one molecule thick, (iii) the molecular mass of the oil is 200   **3.29** 2.3×10^{23} molecules/mole   *Understanding Key Concepts*   **3.30** box (b)   **3.32** $C_2H_4 + 3 O_2 \rightarrow 2 CO_2 + 2 H_2O$

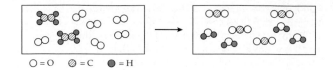

O = O   ⊘ = C   ● = H

3.34 $C_{17}H_{18}F_3NO$; 309.36 amu   **3.36** 0.67 M   *Additional Problems*   **3.38** (a) is not balanced, (b) is balanced.   **3.40** (a) $Mg + 2 HNO_3 \rightarrow H_2 + Mg(NO_3)_2$; (b) $CaC_2 + 2 H_2O \rightarrow Ca(OH)_2 + C_2H_2$; (c) $2 S + 3 O_2 \rightarrow 2 SO_3$; (d) $UO_2 + 4 HF \rightarrow UF_4 + 2 H_2O$ **3.42** Hg_2Cl_2: 472.1 amu; $C_4H_8O_2$: 88.1 amu; CF_2Cl_2: 120.9 amu   **3.44** (a) 47.88 g; (b) 159.81 g; (c) 200.59 g; (d) 18.02 g   **3.46** 5.0 mol   **3.48** 0.867 mol   **3.50** 119 amu   **3.52** 1.97×10^{-3} mol $FeSO_4$; 1.19×10^{21} Fe(II) atoms   **3.54** 6.44×10^{-4} mol; 3.88×10^{20} molecules   **3.56** (a) 0.14 mol; (b) 0.0051 mol; (c) 2.7×10^{-3} mol   **3.58** 167 kg   **3.60** (a) $2 Fe_2O_3 + 3 C \rightarrow 4 Fe + 3 CO_2$; (b) 4.93 mol C; (c) 59.2 g C   **3.62** (a) $2 Mg + O_2 \rightarrow 2 MgO$; (b) 16.5 g O_2, 41.5 g MgO; (c) 38.0 g Mg, 63.0 g MgO   **3.64** (a) $2 HgO \rightarrow 2 Hg + O_2$; (b) 42.1 g Hg, 3.36 g O_2; (c) 451 g HgO   **3.66** AgCl   **3.68** 15.8 g NH_3; 83.3 g N_2 left over   **3.70** 5.22 g $C_2H_4Cl_2$ **3.72** 0.526 L CO_2; $CaCO_3$ is the limiting reactant.   **3.74** 3.2 g   **3.76** 86.8%   **3.78** (a) 0.0420 mol; (b) 0.12 mol   **3.80** 160 mL   **3.82** 0.0685 M   **3.84** Na^+, 0.147 M; Ca^{2+}, 0.002 98 M; K^+, 0.004 02 M; Cl^-, 0.157 M   **3.86** 1.71 M   **3.88** 15.5 g   **3.90** 57.2 mL   **3.92** 20.0% C; 6.72% H; 46.6% N; 26.6% O   **3.94** SnF_2 **3.96** C_7H_8 **3.98** 13,000 amu   **3.100** Disilane is Si_2H_6.   **3.102** (a) 39.99% C; 6.71% H; 53.27% O; (b) 2.06% H; 32.70% S; 65.25% O; (c) 24.75% K; 34.77% Mn; 40.51% O; (d) 45.89% C; 2.75% H; 7.65% N; 26.20% O; 17.51% S   **3.104** (a) $SiCl_4 + 2 H_2O \rightarrow SiO_2 + 4 HCl$; (b) $P_4O_{10} + 6 H_2O \rightarrow 4 H_3PO_4$; (c) $CaCN_2 + 3 H_2O \rightarrow CaCO_3 + 2 NH_3$; (d) $3 NO_2 + H_2O \rightarrow 2 HNO_3 + NO$ **3.106** $C_{10}H_{10}Fe$ **3.108** Na^+, 0.295 M; Li^+, 0.0406 M; SO_4^{2-}, 0.0590 M; PO_4^{3-}, 0.0725 M   **3.110** High resolution mass spectrometry is capable of measuring the mass of molecules with a particular isotopic composition.   **3.112** $C_6H_{12}O_6 + 6 O_2 \rightarrow 6 CO_2 + 6 H_2O$; 97.2 g CO_2, 56.1 L CO_2 **3.114** (a) 79.9% Cu, 20.1% S; (b) Cu_2S; (c) 4.2×10^{22} Cu^+ ions/cm^3 **3.116** 5.32 g PCl_3; 4.68 g PCl_5 **3.118** The mass ratio of NH_4NO_3 to $(NH_4)_2HPO_4$ is 2 to 1.

Chapter 4

4.1 (a) precipitation; (b) redox; (c) acid–base neutralization   **4.2** 0.675 M   **4.3** A_2Y is the strongest electrolyte because it is completely dissociated into ions. A_2X is the weakest electrolyte because it is the least dissociated.   **4.4** (a) $2 Ag^+(aq) + CrO_4^{2-}(aq) \rightarrow Ag_2CrO_4(s)$; (b) $2 H^+(aq) + MgCO_3(s) \rightarrow H_2O(l) + CO_2(g) + Mg^{2+}(aq)$ **4.5** (a) insoluble; (b) insoluble; (c) soluble; (d) insoluble; (e) soluble; (f) soluble   **4.6** (a) $Ni^{2+}(aq) + S^{2-}(aq) \rightarrow NiS(s)$; (b) $Pb^{2+}(aq) + CrO_4^{2-}(aq) \rightarrow PbCrO_4(s)$; (c) $Ag^+(aq) + Br^-(aq) \rightarrow AgBr(s)$ **4.7** $3 CaCl_2(aq) + 2 Na_3PO_4(aq) \rightarrow Ca_3(PO_4)_2(s) + 6 NaCl(aq)$; $3 Ca^{2+}(aq) + 2 PO_4^{3-}(aq) \rightarrow Ca_3(PO_4)_2(s)$ **4.8** The precipitate is either $Mg_3(PO_4)_2$ or $Zn_3(PO_4)_2$.   **4.9** (a) $2 Cs^+(aq) + 2 OH^-(aq) + 2 H^+(aq) + SO_4^{2-}(aq) \rightarrow 2 Cs^+(aq) + SO_4^{2-}(aq) + 2 H_2O(l)$; $H^+(aq) + OH^-(aq) \rightarrow H_2O(l)$; (b) $Ca^{2+}(aq) + 2 OH^-(aq) + 2 CH_3CO_2H(aq) \rightarrow Ca^{2+}(aq) + 2 CH_3CO_2^-(aq) + 2 H_2O(l)$; $CH_3CO_2H(aq) + OH^-(aq) \rightarrow CH_3CO_2^-(aq) + H_2O(l)$

4.10 HY is the strongest acid; HX is the weakest acid. **4.11** (a) Cl -1, Sn $+4$; (b) O -2, Cr $+6$; (c) O -2, Cl -1, V $+5$; (d) O -2, V $+3$; (e) O -2, H $+1$, N $+5$; (f) O -2, S $+6$, Fe $+2$ **4.12** $2\,Cu^{2+}(aq) + 4\,I^-(aq) \rightarrow 2\,CuI(s) + I_2(aq)$; Cu^{2+}: $+2$; I^-: -1; CuI: Cu $+1$, I -1; I_2: 0; oxidizing agent, Cu^{2+}; reducing agent, I^- **4.13** (a) C is oxidized. C is the reducing agent. The Sn in SnO_2 is reduced. SnO_2 is the oxidizing agent. (b) Sn^{2+} is oxidized. Sn^{2+} is the reducing agent. Fe^{3+} is reduced. Fe^{3+} is the oxidizing agent. **4.14** (a) N.R.; (b) N.R. **4.15** Because B will reduce A^+, B is above A in the activity series. Because B will not reduce C^+, C is above B in the activity series. Therefore C must be above A in the activity series and C will reduce A^+. **4.16** $8\,H^+(aq) + Cr_2O_7^{2-}(aq) + I^-(aq) \rightarrow 2\,Cr^{3+}(aq) + IO_3^-(aq) + 4\,H_2O(l)$ **4.17** $H_2O(l) + 2\,MnO_4^-(aq) + Br^-(aq) \rightarrow 2\,MnO_2(s) + BrO_3^-(aq) + 2\,OH^-(aq)$ **4.18** (a) $MnO_4^-(aq) \rightarrow MnO_2(s)$, $IO_3^-(aq) \rightarrow IO_4^-(aq)$; (b) $NO_3^-(aq) \rightarrow NO_2(g)$, $SO_2(aq) \rightarrow SO_4^{2-}(aq)$ **4.19** $2\,NO_3^-(aq) + 8\,H^+(aq) + 3\,Cu(s) \rightarrow 3\,Cu^{2+}(aq) + 2\,NO(g) + 4\,H_2O(l)$ **4.20** $4\,Fe(OH)_2(s) + 2\,H_2O(l) + O_2(g) \rightarrow 4\,Fe(OH)_3(s)$ **4.21** 1.98 M **4.22** $Na_2S_2O_3$ is used to solubilize the remaining unreduced AgBr on the film so that it is no longer sensitive to light. **4.23** Light is passed through the negative image onto special photographic paper that is coated with the same kind of gelatin–AgBr emulsion used on the original film. Developing the photographic paper with hydroquinone and fixing the image with sodium thiosulfate reverses the negative image, and a final, positive image is produced. *Understanding Key Concepts* **4.24** (a) (1); (b) (2); (c) (3) **4.26** (a) (2); (b) (3); (c) (1) **4.28** C > A > D > B *Additional Problems* **4.30** (a) precipitation; (b) redox; (c) acid–base neutralization **4.32** (a) $Hg^{2+}(aq) + 2\,I^-(aq) \rightarrow HgI_2(s)$; (b)

$$2\,HgO(s) \xrightarrow{\text{Heat}} 2\,Hg(l) + O_2(g);$$

(c) $H_3PO_4(aq) + 3\,OH^-(aq) \rightarrow PO_4^{3-}(aq) + 3\,H_2O(l)$ **4.34** $Ba(OH)_2$ dissociates into $Ba^{2+}(aq)$ and $2\,OH^-(aq)$, and conducts electricity. H_2SO_4 dissociates into $H^+(aq)$ and $HSO_4^-(aq)$ and conducts electricity. When equal molar solutions of $Ba(OH)_2$ and H_2SO_4 are mixed, the insoluble $BaSO_4$ is formed along with two H_2O. In water $BaSO_4$ does not produce any appreciable amount of ions and the mixture does not conduct electricity. **4.36** (a) strong; (b) weak; (c) strong; (d) strong; (e) weak **4.38** (a) 2.25 M; (b) 1.42 M **4.40** (a) insoluble; (b) soluble; (c) insoluble; (d) insoluble **4.42** (a) No precipitate will form; (b) $Fe(OH)_2(s)$ will precipitate. (c) No precipitate will form. **4.44** (a) $Pb(NO_3)_2(aq) + Na_2SO_4(aq) \rightarrow PbSO_4(s) + 2\,NaNO_3(aq)$; (b) $3\,MgCl_2(aq) + 2\,K_3PO_4(aq) \rightarrow Mg_3(PO_4)_2(s) + 6\,KCl(aq)$; (c) $ZnSO_4(aq) + Na_2CrO_4(aq) \rightarrow ZnCrO_4(s) + Na_2SO_4(aq)$ **4.46** Add HCl(aq); it will selectively precipitate AgCl(s). **4.48** Cs^+ and/or NH_4^+ **4.50** Add the solution to an active metal, such as magnesium. Bubbles of H_2 gas indicate the presence of an acid. **4.52** (a) $2\,H^+(aq) + 2\,ClO_4^-(aq) + Ca^{2+}(aq) + 2\,OH^-(aq) \rightarrow Ca^{2+}(aq) + 2\,ClO_4^-(aq) + 2\,H_2O(l)$; (b) $CH_3CO_2H(aq) + Na^+(aq) + OH^-(aq) \rightarrow CH_3CO_2^-(aq) + Na^+(aq) + H_2O(l)$ **4.54** (a) $H^+(aq) + OH^-(aq) \rightarrow H_2O(l)$; (b) $H^+(aq) + OH^-(aq) \rightarrow H_2O(l)$ **4.56** best reducing agents, bottom left; best oxidizing agents, top right of the periodic table (excluding noble gases) **4.58** (a) gains electrons; (b) loses electrons; (c) loses electrons; (d) gains electrons **4.60**

(a) NO_2: O -2, N $+4$; (b) SO_3: O -2, S $+6$; (c) $COCl_2$: O -2, Cl -1, C $+4$; (d) CH_2Cl_2: Cl -1, H $+1$, C 0; (e) $KClO_3$: O -2, K $+1$, Cl $+5$; (f) HNO_3: O -2, H $+1$, N $+5$ **4.62** (a) ClO_3^-: O -2, Cl $+5$; (b) SO_3^{2-}: O -2, S $+4$; (c) $C_2O_4^{2-}$: O -2, C $+3$; (d) NO_2^-: O -2, N $+3$; (e) BrO^-: O -2, Br $+1$ **4.64** (a) Ca(s) is oxidized; $Sn^{2+}(aq)$ is reduced; (b) not a redox reaction **4.66** (a) N.R.; (b) N.R.; (c) N.R.; (d) $Au^{3+}(aq) + 3\,Ag(s) \rightarrow 3\,Ag^+(aq) + Au(s)$ **4.68** A > B > C > D **4.70** (a) N.R.; (b) N.R. **4.72** (a) reduction; (b) oxidation; (c) oxidation; (d) reduction **4.74** (a) $3\,e^- + 4\,H^+(aq) + NO_3^-(aq) \rightarrow NO(g) + 2\,H_2O(l)$; (b) $Zn(s) \rightarrow Zn^{2+}(aq) + 2\,e^-$; (c) $Ti^{3+}(aq) + 2\,H_2O(l) \rightarrow TiO_2(s) + 4\,H^+(aq) + e^-$; (d) $Sn^{4+}(aq) + 2\,e^- \rightarrow Sn^{2+}(aq)$ **4.76** (a) oxidation: $Te(s) \rightarrow TeO_2(s)$, reduction: $NO_3^-(aq) \rightarrow NO(g)$; (b) oxidation: $Fe^{2+}(aq) \rightarrow Fe^{3+}(aq)$, reduction: $H_2O_2(aq) \rightarrow H_2O(l)$ **4.78** (a) $14\,H^+(aq) + Cr_2O_7^{2-}(aq) + 6\,e^- \rightarrow 2\,Cr^{3+}(aq) + 7\,H_2O(l)$; (b) $4\,H_2O(l) + CrO_4^{2-}(aq) + 3\,e^- \rightarrow Cr(OH)_4^-(aq) + 4\,OH^-(aq)$; (c) $Bi^{3+}(aq) + 6\,OH^-(aq) \rightarrow BiO_3^-(aq) + 3\,H_2O(l) + 2\,e^-$; (d) $H_2O(l) + ClO^-(aq) + 2\,e^- \rightarrow Cl^-(aq) + 2\,OH^-(aq)$ **4.80** (a) $H_2O(l) + 2\,MnO_4^-(aq) + 3\,IO_3^-(aq) \rightarrow 2\,MnO_2(s) + 3\,IO_4^-(aq) + 2\,OH^-(aq)$; (b) $2\,Cu(OH)_2(s) + N_2H_4(aq) \rightarrow 2\,Cu(s) + 4\,H_2O(l) + N_2(g)$; (c) $3\,Fe(OH)_2(s) + 4\,H_2O(l) + CrO_4^{2-}(aq) \rightarrow 3\,Fe(OH)_3(s) + Cr(OH)_4^-(aq) + OH^-(aq)$; (d) $ClO_4^-(aq) + 2\,H_2O_2(aq) \rightarrow ClO_2^-(aq) + 2\,H_2O(l) + 2\,O_2(g)$ **4.82** (a) $Zn(s) + 2\,VO^{2+}(aq) + 4\,H^+(aq) \rightarrow Zn^{2+}(aq) + 2\,V^{3+}(aq) + 2\,H_2O(l)$; (b) $2\,H^+(aq) + Ag(s) + NO_3^-(aq) \rightarrow Ag^+(aq) + NO_2(g) + H_2O(l)$; (c) $3\,Mg(s) + 16\,H^+(aq) + 2\,VO_4^{3-}(aq) \rightarrow 3\,Mg^{2+}(aq) + 2\,V^{2+}(aq) + 8\,H_2O(l)$; (d) $6\,H^+(aq) + IO_3^-(aq) + 8\,I^-(aq) \rightarrow 3\,I_3^- + 3\,H_2O(l)$ **4.84** 0.670 g I_2 **4.86** 0.134 M **4.88** 80.32% **4.90** 0.101% **4.92** (a) $4\,[Fe(CN)_6]^{3-}(aq) + N_2H_4(aq) + 4\,OH^-(aq) \rightarrow 4\,[Fe(CN)_6]^{4-}(aq) + N_2(g) + 4\,H_2O(l)$; (b) $SeO_3^{2-}(aq) + Cl_2(g) + 2\,OH^-(aq) \rightarrow SeO_4^{2-}(aq) + 2\,Cl^-(aq) + H_2O(l)$; (c) $2\,CoCl_2(aq) + H_2O(l) + HO_2^-(aq) + 3\,OH^-(aq) \rightarrow 2\,Co(OH)_3(s) + 4\,Cl^-(aq)$ **4.94** (a) C_2H_6: H $+1$, C -3; (b) $Na_2B_4O_7$: O -2, Na $+1$, B $+3$; (c) Mg_2SiO_4: O -2, Mg $+2$, Si $+4$ **4.96** Ni^{2+} and Au^{3+} **4.98** $K_{sp} = [Mg^{2+}][F^-]^2 = 7.0 \times 10^{-11}$ **4.100** (a) Add HCl: $Hg_2^{2+}(aq) + 2\,Cl^-(aq) \rightarrow Hg_2Cl_2(s)$; (b) Add H_2SO_4: $Pb^{2+}(aq) + SO_4^{2-}(aq) \rightarrow PbSO_4(s)$; (c) Add Na_2CO_3: $Ca^{2+}(aq) + CO_3^{2-}(aq) \rightarrow CaCO_3(s)$; (d) Add Na_2SO_4: $Ba^{2+}(aq) + SO_4^{2-}(aq) \rightarrow BaSO_4(s)$ **4.102** (a) $2\,Mn(OH)_2(s) + H_2O_2(aq) \rightarrow 2\,Mn(OH)_3(s)$; (b) $4\,H^+(aq) + 3\,MnO_4^{2-}(aq) \rightarrow MnO_2(s) + 2\,MnO_4^-(aq) + 2\,H_2O(l)$; (c) $8\,I^-(aq) + IO_3^-(aq) + 6\,H^+(aq) \rightarrow 3\,I_3^-(aq) + 3\,H_2O(l)$; (d) $2\,H_2O(l) + 2\,P(s) + 3\,PO_4^{3-}(aq) + OH^-(aq) \rightarrow 5\,HPO_3^{2-}(aq)$ **4.104** (a) $14\,H^+(aq) + Cr_2O_7^{2-}(aq) + 6\,Cr^{2+}(aq) \rightarrow 8\,Cr^{3+}(aq) + 7\,H_2O(l)$; (b) K^+, 0.0833 M; NO_3^-, 0.617 M; H^+, 0.183 M; $Cr_2O_7^{2-}$, 0.0250 M; Cr^{3+}, 0.133 M **4.106** (a) (1) $3\,Cu(s) + 8\,H^+(aq) + 2\,NO_3^-(aq) \rightarrow 3\,Cu^{2+}(aq) + 2\,NO(g) + 4\,H_2O(l)$; (2) $2\,Cu^{2+}(aq) + 2\,SCN^-(aq) + H_2O(l) + HSO_3^-(aq) \rightarrow 2\,CuSCN(s) + HSO_4^-(aq) + 2\,H^+(aq)$; (3) $10\,Cu^+(aq) + 12\,H^+(aq) + 2\,IO_3^-(aq) \rightarrow 10\,Cu^{2+}(aq) + I_2(aq) + 6\,H_2O(l)$; (4) $I_2(aq) + 2\,S_2O_3^{2-}(aq) \rightarrow 2\,I^-(aq) + S_4O_6^{2-}(aq)$; (5) $2\,ZnNH_4PO_4 \rightarrow Zn_2P_2O_7 + H_2O + 2\,NH_3$; (b) 77.1% Cu; (c) 19.5% Zn

Chapter 5

5.1 gamma ray, 8.43×10^{18} Hz; radar wave, 2.91×10^9 Hz **5.2** 2.93 m; 3.14×10^{-10} m **5.3** (b) has the higher frequency. (b) represents the more intense beam of light. (b) represents blue light. (a) represents red light. **5.4** 397.0 nm **5.5** 1875 nm **5.6** 820.4 nm **5.7** 1310 kJ/mol

5.8 IR, 77.2 kJ/mol; UV, 479 kJ/mol; X ray, 2.18×10^4 kJ/mol　**5.9** 2.34×10^{-38} m　**5.10** 5×10^{-34} m
5.11

n	l	m_l	Orbital	No. of Orbitals
5	0	0	$5s$	1
	1	−1, 0, +1	$5p$	3
	2	−2, −1, 0, +1, +2	$5d$	5
	3	−3, −2, −1, 0, +1, +2, +3	$5f$	7
	4	−4, −3, −2, −1, 0, +1, +2, +3, +4	$5g$	9

There are 25 possible orbitals in the fifth shell.
5.12 (a) $2p$; (b) $4f$; (c) $3d$　**5.13** (a) $n = 3$, $l = 0$, $m_l = 0$; (b) $n = 2$, $l = 1$, $m_l = −1, 0, +1$; (c) $n = 4$, $l = 2$, $m_l = −2, −1, 0, +1, +2$
5.14 four nodal planes　**5.15** $n = 4$ and $l = 2$　**5.16** 1.31×10^3 kJ/mol
5.17 (a) Ti, $1s^2 2s^2 2p^6 3s^2 3p^6 4s^2 3d^2$ or [Ar] $4s^2 3d^2$

$$[Ar] \quad \frac{\uparrow\downarrow}{4s} \qquad \frac{\uparrow}{} \; \frac{\uparrow}{} \; \frac{}{} \; \frac{}{} \; \frac{}{}_{3d}$$

(b) Zn, $1s^2 2s^2 2p^6 3s^2 3p^6 4s^2 3d^{10}$ or [Ar] $4s^2 3d^{10}$

$$[Ar] \quad \frac{\uparrow\downarrow}{4s} \qquad \frac{\uparrow\downarrow}{} \; \frac{\uparrow\downarrow}{} \; \frac{\uparrow\downarrow}{} \; \frac{\uparrow\downarrow}{} \; \frac{\uparrow\downarrow}{}_{3d}$$

(c) Sn, $1s^2 2s^2 2p^6 3s^2 3p^6 4s^2 3d^{10} 4p^6 5s^2 4d^{10} 5p^2$ or [Kr] $5s^2 4d^{10} 5p^2$

$$[Kr] \quad \frac{\uparrow\downarrow}{5s} \qquad \frac{\uparrow\downarrow}{} \; \frac{\uparrow\downarrow}{} \; \frac{\uparrow\downarrow}{} \; \frac{\uparrow\downarrow}{} \; \frac{\uparrow\downarrow}{}_{4d} \qquad \frac{\uparrow}{} \; \frac{\uparrow}{} \; \frac{}{}_{5p}$$

(d) Pb, [Xe] $6s^2 4f^{14} 5d^{10} 6p^2$　**5.18** $1s^2 2s^2 2p^6$　**5.19** Ni　**5.20** Cr, Cu, Nb, Mo, Ru, Rh, Pd, Ag, La, Ce, Gd, Pt, Au, Ac, Th, Pa, U, Np, Cm　**5.21** (a) Ba; (b) W; (c) Sn; (d) Ce　**5.22** The solar "gas" of energetic protons and electrons that collide with atmospheric gases and thus produce light emission are attracted to the earth's north and south magnetic poles.　*Understanding Key Concepts*
5.24

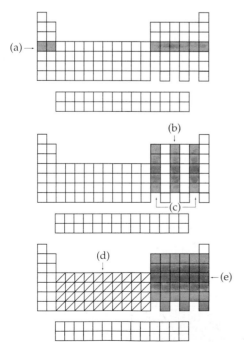

5.26 Ga　**5.28** Sr (215 pm) > Ca (197 pm) > Br (114 pm)
Additional Problems　**5.30** Violet has the higher frequency and energy. Red has the higher wavelength.　**5.32** 5.5×10^{-8} m　**5.34** (a) $\nu = 99.5$ MHz, $E = 3.97 \times 10^{-5}$ kJ/mol, $\nu = 115.0$ kHz, $E = 4.589 \times 10^{-8}$ kJ/mol. The FM radio wave (99.5 MHz) has the higher energy; (b) $\lambda = 3.44 \times 10^{-9}$ m, $E = 3.48 \times 10^4$ kJ/mol, $\lambda = 6.71 \times 10^{-2}$ m, $E = 1.78 \times 10^{-3}$ kJ/mol. The X ray ($\lambda = 3.44 \times 10^{-9}$ m) has the higher energy.　**5.36** (a) 1320 nm, near IR; (b) 0.149 m, radio wave; (c) 65.4 nm, UV　**5.38** 2.45×10^{-12} m, γ ray
5.40 1.06×10^{-34} m　**5.42** 9.14×10^{-24} m/s　**5.44** For $n = 3$, $E = 182.3$ kJ/mol; for $n = 4$, $E = 246.1$ kJ/mol; For $n = 5$, $E = 275.6$ kJ/mol　**5.46** 328.1 kJ/mol　**5.48** 363 kJ/mol
5.50 n is the principal quantum number. The size and energy level of an orbital depends on n. l is the angular-momentum quantum number. l defines the three-dimensional shape of an orbital. m_l is the magnetic quantum number. m_l defines the spatial orientation of an orbital m_s is the spin quantum number. . m_s indicates the spin of the electron and can have either of two values, +1/2 or −1/2.　**5.52** The probability of finding the electron drops off rapidly as distance from the nucleus increases, although it never drops to zero, even at large distances. As a result, there is no definite boundary or size for an orbital. However, we usually imagine the boundary surface of an orbital enclosing the volume where an electron spends 95% of its time.　**5.54** Part of the electron–nucleus attraction is canceled by the electron–electron repulsion, an effect we describe by saying that the electrons are shielded from the nucleus by the other electrons. The net nuclear charge felt by an electron is called the effective nuclear charge, Z_{eff}. $Z_{eff} = Z_{actual} −$ electron shielding　**5.56** (a) $4s$: $n = 4$; $l = 0$; $m_l = 0$; $m_s = \pm 1/2$; (b) $3p$: $n = 3$; $l = 1$; $m_l = −1, 0, +1$; $m_s = \pm 1/2$; (c) $5f$: $n = 5$; $l = 3$; $m_l = −3, −2, −1, 0, +1, +2, +3$; $m_s = \pm 1/2$; (d) $5d$: $n = 5$; $l = 2$; $m_l = −2, −1, 0, +1, +2$; $m_s = \pm 1/2$　**5.58** (a) not allowed for $l = 0$, $m_l = 0$ only; (b) allowed; (c) not allowed for $n = 4$, $l = 0, 1, 2,$ or 3 only　**5.60** The maximum number of electrons will occur when the $5g$ orbital is filled: [Rn] $7s^2 5f^{14} 6d^{10} 7p^6 8s^2 5g^{18} = 138$ electrons　**5.62** 8×10^{-31} m　**5.64** The principal quantum number n increases by 1 from one period to the next. As the principal quantum number increases, the number of orbitals in a shell increases. The progression of elements parallels the number of electrons in a particular shell.　**5.66** (a) $5d$; (b) $4s$; (c) $6s$　**5.68** (a) $3d$; (b) $4p$; (c) $6d$; (d) $6s$
5.70 (a) $1s^2 2s^2 2p^6 3s^2 3p^6 4s^2 3d^2$;
(b) $1s^2 2s^2 2p^6 3s^2 3p^6 4s^2 3d^{10} 4p^6 5s^2 4d^6$;
(c) $1s^2 2s^2 2p^6 3s^2 3p^6 4s^2 3d^{10} 4p^6 5s^2 4d^{10} 5p^2$;
(d) $1s^2 2s^2 2p^6 3s^2 3p^6 4s^2 3d^{10} 4p^6 5s^2$;
(e) $1s^2 2s^2 2p^6 3s^2 3p^6 4s^2 3d^{10} 4p^4$
5.72
(a) Rb,　[Kr] $\dfrac{\uparrow}{5s}$

(b) W,　[Xe] $\dfrac{\uparrow\downarrow}{6s} \quad \dfrac{\uparrow\downarrow}{} \dfrac{\uparrow\downarrow}{} \dfrac{\uparrow\downarrow}{} \dfrac{\uparrow\downarrow}{} \dfrac{\uparrow\downarrow}{} \dfrac{\uparrow\downarrow}{} \dfrac{\uparrow\downarrow}{}_{4f} \quad \dfrac{\uparrow}{} \dfrac{\uparrow}{} \dfrac{\uparrow}{} \dfrac{\uparrow}{}_{5d} \dfrac{}{}$

(c) Ge,　[Ar] $\dfrac{\uparrow\downarrow}{4s} \quad \dfrac{\uparrow\downarrow}{} \dfrac{\uparrow\downarrow}{} \dfrac{\uparrow\downarrow}{} \dfrac{\uparrow\downarrow}{} \dfrac{\uparrow\downarrow}{}_{3d} \quad \dfrac{\uparrow}{} \dfrac{\uparrow}{}_{4p} \dfrac{}{}$

(d) Zr,　[Kr] $\dfrac{\uparrow\downarrow}{5s} \quad \dfrac{\uparrow}{} \dfrac{\uparrow}{}_{4d} \dfrac{}{} \dfrac{}{} \dfrac{}{}$

5.74 $4s > 4d > 4f$　**5.76** [Rn] $7s^2 5f^{14} 6d^{10} 7p^6$　**5.78** (a) 2; (b) 2; (c) 1; (d) 3　**5.80** $Z = 121$　**5.82** Atomic radii increase

down a group because the electron shells are farther away from the nucleus.   **5.84** F < O < S   **5.86** Mg has a higher ionization energy than Na because Mg has a higher Z_{eff} and a smaller size.   **5.88** λ = 410.2 nm; E = 291.6 kJ/mol   **5.90** 2279 nm   **5.92** (a) 0.151 kJ/mol; (b) 2.17×10^{-8} kJ/mol; (c) 2.91 kJ/mol
5.94

(a) Ra, [Rn] $7s^2$ [Rn] $\underset{7s}{\uparrow\downarrow}$

(b) Sc, [Ar] $4s^2\,3d^1$ [Ar] $\underset{4s}{\uparrow\downarrow}$ $\underset{\qquad 3d \qquad}{\uparrow \;\; — \;\; — \;\; — \;\; —}$

(c) Lr, [Rn] $7s^2\,5f^{14}\,6d^1$

[Rn] $\underset{7s}{\uparrow\downarrow}$ $\underset{\qquad\qquad 5f \qquad\qquad}{\uparrow\downarrow\;\uparrow\downarrow\;\uparrow\downarrow\;\uparrow\downarrow\;\uparrow\downarrow\;\uparrow\downarrow\;\uparrow\downarrow}$ $\underset{\qquad\qquad 6d \qquad\qquad}{\uparrow\;\;—\;\;—\;\;—\;\;—}$

(d) B, [He] $2s^2\,2p^1$ [He] $\underset{2s}{\uparrow\downarrow}$ $\underset{\quad 2p \quad}{\uparrow\;\;—\;\;—}$

(e) Te, [Kr] $5s^2\,4d^{10}\,5p^4$

[Kr] $\underset{5s}{\uparrow\downarrow}$ $\underset{\qquad\quad 4d \qquad\quad}{\uparrow\downarrow\;\uparrow\downarrow\;\uparrow\downarrow\;\uparrow\downarrow\;\uparrow\downarrow}$ $\underset{\quad 5p \quad}{\uparrow\downarrow\;\uparrow\;\uparrow}$

5.96 580 nm
5.98

(a) Sr, [Kr] $\underset{5s}{\uparrow\downarrow}$

(b) Cd, [Kr] $\underset{5s}{\uparrow\downarrow}$ $\underset{\qquad 4d \qquad}{\uparrow\downarrow\;\uparrow\downarrow\;\uparrow\downarrow\;\uparrow\downarrow\;\uparrow\downarrow}$

(c) Ti, [Ar] $\underset{4s}{\uparrow\downarrow}$ $\underset{\qquad 3d \qquad}{\uparrow\;\;\uparrow\;\;—\;\;—\;\;—}$

(d) Se, [Ar] $\underset{4s}{\uparrow\downarrow}$ $\underset{\qquad 3d \qquad}{\uparrow\downarrow\;\uparrow\downarrow\;\uparrow\downarrow\;\uparrow\downarrow\;\uparrow\downarrow}$ $\underset{\quad 4p \quad}{\uparrow\downarrow\;\uparrow\;\uparrow}$

5.100 K, Z_{eff} = 2.26; Kr, Z_{eff} = 4.06   **5.102** 8.3×10^{28} photons
5.104 940 kJ/mol
5.106

$$\Delta E = \frac{Z^2 e^2}{2a_0}\left[\frac{1}{n_1^2} - \frac{1}{n_2^2}\right]$$

This equation shows that ΔE is proportional to

$$\left[\frac{1}{n_1^2} - \frac{1}{n_2^2}\right]$$

where n_1 and n_2 are integers with $n_2 > n_1$. This is similar to the Balmer–Rydberg equation where $1/\lambda$ or ν is proportional to

$$\left[\frac{1}{m^2} - \frac{1}{n^2}\right]$$

where m and n are integers with $n > m$.

Chapter 6

6.1 (a) Ra^{2+} [Rn]; (b) La^{3+} [Xe]; (c) Ti^{4+} [Ar]; (d) N^{3-} [Ne]. Each ion has the ground-state configuration of the noble gas closest to it in the periodic table.   **6.2** Zn^{2+} **6.3** (a) O^{2-}; (b) S; (c) Fe; (d) H^- **6.4** K^+, r = 133 pm; Cl^-, r = 184 pm; K, r = 227 pm   **6.5** (a) Br; (b) S; (c) Se; (d) Ne   **6.6** (a) Be; (b) Ga   **6.7** (b) Cl has the highest E_{i1} and smallest E_{i4}.
6.8 Al < Kr < Ca   **6.9** Cr can accept an electron into a $4s$

orbital. Both Mn and Fe accept the added electron into a $3d$ orbital that contains an electron, but Mn has a lower value of Z_{eff}.   **6.10** Kr has the least favorable E_{ea}; Ge has the most favorable E_{ea}.   **6.11** (a) KCl; (b) CaF_2; (c) CaO
6.12 –562 kJ/mol   **6.13** (a) smaller; (b) larger   **6.14** (a) –2; (b) –1; (c) –1/2   **6.15** (a) $2\,Cs(s) + 2\,H_2O(l) \rightarrow 2\,Cs^+(aq) + 2\,OH^-(aq) + H_2(g)$; (b) $Na(s) + N_2(g) \rightarrow$ N.R.; (c) $Rb(s) + O_2(g) \rightarrow RbO_2(s)$; (d) $2\,K(s) + 2\,NH_3(g) \rightarrow 2\,KNH_2(s) + H_2(g)$; (e) $2\,Rb(s) + H_2(g) \rightarrow 2\,RbH(s)$ **6.16** (a) $Be(s) + Br_2(l) \rightarrow BeBr_2(s)$; (b) $Sr(s) + 2\,H_2O(l) \rightarrow Sr(OH)_2(aq) + H_2(g)$; (c) $2\,Mg(s) + O_2(g) \rightarrow 2\,MgO(s)$ **6.17** $BeCl_2(s) + 2\,K(s) \rightarrow Be(s) + 2\,KCl(s)$ **6.18** MgS(s); –2   **6.19** H^+ is the oxidizing agent. Al is the reducing agent.   **6.20** $2\,Al(s) + 3\,S(s) \rightarrow Al_2S_3(s)$ **6.21** (a) $Br_2(l) + Cl_2(g) \rightarrow 2\,BrCl(g)$; (b) $2\,Al(s) + 3\,F_2(g) \rightarrow 2\,AlF_3(s)$; (c) $H_2(g) + I_2(s) \rightarrow 2\,HI(g)$
6.22 $Br_2(l) + 2\,NaI(s) \rightarrow 2\,NaBr(s) + I_2(s)$, Br_2 is the oxidizing agent. I^- is the reducing agent.   **6.23** (a) XeF_2: F –1, Xe +2; (b) XeF_4: F –1, Xe +4; (c) $XeOF_4$: F –1, O –2, Xe +6   **6.24** (a) [Kr]; (b) [Xe]; (c) [Ar]-like (note that Ga^{3+} has ten $3d$ electrons in addition to the two $3s$ and six $3p$ electrons); (d) [Ne]   **6.25** gain 2 electrons   **6.26** evaporation of seawater; mining deposits of halite or rock salt
Understanding Key Concepts   **6.28** (a) ionic compound; (b) covalent compound
6.30

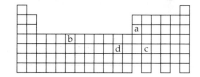

(a) Al^{3+}; (b) Cr^{3+}; (c) Sn^{2+}; (d) Ag^+ **6.32** (a) I_2; (b) Na; (c) NaCl; (d) Cl_2 **6.34** green—CBr_4: C, +4; Br, –1; blue—SrF_2: Sr, +2; F, –1; red—PbS: Pb, +2; S, –2 or PbS_2: Pb, +4; S, –2   ***Additional Problems***   **6.36** (a) La^{3+}, [Xe]; (b) Ag^+, [Kr] $4d^{10}$; (c) Sn^{2+}, [Kr] $5s^2 4d^{10}$ **6.38** Cr^{2+}; Fe^{2+} **6.40** positive sign because energy is required to remove an electron   **6.42** Largest E_{i1} in Group 8A because of the largest values of Z_{eff}. Smallest E_{i1} in Group 1A because of the smallest values of Z_{eff}.   **6.44** (a) Ca; (b) Ca   **6.46** Ar has the highest E_{i2} and the lowest E_{i7}.   **6.48** (a) Lowest K, highest Li; (b) lowest B, highest Cl; (c) lowest Ca, highest Cl   **6.50** They have the same magnitude but opposite sign   **6.52** Na^+ has a more negative electron affinity than either Na or Cl.   **6.54** because of the positive Z_{eff}.   **6.56** (a) F; (b) Na; (c) Br   **6.58** $MgCl_2$ > LiCl > KCl > KBr   **6.60** +195 kJ/mol   **6.62** –325 kJ/mol   **6.64** 808 kJ/mol   **6.66** –537 kJ/mol   **6.68** –176 kJ/mol
6.70

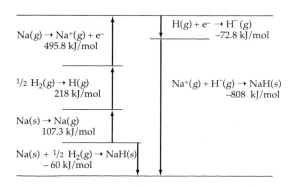

6.72

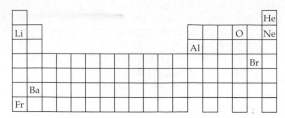

6.74 Solids: I_2; Liquids: Br_2; Gases: F_2, Cl_2, He, Ne, Ar, Kr, Xe   **6.76** (a) solid; (b) dark, like I_2; (c) yes, NaAt

6.78

(a) $2\ NaCl \xrightarrow[\substack{580°C}]{\substack{Electrolysis \\ in\ CaCl_2}} 2\ Na(l) + Cl_2(g)$

(b) $2\ Al_2O_3 \xrightarrow[\substack{980°C}]{\substack{Electrolysis \\ in\ Na_3AlF_6}} 4\ Al(l) + 3\ O_2(g)$

(c) Ar is obtained from the distillation of liquid air; (d) $2\ Br^-(aq) + Cl_2(g) \rightarrow Br_2(l) + 2\ Cl^-(aq)$ **6.80** Main-group elements tend to undergo reactions that leave them with eight valence electrons. The octet rule works because taking electrons away from a filled octet is difficult because they are tightly held by a high Z_{eff}; adding more electrons to a filled octet is difficult because, with s and p sublevels full, there is no low-energy orbital available.   **6.82** (a) $2\ K(s) + H_2(g) \rightarrow 2\ KH(s)$; (b) $2\ K(s) + 2\ H_2O(l) \rightarrow 2\ K^+(aq) + 2\ OH^-(aq) + H_2(g)$; (c) $2\ K(s) + 2\ NH_3(g) \rightarrow 2\ KNH_2(s) + H_2(g)$; (d) $2\ K(s) + Br_2(l) \rightarrow 2\ KBr(s)$; (e) $K(s) + N_2(g) \rightarrow$ N.R.; (f) $K(s) + O_2(g) \rightarrow KO_2(s)$ **6.84** (a) $Cl_2(g) + H_2(g) \rightarrow 2\ HCl(g)$; (b) $Cl_2(g) + Ar(g) \rightarrow$ N.R.; (c) $Cl_2(g) + Br_2(l) \rightarrow 2\ BrCl(g)$; (d) $Cl_2(g) + N_2(g) \rightarrow$ N.R.   **6.86** $AlCl_3 + 3\ Na \rightarrow Al + 3\ NaCl$. Al^{3+} is reduced. Na is oxidized.   **6.88** 590 g   **6.90** 5.56 g; H_2 is the limiting reactant.   **6.92** (a) H^+ is the oxidizing agent. Mg is the reducing agent. (b) F_2 is the oxidizing agent. Kr is the reducing agent. (c) Cl_2 is the oxidizing agent. I_2 is the reducing agent.   **6.94** Cu^{2+} has fewer electrons and a larger effective nuclear charge.   **6.96** MgF, -294 kJ/mol; MgF_2, -1114 kJ/mol   **6.98** (a) Na is used in table salt (NaCl), glass, rubber, and pharmaceutical agents; (b) Mg is used as a structural material when alloyed with Al. (c) F_2 is used in the manufacture of Teflon, $(C_2F_4)_n$, and in toothpaste as SnF_2.   **6.100** (a) $2\ Li(s) + H_2(g) \rightarrow 2\ LiH(s)$; (b) $2\ Li(s) + 2\ H_2O(l) \rightarrow 2\ Li^+(aq) + 2\ OH^-(aq) + H_2(g)$; (c) $2\ Li(s) + 2\ NH_3(g) \rightarrow 2\ LiNH_2(s) + H_2(g)$; (d) $2\ Li(s) + Br_2(l) \rightarrow 2\ LiBr(s)$; (e) $6\ Li(s) + N_2(g) \rightarrow 2\ Li_3N(s)$; (f) $4\ Li(s) + O_2(g) \rightarrow 2\ Li_2O(s)$ **6.102** When moving diagonally down and right on the periodic table, the increase in atomic radius caused by going to a larger shell is offset by a decrease caused by the higher Z_{eff}.   **6.104**

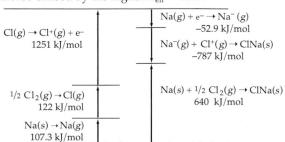

6.106 $E_{ea2} = +744$ kJ/mol. O^{2-} is not stable in the gas phase; it is stable in MgO because of the large lattice energy which results from the $+2$ and -2 charge of the ions and their small size.   **6.108** (a) 64.0%; (b) Ca; (c) $Ca(s) + Cl_2(g) \rightarrow CaCl_2(s)$; $CaCl_2(aq) + 2\ AgNO_3(aq) \rightarrow 2\ AgCl(s) + Ca(NO_3)_2(aq)$; (d) Cl_2 is in excess; 0.47 g Cl_2 unreacted   **6.110** (a) -832 kJ/mol; (b) Sr is the limiting reactant. 36.1 g $SrCl_2$ (c) 190 kJ

Chapter 7

7.1 (a) polar covalent; (b) ionic; (c) polar covalent; (d) polar covalent   **7.2** $CCl_4 \sim ClO_2 < TiCl_3 < BaCl_2$

7.3

(a) $H:\overset{..}{\underset{..}{S}}:H$ (b)
$$\overset{H}{\underset{\overset{|}{:\overset{..}{\underset{..}{Cl}}:}}{\underset{..}{:Cl}:\overset{..}{\underset{..}{C}}:\overset{..}{\underset{..}{Cl}}:}$$

7.4

$H:\overset{..}{\underset{..}{O}}:H + H^+ \longrightarrow \left[H:\overset{..}{\underset{..}{O}}:H \atop {\overset{H}{}}\right]^+$

Hydronium ion

7.5

(a)
$$\begin{array}{ccc} H & H & H \\ | & | & | \\ H-C-C-C-H \\ | & | & | \\ H & H & H \end{array}$$

(b) $H-\overset{..}{\underset{..}{O}}-\overset{..}{\underset{..}{O}}-H$

(c)
$$\begin{array}{c} H \\ | \\ H-C-\overset{..}{N}-H \\ | \quad | \\ H \quad H \end{array}$$

(d)
$$\begin{array}{cc} H & H \\ | & | \\ H-C=C-H \end{array}$$

(e) $H-C\equiv C-H$

(f)
$$\begin{array}{c} :\overset{..}{\underset{}{Cl}}: \\ | \\ :\overset{..}{\underset{..}{Cl}}-C=\overset{..}{\underset{..}{O}}: \end{array}$$

7.6

$$\begin{array}{cc} H & H \\ | & | \\ H-C-C-\overset{..}{\underset{..}{O}}-H \\ | & | \\ H & H \end{array} \quad and \quad \begin{array}{ccc} H & & H \\ | & & | \\ H-C-\overset{..}{\underset{..}{O}}-C-H \\ | & & | \\ H & & H \end{array}$$

7.7 $C_4H_5N_3O$

$$\begin{array}{c} \overset{..}{\underset{..}{O}}: \\ \| \\ \overset{H}{\underset{H}{\diagup}}\overset{..}{N}-C \quad \overset{N}{\diagdown}C \quad N-H \\ \overset{\diagdown}{N}-C \quad \overset{|}{C}=C \\ H \quad H \quad H \end{array}$$

7.8 $:C\equiv O:$

7.9

(a)
$$:\overset{..}{\underset{..}{Cl}}-Al-\overset{..}{\underset{..}{Cl}}: \atop {\underset{:\overset{..}{\underset{..}{Cl}}:}{|}}$$

(b)
$$:\overset{..}{\underset{..}{Cl}}-\overset{..}{\underset{..}{I}}-\overset{..}{\underset{..}{Cl}}: \atop {\underset{:\overset{..}{\underset{..}{Cl}}:}{|}}$$

(c)
$$:\overset{..}{\underset{..}{F}} \quad \overset{:\overset{..}{\underset{}{O}}:}{\underset{Xe}{|}} \quad \overset{..}{\underset{..}{F}}: \atop :\overset{..}{\underset{}{F}} \quad \overset{..}{\underset{}{F}}:$$

(d) $:\overset{..}{\underset{..}{Br}}-\overset{..}{\underset{..}{O}}-H$

7.10

(a) $\left[:\ddot{O}-H\right]^{-}$ (b) $\left[H-\overset{\cdot\cdot}{\underset{\cdot\cdot}{S}}-H\right]^{+}$ with H below   (c) $\left[\begin{array}{c}:\ddot{O}: \\ | \\ C-\ddot{O}-H \\ \| \\ :O:\end{array}\right]^{-}$

7.11 $:\ddot{N}=N=\ddot{O}: \longleftrightarrow :N\equiv N-\ddot{O}:$

7.12

(a) $\ddot{S}$ structures: $:\ddot{O}\diagup^{\overset{\cdot\cdot}{S}}\diagdown\ddot{O}: \longleftrightarrow :O\diagup^{\overset{\cdot\cdot}{S}}\diagdown\ddot{O}:$

(b) $\left[\begin{array}{c}:\ddot{O}: \\ \| \\ C \\ \diagup \diagdown \\ :\ddot{O} \quad \ddot{O}:\end{array}\right]^{2-} \longleftrightarrow \left[\begin{array}{c}:\ddot{O}: \\ | \\ C \\ \diagup \diagdown \\ :\ddot{O} \quad \ddot{O}:\end{array}\right]^{2-} \longleftrightarrow \left[\begin{array}{c}:\ddot{O}: \\ | \\ C \\ \diagup \diagdown \\ :O \quad \ddot{O}:\end{array}\right]^{2-}$

(c) $\left[H-C\begin{array}{c}\ddot{O}: \\ \diagup \\ \diagdown \\ :\ddot{O}:\end{array}\right]^{-} \longleftrightarrow \left[H-C\begin{array}{c}:\ddot{O}: \\ \diagup \\ \diagdown \\ O:\end{array}\right]^{-}$

7.13

Two resonance structures of a benzene-ring-like molecule with $-\ddot{O}-CH_3$ group.

7.14

$\left[\begin{array}{c}:O: \\ \| \\ N^{+} \\ \diagup \diagdown \\ :\ddot{O} \quad \ddot{O}:\end{array}\right]^{-} \longleftrightarrow \left[\begin{array}{c}:\ddot{O}\overset{-}{\cdot}: \\ | \\ N^{+} \\ \diagup \diagdown \\ :O \quad \ddot{O}:\end{array}\right]^{-} \longleftrightarrow \left[\begin{array}{c}:\ddot{O}: \\ | \\ N^{+} \\ \diagup \diagdown \\ :\ddot{O} \quad O:\end{array}\right]^{-}$

7.15

(a) $:\overset{-}{\ddot{N}}=C=\ddot{O}:$ (b) $:\overset{-}{\ddot{O}}-\overset{+}{\ddot{O}}=\ddot{O}:$

7.16 (a) bent; (b) trigonal pyramidal; (c) linear; (d) octahedral; (e) square pyramidal; (f) tetrahedral; (g) tetrahedral; (h) tetrahedral; (i) square planar; (j) trigonal planar

7.17

$\begin{array}{c}H \quad :O: \\ | \quad\quad \| \\ H-C-C-\ddot{O}-H \\ | \\ H\end{array}$ and $\begin{array}{c}H \diagdown\diagdown C-C\diagup^{O} \\ H\diagup \quad \diagdown O-H \\ H\end{array}$

7.18 (a) tetrahedral; (b) see-saw   **7.19** Each C is sp^3 hybridized. The C–C bond is formed by the overlap of one singly occupied sp^3 hybrid orbital from each C. The C–H bonds are formed by the overlap of one singly occupied sp^3 orbital on C with a singly occupied H $1s$ orbital. **7.20** The carbon in formaldehyde is sp^2 hybridized.

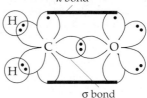

7.21 In HCN the carbon is sp hybridized.

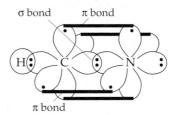

7.22 The hybridization of the central I is sp^3d.

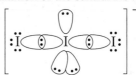

7.23 SF_2 sp^3; SF_4 sp^3d; SF_6 sp^3d^2
7.24 (a) sp; (b) sp^3d
7.25

For He$_2^{+}$ σ^*_{1s} ↑
                    σ_{1s} ↑↓

He$_2^{+}$ should be stable with a bond order of 1/2.
7.26

For B$_2$ σ^*_{2p} —
            π^*_{2p} — —
            σ_{2p} —
            π_{2p} ↑ ↑
            σ^*_{2s} ↑↓
            σ_{2s} ↑↓

Bond order = 1; paramagnetic

For C$_2$ σ^*_{2p} —
            π^*_{2p} — —
            σ_{2p} —
            π_{2p} ↑↓ ↑↓
            σ^*_{2s} ↑↓
            σ_{2s} ↑↓

Bond order = 2; diamagnetic
7.27

$\left[\begin{array}{c}H \\ \diagdown \\ C=\ddot{O}: \\ | \\ :\ddot{O}:\end{array}\right]^{-} \longleftrightarrow \left[\begin{array}{c}H \\ \diagdown \\ C-\ddot{O}: \\ \| \\ :O:\end{array}\right]^{-}$

7.28 Handed biomolecules have specific shapes that only match complementary-shaped receptor sites in living systems. The mirror-image forms of the molecules can't fit into the receptor sites and thus don't elicit the same biological response.   **7.29** (a) has no handedness; (b) has handedness.   ***Understanding Key Concepts***   **7.30** (a) square pyramidal; (b) trigonal pyramidal; (c) square planar; (d) trigonal planar   **7.32** (c) does not have a tetrahedral central atom.   **7.34** (a) $C_8H_9NO_2$; (b) and (c)

*sp*², trigonal planar, although VSEPR theory would predict otherwise.

7.36

All carbons that have only single bonds are *sp*³ hybridized. The three carbons that have double bonds are *sp*² hybridized.   ***Additional Problems***   **7.38** Electronegativity increases from left to right across a period and decreases down a group.   **7.40** K < Li < Mg < Pb < C < Br   **7.42** (a) polar covalent; (b) polar covalent; (c) polar covalent; (d) polar covalent; (e) Na⁺–OH⁻ is ionic; OH⁻ is polar covalent.

7.44

(a) $\overset{\delta-}{C}\!-\!\overset{\delta+}{H}$ $\overset{\delta+}{C}\!-\!\overset{\delta-}{Cl}$ (b) $\overset{\delta-}{Si}\!-\!\overset{\delta+}{Li}$ $\overset{\delta+}{Si}\!-\!\overset{\delta-}{Cl}$

(c) $\overset{\delta-}{N}\!-\!\overset{\delta+}{Cl}$ N–Mg

7.46 The transition metals are characterized by partially filled *d* orbitals which can be used to expand their valence shell beyond the normal octet of electrons.

7.48

7.50

7.52

7.54 (a) yes; (b) yes; (c) yes; (d) yes   **7.56** (a) Al; (b) P

7.58

7.60 :C≡O: C, –1; O, +1

7.62

7.64

Structure (a) is more important because of the octet of electrons around carbon.   **7.66** (a) trigonal planar; (b) trigonal bipyramidal; (c) linear; (d) octahedral   **7.68** (a) 4; (b) 6; (c) 3 or 4; (d) 2 or 5; (e) 6; (f) 4   **7.70** (a) bent; (b) tetrahedral; (c) bent; (d) trigonal planar   **7.72** (a) trigonal bipyramidal; (b) see-saw; (c) trigonal pyramidal; (d) tetrahedral   **7.74** (a) tetrahedral; (b) tetrahedral; (c) tetrahedral; (d) trigonal pyramidal; (e) tetrahedral   **7.76** (a) approximately 109°; (b) approximately 120°; (c) 90°; (d) approximately 120°

7.78

H–C$_a$–H, ~120°; H–C$_a$–C$_b$, ~120°; C$_a$–C$_b$–C$_c$, ~120°; C$_b$–C$_c$–N, 180°; C$_a$–C$_b$–H, ~120°; H–C$_b$–C$_c$, ~120°   **7.80** The geometry about each carbon is tetrahedral with a C–C–C bond angle of approximately 109°. Because the geometry about each carbon is tetrahedral, the cyclohexane ring cannot be flat.   **7.82** In a π bond, the shared electrons occupy a region above and below a line connecting the two nuclei. A σ bond has its shared electrons located along the axis between the two nuclei.   **7.84** (a) *sp*; (b) *sp*³*d*; (c) *sp*³*d*²; (d) *sp*³   **7.86** (a) *sp*³; (b) *sp*³*d*²; (c) *sp*² or *sp*³; (d) *sp* or *sp*³*d*; (e) *sp*³*d*²   **7.88** (a) *sp*²; (b) *sp*³; (c) *sp*³*d*²; (d) *sp*²

7.90 C is sp^2; N atoms are sp^3

$$
\underset{\substack{\sim109^\circ}}{H_2N} \overset{\displaystyle O}{\underset{\displaystyle \underset{H}{\overset{\displaystyle \|}{C}}}{}} \underset{\sim109^\circ}{N} \text{—H}
$$

~120°

7.92

	O_2^+	O_2	O_2^-
σ^*_{2p}	—	—	—
π^*_{2p}	↑ —	↑ ↑	↑↓ ↑
π_{2p}	↑↓ ↑↓	↑↓ ↑↓	↑↓ ↑↓
σ_{2p}	↑↓	↑↓	↑↓
σ^*_{2s}	↑↓	↑↓	↑↓
σ_{2s}	↑↓	↑↓	↑↓

Bond order: O_2^+, 2.5; O_2, 2.0; O_2^-, 1.5. All are stable. All have unpaired electrons.

7.94

p orbitals in allyl cation

allyl cation showing only the σ bonds (each C is sp^2 hybridized)

delocalized MO model for π bonding in the allyl cation

7.96 Every carbon is sp^2 hybridized. There are 18 σ bonds and 5 π bonds.

7.98

$$
\underset{\substack{H \\ |}}{\overset{\substack{H \\ |}}{H-C}}-\ddot{N}=C=\ddot{O}: \longleftrightarrow \underset{\substack{H \\ |}}{\overset{\substack{H \\ |}}{H-C}}-\overset{+}{N}\equiv C-\ddot{\underset{..}{\overset{..}{O}}}:
$$

7.100 (a) B −1, O +1; (b) Reactants: B, trigonal planar, sp^2; O, bent, sp^2. Product: B, tetrahedral, sp^3; O, trigonal pyramidal, sp^3. **7.102** The triply bonded carbon atoms are sp hybridized. The theoretical bond angle for C–C≡C is 180°. Benzyne is so reactive because the C–C≡C bond angle is closer to 120° and is very strained.

7.104

$$
\underset{\substack{:\ddot{Cl}: \\ |}}{:\ddot{Cl}} - \underset{\substack{| \\ :\ddot{Cl}:}}{\overset{\substack{:\ddot{Cl}: \\ |}}{C}} - \underset{\substack{| \\ H}}{\overset{\substack{:\ddot{O}-H \\ |}}{C}} - \ddot{O} - H
$$

7.106

$C_2{}^{2-}$

σ^*_{2p}	—
π^*_{2p}	— —
σ_{2p}	↑↓
π_{2p}	↑↓ ↑↓
σ^*_{2s}	↑↓
σ_{2s}	↑↓

Bond order = 3 **7.108** −109 kJ

7.110

(a)
$$
\ddot{\underset{..}{O}}\overset{\displaystyle }{\underset{\displaystyle }{}}\underset{+}{N}-\underset{\substack{.. \\ ..}}{\ddot{O}}-\underset{+}{N}\underset{:O:}{\overset{O:}{}}
$$

(b), (c), (d), (e), (f), (g), (h)

Structures (a)–(d) make more important contributions to the resonance hybrid because of only −1 and 0 formal charges on the oxygens.

7.112

21 σ bonds
5 π bonds

Each C atom is sp^2 hybridized.

7.114 (a)

$$
\left[:\ddot{O}=Cr-\ddot{O}-Cr=\ddot{O}: \right]^{2-}
$$

(b) Each Cr atom has 6 pairs of electrons around it. The likely geometry about each Cr atom is tetrahedral because each Cr has 4 charge clouds.

Chapter 8

8.1 3.2×10^2 kJ   **8.2** (a) and (b) are state functions; (c) is not.   **8.3** $+1.9 \times 10^4$ J; flows into the system.   **8.4** $w = -0.25$ kJ; the expanding system loses work energy and does work on the surroundings.   **8.5** (a) $P\Delta V$ is negative for this reaction because the system volume is decreased at constant pressure. ΔH is negative. Its value is slightly more negative than ΔE.   **8.6** $w = 0.57$ kJ; $\Delta E = -120$ kJ   **8.7** –45.2 kJ   **8.8** (a) 780 kJ evolved; (b) 1.24 kJ absorbed   **8.9** $q = -32$ kJ   **8.10** 0.13 J/(g · °C)   **8.11** -1.1×10^2 kJ   **8.12** –202 kJ   **8.13** (a) A + 2 B → D; $\Delta H° = -150$ kJ; (b) red arrow: step 1, green arrow: step 2, blue arrow: overall reaction; (c) top energy level represents A + 2 B, middle energy level represents C + B, bottom energy level represents D
8.14

Reactants $CH_4 + 2 Cl_2$

$\Delta H° = -98.3$ kJ

$CH_3Cl + HCl + Cl_2$

$\Delta H° = -202$ kJ

$\Delta H° = -104$ kJ

Products $CH_2Cl_2 + 2 HCl$

8.15 –905.6 kJ   **8.16** +2816 kJ   **8.17** –39 kJ   **8.18** –81 kJ   **8.19** –2635.5 kJ/mol; –45.35 kJ/g; –26.3 kJ/mL   **8.20** $\Delta S°$ is negative because the reaction decreases the number of moles of gaseous molecules.   **8.21** The reaction proceeds from a solid and a gas (reactants) to all gas (product). This is more disordered and $\Delta S°$ is positive.   **8.22** (a) spontaneous; (b) nonspontaneous   **8.23** $\Delta G° = -32.9$ kJ; reaction is spontaneous; $T = 190°C$ *Understanding Key Concepts*   **8.24** (a) yes; $w < 0$; (b) yes; $\Delta H < 0$; exothermic
8.26

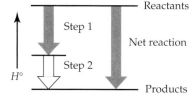

Step 1
Net reaction
Step 2
$H°$
Reactants
Products

8.28 The volume decreases from 5 L to 3 L.

1 atm

$V = 3.00$ L

8.30 $\Delta G < 0$; $\Delta S > 0$; $\Delta H > 0$ *Additional Problems*   **8.32** Heat is the energy transferred from one object to another as the result of a temperature difference between them. Temperature is a measure of the kinetic energy of molecular motion. Energy is the capacity to do work or supply heat. Work is defined as the distance moved times the force that opposes the motion ($w = d \times F$). Kinetic energy

is the energy of motion. Potential energy is stored energy.
8.34 Car: 7.1×10^5 J; truck: 6.7×10^5 J   **8.36** –70 J. The energy change is negative.   **8.38** $\Delta E = q_v$ is the heat of a reaction at constant volume. $\Delta H = q_p$ is the heat of a reaction at constant pressure.   **8.40** ΔH and ΔE are nearly equal when there are no gases involved in a chemical reaction, or, if gases are involved, $\Delta V = 0$ **8.42** –0.30 kJ   **8.44** $w = 0.283$ kJ; $\Delta E = -314$ kJ   **8.46** 25.5 kJ   **8.48** 131 kJ   **8.50** 0.388 kJ evolved; exothermic   **8.52** Heat capacity is the amount of heat required to raise the temperature of a substance a given amount. Specific heat is the amount of heat necessary to raise the temperature of exactly 1 g of a substance by exactly 1°C.   **8.54** 1.23 J/(g · °C)   **8.56** –83.7 kJ   **8.58** $\Delta H = -56$ kJ/mol; same temperature increase because NaOH is still the limiting reactant.   **8.60** The standard state of an element is its most stable form at 1 atm and 25°C.   **8.62** The overall enthalpy change for a reaction is equal to the sum of the enthalpy changes for the individual steps in the reaction. Hess's law works because of the law of conservation of energy.   **8.64** –395.7 kJ/mol   **8.66** –909.3 kJ   **8.68** +104.0 kJ/mol   **8.70** –17.1 kJ/mol   **8.72** –123 kJ   **8.74** –2645 kJ   **8.76** Entropy is a measure of molecular disorder.   **8.78** A reaction can be spontaneous yet endothermic if ΔS is positive and the $T\Delta S$ term is larger than ΔH.   **8.80** (a) positive; (b) negative   **8.82** (a) zero; (b) zero; (c) negative   **8.84** ΔS is positive. The reaction increases the total number of molecules.   **8.86** (a) spontaneous, exothermic; (b) nonspontaneous, exothermic; (c) spontaneous, endothermic; (d) nonspontaneous, endothermic   **8.88** 570 K   **8.90** (a) spontaneous at all temperatures; (b) and (c) have a crossover temperature; (d) nonspontaneous at all temperatures   **8.92** 31.6 J/(K · mol)   **8.94** –468 kJ   **8.96** –171.2 kJ   **8.98** 279 K   **8.100** (a) +34 kJ; (b) –451 kJ; (c) –87 kJ   **8.102** (a) 2 $C_8H_{18}(l)$ + 25 $O_2(g)$ → 16 $CO_2(g)$ + 18 $H_2O(g)$; (b) +132.4 kJ/mol   **8.104** (a) $\Delta G = -T\Delta S_{total}$; (b) $\Delta S_{surr} = -9451$ J/(K · mol)   **8.106** $\Delta H° = +201.9$ kJ   **8.108** (a) 2 K(s) + 2 $H_2O(l)$ → 2 KOH(aq) + $H_2(g)$; (b) –393.2 kJ; (c) 47.7°C; (d) 0.483 M; 174 mL of 0.554 M H_2SO_4

Chapter 9

9.1 1.00 atm = 14.7 psi; 1.00 mm Hg = 1.93×10^{-2} psi   **9.2** 10.3 m   **9.3** 0.650 atm   **9.4** 1000 mm Hg
9.5

(a)   (b)

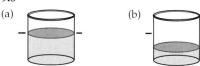

9.6 4.461×10^3 mol; 7.155×10^4 g   **9.7** 5.0 atm   **9.8** 267 mol   **9.9** 28°C   **9.10** 14.8 g; 7.55 L   **9.11** 190 L   **9.12** 34.1 amu; H_2S, hydrogen sulfide   **9.13** $X_{H_2} = 0.7281$; $X_{N_2} = 0.2554$; $X_{NH_3} = 0.0165$ **9.14** $P_{total} = 25.27$ atm; $P_{H_2} = 18.4$ atm; $P_{N_2} = 6.45$ atm; $P_{NH_3} = 0.417$ atm   **9.15** 0.0280 atm   **9.16** $P_{red} = 300$ mm Hg; $P_{yellow} = 100$ mm Hg; $P_{green} = 200$ mm Hg   **9.17** at 37°C, 525 m/s; at –25°C, 470 m/s   **9.18** (a) O_2, 1.62; (b) C_2H_2, 1.04   **9.19** $^{20}Ne(1.05) > ^{21}Ne(1.02) > ^{22}Ne(1.00)$ **9.20** ideal gas law: 20.5 atm; van der Waals equation: 20.3 atm   **9.21** 3.8×10^{-5} m   **9.22** 2.0%   **9.23** (a) 5.9 mm Hg; (b) 0.41 g   *Understanding Key Concepts*   **9.24** (a) The volume will increase by a factor of 1.5; (b) The volume will decrease by a factor of 2; (c) There is no change in volume.

(a)

(b)

(c)

9.26 (c)   **9.28** The gas pressure in the bulb in mm Hg is equal to the difference in the height of the Hg in the two arms of the manometer.   **9.30** (a) yellow; (b) 36 amu *Additional Problems*   **9.32** Temperature is a measure of the average kinetic energy of gas particles.   **9.34** 0.632 atm; 6.40×10^4 Pa   **9.36** 930 mm Hg   **9.38** 1.046×10^5 Pa   **9.40** 28.96 amu   **9.42** (a) P would triple; (b) P would be 1/3 the initial P; (c) P would increase by 1.8 times; (d) P would be 0.17 times the initial P.   **9.44** They all contain the same number of gas molecules. **9.46** 7210 L; 51.5 L   **9.48** 2.1×10^4 mm Hg   **9.50** 1×10^{-17} mm Hg   **9.52** 1.23×10^4 g   **9.54** ice cube   **9.56** Weigh the containers. The heavier container contains O_2.   **9.58** 1.5×10^4 g O_2 **9.60** (a) 0.716 g/L; (b) 1.96 g/L; (c) 1.43 g/L; (d) 15.7 g/L   **9.62** 34.0 amu   **9.64** 0.5469 L   **9.66** (a) 9.44 L; (b) 6.05 g Zn   **9.68** (a) 380 g; (b) 5.4 days   **9.70** $P_{N_2} = 0.7808$ atm; $P_{O_2} = 0.2095$ atm; $P_{Ar} = 0.0093$ atm; $P_{CO_2} = 0.000\ 37$ atm   **9.72** $P_{O_2} = 0.970$ atm; $P_{CO_2} = 0.007\ 11$ atm   **9.74** $X_{HCl} = 0.026$; $X_{H_2} = 0.094$; $X_{Ne} = 0.88$ **9.76** $P_{H_2} = 723$ mm Hg; 3.36 g Mg   **9.78** See list in text Section 9.6.   **9.80** Heat is the energy transferred from one object to another as the result of a temperature difference between them. Temperature is a measure of the kinetic energy of molecular motion.   **9.82** $u = 443$ m/s   **9.84** For H_2, $u = 1360$ m/s; for He, $u = 2010$ m/s   **9.86** 17.2 amu **9.88** Relative rates are HCl(1.05) > F_2(1.02) > Ar(1.00). **9.90** $-272.83°C$ **9.92** Relative rates are $^{35}Cl_2$(1.03) > $^{35}Cl^{37}Cl$(1.01) > $^{37}Cl_2$(1.00).   **9.94** 1.1 L   **9.96** 504.3 g   **9.98** (a) A contains $CO_2(g)$ and $N_2(g)$; B contains $CO_2(g)$, $N_2(g)$, and $H_2O(s)$; (b) 0.0013 mol H_2O; (c) A contains $N_2(g)$, B contains $N_2(g)$ and $H_2O(s)$; C contains $N_2(g)$ and $CO_2(s)$; (d) 0.010 92 mol N_2; (e) 0.0181 mol CO_2 **9.100** ideal gas law: $P = 59.1$ atm at 0°C; 70.0 atm at 50°C; 80.8 atm at 100°C. van der Waals equation: $P = 36.5$ atm at 0°C; 48.5 atm at 50°C; 60.5 atm at 100°C   **9.102** 816 atm   **9.104** (a) 0.901 mol; (b) 1.44 atm; (c) $P_{CH_4} = 1.32$ atm; $P_{C_2H_6} = 0.12$ atm; (d) 771 kJ liberated   **9.106** (a) 2 $C_8H_{18}(l)$ + 25 $O_2(g) \rightarrow$ 16 $CO_2(g)$ + 18 $H_2O(g)$; (b) 1.1×10^{11} kg; (c) 5.7×10^{13} L; (d) 59.5 mol, 1.33×10^3 L **9.108** (a) 0.0290 mol; (b) 0.0100 mol A; A = H_2O; (c) 0.0120 mol B; B = CO_2; (d) 0.001 00 mol C; C = O_2; 0.006 00 mol D; 28.0 g/mol; D = N_2; (e) 4 $C_3H_5N_3O_9(l) \rightarrow$ 10 $H_2O(g)$ + 12 $CO_2(g)$ + $O_2(g)$ + 6 $N_2(g)$

Chapter 10

10.1 41%; HF has more ionic character than HCl.   **10.2** (a) SF_6 is symmetrical (octahedral) and has no dipole moment; (b) $H_2C=CH_2$ is symmetrical; no dipole moment;

(c)

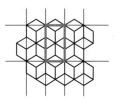

The C–Cl bonds in $CHCl_3$ are polar covalent bonds, and the molecule is polar.

(d)

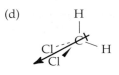

The C–Cl bonds in CH_2Cl_2 are polar covalent bonds, and the molecule is polar.

10.3

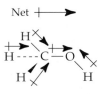

10.4 (a) HNO_3; (b) HNO_3; (c) Ar   **10.5** H_2S, dipole–dipole, dispersion; CH_3OH, hydrogen bonding, dipole–dipole, dispersion; C_2H_6, dispersion; Ar, dispersion; Ar < C_2H_6 < H_2S < CH_3OH **10.6** (a) positive; (b) negative; (c) positive   **10.7** 334 K   **10.8** 47°C   **10.9** 31.4 kJ/mol **10.10** (a) 2 atoms; (b) 4 atoms   **10.11** 167 pm   **10.12** 9.31 g/cm^3 **10.13** There are several possibilities. Here's one.

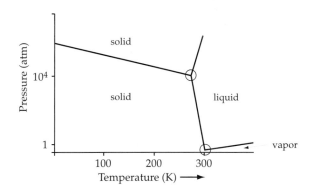

10.14 For CuCl: 4 minuses, 4 pluses; for $BaCl_2$: 8 pluses, 8 minuses   **10.15** (a) 1 Re atom, 3 O atoms; (b) ReO_3; (c) +6; (d) linear; (e) octahedral   **10.16** The triple point pressure of 5.11 atm   **10.17** (a) $CO_2(s) \rightarrow CO_2(g)$; (b) $CO_2(l) \rightarrow CO_2(g)$; (c) $CO_2(g) \rightarrow CO_2(l) \rightarrow$ supercritical CO_2 **10.18** (a)

(b) two; (c) Increasing the pressure favors the liquid phase, giving the solid/liquid boundary a negative slope. At 1 atm pressure the liquid phase is more dense than the solid phase.   **10.19** The molecules in a liquid crystal display can move around, as in viscous liquids, but they have a restricted range of motion, as in solids.   **10.20** Liquid crystal molecules have a rigid rodlike shape with a length four to eight times greater than their diameter. *Understanding Key Concepts*   **10.22** (a) cubic closest-packed; (b) 4 S^{2-}; 4 Zn^{2+} **10.24** (a) normal bp ≈ 300 K;

normal mp ≈ 180 K; (b) (i) solid, (ii) gas, (iii) supercritical fluid   **10.26** Here are two possibilities.

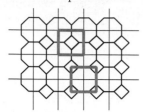

Additional Problems   **10.28** If a molecule has polar covalent bonds, the molecular shape will determine whether the bond dipoles cancel or not.   **10.30** (a) Dipole–dipole forces; dispersion forces are also present. (b) dispersion forces; (c) dispersion forces; (d) Dipole–dipole forces and hydrogen bonding; dispersion forces are also present.   **10.32** For CH_3OH and CH_4, dispersion forces are small. CH_3OH can hydrogen bond; CH_4 cannot. This accounts for the large difference in boiling points. For 1-decanol and decane, dispersion forces are comparable and relatively large. 1-decanol can hydrogen bond; decane cannot. This accounts for the 55°C higher boiling point for 1-decanol.

10.34

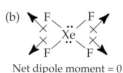

(a), (b), (c), (d)

10.36

SO_2 is bent and the individual bond dipole moments add to give a net dipole moment. CO_2 is linear and the individual bond dipole moments cancel.

10.38

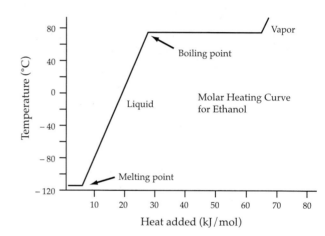

Hydrogen bond

10.40 ΔH_{vap} is usually larger than ΔH_{fusion} because ΔH_{vap} is the heat required to overcome all intermolecular forces.
10.42 (a) $Hg(l) \rightarrow Hg(g)$; (b) no change of state, Hg remains a liquid; (c) $Hg(g) \rightarrow Hg(l) \rightarrow Hg(s)$ **10.44** As the pressure is lowered, more of the liquid H_2O is converted to H_2O vapor. This conversion is endothermic and the temperature decreases. The decrease in pressure and temperature takes the system across the liquid/solid boundary in the phase diagram so the H_2O that remains turns to ice.   **10.46** 2.40 kJ   **10.48** 3.73 kJ

10.50

Molar Heating Curve for Ethanol

10.52 88.2 J/(K · mol)   **10.54** 28.0 kJ/mol   **10.56** 294 mm Hg

10.58

T(K)	P_{vap}(mm Hg)	$\ln P_{vap}$	$1/T$
263	80.1	4.383	0.003 802
273	133.6	4.8949	0.003 663
283	213.3	5.3627	0.003 534
293	329.6	5.7979	0.003 413
303	495.4	6.2054	0.003 300
313	724.4	6.5853	0.003 195

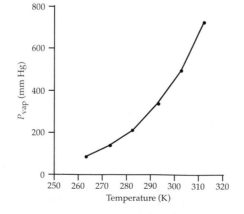

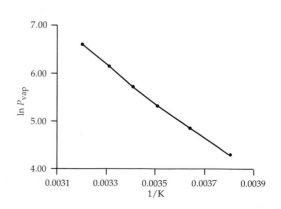

10.60 ΔH_{vap} = 30.1 kJ/mol   **10.62** ΔH_{vap} = 30.1 kJ/mol. The calculated ΔH_{vap} and that obtained from the plot in Problem 10.60 are the same.   **10.64** molecular solid, CO_2, I_2; metallic solid, any metallic element; covalent network solid, diamond; ionic solid, NaCl   **10.66** The unit cell is the smallest repeating unit in a crystal.   **10.68** r = 128 pm; density = 8.90 g/cm^3 **10.70** 404.9 pm   **10.72** 137 pm   **10.74** face-centered cubic   **10.76** Six Na^+ ions touch each H^- ion and six H^- ions touch each Na^+ ion.   **10.78** 244 pm   **10.80** (a) gas; (b) liquid; (c) solid
10.82

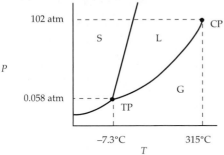

10.84 (a) $Br_2(s)$; (b) $Br_2(l)$ **10.86** Solid O_2 does not melt when pressure is applied because the solid is denser than the liquid and the solid/liquid boundary in the phase diagram slopes to the right.
10.88

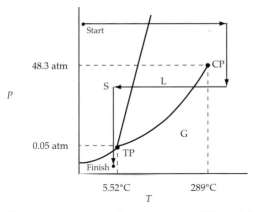

The starting phase is benzene as a solid, and the final phase is benzene as a gas.   **10.90** solid → liquid → supercritical fluid → liquid → solid → gas   **10.92** Because chlorine is larger than fluorine, the charge separation is larger in CH_3Cl compared to CH_3F resulting in CH_3Cl having a slightly larger dipole moment.   **10.94** 0.192 kJ   **10.96** 0.837 atm   **10.98** 650°C   **10.100** –30.7°C   **10.102** 23.3 kJ/mol
10.104

Kr cannot be liquified at room temperature because room temperature is above T_c (–63°C).   **10.106** 68%   **10.108** 6.01×10^{23} atoms/mol   **10.110** (a) 556 pm; (b) 2.26 g/cm^3 **10.112** (a) 1.926 g; (b) Fe_3O_4; (c) 24 Fe atoms; 32 O atoms   **10.114** (a) 231 pm; (b) K; (c) density of solid = 0.857 g/cm^3; density of vapor = 7.29×10^{-6} g/cm^3

Chapter 11

11.1 toluene < Br_2 < KBr   **11.2** (a) Na^+ because Na^+ is smaller than Cs^+; (b) Ba^{2+} because of its higher charge   **11.3** 5.52 mass %   **11.4** 4.6×10^{-5} g   **11.5** 0.614 M   **11.6** molality = 0.0249 m; $X_{C_{27}H_{46}O}$ = 2.96×10^{-3} **11.7** 312 g   **11.8** 0.251 M   **11.9** 0.513 m **11.10** 0.621 m **11.11** 3.2×10^{-2} mol/(L·atm)   **11.12** (a) 0.080 M; (b) 1.3×10^{-5} M   **11.13** 98.6 mm Hg   **11.14** 17.6 g   **11.15** upper curve: pure solvent; lower curve: solution   **11.16** (a) 27.1 mm Hg; (b) 46.6 mm Hg   **11.17** upper and lower curves: two pure liquids; middle curve: the mixture   **11.18** 62.1°C   **11.19** –2.15°C   **11.20** 0.793 m **11.21** (a) 62°C; (b) 2 m **11.22** 9.54 atm   **11.23** 0.156 M   **11.24** 128 g/mol   **11.25** 342 g/mol
11.26 (a)

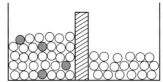

(b) ~50°C; (d) 90% dichloromethane and 10% chloroform   **11.27** Both solvent molecules and small solute particles can pass through a semipermeable dialysis membrane. Only large colloidal particles such as proteins can't pass through. Only solvent molecules can pass through a semipermeable membrane used for osmosis.   *Understanding Key Concepts*   **11.28** (a) < (b) < (c)
11.30

Assume that only the blue (open) spheres (solvent) can pass through the semipermeable membrane. There will be a net transfer of solvent from the right compartment (pure solvent) to the left compartment (solution) to achieve equilibrium.   **11.32** When 100 mL of 9 M H_2SO_4 at 0°C is added to 100 mL of liquid water at 0°C, the temperature rises because ΔH_{soln} for H_2SO_4 is exothermic. When 100 mL of 9 M H_2SO_4 at 0°C is added to 100 g of solid ice at 0°C, some of the ice will melt (an endothermic process) and the temperature will fall because the H_2SO_4 (solute) lowers the freezing point of the ice/water mixture.   **11.34** (b)   *Additional Problems*   **11.36** The larger the surface area, the more solid–solvent interactions, and the more rapidly the solid will dissolve. Powdered NaCl has a much larger surface area than a large block of NaCl.   **11.38** Substances tend to dissolve when the solute and solvent have the same type and magnitude of intermolecular forces. **11.40** Energy is required to overcome intermolecular forces holding solute particles together in the crystal. For an ionic solid, this is the lattice energy. **11.42** Ethyl alcohol and water are both polar with small dispersion forces. They both can hydrogen bond, and are miscible. Pentyl alcohol is slightly polar and can hydrogen bond. It has, however, a relatively large dispersion force because of its size, which limits its water solubility. **11.44** 42.5°C **11.46**

$$molarity = \frac{moles\ of\ solute}{liters\ of\ solution}; \quad molality = \frac{moles\ of\ solute}{kg\ of\ solvent}$$

11.48 (a) Dissolve 1.50 mol of glucose in water; dilute to 1.00 L; (b) Dissolve 1.135 mol of KBr in 1.00 kg of H_2O; (c) Mix together 0.15 mol of CH_3OH with 0.85 mol of H_2O. **11.50** Dissolve 4.42×10^{-3} mol (0.540 g) of $C_7H_6O_2$ in enough $CHCl_3$ to make 165 mL of solution. **11.52** (a) 0.500 M KCl; (b) 1.75 M glucose **11.54** (a) 11.2 mass %; (b) 0.002 70 mass % ; (c) 3.65 mass % **11.56** 0.20 ppm **11.58** (a) 0.196 m; (b) $X_{C_{10}H_{14}N_2} = 0.0145$; $X_{CH_2Cl_2} = 0.985$ **11.60** 1.81 M **11.62** 10.7 m **11.64** 3.7 g **11.66** (a) 0.0187; (b) 16.0%; (c) 1.06 m **11.68** 0.068 M **11.70** 0.06 atm **11.72** 0.010 mol/(L · atm) **11.74** The difference in entropy between the solvent in a solution and a pure solvent **11.76**

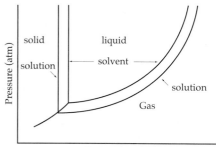

11.78 (a) 70.5 mm Hg; (b) 68.0 mm Hg **11.80** (a) 100.57°C; (b) 101.6°C **11.82** 219 mm Hg **11.84** In the liquid, $X_{acetone} = 0.602$ and $X_{ethyl\ acetate} = 0.398$. In the vapor, $X_{acetone} = 0.785$; $X_{ethyl\ acetate} = 0.215$ **11.86** 3.6 °C · kg/mol **11.88** 0.573 m **11.90** (a) 13.0 atm; (b) 65.2 atm **11.92** 0.197 M **11.94** osmotic pressure because, of the four colligative properties, osmotic pressure gives the largest colligative property change per mole of solute **11.96** 342.5 amu **11.98** HCl is a strong electrolyte in H_2O and completely dissociates into two solute particles per each HCl. HF is a weak electrolyte in H_2O. Only a few percent of the HF molecules dissociates into ions. **11.100** molecular mass = 538 amu; molecular formula is $C_{40}H_{56}$ **11.102** K_f for snow (H_2O) is 1.86 °C · kg/mol. Reasonable amounts of salt are capable of lowering the freezing point (ΔT_f) of the snow below an air temperature of –2°C. Reasonable amounts of salt, however, are not capable of causing a ΔT_f of more than 30°C. **11.104** 2.60×10^3 g **11.106** When solid $CaCl_2$ is added to liquid water, the temperature rises because ΔH_{soln} for $CaCl_2$ is exothermic. When solid $CaCl_2$ is added to ice at 0°C, some of the ice will melt (an endothermic process) and the temperature will fall because the $CaCl_2$ lowers the freezing point of an ice/water mixture. **11.108** 1.3 g **11.110** fp = –2.3°C; bp = 100.63°C **11.112** (a) 11.7 g of rubbing alcohol; (b) 0.75 mol C_3H_8O **11.114** molecular mass =

271 amu; molecular formula is $C_{18}H_{24}O_2$ **11.116** (a) 24.07 g $BaSO_4$; (b) More precipitate will form because of the excess $BaCl_2$ in the solution. **11.118** 10% LiCl; 41% H_2O; 49% CH_3OH **11.120** (a) C_2H_3; (b) 108 amu; (c) C_8H_{12}

Chapter 12

12.1 (a) 1.6×10^{-4} M/s; (b) 3.2×10^{-4} M/s **12.2** Rate of decomposition of $N_2O_5 = 2.2 \times 10^{-5}$ M/s; Rate of formation of $O_2 = 1.1 \times 10^{-5}$ M/s **12.3** Rate = $k[BrO_3^-][Br^-][H^+]^2$; 1st order in BrO_3^-, 1st order in Br^-, 2nd order in H^+, 4th order overall; Rate = $k[H_2][I_2]$, 1st order in H_2, 1st order in I_2, 2nd order overall; Rate = $k[CH_3CHO]^{3/2}$, 3/2 order in CH_3CHO, 3/2 order overall **12.4** (a) Rate = $k[H_2O_2][I^-]$; (b) 1.15×10^{-2} /(M · s); (c) 1.38×10^{-3} M/s **12.5** 1/s; 1/s; $1/(M^3 \cdot s)$; $1/(M \cdot s)$; $1/(M^{1/2} \cdot s)$ **12.6** (a) Zero order in A; second order in B; second order overall; (b) Rate = $k[B]^2$ **12.7** (a) 0.080 M; (b) 61 h **12.8** A plot of ln[cyclopropane] versus time is linear, indicating that the data fit the equation for a first-order reaction. $k = 6.6 \times 10^{-4}$/s (0.040/min) **12.9** (a) 11 h; (b) 0.019 M; (c) 22 h **12.10** (a) 5 min; (b)

○ red ● blue

12.11 (a) A plot of 1/[HI] versus time is linear. The reaction is second order; (b) 0.0308/(M · min); (c) 260 min; (d) 81.2 min **12.12** (a) $2 NO_2(g) + F_2(g) \rightarrow 2 NO_2F(g)$; $F(g)$ is a reaction intermediate; (b) each elementary reaction is bimolecular. **12.13** (a) Rate = $k[O_3][O]$; (b) Rate = $k[Br]^2[Ar]$; (c) Rate = $k[Co(CN)_5(H_2O)^{2-}]$ **12.14**

$$Co(CN)_5(H_2O)^{2-}(aq) \rightarrow Co(CN)_5^{2-}(aq) + H_2O(l) \quad (slow)$$
$$\underline{Co(CN)_5^{2-}(aq) + I^-(aq) \rightarrow Co(CN)_5I^{3-}(aq) \qquad (fast)}$$
Overall reaction $\quad Co(CN)_5(H_2O)^{2-}(aq) + I^-(aq) \rightarrow$
$$Co(CN)_5I^{3-}(aq) + H_2O(l)$$

The rate law for the first (slow) elementary reaction is: Rate = $k[Co(CN)_5(H_2O)^{2-}]$ **12.15** (a) 80 kJ/mol; (b) endothermic; (c)

$$\begin{array}{c} A \text{---} C \\ \vdots \quad \vdots \\ B \text{---} D \end{array}$$

12.16 (a) 104 kJ/mol; (b) 1.4×10^{-4}/s **12.17** (a) first order in A, zero order in B, first order in C; (b) Rate = $k[A][C]$; (c)

$$A + C \rightarrow AC \quad (\text{rate determining step})$$
$$\underline{AC + B \rightarrow AB + C}$$
Overall reaction $\quad A + B \rightarrow AB$

(d) C is a catalyst; it is consumed in the first step and then regenerated in the second step. **12.18** Because the bonds in nitro groups are weak and because the explosion products (CO_2, N_2, H_2O, and O_2) are extremely stable, a great deal of energy is released. **12.19** Secondary explosives should have a higher activation energy. **12.20** –3078 kJ **12.21** 4.40×10^3 L *Understanding Key Concepts* **12.22** 2:1:4:2 **12.24** (a) 2:4:3; (b) the same; (c) rates will double, and half-lives will be unaffected.   **12.26** (a) Because the half-life is inversely proportional to the concentration of A molecules, the reaction is second order in A; (b) Rate = $k[A]^2$; (c)

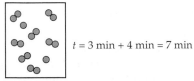

$t = 3 \text{ min} + 4 \text{ min} = 7 \text{ min}$

$t = 7$ min   **12.28** (a) Measure the concentration of AB_2 as a function of time; (b) If an ln $[AB_2]$ versus time plot is linear, the reaction is first order. If a $1/[AB_2]$ versus time plot is linear, the reaction is second order. (c) $k = \text{Rate}/[AB_2]^x$ (x = reaction order)   **12.30** (a) Rate = $k[B_2][C]$; (b) $B_2 + C \rightarrow CB + B$ (slow); $CB + A \rightarrow AB + C$ (fast); (c) C is a catalyst. C does not appear in the chemical equation because it is consumed in the first step and regenerated in the second step.   **12.32** (a)

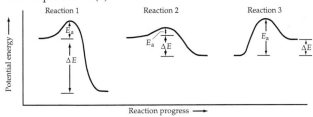

Reaction 1   Reaction 2   Reaction 3

(b) Reaction 2 is the fastest (smallest E_a) and Reaction 3 is the slowest (largest E_a). (c) Reaction 3 is the most endothermic (positive ΔE) and Reaction 1 is the most exothermic (largest negative ΔE).   ***Additional Problems***   **12.34** M/s or mol/(L·s)   **12.36** (a) 6.0×10^{-5} M/s; (b) 3.3×10^{-5} M/s   **12.38** (a) 2.4×10^{-5} M/s; (b) 3.1×10^{-5} M/s   **12.40** (a) 3 times faster; (b) 2 times faster   **12.42**

$$-\frac{\Delta[N_2]}{\Delta t} = -\frac{1}{3}\frac{\Delta[H_2]}{\Delta t} = \frac{1}{2}\frac{\Delta[NH_3]}{\Delta t}$$

12.44 2nd order in NO; 1st order in Br_2; 3rd order overall   **12.46** Rate = $k[H_2][ICl]$; $1/(M \cdot s)$ **12.48** (a) Rate = $k[CH_3Br][OH^-]$; (b) the rate will decrease by a factor of 5; (c) the rate will increase by a factor of 4   **12.50** (a) Rate = $k[CH_3COCH_3]$; (b) 8.7×10^{-3}/s; (c) 1.6×10^{-5} M/s   **12.52** (a) Rate = $k[NH_4^+]^m[NO_2^-]^n$; (b) $3.0 \times 10^{-4}/(M \cdot s)$; (c) 6.1×10^{-6} M/s   **12.54** (a) 0.015 M; (b) 40 min; (c) 7.2 min   **12.56** $t_{1/2} = 17$ min; $t = 69$ min   **12.58** 16.0 h   **12.60** (a) 5.2×10^{-3} M; (b) 3.1 h   **12.62** $t_{1/2} = 21$ min; $t = 42$ min   **12.64** A plot of ln $[N_2O]$ versus time is linear. The reaction is first order in N_2O; $k = 3.79 \times 10^{-5}$/s   **12.66** 2.79×10^{-3}/s   **12.68** An elementary reaction is an individual molecular event that involves the breaking and/or making of chemical bonds. The overall reaction describes only the stoichiometry of the overall process but provides no information about how the reaction occurs.   **12.70** There is no relationship between the coefficients in a balanced chemical equation for an overall reaction and the exponents in the rate law unless the overall reaction occurs in a single elementary step, in which case the coefficients in the balanced equation are the exponents in the rate law.   **12.72** (a) $H_2(g) + 2 ICl(g) \rightarrow I_2(g) + 2 HCl(g)$; (b) $HI(g)$; (c) each elementary reaction is bimolecular   **12.74** (a) bimolecular, Rate = $k[O_3][Cl]$; (b) unimolecular, Rate =

$k[NO_2]$; (c) bimolecular, Rate = $k[ClO][O]$; (d) termolecular, Rate = $k[Cl]^2[N_2]$ **12.76** (a) $2 NO_2Cl(g) \rightarrow 2 NO_2(g) + Cl_2(g)$; (b) 1. unimolecular; 2. bimolecular; (c) Rate = $k[NO_2Cl]$ **12.78** $NO_2(g) + F_2(g) \rightarrow NO_2F(g) + F(g)$; $F(g) + NO_2(g) \rightarrow NO_2F(g)$ **12.80** Very few collisions involve a collision energy greater than or equal to the activation energy, and only a fraction of those have the proper orientation for reaction.   **12.82**

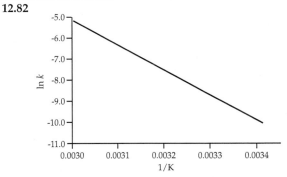

$E_a = 104$ kJ/mol   **12.84** (a) 134 kJ/mol; (b) $6.0/(M \cdot s)$ **12.86** 87 kJ/mol   **12.88** (a)

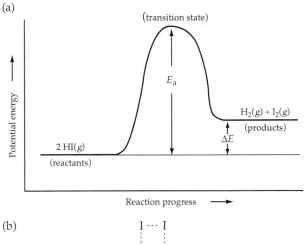

(b)

I --- I
┊   ┊
H --- H

12.90 A catalyst does participate in the reaction, but it is not consumed because it reacts in one step of the reaction and is regenerated in a subsequent step.   **12.92** A catalyst increases the rate of a reaction by changing the reaction mechanism and lowering the activation energy.   **12.94** (a) $O_3(g) + O(g) \rightarrow 2 O_2(g)$; (b) Cl acts as a catalyst; (c) ClO is a reaction intermediate; (d) A catalyst reacts in one step and is regenerated in a subsequent step. A reaction intermediate is produced in one step and consumed in another.   **12.96** (a) $NH_2NO_2(aq) \rightarrow N_2O(g) + H_2O(l)$; (b) OH^- acts as a catalyst; $NHNO_2^-$ is a reaction intermediate; (c) The rate will decrease because added acid decreases the concentration of OH^-, which appears in the rate law since it is a catalyst.   **12.98** The first maximum represents the potential energy of the transition state for the first step. The second maximum represents the potential energy of the transition state for the second step. The saddle point between the two maxima represents the potential energy of the intermediate products.   **12.100** (a) increase; (b) decrease; (c) increase; (d) no change

12.102 (a) Rate = $k[C_2H_4Br_2][I^-]$; (b) $4.98 \times 10^{-3}/M \cdot s$); (c) 1.12×10^{-4} M/s   **12.104** For E_a = 50 kJ/mol, f = 2.0 × 10^{-9}; for E_a = 100 kJ/mol; f = 3.9 × 10^{-18} **12.106** (a) $2\,NO_2(g) + Br_2(g) \rightarrow 2\,NOBr(g)$; (b) $NOBr_2$ is a reaction intermediate; (c) Rate = $k[NO][Br_2]$; (d) It can't be the first step. It must be the second step.   **12.108 (a)** $I^-(aq)$ + $OCl^-(aq) \rightarrow Cl^-(aq) + OI^-(aq)$; (b)

$$\text{Rate} = k\frac{[I^-][OCl^-]}{[OH^-]}$$

k = 60/s; (c) The reaction does not occur by a single-step mechanism because OH^- appears in the rate law but not in the overall reaction. (d)

$OCl^-(aq) + H_2O(l) \rightleftharpoons HOCl(aq) + OH^-(aq)$	(fast)
$HOCl(aq) + I^-(aq) \rightarrow HOI(aq) + Cl^-(aq)$	(slow)
$HOI(aq) + OH^-(aq) \rightarrow H_2O(l) + OI^-(aq)$	(fast)

Overall reaction $I^-(aq) + OCl(aq) \rightarrow$
$$Cl^-(aq) + OI^-(aq)$$

12.110 (a) $\text{Rate}_f = k_f[A]$ and $\text{Rate}_r = k_r[B]$; (b)

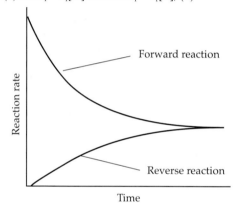

(c) [B]/[A] = 3   **12.112** 1.6×10^4 y   **12.114** (a) 8.7×10^{19} molecules/min; (b) 1.2 atm   **12.116** 0.71 atm   **12.118** –122 J

Chapter 13

13.1

$$\text{(a) } K_c = \frac{[SO_3]^2}{[SO_2]^2[O_2]} \qquad \text{(b) } K_c = \frac{[SO_2]^2[O_2]}{[SO_3]^2}$$

13.2 (a) 7.9×10^4; (b) 1.3×10^{-5} **13.3** The [B]/[A] ratio in the equilibrium mixture is 0.333. Only mixture (3) has this same ratio and is also at equilibrium.   **13.4** 9.48   **13.5** at 500 K: $K_p = 1.7 \times 10^4$; at 1000 K: K_c = 1.1
13.6

$$\text{(a) } K_c = \frac{[H_2]^3}{[H_2O]^3}, \qquad K_p = \frac{(P_{H_2})^3}{(P_{H_2O})^3}$$

(b) $K_c = [H_2]^2[O_2]$, $K_p = (P_{H_2})^2(P_{O_2})$ **13.7** $K_c = 1.2 \times 10^{-42}$. Since K_c is very small, the equilibrium mixture contains mostly H_2 molecules. H is in periodic group 1A. A very small value of K_c is consistent with strong bonding between 2 H atoms, each with one valence electron.   **13.8** (a) Because $Q_c < K_c$, the reaction is not at equilibrium. The reaction will proceed to the right to reach equilibrium; (b) Because $Q_c > K_c$, the reaction is not at equilibrium. The

reaction will proceed to the left to reach equilibrium.
13.9 (a) 2; (b) (1) reverse; (3), forward   **13.10** (a) 3.5 × 10^{-22} M; (b) 210 H atoms; 6.0×10^{22} H$_2$ molecules   **13.11** $[CO_2] = [H_2] = 0.101$ M; $[CO] = [H_2O] = 0.049$ M   **13.12** $[N_2O_4] = 0.0429$ M; $[NO_2] = 0.0141$ M   **13.13** $[N_2O_4] =$ 0.0292 M; $[NO_2] = 0.0116$ M   **13.14** (a) increases; (b) decreases; (c) decreases; (d) increases; At equilibrium,

$$Q_c = K_c = \frac{[CO_2][H_2]}{[CO][H_2O]}$$

If some CO_2 is removed from the equilibrium mixture, the numerator in Q_c is decreased, which means that $Q_c < K_c$ and the reaction will shift to the right, increasing the H$_2$ concentration.   **13.15** (a) remains the same; (b) increases; (c) decreases
13.16

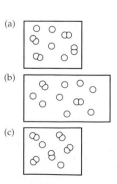

13.17 The equilibrium mixture will contain more NO, the higher the temperature.   **13.18** The amount of ethyl acetate decreases. As the temperature is decreased, the reaction shifts from left to right. The product concentrations increase, and the reactant concentrations decrease. This corresponds to an increase in K_c.   **13.19** There are more $AB(g)$ molecules at the higher temperature. The equilibrium shifted to the right at the higher temperature, which means the reaction is endothermic.   **13.20** (a) remains the same; (b) increases; (c) decreases; (d) remains the same; (e) decreases   **13.21** (a) k_f; (b) 2.5 × 10^{-28} M^{-1} s^{-1}; (c) k_r decreases more than k_f decreases, and therefore K_c increases.   **13.22** The reaction will shift to the left. This will decrease the effectiveness of Hb for carrying O_2.   **13.23** The equilibrium shifts to the left.   **13.24** 26 π electrons.   **13.25** 2.58×10^{21} O_2 molecules   *Understanding Key Concepts*   **13.26** (a) (1) and (3); (b) $K_c = 1.5$; (c) Because the same number of molecules appear on both sides of the equation, the volume terms in K_c all cancel. Therefore, we can calculate K_c without including the volume.   **13.28** (a) (3): (b) reaction (1): reverse direction; reaction (2): forward direction   **13.30** When the stopcock is opened, the reaction will go in the reverse direction because there will be initially an excess of AB molecules.   **13.32** (a) $AB \rightarrow A + B$; (b) The reaction is endothermic because a stress of added heat (higher temperature) shifts the $AB \rightleftharpoons A + B$ equilibrium to the right. (c) If the volume is increased, the pressure is decreased. The stress of decreased pressure will be relieved by a shift in the equilibrium from left to right, thus increasing the number of A atoms.
13.34

Additional Problems   **13.36**

(a) $K_c = \dfrac{[CO][H_2]^3}{[CH_4][H_2O]}$ (b) $K_c = \dfrac{[ClF_3]^2}{[F_2]^3[Cl_2]}$

(c) $K_c = \dfrac{[HF]^2}{[H_2][F_2]}$

13.38

(a) $K_p = \dfrac{(P_{CO})(P_{H_2})^3}{(P_{CH_4})(P_{H_2O})}$, $\Delta n = 2$ and $K_p = K_c(RT)^2$

(b) $K_p = \dfrac{(P_{ClF_3})^2}{(P_{F_2})^3(P_{Cl_2})}$, $\Delta n = -2$ and $K_p = K_c(RT)^{-2}$

(c) $K_p = \dfrac{(P_{HF})^2}{(P_{H_2})(P_{F_2})}$, $\Delta n = 0$ and $K_p = K_c$

13.40

$$K_c = \dfrac{[C_2H_5OC_2H_5][H_2O]}{[C_2H_5OH]^2}$$

13.42

$$K_c = \dfrac{[\text{Isocitrate}]}{[\text{Citrate}]}$$

13.44 1.3×10^8 **13.46** 0.058 **13.48** 29.0 **13.50** (a)

$$K_c = \dfrac{[CH_3CO_2C_2H_5][H_2O]}{[CH_3CO_2H][C_2H_5OH]}$$

(b) $K_c = 3.4$. Because there are the same number of molecules on both sides of the equation, the volume terms in K_c cancel. Therefore, we can calculate K_c without including the volume.   **13.52** 23.6 **13.54** $K_p = 0.0313$ atm; $K_c = 1.28 \times 10^{-3}$

13.56 (a)

$K_c - \dfrac{[CO_2]^3}{[CO]^3}$ $K_p = \dfrac{(P_{CO_2})^3}{(P_{CO})^3}$ (b) $K_c = \dfrac{1}{[O_2]^3}$ $K_p = \dfrac{1}{(P_{O_2})^3}$

(c) $K_c = [SO_3]$, $K_p = P_{SO_3}$; (d) $K_c = [Ba^{2+}][SO_4^{2-}]$ **13.58** (a) mostly product; (b) mostly reactants   **13.60** (c)   **13.62** $K_c = 1.2 \times 10^{82}$ is very large. When equilibrium is reached, very little if any ethanol will remain because the reaction goes to completion.   **13.64** The reaction is not at equilibrium because $Q_c > K_c$. The reaction will proceed from right to left to reach equilibrium.   **13.66** 5.9×10^{-3} M   **13.68** $[NO] = 0.056$ M; $[N_2] = [O_2] = 1.37$ M   **13.70** $[PCl_3] = [Cl_2] = 0.071$ M; $[PCl_5] = 0.089$ M   **13.72** (a) 6.8 mol; (b) 0.03 mol CH_3CO_2H; 9.03 mol C_2H_5OH; 0.97 mol $CH_3CO_2C_2H_5$; 0.97 mol H_2O **13.74** $P_{ClF} = P_{F_2} = 0.389$ atm; $P_{ClF_3} = 1.08$ atm   **13.76** (a) increases; (b) increases; (c) decreases; (d) decreases; Disturbing the equilibrium by decreasing $[Cl^-]$ increases Q_c ($Q_c = 1/[Ag^+]_t[Cl^-]_t$) to a value greater than K_c. To reach a new state of equilibrium, Q_c must decrease, which means that the denominator must increase; that is, the reaction must go from right to left, thus decreasing the amount of solid AgCl.   **13.78** (a) decreases; (b) remains the same; (c) increases   **13.80** $[H_2]$ decreases when the temperature is increased. As the temperature is decreased, the reaction shifts to the right. $[CO_2]$ and $[H_2]$ increase, $[CO]$ and $[H_2O]$ decrease, and K_c

increases.   **13.82** (a) increases; (b) increases; (c) decreases; (d) increases   **13.84** (a) decreases; (b) increases; (c) no change; (d) increases; (e) no change   **13.86**

$$k_f[A][B] = k_r[C]; \quad \dfrac{k_f}{k_r} = \dfrac{[C]}{[A][B]} = K_c$$

13.88 210 **13.90** k_r increases more than k_f, this means that E_a (reverse) is greater than E_a (forward). The reaction is exothermic when E_a (reverse) > E_a (forward).   **13.92** (a) $K_c = 1.51$; $K_p = 49.6$; (b)

13.94 1.5×10^{-6} M   **13.96** $[H_2] = [I_2] = 0.045$ M; $[HI] = 0.31$ M   **13.98** $[CO] = [H_2] = 0.18$ M; $[H_2O] = 1.02$ M   **13.100** A decrease in volume (a) and the addition of reactants (c) will affect the composition of the equilibrium mixture, but leave the value of K_c unchanged. A change in temperature (b) affects the value of K_c. Addition of a catalyst (d) or an inert gas (e) affects neither the composition of the equilibrium mixture nor the value of K_c.   **13.102** (a) 2.5; (b) 0.0038; (c) K_c for the water solution is so much smaller than K_c for the benzene solution because H_2O can hydrogen bond with acetic acid, thus preventing acetic acid dimer formation. Benzene cannot hydrogen bond with acetic acid.   **13.104** (a) $K_c = 0.573$ and $K_p = 23.5$; (b) The reaction proceeds to the right to reach equilibrium. $[PCl_5] = 0.365$ M, $[PCl_3] = 0.285$ M, $[Cl_2] = 0.735$ M   **13.106**

(a) $K_c = \dfrac{[C_2H_6][C_2H_4]}{[C_4H_{10}]}$ $K_p = \dfrac{(P_{C_2H_6})(P_{C_2H_4})}{(P_{C_4H_{10}})}$

(b) 0.19; (c) 38%; $P_{total} = 69$ atm; (d) A decrease in volume would decrease the % conversion of C_4H_{10}.   **13.108** (a) 0.0840; (b) $[(CH_3)_2C=CH_2] = [HCl] = 0.094$ M; $[(CH_3)_3CCl] = 0.106$ M; (c) $P_{t\text{-butyl chloride}} = 0.055$ atm; $P_{isobutylene} = 0.345$ atm; $P_{HCl} = 0.545$ atm   **13.110** The activation energy (E_a) is positive, and for an exothermic reaction, $E_{a,r} > E_{a,f}$.

$$K_c = \dfrac{k_f}{k_r} = \dfrac{A_f\, e^{-E_{a,f}/RT}}{A_r\, e^{-E_{a,r}/RT}} = \dfrac{A_f}{A_r} e^{(E_{a,r} - E_{a,f})/RT}$$

$(E_{a,r} - E_{a,f})$ is positive, so the exponent is always positive. As the temperature increases, the exponent, $(E_{a,r} - E_{a,f})/RT$, decreases and the value for K_c decreases as well.   **13.112** (a) 1.47; (b) 10.3; (c) In agreement with Le Châtelier's principle, the reaction is endothermic because K_p increases with increasing temperature.   **13.114** (a) 1.52; (b) 6.89 kJ   **13.116** 6.0×10^2 O_3 molecules

Chapter 14

14.1 $d_{H_2} = 8.24 \times 10^{-5}$ g/cm^3. Air is 14 times more dense than H_2.   **14.2** (a) 18.2% D; (b) 24.3% ^{13}C; (c) The isotope effect for H is larger than that for C because D is two times the mass of ^{1}H while ^{13}C is only about 8% heavier than ^{12}C.   **14.3** $2\,Ga(s) + 6\,H^+(aq) \rightarrow 3\,H_2(g) + 2\,Ga^{3+}(aq)$ **14.4** (a) $SrH_2(s) + 2\,H_2O(l) \rightarrow 2\,H_2(g) + Sr^{2+}(aq) + 2\,OH^-(aq)$; (b) $KH(s) + H_2O(l) \rightarrow H_2(g) + K^+(aq) + OH^-(aq)$

14.5 1.7×10^2 kg **14.6** (a) (1) HF, covalent, (2) CH_4, covalent, (3) KH, ionic, (4) TiH_x, interstitial; (b) KH and TiH_x are solids at room temperature while HF and CH_4 are gases. HF has stronger intermolecular forces than CH_4, because HF can hydrogen bond and CH_4 cannot. Consequently, CH_4 has the lowest melting point. (c) KH **14.7** $PdH_{0.74}$; $d_H = 0.0841$ g/cm^3; $M_H = 83.4$ M **14.8** 15.5 mL **14.9** (a) Li_2O, Ga_2O_3, CO_2; (b) Li_2O is the most ionic. CO_2 is the most covalent. (c) CO_2 is the most acidic. Li_2O is the most basic. (d) Ga_2O_3 **14.10** (a) $Li_2O(s) + H_2O(l) \rightarrow 2\,Li^+(aq) + 2\,OH^-(aq)$; (b) $SO_3(l) + H_2O(l) \rightarrow H^+(aq) + HSO_4^-(aq)$; (c) $Cr_2O_3(s) + 6\,H^+(aq) \rightarrow 2\,Cr^{3+}(aq) + 3\,H_2O(l)$; (d) $Cr_2O_3(s) + 2\,OH^-(aq) + 3\,H_2O(l) \rightarrow 2\,Cr(OH)_4^-(aq)$ **14.11** (a) –1, peroxide; (b) –2, oxide; (c) –1/2, superoxide; (d) –1, peroxide; (e) –2, oxide **14.12** (a) $Rb_2O_2(s) + H_2O(l) \rightarrow 2\,Rb^+(aq) + HO_2^-(aq) + OH^-(aq)$; (b) $CaO(s) + H_2O(l) \rightarrow Ca^{2+}(aq) + 2\,OH^-(aq)$; (c) $2\,CsO_2(s) + H_2O(l) \rightarrow O_2(g) + 2\,Cs^+(aq) + HO_2^-(aq) + OH^-(aq)$; (d) $SrO_2(s) + H_2O(l) \rightarrow Sr^{2+}(aq) + HO_2^-(aq) + OH^-(aq)$; (e) $CO_2(g) + H_2O(l) \rightarrow H^+(aq) + HCO_3^-(aq)$ **14.13**

O_2^- is paramagnetic with one unpaired electron. O_2^- bond order $= (8 - 5)/2 = 1.5$
14.14

$$H-\overset{..}{\underset{..}{O}}-\overset{..}{\underset{..}{O}}-H$$

The electron dot structure indicates a single bond which is consistent with an O–O bond length of 148 pm. **14.15** $PbS(s) + 4\,H_2O_2(aq) \rightarrow PbSO_4(s) + 4\,H_2O(l)$ **14.16** (a) $2\,Li(s) + 2\,H_2O(l) \rightarrow H_2(g) + 2\,Li^+(aq) + 2\,OH^-(aq)$; (b) $Sr(s) + 2\,H_2O(l) \rightarrow H_2(g) + Sr^{2+}(aq) + 2\,OH^-(aq)$; (c) $Br_2(l) + H_2O(l) \rightleftharpoons HOBr(aq) + H^+(aq) + Br^-(aq)$ **14.17** $NiSO_4 \cdot 7\,H_2O$ **14.18** Hydrogen can be stored as a solid in the form of solid interstitial hydrides or in the recently discovered tube-shaped molecules called carbon nanotubes. **14.19** 1.5×10^7 kJ; 1.0×10^3 kg O_2 ***Understanding Key Concepts*** **14.20** (a) (1) covalent, (2) ionic, (3) covalent, (4) interstitial; (b) (1) H, +1; other element, –3, (2) H, –1; other element, +1 (3) H, +1; other element, –2 **14.22** (a), (b), and (d) are different kinds of water and have similar properties. (c) and (e) are different kinds of hydrogen peroxide and have similar properties. The properties of (a), (b), and (d) are quite different from those of (c) and (e). **14.24** (a) (1) –2, +2; (2) –2, +1; (3) –2, +5; (b) (1) three-dimensional; (2) molecular; (3) molecular; (c) (1) solid; (2) gas or liquid; (3) gas or liquid; (d) (2) hydrogen; (3) nitrogen **14.26** (a) 4; (b) (3); (c) (1) and (4) ***Additional Problems*** **14.28** Quantitative differences in properties that arise from the differences in the masses of the isotopes are known as isotope effects. Examples: H_2 and D_2 have different melting and boiling points. H_2O and D_2O have different dissociation constants. **14.30** % mass difference (^{1}H and ^{2}H) = 98.85%; % mass difference (^{2}H and ^{3}H) =

49.74%. The differences in properties will be larger for H_2O and D_2O rather than for D_2O and T_2O because of the larger relative difference in mass for H and D versus D and T. This is supported by the data in Table 14.1. **14.32** 18:

$H_2^{16}O$	$H_2^{17}O$	$H_2^{18}O$
$D_2^{16}O$	$D_2^{17}O$	$D_2^{18}O$
$T_2^{16}O$	$T_2^{17}O$	$T_2^{18}O$
$HD^{16}O$	$HD^{17}O$	$HD^{18}O$
$HT^{16}O$	$HT^{17}O$	$HT^{18}O$
$DT^{16}O$	$DT^{17}O$	$DT^{18}O$

14.34 (a) $Zn(s) + 2\,H^+(aq) \rightarrow H_2(g) + Zn^{2+}(aq)$; (b) $H_2O(g) + C(s) \rightarrow CO(g) + H_2(g)$; (c) $H_2O(g) + CH_4(g) \rightarrow CO(g) + 3\,H_2(g)$; (d) electrolysis: $2\,H_2O(l) \rightarrow 2\,H_2(g) + O_2(g)$; also (b) and (c) above. **14.36** the steam–hydrocarbon reforming process

$$CH_4(g) + H_2O(g) \xrightarrow[\text{Ni catalyst}]{1100°C} CO(g) + 3\,H_2(g)$$

$$CO(g) + H_2O(g) \xrightarrow{400°C} CO_2(g) + H_2(g)$$

$$CO_2(g) + 2\,OH^-(aq) \longrightarrow CO_3^{2-}(aq) + H_2O(l)$$

14.38 (a) LiH; (b) 12.9 kg **14.40** (a) MgH_2, H^-; (b) PH_3, covalent; (c) KH, H^-; (d) HBr, covalent **14.42** H_2S: covalent hydride, gas, weak acid in H_2O; NaH: ionic hydride, solid (salt like), reacts with H_2O to produce H_2; PdH_x metallic (interstitial) hydride, solid, stores hydrogen **14.44** (a) CH_4, covalent bonding; (b) NaH, ionic bonding **14.46** (a) bent; (b) trigonal pyramidal; (c) tetrahedral **14.48** A nonstoichiometric compound is a compound whose atomic composition cannot be expressed as a ratio of small whole numbers. An example is PdH_x. The lack of stoichiometry results from the hydrogen occupying holes in the solid state structure. **14.50** (a) $d_H = 0.16$ g/cm^3; the density of H in TiH_2 is about 2.25 times the density of liquid H_2. (b) 1.8×10^3 cm^3 **14.52** (a) O_2 is obtained in industry by the fractional distillation of liquid air; (b) In the laboratory, O_2 is prepared by the thermal decomposition of $KClO_3(s)$.

$$2\,KClO_3(s) \xrightarrow[\text{MnO}_2]{\text{Heat}} 2\,KCl(s) + 3\,O_2(g)$$

14.54 7.34 L **14.56** (a) $4\,Li(s) + O_2(g) \rightarrow 2\,Li_2O(s)$; (b) $P_4(s) + 5\,O_2(g) \rightarrow P_4O_{10}(s)$; (c) $4\,Al(s) + 3\,O_2(g) \rightarrow 2\,Al_2O_3(s)$; (d) $Si(s) + O_2(g) \rightarrow SiO_2(s)$ **14.58**

$$:\overset{..}{O}::\overset{..}{O}:$$

The electron dot structure shows an O=O double bond. It also shows all electrons paired. This is not consistent with the fact that O_2 is paramagnetic. **14.60** $Li_2O < BeO < B_2O_3 < CO_2 < N_2O_5$ **14.62** $N_2O_5 < Al_2O_3 < K_2O < Cs_2O$ **14.64** (a) CrO_3; (b) N_2O_5; (c) SO_3 **14.66** (a) $Cl_2O_7(l) + H_2O(l) \rightarrow 2\,H^+(aq) + 2\,ClO_4^-(aq)$; (b) $K_2O(s) + H_2O(l) \rightarrow 2\,K^+(aq) + 2\,OH^-(aq)$; (c) $SO_3(l) + H_2O(l) \rightarrow H^+(aq) + HSO_4^-(aq)$ **14.68** (a) $ZnO(s) + 2\,H^+(aq) \rightarrow Zn^{2+}(aq) + H_2O(l)$; (b) $ZnO(s) + 2\,OH^-(aq) + H_2O(l) \rightarrow Zn(OH)_4^{2-}(aq)$ **14.70** A peroxide has oxygen in the –1 oxidation state, for example, H_2O_2. A superoxide has oxygen in the –1/2 oxidation state, for example, KO_2. **14.72** (a) BaO_2; (b) CaO; (c) CsO_2; (d) Li_2O; (e) Na_2O_2

14.74

	O_2	O_2^-	O_2^{2-}
σ^*_{2p}	—	—	—
π^*_{2p}	↑ ↑	↑↓ ↑	↑↓ ↑↓
π_{2p}	↑↓ ↑↓	↑↓ ↑↓	↑↓ ↑↓
σ_{2p}	↑↓	↑↓	↑↓
Bond order:	2	1.5	1

(a) The O–O bond length increases because the bond order decreases. The bond order decreases because of the increased occupancy of antibonding orbitals; (b) O_2^- has 1 unpaired electron and is paramagnetic. O_2^{2-} has no unpaired electrons and is diamagnetic.   **14.76**
(a) $H_2O_2(aq) + 2\,H^+(aq) + 2\,I^-(aq) \rightarrow I_2(aq) + 2\,H_2O(l)$;
(b) $3\,H_2O_2(aq) + 8\,H^+(aq) + Cr_2O_7^{2-}(aq) \rightarrow 2\,Cr^{3+}(aq) + 3\,O_2(g) + 7\,H_2O(l)$
14.78

14.80

$$3\,O_2(g) \xrightarrow[\text{discharge}]{\text{Electric}} 2\,O_3(g)$$

14.82 (a) $2\,F_2(g) + 2\,H_2O(l) \rightarrow O_2(g) + 4\,HF(aq)$;
(b) $Cl_2(g) + H_2O(l) \rightleftharpoons HOCl(aq) + H^+(aq) + Cl^-(aq)$;
(c) $I_2(s) + H_2O(l) \rightarrow HOI(aq) + H^+(aq) + I^-(aq)$;
(d) $Ba(s) + 2\,H_2O(l) \rightarrow H_2(g) + Ba^{2+}(aq) + 2\,OH^-(aq)$
14.84 $AlCl_3 \cdot 6\,H_2O$

14.86 12.41%   **14.88** The mineral gypsum is $CaSO_4 \cdot 2\,H_2O$; $x = 2$ **14.90** 1.5×10^{11} kg   **14.92** a reducing agent; Ca or Al   **14.94** 2.2×10^3 L   **14.96** (a) B_2O_3, diboron trioxide; (b) H_2O_2, hydrogen peroxide; (c) SrH_2, strontium hydride; (d) CsO_2, cesium superoxide; (e) $HClO_4$, perchloric acid; (f) BaO_2, barium peroxide   **14.98** (a) 6: $^{16}O_2$, $^{17}O_2$, $^{18}O_2$, $^{16}O^{17}O$, $^{16}O^{18}O$, $^{17}O^{18}O$;
(b) 18:

$^{16}O_3$	$^{16}O_2{}^{17}O$	$^{18}O_3$
$^{17}O_2{}^{16}O$	$^{17}O_3$	$^{16}O_2{}^{18}O$
$^{18}O_2{}^{16}O$	$^{18}O_2{}^{17}O$	$^{17}O_2{}^{18}O$
$^{16}O^{17}O^{16}O$	$^{17}O^{18}O^{17}O$	$^{16}O^{17}O^{18}O$
$^{16}O^{18}O^{16}O$	$^{17}O^{16}O^{17}O$	$^{18}O^{16}O^{18}O$
$^{17}O^{18}O^{16}O$	$^{18}O^{16}O^{17}O$	$^{18}O^{17}O^{18}O$

14.100 (a) $2\,H_2(g) + O_2(g) \rightarrow 2\,H_2O(l)$; (b) $O_3(g) + 2\,I^-(aq) + H_2O(l) \rightarrow O_2(g) + I_2(aq) + 2\,OH^-(aq)$; (c) $H_2O_2(aq) + 2\,H^+(aq) + 2\,Br^-(aq) \rightarrow 2\,H_2O(l) + Br_2(aq)$; (d) $2\,Na(l) + H_2(g) \rightarrow 2\,NaH(s)$; (e) $2\,Na(s) + 2\,H_2O(l) \rightarrow H_2(g) + 2\,Na^+(aq) + 2\,OH^-(aq)$ **14.102** K is oxidized by water. F_2 is reduced by water. Cl_2 and Br_2 disproportionate when treated with water.   **14.104** (a) −128.2 kJ; (b) −41.2 kJ; (c) −78.0 kJ; (d) 2816 kJ   **14.106** 496 J   **14.108** M = Sr; SrH_2 **14.110** (a) 0.58 mm Hg; (b) 9.4×10^{18} HD molecules; (c) 7.3×10^{14} D_2 molecules

Chapter 15

15.1 (a) $H_2SO_4(aq) + H_2O(l) \rightleftharpoons H_3O^+(aq) + HSO_4^-(aq)$
conjugate base
(b) $HSO_4^-(aq) + H_2O(l) \rightleftharpoons H_3O^+(aq) + SO_4^{2-}(aq)$
conjugate base
(c) $H_3O^+(aq) + H_2O(l) \rightleftharpoons H_3O^+(aq) + H_2O(l)$
conjugate base
15.2 (a) H_2CO_3; (b) HCO_3^-; (c) H_2O **15.3** acids: HF and H_2S; bases: HS^- and F^-; conjugate acid–base pairs: H_2S and HS^-; HF and F^- **15.4** (a) from right to left; (b) from left to right   **15.5** (a) HY; (b) X^-; (c) to the left   **15.6** $[H_3O^+] = 2.0 \times 10^{-9}$ M, solution is basic.   **15.7** 2.3×10^{-7} M   **15.8** (a) 8.20; (b) 4.22   **15.9** $[H_3O^+] = 4.0 \times 10^{-8}$ M; $[OH^-] = 2.5 \times 10^{-7}$ M   **15.10** (a) 1.30; (b) −0.78; (c) 12.30; (d) 12.30   **15.11** 11.81   **15.12** $K_a = 3.5 \times 10^{-8}$. This value of K_a agrees with the value in Table 15.2.   **15.13** (a) HY < HX < HZ; (b) HZ; (c) HY has the highest pH; HX has the lowest pH.   **15.14** (a) pH = 2.38; $[H_3O^+] = [CH_3CO_2^-] = 4.2 \times 10^{-3}$ M; $[CH_3CO_2H] = 1.00$ M; $[OH^-] = 2.4 \times 10^{-12}$ M; (b) pH = 3.38; $[H_3O^+] = [CH_3CO_2^-] = 4.2 \times 10^{-4}$ M; $[CH_3CO_2H] = 0.0096$ M; $[OH^-] = 2.4 \times 10^{-11}$ M   **15.15** pH = 3.20   **15.16** (a) 8.0%; (b) 2.6%   **15.17** $[H_3O^+] = [HSO_3^-] = 0.032$ M; $[H_2SO_3] = 0.07$ M; $[SO_3^{2-}] = 6.3 \times 10^{-8}$ M; $[OH^-] = 3.1 \times 10^{-13}$ M; pH = 1.49   **15.18** $[H_2SO_4] = 0$ M; $[HSO_4^-] = 0.49$ M; $[SO_4^{2-}] = 0.011$ M; $[H_3O^+] = 0.51$ M; $[OH^-] = 2.0 \times 10^{-14}$ M; pH = 0.29   **15.19** $[NH_4^+] = [OH^-] = 2.7 \times 10^{-3}$ M; $[NH_3] = 0.40$ M; $[H_3O^+] = 3.7 \times 10^{-12}$ M; pH = 11.43   **15.20** pH = 9.45   **15.21** (a) 7.7×10^{-12}; (b) 2.9×10^{-7} **15.22** (a) acidic. pH = 4.92; (b) acidic. pH = 5.00   **15.23** pH = 8.32   **15.24** $K_a = 5.6 \times 10^{-10}$; $K_b = 2.0 \times 10^{-5}$; Because $K_b > K_a$, the solution is basic.   **15.25** (a) neutral; (b) basic; (c) acidic; (d) acidic; (e) acidic   **15.26** (a) H_2Se; (b) HI; (c) HNO_3; (d) H_2SO_3 **15.27** (a) Lewis acid, $AlCl_3$; Lewis base, Cl^-; (b) Lewis acid, Ag^+; Lewis base, NH_3; (c) Lewis acid, SO_2; Lewis base, OH^-; (d) Lewis acid, Cr^{3+}; Lewis base, H_2O
15.28

15.29 The O^{2-} from CaO is the Lewis base, and SO_2 is the Lewis acid.

15.30 4.101   *Understanding Key Concepts*   **15.32** (a) X^-, Y^-, Z^-; (b) HX < HZ < HY; (c) HY; (d) HX; (e) 20%   **15.34** (b)   **15.36** (a) $Y^- < Z^- < X^-$; (b) Y^-; (c) The numbers of HA molecules and OH^- ions are equal because the reaction of A^- with water has a 1:1 stoichiometry: $A^- + H_2O \rightleftharpoons HA + OH^-$ **15.38** (a) Brønsted–Lowry acids: NH_4^+, $H_2PO_4^-$; Brønsted–Lowry bases: SO_3^{2-}, OCl^-, $H_2PO_4^-$; (b) Lewis acids: Fe^{3+}, BCl_3; Lewis bases: SO_3^{2-}, OCl^-, $H_2PO_4^-$ *Additional Problems*   **15.40** NH_3, CN^-, and NO_2^-

15.42 (a) SO_4^{2-}; (b) HSO_3^-; (c) HPO_4^{2-}; (d) NH_3;
(e) OH^-; (f) NH_2^- **15.44**

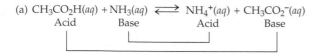

(a) $CH_3CO_2H(aq) + NH_3(aq) \rightleftharpoons NH_4^+(aq) + CH_3CO_2^-(aq)$
    Acid   Base   Acid   Base

(b) $CO_3^{2-}(aq) + H_3O^+(aq) \rightleftharpoons H_2O(l) + HCO_3^-(aq)$
    Base   Acid   Base   Acid

(c) $HSO_3^-(aq) + H_2O(l) \rightleftharpoons H_3O^+(aq) + SO_3^{2-}(aq)$
    Acid   Base   Acid   Base

(d) $HSO_3^-(aq) + H_2O(l) \rightleftharpoons H_2SO_3(aq) + OH^-(aq)$
    Base   Acid   Acid   Base

15.46 Strong acids: HNO_3 and H_2SO_4; Strong bases:
H^- and O^{2-} **15.48** (a) left; (b) left; (c) right; (d) right
15.50 (a) $[OH^-] = 2.9 \times 10^{-6}$ M, basic; (b) $[H_3O^+] = 1.0 \times 10^{-12}$ M, basic; (c) $[H_3O^+] = 1.0 \times 10^{-4}$ M, acidic; (d) $[OH^-] = 1.0 \times 10^{-7}$ M, neutral; (e) $[OH^-] = 1.2 \times 10^{-10}$ M, acidic
15.52 (a) 4.70; (b) 11.6; (c) 8.449; (d) 3; (e) 15.08   **15.54** (a)
8×10^{-5} M; (b) 1.5×10^{-11} M; (c) 1.0 M; (d) 5.6×10^{-15} M;
(e) 10 M   **15.56** (a) 3; (b) 5.00; (c) 0.30   **15.58** (a) 2; (b) 7.4;
(c) 6.9; (d) 4.8 to 8.3   **15.60** (a) 0.40; (b) 10.57; (c) 10.00
15.62 (a) 13.90; (b) 1.19; (c) 2.30; (d) 2.92   **15.64**

(a) $HClO_2(aq) + H_2O(l) \rightleftharpoons H_3O^+(aq) + ClO_2^-(aq)$;
$$K_a = \frac{[H_3O^+][ClO_2^-]}{[HClO_2]}$$

(b) $HOBr(aq) + H_2O(l) \rightleftharpoons H_3O^+(aq) + OBr^-(aq)$;
$$K_a = \frac{[H_3O^+][OBr^-]}{[HOBr]}$$

(c) $HCO_2H(aq) + H_2O(l) \rightleftharpoons H_3O^+(aq) + HCO_2^-(aq)$;
$$K_a = \frac{[H_3O^+][HCO_2^-]}{[HCO_2H]}$$

15.66 (a) $C_6H_5OH < HOCl < CH_3CO_2H < HNO_3$;
(b) $HNO_3 > CH_3CO_2H > HOCl > C_6H_5OH$; 1 M HNO_3,
$[H_3O^+] = 1$ M; 1 M CH_3CO_2H, $[H_3O^+] = 4 \times 10^{-3}$ M; 1 M
$HOCl$, $[H_3O^+] = 2 \times 10^{-4}$ M; 1 M C_6H_5OH, $[H_3O^+] = 1 \times 10^{-5}$ M   **15.68** $K_a = 2.0 \times 10^{-9}$ **15.70** $[H_3O^+] = [C_6H_5O^-] = 3.6 \times 10^{-6}$ M; $[C_6H_5OH] = 0.10$ M; $[OH^-] = 2.8 \times 10^{-9}$ M;
pH = 5.44; % dissociation = 0.0036%   **15.72** pH = 1.59;
% dissociation = 1.7%   **15.74**

$H_2SeO_4(aq) + H_2O(l) \rightleftharpoons H_3O^+(aq) + HSeO_4^-(aq)$;
$$K_{a1} = \frac{[H_3O^+][HSeO_4^-]}{[H_2SeO_4]}$$

$HSeO_4^-(aq) + H_2O(l) \rightleftharpoons H_3O^+(aq) + SeO_4^{2-}(aq)$;
$$K_{a2} = \frac{[H_3O^+][SeO_4^{2-}]}{[HSeO_4^-]}$$

15.76 $[H_3O^+] = [HCO_3^-] = 6.6 \times 10^{-5}$ M; $[H_2CO_3] = 0.010$ M; $[CO_3^{2-}] = 5.6 \times 10^{-11}$ M; $[OH^-] = 1.5 \times 10^{-10}$ M;
pH = 4.18   **15.78** pH = 1.08; $[C_2O_4^{2-}] = 6.4 \times 10^{-5}$ M   **15.80**

(a) $(CH_3)_2NH(aq) + H_2O(l) \rightleftharpoons$
                          $(CH_3)_2NH_2^+(aq) + OH^-(aq)$;
$$K_b = \frac{[(CH_3)_2NH_2^+][OH^-]}{[(CH_3)_2NH]}$$

(b) $C_6H_5NH_2(aq) + H_2O(l) \rightleftharpoons$
                       $C_6H_5NH_3^+(aq) + OH^-(aq)$;
$$K_b = \frac{[C_6H_5NH_3^+][OH^-]}{[C_6H_5NH_2]}$$

(c) $CN^-(aq) + H_2O(l) \rightleftharpoons HCN(aq) + OH^-(aq)$;
$$K_b = \frac{[HCN][OH^-]}{[CN^-]}$$

15.82 $K_b = 1 \times 10^{-6}$ **15.84** (a) 11.96; (b) 8.92; (c) 9.42   **15.86**
(a) 2.0×10^{-11}; (b) 1.1×10^{-6}; (c) 2.3×10^{-5}; (d) 5.6×10^{-6}
15.88

(a) $CH_3NH_3^+(aq) + H_2O(l) \rightleftharpoons H_3O^+(aq) + CH_3NH_2(aq)$
    Acid   Base   Acid   Base

(b) $Cr(H_2O)_6^{3+}(aq) + H_2O(l) \rightleftharpoons H_3O^+(aq) + Cr(H_2O)_5OH^{2+}(aq)$
    Acid   Base   Acid   Base

(c) $CH_3CO_2^-(aq) + H_2O(l) \rightleftharpoons OH^-(aq) + CH_3CO_2H(aq)$
    Base   Acid   Base   Acid

(d) $PO_4^{3-}(aq) + H_2O(l) \rightleftharpoons OH^-(aq) + HPO_4^{2-}(aq)$
    Base   Acid   Base   Acid

15.90 (a) basic; (b) neutral; (c) acidic; (d) neutral; (e) basic;
(f) acidic   **15.92** (a) $[H_3O^+] = [C_2H_5NH_2] = 1.2 \times 10^{-6}$ M;
$[C_2H_5NH_3^+] = 0.10$ M; $[NO_3^-] = 0.10$ M; $[OH^-] = 8.0 \times 10^{-9}$ M;
pH = 5.90; (b) $[CH_3CO_2H] = [OH^-] = 7.5 \times 10^{-6}$ M; $[CH_3CO_2^-] = 0.10$ M; $[Na^+] = 0.10$ M; $[H_3O^+] = 1.3 \times 10^{-9}$ M; pH = 8.89;
(c) $[Na^+] = [NO_3^-] = 0.10$ M; $[H_3O^+] = [OH^-] = 1.0 \times 10^{-7}$ M;
pH = 7.00   **15.94** (a) $PH_3 < H_2S < HCl$; electronegativity
increases from P to Cl; (b) $NH_3 < PH_3 < AsH_3$; X–H bond
strength decreases from N to As (down a group); (c) $HBrO < HBrO_2 < HBrO_3$; acid strength increases with the number of
O atoms   **15.96** (a) HCl; The strength of a binary acid H_nA
increases as A moves from left to right and from top to bottom in the periodic table. (b) $HClO_3$; The strength of an
oxoacid increases with increasing electronegativity and
increasing oxidation state of the central atom. (c) HBr; The
strength of a binary acid H_nA increases as A moves from left
to right and from top to bottom in the periodic table.   **15.98**
(a) H_2Te, weaker X–H bond; (b) H_3PO_4, P has higher electronegativity; (c) $H_2PO_4^-$, lower negative charge; (d) NH_4^+,
higher positive charge and N is more electronegative than C.
15.100 (a) Lewis acid, SiF_4; Lewis base, F^-; (b) Lewis acid,
Zn^{2+}; Lewis base, NH_3; (c) Lewis acid, $HgCl_2$; Lewis base,
Cl^-; (d) Lewis acid, CO_2; Lewis base, H_2O

15.102

(a) $2 \; \ddot{\ddot{F}} \colon^{-} + SiF_4 \longrightarrow SiF_6{}^{2-}$

(b) $4 \; \ddot{N}H_3 + Zn^{2+} \longrightarrow Zn(NH_3)_4{}^{2+}$

(c) $2 \; \ddot{\ddot{Cl}} \colon^{-} + HgCl_2 \longrightarrow HgCl_4{}^{2-}$

(d) $H_2\ddot{O} \colon + CO_2 \longrightarrow H_2CO_3$

15.104 (a) CN^-, Lewis base; (b) H^+, Lewis acid; (c) H_2O, Lewis base; (d) Fe^{3+}, Lewis acid; (e) OH^-, Lewis base; (f) CO_2, Lewis acid; (g) $P(CH_3)_3$, Lewis base; (h) $B(CH_3)_3$, Lewis acid **15.106** H_2S, acid only; HS^-, acid and base; S^{2-}, base only; H_2O, acid and a base; H_3O^+, acid only; OH^-, base only

15.108

$$\left[H-\ddot{O}-H \atop \quad\; | \atop \quad\; H \right]^{+} + H-\ddot{\ddot{O}}-H \longrightarrow \left[H-\ddot{O}-H\cdots\ddot{\ddot{O}}-H \atop \quad\; | \qquad\qquad | \atop \quad\; H \qquad\qquad H \right]^{+}$$

H_3O^+ can hydrogen bond with additional H_2O molecules.
15.110 $HCO_3{}^-(aq) + Al(H_2O)_6{}^{3+}(aq) \rightarrow H_2O(l) + CO_2(g) + Al(H_2O)_5(OH)^{2+}(aq)$ **15.112** fraction dissociated $= 6.09 \times 10^{-10}$; % dissociation $= 6.09 \times 10^{-8}$ % **15.114** $K_{a1} = 1.0 \times 10^{-8}$; $K_{a2} = 7.7 \times 10^{-4}$ **15.116** (a) $A^-(aq) + H_2O(l) \rightleftharpoons HA(aq) + OH^-(aq)$, basic; (b) $M(H_2O)_6{}^{3+}(aq) + H_2O(l) \rightleftharpoons H_3O^+(aq) + M(H_2O)_5(OH)^{2+}(aq)$, acidic; (c) $2 \; H_2O(l) \rightleftharpoons H_3O^+(aq) + OH^-(aq)$, neutral; (d) $M(H_2O)_6{}^{3+}(aq) + A^-(aq) \rightleftharpoons HA(aq) + M(H_2O)_5(OH)^{2+}(aq)$, acidic because K_a for $M(H_2O)_6{}^{3+}$ (10^{-4}) is greater than K_b for A^- (10^{-9}). **15.118** (a) acidic; (b) basic **15.120**

$$\text{Fraction dissociated} = \frac{[HA]_{diss}}{[HA]_{initial}}$$

For a weak acid, $[HA]_{diss} = [H_3O^+] = [A^-]$

$$K_a = \frac{[H_3O^+][A^-]}{[HA]} = \frac{[H_3O^+]^2}{[HA]}; \quad [H_3O^+] = \sqrt{K_a[HA]}$$

$$\text{Fraction dissociated} = \frac{[HA]_{diss}}{[HA]} = \frac{[H_3O^+]}{[HA]}$$
$$= \frac{\sqrt{K_a[HA]}}{[HA]} = \sqrt{\frac{K_a}{[HA]}}$$

$$\text{\% dissociation} = \sqrt{\frac{K_a}{[HA]}} \times 100\%$$

15.122 pH = 2.54 **15.124** $[H_3PO_4] = 0.67$ M; $[H_2PO_4{}^-] = 0.071$ M; $[HPO_4{}^{2-}] = 6.2 \times 10^{-8}$ M; $[PO_4{}^{3-}] = 4.2 \times 10^{-19}$ M; $[H_3O^+] = 0.0708$ M; $[OH^-] = 1.4 \times 10^{-13}$ M; pH = 1.15 **15.126** 0.25 m; $-0.93°C$ **15.128** $H_3O^+(aq) + PO_4{}^{3-}(aq) \rightleftharpoons HPO_4{}^{2-}(aq) + H_2O(l)$; $H_3O^+(aq) + HPO_4{}^{2-}(aq) \rightleftharpoons H_2PO_4{}^-(aq) + H_2O(l)$; $H_3O^+(aq) + H_2PO_4{}^-(aq) \rightleftharpoons H_3PO_4(aq) + H_2O(l)$; pH = 2.02

Chapter 16

16.1 (a) $HNO_2(aq) + OH^-(aq) \rightleftharpoons NO_2{}^-(aq) + H_2O(l)$; pH > 7.00; (b) $H_3O^+(aq) + NH_3(aq) \rightleftharpoons NH_4{}^+(aq) + H_2O(l)$; pH < 7.00; (c) $OH^-(aq) + H_3O^+(aq) \rightleftharpoons 2 \; H_2O(l)$; pH = 7.00

16.2 (a) $HF(aq) + OH^-(aq) \rightleftharpoons H_2O(l) + F^-(aq)$; $K_n = 3.5 \times 10^{10}$; (b) $H_3O^+(aq) + OH^-(aq) \rightleftharpoons 2 \; H_2O(l)$; $K_n = 1.0 \times 10^{14}$; (c) $HF(aq) + NH_3(aq) \rightleftharpoons NH_4{}^+(aq) + F^-(aq)$; $K_n = 6.3 \times 10^5$. The tendency to proceed to completion is: reaction (c) < reaction (a) < reaction (b) **16.3** $[H_3O^+] = 1.2 \times 10^{-9}$ M; $[OH^-] = 8.2 \times 10^{-6}$ M; $[Na^+] = [CN^-] = 0.010$ M; $[HCN] = 0.025$ M; pH = 8.91; % dissociation $= 4.9 \times 10^{-6}$ % **16.4** pH = 9.55 **16.5** Solution (a) has the largest percent dissociation. Solution (c) has the highest pH. **16.6** (a) (1) and (3); (b) (3) **16.7** For the buffer, pH = 3.76; (a) pH = 3.71; (b) pH = 3.87 **16.8** This solution has less buffering capacity than the solution in Problem 16.7 because it contains less HF and F^- per 100 mL. Note that the change in pH is greater than that in Problem 16.7. **16.9** pH = 9.95 **16.10** Make the Na_2CO_3 concentration 1.4 times the concentration of $NaHCO_3$. **16.11** $HOCl$ ($K_a = 3.5 \times 10^{-8}$) and $NaOCl$. **16.12** (a) 9.44; (b) 7.87 **16.13** (a) 2.17; (b) 11.77. The results are consistent with the pH data in Table 16.1. **16.14** (a) 12.00; (b) 4.08; (c) 2.15 **16.15** (a) (3); (b) (1); (c) (4); (d) (2) **16.16** 40 mL of 0.0400 M NaOH are required to reach the equivalence point. (a) 6.98; (b) 7.46; (c) 9.75 **16.17** Use thymolphthalein (pH 9.4–10.6). Bromthymol blue is unacceptable because it changes color halfway to the equivalence point. **16.18** (a) 4.51; (b) 7.20; (c) 7.7 **16.19** (a) 5.97; (b) 9.62; (c) 10.91 **16.20** (a) $K_{sp} = [Ag^+][Cl^-]$; (b) $K_{sp} = [Pb^{2+}][I^-]^2$; (c) $K_{sp} = [Ca^{2+}]^3[PO_4{}^{3-}]^2$ **16.21** 2.1×10^{-33} **16.22** 1.10×10^{-10} **16.23** AgCl, solubility = 0.0019 g/L; Ag_2CrO_4, solubility = 0.022 g/L. Ag_2CrO_4 has the higher solubility, despite its smaller value of K_{sp}. **16.24** (a) AgZ; (b) AgY **16.25** 1.4×10^{-5} M **16.26** AgCN, $Al(OH)_3$, and ZnS **16.27** 1.1×10^{-12} M **16.28** $K = 25.4$; 0.045 mol/L **16.29** (a) IP > K_{sp}; a precipitate of $BaCO_3$ will form. (b) IP < K_{sp}; no precipitate will form. **16.30** For $Mn(OH)_2$, IP < K_{sp}; no precipitate will form. For $Fe(OH)_2$, IP > K_{sp}; a precipitate of $Fe(OH)_2$ will form. **16.31** $Q_c > K_{spa}$ for CdS; CdS will precipitate. $Q_c < K_{spa}$ for ZnS; Zn^{2+} will remain in solution. **16.32** This protein has both acidic and basic sites. H_3PO_4–$H_2PO_4{}^-$, an acidic buffer, protonates the basic sites in the protein making them positive and the protein migrates towards the negative electrode. H_3BO_3–$H_2BO_3{}^-$ is a basic buffer. At basic pH's, the acidic sites in the protein are dissociated making them negative and the protein migrates towards the positive electrode. **16.33** To increase the rate at which the proteins migrate toward the negative electrode, increase the number of basic sites that are protonated by lowering the pH. Decrease the $[HPO_4{}^{2-}]/[H_2PO_4{}^-]$ ratio. ***Understanding Key Concepts*** **16.34** (a) (1), (3), and (4); (b) (4) **16.36** (4) **16.38** (a) (i) (1), (ii) (4), (iii) (3), (iv) (2); (b) less than 7 **16.40** A precipitate will form when IP > K_{sp}. A precipitate will form only in (3). **16.42** (a) The bottom curve represents the titration of a strong acid; the top curve represents the titration of a weak acid; (b) pH = 7 for titration of the strong acid; pH = 10 for titration of the weak acid; (c) pK_a ~ 6.3 ***Additional Problems*** **16.44** (a) pH = 7.00; (b) pH > 7.00; (c) pH < 7.00; (d) pH > 7.00 **16.46** (a) 1.0×10^{14}; (b) 3.5×10^6; (c) 4.3×10^4; (d) 6.5×10^9; (c) < (b) < (d) < (a) **16.48** Solution (a) **16.50** $K_n = 2.3 \times 10^{-5}$; K_n is small so the neutralization reaction does not proceed very far to completion. **16.52** (a) $NaNO_2$; (c) HCl; (d) $Ba(NO_2)_2$ **16.54** (a) pH increases; (b) no change; (c) pH decreases **16.56** pH = 3.06 **16.58** For 0.10 M HN_3: % dissociation = 1.4%. For 0.10 M HN_3 in 0.10 M HCl:

% dissociation = 0.019%. The % dissociation is less because of the common ion (H_3O^+) effect.   **16.60** Solutions (a), (c), and (d)   **16.62** Solution (a)   **16.64** When blood absorbs acid, the equilibrium shifts to the left, decreasing the pH, but not by much because the $[HCO_3^-]/[H_2CO_3]$ ratio remains nearly constant. When blood absorbs base, the equilibrium shifts to the right, increasing the pH, but not by much because the $[HCO_3^-]/[H_2CO_3]$ ratio remains nearly constant.   **16.66** pH = 9.09. The pH of a buffer solution will not change on dilution because the acid and base concentrations will change by the same amount and their ratio will remain the same.   **16.68** pH = 9.25; (a) pH = 9.29; (b) pH = 9.07   **16.70** (a) 9.24; (b) 3.74; (c) 7.46   **16.72** pH = 4.04   **16.74** The volume of the 1.0 M NH_3 solution should be 3.5 times the volume of the 1.0 M NH_4Cl solution   **16.76** $H_2PO_4^-$–HPO_4^{2-} because the pK_a for $H_2PO_4^-$ (7.21) is closest to 7.00.   **16.78** (a) 9.00 mmol; (b) 20.0 mL; (c) 7.00; (d)

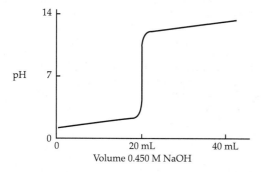

16.80 0.0500 M   **16.82** (a) 0.92; (b) 1.77; (c) 3.5; (d) 7.00; (e) 10.5; (f) 12.60

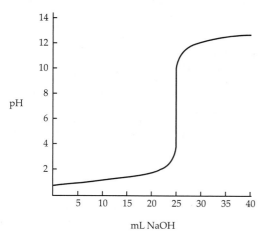

mL NaOH

16.84 50.0 mL; (a) 2.85; (b) 3.46; (c) 8.25; (d) 12.70   **16.86** (a) 1.74; (b) 2.34; (c) 6.02; (d) 9.70; (e) 11.11   **16.88** (a) pH = 8.04, phenol red; (b) pH = 7.00, bromthymol blue or phenol red (Any indicator that changes color in the pH range 4–10 is satisfactory for a strong acid–strong base titration.); (c) pH = 5.92, chlorphenol red   **16.90** (a) $Ag_2CO_3(s) \rightleftharpoons 2 Ag^+(aq) + CO_3^{2-}(aq)$; $K_{sp} = [Ag^+]^2[CO_3^{2-}]$; (b) $PbCrO_4(s) \rightleftharpoons Pb^{2+}(aq) + CrO_4^{2-}(aq)$; $K_{sp} = [Pb^{2+}][CrO_4^{2-}]$; (c) $Al(OH)_3(s) \rightleftharpoons Al^{3+}(aq) + 3 OH^-(aq)$; $K_{sp} = [Al^{3+}][OH^-]^3$; (d) $Hg_2Cl_2(s) \rightleftharpoons Hg_2^{2+}(aq) + 2 Cl^-(aq)$; $K_{sp} = [Hg_2^{2+}][Cl^-]^2$ **16.92** (a) 8.4×10^{-9}; (b) 5.8×10^{-3} M; (c) 0.13 M   **16.94** 8.39×10^{-12} **16.96** (a) 1.1×10^{-5} M; (b) 1.1×10^{-4} M; (c) 1.6×10^{-5} M   **16.98** (a) $AgNO_3$, source of Ag^+; equilib-

rium shifts left; (b) HNO_3, source of H_3O^+, removes CO_3^{2-}; equilibrium shifts right; (c) Na_2CO_3, source of CO_3^{2-}; equilibrium shifts left; (d) NH_3, forms $Ag(NH_3)_2^+$; removes Ag^+; equilibrium shifts right   **16.100** (a) 5.3×10^{-7} M; (b) 2.8×10^{-10} M   **16.102** (b), (c), and (d) are more soluble in acidic solution. (a) $AgBr(s) \rightleftharpoons Ag^+(aq) + Br^-(aq)$; (b) $CaCO_3(s) + H_3O^+(aq) \rightleftharpoons Ca^{2+}(aq) + HCO_3^-(aq) + H_2O(l)$; (c) $Ni(OH)_2(s) + 2 H_3O^+(aq) \rightleftharpoons Ni^{2+}(aq) + 4 H_2O(l)$; (d) $Ca_3(PO_4)_2(s) + 2 H_3O^+(aq) \rightleftharpoons 3 Ca^{2+}(aq) + 2 HPO_4^{2-}(aq) + 2 H_2O(l)$ **16.104** 3.5×10^{-22} M   **16.106** (a) $AgI(s) + 2 CN^-(aq) \rightleftharpoons Ag(CN)_2^-(aq) + I^-(aq)$, $K = 2.6 \times 10^4$; (b) $Al(OH)_3(s) + OH^-(aq) \rightleftharpoons Al(OH)_4^-(aq)$, $K = 6$; (c) $Zn(OH)_2(s) + 4 NH_3(aq) \rightleftharpoons Zn(NH_3)_4^{2+} + 2 OH^-(aq)$, $K = 3.2 \times 10^{-8}$ **16.108** (a) 9.2×10^{-9} M; (b) 0.050 M   **16.110** IP = 4.5×10^{-7}; IP > K_{sp}; $BaSO_4(s)$ will precipitate.   **16.112** For $BaSO_4$, IP = 4.8×10^{-11}; IP < K_{sp}; $BaSO_4$ will not precipitate. For $Fe(OH)_3$, IP = 1.6×10^{-20}; IP > K_{sp}; $Fe(OH)_3(s)$ will precipitate.   **16.114** IP = 9.6×10^{-11}; IP > K_{sp}; $Mg(OH)_2(s)$ will precipitate.   **16.116** Yes. $Q_c = 1.1 \times 10^{-2}$; For FeS, $Q_c < K_{spa}$, and no FeS will precipitate. For SnS, $Q_c > K_{spa}$, and SnS will precipitate.   **16.118** (a) Add Cl^- to precipitate AgCl; (b) add CO_3^{2-} to precipitate $CaCO_3$; (c) add H_2S to precipitate MnS; (d) add NH_3 and NH_4Cl to precipitate $Cr(OH)_3$ (Need buffer to control $[OH^-]$; excess OH^- produces the soluble $Cr(OH)_4^-$.)   **16.120** Prepare aqueous solutions of the three salts. Add a solution of $(NH_4)_2HPO_4$. If a white precipitate forms, the solution contains Mg^{2+}. Perform flame test on the other two solutions. A yellow flame test indicates Na^+. A violet flame test indicates K^+.   **16.122** (a), (b), and (d)   **16.124** 12 mL   **16.126** 5.3×10^{-12} **16.128** $[NH_4^+]$ = 0.50 M; $[NH_3]$ = 1.0 M; pH = 9.55   **16.130** molar solubility = 3.5×10^{-8} M; solubility = 3×10^{-6} g/L   **16.132** (a)
$HA^-(aq) + H_2O(l) \rightleftharpoons H_3O^+(aq) + A^{2-}(aq)$; $K_{a2} = 10^{-10}$;
$HA^-(aq) + H_2O(l) \rightleftharpoons H_2A(aq) + OH^-(aq)$; $K_b = K_w/K_{a1} = 10^{-10}$; $2 HA^-(aq) \rightleftharpoons H_2A(aq) + A^{2-}(aq)$; $K = K_{a2}/K_{a1} = 10^{-6}$
$2 H_2O(l) \rightleftharpoons H_3O^+(aq) + OH^-(aq)$; $K_w = 1.0 \times 10^{-14}$
The principal reaction is the one with the largest K, and that is the third reaction);
(b)

$$K_{a1} = \frac{[H_3O^+][HA^-]}{[H_2A]} \text{ and } K_{a2} = \frac{[H_3O^+][A^{2-}]}{[HA^-]}$$

$$[H_3O^+] = \frac{K_{a1}[H_2A]}{[HA^-]} \text{ and } [H_3O^+] = \frac{K_{a2}[HA^-]}{[HA^{2-}]}$$

$$\frac{K_{a1}[H_2A]}{[HA^-]} \times \frac{K_{a2}[HA^-]}{[A^{2-}]} = [H_3O^+]^2;$$

$$\frac{K_{a1}K_{a2}[H_2A]}{[A^{2-}]} = [H_3O^+]^2$$

Because the principal reaction is
$$2 HA^-(aq) \rightleftharpoons H_2A(aq) + A^{2-}(aq)$$
$[H_2A] = [A^{2-}]$; $K_{a1}K_{a2} = [H_3O^+]^2$;
$\log K_{a1} + \log K_{a2} = 2 \log[H_3O^+]$;

$$\frac{\log K_{a1} + \log K_{a2}}{2} = \log[H_3O^+];$$

$$\frac{-\log K_{a1} + (-\log K_{a2})}{2} = -\log[H_3O^+]$$

$$\frac{pK_{a1} + pK_{a2}}{2} = pH$$

(c) 3×10^{19} A^{2-} ions   **16.134** (a) 2.2×10^{-5} M; (b) [HCl] = 0.124 M; $[H_3PO_4]$ = 0.0960 M; (c) 100%; (d) 0.89;

(e)

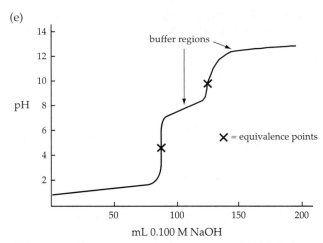

(f) Bromcresol green or methyl orange are suitable indicators for the first equivalence point. Thymolphthalein is a suitable indicator for the second equivalence point.
16.136 4.00×10^{-6}

Chapter 17

17.1 (a) spontaneous; (b), (c), and (d) nonspontaneous
17.2 (a) negative; (b) positive; (c) positive; (d) negative
17.3 (a) $A_2 + 2 B \rightarrow 2 AB$; (b) negative   **17.4** (a) disordered N_2O; (b) silica glass; (c) 1 mole N_2 at STP; (d) 1 mole N_2 at 273 K and 0.25 atm   **17.5** +160.4 J/K   **17.6** ΔS_{total} = -438 J/K. Because ΔS_{total} is negative, the reaction is not spontaneous under standard-state conditions at 25°C.
17.7 (a) nonspontaneous at 25°C; (b) 52°C   **17.8** (a) No; (b) 357°C   **17.9** ΔH = $-$; ΔS = $-$; ΔG = $-$ **17.10** (a) +130.5 kJ; (b) Because $\Delta G > 0$, the reaction is nonspontaneous at 25°C; (c) 839°C   **17.11** (a) $\Delta S°$ is positive; (b) at high temperature   **17.12** (a) -150.2 kJ; Yes, because $\Delta G < 0$; (b) No, because $\Delta G°_f(C_2H_2) > 0$.   **17.13** ΔG = +39.6 kJ. Because $\Delta G > 0$, the reaction is spontaneous in the reverse direction.   **17.14** (a) (3) has the largest ΔG; (1) has the smallest ΔG; (b) 15 kJ   **17.15** 1×10^{-23} **17.16** 0.03 atm   **17.17** 80 kJ   **17.18** No, because the decrease in entropy of the growing person is more than compensated for by an increase in the entropy of the surroundings.   **17.19** Yes. You might, for example, see an explosion run backwards, which is impossible because it would decrease the entropy of the universe.   *Understanding Key Concepts*
17.20 (a)

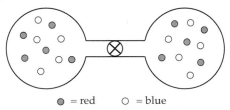

● = red   ○ = blue

(b) $\Delta H = 0$ (no heat is gained or lost in the mixing of ideal gases); ΔS = + (the mixture of the two gases is more disordered); ΔG = $-$ (the mixing of the two gases is spontaneous); (c) For an isolated system, $\Delta S_{surr} = 0$ and $\Delta S_{sys} = \Delta S_{total} > 0$ for the spontaneous process; (d) ΔG = + and the process is nonspontaneous.   **17.22** ΔH = $-$ (heat is lost during condensation); ΔS = $-$ (liquid is more ordered than vapor); ΔG = $-$ (the reaction is spontaneous)

17.24 (a) $2 A_2 + B_2 \rightarrow 2 A_2B$; (b) ΔH = $-$ (because ΔS is negative, ΔH must also be negative in order for ΔG to be negative). ΔS = $-$ (the reaction becomes more ordered in going from reactants [3 molecules] to products [2 molecules]). ΔG = $-$ (the reaction is spontaneous).   **17.26** (a) $\Delta H°$ = +, $\Delta S°$ = +; (b) $\Delta S°$ is for the complete conversion of 1 mole of A_2 in its standard state to 2 moles of A in its standard state. (c) There is not enough information to say anything about the sign of $\Delta G°$. $\Delta G°$ decreases (becomes less positive or more negative) as the temperature increases. (d) K_p increases as the temperature increases. As the temperature increases there will be more A and less A_2. (e) $\Delta G = 0$ at equilibrium   **17.28** (1) $\Delta G°$ = 0; (2) negative; (3) positive   *Additional Problems*   **17.30** A spontaneous process is one that proceeds on its own without any external influence. For example: $H_2O(s) \rightarrow H_2O(l)$ at 25°C. A nonspontaneous process takes place only in the presence of some continuous external influence. For example: $2 NaCl(s) \rightarrow 2 Na(s) + Cl_2(g)$ **17.32** (a) and (d) nonspontaneous; (b) and (c) spontaneous   **17.34** (b) and (d) spontaneous (because of the large positive K_p's)   **17.36** Molecular randomness or disorder is called entropy. $H_2O(s) \rightarrow H_2O(l)$ at 25°C.   **17.38** (a) +; (b) $-$; (c) $-$; (d) +   **17.40** (a) $-$; (b) $-$; (c) +; (d) $-$ **17.42** (a) 2.30×10^{-22} J/K; (b) 2.30×10^{-21} J/K; (c) 11.5 J/K.   **17.44** (a) H_2 at 25°C in 50 L; (b) O_2 at 25°C, 1 atm; (c) H_2 at 100°C, 1 atm; (d) CO_2 at 100°C, 0.1 atm   **17.46** The standard molar entropy of a substance is the entropy of 1 mol of the pure substance at 1 atm pressure and 25°C. $\Delta S°$ = $S°$(products) $-$ $S°$(reactants)   **17.48** (a) $C_2H_6(g)$; (b) $CO_2(g)$; (c) $I_2(g)$; (d) $CH_3OH(g)$ **17.50** (a) $\Delta S°$ = +125 J/K (+, because moles of gas increase); (b) $\Delta S°$ = -181.2 J/K ($-$, because moles of gas decrease); (c) $\Delta S°$ = +137.4 J/K (+, because moles of gas increase); (d) $\Delta S°$ = -626.4 J/K ($-$, because moles of gas decrease)   **17.52** In any spontaneous process, the total entropy of a system and its surroundings always increases.   **17.54** $\Delta S_{surr} = -\Delta H/T$; (a) positive; (b) negative   **17.56** ΔS_{sys} = -297.3 J/K; ΔS_{surr} = -30.7 J/K; ΔS_{total} = -328.0 J/K; Because $\Delta S_{total} < 0$, the reaction is nonspontaneous.   **17.58** (a) ΔS_{surr} = -89.5 J/(K · mol); ΔS_{total} = -2.5 J/(K · mol); (b) ΔS_{surr} = -87.0 J/(K · mol); ΔS_{total} = 0; (c) ΔS_{surr} = -84.6 J/(K · mol); ΔS_{total} = +2.4 J/(K · mol). Benzene does not boil at 70°C because ΔS_{total} is negative. The normal boiling point for benzene is 80°C (353 K), where $\Delta S_{total} = 0$.   **17.60**

ΔH	ΔS	$\Delta G = \Delta H - T\Delta S$	*Reaction Spontaneity*
$-$	+	$-$	Spontaneous at all temperatures
$-$	$-$	$-$ or +	Spontaneous at low temperatures where $|\Delta H| > |T\Delta S|$ Nonspontaneous at high temperatures where $|\Delta H| < |T\Delta S|$
+	$-$	+	Nonspontaneous at all temperatures
+	+	$-$ or +	Spontaneous at high temperatures where $T\Delta S > \Delta H$ Nonspontaneous at low temperatures where $T\Delta S < \Delta H$

17.62 (a) $\Delta H = +$, $\Delta S = +$, $\Delta G = +$; (b) $\Delta H = +$, $\Delta S = +$, $\Delta G = -$ **17.64** (a) $\Delta G_{vap} = +0.9$ kJ/mol; benzene does not boil; (b) $\Delta G_{vap} = 0$; 80°C is the boiling point; (c) $\Delta G_{vap} = -0.9$ kJ/mol; benzene boils   **17.66** 122°C   **17.68** (a) $\Delta G°$ is the change in free energy that occurs when reactants in their standard states are converted to products in their standard states; (b) $\Delta G°_f$ is the free-energy change for formation of one mole of a substance in its standard state from the most stable form of the constituent elements in their standard states.   **17.70** (a) $\Delta H° = 66.4$ kJ; $\Delta S° = -121.5$ J/K; $\Delta G° = +102.6$ kJ; Because $\Delta G°$ is positive, the reaction is nonspontaneous under standard-state conditions at 25°C; (b) $\Delta H° = -78.0$ kJ; $\Delta S° = 494.2$ J/(K · mol); $\Delta G° = -225.3$ kJ; Because $\Delta G°$ is negative, the reaction is spontaneous under standard-state conditions at 25°C; (c) $\Delta H° = -492.6$ kJ; $\Delta S° = -136.1$ J/(K · mol); $\Delta G° = -452.0$ kJ; the reaction is spontaneous   **17.72** (a) +102.6 kJ; (b) −225.8 kJ; (c) −452 kJ   **17.74** (a) stable; (b) stable; (c) unstable; (d) unstable   **17.76** $\Delta G° = -6.1$ kJ. The reaction becomes nonspontaneous at high temperatures because $\Delta S°$ is negative. The reaction becomes nonspontaneous at 72°C.   **17.78** $\Delta G° = -503.1$ kJ. Because $\Delta G°$ is negative, the reaction is possible. Look for a catalyst. Because $\Delta G°_f$ for benzene is positive (+124.5 kJ/mol), the synthesis of benzene from graphite and gaseous H_2 at 25°C and 1 atm pressure is not possible.   **17.80** $\Delta G = \Delta G° + RT \ln Q$ **17.82** (a) −176.0 kJ; (b) −133.8 kJ; (c) −141.8 kJ   **17.84** (a) If $K > 1$, $\Delta G°$ is negative; (b) If $K = 1$, $\Delta G° = 0$; (c) If $K < 1$, $\Delta G°$ is positive.   **17.86** $K_p = 7.1 \times 10^{24}$ **17.88** 0.079 atm   **17.90** $K_p = 2.9 \times 10^{28}$ **17.92** The kinetic parameters [(a), (b), and (h)] are affected by a catalyst. The thermodynamic and equilibrium parameters [(c), (d), (e), (f), and (g)] are not affected by a catalyst.   **17.94** (a) Spontaneous does not mean fast, just possible. (b) For a spontaneous reaction $\Delta S_{total} > 0$. ΔS_{sys} can be positive or negative. (c) An endothermic reaction can be spontaneous if $\Delta S_{sys} > 0$; (d) This statement is true because the sign of ΔG changes when the direction of a reaction is reversed.
17.96

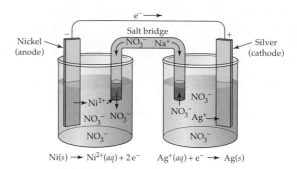

17.98 78°C   **17.100** (a) endothermic; (b) 33.6 kJ   **17.102** (a) $\Delta H° = -1203.4$ kJ; $\Delta S° = -216.6$ J/K; $\Delta G° = -1138.8$ kJ; spontaneous. $\Delta G°$ becomes less negative as the temperature is raised. (b) $\Delta H° = +101$ kJ; $\Delta S° = 174.8$ J/K; $\Delta G° = +49$ kJ; not spontaneous. $\Delta G°$ becomes less positive as the temperature is raised. (c) $\Delta H° = -852$ kJ; $\Delta S° = -38.5$ J/K; $\Delta G° = -840$ kJ; spontaneous. $\Delta G°$ becomes less negative as the temperature is raised. (d) $\Delta H° = +135.6$ kJ; $\Delta S° = +333$ J/K; $\Delta G° = +36.4$ kJ; not spontaneous. $\Delta G°$ becomes less positive as the temperature is raised.   **17.104** (a) −253 J/(K · mol); (b) −376 J/(K · mol); (c) −345 J/(K · mol)   **17.106** $\Delta S° = -82.0$ J/K; $\Delta H° = +2.71$ kJ   **17.108** 26°C

17.110 [N_2] = 0.078 M; [H_2] = 0.235 M; [NH_3] = 0.043 M
17.112 (a) −371.4 kJ; (b) $\Delta G = -350.1$ kJ

Chapter 18

18.1

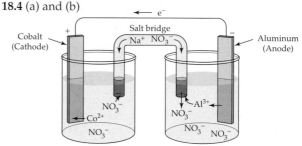

$Ni(s) \longrightarrow Ni^{2+}(aq) + 2 e^-$ $Ag^+(aq) + e^- \longrightarrow Ag(s)$

18.2 $Fe(s) \mid Fe^{2+}(aq) \parallel Sn^{2+}(aq) \mid Sn(s)$ **18.3** $Pb(s) + Br_2(l) \rightarrow Pb^{2+}(aq) + 2 Br^-(aq)$. There is a Pb anode in an aqueous solution of Pb^{2+}. The cathode is a Pt wire that dips into a pool of liquid Br_2 and an aqueous solution that is saturated with Br_2. A salt bridge connects the anode and the cathode compartment. The electrodes are connected through an external circuit.
18.4 (a) and (b)

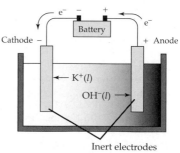

(c) $2 Al(s) + 3 Co^{2+}(aq) \rightarrow 2 Al^{3+}(aq) + 3 Co(s)$; (d) $Al(s) \mid Al^{3+}(aq) \parallel Co^{2+}(aq) \mid Co(s)$ **18.5** −270 kJ   **18.6** −0.74 V   **18.7** Cl_2 is the stronger oxidizing agent. Mg is the stronger reducing agent.   **18.8** (a) Reaction can occur; $E° = 0.23$ V. (b) Reaction cannot occur; $E° = -1.40$ V.   **18.9** (a) D is the strongest reducing agent. A^{3+} is the strongest oxidizing agent. (b) B^{2+} can oxidize C and D. C can reduce A^{3+} and B^{2+}. (c) $A^{3+} + 2 D \rightarrow A^+ + 2 D^+$; 2.85 V   **18.10** 0.25 V   **18.11** (a) −0.03 V; (b) +0.06 V   **18.12** 6.9   **18.13** $K = 10^{31}$ **18.14** −0.140 V   **18.15** (a) $Zn(s) + 2 MnO_2(s) + 2 NH_4^+(aq) \rightarrow Zn^{2+}(aq) + Mn_2O_3(s) + 2 NH_3(aq) + H_2O(l)$; (b) $Zn(s) + 2 MnO_2(s) \rightarrow ZnO(s) + Mn_2O_3(s)$; (c) $Zn(s) + HgO(s) \rightarrow ZnO(s) + Hg(l)$; (d) $Cd(s) + 2 NiO(OH)(s) + 2 H_2O(l) \rightarrow Cd(OH)_2(s) + 2 Ni(OH)_2(s)$
18.16 (a)

(b)
anode $4 OH^-(l) \rightarrow O_2(g) + 2 H_2O(l) + 4 e^-$
cathode $4 K^+(l) + 4 e^- \rightarrow 4 K(l)$

overall $4 K^+(l) + 4 OH^-(l) \rightarrow 4 K(l) + O_2(g) + 2 H_2O(l)$
18.17 (a)
anode $2 Cl^-(aq) \rightarrow Cl_2(g) + 2 e^-$
cathode $2 H_2O(l) + 2 e^- \rightarrow H_2(g) + 2 OH^-(aq)$

overall $2 Cl^-(aq) + 2 H_2O(l) \rightarrow Cl_2(g) + H_2(g) + 2 OH^-(aq)$
(b)
anode $2 H_2O(l) \rightarrow O_2(g) + 4 H^+(aq) + 4 e^-$
cathode $2 Cu^{2+}(aq) + 4 e^- \rightarrow 2 Cu(s)$

overall $2 Cu^{2+}(aq) + 2 H_2O(l) \rightarrow 2 Cu(s) + O_2(g) + 4 H^+(aq)$
18.18 268 kg Al   **18.19** 7.45 h   **18.20** When a beam of
white light strikes the anodized surface, part of the light
is reflected from the outer TiO_2, while part penetrates
through the semitransparent TiO_2 and is reflected from
the inner metal. If the two reflections of a particular
wavelength are out of phase, they interfere destructively
and that wavelength is canceled from the reflected light.
Because $n\lambda = 2d \times \sin \theta$, the canceled wavelength depends
on the thickness of the TiO_2 layer.   **18.21** 62.6 min
Understanding Key Concepts **18.22** (a)–(d)

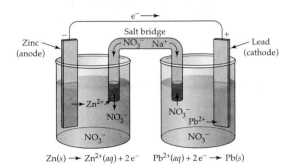

Zn(s) → Zn²⁺(aq) + 2 e⁻   Pb²⁺(aq) + 2 e⁻ → Pb(s)

(e)
anode $Zn(s) \rightarrow Zn^{2+}(aq) + 2 e^-$
cathode $Pb^{2+}(aq) + 2 e^- \rightarrow Pb(s)$

overall $Zn(s) + Pb^{2+}(aq) \rightarrow Zn^{2+}(aq) + Pb(s)$
18.24 (a) The three cell reactions are the same except for
cation concentrations: $Cu(s) + 2 Fe^{3+}(aq) \rightarrow Cu^{2+}(aq) +$
$2 Fe^{2+}(aq)$
(b)

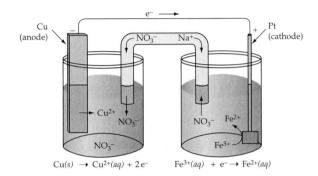

Cu(s) → Cu²⁺(aq) + 2 e⁻   Fe³⁺(aq) + e⁻ → Fe²⁺(aq)

(c) (1) $E = 0.43$ V; (2) $E = 0.39$ V; (3) $E = 0.46$ V. Cell (3) has
the largest potential, while cell (2) has the smallest as cal-

culated from the Nernst equation.   **18.26** (a) electrolytic;
(b)

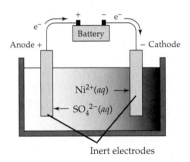

(c)
anode $2 H_2O(l) \rightarrow O_2(g) + 4 H^+(aq) + 4 e^-$
cathode $Ni^{2+}(aq) + 2 e^- \rightarrow Ni(s)$

overall $2 Ni^{2+}(aq) + 2 H_2O(l) \rightarrow 2 Ni(s) + O_2(g) + 4 H^+(aq)$
18.28 (a) oxidizing agents: PbO_2, H^+, $Cr_2O_7^{2-}$; reducing
agents: Al, Fe, Ag. (b) PbO_2 is the strongest oxidizing
agent. H^+ is the weakest oxidizing agent. (c) Al is the
strongest reducing agent. Ag is the weakest reducing
agent. (d) oxidized by Cu^{2+}: Fe and Al; reduced by H_2O_2:
PbO_2 and $Cr_2O_7^{2-}$.   ***Additional Problems*** **18.30** The electrode where oxidation takes place is called the anode (for
example, the lead electrode in the lead storage battery). The
electrode where reduction takes place is called the cathode
(for example, the PbO_2 electrode in the lead storage battery).
18.32 The cathode of a galvanic cell is considered to be the
positive electrode because electrons flow through the external circuit toward the positive electrode (the cathode).
18.34 (a)

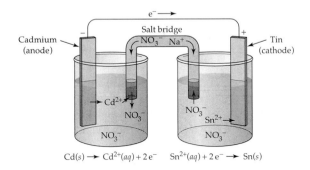

Cd(s) → Cd²⁺(aq) + 2 e⁻   Sn²⁺(aq) + 2 e⁻ → Sn(s)

(b)

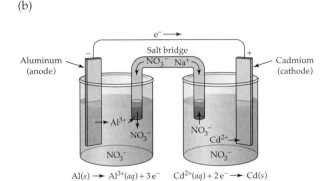

Al(s) → Al³⁺(aq) + 3 e⁻   Cd²⁺(aq) + 2 e⁻ → Cd(s)

(c)

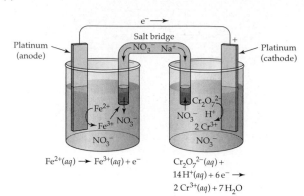

$Fe^{2+}(aq) \longrightarrow Fe^{3+}(aq) + e^-$

$Cr_2O_7^{2-}(aq) +$
$14 H^+(aq) + 6 e^- \longrightarrow$
$2 Cr^{3+}(aq) + 7 H_2O$

18.36 (a) $Cd(s) | Cd^{2+}(aq) \| Sn^{2+}(aq) | Sn(s)$;
(b) $Al(s) | Al^{3+}(aq) \| Cd^{2+}(aq) | Cd(s)$;
(c) $Pt(s) | Fe^{2+}(aq), Fe^{3+}(aq) \| Cr_2O_7^{2-}(aq), Cr^{3+}(aq) | Pt(s)$
18.38 (a)

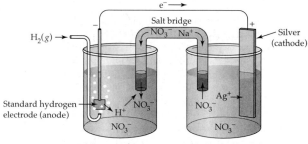

$H_2(g) \longrightarrow 2 H^+(aq) + 2 e^-$ $Ag^+(aq) + e^- \longrightarrow Ag(s)$

(b)
anode $H_2(g) \longrightarrow 2 H^+(aq) + 2 e^-$
cathode $2 Ag^+(aq) + 2 e^- \longrightarrow 2 Ag(s)$

overall $H_2(g) + 2 Ag^+(aq) \longrightarrow 2 H^+(aq) + 2 Ag(s)$
(c) $Pt(s) | H_2(g) | H^+(aq) \| Ag^+(aq) | Ag(s)$
18.40 (a)
anode $Co(s) \longrightarrow Co^{2+}(aq) + 2 e^-$
cathode $Cu^{2+}(aq) + 2 e^- \longrightarrow Cu(s)$

overall $Co(s) + Cu^{2+}(aq) \longrightarrow Co^{2+}(aq) + Cu(s)$

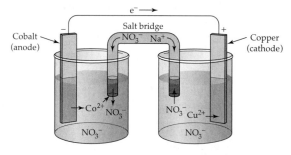

(b)
anode $2 Fe(s) \longrightarrow 2 Fe^{2+}(aq) + 4 e^-$
cathode $O_2(aq) + 4 H^+(aq) + 4 e^- \longrightarrow 2 H_2O(l)$

overall $2 Fe(s) + O_2(g) + 4 H^+(aq) \longrightarrow 2 Fe^{2+}(aq) + 2 H_2O(l)$

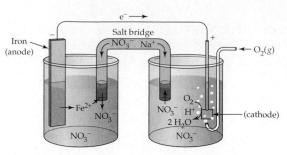

18.42 electrical potential, volt (V); charge, coulomb (C); energy, joule (J). $1 J = 1 C \cdot 1 V$ **18.44** E is the standard cell potential ($E°$) when all reactants and products are in their standard states—solutes at 1 M concentrations, gases at a partial pressure of 1 atm, solids and liquids in pure form, all at 25°C.   **18.46** –309 kJ   **18.48** +1.23 V   **18.50** –0.36 V   **18.52** $Sn^{4+}(aq) < Br_2(l) < MnO_4^-(aq)$ **18.54** $Cr_2O_7^{2-}$ strongest; Fe^{2+} weakest   **18.56** (a) $E° = 0.26$ V, $\Delta G° = -50$ kJ; (b) $E° = 1.26$ V, $\Delta G° = -730$ kJ; (c) $E° = 0.56$ V, $\Delta G° = -324$ kJ   **18.58** (a) $E° = -0.90$ V, nonspontaneous; (b) $E° = 2.11$ V, spontaneous   **18.60** (a) $E° = +0.94$ V, $Sn^{2+}(aq)$ can be oxidized by $Br_2(l)$; (b) $E° = -0.41$ V, $Ni^{2+}(aq)$ cannot be reduced by $Sn^{2+}(aq)$; (c) $E° = -0.93$ V, $Ag(s)$ cannot be oxidized by $Pb^{2+}(aq)$; (d) $E° = +0.20$ V, $I_2(s)$ can be reduced by $Cu(s)$ **18.62** 0.87 V   **18.64** $E = 0.35$ V, $[Cu^{2+}] = 1 \times 10^{-16}$ M   **18.66** (a) 0.64 V; (b) 0.77 V; (c) –0.23 V; (d) –1.05 V   **18.68** 9.0   **18.70** $\Delta G° = -nFE°$. Because n and F are always positive, $\Delta G°$ is negative when $E°$ is positive because of the negative sign in the equation.

$$E° = \frac{0.0592}{n} \log K; \quad \log K = \frac{nE°}{0.0592}; \quad K = 10^{\frac{nE°}{0.0592}}$$

If $E°$ is positive, the exponent is positive (because n is positive) and K is greater than 1.   **18.72** 6×10^{35} **18.74** (a) 6×10^8; (b) 10^{128}; (c) 10^{57} **18.76** 9×10^{-3} **18.78** Rust is a hydrated form of iron(III) oxide ($Fe_2O_3 \cdot H_2O$). Rust forms from the oxidation of Fe in the presence of O_2 and H_2O. Rust can be prevented by coating Fe with Zn (galvanizing).   **18.80** Cathodic protection is the attachment of a more easily oxidized metal to the metal you want to protect. This forces the metal you want to protect to be the cathode, hence the name, cathodic protection. Zn and Al can offer cathodic protection to Fe (Ni and Sn cannot).   **18.82** (a)

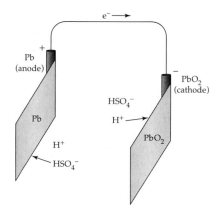

(b)

anode	$Pb(s) + HSO_4^-(aq) \rightarrow PbSO_4(s) + H^+(aq) + 2\ e^-$
cathode	$PbO_2(s) + 3\ H^+(aq) + HSO_4^-(aq) + 2\ e^- \rightarrow$
	$PbSO_4(s) + 2\ H_2O(l)$

overall $Pb(s) + PbO_2(s) + 2\ H^+(aq) + 2\ HSO_4^-(aq) \rightarrow$
$2\ PbSO_4(s) + 2\ H_2O(l)$

(c) $K = 1 \times 10^{65}$; (d) cell voltage = 0   **18.84** 6.62 g
18.86 (a)

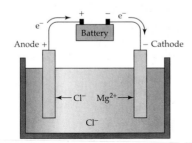

(b)

anode	$2\ Cl^-(l) \rightarrow Cl_2(g) + 2\ e^-$
cathode	$Mg^{2+}(l) + 2\ e^- \rightarrow Mg(l)$

overall $Mg^{2+}(l) + 2\ Cl^-(l) \rightarrow Mg(l) + Cl_2(g)$
18.88 Possible anode reactions:
$2\ Cl^-(aq) \rightarrow Cl_2(g) + 2\ e^-$
$2\ H_2O(l) \rightarrow O_2(g) + 4\ H^+(aq) + 4\ e^-$
Possible cathode reactions:
$2\ H_2O(l) + 2\ e^- \rightarrow H_2(g) + 2\ OH^-(aq)$
$Mg^{2+}(aq) + 2\ e^- \rightarrow Mg(s)$
Actual reactions:
anode $2\ Cl^-(aq) \rightarrow Cl_2(g) + 2\ e^-$
cathode $2\ H_2O(l) + 2\ e^- \rightarrow H_2(g) + 2\ OH^-(aq)$
This anode reaction takes place instead of $2\ H_2O(l) \rightarrow$
$O_2(g) + 4\ H^+(aq) + 4\ e^-$ because of a high overvoltage for
formation of gaseous O_2. This cathode reaction takes
place instead of $Mg^{2+}(aq) + 2\ e^- \rightarrow Mg(s)$ because H_2O is
easier to reduce than Mg^{2+}.   **18.90** (a)

anode	$2\ Br^-(aq) \rightarrow Br_2(l) + 2\ e^-$
cathode	$2\ H_2O(l) + 2\ e^- \rightarrow H_2(g) + 2\ OH^-(aq)$

overall $2\ H_2O(l) + 2\ Br^-(aq) \rightarrow Br_2(l) + H_2(g) + 2\ OH^-(aq)$
(b)

anode	$2\ Cl^-(aq) \rightarrow Cl_2(g) + 2\ e^-$
cathode	$Cu^{2+}(aq) + 2\ e^- \rightarrow Cu(s)$

overall $Cu^{2+}(aq) + 2\ Cl^-(aq) \rightarrow Cu(s) + Cl_2(g)$
(c)

anode	$4\ OH^-(aq) \rightarrow O_2(g) + 2\ H_2O(l) + 4\ e^-$
cathode	$4\ H_2O(l) + 4\ e^- \rightarrow 2\ H_2(g) + 4\ OH^-(aq)$

overall $2\ H_2O(l) \rightarrow O_2(g) + 2\ H_2(g)$
18.92 3.22 g   **18.94** Time = 38.9 h; 4.87×10^5 L Cl_2 **18.96**
84.9 g   **18.98** (a) $2\ MnO_4^-(aq) + 16\ H^+(aq) + 5\ Sn^{2+}(aq) \rightarrow$
$2\ Mn^{2+}(aq) + 5\ Sn^{4+}(aq) + 8\ H_2O(l)$; (b) MnO_4^- is the oxidizing agent; Sn^{2+} is the reducing agent. (c) 1.36 V   **18.100**
(a) Ag^+ is the strongest oxidizing agent; Pb is the
strongest reducing agent.

(b)

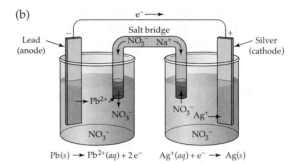

(c) $Pb(s) + 2\ Ag^+(aq) \rightarrow Pb^{2+}(aq) + 2\ Ag(s)$; $E° = 0.93$ V,
$\Delta G° = -180$ kJ, $K = 10^{31}$; (d) $E = 0.87$ V
18.102 (a)

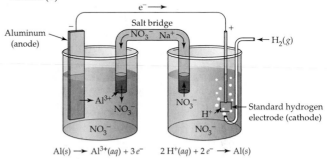

(b) $2\ Al(s) + 6\ H^+(aq) \rightarrow 2\ Al^{3+}(aq) + 3\ H_2(g)$; $E° = 1.66$ V;
(c) 1.59 V; (d) $\Delta G° = -961$ kJ, $K = 10^{168}$; (e) 1.40 g   **18.104**
4.0×10^{10} kWh   **18.106** (a) 1.27 V; (b) 0.95 V   **18.108**
1.7×10^{-10} **18.110** 6×10^{-23} **18.112** (a)

anode	$4[Al(s) \rightarrow Al^{3+}(aq) + 3\ e^-]$
cathode	$3[O_2(g) + 4\ H^+(aq) + 4\ e^- \rightarrow 2\ H_2O(l)]$

overall $4\ Al(s) + 3\ O_2(g) + 12\ H^+(aq) \rightarrow$
$4\ Al^{3+}(aq) + 6\ H_2O(l)$

(b)

$$E = E° - \frac{2.303\ RT}{nF} \log \frac{[Al^{3+}]^4}{(P_{O_2})^3\ [H^+]^{12}}$$

where $T = 310$ K; (c) 2.63 V   **18.114** (a) $4\ CH_2{=}CHCN +$
$2\ H_2O \rightarrow 2\ NC(CH_2)_4CN + O_2$; (b) 60.5 kg; (c) 7030 L
18.116 (a) $Cr_2O_7^{2-}(aq) + 6\ Fe^{2+}(aq) + 14\ H^+(aq) \rightarrow$
$2\ Cr^{3+}(aq) + 6\ Fe^{3+}(aq) + 7\ H_2O(l)$; (b) 1.02 V   **18.118**
(a) $Zn(s) + 2\ Ag^+(aq) + H_2O(l) \rightarrow ZnO(s) + 2\ Ag(s) +$
$2\ H^+(aq)$, $\Delta H° = -273.7$ kJ, $\Delta S° = -128.1$ J/K; $\Delta G° =$
-235.5 kJ; (b) $E° = 1.220$ V; $K = 2 \times 10^{41}$; (c) decreases, $E =$
0.682 V; (d) IP = 1.6×10^{-11}, IP < K_{sp}, AgCl will not precipitate. IP = 1.6×10^{-11}, IP > K_{sp}, AgBr will precipitate.

Chapter 19

19.1 (a) B; (b) Br; (c) Se   **19.2** (a)

Nitrogen can form very strong $p\pi$–$p\pi$ bonds. Phosphorus forms weaker $p\pi$–$p\pi$ bonds, so it tends to form more single bonds. (b) Sulfur can use empty $3d$ orbitals to form octahedral sp^3d^2 hybrid orbitals to bond with six fluorines in SF_6. With just $2s$ and $2p$ valence orbitals, oxygen cannot expand its valence orbitals beyond four.   **19.3** Carbon forms strong π bonds with oxygen. Silicon does not form strong π bonds with oxygen, and what results are chains

of alternating silicon and oxygen singly bonded to each other.   **19.4** H–C≡N: The carbon is sp hybridized.   **19.5** Hb–O_2 + CO ⇌ Hb–CO + O_2. Mild cases of carbon monoxide poisoning can be treated with O_2. Le Châtelier's principle says that adding a product (O_2) will cause the reaction to proceed in the reverse direction, back to Hb–O_2.   **19.6** (a) $Si_3O_9^{6-}$; (b) $Si_4O_{13}^{10-}$

19.7 4 $H_3PO_4(aq) \rightarrow H_6P_4O_{13}(aq)$ + 3 $H_2O(l)$

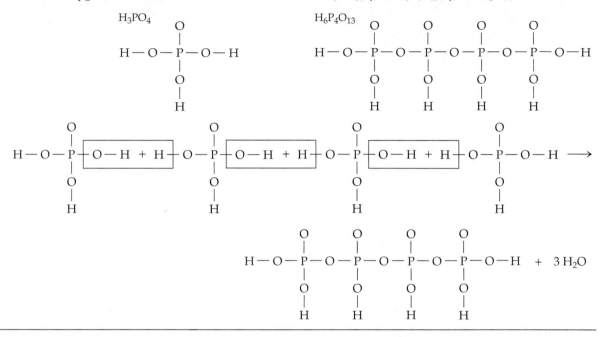

19.8 (a) SO_3^{2-}, HSO_3^-, SO_4^{2-}; (b) HSO_4^-; (c) SO_3^{2-}; (d) HSO_4^-
19.9

(a) bent

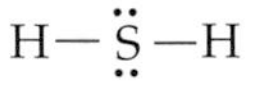

H—S̈—H

(b) bent

:Ö—S̈=Ö: ⟷ :Ö=S̈—Ö:

(c) trigonal planar

:Ö: :Ö: Ö:
| | ‖
:Ö—S̈=Ö: ⟷ :Ö=S̈—Ö: ⟷ :Ö—S̈—Ö:

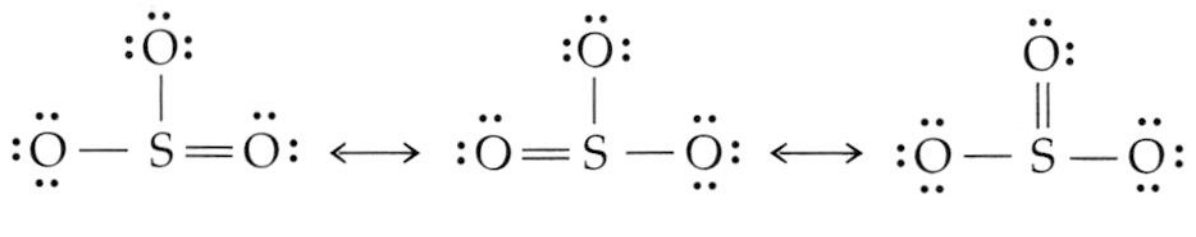

19.10 Because copper is always conducting, a photocopier drum coated with copper would not hold any charge so no document image would adhere to the drum for copying.   *Understanding Key Concepts*   **19.12**

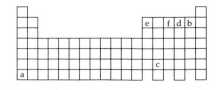

19.14 (a) N_2, O_2, F_2, P_4, S_8, Cl_2

(b) :N≡N: :Ö=Ö: :F̈—F̈:   :C̈l—C̈l:

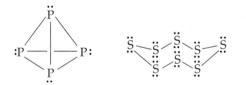

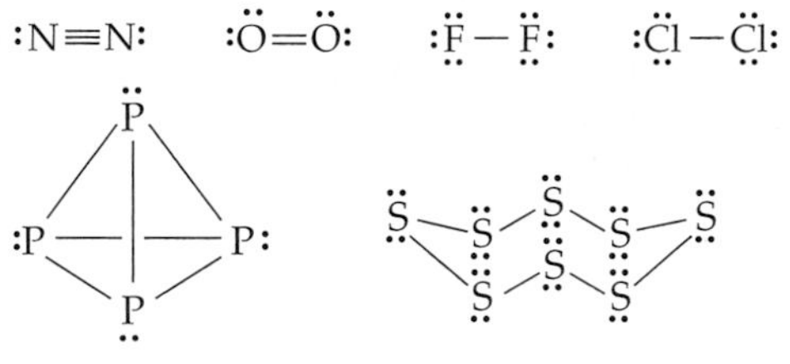

(c) The smaller N and O can form effective π bonds, whereas P and S cannot. In both F_2 and Cl_2, the atoms are joined by a single bond.   **19.16** (a) H_2O, CH_4, HF, B_2H_6, NH_3; (b)

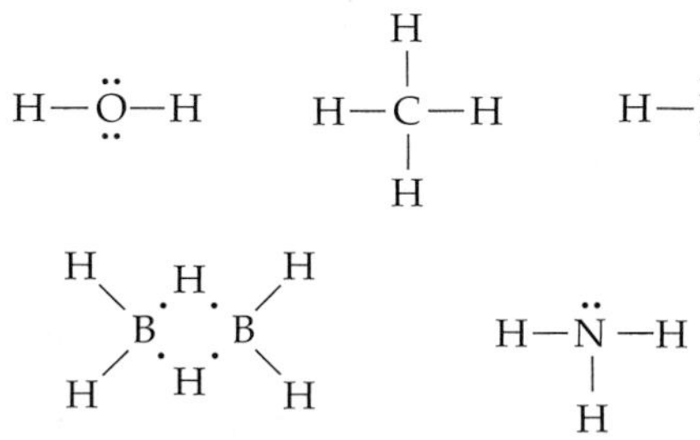

                        H
                        |
H—Ö—H   H—C—H   H—F̈:
                        |
                        H

H   H   H
 \ | /
   B   B
 / | \
H   H   H

H—N̈—H
    |
    H

There is a problem in drawing an electron-dot structure for B_2H_6 because this molecule is electron deficient and has two three-center, two-electron bonds.

Additional Problems **19.18** (a) Cl; (b) Si; (c) O **19.20** (a) Al; (b) P; (c) Pb **19.22** (a) I; (b) N; (c) F **19.24** (a) Sn; (b) Ge; (c) Bi **19.26** (a) CaH_2; (b) Ga_2O_3; (c) KCl **19.28** Molecular: (a) B_2H_6, (c) SO_3, (d) $GeCl_4$; Extended three-dimensional structure: (b) $KAlSi_3O_8$ **19.30** (a) P_4O_{10}; (b) B_2O_3; (c) SO_2 **19.32** (a) Sn; (b) Cl; (c) Sn; (d) Se; (e) B **19.34** Boron has only $2s$ and $2p$ valence orbitals and can form a maximum of four bonds. Aluminum has available $3d$ orbitals and can use octahedral sp^3d^2 hybrid orbitals to bond to six F^- ions. **19.36** In O_2 a π bond is formed by $2p$ orbitals on each O. S does not form strong π bonds with its $3p$ orbitals, which leads to the S_8 ring structure. **19.38** +3 for B, Al, Ga, and In; +1 for Tl **19.40** Boron is a hard semi-conductor with a high melting point. Boron forms only molecular compounds and does not form an aqueous B^{3+} ion. $B(OH)_3$ is an acid. **19.42**

$$2\ BBr_3(g)\ +\ 3\ H_2(g)\ \xrightarrow[1200°C]{Ta\ wire}\ 2\ B(s)\ +\ 6\ HBr(g)$$

19.44 (a) An electron deficient molecule is a molecule that doesn't have enough electrons to form a two-center two-electron bond between each pair of bonded atoms. B_2H_6 is an electron deficient molecule. (b) A three-center two-electron bond has three atoms bonded together using just two electrons. The B–H–B bridging bond in B_2H_6 is a three-center two-electron bond. **19.46** (a) Al; (b) Tl; (c) B; (d) B **19.48** (a) Pb; (b) C; (c) Si; (d) C **19.50** (a) tetrahe-dral, sp^3; (b) linear, sp; (c) trigonal planar, sp^2; (d) octahedral, sp^3d^2 **19.52** Diamond is a very hard, high melting solid. It is an electrical insulator. Diamond has a covalent network structure in which each C atom uses sp^3 hybrid orbitals to form a tetrahedral array of σ bonds. The interlocking, three-dimensional network of strong bonds makes diamond the hardest known substance with the highest melting point for an element. Because the valence electrons are localized in the σ bonds, diamond is an electrical insulator. **19.54** (a) carbon tetrachloride, CCl_4; (b) carbon monoxide, CO; (c) methane, CH_4 **19.56** (1) To provide the "bite" in soft drinks; $CO_2(aq) + H_2O(l) \rightleftharpoons H_2CO_3(aq)$; (2) CO_2 in fire extinguishers; CO_2 is non-flammable and 1.5 times more dense than air; (3) Refrigerant; dry ice, sublimes at –78°C **19.58**

$$SiO_2(l)\ +\ 2\ C(s)\ \xrightarrow{Heat}\ Si(l)\ +\ 2\ CO(g)$$

sand

Purification of Si for semiconductor devices: (1) $Si(s) + 2\ Cl_2(g) \rightarrow SiCl_4(l)$; (2) $SiCl_4$ is purified by distillation; (3) $SiCl_4(g)\ +\ 2\ H_2(g)\ \xrightarrow{Heat}\ Si(s)\ +\ 4\ HCl(g)$; (4) Zone refining

19.60

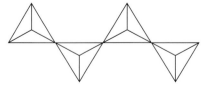

(a) $SiO_4{}^{4-}$ (b) $Si_4O_{13}{}^{10-}$

The charge on the anion is equal to the number of termi-nal O atoms **19.62** (a) $Si_3O_{10}{}^{8-}$; (b) $Ca_2Cu_2Si_3O_{10} \cdot 2\ H_2O$ **19.64** (a) P; (b) Sb and Bi; (c) N; (d) Bi **19.66** (a) N_2O, +1; (b) N_2H_4, –2; (c) Ca_3P_2, –3; (d) H_3PO_3, +3; (e) H_3AsO_4, +5 **19.68** $:N{\equiv}N:$; N_2 is unreactive because of the large

amount of energy necessary to break the $N{\equiv}N$ triple bond. **19.70** (a) bent, sp^2; (b) trigonal pyramidal, sp^3; (c) trigonal bipyramidal, sp^3d; (d) tetrahedral, sp^3 **19.72**

White phosphorus

Red phosphorus

White phosphorus is reactive due to the considerable strain in the P_4 molecule. **19.74** (a) The structure for phosphorous acid is

H — O — P — H with =O above and O—H below

Only the two hydrogens bonded to oxygen are acidic. (b) Nitrogen forms strong π bonds, and in N_2 the nitrogen atoms are triple bonded to each other. Phosphorus does not form strong $p\pi$–$p\pi$ bonds, and so the P atoms are sin-gle bonded to each other in P_4. **19.76** (a) $2\ NO(g) + O_2(g) \rightarrow 2\ NO_2(g)$; (b) $4\ HNO_3(aq) \rightarrow 4\ NO_2(aq) + O_2(g) + 2\ H_2O(l)$; (c) $3\ Ag(s) + 4\ H^+(aq) + NO_3{}^-(aq) \rightarrow 3\ Ag^+(aq) + NO(g) + 2\ H_2O(l)$; (d) $N_2H_4(aq) + 2\ I_2(aq) \rightarrow N_2(g) + 4\ H^+(aq) + 4\ I^-(aq)$ **19.78** (a) O; (b) Te; (c) Po; (d) O **19.80** (a) crown-shaped S_8 rings; (b) crown-shaped S_8 rings; (c) disordered, tangled S_n chains; (d) long polymer chains (S_n, $n > 200,000$) **19.82** (a) hydrogen sulfide, H_2S; lead(II) sul-fide, PbS; (b) sulfur dioxide, SO_2; sulfurous acid, H_2SO_3; (c) sulfur trioxide, SO_3; sulfur hexafluoride, SF_6 **19.84** a three-step reaction sequence in which (1) sulfur burns in air to give SO_2, (2) SO_2 is oxidized to SO_3 in the presence of a vanadium(V) oxide catalyst, and (3) SO_3 reacts with water to give H_2SO_4.
(1) $S(s) + O_2(g) \rightarrow SO_2(g)$

$$(2)\ 2\ SO(g)\ +\ O_2(g)\ \xrightarrow[V_2O_5\ catalyst]{Heat}\ 2\ SO_3(g)$$

(3) $SO_3(g) + H_2O$ (in conc H_2SO_4) $\rightarrow H_2SO_4(l)$
19.86 (a) $Zn(s) + 2\ H_3O^+(aq) \rightarrow Zn^{2+}(aq) + H_2(g) + 2\ H_2O(l)$; (b) $BaSO_3(s) + 2\ H_3O^+(aq) \rightarrow H_2SO_3(aq) + Ba^{2+}(aq) + 2\ H_2O(l)$; (c) $Cu(s) + 2\ H_2SO_4(l) \rightarrow Cu^{2+}(aq) + SO_4{}^{2-}(aq) + SO_2(g) + 2\ H_2O(l)$; (d) $H_2S(aq) + I_2(aq) \rightarrow S(s) + 2\ H^+(aq) + 2\ I^-(aq)$ **19.88** (a) Acid strength increases as the number of O atoms increases. (b) In comparison with S, O is much too electronegative to form compounds of O in the +4 oxidation state. Also, S uses sp^3d hybrid orbitals for bonding in SF_4, but O doesn't have valence d orbitals and so it can't form four bonds to F. (c) Each S is sp^3 hybridized with two lone pairs of electrons. The bond angles are therefore 109.5°. A planar ring would require bond angles of 135°. **19.90** (a) $HBrO_3$, +5; (b) HIO, +1; (c) $NaClO_2$, +3; (d) $NaIO_4$, +7

19.92 (a) iodic acid, (b) chlorous acid; (c) sodium hypobromite; (d) lithium perchlorate   **19.94**

(a)   :Ö—I—Ö—H   trigonal pyramidal
           |
          :Ö:

(b)   [:Ö—Cl—Ö:]⁻   bent

(c)   H—Ö—Cl: bent

(d)   [structure with I center, octahedral]   octahedral

19.96 Oxygen atoms are highly electronegative. Increasing the number of oxygen atoms increases the polarity of the O–H bond and increases the acid strength.   **19.98** (a) $Br_2(l) + 2\ OH^-(aq) \rightarrow OBr^-(aq) + Br^-(aq) + H_2O(l)$; (b) $Cl_2(g) + H_2O(l) \rightarrow HOCl(aq) + H^+(aq) + Cl^-(aq)$; (c) $3\ Cl_2(g) + 6\ OH^-(aq) \rightarrow ClO_3^-(aq) + 5\ Cl^-(aq) + 3\ H_2O(l)$ **19.100** (a) $Na_2B_4O_7 \cdot 10\ H_2O$; (b) $Ca_3(PO_4)_2$; (c) elemental sulfur, FeS_2, PbS, HgS, $CaSO_4 \cdot 2\ H_2O$ **19.102** (a) LiCl; (b) SiO_2; (c) P_4O_{10} **19.104** In silicate and phosphate anions, both Si and P are surrounded by tetrahedra of O atoms, which can link together to form chains and rings.   **19.106** Earth's crust: O, Si, Al, Fe; human body: O, C, H, N   **19.108** C, Si, Ge, and Sn have allotropes with the diamond structure. Sn and Pb have metallic allotropes. C (nonmetal), Si (semimetal), Ge (semimetal), Sn (semimetal and metal), Pb (metal)   **19.110** (a) $H_3PO_4(aq) + H_2O(l) \rightleftharpoons H_3O^+(aq) + H_2PO_4^-(aq)$; H_3PO_4 is a Brønsted–Lowry acid. (b) $B(OH)_3(aq) + 2\ H_2O(l) \rightleftharpoons B(OH)_4^-(aq) + H_3O^+(aq)$; $B(OH)_3$ is a Lewis acid.   **19.112** (a) In diamond each C is covalently bonded to four additional C atoms in a rigid three-dimensional network solid. Graphite is a two-dimensional covalent network solid of carbon sheets that can slide over each other. Both are high melting because melting requires the breaking of C–C bonds. (b) Chlorine does not form perhalic acids of the type H_5XO_6 because its smaller size favors a tetrahedral structure over an octahedral one.   **19.114** The angle required by P_4 is 60°. The strain would not be reduced by using sp^3 hybrid orbitals because their angle is ~109°.   **19.116** (a)

·N̈=Ö: [:Ö—Ö·]⁻ ⟷ [:Ö—Ö:]⁻   [:Ö=N̈—Ö—Ö:]⁻

The O–N–O bond angle should be ~120°. (b) The bond order is 2 1/2 with one unpaired electron.

σ*₂ₚ   —

π*₂ₚ   ↑   —

σ₂ₚ   ↑↓

π₂ₚ   ↑↓   ↑↓

σ*₂ₛ   ↑↓

σ₂ₛ   ↑↓

19.118 $E° = 0.30$ V; $\Delta G° = -58$ kJ; disproportionation of In^+ is spontaneous. $E° = -1.59$ V; $\Delta G° = +307$ kJ; disproportionation of Tl^+ is nonspontaneous.   **19.120** (a) 8.62% $(NH_4)_2HPO_4$; 91.38% NH_4NO_3; (b) 7.50

Chapter 20

20.1 (a) [Ar] $3d^3 4s^2$; (b) [Ar] $3d^7$; (c) [Ar] $3d^3$
20.2

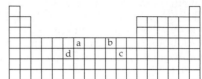

20.3 Z_{eff} increases from left to right. (a) Ti strongest, Zn weakest; (b) FeO_4^{2-} strongest, VO_4^{3-} weakest   **20.4** $[Cr(NH_3)_2(SCN)_4]^-$ **20.5** +2   **20.6** (a) tetraamminecopper(II) sulfate; (b) sodium tetrahydroxochromate(III); (c) triglycinatocobalt(III); (d) pentaaquaisothiocyanatoiron(III) ion   **20.7** (a) $[Zn(NH_3)_4](NO_3)_2$; (b) $Ni(CO)_4$; (c) $K[Pt(NH_3)Cl_3]$; (d) $[Au(CN)_2]^-$ **20.8** (a) Two

[cis and trans structures of Pt complex with NCS, SCN, NH₃]
cis   trans

(b) None; (c) Two

[two Co complex structures with NH₃, NO₂]

(d) None; (e) Two

[trans and cis Cr complex structures with N, Br]
trans   cis

(f) None
20.9 (a) (1), trans; (2), cis; (3), cis; (4), trans; (b) (1) and (4) are identical, and (2) and (3) are identical. The cis and trans isomers are different.   **20.10** (b) foot, (d) corkscrew   **20.11** (a) (2) and (3) are chiral and (1) and (4) are achiral. (b)

[two Rh complex structures with Br, N, Cl]

20.12 (a) $[Fe(C_2O_4)_3]^{3-}$

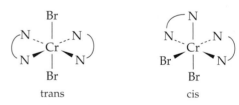

(c) $[Co(NH_3)_2(en)_2]^{3+}$

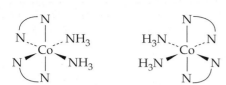

20.13

(a) Fe^{3+}

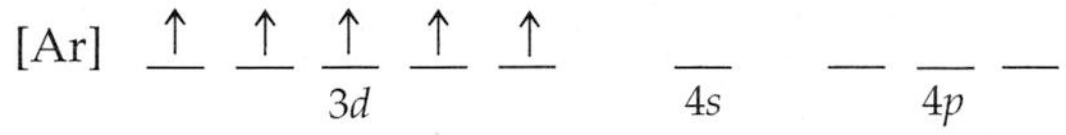

$[Ar]$ ↑ ↑ ↑ ↑ ↑   __   __ __ __
     $3d$ $4s$ $4p$

$[Fe(CN)_6]^{3-}$

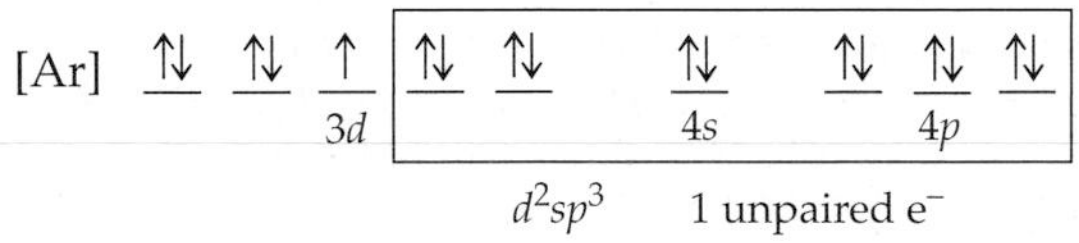

$[Ar]$ ⇅ ⇅ ↑ | ⇅ ⇅   ⇅   ⇅ ⇅ ⇅ |
     $3d$ $4s$ $4p$

       d^2sp^3 1 unpaired e^-

(b) Co^{2+}

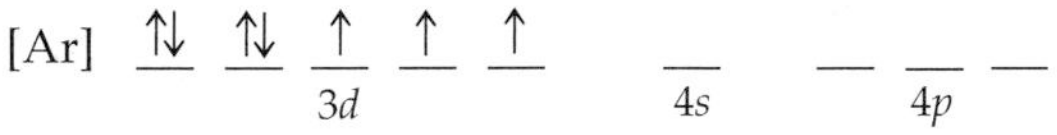

$[Ar]$ ⇅ ⇅ ↑ ↑ ↑   __   __ __ __
     $3d$ $4s$ $4p$

$[Co(H_2O)_6]^{2+}$

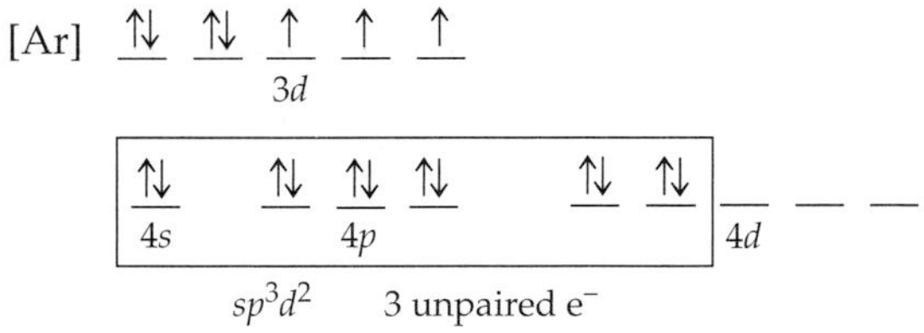

$[Ar]$ ⇅ ⇅ ↑ ↑ ↑
     $3d$

| ⇅   ⇅ ⇅ ⇅   ⇅ ⇅ | __ __ __
$4s$ $4p$ $4d$

    sp^3d^2 3 unpaired e^-

(c) V^{3+}

$[Ar]$ ↑ ↑ __ __ __   __   __ __ __
     $3d$ $4s$ $4p$

$[VCl_4]^-$

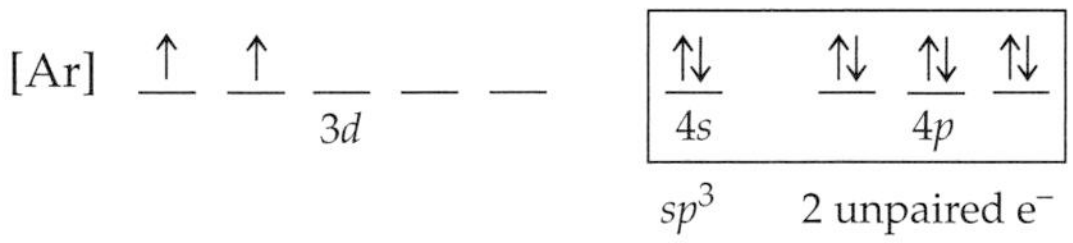

$[Ar]$ ↑ ↑ __ __ __ | ⇅   ⇅ ⇅ ⇅ |
     $3d$ $4s$ $4p$

       sp^3 2 unpaired e^-

(d) Pt^{2+}

$[Xe]$ ⇅ ⇅ ⇅ ↑ ↑   __   __ __ __
     $5d$ $6s$ $6p$

$[PtCl_4]^{2-}$

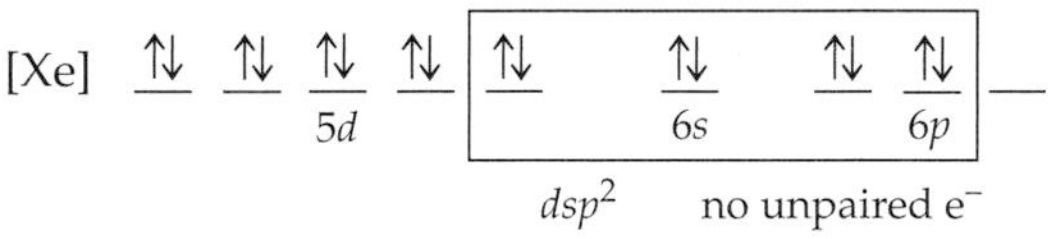

$[Xe]$ ⇅ ⇅ ⇅ ⇅ | ⇅   ⇅   ⇅ ⇅ | __
     $5d$ $6s$ $6p$

       dsp^2 no unpaired e^-

20.14

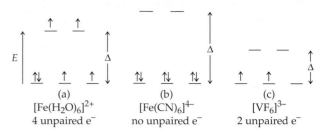

     (a)   (b)   (c)
  $[Fe(H_2O)_6]^{2+}$ $[Fe(CN)_6]^{4-}$ $[VF_6]^{3-}$
  4 unpaired e^- no unpaired e^- 2 unpaired e^-

20.15

(a) $[NiCl_4]^{2-}$ (tetrahedral)   (b) $[Ni(CN)_4]^{2-}$ (square planar)

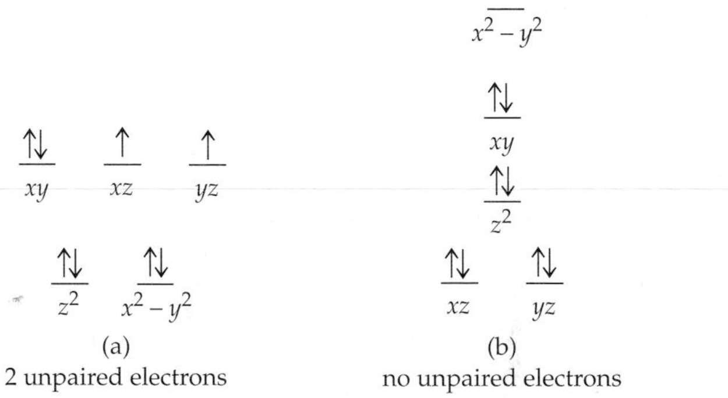

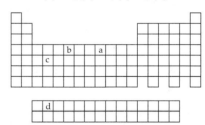

      (a)   (b)
2 unpaired electrons   no unpaired electrons

20.16 Although titanium has a large positive $E°$ for oxidation, the bulk metal is remarkably immune to corrosion because its surface becomes coated with a thin, protective oxide film.   **20.17** 1.00×10^{11} L   *Understanding Key Concepts*   **20.18** (a) Co; (b) Cr; (c) Zr; (d) Pr

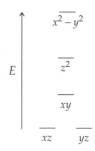

20.20 (a) bidentate; (b) monodentate; (c) tridentate; (d) monodentate. (a) and (c) can form chelate rings. **20.22** (a) (1) cis; (2) trans; (3) trans; (4) cis; (b) (1) and (4) are the same. (2) and (3) are the same. (c) None of the isomers exist as enantiomers because their mirror images are identical.

20.24

Additional Problems   **20.26** (a) [Ar] $3d^8 4s^2$; (b) [Ar] $3d^5$ $4s^1$; (c) [Kr] $4d^2 5s^2$; (d) [Ar] $3d^{10} 4s^2$ **20.28** (a) [Ar] $3d^7$; (b) [Ar] $3d^5$; (c) [Kr] $4d^3$; (d) [Ar] $3d^0$ **20.30** (a) 1; (b) 2; (c) 0; (d) 3   **20.32** Ti is harder than K and Ca largely because the sharing of d, as well as s, electrons results in stronger metallic bonding.   **20.34** (a) The decrease in radii with increasing atomic number is expected because the added d electrons only partially shield the added nuclear charge. As a result, Z_{eff} increases. With increasing Z_{eff}, the electrons are more strongly attracted to the nucleus, and atomic size decreases. (b) The densities of the transition metals are inversely related to their atomic radii.   **20.36** The lanthanide contraction is the general decrease in atomic radii of the f-block lanthanide elements. It is due to the increase in Z_{eff} as the $4f$ subshell is filled.
20.38

Element	$E_{i1} + E_{i2}$ (kJ/mol)
Sc	1866
Ti	1968
V	2064
Cr	2225
Mn	2226
Fe	2320
Co	2404
Ni	2490
Cu	2703
Zn	2639

The general trend is due to the increase in Z_{eff} with increasing atomic number. Higher than expected values for the sum of the first two ionization energies are observed for Cr and Cu because of their anomalous electron configurations (Cr: $3d^5 4s^1$; Cu: $3d^{10} 4s^1$). An increasing Z_{eff} affects $3d$ orbitals more than the $4s$ orbital and the second ionization energy for an electron from the $3d$ orbital is higher than expected.   **20.40** (a) Cr(s) + 2 H$^+$(aq) → Cr^{2+}(aq) + H$_2$(g); (b) Zn(s) + 2 H$^+$(aq) → Zn^{2+}(aq) + H$_2$(g); (c) N.R.; (d) Fe(s) + 2 H$^+$(aq) → Fe^{2+}(aq) + H$_2$(g)   **20.42** (b) Mn; (d) Cu   **20.44** Sc(III), Ti(IV), V(V), Cr(VI), Mn(VII), Fe(VI), Co(III), Ni(II), Cu(II), Zn (II)   **20.46** Cu^{2+} is a stronger oxidizing agent than Cr^{2+} because of a higher Z_{eff}.   **20.48** Cr^{2+} is more easily oxidized than Ni^{2+} because of a smaller Z_{eff}.   **20.50** Mn^{2+} < MnO$_2$ < MnO$_4^-$ because of increasing oxidation state of Mn.   **20.52** (a) Cr$_2$O$_3$(s) + 2 Al(s) → 2 Cr(s) + Al$_2$O$_3$(s); (b) Cu$_2$S(l) + O$_2$(g) → 2 Cu(l) + SO$_2$(g)   **20.54** (a) Cr$_2$O$_7^{2-}$; (b) Cr^{3+}; (c) Cr^{2+}; (d) Fe^{2+}; (e) Cu^{2+} **20.56** Cr(OH)$_2$ < Cr(OH)$_3$ < CrO$_2$(OH)$_2$. Acid strength increases with polarity of the O–H bond, which increases in turn with the oxidation state of Cr.   **20.58** (c) Cr(OH)$_3$ **20.60** (a) Add excess KOH(aq) and Fe^{3+} will precipitate as Fe(OH)$_3$(s). Na$^+$(aq) will remain in solu-

tion. (b) Add excess NaOH(aq) and Fe^{3+} will precipitate as Fe(OH)$_3$(s). Cr(OH)$_4^-$(aq) will remain in solution. (c) Add excess NH$_3$(aq) and Fe^{3+} will precipitate as Fe(OH)$_3$(s). Cu(NH$_3$)$_4^{2+}$(aq) will remain in solution.   **20.62** (a) Cr$_2$O$_7^{2-}$(aq) + 6 Fe^{2+}(aq) + 14 H$^+$(aq) → 2 Cr^{3+}(aq) + 6 Fe^{3+}(aq) + 7 H$_2$O(l); (b) 4 Fe^{2+}(aq) + O$_2$(g) + 4 H$^+$(aq) → 4 Fe^{3+}(aq) + 2 H$_2$O(l); (c) Cu$_2$O(s) + 2 H$^+$(aq) → Cu(s) + Cu^{2+}(aq) + H$_2$O(l); (d) Fe(s) + 2 H$^+$(aq) → Fe^{2+}(aq) + H$_2$(g)   **20.64** (a) 2 CrO$_4^{2-}$(aq) + 2 H$_3$O$^+$(aq) → Cr$_2$O$_7^{2-}$(aq) + 3 H$_2$O(l); (b) [Fe(H$_2$O)$_6$]$^{3+}$(aq) + SCN$^-$(aq) → [Fe(H$_2$O)$_5$(SCN)]$^{2+}$(aq) + H$_2$O(l); (c) 3 Cu(s) + 2 NO$_3^-$(aq) + 8 H$^+$(aq) → 3 Cu^{2+}(aq) + 2 NO(g) + 4 H$_2$O(l); (d) Cr(OH)$_3$(s) + OH$^-$(aq) → Cr(OH)$_4^-$(aq); 2 Cr(OH)$_4^-$(aq) + 3 HO$_2^-$(aq) → 2 CrO$_4^{2-}$(aq) + 5 H$_2$O(l) + OH$^-$(aq)   **20.66** Ni^{2+} accepts six pairs of electrons, two each from the three ethylenediamine ligands. Ni^{2+} is an electron pair acceptor, a Lewis acid. The two nitrogens in each ethylenediamine donate a pair of electrons to the Ni^{2+}. The ethylenediamine is an electron pair donor, a Lewis base.   **20.68** (a) [Ag(NH$_3$)$_2$]$^+$; (b) [Ni(CN)$_4$]$^{2-}$; (c) [Cr(H$_2$O)$_6$]$^{3+}$ **20.70** (a) +1; (b) +3; (c) +2; (d) +4; (e) +3   **20.72** (a) Ni(CO)$_4$; (b) [Ag(NH$_3$)$_2$]$^+$; (c) [Fe(CN)$_6$]$^{3-}$; (d) [Ni(CN)$_4$]$^{2-}$
20.74

The iron is in the +3 oxidation state, and the coordination number is six. The geometry about the Fe is octahedral. There are three chelate rings, one formed by each oxalate ligand.   **20.76** (a) +3; (b) +1; (c) +3; (d) +1   **20.78** (a) tetrachloromanganate(II); (b) hexaamminenickel(II); (c) tricarbonatocobaltate(III); (d) bis(ethylenediamine)-dithiocyanatoplatinum(IV)   **20.80** (a) cesium tetrachloroferrate(III); (b) hexaaquavanadium(III) nitrate; (c) tetraamminedibromocobalt(III) bromide; (d) diglycinatocopper(II)   **20.82** (a) [Pt(NH$_3$)$_4$]Cl$_2$; (b) Na$_3$[Fe(CN)$_6$]; (c) [Pt(en)$_3$](SO$_4$)$_2$; (d) Rh(NH$_3$)$_3$(SCN)$_3$
20.84

[Ru(NH$_3$)$_5$(NO$_2$)]Cl

[Ru(NH$_3$)$_5$(ONO)]Cl

[Ru(NH$_3$)$_5$Cl]NO$_2$

[Ru(NH₃)₅(NO₂)]Cl and [Ru(NH₃)₅(ONO)]Cl are linkage isomers. [Ru(NH₃)₅Cl]NO₂ is an ionization isomer of both [Ru(NH₃)₅(NO₂)]Cl and [Ru(NH₃)₅(ONO)]Cl.   **20.86** (a) [Cr(NH₃)₂Cl₄]⁻ can exist as cis and trans diastereoisomers; (d) [PtCl₂Br₂]²⁻ (square planar) can exist as cis and trans diastereoisomers.   **20.88** (c) *cis*-[Cr(en)₂(H₂O)₂]³⁺; (d) [Cr(C₂O₄)₃]³⁻   **20.90** (a) Ru(NH₃)₄Cl₂ can exist as cis and trans diastereoisomers.

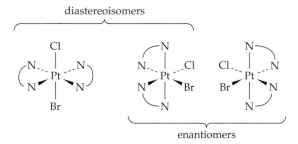

cis   trans

(b) [Pt(en)₃]⁴⁺ can exist as enantiomers.

(c) [Pt(en)₂ClBr]²⁺ can exist as both diastereoisomers and enantiomers.

diastereoisomers

enantiomers

20.92 Plane-polarized light is light in which the electric vibrations of the light wave are restricted to a single plane. The following chromium complex can rotate the plane of plane-polarized light.

[Cr(en)₃]³⁺

20.94 The measure of the amount of light absorbed by a substance is called the absorbance, and a graph of absorbance versus wavelength is called an absorption spectrum; orange   **20.96**

(a) [Ti(H₂O)₆]³⁺

Ti³⁺

[Ar] ↑ __ __ __ __   __   __ __ __
        3d   4s   4p

[Ti(H₂O)₆]³⁺

[Ar] ↑ __ __ | ↑↓ ↑↓   ↑↓   ↑↓ ↑↓ ↑↓ |
        3d   4s   4p
              d²sp³   1 unpaired e⁻

(b) [NiBr₄]²⁻

Ni²⁺

[Ar] ↑↓ ↑↓ ↑↓ ↑ ↑   __   __ __ __
        3d   4s   4p

[NiBr₄]²⁻

[Ar] ↑↓ ↑↓ ↑↓ ↑ ↑ | ↑↓   ↑↓ ↑↓ ↑↓ |
        3d   4s   4p
                    sp³   2 unpaired e⁻

(c) [Fe(CN)₆]³⁻ (low-spin)

Fe³⁺

[Ar] ↑ ↑ ↑ ↑ ↑   __   __ __ __
        3d   4s   4p

[Fe(CN)₆]³⁻

[Ar] ↑↓ ↑↓ ↑ | ↑↓ ↑↓   ↑↓   ↑↓ ↑↓ ↑↓ |
        3d   4s   4p
            d²sp³   1 unpaired e⁻

(d) [MnCl₆]³⁻ (high-spin)

Mn³⁺

[Ar] ↑ ↑ ↑ ↑ __   __   __ __ __
        3d   4s   4p

[MnCl₆]³⁻

[Ar] ↑ ↑ ↑ ↑ __
        3d

| ↑↓   ↑↓ ↑↓ ↑↓   ↑↓ ↑↓ | __ __ __
  4s   4p   4d
      sp³d²   4 unpaired e⁻

20.98 [Ti(H₂O)₆]³⁺, Ti³⁺, 3d¹

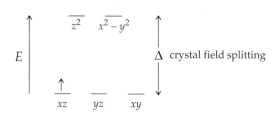

[Ti(H₂O)₆]³⁺ is colored because it can absorb light in the visible region, exciting the electron to the higher-energy set of orbitals.   **20.100** Δ = 220 kJ/mol; NCS⁻ is a weaker-field ligand than H₂O; color of [Ti(NCS)₆]³⁻ should be red.   **20.102**

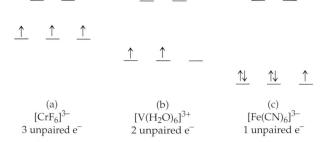

(a)   (b)   (c)
[CrF₆]³⁻   [V(H₂O)₆]³⁺   [Fe(CN)₆]³⁻
3 unpaired e⁻   2 unpaired e⁻   1 unpaired e⁻

20.104

Ni^{2+}(aq)   Zn^{2+}(aq)

↑   ↑   ↑↓   ↑↓

↑↓   ↑↓   ↑↓   ↑↓   ↑↓   ↑↓

Ni^{2+}(aq) is green because the Ni^{2+} ion can absorb light, which promotes electrons from the filled d orbitals to the higher energy half-filled d orbitals. Zn^{2+}(aq) is colorless because the d orbitals are completely filled and no electrons can be promoted, so no light is absorbed. **20.106** Weak-field ligands produce a small Δ. Strong-field ligands produce a large Δ. For a metal complex with weak-field ligands, $\Delta < P$, where P is the pairing energy, and it is easier to place an electron in either d_{z^2} or $d_{x^2-y^2}$ than to pair up electrons; high-spin complexes result. For a metal complex with strong-field ligands, $\Delta > P$ and it is easier to pair up electrons than to place them in either d_{z^2} or $d_{x^2-y^2}$; low-spin complexes result.
20.108

$\overline{x^2 - y^2}$

↑↓
xy
↑↓
z^2

↑↓   ↑↓
xz yz

Square planar geometry is most common for metal ions with d^8 configurations because this configuration favors low-spin complexes in which all four lower energy d orbitals are filled and the higher energy $d_{x^2-y^2}$ orbital is vacant.   **20.110** (a) [Mn(CN)$_6$]$^{3-}$; (d) [FeF$_6$]$^{4-}$ **20.112** (a) 4 Co^{3+}(aq) + 2 H$_2$O(l) → 4 Co^{2+}(aq) + O$_2$(g) + 4 H$^+$(aq); (b) 4 Cr^{2+}(aq) + O$_2$(g) + 4 H$^+$(aq) → 4 Cr^{3+}(aq) + 2 H$_2$O(l); (c) 3 Cu(s) + Cr$_2$O$_7^{2-}$(aq) + 14 H$^+$(aq) → 3 Cu^{2+}(aq) + 2 Cr^{3+}(aq) + 7 H$_2$O(l); (d) 2 CrO$_4^{2-}$(aq) + 2 H$^+$(aq) → Cr$_2$O$_7^{2-}$(aq) + H$_2$O(l)   **20.114** [Fe^{3+}] = 0.200 M, [Cr^{3+}] = 0.0665 M, [Cr$_2$O$_7$]$^{2-}$ = 0.0166 M **20.116** (a) sodium aquabromodioxalatoplatinate(IV); (b) hexaamminechromium(III) trioxalatocobaltate(III; (c) hexaamminecobalt(III) trioxalatochromate(III); (d) diamminebis(ethylenediamine)rhodium(III) sulfate **20.118** Cl$^-$ is a weak-field ligand, whereas CN$^-$ is a strong-field ligand. Δ for [Fe(CN)$_6$]$^{3-}$ is larger than the pairing energy P; Δ for [FeCl$_6$]$^{3-}$ is smaller than P. Fe^{3+} has a 3d^5 electron configuration.

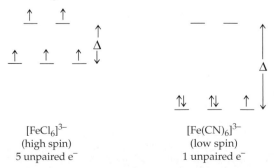

[FeCl$_6$]$^{3-}$ [Fe(CN)$_6$]$^{3-}$
(high spin)   (low spin)
5 unpaired e$^-$ 1 unpaired e$^-$

Because of the difference in Δ, [FeCl$_6$]$^{3-}$ is high-spin with five unpaired electrons, whereas [Fe(CN)$_6$]$^{3-}$ is low-spin with only one unpaired electron.   **20.120** For d^1–d^3 and d^8–d^{10} complexes, only one ground-state electron configuration is possible. In d^1–d^3 complexes, all the electrons occupy the lower-energy d orbitals , independent of the value of Δ. In d^8–d^{10} complexes, the lower-energy set of d orbitals is filled with three pairs of electrons, while the higher-energy set contains two, three, or four electrons, again independent of the value of Δ.   **20.122** [CoCl$_4$]$^{2-}$ is tetrahedral. [Co(H$_2$O)$_6$]$^{2+}$ is octahedral. Because $\Delta_{tet} < \Delta_{oct}$, these complexes have different colors. [CoCl$_4$]$^{2-}$ has absorption bands at longer wavelengths.
20.124

20.126

1 can exist as enantiomers.   **20.128** (a) Reaction (2) should have the larger entropy change because three bidentate en ligands displace six water molecules. (b) $\Delta G° = \Delta H° - T\Delta S°$. Because $\Delta H°_1$ and $\Delta H°_2$ are almost the same, the difference in $\Delta G°$ is determined by the difference in $\Delta S°$. Because $\Delta S°_2$ is larger than $\Delta S°_1$, $\Delta G°_2$ is more negative than $\Delta G°_1$, which is consistent with the greater stability of Ni(en)$_3^{2+}$. (c) 180 J/(K · mol)   **20.130** (a) Cr(s) + 2 H$^+$(aq) → Cr^{2+}(aq) + H$_2$(g); (b) 1.26 L; (c) 2.23;

(d)

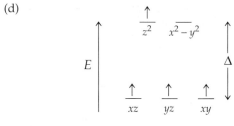

[Cr(H₂O)₆²⁺]

[Ar] ↑ ↑ ↑ ↑ _ _
    3d

| ↑↓ | ↑↓ ↑↓ ↑↓ | ↑↓ ↑↓ | _ _ _ |
| 4s | 4p | | 4d |

sp^3d^2 4 unpaired e⁻

(e) The addition of KCN converts $Cr(H_2O)_6^{2+}(aq)$ to $Cr(CN)_6^{4-}(aq)$. CN^- is a strong field ligand and increases Δ changing the chromium complex from high spin, with 4 unpaired electrons, to low spin, with only 2 unpaired electrons.

Chapter 21

21.1 (a) $Cr_2O_3(s) + 2 Al(s) \rightarrow 2 Cr(s) + Al_2O_3(s)$;
(b) $Cu_2S(s) + O_2(g) \rightarrow 2 Cu(s) + SO_2(g)$; (c) $PbO(s) + C(s) \rightarrow Pb(s) + CO(g)$;

(d) $2 K^+(l) + 2 Cl^-(l) \xrightarrow{\text{Electrolysis}} 2 K(l) + Cl_2(g)$

21.2 The electron configuration for Hg is $[Xe] 4f^{14} 5d^{10} 6s^2$. Assuming the 5d and 6s bands overlap, the composite band can accommodate 12 valence electrons per metal atom. Weak bonding and a low melting point are expected for Hg because both the bonding and antibonding MOs are occupied. **21.3** (a) The composite s–d band can accommodate 12 valence electrons per metal atom. (1) is half-filled, Mo $(5s^1 4d^5)$. (2) is one-fourth full, Y $(5s^2 4d^1)$. (3) is almost completey filled, Ag $(5s^1 4d^{10})$. (b) Strong bonding and the highest melting point are expected for Mo because all the bonding MOs are occupied and all the antibonding MOs are empty. Weak bonding and the lowest melting point are expected for Ag because the bonding MOs are occupied and the antibonding MOs contain 5 electrons per Ag atom. (c) Strong bonding and hardness are expected for Mo. Weak bonding and relative softness are expected for Ag. **21.4** Ge doped with As is an n-type semiconductor because As has an additional valence electron. The extra electrons are in the conduction band. The number of electrons in the conduction band of the doped Ge is much higher than for pure Ge, and the conductivity of the doped semiconductor is higher. **21.5** (a) (1), silicon; (2), white tin; (3), diamond; (4), silicon doped with aluminum; (b) (3) < (1) < (4) < (2). Diamond (3) is an insulator with a large band gap. Silicon (1) is a semiconductor with a smaller band gap than diamond. Silicon doped with aluminum (4) is a p-type semiconductor that has vacancies (positive holes) in the valence bond. White tin (2) has a partially filled s–p composite band and is a metallic conductor.

21.6

8 Cu at corners	$8 \times 1/8 = 1$ Cu
8 Cu on edges	$8 \times 1/4 = \underline{2}$ Cu
	Total $= 3$ Cu

12 O on edges	$12 \times 1/4 = 3$ O
8 O on faces	$8 \times 1/2 = \underline{4}$ O
	Total $= 7$ O

21.7 $Si(OCH_3)_4 + 4 H_2O \rightarrow Si(OH)_4 + 4 HOCH_3$
21.8 $Ba[OCH(CH_3)_2]_2 + Ti[OCH(CH_3)_2]_4 + 6 H_2O \rightarrow BaTi(OH)_6(s) + 6 HOCH(CH_3)_2$;

$$BaTi(OH)_6(s) \xrightarrow{\text{Heat}} BaTiO_3(s)$$

21.9 (a) ceramic–metal; (b) ceramic–ceramic; (c) ceramic–polymer; (d) ceramic–metal **21.10** Diamond is a pure element and not a true ceramic though it shares many properties with ceramics including hardness and a high melting point.
21.11

In diamond each carbon atom is covalently bonded to four other carbons. This network of strong bonds makes the diamond structure very rigid, which is why diamond is so hard. ***Understanding Key Concepts*** **21.12** A—metal oxide; B—metal sulfide; C—metal carbonate; D—free metal **21.14** (a) (1) and (4) are semiconductors; (2) is a metal; (3) is an insulator. (b) (3) < (1) < (4) < (2). The conductivity increases with decreasing band gap. (c) (1) and (4) increases; (2) decreases; (3) not much change.
21.16

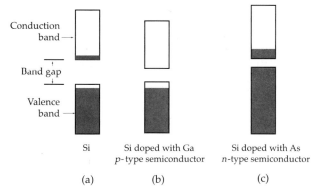

(d) The electrical conductivity of the doped silicon in both cases is higher than for pure silicon. Si doped with Ga is a p-type semiconductor with many more positive holes in the conduction band than in pure Si. This results in a higher conductivity. Si doped with As is an n-type semiconductor with many more electrons in the conduction band than in pure Si. This also results in a higher conductivity.
Additional Problems **21.18** TiO_2, MnO_2, and Fe_2O_3 **21.20** (a) sulfide; (b) oxide; (c) uncombined; (d) sulfide **21.22** The less electronegative early transition metals tend to form ionic compounds by losing electrons to highly electronegative nonmetals such as oxygen. The more electronegative late transition metals tend to form compounds with more covalent character by bonding to the

less electronegative nonmetals such as sulfur.   **21.24**
(a) hematite; (b) galena; (c) rutile; (d) chalcopyrite   **21.26**
The flotation process exploits the differences in the ability
of water and oil to wet the surfaces of the mineral and the
gangue. The gangue, which contains ionic silicates, is
moistened by the polar water molecules and sinks to the
bottom of the tank. The mineral particles, which contain
the less polar metal sulfide, are coated by the oil and
become attached to the soapy air bubbles created by
the detergent. The metal sulfide particles are carried to
the surface in the soapy froth, which is skimmed off at the
top of the tank. This process would not work well for a
metal oxide because it is too polar and will be wet by the
water and sink with the gangue.   **21.28** Since $E° < 0$ for
Zn^{2+}, the reduction of Zn^{2+} is not favored. Since $E° > 0$ for
Hg^{2+}, the reduction of Hg^{2+} is favored. The roasting of
CdS should yield CdO because, like Zn^{2+}, $E° < 0$ for the
reduction of Cd^{2+}.   **21.30** (a) $V_2O_5(s) + 5\,Ca(s) \rightarrow 2\,V(s) +$
$5\,CaO(s)$; (b) $2\,PbS(s) + 3\,O_2(g) \rightarrow 2\,PbO(s) + 2\,SO_2(g)$;
(c) $MoO_3(s) + 3\,H_2(g) \rightarrow Mo(s) + 3\,H_2O(g)$; (d) $3\,MnO_2(s) +$
$4\,Al(s) \rightarrow 3\,Mn(s) + 2\,Al_2O_3(s)$;

(e) $MgCl_2(l) \xrightarrow{\text{Electrolysis}} Mg(l) + Cl_2(g)$

21.32 $\Delta H° = -878.2$ kJ; $\Delta G° = -834.4$ kJ. $\Delta H°$ and $\Delta G°$ are
different because of the entropy change associated with
the reaction. The minus sign for $(\Delta H° - \Delta G°)$ indicates
that the entropy is negative, which is consistent with a
decrease in the number of moles of gas from 3 mol to
2 mol.   **21.34** (a) 110 kg; (b) 1.06×10^5 L   **21.36** 0.460 kg
21.38 $Fe_2O_3(s) + 3\,CO(g) \rightarrow 2\,Fe(l) + 3\,CO_2(g)$. Fe_2O_3 is the
oxidizing agent. CO is the reducing agent.   **21.40** Slag is
a by-product of iron production, consisting mainly of
$CaSiO_3$. It is produced from the gangue in iron ore.   **21.42**
Molten iron from a blast furnace is exposed to a jet of
pure oxygen gas for about 20 minutes. The impurities are
oxidized to yield a molten slag that can be poured off.
$P_4(l) + 5\,O_2(g) \rightarrow P_4O_{10}(l)$; $6\,CaO(s) + P_4O_{10}(l) \rightarrow$
$2\,Ca_3(PO_4)_2(l)$ (slag); $2\,Mn(l) + O_2(g) \rightarrow 2\,MnO(s)$;
$MnO(s) + SiO_2(s) \rightarrow MnSiO_3(l)$ (slag)   **21.44** $SiO_2(s) +$
$2\,C(s) \rightarrow Si(s) + 2\,CO(g)$; $Si(s) + O_2(g) \rightarrow SiO_2(s)$; $CaO(s) +$
$SiO_2(s) \rightarrow CaSiO_3(l)$ (slag)
21.46

Each K has a single valence electron and has eight nearest
neighbor K atoms. The valence electrons can't be localized in an electron-pair bond between any particular pair
of K atoms.   **21.48** Malleability and ductility of metals
follow from the fact that the delocalized bonding extends
in all directions. When a metallic crystal is deformed, no
localized bonds are broken. Instead, the electron sea
simply adjusts to the new distribution of cations, and
the energy of the deformed structure is similar to that of
the original. Thus, the energy required to deform a metal
is relatively small.   **21.50** The energy required to deform
a transition metal like W is greater than that for Cs
because W has more valence electrons and hence more
electrostatic "glue."   **21.52** The difference in energy
between successive MOs in a metal decreases as the
number of metal atoms increases so that the MOs merge

into an almost continuous band of energy levels.
Consequently, MO theory for metals is often called band
theory.   **21.54** The energy levels within a band occur in
degenerate pairs; one set of energy levels applies to
electrons moving to the right, and the other set applies to
electrons moving to the left. In the absence of an electrical
potential, the two sets of levels are equally populated. As
a result there is no net electric current. In the presence of
an electrical potential those electrons moving to the right
are accelerated, those moving to the left are slowed down,
and some change direction. Thus, the two sets of energy
levels are now unequally populated. The number of
electrons moving to the right is now greater than the
number moving to the left, and so there is a net electric
current.   **21.56**

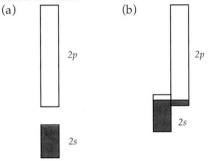

Diagram (b) shows the 2s and 2p bands overlapping in
energy and the resulting composite band is only partially
filled. Thus, Be is a good electrical conductor.   **21.58**
Transition metals have a d band that can overlap the
s band to give a composite band consisting of six MOs per
metal atom. Half of the MOs are bonding and half are
antibonding, and thus one expects maximum bonding for
metals that have six valence electrons per metal atom.
Accordingly, the melting points of the transition metals
go through a maximum at or near group 6B.   **21.60**
A semiconductor is a material that has an electrical
conductivity intermediate between that of a metal and
that of an insulator. Si, Ge, and Sn (gray) are semiconductors.
21.62

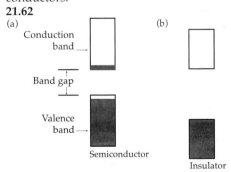

The MOs of a semiconductor are similar to those of an
insulator, but the band gap in a semiconductor is smaller.
As a result, a few electrons have enough energy to jump
the gap and occupy the higher-energy, conduction band.
The conduction band is thus partially filled, and the
valence band is partially empty. When an electrical
potential is applied to a semiconductor, it conducts a
small amount of current because the potential can accelerate the electrons in the partially filled bands.   **21.64** As
the band gap increases, the number of electrons able to
jump the gap and occupy the higher-energy conduction

band decreases, and thus the conductivity decreases. **21.66** An *n*-type semiconductor is a semiconductor doped with a substance with more valence electrons than the semiconductor itself. Si doped with P is an example.

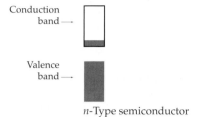

n-Type semiconductor

21.68 In the MO picture, the extra electrons occupy the conduction band. The number of electrons in the conduction band of the doped Ge is much greater than for pure Ge, and the conductivity of the doped semiconductor is correspondingly higher.   **21.70** (a) *p*-type; (b) *n*-type; (c) *n*-type   **21.72** Al_2O_3 < Ge < Ge doped with In < Fe < Cu   **21.74** (1) A superconductor is able to levitate a magnet. (2) In a superconductor, once an electric current is started, it flows indefinitely without loss of energy. A superconductor has no electrical resistance.   **21.76** The fullerides are three-dimensional superconductors, whereas the copper oxide ceramics are two-dimensional superconductors.   **21.78** Ceramics are inorganic, non-molecular solids, including both crystalline and amorphous materials. Ceramics have higher melting points, and they are stiffer, harder, and more resistant to wear and corrosion than are metals.   **21.80** Ceramics have higher melting points, and they are stiffer, harder and more wear resistant than are metals.   **21.82** The brittleness of ceramics is due to strong chemical bonding. In silicon nitride each Si atom is bonded to four N atoms and each N atom is bonded to three Si atoms. The strong, highly directional covalent bonds prevent the planes of atoms from sliding over one another when the solid is subjected to stress. As a result, the solid can't deform to relieve the stress. It maintains its shape up to a point, but then the bonds give way suddenly and the material fails catastrophically when the stress exceeds a certain threshold value. By contrast, metals are able to deform under stress because their planes of metal cations can slide easily in the electron sea.   **21.84** Ceramic processing is the series of steps that leads from raw material to the finished ceramic object.   **21.86** A sol is a colloidal dispersion of tiny particles. A gel is a more rigid gelatin-like material consisting of larger particles.   **21.88** $Zr[OCH(CH_3)_2]_4 + 4 H_2O \rightarrow Zr(OH)_4 + 4 HOCH(CH_3)_2$ **21.90** $(HO)_3Si–O–H + H–O–Si(OH)_3 \rightarrow$ $(HO)_3Si–O–Si(OH)_3 + H_2O$. Further reactions of this sort give a three-dimensional network of Si–O–Si bridges. On heating, SiO_2 is obtained.   **21.92** $3 SiCl_4(g) + 4 NH_3(g) \rightarrow$ $Si_3N_4(s) + 12 HCl(g)$ **21.94** Graphite/epoxy composites are good materials for making tennis rackets and golf clubs because of their high strength-to-weight ratios. **21.96** The chemical composition of the alkaline earth metals is that of metal sulfates and sulfites, MSO_4 and MSO_3. **21.98** Band theory better explains how the number of valence electrons affects properties such as melting point and hardness.   **21.100** Transition metals have a *d* band that can overlap the *s* band to give a composite band consisting of six MOs per metal atom. Half of the MOs are

bonding and half are antibonding. Strong bonding and a high enthalpy of vaporization are expected for V because almost all of the bonding MOs are occupied and all of the antibonding MOs are empty. Weak bonding and a low enthalpy of vaporization are expected for Zn because both the bonding and the antibonding MOs are occupied. **21.102** With a band gap of 130 kJ/mol, GaAs is a semiconductor. Because Ge lies between Ga and As in the periodic table, GaAs is isoelectronic with Ge.

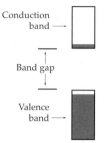

GaAs semiconductor

21.104 36.7 g $Y(OCH_2CH_3)_3$; 74.4 g $Ba(OCH_2CH_3)_2$; 109 g $YBa_2Cu_3O_7$ **21.106** (a)

$$6 Al(OCH_2CH_3)_3 + 2 Si(OCH_2CH_3)_4 + 26 H_2O \rightarrow$$
$$6 Al(OH)_3(s) + 2 Si(OH)_4(s) + 26 HOCH_2CH_3$$
$$\text{sol}$$

(b) H_2O is eliminated from the sol through a series of reactions linking the sol particles together through a three-dimensional network of O bridges to form the gel. $(HO)_2Al–O–H + H–O–Si(OH)_3 \rightarrow (HO)_2Al–O–Si(OH)_3 +$ H_2O; (c) The remaining H_2O and solvent are removed from the gel by heating to produce the ceramic, $3 Al_2O_3 \cdot 2 SiO_2$.   **21.108** (a)

$$H_2C =\!\!= CH + H_2C =\!\!= CH \longrightarrow$$
$$\qquad | \qquad\qquad |$$
$$\qquad CN \qquad\quad\; CN$$

$$-CH_2-CH-CH_2-CH-$$
$$\qquad\quad | \qquad\qquad |$$
$$\qquad\quad CN \qquad\quad\; CN$$

(b) −89 kJ/unit; exothermic   **21.110** (a) $\Delta G° = +12.6$ kJ, $K_p = 0.028$; (b) $\Delta G° = +45.4$ kJ, $K_p = 1.9 \times 10^{-5}$; (c) $\Delta S°$ is large and negative because as the reaction proceeds in the forward direction, the number of moles of gas decreases from four to one. Because $\Delta S°$ is negative, $-T\Delta S°$ is positive, and as T increases, $\Delta G°$ becomes more positive because $\Delta G° = \Delta H° - T\Delta S°$. (d) The reaction is exothermic because $\Delta H°$ is negative. $Ni(s) + 4 CO(g) \rightleftharpoons Ni(CO)_4(g) +$ Heat. Heat is added as the temperature is raised and the reaction proceeds in the reverse direction to relieve this stress, as predicted by Le Châtelier's principle. As the reverse reaction proceeds, the partial pressure of CO increases and the partial pressure of $Ni(CO)_4$ decreases. K_p decreases as calculated because $K_p = P_{Ni(CO)_4}/(P_{CO})^4$. **21.112** (a) $(NH_4)_2Zn(CrO_4)_2(s) \rightarrow ZnCr_2O_4(s) + N_2(g) +$ $4 H_2O(g)$; (b) 7.251 g; (c) 7.35 L; (d) 1/8 of the tetrahedral holes and 1/2 of the octahedral holes are filled; (e) Octahedral Cr^{3+} has three unpaired electrons in the lower-energy *d* orbitals (*xy*, *xz*, *yz*). Cr^{3+} can absorb visible light to promote one of these *d* electrons to one of the higher-energy *d* orbitals making this compound colored. All of the *d* orbitals in Zn^{2+} are filled and no *d* electrons can be promoted; consequently the Zn^{2+} ion does not contribute to the color.

Chapter 22

22.1 (a) $^{106}_{44}Ru \rightarrow ^{0}_{-1}e + ^{106}_{45}Rh$; (b) $^{189}_{83}Bi \rightarrow ^{4}_{2}He + ^{185}_{81}Tl$; (c) $^{204}_{84}Po + ^{0}_{-1}e \rightarrow ^{204}_{83}Bi$ **22.2** $^{4}_{2}He$ **22.3** 64.2 h
22.4 1.21×10^{-4} y^{-1} **22.5** 14.0%   **22.6** 44.5 d   **22.7** 3.0 d
22.8 ^{199}Au decays by beta emission. ^{173}Au decays by alpha emission.   **22.9** The shorter arrow pointing right is for beta emission. The longer arrow pointing left is for alpha emission. A = $^{250}_{97}Bk$; B = $^{250}_{98}Cf$; C = $^{246}_{96}Cm$; D = $^{242}_{94}Pu$; E = $^{238}_{92}U$
22.10 Mass defect = 0.136 99 g/mol; binding energy = 8.00 MeV/nucleon   **22.11** -9.47×10^{-12} kg/mol   **22.12** 1.79×10^{10} kJ/mol   **22.13** 5.31×10^{8} kJ/mol   **22.14** $^{40}_{18}Ar + ^{1}_{1}p \rightarrow ^{40}_{19}K + ^{1}_{0}n$ **22.15** $^{238}_{92}U + ^{2}_{1}H \rightarrow ^{238}_{93}Np + 2 \, ^{1}_{0}n$
22.16 1.53×10^{4} y   **22.17** Elements heavier than iron arise from nuclear reactions occurring as a result of supernova explosions.   *Understanding Key Concepts*   **22.18** two half-lives   **22.20** $^{14}_{6}C$ would decay by beta emission because the n/p ratio is high.   **22.22** ^{160}W is neutron poor and decays by alpha emission. ^{185}W is neutron rich and decays by beta emission.   **22.24** The shorter arrow pointing right is for beta emission. The longer arrow pointing left is for alpha

emission. A = $^{241}_{94}Pu$; B = $^{241}_{95}Am$; C = $^{237}_{93}Np$; D = $^{233}_{91}Pa$; E = $^{233}_{92}U$
Additional Problems   **22.26** Positron emission is the conversion of a proton in the nucleus into a neutron plus an ejected positron. Electron capture is the process in which a proton in the nucleus captures an inner-shell electron and is thereby converted into a neutron.   **22.28** Alpha particles move relatively slowly and can be stopped by the skin. However, inside the body, alpha particles give up their energy to the immediately surrounding tissue. Gamma rays move at the speed of light and are very penetrating. Therefore they are equally hazardous internally and externally.   **22.30** There is no radioactive "neutralization" reaction like there is an acid–base neutralization reaction.
22.32 (a) $^{126}_{50}Sn \rightarrow ^{0}_{-1}e + ^{126}_{51}Sb$; (b) $^{210}_{88}Ra \rightarrow ^{4}_{2}He + ^{206}_{86}Rn$; (c) $^{77}_{37}Rb \rightarrow ^{0}_{1}e + ^{77}_{36}Kr$; (d) $^{76}_{36}Kr + ^{0}_{-1}e \rightarrow ^{76}_{35}Br$ **22.34** (a) $^{0}_{1}e$; (b) $^{4}_{2}He$; (c) $^{0}_{-1}e$ **22.36** (a) $^{162}_{75}Re \rightarrow ^{158}_{73}Ta + ^{4}_{2}He$; (b) $^{138}_{62}Sm + ^{0}_{-1}e \rightarrow ^{138}_{61}Pm$; (c) $^{188}_{74}W \rightarrow ^{188}_{75}Re + ^{0}_{-1}e$; (d) $^{165}_{73}Ta \rightarrow ^{165}_{72}Hf + ^{0}_{1}e$ **22.38** $^{237}_{93}Np$, $^{233}_{91}Pa$, $^{233}_{92}U$, $^{229}_{90}Th$, $^{225}_{88}Ra$, $^{225}_{89}Ac$, $^{221}_{87}Fr$, $^{217}_{85}At$, $^{213}_{83}Bi$, $^{213}_{84}Po$, $^{209}_{82}Pb$, $^{209}_{83}Bi$ **22.40** 6 α, 4 β **22.42** It takes 44.5 days for half of the original amount of ^{59}Fe to decay.   **22.44** 0.247 d^{-1} **22.46** 3.04 d
22.48 After 65 d: 99.97%. After 65 y: 90.10%. After 650 y: 35.27%.   **22.50** 6980 y   **22.52** 23.2%   **22.54** 621 α particles
22.56 1.1×10^{6} y   **22.58** 34.6 d   **22.60** The lost mass is converted into the binding energy that is used to hold the nucleons together.   **22.62** 0.000 828 nm   **22.64** (a) 0.480 59 g/mol; (b) 0.854 99 g/mol   **22.66** (a) 8.76 MeV/nucleon; (b) 8.74 MeV/nucleon   **22.68** 5.42×10^{8} kJ/mol
22.70 1.02×10^{-9} g   **22.72** 9.87×10^{7} kJ/mol   **22.74** (a) $^{133}_{49}In$; (b) $^{13}_{7}N$ **22.76** $^{209}_{83}Bi + ^{58}_{26}Fe \rightarrow ^{266}_{109}Mt + ^{1}_{0}n$ **22.78** $^{12}_{6}C$ **22.80** 4.52×10^{9} kJ/mol   **22.82** 1.18×10^{9} kJ/mol   **22.84** 20.3 s
22.86 (a) 0.467 00 g/mol, 8.72 MeV/nucleon; (b) 0.600 15 g/mol, 8.76 MeV/nucleon   **22.88** 1.77×10^{9} kJ/mol
22.90 $^{238}_{92}U + ^{1}_{0}n \rightarrow ^{239}_{94}Pu + 2 \, ^{0}_{-1}e$ **22.92** 11,000 y   **22.94** 5.73 L

Chapter 23

23.1

23.2

23.3

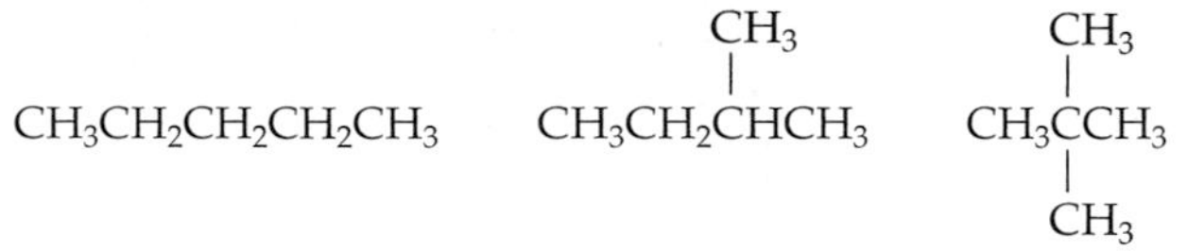

CH₃CH₂CH₂CH₂CH₃

$$CH_3CH_2CHCH_3$$
with CH₃ above CHCH₃

$$CH_3CCH_3$$
with CH₃ above and CH₃ below

CH₃CH₂CH₂CH₂CH₃   CH₃CH₂CHCH₃ (CH₃ above)   CH₃CCH₃ (CH₃ above, CH₃ below)

23.4

C₇H₁₆   CH₃CH₂CH₂CCH₃ (CH₃ above, CH₃ below)

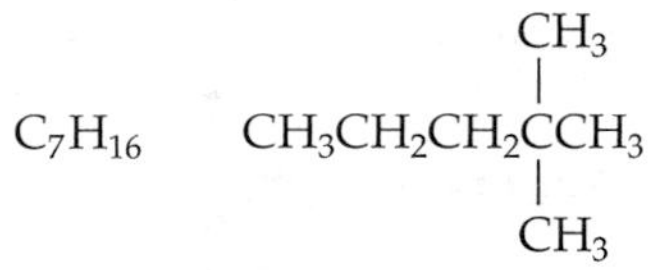

23.5 Structures (a) and (c) are identical.   **23.6** (a) pentane, 2-methylbutane, 2,2-dimethylpropane; (b) 3,4-dimethylhexane; (c) 2,4-dimethylpentane; (d) 2,2,5-trimethylheptane

23.7

(a)   CH₃CH₂CHCHCH₂CH₂CH₂CH₂CH₃
with CH₃ above first CH and CH₃ below second CH

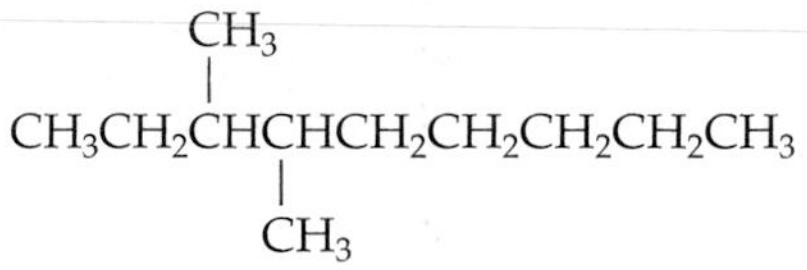

(b)   CH₃CH₂CH—CCH₂CH₂CH₃
with CH₃ above the C; below CH is CH₂ then CH₃; beside C is CH₃

(c)   CH₃CCH₂CHCH₂CH₂CH₂CH₃
with CH₃ and CH₂CH₂CH₃ above, CH₃ below

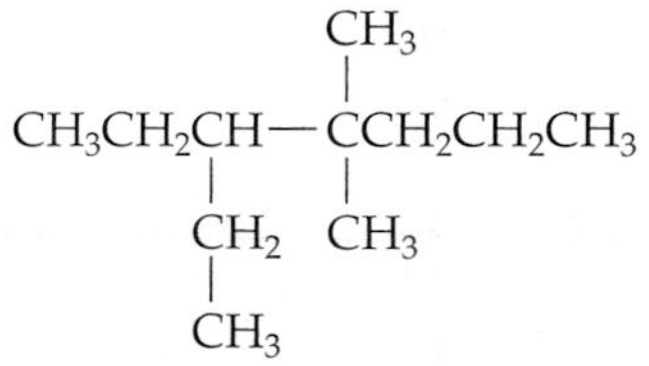

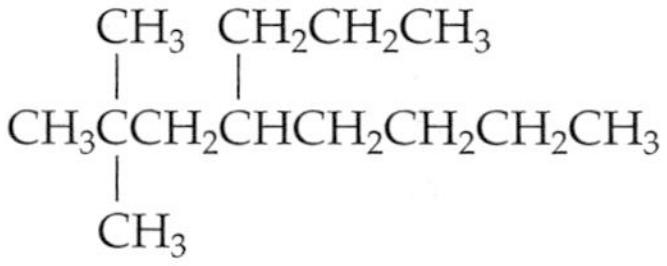

(d)   CH₃CCH₂CHCH₃
with CH₃ and CH₃ above, CH₃ below

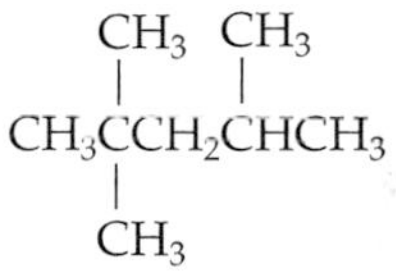

23.8 2,3-dimethylhexane   **23.9** (a) 1,4-dimethylcyclohexane; (b) 1-ethyl-3-methylcyclopentane; (c) isopropylcyclobutane

23.10

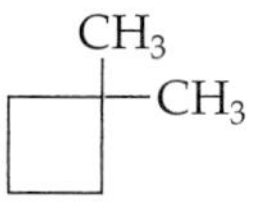

(a) cyclobutane with CH₃ and CH₃ substituents

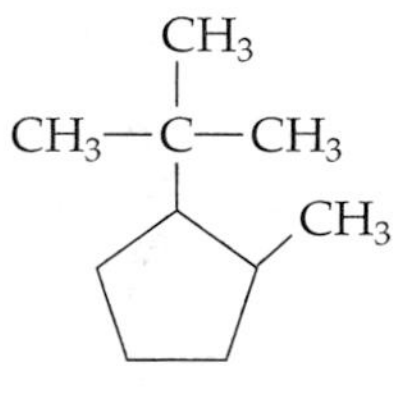

(b) CH₃—C—CH₃ structure attached to cyclopentane with CH₃

(c) cycloheptane ring with two CH₃ groups

23.11

ClCH₂CHCH₂CH₃ (CH₃ above)

CH₃CCH₂CH₃ (CH₃ above, Cl below)

CH₃CHCHCH₃ (CH₃ above, Cl below)

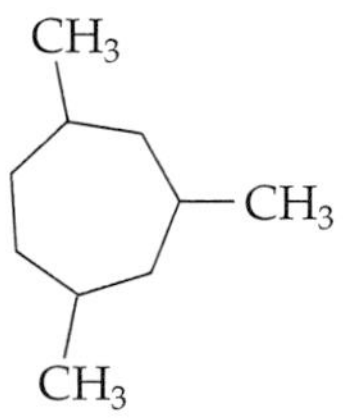

CH₃CHCH₂CH₂Cl (CH₃ above)

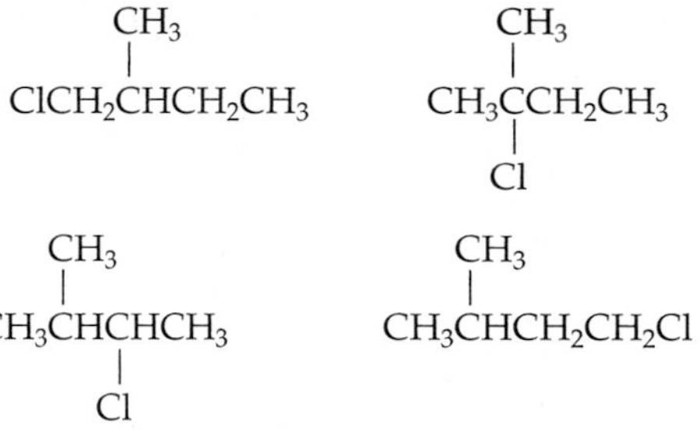

23.12 (a) carboxylic acid, alcohol; (b) alkene, aromatic ring

23.13

(a)   O (double bond)
        CH₃CH

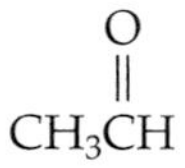

(b)   O (double bond)
        CH₃CH₂COH

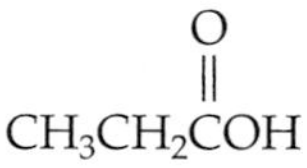

23.14 (a) 3-methyl-1-butene; (b) 4-methyl-3-heptene; (c) 3-ethyl-1-hexyne

23.15

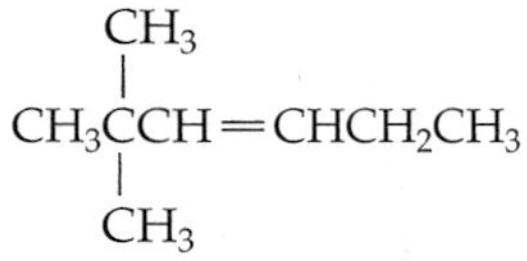

CH₃CCH=CHCH₂CH₃ (CH₃ above, CH₃ below)
(a)

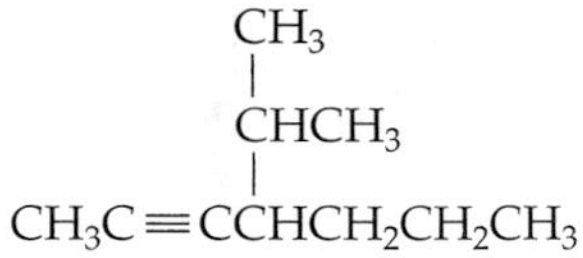

CHCH₃ (CH₃ above)
CH₃C≡CCHCH₂CH₂CH₃
(b)

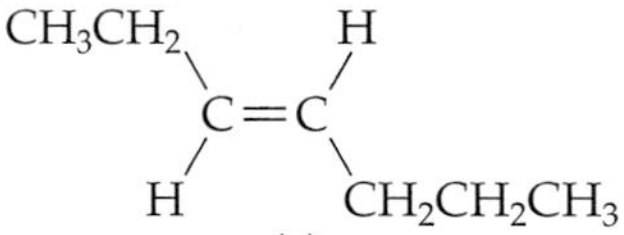

CH₃CH₂ , H
        C=C
H , CH₂CH₂CH₃
(c)

23.16

CH₃CH₂CH₂CH₃   CH₃CHCHCH₃ (Br Br above)   CH₃CH₂CHCH₃ (OH above)
(a)   (b) 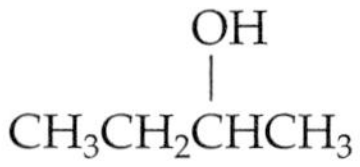   (c)

23.17

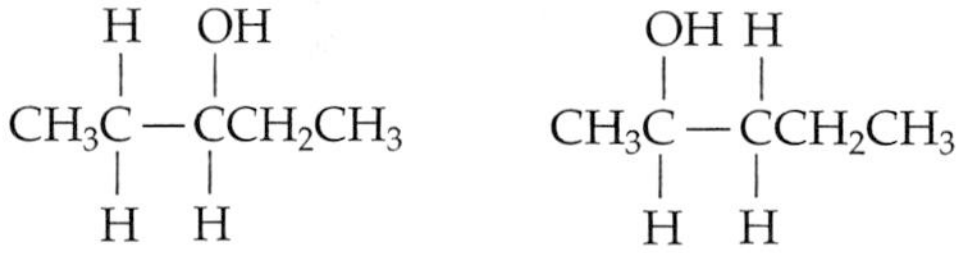

H OH
CH₃C—CCH₂CH₃
H H

OH H
CH₃C—CCH₂CH₃
H H

23.18

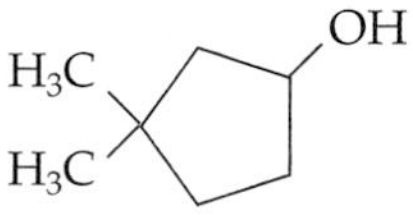

cyclopentane with H₃C, H₃C on one carbon and OH on another

23.19

(a) benzene ring with Br, Br

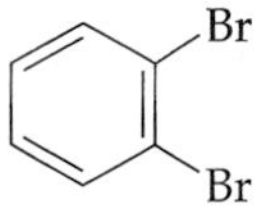

(b) benzene ring with NO₂ and Cl

(c) benzene ring with CH₂CH₃, CH₂CH₃

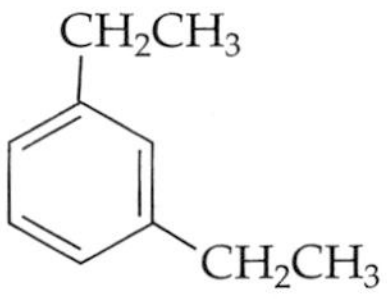

23.20

(a) 2-bromo-4-methyltoluene structure with CH_3, Br, CH_3

(b) CH_3, NO_2, CH_3 substituted benzene

(c) CH_3, SO_3H, CH_3 substituted benzene

23.21

Three structures: (left) CH_3 with Br (ortho); (middle) CH_3 with Br (meta); (right) CH_3 with Br (para)

23.22

(a) benzene ring with $\overset{+}{N}H_2CH_3$ Cl^-

(b) $CH_3CH_2CH_2\overset{+}{N}H_3$ Cl^-

23.23

(a) $CH_3\overset{CH_3}{\underset{|}{CH}}CH_2CH_2\overset{O}{\overset{||}{C}}-OH$

(b) benzene ring $-\overset{O}{\overset{||}{C}}-O-\overset{CH_3}{\underset{|}{CH}}CH_3$

(c) $CH_3CH_2\overset{O}{\overset{||}{C}}-NHCH_2CH_3$

23.24

(a) benzene ring with $\overset{O}{\overset{||}{C}}NH_2$ and CH_3

(b) $CH_3\overset{Cl}{\underset{|}{CH}}CH_2\overset{O}{\overset{||}{C}}\overset{CH_3}{\underset{|}{CH}}CH_2CH_3$

23.25

$CH_3\overset{CH_3}{\underset{|}{CH}}CH\overset{}{}CH_2\overset{O}{\overset{||}{C}}O\overset{}{}\overset{CH_3}{\underset{|}{CH}}CH_3$

$CH_3\overset{CH_3}{\underset{|}{CH}}CHCH_2\overset{O}{\overset{||}{C}}OCH\overset{}{\underset{|}{CH_3}}CH_3$

23.26

(a) CO_2CH_3 | $CH=CH_2$

(b) $HOCH_2CH_2CH_2OH$ + $HO\overset{O}{\overset{||}{C}}CH_2CH_2\overset{O}{\overset{||}{C}}OH$

23.27

Alcohol → HO, H   Cyclic ester
structure with Alcohol → HO, Alcohol → HO, OH ← Alcohol, Alkene

Understanding Key Concepts   **23.28**

(a) $CH_3\underset{\underset{CH_3}{|}}{\overset{CH_3}{\overset{|}{C}}}CH_2CH_3$

(b) $CH_3\underset{\underset{OH}{|}}{\overset{CH_3}{\overset{|}{C}H}}CHCH_3$

23.30 (a) alkene, ketone, ether; (b) alkene, amine, carboxylic acid   **23.32** $CH_2{=}CCl_2$ **23.34** There are many possibilities. Here are two:

$CH_3CH_2\overset{O}{\overset{||}{C}}CH{=}\underset{\underset{CH_3}{|}}{C}CH_3$

$CH_3\overset{O}{\overset{||}{C}}CH_2CH_2\underset{}{\overset{CH_3}{\overset{|}{C}}}{=}CH_2$

Additional Problems   **23.36** A functional group is a part of a larger molecule and is composed of an atom or group of atoms that has a characteristic chemical behavior.
23.38

(a) $CH_3CH_2\overset{O}{\overset{||}{C}}CH_2CH_3$

(b) $CH_3CH_2CH_2\overset{O}{\overset{||}{C}}OCH_2CH_3$

(c) $NH_2CH_2\overset{O}{\overset{||}{C}}OH$

23.40

$CH_3CH_2CH_2OH$ $CH_3\underset{\underset{OH}{|}}{C}HCH_3$ $CH_3CH_2OCH_3$

23.42 (a) alkene and aldehyde; (b) aromatic ring, alcohol, and ketone   **23.44** In a straight-chain alkane, all the carbons are connected in a row. In a branched-chain alkane, there are branching connections of carbons along the carbon chain.
23.46 sp^3 hybrid orbitals   **23.48** C_3H_9 contains one more H than needed for an alkane.   **23.50** (a) 4-ethyl-3-methyloctane; (b) 4-isopropyl-2-methylheptane; (c) 2,2,6-trimethylheptane; (d) 4-ethyl-4-methyloctane   **23.52**

(a) $CH_3CH_2\underset{\underset{CH_2CH_3}{|}}{C}HCH_2CH_2CH_3$

(b) $CH_3\underset{\underset{CH_3}{|}}{\overset{CH_3}{\overset{|}{C}}}{-}CHCH_2CH_3$

(c)

$$CH_3CH_2C \overset{\overset{\displaystyle CH_2CH_3}{|}}{\underset{\underset{\displaystyle CH_3}{|}}{\rule{0pt}{0pt}}} \overset{}{\underset{\underset{\displaystyle CH_3}{|}}{\rule{0pt}{0pt}}} CHCH_2CH_2CH_3$$

(d)

$$CH_3CHCH_2CH_2CHCH_2CH_2CH_3$$
with substituents CH_3 and $\overset{\overset{\displaystyle CH_3}{|}}{CHCH_3}$

23.54 (a) 1,1-dimethylcyclopentane; (b) 1-isopropyl-2-methylcyclohexane; (c) 1,2,4-trimethylcyclooctane
23.56 hexane, 2-methylpentane, 3-methylpentane, 2,2-dimethylbutane, and 2,3-dimethylbutane
23.58

(a)

$$ClCH_2CH_2CH_2CH_2CH_2CH_3 \qquad CH_3\overset{\overset{\displaystyle Cl}{|}}{C}HCH_2CH_2CH_2CH_3$$

$$CH_3CH_2\overset{\overset{\displaystyle Cl}{|}}{C}HCH_2CH_2CH_3$$

(b)

$$ClCH_2CH_2\overset{\overset{\displaystyle CH_3}{|}}{C}HCH_2CH_3 \qquad CH_3\overset{\overset{\displaystyle Cl}{|}}{C}H\overset{\overset{\displaystyle CH_3}{|}}{C}HCH_2CH_3$$

$$CH_3CH_2\overset{\overset{\displaystyle CH_3}{|}}{\underset{\underset{\displaystyle Cl}{|}}{C}}CH_2CH_3 \qquad CH_3CH_2\overset{\overset{\displaystyle CH_2Cl}{|}}{C}HCH_2CH_3$$

(c)

structures with CH₃ and Cl substituents on cyclohexane rings; CH₂Cl cyclohexane; Cl CH₃ cyclohexane

23.60 (a) sp^2; (b) sp; (c) sp^2 **23.62** Today the term "aromatic" refers to the class of compounds containing a six-membered ring with three double bonds, not to the fragrance of a compound.
23.64

$$CH_3CH=CHCH_2CH_3 \qquad HC\equiv CCH_2CH_3$$
(a)   (b)

(c) dimethylbenzene ring with two CH₃ groups (meta)

23.66 (a) 4-methyl-2-pentene; (b) 3-methyl-1-pentene; (c) 1,2-dichlorobenzene, or o-dichlorobenzene; (d) 2-methyl-2-butene; (e) 7-methyl-3-octyne

23.68

$$CH_2=CHCH_2CH_2CH_3 \qquad CH_3CH=CHCH_2CH_3$$
1-pentene   2-pentene

$$CH_2=\overset{\overset{\displaystyle CH_3}{|}}{C}CH_2CH_3 \qquad CH_3\overset{\overset{\displaystyle CH_3}{|}}{C}=CHCH_3 \qquad CH_2=CH\overset{\overset{\displaystyle CH_3}{|}}{C}HCH_3$$
2-methyl-1-butene   2-methyl-2-butene   3-methyl-1-butene

Only 2-pentene can exist as cis–trans isomers.   **23.70** (b), (c)
23.72

(a)

$$\overset{H}{\underset{CH_3CH_2}{}}C=C\overset{H}{\underset{CH_2CH_2CH_3}{}}$$

(b)

$$\overset{H}{\underset{H_3C}{}}C=C\overset{H}{\underset{CHCH_3 \,|\, CH_3}{}}$$

(c)

$$\overset{CH_3CH \,|\, CH_3 \text{ and } H}{\underset{H}{}}C=C\overset{H}{\underset{CHCH_3 \,|\, CH_3}{}}$$

23.74 Alkanes have free rotation about carbon–carbon single bonds and alkynes are linear about the carbon–carbon triple bond.   **23.76**

$$CH_3\overset{\overset{\displaystyle CH_3}{|}}{\underset{\underset{\displaystyle H}{|}}{C}}\overset{\overset{\displaystyle CH_3}{|}}{\underset{\underset{\displaystyle H}{|}}{C}}CH_3 \quad CH_3\overset{\overset{\displaystyle CH_3}{|}}{\underset{\underset{\displaystyle Br}{|}}{C}}\overset{\overset{\displaystyle CH_3}{|}}{\underset{\underset{\displaystyle Br}{|}}{C}}CH_3 \quad CH_3\overset{\overset{\displaystyle CH_3}{|}}{\underset{\underset{\displaystyle OH}{|}}{C}}\overset{\overset{\displaystyle CH_3}{|}}{\underset{\underset{\displaystyle H}{|}}{C}}CH_3$$
(a)   (b)   (c)

23.78

benzene rings with Cl, Br, NO₂, SO₃H, Cl substituents
(a)   (b)   (c)   (d)

23.80

(a)

$$CH_3\overset{\overset{\displaystyle CH_3}{|}}{\underset{\underset{\displaystyle OH}{|}}{C}}CH_2\overset{\overset{\displaystyle CH_3}{|}}{C}HCH_3$$

(b) OH, CH₃, CH₃ substituted cyclohexane

(c)

$$HOCH_2CH_2CH_2CH_2\overset{\overset{\displaystyle CH_2CH_3}{|}}{\underset{\underset{\displaystyle CH_2CH_3}{|}}{C}}CH_2CH_3$$

(d)

$$CH_3CH_2\overset{\overset{\displaystyle CH_2CH_3}{|}}{\underset{\underset{\displaystyle OH}{|}}{C}}CH_2CH_2CH_3$$

23.82 Quinine, a base, will dissolve in aqueous acid, but menthol is insoluble.   **23.84** An aldehyde has a terminal carbonyl group. A ketone has the carbonyl group located between two carbon atoms.   **23.86** oxidation of the related alcohol   **23.88** (a) ketone; (b) aldehyde; (c) ketone; (d) amide; (e) ester   **23.90** 0.81%   **23.92** (a) methyl 4-methylpentanoate; (b) 4,4-dimethylpentanoic acid; (c) 2-methylpentanamide   **23.94**

(a)
$$\text{CH}_3\text{CH}_2\text{CH}_2\text{CH}_2\overset{\displaystyle O}{\overset{\|}{\text{C}}}\text{—OCH}_3$$

(b)
$$\text{CH}_3\text{CH}_2\underset{\underset{\displaystyle CH_3}{|}}{\text{CH}}\overset{\displaystyle O}{\overset{\|}{\text{C}}}\text{—O}\underset{\underset{\displaystyle CH_3}{|}}{\text{CH}}$$

(c)
$$\text{CH}_3\overset{\displaystyle O}{\overset{\|}{\text{C}}}\text{—O—}\bigcirc$$

23.98 amine, aromatic ring, and ester

$$\text{H}_2\text{N}—\bigcirc—\overset{\displaystyle O}{\overset{\|}{\text{C}}}\text{OH}$$
carboxylic acid

$$\text{HOCH}_2\text{CH}_2\underset{\underset{\displaystyle CH_2CH_3}{|}}{\text{N}}\text{CH}_2\text{CH}_3$$
alcohol

23.100 Polymers are large molecules formed by the repetitive bonding together of many smaller molecules, called monomers.

23.102

$$\left(\text{CH}_2\underset{\underset{\displaystyle Cl}{|}}{\text{CH}}\text{CH}_2\underset{\underset{\displaystyle Cl}{|}}{\text{CH}}\text{CH}_2\underset{\underset{\displaystyle Cl}{|}}{\text{CH}}\text{CH}_2\underset{\underset{\displaystyle Cl}{|}}{\text{CH}}\right)_n$$

23.104

(a) $\text{CH}_2\!=\!\underset{\underset{\displaystyle CN}{|}}{\text{CH}}$ (b) $\text{CH}_2\!=\!\underset{\underset{\displaystyle CH_3}{|}}{\text{CH}}$ (c) $\text{CH}_2\!=\!\text{CCl}_2$

23.96

(a)
$$\text{CH}_3\text{CH}_2\text{CH}_2\text{CH}_2\overset{\displaystyle O}{\overset{\|}{\text{C}}}\text{OH} + \text{CH}_3\text{OH} \xrightarrow{\text{H}^+}$$
$$\text{CH}_3\text{CH}_2\text{CH}_2\text{CH}_2\overset{\displaystyle O}{\overset{\|}{\text{C}}}\text{OCH}_3 + \text{H}_2\text{O}$$

(b)
$$\text{CH}_3\text{CH}_2\underset{\underset{\displaystyle CH_3}{|}}{\text{CH}}\overset{\displaystyle O}{\overset{\|}{\text{C}}}\text{OH} + \text{H}\underset{\underset{\displaystyle CH_3}{|}}{\overset{\overset{\displaystyle CH_3}{|}}{\text{C}}}\text{OH} \xrightarrow{\text{H}^+}$$
$$\text{CH}_3\text{CH}_2\underset{\underset{\displaystyle CH_3}{|}}{\text{CH}}\overset{\displaystyle O}{\overset{\|}{\text{C}}}\text{O}\underset{\underset{\displaystyle CH_3}{|}}{\text{CH}} + \text{H}_2\text{O}$$

(c)
$$\text{CH}_3\overset{\displaystyle O}{\overset{\|}{\text{C}}}\text{OH} + \text{HO}—\bigcirc \xrightarrow{\text{H}^+}$$
$$\text{CH}_3\overset{\displaystyle O}{\overset{\|}{\text{C}}}\text{O}—\bigcirc + \text{H}_2\text{O}$$

23.106

repeating unit

23.108

(a)
$$\text{CH}_3\underset{\underset{\displaystyle CH_3}{|}}{\text{CH}}\text{CH}_2\text{CH}_2\text{CH}_2\text{CH}_2\text{CH}_3$$

(b)
$$\text{CH}_3\underset{\underset{\displaystyle CH_3}{|}}{\text{CH}}\text{CH}_2\underset{\underset{\displaystyle CH_2CH_3}{|}}{\text{CH}}\text{CH}_2\text{CH}_3$$

(c)
$$\overset{\sim}{\text{C}}\text{H}_3\text{CH}_2\underset{\underset{\displaystyle CH_3}{|}}{\text{CH}}\text{—}\underset{\underset{\displaystyle CH_3}{|}}{\overset{\overset{\displaystyle CH_2CH_3}{|}}{\text{C}}}\text{CH}_2\text{CH}_2\text{CH}_2\text{CH}_3$$

(d)
$$\text{CH}_3\underset{\underset{\displaystyle CH_3}{|}}{\text{CH}}\text{CH}_2\underset{\underset{\displaystyle CH_3}{|}}{\overset{\overset{\displaystyle CH_3}{|}}{\text{C}}}\text{CH}_2\text{CH}_2\text{CH}_3$$

(e)
$$\overset{\text{CH}_3 \quad \text{CH}_3}{\bigtriangleup\!\!\!\bigtriangledown}$$

(f)
$$\text{CH}_3\text{CH}_2\underset{\underset{\displaystyle H_3C}{|}}{\text{CH}}\underset{\underset{\displaystyle CHCH_3}{|}}{\text{CH}}\text{CH}_2\text{CH}_2\text{CH}_3$$
$$\underset{\displaystyle CH_3}{|}$$

23.110 Cyclohexene will react with Br$_2$ and decolorize it. Cyclohexane will not react.
23.112

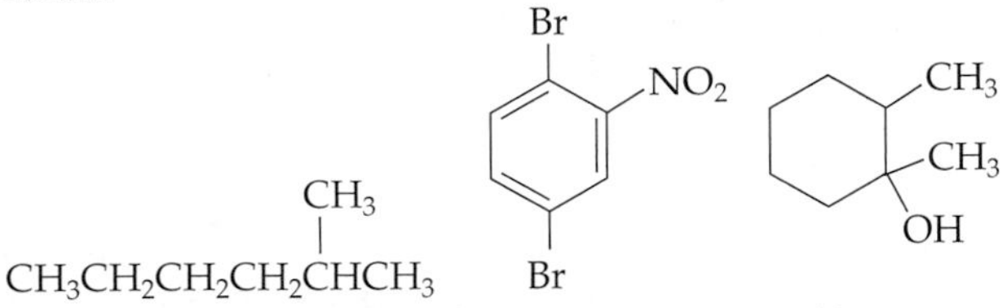

(a)   (b)   (c)

23.114 (a) CHO; (b) 116 amu;

(c)

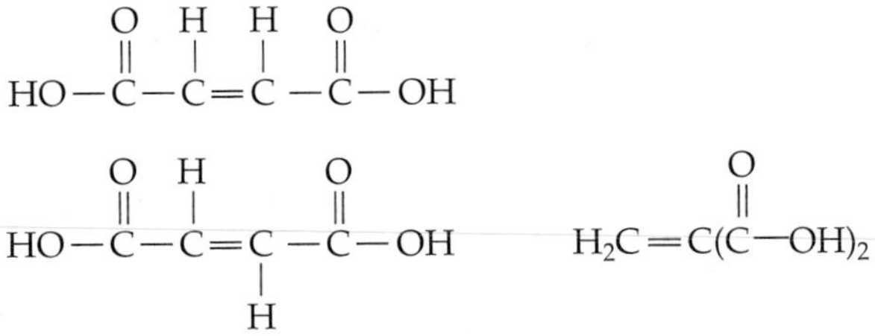

(d)

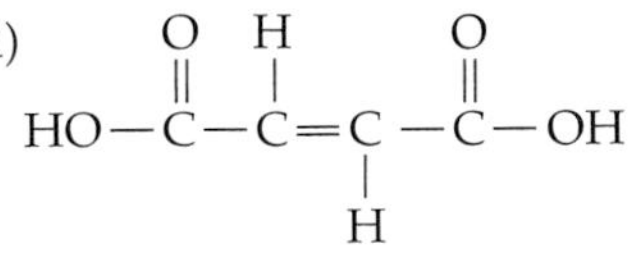

Chapter 24

24.1

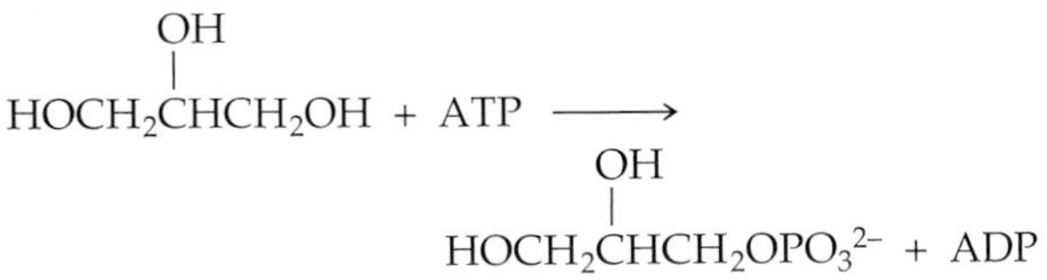

24.2 Amino acids that contain an aromatic ring: phenylalanine, tryptophan, tyrosine. Amino acids that contain sulfur: methionine, cysteine. Amino acids that are alcohols: serine, threonine, tyrosine. Amino acids that have alkyl-group side chains: alanine, valine, isoleucine, leucine. **24.3**

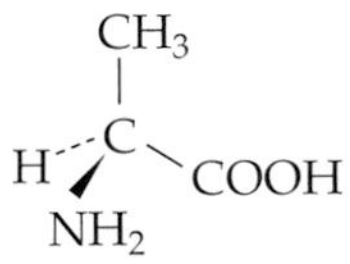

24.4 (a) isoleucine (neutral); (b) histidine (basic) **24.5** (a) and (c) **24.6** No carbon in 2-aminopropane has four different groups attached to it so the molecule is achiral. The second carbon in 2-aminobutane has four different groups attached to it so the molecule is chiral.

24.7 (b) and (c) **24.8** isoleucine and threonine **24.9** (b) and (c) are identical. (a) is an enantiomer.
24.10

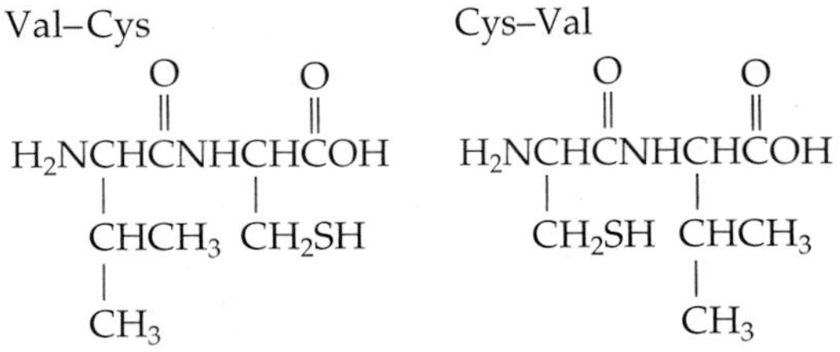

24.11 Val–Tyr–Gly; Val–Gly–Tyr; Tyr–Gly–Val; Tyr–Val–Gly; Gly–Tyr–Val; Gly–Val–Tyr **24.12** (a) aldopentose; (b) ketotriose; (c) aldotetrose
24.13

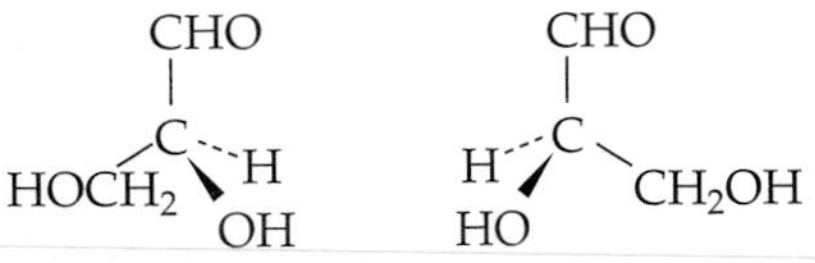

24.14 8 **24.15** aldohexose; 4 chiral carbons

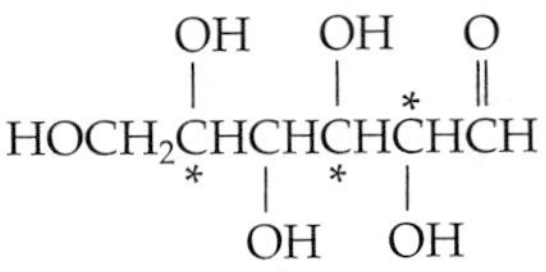

24.16

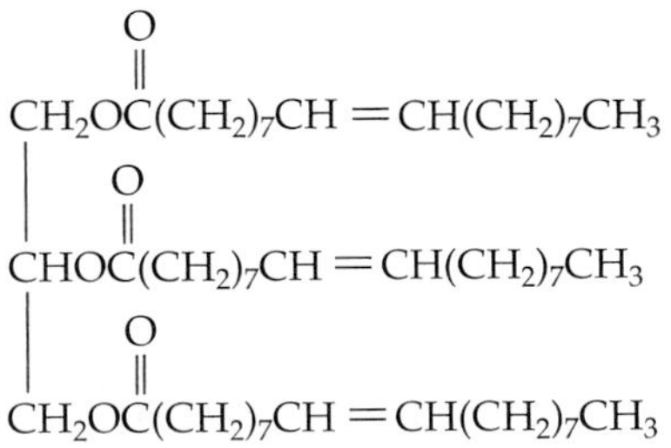

24.17

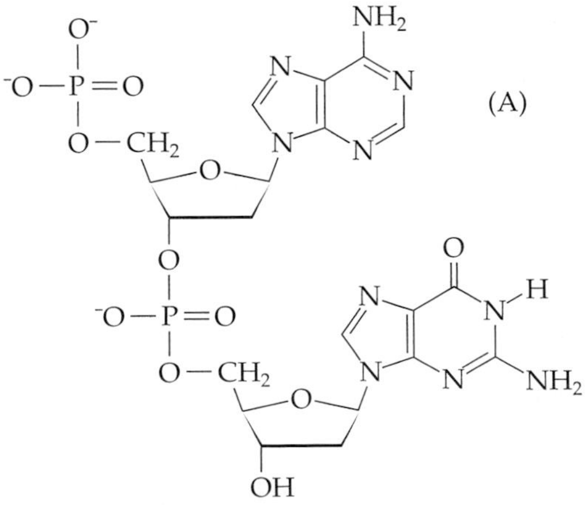

24.18

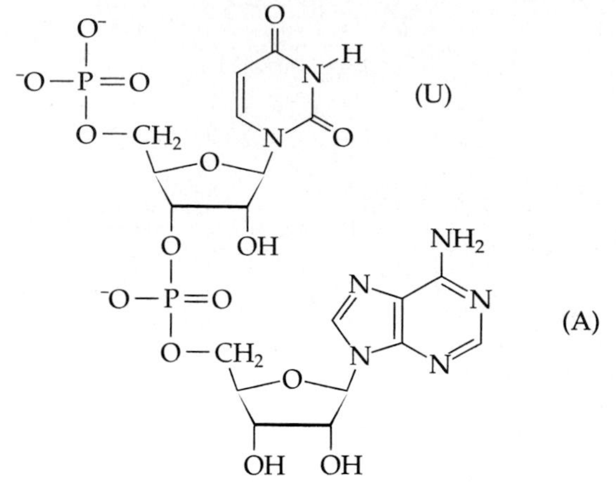

24.19 (a) adenine (DNA, RNA); (b) thymine (DNA)
24.20 Original: G–G–C–C–C–G–T–A–A–T
Complement: C–C–G–G–G–C–A–T–T–A
24.21

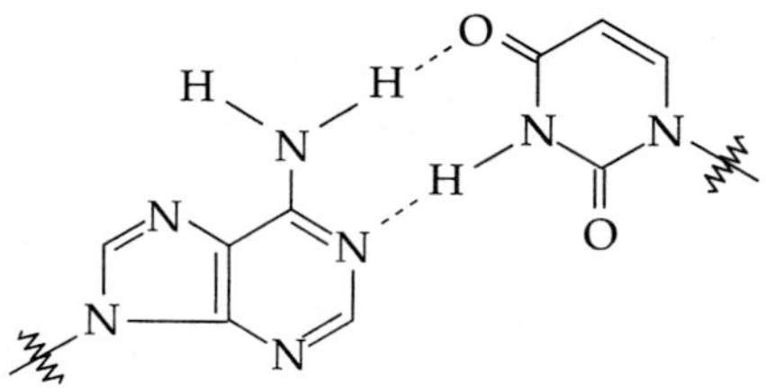

24.22 C–G–T–G–A–T–T–A–C–A   (DNA)
G–C–A–C–U–A–A–U–G–U   (RNA)
24.23 U–G–C–A–U–C–G–A–G–U   (RNA)
A–C–G–T–A–G–C–T–C–A   (DNA)
24.24 Human genes contain short, repeating sequences of noncoding DNA, called short tandem repeat (STR) loci. The base sequences in these STR loci are slightly different for every individual.   **24.25** (1) The sample is cleaved by enzymes that cut out the STR loci. (2) The fragments are isolated and purified by electrophoresis. (3) The fragments are amplified using the polymerase chain reaction (PCR). (4) The sequence of nucleotides in the DNA fragments is determined using the Sanger dideoxy method.
Understanding Key Concepts   **24.26** (a) serine; (b) methionine   **24.28** β form   **24.30** Ser–Val
Additional Problems   **24.32** The amino group is connected to the carbon atom alpha to (next to) the carboxylic acid group.   **24.34** (a) serine; (b) threonine; (c) proline; (d) phenylalanine; (e) cysteine
24.36

(a)   (b)

24.38 Val–Ser–Phe–Met–Thr–Ala   **24.40** (a) The primary structure of a protein is the sequence in which amino acids are linked together. (b) The secondary structure of a protein is the orientation of segments of a protein chain into a regular pattern. (c) The tertiary structure of a protein is the way in which a protein chain folds into a specific three-dimensional shape.   **24.42** hydrophobic

interactions, covalent disulfide bridges, electrostatic salt bridges, and hydrogen bonds   **24.44** Nearby cysteine residues can link together, forming a disulfide bridge.
24.46 Met–Ile–Lys, Met–Lys–Ile, Ile–Met–Lys, Ile–Lys–Met, Lys–Met–Ile, Lys–Ile–Met   **24.48**

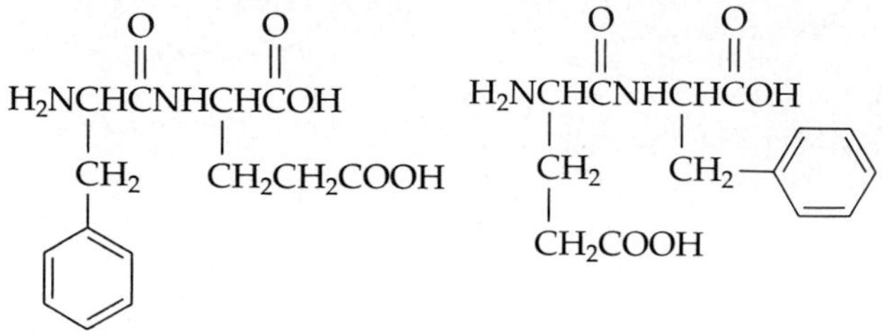

24.50 (a) and (c)   **24.52** (a)
24.54

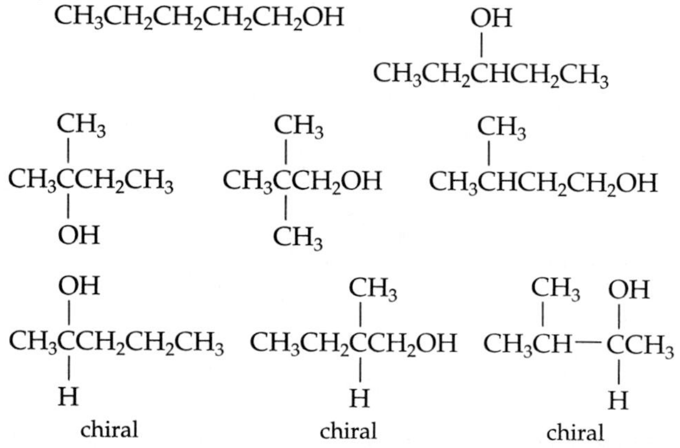

24.56 An aldose contains the aldehyde functional group while a ketose contains the ketone functional group.
24.58 Starch differs from cellulose in that it contains α- rather than β-glucose units.
24.60

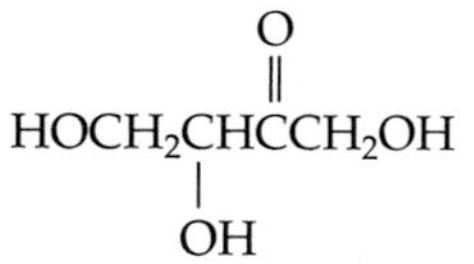

24.62

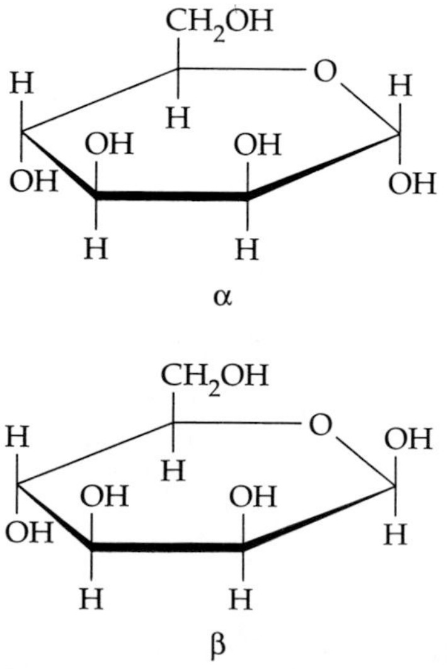

24.64

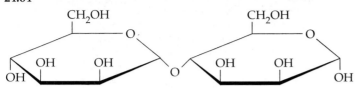

24.66 Long-chain carboxylic acids are called fatty acids.
24.68

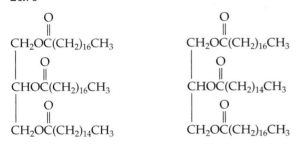

24.70

$$CH_2OC(CH_2)_{16}CH_3$$ with O double bond
$$CHOC(CH_2)_{16}CH_3$$ with O double bond
$$CH_2OC(CH_2)_{14}CH_3$$ with O double bond

$$CH_2OC(CH_2)_{16}CH_3$$ with O double bond
$$CHOC(CH_2)_{14}CH_3$$ with O double bond
$$CH_2OC(CH_2)_{16}CH_3$$ with O double bond

The two fat molecules differ depending on where the palmitic acid chain is located.
24.72

$$CH_3(CH_2)_{16}CO^- \ K^+ \quad CH_3(CH_2)_7CH=CH(CH_2)_7CO^- \ K^+$$
Potassium stearate   Potassium oleate

$$CH_3CH_2CH=CHCH_2CH=CHCH_2CH=CH(CH_2)_7CO^- \ K^+$$
Potassium linolenate

$$HOCH_2CHCH_2OH$$ with OH
Glycerol

24.74

(a)
$$CH_3(CH_2)_7CHCH(CH_2)_7COOH$$ with Br Br

(b) $CH_3(CH_2)_{16}COOH$

(c)
$$CH_3(CH_2)_7CH=CH(CH_2)_7COCH_3$$ with O

24.76 a phosphate group, an aldopentose sugar, and an amine base   **24.78** Most DNA of higher organisms is found in the nucleus of cells.   **24.80** A chromosome is a threadlike strand of DNA in the cell nucleus. A gene is a segment of a DNA chain that contains the instructions necessary to make a specific protein.
24.82

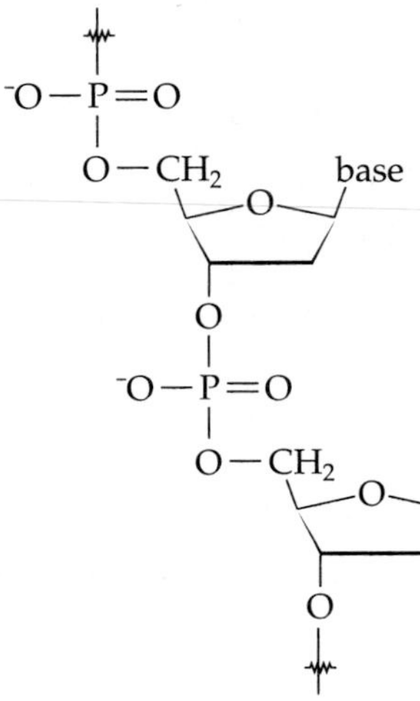

$$^-O-P=O$$ etc.
$$O-CH_2 \quad base$$

$$^-O-P=O$$
$$O-CH_2 \quad base$$

24.84

structure with NH_2, N, O, $^-O-P=O$, $O-CH_2$, OH

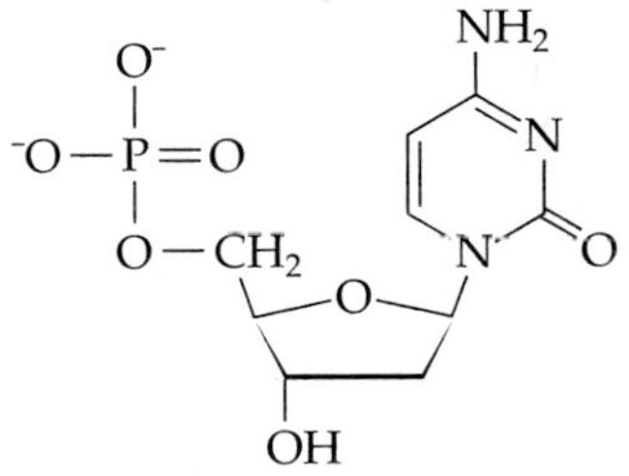

24.86 Original: T–A–C–C–G–A
           Complement: A–T–G–G–C–T
24.88 It takes three nucleotides to code for a specific amino acid. In insulin the 21 amino acid chain would require $(3 \times 21) = 63$ nucleotides and the 30 amino acid chain would require $(3 \times 30) = 90$ nucleotides.
24.90

$$CH_3(CH_2)_{18}CO(CH_2)_{31}CH_3$$ with O

24.92 $\Delta G^\circ = +2870$ kJ/mol
24.94

(a)

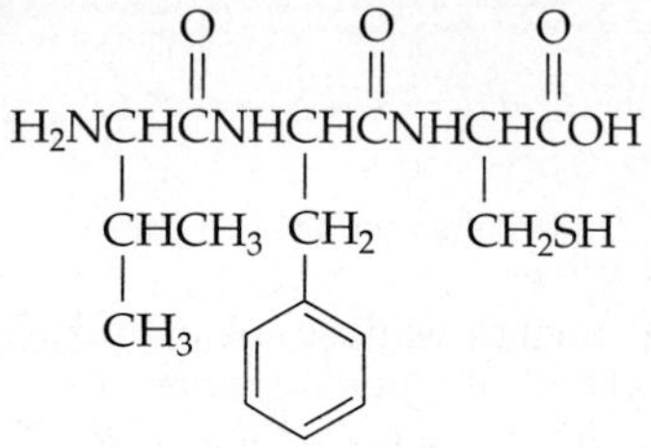

$$H_2NCHCNHCHCNHCHCOH$$

with three C=O groups above the chain, and below: $CHCH_3$, CH_2, CH_2SH; below $CHCH_3$: CH_3; below CH_2: a phenyl ring.

(b)

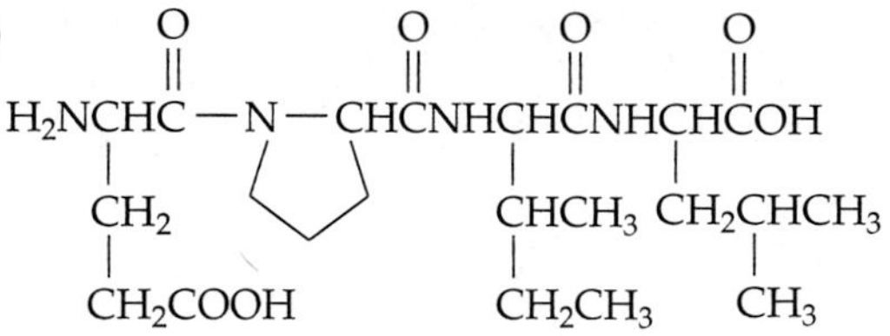

$$H_2NCHC-N-CHCNHCHCNHCHCOH$$

with four C=O groups above, a pyrrolidine ring at the N, and below the chain: CH_2, $CHCH_3$, CH_2CHCH_3; below those: CH_2COOH, CH_2CH_3, CH_3.

24.96

$$CH_3(CH_2)_{16}CO(CH_2)_{21}CH_3$$

with O double-bonded above the C.

24.98 Original: A–G–T–T–C–A–T–C–G
Complement: T–C–A–A–G–T–A–G–C
24.100 (a) 1.27 g; (b) 1.04 g I_2 in excess, 0.23 g I_2 reacted;
(c) 46; (d) 1.4

Glossary

Absorption spectrum a plot of the amount of light absorbed versus wavelength (*Section 20.10*)

Accuracy how close to the true value a given measurement is (*Section 1.11*)

Achiral lacking handedness (*Section 20.9*)

Acid a substance that provides H^+ ions when dissolved in water (*Section 2.9*)

Acid–base indicator a substance that changes color in a specific pH range (*Section 15.6*)

Acid–base neutralization reaction a process in which an acid reacts with a base to yield water plus an ionic compound called a salt (*Sections 4.5, 16.1*)

Acid-dissociation constant (K_a) the equilibrium constant for the dissociation of an acid in water (*Section 15.8*)

Actinide one of the 14 inner-transition metals following actinium in the periodic table (*Section 1.3*)

Activation energy (E_a) the height of the energy barrier between reactants and products (*Section 12.9*)

Active site a small three-dimensional region of an enzyme with the specific shape necessary to bind the substrate and catalyze the appropriate reaction (*Section 24.6*)

Activity series a list of elements in order of their reducing ability in aqueous solution (*Section 4.8*)

Acyclic an open-chain compound; not part of a ring (*Section 23.6*)

Addition reaction an organic reaction in which a reagent adds to the multiple bond of the unsaturated reactant to yield a saturated product (*Section 23.10*)

Adenosine triphosphate (ATP) a molecule that is formed as the final result of food catabolism and plays a pivotal role in the production of biological energy (*Section 24.1*)

Advanced ceramic a ceramic material that has high-tech engineering, electronic, or biomedical applications (*Section 21.7*)

Alcohol an organic molecule that contains an −OH group (*Section 23.12*)

Aldehyde an organic molecule that contains one alkyl group and one hydrogen bonded to a C=O carbon (*Section 23.13*)

Alkali metal an element in group 1A of the periodic table (*Sections 1.4, 6.7*)

Alkaline earth metal an element in group 2A of the periodic table (*Sections 1.4, 6.8*)

Alkane a compound that contains only carbon and hydrogen and has only single bonds (Section 23.2)

Alkene a hydrocarbon that has a carbon–carbon double bond (*Section 23.9*)

Alkyl group the part of an alkane that remains when a hydrogen is removed (*Section 23.5*)

Alkyne a hydrocarbon that has a carbon–carbon triple bond (*Section 23.9*)

Allotropes different structural forms of an element (*Section 10.10*)

Alloy a solid solution of two or more metals (*Section 21.2*)

Alpha- (α-) helix a secondary structure in which a protein chain adopts a spiral arrangement (*Section 24.5*)

Alpha (α) radiation a type of radioactive emission; a helium nucleus (*Sections 2.4, 22.2*)

Aluminosilicate a silicate mineral in which partial substitution of Si^{4+} by Al^{3+} has occurred (*Section 19.7*)

Amide an organic molecule that contains one alkyl group and one nitrogen bonded to a C=O carbon (*Section 23.14*)

Amine an organic derivative of ammonia (*Section 23.12*)

Amino acid the building block from which proteins are made (*Section 24.2*)

Amorphous solid a solid whose constituent particles are randomly arranged and have no ordered, long-range structure (*Section 10.6*)

Amphoteric exhibiting both acidic and basic properties (*Section 14.9*)

Amplitude a wave's height measured from the midpoint between peak and trough (*Section 5.2*)

Anabolism metabolic reaction sequences that put building blocks together to assemble larger molecules (*Section 24.1*)

Angular-momentum quantum number (l) a variable in the solutions to the Schrödinger wave equation that defines the three-dimensional shape of an orbital (*Section 5.7*)

Anion a negatively charged atom or group of atoms (*Section 2.8*)

Anode the electrode at which oxidation takes place (*Section 18.1*)

Anodizing the oxidation of a metal anode to yield a protective metal oxide coat (*Chapter 18 Interlude*)

Antibonding molecular orbital a molecular orbital that is higher in energy than the atomic orbitals it is derived from (*Section 7.13*)

Apoenzyme the protein part of an enzyme (*Section 24.6*)

Aqueous solution a solution with water as solvent (*Chapter 4 Introduction*)

Aromatic the class of compounds related to benzene (*Section 23.11*)

Arrhenius acid a substance that provides H^+ ions when dissolved in water (*Sections 2.9, 4.5*)

Arrhenius base a substance that provides OH^- ions when dissolved in water (*Sections 2.9, 4.5*)

Arrhenius equation an equation relating reaction rate constant, temperature, and activation energy; $k = Ae^{-E_a/RT}$ (*Section 12.9*)

Atom the smallest particle that retains the chemical properties of an element (*Chapter 2*)

Atomic mass the weighted average mass of an element's atoms (*Section 2.6*)

Atomic mass unit (amu) a convenient unit of mass; 1/12th the mass of a $^{12}_{6}C$ atom (*Section 2.6*)

Atomic number (Z) the number of protons in an atom's nucleus (*Section 2.5*)

Aufbau principle a set of rules that guides the filling order of orbitals in atoms (*Section 5.12*)

Avogadro's law The volume of a gas at a fixed pressure and temperature is proportional to its molar amount (*Section 9.2*).

Avogadro's number the number of units in a mole; 6.022×10^{23} (*Section 3.3*)

Balanced said of an equation in which the numbers and kinds of atoms on both sides of the arrow are the same (*Section 3.1*)

Band gap the energy difference between the valence band and the conduction band (*Section 21.5*)

Band theory the molecular orbital theory for metals (*Section 21.4*)

Base a substance that provides OH^- ions when dissolved in water (*Section 2.9*)

Base-dissociation constant (K_b) the equilibrium constant for the reaction of a base with water (*Section 15.12*)

Basic oxygen process a method for purifying iron and converting it to steel (*Section 21.3*)

Battery *see* Galvanic cell (*Section 18.9*)

Bayer process purification of Al_2O_3 by treating bauxite with hot aqueous NaOH (*Section 16.12*)

Beta- (β-) pleated sheet a secondary structure in which a protein chain doubles back on itself after a hairpin bend (*Section 24.5*)

Beta (β) radiation a type of radioactive emission consisting of electrons (*Section 22.2*)

Bidentate ligand a ligand that bonds to a metal using electron pairs on two donor atoms (*Section 20.6*)

Bimolecular reaction an elementary reaction that results from collisions between two reactant molecules (*Section 12.7*)

Binary hydride a compound that contains hydrogen and one other element (*Section 14.5*)

Binding energy the energy that holds nucleons together in the nucleus of an atom (*Section 22.5*)

Biochemistry the chemistry of living organisms (*Chapter 24*)

Bleaching decolorizing a colored material (*Section 4.12*)

Body-centered cubic packing a packing arrangement of spheres into a body-centered cubic unit cell (*Section 10.8*)

Body-centered cubic unit cell a cubic unit cell with an atom at each of its eight corners and an additional atom in the center of the cube (*Section 10.8*)

Boiling point the temperature at which liquid and vapor coexist in equilibrium (*Section 10.4*)

Bond angle the angle at which two adjacent bonds intersect (*Section 7.9*)

Bond dissociation energy (D) the amount of energy necessary to break a chemical bond in an isolated molecule in the gaseous state (*Section 7.2*)

Bond length the optimum distance between nuclei in a covalent bond (*Section 7.1*)

Bond order the number of electron pairs shared between two bonded atoms (*Section 7.5*)

Bonding electron pair a pair of valence electrons in a covalent bond (*Section 7.5*)

Bonding molecular orbital a molecular orbital that is lower in energy than the atomic orbitals it is derived from (*Section 7.13*)

Borane any compound of boron and hydrogen (*Section 19.4*)

Born–Haber cycle a pictorial way of viewing the steps in the formation of an ionic solid from its elements (*Section 6.6*)

Boyle's law the volume of a fixed amount of gas at a constant temperature varies inversely with its pressure (*Section 9.2*)

Bragg equation an equation used in X-ray crystallography for calculating the distance between atoms in a crystal (*Section 10.7*)

Branched-chain alkane an alkane with a branching connection of carbons (*Section 23.2*)

Brønsted–Lowry acid a substance that can transfer H^+ (*Section 15.1*)

Brønsted–Lowry base a substance that can accept H^+ (*Section 15.1*)

Buffer capacity a measure of the amount of acid or base that a buffer can absorb without a significant change in pH (*Section 16.3*)

Buffer solution a solution of a weak acid and its conjugate base that resists drastic changes in pH (*Section 16.3*)

C-Terminal amino acid the amino acid with a free $-CO_2H$ group on the end of a protein chain (*Section 24.4*)

Carbohydrate a large class of organic molecules related to glucose (*Section 24.7*)

Carbonyl group the C=O group (*Section 23.13*)

Carboxylate anion the anion that results from deprotonation of a carboxylic acid (*Section 23.14*)

Carboxylic acid an organic molecule that contains the $-CO_2H$ group (*Section 23.14*)

Catabolism metabolic reaction sequences that break molecules apart (*Section 24.1*)

Catalyst a substance that increases the rate of a reaction without itself being consumed (*Section 12.11*)

Cathode the electrode at which reduction takes place (*Section 18.1*)

Cathode ray the visible glow emitted when an electric potential is applied between two electrodes in an evacuated chamber (*Section 2.3*)

Cathodic protection a technique for protecting a metal from corrosion by connecting it to a second metal that is more easily oxidized (*Section 18.10*)

Cation a positively charged atom or group of atoms (*Section 2.8*)

Cell potential (*E*) *see* Electromotive force (*Section 18.3*)

Cell voltage *see* Electromotive force (*Section 18.3*)

Celsius degree (°C) a common unit of temperature; $0°C = 273.15$ K (*Section 1.8*)

Centimeter (cm) a common unit of length; 1 cm = 0.01 m (*Section 1.7*)

Ceramic an inorganic, nonmetallic, nonmolecular solid (*Section 21.7*)

Ceramic composite a hybrid material made of two ceramics (*Section 21.8*)

Ceramic–metal composite a hybrid material made of a metal reinforced with a ceramic (*Section 21.8*)

Ceramic–polymer composite a hybrid material made of a polymer reinforced with a ceramic (*Section 21.8*)

Chain reaction a reaction whose product initiates further reaction (*Section 22.6*)

Change of state *see* Phase change (*Section 10.4*)

Charles' law the volume of a fixed amount of gas at a constant pressure varies directly with its absolute temperature (*Section 9.2*).

Chelate the complex formed by a metal atom and a polydentate ligand (*Section 20.6*)

Chelating agent a polydentate ligand (*Section 20.6*)

Chemical bond the forces that hold atoms together in chemical compounds (*Section 2.8*)

Chemical energy potential energy stored in the chemical bonds of molecules (*Section 8.2*)

Chemical equation a format for writing a chemical reaction, listing reactants on the left, products on the right, and an arrow between them (*Section 2.7*)

Chemical equilibrium the state reached when the concentrations of reactants and products remain constant in time (*Chapter 13*)

Chemical formula a format for listing the number and kind of constituent elements in a compound (*Section 2.7*)

Chemical kinetics the area of chemistry concerned with reaction rates and the sequence of steps by which reactions occur (*Chapter 12*)

Chemical property a characteristic that results in a change in the chemical makeup of a sample (*Section 1.4*)

Chemical reaction the transformation of one substance into another (*Section 2.7*)

Chemistry the study of the composition, properties, and transformations of matter (*Chapter 1*)

Chiral having handedness (*Section 20.9*)

Chlor-alkali industry the commercial production method for Cl_2 and NaOH by electrolysis of aqueous sodium chloride (*Section 18.12*)

Chromosome a threadlike strand of DNA in the nucleus of cells (*Section 24.14*)

Cis isomer the isomer of a metal complex or alkene in which identical ligands or groups are adjacent rather than opposite (*Sections 20.8, 23.9*)

Clausius–Clapeyron equation a mathematical relationship between vapor pressure and heat of vaporization for a substance (*Section 10.5*)

Coefficient a number placed before a formula in an equation to indicate how many formula units are required to balance the equation (*Section 3.1*)

Coenzyme an inorganic ion or a small organic molecule that acts as an enzyme cofactor (*Section 24.6*)

Cofactor the small, nonprotein part of an enzyme (*Section 24.6*)

Colligative property a property that depends only on the amount of dissolved solute rather than on the chemical identity of the solute (*Section 11.5*)

Collision theory a model in which bimolecular reactions occur when two properly oriented reactant molecules come together in a sufficiently energetic collision (*Section 12.9*)

Colloid a homogeneous mixture containing particles with diameters in the range 2–500 nm (*Section 11.1*)

Combustion a chemical reaction that sustains a flame (*Sections 4.12, 8.8*)

Common-ion effect the shift in the position of an equilibrium on addition of a substance that provides an ion in common with one of the ions already involved in the equilibrium (*Section 16.2*)

Complex ion an ion that contains a metal cation bonded to one or more small molecules or ions (*Section 16.12*)

Compound a chemical substance composed of more than one kind of element (*Section 2.7*)

Condensation the change of a gas to a liquid (*Section 10.4*)

Condensed structure a shorthand method for drawing organic structures in which C−H and C−C single bonds are "understood" rather than shown (*Section 23.3*)

Conduction band the antibonding molecular orbitals in a semiconductor (*Section 21.5*)

Conformation the three-dimensional structure of a molecule (*Section 23.4*)

Conjugate acid the species HA formed by addition of H^+ to the base A^- (*Section 15.1*)

Conjugate acid–base pair chemical species whose formulas differ only by one proton (*Section 15.1*)

Conjugate base the species A^- formed by loss of H^+ from the acid HA (*Section 15.1*)

Constitutional isomers isomers that have different connections among their constituent atoms (Section 20.8)

Contact process the commercial process for making sulfuric acid from sulfur (*Section 19.13*)

Conversion factor an expression that describes the relationship between units (*Section 1.13*)

Coordinate covalent bond a bond formed when one atom donates two electrons to another atom that has a vacant valence orbital (*Section 7.5*)

Coordination compound a compound in which a central metal ion is attached to a group of surrounding molecules or ions by coordinate covalent bonds (*Section 20.5*)

Coordination number the number of nearest-neighbor atoms in a crystal (*Section 10.8*) or the number of ligand donor atoms that surround a central metal ion in a complex (*Section 20.5*)

Core electrons inner-shell electrons (*Section 6.3*)

Corrosion the oxidative deterioration of a metal, such as the conversion of iron to rust (*Section 18.10*)

Cosmic ray energetic particles, primarily protons, coming from interstellar space (*Section 22.8*)

Coulomb's law The force resulting from the interaction of two electric charges is equal to a constant k times the magnitude of the charges divided by the square of the distance between them (*Section 6.6*)

Covalent bond a bond that occurs when two atoms share several (usually two) electrons (*Section 2.8; Chapter 7*)

Covalent hydride a compound in which hydrogen is attached to another element by a covalent bond (*Section 14.5*)

Covalent network solid a solid whose atoms are linked together by covalent bonds into a giant three-dimensional array (*Section 10.6*)

Critical mass the amount of material necessary for a nuclear chain reaction to become self-sustaining (*Section 22.6*)

Critical point a combination of temperature and pressure beyond which a gas cannot be liquefied (*Section 10.11*)

Crystal field splitting the energy splitting between two sets of *d* orbitals in a metal complex (*Section 20.12*)

Crystal field theory a model that views the bonding in metal complexes as arising from electrostatic interactions and considers the effect of the ligand charges on the energies of the metal ion *d* orbitals (*Section 20.12*)

Crystalline solid a solid whose atoms, ions, or molecules have an ordered arrangement extending over a long range (*Section 10.6*)

Cubic centimeter (cm^3) a common unit of volume, equal in size to the milliliter; $1 \text{ cm}^3 = 10^{-6} \text{ m}^3$ (*Section 1.9*)

Cubic closest-packed a packing arrangement of spheres into a face-centered cubic unit cell with three alternating layers (*Section 10.8*)

Cubic meter (m^3) the SI unit of volume (*Section 1.9*)

Cycloalkane an alkane that contains a ring of carbon atoms (*Section 23.6*)

***d*-Block element** a transition metal element, which results from *d*-orbital filling (*Section 5.13; Chapter 20*)

Dalton's law of partial pressures The total pressure exerted by a mixture of gases in a container at constant *V* and *T* is equal to the sum of the pressures exerted by each individual gas in the container (*Section 9.5*).

de Broglie equation an equation that relates mass, wavelength, and velocity, $m = h/\lambda v$ (*Section 5.5*)

Decay constant the first-order rate constant for radioactive decay (*Section 22.3*)

Decay series a sequence of nuclear disintegrations ultimately leading to a nonradioactive product (*Section 22.4*)

Degenerate having the same energy level (*Section 5.12*)

Density an intensive physical property that relates the mass of an object to its volume (*Section 1.10*)

Deoxyribonucleic acid (DNA) an immense biological polymer of deoxyribonucleotide units that contains an organism's genetic information (*Section 24.12*)

Deposition the change of a gas directly to a solid (*Section 10.4*)

Diamagnetic a substance that is weakly repelled by a magnetic field (*Section 7.14*)

Diastereoisomers non-mirror-image stereoisomers (*Section 20.8*)

Diffraction scattering of a light beam by an object containing regularly spaced lines or points (*Section 10.7*)

Diffusion the mixing of different gases by random molecular motion and with frequent collisions (*Section 9.7*)

Dimensional-analysis method a method of problem solving whereby problems are set up so that unwanted units cancel (*Section 1.13*)

Dipole a pair of separated electrical charges (*Section 10.1*)

Dipole–dipole force an intermolecular force resulting from electrical interactions among dipoles on neighboring molecules (*Section 10.2*)

Dipole moment (μ) the measure of net molecular polarity; $\mu = Q \times r$ (*Section 10.1*)

Diprotic acid an acid that has two dissociable protons (*Section 15.7*)

Disproportionation reaction a reaction in which a substance is both oxidized and reduced (*Section 14.10*)

Dissociate splitting apart to give ions when dissolved in water (*Section 4.2*)

Donor atom the atom attached directly to a metal in a coordination compound (*Section 20.6*)

Doping the addition of a small amount of impurities to increase the conductivity of a semiconductor (*Section 21.5*)

Double bond a covalent bond formed by sharing four electrons (*Section 7.5*)

Effective nuclear charge (Z_{eff}) the net nuclear charge actually felt by an electron (*Section 5.11*)

Effusion the escape of gas molecules through a tiny hole in a membrane without collisions (*Section 9.7*)

Electrochemical cell a device for interconverting chemical and electrical energy (*Section 18.1*)

Electrochemistry the area of chemistry concerned with the interconversion of chemical and electrical energy (*Chapter 18*)

Electrode a conductor through which electrical current enters or leaves a cell (*Section 18.1*)

Electrolysis the process of using an electric current to bring about chemical change (*Section 18.11*)

Electrolyte a substance that dissolves in water to produce ions (*Section 4.2*)

Electrolytic cell an electrochemical cell in which an electric current drives a nonspontaneous reaction (*Section 18.11*)

Electromagnetic radiation radiant energy (*Section 5.2*)

Electromagnetic spectrum the range of different kinds of electromagnetic radiation (*Section 5.2*)

Electromotive force (emf) the electrical potential that pushes electrons away from the anode and pulls them toward the cathode (*Section 18.3*)

Electron a negatively charged, fundamental atomic particle (*Section 2.3*)

Electron affinity (E_{ea}) the energy change that occurs when an electron is added to an isolated atom in the gaseous state (*Section 6.5*)

Electron capture a process in which a proton in the nucleus captures an inner-shell electron and is thereby converted into a neutron (*Section 22.2*)

Electron configuration a description of which orbitals in an atom are occupied by electrons (*Section 5.12*)

Electron-dot structure a representation of a molecule that shows valence electrons as dots (*Section 7.5*)

Electron-sea model a model that visualizes metals as a three-dimensional array of metal cations immersed in a sea of delocalized electrons that are free to move about (*Section 21.4*)

Electronegativity (EN) the ability of an atom in a molecule to attract the shared electrons in a bond (*Section 7.4*)

Electroplating the coating of one metal on the surface of another using electrolysis (*Section 18.12*)

Electrorefining the purification of a metal by means of electrolysis (*Section 18.12*)

Element a fundamental substance that can't be chemically changed or broken down into anything simpler (*Sections 1.2, 2.1*)

Elementary reaction a single step in a reaction mechanism (*Section 12.7*)

Empirical formula a formula that gives the ratios of atoms in a compound (*Section 3.11*)

Enantiomers stereoisomers that are nonidentical mirror images of one another (*Section 20.9*)

Endothermic a reaction in which heat is absorbed and the temperature of the surroundings falls (*Section 8.7*)

Energy the capacity to do work or supply heat (*Section 8.1*)

Enthalpy change (ΔH) the heat change at constant pressure; $\Delta H = \Delta E + P\Delta V$ (*Section 8.5*)

Enthalpy (H) the quantity $E + PV$ (*Section 8.5*)

Entropy (S) the amount of molecular disorder or randomness in a system (*Sections 8.13, 17.3*)

Entropy of solution (ΔS$_{soln}$) the entropy change during formation of a solution (*Section 11.2*)

Enzyme a large protein that acts as a catalyst for a biological reaction (*Section 24.6*)

Equilibrium constant (K$_c$) the constant in the equilibrium equation (*Section 13.2*)

Equilibrium constant (K$_p$) the equilibrium constant for reaction of gases, defined using partial pressures (*Section 13.3*)

Equilibrium equation an equation that relates the concentrations in an equilibrium mixture (*Section 13.2*)

Equilibrium mixture a mixture of reactants and products at equilibrium (*Section 13.1*)

Equivalence point the point in a titration at which stoichiometrically equivalent quantities of reactants have been mixed together (*Section 16.5*)

Ester an organic molecule that contains the $-CO_2R$ group (*Section 23.14*)

Ether an organic molecule that contains two alkyl groups bonded to the same oxygen atom (*Section 23.12*)

Exothermic a reaction in which heat is evolved and the temperature of the surroundings rises (*Section 8.7*)

Extensive property a property whose value depends on the sample size (*Section 1.4*)

***f*-Block element** a lanthanide or actinide element, which results from *f*-orbital filling (*Section 5.13*)

Face-centered cubic unit cell a cubic unit cell with an atom at each of its eight corners and an additional atom on each of its six faces (*Section 10.8*)

Faraday the electrical charge on 1 mol of electrons (96,485 C/mol e⁻) (*Section 18.3*)

Fatty acid a long-chain carboxylic acid found as a constituent of fats and oils (*Section 24.11*)

First law of thermodynamics The energy of the universe is constant (*Section 8.3*).

First-order reaction one whose rate depends on the concentration of a single reactant raised to the first power (*Section 12.4*)

Flotation a metallurgical process that exploits differences in the ability of water and oil to wet the surfaces of mineral and gangue (*Section 21.2*)

Formal charge the result of a method of electron bookkeeping that tells whether an atom in a molecule has gained or lost electrons compared to an isolated atom (*Section 7.8*)

Formation constant (K$_f$) the equilibrium constant for formation of a complex ion (*Section 16.12*)

Formula mass the sum of atomic masses of all atoms in one formula unit of a substance (*Section 3.3*)

Formula unit one unit (atom, ion, or molecule) corresponding to a given formula (*Section 3.1*)

Fractional distillation the separation of volatile liquids on the basis of boiling point (*Section 11.10*)

Free-energy change (ΔG) $\Delta G = \Delta H - T\Delta S$ (*Sections 8.14, 17.7*)

Freezing the change of a liquid to a solid (*Section 10.4*)

Frequency (ν) the number of wave maxima that pass by a fixed point per unit time (*Section 5.2*)

Frequency factor the parameter A $(= pZ)$ in the Arrhenius equation (*Section 12.9*)

Fuel cell a galvanic cell in which one of the reactants is a traditional fuel such as methane or hydrogen (*Section 18.9*)

Functional group a part of a larger molecule; composed of an atom or group of atoms that has characteristic chemical behavior (*Section 23.8*)

Fusion melting; the change of a solid to a liquid (*Section 10.4*)

Galvanic cell an electrochemical cell in which a spontaneous chemical reaction generates an electric current (*Section 18.1*)

Galvanizing a process for protecting steel from corrosion by coating it with zinc (*Section 18.10*)

Gamma (γ) radiation a type of radioactive emission consisting of a stream of high-energy photons (*Section 22.2*)

Gangue the economically worthless material consisting of sand, clay, and other impurities that accompanies an ore (*Section 21.2*)

Gas constant (R) the constant in the ideal gas law $PV = nRT$ (*Section 9.3*)

Gas laws relationships among the variables P, V, n, and T for a gas sample (*Section 9.2*)

Gas shift reaction the reaction of CO with H_2O to yield hydrogen (*Section 14.3*)

Gene a segment of a DNA chain that contains the instructions necessary to make a specific protein (*Section 24.14*)

Geometric isomers *see* Diastereoisomers (*Section 20.8*)

Gibbs free-energy change (ΔG) $\Delta G = \Delta H - T\Delta S$ (*Sections 8.14, 17.7*)

Graham's law The rate of effusion of a gas is inversely proportional to the square root of its molar mass (*Section 9.7*).

Gram (g) a common unit of mass; 1 g = 0.001 kg (*Section 1.6*)

Ground-state configuration the lowest-energy electron configuration of an atom (*Section 5.12*)

Group a column of elements in the periodic table (*Section 1.3*)

Half-life ($t_{1/2}$) the time required for a reactant concentration to drop to one-half of its initial value (*Section 12.5*)

Half-reaction the oxidation or reduction part of a redox reaction (*Sections 4.10, 18.1*)

Half-reaction method a method for balancing redox equations (*Section 4.10*)

Hall–Heroult process the commercial production method for aluminum by electrolysis of a molten mixture of aluminum oxide and cryolite (*Section 18.12*)

Halogen an element in group 7A of the periodic table (*Sections 1.4, 6.10*)

Halogenation the addition of halogen (Cl_2 or Br_2) to an alkene (*Section 23.10*)

Hard water water that contains appreciable concentrations of Ca^{2+}, Mg^{2+}, or Fe^{2+} cations (*Section 14.13*)

Heat the energy transferred from one object to another as the result of a temperature difference between them (*Section 8.2*)

Heat capacity (C) the amount of heat required to raise the temperature of an object or substance a given amount (*Section 8.8*)

Heat of combustion the amount of energy released on burning a substance (*Section 8.12*)

Heat of fusion (ΔH_{fusion}) the amount of heat required for melting (*Section 10.4*)

Heat of reaction (ΔH) the enthalpy change for a reaction (*Section 8.5*)

Heat of solution (ΔH_{soln}) the enthalpy change during formation of a solution (*Section 11.2*)

Heat of sublimation (ΔH_{subl}) the amount of heat required for sublimation (*Section 8.7*)

Heat of vaporization (ΔH_{vap}) the amount of heat required for evaporation (*Section 10.4*)

Heisenberg uncertainty principle We can never know both the position and the velocity of an electron beyond a certain level of precision (*Section 5.6*).

Henderson–Hasselbalch equation an equation relating the pH of a solution to the pK_a of the weak acid; pH = pK_a + log ([base]/[acid]) (*Section 16.4*)

Henry's law The solubility of a gas in a liquid at a given temperature is directly proportional to the partial pressure of the gas over the solution (*Section 11.4*).

Hertz (Hz) a unit of frequency; 1 Hz = 1 s^{-1} (*Section 5.2*)

Hess's law The overall enthalpy change for a reaction is equal to the sum of the enthalpy changes for the individual steps in the reaction (*Section 8.9*).

Heterogeneous catalyst a catalyst that exists in a different phase than the reactants (*Section 12.12*)

Heterogeneous equilibria equilibria in which reactants and products are present in more than one phase (*Section 13.4*)

Heterogeneous mixture a mixture having regions with differing compositions (*Section 2.7*)

Hexagonal closest-packed a packing arrangement of spheres into a noncubic unit cell with two alternating layers (*Section 10.8*)

High-spin complex a metal complex in which the d electrons are arranged to give the maximum number of unpaired electrons (*Section 20.11*)

Holoenzyme the assembly of apoenzyme plus cofactor (*Section 24.6*)

Homogeneous catalyst a catalyst that exists in the same phase as the reactants (*Section 12.12*)

Homogeneous equilibria equilibria in which all reactants and products are in a single phase, usually either gaseous or solution (*Section 13.4*)

Homogeneous mixture a mixture having a constant composition throughout (*Section 2.7*)

Hund's rule If two or more degenerate orbitals are available, one electron goes in each until all are half full (*Section 5.12*).

Hybrid atomic orbital a wave function derived by combination of atomic wave functions (*Section 7.11*)

Hydrate a solid compound that contains water molecules (*Section 14.15*)

Hydration the addition of water to an alkene (*Section 23.10*)

Hydrocarbon a compound that contains only carbon and hydrogen (*Section 23.2*)

Hydrogen bond an attractive interaction between a hydrogen atom bonded to an electronegative O, N, or F atom and an unshared electron pair on a nearby electronegative atom (*Section 10.2*)

Hydrogenation the addition of H_2 to an alkene to yield an alkane (*Section 23.10*)

Hydronium ion H_3O^+ (*Sections 4.5, 15.1*)

Hygroscopic absorbing water from the air (*Section 14.15*)

Ideal gas law a description of how the volume of a gas is affected by changes in pressure, temperature, and amount; $PV = nRT$ (*Section 9.3*)

Initial rate the instantaneous rate of a reaction at the beginning (*Section 12.1*)

Inner transition metal element an element in the 14 groups shown separately at the bottom of the periodic table (*Section 1.3*)

Instantaneous rate the rate of a reaction at a particular time (*Section 12.1*)

Integrated rate law the integrated form of a rate law (*Section 12.4*)

Intensive property a property whose value does not depend on the sample size (*Section 1.4*)

Interhalogen compound a compound X–Y, where X and Y are different halogens (*Section 6.10*)

Intermolecular force an attraction between molecules that holds them together (*Section 10.2*)

Internal energy (E) the sum of kinetic and potential energies for each particle in the system (*Section 8.3*)

International System of Units (SI) the seven base units, along with others derived from them, used for all scientific measurements (*Section 1.5*)

Interstitial hydride a metallic hydride that consists of a crystal lattice of metal atoms with the smaller hydrogen atoms occupying holes between the larger metal atoms (*Section 14.5*)

Ion a charged atom or group of atoms (*Section 2.8*)

Ion–dipole force an intermolecular force resulting from electrical interactions between an ion and the partial charges on a polar molecule (*Section 10.2*)

Ion exchange a process for softening hard water in which the Ca^{2+} and Mg^{2+} ions are replaced by Na^+ (*Section 14.13*)

Ion product (IP) a number defined in the same way as K_{sp}, except that the concentrations in the expression for IP are not necessarily equilibrium values (*Section 16.13*)

Ion-product constant for water (K_w) $[H_3O^+][OH^-] = 1.0 \times 10^{-14}$ (*Section 15.4*)

Ionic bond a bond that results from a transfer of one or more electrons between atoms (*Sections 2.8, 6.6*)

Ionic equation a reaction written so that the ions are explicitly shown (*Section 4.3*)

Ionic hydride a saltlike, high-melting, white, crystalline compound formed by the alkali metals and the heavier alkaline earth metals (*Section 14.5*)

Ionic solid a solid whose constituent particles are ions ordered into a regular three-dimensional arrangement held together by ionic bonds (*Sections 2.8, 6.6*)

Ionization energy (E_i) the amount of energy necessary to remove the outermost electron from an isolated neutral atom in the gaseous state (*Section 6.3*)

Ionization isomers isomers that differ in the anion bonded to the metal ion (*Section 20.8*)

Ionizing radiation radiation that knocks an electron from a molecule, thereby ionizing it (*Section 22.8*)

Isomers compounds that have the same formula but a different bonding arrangement of their constituent atoms (*Section 20.8*)

Isotope effect differences in properties that arise from the differences in isotopic mass (*Section 14.2*)

Isotopes atoms with identical atomic numbers but different mass numbers (*Section 2.5*)

Kelvin (K) the SI unit of temperature; 0 K = absolute zero (*Section 1.8*)

Ketone an organic molecule that contains two alkyl groups bonded to a C=O carbon (*Section 23.13*)

kilogram (kg) the SI unit of mass; 1 kg = 2.205 U.S. lb (*Section 1.6*)

Kinetic energy (E_K) the energy of motion; $E_K = (1/2)mv^2$ (*Section 8.1*)

Kinetic–molecular theory a theory describing the behavior of gases (*Section 9.6*)

Lanthanide one of the 14 inner-transition metals following lanthanum in the periodic table (*Section 1.3*)

Lanthanide contraction the decrease in atomic radii across the *f*-block lanthanide elements (*Section 20.2*)

Lattice energy (U) the sum of the electrostatic interactions between ions in a solid that must be overcome to break a crystal into individual ions (*Section 6.6*)

Law of conservation of energy Energy can be neither created nor destroyed (*Section 8.2*).

Law of definite proportions Different samples of a pure chemical substance always contain the same proportion of elements by mass (*Section 2.1*).

Law of mass conservation Mass is neither created nor destroyed in chemical reactions (*Section 2.1*).

Law of multiple proportions The mass ratios are small, whole-number multiples of one another when two elements combine in different ways to form different substances (*Section 2.2*).

Le Châtelier's principle If a stress is applied to a reaction mixture at equilibrium, reaction occurs in the direction that relieves the stress (*Section 13.6*).

Lewis acid an electron-pair acceptor (*Section 15.16*)

Lewis base an electron-pair donor (*Section 15.16*)

Lewis structure a representation of a molecule that shows valence electrons as dots (*Section 7.5*)

Ligand a molecule or ion that bonds to the central metal ion in a complex (*Section 20.5*)

Ligand donor atom an atom attached directly to the metal ion in a metal complex (*Section 20.5*)

Limiting reactant the reactant present in limiting amount that controls the extent to which reaction occurs (*Section 3.6*)

Line spectrum the wavelengths of light emitted by an energetically excited atom (*Section 5.3*)

Line structure a shorthand structure for an organic molecule in which only bonds (lines) are shown; carbon atoms are understood to be at every junction of lines (*Section 23.6*)

Linkage isomers isomers that arise when a ligand bonds to a metal through either of two different donor atoms (*Section 20.8*)

Lipid a naturally occurring organic molecule that dissolves in nonpolar organic solvents when a sample of plant or animal tissue is crushed or ground (*Section 24.11*)

Liquid crystal a state of matter in which molecules move around as if in a viscous liquid, but have a restricted range of motion as if in a solid (*Chapter 10 Interlude*)

Liter (L) a common unit of volume; $1 \text{ L} = 10^{-3} \text{ m}^3$ (*Section 1.9*)

Lock-and-key model a model that pictures an enzyme as a large, irregularly shaped molecule with a cleft into which substrate can fit (*Section 24.6*)

London dispersion force an intermolecular force resulting from the motion of electrons around atoms (*Section 10.2*)

Lone pair electrons a pair of valence electrons not used for bonding (*Section 7.5*)

Low-spin complex a metal complex in which the *d* electrons are paired up to give a maximum number of doubly occupied *d* orbitals and a minimum number of unpaired electrons (*Section 20.11*)

Magnetic quantum number (m_l) a variable in the solutions to the Schrödinger wave equation that defines the spatial orientation of an orbital (*Section 5.7*)

Main group element an element in the two groups on the left and the six groups on the right of the periodic table (*Section 1.3, Chapters 6, 19*)

Mass the amount of matter in an object (*Section 1.6*)

Mass defect the loss in mass that occurs when protons and neutrons combine to form a nucleus (*Section 22.5*)

Mass number (*A*) the total number of protons and neutrons in an atom (*Section 2.5*)

Mass percent a unit of concentration; the mass of one component divided by the total mass of the solution times 100% (*Section 11.3*)

Matter a catchall term used to describe anything you can touch, taste, or smell (*Section 1.6*)

Melting point the temperature at which solid and liquid coexist in equilibrium (*Section 10.4*)

Metabolism the sum of the many organic reactions that go on in cells (*Section 24.1*)

Metal complex see Coordination compound (*Section 20.5*)

Metal an element on the left side of the periodic table, bounded on the right by a zigzag line running from boron to astatine (*Sections 1.4, 21.4*)

Metallic hydride a compound formed by reaction of lanthanide, actinide, or some *d*-block transition metals with variable amounts of hydrogen (*Section 14.5*)

Metallic solid a solid consisting of metal atoms, whose crystals have metallic properties such as electrical conductivity (*Section 10.6*)

Metalloid see Semimetal (*Section 1.4*)

Metallurgy the science and technology of extracting metals from their ores (*Section 21.2*)

Meter (m) the SI unit of length (*Section 1.7*)

Microgram (μg) a common unit of mass; $1 \ \mu g = 0.001 \ mg = 10^{-6} \ g$ (*Section 1.6*)

Micrometer (μm) a common unit of length; $1 \ \mu m = 0.001 \ mm = 10^{-6} \ m$ (*Section 1.7*)

Milligram (mg) a common unit of mass; $1 \ mg = 0.001 \ g = 10^{-6} \ kg$ (*Section 1.6*)

Milliliter (mL) a common unit of volume; $1 \ mL = 1 \ cm^3$ (*Section 1.9*)

Millimeter (mm) a common unit of length; $1 \ mm = 0.001 \ m$ (*Section 1.7*)

Mineral a crystalline, inorganic constituent of the rocks that make up the earth's crust (*Section 21.1*)

Miscible mutually soluble in all proportions (*Section 11.4*)

Mixture a blend of two or more substance in some random proportion (*Section 2.7*)

Molal boiling-point-elevation constant (*K*$_b$) the amount by which the boiling point of a solvent is raised by dissolved substances (*Section 11.7*)

Molal freezing-point-depression constant (*K*$_f$) the amount by which the boiling point of a solvent is lowered by dissolved substances (*Section 11.7*)

Molality (*m*) a unit of concentration; the number of moles of solute per kilogram of solvent (mol/kg) (Section 11.3)

Molar heat capacity (*C*$_m$) the amount of heat necessary to raise the temperature of 1 mol of a substance 1°C (*Section 8.8*)

Molar mass the mass of 1 mol of substance; equal to the molecular or formula mass of the substance in grams (*Section 3.3*)

Molarity (M) a common unit of concentration; the number of moles of solute per liter of solution (*Sections 3.7, 11.3*)

Mole (mol) the SI unit for amount of substance; the quantity of a substance that contains as many molecules or formula units as there are atoms in exactly 12 g of carbon-12 (*Section 3.3*)

Mole fraction (*X*) a unit of concentration; the number of moles of a component divided by the total number of moles in the mixture (*Section 9.5*)

Molecular equation a reaction written using the full formulas of reactants and products (*Section 4.3*)

Molecular formula a formula that tells the actual numbers of atoms in a compound (*Section 3.11*)

Molecular mass the sum of atomic masses of the atoms in a molecule (*Section 3.3*)

Molecular orbital theory a quantum mechanical description of bonding in which electrons occupy molecular orbitals that belong to the entire molecule rather than to an individual atom (*Section 7.13*)

Molecular solid a solid whose constituent particles are molecules held together by intermolecular forces (*Section 10.6*)

Molecularity the number of molecules (or atoms) on the reactant side of the chemical equation for an elementary reaction (*Section 12.7*)

Molecule the unit of matter that results when two or more atoms are joined by covalent bonds (*Section 2.8*)

Mond process a chemical method for purification of nickel from its ore (*Section 21.2*)

Monodentate ligand a ligand that bonds to a metal using the electron pair of a single donor atom (*Section 20.6*)

Monomer a small molecule that, when bonded to itself many times, forms a polymer (*Section 23.15*)

Monoprotic acid an acid that has a single dissociable proton (*Section 15.7*)

Monosaccharide a carbohydrate such as glucose that can't be broken down into smaller molecules by hydrolysis (*Section 24.7*)

N-Terminal amino acid the amino acid with a free $-NH_2$ group on the end of a protein chain (*Section 24.4*)

***n*-Type semiconductor** a semiconductor doped with an impurity that has more electrons than necessary for bonding (*Section 21.5*)

Nanometer (nm) a common unit of length; $1 \ nm = 10^{-9} \ m$ (*Section 1.7*)

Nernst equation an equation for calculating cell potentials under non-standard-state conditions; $E = E^\circ - (RT \ln Q)/(nF)$ (*Section 18.6*)

Net ionic equation a reaction written so that the spectator ions are removed (*Section 4.3*)

Neutralization reaction see Acid–base neutralization reaction (*Sections 4.5, 16.1*)

Neutron a neutral, fundamental atomic particle in the nucleus of atoms (*Section 2.4*)

Newton the SI unit for force (*Section 9.1*)

Noble gas an element in group 8A of the periodic table (*Sections 1.4, 6.11*)

Node a region where a wave has zero amplitude (*Section 5.8*)

Nonelectrolyte a substance that does not produce ions when dissolved in water (*Section 4.2*)

Nonmetal an element on the right side of the periodic table, bounded on the left by a zigzag line running from boron to astatine (*Section 1.4*)

Nonspontaneous process a process that requires a continuous input of energy to proceed (*Section 8.13*)

Nonstoichiometric compound a compound whose atomic composition can't be expressed as a ratio of small whole numbers (*Section 14.5*)

Normal boiling point the temperature at which boiling occurs when there is exactly 1 atm of external pressure (*Sections 10.5, 10.11*)

Normal melting point the temperature at which melting occurs when there is exactly 1 atm of external pressure (*Section 10.11*)

Nuclear chemistry the study of the properties and reactions of atomic nuclei (*Chapter 22*)

Nuclear equation an equation for a nuclear reaction in which the sums of the nucleons are the same on both sides and the sums of the charges on the nuclei and any elementary particles are the same on both sides (*Section 22.2*)

Nuclear fission the fragmenting of heavy nuclei (*Section 22.6*)

Nuclear fusion the joining together of light nuclei (*Section 22.6*)

Nuclear reaction a reaction that changes an atomic nucleus (*Section 22.1*)

Nuclear transmutation the change of one element into another (*Section 22.7*)

Nucleic acid a polymer made up of nucleotide units linked together to form a long chain (*Section 24.12*)

Nucleon a general term for both protons and neutrons (*Section 22.1*)

Nucleoside a constituent of nucleotides; composed of an aldopentose sugar plus an amine base (*Section 24.12*)

Nucleotide a building block from which nucleic acids are made; composed of a nucleoside plus phosphoric acid (*Section 24.12*)

Nucleus the central core of an atom consisting of protons and neutrons (*Section 2.4*)

Nuclide the nucleus of a specific isotope (*Section 22.1*)

Octet rule the statement that main-group elements tend to undergo reactions that leave them with eight valence electrons (*Section 6.12*)

Orbital a solution to the Schrödinger wave equation; describes a region of space where an electron is likely to be found (*Section 5.7*)

Ore a mineral deposit from which a metal can be produced economically (*Section 21.1*)

Organic chemistry the study of carbon compounds (*Chapter 23*)

Osmosis the passage of solvent through a membrane from the less concentrated side to the more concentrated side (*Section 11.8*)

Osmotic pressure the amount of pressure necessary to cause osmosis to stop (*Section 11.8*)

Ostwald process the commercial process for making nitric acid from ammonia (*Section 19.10*)

Overvoltage the additional voltage required above that calculated for an electrolysis reaction (*Section 18.11*)

Oxidation the loss of one or more electrons by a substance (*Section 4.6*)

Oxidation number a value that measures whether an atom in a compound is neutral, electron-rich, or electron-poor compared to an isolated atom (*Section 4.6*)

Oxidation-number method a method for balancing redox equations (*Section 4.9*)

Oxidation–reduction (redox) reaction a process in which one or more electrons are transferred between reaction partners (*Sections 4.1, 4.6*)

Oxide a binary compound with oxygen in the −2 oxidation state (*Section 14.9*)

Oxidizing agent a substance that causes an oxidation by accepting an electron (*Section 4.7*)

Oxoacid an acid that contains oxygen in addition to hydrogen and another element (*Section 2.10*)

Oxoanion an anion in which an atom is combined with oxygen (*Section 2.10*)

***p*-Block element** an element in groups 3A–8A, which results from *p*-orbital filling (*Section 5.13*)

***p*-Type semiconductor** a semiconductor doped with an impurity that has fewer electrons than necessary for bonding (*Section 21.5*)

Paramagnetic a substance that is attracted by a magnetic field (*Section 7.14*)

Pascal the SI unit for pressure (*Section 9.1*)

Pauli exclusion principle No two electrons in an atom can have the same four quantum numbers (*Section 5.10*).

Peptide bond the amide bond linking two amino acids (*Section 24.2*)

Percent composition a list of elements present in a compound and the mass percent of each (*Section 3.11*)

Percent dissociation the concentration of the acid that dissociates divided by the initial concentration of the acid times 100% (*Section 15.10*)

Percent yield the amount of product actually formed in a reaction divided by the amount theoretically possible and multiplied by 100% (*Section 3.5*)

Period a row of elements in the periodic table (*Section 1.3*)

Periodic table a chart of the elements arranged by increasing atomic number so that elements in a given group have similar chemical properties (*Section 1.3*)

Peroxide a binary compound with oxygen in the −1 oxidation state (*Section 14.10*)

Petroleum a complex mixture of organic substances, primarily hydrocarbons (*Section 8.12*)

pH the negative base-10 logarithm of the molar hydronium ion concentration (*Section 15.5*)

pH titration curve a plot of the pH of a solution as a function of the volume of added base or acid (*Section 16.5*)

Phase a state of matter (*Section 10.4*)

Phase change a process in which the physical form but not the chemical identity of a substance changes (*Section 10.4*)

Phase diagram a plot showing the effects of pressure and temperature on the physical state of a substance (*Section 10.11*)

Photoelectric effect the ejection of electrons from a metal on exposure to radiant energy (*Section 5.4*)

Photon the smallest possible amount of radiant energy; a quantum (*Section 5.4*)

Physical property a characteristic that can be determined without changing the chemical makeup of a sample (*Section 1.4*)

Pi (π) bond a bond in which shared electrons occupy a region above and below a line connecting the two nuclei (*Section 7.12*)

Picometer (pm) a common unit of length; 1 pm = 10^{-12} m (*Section 1.7*)

Planck's constant (h) 6.626×10^{-34} J · s; a fundamental physical constant that relates energy and frequency, $E = h\nu$ (*Section 5.4*)

Polar covalent bond a bond in which the bonding electrons are attracted somewhat more strongly by one atom than by the other (*Section 7.4*)

Polarizability the ease with which a molecule's electron cloud can be distorted by a nearby electric field (*Section 10.2*)

Polyatomic ion a charged, covalently bonded group of atoms (*Section 2.8*)

Polydentate ligand a ligand that bonds to a metal through electron pairs on more than one donor atom (*Section 20.6*)

Polymer a large molecule formed by the repetitive bonding together of many smaller molecules (*Section 23.15*)

Polypeptide a chain of up to 100 amino acids (*Section 24.2*)

Polyprotic acid an acid that contains more than one dissociable proton (*Section 15.11*)

Polysaccharide a compound such as cellulose that is made of many simple sugars linked together (*Section 24.7*)

Positron emission the conversion of a proton in the nucleus into a neutron plus an ejected positron (*Section 22.2*)

Potential energy (E_P) energy that is stored, either in an object because of its position or in a molecule because of its chemical composition (*Section 8.1*)

Precipitation reaction a reaction in which an insoluble solid precipitate forms and drops out of solution (*Section 4.4*)

Precision how well a number of independent measurements agree with one another (*Section 1.11*)

Primary protein structure the sequence in which amino acids are linked together (*Section 24.5*)

Primitive-cubic unit cell a cubic unit cell with an atom at each of its eight corners (*Section 10.8*)

Principal quantum number (n) a variable in the solutions to the Schrödinger wave equation on which the size and energy level of an orbital primarily depends (*Section 5.7*)

Property any characteristic that can be used to describe or identify matter (*Section 1.4*)

Protein a biological polymer made up of many amino acids linked together to form a long chain (*Section 24.2*)

Proton a positively charged, fundamental atomic particle in the nucleus of atoms (*Section 2.4*)

Qualitative analysis a procedure for identifying the ions present in an unknown solution (*Section 16.15*)

Quantized changing only in discrete amounts (*Section 5.4*)

Quantum the smallest possible amount of radiant energy (*Section 5.4*)

Quantum mechanical model a model of atomic structure that concentrates on an electron's wavelike properties (*Section 5.6*)

Quantum number a variable in the Schrödinger wave equation; describes the energy level and position in space where an electron is most likely to be found (*Section 5.7*)

Quaternary protein structure the aggregation of several protein chains to form a larger unit (*Section 24.5*)

Racemic mixture a 50:50 mixture of enantiomers (*Section 20.9*)

Radioactivity the spontaneous emission of radiation (*Section 22.2*)

Radiocarbon dating a technique for dating archaeological artifacts by measuring the amount of ^{14}C in the sample (*Section 22.10*)

Radioisotope a radioactive isotope (*Section 22.2*)

Radionuclide a radioactive nucleus (*Section 22.2*)

Raoult's law The vapor pressure of a solution containing a nonvolatile solute is equal to the vapor pressure of pure solvent times the mole fraction of the solvent (*Section 11.6*).

Rate constant the proportionality constant in a rate law (*Section 12.2*)

Rate-determining step the slowest step in a reaction mechanism (*Section 12.8*)

Rate law an equation that tells how reaction rate depends on the concentration of each reactant (*Section 12.2*)

Reaction intermediate a species that is formed in one step of a reaction mechanism and consumed in a subsequent step (*Section 12.7*)

Reaction mechanism the sequence of molecular events that defines the pathway from reactants to products (*Section 12.7*)

Reaction order the value of the exponents of concentration terms in the rate law (*Section 12.2*)

Reaction quotient (Q_c) similar to the equilibrium constant K_c except that the concentrations in the equilibrium constant expression are not necessarily equilibrium values (*Section 13.5*)

Reaction rate the increase in the concentration of a product per unit time or the decrease in the concentration of a reactant per unit time (*Section 12.1*)

Redox reaction an oxidation–reduction reaction (*Section 4.6*)

Redox titration a procedure for determining the concentration of a redox agent (*Section 4.11*)

Reducing agent a substance that causes a reduction by donating an electron (*Section 4.7*)

Reduction the gain of one or more electrons by a substance (*Section 4.6*)

Replication the process by which identical copies of DNA are made (*Section 24.14*)

Resonance hybrid an average of several valid electron-dot structures for a molecule (*Section 7.7*)

Respiration the process of breathing and using oxygen for biological redox reactions (*Section 4.12*)

Reverse osmosis the passage of solvent through a membrane from the more concentrated side to the less concentrated side (*Section 11.9*)

Ribonucleic acid (RNA) a biological polymer of ribonucleotide units that serves to transcribe the genetic information in DNA and uses that information to direct the synthesis of proteins (*Section 24.12*)

Roasting a metallurgical process that involves heating a mineral in air (*Section 21.2*)

Rounding off deleting digits to keep only the correct number of significant figures (*Section 1.12*)

s-Block element an element in groups 1A or 2A, which results from s-orbital filling (*Section 5.13*)

Sacrificial anode an easily oxidized metal that corrodes instead of a less reactive metal to which it is connected (*Section 18.10*)

Salt an ionic compound formed in an acid–base neutralization reaction (*Section 4.1*)

Salt bridge a tube that contains a gel permeated with a solution of an inert electrolyte connecting the two sides of an electrochemical cell (*Section 18.1*)

Saponification the base-catalyzed hydrolysis of an ester to yield a carboxylic acid and an alcohol (*Section 23.14*)

Saturated hydrocarbon a hydrocarbon that contains only single bonds (*Section 23.2*)

Saturated solution a solution containing the maximum possible amount of dissolved solute at equilibrium (*Section 11.4*)

Scientific notation a system in which a large or small number is written as a number between 1 and 10 times a power of 10 (*Section 1.5*)

Second Law of thermodynamics In any spontaneous process, the total entropy of a system and its surroundings always increases (*Section 17.6*).

Second-order reaction one whose rate depends on the concentration of a single reactant raised to the second power or on the concentrations of two different reactants, each raised to the first power (*Section 12.6*)

Secondary protein structure orienting of segments of a protein chain into a regular pattern (*Section 24.5*)

Semiconductor a material that has an electrical conductivity intermediate between that of a metal and that of an insulator (*Section 21.5*)

Semimetal an element adjacent to the zigzag boundary between metals and nonmetals (*Section 1.4*)

Semipermeable membrane a membrane that allows passage of water or other small molecules but not the passage of large solute molecules or ions (*Section 11.8*)

Shell a grouping of orbitals according to principal quantum number (*Section 5.7*)

Side chain the group attached to the α carbon of an amino acid (*Section 24.1*)

Sigma (σ) bond a covalent bond in which the shared electrons are centered about the axis between the two nuclei (*Section 7.10*)

Significant figures the total number of digits in a measurement (*Section 1.11*)

Silicate an ionic compound that contains silicon oxoanions along with cations, such as Na^+, K^+, Mg^{2+}, or Ca^{2+} (*Section 19.7*)

Simple cubic packing a packing arrangement of spheres into a primitive-cubic unit cell (*Section 10.8*)

Single bond a covalent bond formed by sharing two electrons (*Section 7.5*)

Sintering a process in which the particles of a powder are "welded" together without completely melting (*Section 21.7*)

Slag a by-product of iron production, consisting mainly of calcium silicate (*Section 21.3*)

Solubility the amount of a substance that dissolves in a given volume of solvent (*Sections 4.4, 11.4*)

Solubility product constant (K_{sp}) the equilibrium constant for a dissolution reaction (*Section 16.10*)

Solute the dissolved substance in a solution (*Section 11.1*)

Solution a homogeneous mixture containing particles the size of a typical ion or covalent molecule (*Section 11.1*)

Solvent the major component in a solution (*Section 11.1*)

Sol–gel method a method of preparing ceramics, involving synthesis of a metal oxide powder from a metal alkoxide (*Section 21.7*)

sp Hybrid orbital a hybrid orbital formed by combination of one atomic s orbital with one p orbital (*Section 7.12*)

sp^2 Hybrid orbital a hybrid orbital formed by combination of one s and two p atomic orbitals (*Section 7.12*)

sp^3 Hybrid orbital a hybrid orbital formed by combination of one s and three p atomic orbitals (*Section 7.11*)

sp^3d Hybrid orbital a hybrid orbital formed by combination of one s, three p, and one d atomic orbitals (*Section 7.12*)

sp^3d^2 Hybrid orbital a hybrid orbital formed by combination of one s, three p, and two d atomic orbitals (*Section 7.12*)

Specific heat the amount of heat necessary to raise the temperature of 1 gram of a substance 1°C (*Section 8.8*)

Spectator ion an ion that appears on both sides of the reaction arrow (*Section 4.3*)

Spectrochemical series an ordered list of ligands in which crystal field splitting increases (*Section 20.12*)

Spin quantum number (m_s) a variable that describes the spin of an electron, either $+1/2$ or $-1/2$ (*Section 5.10*)

Spontaneous process one that proceeds on its own without any continuous external influence (*Section 8.13*)

Standard cell potential ($E°$) the cell potential when both reactants and products are in their standard states (*Section 18.3*)

Standard electrode potential *see* Standard reduction potential (*Section 18.4*)

Standard enthalpy of reaction ($\Delta H°$) enthalpy change under standard-state conditions (*Section 8.6*)

Standard entropy of reaction ($\Delta S°$) the entropy change for a chemical reaction under standard-state conditions (*Section 17.5*)

Standard free-energy change (ΔG°) the free-energy change that occurs when reactants in their standard states are converted to products in their standard states (*Section 17.8*)

Standard free energy of formation (ΔG°f) the free-energy change for formation of 1 mol of a substance in its standard state from the most stable form of the constituent elements in their standard states (*Section 17.9*)

Standard heat of formation (ΔH°f) the enthalpy change $\Delta H°_f$ for the hypothetical formation of 1 mol of a substance in its standard state from the most stable forms of its constituent elements in their standard states (*Section 8.10*)

Standard hydrogen electrode (S.H.E.) a reference half-cell consisting of a platinum electrode in contact with H_2 gas and aqueous H^+ ions at standard-state conditions (*Section 18.4*)

Standard molar entropy (S°) the entropy of 1 mol of a pure substance at 1 atm pressure and a specified temperature, usually 25°C (*Section 17.5*)

Standard molar volume the volume of 1 mol of a gas at 0°C and 1 atm pressure; 22.414 L (*Section 9.2*)

Standard reduction potential the standard potential for a reduction half-cell (*Section 18.4*)

Standard temperature and pressure (STP) $T = 273.15$ K; $P = 1$ atm (*Section 9.3*)

State function a function or property whose value depends only on the present condition of the system, not on the path used to arrive at that condition (*Section 8.3*)

Steam–hydrocarbon re-forming process an important industrial method for producing hydrogen from methane (*Section 14.3*)

Stereoisomers isomers that have the same connections among atoms but have a different arrangement of the atoms in space (*Section 20.8*)

Steric factor the fraction of collisions with the proper orientation for converting reactants to products (*Section 12.9*)

Steroid a lipid whose structure is similar to that of cholesterol (*Section 24.11*)

Stoichiometry mole/mass relationships between reactants and products (*Section 3.4*)

Straight-chain alkane an alkane that has all its carbons connected in a row (*Section 23.2*)

Strong acid an acid that dissociates completely in water to give H^+ ions and is a strong electrolyte (*Sections 4.5, 15.2*)

Strong base a base that dissociates or reacts completely with water to give OH^- ions and is a strong electrolyte (*Sections 4.5, 15.2*)

Strong electrolyte a compound that dissociates completely into ions when dissolved in water (*Section 4.2*)

Strong-field ligand ligand that has a large crystal field splitting (*Section 20.12*)

Structural formula a representation that shows the specific connections between atoms in a molecule (*Section 2.8*)

Sublimation the direct conversion of a solid to a vapor without going through a liquid state (*Sections 8.7, 10.4*)

Subshell a grouping of orbitals by angular-momentum quantum number (*Section 5.7*)

Subsidiary reaction any proton-transfer process other than the principal one in an acid–base reaction (*Section 15.9*)

Substitution reaction an organic reaction in which one group substitutes for another, particularly on aromatic rings (*Section 23.11*)

Substrate the compound acted on by an enzyme (*Section 24.6*)

Superconducting transition temperature (Tc) the temperature below which a superconductor loses all electrical resistance (*Section 21.6*)

Superconductor a material that loses all electrical resistance below a certain temperature (*Section 21.6*)

Supercritical fluid a state of matter beyond the critical point that is neither liquid nor gas (*Section 10.11*)

Superoxide a binary compound with oxygen in the $-1/2$ oxidation state (*Section 14.10*)

Supersaturated solution a solution containing a greater-than-equilibrium amount of solute (*Section 11.4*)

Surface tension the resistance of a liquid to spreading out and increasing its surface area (*Section 10.3*)

Suspension a homogeneous mixture containing particles greater than about 1000 nm in diameter that are visible with a low-power microscope (*Section 11.1*)

Symmetry plane a plane that cuts through an object so that one half of the object is a mirror image of the other half (*Section 20.9*)

Termolecular reaction an elementary reaction that results from collisions between three reactant molecules (*Section 12.7*)

Temperature a measure of the kinetic energy of molecular motion (*Section 8.2*)

Tertiary protein structure folding of a protein chain into a specific three-dimensional shape (*Section 24.5*)

Theory a consistent explanation of known observations (*Section 1.1*)

Thermochemistry a study of the heat changes that take place during reactions (*Chapter 8*)

Thermodynamic standard state conditions under which thermodynamic measurements are reported; 298.15 K (25°C), 1 atm pressure for each gas, 1 M concentration for solutions (*Section 8.6*)

Thermodynamics the study of the interconversion of heat and other forms of energy (*Chapter 17*)

Third law of thermodynamics The entropy of a perfectly ordered crystalline substance at 0 K is zero (*Section 17.4*).

Three-center, two-electron bond a covalent bond in which three atoms share two electrons (*Section 19.4*)

Titration a procedure for determining the concentration of a solution (*Section 3.10*)

Trans isomer the isomer of a metal complex or alkene in which identical ligands or groups are opposite one another rather than adjacent (*Sections 20.8, 23.9*)

Transcription the process by which information in DNA is transferred to and decoded by RNA (*Section 24.14*)

Transition metal element an element in the 10 groups in the middle of the periodic table (*Section 1.3; Chapter 20*)

Transition state   the configuration of atoms at the maximum in the potential energy profile for a reaction (*Section 12.9*)

Translation   the process by which RNA builds proteins (*Section 24.14*)

Transuranium elements   the 23 artificially produced elements beyond uranium in the periodic table (*Section 22.4*)

Triacylglycerol   a triester of glycerol (1,2,3-propanetriol) with three long-chain carboxylic acids (*Section 24.11*)

Triple bond   a covalent bond formed by sharing six electrons (*Section 7.5*)

Triple point   a unique combination of pressure and temperature at which gas, liquid, and solid phases coexist in equilibrium (*Section 10.11*)

Turnover number   the number of substrate molecules acted on by one molecule of enzyme per unit time (*Section 24.6*)

Unimolecular reaction   an elementary reaction that involves a single reactant molecule (*Section 12.7*)

Unit cell   a small repeating unit that makes up a crystal (*Section 10.8*)

Unsaturated   an organic molecule that has a double or triple bond (*Section 23.9*)

Valence band   the bonding molecular orbitals in a semiconductor (*Section 21.5*)

Valence bond theory   a quantum mechanical description of bonding that pictures covalent bond formation as the overlap of two singly occupied atomic orbitals (*Section 7.10*)

Valence shell   the outermost electron shell (*Section 5.13*)

Valence-shell electron-pair repulsion (VSEPR) model   a model for predicting the approximate geometry of a molecule (*Section 7.9*)

Van der Waals forces   an alternative name for intermolecular forces (*Section 10.2*)

Vapor pressure   the partial pressure of a gas in equilibrium with liquid (*Section 10.5*)

Vaporization   the change of a liquid to a gas (*Section 10.4*)

Vinyl monomer   a simple alkene used to make a polymer (*Section 23.15*)

Viscosity   the measure of a liquid's resistance to flow (*Section 10.3*)

Voltaic cell   *see* Galvanic cell (*Section 18.1*)

Volume   the amount of space occupied by an object (*Section 1.9*)

Water-gas shift reaction   a method for the industrial preparation of H_2 by reaction of CO with H_2O (*Section 14.3*)

Watson–Crick model   a model of DNA structure, consisting of two polynucleotide strands coiled around each other in a double helix (*Section 24.13*)

Wave function   a solution to the Schrödinger wave equation (*Section 5.7*)

Wavelength (λ)   the length of a wave from one maximum to the next (*Section 5.2*)

Weak acid   an acid that dissociates incompletely in water and is a weak electrolyte (*Sections 4.5, 15.2*)

Weak base   a base that dissociates or reacts incompletely with water and is a weak electrolyte (*Sections 4.5, 15.2*)

Weak electrolyte   a compound that dissociates incompletely when dissolved in water (*Section 4.2*)

Weak-field ligand   a ligand that has a small crystal field splitting (*Section 20.12*)

Work (w)   the distance (d) moved times the force (F) that opposes the motion (*Section 8.4*)

Yield   the amount of product formed in a reaction (*Section 3.5*)

Zone refining   a purification technique in which a heater melts a narrow zone at the top of a rod of some material and then sweeps slowly down the rod bringing impurities with it (*Section 19.7*)

Index

Photo Credits

FRONTMATTER p. iii (T) Richard Kaylin/Stone **(B)** Ken Eward/Photo Researchers, Inc. **iv (T)** Dr. Jeremy Burgess/Science Photo Library/Photo Researchers, Inc. **v** Alexis Komenda/Liaison Agency, Inc. **vi** Rolf Hicker/Stone **vii (T)** George Steinmetz/George Steinmetz Photography **viii (T)** Edward Parker/Still Pictures/Peter Arnold, Inc. **ix** Christel Rosenfeld/Stone **x (T)** Tim Davis/Stone **xi** GTE Laboratories Incorporated **xiii** Mark A. Leman/Stone

CHAPTER 1 1 Thomas Riggs/NGS Image Collection **2 (T)** Dr. Jeremy Burgess/Science Photo Library/Photo Researchers, Inc. **(BL)** Patrick Curtet/Point de Vue/Liaison Agency, Inc. **(BR)** Torleif Svensson/The Stock Market **3** McCracken Photographers/Pearson Education/PH College **4 (T)** James Randklev/Stone **(B)** Thomas Zimmermann/Stone **7** McCracken Photographers, Inc./Pearson Education/PH College **8 (B)** McCracken Photographers, Inc./Pearson Education/PH College **9 (T)** Gary Benson/Stone **(BL)** Tom Bochsler/Pearson Education/PH College **(BR)** McCracken Photographers, Inc./Pearson Education/PH College **12** McCracken Photographers/Pearson Education/PH College **13 (T)** McCracken Photographers/Pearson Education/PH College **(B)** Dr. Tony Brain/Science Photo Library/Photo Researchers, Inc. **16** McCracken Photographers/Pearson Education/PH College **17** Tom Stewart/The Stock Market **18** McCracken Photographers, Inc./Pearson Education/PH College **19** McCracken Photographers/Pearson Education/PH College **21** McCracken Photographers, Inc./Pearson Education/PH College **23** Lori Adamski Peek/Stone **26** Levy/Liaison Agency, Inc.

CHAPTER 2 36 William Curtsinger/NGS Image Collection **40** Dr. E. R. Degginger/Color-Pic, Inc. **44** Bob Daemmrich/Stock Boston **47** Will McIntyre/Stone **50** Craig Tuttle/The Stock Market **51 (L)** Paul Silverman/Fundamental Photographs **(R)** Tom Pantages/Tom Pantages **52** UPI/Corbis-Bettmann **55** Richard M. Busch/Richard M. Busch **57** Stephane Alix/Liaison Agency, Inc. **61** Paul Silverman/Fundamental Photographs **64** Tom Pantages/Tom Pantages **65** Courtesy of International Business Machines Corporation. Unauthorized use not permitted. Almaden Research Center.

CHAPTER 3 78 Tom Pantages/Tom Pantages **79** Jeff Greenberg/Stock Boston **80** Jean Marc Barey/Agence Vandystadt/Photo Researchers, Inc. **81** Phil Degginger/Color-Pic, Inc. **83** Bethlehem Steel Corporation **91** NASA/Johnson Space Center **95** Paul Silverman/Fundamental Photographs **97** Ed Degginger/Color-Pic, Inc. **103** McCracken Photographers/Pearson Education/PH College **106 (L)** Library of Congress **(R)** Science Photo Library/Photo Researchers, Inc.

CHAPTER 4 117 Franklin J. Viola/Viola's Photo Visions, Inc. **118** McCracken Photographers/Pearson Education/PH College **123** McCracken Photographers/Pearson Education/PH College **126** McCracken Photographers/Pearson Education/PH College **128** McCracken Photographers/Pearson Education/PH College **131** Courtesy of the Leon Levy Expedition to Ashkelon, Dr. Lawrence Stager, Director **133** McCracken Photographers/Pearson Education/PH College **134** McCracken Photographers/Pearson Education/PH College **135 (T)** Peticolas/Megna/Fundamental Photographs **(B)** Tom Pantages/Tom Pantages **136** Phil Degginger/Color-Pic, Inc. **140** Tom Pantages/Tom Pantages **145** McCracken Photographers/Pearson Education/PH College **146** Tom Pantages/Tom Pantages **148** Hulton Getty/Liaison Agency, Inc.

CHAPTER 5 159 Art Wolfe/Art Wolfe, Inc. **163** Philip Long/Stone **165 (TL)** Pictor **(TC)** Malcolm Boulton/Photo Researchers, Inc. **(TR)** Alexis Komenda/Liaison Agency, Inc. **(B)** Tom Pantages/Tom Pantages **167** Phil

Degginger/Color-Pic, Inc. **168** Yoav Levy/Phototake NYC **172 (T)** Donald Deitz/Stock Boston **(B)** Science Photo Library/Photo Researchers, Inc. **173** Photograph by Dr. Harold E. Edgerton. © Harold & Esther Edgerton Foundation, 1999, courtesy of Palm Press, Inc. **192** Pekka Parviainen/Science Photo Library/Photo Researchers, Inc.

CHAPTER 6 201 Rolf Hicker/Stone **211** Carey Van Loon/Carey Van Loon **212** Ed Degginger/Color-Pic, Inc. **217** McCracken Photographers/Pearson Education/PH College **222 (T)** Karl Hartmann/Traudel Sachs/Phototake NYC **(CL)** Ed Degginger/Color-Pic, Inc. **(CR)** McCracken Photographers/Pearson Education/PH College **(BL)** Hartmann/Sachs/Phototake NYC **(BR)** Jeffrey L. Rotman/Jeffrey L. Rotman **224** McCracken Photographers/Pearson Education/PH College **225** Carey Van Loon/Carey Van Loon **226** Yoav Levy/Phototake NYC **227** Richard T. Nowitz/Phototake NYC **230** Stephen Frink/Stone **234** Tomasz Tomaszewski Photography, Warsaw, Poland **235** Co Rentmeester/The Image Bank

CHAPTER 7 244 Jeff Greenberg/Omni-Photo Communications, Inc. **247** McCracken Photographers/Pearson Education/PH College **258** Phil Degginger/Color-Pic, Inc. **283** McCracken Photographers/Pearson Education/PH College **287** Craig X. Sotres/Pearson Education/PH College **288** Ed Degginger/Color-Pic, Inc.

CHAPTER 8 299 Keith Wood/Stone **300** Jim Caccavo/Stock Boston **302** Ian Murphy/Stone **304** Dorian Hanner/Dorian Hanner **305** Lori Adamski Peek/Stone **308** SuperStock, Inc. **311 (T)** Jason Gurr/Fundamental Photographs **(B)** Tom Pantages/Tom Pantages **315** Paul Harris/Stone **316** McCracken Photographers/Pearson Education/PH College **321** Fred Lyon/Fred Lyon **322** Liane Enkelis/Stock Boston **326 (T)** Francois Gohier/Photo Researchers, Inc. **(B)** Spencer Grant/Stock Boston **327** V. O'Brien/Mauritius GmbH/Phototake NYC **328 (L)** D. Van Kirk/The Image Bank **(C)** Frank Spinelli/Stone **(R)** Ezio Geneletti/The Image Bank **332** Kristen Brochmann/Fundamental Photographs **334** Springer/Corbis

CHAPTER 9 345 George Steinmetz/George Steinmetz Photography **346** Kaz Mori/The Image Bank **347** Paul Silverman/Fundamental Photographs **348** NASA/Stock Boston **349** Keren Su/Stock Boston **352** William Johnson/Stock Boston **353** Reprinted with permission of Chappell Studio/MarathonFoto **355** E. R.Degginger/Color-Pic, Inc. **356** Donald Johnston/Stone **358** E. R.Degginger/Color-Pic, Inc. **360** Denis & Kelly Tapparel/Stone **364** Andrew McClenaghan/Science Photo Library/Photo Researchers, Inc. **368** StockTrek/PhotoDisc, Inc **369** David R. Frazier/Photo Researchers, Inc. **370** NASA/Goddard Space Flight Center **372** Stacy Pick/Stock Boston

CHAPTER 10 384 Stephane Compoint/Corbis Sygma **388** David Taylor/Science Photo Library/Photo Researchers, Inc. **394 (TL)** Index Stock Imagery, Inc. **(TR)** Kristen Brochmann/Fundamental Photographs **(B)** Tom Pantages/Tom Pantages **395** Hermann Eisenbeiss/Photo Researchers, Inc. **396 (T)** Rod Kaye Photography/Aristock, Inc. **(B)** Randy G. Taylor/Liaison Agency, Inc. **397** Ed Degginger/Color-Pic, Inc. **404 (L** Jeffrey A. Scovil/Jeffrey A. Scovil **(R)** Corbis Digital Stock **407** Kenneth Eward/BioGrafx **109 (T)** Tony Mendoza/Stock Boston **(B)** Michael Dalton/Fundamental Photographs **416** General Electric Corporate Research & Development Center **417** Corning Incorporated **420** Paul Silverman/Fundamental Photographs **421** Courtesy of International Business Machines Corporation. Unauthorized use not permitted.

CHAPTER 11 431 David Doubilet/Doubilet Photography, Inc. **432 (T)** Japack/Corbis **(B)** Michael Dalton/Fundamental Photographs **433**

Edward Parker/Still Pictures/Peter Arnold, Inc. **438** Howard N. Kaplan/HNK Architectural Photography, Inc **443** Charles D. Winters/Timeframe Photography Inc./Photo Researchers, Inc. **445** Karl E. Huggins **446** Molkenthin Studio/The Stock Market **456** McCracken Photographers/Pearson Education/PH College **457** Marc Muench/Stone **460** Thomas Kitchin/Tom Stack & Associates **463** Hank Morgan/Science Source/Photo Researchers, Inc.

CHAPTER 12 473 Keith Ciriegio/AP/Wide World Photos **474 (T)** Grant Heilman/Grant Heilman Photography **(BR)** Emory Kristof/National Geographic Society **480** McCracken Photographers/Pearson Education/PH College **483** McCracken Photographers/Pearson Education/PH College **495** Michio Hoshino/Minden Pictures **499** Jeffry W. Myers/West Stock **501** McCracken Photographers/Pearson Education/PH College **508** Tom Pantages/Tom Pantages **509** McCracken Photographers/Pearson Education/PH College **513** AC/General Motors/Peter Arnold, Inc. **514** Bruce Forster/Stone **515** Tom Pantages/Tom Pantages

CHAPTER 13 528 Norbert Wu/Norbert Wu Productions **531** Martin Rogers/Stock Boston **537** Dan McCoy/Rainbow **538** McCracken Photographers/Pearson Education/PH College **555 (T)** McCracken Photographers/Pearson Education/PH College **(B)** Tom Pantages/Tom Pantages **559** Benjamin Rondel/The Stock Market **561** Jon Burbank/The Image Works

CHAPTER 14 573 Ralph A. Clevenger/Corbis **574** National Optical Astronomy Observatories **578** SuperStock, Inc. **583** Runk/Schoenberger/ Grant Heilman Photography, Inc. **584** Heine Schneebeli/Science Photo Library/Photo Researchers, Inc. **586 (L)** McCracken Photographers/Pearson Education/PH College **(R)** Tom Bochsler/Pearson Education/PH College **590** Courtesy of DuPont Nomex(R) **593** Paul Silverman/Fundamental Photographs **598** Martin Bond/Science Photo Library/Photo Researchers, Inc. **599 (L)** NASA Headquarters **(R)** Robert F. Sisson/SIPA Press

CHAPTER 15 608 Christel Rosenfeld/Stone **620 (L)** Runk/Schoenberger/ Grant Heilman Photography, Inc. **621** Jim Strawser/Grant Heilman Photography, Inc. **634** Leonard Lessin/Peter Arnold, Inc. **648** Ben Osborne/Stone **649** Will McIntyre/Photo Researchers, Inc.

CHAPTER 16 661 Tom Bean/Stone **674** Donald Clegg and Roxy Wilson/ Pearson Education/PH College **687 (T)** Manfred Kage/Peter Arnold, Inc. **689** Ed Degginger/Color-Pic, Inc. **690** Sovereign/ISM/Phototake NYC **697** Dan McCoy/Rainbow **704 (T)** Jean-Marc Loubat/Agence Vandystadt/ Photo Researchers, Inc. **(B)** Biology Media/Photo Researchers, Inc.

CHAPTER 17 717 Tim Davis/Stone **719** Terje Rakke/The Image Bank **720 (T)** John Kaprielian/Photo Researchers, Inc. **(B)** Jack Dykinga/Stone **723** Paul Silverman/Fundamental Photographs **731 (L)** James H. Karales/Peter Arnold, Inc. **(R)** Nancy Simmerman/Stone **732** E. R. Degginger/Photo Researchers, Inc. **747** Chris Bjornberg/Photo Researchers, Inc. **748** U.S. Department of Agriculture

CHAPTER 18 759 Phil Degginger/Color-Pic, Inc. **763** Ed Degginger/Color-Pic, Inc. **778** Corning Incorporated **782** SuperStock, Inc. **783** Tony Freeman/PhotoEdit **784 (T)** Tom Pantages/Tom Pantages **(B)** Michael Dalton/Fundamental Photographs **786 (T)** Chicago Transit Authority **(B)** Porter Gifford/Liaison Agency, Inc. **787** Zandria Muench/Stone **788** Merlin Metalworks, Inc. **792** Runk/Schoenberger/Grant Heilman Photography, Inc. **795** Science VU-AMAX/Visuals Unlimited **799 (T)** Ed Degginger/Color-Pic, Inc. **(B)** Bill Seeley/Reactive Metals Studio, Inc.

CHAPTER 19 811 Richard Kaylin/Stone **817 (L)** Paul Silverman/Fundamental Photographs **(R)** Hank Morgan/Science Source/Photo Researchers, Inc. **818** U.S. Borax Inc. **821** Michael J. Bronikowski, Center for Nanoscale Science and Technology, Rice University **823** Peter Turnley/Black Star **825** Photo Courtesy of Texas Instruments Incorporated **826 (T)** Jeffrey A. Scovil/Jeffrey A. Scovil **(BL)** Jeffrey A. Scovil/Jeffrey A.

Scovil **826 (BR)** William E. Ferguson/William E. Ferguson **827 (L)** Runk/Schoenberger/Grant Heilman Photography, Inc. **(R)** Barry L. Runk/Grant Heilman Photography, Inc. **828 (L)** Paul Silverman/ Fundamental Photographs **(R)** Roberto de Gugliemo/Science Photo Library/Photo Researchers, Inc. **828 (B)** Barry L. Runk/Grant Heilman Photography, Inc. **829 (L)** Ed Degginger/Color-Pic, Inc. **(C)** Stephen Frisch/Stephen Frisch **(R)** Stephen Frisch/Stephen Frisch **830 (TL)** Ed Degginger/Color-Pic, Inc. **(BL)** Russ Lappa/Science Source/Photo Researchers, Inc. **(R)** Paul Silverman/Fundamental Photographs **833** Kristen Brochmann/Fundamental Photographs **835** Tom Bochsler/ Pearson Education/PH College **839** U.S. Department of Agriculture **840** Stephen Frisch/Stephen Frisch **841 (T)** Kent Knudson/FPG International LLC **(B)** Specimen from North Museum, Franklin and Marshall College. Photo by Runk/Schoenberger/Grant Heilman Photography, Inc. **842 (T)** Manfred Kage/Peter Arnold, Inc.

CHAPTER 20 859 Rich Chisholm/The Stock Market **863** Paul Silverman/Fundamental Photographs **868** Jason Seley (1919-1983), Herakles in Ithaka I, 1980-81. Photo by Robert Barker, Cornell University Photography. Sculpture, welded steel, 342.9 cm. Gift of the Artist, Herbert F. Johnson Museum of Art, Cornell University. **869 (T)** Donald Clegg and Roxy Wilson/Pearson Education/PH College **872** Dale Boyer/Photo Researchers, Inc. **873** Paul Silverman/Fundamental Photographs **899** Ed Kashi/Phototake NYC

CHAPTER 21 910 Railway Technical Research Institute, Tokyo, Japan **911 (T)** Dan McCoy/Rainbow **(BL)** Barry L. Runk/Grant Heilman Photography, Inc. **(BC)** Karl Hartmann/Traudel Sachs/Phototake NYC **(BR)** Jeffrey A. Scovil/Jeffrey A. Scovil **913** Paul Silverman/Fundamental Photographs **915** Runk/Schoenberger/Grant Heilman Photography, Inc. **916** Erich Hartmann/Magnum Photos, Inc. **917** Science VU/Visuals Unlimited **918** Peter Poulides/Stone **924** Astrid and Hanns-Frieder Michler/Science Photo Library/Photo Researchers, Inc. **925** Chemical Design Ltd./Science Photo Library/Photo Researchers, Inc. **926** Phil Degginger/Color-Pic, Inc. **928** GTE Laboratories Incorporated **930** Corning Incorporated **932** Northrop Grumman Corporation/Military Aircraft Systems Division **933** J. & L. Weber/Peter Arnold

CHAPTER 22 943 Randy Montoya/Sandia National Laboratories **944** Fred Morris/Gnu Images/The Stock Market **951** Inga Spence/Index Stock Imagery, Inc. **961** Olivier Blaise/Liaison Agency, Inc. **962** Joseph Sohm/Photo Researchers, Inc. **965** Fermilab Visual Media Services **966 (T)** Yoav Levy/Phototake NYC **(B)** Rennie Van Munchow/Phototake NYC **967** Kevin Schafer/Peter Arnold, Inc. **969** John D. Cunningham/Visuals Unlimited **970** Yan/Photo Researchers, Inc. **971** Martin Dohrn/Science Photo Library/Photo Researchers, Inc. **972 (L)** Roger Tully/Stone **(R)** Scott Camazine/Photo Researchers, Inc. **973 (T)** NASA Headquarters **(B)** Christopher Burrows, European Space Agency/Space Telescope Science Institute/NASA

CHAPTER 23 981 Geoff Tompkinson/Science Photo Library/Photo Researchers, Inc. **984** Larry Lefever/Grant Heilman Photography, Inc. Ken Graham/Stone **1001** Lance Nelson/The Stock Market **1003** Letraset Phototone **1006 (T)** Hank Morgan/Rainbow **(B)** Kristen Brochmann/Fundamental Photographs **1008** Thomas Kitchin/Tom Stack & Associates **1010** Richard Weiss/Silver Burdett Ginn **1014** Richard Pasley/Stock Boston **1016** Paul S. Howell/Liaison Agency, Inc.

CHAPTER 24 1027 Norbert Wu/Norbert Wu Productions **1028** Mark A. Leman/Stone **1031** Art Wolfe/Stone **1038** Ken Eward/Science Source/Photo Researchers, Inc. **1039** Gregory G. Dimijian/Photo Researchers, Inc. **1040 (T)** Christopher Arnesen/Stone **(B)** James D. Watt/Planet Earth Pictures Ltd **1041 (L)** Ken Eward/Photo Researchers, Inc. **1043** Ken Eward/Biografx/Science Source/Photo Researchers, Inc. **1049 (T)** Jim Anderson/Woodfin Camp & Associates **(B)** Thomas Kitchin/Tom Stack & Associates **1051** Darrell Gulin/Stone **1056** Leonard Lessin/Peter Arnold, Inc. **1060** Telegraph Colour Library/FPG International LLC

Useful Conversion Factors and Relationships

Length

SI unit: meter (m)

$$1 \text{ km} = 0.621\,37 \text{ mi}$$
$$1 \text{ mi} = 5280 \text{ ft}$$
$$= 1.6093 \text{ km}$$
$$1 \text{ m} = 1.0936 \text{ yd}$$
$$1 \text{ in.} = 2.54 \text{ cm (exactly)}$$
$$1 \text{ cm} = 0.393\,70 \text{ in.}$$
$$1 \text{ Å} = 10^{-10} \text{ m}$$

Mass

SI unit: kilogram (kg)

$$1 \text{ kg} = 10^3 \text{ g} = 2.2046 \text{ lb}$$
$$1 \text{ lb} = 16 \text{ oz} = 453.59 \text{ g}$$
$$1 \text{ amu} = 1.660\,54 \times 10^{-27} \text{ kg}$$

Temperature

SI unit: Kelvin (K)

$$0 \text{ K} = -273.15^\circ\text{C}$$
$$= -459.67^\circ\text{F}$$
$$\text{K} = {}^\circ\text{C} + 273.15$$
$$^\circ\text{C} = \tfrac{5}{9}\,(^\circ\text{F} - 32)$$
$$^\circ\text{F} = \tfrac{9}{5}\,(^\circ\text{C}) + 32$$

Energy (derived)

SI unit: Joule (J)

$$1 \text{ J} = 1 \text{ (kg}\cdot\text{m}^2)/\text{s}^2$$
$$1 \text{ J} = 0.239\,01 \text{ cal}$$
$$= 1 \text{ C} \times 1 \text{ V}$$
$$1 \text{ cal} = 4.184 \text{ J}$$
$$1 \text{ eV} = 1.602 \times 10^{-19} \text{ J}$$

Pressure (derived)

SI unit: Pascal (Pa)

$$1 \text{ Pa} = 1 \text{ N/m}^2$$
$$= 1 \text{ kg}/(\text{m}\cdot\text{s}^2)$$
$$1 \text{ atm} = 101,325 \text{ Pa}$$
$$= 760 \text{ mm Hg (torr)}$$
$$= 14.70 \text{ lb/in}^2$$
$$1 \text{ bar} = 10^5 \text{ Pa}$$

Volume (derived)

SI unit: cubic meter (m³)

$$1 \text{ L} = 10^{-3} \text{ m}^3$$
$$= 1 \text{ dm}^3$$
$$= 10^3 \text{ cm}^3$$
$$= 1.0567 \text{ qt}$$
$$1 \text{ gal} = 4 \text{ qt}$$
$$= 3.7854 \text{ L}$$
$$1 \text{ cm}^3 = 1 \text{ mL}$$
$$1 \text{ in}^3 = 16.4 \text{ cm}^3$$

Fundamental Constants

Atomic mass unit	1 amu	$= 1.660\,539 \times 10^{-27}$ kg
	1 g	$= 6.022\,142 \times 10^{23}$ amu
Avogadro's number	N_A	$= 6.022\,142 \times 10^{23}$/mol
Boltzmann's constant	k	$= 1.380\,650 \times 10^{-23}$ J/K
Electron charge	e	$= 1.602\,176 \times 10^{-19}$ C
Faraday's constant	$\mathcal{F}$	$= 9.648\,534 \times 10^4$ C/mol
Gas constant	R	$= 8.314\,472$ J/(mol·K)
		$= 0.082\,058\,2$ (L·atm)/(mol·K)
Mass of electron	m_e	$= 5.485\,799 \times 10^{-4}$ amu
		$= 9.109\,382 \times 10^{-31}$ kg
Mass of neutron	m_n	$= 1.008\,665$ amu
		$= 1.674\,927 \times 10^{-27}$ kg
Mass of proton	m_p	$= 1.007\,276$ amu
		$= 1.672\,622 \times 10^{-27}$ kg
Pi	π	$= 3.141\,592\,653\,6$
Planck's constant	h	$= 6.626\,069 \times 10^{-34}$ J·s
Speed of light	c	$= 2.997\,924\,58 \times 10^8$ m/s